Organic Electronic Spectral Data
Volume XX   1978

Organic Electronic Spectral Data

Volume XX   1978

JOHN P. PHILLIPS, DALLAS BATES
HENRY FEUER & B. S. THYAGARAJAN

EDITORS

CONTRIBUTORS

Dallas Bates

H. Feuer

L. D. Freedman

C. M. Martini

F. C. Nachod

J. P. Phillips

AN INTERSCIENCE ® PUBLICATION
JOHN WILEY & SONS
New York • Chichester • Brisbane • Toronto • Singapore

INTRODUCTION TO THE SERIES

In 1956 a cooperative effort to abstract and publish in formula order all the ultraviolet-visible spectra of organic compounds presented in the journal literature was organized through the enterprise and leadership of M.J. Kamlet and H.E. Ungnade. Organic Electronic Spectral Data was incorporated in 1957 to create a formal structure for the venture, and coverage of the literature from 1946 onward was then carried out by chemists with special interests in spectrophotometry through a page by page search of the major chemical journals. After the first two volumes (covering the literature from 1946 through 1955) were produced, a regular schedule of one volume for each subsequent period of two years was introduced. In 1966 an annual schedule was inaugurated.

Altogether, more than fifty chemists have searched a group of journals totalling more than a hundred titles during the course of this sustained project. Additions and subtractions from both the lists of contributors and of journals have occurred from time to time, and it is estimated that the effort to cover all the literature containing spectra may not be more than 95% successful. However, the total collection is by far the largest ever assembled, amounting to approximately 450,000 spectra in the twenty volumes so far.

Volume XXI is in preparation.

PREFACE

Processing of the data provided by the contributors to Volume XX as to the last several volumes was performed at the University of Louisville.

John P. Phillips
Dallas Bates
Henry Feuer
B.S. Thyagarajan

ORGANIZATION AND USE OF THE DATA

The data in this volume were abstracted from the journals listed in the reference section at the end. Although a few exceptions were made, the data generally had to satisfy the following requirements: the compound had to be pure enough for satisfactory elemental analysis and for a definite empirical formula; solvent and phase had to be given; and sufficient data to calculate molar absorptivities had to be available. Later it was decided to include spectra even if solvent was not mentioned. Experience has shown that the most probable single solvent in such circumstances is ethanol.

All entries in the compilation are organized according to the molecular formula index system used by Chemical Abstracts. Most of the compound names have been made to conform with the Chemical Abstracts system of nomenclature.

Solvent or phase appears in the second column of the data lists, often abbreviated according to standard practice; there is a key to less obvious abbreviations on the next page. Anion and cation are used in this column if the spectra are run in relatively basic or acidic conditions respectively but exact specifications cannot be ascertained.

The numerical data in the third column present wavelength values in nanometers (millimicrons) for all maxima, shoulders and inflections, with the logarithms of the corresponding molar absorptivities in parentheses. Shoulders and inflections are marked with a letter s. In spectra with considerable fine structure in the bands a main maximum is listed and labelled with a letter f. Numerical values are given to the nearest nanometer for wavelength and nearest 0.01 unit for the logarithm of the molar absorptivity. Spectra that change with time or other common conditions are labelled "anom." or "changing", and temperatures are indicated if unusual.

The reference column contains the code number of the journal, the initial page number of the paper, and in the last two digits the year (1978). A letter is added for journals with more than one volume or section for a year. The complete list of all articles and authors thereof appears in the References at the end of the book.

Several journals that were abstracted for previous volumes in this series have been omitted, usually for lack of useful data, and several new ones have been added. Most Russian journals have been abstracted in the form of the English translation editions.

ABBREVIATIONS

s	shoulder or inflection
f	fine structure
n.s.g.	no solvent given in original reference
$C_6H_{11}Me$	methylcyclohexane
C_6H_{12}	cyclohexane
DMF	dimethylformamide
DMSO	dimethylsulfoxide
THF	tetrahydrofuran

Other solvent abbreviations generally follow the practice of Chemical Abstracts.

Underlined data were estimated from graphs.

JOURNALS ABSTRACTED

Journal	No.	Journal	No.
Acta Chem. Scand.	1	Tetrahedron Letters	88
Indian J. Chem.	2	Angew. Chem.	89
J. Heterocyclic Chem.	4	J. Inorg. Nucl. Chem.	90
Ann. Chem. Liebigs	5	J. Applied Chem. U.S.S.R.	93
Australian J. Chem.	12	Chem. Pharm. Bull. Japan	94
Steroids	13	J. Pharm. Soc. Japan	95
Bull. Chem. Soc. Japan	18	The Analyst	96
Bull. Acad. Polon. Sci.	19	Z. Chemie	97
Bull. soc. chim. Belges	20	J. Agr. Food Chem.	98
Bull. soc. chim. France	22	Theor. Exptl. Chem.	99
Can. J. Chem.	23	Lloydia	100
Chem. Ber.	24	J. Organometallic Chem.	101
Chem. and Ind.(London)	25	Phytochemistry	102
Chimia	27	Khim. Geterosikl. Soedin.	103
Compt. rend.	28	Zhur. Organ. Khim.	104
Doklady Akad. Nauk S.S.S.R.	30	Khim. Prirodn. Soedin	105
Experientia	31	Die Pharmazie	106
Gazz. chim. ital.	32	Synthetic Comm.	107
Helv. Chim. Acta	33	Russian J. Phys. Chem.	110
J. Chem. Eng. Data	34	European J. Med. Chem.	111
J. Am. Chem. Soc.	35	Spectroscopy Letters	112
J. Pharm. Sci.	36	Acta Chim. Acad. Sci. Hung	114
J. Chem. Soc., Perkin Trans. II	39B	Egyptian J. Chem.	115
J. Chem. Soc., Perkin Trans. I	39C	Macromolecules	116
Nippon Kagaku Kaishi	40	Org. Preps. and Procedures	117
J. Indian Chem. Soc.	42	Synthesis	118
J. Org. Chem.	44	S. African J. Chem.	119
J. Polymer Sci., Polymer Chem. Ed.	47	Pakistan J. Sci. Ind. Research	120
J. prakt. Chem.	48	J. Macromol. Sci.	121
Monatsh. Chem.	49	Moscow U. Chem. Bull.	123
Naturwiss.	51	Ukrain. Khim. Zhur.	124
Rec. trav. chim.	54	Makromol. Chem.	126
Polish J. Chem.	56	Croatica Chem. Acta	128
J. Chem. Soc., Faraday Trans. II	60	Bioorg. Chem.	130
Ber. Bunsen Gesell. Phys. Chem.	61	Pharm. Acta Helv.	133
Z. phys. Chem.	62	J. Appl. Spectroscopy S.S.S.R.	135
Z. physiol. Chem.	63	Carbohydrate Research	136
Z. Naturforsch.	64	Finnish Chem. Letters	137
Zhur. Obshchei Khim.	65	Chemistry Letters	138
J. Structural Chem.	67	P, S and Related Elements	139
Biochemistry	69	J. Anal. Chem. S.S.S.R.	140
Izvest Akad. Nauk S.S.S.R.	70	Heterocycles	142
Coll. Czech. Chem. Comm.	73	Arzneimittel. Forsch.	145
J. Chem. Soc., Chem. Comm.	77	J. Appl. Chem. Biotech.	146
Tetrahedron	78	J. Luminescence	147
Revue Roumaine Chim.	80	Photochem. Photobiol.	149
Arch. Pharm.	83	J. Chem. Research	150
Talanta	86		
J. Med. Chem.	87		

Organic Electronic Spectral Data
Volume XX   1978

Compound	Solvent	$\lambda_{max}(\log \epsilon)$	Ref.
$CH_3N_3O_4$			
Methanamine, N,N-dinitro-	H_2O	245(3.33)	104-0846-78
$C_2H_2Cl_3NO$			
Acetamide, 2,2,2-trichloro-	EtOH	212(2)(approx.)	135-1225-78
$C_2H_2F_3NO$			
Acetamide, 2,2,2-trifluoro-	EtOH	212(2)(approx.)	135-1225-78
$C_2H_2N_2S$			
1,2,3-Thiadiazole	C_6H_{12}	211(3.64),249(3.16), 294(2.29)	44-2487-78
$C_2H_2N_2S_2$			
1,2,3-Thiadiazole-5-thiol, potassium salt	MeOH	223(3.65),335(3.81)	4-1295-78
$C_2H_2O_4$			
Oxalic acid	M HCl	<u>240s(1.8)</u>	140-0060-78
$C_2H_3ClN_2O_3$			
Acetamide, 2-chloro-N-nitro-	H_2O	266(3.78)	104-2316-78
$C_2H_3Cl_2NO$			
Acetamide, 2,2-dichloro-	EtOH	212(2)(approx.)	135-1225-78
$C_2H_3F_2NO$			
Acetamide, 2,2-difluoro-	EtOH	212(2)(approx.)	135-1225-78
C_2H_3NS			
Methane, isothiocyanato-, iodine complex	CCl_4	450(3.26)	22-0439-78
C_2H_4ClNO			
Acetamide, 2-chloro-	EtOH	212(2)(approx.)	135-1225-78
C_2H_4FNO			
Acetamide, 2-fluoro-	EtOH	212(2)(approx.)	135-1225-78
$C_2H_4N_2O_3$			
Acetamide, N-nitro-	H_2O	266(3.82)	104-2316-78
$C_2H_4N_4O_4S$			
Carbamimidothioic acid, dinitromethyl ester	H_2O MeCN	346(4.08) 337(4.00)	70-2087-78 70-2087-78
$C_2H_4S_2$			
Ethane(dithioic) acid, sodium salt	EtOH	337(4.06),487(1.77)	64-0976-78B
$C_2H_5N_3O_4$			
Ethanamine, N,N-dinitro-	H_2O	245(3.33)	104-0846-78
$C_2H_6N_2O_2$			
Methanamine, N-methyl-N-nitro-	H_2O	<u>238(3.7)</u>	104-0221-78
C_2H_6OS			
Methane, sulfinylbis-	hexane H_2O EtOH	204(3.45),222s(2.95) 193(3.08),208(2.95) 198(3.15),210(3.04)	70-0901-78 70-0901-78 70-0901-78

Compound	Solvent	$\lambda_{max}(\log \epsilon)$	Ref.
$C_3Cl_5N_3S$ 1H-1,2,4,6-Thiatriazine, 1,3-dichloro- 5-(trichloromethyl)-	hexane	278(3.79)	103-0226-78
$C_3H_2F_5NO$ Propanamide, 2,2,3,3,3-pentafluoro-	EtOH	212(2)(approx.)	135-1225-78
$C_3H_2N_2O_2S_2$ 1,2,3-Thiadiazole-4-carboxylic acid, 5-mercapto-, dipotassium salt	H_2O	238(3.73),333(3.82)	4-1295-78
$C_3H_3ClN_4O$ 1,3,5-Triazin-2(1H)-one, 4-amino- 6-chloro-	pH 1 pH 7 pH 11	255(3.87) 250(3.66) 247(3.56)	87-0883-78 87-0883-78 87-0883-78
$C_3H_3N_3O_5S$ 2H-1,2,6-Thiadiazin-3(6H)-one, 4-nitro-, 1,1-dioxide	EtOH	267(3.71),337(4.15)	4-0253-78
$C_3H_4BrNO_2$ 1-Propene, 2-bromo-3-nitro-, anion	pH 13	276(4.12)	44-3116-78
$C_3H_4N_2O$ 2H-Imidazol-2-one, 1,3-dihydro-	pH 7	206(3.92)	73-1511-78
$C_3H_4N_2OS$ 4-Thiazolone, 2-amino-4,5-dihydro-	neutral cation anion	221(4.29),249(3.91) 206(4.27),238(3.79) 232(4.23)	104-1229-78 104-1229-78 104-1229-78
$C_3H_4N_2O_2S$ 2,4-Thiazolidinedione, 2-oxime	n.s.g.	210(3.95),230(3.89)	139-0303-78A
$C_3H_4N_2O_4S$ 2H-1,2,6-Thiadiazine-3,5(4H,6H)-dione, monopotassium salt	H_2O	208(3.76)	4-0221-78
$C_3H_4N_2S$ 1,2,3-Thiadiazole, 4-methyl- 1,2,3-Thiadiazole, 5-methyl-	C_6H_{12} C_6H_{12}	213(3.68),258(3.27), 290(2.34) 217(3.72),253(3.32), 293(2.39)	44-2487-78 44-2487-78
$C_3H_4N_2S_3$ 1,2,4-Thiadiazole-5(2H)-thione, 3-(methylthio)-	n.s.g.	242(4.08),313(4.15)	106-0764-78
$C_3H_4N_4O_4S$ 2H-1,2,6-Thiadiazine-3,4-dione, 5-am- ino-, 4-oxime, 1,1-dioxide	H_2O	203(3.71),254(3.69)	4-0221-78
$C_3H_4OS_2$ 1,3-Oxathiolane-2-thione	EtOH	282(3.68)	118-0286-78
$C_3H_4O_2S$ 1,3-Dioxolane-2-thione	EtOH	235(4.18)	118-0286-78
$C_3H_5NO_2$ 1-Propene, 3-nitro-, anion	pH 13	275(4.32)	44-3116-78

Compound	Solvent	$\lambda_{max}(\log \epsilon)$	Ref.
C_3H_5NS			
Ethane, isothiocyanato-, iodine complex	CCl_4	465(3.15)	22-0439-78
$C_3H_5N_3O_3S$			
2H-1,2,6-Thiadiazin-3(4H)-one, 5-amino-, 1,1-dioxide	H_2O	210(4.15),220s(4.10), 295)2.88)	4-0221-78
C_3H_6ClNO			
Propane, 2-chloro-2-nitroso-	benzene	652(1.38)	39-0012-78B
	MeOH	647(1.15)	39-0012-78B
$C_3H_6N_2O_4S$			
Sulfonium, dimethyl-, dinitromethylide	MeCN	329(3.98)	70-2087-78
$C_3H_6N_3O_2P$			
1,5,2-Diazaphosphorin-6(1H)-one, 2-amino-2,5-dihydro-, 2-oxide	H_2O	235(3.64)	130-0421-78
$C_3H_6N_4$			
1-Triazene-1-carbonitrile, 3,3-dimethyl-	CH_2Cl_2	268(4.14)	89-0203-78
$C_3H_6N_4O_2$			
5H-Tetrazol-5-one, 1,4-dihydro-1,4-dimethyl-, 2-oxide	MeOH	256(3.8)	33-1622-78
$C_3H_6N_4O_3S$			
2H-1,2,6-Thiadiazin-3(6H)-one, 4,5-diamino-, 1,1-dioxide	H_2O	211(3.86),286s(3.23)	4-0221-78
$C_3H_6N_4S$			
5H-Tetrazole-5-thione, 1,4-dihydro-1,4-dimethyl-	EtOH	206(3.77),246(4.13), 275(3.19)	24-0566-78
C_3H_6O			
Acetone	neat	<u>276(1.2)</u>	18-1708-78
(mole per cent)	90%	<u>276(1.2)</u>	18-1708-78
	56%	<u>274(1.2)</u>	18-1708-78
C_3H_6OS			
Ethene, (methylsulfinyl)-	hexane	196(3.49),213s(3.15), 245(3.38)	70-0901-78
	H_2O	194(3.38),208s(3.28), 223(3.26)	70-0901-78
	EtOH	196(3.32),208s(3.26), 231(3.30)	70-0901-78
$C_3H_6OS_2$			
Carbonodithioic acid, O-ethyl ester, sodium salt	n.s.g.	227(3.79),303(4.06)	44-2930-78
$C_3H_6OS_3$			
1,2,3-Trithian-5-ol	MeOH	263(0.51)	138-1219-78
$C_3H_6S_2$			
Propane(dithioic) acid, sodium salt	EtOH	338(4.23),460s(--)	64-0976-78B
$C_3H_7N_3O_4$			
1-Propanamine, N,N-dinitro-	H_2O	246(3.34)	104-0846-78

Compound	Solvent	$\lambda_{max}(\log \epsilon)$	Ref.
$C_4HClN_2O_2$ 2H-Imidazol-2-one, 4-chloro-5-formyl-	MeOH	245(2.8),335(3.5)	103-0208-78
C_4HClO_3 3-Cyclobutene-1,2-dione, 3-chloro-4-hydroxy-	MeCN	224(4.17),237s(4.10)	33-1784-78
$C_4H_2Cl_2N_2$ Pyrimidine, 2,4-dichloro-	EtOH	210(3.87),258(3.59)	104-0377-78
$C_4H_2Cl_4$ 1,3-Butadiene, 1,2,3,4-tetrachloro-, (Z,Z)-	n.s.g.	249(4.62)	2-0151-78
$C_4H_2F_6INO$ Butanamide, 2,2,3,3,4,4-hexafluoro-4-iodo-	EtOH	212(2)(approx.)	135-1225-78
$C_4H_2F_7NO$ Butanamide, 2,2,3,3,4,4,4-heptafluoro-	EtOH	212(2)(approx.)	135-1225-78
$C_4H_2N_2O_5$ Furan, 2,4-dinitro- Meisenheimer methoxide complex	MeOH MeOH	286(3.87) 500(4.14)	44-4303-78 44-4303-78
$C_4H_3NO_2$ 1H-Pyrrole-2,5-dione (maleimide)	pH 2	217(4.22),277(2.78)	39-1277-78B
$C_4H_3N_3O_3S$ 2H-1,2,6-Thiadiazine-4-carbonitrile, 3,6-dihydro-3-oxo-, 1,1-dioxide	H_2O	290(4.04)	4-0253-78
$C_4H_3N_3S_3Si$ Silane, triisothiocyanatomethyl-	hexane	245(3.1)	114-0217-78A
$C_4H_3N_5O_2$ 1H-1,2,3-Triazolo[4,5-d]pyrimidine-5,7(4H,6H)-dione	pH 1 pH 12	265(3.88) 283(3.73)	104-1617-78 104-1617-78
$C_4H_3N_5S$ 1,2,3,4-Thiatriazole, 5-(1H-imidazol-1-yl)-	MeOH	245(3.92),271(3.83)	78-0453-78
C_4H_4ClNO Butanenitrile, 2-chloro-3-oxo-	EtOH	234(4.02)	44-3821-78
$C_4H_4ClN_3$ 2-Pyrimidinamine, 4-chloro- 4-Pyrimidinamine, 2-chloro-	EtOH EtOH	232(4.23),298(3.66) 236(4.10),276(3.64)	104-0377-78 104-0377-78
$C_4H_4N_2O$ 2(1H)-Pyrazinone	H_2O	220(4.00),318(2.74)	142-1733-78A
$C_4H_4N_2S_3$ 3,4,6-Pyridazinetrithiol	pH -1.0 pH 5.0	244(3.99),286(4.23) 210(4.01),296(4.34)	12-0389-78 12-0389-78
$C_4H_4N_4O_2S$ Sulfamide, (2,2-dicyanoethenyl)-	EtOH	276(4.35)	4-0253-78

Compound	Solvent	λ_{max}(log ϵ)	Ref.

$C_4H_4N_4O_3S$
  Imidazo[4,5-c][1,2,6]thiadiazin-
  4(3H)-one, 1,5-dihydro-, 2,2-dioxide
	H_2O	207(4.00),227s(3.76),
		289(3.90)
			4-0221-78

$C_4H_4N_4O_4$
  2,4(1H,3H)-Pyrimidinedione, 6-(nitro-
  amino)-
	pH 3.0	221(4.27),240s(3.78),
		313(4.05)
			24-1006-78

	pH 8.0	220s(4.10),275s(3.30), 325(4.15)	24-1006-78
	pH 14.5	226(3.91),339(4.19)	24-1006-78
$C_4H_5Cl_2NO$ 2-Pyrrolidinone, 3,3-dichloro-	MeOH	206(4.00)	83-0294-78

$C_4H_5NO_4$
  2-Propenoic acid, 3-nitro-, methyl
  ester, (E)-
	C_6H_{12}	465(2.39)	73-2395-78

C_4H_5NS
  1-Propene, 3-isothiocyanato-, iodine
  complex
	CCl_4	460(3.16)	22-0439-78

$C_4H_5N_5$ 1H-Pyrazolo[1,5-d]tetrazole, 1-methyl-	EtOH	276(3.73),320s(3.28)	4-0395-78

$C_4H_5N_7$
  1H-1,2,3-Triazolo[4,5-d]pyrimidine-
  5,7-diamine
	pH 2.2
	pH 8.4
	DMSO
		252(4.08),276s(3.95)
		249(3.76),288(3.88)
		288(3.96)
			33-0108-78
			33-0108-78
			33-0108-78

C_4H_6 1,3-Butadiene	EtOH	217(4.32)	39-1338-78B
$C_4H_6BrN_3$ 1H-Pyrazol-3-amine, 4-bromo-5-methyl-	MeOH	235(4.52)	103-0313-78
$C_4H_6HgO_4$ Mercuric acetate	MeOH	244(2.37)	78-1943-78
$C_4H_6N_2$ 2,3-Diazabicyclo[2.2.0]hex-2-ene	isooctane	334(1.76),341(1.83), 350(1.82)	88-2469-78
$C_4H_6N_2OS$ 4-Thiazolidinone, 2-imino-3-methyl-	neutral cation	214(4.15) 210(4.26),241(3.78)	104-1229-78 104-1229-78

$C_4H_6N_2O_2$
  1,3,4-Oxadiazol-2(5H)-one, 5,5-dimeth-
  yl-
	hexane	214(3.46),<u>362(2.4)</u>,
		373(2.54),<u>380(2.3)</u>
			23-1319-78

$C_4H_6N_2O_2S$
  2,4-Thiazolidinedione, 5-methyl-,
  2-oxime
	n.s.g.	210(3.90),230(3.86)	139-0303-78A

$C_4H_6N_2S$ 1H-Imidazole, 2-(methylthio)-	MeOH EtOH CH_2Cl_2	218(3.85),245(3.67) 220(3.936),251(3.605) 225(3.411),255(3.684)	44-4774-78 28-0201-78B 28-0201-78B

Compound	Solvent	$\lambda_{max}(\log \epsilon)$	Ref.
$C_4H_6N_2S_2$			
1,2,3-Thiadiazole, 5-(ethylthio)-	MeOH	240(3.27),288(3.75), 301(3.63)	4-1295-78
$C_4H_6N_2S_2Si$			
Silane, diisothiocyanatodimethyl-	hexane	<u>242(3.2)</u>	114-0217-78A
$C_4H_6N_2S_3$			
1,2,4-Thiadiazole, 3,5-bis(methylthio)-	n.s.g.	254(4.13),284s(3.64)	106-0764-78
$C_4H_6N_4O_2$			
2,4(1H,3H)-Pyrimidinedione, 6-hydrazino-	0.5M HCl	262(3.90)	56-0037-78
	H_2O	200(3.83),268(4.10)	56-0037-78
1,2,3,4-Tetrazine-5,6-dione, 1,4-di- hydro-1,4-dimethyl-	MeOH	258(3.9)	33-1622-78
1,2,4-Triazine-3,5(2H,4H)-dione, 6-amino-2-methyl-	pH 1	307(3.71)	124-1064-78
	EtOH-pH 7	311(3.74)	124-1064-78
	pH 13	296(3.72)	124-1064-78
1,2,4-Triazine-3,5(2H,4H)-dione, 6-amino-4-methyl-	pH 1	298(3.65)	124-1064-78
	EtOH-pH 7	302(3.65)	124-1064-78
	pH 13	321(3.82)	124-1064-78
1,2,4-Triazine-3,5(2H,4H)-dione, 6-(methylamino)-	pH 1	308(3.66)	103-1273-78
	H_2O	308(3.61)	103-1273-78
	pH 13	298(3.57)	103-1273-78
$C_4H_6N_4O_3$			
1-Imidazolidinecarboxamide, 3-nitroso- 2-oxo-	EtOH	252(4.07),386(1.95), 401(2.13),419(2.12)	94-2635-78
C_4H_6OS			
Ethene, 1,1'-sulfinylbis-	hexane	197(4.03),236(3.45)	70-0901-78
	H_2O	195(4.06),222(3.42)	70-0901-78
	EtOH	196(4.13),226(3.42)	70-0901-78
$C_4H_6O_2$			
2-Propenal, 3-methoxy-	C_6H_{12}	233(4.74),312(2.78)	44-2833-78
$C_4H_6O_2S$			
1,3-Dioxolane-2-thione, 4-methyl-	EtOH	234(4.15)	118-0286-78
Propanedial, (methylthio)-	H_2O	272(4.30)	73-1248-78
$C_4H_7NO_2$			
2-Butene, 1-nitro-	pH 13	274(4.37)	44-3116-78
1-Propene, 2-methyl-1-nitro-	hexane	240(3.95)	44-3116-78
	H_2O	262(3.95)	44-3116-78
	EtOH	250(3.93)	44-3116-78
1-Propene, 2-methyl-3-nitro-	pH 13	276(4.16)	44-3116-78
$C_4H_7N_3O$			
3H-Pyrazol-3-one, 5-(aminomethyl)- 1,2-dihydro-	MeOH	240(3.60)	1-0056-78
$C_4H_7N_{11}$			
2H-Tetrazole, 5,5'-(1-triazene-1,3-di- yl)bis[2-methyl-	EtOH	245(3.96),296(4.38)	103-0800-78
C_4H_8ClNO			
Butane, 2-chloro-2-nitroso-	benzene	653(1.33)	39-0012-78B
	MeOH	651(1.18)	39-0012-78B
	THF	651(1.23)	39-0012-78B
	EtI	654(1.31)	39-0012-78B

Compound	Solvent	$\lambda_{max}(\log \epsilon)$	Ref.
C₄H₈N₂O			
2-Imidazolidinone, 1-methyl-	EtOH	none	28-0521-78A
	CH₂Cl₂	245(1.485)	28-0521-78A
2(1H)-Pyrimidinone, tetrahydro-	EtOH	245(1.725)	28-0521-78A
	CH₂Cl₂	254(1.756)	28-0521-78A
C₄H₈N₂O₃			
Glycine, N-ethyl-N-nitroso-	EtOH	235(3.81),350(1.92)	94-3909-78
Methanol, (methylnitrosoamino)-, acetate	n.s.g.	228(3.82),367(1.94)	94-3905-78
Propanamide, 2-methyl-N-nitro-	H₂O	268(3.84)	104-2316-78
C₄H₈N₂O₄			
Glycine, N-[2-(hydroxyamino)-2-oxo-ethyl]-	acid	196(3.67)	73-2897-78
	anion	196(3.67)	73-2897-78
	dianion	217(3.79)	73-2897-78
C₄H₈N₂S			
1H-Imidazole, 4,5-dihydro-2-(methyl-thio)-	EtOH	224(3.734)	28-0521-78A
	CH₂Cl₂	none	28-0521-78A
C₄H₈N₄			
1,2,4,5-Tetrazine, 1,4-dihydro-1,4-di-methyl-	heptane	242(3.77)	104-0576-78
3H-1,2,4-Triazol-5-amine, 3,3-dimethyl-	EtOH	285(2.375)	23-2194-78
C₄H₈N₄O₂			
1,2,3,4-Tetrazin-5(4H)-one, 1,6-di-hydro-1,4-dimethyl-, 2-oxide	MeOH	292(3.7)	33-1622-78
C₄H₈OS			
Ethene, (ethylsulfinyl)-	hexane	198(3.49),213s(3.20), 248(3.42)	70-0901-78
	H₂O	197(3.40),208s(3.30), 234(3.40)	70-0901-78
C₄H₈S			
1-Propene, 1-(methylthio)-, cis	n.s.g.	225(3.86),245(3.49)	73-2619-78
1-Propene, 1-(methylthio)-, trans	n.s.g.	228(3.51),251(3.34)	73-2619-78
1-Propene, 3-(methylthio)-	n.s.g.	219(3.45),235(3.01)	73-2619-78
C₄H₈S₂			
Butane(dithioic) acid, sodium salt	EtOH	338(4.20),460s(--)	64-0976-78B
Propane(dithioic) acid, 2-methyl-, sodium salt	EtOH	338(4.15),460s(--)	64-0976-78B
C₄H₉NO			
Acetamide, N,N-dimethyl-	H₂O	220s(2.1)	104-0221-78
C₄H₉NOS			
Thionitrous acid, S-butyl ester	hexane-1% CCl₄ at 0°	342(2.89),519(1.11), 550(1.43)	39-0913-78C
Thionitrous acid, S-(1,1-dimethyl-ethyl) ester	hexane-1% CCl₄ at 0°	345(2.83),562(0.85), 552(0.78),598(1.23), 605(1.20)	39-0913-78C
C₄H₉NSSi			
Silane, isothiocyanatotrimethyl-	hexane	200(3.4),243(2.3)	114-0217-78A
	dioxan	236(2.9)	114-0217-78A

Compound	Solvent	$\lambda_{max}(\log \epsilon)$	Ref.
$C_4H_9N_3O_4$ 1-Butanamine, N,N-dinitro-	H_2O	248(3.34)	118-0846-78

$C_4H_{10}AgN_5O_3$
  Silver, (nitrato-O)(1,4,5,6-tetrahydro-
  1,4-dimethyl-1,2,3,4-tetrazine-N^1)-
	MeOH	263(3.9)	33-1622-78

$C_4H_{10}Cl_4Hg_2N_4$
  Mercury, tetrachloro[μ-(1,4,5,6-tetra-
  hydro-1,4-dimethyl-1,2,3,4-tetra-
  zine-N^1,N^4)]di-
	MeOH	268(3.08)	33-1622-78

$C_4H_{10}N_2O_2$ Ethanamine, N-ethyl-N-nitro-	H_2O	240(3.7)	104-0221-78

$C_4H_{10}N_4$
  1,2,3,4-Tetrazine, 1,4,5,6-tetrahydro-
  1,4-dimethyl-
	MeOH	225(3.47),268(3.75)	33-1622-78

$C_4H_{10}N_4O$
  1,2,3,4-Tetrazine, 1,4,5,6-tetrahydro-
  1,4-dimethyl-, 2-oxide
	MeOH	268(4.01)	33-1622-78

$C_4H_{10}OS$ Ethane, 1,1'-sulfinylbis-	hexane H_2O EtOH	208(3.54),224(3.11) 195(3.15),212(2.95) 199(3.28),214s(2.95)	70-0901-78 70-0901-78 70-0901-78
$C_4H_{12}N_2Si$ Diazene, methyl(trimethylsilyl)-	CH_2Cl_2	491(0.78)	60-1909-78
$C_4H_{12}N_4$ 2-Tetrazene, 1,1,4,4-tetramethyl-	EtOH EtOH	277(3.90) 277(3.90)	23-1657-78 35-2444-78
$C_4N_4S_4Si$ Silane, tetraisothiocyanato-	hexane	245(3.7)	114-0217-78A

C_4OS_5
  [1,3]Dithiolo[4,5-d]-1,3-dithiol-2-one,
  5-thioxo-
	$CHCl_3$	250(3.89),290(3.55),
		  400(4.14)
			97-0345B-78

$C_4O_2S_4$
  [1,3]Dithiolo[4,5-d]-1,3-dithiole-
  2,5-dione
	$CHCl_3$	278(3.72)	97-0345B-78

C_4S_6
  [1,3]Dithiolo[4,5-d]-1,3-dithiole-
  2,5-dithione
	$CHCl_3$	282(3.84),410(4.28)	97-0345B-78

Compound	Solvent	$\lambda_{max}(\log \epsilon)$	Ref.
C_5F_5N			
Pyridine, pentafluoro-	n.s.g.	200(2.9),256(3.08)	39-0363-78C
$C_5H_2BrNS_2$			
Thieno[3,2-d]thiazole, 5-bromo-	MeOH	232(4.38),255s(3.67)	4-0081-78
$C_5H_2F_6O_2$			
2,4-Pentanedione, 1,1,1,5,5,5-hexa-fluoro-	C_6F_{14}	272(3.8)	18-1723-78
$C_5H_2N_2O_2S_2$			
Thieno[3,2-d]thiazole, 5-nitro-	MeOH	238(4.03),310s(3.76), 339(3.82),458(3.53)	4-0081-78
$C_5H_2N_2O_3$			
3-Furancarbonitrile, 5-nitro-	MeOH	286(3.94)	44-4303-78
Meisenhemer methoxide complex	MeOH	388(4.28)	44-4303-78
$C_5H_2O_3$			
4-Cyclopentene-1,2,3-trione	MeCN	218(4.11),254s(3.25), 340s(1.08),455(1.11), 541(1.04)	138-0217-78
$C_5H_3ClN_4$			
1,2,4-Triazolo[4,3-a]pyrazine, 8-chloro-	pH 1	216(4.42),255(3.38), 261(3.40),270s(3.44), 299(3.73)	4-0987-78
	H_2O	216(4.42),253(3.37), 262(3.41),270s(3.45), 297(3.74)	4-0987-78
	pH 11	216(4.39),255(3.40), 264(3.44),297(3.74)	4-0987-78
$C_5H_3IN_4O_2$			
1H-Purine-2,6-dione, 3,7-dihydro-8-iodo-	pH 3.0	215(4.15),278(4.10)	44-0544-78
	pH 8.0	213(4.28),280(4.15)	44-0544-78
	pH 13	245s(3.83),289(4.01)	44-0544-78
$C_5H_3NS_2$			
Thieno[3,2-d]thiazole	MeOH	226(4.20),249s(3.47)	4-0081-78
$C_5H_4Cl_2N_2O$			
3(2H)-Pyridazinone, 4,5-dichloro-2-methyl-	MeOH	218(4.34),240(--), 300(3.63)	73-2415-78
4(1H)-Pyridazinone, 5,6-dichloro-1-methyl-	MeOH	211(4.06),285(4.10)	73-2415-78
$C_5H_4N_2O_5$			
Furan, 2-nitro-5-(nitromethyl)-	dioxan	304(4.017)	73-0463-78
$C_5H_4N_4$			
[1,2,3]Triazolo[1,5-a]pyrazine	EtOH	274(3.69),283(3.69), 300s(--)	33-1755-78
[1,2,3]Triazolo[1,5-c]pyrimidine	heptane	266(3.75),273s(3.66), 294(3.20),301(3.21), 309(3.21),323(3.08), 340(2.65)	4-1041-78
	$CHCl_3$	267(3.87),276s(3.79), 296(3.38)	4-1041-78
[1,2,4]Triazolo[1,5-c]pyrimidine	dioxan	254(3.71),256(3.68)	12-2505-78

Compound	Solvent	λ_{max}(log ϵ)	Ref.
1,2,4-Triazolo[4,3-c]pyrimidine	pH 7.0	241s(3.70),249(3.79), 258(3.77),269s(3.63)	12-2505-78
	dioxan	245s(3.68),252(3.73), 262(3.68),276(3.57)	12-2505-78
$C_5H_4N_4O$			
Cyanamide, (1,4-dihydro-4-oxo-2-pyrimi-	pH 1	241(4.21),265s(3.94)	44-1193-78
dinyl)-	pH 7	243(4.19),295(3.84)	44-1193-78
	pH 13	230(3.99),267(3.72)	44-1193-78
1,2,4-Triazolo[4,3-a]pyrazin-8(7H)-one	pH 1	216(4.26),248s(3.57), 257s(3.60),265s(3.64), 284(3.73)	4-0987-78
	H_2O	216(4.30),250s(3.60), 256s(3.62),266s(3.66), 283(3.76)	4-0987-78
	pH 11	216(4.12),276(3.81), 289s(3.80)	4-0987-78
$C_5H_4N_4OS$			
Imidazo[1,2-a]-1,3,5-triazin-2(1H)-one,	pH 1	247(4.02),266(4.00), 298(3.98)	44-4784-78
3,4-dihydro-4-thioxo-	pH 13	247(4.02),271(3.90), 300s(3.52)	44-4784-78
$C_5H_4N_4O_2$			
Imidazo[1,2-a]-1,3,5-triazine-	pH 1	232(3.93),253(3.88)	44-4784-78
2,4(1H,3H)-dione	pH 1	231(3.95),248(3.91)	87-0883-78
	pH 7	234(3.98),251(3.97)	87-0883-78
	pH 13	249(4.00)	44-4784-78
	pH 13	246(3.98)	87-0883-78
$C_5H_4N_4O_2S$			
[1,2,3]Thiadiazolo[4,5-d]pyrimidine-	EtOH	250(3.87),320(3.56)	44-1677-78
5,7(4H,6H)-dione, 6-methyl-			
$C_5H_4N_4S$			
1,3,4-Thiadiazole, 2-(1H-imidazol-1-yl)-	MeOH	257(3.96)	78-0453-78
1,2,4-Triazolo[4,3-a]pyrazine-8(7H)-	pH 1	208(4.22),268(3.84), 343s(4.08),353(4.13), 365s(4.03)	4-0987-78
thione	H_2O	208(4.23),268(3.84), 342s(4.08),353(4.13), 365s(4.03)	4-0987-78
	pH 11	210(4.20),270(3.76), 337(4.05)	4-0987-78
$C_5H_4N_4Se$			
1,2,4-Triazolo[4,3-a]pyrazine-8(7H)-	pH 1	213(4.33),291(3.72), 386(3.84)	4-0987-78
selenone	H_2O	213(4.35),262s(3.58), 312(3.77),332s(3.69), 382s(3.52)	4-0987-78
	pH 11	212(4.32),252s(3.74), 292s(3.70),317s(3.72), 344(3.82)	4-0987-78
$C_5H_4O_3$			
2H-Pyran-2,6(3H)-dione	EtOH	233(3.91),350(4.21)	2-0196-78
	EtOH-HCl	328(4.01)	2-0196-78

Compound	Solvent	λ_{max}(log ϵ)	Ref.
$C_5H_4O_4$			
4-Cyclopentene-1,3-dione, 2,2-dihy- droxy-	MeCN	217(4.11),341(2.28), 407(2.45)	138-0217-78
C_5H_5Cl			
1-Penten-4-yne, 3-chloro-	C_6H_{12}	227(2.81)	150-2340-78
2-Penten-4-yne, 1-chloro-	C_6H_{12}	228(4.08)	150-2340-78
$C_5H_5ClFN_3$			
4-Pyrimidinamine, 2-chloro-5-fluoro- N-methyl-	MeOH	240(4.03),280(3.73)	44-4200-78
$C_5H_5ClN_2OS$			
3(2H)-Pyridazinethione, 4-chloro- 5-hydroxy-2-methyl-	MeOH	257(4.28),294(3.98)	73-2415-78
$C_5H_5ClN_2O_2$			
3(2H)-Pyridazinone, 4-chloro-5-hydroxy- 2-methyl-	MeOH	216(4.48),227(--), 287(3.71)	73-2415-78
2,4(1H,3H)-Pyrimidinedione, 5-chloro- 1-methyl-	pH 2	210(3.94),281(3.95)	35-5170-78
	pH 12	277(3.79)	35-5170-78
$C_5H_5F_3O_2$			
2,4-Pentanedione, 1,1,1-trifluoro-	C_6F_{14}	282(3.8)	18-1723-78
C_5H_5NO			
2-Pyridinol	hexane	230(3.9),299(3.59)	94-1403-78
	MeOH	227(3.90),299(3.73)	94-1403-78
4-Pyridinol	MeOH	256(4.18)	94-1403-78
$C_5H_5NO_2$			
2-Propenoic acid, 3-cyano-, methyl ester, (E)-	C_6H_{12}	465(2.43)	73-2395-78
(Z)-	C_6H_{12}	460(2.65)	73-2395-78
1H-Pyrrole-2,5-dione, 1-methyl-	pH 2	219(4.21),300(2.78)	39-1277-78B
$C_5H_5NO_3$			
1H-Pyrrole-2,5-dione, 1-(hydroxy- methyl)-	pH 2	218(4.18),283(2.70)	39-1277-78B
$C_5H_5NO_3S$			
3-Isothiazolecarboxylic acid, 4-(hy- droxymethyl)-	EtOH	225(3.62),263(3.79)	44-0079-78
$C_5H_5NS_2$			
2(1H)-Pyridinethione, 3-mercapto-	EtOH	234(4.18),293(3.89), 376(3.81)	56-2039-78
$C_5H_5N_3O_4$			
1H-Pyrrole, 2-methyl-1,4-dinitro-	H_2O	210(4.03),229(4.20), 270(4.10),329(3.91), 386(3.23)	78-0505-78
	pH 9.9	299(4.08),345(4.01)	78-0505-78
	EtOH	212(4.02),234(4.19), 275(4.11),336(3.88), 398(3.28)	78-0505-78
$C_5H_5N_5$			
1,2,4-Triazolo[4,3-a]pyrazin-8-amine	pH 1	225(4.40),230s(4.31), 252s(3.65),261s(3.68), 293(3.94)	4-0987-78

Compound	Solvent	λ_{max}(log ϵ)	Ref.
1,2,4-Triazolo[4,3-a]pyrazin-8-amine (cont.)	H₂O	226(4.19),267s(3.79), 276s(3.85),295(3.95)	4-0987-78
	pH 11	226(4.17),267s(3.74), 277s(3.82),296(3.93)	4-0987-78
1H-1,2,3-Triazolo[4,5-b]pyridin-5-amine	EtOH	246s(3.54),314(4.07)	33-0108-78
6H-1,2,3-Triazolo[4,5-d]pyrimidine, 6-methyl-	pH 1.0	253(3.96)	39-0513-78C
	pH 7.0	215(4.31),270(3.80)	39-0513-78C
C₅H₅N₅O			
Imidazo[1,2-a]-1,3,5-triazin-4(1H)-one, 2-amino-	pH 1	242s(3.97),261(4.13)	87-0883-78
	pH 7	252(4.08)	87-0883-78
	pH 11	255(4.03)	87-0883-78
1,2,4-Triazolo[4,3-a]pyrazin-8(7H)-one, oxime	pH 1	202s(3.90),231(4.18), 293(3.86),306s(3.82)	4-0987-78
	H₂O	226(4.14),273(3.90), 306s(3.68)	4-0987-78
	pH 11	228(4.13),240s(4.09), 283(3.89),303s(3.80)	4-0987-78
7H-1,2,3-Triazolo[4,5-b]pyridin-7-one, 5-amino-1,4-dihydro-	pH 1	264(3.85),297(4.17)	44-4910-78
	pH 11	284(4.22)	44-4910-78
C₅H₅N₅O₂			
1H-1,2,3-Triazolo[4,5-d]pyrimidine-5,7(4H,6H)-dione, 1-methyl-	pH 1	274(3.69)	104-1617-78
	pH 12	284(3.67)	104-1617-78
1H-1,2,3-Triazolo[4,5-d]pyrimidine-5,7(4H,6H)-dione, 4-methyl-	pH 1	272(3.82)	104-1617-78
	pH 12	246(3.88),273(3.96)	104-1617-78
1H-1,2,3-Triazolo[4,5-d]pyrimidine-5,7(4H,6H)-dione, 6-methyl-	pH 1	267(3.91)	104-1617-78
	pH 12	246(3.76),273(3.90)	104-1617-78
2H-1,2,3-Triazolo[4,5-d]pyrimidine-5,7(4H,6H)-dione, 2-methyl-	pH 1	272(3.98)	104-1617-78
	pH 12	300(3.72)	104-1617-78
3H-1,2,3-Triazolo[4,5-d]pyrimidine-5,7(4H,6H)-dione, 3-methyl-	pH 1	233(3.92),254(4.05)	104-1617-78
	pH 12	247(4.02),279(3.99)	104-1617-78
C₅H₅N₅S			
7H-1,2,3-Triazolo[4,5-b]pyridine-7-thione, 5-amino-1,4-dihydro-	pH 1	299s(4.12),308(4.15), 316(4.15),330(4.07)	44-4910-78
	pH 11	310(4.25)	44-4910-78
C₅H₆AsN			
4-Arseninamine	EtOH	234(3.94),318(4.16)	88-1175-78
C₅H₆ClN₃			
2-Pyrimidinamine, 4-chloro-N-methyl-	EtOH	206(3.80),240(4.27), 309(3.52)	104-0377-78
4-Pyrimidinamine, 2-chloro-N-methyl-	EtOH	208(3.89),249(4.19), 280(3.64)	104-0377-78
C₅H₆ClN₃OS			
4-Pyrimidinamine, 6-chloro-2-(methyl-sulfinyl)-	MeOH	242(4.14),280(3.57)	24-1006-78
C₅H₆N₂			
Pyridazine, 4-methyl-	EtOH	248(3.30),300(c.2.3)	94-3633-78
C₅H₆N₂O			
4(1H)-Pyrimidinone, 2-methyl-	hexane	220(3.93),273(3.66)	94-1403-78
	MeOH	221(3.89),273(3.65)	94-1403-78
	BuOH	222(3.87),274(3.64)	94-1403-78
4(1H)-Pyrimidinone, 5-methyl-	H₂O	260(3.74)	56-1085-78

Compound	Solvent	$\lambda_{max}(\log \epsilon)$	Ref.
$C_5H_6N_2O_2$			
2H-Imidazol-2-one, 1-acetyl-1,3-di-hydro-	n.s.g.	206(3.53),260(3.71)	73-1511-78
2,4(1H,3H)-Pyrimidinedione, 1-methyl-	pH 2 and 7	266(4.14)	73-3444-78
	pH 12	264(3.99)	73-3444-78
$C_5H_6N_2O_2S_2$			
1,2,3-Thiadiazole-4-carboxylic acid, 5-(ethylthio)-	MeOH	233(3.74),301(3.93)	4-1295-78
1,2,3-Thiadiazole-4-carboxylic acid, 5-mercapto-, ethyl ester, potassium salt	MeOH	333(3.90)	4-1295-78
$C_5H_6N_4O$			
Formamide, N-[2-(1H-1,2,4-triazol-3-yl)ethylidene]-	pH 7.0	266(4.30)	12-2505-78
[1,2,3]Triazolo[1,5-c]pyrimidin-7-ol, 6,7-dihydro-	wet MeOH	249s(4.04),271(4.17)	4-1041-78
$C_5H_6N_4O_3S$			
2H-[1,2,3,5]Thiatriazolo[5,4-c]pyrimidine, 5,7(1H,6H)-dione, 1-oxide	EtOH	260(3.85),400(4.04)	44-1677-78
$C_5H_6N_4O_4$			
2(1H)-Pyrimidinone, 4-methoxy-6-(nitro-amino)-	pH -1.0	224(3.94),258(3.79), 329(4.30)	24-1006-78
	pH 6.0	233(4.02),253s(3.78), 334(4.18)	24-1006-78
	pH 14.0	223(4.16),270(3.89), 325s(3.54)	24-1006-78
C_5H_6O			
Acetaldehyde, cyclopropylidene-	C_6H_{12}	220(3.96),328(1.38)	88-3287-78
$C_5H_6O_2$			
2-Propynoic acid, ethyl ester	heptane	208(3.01)	65-1244-78
	MeCN	206(2.99)	65-1244-78
$C_5H_6O_2S_4$			
3H-1,2-Dithiole-3-thione, 4-(ethylsul-fonyl)-	EtOH	232(4.01),263(3.65), 276(3.61),323(3.53), 415(3.88)	39-1017-78C
$C_5H_6O_3$			
3(2H)-Furanone, 4-methoxy-	ether	277(3.86)	33-0430-78
$C_5H_7ClN_2O_2S$			
2,4-Thiazolidinedione, 5-(2-chloro-ethyl)-, 2-oxime	n.s.g.	208(3.89),232(3.87)	139-0303-78A
$C_5H_7FN_4$			
2,4-Pyrimidinediamine, 5-fluoro-N^4-methyl-, monohydrochloride	pH 1	206(4.31),235s(4.10), 267(3.88)	44-4200-78
$C_5H_7NO_4$			
1,3-Dioxepin, 4,7-dihydro-5-nitro-	MeOH	245(3.84)	128-0259-78
$C_5H_7N_3$			
1H-Imidazo[1,2-a]imidazole, 2,3-di-hydro-	EtOH	215(3.88)	2-0226-78

Compound	Solvent	λ_{max}(log ϵ)	Ref.
$C_5H_7N_3O$			
Ethanone, 1-(2-amino-1H-imidazol-4-yl)-	EtOH	300(4.15)	35-4208-78
	EtOH-HCl	268(4.00)	35-4208-78
	EtOH-NaOH	244(3.52),317(4.08)	35-4208-78
Ethanone, 1-(3-amino-1H-pyrazol-4-yl)-	EtOH-HCl	232s(3.68),281(3.81)	35-4208-78
	EtOH-NaOH	260(4.08),292s(3.79)	35-4208-78
$C_5H_7N_3OS$			
2(1H)-Pyrimidinethione, 6-amino-4-meth-oxy-	pH 1.0	230(4.15),279(4.33)	24-1006-78
	pH 6.0	238(3.99),261(4.09), 301(4.07)	24-1006-78
	pH 11.0	225(4.20),268(4.15)	24-1006-78
4(1H)-Pyrimidinethione, 3-amino-6-meth-yl-2-oxo-3,4-dihydro-	H_2O	329.5(4.14)	94-2765-78
$C_5H_7N_3O_2$			
2,4(1H,3H)-Pyrimidinedione, 3-amino-6-methyl-	EtOH	260(3.88)	94-2765-78
2(1H)-Pyrimidinone, 4-amino-6-methoxy-	pH 2.0	264(4.25)	24-1006-78
	pH 7.0	218s(3.89),271(4.12)	24-1006-78
	pH 13.0	265(3.99)	24-1006-78
$C_5H_7N_3O_2S$			
4-Thiazoleacetic acid, 3-amino-2,3-di-hydro-2-imino-, monohydrochloride	EtOH	262(3.50)	4-0401-78
$C_5H_7N_3S_2$			
2,4(1H,3H)-Pyrimidinedithione, 3-amino-6-methyl-	H_2O	290(4.36)	94-2765-78
$C_5H_7N_5O_2$			
1,3,5-Triazin-2(1H)-one, 4-amino-6-[(2-hydroxyethenyl)amino]-	pH 1	232(4.31)	87-0883-78
	pH 7 and 11	234(3.94)	87-0883-78
C_5H_8			
Cyclopropane, ethenyl-	C_6H_{12}	192(4.00)	39-1338-78B
$C_5H_8ClNO_4$			
1,3-Dioxepane, 5-chloro-6-nitro-	MeOH	216(3.29)	128-0259-78
$C_5H_8N_2$			
2,3-Diazabicyclo[2.2.1]hept-2-ene	hexane	341(2.62)	35-5122-78
	H_2O	334(--)	35-5122-78
	MeOH	334(--)	35-5122-78
	MeCN	340(--)	35-5122-78
3H-Pyrazole, 4-ethylidene-4,5-dihydro-	MeOH	320(2.72)	23-0992-78
$C_5H_8N_2O$			
1H-Pyrazolium, 3-hydroxy-1,1-dimethyl-, hydroxide, inner salt	MeOH	254(3.25)	24-0780-78
$C_5H_8N_2OS$			
4-Thiazolidinone, 3-methyl-2-(methyl-imino)-	neutral	218(4.15)	104-1229-78
	cation	218(4.31),235(--)	104-1229-78
4(5H)-Thiazolone, 2-amino-5,5-dimethyl-	neutral	222(4.33),250(3.90)	104-1229-78
	cation	207(4.28),239(3.79)	104-1229-78
	anion	232(4.29)	104-1229-78
4(5H)-Thiazolone, 2-(dimethylamino)-	neutral	241(4.30)	104-1229-78
	cation	219(4.28),240(--)	104-1229-78

Compound	Solvent	$\lambda_{max}(\log \epsilon)$	Ref.
$C_5H_8N_2O_2$			
3(2H)-Isoxazolone, 5-(1-aminoethyl)-, (S)-	MeOH	210(3.74)	1-0469-78
3(2H)-Isoxazolone, 5-(aminomethyl)-4-methyl-, hydrobromide	MeOH	211(3.87)	1-0469-78
$C_5H_8N_2O_2S$			
2,4-Thiazolidinedione, 5,5-dimethyl-2-oxime	n.s.g.	210(3.94),230(3.94)	139-0303-78A
2,4-Thiazolidinedione, 5-ethyl-2-oxime	n.s.g.	210(3.94),230(3.90)	139-0303-78A
$C_5H_8N_2S$			
1H-Imidazole, 1-methyl-2-(methylthio)-	EtOH	223(3.846),250(3.684)	28-0201-78B
	CH_2Cl_2	226(3.886),255s(3.688)	28-0201-78B
$C_5H_8N_4O$			
1H-Imidazole-4-carboxamide, 5-amino-1-methyl-	pH 1	240(3.88),269(4.00)	94-1929-78
	pH 7	267.5(4.08)	94-1929-78
	pH 13	267.5(4.08)	94-1929-78
	EtOH	268(4.09)	94-1929-78
$C_5H_8N_4O_2$			
1,2,4-Triazine-3,5(2H,4H)-dione, 2-methyl-6-(methylamino)-	pH 1	314(3.72)	124-1064-78
	EtOH-pH 7	318(3.71)	124-1064-78
	pH 13	302(3.70)	124-1064-78
1,2,4-Triazine-3,5(2H,4H)-dione, 4-methyl-6-(methylamino)-	pH 1	305(3.64)	124-1064-78
	EtOH-pH 7	309(3.64)	124-1064-78
	pH 13	324(3.77)	124-1064-78
1,2,4-Triazin-3(2H)-one, 5-methoxy-6-(methylamino)-	pH 1	305(3.70)	124-1064-78
	pH 7	310(3.71)	124-1064-78
	pH 13	306(3.68)	124-1064-78
$C_5H_8N_4O_3$			
1-Imidazolidinecarboxamide, 4(and 5)-methyl-3-nitroso-2-oxo-	EtOH	252(3.79),392(1.57), 407(2.74),424(1.71)	94-2635-78
C_5H_8OS			
3-Buten-2-one, 4-(methylthio)-	H_2O	301(4.11)	70-0133-78
$C_5H_8OS_2$			
1,3-Oxathiolane-2-thione, 4,5-dimethyl-	EtOH	230(3.58),284(3.96)	118-0286-78
$C_5H_8O_2S$			
1,3-Dioxolane-2-thione, 4,5-dimethyl-	EtOH	234(4.07)	118-0286-78
2-Propenal, 3-methoxy-2-(methylthio)-	EtOH	242(4.06),286(3.76)	73-1248-78
2-Propenoic acid, 3-(methylthio)-, methyl ester, (E)-	C_6H_{12}	370(3.16)	73-2395-78
(Z)-	C_6H_{12}	350(3.10)	73-2395-78
Sulfonium, dimethyl-, 1-formyl-2-oxo-ethylide	H_2O	251(4.35)	73-1248-78
	EtOH	251(4.37)	73-1248-78
$C_5H_8O_3$			
2-Propenoic acid, 3-methoxy-, methyl ester	C_6H_{12}	230(2.06)	73-2395-78
$C_5H_8O_3S_2$			
Acetic acid, [(ethoxythioxomethyl)-thio]-	EtOH	221(3.81),278(4.04)	44-2930-78

Compound	Solvent	$\lambda_{max}(\log \epsilon)$	Ref.
$C_5H_9ClN_2O$			
1H-Pyrazolium, 2,3-dihydro-1,1-dimeth- yl-3-oxo-, chloride	MeOH	231(3.04)	24-0780-78
$C_5H_9NO_2$			
2-Butene, 2-methyl-3-nitro-	hexane	251(3.54)	44-3116-78
C_5H_9NS			
Butane, 1-isothiocyanato-, iodine com- plex	CCl_4	460(3.15)	22-0439-78
Propane, 2-isothiocyanato-	hexane	<u>251(3.0)</u>	114-0217-78A
iodine complex	CCl_4	<u>460(3.22)</u>	22-0439-78
$C_5H_9N_3O$			
1H-Imidazole-4-ethanol, 2-amino-, hydrochloride	EtOH	216(3.91)	35-4208-78
3H-Pyrazol-3-one, 5-(1-aminoethyl)- 1,2-dihydro-	MeOH	221(3.69),240(3.49)	1-0056-78
3H-Pyrazol-3-one, 5-(2-aminoethyl)- 1,2-dihydro-	MeOH	226(3.58),240(3.52)	1-0056-78
dihydrochloride	MeOH	218(3.74),246(3.24)	1-0056-78
$C_5H_9N_3O_4$			
α-L-Arabinopyranosyl azide	MeOH	274(1.61)	78-1427-78
β-D-Ribofuranosyl azide	MeOH	275(1.62)	78-1427-78
$C_5H_9N_5$			
1H-Imidazole-4-carboximidamide,	pH 1	283.5(4.04)	94-1929-78
5-amino-1-methyl-	pH 7	285.5(4.07)	94-1929-78
	pH 13	267(3.96)	94-1929-78
$C_5H_9N_5O$			
4(1H)-Pyrimidinone, 2-amino-6-(1-meth- ylhydrazino)-	MeOH	226(4.38),274(4.24)	44-4844-78
$C_5H_9N_5O_2$			
1,2,4-Triazine-3,5(2H,4H)-dione,	pH 1	303(3.66)	103-1273-78
6-[(2-aminoethyl)amino]-	H_2O	297(3.64)	103-1273-78
	pH 13	297(3.58)	103-1273-78
$C_5H_{10}N_2O$			
2-Imidazolidinone, 1-ethyl-	EtOH	237.6(1.683)	28-0521-78A
	CH_2Cl_2	245.5(1.661)	28-0521-78A
2(1H)-Pyrimidinone, tetrahydro-	EtOH	244(1.757)	28-0521-78A
1-methyl-	CH_2Cl_2	254(1.725)	28-0521-78A
$C_5H_{10}N_2O_3$			
β-Alanine, N-ethyl-N-nitroso-	EtOH	233(3.83),353(1.91)	94-3909-78
Butanoic acid, 4-(methylnitrosoamino)-	EtOH	231(3.87),348(1.96)	94-3909-78
Methanol, (ethylnitrosoamino)-, acetate	n.s.g.	231(3.84),372(1.91)	94-3905-78
$C_5H_{10}N_2O_4$			
Glycine, N-[2-(hydroxymethylamino)-	acid	203(3.74)	73-2897-78
2-oxoethyl]-	anion	232(3.63)	73-2897-78
	dianion	233(3.79)	73-2897-78
$C_5H_{10}N_2S$			
1H-Imidazole, 4,5-dihydro-1-methyl-	EtOH	228(3.740)	28-0521-78A
2-(methylthio)-	CH_2Cl_2	230(3.720)	28-0521-78A
2-Imidazolidinethione, 1,3-dimethyl-	EtOH	240(4.331)	28-0521-78A
	CH_2Cl_2	246(4.339)	28-0521-78A

Compound	Solvent	λ_{max}(log ϵ)	Ref.
Pyrimidine, 1,4,5,6-tetrahydro-2-(methylthio)-	EtOH	225(3.748)	28-0521-78A
	CH$_2$Cl$_2$	none	28-0521-78A
2(1H)-Pyrimidinethione, tetrahydro-1-methyl-	EtOH	244.8(4.125)	28-0521-78A
	CH$_2$Cl$_2$	254(4.140)	28-0521-78A
C$_5$H$_{10}$OS			
Propane, 1-(ethenylsulfinyl)-	hexane	198(3.43),213s(3.26), 248(3.42)	70-0901-78
Propane, 2-(ethenylsulfinyl)-	hexane	196(3.53),213s(3.26), 251(3.46)	70-0901-78
	H$_2$O	208s(3.34),229(3.36)	70-0901-78
	EtOH	213s(3.23),236(3.28)	70-0901-78
C$_5$H$_{10}$OS$_3$			
1,3-Oxathiolane, 2,2-bis(methylthio)-	EtOH	229(2.74)	118-0286-78
C$_5$H$_{10}$S			
1-Propene, 1-(ethylthio)-, cis	n.s.g.	225(3.86),246(3.57)	73-2619-78
1-Propene, 1-(ethylthio)-, trans	n.s.g.	229(3.76),250(3.61)	73-2619-78
1-Propene, 3-(ethylthio)-	n.s.g.	214(3.49),235(3.03)	73-2619-78
C$_5$H$_{10}$S$_2$			
Pentane(dithioic) acid, sodium salt	EtOH	338(4.15),460s(--)	64-0976-78B
C$_5$H$_{11}$ClN$_2$O$_4$			
1H-Imidazole, 4,5-dihydro-1,2-dimethyl-, monoperchlorate	MeOH	226(3.73)	39-1453-78C
Pyrimidinium, 1,4,5,6-tetrahydro-2-methyl-, perchlorate	MeOH	208(3.72)(approx.)	39-1453-78C
C$_5$H$_{11}$NO$_2$S$_2$			
Carbamodithioic acid, bis(2-hydroxyethyl)-	MeOH	214(3.81),255(4.48), 295(4.48)	90-0211-78
C$_5$H$_{11}$N$_3$O			
Methanehydrazonamide, N'-formyl-N,N,N'-trimethyl-	heptane	218(3.81),258(3.85)	104-0576-78
C$_5$H$_{12}$N$_2$O$_2$			
1-Butanol, 4-(methylnitrosoamino)-	EtOH	231(3.90),347(1.96)	94-3891-78
1-Propanol, 3-(ethylnitrosoamino)-	EtOH	233(3.85),351(1.93)	94-3891-78
C$_5$H$_{12}$N$_2$S			
Thiourea, tetramethyl-	isooctane	219(3.72),240s(3.95) 261(4.20),330(2.32)	1-0259-78A
C$_5$H$_{12}$OS			
Propane, 1-(ethylsulfinyl)-	hexane	208(3.52),224s(3.11)	70-0901-78
	H$_2$O	196(3.04),212(2.95)	70-0901-78
	EtOH	197(3.30),213s(3.00)	70-0901-78
C$_5$H$_{12}$OSi			
Silane, acetyltrimethyl-	C$_6$H$_{12}$	358(2.00),372(2.10), 388(1.97)	112-0751-78

Compound	Solvent	$\lambda_{max}(\log \epsilon)$	Ref.
$C_6Br_4O_2$ 2,5-Cyclohexadiene-1,4-dione, 2,3,5,6-tetrabromo-	$CHCl_3$	314(4.24),398(2.45)	40-1127-78
$C_6Cl_2F_6N_2$ Pyridazine, 4,5-dichloro-3,6-bis(tri- fluoromethyl)-	CCl_4	304(2.53)	39-0378-78C
$C_6Cl_4O_2$ 2,5-Cyclohexadiene-1,4-dione, 2,3,5,6-tetrachloro-	$CHCl_3$	294(4.29),374(2.39)	40-1127-78
$C_6F_4O_2$ 2,5-Cyclohexadiene-1,4-dione, 2,3,5,6-tetrafluoro-	$CHCl_3$	258(4.33),336(2.28)	40-1127-78
C_6HBrN_4OS [1,2,5]Thiadiazolo[3,4-e]-2,1,3-benz- oxadiazole, 5-bromo-	EtOH	263(4.28),292(3.80), 302(3.78)	94-3896-78
$C_6HBrN_4O_2S$ [1,2,5]Thiadiazolo[3,4-e]-2,1,3-benz- oxadiazole, 5-bromo-, 3-oxide	EtOH	217(4.23),277(4.42), 337(4.07)	94-3896-78
$C_6H_2Br_2N_2O$ 2,5-Cyclohexadien-1-one, 2,6-dibromo- 4-diazo-	MeOH	278(3.91),369(4.64)	70-2455-78
$C_6H_2Br_2N_4OS$ [1,2,5]Thiadiazolo[3,4-e]-2,1,3-benz- oxadiazole, 4,5-dibromo-4,5-dihydro-	EtOH	286(4.15),291(4.14)	94-3896-78
$C_6H_2Cl_2O_2$ 2,5-Cyclohexadiene-1,4-dione, 2,6-di- chloro-	$CHCl_3$	276(4.25),342(1.71)	40-1127-78
$C_6H_2Cl_2O_3$ 2,5-Furandicarbonyl dichloride	hexane	380(3.36)	116-0568-78
$C_6H_2Cl_2O_4$ 2,5-Cyclohexadiene-1,4-dione, 2,5-di- chloro-3,6-dihydroxy-	neutral anion dianion	302(4.31),460(2.30) 310(4.21),530(2.94) 332(4.43),525(2.27)	150-2216-78 150-2216-78 150-2216-78
$C_6H_2N_4OS$ [1,2,5]Thiadiazolo[3,4-e]-2,1,3-benz- oxadiazole	EtOH	253(4.31),259(4.27), 290(3.86),298(3.87)	94-3896-78
$C_6H_2N_4O_2S$ [1,2,5]Thiadiazolo[3,4-e]-2,1,3-benz- oxadiazole, 3-oxide	EtOH	219(3.95),270(4.20), 274s(4.19),328(3.88), 336(3.88)	94-3896-78
$C_6H_2N_4S$ 1,2,5-Thiadiazole-3-carbonitrile, 4-(2-cyanoethenyl)-	EtOH	251(3.93),259(3.96), 294(3.75)	94-3896-78
$C_6H_3ClO_3$ 2H-Pyran-5-carbonyl chloride, 2-oxo-	C_6H_{12}	258(4.14),265(4.03), 290s(3.46)	44-4415-78

Compound	Solvent	$\lambda_{max}(\log \epsilon)$	Ref.
$C_6H_3Cl_2NO_3$ Phenol, 2,6-dichloro-4-nitro-	benzene	300(4.84)	70-1573-78
$C_6H_4BrClO_2S$ Benzenesulfonyl chloride, 4-bromo-	EtOH	237(4.18),269(3.06)	39-0295-78C
$C_6H_4BrNO_3$ Phenol, 2-bromo-4-nitro- Phenol, 2-bromo-6-nitro-	benzene benzene	300(3.93) 363(3.49)	70-1573-78 70-1573-78
C_6H_4ClNOS Thionitrous acid, S-(4-chlorophenyl) ester	hexane-1% CCl_4 at 0°	526(1.42),566(1.72)	39-0913-78C
$C_6H_4ClNO_3$ Phenol, 2-chloro-4-nitro- Phenol, 2-chloro-6-nitro- Phenol, 4-chloro-2-nitro- anion	benzene $C_2H_4Cl_2$ DMSO benzene $C_2H_4Cl_2$ DMSO benzene $C_2H_4Cl_2$ DMSO H_2O	300(3.93) 303(4.02) 322(--),433(--) 364(3.47) 360(3.51) 346(--),459(--) 367(3.53) 367(3.52) 351(--),466(--) 425(3.64)	70-1573-78 70-1573-78 70-2243-78 70-1573-78 70-1573-78 70-2243-78 70-1573-78 70-1573-78 70-2243-78 120-0111-78
$C_6H_4Cl_2N_2O_2$ 2-Pyrimidineacetic acid, 4,6-dichloro- 	neutral anion	254s(3.66),258(3.70), 264s(3.55) 254s(3.65),258(3.68), 264s(3.54)	12-0649-78 12-0649-78
$C_6H_4N_2$ 2-Pyridinecarbonitrile 3-Pyridinecarbonitrile 4-Pyridinecarbonitrile	hexane hexane hexane	216(4.01),263(3.43) 218(4.02),263(3.32) 212(3.69),271(3.18)	110-0044-78 110-0044-78 110-0044-78
$C_6H_4N_2O$ 2,5-Cyclohexadien-1-one, 4-diazo-	MeOH	257(3.66),354(4.74)	70-2455-78
$C_6H_4N_2OS$ Benzo[c]-1,2,5-thiadiazole, 1-oxide	EPA at 85^oK	285s(<u>3.5</u>),447(<u>3.5</u>)	1-0625-78
$C_6H_4N_2O_5$ Phenol, 2,4-dinitro-, anion	H_2O	360(4.17)	120-0111-78
$C_6H_4N_2S$ Isothiazolo[4,5-b]pyridine	EtOH	226(4.24),295(3.53), 306(3.46)	78-0989-78
$C_6H_4N_4OS$ [1,2,5]Thiadiazolo[3,4-e]-2,1,3-benz-oxadiazole, 4,5-dihydro-	EtOH	232(3.40),241s(3.32), 250s(3.04),290(4.19), 294(4.18)	94-3896-78
$C_6H_4N_4O_2$ 6,7-Pteridinedione, 5,8-dihydro- 	pH 4.0 pH 8.0 pH 12.0	220(3.98),249(3.71), 301(4.18) 227(4.03),268(3.71), 319(4.29) 220(4.47),240(4.13),	24-1763-78 24-1763-78 24-1763-78

Compound	Solvent	$\lambda_{max}(\log \epsilon)$	Ref.
6,7-Pteridinedione, 5,8-dihydro- (cont.)		324(4.30),338(4.25)	24-1763-78
$C_6H_4N_4O_2S$ [1,2,5]Thiadiazolo[3,4-e]-2,1,3-benz-oxadiazole, 4,5-dihydro-, 3-oxide	EtOH	215(3.83),262(3.91), 308s(3.09),314(4.11), 320(4.10)	94-3896-78
$C_6H_4N_4O_3$ 2,4(1H,3H)-Pteridinedione, 3-hydroxy-	pH 2	231(4.14),322(3.87)	44-0167-78
	pH 6.8	217(4.20),243(3.94), 327(3.90),338s(3.88)	44-0167-78
	pH 12	217(4.24),261(4.28), 356(3.92)	44-0167-78
4,6,7(1H)-Pteridinetrione, 5,8-dihydro-	pH 13.0	225(4.45),319(4.10), 330(4.11)	24-1763-78
$C_6H_4OS_4$ 1,3-Dithiole, 2-(1,3-dithiol-2-yli-dene)-, 1-oxide	EtOH	208(3.92),265s(3.38), 295(3.50),350s(3.77), 388(3.98)	44-4394-78
$C_6H_5BrN_2O_2$ 2-Pyrimidineacetic acid, 4-bromo-	neutral anion	252(3.61) 256(3.60)	12-0649-78 12-0649-78
C_6H_5Cl Benzene, chloro-	Me_3SiCl	<u>255f(3.2)</u>	65-0891-78
$C_6H_5ClN_2O_2$ 2-Pyrimidineacetic acid, 4-chloro-	neutral anion	252(3.57) 256(3.56)	12-0649-78 12-0649-78
$C_6H_5ClN_2O_3$ Acetamide, 2-chloro-N-(2,5-dihydro-2,5-dioxo-1H-pyrrol-1-yl)-	n.s.g.	215(4.15)	63-0211-78
Acetic acid, chloro-, (5-oxo-2(5H)-furanylidene)hydrazide	n.s.g.	310(4.30)	63-0211-78
$C_6H_5ClN_4$ 1,2,4-Triazolo[4,3-a]pyrazine, 8-chloro-3-methyl-	pH 1	203s(3.92),211s(4.15), 220(4.39),249(3.28), 256(3.33),266(3.34), 276s(3.36),303(3.63)	4-0987-78
	H_2O	219(4.40),249(3.29), 256(3.34),266(3.34), 274s(3.34),304(3.64)	4-0987-78
	pH 11	220(4.39),249(3.29), 256(3.34),265(3.36), 276s(3.37),304(3.68)	4-0987-78
C_6H_5ClS Benzenethiol, 4-chloro-	pH 2	245.0(3.02)	44-2248-78
with hexadecyltrimethylammonium brom-ide	pH 2	247.0(3.22)	44-2248-78
Benzenethiol, 4-chloro-	pH 8.5	270.0(4.18)	44-2248-78
with hexadecyltrimethylammonium brom-ide	pH 8.5	285.0(4.19)	44-2248-78
$C_6H_5Cl_2P$ Phosphonous dichloride, phenyl-	methacrylate	293(1.26)	47-0041-78

Compound	Solvent	$\lambda_{max}(\log \epsilon)$	Ref.
$C_6H_5FN_2O_2$			
2-Pyrimidineacetic acid, 4-fluoro-	anion	242(3.42)	12-0649-78
$C_6H_5F_3$			
Cyclopentadiene, (trifluoromethyl)-	hexane	242(3.67)	1-0293-78
$C_6H_5IN_2O_2$			
2-Pyrimidineacetic acid, 4-iodo-	neutral	262(3.78)	12-0649-78
	cation	300(3.76)(extrapolated)	12-0649-78
	anion	264(3.75)	12-0649-78
C_6H_5NO			
4H-Furo[3,2-b]pyrrole	EtOH	248(4.20)	23-1429-78
2-Pyridinecarboxaldehyde	MeCN	270(3.85)	90-0703-78
C_6H_5NOS			
Benzenamine, N-sulfinyl-	hexane-1%	261(--),380(--),	39-0913-78C
	CCl$_4$ at 0°	530(1.43),570(1.62)	
$C_6H_5NO_2$			
3-Pyridinecarboxylic acid	MeOH	216(3.93),257s(3.37),	102-2047-78
		263(3.48),269s(3.35)	
$C_6H_5NO_3$			
Phenol, 2-nitro-	benzene	357(3.56)	70-1573-78
	$C_2H_4Cl_2$	357(3.60)	70-1573-78
	DMSO	346(--),453(--)	70-2243-78
anion	H_2O	416(3.68)	120-0111-78
Phenol, 4-nitro-	benzene	300(4.03)	70-1573-78
	$C_2H_4Cl_2$	303(4.06)	70-1573-78
	DMSO	323(--),435(--)	70-2243-78
anion	H_2O	400(4.26)	120-0111-78
2H-Pyran-2,6(3H)-dione, 3-(amino-	MeOH	328(4.39)	44-4415-78
methylene)-, (Z)-	C_6H_{12}	324(--)	44-4415-78
	H_2O	332(--)	44-4415-78
(also other solvents)	MeCN	325(--)	44-4415-78
C_6H_5NS			
4H-Thieno[3,2-b]pyrrole	EtOH	217(3.85),258(4.10)	23-1429-78
6H-Thieno[2,3-b]pyrrole	EtOH	212(4.26),250(3.56)	23-1429-78
C_6H_5NSe			
4H-Selenolo[3,2-b]pyrrole	EtOH	219(3.81),266(4.09)	23-1429-78
6H-Selenolo[2,3-b]pyrrole	EtOH	214(4.30),251(3.61)	23-1429-78
$C_6H_5N_3O$			
2(1H)-Pyrimidinone, 4-amino-5-ethynyl-	pH 1	238(4.20),301(4.03)	39-1263-78C
	pH 7	235(4.25),288(3.89)	39-1263-78C
	pH 12	256(4.22),303(4.06)	39-1263-78C
4H-Pyrrolo[2,3-d]pyrimidin-4-one,	MeOH	260(4.01)	24-2925-78
1,7-dihydro-			
$C_6H_5N_3OS$			
4H-Pyrrolo[2,3-d]pyrimidin-4-one,	MeOH	241(3.93),300(4.31)	24-2925-78
1,2,3,7-tetrahydro-2-thioxo-			
$C_6H_5N_3OS_2$			
5H-1,3,4-Thiadiazolo[3,2-a]pyrimidin-	H_2O	252(4.30),287(3.96),	94-2765-78
5-one, 2,3-dihydro-7-methyl-2-thioxo-		316(4.02)	

Compound	Solvent	$\lambda_{max}(\log \epsilon)$	Ref.
$C_6H_5N_3O_4$			
Pyridinium dinitromethylide	MeCN	340(4.24)	70-2087-78
	H_2SO_4	345(--)	70-2087-78
$C_6H_5N_3S_2$			
5H-1,3,4-Thiadiazolo[3,2-a]pyrimidine- 5-thione, 7-methyl-	H_2O	232(3.92),297(3.79), 355(4.14)	94-2765-78
$C_6H_5N_5O$			
6(5H)-Pteridinone, 4-amino-	pH 0	225(4.25),280(4.31), 345s(3.20),360s(3.18)	24-1763-78
	pH 12.0	255(4.25),277s(3.89), 367(3.86),384s(3.76)	24-1763-78
$C_6H_5N_5O_2$			
6,7-Pteridinedione, 4-amino-1,5-di- hydro-	pH -1.0	221(4.26),302(4.10), 310s(4.09)	24-1763-78
	pH 4.0	217(4.39),235s(4.07), 290s(3.89),312(3.95)	24-1763-78
	pH 10.0	218(4.27),305s(3.97), 322(4.06),330(4.06)	24-1763-78
	pH 14.0	227(4.46),240s(4.28), 319(4.18),330(4.23), 344s(4.09)	24-1763-78
$C_6H_6B_2ClN_3O_3$			
Boronic acid, (5-chloro-1,2-dihydro- 2-hydroxypyrimido[5,4-e]-1,2-aza- borin-3-yl)-	pH 1	217(4.54),302(4.15)	44-0950-78
	H_2O	216(4.52),302(4.08)	44-0950-78
	pH 13	317(4.20)	44-0950-78
$C_6H_6Cl_4N_2$			
Pyridazine, 4,4,5,6-tetrachloro- 3,4-dihydro-3,3-dimethyl-	MeOH	294(4.49)	150-0582-78
$C_6H_6N_2$			
7,8-Diazatetracyclo[3.3.0.0^{2,4}.0^{3,6}]- oct-7-ene	hexane	373(2.49)	35-5122-78
	H_2O	362(--)	35-5122-78
	MeOH	367(--)	35-5122-78
	MeCN	369(--)	35-5122-78
Propanedinitrile, (1-methylethenyl)-	C_6H_{12}	232(4.05)	39-0995-78B
Pyrrolo[2,3-b]pyrrole, 1,6-dihydro-	EtOH	243(3.94)	23-1429-78
$C_6H_6N_2O_2$			
2-Pyrimidineacetic acid	neutral	246(3.43)	12-0649-78
	cation	250(3.63)	12-0649-78
	anion	244s(3.36),248(3.42), 256s(3.29)	12-0649-78
$C_6H_6N_4$			
[1,2,4]Triazolo[1,5-c]pyrimidine, 2-methyl-	pH 7.0	256(3.72)	12-2505-78
[1,2,4]Triazolo[1,5-c]pyrimidine, 5-methyl-	pH 7.0	257(3.92)	12-2505-78
[1,2,4]Triazolo[1,5-c]pyrimidine, 7-methyl-	pH 7.0	254(3.73)	12-2505-78
[1,2,4]Triazolo[1,5-c]pyrimidine, 8-methyl-	pH 7.0	256(3.83)	12-2505-78
1,2,4-Triazolo[4,3-c]pyrimidine, 3-methyl-	pH 7.0	243s(3.69),252(3.74), 261(3.62),283s(3.37)	12-2505-78
1,2,4-Triazolo[4,3-c]pyrimidine, 5-methyl-	pH 7.0	242s(3.70),250(3.78), 259(3.73),276s(3.56)	12-2505-78

Compound	Solvent	$\lambda_{max}(\log \epsilon)$	Ref.
1,2,4-Triazolo[4,3-c]pyrimidine, 7-methyl-	pH 7.0	244s(3.74),252(3.81), 259s(3.76)	12-2505-78
1,2,4-Triazolo[4,3-c]pyrimidine, 8-methyl-	pH 7.0	243s(3.76),250(3.83), 259(3.78),275s(3.59)	12-2505-78
$C_6H_6N_4O$			
4,5-Benzofurazandiamine	EtOH	238(4.32),290(3.42), 440(3.60)	94-3896-78
6(5H)-Pteridinone, 1,7-dihydro-	pH 2.4	292(4.01)	24-1763-78
	pH 7.4	293(3.93)	24-1763-78
	pH 13.0	305(4.07)	24-1763-78
1,2,4-Triazolo[4,3-a]pyrazine, 8-methoxy-	pH 1	213(4.26),239s(3.41), 248s(3.50),256s(3.55), 267s(3.62),278(3.70), 287s(3.53),299s(3.48)	4-0987-78
	H_2O	212(4.29),240s(3.39), 246s(3.46),255s(3.55), 266s(3.64),278(3.71), 286s(3.69),301s(3.39)	4-0987-78
	pH 11	213(4.26),239s(3.36), 249s(3.48),255s(3.54), 266s(3.63),278(3.70), 286s(3.68),302s(3.36)	4-0987-78
1,2,4-Triazolo[4,3-a]pyrazin-8(7H)-one, 3-methyl-	pH 1	217(4.15),258(3.65), 285(3.62)	4-0987-78
	H_2O	219(4.27),248(3.54), 255(3.54),265s(3.55), 284(3.65)	4-0987-78
	pH 11	220s(4.05),223(4.05), 268s(3.65),275(3.67), 292(3.63)	4-0987-78
1,2,4-Triazolo[4,3-a]pyrazin-8(7H)-one, 7-methyl-	pH 1	216(4.39),250s(3.76), 257s(3.81),266s(3.84), 282(3.90)	4-0987-78
	H_2O	216(4.08),249s(3.43), 257s(3.48),267s(3.57), 283(3.57)	4-0987-78
	pH 11	217(4.35),251s(3.73), 257s(3.77),267s(3.83), 283(3.86)	4-0987-78
[1,2,4]Triazolo[1,5-a]pyrimidin-7-ol, 5-methyl-	MeOH	212(--),242(3.58), 272(4.02)	106-0051-78
$C_6H_6N_4OS$			
5H-1,3,4-Thiadiazolo[3,2-a]pyrimidin-5-one, 2-amino-7-methyl-	MeOH	218(3.92),255(3.40), 306(3.76)	94-2765-78
$C_6H_6N_4OS_2$			
1,2,3-Thiadiazolo[4,5-d]pyrimidin-5(4H)-one, 6,7-dihydro-4,6-di-methyl-7-thioxo-	EtOH	250(4.11),375(4.03)	44-1677-78
$C_6H_6N_4O_2$			
Imidazo[1,2-a]-1,3,5-triazine-2,4(3H,8H)-dione, 8-methyl-	pH 1	238(4.02),252(3.99)	44-4774-78
4,6-Pteridinedione, 1,5,7,8-tetrahydro-	pH 6.3	274(3.90),318(3.69)	24-1763-78
	pH 12.0	273(3.83),300s(3.67)	24-1763-78
6(5H)-Pteridinone, 7,8-dihydro-7-hydroxy-	pH 1.7	287(4.09)	24-1763-78
	pH 5.2	289(4.00)	24-1763-78
	pH 13.0	224(4.29),256(3.97), 356(3.84)	24-1763-78

Compound	Solvent	$\lambda_{max}(\log \epsilon)$	Ref.
1H-Pyrazolo[3,4-d]pyrimidine-4,6(5H,7H)-dione, 3-methyl-	H_2O and 0.5M NaOH	207(4.24),244(3.89), 260(3.81)	56-0037-78
$C_6H_6N_4O_2S$			
1,2,3-Thiadiazolo[4,5-d]pyrimidine-5,7(4H,6H)-dione, 4,6-dimethyl-	EtOH	242(3.81),327(3.79)	44-1677-78
1,2,3-Thiadiazolo[4,5-d]pyrimidinium, 6,7-dihydro-5-hydroxy-3,6-dimethyl-7-oxo-, hydroxide, inner salt	EtOH	242(5.22),300(2.69), 405(3.42)	44-1677-78
$C_6H_6N_4O_4$			
1,4-Benzenediamine, 2,3-dinitro-	benzene	472(3.62)	39-1194-78C
	EtOH	491(--)	39-1194-78C
	$CHCl_3$	474(--)	39-1194-78C
1,4-Benzenediamine, 2,5-dinitro-	C_6H_{12}	518(3.66)	39-1194-78C
	EtOH	560(--)	39-1194-78C
	CH_2Cl_2	540(--)	39-1194-78C
1,4-Benzenediamine, 2,6-dinitro-	C_6H_{12}	498(3.52)	39-1194-78C
	EtOH	516(--)	39-1194-78C
	$CHCl_3$	512(--)	39-1194-78C
Hydrazine, (2,4-dinitrophenyl)-	toluene	343(4.05)	104-2229-78
	EtOH	260(3.89),350(4.09)	104-2229-78
$C_6H_6N_4S$			
2,1,3-Benzothiadiazole-4,5-diamine	EtOH	264(4.37),324(3.91), 450(3.45)	94-3896-78
1,2,4-Triazolo[4,3-a]pyrazine, 8-(methylthio)-	pH 1	207(4.42),257(3.96), 305s(4.09),314(4.13), 326s(4.00)	4-0987-78
	H_2O	208(4.44),250s(3.90), 257(3.97),262s(3.93), 305s(4.11),313(4.14), 325s(4.00)	4-0987-78
	pH 11	210(4.35),259(4.15), 270s(4.06),291s(4.04), 304s(4.12),312(4.15), 320s(4.06)	4-0987-78
1,2,4-Triazolo[4,3-a]pyrazine-8(7H)-thione, 3-methyl-	pH 1	210(4.25),271(3.86), 355(4.08)	4-0987-78
	H_2O	212(4.26),268(3.86), 309s(3.66),342s(4.08), 354(4.14)	4-0987-78
	pH 11	210(4.31),272(3.70), 337(4.02)	4-0987-78
$C_6H_6N_4Se$			
1,2,4-Triazolo[4,3-a]pyrazine-8(7H)-selenone, 3-methyl-	pH 1	217(4.27),286(3.66), 387(3.84)	4-0987-78
	H_2O	217(4.31),305(3.65), 385(3.80)	4-0987-78
	pH 11	216(4.29),220s(4.24), 231s(3.98),265s(3.79), 275(3.80),287s(3.76), 353(3.54)	4-0987-78
$C_6H_6N_6O$			
6(5H)-Pteridinone, 2,4-diamino-	pH 0.0	245(4.14),280(3.66), 355(3.89),365s(3.84)	24-1763-78
	pH 5.8	247(4.06),380(3.65)	24-1763-78
	pH 11.0	258(4.23),398(3.81)	24-1763-78

Compound	Solvent	$\lambda_{max}(\log \epsilon)$	Ref.
$C_6H_6N_6O_2$			
6,7-Pteridinedione, 2,4-diamino-1,5-dihydro-	pH -6.2	215(4.20),302(4.00), 327(4.11),335s(4.09)	24-1763-78
	pH 0.15	230(4.37),292s(4.02), 302(4.06),335(4.02)	24-1763-78
	pH 4.94	232(4.27),290(3.88), 336(3.96),346(3.96)	24-1763-78
	pH 9.12	232(4.27),290(3.88), 336(3.96),346(3.96)	24-1763-78
	pH 14.0	243(4.33),287(3.82), 293(3.81),345(4.15)	24-1763-78
C_6H_6O			
Phenol, sulfate, potassium salt	90% MeOH	221(2.79),252s(2.30), 259(2.40),264(2.42), 271(2.28)	23-2681-78
Phenol, TiCl$_4$ complex	acetone	410(2.38)	98-0973-78
$C_6H_6O_2$			
1,2-Benzenediol, disulfate, dipotassium salt	80% MeOH	215s(3.01),221(3.04), 266(2.73),271(2.66)	23-2681-78
1,2-Benzenediol, TiCl$_4$ complex	acetone	430(3.28)	98-0973-78
1,3-Benzenediol, disulfate, dipotassium salt	80% MeOH	220(3.05),265(2.60), 272(2.52)	23-2681-78
1,3-Benzenediol, TiCl$_4$ complex	acetone	410(2.20)	98-0973-78
1,4-Benzenediol	EtOH	225(3.71),294(3.45)	2-0731-78
	HClO$_4$	221(3.74),286(3.74)	2-0731-78
	base	227(3.64),284(3.17)	2-0731-78
disulfate, dipotassium salt	80% MeOH	224(3.11),268(2.71)	23-2681-78
1,4-Benzenediol, TiCl$_4$ complex	acetone	440(3.61)	98-0973-78
$C_6H_6O_2S$			
7-Thiabicyclo[2.2.1]heptane-2,5-dione	hexane	239(2.81),258(2.34), 289(2.46),299(2.60), 309(2.62),321(2.37)	18-1178-78
	H$_2$O	240(2.81),262(2.49), 300(2.82)	18-1178-78
	EtOH	239(2.86),262(2.48), 298(2.75),305(2.77)	18-1178-78
	MeCN	239(2.87),262(2.42), 296(2.67),305(2.71), 315(2.53)	18-1178-78
$C_6H_6O_3$			
1,3,5-Benzenetriol, trisulfate, tri-potassium salt	80% MeOH	215(3.10),266(2.26), 270(2.23)	23-2681-78
1,3,5-Benzenetriol, TiCl$_4$ complex	acetone	410(3.04)	98-0973-78
$C_6H_6O_4$			
1,2,3,5-Benzenetetrol	MeOH	239(2.74),266-290(2.75)	23-2681-78
2,5-disulfate, dipotassium salt	MeOH	236(2.94),269(2.73)	88-0847-78
4H-Pyran-4-one, 3-hydroxy-2-(hydroxy-methyl)-	MeOH	268(4.20)	94-0643-78
$C_6H_6O_7$			
L-threo-Hex-2-enaric acid, 1,4-lactone	pH 6.86	266.5(4.01)	136-0251-78A
$C_6H_6O_{10}S$			
L-threo-Hex-2-enaric acid, 1,4-lactone, 2-(hydrogen sulfate), tripotassium salt	pH 6.86	254(4.23)	136-0251-78A

Compound	Solvent	$\lambda_{max}(\log \epsilon)$	Ref.
C_6H_6S			
Benzenethiol	pH 2	237.8(2.43)	44-2248-78
	pH 8.5	262.5(4.10)	44-2248-78
with hexadecyltrimethylammonium	pH 2	238.0(2.40)	44-2248-78
bromide	pH 8.5	275.0(3.98)	44-2248-78
$C_6H_7BrO_4$			
2-Butenedioic acid, 2-(bromomethyl)-,	EtOH	215(4.12)	39-1490-78C
1-methyl ester, (Z)-			
C_6H_7FO			
2-Cyclohexen-1-one, 2-fluoro-	hexane	228(4.42),322(1.56)	44-4512-78
C_6H_7NO			
Pyridine, 2-methoxy-	hexane	215(3.77),272(3.58),	94-1403-78
		280f(3.43)	
	MeOH	215(3.91),271(3.62),	94-1403-78
		280s(3.43)	
Pyridine, 4-methoxy-	hexane	214(3.82),239(2.81)	94-1403-78
	MeOH	219(3.92),245s(2.95)	94-1403-78
3-Pyridinol, 2-methyl-	EtOH	281(3.71)	33-2542-78
	EtOH-NaOH	243(3.97),304(3.71)	33-2542-78
2(1H)-Pyridinone, 1-methyl-	hexane	235(3.75),240f(3.63),	94-1403-78
		308(3.64),318f(3.62),	
		333f(3.30)	
	MeOH	229(3.81),302(3.71)	94-1403-78
4(1H)-Pyridinone, 1-methyl-	MeOH	263.5(4.22)	94-1403-78
$C_6H_7NO_2$			
Butanenitrile, 2-acetyl-3-oxo- (enol)	EtOH	276(4.01)	18-0839-78
2(1H)-Pyridinone, 6-methoxy-	propylene	279(3.8),310(3.5)	35-7055-78
	carbonate		
	+ 2M NaClO₄	308(3.9)	35-7055-78
1H-Pyrrole-2,5-dione, 1-ethyl-	pH 2	219(4.16),300(2.78)	39-1277-78B
$C_6H_7NO_2S$			
3-Isothiazolecarboxylic acid, 4-methyl-,	EtOH	227(3.64),268(3.78)	44-0079-78
methyl ester			
$C_6H_7NO_3$			
4-Isoxazolecarboxaldehyde, 3-methoxy-	MeOH	236(3.81)	1-0027-78
5-methyl-			
$C_6H_7NO_5S$			
Furan, 2-[(methylsulfonyl)methyl]-	MeOH	208(4.01),236s(3.69),	73-0156-78
5-nitro-		308(4.15)	
C_6H_7NS			
Methanamine, N-(2-thienylmethylene)-	C_6H_{12}	255(3.98),273(3.80)	18-2718-78
$C_6H_7N_3$			
1H-Pyrazolo[1,2-a][1,2,3]triazol-4-ium,	EtOH	285(4.02)	88-1291-78
2-methyl-, hydroxide, inner salt			
$C_6H_7N_3O_2$			
1,4-Benzenediamine, 2-nitro-	C_6H_{12}	441(3.59)	39-1194-78C
	EtOH	473(--)	39-1194-78C
	CH_2Cl_2	465(--)	39-1194-78C
$C_6H_7N_3O_4$			
1H-Pyrrole, 2,5-dimethyl-1,3-dinitro-	H_2O	217s(4.07),234(4.24),	78-0505-78

Compound	Solvent	$\lambda_{max}(\log \epsilon)$	Ref.
1H-Pyrrole, 2,5-dimethyl-1,3-dinitro- (cont.)		268(4.03),325(3.82), 378s(3.71)	78-0505-78
	EtOH	220s(4.05),237(4.24), 275(4.05),332(3.80), 390(3.21)	78-0505-78
$C_6H_7N_5$			
1,2,4-Triazolo[4,3-a]pyrazin-8-amine, 3-methyl-	pH 1	228(4.30),257s(3.58), 265(3.62),274s(3.65), 297(3.72),322s(3.42)	4-0987-78
	H_2O	230(4.12),236s(4.04), 259s(3.58),268s(3.65), 277s(3.70),297(3.80)	4-0987-78
	pH 11	204(3.95),230(4.14), 259s(3.65),268s(3.70), 277s(3.74),297(3.84)	4-0987-78
[1,2,4]Triazolo[1,5-a]pyrimidin-7-amine, 5-methyl-	MeOH	219(--),260(3.80), 286(4.13)	106-0051-78
$C_6H_7N_5O$			
6(5H)-Pteridinone, 4-amino-7,8-dihydro-	pH 1.0	225(4.06),285(4.05)	24-1763-78
	pH 8.0	223(4.12),281(3.93)	24-1763-78
	pH 13.0	227(4.10),293(4.03)	24-1763-78
1,2,4-Triazolo[4,3-a]pyrazin-8(7H)-one, 3-methyl-, oxime	pH 1	200(3.83),229(4.21), 249s(3.84),295(3.80)	4-0987-78
	H_2O	234(4.13),273(3.92), 303s(3.70)	4-0987-78
	pH 11	201(3.87),224(4.18), 240s(4.13),280(3.92), 307s(3.76)	4-0987-78
[1,2,4]Triazolo[1,5-a]pyrimidine- 5-methanol, 7-amino-	MeOH	219(--),260(3.74), 288(4.10)	106-0051-78
[1,2,4]Triazolo[1,5-a]pyrimidin-6-ol, 7-amino-5-methyl-	MeOH	221(--),258(3.60), 305(4.09)	106-0051-78
	MeOH-NaOH	232(4.19),275(3.71), 338(3.92)	106-0051-78
$C_6H_7N_5O_2$			
1H-1,2,3-Triazolo[4,5-d]pyrimidine- 5,7(4H,6H)-dione, 1,4-dimethyl-	pH 6 pH 12	279(3.80) 250(3.85),282(3.86)	104-1617-78 104-1617-78
1H-1,2,3-Triazolo[4,5-d]pyrimidine- 5,7(4H,6H)-dione, 1,6-dimethyl-	pH 1 pH 12	280(3.77) 303(3.82)	104-1617-78 104-1617-78
1H-1,2,3-Triazolo[4,5-d]pyrimidine- 5,7(4H,6H)-dione, 4,6-dimethyl-	pH 1 pH 12	270(3.84) 269(3.87)	104-1617-78 104-1617-78
2H-1,2,3-Triazolo[4,5-d]pyrimidine- 5,7(4H,6H)-dione, 2,4-dimethyl-	pH 6 pH 12	278(3.91) 251(3.94),282(3.99)	104-1617-78 104-1617-78
2H-1,2,3-Triazolo[4,5-d]pyrimidine- 5,7(4H,6H)-dione, 2,6-dimethyl-	pH 1 pH 12	274(3.93) 300(3.82)	104-1617-78 104-1617-78
3H-1,2,3-Triazolo[4,5-d]pyrimidine- 5,7(4H,6H)-dione, 3,4-dimethyl-	pH 6	239(3.85),259(3.94)	104-1617-78
3H-1,2,3-Triazolo[4,5-d]pyrimidine- 5,7(4H,6H)-dione, 3,6-dimethyl-	pH 6	249(3.75),278(3.80)	104-1617-78
$C_6H_7N_5O_4S$			
[1,2,4]Triazolo[1,5-a]pyrimidin-6-ol, 7-amino-5-methyl-, hydrogen sulfate	MeOH	221(--),257(3.55), 297(4.01)	106-0051-78
C_6H_7P			
Phosphine, phenyl-	C_6H_{12}	234(3.54),260(2.80), 265(2.85),272(2.72)	65-1394-78

$C_6H_8-C_6H_8N_2O_2$

Compound	Solvent	$\lambda_{max}(\log \epsilon)$	Ref.
C_6H_8 1,3,5-Hexatriene	C_6H_{12}	248(4.76),258(4.90), 268(4.83)	39-1338-78B
C_6H_8BT1 Boratabenzene, 1-methyl-, thallium	MeCN	228(3.94),285s(3.54)	101-0265-78J
$C_6H_8B_2N_4O_3$ Boronic acid, (5-amino-1,2-dihydro- 2-hydroxypyrimido[5,4-e]-1,2-aza- borin-3-yl)-	pH 1 H_2O pH 13	225(4.24),283(4.10), 300(4.02) 203(4.29),233(4.23), 284(3.95),302(4.03) 228(4.29),288(4.03)	44-0950-78 44-0950-78 44-0950-78
$C_6H_8ClN_3$ 2-Pyrimidinamine, 4-chloro-N,N-dimeth- yl- 4-Pyrimidinamine, 2-chloro-N,N-dimeth- yl-	heptane heptane	210(3.79),247(4.35), 315(3.50) 207(3.85),245(4.20), 293(4.68)	104-0377-78 104-0377-78
$C_6H_8N_2$ 1,5-Diazatetracyclo[3.3.0.0^{2,8}.0^{4,6}]- octane	C_6H_{12}	209(2.18)	33-0795-78
$C_6H_8N_2O$ Ethanone, 1-(4,5-dihydro-4-methylene- 1H-pyrazol-3-yl)- Pyrimidine, 4-methoxy-2-methyl- 4-Pyrimidinol, 2,5-dimethyl- 2(1H)-Pyrimidinone, 4,6-dimethyl-, hydrobromide 2(1H)-Pyrimidinone, 5-ethyl- 4(1H)-Pyrimidinone, 1,2-dimethyl- 4(3H)-Pyrimidinone, 2,3-dimethyl- 	EtOH hexane MeOH BuOH hexane MeOH BuOH n.s.g. H_2O hexane-5% MeOH MeOH BuOH hexane-5% MeOH MeOH BuOH	268(3.45),350(3.81) 213(3.80),249(3.56) 215(3.83),251(3.59) 215(3.80),250(3.54) 225(3.84),272(3.70) 228(3.79),274(3.76) 226(3.81),275(3.74) 305(3.79) 259(3.70) 240(4.14) 240(4.15) 241(4.13) 221(3.74),285(3.60) 222(3.83),276(3.68) 223(3.86),278(3.65)	22-0401-78 94-1403-78 94-1403-78 94-1403-78 94-1403-78 94-1403-78 94-1403-78 44-3544-78 56-1085-78 94-1403-78 94-1403-78 94-1403-78 94-1403-78 94-1403-78 94-1403-78
$C_6H_8N_2OS$ 4(1H)-Pyridinethione, 1-amino-3-hy- droxy-2-methyl- 2(1H)-Pyrimidinethione, 4-methoxy- 5-methyl-	MeOH H_2O	254(3.98),275(3.36), 348(4.27) 269(3.65),304(3.70)	18-0179-78 56-1085-78
$C_6H_8N_2O_2$ 1H-Pyrazole-3-carboxylic acid, 4,5-di- hydro-4-methylene-, methyl ester 2,4(1H,3H)-Pyrimidinedione, 1,5-di- methyl- 4-Pyrimidinol, 5-methoxy-2-methyl- 2(1H)-Pyrimidinone, 4-methoxy-1-methyl- 1H-Pyrrolo[1,2-c]imidazole-1,3(2H)-di- one, tetrahydro-	EtOH pH 2 and 7 pH 12 MeOH BuOH pH 2,7,12 MeOH	250(3.26),327(2.78) 272(4.17) 270(4.04) 238(3.86),275(3.87) 238(3.82),278(3.84) 275(3.72) 210(3.57),240(2.65)	22-0401-78 73-3444-78 73-3444-78 94-1403-78 73-3444-78 19-0851-78

Compound	Solvent	$\lambda_{max}(\log \epsilon)$	Ref.
$C_6H_8N_2O_2S$			
Benzenesulfonamide, 4-amino- β-cyclodextrin complex	pH 7.0 pH 7.0	259(4.23) 260(4.22)	94-1162-78 94-1162-78
$C_6H_8N_2O_2S_2$			
Propanoic acid, 3-(1,2,3-thiadiazol- 5-ylthio)-, methyl ester	MeOH	237s(3.28),280(3.56), 299(3.62)	4-1295-78
Propanoic acid, 3-(1,3,4-thiadiazol- 2-ylthio)-, methyl ester	MeOH	266(3.70)	4-1295-78
$C_6H_8N_2O_4$			
2,4,6(1H,3H,5H)-Pyrimidinetrione, 5-ethyl-1-hydroxy-	pH 0.09 pH 7.7	196(4.06),269(4.15) 198(4.14),224(3.72)	12-2517-78 12-2517-78
$C_6H_8N_2O_5S$			
2H-1,2,6-Thiadiazine-4-carboxylic acid, 3,6-dihydro-3-oxo-, ethyl ester, 1,1-dioxide	EtOH	244(4.08),288(3.89)	4-0253-78
$C_6H_8N_4$			
Ethanone, 1-(3-pyridazinyl)-, hydrazone	MeOH	224(3.69),288(4.06)	18-0179-78
3-Pyridazinecarboxaldehyde, 6-methyl-, hydrazone	MeOH	230(3.74),291(4.28)	18-0179-78
$C_6H_8N_4O$			
Acetamide, N-[2-(1H-1,2,4-triazol-3- yl)ethenyl]-	pH 7.0	267(4.31)	12-2505-78
Formamide, N-[2-(5-methyl-1H-1,2,4- triazol-3-yl)ethenyl]-	pH 7.0	266(4.31)	12-2505-78
$C_6H_8N_4O_2$			
Hydrazinecarboxaldehyde, 2-(1,4-dihy- dro-6-methyl-4-oxo-2-pyrimidinyl)-	MeOH	209(4.14),225s(4.05), 282(3.82)	78-2927-78
$C_6H_8N_4O_3$			
Acetic acid, 2-(1,2,3,6-tetrahydro- 2,6-dioxo-4-pyrimidinyl)hydrazide	H_2O 0.5M HCl 0.5M NaOH	200(4.06),267(4.29) 266.5(4.29) 284(4.38)	56-0037-78 56-0037-78 56-0037-78
1,2,4-Triazine-3,5(2H,4H)-dione, 2-acetyl-6-(methylamino)-	EtOH-pH 7	277(3.73)	124-1064-78
$C_6H_8N_4O_3S$			
4-Pyrimidinamine, 6-methoxy-2-(methyl- thio)-N-nitro-	pH 1.0 pH 9.0 MeOH	253(4.31),330(4.04) 255(4.25),280s(4.08), 310s(3.89) 253(4.39),285s(3.81), 330s(3.40)	24-1006-78 24-1006-78 24-1006-78
2H-[1,2,3,5]thiatriazolo[5,4-c]pyrimi- dine-5,7(1H,6H)-dione, 1,6-dimeth- yl-, 3-oxide	EtOH	265(4.25),343(4.24)	44-1677-78
$C_6H_8N_4O_4$			
4-Pyrimidinamine, 2,6-dimethoxy-N-ni- tro-	pH 1.0 pH 6.0 MeOH	245s(3.76),322(4.08) 263(3.87),307(3.91) 223(4.02),265(3.80), 307(4.04)	24-1006-78 24-1006-78 24-1006-78
2(1H)-Pyrimidinone, 4-methoxy-1-methyl- 6-(nitroamino)-	pH 1.0 pH 6.0	229(3.95),264(3.92), 331(4.16) 235s(3.98),263(3.83), 327(4.05)	24-1006-78 24-1006-78

Compound	Solvent	$\lambda_{max}(\log \epsilon)$	Ref.
$C_6H_8N_4O_4S$			
4-Pyrimidinamine, 6-methoxy-2-(methyl-sulfinyl)-N-nitro-	MeOH	238(4.13),270s(3.98), 315(3.81)	24-1006-78
$C_6H_8N_4O_5S$			
4-Pyrimidinamine, 6-methoxy-2-(methyl-sulfonyl)-N-nitro-	MeOH	236(4.24),272(4.01), 305s(3.71)	24-1006-78
$C_6H_8N_6$			
[1,2,4]Triazolo[5,1-c][1,2,4]triazin-7-amine, 3,4-dimethyl-	H_2O	228(4.51),303(3.75), 333s(3.74)	39-0885-78C
$C_6H_8N_6O$			
6(5H)-Pteridinone, 2,4-diamino-7,8-di-hydro-	pH 0.0	245(4.14),280(3.66), 355(3.89),365s(3.84)	24-1763-78
	pH 9.0	222(4.32),282(4.12)	24-1763-78
$C_6H_8N_6OS$			
1,2,3-Thiadiazolo[4,5-d]pyrimidine-5,7(4H,6H)-dione, 4,6-dimethyl-, 7-hydrazone	EtOH	245s(3.37),325(3.23)	44-1677-78
C_6H_8OS			
7-Thiabicyclo[2.2.1]heptan-2-one	hexane	223(2.96),255(2.28), 260(2.27),289(2.23), 298(2.36),308(2.39), 319(2.17)	18-1178-78
	H_2O	258(2.32),300(2.59)	18-1178-78
	EtOH	223(3.03),260(2.37), 304(2.56)	18-1178-78
	MeCN	222(2.86),258(2.44), 296(2.46),304(2.51), 315(2.33)	18-1178-78
$C_6H_8O_2$			
Cyclopentanone, 2-(hydroxymethylene)-	hexane	275.2(3.90)	32-0581-78
	MeOH	270.0(4.04)	32-0581-78
1-Cyclopentenecarboxylic acid	H_2O	224.0(3.90)	23-2003-78
	95.3% H_2SO_4	262.5(4.02)	23-2003-78
2-Cyclopenten-1-one, 2-methoxy-	EtOH	249(3.99),305(1.95)	78-1509-78
$C_6H_8O_3$			
Furan, 3,4-dimethoxy-	EtOH	213(3.82)	33-0430-78
2,5-Furandimethanol	EtOH	225(2.92)	116-0568-78
2H-Pyran-2-one, 5,6-dihydro-4-hydroxy-6-methyl-	H_2O	250(4.59)	4-1153-78
	MeOH	247(4.55)	4-1153-78
	EtOH	243(4.39),278(4.07)	4-1153-78
	$CHCl_3$	244(4.12),284(3.81)	4-1153-78
	MeCN	242(4.20),285(3.56)	4-1153-78
$C_6H_8O_3S$			
7-Thiabicyclo[2.2.1]heptan-2-one, 7,7-dioxide	H_2O	293(1.83)	18-1178-78
	EtOH	288(1.73),297(1.78), 307(1.71),319(1.36)	18-1178-78
$C_6H_8O_6$			
L-Ascorbic acid	pH 2.0	243(1.0)	136-0127-78H
	pH 7.0	265(1.22)	136-0127-78H
2-phosphate	pH 2	238(0.95)	136-0127-78H
	pH 10	264(1.20)	136-0127-78H
2,2'-phosphate	pH 2	236(1.24)	136-0127-78H

Compound	Solvent	$\lambda_{max}(\log \epsilon)$	Ref.
L-Ascorbic acid, 2,2'-phosphate	pH 7	258(1.34)	136-0127-78H
(cont.)	pH 10	259(1.48)	136-0127-78H
L-Ascorbic acid, 2-sulfate	pH 2	232(1.04)	136-0127-78H
	pH 7.0	255(1.21)	136-0127-78H
	pH 10	263(0.90)	136-0127-78H
C_6H_8S			
1-Buten-3-yne, 1-(ethylthio)-, cis	H_2O	273(3.98)	70-0133-78
1-Buten-3-yne, 1-(ethylthio)-, trans	H_2O	272(4.08)	70-0133-78
$C_6H_9BrN_2O_2S$			
2,4-Thiazolidinedione, 5-(2-bromo-propyl)-, 2-oxime	n.s.g.	210(3.91),230(3.89)	139-0303-78A
$C_6H_9FN_4$			
2-Pyrimidinamine, 5-fluoro-1,4-dihydro-1-methyl-4-(methylimino)-, monohydrochloride	pH 1	240(4.06),276(3.92)	44-4200-78
picrate	MeOH	207(4.67),353(4.20)	44-4200-78
2,4-Pyrimidinediamine, 5-fluoro-N,N-dimethyl-, monohydrochloride	H_2O	215(4.31),280s(3.72)	44-4200-78
picrate	MeOH	209(4.60),300(3.86), 354(4.18)	44-4200-78
$C_6H_9NO_4S$			
Furo[2,3-d]oxazole-2(3H)-thione, tetrahydro-6-hydroxy-5-(hydroxymethyl)-, [3aR-(3aα,5β,6α,6aα)]-	MeOH	243(4.28),285s(3.00)	44-4200-78
$C_6H_9N_3$			
Imidazo[1,2-a]pyrimidine, 5,6,7,8-tetrahydro-	EtOH	215(3.90)	2-0226-78
$C_6H_9N_3O_2S$			
4-Pyrimidinamine, 6-methoxy-2-(methylsulfinyl)-	MeOH	213(4.47),240(3.90), 270s(3.43)	24-1006-78
$C_6H_9N_3O_3S$			
4-Pyrimidinamine, 6-methoxy-2-(methylsulfonyl)-	MeOH	215(4.52),243(3.80), 270(3.30)	24-1006-78
$C_6H_9N_3O_4S$			
2-Propenoic acid, 3-[(aminosulfonyl)-amino]-2-cyano-, ethyl ester	EtOH	275(4.27)	4-0253-78
C_6H_{10}			
Cyclopropane, 1-propenyl-	C_6H_{12}	194.5(4.11)	39-1338-78B
$C_6H_{10}BrN_2O_3P$			
1,3,2-Diazaphosphorin-4(1H)-one, 5-bromo-2-ethoxy-2,3-dihydro-6-methyl-, 2-oxide	EtOH	276(3.92)	103-0784-78
$C_6H_{10}ClNO$			
Cyclohexane, 1-chloro-1-nitroso-	MeOH	650(1.26)	39-0012-78B
$C_6H_{10}Cl_3NO$			
Acetamide, 2,2,2-trichloro-N,N-diethyl-	H_2O	<u>240(2.6)</u>	104-0221-78
$C_6H_{10}N_2$			
7-Azabicyclo[4.1.0]hept-3-en-7-amine	C_6H_{12}	216(2.91)	33-0795-78

Compound	Solvent	λ_{max}(log ϵ)	Ref.
2,3-Diazabicyclo[2.2.2]oct-2-ene	hexane	377(2.29)	35-5122-78
	H_2O	364(--)	35-5122-78
	MeOH	373(--)	35-5122-78
	MeCN	376(--)	35-5122-78
1,2-Diazaspiro[2.5]oct-1-ene	hexane	366(2.18)	35-5122-78
	H_2O	352(--)	35-5122-78
	MeOH	352(--)	35-5122-78
	MeCN	352(--)	35-5122-78
3H-Pyrazole, 4-ethylidene-4,5-dihydro-3-methyl-, (E)-	MeOH	324(2.60)	23-0992-78
$C_6H_{10}N_2O$			
1H-Pyrazole-3-ethanol, 5-methyl-	EtOH	221(3.45)	118-0900-78
$C_6H_{10}N_2O_2$			
3(2H)-Isoxazolone, 5-(1-aminoethyl)-4-methyl-, (±)-	MeOH	213(3.74)	1-0469-78
3(2H)-Isoxazolone, 5-(2-aminoethyl)-4-methyl-	MeOH	212(3.83)	1-0469-78
3(2H)-Isoxazolone, 5-(2-amino-1-methylethyl)-, (±)-	MeOH	210(3.87)	1-0469-78
3(2H)-Isoxazolone, 5-(2-aminopropyl)-, monohydrobromide, (±)-	MeOH	210(3.90)	1-0469-78
$C_6H_{10}N_2O_5$			
Glycine, N-(acetoxymethyl)-N-nitroso-, methyl ester	EtOH	230(3.80),369(1.96)	94-3914-78
$C_6H_{10}N_2S$			
1H-Imidazole, 1-ethyl-2-(methylthio)-	EtOH	224(3.834),250(3.647)	28-0201-78B
	CH_2Cl_2	226(3.810),255(3.633)	28-0201-78B
1H-Imidazole, 2-(ethylthio)-1-methyl-	EtOH	223(3.747),250(3.330)	28-0201-78B
	CH_2Cl_2	222(3.491),245s(3.334)	28-0201-78B
$C_6H_{10}N_4O$			
1H-Imidazole-4-carboxamide, 5-amino-1-ethyl-	pH 1	241(3.88),269(4.01)	94-1929-78
	pH 7	268(4.09)	94-1929-78
	pH 13	268(4.09)	94-1929-78
	EtOH	269(4.10)	94-1929-78
$C_6H_{10}N_4O_2$			
2,4(1H,3H)-Pyrimidinedione, 6-hydrazino-1,3-dimethyl-	H_2O	206(4.00),268(4.17)	56-0037-78
	0.5M HCl	265(3.97)	56-0037-78
1,2,4-Triazine-3,5(2H,4H)-dione, 6-(dimethylamino)-2-methyl-	pH 1	318(3.69)	124-1064-78
	EtOH-pH 7	328(3.77)	124-1064-78
	pH 13	302(3.70)	124-1064-78
1,2,4-Triazine-3,5(2H,4H)-dione, 6-(dimethylamino)-4-methyl-	pH 1	310(3.62)	124-1064-78
	EtOH-pH 7	318(3.70)	124-1064-78
	pH 13	324(3.79)	124-1064-78
1,2,4-Triazin-3(2H)-one, 6-(dimethylamino)-5-methoxy-	pH 1	314(3.72)	124-1064-78
	pH 7	316(3.71)	124-1064-78
	pH 13	314(3.60)	124-1064-78
$C_6H_{10}OS$			
3-Buten-2-one, 4-(ethylthio)-	H_2O	301(4.15)	70-0133-78
$C_6H_{10}OS_2$			
1,3-Dithiolo[4,5-b]furan, tetrahydro-3a-methyl-	MeOH	257(2.91)	98-1173-78

Compound	Solvent	$\lambda_{max}(\log \epsilon)$	Ref.
$C_6H_{10}O_2$ 2-Hexenal, 5-hydroxy-, [R-(E)]-	MeOH	223(4.05)	18-3035-78
$C_6H_{10}O_2S_3$ Thiodicarbonic acid, diethyl ester	EtOH	249(3.83),302(4.26)	44-2930-78
$C_6H_{10}O_2S_4$ Carbonodithioic acid, O,O'-1,2-ethane- diyl S,S'-dimethyl ester	EtOH	229(3.75),278(4.26)	118-0286-78
$C_6H_{10}O_2Si$ 2-Propynoic acid, trimethylsilyl ester	heptane MeCN	203(3.33),210(3.30) 203(3.32)	65-1244-78 65-1244-78
$C_6H_{10}O_3$ Ethanone, 1-(4,4-dimethyl-1,2-dioxetan- 3-yl)-	CCl₄ 90% MeCN- 10% CCl₄	290(1.73),350(1.23) 350(1.18)	35-6649-78 35-6649-78
$C_6H_{10}O_3S_2$ Thiodicarbonic acid, diethyl ester EtO.CO.S.CS.OEt	EtOH	273(3.97)	44-2930-78
$C_6H_{10}O_5$ β-L-threo-Pentopyranosid-4-ulose, methyl	H_2O pH 13	260(2.97) 317(3.24),341s(--)	23-1752-78 23-1752-78
$C_6H_{10}S$ Cyclobutanethione, 2,2-dimethyl-	C_6H_{12} MeCN	210(4.01),230(3.96), 495(1.03) 210(3.99),230(3.97), 481(1.04)	54-0121-78 54-0121-78
7-Thiabicyclo[2.2.1]heptane	hexane EtOH	213(3.61),249(1.56) 211(3.47),248(1.61)	18-1178-78 18-1178-78
$C_6H_{11}ClN_2$ 3H-Pyrazole, 3-chloro-4,5-dihydro- 3,5,5-trimethyl-	hexane H_2O MeOH MeCN	332(2.02) 322(--) 328(--) 330(--)	35-5122-78 35-5122-78 35-5122-78 35-5122-78
$C_6H_{11}ClN_2O_4$ 1H-1,4-Diazepinium, 2,3-dihydro- 5-methyl-, perchlorate	MeOH	330(4.13)	39-1453-78C
$C_6H_{11}NO$ 2H-Pyrrole, 3,4-dihydro-2,5-dimethyl-, 1-oxide	n.s.g.	231(3.86)	12-2317-78
$C_6H_{11}NO_2$ Methanaminium, N,N-dimethyl-, 1-formyl 2-oxoethylide	H_2O EtOH	257(4.46) 257(4.46)	73-1261-78 73-1261-78
$C_6H_{11}N_2OPS_2$ 1,3,2-Diazaphosphorine-4(1H)-thione, 2-ethoxy-2,3-dihydro-6-methyl-, 2-sulfide	EtOH	338(4.17)	103-0784-78
$C_6H_{11}N_3O_5$ β-D-Galactopyranosyl azide β-D-Glucopyranosyl azide	MeOH MeOH	275(1.56) 274(1.54)	78-1427-78 78-1427-78

Compound	Solvent	$\lambda_{max}(\log \epsilon)$	Ref.
α-D-Mannopyranosyl azide	MeOH	273(1.47)	78-1427-78
$C_6H_{11}N_5$			
1H-Imidazole-4-carboximidamide,	pH 1	284(4.05)	94-1929-78
5-amino-1-ethyl-	pH 7	286(4.08)	94-1929-78
	pH 13	267(3.96)	94-1929-78
$C_6H_{11}N_5O$			
1H-Imidazole-4-carboximidamide,	pH 1	279(3.97)	94-1929-78
5-amino-1-ethyl-N-hydroxy-	pH 7	257.5(3.97)	94-1929-78
	pH 13	260.5(4.00)	94-1929-78
	EtOH	258.5(4.03)	94-1929-78
$C_6H_{12}B_2F_8S_2$			
1,5-Dithiocane, dication bis(tetra- fluoroborate)	MeCN	212(3.58),233(3.98)	35-6416-78
$C_6H_{12}ClNO$			
Acetamide, 2-chloro-N,N-diethyl-	H_2O	237(2.4)	104-0221-78
Butane, 2-chloro-3,3-dimethyl-2-nitroso-	benzene	668(1.27)	39-0012-78B
	MeOH	666(1.15)	39-0012-78B
$C_6H_{12}N_2$			
2-Butenal, dimethylhydrazone	C_6H_{12}	275(4.40)	28-0047-78A
Diazene, (1,1-dimethyl-2-propenyl)- methyl-	hexane	363(1.23)	35-7009-78
1,2-Diazete, 3,4-dihydro-3,3,4,4-tetra-	hexane	347(2.12)	35-1876-78
methyl-	hexane	347(2.12)	35-5122-78
$C_6H_{12}N_2O$			
2-Imidazolidinone, 1-(1-methylethyl)-	EtOH	240(1.834)	28-0521-78A
	CH_2Cl_2	248(2.028)	28-0521-78A
2(1H)-Pyrimidinone, 1-ethyltetrahydro-	EtOH	none	28-0521-78A
	CH_2Cl_2	254(1.870)	28-0521-78A
$C_6H_{12}N_2O_2$			
Ethenamine, N,N-diethyl-2-nitro-	MeOH	356(4.50)	20-0693-78
2-Propanamine, 2-methyl-N-(2-nitro- ethenyl)-	MeOH	350(4.31)	20-0693-78
1-Propen-1-amine, N-(1-methylethyl)- 2-nitro-	EtOH	230(3.26),353(4.29)	44-0497-78
	EtOH	262(3.00),368(4.21)	44-0497-78
1-Propen-1-amine, 2-nitro-N-propyl-	EtOH	257(2.90),370(3.97)	44-0497-78
$C_6H_{12}N_2O_3$			
β-Alanine, N-nitroso-N-propyl-	EtOH	234(3.86),353(1.94)	94-3909-78
Butanoic acid, 4-(ethylnitrosoamino)-	EtOH	233(3.88),352(1.92)	94-3909-78
Glycine, N-butyl-N-nitroso-	EtOH	237(3.77),350(1.90)	94-3909-78
Methanol, [(1-methylethyl)nitroso- amino]-, acetate	n.s.g.	233(3.83),375(1.95)	94-3905-78
Methanol, (nitrosopropylamino)-, acetate	n.s.g.	232(3.82),372(1.94)	94-3905-78
$C_6H_{12}N_2S$			
1H-Imidazole, 1-ethyl-4,5-dihydro-	EtOH	228(3.763)	28-0521-78A
2-(methylthio)-	CH_2Cl_2	230(3.755)	28-0521-78A
2-Imidazolidinethione, 1-ethyl-	EtOH	240.5(4.304)	28-0521-78A
3-methyl-	CH_2Cl_2	246.5(4.320)	28-0521-78A
Pyrimidine, 1,4,5,6-tetrahydro- 1-methyl-2-(methylthio)-	EtOH	225(3.934)	28-0521-78A
2(1H)-Pyrimidinethione, tetrahydro-	EtOH	240(4.331)	28-0521-78A
1,3-dimethyl-	CH_2Cl_2	246(4.315)	28-0521-78A

Compound	Solvent	$\lambda_{max}(\log \epsilon)$	Ref.
$C_6H_{12}O$			
Butane, 1-(ethenyloxy)-	C_6H_{12}	194(3.99)	99-0194-78
$C_6H_{12}OS$			
Butane, 1-(ethenylsulfinyl)-	hexane	196(3.57),211s(3.28), 248(3.45)	70-0901-78
	H_2O	195(3.46),210s(3.36), 226(3.36)	70-0901-78
Propane, 1-(ethenylsulfinyl)-2-methyl-	hexane	196(3.58),213s(3.28), 248(3.46)	70-0901-78
	H_2O	208s(3.48),224(3.36)	70-0901-78
Propane, 2-(ethenylsulfinyl)-2-methyl-	hexane	195(3.54),217s(3.20), 253(3.51)	70-0901-78
	H_2O	232(3.40)	70-0901-78
	EtOH	228(3.45)	70-0901-78
$C_6H_{12}OS_3$			
1,3-Oxathiane, 2,2-bis(methylthio)-	EtOH	226(2.60)	118-0286-78
1,3-Oxathiolane, 5-methyl-2,2-bis(methylthio)-	EtOH	230(2.78)	118-0286-78
$C_6H_{12}S$			
1-Propene, 3-[(1-methylethyl)thio]-	n.s.g.	214(3.38),235(2.91)	73-2619-78
$C_6H_{12}S_2$			
1,5-Dithiocane	n.s.g.	202(3.48)	42-0146-78
Hexane(dithioic) acid, sodium salt	EtOH	338(4.15),460s(--)	64-0976-78B
$C_6H_{13}NO_3$			
Carbamic acid, (1,1-dimethylethyl)hydroxy-, methyl ester	n.s.g.	225(3.59),245s(3.47), 601(0.45),640s(0.40), 690s(0.20)	39-1066-78C
$C_6H_{13}N_2$			
1H-Imidazolium, 4,5-dihydro-1,2,3-trimethyl-, perchlorate	MeOH	235(3.88)	39-1453-78C
1H-Imidazolium, 4,5-dihydro-2,4,4-trimethyl-, perchlorate	MeOH	216(3.83)	39-1453-78C
$C_6H_{13}N_2S$			
1H-Imidazolium, 4,5-dihydro-1,3-dimethyl-2-(methylthio)-, iodide	EtOH	222-253(4.17-3.85)	28-0521-78A
	CH Cl	243.5(4.295)	28-0521-78A
$C_6H_{13}N_3O$			
Ethanehydrazonamide, N'-ethyl-N,N-dimethyl-2-oxo-	heptane	253(3.52),277(3.23)	104-0576-78
Methanehydrazonamide, N'-ethyl-N'-formyl-N,N-dimethyl-	heptane	218(3.76),261(3.78)	104-0576-78
$C_6H_{14}N_2O_2$			
1-Butanol, 4-(ethylnitrosoamino)-	EtOH	237(3.87),351(1.93)	94-3891-78
Ethanol, 2-(butylnitrosoamino)-	EtOH	235(3.85),351(1.95)	94-3891-78
1-Propanamine, N-nitroso-N-propyl-	H_2O	243(3.8)	104-0221-78
1-Propanol, 3-(nitrosopropylamino)-	EtOH	234(3.86),352(1.94)	94-3891-78
$C_6H_{14}OS$			
Propane, 1,1'-sulfinylbis-	hexane	208(3.56),226(3.11)	70-0901-78
	H_2O	212(2.95)	70-0901-78
	EtOH	201s(3.08),214(2.90)	70-0901-78
Propane, 2,2'-sulfinylbis-	hexane	209s(3.52),229(3.04)	70-0901-78
	H_2O	196(3.20),216(2.95)	70-0901-78

Compound	Solvent	$\lambda_{max}(\log \epsilon)$	Ref.
Propane, 2,2'-sulfinylbis- (cont.)	EtOH	197(3.46),218(2.95)	70-0901-78
$C_6H_{15}Cl_2NPt$ Platinum, dichloro(η^2-ethene)(2-methyl- 2-propanamine)-	EtOH	235(3.34),265(3.21), 303(2.98)	101-0357-78A
$C_6H_{15}N_2P$ Diazene, (1,1-dimethylethyl)(dimethyl- phosphino)-	CH_2Cl_2	425(1.83)	60-1909-78
$C_6H_{18}N_2O_2Si_2$ Hyponitrous acid, bis(trimethylsilyl) ester	hexane	259(3.04)	60-1909-78
$C_6H_{18}N_2Si_2$ Diazene, bis(trimethylsilyl)-	hexane	785(0.70)	60-1909-78
$C_6I_4O_2$ 2,5-Cyclohexadiene-1,4-dione, 2,3,5,6-tetraiodo-	$CHCl_3$	336(3.82)	40-1127-78

Compound	Solvent	λ_{max}(log ϵ)	Ref.
C$_7$H$_3$ClN$_2$O$_2$S			
2,1-Benzisothiazole, 3-chloro-5-nitro-	EtOH	213s(--),270(4.39), 297s(--),308(3.83), 332(3.85),351(3.65)	4-0529-78
2,1-Benzisothiazole, 3-chloro-7-nitro-	EtOH	207(4.30),304(3.71), 315s(--),359(3.78)	4-0529-78
C$_7$H$_3$Cl$_2$NOS			
2,1-Benzisothiazol-3(1H)-one, 5,7-di-chloro-	EtOH	202(4.29),224(4.35), 237s(--),253(3.98), 262s(--),372(3.66)	4-0529-78
C$_7$H$_4$ClNOS			
2,1-Benzisothiazol-3(1H)-one, 5-chloro-	EtOH	222(4.65),248s(--), 256(4.04),264(3.50)	4-0529-78
C$_7$H$_4$ClNS			
2,1-Benzisothiazole, 3-chloro-	EtOH	204(4.35),222(4.27), 291(3.91),298s(--), 303(4.01),328(3.77)	4-0529-78
C$_7$H$_4$ClN$_3$O			
Benzoyl chloride, 4-azido-	CHCl$_3$	298(4.36)	40-1690-78
C$_7$H$_4$FN			
Benzonitrile, 3-fluoro-	heptane	280.5(3.17)(smoothed)	73-2763-78
	MeOH	282(3.19)(smoothed)	73-2763-78
Benzonitrile, 4-fluoro-	heptane	274(2.36)(smoothed)	73-2763-78
C$_7$H$_4$F$_5$N			
3-Azabicyclo[4.2.0]octa-2,4-diene, 1,2,4,5,6-pentafluoro-	n.s.g.	235(3.43)	39-0363-78C
C$_7$H$_4$IN$_3$O$_6$			
Iodonium, (4-nitrophenyl)-, dinitro-methylide	DMSO	347(4.06)	70-2080-78
	MeCN	296(--),330(--)	70-2080-78
C$_7$H$_4$N$_2$O$_2$			
Benzonitrile, 3-nitro-	heptane	294.5(c.2.83)	73-2763-78
C$_7$H$_4$N$_2$O$_3$S			
2,1-Benzisothiazol-3(1H)-one, 5-nitro-	EtOH	204(4.17),222(4.19), 228(4.19),236s(--), 270(4.01),345(4.02)	4-0529-78
C$_7$H$_4$O$_3$			
4,6-Cycloheptadiene-1,2,3-trione	n.s.g.	335(3.54),574(1.2)	88-1299-78
C$_7$H$_4$O$_4$			
1,4-Cyclohexadiene-1-carboxylic acid, 3,6-dioxo-	pH 1	290(4.1),432(2.9), 470(2.5)	24-0832-78
C$_7$H$_5$BrClNO$_2$			
Benzene, 1-(bromochloromethyl)-4-nitro-	MeOH	269(4.08)	12-2477-78
C$_7$H$_5$ClN$_2$			
Benzene, 1-chloro-3-(diazomethyl)-	DMF	477(1.43)	142-0105-78C
Benzene, 1-chloro-4-(diazomethyl)-	DMF	484(1.48)	142-0105-78C
C$_7$H$_5$ClS$_2$			
Benzenecarbodithioic acid, 4-chloro-	EtOH	286(3.98),361(3.91),	64-0976-78B

Compound	Solvent	$\lambda_{max}(\log \epsilon)$	Ref.
sodium salt (cont.)		489(2.25)	64-0976-78B
C₇H₅Cl₃N₂O			
2-Propanone, 1,1,1-trichloro-3-(2-pyr-imidinyl)-	n.s.g.	277(4.2),320s(3.8), 380s(3.3)	103-1016-78
C₇H₅F₃N₂O			
2-Propanone, 1,1,1-trifluoro-3-(2-pyr-imidinyl)-	n.s.g.	260(4.1),365(3.4)	103-1016-78
C₇H₅IN₂O₄			
Iodonium, phenyl-, dinitromethylide	MeCN	340(3.95)	70-2080-78
	DMSO	340(--)	70-2080-78
C₇H₅N			
Benzonitrile	hexane	192(4.72),223(4.11), 270(2.81)	110-0044-78
	heptane	277(2.75)(smoothed)	73-2763-78
	MeOH	278(2.85)(smoothed)	73-2763-78
C₇H₅NO			
Benzonitrile, 3-hydroxy-	heptane	295(3.46)	73-2763-78
Benzonitrile, 4-hydroxy-	heptane	284(3.40)	73-2763-78
3H-Pyrrolizin-3-one	EtOH	292(5.0),416(3.7)	24-2407-78
C₇H₅NOS			
2,1-Benzisothiazol-3(1H)-one	EtOH	217(4.79),230s(--), 243s(--),353(3.67)	4-0529-78
C₇H₅NS₂			
2(3H)-Benzothiazolethione	dioxan	232(3.8),240(3.8), 322(4.4)	65-1352-78
C₇H₅N₃			
Pyrido[3,2-c]pyridazine	EtOH	208(4.62),263(3.59), 306(3.62),318(3.65)	88-2731-78
Pyrido[3,4-c]pyridazine	EtOH	209(4.46),284(3.58)	88-2731-78
C₇H₅N₃O₂			
Benzoic acid, 4-azido-	MeOH	217(3.86),272(4.31)	40-1690-78
Pyrido[2,3-b]pyrazine-2,3-dione, 1,2,3,4-tetrahydro-	pH -2.28	220(4.03),226(3.97), 323(4.32)	24-1763-78
	pH 5.0	220(3.87),227(3.81), 241(3.60),311(4.20), 323(4.13)	24-1763-78
	pH 10.2	227(4.02),256(3.61), 317(4.19),328(4.28), 343(4.09)	24-1763-78
	pH 14.2	224(4.37),264(3.62), 333(4.36),347(3.32)	24-1763-78
Pyrido[3,4-b]pyrazine-2,3-dione, 1,2,3,4-tetrahydro-	pH 1.0	219(4.43),302(4.06)	24-1763-78
	pH 6.11	235(3.99),251(4.00), 300(3.99)	24-1763-78
	pH 9.9	217(4.49),241s(4.08), 304(4.05),316(4.11), 329(3.88)	24-1763-78
	pH 14.2	227(4.58),247s(4.24), 308(4.05),319(4.19), 333(4.05)	24-1763-78

Compound	Solvent	λ_{max}(log ϵ)	Ref.
$C_7H_5N_3O_7$ 2,5-Cyclohexadien-1-one, 4-(methyl-aci-nitro)-2,6-dinitro-	MeOH	307(3.90),406(4.24)	104-2064-78
$C_7H_5N_5$ Imidazo[1,2-a]-1,2,4-triazolo[3,4-c]-pyrazine	pH 1	204s(4.16),214s(4.32), 220(4.38),226(4.37), 231s(4.31),241s(3.93), 262s(3.85),277(3.95), 290s(3.74)	4-0987-78
	H₂O	209s(4.39),225(4.50), 231(3.47),246(3.90), 268s(3.87),277(3.92), 290(3.76)	4-0987-78
	pH 11	212s(4.17),219s(4.37), 225(4.49),231(4.46), 246(3.89),269s(3.86), 279(3.91),290(3.76)	4-0987-78
$C_7H_5N_5O_2$ 1H-Tetrazole, 5-(4-nitrophenyl)-	pH 1.55 pH 5.48	278(4.1) 300(4.1)	104-1582-78 104-1582-78
C_7H_6AsN 1H-1,3-Benzazarsole	MeOH	235(4.48),260s(3.72), 298s(3.51),320(3.78)	101-0001-78K
	MeOH-KOH	235(4.37),256(3.78), 318(3.64)	101-0001-78K
$C_7H_6ClN_3O$ 1H-Benzotriazole, 5-chloro-1-methoxy-	EtOH	273(3.86)	4-1043-78
$C_7H_6Cl_2N_2O_2$ 2-Pyrimidineacetic acid, 4,6-dichloro-, methyl ester (in 5% MeOH)	neutral	254s(3.66),258(3.69), 264s(3.56)	12-0649-78
	cation	258(3.74),266(3.75), 274s(3.48)	12-0649-78
C_7H_6NP 1H-1,3-Benzazaphosphole	MeOH	226(4.45)	88-0441-78
$C_7H_6N_2$ Acetonitrile, 2-azabicyclo[2.2.0]hex-5-en-3-ylidene-	MeOH	274(4.02)	44-0944-78
Benzene, (diazomethyl)-	DMF	484(1.38)	142-0105-78C
$C_7H_6N_2O$ Indazol-3-ol	EtOH	220(4.35),310(3.57)	118-0633-78
Propanedinitrile, (3-hydroxy-2-buten-ylidene)-	CHCl₃	236(3.8),337(4.0)	83-0287-78
$C_7H_6N_2OS$ Thieno[2,3-d]pyrimidin-4(3H)-one, 3-methyl-	EtOH	214(4.29),253(3.60), 262(3.67),293(3.82)	1-0303-78
$C_7H_6N_2O_2$ 1H-Pyrrolo[3,2-c]pyridine-4,6(5H,7H)-dione	pH 1 pH 11 EtOH-pH 7	245(3.70),290(3.60) 267(4.04),315(3.96) 241(3.92),279(3.82)	87-0990-78 87-0990-78 87-0990-78

Compound	Solvent	$\lambda_{max}(\log \epsilon)$	Ref.
$C_7H_6N_2O_2S$			
Benzenesulfonamide, 3-cyano-	MeOH	282.5(2.68)(smoothed)	73-2763-78
Benzenesulfonamide, 4-cyano-	MeOH	284.5(3.08)(smoothed)	73-2763-78
1H-4,1,2-Benzothiadiazine 4,4-dioxide	EtOH	302(3.95)	39-0539-78C
$C_7H_6N_4O$			
4(1H)-Pyridinone, 1-(4H-1,2,4-triazol-4-yl)-	EtOH	210(3.45),265(3.04)	120-0001-78
$C_7H_6N_4OS$			
4(1H)-Pteridinone, 2,3-dihydro-3-methyl-2-thioxo-	pH 5.0	219(3.94),283(4.42), 348(3.96)	24-0971-78
	pH 10.0	228(4.04),248(3.99), 307(4.38),372(3.77)	24-0971-78
$C_7H_6N_4O_3$			
6,7-Pteridinedione, 1,5-dihydro-4-methoxy-	pH 4.0	220(3.94),272(3.89), 297s(4.03),306(4.04)	24-1763-78
	pH 10.0	217(4.42),233s(4.04), 298s(4.04),309(4.17), 322(4.19),338(3.95)	24-1763-78
	pH 13.0	218(4.51),233s(4.19), 312(4.21),324(4.30), 338(4.13)	24-1763-78
$C_7H_6N_4O_4$			
1H-Purine-2,6-dione, 3-acetoxy-3,7-dihydro-, monohydrochloride	pH 4.0	267(3.96)(changing)	44-0544-78
$C_7H_6O_2$			
1,3-Cyclopentadiene-1-carboxaldehyde, 5-(hydroxymethylene)-	MeOH	242(4.204),315(3.897), 381(3.806)	73-0938-78
$C_7H_6O_2S_2$			
3H-1,2-Benzodithiole 1,1-dioxide	EtOH	213(3.76),260(2.76), 266(2.86),273(2.79)	44-3374-78
3H-1,2-Benzodithiole 2,2-dioxide	EtOH	205(4.42),237(3.97), 274(2.76)	44-3374-78
5H-Cyclopenta-1,4-dithiin-5,7(6H)-dione, 2,3-dihydro-	MeOH	249(4.73),368(4.68)	24-3233-78
$C_7H_6O_3$			
Benzoic acid, 2-hydroxy-	H_2O	302(3.57)	36-0334-78
	anion	295(3.55)	36-0334-78
	dianion	309(3.61)	36-0334-78
Benzoic acid, 3-hydroxy-	H_2O	297(3.42)	36-0334-78
	anion	287(3.36)	36-0334-78
	dianion	312(3.48)	36-0334-78
Benzoic acid, 4-hydroxy-	H_2O	256(4.17)	36-0334-78
	anion	246(4.10)	36-0334-78
	dianion	281(4.24)	36-0334-78
TiCl$_4$ complex	acetone	410(2.00)	98-0973-78
2-Furancarboxaldehyde, 5-acetyl-	EtOH	290(4.30)	103-0940-78
$C_7H_6O_4$			
2H-Pyran-6-carboxylic acid, 2-oxo-, methyl ester	MeOH	245(4.03),286(3.65)	44-4415-78
$C_7H_6S_2$			
Benzenecarbodithioic acid, sodium salt	EtOH	283(3.95),360(3.90), 476(2.25)	64-0976-78B

Compound	Solvent	$\lambda_{max}(\log \epsilon)$	Ref.
C$_7$H$_7$BrN$_2$OS$_2$ Hydrazinecarbodithioic acid, [(5-bromo-2-furanyl)methylene]-, methyl ester	dioxan	353(4.46)	73-2643-78
C$_7$H$_7$BrN$_2$O$_2$ 2-Pyrimidineacetic acid, 4-bromo-, methyl ester	neutral	252(3.60)	12-0649-78
C$_7$H$_7$BrOS 4H-Thiopyran-4-one, 2-bromo-3,5-di-methyl-	EtOH	244(3.99),297(4.16), 304(4.15)	44-4966-78
C$_7$H$_7$BrO$_3$ 4H-Pyran-4-one, 3-bromo-2-methoxy-6-methyl-	EtOH	210(4.27),264(3.91)	23-1796-78
C$_7$H$_7$ClN$_2$O Benzenecarboximidamide, 4-chloro-N-hydroxy-	EtOH C$_6$H$_{12}$	227(4.15),268(3.84) 227(--),262(--)	73-2740-78 73-2740-78
C$_7$H$_7$ClN$_2$O$_2$ 2-Pyrimidineacetic acid, 4-chloro-, methyl ester	neutral	252(3.57)	12-0649-78
C$_7$H$_7$ClN$_2$O$_3$ 2-Pyrimidineacetic acid, 4-chloro-6-methoxy- (spectra in 5% MeOH)	neutral cation anion	254(3.60) 257(4.10) 254(3.58)	12-0649-78 12-0649-78 12-0649-78
C$_7$H$_7$ClN$_4$O Furo[2,3-d]pyrimidine-2,4-diamine, 5-(chloromethyl)-	acid pH 4.0 pH 7.0 base	255(4.02),297(3.73) 259(3.93),274(3.79) 215(4.26),278(3.82) 259(3.87),273s(3.79)	44-3937-78 44-3937-78 44-3937-78 44-3937-78
C$_7$H$_7$FN$_2$O$_2$ 2-Pyrimidineacetic acid, 4-fluoro-, methyl ester	neutral	242(3.44)	12-0649-78
C$_7$H$_7$FO 1,3-Cyclopentadiene, 5-(fluoromethoxy-methylene)-	hexane	279(4.23)	1-0293-78
C$_7$H$_7$IN$_2$OS$_2$ Hydrazinecarbodithioic acid, [(5-iodo-2-furanyl)methylene]-, methyl ester	dioxan	356(4.57)	73-2643-78
C$_7$H$_7$IN$_2$O$_2$ 2-Pyrimidineacetic acid, 4-iodo-, methyl ester (in 5% MeOH)	cation anion	300(3.77)(extrapolated) 262(3.79)	12-0649-78 12-0649-78
C$_7$H$_7$NOS Thionitrous acid, S-(4-methylphenyl) ester Thionitrous acid, S-(phenylmethyl) ester	hexane-1% CCl$_4$ at 0° hexane-1% CCl$_4$ at 0°	261(--),380(--), 533(1.42),574(1.71) 340(3.01),530s(1.23), 560(1.42)	39-0913-78C 39-0913-78C
C$_7$H$_7$NO$_2$ Benzamide, N-hydroxy- 2-Pyridinecarboxaldehyde, 3-hydroxy-6-methyl-	H$_2$O or EtOH MeOH	230(3.845) 215(3.96),241(3.52), 325(3.56)	73-1571-78 4-0029-78

Compound	Solvent	λ_{max} (log ϵ)	Ref.
$C_7H_7NO_2S$			
Benzene, 1-(methylthio)-4-nitro-	EtOH	220(3.80),338(4.15)	65-2027-78
$C_7H_7NO_3$			
3-Pyridinecarboxaldehyde, 1,2-dihydro- 4-hydroxy-6-methyl-2-oxo-	EtOH	232(3.79),272(3.08), 337(3.72)	23-0613-78
3-Pyridinecarboxylic acid, 2-hydroxy-, methyl ester	hexane	229(3.83),300(3.86)	94-1403-78
	MeOH	235(3.87),328(3.91)	94-1403-78
	ether	228(3.72),299(2.63), 332(3.44)	94-1415-78
	Pr_2O	228(3.71),299(3.67), 331(3.34)	94-1415-78
	isoPr$_2$O	299(3.64),332(3.36)	94-1415-78
	Bu$_2$O	228(3.68),299(3.67), 331(3.23)	94-1415-78
	dioxan	232(3.73),300s(3.44), 336(3.67)	94-1415-78
3-Pyridinecarboxylic acid, 4-hydroxy-, methyl ester	hexane	233(4.00),282(3.36)	94-1403-78
	MeOH	251(4.09),256(4.09), 281(3.67)	94-1403-78
	ether	232(3.95),278(3.37)	94-1415-78
	Pr_2O	233(3.92),278(3.31)	94-1415-78
	isoPr$_2$O	277.5(3.33)	94-1415-78
	Bu$_2$O	233(3.87),279(3.29)	94-1415-78
	dioxan	233(3.93),257s(3.19), 278(3.36)	94-1415-78
$C_7H_7NO_4$			
2,3,6(1H)-Pyridinetrione, 5-methoxy- 1-methyl-	n.s.g.	256(3.09)	88-1751-78
$C_7H_7NO_5$			
2,3,4(1H)-Pyridinetrione, 6-hydroxy- 5-methoxy-1-methyl-	MeOH	215(3.94),275(3.97), 385(3.28)	142-0191-78C
$C_7H_7N_2S$			
Benzenediazonium, 4-(methylthio)-, tetrafluoroborate	EtOH	374(4.16)	65-2027-78
$C_7H_7N_3O$			
Pyrido[2,3-b]pyrazin-2(1H)-one, 3,4-dihydro-	pH 0.0	275(3.67),325(4.03)	24-1763-78
	pH 7.0	270s(3.43),312(3.88)	24-1763-78
	pH 14.2	225(4.11),280s(3.70), 319(4.00)	24-1763-78
Pyrido[3,4-b]pyrazin-3(2H)-one, 1,4-dihydro-	pH 1.0	211(4.45),233(4.38), 241(4.31),296(3.79)	24-1763-78
	pH 10.0	225(4.47),286(3.75)	24-1763-78
	pH 14.0	232(4.54),299(3.87)	24-1763-78
6H-Pyrrolo[3,2-c]pyridin-6-one, 4-amino-1,5-dihydro-	pH 1	263(3.84),308(3.85)	4-1505-78
	pH 7.1	271(3.89),327(3.84)	4-1505-78
	pH 11	272(3.83),332(3.72)	4-1505-78
$C_7H_7N_3OS$			
4H-Pyrrolo[2,3-d]pyrimidin-4-one, 1,7-dihydro-2-(methylthio)-	MeOH	218(4.27),270(4.09)	24-2925-78
$C_7H_7N_3O_2$			
Pyrrolo[3,2-d]pyridazine-1,4-dione, 7-methyl-	DMSO	281(3.82),300(3.58)	39-0483-78C
1H-Pyrrolo[2,3-d]pyrimidine-2,4(3H,7H)- dione, 5-methyl-	DMSO	271(3.65)	39-0483-78C

Compound	Solvent	$\lambda_{max}(\log \epsilon)$	Ref.
1H-Pyrrolo[3,2-d]pyrimidine-2,4(3H,5H)-dione, 7-methyl-	MeOH	269(4.13)	39-0483-78C
$C_7H_7N_3O_3$ 1H-Pyrrolo[2,3-d]pyrimidine-2,4,5(3H)-trione, 6,7-dihydro-1-methyl-	H_2O	275(4.34)	103-0443-78
1H-Pyrrolo[2,3-d]pyrimidine-2,4,5(3H)-trione, 6,7-dihydro-3-methyl-	H_2O	268(4.23)	103-0443-78
$C_7H_7N_3O_6S$ Methanesulfonamide, N-(2,4-dinitrophenyl)-	dioxan	260(4.1),290s(4.0)	104-2361-78
$C_7H_7N_5O_2$ Pyrimido[4,5-e]-1,2,4-triazine-6,8(5H,7H)-dione, 5,7-dimethyl-	EtOH	240(4.28),312(3.85)	94-0367-78
Pyrimido[5,4-e]-1,2,4-triazine-5,7(6H,8H)-dione, 6,8-dimethyl-	EtOH	237(3.97),275s(2.99), 343(3.16)	44-0469-78
$C_7H_7N_5O_3$ Pyrimido[5,4-e]-1,2,4-triazine-5,7(6H,8H)-dione, 6,8-dimethyl-, 4-oxide	EtOH	240(4.10),304(3.21), 323s(2.78)	44-0469-78
C_7H_8AsCl Arsenin, 4-(2-chloroethyl)-	C_6H_{12}	220(4.00),272(3.97)	101-0247-78E
$C_7H_8BrN_5O$ 6H-Purin-6-one, 2-amino-8-bromo-1,7-dihydro-1,7-dimethyl-	pH 1	217(4.28),253s(3.91), 272(4.11)	44-2325-78
	pH 7	216(4.37),237s(3.94), 276(4.15)	44-2325-78
$C_7H_8ClN_3O_3$ 2,4(1H,3H)-Pyrimidinedione, 6-amino-5-(chloroacetyl)-1-methyl-	EtOH	243(3.96),278(4.10)	103-0443-78
2,4(1H,3H)-Pyrimidinedione, 6-amino-5-(chloroacetyl)-3-methyl-	EtOH	243(3.93),275(4.10)	103-0443-78
$C_7H_8ClN_3S$ Hydrazinecarbothioamide, N-(3-chlorophenyl)-	EtOH	222(4.04),245(4.28), 278s(3.37)	104-0106-78
$C_7H_8F_2N_6S$ 3H-1,2,4-Triazole-3-thione, 4-amino-5-[5-(difluoromethyl)-3-methyl-1H-pyrazol-1-yl]-2,4-dihydro-	EtOH	256(4.41)	103-0458-78
$C_7H_8N_2$ Benzaldehyde, hydrazone	n.s.g.	273(4.17)	104-0717-78
3,4-Diazatricyclo[4.2.1.0^{2,5}]nona-3,7-diene	hexane	354(2.29)	35-5122-78
	H_2O	342(--)	35-5122-78
	MeOH	345(--)	35-5122-78
	MeCN	347(--)	35-5122-78
3,5,6-Methenocyclopentapyrazole, 3,3a,4,5,6,6a-hexahydro-	hexane	342(2.81)	35-5122-78
	H_2O	331(--)	35-5122-78
	MeOH	338(--)	35-5122-78
	MeCN	338(--)	35-5122-78
$C_7H_8N_2O$ Benzenecarboximidamide, N-hydroxy-	EtOH	218(3.99),259(3.75)	73-2740-78

Compound	Solvent	λ_{max}(log ϵ)	Ref.
2-Propanone, 1-(3-pyridazinyl)-	C_6H_{12}	283(3.68),323(3.68), 335(3.61)	94-3633-78
	EtOH	256(3.57),320(3.42)	94-3633-78
3-Pyridinecarboxaldehyde, 2-(methyl- amino)-	EtOH	263(3.81),368(3.79)	104-1218-78
$C_7H_8N_2O_2$			
1H-Azepine-4-carbonitrile, 2,5,6,7- tetrahydro-3-hydroxy-2-oxo-	n.s.g.	258(3.80)	103-1013-78
8H-Isoxazolo[5,4-c]azepin-8-one, 4,5,6,7-tetrahydro-	n.s.g.	238(4.02)	103-1013-78
2-Pyrimidineacetic acid, methyl ester	neutral	240s(3.40),246(3.44), 274s(2.55)	12-0649-78
	cation	250(3.64)	12-0649-78
2-Pyrimidineacetic acid, 4-methyl-	neutral	246(3.57)	12-0649-78
	cation	250(3.73)	12-0649-78
	anion	248(3.55),275s(2.70)	12-0649-78
$C_7H_8N_2O_3$			
2H-Imidazol-2-one, 1,3-diacetyl- 1,3-dihydro-	n.s.g.	239(4.08)	73-1511-78
2-Pyrimidineacetic acid, 4-methoxy-	neutral	250(3.69)(extrapolated)	12-0649-78
	cation	240(3.82)	12-0649-78
	anion	252(3.59)	12-0649-78
$C_7H_8N_2O_3S$			
Acetic acid, (1-methyl-5-oxo-2-thioxo- 4-imidazolidinylidene)-, methyl ester	EtOH	348(4.32)	5-0227-78
$C_7H_8N_2O_4S$			
Thiophene, 2-(1-methyl-1-nitroethyl)- 5-nitro-	MeOH	324(3.91)	12-2463-78
$C_7H_8N_2O_4S_2$			
1,2,3-Thiadiazole-4-carboxylic acid, 5-[(3-methoxy-3-oxopropyl)thio]-	MeOH	231(3.75),301(3.90)	4-1295-78
$C_7H_8N_4$			
7H-Purine, 6,7-dimethyl-	pH 0	209(4.26),259(3.82)	142-0287-78A
	pH 7	202(4.30),264(3.88)	142-0287-78A
[1,2,4]Triazolo[1,5-c]pyrimidine, 2,5-dimethyl-	pH 7.0	256(3.77)	12-2505-78
[1,2,4]Triazolo[1,5-c]pyrimidine, 2,7-dimethyl-	pH 7.0	256(3.75)	12-2505-78
[1,2,4]Triazolo[1,5-c]pyrimidine, 2,8-dimethyl-	pH 7.0	257(3.78)	12-2505-78
[1,2,4]Triazolo[1,5-c]pyrimidine, 5,7-dimethyl-	pH 7.0	256(3.89)	12-2505-78
[1,2,4]Triazolo[1,5-c]pyrimidine, 5,8-dimethyl-	pH 7.0	256(3.92),269s(3.80)	12-2505-78
[1,2,4]Triazolo[1,5-c]pyrimidine, 7,8-dimethyl-	pH 7.0	259(3.86)	12-2505-78
[1,2,4]Triazolo[1,5-c]pyrimidine, 2-ethyl-	pH 7.0	256(3.71)	12-2505-78
1,2,4-Triazolo[4,3-c]pyrimidine, 3,5-dimethyl-	pH 7.0	247s(3.74),255(3.78), 264(3.70),277s(3.51)	12-2505-78
1,2,4-Triazolo[4,3-c]pyrimidine, 3,7-dimethyl-	pH 7.0	246s(3.81),254(3.88), 263s(3.79),278s(3.45)	12-2505-78
1,2,4-Triazolo[4,3-c]pyrimidine, 3,8-dimethyl-	pH 7.0	245s(3.76),253(3.82), 262(3.75),281(3.55)	12-2505-78

Compound	Solvent	$\lambda_{max}(\log \epsilon)$	Ref.
1,2,4-Triazolo[4,3-c]pyrimidine, 5,7-dimethyl-	pH 7.0	245s(3.79),253(3.87), 261s(3.82),277s(3.55)	12-2505-78
1,2,4-Triazolo[4,3-c]pyrimidine, 5,8-dimethyl-	pH 7.0	244s(3.78),253(3.88), 261(3.86),276s(3.73)	12-2505-78
1,2,4-Triazolo[4,3-c]pyrimidine, 7,8-dimethyl-	pH 7.0	245s(3.80),254(3.89), 261(3.85),278s(3.58)	12-2505-78
1,2,4-Triazolo[4,3-c]pyrimidine, 3-ethyl-	pH 7.0	244s(3.71),252(3.76), 262(3.67),275s(3.41)	12-2505-78
$C_7H_8N_4O$			
Furo[2,3-d]pyrimidine-2,4-diamine, 5-methyl-	acid	257(4.13),302(3.87)	44-3937-78
	pH 7.0	215(4.05),258(4.08), 280s(3.88)	44-3937-78
	base	257(4.08),275s(3.91)	44-3937-78
4H-Pyrrolo[2,3-d]pyrimidin-4-one, 2-amino-1,7-dihydro-5-methyl-	acid	227(4.17),265(3.93)	44-3937-78
	pH 7.0	223(4.27),262(3.96), 276s(3.85)	44-3937-78
	base	262(3.90)	44-3937-78
4H-Pyrrolo[2,3-d]pyrimidin-4-one, 2-amino-1,7-dihydro-6-methyl-, hydrochloride	acid	222(4.10),264(4.05)	44-3937-78
	pH 7.0	217(4.29),260(4.08), 280s(3.91)	44-3937-78
	base	260(4.04)	44-3937-78
1,2,4-Triazolo[4,3-a]pyrazine, 8-methoxy-3-methyl-	pH 1	217(4.31),242s(3.56), 249s(3.61),258s(3.65), 269s(3.68),283(3.76), 291s(3.75),301s(3.59)	4-0987-78
	H_2O	217(4.41),247s(3.52), 261s(3.62),271s(3.69), 282(3.76),287s(3.74), 304s(3.47)	4-0987-78
	pH 11	217(4.38),247s(3.51), 258s(3.61),269s(3.68), 281(3.75),289s(3.73), 304s(3.46)	4-0987-78
1,2,4-Triazolo[4,3-a]pyrazin-8(7H)-one, 3,7-dimethyl-	pH 1	219(4.25),251s(3.70), 258s(3.75),267(3.76), 282s(3.75)	4-0987-78
	H_2O	221(4.24),250s(3.57), 258(3.61),267(3.62), 285(3.67)	4-0987-78
	pH 11	220(4.24),249s(3.57), 258(3.62),267(3.63), 286(3.68)	4-0987-78
$C_7H_8N_4O_2$			
6(5H)-Pteridinone, 1,7-dihydro-4-methoxy-	pH -1.0	295(4.06)	24-1763-78
	pH 5.1	213(4.51),279(3.93)	24-1763-78
	pH 14.0	220(4.38),284(3.95), 320s(3.80)	24-1763-78
1H-Pyrazolo[3,4-d]pyrimidine-4,6(5H,7H)-dione, 3,7-dimethyl-	H_2O and 0.5M NaOH	212(4.26),241(3.81), 262(4.10)	56-0037-78
2H-Pyrazolo[3,4-d]pyrimidine-4,6(5H,7H)-dione, 5,7-dimethyl-	EtOH	245(3.99),255(3.96)	4-0359-78
Theophylline	pH 6.0	270(4.02)	24-0982-78
	pH 11.0	274(4.09)	24-0982-78
$C_7H_8N_4O_2S$			
1,2,3-Thiadiazolo[4,5-d]pyrimidine-5,7(4H,6H)-dione, 6-propyl-	EtOH	245s(3.46),275(3.42), 325(3.44)	44-1677-78

Compound	Solvent	$\lambda_{max}(\log \epsilon)$	Ref.
$C_7H_8N_4O_4$			
1,4-Benzenediamine, N-methyl-2,3-di-nitro-	C_6H_{12}	499(3.80)	39-1194-78C
	EtOH	512(--)	39-1194-78C
	CHCl$_3$	507(--)	39-1194-78C
1,4-Benzenediamine, N-methyl-2,5-di-nitro-	C_6H_{12}	558(3.69)	39-1194-78C
	EtOH	592(--)	39-1194-78C
	CHCl$_3$	588(--)	39-1194-78C
1,4-Benzenediamine, N-methyl-2,6-di-nitro-	C_6H_{12}	519(3.83)	39-1194-78C
	EtOH	548(--)	39-1194-78C
	CHCl$_3$	540(--)	39-1194-78C
$C_7H_8N_4S$			
1,2,4-Triazolo[4,3-a]pyrazine, 3-methyl-8-(methylthio)-	pH 1	210(4.42),259(3.97), 305s(4.02),318(4.08), 332s(3.93)	4-0987-78
	H_2O	210(4.49),259(4.01), 291s(3.92),307s(4.06), 317(4.08),331s(3.87)	4-0987-78
	pH 11	212(4.42),258(4.01), 291s(3.92),306s(4.06), 316(4.09),331s(3.87)	4-0987-78
$C_7H_8N_4S_2$			
5H-Dimidazo[2,1-b:2',1'-d][1,3,5]thia-diazine-5-thione, 2,3,8,9-tetrahydro-	EtOH	245(4.30),285(4.45)	88-3823-78
1H-Purine, 2,6-bis(methylthio)-	EtOH	257(4.31),308(4.02)	104-1624-78
C_7H_8O			
Benzene, methoxy-	heptane	271(3.46)	64-0197-78B
Benzenemethanol	C_6H_{12}	260f(2.3)	28-0021-78A
Tricyclo[4.1.0.0^{2,4}]heptan-5-one, anti	EtOH	287(1.45)	44-3848-78
syn	EtOH	283(1.85)	44-3848-78
C_7H_8OS			
Benzene, (methylsulfinyl)-	hexane	212(3.53),249(3.43)	70-0901-78
Benzenethiol, 4-methoxy-	pH 2	239.0(2.88)	44-2248-78
	pH 8.5	262.5(4.15)	44-2248-78
with hexadecyltrimethylammonium brom-ide	pH 2	241(2.81)	44-2248-78
	pH 8.5	270(4.22)	44-2248-78
$C_7H_8O_2$			
1,3-Cyclobutanedione, 2-(1-methyleth-ylidene)-	EtOH	255(4.04)	12-1757-78
2,4-Cycloheptadien-1-one, 6-hydroxy-	MeOH	240(3.38),291(3.66)	18-2131-78
2,6-Cycloheptadien-1-one, 4-hydroxy-	MeOH	233(4.06),240(3.97), 303(3.11),312(3.08)	18-2131-78
Cyclopentadiene-1-carboxylic acid, methyl ester	EtOH	275(3.77)	88-1071-78
2,4-Heptadienal, 6-oxo-	MeOH	274(4.44),352(2.54), 341(2.52)	18-2131-78
8-Oxabicyclo[5.1.0]oct-4-en-3-one	MeOH	228(3.89),275(2.63)	18-2131-78
2-Oxetanone, 4-methylene-3-(1-methyl-ethylidene)-	EtOH	223(3.93),270(3.52)	12-1757-78
$C_7H_8O_2S$			
2H-Pyran-2-thione, 4-hydroxy-3,5-di-methyl-	EtOH	216(4.01),241(4.19), 272s(3.51),324(4.03)	44-4966-78
2H-Thiopyran-2-one, 4-hydroxy-3,5-di-methyl-	isooctane	222(3.95),227(3.13), 324(3.95)	44-4966-78
	EtOH	231(4.34),243s(3.99),	44-4966-78

Compound	Solvent	λ_{max}(log ϵ)	Ref.
2H-Thiopyran-2-one, 4-hydroxy-3,5-di- methyl- (cont.)		273(3.52),291(3.59), 321(3.70)	44-4966-78
$C_7H_8O_3$			
2-Cyclopenten-1-one, 4-acetoxy-	$CHCl_3$	207(4.13)	88-2553-78
4H-Pyran-4-one, 2-methoxy-6-methyl-	EtOH	240(4.10)	23-1796-78
C_7H_8S			
Benzene, (methylthio)-	heptane	255(3.91)	64-0197-78B
Benzenethiol, 4-methyl-	heptane	254(4.02),275(3.20)	65-2027-78
	pH 2	238.0(2.71)	44-2248-78
	pH 8.5	265.0(4.10)	44-2248-78
with hexadecyltrimethylammonium brom-	pH 2	240.0(2.72)	44-2248-78
ide	pH 8.5	276.0(4.10)	44-2248-78
$C_7H_9BF_4N_2O$			
Pyridinium, 3-(aminocarbonyl)-1-meth- yl-, tetrafluoroborate	H_2O	266(2.62)	73-3056-78
C_7H_9BrO			
2,5-Heptadien-4-one, 2-bromo-	EtOH	238(3.90),268(4.05)	70-0932-78
$C_7H_9FN_2O$			
Pyridinium, 3-(aminocarbonyl)-1-meth- yl-, fluoride	H_2O	265(2.63)	73-3056-78
C_7H_9N			
Benzenemethanamine	C_6H_{12}	<u>260f(2.3)</u>	28-0021-78A
	H_2O	<u>258f(2.3)</u>	28-0021-78A
	+ HCl	<u>256f(2.4)</u>	28-0021-78A
	MeOH	<u>260f(2.3)</u>	28-0021-78A
C_7H_9NO			
2H-Pyrrol-2-one, 3-ethenyl-1,5-dihydro- 4-methyl-	MeOH	223(4.18)	24-0486-78
$C_7H_9NO_2$			
1,3-Cyclohexadiene-1-carboxylic acid, 5-amino- (dl-gabaculine)	H_2O	275(3.93)	44-1448-78
1-Propanone, 1-(2-furanyl)-, oxime	n.s.g.	204(4.08),245s(4.08), 267(4.19)	104-2420-78
2-Pyridinemethanol, 3-hydroxy-6-methyl-	MeOH	223(3.85),288(3.73)	4-0029-78
$C_7H_9NO_2S$			
1H-Pyrrole, 2-(2-propenylsulfonyl)-	EtOH	237(4.00)	23-0221-78
3-Thiophenecarboxylic acid, 2-amino-, ethyl ester	MeOH	221(4.44),262(3.57), 295(3.77)	24-0770-78
$C_7H_9NO_3$			
1H-Pyrrole-2-carboxylic acid, 3-hy- droxy-, ethyl ester	EtOH	263(4.23)	94-2224-78
$C_7H_9NO_3S$			
2-Thiophenemethanol, α,α-dimethyl- 5-nitro-	MeOH	327(3.98)	12-2463-78
$C_7H_9NO_5S$			
Furan, 2-[(ethylsulfonyl)methyl]- 5-nitro-	MeOH	213(3.85),234s(3.55), 312(4.07)	73-0156-78

Compound	Solvent	λ_{max}(log ϵ)	Ref.
C₇H₉NS₂			
Pyridine, 2,3-bis(methylthio)-	EtOH	225(4.28),233(4.30), 261(4.15),309(3.98)	56-2039-78
C₇H₉N₂O			
Pyridinium, 3-(aminocarbonyl)-1-meth- yl-, fluoride	H₂O	265(2.63)	73-3056-78
tetrafluoroborate	H₂O	266(2.62)	73-3056-78
C₇H₉N₃O			
Imidazo[1,2-a]pyrimidin-5(1H)-one, 2,3-dihydro-7-methyl-	EtOH	228(4.06),279(3.77)	12-0179-78
Pyrazolo[3,4-c]azepin-8(1H)-one, 4,5,6,7-tetrahydro-	n.s.g.	228(4.09)	103-1009-78
Pyrido[2,3-d]pyrimidin-4(1H)-one, 5,6,7,8-tetrahydro-	MeOH	229(4.10),269(3.84), 278(3.91)	24-2297-78
C₇H₉N₃OS			
4H-Thiazolo[4,5-c]azepin-4-one, 2-amino-5,6,7,8-tetrahydro-	EtOH	284(3.61)	103-0306-78
C₇H₉N₃O₂			
1,4-Benzenediamine, N¹-methyl-2-nitro-	C₆H₁₂	472(3.71)	39-1194-78C
	EtOH	498(--)	39-1194-78C
	CHCl₃	500(--)	39-1194-78C
1,4-Benzenediamine, N⁴-methyl-2-nitro-	C₆H₁₂	459(3.58)	39-1194-78C
	EtOH	498(--)	39-1194-78C
	CHCl₃	485(--)	39-1194-78C
C₇H₉N₃O₃			
2(1H)-Pyridinone, 1-methyl-4-(methyl- amino)-3-nitro-	EtOH	242(4.1),340(3.9)	142-0739-78A
1(2H)-Pyrimidinepropanoic acid, 4-amino-2-oxo-	pH 2	278(4.11)	73-3444-78
2(1H)-Pyrimidinone, 3,4-dihydro-1,3-di- methyl-4-(nitromethylene)-	EtOH	257(3.6),285(3.5), 384(4.5)	142-0739-78A
C₇H₉N₃O₃S			
4-Thiazoleacetic acid, 2-amino-α-(hy- droxyimino)-, ethyl ester, (E)-	EtOH	227(4.11),260s(3.69), 304(3.46)	78-2233-78
(Z)-	EtOH	225(4.20),285s(3.71)	78-2233-78
C₇H₉N₃O₄			
2,4(1H,3H)-Pyrimidinedione, 6-amino- 5-(hydroxyacetyl)-3-methyl-	EtOH	244(3.95),277(4.19)	103-0443-78
C₇H₉N₅			
6H-Purin-6-imine, 3,9-dihydro-3,9-di- methyl-, monohydrochloride	pH 1	270(4.20)	88-5007-78
	pH 7	270(4.19)	88-5007-78
perchlorate	pH 1	270(4.19)	88-5007-78
	pH 7	270(4.19)	88-5007-78
Pyrazolo[5,1-c][1,2,4]triazin-7-amine, 3,4-dimethyl-	EtOH	254(4.52),337(3.46)	39-0885-78C
C₇H₉N₅O			
2H-Purin-2-one, 6-amino-3,9-dihydro- 3,9-dimethyl-	pH 7	279(4.07)	88-1447-78
6H-Purin-6-one, 2-amino-3,7-dihydro- 3,7-dimethyl-	pH 7	270(3.76)	88-1447-78
6H-Purin-6-one, 2-amino-3,9-dihydro- 3,9-dimethyl-	pH 1	248(4.10)	88-2907-78
	pH 7	265(4.04)	88-1447-78

Compound	Solvent	$\lambda_{max}(\log \epsilon)$	Ref.
6H-Purin-6-one, 2-amino-3,9-dihydro-3,9-dimethyl- (cont.)	pH 7	216(4.48),247(3.93), 266(4.06)	88-2907-78
	pH 13	247(3.94),266(4.04)	88-2907-78
$C_7H_9N_5O_2$			
1H-Purine-6,8-dione, 2-amino-7,9-dihydro-1,7-dimethyl-	pH 1	250(4.04),291(4.00)	44-2325-78
	pH 7	250(4.03),293(4.00)	44-2325-78
	pH 13	225(4.57),260s(3.86), 294(4.06)	44-2325-78
1H-1,2,3-Triazolo[4,5-d]pyrimidine-5,7(4H,6H)-dione, 1,4,6-trimethyl-	pH 6	380(3.74)	104-1617-78
2H-1,2,3-Triazolo[4,5-d]pyrimidine-5,7(4H,6H)-dione, 2,4,6-trimethyl-	pH 6	276(3.87)	104-1617-78
$C_7H_9N_5O_3$			
Carbamic azide, [(3-methoxy-4-methyl-5-isoxazolyl)methyl]-	MeOH	212(4.09)	1-0469-78
$C_7H_9N_5S_2$			
1,2-Hydrazinedicarbothioamide, N-2-pyridinyl-	EtOH	290(4.30)	104-0181-78
C_7H_{10}			
1,3-Cycloheptadiene	heptane	246.5(4.03)	70-0091-78
$C_7H_{10}N_2$			
3,4-Diazatricyclo[4.2.1.0^{2,5}]non-3-ene	hexane	353(2.29)	35-5122-78
	H_2O	340(--)	35-5122-78
	MeOH	345(--)	35-5122-78
	MeCN	345(--)	35-5122-78
1H-Pyrazole, 1-ethenyl-3,5-dimethyl-	MeOH	254(4.06)	95-0165-78
$C_7H_{10}N_2O$			
Pyrimidine, 4-methoxy-2,5-dimethyl-	hexane	215(3.87),255(3.63)	94-1403-78
	MeOH	219(3.86),256(3.67)	94-1403-78
	BuOH	218(3.88),255(3.65)	94-1403-78
4(1H)-Pyrimidinone, 1,2,5-trimethyl-	hexane-5% MeOH	246(4.07)	94-1403-78
	MeOH	246(4.14)	94-1403-78
	BuOH	246.5(4.09)	94-1403-78
4(1H)-Pyrimidinone, 2,3,5-trimethyl-	hexane-5% MeOH	226(3.68),283(3.72)	94-1403-78
	MeOH	226(3.76),275(3.77)	94-1403-78
	BuOH	228(3.72),277(3.75)	94-1403-78
2H-Pyrrol-2-one, 5-amino-4-ethylidene-3,4-dihydro-3-methyl-	MeOH	218(3.95),265(4.10)	44-0283-78
$C_7H_{10}N_2OS$			
4(1H)-Pyridinethione, 1-amino-3-methoxy-2-methyl-	MeOH	252(3.89),285(3.21), 349(4.33)	18-0179-78
2(1H)-Pyrimidinone, 4-(propylthio)-	pH 1	300(4.05)	56-1255-78
	pH 2	303(3.97)	56-1255-78
	pH 3-10	300(4.05)	56-1255-78
	pH 11-12	300(4.01)	56-1255-78
	H_2O	300(4.05)	56-1255-78
$C_7H_{10}N_2OS_2$			
3(2H)-Pyridazinone, 4-(ethylthio)-5-mercapto-2-methyl-	MeOH	216(4.52),250(--), 296(3.91)	73-2415-78

Compound	Solvent	$\lambda_{max}(\log \epsilon)$	Ref.
$C_7H_{10}N_2O_2$			
2,4-Hexadienedial, 1-(O-methyloxime) 6-oxime	EtOH	295s(4.55),307(4.69), 320(4.62)	94-2575-78
8-Oxa-1-aza-5-azoniaspiro[4.5]dec-3-ene, 2-oxo-, hydroxide, inner salt	MeOH	256(3.23)	24-0780-78
3H-Pyrazol-3-ol, 2,4-dihydro-3-methyl-4-methylene-, acetate	EtOH	285(3.73)	22-0401-78
7H-Pyrazolo[1,2-d][1,4,5]oxadiazepin-7-one, 1,2,4,5-tetrahydro-	MeOH	254(3.89)	24-0780-78
Pyrimidine, 4,5-dimethoxy-2-methyl-	hexane	223(3.87),265(3.63)	94-1403-78
	MeOH	227(3.92),269(3.72)	94-1403-78
	BuOH	227(3.89),269(3.72)	94-1403-78
2(1H)-Pyrimidinone, 4-methoxy-1,5-di-methyl-	pH 2,7,12	280(3.72)	73-3444-78
4(1H)-Pyrimidinone, 5-methoxy-1,2-di-methyl-	hexane-5% MeOH	263(4.04)	94-1403-78
	MeOH	264(4.10)	94-1403-78
	BuOH	263(4.05)	94-1403-78
4(3H)-Pyrimidinone, 5-methoxy-2,3-di-methyl-	hexane-5% MeOH	235(3.69),282(3.78)	94-1403-78
	MeOH	239(3.81),278(3.85)	94-1403-78
	BuOH	239(3.76),279(3.82)	94-1403-78
$C_7H_{10}N_2O_2S_2$			
Propanoic acid, 3-[(5-methyl-1,3,4-thiadiazol-2-yl)thio]-, methyl ester	MeOH	210(3.47),267(3.82)	4-1295-78
$C_7H_{10}N_2O_3$			
Acetic acid, (4-oxo-2-imidazolidinyli-dene)-, ethyl ester	pH 1	280(2.18)	5-0473-78
	pH 13	260(--),285(2.18)	5-0473-78
2,4-Hexadienal, 6-nitro-, O-methyloxime	M NaOH	244(3.79),354(4.71)	94-2422-78
	EtOH	273(4.37),368(4.11)	94-2422-78
	$CHCl_3$	276(4.53)	94-2422-78
2,4-Hexadienamide, N-hydroxy-6-(meth-oxyimino)-	EtOH	295(4.61)	94-2575-78
sodium salt	EtOH	296(4.60)	94-2575-78
Hydrazinecarboxylic acid, (4-oxo-2-but-enylidene)-, ethyl ester, (?,E)-	EtOH	297(4.40)	94-2205-78
5-Isoxazoleacetamide, 3-methoxy-4-meth-yl-	MeOH	217(3.82)	1-0469-78
1H-Pyrazolium, 2,3-dihydro-5-(methoxy-carbonyl)-1,1-dimethyl-3-oxo-, hydroxide, inner salt	MeOH	211(4.01),299(3.00)	24-0780-78
$C_7H_{10}N_2O_4$			
4-Isoxazoleacetic acid, α-amino-3-meth-oxy-5-methyl-, (±)-	MeOH	none above 210 nm	1-0027-78
$C_7H_{10}N_2S$			
1H-Imidazole, 1-methyl-2-(2-propenyl-thio)-	EtOH	226(3.788),253(3.662)	28-0201-78B
	CH_2Cl_2	225(3.878),247s(3.706)	28-0201-78B
$C_7H_{10}N_2S_3$			
Compound, m. 137-8°	pH 4	290(4.34)	12-0389-78
Pyridazine, 3,4,6-tris(methylthio)-	pH -0.5	282(4.42),326(4.14)	12-0389-78
	pH 6.0	268(4.38)	12-0389-78
3(2H)-Pyridazinethione, 2-methyl-4,5-bis(methylthio)-	pH -4.0	275(4.54),323(4.25)	12-0389-78
3(2H)-Pyridazinethione, 2-methyl-4,6-bis(methylthio)-	pH 4.0	256(4.09),284(4.33)	12-0389-78

Compound	Solvent	$\lambda_{max}(\log \epsilon)$	Ref.
$C_7H_{10}N_4O$			
Acetamide, N-[2-(5-methyl-1H-1,2,4-triazol-3-yl)ethenyl]-	pH 7.0	267(4.34)	12-2505-78
Formamide, N-[2-(5-ethyl-1H-1,2,3-triazol-3-yl)ethenyl]-	pH 7.0	266(4.30)	12-2505-78
Methanehydrazonic acid, N-4-pyrimidinyl-, ethyl ester	pH 7.0	265(3.72)	12-2505-78
Pyrazolo[3,4-c]azepin-8(1H)-one, 3-amino-4,5,6,7-tetrahydro-	n.s.g.	233(4.08),280s(3.01)	103-1013-78
$C_7H_{10}N_4O_3$			
1,2,4-Triazine-3,5(2H,4H)-dione, 2-acetyl-6-(dimethylamino)-	EtOH-pH 7	321(3.77)	124-1064-78
$C_7H_{10}N_4O_3S$			
6H-1,3,4-Thiadiazine-5-acetic acid, 2-amino-α-(hydroxyimino)-, ethyl ester, (E)-	EtOH	268(4.00?),312(3.87?), 337(3.79?)	4-0401-78
$C_7H_{10}N_4S_3$			
1-Imidazolidinecarbodithioic acid, 2-thioxo-, 4,5-dihydro-1H-imidazol-2-yl ester	EtOH	262(3.79),280(3.86)	88-3823-78
$C_7H_{10}N_6OS$			
1,2,3-Thiadiazolo[4,5-d]pyrimidine-5,7(4H,6H)-dione, 4,6-dimethyl-, 7-(methylhydrazone)	EtOH	243s(3.40),327(3.26)	44-1677-78
$C_7H_{10}N_6S$			
3H-1,2,4-Triazole-3-thione, 4-amino-5-(3,5-dimethyl-1H-pyrazol-1-yl)-2,4-dihydro-	EtOH	256(4.28)	103-0458-78
$C_7H_{10}O$			
Bicyclo[2.2.1]heptan-2-one	hexane	293(1.34)	18-1178-78
	EtOH	290(1.42)	18-1178-78
Ethanone, 1-(1-cyclopenten-1-yl)-	EtOH	239(4.00),305(1.66)	44-0700-78
$C_7H_{10}O_2$			
Cyclohexanone, 2-(hydroxymethylene)-	hexane	291.0(3.84)	32-0581-78
	C_6H_{12}	291.5(3.83)	32-0581-78
	MeOH	285.4(3.74)	32-0581-78
	EtOH	283.0(3.84)	32-0581-78
2-Cyclohexen-1-one, 2-methoxy-	C_6H_{12}	253(3.75),315(1.54)	78-1509-78
1,3-Cyclopentanedione, 2,4-dimethyl-	EtOH	271(4.18)	102-2015-78
2-Propynoic acid, 1,1-dimethylethyl ester	heptane	208(2.78)	65-1244-78
$C_7H_{10}O_2S$			
1,3-Benzodioxole-2-thione, hexahydro-	EtOH	243(3.76)	118-0286-78
$C_7H_{10}O_4$			
2-Cyclobuten-1-one, 3,4,4-trimethoxy-	MeOH	245(4.13),277(1.90)	33-1784-78
$C_7H_{10}S$			
1-Buten-3-yne, 1-(propylthio)-, cis	H_2O	275(4.05)	70-0133-78
1-Buten-3-yne, 1-(propylthio)-, trans	H_2O	273(4.16)	70-0133-78
$C_7H_{11}Br$			
2,4-Hepta-iene, 7-bromo-, (E,E)-	EtOH	321(3.26)	105-0101-78

Compound	Solvent	$\lambda_{max}(\log \epsilon)$	Ref.
$C_7H_{11}ClN_2O_2$			
3H-Pyrazol-3-one, 1-[2-(2-chloroeth-oxy)ethyl]-1,2-dihydro-	MeOH	227(3.78),255(2.65)	24-0780-78
$C_7H_{11}NO_2$			
2,4-Azetidinedione, 3,3-diethyl-	CH_2Cl_2	246(2.13),251s(2.05)	33-3050-78
1H-Pyrrole-2,4-dimethanol, 1-methyl-, Ehrlich reaction product	n.s.g.	565(3.76)	39-0896-78C
$C_7H_{11}NS$			
Cyclohexane, isothiocyanato-, iodine complex	CCl_4	460(3.15)	22-0439-78
$C_7H_{11}N_2O_2$			
8-Oxa-1-aza-5-azoniaspiro[4.5]dec-3-ene, 2-oxo-, chloride	MeOH	239(3.00)	24-0780-78
Pyrazolo[1,2-d][1,4,5]oxadiazepin-6-ium, 1,2,4,5-tetrahydro-7-hydroxy-, chloride	MeOH	240(3.74),250s(3.72)	24-0780-78
$C_7H_{11}N_3$			
1H-Imidazo[1,2-a][1,3]diazepine, 5,6,7,8-tetrahydro-	EtOH	218(3.90)	2-0226-78
$C_7H_{11}N_3O_2$			
2-Pyrrolidinecarbonitrile, 2-ethyl-1,4-dihydroxy-5-imino-	H_2O	238(4.03)	1-0118-78
	acid	none	1-0118-78
$C_7H_{11}N_3O_2S$			
1,3,4-Thiadiazine-5-acetic acid, 2-amino-, ethyl ester	EtOH	248(3.76?),345(4.00?)	4-0401-78
4-Thiazoleacetic acid, 3-amino-2,3-di-hydro-2-imino-, ethyl ester	EtOH	290(4.41?)	4-0401-78
$C_7H_{11}N_3O_3$			
2(1H)-Pyrimidinone, 4-amino-1-(2,3-di-hydroxypropyl)-, (±)-	pH 2	282(4.08)	73-2054-78
	pH 7 and 12	273(3.92)	73-2054-78
$C_7H_{11}N_5O_3$			
2,4-Pyrimidinediamine, 6-methoxy-N^2,N^2-dimethyl-N^4-nitro-	pH -2.0	224(4.31),250s(4.17), 316(3.79)	24-1006-78
	pH 3.0	220(4.23),251(4.19), 270s(3.95),345(4.15)	24-1006-78
	pH 9.0	245(4.27),275s(3.97), 340s(3.50)	24-1006-78
	MeOH	225s(4.16),248(4.34), 335(3.85)	24-1006-78
C_7H_{12}			
2-Hexene, 4-methylene-, cis	EtOH	225(3.82)	30-0597-78
2-Hexene, 4-methylene-, trans	EtOH	229(4.30)	30-0597-78
$C_7H_{12}N_2$			
2,3-Diazabicyclo[2.2.2]oct-2-ene, 1-methyl-	hexadecane	380(2.30)	35-1876-78
3H-Pyrazole, 4-ethylidene-4,5-dihydro-3,5-dimethyl-	MeOH	327(2.52)	23-0992-78
$C_7H_{12}N_2O$			
1H-Pyrazole-3-ethanol, α,5-dimethyl-	EtOH	221(3.48)	118-0900-78

Compound	Solvent	$\lambda_{max}(\log \epsilon)$	Ref.
3H-Pyrazol-3-one, 2,4-dihydro-5-methyl-2-propyl-	heptane	258(3.50),313(2.41)	104-0576-78
4H-Pyrazol-4-one, 3,5-dihydro-3,3,5,5-tetramethyl-	hexane	357(2.22)	35-5122-78
	H_2O	347(--)	35-5122-78
	MeOH	350(--)	35-5122-78
	MeCN	355(--)	35-5122-78
$C_7H_{12}N_2O_2$			
Acetic acid, diazo-, 2-methylbutyl ester, (S)-	EtOH	213(3.65),248(4.24)	44-4447-78
5-Isoxazoleethanamine, 3-methoxy-4-methyl-, monohydrochloride	MeOH	216(3.81)	1-0469-78
Pyrrolidine, 1-(2-nitro-1-propenyl)-	EtOH	256(3.20),382(4.17)	44-1238-78
$C_7H_{12}N_2O_5$			
β-Alanine, N-(acetoxymethyl)-N-nitroso-, methyl ester	EtOH	230(3.82),371(1.92)	94-3914-78
Ethanol, 2-[(acetoxymethyl)nitrosoamino]-, acetate	EtOH	231(3.79),372(1.91)	94-3914-78
$C_7H_{12}N_2S$			
1H-Imidazole, 1-(1-methylethyl)-2-(methylthio)-	EtOH	224(3.281),250(3.635)	28-0201-78B
	CH_2Cl_2	227(3.160),255(3.584)	28-0201-78B
1H-Imidazole, 1-methyl-2-[(1-methylethyl)thio]-	EtOH	225(3.800),250(3.646)	28-0201-78B
	CH_2Cl_2	224(3.834),249s(3.709)	28-0201-78B
1H-Imidazole, 1-methyl-2-(propylthio)-	EtOH	222(3.804),250(3.694)	28-0201-78B
	CH_2Cl_2	224(3.853),245s(3.721)	28-0201-78B
$C_7H_{12}N_3O_2$			
1,2,4-Triazolidin-1-yl, 2-(1,1-dimethylethyl)-4-methyl-3,5-dioxo-	benzene	303(3.49),413(2.77)	44-0808-78
$C_7H_{12}N_4O_2$			
1,2,4-Triazine-3,5(2H,4H)-dione, 6-(dimethylamino)-2,4-dimethyl-	pH 1	314(3.67)	124-1064-78
	EtOH-pH 7	321(3.76)	124-1064-78
	pH 13	314(3.73)	124-1064-78
$C_7H_{12}OS$			
3-Buten-2-one, 4-[(1-methylethyl)thio]-	H_2O	303(4.20)	70-0133-78
3-Buten-2-one, 4-(propylthio)-	H_2O	302(4.18)	70-0133-78
5-Thiocanone	CCl_4	238(3.41)	42-0146-78
$C_7H_{12}O_2S$			
2-Butenoic acid, 4-(methylthio)-, ethyl ester	EtOH	210(4.11),267(3.31)	39-0955-78C
2-Propenoic acid, 2-[(methylthio)methyl]-, ethyl ester	EtOH	207(3.87)	39-0955-78C
$C_7H_{12}O_2S_4$			
Carbonodithioic acid, O,O'-(1-methyl-1,2-ethanediyl) S,S'-dimethyl ester	EtOH	230(3.57),278(4.24)	118-0286-78
Carbonodithioic acid, O,O'-1,3-propanediyl S,S'-dimethyl ester	EtOH	228(3.79),278(4.04)	118-0286-78
$C_7H_{12}O_4$			
2-Propenoic acid, 3-methoxy-2-(methoxymethyl)-, methyl ester, cis	MeOH	240(4.20)	94-0038-78
trans	MeOH	237(4.18)	94-0038-78
$C_7H_{12}S$			
1,3-Butadiene, 1-(ethylthio)-3-methyl-	C_6H_{12}	276(4.28)	88-3729-78

Compound	Solvent	$\lambda_{max}(\log \epsilon)$	Ref.
Cyclobutanethione, 2,2,3-trimethyl-	C_6H_{12}	213(3.83),231(3.89), 496(1.03)	54-0121-78
	MeCN	213(3.81),232(3.88), 482(1.05)	54-0121-78
$C_7H_{12}S_2$			
Cyclohexanecarbodithioic acid, sodium salt	EtOH	338(4.18),460s(--)	64-0976-78B
$C_7H_{12}Sn$			
Stannacyclohexa-2,5-diene, 1,1-dimethyl-	C_6H_{12}	217(3.27)	101-0257-78E
$C_7H_{13}NO_2$			
Dimethylethylammonium 1,3-dioxopropanide	H_2O	257(4.45)	73-1261-78
	EtOH	257(4.45)	73-1261-78
$C_7H_{13}NO_3$			
2-Hexenamide, 3,5-dihydroxy-N-methyl-	hexane	340(4.12)	4-1153-78
	EtOH	270(4.18)	4-1153-78
$C_7H_{13}N_2S$			
1H-Imidazolium, 2-(ethylthio)-1,3-dimethyl-, iodide	EtOH	221(4.313),261(3.614)	28-0201-78B
	CH_2Cl_2	240(4.244),271(3.585), 290(3.090)	28-0201-78B
$C_7H_{13}N_3O$			
Imidazole-4-carboxamidine, 5-amino-1-ethyl-N'-methoxy-	pH 1	282(3.99)	94-1929-78
	pH 7	263.5(3.99)	94-1929-78
	pH 13	263.5(3.99)	94-1929-78
$C_7H_{14}B_2F_8S_2$			
1,5-Dithiacyclonane, dication, bis-(tetrafluoroborate)	MeCN	237(3.84)	35-6416-78
$C_7H_{14}N_2$			
3-Buten-2-one, 3-methyl-, dimethylhydrazone	C_6H_{12}	275(4.41)	28-0047-78A
3H-Pyrazole, 4,5-dihydro-3,3,5,5-tetramethyl-	hexane	327(2.31)	35-1876-78
	hexane	327(2.28)	35-5122-78
	C_6H_5Et	328(2.22)	24-0596-78
	H_2O	320(--)	35-5122-78
	MeOH	324(2.20)	35-1876-78
	MeCN	327(--)	35-5122-78
$C_7H_{14}N_2O_2$			
2-Propanone, 1-(butylnitrosoamino)-	EtOH	240(3.76),350(1.93)	94-3891-78
1-Propen-1-amine, N-(1,1-dimethylethyl)-2-nitro-	EtOH	260(3.11),370(4.86)	44-0497-78
1-Propen-1-amine, N-(2-methylpropyl)-2-nitro-	EtOH	243(3.15),369(3.98)	44-0497-78
$C_7H_{14}N_2O_3$			
Butanoic acid, (4-nitrosopropylamino)-	EtOH	234(3.86),352(1.93)	94-3909-78
Glycine, N-nitroso-N-pentyl-	EtOH	236(3.79),351(1.93)	94-3909-78
Methanol, (butylnitrosoamino)-, acetate	n.s.g.	233(3.85),372(1.93)	94-3905-78
Methanol, [(1,1-dimethylethyl)nitrosoamino]-, acetate	n.s.g.	235(3.83),381(1.87)	94-3905-78
Methanol, [(1-methylpropyl)nitrosoamino]-, acetate	n.s.g.	234(3.87),372(1.99)	94-3905-78

Compound	Solvent	$\lambda_{max}(\log \epsilon)$	Ref.
Methanol, [(2-methylpropyl)nitroso-amino]-, acetate	n.s.g.	234(3.77),375(1.87)	94-3905-78
Propanoic acid, 3-(butylnitrosoamino)-	EtOH	234(3.84),353(1.90)	94-3909-78
$C_7H_{14}N_2S$			
1H-Imidazole, 4,5-dihydro-1-(1-methyl-ethyl)-2-(methylthio)-	EtOH	230(3.724)	28-0521-78A
	CH_2Cl_2	231(3.679)	28-0521-78A
2-Imidazolidinethione, 1-methyl-3-(1-methylethyl)-	EtOH	241(4.226)	28-0521-78A
	CH_2Cl_2	247(4.252)	28-0521-78A
Pyrimidine, 1-ethyl-1,4,5,6-tetrahydro-2-(methylthio)-	EtOH	225(4.013)	28-0521-78A
	CH_2Cl_2	none	28-0521-78A
2(1H)-Pyrimidinethione, 1-ethyltetra-hydro-3-methyl-	EtOH	240.5(4.304)	28-0521-78A
	CH_2Cl_2	246.5(4.312)	28-0521-78A
$C_7H_{14}OS_3$			
1,3-Oxathiolane, 4,5-dimethyl-2,2-bis-(methylthio)-	EtOH	228(2.70)	118-0286-78
$C_7H_{14}S$			
1-Propene, 3-[(1,1-dimethylethyl)thio]-	n.s.g.	210(3.46),235(2.91)	73-2619-78
$C_7H_{15}NO$			
Propanamide, N,N-diethyl-	H_2O	<u>221(2.2)</u>	104-0221-78
$C_7H_{15}N_2S$			
1H-Imidazolium, 1-ethyl-4,5-dihydro-3-methyl-2-(methylthio)-, iodide	EtOH	221(4.155),256(3.841)	28-0521-78A
	CH_2Cl_2	244(4.309)	28-0521-78A
Pyrimidinium, 1,4,5,6-tetrahydro-1,3-dimethyl-2-(methylthio)-	EtOH	221-243(4.21-4.00)	28-0521-78A
	CH_2Cl_2	242.5(4.349)	28-0521-78A
$C_7H_{15}N_3O$			
Methanehydrazonamide, N'-formyl-N,N-di-methyl-N'-(1-methylethyl)-	heptane	218(3.64),263(3.39)	104-0576-78
Methanehydrazonamide, N'-formyl-N,N-di-methyl-N'-propyl-	heptane	218(3.56),262(3.55)	104-0576-78
$C_7H_{16}N_2O_2$			
1-Butanol, 4-(nitrosopropylamino)-	EtOH	235(3.85),351(1.93)	94-3891-78
Ethanol, 2-(nitrosopentylamino)-	EtOH	235(3.85),351(1.94)	94-3891-78
1-Propanol, 3-(butylnitrosoamino)-	EtOH	235(3.86),350(1.94)	94-3891-78
2-Propanol, 1-(butylnitrosoamino)-	EtOH	236(3.84),352(1.93)	94-3891-78
$C_7H_{18}GeN_2$			
Diazene, (1,1-dimethylethyl)trimethyl-germyl)-	CH_2Cl_2	480(1.08)	60-1909-78
$C_7H_{18}N_2O_3Si$			
Diazene, (1,1-dimethylethyl)(trimeth-oxysilyl)-	CH_2Cl_2	506(0.85)	60-1909-78
$C_7H_{18}N_2Si$			
Diazene, (1,1-dimethylethyl)(trimethyl-silyl)-	CH_2Cl_2	500(0.95)	60-1909-78

Compound	Solvent	$\lambda_{max}(\log \epsilon)$	Ref.
C_8BrCl_7			
Bicyclo[3.2.0]hepta-2,6-diene, 7-bromo-1,2,3,5,6-pentachloro-4-(dichloro-methylene)-	heptane	213(4.00),290(4.24)	5-0804-78
1,3,5-Cycloheptatriene, 3-bromo-1,2,4,5,6-pentachloro-7-(dichloro-methylene)-	heptane	215(4.27),246(4.23), 262(4.24)	5-0804-78
$C_8Br_2Cl_6$			
Bicyclo[3.2.0]hepta-2,6-diene, 1,2,3,5,6,7-hexachloro-4-(dibromo-methylene)-	heptane	212(4.06),288(4.33)	5-0804-78
$C_8Cl_2N_2O_2$			
1,4-Cyclohexadiene-1,2-dicarbonitrile, 4,5-dichloro-3,6-dioxo-	$CHCl_3$	288(3.75),387(2.90)	40-1127-78
C_8Cl_8			
1,3,5-Cycloheptatriene, 1,2,3,4,5,6-hexachloro-7-(dichloromethylene)-	heptane	215(4.20),241(4.30), 258(4.23)	5-0804-78
C_8Cl_9F			
Bicyclo[3.2.0]hepta-2,6-diene, 1,3,4,5,6,7-hexachloro-4-fluoro-2-(trichloromethyl)-	heptane	212(4.03),229s(3.82)	5-1775-78
C_8Cl_{10}			
Bicyclo[3.2.0]hepta-2,6-diene, 1,3,4,4,5,6,7-heptachloro-2-(trichloromethyl)-	heptane	212(3.93),227s(3.84)	5-1775-78
$C_8H_2Cl_6$			
Bicyclo[3.2.0]hepta-2,6-diene, 1,2,3,5,6,7-hexachloro-4-methylene-	heptane	213(3.99),246(4.01), 253s(3.97),267s(3.81)	5-1406-78
$C_8H_2Cl_8$			
Bicyclo[3.2.0]hepta-2,6-diene, 1,2,3,4,5,6,7-heptachloro-4-(chloromethyl)-	heptane	218(3.98)	5-1406-78
$C_8H_2N_2O_4$			
Propanedinitrile, (3,4-dihydroxy-2,5-dioxo-3-cyclopenten-1-ylidene)-, dipotassium salt	MeOH	444s(4.06),540(4.92)	35-2586-78
$C_8H_3Br_4N$			
1H-Indole, 2,3,5,6-tetrabromo-	EtOH	230(4.67),294(3.98), 301(3.98)	88-4479-78
$C_8H_4Br_2S_2$			
2,2'-Bithiophene, 3,3'-dibromo-	heptane	220(3.74),255(4.01), 294s(3.62),310s(3.48), 327s(3.07)	20-0027-78
2,2'-Bithiophene, 5,5'-dibromo-	heptane	237s(3.76),247(3.76), 255s(3.70),305s(4.13), 318(4.23),322s(4.18), 350s(3.87)	20-0027-78
$C_8H_4ClNO_2$			
2H-Pyrano[2,3-b]pyridin-2-one, 4-chloro-	EtOH	262(3.74),272(3.72), 308(4.03)	103-0507-78

Compound	Solvent	$\lambda_{max}(\log \epsilon)$	Ref.
$C_8H_4ClNO_2S$ Maleimide, 2-chloro-3-(2-thienyl)-	n.s.g.	252(4.05),373(3.93)	88-0125-78
$C_8H_4ClN_5S$ 1H-Purine, 6-chloro-8-(4-thiazolyl)-	pH 2.2 pH 6.4 pH 12.8	298(4.34) 299(4.32) 302(4.30)	87-0344-78 87-0344-78 87-0344-78
$C_8H_4Cl_2N_2$ 1,5-Naphthyridine, 2,6-dichloro-	EtOH	214(5.38),256(3.63), 266(3.53),310(3.76), 323(3.76)	4-0685-78
$C_8H_4Cl_2S_2$ 1,3-Benzodithiole, 2-(dichloromethyl- ene)-	C_6H_{12}	207(3.89),242(4.18), 266(4.00),277(4.00), 314(3.62)	44-0416-78
$C_8H_4F_2INO_2$ Benzene, 1-(1,2-difluoro-2-iodoethenyl)- 4-nitro-, (Z)-	ether	321(4.27)	104-0191-78
$C_8H_4F_3N$ Benzonitrile, 3-(trifluoromethyl)- Benzonitrile, 4-(trifluoromethyl)-	heptane MeOH heptane MeOH	277.5(2.57) 278(2.69) 279(2.79) 280.5(3.09)	73-2763-78 73-2763-78 73-2763-78 73-2763-78
$C_8H_4INO_2$ 1H-Indole-2,3-dione, 5-iodo-	EtOH	430(3.90)	111-0515-78
$C_8H_4INO_2S$ Maleimide, 2-iodo-3-(2-thienyl)-	n.s.g.	253(4.07),385(3.86)	88-0125-78
$C_8H_4N_2$ 1,2-Benzenedicarbonitrile 1,3-Benzenedicarbonitrile 1,4-Benzenedicarbonitrile	hexane hexane heptane MeOH hexane heptane MeOH	206(4.63),236(3.93), 280(2.90) 206(4.75),226(4.16), 281(2.48) 288(2.49)(smoothed) 288(2.75) 235(4.35),281(3.04) 288.5(3.16)(smoothed) 290(3.28)	110-0044-78 110-0044-78 73-2763-78 73-2763-78 110-0044-78 73-2763-78 73-2763-78
$C_8H_4N_2O_4$ 1H-Indole-2,3-dione, 5-nitro- 1H-Isoindole-1,3(2H)-dione, 4-nitro- 1H-Isoindole-1,3(2H)-dione, 5-nitro-	EtOH EtOH EtOH	320(4.09) 226(4.06) 233(4.13)	111-0515-78 94-0530-78 94-0530-78
$C_8H_4N_4O_4$ 4H,8H-Benzo[1,2-c:4,5-c']diisoxazole- 4,8-dione, 3,7-diamino-	MeOH	216(3.8),284(4.1), 326(4.0)	24-3346-78
$C_8H_5Br_5O_2$ 4H-Pyrimidin-4-one, 2,3,5-tribromo- 6-(1,1-dibromopropyl)-	MeOH	238(4.20),247s(--), 275(4.06)	88-3165-78
$C_8H_5ClN_2O$ Quinoxaline, 2-chloro-, 4-oxide	MeOH	243(4.61),320(3.93)	44-4125-78

Compound	Solvent	$\lambda_{max}(\log \epsilon)$	Ref.
$C_8H_5Cl_2N_3S$			
1H-1,2,4,6-Thiatriazine, 1,3-dichloro-5-phenyl-	hexane	259(4.26)	103-0226-78
$C_8H_5F_2I$			
Benzene, (1,2-difluoro-2-iodoethenyl)-, (Z)-	heptane	255(4.21)	104-0191-78
$C_8H_5F_3O_2S$			
1,3-Butanedione, 4,4,4-trifluoro-1-(2-thienyl)-	heptane	315(4.3)	40-0489-78
	MeCN	320(4.2)	40-0489-78
	CHCl$_3$	323(4.2)	40-0489-78
$C_8H_5F_3O_3$			
1,3-Butanedione, 4,4,4-trifluoro-1-(2-furanyl)-	heptane	320(4.3),345(4.3)	40-0489-78
	MeCN	320s(4.3),350(4.3)	40-0489-78
C_8H_5NO			
Benzonitrile, 3-formyl-	heptane	294.5(2.45)(smoothed)	73-2763-78
	MeOH	293(2.84)	73-2763-78
Benzonitrile, 4-formyl-	heptane	297.5(3.16)(smoothed)	73-2763-78
C_8H_5NOS			
2H-Thiopyrano[2,3-b]pyridin-2-one	EtOH	228(4.42),256(3.82), 278(3.46),288(3.43), 313(3.76)	78-0989-78
$C_8H_5NO_2$			
Benzoic acid, 3-cyano-	MeOH	286.5(2.92)(smoothed)	73-2763-78
1H-Indole-2,3-dione	EtOH	285(3.53)	111-0515-78
$C_8H_5NO_3$			
2H-3,1-Benzoxazine-2,4(1H)-dione, anion	MeCN-pH 9.7	350(3.59)	44-0268-78
1H-Isoindole-1,3(2H)-dione, 4-hydroxy-	EtOH	222(4.19)	94-0530-78
1H-Isoindole-1,3(2H)-dione, 5-hydroxy-	EtOH	225(4.23)	94-0530-78
$C_8H_5N_3$			
Benzonitrile, 4-(diazomethyl)-	DMF	459(1.67)	142-0105-78C
$C_8H_5N_3O$			
1H-Benzimidazole-2-carbonitrile, 1-hydroxy-	MeOH	225(4.29),286(4.11)	44-0076-78
Furo[3,2-b]pyridine-3-carbonitrile, 2-amino-, hydrate	MeOH	209(3.83),215s(3.71), 258(4.06),310(3.94)	4-1411-78
$C_8H_5N_3O_2$			
1H-Benzimidazole-2-carbonitrile, 1-hydroxy-, 3-oxide	MeOH	235(4.37),303(3.96), 369(3.65)	44-0076-78
$C_8H_5N_3S_2$			
Isothiazolo[3',4':4,5]thieno[2,3-b]-pyridin-3-amine	EtOH	239(4.03),271(4.08), 280(4.24),292(4.33), 361(3.34)	32-0057-78
$C_8H_5N_5O$			
Quinoxaline, 2-azido-, 1-oxide	MeOH	225(4.26),265(4.47), 335(3.72)	44-0076-78
Tetrazolo[1,5-a]quinoxaline, 5-oxide	MeOH	230(4.30),262s(--), 317(3.97)	44-4125-78

Compound	Solvent	$\lambda_{max}(\log \epsilon)$	Ref.
$C_8H_5N_5O_2$ Quinoxaline, 2-azido-, 1,4-dioxide	MeOH	238(4.28),283(4.40), 395(3.78)	44-0076-78
$C_8H_5N_5O_5$ 6,7-Pteridinedicarboxylic acid, 2-am- ino-1,4-dihydro-4-oxo-	pH -1.0	238(4.02),261(3.99), 328(3.97)	24-3790-78
	pH 3.0	246s(4.05),292(4.08), 350(3.94)	24-3790-78
	pH 6.0	240s(4.07),283(4.12), 345(3.93)	24-3790-78
	pH 11.7 ˙	262(4.32),365(3.97)	24-3790-78
$C_8H_5N_5S$ 1H-Purine, 8-(4-thiazolyl)-	pH 4.8	296(4.46)	87-0344-78
	pH 6.9	296(4.46)	87-0344-78
	pH 12.6	312(4.33)	87-0344-78
$C_8H_5N_5S_2$ 6H-Purine-6-thione, 1,7-dihydro- 8-(4-thiazolyl)-	pH 1.0	303(4.22),317(4.22)	87-0344-78
	pH 3.0	245(4.22),293(4.12), 317(4.21)	87-0344-78
	pH 4.5	245(4.22),252(4.25), 292(4.00),318(4.15)	87-0344-78
$C_8H_6BrClO_2$ 1,4-Benzodioxin, 6-bromo-5-chloro- 2,3-dihydro-	EtOH	210(4.69),228(4.00), 293(3.36)	103-1188-78
1,4-Benzodioxin, 6-bromo-7-chloro- 2,3-dihydro-	EtOH	210(4.68),227(3.95), 296(3.59),330(2.70)	103-1188-78
$C_8H_6BrFO_2$ 1,4-Benzodioxin, 6-bromo-5-fluoro- 2,3-dihydro-	EtOH	209(4.45),227(3.96), 281(3.00)	103-1188-78
1,4-Benzodioxin, 6-bromo-7-fluoro- 2,3-dihydro-	EtOH	205(4.39),224(3.84), 294(3.62)	103-1188-78
C_8H_6BrNOS 2,1-Benzisothiazol-3(1H)-one, 5-bromo- 1-methyl-	EtOH	203(4.17),224(4.42), 225s(--),252s(--), 262(4.16),378(3.67)	4-0529-78
$C_8H_6BrNO_4$ Benzeneacetic acid, 2-bromo-4-nitro-	EtOH	268(3.99)	73-0471-78
1,4-Benzodioxin, 7-bromo-2,3-dihydro- 5-nitro-	EtOH	225(4.22),274(3.70), 338(3.48)	103-1188-78
$C_8H_6Br_2O_4$ 2,5-Cyclohexadiene-1,4-dione, 2,6-di- bromo-3,5-dimethoxy-	$CHCl_3$	312(4.15)	40-1127-78
C_8H_6ClN Benzonitrile, 3-(chloromethyl)-	heptane	281.5(2.83)(smoothed)	73-2763-78
	MeOH	283.5(2.79)	73-2763-78
Benzonitrile, 4-(chloromethyl)-	heptane	282(2.73)	73-2763-78
	MeOH	282(2.90)	73-2763-78
$C_8H_6ClNO_3S_2$ 2,1-Benzisothiazole-5-sulfonyl chlor- ide, 1,3-dihydro-1-methyl-3-oxo-	EtOH	219(4.40),237s(--), 263(4.16),314(3.83), 364(3.43)	4-0529-78

Compound	Solvent	$\lambda_{max}(\log \epsilon)$	Ref.
$C_8H_6ClNO_4$			
1,4-Benzodioxin, 5-chloro-2,3-dihydro-6-nitro-	EtOH	208(4.08),254(3.75), 299(3.70)	103-1188-78
1,4-Benzodioxin, 5-chloro-2,3-dihydro-7-nitro-	EtOH	208(4.22),225(4.20), 244(3.97),315(3.86)	103-1188-78
$C_8H_6ClN_5$			
1,2,4,5-Tetrazin-3-amine, 6-(4-chloro-phenyl)-	dioxan	299(3.39),535(1.73)	30-0338-78
$C_8H_6Cl_2O_4$			
2,5-Cyclohexadiene-1,4-dione, 2,5-di-chloro-3,6-dimethoxy-	$CHCl_3$	304(4.13)	40-1127-78
$C_8H_6FNO_4$			
1,4-Benzodioxin, 5-fluoro-2,3-dihydro-7-nitro-	EtOH	219(3.74),236(3.50), 312(3.57)	103-1188-78
$C_8H_6N_2O_2$			
3H-Indol-3-one, oxime, 1-oxide	MeOH	237(3.36),302(3.33)	39-1117-78C
	MeOH-base	249(3.11),258(3.17), 295(2.96),285(2.43)	39-1117-78C
1H-Isoindole-1,3(2H)-dione, 4-amino-	EtOH	225(4.21),237(4.19)	94-0530-78
1H-Isoindole-1,3(2H)-dione, 5-amino-	EtOH	245(4.36),257(4.30)	94-0530-78
1,5-Naphthyridine-4,8-dione, 1,5-di-hydro-	H_2O	232(4.59),262(3.51), 318(4.29),330(4.44)	44-1331-78
2,3-Quinoxalinedione, 1,4-dihydro-	pH 7.0	228(3.98),235(3.91), 257(3.66),300s(3.97), 312(4.06),324s(4.00), 343s(3.64)	24-1753-78 +24-1763-78
	pH 11.5	214(4.34),235(3.81), 242s(3.77),298s(3.59), 312(3.81),324(3.85), 338s(3.64)	24-1753-78
	2M NaOH	226(4.25),245s(3.85), 262s(3.53),304s(3.55), 314(3.82),327(3.94), 341(3.79)	24-1753-78
$C_8H_6N_2O_3$			
Ethanone, 1-(5-benzofurazanyl)-, N-oxide	MeOH	243(4.33),370(3.78)	87-0483-78
$C_8H_6N_2O_3S$			
2,1-Benzisothiazol-3(1H)-one, 1-methyl-5-nitro-	EtOH	207(4.18),225(4.22), 232(4.23),240s(--), 170(4.03),252(4.05)	4-0529-78
$C_8H_6N_2S_2$			
1,5-Naphthyridine-4,8-dithione, 1,5-dihydro-	MeCN	246s(4.17),260(4.29), 328(3.78),410s(3.87), 450(4.13)(changing)	44-1331-78
after 30 hours	MeCN	222(4.20),238s(4.13), 263s(3.92),318(3.96), 330(4.02)	44-1331-78
$C_8H_6N_4$			
5,5'-Bipyrimidine	$CHCl_3$	242(4.19),255s(4.06)	24-1330-78
1,2,3-Triazolo[4,5-b]indole, 1,4-di-hydro-	EtOH	241(4.38),274(4.19), 279(4.18),302(3.51)	33-0108-78

Compound	Solvent	λ_{max}(log ϵ)	Ref.
C$_8$H$_6$N$_4$OS			
10H-Thiazolo[2,3-b]pteridin-10-one, 7,8-dihydro-	pH -3.0	266(4.12),309(3.97), 327s(3.85),372(3.38)	24-0971-78
	pH 3.0	230(3.94),265(4.11), 278(4.10),330(3.85)	24-0971-78
C$_8$H$_6$N$_4$O$_3$			
4,4'-Bipyrimidine-2,2',6(1H,1'H,3H)-trione	pH 0.9	303(3.97),314(3.92), 336s(3.84)	44-0511-78
	pH 7.2	304(3.93),314(3.92), 336s(3.83)	44-0511-78
	pH 12.9	325(4.05)	44-0511-78
C$_8$H$_6$N$_6$OS			
6H-Purin-6-one, 8-(4-thiazolyl)-, oxime	pH 2.8	237(4.02),303(4.29)	87-0344-78
	pH 6.9	234(4.16),300(4.11)	87-0344-78
C$_8$H$_6$N$_6$S			
1H-Purin-6-amine, 8-(4-thiazolyl)-	pH 1.2	206(4.40)	87-0344-78
	pH 6.9	232(4.29),295(4.32)	87-0344-78
	pH 10.7	233(4.34),300(4.34)	87-0344-78
C$_8$H$_6$N$_6$S$_2$			
6H-Purine-6-thione, 2-amino-1,7-di-hydro-8-(4-thiazolyl)-	pH 0.9	222(4.34),314(4.41)	87-0344-78
	pH 2.8	233(4.21),314(4.39)	87-0344-78
	pH 13.8	222(4.27),329(4.33)	87-0344-78
C$_8$H$_6$S$_2$			
2,2'-Bithiophene	heptane	245(3.33),256s(3.27), 288s(3.49),301(3.58), 311s(3.54),317s(3.47), 330s(3.15)	20-0027-78
C$_8$H$_6$S$_8$			
1,3-Dithiole-2-thione, 4,4'-dithiobis-[5-methyl-	pyridine	369(4.51),375(4.51)	39-1017-78C
C$_8$H$_7$BF$_2$O$_4$			
2H-Pyran-2-one, 3-acetyl-4-[(difluoro-boryl)oxy]-6-methyl-	MeOH	224(3.98),316(4.09)	23-1796-78
C$_8$H$_7$BrO			
Benzene, 1-bromo-4-(ethenyloxy)-	C$_6$H$_{12}$	197(4.43),232(4.19), 280(3.05)	99-0194-78
C$_8$H$_7$ClN$_2$O$_2$			
Benzenamine, N-(2-chloro-2-aci-nitro-ethylidene)-	EtOH	243(4.14),393(4.35)	104-0817-78
C$_8$H$_7$ClN$_4$O			
Acetamide, N-(7-chloro-1H-imidazo-[4,5-b]pyridin-5-yl)-	pH 1	248(3.92),255s(3.80), 298(4.19)	4-0839-78
	pH 11	301(4.14)	4-0839-78
C$_8$H$_7$ClN$_4$O$_3$			
9H-Purine-9-propanoic acid, α-chloro-1,6-dihydro-6-oxo-, (±)-	pH 2	250(4.07)	73-3444-78
C$_8$H$_7$ClO			
Benzene, [(2-chloroethenyl)oxy]-	dioxan	198(4.42),228(4.26), 268(2.95),272(2.60), 273(2.48)	99-0194-78

Compound	Solvent	λ_{max}(log ϵ)	Ref.
$C_8H_7Cl_3N_2O_2$			
2-Propanone, 1,1,1-trichloro-3-(6-meth-oxy-4-pyrimidinyl)-	n.s.g.	<u>300(3.9),360(4.2)</u>	103-1016-78
$C_8H_7F_3$			
2,4-Heptadiyne, 1,1,1-trifluoro-6-methyl-	hexane	255(3.00)	104-1089-78
$C_8H_7F_3N_2O$			
2-Propanone, 1,1,1-trifluoro-3-(6-meth-yl-4-pyrimidinyl)-	n.s.g.	<u>295(3.81),360(4.1)</u>	103-1016-78
C_8H_7N			
Benzonitrile, 3-methyl-	heptane	282.5(2.89)(smoothed)	73-2763-78
	MeOH	283.5(2.92)	73-2763-78
Benzonitrile, 4-methyl-	hexane	230(4.25),268(2.60)	110-0044-78
	heptane	279(2.58)(smoothed)	73-2763-78
	MeOH	279(2.76)	73-2763-78
C_8H_7NO			
Benzonitrile, 3-methoxy-	heptane	296.5(3.42)(smoothed)	73-2763-78
Benzonitrile, 4-methoxy-	heptane	283.5(3.10)(smoothed)	73-2763-78
	71% H_2SO_4	<u>245(4.3)</u>	104-0111-78
Furo[2,3-c]pyridine, 2-methyl-	EtOH	254(4.04),270s(3.88), 279(3.59)	33-2542-78
Furo[3.2-b]pyridine, 2-methyl-	EtOH	243(3.74),287(4.02)	33-2542-78
2H-Pyrano[3,2-b]pyridine	EtOH	256(3.08),317(3.82)	33-2542-78
Pyridine, 3-(2-propynyloxy)-	EtOH	275(3.60),282s(3.45)	33-2542-78
C_8H_7NOS			
2,1-Benzisothiazole, 3-methoxy-	EtOH	212s(--),222(4.18), 285s(--),294(3.54), 342(3.58)	4-0529-78
2,1-Benziisothiazol-3(1H)-one, 1-methyl-	EtOH	222(4.31),234(4.24), 247(3.94),253s(--), 366(3.58)	4-0529-78
2,1-Benzisothiazol-3(1H)-one, 5-methyl-	EtOH	221(4.37),245(3.98), 252s(--),267s(--), 358(3.65)	4-0529-78
$C_8H_7NO_2$			
Furo[3,2-b]pyridine, 2-methyl-, N-oxide	EtOH	221s(3.98),238(4.31), 294s(4.03),302(4.05), 310s(3.99)	33-2542-78
$C_8H_7NO_2S$			
Benzonitrile, 3-(methylsulfonyl)-	heptane	280.5(2.67)	73-2763-78
	MeOH	280.5(2.77)	73-2763-78
Benzonitrile, 4-(methylsulfonyl)-	MeOH	285.5(3.19)	73-2763-78
$C_8H_7NO_2S_2$			
2H-1,4-Benzothiazine-3(4H)-thione, 1,1-dioxide	MeOH	235(4.02),292(3.97), 331(4.28)	2-0678-78
$C_8H_7NO_3$			
Aziridine, 1-[(2-oxo-2H-pyran-5-yl)-carbonyl]-	C_6H_{12}	251(4.22),285(3.72)	44-4415-78
$C_8H_7NO_4$			
Ethanone, 1-(4-hydroxy-3-nitrophenyl)-, anion	H_2O	399(3.66)	120-0111-78

Compound	Solvent	$\lambda_{max}(\log \epsilon)$	Ref.
Ethanone, 2-hydroxy-1-(4-nitrophenyl)-	H_2O	267(3.96)	44-0610-78
3,4-Pyridinedicarboxylic acid, 2-methyl-	EtOH	217(3.84),280(3.79)	106-0782-78
$C_8H_7NO_5$			
Benzaldehyde, 5-hydroxy-4-methoxy- 2-nitro-	MeOH	222(4.00),261(4.06), 310(3.74),336(3.72)	39-0440-78C
	MeOH-base	235(4.09),293(3.97), 438(4.10)	39-0440-78C
$C_8H_7N_3O$			
Imidazo[1,2-a]pyridine, 2-methyl- 3-nitroso-	n.s.g.	249(4.56),278(4.66), 322(4.49),377(4.68)	103-1241-78
Pyridine, 2-(methylfurazanyl)-	MeOH	240(3.85),269(3.72)	4-1093-78
Pyridine, 2-(3-methyl-1,2,4-oxadiazol- 5-yl)-	MeOH	240(4.22),275(4.13)	4-1093-78
Pyrido[2,3-d]pyrimidin-4(1H)-one, 6-methyl-	pH 1	266(3.76),320(3.97)	44-0828-78
	pH 7	261(3.88),305(4.26), 316(3.79)	44-0828-78
	pH 11	275(3.70),318(3.91)	44-0828-78
$C_8H_7N_3O_2$			
1H-Benzimidazole-1-carboxamide, 2,3-dihydro-2-oxo-	EtOH	235(3.70),277(3.53), 285(3.49)	94-2635-78
2H-Benzotriazole-4,7-dione, 2,5-di- methyl-	EtOH	228(3.95),250(3.59), 281(2.88),289(2.74)	87-0578-78
2H-Cyclopenta[d]pyridazine, 2-methyl- 7-nitro-	ether	243(4.38),267(4.08), 272s(4.04),292(3.72), 336(3.92),349s(3.81), 416(4.04)	44-1602-78
Furo[3,2-b]pyridine-3-carboxamide, 2-amino-	MeOH	209(4.08),217(3.90), 256(4.32),306(4.21)	4-1411-78
Pyrido[2,3-b]pyrazine-2,3-dione, 1,4-dihydro-1-methyl-	pH 2.0	208(4.52),221(3.97), 242s(3.57),312(4.20), 323(4.14),340s(3.77)	24-1763-78
	pH 12.0	212(4.46),228(4.08), 255(3.59),318(4.20), 330(4.34),345(4.17)	24-1763-78
$C_8H_7N_3O_7$			
2,5-Cyclohexadien-1-one, 4-(ethyl-aci- nitro)-2,6-dinitro-	MeOH	307(4.08),406(4.19)	104-2064-78
$C_8H_7N_3S$			
5-Azacinnoline, 4-(methylthio)-	heptane	235(3.97),349(3.94)	103-1032-78
$C_8H_7N_3S_2$			
Thieno[2,3-b]pyridine-2-carbothioamide, 3-amino-	EtOH	249(3.72),300(4.26), 316(4.09),327(3.99), 406(3.89)	32-0057-78
$C_8H_7N_5$			
Imidazo[1,2-a]-1,2,4-triazolo[3,4-c]- pyrazine, 3-methyl-	pH 1	221(4.37),226s(4.36), 235s(4.20),244s(4.03), 277(3.91),287s(3.88), 299s(3.69)	4-0987-78
	H_2O	220s(4.38),227(4.51), 233(4.49),247(3.93), 269s(3.84),281(3.89), 293(3.74)	4-0987-78
	pH 11	220s(4.38),226(4.51), 233(4.48),246(3.91),	4-0987-78

Compound	Solvent	$\lambda_{max}(\log \epsilon)$	Ref.
Imidazo[1,2-a]-1,2,4-triazolo[3,4-c]-pyrazine, 3-methyl- (cont.)		269s(3.83),280(3.88), 292(3.73)	4-0987-78
1,2,4,5-Tetrazin-3-amine, 6-phenyl-	dioxan	273(4.38),400(2.24), 543(1.28)	30-0338-78
$C_8H_7N_5O_2$			
1H-Tetrazole, 1-methyl-5-(3-nitro-phenyl)-	pH 7.5	221(4.3)	104-2252-78
2H-Tetrazole, 2-methyl-5-(3-nitro-phenyl)-	pH 7.5	230(4.4)	104-2252-78
$C_8H_7N_5O_3$			
6-Pteridinecarboxylic acid, 2-amino-1,4-dihydro-7-methyl-4-oxo-	pH 0.0	262(3.97),324(4.04)	24-3790-78
	pH 2.3	240s(4.00),292(4.08), 340(3.92),360s(3.72)	24-3790-78
	pH 5.0	238s(4.04),280(4.11), 346(3.88)	24-3790-78
	pH 11.0	257(4.30),360(3.93)	24-3790-78
7-Pteridinecarboxylic acid, 2-amino-1,4-dihydro-6-methyl-4-oxo-	pH -1.0	237(4.08),248s(4.05), 338(3.93)	24-3790-78
	pH 2.1	240s(4.08),272s(3.63), 328(3.98),372s(3.16)	24-3790-78
	pH 5.0	243(4.10),273(4.10), 350(3.88)	24-3790-78
	pH 11.0	255(4.36),367(3.91)	24-3790-78
Pteridinium, 2-amino-6-carboxy-1,4-di-hydro-8-methyl-4-oxo-, hydroxide, inner salt	pH 0.0	273(4.08),300(4.18), 390(4.05)	24-3790-78
	pH 3.5	265(4.11),295(4.17), 390(4.03)	24-3790-78
	pH 8.0	279(4.31),400(3.97)	24-3790-78
$C_8H_7N_5O_4$			
Pyrimido[4,5-c]pyridazine-3-carboxylic acid, 7-amino-1,4,5,6-tetrahydro-1-methyl-4,5-dioxo-, disodium salt	pH 2	267(4.66),315(3.80)	44-4844-78
	M NaOH	251(4.48),306(4.03)	44-4844-78
Pyrimido[4,5-c]pyridazine-4-carboxylic acid, 7-amino-1,2,3,5-tetrahydro-1-methyl-3,5-dioxo-, disodium salt	pH 2	248(4.35),255s(4.32), 380(3.69)	44-4844-78
	M NaOH	254(4.55),276s(3.90), 409(3.73)	44-4844-78
C_8H_8			
Benzocyclobutene	EtOH	259(3.14),265(3.32), 271(3.32)	35-3730-78
Styrene	C_6H_{12}	205(4.40),214(4.10), 248(4.15),274(2.93), 282(2.92),291(2.74)	99-0194-78
C_8H_8AsN			
1H-1,3-Benzazaarsole, 2-methyl-	MeOH	239(4.43),266s(3.85), 286(3.66),325(3.71)	101-0001-78K
	MeOH-KOH	238(4.48),256s(4.18), 283s(3.81),325(3.79)	101-0001-78K
$C_8H_8ClNO_3$			
1H-Pyrrole-3,4-dicarboxaldehyde, 2-chloro-5-methoxy-1-methyl-	EtOH	243(4.30),310(4.19)	104-2041-78
$C_8H_8ClN_3O$			
1H-Benzotriazole, 5-chloro-1-ethoxy-	EtOH	273(4.10)	4-1043-78

Compound	Solvent	$\lambda_{max}(\log \epsilon)$	Ref.
$C_8H_8Cl_2O_4$			
3,5-Cyclohexadiene-1-carboxylic acid, 3,5-dichloro-1,2-dihydroxy-, methyl ester	MeOH	276(3.64)	78-1707-78
$C_8H_8F_4N_6S$			
3H-1,2,4-Triazole-3-thione, 4-amino-2,4-dihydro-5-[3-methyl-5-(1,1,2,2-tetrafluoroethyl)-1H-pyrazol-1-yl]-	EtOH	256(4.35)	103-0458-78
$C_8H_8IN_3$			
Pyrido[2,3-b]pyrazinium, 5-methyl-, iodide, 1:1 indole complex	EtOH	275(4.19),290(4.11), 348(3.98),496(3.64)	104-0398-78
C_8H_8NP			
1H-1,3-Benzazaphosphole, 2-methyl-	MeOH	231(4.43)	88-0441-78
$C_8H_8N_2$			
Benzene, 1-(diazomethyl)-3-methyl-	DMF	490(1.40)	142-0105-78C
Benzene, 1-(diazomethyl)-4-methyl-	DMF	488(1.43)	142-0105-78C
9,10-Diazapentacyclo[4.4.0.0^{2,5}.0^{3,8}-0^{4,7}]dec-9-ene	hexane	401(2.18)	35-5122-78
	H_2O	382(--)	35-5122-78
	MeCN	390(--),402(--)	35-5122-78
1H-Pyrrolo[3,2-b]pyridine, 3-methyl- (4-azaskatole)	MeOH	225(4.34),292(3.84)	95-0898-78
$C_8H_8N_2O$			
Benzene, 1-(diazomethyl)-4-methoxy-	DMF	507(1.46)	142-0105-78C
1H-Indazole, 3-methoxy-	EtOH	216(4.47),252(3.21), 300(3.68)	118-0633-78
3H-Indazol-3-one, 1,2-dihydro-1-methyl-	EtOH	312(3.65)	118-0633-78
3H-Indazol-3-one, 1,2-dihydro-2-methyl-	EtOH	218(4.30),235(4.04), 310(3.70)	118-0633-78
$C_8H_8N_2O_2$			
1H-Pyrrolo[3,2-c]pyridine-4,6(5H,7H)-dione, 5-methyl-	pH 1	245(3.65),286(3.54)	87-0990-78
	pH 7	241(3.62),283(3.45), 350(2.89)	87-0990-78
	pH 11	268(3.56),308(3.57)	87-0990-78
$C_8H_8N_2O_2S$			
Imidazo[1,2-a]pyridine, 2-(methylsul-fonyl)-	EtOH	220(4.55),279(3.64), 290(3.64)	95-0631-78
monohydrobromide	EtOH	220(4.59),278(3.67), 293(3.68)	95-0631-78
$C_8H_8N_2O_3$			
Aziridine, 1-[2-(5-nitro-2-furanyl)-ethenyl]-, (E)-	EtOH	270(4.30),497(4.40)	103-0250-78
[3,4'-Bi-2H-pyrrole]-2,2'-dione, 1,1',5,5'-tetrahydro-4-hydroxy-	EtOH	241(4.11),352(4.50), 367s(4.43)	35-4225-78
	EtOH-H_2SO_4	238(4.07),354(4.60), 368(4.56)	35-4225-78
	EtOH-KOH	255(4.28),287(4.33), 333(4.35)	35-4225-78
$C_8H_8N_2O_3S_2$			
2,1-Benzisothiazole-5-sulfonamide, 1,3-dihydro-1-methyl-3-oxo-	EtOH	225(4.40),237(4.33), 261(4.25),282s(--), 368(3.63)	4-0529-78

Compound	Solvent	λ_{max}(log ϵ)	Ref.
$C_8H_8N_2O_4$			
Benzenecarboximidic acid, 4-hydroxy-3-nitro-, methyl ester, monohydrochloride	pH 1	255(4.3),345(2.5)	63-0407-78
	pH 5.2	313(4.2),397(2.6)	63-0407-78
2,4(1H,3H)-Pyrimidinedione, 1-(2,3-dioxobutyl)-	n.s.g.	260(4.01),410(1.51)	78-2861-78
$C_8H_8N_2S$			
2-Benzothiazolamine, N-methyl-	EtOH	222(4.50),265(4.14), 289s(3.50)	103-0380-78
2(3H)-Benzothiazolimine, 3-methyl-	EtOH	222(4.57),261(4.03), 295(3.77)	103-0380-78
$C_8H_8N_3$			
Pyrido[2,3-b]pyrazinium, 5-methyl-, iodide, 1:1 indole complex	EtOH	275(4.19),290(4.11), 348(3.98),496(3.64)	104-0398-78
$C_8H_8N_4$			
1,5-Naphthyridine-4,8-diamine	H_2O	235(4.57),328(4.13)	44-1331-78
$C_8H_8N_4O$			
Phenol, 2-(2-methyl-2H-tetrazol-5-yl)-	EtOH	241(4.07),295(3.73)	44-1664-78
$C_8H_8N_4OS$			
4(1H)-Pteridinone, 1-methyl-2-(methylthio)-	pH -2.7	223(3.85),263(4.06), 315(4.04)	24-0971-78
	pH 4.0	258(4.19),303s(3.88), 333(4.01),343s(3.94)	24-0971-78
4(3H)-Pteridinone, 3-methyl-2-(methylthio)-	pH -3.4	255(3.91),287(4.25), 385(3.99)	24-0971-78
	pH 3.0	243(4.06),264(4.08), 283(4.11),337(3.81)	24-0971-78
$C_8H_8N_4O_2$			
Acetamide, N-(4,5-dihydro-4-oxo-1H-imidazo[4,5-c]pyridin-6-yl)-	pH 1	278(4.14)	87-1212-78
	pH 7	261(4.09),272s(4.08), 292(3.96)	87-1212-78
	pH 11	276(4.05)	87-1212-78
$C_8H_8N_4O_2S$			
1,3,4-Thiadiazole-2-carboxylic acid, 5-(1H-imidazol-1-yl)-, ethyl ester	MeOH	278(4.06)	78-0453-78
$C_8H_8N_4O_3$			
2,6,7(1H)-Pteridinetrione, 5,8-dihydro-4,8-dimethyl-	pH -3.0	229(4.29),251s(4.00), 273(3.66),287(3.66), 324(4.08),370s(3.13)	24-1763-78
	pH 3.0	230(4.39),254(4.10), 341(3.98)	24-1763-78
	pH 9.0	236(4.44),295(3.72), 357(4.04)	24-1763-78
	pH 13.0	231(4.34),295(3.67), 360(4.13)	24-1763-78
1,2,4-Triazolo[4,3-c]pyrimidine-5,7(1H,6H)-dione, 8-acetyl-3-methyl-	0.5M HCl	208(3.98),252(4.15), 298(4.21)	56-0037-78
	H_2O	208(4.13),255(4.03), 307(--)	56-0037-78
	0.5M NaOH	262(3.90),318(4.27)	56-0037-78
1,2,4-Triazolo[4,3-c]pyrimidin-5(6H)-one, 8-acetyl-7-hydroxy-3-methyl-	0.5M HCl	262(3.75),318(4.11)	56-0037-78
	H_2O	210(4.25),262(3.93), 310(--)	56-0037-78

Compound	Solvent	$\lambda_{max}(\log \epsilon)$	Ref.
$C_8H_8N_4O_4$ 9H-Purine-9-propanoic acid, 1,6-di-hydro-α-hydroxy-6-oxo-, (±)-	pH 2	250(4.03)	73-3444-78
$C_8H_8N_6$ 1,2,4,5-Tetrazin-3(2H)-one, 6-phenyl-, hydrazone	dioxan	285(4.50),402(2.16), 540(1.22)	30-0338-78
$C_8H_8N_6O_2$ 6-Pteridinecarboxylic acid, 2,4-di-amino-7-methyl-	pH 0.0	256(4.27),334(4.15), 341s(4.13)	24-3790-78
	pH 4.0	249(4.22),337(4.09), 345s(4.06)	24-3790-78
	pH 9.0	223(4.08),260(4.34), 284s(3.81),367(3.98)	24-3790-78
C_8H_8O Benzene, (ethenyloxy)-	C_6H_{12}	196(4.42),225(4.15), 263(3.03),269(3.14), 276(3.03)	99-0194-78
	heptane	270(3.10)	64-0197-78B
C_8H_8OS Benzene, (ethenylsulfinyl)-	hexane	198(4.34),217s(4.06), 242(3.61)	70-0901-78
	H_2O	200(4.34),223(4.07), 259(3.11),266(3.04), 274(2.85)	70-0901-78
$C_8H_8OS_2$ Benzenecarbodithioic acid, 4-methoxy-, sodium salt	6:4 C_6H_{12}-ether	321(4.18),360(3.99), 502(2.30)	64-0976-78B
$C_8H_8O_2$ 2H-Cyclopenta[b]furan-2-one, 3,3a,6,6a-tetrahydro-3-methylene-	MeOH	208(4.08)	87-0815-78
1,3-Dioxolane, 2-(2,4-cyclopentadien-1-ylidene)-	EtOH	292(4.42),301(4.40)	1-0293-78
$C_8H_8O_2S$ Benzaldehyde, 2-hydroxy-6-(methylthio)-	MeOH	237(5.00),296(4.43), 353(4.38)	18-2437-78
2-Thiabicyclo[3.3.1]nona-3,7-dien-9-one, 6-hydroxy-isomer	EtOH	230(3.50),318(2.28)	138-0157-78
	EtOH	231(3.62),310s(2.60)	138-0157-78
9-Thiabicyclo[3.3.1]nona-3,7-dien-2-one, 6-hydroxy-, endo	EtOH	232(3.67),298(2.94), 360s(2.49)	138-0157-78
$C_8H_8O_2S_2$ 1,4-Dithiocin-6-ol, acetate	C_6H_{12}	225(3.73),249s(3.48), 255s(3.41),285(3.49)	78-3631-78
$C_8H_8O_3$ Benzaldehyde, 4-hydroxy-3-methoxy-	MeOH	232(4.19),279(4.02), 309(4.01)	95-1607-78
	MeOH-base	251(3.92),349(4.40)	95-1607-78
2,4-Benzofurandione, 3,5,6,7-tetra-hydro-	EtOH	252(4.15)	107-0353-78
Benzoic acid, 2-hydroxy-, methyl ester	H_2O	302(3.49)	36-0334-78
anion	H_2O	333(3.62)	36-0334-78

Compound	Solvent	$\lambda_{max}(\log \epsilon)$	Ref.
Benzoic acid, 3-hydroxy-, methyl ester	H_2O	296(3.45)	36-0334-78
anion	H_2O	322(3.46)	36-0334-78
Benzoic acid, 4-hydroxy-, methyl ester	H_2O	256(4.20)	36-0334-78
	MeOH	210(3.90),257(4.01)	39-0876-78C
anion	H_2O	296(4.39)	36-0334-78
Benzoic acid, 2-methoxy-	H_2O	295(3.52)	36-0334-78
anion	H_2O	280(3.31)	36-0334-78
Benzoic acid, 3-methoxy-	H_2O	301(3.41)	36-0334-78
anion	H_2O	286(3.34)	36-0334-78
Benzoic acid, 4-methoxy-	H_2O	257(4.16)	36-0334-78
anion	H_2O	250(4.11)	36-0334-78
Menisdaurilide	EtOH	256.5(4.24)	94-1677-78
$C_8H_8O_4$			
Benzoic acid, 4-hydroxy-3-methoxy-, TiCl$_4$ complex	acetone	405(2.08)	98-0973-78
2,5-Cyclohexadiene-1,4-dione, 2,6-di-methoxy-	MeOH	286(4.18),385(2.78)	95-1607-78
	MeOH-base	247(4.06),286(3.83)	95-1607-78
$C_8H_8O_5$			
2,5-Furandicarboxaldehyde, 3,4-dimeth-oxy-	ether	237(4.02),299(4.25)	33-0430-78
C_8H_8S			
Benzene, (ethenylthio)-	heptane	265(3.46)	64-0197-78B
$C_8H_8S_2$			
Benzenecarbodithioic acid, 4-methyl-, sodium salt	EtOH	297(4.01),363(3.82), 476(2.20)	64-0976-78B
$C_8H_9BrO_3$			
4H-Pyran-4-one, 3-bromo-2-ethoxy-6-methyl-	EtOH	200(4.25),264(3.90)	23-1796-78
$C_8H_9BrO_4$			
3,5-Cyclohexadiene-1-carboxylic acid, 3-bromo-1,2-dihydroxy-, methyl ester	MeOH	277(3.81)	78-1707-78
3,5-Cyclohexadiene-1-carboxylic acid, 5-bromo-1,2-dihydroxy-, methyl ester	MeOH	266(3.42)	78-1707-78
$C_8H_9Br_2NO_3$			
1H-Pyrrole-2-carboxylic acid, 4,5-di-bromo-3-hydroxy-1-methyl-, ethyl ester	EtOH	275(4.20)	94-3521-78
$C_8H_9ClN_2O_2$			
1H-Pyrrole-3-carboxaldehyde, 2-chloro-4,5-dihydro-1-methyl-4-[(methyl-amino)methylene]-5-oxo-	EtOH	295(4.21),400(4.17)	104-2041-78
$C_8H_9ClN_2O_3$			
2-Pyrimidineacetic acid, 4-chloro-6-methoxy-, methyl ester (spectra in 5% MeOH)	neutral	254(3.61)	12-0649-78
	cation	257(4.03)	12-0649-78
$C_8H_9ClO_4$			
3,5-Cyclohexadiene-1-carboxylic acid, 3-chloro-1,2-dihydroxy-, methyl ester	MeOH	273(3.75)	78-1707-78
3,5-Cyclohexadiene-1-carboxylic acid, 5-chloro-1,2-dihydroxy-, methyl ester	MeOH	264(3.46)	78-1707-78

Compound	Solvent	$\lambda_{max}(\log \epsilon)$	Ref.
$C_8H_9Cl_3O$			
3-Hexyn-2-one, 1,1,1-trichloro-5,5-di-methyl-	hexane	234(3.88)	118-0307-78
$C_8H_9FN_2O_3$			
2,4(1H,3H)-Pyrimidinedione, 5-fluoro-1-(tetrahydro-2-furanyl)-	pH 2	271(3.95)	87-0738-78
	pH 12	270(3.84)	87-0738-78
	EtOH	270.5(3.94)	87-0738-78
2,4(1H,3H)-Pyrimidinedione, 5-fluoro-3-(tetrahydro-2-furanyl)-	pH 2	269(4.11)	87-0738-78
	pH 2	269(3.81)	95-1551-78
	pH 12	301(4.24)	87-0738-78
	pH 12	301(3.94)	95-1551-78
	MeOH	269(4.11)	87-0738-78
	MeOH	269(3.81)	95-1551-78
$C_8H_9F_3O_5S$			
1-Cyclopentene-1-carboxylic acid, 2-[[(trifluoromethyl)sulfonyl]-oxy]-, methyl ester	C_6H_{12}	221(4.08)	44-2291-78
2-Cyclopentene-1-carboxylic acid, 2-[[(trifluoromethyl)sulfonyl]-oxy]-, methyl ester	C_6H_{12}	198(3.62)	44-2291-78
C_8H_9NO			
Benzene, 1,3-dimethyl-4-nitroso-	$CHCl_3$	765(1.51)	44-2932-78
Furo[2,3-b]pyridine, 2,3-dihydro-2-methyl-	EtOH	289(3.86)	33-2542-78
Pyridine, 2-(1-methoxyethenyl)-	MeOH	250(3.8)	103-0530-78
Pyridine, 3-(2-propenyloxy)-	EtOH	277(3.60),285s(3.41)	33-2542-78
2(1H)-Pyridinone, 1-(1-propenyl)-, cis	EtOH	309(3.74)	78-2609-78
trans	EtOH	318(3.80)	78-2609-78
C_8H_9NOS			
3-Pyridinol, 2-(1-propenylthio)-	EtOH	250(3.87),311(3.90)	1-0066-78
2(1H)-Pyridinethione, 3-hydroxy-1-(1-methylethenyl)-	EtOH	280(3.75),373(4.06)	1-0066-78
Thionitrous acid, S-(1-phenylethyl) ester	hexane-1% CCl_4 at 0°	341(3.00),520s(1.30), 555(1.61)	39-0913-78C
$C_8H_9NO_2$			
Acetamide, N-hydroxy-N-phenyl-	H_2O	214(3.84),242(3.84)	73-1571-78
	10% EtOH	214(3.84),245(3.85)	73-1571-78
	30% EtOH	212(3.81),247(3.82)	73-1571-78
	50% EtOH	212(3.81),251(3.83)	73-1571-78
Benzamide, 2-hydroxy-N-methyl-	EtOH	237(3.88),301(3.92), 314s(3.44)	33-0716-78
Benzamide, N-methoxy-	H_2O or EtOH	232(3.865)	73-1571-78
Formamide, N-(4-methoxyphenyl)-	84.7% H_2SO_4	264(4.0)	104-0111-78
Phenol, 3,5-dimethyl-2-nitroso-	DMF	705(1.70)	104-2189-78
$C_8H_9NO_3$			
Acetamide, N-hydroxy-N-(4-hydroxy-phenyl)-	EtOH	254(3.93)	87-0649-78
	pH 5.02	233(3.94)	87-0649-78
3-Pyridinecarboxylic acid, 1,2-dihydro-1-methyl-2-oxo-, methyl ester	hexane	234(3.64),342(3.81)	94-1403-78
	MeOH	236(3.71),331(3.89)	94-1403-78
	ether	233(3.76),340(3.81)	94-1415-78
	Pr_2O	233(3.64),339(3.82)	94-1415-78
	isoPr$_2$O	339(3.74)	94-1415-78
	Bu_2O	232(3.62),340(3.79)	94-1415-78
	dioxan	232(3.71),338(3.85)	94-1415-78

Compound	Solvent	$\lambda_{max}(\log \epsilon)$	Ref.
3-Pyridinecarboxylic acid, 1,4-di- hydro-1-methyl-4-oxo-, methyl ester	hexane-5% MeOH	258(3.94),262(3.94), 288(3.54)	94-1403-78
	MeOH	257(3.93),262(3.93), 286(3.46)	94-1403-78
	ether	251s(3.79),258(3.90), 265(3.87),298(3.46)	94-1415-78
	Pr_2O-5% MeOH	257(3.91),263(3.90), 291(3.45)	94-1415-78
	isoPr$_2$O	258(3.89),264(3.87), 297(3.44)	94-1415-78
	Bu$_2$O	258(3.89),265(3.86), 296(3.44)	94-1415-78
	dioxan	258(3.89),265(3.86), 298(3.41)	94-1415-78
3-Pyridinecarboxylic acid, 2-methoxy-, methyl ester	hexane	227(3.87),288(3.78)	94-1403-78
	MeOH	227(3.87),288(3.77)	94-1403-78
	ether	226(3.88),287(3.79)	94-1415-78
	Pr$_2$O	227(3.77),287(3.75)	94-1415-78
	isoPr$_2$O	287(3.69)	94-1415-78
	Bu$_2$O	226(3.82),287(3.73)	94-1415-78
	dioxan	226(3.82),287(3.74)	94-1415-78
3-Pyridinecarboxylic acid, 4-methoxy-, methyl ester	hexane	228(3.87),269(3.16)	94-1403-78
	MeOH	229(3.89),271(3.24)	94-1403-78
	ether	227(3.88),268(3.21)	94-1415-78
	Pr$_2$O	227(3.83),269(3.17)	94-1415-78
	isoPr$_2$O	268(3.23)	94-1415-78
	Bu$_2$O	228(3.84),268(3.26)	94-1415-78
	dioxan	228(3.87),269(3.18)	94-1415-78
$C_8H_9NO_4$			
4-Oxa-1-azabicyclo[3.2.0]hept-2-ene- 2-carboxylic acid, 3-ethyl-7-oxo-, potassium salt	H$_2$O	260.5(3.76)	77-0469-78
5-Oxa-1-azabicyclo[4.2.0]oct-2-ene- 2-carboxylic acid, 3-methyl-8-oxo-	EtOH	258(3.91)	39-1450-78C
2,5-Pyrrolidinedione, 1-(tetrahydro- 5-oxo-2-furanyl)-	THF	211(2.29),320(1.04)	40-0404-78
$C_8H_9NO_4S$			
3-Isothiazolecarboxylic acid, 4-(acet- oxymethyl)-, methyl ester	EtOH	227(3.62),262(3.80)	44-0079-78
$C_8H_9NO_5$			
4-Isoxazoleacetic acid, 3-methoxy- 5-methyl-α-oxo-, methyl ester, (±)-	MeOH	245(3.93)	1-0027-78
$C_8H_9NO_6S$			
Acetamide, N-hydroxy-N-[4-(sulfooxy)- phenyl]-, monopotassium salt	pH 13	216(3.90),274(3.54)	87-0649-78
	90% EtOH	252(4.06)	87-0649-78
	0.1M HOAc	244(3.86)	87-0649-78
$C_8H_9N_3$			
3-Pyridinecarbonitrile, 2-amino- 4,6-dimethyl-	EtOH	218(4.17),247(3.80), 320(3.60)	103-1139-78
$C_8H_9N_3O$			
Imidazo[1,2-a]pyridine-8-carboxamide, 2,3-dihydro-	CHCl$_3$	268(3.9),426(3.5)	24-2594-78
Pyrido[2,3-b]pyrazin-2(1H)-one, 3,4-di- hydro-1-methyl-	pH 1.0	276s(3.54),324(3.95)	24-1763-78
	pH 7.0	266s(3.28),311(3.81)	24-1763-78

Compound	Solvent	$\lambda_{max}(\log \epsilon)$	Ref.
$C_8H_9N_3OS$			
Benzoic acid, 2-(aminothioxomethyl)hydrazide	MeOH	244(4.32)	80-0397-78
nickel complex	MeOH	258(4.35)	80-0397-78
$C_8H_9N_3O_2$			
Hydrazine, 1-(2-nitroethenyl)-2-phenyl-	EtOH	243(4.13),380(4.33)	20-0693-78
Imidazo[1,2-b]pyridazine-3,6(2H,5H)-dione, 2,2-dimethyl-	EtOH	358(3.29)	33-2116-78
4(1H)-Pyridinimine, 1-[1-(hydroxyimino)-2-oxopropyl]-	n.s.g.	250(3.56),259(4.15)	103-1241-78
1H-Pyrrolo[3,2-d]pyrimidine-2,4(3H,5H)-dione, 3,7-dimethyl-	MeOH	270(4.10)	39-0483-78C
$C_8H_9N_3O_3$			
Acetaldehyde, hydroxy-, (4-nitrophenyl)hydrazone	n.s.g.	358(4.01)	2-0469-78
1(2H)-Pyrimidinebutanenitrile, 3,4-dihydro-β-hydroxy-2,4-dioxo-	H_2O	266.5(4.00)	126-0905-78
1H-Pyrrolo[2,3-d]pyrimidine-2,4,5(3H)-trione, 6,7-dihydro-1,3-dimethyl-	H_2O	250(3.82),278(4.11)	103-0443-78
1H-Pyrrolo[2,3-d]pyrimidine-2,4,5(3H)-trione, 1-ethyl-6,7-dihydro-	H_2O	275(4.35)	103-0443-78
$C_8H_9N_3O_5$			
Ethanol, 2-[(2,4-dinitrophenyl)amino]-	DMSO	362(4.25),390s(--)	18-2601-78
$C_8H_9N_5O_2$			
Carbamic acid, 1H-1,2,3-triazolo[4,5-b]pyridin-5-yl-, ethyl ester	pH 1	240(3.81),251(3.66),302(4.19),308s(4.15)	44-4910-78
	pH 11	303(4.18)	44-4910-78
6,7-Pteridinedione, 5,8-dihydro-8-methyl-2-(methylamino)-	pH 1.1	242(4.49),263(4.29),348(3.78)	24-1763-78
	pH 6.0	242(4.39),290s(3.63),353(3.99)	24-1763-78
	pH 12.0	234(4.33),260s(4.10),296(3.75),360(4.09)	24-1763-78
Pyrimido[4,5-c]pyridazine-4,5(1H,6H)-dione, 7-amino-1,3-dimethyl-	MeOH	255(4.60),300(3.88),310s(3.75)	44-4844-78
$C_8H_9N_5O_3$			
6-Pteridinecarboxylic acid, 2-amino-1,4,7,8-tetrahydro-8-methyl-4-oxo-	pH -1.0	232(4.15),286(4.12),390(3.81)	24-3790-78
	pH 2.0	264(4.34),283s(3.96),403(3.91)	24-3790-78
	pH 6.0	251(4.30),283(4.00),378(3.84)	24-3790-78
	pH 11.0	259(4.26),282s(3.79),387(3.91)	24-3790-78
9H-Purine-9-propanoic acid, 6-amino-α-hydroxy-, (S)-	pH 2	262(4.15)	73-3444-78
Pyrimido[5,4-e]-1,2,4-triazine-3,5,7(6H)-trione, 2,8-dihydro-2,6,8-trimethyl-	H_2O	240(4.27),280(3.30),415(3.45)	44-0469-78
	MeOH	218(4.10),285(3.86),415(2.95)	44-0469-78
$C_8H_9N_5O_4$			
6-Pteridinecarboxylic acid, 2-amino-5-formyl-1,4,5,6,7,8-hexahydro-4-oxo-	pH -5.5	264(4.19)	24-3790-78
	pH 0.0	283(4.18)	24-3790-78
	pH 5.0	222(4.36),285(4.14)	24-3790-78
	pH 13.0	278(4.12)	24-3790-78

Compound	Solvent	λ_{max}(log ϵ)	Ref.
C_8H_{10}			
Benzene, 1,2-dimethyl-	EtOH	262(2.41),270(2.33)	35-3730-78
Bi-1-cyclobuten-1-yl	C_6H_{12}	233s(4.22),240(4.28),	5-1880-78
		249s(4.10)	
Bicyclo[2.2.1]hept-2-ene, 5-methylene-	C_6H_{12}	210s(4.04)	35-1172-78
1,3-Butadiene, 2-(1-cyclobutenyl)-	C_6H_{12}	215(4.02),221(4.03),	5-1880-78
		241(4.12),250s(4.07)	
1,2,6-Heptatriene, 5-methylene-	C_6H_{12}	223(4.29)	5-1880-78
1,5-Hexadiene, 3,4-bis(methylene)-	C_6H_{12}	216.5(4.54)	5-1880-78
Tricyclo[4.1.0.0^{4,7}]heptane, 2-methyl-ene-	C_6H_{12}	210(4.07)	35-1172-78
$C_8H_{10}BrClN_2O$			
3-Pyrrolidinone, 2-bromo-2-chloro-1-(3,4-dihydro-2H-pyrrol-5-yl)-	MeOH	230(3.73)	83-0294-78
$C_8H_{10}BrNO_3$			
1H-Pyrrole-2-carboxylic acid, 4-bromo-3-hydroxy-1-methyl-, ethyl ester	EtOH	265(4.22)	94-3521-78
1H-Pyrrole-2-carboxylic acid, 5-bromo-3-hydroxy-1-methyl-, ethyl ester	EtOH	272(4.27)	94-3521-78
$C_8H_{10}BrN_5O_2$			
1,2-Propanediol, 3-(6-amino-8-bromo-9H-purin-9-yl)-	pH 2	264(4.30)	73-3103-78
	pH 12	266(4.29)	73-3103-78
$C_8H_{10}BrN_7$			
1H-Pyrazol-3-amine, 4-[(4-bromo-5-meth-yl-1H-pyrazol-3-yl)azo]-5-methyl-	MeOH	234(4.15),326(4.11),	103-0313-78
		340(4.09)	
$C_8H_{10}ClN_3O_3$			
2,4(1H,3H)-Pyrimidinedione, 6-amino-5-(chloroacetyl)-1,3-dimethyl-	EtOH	244(3.56),281(4.02)	103-0443-78
2,4(1H,3H)-Pyrimidinedione, 6-amino-5-(chloroacetyl)-1-ethyl-	EtOH	243(3.90),276(4.12)	103-0443-78
$C_8H_{10}Cl_2N_2$			
1H-Pyrrole, 2,3-dichloro-1-(3,4-di-hydro-2H-pyrrol-5-yl)-4,5-dihydro-	MeOH	260(4.25)	83-0294-78
$C_8H_{10}N_2$			
Acetaldehyde, phenylhydrazone	n.s.g.	270(4.23)	104-0717-78
Benzaldehyde, methylhydrazone	n.s.g.	286(4.18)	104-0717-78
7,8-Diazabicyclo[4.2.2]deca-2,4,7-tri-ene (shoulder present)	MeCN	238(3.32),306(2.72)	35-0285-78
Ethanone, 1-phenyl-, hydrazone	n.s.g.	262(4.00)	104-0717-78
$C_8H_{10}N_2O$			
Ethanone, 1-[2-(methylamino)-3-pyridin-yl]-	EtOH	261(3.90),368(3.80)	104-1218-78
Methanone, cyclobutyl-1H-imidazol-2-yl-	EtOH	280(4.16)	33-2831-78
Methanone, cyclobutyl-1H-imidazol-4-yl-	EtOH	259(4.04)	33-2831-78
4-Penten-1-one, 1-(1H-imidazol-2-yl)-	EtOH	279(4.09)	33-2831-78
4-Penten-1-one, 1-(1H-imidazol-4-yl)-	EtOH	257(4.14)	33-2831-78
2-Propanone, 1-(6-methyl-3-pyridazin-yl)-	C_6H_{12}-5% ether	262(3.60),283s(3.36), 323(3.30),333(3.28)	94-3633-78
	EtOH	256(3.61),320(3.30)	94-3633-78
3-Pyridinecarboxaldehyde, 2-(ethyl-amino)-	EtOH	265(3.79),369(3.74)	104-1218-78
3-Pyridinecarboxaldehyde, 6-methyl-2-(methylamino)-	EtOH	267(3.68),367(3.79)	104-1218-78

Compound	Solvent	λ_{max}(log ϵ)	Ref.
$C_8H_{10}N_2OS_2$			
5-Oxa-2,8-dithia-10,13-diazabicyclo-[7.3.1]trideca-1(13),9,11-triene	EtOH	210(3.73),255(4.79), 300(4.11)	44-3362-78
$C_8H_{10}N_2O_2$			
2-Furancarbonitrile, 4-acetyl-5-amino-2,3-dihydro-2-methyl-	n.s.g.	290(4.26)	28-0663-78A
8H-Pyrido[1,2-a]pyrazin-8-one, 1,2,3,4-tetrahydro-9-hydroxy-, dihydrochloride	H_2O	218(4.14),278(4.07), 285(4.09)	12-0187-78
$C_8H_{10}N_2O_2S$			
2-Pyrimidineacetic acid, 4-(ethylthio)-	neutral	258(3.79),286(3.88)	12-0649-78
	cation	225(3.86),255(3.40), 306(4.20)	12-0649-78
	anion	256(3.84),284(3.89)	12-0649-78
$C_8H_{10}N_2O_3$			
2-Pyrimidineacetic acid, 4-methoxy-, methyl ester	neutral	250(3.56)	12-0649-78
	cation	240(3.82)	12-0649-78
5-Pyrimidinecarboxylic acid, 4-hydroxy-2-methyl-, ethyl ester	hexane	228(3.97),271(3.83), 300s(2.87)	94-1403-78
	MeOH	226(3.89),297(3.86)	94-1403-78
	BuOH	226(3.84),299(3.82)	94-1403-78
	ether	223(3.89),302(3.82)	94-1415-78
	Pr_2O	223(3.82),275s(3.53), 302(3.72)	94-1415-78
	isoPr$_2$O	275s(3.57),303(3.77)	94-1415-78
	Bu$_2$O	223(3.88),276s(3.55), 303(3.70)	94-1415-78
	dioxan	302(3.82)	94-1415-78
$C_8H_{10}N_2O_4$			
Diazenecarboxylic acid, (5-oxo-1,3-pentadienyl)-, ethyl ester, 1-oxide, (?,Z,E)-	EtOH	283(4.43)	94-2205-78
2-Pyrimidineacetic acid, 4,6-dimethoxy-	neutral	242(3.60)(extrapolated)	12-0649-78
	cation	252(3.90)	12-0649-78
	anion	248(3.39)	12-0649-78
$C_8H_{10}N_2O_4S$			
2-Pyrimidineacetic acid, 4-(ethylsulfonyl)-	neutral	253s(3.53),256(3.56), 264(3.43)	12-0649-78
	anion	254s(3.44),260(3.49), 265(3.43)	12-0649-78
$C_8H_{10}N_2O_5$			
4-Isoxazoleacetic acid, α-(hydroxyimino)-3-methoxy-5-methyl-, methyl ester	MeOH	241s(3.70)	1-0027-78
isomer	MeOH	212(4.04),243s(3.62)	1-0027-78
1(2H)-Pyrimidinebutanoic acid, 3,4-dihydro-β-hydroxy-2,4-dioxo-	H_2O	266.5(4.00)	126-0905-78
$C_8H_{10}N_2O_5S$			
4H-Pyran-3-carboxamide, N-(aminosulfonyl)-2,6-dimethyl-4-oxo-	EtOH	247(3.93)	4-0477-78
$C_8H_{10}N_2S_2$			
2,4(1H,3H)-Quinazolinedithione, 5,6,7,8-tetrahydro-	EtOH	292(4.37),350(4.02)	106-0185-78
	EtOH-NaOH	258(4.27),282(4.37), 370(4.04)	106-0185-78

Compound	Solvent	λ_{max}(log ϵ)	Ref.
$C_8H_{10}N_2S_3$			
5,8-Ethano-1H,3H-[1,3,4]thiadiazolo-[3,4-a]pyridazine-1,3-dithione, tetrahydro-	MeOH	266(4.2),338(4.2)	44-2224-78
$C_8H_{10}N_4$			
[1,2,4]Triazolo[1,5-c]pyrimidine, 2-ethyl-7-methyl-	pH 7.0	256(3.73)	12-2505-78
[1,2,4]Triazolo[1,5-c]pyrimidine, 2,5,7-trimethyl-	pH 7.0	257(3.82)	12-2505-78
[1,2,4]Triazolo[1,5-c]pyrimidine, 2,5,8-trimethyl-	pH 7.0	257(3.87)	12-2505-78
[1,2,4]Triazolo[1,5-c]pyrimidine, 2,7,8-trimethyl-	pH 7.0	260(3.85)	12-2505-78
[1,2,4]Triazolo[1,5-c]pyrimidine, 5,7,8-trimethyl-	pH 7.0	258(3.95),274s(3.81)	12-2505-78
1,2,4-Triazolo[4,3-c]pyrimidine, 3-ethyl-7-methyl-	pH 7.0	246s(3.79),255(3.85), 263s(3.77)	12-2505-78
1,2,4-Triazolo[4,3-c]pyrimidine, 3,5,7-trimethyl-	pH 7.0	249s(3.84),257(3.89), 266s(3.80),280(3.49)	12-2505-78
1,2,4-Triazolo[4,3-c]pyrimidine, 3,5,8-trimethyl-	pH 7.0	248s(3.75),257(3.83), 264(3.79),278s(3.64)	12-2505-78
1,2,4-Triazolo[4,3-c]pyrimidine, 3,7,8-trimethyl-	pH 7.0	248s(3.75),256(3.81), 264s(3.73),287(3.40)	12-2505-78
1,2,4-Triazolo[4,3-c]pyrimidine, 5,7,8-trimethyl-	pH 7.0	245s(3.84),256(3.96), 263s(3.93),281s(3.69)	12-2505-78
$C_8H_{10}N_4O$			
Furo[2,3-d]pyrimidine-2,4-diamine, 5,6-dimethyl-	acid	261(4.17),307(3.86)	44-3937-78
	pH 7.0	259(4.14),280s(3.93)	44-3937-78
	base	259(4.13)	44-3937-78
4H-Pyrazolo[3,4-d]pyridazin-4-one, 1,5-dihydro-1,3,7-trimethyl-	EtOH	217(4.15),227(3.81), 278(3.72)	4-0813-78
4H-Pyrazolo[3,4-d]pyridazin-4-one, 2,5-dihydro-2,3,7-trimethyl-	EtOH	216(4.09),276(3.76)	4-0813-78
4H-Pyrrolo[2,3-d]pyrimidin-4-one, 2-amino-1,7-dihydro-5,6-dimethyl-	acid	229(4.15),272(4.00)	44-3937-78
	pH 7.0	223(4.26),266(3.99), 283s(3.90)	44-3937-78
	base	266(3.96)	44-3937-78
4H-Pyrrolo[3,2-d]pyrimidin-4-one, 2-(dimethylamino)-1,5-dihydro-	pH 1	238(4.37),272(4.14)	44-2536-78
	pH 7	230(4.45),263(3.94), 282(3.97)	44-2536-78
	pH 13	232(4.44),295(3.71)	44-2536-78
$C_8H_{10}N_4O_2$			
1H-Pyrazolo[3,4-d]pyrimidine-4,6(5H,7H)-dione, 3,5,7-trimethyl-	H_2O and 0.5M NaOH	208(4.30),246(3.93), 258(3.99)	56-0037-78
2H-Pyrazolo[3,4-d]pyrimidine-4,6(5H,7H)-dione, 2,5,7-trimethyl-	EtOH	238(3.66),263(3.81)	4-0359-78
1,2,4-Triazolo[4,3-c]pyrimidine-5,7(1H,6H)-dione, 8-ethyl-3-methyl-	0.5M HCl	257(3.75),294(3.92)	56-0037-78
	H_2O	200(3.92),257(3.81), 300(3.89)	56-0037-78
	0.5M NaOH	263(3.86),295(3.83)	56-0037-78
$C_8H_{10}N_4O_2S$			
6H-Purine-6-thione, 9-(2,3-dihydroxypropyl)-1,9-dihydro-	pH 2	322(4.41)	73-3103-78
	pH 7	322(4.44)	73-3103-78
	pH 12	310(4.37)	73-3103-78
1,2,3-Thiadiazolo[4,5-d]pyrimidine-5,7(4H,6H)-dione, 4,6-diethyl-	EtOH	240(3.35),327(3.37)	44-1677-78

Compound	Solvent	$\lambda_{max}(\log \epsilon)$	Ref.
1,2,3-Thiadiazolo[4,5-d]pyrimidinium, 4,5,6,7-tetrahydro-3-methyl-5,7-di-oxo-6-propyl-, hydroxide, inner salt	EtOH	243(3.68),300(2.27), 410(2.82)	44-1677-78
C₈H₁₀N₄O₃			
2,6-Pteridinedione, 1,5,7,8-tetrahydro-7-hydroxy-4,8-dimethyl-	pH -1.0	227(4.05),257(4.01), 319(4.03)	24-1763-78
	pH 5.0	236(4.23),296(3.97)	24-1763-78
	pH 12.0	214(4.29),235(4.20), 275(4.09),309s(3.93)	24-1763-78
6H-Purin-6-one, 9-(2,3-dihydroxyprop-yl)-, (S)-	pH 2	250(4.03)	73-3103-78
	pH 7	250(4.03)	73-3103-78
	pH 12	252(4.08)	73-3103-78
C₈H₁₀N₄O₄			
Acetic acid, 2-acetyl-1-(1,2,3,6-tetra-hydro-2,6-dioxo-4-pyrimidinyl)hydra-zide	0.5M HCl	267.5(4.18)	56-0037-78
	H₂O	208(4.01),268(4.14)	56-0037-78
	0.5M NaOH	285(4.22)	56-0037-78
1,4-Benzenediamine, N¹,N⁴-dimethyl-2,3-dinitro-	C₆H₁₂	530(3.84)	39-1194-78C
	EtOH	539(--)	39-1194-78C
	CHCl₃	539(--)	39-1194-78C
1,4-Benzenediamine, N¹,N⁴-dimethyl-2,5-dinitro-	C₆H₁₂	596(3.70)	39-1194-78C
	EtOH	624(--)	39-1194-78C
	CHCl₃	632(--)	39-1194-78C
1,4-Benzenediamine, N¹,N⁴-dimethyl-2,6-dinitro-	C₆H₁₂	504(3.71)	39-1194-78C
	EtOH	532(--)	39-1194-78C
	CHCl₃	525(--)	39-1194-78C
2(1H)-Pyrimidinone, 4-methoxy-6-(nitro-amino)-1-(2-propenyl)-	pH 1.0	228(3.93),264(3.92), 332(4.16)	24-1006-78
	pH 6.0	235s(3.98),263(3.86), 336(4.01)	24-1006-78
	MeOH	230s(3.96),262(3.87), 328(4.05)	24-1006-78
C₈H₁₀N₄S₂			
7H-Purine, 7-methyl-2,6-bis(methyl-thio)-	EtOH	239(4.30),258(4.32), 320(3.99)	104-1624-78
9H-Purine, 9-methyl-2,6-bis(methyl-thio)-, hydriodide	EtOH	234(4.22),258(4.37), 308(4.13)	104-1624-78
2H-Purine-2-thione, 3,7-dihydro-3,7-di-methyl-6-(methylthio)-	H₂O	254(4.08),279(4.27), 294(4.26),349(3.97)	104-1624-78
6H-Purine-6-thione, 3,7-dihydro-3,7-di-methyl-2-(methylthio)-	H₂O	228(4.10),276s(3.91), 300(4.09),344(4.42)	104-1624-78
C₈H₁₀N₆O			
Pyrimido[4,5-c]pyridazin-4(1H)-one, 5,7-diamino-1,3-dimethyl-	MeOH	222(4.11),247(4.49), 306(4.06)	44-4844-78
C₈H₁₀N₈O₄			
1,2,4-Triazine-3,5(2H,4H)-dione, 6,6'-(1,2-ethanediyldiimino)bis-	H₂O	307(3.94)	103-1273-78
	pH 13	296(3.87)	103-1273-78
C₈H₁₀O			
Bicyclo[2.2.1]heptan-7-one, 2-methyl-ene-, (1R)-	isopentane	302(2.0)	88-4467-78
9-Oxabicyclo[4.2.1]nona-2,4-diene	EtOH	257(3.73)	49-0575-78
C₈H₁₀OS			
Benzene, 1-methoxy-4-(methylthio)-	heptane	257(4.03),290(3.20)	65-2027-78

Compound	Solvent	$\lambda_{max}(\log \epsilon)$	Ref.
$C_8H_{10}O_2$			
Benzeneethanol, 4-hydroxy-	EtOH	224(3.89),279(3.23)	102-1069-78
4-Cyclopentene-1,3-dione, 2,2,4-tri-	EtOH	231(4.12),272(4.05)	102-2015-78
	EtOH-NaOH	231(4.14),272(4.06)	102-2015-78
2-Cyclopenten-1-one, 4-hydroxy-	CHCl$_3$	270(4.25)	88-2553-78
3-(1-propenyl)-, trans			
Phenol, 4-(methoxymethyl)-	MeOH	229(3.95),280(3.18)	98-0195-78
$C_8H_{10}O_2S$			
Phenol, 2-[(methylsulfinyl)methyl]-	n.s.g.	279(3.53)	39-1580-78C
4H-Pyran-4-one, 3,5-dimethyl-2-(meth-	isooctane	211(4.16),222s(3.92),	44-4966-78
ylthio)-		275(4.09)	
	EtOH	215(4.23),226s(4.05),	44-4966-78
		284(4.14)	
2H-Thiopyran-2-one, 4-methoxy-3,5-di-	isooctane	224(4.20),332(3.49)	44-4966-78
methyl-			
4H-Thiopyran-4-one, 2-methoxy-3,5-di-	isooctane	208(4.13),232(3.85),	44-4966-78
methyl-		289(4.05)	
$C_8H_{10}O_3$			
4H-Pyran-4-one, 2-ethoxy-6-methyl-	EtOH	240(4.06)	23-1796-78
$C_8H_{10}O_3S$			
Benzenesulfonic acid, 4-methyl-,	n.s.g.	262(4.00)	124-0844-78
methyl ester			
9-Thiabicyclo[3.3.1]nonane-2,6-dione,	EtOH	230(2.79),298(2.06),	88-2819-78
9-oxide		306(2.06)	
$C_8H_{10}O_4$			
3,5-Cyclohexadiene-1-carboxylic acid,	MeOH	260.5(3.49)	78-1707-78
1,2-dihydroxy-, methyl ester			
2,4-Hexadienedioic acid, monoethyl	CHCl$_3$	262(4.35)	35-5472-78
ester, cis-cis			
$C_8H_{10}O_5$			
Oxiranecarboxylic acid, 3-(3-methoxy-	EtOH	214(4.23)	24-3665-78
3-oxo-1-propenyl)-, methyl ester			
$C_8H_{10}S$			
Benzene, 1-methyl-4-(methylthio)-	heptane	255(4.03),280(3.10)	65-2027-78
$C_8H_{10}S_2$			
Benzene, 1,4-bis(methylthio)-	heptane	275(4.30),305(3.20)	65-2027-78
$C_8H_{11}ClN_2$			
1H-Pyrrole, 4-chloro-1-(3,4-dihydro-	MeOH	258(4.31)	83-0294-78
2H-pyrrol-5-yl)-2,3-dihydro-			
$C_8H_{11}ClN_2O_3$			
2,4(1H,3H)-Pyrimidinedione, 1-(3-chlo-	H$_2$O	272(3.99)	103-0443-78
ro-2-hydroxypropyl)-5-methyl-			
$C_8H_{11}KSi$			
Potassium, (dimethylphenylsilyl)-	THF	275s(--),340(3.88)	120-0052-78
$C_8H_{11}N$			
Benzenamine, N,N-dimethyl-	C$_6$H$_{12}$	251(4.18)	65-1394-78
$C_8H_{11}NO_2$			
8-Azabicyclo[3.2.1]oct-2-ene-2-carbox-	MeOH	212(3.95)	1-0327-78
ylic acid, hydrobromide, (1R)-			

Compound	Solvent	λ_{max}(log ϵ)	Ref.
2(1H)-Pyridinone, 1-(2-hydroxypropyl)-	EtOH	303(3.75)	78-2609-78
C$_8$H$_{11}$NO$_3$			
Carbamic acid, (5-oxo-1,3-pentadien-yl)-, ethyl ester, (Z,E)-	EtOH	318(4.65)	94-2205-78
1(2H)-Pyridinecarboxylic acid, 3,6-di-hydro-3-oxo-, ethyl ester	EtOH	218(3.96)	107-0099-78
1H-Pyrrole-2-carboxylic acid, 3-hy-droxy-1-methyl-, ethyl ester	EtOH	264(4.22)	94-2224-78
1H-Pyrrole-2-carboxylic acid, 3-hy-droxy-5-methyl-, ethyl ester	MeOH	270(4.36)	35-6491-78
1H-Pyrrole-3-carboxylic acid, 4-hy-droxy-1-methyl-, ethyl ester	EtOH	214(4.04),243(4.10)	39-0896-78C
C$_8$H$_{11}$NO$_4$			
5-Isoxazoleacetic acid, 3-methoxy-α,4-dimethyl-	MeOH	214(3.82)	1-0469-78
Propanedioic acid, (1-aziridinylmeth-ylene)-, dimethyl ester	isooctane	271(4.10)	78-2321-78
C$_8$H$_{11}$NO$_5$			
4-Isoxazoleacetic acid, α-hydroxy-3-methoxy-5-methyl-, methyl ester	MeOH	212(3.82)	1-0027-78
C$_8$H$_{11}$NO$_5$S			
Furan, 2-[[(1-methylethyl)sulfonyl]-methyl]-5-nitro-	MeOH	209(4.05),238s(3.63), 312(4.07)	73-0156-78
Furan, 2-nitro-5-[(propylsulfonyl)-methyl]-	MeOH	213(3.79),234s(3.55), 311(4.07)	73-0156-78
C$_8$H$_{11}$NS			
1H-Pyrrole, 1-methyl-2-(2-propenyl-thio)-	EtOH	222(3.79),248(3.83)	23-0221-78
C$_8$H$_{11}$N$_3$OS			
Pyrido[2.3-d]pyrimidin-4(1H)-one, 2,3,5,6,7,8-hexahydro-3-methyl-2-thioxo-	MeOH	222(4.08),291(4.17)	24-2297-78
C$_8$H$_{11}$N$_3$O$_2$			
1,4-Benzenediamine, N^1,N^1-dimethyl-2-nitro-	C$_6$H$_{12}$	429(3.15)	39-1194-78C
	EtOH	440(--)	39-1194-78C
	CHCl$_3$	461(--)	39-1194-78C
1,4-Benzenediamine, N^1,N^4-dimethyl-2-nitro-	C$_6$H$_{12}$	489(3.68)	39-1194-78C
	EtOH	520(--)	39-1194-78C
	CHCl$_3$	520(--)	39-1194-78C
1,4-Benzenediamine, N^4,N^4-dimethyl-2-nitro-	C$_6$H$_{12}$	461(3.52)	39-1194-78C
	EtOH	484(--)	39-1194-78C
	CHCl$_3$	491(--)	39-1194-78C
Pyrido[2,3-d]pyrimidine-2,4(1H,3H)-dione, 5,6,7,8-tetrahydro-3-methyl-	MeOH	217(4.08),286(4.31)	24-2297-78
C$_8$H$_{11}$N$_3$O$_3$			
Ethanol, 2-[methyl(5-nitro-2-pyridin-yl)amino]-	DMSO	387(4.31)	44-0441-78
C$_8$H$_{11}$N$_3$O$_3$S			
4-Thiazoleacetic acid, 2-amino-α-(meth-oxyimino)-, ethyl ester, (E)-	EtOH	232(4.24),300(3.49)	78-2233-78
(Z)-	EtOH	234(4.15),296(3.76)	78-2233-78

$C_8H_{11}N_3O_4-C_8H_{11}N_5O_2$

Compound	Solvent	$\lambda_{max}(\log \epsilon)$	Ref.
$C_8H_{11}N_3O_4$			
α-Alanine, N-methyl-β-(1-uracilyl)-	0.05M HCl	260(3.99)	126-2195-78
	H_2O	262(3.96)	126-2195-78
	0.05M NaOH	264(3.94)	126-2195-78
2,4(1H,3H)-Pyrimidinedione, 6-amino-5-(hydroxyacetyl)-1,3-dimethyl-	EtOH	247(3.91),276(4.15)	103-0443-78
$C_8H_{11}N_3O_4S$			
L-Cysteine, S-(1,2,3,4-tetrahydro-1-methyl-2,4-dioxo-5-pyrimidinyl)-	pH 2	282(3.86)	35-5170-78
	pH 12	278(3.68)	35-5170-78
$C_8H_{11}N_3O_6$			
1H-Imidazole, 2-nitro-1-β-D-ribofuranosyl-	MeOH-acid	222(3.52),325(3.88)	44-4784-78
	MeOH-base	220(3.60),325(3.87)	44-4784-78
$C_8H_{11}N_3S$			
Hydrazinecarbothioamide, N-methyl-2-phenyl-	EtOH	240(4.36),275s(3.38)	104-0106-78
Hydrazinecarbothioamide, 1-methyl-2-phenyl-	EtOH	240(4.53),280s(3.32)	104-0106-78
Hydrazinecarbothioamide, 2-methyl-2-phenyl-	EtOH	242(4.29),280s(3.38)	104-0106-78
Hydrazinecarbothioamide, 2-(4-methyl-phenyl)-	EtOH	243(4.00),275s(3.00)	104-0106-78
$C_8H_{11}N_3S_2$			
2,4(1H,3H)-Quinazolinedithione, 3-amino-5,6,7,8-tetrahydro-	EtOH	288(4.39),312(4.30)	106-0185-78
	EtOH-NaOH	283(4.39),351(4.08)	106-0185-78
$C_8H_{11}N_4S_2$			
Purinium, 6,9-dihydro-7,9-dimethyl-2-(methylthio)-6-thioxo-, iodide	EtOH	226(4.39),273(4.13),341(4.22)	104-1624-78
$C_8H_{11}N_5$			
6H-Purin-6-imine, 3-ethyl-3,9-dihydro-9-methyl-, monohydrochloride monoperchlorate	pH 1	271(4.20)	88-5007-78
	pH 7	271(4.19)	88-5007-78
	pH 1	271(4.20)	88-5007-78
	pH 7	271(4.19)	88-5007-78
6H-Purin-6-imine, 9-ethyl-3,9-dihydro-3-methyl-, monoperchlorate	pH 1	270(4.20)	88-5007-78
	pH 7	270(4.20)	88-5007-78
[1,2,4]Triazolo[1,5-a]pyrimidin-7-amine, N-ethyl-5-methyl-	MeOH	220(--),264(3.80),291(4.22)	106-0051-78
$C_8H_{11}N_5O$			
1H-Imidazole-4-carboxamide, 5-(cyanomethylamino)-1-ethyl-	pH 1	235s(3.79)	88-2907-78
	pH 7	241(3.88)	88-2907-78
9H-Purin-6-amine, 9-ethyl-N-methoxy-	pH 1	271.5(4.13)	94-1929-78
	pH 7	269(4.17)	94-1929-78
	pH 13	286(4.07)	94-1929-78
9H-Purine-9-propanol, 6-amino-	pH 2	261(4.13)	73-3444-78
[1,2,4]Triazolo[1,5-a]pyrimidine-5-methanol, 7-(ethylamino)-	MeOH	220(--),265(3.73),293(4.17)	106-0051-78
[1,2,4]Triazolo[1,5-a]pyrimidin-6-ol, 7-(ethylamino)-5-methyl-	MeOH	225(--),262(3.64),314(4.15)	106-0051-78
	MeOH-NaOH	238(4.13),280(3.45),348(3.89)	106-0051-78
$C_8H_{11}N_5O_2$			
6(5H)-Pteridinone, 7,8-dihydro-7-hydroxy-8-methyl-2-(methylamino)-	pH 2.0	222(4.32),247(4.39),302s(3.74)	24-1763-78
	pH 7.9	216(4.29),272(4.08),	24-1763-78

Compound	Solvent	λ_{max} (log ϵ)	Ref.
6(5H)-Pteridinone, 7,8-dihydro-7-hy-droxy-8-methyl-2-(methylamino)-(cont.)	pH 14.0	309(3.98) 220(4.33),282(4.13), 322(3.96)	24-1763-78 24-1763-78
Purine, 6-amino-3-(2,3-dihydroxypro-pyl)-	pH 2 pH 7 pH 12	276(4.15) 275(4.13) 274(4.14)	73-3103-78 73-3103-78 73-3103-78
Purine, 6-amino-9-(2,3-dihydroxypro-pyl)-	pH 2 pH 2 pH 7 pH 12	262(4.15) 259(4.11) 261(4.06) 261(4.07)	73-2054-78 73-3103-78 73-3103-78 73-3103-78
6H-Purin-6-one, 2-amino-1,7-dihydro-8-methoxy-1,7-dimethyl-	pH 1 pH 7	212s(4.18),251s(3.91), 274(4.08) 217(4.40),245(3.83), 285(3.98)	44-2325-78 44-2325-78
$C_8H_{11}N_5O_2S$ 8H-Purine-8-thione, 6-amino-9-(2,3-di-hydroxypropyl)-	pH 2 pH 7 pH 12	243(4.29),308(4.51) 238(4.43),304(4.57) 241(4.32),306(4.52)	73-3103-78 73-3103-78 73-3103-78
$C_8H_{11}N_5O_3$ 1,2-Propanediol, 3-(6-amino-9H-purin-9-yl)-, N-oxide	pH 2 pH 7 pH 12	220(4.10),254(4.03), 278(3.83) 224(4.08),254(4.03), 278(3.83) 254(4.03),278(3.83)	73-3103-78 73-3103-78 73-3103-78
6H-Purin-6-one, 2-amino-9-(2,3-di-hydroxypropyl)-1,9-dihydro-	pH 2 pH 7 pH 12	251(3.95) 251(3.95) 282(3.83)	73-3103-78 73-3103-78 73-3103-78
8H-Purin-8-one, 6-amino-9-(2,3-di-hydroxypropyl)-7,9-dihydro-	pH 2 pH 7 pH 12	289(4.19) 286(4.13) 295(4.18)	73-3103-78 73-3103-78 73-3103-78
$C_8H_{11}N_7$ 1H-Pyrazol-3-amine, 5-methyl-4-[(5-methyl-1H-pyrazol-3-yl)azo]-	MeOH DMF	236(4.23),360(4.20) 416(3.29)	103-0313-78 103-0313-78
$C_8H_{11}P$ Phosphine, dimethylphenyl-	C_6H_{12}	251(2.74)	65-1394-78
C_8H_{12} 1,3-Cycloheptadiene, 5-methyl- 2,4,6-Octatriene, trans	heptane EtOH	249(3.94) 263(4.68)	70-0091-78 70-0923-78
$C_8H_{12}NO_2$ N-Oxynorpseudopelletierine	MeOH $CHCl_3$	445(3.56) 462(0.93)	22-0612-78 22-0612-78
$C_8H_{12}N_2$ 7,8-Diazatricyclo[4.2.2.0^{2,5}]dec-7-ene	hexane H_2O MeOH MeCN	383(2.31) 373(--) 382(--) 385(--)	35-5122-78 35-5122-78 35-5122-78 35-5122-78
1H-Imidazole, 1-ethenyl-2-propyl-	EtOH	231.5(4.07)	93-2264-78
$C_8H_{12}N_2O$ 1-Aza-5-azoniaspiro[4.5]dec-3-ene, 2-oxo-, hydroxide, inner salt	MeOH	256(3.26)	24-0780-78
1-Butanone, 1-(4,5-dihydro-4-methylene-1H-pyrazol-3-yl)-	EtOH	318(2.85),335(2.78)	22-0401-78
1H-Imidazole, 1-(1-oxopentyl)-	THF	245(3.52)	33-2823-78

Compound	Solvent	$\lambda_{max}(\log \epsilon)$	Ref.
1-Pentanone, 1-(1H-imidazol-2-yl)-	EtOH	278(4.10)	33-2823-78
	EtOH-HCl	266(3.92),322(1.18)	33-2823-78
	EtOH-NaOH	310(4.17)	33-2823-78
1-Pentanone, 1-(1H-imidazol-4-yl)-	EtOH	254(4.09),310(1.20)	33-2823-78
	EtOH-HCl	234(4.07),310(1.20)	33-2823-78
	EtOH-NaOH	285(4.10)	33-2823-78
1H,5H-Pyrazolo[1,2-a][1,2]diazepin-1-one, 6,7,8,9-tetrahydro-	MeOH	254(3.88)	24-0780-78
2-Pyrrolidinone, 1-(3,4-dihydro-2H-pyrrol-5-yl)-	MeOH	226(4.05)	83-0294-78

$C_8H_{12}N_2OS$

Compound	Solvent	$\lambda_{max}(\log \epsilon)$	Ref.
2(1H)-Pyrimidinone, 4-(butylthio)-	pH 1.00	330(4.08)	56-1255-78
	pH 2.00	303(3.96)	56-1255-78
	pH 3-10	300(4.04)	56-1255-78
	pH 11-13	300(4.00)	56-1255-78
2(1H)-Pyrimidinone, 5-methyl-4-(propyl-thio)-	pH 1.00	330(4.05)	56-1255-78
	pH 2	303(4.17)	56-1255-78
	pH 3-10	300(4.16)	56-1255-78
	pH 11-13	300(4.19)	56-1255-78
2(1H)-Pyrimidinone, 6-methyl-4-(propyl-thio)-	H_2O	305(3.99)	56-1255-78
4(1H)-Pyrimidinone, 1-butyl-2,3-dihydro-2-thioxo-	pH 1-7	269(4.09),290s(4.01)	44-1193-78
	pH 13	236(4.31),270(4.16)	44-1193-78

$C_8H_{12}N_2O_2$

Compound	Solvent	$\lambda_{max}(\log \epsilon)$	Ref.
Propanoic acid, 4,5-dihydro-5-methyl-4-methylene-1H-pyrazol-5-yl ester	EtOH	220(3.40),285(3.64)	22-0401-78

$C_8H_{12}N_2O_3$

Compound	Solvent	$\lambda_{max}(\log \epsilon)$	Ref.
2,4-Heptadienal, 6-aci-nitro-, O-methyloxime	M NaOH	250(3.83),353(4.63)	94-2422-78
	EtOH	251(3.82),330(4.57),343(4.63)	94-2422-78
	$CHCl_3$	283(4.19)	94-2422-78
2,4-Hexadienal, 6-(methyl-aci-nitro)-, O-methyloxime	EtOH	245(3.68),317(4.56),330(4.67),347(4.67)	94-2575-78
Hydrazinecarboxylic acid, (1-methyl-4-oxo-2-butenylidene)-, ethyl ester, (?,E)-	EtOH	294(4.38)	94-2205-78
1H-Pyrazole-4-carboxylic acid, 3-(hydroxymethyl)-5-methyl-, ethyl ester	EtOH	230(3.90)	118-0448-78

$C_8H_{12}N_2O_4$

Compound	Solvent	$\lambda_{max}(\log \epsilon)$	Ref.
Carbamic acid, [2-(2,3-dihydro-3-oxo-5-isoxazolyl)-1-methylethyl]-, methyl ester, (±)-	MeOH	211(3.86)	1-0469-78
Carbamic acid, [(3-methoxy-4-methyl-5-isoxazolyl)methyl]-, methyl ester	MeOH	215(3.87)	1-0469-78
4-Isoxazoleacetic acid, α-amino-3-methoxy-5-methyl-, methyl ester, hydrochloride, (±)-	MeOH	none above 210 nm	1-0027-78

$C_8H_{12}N_4O$

Compound	Solvent	$\lambda_{max}(\log \epsilon)$	Ref.
Ethanehydrazonic acid, N-4-pyrimidinyl-, ethyl ester	pH 7.0	266(3.75)	12-2505-78

$C_8H_{12}N_4O_3$

Compound	Solvent	$\lambda_{max}(\log \epsilon)$	Ref.
Acetic acid, 2-(1,2,3,6-tetrahydro-1,3-dimethyl-2,6-dioxo-4-pyrimidinyl)hydrazide	0.5M HCl	267(4.19)	56-0037-78
	H_2O	202(4.08),268(4.21)	56-0037-78
	0.5M NaOH	279(4.25)	56-0037-78

Compound	Solvent	$\lambda_{max}(\log \epsilon)$	Ref.
1,2,4-Triazine-3,5(2H,4H)-dione,	pH 1	309(3.77)	124-1064-78
2-methyl-6-morpholino-	EtOH-pH 7	314(3.71)	124-1064-78
	pH 13	298(3.67)	124-1064-78
1,2,4-Triazine-3,5(2H,4H)-dione,	pH 1	302(3.65)	124-1064-78
4-methyl-6-morpholino-	EtOH-pH 7	308(3.63)	124-1064-78
	pH 13	321(3.84)	124-1064-78
$C_8H_{12}N_4O_5$			
1,3,5-Triazin-2(1H)-one, 4-amino-	H_2O	241(3.83)(changing)	87-0204-78
1-β-D-ribofuranosyl-	pH 7	241(3.84)(changing)	87-0204-78
$C_8H_{12}N_6O$			
9H-Purine-9-propanol, β,6-diamino-, (±)-	pH 2,7,12	261(4.13)	73-3444-78
$C_8H_{12}N_6O_2$			
1,2-Propanediol, 3-(2,6-diamino-9H-	pH 2	252(3.99),290(4.00)	73-3103-78
purin-9-yl)-	pH 7	255(3.92),280(3.99)	73-3103-78
	pH 12	255(3.92),280(4.02)	73-3103-78
$C_8H_{12}N_6O_4S_2$			
1,2-Ethanediamine, N,N'-bis(2-oximino-	n.s.g.	268(4.44)	139-0303-78A
4-oxothiazolidin-5-yl)-			
$C_8H_{12}O$			
2-Cyclohexen-1-one, 3,4-dimethyl-	n.s.g.	236(4.13)	22-0255-78
2-Cyclopenten-1-one, 3-(1-methylethyl)-	n.s.g.	227(4.16)	22-0255-78
3,5-Heptadien-2-one, 6-methyl-	EtOH	294(4.25)	104-1319-78
$C_8H_{12}OS_2$			
Butanal, 4-(1,3-dithian-2-ylidene)-	n.s.g.	256(3.71)	33-3087-78
$C_8H_{12}O_2$			
1,3-Cyclobutanedione, 2,2,4,4-tetra-	EtOH	204(2.16),226(2.08)	42-0242-78
methyl-			
Cycloheptanone, 2-(hydroxymethylene)-	hexane	283.4(3.90)	32-0581-78
	C_6H_{12}	284.0(3.94)	32-0581-78
	MeOH	281.5(3.91)	32-0581-78
	EtOH	280.5(3.89)	32-0581-78
1,3-Cyclopentanedione, 2,2,4-trimethyl-	EtOH	229(4.31)	102-2015-78
	EtOH-NaOH	228(4.42)	102-2015-78
2-Cyclopenten-1-one, 5-hydroxy-2-pro-	EtOH	231(3.87)	107-0155-78
pyl-			
2-Hexynal, 4-hydroxy-5,5-dimethyl-	hexane	222(3.82),230(3.79)	118-0307-78
2-Propynoic acid, 2,2-dimethylpropyl	heptane	208(3.02)	65-1244-78
ester			
$C_8H_{12}O_2S$			
2H-Thiopyran-3(4H)-one, dihydro-2-(hy-	C_6H_{12}	216(3.80),250(3.14),	24-3246-78
droxymethylene)-4,4-dimethyl-		360(3.58)	
	MeOH	218(3.94),250(3.22),	24-3246-78
		338(3.65)	
$C_8H_{12}O_2S_2$			
1,4-Dithiepane-2-propanal, 3-oxo-	n.s.g.	239(3.70),283(3.30)	33-3087-78
3-Hexene-2,5-dione, 3,4-bis(methyl-	C_6H_{12}	315(3.93)	24-3233-78
thio)-,			
cis-cis	C_6H_{12}	318(3.94)	24-3233-78
$C_8H_{12}O_4S_8Sn$			
Tin, bis(O-methyl carbonodithioato-S)-	C_6H_{12}	218(4.49),273(4.52)	12-1493-78
bis(O-methyl carbonodithioato-S,S')-			

Compound	Solvent	$\lambda_{max}(\log \epsilon)$	Ref.
$C_8H_{12}S$			
1-Buten-3-yne, 1-[(1,1-dimethylethyl)-thio]-, cis	H_2O	274(4.05)	70-0133-78
trans	H_2O	275(4.09)	70-0133-78
$C_8H_{13}AsGe$			
Arsenin, 4-(trimethylgermyl)-	C_6H_{12}	218(3.70),274(3.74)	101-0247-78E
$C_8H_{13}AsSi$			
Arsenin, 4-(trimethylsilyl)-	C_6H_{12}	220(3.68),272(3.71)	101-0247-78E
$C_8H_{13}BrN_2O_3$			
2-Furanmethanaminium, N,N,N-trimethyl-5-nitro-, bromide	H_2O	201(4.01),241(3.80), 292(3.67)	73-2041-78
$C_8H_{13}ClN_2O$			
1-Aza-5-azoniaspiro[4.5]dec-3-ene, 2-oxo-, chloride	MeOH	229(3.11)	24-0780-78
5H-Pyrazolo[1,2-a][1,2]diazepin-4-ium, 6,7,8,9-tetrahydro-3-hydroxy-, chloride	MeOH	234(3.83),254s(3.56)	24-0780-78
3H-Pyrazol-3-one, 1-(5-chloropentyl)-1,2-dihydro-	MeOH	228(3.79),255s(2.54)	24-0780-78
$C_8H_{13}ClN_2O_4$			
Cyclopenta[e]-1,4-diazepine, 1,2,3,6,7,8-hexahydro-, monoperchlorate	MeOH	348(4.08)	39-1453-78C
$C_8H_{13}ClO_4S_2$			
1,2-Dithiol-1-ium, 4-ethyl-3-propyl-, perchlorate	MeOH	211(3.29),255(3.62), 305(3.82)	39-0195-78C
$C_8H_{13}FO$			
Ethanone, 1-cyclohexyl-2-fluoro-	C_6H_{12}	277(1.51)	22-0129-78
$C_8H_{13}NO$			
3-Butyn-2-one, 4-(diethylamino)-	CH_2Cl_2	282(4.18)	33-1609-78
$C_8H_{13}NO_2$			
2,4-Azetidinedione, 3,3-diethyl-1-methyl-	CH_2Cl_2	247(4.13)	33-3050-78
1H-Pyrrole-2,3-dimethanol, 1,5-dimethyl-, Ehrlich reaction product	n.s.g.	528(4.55)	39-0896-78C
1H-Pyrrole-2-ethanol, 3-(hydroxymethyl)-1-methyl-, Ehrlich product	n.s.g.	568(4.96)	39-0896-78C
2-Pyrrolidinone, 1-ethenyl-5-ethyl-5-hydroxy-	THF	228.5(4.20)	40-0404-78
$C_8H_{13}NO_3$			
5-Isoxazolepropanol, 3-methoxy-γ-methyl-, (±)-	MeOH	212(3.88)	1-0469-78
$C_8H_{13}NO_4$			
1,3-Dioxepin, 4,7-dihydro-2-(1-methylethyl)-5-nitro-	MeOH	245(3.71)	128-0259-78
$C_8H_{13}N_2O$			
1-Aza-5-azoniaspiro[4.5]dec-3-ene, 2-oxo-, chloride	MeOH	229(3.11)	24-0780-78

Compound	Solvent	$\lambda_{max}(\log \epsilon)$	Ref.
5H-Pyrazolo[1,2-a][1,2]diazepin-4-ium, 6,7,8,9-tetrahydro-3-hydroxy-, chloride	MeOH	234(3.83),254s(3.56)	24-0780-78
$C_8H_{13}N_2O_3$			
2-Furanmethanaminium, N,N,N-trimethyl-5-nitro-, bromide	H_2O	201(4.01),241(3.80), 292(3.67)	73-2041-78
$C_8H_{13}N_2S$			
1H-Imidazolium, 1,3-dimethyl-2-(2-propenylthio)-, iodide	EtOH	221(4.277),260(3.393)	28-0201-78B
	CH_2Cl_2	242(4.206),262(3.995)	28-0201-78B
$C_8H_{13}N_3O$			
2H-Pyrazolo[4,3-c]pyridin-3-ol, 4,5,6,7-tetrahydro-4,4-dimethyl-	MeOH	243(3.68)	1-0056-78
$C_8H_{13}N_3O_4$			
1H-Imidazol-2-amine, 1-α-D-ribofuranosyl-	pH 1	213(3.90)	44-4784-78
$C_8H_{13}N_5O$			
4(1H)-Pteridinone, 2-amino-5,6,7,8-tetrahydro-1,6-dimethyl-, hydrochloride	pH -2	269(4.11),338(3.04)	12-1081-78
	pH 3.5	269(3.94)	12-1081-78
	pH 8.5	218(4.11),298(3.90)	12-1081-78
4(1H)-Pteridinone, 2-amino-5,6,7,8-tetrahydro-5,6-dimethyl-, hydrochloride	pH -1	264(4.23)	12-1081-78
	pH 3.6	219(4.10),265(4.10)	12-1081-78
	pH 8.5	286(4.04)	12-1081-78
	pH 13	279(3.94)	12-1081-78
4(1H)-Pteridinone, 5,6,7,8-tetrahydro-6-methyl-2-(methylamino)-, dihydrochloride	pH -1	266(4.07)	12-1081-78
	pH 3.5	226(4.09),269(3.88)	12-1081-78
	pH 8.5	227(4.06),306(3.80)	12-1081-78
4(3H)-Pteridinone, 2-amino-5,6,7,8-tetrahydro-3,6-dimethyl-	pH -1	267(4.22)	12-1081-78
1H-1,2,3-Triazolo[4,5-d]pyrimidine, 6,7-dihydro-6-methyl-7-propoxy-, hydrochloride	PrOH	261(3.94)	39-0513-78C
$C_8H_{13}N_5O_2$			
1H-Imidazole-4-carboximidamide, 5-(formylmethylamino)-N-methoxy-1-methyl-	pH 1	253(3.88)	88-5007-78
	pH 7	250(3.75)	88-5007-78
	pH 13	250s(3.75)	88-5007-78
	EtOH	250s(3.74)	88-5007-78
1H-1,2,3-Triazole-4-carboxamide, N-(aminocarbonyl)-1-butyl-	EtOH	227(4.16)	4-1349-78
$C_8H_{13}N_5O_3$			
2,4-Pyrimidinediamine, 6-methoxy-N^2,N^2,N^4-trimethyl-N^4-nitro-	pH -1.0	226(4.36),266s(3.85), 305s(3.64)	24-1006-78
	pH 4.0	244(4.35),313(3.45)	24-1006-78
	MeOH	246(4.36),313(3.52)	24-1006-78
$C_8H_{13}N_7$			
1,2-Propanediamine, 3-(6-amino-9H-purin-9-yl)-, (±)-	pH 2	262(4.14)	73-3444-78
$C_8H_{13}S_2$			
1,2-Dithiol-1-ium, 4-ethyl-3-propyl-, perchlorate	MeOH	211(3.29),255(3.62), 305(3.82)	39-0195-78C
$C_8H_{14}NO$			
9-Azabicyclo[3.3.1]non-9-yloxy	C_6H_{12}	244(3.64),483(0.93)	22-0612-78

Compound	Solvent	λ_{max}(log ϵ)	Ref.
9-Azabicyclo[3.3.1]non-9-yloxy (cont.)	MeOH	244(3.52),428(0.99)	22-0612-78
$C_8H_{14}N_2$			
9-Azabicyclo[6.1.0]non-4-en-9-amine, (1α,4Z,8α)-	C_6H_{12}	221(2.74)	33-0795-78
2,3-Diazabicyclo[2.2.2]oct-2-ene, 1,4-dimethyl-	hexane	383(2.28)	35-5122-78
	H_2O	371(--)	35-5122-78
	MeOH	375(--)	35-5122-78
	MeCN	382(--)	35-5122-78
3H-Pyrazole, 4-(2,2-dimethylpropylidene)-4,5-dihydro-	MeOH	320(2.64)	23-0992-78
2H-Pyrrole, 3,4-dihydro-5-(1-pyrrolidinyl)-	MeOH	211(3.99)	83-0294-78
$C_8H_{14}N_2O_2$			
1,3-Cyclobutanedione, 2,2,4,4-tetramethyl-, dioxime	EtOH	206(3.78)	42-0242-78
3H-Pyrazole-3-carboxylic acid, 4,5-dihydro-3,5,5-trimethyl-, methyl ester	hexane	327(2.22)	35-5122-78
	H_2O	318(--)	35-5122-78
	MeOH	325(--)	35-5122-78
	MeCN	327(--)	35-5122-78
3H-Pyrazol-3-ol, 3,5,5-trimethyl-, acetate	hexane	328(2.33)	35-5122-78
	H_2O	323(--)	35-5122-78
	MeOH	327(--)	35-5122-78
	MeCN	328(--)	35-5122-78
$C_8H_{14}N_2O_5$			
Butanoic acid, 4-[(acetoxymethyl)nitrosoamino]-, methyl ester	EtOH	232(3.87),372(2.00)	94-3914-78
1,3-Dioxepane, 2-(1-methylethyl)-5-nitro-6-nitroso-	MeCN	205(3.58),291(3.43)	128-0259-78
1-Propanol, 3-[(acetoxymethyl)nitrosoamino]-, acetate	EtOH	231(3.85),372(1.94)	94-3914-78
$C_8H_{14}N_2O_6S$			
Propanedioic acid, [[(aminosulfonyl)amino]methylene]-, diethyl ester	EtOH	272(4.26)	4-0253-78
$C_8H_{14}N_2S$			
1H-Imidazole, 2-(butylthio)-1-methyl-	EtOH	222(3.846),250(3.728)	28-0201-78B
	CH_2Cl_2	224(3.883),245s(3.740)	28-0201-78B
1H-Imidazole, 1-(1,1-dimethylethyl)-	EtOH	225(3.792),254(3.635)	28-0201-78B
	CH_2Cl_2	230(3.790),257(3.560)	28-0201-78B
$C_8H_{14}N_4O$			
1,2,3,4-Tetraazatricyclo[7.3.0.0^{4,8}]-dodecane 2-N-oxide	MeOH	287(4.2)	33-1622-78
$C_8H_{14}N_4O_6$			
Urea, [amino(formylamino)methylene]-β-D-ribofuranosyl-	H_2O	238(4.28)(changing)	87-0204-78
	pH 7	238(4.27)(changing)	87-0204-78
$C_8H_{14}N_4S_4$			
1,3-Dithia-5,8-diazaspiro[3.4]octane, 2-[(4,5-dihydro-1H-imidazol-2-yl)-thio]-2-(methylthio)-	EtOH	264s(4.48),272(4.49)	88-3823-78
$C_8H_{14}OS$			
3-Buten-2-one, 4-(butylthio)-	H_2O	302(4.18)	70-0133-78
3-Buten-2-one, 4-[(1,1-dimethylethyl)-thio]-	H_2O	307(4.23)	70-0133-78

Compound	Solvent	$\lambda_{max}(\log \epsilon)$	Ref.
3-Buten-2-one, 4-[(2-methylpropyl)-thio]-	H_2O	303(4.21)	70-0133-78
Cyclohexane, (ethenylsulfinyl)-	hexane	198(3.45),213s(3.28), 251(3.48)	70-0901-78
$C_8H_{14}O_2$			
3-Octenoic acid	hexane	213(3.34)	119-0075-78
$C_8H_{14}O_2S_2$			
Dithiocarbonic acid, O-(2-hydroxycyclo-hexyl) ester	EtOH	228(3.79),278(4.04)	118-0286-78
$C_8H_{14}O_2S_4$			
Dithiocarbonic acid, O,O'-(1,2-dimeth-anediyl) S,S'-dimethyl ester	EtOH	230(3.77),279(4.11)	118-0286-78
$C_8H_{14}S$			
Cyclobutanethione, 2,2,3,3-tetramethyl-	C_6H_{12}	215(3.03),231(3.88), 496(1.06)	54-0121-78
	MeCN	215(3.79),231(3.88), 483(1.07)	54-0121-78
$C_8H_{15}NO_2$			
Acetamide, N-acetyl-N-(1-methylpropyl)-	$CHCl_3$	243(3.07)	20-0621-78
Acetamide, N-acetyl-N-(2-methylpropyl)-	$CHCl_3$	244(2.83)	20-0621-78
Alanine, N-(2-methylpropylidene)-, methyl ester	hexane	216(2.79)	94-0466-78
	MeOH	201(2.79)	94-0466-78
$C_8H_{15}NO_3$			
Dimethyl(3-hydroxypropyl)ammonium 1,3-dioxo-2-propanide	EtOH	257(4.44)	73-1261-78
$C_8H_{15}N_2S$			
1H-Imidazolium, 1,3-dimethyl-2-[(1-methylethyl)thio]-, iodide	EtOH	222(4.288),259(3.610)	28-0201-78B
	CH_2Cl_2	240(4.262),268(3.645), 295(3.148)	28-0201-78B
1H-Imidazolium, 1,3-dimethyl-2-(propyl-thio)-, iodide	EtOH	221(4.235),260(3.560)	28-0201-78B
	CH_2Cl_2	240(4.269),269(3.654), 296(3.026)	28-0201-78
$C_8H_{16}B_2F_8S_2$			
1,6-Dithiecane dication bis(tetra-fluoroborate)	MeCN	232(3.57)	35-6416-78
$C_8H_{16}N_2$			
9-Azabicyclo[6.1.0]nonan-9-amine, cis	C_6H_{12}	218(2.84)	33-0795-78
2-Hexenal, dimethylhydrazone, (E,E)-	C_6H_{12}	275(4.42)	28-0047-78A
Pyridazine, 3,4,5,6-tetrahydro-3,3,6,6-tetramethyl-	hexane	380(2.14)	35-1876-78
$C_8H_{16}N_2O_2$			
2-Butanone, 1-(butylnitrosoamino)-	EtOH	240(3.74),349(1.92)	94-3891-78
2-Butanone, 4-(butylnitrosoamino)-	EtOH	234(3.85),351(1.92)	94-3891-78
1-Buten-1-amine, N-(1,1-dimethyl-ethyl)-2-nitro-	EtOH	259(2.58),370(4.13)	44-0497-78
2-Propanone, 1-(nitrosopentylamino)-	EtOH	241(3.75),351(1.90)	94-3891-78
$C_8H_{16}N_2O_3$			
β-Alanine, N-nitroso-N-pentyl-	EtOH	235(3.85),353(1.93)	94-3909-78
Butanoic acid, 4-(butylnitrosoamino)-	EtOH	235(3.87),352(1.94)	94-3909-78

Compound	Solvent	$\lambda_{max}(\log \epsilon)$	Ref.
Butanoic acid, 4-[(1,1-dimethylethyl)-nitrosoamino]-	EtOH	232(3.85),355(1.77)	94-3909-78
$C_8H_{16}N_2O_4$ Butanoic acid, 4-(butylnitrosoamino)-3-hydroxy-	EtOH	236(3.86),353(1.97)	94-3909-78
$C_8H_{16}N_2S$ 1H-Imidazole, 1-(1,1-dimethylethyl)-4,5-dihydro-2-(methylthio)-	EtOH	225(3.467)	28-0521-78A
	CH_2Cl_2	240(3.416)	28-0521-78A
Pyrimidine, 1,4,5,6-tetrahydro-1-(1-methylethyl)-2-(methylthio)-	EtOH	227(3.892)	28-0521-78A
	CH_2Cl_2	none	28-0521-78A
2(1H)-Pyrimidinethione, tetrahydro-1-methyl-3-(1-methylethyl)-	EtOH	241(4.226)	28-0521-78A
	CH_2Cl_2	246(4.252)	28-0521-78A
$C_8H_{16}N_4$ 1,2,4,5-Tetrazine, 1,4-dihydro-1,4-di-propyl-	heptane	242(3.37)	104-0576-78
$C_8H_{17}AsO_5S$ β-D-Galactopyranose, 1-thio-, 1-(di-methylarsinite)	MeOH	222(3.78)	136-0069-78E
$C_8H_{17}NO$ Butanamide, N,N-diethyl-	H_2O	223(2.3)	104-0221-78
$C_8H_{17}NOS$ Thionitrous acid, S-octyl ester	hexane-1% CCl_4 at 0°	339(2.85),515(0.90), 547(1.23)	39-0913-78C
$C_8H_{17}N_2S$ 1H-Imidazolium, 4,5-dihydro-1-methyl-3-(1-methylethyl)-2-(methylthio)-, iodide	EtOH	221(4.239),256(3.921)	28-0521-78A
	CH_2Cl_2	245(4.286)	28-0521-78A
Pyrimidinium, 1-ethyl-1,4,5,6-tetra-hydro-3-methyl-2-(methylthio)-, iodide	EtOH	221(4.00),245(4.235)	28-0521-78A
	CH_2Cl_2	244(4.317)	28-0521-78A
$C_8H_{17}N_3O$ Acetic acid, [(dimethylamino)methyl-ene]propylhydrazide	heptane	227(3.48),263(3.70)	104-0576-78
Methanehydrazonamide, N'-(1,1-dimeth-ylethyl)-N'-formyl-N,N-dimethyl-	heptane	218(3.57),263(3.32)	104-0576-78
$C_8H_{18}N_2O_2$ 1-Butanamine, N-butyl-N-nitro-	H_2O	245(3.9)	104-0221-78
1-Butanol, 4-(butylnitrosoamino)-	EtOH	235(3.88),352(1.93)	94-3891-78
1-Butanol, 4-[(1,1-dimethylethyl)-nitrosoamino]-	EtOH	232(3.86),353(1.78)	94-3891-78
2-Butanol, 1-(butylnitrosoamino)-	EtOH	237(3.85),351(1.96)	94-3891-78
2-Butanol, 4-(butylnitrosoamino)-	EtOH	234(3.83),350(1.91)	94-3891-78
Diazene, bis(1,1-dimethylethyl)-, 1,2-dioxide	H_2O	286(3.95)	104-0412-78
1-Propanol, 3-(nitrosopentylamino)-	EtOH	235(3.86),351(1.94)	94-3891-78
$C_8H_{18}N_2O_3$ 1,3-Butanediol, 4-(butylnitrosoamino)-	EtOH	237(3.81),352(1.93)	94-3909-78
$C_8H_{18}N_4$ 1,2,3,4-Tetrazine, 1,4-diethyl-1,4,5,6-tetrahydro-5,6-dimethyl-	MeOH	228(3.65),272(4.1)	33-1622-78

Compound	Solvent	$\lambda_{max}(\log \epsilon)$	Ref.
$C_8H_{18}N_4O$			
1,2,3,4-Tetrazine, 1,4-diethyl-1,4,5,6-tetrahydro-5,6-dimethyl-, 2-oxide, trans	MeOH	290(4.04)	33-1622-78
$C_8H_{18}OS$			
Propane, 1,1'-sulfinylbis[2-methyl-	hexane	207s(3.43),224(2.90)	70-0901-78
	H_2O	213(3.00)	70-0901-78
	EtOH	204(3.04),214(3.00)	70-0901-78
Propane, 2,2'-sulfinylbis[2-methyl-	hexane	206(3.57),229s(2.95)	70-0901-78
	H_2O	196(3.15),221(2.85)	70-0901-78
$C_8H_{18}O_2Zn$			
Zinc, bis(3-methoxypropyl-C,O)-, 2,2'-bipyridine complex	benzene	405(2.48)	101-0245-78J
$C_8H_{18}S_2Zn$			
Zinc, bis[3-(methylthio)propyl-C,S]-, 2,2'-bipyridine complex	benzene	375(2.43)	101-0245-78J
$C_8H_{18}Zn$			
Zinc, dibutyl-	benzene	425(2.56)	101-0245-78J
$C_8H_{20}MoO_4$			
Molybdenum, tetraethoxy-	C_6H_{12}	700(2.53)	35-2744-78
$C_8H_{20}N_8O_2$			
Ethanamine, 2,2'-(1,4-dimethyl-2-tetrazene-1,4-diyl)bis[N-methyl-N-nitroso-	MeOH	234(4.23),278(4.08)	33-1622-78
$C_8N_8O_2$			
1,4-Cyclohexadiene-1,4-dicarbonitrile, 2,5-diazido-3,6-dioxo-	MeCN	226(4.18),260(4.24), 334(4.40)	89-0369-78

Compound	Solvent	$\lambda_{max}(\log \epsilon)$	Ref.
$C_9H_3N_3$ 1,2,4-Benzenetricarbonitrile	hexane	214(4.62),247(4.18), 296(3.26)	110-0044-78
$C_9H_4Cl_6$ Benzene, pentachloro(1-chloro-1-propen-yl)-, (Z)-	heptane	217(4.39)	5-1406-78
Bicyclo[3.2.0]hepta-2,6-diene, 1,2,3,5,6,7-hexachloro-4-ethylidene-, (1α,4E,5α)-	heptane	210(4.05),270(3.92)	5-1406-78
$C_9H_5BrN_2O$ Propanedinitrile, [1-(5-bromo-2-furan-yl)ethylidene]-	EtOH	214(4.02),253(3.51), 355(4.44)	73-0870-78
2-Quinoxalinecarboxaldehyde, 3-bromo-	CHCl$_3$	333(4.88)	97-0177-78
$C_9H_5Br_4N$ 1H-Indole, 2,3,5,6-tetrabromo-1-methyl-	EtOH	233(4.69),296(4.00), 303(4.00)	88-4479-78
$C_9H_5ClN_2OS_2$ Methanone, (4-chlorophenyl)(5-mercapto-1,2,3-thiadiazol-4-yl)-, potassium salt	MeOH	257(4.19),360(3.98)	4-1295-78
C_9H_5ClOS 2H-1-Benzothiopyran-2-one, 4-chloro-	EtOH	264(3.62),294(3.64), 303(3.63)	103-0507-78
$C_9H_5NO_3S$ 2H-Thiopyrano[2,3-b]pyridine-3-carbox-ylic acid, 2-oxo-	EtOH	230(4.38),261(3.82), 301(3.64),335(3.86)	78-0989-78
$C_9H_5N_3OS$ Thiazolo[4,5-b]quinoxalin-2(3H)-one	EtOH	225(4.263),266(4.231)	2-0683-78
$C_9H_5N_3S_2$ Pyrido[3',2':4,5]thieno[3,2-d]pyrimi-dine-4(1H)-thione	EtOH	252(3.95),264(4.04), 268(4.03),303(4.00), 313(3.99),321(3.79), 374(3.90),408(3.48)	32-0057-78
$C_9H_5N_7S$ 5H-Tetrazolo[5',1':2,3][1,3,4]thiadia-zino[5,6-b]quinoxaline	EtOH	213(4.36),227(4.44), 270(4.01),302(3.99)	2-0307-78
C_9H_6BrNOS 4H-1-Benzothiopyran-4-one, 2-amino-3-bromo-	EtOH	231(4.41),258s(--), 270s(--),327(4.10)	103-0141-78
$C_9H_6BrNO_2S$ Maleimide, 2-bromo-1-methyl-3-(2-thi-enyl)-	n.s.g.	254(3.92),385(3.71)	88-0125-78
C_9H_6BrNS Thiazole, 5-bromo-2-phenyl-	EtOH	298(4.23)	39-0685-78C
$C_9H_6Br_2N_2O_2$ 2-Pyridinepropanoic acid, α,β-dibromo-3-cyano-	H$_2$O	264(3.97)	103-0049-78

Compound	Solvent	$\lambda_{max}(\log \epsilon)$	Ref.
$C_9H_6Br_3N$			
1H-Indole, 2,3,5-tribromo-1-methyl-	EtOH	228(4.53),282(3.86), 290(3.89),297(3.89)	88-4479-78
1H-Indole, 2,3,6-tribromo-1-methyl-	EtOH	230(4.59),288(4.00), 294(4.00)	88-4479-78
$C_9H_6ClNO_2S$			
1H-Pyrrole-2,5-dione, 3-chloro-4-(5-methyl-2-thienyl)-	MeOH	240(3.94),258(4.00), 390(3.94)	88-0125-78
$C_9H_6ClNO_2S_2$			
1,4,2-Dithiazine, 3-chloro-5-phenyl-, 1,1-dioxide	EtOH	245(3.99)	18-1805-78
C_9H_6ClNS			
Benzene, 1-chloro-4-(2-isothiocyanato-ethenyl)-, trans	dioxan	308(4.70)	73-1917-78
$C_9H_6Cl_2N_2S$			
Cinnoline, 4-chloro-3-[(chloromethyl)-thio]-	EtOH	252(4.38),360(3.30)	4-0115-78
C_9H_6DNS			
Thiazole-2-d, 4-phenyl-	EtOH	253(4.18)	39-0685-78C
Thiazole-5-d, 2-phenyl-	EtOH	288(4.15)	39-0685-78C
$C_9H_6Li_2$			
Lithium, [μ-(3-phenyl-2-propynyli-dene)]di-	ether	382(4.38)	101-0001-78N
$C_9H_6N_2O$			
1,8-Naphthyridine-2-carboxaldehyde	CHCl₃	275(3.82),286s(3.75), 315(3.67),326s(3.60)	97-0030-78
Propanedinitrile, [1-(2-furanyl)ethyli-dene]-	EtOH	208(3.85),249(3.44), 345(4.37)	73-0870-78
$C_9H_6N_2O_2S$			
Pyridinium, 1-(5-formyl-2,3-dihydro-2-oxo-4-thiazolyl)-, hydroxide, inner salt	MeOH	255(3.2),320(3.1), 450(2.7)	103-0208-78
$C_9H_6N_2O_3$			
1(2H)-Isoquinolinone, 5-nitro-	20% MeCN	261(4.06),313(3.83), 372(3.69)	44-1132-78
Isoxazole, 3-(4-nitrophenyl)-	HOAc	279(4.13)	103-0264-78
	HOAc-H₂SO₄	281(4.28)	103-0264-78
Isoxazole, 5-(4-nitrophenyl)-	HOAc	280(4.05)	103-0264-78
	HOAc-H₂SO₄	282(4.33)	103-0264-78
$C_9H_6N_4O$			
Isoxazolo[4,5-d]-1,2,3-triazole, 3-phenyl-	EtOH	266(4.20)	33-0108-78
[1,2,4]Triazolo[4,3-a]quinoxaline, 5-oxide	MeOH	228s(--),263(3.87), 322(4.04)	44-4125-78
$C_9H_6N_4S$			
1H-Isothiazolo[4,5-d]-1,2,3-triazole, 3-phenyl-	EtOH	290s(4.16),299(4.19)	33-0108-78
$C_9H_6N_4S_2$			
3H-[1,3,4]Thiadiazino[5,6-b]quinoxa-	EtOH	215(4.5),295(4.15)	2-0307-78

Compound	Solvent	$\lambda_{max}(\log \epsilon)$	Ref.
line-3-thione, 1,2-dihydro- (cont.)			2-0307-78
$C_9H_6O_2S$			
Benzo[b]thiophene-2-carboxylic acid	EtOH	230(4.24),241s(--), 277s(--),284(4.12)	103-1085-78
$C_9H_6O_3$			
2H-1-Benzopyran-2-one, 4-hydroxy-	MeOH	302(4.04)	49-0123-78
	n.s.g.	210(4.34),269(3.86), 280(3.95),303(3.88), 316(3.66)	78-1221-78
β-cyclodextrin complex	H_2O	286(4.19),295s(4.14)	133-0241-78
2H-1-Benzopyran-2-one, 7-hydroxy-	H_2O	325(4.18)	51-0652-78
	MeOH	324(4.17)	49-0123-78
	EtOH	326(4.18)	51-0652-78
anion	H_2O	369(4.28)	51-0652-78
anion	EtOH	377(4.32)	51-0652-78
$C_9H_6O_4$			
2H-1-Benzopyran-2-one, 4,7-dihydroxy-	MeOH	309(4.25)	49-0123-78
$C_9H_7BrN_2OS$			
4(5H)-Thiazolone, 2-[(4-bromophenyl)- amino]-	dioxan	274(4.05)	103-0148-78
$C_9H_7BrO_3$			
1,4-Benzodioxin-5-carboxaldehyde, 7-bromo-2,3-dihydro-	EtOH	224(4.44),271(3.85), 347(3.48)	103-1188-78
$C_9H_7BrO_3S$			
Benzenesulfonic acid, 4-bromo-, 2-propynyl ester	n.s.g.	234(4.64)	124-0844-78
$C_9H_7BrO_4$			
1,4-Benzodioxin-5-carboxylic acid, 6-bromo-2,3-dihydro-	EtOH	225(3.68),297(3.09)	103-1188-78
1,4-Benzodioxin-5-carboxylic acid, 7-bromo-2,3-dihydro-	EtOH	215(4.42),314(3.34)	103-1188-78
$C_9H_7ClN_2OS$			
4(5H)-Thiazolone, 2-[(3-chlorophenyl)- amino]-	dioxan	272(3.87)	103-0148-78
4(5H)-Thiazolone, 2-[(4-chlorophenyl)- amino]-	dioxan	274(3.99)	103-0148-78
$C_9H_7ClN_2O_2S$			
Benzene, 1-(1-chloro-2-isothiocyanato- ethyl)-4-nitro-	dioxan	264(4.0)	73-1917-78
2,4-Thiazolidinedione, 5-(4-chloro- phenyl)-, 2-oxime, (Z)-	n.s.g.	208(4.17),225(4.27)	139-0303-78A
$C_9H_7ClN_2O_3S$			
Acetyl chloride, diazo[(4-methylphen- yl)sulfonyl]-	C_6H_{12}	366(0.35)	130-0189-78
$C_9H_7ClN_4$			
Formaldehyde, (4-chloro-1(2H)-phthala- zinylidene)hydrazone, (E)-	dioxan	282(4.38),345(3.98)	103-0567-78
	MeCN	212(4.54),281(4.32), 342(3.94)	103-0567-78

Compound	Solvent	$\lambda_{max}(\log \epsilon)$	Ref.
$C_9H_7ClO_3S$			
Benzenesulfonic acid, 3-chloro-, 2-propynyl ester	n.s.g.	221(4.41)	124-0844-78
Benzenesulfonic acid, 4-chloro-, 2-propynyl ester	n.s.g.	229(4.71)	124-0844-78
$C_9H_7Cl_2NS$			
Benzene, 1-chloro-4-(1-chloro-2-iso-thiocyanatoethyl)-	dioxan	none above 250 nm	73-1917-78
C_9H_7Li			
Lithium, (3-phenyl-2-propynyl)-	ether	362(2.81)	101-0001-78N
C_9H_7NO			
Benzonitrile, 3-acetyl-	heptane	293(2.78)(smoothed)	73-2763-78
	MeOH	292.5(2.77)	
Benzonitrile, 4-acetyl-	heptane	295(3.09)(smoothed)	73-2763-78
Isoxazole, 3-phenyl-	EtOH	240(4.05)	78-1571-78
	H_2SO_4	276(4.07)	78-1571-78
	HOAc	251(3.88)	103-0264-78
	HOAc-H_2SO_4	275(4.17)	103-0264-78
Isoxazole, 5-phenyl-	HOAc	262(4.27)	103-0264-78
	HOAc-H_2SO_4	290(4.32)	103-0264-78
5H-Pyrrolo[1,2-a]azepin-5-one	EtOH	242(4.15),290(3.60), 300(3.58),350(2.98), 410(3.36)	24-2407-78
9H-Pyrrolo[1,2-a]azepin-9-one	EtOH	234(4.31),261(4.34), 310(3.77)	24-2407-78
C_9H_7NOS			
Benzo[b]thiophene-2-carboxamide	EtOH	228(4.36),245(4.32), 279s(--),287(4.33)	103-1085-78
4H-1-Benzothiopyran-4-one, 2-amino-	EtOH	233(4.48),252s(--), 258s(--),320(4.18)	103-0141-78
Benzoxazole, 2-(ethenylthio)-	EtOH	245(4.10),253(4.11), 260(4.07),279(4.21), 286(4.18)	121-1477-78
2H-Thiopyrano[2,3-b]pyridin-2-one, 3-methyl-	EtOH	227(4.26),258(3.77), 292(3.60),319(3.75)	78-0989-78
$(C_9H_7NOS)_n$			
Benzoxazole, 2-(ethenylthio)-, homo-polymer	THF	245(4.02),259(3.88), 280(3.97),288(3.93)	121-1477-78
$C_9H_7NO_2$			
Benzonitrile, 3-acetoxy-	heptane	295(2.97)	73-2763-78
Benzonitrile, 3-carbomethoxy-	heptane	288(2.70)(smoothed)	73-2763-78
Benzonitrile, 4-carbomethoxy-	heptane	291.5(3.15)(smoothed)	73-2763-78
1H-Indole-3-carboxaldehyde, 1-hydroxy-	MeOH	213(3.87),245(3.56), 255(3.49),308(3.47), 350(3.35)	39-1117-78C
1H-Indole-2,3-dione, 1-methyl-	MeOH	245(4.29),251(4.22), 300(3.37)	150-1683-78
2,4(1H,3H)-Quinolinedione	EtOH	313(3.80)	64-0332-78B
$C_9H_7NO_3$			
Furo[3,2-b]pyridine-2-carboxylic acid, 5-methyl- (hydrate)	MeOH	254(3.87),302(4.20), 308(4.16)	4-0029-78
hydrochloride	20% HCl	230(3.61),258(3.38), 287(3.37),315(4.39)	4-0029-78

Compound	Solvent	λ_{max}(log ϵ)	Ref.
1H-Isoindole-1,3(2H)-dione, 4-methoxy-	EtOH	227.5(4.21)	94-0530-78
1H-Isoindole-1,3(2H)-dione, 5-methoxy-	EtOH	234.5(4.26)	94-0530-78
C$_9$H$_7$NO$_4$			
Furo[3,4-b]pyridine-7-acetic acid, 5,7-dihydro-5-oxo-	H$_2$O	270(3.60)	103-0049-78
C$_9$H$_7$NO$_5$			
1,4-Benzodioxin-5-carboxaldehyde, 2,3-dihydro-6-nitro-	EtOH	212(3.80),221(3.82), 350(3.61)	103-1188-78
1,4-Benzodioxin-5-carboxaldehyde, 2,3-dihydro-7-nitro-	EtOH	253(4.10),323(3.81)	103-1188-78
1,4-Benzodioxin-6-carboxaldehyde, 2,3-dihydro-7-nitro-	EtOH	259(4.09),330(3.67)	103-1188-78
2H-1,4-Benzoxazine-5-carboxylic acid, 3,4,6,7-tetrahydro-6,7-dioxo-	2:1 dioxan-EtOH	275(3.94),307(4.09), 457(3.08)	150-2201-78
C$_9$H$_7$NO$_5$S			
Benzenesulfonic acid, 3-nitro-, 2-propynyl ester	n.s.g.	248(4.55)	124-0844-78
Benzenesulfonic acid, 4-nitro-, 2-propynyl ester	n.s.g.	250(4.84)	124-0844-78
C$_9$H$_7$NO$_5$S$_2$			
Furan, 2-nitro-5-[(2-thienylsulfonyl)-methyl]-	MeOH	233(4.06),258s(4.90), 313(4.00)	73-0160-78
C$_9$H$_7$NO$_6$			
1,4-Benzodioxin-5-carboxylic acid, 2,3-dihydro-6-nitro-	EtOH	217(4.02),320(3.75)	103-1188-78
C$_9$H$_7$NO$_6$S			
Furan, 2-[(2-furanylsulfonyl)methyl]-5-nitro-	MeOH	218(4.26),258(3.43), 267(3.54),274(3.64), 312(4.11)	73-0160-78
C$_9$H$_7$NS			
Thiazole, 2-phenyl-	C$_6$H$_{12}$	287(4.14)	39-0685-78C
C$_9$H$_7$NS$_2$			
Benzothiazole, 2-(ethenylthio)-	EtOH	203(4.28),226(4.23), 248(3.93),286(4.14), 294(4.14),304(4.08)	121-1477-78
(C$_9$H$_7$NS$_2$)$_n$			
Benzothiazole, 2-(ethenylthio)-, homopolymer	THF	226(4.24),249(3.92), 285(4.01),295(4.00), 303(3.92)	121-1477-78
C$_9$H$_7$N$_3$O			
2-Quinoxalinecarboxaldehyde, 3-amino-	CHCl$_3$	420(3.31)	97-0177-78
C$_9$H$_7$N$_3$O$_2$			
2H-Benzimidazole-2-carbonitrile, 2-methyl-, 1,3-dioxide	MeOH	247(4.22),531(3.69)	44-0076-78
3H-2,1,4-Benzoxadiazine-3-carbonitrile, 3-methyl-, 4-oxide	MeOH	228(4.23),405(3.56)	44-0076-78
C$_9$H$_7$N$_3$O$_3$S			
Pyrimidine, 2-[[(5-nitro-2-furanyl)-methyl]thio]-	MeOH	208(4.11),244(4.32), 321(4.20)	73-0160-78

Compound	Solvent	$\lambda_{max}(\log \epsilon)$	Ref.
4(5H)-Thiazolone, 2-[(4-nitrophenyl)-amino]-	dioxan	316(4.14)	103-0148-78
$C_9H_7N_3O_4S$ 2,4-Thiazolidinedione, 5-(4-nitrophenyl)-, 2-oxime, (Z)-	n.s.g.	215s(--),260(4.13)	139-0303-78A
$C_9H_7N_3O_5S$ Pyrimidine, 2-[[(5-nitro-2-furanyl)-methyl]sulfonyl]-	MeOH	216(4.19),241s(3.90), 311(4.25)	73-0160-78
$C_9H_7N_5$ 1,2,4-Triazolo[3,4-a]phthalazin-3-amine	MeOH	209(4.47),258(4.39), 266(4.42)	4-0311-78
$C_9H_7N_5O$ Tetrazolo[1,5-a]quinoxaline, 4-methyl-, 5-oxide	MeOH	232(4.35),262s(--), 317(4.04)	44-4125-78
$C_9H_7N_5O_2$ Quinoxaline, 2-azido-3-methyl-, 1,4-dioxide	MeOH	237(4.37),280(4.48), 380(3.90)	44-0076-78
$C_9H_7N_5O_2S_2$ 1H-Purine, 6-(methylsulfonyl)-8-(4-thiazolyl)-	pH 0.9 pH 4.8 pH 13	314(4.25) 314(4.24) 326(4.18)	87-0344-78 87-0344-78 87-0344-78
$C_9H_5N_5S$ 1H-1,3,4-Thiadiazino[5,6-b]quinoxalin-3-amine sulfate (1:1)	EtOH EtOH	217(4.35),225(4.15), 303(4.05),313(4.05) 218(4.30),278(4.30), 303(4.01),312(4.01)	2-0307-78 2-0307-78
Thiazolo[4,5-b]quinoxalin-3(2H)-amine, 2-imino- hydrochloride	EtOH EtOH	219(4.57),259(4.4) 217(4.44),236(4.45)	2-0683-78 2-0683-78
C_9H_8 Cyclopropa[3,4]benzocyclobutene	hexane	264s(3.1+),270(3.2+), 277(3.2+)	35-0980-78
$C_9H_8ClFN_2O_4$ 6H-Furo[2',3':4,5]oxazolo[3,2-a]pyrimidin-6-one, 2-(chloromethyl)-7-fluoro-2,3,3a,9a-tetrahydro-3-hydroxy-, [2R-(2α,3β,3aβ,9aβ)]-	H_2O	225(3.89),253(3.99)	73-3268-78
C_9H_8ClN 2H-Azirine, 2-(chloromethyl)-3-phenyl- Benzenecarboximidoyl chloride, N-ethenyl-	EtOH C_6H_{12}	241(4.15) 264(4.11)	35-4481-78 46-4481-78
C_9H_8ClNO 2-Propenal, 3-[(4-chlorophenyl)amino]-	96% H_2SO_4	250(4.0),332(4.3) (changing)	94-0930-78
$C_9H_8ClNO_2$ Furo[3.2-c]pyridin-4(5H)-one, 7-chloro-2,6-dimethyl-	EtOH	256(4.09),264(4.06), 305(3.87)	23-0613-78

$C_9H_8ClN_3S-C_9H_8N_2O_2$

Compound	Solvent	$\lambda_{max}(\log \epsilon)$	Ref.
$C_9H_8ClN_3S$			
Pyrido[2,3-d]pyrimidine, 4-chloro-6-methyl-2-(methylthio)-	pH 1	246(4.21),275(4.29), 375(3.90)	44-0828-78
	pH 7	243(4.35),271(4.31), 354(3.80)	44-0828-78
	pH 11	235(4.19),267(4.31), 346(3.94)	44-0828-78
$C_9H_8Cl_3OP$			
Phosphonic dichloride, [2-chloro-2-(4-methylphenyl)ethenyl]-	$C_2H_4Cl_2$	225(4.08),289(4.34)	65-1817-78
	MeCN	222(4.08),285(4.36)	65-1817-78
$C_9H_8Cl_3O_2P$			
Phosphonic dichloride, [2-chloro-2-(4-methoxyphenyl)ethenyl]-	$C_2H_4Cl_2$	233(4.11),311(4.36)	65-1817-78
	MeCN	230(4.08),300(4.38)	65-1817-78
$C_9H_8Cl_4OP$			
Phosphorus(1+), trichloro[2-chloro-2-(4-methoxyphenyl)ethenyl]-, (T-4)-, hexachlorophosphate	$C_2H_4Cl_2$	255(4.00),270s(3.93), 398(4.45)	65-1817-78
	MeCN	249(3.99),266s(3.92), 369(4.46)	65-1817-78
$C_9H_8Cl_4P$			
Phosphorus(1+), trichloro[2-chloro-2-(4-methylphenyl)ethenyl]-, (T-4)-, hexachlorophosphate	$C_2H_4Cl_2$	269(4.00),349(4.38)	65-1817-78
	MeCN	234(3.89),270s(3.97), 333(4.38)	65-1817-78
	$MeNO_2$	267s(3.93),338(4.36)	65-1817-78
C_9H_8N			
Quinolizinium bromide	H_2O	225(4.31),270(3.57), 281(3.57),300s(--), 310(4.08),317(4.02), 324(4.29)	40-1249-78
C_9H_8NO			
Quinolizinium, 1-hydroxy-, bromide	H_2O	237(4.40),300s(--), 334s(--),347(4.42)	40-1249-78
Quinolizinium, 2-hydroxy-, bromide	H_2O	232(4.15),300(4.05), 335(3.81)	40-1249-78
Quinolizinium, 3-hydroxy-, bromide	H_2O	238(4.34),266(4.17), 268(4.18),328(3.88)	40-1249-78
$C_9H_8N_2$			
3H-1,2-Benzodiazepine	EtOH	225(4.34),261(3.95)	94-1890-78
$C_9H_8N_2O$			
3H-1,2-Benzodiazepine, 1-oxide	EtOH	223(4.36),310(3.23)	94-1896-78
3H-1,2-Benzodiazepine, 2-oxide	EtOH	222(4.20),310(3.11)	94-1896-78
3H-Pyrazol-3-one, 1,2-dihydro-1-phenyl-	MeOH	269(4.24)	24-0780-78
4(1H)-Quinazolinone, 1-methyl-	EtOH	268(3.94),277(3.99), 306(4.23),317(4.14)	102-2125-78
	EtOH-HCl	278(4.09),294(4.04), 305(3.90)	102-2125-78
$C_9H_8N_2OS$			
2H-1-Benzothiopyran-2-one, 3,4-diamino-	EtOH	236(4.64),278(3.96), 365(4.03)	103-0033-78
4(5H)-Thiazolone, 2-(phenylamino)-	dioxan	270(3.93)	103-0148-78
$C_9H_8N_2O_2$			
Ethanone, 1-(2-aminofuro[3,2-b]pyridin-	MeOH	208(4.37),251(3.95),	4-1411-78

Compound	Solvent	λ_{max} (log ϵ)	Ref.
3-yl)- (cont.)		297(4.09)	4-1411-78
1H-Indole-3-carboxamide, 1-hydroxy-	MeOH	218(4.13),243(3.83), 297(3.64),340(3.27)	39-1117-78C
1,3,4-Oxadiazol-2(3H)-one, 5-methyl-3-phenyl-	EtOH	239(3.90),269s(2.86), 277s(2.61)	33-1477-78
2,3-Quinoxalinedione, 1,4-dihydro-1-methyl-	pH 5.0	232(4.09),239(4.02), 258(3.76),266s(3.72), 302s(3.99),314(4.09), 328s(4.00),346s(3.69)	24-1753-78
	pH 12.0	237(4.11),255s(3.76), 300s(3.87),314(4.09), 325(4.13),340s(3.91)	24-1755-78
C$_9$H$_8$N$_2$O$_2$S			
2,4-Thiazolidinedione, 5-phenyl-, 2-oxime, (Z)-	n.s.g.	203(4.29)	139-0303-78A
C$_9$H$_8$N$_2$O$_3$			
5H-Pyrano[2.3-b]pyridin-5-one, 4-hydroxy-7-methyl-, oxime	EtOH	313(4.16)	83-0848-78
2,4(1H,3H)-Pyrimidinedione, 1-(2,5-dihydro-5-methylene-2-furanyl)-, (R)-	pH 7	259(4.11)	44-3324-78
2,4(1H,3H)-Pyrimidinedione, 1-(5-methyl-2-furanyl)-	pH 7	210(4.08),255(3.94)	44-3324-78
5H-Pyrrolo[3,4-b]pyridine-7-acetic acid, 6,7-dihydro-5-oxo-	H$_2$O	262(3.38)	103-0049-78
C$_9$H$_8$N$_2$O$_4$			
Benzoic acid, 2-[[(hydroxyimino)acetyl]amino]-, anion	MeCN-pH 9.7	298(4.10),310s(--)	44-0268-78
C$_9$H$_8$N$_2$O$_4$S			
1H-4,1,2-Benzothiadiazine-3-carboxylic acid, methyl ester, 4,4-dioxide	EtOH	326(4.04)	39-0539-78C
C$_9$H$_8$N$_2$O$_6$			
6,9-Epoxypyrimido[1,6-a]azepine-1,3,5(2H)-trione, 6,7,8,9-tetrahydro-7,8-dihydroxy-	0.2M HCl	312(3.79)	44-0481-78
	dioxan	315(--)	44-0481-78
C$_9$H$_8$N$_2$S			
1H-Benzimidazole, 2-(ethenylthio)-	EtOH	216(4.61),252(3.88), 277(3.81),286(4.03), 293(4.07)	121-1477-78
(C$_9$H$_8$N$_2$S)$_n$			
1H-Benzimidazole, 2-(ethenylthio)-, homopolymer	THF	213(4.44),254(3.84), 260(3.82),287(4.05), 294(3.97)	121-1477-78
C$_9$H$_8$N$_2$S$_2$			
1,2,3-Thiadiazole, 5-[(phenylmethyl)-thio]-	MeOH	282(3.38),301(3.44)	4-1295-78
C$_9$H$_8$N$_4$			
Cyanamide, (1,4-dihydro-2-quinazolin-yl)-	MeOH	234(4.21),261(4.26)	4-1409-78
1H-1,2,4-Triazolo[4,3-a]benzimidazole, 3-methyl-	EtOH-HCl	275(3.60),282(3.68), 293(3.46),298(3.45)	22-0273-78
	CF$_3$COOH	269(3.62),276(3.67), 287(3.46),295s(3.40)	22-0273-78

Compound	Solvent	$\lambda_{max}(\log \epsilon)$	Ref.
$C_9H_8N_4O$			
2H-Indol-2-one, 3-azido-1,3-dihydro-3-methyl-	EtOH	253(3.84),290(3.21)	94-2866-78
$C_9H_8N_4OS$			
Propanamide, 2-cyano-3-(2-pyridinyl-amino)-3-thioxo-	EtOH	310(4.11)	104-0181-78
7H,11H-[1,3]Thiazino[2,3-b]pteridin-11-one, 8,9-dihydro-	pH -3.0	222(3.95),269(4.16),320(4.03),335s(3.93),390(3.02)	24-0971-78
	pH 7.0	243(4.04),263s(3.94),292(4.19),342(3.85)	24-0971-78
$C_9H_8N_4O_3$			
[4,4'-Bipyrimidine]-2,2',6(1H,1'H,3H)-trione, 5-methyl-	pH 0.9	317(3.95)	44-0511-78
	pH 7.2	314(3.94)	44-0511-78
	pH 12.9	303(4.05)	44-0511-78
[4,4'-Bipyrimidine]-2,2',6(1H,1'H,3H)-trione, 5'-methyl-	pH 0.9	320(3.84)	44-0511-78
	pH 7.2	319(3.84)	44-0511-78
	pH 12.9	299(3.99)	44-0511-78
$C_9H_8N_6S$			
3H-1,3,4-Thiadiazino[5,6-b]quinoxalin-3-one, 1,2-dihydro-, hydrazone	EtOH	215(4.443),308(4.32)	2-0307-78
C_9H_8O			
1H-Inden-1-one, 2,3-dihydro- (cation)	H_2SO_4	283(4.27)	65-1644-78
$C_9H_8O_2$			
4,11-Dioxatricyclo[5.1.0]undeca-1,6-diene, transoid	MeCN	230s(2.93)	77-0377-78
2-Propenoic acid, 3-phenyl-	MeOH	273(4.34)	19-0907-78
	EtOH	215(4.28),221(4.18),268(4.31)	133-0056-78
Tricyclo[4.2.1.0^{2,5}]non-7-ene-3,4-dione, endo	C_6H_{12}	271(2.19),526(1.80)	24-2557-78
	MeCN	289(2.14),521(1.72)	24-2557-78
exo	C_6H_{12}	271(3.34),546(2.28)	24-2557-78
	MeCN	278(3.33),533(2.23)	24-2557-78
$C_9H_8O_3$			
2-Oxaspiro[4.5]deca-6,9-diene-1,8-dione	MeOH	243(4.1)	24-1944-78
2-Propenoic acid, 3-(4-hydroxyphenyl)-, TiCl$_4$ complex	acetone	410(2.66)	98-0973-78
$C_9H_8O_3S$			
2-Propyn-1-ol, benzenesulfonate	n.s.g.	265(4.29)	124-0844-78
$C_9H_8O_4$			
1,3-Benzodioxole-4-carboxaldehyde, 6-methoxy-	MeOH	230(3.68),255(4.05),290(3.51),315(2.82)	2-0289-78
2-Propenoic acid, 3-(3,4-dihydroxyphenyl)-, TiCl$_4$ complex	acetone	450(3.30)	98-0973-78
$C_9H_9BrN_4O_2$			
Propanediamide, 2-[(4-bromophenyl)hydrazono]-	EtOH	230(4.27),302(3.72),360(4.41)	104-1956-78
C_9H_9BrO			
2,4,6-Cycloheptatrien-1-one, 4-bromo-2,7-dimethyl-	H_2O	239(4.45),330(3.92),345(3.83)	35-1778-78

Compound	Solvent	$\lambda_{max}(\log \epsilon)$	Ref.
$C_9H_9BrO_2$			
1,4-Benzodioxin, 6-bromo-2,3-dihydro-5-methyl-	EtOH	212(3.80),288(3.32)	103-1188-78
1,4-Benzodioxin, 6-bromo-2,3-dihydro-7-methyl-	EtOH	209(5.06),224(4.49), 295(4.09)	103-1188-78
$C_9H_9BrO_3$			
1,4-Benzodioxin, 6-bromo-2,3-dihydro-7-methoxy-	EtOH	210(4.24),226(3.81), 301(3.64)	103-1188-78
$C_9H_9Br_2N_3Se$			
Selenium, dibromo(1,2-dihydro-1-methyl-4-phenyl-3H-1,2,4-triazolium-3-ylidene)-, hydroxide, inner salt	MeCN	262(4.54),404(2.42)	70-1220-78
$C_9H_9ClN_2$			
1H-Imidazole, 2-(4-chlorophenyl)-4,5-dihydro-	EtOH	247(4.11)	104-0758-78
hydrochloride	EtOH	246(4.19)	104-0758-78
$C_9H_9ClN_2O_2$			
Benzenamine, N-(2-chloro-2-nitroethenyl)-N-methyl-	EtOH	245(4.13),380(4.50)	104-0817-78
$C_9H_9ClN_2O_4$			
6H-Furo[2',3':4,5]oxazolo[3,2-a]pyrimidin-6-one, 2-(chloromethyl)-2,3,3a,9a-tetrahydro-3-hydroxy-, [2R-$(2\alpha,3\beta,3a\beta,9a\beta)$]-	H_2O	225(4.01),249(3.98)	73-3268-78
$C_9H_9ClN_4$			
Phthalazine, 1-chloro-4-(1-methylhydrazino)-	dioxan	260s(3.72),315(3.90)	103-0436-78
	MeCN	213(4.73),259s(3.72), 316(3.89)	103-0436-78
$C_9H_9ClN_4O_2$			
Propanediamide, 2-[(2-chlorophenyl)hydrazono]-	EtOH	240(4.16),358(4.37)	104-1956-78
Propanediamide, 2-[(3-chlorophenyl)hydrazono]-	EtOH	230(4.25),302(3.75), 353(4.34)	104-1956-78
Propanediamide, 2-[(4-chlorophenyl)hydrazono]-	EtOH	238(4.18),303(3.81), 360(4.42)	104-1956-78
$C_9H_9Cl_2FN_2O_4$			
Uridine, 2',5'-dichloro-2',5'-dideoxy-5-fluoro-	H_2O	207(4.12),267(4.07)	73-3268-78
$C_9H_9FN_2O_5$			
2,2'-Anhydro-1-β-D-arabinofuranosyl-5-fluorouracil	EtOH	223(3.86),255(3.96)	94-2990-78
C_9H_9FO			
2-Propanone, 1-fluoro-3-phenyl-	C_6H_{12}	265(3.34),296(2.30)	22-0129-78
$C_9H_9F_3$			
2,4-Heptadiyne, 1,1,1-trifluoro-6,6-dimethyl-	hexane	240(2.65),255(2.51)	104-1089-78
$C_9H_9IN_4O_2$			
Propanediamide, 2-[(4-iodophenyl)hydrazono]-	EtOH	230(4.31),305(3.79), 365(4.46)	104-1956-78

Compound	Solvent	λ_{max}(log ϵ)	Ref.
C$_9$H$_9$NO			
Furo[2,3-c]pyridine, 2,7-dimethyl-	EtOH	210s(3.28),216(3.47), 220(3.58),253(3.88), 264s(3.70),267s(3.67), 272(3.66),282(3.47)	33-2542-78
1H-Indole, 1-methoxy-	MeOH	218(4.37),270(3.61), 288s(3.54),296(3.40)	39-1117-78C
Pyridine, 2-methyl-3-(2-propynyloxy)-	EtOH	285.5(3.83)	33-2542-78
C$_9$H$_9$NO$_2$			
Benzene, [1-(nitromethyl)ethenyl]-	pH 13	288(4.04)	44-3116-78
Furo[3,2-c]pyridin-4(5H)-one, 2,6-di- methyl-	EtOH	254(4.11),262(4.09), 295(3.94)	23-0613-78
3,8(2H,5H)-Isoquinolinedione, 6,7-di- hydro-	MeOH	221(4.13),279(4.22)	44-0966-78
5-Isoxazolol, 4,5-dihydro-3-phenyl-	EtOH	258(4.51)	78-1571-78
C$_9$H$_9$NO$_2$S$_2$			
2H-1,4-Benzothiazine, 3-(methylthio)-, 1,1-dioxide	MeCN	244(4.17),277(3.89)	2-0678-78
C$_9$H$_9$NO$_3$			
Glycine, N-benzoyl-	EtOH	231(3.89)	39-1157-78B
C$_9$H$_9$NO$_3$S			
5H-Thiazolo[3,2-a]pyridine-8-carboxylic acid, 2,3-dihydro-5-oxo-, methyl ester	MeOH	211(4.0),241(3.9), 290(4.2),320(4.0)	28-0385-78B
C$_9$H$_9$NO$_4$			
1,4-Benzodioxin, 2,3-dihydro-5-methyl- 6-nitro-	EtOH	209(3.97),222(3.94), 252(3.80),311(3.72)	103-1188-78
1,4-Benzodioxin, 2,3-dihydro-5-methyl- 7-nitro-	EtOH	214(4.29),246(4.01), 318(3.94)	103-1188-78
1,2-Propanedione, 1-(1,2-dihydro-4-hy- droxy-6-methyl-2-oxo-3-pyridinyl)-	EtOH	207(4.09),230(4.08), 318(4.13)	23-0613-78
C$_9$H$_9$NO$_5$			
Benzaldehyde, 4,5-dimethoxy-2-nitro-	MeOH	222(3.93),252(4.01), 261(4.03),312(3.64), 340(3.67)	39-0440-78C
1,4-Benzodioxin, 2,3-dihydro-5-methoxy- 8-nitro-	EtOH	245(3.89),312(3.93)	103-1188-78
C$_9$H$_9$NS$_2$			
2,1-Benzisothiazole, 3-(ethylthio)-	EtOH	210s(--),228(4.33), 296(3.81),305(3.84), 353(3.77)	4-0529-78
C$_9$H$_9$N$_3$			
Benzenamine, 4-(1H-pyrazol-1-yl)-	isoPrOH dioxan	270(4.24) 274(4.31)	103-1123-78 103-1123-78
Isoquinoline, 3-hydrazino-	EtOH	209(4.37),236(4.51), 286(3.91),291(3.92), 366(3.29)	4-0463-78
2-Quinoxalinamine, 3-methyl-	n.s.g.	350(3.72)	103-1163-78
C$_9$H$_9$N$_3$O			
Imidazo[1,2-a]pyridine, 2,8-dimethyl- 3-nitroso-	n.s.g.	250(4.38),279(4.43), 322(4.33),384(4.38)	103-1241-78

Compound	Solvent	$\lambda_{max}(\log \epsilon)$	Ref.
C₉H₉N₃OS			
Pyrido[2,3-d]pyrimidin-4(1H)-one, 6-methyl-2-(methylthio)-	pH 1	276(4.25),290(4.19), 342(4.17)	44-0828-78
	pH 7	259(4.19),274(4.22), 313(1.85)	44-0828-78
	pH 14	256(4.41),275(4.07), 328(3.88)	44-0828-78
C₉H₉N₃O₂			
2H-Benzotriazole-4,7-dione, 2,5,6-tri-methyl-	EtOH	224(4.30),258(4.22), 333(2.78)	87-0578-78
1H-Imidazole, 4,5-dihydro-2-(4-nitro-phenyl)-	EtOH	266(4.06)	104-0758-78
hydrochloride	EtOH	259(4.16)	104-0758-78
2(1H)-Pyrimidinone, 4-amino-1-(5-meth-yl-2-furanyl)-	pH 7	230(4.09),268(3.88)	44-3324-78
5H-Pyrrolo[3,4-b]pyridine-7-acetamide, 6,7-dihydro-5-oxo-	H₂O	214(4.03),269(3.75)	103-0049-78
C₉H₉N₃O₃			
Diisoxazolo[5,4-b:4',5'-e]pyridine, 4-ethynyl-3a,4,4a,7a,8,8a-hexa-hydro-8-hydroxy-	EtOH	212(3.65)	49-0337-78
C₉H₉N₃O₈			
1,4-Dioxaspiro[4.5]deca-6,9-diene, 8-(methyl-aci-nitro)-6,10-dimethyl-	MeOH	368(4.4)	104-1041-78
1,4-Dioxaspiro[4.5]decadiene, 10-(meth-yl-aci-nitro)-6,8-dimethyl-	MeOH	466(4.3)	104-1041-78
C₉H₉N₅O			
9H-Imidazo[1,2-a]purin-9-one, 1,4-di-hydro-4,6-dimethyl-	pH 1.3	227(4.58),231(4.58), 282(4.02)	44-1644-78
	pH 6.8	230(4.53),265(3.81), 304(3.84)	44-1644-78
	pH 11.6	230(4.59),273(4.02), 300(4.03)	44-1644-78
C₉H₉N₅O₃			
7-Pteridinecarboxylic acid, 2-amino-1,4-dihydro-6-methyl-4-oxo-, methyl ester	pH 0.2	237(4.09),337(3.96)	24-3790-78
	pH 5.2	243(4.08),274(4.10), 356(3.81)	24-3790-78
	pH 9.6	257(4.28),373(3.83)	24-3790-78
C₉H₉N₅O₄			
L-Alanine, N-(4-azido-2-nitrophenyl)-, hydrochloride	H₂O	249(4.30),333(3.23)	69-3321-78
Propanediamide, 2-[(2-nitrophenyl)-hydrazono]-	EtOH	280(3.99),330(4.08), 390(4.17)	104-1956-78
Propanediamide, 2-[(4-nitrophenyl)-hydrazono]-	EtOH	240(4.09),300(3.75), 380(4.60)	104-1956-78
Pyrimido[4,5-c]pyridazine-3-carboxylic acid, 5-amino-1,5,7,8-tetrahydro-1-methyl-4,7-dioxo-, methyl ester	MeOH	257(4.53),314(3.92)	44-4844-78
C₉H₁₀			
Benzene, 2-propenyl-	C₆H₁₂	260f(2.3),270(2.2)	28-0021-78A
1,4-Cycloheptadiene, 6,7-bis(methyl-ene)-	pentane	224(4.5),260s(3.6)	138-0961-78
dimer	hexane	211(4.40),233s(4.10), 263s(3.57)	138-0961-78

Compound	Solvent	$\lambda_{max}(\log \epsilon)$	Ref.
1H-Indene, 2,3-dihydro-	EtOH	259(2.95),266(3.09), 273(3.13)	35-3730-78
$C_9H_{10}BrClN_2O_4$ Uridine, 2'-bromo-5'-chloro-2',5'-di-deoxy-	H_2O	257(4.14)	73-3268-78
$C_9H_{10}BrNO_3$ 3H-Azepine-3-carboxylic acid, 5-bromo-2-methoxy-, methyl ester	EtOH	264(3.86)	39-0191-78C
$C_9H_{10}BrO_4P$ Phosphonic acid, (3-bromobenzoyl)-, dimethyl ester	H_2O dioxan	264(3.34)(hydrated) 260(3.98)	35-7382-78 35-7382-78
$C_9H_{10}ClFN_2O_4$ Uridine, 5'-chloro-2',5'-dideoxy-5-fluoro-	H_2O	269(4.01)	73-3268-78
$C_9H_{10}ClFN_2O_5$ 2,4(1H,3H)-Pyrimidinedione, 1-(5-chloro-5-deoxy-β-D-arabinofuranosyl)-5-fluoro-	H_2O	207(3.97),269(3.95)	73-3268-78
$C_9H_{10}ClNO_3$ 3H-Azepine-3-carboxylic acid, 5-chloro-2-methoxy-, methyl ester 3H-Azepine-3-carboxylic acid, 6-chloro-2-methoxy-, methyl ester	EtOH EtOH	262(4.08) 260(3.95)	39-0191-78C 39-0191-78C
$C_9H_{10}ClNO_4S$ 2,1-Benzisothiazolium, 1,3-dimethyl-, perchlorate	MeCN	300(4.09),350(3.69)	44-1233-78
$C_9H_{10}ClNO_5$ 2,1-Benzisoxazolium, 1,3-dimethyl-, perchlorate	H_2O	201(4.40),335(3.54)	44-1233-78
$C_9H_{10}ClN_3O$ 1H-Benzotriazole, 5-chloro-1-propoxy-	EtOH	273(4.12)	4-1043-78
$C_9H_{10}Cl_2N_2O_4$ Uridine, 2',5'-dichloro-2',5'-dideoxy-	H_2O	259(4.09)	73-3268-78
$C_9H_{10}FN_3O_3S$ 6H-Furo[2',3':4,5]thiazolo[3,2-a]pyr-imidine-2-methanol, 7-fluoro-2,3,3a,9a-tetrahydro-3-hydroxy-6-imino-, monohydrochloride, [2R-(2α,3β,3aβ,9aβ)]-	MeOH	247(4.37),285s(3.71)	44-4200-78
$C_9H_{10}F_4N_6S$ 3H-1,2,4-Triazole-3-thione, 4-amino-5-[3-ethyl-5-(1,1,2,2-tetrafluoro-ethyl)-1H-pyrazol-1-yl]-2,4-dihydro-	EtOH	256(4.30)	103-0458-78
$C_9H_{10}NS_2$ 2,1-Benzisothiazolium, 1-methyl-3-(methylthio)-, iodide	EtOH	222(4.42),304(3.94), 115(4.06)[sic]	4-0529-78

Compound	Solvent	$\lambda_{max}(\log \epsilon)$	Ref.
$C_9H_{10}N_2O$			
1H-Indazole, 3-methoxy-1-methyl-	EtOH	310(3.70)	118-0633-78
3H-Indazol-3-one, 1,2-dihydro-1,2-di-methyl-	EtOH	239(4.04),312(3.70)	118-0633-78
2,7-Naphthyridin-3(2H)-one, 1,4-di-hydro-1-methyl-, (S)- (jasminidine)	EtOH	259(3.31),266(3.22)	102-1069-78
1,2,4-Oxadiazole, 4,5-dihydro-5-methyl-3-phenyl-	C_6H_{12}	220(4.03),275(3.77)	78-2967-78
7H-Pyrid[3,2-c]azepin-7-one, 5,6,8,9-tetrahydro-	H_2O	263(3.59)	103-0049-78
2(1H)-Quinoxalinone, 3,4-dihydro-1-methyl-	MeOH	222(4.52),260s(3.40), 302(3.59)	24-1753-78
2(1H)-Quinoxalinone, 3,4-dihydro-3-methyl-	pH -2.0	243(4.05),273s(3.46)	24-1753-78
	pH 3.0	222(4.52),265(3.51), 297(3.51)	24-1753-78
	pH 14.2	228(4.52),285(3.74), 303(3.74)	24-1753-78
$C_9H_{10}N_2OS$			
Thieno[2,3-d]pyrimidin-4(3H)-one, 2,3,6-trimethyl-	EtOH	217(4.60),265(3.76), 301(3.89)	1-0303-78
$C_9H_{10}N_2O_2$			
Acetamide, N-acetyl-N-3-pyridinyl-	$CHCl_3$	258(3.36)	20-0621-78
Acetamide, 2-cyano-2-(3-oxo-1-cyclo-hexenyl)-	MeOH	370(4.34)	44-0966-78
$C_9H_{10}N_2O_3$			
Acetamide, 2-(hydroxyimino)-N-(4-meth-oxyphenyl)-	MeOH	288(4.01)	73-0309-78
Ethanediamide, (4-methoxyphenyl)-	MeOH	281(4.03)	73-0309-78
2-Propanone, 1-(1-methyl-5-nitro-2(1H)-pyridinylidene)-	$CHCl_3$	279(3.57),405(4.42)	103-0347-78
1H-Pyrrole-3-carboxylic acid, 5-cyano-4-hydroxy-2-methyl-, ethyl ester	EtOH	224(4.50),242s(4.17)	94-3080-78
$C_9H_{10}N_2O_3S$			
1H-2,1,3-Benzothiadiazine, 4-methoxy-1-methyl-, 2,2-dioxide	EtOH	222(5.44),258(4.73), 265(4.68),333(4.46)	4-1521-78
1H-2,1,3-Benzothiadiazin-4(3H)-one, 1,3-dimethyl-, 2,2-dioxide	EtOH	219(5.31),237(4.83), 305(4.48)	4-1521-78
$C_9H_{10}N_2O_4$			
4-Isoxazoleacetonitrile, α-acetoxy-3-methoxy-5-methyl-, (±)-	MeOH	210(3.77)	1-0027-78
2(1H)-Pyridinone, 4-hydroxy-3-[1-(hy-droxyamino)-3-oxo-1-butenyl]-	EtOH	360(4.37)	83-0848-78
2H-Pyrido[1,2-a]pyrazine-3-carboxylic acid, 1,3,4,8-tetrahydro-9-hydroxy-8-oxo-, (S)-	H_2O	199(3.96),217(3.94), 284(3.89)	12-0187-78
2,4(1H,3H)-Pyrimidinedione, 1-(2,3-di-oxobutyl)-3-methyl-	MeCN	260(3.98),415(1.56)	78-2861-78
2,4(1H,3H)-Pyrimidinedione, 1-(3,4-di-oxopentyl)-	MeOH	263(4.00)	78-2861-78
$C_9H_{10}N_2O_5$			
4,2'-Anhydro-5-β-D-arabinofuranosyl-uracil	pH 1	277(3.62)	87-0096-78
	pH 7	277(3.59)	87-0096-78
	pH 10	283(3.76)	87-0096-78
2-Propenoic acid, 2-amino-3-(5-nitro-2-furanyl)-, ethyl ester	EtOH	278(4.03),440(4.19)	4-0555-78

Compound	Solvent	$\lambda_{max}(\log \epsilon)$	Ref.
$C_9H_{10}N_2O_5S$			
5'-Deoxy-5'-thio-S^6,5'-cyclouridine	H_2O	293(4.05)	94-2664-78
	base	289(4.00)	94-2664-78
$C_9H_{10}N_4OS$			
Acetamide, N-[7-(methylthio)-1H-imid-	EtOH-HCl	208(4.08),245(4.22),	87-0112-78
azo[4,5-b]pyridin-5-yl]-		296(4.35)	
	EtOH-pH 7	245(4.24),290(4.26)	87-0112-78
	EtOH-NaOH	243(4.24),301(4.23)	87-0112-78
4(1H)-Pteridinone, 2,3-dihydro-3,6,7-	pH 5.0	219(3.98),287(4.39),	24-0971-78
trimethyl-2-thioxo-		348(4.05)	
	pH 10.0	230s(4.00),252(4.10),	24-0971-78
		307(4.39),366(3.82)	
$C_9H_{10}N_4O_2$			
Propanediamide, 2-(phenylhydrazono)-	EtOH	235(4.13),298(3.5),	104-1956-78
		360(4.36)	
$C_9H_{10}N_4O_3$			
1H-Pyrazolo[3,4-d]pyrimidine-4,6(5H,7H)-	H_2O and 0.5	213(4.25),242(3.81)	56-0037-78
dione, 1-acetyl-3,7-dimethyl-			
1,2,4-Triazolo[4,3-c]pyrimidine-	0.5M HCl	248(3.92),284(4.13)	56-0037-78
5,7(1H,6H)-dione, 1-acetyl-3,6-	H_2O	251(3.53),285(4.12)	56-0037-78
dimethyl-	0.5M NaOH	261(4.00),292(4.05)	56-0037-78
1,2,4-Triazolo[4,3-c]pyrimidin-5(6H)-	0.5M HCl	210(3.98),258(4.08),	56-0037-78
one, 8-acetyl-7-hydroxy-3,6-dimethyl-		296(4.14)	
	H_2O	208(4.13),258(4.06),	56-0037-78
		297(4.10)	
	0.5M NaOH	273(4.19)	56-0037-78
1,2,4-Triazolo[4,3-c]pyrimidin-5(6H)-	H_2O	200(4.17),267(3.90),	56-0037-78
one, 7-(acetyloxy)-3,6-dimethyl-		301(4.16)	
$C_9H_{10}N_5O_3$			
Pteridinium, 2-amino-1,4-dihydro-	pH 1.0	223(4.32),272(4.02),	24-3790-78
6-(methoxycarbonyl)-8-methyl-4-		300(4.13),390(4.00)	
oxo-, salt with 4-methylbenzene-	pH 7.0	286(4.19),387(3.99)	24-3790-78
sulfonic acid			
$C_9H_{10}N_6$			
Benzaldehyde, (2-methyl-2H-tetrazol-	pH 13	305(3.97),356(3.92)	103-0800-78
5-yl)hydrazone	M NaOH	360(4.36)	103-0800-78
	EtOH	307(4.37)	103-0800-78
$C_9H_{10}N_6O_3$			
Pyrimido[4,5-c]pyridazine-3-carboxylic	MeOH	228(4.18),256(4.48),	44-4844-78
acid, 5,7-diamino-1,4-dihydro-1-		261s(4.46),313(3.94)	
methyl-4-oxo-, methyl ester			
$C_9H_{10}O$			
Benzene, 1-ethenyl-4-methoxy-	EtOH	259(4.26),292(3.39),	44-4316-78
		303(3.15)	
Benzene, 1-(ethenyloxy)-3-methyl-	C_6H_{12}	202(4.51),228(4.12),	99-0194-78
		272(3.08),278(--)	
Benzene, 1-(ethenyloxy)-4-methyl-	C_6H_{12}	198(4.46),228(4.17),	99-0194-78
		276(3.17)	
Bicyclo[4.2.1]nona-2,4-dien-7-one	EtOH	210(3.55),257(3.70)	24-3927-78
2,4,6-Cycloheptatrien-1-one, 2,7-di-	H_2O	234(4.40),241(4.33),	35-1778-78
methyl-		325(3.84),336(3.77)	
1H-Inden-1-ol, 2,3-dihydro-, (S)-	MeOH	214(3.67),272(2.96)	35-6035-78
2-Propanone, 1-phenyl-	C_6H_{12}	260f(2.4),290(2.1)	28-0621-78A

Compound	Solvent	$\lambda_{max}(\log \epsilon)$	Ref.
2-Propynal, 3-(1-cyclohexen-1-yl)-	hexane	258s(3.90),266(4.03), 278(4.03)	118-0307-78
Tetracyclo[4.3.0.0^{2,9}.0^{4,8}]nonan-3-one	isooctane	278(1.82)	44-3904-78
	EtOH	271(1.70)	44-3904-78
Tricyclo[6.1.0.0^{2,4}]non-6-en-5-one,	EtOH	263(3.65)	35-5856-78
anti	EtOH	263(3.63)	88-0977-78
syn	EtOH	225(3.72)	35-5856-78
$C_9H_{10}O_2$			
Benzene, 1-(ethenyloxy)-4-methoxy-	C_6H_{12}	230(4.16),286(3.40)	99-0194-78
2H-1-Benzopyran-2-one, 5,6,7,8-tetra-hydro-	EtOH	222(3.5),308(3.8)	23-0419-78
2,4,6-Cycloheptatrien-1-one, 4-hydroxy-2,7-dimethyl-	pH 13	238(4.34),368(4.28)	35-1778-78
2,5-Cyclohexadiene-1,4-dione, trimethyl-	H_2O	260(4.31),352(2.61)	28-0353-78B
Ethanone, 1-[5-(1-hydroxyethylidene)-1,3-cyclopentadien-1-yl]-	MeOH	246(4.342),319(4.139), 380(4.086)	73-0938-78
Tricyclo[4.2.1.0^{2,5}]nonane-3,4-dione,	C_6H_{12}	533(1.96)	24-2557-78
endo	MeCN	525(1.80)	24-2557-78
exo	C_6H_{12}	526(1.61)	24-2557-78
	MeCN	505(1.58)	24-2557-78
$C_9H_{10}O_2S$			
Benzaldehyde, 2-methoxy-6-(methylthio)-	MeOH	236(4.87),298(4.40), 372(4.28)	18-2437-78
2-Propanone, 1-(phenylsulfinyl)-	MeOH	250(3.90)	12-1965-78
$C_9H_{10}O_3$			
1,2-Benzodioxin, 3,8a-dihydro-3-meth-oxy-	CHCl	304(3.9+)	88-3227-78
1,3-Benzodioxole, 2-methoxy-2-methyl-	MeOH	276s(3.40),280(3.43), 285s(3.33)	12-2259-78
2,5-Cyclohexadiene-1,4-dione, hydroxy-trimethyl-	H_2O	265(4.20)	28-0353-78B
Ethanone, 1-(2-hydroxy-5-methoxyphen-yl)-	MeOH	229(4.01),274(4.17), 316(3.84)	95-1607-78
	MeOH-base	231(4.17),275(4.04), 350(3.76)	95-1607-78
2-Oxaspiro[4.5]dec-6-ene-1,8-dione	MeOH	219(4.04)	24-1944-78
$C_9H_{10}O_4$			
Benzaldehyde, 4-hydroxy-3,5-dimethoxy-	MeOH	216(4.27),232(4.23), 308(4.23)	95-1607-78
	MeOH-base	217(4.13),253(4.05), 368(4.40)	95-1607-78
1,2,3-Benzenetriol, 4-(3-hydroxy-1-propenyl)-, (E)-	MeOH	225(4.33),269(4.14)	24-3939-78
Benzoic acid, 3,4-dimethoxy-	EtOH	254(3.97),290(3.64)	105-0377-78
Benzoic acid, 4-hydroxy-3-methoxy-, methyl ester	MeOH	220(4.37),263(4.13), 293(3.86)	95-1607-78
	MeOH-base	235(4.14),314(4.37)	95-1607-78
2H-Pyran-2-one, 3-acetyl-4-methoxy-6-methyl-	MeOH	313(3.90)	118-0144-78
$C_9H_{10}O_5$			
1,3-Benzodioxol-5(7aH)-one, 4,7a-di-methoxy-	EtOH	230s(3.79),321(3.51)	44-3983-78
1,3-Benzodioxol-5(7aH)-one, 6,7a-di-methoxy-	EtOH	252(4.25),306(3.20)	44-3983-78

Compound	Solvent	$\lambda_{max}(\log \epsilon)$	Ref.
$C_9H_{10}O_5S_2$ Thiosulfuric acid, S-[(4-acetoxyphenyl)methyl] ester, sodium salt	EtOH	265(2.64),272(2.51)	44-1197-78
$C_9H_{10}O_6$ 1,5-Cyclohexadiene-1-carboxylic acid, 3-(carboxymethoxy)-4-hydroxy-, trans	H_2O	273(3.34)	77-0869-78
$C_9H_{10}O_6S$ Benzoic acid, 3-methoxy-4-[(methylsulfonyl)oxy]-	EtOH	211(4.01),236(3.80), 289(3.37)	33-1200-78
$C_9H_{11}BrN_2O_3$ Ethanol, 2-[(2-bromo-4-nitrophenyl)methylamino]-	DMSO	390(3.67)	44-0441-78
2,4(1H,3H)-Pyrimidinedione, 1-(3-bromo-4-oxopentyl)-	MeOH	267(4.01)	78-2861-78
$C_9H_{11}BrN_2O_6$ 2,4(1H,3H)-Pyrimidinedione, 1-O-β-D-apio-L-furanosyl-5-bromo-	EtOH	275(3.93)	87-0706-78
$C_9H_{11}Cl$ Bicyclo[5.1.0]octa-2,4-diene, 4-(chloromethyl)-	C_6H_{12}	232(3.49)	44-1118-78
$C_9H_{11}ClN_2$ Benzenecarboximidamide, 4-chloro-N,N'-dimethyl-	EtOH	227(4.12)	104-0758-78
hydrochloride	EtOH	231(4.14)	104-0758-78
$C_9H_{11}ClN_2O_4$ Uridine, 5'-chloro-2',5'-dideoxy-	H_2O	261(4.05)	73-3268-78
$C_9H_{11}ClN_2O_5$ 2,4(1H,3H)-Pyrimidinedione, 1-(5-chloro-5-deoxy-β-D-arabinofuranosyl)-	H_2O	207(3.97),262(4.05)	73-3268-78
2,4(1H,3H)-Pyrimidinedione, 5-(2-chloro-2-deoxy-β-D-ribofuranosyl)-	pH 1	262(3.92)	87-0096-78
	pH 7	262(3.94)	87-0096-78
	pH 10	285(3.81)	87-0096-78
$C_9H_{11}ClO$ Bicyclo[2.2.1]heptan-7-ol, 5-chloro-2,3-bis(methylene)-	isooctane	242s(3.91),249(3.98), 256s(3.84)	33-0732-78
	EtOH	242s(3.92),249(3.97), 256s(3.84)	33-0732-78
Bicyclo[4.2.1]non-3-en-2-one, 3-chloro-	C_6H_{12}	247(3.73)	78-1965-78
Tricyclo[2.2.1.0^{2,6}]heptan-3-ol, 1-(chloromethyl)-7-methylene-	isooctane	208(3.91)	33-0732-78
	EtOH	209(3.91)	33-0732-78
$C_9H_{11}ClO_2S$ Sulfonium, (2-formyl-3-hydroxyphenyl)-dimethyl-, chloride	H_2O	255(4.33),332(4.32), 380s(3.74)	18-2437-78
$C_9H_{11}ClO_3S$ Ethanol, 2-chloro-, 4-methylbenzenesulfonate	n.s.g.	262(4.15)	124-0844-78
$C_9H_{11}Cl_2N_3O_2$ Cytosine, 1-(3,5-dichloro-2,3,5-trideoxy-β-D-threo-pentofuranosyl)-	pH 7	271(3.98)	44-3324-78

Compound	Solvent	$\lambda_{max}(\log \epsilon)$	Ref.
$C_9H_{11}FN_2O_4$			
Uridine, 2',5'-dideoxy-5-fluoro-	H_2O	270(3.98)	73-3268-78
$C_9H_{11}FN_2O_5$			
2,4(1H,3H)-Pyrimidinedione, 1-(5-deoxy-β-D-arabinofuranosyl)-5-fluoro-	H_2O	209(3.92),271(3.94)	73-3268-78
$C_9H_{11}IN_2O_6$			
2,4(1H,3H)-Pyrimidinedione, 1-O-β-apio-L-furanosyl-5-iodo-	EtOH	280(3.82)	87-0706-78
$C_9H_{11}IO_2$			
2(4H)-Benzofuranone, 5,6,7,7a-tetra-hydro-7a-(iodomethyl)-, (±)-	EtOH	215(4.06)	39-0387-78C
$C_9H_{11}N$			
Benzenamine, 2-(2-propenyl)-	MeOH	214(3.93),245(3.97), 294(3.42)	78-1943-78
1H-Inden-1-amine, 2,3-dihydro-	C_6H_{12}	210(3.89),273(2.91), 280s(2.47)	35-6035-78
$C_9H_{11}NO$			
Furo[2,3-c]pyridine, 2,3-dihydro-2,7-dimethyl-	EtOH	276s(3.62),284(3.72)	33-2542-78
4H-Indol-4-one, 1,5,6,7-tetrahydro-1-methyl-	MeOH	210(4.21),251(4.06), 274s(3.98)	24-1780-78
Pyridine, 2-(1-ethoxyethenyl)-	MeOH	258(3.75)	103-0530-78
Pyridine, 2-methyl-3-(2-propenyloxy)-	EtOH	277(3.72),285s(3.56)	33-2542-78
2(1H)-Pyridinone, 1-(2-methyl-1-prop-enyl)-	EtOH	309(3.77)	78-2609-78
2(1H)-Pyridinone, 1-(2-methyl-2-prop-enyl)-	EtOH	302(3.70)	78-2609-78
$C_9H_{11}NOS$			
Ethanethioic acid, S-[3-(2-propenyl)-1H-pyrrol-2-yl] ester	EtOH	226(4.02),241(4.04)	23-0221-78
$C_9H_{11}NO_2$			
Acetic acid, 2-azabicyclo[2.2.0]hex-5-en-3-ylidene-, ethyl ester	MeOH	284(4.15)	44-0944-78
2,4,6-Cycloheptatrien-1-one, 3-(di-methylamino)-	MeOH	226(4.14),271(4.26), 360(4.04)	18-2338-78
Furo[3.2-c]pyridin-4(2H)-one, 3,5-di-hydro-2,6-dimethyl-	EtOH	223(3.82),237s(3.53), 294(3.84)	23-0613-78
Phenol, 2,3,5-trimethyl-6-nitroso-	DMF	695(1.56)	104-2189-78
Phenol, 3,4,5-trimethyl-2-nitroso-	DMF	700(1.62)	104-2189-78
$C_9H_{11}NO_3$			
Acetamide, N-hydroxy-N-(4-methoxyphen-yl)-	pH 1	235(4.01)	87-0649-78
	pH 7.0	235(4.01)	87-0649-78
3H-Azepine-3-carboxylic acid, 2-meth-oxy-, methyl ester	EtOH	257(3.75)	39-0191-78C
2(1H)-Pyridinone, 4-hydroxy-6-methyl-3-(2-oxopropyl)-	EtOH	215(4.12),280(3.65)	23-0613-78
$C_9H_{11}NO_4$			
1-Azabicyclo[3.2.0]hept-2-ene-2-carbox-ylic acid, 6-(1-hydroxyethyl)-7-oxo-, monosodium salt	H_2O	265(3.81)	35-8004-78
4-Oxazolecarboxylic acid, 5-methyl-2-(1-oxopropyl)-, methyl ester	MeOH	227(3.93),265(4.04)	88-2791-78

$C_9H_{11}NO_5S-C_9H_{11}N_5O_3$

Compound	Solvent	$\lambda_{max}(\log \epsilon)$	Ref.
$C_9H_{11}NO_5S$			
Ethanol, 2-nitro-, 4-methylbenzenesulfonate	n.s.g.	242(4.61)	124-0844-78
$C_9H_{11}N_3$			
Benzenamine, 4-(diazomethyl)-N,N-dimethyl-	DMF	523(1.56)	142-0105-78C
1H-Benzotriazole, 1,5,6-trimethyl-	4.2M HCl	286(4.06)	87-0578-78
$C_9H_{11}N_3O$			
1H-Benzotriazole, 1,5,6-trimethyl-,	4.2M HCl	287(4.01)	87-0578-78
3-oxide	EtOH	278(3.57),288(3.61), 324(3.92)	87-0578-78
3H-Pyrazolo[3,4-b]pyridin-3-one,	MeOH	229(4.21),321(3.58)	20-0309-78
1,2-dihydro-1,4,6-trimethyl-	THF	232(4.46),316(3.63)	20-0309-78
2H-Pyrido[1,2-a]pyrimidine-9-carboxamide, 3,4-dihydro-	CH_2Cl_2	263(3.9),397(3.4)	24-2594-78
$C_9H_{11}N_3O_3$			
1(2H)-Pyrimidinebutanenitrile, 3,4-dihydro-β-hydroxy-5-methyl-2,4-dioxo-	H_2O	273(3.98)	126-0905-78
$C_9H_{11}N_3O_4$			
Acetamide, N-[2-[(5-nitro-2-pyridinyl)-oxy]ethyl]-	DMSO	303(4.01)	44-0441-78
4,2'-Anhydro-5-β-D-arabinofuranosylisocytosine	pH 1	278(3.54)	87-0096-78
	pH 7	286(3.76)	87-0096-78
	pH 10	285(3.78)	87-0096-78
Ethanol, 2-[(5-nitro-2-pyridinyl)-amino]-, acetate	DMSO	369(4.25)	44-0441-78
2,6-Pyridinedimethanol, dicarbamate	EtOH	264(4.65)	95-1402-78
2,4(1H,3H)-Pyrimidinedione, 1-[3-(hydroxyimino)-4-oxopentyl]-	MeOH	264(4.02)	78-2861-78
$C_9H_{11}N_3O_5$			
$O^6,5'$-Cyclocytidine	M HCl	276(4.24)	94-2340-78
	H_2O	270(4.05)	94-2340-78
$C_9H_{11}N_5O_2$			
Acetamide, N-(6,7-dihydro-1,7-dimethyl-6-oxo-1H-purin-2-yl)-	pH 1.0	260(4.05)	44-2325-78
	pH 7.0	216(4.36),261(3.99)	44-2325-78
	pH 13	268(4.00)	44-2325-78
6,7-Pteridinedione, 4-(dimethylamino)-5,8-dihydro-8-methyl-	pH -1.0	237(4.21),312s(4.14), 326(4.18),345s(4.05)	24-1763-78
	pH 4.0	234(4.17),254(4.13), 320(4.03)	24-1763-78
	pH 13.0	238(4.12),250s(4.12), 320s(4.08),335(4.16), 345s(4.15)	24-1763-78
$C_9H_{11}N_5O_3$			
Acetamide, N-(6,7,8,9-tetrahydro-1,7-dimethyl-6,8-dioxo-1H-purin-2-yl)-	pH 1	265(3.89),296(3.74)	44-2325-78
	pH 12	228(4.35),284(3.96)	44-2325-78
6-Pteridinecarboxylic acid, 2-amino-1,4,7,8-tetrahydro-8-methyl-4-oxo-, methyl ester	MeOH	217(4.22),265(4.32), 282s(3.86),400(3.95)	24-3790-78
4(1H)-Pteridinone, 2-amino-6-(dimethoxymethyl)-	pH 1	248(4.05),318(3.96), 335s(3.83)	44-0736-78
	pH 13	256(4.41),282s(3.86), 360(3.89)	44-0736-78

Compound	Solvent	$\lambda_{max}(\log \epsilon)$	Ref.
$C_9H_{11}O_2S$			
Sulfonium, (2-formyl-3-hydroxyphenyl)-dimethyl-, chloride	H_2O	255(4.33),332(4.32), 380s(3.74)	18-2437-78
$C_9H_{11}O_4P$			
Phosphonic acid, benzoyl-, dimethyl ester	H_2O dioxan	268(3.74)(hydrate) 262(4.11)	35-7382-78 35-7382-78
$C_9H_{12}BrN_3O_5$			
Cytidine, 5-bromo-	pH 1 pH 7.0	299(4.04) 287(3.86)	94-2340-78 94-2340-78
$C_9H_{12}ClN_3$			
Pyrimidine, 2-chloro-4-piperidino-	heptane	203(4.09),249(4.24), 290(3.65)	104-0377-78
Pyrimidine, 4-chloro-2-piperidino-	heptane	205(3.88),250(4.47), 316(3.43)	104-0377-78
$C_9H_{12}FN_3O_4S$			
Cytidine, 5-fluoro-2-thio-	H_2O	217(3.77),260(4.33)	44-4200-78
$C_9H_{12}IN_3O_4$			
Uridine, 5'-amino-2',5'-dideoxy-5-iodo-	pH 2 pH 12	285(3.87) 279(3.74)	87-0106-78 87-0106-78
$C_9H_{12}N$			
Cycloheptatrienylium, (dimethylamino)-, perchlorate	MeCN	239(4.33),333(4.19)	150-4801-78
$C_9H_{12}N_2$			
Acetaldehyde, methylphenylhydrazone	n.s.g.	277(4.27)	104-0717-78
Benzaldehyde, dimethylhydrazone	n.s.g.	296(4.20)	104-0717-78
1H-Benzimidazole, 2,3-dihydro-1,3-dimethyl-	hexane	217(4.46),264(3.73), 312(3.72)	44-2621-78
Ethanone, 1-phenyl-, methylhydrazone	n.s.g.	278(3.95)	104-0717-78
2-Propanone, phenylhydrazone	n.s.g.	272(4.18)	104-0717-78
$C_9H_{12}N_2O$			
Ethanone, 1-[2-(ethylamino)-3-pyridinyl]-	EtOH	261(3.88),369(3.74)	104-1218-78
4H-Indol-4-one, 1,5,6,7-tetrahydro-1-methyl-, oxime, (Z)-	MeOH	211(4.06),263(4.05)	24-1780-78
Pyrrolo[3,2-b]azepin-5(1H)-one, 4,6,7,8-tetrahydro-1-methyl-	MeOH	214(4.09),241s(3.73), 254s(3.69)	24-1780-78
Pyrrolo[3,2-c]azepin-4(1H)-one, 5,6,7,8-tetrahydro-1-methyl-	MeOH	209(4.26),240s(3.87), 258s(3.76)	24-1780-78
$C_9H_{12}N_2O_2$			
3H-Azepine-3-carboxamide, 2-ethoxy-	EtOH	255(3.65)	39-0191-78C
1H-1,2-Diazepine-1-carboxylic acid, 4-methyl-, ethyl ester	MeOH	224(3.93),240s(3.78), 350s(2.42)	33-2887-78
1H-1,2-Diazepine-1-carboxylic acid, 6-methyl-, ethyl ester	MeOH	250(--),350s(2.38)	33-2887-78
Pyridinium, 1-[(ethoxycarbonyl)amino]-3-methyl-, hydroxide, inner salt	$CHCl_3$	336(3.90)	33-2887-78
8H-Pyrido[1,2-a]pyrazin-8-one, 1,2,3,4-tetrahydro-9-hydroxy-1-methyl-	EtOH	219(3.89),287(3.93)	12-0187-78
2-Pyrimidineacetic acid, 4-methyl-, ethyl ester	neutral	246(3.55),255s(3.39), 275(2.46)	12-0649-78
	cation	250(3.74)	12-0649-78

Compound	Solvent	λ_{max} (log ϵ)	Ref.
2-Pyrimidineacetic acid, 4-(1-methyl-	neutral	248(3.59)(extrapolated)	12-0649-78
ethyl)-	cation	250(3.75)	12-0649-78
	anion	250(3.54),275s(2.49)	12-0649-78
$C_9H_{12}N_2O_2S$			
2-Pyrimidineacetic acid, 4-(ethylthio)-,	neutral	258(3.85),285(3.89),	12-0649-78
methyl ester		300(3.24)	
	cation	225(3.90),255(3.39),	12-0649-78
		306(4.21)	
$C_9H_{12}N_2O_3$			
3H-Azepine-3-carboxamide, 2,4-dimeth-	EtOH	262(3.85)	39-0191-78C
oxy-			
5-Pyrimidinecarboxylic acid, 1,4-di-	hexane-5%	240(4.10),283(3.70)	94-1403-78
hydro-1,2-dimethyl-4-oxo-, ethyl	MeOH		
ester	MeOH	240(4.12),283(3.74)	94-1403-78
	BuOH	241(4.07),284(3.68)	94-1403-78
	ether	237(4.15),285(3.79)	94-1415-78
	Pr_2O-10%	241(4.04),284(3.73)	94-1415-78
	MeOH		
5% methanol	$isoPr_2O$	284.5(3.73)	94-1415-78
5% methanol	Bu_2O	241(4.02),285(3.65)	94-1415-78
	dioxan	240(4.05),287(3.67)	94-1415-78
5-Pyrimidinecarboxylic acid, 1,6-di-	hexane-5%	224(3.64),304(3.77)	94-1403-78
hydro-1,2-dimethyl-6-oxo-, ethyl	MeOH		
ester	MeOH	226(3.77),302(3.90)	94-1403-78
	BuOH	225(3.77),303(3.85)	94-1403-78
	ether	222(3.79),303(3.88)	94-1415-78
	Pr_2O	224(3.74),303(3.83)	94-1415-78
	$isoPr_2O$	303(3.86)	94-1415-78
	Bu_2O	222(3.74),305(3.81)	94-1415-78
	dioxan	305(3.81)	94-1415-78
5-Pyrimidinecarboxylic acid, 4-methoxy-	hexane	226(3.98),263(3.77)	94-1403-78
2-methyl-, ethyl ester	MeOH	228(4.01),263(3.79)	94-1403-78
	BuOH	228(3.98),263(3.78)	94-1403-78
	ether	226(4.05),262(3.83)	94-1415-78
	Pr_2O	226(4.03),263(3.81)	94-1415-78
	$isoPr_2O$	262(3.80)	94-1415-78
	Bu_2O	226(4.01),263(2.80)	94-1415-78
	dioxan	263(3.80)	94-1415-78
$C_9H_{12}N_2O_4$			
Butanedioic acid, diazo(1-methylethyli-	C_6H_{12}	256(4.12)	78-2797-78
dene)-, dimethyl ester			
Diazenecarboxylic acid, (1-methyl-5-	EtOH	286(4.46)	94-2205-78
oxo-1,3-pentadienyl)-, ethyl ester,			
1-oxide, (?,Z,E)-			
2,4-Hexadienamide, N-acetoxy-6-(meth-	EtOH	297.5(4.63)	94-2575-78
oxyimino)-			
8-Oxa-1-aza-5-azoniaspiro[4.5]dec-3-ene,	MeOH	210(4.00),298(3.15)	24-0780-78
4-(methoxycarbonyl)-2-oxo-, hydrox-			
ide, inner salt			
2-Pyrimidineacetic acid, 4,6-dimeth-	neutral	242(3.58)	12-0649-78
oxy-, methyl ester	cation	252(3.92)	12-0649-78
$C_9H_{12}N_2O_4S$			
2-Pyrimidineacetic acid, 4-(ethylsul-	neutral	253s(3.53),256(3.56),	12-0649-78
fonyl)-, methyl ester		264(3.44)	
$C_9H_{12}N_2O_4S_2$			
1,2,3-Thiadiazole-4-carboxylic acid,	MeOH	233(3.78),299(3.95)	4-1295-78

Compound	Solvent	$\lambda_{max}(\log \epsilon)$	Ref.
5-[(3-methoxy-3-oxopropyl)thio]-, ethyl ester (cont.)			4-1295-78
$C_9H_{12}N_2O_5$			
D-Arabinitol, 1,4-anhydro-2-deoxy- 2-(3,4-dihydro-2,4-dioxo-1(2H)- pyrimidinyl)-	pH 1 and 7 pH 13	267(3.99) 266(3.87)	44-0541-78 44-0541-78
1(2H)-Pyrimidinebutanoic acid, 3,4-di- hydro-β-hydroxy-5-methyl-2,4-dioxo-	H_2O	273(4.00)	126-0905-78
2,4(1H,3H)-Pyrimidinedione, 1-(5-deoxy- β-D-arabinofuranosyl)-	H_2O	210(3.92),264(4.00)	73-3268-78
$C_9H_{12}N_2O_6$			
2,4(1H,3H)-Pyrimidinedione, 1-O-β-D- apio-L-furanosyl-	EtOH	261(4.03)	87-0706-78
2,4(1H,3H)-Pyrimidinedione, 5-β-D- arabinofuranosyl-	pH 1 pH 7 pH 10	262(3.88) 263(3.86) 268s(3.69),290(3.58)	87-0096-78 87-0096-78 87-0096-78
Uridine	pH 7	262(4.00)	39-0762-78C
$C_9H_{12}N_2S$			
Ethanethioic acid, 2-methyl-2-phenyl- hydrazide	MeOH	220(3.85),237(4.09), 250(3.94),275(4.15)	123-0701-78
$C_9H_{12}N_2S_2$			
2,4(1H,3H)-Quinazolinedithione, 5,6,7,8-tetrahydro-3-methyl-	EtOH EtOH-NaOH	298(4.43),350(4.04) 259(4.23),287(4.32), 365(4.23)	106-0185-78 106-0185-78
$C_9H_{12}N_4$			
[1,2,4]Triazolo[1,5-c]pyrimidine, 2,5,7,8-tetramethyl-	pH 7.0	260(3.89)	12-2505-78
[1,2,4]Triazolo[4,3-c]pyrimidine, 3,5,7,8-tetramethyl-	pH 7.0	250s(3.83),259(3.93), 266s(3.88),284s(3.62)	12-2505-78
$C_9H_{12}N_4O$			
4H-Pyrrolo[2,3-d]pyrimidin-4-one, 1,7- dihydro-5-[(dimethylamino)methyl]-	MeOH	262(3.94)	24-2925-78
$C_9H_{12}N_4O_3$			
2,4,6(3H)-Pteridinetrione, 1,5,7,8- tetrahydro-1,3,8-trimethyl-	EtOH	245(6.1),262s(5.9), 313(5.9)	35-7661-78
1H-Pyrimido[4,5-b][1,4]diazepine- 2,4,6(3H)-trione, 5,7,8,9-tetra- hydro-1,3-dimethyl-	EtOH	267(4.9),297(5.0)	35-7661-78
Theophylline, 8-(methoxymethyl)-	EtOH	207(4.46),275(4.01)	73-3414-78
$C_9H_{12}N_4O_4$			
1,2,4-Triazine-3,5(2H,4H)-dione, 2-acetyl-6-morpholino-	EtOH-pH 7	312(3.78)	124-1064-78
$C_9H_{12}N_6O_2$			
2,4-Pteridinediamine, 6-(dimethoxy- methyl)-	MeOH	261(4.37),284s(3.73), 368(3.85)	44-0736-78
9H-Purine-9-propanoic acid, 6-amino- α-methylamino-	0.05M HCl H_2O ?	257(4.28) 260(4.14) 260(3.99)	126-2195-78 126-2195-78 126-2195-78
$C_9H_{12}O$			
Bicyclo[3.3.1]non-8-en-3-one	C_6H_{12}	298(2.01),307(2.01), 318(2.00),329(1.86)	44-3653-78

Compound	Solvent	$\lambda_{max}(\log \epsilon)$	Ref.
Bicyclo[4.2.1]non-3-en-2-one	EtOH	229(3.94)	78-1965-78
Bicyclo[2.2.2]octanone, 6-methylene-	EtOH	283(2.38)	44-3653-78
2-Cyclohexen-1-one, 5-ethenyl-3-methyl-	EtOH	237(3.99)	44-3653-78
2-Cyclohexen-1-one, 4-(1-methylethyli-dene)-	n.s.g.	302(4.19)	22-0255-78
2,4,7-Cyclononatrien-1-ol, (Z,Z,Z)-	C_6H_{12}	223(3.64)	35-5856-78
3-Nortricyclanol, 6-methyl-5-methyl-ene-, anti	EtOH	205.5(3.90)	33-0732-78
$C_9H_{12}OS$			
Benzene, [(1-methylethyl)sulfinyl]-	hexane	209(3.83),255(3.63)	70-0901-78
	H_2O	210(3.85),240(3.66), 267(2.85),274(2.70)	70-0901-78
2-Propanol, 1-(phenylthio)-, (S)-(+)-	MeOH	254(3.9)	12-1965-78
$C_9H_{12}O_2$			
1,4-Benzenediol, 2,3,5-trimethyl-	H_2O	286(3.44)	28-0353-78B
3H-2-Benzopyran-3-one, 1,4,5,6,7,8-hexahydro-	MeOH	223(4.00)	44-1248-78
1,3-Cyclobutanedione, 2,2-dimethyl-4-(1-methylethylidene)-	EtOH	218(3.76),263(3.95)	12-1757-78
1,3,6-Cycloheptatriene-1,2-dimethanol	EtOH	261(3.40)	138-0961-78
1,3(and 4)-Cyclopentadiene-1-acetic acid, ethyl ester	n.s.g.	247(3.50)	104-0264-78
4,6-Heptadienoic acid, 2-ethylidene-	EtOH	209(3.69)	18-2375-78
2-Oxetanone, 3,4-bis(1-methylethyli-dene)-	EtOH	224(4.14),291(3.43)	12-1757-78
Phenol, 4-(ethoxymethyl)-	EtOH	230(3.95),280(3.18)	98-0195-78
$C_9H_{12}O_2S$			
Benzenesulfinic acid, 4-methyl-, ethyl ester	hexane	203(3.96),223(3.96), 254(3.50)	118-0441-78
$C_9H_{12}O_3$			
Benzeneethanol, 3-hydroxy-4-methoxy-	EtOH	222(3.97),280(3.62)	94-2111-78
Benzenemethanol, 3,4-dimethoxy-	hexane	234(4.50),292(3.96)	102-1791-78
2-Butynal, 4-[(tetrahydro-2H-pyran-2-yl)oxy]-	hexane	221(3.86),232(3.83)	118-0307-78
2-Furancarboxylic acid, 5-butyl-	EtOH	257(4.03)	44-4081-78
6-Heptynoic acid, 3-oxo-, ethyl ester	EtOH-NaOH	274(4.36)	22-0131-78
2,4-Hexadienal, 2-(1-hydroxy-2-oxo-propyl)-, [R-(E,E)]- (avellaneol)	n.s.g.	278(4.29)	88-3527-78
Phenol, 2-methoxy-4-(methoxymethyl)-	MeOH	239(3.82),280(3.43)	98-0195-78
$C_9H_{12}O_3S$			
Benzene, [2-(methylsulfonyl)ethoxy]-	n.s.g.	270(3.35),277(3.29)	39-1580-78C
Benzenesulfonic acid, 4-methyl-, ethyl ester	n.s.g.	262(4.02)	124-0844-78
2-Propanol, 1-(phenylsulfonyl)-, (S)-(+)-	MeOH	220(3.9),242(2.8), 247(2.9),253(2.9), 259(3.0),266(3.1), 272(3.0)	12-1965-78
$C_9H_{12}O_4$			
Acetic acid, (5-methoxy-2(5H)-furanyli-dene)-, ethyl ester, (E)-	EtOH	280(3.96)	35-7934-78
(Z)-	EtOH	226(3.81),278(4.08)	35-7934-78
3,5-Cyclohexadiene-1-carboxylic acid, 1,2-dihydroxy-3-methyl-, methyl ester	MeOH	268(3.72)	78-1707-78
2,5-Cyclohexadien-1-one, 3,4,4-trimeth-oxy-	EtOH	224(4.03),290(3.67)	44-3983-78

Compound	Solvent	λ_{max}(log ϵ)	Ref.
2-Furanacetic acid, α-methoxy-, ethyl ester	EtOH	220(3.80)	35-7934-78
3-Furancarboxylic acid, 4,5-dihydro-2,5-dimethyl-4-oxo-, ethyl ester	EtOH	213(3.98),262(4.06)	118-0448-78
$C_9H_{12}O_5$			
2,4-Hexadienedioic acid, 2-hydroxy-, 1-(1-methylethyl) ester	EtOH	303(4.4)	5-1734-78
$C_9H_{13}BrO$			
Bicyclo[3.3.1]nonan-3-one, 1-bromo-	EtOH	224(2.63),283(1.81)	44-3653-78
$C_9H_{13}BrO_2$			
2-Cyclohexen-1-one, 2-bromo-3-hydroxy-5-propyl-	EtOH	293(4.36)	2-0970-78
$C_9H_{13}ClN_2O_4$			
1H-Pyrazole-3-carboxylic acid, 2-[2-(2-chloroethoxy)ethyl]-2,5-dihydro-5-oxo-, methyl ester	MeOH	229(4.06),271(3.58)	24-0780-78
$C_9H_{13}ClN_4O_3$			
D-Arabinitol, 2-[(5-amino-6-chloro-4-pyrimidinyl)amino]-1,4-anhydro-2-deoxy-	pH 1 pH 7, 13	305(4.12) 262(3.93),292(3.99)	44-0541-78 44-0541-78
$C_9H_{13}ClO$			
Bicyclo[3.3.1]nonan-3-one, 1-chloro-	EtOH	283(1.34)	44-3653-78
$C_9H_{13}IN_4O_3$			
Cytidine, 5'-amino-2',5'-dideoxy-5-iodo-	pH 2 pH 12	299(3.89) 291(3.80)	87-0106-78 87-0106-78
$C_9H_{13}NO$			
Ethanone, 1-(3-propyl-1H-pyrrol-2-yl)-	EtOH	291(4.10)	23-0221-78
3-Pyridinol, 6-(1,1-dimethylethyl)-	EtOH	219(--),283(3.20)	150-3551-78
2(1H)-Pyridinone, 1-methyl-3-(1-methylethyl)-	EtOH	234(3.74),303(3.83)	142-0023-78B
2(1H)-Pyridinone, 1-methyl-5-(1-methylethyl)-	EtOH	231(3.92),310(3.72)	142-0023-78B
$C_9H_{13}NO_2$			
2(1H)-Pyridinone, 1-(2-hydroxy-2-methylpropyl)-	EtOH	304(3.76)	78-2609-78
$C_9H_{13}NO_3$			
2(1H)-Pyridinone, 4-hydroxy-3-(2-hydroxypropyl)-6-methyl-	EtOH	215(4.15),288(3.75)	23-0613-78
1H-Pyrrole-2-carboxylic acid, 3-hydroxy-, 1,1-dimethylethyl ester	EtOH	265(4.21)	94-2224-78
$C_9H_{13}NO_4$			
Propanedioic acid, (1-azetidinylmethylene)-, dimethyl ester	isooctane	221(3.72),291(4.30)	78-2321-78
2(1H)-Pyridinone, 3-(1,2-dihydroxypropyl)-4-hydroxy-6-methyl-	EtOH	218(4.18),280(3.56)	23-0613-78
$C_9H_{13}NO_5S$			
Furan, 2-[(butylsulfonyl)methyl]-5-nitro-	MeOH	210(3.83),240s(3.55), 313(3.96)	73-0156-78

Compound	Solvent	$\lambda_{max}(\log \epsilon)$	Ref.
$C_9H_{13}NS$			
Benzenamine, N,N-dimethyl-4-(methyl-thio)-	EtOH	275(4.32),310(3.36)	65-2027-78
$C_9H_{13}N_3O$			
Imidazo[1,2-a]pyrimidin-5(1H)-one, 6-ethyl-2,3-dihydro-7-methyl-	EtOH	232(3.96),290(3.90)	12-0179-78
$C_9H_{13}N_3O_2$			
1,4-Benzenediamine, N^1,N^4,N^4-trimethyl-2-nitro-	C_6H_{12}	489(3.65)	39-1194-78C
	EtOH	505(--)	39-1194-78C
	$CHCl_3$	522(--)	39-1194-78C
1,4-Benzenediamine, N^1,N^4,N^4-trimethyl-3-nitro-	C_6H_{12}	435(3.04)	39-1194-78C
	EtOH	451(--)	39-1194-78C
	$CHCl_3$	479(--)	39-1194-78C
$C_9H_{13}N_3O_3$			
Pyrido[2,3-d]pyrimidine-2,4(1H,3H)-di-one, 5,6,7,8-tetrahydro-3-(methoxy-methyl)-	MeOH	220(3.99),290(4.26)	24-2297-78
$C_9H_{13}N_3O_3S$			
4-Thiazoleacetic acid, 2-amino-α-(eth-oxyimino)-, ethyl ester, (Z)-	EtOH	234(4.16),294(4.21)	78-2233-78
$C_9H_{13}N_3O_5$			
Cytidine	pH 7	271(3.96)	39-0762-78C
2(1H)-Pyrimidinone, 4-amino-5-β-D-ara-binofuranosyl-	pH 1	283(3.96)	87-0096-78
	pH 7	272(3.74)	87-0096-78
	pH 10	272(3.70)	87-0096-78
4(1H)-Pyrimidinone, 2-amino-5-α-D-ara-binofuranosyl-	pH 1	262(3.78)	87-0096-78
	pH 7	290(3.68)	87-0096-78
	pH 10	279(3.70)	87-0096-78
β-	pH 1	262(3.88)	87-0096-78
	pH 7	266(3.58),290s(3.67)	87-0096-78
	pH 10	277(3.81)	87-0096-78
1,2,4-Triazine-3,5(2H,4H)-dione, 2-(2-deoxy-β-D-erythro-pentofuranosyl)-6-methyl-	pH 1.5	264(3.78)	136-0175-78C
	pH 12.7	254(3.89)	136-0175-78C
	EtOH	267(3.89)	136-0175-78C
$C_9H_{13}N_3O_6$			
1H-1,2,3-Triazole-4-carboxylic acid, 1-β-D-ribofuranosyl-, methyl ester	MeOH	215(3.98)	35-2248-78
$C_9H_{13}N_3S_2$			
2,5-Pyrrolidinedithione, 1-methyl-3,4-bis[(methylamino)methylene]-	EtOH	290(4.01),360(4.50), 470(3.91)	104-2041-78
$C_9H_{13}N_4S_2$			
7H-Purinium, 3,7-dimethyl-2,6-bis(meth-ylthio)-, iodide	H_2O	226(4.30),244(4.24), 294(4.29),334(4.26)	104-1624-78
7H-Purinium, 7,9-dimethyl-2,6-bis(meth-ylthio)-, iodide	H_2O	224(4.42),264(4.32), 320(4.03)	104-1624-78
$C_9H_{13}N_5$			
6H-Purin-6-imine, 3,9-diethyl-3,9-di-hydro-, monoperchlorate	pH 1	271(4.20)	88-5007-78
	pH 7	271(4.20)	88-5007-78
$C_9H_{13}N_5O_2$			
2,3-Butanediol, 1-(6-amino-9H-purin-9-yl)-, (R*,S*)-(±)-	pH 2	260(4.13)	73-3444-78

Compound	Solvent	$\lambda_{max}(\log \epsilon)$	Ref.
1,2-Propanediol, 3-[6-(methylamino)- 9H-purin-9-yl]-	pH 2	264(4.17)	73-3103-78
	pH 7	268(4.14)	73-3103-78
	pH 12	268(4.14)	73-3103-78
2-Propanol, 1-[(6-amino-9H-purin-9-yl)- methoxy]-	pH 2 and 12	262(4.11)	73-3444-78
6(5H)-Pteridinone, 4-(dimethylamino)- 7,8-dihydro-7-hydroxy-8-methyl-	pH 0.0	250(4.26),307(4.22)	24-1763-78
	pH 6.0	250(4.28),293(4.19)	24-1763-78
	pH 13.0	239(4.26),303(4.20)	24-1763-78
$C_9H_{13}N_5O_2S$			
1,2-Propanediol, 3-[6-amino-2-(methyl- thio)-9H-purin-9-yl]-	pH 2	270(4.25)	73-3103-78
	pH 12	233(4.42),277(4.25)	73-3103-78
Propanoic acid, 2-[(2-amino-1,6-di- hydro-6-thioxo-4-pyrimidinyl)methyl- hydrazono]-, methyl ester	MeOH	240(4.12),290(4.09), 317(4.13),351(4.18)	44-4844-78
$C_9H_{14}ClN_2O_3PS_2$			
Phosphorothioic acid, O-(5-chloro- 1,6-dihydro-1-methyl-6-thioxo-4- pyridazinyl) O,O-diethyl ester	MeOH	218(4.22),256(3.60), 292(3.70)	73-2415-78
$C_9H_{14}FN_4O_4$			
Pyrimidinium, 1-β-D-arabinofuranosyl- 2,4-diamino-5-fluoro-, chloride	H_2O	206(4.26),236(4.04), 275(3.87)	44-4200-78
$C_9H_{14}N_2$			
1H-1,5-Benzodiazepine, 5a,6,7,8,9,9a- hexahydro-, monoperchlorate	MeOH	330(4.11)	39-1453-78C
$C_9H_{14}N_2O$			
3-Butyn-2-one, 4-(4-methyl-1-piperazin- yl)-	CH_2Cl_2	277(4.27)	33-1609-78
$C_9H_{14}N_2OS$			
2(1H)-Pyrimidinone, 4-(butylthio)- 5?-methyl-	pH 1.00	330(4.05)	56-1255-78
	pH 2.00	303(4.16)	56-1255-78
	pH 3-10	300(4.14)	56-1255-78
	pH 11-13	300(4.18)	56-1255-78
2(1H)-Pyrimidinone, 4-(butylthio)- 6?-methyl-	H_2O	305(3.95)	56-1255-78
1H-Pyrrolo[1,2-c]imidazol-1-one, hexahydro-2-propyl-3-thioxo-	MeOH	202(3.90),246(4.15), 272(4.23)	19-0851-78
$C_9H_{14}N_2O_2$			
2H-Benzimidazole, 4,5,6,7-tetrahydro- 2,2-dimethyl-, 1,3-dioxide	EtOH	212(3.90),350(3.90)	70-0844-78
1H-Pyrazol-5-ol, 4-ethylidene-4,5-di- hydro-5-methyl-, propanoate	EtOH	281(3.74)	22-0401-78
$C_9H_{14}N_2O_2S$			
2,4-Thiazolidinedione, 5-cyclohexyl-, 2-oxime, (Z)-	n.s.g.	213(3.93),232(3.90)	139-0303-78A
$C_9H_{14}N_2O_3$			
2,4-Octadienal, 6-nitro-, O-methyloxime	M NaOH	252(3.92),356(4.62)	94-2422-78
	EtOH	248(3.75),330(4.60), 345(4.66)	94-2422-78
	$CHCl_3$	282(4.19)	94-2422-78
1H-Pyrazole-4-carboxylic acid, 3-(1-hy- droxyethyl)-5-methyl-, ethyl ester	EtOH	229(3.93)	118-0448-78

Compound	Solvent	λ_{max}(log ϵ)	Ref.
1H-Pyrazole-4-carboxylic acid, 3-(2-hydroxyethyl)-5-methyl-, ethyl ester	EtOH	227(3.93)	118-0900-78
$C_9H_{14}N_2O_4$			
Carbamic acid, [2-(3-methoxy-5-isoxazolyl)propyl]-, methyl ester, (±)-	MeOH	210(3.89)	1-0469-78
Carbamic acid, [1-(3-methoxy-4-methyl-5-isoxazolyl)ethyl]-, methyl ester	MeOH	212(3.81)	1-0469-78
2,4(1H,3H)-Pyrimidinedione, 1-(3,5-dihydroxypentyl)-, (±)-	pH 2	262(3.98)	73-3444-78
2,4,6(1H,3H,5H)-Pyrimidinetrione, 5,5-diethyl-1-hydroxy-3-methyl-	pH 1.3	199(4.96),233(3.80)	12-2517-78
	pH 9.32	205(4.24),293(3.54)	12-2517-78
2,4,6(1H,3H,5H)-Pyrimidinetrione, 5,5-diethyl-1-methoxy-	pH 1.41	198(4.88),221(3.86)	12-2517-78
	pH 9.86	200(4.41),243(3.98)	12-2517-78
$C_9H_{14}N_2O_4S$			
1H-Imidazole, 2-(methylthio)-1-β-D-ribofuranosyl-	MeOH	224(3.79),249(3.66)	44-4774-78
$C_9H_{14}N_2O_5S_3$			
β-D-Glucopyranoside, 3-(methylthio)-1,2,4-thiadiazol-5-yl 1-thio-	n.s.g.	253(4.08),286s(3.52)	106-0764-78
$C_9H_{14}N_2O_6S$			
1H-Imidazole, 2-(methylsulfonyl)-1-β-D-ribofuranosyl-	MeOH	237(3.97)	44-4774-78
$C_9H_{14}N_2S$			
1H-Imidazole, 2-(cyclopentylthio)-1-methyl-	EtOH	226(3.843),250(3.708)	28-0201-78B
	CH_2Cl_2	226(3.890),248s(3.740)	28-0201-78B
$C_9H_{14}N_2S_2$			
Pyrimidine, 2,4-bis(ethylthio)-6-methyl-	H_2O	255(3.24),300(3.15)	56-1255-78
$C_9H_{14}N_2Si$			
Diazene, phenyl(trimethylsilyl)-	CH_2Cl_2	575(1.42)	60-1909-78
$C_9H_{14}N_3O_5PS$			
Phosphorohydrazidothioic acid, [(5-nitro-2-furanyl)methylene]-, 0,0-diethyl ester	dioxan	370(4.48)	73-2643-78
$C_9H_{14}N_4O$			
Propanehydrazonic acid, N-4-pyrimidinyl-, ethyl ester, (E)-	pH 7.0	271(4.16)	12-2505-78
(Z)-	pH 7.0	267(4.04)	12-2505-78
1-Propen-2-amine, N,N-dimethyl-1-(6-methyl-1,2,4-triazin-5-yl)-, N-oxide	EtOH	251(3.98),278(4.16), 360(3.68),430(4.20)	24-0240-78
$C_9H_{14}N_4OS_2$			
Formamide, N-methyl-N-[4-(methylamino)-2,6-bis(methylthio)-5-pyrimidinyl]-	EtOH	242(4.54),294(3.90)	104-1624-78
$C_9H_{14}N_4O_2$			
2,4(1H,3H)-Pteridinedione, 5,6,7,8-tetrahydro-1,3,6-trimethyl-, hydrochloride	pH -1	268(4.24)	12-1081-78
	pH 3.4	267(4.16)	12-1081-78
	pH 8.5	304(3.9)	12-1081-78
	pH 12.7	286(3.51)	12-1081-78
1H-Pyrazole-4-carboxylic acid, 5-[[(dimethylamino)methylenelamino]-, ethyl ester	EtOH	287(4.05)	103-0310-78

Compound	Solvent	$\lambda_{max}(\log \epsilon)$	Ref.
$C_9H_{14}N_4O_4$			
Acetamide, N-(6-amino-1,2,3,4-tetra-hydro-1,3-dimethyl-2,4-dioxo-5-py-rimidinyl)-2-methoxy-	EtOH	197(4.17),268(4.16)	73-3414-78
2,4-Pyrimidinediamine, 5-α-D-arabino-furanosyl-	pH 1	270(3.66)	87-0096-78
	pH 7	282(3.72)	87-0096-78
	pH 10	285(3.79)	87-0096-78
$C_9H_{14}O$			
2-Cyclohexen-1-one, 5-ethyl-3-methyl-	EtOH	240(3.66)	78-2783-78
2-Cyclopenten-1-one, 3-methyl-5-(1-methylethyl)-	n.s.g.	227(4.10)	22-0255-78
2-Cyclopenten-1-one, 2,4,4,5-tetra-methyl-	EtOH	231(3.75)	35-1799-78
Ethanone, 1-bicyclo[2.2.1]hept-2-yl-, endo	EtOH	280(1.38)	44-4215-78
exo	EtOH	280(1.43)	44-4215-78
$C_9H_{14}O_2$			
Bicyclo[3.3.1]nonan-3-one, 1-hydroxy-	EtOH	280(1.26)	44-3653-78
Cyclohexanol, 3-(2-hydroxyethylidene)-4-methylene-, (Z)-(±)-	EtOH	219(3.81)	39-0387-78C
Cyclooctanone, 2-(hydroxymethylene)-	hexane	291.2(3.81)	32-0581-78
	C_6H_{12}	291.4(3.86)	32-0581-78
	MeOH	290.0(3.79)	32-0581-78
	EtOH	288.1(3.78)	32-0581-78
2-Cyclopenten-1-one, 4-(1-ethoxyethyl)-	EtOH	215(4.02),307(1.60)	118-0543-78
2-Cyclopenten-1-one, 5-hydroxy-2-(2-methylpropyl)-	EtOH	232(3.82)	107-0155-78
3,5-Octadienoic acid, methyl ester	hexane	228(4.26)	119-0075-78
3,6-Octadienoic acid, methyl ester	hexane	213(3.30)	119-0075-78
4,6-Octadienoic acid, methyl ester	hexane	228(4.26)	119-0075-78
$C_9H_{14}O_2S$			
2-Cyclohexen-1-one, 3-hydroxy-5,5-di-methyl-2-(methylthio)-	MeOH	248(3.92),281(3.95) (changing)	44-2676-78
$C_9H_{14}O_2S_2$			
1,4-Dithiepane-2-butanal, 3-oxo-	n.s.g.	240(3.56),280s(--), 313(2.11)	33-3087-78
$C_9H_{14}O_3$			
3,5-Heptadien-2-one, 7,7-dimethoxy-, (E,E)-	MeOH	266(4.48),330(2.70)	18-2131-78
5-Heptenoic acid, 2-ethylidene-4-hy-droxy-, (E,E)-	EtOH	213(4.02)	18-2375-78
2,4-Octadienoic acid, 7-hydroxy-, methyl ester, [R-(E,E)]-	MeOH	260(4.44)	18-3035-78
[R-(E,Z)]-	MeOH	265(4.32)	18-3035-78
$C_9H_{14}SSi$			
Silane, trimethyl(phenylthio)-	heptane	243(3.60),265(3.00)	65-2027-78
$C_9H_{15}BrN_2O_4$			
2-Furanmethanaminium, N-(2-hydroxy-ethyl)-N,N-dimethyl-5-nitro-, bromide	H_2O	203(4.07),295(3.70)	73-2041-78
$C_9H_{15}BrO_2Sn$			
Stannane, [(3-bromo-1-oxo-2-propynyl)-oxy]triethyl-	MeCN	212(3.63)	65-1244-78

Compound	Solvent	$\lambda_{max}(\log \epsilon)$	Ref.
$C_9H_{15}ClO_2Sn$ Stannane, [(3-chloro-1-oxo-2-propynyl)-oxy]triethyl-	MeCN	212(3.49)	65-1244-78
$C_9H_{15}IO_2Sn$ Stannane, triethyl[(3-iodo-1-oxo-2-propynyl)oxy]-	MeCN	214(3.75),232s(--)	65-1244-78
$C_9H_{15}NO$ Cyclohexanone, 2-[(dimethylamino)methylene]-	EtOH	340(4.33)	70-0102-78
$C_9H_{15}NO_2$ 2,4-Azetidinedione, 3,3-bis(1-methylethyl)-	CH_2Cl_2	247(2.12),253s(2.08)	33-3050-78
$C_9H_{15}NO_4$ (2-Acetoxyethyl)dimethylaminium 1,3-dioxo-2-propanide	EtOH	257(4.46)	73-1261-78
$C_9H_{15}N_3O_2S$ 2,4-Thiazolidinedione, 5-[2-(1-pyrrolidinyl)ethyl]-, 2-oxime	n.s.g.	263(4.23)	139-0303-78A
$C_9H_{15}N_3O_3S$ 2,4-Thiazolidinedione, 5-(2-morpholino-ethyl)-, 2-oxime	n.s.g.	263(4.22)	139-0303-78A
$C_9H_{15}N_5$ 2-Pentenedinitrile, 2-amino-3,4-bis(dimethylamino)-	MeOH	267(4.10)	35-0926-78
$C_9H_{15}N_5O$ 4(1H)-Pteridinone, 2-amino-5,6,7,8-tetrahydro-1,5,6-trimethyl-	pH -2 pH 3.5 pH 8.5-13	267(4.13) 266(4.00) 286(3.98)	12-1081-78 12-1081-78 12-1081-78
4(3H)-Pteridinone, 2-amino-5,6,7,8-tetrahydro-3,5,6-trimethyl-, hydrochloride	pH -2 pH 3.5 pH 8.5-13	265(4.11) 264(3.93) 284(3.89)	12-1081-78 12-1081-78 12-1081-78
4(3H)-Pteridinone, 2,3,5,8-tetrahydro-2-imino-1,3,6-trimethyl-, dihydro-chloride	pH -1 pH 3-12	264(4.13) 218(4.28),263(3.89)	12-1081-78 12-1081-78
$C_9H_{15}N_5O_2$ 1H-1,2,3-Triazole-4-carboxamide, 1-butyl-N-[(methylamino)carbonyl]-	EtOH	232(3.81)	4-1349-78
$C_9H_{15}N_5O_4$ 1H-Imidazole-4-carboximidamide, 5-amino-1-β-D-ribofuranosyl-	pH 1 pH 7 pH 13	284(4.05) 286(4.08) 267.5(3.95)	94-1929-78 94-1929-78 94-1929-78
$C_9H_{16}GeO_2$ Germane, triethyl[(1-oxo-2-propynyl)-oxy]-	heptane	208(3.04)	65-1244-78
$C_9H_{16}N_2$ 1-Cyclohexene-1-carboxaldehyde, dimethylhydrazone, (E)-	C_6H_{12}	275(4.39)	28-0047-78A

Compound	Solvent	$\lambda_{max}(\log \epsilon)$	Ref.
$C_9H_{16}N_2O$			
3H-Pyrrol-3-one, 4,5-dihydro-5,5-di-	EtOH	230(3.98)	22-0621-78
methyl-2-(1-methylethyl)-, oxime	EtOH-NaOH	275(3.94)	22-0621-78
after acidification	EtOH-NaOH	230(3.97)	22-0621-78
$C_9H_{16}N_2O_2$			
Cyclohexanamine, N-(2-nitro-1-propenyl)-	EtOH	258(2.86),370(4.09)	44-0497-78
3,4-Piperidinedione, 2,2,6,6-tetra-	EtOH	231(4.09)	22-0621-78
methyl-, 4-oxime	EtOH-NaOH	276(4.19)	22-0621-78
after acidifcation	EtOH-NaOH	263(4.08)	22-0621-78
$C_9H_{16}N_2O_5$			
1-Butanol, 4-[(acetoxymethyl)nitroso-	EtOH	232(3.82),372(1.91)	94-3914-78
amino]-, acetate			
Glycine, N-(1-acetoxybutyl)-N-nitroso-,	EtOH	236(3.80),366(1.88)	94-3914-78
methyl ester			
$C_9H_{16}N_6O_2$			
1H-1,2,3-Triazole-4-carboxamide,	EtOH	226(4.01)	4-1349-78
N-(aminocarbonyl)-1-[3-(dimethyl-			
amino)propyl]-			
$C_9H_{16}OS_2$			
1,5-Dithiaspiro[5.5]undecan-7-ol	n.s.g.	227(2.88),246(3.02), 250s(--)	33-3087-78
$C_9H_{16}OS_3$			
1,3-Benzoxathiole, hexahydro-2,2-	EtOH	227(2.95)	118-0286-78
bis(methylthio)-			
$C_9H_{16}OSn$			
2-Cyclohexen-1-one, 3-(trimethyl-	MeOH	236(4.08)	77-1033-78
stannyl)-			
$C_9H_{16}O_2$			
2,5-Nonanedione	MeOH	274(2.47)	44-4081-78
$C_9H_{16}O_2Sn$			
Stannane, triethyl[(1-oxo-2-propynyl)-	MeCN	207(3.40)	65-1244-78
oxy]-			
$C_9H_{16}O_4$			
Pentanedioic acid, 2,2,4,4-tetramethyl-	HCl	210(2.18)	137-0151-78
	NaOH	218(1.78)	137-0151-78
$C_9H_{17}NO_2$			
L-Alanine, N-(2-methylpropylidene)-,	hexane	216(2.80)	94-0466-78
ethyl ester, (E)-	MeOH	201(2.83)	94-0466-78
Butyldimethylammonium 1,3-dioxo-2-prop-	EtOH	257(4.45)	73-1261-78
anide			
2-Propenal, 3-butoxy-2-(dimethylamino)-	EtOH	231(3.86),286(3.71)	73-1261-78
$C_9H_{17}N_2S$			
1H-Imidazolium, 2-(butylthio)-1,3-di-	EtOH	221(4.300),261(4.625)	28-0201-78B
methyl-, iodide	CH_2Cl_2	239(4.228),265(3.730)	28-0201-78B
$C_9H_{17}N_3$			
1H-Pyrazol-3-amine, N,N,5-trimethyl-	heptane	220(3.20),252(3.42)	104-0576-78
1-propyl-			

C$_9$H$_{18}$N$_2$–C$_9$H$_{22}$OSi$_3$

Compound	Solvent	λ_{max}(log ϵ)	Ref.
C$_9$H$_{18}$N$_2$			
2-Heptenal, dimethylhydrazone, (E,E)-	C$_6$H$_{12}$	275(4.37)	28-0047-78A
2-Pentenal, 4,4-dimethyl-, dimethyl-hydrazone, (E,E)-	C$_6$H$_{12}$	275(4.40)	28-0047-78A
C$_9$H$_{18}$N$_2$O$_3$			
Butanoic acid, 4-(nitrosopentylamino)-	EtOH	235(3.85),353(1.93)	94-3909-78
C$_9$H$_{18}$OSi			
Silane, trimethyl(1-oxo-2-hexenyl)-	C$_6$H$_{12}$	425(1.72),440(1.66), 445(1.53)	101-0C01-78B
C$_9$H$_{18}$S			
1-Butene, 1-(butylthio)-2-methyl-,(E)-	C$_6$H$_{12}$	229(3.97),249(3.75)	73-2635-78
1-Butene, 1-(butylthio)-2-methyl-,(Z)-	C$_6$H$_{12}$	252(3.88)	73-2635-78
1-Butene, 1-(butylthio)-3-methyl-,(E)-	C$_6$H$_{12}$	231(4.02),246(3.88)	73-2635-78
1-Butene, 1-(butylthio)-3-methyl-,(Z)-	C$_6$H$_{12}$	229(3.98),246(3.77)	73-2635-78
3-Pentanethione, 2,2,4,4-tetramethyl-	n.s.g.	237(3.89),539(1.0)	88-3463-78
C$_9$H$_{19}$AsO$_5$S			
D-Galactopyranoside, methyl 6-thio-, 6-(dimethylarsinite)	MeOH	223(3.64)	136-0069-78E
C$_9$H$_{19}$NO			
2-Piperidinemethanol, α-(1-methylethyl)-	n.s.g.	221(3.33),259(3.05)	42-0916-78
2-Piperidinemethanol, α-propyl-	n.s.g.	223(3.33),260(3.07)	42-0916-78
C$_9$H$_{19}$N$_2$S			
Pyrimidinium, 1,4,5,6-tetrahydro-1-methyl-3-(1-methylethyl)-2-(methylthio)-, iodide	EtOH CH$_2$Cl$_2$	221(4.006),245(4.228) 253(4.316)	28-0521-78A 28-0521-78A
C$_9$H$_{19}$N$_3$O			
Acetic acid, [1-(dimethylamino)ethylidene)propylhydrazide	heptane	227(3.75),249s(3.00)	104-0576-78
Butanehydrazonamide, N,N-dimethyl-3-oxo-N'-propyl-	heptane	220(3.45),287(3.75)	104-0576-78
C$_9$H$_{20}$N$_2$O$_2$			
1-Butanol, 4-(nitrosopentylamino)-	EtOH	234(3.85),350(1.92)	94-3891-78
C$_9$H$_{22}$OSi$_3$			
1,2,4-Trisilacyclohexan-3-one, 1,1,2,2,4,4-hexamethyl-	C$_6$H$_{12}$	234(3.88)	101-0C25-78G

Compound	Solvent	$\lambda_{max}(\log \epsilon)$	Ref.
$C_{10}Cl_6N_2$ Propanedinitrile, (1,3,4,5,6,7-hexachlorobicyclo[3.2.0]hepta-3,6-dien-2-ylidene)-	heptane	210(4.06),299s(4.25), 306(4.25),330s(4.05)	24-2738-78
$C_{10}F_8O$ 1H-Inden-1-one, 3,4,5,6,7-pentafluoro-2-(trifluoromethyl)-	heptane	319(3.41),332(3.45), 344s(3.30)	78-3215-78
$C_{10}H_2N_4$ 1,2,4,5-Benzenetetracarbonitrile	hexane	224(4.65),256(4.14), 314(3.40)	110-0044-78
$C_{10}H_4Cl_2O_4$ 1,4-Naphthalenedione, 2,3-dichloro-5,8-dihydroxy-	C_6H_{12}	491(3.49),530(3.58), 554(3.30),570(3.45)	150-2319-78
	EtOH	495(--),525(--), 567(--)	150-2319-78
$C_{10}H_4N_4S_2$ Naphtho[1,8-cd:4,5-c'd']bis[1,2,6]thiadiazine	MeCN	255(4.38),273(4.41), 335s(3.65),352s(3.60), 459(4.15),489(4.29)	35-1235-78
$C_{10}H_4O_8S_4$ 1,3-Dithiole-4,5-dicarboxylic acid, 2-(4,5-dicarboxy-1,3-dithiol-2-ylidene)-	DMF	302(4.10),311(4.17), 477(3.33)	44-0595-78
$C_{10}H_5BrN_2O_6$ Furan, 2-[2-(5-bromo-2-furanyl)-1-nitroethenyl]-5-nitro-, (E)-	dioxan	302(4.01),420(4.32)	73-3252-78
$C_{10}H_5BrO_3$ 1,4-Naphthalenedione, 2-bromo-6-hydroxy-	EtOH	275(4.30),285(4.18), 410(3.18)	12-1335-78
$C_{10}H_5Cl_2NO_2$ 1,4-Naphthalenedione, 5-amino-2,3-dichloro-	C_6H_{12} EtOH	498(3.72) 531(--)	150-2319-78 150-2319-78
$C_{10}H_5Cl_2NO_3$ 1,4-Naphthalenedione, 5-amino-2,3-dichloro-8-hydroxy-	C_6H_{12}	538(3.88),576(4.07), 624(3.99)	150-2319-78
	EtOH	576(--),619(--)	150-2319-78
$C_{10}H_5Cl_3O$ 3-Butyn-2-one, 1,1,1-trichloro-4-phenyl-	hexane	285(4.13),301(4.15)	118-0307-78
$C_{10}H_5Cl_5$ 1,3-Cyclopentadiene, 1,2,3,4-tetrachloro-5-(4-chloro-2-cyclopenten-1-ylidene)-	hexane	250s(4.39),327(4.50), 341(4.44),356s(4.10), 450(2.60)	1-0149-78
$C_{10}H_5IN_2O_6$ Furan, 2-[2-(5-iodo-2-furanyl)-1-nitroethenyl]-5-nitro-	dioxan	308(4.03),425(4.35)	73-3252-78
$C_{10}H_5N_3O_8$ Furan, 2,2'-(1-nitro-1,2-ethenediyl)-	dioxan	310(4.31),372(4.15)	73-3252-78

Compound	Solvent	$\lambda_{max}(\log \epsilon)$	Ref.
bis[5-nitro-, (E)- (cont.)			73-3252-78
(Z)-	dioxan	310(4.23),392(4.18)	73-3252-78
$C_{10}H_6BrClO$			
2(5H)-Furanone, 5-bromo-3-(4-chloro-phenyl)-	EtOH	283(4.09)	44-4115-78
$C_{10}H_6BrFO_2$			
2(5H)-Furanone, 5-bromo-3-(4-fluoro-phenyl)-	EtOH	282(3.96)	44-4115-78
$C_{10}H_6BrNO$			
2-Quinolinecarboxaldehyde, 4-bromo-	DMF	301(3.81)	97-0138-78
$C_{10}H_6Br_2N_2$			
4,4'-Bipyridine, 3,3'-dibromo-	EtOH	272(3.70)	39-1126-78C
$C_{10}H_6Br_2O_2$			
1,2-Naphthalenediol, 3,6-dibromo-	MeOH	266(4.31),348(3.62), 392s(3.33)	12-2259-78
$C_{10}H_6ClNO$			
2-Quinolinecarboxaldehyde, 4-chloro-	DMF	303(4.18)	97-0138-78
$C_{10}H_6ClN_3O_2S$			
2-Naphthalenesulfonyl chloride, 5-azido-	CHCl$_3$	255(4.51),314(3.78), 349(3.15)	40-1690-78
$C_{10}H_6Co_2O_6S_2$			
Cobalt, [μ-[(1,2-η;1,2-η)-bis(methyl-thio)ethyne]]hexacarbonyldi-, (Co-Co)-	EtOH	290(4.66),317(4.51), 468(2.54),548(2.15)	101-0159-78Q
$C_{10}H_6N_2O$			
Benzofuro[2,3-d]pyridazine	H$_2$O	207(4.20),223(4.21), 273(4.09)	4-1387-78
$C_{10}H_6N_2OS$			
Thiazolo[3,2-a]quinoxalin-10-ium, 1-hydroxy-, hydroxide, inner salt	H$_2$O	240(4.12),263(3.75), 292(3.78),438(4.11)	2-0678-78
$C_{10}H_6N_2O_2$			
1,8-Naphthyridine-2,7-dicarboxaldehyde	MeOH	256(3.53),302s(3.67), 308(3.69),314(3.72)	97-0020-78
$C_{10}H_6N_2O_3$			
2H-Pyrano[3,2-c]pyridine-3-carboni-trile, 5,6-dihydro-7-methyl-2,5-dioxo-	MeOH	388(4.26)	49-1075-78
2-Quinolinecarboxaldehyde, 4-nitro-	DMF	333(3.55)	97-0138-78
$C_{10}H_6N_2O_6$			
Furan, 2-[2-(2-furanyl-1-nitroethenyl]-5-nitro-, (E)-	dioxan	314(4.19),357(4.24)	73-3252-78
$C_{10}H_6N_2S$			
Naphtho[1,8-cd]thiadiazine	film	234(4.73),255(4.07), 330(3.92),345(3.93), 645(2.74)	89-0468-78
$C_{10}H_6N_2S_2$			
1,6-Diazathianthrene	H$_2$O	206(4.18),251(4.22), 295(3.77)	56-2039-78

Compound	Solvent	$\lambda_{max}(\log \epsilon)$	Ref.
$C_{10}H_6N_2S_3$			
[1,3]Thiazino[5,4-b]indole-2,4-dithione, 1,5-dihydro-	EtOH	<u>310(4.5),375(4.1), 480(4.0)</u>	103-1217-78
$C_{10}H_6N_4$			
2,7,9,10-Tetraazaphenanthrene	dioxan	237(4.76),261s(4.09), 295(4.17)	39-1126-78C
4,5,9,10-Tetraazaphenanthrene	EtOH	232(4.52),259(4.43), 296s(3.78),329(3.49), 345(3.42)	39-1126-78C
$C_{10}H_6N_4O$			
2-Quinolinecarboxaldehyde, 4-azido-	DMF	302(3.92)	97-0138-78
2,7,9,10-Tetraazaphenanthrene 9-oxide	dioxan	242(4.63),258s(4.37), 334(4.13)	39-1126-78C
$C_{10}H_6N_4O_4$			
4,4'-Bipyridine, 3,3'-dinitro-	MeOH	245(3.84),274s(3.69)	39-1126-78C
$C_{10}H_6O_3$			
4H-1-Benzopyran-2-carboxaldehyde, 4-oxo-	EtOH	222(4.25),299(3.83)	118-0208-78
1,4-Naphthalenedione, 6-hydroxy-	EtOH	225s(4.30),270(4.45), 395(3.52)	12-1335-78
$C_{10}H_6O_4$			
1,4-Naphthalenedione, 5,8-dihydroxy-	C_6H_{12}	490(3.80),528(3.85), 550(3.68),568(3.71)	150-2319-78
	EtOH	486(--),518(--), 560(--)	150-2319-78
$C_{10}H_6O_4S_2$			
Naphtho[1,8-cd]-1,2-dithiole, 1,1,2,2-tetraoxide	60% dioxan	302(3.81)	44-0914-78
$C_{10}H_6O_5$			
2H-1-Benzopyran-2,5,8-trione, 7-methoxy-	EtOH	263(3.98),289(4.10), 300(4.09)	102-0505-78
$C_{10}H_6S_2$			
Naphtho[1,8-cd]-1,2-dithiole	$C_6H_{11}Me$	206(4.52),216s(4.15), 223s(4.04),253(4.42), 286(2.70),292s(2.60), 357(4.08),371(4.18), 376s(4.04),460s(1.85)	35-0202-78
$C_{10}H_7BrN_2O$			
Pyrimidine, 5-bromo-4-phenyl-, 1-oxide	EtOH	210(4.05),229(4.03), 296(4.23),326s(4.04)	103-1132-78
$C_{10}H_7BrO_2$			
2(5H)-Furanone, 5-bromo-3-phenyl-	EtOH	273(4.00)	44-4115-78
$C_{10}H_7BrO_3$			
2(5H)-Furanone, 5-bromo-5-hydroxy-3-phenyl-	EtOH	286(4.18)	44-4115-78
$C_{10}H_7ClN_2$			
Pyrazine, 2-chloro-3-phenyl-	EtOH	237(3.89),249(3.86), 287(3.89)	44-3367-78

Compound	Solvent	$\lambda_{max}(\log \epsilon)$	Ref.
Pyrazine, 2-chloro-5-phenyl-	EtOH	254(4.23),291(3.99), 311(4.03)	44-3367-78
Pyrazine, 2-chloro-6-phenyl-	EtOH	252(4.03),290(3.98), 310(4.03)	44-3367-78
$C_{10}H_7F_3O_2$ 1,3-Butanedione, 4,4,4-trifluoro- 1-phenyl-	heptane CHCl$_3$ MeCN	325(4.3) 332(4.3) 335(4.3)	40-0489-78 40-0489-78 40-0489-78
$C_{10}H_7NO_2$ 1,4-Naphthalenedione, 5-amino-	C_6H_{12} EtOH	484(3.73) 516(--)	150-2319-78 150-2319-78
1,4-Naphthalenedione, 6-amino-	C_6H_{12} EtOH	410(3.38) 479(--)	150-2319-78 150-2319-78
$C_{10}H_7NO_3$ 1,4-Naphthalenedione, 5-amino-8-hydr-oxy-	C_6H_{12} EtOH	525(3.82),564(3.97), 589(3.77),609(3.88) 572(--),617(--)	150-2319-78 150-2319-78
$C_{10}H_7NO_3S$ 2-Thiazolecarboxylic acid, 4-hydroxy-5-phenyl- (hydrate)	MeOH	257(3.90),352(4.18)	5-0473-78
$C_{10}H_7NO_3S_2$ Thiazolo[2,3-c][1,4]benzothiazin-1(2H)-one, 5,5-dioxide	MeOH	280(3.82),352(4.02)	2-0678-78
$C_{10}H_7NO_4$ 1H-Indole-3-acetic acid, 1-hydroxy-α-oxo-	MeOH	222(4.32),245(3.85), 255(3.83),313(3.73), 395(3.31)	39-1117-78C
$C_{10}H_7NO_4S$ 1,2-Benzisothiazol-3(2H)-one, 2-(1-oxo-2-propenyl)-, 1,1-dioxide	THF	253s(4.21),273(4.18)	40-0404-78
$C_{10}H_7NO_5$ Ethanone, 1-(2-furanyl)-2-(5-nitro-2-furanyl)- (and other pH values)	pH 1 pH 2 pH 3 pH 6 pH 9	none 490(3.3) 465(3.4) 463(3.9) 457(4.6)	95-0286-78 95-0286-78 95-0286-78 95-0286-78 95-0286-78
$C_{10}H_7NS$ Pyrrolo[2,1-b]benzothiazole	EtOH	221s(3.909),228(3.953), 248(4.079),309(3.532)	39-1198-78C
$C_{10}H_7N_3$ 4-Isoquinolinecarbonitrile, 3-amino-	EtOH	241(4.76),276(3.77), 286(3.70),384(3.67)	138-0677-78
$C_{10}H_7N_3O_2$ Propanedinitrile, (3-acetoxy-2-pyri-dinyl)-	MeOH MeOH-NaOH	227s(3.78),292(4.13), 366(3.68) 226(3.94),291(4.16), 348(3.68)	4-1411-78 4-1411-78
$C_{10}H_7N_3O_2S$ 1H-Pyrimido[5,4-b][1,4]benzothiazine-	12M HCl	285(3.68),368(3.28)	5-0193-78

Compound	Solvent	$\lambda_{max}(\log \epsilon)$	Ref.
2,4(3H,10H)-dione (cont.)	pH 13	278s(3.91),320s(3.45)	5-0193-78
	MeOH-DMF	253(4.31),278(3.91),	5-0193-78
		324(3.32),385s(3.08)	
$C_{10}H_7N_3O_3S$			
2-Naphthalenesulfonic acid, 5-azido-, sodium salt	MeOH	238(4.58),305(3.93), 325(3.23),334(3.28), 415(2.36)	40-1690-78
Pyrimidine, 2-[[5-(2-nitroethenyl)-2-furanyl]thio]-	EtOH	237(4.06),354(4.11)	73-2037-78
$C_{10}H_7N_3O_6$			
2-Furanmethanamine, 5-nitro-α-[(5-nitro-2-furanyl)methylene]-	EtOH	284(4.18),485(4.38)	4-0555-78
$C_{10}H_7N_3S_2$			
Pyrido[3',2':4,5]thieno[3,2-d]pyrimidine, 4-(methylthio)-	EtOH	236(3.95),247(3.70), 270(3.98),298(4.17), 332(3.76),335(3.75), 370(2.30),385(2.18)	32-0057-78
$C_{10}H_7N_7S$			
5H-Tetrazolo[5',1':2,3][1,3,4]thiadiazino[5,6-b]quinoxaline, 5-methyl-	EtOH	211(4.31),227(4.3), 294(3.83)	2-0307-78
$C_{10}H_8$			
Benzene, 1-buten-3-ynyl-	hexane	226s(3.98),273(4.31), 302s(3.68)	24-2563-78
$C_{10}H_8BrNO_2$			
2H-Azirine-2-carboxylic acid, 3-(4-bromophenyl)-, methyl ester	EtOH	259(4.25)	103-0843-78
Isoxazole, 3-(4-bromophenyl)-5-methoxy-	EtOH	248(4.28)	103-0843-78
Oxazole, 2-(4-bromophenyl)-5-methoxy-	EtOH	298(4.24)	104-1353-78
$C_{10}H_8BrNO_2S$			
1H-Pyrrole-2,5-dione, 3-bromo-1-methyl-4-(5-methyl-2-thienyl)-	MeCN	243(4.06),265(4.07), 402(3.96)	88-0125-78
$C_{10}H_8BrN_3O_3$			
1H-1,5-Benzodiazepin-2-one, 8-bromo-1,3-dihydro-4-methyl-7-nitro-	EtOH	270(4.13),300s(3.86)	103-0455-78
	50% H_2SO_4	202(4.12),265(4.13)	103-0455-78
$C_{10}H_8ClNO_2$			
Acetamide, N-(2-chloro-3-benzofuranyl)-	EtOH	254(4.06),275(3.60), 279(3.61),282(3.32), 284(3.49)	39-0419-78C
Oxazole, 2-(3-chlorophenyl)-5-methoxy-	EtOH	296(4.21)	104-1353-78
Oxazole, 2-(4-chlorophenyl)-5-methoxy-	EtOH	296(4.23)	104-1353-78
$C_{10}H_8ClNO_2S_3$			
1,4,2-Dithiazole, 5-[(4-chlorophenyl)methylene]-3-(methylthio)-, 1,1-dioxide	EtOH	320(4.34)	18-1805-78
$C_{10}H_8ClNO_3$			
Furo[2,3-b]pyridine-2-carboxylic acid, 6-chloro-, ethyl ester	MeOH	258(4.06),268(4.08), 298(4.28),305(4.26)	4-0031-78
$C_{10}H_8ClN_3O_3$			
2H-1,5-Benzodiazepin-2-one, 8-chloro-	EtOH	270(4.25),300s(3.94)	103-0455-78

Compound	Solvent	$\lambda_{max}(\log \epsilon)$	Ref.
1,3-dihydro-4-methyl-7-nitro- (cont.)	50% H_2SO_4	196(4.26),260(4.06), 300(3.69)	103-0455-78
$C_{10}H_8Cl_2$ Bicyclo[4.2.2]deca-2,4,7,9-tetraene, 2,5-dichloro-	C_6H_{12}	274(3.98),286(4.02), 298(3.82)	5-2074-78
$C_{10}H_8Cl_2FeSi$ Ferrocene, 1,1'-(dichlorosilylene)-	isooctane	324(2.21),470(2.49)	35-7264-78
$C_{10}H_8Cl_3OP$ Phosphonic dichloride, (2-chloro-4-phenyl-1,3-butadienyl)-	$C_2H_4Cl_2$ MeCN	235(4.00),319(4.56) 228(3.99),316(4.56)	65-1817-78 65-1817-78
$C_{10}H_8Cl_4P$ Phosphonium, trichloro(2-chloro-4-phenyl-1,3-butadienyl)-	$C_2H_4Cl_2$	253(3.80),273s(3.89), 328s(4.04),389(4.36)	65-1817-78
$C_{10}H_8F_8N_6S$ 3H-1,2,4-Triazole-3-thione, 4-amino-2,4-dihydro-5-[3-methyl-5-(1,1,2,2-3,3,4,4-octafluorobutyl)-1H-pyrazol-1-yl]-	EtOH	256(4.36)	103-0458-78
$C_{10}H_8Li_2$ Lithium, [μ-(1-methyl-3-phenyl-2-propynylidene)]di-	ether	382(4.00)	101-0001-78N
$C_{10}H_8N_2$ 1H-Indole-2-carbonitrile, 1-methyl- 7H-Pyrrolo[3,4-e]isoindole	EtOH CH_2Cl_2	284(4.504) 257(4.17),261(4.18), 273(4.10),290(4.07), 300(4.03),327s(2.67)	103-1204-78 89-0068-78
$C_{10}H_8N_2O$ Cycloprop[cd]azulen-2(1H)-one, 1-diazo-2a,2b,6a,6b-tetrahydro- Spiro[cycloprop[cd]azulene-1(2H),3'-[3H]diazirin]-2-one, 2a,2b,6a,6b-tetrahydro-	EtOH EtOH	249(4.09),300(3.57), 392s(1.57) 253(4.62),315(2.26), 330(2.00)	77-0442-78 77-0442-78
$C_{10}H_8N_2O_2$ [1,1'(2H,4'H)-Bipyridine]-2,4'-dione [1,1'(4H,4'H)-Bipyridine]-4,4'-dione monohydrochloride 4-Isoquinolinecarbonitrile, 2,3,5,6,7,8-hexahydro-3,8-dioxo- 2-Propenoic acid, 3-(1H-indazol-3-yl)- 2-Quinoxalinecarboxaldehyde, 3-methoxy-	EtOH EtOH EtOH MeOH EtOH $CHCl_3$	225(3.70),265(4.11), 300(3.76) 220(3.85),257(4.08), 275(4.11),330(3.60) 220(3.78),255(3.94), 275(3.96),325(3.48) 227(4.25),232s(4.20), 279(4.11),324(3.83) 216(4.32),252(3.9), 310(4.17) 312(3.83)	120-0001-78 120-0001-78 120-0001-78 44-0966-78 103-0771-78 97-0177-78
$C_{10}H_8N_2O_2S$ Acetic acid, (1,8-naphthyridin-2-yl-thio)- Acetic acid, (2-quinoxalinylthio)-	MeOH MeOH	244(4.28),342(4.07) 242(4.18),264(4.18), 353(3.92)	2-0678-78 2-0678-78

Compound	Solvent	λ_{max} (log ϵ)	Ref.
$C_{10}H_8N_2O_2S_2$ 1,2,3-Thiadiazole-4-carboxylic acid, 5-[(phenylmethyl)thio]-	MeOH	240(3.82),302(3.96)	4-1295-78
$C_{10}H_8N_2O_3$ Acetamide, N-(2,3-dihydro-1,3-dioxo-1H-isoindol-4-yl)-	EtOH	227(4.28),234(4.24)	94-0530-78
Acetamide, N-(2,3-dihydro-1,3-dioxo-1H-isoindol-5-yl)-	EtOH	249(4.34),255(4.29)	94-0530-78
$C_{10}H_8N_2O_3S$ 5H-Thiazolo[3,2-a]pyridine-7-carboxylic acid, 8-cyano-2,3-dihydro-5-oxo-, methyl ester	MeOH	210(4.3),238(4.2), 282(4.6),353(3.8)	28-0385-78B
$C_{10}H_8N_2O_4$ Oxazole, 5-methoxy-2-(4-nitrophenyl)-	EtOH	239(3.67),347(3.86)	104-1353-78
$C_{10}H_8N_2O_4S$ Thieno[2,3-d]pyridazine-4,7-dicarboxylic acid, dimethyl ester	CH_2Cl_2	252(4.34),316(2.49)	83-0728-78
$C_{20}H_8N_2Se$ Pyridine, 2,2'-selenobis-	EtOH	246(3.89),289(3.77)	64-0118-78B
$C_{10}H_8N_4O$ Pyridine, 3,3'-azoxybis-	n.s.g.	244(3.94),314(4.21)	107-0109-78
$C_{10}H_8N_4O_2$ 6H-Purin-6-one, 1,9-dihydro-9-(5-methyl-2-furanyl)-	pH 7	242(4.37)	44-3324-78
$C_{10}H_8N_4O_2S$ 4(1H)-Pyrimidinone, 6-amino-2,3-dihydro-5-nitroso-1-phenyl-2-thioxo-	EtOH	346(4.44)	95-1072-78
$C_{10}H_8N_4O_3$ 2,4(1H,3H)-Pyrimidinedione, 6-amino-5-nitroso-1-phenyl-	EtOH	228(4.30),321(4.20)	95-1072-78
$C_{10}H_8N_4O_3S$ Benzenesulfonic acid, 4-[(1,1-dicyanoethyl)azo]-, sodium salt	H_2O	280(4.08)	126-2845-78
$C_{10}H_8N_4O_4S_4$ 1,3-Dithiole-4,5-dicarboxamide, 2-[4,5-bis(aminocarbonyl)-1,3-dithiol-2-ylidene]-	DMF	319(4.18),449(3.31)	44-0595-78
$C_{10}H_8N_4O_6$ 2H-Pyrrol-2-one, 1-[(2,4-dinitrophenyl)amino]-1,5-dihydro-5-hydroxy-	n.s.g.	332(4.19)	39-0401-78C
$C_{10}H_8N_4S$ Thiazolo[4,5-b]quinoxalin-2-amine, N-methyl-	EtOH	227(4.53),250(4.21), 226(4.33)[sic]	2-0683-78
$C_{10}H_8O$ Cycloprop[a]inden-6(1H)-one, 1a,6a-dihydro-	EtOH	255(3.81),298(3.18), 305s(3.13)	44-4316-78

Compound	Solvent	λ_{max} (log ϵ)	Ref.
2-Decene-4,6,8-triyn-1-ol, (E)-	EtOH	230(4.92),242(5.14), 257(3.57),272(3.88), 288(4.16),307(4.28), 328(4.11)	39-1487-78C
$C_{10}H_8O_3$ 2H-1-Benzopyran-2-one, 4-methoxy-	n.s.g.	208(4.39),214(4.43), 265(4.06),275(4.05), 302(3.84),314(3.67)	78-1221-78
4H-1-Benzopyran-4-one, 2-methoxy-	EtOH	220(4.26),265(3.98), 284(3.90)	23-1796-78
$C_{10}H_8O_4$ 2(3H)-Benzofuranone, 4-acetyl-5-hydroxy-	EtOH	234(4.16),250s(3.97), 345(3.58)	12-2099-78
$C_{10}H_9Br$ Bicyclo[4.2.2]deca-2,4,7,9-tetraene, 2(and 1)-bromo-	C_6H_{12}	269(3.67),280(3.66), 291(3.43)	5-2074-78
$C_{10}H_9BrN_2O_2S_2$ 1,4,2-Dithiazin-3-amine, 5-(4-bromophenyl)-N-methyl-, 1,1-dioxide	EtOH	251(4.27)	18-1805-78
$C_{10}H_9BrN_2O_3$ Pyridinium, 1-[(5-nitro-2-furanyl)-methyl-, bromide	H_2O	202(4.11),257(3.70), 298(3.62)	73-2041-78
$C_{10}H_9BrN_2O_3S$ 1,4,3-Oxathiazin-2-amine, 6-(4-bromophenyl)-N-methyl-, 4,4-dioxide	EtOH	270(4.21)	18-1805-78
$C_{10}H_9BrN_4O$ 1H-Pyrazole-4,5-dione, 3-methyl-, 4-[(4-bromophenyl)hydrazone]	EtOH	253(4.04),260(4.00), 335(3.96),411(4.14)	80-0079-78
	acetone	406(4.32)	80-0079-78
	CHCl$_3$	417(4.36)	80-0079-78
$C_{10}H_9BrO_3$ 2(3H)-Benzofuranone, 4-bromo-5-hydroxy-6,7-dimethyl-	CHCl$_3$	249(3.37),298(3.46)	12-1353-78
$C_{10}H_9Cl$ Bicyclo[4.2.2]deca-2,4,7,9-tetraene, 2(and 1)-chloro-	C_6H_{12}	267(3.70),277(3.70), 287(3.46)	5-2074-78
$C_{10}H_9ClN_2O_2S_2$ 1,4,2-Dithiazin-3-amine, 5-(4-chlorophenyl)-N-methyl-, 1,1-dioxide	EtOH	248(4.27)	18-1805-78
$C_{10}H_9ClN_2O_3S$ 1,4,3-Oxathiazin-2-amine, 6-(4-chlorophenyl)-N-methyl-, 4,4-dioxide	EtOH	266(4.17)	18-1805-78
$C_{10}H_9ClN_4$ Acetaldehyde, (4-chloro-1(2H)-phthalazinylidene)hydrazone	dioxan	282(4.42),345(3.96)	103-0567-78
	MeCN	212(4.51),281(4.29), 343(3.93)	103-0567-78
Formaldehyde, (4-chloro-2-methyl-1(2H)-phthalazinylidene)hydrazone	dioxan	287(4.09),355(3.95)	103-0567-78

Compound	Solvent	$\lambda_{max}(\log \epsilon)$	Ref.
Formaldehyde, (4-chloro-1-phthalazinyl)-methylhydrazone	dioxan	245s(3.95),325(3.92)	103-0436-78
	MeCN	217(4.56),250s(3.95), 323(3.92)	103-0436-78
	crystal	212(--),265(--), 332(--)	103-0436-78
$C_{10}H_9ClN_4O$			
1H-Pyrazole-4,5-dione, 3-methyl-, 4-[(2-chlorophenyl)hydrazone]	EtOH	254(4.15),262(4.11), 338(4.01),404(4.17)	80-0079-78
	acetone	396(4.20)	80-0079-78
	$CHCl_3$	414(4.34)	80-0079-78
	CCl_4	410(4.26)	80-0079-78
1H-Pyrazole-4,5-dione, 3-methyl-, 4-[(4-chlorophenyl)hydrazone]	EtOH	253(4.18),260(4.15), 336(4.11),412(4.32)	80-0079-78
$C_{10}H_9ClO_3$			
Ethanone, 1-(5-chloro-2,3-dihydro-1,4-benzodioxin-6-yl)-	EtOH	212(4.04),225(4.13), 235(4.04),275(3.77), 310(3.52)	103-1188-78
Ethanone, 1-(7-chloro-2,3-dihydro-1,4-benzodioxin-6-yl)-	EtOH	225(4.22),236(4.26), 275(3.80),312(3.49)	103-1188-78
Ethanone, 1-(8-chloro-2,3-dihydro-1,4-benzodioxin-6-yl)-	EtOH	224(4.33),279(4.03)	103-1188-78
$C_{10}H_9Cl_2NO$			
Benzonitrile, 3,5-dichloro-2,4,6-tri-methyl-, N-oxide	MeCN	270(4.14)	39-0607-78B
$C_{10}H_9Cl_3FeSi$			
Ferrocene, (trichlorosilyl)-	isooctane	324(2.04),440(2.15)	35-7264-78
$C_{10}H_9Li$			
Lithium, (1-methyl-3-phenyl-2-propynyl)-	ether	362(3.16)	101-0001-78N
$C_{10}H_9N$			
2-Azabicyclo[4.4.1]undeca-1,3,5,7,9-pentaene	C_6H_{12}	240(4.45),258(4.07), 297s(3.42),364(3.34)	89-0853-78
Benzeneacetonitrile, α-ethylidene-, (Z)-	EtOH	217(4.11),222(4.07), 266(4.31)	23-0041-78
Quinoline, 6-methyl-	81% H_2SO_4	243(4.6),320(3.8)	94-0930-78
$C_{10}H_9NO$			
2-Azabicyclo[4.4.1]undeca-4,6,8,10-tetraen-3-one	MeCN	223(4.10),270(4.46), 302(3.67),373(3.54)	89-0853-78
1H-Indole-7-carboxaldehyde, 1-methyl-	MeOH	226(4.38),247(4.34), 340(3.92)	103-1278-78
$C_{10}H_9NO_2$			
Acetamide, N-3-benzofuranyl-	EtOH	237(3.85),245(3.87)	39-0419-78C
Benzamide, 2-(2-propynyloxy)-	EtOH	230(3.93),286(3.46), 296s(3.40)	33-0716-78
Benzene, (4-nitro-1-butynyl)-	EtOH	239(4.50),249(4.45)	104-0676-78
1,4-Benzoxazepin-5(2H)-one, 3,4-di-hydro-3-methylene-	EtOH	237s(4.00),240(4.01), 247s(3.99),259s(3.90), 265s(3.93),271(3.95), 285s(3.89),300s(3.64)	33-0716-78
1,4-Benzoxazepin-5(4H)-one, 3-methyl-	EtOH	244s(3.68),248s(3.84), 254s(4.01),259(4.09), 265(4.18),271(4.11), 278(4.14),298s(3.91), 307(4.27),318(3.98)	33-0716-78

Compound	Solvent	λ_{max}(log ϵ)	Ref.
4H-1,3-Benzoxazin-4-one, 2-ethenyl-2,3-dihydro-	EtOH	232(3.86),296(3.35)	33-0716-78
1H-Indole-3-carboxaldehyde, 1-methoxy-	MeOH	214(4.19),245(4.03), 300(4.02)	39-1117-78C
1H-Indole-2-carboxylic acid, 1-methyl-	EtOH	296(4.16)	103-1204-78
Oxazole, 5-methoxy-2-phenyl-	EtOH	289(4.18)	104-1353-78
2(1H)-Quinolinone, 4-hydroxy-1-methyl-	MeOH	317(3.77),328(3.67)	49-0123-78
$C_{10}H_9NO_2S$			
4(1H)-Quinolinone, 3-(methylsulfinyl)-	n.s.g.	244(4.30),251(4.30), 294(3.85),317(3.95), 329(3.95)	4-0113-78
Thieno[3,2-b]pyridine-2-carboxylic acid, ethyl ester	EtOH	239(4.21),286(3.10)	78-0989-78
$C_{10}H_9NO_2S_3$			
1,4,2-Dithiazine, 3-(methylthio)-5-phenyl-	EtOH	252(4.28)	18-1805-78
1,4,2-Dithiazole, 3-(methylthio)-5-(phenylmethylene)-, 1,1-dioxide	EtOH	296(4.36)	18-1805-78
$C_{10}H_9NO_4$			
Furo[3.4-b]pyridine-7-acetic acid, 5,7-dihydro-2-methyl-5-oxo-	H_2O	225(3.97),275(3.87)	103-0049-78
$C_{10}H_9NO_4S_2$			
Acetic acid, (4H-1,4-benzothiazin-3-ylthio)-, S,S-dioxide	MeOH	248(4.08),284(3.95)	2-0678-78
$C_{10}H_9NO_6$			
2-Propenoic acid, 3-(5-hydroxy-4-methoxy-2-nitrophenyl)-, (E)-	MeOH	230(4.05),273(4.32), 345(3.81)	39-0440-78C
	MeOH-base	246(4.17),296(4.09), 432(4.11)	39-0440-78C
$C_{10}H_9NS$			
Benzene, 1-(2-isothiocyanatoethenyl)-4-methyl- (mostly trans)	dioxan	307(4.19)	73-1917-78
$C_{10}H_9NS_2$			
2(3H)-Thiazolethione, 4-(4-methylphenyl)-	EtOH	242(4.22),292s(4.02), 322(4.17)	39-1017-78C
$C_{10}H_9NS_3$			
Benzenecarbothioamide, N-1,3-dithiolan-2-ylidene-	CH_2Cl_2	239(4.13),327(4.41), 370s(4.00),560(2.21)	18-0301-78
$C_{10}H_9N_2O_3$			
Pyridinium, 1-[(5-nitro-2-furanyl)-methyl]-, bromide	H_2O	202(4.11),257(3.70), 298(3.62)	73-2041-78
$C_{10}H_9N_3$			
[2,2'-Bipyridin]-4-amine	neutral	232(4.27),275(3.98)	39-1215-78B
	cation	240(4.23),275(4.09)	39-1215-78B
1H-Pyrazolo[1,5-a]benzimidazole, 2-methyl-	2N H_2SO_4	294(4.04),301(4.15)	4-0715-78
	EtOH	229(4.36),306(4.00)	4-0715-78
$C_{10}H_9N_3O$			
2-Propenamide, 3-(3-cyano-6-methyl-2-pyridinyl)-, (E)-	H_2O	268(4.35)	103-0049-78

Compound	Solvent	$\lambda_{max}(\log \epsilon)$	Ref.
1,2,4-Triazin-6(1H)-one, 3-methyl-5-phenyl-	MeOH	226(3.78),277(3.75), 333(3.98)	4-1271-78
1,2,4-Triazin-6(1H)-one, 5-methyl-3-phenyl-	MeOH	263(4.32),315(3.35)	4-1271-78
$C_{10}H_9N_3OS$			
4(5H)-Isoxazolone, 3-methyl-5-thioxo-, 4-(phenylhydrazone)	dioxan	410(4.06)	64-0075-78B cf64-1056-78B
4(1H)-Pyrimidinone, 6-amino-2,3-di-hydro-1-phenyl-2-thioxo-	EtOH	285(4.30)	95-1072-78
3H-1,4,8b-Triazaacenaphthylen-3-one, 4,5-dihydro-2-(methylthio)-	EtOH	230(4.01),269(4.12), 322(3.50)	95-0631-78
$C_{10}H_9N_3O_2$			
2,4(1H,3H)-Pyrimidinedione, 6-amino-1-phenyl-	n.s.g.	270(4.41)	95-1072-78
$C_{10}H_9N_3O_3$			
Pyridazine, 1-acetyl-4-(acetoxycyano-methylene)-1,4-dihydro-	EtOH	342(4.32)	4-0637-78
$C_{10}H_9N_3O_3S$			
1,3,4-Thiadiazol-2(3H)-one, 5-ethyl-3-(4-nitrophenyl)-	EtOH	250s(3.757),310(4.183)	146-0864-78
$C_{10}H_9N_3O_4$			
2H-Benzotriazole-2-acetic acid, 4,7-di-hydro-5-methyl-4,7-dioxo-, methyl ester	EtOH	227(4.22),250(4.13), 320(3.04)	87-0578-78
$C_{10}H_9N_3S$			
Pyrido[3,2-c]pyridazine, 4-(2-propenyl-thio)-	heptane	235(3.77),332(3.72), 344(3.71)	103-1032-78
$C_{10}H_9N_3S_2$			
Isothiazolo[3',4':4,5]thieno[2,3-b]pyr-idin-3-amine, 6,8-dimethyl-	EtOH	241(4.08),270(4.17), 280(4.32),291(4.41), 351(3.43)	32-0057-78
$C_{10}H_9N_5$			
1H-Pyrazole, 3-azido-1-methyl-4-phenyl-	EtOH	250s(4.09),267(4.20)	4-0395-78
1H-Pyrazolo[1,5-d]tetrazole, 1-methyl-7-phenyl-	EtOH	257(4.15),298s(3.46)	4-0395-78
3H-Pyrazolo[1,5-d]tetrazole, 3-methyl-7-phenyl-	EtOH	258(3.85),264(2.86), 268(3.85),346(3.15)	4-0395-78
Pyrazolo[3,4-d]-1,2,3-triazole, 1,4-di-hydro-6-methyl-4-phenyl- (spectra in 50% MeOH)	pH 1.3 pH 7.0 pH 12.0	237(4.34),293(3.95) 238(4.35),288(4.00) 240(4.29),283(4.25)	33-0108-78 33-0108-78 33-0108-78
$C_{10}H_9N_5O$			
9H-Purin-6-amine, 9-(5-methyl-2-furan-yl)-	pH 1 pH 7 pH 11	252(4.33) 252(4.29) 251(4.27)	44-0998-78 44-0998-78 44-0998-78
$C_{10}H_9N_5O_3$			
1H-Pyrazole-4,5-dione, 3-methyl-, 4-[(2-nitrophenyl)hydrazone]	EtOH	250(4.30),275(4.22), 350(4.29),420(4.30)	80-0079-78
	acetone	418(3.96)	80-0079-78
	CHCl$_3$	425(4.18)	80-0079-78
	CCl$_4$	419(4.39)	80-0079-78

Compound	Solvent	λ_{max}(log ϵ)	Ref.
1H-Pyrazole-4,5-dione, 3-methyl-, 4-[(3-nitrophenyl)hydrazone]	EtOH	252(4.16),345(4.23), 400(4.15)	80-0079-78
	acetone	400(4.10)	80-0079-78
1H-Pyrazole-4,5-dione, 3-methyl-, 4-[(4-nitrophenyl)hydrazone]	EtOH	253(4.24),344(4.20), 407(4.32)	80-0079-78
	acetone	410(4.32)	80-0079-78
$C_{10}H_9N_5O_3S$ 1,3,4-Thiadiazol-2(3H)-imine, 5-ethyl-3-(4-nitrophenyl)-N-nitroso-	EtOH	288(4.106),348(4.221), 430(1.762)	146-0864-78
$C_{10}H_9N_5O_4$ 7-Pteridinecarboxylic acid, 2-(acetyl-amino)-1,4-dihydro-6-methyl-4-oxo-	pH 0.0	243(4.07),282(4.16), 347(3.88)	24-3790-78
	pH 4.0	249(4.03),280(4.15), 339(3.96)	24-3790-78
	pH 11.0	257(4.40),348(3.91)	24-3790-78
$C_{10}H_9N_5O_5$ 6,7-Pteridinedicarboxylic acid, 2-ami-no-1,4-dihydro-4-oxo-, dimethyl ester	pH -1.0	238(4.07),262(4.06), 327(4.01)	24-3790-78
	pH 4.0	243(4.04),303(4.17), 363(3.91)	24-3790-78
	pH 9.0	278(4.28),376(3.98)	24-3790-78
$C_{10}H_{10}$ Benzo[1,2:3,4]dicyclobutene	EtOH	266(3.13),269(3.14), 275(3.19)	35-3730-78
Benzo[1,2:4,5]dicyclobutene	EtOH	276(3.66),280(3.71), 286(3.59)	35-3730-78
$C_{10}H_{10}BrNO_2$ Benzene, 1-(1-bromo-2-methyl-1-propen-yl)-4-nitro-	MeOH	265(3.92),304(3.81)	12-2477-78
$C_{10}H_{10}BrN_3O_2S$ Thiourea, (6-bromo-3,4-dihydro-3-meth-yl-2-oxo-2H-1-benzoxazin-4-yl)-	EtOH	225(4.01),267(4.23), 292(3.68)	4-1193-78
1,3,5-Triazin-2(1H)-one, 6-(5-bromo-2-hydroxyphenyl)tetrahydro-1-methyl-4-thioxo-	EtOH	226s(4.05),270(4.30), 295s(3.52)	4-1193-78
$C_{10}H_{10}ClNO_2$ Acetamide, N-acetyl-N-(4-chlorophenyl)-	CHCl$_3$	245(3.34)	20-0621-78
Benzene, 1-(1-chloro-2-methyl-1-propen-yl)-4-nitro-	hexane	252(3.86),299(3.94)	12-2477-78
4-Isoxazolol, 5-(4-chlorophenyl)-4,5-dihydro-3-methyl-	MeOH	263(2.43),269(2.49), 276(2.36)	142-0187-78C
$C_{10}H_{10}ClNS$ Benzene, 1-(1-chloro-2-isothiocyanato-ethyl)-4-methyl-	dioxan	none above 250 nm	73-1917-78
$C_{10}H_{10}ClN_3$ 1-Phthalazinamine, 4-chloro-N,N-dimeth-yl-	dioxan	262s(3.56),320(3.76)	103-0436-78
	MeCN	214(4.56),262s(3.43), 322(3.66)	103-0436-78
	solid	212(--),264(--), 322(--)	103-0436-78

Compound	Solvent	λ_{max}(log ϵ)	Ref.
$C_{10}H_{10}ClN_3O_2S$ 1,3,5-Triazin-2(1H)-one, 6-(5-chloro-2-hydroxyphenyl)tetrahydro-1-methyl-4-thioxo-	EtOH	227(4.31),295s(3.54)	4-1193-78
$C_{10}H_{10}ClN_5$ 1,2,4,5-Tetrazin-3-amine, 6-(4-chlorophenyl)-N,N-dimethyl-	dioxan	301(4.49),416(3.41), 538(2.78)	30-0338-78
$C_{10}H_{10}Cl_2N_4O_2$ 6H-Purin-6-one, 9-[4-chloro-5-(chloromethyl)tetrahydro-2-furanyl]-1,9-dihydro-, [2R-(2α,4α,5α)]-	pH 7	249(4.15)	44-3324-78
$C_{10}H_{10}Cl_2N_4O_4$ 1H-Imidazo[4,5-d]pyridazine, 4,7-dichloro-1-β-D-ribofuranosyl-	pH 1,7,11	250(3.78)	4-0001-78
$C_{10}H_{10}F_2IN$ Benzenamine, 4-(1,2-difluoro-2-iodoethenyl)-N,N-dimethyl-, (Z)-	heptane	300(4.24)	104-0191-78
$C_{10}H_{10}F_6$ Cyclobutene, 3-methyl-4-(1-methylethylidene)-1,2-bis(trifluoromethyl)-	hexane	260(3.82)	39-1161-78C
$C_{10}H_{10}Fe$ Ferrocene ferricinium ion	CH_2Cl_2 EtOH	440(2.00) 617(2.63)	101-0077-78L 101-0077-78L
$C_{10}H_{10}NO$ Quinolinium, 8-hydroxy-1-methyl-, chloride	pH 13	273(4.52),346(3.11), 446(3.21)	18-3489-78
	MeOH-KOH	278(4.52),353(3.30), 474(3.19)	18-3489-78
$C_{10}H_{10}N_2$ 3H-1,2-Benzodiazepine, 5-methyl-	EtOH	225(4.40),261(3.00)	94-1890-78
1H-Imidazole, 2-methyl-4-phenyl-	EtOH	203(4.21),267(4.18)	44-2289-78
1H-Pyrazole, 3-methyl-5-phenyl-	pH 7	248(4.1)	78-2259-78
	EtOH	251(4.2)	78-2259-78
$C_{10}H_{10}N_2O$ 1H-Benzimidazole, 1-acetyl-2-methyl-	EtOH	230(4.06),277(3.74)	12-2675-78
3H-1,2-Benzodiazepine, 5-methyl-, 1-oxide	EtOH	234(4.49),306(3.75)	94-1896-78
3H-1,2-Benzodiazepine, 5-methyl-, 2-oxide	EtOH	239(4.46),303(3.76)	94-1896-78
Ethanone, 1-(2-methyl-1H-benzimidazol-4-yl)-	EtOH	215(4.15),280(3.98), 315(3.83)	12-2675-78
Ethanone, 1-(2-methyl-1H-benzimidazol-5-yl)-	EtOH	230(4.18),277(3.92)	12-2675-78
1H-Pyrazole, 3-methoxy-1-phenyl-	MeOH	265(4.27)	24-0780-78
1H-Pyrazolium, 2,3-dihydro-1-methyl-3-oxo-1-phenyl-, hydroxide, inner salt	MeOH	257(3.28)	24-0780-78
1H-Pyrazol-3-ol, 1-methyl-5-phenyl-	MeOH	205(4.35),240(4.12)	24-0791-78
$C_{10}H_{10}N_2OS$ 4-Thiazolidinone, 3-methyl-2-(phenylimino)-	dioxan	272(3.91)	103-0148-78

Compound	Solvent	λ_{max}(log ϵ)	Ref.
4(5H)-Thiazolone, 2-(methylphenyl-amino)-	dioxan	256(4.25)	103-0148-78
4(5H)-Thiazolone, 2-[(4-methylphenyl)-amino]-	dioxan	274(3.99)	103-0148-78
$C_{10}H_{10}N_2O_2$			
1,5-Naphthyridine, 4,8-dimethoxy-	H_2O	222(4.71),282(4.00)	44-1331-78
1,5-Naphthyridine-4,8-dione, 1,5-di-hydro-1,5-dimethyl-	H_2O	232(4.49),267(3.48), 335(4.29),348(4.33)	44-1331-78
1,5-Naphthyridine-4,8-dione, 1,5-di-hydro-1,7-dimethyl-	H_2O	232(4.53),267(3.54), 297s(3.70),325(4.30), 338(4.40)	44-1331-78
1,5-Naphthyridine-4,8-dione, 1,5-di-hydro-3,7-dimethyl-	H_2O	233s(4.57),237s(4.66), 240(4.71),273(3.63), 332s(4.24),337s(4.26), 343(4.37)	44-1331-78
2-Propen-1-amine, N-[(4-nitrophenyl)-methylene]-	$CHCl_3$	204(4.21)	70-0550-78
2,3-Quinoxalinedione, 1,4-dihydro-1,4-dimethyl-	pH 7.0	233(4.10),240(4.04), 255(3.80),264s(3.71), 302s(3.98),314(4.07), 326s(4.02),342(3.68)	24-1753-78
$C_{10}H_{10}N_2O_2S$			
2,4-Thiazolidinedione, 5-(phenylmeth-yl)-, 2-oxime, (Z)-	n.s.g.	208(4.20),230s(--)	139-0303-78A
4(5H)-Thiazolone, 2-[(4-methoxyphenyl)-amino]-	dioxan	283(4.01)	103-0148-78
$C_{10}H_{10}N_2O_2S_2$			
1,4,2-Dithiazin-3-amine, N-methyl-5-phenyl-, 1,1-dioxide	EtOH	244(4.24)	18-1805-78
1,4,2-Dithiazol-3-amine, N-methyl-5-(phenylmethylene)-, 1,1-dioxide	EtOH	296(4.32)	18-1805-78
$C_{10}H_{10}N_2O_3$			
2H-1-Benzazepin-2-one, 1,3,4,5-tetra-hydro-8-nitro-	MeOH	243(4.48),275(3.88)	24-1780-78
Furo[3,2-b]pyridine-3-carboxylic acid, 2-amino-, ethyl ester	MeOH	221(3.61),259(4.04), 306(4.06)	4-1411-78
1(2H)-Naphthalenone, 3,4-dihydro-7-nitro-, oxime	MeOH	259(4.40)	24-1780-78
2,4(1H,3H)-Pyrimidinedione, 1-(2,5-di-hydro-5-methylene-2-furanyl)-5-meth-yl-, (R)-	pH 7	264(4.09)	44-3324-78
2,4(1H,3H)-Pyrimidinedione, 5-methyl-1-(5-methyl-2-furanyl)-	pH 7	209(4.14),264(3.97)	44-3324-78
5H-Pyrrolo[3,4-b]pyridine-7-acetic acid, 6,7-dihydro-2-methyl-5-oxo-	H_2O	270(3.65)	103-0049-78
$C_{10}H_{10}N_2O_3S$			
1,4,3-Oxathiazin-2-amine, N-methyl-6-phenyl-	EtOH	260(4.17)	18-1805-78
Quinoxaline, 2-methyl-3-(methylsulfon-yl)-, 1-oxide	MeOH	274(4.60),330(3.91)	44-4125-78
$C_{10}H_{10}N_2O_4$			
5H-Pyrano[2,3-b]pyridin-5-one, 4-hy-droxy-7-methyl-, O-methyloxime, 8-oxide	EtOH	316(4.03)	83-0848-78

Compound	Solvent	$\lambda_{max}(\log \epsilon)$	Ref.
$C_{10}H_{10}N_2O_4S$			
1H-4,1,2-Benzothiadiazine-3-carboxylic acid, ethyl ester, 4,4-dioxide	EtOH	328(4.11)	39-0539-78C
$C_{10}H_{10}N_2O_5$			
2,4(1H,3H)-Pyrimidinedione, 1-(5,5,6,6-tetradehydro-5,6-dideoxy-β-D-ribo-hexofuranosyl)-	MeOH	260(4.03)	44-0367-78
$C_{10}H_{10}N_2O_7$			
Furan, 3,4-dimethoxy-2,5-bis(2-nitro-ethenyl)-	EtOH	326(4.13),423(4.30)	33-0430-78
$C_{10}H_{10}N_2S$			
1H-Imidazole, 2-(methylthio)-1-phenyl-	EtOH	260(3.678)	28-0201-78B
	CH_2Cl_2	226(4.060),263(3.662)	28-0201-78B
$C_{10}H_{10}N_4$			
[4,4'-Bipyridine]-3,3'-diamine	MeOH	235s(5.23),318(4.85)	39-1126-78C
1H-Indole, 2-(azidomethyl)-3-methyl-	EtOH	225(4.38),280(3.79), 284(3.80),292(3.73)	94-2874-78
1H-Tetrazole, 5-phenyl-1-(2-propenyl)-	EtOH	231(4.02)	44-1664-78
2H-Tetrazole, 5-phenyl-2-(2-propenyl)-	EtOH	239(4.22)	44-1664-78
1H-1,2,4-Triazolo[4,3-a]benzimidazole, 1,3-dimethyl-	EtOH-HCl	267s(3.33),275(3.54), 283(3.57),292s(2.63)	22-0273-78
	CF_3COOH	264s(3.33),272(3.50), 278(3.54),287s(2.67)	22-0273-78
9H-1,2,4-Triazolo[4,3-a]benzimidazole, 3,9-dimethyl-	EtOH-HCl	276(3.42),283(3.55), 297(3.48)	22-0273-78
	CF_3COOH	273(3.45),279(3.55), 295(3.34)	22-0273-78
$C_{10}H_{10}N_4O$			
1H-Pyrazole-4,5-dione, 3-methyl-, 4-(phenylhydrazone)	EtOH	251(4.13),258(4.11), 335(3.95),409(4.16)	80-0079-78
$C_{10}H_{10}N_4OS$			
5H-Thiazolo[3,2-a]pteridin-5-one, 8,9-dihydro-2,3-dimethyl-	pH -2.0	227(4.13),254s(3.83), 279(4.03),310s(4.02), 324(4.13),336s(4.08)	24-0971-78
	pH 3.0	234(4.14),261(4.25), 319s(3.95),335(4.09), 347s(4.06)	24-0971-78
10H-Thiazolo[2,3-b]pteridin-10-one, 7,8-dihydro-2,3-dimethyl-	pH -3.0	225s(3.87),260s(4.00), 282(4.14),302s(4.01), 317(3.87),330(3.87), 372(3.90)	24-0971-78
	pH 3.0	235s(4.06),247(4.09), 284(4.15),333(3.97)	24-0971-78
$C_{10}H_{10}N_4O_2$			
Butanenitrile, 2-[(4-nitrophenyl)hydra-zono]-	EtOH	370(4.434)	146-0864-78
3(2H)-Furazanone, 4-benzoyl-, 3-(methylhydrazone)	EtOH	204(3.26),255(2.87), 290(2.95),358(2.54)	103-0503-78
1H-Pyrazole-4,5-dione, 3-methyl-, 4-[(4-hydroxyphenyl)hydrazone]	EtOH	265(3.98),438(4.20)	80-0079-78
	acetone	430(4.30)	80-0079-78
$C_{10}H_{10}N_4O_2S$			
Benzenesulfonamide, 4-amino-N-2-pyrimi-dinyl-	pH 7.0	241(4.29),258(4.31)	94-1162-78

$C_{10}H_{10}N_4O_2S-C_{10}H_{10}O$

Compound	Solvent	$\lambda_{max}(\log \epsilon)$	Ref.
Benzenesulfonamide, 4-amino-N-2-pyrimidinyl-, β-cyclodextrin complex	pH 7.0	241(4.22),260(4.26)	94-1162-78
1,3,4-Thiadiazol-2(3H)-imine, 5-ethyl-3-(4-nitrophenyl)-	EtOH	240(4.171),350(4.206)	146-0864-78
$C_{10}H_{10}N_4O_3$			
[4,5'-Bipyrimidine]-2',4'(1'H,3'H)-dione, 2-ethoxy-	EtOH	315(4.26)	56-1579-78
[4,4'-Bipyrimidine]-2,2',6(1H,1'H.3H)-trione, 5,5'-dimethyl-	pH 0.9	255(3.81),322(3.85)	44-0511-78
	pH 7.2	258(3.79),381(3.85)	44-0511-78
	pH 12.9	302(4.11)	44-0511-78
$C_{10}H_{10}N_4O_4$			
1H-Imidazo[4,5-d]pyridazine-4,7-dicarboxylic acid, 1-methyl-, dimethyl ester	CH_2Cl_2	281(3.81)	83-0728-78
$C_{10}H_{10}N_4O_5$			
8,5'-Cycloinosine, (R)-	0.5M HCl	254(4.13)	78-2633-78
	H_2O	254(4.15)	78-2633-78
	0.5M NaOH	259(4.17)	78-2633-78
8,5'-Cycloinosine, (S)-	0.5M HCl	254(4.12)	78-2633-78
	H_2O	253(4.15)	78-2633-78
	0.5M NaOH	260(4.16)	78-2633-78
$C_{10}H_{10}N_6$			
1H-Cyclohepta[1,2-d:3,4-d']diimidazole-2,8-diamine, 4-methyl-	MeOH	245(3.78),296(4.44), 363(3.69),402(3.82)	35-4208-78
	MeOH-HCl	230(3.98),288(4.47), 357(--),395(3.96)	35-4208-78
	MeOH-NaOH	227(3.92),296(4.52), 364(3.66),404(3.87)	35-4208-78
1H-Cyclohepta[1,2-d:4,5-d']diimidazole-2,6-diamine, 4-methyl- (parazoanthoxanthin A)	MeOH	285(4.50),295(4.57), 405(4.18)	35-4208-78
	MeOH-HCl	293(4.64),390(4.05)	35-4208-78
	MeOH-NaOH	295(4.82),402(4.25)	35-4208-78
$C_{10}H_{10}N_6S$			
3H-1,3,4-Thiadiazino[5,6-b]quinoxalin-3-one, 1,2-dihydro-1-methyl-, hydrazone	EtOH	210(4.4),250(4.28), 253(4.28),294(4.3)	2-0307-78
$C_{10}H_{10}O$			
Bicyclo[4.2.2]deca-2,4,7-trien-9-one	C_6H_{12}	207(3.69),216(3.67), 258(3.63),267(3.62), 298(2.16),309(2.04), 320s(1.78)	5-2074-78
Bicyclo[4.2.2]deca-3,7,9-trien-2-one	C_6H_{12}	226(3.59),322(1.90), 332(1.95),345(1.91), 360(1.77),309s(1.81), 377(1.34)	5-2074-78
2-Buten-1-one, 1-phenyl-	EtOH	254(4.24),330(1.77)	44-2056-78
1H-Indene-4-carboxaldehyde, 2,3-dihydro-	EtOH	252(4.04)	44-2167-78
1H-Indene-5-carboxaldehyde, 2,3-dihydro-	EtOH	248(4.10)	44-2167-78
1H-Inden-1-one, 2,3-dihydro-3-methyl-	EtOH	215(3.58),245(3.74), 290(3.10)	12-1113-78
1(2H)-Naphthalenone, 3,4-dihydro-	H_2SO_4	296(4.25)	65-1644-78
2,4,6-Nonatrien-8-ynal, 7-methyl-, (E,E,E)-	ether	234(4.16),305s(4.52), 324(4.68),337(4.65)	18-2112-78

Compound	Solvent	$\lambda_{max}(\log \epsilon)$	Ref.
9-Oxatricyclo[6.3.0.0^{2,7}]undeca-3,5,10-triene	hexane	220(3.80),283(3.24)	44-0315-78
Tricyclo[4.3.1.0^{7,9}]deca-2,4-dien-10-one	EtOH	257(3.41),266(3.60), 277(3.58),318(2.74)	88-2387-78
$C_{10}H_{10}O_2$			
1,3,5-Cycloheptatriene-1-carboxylic acid, 6-ethenyl-	C_6H_{12}	238(4.52),292(3.42), 343(3.70)	89-0853-78
Cyclopropanecarboxylic acid, 2-phenyl-, cis	EtOH	210(3.91),261(3.89), 263(2.32)	44-4447-78
trans	EtOH	220(4.01),260(2.50)	44-4447-78
1H-Indene-1-carboxylic acid, 2,3-dihydro-, (S)-	MeOH	213(3.71),273(2.93)	44-2167-78
potassium salt	MeOH	215(3.73),274(3.08)	44-2167-78
$C_{10}H_{10}O_3$			
6-Benzofuranol, 4-methoxy-3-methyl-	EtOH	258(4.20),287s(3.20)	12-1533-78
$C_{10}H_{10}O_3S$			
2-Propyn-1-ol, 4-methylbenzenesulfonate	hexane	262(3.88)	124-0844-78
	MeOH	262(4.14)	124-0844-78
	EtOH	262(4.11)	124-0844-78
	dioxan	262(4.07)	124-0844-78
	CHCl$_3$	262(4.09)	124-0844-78
	$C_2H_4Cl_2$	262(4.12)	124-0844-78
2-Thiabicyclo[3.3.1]nona-3,7-dien-6-one, 9-acetoxy-	EtOH	217(3.78),245s(3.00), 252s(2.99),291(2.90), 350s(2.60)	138-0157-78
2-Thiabicyclo[3.3.1]nona-3,7-dien-9-one, 6-acetoxy-	EtOH	229(3.89),300s(2.77), 320s(2.58)	138-0157-78
9-Thiabicyclo[3.3.1]nona-3,7-dien-2-one, 6-acetoxy-	EtOH	234(3.67),295(2.78), 358s(2.48)	138-0157-78
$C_{10}H_{10}O_4$			
2,4-Benzofurandione, 3-acetyl-3,5,6,7-tetrahydro-	EtOH	307(4.15)	107-0353-78
1H-2-Benzopyran-1-one, 3,4-dihydro-5,8-dihydroxy-3-methyl-, (R)-	EtOH	222(4.28),250(3.78), 350(3.65)	102-0511-78
	EtOH-AlCl$_3$	222(4.29),252(3.84), 352(3.60)	102-0511-78
$C_{10}H_{10}O_4S$			
Benzenesulfonic acid, 4-methoxy-, 2-propynyl ester	n.s.g.	240(4.57)	124-0844-78
$C_{10}H_{10}O_5$			
4H-1-Benzopyran-4-one, 2,3-dihydro-5,7-dihydroxy-8-methoxy-	EtOH	212(4.36),236s(3.95), 292(4.24),334(3.90)	33-1257-78
	EtOH-NaOAc	212(4.19),248(3.54), 330(4.39)	33-1257-78
$C_{10}H_{10}O_6$			
2H-Pyran-2-one, 3-acetoxy-6-(acetoxymethyl)-	MeOH	292(3.76)	136-0267-78A
$C_{10}H_{10}Ru$			
Ruthenocene	EtOH	274(2.46),324(2.43)	18-0909-78
$C_{10}H_{10}S_4$			
1,3-Benzodithiole, 2-(1,3-dithiol-2-ylidene)-4,5,6,7-tetrahydro-	n.s.g.	286(4.01),298(4.05), 314(4.02),323(3.98),	44-0369-78

Compound	Solvent	$\lambda_{max}(\log \epsilon)$	Ref.
(cont.)		478(2.33)	44-0369-78
$C_{10}H_{11}BFeO_2$			
Ferrocene, borono-	MeOH	327(2.26),444(2.22)	101-0101-78G
	EtOH	328(2.23),444(2.22)	101-0101-78G
	75% EtOH	325(2.24),441(2.22)	101-0101-78G
	50% EtOH	327(2.29),440(2.22)	101-0101-78G
	DMF	327(2.26),444(2.22)	101-0101-78G
	CH_2Cl_2	325s(2.23),440(2.16)	101-0101-78G
$C_{10}H_{11}BrN_2O_3$			
1H-Pyrrole-2-carboxylic acid, 4-bromo-5-(cyanomethyl)-3-hydroxy-1-methyl-, ethyl ester	EtOH	269(4.16)	94-3521-78
$C_{10}H_{11}BrN_2O_4$			
Acetamide, N-[2-(2-bromo-4-nitrophenoxy)ethyl]-	DMSO	317(3.96)	44-0441-78
Benzene, 1-(1-bromo-2-methyl-2-nitropropyl)-4-nitro-	C_6H_{12}	262(4.11)	12-2477-78
Ethanol, 2-[(2-bromo-4-nitrophenyl)-amino]-, acetate	DMSO	386(4.26)	44-0441-78
$C_{10}H_{11}BrN_5O_6P$			
Adenosine, 2-bromo-, cyclic 3',5'-(hydrogen phosphate)	pH 1	265.0(4.16)	94-2391-78
	pH 13	265.5(4.17)	94-2391-78
$C_{10}H_{11}BrO$			
Bicyclo[4.2.2]deca-7,9-dien-2-one, 4-bromo-	C_6H_{12}	299(2.02),308(2.08), 318(2.06),330s(1.76)	5-2074-78
$C_{10}H_{11}Br_2NO_2$			
Benzene, 1-(1,2-dibromo-2-methylpropyl)-4-nitro-	C_6H_{12}	268(4.11)	12-2477-78
$C_{10}H_{11}ClN_2$			
Pyrimidine, 2-(4-chlorophenyl)-1,4,5,6-tetrahydro-	EtOH	237(4.12)	104-0758-78
hydrochloride	EtOH	237(4.16)	104-0758-78
$C_{10}H_{11}ClN_2O_3S$			
2H-1,2,4-Benzothiadiazine-3-methanol, 6-chloro-α,α-dimethyl-, 1,1-dioxide	MeOH	267(3.88)	4-0063-78
$C_{10}H_{11}ClN_2O_4$			
Benzene, 1-(1-chloro-2-methyl-2-nitropropyl)-2-nitro-	MeOH	257(3.58)	12-2477-78
$C_{10}H_{11}ClN_4O_2$			
Propanediamide, 2-[(5-chloro-2-methylphenyl)hydrazono]-	EtOH	245(4.09),290(3.79), 365(4.26)	104-1956-78
$C_{10}H_{11}ClN_4O_3$			
D-Arabinitol, 1,4-anhydro-2-(6-chloro-9H-purin-9-yl)-2-deoxy-	pH 1 and 7	265(3.97)	44-0541-78
	pH 13	257(3.93)	44-0541-78
$C_{10}H_{11}ClN_5O_6P$			
Adenosine, 2-chloro-, cyclic 3',5'-(hydrogen phosphate)	pH 1	264.5(4.17)	94-2391-78
	pH 13	264.5(4.20)	94-2391-78

Compound	Solvent	$\lambda_{max}(\log \epsilon)$	Ref.
$C_{10}H_{11}FN_5O_6P$			
Adenosine, 2-fluoro-, cyclic 3',5'-(hydrogen phosphate)	pH 1	262(4.13),269s(4.07)	94-2391-78
	pH 13	262(4.18),269s(4.07)	94-2391-78
$C_{10}H_{11}F_3N_2O_3S$			
Acetamide, N-[4-methyl-3-[[(trifluoromethyl)sulfonyl]amino]phenyl]-	pH 8.5	280(4.2)	98-1316-78
$C_{10}H_{11}IN_5O_6P$			
Adenosine, 2-iodo-, cyclic 3',5'-(hydrogen phosphate)	pH 1	265.5(4.16)	94-2391-78
	pH 13	267.0(4.17)	94-2391-78
$C_{10}H_{11}N$			
Benzonitrile, 2,4,5-trimethyl-	hexane	205(4.85),235(4.23), 284(3.20)	110-0044-78
Benzonitrile, 2,4,6-trimethyl-	75.3% H_2SO_4	243(4.2),292(3.4)	104-0111-78
Cycloprop[a]inden-1a(1H)-amine, 6,6a-dihydro-	MeOH	247(2.74)	35-2181-78
2H-Isoindole, 4,6-dimethyl-	CH_2Cl_2	281(3.25),292s(3.41), 308s(3.71),321(3.84), 334(3.74)	142-0409-78C
2H-Isoindole, 5,6-dimethyl-	CH_2Cl_2	250(3.76),281(3.33), 294(3.29),313(3.37), 324(3.40),337s(3.26)	142-0409-78C
1-Propen-1-amine, N-(phenylmethylene)-	MeOH	280(4.2)	78-0833-78
$C_{10}H_{11}NO$			
Acetamide, N-(2-phenylethenyl)-, trans	MeOH	278(4.37)	56-2233-78
2H-Azirine, 2-(methoxymethyl)-3-phenyl-	C_6H_{12}	246(4.20)	35-4481-78
Benzo[b]cyclopropa[d]pyran-7b(1H)-amine, 1a,2-dihydro-	MeOH	273(3.66),280(3.63)	35-3494-78
1H-Isoindol-1-one, 2,3-dihydro-2,3-dimethyl-, (-)-	EtOH	220(4.06),229(4.05), 237(3.86),269(3.40), 278(3.15)	103-0538-78
2-Propenal, 3-[(4-methylphenyl)amino]-	81.3% H_2SO_4	250(4.1),335(4.3)	94-0930-78
$C_{10}H_{11}NO_2$			
Acetamide, N-acetyl-N-phenyl-	CHCl	246(2.75)	20-0621-78
Acetamide, N-(2,3-dihydro-3-benzofuranyl-	EtOH	243(2.36),280(3.48), 285(3.41),287(3.42)	39-0419-78C
1H-Indole, 1,5-dimethoxy-	MeOH	213(4.07),273(3.75), 303(3.47),314s(3.50)	39-1117-7-C
8(5H)-Isoquinolinone, 6,7-dihydro-3-methoxy-	MeOH	268(4.12)	44-0966-78
4-Isoxazolol, 4,5-dihydro-3-methyl-5-phenyl-	MeOH	254(2.20),260(2.31), 266(2.21)	142-0187-78C
2(1H)-Pyridinone, 1-(1-methyl-3-oxo-1-butenyl)-	EtOH	312(3.71)	78-2609-78
2(1H)-Pyridinone, 1-(2-oxo-3-pentenyl)-	EtOH	312(3.71)	78-2609-78
$C_{10}H_{11}NO_3$			
Acetic acid, cyano(3-hydroxy-2-cyclohexen-1-ylidene)-, methyl ester	MeOH	338(4.43)	44-0966-78
Alanine, N-benzoyl-	EtOH	230(3.97)	39-1157-78B
Glycine, N-benzoyl-N-methyl-	EtOH	211(3.89)	39-1157-78B
2-Propenoic acid, 2-methyl-3-(2-oxo-1(2H)-pyridinyl)-, methyl ester, (E)-	EtOH	317(3.84)	78-2609-78
(Z)-	EtOH	315(3.77)	78-2609-78
2H-Pyran-2,6(3H)-dione, 3-(1-pyrrolidinylmethylene)-, (Z)-	MeOH	287(3.78),357(4.50)	44-4415-78

Compound	Solvent	$\lambda_{max}(\log \epsilon)$	Ref.
2(1H)-Quinolinone, 3,4-dihydro-6-hydroxy-7-methoxy-	MeOH	220(4.17),266(3.95), 292(3.77)	39-0440-78C
$C_{10}H_{11}NO_3S$			
2H-Azirine-2-methanol, 3-phenyl-	C_6H_{12}	242(4.08)	35-4481-78
Propanenitrile, 3-hydroxy-, 4-methyl-benzenesulfonate	n.s.g.	228(4.27)	124-0844-78
5H-Thiazolo[3,2-a]pyridine-8-carboxylic acid, 2,3-dihydro-5-oxo-, ethyl ester	MeOH	209(4.1),237(3.9), 290(4.2),323(3.9)	28-0385-78B
$C_{10}H_{11}NO_4$			
2-Butenedioic acid, 2-(1H-pyrrol-1-yl)-, dimethyl ester, (E)-	MeOH	283(4.17),328s(3.17)	44-3727-78
2-Butenedioic acid, 2-(1H-pyrrol-2-yl)-, dimethyl ester, (E)-	MeOH	355(4.06)	44-3727-78
(Z)-	MeOH	346(4.17)	44-3727-78
3-Furancarboxylic acid, 2-(2-cyanoethyl)-2,5-dihydro-2,4-dimethyl-5-oxo-	EtOH	227.5(4.01)	78-0955-78
2-Furanpropanenitrile, 5-formyl-3,4-dimethoxy-	EtOH	291(4.23)	33-0430-78
1-Oxa-2-azaspiro[4.5]deca-2,6,9-triene-3-carboxylic acid, 8-hydroxy-, methyl ester	EtOH	231(3.71),242(3.61), 280(3.06)	5-0066-78
1-Oxadethiaceph-3-em-4-carboxylic acid, 2-ethylidene-3-methyl-, (Z)-	EtOH	296(4.16)	39-1450-78C
1-Propanone, 2-methyl-2-(nitrooxy)-1-phenyl-	EtOH	244(4.28),276s(3.15)	33-0589-78
$C_{10}H_{11}NO_4S$			
1-Propene-2-sulfonic acid, 1-[(4-methylphenyl)amino]-3-oxo-, sodium salt	100% H_2SO_4	250(4.17),340(4.23)	94-0930-78
$C_{10}H_{11}NO_6$			
Phenol, 5-(1,3-dioxolan-2-yl)-2-methoxy-4-nitro-	MeOH	246(3.97),286(3.66), 336(3.67)	39-0440-78C
	MeOH-base	270(3.92),426(4.10)	39-0440-78C
$C_{10}H_{11}NO_6S$			
Benzoic acid, 2-[[(methoxycarbonyl)-amino]sulfonyl]-, methyl ester	EtOH	222(3.91),270(3.14), 277(3.16)	40-0582-78
$C_{10}H_{11}NS$			
Benzene, (2-isothiocyanato-1-methyl-ethyl)-, (R)-	isooctane	250(3.38)	30-0350-78
Benzene, (2-isothiocyanatopropyl)-, (S)-(+)-	isooctane	252(3.17)	30-0350-78
$C_{10}H_{11}N_2$			
Quinolinium, 1-amino-2-methyl-, sulfate (1:1)	pH 1.8	231(4.65),315(3.94)	94-1015-78
	pH 5.8	231(4.65),315(3.95)	94-1015-78
	pH 12.3	232(4.65),315(3.93)	94-1015-78
$C_{10}H_{11}N_2OPS_2$			
1,3,2-Diazaphosphorine-4(1H)-thione, 2,3-dihydro-6-methyl-2-phenoxy-, 2-sulfide	EtOH	333(4.31)	103-0784-78
$C_{10}H_{11}N_2O_3P$			
1,3,2-Diazaphosphorin-4(1H)-one, 2,3-dihydro-6-methyl-2-phenoxy-, 2-oxide	EtOH	259(5.14)	103-0784-78

Compound	Solvent	$\lambda_{max}(\log \epsilon)$	Ref.
$C_{10}H_{11}N_3$			
1H-Pyrazol-5-amine, 3-methyl-1-phenyl-	EtOH	243(4.11)	4-0715-78
	2N H_2SO_4	236(4.11)	4-0715-78
$C_{10}H_{11}N_3O$			
1,2,4-Triazin-6(1H)-one, 4,5-dihydro-3-methyl-5-phenyl-	MeOH	277(3.56)	4-1271-78
1,2,4-Triazin-6(1 H)-one, 4,5-dihydro-5-methyl-3-phenyl-	MeOH	225(4.12),298(3.87)	4-1271-78
$C_{10}H_{11}N_3OS$			
Pyrido[2,3-d]pyrimidine, 4-methoxy-6-methyl-2-(methylthio)-	pH 1	259(4.33),355(4.07)	44-0828-78
	pH 7	242(4.23),266(4.26), 330(3.88)	44-0828-78
	pH 11	242(4.22),266(4.26), 329(3.88)	44-0828-78
Pyrido[2,3-d]pyrimidin-4(8H)-one, 6,8-dimethyl-2-(methylthio)-	pH 1	278(4.20),298(4.15), 345(4.18)	44-0828-78
	pH 7	274(4.41),376(4.11)	44-0828-78
	pH 11	274(4.41),376(4.11)	44-0828-78
$C_{10}H_{11}N_3O_2$			
1H-Indazole-3-propanoic acid, β-amino-	EtOH	209(4.65),244(3.76), 293(3.82)	103-0771-78
5H-Pyrrolo[3,4-b]pyridine-7-acetamide, 6,7-dihydro-2-methyl-5-oxo-	H_2O	219(4.01),275(3.81)	103-0049-78
$C_{10}H_{11}N_3O_2S$			
1,3,5-Triazin-2(1H)-one, tetrahydro-6-(2-hydroxyphenyl)-1-methyl-4-thioxo-	EtOH	268(4.31)	4-1193-78
$C_{10}H_{11}N_3O_3$			
Benzonitrile, 2-[(2-hydroxyethyl)methylamino]-5-nitro-	DMSO	388(4.25)	44-0441-78
1,3,5-Triazine-2,4(1H,3H)-dione, dihydro-6-(2-hydroxyphenyl)-1-methyl-	EtOH	217s(3.94),278(3.50)	4-1193-78
Urea, (3,4-dihydro-3-methyl-2-oxo-2H-1,3-benzoxazin-4-yl)-	EtOH	226(4.05),266s(3.67), 274(3.73),288(3.63)	4-1193-78
$C_{10}H_{11}N_3O_4$			
1H-Pyrrolo[3,2-d]pyrimidine-7-carboxylic acid, 2,3,4,5-tetrahydro-1,3-dimethyl-2,4-dioxo-, methyl ester	EtOH	230(4.54),273(3.91)	142-0793-78A
$C_{10}H_{11}N_3O_5$			
Uridine, 5-cyano-2'-deoxy-	H_2O	217(4.02),278(4.09)	94-2657-78
	base	277(3.94)	94-2657-78
$C_{10}H_{11}N_3O_6$			
Acetamide, N-[2-(2,4-dinitrophenoxy)-ethyl]-	DMSO	298(4.05)	18-2601-78
	DMSO-NaOMe	346(4.16),359(4.14), 506(4.45)	18-2601-78
Ethanol, 2-[(2,4-dinitrophenyl)amino]-, acetate	DMSO	358(4.24),410s(--)	18-2601-78
	DMSO-NaOMe	432(4.32),490s(4.20)	18-2601-78
5-Pyrimidinecarbonitrile, 1-β-D-arabinofuranosyl-1,2,3,4-tetrahydro-2,4-dioxo-	H_2O	218(4.02),278(4.11)	94-2657-78
	base	277(3.98)	94-2657-78
$C_{10}H_{11}N_3S$			
1H-Imidazol-1-amine, 2-(methylthio)-4-phenyl-	MeOH	271(4.23)	18-1846-78

Compound	Solvent	$\lambda_{max}(\log \epsilon)$	Ref.
1H-Imidazol-1-amine, 2-(methylthio)-5-phenyl-	MeOH	278(4.23)	18-1846-78
$C_{10}H_{11}N_3S_2$			
Thieno[2,3-b]pyrimidine-2-carbothio-amide, 3-amino-4,6-dimethyl-	EtOH	260(3.77),301(4.23), 311(4.15),323(4.05), 401(3.94)	32-0057-78
$C_{10}H_{11}N_5O$			
9H-Imidazo[1,2-a]purin-9-one, 1,4-di-hydro-1,4,6-trimethyl-	pH 7	233(4.48),265(3.72), 311(3.79)	88-1447-78
9H-Imidazo[1,2-a]purin-9-one, 3,4-di-hydro-3,4,6-trimethyl-	pH 7	233(4.54),262(3.62), 295(3.88)	88-1447-78
	pH 7	234(4.60),264(3.60), 297(3.89)	88-2907-78
	pH 13	234(4.60),264(3.60), 297(3.88)	88-2907-78
	pH 1	228(4.63),278(4.06)	88-2907-78
$C_{10}H_{11}N_5O_4$			
8,5'-Cycloadenosine, (R)-	0.5M HCl	262(4.22)	78-2449-78
	H_2O	266(4.20)	78-2449-78
	0.5M NaOH	267(4.22)	78-2449-78
8,5'-Cycloadenosine, (S)-	0.5M HCl	262(4.21)	78-2449-78
	H_2O	265(4.21)	78-2449-78
	0.5M NaOH	267(4.22)	78-2449-78
8,5'-Imino-9-(5-deoxy-β-D-ribofurano-syl)hypoxanthine	MeOH	261(4.31)	44-2320-78
$C_{10}H_{11}N_5O_5$			
6,7-Pteridinedicarboxylic acid, 2-am-ino-1,4,7,8-tetrahydro-4-oxo-, dimethyl ester	pH -1.0	276(4.04),371(3.88)	24-3790-78
	pH 5.0	262(4.11),281s(3.79), 383(4.03)	24-3790-78
	pH 11.0	264(4.14),395(4.04)	24-3790-78
$C_{10}H_{11}N_7O_3S$			
Inosine, 2'-azido-2'-deoxy-6-thio-	pH 2	324(4.36)	94-0985-78
	pH 12	315(4.36)	94-0985-78
	50% EtOH	323.5(4.36)	94-0985-78
$C_{10}H_{11}O_4P$			
Acetic acid, 2,2'-(phenylphosphini-dene)bis-	H_2O	206(3.98),218s(3.93), 252(3.59),261s(--), 268s(--),275s(--)	73-0057-78
	dianion	206(3.96),218(3.91), 255(3.61),261s(--), 268s(--),275s(--)	73-0057-78
$C_{10}H_{12}$			
Cyclohexene, 3-methylene-4-(1,2-propa-dienyl)-	EtOH	224(4.50),273s(3.61)	24-3112-78
Cyclohexene, 3-methylene-5-(2-propen-ylidene)-, (E)-	EtOH	226(4.67)	24-3112-78
(Z)-	EtOH	223(4.67),228s(4.65), 243s(4.47)	24-3112-78
Dispiro[2.2.2.2]deca-4,9-diene	hexane	222(4.30)	35-1806-78
1H-Indene, 2,3-dihydro-1-methyl-, (S)-	gas	218(2.96),271(2.88)	35-6035-78
	hexane	273(3.26)	35-6035-78
	MeOH	273(3.11)	35-6035-78

Compound	Solvent	$\lambda_{max}(\log \epsilon)$	Ref.
$C_{10}H_{12}BrN_5O_3$			
Adenosine, 2'bromo-2'-deoxy-	pH 2	257(4.18)	94-2449-78
	H_2O	259.5(4.18)	94-2449-78
	pH 12	260(4.18)	94-2449-78
$C_{10}H_{12}ClNO_4$			
2,5-Cyclohexadien-1-ol, 3-chloro-4,5-dimethyl-4-nitro-, acetate	MeOH	199.5(4.16)	23-1063-78
isomer 1b	MeOH	201(4.10)	23-1063-78
$C_{10}H_{12}ClN_3O$			
1H-Benzotriazole, 1-butoxy-5-chloro-	EtOH	273(4.06)	4-1043-78
$C_{10}H_{12}ClN_5O_3$			
Adenosine, 2'-chloro-2'-deoxy-	pH 2	256.5(4.17)	94-2449-78
	H_2O	259.1(4.16)	94-2449-78
	pH 12	259.5(4.18)	94-2449-78
$C_{10}H_{12}ClN_5O_4$			
3H-1,2,3-Triazolo[4,5-b]pyridin-5-amine, 7-chloro-3-β-D-ribofuranosyl-	pH 1	255(3.75),317(4.14)	44-4910-78
	pH 11	255(3.75),316(4.13)	44-4910-78
$C_{10}H_{12}Cl_3NO$			
4-Pyridineethanol, 3-ethyl-α-(trichloromethyl)-, (R)-	EtOH	261(3.46),268(3.37)	35-0581-78
$C_{10}H_{12}D_2S$			
Tricyclo[3.3.1.1^{3,7}]decanethione-4,4-d_2, (1S)-	isoctane	507(1.05)	88-4857-78
$C_{10}H_{12}FN_5O_3$			
Adenosine, 2'-deoxy-2'-fluoro-	pH 2	259(4.16)	94-2449-78
	H_2O	259.5(4.15)	94-2449-78
	pH 12	260(4.15)	94-2449-78
$C_{10}H_{12}F_6N_2$			
Pyridazine, 4,5-dihydro-4,4,5,5-tetramethyl-3,6-bis(trifluoromethyl)-	ether	232(3.10),263(2.85)	39-0378-78C
$C_{10}H_{12}IN_5O_3$			
Adenosine, 2'-deoxy-2'-iodo-	pH 2	256.5(4.15)	94-2449-78
	H_2O	259(4.15)	94-2449-78
	pH 12	259(4.16)	94-2449-78
$C_{10}H_{12}NO_4$			
Pyridinium, 3,5-bis(methoxycarbonyl)-1-methyl-, (OC-6-11)-, hexachloroplatinate	H_2O	205(4.65),260(4.56)	88-0303-78
$C_{10}H_{12}N_2$			
1,4-Ethanoquinoxaline, 2,3-dihydro-	hexane	313(2.18)	44-2621-78
1H-Inden-1-one, 2,3-dihydro-4-methyl-, hydrazone	EtOH	213(4.27),266(4.09), 272s(4.07),292s(3.90), 302(3.83)	5-0440-78
1H-Inden-1-one, 2,3-dihydro-5-methyl-, hydrazone	EtOH	210(4.22),219s(4.13), 265(4.13),299s(3.96), 302s(3.93)	5-0440-78
1H-Inden-1-one, 2,3-dihydro-6-methyl-, hydrazone	EtOH	211(4.26),255s(3.98), 263(4.03),299s(3.90), 309s(3.88)	5-0440-78

Compound	Solvent	λ_{max} (log ϵ)	Ref.
1H-Inden-1-one, 2,3-dihydro-7-methyl-, hydrazone	EtOH	214(4.35),255(4.09), 259(4.11),287s(3.73), 299(3.60)	5-0440-78
$C_{10}H_{12}N_2O$			
7H-Pyrid[3,2-c]azepin-7-one, 5,6,8,9-tetrahydro-2-methyl-	H_2O	269(3.73)	103-0049-78
2(1H)-Quinoxalinone, 3,4-dihydro-1,3-dimethyl-	MeOH	222(4.52),260s(3.42), 305(3.95)	24-1753-78
2(1H)-Quinoxalinone, 3,4-dihydro-1,4-dimethyl-	pH -2.0	238s(4.03),244(4.06), 273s(3.47)	24-1753-78
	pH 3.0	223(4.54),264(3.55), 300(3.59)	24-1753-78
$C_{10}H_{12}N_2OS$			
Thieno[2,3-d]pyrimidin-4(3H)-one, 6-ethyl-2,3-dimethyl-	EtOH	216(4.37),266(3.92), 300(4.00)	1-0303-78
$C_{10}H_{12}N_2O_2$			
1H,5H-Pyrazolo[1,2-a]pyrazole-1,5-dione, 2,3,6,7-tetramethyl-, anti	dioxan	322(4.18)	35-6516-78
syn	dioxan	235(4.16),255s(3.72), 359(3.81)	35-6516-78
4-Pyridineacetic acid, α-(1-aminoethylidene)-, methyl ester	EtOH	296(4.2)	4-1425-78
$C_{10}H_{12}N_2O_3$			
Glycine, N-[4-(methylamino)benzoyl]-	EtOH	296(4.22)	87-1162-78
Pyrano[2,3-c]pyrazol-6(1H)-one, 5-(2-hydroxyethyl)-3,4-dimethyl-	EtOH	306(5.18)	95-0335-78
3H-Pyrazol-3-one, 4-[1-(dihydro-2-oxo-3(2H)-furanylidene)ethyl]-2,4-dihydro-5-methyl-	EtOH	291.5(4.90)	95-0335-78
1H-Pyrrole-2-carboxylic acid, 4-(cyanomethyl)-3-hydroxy-1-methyl-, ethyl ester	EtOH	265(4.30)	94-3521-78
Pyrrolidine, 1-[2-(5-nitro-2-furanyl)-ethenyl]-	EtOH	280(3.90),510(4.08)	103-0250-78
$C_{10}H_{12}N_2O_3S_2$			
2,1-Benzisothiazole-5-sulfonamide, N-ethyl-1,3-dihydro-1-methyl-3-oxo-	EtOH	224(4.35),237(4.25), 262(4.20),283s(--), 369(3.42)	4-0529-78
$C_{10}H_{12}N_2O_4$			
Imidazo[1,2-a]pyridine-8-carboxylic acid, 1,2,3,5,6,7-hexahydro-2,3-dioxo-, ethyl ester	MeOH	223(3.86),310(4.47), 318s(4.25)	24-2813-78
Morpholine, 4-[2-(5-nitro-2-furanyl)-ethenyl]-, (E)-	EtOH	277(4.04),495(4.20)	103-0250-78
2,4,6(1H,3H,5H)-Pyrimidinetrione, 1,3-dimethyl-5-(3-oxo-1-butenyl)-	$CHCl_3$	383(4.3)	83-0287-78
$C_{10}H_{12}N_2O_5$			
2,2'-Anhydro-3-β-D-arabinofuranosyl-6-methyluracil	H_2O	218(3.75),272(3.80)	103-0901-78
2,4(1H,3H)-Pyrimidinedione, 1-[2-(2-methyl-1,3-dioxolan-2-yl)-2-oxo-ethyl]-	MeOH	265(4.02)	78-2861-78

Compound	Solvent	$\lambda_{max}(\log \epsilon)$	Ref.
$C_{10}H_{12}N_2O_7$			
2(1H)-Pyridinone, 5-nitro-1-β-D-ribo-furanosyl-	pH 1	302(4.05)	87-0427-78
	pH 11	302(4.02)	87-0427-78
$C_{10}H_{12}N_2S$			
2-Imidazolidinethione, 1-methyl-3-phenyl-	EtOH	250(4.098)	28-0521-78A
	CH_2Cl_2	254(4.109)	28-0521-78A
$C_{10}H_{12}N_3OPS$			
1,3,2-Diazaphosphorin-4-amine, 1,2-di-hydro-6-methyl-2-phenoxy-, 2-sulfide	EtOH	243(3.87),283(3.68)	103-0784-78
$C_{10}H_{12}N_3O_2P$			
1,5,2-Diazaphosphorin-6(1H)-one, 2-amino-2,5-dihydro-5-(phenyl-methyl)-, 2-oxide	EtOH	222(3.81),243(3.96)	130-0421-78
$C_{10}H_{12}N_4O$			
Benzofuro[2,3-d]pyrimidine-2,4-diamine, 5,6,7,8-tetrahydro-	acid	217(4.15),265(4.14)	44-3937-78
	pH 7.0	216(4.32),265(4.13)	44-3937-78
	base	261(4.01)	44-3937-78
3-Cyclohexene-1-carboxamide, 4-amino-1,3-dicyano-N-methyl-	MeOH	264(4.11)	42-0281-78
$C_{10}H_{12}N_4OS$			
4(1H)-Pteridinone, 1,6,7-trimethyl-2-(methylthio)-	pH -2.4	225(4.06),253(3.97), 272(4.00),322(4.12), 333(4.08)	24-0971-78
	pH 4.0	231(4.13),258(4.28), 289(3.80),333(4.12), 344(4.08)	24-0971-78
4(3H)-Pteridinone, 3,6,7-trimethyl-2-(methylthio)-	pH -2.4	263s(4.02),286(4.23), 296s(4.14),382(4.09)	24-0971-78
	pH 2.0	246(4.19),286(4.16), 335(3.91)	24-0971-78
$C_{10}H_{12}N_4O_2$			
Propanediamide, 2-[(4-methylphenyl)hy-drazono]-	EtOH	235(4.52),300(3.98), 350(4.73)	104-1956-78
$C_{10}H_{12}N_4O_3$			
Propanediamide, 2-[(4-methoxyphenyl)hy-drazono]-	EtOH	235(4.30),300(3.86), 375(4.46)	104-1956-78
1H-Purine-2,6-dione, 3,7-dihydro-1,3-dimethyl-7-(oxiranylmethyl)-	EtOH	206(4.37),275(3.90)	73-3414-78
1H-Pyrazolo[3,4-d]pyrimidine-4,6(5H,7H)-dione, 1-acetyl-3,5,7-trimethyl-	H_2O and 0.5M NaOH	205(4.29),247(3.83)	56-0037-78
1H-Pyrrolo[2,3-d]pyrimidine-6-acetic acid, 2-amino-4,7-dihydro-4-oxo-, ethyl ester	acid	221(4.16),263(4.08)	44-3937-78
	pH 7.0	216(4.27),261(4.14), 280s(3.96)	44-3937-78
	base	262(4.06)	44-3937-78
$C_{10}H_{12}N_4O_4$			
[4,4'-Bipyrimidine]-2,2',6(1H,1'H,3H)-trione, 4,5-dihydro-5-hydroxy-5,5'-dimethyl-	pH 6	265(3.90)	77-0284-78
1H-Imidazo[4,5-d]pyridazine, 1-β-D-ribofuranosyl-	pH 1	252(3.69)	4-0001-78
	pH 7	242(3.66)	4-0001-78
	pH 11	243(3.69)	4-0001-78

Compound	Solvent	$\lambda_{max}(\log \epsilon)$	Ref.
$C_{10}H_{12}N_4O_4S$			
6H-Purine-6-thione, 1,7-dihydro-1-α-D-	pH 1	324(3.87)	4-0929-78
ribofuranosyl-	pH 7	237(3.64),325(3.85)	4-0929-78
	pH 13	239(3.61),328(3.92)	4-0929-78
6H-Purine-6-thione, 1,7-dihydro-1-β-D-	pH 1	323(3.88)	4-0929-78
ribofuranosyl-	pH 7	236(3.62),325(3.87)	4-0929-78
	pH 13	238(3.60),328(3.93)	4-0929-78
$C_{10}H_{12}N_4O_4S_2$			
Imidazo[1,2-a]-1,3,5-triazine-2,4-	pH 1	293(4.54),310s(4.40)	44-4784-78
(3H,8H)-dithione, 8-α-D-ribofuranosyl-	pH 13	289(4.43),306(4.47)	44-4784-78
Imidazo[1,2-a]-1,3,5-triazine-2,4-	pH 1	294(4.55),310s(4.41)	44-4784-78
(3H,8H)-dithione, 8-β-D-ribofuranosyl-	pH 13	291s(4.41),306(4.45)	44-4784-78
$C_{10}H_{12}N_4O_5$			
Acetic acid, 1,2-diacetyl-2-(1,2,3,6-	0.5M HCl	265(4.27)	56-0037-78
tetrahydro-2,6-dioxo-4-pyrimidinyl)-	H_2O	206(3.98),266(4.26)	56-0037-78
hydrazide	0.5M NaOH	286(4.36)	56-0037-78
Cytidine, 6-cyano-	M HCl	300(4.00)	94-2340-78
	H_2O	290(3.94)	94-2340-78
4H-Imidazo[4,5-d]pyridazin-4-one,	pH 1,7,11	258(3.74)	4-0001-78
3,5-dihydro-3-β-D-ribofuranosyl-			
4H-Imidazo[4,5-d]pyridazin-4-one,	pH 1	280(3.67)	4-0001-78
5,6-dihydro-6-β-D-ribofuranosyl-	pH 7	302(3.74)	4-0001-78
	pH 11	312(3.94)	4-0001-78
Pyrazolo[4,3-d]pyrimidin-7-ol, 2-β-D-	MeOH	218(4.21),262s(3.70),	104-0601-78
ribofuranosyl-		276(3.81),282s(3.81)	
$C_{10}H_{12}N_4O_5S$			
Imidazo[1,2-a]-1,3,5-triazin-2(3H)-one,	pH 1	252(4.13),268(4.18),	44-4784-78
4,8-dihydro-8-α-D-ribofuranosyl-		299(4.06)	
4-thioxo-			
Imidazo[1,2-a]-1,3,5-triazin-2(3H)-one,	pH 1	251(4.10),269(4.14),	44-4784-78
4,8-dihydro-8-β-D-ribofuranosyl-		299(4.06)	
4-thioxo-	pH 13	248(4.22),280(4.19)	44-4784-78
Imidazo[1,2-a]-1,3,5-triazin-2(1H)-one,	pH 1	214(4.22),276(4.05),	44-4774-78
3,4-dihydro-1-β-D-ribofuranosyl-		294s(3.95)	
	pH 13	240s(3.63),288(4.16)	44-4774-78
$C_{10}H_{12}N_4O_6$			
1H-Imidazo[4,5-d]pyridazine-4,7-dione,	pH 1	225s(4.18),233(3.91),	4-0001-78
5,6-dihydro-1-β-D-ribofuranosyl-		255(3.56)	
	pH 7	228(4.21),234s(4.08),	4-0001-78
		279(3.63)	
	pH 11	228(4.16),234s(4.05),	4-0001-78
		279(3.63)	
Imidazo[1,2-a]-1,3,5-triazine-2,4-	pH 1	233(3.79)	44-4774-78
(1H,3H)-dione, 1-β-D-ribofuranosyl-	pH 13	224s(3.86),240s(3.68)	44-4774-78
Imidazo[1,2-a]-1,3,5-triazine-2,4-	pH 1	235(4.03),256(4.03)	44-4784-78
(3H,8H)-dione, 8-α-D-ribofuranosyl-			
Imidazo[1,2-a]-1,3,5-triazine-2,4-	pH 1	234(4.03),256(4.01)	44-4784-78
(3H,8H)-dione, 8-β-D-ribofuranosyl-	pH 1 and 7	233(4.03),256(4.01)	87-0883-78
	pH 11	247(4.08)	87-0883-78
	pH 13	239(4.06),249(4.10)	44-4784-78
$C_{10}H_{12}N_6$			
Methanimidamide, 1-cyano-N'-[1,2-di-	MeOH	277(4.279),315(4.076),	35-0926-78
cyano-2-(dimethylamino)ethenyl]-		393(4.472)	
N,N-dimethyl-			

Compound	Solvent	$\lambda_{max}(\log \epsilon)$	Ref.
$C_{10}H_{12}N_6O_3$			
8,11-Epoxy[1,3]diazocino[1,2-e]purine-9,10-diol, 4-amino-6,7,8,9,10,11-hexahydro-, [8S-(8α,9α,10α,11α)]-	MeOH	210(4.40),272(4.25)	44-2320-78
$C_{10}H_{12}N_6O_4$			
8,11-Epoxy[1,3]diazocino[1,2-e]purin-4(1H)-one, 6-amino-6,7,8,9,10,11-hexahydro-, monohydrochloride, [8S-(8α,9α,10α,11α)]-	MeOH	260(4.18),290s(--)	44-2320-78
$C_{10}H_{12}N_8O_2$			
Adenine, 9-(3-azido-2,3-dideoxy-α-D-ribofuranosyl)-	pH 1	257.5(4.18)	44-3044-78
	H_2O	260(4.19)	44-3044-78
Adenine, 9-(3-azido-2,3-dideoxy-β-D-ribofuranosyl)-	pH 1	257(4.17)	44-3044-78
	H_2O	259.5(4.18)	44-3044-78
$C_{10}H_{12}N_8O_3$			
Adenosine, 2'-azido-2'-deoxy-	pH 1	257(4.18)	78-1133-78
	neutral	259(4.18)	78-1133-78
	pH 13	259(4.18)	78-1133-78
9H-Purin-6-amine, 9-(2-azido-2-deoxy-β-D-arabinofuranosyl)-	MeOH	259(4.18)	88-3653-78
6H-Purin-6-one, 2-amino-7-(3-azido-2,3-dideoxy-D-ribofuranosyl)-	pH 1	249(4.00),269s(3.82)	44-3044-78
	H_2O	241s(3.81),285(3.87)	44-3044-78
	pH 13	281.5(3.82)	44-3044-78
6H-Purin-6-one, 2-amino-9-(3-azido-2,3-dideoxy-α-D-ribofuranosyl)-	pH 1	255(4.09),275s(3.92)	44-3044-78
	H_2O	253(4.14),271s(3.99)	44-3044-78
	pH 13	266(4.06)	
β-anomer	pH 1	256(4.09),275s(3.93)	44-3044-78
	H_2O	253(4.14),270s(3.99)	44-3044-78
	pH 13	265(4.06)	44-3044-78
$C_{10}H_{12}O$			
Bicyclo[5.1.0]octa-2,5-dien-4-one, 3,5-dimethyl-	EtOH	216(3.77),281(3.74)	35-1778-78
cation	H_2SO_4	207(4.08),250(4.25),328(3.67)	35-1778-78
2,4,6-Cycloheptatrien-1-one, 3-(1-methylethyl)-	EtOH	236(4.23),303(3.67)	35-1778-78
2,4,6-Cycloheptatrien-1-one, 4-(1-methylethyl)- (nezukone)	EtOH	237(3.72)	35-1778-78
2,4,6-Cycloheptatrien-1-one, 2,4,7-tri-methyl-	H_2O	237(4.19),333(3.89),342(3.86)	35-1778-78
Ditwist-brendan-5-one, (-)-(1R,2R,4S,6S,7R,8S)-	isooctane	291(1.28)	44-0689-78
5H-Inden-5-one, 1,6,7,7a-tetrahydro-2-methyl-	EtOH	289(4.27)	88-2445-78
Tetracyclo[7.1.0.0^{2,4}.0^{5,7}]decan-8-one, anti-anti	EtOH	280(1.83)	88-0977-78
$C_{10}H_{12}O_2$			
Benzaldehyde, 2-hydroxy-3-(1-methylethyl)-	EtOH	219(4.22),262(4.08),335(3.46)	32-0079-78
Benzaldehyde, 2-hydroxy-5-(1-methylethyl)-	EtOH	220(4.02),270(3.98),335(3.56)	32-0079-78
Benzene, [2-(ethenyloxy)ethoxy]-	n.s.g.	219(3.02),265(3.11),270(3.24),277(3.17)	47-1343-78
Bicyclo[3.3.1]non-3-ene-2,9-dione, 5-methyl-	EtOH	225(3.8)	23-0419-78

Compound	Solvent	$\lambda_{max}(\log \epsilon)$	Ref.
2,5-Cyclohexadiene-1,4-dione, 2,3,5,6-tetramethyl-	$CHCl_3$	270(4.17),340(2.22)	40-1127-78
4-Cyclopentene-1,3-dione, 2,2-dimethyl-4-(1-methylethenyl)-	EtOH	284(4.02),384s(1.72)	78-1567-78
Ethanone, 1-(2-hydroxy-4,5-dimethyl-phenyl)-	EtOH	263(4.11),333(3.60)	23-0517-78
Ethanone, 1-(2-methoxy-4-methylphenyl)-	EtOH	216(4.31),256(4.05), 309(3.70)	95-0503-78
1H-Indene-1,6(2H)-dione, 3,3a,4,5-tetra-hydro-3a-methyl-	EtOH	255(4.10)	33-0626-78
1H-Indene-2,7(4H,7aH)-dione, 5,6-di-hydro-7a-methyl-	EtOH	239(4.40)	78-2201-78
Tricyclo[3.1.0.0^{2,6}]hexane-3,4-dione, 1,2,5,6-tetramethyl-	EtOH	272(1.1),436(1.4)	44-1912-78
$C_{10}H_{12}O_2S$			
Benzenesulfinic acid, 4-methyl-, 2-propenyl ester	hexane	203(4.01),225(3.95), 257(3.51)	118-0441-78
Phenol, 2-(tetrahydro-2-thienyl)-, S-oxide	MeOH-EtOAc	280(3.52)	39-1580-78C
geometric isomer	MeOH-EtOAc	280(3.45)	39-1580-78C
$C_{10}H_{12}O_2S_2$			
Thieno[3,2-b]thiophene-3,6(2H,5H)-di-one, 2,2,5,5-tetramethyl-	C_6H_{12}	278(2.67),438(3.67), 460(3.66),467(3.66)	24-3233-78
	benzene	280s(3.06),445(3.67), 467(3.67)	24-3233-78
	EtOH	277(2.89),460(3.64)	24-3233-78
	HOAc	279(2.60),450s(3.64), 461(3.64)	24-3233-78
	$CHCl_3$	278(2.90),454(3.67), 467(3.66)	24-3233-78
	CCl_4	278(2.99),444(3.68), 466(3.68)	24-3233-78
	H_2SO_4	580(3.51)	24-3233-78
	CF_3COOH	488(3.68)	24-3233-78
$C_{10}H_{12}O_3$			
2-Cyclopenten-1-one, 4-(acetoxymeth-ylene)-2,3-dimethyl-, (E)-	n.s.g.	287(4.22)	33-0266-78
2-Cyclopenten-1-one, 4-acetoxy-3-(1-propenyl)-, [R-(E)]-	$CHCl_3$	270(4.42)	88-2553-78
Ethanone, 1-(3,4-dimethoxyphenyl)-	MeOH	227(4.27),272(4.06), 302(3.92)	33-1200-78
$C_{10}H_{12}O_3S$			
Benzenesulfonic acid, 4-methyl-, 2-propenyl ester	n.s.g.	262(4.11)	124-0844-78
$C_{10}H_{12}O_4$			
2-Cyclopenten-1-one, 5-acetoxy-4-hy-droxy-3-(1-propenyl)-, [4S-[3(E),4α,5β]]-	$CHCl_3$	278(4.46)	88-2553-78
$C_{10}H_{12}O_5$			
Acetic acid, (5-acetoxy-2(5H)-furanyli-dene)-, ethyl ester, (Z)-	EtOH	272(4.21)	35-7934-78
Benzaldehyde, 6-hydroxy-2,3,4-trimeth-oxy-	MeOH	237(4.19),287(4.35), 337(3.76)	95-1607-78
	MeOH-base	244(4.33),285(4.22), 373(3.95)	95-1607-78

Compound	Solvent	λ_{max}(log ϵ)	Ref.
Benzoic acid, 4-hydroxy-3,5-dimethoxy-, methyl ester	MeOH	219(4.40),276(4.05)	95-1607-78
	MeOH-base	239(4.14),325(4.30)	95-1607-78
2-Furanacetic acid, α-acetoxy-, ethyl ester	EtOH	218(3.80)	35-7934-78
2H-Pyran-5-carboxylic acid, 4-methoxy-6-methyl-2-oxo-, ethyl ester	MeOH	311(3.85)	118-0144-78
1H,6H-Pyrano[3,4-c]pyran-1,6-dione, 5-ethyl-3,4,5,8-tetrahydro-5-hydroxy-, ($\pm$)-	MeOH	228(3.74)	102-0135-78
	MeOH-HCl	228(3.74)	102-0135-78
	MeCN	227(3.75)	102-0135-78
$C_{10}H_{12}O_6$			
1,4-Cyclohexadiene-1,4-dicarboxylic acid, 2,5-dihydroxy-, dimethyl ester	C_6H_{12}	245(4.39)	48-0991-78
	EtOH	243(4.35)	48-0991-78
	$CHCl_3$	245(4.34)	48-0991-78
2,5-Cyclohexadiene-1,4-dione, 2,3,5,6-tetramethoxy-	$CHCl_3$	282(4.23),400(2.42)	40-1127-78
D-glycero-Hex-1-en-3-ulose, 2,6-diacetoxy-1,5-anhydro-5-deoxy-	MeOH	268(3.85)	136-0433-78H
$C_{10}H_{13}BrN_2$			
1H-Imidazo[1,2-a]pyridin-4-ium, 1,2,3-trimethyl-, bromide	EtOH	204(4.21),229(4.33), 290(3.90)	4-1149-78
$C_{10}H_{13}BrN_2O_7$			
1H-Imidazole-4-carboxylic acid, 5-bromo-2,3-dihydro-2-oxo-1-β-D-ribofuranosyl-, methyl ester	MeOH	277(4.02)	94-3322-78
$C_{10}H_{13}BrO_2$			
2,3-Bornanedione, 4-bromo-	n.s.g.	456(1.70)	78-1845-78
$C_{10}H_{13}BrO_3$			
Cyclopenta[c]pyran-4-carboxylic acid, 1-bromo-1,4aα,5,6,7,7aα-hexahydro-, methyl ester, ($\pm$)-	MeOH	238(4.08)	24-2423-78
$C_{10}H_{13}ClN_2O_6$			
1H-Imidazole-4-carboxylic acid, 2-chloro-1-β-D-ribofuranosyl-, methyl ester	MeOH	234(4.02)	94-3322-78
$C_{10}H_{13}ClN_5O_6PS$			
Adenosine, 2-chloro-, 5'-(dihydrogen phosphorothioate)	pH 1	264(4.10)	87-0520-78
$C_{10}H_{13}ClN_6O_6S$			
Adenosine, 2-chloro-, 5'-sulfamate	pH 1	263.5(4.16)	87-0520-78
$C_{10}H_{13}ClO_5$			
2H-Pyran-2-one, 5-acetoxy-6-(2-chloro-1-hydroxypropyl)-5,6-dihydro-, [5S-[5α,6α(1S*,2R*)]]-	EtOH	205(4.01)	18-3175-78
$C_{10}H_{13}Cl_3O_2$			
3-Heptyn-2-one, 1,1,1-trichloro-5-hydroxy-5,6,6-trimethyl-	hexane	234(3.82)	118-0307-78
$C_{10}H_{13}IO_2S$			
1-Butanesulfinic acid, 4-iodo-, phenyl ester	n.s.g.	211(3.73),272(3.17)	39-1580-78C

Compound	Solvent	λ_{max}(log ϵ)	Ref.
$C_{10}H_{13}N$			
5H-2-Pyrindine, 6,7-dihydro-1,3-di-methyl-	EtOH	267(3.43)	39-1293-78C
2H-Pyrrole, 5-(1-buten-3-ynyl)-3,4-di-hydro-3,3-dimethyl-, (E)-	MeOH	253(4.39),263s(4.30)	83-0977-78
$C_{10}H_{13}NO$			
Formamide, N-(2,4,6-trimethylphenyl)-	75.3% H_2SO_4	240s(3.3)	104-0111-78
1H-Indol-5-ol, 2,3-dihydro-1,2-di-methyl-	MeOH	243(3.19),277(3.18)	94-2027-78
hydrobromide	MeOH	242(3.51),275(3.34)	94-2027-78
4H-Indol-4-one, 1,5,6,7-tetrahydro-1,3-dimethyl-	MeOH	213(4.15),256(4.20)	24-1780-78
1(2H)-Isoquinolinone, 5,6,7,8-tetra-hydro-2-methyl-	EtOH	236(3.60),295(3.83)	44-4069-78
Pyridine, 2-[2-(1-methylethoxy)ethen-yl]-, (E)-	MeOH	265(3.82)	103-0530-78
$C_{10}H_{13}NOS$			
Ethanethioic acid, S-[1-methyl-3-(2-propenyl)-1H-pyrrol-2-yl] ester	EtOH	227(4.03),245(4.07)	23-0221-78
3-Pyridinol, 2-(ethenylthio)-6-(1-meth-ylethyl)-	EtOH	248(3.99),312(4.00)	1-0066-78
$C_{10}H_{13}NO_2$			
Benzene, 1-(1,1-dimethylethyl)-2-nitro-	MeOH	250s(3.06),280s(2.80), 380s(2.60)	104-1936-78
Benzene, 1-(1,1-dimethylethyl)-4-nitro-	MeOH	216(3.82),276(3.95)	104-1936-78
1-Propanol, 2-[(phenylmethylene)-amino]-, N-oxide	EtOH	298(c.4.18)	44-1900-78
4-Pyridineacetic acid, 3-ethyl-, methyl ester	MeOH	263(3.47),270s(3.38)	35-0571-78
$C_{10}H_{13}NO_3$			
Acetic acid, (1-ethenyl-5-oxo-2-pyrro-lidinylidene)-, ethyl ester	THF	266.5(4.32)	40-0404-78
2H-Pyran-2,6(3H)-dione, 3-[(diethyl-amino)methylene]-, (Z)-	MeOH	282(3.85),356(4.58)	44-4415-78
2H-Pyran-2,6(3H)-dione, 3-[[(1,1-di-methylethyl)amino]methylene]-, (Z)-	MeOH	337(4.62),345s(3.59)	44-4415-78
2-Pyridinecarboxylic acid, 5-butoxy-	M HCl	254(3.89),295(4.03)	103-0183-78
	H_2O	245(3.95),290(3.86)	103-0183-78
	M KOH	242(4.04),282(3.80)	103-0183-78
	EtOH	248(4.10),282(3.85)	103-0183-78
	25% EtOH	245(3.98),289(3.82)	103-0183-78
3-Pyridinecarboxylic acid, 6-butoxy-	M HCl	236(4.01),284(3.95)	103-0183-78
	H_2O	240(4.04),275(3.78)	103-0183-78
	M KOH	238(4.04),275(3.75)	103-0183-78
	EtOH	243(4.11),274(3.80)	103-0183-78
	25% EtOH	241(4.06),275(3.80)	103-0183-78
2(1H)-Pyridinone, 1-(2-acetoxypropyl)-	EtOH	303(3.70)	78-2609-78
$C_{10}H_{13}NO_4$			
3H-Azepine-3-carboxylic acid, 2,5-di-methoxy-, methyl ester	EtOH	268(4.09)	39-0191-78C
3H-Azepine-3-carboxylic acid, 2,6-di-methoxy-, methyl ester	EtOH	264(3.90)	39-0191-78C
2,5-Cyclohexadienepropanoic acid, 4-hydroxy-α-(hydroxyimino)-, methyl ester	EtOH	219(3.43),250(3.83)	5-0066-78

Compound	Solvent	$\lambda_{max}(\log \epsilon)$	Ref.
1-Oxadethiaceph-3-em-4-carboxylic acid, 2-ethyl-3-methyl-	EtOH	257(3.83)	39-1450-78C
1H-Pyrrole-3-carboxylic acid, 4-acetoxy-1-methyl-, ethyl ester	EtOH	210(4.16),234(4.00)	39-0896-78C
$C_{10}H_{13}NO_4S$			
2-Pentenedioic acid, 4-(2-thiazolidinylidene)-, dimethyl ester	MeOH	255(3.8),291(4.1), 338(4.5)	28-0385-78B
$C_{10}H_{13}NO_4S_2$			
1H-Pyrrole, 2,5-bis(2-propenylsulfonyl)-	EtOH	237(4.24)	23-0221-78
$C_{10}H_{13}NO_5$			
1H-Pyrrole-2,4-dicarboxylic acid, 3-hydroxy-, diethyl ester	EtOH	226(4.47),258(4.24)	94-2224-78
$C_{10}H_{13}NS_2$			
1H-Pyrrole, 2,5-bis(2-propenylthio)-	EtOH	220(3.93),264(4.05)	23-0221-78
$C_{10}H_{13}N_2$			
Benzenaminium, 3-cyano-N,N,N-trimethyl-, iodide	MeOH	277(2.64)(smoothed)	73-2763-78
Benzenaminium, 4-cyano-N,N,N-trimethyl-, iodide	MeOH	277.5(2.88)	73-2763-78
1H-Imidazo[1,2-a]pyridin-4-ium, 1,2,3-trimethyl-, bromide (hydrate)	EtOH	204(4.21),229(4.33), 290(3.90)	4-1149-78
$C_{10}H_{13}N_3O$			
Methanehydrazonamide, N'-formyl-N,N-dimethyl-N'-phenyl-, (E)-	heptane	218(3.84),258(4.04)	104-0576-78
1H-Pyrazolo[3,4-b]pyridine, 3-methoxy-1,4,6-trimethyl-	MeOH	230(4.23),318(3.62)	20-0309-78
	THF	231(4.43),314(3.67)	20-0309-78
3H-Pyrazolo[3,4-b]pyridin-3-one, 1,2-dihydro-1,2,4,6-tetramethyl-	MeOH	219(4.08),247(4.08), 320(3.57)	20-0309-78
	THF	224(4.21),253(4.20), 316(3.54)	20-0309-78
Pyrido[2,3-d]pyrrolo[1,2-a]pyrimidin-5(1H)-one, 2,3,4,7,8,9-hexahydro-	MeOH	225(4.19),271(3.95)	24-2297-78
$C_{10}H_{13}N_3O_3$			
3-Pyridazineacetic acid, α-(1-aminoethylidene)-6-methoxy-, methyl ester	EtOH	278(4.3)	4-1425-78
Pyrimidinebutanenitrile, 4-ethoxy-β-hydroxy-2-oxo-	EtOH	276(3.79)	126-0905-78
$C_{10}H_{13}N_3O_3S$			
4-Thiazoleacetic acid, 2-amino-α-[(2-propenyloxy)imino]-, ethyl ester, (Z)-	EtOH	234(4.19),295(3.81)	78-2233-78
$C_{10}H_{13}N_3O_4$			
Benzenamine, N-butyl-2,4-dinitro-	dioxan	350(4.2)	104-2361-78
1H-Cyclopentapyrimidine-1-propanoic acid, α-amino-2,3,4,5,6,7-hexahydro-2,4-dioxo-	0.05M HCl	272(4.00)	126-2195-78
	H_2O	274(3.99)	126-2195-78
	0.05M NaOH	276(3.95)	126-2195-78
2,6-Pyridinedimethanol, carbamate methylcarbamate	EtOH	264(3.68)	95-1402-78
$C_{10}H_{13}N_3O_7$			
Cytidine-6-carboxylic acid	pH 2	283(4.01)	94-2340-78
	pH 7	274(3.84)	94-2340-78

Compound	Solvent	$\lambda_{max}(\log \epsilon)$	Ref.
$C_{10}H_{13}N_5O_2$			
6,7-Pteridinedione, 2-(dimethylamino)-8-ethyl-5,8-dihydro-	pH 1.0	230s(4.28),253(4.49), 306(3.63)	24-1763-78
	pH 6.0	253(4.46),300s(3.66), 367(3.95)	24-1763-78
	pH 13.0	225(4.26),285(4.20), 320(3.94)	24-1763-78
$C_{10}H_{13}N_5O_3$			
Acetamide, N-(6,7-dihydro-8-methoxy-1,7-dimethyl-6-oxo-1H-purin-2-yl)-	pH 1	217(4.42),261(4.03), 278s(3.90)	44-2325-78
	pH 7	220(4.52),263(4.03), 277s(3.88)	44-2325-78
	pH 12	218(4.45),269(4.03)	44-2325-78
D-Arabinitol, 2-(6-amino-9H-purin-9-yl)-1,4-anhydro-2-deoxy-	pH 1	258(4.15)	44-0541-78
	pH 7,13	260(4.16)	44-0541-78
9H-Purin-6-amine, 9-(5-deoxy-α-D-arabinofuranosyl)-	H_2O	259(4.17)	44-0161-78
$C_{10}H_{13}N_5O_4$			
Adenosine	pH 1.5	205(4.326),257(4.164)	5-1796-78
	pH 7.0	206(4.326),260(4.173)	5-1796-78
	pH 7	259(4.15)	39-0762-78C
Imidazo[1,2-a]-1,3,5-triazin-4(8H)-imine, 8-α-D-ribofuranosyl-, hydrochloride	H_2O	210(4.20),248(3.73), 288(4.01)	44-4784-78
	pH 13	240s(3.95),247(4.03), 254s(3.93),300(3.96)	44-4784-78
β-anomer	H_2O	210(4.19),248(3.72), 288(4.01)	44-4784-78
	pH 13	241s(3.92),247(3.99), 253s(3.91),299(4.01)	44-4784-78
Inosine, 2'-amino-2'-deoxy-	pH 1	249(4.07)	94-0985-78
	H_2O	248.5(4.05)	94-0985-78
	pH 13	254(4.09)	94-0985-78
1,2-Propanediol, 3-(8-acetoxy-6-amino-9H-purin-9-yl)-, (±)-	pH 2	279(4.10)	73-3103-78
	pH 7	276(4.18)	73-3103-78
	pH 12	273(4.10)	73-3103-78
1H-Pyrazolo[4,3-d]pyrimidin-7-amine, 1-β-D-ribofuranosyl-	MeOH	211(4.16),228s(3.79), 290s(3.88),295(3.89), 307s(3.64)	104-0601-78
2H-Pyrazolo[4,3-d]pyrimidin-7-amine, 2-β-D-ribofuranosyl-	MeOH	213(4.28),235s(3.82), 246s(3.53),293(3.91), 301(3.94),313s(3.73)	104-0601-78
2,4(1H,3H)-Pyrimidinedione, 1-(3-azido-2,3-dideoxy-α-D-ribofuranosyl)-	H_2O	268(4.01)	44-3044-78
	pH 13	267(3.90)	44-3044-78
$C_{10}H_{13}N_5O_4S$			
Imidazo[1,2-a]-1,3,5-triazine-2(1H)-thione, 4-amino-1-β-D-ribofuranosyl-	pH 1	240s(3.98),259(4.16), 275s(4.10),306(4.20)	44-4774-78
	pH 13	256(4.24),283(4.38)	44-4774-78
D-Ribofuranosylamine, N-(6-amino-1,2,3-thiadiazolo[4,5-c]pyridin-4-yl)-	pH 1	279(3.91),327(4.19)	44-4910-78
	pH 11	233(4.12),244(4.09), 315(4.06),351s(3.76)	44-4910-78
	EtOH	236(4.03),256(3.99), 319(3.96),351(3.75)	44-4910-78
7H-1,2,3-Triazolo[4,5-b]pyridine-7-thione, 5-amino-3,4-dihydro-3-β-D-ribofuranosyl-	pH 1	287(4.00),310(4.08), 320(4.10),329(4.09)	44-4910-78
	pH 11	248(4.04),310(4.25)	44-4910-78
$C_{10}H_{13}N_5O_5$			
Guanosine	pH 7.0	254(4.13)	39-0762-78C

Compound	Solvent	$\lambda_{max}(\log \epsilon)$	Ref.
Imidazo[1,2-a]-1,3,5-triazin-2(1H)-one,	pH 1	232(4.04),258s(3.54)	44-4774-78
4-amino-1-β-D-ribofuranosyl-	pH 13	240(4.16)	44-4774-78
Imidazo[1,2-a]-1,3,5-triazin-2(8H)-one,	pH 1	243(4.09),265s(3.83)	44-4784-78
4-amino-8-α-D-ribofuranosyl-	pH 13	247(4.10)	44-4784-78
β-anomer	pH 1	242(4.09),265s(3.83)	44-4784-78
	pH 13	247(4.18)	44-4784-78
Imidazo[1,2-a]-1,3,5-triazin-4(1H)-one,	pH 2	238s(3.94),262(4.11)	44-4774-78
2-amino-1-β-D-ribofuranosyl-	pH 13	223(4.35),267s(3.43)	44-4774-78
Imidazo[1,2-a]-1,3,5-triazin-4(1H)-one,	pH 1	238s(3.94),264(4.17)	87-0883-78
2-amino-8-β-D-ribofuranosyl-	pH 7	210(4.46),256(4.14)	87-0883-78
	pH 11	217(3.67),256(4.14)	87-0883-78
7H-1,2,3-Triazolo[4,5-b]pyridin-7-one,	pH 1	250(3.90),296(4.25)	44-4910-78
5-amino-3,4-dihydro-3-β-D-ribo-	pH 11	284(4.26)	44-4910-78
furanosyl-			
$C_{10}H_{13}N_7O_3$			
8,5'-Aminimino-9-(5-deoxy-β-D-ribo-	MeOH	213(4.44),272(4.32)	44-2320-78
furanosyl)adenine			
$C_{10}H_{13}O_4P$			
Phosphonic acid, (4-methylbenzoyl)-,	H_2O	282(3.93)(hydrate)	35-7382-78
dimethyl ester	dioxan	272(4.12)	35-7382-78
$C_{10}H_{13}O_5P$			
Phosphonic acid, (4-methoxybenzoyl)-,	H_2O	310(4.14)(hydrate)	35-7382-78
dimethyl ester	dioxan	298(4.20)	35-7382-78
$C_{10}H_{14}$			
Benzene, 1,2,3,4-tetramethyl-	EtOH	267(2.48),272(2.39),	35-3730-78
		276(2.37)	
1,2:3,4-Dimethanocyclohexane, 5-ethenyl-	EtOH	254(4.18)	88-0977-78
$C_{10}H_{14}BrNO_2S$			
Benzenesulfonamide, N-(3-bromopropyl)-	$CHCl_3$	258(2.03),264(2.04)	70-0149-78
4-methyl-			
$C_{10}H_{14}ClN_3O_4$			
1H-1,2,5-Triazapentadienium, 1,1-di-	$CHCl_3$	377(4.52)	39-1023-78C
methyl-5-phenyl-, perchlorate			
1H-1,2,5-Triazapentadienium, 5,5-di-	$CHCl_3$	405(4.54)	39-1023-78C
methyl-1-phenyl-, perchlorate			
$C_{10}H_{14}N_2$			
Ethanone, 1-phenyl-, dimethylhydrazone	n.s.g.	312(3.27)	104-0717-78
2-Propanone, methylphenylhydrazone	n.s.g.	287(3.49)	104-0717-78
Quinoxaline, 1,2,3,4-tetrahydro-1,4-di-	hexane	229(4.53),263(3.51),	44-2621-78
methyl-		311(3.79)	
$C_{10}H_{14}N_2O$			
Benzenamine, N,N,3,5-tetramethyl-	MeOH	234(3.63),274(3.73),	87-1044-78
4-nitroso-		279s(3.72),335s(3.20),	
		413(4.53)	
Ethanone, 2-cyclopentyl-1-(1H-imidazol-	EtOH	280(4.11)	33-2831-78
2-yl)-			
Ethanone, 2-cyclopentyl-1-(1H-imidazol-	EtOH	257(4.05)	33-2831-78
4-yl)-			
4H-Indol-4-one, 1,5,6,7-tetrahydro-	MeOH	216(4.11),252(4.06)	24-1780-78
1,3-dimethyl-, oxime, (E)-			
Pyrrolo[3,2-b]azepin-5(1H)-one,	MeOH	217(3.99),243s(4.12)	24-1780-78
4,6,7,8-tetrahydro-1,3-dimethyl-			

Compound	Solvent	λ_{max}(log ϵ)	Ref.
Pyrrolo[3,2-c]azepin-4(1H)-one, 5,6,7,8-tetrahydro-1,3-dimethyl-	MeOH	212(4.27),243s(4.12)	24-1780-78
$C_{10}H_{14}N_2O_2$			
Benzenamine, N-ethyl-3,5-dimethyl-2-nitro-	EtOH	238(4.19),292(3.56), 423(3.57)	78-2213-78
Benzenamine, N-ethyl-3,5-dimethyl-4-nitro-	EtOH	246(3.92),305(3.45), 392(3.79)	78-2213-78
Benzenamine, N-methyl-4-nitro-2-propyl-	EtOH	228(3.95),260s(3.65), 387(4.30)	104-1544-78
Benzenamine, N,N,3,5-tetramethyl-4-nitro-	EtOH	249(3.98),308(3.47), 393(3.79)	78-2213-78
8,11-Diazatricyclo[5.2.2.0^{1,5}]undecane, 7-methyl-9,10-dioxo-	EtOH	267(2.00)	39-1293-78C
Diazene, (ethoxymethoxymethyl)phenyl-, (E)-	EtOH	215(3.95),269(3.92)	44-1459-78
Diazene, phenyl(propoxymethyl)-, 1-oxide, (Z)-	EtOH	247(4.02)	44-1459-78
2-Pyrimidineacetic acid, 4-(1,1-dimethylethyl)-	neutral	250(3.59)(extrapolated)	12-0649-78
	cation	250(3.75)	12-0649-78
	anion	250(3.58),275s(2.60)	12-0649-78
2-Pyrimidineacetic acid, 4-(1-methylethyl)-, methyl ester	neutral	248(3.57),275(2.54)	12-0649-78
	cation	250(3.75)	12-0649-78
4(1H)-Pyrimidinone, 6-hydroxy-2-methyl-5-(4-pentenyl)-	EtOH	248(3.65),263(3.85), 274(3.70)	39-1293-78C
$C_{10}H_{14}N_2O_2S_2$			
2-Pyrimidineacetic acid, 4,6-bis(ethylthio)-	neutral	254(4.36),300(4.16)	12-0649-78
	cation	252(3.66),266(3.70), 285(3.67),326(4.09)	12-0649-78
	anion	254(4.35),300(4.14)	12-0649-78
$C_{10}H_{14}N_2O_3$			
Ethenamine, N,N-diethyl-2-(5-nitro-2-furanyl)-, (E)-	EtOH	280(4.26),511(4.52)	103-0250-78
1H-Pyrazole-4-carboxylic acid, 3-acetyl-1,5-dimethyl-, ethyl ester	n.s.g.	213(3.80),245(3.72)	44-2665-78
1H-Pyrazole-4-carboxylic acid, 5-acetyl-1,3-dimethyl-, ethyl ester	n.s.g.	254(3.73)	44-2665-78
$C_{10}H_{14}N_2O_4$			
Pyrazinecarboxylic acid, 1-acetyl-1,2,5,6-tetrahydro-2,3,5-trimethyl-6-oxo-	n.s.g.	259(2.62),302(2.20)	70-0963-78
2,5,8,11-Tetraoxa-13,16-diazabicyclo[10.3.1]hexadeca-1(16),12,14-triene	EtOH	230(5.94),270(4.79)	44-3362-78
$C_{10}H_{14}N_2O_4S$			
Thiophene, 2-nitro-5-(1,1,2-trimethyl-2-nitropropyl)-	MeOH	325(4.02)	12-2463-78
$C_{10}H_{14}N_2O_5$			
2,2'-Anhydro-3-β-D-arabinofuranosyl-6-hydroxy-6-methyl-5,6-dihydrouracil	H_2O	210(3.5)	103-0901-78
2,4(1H,3H)-Pyrimidinedione, 1-(3,3-dimethoxy-2-oxobutyl)-	n.s.g.	260(4.01)	78-2861-78
$C_{10}H_{14}N_2O_6$			
2,4(1H,3H)-Pyrimidinedione, 1-O-β-D-apio-L-furanosyl-5-methyl-	EtOH	266(3.99)	87-0706-78

Compound	Solvent	$\lambda_{max}(\log \epsilon)$	Ref.
2,4(1H,3H)-Pyrimidinedione, 3-β-D-ara-binofuranosyl-6-methyl-	H_2O	208(3.90),262(3.96)	103-0901-78
	pH 13	286(4.0)	103-0901-78
Uridine, 3-methyl-	pH 7	262(3.95)	39-0762-78C
Uridine, 5'-methyl-	pH 1	263(3.91)	88-4403-78
	pH 13	217(4.04),265(3.84)	88-4403-78
	MeOH	264(3.75)	88-4403-78
$C_{10}H_{14}N_2O_6S$			
1H-Imidazole-4-carboxylic acid, 2,3-di-hydro-1-β-D-ribofuranosyl-2-thioxo-, methyl ester	MeOH	266(4.10),292(3.96)	94-3322-78
$C_{10}H_{14}N_2O_7$			
1H-Imidazole-4-carboxylic acid, 2,3-di-hydro-2-oxo-1-β-D-ribofuranosyl-, methyl ester	MeOH	268.5(4.10)	94-3322-78
$C_{10}H_{14}N_2S$			
Propanethioic acid, 2-methyl-2-phenyl-hydrazide	MeOH	222(3.76),238(4.02), 251(3.86),276(4.08)	123-0701-78
$C_{10}H_{14}N_4$			
[1,2,4]Triazolo[1,5-c]pyrimidine, 2-ethyl-5,7,8-trimethyl-	pH 7.0	260(3.91)	12-2505-78
1,2,4-Triazolo[4,3-c]pyrimidine, 3-ethyl-5,7,8-trimethyl-	pH 7.0	251s(3.87),259(3.95), 267s(3.90),285s(3.64)	12-2505-78
$C_{10}H_{14}N_4O$			
Imidazo[1,2-b]pyridazin-4-ium, 3-(di-methylamino)-5,6-dihydro-2,2-dimeth-yl-6-oxo-, hydroxide, inner salt	MeCN	273s(3.63),425(3.40)	33-2116-78
$C_{10}H_{14}N_4OS$			
4H-Pyrrolo[2,3-d]pyrimidin-4-one, 5-[(dimethylamino)methyl]-1,7-dihydro-2-(methylthio)-	MeOH	223(4.25),276(4.10)	24-2925-78
$C_{10}H_{14}N_4O_2$			
Hydrazinecarboxamide, 2-nitroso-N-phenyl-2-propyl-	EtOH	236(4.35),350(2.02)	104-2319-78
$C_{10}H_{14}N_4O_3$			
Pyrido[2,3-d]pyrimidine-8(2H)-carbox-amide, 1,3,4,5,6,7-hexahydro-N,3-di-methyl-2,4-dioxo-	MeOH	212(4.04),288(4.17)	24-2297-78
$C_{10}H_{14}N_4O_3S$			
Thiazolo[2,3-c]-1,2,4-triazole-5-acetic acid, 2,3-dihydro-α-(hydroxyimino)-3,3-dimethyl-, ethyl ester	EtOH	257(4.10?),346(4.29?)	4-0401-78
$C_{10}H_{14}N_4O_4$			
1H-Purine-2,6-dione, 7-(2,3-dihydroxy-propyl)-3,7-dihydro-1,3-dimethyl-, (±)-	pH 2,7,12	274(4.00)	73-2054-78
2,4(1H,3H)-Pyrimidinedione, 1,3-dimeth-yl-6-(N,N'-diacetylhydrazino)-	0.5M HCl	276(3.98)	56-0037-78
	H_2O	213(4.08),276(3.97)	56-0037-78
	0.5M NaOH	278(3.94)	56-0037-78
1(2H)-Pyrimidinepropanoic acid, 3,4-di-hydro-5-methyl-2,4-dioxo-, 2-acetyl-hydrazide	H_2O	265(3.99)	78-2861-78

Compound	Solvent	$\lambda_{max}(\log \epsilon)$	Ref.
$C_{10}H_{14}N_5O_7P$			
5'-Adenylic acid	pH 2.0	205(4.326),257(4.164)	5-1796-78
	pH 7.9	207(4.320),260(4.176)	5-1796-78
$C_{10}H_{14}N_5O_8P$			
Imidazo[1,2-a]-1,3,5-triazin-4(8H)-one,	pH 1	239s(3.85),265(4.07)	87-0883-78
2-amino-8-(5-O-phosphono-β-D-ribo-	pH 7	211(4.35),257(4.04)	87-0883-78
furanosyl)-	pH 11	257(4.04)	87-0883-78
$C_{10}H_{14}N_6$			
2-Propene-1,1,3-tricarbonitrile,	MeOH	273(3.93)	35-0926-78
2-amino-1,3-bis(dimethylamino)-			
$C_{10}H_{14}N_6O_2$			
Adenosine, 3'-amino-2',3'-dideoxy-	pH 1	257(4.17)	44-3044-78
	H_2O	259(4.17)	44-3044-78
$C_{10}H_{14}N_6O_3$			
Adenosine, 2'-amino-2'-deoxy-, dihydro-	pH 1	256.5(4.17)	78-1133-78
chloride	H_2O	258.5(4.18)	78-1133-78
	pH 13	259(4.18)	78-1133-78
Guanosine, 3'-amino-2',3'-dideoxy-	pH 1	256(4.08),275s(3.93)	44-3044-78
	H_2O	253(4.12),270s(3.97)	44-3044-78
	pH 13	265(4.05)	44-3044-78
9H-Purin-6-amine, 9-(2-amino-2-deoxy-	pH 1	262(4.16)	88-3653-78
β-D-arabinofuranosyl)-	pH 6.0	259.5(4.17)	88-3653-78
	pH 13	260.5(4.19)	88-3653-78
$C_{10}H_{14}N_6O_4$			
Guanosine, 2'-amino-2'-deoxy-	pH 1	256(4.05),280s(3.90)	94-0240-78
	H_2O	252(4.11),270s(3.96)	94-0240-78
	pH 13	259(4.04)	94-0240-78
$C_{10}H_{14}N_6O_5$			
5,8-Epoxyimidazo[1,2-a][1,3]diazocine-	MeOH-HCl	216(4.32),277(4.09)	44-2320-78
2-carboxamide, 10-amino-3-(formylam-			
ino)-5,6,7,8,9,10-hexahydro-6,7-di-			
hydroxy-, [5R-(5α,6α,7α,8α)]-			
Guanosine, N-1-amino-	pH 13	254(4.13)	64-0326-78C
$C_{10}H_{14}N_6O_6S$			
Adenosine, 5'-deoxy-5'-(sulfoamino)-	H_2O	259(4.14)	87-0704-78
$C_{10}H_{14}N_7O_{13}P_3$			
6-Azidopurine ribonucleoside 5'-tri-	pH 2	252(3.64),260(3.65),	69-0094-78
phosphate, tetralithium salt		287(3.90)	
$C_{10}H_{14}O$			
Bicyclo[3.1.1]hept-2-ene-6-carboxalde-	C_6H_{12}	290(1.30)	78-1957-78
hyde, 2,6-dimethyl-			
Bicyclo[5.1.0]octa-2,4-diene, 4-(meth-	C_6H_{12}	251(3.46)	44-1118-78
oxymethyl)-			
2,6-Cycloheptadien-1-one, 2-(1-methyl-	EtOH	237(3.78)	35-1778-78
ethyl)-			
2,5-Cyclohexadien-1-one, 4-methyl-	MeOH	238(4.1)	24-1944-78
4-propyl-			
1,2,5-Cyclononatriene, 7-methoxy-	C_6H_{12}	222(3.57)	35-5856-78
Ethanone, 1-(5-methyl-1,5-cyclohepta-	EtOH	233(3.74)	88-3603-78
dien-1-yl)-			
2(1H)-Pentalenone, 4,5,6,6a-tetrahydro-	EtOH	240(4.20)	35-1799-78
1,3-dimethyl-, cis			

Compound	Solvent	$\lambda_{max}(\log \epsilon)$	Ref.
2(1H)-Pentalenone, 4,5,6,6a-tetrahydro-1,3-dimethyl-, trans	EtOH	240(3.99)	35-1799-78
2(1H)-Pentalenone, 4,5,6,6a-tetrahydro-4,6a-dimethyl-	MeOH	232(4.08),290(2.20)	44-2557-78
$C_{10}H_{14}OS$			
2-Propanol, 1-[(phenylmethyl)thio]-, (S)-	MeOH	261(2.6),267(2.4)	12-1965-78
$C_{10}H_{14}OSi$			
Silane, benzoyltrimethyl-	C_6H_{12}	405(1.97),424(2.05), 445(1.81)	112-0741-78
$C_{10}H_{14}O_2$			
Benzene, (2-ethoxyethoxy)-	n.s.g.	195(4.53),221(3.94), 271(3.26),278(3.18)	47-1343-78
3H-2-Benzopyran-3-one, 1,5,6,7,8,8a-hexahydro-4-methyl-	MeOH	232(4.00)	44-1248-78
2-Cyclohexen-1-one, 4-(3-hydroxy-1-methylpropylidene)-	EtOH	222(3.68),303(4.15)	39-0024-78C
2-Cyclopentene-1-acetaldehyde, 2-formyl-α,3-dimethyl-	EtOH	253(4.08)	138-0433-78
Rhododendrol, (+)-	EtOH	221(3.84),276(3.27), 283s(--)	95-0041-78
$C_{10}H_{14}O_2S$			
2-Propanol, 1-[(phenylmethyl)sulfinyl]-(S_S,S_C)-(+)-	MeOH	225(2.50),262(2.57), 267(2.43),272(2.33)	12-1965-78
(-)-	MeOH	224(2.48),265(2.48), 270(2.30)	12-1965-78
$C_{10}H_{14}O_3$			
Acetic acid, (5-hydroxy-2-methylene-cyclohexylidene)-, methyl ester, (Z)-(±)-	EtOH	219(3.81)	39-0387-78C
2-Cyclopenten-1-one, 5-acetoxy-2-propyl-	EtOH	231(3.91)	107-0155-78
3-Hexynoic acid, 5,5-dimethyl-2-oxo-, ethyl ester	hexane	238(3.82)	118-0307-78
1-Oxaspiro[4.5]decane-2,4-dione, 3-methyl-	EtOH	234(4.01)	4-1493-78
Phenol, 2,5-dimethoxy-3,4-dimethyl-	EtOH	284(3.39)	23-0517-78
Phenol, 4-(ethoxymethyl)-2-methoxy-	EtOH	233(3.83),282(3.45)	98-0195-78
$C_{10}H_{14}O_3S$			
Benzenesulfonic acid, 4-methyl-, 1-methylethyl ester	n.s.g.	225(4.44)	124-0844-78
Benzenesulfonic acid, 4-methyl-, propyl ester	n.s.g.	262(4.11)	124-0844-78
2-Propanol, 1-[(phenylmethyl)sulfonyl]-, (S)-	MeOH	255(2.4),259(2.4), 265(2.4),270(2.3)	12-1965-78
$C_{10}H_{14}O_3S_2$			
5H-1,4-Dithiepin-2-propanal, 3-acetoxy-6,7-dihydro-	n.s.g.	257(3.83)	33-3087-78
$C_{10}H_{14}O_4$			
Acetic acid, (5-ethoxy-2(5H)-furanylidene)-, ethyl ester	EtOH	278(3.93)	35-7934-78
Cyclopenta[c]pyran-4-carboxylic acid, 1,4aα,5,6,7,7aα-hexahydro-1-hydroxy-,	MeOH pH 12	238(4.05) 275(--)	24-2423-78 24-2423-78

Compound	Solvent	$\lambda_{max}(\log \epsilon)$	Ref.
methyl ester (cont.)			24-2423-78
Cyclopenta[c]pyran-4-carboxylic acid, 1,4aα,5,6,7,7aα-hexahydro-1α-methoxy-, (±)-	MeOH	236(4.03)	24-2423-78
Cyclopropanecarboxylic acid, 3-(2-carboxy-1-propenyl)-2,2-dimethyl-, (E,E)-	EtOH	237(4.20)	24-2206-78
3-Furancarboxylic acid, 4,5-dihydro-4-oxo-2-propyl-, ethyl ester	EtOH	214(3.82),262(4.02)	118-0448-78
2,5-Heptadienoic acid, 2,4,4-trimethyl-, (E,E)-	EtOH	214(4.18)	24-2206-78
2H-Pyran-2-acetic acid, 3,6-dihydro-3,3,5-trimethyl-6-oxo-	EtOH	217(3.74)	24-2206-78
$C_{10}H_{14}O_4S$			
Ethanol, 2-methoxy-, 4-methylbenzenesulfonate	n.s.g.	224(4.87)	124-0844-78
$C_{10}H_{14}O_5$			
2,5-Cyclohexadien-1-one, 2,4,4,5-tetramethoxy-	EtOH	249(4.21),304(3.26)	44-3983-78
3-Furancarboxylic acid, 4,5-dihydro-5-methoxy-2,5-dimethyl-4-oxo-, ethyl ester	EtOH	216(4.07),270(4.08)	44-2665-78
$C_{10}H_{14}O_6$			
2H-Pyran-2-one, 5-acetoxy-6-(acetoxymethyl)tetrahydro-, (5S-trans)-	EtOH	205(3.96)	18-3175-78
$C_{10}H_{14}S$			
Benzene, (butylthio)-	heptane	255(3.95),275(3.40)	65-2027-78
$C_{10}H_{15}BrN_2O$			
Imidazo[1,2-a]pyridinium, 2,3-dihydro-2-hydroxy-1,2,3-trimethyl-, bromide	EtOH	202(4.11),238(4.04), 326(3.65)	4-1149-78
$C_{10}H_{15}BrO_2$			
6-Oxabicyclo[3.2.1]octan-3-one, 4α-bromo-4β,7,7-trimethyl-	MeOH	307(1.92)	44-0740-78
$C_{10}H_{15}ClO_2$			
6-Oxabicyclo[3.2.1]octan-3-one, 4α-chloro-4β,7,7-trimethyl-	MeOH	298(1.60)	44-0740-78
$C_{10}H_{15}Cl_2NPt$			
Platinum, dichloro(η^2-ethene)(2,4,6-trimethylpyridine)-	EtOH	235(3.66),265(3.74), 300(2.91)	101-0357-78A
$C_{10}H_{15}IO_2$			
2-Cyclohexen-1-one, 3-ethoxy-2-iodo-5,5-dimethyl-	MeOH	283(3.97)	44-2676-78
$C_{10}H_{15}N$			
Benzenamine, N-(1,1-dimethylethyl)-	MeOH	244(3.80),290(2.87)	104-1936-78
hydrochloride	MeOH	254(2.30)	104-1936-78
$C_{10}H_{15}NO$			
2(1H)-Pyridinone, 3-(1,1-dimethylethyl)-1-methyl-	EtOH	233(3.75),302(3.83)	142-0023-78B
2(1H)-Pyridinone, 5-(1,1-dimethylethyl)-1-methyl-	EtOH	230(3.96),309(3.75)	142-0023-78B

Compound	Solvent	$\lambda_{max}(\log \epsilon)$	Ref.
2H-Pyrrole-5-ethanol, α-ethynyl-3,4-dihydro-3,3-dimethyl-	MeOH	203(3.20),250s(2.71)	83-0977-78
$C_{10}H_{15}NO_2$			
1-Oxaspiro[4.5]dec-3-en-2-one, 4-amino-3-methyl-	EtOH	260(4.25)	4-1493-78
$C_{10}H_{15}NO_2SSi$			
Silane, trimethyl[[(4-nitrophenyl)-thio]methyl]-	EtOH	227(3.77),343(4.17)	65-2027-78
$C_{10}H_{15}NO_3$			
3,5-Hexadien-2-one, 3-acetoxy-6-(dimethylamino)-	hexane	244(3.63),380(4.62)	70-0107-78
	H_2O	278(3.77),400(4.77)	70-0107-78
1H-Pyrrole-2-carboxylic acid, 3-hydroxy-1-methyl-, 1,1-dimethylethyl ester	EtOH	264(4.18)	94-2224-78
$C_{10}H_{15}NO_4$			
Propanedioic acid, (1-pyrrolidinylmethylene)-, dimethyl ester	isooctane	225(3.72),288(4.35)	78-2321-78
$C_{10}H_{15}NO_5S$			
2-Thiophenepentanoic acid, 2,5-dihydro-4-hydroxy-3-nitro-, methyl ester, compd. with α-methylbenzenemethanamine, (-)-	n.s.g.	348(4.20)	35-7423-78
$C_{10}H_{15}N_2O$			
Imidazo[1,2-a]pyridinium, 2,3-dihydro-2-hydroxy-1,2,3-trimethyl-, bromide	EtOH	202(4.11),238(4.04), 326(3.65)	4-1149-78
$C_{10}H_{15}N_2SSi$			
Benzenediazonium, 4-[[(trimethylsilyl)-methyl]thio]-, tetrafluoroborate	EtOH	384(4.14)	65-2027-78
$C_{10}H_{15}N_3$			
1-Propanone, 2-amino-2-methyl-1-phenyl-, hydrazone	hexane	211(4.15),278(3.34)	103-0278-78
$C_{10}H_{15}N_3O_2$			
1,4-Benzenediamine, N,N,N',N'-tetramethyl-2-nitro-	C_6H_{12}	423(3.08)	39-1194-78C
	EtOH	445(--)	39-1194-78C
	$CHCl_3$	468(--)	39-1194-78C
$C_{10}H_{15}N_3O_3S$			
4-Thiazoleacetic acid, 2-amino-α-[(1-methylethoxy)imino]-, ethyl ester, (Z)-	EtOH	243(4.17),292(3.81)	78-2233-78
$C_{10}H_{15}N_3O_4$			
Thymidine, 3'-amino-3'-deoxy-	pH 1	266(3.96)	87-0109-78
	pH 13	266.5(3.87)	87-0109-78
$C_{10}H_{15}N_3O_5$			
Cytidine, N-methyl-	pH 1	222(4.08),265(3.86)	44-1193-78
	pH 7	222(4.16),293(3.74)	44-1193-78
	pH 13	233(4.03),281(3.79)	44-1193-78
α-anomer	pH 1	223(4.08),265(3.84)	44-1193-78
	pH 7	220(4.21),295(3.64)	44-1193-78
	pH 13	233(4.03),282(3.81)	44-1193-78

Compound	Solvent	λ_{max}(log ϵ)	Ref.
Cytidine, 2'-O-methyl-	pH 7	270(3.94)	39-0762-78C
Cytidine, 3'-O-methyl-	pH 7	270(3.95)	39-0762-78C
$C_{10}H_{15}N_3O_6$			
2,4(1H,3H)-Pyrimidinedione, 1-(2-amino-2-deoxy-β-D-glucopyranosyl)-	H_2O	257(4.09)	136-0185-78C
$C_{10}H_{15}N_3O_8S$			
Thymidine, 5'-deoxy-5'-(sulfoamino)-	H_2O	267(3.96)	87-0704-78
$C_{10}H_{15}N_5$			
[1,2,4]Triazolo[1,5-a]pyrimidin-6-ol, 7-(diethylamino)-5-methyl-	MeOH	221(--),271(3.80), 306(4.26)	106-0051-78
$C_{10}H_{15}N_5O$			
Ethanol, 2-[ethyl(5-methyl[1,2,4]triazolo[1,5-a]pyrimidin-7-yl)amino]-	MeOH	221(--),271(3.78), 306(4.27)	106-0051-78
[1,2,4]Triazolo[1,5-a]pyrimidin-6-ol, 7-(diethylamino)-5-methyl-	MeOH	229(--),271(3.55), 332(4.16)	106-0051-78
	MeOH-NaOH	239(4.30),358(3.96)	106-0051-78
$C_{10}H_{15}N_5O_2$			
1,3-Pentanediol, 5-(6-amino-9H-purin-9-yl)-, (±)-	pH 2	260(4.13)	73-3444-78
1,2-Propanediol, 3-[6-(dimethylamino)-9H-purin-9-yl]-	pH 2	270(4.20)	73-3103-78
	pH 7	278(4.20)	73-3103-78
	pH 12	278(4.21)	73-3103-78
6(5H)-Pteridinone, 2-(dimethylamino)-8-ethyl-7,8-dihydro-7-hydroxy-	pH 0.0	230s(4.28),253(4.49), 310s(3.61)	24-1763-78
	pH 9.0	225(4.23),277(4.14), 320s(3.95)	24-1763-78
	pH 13.0	225(4.26),284(4.20), 335s(3.82)	24-1763-78
9H-Purin-6-amine, 9-(2,3-dimethoxypropyl)-, (±)-	pH 2,7,12	262(4.13)	73-3444-78
$C_{10}H_{16}$			
1,3-Cyclohexadiene, 5-ethyl-1,3-dimethyl-	EtOH	267(3.36)	78-2783-78
Cyclooctane, 1,2-bis(methylene)-	C_6H_{12}	229(3.57)	54-0105-78
1,3,6-Octatriene, 2,6-dimethyl-	n.s.g.	238(4.32)	88-4073-78
$C_{10}H_{16}N_2$			
1,2-Benzenediamine, N,N,N',N'-tetramethyl-	hexane	233(4.32),268(3.93), 294(3.45)	44-2621-78
$C_{10}H_{16}N_2O_2$			
1H-Pyrazole-5-carboxylic acid, 4,5-dihydro-5-methyl-4-(1-methylethylidene)-, ethyl ester	EtOH	283(3.72)	22-0401-78
$C_{10}H_{16}N_2O_3$			
1H-Pyrazole-4-carboxylic acid, 3-(hydroxymethyl)-5-propyl-, ethyl ester	EtOH	227(3.98)	118-0448-78
1H-Pyrazole-4-carboxylic acid, 3-(2-hydroxypropyl)-5-methyl-	EtOH	227(3.93)	118-0900-78
$C_{10}H_{16}N_2O_5$			
Glycine, N-[5-hydroxy-5-(hydroxymethyl)-3-imino-2-methoxy-1-cyclohexen-1-yl]-	H_2O	320(4.55)	88-1401-78

Compound	Solvent	$\lambda_{max}(\log \epsilon)$	Ref.
Palythine	H_2O	320(4.56)	88-2299-78
$C_{10}H_{16}N_2S_2$			
Pyrimidine, 2,4-bis(propylthio)-	H_2O	255(3.20),305(3.14)	56-1255-78
$C_{10}H_{16}N_4O_5$			
1H-Imidazole-4-carboxamide, 5-(methyl-	pH 1	259(3.88)	88-4047-78
amino)-1-β-D-ribofuranosyl-	pH 7	267(3.92)	88-4047-78
	pH 13	268(3.93)	88-4047-78
$C_{10}H_{16}N_6O_5$			
Urea, N'-(aminoiminomethyl)-N-1H-imida-	MeOH-HCl	219(4.02)	44-4774-78
zol-2-yl-N-β-D-ribofuranosyl-	MeOH-NaOH	227(4.39)	44-4774-78
$C_{10}H_{16}N_6O_6$			
1H-Imidazole-4,5-dicarboxylic acid,	pH 1	262(3.99)	4-0001-78
1-β-D-ribofuranosyl-, hydrazide	pH 7	241(4.01)	4-0001-78
	pH 11	236s(4.01)	4-0001-78
$C_{10}H_{16}N_6S_4$			
1,3-Dithia-5,8-diazaspiro[3.4]octane,	EtOH	262s(4.43),280(4.50)	88-3823-78
2,2-bis[(4,5-dihydro-1H-imidazol-			
2-yl)thio]-			
$C_{10}H_{16}O$			
4(1H)-Azulenone, octahydro-, cis	EtOH	234(4.00),369(4.11)	44-2153-78
4(1H)-Azulenone, octahydro-, trans	EtOH	285(1.46)	44-2153-78
Camphor	THF-LiClO₄	290(1.43)(changing)	44-1126-78
after time	THF-LiClO₄	286.5(1.50)	44-1126-78
Cycloheptanone, 2-(1-methylethylidene)-	hexane	238(3.86),323(1.85)	44-0604-78
2-Cyclohexen-1-one, 3-ethyl-2,5-dimeth-	n.s.g.	250(4.12)	22-0255-78
yl-			
2-Cyclohexen-1-one, 3-methyl-6-(1-meth-	MeOH	237(4.11)	22-0255-78
ylethyl)-			
2-Cyclohexen-1-one, 4-methyl-4-propyl-	MeOH	229(4.08)	24-1944-78
3-Cyclohexen-1-one, 2-methyl-4-(1-meth-	EtOH	290(1.96)	44-0519-78
ylethyl)-			
Cyclooctanone, 2-ethylidene-, (E)-	hexane	237(3.88)	54-0305-78
1-Cyclopentene-1-carboxaldehyde,	EtOH	250(4.08)	23-1628-78
2-methyl-5-(1-methylethyl)-, (S)-			
1-Cyclopentene-1-methanol, α-methyl-	EtOH	210(2.01)(end absorption)	44-2153-78
α-2-propenyl-			
2-Cyclopenten-1-one, 2,4-dimethyl-	n.s.g.	235(4.10)	22-0255-78
3-(1-methylethyl)-			
2-Cyclopenten-1-one, 3-ethyl-2,4,5-tri-	EtOH	239(4.09)	35-1799-78
methyl-, cis			
trans	EtOH	241(4.16)	35-1799-78
2-Cyclopenten-1-one, 5-methyl-3-(2-	n.s.g.	229(4.14)	22-0255-78
methylpropyl)-			
Ethanone, 1-[2-(2-propenyl)cyclopent-	EtOH	281(1.34)	44-2153-78
yl]-			
$C_{10}H_{16}OS$			
Bicyclo[3.3.1]nonan-3-one, 1-(methyl-	EtOH	240s(2.26),287(1.48)	44-3653-78
thio)-			
$C_{10}H_{16}OS_2$			
Cycloheptanone, 2-[bis(methylthio)meth-	hexane	223s(3.52),270(3.54),	44-0604-78
ylene]-		297s(3.43)	

Compound	Solvent	$\lambda_{max}(\log \epsilon)$	Ref.
$C_{10}H_{16}O_2$			
Bicyclo[3.3.1]nonan-3-one, 1-methoxy-	EtOH	278(1.32)	44-3653-78
Cycloheptanone, 3-(2-oxopropyl)-	EtOH	280(1.67)	44-3653-78
2-Cyclohepten-1-one, 6-hydroxy-2-(1-methylethyl)-	EtOH	238(3.88)	35-1778-78
2-Cyclohexen-1-one, 4-hydroxy-4,5,6,6-tetramethyl-	n.s.g.	217(3.9)	39-0159-78C
2,5-Heptadienoic acid, 4,4,6-trimethyl-, (E)-	EtOH	213(4.01)	78-3331-78
3-Nonene-2,7-dione, 8-methyl-	pentane	218(4.13)	142-0083-78C
2-Propenoic acid, 3-(2,2,3,3-tetramethylcyclopropyl)-, (E)-	EtOH	245(4.20)	78-3331-78
(Z)-	EtOH	253(4.05)	78-3331-78
$C_{10}H_{16}O_2S_2$			
1,4-Dithiepane-2-pentanal, 3-oxo-	n.s.g.	239(3.56),283s(2.99), 318s(2.30)	33-3087-78
$C_{10}H_{16}O_2S_4$			
Carbonodithioic acid, O,O'-1,2-cyclohexanediyl S,S'-dimethyl ester, trans	EtOH	230(3.51),279(4.00)	118-0286-78
$C_{10}H_{16}O_3$			
2-Butenoic acid, 2-acetyl-, 1,1-dimethylethyl ester	EtOH	210(3.84)(end absorption)	44-0700-78
isomer	EtOH	217(3.92)	44-0700-78
$C_{10}H_{16}O_4$			
2-Cyclobuten-1-one, 3,4,4-triethoxy-	EtOH	248(4.14),289(1.90)	33-1784-78
$C_{10}H_{16}S$			
Bicyclo[4.2.0]octane-7-thione, 8,8-dimethyl-	C_6H_{12}	213(3.77),232(3.75), 491(1.02)	54-0121-78
	MeCN	213(3.83),232(3.78), 478(1.02)	54-0121-78
$C_{10}H_{16}SSi$			
Silane, trimethyl[(phenylthio)methyl]-	heptane	255(4.07),275(3.30)	65-2027-78
$C_{10}H_{16}S_2$			
1,6-Cyclodecanedithione	$CHCl_3$	310(2.81)	42-0146-78
	CCl_4	265(3.43)	42-0146-78
	THF	275(2.95)	42-0146-78
$C_{10}H_{17}NO$			
Cyclohexanamine, N-(2-methoxy-1-propen-1-ylidene)-	C_6H_{12}	207(3.57),245(3.23), 360(1.48)	24-1223-78
$C_{10}H_{17}NO_2$			
Oxazole, 2-hexyl-5-methoxy-	EtOH	225(3.69)	104-1353-78
$C_{10}H_{17}NO_4$			
Propanedioic acid, [(diethylamino)methylene]-, dimethyl ester	isooctane	227(3.56),282(4.27)	78-2321-78
$C_{10}H_{17}N_2S$			
1H-Imidazolium, 2-(cyclopentylthio)-1,3-dimethyl-, iodide	EtOH	222(4.237),264(3.628)	28-0201-78B
	CH_2Cl_2	240(4.297),265(3.703)	28-0201-78B
$C_{10}H_{17}N_3$			
Pyrazinamine, 3,6-diethyl-N,N-dimethyl-	EtOH	258(3.87),285(3.51),	94-1322-78

Compound	Solvent	λ_{max}(log ϵ)	Ref.
(cont.)		328(3.73)	94-1322-78
$C_{10}H_{17}N_3O_3S$ 4(1H)-Pyrimidinone, 6-amino-5-(2,2-di- ethoxyethyl)-2,3-dihydro-2-thioxo-	MeOH	242(3.85),288(4.28)	24-2925-78
$C_{10}H_{17}N_5$ 1H-Imidazole-4-carbonitrile, 5-amino- 1-[2-(diethylamino)ethyl]-	EtOH	244(4.08)	2-0786-78
$C_{10}H_{17}N_5O$ 4(1H)-Pteridinone, 2,3,5,6,7,8-hexahy- dro-2-imino-1,3,5,6-tetramethyl-, dihydrochloride 4(1H)-Pteridinone, 2,3,5,6,7,8-hexahy- dro-1,3,6-trimethyl-2-(methylimino)-, dihydrochloride	pH -1 pH 3.4-12.7 pH -2 pH 3.5-13	264(4.18) 220(4.43),264(4.04) 265(4.26) 225(4.35),268(4.07)	12-1081-78 12-1081-78 12-1081-78 12-1081-78
$C_{10}H_{17}S_2$ 1,2-Dithiol-1-ium, 4-(1-methylethyl)- 3-(2-methylpropyl)-, perchlorate	MeOH	211(3.31),258(3.62), 306(3.83)	39-0195-78C
$C_{10}H_{18}GeO_2$ Germane, triethyl[(1-oxo-2-butynyl)- oxy]-	heptane	210(3.23)	65-1244-78
$C_{10}H_{18}N_2$ Diazene, bis(1,1-dimethyl-2-propenyl)- 3H-Pyrazole, 4,5-dihydro-3,3,5,5-tetra- methyl-4-(1-methylethylidene)-	hexane hexane H_2O MeOH MeCN	366(1.47) 327(2.20) 317(--) 322(--) 327(--)	35-7009-78 35-5122-78 35-5122-78 35-5122-78 35-5122-78
$C_{10}H_{18}N_2O_2$ 6-Oxa-3-azabicyclo[3.2.2]nonane, 5,7,7-trimethyl-N-nitroso-	EtOH	238(3.86)	39-0804-78C
$C_{10}H_{18}N_2O_5$ β-Alanine, N-(1-acetoxybutyl)-N-nitro- so-, methyl ester	EtOH	231(3.82),369(1.90)	94-3914-78
$C_{10}H_{18}N_3O_4P$ 1,3,2-Diazaphosphorin-4(1H)-one, 2-eth- oxy-2,3-dihydro-6-methyl-5-morpho- lino-, 2-oxide	EtOH	263(3.90)	103-0784-78
$C_{10}H_{18}N_4$ 1,2,3,4-Tetraazatricyclo[8.4.0.0^{4,9}]- tetradecane	MeOH	227(3.36),280(3.74)	33-1622-78
$C_{10}H_{18}N_4O$ 1,2,3,4-Tetraazatricyclo[8.4.0.0^{4,9}]- tetradecane 2-N-oxide	MeOH	294(4.06)	33-1622-78
$C_{10}H_{18}O$ 2,4-Hexadien-1-ol, 5-methyl-4-(1-meth- ylethyl)-, (E)-	pentane	239(3.94)	142-0083-78C
$C_{10}H_{18}OS_2$ 1,5-Dithiaspiro[5.6]dodecan-7-ol	n.s.g.	222s(--),248(3.00)	33-3087-78

$C_{10}H_{18}O_2Si-C_{10}H_{21}NO$

Compound	Solvent	$\lambda_{max}(\log \epsilon)$	Ref.
$C_{10}H_{18}O_2Si$ 2-Butynoic acid, triethylsilyl ester	heptane	209(3.39),215(3.35)	65-1244-78
$C_{10}H_{18}O_2Sn$ Stannane, triethyl[(1-oxo-2-butynyl)-oxy]-	MeCN	209(3.63)	65-1244-78
$C_{10}H_{18}O_3$ Heptanoic acid, 2,2-dimethyl-6-oxo-, methyl ester	EtOH	281(1.36)	33-2681-78
$C_{10}H_{19}NO_2$ L-Valine, N-(2-methylpropylidene)-, methyl ester, (E)-	hexane MeOH	216(2.83) 201(2.80)	94-0466-78 94-0466-78
$C_{10}H_{20}N_2$ 1,2-Diazocine, 3,4,5,6,7,8-hexahydro-3,3,8,8-tetramethyl-	PhEt	370(1.23)	24-0596-78
$C_{10}H_{20}N_2O_2$ 1-Hexanamine, N-(2-nitro-1-butenyl)-	EtOH	244(3.00),370(3.85)	44-0497-78
$C_{10}H_{20}N_2O_3$ 1-Butanol, 1-(butylnitrosoamino)-, acetate	n.s.g.	232(3.83),369(1.86)	94-3905-78
$C_{10}H_{20}N_2S_2$ Piperidine, 1,1'-dithiobis-	n.s.g.	253(4.0)	23-1080-78
$C_{10}H_{20}N_4$ Piperidine, 1,1'-azobis-	dioxan	284(3.97)	60-1909-78
$C_{10}H_{20}N_6$ Dipyrazino[1,2-d:2',1'-f][1,2,3,4]tetra-zine, 1,2,3,4,9,10,11,12,12a,12b-deca-hydro-2,11-dimethyl-, trans	MeOH	216(3.65),274(3.52)	33-1622-78
$C_{10}H_{20}N_6O$ Dipyrazino[1,2-d:2',1'-f][1,2,3,4]tetra-zine, 1,2,3,4,9,10,11,12,12a,12b-deca-hydro-2,11-dimethyl-, 6-oxide, trans	MeOH	288(4.0)	33-1622-78
$C_{10}H_{20}O_6$ 2H-Pyran, tetrahydro-2,3,3,5,6-penta-methoxy-	EtOH	210(2.85)(end absorption)	12-0627-78
$C_{10}H_{20}Si_2$ Silane, 1-buten-3-yne-1,4-diylbis[tri-methyl-	EtOH	286(3.00)	70-2160-78
$C_{10}H_{21}FOSi$ Silane, acetylbis(1,1-dimethylethyl)-fluoro-	C_6H_{12}	361(2.17),374(2.32), 388(2.20)	112-0751-78
$C_{10}H_{21}NO$ Acetamide, N,N-dibutyl- 2-Piperidinemethanol, α-butyl- 2-Piperidinemethanol, α-(2-methylpro-pyl)-	H_2O n.s.g. n.s.g.	223(4.2) 224(3.65),262(3.38) 227(3.66),266(3.38)	104-0221-78 42-0916-78 42-0916-78

Compound	Solvent	$\lambda_{max}(\log \epsilon)$	Ref.
$C_{10}H_{21}NOS$ Thionitrous acid, S-decyl ester	hexane-1% CCl$_4$ at 0°	339(c.2.85),515(1.08), 547(1.30)	39-0913-78C
$C_{10}H_{21}N_3O_2$ 1,1-Ethenediamine, N,N'-bis(1,1-dimethylethyl)-2-nitro-	MeOH	330(4.03)	20-0693-78
$C_{10}H_{22}OS$ Pentane, 1,1'-sulfinylbis-	hexane H$_2$O EtOH	207(3.53),225s(3.08) 196(3.11),213(2.95) 202(3.15),214s(2.95)	70-0901-78 70-0901-78 70-0901-78
$C_{10}H_{22}OSi$ Silane, acetylbis(1,1-dimethylethyl)-	C$_6$H$_{12}$	361(2.12),374(2.32), 388(2.20)	112-0751-78
$C_{10}H_{22}O_2Zn$ Zinc, bis(4-methoxybutyl)-, 2,2'-bipyridine complex	benzene	418(2.32)	101-0245-78J
$C_{10}H_{24}N_2Zn$ Zinc, bis[3-(dimethylamino)propyl]-, 2,2'-bipyridine complex	benzene	530(2.40)	101-0245-78J

Compound	Solvent	$\lambda_{max}(\log \epsilon)$	Ref.
$C_{11}H_2N_4O_3$			
Propanedinitrile, 2,2'-(4,5-dihydroxy-2-oxo-4-cyclopentene-1,3-diylidene)-bis-	H_2O	228s(4.03),238s(4.12), 254(4.25),281s(3.91), 314(3.77),447s(4.11), 534(4.95)	35-2586-78
dipotassium salt	H_2O	227s(3.99),238s(4.16), 252(4.24),282s(3.90), 313(3.76),448s(4.12), 534(4.94)	35-2586-78
$C_{11}H_5BrCl_3NO_6S$			
Furan, 2-[2-(5-bromo-2-furanyl)-1-[(trichloromethyl)sulfonyl]ethenyl]-5-nitro-, (E)-	EtOH	210(4.06),309(4.43), 367(4.06)	73-1618-78
$C_{11}H_5BrOS$			
3H-Naphtho[1,8-bc]thiophen-3-one, 4-bromo-	C_6H_{12}	214s(--),257(4.216), 267(4.293),316(3.725), 356s(--),378(3.929), 390s(--)	5-0627-78
$C_{11}H_5Cl_3INO_6S$			
Furan, 2-[2-(5-iodo-2-furanyl)-1-(trichloromethyl)sulfonyl]ethenyl]-5-nitro-, (E)-	EtOH	210(4.11),222(4.09), 255s(3.90),318(4.39), 374(4.15)	73-1618-78
$C_{11}H_5Cl_3N_2O_8S$			
Furan, 2,2'-[1-[(trichloromethyl)sulfonyl]-1,2-ethenediyl]bis[5-nitro-	EtOH	210(4.12),244(4.06), 309(4.37),386(3.88)	73-1618-78
$C_{11}H_6Cl_3NO_5S_2$			
Furan, 2-nitro-5-[2-(2-thienyl)-1-[(trichloromethyl)sulfonyl]ethenyl]-, (E)-	EtOH	208(4.00),305(4.27), 350(4.02)	73-1618-78
$C_{11}H_6Cl_3NO_6S$			
Furan, 2-[2-(2-furanyl)-1-[(trichloromethyl)sulfonyl]ethenyl]-5-nitro-	EtOH	210(3.83),223s(3.78), 244(3.72),303(4.29), 356(3.88)	73-1618-78
$C_{11}H_6INO_4$			
1-Naphthalenecarboxylic acid, 8-iodo-3-nitro-	MeOH	226(4.43),325(3.90)	95-0358-78
$C_{11}H_6N_2O$			
4-Quinolinecarbonitrile, 2-formyl-	DMF	322(3.61)	97-0138-78
$C_{11}H_6N_2O_3$			
Benz[cd]indol-2(1H)-one, 4-nitro-	MeOH	231(5.58),300(4.90)	95-0358-78
$C_{11}H_6N_2O_3S$			
1-Azaphenoxathiin, 7-nitro-	EtOH	214(3.88),224(3.80), 246(3.62),261(3.57), 311(3.38),370(3.32)	4-0609-78
1-Azaphenoxathiin, 9-nitro-	EtOH	224(4.24),291(3.83), 398(3.34)	4-0609-78
$C_{11}H_6OS$			
3H-Naphtho[1,8-bc]thiophen-3-one	MeOH	212s(--),259(4.165), 308(3.720),376(3.897)	5-0627-78

Compound	Solvent	$\lambda_{max}(\log \epsilon)$	Ref.
$C_{11}H_6OS_2$			
2-Cyclopropen-1-one, 2,3-di-2-thienyl-	EtOH	252(4.34),271(4.26), 325s(4.44),338(4.58), 356(4.51)	88-0803-78
$C_{11}H_6O_3$			
5H-Benzocycloheptene-5,6,9-trione	CH_2Cl_2	249(4.16),318(3.47), 450s(1.7)	88-2289-78
Furo[2,3-f]coumarin	EtOH	214(4.253),220(4.231), 243(4.339),296(4.077), 345s(--)	111-0213-78
Furo[2,3-h]coumarin	EtOH	205(4.388),215s(--), 247(4.447),299(4.088), 330s(--)	111-0213-78
Furo[3,2-f]coumarin	EtOH	211(4.606),245s(--), 314(4.208)	111-0213-78
Furo[3,2-h]coumarin	EtOH	210(4.449),215(4.450), 240(4.343),301(4.097), 335(4.070)	111-0213-78
Psoralen	C_6H_{12}	241(4.53),248(4.58), 262(3.97),275s(3.97), 285(4.10),296(4.05), 332(3.88)	94-3433-78
	benzene	287(3.97),298(3.93), 331(3.78)	94-3433-78
	MeOH	242s(4.37),246(4.39), 290(3.99),326(3.76)	94-3433-78
	dioxan	242(4.56),248(4.58), 262(3.94),286(4.10), 331(3.89)	94-3433-78
	MeCN	242s(4.37),247(4.39), 288(3.97),327(3.74)	94-3433-78
	CCl_4	287(3.96),297(3.91), 331(3.75)	94-3433-78
	$CF_3CHOHCF_3$	200(4.51),210s(4.42), 242(4.34),304(4.10)	94-3433-78
$C_{11}H_7BrN_2O$			
1H-Pyrido[3,2-b][1,4]benzoxazine, 3-bromo-	MeOH	232(4.71),349(4.10)	94-1375-78
$C_{11}H_7BrO_3$			
1,4-Naphthalenedione, 2-bromo-6-meth-oxy-	EtOH	271(4.28),289(4.17), 337(3.32),400(3.26)	12-1335-78
1,4-Naphthalenedione, 2-bromo-7-meth-oxy-	EtOH	267(4.27),289(4.04), 345s(3.15),400(3.31)	12-1335-78
$C_{11}H_7ClN_2O$			
1H-Pyrido[3,2-b][1,4]benzoxazine, 3-chloro-	MeOH	232(4.66),350(4.05)	94-1375-78
$C_{11}H_7ClN_2O_4$			
1,4-Naphthalenedione, 2-chloro-3-(meth-ylamino)-5-nitro-	C_6H_{12} EtOH	481(3.51) 490(--)	39-1083-78C 39-1083-78C
1,4-Naphthalenedione, 3-chloro-2-(meth-ylamino)-5-nitro-	C_6H_{12} EtOH	482(3.56) 490(--)	39-1083-78C 39-1083-78C
$C_{11}H_7ClO_3$			
1,4-Naphthalenedione, 2-chloro-3-meth-oxy-	C_6H_{12} EtOH	334(3.49) 338(--)	39-1083-78C 39-1083-78C

Compound	Solvent	λ_{max}(log ϵ)	Ref.
$C_{11}H_7ClO_4$ 1,4-Naphthalenedione, 2-chloro-8-hy-droxy-6-methoxy-	EtOH	220(4.59),273(4.15), 283s(4.03),445(3.81)	44-1435-78
$C_{11}H_7Cl_2NO_3$ 1,4-Naphthalenedione, 5-amino-2,3-di-chloro-8-methoxy-	C_6H_{12} EtOH	540(3.80) 582(--)	150-2319-78 150-2319-78
$C_{11}H_7Cl_3N_2O_5S$ 1H-Pyrrole, 2-[2-(5-nitro-2-furanyl)-2-[(trichloromethyl)sulfonyl]eth-enyl]-	EtOH	207(4.00),228(3.90), 312(4.27),345(4.32), 400s(4.04)	73-1618-78
$C_{11}H_7F_9N_6OS$ Pyrazole, 1-(4-amino-5-thioxo-1,2,4-triazol-3-yl)-5-[(heptafluoro-2-fur-anyl)difluoromethyl]-3-methyl-	EtOH	256(4.28)	103-0458-78
$C_{11}H_7NOS$ Furo[3,2-d]isothiazole, 5-phenyl-	n.s.g.	218(4.31),287(4.40)	103-0465-78
$C_{11}H_7NOS_2$ 4H-1-Benzothiopyran-3-carbonitrile, 2-(methylthio)-4-oxo-	EtOH	245(3.75),284(4.01), 317s(3.62),330(3.71)	78-0725-78
$C_{11}H_7NO_3$ 5-Azulenecarboxaldehyde, 7-nitro-	MeOH	294(4.52),414(4.04), 560(3.08)	118-0592-78
2-Propenal, 3-(1,3-dihydro-1,3-dioxo-2H-isoindol-2-yl)-	THF	240(4.28),260(4.64), 270(4.46)	40-0404-78
$C_{11}H_7NO_3S$ Methanone, (4-nitrophenyl)-2-thienyl-	C_6H_{12}	252(4.33),294(3.48)	18-2718-78
$C_{11}H_7NS_2$ Benzothiazole, 2-(3-thienyl)-	MeOH	215s(4.255),225(4.301), 246s(3.931),256(3.854), 299(4.212)	39-1198-78C
Thieno[3,2-d]isothiazole, 5-phenyl-	n.s.g.	245(4.35),290(4.22)	103-0465-78
$C_{11}H_7N_3O$ Naphtho[1,2-e][1,2,4]triazin-2(3H)-one	MeOH	411s(3.43),422(3.59), 448s(3.29)	39-0789-78C
	dioxan	415s(3.68),430(3.49), 452s(3.32)	39-0789-78C
$C_{11}H_7N_3O_2$ Naphtho[1,2-e][1,2,4]triazin-2(3H)-one, 1-oxide	MeOH	373s(3.55),387(3.60), 415(3.60),435s(3.55)	39-0789-78C
	dioxan	372s(3.59),389(3.64), 405s(3.61),435s(3.48)	39-0789-78C
$C_{11}H_7N_5O_3$ 1H-Purine, 6-(4-nitrophenoxy)-	EtOH	271(4.18)	44-2587-78
$C_{11}H_8BrNO_2S$ Acetamide, N-(3-bromo-4-oxo-4H-1-benzo-thiopyran-2-yl)-	EtOH	230(4.18),263(4.37), 271(4.44),306(3.64), 347(4.12)	103-0141-78
4-Oxazolidinone, 5-[(4-bromophenyl)-methylene]-3-methyl-2-thioxo-	n.s.g.	256(4.17),292(4.09), 343(4.51)	104-1225-78

Compound	Solvent	$\lambda_{max}(\log \epsilon)$	Ref.
4(5H)-Oxazolone, 5-[(4-bromophenyl)-methylene]-2-(methylthio)-	n.s.g.	242(4.28),328(4.48)	104-1225-78
$C_{11}H_8Br_2N_2O$			
Pyrimidine, 5-bromo-4-(bromomethyl)-6-phenyl-, 3-oxide	EtOH	223(4.10),242(4.09), 298(4.20),346(3.72)	103-1132-78
Pyrimidine, 2-(dibromomethyl)-4-phenyl-, 1-oxide	EtOH	207(4.17),220(4.23), 313(4.35)	103-1132-78
Pyrimidine, 4-(dibromomethyl)-6-phenyl-, 3-oxide	EtOH	207(4.05),220(4.12), 308(4.27)	103-1132-78
$C_{11}H_8Br_5NO_2$			
Bicyclo[4.2.0]oct-2-ene-7,8-dicarbox-imide, 4,4,5,7,8-pentabromo-, anti-anti	MeOH	none	24-2677-78
$C_{11}H_8Cl$			
Benzocycloheptenylium, 7-chloro-, hexachloroantimonate	H_2SO_4	240(4.36),295(4.99), 350(3.83),364(4.04), 428(3.42)	78-0533-78
$C_{11}H_8ClNO_2$			
1,4-Naphthalenedione, 2-chloro-3-(meth-ylamino)-	C_6H_{12} EtOH	454(3.04) 477(--)	39-1083-78C 39-1083-78C
$C_{11}H_8ClNS$			
Benzenamine, 2-chloro-N-(2-thienylmeth-ylene)-	C_6H_{12}	266(4.12),288(4.14), 324(3.88)	18-2718-78
Benzenamine, 4-chloro-N-(2-thienylmeth-ylene)-	C_6H_{12}	226(4.00),264(4.05), 294(4.11),330(4.10)	18-2718-78
$C_{11}H_8ClN_5O$			
6H-Imidazo[4,5-e]-1,2,4-triazin-6-one, 3-(4-chlorophenyl)-1,5-dihydro-5-methyl-	EtOH	247(4.02),278(3.84), 289s(3.83),308(3.78)	94-3154-78
$C_{11}H_8F_3NO_2$			
Acetamide, N-(1-benzoylethenyl)-2,2,2-trifluoro-	EtOH	252(3.88)	35-4481-78
Acetic acid, trifluoro-, (3-phenyl-2H-azirin-2-yl)methyl ester	C_6H_{12}	243(4.15)	35-4481-78
4-Oxazolemethanol, 5-phenyl-2-(tri-fluoromethyl)-	EtOH	255(4.25)	35-4481-78
$C_{11}H_8FeN_2O_3$			
Iron, tricarbonyl[(5,6,7,8-η)-1,4-di-hydrocycloheptapyrazole]-	EtOH	220s(4.34),298s(3.68)	142-0293-78C
$C_{11}H_8N_2O$			
1H-Pyrido[3,2-b][1,4]benzoxazine	MeOH	230(4.29),339(3.69)	94-1375-78
5H-Pyrido[2,3-b][1,4]benzoxazine	EtOH	230(4.41),323(3.92)	94-1375-78
5H-Pyrido[3,4-b][1,4]benzoxazine	EtOH	241(4.54),319(3.70)	94-1375-78
10H-Pyrido[4,3-b][1,4]benzoxazine	EtOH	238(4.57),312(3.88)	94-1375-78
$C_{11}H_8N_2O_2$			
4-Quinolinecarboxamide, 2-formyl-	DMF	302(3.72)	97-0138-78
$C_{11}H_8N_2O_2S$			
Benzenamine, 2-nitro-N-(2-thienyl-methylene)-	C_6H_{12}	228(4.21),252(4.21), 270(4.09),366(3.60)	18-2718-78

Compound	Solvent	λ_{max}(log ϵ)	Ref.
Benzenamine, 4-nitro-N-(2-thienylmethylene)-	C_6H_{12}	256(3.83),282(3.89), 320(4.03)	18-2718-78
$C_{11}H_8N_2O_3$			
2H-Pyrano[3,2-c]pyridine-3-carbonitrile, 5,6-dihydro-6,7-dimethyl-2,5-dioxo-	MeOH	381(4.24)	49-1075-78
$C_{11}H_8N_2O_4S$			
Benzoic acid, 2-[(2-nitro-3-thienyl)- amino]-	MeOH	214(4.25),256(4.15), 311(3.95),422(4.25)	150-5101-78
4-Oxazolidinone, 3-methyl-5-[(4-nitrophenyl)methylene]-2-thioxo-	n.s.g.	233(4.19),295(4.28), 348(4.56)	104-1225-78
4(5H)-Oxazolone, 2-(methylthio)-5-[(4-nitrophenyl)methylene]-	n.s.g.	259(4.17),328(4.30)	104-1225-78
$C_{11}H_8N_2O_6$			
Furan, 2-[2-(5-methyl-2-furanyl)- 1-nitroethenyl]-5-nitro-	dioxan	317(4.13),381(4.20)	73-3252-78
$C_{11}H_8N_4O$			
1H-Purine, 6-phenoxy-	EtOH	254(4.23)	44-2587-78
$C_{11}H_8N_4O_2$			
Benzo[g]pteridine-2,4(3H,10H)-dione, 10-methyl-	H_2O	215(4.45),266(4.60), 330(3.87),435(4.00)	150-1325-78
$C_{11}H_8N_4O_2S$			
1,2,3-Thiadiazolo[4,5-d]pyrimidine- 5,7(4H,6H)-dione, 6-(phenylmethyl)-	EtOH	230s(3.77),260(3.53), 398(3.80)	44-1677-78
1,2,3-Thiadiazolo[4,5-d]pyrimidinium, 4,5,6,7-tetrahydro-6-methyl-5,7-di- oxo-2-phenyl-, hydroxide, inner salt	EtOH	225(4.39),282(4.42)	44-1677-78
$C_{11}H_8O$			
2H-Naphtho[1,8-bc]furan	EtOH	202(4.15),251(3.89), 312(3.94),340(3.62)	18-2068-78
$C_{11}H_8OS$			
Methanone, phenyl-2-thienyl-	C_6H_{12}	252(4.15),279(4.10)	18-2718-78
$C_{11}H_8O_3$			
5H-Benzocyclohepten-5-one, 6,9-di- hydroxy-	n.s.g.	242(4.34),261s(3.96), 287s(3.51),300s(3.35), 362(3.41),408(3.48), 486(2.79)	88-2289-78
1,4-Naphthalenedione, 2-hydroxy- 3-methyl-	EtOH	206(4.28),250(4.35), 274(4.38),328(3.45), 402(3.38),491(3.26)	104-1962-78
1,4-Naphthalenedione, 5-hydroxy- 7-methyl-	EtOH	215(4.54),253(4.13), 428(3.61)	98-0869-78
1,4-Naphthalenedione, 2-methoxy-	C_6H_{12} EtOH	328(3.53) 333(--)	39-1083-78C 39-1083-78C
1,4-Naphthalenedione, 5-methoxy-	C_6H_{12} EtOH	387(3.56) 400(--)	150-2319-78 150-2319-78
$C_{11}H_8O_3S$			
Spiro[1H-indene-1,2'-[1,3]oxathiolane]- 2,3-dione	MeCN	215(4.08),236s(3.87), 279(3.87),307s(2.61), 517(1.86)	24-3058-78

Compound	Solvent	$\lambda_{max}(\log \epsilon)$	Ref.
$C_{11}H_8O_4$			
2H-1-Benzopyran-2-one, 3-acetyl-7-hy-	H_2O	361(4.36)	51-0652-78
droxy-	EtOH	364(4.38)	51-0652-78
anion	H_2O	414(4.64)	51-0652-78
	EtOH	431(4.66)	51-0652-78
Spiro[1,3-dioxolane-2,2'-[2H]indene]-	MeOH	231(4.57),249s(4.07),	24-3058-78
1',3'-dione		283(3.02),291(2.99),	
		305s(2.40),365(1.76)	
$C_{11}H_8S_2$			
1-Naphthalenecarbodithioic acid,	6:4 C_6H_{12}-	325(3.90),347(4.07),	64-0976-78B
sodium salt	ether	475(2.20)	
$C_{11}H_9AsO$			
4-Arseninol, 2-phenyl-	EtOH	262(4.20),310(3.85)	88-1471-78
$C_{11}H_9BrN_2O$			
Pyrimidine, 2-(bromomethyl)-4-phenyl-,	EtOH	207(4.06),220(4.13),	103-1132-78
1-oxide		311(4.28)	
Pyrimidine, 4-(bromomethyl)-6-phenyl-,	EtOH	217(4.18),305(4.27),	103-1132-78
1-oxide		341s(4.00)	
Pyrimidine, 5-bromo-4-methyl-6-phenyl-,	EtOH	208(4.07),229(4.13)	103-1132-78
1-oxide		294(4.16)	
$C_{11}H_9BrN_2O_3$			
Pyridinium, 1-[2-(5-nitro-2-furanyl)-	EtOH	225(4.10),261(3.80),	103-0250-78
ethenyl]-, bromide		345(4.10)	
$C_{11}H_9Br_4NO_2$			
Bicyclo[4.2.0]oct-2-ene-7,8-dicarbox-	MeOH	none	24-2677-78
imide, 4,5,7,8-tetrabromo-N-methyl-,			
anti-anti			
$C_{11}H_9Br_4N_3$			
3,5-Pyridinedicarbonitrile, 2,6-bis(di-	EtOH	208(4.10),267(3.91),	103-1236-78
bromomethyl)-1,4-dihydro-4,4-dimethyl-		361(3.51)	
$C_{11}H_9Cl$			
5H-Benzocycloheptene, 7-chloro-	MeOH	280(3.83)	78-0533-78
$C_{11}H_9ClN_2$			
1H-Pyrazole, 5-chloro-1-ethenyl-	MeOH	271(4.31),276(4.29),	95-0095-78
3-phenyl-		282(4.24)	
$C_{11}H_9ClN_2O$			
Pyridazine, 3-(4-chlorophenyl)-6-meth-	MeOH	256(--)	2-1000-78
oxy-			
$C_{11}H_9ClN_2O_2$			
1H-Pyrrole-2-carboxylic acid, 4-(2-am-	MeOH	221(4.23),302(3.48)	88-3097-78
ino-3-chlorophenyl)-	MeOH-HCl	217(4.02),270(3.86)	88-3097-78
	MeOH-NaOH	217(4.33),265(3.86),	88-3097-78
		304(3.68)	
Sydnone, 4-[2-(4-chlorophenyl)ethenyl]-	benzene	355(4.47)	48-0071-78
3-methyl-, (E)-	EtOH	349(4.48)	48-0071-78
(Z)-	benzene	346(4.14)	48-0071-78
	EtOH	334(4.18)	48-0071-78
$C_{11}H_9FO_3$			
Benzeneacetic acid, 4-fluoro-α-(2-oxo-	EtOH	288(3.86)	44-4115-78
ethylidene)-, methyl ester			

Compound	Solvent	λ_{max}(log ϵ)	Ref.
$C_{11}H_9N$			
Pyridine, 3-phenyl-	MeOH	245(4.15),275s(--)	95-0910-78
	MeOH-HCl	232(4.11),256(4.05), 290s(--)	95-0910-78
$C_{11}H_9NO$			
2H-Azirine, 3-[2-(2-propynyloxy)phenyl]-	MeOH	245(4.36),302(3.97)	44-0066-78
1,4-Ethenoisoquinolin-3(2H)-one, 1,4-dihydro-	EtOH	254(3.88)	78-2617-78
Phenol, 4-(3-pyridinyl)-	MeOH	270(4.15)	95-0914-78
	MeOH-HCl	246(4.02),289(4.10)	95-0914-78
	MeOH-NaOH	304(4.22)	95-0914-78
Pyridinium, 3-hydroxy-1-phenyl-, hydroxide, inner salt	EtOH-NaOH	222(4.10),243(4.03), 341(3.55)	1-0068-78
2(1H)-Pyridinone, 1-phenyl-	EtOH	221(3.46),231(3.44), 310(3.60)	1-0068-78
$C_{11}H_9NOS$			
2(3H)-Furanthione, 3-(aminomethylene)-5-phenyl-	n.s.g.	228(3.48),268(3.80), 315(3.59),380(3.57)	103-0465-78
$C_{11}H_9NO_2$			
1,4-Naphthalenedione, 2-(methylamino)-	C_6H_{12}	423(3.45)	39-1083-78C
	EtOH	452(--)	39-1083-78C
1,4-Naphthalenedione, 5-(methylamino)-	C_6H_{12}	529(3.76)	150-2319-78
	EtOH	552(--)	150-2319-78
2-Quinolinecarboxaldehyde, 4-methoxy-	DMF	336(3.41)	97-0138-78
3-Quinolinecarboxaldehyde, 2-methyl-, 1-oxide	EtOH	240(4.66),320(3.86)	95-0802-78
$C_{11}H_9NO_2S$			
4-Oxazolidinone, 3-methyl-5-(phenylmethylene)-2-thioxo-	n.s.g.	239(4.19),290(4.15), 340(4.45)	104-1225-78
4(5H)-Oxazolone, 2-(methylthio)-5-(phenylmethylene)-	n.s.g.	230(4.32),289s(4.64), 326(4.70)	104-1225-78
Phenol, 3-methyl-4-nitroso-5-(2-thienyl)-	DMF	695(1.36)	104-2189-78
Pyridine, 2-(phenylsulfonyl)-	EtOH	229(4.01),257s(3.65), 260(3.66)	95-0095-78
Spiro[2H-indene-2,2'-[1,3]thiazolidine]-1,3-dione	MeOH	227(4.65),247s(4.07), 280(3.18),286s(3.17), 302s(3.00),353(2.55)	24-3058-78
$C_{11}H_9NO_3$			
1,4-Naphthalenedione, 5-amino-8-methoxy-	C_6H_{12}	512(3.60)	150-2319-78
	EtOH	566(--)	150-2319-78
1,4-Naphthalenedione, 5-hydroxy-8-(methylamino)-	C_6H_{12}	557(3.73),598(3.96), 650(3.94)	150-2319-78
	EtOH	597(--),646(--)	150-2319-78
$C_{11}H_9NO_3S$			
2H-Thiopyrano[5,6-b]pyridine-3-carboxylic acid, 2-oxo-, ethyl ester	EtOH	232(4.38),257(3.88), 292(3.60),342(3.76)	78-0989-78
$C_{11}H_9NO_4$			
5H-1,3-Dioxolo[4,5-f]indole-6,7-dione, 5-ethyl-	5% MeOH	204(4.09),276(4.27), 329(3.69),385(2.92)	4-0333-78
1H-Indole-3-acetic acid, 1-methoxy-α-oxo-	MeOH	212(4.14),253(3.69), 310(3.72)	39-1117-78C
3,5,8(2H)-Isoquinolinetrione, 7-methoxy-2-methyl-	EtOH	212(4.30),238(4.01), 312(3.94)	4-0569-78

Compound	Solvent	$\lambda_{max}(\log \epsilon)$	Ref.
$C_{11}H_9NO_5S$			
1,2-Benzisothiazol-3(2H)-one, 2-(tetra-hydro-5-oxo-2-furanyl)-, 1,1-dioxide	THF	214(4.19),278(2.99), 287(2.89)	40-0404-78
3(2H)-Benzofuranone, 6-hydroxy-2-[1-(methylthio)-2-nitroethylidene]-	EtOH	252(3.92),362(4.37)	142-0399-78A
$C_{11}H_9NS$			
Benzenamine, N-(2-thienylmethylene)-	C_6H_{12}	225(3.95),264(3.94), 288(4.00),320(3.93)	18-2718-78
Pyrrolo[2,1-b]benzothiazole, 1-methyl-	EtOH	220(3.941),251(4.053), 317(3.434)	39-1198-78C
Pyrrolo[2,1-b]benzothiazole, 3-methyl-	EtOH	220s(3.966),232(4.027), 249(3.982),309(3.430)	39-1198-78C
$C_{11}H_9NS_2$			
2(3H)-Thiophenethione, 3-(aminomethyl-ene)-5-phenyl-	n.s.g.	235(4.14),320(4.21), 409(4.03)	103-0465-78
$C_{11}H_9N_2O_3$			
Pyridinium, 1-[2-(5-nitro-2-furanyl)-ethenyl]-, bromide	EtOH	225(4.10),261(3.80), 345(4.10)	103-0250-78
$C_{11}H_9N_3$			
Cyclohepta[b]pyrrole-3-carbonitrile, 1,2-dihydro-2-imino-1-methyl-	EtOH	228(4.11),283(4.38), 435(4.05)	138-0677-78
1H-Perimidin-2-amine, hydrochloride	EtOH	240(4.3),325(3.9)	103-0694-78
Pyrazine, [2-(2-pyridinyl)ethenyl]-	heptane	190(4.30),220(3.85), 260(4.10),320(4.31)	61-0396-78
Pyrazine, [2-(3-pyridinyl)ethenyl]-	heptane	183(4.27),219(3.86), 318(4.17),316(4.28)	61-0396-78
Pyrazine, [2-(4-pyridinyl)ethenyl]-	heptane	189(4.30),215(4.04), 270(4.33),310(4.33)	61-0396-78
Pyridazine, 3-[2-(2-pyridinyl)ethenyl]-	heptane	193(4.32),219(3.90), 269(4.22),312(4.29)	61-0396-78
Pyridazine, 3-[2-(3-pyridinyl)ethenyl]-	heptane	191(4.25),218s(3.90), 303(4.26)	61-0396-78
Pyridazine, 3-[2-(4-pyridinyl)ethenyl]-	heptane	196(4.08),223(3.78), 287(4.23)	61-0396-78
Pyridazine, 4-[2-(2-pyridinyl)ethenyl]-	heptane	190(4.26),220(3.92), 274s(4.02),308(4.27)	61-0396-78
Pyridazine, 4-[2-(3-pyridinyl)ethenyl]-	heptane	190(4.17),216(3.93), 301(4.20)	61-0396-78
Pyridazine, 4-[2-(4-pyridinyl)ethenyl]-	heptane	195(4.35),224(3.86), 286(4.28)	61-0396-78
Pyrimidine, 2-[2-(2-pyridinyl)ethenyl]-	heptane	190(4.35),218s(3.93), 268s(4.04),305(4.44)	61-0396-78
Pyrimidine, 2-[2-(3-pyridinyl)ethenyl]-	heptane	186(4.27),208(3.89), 300(4.36)	61-0396-78
Pyrimidine, 2-[2-(4-pyridinyl)ethenyl]-	heptane	195(4.44),220(4.09), 286(4.46)	61-0396-78
Pyrimidine, 4-[2-(2-pyridinyl)ethenyl]-	heptane	191(4.40),217(4.08), 264(4.01),310(4.33)	61-0396-78
Pyrimidine, 4-[2-(3-pyridinyl)ethenyl]-	heptane	189(4.22),217(3.97), 270s(4.00),304(4.41)	61-0396-78
Pyrimidine, 4-[2-(4-pyridinyl)ethenyl]-	heptane	190(4.28),218(3.90), 299(4.38)	61-0396-78
Pyrimidine, 5-[2-(2-pyridinyl)ethenyl]-	heptane	182(4.26),217(3.86), 274s(4.00),302(4.26)	61-0396-78
Pyrimidine, 5-[2-(3-pyridinyl)ethenyl]-	heptane	191(4.26),218s(3.90), 278(4.26)	61-0396-78
Pyrimidine, 5-[2-(4-pyridinyl)ethenyl]-	heptane	193(4.18),222(3.76),	61-0396-78

Compound	Solvent	λ_{max}(log ϵ)	Ref.
(cont.) 1,3a,6a-Triazapentalene, 2-phenyl-	EtOH	294(4.26) 249(4.40),278(4.32), 329(3.75)	61-0396-78 88-1291-78
$C_{11}H_9N_3O$ 1,9,9b-Triazaphenalen-2(3H)-one, 8-methyl-, monohydrobromide	EtOH	234(4.30),288(4.29), 322(3.93),338(4.03), 490(3.40),527(3.52), 565(3.54),610(3.40)	95-0623-78
[1,2,3]Triazolo[5,1-c][1,4]benzoxaze- pine, 4,10-dihydro-10-methylene-	MeOH	290(3.85)	44-0066-78
$C_{11}H_9N_3OS$ 1H-Pyrido[1,2-c]pyrimidine-4-carbo- nitrile, 8-methyl-3-(methylthio)- 1-oxo-	EtOH	255(4.39),320(4.04), 330(4.09),396(4.13)	95-0623-78
$C_{11}H_9N_3O_2$ 2-Pyrimidinecarboxaldehyde, 4-phenyl-, oxime, 1-oxide	EtOH	204(4.05),256(4.26), 317(4.36)	103-1137-78
4-Pyrimidinecarboxaldehyde, 6-phenyl-, oxime, 3-oxide	EtOH	204(3.96),222(3.92), 290(4.32),363(3.66)	103-1137-78
$C_{11}H_9N_3O_2S$ 1H-Pyrimido[5,4-b][1,4]benzothiazine- 2,4(3H,10H)-dione, 10-methyl-	MeOH	248(4.26),281(3.95), 360s(3.25)	5-0193-78
	pH 13	255(4.33),280s(3.91), 320s(3.36)	5-0193-78
	6M NaOH	249(4.27),278s(3.85), 315s(3.40)	5-0193-78
	12M HCl	251(4.21),291(3.83), 374(3.53)	5-0193-78
$C_{11}H_9N_3O_4$ Sydnone, 3-methyl-4-[2-(4-nitrophenyl)- ethenyl]-, (E)-	benzene EtOH	393(4.49) 383(4.48)	48-0071-78 48-0071-78
(Z)-	benzene EtOH	393(4.19) 366(4.18)	48-0071-78 48-0071-78
$C_{11}H_9N_3S$ 6H-Pyrazolo[4,3-d]isothiazole, 4-methyl-6-phenyl-	n.s.g.	237(4.25),275(4.17)	103-0465-78
1,9,9b-Triazaphenalene, 2-(methyl- thio)-, monohydrobromide	EtOH	238(4.10),290(4.03), 320(4.06),354(3.84), 376(3.79),380(3.74), 418(3.74),540(3.14), 560(3.18),630(3.05)	95-0623-78
$C_{11}H_9N_3S_2$ Pyrido[3',2':4,5]thieno[3,2-d]pyrimi- dine-4(1H)-thione, 7,9-dimethyl-	EtOH	252(4.03),267(4.10), 270(4.09),298(4.10), 307(4.11),318(3.96), 372(4.01),405(3.54)	32-0057-78
$C_{11}H_9N_5O$ 6H-Imidazo[4,5-e]-1,2,4-triazin-6-one, 1,5-dihydro-5-methyl-3-phenyl-	EtOH	243(3.97),271s(3.74), 287s(3.69),306(3.68)	94-3154-78
$C_{11}H_9N_5O_2$ Pyridine, 2,6-bis(3-methyl-1,2,4-oxadi- azol-5-yl)-	MeOH	229(4.33),288(4.01)	4-1093-78

Compound	Solvent	$\lambda_{max}(\log \epsilon)$	Ref.
Pyridine, 2,6-bis(5-methyl-1,2,4-oxadi-azol-3-yl)-	MeOH	236(3.99),282(3.88)	4-1093-78
Pyridine, 2,6-bis(5-methyl-1,3,4-oxadi-azol-2-yl)-	MeOH	236(4.36),293(3.99)	4-1093-78
$C_{11}H_9N_5O_2S$			
Pyrimido[4,5-e]-1,2,4-triazine-6,8-di-one, 5,7-dihydro-5,7-dimethyl-3-(2-thienyl)-	EtOH	246(4.50),354(4.54)	94-0367-78
$C_{11}H_9N_5O_3$			
Pyrimido[4,5-e]-1,2,4-triazine-6,8-di-one, 3-(2-furanyl)-5,7-dihydro-5,7-dimethyl-	EtOH	244(4.41),351(4.42)	94-0367-78
$C_{11}H_{10}$			
4H-Cyclopentacyclooctene	EtOH	232(4.10),255s(3.86), 323(3.81),420s(2.43)	35-6535-78
$C_{11}H_{10}AsN$			
4-Arseninamine, 2-phenyl-	EtOH	273(4.43),334(4.17)	88-1175-78
$C_{11}H_{10}BTl$			
Boratabenzene, 1-phenyl-, thallium	MeCN	277(4.13)	101-0265-78J
$C_{11}H_{10}BrNS$			
Thiopyrano[2,3-b]indole, 7-bromo-2,3,4,9-tetrahydro-	EtOH	229(4.28),245(4.55), 280s(3.83),292(3.98), 313(4.09)	94-3695-78
$C_{11}H_{10}BrN_3O$			
1,9,9b-Triazaphenalen-2(3H)-one, 8-methyl-, monohydrobromide	EtOH	234(4.30),288(4.29), 322(3.93),338(4.03), 490(3.40),527(3.52), 565(3.54),610(3.40)	95-0623-78
$C_{11}H_{10}BrN_3O_3$			
1H-Pyrazole-5-carboxylic acid, 4-(5-bromo-2-furanyl)-5-cyano-4,5-di-hydro-4-methyl-, methyl ester	EtOH	231(4.01)	73-0870-78
3H-Pyrazole-3-carboxylic acid, 4-(5-bromo-2-furanyl)-3-cyano-4,5-di-hydro-4-methyl-, methyl ester	EtOH	288(4.12),388(3.42)	73-0870-78
$C_{11}H_{10}BrN_3S$			
1,9,9b-Triazaphenalene, 2-(methylthio)-, monohydrobromide	EtOH	238(4.10),290(4.03), 320(4.06),354(3.84), 376(3.79),380(3.74), 418(3.74),540(3.14), 560(3.18),630(3.05)	95-0623-78
$C_{11}H_{10}Br_2N_2Se$			
2H-Dipyrido[2,1-b:1',2'-e][1,3,5]selen-adiazinediium dibromide	aq HBr	233(4.22),277(4.04), 377(3.58)	64-0118-78B
$C_{11}H_{10}ClNO$			
2-Propyn-1-one, 1-(4-chlorophenyl)-3-(dimethylamino)-	MeCN	257(4.15),320(4.22)	33-1609-78
$C_{11}H_{10}ClNOS$			
4(1H)-Quinolinone, 3-[(chloromethyl)-	EtOH	254(4.20),298(3.78),	4-0115-78

Compound	Solvent	$\lambda_{max}(\log \epsilon)$	Ref.
thio]-1-methyl- (cont.)		330(4.00),338(4.02)	4-0115-78
$C_{11}H_{10}ClNO_3$			
Ethanone, 2-[3-(chloromethyl)-1,4,2-dioxazol-5-yl]-1-phenyl-	MeOH	248(4.07),270(3.93)	97-0057-78
$C_{11}H_{10}ClN_3$			
1H-Perimidin-2-amine, hydrochloride	EtOH	<u>240(4.3),325(3.9)</u>	103-0694-78
$C_{11}H_{10}ClN_3O$			
Imidazo[1,5,4-ef][1,5]benzodiazepin-6(7H)-one, 9-chloro-4,5-dihydro-2-methyl-	MeOH	269s(4.16),274(4.19), 292(3.93),302(3.89)	111-0053-78
Imidazo[4,5,1-jk][1,4]benzodiazepin-7(4H)-one, 9-chloro-5,6-dihydro-2-methyl-	MeOH	275(3.86),313(3.78)	111-0053-78
1,2,4-Triazin-5(2H)-one, 3-(4-chlorophenyl)-6-ethyl-	EtOH	251(4.26)	114-0091-78B
1,2,4-Triazin-6(1H)-one, 3-(4-chlorophenyl)-5-ethyl-	EtOH	267(4.26)	114-0091-78B
$C_{11}H_{10}Cl_2O_3S$			
4H-1-Benzopyran-4-one, 2,3-dichloro-2,3-dihydro-8-methoxy-3-(methylthio)-	n.s.g.	264(3.90),330(3.30)	4-0115-78
$C_{11}H_{10}NO_2$			
Pyridinium, 1-(2,3-dihydroxyphenyl)-, iodide	EtOH EtOH-NaOH	350(4.26) 450(3.11)	2-0165-78 2-0165-78
Pyridinium, 1-(2,5-dihydroxyphenyl)-, iodide	EtOH EtOH-NaOH	370(3.10) 485(3.11)	2-0165-78 2-0165-78
$C_{11}H_{10}N_2$			
1H-Perimidine, 2,3-dihydro-	MeOH	213(4.46),236(4.32), 347(3.70)	103-1156-78
2-Pyridinamine, 5-phenyl-	MeOH MeOH-HCl	268(4.33),315s(--) 262(4.32),319(3.79)	95-0910-78 95-0910-78
$C_{11}H_{10}N_2O$			
[1]Benzopyrano[4,3-c]pyrazole, 2,4-dihydro-2-methyl-	EtOH	254(3.97),263(3.95), 298(3.79)	44-1664-78
3-Buten-2-one, 4-(1H-indazol-3-yl)-	EtOH	207(4.24),253(3.89), 330(4.23)	103-0771-78
Ethanone, 1-(1-phenyl-1H-imidazol-2-yl)-	EtOH	279(3.93)	77-0124-78
1H-Pyrazol-5-ol, 1-ethenyl-3-phenyl-	MeOH	266.5(4.09)	95-0095-78
Pyrimidine, 2-methyl-4-phenyl-, 1-oxide	EtOH	211(4.25),294(4.16), 331(4.23)	103-1132-78
$C_{11}H_{10}N_2OS$			
2-Quinoxalinecarboxaldehyde, 3-(ethylthio)-	EtOH	362(3.88)	97-0177-78
$C_{11}H_{10}N_2O_2$			
1H-Imidazole-2-carboxaldehyde, 4-methyl-1-phenyl-, 3-oxide	MeOH	291(4.00),356(3.71)	77-0124-78
Pyridinium, 1,1'-methylenebis[3-hydroxy-, dihydroxide bis(inner salt)	EtOH	228(4.20),263(3.85), 342(3.71)	150-1182-78
Pyrimidine, 5-methoxy-4-phenyl-, 1-oxide	EtOH	207(4.23),226(4.18), 344(4.37)	103-1132-78
2-Quinoxalinecarboxaldehyde, 3-ethoxy-	EtOH	338(3.74)	97-0177-78
Sydnone, 3-methyl-4-(2-phenylethenyl)-,	benzene	352(4.42)	48-0071-78

Compound	Solvent	$\lambda_{max}(\log \epsilon)$	Ref.
(E)- (cont.)	EtOH	347(4.43)	48-0071-78
Sydnone, 3-methyl-4-(2-phenylethenyl)-,	benzene	342(4.06)	48-0071-78
(Z)-	EtOH	330(4.08)	48-0071-78
$C_{11}H_{10}N_2O_3$			
1H-Indole, 1-methoxy-3-(2-nitroethen-yl)-	MeOH	225(3.98),282(3.64), 394(4.02)	39-1117-78C
2-Propyn-1-one, 3-(dimethylamino)-1-(4-nitrophenyl)-	MeCN	266(4.27),346(4.07)	33-1609-78
Pyridine, 1-acetyl-4-[(acetyloxy)cyano-methylene]-1,4-dihydro-	EtOH	347(4.36)	4-0637-78
2-Pyridineacetonitrile, α-acetyl-3-(acetyloxy)-	MeOH	226(3.58),294(3.80), 374(3.61)	4-1411-78
	MeOH-NaOH	225s(3.59),242(3.63), 292(3.52),329(3.41), 384(3.43),396s(3.39)	4-1411-78
$C_{11}H_{10}N_2O_4S_2$			
Thiophene, 2,2'-(1-methylethylidene)-bis[5-nitro-	MeOH	328(4.29)	12-2463-78
$C_{11}H_{10}N_2Se$			
2H-Dipyrido[2,1-b:1',2'-e][1,3,5]selena-diazinediium dibromide	aq HBr	233(4.22),277(4.04), 377(3.58)	64-0118-78B
$C_{11}H_{10}N_4$			
Propanedinitrile, methyl[(3-methylphen-yl)azo]-	CHCl$_3$	290(4.09)	126-1803-78
Propanedinitrile, methyl[(4-methylphen-yl)azo]-	toluene	297(4.18)	126-1803-78
$C_{11}H_{10}N_4O$			
2H-Tetrazole, 2-methyl-5-[2-(2-propyn-yloxy)phenyl]-	EtOH	236(4.05),287(3.59)	44-1664-78
$C_{11}H_{10}N_4O_2$			
1H-Pyrazole-4-carboxylic acid, 3-(3-cy-ano-2-pyridinyl)-4,5-dihydro-, methyl ester	H$_2$O	303(4.02)	103-0049-78
$C_{11}H_{10}N_4O_3$			
Imidazo[1,5,4-ef][1,5]benzodiazepin-6(7H)-one, 4,5-dihydro-2-methyl-8-nitro-	MeOH	227s(4.11),353(4.22)	111-0053-78
Imidazo[4,5,1-jk][1,4]benzodiazepin-7(4H)-one, 5,6-dihydro-2-methyl-9-nitro-	MeOH	260(4.34),310(3.92)	111-0053-78
$C_{11}H_{10}N_4O_3S$			
Benzenesulfonic acid, 4-[(1,1-dicyano-ethyl)azo]-3-methyl-, sodium salt	H$_2$O	280(4.08)	126-2845-78
$C_{11}H_{10}N_4O_5$			
1H-Pyrazole-5-carboxylic acid, 5-cyano-4,5-dihydro-4-methyl-4-(5-methyl-2-furanyl)-, methyl ester	EtOH	221(3.85),227(3.88), 307(3.77)	73-0870-78
$C_{11}H_{10}N_4O_6$			
2H-Pyrrol-2-one, 1-[(2,4-dinitrophen-yl)amino]-1,5-dihydro-5-hydroxy-4-methyl-	n.s.g.	255(4.01),330(4.17)	39-0401-78C

Compound	Solvent	$\lambda_{max}(\log \epsilon)$	Ref.
2H-Pyrrol-2-one, 1-[(2,4-dinitrophenyl)-amino]-1,5-dihydro-5-hydroxy-5-methyl-	n.s.g.	260(3.88),328(4.16)	39-0401-78C
$C_{11}H_{10}N_4S$ 1H-Pyrido[1,2-c]pyrimidine-4-carbonitrile, 1-imino-8-methyl-3-(methylthio)-	EtOH	261(4.25),316(3.98), 377s(3.86),440(4.26)	95-0623-78
$C_{11}H_{10}O$ 5H-Benzocyclohepten-5-one, 8,9-dihydro-	C_6H_{12}	229(3.77),266(3.74), 293s(2.85)	35-5141-78
$C_{11}H_{10}ORu$ Ruthenocene, formyl-	EtOH	251(4.12),344(3.16)	18-0909-78
$C_{11}H_{10}O_2$ Bicyclo[4.2.2]deca-2,4,7,9-tetraene-2-carboxylic acid	C_6H_{12}	295(3.85)	5-2074-78
$C_{11}H_{10}O_2S_4$ 3H-1,2-Dithiole-3-thione, 5-(4-methylphenyl)-4-(methylsulfonyl)-	EtOH	236(4.41),283(4.04), 322(4.01),415(3.95)	39-1017-78C
$C_{11}H_{10}O_3$ 2H-1-Benzopyran-2-one, 4-methoxy-5-methyl-	n.s.g.	207(4.45),215(4.37), 275(4.07),283(4.05), 305(3.67),315(3.48)	78-1221-78
2H-1-Benzopyran-2-one, 4-methoxy-6-methyl-	n.s.g.	206(4.31),217(4.30), 267(3.91),278(3.87), 309(3.60),322(3.42)	78-1221-78
2H-1-Benzopyran-2-one, 4-methoxy-7-methyl-	n.s.g.	206(4.43),216(4.36), 267(3.94),277(4.03), 303(3.91),315(3.78)	78-1221-78
2H-1-Benzopyran-2-one, 4-methoxy-8-methyl-	n.s.g.	206(4.40),216(4.35), 268(4.06),278(4.03), 306(3.65),315(3.52)	78-1221-78
2H-1-Benzopyran-2-one, 5-methoxy-4-methyl-	n.s.g.	209(4.28),225(3.81), 245(3.70),296(4.01), 321(3.70)	78-1221-78
2H-1-Benzopyran-2-one, 6-methoxy-4-methyl-	n.s.g.	205(4.24),226(4.40), 274(4.02),340(3.64)	78-1221-78
2H-1-Benzopyran-2-one, 7-methoxy-4-methyl-	n.s.g.	206(4.31),219(4.24), 240(3.57),319(4.14), 336(3.87)	78-1221-78
2H-1-Benzopyran-2-one, 8-methoxy-4-methyl-	n.s.g.	207(4.28),251(4.03), 285(4.03)	78-1221-78
4H-1-Benzopyran-4-one, 2-ethoxy-	EtOH	220(4.27),267(4.03), 285(3.91)	23-1796-78
Cyclopenta[b]pyran-6(2H)-one, 7,7a-dihydro-5-hydroxy-7-(2-propenylidene)-	MeOH MeOH-KOH	248(4.34),340(4.08) 263(4.48),423(4.00)	88-0961-78 88-0961-78
geometric isomer	MeOH	227(4.18),301(4.23)	88-0961-78
Ethanone, 1-(4-hydroxy-2-methyl-5-benzofuranyl)-	MeOH	231(4.2),282(4.31), 309(4.27)	107-0251-78
Ethanone, 1-(6-hydroxy-2-methyl-5-benzofuranyl)-	MeOH	281(3.92),312(4.10)	107-0251-78
1,2,4-Ethanylylidene-6H-5-oxacyclobut[cd]indene-6,8-dione, octahydro-	MeCN	199(3.49)	88-1223-78
$C_{11}H_{10}O_3S$ 4H-1-Benzopyran-4-one, 8-methoxy-3-(methylthio)-	n.s.g.	252(4.35),320(3.64)	4-0115-78

Compound	Solvent	$\lambda_{max}(\log \epsilon)$	Ref.
4H-1-Benzopyran-4-one, 2-[(methylsulfin-yl)methyl]-	EtOH	224(4.24),266(3.90), 295(3.89)	118-0208-78
$C_{11}H_{10}O_3S_2$			
3(2H)-Benzofuranone, 2-[bis(methyl-thio)methylene]-6-hydroxy-	EtOH	226(4.00),254(3.86), 364(4.29)	142-0399-78A
$C_{11}H_{10}O_4$			
1H-2-Benzopyran-5-carboxaldehyde, 3,4-dihydro-8-hydroxy-3-methyl-1-oxo-,	MeOH	237(4.34),277(3.96), 316(3.61)	102-0511-78
(R)-	MeOH-AlCl₃	247(4.36),350(4.00), ?(3.71)(2λ,3ε)	102-0511-78
2H-1-Benzopyran-6-carboxaldehyde, 3,4-dihydro-5-hydroxy-2-methyl-4-oxo-	EtOH	242(3.78),267(3.72), 284s(3.61),350(3.17)	102-0511-78
	EtOH-AlCl₃	243(3.73),285(3.69), 303s(3.54),400(3.18)	102-0511-78
2H-1-Benzopyran-6-carboxaldehyde, 3,4-dihydro-7-hydroxy-2-methyl-4-oxo-	EtOH	258(4.25),285(3.99), 312(3.98),370(3.61)	102-0511-78
	EtOH-AlCl₃	270(4.43),360(3.31)	102-0511-78
1H-2-Benzopyran-1-one, 6,7-dimethoxy-	MeOH	240(3.58),280(3.05), 325(2.92)	4-0257-78
2-Butenoic acid, 4-formyl-3-hydroxy-, phenyl ester	EtOH	262(4.25),322(3.66)	102-0511-78
	EtOH-AlCl₃	234s(4.34),276(4.25), 375(3.57)	102-0511-78
$C_{11}H_{10}O_4S$			
4H-1-Benzopyran-4-one, 8-methoxy-3-(methylsulfinyl)-	n.s.g.	240(4.23),312(3.60)	4-0115-78
$C_{11}H_{10}O_5$			
1H-2-Benzopyran-5-carboxylic acid, 3,4-dihydro-8-hydroxy-3-methyl-1-oxo-, (R)-	EtOH	222(4.46),242(3.97), 322(3.94)	102-0511-78
	EtOH-AlCl₃	230(4.44),247(3.99), 312(4.25)	102-0511-78
5-Isobenzofurancarboxaldehyde, 1,3-di-hydro-4-hydroxy-6-methoxy-7-methyl-3-oxo-	EtOH	265(4.15),299(3.89), 368(--)	102-0511-78
	EtOH-AlCl₃	263(4.15),306(3.72), 368(3.49)	102-0511-78
$C_{11}H_{10}O_5S$			
4H-1-Benzopyran-4-one, 8-methoxy-3-(methylsulfonyl)-	n.s.g.	252(4.08),308(3.60)	4-0113-78
$C_{11}H_{11}BrN_2O_3$			
Pyridinium, 2-methyl-1-[(5-nitro-2-furanyl)methyl]-, bromide	H₂O	203(4.27),269(3.93), 299(3.71)	73-2041-78
Pyridinium, 3-methyl-1-[(5-nitro-2-furanyl)methyl]-, bromide	H₂O	203(4.26),269(3.84), 298(3.74)	73-2041-78
Pyridinium, 4-methyl-1-[(5-nitro-2-furanyl)methyl]-, bromide	H₂O	203(4.28),255(3.79), 299(3.79)	73-2041-78
$C_{11}H_{11}BrN_2O_4$			
Glycine, N-methyl-N-nitroso-, 2-(4-bromophenyl)-2-oxoethyl ester	EtOH	257(4.31)	94-3909-78
$C_{11}H_{11}Br_2N$			
Bicyclo[6.1.0]nona-2,6-diene-9-carbo-nitrile, 4,5-dibromo-9-methyl-, (1α,4α,5β,8α,9α)-	dioxan	212(3.84)	24-0282-78
(1α,4α,5β,8α,9β)-	dioxan	212(3.89)	24-0282-78

Compound	Solvent	$\lambda_{max}(\log \epsilon)$	Ref.
$C_{11}H_{11}Br_2N_3$			
3,5-Pyridinedicarbonitrile, 2,6-bis(dibromomethyl)-1,4-dihydro-4,4-dimethyl-	EtOH	208(4.61),245(4.65), 369(4.30)	103-1236-78
$C_{11}H_{11}ClN_4$			
Acetaldehyde, (4-chloro-2-methyl-1(2H)-phthalazinylidene)hydrazone	dioxan	287(4.22),358(4.00)	103-0567-78
	MeCN	213(4.56),285(4.24), 350(4.02)	103-0567-78
Acetaldehyde, (4-chloro-1-phthalazinyl)methylhydrazone	dioxan	245s(4.13),332(4.02)	103-0436-78
	MeCN	217(4.63),241s(4.02), 333(3.98)	103-0436-78
	crystal	208(--),275(--), 355(--)	103-0436-78
1(2H)-Phthalazinone, 1-(methylethylidene)hydrazone	MeOH	214(4.57),282(4.30), 355(4.04)	103-0567-78
	dioxan	282(4.48),350(4.08)	103-0567-78
	MeCN	212(4.57),281(4.30), 355(4.02)	103-0567-78
	CCl_4	284(--),350(--)	103-0567-78
Pyrimido[4,5-b][1,8]naphthyridine, 5-chloro-6,7,8,9-tetrahydro-5-chloro-	EtOH	246(4.20),285(3.78), 358(4.02)	103-1261-78
$C_{11}H_{11}ClN_4O$			
Benzamide, 4-chloro-N-(4,5-dimethyl-1H-1,2,3-triazol-1-yl)-	MeOH	238(4.17)	4-1255-78
$C_{11}H_{11}ClO$			
3-Buten-2-one, 4-(3-chlorophenyl)-3-methyl-, (E)-	MeOH	274(4.24)	56-2233-78
3-Buten-2-one, 4-(4-chlorophenyl)-3-methyl-, (E)-	MeOH	282(4.35)	56-2233-78
4-Penten-1-one, 1-(4-chlorophenyl)-	EtOH	250(4.30)	35-0883-78
$C_{11}H_{11}Cl_2NO$			
2H-Indol-2-one, 3-chloro-3-(3-chloropropyl)-1,3-dihydro-	EtOH	217(4.25),259(3.48), 305s(2.95)	94-3695-78
$C_{11}H_{11}Cl_2N_3O_4$			
3H-Imidazo[4,5-b]pyridine, 5,7-dichloro-3-β-D-ribofuranosyl-	pH 1	254(3.78),287(3.95)	4-0839-78
	pH 11	254s(3.76),259(3.77), 289(3.95)	4-0839-78
$C_{11}H_{11}FN_2OS$			
Benzamide, N-(5,6-dihydro-4H-1,3-thiazin-2-yl)-4-fluoro-	EtOH	253(4.00),306(3.94)	87-0895-78
$C_{11}H_{11}I$			
Bicyclo[5.3.1]undeca-1,3,5,9-tetraene, 7-iodo-	hexane	223(4.26),319(3.66)	35-6535-78
$C_{11}H_{11}I_2N_3$			
3,5-Pyridinedicarbonitrile, 1,4-dihydro-2,6-bis(iodomethyl)-4,4-dimethyl-	EtOH	210(4.77),250(4.62), 370(4.25)	103-1226-78
$C_{11}H_{11}N$			
Bicyclo[6.1.0]nona-2,4,6-triene-9-carbonitrile, 9-methyl-, (1α,8α,9α)-	C_6H_{12}	209(4.05),245(3.54)	24-0282-78
(1α,8α,9β)-	C_6H_{12}	209(4.09),248(3.57)	24-0282-78
Cyclobutanecarbonitrile, 1-phenyl-	EtOH	259(2.11)	44-0700-78

Compound	Solvent	$\lambda_{max}(\log \epsilon)$	Ref.
$C_{11}H_{11}NO$			
2-Azabicyclo[4.4.1]undeca-2,4,6,8,10-pentaene, 3-methoxy-	C_6H_{12}	236(4.33),264(4.42), 294s(3.72),372(3.67)	89-0853-78
3-Azabicyclo[4.4.1]undeca-1,3,5,7,9-pentaene, 4-methoxy-	MeOH	258(4.33),320(3.63), 385(3.29)	44-3813-78
2H-Azirine, 3-[2-(2-propenyloxy)phenyl]-	MeOH	248(3.73),380(4.42)	35-3494-78
2H-Azirine-2-acetaldehyde, 2-methyl-3-phenyl-	C_6H_{12}	243(4.10),278(3.11), 288(3.00)	44-3757-78
[1]Benzopyrano[4,3-b]pyrrole, 2,3,3a,4-tetrahydro-	MeOH	253(--),310(4.30)	35-3494-78
10bH-Oxazolo[2,3-a]isoquinoline, 2,3-dihydro-	EtOH	208(3.97),231(3.67), 246(3.45)	44-0672-78
2-Propyn-1-one, 3-(dimethylamino)-1-phenyl-	MeCN	248(4.02),315(4.18)	33-1609-78
Pyrano[2,3-b]indole, 2,3,4,9-tetra-hydro-	EtOH	219s(4.35),231(4.46), 275(3.88),294(3.70)	94-3695-78
1H-Pyrrol-3-ol, 2-methyl-5-phenyl-	MeOH	245(4.13),345(3.82)	49-0137-78
$C_{11}H_{11}NOS$			
2H-Indol-2-one, 3-[(ethylthio)methylene]-1,3-dihydro-, (E)-	MeOH	243(4.09),252(4.10), 278s(3.87),284(3.91), 331(4.22)	138-1281-78
(Z)-	MeOH	235(3.82),255(3.85), 277(4.09),282(4.11), 334(4.25)	138-1281-78
Spiro[3H-indole-3,2'(3'H)-thiophen]-2(1H)-one, 4',5'-dihydro-	EtOH	254(3.84),289(3.15)	94-3695-78
$C_{11}H_{11}NO_2$			
2-Azabutadiene, 1-acetoxy-1-phenyl-	C_6H_{12}	223(3.90)	35-4481-78
2,4-Azetidinedione, 3-ethyl-3-phenyl-	CH_2Cl_2	247(2.73),251s(2.71), 256s(2.64),262s(2.32)	33-3050-78
2H-Azirine-2-methanol, 3-phenyl-, acetate	C_6H_{12}	241(4.04)	35-4481-78
Benzamide, N-methyl-2-(1,2-propadienyloxy)-	EtOH	284(3.47),293s(3.38)	33-0716-78
Benzamide, N-methyl-2-(2-propynyloxy)-	EtOH	232(3.95),285(3.44), 293s(3.35)	33-0716-78
1,4-Benzoxazepin-5(2H)-one, 3,4-di-hydro-4-methyl-3-methylene-	EtOH	239s(3.92),265(3.82), 312s(3.34)	33-0716-78
4H-1,3-Benzoxazin-4-one, 2,3-dihydro-2-ethenyl-3-methyl-	EtOH	234s(3.83),252s(3.67), 298(3.38)	33-0716-78
Isoxazole, 5-(methoxymethyl)-3-phenyl-	EtOH	242(4.10)	22-0415-78
Oxazole, 5-methoxy-2-(3-methylphenyl)-	EtOH	289(4.20)	104-1353-78
Oxazole, 5-methoxy-2-(4-methylphenyl)-	EtOH	289(4.20)	104-1353-78
Oxazole, 5-methoxy-2-(phenylmethyl)-	EtOH	229(3.84)	104-1353-78
2(1H)-Pyridinone, 1-(3-oxo-1-cyclohex-en-1-yl)-	EtOH	315(3.77)	78-2609-78
Spiro[furan-2(3H),3'-[3H]indol]-2'(1'H)-one, 4,5-dihydro-	EtOH	250(3.71),268s(3.45), 295(3.00)	94-3695-78
$C_{11}H_{11}NO_2S$			
4(1H)-Quinolinone, 1-methyl-3-(methyl-sulfinyl)-	EtOH	246(4.28),254(4.28), 296(3.85),322(4.04), 377(4.04)	4-0115-78
$C_{11}H_{11}NO_2S_3$			
1,4,2-Dithiazole, 5-[(4-methylphenyl)-methylene]-3-(methylthio)-	EtOH	322(4.17)	18-1805-78

Compound	Solvent	$\lambda_{max}(\log \epsilon)$	Ref.
$C_{11}H_{11}NO_3$			
2-Azetidinecarboxylic acid, 1-benzoyl-, (S)-	EtOH	234(3.82)	39-1157-78B
2H-1-Benzopyran-2-one, 7-(dimethyl-amino)-4-hydroxy-	MeOH	350(4.38)	49-0123-78
Oxazole, 5-methoxy-2-(4-methoxyphenyl)-	EtOH	290(4.26)	104-1353-78
2-Propanone, 1-(3-phenyl-1,4,2-diox-azol-5-yl)-	MeOH	263(4.44)	97-0057-78
$C_{11}H_{11}NO_5S$			
5H-Thiazolo[3,2-a]pyridine-7,8-dicarb-oxylic acid, 2,3-dihydro-5-oxo-, dimethyl ester	MeOH	210(4.3),230(4.0), 290(4.1),328(3.9)	28-0385-78B
$C_{11}H_{11}NS$			
Benzene, (1-isothiocyanato-2-methyl-1-propenyl)-	hexane	265s(3.99),274(4.03)	24-3750-78
Thiopyrano[2,3-b]indole, 2,3,4,9-tetra-hydro-	EtOH	224s(4.32),239(4.45), 287(3.97),302(4.04)	94-3695-78
$C_{11}H_{11}N_2O_2$			
1-Pyrazolidinyl, 4,4-dimethyl-3,5-di-oxo-2-phenyl-	benzene	370s(1.88),480(1.20) (anom.)	44-0808-78
$C_{11}H_{11}N_2O_3$			
Pyridinium, 2-methyl-1-[(5-nitro-2-furanyl)methyl]-, bromide	H_2O	203(4.27),269(3.93), 299(3.71)	73-2041-78
Pyridinium, 3-methyl-1-[(5-nitro-2-furanyl)methyl]-, bromide	H_2O	203(4.26),269(3.84), 298(3.74)	73-2041-78
Pyridinium, 4-methyl-1-[(5-nitro-2-furanyl)methyl]-, bromide	H_2O	203(4.28),255(3.79), 299(3.79)	73-2041-78
$C_{11}H_{11}N_3$			
[2,2'-Bipyridin]-4-amine, N-methyl-	neutral	228(4.34),272(4.20)	39-1215-78B
	cation	235(4.32),263(4.25)	39-1215-78B
4H-Pyrazolo[1,5-a]benzimidazole, 2,4-dimethyl-	2N H_2SO_4	294(4.04),306(4.78)	4-0715-78
	EtOH	232(4.54),308(4.20)	4-0715-78
$C_{11}H_{11}N_3O$			
Imidazo[1,5,4-ef][1,5]benzodiazepin-6(7H)-one, 4,5-dihydro-2-methyl-	MeOH	272(4.18),283s(4.06), 294(3.93)	111-0053-78
Imidazo[4,5,1-jk][1,4]benzodiazepin-7(4H)-one, 5,6-dihydro-2-methyl-	MeOH	275(3.86),299(3.77)	111-0053-78
2-Propanone, 1-(5-phenyl-1H-1,2,4-tria-zol-3-yl)-	EtOH	244(4.15)	94-1825-78
1,2,4-Triazin-5(2H)-one, 6-ethyl-3-phenyl-	EtOH	244(4.28)	114-0091-78B
1,2,4-Triazin-6(1H)-one, 1,3-dimethyl-5-phenyl-	MeOH	225(3.92),278(3.87), 337(4.13)	4-1271-78
1,2,4-Triazin-6(1H)-one, 1,5-dimethyl-3-phenyl-	MeOH	265(3.33),320(3.48)	4-1271-78
1,2,4-Triazin-6(1H)-one, 5-ethyl-3-phenyl-	EtOH	213s(4.01),262(4.27)	114-0091-78B
$C_{11}H_{11}N_3OS$			
3H-1,4,8b-Triazaacenaphthylen-3-one, 4,5-dihydro-4-methyl-2-(methylthio)-	EtOH	240(4.00),271(4.09), 326(3.49)	95-0623-78
$C_{11}H_{11}N_3O_2$			
1H-1,2,3-Triazole-4-carboxylic acid, 5-methyl-1-phenyl-, methyl ester	EtOH	218(4.36)	22-0415-78

Compound	Solvent	$\lambda_{max}(\log \epsilon)$	Ref.
1H-1,2,3-Triazole-5-carboxylic acid, 4-methyl-1-phenyl-, methyl ester	EtOH	235(4.04)	22-0415-78
$C_{11}H_{11}N_3O_2S$			
Benzenesulfonamide, 4-amino-N-2-pyridin-yl-	pH 7.0	243(4.20),261(4.23), 312(3.95)	94-1162-78
β-cyclodextrin complex	pH 7.0	245(4.15),266(4.21), 315(3.88)	94-1162-78
Propanoic acid, 2-cyano-3-(2-pyridinyl-amino)-3-thioxo-, ethyl ester	EtOH	307(4.29)	104-0181-78
$C_{11}H_{11}N_3O_3$			
2H-Isoindole-2-propanoic acid, 1,3-di-hydro-1,3-dioxo-, hydrazide	MeOH	220(4.60),232(4.20), 241(4.03),293(3.22)	1-0056-78
1H-Pyrazole-5-carboxylic acid, 5-cyano-4-(2-furanyl)-4,5-dihydro-4-methyl-, methyl ester	EtOH	222(3.99)	73-0870-78
3H-Pyrazole-3-carboxylic acid, 3-cyano-4-(2-furanyl)-4,5-dihydro-4-methyl-, methyl ester	EtOH	222(4.03),331(3.30)	73-0870-78
1H-Pyrrole-2-carboxylic acid, 4-cyano-5-(cyanomethyl)-3-hydroxy-1-methyl-, ethyl ester	EtOH	213(3.99),267(4.21), 294(3.98)	94-3521-78
$C_{11}H_{11}N_3O_3S$			
1,3,4-Thiadiazol-2(3H)-one, 5-(1-meth-ylethyl)-3-(4-nitrophenyl)-	EtOH	250s(3.81),310(4.18)	146-0864-78
$C_{11}H_{11}N_3O_4$			
Acetamide, N-[2-(2-cyano-4-nitrophen-oxy)ethyl]-	DMSO	303(4.06)	44-0441-78
2H-1,5-Benzodiazepin-2-one, 1,3-di-hydro-8-methoxy-4-methyl-7-nitro-	EtOH	218(4.46),265(4.23), 310(3.81)	103-0455-78
	50% H_2SO_4	212(4.26),265(3.96)	103-0455-78
2H-1,5-Benzodiazepin-2-one, 1,3-di-hydro-8-methoxy-4-methyl-9-nitro-	EtOH	214(4.64),276(3.95)	103-0455-78
	50% H_2SO_4	204(4.49),230(4.20), 290(3.95)	103-0455-78
Benzonitrile, 2-[(2-acetoxyethyl)ami-no]-5-nitro-	DMSO	372(4.23)	44-0441-78
2H-Benzotriazole-2-acetic acid, 4,7-di-hydro-5,6-dimethyl-4,7-dioxo-, methyl ester	EtOH	226(4.32),255(4.23), 330(2.90)	87-0578-78
$C_{11}H_{11}N_5O_3$			
Benzamide, N-(4,5-dimethyl-1H-1,2,3-triazol-1-yl)-4-nitro-	MeOH	261(3.90)	4-1255-78
$C_{11}H_{11}N_5O_3S$			
1,3,4-Thiadiazol-2(3H)-imine, 5-(1-methylethyl)-3-(4-nitrophenyl)-N-nitroso-	EtOH	292(4.074),340(4.128), 430(1.80)	146-0864-78
$C_{11}H_{11}N_5O_5$			
Butanoic acid, 2-[(5-azido-2-nitrobenz-oyl)amino]-	EtOH	260(3.66),307(4.06)	63-1659-78
$C_{11}H_{11}N_9$			
3,5-Pyridinedicarbonitrile, 2,6-bis-(azidomethyl)-1,4-dihydro-4,4-di-methyl-	EtOH	218(4.81),272(4.80), 349(4.35)	103-1236-78

Compound	Solvent	$\lambda_{max}(\log \epsilon)$	Ref.
$C_{11}H_{12}$			
Benzene, 3-cyclopenten-1-yl-	C_6H_{12}	248(2.70),255(2.41), 258(2.71),262(2.71), 268(2.61),284(2.21), 308(1.73)	35-0877-78
Benzene, (2-ethenylcyclopropyl)-, cis	C_6H_{12}	248(2.57),255(2.52), 260(2.50),265(2.48), 274(2.24),290(1.69)	35-0877-78
Benzene, (2-ethenylcyclopropyl)-, trans	C_6H_{12}	262(2.82),268(2.83), 276(2.73)	35-0877-78
Benzene, (1-ethyl-2-propynyl)- (contains some diene)	n.s.g.	247(4.06),280(3.04), 288(2.80)	104-1908-78
Benzene, 1,3-pentadienyl-	EtOH	210(4.21),221(4.09), 227(4.09),235(3.97), 279(4.44),286(4.45), 308(4.02)	35-0877-78
Bicyclo[4.2.2]deca-2,4,7,9-tetraene, 2(and 1)-methyl-	C_6H_{12}	263(3.66),271(3.69), 281s(3.51)	5-2074-78
1H-Cyclobut[f]indene, 2,4,5,6-tetra-hydro-	EtOH	276(3.61),280(3.64), 286(3.57)	35-3730-78
3,4-Cyclopentenobenzocyclobutene	EtOH	267(2.99),271(2.94), 276(3.02)	35-3730-78
$C_{11}H_{12}BrNO_7$			
3-Pyridinecarboxylic acid, 5-bromo-1,6-dihydro-6-oxo-1-β-D-ribofurano-syl-	pH 1	213(3.87),257(3.95)	87-0427-78
	pH 7	212(4.26),257(3.95), 213[sic](3.87)	87-0427-78
	pH 11	214(4.27),265(4.06), 307(3.86)	87-0427-78
$C_{11}H_{12}BrN_5$			
1H-Pyrazol-3-amine, 4-[(2-bromo-4-meth-ylphenyl)azo]-5-methyl-	$CHCl_3$	375(4.06),383(4.03)	103-0885-78
$C_{11}H_{12}ClNO$			
Acetamide, N-[2-(3-chlorophenyl)-1-methylethenyl]-, cis	MeOH	263(4.16)	56-2233-78
trans	MeOH	265(4.01)	56-2233-78
Acetamide, N-[2-(4-chlorophenyl)-1-methylethenyl]-, cis	MeOH	266(4.22)	56-2233-78
trans	MeOH	274(4.26)	56-2233-78
3-Buten-2-one, 4-(3-chlorophenyl)-3-methyl-, oxime	MeOH	269(4.33)	56-2233-78
3-Buten-2-one, 4-(4-chlorophenyl)-3-methyl-, oxime	MeOH	275(4.48)	56-2233-78
$C_{11}H_{12}ClNO_2$			
2,4,6-Cycloheptatrien-1-one, 5-chloro-2-hydroxy-3-pyrrolidino-	MeOH	277(4.38),368(4.19), 430(3.82)	18-2338-78
$C_{11}H_{12}ClN_3$			
Imidazo[4,5,1-jk][1,4]benzodiazepine, 9-chloro-4,5,6,7-tetrahydro-2-methyl-	MeOH	255s(3.81),260(3.76), 284(3.67),293(3.66)	111-0053-78
$C_{11}H_{12}ClN_3O$			
Benzotriazole, 5-chloro-1-cyclopentoxy-	EtOH	273(3.88)	4-1043-78
$C_{11}H_{12}N_2$			
2H-Imidazole, 2,2-dimethyl-4-phenyl-	C_6H_{12}	208(4.47),221(4.09), 258(3.97),278s(3.43), 289(2.98)	44-2289-78

Compound	Solvent	$\lambda_{max}(\log \epsilon)$	Ref.
$C_{11}H_{12}N_2O$			
Benz[cd]indazole, 1,3,4,5-tetrahydro-8-methoxy-	EtOH	258(3.67),266(3.65), 297(3.73)	44-2508-78
[1]Benzopyrano[4,3-c]pyrazole, 2,3,3a,4-tetrahydro-2-methyl-	EtOH	297(3.84),325(4.00)	44-1664-78
1H-Indole-1-carboxamide, 2,3-dimethyl-	EtOH	210(4.21),235(4.25), 270(3.94)	118-0374-78
1H-Pyrazole, 3-methoxy-1-methyl-5-phenyl-	MeOH	238(4.08)	24-0791-78
Pyrido[2,3-g]indolizin-5-one, 1,2,3,5,6,10b-hexahydro-	EtOH	254s(3.66),260(3.75), 264s(3.70)	44-2125-78
$C_{11}H_{12}N_2OS$			
2,1-Benzisothiazole, 3-morpholino-	EtOH	232(4.53),280s(3.46), 291(3.55),304(3.49), 370(3.86)	48-0313-78
4(5H)-Thiazolone, 2-[methyl(4-methylphenyl)amino]-	dioxan	256(4.28)	103-0148-78
$C_{11}H_{12}N_2O$			
Benzoxazole, 7-butyl-6-nitroso-	CHCl$_3$	765(1.57)	39-0692-78C
1H-Pyrido[3,2-b]indole-2,9-dione, 2,3,4,6,7,8-hexahydro-	EtOH	249(4.18),313(3.51)	94-3080-78
$C_{11}H_{12}N_2O_2S$			
Benzo[1,2-c:3,4-c']dipyrrole, 1,2,3,7-tetrahydro-2-(methylsulfonyl)-	CH$_2$Cl$_2$ at -20°	265(3.11),281(3.08), 291(3.21),313s(3.63), 325(3.72),337s(3.60)	89-0068-78
4(5H)-Thiazolone, 2-[(4-methoxyphenyl)-methylamino]-	dioxan	255(3.28)	103-0148-78
$C_{11}H_{12}N_2O_2S_2$			
1,4,2-Dithiazin-3-amine, N-ethyl-5-phenyl-, 1,1-dioxide	EtOH	244(4.26)	18-1805-78
1,4,2-Dithiazin-3-amine, N-methyl-5-(4-methylphenyl)-, 1,1-dioxide	EtOH	249(4.20)	18-1805-78
1,4,2-Dithiazol-3-amine, N-ethyl-5-(phenylmethylene)-, 1,1-dioxide	EtOH	296(4.37)	18-1805-78
$C_{11}H_{12}N_2O_3$			
2,7-Naphthyridine-4-carboxylic acid, 5,6,7,8-tetrahydro-8-methyl-6-oxo-, methyl ester, (S)- (jasminine)	EtOH	214(3.81),270(3.24)	102-1069-78
1H-Pyrrolo[3,2-b]pyridine-3-carboxylic acid, 4,5-dihydro-2-methyl-5-oxo-, ethyl ester	EtOH	225(4.20),254s(3.85), 259(3.85),333(4.20)	94-3080-78
$C_{11}H_{12}N_2O_3S$			
Morpholine, 4-[(2-nitrophenyl)thioxomethyl]-	EtOH	254(4.13),274(4.14), 359s(3.05)	48-0313-78
1,4,3-Oxathiazin-2-amine, N-ethyl-6-phenyl-, 4,4-dioxide	EtOH	260(4.15)	18-1805-78
$C_{11}H_{12}N_2O_3S_2$			
Ethanethioic acid, diazo[(4-methylphenyl)sulfonyl]-, S-ethyl ester	C$_6$H$_{12}$	243(0.63),390(0.32)	130-0189-78
$C_{11}H_{12}N_2O_4$			
2,4(1H,3H)-Pyrimidinedione, 1-(5-ethynyltetrahydro-4-hydroxy-2-furanyl)-5-methyl-, [2R-(2α,4β,5α)]-	MeOH	265(4.09)	44-0367-78

Compound	Solvent	$\lambda_{max}(\log \epsilon)$	Ref.
$C_{11}H_{12}N_2O_4S$			
Acetic acid, diazo[(4-methylphenyl)sulfonyl]-, ethyl ester	n.s.g.	<u>245(4.2)</u>,370(2.15)	35-0309-78
1H-4,1,2-Benzothiadiazine-3-carboxylic acid, 1-methyl-, ethyl ester, 4,4-dioxide	EtOH	331(4.15)	39-0539-78C
Imidazo[1,2-a]pyridine-3-carboxylic acid, 2-(methylsulfonyl)-, ethyl ester	EtOH	240(4.43),245(4.36), 290(3.92),308(3.86), 320s(3.71)	95-0631-78
$C_{11}H_{12}N_2O_5$			
Uridine, 2'-deoxy-5-ethynyl-	EtOH	228(4.00),289(4.06)	39-1263-78C
$C_{11}H_{12}N_2O_6$			
Uridine, 5-ethynyl-	pH 7	230(3.97),290(4.03)	39-1263-78C
	pH 12	288(3.94)	39-1263-78C
$C_{11}H_{12}N_2S$			
1H-Imidazole, 1-methyl-2-[(phenylmethyl)thio]-	EtOH	215(4.041),257(3.740)	28-0201-78B
	CH_2Cl_2	216(4.103),253s(3.738)	28-0201-78B
2-Thiazolamine, N,N-dimethyl-4-phenyl-	EtOH	237(4.42),265s(3.98), 289(3.82)	94-3017-78
$C_{11}H_{12}N_3O_4P$			
Carbamic acid, (5,6-dihydro-6-oxo-1,5,2-diazaphosphorin-2(1H)-yl)-, phenylmethyl ester, P-oxide	EtOH	204(3.83),238(3.81)	130-0421-78
$C_{11}H_{12}N_4$			
Cyanamide, (3,4-dihydro-1,3-dimethyl-2(1H)-quinazolinylidene)-	MeOH	240(4.22),260(4.13)	4-1409-78
1H-Indole, 2-(1-azidoethyl)-3-methyl-	EtOH	225(4.47),278(3.90), 284(3.91),293(3.83)	94-2874-78
1(2H)-Phthalazinone, (1-methylethylidene)hydrazone	dioxan	280(4.53),350(4.08)	103-0567-78
	MeCN	207(4.56),279(4.46), 344(3.94)	103-0567-78
1H-Tetrazole, 1-(3-butenyl)-5-phenyl-	EtOH	231(4.02)	44-1664-78
2H-Tetrazole, 2-(3-butenyl)-5-phenyl-	EtOH	238(4.20)	44-1664-78
1H-1,2,4-Triazolo[4,3-a]benzimidazole, 1-ethyl-3-methyl-	EtOH	233(4.03),253(3.88), 259(3.87),291(3.48), 299(3.45)	4-1027-78
9H-1,2,4-Triazolo[4,3-a]benzimidazole, 9-ethyl-3-methyl-	EtOH	232(3.95),240s(3.73), 291(3.70),297(3.71)	4-1027-78
$C_{11}H_{12}N_4O$			
1H-Imidazole-4-carboxamide, 5-amino-1-(phenylmethyl)-	pH 1	244(3.93),269(4.05)	94-1929-78
	pH 7	268(4.12)	94-1929-78
	pH 13	268(4.12)	94-1929-78
	EtOH	270(4.14)	94-1929-78
1H-Pyrazole-4,5-dione, 4-[(2-methylphenyl)hydrazone]	EtOH	252(4.12),260(4.11), 336(4.00),416(4.21)	80-0079-78
	acetone	413(4.23)	80-0079-78
	$CHCl_3$	422(4.27)	80-0079-78
	CCl_4	420(4.26)	80-0079-78
1H-Pyrazole-4,5-dione, 4-[(4-methylphenyl)hydrazone]	EtOH	254(4.08),262(4.07), 340(4.04),418(4.26)	80-0079-78
	acetone	414(4.36)	80-0079-78
	$CHCl_3$	420(4.32)	80-0079-78
	CCl_4	419(4.42)	80-0079-78
2H-Tetrazole, 2-methyl-5-[2-(2-propenyloxy)phenyl]-	EtOH	235(4.00),288(3.65)	44-1664-78

Compound	Solvent	$\lambda_{max}(\log \epsilon)$	Ref.
$C_{11}H_{12}N_4O_2$			
Butanenitrile, 3-methyl-2-[(4-nitro-phenyl)hydrazono]-	EtOH	365(4.430)	146-0864-78
1H-Pyrazole-4,5-dione, 4-[(2-methoxy-phenyl)hydrazono]-	EtOH	255(4.07),430(4.25)	80-0079-78
1H-Pyrazole-4,5-dione, 4-[(4-methoxy-phenyl)hydrazono]-	EtOH	261(4.04),432(4.34)	80-0079-78
$C_{11}H_{12}N_4O_2S$			
Benzenesulfonamide, 4-amino-N-(4-meth-yl-2-pyrimidinyl)-	pH 7.0	242(4.26),259(4.28)	94-1162-78
β-cyclodextrin complex	pH 7.0	242(4.20),262(4.25)	94-1162-78
1,3,4-Thiadiazol-2(3H)-imine, 5-(1-methylethyl)-3-(4-nitrophenyl)-	EtOH	240(4.159),340(4.179)	146-0864-78
$C_{11}H_{12}N_4O_3$			
Pyrimido[5',4':4,5]pyrrolo[2,3-c]azep-ine-2,4,9(3H)-trione, 1,5,6,7,8,10-hexahydro-1-methyl-	EtOH	241(4.18),305(4.03)	103-0306-78
Pyrimido[5',4':4,5]pyrrolo[2,3-c]azep-ine-2,4,9(3H)-trione, 1,5,6,7,8,10-hexahydro-3-methyl-	EtOH	222(4.88),286(3.26)	103-0306-78
$C_{11}H_{12}N_4O_3S$			
Benzenesulfonamide, 4-amino-N-(6-meth-oxy-4-pyrimidinyl)-	pH 7.0	250s(4.35),263(4.41)	94-1162-78
β-cyclodextrin complex	pH 7.0	250s(4.32),264(4.38)	94-1162-78
$C_{11}H_{12}N_4O_6S$			
β-D-Glucopyranuronic acid, 1-deoxy-1-(1,6-dihydro-6-thioxo-9H-purin-9-yl)-, sodium salt	H_2O	285(4.11)	106-0250-78
	MeOH	279(4.16)	106-0250-78
$C_{11}H_{12}N_5O_5$			
Pteridinium, 2-amino-1,4-dihydro-6,7-bis(methoxycarbonyl)-8-methyl-4-oxo-, chloride	pH -1.0	282s(4.13),305(4.22), 397(4.07)	24-3790-78
	pH 5.0	262(4.24),382(4.22)	24-3790-78
	pH 10.0	260(4.21),305s(3.27), 398(4.30)	24-3790-78
$C_{11}H_{12}N_6O_5$			
[1,2,4]Oxadiazolo[3,2-i]purin-2(7H)-imine, 7-β-D-ribofuranosyl-, hydrobromide	pH 1	224(4.40),283(4.32)	94-2122-78
	H_2O	224(4.35),283(4.31)	94-2122-78
	pH 13	247(4.48),294(4.22)	94-2122-78
9H-[1,2,4]Triazolo[1,5-a]purin-9-one, 3,4-dihydro-3-β-D-ribofuranosyl-	pH 1	277(4.07)	64-0326-78C
	H_2O	285(4.15)	64-0326-78C
	pH 13	285(4.17)	64-0326-78C
$C_{11}H_{12}N_8O_7$			
Imidazole-5-carboxamidine, 5-amino-1-methyl-, picrate	pH 1	282(4.15)	94-1929-78
	pH 7	285.5(4.15)	94-1929-78
	EtOH	292(4.24)	94-1929-78
$C_{11}H_{12}O$			
5H-Benzocyclohepten-1-ol, 6,9-dihydro-	EtOH	220s(--),273(3.35)	35-0877-78
	EtOH-base	240(3.92),294(3.44)	35-0877-78
Benzofuran, 3,5,6-trimethyl-	EtOH	248(4.01),278(3.48), 284(3.48),290(3.56)	23-0517-78
Bicyclo[4.2.2]deca-2,4,7,9-tetraene, 2-methoxy-	C_6H_{12}	268(3.76),276s(3.74)	5-2074-78
1-Butanone, 2-methylene-1-phenyl-	EtOH	246(3.98),336(1.97)	44-4316-78

Compound	Solvent	$\lambda_{max}(\log \epsilon)$	Ref.
3-Buten-2-one, 3-methyl-4-phenyl-	MeOH	278(4.31)	56-2233-78
5H-Cycloprop[a]inden-6-ol, 1,1a,6,6a-tetrahydro-6-methyl-	EtOH	264(2.83),270(2.95), 278(2.90),296(2.13), 307(2.04)	44-4316-78
1H-Inden-1-one, 2-ethyl-2,3-dihydro-	EtOH	245(4.08),292(3.34)	44-4316-78
1H-Inden-1-one, 3-ethyl-2,3-dihydro-	EtOH	245(4.06),288(3.39), 293(3.39)	44-4316-78
1-Naphthalenecarboxaldehyde, 5,6,7,8-tetrahydro-	EtOH	254(4.03)	44-2167-78
2-Naphthalenecarboxaldehyde, 5,6,7,8-tetrahydro-	EtOH	266(4.14)	44-2167-78
1(2H)-Naphthalenone, 3,4-dihydro-2-methyl-	EtOH	210(4.15),248(4.06), 292(3.19)	44-4316-78
Trishomocuban-4-one, (+)-D	isooctane	272s(1.10),286(1.24)	44-0689-78
$C_{11}H_{12}OS_2$			
1,4-Dithiepan-2-one, 3-phenyl-	n.s.g.	247(4.19),270s(4.03), 330(2.89)	33-3087-78
$C_{11}H_{12}O_2$			
2H-1-Benzopyran-3(4H)-one, 2,7-dimethyl-	MeOH	227(3.34),305(2.87)	44-0303-78
Cyclobutanecarboxylic acid, 1-phenyl-	EtOH	218s(3.67),256(2.16)	44-0700-78
Cyclopropanecarboxylic acid, 2-phenyl-, methyl ester, cis	EtOH	210(3.92),216(3.91), 255(2.25),260(2.31), 265s(2.22)	44-4447-78
trans	EtOH	220(4.03),253s(2.37), 259(2.50),265(2.55), 272(2.37)	44-4447-78
$C_{11}H_{12}O_3$			
Benzaldehyde, 2-[2-(2-ethenyloxy)ethoxy]-	n.s.g.	213(4.27),252(3.95), 317(3.60)	47-1343-78
Benzaldehyde, 3-[2-(2-ethenyloxy)ethoxy]-	n.s.g.	217(4.34),251(3.92)	47-1343-78
Benzaldehyde, 4-[2-(2-ethenyloxy)ethoxy]-	n.s.g.	205(4.09),218(4.14), 275(4.26)	47-1343-78
4,7-Benzofurandione, 2,3-dihydro-3,5,6-trimethyl-	EtOH	285(4.28),420(2.60)	23-0517-78
1H-2-Benzopyran-1-one, 3,4-dihydro-8-hydroxy-3,5-dimethyl-, (R)-	EtOH	247(3.82),322(3.63)	102-0511-78
	EtOH-AlCl₃	260(3.66),360(3.51)	102-0511-78
1H-2-Benzopyran-1-one, 3,4-dihydro-8-hydroxy-3,7-dimethyl-, (R)-	EtOH	250(3.85),320(3.60)	102-0511-78
	EtOH-AlCl₃	264(3.70),360(3.49)	102-0511-78
4H-1-Benzopyran-4-one, 2,3-dihydro-7-hydroxy-2,6-dimethyl-	EtOH	237(3.68),279(3.72), 329(3.59)	102-0511-78
2-Butenoic acid, 3-(4-methoxyphenyl)-	n.s.g.	217(4.23),274(4.38)	2-0200-78
Cyclopropanecarboxylic acid, 2-(4-methoxyphenyl)-, trans	EtOH	231(4.15),279(3.22), 282(3.21)	44-4316-78
Ethanone, 1-(2,3-dihydro-5-methyl-1,4-benzodioxin-6-yl)-	EtOH	216(4.26),232(4.14), 274(4.04),306(3.63)	103-1188-78
Ethanone, 1-(2,3-dihydro-7-methyl-1,4-benzodioxin-6-yl)-	EtOH	231(4.29),271(4.03), 310(3.69)	103-1188-78
Ethanone, 1-(2,3-dihydro-8-methyl-1,4-benzodioxin-6-yl)-	EtOH	215(4.16),231(4.11), 280(3.94)	103-1188-78
1(2H)-Naphthalenone, 3,4-dihydro-4,8-dihydroxy-6-methyl-	EtOH	218(4.22),229s(--), 268(4.05),333(3.53)	98-0869-78
1,4-Pentanedione, 3-hydroxy-1-phenyl-	MeOH	242(3.68),277(2.70)	49-0137-78
Spiro[2H-1,5-benzodioxepin-3(4H),3'-oxetane]	EtOH	274(3.11)	56-1913-78

Compound	Solvent	$\lambda_{max}(\log \epsilon)$	Ref.
$C_{11}H_{12}O_3S$			
2-Propen-1-one, 3-(ethylsulfonyl)-1-phenyl-, (E)-	EtOH	205(4.08),221(4.02), 267(3.96)	139-0191-78B
(Z)-	EtOH	205(4.09),254(3.98)	139-0191-78B
$C_{11}H_{12}O_4$			
1,3-Benzodioxol-5(7aH)-one, 7a-methoxy-4-(2-propenyl)-	EtOH	225(3.86),305(3.58)	44-3983-78
1,3-Benzodioxol-5(7aH)-one, 7a-methoxy-6-(2-propenyl)-	EtOH	237(3.97),295(3.51)	44-3983-78
1H-2-Benzopyran-1-one, 3,4-dihydro-8-hydroxy-5-methoxy-3-methyl-, (R)-	EtOH EtOH-AlCl$_3$	240s(3.68),343(3.56) 385(3.62)	102-0511-78 102-0511-78
2H-6,9a-Methanocycloocta[b]furan-2,4(5H)-dione, 6,7,8,9-tetra-hydro-3-hydroxy-	MeCN	270(3.86),335(3.51)	44-3653-78
$C_{11}H_{12}O_4S$			
4H-1-Benzopyran-4-one, 2,3-dihydro-8-methoxy-3-(methylsulfinyl)-	n.s.g.	272(3.95),344(3.30)	4-0115-78
$C_{11}H_{12}O_5$			
Benzeneacetic acid, 2-acetyl-3,6-di-hydroxy-, methyl ester	EtOH	328(3.48)	12-2099-78
1H-2-Benzopyran-1-one, 3,4-dihydro-8-hydroxy-7-(hydroxymethyl)-6-methoxy- (stellatin)	MeOH MeOH-KOH	265(4.23),300(3.77) 338(3.78)	77-0627-78 77-0627-78
2H-Cyclopenta[b]furan-3-acetic acid, 3,3a,6,6a-tetrahydro-α,2-dioxo-, ethyl ester	EtOH	212(3.20),278(3.53)	87-0815-78
Ethanone, 1,1'-(2,4,6-trihydroxy-5-methyl-1,3-phenylene)bis-	EtOH	274(4.38),295s(4.25), 336(3.68),382s(3.42)	94-3585-78
2-Propenoic acid, 3-(4-hydroxy-3,5-di-methoxyphenyl)-, TiCl$_4$ complex	acetone	405(2.95)	98-0973-78
$C_{11}H_{12}S_2$			
1,3-Dithiane, 2-(phenylmethylene)-	n.s.g.	232(3.91),307(4.15)	33-3087-78
$C_{11}H_{13}ClN_2O_4$			
Benzene, 1-(1-chloro-2-methyl-2-nitro-butyl)-4-nitro-	MeOH	265(4.05)	12-2477-78
more polar isomer	MeOH	265(4.06)	12-2477-78
1,4-Diazepinium, 2,3-dihydro-6-phenyl-, perchlorate	MeOH	246(4.04),352(3.87)	39-1453-78C
$C_{11}H_{13}ClN_4O_4$			
3H-Imidazo[4,5-b]pyridin-5-amine, 7-chloro-3-β-D-ribofuranosyl-	pH 1 pH 1 pH 7	240(3.75),319(4.04) 218(4.34),318(4.04) 220(4.29),249(3.80), 256s(3.71),311(4.03)	4-0839-78 87-0112-78 87-0112-78
	pH 11	250(3.81),255(3.81), 311(4.05)	4-0839-78
	pH 13	249(3.77),256s(3.71), 311(4.04)	87-0112-78
$C_{11}H_{13}ClN_4O_5$			
9H-Purine, 6-chloro-9-β-D-glucopyrano-syl-	EtOH	264(3.96)	12-1095-78
$C_{11}H_{13}ClO$			
1H-Indene-3-carboxaldehyde, 6-chloro-2,4,5,7a-tetrahydro-7a-methyl-	EtOH	234(4.18)	77-0528-78

Compound	Solvent	$\lambda_{max}(\log \epsilon)$	Ref.
$C_{11}H_{13}FN_2O_5$			
2,2'-Anhydro-1-(6-deoxy-6-fluoro-β-D-glucopyranosyl)thymine	H_2O	227(3.98),235s(3.98), 249(3.87)	48-0157-78
$C_{11}H_{13}F_3N_5O_7P$			
5'-Adenylic acid, 2-(trifluoromethyl)-	pH 1	260(4.09)	87-0520-78
$C_{11}H_{13}N$			
Cycloprop[a]inden-1a(1H)-amine, 6,6a-dihydro-1-methyl-, endo	MeOH	265(3.26),270(3.26)	35-2181-78
exo	MeOH	270(3.40),277(3.34)	35-2181-78
1-Propen-1-amine, 2-methyl-N-(phenylmethylene)-, (E)-	MeOH	290(4.2)	78-0833-78
$C_{11}H_{13}NO$			
Acetamide, N-(1-methyl-2-phenylethenyl)-, cis	MeOH	259(4.16)	56-2233-78
trans	MeOH	268(4.21)	56-2233-78
1H-2-Benzazepin-1-one, 2,3,4,5-tetrahydro-5-methyl-, (+)-	EtOH	220(4.00),269(3.03), 276(2.95)	103-0538-78
Benzo[b]cyclopropa[d]pyran-7b(1H)-amine, 1a,2-dihydro-1-methyl-	C_6H_{12}	227(3.86),277(3.28), 283(3.26)	35-3494-78
Benzonitrile, 3-(1,1-dimethylethyl)-4-hydroxy-	EtOH	254(4.15),275s(3.49), 286(3.19)	12-0907-78
3-Buten-2-one, 3-methyl-4-phenyl-, oxime	MeOH	268(4.36)	56-2233-78
2-Propen-1-one, 3-(dimethylamino)-1-phenyl-	C_6H_{12}	324(4.25)	48-0945-78
	EtOH	341(4.36)	48-0945-78
$C_{11}H_{13}NO_2$			
Acetamide, N-acetyl-N-(4-methylphenyl)-	$CHCl_3$	242(3.15)	20-0621-78
Acetamide, N-acetyl-N-(phenylmethyl)-	$CHCl_3$	245(2.71)	20-0621-78
2,4,6-Cycloheptatrien-1-one, 2-hydroxy-3-pyrrolidino-	MeOH	269(4.34),361(4.19), 422(3.79)	18-2338-78
2,4,6-Cycloheptatrien-1-one, 2-hydroxy-5-pyrrolidino-	MeOH	240(4.32),286(3.51), 374(4.29),394s(4.23)	18-2338-78
2H-Indol-2-one, 1,3-dihydro-3-(3-hydroxypropyl)-	EtOH	251(3.93),282s(3.14), 293s(2.86)	94-3695-78
4-Isoxazolol, 4,5-dihydro-3-methyl-5-(4-methylphenyl)-	MeOH	261(2.47),266(2.55), 274(2.47)	142-0187-78C
Pyrano[2,3-b]indol-4a(2H)-ol, 3,4,9,9a-tetrahydro-	EtOH	241(3.87),294(3.37)	94-3695-78
$C_{11}H_{13}NO_3$			
Alanine, N-benzoyl-N-methyl-	EtOH	211(3.88)	39-1157-78B
Benzoic acid, 4-morpholino-	EtOH	211(4.34),300(4.34)	104-1910-78
2,4,6-Cycloheptatrien-1-one, 2-hydroxy-3-morpholino-	MeOH	226(4.24),273(4.19), 355(4.03)	18-2338-78
1H-Indole-2-carboxylic acid, 4,5,6,7-tetrahydro-1,3-dimethyl-4-oxo-	MeOH	236(4.41),258s(4.10), 284(4.30)	24-1780-78
4-Isoxazolol, 4,5-dihydro-5-(4-methoxyphenyl)-3-methyl-	MeOH	277(3.19),284(3.12)	142-0187-78C
2,4-Pyrrolidinedione, 1-methyl-3-(1-oxo-2,4-hexadienyl)-	EtOH	238(4.12),358(4.44), 374s(4.34)	35-4225-78
	EtOH-H_2SO_4	237(4.12),353(4.46), 373s(4.46)	35-4225-78
	EtOH-KOH	259(4.27),280(4.26), 337(4.23)	35-4225-78
$C_{11}H_{13}NO_3S$			
2H-Indol-2-one, 1,3-dihydro-4,5-dimeth-	MeOH	258(3.89),310(3.40)	39-1476-78C

Compound	Solvent	$\lambda_{max}(\log \epsilon)$	Ref.
oxy-3-(methylthio)- (cont.)			39-1476-78C
2H-Indol-2-one, 1,3-dihydro-4,7-dimeth-	MeOH	251s(3.46),312(3.71)	39-1476-78C
oxy-3-(methylthio)-			
2H-Indol-2-one, 1,3-dihydro-5,6-dimeth-	MeOH	220(4.43),270(3.83),	39-1476-78C
oxy-3-(methylthio)-		300(3.56)	
2H-Indol-2-one, 1,3-dihydro-6,7-dimeth-	MeOH	225(4.01),256s(3.38),	39-1476-78C
oxy-3-(methylthio)-		310s(2.36)	
4a,8a-Methanoisoquinolin-3(4H)-one,	MeOH	272(3.26)	44-3813-78
1,2-dihydro-4-(methylsulfonyl)-			

$C_{11}H_{13}NO_4$

Compound	Solvent	$\lambda_{max}(\log \epsilon)$	Ref.
Acetamide, N-(2-formyl-4,5-dimethoxy-	MeOH	250(4.4),289(3.3),	5-0608-78
phenyl)-		343(3.8)	
β-Alanine, N-(4-methyl-3,6-dioxo-1,4-	MeOH	271(4.08),472(3.44)	64-0912-78C
cyclohexadien-1-yl)-, methyl ester			
1,4-Dioxino[2',3':4,5]furo[3,2-c]pyri-	EtOH	215(4.42),290(3.81)	23-0613-78
din-9(4aH)-one, 2,3,8,9b-tetrahydro-			
4a,7-dimethyl-			

$C_{11}H_{13}NO_5$

Compound	Solvent	$\lambda_{max}(\log \epsilon)$	Ref.
3H-Azepine-3,6-dicarboxylic acid,	EtOH	276(3.96)	39-0191-78C
2-methoxy-, dimethyl ester			

$C_{11}H_{13}NO_5S_2$

Compound	Solvent	$\lambda_{max}(\log \epsilon)$	Ref.
Benzenecarbothioic acid, 2-[[(methoxy-	EtOH	233s(3.87),245s(3.78),	40-0582-78
carbonyl)amino]sulfonyl]-, S-ethyl		271s(3.44),279s(3.37)	
ester			

$C_{11}H_{13}NO_6$

Compound	Solvent	$\lambda_{max}(\log \epsilon)$	Ref.
1,3-Dioxolane, 2-(4,5-dimethoxy-	MeOH	220(4.02),240(4.04),	39-0440-78C
2-nitrophenyl)-		295(3.63)	

$C_{11}H_{13}NO_6S$

Compound	Solvent	$\lambda_{max}(\log \epsilon)$	Ref.
Benzoic acid, 2-[[(methoxycarbonyl)-	EtOH	226(3.91),270(3.19),	40-0582-78
methylamino]sulfonyl]-, methyl ester		277(3.22)	

$C_{11}H_{13}NO_7$

Compound	Solvent	$\lambda_{max}(\log \epsilon)$	Ref.
3-Pyridinecarboxylic acid, 1,6-dihydro-	pH 1	259(4.20)	87-0427-78
6-oxo-1-β-D-ribofuranosyl-	pH 11	252(4.10),300(3.75)	87-0427-78

$C_{11}H_{13}NS$

Compound	Solvent	$\lambda_{max}(\log \epsilon)$	Ref.
Benzene, (3-isothiocyanatobutyl)-, (R)-	isooctane	243(3.43)	30-0350-78
Benzene, (3-isothiocyanato-1-methyl-	isooctane	249(3.38)	30-0350-78
propyl)-, (S)-			
Benzene, (3-isothiocyanato-2-methyl-	isooctane	249(3.39)	30-0350-78
propyl)-, (R)-			
Benzothiazole, 2-(1,1-dimethylethyl)-	CCl₄	258(4.00),284(3.23),	44-0731-78
		295(3.23)	

$C_{11}H_{13}N_2OPS_2$

Compound	Solvent	$\lambda_{max}(\log \epsilon)$	Ref.
1,3,2-Diazaphosphorine, 1,2-dihydro-	EtOH	278(3.90),318(3.82)	103-0784-78
6-methyl-4-(methylthio)-2-phenoxy-,			
2-sulfide			

$C_{11}H_{13}N_3$

Compound	Solvent	$\lambda_{max}(\log \epsilon)$	Ref.
1H-Benzotriazole, 1-cyclopentyl-	n.s.g.	255(4.80),263(4.70),	4-0127-78
		279(4.72)	
2H-Benzotriazole, 2-cyclopentyl-	n.s.g.	273(5.13),279(5.15),	4-0127-78
		286(5.08)	
Imidazo[1,5,4-ef][1,5]benzodiazepine,	MeOH	275(3.92),290s(3.80)	111-0053-78
4,5,6,7-tetrahydro-2-methyl-			

Compound	Solvent	λ_{max}(log ϵ)	Ref.
Imidazo[4,5,1-jk][1,4]benzodiazepine, 4,5,6,7-tetrahydro-2-methyl-, hydrochloride	MeOH	252s(3.87),256(3.88), 276(3.69),286(3.67)	111-0053-78
3-Quinolinecarbonitrile, 1,2,5,6,7,8-hexahydro-2-imino-1-methyl-	EtOH	220(3.98),256(3.89), 346(3.54),398(3.58)	88-3469-78
$C_{11}H_{13}N_3O$			
1,3,4-Oxadiazole-2-methanamine, α,α-dimethyl-5-phenyl-	EtOH	251(4.29)	33-2419-78
1,2,4-Triazin-6(1H)-one, 2,5-dihydro-1,3-dimethyl-5-phenyl-	MeOH	277(3.44)	4-1271-78
1,2,4-Triazin-6(1H)-one, 2,5-dihydro-1,5-dimethyl-3-phenyl-	MeOH	225(4.21),301(3.89)	4-1271-78
1,2,4-Triazin-6(1H)-one, 2,5-dihydro-2,3-dimethyl-5-phenyl-	H_2O	283(3.67)	4-1271-78
	pH 12	283(3.57)	4-1271-78
	MeOH	290(3.71)	4-1271-78
	MeOH-H_2SO_4	265(3.58)	4-1271-78
	MeCN	246(3.56),303(3.68)	4-1271-78
	MeCN-H_2SO_4	262(3.60)	4-1271-78
	$CHCl_3$	304(3.66)	4-1271-78
1,2,4-Triazin-6(1H)-one, 2,5-dihydro-2,5-dimethyl-3-phenyl-	H_2O	227(4.07),300(3.76)	4-1271-78
	pH 12	227s(4.10),300(3.64)	4-1271-78
	MeOH	223(4.03),308(3.75)	4-1271-78
	MeOH-H_2SO_4	230(4.00),277(3.73)	4-1271-78
	MeCN	224(4.09),327(3.58)	4-1271-78
	MeCN-H_2SO_4	231(4.04),275(3.76)	4-1271-78
	$CHCl_3$	328(3.48)	4-1271-78
$C_{11}H_{13}N_3OS$			
Benzoic acid, 2-[(2-propenylamino)thioxomethyl]hydrazide	MeOH	244(4.29),316(3.08)	80-0397-78
nickel complex	MeOH	260(4.64),322(4.27)	80-0397-78
4(5H)-Thiazolone, 2-[[4-(dimethylamino)phenyl]amino]-	dioxan	272(4.02),308(4.09)	103-0148-78
$C_{11}H_{13}N_3O_2$			
1H-Benzotriazole-1-acetic acid, 5,6-dimethyl-, methyl ester	4.2M HCl	287(3.90)	87-0578-78
Pyrrolo[3,4-d][1,2]diazepine-3(7H)-carboxylic acid, 1-methyl-, ethyl ester	EtOH	225(4.17),254(4.31)	88-4093-78
$C_{11}H_{13}N_3O_2S$			
1,3,5-Triazin-2(1H)-one, tetrahydro-6-(2-hydroxyphenyl)-1,3-dimethyl-4-thioxo-	EtOH	270(4.29)	4-1193-78
$C_{11}H_{13}N_3O_3$			
1H-Benzotriazoleacetic acid, 5,6-dimethyl-, methyl ester, 3-oxide	EtOH	277(3.60),286(3.60), 323(3.87)	87-0578-78
	4.2M HCl	287(3.87)	87-0578-78
3H-Pyrrolo[3,2-d]pyrimidine-7-carboxylic acid, 4,5-dihydro-3,6-dimethyl-4-oxo-, ethyl ester	EtOH	235(4.70),256s(3.92)	94-3080-78
$C_{11}H_{13}N_3O_4$			
Cytidine, 2'-deoxy-5-ethynyl-	pH 2	238(4.14),307(3.99)	39-1263-78C
	pH 6	236(4.19),295(3.98)	39-1263-78C
	pH 12	266(3.67)	39-1263-78C
Propanamide, 2-(acetylamino)-N-(4-nitrophenyl)-	EtOH	303(4.085)	146-0864-78

Compound	Solvent	$\lambda_{max}(\log \epsilon)$	Ref.
1H-Pyrrolo[3,2-d]pyrimidine-7-carbox-	EtOH	230(4.48),270(4.01)	94-3080-78
ylic acid, 2,3,4,5-tetrahydro-3,6-	pH 1	231(--),274(--)	94-3080-78
dimethyl-2,4-dioxo-, ethyl ester	pH 13	242(--),285(--)	94-3080-78
1H-Pyrrolo[3,2-d]pyrimidine-7-carbox-	EtOH	235(4.23),273(3.50)	142-0793-78A
ylic acid, 2,3,4,5-tetrahydro-1,3,5-			
trimethyl-2,4-dioxo-, methyl ester			
1H-Pyrrolo[3,2-d]pyrimidine-7-carbox-	EtOH	233(4.24),275(3.73)	142-0793-78A
ylic acid, 2,3,4,5-tetrahydro-1,3,6-			
trimethyl-2,4-dioxo-, methyl ester			
1,3,5-Triazine-2,4(1H,3H)-dione, 6-(2-	EtOH	282(3.51)	4-1193-78
hydroxy-3-methoxyphenyl)-1-methyl-			
$C_{11}H_{13}N_3S$			
Pyrido[3,2-c]pyridazine, 4-(butylthio)-	heptane	246(4.17),259(4.23),	103-1032-78
		339(4.13),348(4.13)	
Pyrido[3,2-c]pyridazine, 4-[(1-methyl-	heptane	235(3.90),255(3.97),	103-1032-78
propyl)thio]-		340(3.90),350(3.90)	
Pyrido[3,2-c]pyridazine, 4-[(2-methyl-	heptane	235(4.00),255(4.60),	103-1032-78
propyl)thio]-		339(3.98),348(3.98)	
$C_{11}H_{13}N_4$			
1H-1,2,4-Triazolo[4,3-a]benzimidazolium,	EtOH	263s(3.42),277(3.65),	22-0273-78
1,3,9-trimethyl-		284(3.67),297s(2.72)	
	CF_3COOH	265s(3.53),274(3.68),	22-0273-78
		281(3.69),303s(2.74)	
$C_{11}H_{13}N_4O_8P$			
2,4(1H,3H)-Pteridinedione, 1-(2-deoxy-	pH 1.0	229(4.104),316(3.845)	5-1780-78
5-O-phosphono-β-D-erythro-pentofur-	pH 7.0	230(4.104),317(3.839)	5-1780-78
anosyl)-	pH 13.0	238(4.176),277(3.556),	5-1780-78
		323(3.839)	
$C_{11}H_{13}N_4O_9P$			
2,4(1H,3H)-Pteridinedione, 1-(5-O-phos-	pH 1.0	228(4.114),314(3.863)	5-1780-78
phono-β-D-ribofuranosyl)-	pH 7.0	229(4.107),315(3.845)	5-1780-78
	pH 13.0	237(4.164),278s(3.591),	5-1780-78
		324(3.857)	
$C_{11}H_{13}N_5$			
1H-Imidazole-4-carboximidamide,	pH 1	284(4.07)	94-1929-78
5-amino-1-(phenylmethyl)-	pH 7	285(4.10)	94-1929-78
	pH 13	267(4.00)	94-1929-78
2-Propanone, 1-(5-phenyl-1H-1,2,4-tria-	EtOH	243(4.21),310(3.44)	94-1825-78
zol-3-yl)-, hydrazone			
$C_{11}H_{13}N_5O_3$			
9H-Purin-6-amine, 9-(5,6-dideoxy-α-D-	M HCl	256.5(4.15)	78-2627-78
arabino-hex-5-enofuranosyl)-	H_2O	259(4.17)	78-2627-78
	pH 13	232(3.58)	78-2627-78
$C_{11}H_{13}N_5O_5$			
6,7-Pteridinedicarboxylic acid, 2-am-	pH -2.0	232(4.18),282(4.12),	24-3790-78
ino-1,4,7,8-tetrahydro-8-methyl-		377(3.97)	
4-oxo-, dimethyl ester	pH 4.0	219(4.33),265(4.25),	24-3790-78
		390(3.90)	
	pH 11.0	263(4.24),404(4.13)	24-3790-78
$C_{11}H_{13}N_5O_6$			
6,7-Pteridinedicarboxylic acid, 2-am-	pH -4.0	263(4.15)	24-3790-78
ino-5-formyl-1,4,5,6,7,8-hexahydro-	pH 0.0	282(4.11)	24-3790-78
4-oxo-, dimethyl ester	pH 5.0	220(4.41),268(4.12)	24-3790-78

Compound	Solvent	$\lambda_{max}(\log \epsilon)$	Ref.
(cont.)	pH 12.0	280(4.12)	24-3790-78
$C_{11}H_{13}OP$			
4-Phosphorinanone, 1-phenyl-	MeOH	250(3.95)	70-1339-78
hydrochloride	MeOH	250(3.65)	70-1339-78
hydriodide	MeOH	265(3.00)	70-1339-78
$C_{11}H_{13}OPS$			
4-Phosphorinanone, 1-phenyl-, P-sulfide	MeOH	247(3.30)	70-1339-78
$C_{11}H_{13}OPSe$			
4-Phosphorinanone, 1-phenyl-, P-selenide	MeOH	255(3.30)	70-1339-78
$C_{11}H_{13}O_2P$			
4-Phosphorinanone, 1-phenyl-, P-oxide	MeOH	265(2.85)	70-1339-78
$C_{11}H_{14}$			
Naphthalene, 1,2,3,4,4a,8a-hexahydro-2-methylene-	ether	263(3.60)	5-2074-78
$C_{11}H_{14}BrFN_2O_5$			
2,4(1H,3H)-Pyrimidinedione, 1-(2-bromo-2,6-dideoxy-6-fluoro-β-D-glucopyranosyl)-5-methyl-	H_2O	264(4.09)	48-0157-78
$C_{11}H_{14}BrNO_5$			
1H-Pyrrole-2,4-dicarboxylic acid, 5-bromo-3-hydroxy-1-methyl-, diethyl ester	EtOH	233(4.36),267(4.15)	94-3521-78
$C_{11}H_{14}ClFN_2O_5$			
2,4(1H,3H)-Pyrimidinedione, 1-(2-chloro-2,6-dideoxy-6-fluoro-β-D-glucopyranosyl)-5-methyl-	H_2O	264(4.01)	48-0157-78
$C_{11}H_{14}ClNO_2$			
Isoquinolinium, 3,4-dihydro-7-hydroxy-6-methoxy-2-methyl-, chloride	EtOH	251(4.24),312(4.00), 370(3.86)	95-1658-78
$C_{11}H_{14}ClN_3O$			
1H-Benzotriazole, 5-chloro-1-(pentyloxy)-	EtOH	273(3.98)	4-1043-78
$C_{11}H_{14}Cl_2O_3$			
3-Furancarboxylic acid, 5-(2,2-dichloroethenyl)-4,5-dihydro-2,4,4-trimethyl-, methyl ester	MeOH	254(4.05)	25-0274-78
$C_{11}H_{14}FIN_2O_5$			
2,4(1H,3H)-Pyrimidinedione, 1-(2,6-dideoxy-6-fluoro-2-iodo-β-D-glucopyranosyl)-5-methyl-	MeOH	264(4.11)	48-0157-78
$C_{11}H_{14}FN$			
Piperidine, 1-(2,4-cyclopentadien-1-ylidenefluoromethyl)-	hexane	318(4.48)	1-0293-78
$C_{11}H_{14}F_2N_2O_5$			
2,4(1H,3H)-Pyrimidinedione, 1-(2,6-dideoxy-2,6-difluoro-β-D-gluco-	H_2O	265(4.01)	48-0157-78

Compound	Solvent	$\lambda_{max}(\log \epsilon)$	Ref.
pyranosyl)-5-methyl- (cont.)			48-0157-78
$C_{11}H_{14}NO_2$			
Isoquinolinium, 3,4-dihydro-7-hydroxy-6-methoxy-2-methyl-, chloride	EtOH	251(4.24),312(4.00), 370(3.86)	95-1658-78
$C_{11}H_{14}N_2$			
Pyrido[2,3-g]indolizine, 1,2,3,5,6,10b-hexahydro-	EtOH	255s(3.44),263(3.56), 268s(3.52),276s(3.40)	44-2125-78
$C_{11}H_{14}N_2O$			
1H-Indole-3-propanol, β-amino-, (S)-	MeOH	229(3.89),276(3.63), 283(3.66),292(3.60)	35-0938-78
2(1H)-Quinoxalinone, 3,4-dihydro-1,3,4-trimethyl-	MeOH	225(4.56),262(3.50), 308(3.61)	24-1753-78
$C_{11}H_{14}N_2OS_2$			
Benzenamine, N-[(4-morpholinylthioxomethyl)thio]-	CH_2Cl_2	252(4.02),290(4.33), 398(3.20)	48-0313-78
$C_{11}H_{14}N_2O_2$			
Diazene, [methoxy(2-propenyloxy)methyl]phenyl-, (E)-	EtOH	214(4.02),269(3.92)	44-1459-78
Propanoic acid, 2-(phenylhydrazono)-, ethyl ester	EtOH	325(4.32)	104-1402-78
$C_{11}H_{14}N_2O_3$			
DL-Alanine, N-[4-(methylamino)benzoyl]-	MeOH	297(4.07)	87-1162-78
β-Alanine, N-[4-(methylamino)benzoyl]-	MeOH	294(3.90)	87-1162-78
Glycine, N-methyl-N-[4-(methylamino)-benzoyl]-	EtOH	287(4.13)	87-1162-78
1,8-Naphthyridine-4a(2H)-carboxylic acid, 1,5,6,7-tetrahydro-2-oxo-, ethyl ester	MeOH	214(4.17),266(3.79)	24-2813-78
Piperidine, 1-[2-(5-nitro-2-furanyl)-ethenyl]-	EtOH	280(4.41),510(4.64)	103-0250-78
2,4(1H,3H)-Pyrimidinedione, 5-(3,4-dihydro-2H-pyran-2-yl)-1,3-dimethyl-	MeOH	270(3.90)	44-4110-78
2,4(1H,3H)-Pyrimidinedione, 5-(5,6-dihydro-2H-pyran-2-yl)-1,3-dimethyl-	MeOH	270(3.91)	44-4110-78
2,4(1H,3H)-Pyrimidinedione, 5-(5,6-dihydro-2H-pyran-6-yl)-1,3-dimethyl-	MeOH	270(3.90)	44-4110-78
1H-Pyrrolo[3,2-b]pyridine-3-carboxylic acid, 4,5,6,7-tetrahydro-2-methyl-5-oxo-, ethyl ester	EtOH	229(4.20),291(3.68)	94-3080-78
$C_{11}H_{14}N_2O_4$			
1H-1,2-Diazepine-1,4-dicarboxylic acid, diethyl ester	MeOH	222(3.99),300s(2.95)	33-2887-78
1,6-Naphthyridine-6(2H)-carboxylic acid, 1,5,7,8-tetrahydro-4-hydroxy-2-oxo-, ethyl ester	EtOH	285(3.80)	142-0267-78C
Pyridinium, 3-(ethoxycarbonyl)-1-[(ethoxycarbonyl)amino]-, hydroxide, inner salt	$CHCl_3$	324(3.70)	33-2887-78
2,4,6(1H,3H,5H)-Pyrimidinetrione, 5-(3-methoxy-2-butenylidene)-1,3-dimethyl-	MeOH	368(4.5)	83-0287-78
$C_{11}H_{14}N_2O_6$			
3-Pyridinecarboxamide, 1,6-dihydro-	pH 1	257(4.14),295(3.72)	87-0427-78

Compound	Solvent	λ_{max}(log ϵ)	Ref.
6-oxo-1-β-D-ribofuranosyl- (cont.)	pH 11	257(4.14),295(3.72)	87-0427-78
$C_{11}H_{14}N_2S_2$			
2-Benzothiazolesulfenamide, N,N-di-ethyl-	dioxan	<u>273(3.6)</u>	65-1352-78
$C_{11}H_{14}N_4O$			
3-Cyclohexene-1-carboxamide, 4-amino-1,3-dicyano-N-ethyl-	MeOH	265(4.15)	42-0281-78
$C_{11}H_{14}N_4O$			
Phthalazine, 1-ethoxy-4-(1-methylhydra-zino)-	dioxan	250s(3.70),303(3.72)	103-0436-78
$C_{11}H_{14}N_4O_2$			
4H-Imidazo[4,5-c]pyridin-4-one, 6-am-ino-1,5-dihydro-1-(tetrahydro-2H-pyran-2-yl)-	pH 1	285(4.13),313(3.84)	87-1212-78
	pH 7	270(4.10),300(3.97)	87-1212-78
	pH 11	272(4.09),295(3.96)	87-1212-78
4H-Imidazo[4,5-c]pyridin-4-one, 6-am-ino-3,5-dihydro-3-(tetrahydro-2H-pyran-2-yl)-	pH 1	275(4.09),317(3.79)	87-1212-78
	pH 7	257(3.83),315(3.89)	87-1212-78
	pH 11	257(3.82),315(3.87)	87-1212-78
Propanediamide, 2-[(4-ethylphenyl)hydra-zono]-	EtOH	235(4.10),365(4.31)	104-1956-78
$C_{11}H_{14}N_4O_3$			
1H-Purine-2,6-dione, 3,7-dihydro-1,3-dimethyl-8-(tetrahydro-2-furanyl)-	pH 6.0	275(4.07)	24-0982-78
	pH 12.0	278(4.16)	24-0982-78
$C_{11}H_{14}N_4O_4$			
1H-Imidazole-4-carboxamide, 5-(cyano-methyl)-1-(5-deoxy-β-D-ribofurano-syl)-	pH 1	218(4.00)	87-1212-78
	pH 7	233(4.01)	87-1212-78
	pH 11	234(3.99)	87-1212-78
1H-Imidazo[4,5-c]pyridin-6-amine, 1-β-D-ribofuranosyl-	pH 1	226(4.64),254s(3.59), 261(3.68),268(3.59), 320(3.56)	44-0289-78
	pH 7	218(4.49),256(3.70), 298(3.49)	44-0289-78
	pH 11	222(4.69),255(3.75), 297(3.56)	44-0289-78
3H-Imidazo[4,5-c]pyridin-6-amine, 3-β-D-ribofuranosyl-	pH 1	222(4.54),224(3.67), 339(3.78)	44-0289-78
	H_2O	215(4.49),253s(3.46), 314(3.53)	44-0289-78
	pH 11	221(4.22),253s(3.46), 312(3.56)	44-0289-78
4H-Imidazo[4,5-c]pyridin-4-one, 6-am-ino-1-(5-deoxy-β-D-ribofuranosyl)-1,3-dihydro-	pH 1	284(4.11),308s(3.84)	87-1212-78
	pH 7	272(4.09),299(3.96)	87-1212-78
	pH 11	272(4.09),295s(3.96)	87-1212-78
4H-Imidazo[4,5-c]pyridin-4-one, 6-am-ino-3-(5-deoxy-β-D-ribofuranosyl)-3,5-dihydro-	pH 1	277(4.07),317(3.78)	87-1212-78
	pH 7	258(3.81),317(3.88)	87-1212-78
	pH 11	258(3.81),317(3.86)	87-1212-78
1H,6H-Oxazepino[3,4-f]purine-2,4-(3H,10H)-dione, 7,8-dihydro-7-hydroxy-1,3-dimethyl-	EtOH	209(4.44),273(3.96)	73-3414-78
Theophylline, 7-(oxiranylmethyl)-8-(hydroxymethyl)-	EtOH	208(4.42),278(3.97)	73-3414-78
$C_{11}H_{14}N_4O_4S$			
3H-Imidazo[4,5-b]pyridine-7-thiol, 5-amino-3-β-D-ribofuranosyl-	pH 1	223(4.16),226(3.86), 327(3.90)	87-0112-78

Compound	Solvent	$\lambda_{max}(\log \epsilon)$	Ref.
3H-Imidazo[4,5-b]pyridine-7-thiol, 5-amino-3-β-D-ribofuranosyl- (cont.)	pH 7	213(4.19),228(4.18), 261(3.95),268s(3.94), 317(3.92)	87-0112-78
	pH 13	243(4.05),299(4.21)	87-0112-78
4H-Imidazo[4,5-c]pyridine-4-thione, 6-amino-1,5-dihydro-1-β-D-ribo-furanosyl-	pH 1	223s(4.23),245s(3.82), 291(3.86),374(4.23)	44-0289-78
	pH 7	228(4.25),253(3.82), 283(3.93),353(4.26)	44-0289-78
	pH 11	226(4.19),245s(3.98), 283(3.82),323(4.13)	44-0289-78
4H-Imidazo[4,5-c]pyridine-4-thione, 6-amino-3,5-dihydro-3-β-D-ribo-furanosyl-	pH 1	227(4.20),255(3.70), 283(3.81),378(4.14)	44-0289-78
	pH 7	231(4.30),264(3.94), 373(4.14)	44-0289-78
	pH 11	227(4.24),261(3.79), 343(3.92)	44-0289-78
1H-Pyrazolo[4,3-d]pyrimidine, 7-(methyl-thio)-1-β-D-ribofuranosyl-	MeOH	198(4.03),219(4.20), 255(3.61),259s(3.60), 303s(3.94),311(4.01), 322s(3.89)	104-0601-78
2H-Pyrazolo[4,3-d]pyrimidine, 7-(methyl-thio)-2-α-D-ribofuranosyl-	MeOH	201(4.08),218(4.20), 261(3.66),266s(3.66), 310(3.98),320(4.05), 333(3.90)	104-0601-78
2H-Pyrazolo[4,3-d]pyrimidine, 7-(methyl-thio)-2-β-D-ribofuranosyl-	MeOH	218(4.31),261(3.79), 268s(3.79),309(4.10), 320(4.16),330(4.02)	104-0601-78
$C_{11}H_{14}N_4O_5$			
Acetic acid, 1,2-diacetyl-2-(1,2,3,6-tetrahydro-1-methyl-2,6-dioxo-4-pyrimidinyl)hydrazide	0.5M HCl	265(4.28)	56-0037-78
	H_2O	206(4.04),266(4.27)	56-0037-78
	0.5M NaOH	284(4.39)	56-0037-78
Inosine, 3-methyl-	pH 7	259(4.14)	88-4047-78
$C_{11}H_{14}N_4O_5S$			
6H-Purine-6-thione, 7-β-D-gluco-furanosyl-	EtOH	333(4.26)	12-1095-78
	HOAc	331(4.30)	12-1095-78
	NH_3	319(4.27)	12-1095-78
$C_{11}H_{14}N_4O_{12}P_2$			
2,4(1H,3H)-Pteridinedione, 1-[5-O-[hydroxy(phosphonooxy)phosphinyl]-β-D-ribofuranosyl]-	pH 1.0	228(4.116),314(3.865)	5-1788-78
	pH 7.0	229(4.109),315(3.851)	5-1788-78
	pH 13.0	238(4.163),278s(3.600), 324(3.861)	5-1788-78
$C_{11}H_{14}N_5O_9P$			
4,7(1H,8H)-Pteridinedione, 2-amino-8-(5-O-phosphono-β-D-ribofuranosyl)-	pH 1.0	292(3.977),348(4.083)	5-1780-78
	pH 7.0	292(3.971),350(4.097)	5-1780-78
	pH 13.0	260(4.061),283s(3.613), 363(4.114)	5-1780-78
$C_{11}H_{14}N_6O_5$			
Adenosine, N-(aminocarbonyl)-	pH 1	211(4.38),275(4.32)	94-2122-78
	H_2O	210(4.39),267(4.33), 274s(4.25)	94-2122-78
	pH 13	279s(4.16),293(4.33)	94-2122-78
	EtOH	268(4.30),275s(4.20)	39-0131-78C
$C_{11}H_{14}O$			
Benzene, 2-(ethenyloxy)-1,3,5-trimeth-yl-	heptane	276(2.66)	64-0197-78B

Compound	Solvent	$\lambda_{max}(\log \epsilon)$	Ref.
Benzofuran, 2,3-dihydro-2,2,4-trimethyl-	EtOH	280(3.34)	77-0526-78
Benzofuran, 2,3-dihydro-3,5,6-trimethyl-	EtOH	287(3.58)	23-0517-78
2,4,6-Cycloheptatrien-1-one, 2,4,5,7-tetramethyl-	H_2O	239(4.25),339(3.76), 346(3.76)	35-1778-78
2,5-Cyclohexadien-1-one, 4-methyl-4-(2-methyl-2-propenyl)-	n.s.g.	235(4.06)	33-0401-78
1-Propanone, 2,2-dimethyl-1-phenyl-	C_6H_{12}	322(1.90),359(1.23)	112-0751-78
Tricyclo[3.1.0.0^{2,6}]hexan-3-one, 1,2,5,6-tetramethyl-	EtOH	215(3.7),314(1.3)	44-1912-78
$C_{11}H_{14}O_2$			
Benzaldehyde, 3-(1,1-dimethylethyl)-2-hydroxy-	EtOH	218(4.23),263(4.06), 337(3.54)	32-0079-78
Benzaldehyde, 3-(1,1-dimethylethyl)-4-hydroxy-	EtOH	228(4.13),295(4.20)	12-0907-78
Benzenepropanoic acid, β-ethyl-	EtOH	237-268f(2.4)	44-4316-78
Bicyclo[3.2.1]oct-3-ene-2,7-dione, 1,3,4-trimethyl-	EtOH	256(3.80)	78-0951-78
Ethanone, 1-[6-(2-methyl-1-propenyl)-2-oxabicyclo[3.1.0]hex-3-en-6-yl]-, (1α,5α,6β)-	MeCN	270(4.10),300s(4.03)	78-2797-78
$C_{11}H_{14}O_3$			
Benzaldehyde, 2-(2-ethoxyethoxy)-	n.s.g.	214(4.36),251(4.06), 318(3.71)	47-1343-78
Benzaldehyde, 3-(2-ethoxyethoxy)-	n.s.g.	219(4.41),252(4.00), 310(3.50)	47-1343-78
Benzaldehyde, 4-(2-ethoxyethoxy)-	n.s.g.	197(4.24),219(4.15), 275(4.30)	47-1343-78
Benzaldehyde, 2-hydroxy-5-(3-hydroxybutyl)-	MeOH	223(4.39),260(4.14), 339(3.67)	95-1607-78
	MeOH-base	236(4.39),266(4.00), 395(3.93)	95-1607-78
Benzoic acid, 3-(1,1-dimethylethyl)-4-hydroxy-	EtOH	261(4.13)	12-0907-78
2-Cyclopenten-1-one, 2,3-dimethyl-4-(1-oxopropoxy)methylene]-, (E)-	n.s.g.	287(4.25)	33-0266-78
Ethanone, 1-(2,5-dihydroxy-3,4,6-trimethylphenyl)- (plus shoulders)	EtOH	277(3.82),365(3.30)	44-3723-78
Ethanone, 1-(2,6-dihydroxy-3,4,5-trimethylphenyl)-	EtOH	210s(4.23),224s(4.04), 280(4.17),360(3.45)	44-3723-78
Ethanone, 1-(2-ethoxy-4-methoxyphenyl)-	EtOH	229(4.14),269(4.07), 306(3.89)	95-0503-78
3a(1H)-Pentalenecarboxylic acid, 2,3,4,5-tetrahydro-5-oxo-, ethyl ester	MeOH	232(3.99),295(1.54)	44-2557-78
$C_{11}H_{14}O_3S$			
2-Butanone, 3-[(phenylmethyl)sulfonyl]-	MeOH	219(3.9),253(2.3), 259(2.3),265(2.4), 270(2.3)	12-1965-78
$C_{11}H_{14}O_4$			
Benzaldehyde, 3,6-dihydroxy-2-methoxy-4-propyl-	EtOH	246(4.12),290(4.13), 366(3.76)	73-1438-78
Benzenepropanoic acid, 2-hydroxy-4-methoxy-, methyl ester	EtOH	211(3.95),222s(3.82), 281(3.34)	102-1383-78
Bornane-4-carboxylic acid, 2,3-dioxo-	n.s.g.	459(1.44)	78-1845-78

Compound	Solvent	$\lambda_{max}(\log \epsilon)$	Ref.
1-Butanone, 1-(2,4,6-trihydroxy-3-methylphenyl)-	EtOH	286(4.24)	142-1355-78A
Ethanone, 1,1',1'',1'''-cyclopropanedi-ylidenetetrakis-	EtOH	280(2.18)	44-3070-78
Ethanone, 1-[5-(4-methyl-1,3-dioxan-2-yl)-2-furanyl]-	EtOH	278(4.18)	103-0940-78
2,4-Hexadienal, 2-(1-acetoxy-2-oxo-propyl)-, [R-(E,E)]-	EtOH	278(4.28)	88-3527-78
2H-Pyran-6-carboxylic acid, 4-(1,1-di-methylethyl)-2-oxo-, methyl ester	MeOH	207(4.29),298(3.80)	35-5472-78
1H,3H-Pyrano[3.4-c]pyran-1-one, 5-eth-enyl-4,4a,5,6-tetrahydro-6-methoxy-, (±)-	MeOH	243.5(3.95)	33-1221-78
$C_{11}H_{14}O_4S$			
Cyclopenta[cd]pentalen-1(2H)-one, 2a,3,4,4a,5,6b-hexahydro-3-[(meth-ylsulfonyl)oxy]-	CHCl$_3$	243(3.97)	23-1687-78
$C_{11}H_{14}O_6$			
3-Furancarboxylic acid, 5-acetoxy-4,5-dihydro-2,5-dimethyl-4-oxo-, ethyl ester	EtOH	212(3.9),262(3.93)	118-0291-78
2,4-Hexadienedioic acid, 2-acetoxy-, 1-(1-methylethyl) ester, (Z,E)-	EtOH	272(4.5)	5-1734-78
$C_{11}H_{14}S$			
Benzene, 2-(ethenylthio)-1,3,5-tri-methyl-	heptane	271(3.43)	64-0197-78B
$C_{11}H_{15}BrN_2O_7$			
1H-Imidazole-4-carboxylic acid, 5-bro-mo-2,3-dihydro-3-methyl-2-oxo-1-β-D-ribofuranosyl-	MeOH	280(4.02)	94-3322-78
$C_{11}H_{15}ClN_2O_2$			
1H-Pyrrole-3-carboxaldehyde, 2-chloro-4-[(diethylamino)methylene]-4,5-di-hydro-1-methyl-5-oxo-	EtOH	298(4.38),405(4.34)	104-2041-78
$C_{11}H_{15}ClN_5O_7P$			
5'-Adenylic acid, 2-chloro-N-methyl-	pH 1	272(4.18)	87-0520-78
$C_{11}H_{15}FN_2O_5$			
2,4(1H,3H)-Pyrimidinedione, 1-(2,6-di-deoxy-6-fluoro-β-D-glucopyranosyl)-5-methyl-	H$_2$O	265(4.03)	48-0157-78
$C_{11}H_{15}FN_2O_6$			
2,4(1H,3H)-Pyrimidinedione, 1-(6-deoxy-6-fluoro-β-D-mannopyranosyl)-5-meth-yl-	H$_2$O	266.5(3.95)	48-0157-78
$C_{11}H_{15}NO$			
Benzene, 2-butyl-4-methyl-1-nitroso-	CHCl$_3$	765(1.51)	44-2932-78
Phenol, 4-(dimethylamino)-3-(2-prop-enyl)-	MeOH	238(3.65)	94-2027-78
	MeOH-NaOH	244(--),300(--)	94-2027-78
Propanamide, 2-methyl-N-(4-methyl-phenyl)-	MeOH	230(3.90)	2-1067-78

Compound	Solvent	$\lambda_{max}(\log \epsilon)$	Ref.
$C_{11}H_{15}NO_2$			
Benzeneacetic acid, 4-(dimethylamino)-, methyl ester	EtOH	258(4.20)	78-1889-78
Propanamide, N-(3-methoxyphenyl)-2-methyl-	MeOH	280(3.63)	2-1067-78
1-Propanol, 2-methyl-2-[(phenylmethylene)amino]-, N-oxide	hexane	305(c.4.18)	44-1900-78
$C_{11}H_{15}NO_3$			
2(1H)-Pyridinone, 1-(2-acetoxy-2-methylpropyl)-	EtOH	300(3.62)	78-2609-78
1H-Pyrrole-2-carboxylic acid, 2,3-dihydro-1-methyl-3-oxo-2-(2-propenyl)-, ethyl ester	EtOH	332(3.88)	94-3521-78
1H-Pyrrole-2-carboxylic acid, 1-methyl-3-(2-propenyloxy)-, ethyl ester	EtOH	267(4.05)	94-3521-78
$C_{11}H_{15}NO_4$			
1,3-Propanediol, 2-(hydroxymethyl)-2-[(phenylmethylene)amino]-, N-oxide	EtOH	298(c.4.18)	44-1900-78
$C_{11}H_{15}NO_5$			
1H-Pyrrole-2,4-dicarboxylic acid, 3-hydroxy-1-methyl-, diethyl ester	EtOH	229(4.44),262(4.10)	94-2224-78
$C_{11}H_{15}NO_6$			
1H-Pyrrole-2-acetic acid, α-(1-hydroxyethyl)-4-hydroxy-5-(methoxycarbonyl)-, methyl ester	MeOH	270(4.23)	35-6491-78
$C_{11}H_{15}N_2S$			
1H-Imidazolium, 4,5-dihydro-1-methyl-2-(methylthio)-3-phenyl-, iodide	EtOH CH$_2$Cl$_2$	218(3.233),258(3.932) 244(3.320)	28-0521-78A 28-0521-78A
$C_{11}H_{15}N_3O$			
5H-Dipyrido[1,2-a:2',3'-d]pyrimidin-5-one, 1,2,3,4,7,8,9,10-octahydro-	MeOH	230(4.11),269(4.06)	24-2297-78
Ethanehydrazonamide, N,N,N'-trimethyl-2-oxo-N'-phenyl-	heptane	227(3.75),340(4.44)	104-0576-78
$C_{11}H_{15}N_3OS$			
5H-Pyrido[2',3':4,5]pyrimido[1,2-d]-[1,4]thiazepin-5-one, 1,2,3,4,7,8-10,11-octahydro-	MeOH	227(4.17),279(4.12)	24-2297-78
$C_{11}H_{15}N_3O_2$			
Hydrazinecarboxamide, 2-[2-(4-methoxyphenyl)-1-methylethylidene]-	EtOH	266(4.18)	95-0503-78
5-Oxazolamine, N-(2,4-dimethyl-5-oxazolyl)-N,2,4-trimethyl-	n.s.g.	233(3.81)	70-0963-78
$C_{11}H_{15}N_3O_4$			
2,6-Pyridinedimethanol, bis(methylcarbamate)	EtOH	263(3.73)	95-1402-78
$C_{11}H_{15}N_3O_6$			
3-Pyridinecarboxylic acid, 1,6-dihydro-6-oxo-1-β-D-ribofuranosyl-	pH 1 pH 11	261(4.17),295(3.75) 255(4.15),295(3.75)	87-0427-78 87-0427-78
$C_{11}H_{15}N_5O_4$			
Adenosine, 2-methyl-	pH 1.0	204(4.328),257(4.127)	5-1796-78

Compound	Solvent	λ_{max}(log ϵ)	Ref.
Adenosine, 2-methyl- (cont.)	pH 7.0	209(4.380),262(4.149)	5-1796-78
Adenosine, 2'-O-methyl-	pH 7	258(4.17)	39-0762-78C
Adenosine, 3'-O-methyl-	pH 7	258(4.17)	39-0762-78C
1H-Imidazo[4,5-c]pyridine-4,6-diamine, 1-β-D-ribofuranosyl-	pH 1	217(4.47),271(4.06), 315(3.95)	44-0289-78
	pH 7	217(4.47),272(4.05), 295s(3.85)	44-0289-78
	pH 14	218(4.27),273(4.02), 288s(3.95)	44-0289-78
3H-Imidazo[4,5-c]pyridine-4,6-diamine, 3-β-D-ribofuranosyl-	pH 1	214(4.30),273(3.84)	44-0289-78
	pH 7	218(3.36),248(3.63), 318(3.69)	44-0289-78
	pH 11	222(4.19),248(3.63), 313(3.71)	44-0289-78
9H-Purin-6-amine, 9-(6-deoxy-α-D-altro-furanosyl)-	H_2O	260(4.13)	44-0962-78
9H-Purin-6-amine, 9-(6-deoxy-β-D-galactofuranosyl)-L-	H_2O	259(4.18)	78-2627-78
	H_2O	259(4.17)	44-0962-78
9H-Purin-6-amine, 9-(5-deoxy-α-L-arabino-hexofuranosyl)-	MeOH	260(4.16)	136-0089-78C
9H-Purin-6-amine, 9-(5-deoxy-β-D-xylo-hexofuranosyl)-	MeOH	260(4.19)	136-0089-78C
$C_{11}H_{15}N_5O_4S$			
3H-1,2,3-Triazolo[4,5-b]pyridin-5-amine, 7-(methylthio)-3-β-D-ribofuranosyl-	pH 1	235s(3.87),297(4.17), 310s(4.22),324s(4.06)	44-4910-78
	pH 11	240(4.00),296(4.22), 310s(4.17)	44-4910-78
$C_{11}H_{15}N_5O_5$			
Guanosine, 1-methyl-	pH 7	257(4.04)	39-0762-78C
Guanosine, 2'-O-methyl-	pH 7.0	254(4.15)	39-0762-78C
Guanosine, 3-methyl-	H_2O	216(4.28),243(3.86), 263(3.87)	88-2579-78
	pH 7	217(4.44),250(4.00), 265(4.07)	88-2907-78
	pH 13	250(4.00),265(4.07)	88-2907-78
Guanosine, O^6-methyl-	n.s.g.	248(4.09),278(4.08)	39-0762-78C
7H-Purin-6-amine, 7-β-D-glucopyranosyl-	pH 1	273(4.16)	12-1095-78
	H_2O	271(3.99)	12-1095-78
	pH 13	247s(--),271(4.01)	12-1095-78
9H-Purin-6-amine, 9-α-D-mannofuranosyl-	MeOH	259(4.16)	39-1381-78C
9H-Purin-6-amine, 9-β-D-mannofuranosyl-	MeOH	260(4.17)	39-1381-78C
$C_{11}H_{15}N_5O_6S$			
9H-Purin-6-amine, 9-[2-O-(methylsulfon-yl)-β-D-arabinofuranosyl]-	pH 1	256(4.18)	78-1133-78
	neutral	258.5(4.18)	78-1133-78
	pH 13	259(4.18)	78-1133-78
$C_{11}H_{15}N_5O_7S$			
6H-Purin-6-one, 2-amino-1,9-dihydro-9-[2-O-(methylsulfonyl)-β-D-arabino-furanosyl]-	pH 1	256(4.12),280s(3.93)	94-0240-78
	H_2O	252(4.15),270s(4.00)	94-0240-78
	pH 13	262(4.09)	94-0240-78
$C_{11}H_{15}N_5O_{12}P_2$			
4,7(1H,8H)-Pteridinedione, 2-amino-8-[5-O-[hydroxy(phosphonooxy)phos-phinyl]-β-D-ribofuranosyl]-	pH 1.0	214(4.581),292(4.029), 349(4.149)	5-1788-78
	pH 7.0	214(4.525),292(4.004), 351(4.152)	5-1788-78
	pH 13.0	260(4.086),283s(3.625),	5-1788-78

Compound	Solvent	$\lambda_{max}(\log \epsilon)$	Ref.
(cont.)		364(4.167)	5-1788-78
$C_{11}H_{16}$			
Cyclopropane, 1-ethenyl-2-(hexa-trans-1,cis-3-dienyl)-, trans	EtOH	245(4.46)	39-1338-78B
$C_{11}H_{16}Cl_2N_2$			
Pyrazine, 2,5-dichloro-3-(1-methylethyl)-6-(2-methylpropyl)-	EtOH	285(3.87),299(3.92)	94-2046-78
$C_{11}H_{16}NO$			
Pyrylium, 4-[2-(dimethylamino)ethenyl]-2,6-dimethyl-, perchlorate	HOAc-HClO$_4$	402(4.665)	83-0236-78
$C_{11}H_{16}N_2$			
1H-1,5-Benzodiazepine, 2,3,4,5-tetra-1,5-dimethyl-	hexane	236(4.42),266(3.97),303s(3.59)	44-2621-78
$C_{11}H_{16}N_2O$			
Benzamide, N-(4-aminobutyl)-	EtOH	224(4.00)	105-0637-78
Ethanone, 2-cyclohexyl-1-(1H-imidazol-2-yl)-	EtOH	280(4.09)	33-2831-78
Ethanone, 2-cyclohexyl-1-(1H-imidazol-4-yl)-	EtOH	258(4.14)	33-2831-78
5-Hexen-1-one, 1-(1H-imidazol-2-yl)-4,4-dimethyl-	EtOH	277(4.12)	33-2831-78
5-Hexen-1-one, 1-(1H-imidazol-4-yl)-4,4-dimethyl-	EtOH	257(4.05)	33-2831-78
1H-Isoindol-1-one, 3-(dimethylamino)-3a,4,5,7a-tetrahydro-3a-methyl-	n.s.g.	244.5(4.36)	33-0815-78
Methanone, (3,3-dimethylcyclopentyl)-1H-imidazol-2-yl-	EtOH	277(4.12)	33-2831-78
Methanone, (3,3-dimethylcyclopentyl)-1H-imidazol-4-yl-	EtOH	255(4.12)	33-2831-78
2,4-Pentadienamide, N-[1-(dimethylamino)-2-methyl-2-propenylidene]-, (E,E)-	n.s.g.	253(4.33),290(4.22)	33-0815-78
Pyridine, 2-(2,5-dimethyl-1-pyrrolidinyl)-, 1-oxide, (2R-trans)-	MeOH	245(4.32),277(3.90),335(3.58)	78-2533-78
4(1H)-Pyrimidinone, 2,6-dimethyl-5-(4-pentenyl)-	EtOH	230(3.73),274(3.70)	39-1293-78C
$C_{11}H_{16}N_2OS_2$			
4(1H)-Quinazolinone, 2,3,5,6,7,8-hexahydro-3-(2-mercaptopropyl)-2-thioxo-	MeOH	222(4.12),287(4.25)	106-0185-78
	MeOH-NaOH	267(4.05),318(4.14)	106-0185-78
$C_{11}H_{16}N_2O_2$			
Diazene, (methoxypropoxymethyl)phenyl-, (E)-	EtOH	215(4.01),268(3.96)	44-1459-78
3-Pyridinecarboxylic acid, 4,6-dimethyl-2-(methylamino)-, ethyl ester	EtOH	215(4.19),257(3.89),341(3.75)	88-4135-78 +103-1139-78
2-Pyrimidineacetic acid, 4-(1,1-dimethylethyl)-, methyl ester	neutral cation	248(3.60),275(2.65) 250(3.76)	12-0649-78 12-0649-78
$C_{11}H_{16}N_2O_2S_2$			
2-Pyrimidineacetic acid, 4,6-bis(ethylthio)-, methyl ester	neutral cation	254(4.34),300(4.16) 252(3.66),268(3.71),286(3.69),326(4.11)	12-0649-78 12-0649-78
$C_{11}H_{16}N_2O_3$			
2,4(1H,3H)-Pyrimidinedione, 1,3-dimeth-	MeOH	270(3.90)	44-4110-78

Compound	Solvent	$\lambda_{max}(\log \epsilon)$	Ref.
yl-5-(tetrahydro-2H-pyran-2-yl)-			44-4110-78
$C_{11}H_{16}N_2O_4$			
3-Piperidinecarbonitrile, 2-methoxy-3-(methoxymethyl)-5-(methoxymethylene)-6-oxo-	MeOH	245(4.15)	94-0038-78
$C_{11}H_{16}N_2O_5$			
2,4(1H,3H)-Pyrimidinedione, 1-(3,3-dimethoxy-2-oxobutyl)-3-methyl-	MeCN	260(4.00)	78-2861-78
2,4(1H,3H)-Pyrimidinedione, 1-(3,3-dimethoxy-2-oxobutyl)-5-methyl-	MeCN	265(3.98)	78-2861-78
$C_{11}H_{16}N_2O_6$			
Uridine, 5',5'-dimethyl-	pH 1	264(3.92)	88-4403-78
	pH 13	218(3.92),287(3.80)	88-4403-78
	MeOH	212(3.90),263(3.82)	88-4403-78
Uridine, 3-methyl-2'-O-methyl-	pH 7	262(3.94)	39-0762-78C
Uridine, 3-methyl-3'-O-methyl-	pH 7	262(3.94)	39-0762-78C
$C_{11}H_{16}N_2O_7$			
1H-Imidazole-4-carboxylic acid, 2,3-dihydro-3-methyl-2-oxo-1-β-D-ribofuranosyl-, methyl ester	MeOH	270(4.10)	94-3322-78
$C_{11}H_{16}N_2S$			
Butanethioic acid, 2-methyl-2-phenylhydrazide	MeOH	212(4.14),227(4.02), 247(4.12),280(3.91)	123-0701-78
$C_{11}H_{16}N_4O$			
Hydrazinecarboxamide, 2-(2-amino-2-methyl-1-phenylpropylidene)-	hexane	207(4.08),230(3.81)	103-0278-78
$C_{11}H_{16}N_4O_2$			
Hydrazinecarboxamide, 2-butyl-2-nitroso-N-phenyl-	H_2O	237(4.35),350(2.07)	104-2319-78
$C_{11}H_{16}N_4O_5$			
Inosine, 1,2-dihydro-3-methyl-	pH 1	261(3.69)	88-4047-78
	pH 7	267(3.69)	88-4047-78
	pH 13	268(3.69)	88-4047-78
2H-Pyran-3(6H)-one, 6-(acetoxymethyl)-4-azido-2-(1-methylethoxy)-, 3-oxime, [2S-(2α,3Z,6β)]-	MeOH	251(4.1)	24-3912-78
1H-1,2,4-Triazole-5-carboxamide, 1-(2,3-O-(1-methylethylidene)-β-D-ribofuranosyl-	MeOH	211(4.15)	35-2248-78
$C_{11}H_{16}N_4O_6$			
4-Pyrimidinecarboximidic acid, 6-amino-2,3-dihydro-2-oxo-3-β-D-ribofuranosyl-, methyl ester	H_2O	272(3.90)	94-2340-78
$C_{11}H_{16}N_5O_6PS_2$			
Adenosine, 2-(methylthio)-, 5'-(dihydrogen phosphorothioate)	pH 1	268(4.15)	87-0520-78
$C_{11}H_{16}N_5O_7P$			
5'-Adenylic acid, 2-methyl-	pH 2.0	204(4.316),256(4.107)	5-1796-78
	pH 7.0	208(4.364),261(4.137)	5-1796-78
polymer	pH 7.3	257(3.959)	5-1796-78

Compound	Solvent	$\lambda_{max}(\log \epsilon)$	Ref.
$C_{11}H_{16}N_5O_7PS$			
5'-Adenylic acid, 2-(methylthio)-	pH 1	268(4.22)	87-0520-78
$C_{11}H_{16}N_5O_8P$			
5'-Adenylic acid, 2-methoxy-	pH 1	251(3.82),273(4.05)	87-0520-78
$C_{11}H_{16}N_6O_2$			
Cyclopentanemethanol, 2-amino-4-(6-am-	pH 1	258(4.16)	44-2311-78
ino-9H-purin-9-yl)-3-hydroxy-	H_2O	260(4.17)	44-2311-78
	pH 13	260(4.17)	44-2311-78
$C_{11}H_{16}N_6O_3$			
Adenosine, 2'-amino-2'-deoxy-N-methyl-,	pH 2	262(4.26)	94-0985-78
dihydrochloride	H_2O	266(4.21)	94-0985-78
	pH 12	266(4.21)	94-0985-78
$C_{11}H_{16}N_6O_6S_2$			
Adenosine, 2-(methylthio)-, 5'-sulfam-ate	pH 1	273(4.28)	87-0520-78
$C_{11}H_{16}O$			
Benzenemethanol, α-(1,1-dimethylethyl)-	MeOH	210(3.88),258(2.30)	104-1942-78
2-Buten-1-one, 1-(2-methyl-2-cyclo-hexen-1-yl)-	n.s.g.	243(3.87)	70-0215-78
2-Cyclopenten-1-one, 3-(3-hexenyl)-	EtOH	228(4.18)	33-0990-78
2-Cyclopenten-1-one, 3-methyl-2-(2-pen-tenyl)-, (Z)- (jasmone)	EtOH	234(4.10)	33-0990-78
2H-Inden-2-one, 1,4,5,6,7,7a-hexahydro-1,3-dimethyl-, cis	EtOH	240(4.12)	35-1799-78
trans	EtOH	241(4.10)	35-1799-78
5H-Inden-5-one, 1,2,3,6,7,7a-hexahydro-4,7a-dimethyl-	EtOH	250(4.02)	39-1606-78C
$C_{11}H_{16}OS_2$			
Cyclohexanecarboxaldehyde, 4-(1,3-di-thian-5-ylidene)-	n.s.g.	259(4.04),310(2.48)	33-3087-78
$C_{11}H_{16}O_2$			
3H-2-Benzopyran-3-one, 4-ethyl-1,5,6,7,8,8a-hexahydro-	MeOH	234(3.98)	44-1248-78
4,9-Decadienoic acid, 8-methylene-, cis	EtOH	224(4.27)	149-0007-78
trans	EtOH	224(4.18)	149-0007-78
Limonenecarboxylic acid	EtOH	225(3.97)	78-0599-78
$C_{11}H_{16}O_2S$			
Benzenesulfinic acid, 4-methyl-, butyl ester	hexane	202(3.76),224(3.69), 255(3.24)	118-0441-78
Benzenesulfinic acid, 4-methyl-, 1,1-dimethylethyl ester	hexane	203(4.15),224(4.11), 252(3.75)	118-0441-78
$C_{11}H_{16}O_3$			
Bicyclo[3.3.1]nonan-3-one, 1-acetoxy-	EtOH	250s(1.65)	44-3653-78
2-Cyclohepten-1-one, 6-acetoxy-2,7-di-methyl-	EtOH	240(3.85)	35-1778-78
1-Cyclohexene-1-acetic acid, 2-formyl-, ethyl ester	EtOH	247(4.1)	23-0424-78
2-Cyclopenten-1-one, 5-acetoxy-2-(2-methylpropyl)-	EtOH	232(3.89)	107-0155-78
2-Hepten-6-ynoic acid, 3-ethoxy-, ethyl ester	EtOH	236(4.17)	22-0131-78

Compound	Solvent	$\lambda_{max}(\log \epsilon)$	Ref.
1-Oxaspiro[4.5]dec-3-en-2-one, 4-methoxy-3-methyl-	EtOH	229(4.10)	4-1493-78
2-Propenoic acid, 3-(1-methyl-2-oxocyclohexyl)-, methyl ester, cis	EtOH	210(3.8)	23-0419-78
$C_{11}H_{16}O_3S$			
Benzenesulfonic acid, 4-methyl-, butyl ester	n.s.g.	261(4.08)	124-0844-78
Benzenesulfonic acid, 4-methyl-, 2-methylpropyl ester	n.s.g.	225(4.63)	124-0844-78
2-Butanol, 3-[(phenylmethyl)sulfonyl]-, [R-(R*,S*)]-	MeOH	218(4.09),254(1.36), 259(1.41)	12-1965-78
$C_{11}H_{16}O_3S_2$			
5H-1,4-Dithiepin-2-butanal, 3-acetoxy-6,7-dihydro-	n.s.g.	220s(--),259(3.90)	33-3087-78
$C_{11}H_{16}O_4$			
Cyclopenta[b]pyran-3-carboxylic acid, 2,4aα,5,6,7,7aα-hexahydro-2α-methoxy-, methyl ester, (±)-	MeOH	211(3.85)	24-2423-78
Cyclopenta[c]pyran-4-carboxylic acid, 1,4a,5,6,7,7a-hexahydro-1α-methoxy-1β-	MeOH	238(4.05)	24-2423-78
	MeOH	238(3.98)	24-2423-78
2-Furanacetic acid, α-(1-methylethoxy)-, ethyl ester	EtOH	270(4.07)	35-7934-78
$C_{11}H_{16}O_4S$			
Ethanol, 2-ethoxy-, 4-methylbenzenesulfonate	n.s.g.	224(4.88)	124-0844-78
$C_{11}H_{16}O_6$			
1,5-Cyclohexadiene-1-carboxylic acid, 3-hydroxy-4-[(2-methoxyethoxy)methoxy]-, trans	EtOH	277(3.34)	77-0869-78
$C_{11}H_{17}BrN_2O$			
2H-Azepin-2-one, 4-bromo-1,5,6,7-tetrahydro-3-piperidino-	EtOH	308(3.43)	103-0306-78
$C_{11}H_{17}BrO$			
1,5-Decadien-4-one, 6-bromo-2-methyl-	EtOH	247(3.86)	70-1361-78
$C_{11}H_{17}BrO_2$			
2-Cyclohexen-1-one, 2-bromo-3-hydroxy-5-pentyl-	EtOH	292(4.36)	2-0970-78
$C_{11}H_{17}ClN_2$			
Pyrazine, 3-chloro-2-(1-methylethyl)-5-(2-methylpropyl)-	EtOH	282(3.92),302s(3.53)	94-2027-78
Pyrazine, 3-chloro-5-(1-methylethyl)-2-(2-methylpropyl)-	EtOH	282(3.98),301s(3.63)	94-2046-78
$C_{11}H_{17}ClN_2O$			
Pyrazine, 3-chloro-2-(1-methylethyl)-5-(2-methylpropyl)-, 1-oxide	EtOH	236(4.31),276(4.05), 305(3.58),316(3.52)	94-2046-78
Pyrazine, 3-chloro-5-(1-methylethyl)-2-(2-methylpropyl)-, 1-oxide	EtOH	236(4.25),275(3.99), 305(3.55),315(3.51)	94-2046-78
$C_{11}H_{17}ClN_2O_2$			
Pyrazine, 3-chloro-2-(1-methylethyl)-	EtOH	248(4.39),316(4.23)	94-2046-78

Compound	Solvent	$\lambda_{max}(\log \epsilon)$	Ref.
5-(2-methylpropyl)-, 1,4-dioxide			94-2046-78
Pyrazine, 3-chloro-5-(1-methylethyl)-2-(2-methylpropyl)-, 1,4-dioxide	EtOH	248(4.43),315(4.28)	94-2046-78
$C_{11}H_{17}ClO$			
2-Buten-1-one, 3-chloro-1-(2-methyl-cyclohexyl)-, cis-trans mixture	n.s.g.	243(3.85)	70-0215-78
isomer	n.s.g.	222(4.09)	70-0215-78
1-Propanone, 1-(2-chloro-2-cyclohepten-1-yl)-2-methyl-	hexane	276(2.48)	44-0604-78
$C_{11}H_{17}FN_4O_4$			
4-Pyrimidinamine, 1-β-D-arabinofurano-syl-5-fluoro-1,2-dihydro-N-methyl-2-(methylimino)-, hydrochloride	MeOH	213(4.19),248(4.17),280(3.88)	44-4200-78
picrate	MeOH	213(4.54),284(3.97),355(4.17)	44-4200-78
$C_{11}H_{17}FO$			
2-Buten-1-one, 3-fluoro-1-(2-methyl-cyclohexyl)-	n.s.g.	229(3.93)	70-0215-78
$C_{11}H_{17}N$			
Quinoline, 2,3,4,4a,5,6-hexahydro-6,6-dimethyl-	C_6H_{12}	223(4.23),282(2.23)	33-1025-78
$C_{11}H_{17}NO_3$			
3,5-Hexadien-2-one, 6-(dimethylamino)-3-acetoxy-5-methyl-	H_2O	290(3.52),400(4.69)	70-0107-78
2H-Pyran-5-carboxylic acid, 2-(dimeth-ylamino)-3,6-dimethyl-, methyl ester	hexane	210(3.98),235(3.98),294(3.68)	70-0107-78
$C_{11}H_{17}NO_4$			
Propanedioic acid, [3-(dimethylamino)-2-methyl-2-propenylidene]-, dimethyl ester	hexane	350(4.52)	70-0107-78
Propanedioic acid, (1-piperidinylmeth-ylene)-, dimethyl ester	isooctane	224(3.61),286(4.23)	78-2321-78
$C_{11}H_{17}NO_5$			
Butanedioic acid, (2-methyl-2-pyrroli-dinyl)oxo-, dimethyl ester	MeOH	277(4.28)	83-0977-78
$C_{11}H_{17}NO_5S$			
Furan, 2-[(hexylsulfonyl)methyl]-5-nitro-	MeOH	210(3.91),237s(3.60),313(4.00)	73-0156-78
$C_{11}H_{17}N_3$			
1-Propanone, 2-amino-2-methyl-1-phen-yl-, methylhydrazone	hexane	210(4.00),278(3.26)	103-0278-78
$C_{11}H_{17}N_3O_3$			
Carbamic acid, [2-[3-(aminocarbonyl)-1(4H)-pyridinyl]ethyl]-, ethyl ester	$CHCl_3$	243(2.7),353(3.7)	24-2594-78
$C_{11}H_{17}N_4O_{13}P_3S$			
9H-Purine, 9-[5-O-[hydroxy[[hydroxy-(phosphonooxy)phosphinyl]oxy]phos-phinyl]-β-D-ribofuranosyl]-6-(meth-ylthio)-	EtOH	287s(--),290(4.28)	69-0094-78

Compound	Solvent	$\lambda_{max}(\log \epsilon)$	Ref.
$C_{11}H_{17}N_5OS$ 6H-Purin-6-one, 9-[2-(diethylamino)ethyl]-1,2,3,9-tetrahydro-2-thioxo-	EtOH	288(4.21),296(4.21)	2-0786-78
$C_{11}H_{17}N_5S_2$ 1H-Purine-2,6-dithione, 9-[2-(diethylamino)ethyl]-3,9-dihydro-	pH 1	254(3.87),294(4.42), 343(4.08)	2-0786-78
	pH 13	280(4.40),327(4.06)	2-0786-78
$C_{11}H_{17}N_6O_7P$ 5'-Adenylic acid, 2-(methylamino)-	pH 1	255(4.08),298(3.93)	87-0520-78
$C_{11}H_{18}$ Cyclopropane, 1-ethenyl-2-(trans-1-hexenyl)-, trans	EtOH	207(4.20)	39-1338-78B
$C_{11}H_{18}NO$ 1-Azetidinyloxy, 2,2-dimethyl-4,4-di-2-propenyl-	C_6H_{12}	238(3.40),465(1.02)	88-0795-78
	MeOH	230(3.43),420(1.10)	88-0795-78
$C_{11}H_{18}N_2$ Pyrazine, 2-(1-methylethyl)-5-(2-methylpropyl)-	EtOH	273(3.96),277(3.95)	94-2046-78
$C_{11}H_{18}N_2O$ 1,4,6-Heptatrien-3-one, 1,7-bis(dimethylamino)-	EtOH	430(4.86)	70-0102-78
Pyrazine, 2-(1-methylethyl)-5-(2-methylpropyl)-, 1-oxide	EtOH	226(4.35),270(4.11), 299s(3.65),307s(3.55)	94-2046-78
Pyrazine, 2-(1-methylethyl)-5-(2-methylpropyl)-, 4-oxide	EtOH	227(4.26),270(4.02), 298s(3.57),307s(3.49)	94-2046-78
2(1H)-Pyrazinone, 3-(1-methylethyl)-6-(2-methylpropyl)-	EtOH	228(3.88),324(3.89)	94-2046-78
2(1H)-Pyrazinone, 6-(1-methylethyl)-3-(2-methylpropyl)-	EtOH	227(3.91),325(3.92)	94-2046-78
$C_{11}H_{18}N_2O_2$ Pyrazine, 2-(1-methylethyl)-5-(2-methylpropyl)-, 1,4-dioxide	EtOH	239(3.54),311(3.54)	94-2046-78
2(1H)-Pyrazinone, 1-hydroxy-3-(1-methylethyl)-6-(2-methylpropyl)-	EtOH	238(4.18),341(3.93)	94-2046-78
2(1H)-Pyrazinone, 1-hydroxy-6-(1-methylethyl)-3-(2-methylpropyl)-	EtOH	238(4.15),341(3.90)	94-2046-78
$C_{11}H_{18}N_2O_3$ 2(1H)-Pyrazinone, 1-hydroxy-3-(1-methylethyl)-6-(2-methylpropyl)-, 4-oxide	EtOH	231(4.20),239s(4.17), 261(4.08),294(3.83), 362(3.89)	94-2046-78
2(1H)-Pyrazinone, 1-hydroxy-6-(1-methylethyl)-3-(2-methylpropyl)-, 4-oxide	EtOH	229(4.22),240s(4.17), 260(4.21),298(3.89), 368(4.04)	94-2046-78
$C_{11}H_{18}N_2O_5$ Palythine, methyl ester	H_2O	318(4.56)	88-2299-78
$C_{11}H_{18}N_2S_2$ Pyrimidine, 4-methyl-2,6-bis(propylthio)-	H_2O	255(3.10),300(3.05)	56-1255-78
$C_{11}H_{18}O$ Cyclooctanone, 2-propylidene-	hexane	238(3.90)	54-0305-78

Compound	Solvent	$\lambda_{max}(\log \epsilon)$	Ref.
$C_{11}H_{18}OSSi$			
Silane, [[(4-methoxyphenyl)thio]methyl]-trimethyl-	heptane	258(4.04),290(3.15)	65-2027-78
$C_{11}H_{18}OS_2$			
1,2-Dithiol-1-ium, 3,5-bis(1,1-dimethylethyl)-4-hydroxy-, hydroxide, inner salt	dioxan	463(3.77)	139-0251-78B
Heptanal, 7-(1,3-dithian-2-ylidene)-	n.s.g.	257(3.84)	33-3087-78
Spiro[bicyclo[2.2.2]octane-2,2'-[1,3]-dithian]-3-ol	n.s.g.	224s(--),250(3.02)	33-3087-78
$C_{11}H_{18}O_2$			
2-Cyclohexen-1-one, 3-hydroxy-5-pentyl-	EtOH	283(4.31)	2-0970-78
$C_{11}H_{18}O_2S_2$			
1,4-Dithiepane-2-hexanal, 3-oxo-	n.s.g.	220(3.34),240(3.57), 285(2.95),320(1.90)	33-3087-78
$C_{11}H_{18}O_3$			
2-Heptenoic acid, 3-(1-methylethyl)-6-oxo-, methyl ester	EtOH	217(4.09)	130-0289-78
$C_{11}H_{18}O_4$			
2-Pentenoic acid, 3-methyl-5-[(tetrahydro-2H-pyran-2-yl)oxy]-, (Z)-	EtOH	216(4.12)	33-1975-78
$C_{11}H_{18}SSi$			
Silane, trimethyl[[(4-methylphenyl)-thio]methyl]-	heptane	256(4.08),280(3.27)	65-2027-78
$C_{11}H_{19}NO$			
Bicyclo[3.3.1]nonan-3-one, 1-(dimethyl-amino)-	EtOH	215(2.85),279s(1.78)	44-3653-78
$C_{11}H_{19}NSSi$			
Benzenamine, N,N-dimethyl-4-[(trimethylsilyl)thio]-	heptane	268(4.24),307(3.37)	65-2027-78
$C_{11}H_{19}N_2O_3$			
2-Furanmethanaminium, N,N,N-trimethyl-5-nitro-, bromide	H_2O	202(4.01),229(3.54), 297(4.03)	73-2041-78
$C_{11}H_{19}N_3O$			
Pyrazinamine, 1-acetyl-N-ethyl-1,6-di-hydro-3,5,6-trimethyl-	n.s.g.	200(4.00),223(3.76)	70-0963-78
$C_{11}H_{19}N_3O_3P$			
Phosphorus(1+), (4-diazenyl-N,N-dimethylbenzenaminato)trimethoxy-, (T-4)-, tetrafluoroborate	n.s.g.	275(3.70),457(4.55)	123-0190-78
$C_{11}H_{19}N_3O_3S$			
4(1H)-Pyrimidinone, 6-amino-5-(2,2-diethoxyethyl)-2-(methylthio)-	MeOH	218(4.31),282(3.97)	24-2925-78
$C_{11}H_{19}N_7$			
9H-Purine-2,6-diamine, 9-[2-(diethyl-amino)ethyl]-	EtOH	251s(3.90),255(3.92), 281(4.00)	2-0786-78

Compound	Solvent	$\lambda_{max}(\log \epsilon)$	Ref.
$C_{11}H_{20}N_2O_5$			
Butanoic acid, 4-[(1-acetoxybutyl)ni-trosoamino]-, methyl ester	EtOH	232(3.83),365(1.88)	94-3914-78
$C_{11}H_{20}N_3O_3P$			
1,3,2-Diazaphosphorin-4(1H)-one, 2-eth-oxy-2,3-dihydro-6-methyl-5-piperi-dino-, 2-oxide	EtOH	263(3.90)	103-0784-78
$C_{11}H_{20}N_4O$			
1H-1,2,3-Triazole-4-carboxamide, N,1-dibutyl-	EtOH	217(4.21)	4-1349-78
$C_{11}H_{20}N_4O_4S$			
Tetramethylammonium 5-(1-methyl-1-nitro-ethyl)-2,4-dinitro-2,5-dihydrothio-phen-2-ide	DMSO	580(4.55)	12-2463-78
$C_{11}H_{20}OS_2$			
1,5-Dithiaspiro[5.7]tridecan-7-ol	n.s.g.	224s(--),246(3.04), 300(2.60)	33-3087-78
$C_{11}H_{20}SSi_2$			
Disilane, pentamethyl(phenylthio)-	heptane	250(3.80)	65-2027-78
$C_{11}H_{21}NO_2$			
Valine, N-(2-methylpropylidene)-, ethyl ester	hexane MeOH	217(2.84) 201(2.81)	94-0466-78 94-0466-78
$C_{11}H_{21}N_2O$			
Methanaminium, N-[5-(dimethylamino)-3-ethoxy-2,4-pentadienylidene]-N-methyl-, tetrafluoroborate	CHCl	260(4.3),280(4.0), 414(4.6)	83-0287-78
$C_{11}H_{22}NO$			
1-Azetidinyloxy, 2,2-dimethyl-4,4-di-propyl-	C_6H_{12} MeOH	236(3.40),460(1.04) 243(3.40),425(1.04)	88-0795-78 88-0795-78
$C_{11}H_{22}N_2O_2$			
1-Hepten-1-amine, N-(1,1-dimethyleth-yl)-2-nitro-	EtOH	260(3.04),370(4.20)	44-0497-78
$C_{11}H_{24}SiSn$			
Silane, trimethyl[5-(trimethylstannyl)-1,4-pentadienyl]-, (Z,Z)-	C_6H_{12}	217(3.42)	101-0257-78E

Compound	Solvent	$\lambda_{max}(\log \epsilon)$	Ref.
$C_{12}Cl_6O_4$ Dibenzo[b,e][1,4]dioxin-2,3-dione, 1,4,6,7,8,9-hexachloro-	benzene	400(4.18)	70-0178-78
$C_{12}F_{10}N_2$ Diazene, bis(pentafluorophenyl)-, (E)-	hexane	225(3.95),308(4.32), 430(2.98)	2-0910-78
$C_{12}H_5BrO_3S$ 3H-Naphtho[1,8-bc]thiophene-2-carbox- ylic acid, 4-bromo-3-oxo-	MeOH	230s(--),270(4.186), 349(3.727),387(3.698)	5-0627-78
$C_{12}H_5F_6NS$ Pyrrolo[2,1-b]benzothiazole, 2,3-bis- (trifluoromethyl)-	MeOH	220(4.84),292(3.70), 300(3.77)	44-2697-78
$C_{12}H_5N_3O_2$ 2,3-Naphthalenedicarbonitrile, 5-amino- 1,4-dihydro-1,4-dioxo-	benzene acetone	585(3.61) 598(--)	150-2319-78 150-2319-78
$C_{12}H_6Cl_3NO_8S$ 2-Furancarboxylic acid, 5-[2-(5-nitro- 2-furanyl)-2-[(trichloromethyl)sul- fonyl]ethenyl]-	EtOH	214(4.13),307(4.40), 358(4.07)	73-1618-78
$C_{12}H_6F_6$ Naphthalene, 2,3-bis(trifluoromethyl)-	C_6H_{12}	228(3.32),265(3.25), 275(3.25),305(2.62), 312(2.54),320(2.69)	88-1099-78
$C_{12}H_6FeO_4$ Iron, [(1,7,8,9-η)-bicyclo[5.2.0]nona- 2,5,7,9-tetraen-4-one]tricarbonyl-	CH_2Cl_2 CH_2Cl_2 + CF_3COOH	263(4.08),400s(3.52) 250(4.12),292(4.11), 318(4.08),410s(3.56)	12-1607-78 12-1607-78
$C_{12}H_6N_2O_4$ 3H-Phenoxazin-3-one, 7-nitro-	C_6H_{12}	250(4.25),325(4.15), 338(4.13),356(3.97), 426(3.98)	24-3012-78
$C_{12}H_6O_3S$ 3H-Naphtho[1,8-bc]thiophene-2-carbox- ylic acid, 3-oxo-	MeOH	220s(--),268(4.289), 332(3.827),373(3.881)	5-0627-78
$C_{12}H_6S_3$ Benzo[1,2-c:3,4-c':5,6-c'']trithiophene	n.s.g.	220(4.2),225(4.1), 250(4.7),259(4.8), 292(3.8),310(3.9), 320(3.8)	35-4326-78
$C_{12}H_7BrO$ 1(2H)-Acenaphthylenone, 5-bromo-	dioxan	343(4.13)	104-1971-78
$C_{12}H_7ClF_4$ Spirohex-4-ene, 1-chloro-1,2,2,4-tetra- fluoro-5-phenyl-	EtOH	257(4.43)	44-4873-78
$C_{12}H_7ClN_2O_5$ Furan, 2-[2-(4-chlorophenyl)-1-nitro- ethenyl]-5-nitro-, (E)-	dioxan	230(4.17),312(4.33)	73-0463-78

Compound	Solvent	$\lambda_{max}(\log \epsilon)$	Ref.
$C_{12}H_7Cl_2F_3$ Spirohex-4-ene, 1,1-dichloro-2,2,4-tri- fluoro-5-phenyl-	EtOH	212(4.25),218(4.11), 263(3.40)	44-4873-78
$C_{12}H_7Cl_2HgNO_3$ Mercury, (2,6-dichloro-4-nitrophenol- ato-O^1)phenyl-	benzene	340(4.11)	70-1573-78
$C_{12}H_7Cl_3N_2O_5S$ Pyridine, 3-[2-(5-nitro-2-furanyl)- 2-[(trichloromethyl)sulfonyl]eth- enyl]-	EtOH	205(4.23),234s(4.00), 268(3.90),306(4.07)	73-1618-78
$C_{12}H_7FN_2O_5$ Furan, 2-[2-(4-fluorophenyl)-1-nitro- ethenyl]-5-nitro-, (E)-	dioxan	307(4.26)	73-0463-78
$C_{12}H_7F_6FePO_3$ Iron, [(1,7,8,9-η)-bicyclo[5.2.0]nona- 1(7),3,5,8-tetraenyl hexafluorophos- phate]tricarbonyl-	CH_2Cl_2	262(4.21),349(3.95), 550(2.96)	12-1607-78
$C_{12}H_7NOS$ Naphtho[2',1':4,5]thiazole-2-carbox- aldehyde	n.s.g.	288(3.76),302(3.61), 317(3.11),325(3.14), 332(3.24)	42-0702-78
$C_{12}H_7NO_3$ 1(2H)-Acenaphthylenone, 5-nitro-	dioxan	349(4.37)	104-1971-78
$C_{12}H_7NO_4$ Pyrido[1,2-a]indole-6,10-dione, 7,9-dihydroxy-	MeOH	203(4.35),229(4.27), 252(4.34),409(4.09), 426(4.09)	83-0960-78
$C_{12}H_7N_3$ Propanedinitrile, 2(1H)-quinolinyli- dene)-	pH 13	276(--),293(--), 303(--),379(--)	94-2435-78
	EtOH	286(4.42),398(4.19)	94-2435-78
$C_{12}H_7N_3O_7$ Furan, 2-nitro-5-(1-nitro-2-(2-nitro- phenyl)ethenyl]-, (E)-	dioxan	312(4.11)	73-0463-78
Furan, 2-nitro-5-(1-nitro-2-(4-nitro- phenyl)ethenyl]-, (E)-	dioxan	216(4.08),298(4.24)	73-0463-78
$C_{12}H_7N_5$ 1H-Pyrazolo[3,4-d]pyrimidine-4-carbo- nitrile, 1-phenyl-	EtOH	250(4.59),335(3.29)	95-0089-78
$C_{12}H_7N_5O_3S$ Methanone, [5-(1H-imidazol-1-yl)-1,3,4- thiadiazol-2-yl](4-nitrophenyl)-	MeOH	278(4.24),307(4.22)	78-0453-78
$C_{12}H_8BrHgNO_3$ Mercury, (2-bromo-4-nitrophenolato- O^1)phenyl-	benzene	340(4.12)	70-1573-78
Mercury, (2-bromo-6-nitrophenolato- O^1)phenyl-	benzene	361(3.45)	70-1573-78

Compound	Solvent	$\lambda_{max}(\log \epsilon)$	Ref.
$C_{12}H_8BrNO_3S$ Furan, 2-[(4-bromophenyl)thio]-5-(2-nitroethenyl)-	EtOH	205(4.25),233(4.13), 251(4.18),372(4.15)	73-2037-78
$C_{12}H_8Br_2$ Cyclobuta[b]naphthalene, 1,2-dibromo-1,2-dihydro-, trans	ether	242(4.48),294(3.33), 298(3.32),313(3.21), 328(3.13)	78-0113-78
$C_{12}H_8ClHgNO_3$ Mercury, (2-chloro-4-nitrophenolato-O^1)phenyl-	benzene $C_2H_4Cl_2$ DMSO	339(4.12) 350(4.24) 367(--),434(--)	70-1573-78 70-1573-78 70-2243-78
Mercury, (4-chloro-2-nitrophenolato-O^1)phenyl-	benzene DMSO $C_2H_4Cl_2$	370(3.40) 397(--),466(--) 373(3.46)	70-2243-78 70-2243-78 70-1573-78
Mercury, (6-chloro-2-nitrophenolato-O^1)phenyl-	benzene $C_2H_4Cl_2$	364(3.43) 373(3.46)	70-1573-78 70-1573-78
$C_{12}H_8ClNO_3S$ Furan, 2-[(4-chlorophenyl)thio]-5-(2-nitroethenyl)-	EtOH	205(4.39),229(4.25), 247(4.30),370(4.32)	73-2037-78
$C_{12}H_8Cl_2O_4$ 1,4-Naphthalenedione, 2,3-dichloro-5,8-dimethoxy-	C_6H_{12} EtOH	431(3.68) 476(--)	150-2319-78 150-2319-78
1,4-Naphthalenedione, 6,7-dichloro-5,8-dimethoxy-	C_6H_{12} EtOH	372(3.65) 372(--)	150-2319-78 150-2319-78
$C_{12}H_8Cl_3NO_6S$ Furan, 2-[2-(5-methyl-2-furanyl)-1-[(trichloromethyl)sulfonyl]ethenyl]-5-nitro-, (E)-	EtOH	203(4.04),224(3.96), 249s(3.87),312(4.43), 361(4.17)	73-1618-78
$C_{12}H_8N_2O$ Propanedinitrile, [(3-phenyloxiranyl)methylene]-	EtOH	217(3.93),253(3.17)	24-3665-78
11H-Pyrido[2,1-b]quinazolin-11-one	MeCN	217(4.380),227(4.431), 257(4.210),285(3.470), 349(4.086),377(3.868), 397(3.924)	104-2218-78
	CH_2Cl_2	256(4.278),287(3.566), 352(4.146),380(3.894), 401(3.531)	104-2218-78
$C_{12}H_8N_2O_2$ 2-Naphthalenecarbonitrile, 1,4-dihydro-3-(methylamino)-1,4-dioxo-	C_6H_{12} EtOH	431(3.48) 432(--)	39-1083-78C 39-1083-78C
$C_{12}H_8N_2O_2S$ 2H-1-Benzothiopyrano[2,3-d]pyrimidine-2,4(3H)-dione, 3-methyl-	EtOH	283(3.98),346(2.98), 422(2.92)	88-2803-78
$C_{12}H_8N_2O_3SSe$ Benzoselenazole, 2-[[(5-nitro-2-furanyl)methyl]thio]-	MeOH	232(4.45),248(4.19), 287(4.23),299(4.25), 310(4.27)	73-0160-78
$C_{12}H_8N_2O_3S_2$ Benzothiazole, 2-[[(5-nitro-2-furanyl)methyl]thio]-	MeOH	223(4.51),244s(4.18), 281(4.26),292(4.28),	73-0160-78

Compound	Solvent	$\lambda_{max}(\log \epsilon)$	Ref.
(cont.)		302(4.31),321s(4.27)	73-0160-78
$C_{12}H_8N_2O_4S$			
Benzoxazole, 2-[[(5-nitro-2-furanyl)-methyl]thio]-	MeOH	211(4.41),243(4.24), 278(4.22),285(4.25), 316(4.21)	73-0160-78
$C_{12}H_8N_2O_5$			
Furan, 2-nitro-5-(1-nitro-2-phenyleth-enyl)-, (E)-	dioxan	228(4.00),310(4.30)	73-0463-78
$C_{12}H_8N_2O_5S$			
Furan, 2-(2-nitroethenyl)-5-[(4-nitro-phenyl)thio]-	EtOH	207s(4.18),232(4.02), 345(4.24)	73-2037-78
$C_{12}H_8N_2O_5S_2$			
Benzothiazole, 2-[[(5-nitro-2-furanyl)-methyl]sulfonyl]-	MeOH	218(4.36),239(4.15), 284(4.24),308(4.24)	73-0160-78
$C_{12}H_8N_2O_6$			
Propanal, 2-(6,8-dinitro-2H-1-benzo-pyran-2-ylidene)-	EtOH	378(4.00)	48-0497-78
$C_{12}H_8N_2S$			
Acenaphtho[5,6-cd][1,2,6]thiadiazin-2-S^{IV}, 6,7-dihydro-	film	234(4.71),260(4.22), 325(3.92),358(3.96), 404(1.90),680(2.62)	89-0468-78
$C_{12}H_8N_2Se$			
2,4-Pentadienenitrile, 5-(1,2-benziso-selenazol-3-yl)-, (Z,Z)-	EtOH	239(3.50),276(3.30), 357(3.11)	39-1006-78C
$C_{12}H_8N_4OS$			
Methanone, [5-(1H-imidazol-1-yl)-1,3,4-thiadiazol-2-yl]phenyl-	MeOH	300(4.32)	78-0453-78
$C_{12}H_8N_4O_2$			
Phenazine, 6-amino-2-nitro-	EtOH	558(3.33)	39-0299-78C
1H-Pyrazolo[3,4-d]pyrimidine-4-carbox-ylic acid, 1-phenyl-	EtOH	250(4.34),318(3.20)	95-0891-78
Pyrimido[5,4-c]pyridazine-6,8(2H,7H)-dione, 2-phenyl-	pH 1	203(4.29),306s(3.96), 330(4.02)	44-2536-78
	pH 7	205(4.31),263(3.92), 324(4.10),363(4.17)	44-2536-78
	pH 13	218(4.00),291(4.25), 366(4.10)	44-2536-78
$C_{12}H_8OS$			
3H-Naphtho[1,8-bc]thiophen-3-one, 2-methyl-	MeOH	213s(--),261(4.193), 304(3.723),370(3.919)	5-0627-78
$C_{12}H_8OSe$			
Phenoxaselenin	heptane	202(4.34),218(4.13), 238(4.31),241(4.30), 294(3.54)	101-0235-78C
$C_{12}H_8OTe$			
Phenoxatellurin	heptane	202(4.57),230s(4.07), 257(3.91),290s(3.62), 356(2.55)	101-0235-78C

Compound	Solvent	$\lambda_{max}(\log \epsilon)$	Ref.
$C_{12}H_8O_3$			
1,4-Methano-1H-benzocycloheptene-5,6,9(4H)-trione	CH_2Cl_2	267(3.95),343(3.38), 500s(1.57)	88-2289-78
$C_{12}H_8O_4$			
Psoralen, 5-methoxy-	C_6H_{12}	222(4.26),242s(--), 250(4.33),267(4.16), 301(4.12)	94-3433-78
	MeOH	222(4.38),242s(4.19), 250(4.24),259(4.20), 268(4.24),310(4.16)	94-3433-78
	dioxan	244s(4.26),250(4.30), 260(4.20),268(4.19), 305(4.15)	94-3433-78
	MeCN	222(4.34),243s(4.19), 250(4.25),258(4.19), 268(4.21),310(4.16)	94-3433-78
	CCl_4	306(4.10)	94-3433-78
	$CF_3CHOHCF_3$	200(4.32),218(4.39), 242(4.09),247(4.10), 263(4.18),269(4.18), 320(4.16)	94-3433-78
Psoralen, 8-methoxy-	C_6H_{12}	220(4.27),245(4.33), 250(4.35),265s(4.02), 293(4.00),305s(3.91), 340(3.37)	94-3433-78
	dioxan	244s(4.36),250(4.41), 265(4.13),295(4.05), 340(3.42)	94-3433-78
	MeCN	219(4.38),243s(4.38), 249(4.42),264s(4.11), 297(4.08)	94-3433-78
	CCl_4	297(4.02),340(4.41)	94-3433-78
	$CF_3CHOHCF_3$	200(4.39),216(4.45), 243(4.27),263(4.15), 310(4.10)	94-3433-78
$C_{12}H_8O_4S_2$			
Dibenzo[c,e][1,2]dithiin, 5,5,6,6-tetra-oxide	60% dioxan	313(3.81)	44-0914-78
$C_{12}H_8S_4$			
1,3-Dithiole, 2-(1,3-dithiol-2-yli-dene)-4-phenyl-	n.s.g.	228(4.06),277(4.14), 302(4.15),322(4.10), 400(3.45)	44-4642-78
$C_{12}H_9BrN_2$			
Propanedinitrile, [2-(4-bromophenyl)-1-methylethylidene]-	C_6H_{12}	274(3.56),281(3.42)	39-0995-78B
$C_{12}H_9BrN_2O$			
1H-1,2-Diazepine, 1-benzoyl-4-bromo-	MeOH	226(4.15),286s(3.75), 358(2.60)	33-2887-78
1H-1,2-Diazepine, 1-benzoyl-6-bromo-	MeOH	227(4.08),284s(3.73), 360(2.60)	33-2887-78
1H-Pyrido[3,2-b][1,4]benzoxazine, 3-bromo-1-methyl-	EtOH	236(4.34),239s(4.63), 387(3.87),410s(3.81), 433s(3.64)	95-0585-78
monohydriodide	EtOH	234(4.26),393(3.94)	95-0585-78
10H-Pyrido[3,2-b][1,4]benzoxazine, 3-bromo-10-methyl-	MeOH	235(4.48),350(3.91)	95-0585-78

Compound	Solvent	$\lambda_{max}(\log \epsilon)$	Ref.
Pyridinium, 1-(benzoylamino)-3-bromo-, hydroxide, inner salt	EtOH	337(3.79)	33-2887-78
$C_{12}H_9BrN_2O_3$ Benzenemethanamine, 4-bromo-α-[(5-nitro-2-furanyl)methylene]-	EtOH	263(4.13),484(4.34)	4-0555-78
$C_{12}H_9BrN_2O_3S$ Acetic acid, [1-(4-bromophenyl)-5-oxo-2-thioxo-4-imidazolidinylidene]-, methyl ester, (Z)-	EtOH	348(4.18)	5-0227-78
$C_{12}H_9BrO_2$ Propanal, 2-(6-bromo-2H-1-benzopyran-2-ylidene)-, (E)-	EtOH	374(3.77)	48-0497-78
$C_{12}H_9BrO_3$ 1,4-Naphthalenedione, 2-bromo-7-ethoxy-	EtOH	270(4.34),293s(4.11), 350s(3.21),402(3.28)	12-1335-78
1,4-Naphthalenedione, 2-bromo-6-methoxy-8-methyl-	EtOH	272(4.32),290s(4.18), 350(3.41),400(3.37)	12-1363-78
1,4-Naphthalenedione, 2-bromo-7-methoxy-5-methyl-	EtOH	270(4.19),290(4.03), 350(3.32),400(3.27)	12-1363-78
$C_{12}H_9BrO_4$ 2(5H)-Furanone, 5-acetoxy-5-bromo-3-phenyl-	EtOH	272(4.02)	44-4115-78
$C_{12}H_9ClN_2$ Propanedinitrile, [2-(4-chlorophenyl)-1-methylethylidene]-	C_6H_{12}	273(3.51),280(3.32)	39-0995-78B
$C_{12}H_9ClN_2O$ 1H-1,2-Diazepine, 1-benzoyl-4-chloro-	MeOH	230(4.03),262(3.79), 358(2.66)	33-2887-78
1H-1,2-Diazepine, 1-benzoyl-6-chloro-	MeOH	227(4.11),272(3.84), 358(2.49)	33-2887-78
Pyridinium, 1-(benzoylamino)-3-chloro-, hydroxide, inner salt	EtOH	338(3.76)	33-2887-78
1H-Pyrido[3,2-b][1,4]benzoxazine, 3-chloro-1-methyl-	MeOH	235(4.66),239s(4.32), 388(4.21),410s(4.14), 435s(3.95)	95-0585-78
monohydriodide	MeOH	232(4.47),388(3.95)	95-0585-78
10H-Pyrido[3,2-b][1,4]benzoxazine, 3-chloro-10-methyl-	MeOH	234(4.61),345(4.04)	95-0585-78
$C_{12}H_9ClN_2O_2$ 2,5-Cyclohexadien-1-one, 3-amino-4-[(3-chloro-4-hydroxyphenyl)imino]-	neutral anion	464(3.94) 565(4.21)	39-1292-78B 39-1292-78B
2,5-Cyclohexadien-1-one, 5-amino-2-chloro-4-[(4-hydroxyphenyl)imino]-	neutral anion	480(3.91) 585(4.27)	39-1292-78B 39-1292-78B
$C_{12}H_9ClN_2O_3$ Benzenemethanamine, 4-chloro-α-[(5-nitro-2-furanyl)methylene]-	EtOH	265(4.10),485(4.31)	4-0555-78
$C_{12}H_9ClO_2$ Benzo[1,2-b:4,3-b']difuran, 4-chloro-2,7-dimethyl-	MeOH	225(4.35),270s(4.32), 278(4.49),289(4.28), 301(3.92)	4-0043-78

Compound	Solvent	$\lambda_{max}(\log \epsilon)$	Ref.
$C_{12}H_9ClO_3$			
1,4-Naphthalenedione, 2-chloro-6-meth-oxy-8-methyl-	EtOH	209(4.46),227s(4.09), 269(4.27),400(3.38)	39-1041-78C
$C_{12}H_9ClO_4$			
2(5H)-Furanone, 4-acetyl-5-(2-chloro-phenyl)-3-hydroxy-	H_2O	318(4.13)	4-0737-78
	EtOH	268(3.97),320(3.66)	4-0737-78
	EtOH-NaHCO$_3$	320(4.12)	4-0737-78
	CHCl$_3$	259(3.99)	4-0737-78
$C_{12}H_9Cl_2NO_3$			
1,4-Naphthalenedione, 2,3-dichloro-5-methoxy-8-(methylamino)-	C_6H_{12}	580(3.86)	150-2319-78
	EtOH	612(--),667(--)	150-2319-78
$C_{12}H_9FN_2O$			
1H-1,2-Diazepine, 1-benzoyl-4-fluoro-	MeOH	227(3.93),278s(3.64), 362(2.30)	33-2887-78
1H-1,2-Diazepine, 1-benzoyl-6-fluoro-	MeOH	225(4.03),278s(3.66), 360(2.40)	33-2887-78
Pyridinium, 1-(benzoylamino)-3-fluoro-, hydroxide, inner salt	EtOH	335(3.79)	33-2887-78
$C_{12}H_9FN_2O_3S$			
Acetic acid, [1-(4-fluorophenyl)-5-oxo-2-thioxo-4-imidazolidinylidene]-, methyl ester, (Z)-	EtOH	348(4.36)	5-0227-78
$C_{12}H_9HgNO_3$			
Mercury, (2-nitrophenolato-O^1)phenyl-	benzene	363(3.07)	70-1573-78
	benzene	359(3.07)	70-2243-78
	$C_2H_4Cl_2$	357(3.15)	70-1573-78
	DMSO	377(--),455(--)	70-2243-78
Mercury, (4-nitrophenolato-O^1)phenyl-	benzene	337(4.10)	70-1573-78
	$C_2H_4Cl_2$	345(3.97)	70-1573-78
	DMSO	367(--),434(--)	70-2243-78
$C_{12}H_9IN_2O$			
1H-1,2-Diazepine, 1-benzoyl-4-iodo-	MeOH	227(4.10),260(3.99), 363(2.84)	33-2887-78
1H-1,2-Diazepine, 1-benzoyl-6-iodo-	MeOH	228(4.19),260(4.04), 362(2.51)	33-2887-78
Pyridinium, 1-(benzoylamino)-3-iodo-, hydroxide, inner salt	EtOH	336(3.60)	33-2887-78
$C_{12}H_9IN_2O_3$			
Benzenemethanamine, 4-iodo-α-[(5-nitro-2-furanyl)methylene]-	EtOH	263(4.17),487(4.37)	4-0555-78
$C_{12}H_9N$			
1H-Cyclopenta[c]quinoline	C_6H_{12}	228(4.52),236s(4.43), 243s(4.28),298(3.80), 315s(3.70),325s(3.57)	44-2190-78
	EtOH	210s(4.34),230(4.52), 274(4.13),287s(4.04), 302s(3.95),340s(3.49), 430s(--)	44-2190-78
	ether	210s(4.33),230(4.52), 274(4.29),287s(4.16), 305s(3.93),340s(3.58), 430s(c.3.0)	44-2190-78

Compound	Solvent	$\lambda_{max}(\log \epsilon)$	Ref.
$C_{12}H_9NO$			
1(2H)-Acenaphthylenone, 5-amino-	dioxan	396(3.88)	104-1971-78
Benz[cd]indole, 2-methoxy-	MeOH	250(4.02),280s(3.88),	104-2070-78
		290(3.76),304(3.83),	
		327(3.75),343(3.85),	
		370s(3.25)	
$C_{12}H_9NOS_2$			
4H-1-Benzothiopyran-3-carbonitrile,	EtOH	245(4.04),285(4.27),	78-0725-78
2-(ethylthio)-4-oxo-		317s(3.93),330(4.0)	
$C_{12}H_9NO_2$			
2,5-Cyclohexadien-1-one, 4-[(4-hydroxy-	C_6H_{12}	250(3.85),300(3.68),	48-0557-78
phenyl)imino]-		560(3.57)	
	EtOH	255(4.26),295(4.10),	48-0557-78
		595(3.98)	
	ether	250(4.04),320(3.77),	48-0557-78
		570(3.76)	
	CCl4	320(3.68),570(3.63)	48-0557-78
$C_{12}H_9NO_3$			
2H-Pyran-5-carboxamide, 2-oxo-N-phenyl-	MeOH	257(4.13),280s(4.06)	44-4415-78
2H-Pyran-2,6(3H)-dione, 3-[(phenyl-	MeOH	290(3.70),370(4.63)	44-4415-78
amino)methylene]-, (Z)-			
Pyrrolo[1,2-b]isoquinoline-1,5-dione,	EtOH	375(4.28)	44-2026-78
2,3-dihydro-10-hydroxy-			
$C_{12}H_9NO_3S$			
Furan, 2-(2-nitroethenyl)-5-(phenyl-	EtOH	206(4.39),242(4.23),	73-2037-78
thio)-		371(4.37)	
$C_{12}H_9NO_4$			
1,3-Dioxolo[4,5-g]quinoline-7-carbox-	EtOH	253(4.53),320(3.89),	95-0802-78
aldehyde, 6-methyl-, 5-oxide		334(3.94),350(4.00)	
1H-Indole-2-acetic acid, 3-formyl-	MeOH	252(4.05),263(3.88),	150-1683-78
α-oxo-, methyl ester		305(4.03)	
1H-Isoindole-1,3(2H)-dione, 2-(tetra-	THF	219(4.67),293(3.26)	40-0404-78
hydro-5-oxo-2-furanyl)-			
Propanal, 2-(6-nitro-2H-1-benzopyran-	EtOH	366(4.38)	48-0497-78
2-ylidene)-			
1H-Pyrrole-3,4-dicarboxylic acid,	EtOH	286(4.19)	39-1588-78C
2-phenyl-	EtOH-base	285(4.0)	39-1588-78C
$C_{12}H_9NO_6$			
2(5H)-Furanone, 4-acetyl-3-hydroxy-	H_2O	257(3.76),316(4.12)	4-0737-78
5-(2-nitrophenyl)-	EtOH	266(4.05),318(3.70)	4-0737-78
	EtOH-NaHCO₃	253(3.78),320(4.16)	4-0737-78
	CHCl₃	261(4.17)	4-0737-78
			+95-0802-78
2(5H)-Furanone, 4-acetyl-3-hydroxy-	H_2O	265(3.96),314(4.17)	4-0737-78
5-(3-nitrophenyl)-	EtOH	265(4.26),320(3.73)	4-0737-78
	EtOH-NaHCO₃	260(3.99),320(4.27)	4-0737-78
	CHCl₃	257(4.28)	4-0737-78
1,4-Naphthalenedione, 2,3-dimethoxy-	C_6H_{12}	327(3.42)	39-1083-78C
	EtOH	386(--)	39-1083-78C
$C_{12}H_9NO_7$			
2(5H)-Furanone, 4-acetyl-3-hydroxy-	EtOH	233(4.14),260(3.93),	95-0802-78
5-(5-hydroxy-2-nitrophenyl)-		307(4.06),316(4.03)	

Compound	Solvent	$\lambda_{max}(\log \epsilon)$	Ref.
$C_{12}H_9NS_2$			
Benzothiazole, 2-(5-methyl-3-thienyl)-	EtOH	227(4.325),247(4.006), 256(3.886),302(4.245)	39-1198-78C
$C_{12}H_9N_2O_6P$			
Phosphinic acid, bis(3-nitrophenyl)-	EtOH	263(4.15)	65-1205-78
$C_{12}H_9N_3$			
1H-Benzimidazole, 2-(2-pyridinyl)-	C_6H_{12}	311.5(4.41)	18-3027-78
	benzene	312.0(4.36)	18-3027-78
	MeOH	308(4.32)	18-3027-78
	EtOH	309.0(4.31)	18-3027-78
	PrOH	310(4.32)	18-3027-78
	BuOH	311(4.32)	18-3027-78
	ether	310.0(4.37)	18-3027-78
	$C_2H_4Cl_2$	310.5(4.35)	18-3027-78
1H-Benzimidazole, 2-(3-pyridinyl)-	C_6H_{12}	314.5(4.31)	18-3027-78
	benzene	311.0(4.29)	18-3027-78
	EtOH	312.0(4.30)	18-3027-78
	ether	309.0(4.33)	18-3027-78
	$C_2H_4Cl_2$	310.0(4.30)	18-3027-78
1H-Benzimidazole, 2-(4-pyridinyl)-	C_6H_{12}	310.0(4.32)	18-3027-78
	benzene	309.0(4.32)	18-3027-78
	EtOH	306.5(4.32)	18-3027-78
	ether	306.0(4.37)	18-3027-78
	$C_2H_4Cl_2$	306.5(4.32)	18-3027-78
1,10-Phenanthrolin-4-amine	H_2O	210(4.01),242(4.30), 265(4.35),318(4.13), 357s(4.06)	39-1215-78B
monocation	H_2O	212(3.80),240(4.00), 268(4.42),284s(4.24), 320(4.16)	39-1215-78B
$C_{12}H_9N_3O$			
Naphtho[1,2-e][1,2,4]triazin-2(3H)-one, 3-methyl-	MeOH	412s(3.49),427(3.55), 446s(3.47)	39-0789-78C
	dioxan	413s(3.54),430(3.62), 450s(3.49)	39-0789-78C
3-Pyridinol, 2-(1H-benzimidazol-2-yl)-	C_6H_{12}	344.5(4.48)	18-3027-78
	benzene	342.5(4.45)	18-3027-78
	EtOH	342.0(4.41)	18-3027-78
	ether	345.9(4.43)	18-3027-78
	$C_2H_4Cl_2$	343.0(4.40)	18-3027-78
$C_{12}H_9N_3O_2$			
Naphtho[1,2-e][1,2,4]triazin-2(3H)-one, 3-methyl-, 1-oxide	MeOH	372s(3.63),387(3.66), 417(3.65),437s(3.59)	39-0789-78C
	dioxan	365s(3.62),391(3.69), 412(3.69),437s(3.59)	39-0789-78C
Pyrimido[4,5-b]quinoline-2,4(3H,10H)-dione, 10-methyl-	H_2O	218(4.63),256(4.73), 323(4.08),392(4.08)	150-1325-78
$C_{12}H_9N_3O_3$			
Phenol, 4-[(3-nitrophenyl)azo]-	EtOH	248(4.4),355(4.6)	19-0655-78
1H-Pyrrolo[2,3-d]pyrimidine-2,4,5(3H)-trione, 6,7-dihydro-1-phenyl-	H_2O	244(3.80),278(4.09)	103-0443-78
$C_{12}H_9N_3O_3S$			
1H-Benzimidazole, 2-[[(5-nitro-2-furanyl)methyl]thio]-	MeOH	213(4.53),247(4.27), 296s(4.44),306(4.50)	73-0160-78

Compound	Solvent	$\lambda_{max}(\log \epsilon)$	Ref.
$C_{12}H_9N_3O_5$ Benzenemethanamine, 4-nitro-α-[(5-nitro-2-furanyl)methylene]-	EtOH	270(4.36),480(4.37)	4-0555-78
$C_{12}H_9N_5O$ Pyrazolo[3,4-d]pyrimidine-4-carboxamide, 1-phenyl-	EtOH	204(4.39),249(4.45), 324(3.24)	95-0891-78
Pyrimido[5,4-c]pyridazin-8(2H)-one, 6-amino-2-phenyl-	pH 1	203(4.37),260(4.00), 328s(4.22),362(4.35)	44-2536-78
	pH 7	203(4.32),292(4.26), 376(4.18)	44-2536-78
	pH 13	218(4.18),292(4.35), 376(4.26)	44-2536-78
$C_{12}H_9N_5O_2$ 1H-Pyrazolo[3,4-d]pyrimidine-4-carboxamide, N-hydroxy-1-phenyl-	EtOH	249(4.50),270s(4.00), 325s(3.46)	95-1274-78
$C_{12}H_9N_5O_3$ 2,4(1H,3H)-Pyrimidinedione, 5-[(4-oxopyrido[2,3-d]pyrimidin-3-yl)methyl]-	pH 1 pH 7	264(4.06),318(3.89) 264(3.06),299(3.77), 310(3.63)	44-0828-78 44-0828-78
	pH 11	288(4.13),275(4.09), 310(3.71)[sic]	44-0828-78
2,4(1H,3H)-Pyrimidinedione, 5-[(4-oxopyrido[2,3-d]pyrimidin-8-yl)methyl]-	pH 1 pH 7	261(4.01),327(4.03) 246(4.14),261(3.94), 361(3.96)	44-0828-78 44-0828-78
$C_{12}H_{10}$ Tricyclo[4.1.0.0^{2,7}]hept-3-ene, 5-(2,4-cyclopentadien-1-ylidene)-	pentane	312s(4.31),325(4.44), 337(4.41),351s(4.10), 396(2.53)	88-0645-78
$C_{12}H_{10}BrNO_3$ 6H-1,3-Oxazinium, 5-(4-bromophenyl)-4-hydroxy-2,3-dimethyl-6-oxo-, hydroxide, inner salt	MeCN	226(3.78),258(4.08), 324(3.73)	5-1655-78
$C_{12}H_{10}BrN_3O$ 4H-Pyrimido[2,1-b]quinazolin-4-one, 3-bromo-1,6-dihydro-2-methyl-	MeOH	263(4.12),327(4.20)	4-0077-78
$C_{12}H_{10}ClN$ 1H-Pyrrole, 2-(4-chlorophenyl)-1-ethenyl-	C_6H_{12}	206(4.16),250(4.06), 270(4.10)	103-0399-78
$C_{12}H_{10}ClNO_3$ 6H-1,3-Oxazinium, 5-(4-chlorophenyl)-4-hydroxy-2,3-dimethyl-6-oxo-, hydroxide, inner salt	MeCN	220s(3.97),257(4.19), 323(3.83)	5-1655-78
$C_{12}H_{10}ClNO_4$ 3H-Indole-2,3-dicarboxylic acid, 3-chloro-, dimethyl ester	CH_2Cl_2	240(3.92),300(3.49)	94-2866-78
$C_{12}H_{10}ClN_3O$ 2,1-Benzisoxazole, 5-chloro-3-(3,5-dimethyl-1H-pyrazol-4-yl)-	EtOH	248(4.02),348(3.82)	94-1141-78
4H-Pyrimido[2,1-b]quinazolin-4-one, 8-chloro-1,6-dihydro-2-methyl-	MeOH	260(4.20),313(4.19)	4-0077-78

Compound	Solvent	$\lambda_{max}(\log \epsilon)$	Ref.
$C_{12}H_{10}ClN_3O_3$			
2,4(1H,3H)-Pyrimidinedione, 6-amino-5-(chloroacetyl)-1-phenyl-	EtOH	243(4.02),278(4.18)	103-0443-78
$C_{12}H_{10}ClN_5O$			
6H-Imidazo[4,5-e]-1,2,4-triazin-6-one, 3-(4-chlorophenyl)-5,7-dihydro-5,7-dimethyl-	EtOH	249(4.39),275(4.25), 309(4.07)	94-3154-78
$C_{12}H_{10}ClP$			
Phosphinous chloride, diphenyl-	Me methacrylate	293(2.75)	47-0041-78
$C_{12}H_{10}Cl_5N_3$			
1H-Benzotriazole, 4,5,6,7-tetrachloro-1-(2-chlorocyclohexyl)-, trans	EtOH	221(4.54),275(3.90), 283(3.92),306(3.71)	39-0909-78C
2H-Benzotriazole, 4,5,6,7-tetrachloro-2-(2-chlorocyclohexyl)-, trans	EtOH	222(4.55),294(4.10), 303(4.09)	39-0909-78C
$C_{12}H_{10}F_3NO$			
Ethanone, 1-(1,3-dimethyl-1H-indol-2-yl)-2,2,2-trifluoro-	EtOH	208(4.36),245(4.09), 330(4.38)	39-1471-78C
$C_{12}H_{10}N_2$			
Diazene, diphenyl-	hexane	228(4.11),314(4.29), 440(2.78)	2-0910-78
Propanedinitrile, (1-methyl-2-phenylethylidene)-	C_6H_{12}	245s(4.04),260s(3.68), 266(3.60),273(3.45)	39-0995-78B
$C_{12}H_{10}N_2O$			
5H-Cyclopenta[2,1-b:3,4-b']dipyridin-5-ol, 5-methyl-	EtOH	258(3.49),307(4.13), 313(4.13),318(4.22)	64-0080-78B
Ethanone, 1-phenyl-2-(3-pyridazinyl)-	C_6H_{12}	238(3.75),342(3.92)	94-3633-78
	EtOH	243(3.98),343(3.76)	94-3633-78
Ethanone, 1-phenyl-2-(4-pyridazinyl)-	EtOH	245(4.28),281(3.30)	94-3633-78
Phenol, 4-(phenylazo)-	MeOH	240(4.0),346(4.4), 420s(3.1)	19-0655-78
1H-Pyrido[3,2-b][1,4]benzoxazine, 1-methyl-	MeOH	234(4.14),380(3.75), 399s(3.73),421s(3.56)	95-0585-78
monohydriodide	MeOH	225(4.81),376(4.51)	95-0585-78
1H-Pyrido[3,2-b][1,4]benzoxazine, 3-methyl-	MeOH	230(4.53),343(3.98)	94-1375-78
1H-Pyrido[3,2-b][1,4]benzoxazine, 8-methyl-	MeOH	231(4.45),338(3.91)	94-1375-78
10H-Pyrido[3,2-b][1,4]benzoxazine, 10-methyl-	MeOH	231(4.49),338(3.92)	95-0585-78
9H-Pyrido[3,4-b]indole-3-methanol	EtOH	215(4.06),235(4.20), 270(3.69),290(4.02), 303(2.55),340(3.60)	95-1635-78
9H-Pyrido[3,4-b]indol-6-ol, 1-methyl-(6λ,5ϵ)	MeOH	230(4.51),244s(4.38), 256s(4.23),288s(4.08), 295(3.66),360(?)	100-0383-78
$C_{12}H_{10}N_2O_2$			
2,5-Cyclohexadien-1-one, 2-amino-4-[(4-hydroxyphenyl)imino]-	neutral	470(3.92)	39-1292-78B
	anion	574(4.24)	39-1292-78B
Pyridinium, 1-(benzoylamino)-3-hydroxy-, hydroxide, inner salt	EtOH	322(3.61)	33-2887-78
2-Pyrimidineacetic acid, 4-phenyl-	neutral	248s(3.76),276(4.01)	12-0649-78
	cation	310(4.17)	12-0649-78
	anion	250(3.80),276(4.01)	12-0649-78

Compound	Solvent	$\lambda_{max}(\log \epsilon)$	Ref.
$C_{12}H_{10}N_2O_2S$			
Benzenamine, 4-[(4-nitrophenyl)thio]-	EtOH	209(4.41),245(4.10), 266(4.18),342(4.09)	111-0093-78
2H-[1]Benzothiopyrano[2,3-d]pyrimidine-2,4(3H)-dione, 1,5-dihydro-3-methyl-	EtOH	225(4.07),249(3.97), 296(3.78)	88-2803-78
$C_{12}H_{10}N_2O_3$			
Benzenemethanamine, α-[(5-nitro-2-fur-anyl)methylene]-	EtOH	265(4.15),488(4.36)	4-0555-78
Methanone, 2-furanyl[4-(2-furanyl)-4,5-dihydro-1H-pyrazol-3-yl]-	EtOH	217(4.08),287(3.94), 354(4.15)	104-1830-78
2-Propenenitrile, 3,3'-(3,4-dimethoxy-2,5-furandiyl)bis-, (E,E)-	EtOH	256(3.47),275s(4.09), 375(4.40)	33-0430-78
(E,Z)-	EtOH	256(4.40),275s(4.05), 375(4.30)	33-0430-78
$C_{12}H_{10}N_2O_4$			
2,3-Quinoxalinedicarboxylic acid, dimethyl ester	EtOH	246(4.57),317(3.69)	94-2866-78
$C_{12}H_{10}N_2O_5$			
1,4-Naphthalenedione, 2-methoxy-3-(methylamino)-5-nitro-	C_6H_{12}	498(3.46)	39-1083-78C
	EtOH	526(--)	39-1083-78C
$C_{12}H_{10}N_2O_6$			
4H,8H-Benzo[1,2-c:4,5-c']diisoxazole-4,8-dione, 3,7-diethoxy-	MeCN	213(3.9),273(3.9), 290(4.0)	24-3346-78
$C_{12}H_{10}N_2S$			
3H-Naphtho[1,8-bc]thiophen-3-one, 2-methyl-, hydrazone	MeOH	214(4.144),280s(--), 328s(--)	5-0627-78
6H-Pyrrolo[3,2-d]isothiazole, 6-methyl-5-phenyl-	n.s.g.	245(4.29),275(4.08)	103-0465-78
$C_{12}H_{10}N_4$			
Propanedinitrile, [(3-ethenylphenyl)-azo]methyl-	CHCl₃	290(4.09)	126-1803-78
Propanedinitrile, [(4-ethenylphenyl)-azo]methyl-	toluene	327(4.26)	126-1803-78
$C_{12}H_{10}N_4O$			
3H-Pyrazol-3-one, 5-amino-2,4-dihydro-4-(1H-indol-3-ylmethylene)-	0.1% NaOH	446(3.430)	103-1088-78
	EtOH	400(3.380)	103-1088-78
	EtOH-NaOEt	446(4.462)	103-1088-78
sodium salt	n.s.g.	447(3.416)	103-1088-78
$C_{12}H_{10}N_4O_2$			
Benzo[g]pteridine-2,4(1H,3H)-dione, 1,3-dimethyl-	MeOH	320(3.88),378(3.89)	83-0115-78
	6M HCl	380(4.03)	83-0115-78
L-Lathyrine	pH 1	224(4.2),298(3.7)	102-2027-78
	pH 13	226(4.1),292(3.6)	102-2027-78
1H-Purine, 6-(4-methoxyphenoxy)-	EtOH	257(4.05)	44-2587-78
$C_{12}H_{10}N_4O_2S$			
1,2,3-Thiadiazolo[4,5-d]pyrimidinium, 4,5,6,7-tetrahydro-3-methyl-5,7-di-oxo-6-(phenylmethyl)-, hydroxide, inner salt	EtOH	245(4.36),300(2.69), 410(3.59)	44-1677-78
$C_{12}H_{10}N_4O_3$			
1-Triazene, 3-hydroxy-1-(2-nitrophen-	EtOH-pH 8	330(4.1),405(3.9)	140-0032-78

Compound	Solvent	$\lambda_{max}(\log \epsilon)$	Ref.
y1)-3-phenyl- (cont.)	EtOH-NaOH	<u>380(4.2),490(4.1)</u>	140-0032-78
$C_{12}H_{10}N_4O_4$			
3H-Indole-2,3-dicarboxylic acid, 3-azido-	EtOH	236(4.09),239(4.09), 305(3.78)	94-2866-78
$C_{12}H_{10}N_6O$			
1H-Pyrazolo[3,4-d]pyrimidine-4-carbox-imidamide, N-hydroxy-1-phenyl-	EtOH	246(4.48),322(3.89)	95-0891-78
1H-Pyrazolo[3,4-d]pyrimidine-4-carbox-ylic acid, 1-phenyl-, hydrazide	EtOH	250(4.48),275s(3.98), 320s(3.55)	95-1274-78
$C_{12}H_{10}N_6O_2$			
Pyrimido[4,5-e]-1,2,4-triazine-6,8(5H,7H)-dione, 5,7-dimethyl-3-(3-pyridinyl)-	EtOH	240(4.43),327(4.10)	94-0367-78
Pyrimido[4,5-e]-1,2,4-triazine-6,8(5H,7H)-dione, 5,7-dimethyl-3-(4-pyridinyl)-	EtOH	235(4.62),324(4.24)	94-0367-78
$C_{12}H_{10}O$			
6(5H)-Benzocyclooctenone	C_6H_{12}	223(4.32),256s(4.21), 288(3.71)	88-0667-78
$C_{12}H_{10}OS$			
Methanone, (2-methyl-3-thienyl)phenyl-	hexane	<u>250(4.2),335(4.3)</u>	23-1970-78
	MeOH	<u>256(4.2)</u>	23-1970-78
Methanone, (4-methyl-3-thienyl)phenyl-	hexane	<u>245(4.1),340(4.3)</u>	23-1970-78
2H-Naphtho[1,8-bc]thiophene-3-ol, 2-methyl-	MeOH	224(4.423),246(4.303), 304s(--),312(3.783), 340s(--),349(3.725)	24-1824-78
$C_{12}H_{10}O_2$			
Benzo[1,2-b:4,3-b']difuran, 2,7-di-methyl-	MeOH	221(4.04),265(4.16), 273(4.29),285(4.17), 295(3.60)	4-0043-78
Benzo[1,2-b:4,5-b']difuran, 2,6-di-methyl-	EtOH	230(4.26),268s(4.19), 278(4.40),287(4.07), 293(4.17),299(4.20), 306(4.18)	4-0043-78
5,8-Methano-9H-benzocyclohepten-9-one, 5,8-dihydro-8-hydroxy-	EtOH	246(2.91),295(2.85)	44-3478-78
2H-Naphtho[1,8-bc]furan, 6-methoxy-	EtOH	207(4.59),242(4.38), 246(4.37),325(3.86), 339(3.88)	18-2068-78
2H-Naphtho[1,8-bc]furan, 8-methoxy-	EtOH	213(4.33),238(4.64), 303(3.56),334(3.51)	18-2068-78
3H-Naphtho[1,8-bc]furan-2-carboxalde-hyde, 4,5-dihydro-	EtOH	203(4.16),233s(3.96), 238(4.00),304(4.35)	18-2068-78
Propanal, 2-(2H-1-benzopyran-2-yli-dene)-	·EtOH	371(4.30)	48-0497-78
$C_{12}H_{10}O_2S$			
2-Propenoic acid, 3-benzo[b]thien-2-yl-, methyl ester, (E)-	MeOH	256(3.77),315(4.50)	44-2493-78
Thiophene, 2-(4-methoxybenzoyl)-	C_6H_{12}	220(4.05),264(4.15), 288(4.33)	18-2718-78
$C_{12}H_{10}O_3$			
Furo[2,3-f]-1,3-benzodioxole, 6-(1-methylethenyl)-	EtOH	222(4.10),276(4.06), 289s(3.92),322(4.24),	117-0137-78

Compound	Solvent	$\lambda_{max}(\log \epsilon)$	Ref.
(cont.)		325(4.24),333(4.20)	117-0137-78
1,4-Methano-6H-benzocyclohepten-6-one, 1,4-dihydro-5,9-dihydroxy-	n.s.g.	244(4.45),262s(4.22), 295s(3.80),331(4.03), 342s(3.97),376s(4.05), 393(4.12),424s(3.24)	88-2289-78
2-Naphthalenecarboxaldehyde, 1,4-dihydroxy-3-methyl-	EtOH	220(4.39),270(4.32), 412(3.90)	104-1962-78
1,2-Naphthalenedione, 7-methoxy-5-methyl-	EtOH	277(4.22),354(3.16), 485(3.21)	12-1363-78
1,4-Naphthalenedione, 6-ethoxy-	CHCl$_3$	264(4.31),273s(4.18), 350s(3.23),396(3.40)	12-1335-78
1,4-Naphthalenedione, 7-methoxy-5-methyl-	EtOH	265(4.18),340(3.34), 370s(3.32)	12-1363-78
$C_{12}H_{10}O_4$			
2,8-Decadiene-4,6-diynedioic acid, dimethyl ester, (E,E)-	MeOH	268(4.41),296(4.24), 315(4.43),338(4.39)	78-1323-78
2(5H)-Furanone, 4-acetyl-3-hydroxy-5-phenyl-	H$_2$O	318(4.20)	4-0737-78
	EtOH	267(3.97),323(3.67)	4-0737-78
	EtOH-NaHCO$_3$	323(4.22)	4-0737-78
	CHCl$_3$	256(3.95)	4-0737-78
1,4-Naphthalenedione, 2,3-dimethoxy-	C$_6$H$_{12}$	329(3.46)	39-1083-78C
	EtOH	335(--)	39-1083-78C
1,4-Naphthalenedione, 5,7-dimethoxy-	EtOH	265(4.07),360s(3.27), 415(3.52)	12-1353-78
1,4-Naphthalenedione, 5,8-dimethoxy-	C$_6$H$_{12}$	414(3.63)	150-2319-78
	EtOH	456(--)	150-2319-78
2H-Pyran-2,6(3H)-dione, 4-(4-methoxyphenyl)-	EtOH	274(4.28),373(3.87)	2-0196-78
	EtOH-HCl	232(4.00),315(4.26)	2-0196-78
	EtOH-NaOH	274(4.25),373(3.87)	2-0196-78
	dioxan	275(4.22),376(3.87)	2-0196-78
	dioxan-NaOH	275(4.22),376(3.70)	2-0196-78
	EtOAc	300(4.09)	2-0196-78
	CHCl$_3$	321(4.32)	2-0196-78
Spiro[1,3-dioxane-2,2'-[2H]indene]-1',3'-dione	MeOH	230(4.64),249(4.04), 282(2.95),290(2.92), 305(2.49),360(1.76)	24-3058-78
$C_{12}H_{10}O_6$			
1H-2-Benzopyran-4-carboxylic acid, 6,7-dimethoxy-1-oxo-	EtOH	245(3.79),280(2.92), 325(2.94)	4-0257-78
$C_{12}H_{10}S_4$			
1,3-Benzodithiole, 2-(4,5-dimethyl-1,3-dithiol-2-ylidene)-	n.s.g.	247(3.84),253(3.94), 260(4.00),292(4.15), 319(4.11),454(2.5)	44-0369-78
$C_{12}H_{11}AsO$			
4-Arseninol, 2-(4-methylphenyl)-	EtOH	260(4.38),304(3.90)	88-1471-78
$C_{12}H_{11}BrO$			
1,4-Hexadien-3-one, 1-bromo-1-phenyl-	EtOH	205(4.39),233(4.01), 293(4.05)	70-0932-78
1,4-Pentadien-3-one, 1-bromo-4-methyl-1-phenyl-, (Z)-	EtOH	238(3.93),295(3.85)	70-1361-78
$C_{12}H_{11}BrO_2$			
2-Cyclohexen-1-one, 2-bromo-3-hydroxy-5-phenyl-	EtOH	295(4.32)	2-0970-78

Compound	Solvent	$\lambda_{max}(\log \epsilon)$	Ref.
$C_{12}H_{11}ClN_5O_6P$ 3H-Imidazo[2,1-i]purine, 5-chloro- 3-(3,5-O-phosphinico-β-D-ribo- furanosyl)-	pH 1	225(4.48),279(4.08)	94-2391-78
$C_{12}H_{11}ClO_2$ 2-Cyclopropene-1-carboxylic acid, 2-(3-chlorophenyl)-3-methyl-, methyl ester	EtOH	258(4.92)	104-2144-78
2-Cyclopropene-1-carboxylic acid, 2-(4-chlorophenyl)-3-methyl-, methyl ester	EtOH	260(3.91)	104-2144-78
$C_{12}H_{11}Cl_3N_2O$ Acetamide, 2,2,2-trichloro-N-[(2-meth- yl-1H-indol-5-yl)methyl]-	MeOH	225(4.65),277(3.87), 283(3.80),294(3.73)	103-0856-78
$C_{12}H_{11}F_5$ Benzene, cyclohexylpentafluoro-	EtOH	207(3.70),224(3.30), 259(2.58)	104-1537-78
Benzene, pentafluoro(1,1,2-trimethyl- 2-propenyl)-	EtOH	262(2.48)	39-0232-78C
$C_{12}H_{11}IN_5O_6P$ 3H-Imidazo[2,1-i]purine, 5-iodo- 3-(3,5-O-phosphinico-β-D-ribo- furanosyl)-	pH 1	224(4.37),233s(4.35), 241s(4.34),285(4.10)	94-2391-78
$C_{12}H_{11}N$ 2H-Azirine, 3-phenyl-2-(2-propynyl)-	C_6H_{12}	241(4.06)	44-3757-78
1H-Cyclopenta[c]quinoline, 2,3-dihydro-	C_6H_{12}	228(4.60),270(3.50), 315(3.40)	44-2190-78
1H-Pyrrole, 1-ethenyl-2-phenyl-	C_6H_{12}	208(4.11),250(4.10), 270(4.12)	103-0399-78 +104-1614-78
$C_{12}H_{11}NO$ [1,1'-Biphenyl]-2-ol, 2'-amino-	MeOH	286(3.77)	94-2508-78
	MeOH-KOH	306(--)	94-2508-78
Pyridinium, 5-hydroxy-2-methyl-1-phen- yl-, hydroxide, inner salt	EtOH-NaOH	223(4.24),260(3.90), 343(3.65)	1-0068-78
2(1H)-Pyridinone, 5-methyl-1-phenyl-	EtOH	227(3.68),316(3.64)	1-0068-78
$C_{12}H_{11}NOS$ Benzenamine, 2-methoxy-N-(2-thienyl- methylene)-	C_6H_{12}	255(4.07),285(4.07), 340(3.77)	18-2718-78
Benzenamine, 4-methoxy-N-(2-thienyl- methylene)-	C_6H_{12}	255(4.24),282(4.25), 336(4.23)	18-2718-78
$C_{12}H_{11}NOS_2$ Carbamodithioic acid, phenyl-, 2-furan- ylmethyl ester	n.s.g.	350s(2.01)	97-0381-78
$C_{12}H_{11}NO_2$ 1H-Benzofuro[2,3-c]azepin-1-one, 2,3,4,5-tetrahydro-	n.s.g.	279(3.73)	103-1223-78
5H-[1]Benzopyrano[3,4-b]pyridin-5-one, 1,2,3,4-tetrahydro-	n.s.g.	250(3.92),344(4.23)	103-1223-78
1-Isoquinolineacetic acid, methyl ester	EtOH	220(4.79),265s(3.64), 273(3.71),285(3.64), 310(3.50),323(3.59), 376(2.54),395(2.67),	4-1425-78

Compound	Solvent	$\lambda_{max}(\log \epsilon)$	Ref.
(cont.)		417(2.51)	4-1425-78
1,4-Naphthalenedione, 5-amino-2,3-di-methyl-	C_6H_{12}	472(3.67)	150-2319-78
	EtOH	500(--)	150-2319-78
1,4-Naphthalenedione, 2-(dimethyl-amino)-	C_6H_{12}	444(3.53)	39-1083-78C
	EtOH	471(3.53)	39-1083-78C
2,5-Pyrrolidinedione, 1-(2-phenyleth-enyl)-	THF	219(4.27),273(4.38), 300(3.05)	40-0404-78
$C_{12}H_{11}NO_2S$			
4-Oxazolidinone, 3-methyl-5-[(4-methyl-phenyl)methylene]-2-thioxo-	n.s.g.	270(4.77),359(4.55)	104-1225-78
Pyridine, 2-[(4-methylphenyl)sulfonyl]-	EtOH	237(4.01),255s(3.75)	95-0095-78
Pyridine, 4-[(4-methylphenyl)sulfonyl]-	EtOH	243(4.05)	95-0095-78
$C_{12}H_{11}NO_2S_4$			
2H-1,3-Thiazine-2,6(3H)-dithione, 4-(4-methylphenyl)-5-(methylsulfonyl)-	n.s.g.	236s(4.00),316(4.43), 372(3.63),442s(3.91), 460(4.01)	39-1017-78C
$C_{12}H_{11}NO_3$			
2-Benzofuranacetaldehyde, 7-ethyl-α-oxo-, aldoxime, (Z)-	n.s.g.	237(4.11),274(3.56), 323(4.27)	39-0928-78C
1H-Indole-2-acetic acid, 3-formyl-, methyl ester	MeOH	221(4.11),250(3.95), 265(3.99),301(4.07)	150-1683-78
4-Isoxazolecarboxylic acid, 5-methyl-3-phenyl-, methyl ester	EtOH	220(4.21)	22-0415-78
5-Isoxazolemethanol, 3-phenyl-, acetate	EtOH	233(4.10)	22-0415-78
1,4-Naphthalenedione, 2-methoxy-3-(methylamino)-	C_6H_{12}	484(3.45)	39-1083-78C
	EtOH	492(--)	39-1083-78C
1,4-Naphthalenedione, 5-methoxy-8-(methylamino)	C_6H_{12}	556(3.72)	150-2319-78
	EtOH	595(--)	150-2319-78
6H-1,3-Oxazinium, 4-hydroxy-2,3-di-methyl-6-oxo-5-phenyl-, hydroxide, inner salt	MeCN	255(3.92),323(3.53)	5-1655-78
2-Propenoic acid, 3-(1-methoxy-1H-indol-3-yl)-, (E)-	MeOH	208s(3.71),231(3.88), 273(3.64),321(3.85)	39-1117-78C
2-Quinolinecarboxylic acid, 1,4-di-hydro-1-methyl-4-oxo-, methyl ester	n.s.g.	218(4.37),242(4.26), 340(3.99)	4-0551-78
$C_{12}H_{11}NO_3S$			
4H-1,3-Benzothiazine-2-acetic acid, 4-oxo-, ethyl ester	MeOH	231(4.13),243(4.18), 297(4.42)	5-0473-78
	pH 13	306(4.48),357(4.01)	5-0473-78
4-Oxazolidinone, 5-[(4-methoxyphenyl)-methylene]-3-methyl-2-thioxo-	n.s.g.	256(4.26),293(4.12), 370(4.60)	104-1225-78
4(5H)-Oxazolone, 5-[(4-methoxyphenyl)-methylene]-2-(methylthio)-	n.s.g.	282(4.23),348(4.54)	104-1225-78
2-Thiazolecarboxylic acid, 4-hydroxy-5-phenyl-, ethyl ester	MeOH	246(3.90),257(3.90), 362(4.23)	5-0473-78
$C_{12}H_{11}NO_3S_3$			
6H-1,3-Thiazin-6-one, 2,3-dihydro-4-(4-methylphenyl)-5-(methylsul-fonyl)-2-thioxo-	EtOH	241(4.08),309(4.30), 394(3.96),460(3.62)	39-1017-78C
$C_{12}H_{11}NO_4$			
Acetic acid, (5,8-dihydro-3-methoxy-8-oxo-7(6H)-isoquinolinylidene)-	MeOH	235(4.05),255(4.03), 297(4.10)	44-3778-78
3-Quinolineacetic acid, α-hydroxy-2-methyl-, 1-oxide	EtOH	240(4.18),320(3.85)	95-0802-78

Compound	Solvent	$\lambda_{max}(\log \epsilon)$	Ref.
$C_{12}H_{11}NO_5$			
3-Furancarboxylic acid, 2,3-dihydro-2-(4-nitrophenyl)-, methyl ester, cis	EtOH	267(3.98)	24-3665-78
2-Propenoic acid, 3-[3-(4-nitrophenyl)-oxiranyl]-, methyl ester, cis	EtOH	234(3.97),273(4.03)	24-3665-78
trans	EtOH	217(4.29),275(4.20)	24-3665-78
2H-Pyrano[3,2-c]pyridine-3-carboxylic acid, 5,6-dihydro-7-methyl-2,5-dioxo-, ethyl ester	MeOH	392(4.23)	49-1075-78
$C_{12}H_{11}NO_5S$			
3(2H)-Benzofuranone, 6-methoxy-2-[1-(methylthio)-2-nitroethylidene]-	EtOH	275(3.86),325(3.78)	142-0399-78A
$C_{12}H_{11}NS$			
Pyrrolo[2,1-b]benzothiazole, 1,3-di-methyl-	EtOH	225(3.785),251(3.826), 313(3.243)	39-1198-78C
$C_{12}H_{11}N_3$			
Cyclohepta[b]pyrrole-3-carbonitrile, 1-ethyl-1,2-dihydro-2-imino-	EtOH	230(4.14),283(4.43), 434(4.10)	138-0677-78
4-Isoquinolinecarbonitrile, 3-(dimethylamino)-	EtOH	253(4.61),293(4.09), 398(3.65)	138-0677-78
Propanedinitrile, [[5-(dimethylamino)-methylene]-1,3-cyclopentadien-1-yl]-methylene]-	MeOH	285(4.31),385s(4.09), 460(4.59)	83-0369-78
$C_{12}H_{11}N_3O$			
Acetamide, N-[2,2'-bipyridin]-4-yl-	neutral	230(4.33),280(4.27), 304(4.21)	39-1215-78B
	cation	230(4.50),285(4.15), 327s(3.76),341s(3.71)	39-1215-78B
Benzamide, N-(4-ethenyl-1H-imidazol-2-yl)-	EtOH	228(4.11),259(3.99), 283s(3.97),296(3.99)	35-4208-78
Imidazo[1,2-a]pyrimidin-5(1H)-one, 2,3-dihydro-7-phenyl-	EtOH	242(4.49),315(3.92)	12-0179-78
Pyridinium, 3-amino-1-(benzoylamino)-, hydroxide, inner salt	EtOH	321(3.61)	33-2887-78
4H-Pyrimido[2,1-b]quinazolin-4-one, 1,6-dihydro-2-methyl-	MeOH	257(4.10),312(4.10)	4-0077-78
$C_{12}H_{11}N_3O_2$			
Benzamide, N-(4-acetyl-1H-imidazol-2-yl)-	MeOH	225(4.00),285(4.23)	35-4208-78
Benzamide, N-(4-acetyl-1H-pyrazol-3-yl)-	MeOH	234(4.23),283(4.04), 290s(4.03)	35-4208-78
3-Buten-2-one, 4-[(5-phenyl-1,2,4-oxa-diazol-3-yl)amino]-	MeOH	249(4.20),256(4.21), 288(4.43)	35-4208-78
Imidazo[2,1-a]phthalazine-3,6(2H,5H)-dione, 2,2-dimethyl-	EtOH	229(3.81),241(3.78), 246s(3.76),260(3.64), 270(3.61),281(2.95), 320-345(3.04)	33-2116-78
1H-Pyrido[1,2-c]pyrimidine-4-carbo-nitrile, 3-ethoxy-8-methyl-1-oxo-	EtOH	254(4.04),263(4.01), 282(3.77),300(3.93), 310(3.96),387(4.05)	95-0623-78
$C_{12}H_{11}N_3O_2S$			
1H-Pyrimido[5,4-b][1,4]benzothiazine-2,4(1H,3H)-dione, 1,3-dimethyl-	pH 13	233s(4.19),262(4.04), 281s(3.95),320s(3.58)	5-0193-73
	MeOH	249(4.31),281(3.86), 325s(3.30),380s(3.00)	5-0193-78

Compound	Solvent	$\lambda_{max}(\log \epsilon)$	Ref.
(cont.)	12M HCl	254(4.25),293(3.78), 376(3.41)	5-0193-78
1H-Pyrimido[5,4-b][1,4]benzothiazine- 2,4(1H,3H)-dione, 3,10-dimethyl-	pH 13	256(4.37),281s(3.93), 320s(3.43)	5-0193-78
	MeOH	250(4.24),281(3.93), 360s(3.15)	5-0193-78
	12M HCl	249(4.16),291(3.80), 369(3.45)	5-0193-78
$C_{12}H_{11}N_3O_3$			
Diazene, (2,5-dimethyl-3-furanyl)(4-ni- trophenyl)-	n.s.g.	378(4.00)	39-0401-78C
Ethanone, 1-[5-methyl-1-(4-nitrophen- yl)-1H-pyrazol-3-yl]-	n.s.g.	295(4.14)	39-0401-78C
$C_{12}H_{11}N_3O_3S$			
1H-[1,2]Oxathiolo[3,2-e][1,2,3]thiadia- zole-7-S^{IV}, 3,4-dimethyl-1-(4-nitro- phenyl)-	C_6H_{12}	243s(3.91),325(3.75), 460(4.53)	39-0195-78C
$C_{12}H_{11}N_3O_4$			
2-Quinoxalinecarboxamide, 6-acetyl- 3-methyl-, 1,4-dioxide	MeOH	238(4.38),280(4.37), 385(3.92)	87-0483-78
2-Quinoxalinecarboxamide, 7-acetyl- 3-methyl-, 1,4-dioxide	MeOH	237(4.33),278(4.35), 380(3.80)	87-0483-78
$C_{12}H_{11}N_3O_6$			
2,5-Furandione, 3,4-dimethoxy-, mono- [(4-nitrophenyl)hydrazone]	dioxan	220(4.05),297(3.63), 392(4.61)	33-1033-78
$C_{12}H_{11}N_3S_2$			
Pyrido[3',2':4,5]thieno[3,2-d]pyrimi- dine, 7,9-dimethyl-4-(methylthio)-	EtOH	243(4.34),259(4.09), 274(4.09),298(4.31), 326(3.88),338(3.88), 372(3.11),385(3.00)	32-0057-78
$C_{12}H_{11}N_5O$			
6H-Imidazo[4,5-e]-1,2,4-triazin-6-one, 5,7-dihydro-5,7-dimethyl-3-phenyl-	EtOH	245(4.32),271(4.08), 310(3.98)	94-3154-78
6H-Imidazo[4,5-e]-1,2,4-triazin-6-one, 1,5-dihydro-5-methyl-3-(4-methyl- phenyl)-	EtOH	241(4.11),310s(3.60)	94-3154-78
1H-1,2,3-Triazolo[4,5-b]pyridin-5-am- ine, 7-(phenylmethoxy)-	pH 1 pH 11	257(4.03),296(4.17) 271(3.92),292(4.02)	44-4910-78 44-4910-78
$C_{12}H_{11}N_5O_2$			
6H-Imidazo[4,5-e]-1,2,4-triazin-6-one, 1,5-dihydro-3-(4-methoxyphenyl)-5- methyl-	EtOH	253(4.19),279(4.25), 320s(3.86)	94-3154-78
$C_{12}H_{11}N_5O_6$			
6,7-Pteridinedicarboxylic acid, 2-(ace- tylamino)-1,4-dihydro-4-oxo-, dimeth- yl ester	pH 2.0 pH 9.0	242(4.14),300(4.21), 347(4.05) 273(4.43),366(3.99)	24-3790-78 24-3790-78
$C_{12}H_{12}$			
Azulene, 1,2-dimethyl-	C_6H_{12}	240(4.18),278(4.74), 287(4.78),305(3.84), 320s(3.18),335(3.49), 345s(3.40),352(3.64), 368(2.96),557(2.30),	5-1379-78

Compound	Solvent	$\lambda_{max}(\log \epsilon)$	Ref.
Azulene, 1,2-dimethyl- (cont.)		573(2.35),590(2.40), 623(2.34),647(2.34), 687(1.98),716(1.94)	5-1379-78
Benzobicyclo[3.2.1]octa-2,5-diene	hexane	255s(2.26),263(3.48), 268(2.67),275(2.71)	35-2959-78
Bicyclo[6.2.0]deca-1,3,5,7,9-pentaene, 9,10-dimethyl-	C_6H_{12}	271(4.56),280(4.64), 365s(2.79),390s(2.84), 410s(2.83),435s(2.76), 464s(2.56)	5-1379-78
Cyclohexane, hexakis(methylene)-	hexane	220(--)	33-0844-78
1,4-Methanonaphthalene, 1,2,3,4-tetra-hydro-2-methylene-	hexane	260(2.9),267(3.1), 272(3.1)	35-7364-78
$C_{12}H_{12}AsN$ 4-Arseninamine, 2-(phenylmethyl)-	EtOH	245(4.13),330(4.09)	88-1175-78
$C_{12}H_{12}BrFO$ 1-Penten-3-one, 1-bromo-4-fluoro-4-methyl-1-phenyl-, (Z)-	EtOH	205(4.08),228(3.83), 297(4.08)	70-1361-78
$C_{12}H_{12}BrNO_2$ 2(1H)-Pyridinone, 1-(4-bromophenyl)-5,6-dihydro-4-hydroxy-6-methyl-	EtOH	223(3.98),302(4.34)	4-1153-78
$C_{12}H_{12}BrN_3O_4$ 4,11-Epoxy-5H-1,3-dioxolo[4,5-e]imid-azo[2,1-b][1,3]oxazocine-8-carbo-nitrile, 9-bromo-3a,4,11,11a-tetrahydro-2,2-dimethyl-	EtOH	238(3.89)	94-3322-78
$C_{12}H_{12}ClNOS$ Morpholine, 4-(5-chlorobenzo[b]thien-2-yl)-	MeOH	237(4.48),255s(4.05), 294(4.29)	73-1276-78
$C_{12}H_{12}ClN_2S$ Methanaminium, N-[3-(4-chlorophenyl)-3-isothiocyanato-2-propenylidene]-N-methyl-, perchlorate	MeCN	232(4.13),244(4.12), 321s(4.24),360(4.44)	97-0334-78
1,3-Thiazin-1-ium, 4-(4-chlorophenyl)-2-(dimethylamino)-, perchlorate	MeCN	227(4.13),259(4.14), 336(4.41),396s(3.49)	97-0334-78
$C_{12}H_{12}ClN_5$ 1,2,4,5-Tetrazine, 3-(4-chlorophenyl)-6-pyrrolidino-	dioxan	300(4.50),315(3.03), 551(2.64)	30-0338-78
$C_{12}H_{12}ClN_5O$ 1,2,4,5-Tetrazine, 3-(4-chlorophenyl)-6-morpholino-	dioxan	308(4.13),417(2.75), 543(2.34)	30-0338-78
$C_{12}H_{12}Cl_2N_2O$ 1,2,4-Oxadiazole, 3-(3,5-dichloro-2,4,6-trimethylphenyl)-5-methyl-	MeCN	287(2.78)	39-0607-78B
$C_{12}H_{12}F_6$ Benzene, pentafluoro(2-fluoro-1,1,2-trimethylpropyl)-	EtOH	262(2.79)	39-0232-78C
$(C_{12}H_{12}Fe)_n$ Ferrocene, ethenyl-, homopolymer	benzene	445(2.0)	126-0131-78

Compound	Solvent	$\lambda_{max}(\log \epsilon)$	Ref.
$C_{12}H_{12}FeO_4$			
Iron, tetracarbonyl[(5,6-η)-cycloocta-1Z,5E-diene]-	hexane	208(4.51),267s(3.76)	33-1695-78
$C_{12}H_{12}FeO_4S_4$			
Iron, tricarbonyl[(2,3,4,5-η)-2,3,4,5-tetrakis(methylthio)-2,4-cyclopentadien-1-one]-	EtOH	263(3.11),340(2.75)	101-0159-78Q
$C_{12}H_{12}NO_2$			
Pyridinium, 1-(2,3-dihydroxyphenyl)-2-methyl-, iodide	EtOH	350(3.88)	2-0165-78
	EtOH-NaOH	420(4.07)	2-0165-78
Pyridinium, 1-(2,3-dihydroxyphenyl)-3-methyl-, iodide	EtOH	355(4.30)	2-0165-78
	EtOH-NaOH	440(4.32)	2-0165-78
Pyridinium, 1-(2,3-dihydroxyphenyl)-4-methyl-, iodide	EtOH	359(4.16)	2-0165-78
	EtOH-NaOH	435(4.31)	2-0165-78
Pyridinium, 1-(2,5-dihydroxyphenyl)-2-methyl-, iodide	EtOH	360(3.46)	2-0165-78
	EtOH-NaOH	467(3.43)	2-0165-78
Pyridinium, 1-(2,5-dihydroxyphenyl)-3-methyl-, iodide	EtOH	368(3.45)	2-0165-78
	EtOH-NaOH	460(3.48)	2-0165-78
Pyridinium, 1-(2,5-dihydroxyphenyl)-4-methyl-, iodide	EtOH	380(3.36)	2-0165-78
	EtOH-NaOH	475(3.42)	2-0165-78
$C_{12}H_{12}N_2$			
2-Penten-4-ynal, 3-methyl-, (3-methyl-2-penten-4-ynylidene)hydrazone	EtOH	302s(4.30),315(4.43), 328(4.43),344s(4.26)	18-2112-78
4H-Pyrrolo[1,2-b]pyrazole, 5,6-dihydro-2-phenyl-	EtOH	254(4.26)	44-1664-78
$C_{12}H_{12}N_2OS$			
Azepino[4,3-b][1,4]benzothiazin-1(2H)-one, 3,4,5,11-tetrahydro-	EtOH	235s(4.14),315(3.30)	103-0306-78
1H-[1,2]Oxathiolo[3,2-e][1,2,3]thiadiazole-7-S^{IV}, 3,4-dimethyl-1-phenyl-	C_6H_{12}	236(3.98),275(3.83), 447(4.32)	39-0195-78C
1H-Pyrrolo[1,2-c]imidazol-1-one, hexahydro-2-phenyl-3-thioxo-	MeOH	204(4.34),227(4.03), 238(4.04),272(4.20), 321(2.00)	19-0851-78
$C_{12}H_{12}N_2OS_2$			
Carbamodithioic acid, (2-furanylmethyl)-, 2-pyridinylmethyl ester	n.s.g.	332(2.02)	97-0381-78
$C_{12}H_{12}N_2O_2$			
3(2H)-Furanone, 2,5-dimethyl-2-(phenylazo)-	MeOH	275(4.18),311s(3.70)	39-0681-78C
1,4-Naphthalenedione, 2,3-bis(methylamino)-	C_6H_{12}	525(3.34)	39-1083-78C
	EtOH	544(--)	39-1083-78C
1H-Pyrazol-3-ol, 1-methyl-5-phenyl-, acetate	MeOH	205(4.37),238(4.15)	24-0791-78
Pyridinium, 1,1'-(1,2-ethanediyl)bis[3-hydroxy-, dihydroxide, bis(inner salt)	EtOH	229(4.19),259(4.03), 336(3.80)	150-1182-78
1H-Pyrrolo[1,2-c]imidazole-1,3(2H)-dione, tetrahydro-2-phenyl-	MeOH	204(4.20),216(4.11), 240(3.48),267(2.68)	19-0851-78
2-Quinoxalinecarboxylic acid, 3-methyl-, ethyl ester	EtOH	240(4.46),316(3.71)	94-2866-78
$C_{12}H_{12}N_2O_3$			
2-Furanpropanenitrile, 5-(2-cyanoethenyl)-3,4-dimethoxy-	EtOH	321(4.37),327(4.36)	33-0430-78
5,8-Isoquinolinedione, 3-(dimethylam-	EtOH	248(4.42),317(4.17)	4-0569-78

Compound	Solvent	$\lambda_{max}(\log \epsilon)$	Ref.
ino)-7-methoxy- (cont.)			4-0569-78
$C_{12}H_{12}N_2O_4$			
1,5-Naphthyridine-3-carboxylic acid, 1,4-dihydro-8-methoxy-4-oxo-, ethyl ester	H_2O	221(4.39),245(4.23), 304(4.18),317(4.13)	44-1331-78
1,5-Naphthyridine-3-carboxylic acid, 1,4,5,8-tetrahydro-5-methyl-4,8-dioxo-, ethyl ester	H_2O	232(4.47),263(3.58), 319(4.29),332(4.29)	44-1331-78
2-Pyridineacetic acid, 3-acetoxy-α-cyano-, ethyl ester	MeOH	224(3.99),293(4.21), 375(3.98)	4-1411-78
	MeOH-HCl (pH 3)	227(3.97),289(4.14), 370(4.00)	4-1411-78
	MeOH-NaOH (pH 12)	224(3.98),243(3.98), 293(3.92),384(3.96), 396(3.92)	4-1411-78
$C_{12}H_{12}N_2S$			
Benzenamine, 4,4'-thiobis-	EtOH	208(--),210(4.41), 251s(4.28),265(4.40)	111-0093-78
2H-Pyrrole-2-thione, 3-(aminomethyl-ene)-1,3-dihydro-1-methyl-5-phenyl-	n.s.g.	258(3.76),380(3.19)	103-0465-78
$C_{12}H_{12}N_2S_2$			
Benzenamine, 2,2'-dithiobis-	EtOH	222(4.65),340(4.05)	103-0380-78
$C_{12}H_{12}N_4$			
1H-Carbazole, 1-azido-2,3,4,9-tetra-hydro-	EtOH	224(4.30),277(3.70), 284(3.71),293(3.62)	94-2874-78
1H-Carbazole, 4a-azido-2,3,4,4a-tetra-hydro-	EtOH	222(4.19),226(4.09), 260(3.43),284(3.27)	94-2874-78
1H-Tetrazole, 1-(4-pentynyl)-5-phenyl-	EtOH	232(3.95)	44-1664-78
2H-Tetrazole, 2-(4-pentenyl)-5-phenyl-	EtOH	239(4.22)	44-1664-78
$C_{12}H_{12}N_4O_2$			
3H-Indole-2-carboxylic acid, 3-azido-3-methyl-, ethyl ester	EtOH	233(4.08),297(3.82)	94-2866-78
$C_{12}H_{12}N_4O_2S$			
Benzenesulfonic acid, [(4-ethenyl-1H-pyrazol-3-yl)methylene]hydrazide	EtOH	263(4.23)	39-1297-78C
$C_{12}H_{12}N_4O_3S$			
Acetamide, N-[5-ethyl-3-(4-nitrophen-yl)-1,3,4-thiadiazol-2(3H)-ylidene]-	EtOH	290(3.721),330(3.786)	146-0864-78
$C_{12}H_{12}N_4O_4S$			
Thieno[2,3-d:5,4-d']dipyrimidine-2,4,5,7(3H,6H)-tetrone, 1,8-di-hydro-1,3,6,8-tetramethyl-	EtOH	245(2.95),304(2.80)	44-1677-78
$C_{12}H_{12}N_4O_4S_2$			
[1,4]Dithiino[2,3-d:6,5-d']dipyrimid-ine-2,4,6,8(1H,3H,7H,9H)-tetrone, 1,3,7,9-tetramethyl-	EtOH	230(3.64),280(3.27), 310(2.98)	44-1677-78
$C_{12}H_{12}N_4O_5S$			
α-L-threo-Hex-4-enopyranuronic acid, 1,4-dideoxy-1-(1,6-dihydro-6-thioxo-9H-purin-9-yl)-, methyl ester	n.s.g.	250(4.01),279(4.26)	106-0250-78

Compound	Solvent	$\lambda_{max}(\log \epsilon)$	Ref.
$C_{12}H_{12}N_6OS$			
1,2,3-Thiadiazolo[4,5-d]pyrimidine-5,7(4H,6H)-dione, 4,6-dimethyl-, 7-(phenylhydrazone)	EtOH	247(3.74),325(2.61)	44-1677-78
$C_{12}H_{12}O$			
5(8H)-Benzocyclooctenone, 9,10-dihydro-	C_6H_{12}	214(3.79),242(3.91), 250s(3.85),287s(3.08)	35-5141-78
Bicyclo[5.3.1]undeca-1,3,5,7,9-pentaene, 5-methoxy-	hexane	280(4.31),300s(3.73), 490(2.70)	35-4320-78
2-Cyclopenten-1-one, 3-methyl-2-phenyl-	EtOH	221(4.15),248(3.92)	44-1481-78
4-Dodecene-6,8,10-triyn-3-ol, (E)-	ether	232(4.71),244(4.92), 259(3.46),274(3.76), 290(4.07),309(4.18), 331(4.01)	39-1487-78C
Ethanone, 1-(3,4-dihydro-1-naphthalenyl)-	MeOH	205(4.47),234(3.98), 280(3.58)	56-0529-78
1,4-Methano-7H-benzocyclohepten-7-one, 1,2,3,4-tetrahydro-	hexane	235(4.26),312(3.97), 324(3.96),400s(2.37)	88-0569-78
	MeOH	240(4.36),327(4.17)	88-0569-78
$C_{12}H_{12}O_2$			
3-Benzofurancarboxaldehyde, 2-ethyl-5-methyl-	MeOH	235(4.28),270(3.93)	83-0714-78
5-Benzofurancarboxaldehyde, 2,3-dihydro-2-(1-methylethenyl)-	MeOH	231(4.35),285s(4.34), 296(4.35)	150-5451-78
Bicyclo[4.2.2]deca-2,4,7,9-tetraene-2-carboxylic acid, methyl ester	C_6H_{12}	290(3.83)	5-2074-78
2(5H)-Furanone, 5,5-dimethyl-3-phenyl-	C_6H_{12}	224(4.00),257(4.04), 295(3.04)	35-8247-78
5,8-Methano-9H-benzocyclohepten-9-one, 5,6,7,8-tetrahydro-8-hydroxy-	EtOH	220(2.80),246(3.90), 295(2.85)	44-3478-78
1-Naphthalenol, 7-methoxy-5-methyl-	EtOH	230s(4.46),243(4.50), 280s(3.62),293(3.70), 323(3.33),347(3.35)	12-1363-78
$C_{12}H_{12}O_3$			
Cyclopenta[cd]pentalen-1(2H)-one, 6-acetoxy-2a,4a,5,6b-tetrahydro-	EtOH	212(3.44),293(2.40)	23-1687-78
Oxiranecarboxylic acid, 3-(2-phenylethenyl)-, methyl ester	EtOH	258(4.36),283s(3.63), 294(3.34)	24-3665-78
2-Propenoic acid, 3-(3-phenyloxiranyl)-, methyl ester, cis	EtOH	234s(3.99)	24-3665-78
trans	EtOH	234(4.20),271s(3.32)	24-3665-78
$C_{12}H_{12}O_3S_2$			
3(2H)-Benzofuranone, 2-[bis(methylthio)methylene]-6-methoxy-	EtOH	225(3.99),256(3.94), 352(4.30)	142-0399-78A
$C_{12}H_{12}O_4$			
2H-1-Benzopyran-8-carboxaldehyde, 3,4-dihydro-5-methoxy-2-methyl-4-oxo-	EtOH	245(4.00),260(4.00), 282s(3.76),340(3.49)	102-0511-78
2H-1-Benzopyran-2-one, 4,7-dimethoxy-5-methyl-	n.s.g.	226(3.95),288(4.06), 306(4.17),317(4.10)	78-1221-78
2H-1-Benzopyran-2-one, 5,7-dimethoxy-4-methyl-	MeOH	209(4.24),241(3.70), 251(3.65),318(4.06)	119-0143-78
4H-1-Benzopyran-4-one, 5,7-dihydroxy-2,6,8-trimethyl-	MeOH	212(4.53),228(4.33), 257(4.40),262(4.45), 320(4.00),335(3.69)	95-1285-78
4H-1-Benzopyran-4-one, 2-(dimethoxymethyl)-	EtOH	221(4.28),298(3.81)	118-0208-78

Compound	Solvent	$\lambda_{max}(\log \epsilon)$	Ref.
1-Benzoxepin-5(2H)-one, 3,7-bis(hy-droxymethyl)-	EtOH	218(4.31),266(4.05), 327(3.23)	39-1490-78C
	EtOH-NaOH	206(--),234s(--), 250s(--),327(--)	39-1490-78C
$C_{12}H_{12}O_5$			
2H-1-Benzopyran-2-one, 5,7,8-trimeth-oxy-	EtOH	254s(--),260(3.82), 326(3.89)	102-0505-78
2-Oxaspiro[4.5]deca-6,9-diene-4-carbox-ylic acid, 2,8-dioxo-, ethyl ester	MeOH	239(4.0)	24-1944-78
$C_{12}H_{12}O_6$			
1,3,5-Benzenetriol, triacetate	MeCN	208(4.40),221s(4.20), 267(3.13)	102-0579-78
$C_{12}H_{12}O_8S_2$			
1,4-Dithiin-2,3,5,6-tetracarboxylic acid, tetramethyl ester	EtOH	250s(3.83),314(3.82), 356s(2.47)	1-0152-78
$C_{12}H_{13}BrN_2O_2S$			
Imidazo[1,2-a]pyridine-3-carboxylic acid, 5-(bromomethyl)-2-(methyl-thio)-, ethyl ester	EtOH	265(4.49),310s(3.79)	95-0631-78
$C_{12}H_{13}BrO_3$			
5-Benzofuranol, 4-bromo-2-ethoxy-6,7-dimethyl-	EtOH	264(4.08),296(3.58)	12-1353-78
$C_{12}H_{13}BrO_4$			
1,4-Cyclohexadiene-1-acetic acid, 2-bromo-4,5-dimethyl-3,6-dioxo-, ethyl ester	hexane	278(4.13),345(2.52), 420s(1.60)	12-1353-78
$C_{12}H_{13}Cl$			
Benzene, 1-chloro-4-(1-methylene-4-pentenyl)-	EtOH	246(4.02)	35-0883-78
$C_{12}H_{13}ClN_2$			
1H-Pyrido[3,4-b]indole, 4a-chloro-2,3,4,4a-tetrahydro-2-methyl-	EtOH	226(3.94),286(3.04)	44-3705-78
$C_{12}H_{13}ClN_2O_4$			
2(5H)-Furanone, 5-hydroxy-3,4-dimeth-oxy-, (4-chlorophenyl)hydrazone	dioxan	242(4.27),290s(3.32)	33-1033-78
$C_{12}H_{13}ClN_2O_4S$			
Oxazolo[2,3-c][1,2,4]benzothiadiazin-1(2H)-one, 8-chloro-3a,4-dihydro-2,2,3a-trimethyl-, 5,5-dioxide	MeOH	245(3.98)	4-0063-78
$C_{12}H_{13}ClN_4$			
1(2H)-Phthalazinone, 4-chloro-2-meth-yl-, (1-methylethylidene)hydrazone, (Z)-	MeOH	286(--),350(--)	103-0567-78
	dioxan	288(4.28),359(4.02)	103-0567-78
	MeCN	212(4.55),281(4.38), 349(3.95)	103-0567-78
	CCl₄	360(--),389(--)	103-0567-78
2-Propanone, (4-chloro-1-phthalazinyl)-methylhydrazone	dioxan	336(3.88)	103-0436-78
	MeCN	215(4.70),336(3.93)	103-0436-78
	crystal	216(--),338(--)	103-0436-78

Compound	Solvent	λ_{max} (log ϵ)	Ref.

$C_{12}H_{13}ClO_4$
  1,3-Benzodioxol-5(7aH)-one, 7a-(2-chloroethoxy)-6-(2-propenyl)-   EtOH   239(3.97),296(3.48)   44-3983-78

$C_{12}H_{13}FO_6$
  2,4-Cyclohexadiene-1-carboxylic acid, 1,6-diacetoxy-2-fluoro-, methyl ester, cis   MeOH   266(3.56)   78-1707-78
  2,4-Cyclohexadiene-1-carboxylic acid, 1,6-diacetoxy-3-fluoro-, methyl ester, cis   MeOH   258(3.58)   78-1707-78
  2,4-Cyclohexadiene-1-carboxylic acid, 1,6-diacetoxy-4-fluoro-, methyl ester, cis   MeOH   264(3.55)   78-1707-78
  2,4-Cyclohexadiene-1-carboxylic acid, 1,6-diacetoxy-5-fluoro-, methyl ester, cis   MeOH   267.5(3.55)   78-1707-78

$C_{12}H_{13}F_6NO_4S_2$
  Benzenamine, N-butyl-2,4-bis(trifluoromethylsulfonyl)-   dioxan   337(3.75)   104-2208-78

$C_{12}H_{13}N$
  3-Azabicyclo[3.1.0]hex-2-ene, 1-methyl-2-phenyl- C_6H_{12} 239(4.04)   44-2029-78
  3-Azabicyclo[3.1.0]hex-2-ene, 2-methyl-1-phenyl- C_6H_{12} 250(2.97)   44-2029-78
  2H-Azirine, 3-(2-ethenylphenyl)-2,2-dimethyl-   MeOH   232(4.61),261(4.49)   35-2181-78
  2H-Azirine, 3-methyl-2-phenyl-2-(2-propenyl)- C_6H_{12} 222(4.45),268(3.74), 274(3.79),283(3.77), 290(3.67)   44-2029-78
  2H-Azirine, 2-methyl-3-[2-(1-propenyl)-phenyl]-   MeOH   236(3.78),262(4.56)   35-2181-78
  2H-Azirine, 2-methyl-3-[2-(2-propenyl)-phenyl]-   MeOH   247(4.56)   35-2181-78
  1H-Benz[g]indole, 6,7,8,9-tetrahydro-   EtOH   238(3.59),281(3.89), 271[sic](3.99)   103-0518-78
  3H-Benz[e]indole, 6,7,8,9-tetrahydro-   EtOH   272(4.00),287(3.85), 288(3.60)   103-0518-78
  Bicyclo[6.1.0]nona-2,4,6-triene-9-carbonitrile, 4,9-dimethyl-, (1α,8α,9α)- C_6H_{12} 208(4.11),243(3.59)   24-0282-78
  (1α,8α,9β)- C_6H_{12} 208(4.17),248(3.59)   24-0282-78
  2H-Pyrrole, 3,4-dihydro-5-(2-phenylethenyl)-   EtOH   216(4.28),222(4.32), 228(4.20),280(4.49)   4-0097-78

$C_{12}H_{13}NO$
  3-Azabicyclo[4.4.1]undeca-2,4,6,8,10-pentaene, 4-ethoxy-   MeOH   258(4.31),315(3.56)   89-0855-78
  2H-Azirine, 3-[2-(2-butenyloxy)phenyl]-, (E)- C_6H_{12} 246(3.96),303(3.65)   35-3494-78
  2H-Azirine, 2-methyl-3-[2-(2-propenyloxy)phenyl]- C_6H_{12} 247(4.05),302(3.73)   35-3494-78
  [1]Benzopyrano[4,3-b]pyrrole, 2,3,3a,4-tetrahydro-3-methyl- C_6H_{12} 249(4.32),304(3.59), 316(3.57)   35-3494-78
  5H-Cyclohepta[c]pyridine, 3-ethoxy-   MeOH   224(3.80),293(3.47)   89-0855-78
  1-Naphthalenamine, 7-methoxy-5-methyl-, hydrochloride   EtOH   250(4.19),280s(4.58), 300(3.66),325s(3.60)   12-1363-78
  Pyrano[4,3-b]indole, 1,3,4,5-tetrahydro-5-methyl-   EtOH   227(4.64),283(3.95), 290s(3.91)   39-1471-78C

Compound	Solvent	$\lambda_{max}(\log \epsilon)$	Ref.
1H-Pyrrole, 3-methoxy-2-methyl-5-phenyl-	MeOH	312(4.18)	49-0137-78
2H-Pyrrol-2-one, 1,5-dihydro-5,5-dimethyl-4-phenyl-	EtOH	260(4.09)	78-3291-78
$C_{12}H_{13}NOS$			
Benzo[b]thiophene, 4-butyl-5-nitroso-	$CHCl_3$	755(1.15)	39-0692-78C
$C_{12}H_{13}NO_2$			
2H-1-Benzopyran-2-one, 7-(dimethylamino)-4-methyl-	MeOH	369(4.31)	49-0123-78
1H-Isoindol-1-one, 2-ethenyl-3-ethyl-2,3-dihydro-3-hydroxy-	EtOH	207(4.30),226(4.28), 277(3.96)	40-0404-78
Isoxazole, 5-(ethoxymethyl)-3-phenyl-	EtOH	242(4.19),268(3.46), 276(3.36),283(2.88)	22-0415-78
2(1H)-Pyridinone, 5,6-dihydro-4-hydroxy-6-methyl-1-phenyl-	EtOH	296(4.14)	4-1153-78
2-Pyrrolidinone, 1-ethenyl-5-hydroxy-5-phenyl-	THF	237(4.18),277(2.87), 287s(2.49)	40-0404-78
$C_{12}H_{13}NO_3$			
1H-Benzofuro[2,3-c]azepin-1-one, 2,3,4,5,5a,10a-hexahydro-10a-hydroxy-	n.s.g.	277(4.12)	103-1223-78
2H-1-Benzopyran-2-one, 7-(dimethylamino)-4-hydroxy-3-methyl-	M HCl	299(4.2)	49-0905-78
	pH 4-6	356(4.1)	49-0905-78
	pH 7	339(4.1)	49-0905-78
1,4-Epoxyisoquinoline-2(1H)-carboxylic acid, 3,4-dihydro-, ethyl ester	MeCN	246(4.17)	78-2617-78
1(2H)-Isoquinolinone, 6,7-dimethoxy-2-methyl-	MeOH	245(3.80),270(3.11), 280(3.18),290(3.18), 335(2.90)	4-0257-78
L-Proline, N-benzoyl-	EtOH	220(3.83)	39-1157-78B
$C_{12}H_{13}NO_4$			
Benzene, 1-[2-(ethenyloxy)ethoxy]-2-(2-nitroethenyl)-	n.s.g.	199(4.50),240(3.92), 300(4.03),346(4.03)	47-1343-78
Benzene, 1-[2-(ethenyloxy)ethoxy]-3-(2-nitroethenyl)-	n.s.g.	246(4.53),304(4.10)	47-1343-78
Benzene, 1-[2-(ethenyloxy)ethoxy]-4-(2-nitroethenyl)-	n.s.g.	196(4.53),236(4.02), 346(4.32)	47-1343-78
$C_{12}H_{13}NO_4S_2$			
2(3H)-Benzothiazolethione, 3-β-D-ribofuranosyl-	EtOH	227(4.13),239(4.04), 326(4.36)	4-0657-78
$C_{12}H_{13}NO_5$			
Benzoic acid, 2-[(4-methoxy-1,4-dioxobutyl)amino]-	EtOH	223(4.41),252(4.10), 309(3.86)	95-1376-78
$C_{12}H_{13}NO_5S$			
2(3H)-Benzoxazolethione, 3-β-D-ribofuranosyl-	EtOH	256(3.88),262(3.85), 298(4.41)	4-0657-78
5H-Thiazolo[3,2-a]pyridine-7,8-dicarboxylic acid, 2,3-dihydro-5-oxo-, 8-ethyl 7-methyl ester	MeOH	206(4.3),231(4.2), 289(4.3),325(4.0)	28-0385-78B
$C_{12}H_{13}NS_2$			
Cyclopentanecarbodithioic acid, 2-(phenylimino)-	EtOH	251(4.02),317(3.84), 418(4.33)	39-0558-78C

Compound	Solvent	$\lambda_{max}(\log \epsilon)$	Ref.
$C_{12}H_{13}N_2S$			
Methanaminium, N-(3-isothiocyanato-3-phenyl-2-propenylidene)-N-methyl-, perchlorate	MeCN	223s(4.10),244(4.02), 321s(4.21),356(4.37)	97-0334-78
$C_{12}H_{13}N_3$			
[2,2'-Bipyridin]-4-amine, N,N-dimethyl-	neutral	239(4.16),275(4.19)	39-1215-78B
	monocation	242(4.21),277(4.04), 315s(3.88)	39-1215-78B
1H-Indole-3-butanenitrile, β-amino-, (S)-	MeOH	273(3.62),281(3.64), 289(3.59)	35-0938-78
Pyrazinamine, N,N-dimethyl-5-phenyl-	EtOH	223(3.95),293(4.35), 362(3.90)	94-1322-78
1H-Pyrazolo[1,5-a]benzimidazole, 1-ethyl-2-methyl-	EtOH	230(4.58),308(4.20), 321(4.18)	4-0715-78
	2N H_2SO_4	291(4.08),301(4.20)	4-0715-78
$C_{12}H_{13}N_3O$			
2(1H)-Cycloheptimidazolone, 4(or 5)-pyrrolidino-	MeOH	245(4.24),279(4.38), 367(4.15),457(4.09)	18-3087-78
$C_{12}H_{13}N_3O_2$			
Benzamide, N-[4-(1-hydroxyethyl)-1H-imidazol-2-yl]-	EtOH	226(4.08),281(3.98), 295s(3.93)	35-4208-78
	EtOH-HCl	240(4.03),265(4.14)	35-4208-78
2(1H)-Cycloheptimidazolone, 6-morpholino-	MeOH	255(4.37),292s(3.78), 420(4.42)	18-3087-78
3-Isoxazolamine, N-(1-methylethyl)-	EtOH	760(1.66)	103-0503-78
1,2,4-Triazin-5(2H)-one, 6-ethyl-3-(4-methoxyphenyl)-	EtOH	215s(4.21),281(4.37)	114-0091-78B
1,2,4-Triazin-6(1H)-one, 5-ethyl-3-(4-methoxyphenyl)-	EtOH	209s(4.23),281(4.55)	114-0091-78B
$C_{12}H_{13}N_3O_3S$			
8-Oxa-1,2-diaza-5-azoniaspiro[4.5]deca-1,3-diene, 4-hydroxy-3-(phenylthio)-, hydroxide, inner salt, 2-oxide	CH_2Cl_2	235(4.24),329(3.75)	88-2331-78
$C_{12}H_{13}N_3O_4$			
2H-Benzotriazole-2-acetic acid, 4,7-di-hydro- ,5,6-trimethyl-4,7-dioxo-, methyl ester	EtOH	229(4.13),256(4.04), 340(2.85)	87-0578-78
4,11-Epoxy-5H-1,3-dioxolo[4,5-e]imid-azo[2,1-b][1,3]oxazocine-8-carbo-nitrile, 3a,4,11,11a-tetrahydro-2,2-dimethyl-	EtOH	225(3.97)	94-3322-78
$C_{12}H_{13}N_3O_6$			
2(5H)-Furanone, 5-hydroxy-3,4-dimeth-oxy-, (4-nitrophenyl)hydrazone	pH 13	369(4.09),469(3.93)	33-1033-78
	dioxan	232(4.01),337(4.18)	33-1033-78
$C_{12}H_{13}N_5$			
1,2,4,5-Tetrazine, 3-phenyl-6-pyrroli-dino-	dioxan	298(4.36),418(2.96), 543(2.57)	30-0338-78
$C_{12}H_{13}N_5O$			
1,2,4,5-Tetrazine, 3-morpholino-6-phen-yl-	dioxan	294(3.93),412(2.48), 594(2.08)	30-0338-78
$C_{12}H_{14}$			
Bicyclo[4.2.2]deca-2,4,7,9-tetraene,	C_6H_{12}	269s(3.82),275(3.88),	5-2074-78

Compound	Solvent	$\lambda_{max}(\log \epsilon)$	Ref.
2,5-dimethyl- (cont.)		383(3.72)	5-2074-78
Cyclobutene, 3,4-diethynyl-1,2,3,4-tetramethyl-, trans	hexane	none above 210 nm	5-1675-78
as-Indacene, 1,2,3,6,7,8-hexahydro-	EtOH	268(3.00),272(2.94),277(3.03)	35-3730-78
s-Indacene, 1,2,3,5,6,7-hexahydro-	EtOH	277(3.56),281(3.58),287(3.58)	35-3730-78
$C_{12}H_{14}BrN_3O_5$			
4,11-Epoxy-5H-1,3-dioxolo[4,5-e]imidazo[2,1-b][1,3]oxazocine-8-carboxamide, 9-bromo-3a,4,11,11a-tetrahydro-2,2-dimethyl-	EtOH	247(3.89)	94-3322-78
$C_{12}H_{14}BrN_5$			
1H-Pyrazol-5-amine, 4-[(2-bromo-4-methylphenyl)azo]-1,3-dimethyl-	CHCl$_3$	385(4.11),415s(3.91)	103-0885-78
$C_{12}H_{14}ClNO_6$			
4H-1,3-Benzoxazinium, 2-methyl-3-(1-methylethyl)-4-oxo-, perchlorate	MeCN	240(4.10),290(3.52)	103-0122-78
$C_{12}H_{14}ClN_3O$			
1H-Benzotriazole, 5-chloro-1-(cyclohexyloxy)-	EtOH	273(3.94)	4-1043-78
$C_{12}H_{14}Cl_2O$			
Cyclobutanone, 2,2-dichloro-3-(3-methyl-1,3-butadienyl)-4-(1-methylethylidene)-	C$_6$H$_{12}$	249(4.08)	39-1568-78C
$C_{12}H_{14}Cl_2O_2$			
2-Naphthalenol, 5,7-dichloro-1,2,3,4-tetrahydro-6-methoxy-1-methyl-	EtOH	276(4.15)	39-0110-78C
$C_{12}H_{14}F_3N_3O_7$			
2,4(1H,3H)-Pyrimidinedione, 1-[2-deoxy-2-(trifluoroacetamido)-β-D-glucopyranosyl]-	EtOH	257(4.11)	136-0185-78C
$C_{12}H_{14}GeO_2$			
Germane, trimethyl[(1-oxo-3-phenyl-2-propynyl)oxy]-	heptane	207(4.09),250(4.14),270(4.00)	65-1244-78
$C_{12}H_{14}N_2$			
4,4'-Bipyridinium, 1,1'-dimethyl-, cation radical	n.s.g.	395(4.58),605(4.00)	77-0628-78
3H-Pyrrolo[1,2-b]pyrazole, 3a,4,5,6-tetrahydro-2-phenyl-	EtOH	219(3.80),285(3.88)	44-1664-78
$C_{12}H_{14}N_2O$			
1H-Indole, 4-butyl-5-nitroso-	CHCl$_3$	730(1.34)	39-0692-78C
2H-Indol-2-one, 3-[2-(dimethylamino)-ethylidene]-1,3-dihydro-, perchlorate	EtOH	253(4.43),292(3.72)	42-1122-78
1-Propanone, 1-(2-ethyl-1H-benzimidazol-4-yl)-	EtOH	215(4.13),280(3.96),313(3.81)	12-2675-78
1-Propanone, 1-(2-ethyl-1H-benzimidazol-5-yl)-	EtOH	233(4.24),277(3.97)	12-2675-78
2H-Pyrrol-2-one, 4-ethyl-1,5-dihydro-3-methyl-5-(1H-pyrrol-2-ylmethylene)-	EtOH	226(3.79),260(3.73),383(4.40)	49-0183-78

Compound	Solvent	$\lambda_{max}(\log \epsilon)$	Ref.
4(1H)-Quinazolinone, 6-methyl-2-(1-methylethyl)-	MeOH	223(4.49),265(4.09), 310(3.84)	2-1067-78
4(1H)-Quinazolinone, 7-methyl-2-(1-methylethyl)-	MeOH	230(4.62),260s(3.96), 300(3.62)	2-1067-78
4(1H)-Quinazolinone, 8-methyl-2-(1-methylethyl)-	MeOH	230(4.38),267(3.85), 310(3.60)	2-1067-78
2-Quinoxalinemethanol, α,α,3-trimethyl-	EtOH	236(4.55),306(3.88), 317(3.95)	22-0621-78

$C_{12}H_{14}N_2O_2$

Compound	Solvent	$\lambda_{max}(\log \epsilon)$	Ref.
1H-Azepine-2,3-dione, tetrahydro-, 3-(O-phenyloxime)	n.s.g.	245(3.87)	103-1223-78
Cyclobuta[1,2-b:4,3-b']dipyridine-2,7-dione, 1,4a,4b,8,8a,8b-hexahydro-1,8-dimethyl-, cis-syn	n.s.g.	257(3.35)	77-0620-78
Cyclobuta[1,2-b:4,3-c']dipyridine-2,8(1H,4aH)-dione, 4b,7,8a,8b-tetrahydro-1,7-dimethyl-, cis-syn	n.s.g.	262(3.51)	77-0620-78
Cyclobuta[1,2-c:4,3-c']dipyridine-1,8-dione, 2,4a,4b,7,8a,8b-hexahydro-2,7-dimethyl-	n.s.g.	263(3.87)	77-0620-78
9,10-Diazatricyclo[6.3.0.0^{2,7}]undeca-3,5,10-triene-9-carboxylic acid, ethyl ester, cis-cis-trans-trans	MeCN	246(4.01),280s(3.29)	44-0315-78
2H-Indol-2-one, 3-[2-(dimethylamino)-1-hydroxyethylidene]-1,3-dihydro-	EtOH	264(4.25),310(4.25)	42-1122-78
1H-Pyrido[3,2-b]indole-2,9-dione, 3,4,5,6,7,8-hexahydro-1-methyl-	EtOH	247(4.06),306(3.49)	94-3080-78
1H-Pyrido[3,2-b]indole-2,9-dione, 3,4,5,6,7,8-hexahydro-5-methyl-	EtOH	252(4.19),313(3.51)	94-3080-78
4(1H)-Quinazolinone, 6-methoxy-2-(1-methylethyl)-	MeOH	225(4.46),270(4.11), 322(3.71)	2-1067-78
4(1H)-Quinazolinone, 7-methoxy-2-(1-methylethyl)-	MeOH	240(4.83),295(3.76), 307(3.64)	2-1067-78
4(1H)-Quinazolinone, 8-methoxy-2-(1-methylethyl)-	MeOH	235(4.26),283(3.85), 315(3.78)	2-1067-78
Sydnone, 3-(1,1-dimethylethyl)-4-phenyl-	EtOH	240(3.35),299(3.95)	33-1477-78

$C_{12}H_{14}N_2O_2S$

Compound	Solvent	$\lambda_{max}(\log \epsilon)$	Ref.
Imidazo[1,2-a]pyridine-3-carboxylic acid, 5-methyl-2-(methylthio)-, ethyl ester	EtOH	259(4.29),306(3.66)	95-0631-78

$C_{12}H_{14}N_2O_3$

Compound	Solvent	$\lambda_{max}(\log \epsilon)$	Ref.
Acetic acid, 1,2-diacetyl-2-phenylhydrazide	EtOH	230(3.80)	78-1581-78

$C_{12}H_{14}N_2O_4$

Compound	Solvent	$\lambda_{max}(\log \epsilon)$	Ref.
1H-Benzimidazole, 1-α-D-ribofuranosyl-	MeOH	245(3.83),250s(3.81), 265s(3.50),273(3.57), 281(3.57)	24-0996-78
1H-Benzimidazole, 1-β-D-ribofuranosyl-	MeOH	245(3.83),250s(3.81), 265s(3.51),273(3.57), 281(3.55)	24-0996-78

$C_{12}H_{14}N_2O_4S$

Compound	Solvent	$\lambda_{max}(\log \epsilon)$	Ref.
2H-Benzimidazole-2-thione, 1,3-dihydro-1-β-D-ribofuranosyl-	pH 7.0	217(4.25),245(4.08), 296s(4.37),303(4.44)	24-0996-78
	pH 12.0	223(4.39),258(4.02), 265(4.00),300(4.30)	24-0996-78

Compound	Solvent	$\lambda_{max}(\log \epsilon)$	Ref.
2H-Benzimidazole-2-thione, 1,3-dihydro-1-β-D-ribofuranosyl- (cont.)	MeOH	222(4.27),248(4.03), 300s(4.36),307(4.46)	24-0996-78
	MeOH	249(3.98),296s(4.28), 306(4.39)	4-0657-78
Imidazo[1,2-a]pyridine-3-carboxylic acid, 5-methyl-2-(methylsulfonyl)-, ethyl ester	EtOH	220(4.59),288(3.87)	95-0631-78
5H-Thieno[2,3-d]pyridazine-4,7-dicarboxylic acid, dihydro-2,6-dimethyl-, dimethyl ester	CH_2Cl_2	300(4.0),329(3.88), 411(3.57)	83-0728-78
$C_{12}H_{14}N_2O_5$			
1H-Indole-1-ethanol, β-(2-hydroxyethoxy)-6-nitro-	EtOH	250(3.89),319(3.79), 352(3.68)	136-0017-78E
3,6-Pyridazinedicarboxylic acid, 4-methyl-5-(2-oxopropyl)-, dimethyl ester	CH_2Cl_2	232(3.76),311(2.65)	83-0728-78
$C_{12}H_{14}N_2O_5S$			
5'-Deoxy-5'-thio-2',3'-O-isopropylidene-S,5'-cyclouridine	EtOH	292(4.02)	94-2664-78
$C_{12}H_{14}N_2Se$			
2-Selenazolamine, 4,5-dihydro-4,4-dimethyl-5-methylene-N-phenyl-	EtOH	204(3.54),246(3.40)	103-0738-78
$C_{12}H_{14}N_3O_2$			
1,2,4-Triazolidin-1-yl, 2-(1,1-dimethyl)-3,5-dioxo-4-phenyl-	benzene	313(3.45),440(2.43) (anom.)	44-0808-78
1,2,4-Triazolidin-1-yl, 4-methyl-2-(1-methyl-1-phenylethyl)-3,5-dioxo-	benzene	300(3.46),380(3.08) (anom.)	44-0808-78
$C_{12}H_{14}N_4$			
1H-Tetrazole, 1-(4-pentenyl)-5-phenyl-	EtOH	231(3.95)	44-1664-78
2H-Tetrazole, 2-(4-pentenyl)-5-phenyl-	EtOH	239(4.23)	44-1664-78
$C_{12}H_{14}N_4O$			
Benzamide, N-(4,5-dimethyl-1H-1,2,3-triazol-1-yl)-4-methyl-	MeOH	239(4.14)	4-1255-78
1H-1,2,4-Triazolo[4,3-a]benzimidazole-1-propanol, 3-methyl-	EtOH	224(3.70),236s(4.01), 255(3.86),261(3.88), 270(3.64),291(3.46), 299(3.44)	4-1027-78
$C_{12}H_{14}N_4O_2$			
Benzamide, N-(4,5-dimethyl-1H-1,2,3-triazol-1-yl)-4-methoxy-	MeOH	259(4.16)	4-1255-78
$C_{12}H_{14}N_4O_2S$			
Benzenesulfonamide, 4-amino-N-(2,6-dimethyl-4-pyrimidinyl)-	pH 7.0	259(4.34),278(4.35)	94-1162-78
compound with β-cyclodextrin	pH 7.0	260(4.33),279(4.34)	94-1162-78
Thiourea, (1-methyl-2-pyrrolidinylidene)(4-nitrophenyl)-	MeOH	220(4.17),256(4.03), 332s(4.34),358(4.40)	87-1044-78
$C_{12}H_{14}N_4O_3$			
[4,4'-Bipyrimidin]-2(1H)-one, 2',6'-diethoxy-	pH 0.9	293(3.86),305s(3.85), 333(3.85)	44-0511-78
	pH 8.9	317(3.94)	44-0511-78
	pH 13	299(3.82),318(3.82)	44-0511-78

Compound	Solvent	$\lambda_{max}(\log \epsilon)$	Ref.
$C_{12}H_{14}N_4O_3S$			
Benzenesulfonamide, 4-amino-N-(6-methoxy-2-methyl-4-pyrimidinyl)-	pH 7.0	252s(4.26),266(4.31)	94-1162-78
compd. with β-cyclodextrin	pH 7.0	253(4.23),268(4.31)	94-1162-78
$C_{12}H_{14}N_4O_4$			
[5,5'-Bipyrimidine]-2,2',4,4'-(1H,1'H,3H,3'H)-tetrone, 1,1',3,3'-tetramethyl-	EtOH	238s(3.77),294(3.89)	44-1677-78
Propanediamide, 2-[[4-(1-oxopropoxy)phenyl]hydrazono]-	EtOH	262(3.85),365(4.16)	104-1956-78
$C_{12}H_{14}N_4O_4S$			
Benzenesulfonamide, 4-amino-N-(2,6-dimethoxy-4-pyrimidinyl)-	pH 7.0	255s(4.35),268(4.44)	94-1162-78
compd. with β-cyclodextrin	pH 7.0	255s(4.35),268(4.43)	94-1162-78
$C_{12}H_{14}N_4O_4S_2$			
2,4(1H,3H)-Pyrimidinedione, 5,5'-dithiobis[1,3-dimethyl-	EtOH	220s(3.93),285(4.01)	44-1677-78
$C_{12}H_{14}N_4O_6S$			
β-D-Glucopyranuronic acid, 1-deoxy-1-(1,6-dihydro-6-thioxo-9H-purin-9-yl)-, methyl ester	MeOH	278(4.22)	106-0250-78
$C_{12}H_{14}N_5O_5$			
Pteridinium, 2-amino-3,4-dihydro-6,7-bis(methoxycarbonyl)-3,8-dimethyl-4-oxo-, chloride	pH -1.0	281(4.23),401(4.06)	24-3790-78
	pH 6.0	263(4.21),380(4.22)	24-3790-78
$C_{12}H_{14}N_6O_5$			
Adenosine, N-cyano-N,6-didehydro-1,6-dihydro-1-methoxy-	pH 1	221(4.23),287(4.30)	94-2122-78
	H_2O	220(4.25),287(4.32)	94-2122-78
	pH 13	244.5(4.28)	94-2122-78
$C_{12}H_{14}N_8O_7$			
1H-Imidazole-4-carboxamidine, 5-amino-1-ethyl-, picrate	pH 1	282(4.15)	94-1929-78
	pH 7	286(4.16)	94-1929-78
	EtOH	292(4.24)	94-1929-78
$C_{12}H_{14}O$			
5H-Benzocycloheptene-1-carboxaldehyde	EtOH	254(3.94)	44-2167-78
5H-Benzocycloheptene-2-carboxaldehyde	EtOH	260(4.19)	44-2167-78
3-Buten-2-one, 3-methyl-4-(3-methylphenyl)-	MeOH	280(4.24)	56-2233-78
3-Buten-2-one, 3-methyl-4-(4-methylphenyl)-	MeOH	288(4.37)	56-2233-78
Cyclohexanone, 2-phenyl-	EtOH	288f(2.12)	44-0700-78
Ethanone, 1-(1-phenylcyclobutyl)-	EtOH	218s(3.77),266(2.32), 291(2.40)	44-0700-78
2-Pentanone, 3-(phenylmethylene)-	MeOH	278(4.26)	56-2233-78
4-Penten-1-one, 2-methyl-1-phenyl-	EtOH	242(3.99)	35-0883-78
4-Penten-1-one, 1-(4-methylphenyl)-	EtOH	242(3.98)	35-0883-78
2H-Pyran, 3,4-dihydro-6-methyl-5-phenyl-	EtOH	256(3.76)	44-0700-78
Tricyclo[3.3.2.0^{2,8}]deca-6,9-dien-4-ol, 4-ethenyl-	C_6H_{12}	230s(3.54)	35-1008-78
$C_{12}H_{14}OS$			
2-Buten-1-one, 3-(ethylthio)-1-phenyl-,	EtOH	251(3.93),320(4.36)	139-0191-78B

Compound	Solvent	$\lambda_{max}(\log \epsilon)$	Ref.
(E)- (cont.)			
(Z)-	EtOH	215(3.97),254(3.85), 332(4.18)	139-0191-78B
$C_{12}H_{14}O_2$			
Benzene, 1-ethenyl-2-[2-(ethenyloxy)-ethoxy]-	n.s.g.	207(4.44),244(4.10), 299(3.59)	47-1367-78
Benzene, 1-ethenyl-3-[2-(ethenyloxy)-ethoxy]-	n.s.g.	214(4.47),249(4.06), 294(3.36)	47-1367-78
Benzene, 1-ethenyl-4-[2-(ethenyloxy)-ethoxy]-	n.s.g.	208(4.24),262(4.18), 280(3.89),288(3.80)	47-1367-78
3-Buten-2-one, 4-(3-methoxyphenyl)-3-methyl-	MeOH	278(4.25)	56-2233-78
3-Buten-2-one, 4-(4-methoxyphenyl)-3-methyl-	MeOH	308(4.46)	56-2233-78
2-Cyclohexen-1-one, 4-(4-oxocyclohex-ylidene)-	EtOH	299.5(4.15)	39-0024-78C
Cyclopropanecarboxylic acid, 2-phenyl-, ethyl ester	EtOH	253-273f(2.7)	44-4316-78
4(1H)-Dibenzofuranone, 2,3,6,7,8,9-hexahydro-	MeOH	213(4.31),280(3.64)	24-0853-78
2H-Inden-2-one, 4-ethynyl-3,3a,4,5,6,7-hexahydro-4-hydroxy-3a-methyl-, cis	EtOH	232(4.41)	78-2201-78
1(2H)-Naphthalenone, 3,4-dihydro-6-methoxy-2-methyl-	EtOH	273(4.15)	39-0110-78C
2(1H)-Naphthalenone, 3,4-dihydro-6-methoxy-1-methyl-	EtOH	276(3.20)	39-0110-78C
1(2H)-Naphthalenone, 3,4-dihydro-7-methoxy-5-methyl-	EtOH	225(4.10),260(3.83), 328(3.42)	12-1363-78
3-Penten-2-one, 4-(4-methoxyphenyl)-	n.s.g.	232(3.95),312(4.21)	2-0200-78
4-Penten-1-one, 1-(4-methoxyphenyl)-	EtOH	272(4.21)	35-0883-78
Spiro[2H-1,5-benzodioxepin-3(4H),1'-cyclobutane]	EtOH	275(3.17)	56-1913-78
Spiro[1,3-dioxolane-2,10'-tricyclo-[4.3.1.0^{7,9}]deca[2,4]diene]	EtOH	260s(3.45),266(3.52), 276(3.49),287(3.20)	88-2387-78
$C_{12}H_{14}O_2S$			
2-Propen-1-one, 1-phenyl-3-(propylsul-finyl)-, (E)-	EtOH	210(3.95),228s(--), 271(4.00)	139-0191-78B
(Z)-	EtOH	210(3.98),228(3.90), 269(3.93)	139-0191-78B
$C_{12}H_{14}O_2S_2$			
1,4-Naphthalenedione, 4a,5,8,8a-tetra-hydro-2,3-bis(methylthio)-	C_6H_{12}	260(--),386(3.83)	24-3233-78
$C_{12}H_{14}O_2Si$			
2-Propynoic acid, 3-phenyl-, trimethyl-silyl ester	heptane	208(3.96),247(4.00), 258(4.05)	65-1244-78
$C_{12}H_{14}O_2Sn$			
2-Propynoic acid, 3-phenyl-, trimethyl-stannyl ester	MeCN	206(4.20),247(4.20)	65-1244-78
$C_{12}H_{14}O_3$			
4H-1-Benzopyran-4-one, 2,3-dihydro-2-(hydroxymethyl)-2,3-dimethyl-	MeOH	223(2.24),252(2.78), 324(1.40)	94-2277-78
1H-Inden-1-one, 3-(acetoxymethylene)-2,3,4,5,6,7-hexahydro-, (E)-	n.s.g.	287(4.25)	33-0266-78
1,4-Pentanedione, 3-methoxy-1-phenyl-	MeOH	242(4.04),273(3.31)	49-0137-78

Compound	Solvent	$\lambda_{max}(\log \epsilon)$	Ref.
$C_{12}H_{14}O_4$			
Acetic acid, (2-acetyl-4,5-dimethyl-phenoxy)-	EtOH	255(3.99),312(3.64)	23-0517-78
1,3-Benzenedicarboxylic acid, 2-butyl-	EtOH	281(3.11)	44-0727-78
2H-1-Benzopyran-3-carboxylic acid, 5,6,7,8-tetrahydro-2-oxo-, ethyl ester	EtOH	335(4)	23-0424-78
1H-2-Benzopyran-1-one, 3,4-dihydro-8-hydroxy-6-methoxy-3,5-dimethyl-	EtOH
EtOH-AlCl$_3$	267(4.07),310(3.78)
280(4.07),342(3.72)	102-0511-78
102-0511-78
4H-1-Benzopyran-4-one, 2,3-dihydro-6,7-dihydroxy-2,2,5-trimethyl-	EtOH	241(4.18),282(4.14),
345(3.76)	142-0437-78C
Bicyclo[4.2.0]octa-2,4-diene-7,8-di-carboxylic acid, dimethyl ester	MeOH	272(3.48)	78-1323-78
2,4,6,8-Decatetraenedioic acid, dimethyl ester	MeOH	245(--),317s(--),
328(4.57),344(4.49)	78-1323-78
Propanoic acid, 2-(2-acetyl-5-methoxy-phenyl)-	EtOH	269(3.92)	39-0110-78C
Spiro[2H-1,5-benzodioxepin-3(4H),5'-[1,3]dioxane]	CHCl$_3$	275(3.74),283s(--)	56-1913-78
$C_{12}H_{14}O_5$			
1,2,4-Benzenetriol, 5,6-dimethyl-, 1,4-diacetate	EtOH	278(3.32)	23-0517-78
Ethanone, 1-(2,3-dihydro-5-hydroxy-2,2-dimethoxy-4-benzofuranyl)-	EtOH	220s(4.07),234(4.10),
256(3.90),357(3.54)	12-2099-78
$(C_{12}H_{14}O_5)_n$			
Poly(2,5-furandiylcarbonyloxy-1,6-hex-anediyloxycarbonyl)	CHCl$_3$	264(3.18)	116-0568-78
$C_{12}H_{14}O_6$			
3,5-Cyclohexadiene-1-carboxylic acid, 1,2-diacetoxy-, methyl ester	MeOH	260.5(3.59)	78-1707-78
$C_{12}H_{15}BrN_4S_2$			
2(3H)-Thiazolethione, 3-[(4-amino-2-methyl-5-pyrimidinyl)methyl]-5-(2-bromoethyl)-4-methyl-	EtOH	233(4.13),279(3.81),
325(4.18)	94-3675-78
$C_{12}H_{15}BrO_4$			
Benzeneacetic acid, 2-bromo-3,6-di-hydroxy-4,5-dimethyl-, ethyl ester	hexane	219s(4.02),301(3.15)	12-1353-78
$C_{12}H_{15}ClN$			
Methanaminium, N-[3-chloro-3-(4-methyl-phenyl)-2-propenylidene]-N-methyl-, perchlorate	CH$_2$Cl$_2$	267(4.19),364(4.39),
400s(3.73),569(4.60)	97-0334-78
$C_{12}H_{15}ClN_2O_2$			
1H-Pyrrole-3-carboxaldehyde, 2-chloro-4,5-dihydro-1-methyl-5-oxo-4-(piper-idomethylene)-	EtOH	300(4.22),400(4.24)	104-2041-78
$C_{12}H_{15}ClN_2O_3$			
7H-Pyrido[2,3-d]azepine-7-carboxylic acid, 4-chloro-1,2,5,6,8,9-hexahydro-2-oxo-, ethyl ester	EtOH	238(3.83),318(3.91)	142-0267-78C
$C_{12}H_{15}ClN_2O_5S$			
Propanamide, 2-acetoxy-N-[2-(aminosul-fonyl)-5-chlorophenyl]-2-methyl-	MeOH	250(4.14),286(3.36)	4-0063-78

Compound	Solvent	λ_{max}(log ϵ)	Ref.
$C_{12}H_{15}FN_2O_4$			
2,4(1H,3H)-Pyrimidinedione, 5-fluoro- 1,3-bis(tetrahydro-2-furanyl)-	pH 2 pH 10 EtOH	275(3.95) 275(3.95) 274(3.92)	87-0738-78 87-0738-78 87-0738-78
$C_{12}H_{15}FN_2O_6$			
Uridine, 5-fluoro-2',3'-O-(1-methyl- ethylidene)-	H_2O	209(3.91),270(3.93)	73-3268-78
$C_{12}H_{15}N$			
1H-Indole, 1,2,3,4-tetramethyl-	MeOH	232(4.51),282s(3.79), 289(3.82),297s(3.77)	44-3727-78
2H-Isoindole, 4,5,6,7-tetramethyl-	CH_2Cl_2	255(3.77),295(3.44), 309(3.57),322(3.62), 334(3.50)	142-0409-78C
$C_{12}H_{15}NO$			
Acetamide, N-[1-methyl-2-(3-methylphen- yl)ethenyl]-, cis	MeOH	262(4.18)	56-2233-78
trans	MeOH	268(4.24)	56-2233-78
Acetamide, N-[1-methyl-2-(4-methylphen- yl)ethenyl]-, cis	MeOH	263(4.21)	56-2233-78
trans	MeOH	268(4.31)	56-2233-78
Acetamide, N-[1-(phenylmethylene)pro- pyl]-, trans	MeOH	266(4.09)	56-2233-78
Benzonitrile, 3-(1,1-dimethylethyl)- 4-methoxy-	EtOH	253(4.21),273s(3.41), 284(3.02)	12-0907-78
3-Buten-2-one, 3-methyl-4-(3-methyl- phenyl)-, oxime	MeOH	271(4.35)	56-2233-78
3-Buten-2-one, 3-methyl-4-(4-methyl- phenyl)-, oxime	MeOH	274(4.38)	56-2233-78
2-Pentanone, 3-(phenylmethylene)-, oxime	MeOH	269(4.33)	56-2233-78
$C_{12}H_{15}NOS_2$			
4a,8a-Methanoisoquinolin-3(4H)-one, 1,2-dihydro-4,4-bis(methylthio)-	$CHCl_3$	280(3.41)	44-3813-78
$C_{12}H_{15}NO_2$			
Acetamide, N-[2-(3-methoxyphenyl)- 1-methylethenyl]-, cis	MeOH	261(4.10)	56-2233-78
trans	MeOH	269(4.18)	56-2233-78
Acetamide, N-[2-(4-methoxyphenyl)- 1-methylethenyl]-, cis	MeOH	267(4.29)	56-2233-78
trans	MeOH	273(4.32)	56-2233-78
2-Azabicyclo[2.2.2]oct-5-en-3-one, trans-endo	EtOH	224(3.89)	78-2617-78
trans-exo	EtOH	223(3.92)	78-2617-78
Azulen-4,8-imin-5(3H)-one, 3a,4,8,8a- tetrahydro-7-methoxy-9-methyl-	n.s.g.	251(3.94)	88-1751-78
3-Buten-2-one, 4-(3-methoxyphenyl)- 3-methyl-, oxime	MeOH	268(4.33)	56-2233-78
3-Buten-2-one, 4-(4-methoxyphenyl)- 3-methyl-, oxime	MeOH	285(4.48)	56-2233-78
1H-Indol-5-ol, 2,3-dihydro-1,2-dimeth- yl-, acetate	MeOH	258(3.75),306(3.18)	94-2027-78
4,7-Methano-1H-indene-2-carboxamide, 2,3,3a,4,7,7a-hexahydro-2-methyl- 1-oxo-, (2α,3aβ,4α,7α,7aβ)-	n.s.g.	217(3.18)	33-1427-78
1(2H)-Naphthalenone, 3,4-dihydro- 7-methoxy-5-methyl-, oxime	EtOH	222(4.26),260(4.01), 308(3.59)	12-1363-78

Compound	Solvent	$\lambda_{max}(\log \epsilon)$	Ref.
$C_{12}H_{15}NO_3$			
Benzeneacetic acid, 4-morpholino-	EtOH	206(4.34),250(4.17)	104-1910-78
1H-Indole-2-carboxylic acid, 4,5,6,7-tetrahydro-3-methyl-4-oxo-, ethyl ester	MeOH	235(4.40),258s(4.06), 284(4.16)	24-1780-78
Norvaline, N-benzoyl-	EtOH	232(3.93)	39-1157-78B
$C_{12}H_{15}NO_3S$			
Ethanone, 1-[4-(tetrahydro-2H-1,2-thiazin-2-yl)phenyl]-, S,S-dioxide	CH_2Cl_2	267(4.39)	117-0048-78
4a,8a-Methanoisoquinoline, 1,4-dihydro-3-methoxy-4-(methylsulfonyl)-	MeOH	272(3.12)	44-3813-78
$C_{12}H_{15}NO_4$			
Benzene, 1-(2-ethoxyethoxy)-2-(2-nitroethenyl)-	n.s.g.	201(4.45),243(3.97), 303(4.08),350(4.3)	47-1343-78
Benzene, 1-(2-ethoxyethoxy)-3-(2-nitroethenyl)-	n.s.g.	203(4.39),222(3.49), 248(3.88)	47-1343-78
Benzene, 1-(2-ethoxyethoxy)-4-(2-nitroethenyl)-	n.s.g.	197(4.37),239(4.01), 350(4.32)	47-1343-78
3-Furancarboxylic acid, 2-(2-cyanoethyl)-2,5-dihydro-2,4-dimethyl-5-oxo-, ethyl ester	EtOH	229(3.98)	78-0955-78
1,2-Pentanedione, 1-(1,2-dihydro-4-hydroxy-6-methyl-2-oxo-3-pyridinyl)-4-methyl-	EtOH	207(3.93),230(3.90), 317(3.95)	23-0613-78
$C_{12}H_{15}NO_7$			
Buchanamine	EtOH	221(3.93),258s(3.43), 262(3.45),271s(3.37)	102-2047-78
3-Pyridinecarboxylic acid, 1,6-dihydro-6-oxo-1-β-D-ribofuranosyl-, methyl ester	pH 1 pH 11	260(4.21) 260(4.20)	87-0427-78 87-0427-78
$C_{12}H_{15}N_2S$			
1H-Imidazolium, 1,3-dimethyl-2-[(phenylmethyl)thio]-, iodide	EtOH CH_2Cl_2	219(4.318),265(3.607) 244(2.57),263(4.16)	28-0201-78B 28-0201-78B
$C_{12}H_{15}N_3$			
1H-Benzotriazole, 1-cyclohexyl-	n.s.g.	255(4.24),263(4.27), 279(4.23)	4-0127-78
2H-Benzotriazole, 2-cyclohexyl-	n.s.g.	273(4.54),279(4.53), 286(4.51)	4-0127-78
$C_{12}H_{15}N_3O$			
Propanedinitrile, [5-(dimethylamino)-3-ethoxy-2,4-pentadienylidene]-	CHCl$_3$	214(3.9),276(4.0), 310(3.8),462(4.7)	83-0287-78
	50% H_2SO_4	286(4.3)	83-0287-78
1,2,4-Triazine, 2,5-dihydro-2,3-dimethyl-6-methoxy-5-phenyl-	MeOH	265(3.34),273(3.34)	4-1271-78
1,2,4-Triazine, 2,5-dihydro-2,5-dimethyl-6-methoxy-3-phenyl-	MeOH	220(4.02),276(3.49)	4-1271-78
1,2,4-Triazin-6(1H)-one, 2,5-dihydro-1,2,3-trimethyl-5-phenyl-	MeOH MeOH-H_2SO_4	225s(3.84),250s(3.49) 264(3.52)	4-1271-78 4-1271-78
1,2,4-Triazin-6(1H)-one, 2,5-dihydro-1,2,5-trimethyl-3-phenyl-	MeOH MeOH-H_2SO_4	232(4.20),269s(4.32) 237(4.07),278(3.67)	4-1271-78 4-1271-78
$C_{12}H_{15}N_3O_2$			
1H-Benzotriazole-1-acetic acid, α,5,6-trimethyl-, methyl ester	4.2M HCl	288(3.93)	87-0578-78
Furazanamine, 4-butyl-N-phenyl-, 5-oxide	n.s.g.	245(4.36),298(3.74)	150-4356-78

Compound	Solvent	$\lambda_{max}(\log \epsilon)$	Ref.
1(2H)-Phthalazinone, 4-[(2-hydroxy-propyl)amino]-2-methyl-	EtOH	235s(3.94),244(3.71), 267s(3.12),318(3.94)	33-2116-78
$C_{12}H_{15}N_3O_3$			
1H-Benzotriazole-1-acetic acid, α,5,6-trimethyl-, methyl ester, 3-oxide	EtOH	276(3.59),286(3.60), 324(3.88)	87-0578-78
	4.2M HCl	287(3.84)	87-0578-78
1H-Imidazole-4-carboxylic acid, 5-(cya-nomethyl)-1-(tetrahydro-2H-pyran-2-yl)-, methyl ester	pH 1	220(4.04)	87-1212-78
	pH 7	240(4.00)	87-1212-78
	pH 11	242(3.99)	87-1212-78
1H-Imidazole-5-carboxylic acid, 4-(cya-nomethyl)-1-(tetrahydro-2H-pyran-2-yl)-, methyl ester	pH 1	222(4.04)	87-1212-78
	pH 5	242(4.08)	87-1212-78
	pH 11	242(4.08)	87-1212-78
1,3,5-Triazine-2,4-dione, 6-(2-hydroxy-phenyl)-1,3,5-trimethyl-	EtOH	218s(3.95),278(3.53), 284s(3.51)	4-1193-78
$C_{12}H_{15}N_3O_4$			
Propanoic acid, 2-methyl-, 2-acetyl-2-(4-nitrophenyl)hydrazide	EtOH	303(4.055)	146-0864-78
1H-Pyrrolo[3,2-d]pyrimidine-7-carbox-ylic acid, 6-ethyl-2,3,4,5-tetrahy-dro-1,3-dimethyl-2,4-dioxo-, methyl ester	EtOH	235(4.23),275(3.60)	142-0793-78A
$C_{12}H_{15}N_3O_5$			
5'-Amino-5'-deoxy-2',3'-O-isopropyli-dene-N⁶,5'-cyclouridine	MeOH	275(4.32)	35-2248-78
	EtOH	276(4.35)	94-2664-78
4,11-Epoxy-5H-1,3-dioxolo[4,5-e]imida-zo[2,1-b][1,3]oxazocine-8-carboxam-ide, 3a,4,11,11a-tetrahydro-2,2-di-methyl-	EtOH	238.5(3.95)	94-3322-78
1,2-Hydrazinedicarboxylic acid, 1-[2-methoxy-2-(2-pyridinyl)ethenyl]-, dimethyl ester, cis	MeOH	280(3.84)	103-0534-78
trans	MeOH	280(3.80)	103-0534-78
$C_{12}H_{15}N_3S$			
Pyrido[3,2-c]pyridazine, 4-(pentyl-thio)-	heptane	236(3.52),256(3.60), 317(3.28),336(3.48), 346(3.47)	103-1032-78
Thiourea, (1-methyl-2-pyrrolidinyli-dene)phenyl-	MeOH	296(4.39)	87-1044-78
$C_{12}H_{15}N_5O_2$			
[4,4'-Bipyrimidin]-2-amine, 2',6'-di-ethoxy-	pH 0.9	302s(3.84),314(3.89), 330s(3.85)	44-0511-78
	pH 9.1	303(3.84)	44-0511-78
$C_{12}H_{15}N_5O_5$			
6,7-Pteridinedicarboxylic acid, 2-am-ino-3,4,7,8-tetrahydro-3,8-dimethyl-4-oxo-, dimethyl ester	pH -2.0	233(4.17),280(4.06), 377(3.96)	24-3790-78
	pH 4.0	218(4.35),265(4.25), 393(4.09)	24-3790-78
9H-Purin-6-amine, 9-(5-acetoxy-β-D-arabino-furanosyl)-	MeOH	258(4.18)	87-1218-78
Thymidine, 3'-azido-3'-deoxy-, 5'-acetate	EtOH	264.5(3.98)	44-3044-78
$C_{12}H_{15}N_{11}$			
1H-Pyrazol-3-amine, 5-methyl-4-[[5-methyl-4-[(5-methyl-1H-pyrazol-3-	DMF	471(3.73)	103-0313-78

Compound	Solvent	λ_{max} (log ϵ)	Ref.
yl)azo]-1H-pyrazol-3-yl]azo]- (cont.)			103-0313-78
$C_{12}H_{15}O_{14}P$			
L-Ascorbic acid, 2,2'-(hydrogen phosphate), barium salt	pH 1.0	235(1.26)	136-0127-78H
	pH 7.0	258(1.41)	136-0127-78H
	pH 12	258(1.49)	136-0127-78H
$C_{12}H_{16}$			
Bicyclo[2.2.1]hept-2-ene, 5-(3-methyl-2-butenylidene)-	C_6H_{12}	252(4.45)	35-1172-78
Tricyclo[3.2.0.0^{2,7}]heptane, 3-(3-methyl-2-butenylidene)-	C_6H_{12}	259(4.4)	35-1172-78
$C_{12}H_{16}AsN$			
3H-1,3-Bezazarsole, 2-(1,1-dimethylethyl)-3-methyl-	MeOH	219(4.25),241(3.96), 277(3.61),295(3.60), 312s(3.43)	101-0001-78K
	MeOH-HClO$_4$	235(3.91),307(3.83)	101-0001-78K
$C_{12}H_{16}ClNO_2$			
Cyclohepta[b]pyran-2(5H)-one, 3-chloro-4-(dimethylamino)-6,7,8,9-tetrahydro-	EtOH	244(3.90),314s(3.97), 324(3.97)	4-0511-78
$C_{12}H_{16}ClNO_3S$			
1-Butanesulfonamide, N-(4-acetylphenyl)-4-chloro-	CH_2Cl_2	253(3.74),258(4.03), 269(4.52)	117-0048-78
$C_{12}H_{16}ClN_3O$			
1H-Benzotriazole, 5-chloro-1-(hexyloxy)-	EtOH	273(4.02)	4-1043-78
$C_{12}H_{16}NO$			
2,1-Benzisoxazolium, 1-(1,1-dimethylethyl)-3-methyl-, perchlorate	H_2O	205(4.43),268(3.68), 335(3.72)	44-1233-78
$C_{12}H_{16}N_2$			
Acetaldehyde, 2-butenylphenylhydrazone	n.s.g.	279(4.25)	39-0543-78C
Benzenecarboximidamide, N,N-dimethyl-N'-1-propenyl-	MeOH	255(3.1)	78-0833-78
Benzenecarboximidamide, N,N-dimethyl-N'-2-propenyl-, (E)-	MeOH	220(4.2)	78-0833-78
Diazene, [1-(1-methylethyl)-1-propenyl]phenyl-, (Z,E)-	hexane	227(4.02),306(4.25), 460(2.36)	49-1081-78
Diazene, (1-methyl-4-pentenyl)phenyl-	n.s.g.	264(3.95),400(2.18)	39-0543-78C
1,4-Ethanoquinoxaline, 2,3-dihydro-5,8-dimethyl-	hexane	265(3.18),274(3.26), 314(1.64),320(1.70)	44-2621-78
1,4-Ethanoquinoxaline, 2,3-dihydro-6,7-dimethyl-	hexane	307(1.38),320(1.32)	44-2621-78
1H-Indole-3-ethanamine, N,N-dimethyl-	EtOH	222(4.58),275(3.77), 282(3.80),292(3.74)	42-1122-78
	EtOH	223(4.39),283(3.76), 291(3.69)	94-2411-78
1(2H)-Pyridinamine, 3,6-dihydro-2-methyl-N-phenyl-	n.s.g.	244(4.07),291(3.21)	39-0543-78C
$C_{12}H_{16}N_2O$			
Azepino[4,3-b]indol-1(2H)-one, 3,4,5,6,7,8,9,10-octahydro-	MeOH	211(4.16),242s(3.81), 270(3.52)	24-0853-78
1H-Indol-5-ol, 3-[2-(dimethylamino)-ethyl]-	EtOH	222(4.21),278(3.68), 302(3.55),313(3.43)	94-2411-78

Compound	Solvent	$\lambda_{max}(\log \epsilon)$	Ref.
2H-Indol-2-one, 3-[2-(dimethylamino)-ethyl]-1,3-dihydro-, monoperchlorate	EtOH	250(3.86),280(3.10)	42-1122-78
Proline N-methylanilide	EtOH	223(3.73),261(2.63)	33-3108-78
$C_{12}H_{16}N_2O_2$			
1H-Indol-5-ol, 3-[2-(dimethylamino)-ethyl]-, N-oxide	EtOH	223(4.52),278(3.99), 302(3.85),314(3.71)	94-2411-78
2H-Indol-2-one, 3-[2-(dimethylamino)-ethyl-1-hydroxyethyl]-1,3-dihydro-, perchlorate	EtOH	252(4.01),282(3.45)	42-1122-78
$C_{12}H_{16}N_2O_3$			
Butanoic acid, 2-[[4-(methylamino)benz-oyl]amino]-, (±)-	MeOH	297(4.21)	87-1162-78
Butanoic acid, 4-[[4-(methylamino)benz-oyl]amino]-	MeOH	295(4.06)	87-1162-78
7H-Pyrido[2,3-d]azepine-7-carboxylic acid, 1,2,5,6,8,9-hexahydro-2-oxo-, ethyl ester	EtOH	235(4.02),315(3.94)	142-0267-78C
4H-Pyrido[1,2-a]pyrimidine-9-carboxylic acid, 1,6,7,8-tetrahydro-2-methyl-4-oxo-, ethyl ester	MeOH	225(4.01),248(3.89), 290s(4.30),303(4.39)	24-2813-78
$C_{12}H_{16}N_2O_3S$			
Acetic acid, (1-cyclohexyl-5-oxo-2-thi-oxo-4-imidazolidinylidene)-, methyl ester, (Z)-	EtOH	348(4.35)	5-0227-78
Ethanone, 1-[4-(tetrahydro-2H-1,2-thia-zin-2-yl)phenyl]-, oxime, S,S-dioxide	CH_2Cl_2	258(4.18)	117-0048-78
$C_{12}H_{16}N_2O_4$			
7H-Pyrido[2,3-d]azepine-7-carboxylic acid, 1,2,5,6,8,9-hexahydro-4-hy-droxy-2-oxo-, ethyl ester	EtOH	242(3.76),288(3.90)	142-0267-78C
$C_{12}H_{16}N_2O_5$			
2,4(1H,3H)-Pyrimidinedione, 1,3-dimeth-yl-6-(triacetylhydrazino)-	H_2O	222(4.23),268(4.14)	56-0037-78
Uridine, 2'-deoxy-5-(1-propenyl)-	pH 1.2	237(4.10),293(3.90)	44-2870-78
	pH 7	237(4.10),293(3.90)	44-2870-78
	pH 12.6	237s(4.15),288(3.84)	44-2870-78
Uridine, 2'-deoxy-5-(2-propenyl)-	pH 1.2	267(3.98)	44-2870-78
	pH 7	267(3.98)	44-2870-78
	pH 12.6	266(3.87)	44-2870-78
$C_{12}H_{16}N_2O_5S$			
Thiourea, N-(2-hydroxyphenyl)-N-β-D-ribofuranosyl-	EtOH	250(4.13),276s(3.82)	4-0657-78
$C_{12}H_{16}N_2O_6$			
Acetamide, N-(1,6-dihydro-6-oxo-1-β-D-ribofuranosyl-3-pyridinyl)-	pH 1	248(3.96),314(3.63)	87-0427-78
	pH 11	248(3.96),314(3.63)	87-0427-78
Uridine, 5-(2-propenyl)-	pH 1.2	267(3.99)	44-2870-78
	pH 7.0	267(3.99)	44-2870-78
	pH 12.6	266(3.88)	44-2870-78
$C_{12}H_{16}N_2O_8$			
1H-Imidazole-4,5-dicarboxylic acid, 1-β-D-ribofuranosyl-, dimethyl ester	pH 1	242(3.83)	4-0001-78
	pH 7	252(3.87)	4-0001-78
	pH 11	250(3.88)	4-0001-78

Compound	Solvent	$\lambda_{max}(\log \epsilon)$	Ref.
$C_{12}H_{16}N_4O_2$			
Benzo[1,2-c:4,5-c']dipyrazole-3,7-dione, 4,8-dihydro-1,2,5,6-tetramethyl-	EtOH	230(3.95)	48-0991-78
Propanedinitrile, (dimorpholinomethylene)-	EtOH	254(4.11),290(4.33)	95-1412-78
$C_{12}H_{16}N_4O_4$			
D-Arabinitol, 1,4-anhydro-2-deoxy-2-(6-ethoxy-9H-purin-9-yl)-	pH 1,7,13	252(4.07)	44-0541-78
1H-Purine-2,6-dione, 3,7-dihydro-8-(methoxymethyl)-1,3-dimethyl-7-(oxiranylmethyl)-	EtOH	206(4.37),275(3.90)	73-3414-78
Pyrazolidine, 2-(2,4-dinitrophenyl)-1,3,4-trimethyl-, cis	MeCN	235(3.83),392(3.78)	78-0903-78
trans	MeCN	235(3.83),392(3.78)	78-0903-78
2,4(1H,3H)-Pyrimidinedione, 6-amino-1,3-dimethyl-5-(2,5,6,7-tetrahydro-3-hydroxy-2-oxo-1H-azepin-4-yl)-, monohydrobromide	EtOH	272(3.13)	103-0306-78
$C_{12}H_{16}N_4O_4S$			
3H-Imidazo[4,5-b]pyridin-5-amine, 7-(methylthio)-3-β-D-ribofuranosyl-	pH 1	209(4.17),237(4.01), 243s(3.96),284(4.14), 316(4.04)	87-0112-78
	pH 7	214(4.20),241(4.20), 273(4.11),308(3.99)	87-0112-78
	pH 13	241(4.19),273(4.12), 308(4.00)	87-0112-78
$C_{12}H_{16}N_4O_5$			
2,6-Pyridinedicarboxylic acid, 4-[[(aminocarbonyl)hydrazono]ethylidene]-1,2,3,4-tetrahydro-, dimethyl ester	EtOH	265(4.02),375(4.53)	44-4765-78
$C_{12}H_{16}N_4O_5S$			
7H-Purine, 7-β-D-glucofuranosyl-6-(methylthio)-	acid	300(4.00)	12-1095-78
	base	253(3.52),290(4.04)	12-1095-78
	EtOH	228(4.03),253(3.52)	12-1095-78
$C_{12}H_{16}N_5O_9P$			
7(8H)-Pteridinone, 2-amino-4-methoxy-8-(5-O-phosphono-β-D-ribofuranosyl)-	pH 1.0	231(4.072),280(3.681), 348(4.179)	5-1780-78
	pH 7.0	231(4.079),280(3.672), 349(4.185)	5-1780-78
	pH 13	235s(4.065),280(3.909), 348(4.176)	5-1780-78
$C_{12}H_{16}N_6O_3$			
Adenosine, 3'-(acetylamino)-2',3'-dideoxy-	pH 1	257(4.17)	44-3044-78
	H_2O	259.5(4.19)	44-3044-78
$C_{12}H_{16}N_6O_3S_3$			
Benzenesulfonic acid, 4-[1,2-bis-[[(methylamino)thioxomethyl]hydrazono]ethyl]-	EtOH	337(4.36)	87-0804-78
copper chelate	EtOH	310(4.29),495(3.48)	87-0804-78
$C_{12}H_{16}N_6O_5$			
Adenosine, N-[(methylamino)carbonyl]-	pH 1	277(4.40)	36-0569-78
	pH 13	278(4.26),298(4.27)	36-0569-78
	70% EtOH	269(4.38),276(--)	36-0569-78

Compound	Solvent	$\lambda_{max}(\log \epsilon)$	Ref.
$C_{12}H_{16}O$			
Benzene, 1-(1,1-dimethylethyl)-4-(eth-enyloxy)-	C_6H_{12}	197(4.50),228(4.21), 276(3.10),280(2.99)	99-0194-78
1,3-Bishomoadamant-7-en-4-one	MeOH	225(2.42),296(1.65)	78-0067-78
2-Cyclohexen-1-one, 4-cyclohexylidene-	EtOH	305(4.24)	39-0024-78C
2(3H)-Naphthalenone, 4,4a,5,6-tetra-hydro-7,8-dimethyl-	EtOH	291(4.29)	78-2439-78
$C_{12}H_{16}O_2$			
Benzaldehyde, 3-(1,1-dimethylethyl)-2-hydroxy-6-methyl-	EtOH	220(4.05),265(3.98), 335(3.87)	32-0079-78
Benzene, 1-ethenyl-2-(2-ethoxyethoxy)-	n.s.g.	210(4.60),246(4.12), 300(3.65)	47-1367-78
Benzene, 1-ethenyl-4-(2-ethoxyethoxy)-	n.s.g.	205(4.34),259(4.37), 290(3.49),303(3.23)	47-1367-78
1-Butanone, 2-(methoxymethyl)-1-phenyl-	EtOH	244(4.10),279(3.03), 320(1.90)	44-4316-78
2-Cyclohexen-1-one, 4-(4-hydroxycyclo-hexylidene)-	EtOH	230s(3.81),302(4.24)	39-0024-78C
Cyclopenta[1,3]cyclopropa[1,2]benzen-3(3H)-one, 3a,3b,4,5,6,7-hexahydro-2-methoxy-3b-methyl- (±)-	EtOH	237(3.90),285s(3.34)	44-1114-78
Cycloprop[cd]azulen-2(2aH)-one, 2b,3,4,5,6,6b-hexahydro-2a-methoxy-2b-methyl-, (2aα,2bα,6bα)-	EtOH	241(3.73)	44-1114-78
2H-Inden-2-one, 7-ethenyl-1,4,5,6,7,7a-hexahydro-7-hydroxy-7a-methyl-	EtOH	234(4.40)	78-2201-78
2-Naphthalenol, 1,2,3,4-tetrahydro-6-methoxy-1-methyl-	EtOH	277(3.16)	39-0110-78C
2(4aH)-Naphthalenone, 5,6,7,8-tetra-hydro-3-methoxy-4a-methyl-	EtOH	250(4.00)	44-1114-78
2-Propenoic acid, 3-bicyclo[2.2.1]hept-5-en-2-yl-2-methyl-, methyl ester	n.s.g.	228(3.80)	33-1427-78
1-Propen-2-ol, 1-bicyclo[2.2.1]hept-5-en-2-yl-, acetate	n.s.g.	208(3.53)	33-1427-78
$C_{12}H_{16}O_2S_2$			
3(2H)-Thiophenone, 2-(dihydro-4,4-di-methyl-3-oxo-2(3H)-thienylidene)di-hydro-4,4-dimethyl-, cis	C_6H_{12} CHCl$_3$	282(3.68),394(4.05) 406(4.08)	24-3233-78 24-3233-78
trans	C_6H_{12} CHCl$_3$	450(4.13) 458(4.07)	24-3233-78 24-3233-78
$C_{12}H_{16}O_2Se_2$			
3(2H)-Selenophenone, 2-(dihydro-4,4-di-methyl-3-oxoselenophene-2(3H)-yli-dene)dihydro-4,4-dimethyl-	CHCl$_3$	482(4.04)	24-3233-78
$C_{12}H_{16}O_3$			
Benzoic acid, 3-(1,1-dimethylethyl)-4-hydroxy-, methyl ester	EtOH	216(4.16),263(4.15)	12-0907-78
1,3-Dioxolane, 2-[(4-methoxyphenyl)-methyl]-2-methyl-	EtOH	226(3.99),277(3.19), 283(3.12)	39-0024-78C
$C_{12}H_{16}O_3S$			
Benzene, [(2-butoxyethenyl)sulfonyl]-	C_6H_{12}	208(4.26),234(3.21), 266(3.00),273(2.85)	99-0194-78
$C_{12}H_{16}O_4$			
Benzeneacetic acid, 2,5-dihydroxy-3,4-dimethyl-, ethyl ester	EtOH	217s(3.94),293(3.55)	12-1353-78

Compound	Solvent	$\lambda_{max}(\log \epsilon)$	Ref.
1,3-Benzodioxole-3a(7aH)-carboxylic acid, 2,2,7-trimethyl-, methyl ester, cis	MeOH	267(3.64)	78-1707-78
4H-1-Benzopyran-4-one, 6-ethyl-5,6,7,8-tetrahydro-5-hydroxy-3-(hydroxymethyl)-, trans (diplosporin)	MeOH	214(4.04),252(4.08), 256(4.07)	119-0111-78
2,5-Cyclohexadien-1-one, 3,4,4-trimethoxy-2-(2-propenyl)-	EtOH	227s(3.96),315(3.59)	44-3717-78
	EtOH	227(3.96),315(3.59)	44-3983-78
Ethanone, 1-[5-(5,5-dimethyl-1,3-dioxan-2-yl)-2-furanyl]-	EtOH	276(4.18)	103-0940-78
$C_{12}H_{16}O_5$			
2H-Pyran-5-carboxylic acid, 3-ethenyl-3,4-dihydro-2-methoxy-4-(2-oxoethyl)-, methyl ester	MeOH	235(4.08)	33-1221-78
epimer	MeOH	235(4.08)	33-1221-78
$C_{12}H_{16}O_6$			
2-Furanpropanoic acid, 3-carboxy-2,5-dihydro-2,4-dimethyl-5-oxo-, α-ethyl ester	EtOH	227.5(4.08)	78-0955-78
2,4-Hexadienedioic acid, 2-acetoxy-, 6-methyl 1-(1-methylethyl) ester, (Z,E)-	EtOH	272(4.2)	5-1734-78
$C_{12}H_{16}O_9$			
4H-Pyran-4-one, 3-(β-D-glucopyranosyloxy)-2-(hydroxymethyl)-	MeOH	260(3.86)	94-0643-78
$C_{12}H_{16}Sn$			
Stannane, 1H-indenyltrimethyl-	heptane	235(4.38),288(3.44), 298(3.39)	104-0207-78
	HMPTA	288(3.40),300(3.33)	104-0207-78
$C_{12}H_{17}ClN_2O_4$			
1-Propen-1-aminium, N,N,N-trimethyl-3-(phenylamino)-, perchlorate	EtOH	226(4.03)	48-0659-78
$C_{12}H_{17}ClN_4O_6$			
1-Propen-1-aminium, N,N,N-trimethyl-3-[(4-nitrophenyl)hydrazono]-, perchlorate	EtOH	294(3.70),384(4.56)	48-0659-78
$C_{12}H_{17}FN_2O_8S$			
2,4(1H,3H)-Pyrimidinedione, 1-[6-deoxy-6-fluoro-2-O-(methylsulfonyl)-β-D-glucopyranosyl)-5-methyl-	MeOH	265(3.97)	48-0157-78
$C_{12}H_{17}N$			
Benzenamine, N,N-dimethyl-2-(1-methyl-2-propenyl)-	MeOH	246(3.67)	94-2027-78
Benzenamine, N-(1,2,2-trimethylpropylidene)-	CCl₄	258(3.43),277(3.43)	44-0731-78
$C_{12}H_{17}NO$			
Phenol, 4-(dimethylamino)-3-(2-methyl-2-propenyl)-	MeOH	288(3.28)	94-2027-78
2-Piperidinemethanol, α-phenyl-	n.s.g.	223(3.65),262(3.38)	42-0916-78
$C_{12}H_{17}NOS_2$			
Carbamodithioic acid, cyclohexyl-,	n.s.g.	255(4.19),270(4.14),	97-0381-78

Compound	Solvent	$\lambda_{max}(\log \epsilon)$	Ref.
2-furanylmethyl ester (cont.)		332s(1.98)	97-0381-78
$C_{12}H_{17}NO_2$			
1H-2-Benzopyran-4-carbonitrile, 4a,5,6,7,8,8a-hexahydro-1-hydroxy-1,3-dimethyl-	EtOH	236(4.23)	18-0839-78
Isoquinoline, 1,2,3,4-tetrahydro-6,7-dimethoxy-1-methyl-	EtOH	235(3.72),285(3.44)	100-0169-78
4,7-Methano-1H-indene-2-carboxamide, 2,3,3a,4,7,7a-hexahydro-1-hydroxy-2-methyl-	n.s.g.	209(3.21)	33-1427-78
Propanamide, N-(4-ethoxyphenyl)-2-methyl-	MeOH	245(4.00),275(3.79)	2-1067-78
2-Propenamide, N-[3-(4-oxo-1-cyclohexen-1-yl)propyl]-	EtOH	222(3.68)	94-0972-78
2-Propenamide, N-[3-(4-oxo-2-cyclohexen-1-yl)propyl]-	EtOH	227(4.11)	94-0972-78
$C_{12}H_{17}NO_3$			
Acetamide, 2-(4,4-dimethoxy-2,5-cyclohexadien-1-ylidene)-N,N-dimethyl-	MeOH	232(4.50),271(4.38)	35-1548-78
Benzeneacetamide, α,4-dimethoxy-N,N-dimethyl-	MeOH	232(4.11),275(3.20), 282(3.18)	35-1548-78
2-Pyridinecarboxylic acid, 5-(hexyloxy)-	M HCl	255(3.90),296(4.07)	103-0183-78
	H_2O	245(3.95),290(3.86)	103-0183-78
	M KOH	243(4.03),282(3.78)	103-0183-78
	EtOH	249(4.10),282(3.86)	103-0183-78
	25% EtOH	245(3.98),289(3.82)	103-0183-78
3-Pyridinecarboxylic acid, 6-(hexyloxy)-	M HCl	236(4.04),284(3.97)	103-0183-78
	H_2O	240(4.01),275(3.77)	103-0183-78
	M KOH	238(4.06),275(3.78)	103-0183-78
	EtOH	244(4.11),274(3.80)	103-0183-78
	25% EtOH	240(4.07),275(3.80)	103-0183-78
2(1H)-Pyridinone, 4-hydroxy-6-methyl-3-(4-methyl-2-oxopentyl)-	MeOH	284(3.87)	142-0417-78A
$C_{12}H_{17}NO_4$			
2(1H)-Pyridinone, 4-hydroxy-3-(2-hydroxy-4-methyl-1-oxopentyl)-6-methyl-	MeOH	230(4.07),268(3.50), 328(4.06)	142-0417-78A
$C_{12}H_{17}NO_4S$			
2-Pentenedioic acid, 4-(2-thiazolidinylidene)-, diethyl ester	MeOH	258(3.9),291(4.2), 339(4.4)	28-0385-78B
$C_{12}H_{17}NO_5$			
Acetonitrile, (6-acetoxy-2-hydroxy-3,4-dimethoxycyclohexylidene)-, $(1Z,2\alpha,3\beta,4\beta,6\beta)$-	EtOH	226(3.8)	102-1731-78
1H-Pyrrole-2,4-dicarboxylic acid, 3-hydroxy-, 2-(1,1-dimethylethyl) 4-ethyl ester	EtOH	226(4.49),259(4.26)	94-2224-78
$C_{12}H_{17}N_2$			
1-Propen-1-aminium, N,N,N-trimethyl-3-(phenylamino)-, perchlorate	EtOH	226(4.03)	48-0659-78
$C_{12}H_{17}N_3$			
1,2,4-Triazine, 2,3,4,5-tetrahydro-3,5,5-trimethyl-6-phenyl-	hexane	211(4.09),265(3.80)	103-0278-78

Compound	Solvent	$\lambda_{max}(\log \epsilon)$	Ref.
$C_{12}H_{17}N_3O$			
Pyrido[2',3':4,5]pyrimido[1,2-a]azepin-5(1H)-one, 2,3,4,7,8,9,10,11-octahydro-	MeOH	233(4.21),277(4.10)	24-2297-78
$C_{12}H_{17}N_3O_3$			
2,4,6(1H,3H,5H)-Pyrimidinetrione, 5-[3-(dimethylamino)-2-butenylidene]-1,3-dimethyl-	$CHCl_3$	240(4.1),420(4.8)	83-0287-78
$C_{12}H_{17}N_3O_4$			
Cytidine, 2'-deoxy-5-(1-propenyl)-	pH 1.2	233(4.07),298(3.82)	44-2870-78
	pH 7	233s(4.12),288(3.71)	44-2870-78
	pH 12.6	233s(4.12),288(3.74)	44-2870-78
Cytidine, 2'-deoxy-5-(2-propenyl)-	pH 1.2	288(4.08)	44-2870-78
	pH 7	278(3.91)	44-2870-78
	pH 12.6	278(3.92)	44-2870-78
2,3-Pyrrolidinedicarboxamide, 1-cyclohexyl-4,5-dioxo-	EtOH	245(3.95)	4-1463-78
$C_{12}H_{17}N_3O_5$			
Cytidine, 5-(2-propenyl)-	pH 1.2	288(4.07)	44-2870-78
	pH 7	278(3.91)	44-2870-78
	pH 12.6	278(3.92)	44-2870-78
$C_{12}H_{17}N_3O_6$			
1H-Imidazole-4-carboxamide, 2,3-dihydro-1-[2,3-O-(1-methylethylidene)-β-D-ribofuranosyl]-2-oxo-	H_2O	263.5(4.02)	94-3322-78
	M NaOH	282.5(3.91)	94-3322-78
β-D-Ribofuranose, 5-deoxy-5-[4-(methoxycarbonyl)-1H-1,2,3-triazol-1-yl]-2,3-O-(1-methylethylidene)-	MeOH	214(3.93)	35-2248-78
$C_{12}H_{17}N_3O_7$			
D-Glucose, (4-nitrophenyl)hydrazone	n.s.g.	380(4.17)	2-0469-78
$C_{12}H_{17}N_4O_2$			
1-Propen-1-aminium, N,N,N-trimethyl-3-[(4-nitrophenyl)hydrazono]-, perchlorate	EtOH	294(3.70),384(4.56)	48-0659-78
$C_{12}H_{17}N_5O_4$			
Adenosine, 2',3'-di-O-methyl-	pH 7.0	260(4.16)	39-0762-78C
Adenosine, 2-ethyl-	pH 1.0	205(4.403),257(4.117)	5-1796-78
	pH 7.0	210(4.384),262(4.143)	5-1796-78
$C_{12}H_{17}N_5O_5$			
L-Lathyrine, γ-L-glutamyl-	pH 1	222(4.1),300(3.6)	102-2027-78
	pH 13	228(4.0),293(3.6)	102-2027-78
3H-1,2,3-Triazolo[4,5-b]pyridin-5-amine, 7-ethoxy-3-β-D-ribofuranosyl-	pH 1	252(3.94),294(4.21)	44-4910-78
	pH 11	268s(4.03),295(4.22)	44-4910-78
$C_{12}H_{17}N_5O_6$			
Thymidine, 3'-deoxy-3'-[[(methylnitrosoamino)carbonyl]amino]-	EtOH	265(4.08)	87-0130-78
Thymidine, 5'-deoxy-5'-[[(methylnitrosoamino)carbonyl]amino]-	EtOH	264(4.04)	87-0130-78
$C_{12}H_{17}N_5O_{12}P_2$			
7(8H)-Pteridinone, 2-amino-8-[5-O-[hydroxy(phosphonooxy)phosphinyl]-β-D-	pH 1.0	231(4.057),280(3.650), 348(4.170)	5-1788-78

Compound	Solvent	$\lambda_{max}(\log \epsilon)$	Ref.
ribofuranosyl]-4-methoxy- (cont.)	pH 7.0	231(4.057),280(3.661), 349(4.152)	5-1788-78
	pH 13.0	245s(4.025),280(3.021), 344(4.137)	5-1788-78
$C_{12}H_{18}$			
1,3,7-Cyclodecatriene, 1,7-dimethyl-, (Z,E,E)- (pregeijerene)	EtOH	252(3.85)	12-2527-78
Cyclohexane, 1-(3-methyl-2-butenyli-dene)-3-methylene-	C_6H_{12}	286(4.43)	35-1172-78
$C_{12}H_{18}Cl_2N_2$			
1H-Pyrrole, 4,5-dichloro-1-(3,4-di-hydro-2,2-dimethyl-2H-pyrrol-5-yl)-2,3-dihydro-2,2-dimethyl-	MeOH	268(4.46)	83-0294-78
1H-Pyrrole, 4,5-dichloro-1-(3,4-di-hydro-3,3-dimethyl-2H-pyrrol-5-yl)-2,3-dihydro-3,3-dimethyl-	MeOH	263(4.26)	83-0294-78
$C_{12}H_{18}N_2$			
1,6-Benzodiazocine, 1,2,3,4,5,6-hexa-hydro-1,6-dimethyl-	hexane	234(4.46),270(3.88), 308(3.70)	44-2621-78
11,12-Diazatetracyclo[4.4.2.0^{2,5}.0^{7,10}]-dodec-11-ene, 1,6-dimethyl-	MeOH	393(1.78)	88-1183-78
Quinoxaline, 1,2,3,4-tetrahydro-1,4,5,8-tetramethyl-	hexane	234(4.38),264s(--), 300(3.15)	44-2621-78
$C_{12}H_{18}N_2O$			
Ethanone, 2-cyclohexyl-1-(1-methyl-1H-imidazol-2-yl)-	EtOH	281(4.08)	33-2831-78
Ethanone, 2-cyclohexyl-1-(1-methyl-1H-imidazol-4-yl)-	EtOH	257(4.08)	33-2831-78
2,4-Hexadienamide, N-[1-(dimethylami-no)-2-methyl-2-propenylidene]-	n.s.g.	280(4.34)	33-0815-78
1H-Isoindol-1-one, 3-(dimethylamino)-3a,4,5,7a-tetrahydro-3a,5-dimethyl-	n.s.g.	245(4.33)	33-0815-78
1H-Isoindol-1-one, 3-(dimethylamino)-3a,4,5,7a-tetrahydro-3a,6-dimethyl-	n.s.g.	244(4.35)	33-0815-78
2,4-Pentadienamide, N-[1-(dimethyl-amino)-2-methyl-2-propenylidene]-4-methyl-, (E,E)-	n.s.g.	261(4.26),287(4.23)	33-0815-78
11H-Pyrido[2,1-b]quinazolin-11-one, 1,2,3,4,4a,6,7,8,9,11a-decahydro-	EtOH	239(3.86),308(4.23)	142-1717-78A
4(1H)-Pyrimidinone, 5-(5-hexenyl)-2,6-dimethyl-	EtOH	230(3.72),274(3.70)	39-1293-78C
$C_{12}H_{18}N_2O_2$			
Acetic acid, diazo-, 1,7,7-trimethyl-bicyclo[2.2.1]hept-2-yl ester, (-)-	hexane	220s(3.88),245(3.09)	44-4447-78
	EtOH	215(3.68),249(4.24)	44-4447-78
$C_{12}H_{18}N_2O_4$			
Propanedioic acid, [3,3-bis(dimethyl-amino)-1,2-propadienylidene]-, dimethyl ester	CHCl$_3$	270(3.9),377(4.7)	88-4263-78
1H-Pyrazole-3-carboxylic acid, 4,5-di-hydro-4-(3-methoxy-2-methyl-3-oxo-1-propenyl)-5,5-dimethyl-, methyl ester, (E)-(±)-	MeCN	216s(--),294(4.01)	24-2206-78
(Z)-(±)-	MeCN	216s(--),290(4.00)	24-2206-78
3-Pyridinecarboxylic acid, 1-acetyl-2-(acetylamino)-1,4,5,6-tetrahydro-,	MeOH	228(3.97),291(4.12)	24-2297-78

Compound	Solvent	$\lambda_{max}(\log \epsilon)$	Ref.
ethyl ester (cont.)			24-2297-78
$C_{12}H_{18}N_2O_4S$			
1H-Imidazole, 1-[2,3-0-(1-methylethylidene)-β-D-ribofuranosyl]-2-(methylthio)-	n.s.g.	224(3.81),248(3.68)	44-4774-78
$C_{12}H_{18}N_2O_5$			
2,5,8,11,14-Pentaoxa-16,19-diazabicyclo-[13.3.1]nonadeca-1(19),15,17-triene	EtOH	230(4.95),276(4.90)	44-3362-78
Uridine, 2'-deoxy-5-propyl-	pH 1.2	267(3.95)	44-2870-78
	pH 7	267(3.95)	44-2870-78
	pH 12.6	266(3.86)	44-2870-78
$C_{12}H_{18}N_2O_6$			
Uridine, 5-propyl-	pH 1.2	267(3.95)	44-2870-78
	pH 7	267(3.96)	44-2870-78
	pH 12.6	266(3.83)	44-2870-78
$C_{12}H_{18}N_2S$			
Pentanethioic acid, 2-methyl-2-phenyl-hydrazide	MeOH	220(3.72),240(4.03),253(3.93),267(3.9)	123-0701-78
$C_{12}H_{18}N_4O_2$			
Hydrazinecarboxamide, 2-nitroso-2-pentyl-N-phenyl-	H_2O	237(4.34),350(2.01)	104-2319-78
$C_{12}H_{18}N_5O_6PS_2$			
Adenosine, 2-(ethylthio)-, 5'-(dihydrogen phosphorothioate)	pH 1	270(4.22)	87-0520-78
$C_{12}H_{18}N_5O_7P$			
5'-Adenylic acid, 2-ethyl-	pH 2.0	204(4.354),257(4.124)	5-1796-78
	pH 7.0	209(4.362),261(4.146)	5-1796-78
polymer	pH 7.3	205(4.193),256(3.949)	5-1796-78
$C_{12}H_{18}N_5O_7PS$			
5'-Adenylic acid, 2-(ethylthio)-	pH 1	270(4.23)	87-0520-78
5'-Adenylic acid, N-methyl-2-(methylthio)-	pH 1	271(4.21)	87-0520-78
$C_{12}H_{18}N_5O_8P$			
5'-Adenylic acid, 2-ethoxy-	pH 1	251(3.84),274(4.06)	87-0520-78
$C_{12}H_{18}N_6O_3$			
1,2,4-Cyclopentanetriol, 3-amino-5-[6-(dimethylamino)-9H-purin-9-yl]-,	pH 1	269(4.23)	18-0855-78
	H_2O	277(4.23)	18-0855-78
(1α,2β,3α,4α,5β)-(±)-, HCl salt	pH 13	277(4.23)	18-0855-78
(1α,2β,3β,4β,5α)-(±)-, dihydrochloride	pH 1	268(4.32)	18-0855-78
	H_2O	276(4.32)	18-0855-78
	pH 13	276(4.32)	18-0855-78
$C_{12}H_{18}N_6O_5$			
Urea, N-[5-[4-(aminocarbonyl)-1H-1,2,3-triazol-1-yl]-5-deoxy-2,3-0-(1-methylethylidene)-β-D-ribofuranosyl]-	MeOH	210(4.14)	35-2248-78
$C_{12}H_{18}O$			
2(1H)-Azulenone, 4,5,6,7,8,8a-hexahydro-, cis	EtOH	242(4.14)	35-1799-78
trans	EtOH	242(4.14)	35-1799-78

Compound	Solvent	λ_{max}(log ϵ)	Ref.
Bicyclo[3.2.1]oct-6-en-8-one, 6-butyl-	hexane	278(2.05)	70-0352-78
Bicyclo[4.2.0]oct-7-en-2-one, 7-butyl-	hexane	251s(2.05),292s(2.02), 303s(3.93)	70-0352-78
Bicyclo[4.2.0]oct-7-en-2-one, 8-butyl-	hexane	278s(2.07),288s(2.05), 296s(1.99),305s(1.90)	70-0352-78
2-Cyclohexen-1-ol, 4-cyclohexylidene-	EtOH	248(4.36),242s(4.31), 255s(4.23)	39-0024-78C
2-Cyclohexen-1-one, 4-cyclohexyl-	EtOH	226(4.01)	39-0024-78C
2(3H)-Naphthalenone, 4,4a,5,6,7,8-hexa-hydro-4a,8-dimethyl-	EtOH	245(4.17)	12-2527-78
Spiro[4.5]dec-3-en-2-one, 1,3-dimethyl-	EtOH	235(3.76)	35-1799-78
$C_{12}H_{18}O_2$			
1,3-Benzenedimethanol, 2-butyl-	EtOH	213(4.04),238(1.85), 243(2.00),247(2.20), 252(2.36),259(2.43), 263(2.43),278(2.28)	44-0727-78
3H-2-Benzopyran-3-one, 1,5,6,7,8,8a-hexahydro-4-(1-methylethyl)-	MeOH	232(3.85)	44-1248-78
3H-2-Benzopyran-3-one, 1,5,6,7,8,8a-hexahydro-4-propyl-	MeOH	234(3.88)	44-1248-78
Cyclohexanone, 2-(1-cyclohexen-1-yl)-2-hydroxy-	C_6H_{12}	287(2.22)	104-0067-78
2-Cyclohexen-1-one, 2,3-dimethyl-6-(3-oxobutyl)-	EtOH	245(3.91)	78-2439-78
2-Cyclohexen-1-one, 4-methyl-4-(4-oxo-pentyl)-	MeOH	226(3.84)	44-2562-78
1,3(and 4)-Cyclopentadiene-1-pentanoic acid, ethyl ester	n.s.g.	247(3.53)	104-0264-78
2(3H)-Naphthalenone, 4,4a,5,6,7,8-hexa-hydro-5-hydroxy-1,4a-dimethyl-	EtOH	248(4.29)	39-1461-78C
2(3H)-Naphthalenone, 4,4a,5,6,7,8-hexa-hydro-1-methoxy-4a-methyl-	n.s.g.	258(4.35)	78-1509-78
$C_{12}H_{18}O_3$			
Bicyclo[3.1.0]hexan-2-one, 6-(5,5-di-methyl-1,3-dioxan-2-yl)-, (1α,5α,6β)-(±)-	EtOH	278(1.61)	44-2093-78
Bicyclo[3.1.0]hexan-3-one, 6-(5,5-di-methyl-1,3-dioxan-2-yl)-, (1α,5α,6β)-	EtOH	270(1.45)	44-2093-78
2-Cyclohepten-1-one, 6-(1-acetoxy-1-methylethyl)-	EtOH	226(3.79)	35-1778-78
2-Cyclohexene-1-butanoic acid, 1-meth-yl-4-oxo-, methyl ester	MeOH	227(3.93)	44-2562-78
2-Cyclopentene-1-carboxylic acid, 2-(1-oxohexyl)-	EtOH	235(3.88)	70-1674-78
2,4,6-Dodecatrienoic acid, 8-hydroxy-, (E,E,E)-	EtOH	296(4.46)	18-2077-78
$C_{12}H_{18}O_3S$			
Benzenesulfonic acid, 4-methyl-, pentyl ester	n.s.g.	262(4.10)	124-0844-77
$C_{12}H_{18}O_4$			
2-Butynedioic acid, bis(1,1-dimethyl-ethyl) ester	EtOH	215(2.60),225(3.30), 254(2.30),281(2.00)	24-0523-78
2,5-Cyclohexadien-1-one, 4,4,5-trimeth-oxy-2-propyl-	EtOH	235(4.09),293(3.53)	44-3983-78
2,5-Heptadienoic acid, 2,4,4-trimethyl-, dimethyl ester, (E,E)-	EtOH	207(4.28),219(4.27)	24-2206-78

Compound	Solvent	$\lambda_{max}(\log \epsilon)$	Ref.
2,5-Heptadienoic acid, 2,4,4-trimethyl-, dimethyl ester, (E,Z)-	EtOH	210(4.20)	24-2206-78
3-Heptynoic acid, 5-hydroxy-5,6,6-tri-methyl-2-oxo-, ethyl ester	hexane	239(3.74),282(2.50)	118-0307-78
2-Propenoic acid, 3-(2,3-dihydro-5-methoxy-2,2-dimethyl-3-furanyl)-2-methyl-, methyl ester, (E)-(±)-	EtOH	226(4.18)	24-2206-78
(Z)-(±)-	EtOH	225(4.14)	24-2206-78
$C_{12}H_{18}O_4S$			
2-Butenoic acid, 4,4'-thiobis-, ethyl ester	EtOH	212(4.34)	39-0955-78C
2-Propenoic acid, 2,2'-[thiobis(meth-ylene)bis-, diethyl ester	EtOH	209(4.01)	39-0955-78C
$C_{12}H_{19}BrO_2$			
2-Cyclohexen-1-one, 2-bromo-5-hexyl-3-hydroxy-	EtOH	293(4.33)	2-0970-78
$C_{12}H_{19}N$			
Benzenamine, N,N-dimethyl-2-(1-methyl-propyl)-	MeOH	245(3.61),280(2.95)	94-2027-78
Quinoline, 2,3,4,4a,5,6-hexahydro-4a,6,6-trimethyl-	C_6H_{12}	221(4.32),277(2.20)	33-1025-78
$C_{12}H_{19}NO$			
Cycloheptanone, 2-[3-(dimethylamino)-2-propenylidene]-, (E,E)-	EtOH	400(4.46)	70-0102-78
Ethanone, 1-(3,4-dipropyl-1H-pyrrol-2-yl)-	EtOH	297(4.29)	23-0221-78
4(1H)-Pyrindinone, 1,2,3,5,6,7-hexa-hydro-1-(1-methylpropyl)-	heptane	310(4.11)	103-0993-78
	EtOH	330(4.18)	103-0993-78
	MeCN	322(4.42)	103-0993-78
	CF_3COOH	315(4.18)	103-0993-78
$C_{12}H_{19}NO_2$			
3-Cyclopenten-1-one, 2,2,5-trimethyl-3-morpholino-	C_6H_{12}	278(4.02)	102-2015-78
4H-Pyran-4-one, 5-[3-(dimethylamino)-2-propenylidene]tetrahydro-2,2-di-methyl-	EtOH	418(4.68)	70-0102-78
2-Pyrrolidinone, 5-(2-oxooctylidene)-geometric isomer	n.s.g.	284(4.435)	107-0219-78
	n.s.g.	280(4.487)	107-0219-78
$C_{12}H_{19}NO_6$			
Propanedioic acid, [[(2-ethoxy-2-oxo-ethyl)amino]methylene]-, diethyl ester	EtOH	220(4.17),278(4.29)	94-2224-78
$C_{12}H_{19}N_3$			
1-Propanone, 2-amino-2-methyl-1-phen-yl-, dimethylhydrazone	hexane	208(4.00),280(3.04)	103-0278-78
$C_{12}H_{19}N_3O_3$			
Carbamic acid, [3-[3-(aminocarbonyl)-1(4H)-pyridinyl]propyl]-, ethyl ester	CHCl$_3$	240(3.5),356(3.6)	24-2594-78
$C_{12}H_{19}N_3O_4$			
Cytidine, 2'-deoxy-5-propyl-	pH 1.2	288(4.08)	44-2870-78
	pH 7	278(3.92)	44-2870-78

Compound	Solvent	$\lambda_{max}(\log \epsilon)$	Ref.
Cytidine, 2'-deoxy-5-propyl- (cont.)	pH 12.6	278(3.93)	44-2870-78
$C_{12}H_{19}N_5O$			
6H-Purin-6-one, 9-[3-(diethylamino)pro-pyl]-1,3-dihydro-	EtOH	245(4.06),249s(4.04)	2-0786-78
$C_{12}H_{19}N_5OS$			
6H-Purin-6-one, 9-[3-(diethylamino)pro-pyl]-1,2,3,9-tetrahydro-2-thioxo-	EtOH	286s(4.17),294(4.18)	2-0786-78
$C_{12}H_{19}N_5O_6$			
Formamide, N-[2-amino-1,6-dihydro-1-methyl-6-oxo-4-(β-D-ribofuranosyl-amino)-5-pyrimidinyl]-N-methyl-	pH 7.0	272(4.32)	39-0762-78C
$C_{12}H_{19}N_5S_2$			
1H-Purine-2,6-dithione, 9-[3-(diethyl-amino)propyl]-3,9-dihydro-	pH 1	256(3.86),295(4.38), 347(4.08)	2-0786-78
	H_2O	246(4.10),292(4.35), 347(4.18)	2-0786-78
	pH 13	280(4.54),326(4.10)	2-0786-78
$C_{12}H_{19}N_6O_7P$			
5'-Adenylic acid, 2-(ethylamino)-	pH 1	255(4.14),298(3.94)	87-0520-78
$C_{12}H_{20}ClN_3$			
Pyrazinamine, 5-chloro-N,N-dimethyl-3,6-bis(1-methylethyl)-	EtOH	261(3.94),330(3.73)	94-1322-78
Pyrazinamine, 5-chloro-N,N-dimethyl-3,6-dipropyl-	EtOH	262(3.97),333(3.75)	94-1322-78
$C_{12}H_{20}N_2$			
1,2-Benzenediamine, N,N,N',N',3,6-hexa-methyl-	hexane	215(4.04),253(3.72), 269(3.51)	44-2621-78
$C_{12}H_{20}N_2O$			
Cyclohexanone, 2,6-bis[(dimethylamino)-methylene]-	EtOH	395(4.74)	70-0102-78
1,4,6-Heptatrien-3-one, 1,7-bis(di-methylamino)-2-methyl-	EtOH	410(4.70)	70-0102-78
1,4,6-Heptatrien-3-one, 1,7-bis(di-methylamino)-6-methyl-	EtOH	440(4.85)	70-0107-78
Pyrazine, 3-methoxy-2-(1-methylethyl)-5-(2-methylpropyl)-	EtOH	280s(3.71),297(3.81)	94-2046-78
Pyrazine, 3-methoxy-5-(1-methylethyl)-2-(2-methylpropyl)-	EtOH	282s(3.85),298(3.96)	94-2046-78
$C_{12}H_{20}N_2O_2$			
2(1H)-Pyrazinone, 1-hydroxy-3,6-bis(2-methylpropyl)- (neoaspergillic acid)	EtOH	234(4.02),330(3.91)	94-1320-78
$C_{12}H_{20}N_2O_3$			
2(1H)-Pyrazinone, 1-hydroxy-3,6-bis(2-methylpropyl)-, 4-oxide	EtOH	237(4.21),333(4.10)	94-1320-78
$C_{12}H_{20}N_2O_4$			
1,2-Hydrazinedicarboxylic acid, 1-(1-ethylidene-2-methyl-2-propenyl)-, diethyl ester, (E)-	hexane	230(3.98)	39-1161-78C

Compound	Solvent	$\lambda_{max}(\log \epsilon)$	Ref.
$C_{12}H_{20}N_2S_2$			
Pyrimidine, 2,4-bis(butylthio)-	H_2O	255(3.18),305(3.13)	56-1255-78
$C_{12}H_{20}N_4O_4$			
3-Pyridinecarboxylic acid, 1-[(methyl-amino)carbonyl]-2-[[(methylamino)-carbonyl]amino]-1,4,5,6-tetrahydro-, ethyl ester	MeOH	235(3.80),298(4.28)	24-2297-78
$C_{12}H_{20}O$			
Cycloheptanone, 2-(2,3-dimethylpropyli-dene)-	hexane	240(3.66),316(1.85)	44-0604-78
2-Cyclopenten-1-one, 2,3,5-triethyl-4-methyl-, cis	EtOH	239(4.09)	35-1799-78
trans	EtOH	239(4.08)	35-1799-78
$C_{12}H_{20}O_2$			
2-Cyclopentene-1-propanol, 1-(hydroxy-methyl)-2-(1-methylethenyl)-, (±)-	EtOH	238(3.75)	39-1606-78C
2-Hexynal, 4-(1,1-dimethylethyl)-4-hydroxy-5,5-dimethyl-	hexane	225(3.91),232(3.88)	118-0307-78
$C_{12}H_{20}O_3$			
Cyclopentanone, 2-(5-acetoxypentyl)-	EtOH	226(4.01)	5-1739-78
2,5,8-Dodecanetrione	MeOH	217(1.91)	44-4081-78
2-Octen-4-one, 8-acetoxy-2,6-dimethyl-	EtOH	240(4.11)	2-0188-78
$C_{12}H_{20}O_4S_8Sn$			
Tin, bis(O-ethyl carbonodithioato-S)-bis(O-ethyl carbonodithioate-S,S')-, (OC-6-21)-	C_6H_{12}	217(4.55),272(4.62)	12-1493-78
$C_{12}H_{20}O_5$			
Pentopyranosid-2-ulose, 1,1-dimethyl-ethyl 3,4-O-(1-methylethylidene)-, α-L-erythro-	EtOH	306(1.30)	150-5344-78
$C_{12}H_{21}NO$			
5,7,9-Dodecatrien-3-ol, 2-amino-	n.s.g.	265(4.60)	102-0831-78
$C_{12}H_{21}NOS$			
4H-Thiopyran-4-one, 3-[[bis(1-methyl-ethyl)amino]methylene]tetrahydro-	EtOH	334(4.29)	4-0181-78
$C_{12}H_{21}NO_3$			
4-Dodecenoic acid, 4-amino-6-oxo-	n.s.g.	293(4.37)	107-0219-78
$C_{12}H_{21}NO_4$			
Propanedioic acid, [[bis(1-methyleth-yl)amino]methylene]-, dimethyl ester	isooctane	235(3.57),284(4.27)	78-2321-78
$C_{12}H_{21}NSSi$			
Benzenamine, N,N-dimethyl-4-[[(trimeth-ylsilyl)methyl]thio]-	EtOH	276(4.35),310(3.40)	65-2027-78
$C_{12}H_{21}N_3O_{10}$			
β-D-Lactosyl azide	MeOH	275(1.56)	78-1427-78
β-D-Maltosyl azide	MeOH	274(1.55)	78-1427-78
$C_{12}H_{21}N_7$			
9H-Purine-2,6-diamine, 9-[3-(diethyl-	EtOH	251s(3.85),255(3.89),	2-0786-78

Compound	Solvent	$\lambda_{max}(\log \epsilon)$	Ref.
amino)propyl]- (cont.)		282(3.98)	2-0786-78
$C_{12}H_{22}$			
4,6-Decadiene, 3,8-dimethyl-, (E,E)-	heptane	210s(4.07),220s(4.36), 225s(4.45),231(4.49), 238s(4.29)	78-2015-78
4,6-Decadiene, 3,8-dimethyl-, (E,Z)-	heptane	211s(3.97),220s(4.26), 228s(4.38),233(4.42), 240s(4.26)	78-2015-78
4,6-Decadiene, 3,8-dimethyl-, (Z,Z)-	heptane	212s(3.99),220s(4.23), 229s(4.40),235(4.45), 241s(4.29)	78-2015-78
5,7-Dodecadiene, (E,Z)-	C_6H_{12}	234(4.41)	101-0159-78M
$C_{12}H_{22}NO$			
Bicyclo[3.3.1]nonan-1-aminium, N,N,N-trimethyl-3-oxo-, iodide	EtOH	285s(1.40)	44-3653-78
$C_{12}H_{22}N_4O$			
[1,2,3,4]Tetrazino[1,6-a:4,5-a']bisazepine, 1,2,3,4,5,10,11,12,13,14,14a-14b-dodecahydro-, 7-oxide, trans	MeOH	289(4.04)	33-1622-78
$C_{12}H_{22}S_2Si_2$			
Silane, [1,4-phenylenebis(thio)]bis-[trimethyl-	heptane	256(4.16)	65-2027-78
$C_{12}H_{23}N_2O$			
1,4-Diazaspiro[4.5]dec-1-yloxy, 2,2,3,3-tetramethyl-	EtOH	237(3.43),450(0.81)	35-0934-78
$C_{12}H_{23}N_2O_2$			
1,4-Diazaspiro[4.5]dec-1-yloxy, 4-hydroxy-2,2,3,3-tetramethyl-	EtOH	245(3.39)	35-0934-78
$C_{12}H_{24}N_2$			
2-Decenal, dimethylhydrazone, (E,E)-	C_6H_{12}	275(4.46)	28-0047-78A
$C_{12}H_{24}O_2SiSn$			
Silane, trimethyl[3-oxo-3-[(triethylstannyl)oxy]-1-propynyl]-	MeCN	211(3.65),224s(--)	65-1244-78
$C_{12}H_{27}NSi$			
2-Propanamine, 2-methyl-N-[1-(trimethylsilyl)pentylidene]-	C_6H_{12}	210(3.30),225(3.18), 302(1.91)	23-2286-78
$C_{12}H_{28}MoO_4$			
Molybdenum, tetraisopropoxide	C_6H_{12}	640(2.85)	35-2744-78
$C_{12}H_{36}N_4Si_4$			
2-Tetrazene, 1,1,4,4-tetrakis(trimethylsilyl)-	hexane	292(3.76)	60-1909-78

Compound	Solvent	$\lambda_{max}(\log \epsilon)$	Ref.
$C_{13}F_{10}$ 9H-Fluorene, 1,2,3,4,5,6,7,8,9,9-deca-fluoro-	EtOH	264(3.75),267(3.76), 309(3.53),313(3.56)	104-0559-78
$C_{13}H_4F_8$ Benzene, 1,1'-methylenebis[2,3,4,5-tetrafluoro-	EtOH	262(3.21)	104-1975-78
$C_{13}H_4F_8O$ Benzenemethanol, 2,3,4,5-tetrafluoro-α-(2,3,4,5-tetrafluorophenyl)-	EtOH	262(3.26)	104-1975-78
$C_{13}H_5ClF_2OS$ 9H-Thioxanthen-9-one, 2-chloro-3,6-di-fluoro-	MeOH	257(4.66),367(3.73)	73-2656-78
$C_{13}H_6ClFOS$ 9H-Thioxanthen-9-one, 2-chloro-3-flu-oro-	MeOH	258(4.66),245s(3.69), 298s(3.51),363(3.78)	73-2656-78
$C_{13}H_6N_2O_3$ 2H-Pyrano[3,2-c]quinoline-3-carboni-trile, 5,6-dihydro-2,5-dioxo-	MeOH	352(3.97),394(4.01)	49-1075-78
$C_{13}H_6N_4$ Ethenetricarbonitrile, 1-indolizinyl-	EtOH	216(4.17),238(3.81), 278(3.56),340(3.11), 491(4.62)	4-1471-78
Ethenetricarbonitrile, 3-indolizinyl-	CHCl$_3$	310(3.52),327(3.54), 336(3.52),378(3.63), 507(4.40),533(4.52)	4-1471-78
$C_{13}H_7ClF_2O_2S$ Benzoic acid, 2-[(4-chloro-3-fluoro-phenyl)thio]-	MeOH	260(4.05),295s(3.92)	73-2656-78
$C_{13}H_7Cl_3FNO_5S$ Furan, 2-[2-(3-fluorophenyl)-1-[(tri-chloromethyl)sulfonyl]ethenyl]-5-nitro-	EtOH	205(4.24),218(4.17), 285(4.25),322s(4.06)	73-1618-78
$C_{13}H_7Cl_3N_2O_7S$ Furan, 2-nitro-5-[2-(4-nitrophenyl)-1-[(trichloromethyl)sulfonyl]eth-enyl]-	EtOH	204(4.30),215s(4.24), 286(4.30),313s(4.12)	73-1618-78
$C_{13}H_7Cl_4NO_5S$ Furan, 2-[2-(3-chlorophenyl)-1-[(tri-chloromethyl)sulfonyl]ethenyl]-5-nitro-	EtOH	213(4.35),228s(4.15), 285(4.25),324s(4.03)	73-1618-78
Furan, 2-[2-(4-chlorophenyl)-1-[(tri-chloromethyl)sulfonyl]ethenyl]-5-nitro-	EtOH	203(4.19),223(4.16), 297(4.36),340s(4.01)	73-1618-78
$C_{13}H_7N_3O_5$ Benzonitrile, 4-[2-nitro-2-(5-nitro-2-furanyl)ethenyl]-, (E)-	dioxan	234(4.21),298(4.29)	73-0463-78
$C_{13}H_7S_2$ Phenaleno[1,9-cd]-1,2-dithiol-1-ium, hexafluorophosphate	MeCN	213(4.51),226(4.50), 258s(3.87),281(3.97),	35-7629-78

Compound	Solvent	λ_{max}(log ϵ)	Ref.
(cont.)		343(3.75),420s(4.33), 439(4.49)	35-7629-78
radical	MeCN	416(4.39)	35-7629-78
$C_{13}H_8ClNO_4S$ Benzoic acid, 2-[(4-chlorophenyl)thio]- 5-nitro-	MeOH	217(4.43),259(3.89), 338(4.19)	73-0471-78
$C_{13}H_8Cl_2N_2$ Benzene, 1,1'-(diazomethylene)bis[4- chloro-	benzene	524(2.08)	39-1283-78B
$C_{13}H_8Cl_3NO_5S$ Furan, 2-nitro-5-[2-phenyl-1-[(tri- chloromethyl)sulfonyl]ethenyl]-	EtOH	210(4.15),220(4.15), 287(4.35),340s(4.00)	73-1618-78
$C_{13}H_8Cl_3NO_8S$ 2-Furancarboxylic acid, 5-[2-(5-nitro- 2-furanyl)-2-[(trichloromethyl)sul- fonyl]ethenyl]-, methyl ester	EtOH	213(4.09),306(4.36), 356(3.92)	73-1618-78
$C_{13}H_8F_2N_4S$ 1H-Tetrazolium, 2,3-bis(2-fluorophen- yl)-2,5-dihydro-5-thioxo-, hydroxide, inner salt	MeOH	423(2.99)	4-0133-78
	EtOH	428(3.02)	4-0133-78
	PrOH	435(3.00)	4-0133-78
	acetone	473(2.97)	4-0133-78
	MeCN	461(3.03)	4-0133-78
	$MeNO_2$	458(2.99)	4-0133-78
	$CHCl_3$	467(2.98)	4-0133-78
	90% EtOH	421(3.03)	4-0133-78
	50% EtOH	409(3.05)	4-0133-78
	10% EtOH	398(3.10)	4-0133-78
(other solvent mixtures)	10% acetone	402(3.08)	4-0133-78
1H-Tetrazolium, 2,3-bis(4-fluorophen- yl)-2,5-dihydro-5-thioxo-, hydroxide, inner salt	M HCl	390(2.97)	4-0133-78
	H_2O	390(3.02)	4-0133-78
	MeOH	420(3.97)	4-0133-78
	EtOH	424(2.98)	4-0133-78
	PrOH	435(2.98)	4-0133-78
	acetone	468(3.94)	4-0133-78
	MeCN	456(2.98)	4-0133-78
	$MeNO_2$	453(2.96)	4-0133-78
	$CHCl_3$	458(2.95)	4-0133-78
	90% EtOH	420(2.99)	4-0133-78
	50% EtOH	409(3.01)	4-0133-78
	10% EtOH	397(3.06)	4-0133-78
(other solvent mixtures)	10% acetone	403(2.98)	4-0133-78
$C_{13}H_8N_2O_3SSe$ Benzoselenazole, 2-[[5-(2-nitroethen- yl)-2-furanyl]thio]-	EtOH	232s(4.15),246(4.06), 278(3.96),334(3.87)	73-2037-78
$C_{13}H_8N_2O_3S_2$ Benzothiazole, 2-[[5-(2-nitroethenyl)- 2-furanyl]thio]-	EtOH	243(4.05),273s(4.09), 300(3.94),348(4.08)	73-2037-78
$C_{13}H_8N_2O_4$ 1H-Naphtho[2,3-b][1,4]diazepine- 2,4,6,11(3H,5H)-tetrone	$CHCl_3$	266(4.34),296(3.98), 343(3.40),428(3.23)	104-0159-78
$C_{13}H_8N_2O_4S$ Benzoxazole, 2-[[5-(2-nitroethenyl)-	EtOH	245(4.16),276(4.06),	73-2037-78

Compound	Solvent	$\lambda_{max}(\log \epsilon)$	Ref.
2-furanyl]thio]- (cont.)		283(4.05),346(4.12)	73-2037-78
$C_{13}H_8N_2Se$			
Propanedinitrile, (2-methyl-4H-1-benzo-selenin-4-ylidene)-	EtOH	427(4.189)	83-0170-78
$C_{13}H_8O$			
6H-Benzo[3,4]cyclobuta[1,2]cyclohepten-6-one	C_6H_{12}	254(4.42),265(4.54), 272(4.61),284(4.69), 340(3.89),350(3.88), 367s(3.72)	12-1585-78
9H-Fluoren-9-one	CH_2Cl_2	251(4.91),259(5.15), 285s(3.84),296(3.90), 312s(3.72),328s(3.57), 358s(2.43),378(2.50), 398s(2.46)	18-2674-78
$C_{13}H_8OS$			
Bicyclo[3.2.0]hepta-2,4,7-trien-6-one, 7-(phenylthio)-	EtOH	232(4.29),325(4.18)	42-1224-78
9H-Thioxanthen-9-one	CH_2Cl_2	258(4.74),289s(3.89), 301(3.86),364(3.90), 382(3.98)	18-2674-78
$C_{13}H_8OS_2$			
2(1H)-Naphthalenone, 1-(1,3-dithiol-2-ylidene)-	MeCN	278(4.44),318(4.06), 474(4.29)	97-0385-78
$C_{13}H_8O_2$			
6H-Benzo[3,4]cyclobuta[1,2]cyclohepten-6-one, 7-hydroxy-	EtOH	228(4.09),249(4.13), 290(4.52),372(3.77), 390(3.83),422(3.71)	88-0143-78
	H_2SO_4	226(4.33),257(4.13), 326(4.52),395(3.70), 413(3.90),490(3.24)	88-0143-78
9H-Xanthen-9-one	CH_2Cl_2	239(4.70),262(4.15), 277s(3.68),287s(3.69), 327s(3.81),342(3.89)	18-2674-78
$C_{13}H_8O_2S$			
6H-Thiopyrano[2,3-b]furan-6-one, 2-phenyl-	EtOH	298(4.44),385(3.84)	104-0589-78
9H-Thioxanthen-9-one, 10-oxide	CH_2Cl_2	284s(4.17),340(3.67)	18-2674-78
$C_{13}H_8O_3S$			
3H-Naphtho[1,8-bc]thiophene-2-carboxylic acid, 3-oxo-, methyl ester	MeOH	220s(--),265(4.316), 312(3.782),366(3.897)	5-0627-78
9H-Thioxanthen-9-one, 10,10-dioxide	CH_2Cl_2	286s(3.96)	18-2674-78
	MeCN	230(4.32),237s(4.23), 254s(3.88),263(3.89), 283(3.89)	18-2674-78
$C_{13}H_8O_4$			
2H-Furo[2,3-h]-1-benzopyran-2-one, 3-acetyl-	EtOH	215(4.418),243(4.317), 329(4.165),370(4.242)	111-0435-78
7H-Furo[2,3-f][1]benzopyran-7-one, 8-acetyl-	EtOH	204(4.290),215(4.273), 247(4.268),325(4.130)	111-0435-78
7H-Furo[3,2-f][1]benzopyran-7-one, 8-acetyl-	EtOH	209(4.478),225s(--), 255s(--),350(4.320)	111-0435-78
8H-Furo[3,2-h][1]benzopyran-8-one, 7-acetyl-	EtOH	205(4.291),220s(--), 249(4.326),255s(--),	111-0435-78

Compound	Solvent	$\lambda_{max}(\log \epsilon)$	Ref.
(cont.)		263s(--),330(4.107)	111-0435-78
$C_{13}H_9$ 1H-Benzo[3,4]cyclobuta[1,2]cyclohepten-ylium, hexafluorophosphate	MeCN	226(4.24),233(4.18), 279(4.28),328(4.30), 388(3.67),399(3.67), 510(2.86)	12-1569-78
$C_{13}H_9BrN_2$ Benzene, 1-bromo-4-(diazophenylmethyl)-	benzene	524(1.91)	39-1283-78B
$C_{13}H_9BrO_3S$ 2-Propenoic acid, 3-[5-[(4-bromophen-yl)thio]-2-furanyl]-	EtOH	206(4.27),240(4.03), 296(4.19)	73-0621-78
$C_{13}H_9BrO_5S$ 2-Propenoic acid, 3-[5-[(4-bromophen-yl)sulfonyl]-2-furanyl]-	EtOH	206(4.24),243(4.26), 305(4.39)	73-0621-78
$C_{13}H_9Br_2ClN_2$ Benzenecarbohydrazonoyl chloride, N-(2,4-dibromophenyl)-	dioxan	233(4.51),313(4.57), 338(4.36)	104-2203-78
$C_{13}H_9Br_2NO_5$ 1H-Pyrrole-3,4-dicarboxylic acid, 2-(3,5-dibromo-2-methoxyphenyl)-	EtOH	275(3.89)	39-1588-78C
$C_{13}H_9ClN_2$ Benzene, 1-chloro-3-(diazophenylmeth-yl)-	benzene	520(1.92)	39-1283-78B
Benzene, 1-chloro-4-(diazophenylmeth-yl)-	benzene	526(2.02)	39-1283-78B
Pyrrolo[1,2-a]pyrimidine, 2-chloro-7-phenyl-	n.s.g.	254(4.60),325(3.88), 370(3.47)	44-3544-78
$C_{13}H_9ClN_2S$ 2-Benzothiazolamine, 6-chloro-N-phenyl-	EtOH	221(4.26),240(4.25), 306(4.41)	103-0380-78
$C_{13}H_9ClN_4$ Pyrido[3,2-c]pyridazin-4-amine, N-(4-chlorophenyl)-	EtOH	254(4.30),372(4.08)	103-0663-78
$C_{13}H_9ClO_5S$ 2-Propenoic acid, 3-[5-[(4-chlorophen-yl)sulfonyl]-2-furanyl]-	EtOH	206(4.25),238(4.24), 304(4.41)	73-0621-78
$C_{13}H_9Cl_2NO_2$ 1H-Cyclobuta[a]indene-1,2-dicarboxim-ide, 1,2-dichloro-2,2a,3,7b-tetra-hydro-, anti-anti	MeCN	222(3.57),265(3.04), 273(3.06)	24-2677-78
1H-Cyclobuta[a]indene-1,2-dicarboxim-ide, 1,2-dichloro-2,2a,3,7b-tetra-hydro-, syn-syn	MeCN	222(3.58),267(3.09), 274(3.12)	24-2677-78
1H-Cyclopenta[b]quinoline-9-carboxylic acid, 5,7-dichloro-2,3-dihydro-	pH 13	222(3.40),241(3.43), 331(2.63)	34-0261-78
	EtOH	223(4.50),241(4.32), 330(2.50)	34-0261-78
	dioxan	250(4.21),335(3.69)	34-0261-78
	H_2O	220(--),241(--), 339(--)	34-0261-78

Compound	Solvent	$\lambda_{max}(\log \epsilon)$	Ref.
$C_{13}H_9Cl_2NO_5$			
1H-Pyrrole-3,5-dicarboxylic acid, 2-(3,5-dichloro-2-methoxyphenyl)-	EtOH	275(3.62)	39-1588-78C
$C_{13}H_9FN_2$			
Benzene, 1-(diazophenylmethyl)-4-fluoro-	benzene	529(1.93)	39-1283-78B
$C_{13}H_9IN_2$			
Benzene, 1-(diazophenylmethyl)-4-iodo-	benzene	525(1.89)	39-1283-78B
$C_{13}H_9K$			
Potassium, 9H-fluoren-9-yl-	THF	470(3.18),500(3.03), 520s(--)	120-0052-78
$C_{13}H_9N$			
5H-Indeno[1,2-b]pyridine, 5-methylene-, dimer	EtOH	206(4.5),256(4.0), 312(4.3)	103-0997-78
$C_{13}H_9NO$			
Benzo[b]cyclohept[e][1,4]oxazine	MeOH	215(4.33),262(4.38), 270s(4.35),410(4.05)	18-2185-78
	MeOH-HCl	225(4.34),265s(4.35), 273(4.38),320(3.87), 435(3.93)	18-2185-78
6(5H)-Phenanthridinone	MeOH	224s(4.59),230(4.67), 237(4.63),249(4.19), 258(4.28),309(3.78), 323(3.95),337(3.80)	94-2508-78
$C_{13}H_9NOS$			
Ethanone, 1-naphtho[2,1-d]thiazol-2-yl-	n.s.g.	286(3.83),315(3.94), 358(3.90)	42-0702-78
$C_{13}H_9NO_2$			
1H-Benz[f]isoindole-1,3(2H)-dione, 2-methyl-	ether	215(4.46),256(4.96), 274(3.94),280(3.97), 289(3.99),321(3.20), 337(3.54),353(3.76)	78-2263-78
$C_{13}H_9NO_2S_2$			
2,5-Pyrroledione, 2,5-dihydro-1-methyl-3,4-bis(2-thienyl)-	n.s.g.	268(3.98),414(3.69)	88-0125-78
$C_{13}H_9NO_4$			
[1]Benzopyrano[2,3-b]pyrrole-2,4-dione, 1,3-dihydro-3-(hydroxymethylene)-1-methyl-	MeOH	223(4.28),291(3.79), 334(4.12)	83-0018-78
Oxazolo[3,2-a]quinolinium, 5-acetoxy-2-hydroxy-, hydroxide, inner salt	EtOH	234(4.501),267(4.107), 328(3.52)	88-1887-78
Pyrido[1,2-a]indole-6,10-dione, 9-hydroxy-7-methoxy-	MeOH	204(4.34),230(4.28), 252(4.33),407(4.10), 423(4.12)	83-0960-78
$C_{13}H_9NO_5S$			
2-Propenoic acid, 3-[5-[(4-nitrophenyl)thio]-2-furanyl]-	EtOH	206(4.25),315(4.36)	73-0621-78
$C_{13}H_9NO_7S$			
2-Propenoic acid, 3-[5-[(4-nitrophenyl)sulfonyl]-2-furanyl]-	EtOH	205(4.23),281(4.21), 291(4.26)	73-0621-78

Compound	Solvent	$\lambda_{max}(\log \epsilon)$	Ref.
$C_{13}H_9NO_8$ 2(5H)-Furanone, 4-acetyl-3-hydroxy- 5-(6-nitro-1,3-benzodioxol-5-yl)-	EtOH	270(4.17),268(4.13), 315(3.89),345(3.67), 381(3.68)	95-0802-78
$C_{13}H_9N_3$ Propanedinitrile, [2-(4-cyanophenyl)- 1-methylethylidene]-	C_6H_{12}	281(2.86)	39-0995-78B
$C_{13}H_9N_3O$ 1H-1,2-Diazepine-4-carbonitrile, 1-benzoyl-	MeOH	232(3.90),272s(3.69), 360(2.67)	33-2887-78
Pyridinium, 1-(benzoylamino)-3-cyano-, hydroxide, inner salt	EtOH	346(3.72)	33-2887-78
$C_{13}H_9N_3O_2$ 1H-Pyrazolo[4',3':3,4]pyrido[2,1-a]iso- indole-1,7(2H)-dione, 4,5-dihydro-	MeOH	232(4.34),306(4.21)	1-0056-78
2H-Pyrido[1,2-a]-1,3,5-triazine- 2,4(3H)-dione, 3-phenyl-	MeOH	245(4.20),333(3.74)	78-0101-78
$C_{13}H_9N_3O_3$ 1H-Pyrazole, 4-(2-furanyl)-3-(4-nitro- phenyl)-	EtOH	206(4.17),220(4.13), 253(4.11),296(3.94)	73-0870-78
$C_{13}H_9N_3O_3S$ 1H-Benzimidazole, 2-[[5-(2-nitroethen- yl)-2-furanyl]thio]-	EtOH	249(4.20),282(4.11), 288(4.10),354(4.29)	73-2037-78
$C_{13}H_9N_3O_3S_3$ 1,3,4-Thiadiazole-2(3H)-thione, 5-[[(5- nitro-2-furanyl)methyl]thio]- 3-phenyl-	MeOH	209(4.62),233s(4.45), 328(4.48)	73-0160-78
$C_{13}H_9N_3O_4S$ 10H-Phenothiazine, 10-methyl-1,3-di- nitro-	dioxan	239(4.44),274(4.08), 299(4.15),438(3.74)	104-0832-78
10H-Phenothiazine, 10-methyl-2,4-di- nitro-	dioxan	239(4.58),295(4.21), 470(3.56)	104-0832-78
$C_{13}H_9N_3S$ Pyrido[3,2-c]pyridazine, 4-(phenyl- thio)-	heptane	225(3.85),340(3.51)	103-1032-78
$C_{13}H_{10}$ 6H-Benzo[3,4]cyclobuta[1,2]cycloheptene	C_6H_{12}	237-256(4.42),315(2.64), 328(2.72),343(2.72), 358(2.46),376(1.99)	12-1569-78
9H-Fluorene, potassium derivative	THF	470(3.18),500(3.03), 520s(--)	120-0052-78
$C_{13}H_{10}AsN$ 1H-1,3-Benzazarsole, 2-phenyl-	MeOH	244(4.24),267(4.43), 308s(3.78),320(3.79), 372(3.83)	101-0001-78K
	MeOH-KOH	260(4.26),298s(3.80), 353(3.69)	101-0001-78K
$C_{13}H_{10}BrClN_2$ Benzenecarbohydrazonoyl chloride, N-(4-bromophenyl)-	dioxan	242(4.20),310(4.19), 340(4.39)	104-2203-78

Compound	Solvent	$\lambda_{max}(\log \epsilon)$	Ref.
$C_{13}H_{10}BrN_2O_4P$			
Phenophosphazine, 2-bromo-5,10-dihydro-10-hydroxy-5-methyl-8-nitro-, 10-oxide	EtOH	222(4.45),261(4.10), 293(3.49),340s(4.00), 375(4.15)	65-1201-78
$C_{13}H_{10}Br_2O_3$			
Naphtho[1,2-d]-1,3-dioxole, 4,7-dibromo-2-methoxy-2-methyl-	MeOH	249(4.83),284(3.54), 296(3.55),309(3.46), 341(3.41),351(3.46)	12-2259-78
$C_{13}H_{10}ClNO_2$			
[1,1'-Biphenyl]-3-ol, 4'-chloro-5-methyl-6-nitro-	DMF	730(1.23)	104-2189-78
$C_{13}H_{10}ClNO_2S$			
Benzoic acid, 4-amino-2-[(4-chlorophenyl)thio]-	MeOH	226(4.27),253(4.30), 289(4.25)	73-0471-78
$C_{13}H_{10}ClNO_3S$			
Benzenemethanol, 2-[(4-chlorophenyl)-thio]-4-nitro-	MeOH	251(4.29),270s(4.15), 338s(3.13)	73-0471-78
$C_{13}H_{10}Cl_2N_2OS_2$			
Hydrazinecarbodithioic acid, [[5-(3,4-dichlorophenyl)-2-furanyl]methylene]-, methyl ester	dioxan	379(4.08)	73-2643-78
$C_{13}H_{10}Cl_2N_4$			
Formazan, 1,5-bis(4-chlorophenyl)-	EtOH	240(4.38),294(4.21), 367(4.48),446(4.00)	104-1956-78
$C_{13}H_{10}Cl_3NO_4$			
1H-Pyrrole-2-carboxylic acid, 3,4-dichloro-2,5-dihydro-2-hydroxy-5-oxo-1-phenyl-, 2-chloroethyl ester	EtOH	226(3.83),271(3.37)	78-0591-78
$C_{13}H_{10}NP$			
1H-1,3-Benzazaphosphole, 2-phenyl-	MeOH	256(4.34)	88-0441-78
$C_{13}H_{10}N_2$			
Benzene, 1,1'-(diazomethylene)bis-	benzene	527(2.01)	39-1283-78B
$C_{13}H_{10}N_2O$			
Phenazinium, 1-hydroxy-5-methyl-, hydroxide, inner salt	DMSO-Pr_4N^+ ClO_4^-	700(3.7)	35-0211-78
zinc complex	same	397(3.9),700(3.5)	35-0211-78
Pyrrolo[1,2-a]pyrimidin-2(1H)-one, 7-phenyl-	n.s.g.	243(4.50),249s(4.47), 290(4.12),299(4.13), 329s(3.54)	44-3544-78
$C_{13}H_{10}N_2O_2$			
Benzaldehyde, 3-[(4-hydroxyphenyl)azo]-	EtOH	232(4.4),250(4.4), 352(4.5)	19-0655-78
Benzo[b][1,6]naphthyridine-1,3(2H,5H)-dione, 5-methyl-	pH 7	282(4.33),303(4.48), 333(3.52),470(3.95)	150-1325-78
	pH 13	286(4.67),331(3.67), 466(3.96)	150-1325-78
Spiro[5H-cyclopenta[2,1-b:3,4-b']dipyridine-5,2'-[1,3]dioxolane]	EtOH	265(3.62),307(4.16), 313(4.13),320(4.27)	64-0080-78B

Compound	Solvent	$\lambda_{max}(\log \epsilon)$	Ref.
$C_{13}H_{10}N_2O_2S$			
5H-Thieno[2,3-b][1,4]benzodiazepin-5-one, 2-acetyl-4,10-dihydro-	MeOH	206(4.40),229(4.25), 243s(4.20),278(4.00), 358(3.98)	150-5101-78
$C_{13}H_{10}N_2O_2S_2$			
5(4H)-Thiazolone, 2-(6-hydroxy-2-benzothiazolyl)-4-(1-methylethylidene)-	EtOH	257(4.02),265(4.04), 280s(3.85),315(3.81), 373(4.24)	44-2366-78
	EtOH-base	250(3.97),285(3.91), 310s(3.82),442(4.32)	44-2366-78
	DMSO-KOPh	320(3.97),473(4.55)	44-2366-78
$C_{13}H_{10}N_2O_3S$			
Pyrrolo[1,2-a]quinazolin-1(3aH)-one, 3-acetyl-4,5-dihydro-2-hydroxy-5-thioxo-	EtOH	265(4.20),321(4.09)	142-1729-78A
$C_{13}H_{10}N_2O_4$			
2-Propen-1-one, 2-amino-3-(5-nitro-2-furanyl)-1-phenyl-	EtOH	259(4.09),449(4.29)	4-0555-78
Pyrrolo[1,2-a]quinazoline-1,5-dione, 3-acetyl-3a,4-dihydro-2-hydroxy-	EtOH	247(4.22),308(3.99), 333(3.87)	142-1729-78A
$C_{13}H_{10}N_2O_4Se$			
Selenonium, diphenyl-, dinitromethylide	H$_2$O	344(4.02)	70-0945-78
	MeOH	339(3.95)	70-0945-78
	MeCN	340(3.94)	70-2087-78
	DMSO	346(3.96)	70-2087-78
$C_{13}H_{10}N_2O_5$			
Furan, 2-[2-(4-methylphenyl)-1-nitroethenyl]-5-nitro-, (E)-	dioxan	231(4.04),318(4.19)	73-0463-78
$C_{13}H_{10}N_2O_5S$			
Benzoic acid, 2-[(5-acetyl-3-nitro-2-thienyl)amino]-	MeOH	219(4.35),262(4.05), 351(4.30)	150-5101-78
$C_{13}H_{10}N_2O_6$			
Furan, 2-[2-(2-methoxyphenyl)-1-nitroethenyl]-5-nitro-, (E)-	dioxan	304(4.02),362(4.10)	73-0463-78
Furan, 2-[2-(4-methoxyphenyl)-1-nitroethenyl]-5-nitro-, (E)-	dioxan	238(4.09),310(4.13), 363(4.22)	73-0463-78
$C_{13}H_{10}N_2S$			
2(3H)-Benzothiazolimine, N-phenyl-	EtOH	219(4.30),236(4.23), 301(4.36)	103-0380-78
Imidazo[2,1-b]naphtho[1,2-d]thiazole, 9,10-dihydro-	MeOH	223(4.51),260(4.45), 345(3.83)	83-0267-78
$C_{13}H_{10}N_3O_6P$			
Phenophosphazine, 5,10-dihydro-10-hydroxy-5-methyl-2,8-dinitro-, 10-oxide	EtOH	220(4.47),270s(3.83), 328(3.87),390(4.38)	65-1201-78
$C_{13}H_{10}N_4$			
Pyrido[3,2-c]pyridazin-4-amine, N-phenyl-	EtOH	255(4.15),370(3.96)	103-0663-78
2H-Tetrazole, 2,5-diphenyl-	EtOH	240s(4.11),271(4.24)	33-1477-78
$C_{13}H_{10}N_4O$			
1,3,5-Heptatriene-1,1,7,7-tetracarbo-	MeOH	304(5.2),333(4.8),	83-0287-78

Compound	Solvent	$\lambda_{max}(\log \epsilon)$	Ref.
nitrile, 4-ethoxy- (cont.)		344(4.9),525(5.9)	83-0287-78
1,9,9b-Triazaphenalene-3-carbonitrile, 2-ethoxy-	EtOH	249(4.25),282(4.28), 341(4.21),387(4.20), 392(4.21),405(4.10), 413(4.22),620(3.04), 660(3.17)	95-0623-78
$C_{13}H_{10}N_4OS$			
1H-Pyrazolo[3,4-d]pyrimidine-4-carbo-thioic acid, S-methyl ester	EtOH	253(4.46),295s(3.84), 350s(3.39)	95-1274-78
$C_{13}H_{10}N_4O_2$			
1-Phenazinamine, N-methyl-3-nitro-	EtOH	548(3.46)	39-0299-78C
1-Phenazinamine, N-methyl-4-nitro-	EtOH	511(3.86)	39-0299-78C
1-Phenazinamine, N-methyl-6-nitro-	EtOH	578(3.40)	39-0299-78C
1-Phenazinamine, N-methyl-7-nitro-	EtOH	560(3.38)	39-0299-78C
1-Phenazinamine, N-methyl-8-nitro-	EtOH	592(3.36)	39-0299-78C
1-Phenazinamine, N-methyl-9-nitro-	EtOH	578(3.40)	39-0299-78C
1H-Pyrazolo[3,4-d]pyrimidine-4-carbox-ylic acid, 1-phenyl-, methyl ester	EtOH	251(4.38),330(3.10)	95-1274-78
$C_{13}H_{10}N_4O_3$			
Benzo[g]pteridine-2,4(3H,10H)-dione, 7-acetyl-10-methyl-	pH 7.0	282(4.51),348(3.74), 427(4.04)	35-7670-78
$C_{13}H_{10}N_4O_5$			
2,4-Dinitrobenzeneazo-3'-(2',5'-dimeth-ylfuran)	n.s.g.	340(4.01),470(4.57)	39-0401-78C
$C_{13}H_{10}O$			
6H-Benzo[3,4]cyclobuta[1,2]cyclohepten-6-one, 7,8-dihydro-	C_6H_{12}	246(4.66),268(4.08), 281(4.11),292(4.09), 304(4.01),317(3.63), 344(3.64),357s(3.47)	12-1585-78
6H-Benzo[3,4]cyclobuta[1,2]cyclohepten-6-one, 9,9a-dihydro-	C_6H_{12}	231(4.01),236(4.00), 251s(3.93),265s(3.87), 278(3.99),290(4.10), 305(4.18),317(4.25)	12-1585-78
7H-Benzo[3,4]cyclobuta[1,2]cyclohepten-7-one, 6,8-dihydro-	C_6H_{12}	235(4.59),309(3.79), 355(3.75)	12-1585-78
Methanone, diphenyl-	ether-LiClO₄	247.5(4.27)(changing)	44-1126-78
after some time	ether-LiClO₄	258(4.15)	44-1126-78
$C_{13}H_{10}OS$			
3H-Naphtho[1,8-bc]thiophen-3-one, 2-ethyl-	MeOH	221(4.375),256(4.161), 310(3.710),350(3.987), 375s(--)	5-0627-78
$C_{13}H_{10}O_3S$			
2-Propenoic acid, 3-[5-(phenylthio)-2-furanyl]-	EtOH	206(4.24),243(4.06), 309(4.13)	73-0621-78
$C_{13}H_{10}O_5S$			
2-Propenoic acid, 3-[5-(phenylsulfon-yl)-2-furanyl]-	EtOH	206(4.13),227(4.05), 302(4.38)	73-0621-78
$C_{13}H_{10}O_6$			
2-Propenoic acid, 3-(8-hydroxy-6-meth-oxy-1-oxo-1H-2-benzopyran-3-yl)-	MeOH	260(4.67),305s(4.07), 313(4.09),342(4.19), 354s(4.05)	44-2339-78

Compound	Solvent	$\lambda_{max}(\log \epsilon)$	Ref.
$C_{13}H_{11}$			
Cycloheptatrienylium, phenyl-	MeCN	366(4.18)	138-0237-78
$C_{13}H_{11}BrNO_2P$			
Phenophosphazine, 2-bromo-5,10-dihydro-10-hydroxy-5-methyl-, 10-oxide	EtOH	222(4.47),280(4.33), 304s(4.04),338(3.70)	65-1201-78
$C_{13}H_{11}BrN_2O$			
Ethanol, 1-[2-[(4-bromophenyl)azo]-2,4-cyclopentadien-1-ylidene]-	MeOH	258(4.459),425(4.260), 468(4.301)	73-0938-78
$C_{13}H_{11}BrN_2OS_2$			
Hydrazinecarbodithioic acid, [(5-bromo-2-furanyl)methylene]-, phenylmethyl ester	dioxan	356(4.58)	73-2643-78
Hydrazinecarbodithioic acid, [[5-(4-bromophenyl)-2-furanyl]methylene]-, methyl ester	dioxan	380(4.66)	73-2643-78
$C_{13}H_{11}Br_4N_3$			
8-Azaspiro[4.5]deca-6,9-diene-6,10-dicarbonitrile, 7,9-bis(dibromomethyl)-	EtOH	208(4.42),250(4.32), 371(4.05)	103-1236-78
$C_{13}H_{11}ClNOP$			
Phenophosphazine, 10-chloro-5,10-dihydro-5-methyl-, 10-oxide	dioxan	231(4.44),273(4.23), 305(3.95),346(3.81)	65-1205-78
$C_{13}H_{11}ClN_2O$			
Ethanol, 1-[2-(4-chlorophenyl)azo]-2,4-cyclopentadien-1-ylidene]-	MeOH	258(4.505),410(4.292), 465(4.309)	73-0938-78
$C_{13}H_{11}ClN_2OS$			
1H-Pyrrole-5-carboxaldehyde, 2-chloro-4,5-dihydro-1-methyl-4-[(phenylamino)methylene]-5-thioxo-	octane	260(4.03),320(4.53), 490(4.12)	104-2041-78
	EtOH	247(4.15),325(4.40), 470(4.03)	104-2041-78
$C_{13}H_{11}ClN_2OS_2$			
Hydrazinecarbodithioic acid, [[5-(2-chlorophenyl)-2-furanyl]methylene]-, methyl ester	dioxan	370(4.50)	73-2643-78
Hydrazinecarbodithioic acid, [[5-(4-chlorophenyl)-2-furanyl]methylene]-, methyl ester	dioxan	379(4.64)	73-2643-78
$C_{13}H_{11}ClN_2OSe$			
1H-Pyrrole-3-carboxaldehyde, 2-chloro-4,5-dihydro-1-methyl-4-[(phenylamino)methylene]-5-selenoxo-	octane	255(3.92),335(4.38), 515(3.92)	104-2041-78
	EtOH	235(4.28),330(4.40), 485(3.81)	104-2041-78
$C_{13}H_{11}ClN_2O_2$			
1H-Pyrrole-3-carboxaldehyde, 2-chloro-4,5-dihydro-1-methyl-5-oxo-4-[(phenylamino)methylene]-	octane	245(3.96),330(4.23), 435(4.34)	104-2041-78
	EtOH	250(4.05),330(3.96), 430(4.22)	104-2041-78
$C_{13}H_{11}ClN_4$			
Indazolo[3,2-c][1,2,4]benzotriazine, 9-chloro-1,2,3,4-tetrahydro-	C_6H_{12}	353(3.96),390(--)	24-2258-78

Compound	Solvent	$\lambda_{max}(\log \epsilon)$	Ref.
$C_{13}H_{11}ClN_4O_2$			
[1,2,4]Triazino[4,3-b]indazole-4-carb-oxylic acid, 8-chloro-3-methyl-, ethyl ester	C_6H_{12}	360(3.90)	24-2258-78
$C_{13}H_{11}ClO_3$			
3H-Naphtho[1,8-bc]furan-2-carboxalde-hyde, 6-chloro-4,5-dihydro-8-methoxy-	EtOH	206(4.46),247(4.23), 301(4.29)	18-2068-78
$C_{13}H_{11}ClO_4$			
3-Furancarboxylic acid, 2-(4-chloro-phenyl)-4,5-dihydro-4-oxo-, ethyl ester	EtOH	218(4.05),260(3.85), 301(4.24)	118-0448-78
2(5H)-Furanone, 4-acetyl-5-(2-chloro-phenyl)-3-methoxy-	EtOH	269(4.05)	4-0737-78
2(5H)-Furanone, 3-acetyl-5-[(4-chloro-phenyl)methyl]-4-hydroxy-	EtOH	227(3.83),265(3.86)	4-0327-78
$C_{13}H_{11}ClO_6$			
1H-2-Benzopyran-4-carboxaldehyde, 3-chloro-5,6,7-trimethoxy-1-oxo-	MeOH	215(3.62),250(3.54)	4-0257-78
$C_{13}H_{11}FN_2OS$			
2H-Benzimidazol-2-one, 1-(5-ethyl-2-thienyl)-5-fluoro-1,3-dihydro-	MeOH	227(4.10),281(3.92)	39-0937-78C
4H-Thieno[2,3-b][1,5]benzodiazepin-4-one, 2-ethyl-7-fluoro-5,10-dihydro-	MeOH	228(4.45),248(4.27)	39-0937-78C
$C_{13}H_{11}IN_2OS_2$			
Hydrazinecarbodithioic acid, [(5-iodo-2-furanyl)methylene]-, phenylmethyl ester	dioxan	360(4.65)	73-2643-78
$C_{13}H_{11}NO$			
2,5-Cyclohexadien-1-one, 4-[(4-methyl-phenyl)imino]-	C_6H_{12}	260(4.22),295(4.12), 460(3.57)	48-0557-78
	EtOH	262(4.17),300(4.12), 465(3.57)	48-0557-78 +80-0617-78
	ether	260(4.32),290(4.24), 462(3.65)	48-0557-78
	$CHCl_3$	300(4.09),462(3.67)	48-0557-78
	CCl_4	300(4.17),460(3.68)	48-0557-78
5H-Indeno[1,2-b]pyridin-5-ol, 5-methyl-	EtOH	212(5.12),284(4.7), 310(4.84)	103-0997-78
$C_{13}H_{11}NO_2$			
Benzamide, N-hydroxy-N-phenyl-	H_2O	219(3.86),255(3.81)	73-1571-78
	10% EtOH	219(3.86),257(3.82)	73-1571-78
	30% EtOH	219(3.86),260(3.83)	73-1571-78
	50% EtOH	219(3.86),264(3.82)	73-1571-78
[1,1'-Biphenyl]-3-ol, 5-methyl-6-ni-troso-	DMF	740(1.20)	104-2189-78
2,4,6-Cycloheptatrien-1-one, 2-[(2-hy-droxyphenyl)amino]-	MeOH	205(4.40),231(4.40), 342(4.00),404(4.18)	18-2185-78
	MeOH-HCl	239(4.36),382(4.03)	18-2185-78
	MeOH-NaOH	231(4.48),340(3.95), 411(4.06)	18-2185-78
2,5-Cyclohexadien-1-one, 4-[(4-meth-oxyphenyl)imino]-	C_6H_{12}	260(4.32),310(4.10), 480(3.87)	48-0557-78
	EtOH	265(4.24),315(4.03), 439(3.98)	48-0557-78 +80-0617-78

Compound	Solvent	$\lambda_{max}(\log \epsilon)$	Ref.
2,5-Cyclohexadien-1-one, 4-[(4-meth-oxyphenyl)imino]- (cont.)	ether	260(4.30),310(4.09), 483(3.88)	48-0557-78
	CHCl$_3$	310(4.12),488(3.96)	48-0557-78
	CCl$_4$	310(4.09),485(3.89)	48-0557-78
1,4-Naphthalenedione, 2-(1-azetidinyl)-	EtOH	239(4.18),277(4.40), 437(3.68)	119-0037-78
5H-Oxazolo[3,2-a]quinolin-5-one, 1,2-dimethyl-	EtOH	259(4.179),329(4.052)	88-1887-78
9H-Pyrrolo[1,2-a]indol-9-one, 7-meth-oxy-6-methyl-	MeOH	258(4.18),290s(3.94), 335(3.78)	39-0460-78C
$C_{13}H_{11}NO_2S$			
Benzamide, 2-(phenylsulfinyl)-	MeOH	212(4.24)	2-0143-78
$C_{13}H_{11}NO_3$			
Furo[2,3-b]quinoline, 4,7-dimethoxy-	EtOH	248(4.75),308(3.92)	102-2145-78
2H-Pyran-2,6(3H)-dione, 3-[[(phenyl-methyl)amino]methylene]-, (Z)-	MeOH	338(4.54),347s(4.52)	44-4415-78
1H-Pyrrole-2-carboxaldehyde, 3-acetoxy-5-phenyl-	MeOH	232(4.01),250(3.87), 325(4.27)	49-0137-78
$C_{13}H_{11}NO_3S$			
Furan, 2-[(4-methylphenyl)thio]-5-(2-nitroethenyl)-	EtOH	205(4.20),243(4.04), 375(4.06)	73-2037-78
$C_{13}H_{11}NO_4$			
1H-Carbazole-6-carboxylic acid, 2,3,4,9-tetrahydro-8-hydroxy-1-oxo-	EtOH	232(4.42),285(4.16)	102-0834-78
3-Quinolineacetic acid, 2-methyl-α-oxo-, methyl ester, 1-oxide	EtOH	240(4.55),320(3.78)	95-0802-78
$C_{13}H_{11}NO_5$			
1,3-Dioxolo[4,5-g]quinoline-7-carbox-ylic acid, 5-ethyl-5,8-dihydro-8-oxo-	aq KOH	257(4.57),265(4.58), 296(3.86),310(3.97), 324(4.08),338(4.12)	87-0485-78
2H-Isoindolepropanoic acid, β,1,3-tri-oxo-, ethyl ester	MeOH	220(4.56),232(4.13), 241(3.99),290(3.26)	1-0056-78
1H-Pyrrole-3,4-dicarboxylic acid, 2-(2-methoxyphenyl)-	EtOH	283(4.39),301(4.39)	39-1588-78C
	base	280(4.05)	39-1588-78C
$C_{13}H_{11}NO_6$			
1,3-Dioxolo[4,5-g]quinoline-7-carbox-ylic acid, 5-ethyl-5,6-dihydro-8-hy-droxy-6-oxo-	aq KOH	226(4.61),302(4.06), 322(4.17),336(4.09)	87-0485-78
2(5H)-Furanone, 4-acetyl-3-methoxy-5-(2-nitrophenyl)-	EtOH	266(4.20)	4-0737-78
	EtOH	266(4.05)	95-0802-78
2(5H)-Furanone, 4-acetyl-3-methoxy-5-(3-nitrophenyl)-	EtOH	267(4.08)	4-0737-78
$C_{13}H_{11}N_2O_4P$			
Phenophosphazine, 5,10-dihydro-10-hy-droxy-5-methyl-2-nitro-, 10-oxide	EtOH	222(4.45),253(4.05), 289(3.58),330s(3.88), 375(4.19)	65-1201-78
$C_{13}H_{11}N_3$			
3H-Pyrido[3,4-b]indole-3-acetonitrile, 4,9-dihydro-, (S)-	MeOH	235(4.01),240(3.98), 316(3.98)	35-0938-78
$C_{13}H_{11}N_3O$			
Diazene, [(hydroxyimino)phenylmethyl]-phenyl-	EtOH	430(2.57)	96-0879-78

Compound	Solvent	$\lambda_{max}(\log \epsilon)$	Ref.
Phenol, 2-(2H-benzotriazol-2-yl)-4-methyl-	isoPrOH	337(4.14)	117-0097-78
$C_{13}H_{11}N_3O_2$			
1H-1,2-Diazepine-4-carboxamide, 1-benzoyl-	MeOH	229(4.23),270s(3.93), 370(3.34)	33-2887-78
Pyridinium, 3-(aminocarbonyl)-1-(benzoylamino)-, hydroxide, inner salt	EtOH	331(3.62)	33-2887-78
Pyrimido[4,5-b]quinoline-2,4(1H,3H)-dione, 1,3-dimethyl-	MeOH	315(4.02),368(4.01), 389s(--)	83-0115-78
	6M HCl	345(4.28),406(4.16), 424s(--)	83-0115-78
Pyrimido[5,4-b]quinoline-2,4(1H,3H)-dione, 1,3-dimethyl-	MeOH	308(4.06),353(3.85)	83-0115-78
	6M HCl	334(4.33)	83-0115-78
Pyrrolo[3,2-d]pyrimidine-2,4(1H,3H)-dione, 7-methyl-3-phenyl-	MeOH	271(4.24)	39-0483-78C
$C_{13}H_{11}N_3O_2S$			
1H-Pyrimido[5,4-b][1,4]benzothiazine-2,4(3H,10H)-dione, 10-(2-propenyl)-	pH 13	259(4.32),281s(3.92), 325s(3.30)	83-0303-78
	MeOH	248(4.27),282(4.05), 355s(3.23)	83-0303-78
	12M HCl	250(4.23),291(3.90), 368(3.49)	83-0303-78
1,9,9b-Triazaphenalene-3-carboxylic acid, 2-(methylthio)-, methyl ester	EtOH	257(4.36),320(4.38), 355(4.18),403(4.24), 404(4.29)	95-0623-78
$C_{13}H_{11}N_3O_3$			
1H-Pyrrolo[2,3-d]pyrimidine-2,4,5(3H)-trione, 6,7-dihydro-3-methyl-1-phenyl-	H_2O	250(4.01),278(4.09)	103-0443-78
$C_{13}H_{11}N_3O_3S$			
[1,2]Oxathiolo[4,3,2-hi][1,2,3]benzothiadiazole-3-S^{IV}, 2,6,7,8-tetrahydro-2-(4-nitrophenyl)-	C_6H_{12}	247s(3.85),332s(3.77), 459(4.53),485(4.60)	39-0195-78C
$C_{13}H_{11}N_3O_3S_2$			
Hydrazinecarbodithioic acid, [[5-(2-nitrophenyl)-2-furanyl]methylene]-, methyl ester	dioxan	370(4.52)	73-2643-78
Hydrazinecarbodithioic acid, [[5-(3-nitrophenyl)-2-furanyl]methylene]-, methyl ester	dioxan	377(4.51)	73-2643-78
Hydrazinecarbodithioic acid, [[5-(4-nitrophenyl)-2-furanyl]methylene]-, methyl ester	dioxan	408(4.57)	73-2643-78
$C_{13}H_{11}N_5O$			
Formamide, N-[(1-phenyl-1H-pyrazolo[3,4-d]pyrimidin-4-yl)methyl]-	EtOH	243(4.49),300(3.49)	95-0891-78
$C_{13}H_{11}N_5O_2$			
7-Azalumazine, 1,3-dimethyl-6-phenyl-	EtOH	278(4.45),375(3.55)	94-3154-78
7H-Purine-7-carboxylic acid, 6-amino-, phenylmethyl ester	pH 1	273(4.01)	35-5232-78
	MeOH	288(3.80)	35-5232-78
	MeCN	289(3.75)	35-5232-78
9H-Purine-9-carboxylic acid, 6-amino-, phenylmethyl ester	pH 1	253(4.17)	35-5232-78
	MeOH	253(4.16)	35-5232-78
	MeCN	252(4.12)	35-5232-78

Compound	Solvent	$\lambda_{max}(\log \epsilon)$	Ref.
Pyrimido[4,5-e]-1,2,4-triazine-6,8(5H,7H)-dione, 5,7-dimethyl-3-phenyl-	EtOH	247(4.46),331(4.25)	94-0367-78
$C_{13}H_{11}N_5O_3$			
Pyrido[2,3-d]pyrimidin-4(8H)-one, 6-methyl-8-[(1,2,3,4-tetrahydro-2,4-dioxo-5-pyrimidinyl)methyl]-	pH 1	263(3.26),331(3.99)	44-0828-78
	pH 7	251(4.12),263(4.01), 365(3.86)	44-0828-78
	pH 11	245(4.01),284(3.08), 364(3.94)	44-0828-78
2,4(1H,3H)-Pyrimidinedione, 5-[(6-methyl-4-oxopyrido[2,3-d]pyrimidin-3(4H)-yl)methyl]-	pH 1	264(4.08),327(3.93)	44-0828-78
	pH 7	265(4.18),306(3.77), 317(3.62)	44-0828-78
	pH 11	275(4.12),288(4.11), 317(3.65)	44-0828-78
$C_{13}H_{11}N_5O_4$			
Pyrido[2,3-d]pyrimidine-2,4(1H,3H)-dione, 6-methyl-3-[(1,2,3,4-tetrahydro-2,4-dioxo-5-pyrimidinyl)methyl]-	pH 1	247(4.07),263(3.90), 315(3.82)	44-0828-78
	pH 7	247(4.07),263(3.90), 315(3.79)	44-0828-78
	pH 11	271(3.15),345(3.64)	44-0828-78
$C_{13}H_{11}N_5S_2$			
Thiocyanic acid, (3,5-dicyano-1,4-dihydro-4,4-dimethyl-2,6-pyridinediyl)bis(methylene) ester	EtOH	222(4.79),270(4.63), 356(4.32)	103-1236-78
$C_{13}H_{12}$			
7,8-Benzobicyclo[4.2.1]nona-2,4,7-triene	n.s.g.	253(3.51),261(3.65), 270(3.71),282(3.51)	35-2959-78
6H-Benzo[3,4]cyclobuta[1,2]cycloheptene, 7,8-dihydro-	C_6H_{12}	237(4.56),288(3.53), 311(3.49),318s(3.71), 321(3.72),327(3.51), 336(3.88)	12-1569-78
$C_{13}H_{12}AsNO$			
Acetamide, N-(2-phenyl-4-arseninyl)-	EtOH	273(4.34),321(4.04)	88-1175-78
$C_{13}H_{12}BrNOS$			
Thiopyrano[2,3-b]indole, 9-acetyl-7-bromo-2,3,4,9-tetrahydro-	EtOH	212(4.42),239(4.24), 268s(3.88),306(4.22)	94-3695-78
$C_{13}H_{12}BrNO_2$			
Methanone, (2-bromo-5-methoxy-4-methylphenyl)-1H-pyrrol-2-yl-	MeOH	227(3.82),302(3.94)	39-0460-78C
$C_{13}H_{12}BrN_2O$			
10H-Pyrido[3,2-b][1,4]benzoxazinium, 3-bromo-1,10-dimethyl-, iodide	MeOH	229(4.95),380(4.33)	95-0585-78
$C_{13}H_{12}BrN_3O$			
4H-Pyrimido[2,1-b]quinazolin-4-one, 3-bromo-6,11-dihydro-2,11-dimethyl-	MeOH	260(4.12),323(4.16)	4-0077-78
$C_{13}H_{12}BrN_3O_3S$			
4(3H)-Pyrimidinone, 5-bromo-6-methyl-2-(methylthio)-1-[(2-nitrophenyl)methyl]-	MeOH	250(4.09),301(4.19)	4-0077-78

Compound	Solvent	$\lambda_{max}(\log \epsilon)$	Ref.
$C_{13}H_{12}BrN_5O_5$ Isoxazolo[2,3-a]pyridinium, 7-[(2,4-di- nitrophenyl)hydrazono]-4,5,6,7-tetra- hydro-, bromide	EtOH	209(3.25),238(3.14), 253(3.28)	78-1581-78
$C_{13}H_{12}ClN_3O$ 1H-Azepine-4-carbonitrile, 3-[(4-chlo- rophenyl)amino]-2,5,6,7-tetrahydro- 2-oxo-	n.s.g.	253(4.13),336(4.10)	103-1013-78
$C_{13}H_{12}ClN_3O_2$ Pyrido[2,3-d]pyrimidine-2,4(1H,3H)-di- one, 3-(4-chlorophenyl)-5,6,7,8- tetrahydro-	MeOH	221(4.04),291(4.28)	24-2297-78
$C_{13}H_{12}ClN_3O_3$ 2,4(1H,3H)-Pyrimidinedione, 6-amino- 5-(chloroacetyl)-3-methyl-1-phenyl-	EtOH	244(3.60),277(3.82)	103-0443-78
$C_{13}H_{12}ClN_3O_3S$ 4(3H)-Pyrimidinone, 3-[(5-chloro-2-ni- trophenyl)methyl]-6-methyl-2-(meth- ylthio)-	MeOH	284(4.17)	4-0077-78
$C_{13}H_{12}ClN_3O_4$ 1H-Benzimidazolium, 1-methyl-3-(2-pyri- dinyl)-, perchlorate	EtOH	220(4.42),270(4.32)	73-2046-78
$C_{13}H_{12}ClN_5O$ 6H-Imidazo[4,5-e]-1,2,4-triazin-6-one, 3-(4-chlorophenyl)-7-ethyl-5,7-di- hydro-5-methyl-	EtOH	249(4.36),289(4.25), 309(4.11)	94-3154-78
$C_{13}H_{12}F_3NO_4$ Acetamide, N-[1-(acetoxymethyl)-2-oxo- 2-phenylethyl]-2,2,2-trifluoro-	EtOH	247(4.12)	35-4481-78
$C_{13}H_{12}NO_2P$ Phenophosphazine, 5,10-dihydro-10-hy- droxy-5-methyl-, 10-oxide	EtOH	222(4.50),274(4.28), 295s(3.95),333(3.78)	65-1201-78
$C_{13}H_{12}N_2$ 1H-Azuleno[8,1-cd]pyridazine, 1,3-di- methyl-	hexane	241(4.52),281(4.00), 347(4.02),362(4.06), 387(2.93),403(2.61), 410(2.84),587(2.69), 635(2.87),700(2.89), 760(2.46),785(2.64)	142-0387-78C
Benzaldehyde, phenylhydrazone	n.s.g.	344(4.34)	104-0717-78
2-Butenenitrile, 4-(3-methyl-2-phenyl- 2H-azirin-2-yl)-, (E)-	C_6H_{12}	255(3.32)	44-3757-78
(Z)-	C_6H_{12}	255(3.24)	44-3757-78
Propanedinitrile, [1-methyl-2-(4-meth- ylphenyl)ethylidene]-	C_6H_{12}	271(3.58),277(3.56)	39-0995-78B
$C_{13}H_{12}N_2O$ Benzoic acid, N'-phenylhydrazide	EtOH	232(4.24),273(3.52)	33-1477-78
Ethanone, 2-(6-methyl-3-pyridazinyl)- 1-phenyl-	C_6H_{12}	238(3.99),340(4.11)	94-3633-78
	EtOH	238(3.98),341(3.82)	94-3633-78
Propanedinitrile, [2-(3-methoxyphenyl)- 1-methylethylidene]-	C_6H_{12}	281(3.32)	39-0995-78B

Compound	Solvent	$\lambda_{max}(\log \epsilon)$	Ref.
Propanedinitrile, [2-(4-methoxyphenyl)-1-methylethylidene]-	C_6H_{12}	278(3.57),286(3.58), 291s(3.58)	39-0995-78B
2-Propanone, 1-phenyl-3-(3-pyridazinyl)-	EtOH	254(3.23),323(3.49)	94-3633-78
1H-Pyrido[3,2-b][1,4]benzoxazine, 1,3-dimethyl-	MeOH	231(4.25),382(3.87), 401s(3.86),425s(3.71)	95-0585-78
hydriodide	MeOH	226(4.72),377(4.09)	95-0585-78
10H-Pyrido[3,2-b][1,4]benzoxazine, 3,10-dimethyl-	MeOH	231(4.53),341(4.09)	95-0585-78
Urea, N,N'-diphenyl-	MeOH	254(4.54)	62-0762-78A
$C_{13}H_{12}N_2OS$			
[1,2]Oxathiolo[4,3,2-hi][1,2,3]benzo-thiadiazole-3-S^{IV}, 2,6,7,8-tetrahy-dro-2-phenyl-	C_6H_{12}	238(3.94),277(3.88), 456(4.37),467(4.33)	39-0195-78C
4H-Thieno[2,3-b][1,5]benzodiazepin-4-one, 2-ethyl-5,10-dihydro-	MeOH	227(4.33),257(4.19)	39-0937-78C
$C_{13}H_{12}N_2O_2$			
Benzo[b][1,6]naphthyridine-1,3(2H,4H)-dione, 5,10-dihydro-5-methyl-	pH 3 or 5	347(4.00)	150-1325-78
2,5-Cyclohexadien-1-one, 3-amino-4-[(4-hydroxy-2-methylphenyl)imino]-	neutral	480(3.84)	39-1292-78B
	anion	600(4.10)	39-1292-78B
2,5-Cyclohexadien-1-one, 3-amino-4-[(4-hydroxy-3-methylphenyl)imino]-	neutral	482(3.90)	39-1292-78B
	anion	605(4.25)	39-1292-78B
2,5-Cyclohexadien-1-one, 3-amino-4-[(4-hydroxyphenyl)imino]-2-methyl-	neutral	470(3.85)	39-1292-78B
	anion	563(4.25)	39-1292-78B
2,5-Cyclohexadien-1-one, 5-amino-4-[(4-hydroxyphenyl)imino]-2-methyl-	neutral	464(3.92)	39-1292-78B
	anion	556(4.22)	39-1292-78B
2,3-Diazabicyclo[3.2.0]hepta-3,6-diene, 2-benzoyl-5-methoxy-	EtOH	230(3.97),265(4.01), 301(3.28)	33-2887-78
Pyridinium, 1-(benzoylamino)-3-methoxy-, hydroxide, inner salt	EtOH	322(3.78)	33-2887-78
2-Pyrimidineacetic acid, 4-phenyl-, methyl ester	neutral	276(4.07),250s(3.82)	12-0649-78
	cation	310(4.10)	12-0649-78
4(1H)-Pyrimidinone, 2-methyl-1-(2-oxo-2-phenylethyl)-	n.s.g.	248(4.48)	44-3544-78
hydrate	n.s.g.	248(4.47)	44-3544-78
$C_{13}H_{12}N_2O_3$			
Methanone, [4-(2-furanyl)-4,5-dihydro-1H-pyrazol-3-yl](5-methyl-2-furanyl)-	EtOH	217(4.07),296(3.94), 356(4.20)	104-1830-78
$C_{13}H_{12}N_2O_4$			
Benzenemethanamine, 4-methoxy-α-[(5-ni-tro-2-furanyl)methylene]-	EtOH	275(4.12),503(4.42)	4-0555-78
Propanoic acid, 3-(1,2-dihydro-2-oxo-3H-indol-3-ylidene)-2-(hydroxyimi-no)-, ethyl ester	EtOH	256(4.62),312(4.60)	104-1463-78
1H-Pyrazole-5-carboxylic acid, 3-(benz-oyloxy)-1-methyl-, methyl ester	MeOH	232(4.39)	24-0780-78
$C_{13}H_{12}N_2O_4S$			
4(1H)-Cinnolinone, 1-acetyl-3-[(acet-yloxymethyl)thio]-	EtOH	256(4.36),262(4.32), 360(4.04)	4-0115-78
$C_{13}H_{12}N_2O_6$			
5H-Pyrano[2,3-b]pyridin-5-one, 4-acet-oxy-7-methyl-, O-acetyloxime, 8-oxide	EtOH	330(4.07)	83-0848-78
$C_{13}H_{12}N_2S$			
2-Imidazolidinethione, 1-(1-naphthyl)-	MeOH	223(4.91),283(3.65)	83-0267-78

Compound	Solvent	$\lambda_{max}(\log \epsilon)$	Ref.
Thiourea, diphenyl-, iodine complex	$CHCl_3$	294(3.81)	62-1083-78A
$C_{13}H_{12}N_3$			
1H-Benzimidazolium, 1-methyl-3-(2-pyridinyl)-, perchlorate	EtOH	220(4.42),270(4.32)	73-2046-78
$C_{13}H_{12}N_3O_4P$			
2-Phenophosphazinamine, 5,10-dihydro-10-hydroxy-5-methyl-8-nitro-, 10-oxide	EtOH	265(3.59),293s(3.20), 335(3.34),380(3.60)	65-1201-78
$C_{13}H_{12}N_4$			
Indazolo[3,2-c][1,2,4]benzotriazine, 1,2,3,4-tetrahydro-	C_6H_{12}	351(3.84)	24-2258-78
$C_{13}H_{12}N_4O$			
4H-Pyrazolo[3,4-d]pyridazin-4-one, 1,5-dihydro-1,7-dimethyl-3-phenyl-	EtOH	214(4.01),260(4.16)	4-0813-78
4H-Pyrazolo[3,4-d]pyridazin-4-one, 2,5-dihydro-2,7-dimethyl-3-phenyl-	EtOH	214(4.13),253(4.10), 268(4.02)	4-0813-78
4H-Pyrrolo[2,3-d]pyrimidin-4-one, 2-amino-1,7-dihydro-5-methyl-6-phenyl-	acid	230(4.32),290(4.34)	44-3937-78
	pH 7.0	227(4.36),297(4.34)	44-3937-78
	base	316(4.34)	44-3937-78
4H-Pyrrolo[2,3-d]pyrimidin-4-one, 2-amino-1,7-dihydro-6-methyl-5-phenyl-	acid	235(4.22),254s(4.13), 275s(4.10)	44-3937-78
	pH 7.0	234(4.26),264(4.09), 297(4.04)	44-3937-78
	base	266(4.07)	44-3937-78
1H-Pyrrolo[2,3-b]quinoxaline-3-carboxamide, N,1-dimethyl-	DMSO	259(4.8),279(3.9), 310(3.9),372(4.0), 391(5.3)	24-2376-78
$C_{13}H_{12}N_4OS$			
Butanamide, N-thiazolo[4,5-b]quinoxalin-2(3H)-ylidene-	EtOH	212(4.52),278(4.45)	2-0683-78
$C_{13}H_{12}N_4O_2$			
[1,2,4]Triazino[4,3-b]indazole-4-carboxylic acid, 3-methyl-, ethyl ester	C_6H_{12}	370(3.84)	24-2258-78
$C_{13}H_{12}N_4O_5S$			
Benzenesulfonamide, N-[4-cyano-5-(methoxyimino)-1,3-pentadienyl]-3-nitro-	MeOH	379(4.58)	83-0433-78
Benzenesulfonamide, N-[4-cyano-5-(methoxyimino)-1,3-pentadienyl]-4-nitro-	MeOH	376(4.54)	83-0433-78
$C_{13}H_{12}N_4S$			
Hydrazinecarbothioamide, 2-(phenylmethylene)-N-2-pyridinyl-	EtOH	233(4.20),325(4.48)	104-0181-78
$C_{13}H_{12}N_5O_5$			
Isoxazolo[2,3-a]pyridinium, 7-[(2,4-dinitrophenyl)hydrazono]-4,5,6,7-tetrahydro-, bromide	EtOH	209(3.25),238(3.14), 253(3.28)	78-1581-78
$C_{13}H_{12}O$			
1,4-Ethano-7H-benzocyclohepten-7-one, 1,4-dihydro-	hexane	234(4.29),308(3.92), 320(3.89)	88-0569-78
	MeOH	237(4.34),323(4.09)	88-0569-78

Compound	Solvent	$\lambda_{max}(\log \epsilon)$	Ref.
$C_{13}H_{12}ORu$			
Ruthenocene, 1,1'-(1-oxo-1,3-propane-diyl)-	EtOH	251(3.69),292s(3.01), 332s(2.84)	18-0909-78
$C_{13}H_{12}OS_4$			
Dispiro[1,3-dithiolane-2,1'-[1H]indene-3'(2'H),2"-[1,3]dithiolan]-2'-one	MeOH	229(4.15),293(3.29)	24-3058-78
$C_{13}H_{12}O_2$			
Benzo[1,2-b:4,5-b']difuran, 2,4,6-tri-methyl-	MeOH	229(4.29),279(4.41), 292s(4.29),298s(4.20), 303(4.04)	4-0043-78
2-Oxetanone, 3-(1-methylethylidene)-4-(phenylmethylene)-	EtOH	224(4.01),254(4.23), 261(4.29),272(4.16), 313(3.95)	12-1757-78
Propanal, 2-(8-methyl-2H-1-benzopyran-2-ylidene)-, (E)-	EtOH	366(4.30)	48-0497-78
$C_{13}H_{12}O_2Ru$			
Ruthenocene, (2-carboxyethenyl)-	EtOH	226s(3.70),268(3.90), 346(3.14)	18-0909-78
$C_{13}H_{12}O_3$			
2-Naphthalenecarboxaldehyde, 6,7-di-methoxy-	MeOH	261(4.68),309(4.25)	83-0328-78
3H-Naphtho[1,8-bc]furan-2-carboxalde-hyde, 4,5-dihydro-6-methoxy-	EtOH	206(4.29),224s(4.01), 306(4.38)	18-2068-78
Propanal, 2-(6-methoxy-2H-1-benzopyran-2-ylidene)-	EtOH	395(4.30)	48-0497-78
Propanal, 2-(8-methoxy-2H-1-benzopyran-2-ylidene)-, (E)-	EtOH	379(4.00)	48-0497-78
2H-Pyran-2-one, 4-(4-methoxyphenyl)-6-methyl-	n.s.g.	233(4.16),307(4.27)	2-0200-78
$C_{13}H_{12}O_4$			
Cyclopenta[b]pyran-6(2H)-one, 5-acet-oxy-7,7a-dihydro-7-(2-propenylidene)-	MeOH	229(4.18),314(4.23)	88-0961-78
geometric isomer	MeOH	276(4.13)	88-0961-78
3-Furancarboxylic acid, 2,5-dihydro-5-methyl-2-oxo-4-phenyl-, methyl ester	MeOH	277(4.10)	1-0665-78
3-Furancarboxylic acid, 4,5-dihydro-4-oxo-2-phenyl-, ethyl ester	EtOH	215(3.92),248(3.65), 294(4.10)	118-0448-78
2(5H)-Furanone, 4-acetyl-3-hydroxy-5-(4-methylphenyl)-	H_2O	320(4.20)	4-0737-78
	EtOH	265(3.92),321(3.79)	4-0737-78
	EtOH-NaHCO$_3$	321(4.20)	4-0737-78
	CHCl$_3$	259(3.98)	4-0737-78
2(5H)-Furanone, 3-acetyl-4-hydroxy-5-(phenylmethyl)-	EtOH	218(4.08),234(3.98), 266(3.83)	4-0327-78
2(5H)-Furanone, 4-acetyl-3-methoxy-5-phenyl-	EtOH	267(4.20)	4-0737-78
$C_{13}H_{12}O_4S$			
4H-1-Benzopyran-4-one, 2-[acetoxy(meth-ylthio)methyl]-	EtOH	222(4.27),295(3.83)	118-0208-78
$C_{13}H_{12}O_5$			
3-Furancarboxylic acid, 2,5-dihydro-5-hydroxy-5-methyl-2-oxo-4-phenyl-, methyl ester	MeOH	280(4.05)	1-0665-78

Compound	Solvent	$\lambda_{max}(\log \epsilon)$	Ref.
2(5H)-Furanone, 4-acetyl-3-hydroxy-5-(4-methoxyphenyl)-	H_2O	318(4.22)	4-0737-78
	EtOH	267(3.83),322(3.96)	4-0737-78
	EtOH-NaHCO$_3$	322(4.17)	4-0737-78
	CHCl$_3$	257(4.08)	4-0737-78
2,4-Pentadienoic acid, 5-(6-methoxy-1,3-benzodioxol-5-yl)-, (E,E)-	EtOH	245(4.05),260(4.03),270(4.01),299(4.03),308(4.02),368(4.18)	5-0573-78
(E,Z)-	n.s.g.	370(4.30)	78-1979-78
	n.s.g.	364(4.17)	78-1979-78
$C_{13}H_{12}O_6$ 1H-2-Benzopyran-4-carboxylic acid, 6,7-dimethoxy-1-oxo-, methyl ester	MeOH	245(3.45),280(2.38),320(2.17)	4-0257-78
$C_{13}H_{12}O_7$ 2H-1-Benzopyran-5-carboxaldehyde, 6-hydroxy-3,4,8-trimethoxy-2-oxo-	EtOH	215(4.35),240(4.11),307(4.27)	102-0583-78
$C_{13}H_{13}BrO$ 1,5-Hexadien-3-one, 1-bromo-5-methyl-1-phenyl-, (Z)-	EtOH	227(3.96),297(4.23)	70-1361-78
$C_{13}H_{13}BrO_3$ 2-Cyclohexen-1-one, 2-bromo-3-hydroxy-5-(4-methoxyphenyl)-	EtOH	293(4.32)	2-0970-78
$C_{13}H_{13}ClN_2O_3$ 1H-Pyrazole-4-carboxylic acid, 3-(4-chlorophenyl)-5-(hydroxymethyl)-, ethyl ester	EtOH	220(4.15),246(4.02)	118-0448-78
$C_{13}H_{13}ClN_4O$ [1,2,4]Triazino[4,3-b]indazole, 8-chloro-3-ethoxy-4-ethyl-	C_6H_{12}	379(3.95)	24-2258-78
$C_{13}H_{13}ClO_2$ Naphthalene, 6-(chloromethyl)-2,3-dimethoxy-	MeOH	238(4.80),268(3.84),279(4.81),313(3.34),320(3.21),325(3.35)	83-0328-78
$C_{13}H_{13}ClO_6$ 1H-2-Benzopyran-1-one, 3-(3-chloro-1,2-dihydroxypropyl)-8-hydroxy-6-methoxy-	MeOH	236s(4.68),244(4.73),257s(4.07),276(3.82),286s(3.65),325(4.07)	44-2339-78
$C_{13}H_{13}Cl_2N_3O_4$ 1H-Indole-2,3-dione, 1-[[bis(2-chloroethyl)amino]methyl]-5-nitro-	EtOH	321(4.00)	111-0515-78
$C_{13}H_{13}Cl_3N_2O$ Acetamide, 2,2,2-trichloro-N-[(2,3-dimethyl-1H-indol-5-yl)methyl]-	MeOH	233(4.68),286(3.92)	103-0856-78
$C_{13}H_{13}F_6NO_4S_2$ Benzene, 1-piperidino-2,4-bis(trifluoromethylsulfonyl)-	dioxan	312(3.97)	104-2208-78
$C_{13}H_{13}IN_2O$ 10H-Pyrido[3,2-b][1,4]benzoxazinium, 1,10-dimethyl-, iodide	MeOH	223(4.55),385(3.85)	95-0585-78

Compound	Solvent	λ_{max}(log ϵ)	Ref.
$C_{13}H_{13}KSi$			
Potassium, (methyldiphenylsilyl)-	THF	255s(--),363(4.03)	120-0052-78
$C_{13}H_{13}N$			
2H-Azirine, 2-(2-butynyl)-2-methyl-3-phenyl-	C_6H_{12}	242(4.15)	44-2029-78
Pyridine, 2,5-dimethyl-6-phenyl-	C_6H_{12}	238(3.86),281(3.70)	44-2029-78
1H-Pyrrole, 1-ethenyl-3-methyl-2-phenyl-	C_6H_{12}	200(4.30),254(4.10), 270(4.13)	104-1614-78
$C_{13}H_{13}NO$			
1H-Benzo[b]cycloprop[d]azocine, 1a,9b-dihydro-4-methoxy-	isooctane	285s(3.20)	44-4712-78
1H-Benzo[b]cycloprop[f]azocine, 1a,9a-dihydro-2-methoxy-, cis	isooctane	243s(3.74),284s(3.20)	44-4712-78
1H-Benzo[d]cycloprop[b]azocine, 1a,9b-dihydro-3-methoxy-, cis	isooctane	241(3.74)	44-4712-78
1H-Benzo[e]cycloprop[c]azocine, 1a,9b-dihydro-2-methoxy-, cis	isooctane	258(3.75)	44-4712-78
Cyclopent[b]indol-2(1H)-one, 3,4-dihydro-3,3-dimethyl-	EtOH	217(4.38),279(3.94)	103-0109-78
Phenol, 2-[(4-aminophenyl)methyl]-	MeOH	230(4.15),280(3.88)	95-0914-78
	MeOH-HCl	279(3.70)	95-0914-78
	MeOH-NaOH	237(4.16),283(3.88)	95-0914-78
$C_{13}H_{13}NOS_2$			
Pyrrolidinium, 1-(4-hydroxy-5-phenyl-1,3-dithiol-2-ylidene)-, hydroxide, inner salt	CH_2Cl_2	230(4.13),246(4.17), 315(4.21),456(3.79)	24-2021-78
$C_{13}H_{13}NO_2$			
1H-1-Benzazepine-1-carboxylic acid, ethyl ester	hexane	204(4.37),228(4.07), 241(4.02),245s(3.97), 289(3.24),306s(3.19)	44-0315-78
1H-Benz[g]indole-2-carboxylic acid, 6,7,8,9-tetrahydro-	EtOH	293(4.03),302(3.98)	103-0518-78
3H-Benz[e]indole-2-carboxylic acid, 6,7,8,9-tetrahydro-	EtOH	225(4.35),298(4.03)	103-0518-78
3H-Cyclobut[b]indole-3-carboxylic acid, 2a,7b-dihydro-, ethyl ester	hexane	208(4.49),212s(4.43), 243s(4.05),247(4.08), 255s(3.99),278s(3.31), 283(3.41),292(3.43)	44-0315-78
Cyclopropanecarboxylic acid, 1-cyano-2-methyl-2-phenyl-	neutral	210(4.0),252(2.8)	44-0626-78
1,4-Naphthalenedione, 2,3-dimethyl-5-(methylamino)-	C_6H_{12}	513(3.48)	150-2319-78
	EtOH	530(--)	150-2319-78
2-Pentenoic acid, 2-cyano-4-phenyl-, methyl ester	acid	210(3.9),241(3.7)	44-0626-78
	neutral	215(3.9)	44-0626-78
	base	210(3.9),245(3.9), 340(3.85)	44-0626-78
2(1H)-Pyridinone, 4-(4-methoxyphenyl)-6-methyl-	EtOH	240(4.20),290(4.26)	2-0200-78
1H-Pyrrol-3-ol, 2-methyl-5-phenyl-, acetate	MeOH	298(4.27)	49-0137-78
$C_{13}H_{13}NO_2S_2$			
1,3-Dithiol-1-ium, 4-hydroxy-2-morpholino-5-phenyl-, hydroxide, inner salt	dioxan	255(4.08),312(4.11), 485(3.35)	24-2021-78
Thiophene, 2-[1-methyl-1-(phenylthio)-ethyl]-5-nitro-	MeOH	243(3.75),323(3.80)	12-2463-78

Compound	Solvent	$\lambda_{max}(\log \epsilon)$	Ref.
$C_{13}H_{13}NO_3$			
1H-Isoindole-1,3(2H)-dione, 2-(4-hydroxy-3-methyl-2-butenyl)-, (E)-	EtOH	221(4.65),241(4.00), 294(3.28)	12-1291-78
7-Isoquinolineacetic acid, 3-methoxy-, methyl ester	MeOH	226(4.85)	44-3778-78
1H-Pyrrole-2-carboxylic acid, 3-hydroxy-1-phenyl-, ethyl ester	EtOH	262(4.10)	94-2224-78
2-Quinolinecarboxylic acid, 1-ethyl-1,4-dihydro-4-oxo-, methyl ester	n.s.g.	216(4.41),242(4.32), 338(4.05)	4-0551-78
$C_{13}H_{13}NO_3S$			
Acetic acid, (3,4-dihydro-3-methyl-4-oxo-2H-1,3-benzothiazin-2-ylidene)-, ethyl ester, (E)-	MeOH	233(4.30),246(4.24), 295(4.39)	5-0473-78
Acetic acid, (4-oxo-5-phenyl-2-thiazolidinylidene)-, ethyl ester	MeOH	234(4.01),284(4.37)	5-0473-78
	pH 13	306(4.51)	5-0473-78
4(1H)-Quinolinone, 3-[(acetoxymethyl)-thio]-1-methyl-	EtOH	254(4.20),298(3.74), 328(4.00),336(4.02)	4-0115-78
2-Thiazolecarboxylic acid, 4-methoxy-5-phenyl-, ethyl ester	MeOH	250(3.87),358(4.18)	5-0473-78
$C_{13}H_{13}NO_4$			
Acetic acid, (5,8-dihydro-3-methoxy-8-oxo-7(6H)-isoquinolinylidene)-, methyl ester	MeOH	241(4.13),252(4.12), 301(4.09)	44-3778-78
6H-1,3-Oxazinium, 4-hydroxy-2-(4-methoxyphenyl)-3,5-dimethyl-6-oxo-, hydroxide, inner salt	MeCN	217(3.95),260(4.19), 282(4.04),353(3.12)	5-1655-78
2,4-Pentadienamide, 5-(6-methoxy-1,3-benzodioxol-5-yl)-, (E,E)-	n.s.g.	368(4.33)	78-1979-78
3-Quinolineacetic acid, α-hydroxy-2-methyl-, methyl ester, 1-oxide	EtOH	239(4.15),322(3.87)	95-0802-78
3-Quinolinecarboxaldehyde, 6,7-dimethoxy-2-methyl-, 1-oxide	EtOH	253(3.70),333(3.88), 348(3.93)	95-0802-78
$C_{13}H_{13}NO_5$			
4-Isoquinolinecarboxylic acid, 1,2-dihydro-6,7-dimethoxy-2-methyl-1-oxo-	MeOH	225(3.57),250(3.83), 295(3.00),335(3.02)	4-0257-78
2H-Pyrano[3,2-c]pyridine-3-carboxylic acid, 5,6-dihydro-6,7-dimethyl-2,5-dioxo-, ethyl ester	MeOH	383(4.24)	49-1075-78
$C_{13}H_{13}N_2O$			
10H-Pyrido[3,2-b][1,4]benzoxazinium, 1,10-dimethyl-, iodide	MeOH	223(4.55),385(3.85)	95-0585-78
$C_{13}H_{13}N_3$			
1H-Perimidin-2-amine, N,N-dimethyl-	EtOH	<u>230(4.6),315(4.0), 330(4.0)</u>	103-0694-78
Propanedinitrile, [3-[1-(dimethylamino)ethylidene]-1,4-cyclopentadienylmethylene]-	MeOH	255(3.80),298(4.07), 441(4.67)	83-0369-78
$C_{13}H_{13}N_3O$			
1H-Azepine-4-carbonitrile, 2,5,6,7-tetrahydro-2-oxo-3-(phenylamino)-	n.s.g.	246(4.07),336(4.03)	103-1013-78
2-Butenenitrile, 4-(methoxyimino)-2-[2-(phenylamino)ethenyl]-	MeOH	391.9(4.42)	83-0433-78
Formamide, N-[2-cyano-1-(1H-indol-3-ylmethyl)ethyl]-, (S)-	MeOH	273(3.87),281(3.89), 289(3.84)	35-0938-78

Compound	Solvent	$\lambda_{max}(\log \epsilon)$	Ref.
2,4-Pentadienenitrile, 2-[(methoxy-imino)methyl]-5-(phenylamino)-	MeOH	391.5(4.58)	83-0433-78
1,2-Propanedione, 1-(1H-pyrrol-2-yl)-, 1-(phenylhydrazone)	EtOH	241(4.05),289(3.66), 372(4.22)	4-1485-78
Pyrazolo[3,4-c]azepin-8(2H)-one, 4,5,6,7-tetrahydro-2-phenyl-	n.s.g.	274(4.19),364(2.56)	103-1009-78
Pyrazolo[5,4-c]azepin-8(2H)-one, 4,5,6,7-tetrahydro-1-phenyl-	n.s.g.	235(4.03)	103-1009-78
Pyrido[2,3-d]pyrimidin-4(1H)-one, 5,6,7,8-tetrahydro-2-phenyl-	MeOH	238(4.16),265(4.12), 295(4.01),312(3.97)	24-2297-78
4H-Pyrimido[2,1-b]quinazolin-4-one, 1,6-dihydro-1,2-dimethyl-	MeOH	218(4.28),290(4.20)	4-0077-78
4H-Pyrimido[2,1-b]quinazolin-4-one, 6,11-dihydro-2,8-dimethyl-	MeOH	258(4.14),314(4.09)	4-0077-78
4H-Pyrimido[2,1-b]quinazolin-4-one, 6,11-dihydro-2,11-dimethyl-	MeOH	255(4.09),310(4.10)	4-0077-78
4H,10H-[1,2,3]Triazolo[5,1-c][1,4]-benzoxazepine, 10-(1-methylethylidene)-	MeOH	287(3.08)	44-0066-78
$C_{13}H_{13}N_3OS$			
Pyrido[2,3-d]pyrimidin-4(1H)-one, 2,3,5,6,7,8-hexahydro-3-phenyl-2-thioxo-	MeOH	218(3.96),289(4.19)	24-2297-78
$C_{13}H_{13}N_3OS_2$			
2H-1,3,5-Thiadiazine-2-thione, 6-mor-pholino-4-phenyl-	CH_2Cl_2	244(4.21),306(4.49), 417(4.06)	48-0647-78
$C_{13}H_{13}N_3O_2$			
Imidazo[2,1-a]phthalazin-3(2H)-one, 6-methoxy-2,2-dimethyl-	EtOH	239(4.17),247(4.13), 259(4.07),269(4.21), 320(3.54)	33-2116-78
	MeCN	229(4.18),237s(4.14), 258(4.07),268(4.23), 322(3.50)	33-2116-78
Pyrido[2,3-d]pyrimidine-2,4(1H,3H)-di-one, 5,6,7,8-tetrahydro-3-phenyl-	MeOH	219(4.12),287(4.21)	24-2297-78
2-Quinoxalinecarboxaldehyde, 3-morpho-lino-	$CHCl_3$	439(2.89)	97-0177-78
$C_{13}H_{13}N_3O_2S$			
10H-Pyrimido[5,4-b][1,4]benzothiazine, 2,4-dimethoxy-10-methyl-	MeOH	210(4.25),248(4.47), 295(3.78)	5-0193-78
	12M HCl	244(4.12),314(3.62), 364(3.53)	5-0193-78
1H-Pyrimido[5,4-b][1,4]benzothiazine-2,4(3H,10H)-dione, 1,3,10-trimethyl-	MeOH	219(4.28),249(4.32), 287(3.98),350s(3.38)	5-0193-78
	12M HCl	251(4.23),291(3.84), 366(3.51)	5-0193-78
1H-Pyrimido[5,4-b][1,4]benzothiazinium, 2,3,4,10-tetrahydro-3,5,10-trimethyl-2,4-dioxo-, hydroxide, inner salt	M HCl	281(4.30),305(4.28), 359(3.79)	5-0193-78
	pH 1	254(4.82),266(4.79)	5-0193-78
	MeOH	227(4.28),257(4.38), 271(4.27),315s(3.52)	5-0193-78
3H-Pyrimido[5,4-b][1,4]benzothiazin-4(10H)-one, 2-methoxy-3,10-dimethyl-	MeOH	256(4.51),320s(3.34), 390s(3.00)	5-0193-78
	12M HCl	260s(4.18),266(4.21), 295s(3.55),369(3.31)	5-0193-78
$C_{13}H_{13}N_3O_3S$			
4(3H)-Pyrimidinone, 6-methyl-2-(methyl-	MeOH	240s(3.96),287(4.08)	4-0077-78

Compound	Solvent	$\lambda_{max}(\log \epsilon)$	Ref.
thio)-3-[(2-nitrophenyl)methyl]- (cont.)			4-0077-78
$C_{13}H_{13}N_3O_4$ 2,4(1H,3H)-Pyrimidinedione, 6-amino- 5-(hydroxyacetyl)-3-methyl-1-phenyl-	EtOH	248(4.03),276(4.11)	103-0443-78
$C_{13}H_{13}N_3S$ Hydrazinecarbothioamide, N,2-diphenyl-	EtOH	238(4.27)	104-0106-78
$C_{13}H_{13}N_5$ Adenine, 3-methyl-9-(phenylmethyl)-, perchlorate	pH 1 and 7	271.5(4.25)	88-5007-78
Adenine, 9-methyl-3-(phenylmethyl)-, perchlorate	pH 1 and 7	272(4.20)	88-5007-78
$C_{13}H_{13}N_5O$ Guanine, 3-methyl-7-(phenylmethyl)-	pH 1	245(3.97),267(4.08)	44-1644-78
	pH 7	248(3.91),268(4.02)	44-1644-78
	pH 12	242(3.96),265(4.06)	44-1644-78
6H-Imidazo[4,5-e]-1,2,4-triazin-6-one, 7-ethyl-5,7-dihydro-5-methyl-3-phen- yl-	EtOH	245(4.15),270(3.88), 309(3.81)	94-3154-78
$C_{13}H_{13}N_5O_4$ 2',3'-O-Ispropylidene-5'-oxo-8,5'-cy- cloadenosine	pH 1.35	261(4.18),311(2.11)	78-2633-78
	pH 5.7	265(4.05),336(3.56)	78-2633-78
	pH 12.1	266(4.20)	78-2633-78
$C_{13}H_{13}N_5S_2$ 1,2-Hydrazinedicarbothioamide, N-phen- yl-N'-(2-pyridinyl)-	EtOH	267(4.42)	104-0181-78
$C_{13}H_{14}BrFO$ 1-Hexen-3-one, 1-bromo-5-fluoro-5-meth- yl-1-phenyl-, (Z)-	EtOH	237(3.86),290(4.10)	70-1361-78
$C_{13}H_{14}Cl_2N_2O_2$ 1H-Indole-2,3-dione, 1-[[bis(2-chloro- ethyl)amino]methyl]-	EtOH	283(3.66)	111-0515-78
$C_{13}H_{14}Fe$ Ferrocene, 1,1'-(1,3-propanediyl)-	CH_2Cl_2	441(2.31)	101-0077-78L
$C_{13}H_{14}I_2O$ Benzene, 1-[(2,3-diiodo-2-propenyl)- oxy]-2-(2-methyl-1-propenyl)-	MeOH	240(3.68),280(3.13)	44-0066-78
$C_{13}H_{14}N_2$ Isoquinoline, 3-pyrrolidino-	EtOH	246(4.60),294(4.19), 303(4.28),393(3.41)	138-0677-78
$C_{13}H_{14}N_2O$ 9H-Carbazole-9-carboxamide, 1,2,3,4- tetrahydro-	EtOH	210(4.18),235(4.26), 270(3.94)	118-0374-78
$C_{13}H_{14}N_2OS$ Morpholine, 4-(4-phenyl-2-thiazolyl)-	EtOH	243(4.41),266s(4.03), 283(3.84)	94-3017-78
$C_{13}H_{14}N_2OS_2$ Carbamodithioic acid, (2-furanylmeth-	n.s.g.	336s(1.87)	97-0381-78

Compound	Solvent	$\lambda_{max}(\log \epsilon)$	Ref.
yl)-, 2-(2-pyridinyl)ethyl ester (cont.)			97-0381-78
$C_{13}H_{14}N_2O_2$			
Pyridinium, 1,1'-(1,3-propanediyl)bis-[3-hydroxy-, dihydroxide, bis(inner salt)	EtOH	223(4.37),258(4.13), 335(3.92)	150-1182-78
$C_{13}H_{14}N_2O_2S$			
1H-[1,2]Oxathiolo[3,2-e][1,2,3]thiadiazole-7-S^{IV}, 1-(4-methoxyphenyl)-3,4-dimethyl-	C_6H_{12}	236(4.02),280(3.90), 453(4.32)	39-0195-78C
4-Oxazolidinone, 5-[[4-(dimethylamino)-phenyl]methylene]-3-methyl-2-thioxo-	n.s.g.	276(4.40),295s(4.19), 455(4.66)	104-1225-78
4(5H)-Oxazolone, 5-[[4-(dimethylamino)-phenyl]methylene]-2-(methylthio)-	n.s.g.	273(4.59)	104-1225-78
$C_{13}H_{14}N_2O_3$			
2(1H)-Phthalazineacetic acid, 4-methyl-1-oxo-, ethyl ester	n.s.g.	255.5(4.439)	2-0689-78
1H-Pyrazole-4-carboxylic acid, 3-(hydroxymethyl)-5-phenyl-, ethyl ester	EtOH	218(3.95),243(4.00)	118-0448-78
3H-Pyrazolo[5,1-c][1,4]benzoxazine-2-carboxylic acid, 3a,4-dihydro-, ethyl ester	EtOH	240(3.70),345(4.05)	44-1664-78
1H-Pyrido[3,2-b]indole-2,9-dione, 5-acetyl-3,4,6,7,8,9-hexahydro-	EtOH	255(4.38),290s(3.82), 330s(2.37)	94-3080-78
$C_{13}H_{14}N_2O_3S$			
Acetic acid, [3-(1-ethynyl-1-cyclohexyl)-4-oxo-2-thioxo-5-imidazolidin-ylidene]-	EtOH	320(3.78)	5-0227-78
2H-1-Benzothiopyran-2-one, 4-(butyl-amino)-3-nitro-	EtOH	235(4.47),261(4.09), 317(4.07)	103-0033-78
2H-1-Benzothiopyran-2-one, 4-(diethyl-amino)-3-nitro-	EtOH	238(4.40),266(3.88), 295(3.87),353(3.87)	103-0033-78
$C_{13}H_{14}N_2O_4$			
Benzoic acid, 3-hydroxy-4-[(2-oxo-cyclohexylidene)hydrazino]-	EtOH	225(4.7),270(4.6)	102-0834-78
1,5-Naphthyridine-3-carboxylic acid, 1,4,5,8-tetrahydro-1,5-dimethyl-4,8-dioxo-, ethyl ester	H_2O	232(4.56),265(3.65), 319(4.36),333(4.36)	44-1331-78
2,4,6(1H,3H,5H)-Pyrimidinetrione, 5-ethyl-1-(phenylmethoxy)-	pH 1.17 pH 6.04	215(4.08) 200(4.94),215(4.14), 271(4.27)	12-2517-78 12-2517-78
$C_{13}H_{14}N_2O_6$			
1H-Indole, 1-α-L-arabino-furanosyl-6-nitro-	EtOH	212(4.12),249(3.72), 318(3.64),358(3.55)	136-0017-78E
1H-Indole, 1-α-D-arabino-pyranosyl-6-nitro-	EtOH	212(4.29),249(3.98), 320(3.92),357(3.86)	136-0017-78E
$C_{13}H_{14}N_4O$			
[1,2,4]Triazino[4,3-b]indazole, 3-ethoxy-4-ethyl-	C_6H_{12}	370(3.85)	24-2258-78
$C_{13}H_{14}N_4O_2$			
1H-Pyrazole-5-carbonitrile, 4,5-dihydro-3-(1-methylethyl)-1-(4-nitro-phenyl)-	EtOH	240(3.998),365(4.359)	146-0864-78

Compound	Solvent	$\lambda_{max}(\log \epsilon)$	Ref.
$C_{13}H_{14}N_4O_2S$			
Benzenesulfonic acid, [1-(4-ethenyl-1H-pyrazol-3-yl)ethylidene]hydrazide	EtOH	262(4.11)	39-1297-78C
Benzenesulfonic acid, 4-methyl-, [(4-ethenyl-1H-pyrazol-3-yl)methylene]-hydrazide, (E)-	EtOH	284(4.41)	39-1297-78C
(Z)-	EtOH	268(4.24)	39-1297-78C
$C_{13}H_{14}N_4O_3S$			
Acetamide, N-[5-(1-methylethyl)-3-(4-nitrophenyl)-1,3,4-thiadiazol-2(3H)-ylidene]-	EtOH	275(4.135),325(4.270)	146-0864-78
Benzenesulfonic acid, 4-[(1,1-dicyano-3-methylbutyl)azo]-, sodium salt	H_2O	280(4.08)	126-2845-78
$C_{13}H_{14}N_4O_5$			
1H-Pyrrole-2-carboxylic acid, 1-methyl-4-[[(1-methyl-4-nitro-1H-pyrrol-2-yl)carbonyl]amino]-, methyl ester	DMF	288(4.39)	78-2389-78
$C_{13}H_{14}N_6O$			
4H,10H-[1,2,3]Triazolo[5,1-c][1,4]benz-oxazepine, 10-(1-azido-1-methylethyl)-	MeOH	265(2.30)	44-0066-78
$C_{13}H_{14}O$			
2-Cyclopenten-1-one, 2,5-dimethyl-3-phenyl-	EtOH	220(3.72),279(4.05)	35-1799-78
1,4-Ethano-7H-benzocyclohepten-7-one, 1,2,3,4-tetrahydro-	hexane	233(4.31),236(4.31), 306(3.98),318(3.96)	88-0569-78
	MeOH	234(4.42),237(4.42), 319(4.16)	88-0569-78
$C_{13}H_{14}OS_2$			
Pentacyclo[6.2.1.0^{2,7}.0^4,^{10}.0^{5,9}]undec-ane-3,6-dione mono(ethylene thioacet-al)	EtOH	253(2.51),300(1.64)	104-2336-78
$C_{13}H_{14}O_2$			
2-Cyclohexen-1-one, 3-(phenylmethoxy)-	EtOH	249(4.32)	78-1567-78
3-Cyclohexen-1-one, 4-(4-methoxyphenyl)-	EtOH	255(4.16)	39-0024-78C
2-Cyclopropene-1-carboxylic acid, 2-methyl-3-(4-methylphenyl)-, methyl ester	EtOH	260(5.08)	104-2144-78
5,8-Methano-2H-indeno[1,2-b]furan-2-one, 3,3a,4,4a,5,8,8a,8b-octa-hydro-3-methylene-	MeOH	208(4.02)	87-0815-78
4a,8a-Propanonaphthalene-9,11-dione, 1,4,5,8-tetrahydro-	EtOH	244(4.16)	88-2187-78
$C_{13}H_{14}O_2Ru$			
Ruthenocene, (2-carboxyethyl)-	EtOH	300(2.86)	18-0909-78
$C_{13}H_{14}O_2S_2$			
5H-1,4-Dithiepin-2-ol, 6,7-dihydro-3-phenyl-, acetate	n.s.g.	226(4.07),296(4.15)	33-3087-78
$C_{13}H_{14}O_3$			
Benzofuran, 5,6-dimethoxy-2-(1-methyl-ethenyl)-	MeOH	216(4.24),279(4.09), 289(4.08),315(4.27), 326s(4.20)	117-0137-78

Compound	Solvent	$\lambda_{max}(\log \epsilon)$	Ref.
4H-1-Benzopyran-4-one, 6-hydroxy-2,5,7,8-tetramethyl-	EtOH	206(4.36),238(4.26), 253s(4.17),330(3.74)	44-3723-78
4a,7-Methano-4aH-benzocycloheptene-4,6(5H,7H)-dione, 8,9-dihydro-1-methoxy-	n.s.g.	342(3.32)	12-1561-78
2-Naphthalenemethanol, 6,7-dimethoxy-	MeOH	234(5.00),266(3.87), 311(3.46),317(3.34), 324(3.63)	83-0328-78
2-Propenoic acid, 3-(2-methyl-3-phenyloxiranyl)-, methyl ester, cis	EtOH	234s(3.84)	24-3665-78
trans	EtOH	234(4.18),288s(2.60)	24-3665-78
$C_{13}H_{14}O_4$			
Ethyne, 1-phenyl-2-α-D-ribofuranosyl-	EtOH	242(4.27),252(4.17)	94-0898-78
Ethyne, 1-phenyl-2-β-D-ribofuranosyl-	EtOH	242(4.34),252(4.25)	94-0898-78
Euparine, 2,3-dihydro-12-hydroxy-	ether	273(4.31),325(4.11)	102-1161-78
2-Naphthalenecarboxylic acid, 1,2,3,4-tetrahydro-6-methoxy-1-oxo-, methyl ester	EtOH	276(4.15)	39-0110-78C
$C_{13}H_{14}O_5$			
2-Propenal, 3-(4-acetoxy-3,5-dimethoxy-phenyl)-	EtOH	322(3.78)	42-1152-78
$C_{13}H_{14}Ru$			
Ruthenocene, 1,1'-(1,3-propanediyl)-	EtOH	265s(2.58),325(2.78)	18-0909-78
$C_{13}H_{15}BF_2N_2O$			
Boron, [2-[(3-ethyl-5-methoxy-4-methyl-2H-pyrrol-2-ylidene)methyl]-1H-pyr-rolato-N^1,N^2]difluoro-, (T-4)-	EtOH	475(4.39)	49-0883-78
$C_{13}H_{15}BrN_2O_3$			
2-Furanmethanaminium, N,N-dimethyl-5-nitro-N-phenyl-, bromide	H_2O	203(4.27),232(3.62), 296(4.05)	73-2041-78
$C_{13}H_{15}BrN_2O_4$			
β-Alanine, N-ethyl-N-nitroso-, 2-(4-bromophenyl)-2-oxoethyl ester	EtOH	256.5(4.21)	94-3909-78
Butanoic acid, 4-(methylnitrosoamino)-, 2-(4-bromophenyl)-2-oxoethyl ester	EtOH	256(4.31)	94-3909-78
$C_{13}H_{15}Br_2N_3$			
3,5-Pyridinedicarbonitrile, 2,6-bis-(bromomethyl)-4,4-diethyl-1,4-di-hydro-	EtOH	210(4.67),250(4.54), 370(4.02)	103-1236-78
3,5-Pyridinedicarbonitrile, 2,6-bis-(bromomethyl)-1,4-dihydro-4-methyl-4-propyl-	EtOH	210(4.55),251(4.53), 350(4.21)	103-1236-78
$C_{13}H_{15}ClN_4O_5$			
Acetamide, N-(7-chloro-3-β-D-ribo-furanosyl-3H-imidazo[4,5-b]pyri-din-5-yl)-	pH 1	253(4.05),257s(4.02), 299(4.28)	4-0839-78
	pH 11	257(4.03),264(4.02), 297(4.24)	4-0839-78
$C_{13}H_{15}IN_4O_5$			
7,11-Epoxy-1H,6H,9H-8,10-dioxa-2,5a,11a-triaza-3a-azoniacyclopenta[5,6]cyclo-oct[1,2,3-cd]indene, 2,3,7,7a,10a,11-hexahydro-9,9-dimethyl-1,3-dioxo-,	pH 2	219(4.26),249(3.80)	44-4774-78

Compound	Solvent	$\lambda_{max}(\log \epsilon)$	Ref.
iodide, [7R-(7α,7aβ,10aβ,11α)]- (cont.)			44-4774-78
$C_{13}H_{15}N$			
3-Azabicyclo[3.1.0]hex-2-ene, 1,4-di-methyl-2-phenyl-, exo	C_6H_{12}	238(4.01)	44-2029-78
3-Azabicyclo[3.1.0]hex-2-ene, 1,6-di-methyl-2-phenyl-, endo	C_6H_{12}	237(4.08)	44-2029-78
exo	C_6H_{12}	236(4.08)	44-2029-78
2H-Azirine, 2,2-dimethyl-3-[2-(2-pro-penyl)phenyl]-	MeOH	248(4.54)	35-2181-78
2H-Azirine, 3-methyl-2-(1-methyl-2-propenyl)-2-phenyl-	C_6H_{12}	225(3.85),256(3.30), 272(2.90)	44-2029-78
1H-Indole, 2-methyl-3-(1-methyl-2-propenyl)-	MeCN	223(4.37),282(3.62), 290(3.56)	44-2029-78
1H-Indole, 2-methyl-3-(2-methyl-2-propenyl)-	C_6H_{12}	222(4.40),273(3.73), 278(3.73),282(3.71), 290(3.60)	44-2029-78
1H-Pyrrole, 2-ethenyl-2,5-dihydro-5-methyl-4-phenyl-	MeCN	236(3.19)	44-2029-78
2H-Pyrrole, 2-ethenyl-3,4-dihydro-4-methyl-5-phenyl-	MeOH	243(4.12)	44-2029-78
2H-Pyrrole, 2-ethenyl-3,4-dihydro-5-methyl-4-phenyl-	MeCN	245(2.90)	44-2029-78
$C_{13}H_{15}NO$			
2H-Azirine, 2,2-dimethyl-3-[2-(2-pro-penyloxy)phenyl]-	MeOH	250(4.02),310(3.71)	35-3494-78
3H-1-Benzazonine, 4,7-dihydro-2-meth-oxy-	isooctane	229(3.74),276(3.08)	44-4712-78
Benzenepropanal, 2-(2,2-dimethyl-2H-azirin-3-yl)-	MeOH	245(4.59),275(4.21)	35-2181-78
Cyclohexanone, 2-[(phenylamino)meth-ylene]-	EtOH	358(4.41)	97-0256-78
Naphth[1,2-d]oxazole, 2,3a,4,5-tetra-hydro-2,2-dimethyl-	MeOH	240(4.10),285(3.62)	35-2181-78
2-Propyn-1-one, 3-(diethylamino)-1-phenyl-	MeCN	248(4.02),319(4.20)	33-1609-78
2H-Pyrrol-2-one, 1,3-dihydro-3,3,5-trimethyl-4-phenyl-	EtOH	269(3.78)	78-3291-78
2H-Pyrrol-2-one, 1,5-dihydro-3,5,5-trimethyl-4-phenyl-	EtOH	238(3.94)	78-3291-78
$C_{13}H_{15}NO_2$			
2-Azabicyclo[5.4.0]undeca-3,5,8,10-tetraene-N-carboxylic acid, ethyl ester, cis	hexane	248(4.31),267s(4.21)	44-0315-78
trans	hexane	227(3.40),254(3.39)	44-0315-78
9-Azatricyclo[6.3.0.0^{2,7}]undeca-3,5,10-triene-N-carboxylic acid, ethyl ester, cis-trans-cis-trans	hexane	238(4.28),282(3.24)	44-0315-78
1H-1-Benzazepine-1-carboxylic acid, 2,3-dihydro-, ethyl ester	hexane	227(4.36),252(3.96), 285(2.97),294s(2.83)	44-0315-78
Benzene, [3-(nitromethyl)-1-hexynyl]-	EtOH	240(4.20),251(4.13)	104-0676-78
3H-Cyclobut[b]indole-3-carboxylic acid, 2a,3a,7a,7b-tetrahydro-, ethyl ester	hexane	244(3.64),251(3.63), 261s(3.58),271s(3.42)	44-0315-78
2-Propen-1-one, 3-morpholino-1-phenyl-	C_6H_{12}	327(4.34)	48-0945-78
	EtOH	346(4.40)	48-0945-78
2(1H)-Pyridinone, 5,6-dihydro-4-hy-droxy-6-methyl-1-(4-methylphenyl)-	EtOH	220(3.93),295(4.30)	4-1153-78
2(1H)-Pyridinone, 5,6-dihydro-4-hy-droxy-6-methyl-1-(phenylmethyl)-	EtOH	270(3.72)	4-1153-78

Compound	Solvent	$\lambda_{max}(\log \epsilon)$	Ref.
2(1H)-Pyridinone, 1-(5,5-dimethyl-3-oxo-1-cyclohexen-1-yl)-	EtOH	315(3.74)	78-2609-78
1H-Pyrrolo[1,2-a]indole-5,8-dione, 2,3,6,7-tetrahydro-6,9-dimethyl-	EtOH	222(4.07),236(4.08), 282(3.90),333(4.03)	44-4472-78
$C_{13}H_{15}NO_3$			
Cyclopentanone, 2-(2-nitro-1-phenyl-ethyl)-	EtOH	211(3.99)	2-0405-78
Pipecolic acid, N-benzoyl-	EtOH	212(3.90)	39-1157-78B
2-Propenoic acid, 3-(4-morpholinomethyl)-	EtOH	209(4.35),340(4.20)	104-1910-78
$C_{13}H_{15}NO_4$			
Acetic acid, (5,8-dihydro-8-hydroxy-3-methoxy-7(6H)-isoquinolinylidene)-	MeOH	220(4.36),272(3.56)	44-3778-78
$C_{13}H_{15}NO_6$			
2,4,5-Pyridinetricarboxylic acid, 6-methyl-, 4,5-diethyl ester	EtOH	235(3.77),288(5.75)	106-0782-78
$C_{13}H_{15}NO_6S$			
2-Thiabicyclo[3.2.0]hepta-3,6-diene-4,6,7-tricarboxylic acid, 3-amino-, 6,7-dimethyl 4-ethyl ester	MeOH	219(4.20),297(4.09)	24-0770-78
$C_{13}H_{15}NO_8$			
α-D-Xylofuranoside, methyl 2-(4-nitro-benzoate)	EtOH	278(4.14)	136-0257-78H
$C_{13}H_{15}NS_2$			
Benzenamine, N-(hexahydro-1,3-benzodi-thiol-2-ylidene)-, trans	EtOH	238(4.04),280(3.78)	39-1208-78C
Cyclopentanecarbodithioic acid, 2-(phenylimino)-, methyl ester	EtOH	252(4.14),325(4.10), 412(4.43)	39-0558-78C
$C_{13}H_{15}N_2O_2$			
1-Pyrazolidinyl, 4,4-diethyl-3,5-dioxo-2-phenyl-	benzene	355(1.63),487(1.18) (anom.)	44-0808-78
$C_{13}H_{15}N_2S$			
Methanaminium, N-[3-isothiocyanato-3-(4-methylphenyl)-2-propenyli-dene]-N-methyl-, perchlorate	MeCN	229(4.04),245(4.03), 328s(4.20),364(4.34)	97-0334-78
$C_{13}H_{15}N_3O$			
3H-Pyrazol-3-one, 4-[(dimethylamino)-methylene]-2,4-dihydro-5-methyl-2-phenyl-	$C_6H_{11}Me$	262(4.13),300(4.33), 340(3.38)	48-0521-78
	PrOH	263(4.22),298(4.37), 333(3.91)	48-0521-78
	PrCN	271s(4.14),295(4.31), 330(3.71)	48-0521-78
$C_{13}H_{15}N_3O_2$			
Acetamide, N-[2-(2-cyanoethyl)-4,5,6,7-tetrahydro-4-oxo-1H-indol-3-yl]-	MeOH	246(3.91),280s(3.67)	94-3080-78
1H-Imidazole-4-carboxylic acid, 5-ami-no-1-(phenylmethyl)-, ethyl ester	EtOH	212(4.17),227s(--), 292(4.24)	118-0741B-78
1H-Indole-3-propanamide, α-(acetyl-amino)-	n.s.g.	280(3.74)	63-0813-78
7,8-Isoquinolinedione, 3,5-bis(diethyl-amino)-	EtOH	212(4.22),246(4.28), 289(3.92),329(3.96)	4-0569-78

Compound	Solvent	$\lambda_{max}(\log \epsilon)$	Ref.
3-Isoxazolamine, N-(1,1-dimethylethyl)- 4-nitroso-5-phenyl-	EtOH	760(1.65)	103-0503-78
$C_{13}H_{15}N_3O_2S$ 1,2-Diaza-5-azoniaspiro[4.5]deca-1,3- diene, 4-hydroxy-3-(phenylthio)-, hydroxide, inner salt, 2-oxide	CH_2Cl_2	236(4.26),328(3.77)	88-2331-78
$C_{13}H_{15}N_3O_5$ 1,3,5-Triazine-1(2H)-acetic acid, tetrahydro-2-(2-hydroxyphenyl)- 3,5-dimethyl-4,6-dioxo-	EtOH	216s(3.95),278(3.51), 283s(3.49)	4-1193-78
$C_{13}H_{15}N_3O_6$ 4,2'-Anhydro-5-(3,5-di-0-acetyl-β-D- arabinofuranosyl)isocytosine, hydro- chloride 2(5H)-Furanone, 3,4,5-trimethoxy-, (4-nitrophenyl)hydrazone	pH 1 pH 7 pH 10 dioxan	277(3.58) 285(3.76) 285(3.74) 275(4.01),394(4.47)	87-0096-78 87-0096-78 87-0096-78 33-1033-78
$C_{13}H_{15}N_3S$ 1H-Imidazol-1-amine, 2-(methylthio)- 4-phenyl-N-propylidene-	MeOH	276(4.15)	18-1846-78
$C_{13}H_{15}N_4O_5$ N^8,5'-Anhydro-1-(2,3-0-isopropylidene- β-D-ribofuranosyl)imidazo[1,2-a]- 1,3,5-triazine-2,4(1H,3H)-dione, iodide	pH 2	219(4.26),249(3.80)	44-4774-78
$C_{13}H_{15}N_5O_3$ Benzoic acid, 2-[[[2-(2,5-dihydro- 5-oxo-1H-pyrazol-3-yl)ethyl]amino]- carbonyl]-, hydrazide, monohydro- chloride	MeOH	219(4.43),249(4.01), 299(3.83)	1-0056-78
$C_{13}H_{15}N_5O_4$ 8,5'-Imino-9-(5-deoxy-2,3-0-isopropyli- dene-β-D-ribofuranosyl)hypoxanthine 2',3'-0-Isopropylidene-5'-deoxy-8,5'- cycloguanosine	MeOH 0.5M HCl H_2O 0.5M NaOH	261(4.29),288s(3.96) 255(4.14),280s(3.96) 253(4.17),272s(4.02) 260s(4.09),270(4.11)	44-2320-78 78-2633-78 78-2633-78 78-2633-78
$C_{13}H_{15}N_5O_5$ 2',3'-0-Isopropylidene-8,5'-cycloguano- sine, (S)-	0.5M HCl H_2O 0.5M NaOH	258(4.13),276s(3.95) 257(4.20),270s(4.08) 261s(4.10),271(4.13)	78-2633-78 78-2633-78 78-2633-78
$C_{13}H_{15}N_7S_2$ Hydrazinecarbothioamide, 2,2'-[1-(4-cy- anophenyl)-1,2-ethanediylidene]bis- [N-methyl- copper chelate	EtOH EtOH	336(4.24) 322(4.41),463(3.72)	87-0804-78 87-0804-78
$C_{13}H_{15}O_3PS$ 3-Phosphorinanecarboxylic acid, 4-oxo- 1-phenyl-, methyl ester, 1-sulfide	MeOH CD_3COCD_3	245(3.85) 245(3.90)	70-1339-78 70-1339-78
$C_{13}H_{15}O_3PSe$ 3-Phosphorinanecarboxylic acid, 4-oxo- 1-phenyl-, methyl ester, 1-selenide	CD_3COCD_3	245(3.90)	70-1339-78

Compound	Solvent	$\lambda_{max}(\log \epsilon)$	Ref.
$C_{13}H_{16}$			
Benzene, 1-(1,1-dimethylethyl)-4-(1-propynyl)-	EtOH	244(4.09),255(4.10)	104-2144-78
Benzene, (1-ethylidene-4-pentenyl)-	EtOH	240(4.03)	35-0883-78
Benzene, (2-methyl-1-methylene-4-pentenyl)-	EtOH	233(4.00)	35-0883-78
5H-Benzocycloheptene, 4a,9a-dihydro-4a,6-dimethyl-	C_6H_{12}	236(4.00),257s(3.82)	89-0943-78
Bicyclo[4.4.1]undeca-2,4,7,9-tetraene, 1,6-dimethyl-	hexane	245(3.62),255(3.70),265(3.56)	89-0943-78
1,5-Hexadiene, 2-(4-methylphenyl)-3,3-d₂ has same spectrum)	EtOH	243(4.04)	35-0883-78
5,10-Methanobenzocyclooctene, 5,6,7,8,9,10-hexahydro-	EtOH	260(2.89),270(3.09),273(3.17)	35-2959-78
$C_{13}H_{16}BrN_5$			
1H-Pyrazol-5-amine, 4-[(2-bromo-4-methylphenyl)azo]-1-ethyl-3-methyl-	CHCl₃	385(4.18),420(4.02)	103-0885-78
$C_{13}H_{16}Br_2O_4$			
1,3-Benzodioxol-5(7aH)-one, 6-(2,3-dibromo-3-methylbutyl)-7a-methoxy-	EtOH	238(3.96),297(3.58)	44-3983-78
$C_{13}H_{16}ClN_3O$			
1H-Benzotriazole, 5-chloro-1-(cycloheptyloxy)-	EtOH	273(4.06)	4-1043-78
$C_{13}H_{16}ClN_5O_6$			
Carbamic acid, (7-chloro-2-β-D-ribofuranosyl-2H-1,2,3-triazolo[4,5-b]-pyridin-5-yl)-, ethyl ester	pH 1	262(3.86),310(3.94)	44-4910-78
	pH 11	260(3.94),311(3.91)	44-4910-78
Carbamic acid, (7-chloro-3-β-D-ribofuranosyl-3H-1,2,3-triazolo[4,5-b]-pyridin-5-yl)-, ethyl ester	pH 1	265(4.05),274(4.04),302(4.32),310s(4.23)	44-4910-78
	pH 11	266(4.00),274(3.99),303(4.27),310s(4.25)	44-4910-78
$C_{13}H_{16}N_2O$			
2H-Pyrrol-2-one, 4-ethyl-1,5-dihydro-3-methyl-5-[(1-methyl-1H-pyrrol-2-yl)methylene]-, (Z)-	EtOH	235(3.71),260(3.66),383(4.37),392s(4.37)	49-0183-78
4(3H)-Quinazolinone, 3,6-dimethyl-2-(1-methylethyl)-	MeOH	225(4.49),270(4.01),312(3.64)	2-1067-78
4(3H)-Quinazolinone, 3,7-dimethyl-2-(1-methylethyl)-	MeOH	230(4.60),270(3.93),302(3.65)	2-1067-78
4(3H)-Quinazolinone, 3,8-dimethyl-2-(1-methylethyl)-	MeOH	230(4.62),270(4.17),310(3.90)	2-1067-78
$C_{13}H_{16}N_2OS$			
2H-1-Benzothiopyran-2-one, 3-amino-4-(butylamino)-	EtOH	237(4.40),257(4.01),278(4.00),371(4.02)	103-0033-78
2H-1-Benzothiopyran-2-one, 3-amino-4-(diethylamino)-	EtOH	227(4.47),273(3.99),364(3.97)	103-0033-78
2H-1-Benzothiopyran-2-one, 3-amino-4-[(1,1-dimethylethyl)amino]-	EtOH	235(4.54),277(4.15),373(4.06)	103-0033-78
$C_{13}H_{16}N_2O_2$			
Propanoic acid, 2-[(5,6,7,8-tetrahydro-1-naphthalenyl)hydrazono]-, anti	EtOH	230(4.11),318s(4.08)	103-0518-78
syn	EtOH	240(3.95),340(4.09)	103-0518-78
Propanoic acid, 2-[(5,6,7,8-tetrahydro-2-naphthalenyl)hydrazono]-, anti	EtOH	230(3.95),330(4.21)	103-0518-78

Compound	Solvent	$\lambda_{max}(\log \epsilon)$	Ref.
syn (cont.)	EtOH	245s(3.95),344(4.21)	103-0518-78
1H-Pyrido[3,2-b]indole-2,9-dione, 5-ethyl-3,4,5,6,7,8-hexahydro-	EtOH	213(4.18),252(4.21), 312(3.51)	94-3080-78
1H-Pyrido[3,2-b]indole-2,9-dione, 3,4,5,6,7,8-hexahydro-7,7-dimethyl-	EtOH	250(4.13),313(3.51)	94-3080-78
4(1H)-Quinazolinone, 6-ethoxy-2-(1-methylethyl)-	MeOH	225(4.53),270(4.16), 323(3.76)	2-1067-78
4(1H)-Quinazolinone, 7-ethoxy-2-(1-methylethyl)-	MeOH	240(4.70),295(3.63), 307(3.64)	2-1067-78
4(1H)-Quinazolinone, 8-ethoxy-2-(1-methylethyl)-	MeOH	236(4.44),275(4.03), 316(3.90)	2-1067-78
4(3H)-Quinazolinone, 6-methoxy-3-methyl-2-(1-methylethyl)-	MeOH	226(4.63),273s(4.17), 325(3.76)	2-1067-78
4(3H)-Quinazolinone, 7-methoxy-3-methyl-2-(1-methylethyl)-	MeOH	242(4.71),297(3.48), 310(3.36)	2-1067-78
4(3H)-Quinazolinone, 8-methoxy-3-methyl-2-(1-methylethyl)-	EtOH	235(4.28),285s(3.88), 317(3.80)	2-1067-78
$C_{13}H_{16}N_2O_2S$			
Benzenecarbothioamide, N-cyclohexyl-4-nitro-	CH_2Cl_2	260(4.30),301s(4.01), 347s(3.63),406s(3.09)	48-0313-78
$C_{13}H_{16}N_2O_3$			
2H-Indol-2-one, 3-[2-(dimethylamino)-1-hydroxyethylidene]-1,3-dihydro-5-methoxy-	EtOH	282(4.19),320(4.00)	42-1122-78
$C_{13}H_{16}N_2O_4$			
Acetamide, N-[3-[2-(formylamino)-5-methoxyphenyl]-3-oxopropyl]-	EtOH	237(4.40),267(3.98), 350(3.65)	142-0385-78A
Butanedioic acid, (methylphenylhydrazono)-, dimethyl ester	MeOH	236(4.00),292s(3.85), 321(3.93)	24-0780-78
Pyrano[2,3-c]pyrazol-6(1H)-one, 5-(2-acetoxyethyl)-1,3,4-trimethyl-	EtOH	227(3.67),316(4.08)	95-0335-78
Pyrano[2,3-c]pyrazol-6(2H)-one, 5-(2-acetoxyethyl)-2,3,4-trimethyl-	EtOH	226(3.45),308(4.19)	95-0335-78
$C_{13}H_{16}N_2O_4S$			
1H-Benzimidazole, 2-(methylthio)-1-β-D-ribofuranosyl-	MeOH	214(4.47),251(3.95), 257s(3.93),283(4.07), 291(4.09)	24-0996-78
$C_{13}H_{16}N_2O_5$			
1H-Indole-1-ethanol, β-(2-hydroxy-1-methylethoxy)-6-nitro-, (R)-	EtOH	252(3.98),313(3.82), 360(3.76)	136-0017-78E
(S)-	EtOH	250(3.96),320(3.88), 365(3.68)	136-0017-78E
1,8-Naphthyridine-4,4a(2H)-dicarboxylic acid, 1,5,6,7-tetrahydro-2-oxo-, 4a-ethyl 4-methyl ester	MeOH	214(4.06),262(4.02)	24-2813-78
3,4-Pyridinedicarboxylic acid, 6-(aminocarbonyl)-2-methyl-, diethyl ester	EtOH	233(4.17),273s(3.42), 280s(3.35)	106-0782-78
3,4-Pyridinedicarboxylic acid, 6-[(hydroxyimino)methyl]-2-methyl-, diethyl ester	EtOH	257(5.00),307(4.97)	106-0782-78
$C_{13}H_{16}N_2O_6$			
1H-Indole, 1-α-D-arabinopyranosyl-2,3-dihydro-6-nitro-	EtOH	252(4.12)	136-0017-78E
1,3-Propanediol, 2-[2-hydroxy-1-(6-nitro-1H-indol-1-yl)ethoxy]-, (R)-	EtOH	250(3.93),321(3.85), 365(3.76)	136-0017-78E

Compound	Solvent	$\lambda_{max}(\log \epsilon)$	Ref.
$C_{13}H_{16}N_2O_8$			
2-Propenoic acid, 3-(1,2,3,6-tetrahydro-2,4-dioxo-1-β-D-ribofuranosyl-5-pyrimidinyl)-, methyl ester, (E)-	H_2O	300(4.30)	35-8106-78
$C_{13}H_{16}N_2S$			
Methanamine, N-[2-(1,1-dimethylethyl)-4H-3,1-benzothiazin-4-ylidene]-	EtOH	<u>250(4.2),260(4.2),</u> <u>319(3.9)</u>	139-0209-78B
$C_{13}H_{16}N_2Se$			
Benzenamine, N-(3,4,4-trimethyl-5-methylene-2-selenazolidinylidene)-	EtOH	205(3.61),237(3.43)	103-0738-78
2-Selenazolamine, 4,5-dihydro-4,4-dimethyl-5-methylene-N-(phenylmethyl)-	hexane	201(3.72),248(3.53), 308(3.20)	103-0738-78
	EtOH	205(3.56),238(3.43), 299(3.12)	103-0738-78
2-Selenazolamine, 4-ethyl-4,5-dihydro-4-methyl-5-methylene-N-phenyl-	hexane	202(3.53),247(3.44)	103-0738-78
	EtOH	204(3.53),246(3.42)	103-0738-78
$C_{13}H_{16}N_4O$			
Acetaldehyde, (4-ethoxy-1-phthalazinyl)methylhydrazone	dioxan	245s(4.15),315(3.93)	103-0436-78
$C_{13}H_{16}N_4O_2S$			
Benzenesulfonic acid, 4-methyl-, [(4-ethyl-1H-pyrazol-3-yl)methylene]hydrazide	EtOH	273(4.13)	39-1297-78C
$C_{13}H_{16}N_4O_3$			
[4,4'-Bipyrimidin]-2(1H)-one, 2',6'-diethoxy-5-methyl-	pH 1	276(3.62),336(3.82)	44-0511-78
	pH 8.9	272s(3.60),321(3.81)	44-0511-78
	pH 13	268(3.73),316(3.76)	44-0511-78
[4,4'-Bipyrimidin]-2(1H)-one, 2',6'-diethoxy-5'-methyl-	pH 0.9	306(3.89)	44-0511-78
	pH 8.8	310(3.93)	44-0511-78
	pH 12.9	275(3.82),300(3.90)	44-0511-78
$C_{13}H_{16}N_4O_4$			
Pyrazolidine, 2-(2,4-dinitrophenyl)-3-ethenyl-1,3-dimethyl-	MeCN	240(4.20),394(4.21)	78-0903-78
Pyrazolidine, 2-(2,4-dinitrophenyl)-1-methyl-3-(1-propenyl)-	MeCN	230(4.34),393(4.33)	78-0903-78
$C_{13}H_{16}N_4O_5$			
4H-Imidazo[4,5-d]pyridazin-4-one, 3,5-dihydro-3-[2,3-O-(1-methylethylidene)-β-D-ribofuranosyl]-	pH 1 or 7	258(3.71)	4-0001-78
	pH 11	260(3.76)	4-0001-78
6H-Purin-6-one, 1,7-dihydro-1-[2,3-O-(1-methylethylidene)-α-D-ribofuranosyl]-	pH 1	249(3.99)	4-0929-78
	pH 7	250(4.02)	4-0929-78
	pH 13	250(4.03)	4-0929-78
β-anomer	pH 1	249(3.98)	4-0929-78
	pH 7	249(4.00)	4-0929-78
	pH 13	260(4.02)	4-0929-78
$C_{13}H_{16}N_4O_6$			
Imidazo[1,2-a]-1,3,5-triazine-2,4-(1H,3H)-dione, 1-[2,3-O-(1-methylethylidene)-β-D-ribofuranosyl]-	pH 1	234(3.76)	44-4774-78
Imidazo[1,2-a]-1,3,5-triazine-2,4-(3H,8H)-dione, 8-[2,3-O-(1-methylethylidene)-α-D-ribofuranosyl]-	pH 1	234(4.04),256(4.03)	44-4784-78

Compound	Solvent	$\lambda_{max}(\log \epsilon)$	Ref.
$C_{13}H_{16}N_6$			
1H-Cyclohepta[1,2-d:3,4-d']diimidazole-2,8-diamine, N^8,N^8,1,4-tetramethyl-	n.s.g.	226(4.02),261(4.02), 306(4.59),317s(4.49), 377(4.00),416(4.18)	88-2235-78
$C_{13}H_{16}N_6O_2S_2$			
Benzoic acid, 4-[1,2-bis[[(methylamino)thioxomethyl]hydrazono]ethyl]-	EtOH	340(4.39)	87-0804-78
copper chelate	EtOH	315(4.21),501(3.43)	87-0804-78
$C_{13}H_{16}N_6O_3$			
4,13-Epoxy-1,3-dioxolo[6,7][1,3]diazocino[1,2-e]purin-8-amine, 3a,4,5,6-13,13a-hexahydro-2,2-dimethyl-, [3aR-(3aα,4β,13β,13aα)]-	MeOH	211(4.50),272(4.33)	44-2320-78
$C_{13}H_{16}N_6O_4$			
8,5'-Aminimino-9-(5-deoxy-2,3-O-isopropylidene-β-D-ribofuranosyl)hypoxanthine-1,4-phthalazinedione complex	MeOH	260(4.28),283s(4.03)	44-2320-78
$C_{13}H_{16}O$			
Benzene, 1-methoxy-4-(1-methylene-4-pentenyl)-	EtOH	272(4.21)	35-0883-78
3,3-d$_2$ compd. has same spectrum			
4-Penten-2-one, 3,3-dimethyl-5-phenyl-, (E)-	C_6H_{12}	253(4.26),284(3.42), 293(3.35)	39-0155-78B
$C_{13}H_{16}OS$			
2-Buten-1-one, 1-phenyl-3-(propylthio)-, (E)-	EtOH	252(4.00),320(4.38)	139-0191-78B
(Z)-	EtOH	214(4.04),253(3.92), 333(4.28)	139-0191-78B
Spiro[6H-cyclohepta[b]thiophene-6,1'-cyclopentan]-4(5H)-one, 7,8-dihydro-	n.s.g.	255(3.9)	42-0847-78
$C_{13}H_{16}OS_2$			
Tetracyclo[6.3.0.0^{5,6}.0^{9,10}]undecane-3,8-dione mono(ethylene thioacetal)	EtOH	250(2.49),300(1.70)	104-2336-78
$C_{13}H_{16}O_2$			
3-Cyclohexen-1-ol, 4-(4-methoxyphenyl)-	EtOH	253(4.18),288(3.26)	39-0024-78C
$C_{13}H_{16}O_3$			
Cyclohexanone, 4-hydroxy-4-(4-methoxyphenyl)-	EtOH	224(4.06),274(3.26)	39-0024-78C
Cyclopropanecarboxylic acid, 2-(4-methoxyphenyl)-, ethyl ester	EtOH	232(4.17),280(3.22), 282(3.22),289s(3.08)	44-4316-78
2,4,6,8-Decatetraenoic acid, 4,9-dimethyl-10-oxo-, methyl ester	hexane	333(4.67),347(4.89), 366(4.88)	39-1511-78C
$C_{13}H_{16}O_3S$			
Cyclohexaneacetic acid, 1-(2-thienylcarbonyl)-	n.s.g.	262(3.8)	42-0847-78
$C_{13}H_{16}O_4$			
1,3-Butanedione, 1-(2,4-dimethoxy-6-methylphenyl)-	EtOH	222(3.92),282(4.04)	18-0842-78
$C_{13}H_{16}O_5$			
1H-2-Benzopyran-1-one, 3,4-dihydro-	EtOH	319(3.55)	102-0511-78

Compound	Solvent	$\lambda_{max}(\log \epsilon)$	Ref.
8-hydroxy-6,7-dimethoxy-3,5-dimethyl-, (R)- (cont.)			102-0511-78
$C_{13}H_{17}BrO$			
2,4,6-Cycloheptatrien-1-one, 4-bromo-2,7-bis(1-methylethyl)-	MeOH	239(4.30),319(3.85)	35-1778-78
$C_{13}H_{17}NO$			
2-Propen-1-one, 3-(diethylamino)-1-phenyl-, (E)-	C_6H_{12}	327(4.36)	48-0945-78
	EtOH	348(4.41)	48-0945-78
$C_{13}H_{17}NOS$			
Ethanol, 2-[(2-propyl-2H-isoindol-1-yl)thio]-	isooctane	330(3.92),345(--)	44-2886-78
	EtOH	332(3.88),345s(--)	44-2886-78
$C_{13}H_{17}NO_2$			
1H-1-Benzazepine-1-carboxylic acid, 2,3,4,5-tetrahydro-, ethyl ester	hexane	204(4.17),229(3.59), 261(2.51)	44-0315-78
Isoquinoline, 6,7-diethoxy-3,4-dihydro-hydrochloride	EtOH	234(4.27),285(3.85), 312(3.79)	114-0045-78A
	H_2O	246(3.92),307(3.67), 353(3.61)	114-0045-78A
Phenol, 4-(dimethylamino)-2-(2-propenyl)-, acetate, hydrochloride	MeOH	256(3.79),309(2.95)	94-2027-78
Phenol, 2-methoxy-4-[2-(2-pyrrolidinyl)ethenyl]-, (E)-(±)-	EtOH	213(4.12),267(4.07), 290s(3.80)	33-1200-78
	EtOH-NaOH	212(4.12),290(4.06), 314(4.05)	33-1200-78
$C_{13}H_{17}NO_3$			
Ethanone, 1-(4-hydroxy-3-methoxyphenyl)-2-(2-pyrrolidinyl)-	EtOH	212(4.08),235(4.10), 282(3.97),309(3.90), 353s(3.37)	33-1200-78
	EtOH-NaOH	212(4.14),252(3.91), 302s(3.55),352(4.34)	33-1200-78
1H-Indole-2-carboxylic acid, 4,5,6,7-tetrahydro-1,3-dimethyl-4-oxo-, ethyl ester	MeOH	235(4.59),260s(4.26), 284(4.27)	24-1780-78
Isoleucine, N-benzoyl-	EtOH	232(3.88)	39-1157-78B
1(2H)-Isoquinolinone, 7-ethoxy-3,4-dihydro-6-methoxy-2-methyl-	MeOH	208s(4.43),220(4.52), 261(3.91),270(3.86), 296(3.78)	100-0271-78
Leucine, N-benzoyl-	EtOH	235(3.88)	39-1157-78B
Norvaline, N-benzoyl-N-methyl-	EtOH	209(3.91)	39-1157-78B
Valine, N-benzoyl-, methyl ester	EtOH	236(3.93)	39-1157-78B
$C_{13}H_{17}NO_4$			
Carbamic acid, [2-(acetyloxy)phenyl]-, butyl ester	hexane	238(4.26)	44-0775-78
Carbonic acid, 2-(acetylamino)phenyl-, butyl ester	hexane	233(4.20)	44-0775-78
$C_{13}H_{17}NO_6S$			
Benzoic acid, 2-[[(methoxycarbonyl)-amino]sulfonyl]-, 1,1-dimethylethyl ester	EtOH	223(3.91),271(3.17), 278(3.17)	40-0582-78
$C_{13}H_{17}NS$			
2H-Isoindole-1-ethanethiol, 2-propyl-	n.s.g.	332(3.87),345s(--)	44-2886-78
2H-Isoindole, 1-(ethylthio)-2-propyl-	isooctane	318s(--),331(3.93), 346(--)	44-2886-78

Compound	Solvent	$\lambda_{max}(\log \epsilon)$	Ref.
(cont.)	EtOH	333(3.88),345s(--)	44-2886-78
$C_{13}H_{17}N_3$			
1H-Benzotriazole, 1-cycloheptyl-	n.s.g.	255(4.39),278(4.44), 362(4.58)	4-0127-78
2H-Benzotriazole, 2-cycloheptyl-	n.s.g.	273(4.11),278(4.15), 286(4.15)	4-0127-78
$C_{13}H_{17}N_3O$			
1,3,4-Oxadiazole-2-methanamine, N,N,α,α-tetramethyl-5-phenyl-	EtOH	251(4.26)	33-2419-78
3-Quinolinecarbonitrile, 1,2,5,6,7,8-hexahydro-2-imino-1-(2-methoxyethyl)-	EtOH	223(4.00),258(3.97), 399(3.77)	88-3469-78
$C_{13}H_{17}N_3O_2$			
3,5-Pyridinedicarbonitrile, 1,4-dihydro-2,6-bis(methoxymethyl)-4,4-dimethyl-	EtOH	218(4.91),277(4.50), 350(4.43)	103-1236-78
$C_{13}H_{17}N_3O_4$			
Pyrido[3,2-c]pyridazine-1,2-dicarboxylic acid, 3,4-dihydro-, diethyl ester	EtOH	233(3.9),269(3.58)	88-2731-78
1H-Pyrido[1,2-c][1,2,3]triazine-1,2(3H)-dicarboxylic acid, diethyl ester	EtOH	211(4.17),260(3.5)	88-2731-78
$C_{13}H_{17}N_3O_4S$			
4H-Pyrazolo[1,5-a]benzimidazolium, 1,2,4-trimethyl-, methyl sulfate	2N H_2SO_4 95% EtOH	294(4.11),304(4.28) 222(4.70),294(4.15), 305(4.30)	4-0715-78 4-0715-78
$C_{13}H_{17}N_3O_5$			
1,2-Hydrazinedicarboxylic acid, 1-[2-ethoxy-2-(2-pyridinyl)ethenyl]-, dimethyl ester, cis	MeOH	280(3.82)	103-0534-78
trans	MeOH	280(3.82)	103-0534-78
$C_{13}H_{17}N_3O_7$			
4-Pyrimidinecarboxamide, 1,2,3,6-tetrahydro-3-[2,3-O-(1-methylethylidene)-β-D-ribofuranosyl]-2,6-dioxo-	pH 6.4 pH 12	267(3.94) 265(3.82)	94-2657-78 94-2657-78
$C_{13}H_{17}N_4O_8P$			
2,4(1H,3H)-Pteridinedione, 1-(2-deoxy-5-O-phosphono-β-D-erythro-pentofuranosyl)-6,7-dimethyl-	pH 1.0	230s(4.093),245s(3.982), 325(3.969)	5-1780-78
	pH 7.0	230s(4.072),245s(3.987), 324(3.964)	5-1780-78
	pH 13.0	241(4.193),273(3.532), 332(3.974)	5-1780-78
$C_{13}H_{17}N_4O_9P$			
2,4(1H,3H)-Pteridinedione, 6,7-dimethyl-1-(5-O-phosphono-β-D-ribofuranosyl)-	pH 1.0	230s(4.057),245s(3.964), 323(3.964)	5-1780-78
	pH 7.0	230s(4.076),245s(3.982), 323(3.969)	5-1780-78
	pH 13.0	240(4.199),272s(3.505), 330(3.991)	5-1780-78
$C_{13}H_{17}N_5O$			
1H-Imidazole-4-carboximidamide, 5-ami-	pH 1	283.5(4.03)	94-1929-78

Compound	Solvent	$\lambda_{max}(\log \epsilon)$	Ref.
no-N'-ethoxy-1-(phenylmethyl)-	pH 7	264.5(4.04)	94-1929-78
(cont.)	pH 13	264.5(4.04)	94-1929-78
	EtOH	266(4.08)	94-1929-78
1H-Imidazole-4-carboximidamide, 5-ami-	pH 1	281.5(3.98)	94-1929-78
no-1-ethyl-N'-(phenylmethoxy)-	pH 7	264(4.02)	94-1929-78
	pH 13	264(4.02)	94-1929-78
	EtOH	265.5(4.09)	94-1929-78
$C_{13}H_{17}N_5O_2$			
[4,4'-Bipyrimidin]-2-amine, 2',6'-di-	pH 0.9	292s(3.69),325(3.75)	44-0511-78
ethoxy-5-methyl-	pH 8.9	271(3.71),313(3.71)	44-0511-78
[4,4'-Bipyrimidin]-2-amine, 2',6'-di-	pH 0.9	278s(3.70),312(3.87)	44-0511-78
ethoxy-5'-methyl-	pH 8.9	278s(3.77),300(3.84)	44-0511-78
$C_{13}H_{17}N_5O_4$			
Acetamide, N-[7-(2,2-dimethyl-1,3-diox-	pH 2	263(4.21)	73-3103-78
olan-4-yl)methyl]-6,7-dihydro-6-oxo-	pH 7	264(4.15)	73-3103-78
1H-purin-2-yl]-, (±)-	pH 12	268(4.06)	73-3103-78
Acetamide, N-[9-(2,2-dimethyl-1,3-diox-	pH 2	263(4.21)	73-3103-78
olan-4-yl)methyl]-6,9-dihydro-6-oxo-	pH 7	260(4.17)	73-3103-78
1H-purin-2-y-]-, (±)-	pH 12	263(4.07)	73-3103-78
$C_{13}H_{17}N_5O_5$			
Imidazo[1,2-a]-1,3,5-triazin-4(8H)-one,	pH 1	237s(3.85),265(4.06)	87-0883-78
2-amino-8-[2,3-O-(1-methylethylidene)-	pH 7	210(4.35),256(4.03)	87-0883-78
β-D-ribofuranosyl]-	pH 11	216(3.36),255(4.04)	87-0883-78
9H-Purin-6-amine, 9-[5-O-(1-oxopropyl)-	MeOH	258(4.17)	87-1218-78
β-D-arabinofuranosyl]-			
$C_{13}H_{17}N_5O_6$			
Carbamic acid, (3-β-D-ribofuranosyl-	pH 1	256(3.80),302(4.26),	44-4910-78
3H-1,2,3-triazolo[4,5-b]pyridin-		311s(4.18)	
5-yl)-, ethyl ester	pH 11	254(3.84),302(4.26),	44-4910-78
		311s(4.18)	
$C_{13}H_{17}N_7OS_2$			
Benzamide, 4-[1,2-bis[[(methylamino)-	EtOH	341(4.29)	87-0804-78
thioxomethyl]hydrazono]ethyl]-			
copper chelate	EtOH	318(4.44),507(3.68)	87-0804-78
$C_{13}H_{17}N_7O_3$			
4,13-Epoxy-1,3-dioxolo[6,7][1,3]diazo-	MeOH	215(4.43),272(4.34)	44-2320-78
cino[1,2-e]purine-6,8-diamine,			
3a,4,5,6,13,13a-hexahydro-2,2-			
dimethyl-, [3aR-(3aα,4β,13β,13aα)]-			
$C_{13}H_{18}$			
1H-Indene, 1-(1,1-dimethylethyl)-2,3-	MeOH	212(3.77),272(3.11)	35-6035-78
dihydro-, (R)-			
$C_{13}H_{18}AsN$			
3H-1,3-Benzazarsole, 2-(1,1-dimethyl-	MeOH	246(3.93),281(3.56),	101-0001-78K
ethyl)-3-ethyl-		297(3.57)	
$C_{13}H_{18}ClNO_2$			
2H-Cycloocta[b]pyran-2-one, 3-chloro-	EtOH	242(3.84),314s(3.97),	4-0511-78
4-(dimethylamino)-5,6,7,8,9,10-		324(3.98)	
hexahydro-			
$C_{13}H_{18}ClN_5O_6$			
Thymidine, 3'-[[[(2-chloroethyl)nitro-	EtOH	266(4.09)	87-0130-78

Compound	Solvent	λ_{max} (log ϵ)	Ref.
soamino]carbonyl]amino]-3'-deoxy- Thymidine, 5'-[[[(2-chloroethyl)nitro- soamino]carbonyl]amino]-5'-deoxy-	EtOH	265(4.11)	87-0130-78 87-0130-78
C$_{13}$H$_{18}$NO$_4$ 3,5-Pyridinedicarboxylic acid, 1,4-di- hydro-1,2-dimethyl-, diethyl ester, anion	DMSO	481(3.83)	103-1226-78
C$_{13}$H$_{18}$N$_2$ 2-Propanone, 2-butenylphenylhydrazone	n.s.g.	252(3.95),280(3.47)	39-0543-78C
Spiro[2H-benzimidazole-2,1'-cyclopent- ane], 1,3-dihydro-1,3-dimethyl-	hexane	224(4.57),314(3.87)	44-2621-78
C$_{13}$H$_{18}$N$_2$O Acetic acid, 1-(1-butenyl)-2-methyl- 2-phenylhydrazide	decalin	238(4.35),288(3.34)	33-1364-78
Azepino[3,2-b]indol-2(1H)-one, 3,4,5,6,7,8,9,10-octahydro-6-methyl-	MeOH	218(4.10),256s(3.86), 295s(3.63)	24-0853-78
Azepino[4,3-b]indol-1(2H)-one, 3,4,5,6,7,8,9,10-octahydro-6-methyl-	MeOH	211(4.49),246s(4.01), 276s(3.68)	24-0853-78
4H-Carbazol-4-one, 1,2,3,5,6,7,8,9- octahydro-9-methyl-, oxime	MeOH	218(4.12),251(4.06)	24-0853-78
Homoproline N-methylanilide, (±)-	EtOH	226(3.74)	33-3108-78
C$_{13}$H$_{18}$N$_2$O$_2$ Benzoic acid, 2-acetyl-2-(1,1-dimethyl- ethyl)hydrazide	EtOH	229(4.21),256s(3.52), 266s(3.36)	33-1477-78
2H-Indol-2-one, 3-[2-(dimethylamino)- ethyl]-1,3-dihydro-5-methoxy-, monoperchlorate	EtOH	248(4.01),302(3.28)	42-1122-78
Piperidine, 1-(3,5-dimethyl-4-nitro- phenyl)-	EtOH	253(4.00),305(3.51), 391(3.76)	78-2213-78
C$_{13}$H$_{18}$N$_2$O$_2$S 2-Propanol, 1-(4-benzothiazolyloxy)- 3-[(1-methylethyl)amino]-	H$_2$O	255(3.73)	94-1443-78
2-Propanol, 1-(5-benzothiazolyloxy)- 3-[(1-methylethyl)amino]-	H$_2$O	255(3.73)	94-1443-78
C$_{13}$H$_{18}$N$_2$O$_3$ Acetic acid, cyano[(3-oxo-1-cyclohexen- 1-yl)amino]-, 1,1-dimethylethyl ester	EtOH	276(4.39)	94-3080-78
1H-Indole-2-carboxylic acid, 3-amino- 4,5,6,7-tetrahydro-4-oxo-, 1,1-di- methylethyl ester	EtOH	255(4.48),283(4.06)	94-3080-78
1H-Indole-2-carboxylic acid, 4,5,6,7- tetrahydro-4-(hydroxyimino)-1,3-di- methyl-, (E)-	MeOH	238(4.42),294(4.12)	24-1780-78
Valine, N-[4-(methylamino)benzoyl]-	MeOH	294(3.98)	87-1162-78
C$_{13}$H$_{18}$N$_2$O$_4$S Phenol, 2-methoxy-6-[2-(3-methyl-2-thi- azolidinyl)ethyl]-4-nitro-	EtOH	210(4.38),249(3.95), 338(4.11),378(3.60)	4-1439-78
C$_{13}$H$_{18}$N$_2$O$_5$S Carbamic acid, [[2-[(butylamino)carbo- nyl]phenyl]sulfonyl]-, methyl ester	EtOH	220(4.01),272(3.17), 278(3.14)	40-0582-78
C$_{13}$H$_{18}$N$_2$O$_6$S 4H-Pyran-3-carboxamide, 2,6-dimethyl-	EtOH	245(3.90)	4-0477-78

Compound	Solvent	$\lambda_{max}(\log \epsilon)$	Ref.
4-oxo-N-[[(tetrahydro-2H-pyran-2-yl)-amino]sulfonyl]- (cont.)			4-0477-78
$C_{13}H_{18}N_2O_7$ 1H-Imidazole-4-carboxylic acid, 2,3-di-hydro-1-[2,3-O-(1-methylethylidene)-β-D-ribofuranosyl]-2-oxo-, methyl ester	MeOH	267.5(4.10)	94-3322-78
$C_{13}H_{18}N_2S$ 2,4-Cyclohexadiene-1-thione, 6-(1,3-di-ethyl-2-imidazolidinylidene)-	benzene	307(4.27)	142-1741-78A
	H_2O	252(4.10)	142-1741-78A
	EtOH	278(4.18)	142-1741-78A
	MeCN	299(4.26)	142-1741-78A
	CH_2Cl_2	299(4.06)	142-1741-78A
$C_{13}H_{18}N_3O_5P$ Carbamic acid, [[(aminocarbonyl)amino]-(2-ethoxyethenyl)phosphinyl]-, phenylmethyl ester, (E)-	EtOH	208(4.34)	130-0421-78
$C_{13}H_{18}N_4O_3$ Morpholine, 4-[1-[(4-nitrophenyl)hydra-zono]propyl]-	EtOH	267(4.078),430(4.411)	146-0864-78
$C_{13}H_{18}N_4O_6$ β-D-Ribofuranose, 5-[4-(aminocarbonyl)-1H-1,2,3-triazol-1-yl]-5-deoxy-2,3-O-(1-methylethylidene)-, 1-acetate	MeOH	210(4.18)	35-2248-78
$C_{13}H_{18}N_6O_2S_3$ Hydrazinecarbothioamide, 2,2'-[1-[4-(methylsulfonyl)phenyl]-1,2-ethane-diylidene]bis[N-methyl-	EtOH	343(4.30)	87-0804-78
copper chelate	EtOH	319(4.47),512(3.75)	87-0804-78
$C_{13}H_{18}N_6O_3$ Acetamide, N-[3-(6-amino-9H-purin-9-yl)-2-hydroxy-5-(hydroxymethyl)cy-clopentyl]-, (1α,2α,3β,5β)-(±)-(1α,2β,3β,5β)-	pH 1	258(4.16)	44-2311-78
	H_2O	260(4.16)	44-2311-78
	pH 13	260(4.17)	44-2311-78
	pH 1	258(4.15)	44-2311-78
	H_2O	260(4.16)	44-2311-78
	pH 13	260(4.17)	44-2311-78
$C_{13}H_{18}N_6O_5$ Adenosine, N-[(ethylamino)carbonyl]-	pH 1	277(4.38)	36-0569-78
	pH 13	270(4.14),278(4.16), 298(4.21)	36-0569-78
	70% EtOH	269(4.37),276(4.29)	36-0569-78
$C_{13}H_{18}N_6S_3$ Hydrazinecarbothioamide, 2,2'-[1-[4-(methylthio)phenyl]-1,2-ethanediyl-ylidene]bis[N-methyl-	EtOH	345(4.34)	87-0804-78
copper chelate	EtOH	312(4.40),540(3.72)	87-0804-78
$C_{13}H_{18}O$ 2,4,6-Cycloheptatrien-1-one, 2,7-bis(1-methylethyl)-	H_2O	238(4.24),243(4.20), 325(3.65),340(3.55)	35-1778-78
2-Cyclohexen-1-one, 3-(1,3-butadienyl)-2,4,4-trimethyl-, (E)-	EtOH	223(4.11),294(4.11)	33-2328-78

Compound	Solvent	$\lambda_{max}(\log \epsilon)$	Ref.
$C_{13}H_{18}OS$ 2-Heptanone, 1-(phenylthio)-	MeOH	250(2.83)	12-1965-78
$C_{13}H_{18}O_2$ Bicyclo[4.2.1]nonane-2,9-dione, 6-methyl-3-(1-methylethylidene)-	EtOH	253(3.74)	77-0529-78
2,4,6-Cycloheptatrien-1-one, 4-hydroxy-2,7-bis(1-methylethyl)-	pH 13	241(4.31),371(4.29)	35-1778-78
2-Cyclohexen-1-one, 4-(4-hydroxy-4-methylcyclohexylidene)-	EtOH	305(4.29)	39-0024-78C
3H-Cyclopenta[1,3]cyclopropa[1,2]benzen-3-one, 3a,3b,4,5,6,7-hexahydro-2-methoxy-3b,7-dimethyl-, $(3a\alpha,3b\beta,7\beta,7aR*)-(\pm)-$	EtOH	234(3.89),285s(3.34)	44-1114-78
Cycloprop[cd]azulen-2(2aH)-one, 2b,3,4,5,6,6b-hexahydro-2a-methoxy-2b,6-dimethyl-	EtOH	240(3.77)	44-1114-78
$C_{13}H_{18}O_2S$ 2-Heptanone, 1-(phenylsulfinyl)-	MeOH	245(3.57),255(3.26),263(3.04)	12-1965-78
$C_{13}H_{18}O_3$ 1,4-Cyclohexanediol, 1-(4-methoxyphenyl)-, cis	EtOH	224(4.04),268s(3.11),274(3.23),281(3.18)	39-0024-78C
trans	EtOH	224(3.96),268(3.00),274(3.13),281(3.08)	39-0024-78C
2-Propenoic acid, 3-(4,6,6-trimethyl-2-oxo-3-cyclohexen-1-yl)-, methyl ester	EtOH	242(4.13)	78-2439-78
$C_{13}H_{18}O_3S$ 2-Heptanone, 1-(phenylsulfonyl)-	MeOH	220(4.01),254(2.69),260(2.90),266(3.06),273(3.00)	12-1965-78
$C_{13}H_{18}O_4$ 1-Butanone, 1-(2-hydroxy-4,6-dimethoxyphenyl)-3-methyl-	EtOH	284(4.16)	142-1355-78A
1H-Indene-4-carboxylic acid, 2,3,5,6-7,7a-hexahydro-1-hydroxy-7a-methyl-5-oxo-, ethyl ester. trans-(±)-	EtOH	224(4.11),293(3.80)	22-0343-78
$C_{13}H_{18}O_5$ Propanedioic acid, (3-oxo-1-cyclohexen-1-yl)-, diethyl ester	MeOH	232(4.13)	44-0966-78
1,2-Propanediol, 1-(2,3-dihydro-4-hydroxy-6-methoxy-3-benzofuranyl)-2-methyl-, [R-(R*,S*)]-	MeOH	220(3.94),228s(3.87),274(2.85),281(2.80)	119-0047-78
4,7-Undecadienedioic acid, 6-oxo-, dimethyl ester	MeOH	240(4.12)	49-0557-78
$C_{13}H_{19}NO$ 1-Piperidineethanol, α-phenyl-	n.s.g.	220(3.11),260(3.79)	42-0916-78
$C_{13}H_{19}NO_2$ Carbamic acid, (3-phenylbutyl)-, ethyl ester, (S)-	EtOH	248(2.15),252(2.27),258(2.40),264(2.25),268(2.22)	103-0538-78
Phenol, 4-(dimethylamino)-2-propyl-, acetate	MeOH	256(3.78),305(3.04)	94-2027-78

Compound	Solvent	$\lambda_{max}(\log \epsilon)$	Ref.
$C_{13}H_{19}NO_5$ 1H-Pyrrole-2,4-dicarboxylic acid, 3-hydroxy-1-methyl-, 2-(1,1-dimethylethyl) 4-ethyl ester	EtOH	229(4.49),262(4.16)	94-2224-78
$C_{13}H_{19}NO_5Se$ α-D-Galactopyranose, 6-deoxy-1,2:3,4-bis-O-(1-methylethylidene)-6-selenocyanato-	MeOH	223(3.22)	136-0069-78E
$C_{13}H_{19}NS_2$ Spiro[cyclohexane-1,2'(4'H)-cyclopenta-[d][1,3]thiazine]-4'-thione, 1',5',6',7'-tetrahydro-2-methyl-	EtOH	245s(3.64),337(3.81), 407(4.33)	39-0558-78C
Spiro[cyclohexane-1,2'(4'H)-cyclopenta-[d][1,3]thiazine]-4'-thione, 1',5',6',7'-tetrahydro-3-methyl-	EtOH	240s(3.66),337(3.86), 407(4.37)	39-0558-78C
Spiro[cyclohexane-1,2'(4'H)-cyclopenta-[d][1,3]thiazine]-4'-thione, 1',5',6',7'-tetrahydro-4-methyl-	EtOH	240s(3.52),336(3.70), 408(4.20)	39-0558-78C
$C_{13}H_{19}NSi$ 3H-1,3-Benzazasilole, 2-(1,1-dimethylethyl)-3,3-dimethyl-	CCl₄	253(3.66),265(3.66), 293(3.53)	44-0731-78
$C_{13}H_{19}NSn$ 3H-1,3-Benzazastannole, 2-(1,1-dimethylethyl)-3,3-dimethyl-	CCl₄	252(3.79),256(3.79), 294(3.58)	44-0731-78
$C_{13}H_{19}N_3$ 1-Propanone, 2-amino-2-methyl-1-phenyl-, (1-methylethylidene)hydrazone	hexane	212(4.34),278(3.38)	103-0278-78
$C_{13}H_{19}N_3O_2$ 5-Oxazolamine, N-(2,4-dimethyl-5-oxazolyl)-2,4-dimethyl-N-propyl-	n.s.g.	231(3.86)	70-0963-78
5-Oxazolamine, 4-ethyl-N-(4-ethyl-N-(4-ethyl-2-methyl-5-oxazolyl)-N,2-dimethyl-	n.s.g.	242(4.17)	70-0963-78
$C_{13}H_{19}N_3O_6$ Butanamide, N-(1-β-D-arabinofuranosyl-1,2-dihydro-2-oxo-4-pyrimidinyl)-	isoPrOH	216(4.17),248(4.14), 303(3.88)	94-0981-78
6,7,8-Triazabicyclo[3.2.1]oct-2-ene-6,7-dicarboxylic acid, 2-methoxy-8-methyl-4-oxo-, diethyl ester	n.s.g.	238(3.79)	88-1751-78
$C_{13}H_{19}N_5O_4$ Adenosine, 2-(1-methylethyl)-	pH 1.0 pH 7.0	205(4.360),256(4.149) 210(4.420),262(4.170)	5-1796-78 5-1796-78
$C_{13}H_{19}N_5O_6$ Urea, N-[5-deoxy-5-[4-(methoxycarbonyl)-1H-1,2,3-triazol-1-yl]-2,3-O-(1-methylethylidene)-β-D-ribofuranosyl]-	MeOH	214(4.00)	35-2248-78
$C_{13}H_{19}N_7O_4$ Adenosine, 8-[[(dimethylamino)methylene]amino]-	H₂O	234(4.24),306(4.34)	5-1365-78

Compound	Solvent	$\lambda_{max}(\log \epsilon)$	Ref.
$C_{13}H_{20}NO_2$			
Morpholinium, 4-ethyl-2-hydroxy- 4-methyl-2-phenyl-, iodide	EtOH	255(2.35)	145-0595-78
$C_{13}H_{20}N_2$			
1H-1,7-Benzodiazonine, 2,3,4,5,6,7- hexahydro-1,7-dimethyl-	hexane	234(4.38),273(3.99), 301(3.62)	44-2621-78
$C_{13}H_{20}N_2O$			
Azepino[3,2-c]quinolizin-6(1H)-one, 2,3,4,5,7,7a,8,9,10,11-decahydro-	EtOH	228(3.93),318(4.34)	142-1717-78A
$C_{13}H_{20}N_2OS_2$			
3,9-Dithia-12-azabicyclo[9.2.2]penta- deca-11,13,14-trien-15-ol, 14-(ami- nomethyl)-	MeOH	230(3.87),250s(3.54), 305(3.62),330s(3.46)	88-3563-78
$C_{13}H_{20}N_2O_4$			
1H-Cyclopentapyrimidine-2,4(3H,5H)-di- one, 1-(2,2-diethoxyethyl)-6,7-di- hydro-	0.05M HCl H_2O 0.05M NaOH	275(3.97) 275(3.96) 273(3.88)	126-2195-78 126-2195-78 126-2195-78
1H-Pyrrole-2,4-dicarboxylic acid, 3-amino-5-methyl-, 2-(1,1-dimethyl- ethyl) 4-ethyl ester	EtOH	237(4.54),280(4.13)	94-3080-78
$C_{13}H_{20}N_2O_5$			
Palythene	H_2O	360(4.70)	88-4909-78
$C_{13}H_{20}N_4O_2S$			
Formamide, N-[(4-amino-2-methyl-5-pyri- midinyl)methyl]-N-[4-hydroxy-1-meth- yl-2-(methylthio)-1-butenyl]-	EtOH	238(4.02),273s(3.78)	94-0722-78
$C_{13}H_{20}N_5O_7P$			
5'-Adenylic acid, 2-(1-methylethyl)-	pH 2.0 pH 7.0	205(4.328),257(4.093) 210(4.375),261(4.134)	5-1796-78 5-1796-78
polymer	pH 7.3	208(4.201),256(3.968)	5-1796-78
$C_{13}H_{20}O$			
Bicyclo[4.2.0]oct-7-en-2-one, 7-butyl- 6-methyl-	heptane	278s(2.42),295s(1.95)	70-0352-78
Bicyclo[4.2.0]oct-7-en-2-one, 8-butyl- 6-methyl-	heptane	278s(2.31),298s(1.96), 307s(1.93)	70-0352-78
2-Cyclohexen-1-one, 3-(2-butenyl)- 2,4,4-trimethyl-	EtOH	248(4.14)	33-2328-78
2H-Inden-2-one, 1,3-diethyl-1,4,5,6,7- 7a-hexahydro-, cis	EtOH	239(4.12)	35-1799-78
trans	EtOH	241(4.17)	35-1799-78
3,5,7-Nonatrien-2-one, 8-methyl- 7-(1-methylethyl)-, (E,E)-	pentane	313(4.41)	142-0083-78C
$C_{13}H_{20}OSi$			
Silane, benzoyltriethyl-	C_6H_{12}	407(1.92),427(2.00), 446(1.76)	112-0751-78
$C_{13}H_{20}O_2$			
3H-2-Benzopyran-3-one, 4-butyl- 1,5,6,7,8,8a-hexahydro-	MeOH	235(3.84)	44-1248-78
Bicyclo[2.2.1]heptan-7-ol, 5-butoxy- 2,3-bis(methylene)-, exo,anti	isooctane	244s(3.94),250(3.99), 268s(3.81)	33-0732-78

Compound	Solvent	$\lambda_{max}(\log \epsilon)$	Ref.
cont.	EtOH	243s(3.89),249(3.92), 257s(3.75)	33-0732-78
3-Buten-2-one, 4-[2,5-dihydro-5,5-di-methyl-4-(1-methylethyl)-2-furanyl]-, (E)-	pentane	218(4.22)	142-0083-78C
1,3-Cyclopentadiene-1-hexanoic acid, ethyl ester (plus isomer)	n.s.g.	247(3.51)	104-0264-78
3,5-Hexadien-2-one, 6-[3,3-dimethyl-2-(1-methylethyl)oxiranyl]-, (E,E)-	pentane	270(4.41)	142-0083-78C
(E,Z)-	pentane	275(4.37)	142-0083-78C
(Z,Z)-	pentane	274(4.23)	142-0083-78C
3,5-Nonadiene-2,8-dione, 7-methyl-7-(1-methylethyl)-, (E,E)-	pentane	261s(4.42),268(4.46)	142-0083-78C
(E,Z)-	pentane	270(4.32),278(4.32)	142-0083-78C
3-Octanone, 1-(2-oxocyclopentylidene)-, (E)-	hexane	240(--)	39-0209-78C
	MeOH	243(3.81),334(3.05)	39-0209-78C
	MeOH-KOH	415(4.53)	39-0209-78C
(Z)-	MeOH	243(3.79),334(3.03)	39-0209-78C
	MeOH-KOH	415(4.42)	39-0209-78C
1-Pentanone, 1-(5-butyl-2-furanyl)-	MeOH	221(3.48),286(4.2)	44-4081-78
$C_{13}H_{20}O_2S$			
2-Heptanol, 1-(phenylsulfinyl)-, [S-(R*,S*)]-	MeOH	243(3.61),265(3.12), 272(2.89)	12-1965-78
$C_{13}H_{20}O_3$			
2-Cyclohexene-1-butanoic acid, 1,3-di-methyl-4-oxo-, methyl ester	MeOH	236(3.74)	44-2562-78
2-Cyclohexen-1-one, 4-(4-acetoxybutyl)-4-methyl-	MeOH	227(3.98)	44-2562-78
Cyclooctanecarboxylic acid, 1-methyl-4-(1-methylethylidene)-5-oxo-	EtOH	252(3.24)	77-0529-78
Cyclopentanebutanoic acid, 1-methyl-3-(1-methylethylidene)-2-oxo-	EtOH	253(4.00)	77-0529-78
2,4,6-Dodecatrienoic acid, 8-hydroxy-6-methyl-, (E,E,E)-	EtOH	296(4.42)	18-2077-78
1-Naphthalenecarboxylic acid, 3,4,4a,5-6,7,8,8a-octahydro-2-hydroxy-, ethyl ester, cis	EtOH	256(3.65)	32-0693-78
4H-Pyran-4-one, 2,3-dihydro-2,3,5-tri-methyl-6-(1-methyl-2-oxobutyl)-	hexane	266(3.93)	78-1769-78
	EtOH	272(3.92)	78-1769-78
$C_{13}H_{20}O_3S$			
2-Heptanol, 1-(phenylsulfonyl)-	MeOH	218(3.97),252(2.71), 258(2.89),265(3.02), 271(2.97)	12-1965-78
$C_{13}H_{20}O_3S_2$			
5H-1,4-Dithiepin-2-hexanal, 3-acetoxy-6,7-dihydro-	n.s.g.	257(3.88)	33-3087-78
$C_{13}H_{20}O_5$			
2H-Pyran-6-carboxylic acid, 4-(1,1-di-methylethyl)-2,2-dimethoxy-, methyl ester	MeOH	286(3.95)	35-5472-78
4-Undecenedioic acid, 6-oxo-, dimethyl ester, (E)-	MeOH	222(4.14)	49-0557-78
$C_{13}H_{21}BrO_2$			
2-Cyclohexen-1-one, 2-bromo-5-heptyl-3-hydroxy-	EtOH	293(4.33)	2-0970-78

Compound	Solvent	λ_{max}(log ϵ)	Ref.
$C_{13}H_{21}BrO_5$ Undecanedioic acid, 5-bromo-6-oxo-, dimethyl ester	MeOH	294(2.08)	49-0557-78
$C_{13}H_{21}FN_4O_4$ 4-Pyrimidinamine, 1-β-D-arabino-furano- syl-N-ethyl-2-(ethylimino)-5-fluoro- 1,2-dihydro-, monohydrochloride	H_2O	217(4.24),251(4.21), 285s(3.92)	44-4200-78
$C_{13}H_{21}NO$ 4(1H)-Quinolinone, 2,3,5,6,7,8-hexahy- dro-1-(1-methylpropyl)-	heptane EtOH MeCN CF_3COOH	316(4.07) 336(4.25) 330(4.21) 322(4.08)	103-0993-78 103-0993-78 103-0993-78 103-0993-78
$C_{13}H_{21}NO_3$ 2,5-Cyclohexadien-1-ol, 1-(1,1-dimeth- ylethyl)-3,4,5-trimethyl-4-nitro- Procerine	MeOH MeOH	225(3.15) 220(4.02)	23-1758-78 73-2312-78
$C_{13}H_{21}NO_6$ Propanedioic acid, [[(2-ethoxy-2-oxo- ethyl)methylamino]methylene]-, diethyl ester	EtOH	282(4.33)	94-2224-78
$C_{13}H_{21}N_2O_4$ 1H-Pyrrol-1-yloxy, 3-[[(1-carboxypro- pyl)amino]carbonyl]-2,5-dihydro- 2,2,5,5-tetramethyl-, (±)-	EtOH	210(3.95),240(3.60)	63-1659-78
$C_{13}H_{21}O_2$ Pyrylium, 2,6-bis(1,1-dimethylethyl)- 4-hydroxy-	H_2SO_4	240(3.87),265(3.90)	88-5071-78
$C_{13}H_{22}$ Cyclohexane, (5-methyl-1,3-hexadienyl)-	C_6H_{12}	232(4.44)	101-0159-78M
$C_{13}H_{22}N_2O_4$ 1,3-Propanediaminium, N,N'-bis(1-form- yl-2-hydroxyethenyl)-N,N,N',N'-tetra- methyl-, dihydroxide, bis(inner salt)	EtOH	255(4.72)	73-1261-78
$C_{13}H_{22}N_2O_6$ Glycine, N-[5-hydroxy-5-hydroxymethyl- 3-[(2-hydroxy-1-methylethyl)imino]- 2-methoxy-1-cyclohexen-1-yl]-	H_2O	332(4.64)	88-4909-78
$C_{13}H_{22}N_2S_2$ Pyrimidine, 2,4-bis(butylthio)- 5-methyl-	H_2O	255(3.08),300(3.02)	56-1255-78
$C_{13}H_{22}O$ 2-Cyclohexen-1-one, 3-butyl-2-propyl- Cyclooctanone, 2-(2,2-dimethylpropyli- dene)- 2-Cyclopenten-1-one, 3-pentyl-2-propyl-	C_6H_{12} MeOH hexane EtOH	238(4.08) 247(3.95) 237(3.91) 237(4.19)	44-4081-78 44-4081-78 54-0305-78 44-4081-78
$C_{13}H_{22}O_2$ 2-Cyclohexen-1-one, 3-(3-hydroxybutyl)- 2,4,4-trimethyl-	EtOH	248(4.16)	33-2328-78

Compound	Solvent	λ_{max}(log ϵ)	Ref.
2-Cyclopenten-1-one, 2-(2-ethylhexyl)-5-hydroxy-	EtOH	230?(3.95)	107-0155-78
2-Cyclopenten-1-one, 2-hydroxy-3-octyl-	EtOH	261(4.13)	107-0155-78
2-Cyclopenten-1-one, 5-hydroxy-2-octyl-	EtOH	231(3.95)	107-0155-78
$C_{13}H_{22}O_3$ 1-Propanone, 1-(1,4-dioxaspiro[4.6]undec-6-yl)-2-methyl-	hexane	286(1.83)	44-0604-78
$C_{13}H_{22}O_4Si$ Cyclopenta[c]pyran-4-carboxylic acid, 1,4a,5,6,7,7a-hexahydro-1-[(trimethylsilyl)oxy]-, (1α,4aα,7aα)-(±)-	MeOH	238(4.10)	24-2423-78
$C_{13}H_{22}O_5$ β-D-lyxo-Hexopyranosid-2-ulose, 1,1-dimethylethyl 6-deoxy-3,4-O-(1-methylethylidene)-	EtOH	306(1.53)	150-5344-78
$C_{13}H_{22}S_4$ Spiro[2H-1,5-benzodithiepin-3(4H),6'-[1,4]dithiepane], hexahydro-, cis	CH$_2$Cl$_2$	232(3.07)	49-1017-78
trans	CH$_2$Cl$_2$	232(2.99)	49-1017-78
$C_{13}H_{23}BrO$ 4-Nonen-3-one, 9-bromo-2,2,6,6-tetramethyl-, (E)-	EtOH	229(4.31)	44-4369-78
$C_{13}H_{23}NO$ 5,7,9-Dodecatrien-3-ol, 2-(methylamino)- (dicarprine B)	n.s.g.	275(4.70)	102-0831-78
$C_{13}H_{23}N_2$ Piperidinium, 1-[3-(1-piperidinyl)-2-propenylidene)-, perchlorate	MeOH	315(4.78)	39-1453-78C
$C_{13}H_{23}N_2O$ Methanaminium, N-[7-(dimethylamino)-3-ethoxy-2,4,6-heptatrienylidene]-N-methyl-, tetrafluoroborate	CH$_2$Cl$_2$	510(5.00)	70-0339-78
1,8-Naphthyridin-1(2H)-yloxy, 3,4,4a,5,6,7-hexahydro-2,2,4a,7,7-pentamethyl-	hexane	226(3.40),265(3.82), 327(2.32),550(0.8)	33-2851-78
$C_{13}H_{23}N_2O_2$ 1,8-Naphthyridin-1(2H)-yloxy, 3,4,4a,5-6,7-hexahydro-2,2,4a,7,7-pentamethyl-, 8-oxide	EtOH	224(3.93),326(3.77), 526s(2.41),591(2.52), c.656(--)	33-2851-78
$C_{13}H_{23}N_3O_2$ 1H-1,4-Diazepine-5,7(2H,6H)-dione, 3-(dimethylamino)-6,6-diethyl-2,2-dimethyl-	CH$_2$Cl$_2$	259(4.00)	33-3050-78
$C_{13}H_{24}N_2$ 1,8-Naphthyridine, 1,2,3,4,4a,5,6,7-octahydro-2,2,4a,7,7-pentamethyl-	EtOH	219(4.06)	33-2851-78
1H-Pyrazole, 3-(1,1-dimethylethyl)-1-(1,2,2-trimethylpropyl)-	EtOH	219(3.74)	138-0263-78
1H-Pyrazole, 5-(1,1-dimethylethyl)-1-(1,2,2-trimethylpropyl)-	EtOH	217(3.49)	138-0263-78

Compound	Solvent	$\lambda_{max}(\log \epsilon)$	Ref.
$C_{13}H_{24}N_2O$			
1,8-Naphthyridine, 1,2,3,4,4a,5,6,7-octahydro-2,2,4a,7,7-pentamethyl-, 8-oxide	EtOH	250(3.99)	33-2851-78
$C_{13}H_{24}N_2O_2$			
1,8-Naphthyridine, 1,2,3,4,4a,5,6,7-octahydro-1-hydroxy-2,2,4a,7,7-pentamethyl-, 8-oxide	EtOH	285(3.88)	33-2851-78
$C_{13}H_{24}O_2$			
5,8-Tridecanedione	C_6H_{12}	268.5(1.72)	44-4081-78
	MeOH	274(1.72)	44-4081-78
$C_{13}H_{25}NS$			
Dodecane, 1-isothiocyanato-, iodine complex	CCl_4	455(3.33)	22-0439-78
$C_{13}H_{25}NSSi_2$			
Benzenamine, N,N-dimethyl-4-[(pentamethyldisilanyl)thio]-	heptane	272(4.22),305(3.37)	65-2027-78

Compound	Solvent	$\lambda_{max}(\log \epsilon)$	Ref.

$C_{14}H_4Cl_2N_2S$
  Propanedinitrile, (6,8-dichloro-3H-
  naphtho[1,8-bc]thien-3-ylidene)-
	MeCN	246(4.11),272(4.34),
		325(3.33),459(4.29),
		480(4.26)
			89-0369-78

$C_{14}H_4Cl_4N_2$
  Benzo[3,4]cyclobuta[1,2-b]quinoxaline,
  6,7,8,9-tetrachloro-
	EtOH	218(4.07),275(4.29),
		287(4.34),309(4.00),
		346(3.51),365(3.63),
		382(3.94),401(3.97)
			78-0495-78

$C_{14}H_4N_2O_8$
  5H,9H-[2]Benzopyrano[5,4,3-cde][1]ben-
  zopyran-5,9-dione, 2,7-dinitro-
	dioxan	330(4.295)	104-0619-78

$C_{14}H_5Cl_5$ Anthracene, 1,2,3,4,9-pentachloro-	heptane	271(5.20),350(3.57), 370(3.82),388(4.01), 410(3.92)	5-1406-78

$C_{14}H_6ClN_7O_6$
  Phthalazine, 1-chloro-4-[(2,4,6-tri-
  nitrophenyl)azo]-
	dioxan	220(4.62),262(4.30),
		335s(3.90),475s(2.70)
			103-0575-78

$C_{14}H_6Cl_6$
  Benzene, pentachloro(1-chloro-2-phenyl-
  ethenyl)-, (Z)-
	heptane	219(4.62),243(4.38)	5-1406-78

Bicyclo[3.2.0]hepta-2,6-diene,
  1,2,3,5,6,7-hexachloro-4-(phenyl-
  methylene)-, (E)-
	heptane	210(4.95),235(4.60),
		241(4.6),320(4.30)
			5-1406-78

(Z)-	heptane	210(4.20),235(4.10), 240(4.10),305(4.25)	5-1406-78

$C_{14}H_6N_2S$
  Propanedinitrile, 3H-naphtho[1,8-bc]-
  thien-3-ylidene-
	MeCN	235(4.02),264(4.26),
		323(3.49),458(4.21)
			89-0369-78

Propanedinitrile, 5H-naphtho[1,8-bc]- thien-5-ylidene-	MeCN	242(4.00),273(3.76), 456(4.39)	89-0369-78

$C_{14}H_6OS_3$
  5H-Benzo[cd][1,2]benzodithiolo[4,3,2-
  ghi][1,2]benzodithiol-10-S^{IV}-5-one
	EtOH	252(4.89),348(3.83),
		468(3.94),612(4.03)
			2-0673-78

$C_{14}H_6OSe_3$
  5H-Benzo[cd][1,2]benzodiselenolo[4,3,2-
  ghi][1,2]benzodiselenol-10-Se^{IV}-5-one
	EtOH	220(4.46),240(4.36),
		320(3.7),500(3.68)
			2-0673-78

$C_{14}H_7BrO_2$ 9,10-Anthracenedione, 1-bromo-	toluene	377(--),397(--), 419(2.04)	18-2264-78
9,10-Anthracenedione, 2-bromo-	toluene	377(--),379(--), 398(--),406(--), 424(1.85)	18-2264-78

$C_{14}H_7Br_2NO_2$
  9,10-Anthracenedione, 1-amino-2,4-di-
  bromo-
	benzene
	EtOH
		<u>475(3.9)</u>
		<u>470(3.9)</u>
			18-1793-78
			18-1793-78

$C_{14}H_7Br_3N_2O_2$
  4(1H)-Quinazolinone, 2-(2,4,6-tri-
  bromophenoxy)-
	dioxan	250(3.87),259(3.89),
		265(3.83),296(3.51),
			4-1033-78

Compound	Solvent	$\lambda_{max}(\log \epsilon)$	Ref.
(cont.)		305(3.62),318(3.52)	4-1033-78
$C_{14}H_7ClN_6O_4$ Phthalazine, 1-chloro-4-[(2,4-dinitro- phenyl)azo]-	dioxan	216(4.63),280(4.27), 332s(4.02),455s(2.90)	103-0575-78
$C_{14}H_7ClO_2$ 9,10-Anthracenedione, 1-chloro-	heptane	377s(<u>2.1</u>),398(<u>2.0</u>), 419(1.88)	18-2264-78
$C_{14}H_7ClO_4$ 9H-Xanthene-2-carboxylic acid, 7-chlo- ro-9-oxo-	MeOH	209(4.32),248(4.98), 283s(3.52),297(3.44), 330s(3.69),346(3.74)	39-0876-78C
$C_{14}H_7Cl_2NS$ Propanedinitrile, 2-(6,8-dichloro-3H- naphtho[1,8-bc]thien-3-ylidene)-	MeOH	225(4.0),246(4.2), 264(4.2),273(4.2), 399(4.2)	5-1974-78
$C_{14}H_7Cl_3N_2O_5S$ Benzonitrile, 4-[2-(5-nitro-2-furanyl)- 2-[(trichloromethyl)sulfonyl]ethenyl]-	EtOH	203(4.27),232(4.07), 258(3.79),308(4.13)	73-1618-78
$C_{14}H_7Cl_3O_5$ 3H-Xanthen-3-one, 2,6,7-trihydroxy- 9-(trichloromethyl)-	pH 8.5	501(3.76)	140-0699-78
9H-Xanthen-9-one, 2,4,5-trichloro- 1,3,6-trihydroxy-8-methyl-	MeOH	209(4.32),252(4.40), 284s(4.00),323(4.00), 353(4.00)	78-2491-78
	MeOH-AlCl₃	209(4.62),253(4.62), 276s(4.52),350(4.52), 400s(4.32)	78-2491-78
$C_{14}H_7FO_2$ 9,10-Anthracenedione, 1-fluoro-	heptane	376(<u>2.0</u>),393(<u>1.9</u>), 419(1.83)	18-2264-78
$C_{14}H_7N_3O_2$ Benzo[3,4]cyclobuta[1,2-b]quinoxaline, 7-nitro-	EtOH	233(4.29),271(4.51), 274s(4.50),310(4.13), 364(3.91),378(4.13), 396(4.10)	78-0495-78
$C_{14}H_7N_3O_3$ Acetamide, N-(6,7-dicyano-5,8-dihydro- 5,8-dioxo-1-naphthalenyl)-	acetone CH₂Cl₂	495(3.69) 519(--)	150-2319-78 150-2319-78
$C_{14}H_7N_3O_6$ 1H-Isoindole-1,3(2H)-dione, 2-(2,4-di- nitrophenyl)-	EtOH	210(4.17),227(4.16), 257(4.11),335(4.29), 380(3.99)	115-0073-78
	CHCl₃ CCl₄	325(4.37),370(3.96) 317(3.82),365(3.36)	115-0073-78 115-0073-78
$C_{14}H_8$ Naphthalene, 1,4-diethynyl- photopolymer	CH₂Cl₂ CH₂Cl₂	<u>320(4.2),333(4.3)</u> 312(3.89),480(3.60)	126-1999-78 126-1999-78
$C_{14}H_8BrNOS$ [1,1'-Biphenyl]-4-carbonyl isothiocyan-	C₆H₁₂	222(4.46),306(4.59)	73-0257-78

Compound	Solvent	$\lambda_{max}(\log \epsilon)$	Ref.
ate, 4'-bromo- (cont.) Methanone, (4-bromophenyl)thieno[2,3-b]pyridin-2-yl-	EtOH	225(4.34),302(4.39)	73-0257-78 78-0989-78
$C_{14}H_8Br_2$ Anthracene, 9,10-dibromo-	C_6H_{12}	379(4.14)	61-1068-78
$C_{14}H_8Br_2O_2$ Benzo[b]benzo[3,4]cyclobuta[1,2-e]-[1,4]dioxin, 7,8-dibromo-4b,10a-dihydro-, cis	ether	259(3.29),265(3.41), 272(3.45),295(3.55)	78-0073-78
$C_{14}H_8ClNOS$ [1,1'-Biphenyl]-4-carbonyl isothiocyanate, 4'-chloro-	C_6H_{12}	220(4.34),302(4.43)	73-0257-78
$C_{14}H_8ClNO_2$ 1H-Isoindole-1,3(2H)-dione, 2-(2-chlorophenyl)-	EtOH	218(3.74),285(3.40)	115-0073-78
	CHCl$_3$	288(3.26)	115-0073-78
	CCl$_4$	290(3.26)	115-0073-78
1H-Isoindole-1,3(2H)-dione, 2-(4-chlorophenyl)-	EtOH	215(4.49),240(4.02), 285(3.20)	115-0073-78
	ether	215(--),235(4.38), 288(--)	115-0073-78
	CHCl$_3$	288(3.34)	115-0073-78
	CCl$_4$	290(3.30)	115-0073-78
1H-Pyrrole-2,5-dione, 3-chloro-4-(1-naphthalenyl)-	MeCN	223(4.25),273(3.33), 381(3.00)	24-2677-78
$C_{14}H_8ClNO_3S$ Dibenzo[b,f]thiepin-10(11H)-one, 8-chloro-3-nitro-	MeOH	227(4.34),264(4.32), 335(3.68)	73-0471-78
$C_{14}H_8ClN_3O_3S$ Ethanone, 1-[2-[(4-chlorophenyl)thio]-4-nitrophenyl]-2-diazo-	MeOH	248(4.32),285s(4.09), 305s(4.01),362(3.59)	73-0471-78
$C_{14}H_8ClN_5O_2$ Phthalazine, 1-chloro-4-[(2-nitrophenyl)azo]-	dioxan	215(4.57),243s(4.36), 290(4.11),450s(2.99)	103-0575-78
Phthalazine, 1-chloro-4-[(4-nitrophenyl)azo]-	dioxan	215(4.58),294(4.28), 330s(4.14),460s(2.93)	103-0575-78
$C_{14}H_8ClN_7O_6$ 1(2H)-Phthalazinone, (2,4,6-trinitrophenyl)hydrazone	dioxan	277(4.13),290s(4.00), 420(4.32)	103-0575-78
	CHCl$_3$	270(--),288(--), 400(--)	103-0575-78
	DMSO	280(--),445(--)	103-0575-78
$C_{14}H_8Cl_2$ Anthracene, 9,10-dichloro-	C_6H_{12}	379(4.12)	61-1068-78
$C_{14}H_8Cl_2O_5$ 9H-Xanthen-9-one, 2,4-dichloro-1,3,6-trihydroxy-8-methyl-	MeOH	211(4.39),247(4.60), 274s(4.25),318(4.30), 350(4.14)	78-2491-78
	MeOH-AlCl$_3$	211(4.62),238s(4.64), 270(4.56),283s(4.54), 347(4.61),397(4.27)	78-2491-78

Compound	Solvent	$\lambda_{max}(\log \epsilon)$	Ref.
9H-Xanthen-9-one, 2,5-dichloro-1,3,6-trihydroxy-8-methyl-	MeOH	212(4.61),250(4.85), 317(4.54),355(4.12)	78-2491-78
	MeOH-AlCl$_3$	212(4.69),252(4.77), 270s(4.61),348(4.67), 400s(4.19)	78-2491-78
9H-Xanthen-9-one, 2,7-dichloro-1,3,6-trihydroxy-8-methyl-	EtOH	245(4.59),260(4.38), 270(4.24),319(4.10), 358(4.28)	78-0577-78
9H-Xanthen-9-one, 4,7-dichloro-1,3,6-trihydroxy-8-methyl-	EtOH	243(4.58),255(4.37), 270(4.25),313(4.09), 360(4.23)	78-0577-78
$C_{14}H_8Cl_3NO_7S$ 1,3-Benzodioxole, 5-[2-(5-nitro-2-furanyl)-2-[(trichloromethyl)sulfonyl]-ethenyl]-	EtOH	206(4.29),236(4.13), 302(4.25),323s(4.20), 372(4.17)	73-1618-78
$C_{14}H_8Cl_6O$ Bicyclo[3.2.0]hepta-3,6-dien-2-ol, 1,3,4,5,6,7-hexachloro-2-(phenylmethyl)-	heptane	210(4.05),270(3.92)	5-1406-78
$C_{14}H_8N_2O$ 6H-Benzo[b]quinolizine-11-carbonitrile, 6-oxo-	EtOH	235(4.51),238(4.52), 265(4.21),296(3.73), 309(3.78),363s(4.06), 384(4.17),394s(4.04), 427(3.93),452(3.70)	44-3536-78
Dibenz[d,f][1,3]oxazepine-6-carbonitrile	MeCN	244(4.39),314(3.67)	94-2508-78
5(6H)-Phenanthridinecarbonitrile, 6-oxo-	MeOH	227(4.60),233(4.69), 255(4.09),263(4.11), 275(4.01),291(3.66), 302(3.76),321(3.87), 330s(3.79)	94-2508-78
6-Phenanthridinecarbonitrile, 5-oxide	MeOH	240(4.50),257(4.54), 287(4.05),298(3.96), 344(4.03),354(4.05)	94-2508-78
$C_{14}H_8N_2O_2$ 6H-Indolo[3,2,1-de][1,5]naphthyridin-6-one, 8-hydroxy-	MeOH	253(4.11),270(4.07), 294s(3.52),330(3.88), 340(3.86),399(3.87)	100-0166-78
	MeOH-NaOH	249(3.99),261s(3.90), 267(3.88),293(3.68), 336(3.87),350(3.86), 448(3.82)	100-0166-78
$C_{14}H_8N_2O_3$ 2H-Pyrano[3,2-c]quinoline-2-carbonitrile, 5,6-dihydro-6-methyl-2,5-dioxo-	MeOH	349(4.04),384(4.07)	49-1075-78
$C_{14}H_8N_2O_3S$ 5H-[2,1]Benzisothiazolo[4,3,2-ghi]benzo[cd][1,2]benzisothiazol-10-SIV-5-one, 1,9-dihydro-4,6-dihydroxy-	EtOH	238(3.54),600(3.07)	2-0673-78
[1,1'-Biphenyl]-4-carbonyl isothiocyanate, 4'-nitro-	C_6H_{12}	305(4.47)	73-0257-78

Compound	Solvent	$\lambda_{max}(\log \epsilon)$	Ref.
$C_{14}H_8N_2O_4$			
1H-Isoindole-1,3(2H)-dione, 2-(2-nitro-phenyl)-	EtOH	223(3.98),280(2.00)	115-0073-78
	CHCl$_3$	285(3.48)	115-0073-78
	CCl$_4$	290(3.36)	115-0073-78
1H-Isoindole-1,3(2H)-dione, 2-(4-nitro-phenyl)-	EtOH	218(3.86),280(3.53), 295(3.53)	115-0073-78
	CHCl$_3$	285(3.38)	115-0073-78
	CCl$_4$	295(3.04)	115-0073-78
2,3-Naphthalenedicarbonitrile, 1,4-di-hydro-5,8-dimethoxy-1,4-dioxo-	benzene	516(--)	150-2319-78
	CH$_2$Cl$_2$	543(3.65)	150-2319-78
$C_{14}H_8N_2O_5$			
Benzofuram. 3-(2,4-dinitrophenyl)-	n.s.g.	366(4.23)	39-0401-78C
$C_{14}H_8N_2S_4$			
Benzothiazole, 2,2'-dithiobis-	dioxan	275(3.5)	65-1352-78
$C_{14}H_8N_4$			
Ethenetricarbonitrile, (2-methyl-1-indolizinyl)-	EtOH	218(4.25),242(3.94), 278(3.59),285(3.58), 362(3.20),515(4.25)	4-1471-78
Ethenetricarbonitrile, (2-methyl-3-indolizinyl)-	EtOH	232(4.23),300(3.71), 331(3.54),393(3.81), 535(4.28)	4-1471-78
$C_{14}H_8N_4O$			
Ethenetricarbonitrile, (1-methoxy-1H-indol-3-yl)-	MeOH	213(4.00),286(3.41), 456(3.79)	39-1117-78C
$C_{14}H_8N_6O_4$			
1,2,4,5-Tetrazine, 3,6-bis(3-nitro-phenyl)-	MeCN	260(4.65),540(2.74)	70-2227-78
$C_{14}H_8OS_4$			
1,3-Benzodithiole, 2-(1,3-benzodithiol-2-ylidene)-, 1-oxide	EtOH	208(3.54),220s(4.36), 296(3.95),406(4.19)	44-4394-78
$C_{14}H_8O_4$			
9,10-Anthracenedione, 1,2-dihydroxy-	pH 1	435(3.56)	140-1305-78
	pH 9.0	530(3.73)	140-1305-78
	pH 14	578(3.97)	140-1305-78
9H-Xanthene-2-carboxylic acid, 9-oxo-	MeOH	217(3.95),246(4.45), 290s(3.14),336(3.64)	39-0876-78C
$C_{14}H_8O_5$			
9,10-Anthracenedione, 1,3,5-trihydroxy-	MeOH-HCOOH	285(4.08),424(4.01)	12-2271-78
9,10-Anthracenedione, 1,3,6-trihydroxy-	EtOH-1% HCOOH	247s(4.01),270s(4.41), 277(4.46),285(4.47), 300s(4.20),343(3.70), 434(3.77)	12-1335-78
9,10-Anthracenedione, 1,3,7-trihydroxy-	EtOH-1% HCOOH	283(4.40),303s(4.26), 347(3.79),389(3.80)	12-1335-78
9,10-Anthracenedione, 1,4,5-trihydroxy-	MeOH-DMSO-HCOOH	283(3.99),462s(4.04), 477s(4.09),489(4.11), 510(3.98),522s(3.89)	12-2271-78
Dibenz[b,e]oxepin-6,11-dione, 1,10-di-hydroxy-	EtOH	276(3.8)	73-1808-78
$C_{14}H_8O_5S$			
1-Anthracenesulfonic acid, 9,10-dihy-dro-9,10-dioxo-, sodium salt	pH 7.0	257(4.55),277s(3.94), 329(3.54)	95-0929-78

Compound	Solvent	$\lambda_{max}(\log \epsilon)$	Ref.
2-Anthracenesulfonic acid, 9,10-di-hydro-9,10-dioxo-, sodium salt	pH 7.0	257(4.65),276(4.13), 330(3.67)	95-0929-78
$C_{14}H_8O_6$			
9,10-Anthracenedione, 1,2,4,8-tetra-hydroxy-	MeOH-DMSO-HCOOH	253(4.46),302(3.96), 462s(3.99),495(4.09), 528s(3.96)	12-2271-78
9,10-Anthracenedione, 1,3,5,8-tetra-hydroxy-	MeOH-1% HCOOH	253(4.19),279(4.30), 302s(4.01),461s(4.11), 476s(4.17),487(4.19), 506s(4.10),518s(3.99)	12-2271-78
1,3-Benzodioxol-5(6H)-one, 6-(6-oxo-1,3-benzodioxol-5(6H)-ylidene)-	$CHCl_3$	242(4.07),254(4.05), 375(3.92),550(3.92)	78-1595-78
$C_{14}H_8O_8$			
9,10-Anthracenedione, 1,2,4,5,6,8-hexa-hydroxy-	MeOH-DMSO-HCOOH	258(4.43),262s(4.36), 282s(4.14),291s(4.08), 327(3.96),467(3.86), 476(3.94),500(4.09), 510(4.16),536(4.17), 548(4.21)	12-2271-78
$C_{14}H_8O_8S_2$			
1,5-Anthracenedisulfonic acid, 9,10-di-hydro-9,10-dioxo-, disodium salt	pH 7.0	259(4.56),325(3.51)	95-0929-78
1,8-Anthracenedisulfonic acid, 9,10-di-hydro-9,10-dioxo-, dipotassium salt	pH 7.0	259(4.46),324(3.42)	95-0929-78
$C_{14}H_8O_9S$			
2-Anthracenesulfonic acid, 9,10-di-hydro-1,3,5,7-tetrahydroxy-9,10-dioxo-	MeOH-HCOOH	251s(4.24),258(4.11), 290(4.38),325s(3.90), 420(3.94)	12-2271-78
$C_{14}H_8O_{12}S_2$			
2,6-Anthracenedisulfonic acid, 9,10-di-hydro-1,3,5,7-tetrahydroxy-9,10-di-oxo-	MeOH-HCOOH	251(4.21),258(4.20), 263s(4.16),293(4.32), 325s(3.97),422(3.99)	12-2271-78
$C_{14}H_9Br$			
Anthracene, 9-bromo-	C_6H_{12}	369(4.01)	61-1068-78
$C_{14}H_9BrO$			
9H-Fluoren-9-one, 4-(bromomethyl)-	EtOH	262(4.71)	94-0635-78
$C_{14}H_9Cl$			
Anthracene, 9-chloro-	C_6H_{12}	368(3.99)	61-1068-78
$C_{14}H_9ClF_3NO_2$			
Benzoic acid, 3-chloro-2-[[3-(trifluoromethyl)phenyl]amino]-, monosodium salt	EtOH	278(4.00),326(3.67)	83-0161-78
Benzoic acid, 4-chloro-2-[[3-(trifluoromethyl)phenyl]amino]-, monosodium salt	EtOH	292(4.24),339(3.83)	83-0161-78
$C_{14}H_9ClFeO_3$			
Iron, tricarbonyl[(1,8,9,10-η)-9-chloro-10-methylbicyclo[6.2.0]deca-1,3,5,7,9-pentaene]	C_6H_{12}	278(4.14),315s(3.72), 430s(2.59)	5-1379-78

Compound	Solvent	$\lambda_{max}(\log \epsilon)$	Ref.
$C_{14}H_9ClN_2O_2$ 4(3H)-Quinazolinone, 2-(4-chlorophen-oxy)-	dioxan	243(3.84),260(3.97), 265(3.94),300(3.51), 307(3.63),320(3.53)	4-1033-78
$C_{14}H_9ClN_3O_6Rh$ Rhodium, dicarbonylchloro[2-[(2,4-di-nitrophenyl)methyl]pyridine-N^1]-, (SP-4-3)-	EtOH	257(4.25),333(3.36)	104-0189-78
$C_{14}H_9ClN_6O$ 1,3,5-Triazine, 2-azido-4-(4-chloro-1-naphthalenyl)-6-methoxy-	C_6H_{12}	327(4.15)	44-1361-78
$C_{14}H_9ClN_6O_4$ 1(2H)-Phthalazinone, 4-chloro-, 2,4-dinitrophenylhydrazone	dioxan	270(4.17),290s(4.00), 405(4.30)	103-0575-78
	CHCl$_3$	270(--),290(--), 405(--)	103-0575-78
$C_{14}H_9ClO_5$ 9H-Xanthen-9-one, 2-chloro-3,6,8-tri-hydroxy-1-methyl-	EtOH	243(4.58),254(4.39), 268(4.13),313(4.18), 352(4.19)	78-0577-78
	MeOH	210(4.56),243(4.76), 254s(4.65),270s(4.30), 312(4.51),345(4.25)	78-2491-78
	MeOH-AlCl$_3$	210(4.78),235(4.79), 265(4.77),342(4.76), 392(4.47)	78-2491-78
9H-Xanthen-9-one, 4-chloro-1,3,6-tri-hydroxy-8-methyl-	EtOH	242(4.54),253(4.34), 312(4.20),346(4.10)	78-0577-78
$C_{14}H_9ClO_7S$ Thiopyrano[2,3-b]furylium, 5-carboxy-2-phenyl-, perchlorate	49% H$_2$SO$_4$	290(4.42),405(3.87)	104-0589-78
$C_{14}H_9Cl_2NO_2$ Cyclobuta[a]naphthalene-1,2-dicarbox-imide, 1,2-dichloro-1,2,2a,8b-tetra-hydro-, antianti	MeOH	255(3.72)	24-2677-78
syn-syn	MeOH	256(4.07)	24-2677-78
$C_{14}H_9F$ Anthracene, 9-fluoro-	C_6H_{12}	364(3.92)	61-1068-78
$C_{14}H_9F_2NO_2$ Benzene, 1-(1,2-difluoro-2-phenyleth-enyl)-4-nitro-, (E)-	ether	336(4.31)	104-0939-78
$C_{14}H_9NOS$ [1,1'-Biphenyl]-4-carbonyl isothiocyan-ate	C_6H_{12}	220(4.46),304(4.48)	73-0257-78
Methanone, phenylthieno[2,3-b]pyridin-2-yl-	EtOH	225(4.14),253(3.99), 300(4.25)	78-0989-78
6H-Thiopyrano[2,3-b]furan-5-carboni-trile, 2-phenyl-	EtOH	225(4.05),272(4.30), 310(4.29),390(3.53)	104-0589-78
$C_{14}H_9NO_2$ 1H-Isoindole-1,3(2H)-dione, 2-phenyl-	EtOH	214(4.00),240(3.90), 291(2.60)	115-0073-78

Compound	Solvent	$\lambda_{max}(\log \epsilon)$	Ref.
1H-Isoindole-1,3(2H)-dione, 2-phenyl- (cont.)	ether	214(--),235(3.91), 291(--)	115-0073-78
	CHCl$_3$	293(3.00)	115-0073-78
	CCl$_4$	294(3.26)	115-0073-78
$C_{14}H_9NO_3$			
1,4-Acridinedione, 2-hydroxy-9-methyl-	dioxan	225(4.47),289(4.52), 360s(3.5)	150-4901-78
1H-Isoindole-1,3(2H)-dione, 2-(3-hy-droxyphenyl)-	EtOH	213(4.09),250(3.68), 300(3.85)	115-0073-78
	CHCl$_3$	315(3.91)	115-0073-78
	CCl$_4$	310(3.90)	115-0073-78
2H-Pyrido[2,1-b][1,3]oxazinium, 4-hy-droxy-2-oxo-3-phenyl-, hydroxide, inner salt	MeCN	216s(4.10),249(4.31), 289s(3.30),367(3.72)	5-1655-78
$C_{14}H_9NS$			
Propanenitrile, 2-(5H-naphtho[1,8-bc]-thien-5-ylidene)-	MeOH	237(4.1),266(4.1), 273(4.1),386(4.4)	5-1974-78
Pyrrolo[3,2,1-kl]phenothiazine	EtOH	226(4.49),253(4.48), 268(4.08),277(4.01), 299(3.63),311(3.74), 342(3.97)	4-0711-78
$C_{14}H_9N_3$			
6H-Benzo[b]quinolizine-11-carbonitrile, 6-imino-	EtOH	233(4.47),242s(4.45), 260(4.11),269(4.09), 309(3.74),322(3.81), 378s(3.99),398(4.05), 426(3.93),454s(3.90), 478s(3.62)	44-3536-78
$C_{14}H_9N_3O$			
6H-Pyrazino[1,2-b]isoquinoline-11-car-bonitrile, 1-methyl-6-oxo-	EtOH	236(4.44),270s(4.08), 364s(4.00),383(4.05), 424s(3.97),438s(3.88)	44-3536-78
13H-Pyrido[1',2':3,4]imidazo[2,1-b]-quinazolin-13-one	MeOH	281s(3.96),289(4.1), 357(3.7),373(3.76), 413(3.4),434(3.35), 460(3.0)	5-0398-78
$C_{14}H_9N_5O_2$			
Phthalazine, 1-[(4-nitrophenyl)azo]-	dioxan	214(4.60),292(4.28), 325(4.13),422s(2.78)	103-0575-78
$C_{14}H_{10}$			
Anthracene	C_6H_{12}	356(3.97)	61-1068-78
$C_{14}H_{10}BF_4N_3$			
6H-Benzo[b]quinolizine-11-carbonitrile, 6-imino-, tetrafluoroborate	MeCN	233(5.02),257s(4.38), 268s(4.34),362s(4.03), 379(4.16),412(4.01), 426(3.81),450s(2.93)	44-3536-78
$C_{14}H_{10}BrNO_2$			
1H-Pyrrole-2,5-dione, 3-bromo-4-(1H-in-den-2-yl)-1-methyl-	MeOH	210(3.66),265(3.94), 407(3.86)	24-2677-78
$C_{14}H_{10}Br_2N_4$			
2H-Cyclopenta[d]pyridazine, 5,7-di-bromo-2-methyl-6-(phenylazo)-	ether	263s(4.20),273(4.26), 297s(4.08),312s(4.04),	44-0664-78

Compound	Solvent	$\lambda_{max}(\log \epsilon)$	Ref.
(cont.)		384(4.04),476(4.04)	44-0664-78
$C_{14}H_{10}ClNOS$			
Dibenzo[b,f]thiepin-10(11H)-one, 2-amino-8-chloro-	MeOH	229(4.44),263(4.27), 355(3.50)	73-0471-78
Dibenzo[b,f]thiepin-10(11H)-one, 8-chloro-, oxime	MeOH	210(4.43),245(4.22), 276(4.06),315(3.45)	73-2427-78
$C_{14}H_{10}ClNO_3S$			
Dibenzo[b,f]thiepin-10-ol, 8-chloro-10,11-dihydro-3-nitro-	MeOH	266(4.30),330(3.09)	73-0471-78
$C_{14}H_{10}ClNO_4S$			
Benzeneacetic acid, 2-[(4-chlorophenyl)thio]-4-nitro-	MeOH	253(4.28),272(4.17), 335(3.32)	73-0471-78
$C_{14}H_{10}ClN_3$			
4H-1,2,4-Triazole, 3-(4-chlorophenyl)-4-phenyl-	EtOH	246(4.28)	88-4629-78
$C_{14}H_{10}ClN_5O_2$			
1(2H)-Phthalazinone, 4-chloro-, (2-nitrophenyl)hydrazone	dioxan	256s(4.34),292(4.23), 356(4.00),470(3.93)	103-0575-78
	CHCl$_3$	262(--),292(--), 350(--),475(--)	103-0575-78
	DMSO	296(--),362(--), 505(--)	103-0575-78
1(2H)-Phthalazinone, 4-chloro-, (4-nitrophenyl)hydrazone	EtOH	210(4.56),280(4.04), 440(4.28)	103-0575-78
	dioxan	260s(4.11),285(3.95), 420(4.26)	103-0575-78
$C_{14}H_{10}Cl_3NO_5S$			
Furan, 2-[2-(4-methylphenyl)-1-(trichloromethyl)sulfonyl]ethenyl]-5-nitro-	EtOH	210(4.17),225(4.15), 299(4.42),345s(4.08)	73-1618-78
$C_{14}H_{10}Cl_3NO_6S$			
Furan, 2-[2-(4-methoxyphenyl)-1-[(trichloromethyl)sulfonyl]ethenyl]-5-nitro-	EtOH	209(4.10),235(4.10), 312(4.43),356s(4.22)	73-1618-78
$C_{14}H_{10}Cl_3NO_7S$			
Phenol, 2-methoxy-4-[2-(5-nitro-2-furanyl)-2-[(trichloromethyl)sulfonyl]-ethenyl]-	EtOH	205(4.28),243(4.09), 310(4.28),378(4.19)	73-1618-78
$C_{14}H_{10}Cs_2$			
Anthracene, 9,10-dihydro-, dianion, cesium salt	n.s.g.	633(4.26)	35-8271-78
$C_{14}H_{10}F_3NO_2$			
Benzoic acid, 2-[[2-(trifluoromethyl)-phenyl]amino]-, sodium salt	EtOH	287(4.22),338(3.89)	83-0161-78
$C_{14}H_{10}N_2O$			
3H-Indol-3-one, 2-(phenylamino)-	EtOH	203(4.21),225(4.12), 258(4.39),303(3.60), 470(3.60)	103-0757-78

Compound	Solvent	$\lambda_{max}(\log \epsilon)$	Ref.
$C_{14}H_{10}N_2O_2$			
1,8-Naphthyridin-2(1H)-one, 4-hydroxy-3-phenyl-	MeOH	231(4.01),257(3.81), 301(4.11),319(4.28)	24-2813-78
1,3,4-Oxadiazol-2(3H)-one, 3,5-diphenyl-	EtOH	238(4.01),269(4.21), 274(4.20),285(4.22)	33-1477-78
Sydnone, 3,4-diphenyl-	EtOH	239(4.03),333(3.99)	33-1477-78
$C_{14}H_{10}N_2O_3Se$			
2,4,6(1H,3H,5H)-Pyrimidinetrione, 5-(2-methyl-4H-1-benzoselenin-4-ylidene)-	HOAc	650(3.921)	83-0170-78
$C_{14}H_{10}N_2O_4$			
3H-Phenoxazin-3-one, 1,9-dimethyl-7-nitro-	C_6H_{12}	251(4.31),339(4.24), 354(4.27),373(4.15), 431(3.97)	24-3012-78
$C_{14}H_{10}N_2O_5$			
Benzoic acid, 4,4'-azoxybis-	MeOH	263(3.96),334(4.24)	39-0171-78C
$C_{14}H_{10}N_4$			
2,4'-Bi-1H-benzimidazole	EtOH	210(4.60),310(4.43), 320(4.51),335(4.42)	12-2675-78
2,5'-Bi-1H-benzimidazole	EtOH	220(4.38),230s(--), 303s(--),312(4.26), 326s(--)	12-2675-78
Naphtho[2,1-e]pyrazolo[5,1-e][1,2,4]-triazine, 2-methyl-	EtOH	410(3.96)	103-0313-78
6H-Pyrazino[1,2-b]isoquinoline-11-carbonitrile, 6-imino-1-methyl-	EtOH	237(4.44),291s(3.77), 303s(3.71),373(3.88), 392(3.91),437(3.90), 458s(3.98)	44-3536-78
1,2,4,5-Tetrazine, 3,6-diphenyl-	MeCN	222(4.57),292(4.61), 550(2.70)	70-2227-78
$C_{14}H_{10}N_4O_2$			
2-Quinoxalinamine, N-(2-nitrophenyl)-	EtOH	264(4.41),399(4.03)	78-0981-78
$C_{14}H_{10}N_4O_3$			
3H-Pyrazole-3-carbonitrile, 4-(2-furanyl)-4,5-dihydro-3-(4-nitrophenyl)-	EtOH	207(4.16),217(4.17), 262(3.97),291(3.92)	73-0870-78
$C_{14}H_{10}N_6O$			
1,3,5-Triazine, 2-azido-4-methoxy-6-(1-naphthalenyl)-	C_6H_{12}	324(4.11)	44-1361-78
$C_{14}H_{10}N_6O_2$			
1-Naphthalenol, 4-(4-azido-6-methoxy-1,3,5-triazin-2-yl)-	C_6H_{12}	346(4.23)	44-1361-78
2-Naphthalenol, 1-(4-azido-6-methoxy-1,3,5-triazin-2-yl)-	C_6H_{12}	375(4.08)	44-1361-78
Pyrido[2,3-b]pyrazin-2(1H)-one, 3-(1,4-dihydro-2-oxopyrido[2,3-b]pyrazin-3(2H)-ylidene)-3,4-dihydro-	pH 1.0	275(3.67),287s(3.72), 301(3.77),430s(4.25), 457(4.46),487(4.48)	24-1763-78
$C_{14}H_{10}N_6O_4$			
1(2H)-Phthalazinone, 2,4-dinitrophenylhydrazone	dioxan	285(3.95),410(4.31)	103-0575-78
$C_{14}H_{10}O$			
Cyclobut[a]acenaphthylen-3-ol, 6b,8a-	50% MeOH	218(4.28),235(4.20),	35-2464-78

Compound	Solvent	$\lambda_{max}(\log \epsilon)$	Ref.
dihydro- (cont.)		306(3.65),330(3.55)	35-2464-78
	50% MeOH-NaOH	251(4.13),343(3.76)	35-2464-78
$C_{14}H_{10}O_3$			
2H-Naphtho[2,3-b]pyran-5,10-dione, 2-methyl-	MeOH	258(4.2),275(4.2)	24-1284-78
$C_{14}H_{10}O_3S$			
3H-Naphtho[1,8-bc]thiophene-2-carboxylic acid, 3-oxo-, ethyl ester	MeOH	220s(--),265(4.328), 313(3.577),365(4.000)	5-0627-78
6H-Thiopyrano[2,3-b]furan-5-carboxylic acid, 2-phenyl-	EtOH	225(4.11),270(4.27), 316(4.36),390(3.57)	104-0589-78
$C_{14}H_{10}O_4$			
9(10H)-Anthracenone, 1,8,10-trihydroxy-	dioxan	224(4.22),265(3.87), 293(3.95),368(4.01)	87-0026-78
2,4,5,6-Phenanthrenetetrol	MeOH	266(4.87)	2-0643-78
$C_{14}H_{10}O_5$			
Benzoic acid, 4-hydroxy-3-(3-hydroxy-benzoyl)-	MeOH	238(4.52),327(3.66)	39-0876-78C
4,9-Ethanonaphtho[2,3-c]furan-1,3,10-trione, 3a,4,9,9a-tetrahydro-9-hydroxy-	MeCN	226(3.64),255(2.54), 292(2.58)	44-3478-78
4,10-Methano-1H-benzo[4,5]cyclohepta-[1,2-c]furan-1,3,9-trione, 3a,4,10-10a-tetrahydro-10-hydroxy-	EtOH	251(4.10),290(3.25), 297s(3.24)	44-3478-78
3H-Xanthen-3-one, 2,6,7-trihydroxy-9-methyl-	pH 9	499(3.60)	140-0699-78
$C_{14}H_{11}BF_4S_2$			
1,3-Benzodithiol-1-ium, 2-(phenylmeth-yl)-, tetrafluoroborate	MeCN-H_2SO_4	253(3.74),257s(3.83), 258(3.85),262(3.86), 320(3.79)	39-1133-78C
$C_{14}H_{11}BrClN$			
Benzenethanimidoyl chloride, α-bromo-N-phenyl-	heptane	225(4.20),272(4.22)	103-0037-78
$C_{14}H_{11}BrFN$			
Benzenethanimidoyl fluoride, α-bromo-N-phenyl-	heptane	224(3.96),255s(3.84)	103-0037-78
$C_{14}H_{11}BrN_2O_2$			
Benzamidoxime, O-benzoyl-4-bromo-	EtOH	231(4.36),260(4.19)	73-2740-78
Benzamidoxime, O-(4-bromobenzoyl)-	EtOH	247(4.35),270s(4.18)	73-2740-78
Methanone, [4-(4-bromophenyl)-4,5-di-hydro-1H-pyrazol-3-yl]-2-furanyl-	EtOH	222(4.2),288(3.91), 357(4.17)	104-1830-78
$C_{14}H_{11}BrN_2O_3$			
Isoquinolinium, 2-[(5-nitro-2-furanyl)-methyl]-, bromide	H_2O	216(4.73),233(4.52), 305(3.90)	73-2041-78
Quinolinium, 1-[(5-nitro-2-furanyl)-methyl]-, bromide	H_2O	202(4.71),224(3.52), 239(3.38),312(4.02)	73-2041-78
$C_{14}H_{11}BrN_4$			
2H-Cyclopenta[d]pyridazine, 5-bromo-2-methyl-7-(phenylazo)-	ether	259(4.26),297(4.08), 391(3.97),473(3.97)	44-0664-78

Compound	Solvent	$\lambda_{max}(\log \epsilon)$	Ref.
$C_{14}H_{11}BrO_3S$			
Benzene, 1-bromo-4-[[2-(phenylsulfon-yl)ethenyl]oxy]-	dioxan	227(4.24),247(4.39)	99-0194-78
$C_{14}H_{11}BrO_4$			
Ethanone, 1-(8-bromo-4-methoxy-5-meth-ylbenzo[1,2-b:5,4-b']difuran-2-yl)-	EtOH	269(4.39),313(4.22), 346s(--)	12-1533-78
$C_{14}H_{11}Br_2NO_2$			
Indeno[1',2':3,4]cyclobuta[1,2-c]pyr-role-1,3(2H,3aH)-dione, 3a,8b-di-bromo-3b,8,8a,8b-tetrahydro-2-methyl-, anti-anti	MeOH	242(2.79),266(3.13), 273(3.13)	24-2677-78
syn-syn	MeCN	241(3.01),268(3.03), 276(3.08)	24-2677-78
$C_{14}H_{11}ClFN$			
Benzeneethanimidoyl fluoride, α-chloro-N-phenyl-	heptane	224(4.12),255s(3.91)	103-0037-78
$C_{14}H_{11}ClN_2$			
Pyrrolo[1,2-c]pyrimidine, 1-chloro-3-methyl-6-phenyl-	n.s.g.	254(4.71),256s(4.71), 283s(3.95),300s(3.57), 358(3.45)	44-3544-78
$C_{14}H_{11}ClN_2O$			
Benzoic acid, [(4-chlorophenyl)methyl-ene]hydrazide	EtOH	299s(4.35),305(4.36)	34-0345-78
$C_{14}H_{11}ClN_2O_2$			
Benzamidoxime, O-benzoyl-4-chloro-	EtOH	230(4.39),266(4.15)	73-2740-78
Benzamidoxime, O-(4-chlorobenzoyl)-	EtOH	241(4.24),270s(4.06)	73-2740-78
11H-Dibenzo[b,e][1,4]diazepin-11-one, 8-chloro-5,10-dihydro-2-methoxy-	MeOH	230(4.51),290s(3.83)	73-0309-78
$C_{14}H_{11}ClN_2O_3S$			
Acetamide, 2-chloro-N-[4-[(4-nitrophen-yl)thio]phenyl]-	EtOH	207(4.45),250(4.26), 273(4.14),337(4.14)	111-0093-78
$C_{14}H_{11}ClN_2O_5$			
Benzoic acid, 2-[(4-chloro-2-nitro-phenyl)amino]-5-methoxy-	MeOH	225s(4.22),268(4.18), 302(4.07),340s(3.69), 452(3.84)	73-0309-78
$C_{14}H_{11}ClOS$			
Ethanone, 1-[5-chloro-2-(phenylthio)-phenyl]-	MeOH	268(3.94),343(3.62)	73-1276-78
$C_{14}H_{11}ClO_3S$			
Benzene, 1-chloro-4-[[2-(phenylsulfon-yl)ethenyl]oxy]-	dioxan	222(4.12),247(4.32)	99-0194-78
$C_{14}H_{11}ClO_5$			
2-Naphthalenecarboxylic acid, 7-chloro-5,8-dihydro-3-methoxy-1-methyl-5,8-dioxo-, methyl ester	CHCl₃	272(4.42),284s(4.18), 390(3.40)	39-1041-78C
$C_{14}H_{11}Cl_2NS$			
2H-1,6-Thiazecine, 3,4,8,9-tetradehy-dro-6-(3,4-dichlorophenyl)-5,6,7,10-tetrahydro-	EtOH	213(4.41),260(4.31), 304(3.30)	42-1228-78

Compound	Solvent	$\lambda_{max}(\log \epsilon)$	Ref.
$C_{14}H_{11}Cl_2N_5O$			
3-Formazancarboxamide, 1,5-bis(2-chlorophenyl)-	EtOH	270(3.83),286(3.82), 437(4.24)	104-1956-78
3-Formazancarboxamide, 1,5-bis(4-chlorophenyl)-	EtOH	272(3.84),302(3.96), 358(4.17),455(3.93)	104-1956-78
$C_{14}H_{11}Cl_4N$			
Benzenamine, 3,4-dichloro-N,N-bis(4-chloro-2-butynyl)-	EtOH	212(4.50),258(4.31), 286(3.36)	42-1228-78
$C_{14}H_{11}Cs$			
Anthracene, 9,10-dihydro-, anion, cesium salt	n.s.g.	444(4.38)	35-8271-78
$C_{14}H_{11}D$			
4a,9a-Methano-9H-fluorene-9-d, (4aα,9β,9aα)-	THF	236(3.68),273(3.55)	44-1586-78
$C_{14}H_{11}F_3N_2O_3S$			
Acetic acid, [5-oxo-2-thioxo-1-[3-(trifluoromethyl)phenyl]-4-imidazolidinylidene]-, ethyl ester, (Z)-	EtOH	348(4.38)	5-0227-78
$C_{14}H_{11}IN_4$			
2H-Cyclopenta[d]pyridazine, 5-iodo-2-methyl-7-(phenylazo)-	ether	259(4.15),307(4.08), 391(3.94),474(3.93)	44-0664-78
$C_{14}H_{11}N$			
1H-Indole, 5-phenyl-	EtOH	207(4.39),252(4.56), 283s(3.97)	103-0173-78
1H-Phenanthro[9,10-b]azirine, 1a,9b-dihydro-	CHCl₃	273(4.12),277(4.15), 281(4.17),288(4.02), 294(3.90),305(3.59)	44-4271-78
$C_{14}H_{11}NO$			
9(10H)-Acridinone, 10-methyl-	CH₂Cl₂	258(4.75),268s(4.50), 277(4.41),294(3.98), 306(3.77),361s(3.70), 380(4.01),400(4.13)	18-2674-78
Phenol, 3-(1H-indol-2-yl)-	EtOH	221(4.20),317(4.05), 335s(3.73)	120-0096-78
$C_{14}H_{11}NOS$			
2,1-Benzisothiazol-3(1H)-one, 1-(phenylmethyl)-	EtOH	217(4.33),247(4.05), 254s(--),365(3.74)	4-0529-78
1-Propanone, 1-(naphtho[2',1':4,5]thiazol-2-yl)-	n.s.g.	288(3.95),318(3.97), 359(3.93)	42-0702-78
Thiopyrano[2,3-b]pyrrol-2(7H)-one, 7-methyl-6-phenyl-	EtOH	275(4.12),392(3.94)	104-0589-78
$C_{14}H_{11}NO_2$			
7-Azabicyclo[4.2.0]octa-1,3,5-triene, 3,4-(methylenedioxy)-7-phenyl-	MeOH	207(4.4),279(4.2), 333(3.9)	5-0608-78
6(5H)-Phenanthridinone, 5-methoxy-	MeOH	231(4.65),236(4.63), 253s(4.16),263(4.20), 325s(3.85),340(3.95), 356s(3.85)	94-2508-78
Phenol, 2-(3-amino-2-benzofuranyl)-	n.s.g.	320(4.3)	103-0702-78
$C_{14}H_{11}NO_2S$			
Naphtho[2,1-d]thiazole-2-carboxylic	n.s.g.	288(3.98),314(4.01),	42-0702-78

Compound	Solvent	$\lambda_{max}(\log \epsilon)$	Ref.
acid, ethyl ester (cont.)		347(3.67)	42-0702-78
Phenanthridine, 6-(methylsulfonyl)-	EtOH	223(4.37),230s(4.35), 251(4.56),296(3.80), 306(3.82),320s(3.52), 337(3.46),355(3.08)	95-0136-78
$C_{14}H_{11}NO_3$			
9H-Carbazole-3-carboxylic acid, 2-hydroxy-, methyl ester (mukonidine)	EtOH	230(4.8),248(4.57), 266(4.64),272(4.73), 308(3.88),316(3.78), 323(3.18)	42-1114-78
3H-Phenoxazine-2,3(10H)-dione, 10-ethyl-	MeOH	235(4.13),266(3.91), 418(3.82),483s(3.73)	64-0912-78C
$C_{14}H_{11}NO_4$			
Acetic acid, 7H-benzocyclohepten-7-ylidenenitro-, methyl ester	MeCN	228(4.41),295(4.35), 303(4.33),377(4.16)	78-0533-78
1H-Isoindole-1,3(2H)-dione, 2-[(2,5-dioxocyclopentyl)methyl]-	EtOH	212(4.32),228(4.20), 238(4.05),295(3.65)	150-5001-78
Pyrido[1,2-a]indole-6,10-dione, 7-ethoxy-9-hydroxy-	MeOH	204(4.29),230(4.21), 252(4.25),408(4.03), 424(4.05)	83-0960-78
$C_{14}H_{11}NO_4S$			
Pyrrolo[2,1-b]benzothiazole-2,3-dicarboxylic acid, dimethyl ester	MeOH	217(5.30),290(4.06)	44-2697-78
$C_{14}H_{11}NO_6$			
1,3-Dioxolo[4,5-g]quinoline-7-acetic acid, 6-methyl-α-oxo-, methyl ester, 5-oxide	EtOH	268(4.52),304(3.88), 332(3.87),347(3.91), 363(3.98)	95-0802-78
$C_{14}H_{11}NO_8$			
2(5H)-Furanone, 4-acetyl-3-methoxy-5-(6-nitro-1,3-benzodioxol-5-yl)-	EtOH	244(4.12),283(4.02), 330(3.72)	95-0802-78
$C_{14}H_{11}N_2O_3$			
Isoquinolinium, 2-[(5-nitro-2-furanyl)methyl]-, bromide	H_2O	216(4.73),233(4.52), 305(3.90)	73-2041-78
Quinolinium, 1-[(5-nitro-2-furanyl)methyl]-, bromide	H_2O	202(4.71),224(3.52), 239(3.38),312(4.02)	73-2041-78
$C_{14}H_{11}N_3$			
Benzenamine, N-(3-imino-1H-indol-2(3H)-ylidene)-	EtOH	204(4.30),225(4.05), 259(4.44),303(3.52), 435(3.52)	103-0757-78
Benzenamine, N-[(1H-indazol-3-yl)methylene]-	EtOH	210(4.02),244(3.49), 294(3.40)	103-0771-78
4H-1,2,4-Triazole, 3,4-diphenyl-	EtOH	236(4.18)	88-4629-78
$C_{14}H_{11}N_3O$			
Acetamide, N-1,10-phenanthrolin-4-yl-	neutral	212(3.81),251(4.20), 291(3.15),310s(3.10)	39-1215-78B
	monocation	218(4.43),240(4.38), 265(4.41),283(4.20), 317(4.15),348(4.07)	39-1215-78B
3a,7a-Epoxy-1H-pentaleno[1,6-de]-1,2,3-triazine, 2,3-dihydro-2-phenyl-, (3aS-cis)-	MeOH	246(4.19),258(4.05), 263s(4.01),400(4.19)	56-0511-78

Compound	Solvent	$\lambda_{max}(\log \epsilon)$	Ref.
$C_{14}H_{11}N_3O_3$			
Benzoic acid, [(2-nitrophenyl)methyl-ene]hydrazide	pH 2.57	220(4.81),273(4.29)	48-1029-78
	pH 7.31	225(4.09),282(4.34)	48-1029-78
	pH 11.80	225(4.02),277(4.11), 313(4.11)	48-1029-78
Phenol, 4-[[3-(2-nitroethenyl)phenyl]-azo]-	EtOH	<u>248(4.4),351(4.7)</u>	19-0655-78
$C_{14}H_{11}N_3O_4$			
Methanone, [4,5-dihydro-4-(3-nitrophen-yl)-1H-pyrazol-3-yl]-2-furanyl-	EtOH	215(3.9),280(4.07), 357(3.7)	104-1830-78
$C_{14}H_{11}N_3O_6$			
Acetamide, N-[4-[2-nitro-2-(5-nitro-2-furanyl)ethenyl]phenyl]-, (E)-	dioxan	246(4.09),310(4.11), 370(4.29)	73-0463-78
$C_{14}H_{11}N_3S$			
Pyrido[3,2-c]pyridazine, 4-[(phenyl-methyl)thio]-	heptane	235(4.10),336(4.10), 345(4.10),353(4.10)	103-1032-78
$C_{14}H_{11}N_5$			
Naphtho[2,1-e]pyrazolo[5,1-c][1,2,4]-triazin-1-amine, 2-methyl-	EtOH	500(4.29)	103-0313-78
$C_{14}H_{11}N_5O_2$			
1(2H)-Phthalazinone, (2-nitrophenyl)hy-drazone	dioxan	252s(4.30),290(4.00), 328s(3.89),418(4.04)	103-0575-78
1(2H)-Phthalazinone, (4-nitrophenyl)hy-drazone	dioxan	244s(4.18),280(4.04), 422(4.32)	103-0575-78
$C_{14}H_{11}N_7O_5$			
3-Formazancarboxamide, 1,5-bis(4-nitro-phenyl)-	EtOH	298(3.65),378(4.11), 442(3.56)	104-1956-78
$C_{14}H_{11}S_2$			
1,3-Benzodithiol-1-ium, 2-(phenylmeth-yl)-, tetrafluoroborate	MeCN-H_2SO_4	253(3.74),257s(3.83), 258(3.85),262(3.86), 320(3.79)	39-1133-78C
$C_{14}H_{12}$			
1,3,5-Cycloheptatriene, 7-(2,4-cyclo-pentadien-1-ylideneethylidene)-	pentane	414(4.61),424(4.61), 450s(4.32)	88-0645-78
	EtOH	417(4.63)	88-0645-78
	MeCN	430(4.62)	88-0645-78
Dicyclobuta[b,g]naphthalene, 1,2,5,6-tetrahydro-	isooctane	221(5.16),238s(3.49), 248(3.10),258(3.42), 268(3.66),279(3.78), 289s(3.69),291(3.71), 302(3.26),308(3.51), 315(3.51),322(3.77)	35-6171-78
Ethene, 1,1-diphenyl-	15M H_2SO_4	315(4.10),427(4.61)	39-0892-78B
	$MeSO_3H$	315(4.04),425(4.57)	39-0892-78B
4a,9a-Methano-9H-fluorene	THF	236(3.71),268s(3.63), 273(3.64),300s(3.17)	44-1586-78
anion	BuLi-THF	250(4.84),297(4.94), 340s(4.31)	44-1586-78
Stilbene, trans	MeOH	203(4.38),229(4.21), 296(4.46),308(4.46)	35-3819-78
$C_{14}H_{12}AsN$			
3H-1,3-Benzazarsole, 3-methyl-2-phenyl-	MeOH	228(4.23),239(4.24),	101-0001-78K

Compound	Solvent	$\lambda_{max}(\log \epsilon)$	Ref.
3H-1,3-Benzazarsole, 3-methyl-2-phenyl- (cont.)		245(4.24),318(4.12), 355s(3.71)	101-0001-78K
	MeOH-HClO₄	219s(4.20),255(3.94), 270(3.76),350(4.16)	101-0001-78K
$C_{14}H_{12}BrCl_2N$ Benzenamine, 4-bromo-N,N-bis(4-chloro- 2-butynyl)-	EtOH	208(4.36),255(4.13), 295s(--)	42-1228-78
$C_{14}H_{12}BrNS$ 2H-1,6-Thiazecine, 6-(4-bromophenyl)- 3,4,8,9-tetradehydro-5,6,7,10-tetra- hydro-	EtOH	206(4.33),258(4.28), 295(3.25)	42-1228-78
$C_{14}H_{12}BrN_3OS$ Benzoic acid, 2-[[(4-bromophenyl)ami- no]thioxomethyl]hydrazide	MeOH	232(4.36)	80-0397-78
$C_{14}H_{12}ClNO_2$ [1,1'-Biphenyl]-3-ol, 4'-chloro- 2,5-dimethyl-6-nitroso-	DMF	725(1.15)	104-2189-78
$C_{14}H_{12}ClN_3O_2$ 1H-Pyrazole, 1-acetyl-4-(5-chloro- 2,1-benzisoxazol-3-yl)-3,5-dimethyl-	EtOH	261(3.86),342(4.06)	94-1141-78
$C_{14}H_{12}ClN_3O_3$ Acetic acid, 2-(4-chlorophenyl)-2-(4- nitrophenyl)hydrazide	EtOH xylene	250(4.02),366(4.28) 360(4.05)	18-0512-78 18-0512-78
Ethanehydrazonic acid, N-(4-nitrophen- yl)-, 4-chlorophenyl ester	EtOH xylene	248(4.17),390(4.43) 370(4.37)	18-0512-78 18-0512-78
$C_{14}H_{12}N_2$ Benzene, 1-(diazophenylmethyl)-3-meth- yl-	benzene	528(1.99)	39-1283-78B
Benzene, 1-(diazophenylmethyl)-4-meth- yl-	benzene	532(2.01)	39-1283-78B
Cycloheptapyrazole, 1,4-dihydro-3-phen- yl-	n.s.g.	246(4.08)	142-0293-78C
Diazene, phenyl(2-phenylethenyl)-, (E,E)-	hexane	235(4.04),344(4.58), 444(2.80)	49-1081-78
$C_{14}H_{12}N_2O$ 6-Azaindolizin-5(6H)-one, 7-methyl- 2-phenyl-	n.s.g.	253(4.69),277s(4.09), 305s(3.72)	44-3544-78
Benzene, 1-(diazophenylmethyl)-4-meth- oxy-	benzene	527(1.99)	39-1283-78B
Benzoic acid, (phenylmethylene)hydra- zide	pH 2.42	225(4.69),253(4.16), 295(4.20)	48-1029-78
	pH 6.3	230(4.23),297(4.49)	48-1029-78
	pH 11.03	230(4.16),309(4.28), 342(4.06)	48-1029-78
	EtOH	294(4.31),330-397s(4.31)	34-0345-78
6(5H)-Phenanthridinone, 5-(methyl- amino)-	MeOH	233(4.62),238(4.64), 251s(4.30),260(4.28), 295s(3.63),312s(3.73), 326(3.90),340(3.83)	94-2508-78
$C_{14}H_{12}N_2OS$ Thieno[2,3-d]pyrimidin-4(3H)-one, 2,3-dimethyl-5-phenyl-	EtOH	222(4.27),242(4.17), 302(3.89)	1-0303-78

Compound	Solvent	$\lambda_{max}(\log \epsilon)$	Ref.
Thieno[2,3-d]pyrimidin-4(3H)-one, 2,3-dimethyl-6-phenyl-	EtOH	204(4.36),219(4.48), 328(4.32)	1-0303-78
Thieno[2,3-d]pyrimidin-4(3H)-one, 3,6-dimethyl-2-phenyl-	EtOH	217(4.72),269(3.75), 311(4.47)	1-0303-78
$C_{14}H_{12}N_2O_2$			
Benzenamine, 2-[2-(2-nitrophenyl)ethenyl]-	benzene	293(4.03),370(3.82)	33-2813-78
Benzoic acid, 2-benzoylhydrazide	EtOH	229(4.38),272s(3.76)	4-0385-78
Benzoic acid, [(2-hydroxyphenyl)methylene]hydrazide	pH 2.43	230(4.27),285(4.35), 296(4.35),325(4.18)	48-1029-78
	pH 7.24	233(4.21),288(4.37), 298(4.37),326(4.19)	48-1029-78
	pH 11.91	238(4.15),297(4.15), 338(4.02)	48-1029-78
Benzoic acid, 2-hydroxy-, (phenylmethylene)hydrazide	pH 2.42	220(4.70),246(4.07), 298(4.23),310(4.28)	48-1029-78
	pH 7.13	225(4.48),250(4.30), 298(4.46),315(4.43)	48-1029-78
	pH 12.2	240(4.26),297(4.42), 309(4.40),350(4.19)	48-1029-78
Benzo[b][1,6]naphthyridine-1,3(2H,5H)-dione, 2,5-dimethyl-	pH 7.0	252(4.18),263(4.24), 283(3.54),303(4.36), 474(3.96)	35-7375-78
	pH 7	282(4.33),303(4.48), 333(3.52),470(3.95)	150-1325-78
1H-Isoindole-1,3(2H)-dione, 2-(7-azabicyclo[2.2.1]hept-3-en-7-yl)-	EtOH	215(4.15),237(4.36), 292(3.16),304(2.99), 322s(2.65)	33-0795-78
Methanone, (4,5-dihydro-4-phenyl-1H-pyrazol-3-yl)-2-furanyl-	EtOH	209(4.12),288(3.89), 355(4.15)	104-1830-78
Quinolinium 1-cyano-2-ethoxy-2-oxoethylide	M HCl	250(4.2),330(3.9)	94-3504-78
	EtOH	255(3.9),320(3.8), 480(3.0)	94-3504-78
$C_{14}H_{12}N_2O_2S$			
Benzenecarbothioamide, 4-nitro-N-(phenylmethyl)-	CH_2Cl_2	257(4.39),299(4.02), 340(3.64),404s(3.06)	48-0313-78
$C_{14}H_{12}N_2O_3$			
1H-Indole, 2,3-dihydro-1-[2-(5-nitro-2-furanyl)ethenyl]-, (E)-	EtOH	279(4.10),505(4.19)	103-0250-78
$C_{14}H_{12}N_2O_4$			
Benzoic acid, 4-[[4-(hydroxymethyl)-phenyl]-ONN-azoxy]-	MeOH	232(3.90),265(3.23), 332(4.23)	39-0171-78C
1H-Pyrazole, 1-acetyl-4-(2-furanyl)-3-(2-furanylcarbonyl)-4,5-dihydro-	EtOH	222(4.29),293(4.19), 334(4.18)	104-1830-78
Pyrrolo[1,2-a]quinazoline-1,5-dione, 3-acetyl-3a,4-dihydro-2-hydroxy-4-methyl-	EtOH	246(4.25),304(3.95), 327(3.81)	142-1729-78A
Pyrrolo[1,2-a]quinazoline-1,5-dione, 3-acetyl-3a,4-dihydro-2-methoxy-	EtOH	305(3.95),317(3.97), 330(3.95)	142-1729-78A
$C_{14}H_{12}N_2O_4S$			
Acetamide, N-[4-[[5-(2-nitroethenyl)-2-furanyl]thio]phenyl]-	EtOH	206(4.49),260(4.40), 382(4.24)	73-2037-78
$C_{14}H_{12}N_2O_5$			
1-Azabicyclo[3.2.0]hept-2-ene-2-carboxylic acid, 7-oxo-, (2-nitrophen-	EtOH	265(3.96)	35-8006-78

Compound	Solvent	$\lambda_{max}(\log \epsilon)$	Ref.
yl)methyl ester (cont.)			35-8006-78
$C_{14}H_{12}N_2S$			
Benzenamine, 4-(6-methyl-2-benzothia-zolyl)-	EtOH	222(4.55),233s(4.50), 292s(4.62),298(4.43), 304s(4.62)	103-0380-78
Benzenamine, N-(3-methyl-2(3H)-benzo-thiazolylidene)-	EtOH	223(4.64),302(4.21)	103-0380-78
$C_{14}H_{12}N_4$			
2H-Cyclopenta[d]pyridazine, 2-methyl-5-(phenylazo)-	ether	233(4.11),257(4.23), 316(4.03),412(4.42)	44-0664-78
2H-Cyclopenta[d]pyridazine, 2-methyl-7-(phenylazo)-	ether	258(4.28),293(4.07), 392(4.18),468(4.16)	44-0664-78
Pyrido[3,2-c]pyridazin-4-amine, N-(4-methylphenyl)-	EtOH	256(4.23),373(4.02)	103-0663-78
$C_{14}H_{12}N_4O$			
2-Naphthalenol, 1-[(5-methyl-1H-pyra-zol-3-yl)azo]-	MeOH	254(4.49),300(4.88), 355(3.71),410(3.78)	103-0313-78
Pyrazolo[3,4-d][1,2]diazepine, 6-benz-oyl-2,6-dihydro-8-methyl-	EtOH	237(4.32)	39-1297-78C
Pyrido[3,2-c]pyridazin-4-amine, N-(4-methoxyphenyl)-	EtOH	256(4.27),369(4.08)	103-0663-78
1,9,9b-Triazaphenalene-3-carbonitrile, 2-ethoxy-8-methyl-	EtOH	239(4.19),250(4.26), 297(4.28),341(4.25), 390(4.18),412(4.16), 550(3.24),640(3.29)	95-0623-78
$C_{14}H_{12}N_4OS$			
Phenol, 4-[[(1-phenyl-1H-tetrazol-5-yl)thio]methyl]-	50% tert-BuOH	228(4.23)	44-1197-78
$C_{14}H_{12}N_4O_2$			
1H-Pyrazolo[3,4-d]pyrimidine-4-carbox-ylic acid, 1-phenyl-, ethyl ester	EtOH	251(4.45),330(3.16)	95-1274-78
Pyrimido[4,5-c]pyridazine-5,7(6H,8H)-dione, 6,8-dimethyl-3-phenyl-	EtOH	255s(4.01),270(4.05), 350(3.22)	4-0781-78 +142-0011-78A
$C_{14}H_{12}N_4O_5$			
Acetic acid, 2,2-bis(4-nitrophenyl)hy-drazide	EtOH	258s(3.92),325s(4.00), 380(4.31)	18-0512-78
	xylene	384(4.21)	18-0512-78
Ethanehydrazonic acid, N-(4-nitrophen-yl)-, 4-nitrophenyl ester	EtOH	225(4.18),382(4.45)	18-0512-78
	xylene	362(4.31)	18-0512-78
$C_{14}H_{12}N_6O_3$			
3-Formazancarboxamide, 1-(4-nitrophen-yl)-5-phenyl-	EtOH	450(3.85)	104-1956-78
$C_{14}H_{12}N_6O_4$			
3H-1,2,4-Triazole-3,5(4H)-dione, 4,4'-[(4,6-dimethyl-1,3-phenylene)bis-(methylene)]bis-	CHCl$_3$	537(2.44),562(2.31)	24-3519-78
$C_{14}H_{12}O$			
15-Oxatetracyclo[6.4.2.1^{2,5}.0^{1,6}]penta-deca-3,6,9,11,13-pentaene	C_6H_{12}	254s(3.41),262(3.48), 272(--),280(3.31)	5-2074-78
$C_{14}H_{12}O_2$			
1,8-Anthracenediol, 9,10-dihydro-	EtOH	253s(4.10),259(4.33),	87-0026-78

Compound	Solvent	λ_{max} (log ϵ)	Ref.
(cont.)		272(3.65),280(3.62)	87-0026-78
3,6-Ethanodicyclopenta[cd,gh]pentalene-	C_6H_{12}	271.5(3.14)	35-5589-78
7,8-dione, 2a,3,3a,5a,6,6a,6b,6c-	EtOH	403.7(1.18)	35-5589-78
octahydro-	CH_2Cl_2	408.5(1.20)	35-5589-78
Ethanone, 1-(3-hydroxy[1,1'-biphenyl]-	EtOH	288(4.35),334(3.87)	95-0503-78
4-yl)-			
2,4-Pentanedione, 3-(3-phenyl-2-prop-	EtOH	230(4.02),324(4.34)	95-0503-78
ynylidene)-			
Phenol, 4,4'-(1,2-ethenediyl)bis-, (E)-	MeOH	228(4.10),300(4.45),	95-0914-78
		307(4.45),325(3.40)	
1H-Xanthene-4-carboxaldehyde, 2,3-di-	EtOH	396(4.47)	48-0497-78
hydro-			
$C_{14}H_{12}O_2S$			
Benzene, [(2-phenylethenyl)sulfonyl]-	dioxan	211(4.35),216(4.30),	99-0194-78
		270(4.43)	
2-Butynoic acid, 2-benzo[b]thien-	MeOH	257(3.94),288s(3.25)	44-2493-78
2-ylethyl ester			
2-Butynoic acid, 2-benzo[b]thien-	MeOH	259(3.68),286(3.35)	44-2493-78
3-ylethyl ester			
4a,9a-(1-Methyletheno)-1H-3,4-dihydro-	MeOH	247(3.80),286(3.17)	44-2493-78
[1]benzothieno[2,3-c]pyran-1-one			
$C_{14}H_{12}O_3$			
1,3-Benzenediol, 5-[2-(4-hydroxyphen-	EtOH	218(4.33),227s(4.17),	94-3050-78
yl)ethenyl]-		307(4.44),320(4.43)	
2(5H)-Furanone, 4-methoxy-5-(3-phenyl-	MeOH	247(3.94),340(4.54)	64-1020-78B
2-propenylidene)-			
Methanone, (2-hydroxy-4-methoxyphenyl)-	isoPrOH	325(3.93)	117-0097-78
phenyl-			
7-Oxabicyclo[4.1.0]hepta-2,4-diene-	isooctane	228(4.17),267(3.65),	35-6483-78
2-methanol, benzoate		273(--),281s(3.57)	
	EtOH	231(4.13),267(3.71)	35-6483-78
$C_{14}H_{12}O_3S$			
Benzene, [(2-phenoxyethenyl)sulfonyl]-	dioxan	244(4.29)	99-0194-78
2-Propenoic acid, 3-[5-[(4-methylphen-	EtOH	206(4.31),245(4.04),	73-0621-78
yl)thio]-2-furanyl]-		313(4.13)	
$C_{14}H_{12}O_4$			
1,2-Benzenediol, 4-[2-(3,5-dihydroxy-	EtOH	221(4.37),240s(4.17),	94-3050-78
phenyl)ethenyl]-		293s(4.21),306s(4.32),	
		326(4.42)	
1,3-Benzenediol, 4-[2-(3,5-dihydroxy-	EtOH	218(4.40),235s(4.23),	138-1241-78
phenyl)ethenyl]-		290s(4.26),301(4.30),	
		330(4.43)	
2H,8H-Benzo[1,2-b:3,4-b']dipyran-	EtOH	211(4.31),248(3.68),	2-0179-78
2,9(10H)-dione, 8,8-dimethyl-		256(3.59),327(4.11)	
2H,8H-Benzo[1,2-b:5,4-b']dipyran-2-one,	dioxan	260(4.23),330(4.11)	2-0375-78
4-hydroxy-8,8-dimethyl-			
7H-Furo[3,2-g][1]benzopyran-7-one,	MeOH	265(3.86),280(3.65),	102-2135-78
2,3-dihydro-4-hydroxy-2-(1-meth-		335(4.30)	
ylethenyl)-			
(-)-	MeOH	265(3.98),330(4.38)	2-0563-78
	MeOH-NaOMe	280(--),340(--)	2-0563-78
7H-Furo[3,2-g][1]benzopyran-7-one,	MeOH	225(4.22),250(4.12),	2-0563-78
4-hydroxy-2-(1-methylethyl)-		265(4.11),315(4.08)	
Lomatin	EtOH	211(4.31),248(3.68),	2-0184-78
		256(3.59),327(4.11)	

Compound	Solvent	$\lambda_{max}(\log \epsilon)$	Ref.
$C_{14}H_{12}O_4S$			
2-Propenoic acid, 3-[5-[(4-methoxyphenyl)thio]-2-furanyl]-	EtOH	206(4.35),234(4.13), 314(4.23)	73-0621-78
$C_{14}H_{12}O_5$			
9,10-Anthracenedione, 1,4,4a,9a-tetrahydro-1,4,5,8-tetrahydroxy-, (1α,4aα,9aα)-	MeOH	205(4.20),218(4.20), 246(4.02),392(3.60)	150-3831-78
2H-Pyran-5-carboxylic acid, 4-(4-methoxyphenyl)-6-methyl-2-oxo-	n.s.g.	232(4.18),307(4.16)	2-0200-78
2H-Pyran-2,6(3H)-dione, 3-acetyl-4-(4-methoxyphenyl)-	EtOH and EtOH-NaOH	275(4.19),362(4.21)	2-0196-78
	EtOH-HCl	325(4.23)	2-0196-78
	dioxan	278(4.21),360(4.11)	2-0196-78
	dioxan-NaOH	278(4.22),360(4.11)	2-0196-78
	EtOAc	330(4.07)	2-0196-78
	$CHCl_3$	326(4.19)	2-0196-78
$C_{14}H_{12}O_5S$			
2-Propenoic acid, 3-[5-[(4-methylphenyl)sulfonyl]-2-furanyl]-	EtOH	207(4.16),237(4.13), 302(4.36)	73-0621-78
$C_{14}H_{12}O_6$			
Anthra[2,3-b]oxirene-3,8-dione, 1a,2,2a,8a,9,9a-hexahydro-2,4,7-trihydroxy- (1α,2β,2aβ,8aβ,9aα)-	MeOH	220(4.28),252(3.72), 391(3.49)	150-3831-78
1-Benzoxepin-3,7-dicarboxylic acid, 2,5-dihydro-5-oxo-, dimethyl ester	EtOH	237(4.44)	39-1490-78C
$C_{14}H_{12}O_6S$			
2-Propenoic acid, 3-[5-[(4-methoxyphenyl)sulfonyl]-2-furanyl]-	EtOH	206(4.24),252(4.25), 304(4.38)	73-0621-78
$C_{14}H_{12}O_7$			
9,10-Anthracenedione, 1,2,3,4-tetrahydro-1,2,3,5,8-pentahydroxy-, (1α,2α,3α)-	MeOH	216(4.49),279(3.83), 483(3.70),510(3.73), 548(3.52)	150-3831-78
$C_{14}H_{12}O_9S_4$			
1,3-Dithiole-4,5-dicarboxylic acid, 2-[4,5-bis(methoxycarbonyl)-1,3-dithiol-2-ylidene]-, dimethyl ester, 1-oxide	EtOH	210(4.65),236(4.52), 303(4.07),370(4.17)	44-4394-78
$C_{14}H_{12}S_4$			
1,3-Benzodithiole, 2-(1,3-benzodithiol-2-ylidene)-4,5,6,7-tetrahydro-	n.s.g.	258(3.83),265(3.89), 271(3.91),281(3.98), 293(4.04),307(4.00), 319(3.99),456(2.28)	44-0369-78
$C_{14}H_{13}AsO_3$			
1(4H)-Arseninacetic acid, 4-oxo-2-phenyl-, methyl ester	EtOH	230(4.32),306(4.39)	88-1471-78
$C_{14}H_{13}BF_4N_6S$			
1H-1,2,3-Triazolo[4,5-b]pyridinium, 5-amino-1-(3-ethyl-2(3H)-benzothiazolylidene)-, tetrafluoroborate	HCOOH	269(3.76),380(4.61), 400s(4.51)	33-0108-78
$C_{14}H_{13}BF_4OS$			
Naphtho[1,8-bc]thiolium, 3-ethoxy-	CF_3COOH	366(4.269),383(4.269)	5-0627-78

Compound	Solvent	$\lambda_{max}(\log \epsilon)$	Ref.
2-methyl-, tetrafluoroborate (cont.)		430s(--)	5-0627-78
$C_{14}H_{13}BrN_2$ Imidazo[1,2-a]pyridin-4-ium, 1-methyl- 2-phenyl-, bromide	EtOH	203(4.48),223(4.31), 288(4.10)	4-1149-78
$C_{14}H_{13}BrN_2O$ Ethanone, 1-(4-bromophenyl)-2-(phenyl- amino)-, oxime, (Z)-	MeOH	247(4.52)	56-1827-78
$C_{14}H_{13}BrN_4$ Ethanedial, (4-bromophenyl)hydrazone phenylhydrazone	EtOH	205(4.1),225s(3.89), 256(3.84),309(3.79), 373(4.41)	136-0089-78E
$C_{14}H_{13}BrO_4$ Ethanone, 1-(8-bromo-5,6-dihydro-4- methoxy-5-methylbenzo[1,2-b:5,4-b']- difuran-3-yl)-	EtOH	250(4.11),327(4.39)	12-1533-78
$C_{14}H_{13}Br_4N_3$ 3-Azaspiro[5.5]undeca-1,4-diene-1,5-di- carbonitrile, 2,4-bis(dibromomethyl)-	EtOH	208(4.33),270(4.94), 360(4.11)	103-1236-78
$C_{14}H_{13}ClN_2O$ Ethanone, 2-[(4-chlorophenyl)amino]- 1-phenyl-, oxime	MeOH	254(4.46)	56-1827-78
$C_{14}H_{13}ClN_2O_2$ 1H-Pyrrole-3-carboxaldehyde, 2-chloro- 4,5-dihydro-1-methyl-4-[(methylphen- ylamino)methylene]-5-oxo-	EtOH	278(4.11),385(4.3)	104-2041-78
$C_{14}H_{13}ClN_2O_3$ Benzoic acid, 2-[(2-amino-4-chlorophen- yl)amino]-5-methoxy-	MeOH	226(4.31),240s(4.10), 282(3.76),311(3.71), 358(3.72)	73-0309-78
2,4-Hexadienimidoyl chloride, N-(benz- oyloxy)-6-(methoxyimino)-	EtOH	235(3.85),313(4.72), 323s(4.69)	94-2575-78
$C_{14}H_{13}ClO$ Spiro[cyclopropane-1,1'(5'H)-[4,8]eth- enoazulen]-5'-one, 4'-chloro- 3'a,4',8',8'a-tetrahydro-	MeOH	237(3.49),273(3.34)	18-1257-78
$C_{14}H_{13}ClO_3S$ Benzenemethanol, α-[chloro(phenylsul- fonyl)methyl]-	dioxan	217(4.27),260(3.13), 266(3.29),273(3.23)	99-0194-78
$C_{14}H_{13}ClO_4$ Propanedioic acid, [3-(4-chlorophenyl)- 2-methyl-2-propenylidene]-, mono- methyl ester, (E,E)-	EtOH	266(3.60),310(4.39)	22-0457-78
(Z,E)-	EtOH	250(3.71),315(4.40)	22-0457-78
$C_{14}H_{13}Cl_2N$ Benzenamine, N,N-bis(4-chloro-2-butyn- yl)-	EtOH	206(4.73),230(4.37), 245(4.27)	42-1228-78
$C_{14}H_{13}Cl_2N_3O_2$ 1,2,4-Triazine-6-acetic acid, 5-chloro-	EtOH	210s(4.11),270(4.37)	114-0091-78B

Compound	Solvent	$\lambda_{max}(\log \epsilon)$	Ref.
3-(4-chlorophenyl)- -methyl-, ethyl ester (cont.)			114-0091-78B
$C_{14}H_{13}Cl_3N_2O$			
2-Azabicyclo[4.1.0]hept-3-ene-4-carbox-amide, 5,7,7-trichloro-2-(phenylmeth-yl)-	EtOH	263(3.64),313(3.64)	18-2698-78
3-Pyridinecarboxamide, 1,2-dihydro-1-(phenylmethyl)-2-(trichloromethyl)-	EtOH	360(3.74)	18-2698-78
$C_{14}H_{13}F$			
Spirohex-4-ene, 1-ethenyl-4-fluoro-5-phenyl-	EtOH	255(4.39)	44-4873-78
$C_{14}H_{13}IN_2O$			
Pyridinium, 2-(2-benzoxazolyl)-1,4-di-methyl-, iodide	EtOH	332(4.31)	4-0017-78
Pyridinium, 2-(2-benzoxazolyl)-1,6-di-methyl-, iodide	EtOH	335(4.26)	4-0017-78
Pyridinium, 4-(2-benzoxazolyl)-1,2-di-methyl-, iodide	EtOH	336(4.33)	4-0017-78
$C_{14}H_{13}IN_2S$			
Benzothiazolium, 3-methyl-2-(4-methyl-2-pyridinyl)-, iodide	EtOH	313(4.12)	4-0017-78
Benzothiazolium, 3-methyl-2-(6-methyl-2-pyridinyl)-, iodide	EtOH	302(4.12)	4-0017-78
Pyridinium, 4-(2-benzothiazolyl)-1,2-dimethyl-, iodide	EtOH	340(4.30)	4-0017-78
$C_{14}H_{13}N$			
Anthracen-1,4-imine, 1,2,3,4-tetra-hydro-	EtOH	265(4.71),274(4.74), 284(4.55),306(3.75), 319(3.76)	44-4469-78
$C_{14}H_{13}NO$			
2H-Benzo[a]quinolizin-2-one, 6,7-di-hydro-4-methyl-	EtOH	253(4.51),276s(4.21), 294s(4.12)	95-0198-78
2,4,6-Cycloheptatrien-1-one, 2-[(2-methylphenyl)amino]-	MeOH	238(4.36),341(4.09), 404(4.19)	18-2338-78
Ethanone, 2-amino-1,2-diphenyl-	EtOH	246(4.20),286(3.31)	44-3394-78
2H-Pyrrol-2-one, 3-ethenyl-1,5-dihydro-4-methyl-5-(phenylmethylene)-, (Z)-	MeOH	342(4.46)	4-1117-78
$C_{14}H_{13}NOS$			
2(1H)-Pyridinethione, 3-hydroxy-6-meth-yl-1-(1-phenylethenyl)-	EtOH	246(4.14),366(4.03)	1-0066-78
$C_{14}H_{13}NO_2$			
Benzoic acid, 2-[(4-methylphenyl)-amino]-	EtOH	223(4.23),286(4.17), 352(3.86)	35-5604-78
[1,1'-Biphenyl]-3-ol, 2,5-dimethyl-6-nitroso-	DMF	710(1.08)	104-2189-78
[1,1'-Biphenyl]-3-ol, 4',5-dimethyl-6-nitroso-	DMF	740(1.26)	104-2189-78
2,4,6-Cycloheptatrien-1-one, 2-[(2-methoxyphenyl)amino]-	MeOH	205(4.70),231(4.69), 344(4.38),404(4.56)	18-2185-78
	MeOH-HCl	205(4.70),240(4.71), 382(4.47)	18-2185-78
	MeOH	234(4.33),345(4.04), 406(4.21)	18-2338-78

Compound	Solvent	$\lambda_{max}(\log \epsilon)$	Ref.
1,4-Naphthalenedione, 2-pyrrolidino-	EtOH	238(4.22),277(4.40), 474(3.74)	119-0037-78
1H,8H-3a,8a-Propanoindeno[1,2-c]pyr-role-1,3(2H)-dione	EtOH	214(3.97),256s(3.00), 263(3.10),269(3.23), 275(3.24)	4-0167-78
5H-Pyrano[3,2-c]quinolin-5-one, 2,6-dihydro-2,2-dimethyl-	EtOH	275(3.80),319s(3.61), 335s(3.82),347(3.92), 361(4.02),378(3.85)	100-0184-78
2-Pyridinecarboxylic acid, 5-methyl-6-phenyl-, methyl ester	C_6H_{12}	238(4.02),278(3.81), 342(2.80)	44-2029-78
3-Pyridinecarboxylic acid, 6-methyl-2-phenyl-, methyl ester	C_6H_{12}	255(4.36)	44-3757-78
$C_{14}H_{13}NO_3$			
Acetic acid, (2-ethenyl-2,3-dihydro-3-oxo-1H-isoindol-1-ylidene)-, ethyl ester	THF	232(4.46),276(4.16), 318(4.89)	40-0404-78
[1,1'-Biphenyl]-3-ol, 4'-methoxy-5-methyl-6-nitroso-	DMF	730(1.23)	104-2189-78
Butanoic acid, 3-oxo-2-(2(1H)-quinolin-ylidene)-, methyl ester	EtOH	220(4.68),263s(4.10), 287(4.28),397(4.26)	4-1425-78
$C_{14}H_{13}NO_4$			
Furo[2,3-b]quinoline, 4,6,7-trimethoxy-(kokusaginin) (same in acid or base)	EtOH	209(4.23),250(4.95), 320(3.97)	83-0135-78
Furo[2,3-b]quinolin-4(9H)-one, 5,7-di-methoxy-9-methyl- (glycarpine)	EtOH	244(4.96),249(4.95), 306(4.29),334(4.19)	102-2145-78
1H-Pyrrole-3,4-dicarboxylic acid, 2-phenyl-, dimethyl ester	EtOH	278(4.11)	39-1588-78C
$C_{14}H_{13}NO_5$			
1,3-Dioxolo[4,5-g]quinoline-7-carbox-ylic acid, 5-ethyl-5,8-dihydro-6-methyl-8-oxo-	aq KOH	229(4.34),255(4.69), 305(4.16),324(4.20), 338(4.19)	87-0485-78
$C_{14}H_{13}NO_6$			
1H-Indole-2,3,4-tricarboxylic acid, trimethyl ester	MeOH	224(4.42),245s(3.86), 252s(3.76),314(4.25)	44-3727-78 44-3727-78
Propanedioic acid, [2-methyl-3-(4-ni-trophenyl)-2-propenylidene]-, mono-methyl ester, (E,E)-	EtOH	266(3.86),296(4.05), 340(4.27)	22-0457-78
$C_{14}H_{13}NO_7$			
2(5H)-Furanone, 4-acetyl-3-methoxy-5-(5-methoxy-2-nitrophenyl)-	EtOH	278(4.16)	95-0802-78
$C_{14}H_{13}NO_8$			
2(5H)-Furanone, 4-acetyl-5-(4,5-dimeth-oxy-2-nitrophenyl)-3-hydroxy-	EtOH	242(4.25),264(4.16), 318(4.08),378(3.78)	95-0802-78
$C_{14}H_{13}NS$			
2H-1,6-Thiazecine, 3,4,8,9-tetradehy-dro-5,6,7,10-tetrahydro-6-phenyl-	EtOH	206(4.15),250(4.00), 287(3.09)	42-1228-78
$C_{14}H_{13}N_2$			
Imidazo[1,2-a]pyridin-4-ium, 1-methyl-2-phenyl-, bromide	EtOH	203(4.48),223(4.31), 288(4.10)	4-1149-78
$C_{14}H_{13}N_2O$			
Pyridinium, 2-(2-benzoxazolyl)-1,4-di-methyl-, iodide	EtOH	332(4.31)	4-0017-78

Compound	Solvent	$\lambda_{max}(\log \epsilon)$	Ref.
Pyridinium, 2-(2-benzoxazolyl)-1,6-di-methyl-, iodide	EtOH	335(4.26)	4-0017-78
Pyridinium, 4-(2-benzoxazolyl)-1,2-di-methyl-, iodide	EtOH	336(4.33)	4-0017-78
$C_{14}H_{13}N_2O_2$			
Cycloheptatrienylium, [methyl(4-nitro-phenyl)amino]-, perchlorate	MeCN	241(4.38),341(4.30)	150-4801-78
$C_{14}H_{13}N_2S$			
Benzothiazolium, 3-methyl-2-(4-methyl-2-pyridinyl)-, iodide	EtOH	313(4.12)	4-0017-78
Benzothiazolium, 3-methyl-2-(6-methyl-2-pyridinyl)-, iodide	EtOH	302(4.12)	4-0017-78
Pyridinium, 4-(2-benzothiazolyl)-1,2-dimethyl-, iodide	EtOH	340(4.30)	4-0017-78
$C_{14}H_{13}N_3$			
Acetaldehyde, (phenylimino)-, phenyl-hydrazone, monoperchlorate	CHCl₃	450(4.64)	39-1023-78C
4-Isoquinolinecarbonitrile, 3-pyrroli-dino-	EtOH	255(4.65),295(4.17),405(3.68)	138-0677-78
Pyrrolo[1,2-a]pyrimidin-2-amine, 4-methyl-7-phenyl-	n.s.g.	257(4.61),301(3.82),331(3.49)	44-3544-78
1,3,10-Triazaphenanthrene, 6,8,9-tri-methyl-	EtOH	270(4.41),303(3.91),337(3.80),353(3.81)	94-0245-78
$C_{14}H_{13}N_3O$			
Benzeneethanol, β-azido-α-phenyl-, erythro-	EtOH	226(3.20),252(2.79),258(2.82),264(2.73),268(2.60)	44-4271-78
threo-	EtOH	226(3/21),247(2.87),252(2.89),258(2.88),264(2.76)	44-4271-78
4-Isoquinolinecarbonitrile, 3-morpho-lino-	EtOH	254(4.51),295(4.10),390(3.57)	138-0677-78
$C_{14}H_{13}N_3OS$			
Benzoic acid, 2-[(phenylamino)thioxo-methyl]hydrazide	MeOH	230(4.36)	80-0397-78
nickel chelate	MeOH	275(4.47)	80-0397-78
Sulfonium, dimethyl-, 1-cyano-2-oxo-2-(2-quinolinylamino)ethylide	pH 1	253(4.49),338(4.27),348s(4.25)	94-2435-78
	EtOH	213(4.40),243(4.41),266(4.46),324(4.06),334(4.03)	94-2435-78
$C_{14}H_{13}N_3O_2$			
1H-Pyrrolo[3,2-d]pyrimidine-2,4(3H,5H)-dione, 7-methyl-3-(phenylmethyl)-	MeOH	271(4.15)	39-0483-78C
$C_{14}H_{13}N_3O_2S$			
Thiazolo[3,2-b][1,2,4]triazole-5-acetic acid, 3-phenyl-, ethyl ester	EtOH	268(4.30?)	4-0401-78
1,9,9b-Triazaphenalene-3-carboxylic acid, 8-methyl-2-(methylthio)-, methyl ester	EtOH	243s(4.63),259(4.78),317(4.77),354(4.58),415(4.59),425(4.62)	95-0623-78
$C_{14}H_{13}N_3O_3$			
Acetic acid, 2-(4-nitrophenyl)-2-phen-ylhydrazide	EtOH	250(3.85),365(4.23)	18-0512-78
	xylene	360(4.13)	18-0512-78

Compound	Solvent	$\lambda_{max}(\log \epsilon)$	Ref.
Benzoic acid, 3-[[4-cyano-5-(methoxy-imino)-1,3-pentadienyl]amino]-	MeOH	396(4.40)	83-0433-78
Benzoic acid, 3-[[5-cyano-5-(methoxy-imino)-1,3-pentadienyl]amino]-	MeOH	401.5(4.62)	83-0433-78
Benzoic acid, 4-[[3-cyano-5-(methoxy-imino)-1,3-pentadienyl]amino]-	MeOH	393(4.65)	83-0433-78
Benzoic acid, 4-[[4-cyano-5-(methoxy-imino)-1,3-pentadienyl]amino]-	MeOH	402(4.46)	83-0433-78
Benzoic acid, 4-[[5-cyano-5-(methoxy-imino)-1,3-pentadienyl]amino]-	MeOH	404(4.43)	83-0433-78
Ethanehydrazonic acid, N-(4-nitrophen-yl)-, phenyl ester	EtOH	250(4.11),395(4.35)	18-0512-78
	xylene	370(4.40)	18-0512-78
Ethanone, 2-[(4-nitrophenyl)amino]-1-phenyl-	MeOH	247(4.09)	56-1827-78
$C_{14}H_{13}N_3O_5$			
Benzenamine, N,N-dimethyl-4-[2-nitro-2-(5-nitro-2-furanyl)ethenyl]-, (E)-	dioxan	275(4.26),310(4.22), 460(4.48)	73-0463-78
$C_{14}H_{13}N_3S$			
Thieno[2,3-d]pyrimidin-4-amine, N,6-di-methyl-2-phenyl-	EtOH	204(4.43),220(4.44), 248(4.44),277(3.98), 284(3.99),294(3.99), 325(4.12)	1-0303-78
$C_{14}H_{13}N_5O$			
Acetamide, N-[(1-phenyl-1H-pyrazolo-[3,4-d]pyrimidin-4-yl)methyl]-	EtOH	243(4.49),300(3.50)	95-0891-78
3-Formazancarboxamide, 1,5-diphenyl-	EtOH	253(3.99),296(3.96), 457(4.28)	104-1956-78
Guanine, O^6-cinnamyl-	EtOH	237(4.39),266s(4.04), 272(4.05),282s(3.93)	44-0516-78
	EtOH-HCl	239(4.34),274(4.04), 281s(4.03)	44-0516-78
	EtOH-NaOH	237(4.33),266s(4.01), 273(4.04),281s(3.99)	44-0516-78
4-Pyridinecarboxylic acid, [(phenylhy-drazono)ethylidene]hydrazide	EtOH	204(4.08),245(3.94), 273(3.91),385(4.43)	136-0089-78E
Pyrimido[5,4-c]pyridazin-8(2H)-one, 6-(dimethylamino)-2-phenyl-	pH 1	202(4.32),272(3.98), 344s(4.25),379(4.34)	44-2536-78
	pH 7	204(4.39),309(4.11), 390(4.30)	44-2536-78
	pH 13	219(4.24),309(4.10), 390(4.28)	44-2536-78
$C_{14}H_{13}N_5O_3S$			
2,4(1H,3H)-Pyrimidinedione, 5-[[6-meth-yl-2-(methylthio)-4-oxopyrido[2,3-d]-pyrimidin-3(4H)-yl]methyl]-	pH 1	266(4.30),285(4.22), 295(3.27),345(4.12)	44-0828-78
	pH 7	276(4.33),320(3.80)	44-0828-78
	pH 11	283(4.36),322(3.78)	44-0828-78
2,4(1H,3H)-Pyrimidinedione, 5-[[6-meth-yl-2-(methylthio)-4-oxopyrido[2,3-d]-pyrimidin-8(4H)-yl]methyl]-	pH 1	269(4.29),295(4.11), 351(3.27)	44-0828-78
	pH 7	274(4.43),371(4.12)	44-0828-78
	pH 11	275(4.49),370(4.13)	44-0828-78
$C_{14}H_{13}N_6S$			
1H-1,2,3-Triazolo[4,5-b]pyridinium, 5-amino-1-(3-ethyl-2(3H)-benzothia-zolylidene)-, tetrafluoroborate	HCOOH	269(3.76),380(4.61), 400s(4.51)	33-0108-78

Compound	Solvent	$\lambda_{max}(\log \epsilon)$	Ref.
$C_{14}H_{13}OS$			
Naphtho[1,8-bc]thiolium, 3-ethoxy-2-methyl-, tetrafluoroborate	CF_3COOH	366(4.269),383(4.269), 430s(--)	5-0627-78
$C_{14}H_{13}PS$			
9-Phosphapentacyclo[4.3.0.0^{2,5}.0^{3,8}-0^{4,7}]nonane, 9-phenyl-, 9-sulfide	EtOH	253(3.52)	44-4338-78
$C_{14}H_{14}$			
1,3,5-Cycloheptatriene, 1-(4-methyl-phenyl)-	EtOH	238(4.04),300(3.86)	88-0345-78
1,3,5-Cycloheptatriene, 2-(4-methyl-phenyl)-	EtOH	243(4.21)	88-0345-78
1,3,5-Cycloheptatriene, 3-(4-methyl-phenyl)-	EtOH	233(3.98),283(3.82)	88-0345-78
1,3,5-Cycloheptatriene, 7-(4-methyl-phenyl)-	EtOH	263(3.56),273(3.46)	88-0345-78
$C_{14}H_{14}BrNO$			
2H-Pyrrol-2-one, 5-[(4-bromophenyl)-methylene]-4-ethyl-1,5-dihydro-3-methyl-, (Z)-	EtOH	233(4.00),357s(4.19), 325s(4.40),336(4.41)	49-0183-78
$C_{14}H_{14}BrN_2O_2$			
Pyridinium, 1-(2-bromoethyl)-4-[(4-ni-trophenyl)methyl]-, bromide	H_2O	215s(4.12),260s(4.13), 265(4.15),274s(4.09)	48-0133-78
deprotonated	H_2O	534(4.51)	48-0133-78
$C_{14}H_{14}Br_2N_2O$			
4H-7a,9a-Diazoniacyclopenta[def]phen-anthrene, 8,9-dihydro-4-hydroxy-4-methyl-	pH 7.0	265(3.61),335(4.30), 348(4.39)	64-0080-78B
$C_{14}H_{14}ClN_3O_3$			
1,2,4-Triazine-5-acetic acid, 3-(4-chlorophenyl)-1,6-dihydro-α-methyl-6-oxo-, ethyl ester	EtOH	211s(4.12),267(4.33)	114-0091-78B
1,2,4-Triazine-6-acetic acid, 3-(4-chlorophenyl)-2,5-dihydro-α-methyl-5-oxo-, ethyl ester	EtOH	250(4.34)	114-0091-78B
$C_{14}H_{14}ClN_7$			
2,4-Pteridinediamine, 6-[[(4-chloro-phenyl)methylamino]methyl]-	MeOH-HCl	250(4.38),337(4.01)	87-0331-78
	MeOH-KOH	260(4.60),375(3.88)	87-0331-78
$C_{14}H_{14}ClN_7O$			
2,4-Pteridinediamine, 6-[[(4-chloro-phenyl)methylamino]methyl]-, 8-oxide	MeOH-HCl	259(4.60),360(3.85)	87-0331-78
	MeOH-KOH	268(4.61),400(3.89)	87-0331-78
$C_{14}H_{14}Cl_2N_4$			
2-Tetrazene, 1,4-bis(4-chlorophenyl)-1,4-dimethyl-	MeCN	352(4.48)	40-0150-78
$C_{14}H_{14}Cl_2O_4$			
6H-Furo[3,2-h][1]benzopyran-5,7-diol, 3,4-dichloro-7,8-dihydro-2,8,8-tri-methyl-, (+)-	EtOH	219(4.48),248s(3.96), 257(4.13),266(4.14), 293(3.46),302s(3.42)	102-1359-78
$C_{14}H_{14}N$			
Cycloheptatrienylium, (methylphenyl-amino)-, perchlorate	MeCN	240(4.35),341(4.23)	150-4801-78

Compound	Solvent	$\lambda_{max}(\log \epsilon)$	Ref.
$C_{14}H_{14}NOP$			
Phenophosphazine, 5,10-dihydro-5,10-di-methyl-, 10-oxide	EtOH	273(4.25),301(3.92), 327(3.76)	65-1205-78
$C_{14}H_{14}NO_2P$			
Phenophosphazine, 5,10-dihydro-10-meth-oxy-5-methyl-, 10-oxide	EtOH	223(4.21),272(4.31), 299(4.00),337(3.80)	65-1205-78
$C_{14}H_{14}NO_3P$			
1H-Pyrrolo[1,2-c][1,3,2]oxazaphosphor-ine, 3,4-dimethyl-1-phenoxy-, 1-ox-ide, (±)-	MeCN	270(4.44)	44-4996-78
$C_{14}H_{14}N_2$			
1H-Azuleno[8,1-cd]pyridazine, 1,3,8-trimethyl-	hexane	229(4.53),250(4.56), 282(4.13),349(4.01), 362(4.05),366(4.05), 404(2.87),578s(2.66), 618(2.81),677(2.81), 753(2.51)	142-0387-78C
9H-Pyrido[3,4-b]indole, 1-propyl-	MeOH	236(4.61),241(4.60), 251s(4.44),?(4.05), 289(4.29),338(3.74), 351(3.75)	95-0898-78
$C_{14}H_{14}N_2O$			
2,5-Cyclohexadien-1-one, 4-[4-(dimeth-ylamino)phenyl]imino]-	C_6H_{12}	270(4.24),295(4.04), 340(3.83),545(4.08)	48-0557-78
	EtOH	275(4.17),300(4.04), 350(3.81),610(4.40)	48-0557-78 +80-0617-78
	ether	275(4.21),295(4.04), 335(3.79),568(4.18)	48-0557-78
	CCl_4	295(4.02),340(3.76), 555(4.14)	48-0557-78
Ethanone, 1-phenyl-2-(phenylamino)-, oxime, (Z)-	MeOH	245(4.20)	56-1827-78
2-Propanone, 1-(6-methyl-3-pyridazin-yl)-3-phenyl-	C_6H_{12}	283(3.75),325(3.88), 337(3.83)	94-3633-78
	EtOH	268s(3.40),322(3.70)	94-3633-78
10H-Pyrido[3,2-b][1,4]benzoxazine, 10-(1-methylethyl)-	C_6H_{12}	233(4.49),333(4.03)	95-0585-78
$C_{14}H_{14}N_2OS_2$			
Hydrazinecarbodithioic acid, [[5-(4-methylphenyl)-2-furanyl]methylene]-, methyl ester	dioxan	382(4.62)	73-2643-78
$C_{14}H_{14}N_2O_2$			
3H-Azepine-3-carboxamide, 2-methoxy-N-phenyl-	EtOH	256(3.63)	39-0191-78C
Benzo[b][1,6]naphthyridine-1,3(2H,4H)-dione, 5,10-dihydro-2,5-dimethyl-	pH 3	350(4.09)	150-1325-78
	pH 13	290(4.39)	150-1325-78
2,5-Cyclohexadien-1-one, 3-amino-4-[(4-hydroxy-2-methylphenyl)-imino]-2-methyl-	neutral	476(3.78)	39-1292-78B
	anion	583(4.11)	39-1292-78B
2,5-Cyclohexadien-1-one, 3-amino-4-[(4-hydroxy-3-methylphenyl)-imino]-2-methyl-	neutral	482(3.86)	39-1292-78B
	anion	592(4.28)	39-1292-78B
2,5-Cyclohexadien-1-one, 5-amino-4-[(4-hydroxy-2-methylphenyl)-imino]-2-methyl-	neutral	470(3.79)	39-1292-78B
	anion	578(4.12)	39-1292-78B

Compound	Solvent	$\lambda_{max}(\log \epsilon)$	Ref.
2,5-Cyclohexadien-1-one, 5-amino-4-[(4-hydroxy-3-methylphenyl)-imino]-2-methyl-	neutral anion	469(3.92) 581(4.25)	39-1292-78B 39-1292-78B
1-Isoquinolineacetic acid, α-(1-amino-ethylidene)-, methyl ester	EtOH	217(4.8),281(4.2), 326(3.6)	4-1425-78
1,8-Naphthyridin-2(1H)-one, 5,6,7,8-tetrahydro-4-hydroxy-3-phenyl-	MeOH	221(4.09),248(4.01), 306s(4.09),311(4.22)	24-2813-78
2(1H)-Pyrimidinone, 4,6-dimethyl-1-(2-oxo-2-phenylethyl)-	n.s.g.	243(4.17),305(3.89)	44-3544-78
2-Quinolineacetic acid, α-(1-aminoeth-ethylidene)-, methyl ester	EtOH	215(4.6),225s(4.5), 238(4.4),282(4.3), 325(3.8)	4-1425-78
5,6-Quinolinedione, 8-piperidino-	EtOH	240(4.61),286(4.11), 314(4.17),495(3.86)	65-2318-78
5,8-Quinolinedione, 6-piperidino-	EtOH	234(4.37),274(4.17), 476(3.72)	65-2318-78
copper chelate dichloride	EtOH	238(4.26),278(3.95), 492(3.75)	65-2318-78
7,8-Quinolinedione, 5-piperidino-	EtOH	239(4.16),278(3.84), 320(3.74),476(3.44)	65-2318-78
copper chelate	EtOH	254(4.32),369(4.02), 689(2.37)	65-2318-78
$C_{14}H_{14}N_2O_2S$			
Ethanone, 1-[4-(3,4-dimethyl-1H-[1,2]-oxathiolo[3,2-e][1,2,3]thiadiazol-7-S^{IV}-1-yl)phenyl]-	C_6H_{12}	233(4.00),299(4.15), 458(4.47)	39-0195-78C
[1,2]Oxathiolo[4,3,2-hi][1,2,3]benzo-thiadiazole-3-S^{IV}, 2,6,7,8-tetrahy-dro-2-(4-methoxyphenyl)-	C_6H_{12}	237(4.00),284(3.89), 466(4.34)	39-0195-78C
$C_{14}H_{14}N_2O_3$			
Benzenemethanol, 4,4'-azoxybis-	MeOH	233(3.90),330(4.04)	39-0171-78C
$C_{14}H_{14}N_2O_3S$			
2H-1-Benzothiopyran-2-one, 3-nitro-4-piperidino-	EtOH	235(4.32),265(3.87), 293(3.84),352(3.87)	103-0033-78
$C_{14}H_{14}N_2O_4$			
11,12-Diazatetracyclo[4.4.2.0^{2,5}.-0^{7,10}]dodeca-3,8,11-triene-1,6-dicarboxylic acid, dimethyl ester	MeOH	366(1.88)	88-1183-78
2,4-Hexadienamide, N-(benzoyloxy)-6-(methoxyimino)-	EtOH	232(4.27),298(4.70)	94-2575-78
5,8-Isoquinolinedione, 7-hydroxy-6-methyl-3-morpholino-	EtOH	250(4.36),325(4.17)	94-2175-78
5,8-Isoquinolinedione, 7-methoxy-3-morpholino-	EtOH	250(4.39),317(4.16)	4-0569-78
$C_{14}H_{14}N_2O_4S$			
Thiocyanic acid, 1-α-L-arabino-furanosyl-1H-indol-5-yl ester	EtOH	240(4.58),275(3.64)	104-2011-78
Thiocyanic acid, 1-α-L-arabino-pyranosyl-1H-indol-5-yl ester	EtOH	240(4.58),285(3.50)	104-2011-78
$C_{14}H_{14}N_2S_2$			
Carbamodithioic acid, methylphenyl-, 2-pyridinylmethyl ester	n.s.g.	338s(1.93)	97-0381-78
2,4(1H,3H)-Quinazolinedithione, 5,6,7,8-tetrahydro-3-phenyl-	EtOH EtOH-NaOH	298(4.39),357(4.04) 258(4.30),293(4.28), 375(4.23)	106-0185-78 106-0185-78

Compound	Solvent	$\lambda_{max}(\log \epsilon)$	Ref.
$C_{14}H_{14}N_4$			
Benzotriazole, 4-amino-5,7-dimethyl-2-phenyl-	EtOH	220(4.163),257(4.38), 309f(4.26),365(3.42)	33-2628-78
$C_{14}H_{14}N_4O$			
Pyrazolo[3,4-d][1,2]diazepine, 6-benz-oyl-2,4,5,6-tetrahydro-8-methyl-	EtOH	295(3.95)	39-1297-78C
4H-Pyrazolo[3,4-d]pyridazin-4-one, 1,5-dihydro-1,3-dimethyl-7-(phenyl-methyl)-	EtOH	215(4.31),280(3.76)	4-0813-78
4H-Pyrazolo[3,4-d]pyridazin-4-one, 1,5-dihydro-3,7-dimethyl-1-(phenyl-methyl)-	EtOH	216(4.28),280(3.79)	4-0813-78
4H-Pyrazolo[3,4-d]pyridazin-4-one, 2,5-dihydro-2,3-dimethyl-7-(phenyl-methyl)-	EtOH	216(4.10),278(3.70)	4-0813-78
4H-Pyrazolo[3,4-d]pyridazin-4-one, 2,5-dihydro-3,7-dimethyl-2-(phenyl-methyl)-	EtOH	215(4.21),277(3.79)	4-0813-78
4H-Pyrazolo[3,4-d]pyridazin-4-one, 1,5-dihydro-1,5,7-trimethyl-3-phenyl-	EtOH	214(4.11),261(4.15)	4-0813-78
4H-Pyrazolo[3,4-d]pyridazin-4-one, 2,5-dihydro-2,5,7-trimethyl-3-phenyl-	EtOH	216(4.20),254(4.08), 272(3.98)	4-0813-78
4H-Pyrrolo[2,3-d]pyrimidin-4-one, 2-amino-3,7-dihydro-6-methyl-5-(phenylmethyl)-	acid	229(4.21),269(4.07)	44-3937-78
	pH 7.0	223(4.32),266(4.06), 286s(3.94)	44-3937-78
	base	267(4.04)	44-3937-78
$C_{14}H_{14}N_4O_2$			
Benzenamine, N,N-dimethyl-4-(3-nitro-phenyl)azo]-	EtOH	260(4.3),435(4.5)	19-0655-78
Hydrazinecarboxamide, 2-nitroso-N-phen-yl-2-(phenylmethyl)-	H_2O	237(4.41),355(1.98)	104-2319-78
Pyrimido[4,5-c]pyridazine-5,7(1H,6H)-dione, 2,8-dihydro-6,8-dimethyl-3-phenyl-	EtOH	257(3.92),345(3.47)	4-0781-78 +142-0011-78A
$C_{14}H_{14}N_4O_3S$			
Pyrazolo[3,4-d][1,2]diazepine, 2-acet-yl-2,3,3a,6-tetrahydro-6-(phenylsul-fonyl)-	EtOH	328(4.19)	39-1297-78C
$C_{14}H_{14}N_4O_4S$			
6H-1,3,4-Thiadiazine-5-acetic acid, 2-amino- -[(benzoyloxy)imino]-, ethyl ester	EtOH	251(4.00?),280(4.00?), 372(3.80?)	4-0401-78
4-Thiazoleacetic acid, 2-(2-benzoylhy-drazino)- -(hydroxyimino)-, ethyl ester, (Z)-	EtOH	237(4.17?),327(4.24?)	4-0401-78
$C_{14}H_{14}N_6O$			
Pyrazinecarbonitrile, 3-amino-6-[[[4-(dimethylamino)phenyl]imino]methyl]-, N-oxide	MeCN	237(4.22),258s(3.89), 338(4.27),376(4.38)	44-0736-78
$C_{14}H_{14}N_6O_4$			
2-Tetrazene, 1,4-dimethyl-1,4-bis(4-nitrophenyl)-	MeCN	436(4.68)	40-0150-78
$C_{14}H_{14}N_8O_4$			
[8,8'-Bi-1H-purine]-2,2',6,6'-tetrone,	pH 7.0	214(4.40),229(4.41),	24-0982-78

Compound	Solvent	$\lambda_{max}(\log \epsilon)$	Ref.
3,3',7,7'-tetrahydro-1,1',3,3'-tetra-methyl- (cont.)		272(4.05),344(4.55), 360s(4.40)	24-0982-78
	pH 13.0	230(4.43),273(3.98), 339(4.57),354s(4.41)	24-0982-78
	MeOH	215s(4.34),229(4.37), 272(4.13),344(4.55), 360s(4.42)	24-0982-78
$C_{14}H_{14}O$			
3-Buten-2-ol, 4-(1-naphthalenyl)-	MeOH	225(4.79),255(3.66), 296(4.05)	83-0328-78
1,3,5-Cycloheptatriene, 1-(4-methoxy-phenyl)-	EtOH	242(4.08),310(4.11)	88-0345-78
1,3,5-Cycloheptatriene, 2-(4-methoxy-phenyl)-	EtOH	250(4.40)	88-0345-78
1,3,5-Cycloheptatriene, 3-(4-methoxy-phenyl)-	EtOH	238(4.21),288(4.19)	88-0345-78
1,3,5-Cycloheptatriene, 7-(4-methoxy-phenyl)-	EtOH	275(3.75),283(3.63)	88-0345-78
$C_{14}H_{14}ORu$			
Ruthenocene, 1,1'-(2-oxo-1,4-butane-diyl)-	EtOH	269s(3.05)	18-0909-78
$C_{14}H_{14}OS$			
4(1H)-Dibenzothiophenone, 2,3-dihydro-2,2-dimethyl-	EtOH	240(4.06),250(4.06), 297(4.21),330s(3.67)	42-1232-78
$C_{14}H_{14}O_2$			
1,2-Ethanediol, 1,2-diphenyl-, (R*,S*)-	EtOH	248s(2.3),254(2.5), 260(2.6),266(2.5)	77-0850-78
3,6-Ethanodicyclopenta[cd,gh]pentalene-7,8-dione, 1,2,2a,3,3a,5a,6,6a,6b,6c-decahydro-	C_6H_{12}	255.7(3.54)	35-5589-78
	EtOH	417.6(1.33)	35-5589-78
	CH_2Cl_2	421.0(1.34)	35-5589-78
4,7-Ethano-1H-indene-8,9-dione, 3a,4,7,7a-tetrahydro-1-(1-methyl-ethylidene)-, (3aα,4α,7α,7aα)-	benzene	455.4(1.92)	5-0440-78
Phenol, 4,4'-(1,2-ethanediyl)bis-	MeOH	279(3.64),284s(3.60), 305s(3.16),325s(3.06)	95-0914-78
	M NaOH	280(3.55),286(3.56), 295s(3.42)	95-0914-78
Phenol, 4-[(4-methylphenoxy)methyl]-	MeOH	281(3.67)	95-0914-78
	M NaOH	282(3.67)	95-0914-78
Spiro[1,4-methanonaphthalen-2(1H),2'-oxiran]-3(4H)-one, 3',3'-dimethyl-, (1α,2α,4α)-	EtOH	225(3.49),262s(2.46), 268(2.68),275(2.79), 325(2.59)	12-1113-78
$C_{14}H_{14}O_2S$			
Benzenesulfinic acid, 4-methyl-, phenylmethyl ester	hexane	199(4.10),225(3.69), 254(3.05)	118-0441-78
$C_{14}H_{14}O_3$			
2H,8H-Benzo[1,2-b:3,4-b']dipyran-2-one, 3,4-dihydro-8,8-dimethyl-	EtOH	231(4.47),272(3.83), 282(3.76),315(3.35)	2-0179-78
2H-Furo[2,3-h]-1-benzopyran-2-one, 3,4-dihydro-8-(1-methylethyl)-	EtOH	222(4.27),256(4.24)	2-0179-78
Oxiranecarboxaldehyde, 2-(2,3-dihydro-3-oxo-1H-inden-1-yl)-3,3-dimethyl-, (R*,S*)-	n.s.g.	225(3.68),248(3.86), 290(3.27)	12-1113-78

Compound	Solvent	$\lambda_{max}(\log \epsilon)$	Ref.
$C_{14}H_{14}O_4$			
2H,6H-Benzo[1,2-b:5,4-b']dipyran-2-one, 7,8-dihydro-4-hydroxy-8,8-dimethyl-	dioxan	280(3.98),320(4.29)	2-0375-78
2H,6H-Benzo[1,2-b:5,4-b']dipyran-2-one, 7,8-dihydro-7-hydroxy-8,8-dimethyl-	EtOH	336(4.47)	102-0328-78
	EtOH-base	320(4.55)	102-0328-78
1-Cyclopentene-1-acetic acid, 2-(4-methoxyphenyl)-5-oxo-	EtOH	226(4.06),300(4.43)	42-0580-78
hydrate	EtOH	226(3.98),300(4.31)	42-0580-78
3-Furancarboxylic acid, 4,5-dihydro-5-methyl-4-oxo-2-phenyl-, ethyl ester	EtOH	216(4.05),248(3.85), 296(4.10)	118-0448-78
2(5H)-Furanone, 3-acetyl-4-hydroxy-5-[(4-methylphenyl)methyl]-	EtOH	224(3.90),230(3.96), 267(3.84)	4-0327-78
2(5H)-Furanone, 4-acetyl-3-methoxy-5-(4-methylphenyl)-	EtOH	264(4.04)	4-0737-78
1,4-Naphthalenedione, 5,7-diethoxy-	EtOH	265(4.15),354s(3.33), 417(3.57)	12-1353-78
1,4-Naphthalenedione, 5,7-dimethoxy-2,3-dimethyl-	EtOH	264(4.24),278s(4.17), 411(3.57)	12-1353-78
2-Pentenedioic acid, 4-(phenylmethylene)-, dimethyl ester, (E,E)-	EtOH	234(4.22),305(4.30)	24-3665-78
(E,Z)-	EtOH	229(3.94),305(4.49)	24-3665-78
Propanedioic acid, (2-methyl-3-phenyl-2-propenylidene)-, monomethyl ester, (E,E)-	EtOH	224(3.90),310(4.45)	22-0457-78
(Z,E)-	EtOH	220(3.60),300(4.40)	22-0457-78
$C_{14}H_{14}O_4S_2$			
4H-Pyran-4-one, 2,2'-dithiobis[3,5-di-methyl-	isooctane	223(4.12),257(4.04)	44-4966-78
$C_{14}H_{14}O_5$			
1,2-Benzenediol, 3,3'-oxybis[5-methyl-(aspermutarubrol)	EtOH	229(4.12),277(3.30)	138-0797-78
	EtOH-base	245(4.09),287(3.66)	138-0797-78
3-Furancarboxylic acid, 4,5-dihydro-2-(4-methoxyphenyl)-4-oxo-, ethyl ester	EtOH	212(3.92),226(4.08), 265(3.67),322(4.41)	118-0448-78
2(5H)-Furanone, 3-acetyl-4-hydroxy-5-[(4-methoxyphenyl)methyl]-	EtOH	230(3.79),266(3.82)	4-0327-78
2(5H)-Furanone, 4-acetyl-3-methoxy-5-(4-methoxyphenyl)-	EtOH	266(4.06)	4-0737-78
7H-Furo[3,2-g][1]benzopyran-7-one, 2,3-dihydro-4-hydroxy-2-(1-hydroxy-1-methylethyl)- (as glucoside)	MeOH	265(3.70),335(4.14)	2-0563-78
	MeOH-NaOAc	280(--),340(--)	2-0563-78
	MeOH-NaOMe	280(--),340(--)	2-0563-78
2-Naphthaleneacetic acid, 1,2,3,4-tet-rahydro-6-methoxy-α,1-dioxo-, methyl ester	EtOH	351(4.15)	39-0110-78C
1,2-Naphthalenedicarboxylic acid, 5,6,7,8-tetrahydro-5-oxo-, dimethyl ester	MeCN	205(4.14),255(4.10), 260s(4.06),302(3.49), 312s(3.42)	88-2145-78
1,2-Naphthalenedicarboxylic acid, 5,6,7,8-tetrahydro-8-oxo-, dimethyl ester	MeCN	225(4.32),248s(3.93), 290(2.93),302(2.93)	88-2145-78
1,4-Naphthalenedione, 6-ethyl-5-hy-droxy-2,7-dimethoxy-	MeOH	221(4.47),258s(4.22), 264(4.24),306(4.00), 424(3.64)	31-1257-78
Oxiraneacetic acid, 3-(methoxycarbo-nyl)-α-(phenylmethylene)-, methyl ester, [2α(E),3β]-	EtOH	274(4.22)	24-3665-78
Oxiranecarboxylic acid, 2-(3-methoxy-3-oxo-1-propenyl)-3-phenyl-, methyl ester, [2α,2(E),3β]-	EtOH	235s(3.82),285s(2.84)	24-3665-78

Compound	Solvent	$\lambda_{max}(\log \epsilon)$	Ref.
2,4-Pentadienoic acid, 5-(6-methoxy-1,3-benzodioxol-5-yl)-, methyl ester, (E,E)-	n.s.g.	374(4.36)	78-1979-78
(E,Z)-	n.s.g.	370(4.18)	78-1979-78
Rutaretin	MeOH	265(3.86),280(3.60), 335(4.30)	102-2135-78
$C_{14}H_{14}O_6$			
1H-2-Benzopyran-4-carboxylic acid, 6,7-dimethoxy-1-oxo-, ethyl ester	MeOH	250(3.93),280(3.03), 320(2.38)	4-0257-78
3,8-Dioxatricyclo[5.1.0.0^{2,4}]octane-5,6-diol, 4-[(benzoyloxy)methyl]-(dideacetylcrotepoxide)	MeOH	230(4.14),273(2.99), 281(3.90)	44-0171-78
1,2-Naphthalenedicarboxylic acid, 5,6,7,8-tetrahydro-3-hydroxy-5-oxo-, dimethyl ester	EtOH	220(4.36),257(4.05), 341(3.66)	142-0327-78C
1,2-Naphthalenedicarboxylic acid, 5,6,7,8-tetrahydro-4-hydroxy-8-oxo-, dimethyl ester	EtOH	234(4.27),322(3.20)	142-0327-78C
$C_{14}H_{14}O_7$			
2-Butenedioic acid, 2-[[4-(methoxycarbonyl)phenoxy]methyl]-, 1-methyl ester, (E)-	EtOH	256(4.37)	39-1490-78C
$C_{14}H_{14}O_8$			
1,2,3,4-Benzenetetracarboxylic acid, tetramethyl ester	EtOH	285(3.09)	78-2617-78
1,2,4,5-Benzenetetracarboxylic acid, tetramethyl ester	EtOH	290(3.41)	78-2617-78
$C_{14}H_{14}S$			
Cyclopenta[b]thiopyran, 4,5,6,7-tetrahydro-2-phenyl-	hexane	198(4.5),235(4.1), 297(3.0)	103-0605-78
$C_{14}H_{15}BrN_2O$			
1H-Imidazo[1,2-a]pyridin-4-ium, 2-hydroxy-1-methyl-2-phenyl-2,3-dihydro-, bromide	EtOH	205(4.37),238(4.16), 328(3.61)	4-1149-78
$C_{14}H_{15}BrN_2O_2$			
Pyridinium, 1-ethyl-4-[(4-nitrophenyl)-methyl]-, bromide	H_2O	556(4.48)	48-0133-78
$C_{14}H_{15}Br_2NO_4$			
1H-Pyrrolo[1,2-a]indole-5,8-dione, 6,9-dibromo-2,3,6,7-tetrahydro-7,7-dimethoxy-6-methyl-	EtOH	254(3.82),294(3.60), 346(3.84)	94-3815-78
$C_{14}H_{15}Br_2N_3$			
3-Azaspiro[5.5]undeca-1,4-diene-1,5-di-carbonitrile, 2,4-bis(bromomethyl)-	EtOH	208(4.10),267(3.91), 361(3.51)	103-1236-78
$C_{14}H_{15}ClN_2O$			
2-Propyn-1-one, 1-(4-chlorophenyl)-3-(4-methyl-1-piperazinyl)-	MeCN	257(4.13),320(4.25)	33-1609-78
$C_{14}H_{15}ClO_2S$			
2-Cyclohexen-1-one, 2-[(4-chlorophenyl)thio]-3-hydroxy-5,5-dimethyl-	MeOH	262(3.87)	44-2676-78

Compound	Solvent	$\lambda_{max}(\log \epsilon)$	Ref.
$C_{14}H_{15}ClO_4$			
6H-Furo[3,2-h][1]benzopyran-5,7-diol, 3-chloro-7,8-dihydro-2,8,8-trimethyl-	EtOH	216(4.34),245s(3.80), 254(3.91),262s(3.86), 289(3.28),298s(3.24)	102-1359-78
$C_{14}H_{15}Cl_2N_2O$			
Quinolinium, 3-(aminocarbonyl)-2-(dichloromethyl)-1-propyl-, chloride	CH_2Cl_2	240(3.5),314(3.7)	5-1536-78
$C_{14}H_{15}Cl_2N_3O_8$			
1H-Benzimidazolium, 1-methyl-3-(1-methylpyridinium-2-yl]-, diperchlorate	EtOH	235(4.24),313(4.08)	73-2046-78
$C_{14}H_{15}Cl_3N_2O$			
Acetamide, 2,2,2-trichloro-N-[(2,3,5-trimethyl-1H-indol-6-yl)methyl]-	$CHCl_3$	235(3.71),291(3.13)	103-0856-78
3-Quinolinecarboxamide, 1,2-dihydro-1-propyl-2-(trichloromethyl)-	CH_2Cl_2	254(4.1),284s(3.5), 383(3.3)	5-1536-78
$C_{14}H_{15}F$			
Spirohex-4-ene, 4-fluoro-1,2-dimethyl-5-phenyl-, cis	EtOH	233(3.84),267(4.10)	44-4873-78
trans	EtOH	267(4.10)	44-4873-78
$C_{14}H_{15}IN_2O$			
10H-Pyrido[3,2-b][1,4]benzoxazinium, 1,3,10-trimethyl-, iodide	MeOH	223(4.78),389(4.11)	95-0585-78
$C_{14}H_{15}N$			
Benzenamine, 2,6-dimethyl-N-phenyl-	MeOH	211(4.25),243(4.08), 271(3.97)	44-4975-78
Benzenamine, 2-methyl-N-(2-methylphenyl)-	MeOH	212(4.28),237(3.91), 282(4.11)	44-4975-78
Phenanthridine, 7,8,9,10-tetrahydro-5-methyl-, perchlorate	MeOH	236(4.70),318(4.00)	65-1707-78
1H-Pyrrole, 1-ethenyl-2-(4-ethylphenyl)-	C_6H_{12}	210(4.07),250(4.11), 265(4.11)	103-0399-78
$C_{14}H_{15}NO$			
9H-Cyclooct[b]indol-9-one, 5,6,7,8,10-11-hexahydro-	MeOH	224(4.71),284(4.09), 291(4.06)	150-1683-78
$C_{14}H_{15}NOS_2$			
Piperidinium, 1-(4-hydroxy-5-phenyl-1,3-dithiol-2-ylidene)-, hydroxide, inner salt	dioxan	224(4.16),252(4.10), 314(4.10),478(3.62)	24-2021-78
$C_{14}H_{15}NO_2$			
1H-Benz[g]indole-2-carboxylic acid, 6,7,8,9-tetrahydro-, methyl ester	EtOH	242(4.19),302(4.13)	103-0518-78
3H-Benz[e]indole-2-carboxylic acid, 6,7,8,9-tetrahydro-, methyl ester	EtOH	238(4.23),302(4.26)	103-0518-78
2-Butenoic acid, 4-(2-methyl-3-phenyl-2H-azirin-2-yl)-, methyl ester, (E)-	C_6H_{12}	243(4.15)	44-3757-78
(Z)-	C_6H_{12}	243(4.22)	44-3757-78
2-Butenoic acid, 4-(3-methyl-2-phenyl-2H-azirin-2-yl)-, methyl ester, (E)-	C_6H_{12}	253(4.10)	44-3757-78
Naphthalene, 1,2,3,4-tetrahydro-5-nitro-	C_6H_{12}	233(4.62),260(4.00)	18-0331-78
Naphthalene, 1,2,3,4-tetrahydro-6-nitro-	EtOH	222(4.63),280(4.46), 326(3.94)	18-0331-78

Compound	Solvent	$\lambda_{max}(\log \epsilon)$	Ref.
1H-Pyrrole-2-acetic acid, 4-methyl-5-phenyl-, methyl ester	C_6H_{12}	236(3.89),278(3.99), 350(2.92)	44-2029-78
1H-Pyrrole-3-acetic acid, 5-methyl-2-phenyl-, methyl ester	EtOH	217(3.43),293(3.76)	44-3757-78
1H-Pyrrol-3-ol, 2-methyl-5-(4-methyl-phenyl)-, acetate	MeOH	295(4.26)	49-0137-78
$C_{14}H_{15}NO_3$			
2H,6H-Benzo[1,2-b:5,4-b']dipyran-2-one, 3,4,7,8-tetrahydro-4-imino-8,8-di-methyl-	dioxan	255(3.96),290(4.02), 320(4.3)	2-0375-78
7-Isoquinolineacetic acid, 3-methoxy-α-methyl-, methyl ester	MeOH	227(4.85)	44-3778-78
1H-Pyrrole-2-carboxylic acid, 3-hy-droxy-1-(phenylmethyl)-, ethyl ester	EtOH	264(4.17)	94-2224-78
$C_{14}H_{15}NO_3S$			
Acetic acid, (3-methyl-4-oxo-5-phenyl-2-thiazolidinylidene)-, ethyl ester, (Z)-	MeOH	243(3.97),285(4.32)	5-0473-78
$C_{14}H_{15}NO_4$			
6H-1,3-Oxazinium, 5-ethyl-4-hydroxy-2-(4-methoxyphenyl)-3-methyl-6-oxo-, hydroxide, inner salt	MeCN	211s(4.22),260(4.25), 285(4.11),350(3.47)	5-1655-78
2,4-Pentadienamide, 5-(6-methoxy-1,3-benzodioxol-5-yl)-N-methyl-	n.s.g.	370(4.33)	78-1979-78
9H-Pyrrolo[1,2-a]indol-9-one, 3,9a-di-hydro-3α-hydroxy-7,9aβ-dimethoxy-6-methyl-, (±)-	MeOH	235(4.13),269s(3.74), 337(3.18)	39-0460-78C
3-Quinolineacetic acid, 1,4-dihydro-α-methoxy-2-methyl-4-oxo-, methyl ester	EtOH	243(4.34),253(4.30), 280s(3.45),293(3.66), 318(4.00),330(4.00)	95-0802-78
$C_{14}H_{15}NO_5$			
2-Azabicyclo[2.2.2]octa-5,7-diene-5,6-dicarboxylic acid, 3-oxo-2-(1-propenyl)-, dimethyl ester, (E)-	EtOH	247(3.72)	78-2617-78
1H-2-Benzopyran-1,3(4H)-dione, 4-[(di-methylamino)methylene]-6,7-dimethoxy-	dioxan	235(3.67),265(3.44)	4-0257-78
1H-Indole-3-carboxylic acid, 4,7-di-hydro-5-methoxy-2,6-dimethyl-4,7-dioxo-, ethyl ester	MeOH	210(4.27),234(4.11), 286(4.01),330(3.64)	88-2251-78
$C_{14}H_{15}NO_6$			
4-Isoquinolinecarboxylic acid, 1,2-di-hydro-5,6,7-trimethoxy-2-methyl-1-oxo-	MeOH	235(3.49),250(3.61), 295(3.01),340(2.87)	4-0257-78
$C_{14}H_{15}N_2O$			
1H-Imidazo[1,2-a]pyridin-4-ium, 2,3-di-hydro-2-hydroxy-1-methyl-2-phenyl-, bromide	EtOH	205(4.37),238(4.16), 328(3.61)	4-1149-78
10H-Pyrido[3,2-b][1,4]benzoxazinium, 1,3,10-trimethyl-, iodide	MeOH	223(4.78),389(4.11)	95-0585-78
$C_{14}H_{15}N_2OS$			
1,3-Thiazin-1-ium, 2-morpholino-4-phenyl-, perchlorate	MeCN	219(3.99),263(4.12), 330(4.30),384s(3.55)	97-0334-78
$C_{14}H_{15}N_2O_2$			
Pyridinium, 1-ethyl-4-[(4-nitrophenyl)-	H_2O	556(4.48)	48-0133-78

Compound	Solvent	λ_{max}(log ϵ)	Ref.
methyl]-, bromide (cont.)			48-0133-78
$C_{14}H_{15}N_3$			
Benzenamine, N,N-dimethyl-4-(phenyl-azo)-	MeOH	<u>260(4.0)</u>,405(4.4)	19-0655-78
$C_{14}H_{15}N_3$			
2H-Pyrrolo[3,4-b]quinoxaline, 2-(1-methylethyl)-	MeOH	254(4.71),322(3.66), 329(3.79),336(3.94), 344(4.03),352(4.11), 357(4.05),510(3.31)	88-4671-78
monoperchlorate	MeOH-HClO$_4$	242(4.45),263(4.48), 362(4.24),371(4.27), 590(3.38)	88-4671-78
$C_{14}H_{15}N_3O$			
2-Butenenitrile, 4-(methoxyimino)-2-[2-[(4-methylphenyl)amino]-ethenyl]-, (?,Z,E)-	MeOH	396.9(4.39)	83-0433-78
3,5-Hexadienenitrile, 2-(methoxyimino)-6-[(4-methylphenyl)amino]-, (?,Z,E)-	MeOH	410(4.56)	83-0433-78
2,4-Pentadienenitrile, 2-[(methoxyimino)methyl]-5-[(4-methylphenyl)amino]-, (?,Z,E)-	MeOH	397.5(4.31)	83-0433-78
Pyridinium, 1-(benzoylimino)-3-(dimethylamino)-, hydroxide, inner salt	EtOH	355(3.60)	33-2887-78
4H-Pyrimido[2,1-b]quinazolin-4-one, 11-ethyl-6,11-dihydro-2-methyl-	MeOH	258(4.10),315(4.14)	4-0077-78
2-Quinoxalinecarboxaldehyde, 3-piperidino-	CHCl$_3$	453(2.74)	97-0177-78
$C_{14}H_{15}N_3OS$			
Pyrido[2,3-d]pyrimidin-4(1H)-one, 2,3,5,6,7,8-hexahydro-3-(phenylmethyl)-2-thioxo-	MeOH	220(4.11),288(4.12)	24-2297-78
$C_{14}H_{15}N_3O_2$			
1H-Azepine-4-carbonitrile, 2,5,6,7-tetrahydro-3-[(4-methoxyphenyl)-amino]-2-oxo-	n.s.g.	247(4.05),336(3.90)	103-1013-78
Pyrido[2,3-d]pyrimidine-2,4(1H,3H)-dione, 5,6,7,8-tetrahydro-3-(phenylmethyl)-	MeOH	220(4.10),289(4.26)	24-2297-78
$C_{14}H_{15}N_3O_2S$			
4-Thiazoleacetic acid, 2-[(phenylmethylene)hydrazino]-, ethyl ester	EtOH	248(4.15?),344(4.27?)	4-0401-78
$C_{14}H_{15}N_3O_3$			
1,2,4-Triazine-5-acetic acid, 1,6-dihydro-α-methyl-6-oxo-3-phenyl-, ethyl ester	EtOH	215s(3.98),262(4.27)	114-0091-78B
1,2,4-Triazine-6-acetic acid, 2,5-dihydro-α-methyl-5-oxo-3-phenyl-, ethyl ester	EtOH	244(4.38)	114-0091-78B
$C_{14}H_{15}N_3O_3S$			
Acetic acid, [2-(benzoylamino)-4H-1,3,4-thiadiazin-5(6H)-ylidene]-, ethyl ester, (Z)-	EtOH	255(3.82?),360(4.05?)	4-0401-78

Compound	Solvent	$\lambda_{max}(\log \epsilon)$	Ref.
Acetic acid, [[(5-methyl-2-phenyl-2H-1,2,3-triazol-4-yl)carbonyl]thio]-, ethyl ester	dioxan	305(4.23)	64-0075-78B
4(3H)-Pyrimidinone, 6-methyl-3-[(5-methyl-2-nitrophenyl)methyl]-2-(methylthio)-	MeOH	285(4.15)	4-0077-78
4-Thiazoleacetic acid, 2-(2-benzoyl-hydrazino)-, ethyl ester	EtOH	240(4.43?),263(4.14?)	4-0401-78
$C_{14}H_{15}N_3O_5$			
Carbonic acid, 1-methyl-1-(1-methyl-2-nitro-1H-imidazol-5-yl)ethyl phenyl ester	CHCl$_3$	319(3.95)	87-0781-78
2-Quinoxalinecarboxamide, 3-methyl-6(and 7)-(2-methyl-1,3-dioxolan-2-yl)-, 1,4-dioxide	MeOH	236(4.34),269(4.42), 382(4.04)	87-0483-78
$C_{14}H_{15}N_3S$			
Hydrazinecarbothioamide, N-methyl-1,2-diphenyl-	EtOH	240(4.38),284s(3.68)	104-0106-78
Hydrazinecarbothioamide, 1-methyl-N,2-diphenyl-	EtOH	240(4.41)	104-0106-78
Hydrazinecarbothioamide, 2-methyl-N,2-diphenyl-	EtOH	238(4.42)	104-0106-78
Hydrazinecarbothioamide, 2-(4-methyl-phenyl)-N-phenyl-	EtOH	245(4.72)	104-0106-78
$C_{14}H_{15}N_3S_2$			
2,4(1H,3H)-Quinazolinedithione, 5,6,7,8-tetrahydro-3-(phenylamino)-	EtOH EtOH-NaOH	294(4.26),360(3.86) 263(4.18),285(4.18), 373(4.05)	106-0185-78 106-0185-78
$C_{14}H_{15}N_5$			
6H-Purin-6-imine, 3-ethyl-3,9-dihydro-9-(phenylmethyl)-, monoperchlorate	pH 1 pH 7	272(4.24) 272(4.24)	88-5007-78 88-5007-78
6H-Purin-6-imine, 9-ethyl-3,9-dihydro-3-(phenylmethyl)-, monoperchlorate	pH 1 pH 7	272(4.21) 272(4.20)	88-5007-78 88-5007-78
$C_{14}H_{15}N_5O$			
9H-Purin-6-amine, N-ethoxy-9-(phenyl-methyl)-	pH 1 pH 7 pH 13 EtOH	270(4.24) 270.5(4.21) 287(4.09) 271(4.16)	94-1929-78 94-1929-78 94-1929-78 94-1929-78
9H-Purin-6-amine, 9-ethyl-N-(phenyl-methoxy)-	pH 1 pH 7 pH 13 EtOH	273.5(4.15) 270.5(4.22) 286.5(4.11) 270.5(4.19)	94-1929-78 94-1929-78 94-1929-78 94-1929-78
6H-Purin-6-one, 3,7-dihydro-3-methyl-2-(methylamino)-7-(phenylmethyl)-	pH 1 pH 7 pH 12	238(3.96),262(4.06) 240(3.97),261(4.07) 236(3.96),265(4.07)	44-1644-78 44-1644-78 44-1644-78
$C_{14}H_{15}N_5O_5S_2$			
5-Thia-1-azabicyclo[4.2.0]oct-2-ene-2-carboxylic acid, 7-[[(2-amino-4-thiazolyl)(methoxyimino)acetyl]-amino]-3-methyl-8-oxo-, (E)-	EtOH	237(4.24),305(3.46)	78-2233-78
(Z)-	EtOH	253(4.20),295(3.80)	78-2233-78
$C_{14}H_{15}N_5O_6$			
Furo[2',3':4,5]oxazolo[3,2-e]purine-8-methanol, 7-acetoxy-4-amino-	pH 1 neutral	257(4.18),280s(3.88) 255(4.20)	78-1133-78 78-1133-78

Compound	Solvent	$\lambda_{max}(\log \epsilon)$	Ref.
6a,7,8,9a-tetrahydro-, acetate (cont.)			78-1133-78
$C_{14}H_{16}$			
Anthracene, 1,4,5,8,9,10-hexahydro-	heptane	208(3.67)	5-1406-78
Benzocyclooctene, 9-ethenyl-5,6,7,8-tetrahydro-	EtOH	216(4.23),261(4.28)	44-1050-78
$C_{14}H_{16}ClN_3O_3$			
1,3,5-Triazine-2,4(1H,3H)-dione, 6-(5-chloro-2-hydroxyphenyl)dihydro-1,3-dimethyl-5-(2-propenyl)-	EtOH	229(3.98),290(3.53)	4-1193-78
$C_{14}H_{16}ClN_5$			
[1,2,4]Triazino[4,3-b]indazol-4-amine, 8-chloro-N,N-diethyl-3-methyl-	C_6H_{12}	399.5(3.96)	24-2258-78
$C_{14}H_{16}Cl_2$			
Bicyclo[8.2.2]tetradeca-5,10,12,13-tetraene, 5,6-dichloro-, (E)-	EtOH	221(3.86),265s(2.45), 271(2.53),279(2.48)	35-1806-78
$C_{14}H_{16}Cl_2N_2$			
1,1'-Bipyridinium, 4,4'-dichloro-2,2',6,6'-tetramethyl-, bis-(tetrafluoroborate)	EtOH	215(3.45),270(3.96)	120-0001-78
dichloride	EtOH	220(3.73),275(4.03)	120-0001-78
$C_{14}H_{16}NO_2$			
2H-[1]Benzopyrano[3,4-b]pyridinium, 1,3,4,5-tetrahydro-4,4-dimethyl-5-oxo-, iodide	n.s.g.	280(3.98),316(3.73)	103-1223-78
$C_{14}H_{16}N_2$			
2-Pyridinamine, N,4,6-trimethyl-3-phenyl-	EtOH	259(3.59),310(3.0)	88-4135-78 +103-1139-78
$C_{14}H_{16}N_2O$			
Acetamide, N-[8-(dimethylamino)-1-naphthalenyl]-	MeOH	210(4.25),234(4.63), 320(4.03)	103-1152-78
$C_{14}H_{16}N_2OS$			
Azepino[4,3-b][1,4]benzothiazin-1(2H)-one, 11-ethyl-3,4,5,11-tetrahydro-	EtOH	254(4.24),285(3.84)	103-0306-78
2H-1-Benzothiopyran-2-one, 3-amino-4-piperidino-	EtOH	225(4.55),274(4.03), 366(3.99)	103-0033-78
1H-[1,2]Oxathiolo[3,2-e][1,2,3]thiadiazole-7-S^{IV}, 3,4-diethyl-1-phenyl-	C_6H_{12}	235(4.03),275(3.91), 446(4.34)	39-0195-78C
$C_{14}H_{16}N_2O_2S$			
[1]Benzothieno[3,2-e]pyrrolo[1,2-a]-pyrimidine-1,5-dione, 2,3,3a,4,6-7,8,9-octahydro-3a-methyl-	EtOH	225(4.59),257s(4.12), 318(4.12)	4-0949-78
2-Propenoic acid, 3-[5-methyl-2-(methylthio)imidazol[1,2-a]pyridin-3-yl]-, ethyl ester	EtOH	254(4.18),275(4.22), 290(4.11),364(4.19)	95-0631-78
$C_{14}H_{16}N_2O_3$			
1H-Pyrazole-4-carboxylic acid, 3-(1-hydroxyethyl)-5-phenyl-, ethyl ester	EtOH	216(4.13),240(3.98)	118-0448-78
1H-Pyrazole-4-carboxylic acid, 3-(hydroxymethyl)-5-(4-methylphenyl)-, ethyl ester	EtOH	215(4.18),243(4.00)	118-0448-78

Compound	Solvent	$\lambda_{max}(\log \epsilon)$	Ref.
$C_{14}H_{16}N_2O_4$			
1H-Pyrazole-4-carboxylic acid, 3-(hydroxymethyl)-5-(4-methoxyphenyl)-, ethyl ester	EtOH	221(4.08),262(3.95)	118-0448-78
Pyrrolo[2,3-b]indole-1,2(2H)-dicarboxylic acid, 3,3a,8,8a-tetrahydro-, dimethyl ester	EtOH	243(3.85),299(3.38)	35-5564-78
D-Ribitol, 1,4-anhydro-1-C-(3-phenyl-1H-pyrazol-4-yl)-, (α)-	EtOH	242.5(3.92)	94-0898-78
(β)-	EtOH	242.5(3.77)	94-0898-78
$C_{14}H_{16}N_2O_6$			
1H-Indole, 1-β-D-fucopyranosyl-6-nitro-	EtOH	212(4.28),246(3.95), 318(3.88),357(3.82)	136-0017-78E
1H-Indole, 1-β-L-rhamnopyranosyl-6-nitro-	EtOH	211(4.36),248(4.01), 318(3.90),359(3.81)	136-0017-78E
$C_{14}H_{16}N_2O_6S$			
2,4,6(1H,3H,5H)-Pyrimidinetrione, 5,5-diethyl-1-[(phenylsulfonyl)oxy]-	pH 0.5	204(4.30),216(4.13)	12-2517-78
$C_{14}H_{16}N_2S$			
1H-Imidazole, 4-phenyl-1-[2-(2-propenylthio)ethyl]-	EtOH	258(4.16)	4-0307-78
Piperidine, 1-(4-phenyl-2-thiazolyl)-	EtOH	239(4.41),265s(4.06), 287(3.87)	94-3017-78
$C_{14}H_{16}N_2S_2$			
1H-[1,2]Dithiolo[5,1-e][1,2,3]thiadiazole-7-S^{IV}, 3,4-diethyl-1-phenyl-	C_6H_{12}	238(4.52),253s(4.20), 293(4.04),489(4.12)	39-0195-78C
$C_{14}H_{16}N_4$			
2-Tetrazene, 1,4-dimethyl-1,4-diphenyl-	MeCN	345(4.43)	40-0150-78
$C_{14}H_{16}N_4O$			
2H-Imidazo[2,1-a]phthalazin-4-ium, 3-(dimethylamino)-5,6-dihydro-2,2-dimethyl-6-oxo-, hydroxide, inner salt	MeCN	234(4.22),261(4.28), 315s(3.40),406(3.59)	33-2116-78
$C_{14}H_{16}N_4O_3S$			
Benzenesulfonic acid, 4-[(1,1-dicyano-3-methylbutyl)azo]-3-methyl-, sodium salt	H_2O	280(4.08)	126-2845-78
$C_{14}H_{16}N_4O_3S_2$			
11,14,17-Trioxa-2,5-dithia-7,21,22,23-tetraazatricyclo[16.3.1.1^{6,10}]tricosa-1(22),6,8,10(23),18,20-hexaene	EtOH	210(3.80),260(4.78), 305(4.20)	44-3362-78
$C_{14}H_{16}N_4O_4$			
1H-Indazole, 1-(2,4-dinitrophenyl)-2,3,3a,4,5,7a-hexahydro-2-methyl-	MeCN	236(4.02),393(4.12)	78-0903-78
$C_{14}H_{16}N_6O_3S$			
Benzenesulfonamide, N-(aminoiminomethyl)-4-[[4-cyano-5-(methoxyimino)-1,3-pentadienyl]amino]-, (?,Z,E)-	MeOH	396.5(4.73)	83-0433-78
$C_{14}H_{16}N_6O_5$			
9H-Purin-6-amine, 9-[3-(acetylamino)-2,3^2-anhydro-3-C-(carboxymethyl)-3-	H_2O	205(4.37),258(4.15)	136-0039-78A

Compound	Solvent	$\lambda_{max}(\log \epsilon)$	Ref.
deoxy-α-D-xylofuranosyl]- (cont.)			136-0039-78A
β-	H_2O	205(4.17),258(4.05)	136-0039-78A
$C_{14}H_{16}O$			
Benzene, 1-methoxy-4-(5-methyl-1,4-cyclohexadien-1-yl)-	n.s.g.	274(4.04)	88-4559-78
1-Naphthalenepropanol, γ-methyl-, (R)-	MeOH	224(4.93),272(3.73), 282(3.82)	83-0328-78
$C_{14}H_{16}OS_2$			
Pentacyclo[6.2.2.0^{2,7}.0^4,^{10}.0^5,^{9}]dodecane-3,6-dione, monoethylene thioacetal	EtOH	255(2.42),300(1.83)	104-2336-78
$C_{14}H_{16}OSe$			
2-Cyclohexen-1-one, 4,4-dimethyl-6-(phenylseleno)-	MeOH	217(4.15),233(4.0)	24-1944-78
$C_{14}H_{16}O_2$			
Benz[d]indene-2,8(1H,3H)-dione, 3a,4,5,6-tetrahydro-3a-methyl-, (3aR*,10aR*)-	EtOH	245(4.19)	88-2187-78
3,6-Ethanodicyclopenta[cd,gh]pentalene-7,8-dione, dodecahydro-	EtOH	424.2(1.48)	35-5589-78
	CH_2Cl_2	427.2(1.49)	35-5589-78
5,9-Methano-6H-benzocyclohepten-6-one, 5,7,8,9-tetrahydro-8-hydroxy-7,7-dimethyl-, exo	n.s.g.	230(3.47),263s(2.77), 273(2.92),278(2.94), 310(2.64)	12-1129-78
5,9-Methano-7H-benzocyclohepten-7-one, 5,6,8,9-tetrahydro-8-hydroxy-6,6-dimethyl-, exo	n.s.g.	230(3.11),255s(2.64), 263(2.82),270(3.01), 276(3.05)	12-1129-78
Spiro[1,4-methanonaphthalen-2,2'(1H)-oxiran]-3-ol, 3,4-dihydro-3',3'-dimethyl-	EtOH	217(3.44),260(2.51), 267(2.68),273(2.71)	12-1113-78
$C_{14}H_{16}O_2S$			
2-Cyclohexen-1-one, 3-hydroxy-5,5-dimethyl-2-(phenylthio)-	MeOH	248(4.30),280s(3.85)	44-2676-78
$C_{14}H_{16}O_3$			
1H-Indene-5-acetic acid, 2,3-dihydro-2,4,6-trimethyl-3-oxo-	EtOH	217(4.53),260(4.20), 303(3.40)	94-2365-78
4a,7-Methano-4aH-benzocycloheptene-4,6(5H,7H)-dione, 1-ethoxy-8,9-dihydro-	n.s.g.	211(3.67),341(3.43)	12-1561-78
4a,7-Methano-4aH-benzocyclohepten-8(5H)-one, 6,7-dihydro-1,4-dimethoxy-	n.s.g.	225(4.06),250s(3.97), 286(3.67),418(3.79)	12-1561-78
$C_{14}H_{16}O_3S$			
Ethanone, 1-[3-methyl-3-(phenylsulfonyl)-1-cyclopenten-1-yl]-	EtOH	242(4)	88-2675-78
$C_{14}H_{16}O_4$			
2H,8H-Benzo[1,2-b:3,4-b']dipyran-2-one, 3,4,9,10-tetrahydro-9-hydroxy-8,8-dimethyl-	EtOH	225(4.0),279(3.2), 287(3.2)	2-0179-78
2-Naphthalenecarboxylic acid, 1,2,3,4-tetrahydro-6-methoxy-2-methyl-1-oxo-, methyl ester	EtOH	277(4.24)	39-0110-78C
$C_{14}H_{16}O_5$			
Cyclopenta[cd]pentalen-1(2H)-one,	CHCl	241(2.94),284(2.77)	23-1687-78

Compound	Solvent	$\lambda_{max}(\log \epsilon)$	Ref.
4,6-diacetoxy-2a,3,4,4a,5,6b-hexahydro- (cont.)			23-1687-78
3(2H)-Pyranone, 4,5-(1-methylethylidenedioxy)-2-phenoxy-	EtOH	231(2.98),259(3.08), 266(3.04)	150-5344-78
	MeCN	310(1.70)	150-5344-78
$C_{14}H_{16}O_6$			
Cyclobuta[1,2:3,4]dicyclopentane-1,4-dione, 3,6-diacetoxy-	EtOH	294(1.60)	118-0543-78
1,2-Naphthalenedicarboxylic acid, 5,6,7,8-tetrahydro-3,5-dihydroxy-, dimethyl ester	EtOH	215(4.41),250(3.92), 320(3.70)	142-0327-78C
2-Oxaspiro[4.5]deca-6,9-diene-3-carboxylic acid, 3-ethoxy-1,8-dioxo-, ethyl ester	MeOH	229(3.9)	88-0475-78
$C_{14}H_{16}O_7$			
2-Propenoic acid, 3,3'-(3,4-dimethoxy-2,5-furandiyl)bis-, dimethyl ester	EtOH	260(4.17),370(4.33)	33-0430-78
$C_{14}H_{16}Ru$			
Ruthenocene, 1,1'-(1,4-butanediyl)-	EtOH	317(2.48)	18-0909-78
$C_{14}H_{16}S_2$			
15,16-Dithiatricyclo[10.2.1.1^{5,8}]hexadeca-5,7,12,14-tetraene	benzene	526(1.22)	18-0309-78
	EtOH	225(4.23),329(4.44), 468(2.69)	18-0309-78
$C_{14}H_{17}BrN_2O_2$			
1-Cyclopentene-1-carboxylic acid, 2-[(2-amino-4-bromophenyl)amino]-, ethyl ester	EtOH	235(3.91),300(4.05)	104-0286-78
$C_{14}H_{17}BrN_2O_4$			
β-Alanine, N-nitroso-N-propyl-, 2-(4-bromophenyl)-2-oxoethyl ester	EtOH	256(4.35)	94-3909-78
Butanoic acid, 4-(ethylnitrosoamino)-, 2-(4-bromophenyl)-2-oxoethyl ester	EtOH	256(4.45)	94-3909-78
Glycine, N-butyl-N-nitroso-, 2-(4-bromophenyl)-2-oxoethyl ester	EtOH	257(4.31)	94-3909-78
$C_{14}H_{17}BrN_2O_6S$			
Uridine, 5-bromo-2',3'-O-(1-methylethylidene)-5'-thio-, 5'-acetate	EtOH	276(3.96)	94-2664-78
$C_{14}H_{17}BrN_2O_7$			
Uridine, 5-bromo-2',3'-O-(1-methylethylidene)-, 5'-acetate	EtOH	276(3.98)	94-2657-78
$C_{14}H_{17}ClN_2O_2$			
Acetamide, 2-chloro-N-[(5-methoxy-2,3-dimethyl-1H-indol-6-yl)methyl]-	CHCl$_3$	230(4.29),287(3.98)	103-0856-78
1-Cyclopentene-1-carboxylic acid, 2-[(2-amino-4-chlorophenyl)amino]-, ethyl ester	EtOH	250(3.85),300(3.92)	104-0286-78
$C_{14}H_{17}ClN_4O_6$			
Carbamic acid, (7-chloro-3-β-D-ribofuranosyl-3H-imidazo[4,5-b]pyridin-5-yl)-, ethyl ester	pH 1	245(3.98),250(4.00), 259s(3.85),301(4.23)	4-0839-78
	pH 11	247s(3.91),253(4.00), 260(3.96),299(4.20)	4-0839-78

Compound	Solvent	$\lambda_{max}(\log \epsilon)$	Ref.
$C_{14}H_{17}ClO$ 1H-Inden-1-one, 6-(2-chloroethyl)- 2,3-dihydro-2,5,7-trimethyl-	EtOH	219(4.66),259(4.24), 303(3.49)	94-2365-78
$C_{14}H_{17}ClO_2$ 1H-Inden-1-one, 6-(2-chloroethyl)- 2,3-dihydro-3-hydroxy-2,5,7-tri- methyl-, (2S-trans)-	EtOH	218(4.46),257(4.17), 299(3.37)	94-2365-78
$C_{14}H_{17}Cl_2N_5OS$ Hydrazinecarbothioamide, 2-[1-[[bis(2- chloroethyl)amino]methyl]-1,2-dihydro- 2-oxo-3H-indol-3-ylidene]-	EtOH	353(4.24)	111-0515-78
$C_{14}H_{17}Cl_2N_5O_3$ 1H-Indole-2,3-dione, 1-[[bis(2-chloro- ethyl)amino]methyl]-5-nitro-, 3-(methylhydrazone)	EtOH	338(4.12)	111-0515-78
$C_{14}H_{17}IN_2$ 1H-Perimidinium, 2,3-dihydro-1,1,3-tri- methyl-, iodide	MeOH	248(4.36),333(3.81)	103-1145-78
$C_{14}H_{17}N$ 2H-Azirine, 3-[2-(2-butenyl)phenyl]- 2,2-dimethyl-, (E)-	MeOH	248(4.20)	35-2181-78
2H-Azirine, 3-[2-(3-butenyl)phenyl]- 2,2-dimethyl-	MeOH	250(4.67)	35-2181-78
$C_{14}H_{17}NO$ Cyclohexanone, 2-[(methylphenylamino)- methylene]-	EtOH	348(4.30)	97-0256-78
2-Cyclohexen-1-one, 3-methyl-2-(methyl- phenylamino)-	MeOH	295(3.29),370(2.37)	77-0029-78
2(1H)-Pyridinone, 5-ethyl-5,6-dihydro- 1-(phenylmethyl)-	EtOH	253.5(3.49)	94-0645-78
2-Pyrrolidinone, 3,3,4-trimethyl- 5-methylene-4-phenyl-	EtOH	238(3.92)	78-3291-78
3H-Pyrrol-3-one, 1,2-dihydro-1,2,2,5- tetramethyl-4-phenyl-	MeCN	228(3.98),343(3.95)	5-1203-78
$C_{14}H_{17}NOS$ 7-Oxa-3-thia-1-azaspiro[5.5]undec-1-ene, 2-phenyl-, (±)-, hydrochloride	EtOH	239(4.12)	87-0895-78
$C_{14}H_{17}NO_2$ 2-Propen-1-one, 2-(morpholinomethyl)- 1-phenyl-	hexane EtOH	323(4.34) 340(4.20)	48-0945-78 48-0945-78
$C_{14}H_{17}NO_2S$ Acetic acid, [(2-propyl-2H-isoindol- 1-yl)thio]-, methyl ester	EtOH	332(3.93),345s(--)	44-2886-78
$C_{14}H_{17}NO_2S_2$ Ethanethioic acid, S,S'-(3,4-di-2-prop- enyl-1H-pyrrole-2,5-diyl) ester	EtOH	256(4.19)	23-0221-78
$C_{14}H_{17}NO_3$ Acetamide, N-(7-methoxy-2,2-dimethyl- 2H-1-benzopyran-5-yl)- (spectrum of free base, not acetyl deriv.)	MeOH	220(4.29),240(4.06), 288(3.47)	24-0439-78

Compound	Solvent	$\lambda_{max}(\log \epsilon)$	Ref.
Cyclohexanone, 2-(2-nitro-1-phenyl-ethyl)-	EtOH	214(3.87)	2-0405-78
[1,3]Dioxepino[5,6-c]pyridine, 1,5-di-hydro-3,3,8-trimethyl-9-(2-propynyl-oxy)-	EtOH	269(3.56),275s(3.53)	33-2542-78
7,7,12-Trimethyl-13-methylene-2-aza-6,8-dioxatetracyclo[8.2.1.0^{2,12}.-0^{4,10}]tridec-3-en-11-one	EtOH	226s(3.45),262(3.02), 270(2.94),309(2.32)	33-2542-78
$C_{14}H_{17}NO_4$			
Cyclopentanone, 2-[1-(4-methoxyphenyl)-2-nitroethyl]-	EtOH	211(4.00),225(4.01), 277(3.17),283(3.11)	2-0405-78
$C_{14}H_{17}NO_4S$			
1H-Indole, 1-α-L-arabinopyranosyl-5-(methylthio)-	EtOH	228(4.40),242(4.24), 250(4.14)	104-2011-78
$C_{14}H_{17}NO_5$			
Cyclopentanone, 2-[1-(4-hydroxy-3-meth-oxyphenyl)-2-nitroethyl]-	EtOH	211(4.26),284(3.48)	2-0405-78
$C_{14}H_{17}NO_6$			
1,3-Benzenedicarboxylic acid, 2-hy-droxy-4,6-dimethyl-5-nitroso-, diethyl ester	DMF	680(1.86)	104-2189-78
Furo[3,2-b]pyridine-2,2(3H)-dicarbox-ylic acid, 3-hydroxy-5-methyl-, diethyl ester	MeOH	222(3.74),291(3.81)	4-0029-78
$C_{14}H_{17}N_2$			
1H-Perimidinium, 2,3-dihydro-1,1,3-tri-methyl-, iodide	MeOH	248(4.36),333(3.81)	103-1145-78
$C_{14}H_{17}N_3OS$			
Pyrido[2,3-d]pyrimidin-4(8H)-one, 6-methyl-8-(3-methyl-2-butenyl)-2-(methylthio)-	pH 1	276(4.16),284(3.28), 294(4.12),347(3.19)	44-0828-78
	pH 7	276(4.38),368(4.10)	44-0828-78
	pH 11	276(4.38),368(4.10)	44-0828-78
$C_{14}H_{17}N_3O_3$			
1,3,5-Triazine-2,4(1H,3H)-dione, dihy-dro-6-(2-hydroxyphenyl)-1,3-dimethyl-5-(2-propenyl)-	EtOH	220s(3.95),280(3.53)	4-1193-78
$C_{14}H_{17}N_3O_4$			
Cyclopentanecarboxylic acid, 2-[(2-ami-no-5-nitrophenyl)imino]-, ethyl ester (spectra in ethanol)	pH 1.15 pH 7	230(3.20),308(3.56) 312(4.09)	104-0286-78 104-0286-78
1-Cyclopentene-1-carboxylic acid, 2-[(2-amino-5-nitrophenyl)amino]-, ethyl ester	EtOH	215(3.79),300(4.29)	104-0286-78
1,3,5-Triazine-2,4(1H,3H)-dione, 6-(2-acetoxyphenyl)dihydro-1,3,5-trimethyl-	EtOH	264(2.66),270(2.60)	4-1193-78
$C_{14}H_{17}N_3O_4S$			
Benzenesulfonamide, N-[3-(3,4-dihydro-2,4-dioxo-1(2H)-pyrimidinyl)propyl]-4-methyl-	CHCl$_3$	267(3.83)	70-0149-78
$C_{14}H_{17}N_3O_6$			
2(5H)-Furanone, 5-ethoxy-3,4-dimeth-	dioxan	276(4.03),397(4.46)	33-1033-78

Compound	Solvent	$\lambda_{max}(\log \epsilon)$	Ref.
oxy-, (4-nitrophenyl)hydrazone (cont.)			33-1033-78
3-Isoxazolecarboxylic acid, 4,5-dihydro- 4-[(2-hydroxy-4,4-dimethyl-6-oxo-1- cyclohexen-1-yl)hydrazono]-5-oxo-, ethyl ester	MeOH	290(3.71),445(4.19)	142-0199-78B
Pentanedioic acid, 2-[[(4-aminobenzoyl)- glycyl]amino]-	MeOH	285(3.32)	87-1165-78
$C_{14}H_{17}N_3S$			
6H-1,3,4-Thiadiazine, 5-phenyl-2-piperi- dino-	EtOH	226(4.08),260(3.99), 286(3.90),347(3.96)	94-3017-78
$C_{14}H_{17}N_4$			
Pyridinium, 1-methyl-3-[4-(dimethyl- amino)phenylazo]-	EtOH	458(4.47)	27-0323-78
Pyridinium, 1-methyl-4-[4-(dimethyl- amino)phenylazo]-	EtOH	555(4.48)	27-0323-78
$C_{14}H_{17}N_5$			
[1,2,4]Triazino[4,3-b]indazol-4-amine, N,N-diethyl-3-methyl-	C_6H_{12}	370(3.83)	24-2258-78
$C_{14}H_{17}N_5O_5$			
Wyosine	H_2O	235(4.25),294(3.62)	88-2579-78
$C_{14}H_{17}N_5O_6$			
9H-Purin-6-amine, 9-(3,5-di-O-acetyl- 9-β-D-arabinofuranosyl)-	pH 1	257(4.19)	78-1133-78
	neutral	258.5(4.19)	78-1133-78
	pH 13	259(4.19)	78-1133-78
$C_{14}H_{17}N_5O_6S$			
9H-Purine-8-thione, 6-amino-9-(3,5-di- acetoxy-β-D-arabinofuranosyl)-7,9- dihydro-	pH 1	243(4.04),298s(4.29), 307(4.35)	78-1133-78
	neutral	230(4.20),298(4.38), 305(4.39)	78-1133-78
	pH 13	298(4.32)	78-1133-78
$C_{14}H_{18}$			
Bicyclo[8.2.2]tetradeca-5,10,12,13- tetraene, cis	hexane	225(3.88),277(2.56), 285(2.51)	35-1806-78
trans	hexane	224(3.85),265s(2.53), 271(2.63),278(2.59)	35-1806-78
Cyclobutene, 1,2,3,4-tetramethyl-3,4- di-1-propynyl-, trans	hexane	none above 210 nm	5-1675-78
4,6-Decadiene-2,8-diyne, 4,5,6,7-tetra- methyl-, (E,E)-	hexane	238.5(4.29)	5-1675-78
$C_{14}H_{18}BF_2NO_2$			
Benzenepropanamide, α-[(difluoroboryl)- oxy]-N,N-diethyl-β-methylene-	EtOH	240(4.05)	44-0954-78
$C_{14}H_{18}B_2F_8N_2O_2$			
1,1'-Bipyridinium, 4,4'-dihydroxy- 2,2',6,6'-tetramethyl-, bis(tet- rafluoroborate)	EtOH	220(3.79),275(4.11)	120-0001-78
$C_{14}H_{18}BrN_3O_6$			
Cytidine, N-acetyl-5-bromo-2',3'-O-(1- methylethylidene)-	H_2O	221(4.27),313(3.81)	94-2340-78

Compound	Solvent	$\lambda_{max}(\log \epsilon)$	Ref.
$C_{14}H_{18}BrN_5$			
1H-Pyrazol-5-amine, 4-[(2-bromo-4-methylphenyl)azo]-3-methyl-1-(1-methylethyl)-	MeOH	240(4.02),318(3.91), 385(4.26),410s(4.20)	103-0885-78
1H-Pyrazol-5-amine, 4-[(2-bromo-4-methylphenyl)azo]-3-methyl-1-propyl-	$CHCl_3$	385(4.46)	103-0885-78
$C_{14}H_{18}Cl_2N_4O$			
1H-Indole-2,3-dione, 1-[[bis(2-chloroethyl)amino]methyl]-, 3-(methylhydrazone)	EtOH	335(4.17)	111-0515-78
$C_{14}H_{18}N_2$			
Benzenamine, N-(3-piperidino-2-propenylidene)-, perchlorate	MeOH	234(3.88),348(4.64)	39-1453-78C
$C_{14}H_{18}N_2O$			
Acetic acid, 1-(1-cyclopenten-1-yl)-2-methyl-2-phenylhydrazide	decalin	240(4.30),288(3.34)	33-1364-78
1-Butanone, 1-(2-propyl-1H-benzimidazol-4-yl)-	EtOH	220(4.23),280(4.02), 313(3.86)	12-2675-78
$C_{14}H_{18}N_2OS$			
7-Oxa-3-thia-1-azaspiro[5.5]undec-1-en-2-amine, N-phenyl-, (±)-	EtOH	261(4.10)	87-0895-78
$C_{14}H_{18}N_2O_2$			
1-Cyclopentene-1-carboxylic acid, 2-[(2-aminophenyl)amino]-, ethyl ester	EtOH	240(3.98),303(4.19)	104-0286-78
Propanoic acid, 2-[(5,6,7,8-tetrahydro-1-naphthalenyl)hydrazono]-, methyl ester, anti	EtOH	237(3.91),323(4.07)	103-0518-78
syn	EtOH	243(3.88),359s(4.01)	103-0518-78
Propanoic acid, 2-[(5,6,7,8-tetrahydro-2-naphthalenyl)hydrazono]-, methyl ester, anti	EtOH	238(3.85),335(4.13)	103-0518-78
syn	EtOH	244(3.81),356(4.06)	103-0518-78
4(3H)-Quinazolinone, 6-ethoxy-3-methyl-2-(1-methylethyl)-	MeOH	225(4.52),272(4.15), 325(3.66)	2-1067-78
4(3H)-Quinazolinone, 7-ethoxy-3-methyl-2-(1-methylethyl)-	MeOH	242(4.75),297s(3.71), 310(3.58)	2-1067-78
4(3H)-Quinazolinone, 8-ethoxy-3-methyl-2-(1-methylethyl)-	MeOH	236(4.42),277(4.01), 317(3.83)	2-1067-78
$C_{14}H_{18}N_2O_3$			
Cyclopenta[d]pyrido[1,2-a]pyrimidine-5-carboxylic acid, 1,2,3,4,6,7,8,10-octahydro-10-oxo-, ethyl ester	MeOH	221(4.21),252(4.11), 309(4.19)	24-2813-78
Pentanoic acid, 2-methyl-3-oxo-2-(phenylazo)-, ethyl ester	EtOH	273(3.84)	104-1402-78
$C_{14}H_{18}N_2O_3S$			
Acetic acid, (dihydro-2'-methylene-5'-oxospiro[cyclohexane-1,3'(2'H)-imidazo[2,1-b]thiazol]-6'(5'H)-ylidene)-, methyl ester, (Z)-	EtOH	348(4.12)	5-0227-78
$C_{14}H_{18}N_2O_4$			
11,12-Diazatetracyclo[4.4.2.0^{2,5}.0^{7,10}]-dodec-11-ene-1,6-dicarboxylic acid,	MeOH	379(1.86)	88-1183-78

Compound	Solvent	$\lambda_{max}(\log \epsilon)$	Ref.
dimethyl ester (cont.) 1H-Indole-2-acetic acid, 3-(acetyl-amino)-4,5,6,7-tetrahydro-4-oxo-, ethyl ester	EtOH	245(3.99),280s(3.74)	88-1183-78 94-3080-78
2-Pyrrolidineacetamide, N,1-dimethyl-3,5-dioxo-4-(1-oxo-2,4-hexadienyl)-	EtOH	285(4.11),339(4.22)	35-4225-78
	EtOH-H_2SO_4	241(4.02),343s(4.33), 356(4.35),373s(4.31)	35-4225-78
	EtOH-KOH	262(4.21),284(4.20), 338(4.18)	35-4225-78
$C_{14}H_{18}N_2O_5$			
1,8-Naphthyridine-4,4a-dicarboxylic acid, 1,2,4a,5,6,7-hexahydro-2-oxo-, diethyl ester	MeOH	216(4.09),259(3.89)	24-2813-78
1H-Pyrrole-2-propanoic acid, 2-cyano-4-(ethoxycarbonyl)-2,3-dihydro-5-methyl-3-oxo-, ethyl ester	EtOH	236(4.13),299(3.93)	94-3080-78
$C_{14}H_{18}N_2O_6$			
1H-Indole, 2,3-dihydro-1-β-D-fuco-pyranosyl-6-nitro-	EtOH	255(4.15)	136-0017-78E
1,6-Naphthyridine-3,6(2H)-dicarboxylic acid, 1,5,7,8-tetrahydro-4-hydroxy-2-oxo-, diethyl ester	EtOH	289(3.82),313(3.68)	142-0267-78C
$C_{14}H_{18}N_2O_8$			
Uridine, 2',3'-O-(3-carboxy-1-methyl-propylidene)-, (R)-	MeOH	261(3.94)	64-0056-78C
$C_{14}H_{18}N_2S$			
1H-Imidazole, 1-[2-(1-methylethyl)-thio]ethyl-4-phenyl-	EtOH	256(4.17)	4-0307-78
$C_{14}H_{18}N_2Se$			
Benzenamine, N-(4-ethyl-3,4-dimethyl-5-methylene-2-selenazolidinylidene)-	EtOH	204(3.60),232(3.39)	103-0738-78
Benzenemethanamine, N-(3,4,4-trimethyl-5-methylene-2-selenazolidinylidene)-	EtOH	211(3.35)	103-0738-78
2-Selenazolamine, 4-ethyl-4,5-dihydro-4-methyl-5-methylene-N-(4-methyl-phenyl)-	hexane	204(3.69),248(3.60)	103-0738-78
	EtOH	206(3.54),247(3.44)	103-0738-78
	EtOH-H_2SO_4	224(3.59),238(3.42)	103-0738-78
2-Selenazolamine, 4-ethyl-4,5-dihydro-4-methyl-5-methylene-N-(phenylmethyl)-	hexane	200(3.64),248(3.43), 308(3.20)	103-0738-78
	EtOH	208(3.54),238(3.35), 299(3.06)	103-0738-78
	EtOH-H_2SO_4	229(3.36),318(3.13)	103-0738-78
$C_{14}H_{18}N_4O_2$			
Propanamide, 2-[(3,4-dihydro-4-oxo-1-phthalazinyl)amino]-N,N,2-trimethyl-	EtOH	235s(3.88),246s(3.61), 319(3.70)	33-2116-78
$C_{14}H_{18}N_4O_3$			
[4,4'-Bipyrimidin]-2(1H)-one, 2',6'-diethoxy-5,5'-dimethyl-	pH 0.9	267(3.83),326(3.81)	44-0511-78
	pH 8.9	268(3.71),314(3.83)	44-0511-78
	pH 12.6	270(3.86),308(3.83)	44-0511-78
$C_{14}H_{18}N_4O_4$			
1H-Indazole, 1-(2,4-dinitrophenyl)-octahydro-2-methyl-, trans	MeCN	237(4.20),399(4.14)	78-0903-78
Pyrazolidine, 2-(2,4-dinitrophenyl)-1,3-dimethyl-3-(1-methylethenyl)-	MeCN	230(4.34),393(4.33)	78-0903-78

Compound	Solvent	$\lambda_{max}(\log \epsilon)$	Ref.
$C_{14}H_{18}N_4O_8$			
Pentanoic acid, 4-[(2,4-dinitrophenyl)-hydrazono]-5,5-dimethoxy-, methyl ester	EtOH	228(4.27),251s(4.15), 358(4.42)	12-0627-78
$C_{14}H_{18}N_6O_3S_2$			
Acetic acid, [4-[1,2-bis[[(methylamino)thioxomethyl]hydrazono]ethyl]-phenoxy]-	EtOH	341(4.43)	87-0804-78
copper chelate	EtOH	308(4.31),502(3.66)	87-0804-78
$C_{14}H_{18}N_6O_7$			
Glycine, N-[[(9-β-D-ribofuranosyl-9H-purin-6-yl)amino]carbonyl]-, methyl ester	EtOH	269(4.32),275s(4.24)	39-0131-78C
$C_{14}H_{18}O$			
Bicyclo[4.2.2]deca-2,4,7,9-tetraene, 2-(1,1-dimethylethoxy)-	C_6H_{12}	207(3.86),213(3.88), 230s(3.42),277(3.83)	5-2074-78
1H-Indene-5-ethanol, 2,4,6-trimethyl-	EtOH	220(4.33),226(4.35), 265(4.03)	94-2365-78
1H-Inden-1-one, 6-ethyl-2,3-dihydro-2,5,7-trimethyl-	EtOH	218(4.48),260(4.25), 305(3.47)	94-2365-78
$C_{14}H_{18}OS$			
Spiro[6H-cyclohepta[b]thiophene-6,1'-cyclopentan]-4(5H)-one, 7,8-dihydro-2-methyl-	n.s.g.	257(4.1)	42-0847-78
$C_{14}H_{18}O_2$			
6H-Cyclopropa[a]naphthalen-6-one, 1,1a,2,3,3a,7b-hexahydro-1a-hydroxy-1,1,3a-trimethyl-	EtOH	243(4.00),303(3.93)	44-0343-78
4(1H)-Dibenzofuranone, 2,3,6,7,8,9-hexahydro-2,2-dimethyl-	MeOH	212(4.22),282(3.56)	24-0853-78
1H-Inden-1-one, 2,3-dihydro-5-(2-hydroxyethyl)-2,4,6-trimethyl-, (R)-	EtOH	217(4.57),260(4.21), 303(3.40)	94-2365-78
$C_{14}H_{18}O_3$			
Cyclohexaneacetic acid, 4-(4-oxo-2-cyclohexen-1-ylidene)-	EtOH	230s(3.77),304(4.27)	39-0024-78C
2-Cyclohexen-1-one, 4-(4-acetoxycyclohexylidene)-	EtOH	222(3.83),300(4.19)	39-0024-78C
1H-Inden-1-one, 2,3-dihydro-6-(2-hydroxyethyl)-2-(hydroxymethyl)-5,7-dimethyl-	EtOH	218(4.46),257(4.17), 299(3.37)	94-2365-78
1H-Inden-1-one, 2,3-dihydro-2-hydroxy-6-(2-hydroxyethyl)-2,5,7-trimethyl-	EtOH	218(4.49),262(4.18), 308(3.39)	94-2365-78
1H-Inden-1-one, 2,3-dihydro-3-hydroxy-6-(2-hydroxyethyl)-2,5,7-trimethyl-	EtOH	217(4.51),259(4.12), 301(3.23)	94-2365-78
$C_{14}H_{18}O_4$			
Acetic acid, (2-acetyl-4,5-dimethyl-phenoxy)-, ethyl ester	EtOH	255(4.07),310(3.68)	23-0517-78
Benzoic acid, 3,4-dimethoxy-, 3-methyl-2-butenyl ester (trichocolein)	n.s.g.	217(--),259(--), 290(--)	31-0155-78
2H-Furo[3,2-h][1]benzopyran-5,7-diol, 3,6,7,8-tetrahydro-2,8,8-trimethyl-	EtOH	227s(3.91),294(3.56)	102-1359-78
$C_{14}H_{18}O_5$			
2H-1-Benzopyran-5,8-dione, 3,4-dihydro-	EtOH	265(4.07),393(2.97)	102-1359-78

Compound	Solvent	$\lambda_{max}(\log \epsilon)$	Ref.
3-hydroxy-7-(2-hydroxypropyl)- 2,2-dimethyl- (cont.)			102-1359-78
$C_{14}H_{18}O_8S$			
2-Butenedioic acid, 2,2'-[thiobis(methylene)]bis-, tetramethyl ester	EtOH	222(4.19)	39-0955-78C
$C_{14}H_{18}S$			
Benzo[b]cycloocta[d]thiophene, 1,4,4a,7,8,11,11a,11b-octahydro-	EtOH	244(3.83)	70-1287-78
Cyclopenta[b]thiopyran, octahydro- 2-phenyl-	hexane	207(4.1),246s(2.4), 253(2.3),259(2.3), 265(2.2)	103-0605-78
Thiophene, 2-(3,7-cyclododecadien- 1-yl)-, (E,Z)-	EtOH	240(3.80)	70-1287-78
Thiophene, 2-(3,4-diethenylcyclohexyl)-	EtOH	250(3.77)	70-1287-78
$C_{14}H_{19}BrO_4$			
Benzeneacetic acid, 2-bromo-3-ethoxy- 6-hydroxy-4,5-dimethyl-, ethyl ester	EtOH	289(3.49)	12-1353-78
5-Benzofuranol, 4-bromo-2,2-diethoxy- 2,3-dihydro-6,7-dimethyl-	hexane	223(3.90),303(3.65)	12-1353-78
$C_{14}H_{19}Cl_2NO_2$			
Benzenamine, N,N-bis(2-chloroethyl)- 4-(1,3-dioxan-2-yl)-	EtOH	262(4.31),290(3.21)	104-1004-78
$C_{14}H_{19}FN_2O_6$			
2,4(1H,3H)-Pyrimidinedione, 1-[6-deoxy- 6-fluoro-2,3-O-(1-methylethylidene)- β-D-mannopyranosyl]-5-methyl-	MeOH	264(3.97)	48-0157-78
$C_{14}H_{19}NO$			
Azocine, 3-ethenyl-2-(2-furanyl)- 1,2,3,4,5,8-hexahydro-1-methyl-	heptane	225(4.32)	70-2284-78
$C_{14}H_{19}NO_2$			
Benzenepropanamide, N,N-diethyl-α-hydroxy-β-methylene-	EtOH	236(4.03)	44-0954-78
$C_{14}H_{19}NO_3$			
Ethanone, 1-(3,4-dimethoxyphenyl)- 2-pyrrolidino- (ruspolinone)	EtOH	227(4.24),272(4.05), 302(3.92)	33-1200-78
Isoleucine, N-benzyl-N-methyl-	EtOH	211(3.86)	39-1157-78B
Leucine, N-benzoyl-N-methyl-	EtOH	211(3.91)	39-1157-78B
$C_{14}H_{19}NO_4$			
1(2H)-Isoquinolinone, 5-ethoxy-3,4-dihydro-6,7-dimethoxy-2-methyl-	MeOH	216(4.52),253s(3.92), 260(3.96),270s(3.88), 296(3.48)	44-0580-78
$C_{14}H_{19}N_3O_3$			
Morpholine, 4-[3-(2,3-dihydro-6-methyl- pyrazolo[5,1-b]oxazol-7-yl)-1-oxo-2- butenyl]-	EtOH	222(4.04),277(4.12)	95-0335-78
$C_{14}H_{19}N_3O_4$			
2,4,6(1H,3H,5H)-Pyrimidinetrione, 5-[5-(dimethylamino)-3-methoxy- 2,4-pentadienylidene]-1,3-dimethyl-	CHCl$_3$	248(4.2),325(3.9), 497(5.3)	83-0287-78

Compound	Solvent	$\lambda_{max}(\log \epsilon)$	Ref.
$C_{14}H_{19}N_3O_5$			
1,2-Hydrazinedicarboxylic acid, 1-[2-(1-methylethoxy)-2-(2-pyridinyl)-ethenyl]-, dimethyl ester, cis	MeOH	275(3.80)	103-0534-78
trans	MeOH	275(3.82)	103-0534-78
$C_{14}H_{19}N_3O_7$			
Cytidine, 2',3'-O-(3-carboxy-1-methyl-propylidene)-, (R)-	MeOH	235(3.86),269(3.88)	64-0056-78C
β-D-Ribofuranose, 5-deoxy-5-[4-(meth-oxycarbonyl)-1H-1,2,3-triazol-1-yl]-2,3-O-(1-methylethylidene)-, 1-acetate	MeOH	214(3.94)	35-2248-78
$C_{14}H_{19}N_3O_9$			
β-D-Galactopyranosyl azide, 2,3,4,6-tetraacetate	MeOH	273(1.62)	78-1427-78
β-D-Glucopyranosyl azide, 2,3,4,6-tetraacetate	MeOH	272(1.56)	78-1427-78
$C_{14}H_{19}N_3S$			
2H-Imidazole-2-thione, 4-(dimethyl-amino)-5-ethyl-1,5-dihydro-1-methyl-5-phenyl-	n.s.g.	285(4.50)	5-1568-78
$C_{14}H_{19}N_5O_2$			
[4,4'-Bipyrimidin]-2-amine, 2',6'-di-ethoxy-5,5'-dimethyl-	pH 0.9	264(3.77),320(3.71)	44-0511-78
	pH 9.24	272(3.80),305(3.72)	44-0511-78
$C_{14}H_{19}N_5O_5$			
Methanimidamide, N'-(4,5-dihydro-4-oxo-1-β-D-ribofuranosyl-1H-imidazo-[4,5-c]pyridin-6-yl)-N,N-dimethyl-	pH 1	275s(4.12),290(4.15)	87-1212-78
	pH 7	230(4.15),312(4.27)	87-1212-78
	pH 11	233(4.13),312(4.27)	87-1212-78
9H-Purin-6-amine, 9-[2,3-O-(1-methyl-ethylidene)-α-D-mannofuranosyl]-β-	MeOH	260(4.15)	39-1381-78C
	MeOH	259(4.16)	39-1381-78C
9H-Purin-6-amine, 9-[5-O-(1-oxobutyl)-β-D-arabinofuranosyl]-	MeOH	259(4.18)	87-1218-78
$C_{14}H_{19}N_7OS_2$			
Acetamide, N-[4-[1,2-bis[[(methylami-no)thioxomethyl]hydrazono]ethyl]-phenyl]-	MeOH	348(4.61)	87-0804-78
copper chelate	MeOH	312(4.49),499(3.85)	87-0804-78
$C_{14}H_{20}$			
Cyclohexane, 1-[2-(3-cyclohexen-1-yl)-ethenyl]-, (E)-(±)-	EtOH	260(--),269(4.63),281(--)	39-0829-78C
$C_{14}H_{20}BrNO_2$			
8-Oxa-5-azoniaspiro[4.5]decane, 7-hy-droxy-7-phenyl-	EtOH	256(2.36)	145-0595-78
$C_{14}H_{20}ClNO_2S$			
2H,5H-Thiopyrano[4,3-b]pyran-2-one, 4-[bis(1-methylethyl)amino]-3-chloro-7,8-dihydro-	EtOH	238s(3.30),316(3.79)	4-0181-78
$C_{14}H_{20}NO_4$			
Pyridinium, 3,5-bis(ethoxycarbonyl)-2,4,6-trimethyl- (anion)	DMSO	454(3.79)	103-1226-78

Compound	Solvent	$\lambda_{max}(\log \epsilon)$	Ref.
$C_{14}H_{20}N_2$			
Benzenecarboximidamide, N,N-diethyl- N'-1-propenyl-	MeOH	260(4.1)	78-0833-78
$C_{14}H_{20}N_2O$			
Acetic acid, 1-(1-ethyl-1-propenyl)- 2-methyl-2-phenylhydrazide	decalin	242(4.27),285(3.34)	33-1364-78
2,5-Cyclohexadien-1-one, 4-diazo- 2,6-bis(1,1-dimethylethyl)-	MeOH	264(3.76),358(4.47)	70-2455-78
Methanone, 1H-imidazol-2-yl(octahydro- 4a(2H)-naphthalenyl)-, trans	EtOH	281(4.07)	33-2831-78
Methanone, 1H-imidazol-4-yl(octahydro- 4a(2H)-naphthalenyl)-, trans	EtOH	264(4.07)	33-2831-78
$C_{14}H_{20}N_2O_2$			
Benzoic acid, 2-(1,1-dimethylethyl)- 2-(oxopropyl)hydrazide	EtOH	228(4.10),270s(3.56)	33-1477-78
$C_{14}H_{20}N_2O_3$			
L-Leucine, N-[4-(methylamino)benzoyl]-	MeOH	297(4.06)	87-1162-78
$C_{14}H_{20}N_2O_4$			
1,3-Dioxepan-5-amine, 2-(1-methyleth- yl)-6-nitro-N-phenyl-	MeOH	209(3.79),242(4.22), 290(3.42)	128-0259-78
1H,5H-Pyrazolo[1,2-a]pyrazole-1,3,5,7- (2H,6H)-tetrone, 2,2,6,6-tetraethyl-	H_2O	232(4.08),262s(--)	88-0543-78
enol methyl ether	H_2O	227(4.06),274(4.19)	88-0543-78
$C_{14}H_{20}N_2O_5S$			
1-Azabicyclo[3,2,0]hept-2-ene-2-carbox- ylic acid, 3-[(2-acetylamino)ethyl]- thio-6-(1-hydroxyethyl)-7-oxo-, methyl ester	MeOH	315(4.04)	35-6491-78
after 24 hours	MeOH-H_3PO_4	270(4.22)	35-6491-78
$C_{14}H_{20}N_2O_6$			
2,4(1H,3H)-Pyrimidinedione, 5-(6-deoxy- 5-C-methyl-2,3-O-(1-methylethyli- dene)-β-DL-ribo-hexofuranosyl]-	pH 13 MeOH	286(3.73) 263(3.72)	88-4403-78 88-4403-78
$C_{14}H_{20}N_4O$			
1(2H)-Phthalazinone, 4-[[2-(dimethyl- amino)-1,1-dimethylethyl]amino]-	EtOH	320(3.70)	33-2116-78
$C_{14}H_{20}N_4O_3$			
Morpholine, 4-[2-methyl-1-[(4-nitro- phenyl)hydrazono]propyl]-	EtOH	260(4.015),405(4.342)	146-0864-78
$C_{14}H_{20}N_4O_3S$			
2H-1,4-Thiazin-3(4H)-one, 4-[(4-amino- 2-methyl-5-pyrimidinyl)methyl]-2-hy- droxy-6-(2-hydroxyethyl)-2,5-dimethyl-	EtOH	223(4.14),267(3.85)	94-0722-78
$C_{14}H_{20}N_4O_8$			
1,1-Hydrazinedicarboxylic acid, 2,2'- (1,2-cyclohexanediylidene)bis-, tetramethyl ester	EtOH	372(3.99)	104-0566-78
$C_{14}H_{20}N_6O_4$			
Acetamide, N-[3-[6-(dimethylamino)-9H- purin-9-yl]-2,4,5-trihydroxycyclo-	pH 1 H_2O	269(4.31) 277(4.31)	18-0855-78 18-0855-78

Compound	Solvent	$\lambda_{max}(\log \epsilon)$	Ref.
pentyl]-, (±)- (cont.)	pH 13	276(4.37)	18-0855-78
2-Propanone, (5-amino-3-β-D-ribofurano-	pH 1	259(4.08),298(4.12)	4-0839-78
syl-3H-imidazo[4,5-b]pyridin-7-yl)hy-	pH 11	248(4.16),287(4.38)	4-0839-78
drazone			
$C_{14}H_{20}N_6O_5$			
Adenosine, N-[(propylamino)carbonyl]-	pH 1	277(4.38)	36-0569-78
	pH 13	269(4.32),276(4.26),	36-0569-78
		298(4.20)	
	70% EtOH	269(4.35),276(4.26)	36-0569-78
$C_{14}H_{20}O$			
Benzene, 2-(ethenyloxy)-1,3-bis(1-meth-	heptane	267(2.57)	64-0197-78B
ylethyl)-			
2-Cyclohexen-1-one, 3-(1,3-butadienyl)-	EtOH	223(4.13),290(4.07)	33-2328-78
2,4,4,5-tetramethyl-			
6H-Cyclopropa[a]naphthalen-6-one,	EtOH	277(4.26)	44-0343-78
1,1a,2,3,3a,4,5,7b-octahydro-1,1,3a-			
trimethyl-, [1aR-(1aα,3β,7bα)]-			
1H-Indene-5-ethanol, 2,3-dihydro-	EtOH	242(2.77),248(2.48),	94-2365-78
2,4,6-trimethyl-		253(2.67),260(2.78),	
		270(2.98),275(2.92),	
		280(3.02)	
1H-Inden-1-ol, 6-ethyl-2,3-dihydro-	EtOH	242(2.33),248(2.50),	94-2365-78
2,5,7-trimethyl-		254(2.66),260(2.73),	
		270(2.98),275(2.78),	
		279(2.87)	
isomer	EtOH	242(2.51),248(2.78),	94-2365-78
		253(2.92),270(2.76),	
		275(2.77),279(2.78)	
1-Penten-3-one, 5-[2-methyl-5-(1-meth-	EtOH	214(4.00)	88-2127-78
ylethenyl)-1-cyclopenten-1-yl]-, (±)-			
Spiro[5.5]undeca-1,7-dien-3-one,	EtOH	230(3.93)	88-2127-78
8,10,10-trimethyl-			
isomer	EtOH	240(3.91)	88-2127-78
$C_{14}H_{20}O_2$			
2,5-Cyclohexadiene-1,4-dione, 2,6-bis-	decane	255(4.39),306(2.85),	70-2215-78
(1,1-dimethylethyl)-		462(1.59)	
3,5-Cyclohexadiene-1,2-dione, 3,5-bis-	decane	387(3.39)	70-2215-78
(1,1-dimethylethyl)-			
3,5-Cyclohexadiene-1,2-dione, 3,6-bis-	decane	403(3.46)	70-2215-78
(1,1-dimethylethyl)-			
2-Cyclohexen-1-one, 2,4,4,5-tetrameth-	EtOH	228(3.97),272(3.94)	33-2328-78
yl-3-(3-oxo-1-butenyl)-			
1,4-Dioxaspiro[4.5]dec-7-ene, 8-(1-cy-	EtOH	229s(4.17),236(4.20),	39-0024-78C
clohexen-1-yl)-		243(4.04)	
Dispiro[5.1.5.1]tetradecane-7,14-dione	EtOH	204(2.91),234(2.26)	42-0242-78
5H-Inden-5-one, 1,2,3,6,7,7a-hexahydro-	EtOH	327(4.43),402(3.36)	33-0626-78
7a-methyl-3-(2-methyl-1-oxopropyl)-	EtOH-KOH	404(4.80)	33-0626-78
Naphth[1,2-b]oxiren-6(2H)-one,	pentane	296(4.25)	33-0626-78
1a,3,3a,4,5,7b-hexahydro-3a-methyl-	MeCN	350(1.70)	33-0626-78
1a-(1-methylethyl)-			
Tricyclo[4.4.0.0^{1,5}]decane-4,9-dione,	pentane	280(2.53)	33-0626-78
6-methyl-10-(1-methylethyl)-			
$C_{14}H_{20}O_2S_2$			
2H-Thiopyran-3(4H)-one, dihydro-4,4-di-	C_6H_{12}	263(3.57),358(3.76)	24-3246-78
methyl-2-(tetrahydro-4,4-dimethyl-3-	MeOH	264(3.58),363(3.77)	24-3246-78
oxo-2H-thiopyran-2-ylidene)-	CHCl$_3$	264(3.59),367(3.79)	24-3246-78

$C_{14}H_{20}O_2S_3-C_{14}H_{21}NO_2$

Compound	Solvent	$\lambda_{max}(\log \epsilon)$	Ref.
$C_{14}H_{20}O_2S_3$			
1,8,13-Trithiadispiro[5.0.5.1]tridec-ane-5,12-dione, 4,4,11,11-tetra-methyl-	EtOH	243(3.76),368(2.62)	24-3246-78
$C_{14}H_{20}O_3$			
1,4-Cyclohexanediol, 1-(4-methoxyphen-yl)-4-methyl-, cis	EtOH	224(4.03),269s(3.11), 274(3.20),281(3.15)	39-0024-78C
trans	EtOH	224(4.01),268s(3.11), 274(3.23),281(3.18)	39-0024-78C
2,4-Hexadienoic acid, 3-methyl-6-(3-oxocyclopentyl)-, ethyl ester	EtOH	264(4.20)	56-0347-78
1-Naphthalenecarboxylic acid, 2,3,4,4a,5,6,7,8-octahydro-6-methyl-2-oxo-, ethyl ester	EtOH	237(3.85)	32-0693-78
Spiro[1,3-dioxolane-2,1'(2'H)-naphtha-len]-6'(7'H)-one, 3',4',8',8'a-tetrahydro-5',8'a-dimethyl-	EtOH	242(4.14)	39-1461-78C
$C_{14}H_{20}O_4$			
5-Benzofuranol, 2,2-diethoxy-2,3-di-hydro-6,7-dimethyl-	EtOH	248(3.68),298(3.45)	12-1353-78
Cyclopentanecarboxylic acid, 2-(1-meth-ylethylidene)-3-oxo-1-(oxopropyl)-, ethyl ester	EtOH	252(3.61)	39-1606-78C
Spiro[1,3-dioxolane-2,1'(2'H)-naphtha-len]-6'(7'H)-one, 3',4',8',8'a-tet-rahydro-4'-hydroxy-5',8'a-dimethyl-, cis-(±)-	EtOH	247(4.07)	39-1461-78C
$C_{14}H_{20}O_5$			
2-Cyclohexene-1-butanoic acid, 1-(eth-oxycarbonyl)-4-oxo-, methyl ester	MeOH	223(3.86)	44-2562-78
3-Furancarboxylic acid, 2,5-dihydro-2,4-dimethyl-2-(4-methyl-3-oxo-hexyl)-5-oxo-	EtOH	227(4.05)	78-0955-78
$C_{14}H_{20}O_6$			
2-Furanpropanoic acid, 3-(ethoxycarbo-nyl)-2,5-dihydro-2,4-dimethyl-5-oxo-, ethyl ester	EtOH	229(4.02)	78-0955-78
$C_{14}H_{20}O_8$			
Cornoside	EtOH	226(3.99)	94-2111-78
$C_{14}H_{21}Cl_2NO_2S$			
2H,5H-Thiopyrano[4,3-b]pyran-2-one, 4-[bis(1-methylethyl)amino]-3,3-di-chloro-3,4,7,8-tetrahydro-	EtOH	230(3.52)	4-0181-78
$C_{14}H_{21}FN_2O_{12}S_3$			
2,4(1H,3H)-Pyrimidinedione, 1-[6-deoxy-6-fluoro-2,3,4-tris-O-(methylsulfon-yl)-β-D-glucopyranosyl]-5-methyl-	MeOH	265(3.99)	48-0157-78
$C_{14}H_{21}NO$			
Benzene, 1,4-bis(1,1-dimethylethyl)-2-nitroso-	C_6H_{12}	296(3.92),317s(3.71), 780(1.6)	23-2665-78
$C_{14}H_{21}NO_2$			
Benzenamine, 4-(5,5-dimethyl-1,3-diox-	EtOH	262(4.27),293(3.29)	104-1004-78

Compound	Solvent	$\lambda_{max}(\log \epsilon)$	Ref.
an-2-yl)-N,N-dimethyl- (cont,)			104-1004-78
$C_{14}H_{21}NO_5$ 1H-Pyrrole-2-propanoic acid, 2-[(1,1-dimethylethoxy)carbonyl]-2,3-dihydro-3-oxo-, ethyl ester	EtOH	318(3.85)	94-3521-78
$C_{14}H_{21}N_3O_3$ 3a-Indene-3a-carboxylic acid, 6-[(aminocarbonyl)hydrazono]-1,2,3,4,5,6-hexahydro-7-methyl-, ethyl ester, (±)-	EtOH	245(3.98)	39-1606-78C
$C_{14}H_{21}N_3O_4$ 2,3-Pyrrolidinedicarboxamide, 1-cyclohexyl-N,N'-dimethyl-4,5-dioxo-	EtOH	235(3.93),273s(3.83)	4-1463-78
$C_{14}H_{21}N_3O_5$ 4(1H)-Pyrimidinone, 2-amino-5-[6-deoxy-5-C-methyl-2,3-O-(1-methylethylidene)-β-DL-ribo-hexofuranosyl]-	pH 13 MeOH	233(3.87),276(3.77) 227(3.73),290(3.73)	138-1297-78 138-1297-78
$C_{14}H_{21}N_3O_6$ Pentanamide, N-(1-β-D-arabinofuranosyl-1,2-dihydro-2-oxo-4-pyrimidinyl)-	isoPrOH	216(4.35),248(4.13), 303(3.86)	94-0981-78
$C_{14}H_{22}$ Cyclohexane, 1,1'-(1,2-ethanediylidene)bis-	EtOH	242(4.50),250(4.58), 259(4.42)	39-0829-78C
$C_{14}H_{22}Cl_2O_4Pd_2$ Palladium, [(1,2,3-η)-4-acetoxy-2-methylbut-2-enyl]di-μ-chlorodi-	EtOH	330(3.31)	44-4769-78
$C_{14}H_{22}FN$ Cyclohexanamine, N-(6-fluoro-4,4-dimethyl-2-cyclohexen-1-ylidene)-	C_6H_{12}	227(4.40),295(2.45)	33-1025-78
$C_{14}H_{22}N_2O$ Cyclohexanone, 2-[(dimethylamino)methylene]-6-[3-(dimethylamino)-2-propenylidene]-	EtOH	230(4.85),260(4.57), 440(4.75)	70-0102-78
$C_{14}H_{22}N_2O_2$ Dispiro[5.1.5.1]tetradecane-7,14-dione, dioxime	EtOH	212(3.34)	42-0242-78
$C_{14}H_{22}N_2O_3S$ Acetic acid, [3-ethyl-4-oxo-5-piperidino-2-thiazolidinylidene]-, ethyl ester	EtOH	245(4.0),285(4.33)	5-0473-78
$C_{14}H_{22}N_2O_6$ 2,5,8,11,14,17-Hexaoxa-19,22-diazabicyclo[16.3.1]docosa-1(22),18,20-triene	EtOH	242(4.78),282(5.15)	44-3362-78
1,1-Hydrazinedicarboxylic acid, 2-(dihydro-2,2,5,5-tetramethyl-4-oxo-3(2H)-furanylidene)-, diethyl ester	heptane	<u>305(3.6)</u>	104-0566-78
$C_{14}H_{22}N_2O_7$ Uridine, 2'-deoxy-5-(4,5-dihydroxypen-	H_2O	267(3.96)	35-8106-78

Compound	Solvent	λ_{max}(log ϵ)	Ref.
tyl)- (cont.)			35-8106-78
$C_{14}H_{22}O$			
Cyclohexanone, 2,3,3-trimethyl-2-(3-methyl-1,3-butadienyl)-, (E)-	pentane	235(4.39),241s(4.27), 292s(2.65),299(2.66), 307s(2.59),319s(2.32)	33-2681-78
(Z)-	pentane	221(3.77),290(2.11), 298(2.13),307(2.06), 317s(1.78)	33-2681-78
2-Cyclohexen-1-one, 3-(2-butenyl)-, (E)-	EtOH	248(4.08)	33-2328-78
(Z)-	EtOH	247(4.11)	33-2328-78
2(3H)-Naphthalenone, 4,4a,5,6,7,8-hexa-hydro-5-methyl-8-(1-methylethyl)-	MeOH	242(4.19)	2-0853-78
1-Penten-3-one, 5-[2-methyl-5-(1-meth-ylethyl)-1-cyclopenten-1-yl]-, (±)-	EtOH	213(3.89)	88-2127-78
Spiro[4.5]dec-6-en-8-one, 1-methyl-4-(1-methylethyl)-	EtOH	238(4.04)	88-2127-78
$C_{14}H_{22}OS_2$			
5-Nonen-2-one, 9-(1,3-dithian-2-yli-dene)-6-methyl-	n.s.g.	257(3.87)	88-3603-78
$C_{14}H_{22}O_2$			
Bicyclo[4.1.0]heptan-2-one, 3,7,7-tri-methyl-3-(3-oxobutyl)-, [1S-(1α,3α-6α)]-	EtOH	220(3.55)	44-0343-78
3-Buten-2-one, 4-(3-hydroxy-2,5,6,6-tetramethyl-1-cyclohexen-1-yl)-	EtOH	221(3.92),284(3.80)	33-2328-78
2-Cyclohexen-1-one, 3-(3-hydroxy-1-butenyl)-2,4,4,5-tetramethyl-	EtOH	263(3.97)	33-2328-78
1,3(or 4)-Cyclopentadiene-1-heptanoic acid, ethyl ester	n.s.g.	247(3.54)	104-0264-78
2(3H)-Naphthalenone, 4,4a,5,6,7,8-hexa-hydro-4a-(hydroxymethyl)-1,7,7-tri-methyl-, (±)-	EtOH	251(4.24)	23-1646-78
$C_{14}H_{22}O_2S_2$			
[2,2'-Bi-2H-thiopyran]-3,3'(4H,4'H)-di-one, tetrahydro-4,4,4',4'-tetrameth-yl-	C_6H_{12}	254(2.82),258(2.78), 309(2.77)	24-3246-78
	EtOH	252(2.83),307(2.83)	24-3246-78
$C_{14}H_{22}O_3$			
2-Cyclohexen-1-one, 4-(4-acetoxybutyl)-2,4-dimethyl-	MeOH	236(3.94)	44-2562-78
2-Cyclohexen-1-one, 4-(4-acetoxypent-yl)-4-methyl-, acetate	MeOH	227(4.08)	44-2562-78
$C_{14}H_{22}O_4$			
2-Butynedioic acid, bis(2,2-dimethyl-propyl) ester	EtOH	214(3.78),224(3.53), 245(2.77),265(2.54)	24-0523-78
$C_{14}H_{23}N$			
Cyclohexanamine, N-(4,4-dimethyl-2-cyclohexen-1-ylidene)-	C_6H_{12}	223(4.31),282(2.30)	33-1025-78
$C_{14}H_{23}NO$			
4H-Cyclohepta[b]pyridin-4-one, 1,2,3,5,6,7,8,9-octahydro-1-(1-methylpropyl)-	heptane	318(4.17)	103-0993-78
	EtOH	338(4.27)	103-0993-78
	MeCN	331(4.17)	103-0993-78
	CF_3COOH	317(4.03)	103-0993-78

Compound	Solvent	$\lambda_{max}(\log \epsilon)$	Ref.
$C_{14}H_{23}NSi$			
Benzenamine, 2,6-dimethyl-N-[1-(tri-methylsilyl)propylidene]-	C_6H_{12}	211(4.19),233(4.00), 275(3.05),324(2.56), 348s(--)	23-2286-78
$C_{14}H_{23}N_3O$			
2-Pyrrolidineacetonitrile, α-(5,5-di-methyl-2-pyrrolidinylidene)-1-hy-droxy-5,5-dimethyl-	MeOH	268(4.23)	78-1241-78
$C_{14}H_{23}N_3O_4$			
Carbamic acid, ethyl-, [2-[2-[[(ethyl-amino)carbonyl]oxy]ethyl]-1-methyl-1H-pyrrol-3-yl]methyl ester, Ehrlich reaction product	n.s.g.	568(5.00)	39-0896-78C
$C_{14}H_{23}P$			
Phosphine, bis(1,1-dimethylethyl)phenyl-	C_6H_{12}	260(3.23)	65-1394-78
$C_{14}H_{24}$			
Cyclododecane, 1,2-bis(methylene)-	C_6H_{12}	230(3.85)	54-0105-78
Cyclohexane, 1,3-octadienyl)-, (E,Z)-	C_6H_{12}	235(4.41)	101-0159-78M
$C_{14}H_{24}ClN_3$			
Pyrazinamine, 5-chloro-N,N-dimethyl-3,6-bis(2-methylpropyl)-	EtOH	264(4.00),336(3.76)	94-1322-78
$C_{14}H_{24}N_2O_4$			
1,4-Butanediaminium, N,N'-bis(1-formyl-2-hydroxyethenyl)-N,N,N',N'-tetra-methyl-, dihydroxide, bis(inner salt)	EtOH	257(4.66)	73-1261-78
$C_{14}H_{24}N_4O_4$			
1,4-Cyclohexadiene-1,4-dicarboxylic acid, 2,5-bis(2,2-dimethylhydra-zino)-, dimethyl ester	EtOH	287(4.33)	48-0991-78
$C_{14}H_{24}O$			
5H-Inden-5-one, octahydro-7a-methyl-3-(2-methylpropyl)-	pentane	299(1.34)	33-0626-78
1-Propanone, 2-methyl-1-(octahydro-3a-methyl-1H-inden-1-yl)-	pentane	285(1.69)	33-0626-78
4,8-Undecadien-3-one, 7-ethyl-5-methyl-, (E,E)-(+)-	MeOH	237(4.27)	44-3454-78
$C_{14}H_{24}OS_2$			
Spiro[5H-1-benzopyran-5,2'-[1,3]dithi-ane], octahydro-2,8a-dimethyl-	n.s.g.	250(2.84)	88-3603-78
$C_{14}H_{24}O_2$			
2-Cyclohexen-1-one, 3-(3-hydroxybutyl)-2,4,4,5-tetramethyl-	EtOH	249(4.10)	33-2328-78
2-Cyclopenten-1-one, 2-methoxy-3-octyl-	EtOH	249(3.93)	107-0155-78
2,6-Nonadienoic acid, 3,8,8-trimethyl-, ethyl ester, (E,Z)-	EtOH	220(4.13)	56-0581-78
$C_{14}H_{24}S_2$			
1,4-Cyclohexadiene, 1,4-bis(ethylthio)-3,3,6,6-tetramethyl-	MeCN	267(3.26)	88-3729-78
2,4,6-Octatriene, 4,5-bis(ethylthio)-2,7-dimethyl-	MeCN	278(3.33)	88-3729-78

Compound	Solvent	$\lambda_{max}(\log \epsilon)$	Ref.
$C_{14}H_{25}AsO_5S$ α-D-Galactopyranose, 1,2:3,4-bis-O-(1-methylethylidene)-6-thio-, dimethyl-arsinite	MeOH	223(3.99)	136-0069-78E
$C_{14}H_{25}AsO_5Se$ α-D-Galactopyranose, 1,2:3,4-bis-O-(1-methylethylidene)-6-seleno-, dimethylarsinite	MeOH	232(3.59)	136-0069-78E
$C_{14}H_{25}NO$ 5,7,9-Dodecatrien-2-amine, 3-methoxy-N-methyl-	n.s.g.	275(4.70)	102-0831-78
$C_{14}H_{25}N_2$ Methanaminium, N-[7-(dimethylamino)-1,4-dimethyl-2,4,6-octatrienyli-dene]-N-methyl-, iodide	CHCl₃	562(5.19)	70-0107-78
$C_{14}H_{25}N_3$ Pyrazinamine, N,N-dimethyl-3,6-bis(2-methylpropyl)-	EtOH	258(3.85),328(3.73)	94-1322-78
$C_{14}H_{26}$ 3,5-Dodecadiene, 2,2-dimethyl-, (E,Z)- 5,7-Tetradecadiene, (E,Z)-	C₆H₁₂ C₆H₁₂	233(4.41) 233(4.38)	101-0159-78M 101-0159-78M
$C_{14}H_{26}B_2N_4$ Boron, tetraethylbis[μ-1H-pyrazolato-$N^1:N^2$)]di-	hexane	210(4.22),250(4.11)	64-0220-78B
$C_{14}H_{26}N_4$ 1,2,3,4-Tetrazine, 1,4-dicyclohexyl-1,4,5,6-tetrahydro-	MeOH	225(3.45),273(3.66)	33-1622-78
$C_{14}H_{26}N_4O$ 1,2,3,4-Tetrazine, 1,4-dicyclohexyl-1,4,5,6-tetrahydro-, 2-oxide	MeOH	293(4.09)	33-1622-78
$C_{14}H_{26}N_8$ [3,3'-Bi-1,2,4-triazine]-6,6'-diamine, 2,2',5,5'-tetrahydro-N,N,N',N',5,5-5',5'-octamethyl-	EtOH	333(4.2)	33-2419-78
$C_{14}H_{26}O$ Cyclohexanone, 2,3,3-trimethyl-2-(3-methylbutyl)- 4-Decen-3-one, 2,2,6,6-tetramethyl-	EtOH EtOH	298(1.36),307s(1.30) 229(3.93)	33-2681-78 44-4369-78
$C_{14}H_{26}S_2Si_2$ Silane, [1,4-phenylenebis(thiomethyl-ene)]bis[trimethyl-	heptane	277(4.25),307(3.22)	65-2027-78
$C_{14}H_{30}O_2S$ 1-Dodecanesulfinic acid, ethyl ester	hexane	196(3.26),223(3.23)	118-0441-78

Compound	Solvent	$\lambda_{max}(\log \epsilon)$	Ref.
$C_{15}H_8ClN$			
9-Anthracenecarbonitrile, 10-chloro-	C_6H_{12}	386(4.00)	61-1068-78
$C_{15}H_8Cl_2N_2O_2$			
4-Quinolinecarboxylic acid, 6,8-di-chloro-2-(2-pyridinyl)-	pH 13	218(4.57),253(4.72), 277(4.61),345(4.16)	34-0261-78
	EtOH	212(3.57),252(3.80), 278(3.72),345(3.34)	34-0261-78
	dioxan	242(4.11),263?(4.24), 282(4.22),350(3.78)	34-0261-78
	H_2O	210(--),253(--), 277(--),345?(--)	34-0261-78
4-Quinolinecarboxylic acid, 6,8-di-chloro-2-(3-pyridinyl)-	pH 13	219(4.09),261(4.25), 345(3.52)	34-0261-78
	EtOH	214(3.90),263(4.19), 347(3.37)	34-0261-78
	dioxan	267(3.39),352(3.85)	34-0261-78
	H_2O	203(--),214(--), 260(--)	34-0261-78
4-Quinolinecarboxylic acid, 6,8-di-chloro-2-(4-pyridinyl)-	pH 13	222(4.20),261(4.53), 350(3.03)	34-0261-78
	EtOH	211(4.52),277(4.60), 350(3.03)	34-0261-78
	dioxan	235(3.15),270(3.54), 350(3.03)	34-0261-78
	H_2O	204(--),222(--), 261(--)	34-0261-78
$C_{15}H_8FN$			
9-Anthracenecarbonitrile, 10-fluoro-	C_6H_{12}	384(3.87)	61-1068-78
$C_{15}H_8F_3N_3O$			
13H-Pyrido[1',2':3,4]imidazo[2,1-b]-quinazolin-13-one, 2-(trifluoro-methyl)-	MeOH	281(3.84),290(3.97), 367(3.77),378(3.82), 410(3.4),433(3.4), 458(3.1)	5-0398-78
$C_{15}H_8N_2O_2$			
Indolo[2,1-b]quinazoline-6,12-dione	MeOH	245(4.55),310(3.8), 390(3.7)	88-3007-78
$C_{15}H_8N_2S$			
Propanedinitrile, (2-methyl-3H-naphtho-[1,8-bc]thien-3-ylidene)-	MeCN	248(4.27),267(4.32), 324(3.73),457(4.28)	89-0369-78
$C_{15}H_8O_4$			
2-Anthracenecarboxylic acid, 9,10-di-hydro-9,10-dioxo-	pH 7.0	259(4.68),280s(4.13), 336(3.73)	95-0929-78
$C_{15}H_9BrN_2O_2$			
Isoquinoline, 1-bromo-4-(4-nitrophen-yl)-	EtOH	219(4.47)	44-0672-78
$C_{15}H_9ClN_2O_2$			
Cyclohept[hi]indolizine-4-carboxylic acid, 2-chloro-1-cyano-, methyl ester	CHCl$_3$	265(4.43),273(4.45), 369(4.03),393(4.08), 417(4.09),463(2.96), 499(3.03),540(3.04), 591(2.94),650(2.65)	18-1573-78

Compound	Solvent	$\lambda_{max}(\log \epsilon)$	Ref.
$C_{15}H_9ClO$			
1H-Inden-1-one, 4-chloro-2-phenyl-	n.s.g.	217(4.32),235(4.26), 285(4.26),315(4.24)	80-1465-78
$C_{15}H_9ClO_2S$			
9H-Thioxanthen-9-one, 2-acetyl-7-chloro-	MeOH	261(4.58),290(4.19), 322(4.23),370s(3.72), 382(3.75)	73-0471-78
$C_{15}H_9ClO_3S$			
1H-2-Benzopyran-1-one, 3-[(4-chloro- phenyl)thio]-4-hydroxy-	MeOH MeOH-base	293(3.96) 327(3.82)	24-2859-78 24-2859-78
$C_{15}H_9ClO_5$			
9,10-Anthracenedione, 1-chloro- 4,5,7-trihydroxy-2-methyl-	EtOH-HCOOH	253(4.26),271(4.25), 291(4.25),450(4.02)	12-1335-78
$C_{15}H_9ClO_5S$			
1H-2-Benzopyran-1,4(3H)-dione, 3-[(4- chlorophenyl)sulfonyl]-	MeOH MeOH-base	297(3.69),305(--) 335(3.89+)(changing)	24-2859-78 24-2859-78
$C_{15}H_9Cl_3N_2O_5S$			
1H-Indole, 3-[2-(5-nitro-2-furanyl)- 2-[(trichloromethyl)sulfonyl]eth- enyl]-	EtOH	216(4.49),254(4.14), 274(4.20),347(4.20), 400(4.09)	73-1618-78
$C_{15}H_9N$			
9-Anthracenecarbonitrile	C_6H_{12}	378(3.98)	61-1068-78
$C_{15}H_9NO$			
Phenanthro[9,10-d]oxazole	C_6H_{12}	252(4.74),277(4.06), 287(3.91),300(4.01)	44-0381-78
$C_{15}H_9NOS$			
Pyrrolo[3,2,1-kl]phenothiazine-2-carb- oxaldehyde	EtOH	248(4.50),373(4.03)	4-0711-78
Thiazolo[3,2-f]phenanthridinium, 3-hy- droxy-, hydroxide, inner salt	CHCl$_3$	316(3.69),330(3.63), 489(3.89)	2-0678-78
$C_{15}H_9NOS_2$			
Thiazolo[2,3-b]benzothiazolium, 3-hy- droxy-2-phenyl-, hydroxide, inner salt	MeOH	282(4.21),420(4.04)	44-2697-78
$C_{15}H_9NO_5$			
2H-1-Benzopyran-2-one, 3-(3-nitrophen- oxy)-	MeOH	220(4.289),290(4.215)	25-0628-78
$C_{15}H_9NO_7S$			
1H-2-Benzopyran-1,4(3H)-dione, 3-[(4- nitrophenyl)sulfonyl]-	MeOH MeOH-base	297(3.72),308(--) 386(3.64)(changing)	24-2859-78 24-2859-78
$C_{15}H_9NS$			
Phenanthro[9,10-d]thiazole	EtOH	233s(4.44),254(4.77), 260s(4.64),283(4.03), 300(3.92),312(3.89)	39-0685-78C
$C_{15}H_9S_3$			
Cyclopropenylium, tri-2-thienyl-	MeCN	271(4.27),296s(4.09), 355(4.65),375(4.65)	88-0803-78

Compound	Solvent	λ_{max}(log ϵ)	Ref.
$C_{15}H_{10}BrNO_2$			
1H-Pyrrole-2,5-dione, 3-bromo-1-methyl-3-(1-naphthalenyl)-	MeCN	217(4.98),278(3.83), 367(3.40)	24-2677-78
$C_{15}H_{10}BrN_3O_3$			
2H-1,5-Benzodiazepin-2-one, 8-bromo-1,3-dihydro-7-nitro-4-phenyl-	EtOH 70% H_2SO_4	273(4.35),320s(4.08) 208(4.28),335(4.29)	103-0455-78 103-0455-78
$C_{15}H_{10}ClNO$			
Isoxazole, 3-(4-chlorophenyl)-5-phenyl-	HOAc HOAc-H_2SO_4	260(4.45) 317(4.51)	103-0264-78 103-0264-78
Isoxazole, 5-(4-chlorophenyl)-3-phenyl-	HOAc HOAc-H_2SO_4	274(4.47) 317(4.48)	103-0264-78 103-0264-78
$C_{15}H_{10}ClN_3O_2$			
Cyclohepta[b]pyrrole-3-carboxylic acid, 2-chloro-6-(dicyanomethylene)-1,6-dihydro-, ethyl ester	MeOH	260(4.29),282(4.02), 290s(4.02),433(4.51), 448s(4.48),472s(4.17)	18-0667-78
$C_{15}H_{10}ClN_3O_3$			
2H-1,5-Benzodiazepin-2-one, 8-chloro-1,3-dihydro-7-nitro-4-phenyl-	EtOH 70% H_2SO_4	263(4.40),320s(4.08) 205(4.26),335(4.29)	103-0455-78 103-0455-78
$C_{15}H_{10}ClN_5O_2$			
Benzaldehyde, 2-nitro-, (4-chloro-1(2H)-phthalazinylidene)hydrazone, (?,E)-	dioxan	286(4.18),390(4.11)	103-0567-78
Benzaldehyde, 4-nitro-, (4-chloro-1(2H)-phthalazinylidene)hydrazone, (?,E)-	dioxan	285(4.26),418(4.36)	103-0567-78
$C_{15}H_{10}ClN_7O_6$			
Phthalazine, 1-chloro-4-[1-methyl-2-(2,4,6-trinitrophenyl)hydrazino]-	dioxan	218(4.64),314(4.11)	103-0575-78
1(2H)-Phthalazinone, 4-chloro-2-methyl-, (2,4,6-trinitrophenyl)hydrazone, (Z)-	dioxan	297(3.85),432(4.26)	103-0575-78
$C_{15}H_{10}Cl_2O_5$			
9H-Xanthen-9-one, 4,7-dichloro-1,6-dihydroxy-3-methoxy-8-methyl-	EtOH	241(4.60),256(4.35), 270(4.19),311(3.99), 365(4.34)	78-0577-78
$C_{15}H_{10}Cl_3N_5O_3S$			
Acetamide, 2-chloro-N-(chloroacetyl)-N-[2-[(chloroacetyl)imino]thiazolo-[4,5-b]quinoxalin-3(2H)-yl]-	EtOH	213?(4.46),244(4.43), 270(4.7)	2-0683-78
$C_{15}H_{10}N_2$			
Propanedinitrile, (1,4-dihydro-1,4-methano-7H-benzocyclohepten-7-ylidene)-	hexane	214(4.14),222s(4.14), 266(3.93),393s(4.34), 399(4.40),407(4.45), 415(4.46),426(4.38), 436(4.34)	88-0569-78
	MeOH	266(4.04),410s(4.49), 415(4.53),426(4.54), 434(4.54)	88-0569-78
3H-Pyrrolo[2,3-c]acridine	EtOH	216(4.39),224(4.39), 238(4.44),280(4.71), 372(3.96)	103-1036-78

Compound	Solvent	$\lambda_{max}(\log \epsilon)$	Ref.
10H-Quindoline	MeOH	227(4.30),269s(4.45), 274(4.46),330s(3.82), 345(4.03)	36-0433-78
	EtOH-HCl	224(4.27),242s(3.94), 273(4.36),280(4.38), 350s(4.03),368(4.26)	36-0433-78
$C_{15}H_{10}N_2O$ Benzo[3,4]cyclobuta[1,2-b]quinoxaline, 7-methoxy-	EtOH	231(4.20),244(4.26), 249(4.29),277(4.37), 296s(3.97),392(4.03), 409(4.03)	78-0495-78
6H-Benzo[b]quinolizine-11-carbonitrile, 2-methyl-6-oxo-	EtOH	232(4.57),262s(4.24), 297(3.67),309(3.72), 364s(4.18),382(4.21), 420s(3.87),452s(3.76)	44-3536-78
$C_{15}H_{10}N_2O_2$ 2-Quinoxalinecarboxaldehyde, 3-phenoxy-	CHCl$_3$	318(3.69)	97-0177-78
$C_{15}H_{10}N_2O_2S$ Pyrrolo[2,1-a]isoquinoline-3-carboxylic acid, 1-cyano-2-(methylthio)-	EtOH	277(4.67),338(3.96), 354(3.88)	95-1412-78
$C_{15}H_{10}N_2O_3$ Isoxazole, 3-(3-nitrophenyl)-5-phenyl-	HOAc	265(4.40)	103-0264-78
	HOAc-H$_2$SO$_4$	315(4.40)	103-0264-78
Isoxazole, 3-(4-nitrophenyl)-5-phenyl-	HOAc	280(4.50)	103-0264-78
	HOAc-H$_2$SO$_4$	320(4.40)	103-0264-78
Isoxazole, 5-(3-nitrophenyl)-3-phenyl-	HOAc	258(4.41)	103-0264-78
	HOAc-H$_2$SO$_4$	300(4.37)	103-0264-78
Isoxazole, 5-(4-nitrophenyl)-3-phenyl-	HOAc	305(4.27)	103-0264-78
	HOAc-H$_2$SO$_4$	317(4.45)	103-0264-78
$C_{15}H_{10}N_2O_7S$ 2H-Thiopyran-3(4H)-one, 5,6-dihydro-2,4-bis[(5-nitro-2-furanyl)methylene]-	EtOH	238(3.75),377(4.04), 435(4.90)	133-0189-78
$C_{15}H_{10}N_4$ 1H-Benzocycloheptene-2,2,3,3-tetracarbonitrile	EtOH	256(3.41)	138-0961-78
Ethenetricarbonitrile, (1,2-dimethyl-3-indolizinyl)-	EtOH	205(4.26),215(4.23), 243(4.09),280(3.79), 307(3.53),344(3.58), 397(3.74),555(4.22)	4-1471-78
Pyrido[3,2-c]pyridazine, 4-(1H-indol-3-yl)-	EtOH	214(4.84),268(4.21), 283(4.15)	103-1267-78
$C_{15}H_{10}N_4O_2$ 1H-Pyrido[3,2-c][1,2,4]triazolo[1,2-a]-pyridazine-1,3(2H)-dione, 2-phenyl-	MeOH	201(3.9),242(3.8), 285s(--)	103-0534-78
$C_{15}H_{10}N_4O_3$ 2-Quinoxalinecarboxaldehyde, 3-[(4-nitrophenyl)amino]-	CHCl$_3$	445(3.47)	97-0177-78
$C_{15}H_{10}N_6$ 1,2,5,6,6a,12c-Hexaazabenz[j]aceanthrylene, 4-methyl-	pH 1	244s(4.2),250(4.28), 259(4.23),263(4.24), 269(4.21),298(4.35),	88-3687-78

Compound	Solvent	$\lambda_{max}(\log \epsilon)$	Ref.
1,2,5,6,6a,12c-Hexaazabenz[j]aceanthrylene, 4-methyl- (cont.)		305(4.39),312s(4.1), 357(3.85),375(3.86), 440(2.97)	88-3687-78
	$CHCl_3$	251(4.25),256(4.28), 275(4.14),321s(4.28), 337(4.48),352(4.59), 400s(2.81),428(2.79), 588s(2.96),627(3.02), 666s(2.96)	88-3687-78
$C_{15}H_{10}N_8$			
Tetrazolo[1,5-a][1,2,4]triazolo[4,3-c]-pyrimidine, 5-methyl-9-(2-quinolinyl)-	$CHCl_3$	241(4.21),244(4.21), 272(4.26),277(4.27), 281(4.25),315(4.06)	88-3687-78
$C_{15}H_{10}O$			
5H-Dibenzo[a,d]cyclohepten-5-one	CH_2Cl_2	257(4.92),310(4.47), 361(3.89)	18-2674-78
$C_{15}H_{10}O_2$			
9,10-Anthracenedione, 2-methyl-	MeOH	258(4.72),330(3.67)	2-1062-78
$C_{15}H_{10}O_3$			
4H-1-Benzopyran-4-one, 2-(2-hydroxyphenyl)-	EtOH	243(4.57),304(4.18)	22-0043-78
4H-1-Benzopyran-4-one, 2-(4-hydroxyphenyl)-	EtOH	252(4.15),300(4.30)	22-0043-78
4H-1-Benzopyran-4-one, 5-hydroxy-2-phenyl-	$CHCl_3$	335(3.66)	18-2425-78
Co chelate	$CHCl_3$	413(3.64)	18-2425-78
Cu chelate	$CHCl_3$	415(2.99)	18-2425-78
Fe(III) chelate (also others)	$CHCl_3$	425(4.07),520s(--)	18-2425-78
$C_{15}H_{10}O_3S$			
1H-2-Benzopyran-1-one, 4-hydroxy-3-(phenylthio)-	MeOH	288(3.96)	24-2859-78
	MeOH-base	329(3.80)	24-2859-78
6H-Thiopyrano[2,3-b]furan-6-one, 5-acetyl-2-phenyl-	EtOH	285(4.49),420(3.97)	104-0589-78
$C_{15}H_{10}O_4$			
9,10-Anthracenedione, 1,4-dihydroxy-6-methyl-	MeOH	209(4.49),219s(--), 258(4.43),279(3.99), 330(3.41),476(3.87), 507(3.61)	150-3831-78
4H-1-Benzopyran-4-one, 5,7-dihydroxy-2-phenyl-	EtOH	268(4.37),318(4.08)	22-0043-78
$C_{15}H_{10}O_4S$			
9H-Xanthene-2-carboxylic acid, 7-(methylthio)-9-oxo-	MeOH	211(4.41),230s(4.40), 251(4.49),258(4.49), 278s(4.29),343(3.49)	39-0876-78C
$C_{15}H_{10}O_5$			
9,10-Anthracenedione, 1,5-dihydroxy-3-(hydroxymethyl)-	MeOH-HCOOH	281(4.00),288(3.99), 424(3.97)	12-2271-78
9,10-Anthracenedione, 1,3,6-trihydroxy-8-methyl-	MeOH-HCOOH	275s(4.22),288(4.35), 307(3.92),344(3.50), 436(3.60)	12-1363-78
9,10-Anthracenedione, 1,3,7-trihydroxy-5-methyl-	MeOH-HCOOH	260s(3.92),284(4.22), 305(3.95),384(4.65)	12-1363-78

Compound	Solvent	$\lambda_{max}(\log \epsilon)$	Ref.
1,3-Benzenediol, 4-furo[2,3-f]-1,3-benzodioxol-6-yl-	EtOH	230s(4.33),243s(4.19), 281(4.15),330(4.53), 346(4.56)	94-1274-78
	EtOH-KOH	243s(4.25),290(4.15), 348(4.54),358(4.51)	94-1274-78
4H-1-Benzopyran-4-one, 2-(3,4,5-trihydroxyphenyl)-	EtOH	228(4.21),306(4.13)	22-0043-78
4H-1-Benzopyran-4-one, 3-(3,4-dihydroxyphenyl)-7-hydroxy-	EtOH	249(4.29),292(4.12)	102-1405-78
4H-1-Benzopyran-4-one, 5,7-dihydroxy-2-(4-hydroxyphenyl)-	EtOH	268(4.21),340(4.30)	22-0043-78
$C_{15}H_{10}O_5S$			
1H-2-Benzopyran-1,4(3H)-dione, 3-(phenylsulfonyl)-	MeOH	298(3.61),305(--)	24-2859-78
	MeOH-base	336(3.89+)(changing)	24-2859-78
$C_{15}H_{10}O_6$			
4H-1-Benzopyran-4-one, 7-hydroxy-2-(3,4,5-trihydroxyphenyl)-	EtOH	224(4.41),308(4.15)	22-0043-78
$C_{15}H_{10}O_7$			
4H-1-Benzopyran-4-one, 5,7-dihydroxy-2-(3,4,5-trihydroxyphenyl)-	EtOH	266(4.15),354(4.30)	22-0043-78
$C_{15}H_{10}O_8$			
9,10-Anthracenedione, 1,2,4,5,6,8-hexahydroxy-3-methyl-	MeOH-DMSO-HCOOH	263(4.48),280s(4.21), 290(4.18),333(3.98), 466s(3.90),478(4.01), 502(4.23),510(4.30), 536(4.31),548(4.37)	12-2271-78
$C_{15}H_{10}O_8S$			
2-Anthracenesulfonic acid, 9,10-dihydro-1,3,8-trihydroxy-6-methyl-9,10-dioxo-	MeOH-HCOOH	294(4.22),443(3.91)	12-2271-78
$C_{15}H_{11}BrN_2$			
Benzenamine, 4-(1-bromo-4-isoquinolinyl)-	EtOH	210(4.46),221(4.56), 253(4.33)	44-0672-78
1H-Pyrazole, 3-(3-bromophenyl)-5-phenyl-	EtOH	254(4.41)	94-1298-78
Quinazoline, 7-bromo-4-methyl-2-phenyl-	EtOH	260(4.45),265(4.47), 290(4.00),303(3.94), 322(3.87),334(3.72)	94-2866-78
Quinoxaline, 6-bromo-2-methyl-3-phenyl-	EtOH	243(4.49),331(3.95)	94-2866-78
$C_{15}H_{11}BrN_2O_3S$			
Benzenamine, N-[6-(4-bromophenyl)-2,3-dihydro-1,4,3-oxathiazin-2-ylidene]-, S,S-dioxide	EtOH	263(4.35)	18-1805-78
1,4,3-Oxathiazin-2-amine, 6-(4-bromophenyl)-N-phenyl-, 4,4-dioxide	EtOH	258(4.38)	18-1805-78
$C_{15}H_{11}BrN_4$			
3H-Indole, 3-azido-6-bromo-3-methyl-2-phenyl-	EtOH	226(3.98),233(4.01), 248(4.21),254(4.22), 306(4.01),316(4.00)	94-2866-78
$C_{15}H_{11}Br_2NO_2$			
7H-Naphtho[1',2':3,4]cyclobuta[1,2-c]-pyrrole-7,9(8H)-dione, 6b,9a-dibromo-6a,6b,9a,9b-tetrahydro-8-methyl-	MeCN	218(4.52),267(3.80)	24-2677-78

Compound	Solvent	$\lambda_{max}(\log \epsilon)$	Ref.
syn-syn (cont.)	MeCN	218(4.58),268(3.85)	24-2677-78
$C_{15}H_{11}ClFN_3$			
Benzenepropanenitrile, α-[(4-chloro-phenyl)hydrazono]-2-fluoro-, anti	EtOH	323(4.21)	118-0445-78
syn	EtOH	338(4.10)	118-0445-78
$C_{15}H_{11}ClN_2$			
Quinazoline, 6-chloro-4-methyl-2-phenyl-	EtOH	253(4.41),262(4.40), 288(4.14),301(4.04), 328(3.53),344(3.38)	94-2866-78
Quinoxaline, 6-chloro-3-methyl-2-phenyl-	EtOH	244(4.44),326(3.99)	94-2866-78
$C_{15}H_{11}ClN_2O$			
Pyrrolo[1,2-c]pyrimidine-5-carboxalde-hyde, 1-chloro-3-methyl-6-phenyl-	n.s.g.	243s(4.27),249(4.29), 276s(3.71),339(3.92)	44-3544-78
$C_{15}H_{11}ClN_2O_2$			
4(1H)-Quinazolinone, 2-(4-chlorophen-oxy)-1-methyl-	dioxan	243(3.99),259(3.78), 292(3.62),301(3.80), 312(3.74)	4-1033-78
4(3H)-Quinazolinone, 2-(4-chlorophen-oxy)-3-methyl-	dioxan	242(3.80),263(3.94), 268(3.92),298(3.49), 309(3.62),322(3.55)	4-1033-78
$C_{15}H_{11}ClN_2O_3S$			
Benzenamine, N-[6-(4-chlorophenyl)-2,3-dihydro-1,4,3-oxathiazin-2-yli-dene]-, S,S-dioxide	EtOH	262(4.33)	18-1805-78
1,4,3-Oxathiazin-2-amine, 6-(4-chloro-phenyl)-N-phenyl-, 4,4-dioxide	EtOH	263(4.25)	18-1805-78
2H-1,2,4-Thiadiazin-3(4H)-one, 5-(4-chlorophenyl)-4-phenyl-, 1,1-dioxide	EtOH	247(4.07)	18-1805-78
$C_{15}H_{11}ClN_4$			
Benzaldehyde, (4-chloro-1(2H)-phthala-zinylidene)hydrazone	dioxan MeCN	290(4.28),370(4.40) 211(4.53),292(4.30), 371(4.34)	103-0567-78 103-0567-78
Cyanamide, (6-chloro-1,4-dihydro-4-phen-yl-2-quinazolinyl)-	MeOH	220(4.40),241s(4.16), 268(4.35)	4-1409-78
1H-Indole, 2-(azidomethyl)-5-chloro-3-phenyl-	EtOH	228(4.48),236(4.44), 266(3.97),295(3.85), 303(3.79)	94-2874-78
3H-Indole, 3-azido-5-chloro-3-methyl-2-phenyl-	EtOH	224(4.21),231(4.19), 277(4.04),287(4.09), 297(4.13),306(4.15)	94-2866-78
$C_{15}H_{11}ClN_4O_2$			
Benzenepropanenitrile, 2-chloro-α-[(4-nitrophenyl)hydrazono]-, anti	EtOH	363(4.52)	118-0445-78
syn	EtOH	367(4.71)	118-0445-78
$C_{15}H_{11}ClOS$			
Dibenzo[b,f]thiepin, 2-chloro-4-meth-oxy-	MeOH	225(4.53),262(4.35), 309(3.90)	73-1747-78
$C_{15}H_{11}ClO_2S$			
Dibenzo[b,f]thiepin-10(11H)-one, 8-chloro-6-methoxy-	MeOH	247(4.32),272s(3.91), 348(3.74)	73-1747-78

Compound	Solvent	λ_{max}(log ϵ)	Ref.
$C_{15}H_{11}ClO_3S$			
Benzoic acid, 2-[(4-acetylphenyl)thio]-	MeOH	242(4.18),260(4.13), 314(4.08)	73-0471-78
$C_{15}H_{11}ClO_5$			
Benzoic acid, 3-(5-chloro-2-methoxy-benzoyl)-4-hydroxy-	MeOH	205(4.18),229(4.48), 260s(3.99),337(3.56)	39-0876-78C
9H-Xanthen-9-one, 2-chloro-3,8-di-hydroxy-6-methoxy-1-methyl-	EtOH	241(4.61),254(4.41), 266(4.19),311(4.12), 357(4.29)	78-0577-78
9H-Xanthen-9-one, 4-chloro-1,6-di-hydroxy-3-methoxy-8-methyl-	EtOH	240(4.52),253(4.34), 271(4.02),310(4.23), 350(4.02)	78-0577-78
$C_{15}H_{11}Cl_2N_3$			
Benzenepropanenitrile, 2-chloro-α-[(4-chlorophenyl)hydrazono]-, anti	EtOH	323(4.25)	118-0445-78
syn	EtOH	337(4.45)	118-0445-78
$C_{15}H_{11}Cl_3N_2O_6S$			
Acetamide, N-[4-[2-(5-nitro-2-furanyl)-2-[(trichloromethyl)sulfonyl]ethen-yl]phenyl]-	EtOH	206(4.21),232(4.19), 315(4.43),365(4.19)	73-1618-78
$C_{15}H_{11}N$			
Benzeneacetonitrile, α-(phenylmethyl-ene)-, (Z)-	EtOH	229(4.08),314(4.36)	23-0041-78
$C_{15}H_{11}NO$			
Benzenepropanenitrile, β-oxo-α-phenyl-	EtOH	292(4.10)	18-3389-78
Benzoxazole, 2-(2-phenylethenyl)-	EtOH	223(4.13),252(3.11), 319(4.16)	56-2479-78
[1,1'-Biphenyl]-2-carboxaldehyde, 2'-(2H-azirin-3-yl)-	C_6H_{12}	245(4.30),295(3.53)	44-0381-78
Isoxazole, 3,5-diphenyl-	EtOH	267(4.37)	78-1571-78
	H_2SO_4	312(4.24)	78-1571-78
	HOAc	268(4.39)	103-0264-78
	HOAc-H_2SO_4	312(4.47)	103-0264-78
$C_{15}H_{11}NOS$			
Benzenamine, N-(4-phenyl-1,3-oxathiol-2-ylidene)-	EtOH	212(4.33),283s(4.31), 297s(4.36)	94-3017-78
[1,1'-Biphenyl]-4-carbonyl isothio-cyanate, 4'-methyl-	C_6H_{12}	222(4.38),312(4.45)	73-0257-78
$C_{15}H_{11}NO_2$			
Benzenepropanenitrile, 2-hydroxy-β-oxo-α-phenyl-	EtOH	262(4.00)	114-0069-78B
1H-Indole-2-carboxylic acid, 5-phenyl-	EtOH	206(4.40),261(4.61), 300s(4.06)	103-0173-78
1H-Isoindole-1,3(2H)-dione, 2-(2-meth-ylphenyl)-	EtOH	220(4.60),290(2.30)	115-0073-78
	CHCl$_3$	290(3.23)	115-0073-78
	CCl$_4$	295(3.28)	115-0073-78
1H-Isoindole-1,3(2H)-dione, 2-(4-meth-ylphenyl)-	EtOH	215(4.27),230s(3.28), 295(3.00)	115-0073-78
	CHCl$_3$	296(3.34)	115-0073-78
	CCl$_4$	300(3.26)	115-0073-78
Phenol, 2-(4-phenyl-5-isoxazolyl)-	EtOH	228(4.28),263(4.03)	114-0069-78
4(1H)-Quinolinone, 2-hydroxy-1-phenyl-	EtOH	316(3.70)	64-0332-78
2(1H)-Quinolinone, 4-(phenoxy)-	EtOH	319(3.75)	64-0332-78

Compound	Solvent	$\lambda_{max}(\log \epsilon)$	Ref.
$C_{15}H_{11}NO_2S$			
Acetic acid, (6-phenanthridinylthio)-	MeOH	243(4.60),305(3.93), 333(3.79),349(3.79)	2-0678-78
[1,1'-Biphenyl]-4-carbonyl isothiocyanate, 4'-methoxy-	C_6H_{12}	227(4.34),327(4.42)	73-0257-78
5(4H)-Oxazolone, 2-phenyl-4-[1-(2-thienyl)ethylidene]-, (Z)-	EtOH	202(4.23),266(4.19), 394(4.44)	118-0832-78
$C_{15}H_{11}NO_3$			
1,4-Acridinedione, 2-methoxy-9-methyl-	dioxan	227(4.47),282(4.41), 373s(3.3)	150-4901-78
4H-1-Benzopyran-4-one, 2-(2-pyridinyl-methyl)-	EtOH	212(4.61),265(4.24), 305(3.72)	118-0208-78
5(4H)-Oxazolone, 4-[1-(2-furanyl)ethylidene]-2-phenyl-, (Z)-	EtOH	202(4.16),263(4.03), 390(4.51)	118-0832-78
2H-Pyrido[2,1-b][1,3]oxazinium, 4-hydroxy-2-oxo-3-(phenylmethyl)-, hydroxide, inner salt	MeCN	268s(3.59),290(3.22), 298s(3.19),360(3.59)	5-1655-78
$C_{15}H_{11}NS$			
Isothiazole, 3,4-diphenyl-	EtOH	277(4.07)	39-0685-78C
Isothiazole, 3,5-diphenyl-	EtOH	252(4.43),278(4.34)	39-0685-78C
Propanenitrile, 2-(2-methyl-3H-naphtho[1,8-bc]thien-3-ylidene)-	C_6H_{12}	248(4.443),268(4.426), 385(4.404)	24-1824-78
$C_{15}H_{11}NS_2$			
Benzenamine, N-(5-phenyl-3H-1,2-dithiol-3-ylidene)-	EtOH	275(4.28),364(3.85)	94-1629-78
$C_{15}H_{11}N_2O_3$			
Isoquinolinium, 2-[2-(5-nitro-2-furanyl)ethenyl]-, bromide, (E)-	EtOH	218(4.05),238(4.02), 345(4.20)	103-0250-78
$C_{15}H_{11}N_3$			
6H-Benzo[b]quinolizine-11-carbonitrile, 6-imino-2-methyl-	EtOH	232(4.43),236s(4.41), 256s(3.96),269s(3.95), 309(3.62),322(3.71), 394(4.01),443s(3.78), 473s(3.48)	44-3536-78
Indolo[2,3-b]quinoxaline, 8-methyl-	EtOH	269s(4.73),273(4.76), 348s(4.29),359(4.35), 386s(3.75)	78-0981-78
1,7,7b-Triazacyclopent[cd]indene, 6-methyl-3-phenyl-	n.s.g.	238s(4.34),247(4.43), 332(4.30),404(3.72), 416(3.69)	44-3544-78
$C_{15}H_{11}N_3O$			
6H-Indolo[2,3-b]quinoxaline, 8-methoxy-	EtOH	236(4.47),277(4.63), 365(4.43)	78-0981-78
2-Quinoxalinecarboxaldehyde, 3-(phenylamino)-	$CHCl_3$	453(3.40)	97-0177-78
$C_{15}H_{11}N_3O_2$			
Benzeneacetonitrile, α-[[(4-nitrophenyl)amino]methylene]-	EtOH	390(4.49)	78-2229-78
Benzo[b][1,6]naphthyridine-7-carbonitrile, 1,2,3,5-tetrahydro-2,5-dimethyl-1,3-dioxo-	pH 7.0	282(4.60),312(4.36), 498(3.91)	35-7375-78
Benzo[b][1,6]naphthyridine-10-carbonitrile, 1,2,3,5-tetrahydro-2,5-dimethyl-1,3-dioxo-	MeOH	229(4.15),291(4.23), 350(3.20),525(3.40)	150-1325-78

Compound	Solvent	$\lambda_{max}(\log \epsilon)$	Ref.
Cyclohepta[b]pyrrole-3-carboxylic acid, 2-(dicyanomethylene)-1,2-dihydro-, ethyl ester	MeOH	257(4.26),290(4.32), 336(3.78),438(4.29)	18-0667-78
1H-Pyrazole, 3-(3-nitrophenyl)-5-phenyl-	EtOH	252(4.40)	94-1298-78
Pyridinium, 1-(5-formyl-2,3-dihydro-2-oxo-1-phenyl-1H-imidazol-4-yl)-, hydroxide, inner salt	MeOH	<u>255(3.4),320(3.5), 400(2.6)</u>	103-0208-78
Quinoxaline, 3-methyl-6-nitro-2-phenyl-	EtOH	255(4.28),318(4.11)	94-2866-78
$C_{15}H_{11}N_3O_4S$			
1-Naphthalenesulfonic acid, 4-hydroxy-3-(2-pyridinylazo)-	pH -0.2	370(4.04),458(4.10)	73-2711-78
	pH 1.0	364(4.08),463(4.10)	73-2711-78
	pH 4-8	359(3.98),478(4.16)	73-2711-78
(also metal chelates)	pH 8-13	498(4.23)	73-2711-78
$C_{15}H_{11}N_3O_6$			
3-Isoxazolecarboxylic acid, 4,5-dihydro-4-[(3-hydroxy-1-oxo-1H-inden-2-yl)hydrazono]-5-oxo-, ethyl ester	MeOH	240(4.29),505(4.34)	142-0199-78B
$C_{15}H_{11}N_5O_2$			
3H-Indole, 3-azido-3-methyl-5-nitro-2-phenyl-	EtOH	232(4.12),335(4.25)	94-2866-78
$C_{15}H_{11}N_7O_7$			
1(2H)-Phthalazinone, 4-[1-methyl-2-(2,4,6-trinitrophenyl)hydrazino]-	dioxan	262(4.20),313(4.20)	103-0575-78
$C_{15}H_{12}$			
Anthracene, 9-methyl-	C_6H_{12}	366(4.02)	61-1068-78
$C_{15}H_{12}BrN_3$			
Benzenepropanenitrile, α-[(4-bromophenyl)hydrazono]-, anti	EtOH	324(4.35)	118-0445-78
syn	EtOH	333(4.40)	118-0445-78
$C_{15}H_{12}ClN$			
3H-Indole, 3-chloro-3-methyl-2-phenyl-	CH_2Cl_2	248(4.43),324(4.04)	94-2866-78
$C_{15}H_{12}ClNO_2$			
9(10H)-Acridinone, 6-chloro-2-methoxy-10-methyl-	EtOH	260(4.62),279(4.72), 405(3.86),424(3.90)	35-5604-78
Benzenepropanamide, 4-chloro-β-oxo-α-phenyl-	EtOH	205(4.41),230(4.37)	18-3389-78
4-Isoxazolol, 3-(4-chlorophenyl)-4,5-dihydro-5-phenyl-	MeOH	269(4.29)	142-0187-78C
$C_{15}H_{12}ClNO_4$			
2H-Cyclohepta[gh]pyrrolizine-1,2-dicarboxylic acid, 4-chloro-, dimethyl ester	CHCl_3	255(4.17),273(3.83), 283(3.74),373(3.79), 394(3.76),415(3.42), 431(3.45),459(3.41), 492(3.23),530(2.78)	39-0429-78C
$C_{15}H_{12}ClNO_6S$			
Thiopyrano[2,3-b]pyrrolium, 3-carboxy-7-methyl-6-phenyl-, perchlorate	49% H_2SO_4	255(4.29),280(4.21), 360(3.89)	104-0589-78

Compound	Solvent	$\lambda_{max}(\log \epsilon)$	Ref.
$C_{15}H_{12}ClN_3$			
Benzenepropanenitrile, α-[(3-chloro-phenyl)hydrazono]-, anti	EtOH	318(4.30)	118-0445-78
syn	EtOH	332(4.33)	118-0445-78
Benzenepropanenitrile, α-[(4-chloro-phenyl)hydrazono]-, anti	EtOH	309(4.27)	118-0445-78
syn	EtOH	333(4.20)	118-0445-78
$C_{15}H_{12}ClN_5O_2$			
1(2H)-Phthalazinone, 4-chloro-2-meth-yl-, (4-nitrophenyl)hydrazone, (Z)-	dioxan	294(4.13),427(4.36)	103-0575-78
$C_{15}H_{12}Cl_2$			
Stilbene, 2,6-dichloro-2'-methyl-, cis	MeOH	224s(4.24),258(4.00)	35-3819-78
trans	MeOH	223s(4.28),283(4.21)	35-3819-78
$C_{15}H_{12}F_2$			
Benzene, 1-(1,2-difluoro-2-phenyleth-enyl)-4-methyl-, (E)-	EtOH	300(4.54)	104-0939-78
$C_{15}H_{12}F_3NO_2$			
Benzoic acid, 3-methyl-2-[[3-(triflu-oromethyl)phenyl]amino]-	EtOH	270(3.96),314(3.75)	83-0161-78
Benzoic acid, 5-methyl-2-[[3-(triflu-oromethyl)phenyl]amino]-	EtOH	290(4.23),347(3.86)	83-0161-78
$C_{15}H_{12}F_3N_5O$			
1H-Purin-2-amine, 6-[[3-[3-(trifluoro-methyl)phenyl]-2-propenyl]oxy]-	EtOH	245(4.35),282(4.02)	44-0516-78
	EtOH-HCl	246(4.29),285(4.04),294s(3.99)	44-0516-78
	EtOH-NaOH	227(4.17),247(4.26),283(3.99)	44-0516-78
$C_{15}H_{12}FeO_3$			
Iron, tricarbonyl[(1,8,9,10-η)-(9,10-dimethylbicyclo[6.2.0]deca-1,3,5,7,9-pentaene]-	EtOH	283(4.30),310(4.22),400s(3.26)	5-1379-78
$C_{15}H_{12}NO_2$			
Isoquinolinium, 2-(2,3-dihydroxyphen-yl)-, iodide	EtOH	350(4.31)	2-0165-78
	EtOH-NaOH	470(4.23)	2-0165-78
Isoquinolinium, 2-(2,5-dihydroxyphen-yl)-, iodide	EtOH	365(3.20)	2-0165-78
	EtOH-NaOH	470(3.25)	2-0165-78
Quinolinium, 1-(2,3-dihydroxyphenyl)-, iodide	EtOH	365(4.21)	2-0165-78
	EtOH-NaOH	490(4.31)	2-0165-78
Quinolinium, 1-(2,5-dihydroxyphenyl)-, iodide	EtOH	390(3.59)	2-0165-78
	EtOH-NaOH	530(3.55)	2-0165-78
Thiopyrano[2,3-b]pyrrolium, 3-carboxy-7-methyl-6-phenyl-, perchlorate	49% H_2SO_4	255(4.29),280(4.21),360(3.89)	104-0589-78
$C_{15}H_{12}N_2$			
Benzenamine, 4-(4-isoquinolinyl)-	EtOH	219(4.73),251(4.18)	44-0672-78
1H-1,5-Benzodiazepine, 4-phenyl-, monoperchlorate	EtOH	220s(4.11),267s(4.29),272(4.33),297(4.24),338(3.81),548(3.07)	48-0659-78
Propanedinitrile, (1,2,3,4-tetrahydro-1,4-methano-7H-benzocyclohepten-7-ylidene)-	hexane	257(4.00),396s(4.41),403(4.45),410(4.45),420(4.36),430(4.32)	88-0569-78
	MeOH	261(4.09),411(4.52),420(4.53),427s(4.52)	88-0569-78
1H-Pyrazole, 3,5-diphenyl-	EtOH	256(4.50)	78-1571-78

Compound	Solvent	$\lambda_{max}(\log \epsilon)$	Ref.
1H-Pyrazole, 3,5-diphenyl- (cont.)	H_2SO_4	276(4.34)	78-1571-78
Quinazoline, 4-methyl-2-phenyl-	EtOH	253(4.38),261(4.43), 284(3.93),320(3.43)	94-2866-78
Quinoxaline, 2-methyl-3-phenyl-	EtOH	240(4.22),324(3.77)	94-2866-78
$C_{15}H_{12}N_2O$			
8H-Cyclohepta[b]quinoxalin-8-one, 7,9-dimethyl-	EtOH	230(4.35),235(4.38), 272(4.66),361(4.05), 378(4.20),400(4.22)	56-2045-78
Phenol, 2-(4-phenyl-1H-pyrazol-3-yl)-	EtOH	254(4.08),261(4.05), 294(3.78)	114-0457-78C
Propanedinitrile, [(2-ethyl-5-methyl-3-benzofuranyl)methylene]-	MeOH	252(4.17),283(3.68), 347(4.14)	83-0714-78
$C_{15}H_{12}N_2OS$			
1H-4,1,2-Benzothiadiazine, 3-acetyl-1-phenyl-	MeOH	225(4.23),283(4.20)	39-0539-78C
$C_{15}H_{12}N_2O_2$			
Benzoxazole, 6-nitroso-7-(2-phenyl-ethyl)-	$CHCl_3$	760(1.53)	39-0692-78C
8H-Cyclohepta[b]quinoxalin-8-one, 7,9-dimethyl-, 5-oxide	EtOH	223(4.16),302(4.65), 383(3.95),415(3.89)	56-2045-78
Ethenamine, N-(diphenylmethylene)-2-nitro-	MeOH	252(3.99),332(4.20)	20-0693-78
1H-Isoindole-1,3(2H)-dione, 2-(3-azatricyclo[3.2.1.0^{2,4}]oct-6-en-3-yl)-, endo	EtOH	213(4.65),230(4.57), 286(3.00),296(3.03), 307s(2.91),340(2.56)	33-0795-78
exo	EtOH	212(4.32),239(4.37), 290s(3.02),296(3.06), 306s(2.95),341(2.67)	33-0795-78
1,8-Naphthyridin-2(1H)-one, 4-hydroxy-3-(2-methylphenyl)-	MeOH	223(4.11),262(4.00), 298(4.08),320(4.21)	24-2813-78
1,8-Naphthyridin-2(1H)-one, 4-hydroxy-3-(3-methylphenyl)-	MeOH	228(4.11),259(4.02), 303(4.12),325(4.26)	24-2813-78
1,8-Naphthyridin-2(1H)-one, 4-hydroxy-3-(4-methylphenyl)-	MeOH	225(4.12),255(3.98), 299(4.04),322(4.19)	24-2813-78
Pyrrolo[1,2-c]pyrimidine-7-carboxalde-hyde, 1,2-dihydro-3-methyl-1-oxo-6-phenyl-	n.s.g.	225(4.12),272(4.28), 293s(3.86),365(4.19)	44-3544-78
Sydnone, 3-methyl-4-[2-(1-naphthalen-yl)ethenyl]-, (E)-	benzene	371(4.37)	48-0071-78
	EtOH	363(4.39)	48-0071-78
(Z)-	benzene	350(4.03)	48-0071-78
	EtOH	340(4.10)	48-0071-78
Sydnone, 3-(4-methylphenyl)-4-phenyl-	EtOH	237(4.12),332(3.96)	33-1477-78
$C_{15}H_{12}N_2O_2S_2$			
1,4,2-Dithiazin-3-amine, N,5-diphenyl-, 1,1-dioxide	EtOH	274(4.35)	18-1805-78
1,4,2-Dithiazol-3-amine, N-phenyl-5-(phenylmethylene)-, 1,1-dioxide	EtOH	304(4.51)	18-1805-78
$C_{15}H_{12}N_2O_3$			
1H-Imidazole, 2,4,5-tri-2-furanyl-4,5-dihydro-	EtOH	250(4.78)	103-1184-78
Methanediamine, 1-(2-furanyl)-N,N'-bis(2-furanylmethylene)-	EtOH	280(4.64)	103-1184-78
1,8-Naphthyridin-2(1H)-one, 4-hydroxy-3-(4-methoxyphenyl)-	MeOH	229(4.10),263(3.92), 304(4.09),327(4.35)	24-2813-78
2-Propenal, 3-[4-[(4-nitrophenyl)meth-ylene]-1(4H)-pyridinyl]-, (E)-	dioxan	260(4.07),350(3.94), 468(4.51)	48-0659-78

Compound	Solvent	λ_{max} (log ϵ)	Ref.
9H-Pyrido[3,4-b]indole-3-carboxylic acid, 1-acetyl-, methyl ester	EtOH	270s(4.47),286(4.66), 374(3.86)	102-0338-78
$C_{15}H_{12}N_2O_3S$			
1,4,3-Oxathiazin-2-amine, N,6-diphenyl-, 4,4-dioxide	EtOH	257(4.28)	18-1805-78
1,4,3-Oxathiazin-2-imine, 2,3-dihydro-N,6-diphenyl-, 4,4-dioxide	EtOH	257(4.29)	18-1805-78
2H-1,2,4-Thiadiazin-3(4H)-one, 4,5-diphenyl-, 1,1-dioxide	EtOH	247(4.04)	18-1805-78
$C_{15}H_{12}N_2S$			
Thiopyrano[2,3-b]pyrrole-3-carbonitrile, 2,7-dihydro-7-methyl-6-phenyl-	EtOH	397(3.71)	104-0589-78
$C_{15}H_{12}N_4$			
Benzaldehyde, 1-phthalazinylhydrazone	dioxan	292(4.13),377(4.23)	103-0567-78
	MeCN	210(4.43),293(4.15), 372(4.24)	103-0567-78
1H-Indole, 2-(azidomethyl)-3-phenyl-	EtOH	225(4.50),272(4.09), 276(4.10),281(4.10), 290(4.04)	94-2874-78
3H-Indole, 3-azido-3-methyl-2-phenyl-	EtOH	227(4.09),233(4.11), 241(4.12),248(4.10), 314(4.05)	94-2866-78
$C_{15}H_{12}N_4O_2$			
Benzenepropanenitrile, α-[(4-nitrophenyl)hydrazono]-, anti	EtOH	364(4.38)	118-0445-78
syn	EtOH	368(4.35)	118-0445-78
$C_{15}H_{12}N_4O_3$			
2,1-Benzisoxazole, 1,3-dihydro-1-methyl-3-[[(4-nitrophenyl)azo]methylene]-	MeCN	395(4.15),555(4.06)	44-1233-78
$C_{15}H_{12}N_6O$			
1,3,5-Triazine, 2-azido-4-methoxy-6-(2-methyl-1-naphthalenyl)-	C_6H_{12}	308(3.59)	44-1361-78
1,3,5-Triazine, 2-azido-4-methoxy-6-(4-methyl-1-naphthalenyl)-	C_6H_{12}	332(4.15)	44-1361-78
$C_{15}H_{12}N_6O_2$			
1,3,5-Triazine, 2-azido-4-methoxy-6-(2-methoxy-1-naphthalenyl)-	C_6H_{12}	337(3.74)	44-1361-78
1,3,5-Triazine, 2-azido-4-methoxy-6-(4-methoxy-1-naphthalenyl)-	C_6H_{12}	346(4.30)	44-1361-78
$C_{15}H_{12}O$			
Anthracene, 9-methoxy-	C_6H_{12}	368(3.98)	61-1068-78
2-Propen-1-one, 1,3-diphenyl-	hexane	190s(4.52),224(4.10), 260(3.85),298(4.39)	65-0340-78
$(C_{15}H_{12}O)_n$			
Methanone, (4-ethenylphenyl)phenyl-, homopolymer	benzene	340(2.3)	126-0131-78
$C_{15}H_{12}OS_4$			
2(1H)-Naphthalenone, 1-[4,5-bis(methylthio)-1,3-dithiol-2-ylidene]-	MeCN	324(4.41),380(4.64), 485(4.66)	97-0385-78

Compound	Solvent	$\lambda_{max}(\log \epsilon)$	Ref.
$C_{15}H_{12}O_2$			
2H-1-Benzopyran-3(4H)-one, 2-phenyl-	MeOH	305(2.81)	44-0303-78
$C_{15}H_{12}O_3$			
Cyclopenta[b]furo[3,2-g]benzopyran-4-one, 1,2,3,4-tetrahydro-9-methyl-	$CHCl_3$	246(4.76),278(4.17), 314(3.99)	42-0468-78
Methanone, (2,4-dihydroxyphenyl)(4-ethenylphenyl)-	MeOH	291(4.19),329(4.11)	116-0312-78
$(C_{15}H_{12}O_3)_n$			
Methanone, (2,4-dihydroxyphenyl)(4-ethenylphenyl)-, homopolymer	MeOH	248(3.85),292(4.04), 324(3.96)	116-0312-78
$C_{15}H_{12}O_3S$			
6H-Thiopyrano[2,3-b]furan-5-carboxylic acid, 2-phenyl-, methyl ester	EtOH	250(4.32),330(4.30), 435(4.26)	104-0589-78
$C_{15}H_{12}O_4$			
4H-Pyran-4-one, 3,5-bis(2-furanylmethylene)tetrahydro-	EtOH	250(3.93),380(4.46)	133-0189-78
$C_{15}H_{12}O_5$			
Benzoic acid, 4-hydroxy-3-(2-methoxybenzoyl)-	MeOH	234(4.34),328(3.57)	39-0876-78C
Benzoic acid, 4-hydroxy-3-(3-methoxybenzoyl)-	MeOH	236(4.37),325(3.66)	39-0876-78C
$C_{15}H_{12}O_6$			
4H-1-Benzopyran-4-one, 2,3-dihydro-5,6,7-trihydroxy-2-(4-hydroxyphenyl)-	EtOH	300(4.13),368(3.18)	18-3627-78
4H-1-Benzopyran-4-one, 2,3-dihydro-5,7,8-trihydroxy-2-(4-hydroxyphenyl)-	EtOH	299(4.17),370(3.52)	18-3627-78
2-Propen-1-one, 3-(4-hydroxyphenyl)-1-(2,3,4,6-tetrahydroxyphenyl)-	EtOH	375(4.47)	18-3627-78
$C_{15}H_{13}Br$			
1-Propene, 1-bromo-1,2-diphenyl-, isomer m. 159°	hexane	260(3.81)	44-0034-78
isomer m. 42°	hexane	224(4.27),274(3.85)	44-0034-78
$C_{15}H_{13}BrN$			
Cycloheptatrienylium, (5-bromo-2,3-dihydro-1H-indol-1-yl)-, perchlorate	MeCN	251(4.26),405(4.40)	150-4801-78
$C_{15}H_{13}BrN_2O$			
4H-1,2,5-Oxadiazine, 3-(4-bromophenyl)-5,6-dihydro-5-phenyl-	MeOH	248(4.38)	56-1827-78
$C_{15}H_{13}BrN_2O_2$			
2-Pyrazoline, 3-(5-methyl-2-furoyl)-4-(4-bromophenyl)-	EtOH	220(4.2),296(3.9), 359(4.19)	104-1830-78
$C_{15}H_{13}BrN_4O$			
Benzoic acid, [[(4-bromophenyl)hydrazono]ethylidene]hydrazide	EtOH	205(4.26),238(4.12), 268s(4.06),317s(3.96), 372(4.63)	136-0089-78E
$C_{15}H_{13}BrO$			
Benzene, 1-bromo-4-[2-(4-methoxyphenyl)ethenyl]-, (E)-	EtOH	231(4.12),293s(4.13), 308(4.43),325(4.47)	87-0889-78

Compound	Solvent	$\lambda_{max}(\log \epsilon)$	Ref.
$C_{15}H_{13}BrO_3$ Benzo[1,2-b:5,4-b']difuran, 8-bromo-4-methoxy-5-methyl-2-(1-methyl-ethenyl)-	EtOH	259(4.40),268(4.41), 295(4.24),304(4.27)	12-1533-78
$C_{15}H_{13}Cl$ Benzene, 1-chloro-2-[2-(2-methylphenyl)ethenyl]-, (E)-	MeOH	227(4.19),294(4.35)	35-3819-78
$C_{15}H_{13}ClN_2O$ Benzeneacetic acid, [(4-chlorophenyl)-methylene]hydrazide	EtOH	288(4.32),296s(4.28), 308s(4.11)	34-0345-78
Benzoic acid, [1-(4-chlorophenyl)eth-ylidene]hydrazide	EtOH	273s(4.27),281(4.27), 292s(4.18)	34-0345-78
4H-1,2,5-Oxadiazine, 5-(4-chlorophenyl)-5,6-dihydro-	MeOH	253(4.47)	56-1827-78
$C_{15}H_{13}ClN_2S$ Quinazoline, 6-chloro-1,4-dihydro-2-(methylthio)-4-phenyl-	MeOH	226(4.30),295(4.07)	4-1409-78
$C_{15}H_{13}ClN_4$ 1(2H)-Phthalazinone, 4-chloro-, methyl-phenylhydrazone	dioxan CHCl$_3$	272(4.43),315(3.85) 272(--),315(--)	103-0575-78 103-0575-78
$C_{15}H_{13}ClN_4O$ 3-Cyclohexene-1-carboxamide, 4-amino-N-(4-chlorophenyl)-1,3-dicyano-	MeOH	262(4.27)	42-0281-78
$C_{15}H_{13}ClN_4O_7$ 2,1-Benzisoxazolium, 1-methyl-3-[[(4-nitrophenyl)hydrazono]methyl]-, perchlorate	MeCN	485(4.48)	44-1233-78
$C_{15}H_{13}ClO_4$ Benzo[1,2-b:4,3-b']difuran-1-carboxylic acid, 5-chloro-2,7-dimethyl-, ethyl ester	MeOH	220(3.90),278(3.86), 289(3.78),299(3.68)	4-0043-78
$C_{15}H_{13}Cl_3N_2O_5S$ Benzenamine, N,N-dimethyl-4-[2-(5-ni-tro-2-furanyl)-2-[(trichloromethyl)-sulfonyl]ethenyl]-	EtOH	208(4.04),250(3.88), 319(3.97),384s(4.10), 428(4.26)	73-1618-78
$C_{15}H_{13}FO_3$ [1,1'-Biphenyl]-4-carboxylic acid, 4'-fluoro-3-hydroxy-, ethyl ester	MeOH	272(4.34),316(3.88)	87-1093-78
$C_{15}H_{13}NO$ 9(10H)-Acridinone, 2,10-dimethyl-	EtOH	254(4.29),297(3.03), 390(3.54),408(3.60)	35-5604-78
9(10H)-Acridinone, 10-ethyl-	CH$_2$Cl$_2$	258(4.84),268s(4.54), 277(4.45),294(3.87), 306(3.76),363s(3.81), 381s(4.12),401(4.26)	18-2674-78
2-Propen-1-one, 1-phenyl-3-(phenyl-amino)-	hexane MeOH	368(4.34) 370(4.44)	48-0945-78 48-0945-78
$C_{15}H_{13}NOS$ 2H-Benzo[a]quinolizin-2-one, 4-methyl-3-(methylthio)-	EtOH	232(4.25),247(4.30), 289(4.42),322(4.0),	95-0198-78

Compound	Solvent	$\lambda_{max}(\log \epsilon)$	Ref.
(cont.)		338(4.0),356(3.78), 376(3.64)	95-0198-78
$C_{15}H_{13}NO_2$			
9(10H)-Acridinone, 2-methoxy-10-methyl-	MeOH	300(3.310)	4-0149-78
Benzenepropanamide, β-oxo-α-phenyl-	EtOH	206(4.30),226(4.30)	18-3389-78
4-Isoxazolol, 4,5-dihydro-3,5-diphenyl-	MeOH	264(4.15)	142-0187-78C
cis	EtOH	266(4.11)	78-1571-78
	H_2SO_4	302(4.16)	78-1571-78
trans	EtOH	260(4.22)	78-1571-78
	H_2SO_4	306(4.14)	78-1571-78
6(5H)-Phenanthridinone, 5-ethoxy-	MeOH	231(4.66),236(4.65), 253s(4.19),263(4.21), 325s(3.87),340(3.95), 356s(3.84)	94-2508-78
2-Propen-1-one, 1-(2-hydroxyphenyl)- 3-phenyl-	EtOH	292(4.22)	56-2479-78
$C_{15}H_{13}NO_2S$			
Phenanthridine, 6-(ethylsulfonyl)-	EtOH	223(4.31),252(4.59), 296(3.80),306(3.80), 337(3.32),355(3.14)	95-0136-78
Thiopyrano[2,3-b]pyrrole-3-carboxylic acid, 2,7-dihydro-7-methyl-6-phenyl-	EtOH	225(4.095),282(4.38), 395(3.68)	104-0589-78
$C_{15}H_{13}NO_3$			
9(10H)-Acridinone, 1-hydroxy-3-methoxy- 10-methyl-	EtOH	213(4.20),223(4.22), 270(4.70),295(4.01), 325(3.85)	83-0135-78
	EtOH-NaOH	271(4.63),305(4.07), 400(3.88)	83-0135-78
Benzamide, N-(2-acetylphenyl)-4-hy- droxy-	EtOH	218(4.48),248(4.37), 279s(4.04),341(4.03)	120-0062-78
Benzamide, N-(2-hydroxy-3-oxo-1,4,6-cy- cloheptatrien-1-yl)-N-methyl-	MeOH	331(3.79),368(3.78)	18-2338-78
9H-Carbazole-3-carboxylic acid, 1-meth- oxy-, methyl ester (mukonine)	EtOH	236(4.6),245(4.57), 266(4.64),274(4.73), 306(3.88),320(3.78)	102-0834-78
9H-Carbazole-3-carboxylic acid, 2-meth- oxy-, methyl ester	EtOH	227(4.5),250(4.28), 256(4.1),285(4.13)	42-1114-78
$C_{15}H_{13}NO_4$			
1,3-Dioxane-4,6-dione, 5-(1H-indol- 3-ylmethylene)-2,2-dimethyl-	EtOH	412(4.704)	103-1088-78
	EtOH-NaOEt	420(4.358)	103-1088-78
$C_{15}H_{13}NO_4S$			
2-Propenoic acid, 3-[[4-(acetylamino)- phenyl]thio]-2-furanyl]-	EtOH	207(4.45),267(4.35), 305(4.29)	73-0621-78
$C_{15}H_{13}NO_4Se$			
Selenonium, diphenyl-, 2-methoxy- 1-nitro-2-oxoethylide	MeCN	314(3.95)	70-0945-78
$C_{15}H_{13}NS_2$			
2-Propenethioamide, 3-mercapto-N,3-di- phenyl-	EtOH	229s(--),265s(--), 347(4.19)	94-1021-78
	DMSO	229s(4.25),256s(4.03), 347(4.19)	94-1026-78
	acid	332(--)	94-1021-78
(also palladium chelate)	base	383(--)	94-1021-78

Compound	Solvent	$\lambda_{max}(\log \epsilon)$	Ref.
$C_{15}H_{13}N_2O_2$			
Cycloheptatrienylium, (2,3-dihydro-5-nitro-1H-indol-1-yl)-, perchlorate	MeCN	262(3.96),416(4.40), 299(3.61)	150-4801-78
$C_{15}H_{13}N_3$			
Benzenepropanenitrile, α-(phenylhydrazono)-, anti	EtOH	319(4.36)	118-0445-78
syn	EtOH	332(4.25)	118-0445-78
2-Quinoxalinamine, N-(3-methylphenyl)-	EtOH	277s(4.40),284(4.41), 383(4.00)	78-0981-78
2-Quinoxalinamine, N-(4-methylphenyl)-	EtOH	274(4.40),286(4.42), 385(3.99)	78-0981-78
4H-1,2,4-Triazole, 3-(4-methylphenyl)-4-phenyl-	EtOH	244(4.23)	88-4629-78
4H-1,2,4-Triazole, 4-(4-methylphenyl)-3-phenyl-	EtOH	236(3.62)	88-4629-78
$C_{15}H_{13}N_3O$			
Benzenamine, N-[(1H-indazol-3-yl)methylene]-4-methoxy-	EtOH	209(4.29),259(3.88), 339(4.24)	103-0771-78
Methanimidamide, 1-cyano-N-(2'-hydroxy-[1,1'-biphenyl]-2-yl)-N'-methyl-	MeOH	244(4.28),285(3.92)	94-2508-78
2-Quinoxalinamine, N-(3-methoxyphenyl)-	EtOH	274(4.42),383(4.00)	78-0981-78
4H-1,2,4-Triazole, 3-(4-methoxyphenyl)-4-phenyl-	EtOH	256(4.21)	88-4629-78
4H-1,2,4-Triazole, 4-(4-methoxyphenyl)-3-phenyl-	EtOH	229(4.33)	88-4629-78
$C_{15}H_{13}N_3OS$			
1H-Benzimidazole-1-carboximidic acid, 2,3-dihydro-2-thioxo-, 4-methylphenyl ester	dioxan	312.5(4.272)	48-0677-78
$C_{15}H_{13}N_3O_2$			
Benzo[b][1,6]naphthyridine-10-carbonitrile, 1,2,3,4,5,10-hexahydro-2,5-dimethyl-1,3-dioxo-	pH 4 / pH 10	342(4.9) / 392(4.36)	150-1325-78 / 150-1325-78
Benzotriazole-4,7-dione, 5,6-dimethyl-2-(phenylmethyl)-	EtOH	233(4.32),251(4.18), 297(3.04),340(2.85)	87-0578-78
$C_{15}H_{13}N_3O_3$			
4H-1,2,5-Oxadiazine, 5,6-dihydro-5-(4-nitrophenyl)-3-phenyl-	MeOH	248(4.27)	56-1827-78
$C_{15}H_{13}N_4O_3$			
2,1-Benzisoxazolium, 1-methyl-3-[[(4-nitrophenyl)hydrazono]methyl]-, perchlorate	MeCN	485(4.48)	44-1233-78
$C_{15}H_{13}N_4S$			
Acridinium, 3,6-diamino-10-methyl-, thiocyanate	MeOH	211(4.1),232(4.1), 261(4.8),281(4.6), 299(4.6),463(4.8)	97-0108-78
$C_{15}H_{13}N_5O_3S$			
Acetamide, N-acetyl-N-[2-(acetylimino)-thiazolo[4,5-b]quinoxalin-3(2H)-yl]-	EtOH	212(4.72),257(4.56), 313(3.73)	2-0683-78
$C_{15}H_{13}N_5O_6$			
2H-Pyrrolo[3,2-b]pyridin-2-one, 4-acetyl-1-[(2,4-dinitrophenyl)amino]-	n.s.g.	216(4.20),269(4.23), 332(3.14)	78-1581-78

Compound	Solvent	$\lambda_{max}(\log \epsilon)$	Ref.
1,4,5,6-tetrahydro- (cont.)			78-1581-78
$C_{15}H_{14}$			
Benzene, 1-methyl-2-(2-phenylethenyl)-, trans	MeOH	229(4.23),295(4.47), 307(4.45)	35-3819-78
4a,9a-Methano-9H-fluorene, 9-methyl-, anti	THF	236(3.74),273(3.61), 301s(3.22)	44-1586-78
$C_{15}H_{14}AsN$			
1H-1,3-Benzazarsole, 1-ethyl-2-phenyl-	MeOH	247(4.21),319(4.17)	101-0001-78K
$C_{15}H_{14}AsNO_2$			
Acetamide, N-acetyl-N-(2-phenyl-4-arseninyl)-	EtOH	259(4.40),299(3.91)	88-1175-78
$C_{15}H_{14}ClNO_2$			
Cyclopenta[b]pyran-2(5H)-one, 3-chloro-6,7-dihydro-4-(methylphenylamino)-	EtOH	242(4.00),346(4.10)	4-0511-78
$C_{15}H_{14}ClNO_2S$			
2H,5H-Thiopyrano[4,3-b]pyran-2-one, 3-chloro-7,8-dihydro-4-(methylphenylamino)-	EtOH	242(4.11),318(3.87), 360(3.79)	4-0181-78
$C_{15}H_{14}ClN_3OS$			
Hydrazinecarboxamide, 2-[1-[5-chloro-2-(phenylthio)phenyl]ethylidene]-	MeOH	253(4.37),280s(4.10)	73-1276-78
$C_{15}H_{14}ClN_3O_3$			
Propanehydrazonic acid, N-(4-nitrophenyl)-, 4-chlorophenyl ester	xylene	370(4.37)	18-0512-78
	EtOH	256s(3.85),355s(3.89), 384(4.32)	18-0512-78
Propanoic acid, 2-(4-chlorophenyl)-2-(4-nitrophenyl)hydrazide	xylene	360(4.07)	18-0512-78
	EtOH	260s(3.92),361(4.26)	18-0512-78
1H-Pyrazole-1-carboxylic acid, 4-(5-chloro-2,1-benzisoxazol-3-yl)-3,5-dimethyl-, ethyl ester	EtOH	253(3.92),337(4.00)	94-1141-78
$C_{15}H_{14}ClN_5O$			
3-Formazancarboxamide, 5-(4-chlorophenyl)-1-(4-methylphenyl)-	EtOH	452(4.01)	104-1956-78
$C_{15}H_{14}Cl_2N_4O_2$			
Formazan, 1,5-bis(5-chloro-2-methoxyphenyl)-	EtOH	248(4.17),296(3.87), 376(4.17),482(3.74)	104-1956-78
$C_{15}H_{14}F_2N_2O_4S$			
3-Thiophenecarboxylic acid, 2-[(2,5-difluoro-4-nitrophenyl)amino]-5-ethyl-, ethyl ester	MeOH	391(4.35)	39-0937-78C
3-Thiophenecarboxylic acid, 2-[(3,5-difluoro-2-nitrophenyl)amino]-5-ethyl-, ethyl ester	MeOH	330(4.08),404(3.64)	39-0937-78C
3-Thiophenecarboxylic acid, 2-[(4,5-difluoro-2-nitrophenyl)amino]-5-ethyl-, ethyl ester	MeOH	333(3.99),429(3.80)	39-0937-78C
$C_{15}H_{14}FeO_4$			
Ferrocene, (2,3-dicarboxy-1-propenyl)-, (E)-	EtOH	242(4.08),290(4.12)	128-0325-78

Compound	Solvent	$\lambda_{max}(\log \epsilon)$	Ref.
$C_{15}H_{14}N$			
Cycloheptatrienylium, (2,3-dihydro-1H-indol-1-yl)-, perchlorate	MeCN	246(4.18),400(4.29)	150-4801-78
$C_{15}H_{14}N_2$			
Benzene, 1,1'-(diazomethylene)bis[4-methyl-	benzene	538(1.95)	39-1283-78B
Phenazine, 2,7,8-trimethyl-	MeOH	256(5.22),377(4.22)	5-0440-78
1-Propene, 1-phenyl-1-(phenylazo)-, (E,Z)-	hexane	227(4.16),302(4.24), 437(2.41)	49-1081-78
1-Propene, 1-phenyl-2-(phenylazo)-	hexane	237(4.11),338(4.34), 454(2.66)	49-1081-78
$C_{15}H_{14}N_2O$			
Benzene, 1-[diazo(4-methoxyphenyl)methyl]-4-methyl-	benzene	544(1.96)	39-1283-78B
Benzeneacetic acid, (phenylmethylene)-hydrazide	EtOH	284(4.29),293s(4.25), 306s(4.05)	34-0345-78
Benzo[g]quinazolin-4(1H)-one, 2-(1-methylethyl)-	MeOH	260(4.66),270(4.61), 330(3.65),348(3.70)	2-1067-78
Benzo[h]quinazolin-4(1H)-one, 2-(1-methylethyl)-	EtOH	248(4.78),295(4.19), 307(4.16),330(3.74), 348(3.79)	2-1067-78
4H-1,2,5-Oxadiazine, 5,6-dihydro-3,5-diphenyl-	MeOH	246(4.19)	56-1827-78
6(5H)-Phenanthridinone, 5-(dimethyl-amino)-	MeOH	231(4.66),237(4.63), 250(4.17),259(4.27), 300s(3.67),310s(3.73), 323(3.89),337(3.82)	94-2508-78
2-Propen-1-one, 3-[(2-aminophenyl)amino]-1-phenyl-	MeCN	213(4.28),247(4.19), 272s(3.88),383(4.33)	48-0659-78
1H-Pyrazol-4-ol, 4,5-dihydro-3,5-diphenyl-, trans	EtOH	292(4.11)	78-1571-78
Pyrrolo[1,2-c]pyrimidine, 1-methoxy-3-methyl-6-phenyl-	n.s.g.	253(4.67),276s(4.02), 289s(3.78),322(3.46)	44-3544-78
$C_{15}H_{14}N_2O_2$			
Benzoic acid, 2-acetyl-2-phenylhydrazide	EtOH	229(4.24)	33-1477-78
Benzoic acid, [(4-methoxyphenyl)methylene]hydrazide	pH 2.4	225(4.87),294(4.34), 315(4.28)	48-1029-78
	pH 7.24	230(4.13),296(4.39), 315(4.51)	48-1029-78
	pH 11.85	230(4.17),296(4.31), 345(4.23)	48-1029-78
	EtOH	290s(4.32),304(4.39)	34-0345-78
Benzo[a]pyrrolo[3,4-f]quinolizine-1,12-dione, 2,3,5,6,10b,11-hexahydro-	MeOH	243(4.10),305(4.08)	30-0232-78
1H-Indazol-3-ol, 5-methoxy-1-(phenylmethyl)-	EtOH	307(3.73)	118-0633-78
Methanone, (4,5-dihydro-4-phenyl-1H-pyrazol-3-yl)(5-methyl-2-furanyl)-	EtOH	209(4.08),300(3.88), 362(3.91)	104-1830-78
Pyrano[2,3-c]pyrazol-6(1H)-one, 3,4-dimethyl-5-(phenylmethyl)-	EtOH	306(4.18)	95-0335-78
$C_{15}H_{14}N_2O_2S$			
Ethanone, 1-[4-(7,8-dihydro[1,2]oxathiolo[4,3,2-hi][1,2,3]benzothiadiazol-3-S^{IV}-2(6H)-yl)phenyl]-	C_6H_{12}	235(3.95),306(4.16), 458(4.48),475(4.44), 483(4.45)	39-0195-78C

Compound	Solvent	$\lambda_{max}(\log \epsilon)$	Ref.
$C_{15}H_{14}N_2O_3$			
[1]Benzopyrano[2,3-b]pyrrole-2,4-dione, 3-[(dimethylamino)methylene]-1,3-dihydro-1-methyl-	MeOH	235(4.13),263(3.88), 292(3.57),355(4.14)	83-0018-78
Isoquinoline, 1,2,3,4-tetrahydro-2-[2-(5-nitro-2-furanyl)ethenyl]-, (E)-	EtOH	298(4.45),313(4.43), 497(4.63)	103-0250-78
Methanone, [4,5-dihydro-4-(4-methoxyphenyl)-1H-pyrazol-3-yl]-2-furanyl-	EtOH	227(4.14),286(3.94), 356(4.13)	104-1830-78
Quinoline, 1,2,3,4-tetrahydro-1-[2-(5-nitro-2-furanyl)ethenyl]-, (E)-	EtOH	298(4.36),324(4.34), 508(4.47)	103-0250-78
$C_{15}H_{14}N_2O_4$			
Benzoic acid, 4-[[4-(hydroxymethyl)-phenyl]-ONN-azoxy]-, methyl ester	MeOH	233(3.95),265(3.93), 335(4.28)	39-0171-78C
$C_{15}H_{14}N_2O_5$			
Propanoic acid, 3-[(3-amino-1,4-dihydro-1,4-dioxo-2-naphthalenyl)amino]-3-oxo-, ethyl ester	EtOH	<u>265(4.5),337s(3.4), 450(3.5)</u>	104-0159-78
2,4,6(1H,3H.5H)-Pyrimidinetrione, 5-[3-hydroxy-3-(2-hydroxyphenyl)-2-propenylidene]-1,3-dimethyl-	MeOH	420(4.3)	83-0503-78
$C_{15}H_{14}N_2O_5S$			
Benzoic acid, 2-[(5-acetyl-3-nitro-2-thienyl)amino]-, ethyl ester	MeOH	217(4.34),263(3.94), 350(4.22)	150-5101-78
Carbamic acid, [[2-[(phenylamino)carbonyl]phenyl]sulfonyl]-, methyl ester	EtOH	228(4.10),272s(3.81), 277s(3.77)	40-0582-78
$C_{15}H_{14}N_4$			
1H-1,2,4-Triazol-3-amine, 5-phenyl-N-(phenylmethyl)-	EtOH	265(3.83)	25-0092-78
	EtOH-acid	259(4.15)	25-0092-78
	EtOH-base	286(4.06)	25-0092-78
4H-1,2,4-Triazol-3-amine, 5-phenyl-4-(phenylmethyl)-	EtOH	259(3.93)	25-0092-78
	EtOH-acid	241(3.95)	25-0092-78
$C_{15}H_{14}N_4O$			
3-Cyclohexene-1-carboxamide, 4-amino-1,3-dicyano-N-phenyl-	MeOH	260(4.16)	42-0281-78
$C_{15}H_{14}N_4O_2$			
2H-Pyrazolo[3,4-d]pyrimidine-4,6(5H,7H)-dione, 5,7-dimethyl-2-(1-phenylethenyl)-	n.s.g.	258(3.78),287(3.81)	4-0359-78
$C_{15}H_{14}N_4O_2S$			
1H-Tetrazolium, 2,5-dihydro-2,3-bis(2-methoxyphenyl)-5-thioxo-, hydroxide, inner salt	MeOH	399(3.00)	4-0133-78
	EtOH	405(3.03)	4-0133-78
	PrOH	410(3.00)	4-0133-78
	$HCONH_2$	405(3.08)	4-0133-78
	MeCN	434(3.05)	4-0133-78
	$MeNO_2$	430(3.03)	4-0133-78
	pyridine	453(3.05)	4-0133-78
	acetone	448(3.01)	4-0133-78
(also solvent mixtures)	$CHCl_3$	437(3.01)	4-0133-78
$C_{15}H_{14}N_4O_4$			
Tetracyclo[4.3.0.0^{2,9}.0^{4,8}]nonan-3-one, 2,4-dinitrophenylhydrazone	EtOH	234(4.26),271s(4.01), 280s(3.89),372(4.37)	44-3904-78

Compound	Solvent	$\lambda_{max}(\log \epsilon)$	Ref.
$C_{15}H_{14}N_4O_5$			
Propanehydrazonic acid, N-(4-nitrophen-yl)-, 4-nitrophenyl ester	xylene	361(4.30)	18-0512-78
	EtOH	250(4.00),390(4.25)	18-0512-78
Propanoic acid, 2,2-bis(4-nitrophenyl)-hydrazide	EtOH	255s(3.89),325s(3.93), 382(4.39)	18-0512-78
$C_{15}H_{14}N_4S$			
1H-Tetrazolium, 2,5-dihydro-2,3-bis(2-methylphenyl)-5-thioxo-, hydroxide, inner salt	MeOH	402(3.01)	4-0133-78
	EtOH	406(3.03)	4-0133-78
	PrOH	413(3.00)	4-0133-78
	acetone	445(3.02)	4-0133-78
	EtOAc	471(3.00)	4-0133-78
	$HCONH_2$	406(3.03)	4-0133-78
	MeCN	438(2.96)	4-0133-78
	$MeNO_2$	442(3.03)	4-0133-78
	pyridine	460(3.02)	4-0133-78
(also solvent mixtures)	$CHCl_3$	456(3.01)	4-0133-78
1H-Tetrazolium, 2,5-dihydro-2,3-bis(4-methylphenyl)-5-thioxo-, hydroxide, inner salt	MeOH	400(3.14)	4-0133-78
	EtOH	405(3.12)	4-0133-78
	PrOH	411(3.06)	4-0133-78
	acetone	456(3.06)	4-0133-78
	$HCONH_2$	405(3.17)	4-0133-78
	MeCN	440(3.10)	4-0133-78
	$MeNO_2$	440(3.01)	4-0133-78
	pyridine	462(3.13)	4-0133-78
(also solvent mixtures)	$CHCl_3$	453(3.12)	4-0133-78
$C_{15}H_{14}N_8$			
Formazan, 1-(2-methyl-2H-tetrazol-5-yl)-3,5-diphenyl-	pH 13	257(4.05),462(4.76)	103-0800-78
	EtOH	280(4.28),427(4.07)	103-0800-78
	dioxan	275(4.29),430(4.11)	103-0800-78
$C_{15}H_{14}O$			
Obtusastyrene	EtOH	256(4.13),285(3.64), 294(3.48)	102-1395-78
$C_{15}H_{14}O_2$			
Ethanone, 1-(3-methoxy[1,1'-biphenyl]-4-yl)-	EtOH	214(4.43),278(4.28)	95-0503-78
$C_{15}H_{14}O_2S$			
Spiro[2H-naphtho[2,3-b][1,4]dioxepin-3(4H),3'-thietane]	$CHCl_3$	246(4.16),283(3.72), 314(3.14),323(3.15), 328(3.17)	56-1913-78
$C_{15}H_{14}O_3$			
2,5-Cyclohexadiene-1,4-dione, 2,3-di-methyl-5-(phenylmethoxy)-	EtOH	270(4.33),370(2.65)	23-0517-78
Methanone, (2,4-dihydroxyphenyl)(4-eth-ylphenyl)-	MeOH	246(3.97),289(4.16), 326(4.06)	116-0312-78
Spiro[2H-naphtho[2,3-b][1,4]dioxepin-3(4H),3'-oxetane]	$CHCl_3$	246(4.15),284(3.72), 314(3.14),322(3.11), 328(3.16)	56-1913-78
$C_{15}H_{14}O_3S$			
Benzene, 1-methyl-2-[[2-(phenylsulfon-yl)ethenyl]oxy]-	dioxan	242(4.16)	99-0194-78
Benzene, 1-methyl-3-[[2-(phenylsulfon-yl)ethenyl]oxy]-	dioxan	245.4(4.29)	99-0194-78
Benzene, 1-methyl-4-[[2-(phenylsulfon-yl)ethenyl]oxy]-	dioxan	246.8(4.24)	99-0194-78

Compound	Solvent	$\lambda_{max}(\log \epsilon)$	Ref.
Ethanone, 1-phenyl-2-[(phenylmethyl)-sulfonyl]-	MeOH	253(4.18)	12-1965-78
$C_{15}H_{14}O_3Se$			
2-Oxaspiro[4.5]dec-6-ene-1,8-dione, 9-(phenylseleno)-	MeOH	217(4.2),236(4.0)	24-1944-78
$C_{15}H_{14}O_4$			
Apiumetin methyl ether	MeOH	260(3.48),330(3.96)	102-2135-78
Benzo[1,2-b:4,3-b']difuran-1-carboxylic acid, 2,7-dimethyl-, ethyl ester	MeOH	220(4.71),269s(3.34), 278(4.12),288s(3.08)	4-0043-78
2H,8H-Benzo[1,2-b:3,4-b']dipyran-2-one, 9-methoxy-8,8-dimethyl-	EtOH	233(4.26),293(4.30), 330(4.10)	2-0179-78
β-Lapachone, 4'-hydroxy-	MeOH	253(4.40),335(3.36), 385(3.26),415(3.33)	2-0035-78
Naphtho[2,3-b]furan-4,9-dione, 2,3-di-hydro-2-(1-hydroxy-1-methylethyl)-, (±)-	EtOH	208(4.39),246(4.25), 252(4.26),288(3.97), 338(3.36),398(3.06)	150-0189-78
2H-Naphtho[2,3-b]pyran-5,10-dione, 3,4-dihydro-4-hydroxy-2,2-dimethyl- (4-hydroxy-α-lapachone)	EtOH	245(4.60),250(4.62), 281(4.47),331(3.70)	2-0035-78
$C_{15}H_{14}O_4S$			
Benzene, 1-methoxy-4-[[2-(phenylsulfon-yl)ethenyl]oxy]-	dioxan	206(4.45),225(4.27), 247(4.13)	99-0194-78
$C_{15}H_{14}O_5$			
2H,8H-Benzo[1,2-b:3,4-b']dipyran-6-car-boxaldehyde, 3,4,9,10-tetrahydro-2,8-dimethyl-4,10-dioxo-	EtOH	250(4.57),288s(4.14), 346(3.87)	102-0511-78
$C_{15}H_{14}O_6$			
9,10-Anthracenedione, 1,2,3,4-tetrahy-dro-2,3,5,8-tetrahydroxy-2-methyl-	MeOH	205s(--),217(4.40), 281(3.86),472(3.75), 503(3.75)	150-3831-78
$C_{15}H_{14}O_7$			
9,10-Anthracenedione, 1,2,3,4-tetrahy-dro-1,2,3,5,8-pentahydroxy-3-methyl-, (1α,2β,3α)-	MeOH	216(4.46),280(3.77), 479(3.65),509(3.70), 545(3.48)	150-3831-78
$C_{15}H_{14}O_7S$			
Propanedioic acid, [(6-hydroxy-3-oxo-2(3H)-benzofuranylidene)(methylthio)-methyl]-, dimethyl ester	EtOH	257(3.91),356(4.29)	142-0399-78A
$C_{15}H_{15}$			
Cycloheptatrienylium, (2,5-dimethyl-phenyl)-	MeCN	364(3.82)	138-0237-78
$C_{15}H_{15}BF_4O_2$			
Methylium, bis(4-methoxyphenyl)-, tetrafluoroborate	$C_2H_4Cl_2$	507(4.98)	104-1197-78
$C_{15}H_{15}BrN_2O_2$			
Pyridinium, 4-[(4-nitrophenyl)methyl]-1-(2-propenyl)-, bromide	H_2O	215s(4.14),260s(4.12), 265(4.14),274s(4.08)	48-0133-78
deprotonated	H_2O	548(4.52)	48-0133-78
$C_{15}H_{15}BrN_2O_4$			
1H-Pyrrolo[1,2-a]indole-9-carbonitrile,	EtOH	244(4.38),285(3.82),	94-3815-78

Compound	Solvent	$\lambda_{max}(\log \epsilon)$	Ref.
6-bromo-2,3,5,6,7,8-hexahydro-7,7-di-methoxy-6-methyl-5,8-dioxo- (cont.)		325(4.14)	94-3815-78
$C_{15}H_{15}BrO_6$			
1,2,3-Benzenetriol, 4-(3-bromo-1-prop-enyl)-, triacetate, (E)-	MeOH	212(4.34),262(4.33)	24-3939-78
$C_{15}H_{15}ClN_4$			
Cycloocta[5,6][1,2,4]triazino[4,3-b]in-dazole, 11-chloro-1,2,3,4,5,6-hexahy-dro-	C_6H_{12}	354(3.81)	24-2258-78
$C_{15}H_{15}ClN_4O_6$			
Riboflavin, 7-chloro-7,8-didemethyl-	n.s.g.	334(3.86),447(3.99)	69-1942-78
Riboflavin, 8-chloro-7,8-didemethyl-	n.s.g.	356(3.93),432(4.08)	69-1942-78
$C_{15}H_{15}ClO_4$			
3-Furancarboxylic acid, 5-[(4-chloro-phenyl)methyl]-4,5-dihydro-2-methyl-4-oxo-, ethyl ester	EtOH	218(4.17),262(3.96)	4-0327-78
Propanedioic acid, [3-(4-chlorophenyl)-2-methyl-2-propenylidene]-, dimethyl ester, (E)-	EtOH	230(4.00),315(4.42)	22-0457-78
$C_{15}H_{15}Cl_2N$			
Benzenamine, N,N-bis(4-chloro-2-butyn-yl)-3-methyl-	EtOH	208(4.43),247(4.01), 286(3.24)	42-1228-78
Benzenamine, N,N-bis(4-chloro-2-butyn-yl)-4-methyl-	EtOH	209(4.25),248(4.06), 295(3.13)	42-1228-78
$C_{15}H_{15}Cl_2NO$			
Benzenamine, N,N-bis(4-chloro-2-butyn-yl)-4-methoxy-	EtOH	205(4.42),244(3.98), 290(2.93)	42-1228-78
$C_{15}H_{15}Cl_2NO_2S$			
2H,5H-Thiopyrano[4,3-b]pyran-2-one, 3,3-dichloro-3,4,7,8-tetrahydro-4-(methylphenylamino)-	EtOH	245(4.15),292(3.32)	4-0181-78
$C_{15}H_{15}Cl_2N_3O_2$			
Pyridinium, 1-[(2,6-dichlorophenyl)-methyl]-3-[[[(methoxymethyl)amino]-carbonyl]amino]-, hydroxide, inner salt	MeOH CH_2Cl_2	315(3.41) 308(3.77)	64-0084-78B 64-0084-78B
$C_{15}H_{15}Cl_3N_2O$			
Acetamide, 2,2,2-trichloro-N-[(2,3,4,9-tetrahydro-1H-carbazol-6-yl)methyl]-	MeOH	232(4.66),286(3.91)	103-0856-78
$C_{15}H_{15}Cs$			
Benzene, 1,1'-methylenebis[3-methyl-, cesium deriv.	n.s.g.	447(4.68)	35-8271-78
Benzene, 1,1'-methylenebis[4-methyl-, cesium deriv.	n.s.g.	446(4.64)	35-8271-78
$C_{15}H_{15}FN_2O_4S$			
3-Thiophenecarboxylic acid, 5-ethyl-2-[(4-fluoro-2-nitrophenyl)amino]-, ethyl ester	MeOH	332(4.02),450(3.70)	39-0937-78C

Compound	Solvent	$\lambda_{max}(\log \epsilon)$	Ref.
$C_{15}H_{15}FO_4$			
Unknown compd., m. 72-4°	MeOH	265(3.97),331(4.15)	87-1093-78
	MeOH-base	265(4.05),420(4.44)	87-1093-78
$C_{15}H_{15}F_3N_2OS$			
Acetamide, 2,2,2-trifluoro-N-[4-(2-methylpropyl)-2-phenyl-5-thiazolyl]-	ether	308(4.08)	78-0611-78
$C_{15}H_{15}N$			
3-Azabicyclo[4.3.0]nona-1,3,5-triene, 4-methyl-2-phenyl-	EtOH	212(3.93),239(3.78), 282(3.72)	39-1293-78C
$C_{15}H_{15}NO$			
2-Aza-9-fluorenol, 9-ethyl-3-methyl-	EtOH	212(5.02),280(4.78), 308(4.32)	103-0997-78
4H-Indol-4-one, 1,5,6,7-tetrahydro-1-methyl-3-phenyl-	MeOH	222(4.59),270(4.27)	24-1780-78
$C_{15}H_{15}NOS$			
2H-1,6-Thiazecine, 3,4,8,9-tetradehydro-5,6,7,10-tetrahydro-6-(4-methoxyphenyl)-	EtOH	206(4.33),249(4.28), 395(3.36)	42-1228-78
$C_{15}H_{15}NO_2$			
8-Azabicyclo[3.2.1]octa-3,6-dien-2-one, 4-methoxy-N-methyl-6-phenyl-	n.s.g.	206(4.14),251(4.32)	88-1751-78
7-Azabicyclo[4.2.0]octa-1,3,5-triene, 3,4-dimethoxy-7-phenyl-	EtOH	278(4.2),318(3.9)	5-0608-78
[1,1'-Biphenyl]-3-ol, 2,4',5-trimethyl-6-nitroso-	DMF	720(1.04)	104-2189-78
Diphenylamine-2-carboxylic acid, 4'-ethyl-	EtOH	287(4.19),352(3.88)	35-5604-78
1,2-Naphthalenedione, 4-piperidino-	EtOH	244(4.55),280(4.29), 472(3.75)	65-2318-78
$C_{15}H_{15}NO_3$			
[1,1'-Biphenyl]-3-ol, 4'-methoxy-2,5-dimethyl-6-nitroso-	DMF	710(1.08)	104-2189-78
1H,8H-3a,8a-Propanoindeno[1,2-c]pyrrole-1,3(2H)-dione, 8-hydroxy-2-methyl-	EtOH	213(3.89),256(2.88), 266s(2.85),274(2.70)	4-0167-78
2(1H)-Pyridinone, 1-(2-acetoxy-2-phenylethyl)-	EtOH	303(3.39)	78-2609-78
2H-Pyrrol-2-one, 1,5-dihydro-4-hydroxy-3-(1-oxo-2-butenyl)-5-(phenylmethyl)-, [S-(E)]-	EtOH-H$_2$SO$_4$, EtOH-KOH	226(4.03),318(4.27) 241(4.22),313(4.17)	88-3173-78 88-3173-78
$C_{15}H_{15}NO_4$			
3H-3-Benzazepine-3,7-dicarboxylic acid, 3-ethyl 7-methyl ester	C_6H_{12}	264(4.68),339(3.21), 354s(3.15),435s(2.70), 467s(2.36)	142-0401-78C
1-Oxadethiaceph-3-em-4-carboxylic acid, 3-methyl-, phenylmethyl ester	EtOH	262(3.72)	39-1450-78C
$C_{15}H_{15}NO_4S$			
1H-Azepine-2-carboxylic acid, 1-[(4-methylphenyl)sulfonyl]-, methyl ester	EtOH	330(3.05)	2-0547-78
Benzoic acid, 2-[[(4-methylphenyl)sulfonyl]amino]-, methyl ester	EtOH	303(3.58)	2-0547-78
Phenanthridinium, 5-methyl-, methyl sulfate	EtOH	246(4.64),320(3.89), 360(3.52)	87-0199-78

Compound	Solvent	$\lambda_{max}(\log \epsilon)$	Ref.
$C_{15}H_{15}NO_5$			
Acetamide, N-[(4-acetyl-2,5-dihydro-3-hydroxy-5-oxo-2-furanyl)methyl]-N-phenyl-	EtOH	230(4.20),265(4.20)	4-0327-78
2-Piperidinone, 4-(5,7-dihydroxy-2-methyl-4-oxo-4H-1-benzopyran-6-yl)-	MeOH	205(4.35),225(4.18), 251(4.25),257(4.27), 295(3.78),318(3.68)	88-2911-78
$C_{15}H_{15}NO_5S$			
2-Propenoic acid, 3-[5-[[4-(dimethylamino)phenyl]sulfonyl]-2-furanyl]-	EtOH	206(4.26),282(4.46), 330(4.26)	73-0621-78
$C_{15}H_{15}NO_6$			
3-Quinolineacetic acid, 6,7-dimethoxy-2-methyl-α-oxo-, methyl ester, 1-oxide	EtOH	256(4.59),334(3.91), 349(3.96)	95-0802-78
$C_{15}H_{15}NO_8$			
2(5H)-Furanone, 4-acetyl-5-(4,5-dimethoxy-2-nitrophenyl)-3-methoxy-	EtOH	245(4.11),280(4.03), 330(3.70)	95-0802-78
$C_{15}H_{15}NS$			
2H-1,6-Thiazecine, 3,4,8,9-tetradehydro-5,6,7,10-tetrahydro-6-(3-methylphenyl)-	EtOH	210(4.36),252(4.14), 287(3.24)	42-1228-78
2H-1,6-Thiazecine, 3,4,8,9-tetradehydro-5,6,7,10-tetrahydro-6-(4-methylphenyl)-	EtOH	206(4.36),252(4.14), 294(3.16)	42-1228-78
$C_{15}H_{15}N_2$			
1H-Imidazo[1,2-a]pyridin-4-ium, 2-methyl-1-(phenylmethyl)-, bromide	EtOH	204(4.45),277s(4.35), 286(4.00)	4-1149-78
$C_{15}H_{15}N_2O_2$			
Pyridinium, 4-[(4-nitrophenyl)methyl]-1-(2-propenyl)-, bromide	H_2O	215s(4.14),260s(4.12), 265(4.14),274s(4.08)	48-0133-78
$C_{15}H_{15}N_3$			
1H-Benzotriazole, 5,6-dimethyl-1-(phenylmethyl)-	4.2M HCl	287(4.00)	87-0578-78
4-Isoquinolinecarbonitrile, 3-piperidino-	EtOH	257(4.56),298(4.17), 402(3.61)	138-0677-78
$C_{15}H_{15}N_3O$			
Anantine, (±)-	EtOH	277(4.48)	88-1801-78
Benzaldehyde, 3-[[4-(dimethylamino)phenyl]azo]-	EtOH	243(4.3),415(4.4)	19-0655-78
Benzamide, 4-[[(methylphenylamino)methylene]amino]-	isoPrOH	316(4.36)	117-0097-78
1H-Benzotriazole, 5,6-dimethyl-1-(phenylmethyl)-, 3-oxide	EtOH	277(3.50),282(3.56), 324(3.89)	87-0578-78
	4.2M HCl	288(3.97)	87-0578-78
Isoanantine, (±)-	EtOH	280(4.48)	88-1801-78
$C_{15}H_{15}N_3OS$			
Hydrazinecarbothioamide, 2-benzoyl-N-(4-methylphenyl)-	MeOH	238(4.34)	80-0397-78
nickel complex	MeOH	270(4.56)	80-0397-78
$C_{15}H_{15}N_3O_2$			
Acetamide, N-phenyl-N-(2,4,5,6-tetra-	n.s.g.	272(4.11),344(3.51)	78-1581-78

Compound	Solvent	$\lambda_{max}(\log \epsilon)$	Ref.
hydro-2-oxo-1H-pyrrolo[3,2-b]pyridin-1-yl)- (cont.)			78-1581-78
2-Butenenitrile, 2-[2-[(4-acetylphenyl)amino]ethenyl]-4-(methoxyimino)-, (?,Z,E)-	MeOH	416(4.27)	83-0433-78
2H-Pyrazolo[4,3-b]pyridine-3-carboxaldehyde, 4-acetyl-4,5,6,7-tetrahydro-2-phenyl-	n.s.g.	223(3.31),275(2.79), 323(2.82)	78-1581-78
$C_{15}H_{15}N_3O_2S$			
Hydrazinecarbothioamide, 2-benzoyl-N-(4-methoxyphenyl)-	MeOH	236(4.33)	80-0397-78
nickel complex	MeOH	267(4.54)	80-0397-78
2H-Pyrimido[4,5-b][1,4]thiazine-2,4(3H)-dione, 1,7-dihydro-1,3,7-trimethyl-6-phenyl- (relative E given in place of log ε)	MeOH	257(1.7),280s(--), 367(1.0)	83-0153-78
	12M HCl	225(0.8),280(1.2), 405(0.8)	83-0153-78
$C_{15}H_{15}N_3O_3$			
Acetic acid, 2-(4-methylphenyl)-2-(4-nitrophenyl)hydrazide	xylene	360(4.23)	18-0512-78
	EtOH	250(3.89),366(4.21)	18-0512-78
Ethanehydrazonic acid, N-(4-nitrophenyl)-, 4-methylphenyl ester	xylene	373(4.40)	18-0512-78
	EtOH	250(4.17),392(4.24)	18-0512-78
Propanehydrazonic acid, N-(4-nitrophenyl)-, phenyl ester	xylene	370(4.39)	18-0512-78
	EtOH	250(4.18),390(4.45)	18-0512-78
Propanoic acid, 2-(4-nitrophenyl)-2-phenylhydrazide	xylene	360(4.15)	18-0512-78
	EtOH	250s(3.87),361(4.21)	18-0512-78
Pyrimido[4,5-b]quinoline-2,4(3H,10H)-dione, 10-(2-hydroxyethyl)-7,8-dimethyl-	n.s.g.	338(4.09),397(4.10)	69-1942-78
$C_{15}H_{15}N_3O_6$			
7H-Pyridazino[4,5-d]azepine-1,4,7-tricarboxylic acid, 7-ethyl 1,4-dimethyl ester	CH_2Cl_2	262(4.43),327(3.64)	83-0786-78
$C_{15}H_{15}N_3O_6S$			
3-Thiophenecarboxylic acid, 2-[(2,4-dinitrophenyl)amino]-5-ethyl-, ethyl ester	MeOH	390(4.16)	39-0937-78C
$C_{15}H_{15}N_5$			
Pyrrolo[3,2-c]tetrazolo[1,5-a]azepine, 5,6,7,8-tetrahydro-8-methyl-10-phenyl-	MeOH	221(4.42),265(4.18)	24-1780-78
$C_{15}H_{15}N_5O_2$			
Acetamide, N-[(4,5-dihydro-3-methyl-4-oxo-1H-pyrazolo[3,4-d]pyridazin-7-yl)methyl]-N-phenyl-	EtOH	216(4.16),270(3.74)	4-0155-78
$C_{15}H_{15}N_5O_7S_2$			
5-Thia-1-azabicyclo[4.2.0]oct-2-ene-2-carboxylic acid, 3-[(acetyloxy)methyl]-7-[[(2-amino-4-thiazolyl)(hydroxyimino)acetyl]amino]-8-oxo-, (E)-	EtOH	230(4.26),250s(4.14), 306(3.43)	78-2233-78
$C_{15}H_{15}OP$			
9-Phosphapentacyclo[4.3.0.0^{2,5}.0^{3,8}-0^{4,7}]nonane, 1-methyl-9-phenyl-, 9-oxide	EtOH	258(2.79),264(2.86), 272(2.77)	44-4338-78

Compound	Solvent	$\lambda_{max}(\log \epsilon)$	Ref.
$C_{15}H_{15}O_2$ Methylium, bis(4-methoxyphenyl)-, tetrafluoroborate	$C_2H_4Cl_2$	507(4.98)	104-1197-78
$C_{15}H_{16}$ 1,1'-Biphenyl, 3,4',5-trimethyl-	dioxan	258(4.17)	24-0205-78
$C_{15}H_{16}BrClO_2$ 2,6-Methanofuro[3,2-b]furan, 5-(1-bromopropylidene)-7-(1-chloro-2-penten-4-ynyl)hexahydro- (cis-maneonene A)	hexane	225(4.17)	44-3194-78
cis-maneonene B	EtOH	227(4.09)	44-3194-78
trans-maneonene B	EtOH	232(4.18)	44-3194-78
cis-maneonene C	EtOH	222(4.34)	44-3194-78
$C_{15}H_{16}BrNO_5$ 1H-Pyrrolo[1,2-a]indole-9-carboxaldehyde, 6-bromo-2,3,5,6,7,8-hexahydro-7,7-dimethoxy-6-methyl-5,8-dioxo-	EtOH	254(4.23),290(3.76), 336(4.02)	94-3815-78
$C_{15}H_{16}Br_2N_2O_2$ Pyridinium, 1-(3-bromopropyl)-4-[(4-nitrophenyl)methyl]-, bromide	H_2O	215s(4.11),260s(4.10), 265(4.12),274s(4.07)	48-0133-78
deprotonated	H_2O	548(4.49)	48-0133-78
$C_{15}H_{16}Br_2O_2$ Isomaneonene A	EtOH	229(4.16)	44-3194-78
Isomaneonene B	EtOH	228(4.16)	44-3194-78
$C_{15}H_{16}ClN_3O_3$ 1,2,4-Triazine-6-acetic acid, 3-(4-chlorophenyl)-5-methoxy-α-methyl-, ethyl ester	EtOH	263(4.21),278(4.21)	114-0091-78B
$C_{15}H_{16}ClN_3O_5S$ 4-Thia-1-azabicyclo[3.2.0]heptane-2-carboxylic acid, 6-amino-3-(chloromethyl)-3-methyl-7-oxo-, (4-nitrophenyl)methyl ester, hydrochloride	EtOH	264(4.01)	88-4703-78
$C_{15}H_{16}NO_5PS_3$ Phosphorothioic acid, O-ethyl O-phenyl S-[(2,3,5,7-tetrahydro-5,7-dioxo-6H-1,4-dithiino[2,3-c]pyrrol-6-yl)methyl] ester	EtOH	209(4.20),264(4.06), 422(3.57)	73-1093-78
$C_{15}H_{16}N_2$ Ethanone, 1-phenyl-, methylphenylhydrazone	n.s.g.	350(3.55)	104-0717-78
9H-Pyrido[3,4-b]indole, 1-butyl-	MeOH	236(4.57),241(4.55), 251s(4.37),283s(4.00), 289(4.24),338(3.68), 351(3.70)	95-0898-78
$C_{15}H_{16}N_2O$ Benzoic acid, 4-methyl-, 2-(4-methylphenyl)hydrazide	EtOH	238(4.37),273s(3.57)	33-1477-78
2,5-Cyclohexadien-1-one, 4-[[4-(dimethylamino)phenyl]imino]-2-methyl-	C_6H_{12}	275(4.22),295(4.05), 335(3.76),540(4.03)	48-0557-78
	EtOH	275(4.32),307(4.00), 335(3.89),600(4.30)	48-0557-78 +80-0617-78

Compound	Solvent	$\lambda_{max}(\log \epsilon)$	Ref.
2,5-Cyclohexadien-1-one, 4-[[4-(dimeth-ylamino)phenyl]imino]-2-methyl-(cont.)	ether	275(3.94),335(3.53), 560(3.85)	48-0557-78
	CCl_4	280(4.04),335(3.72), 550(4.02)	48-0557-78
2,5-Cyclohexadien-1-one, 4-[[4-(dimeth-ylamino)phenyl]imino]-3-methyl-	C_6H_{12}	270(4.27),290(4.11), 338(3.79),540(4.03)	48-0557-78
	EtOH	275(4.22),295(4.10), 350(3.81),608(4.27)	48-0557-78 +80-0617-78
	ether	270(4.29),302(4.01), 340(3.84),560(4.13)	48-0557-78
	CCl_4	335(3.88),550(4.07)	48-0557-78
Ethanone, 2-[(4-methylphenyl)amino]-1-phenyl-, oxime	MeOH	243(4.16)	56-1827-78
4H-Indol-4-one, 1,5,6,7-tetrahydro-1-methyl-3-phenyl-, oxime, (E)-	MeOH	207(4.31),233(4.33), 274(4.01)	24-1780-78
10H-Pyrido[3,2-b][1,4]benzoxazine, 10-butyl-	C_6H_{12}	233(4.45),237(4.45), 342(3.91)	95-0585-78
10H-Pyrido[3,2-b][1,4]benzoxazine, 10-(1-methylpropyl)-	C_6H_{12}	233(4.24),335(3.78)	95-0585-78
10H-Pyrido[3,2-b][1,4]benzoxazine, 10-(2-methylpropyl)-	C_6H_{12}	233(4.85),237(4.86), 343(4.32)	95-0585-78
Pyrrolo[3,2-b]azepin-5(1H)-one, 4,6,7,8-tetrahydro-1-methyl-3-phenyl-	MeOH	207(4.30),241(4.18), 270s(4.02)	24-1780-78
Pyrrolo[3,2-c]azepin-4(1H)-one, 5,6,7,8-tetrahydro-1-methyl-3-phenyl-	MeOH	208(4.49),231s(4.36), 270(4.27)	24-1780-78
$C_{15}H_{16}N_2OS$			
Ethanethioic acid, phenoxy-, 2-methyl-2-phenylhydrazide	MeOH	230(4.08),242(4.28), 256(4.12),264(4.15)	123-0701-78
$C_{15}H_{16}N_2O_2$			
Acetic acid, phenoxy-, 2-methyl-2-phen-ylhydrazide	MeOH	208(4.05),216(3.97), 220(3.98),228(3.91), 240(4.06)	123-0701-78
8,11-Diazatricyclo[5.2.2.0^{1,5}]undecane-9,10-dione, 7-phenyl-	EtOH	224(3.61),269(2.43)	39-1293-78C
Ethanone, 1-(4-acetyl-1,3-dimethyl-1H-pyrazol-5-yl)-2-phenyl-	n.s.g.	217(3.87),259(3.80)	44-2665-78
Ethanone, 1-(4-acetyl-1,5-dimethyl-1H-pyrazol-3-yl)-2-phenyl-	n.s.g.	218(3.96),258(3.81)	44-2665-78
Ethanone, 2-[(4-methoxyphenyl)amino]-1-phenyl-, oxime, (Z)-	MeOH	243(4.19)	56-1827-78
1,6-Naphthyridin-2(1H)-one, 5,6,7,8-tetrahydro-4-hydroxy-6-(phenylmethyl)-	EtOH	286(3.85)	142-0267-78
1,8-Naphthyridin-2(1H)-one, 5,6,7,8-tetrahydro-4-hydroxy-3-(2-methylphen-yl)-	MeOH	224(4.08),259(4.02), 308s(4.10),312(4.18)	24-2813-78
1,8-Naphthyridin-2(1H)-one, 5,6,7,8-tetrahydro-4-hydroxy-3-(3-methylphen-yl)-	MeOH	223(4.12),252(4.03), 305s(4.04),315(4.23)	24-2813-78
1,8-Naphthyridin-2(1H)-one, 5,6,7,8-tetrahydro-4-hydroxy-3-(4-methylphen-yl)-	MeOH	220(4.17),265(3.70), 304s(4.14),319(4.27)	24-2813-78
8H-Pyrido[1,2-a]pyrazin-8-one, 1,2,3,4-tetrahydro-9-hydroxy-1-(phenylmethyl)-	EtOH	220(3.94),288(3.86)	12-0187-78
4(1H)-Pyrimidinone, 6-hydroxy-5-(4-pen-tenyl)-2-phenyl-	EtOH	232(4.40),306(3.99)	39-1293-78C
$C_{15}H_{16}N_2O_3$			
5,8-Isoquinolinedione, 7-hydroxy-6-methyl-3-piperidino-	EtOH	252(4.39),329(4.20), 482(3.77)	94-2175-78

Compound	Solvent	$\lambda_{max}(\log \epsilon)$	Ref.
1,8-Naphthyridin-2(1H)-one, 5,6,7,8-tetrahydro-4-hydroxy-3-(4-methoxyphenyl)-	MeOH	219(3.98),258(3.65), 300(4.18),323(4.31)	24-2813-78
1H-Pyrazole-4-carboxylic acid, 3-acetyl-1-methyl-5-phenyl-, ethyl ester	n.s.g.	239(4.16)	44-2665-78
1H-Pyrazole-4-carboxylic acid, 5-acetyl-1-methyl-3-phenyl-, ethyl ester	n.s.g.	239(4.24)	44-2665-78
$C_{15}H_{16}N_2O_4$			
1,3-Cyclohexanedione, 5,5-dimethyl-2-[[(2-nitrophenyl)amino]methylene]-	EtOH	379(4.16)	97-0256-78
1,3-Cyclohexanedione, 5,5-dimethyl-2-[[(3-nitrophenyl)amino]methylene]-	EtOH	347(4.38)	97-0256-78
1,3-Cyclohexanedione, 5,5-dimethyl-2-[[(4-nitrophenyl)amino]methylene]-	EtOH	372(4.44)	97-0256-78
5,8-Isoquinolinedione, 7-methoxy-6-methyl-3-morpholino-	EtOH	252(4.40),322(4.11), 470(3.71)	94-2175-78
3-Pyridinecarboxamide, N-ethyl-1,2-dihydro-4-hydroxy-6-methoxy-2-oxo-1-phenyl-	MeOH	305(4.2)	120-0101-78
Pyridinium, 1-(3-cyano-4-ethoxy-1-(ethoxycarbonyl)-4-oxo-2-butenyl]-, hydroxide, inner salt	EtOH	226(4.07),262(3.77), 338(4.63)	95-1412-78
1H-Pyrido[3,2-b]indole-2,9-dione, 1,5-diacetyl-3,4,5,6,7,8-hexahydro-	EtOH	249(3.95),300s(3.00)	94-3080-78
$C_{15}H_{16}N_2O_4S$			
3-Thiophenecarboxylic acid, 5-ethyl-2-[(2-nitrophenyl)amino]-, ethyl ester	MeOH	331(4.05),434(3.71)	39-0937-78C
$C_{15}H_{16}N_2O_5$			
2,4,6(1H,3H,5H)-Pyrimidinetrione, 5-[3-(2-hydroxyphenyl)-3-oxopropyl]-1,3-dimethyl-	MeOH	326(3.5)	83-0503-78
$C_{15}H_{16}N_2O_5S$			
3-Thiophenecarboxylic acid, 5-ethyl-2-[(4-methoxy-2-nitrophenyl)amino]-, methyl ester	MeOH	336(4.34)	39-0937-78C
$C_{15}H_{16}N_2S$			
Benzeneethanethioic acid, 2-methyl-2-phenylhydrazide	MeOH	222(3.91),240(4.07), 260(3.91),265(3.92)	123-0701-78
Thiourea, N,N'-bis(phenylmethyl)-, iodine complex	CHCl₃	315(3.63)	62-1083-78A
$C_{15}H_{16}N_2S_2$			
Carbamodithioic acid, ethylphenyl-, 2-pyridinylmethyl ester	n.s.g.	336s(1.95)	97-0381-78
Carbamodithioic acid, methylphenyl-, 2-(2-pyridinyl)ethyl ester	n.s.g.	340s(2.04)	97-0381-78
$C_{15}H_{16}N_4$			
Cycloocta[5,6][1,2,4]triazino[4,3-b]indazole, 1,2,3,4,5,6-hexahydro-	C₆H₁₂	360(3.59)	24-2258-78
$C_{15}H_{16}N_4O$			
4H-Pyrazolo[3,4-d]pyridazin-4-one, 1,5-dihydro-1,3,5-trimethyl-7-(phenylmethyl)-	EtOH	216(4.34),286(3.85)	4-0813-78

Compound	Solvent	λ_{max} (log ϵ)	Ref.
4H-Pyrazolo[3,4-d]pyridazin-4-one, 2,5-dihydro-2,3,5-trimethyl-7-(phenyl-methyl)-	EtOH	217(4.19),284(3.80)	4-0813-78
$C_{15}H_{16}N_4O_3S$			
Pyrazolo[3,4-d][1,2]diazepine, 2-acet-yl-2,3,3a,6-tetrahydro-6-[(4-methyl-phenyl)sulfonyl]-	EtOH	328(4.22)	39-1297-78C
Pyrazolo[3,4-d][1,2]diazepine, 2-acet-yl-2,3,3a,6-tetrahydro-8-methyl-6-(phenylsulfonyl)-	EtOH	323(4.21)	39-1297-78C
$C_{15}H_{16}N_4O_4S$			
Acetamide, N-[4-[[[4-cyano-5-(methoxy-imino)-1,3-pentadienyl]amino]sulfon-yl]phenyl]-, (?,Z,E)-	MeOH	363(4.43)	83-0433-78
$C_{15}H_{16}N_8O_4$			
1H-Purine-2,6-dione, 8,8'-methylenebis-[3,7-dihydro-1,3-dimethyl-	pH 7.0	207(4.71),277(4.39)	24-0982-78
	pH 13.0	217(4.53),283(4.46)	24-0982-78
	MeOH	208(4.61),275(4.38)	24-0982-78
$C_{15}H_{16}N_{10}O_2$			
2H-Pyrimido[4",5":2',3'][1,4]diazepino-6',5':3,4]pyrrolo[1,2-f]pteridine-1,12-dione, 3,10-diamino-5,6,6a,6b-7,8-hexahydro-14-methyl-, trans (drosopterin)	pH -1.5	262(4.24),295s(3.94), 455(4.34)	24-3385-78
	pH 0.8	219(4.45),267(4.12), 325s(3.64),470(4.36)	24-3385-78
	pH 5.0	220(4.55),273(4.12), 324s(3.72),478(4.42)	24-3385-78
	pH 8.8	216(4.51),265(4.13), 330s(3.64),485(4.42)	24-3385-78
	pH 10.0	210(4.55),264(4.16), 330s(3.64),493(4.43)	24-3385-78
	pH 13	210(4.87),265(4.24), 337(3.79),502(4.24)	24-3385-78
$C_{15}H_{16}O$			
Benzene, 1,2-dimethyl-4-(phenylmeth-oxy)-	EtOH	278(3.26),283(3.23)	23-0517-78
Naphtho[2,1-b]furan, 4,5-dihydro-1,5,8-trimethyl- (laevigatin)	EtOH	276(3.80),290s(3.77), 302s(3.36)	88-2653-78
$C_{15}H_{16}OSi$			
Silane, benzoyldimethylphenyl-	C_6H_{12}	404(2.26),423(2.37), 445(2.14)	112-0751-78
$C_{15}H_{16}O_2$			
2H-Benz[e]inden-2-one, 1,4,5,9b-tetra-hydro-7-methoxy-9b-methyl-	EtOH	226(4.28)	39-0110-78C
2H-Benz[e]inden-2-one, 3,3a,4,5-tetra-hydro-7-methoxy-3a-methyl-	EtOH	317(4.18)	39-0110-78C
Cycloprop[2,3]indeno[5,6-b]furan-2(4H)-one, 4a,5,5a,6,6a,6b-hexahydro-3,6b-dimethyl-5-methylene-	MeOH	279(4.10)	142-0139-78A
4,7-Ethano-1H-indene-8,9-dione, 3a,4,7,7a-tetrahydro-4-methyl-1-(1-methylethylidene)-	benzene	457.5(1.98)	5-0440-78
4,7-Ethano-1H-indene-8,9-dione, 3a,4,7,7a-tetrahydro-5-methyl-1-(1-methylethylidene)-	benzene	457.5(1.97)	5-0440-78

Compound	Solvent	$\lambda_{max}(\log \epsilon)$	Ref.
1(2H)-Naphthalenone, 3,4-dihydro-6-methoxy-2-methyl-2-(2-propynyl)-	EtOH	275(4.18)	39-0110-78C
2(1H)-Naphthalenone, 3,4-dihydro-6-methoxy-1-methyl-1-(2-propynyl)-	EtOH	277(4.12)	39-0110-78C
Phenol, 2-[(2-hydroxy-3-methylphenyl)methyl]-6-methyl-	EtOH	274(4.07),280(4.02)	32-0079-78
Spiro[cyclopropane-1,1'(5'H)-[4,8]ethenoazulen]-5'-one, 3'a,4',8',8'a-tetrahydro-4'-methoxy-	MeOH	225(3.89),250(3.81)	18-1257-78
Spiro[cyclopropane-1,1'(7'H)-[4,8]ethenoazulen]-7'-one, 3'a,4',8',8'a-tetrahydro-6'-methoxy-	MeOH	234(3.96),282(3.91)	18-1257-78

$C_{15}H_{16}O_3$

Compound	Solvent	$\lambda_{max}(\log \epsilon)$	Ref.
2H-1-Benzopyran-2-one, 7-(1,1-dimethyl-2-propenyl)-6-hydroxy-4-methyl-	n.s.g.	225(4.12),275(3.95), 340(3.68)	2-1039-78
3H-Cycloprop[2,3]oxireno[4,5]indeno-[5,6-b]furan-3-one, 5,5a,6,6a,7,7a-7b,7c-octahydro-4,7b-dimethyl-6-methylene-	MeOH	226(4.11)	142-0139-78A
2H-Furo[2,3-h]-1-benzopyran-2-one, 8,9-dihydro-4,8,9,9-tetramethyl-	n.s.g.	220(3.96),308(3.95), 320(3.99)	2-1039-78
7H-Furo[3,2-g][1]benzopyran-7-one, 2,3-dihydro-2,3,3,5-tetramethyl-	n.s.g.	219(3.98),310(3.97), 319(3.99)	2-1039-78
Naphtho[1,2-b]furan-8(7H)-one, 6,9-dihydro-4-hydroxy-3,5,9-trimethyl-, (±)-	EtOH	253(4.05),285(3.37), 296(3.21)	94-3704-78
Phenol, 3-[2-(2-hydroxyphenyl)ethyl]-5-methoxy-	EtOH	226s(4.21),276(3.70), 283s(3.66)	102-1179-78
	EtOH-KOH	214(4.02),243s(3.96), 277s(3.98),283(4.00)	102-1179-78
Phenol, 3-[2-(4-hydroxyphenyl)ethyl]-5-methoxy-	EtOH	225(4.30),277s(3.54), 280(3.56)	102-1179-78

$C_{15}H_{16}O_4$

Compound	Solvent	$\lambda_{max}(\log \epsilon)$	Ref.
Benzenemethanol, 3-[4-(hydroxymethyl)phenoxy]-4-methoxy-	EtOH	205(4.33),218(4.27), 270(3.46)	12-0321-78
2H,6H-Benzo[1,2-b:5,4-b']dipyran-2-one, 7,8-dihydro-4-methoxy-8,8-dimethyl-	dioxan	280(3.97),320(4.23)	2-0375-78
4H-1-Benzopyran-4-one, 6-acetoxy-2,5,7,8-tetramethyl-	EtOH	228(4.39),234(4.39), 246s(4.16),252s(4.14), 265s(3.90),275s(3.73), 308(3.79)	44-3723-78
4H-Furo[3,2-g][1]benzopyran-4,9(5H)-dione, 6,7-dihydro-3-methyl-6-(1-methylethyl)-, (R)-	EtOH	222(4.30),270(3.98), 325(3.84),435(2.81)	12-1553-78
3-Furancarboxylic acid, 4,5-dihydro-2-methyl-4-oxo-5-(phenylmethyl)-, ethyl ester	EtOH	214(4.00),264(3.90)	4-0327-78
Propanedioic acid, [2-methyl-3-(4-methylphenyl)-2-propenylidene]-, monomethyl ester, (E,E)-	EtOH	264(3.30),320(4.48)	22-0457-78
Propanedioic acid, (2-methyl-3-phenyl-2-propenylidene)-, dimethyl ester, (E)-	EtOH	226(3.88),310(4.46)	22-0457-78

$C_{15}H_{16}O_4S_2$

Compound	Solvent	$\lambda_{max}(\log \epsilon)$	Ref.
2(3H)-Furanone, 4-[2-(1,3-benzodioxol-5-yl)-1,3-dithian-2-yl]dihydro-, (±)-	EtOH	207(4.23),248(3.51), 290(3.46)	78-1011-78

$C_{15}H_{16}O_5$

Compound	Solvent	$\lambda_{max}(\log \epsilon)$	Ref.
2H,6H-Benzo[1,2-b:5,4-b']dipyran-2-one,	EtOH	222s(4.22),248(3.66),	102-1805-78

Compound	Solvent	$\lambda_{max}(\log \epsilon)$	Ref.
7,8-dihydro-7-hydroxy-6-methoxy-8,8-dimethyl-, (6R-cis)- (cont.)		258(3.57),328(4.27)	102-1805-78
3-Furancarboxylic acid, 4,5-dihydro-5-methoxy-5-methyl-4-oxo-2-phenyl-, ethyl ester	EtOH	212(3.96),254(3.74), 306(4.18)	44-2665-78
Propanedioic acid, [3-(4-methoxyphenyl)-2-methyl-2-propenylidene]-, monomethyl ester, (E,E)-	EtOH	248(3.70),334(4.46)	22-0457-78
Rutaretin methyl ether	MeOH	260(3.80),335(4.30)	102-2135-78
$C_{15}H_{16}O_6$			
1H-2-Benzopyran-1-one, 8-hydroxy-3-(2-hydroxy-4-oxopentyl)-6-methoxy-	MeOH	237s(4.70),243(4.75), 256s(4.14),277(3.87), 286s(3.69),325(3.85)	44-2339-78
2(5H)-Furanone, 3-acetyl-5-[(3,4-dimethoxyphenyl)methyl]-4-hydroxy-	EtOH	210(3.92),231(3.69), 266(3.72)	4-0327-78
$C_{15}H_{16}O_7$			
Benzeneacetic acid, 3,6-diacetoxy-2-acetyl-, methyl ester	EtOH	280(3.19)	12-2099-78
1H-Naphtho[2,3-c]pyran-5,10-dione, 3,4,4a,10a-tetrahydro-3,6,9-tri-hydroxy-7-methoxy-3-methyl-, (3α,4aα,10aα)-(+)-	EtOH	213(4.17),243(4.30), 273(3.87),300(3.71), 391(3.94)	23-1593-78
(3α,4aα,10aβ)-(+)-	EtOH	213(4.15),243(4.31), 273(3.89),300(3.70), 391(3.96)	23-1593-78
$C_{15}H_{16}O_7S$			
4H-1-Benzopyran-4-one, 2,3-diacetoxy-2,3-dihydro-8-methoxy-3-(methylthio)-	n.s.g.	264(3.95),330(3.30)	4-0115-78
$C_{15}H_{16}S$			
4H-1-Benzothiopyran, 5,6,7,8-tetrahydro-2-phenyl-	hexane	198(4.6),240(4.2), 295s(3.2)	103-0605-78
$C_{15}H_{17}BrN_2O_2$			
Pyridinium, 4-[(4-nitrophenyl)methyl]-1-propyl-, bromide	H_2O	260s(4.10),264(4.12), 273(4.08)	48-0133-78
deprotonated	H_2O	564(4.48)	48-0133-78
$C_{15}H_{17}BrO_2$			
2,6-Methanofuro[3,2-b]furan, 5-(1-bromopropylidene)hexahydro-7-(1,3,4-pentatrienyl)-	EtOH	220(4.38)	44-3194-78
$C_{15}H_{17}ClN_2O$			
1H-Imidazo[1,2-a]pyridin-4-ium, 2,3-dihydro-2-hydroxy-2-methyl-1-(phenylmethyl)-, chloride	EtOH	203(4.07),239(4.01), 329(3.67)	4-1149-78
$C_{15}H_{17}ClN_2O_4$			
12H-Indolo[2,3-a]quinolizin-5-ium, 1,2,3,4,6,7-hexahydro-, perchlorate	EtOH-HCl	246(4.02),349(4.34)	22-0355-78
$C_{15}H_{17}ClO_2$			
1(2H)-Naphthalenone, 2-(2-chloro-2-propenyl)-3,4-dihydro-6-methoxy-2-methyl-	EtOH	275(4.08)	39-0110-78C
2(1H)-Naphthalenone, 1-(2-chloro-2-propenyl)-3,4-dihydro-6-methoxy-1-methyl-	EtOH	277(3.27)	39-0110-78C

Compound	Solvent	$\lambda_{max}(\log \epsilon)$	Ref.
$C_{15}H_{17}ClO_3$ Santonin, 14-chloro-	EtOH	242(4.03)	94-2729-78
$C_{15}H_{17}ClO_6$ Pyrylium, 4-[(2,6-dimethyl-4H-pyran-4-ylidene)methyl]-2,6-dimethyl-, perchlorate	HOAc-HClO₄	470(4.529)	83-0236-78
$C_{15}H_{17}Cl_2N_2O_3PS$ Phosphorohydrazidothioic acid, [[5-(3,4-dichlorophenyl)-2-furanyl]methylene]-, O,O-diethyl ester	dioxan	337(4.56)	73-2643-78
$C_{15}H_{17}Cl_3N_2O$ Acetamide, 2,2,2-trichloro-N-[(1,2,3,7-tetramethyl-1H-indol-6-yl)methyl]-	CHCl₃	237(4.58),294(3.83)	103-0856-78
$C_{15}H_{17}CuNO_2$ Copper, 4-(1,1-dimethylethyl)catecholato(pyridine)-	MeOH	244(--),250(3.65), 256(--),358(0.92), 490(0.65)	35-5472-78
	ether	235(--),240(3.57), 246(--),252(--), 284(3.45),371(--), 470(--),676(--)	35-5472-78
$C_{15}H_{17}FN_2O_2S$ 3-Thiophenecarboxylic acid, 2-[(2-amino-4-fluorophenyl)amino]-5-ethyl-, ethyl ester	MeOH	318(3.99)	39-0937-78C
$C_{15}H_{17}FN_2O_7$ 2H,7H-Pyrano[2',3':4,5]oxazolo[3,2-a]pyrimidin-7-one, 3,4-diacetoxy-2-(fluoromethyl)-3,4,4a,10a-tetrahydro-8-methyl-, [2R-(2α,3β,4α,4aβ,10aβ)]-	MeOH	227(3.81),235s(3.82), 249(3.87)	48-0157-78
$C_{15}H_{17}FN_2O_9$ Uridine, 5-fluoro-, 2',3',5'-triacetate	EtOH	270(3.94)	94-2990-78
$C_{15}H_{17}N$ Benzenamine, 4-(1,3,5-cycloheptatrien-1-yl)-N,N-dimethyl-	EtOH	342(4.30)	88-0345-78
Benzenamine, 4-(1,3,5-cycloheptatrien-2-yl)-N,N-dimethyl-	EtOH	276(4.41)	88-0345-78
Benzenamine, 4-(1,3,5-cycloheptatrien-3-yl)-N,N-dimethyl-	EtOH	253(4.06),313(4.27)	88-0345-78
Benzenamine, 4-(1,3,5-cycloheptatrien-7-yl)-N,N-dimethyl-	EtOH	253(4.34),300s(--)	88-0345-78
Bicyclo[8.2.2]tetradeca-5,10,12,13-tetraene-4-carbonitrile, (E)-	EtOH	223(3.83),270(2.62), 278(2.59)	35-1806-78
$C_{15}H_{17}NO$ 1H,8H-3a,8a-Propanoindeno[1,2-c]pyrrol-8-one, 2,3-dihydro-2-methyl-, hydrochloride	EtOH	210(4.23),251(3.92), 295(3.16)	4-0167-78
2H-Pyrrol-2-one, 4-ethyl-1,5-dihydro-3-methyl-5-[(4-methylphenyl)methylene]-, (Z)-	EtOH	231(3.95),237(3.92), 245(3.79),325s(4.37), 337(4.40),355s(4.20)	49-0183-78

Compound	Solvent	$\lambda_{max}(\log \epsilon)$	Ref.
$C_{15}H_{17}NO_2$			
8-Azabicyclo[3.2.1]oct-3-en-2-one, 4-methoxy-8-methyl-6-phenyl-	n.s.g.	203(3.97),250(4.01)	88-1751-78
2-Butenoic acid, 4-[2-(2,2-dimethyl-2H-azirin-3-yl)phenyl]-, methyl ester, (E)-	MeOH	243(4.10)	35-2181-78
1,3-Cyclohexanedione, 5,5-dimethyl-2-[(phenylamino)methylene]-	EtOH	350(4.41)	97-0256-78
2H-Furo[2,3-b]indol-2-one, 3,3a,8,8a-tetrahydro-3a-(3-methyl-2-butenyl)-	EtOH	242(3.25),293(3.13), 319(4.30)	78-0929-78
Indeno[1,2-b]pyrrole-3-carboxylic acid, 2,3,3a,4-tetrahydro-2,2-dimethyl-, methyl ester	MeOH	245(3.96)	35-2181-78
1H-Indole-3-acetic acid, 2-(3-methyl-2-butenyl)-	EtOH	227(4.52),284(3.87), 291(3.82)	78-0929-78
$C_{15}H_{17}NO_3$			
3-Carbazolecarboxylic acid, 5,6,7,8-tetrahydro-1-methoxy-, methyl ester	EtOH	240(4.54),285(4.12), 300(3.85)	102-0834-78
5-Isoxazolecarboxylic acid, 4-ethylidene-4,5-dihydro-5-methyl-3-phenyl-, ethyl ester	EtOH	230(3.98),268(3.93)	22-0415-78
1H-Pyrrole-2-carboxylic acid, 3-hydroxy-1-phenyl-, 1,1-dimethylethyl ester	EtOH	264(4.10)	94-2224-78
$C_{15}H_{17}NO_4$			
1H-Pyrrole-3-carboxylic acid, 4,5-dihydro-5-hydroxy-2,5-dimethyl-4-oxo-1-phenyl-, ethyl ester	EtOH	207(3.82),245(4.20), 320(4.07)	118-0291-78
1H-Pyrrole-3-carboxylic acid, 4,5-dihydro-5-hydroxy-2-methyl-4-oxo-5-(phenylmethyl)-, ethyl ester	EtOH	211(4.05),242(4.23), 307(3.83)	4-1215-78
$C_{15}H_{17}NO_6$			
1H-2-Benzopyran-1,3(4H)-dione, 4-[(dimethylamino)methylene]-5,6,7-trimethoxy-	MeOH	205(3.67),260(3.64)	4-0257-78
$C_{15}H_{17}NS_3$			
Spiro[1,3-benzodithiole-2,4'(5'H)-thiazole], 3a,4,5,6,7,7a-hexahydro-2'-phenyl-	EtOH	248(4.25),285s(3.80)	18-0301-78
$C_{15}H_{17}N_2$			
12H-Indolo[2,3-a]quinolizin-5-ium, 1,2,3,4,6,7-hexahydro-, perchlorate	EtOH-HCl	246(4.02),349(4.34)	22-0355-78
$C_{15}H_{17}N_2O$			
1H-Imidazo[1,2-a]pyridin-4-ium, 2,3-dihydro-2-hydroxy-2-methyl-1-(phenylmethyl)-, chloride	EtOH	203(4.07),239(4.01), 329(3.67)	4-1149-78
$C_{15}H_{17}N_2OS$			
Pyrrolidinium, 1-[3-isothiocyanato-3-(4-methoxyphenyl)-2-propenylidene]-, perchlorate	MeCN	243(4.09),341(4.23), 393(4.39)	97-0334-78
1,3-Thiazin-1-ium, 4-(4-methoxyphenyl)-2-pyrrolidino-, perchlorate	MeCN	259(4.08),298s(3.45), 379(4.42)	97-0334-78
1,3-Thiazin-1-ium, 4-(4-methylphenyl)-2-morpholino-, perchlorate	MeCN	224(4.05),260(4.14), 352(4.42)	97-0334-78

Compound	Solvent	$\lambda_{max}(\log \epsilon)$	Ref.
$C_{15}H_{17}N_2O_2$			
Pyridinium, 4-[(4-nitrophenyl)methyl]-1-propyl-, bromide	H_2O	260s(4.10),264(4.12), 273(4.08)	48-0133-78
$C_{15}H_{17}N$			
1H-Perimidin-2-amine, N-butyl-	EtOH	$\underline{230(5.0),320(4.3)}$	103-0694-78
$C_{15}H_{17}N_3O$			
10H-Pyrido[3,2-b][1,4]benzoxazine-10-ethanamine, N,N-dimethyl-	C_6H_{12}	233(3.64),237(3.64), 342(3.10)	95-0585-78
$C_{15}H_{17}N_3O_3$			
Pyrido[2,3-d]pyrimidine-2,4(1H,3H)-dione, 3-(4-ethoxyphenyl)-5,6,7,8-tetrahydro-	MeOH	224(4.01),293(4.19)	24-2297-78
1,2,4-Triazine-6-acetic acid, 5-methoxy-α-methyl-3-phenyl-, ethyl ester	EtOH	256(4.15),274(4.08)	114-0091-78B
$C_{15}H_{17}N_3O_4$			
1,2,4-Triazine-5-acetic acid, 1,6-dihydro-3-(4-methoxyphenyl)-α-methyl-6-oxo-, ethyl ester	EtOH	210s(4.13),281(4.38)	114-0091-78B
1,2,4-Triazine-6-acetic acid, 2,5-dihydro-3-(4-methoxyphenyl)-α-methyl-5-oxo-, ethyl ester	EtOH	214s(4.15),258s(4.22), 283(4.31)	114-0091-78B
$C_{15}H_{17}N_3O_6$			
7H-Pyridazino[4,5-d]azepine-1,4,7-tricarboxylic acid, 2,4a-dihydro-, 7-ethyl 1,4-dimethyl ester	CH_2Cl_2	292(4.29),413(3.87)	83-0786-78
$C_{15}H_{17}N_3O_7$			
Uridine, 5-cyano-2',3'-O-(1-methylethylidene)-, 5'-acetate	EtOH	236(3.27),275(4.09)	94-2657-78
Uridine, 6-cyano-2',3'-O-(1-methylethylidene)-, 5'-acetate	EtOH	229(3.30),281(3.93)	94-2657-78
$C_{15}H_{17}N_5O$			
7H-1,2,3-Triazolo[4,5-d]pyrimidin-7-one, 6-butyl-3,6-dihydro-3-(phenylmethyl)-	EtOH	259(3.90)	39-0513-78C
$C_{15}H_{17}N_5O_5$			
4,12-Epoxy-4H-1,3-dioxolo[5,6]azepino[1,2-e]purin-11-ol, 9-amino-3a,11,12-12a-tetrahydro-2,2-dimethyl-, acetate, (R)-	H_2O	268(4.17)	78-2449-78
(S)-	H_2O	266(4.18)	78-2449-78
$C_{15}H_{17}O_2$			
Pyrylium, 4-[(2,6-dimethyl-4H-pyran-4-ylidene)methyl]-2,6-dimethyl-, perchlorate	HOAc-HClO₄	470(4.529)	83-0236-78
	CH_2Cl_2	472(5.02)	103-0601-78
	MeNO₂	472(5.00)	103-0601-78
$C_{15}H_{18}$			
Benzene, (3-cyclohexylidene-1-propenyl)-, (E)-	EtOH	290(4.49),316(4.28)	39-0829-78C
$C_{15}H_{18}BrNO_4$			
1H-Pyrrolo[1,2-a]indole-5,8-dione, 6-bromo-2,3,6,7-tetrahydro-7,7-	EtOH	254(4.08),294(3.88), 352(4.08)	94-3815-78

Compound	Solvent	$\lambda_{max}(\log \epsilon)$	Ref.
dimethoxy-6,9-dimethyl- (cont.)			94-3815-78
$C_{15}H_{18}BrN_2O_3PS$			
Phosphorohydrazidothioic acid, [[5-(4-bromophenyl)-2-furanyl]methylene]-, O,O-diethyl ester	dioxan	339(4.59)	73-2643-78
$C_{15}H_{18}Br_2O$			
Spiro[5.5]undec-7-en-3-one, 2-bromo-9-(bromomethylene)-1,1-dimethyl-5-methylene-	EtOH	243(4.32)	88-4805-78
$C_{15}H_{18}ClNO_4$			
Phenanthridinium, 5-ethyl-7,8,9,10-tetrahydro-, perchlorate	MeOH	236(5.06),318(4.24)	65-1707-78
$C_{15}H_{18}ClN_2O_3PS$			
Phosphorohydrazidothioic acid, [[5-(2-chlorophenyl)-2-furanyl]methylene]-, O,O-diethyl ester	dioxan	333(4.51)	73-2643-78
Phosphorohydrazidothioic acid, [[5-(4-chlorophenyl)-2-furanyl]methylene]-, O,O-diethyl ester	dioxan	330(4.57)	73-2643-78
$C_{15}H_{18}Cl_2N_2O_4$			
7H-Pyrido[2,3-d]azepine-3,7-dicarboxylic acid, 2,4-dichloro-5,6,8,9-tetrahydro-, diethyl ester	EtOH	280(3.70)	142-0267-78C
$C_{15}H_{18}Cl_2N_4O_2$			
Acetic acid, [1-[[bis(2-chloroethyl)-amino]methyl]-1,2-dihydro-2-oxo-3H-indol-3-ylidene]hydrazide	EtOH	312(4.19)	111-0515-78
$C_{15}H_{18}N_2O$			
Azepino[3,2-b]indol-2(1H)-one, 3,4,5,6-tetrahydro-4,4,6-trimethyl-	MeOH	233(4.53),277s(3.97),300s(3.88)	24-0853-78
Azepino[4,3-b]indol-1(2H)-one, 3,4,5,6-tetrahydro-4,4,6-trimethyl-	MeOH	218(4.58),233s(4.48),254s(3.99),284s(4.05),291(4.07)	24-0853-78
4H-Carbazol-4-one, 1,2,3,9-tetrahydro-2,2,9-trimethyl-, oxime	MeOH	230(4.30),255s(4.10),264(4.17),290(3.98),302s(3.92)	24-0853-78
Δ^7-Dehydrosophoramine	EtOH	246(3.71),340(3.98)	94-1832-78
2H-Indol-2-one, 1,3-dihydro-3-(2-piperidinoethylidene)-, monoperchlorate	EtOH	250(4.39),258(4.38),288(3.79)	42-1122-78
$C_{15}H_{18}N_2O_2$			
Spiro[imidazolidine-2,2'-[2H]indene]-1',3'-dione, 1,3-diethyl-	MeOH	230(4.68),251(4.05),290(3.27),408(2.24)	24-3058-78
$C_{15}H_{18}N_2O_2S$			
1H-[1]Benzothieno[3,2-e]pyrido[1,2-a]-pyrimidine-1,6(2H)-dione, 3,4,4a,5-7,8,9,10-octahydro-4a-methyl-	EtOH	225(4.45),262s(3.93),317(4.00)	4-0949-78
3-Thiophenecarboxylic acid, 2-[(2-aminophenyl)amino]-5-ethyl-, ethyl ester	MeOH	326(4.00)	39-0937-78C
$C_{15}H_{18}N_2O_3$			
1H-Pyrazole-4-carboxylic acid, 3-(2-hydroxypropyl)-5-phenyl-, ethyl ester	EtOH	216(4.10),242(3.97)	118-0900-78

Compound	Solvent	λ_{max}(log ϵ)	Ref.
$C_{15}H_{18}N_2O_4$			
2-Propenoic acid, 2-acetoxy-3-phenyl-, 1-acetyl-2,2-dimethylhydrazide	MeOH	218(4.12),308(4.25)	24-0791-78
5H-Pyrido[3,2-b]indole-5-acetic acid, 1,2,3,4,6,7,8,9-octahydro-2,9-dioxo-, ethyl ester	EtOH	250(4.23),310(3.48)	94-3080-78
2,4,6(1H,3H,5H)-Pyrimidinetrione, 5,5-diethyl-N-(phenylmethoxy)-	pH 1.20	200(4.63),215(4.10)	12-2517-78
	pH 10.69	196(4.61),244(3.92)	12-2517-78
$C_{15}H_{18}N_2O_4S$			
1H-Benzimidazole, 2-(2-propenyl)-1-β-D-ribofuranosyl-	MeOH	212(4.48),252(3.91), 258s(3.90),284(4.09), 291(4.09)	24-0996-78
$C_{15}H_{18}N_2O_{10}$			
1H-Imidazole-4-carboxylic acid, 2,3-dihydro-2-oxo-1-(2,3,5-tri-O-acetyl-β-D-ribofuranosyl)-	MeOH	263(3.98)	94-3322-78
$C_{15}H_{18}N_2Se$			
3-Selena-1-azaspiro[4.5]dec-1-en-2-amine, 4-methylene-N-phenyl-	EtOH	204(3.66),238(3.51)	103-0738-78
$C_{15}H_{18}N_3O_5PS$			
Phosphorohydrazidothioic acid, [[5-(2-nitrophenyl)-2-furanyl]methylene]-, O,O-diethyl ester	dioxan	360(4.40)	73-2643-78
Phosphorohydrazidothioic acid, [[5-(3-nitrophenyl)-2-furanyl]methylene]-, O,O-diethyl ester	dioxan	334(4.51)	73-2643-78
Phosphorohydrazidothioic acid, [[5-(4-nitrophenyl)-2-furanyl]methylene]-, O,O-diethyl ester	dioxan	386(4.39)	73-2643-78
$C_{15}H_{18}N_4O_2S$			
1H-Pyrimido[5,4-b][1,4]benzothiazine-2,4(3H,10H)-dione, 10-[3-(dimethylamino)propyl]-, hydrochloride	pH 13	254(4.35),281s(3.94), 320s(3.39)	83-0303-78
	MeOH	250(4.27),281(3.92), 360s(3.20)	83-0303-78
	12M HCl	292(3.81),370(3.42)	83-0303-78
$C_{15}H_{18}N_4O_3S$			
Pyrazolo[3,4-d][1,2]diazepine, 2,3,3a,6-tetrahydro-2-(methoxymethyl)-6-[(4-methylphenyl)sulfonyl]-	EtOH	336(4.10)	39-1297-78C
$C_{15}H_{18}N_4O_4$			
4,7-Methano-1H-indazole, 1-(2,4-dinitrophenyl)octahydro-2-methyl-, (3aα,4β,7β,7aα)-	MeCN	391(3.92)	78-0903-78
2,4(1H,3H)-Quinazolinedione, 1-[3-(3,4-dihydro-2,4-dioxo-1(2H)-pyrimidinyl)-propyl]-5,6,7,8-tetrahydro-	H_2O	272(4.27)	56-1035-78
$C_{15}H_{18}N_4O_7$			
Inosine, 2',3'-O-(3-carboxy-1-methyl-propylidene)-	MeOH	245(4.00)	136-0155-78C
$C_{15}H_{18}N_6O_5$			
Acetamide, N-(3a,4,8,9,13,13a-hexahydro-2,2-dimethyl-8-oxo-4,13-epoxy-	MeOH	259(4.28)	44-2320-78

Compound	Solvent	λ_{max}(log ϵ)	Ref.
1,3-dioxolo[6,7][1,3]diazocino[1,2-e]purin-6(5H)-yl)-, [3aR-(3aα,4β,13β,13aα)]- (cont.)	MeOH	259(4.28)	44-2320-78
$C_{15}H_{18}O$			
Azuleno[5,6-c]furan, 5,6,7,7a,8,9-hexahydro-6,6-dimethyl-8-methylene-, (±)-	EtOH	241(4.20)	78-2027-78
Cyclooctanone, 2-(phenylmethylene)-	hexane	276(4.18)	54-0305-78
2-Cyclopenten-1-one, 2,5-diethyl-3-phenyl-	EtOH	221(3.86),278(4.23)	35-1799-78
3-Cyclopenten-1-one, 2,2,5,5-tetramethyl-3-phenyl-	EtOH	243(4.16),281(3.39)	35-1791-78
Ethanone, 1-(3-phenylbicyclo[2.2.1]-hept-2-yl)-, endo-endo	EtOH	260(2.60)	44-4215-78
endo-exo	EtOH	258(2.39),280(1.76)	44-4215-78
exo-endo	EtOH	258(2.40),280(1.72)	44-4215-78
exo-exo	EtOH	260(2.28),280(1.30)	44-4215-78
$C_{15}H_{18}OS$			
Bicyclo[3.3.1]nonan-3-one, 1-(phenylthio)-	EtOH	222(4.04),267(3.11)	44-3653-78
$C_{15}H_{18}OSe$			
Bicyclo[3.3.1]nonan-3-one, 1-(phenylseleno)-	EtOH	220(4.08),280(2.79)	44-3653-78
$C_{15}H_{18}O_2$			
1H-Benz[e]inden-2-ol, 2,4,5,9b-tetrahydro-7-methoxy-9b-methyl-	EtOH	279(3.28)	39-0110-78C
2H-Benz[e]inden-2-ol, 3,3a,4,5-tetrahydro-7-methoxy-3a-methyl-	EtOH	267(4.22)	39-0110-78C
Cyclodeca[b]furan-2(4H)-one, 7,8-dihydro-3,6,10-trimethyl-, (E,E)-	EtOH	269(3.93)	83-0754-78
2-Cyclohexen-1-one, 5,5-dimethyl-3-(phenylmethoxy)-	EtOH	250.5(4.30)	78-1567-78
2-Cyclopropene-1-carboxylic acid, 2-[4-(1,1-dimethylethyl)phenyl]-3-methyl-	EtOH	263(4.22)	104-2144-78
Furanoligularenone	MeOH	226(4.39)	73-1113-78
	dioxan	257(0.14),320(1.63),325(1.66),329(1.62)	73-1113-78
$C_{15}H_{18}O_2S$			
Bicyclo[3.3.1]nonan-3-one, 1-(phenylsulfinyl)-	EtOH	251(3.67)	44-3653-78
2-Cyclohexen-1-one, 3-hydroxy-5,5-dimethyl-2-[(4-methylphenyl)thio]-	MeOH	254(4.02)	44-2676-78
$C_{15}H_{18}O_2Se$			
Bicyclo[3.3.1]nonan-3-one, 1-hydroxy-2-(phenylseleno)-	EtOH	227(3.91),325(2.98)	44-3653-78
$C_{15}H_{18}O_3$			
Cyclodeca[b]furan-6-carboxaldehyde, 2,3,3a,4,5,8,9,11a-octahydro-10-methyl-3-methylene-2-oxo-	isoPrOH	212(3.78),226(3.81)	42-1142-78
3-Cyclohexen-1-ol, 4-(4-methoxyphenyl)-, acetate	EtOH	253(4.19),288s(3.26)	39-0024-78C
10αH-Furanoeremophilane-3,9-dione	EtOH	280(4.16)	94-3704-78
10βH-Furanoeremophilane-3,9-dione	EtOH	281.5(4.15)	94-3704-78

Compound	Solvent	$\lambda_{max}(\log \epsilon)$	Ref.
2(1H)-Naphthalenone, 3,4-dihydro-6-methoxy-1-methyl-1-(2-oxopropyl)-	EtOH	277(3.23)	39-0110-78C
Naphtho[1,2-b]furan-2,8(3H,4H)-dione, 3a,5,5a,6,7,9b-hexahydro-5a,9-dimethyl-3-methylene-	EtOH	246(4.20)	2-0016-78
Naphtho[1,2-b]furan-2,8(3H,4H)-dione, 3a,6,7,9b-tetrahydro-3,6,9-trimethyl-	EtOH	233(3.58),288(4.05)	44-0536-78
Santonin	EtOH	242(4.12),258(4.01)	2-0016-78
$C_{15}H_{18}O_3S$			
2-Cyclohexen-1-one, 3-hydroxy-2-[(4-methoxyphenyl)thio]-5,5-dimethyl-	MeOH	254(3.99)	44-2676-78
$C_{15}H_{18}O_4$			
Benzenepropanoic acid, α-acetyl-4-methoxy-β-methylene-, ethyl ester	n.s.g.	295(4.00)	2-0200-78
Eremophila-1,7-dien-8,12-olide, 8α-hydroxy-3-oxo-	n.s.g.	225(4.36),325(1.87)	73-1113-78
Furanoeremophilane, 10β-hydroxy-3,9-dioxo-	EtOH	285(3.97)	18-3335-78
2-Naphthaleneacetic acid, 1,2,3,4-tetrahydro-6-methoxy-2-methyl-1-oxo-, methyl ester	EtOH	274(4.25)	39-0110-78C
Naphtho[2,3-b]furan-4,6-dione, 4a,5,7-8,8a,9-hexahydro-9-hydroxy-3,4a,5-trimethyl-	EtOH	269(3.49)	94-3704-78
Naphtho[2,3-b]furan-4,9-dione, 4a,5,6-7,8,8a-hexahydro-6-hydroxy-3,4a,5-trimethyl-	EtOH	238(3.51),283(4.13)	94-3704-78
Scabequinone, dihydro-	EtOH	311(4.00),447(2.35)	12-1553-78
$C_{15}H_{18}O_5$			
Ethanone, 1-(2,5-diacetoxy-3,4,6-trimethylphenyl)-	EtOH	210s(4.22),243s(3.54), 280s(2.90)	44-3723-78
Ethanone, 1-(2,6-diacetoxy-3,4,5-trimethylphenyl)-	EtOH	209s(4.28),247(3.75), 285s(3.10)	44-3723-78
$C_{15}H_{18}O_6$			
Diversonol	MeOH	273(4.09),350(3.50)	39-1621-78C
$C_{15}H_{19}BrN_2O_3$			
2-Furanmethanaminium, N,N-diethyl-5-nitro-N-phenyl-, bromide	H_2O	204(4.42),261(4.00), 300(3.86)	73-2041-78
$C_{15}H_{19}BrN_2O_4$			
β-Alanine, N-butyl-N-nitroso-, 2-(4-bromophenyl)-2-oxoethyl ester	EtOH	256.5(4.24)	94-3909-78
Butanoic acid, 4-(nitrosopropylamino)-, 2-(4-bromophenyl)-2-oxoethyl ester	EtOH	256(4.36)	94-3909-78
Glycine, N-nitroso-N-pentyl-, 2-(4-bromophenyl)-2-oxoethyl ester	EtOH	257(4.39)	94-3909-78
$C_{15}H_{19}BrO$			
Spiro[5.5]undeca-1,7-dien-3-one, 9-(bromomethylene)-1,5,5-trimethyl-	EtOH	240(4.48)	88-4805-78
$C_{15}H_{19}ClN_2O_5$			
7H-Pyrido[2,3-d]azepine-3,7-dicarboxylic acid, 4-chloro-1,2,5,6,8,9-hexahydro-2-oxo-, diethyl ester	EtOH	237(3.76),326(4.05)	142-0267-78C

Compound	Solvent	$\lambda_{max}(\log \epsilon)$	Ref.
$C_{15}H_{19}ClN_2O_7$			
Uridine, 6-(chloromethyl)-2',3'-O-(1-methylethylidene)-, 5'-acetate	MeOH	265(4.00)	94-2657-78
$C_{15}H_{19}ClO_2$			
1H-Inden-1-one, 6-(2-chloroethyl)-2,3-dihydro-2-(hydroxymethyl)-2,5,7-trimethyl- (pterosin K)	EtOH	218(4.61),259(4.21), 303(3.41)	94-2365-78
$C_{15}H_{19}IN_4$			
Piperidine, 1-[(cyanoimino)[3-(dimethylamino)methylene]-5-iodo-1,4-cyclopentadien-1-yl]methyl]-	MeOH	239(4.07),275(3.95), 343(4.42)	83-0369-78
$C_{15}H_{19}NO_3$			
2-Aza-6,8-dioxatetracyclo[8.2.1.0^{2,12}-0^{4,10}]tridec-3-en-11-one, 12-methyl-13-methylene-7-(1-methylethyl)-	EtOH	222(3.57),258s(2.90), 261(2.92),265(2.92), 267(2.90),272(2.82), 313(2.36)	33-2542-78
Cyclohexanone, 3-methyl-2-(2-nitro-1-phenylethyl)-, [2α(S*),3β]-	EtOH	211(4.10)	2-0405-78
Cyclohexanone, 4-methyl-2-(2-nitro-1-phenylethyl)-	EtOH	210(4.21),310(3.81)	2-0405-78
Cyclohexanone, 5-methyl-2-(2-nitro-1-phenylethyl)-	EtOH	211(4.10)	2-0405-78
[1,3]Dioxepino[5,6-c]pyridine, 1,5-dihydro-8-methyl-3-(1-methylethyl)-9-(2-propynyloxy)-	EtOH	261s(3.36),268(3.54), 274(3.61)	33-2542-78
7H-Furo[3',4':3,4]pyrido[2,1-i]indol-1(3H)-one, 4,5,8,10,11,12-hexahydro-11-methoxy-, [11S-(11R*,12aR*)]- (cocculolidine)	MeOH	217(4.07)	95-0886-78
Naphtho[1,2-b]furan-2,8(3H,4H)-dione, 3a,6,7,9b-tetrahydro-3,6,9-trimethyl-, 8-oxime	EtOH	276(4.39)	44-0536-78
$C_{15}H_{19}NO_4$			
Cyclohexanone, 2-[1-(2-methoxyphenyl)-2-nitroethyl]-	EtOH	209(4.09),273(3.28), 278(3.26)	2-0405-78
Cyclohexanone, 2-[1-(4-methoxyphenyl)-2-nitroethyl]-	EtOH	226(4.02),277(3.17), 283(3.12)	2-0405-78
$C_{15}H_{19}NO_5$			
Cyclohexanone, 2-[1-(4-hydroxy-3-methoxyphenyl)-2-nitroethyl]-	EtOH	210(4.13),282(3.47)	2-0405-78
Cyclopentanone, 2-[1-(3,4-dimethoxyphenyl)-2-nitroethyl]-, (R*,S*)-	EtOH	211(4.16),282(3.80)	2-0405-78
$C_{15}H_{19}N_3OS_2$			
Thiourea, N-7-oxa-3-thia-1-azaspiro-[5.5]undec-1-en-2-yl-N'-phenyl-, (±)-	EtOH	235(4.07),307(4.41)	87-0895-78
$C_{15}H_{19}N_3O_2$			
3-Pyridinecarboxylic acid, 5-cyano-1-cyclohexyl-1,4-dihydro-4-imino-6-methyl-, methyl ester	EtOH	215(3.80),262(3.66), 376(4.2)	73-2024-78
$C_{15}H_{19}N_3O_2S$			
Urea, N-7-oxa-3-thia-1-azaspiro[5.5]undec-1-en-2-yl-N'-phenyl-, (±)-	EtOH	277(4.30)	87-0895-78

Compound	Solvent	$\lambda_{max}(\log \epsilon)$	Ref.
$C_{15}H_{19}N_3O_4$			
1,3,5-Triazine-2,4(1H,3H)-dione, di-hydro-6-(2-hydroxy-3-methoxyphenyl)-1,3-dimethyl-5-(2-propenyl)-	EtOH	225s(4.00),283(3.53)	4-1193-78
$C_{15}H_{19}N_3O_4S$			
Benzenesulfonamide, N-[3-(3,4-dihydro-3-methyl-2,4-dioxo-1(2H)-pyrimidin-yl)propyl]-4-methyl-	$CHCl_3$	266(3.98)	70-0149-78
Benzenesulfonamide, N-[3-(3,6-dihydro-3-methyl-2,6-dioxo-1(2H)-pyrimidin-yl)propyl]-4-methyl-	$CHCl_3$	268.5(3.82)	70-0149-78
Benzenesulfonamide, N-[3-(3,6-dihydro-4-methyl-2,6-dioxo-1(2H)-pyrimidin-yl)propyl]-4-methyl-	$CHCl_3$	260(3.84)	70-0149-78
$C_{15}H_{19}N_3O_6$			
Glutamic acid, N-[4-(methylamino)benz-oylglycyl]-	MeOH	300(3.99)	87-1165-78
$C_{15}H_{19}N_3O_7S$			
Uridine, 6-(aminothioxomethyl)-2',3'-O-(1-methylethylidene)-, 5'-acetate	EtOH	276(4.10)	94-2657-78
$C_{15}H_{19}N_3S_2$			
2H-1,3,5-Thiadiazine-2-thione, 6-[bis-(1-methylethyl)amino]-4-phenyl-	CH_2Cl_2	243(4.24),303(4.55), 415(4.10)	48-0647-78
$C_{15}H_{19}N_5O_4$			
2-Pyrimidinamine, 1-β-D-arabinofurano-syl-1,4-dihydro-4-imino-N-(4-pyri-dinylmethyl)-, monohydrochloride	pH 2	218(4.38),240s(4.26), 273s(3.95)	94-3244-78
$C_{15}H_{19}N_5O_8S$			
9H-Purin-6-amine, 9-[3,5-diacetoxy-2-O-(methylsulfonyl)-β-D-arabino-furanosyl]-	pH 1	256(4.19)	78-1133-78
	neutral	258.5(4.19)	78-1133-78
	pH 13	258.5(4.18)	78-1133-78
$C_{15}H_{19}N_7O_4$			
Acetamide, N-(8-amino-3,4,13,13a-tetra-hydro-2,2-dimethyl-4,13-epoxy-1,3-di-oxolo[6,7][1,3]diazocino[1,2-e]purin-6(5H)-yl)-, [3aR-(3aα,4β,13β,13aα)]-	MeOH	212(4.45),270(4.31)	44-2320-78
$C_{15}H_{20}$			
Benzene, [2-(2-methylcyclohexylidene)-ethyl]-, (E)-	n.s.g.	256(2.45),262(2.54), 269(2.48)	39-0730-78C
(Z)-	n.s.g.	256(2.56),262(2.65), 269(2.57)	39-0730-78C
Bicyclo[8.2.2]tetradeca-5,10,12,13-tetraene, 4-methyl-, (E)-	EtOH	224(3.84),271(2.63), 278(2.58)	35-1806-78
1H-Indene, 2,3-dihydro-5-(1-methyleth-enyl)-2-(1-methylethyl)-	EtOH	248(4.07)	94-1486-78
$C_{15}H_{20}BrClO$			
Intricenyne	EtOH	223(4.12),231s(3.93)	102-0939-78
$C_{15}H_{20}BrN_5S_2$			
Methanimidamide, N'-[5-[[5-(2-bromo-ethyl)-4-methyl-2-thioxo-3(2H)-thi-azolyl]methyl]-2-methyl-4-pyrimidin-	EtOH	269(4.13),318(4.52)	94-3675-78

Compound	Solvent	λ_{max}(log ϵ)	Ref.
yl]-N,N-dimethyl- (cont.)			94-3675-78
$C_{15}H_{20}Br_2O$ Spiro[5.5]undec-7-en-3-ol, 2-bromo- 9-(bromomethylene)-1,1-dimethyl- 5-methylene-	EtOH	243(4.34),250s(4.32)	88-4805-78
$C_{15}H_{20}CuF_{12}N_2O_3$ Copper, [1,1,1,7,7,7-hexafluoro-2,6-di- hydroxy-2,6-bis(trifluoromethyl)-4- heptanonato(2-)-O^2,O^4](N,N,N',N'- tetramethyl-1,2-ethanediamine-N,N')-	MeOH	613(2.03)	23-2369-78
$C_{15}H_{20}F_4N$ Pyrrolidinium, 3,3,4,4-tetrafluoro- 1-(4,4a,5,6,7,8-hexahydro-4a-methyl- 2(3H)-naphthalenylidene)-	H_2O	286(4.46)	88-2557-78
$C_{15}H_{20}F_{12}N_2NiO_3$ Nickel, [1,1,1,7,7,7-hexafluoro-2,6-di- hydroxy-2,6-bis(trifluoromethyl)-4- heptanonato(2-)-O^2,O^4](N,N,N',N'- tetramethyl-1,2-ethanediamine-N,N')-	CH_2Cl_2	518(2.00)	23-2369-78
$C_{15}H_{20}N_2$ Pyrrolo[2,3-b]indole, 1,2,3,3a,8,8a- hexahydro-3a-(3-methyl-2-butenyl)-	EtOH	207(4.02),243s(3.49), 282(3.34)	78-0929-78
$C_{15}H_{20}N_2O$ Acetic acid, 1-(1-cyclohexen-1-yl)- 2-methyl-2-phenylhydrazide	decalin	240(4.21),286(3.30)	33-1364-78
Cyclopenta[3,4]pyrido[1,2-a][1,8]naph- thyridin-12(6H)-one, 1,2,3,4,7,8,9- 10,10b,11-decahydro-	EtOH	235(3.85),308(4.20)	142-1717-78A
2H-Indol-2-one, 1,3-dihydro-3-(2-piper- idinoethyl)-	EtOH	249(3.92)	42-1122-78
8-Isoquinolineethanamine, 5,6-dihydro- 3-methoxy-N,N-dimethyl-β-methylene-	MeOH	262(4.11)	44-0966-78
2,4-Pentadien-1-one, 5-(dimethylamino)- 1-[2-(dimethylamino)phenyl]-	EtOH	403(4.59)	70-0102-78
$C_{15}H_{20}N_2OS$ 6H-[1]Benzothieno[3,2-e]pyrido[1,2-a]- pyrimidin-6-one, 1,2,3,4,4a,5,7,8- 9,10-decahydro-4a-methyl-	EtOH	231(4.43),265s(3.75), 332(3.77)	4-0949-78
$C_{15}H_{20}N_2OS_4$ Urea, bis[2,3-bis(ethylthio)-2-cyclo- propen-1-ylidene]-	EtOH	319.5(4.35)	18-3653-78
$C_{15}H_{20}N_2O_2$ 1-Cyclopentene-1-carboxylic acid, 5-[(aminomethylphenyl)amino]-, ethyl ester	EtOH	250(3.58),306(3.73)	104-0286-78
2H-Indol-2-one, 1,3-dihydro-3-(1-hy- droxy-2-piperidinoethyl)-, mono- perchlorate	EtOH	250(4.18),280(3.41)	42-1122-78
Tryptophan, N^{in}-(1,1-dimethylethyl)-	H_2O	198(4.33),225(3.51), 277s(--),286(3.73), 291s(--)	63-1617-78

Compound	Solvent	$\lambda_{max}(\log \epsilon)$	Ref.
$C_{15}H_{20}N_2O_3$			
1H-Pyrido[2,1-b]quinazoline-6-carboxylic acid, 2,3,4,5,7,8,9,11-octahydro-11-oxo-, ethyl ester	MeOH	219(4.18),265(4.10), 302(4.09)	24-2813-78
$C_{15}H_{20}N_2O_3S$			
1,4,3-Oxathiazin-2-amine, N-hexyl-6-phenyl-, 4,4-dioxide	EtOH	266(4.10)	18-1805-78
$C_{15}H_{20}N_2O_4$			
1H-Cyclohepta[d]pyridazine-2,3-dicarboxylic acid, 4,7-dihydro-, diethyl ester	EtOH	258(3.56)	138-0961-78
1H-Indole-2-propanoic acid, 3-(acetylamino)-4,5,6,7-tetrahydro-4-oxo-, ethyl ester	EtOH	247(3.94),282(3.56)	94-3080-78
7H-Pyrido[2,3-d]azepine-3,7-dicarboxylic acid, 5,6,8,9-tetrahydro-, diethyl ester	EtOH	285(3.79)	142-0267-78C
$C_{15}H_{20}N_2O_5$			
1H-Cyclopenta[d]pyridazine-1,4-dicarboxylic acid, 2,5-dihydro-1-methoxy-5-(1-methylethylidene)-, dimethyl ester	MeOH	208s(3.55),240(3.95), 268s(3.33)	78-2509-78
7H-Pyrido[2,3-d]azepine-3,7-dicarboxylic acid, 1,2,5,6,8,9-hexahydro-2-oxo-, diethyl ester	EtOH	242(3.91),345(4.03)	142-0267-78C
$C_{15}H_{20}N_2O_6$			
7H-Pyrido[2,3-d]azepine-3,7-dicarboxylic acid, 1,2,5,6,8,9-hexahydro-4-hydroxy-2-oxo-, diethyl ester	EtOH	227(4.32),323(4.00)	142-0267-78C
$C_{15}H_{20}N_2O_7$			
Uridine, 6-methyl-2',3'-O-(1-methylethylidene)-, 5'-acetate	EtOH	258(4.05)	94-2657-78
$C_{15}H_{20}N_2O_8$			
Uridine, 6-(hydroxymethyl)-2',3'-O-(1-methylethylidene)-, 5'-acetate	EtOH	260(4.00)	94-2657-78
$C_{15}H_{20}N_2S$			
1H-Imidazole, 1-[2-(butylthio)ethyl]-4-phenyl-	EtOH	256(4.15)	4-0307-78
$C_{15}H_{20}N_2Se$			
Benzenemethanamine, N-(4-ethyl-3,4-dimethyl-5-methylene-2-selenazolidinylidene)-	EtOH	211(3.39)	103-0738-78
$C_{15}H_{20}N_3O_2$			
1,2,4-Triazolidin-1-yl, 4-(1,1-dimethylethyl)-2-(1-methyl-1-phenylethyl)-3,5-dioxo-	benzene	297(3.40),370s(2.99), 410s(2.78)(anom.)	44-0808-78
$C_{15}H_{20}N_3O_5P$			
1H-Imidazole-4-carboxylic acid, 5-[(dimethoxyphosphinyl)amino]-1-(phenylmethyl)-, ethyl ester	EtOH	213(3.93),244(3.95)	130-0421-78

Compound	Solvent	λ_{max}(log ϵ)	Ref.
$C_{15}H_{20}N_4O_{10}$			
1H-Imidazo[4,5-d]pyridazine-4,7-dione,	pH 1	230s(4.08),275(3.70)	4-0001-78
5,6-dihydro-1,5(or 6)-di-β-D-ribo-	pH 7	232(4.18),238(4.04),	4-0001-78
furanosyl-		290(3.70)	
	pH 11	233(4.15),238(4.04),	4-0001-78
		290(3.70)	
$C_{15}H_{20}O$			
5H-Benzocycloheptene, 6,7,8,9-tetrahy-	EtOH	244(3.91)	44-1569-78
dro-2-methoxy-9,9-dimethyl-5-methyl-			
ene-			
Cyclopenta[2,3]cyclopropa[1,2-a]cyclo-	EtOH	220(3.74),280(3.40)	44-0343-78
propa[c]benzen-6(1H)-one, 2,2a,3,3a-			
6a,6b-hexahydro-3,3,6a,6b-tetrameth-			
yl-, [2aR-(2aα,3aα,3bR*,6aβ,6bα)]-			
5-Hexen-3-one, 2,2,4-trimethyl-6-phen-	C_6H_{12}	253(4.24),285(3.40),	39-0155-78B
yl-, (E)-		293(3.34)	
5-Hexen-3-one, 2,4,4-trimethyl-5-phen-	EtOH	254(4.10),287(3.42),	35-1791-78
yl-		295(3.40)	
5-Hexen-3-one, 2,4,4-trimethyl-6-phen-	EtOH	250(4.03),285(3.17),	35-1778-78
yl-, cis		294(3.11)	
$C_{15}H_{20}OS$			
Spiro[6H-cyclohepta[b]thiophene-6,1'-	n.s.g.	258(3.98)	42-0847-78
cyclopentan]-4(5H)-one, 2-ethyl-			
7,8-dihydro-			
$C_{15}H_{20}O_2$			
Azulene-6,7-dicarbolactone, 1,2,3,3a-	EtOH	284(3.86)	35-6728-78
4,8a-hexahydro-2,2,4-trimethyl-			
Azulene-6,7-dicarbolactone, 1,2,3,3a-	EtOH	281(4.14)	35-6728-78
8,8a-hexahydro-2,2,4-trimethyl-			
Costunolide	MeOH	210(4.05)(end abs.)	36-0347-78
2,5-Cyclohexadiene-1,4-dione, 2-(1,5-	MeOH	253(4.01)	78-1661-78
dimethyl-4-hexenyl)-5-methyl-, (R)-			
(curcuquinone)			
2-Cyclohexen-1-one, 4-[4-(2-oxopropyl)-	EtOH	303(4.27)	39-0024-78C
cyclohexylidene]-			
4H-Cyclopropa[3,4]cyclohepta[1,2-c]pyr-	n.s.g.	217(3.31)	88-1553-78
an-4-one, 4a,5,6,7,7a,8,8a,8b-octahy-			
dro-1,8,8-trimethyl-5-methylene-			
(plagiochilide)			
α-Cyperone, 2-oxo-	EtOH	259.5(4.26)	138-1005-78
Furanoeremophilone, (±)-	EtOH	281(4.17)	94-3704-78
10αH-Furanoeremophilone, (±)-	EtOH	279(4.19)	94-3704-78
1,4-Hexanedione, 3,3,5-trimethyl-1-	EtOH	247(3.87)	35-1791-78
phenyl-			
1H-Indene-5-acetic acid, 2,3-dihydro-	MeOH	264(2.99),271(3.17),	94-1486-78
α-methyl-2-(1-methylethyl)-		277(3.24)	
1H-Inden-1-one, 2,3-dihydro-6-(2-hy-	EtOH	217(4.48),260(4.12),	94-2365-78
droxyethyl)-2,2,5,7-tetramethyl-		304(3.36)	
(pterosin Z)	MeOH	215(3.9),259(3.6),	102-0275-78
		305(2.9)	
1H-Inden-1-one, 2,3-dihydro-6-(2-meth-	EtOH	218(4.52),260(4.20),	94-2365-78
oxyethyl)-2,5,7-trimethyl-, (R)-		305(3.42)	
(pterosin O)			
Inunolide	EtOH	210(4.16)	2-0027-78
$C_{15}H_{20}O_3$			
Costunolide 1,10-epoxide	MeOH	end absorption	36-0347-78
Cyclodeca[b]furan-2(3H)-one, 3a,4,5,8-	isoPrOH	211(3.92)	42-1142-78

Compound	Solvent	$\lambda_{max}(\log \epsilon)$	Ref.
9,11a-hexahydro-6-(hydroxymethyl)-10-methyl-3-methylene- (cont.)			42-1142-78
Cyclohexaneacetic acid, 4-(4-oxo-2-cyclohexen-1-ylidene)-, methyl ester	EtOH	225(3.45),305(3.78)	39-0024-78C
Diplosporin monopropanoate	MeOH	217(4.02),250(4.08)	119-0111-78
3-Epieuryopsonol, (±)-	EtOH	279(4.16)	94-3704-78
Furanoeremophilane, 3β-hydroxy-9-oxo-	EtOH	281(4.21)	94-3704-78
Furanoeremophilane, 10β-hydroxy-3-oxo-	EtOH	219(3.65)	18-3335-78
1H-Inden-1-one, 2,3-dihydro-6-(2-hydroxyethyl)-2-(hydroxymethyl)-2,5,7-trimethyl- (pterosin A)	EtOH	217(4.54),261(4.19), 305(3.39)	94-2365-78
1H-Inden-1-one, 2,3-dihydro-3-hydroxy-6-(2-hydroxyethyl)-2,2,5,7-tetramethyl-	EtOH	217(4.51),260(4.23), 302(3.28)	94-2365-78
Naphtho[1,2-b]furan-2,8(3H,4H)-dione, 3a,5,5a,6,7,9b-hexahydro-3,5a,9-trimethyl-	EtOH	247(4.23)	2-0016-78
Parthenolide	MeOH	210(3.97)(end abs.)	36-0347-78
	EtOH	215(4.02)	42-1152-78
Reynosin	MeOH	210(3.96)(end abs.)	36-0347-78
Santamarine	MeOH	210(4.30)(end abs.)	36-0347-78
Spiro[cyclohexane-1,1'-[1H]indene]-4,7'-diol, 2',3'-dihydro-5'-methoxy-	MeOH	277(2.93),280(2.93), 284(2.95)	94-3641-78
$C_{15}H_{20}O_4$			
Costunolide diepoxide	MeOH	210(3.89)(end abs.)	36-0347-78
Cyclohexaneacetic acid, 4-hydroxy-4-(4-methoxyphenyl)-	EtOH	224(4.05),269s(3.11), 274(3.20),281(3.15)	39-0024-78C
Epipterosin L	EtOH	216(4.54),258(4.18), 301(3.31)	94-2365-78
Epoxysantamarine	MeOH	210(4.16)(end abs.)	36-0347-78
Furanoeremophilane, 3β,10β-dihydroxy-9-oxo-	MeOH	283(3.94)	18-3335-78
2-Naphthaleneacetic acid, 3,4,4a,5,6-7,8,8a-octahydro-7-hydroxy-8,8a-dimethyl-α-methylene-3-oxo-	MeOH	218(3.92),243(3.96)	138-0301-78
1-Propanone, 1-(3,4-dihydro-5,7-dihydroxy-2,2-dimethyl-2H-1-benzopyran-6-yl)-2-methyl-	EtOH-acid EtOH-base	228(4.16),293(4.31) 244s(--),303(4.28), 375(3.70)	39-1303-78C 39-1303-78C
1-Propanone, 1-(3,4-dihydro-5,7-dihydroxy-2,2-dimethyl-2H-1-benzopyran-8-yl)-2-methyl-	EtOH-acid EtOH-base	228s(--),294(4.23) 245(3.91),332(4.47)	39-1303-78C 39-1303-78C
Pterosin L	EtOH	218(4.51),259(4.15), 301(3.30)	94-2365-78
$C_{15}H_{20}O_6$			
Ethanone, 1-[3-(1,2-dihydroxy-2-methylpropyl)-2,3-dihydro-4-hydroxy-6-methoxy-5-benzofuranyl]-, [R-(R*,S*)]-	MeOH	223(4.03),243(4.08), 303(3.99),364(3.89)	119-0047-78
$C_{15}H_{20}O_9$			
Benzaldehyde, 4-(β-D-glucopyranosyloxy)-3,5-dimethoxy-	EtOH	285(4.34)	100-0056-78
$C_{15}H_{20}S$			
2H-1-Benzothiopyran, octahydro-2-phenyl-	hexane	191(4.7),206(4.1), 246s(2.4),253(2.4), 259(2.4),265(2.2)	103-0605-78
$C_{15}H_{21}BrNO_2$			
3-Oxa-6-azoniaspiro[5.5]undecane, 2-(4-	EtOH	262(2.81)	145-0595-78

Compound	Solvent	$\lambda_{max}(\log \epsilon)$	Ref.
bromophenyl)-2-hydroxy-, bromide (cont.)			145-0595-78
$C_{15}H_{21}BrN_2O_4$ 3-Oxa-6-azoniaspiro[5.5]undecane, 2-hy- droxy-2-(4-nitrophenyl)-, bromide	EtOH	259(3.00)	145-0595-78
$C_{15}H_{21}Br_2ClO$ 2H-Benzocyclohepten-2-one, 3,6-dibromo- 7-chloro-3,4,4a,5,6,7,8,9-octahydro- 1,4,4a,7-tetramethyl-	n.s.g.	250(4.20)	88-3931-78
$C_{15}H_{21}ClO_2$ 8H-Indeno[1,2-c]furan-8-one, 1-chloro- 1,3,3a,3b,4,5,6,8a-octahydro- 3,3,3b,4-tetramethyl-	pet ether	248(4.00),343(1.90)	83-0511-78
$C_{15}H_{21}FN_2O_8S$ 2,4(1H,3H)-Pyrimidinedione, 1-[6-deoxy- 6-fluoro-2,3-O-(1-methylethylidene)- 4-O-(methylsulfonyl)-β-D-manno- pyranosyl]-5-methyl-	MeOH	264(3.98)	48-0157-78
$C_{15}H_{21}NO_2$ Benzo[f]quinoline-3-methanol, 1,2,3,4- 4a,5,6,10b-octahydro-8-methoxy-, [3S- (3α,4aα,10bα)]-	EtOH	221(3.93),278(3.32), 287(3.30)	78-1023-78
$C_{15}H_{21}NO_4$ Cyclopentaneacetic acid, 2-(1-cyano- 2-ethoxy-2-oxoethylidene)-α-methyl-, ethyl ester	EtOH	240(4.12)	2-0860-78
5,1-(Epoxyimino)tetrahydrosantonin, (1S,4R,5R)-	EtOH	294(1.23)	44-0536-78
$C_{15}H_{21}NS$ 2H-Isoindole, 1-[(1,1-dimethylethyl)- thio]-2-propyl-	isooctane	320s(--),333(3.97), 348s(--)	44-2886-78
$C_{15}H_{21}N_2O_4$ 3-Oxa-6-azoniaspiro[5.5]undecane, 2-hy- droxy-2-(4-nitrophenyl)-, bromide	EtOH	259(3.00)	145-0595-78
$C_{15}H_{21}N_3O$ 3-Piperidinone, 4-[(2-aminophenyl)imi- no]-2,2,6,6-tetramethyl-	EtOH	238(4.50),308(3.85), 322(3.91)	22-0621-78
	EtOH-NaOH	236(4.54),308(3.85), 332(3.95)	22-0621-78
$C_{15}H_{21}N_3O_5$ 5,1-(Epoxynitrosoimino)tetrahydrosanto- nin oxime, (1S,4R,5R)-	EtOH	246(3.88)	44-0536-78
(1S,4S,5R)-	EtOH	245(3.86)	44-0536-78
$C_{15}H_{21}N_3S$ Pyrido[3,2-c]pyridazine, 4-[(1,1,3,3- tetramethylbutyl)thio]-	heptane	235(4.87),258(4.94), 280(4.52),320(4.61), 350(4.81)	103-1032-78
$C_{15}H_{21}N_5O$ Pyrimido[4,5-b][1,8]naphthyridin-5(7H)- one, 10-[2-(dimethylamino)ethyl]-	EtOH	240(4.37),275(3.37), 362(4.03)	103-1261-78

Compound	Solvent	$\lambda_{max}(\log \epsilon)$	Ref.
6,8,9,10-tetrahydro-9-methyl- (cont.)			103-1261-78
$C_{15}H_{21}N_5O_5$			
9H-Purin-6-amine, 9-[5-O-(2,2-dimethyl-1-oxopropyl)-β-D-arabinofuranosyl]-	MeOH	259(4.17)	87-1218-78
9H-Purin-6-amine, 9-[5-O-(3-methyl-1-oxobutyl)-β-D-arabinofuranosyl]-	MeOH	258(4.17)	87-1218-78
$C_{15}H_{22}$			
Anastreptene	MeCN	190(3.94),215s(3.65), 235(3.43)	78-0041-78
Calamenene, (+)-(1R,4R)-	n.s.g.	224(3.11),270(2.79), 279(2.83)	39-1267-78C
1H-Indene, 2,3-dihydro-2,3-bis(1-methylethyl)-	MeOH	263(3.03),271(3.21), 277(3.28)	94-1486-78
Isopatchoula-3,5-diene	n.s.g.	255(3.96)	2-0148-78
$C_{15}H_{22}NO_2$			
3-Oxa-6-azoniaspiro[5.5]undecane, 2-hydroxy-2-phenyl-, bromide	EtOH	255(2.34)	145-0595-78
$C_{15}H_{22}N_2O$			
Azepino[3,2-b]indol-2(1H)-one, 3,4,5,6-7,8,9,10-octahydro-4,4,6-trimethyl-	MeOH	222(4.16),258s(3.87), 298s(3.59)	24-0853-78
Azepino[4,3-b]indol-1(2H)-one, 3,4,5,6-7,8,9,10-octahydro-4,4,6-trimethyl-	MeOH	211(4.28),245s(3.90), 277s(3.61)	24-0853-78
4H-Carbazol-4-one, 1,2,3,5,6,7,8,9-octahydro-2,2,9-trimethyl-, oxime	MeOH	217(4.14),252(4.06)	24-0853-78
13,14-Dehydrosophorodine	MeOH	253(3)	105-0190-78
Sophocarpine stereoisomer	EtOH	256(3.46)	94-2483-78
$C_{15}H_{22}N_2O_2$			
5,17-Dehydromatrine, N-oxide, (+)-	EtOH	237(4.18)	102-2021-78
8-Isoquinolinol, 8-[1-[(dimethylamino)-methyl]ethenyl]-5,6,7,8-tetrahydro-3-methoxy-	MeOH	276(3.59)	44-0966-78
$C_{15}H_{22}N_2O_4$			
Acetamide, N-[[2-(acetylamino)-4,5-dimethoxyphenyl]methyl]-N-ethyl-	MeOH	250s(--),282(3.7)	5-0608-78
$C_{15}H_{22}N_2O_4S$			
Phenol, 2-methoxy-4-nitro-6-[2-(3,4,4-trimethyl-2-thiazolidinyl)ethyl]-	EtOH	215(4.07),248(4.19), 279s(3.60),340(4.22)	4-1439-78
$C_{15}H_{22}N_6O_3$			
Acetamide, N-[3-[6-(dimethylamino)-9H-purin-9-yl]-2-hydroxy-5-(hydroxymethyl)cyclopentyl]-	pH 1	268(4.26)	44-2311-78
	H_2O	276(4.26)	44-2311-78
	pH 13	276(4.27)	44-2311-78
$C_{15}H_{22}N_6O_5$			
Adenosine, N-[(butylamino)carbonyl]-	pH 1	277(4.40)	36-0569-78
	pH 13	269(4.34),277(4.29), 298(4.22)	36-0569-78
	70% EtOH	269(4.38),276(4.30)	36-0569-78
$C_{15}H_{22}O$			
1(2H)-Azulenone, 3,3a,4,5,6,7-hexahydro-3a,8-dimethyl-5-(1-methylethylidene)-	EtOH	248(4.14)	88-4205-78
Bicyclo[7.2.0]undec-4-en-3-one, 4,11,11-	MeOH	235(3.67)	94-2543-78

Compound	Solvent	$\lambda_{max}(\log \epsilon)$	Ref.
trimethyl-8-methylene- (cont.)			94-2543-78
Buddledin C	MeOH	236(4.04)	94-2543-78
Calamenene, 7-hydroxy-, (1R,4R)-	n.s.g.	280(3.40)	39-1267-78C
2,4,6-Cycloheptatrien-1-one, 4,5-di-methyl-2,7-bis(1-methylethyl)-	MeOH	240(4.32),334(3.86)	35-1778-78
2,5-Cyclohexadien-1-one, 2,6-bis(1,1-dimethylethyl)-4-methylene-	hexane	286(4.79)	70-0448-78
Phenol, 2-(1,5-dimethyl-4-hexenyl)-5-methyl- (curcuphenol)	MeOH	217(3.67),276(3.38)	78-1661-78
	MeOH-base	237(3.72),290(3.55)	78-1661-78
Vulgarone A	EtOH	292(1.94)	78-2893-78
Vulgarone B	EtOH	252(3.91)	78-2893-78
$C_{15}H_{22}O_2$			
Azuleno[4,5-b]furan-2(3H)-one, decahydro-6,9-dimethyl-3-methylene-	EtOH	211(4.08)	2-0539-78
2(3H)-Benzofuranone, 6-ethylhexahydro-6-methyl-3-methylene-7-(1-methylethenyl)-	EtOH	209(4.11)	2-0539-78
Bicyclo[7.2.0]undec-4-en-3-one, 2-hydroxy-4,11,11-trimethyl-8-methylene- (buddledin B)	MeOH	237(3.97)	94-2535-78
1,4-Butanediol, 2-(4,8-dimethyl-3,7-nonadien-5-ynylidene)-, (Z,E)-	EtOH	255s(4.26),267(4.36), 283(4.26)	88-3593-78
2-Cyclohexen-1-one, 4-[4-(2-hydroxy-propyl)cyclohexylidene]-	EtOH	234(3.70),305(4.25)	39-0024-78C
5α-Eremophil-7-ene-2,8-dione, (4βH-10βH)-	EtOH	252.0(4.61)	95-1441-78
1H-Indene-5-ethanol, 2,3-dihydro-2-(hydroxymethyl)-2,4,6-trimethyl-	EtOH	251(2.27),256(2.49), 263(2.74),268(2.84), 272(2.98),277(2.93), 282(3.03)	94-2365-78
Inunolide, dihydro-	EtOH	209(3.88)	2-0027-78
Ligularenolide, tetrahydro-	MeOH	222(4.38)	2-0027-78
8-epi-	MeOH	220(4.08)	2-0027-78
Naphtho[1,2-b]furan-2(3H)-one, decahydro-5a,9-dimethyl-3-methylene-	EtOH	212(4.10)	2-0539-78
Naphtho[1,2-b]furan-2(4H)-one, 5,5a,6-7,8,9,9a,9b-octahydro-3,5a,9-trimethyl-	EtOH	216(4.22)	2-0539-78
6-epi-	EtOH	220(4.08)	2-0539-78
Neoalantolactone	EtOH	277(4.33)	2-0027-78
epimer	EtOH	221(4.17)	2-0027-78
10-Oxabicyclo[7.2.1]dodeca-1(12),4-dien-11-one, 5-methyl-8-(1-methylethyl)-	EtOH	217.5(3.94)	88-4749-78
2-Penten-4-yn-1-ol, 5-(2,2,6-trimethyl-7-oxabicyclo[4.1.0]hept-1-yl)-, (Z)-(±)-	EtOH	231.5(4.00)	39-1511-78C
$C_{15}H_{22}O_3$			
Cyclodeca[b]furan-2(4H)-one, 7,8,9,10-11,11a-hexahydro-11a-hydroxy-3,6,10-trimethyl-	EtOH	218(4.07)	83-0754-78
Deodardione	EtOH	269(3.85)	78-0599-78
Furanoeremophilane, 3β,10β-dihydroxy-	MeOH	220(3.76)	18-3335-78
8H-Indeno[1,2-c]furan-8-one, 1,3,3a,3b-4,5,6,8a-octahydro-1-hydroxy-3,3,3b,4-tetramethyl-	pet ether	247(3.81),342(1.52)	83-0511-78
1-Naphthalenecarboxylic acid, 2,3,4,4a-5,6,7,8-octahydro-6,6-dimethyl-2-oxo-, ethyl ester	EtOH	238(3.82)	32-0693-78

Compound	Solvent	$\lambda_{max}(\log \epsilon)$	Ref.
Oxireno[8,9]cyclodeca[1,2-b]furan-8(2H)-one, 1a,3,6,6a,7,9a,10,10a-octahydro-4,7,10a-trimethyl-	EtOH	214(2.80)	2-0027-78
Propanoic acid, 2,2-dimethyl-, 3-(1,1-dimethylethyl)-4-hydroxyphenyl ester	EtOH	202(1.29),228(0.74), 279(0.41)	117-0079-78
Spiro[1,3-dioxane-2,4'-[4H]inden]-2'(5'H)-one, 3',3'a,6',7'-tetrahydro-3'a,5,5-trimethyl-	EtOH	235(4.42)	78-2201-78
$C_{15}H_{22}O_4$			
Bisoxireno[4,5:8,9]cyclodeca[1,2-b]-furan-4(1aH)-one, decahydro-1a,5,7a-trimethyl-	EtOH	212(2.39)	2-0027-78
2-Cyclohexene-1-carboxylic acid, 2-(3-hydroxy-1-butenyl)-1,3-dimethyl-4-oxo-, ethyl ester	EtOH	292(4.06)	104-0660-78
2-Cyclopentene-1-propanoic acid, 1-(ethoxycarbonyl)-2-(1-methyl-ethenyl)-, methyl ester	EtOH	236(4.11)	39-1606-78C
Furanoeremophilane, 3β,6β,14-trihydroxy-	EtOH	218(3.93)	138-1313-78
Furanoeremophilane, 3β,9β,10β-trihydroxy-	MeOH	224(3.62)	18-3335-78
$C_{15}H_{23}FOSi$			
Silane, benzoylbis(1,1-dimethylethyl)-fluoro-	C_6H_{12}	402(2.10),422(2.27), 443(2.12)	112-0751-78
$C_{15}H_{23}NO_4$			
2,5-Cyclohexadien-1-ol, 1-(1,1-dimethylethyl)-3,4,5-trimethyl-4-nitro-, acetate	MeOH	225(3.20)	23-1758-78
isomer	MeOH	225(3.11)	23-1758-78
3,5-Pyridinedicarboxylic acid, 1-ethyl-1,4-dihydro-2,6-dimethyl-, diethyl ester	EtOH	232(4.31),267(4.13), 360(3.95)	103-1226-78
$C_{15}H_{23}NO_5$			
DL-allo-Octonic acid, 3,6-anhydro-2,8-dideoxy-2-[(dimethylamino)methylene]-7-C-methyl-4,5-O-(1-methylethylidene)-, ζ-lactone	MeOH	300(4.20)	88-4403-78
$C_{15}H_{23}NO_8$			
Propanedioic acid, [[(ethoxycarbonyl)-(2-ethoxy-2-oxoethyl)amino]methylene]-, diethyl ester	EtOH	216(3.90),262(4.05)	94-2224-78
$C_{15}H_{23}N_3O_2$			
5-Oxazolamine, 4-ethyl-N-(4-ethyl-2-methyl-5-oxazolyl)-2-methyl-N-propyl-	n.s.g.	227(4.11)	70-0963-78
$C_{15}H_{23}N_3O_6$			
Hexanamide, N-(1-β-D-arabinofuranosyl)-1,2-dihydro-2-oxo-4-pyrimidinyl)-	isoPrOH	216(4.22),248(4.18), 303(3.91)	94-0981-78
$C_{15}H_{23}N_3O_7$			
α-D-Glucofuranose, 3-amino-3-deoxy-3-C-(1-diazo-2-methoxy-2-oxoethyl)-1,2:5,6-bis-O-(1-methylethylidene)-	MeOH	213(3.73),269(3.88)	136-0039-78A

Compound	Solvent	λ_{max}(log ϵ)	Ref.
C₁₅H₂₃N₆O₈P			
5'-Adenylic acid, N-[(butylamino)carbo-nyl]-, barium salt	pH 1	270(4.31),277(4.29)	36-0569-78
	pH 13	280(sic)(4.22),278(4.21), 298(4.00)	36-0569-78
	70% EtOH	269(4.32),277(4.24)	36-0569-78
C₁₅H₂₃N₇O₄			
Adenosine, 8-[[(diethylamino)methyl-ene]amino]-	H₂O	234(4.24),306(4.34)	5-1365-78
C₁₅H₂₃O			
Phenoxy, 2,6-bis(1,1-dimethylethyl)-4-methyl-	MeCN	400(3.3),620(2.7)	70-1304-78
C₁₅H₂₄			
1H-Indene, 5-ethenylhexahydro-1,3a,4,7a-tetramethyl- (α-pinguisene)	EtOH	237(3.85)	102-0457-78
Naphthalene, 3,4,4a,5,6,7-hexahydro-4a,8-dimethyl-2-(1-methylethyl)- (δ-selinene)	EtOH	248(4.16)	44-1613-78
Naphthalene, 1,2,3,4,4a,5,6,7-octahy-dro-4a,8-dimethyl-1-(1-methylethyl-idene)- (+-selina-4,7(11)-diene)	EtOH	218(3.68)	44-1613-78
1,3-Pentadiene trimer	n.s.g.	232(5.17)	70-1241-78
C₁₅H₂₄NO₅PS₃			
Phosphorothioic acid, O,O-bis(1-methyl-propyl) S-[(2,3,5,7-tetrahydro-5,7-dioxo-6H-1,4-dithiino[2,3-c]pyrrol-6-yl)methyl] ester	EtOH	214(3.60),265(3.92), 420(3.38)	73-1093-78
Phosphorothioic acid, O,O-dibutyl S-[(2,3,5,7-tetrahydro-5,7-dioxo-6H-1,4-dithiino[2,3-c]pyrrol-6-yl)-methyl] ester	EtOH	213(3.70),264(3.96), 420(3.46)	73-1093-78
C₁₅H₂₄N₂O			
1,3,6,8-Nonatetraen-5-one, 1,9-bis(di-methylamino)-2,8-dimethyl-	EtOH	500(4.70)	70-0107-78
C₁₅H₂₄N₃PS₂			
Phosphorodiamidothious acid, tetraeth-yl-, 2-benzothiazolyl ester	dioxan	<u>280(4.0)</u>	65-1352-78
C₁₅H₂₄N₄O₄			
[4,4'-Bi-3H-pyrazole]-3,3'-dicarboxylic acid, 4,4',5,5'-tetrahydro-3,5,5,5'-5'-pentamethyl-, dimethyl ester	EtOH	326(2.63)	24-2206-78
C₁₅H₂₄O			
Cyclohexanol, 1,3,3-trimethyl-2-(3-methyl-2-methylene-3-butenylidene)-, (E)-	EtOH	232(4.22)	33-2681-78
(Z)-	EtOH	230(4.17)	33-2681-78
2H-Inden-2-one, 1,4,5,6,7,7a-hexahydro-1,3-bis(1-methylethyl)-	EtOH	241(4.11)	35-1799-78
4H-Inden-4-one, 1,2,3,5,6,7-hexahydro-3,5,5-trimethyl-7-(1-methylethyl)-, (3R-trans)- (brasilenone)	ether	242(4.02)	78-2077-78
Spiro[5.5]undec-1-en-3-one, 7-(1-meth-ylethyl)-10-methyl-	EtOH	239(3.94)	88-2127-78
isomer	EtOH	241(4.05)	88-2127-78

Compound	Solvent	λ_{max} (log ϵ)	Ref.
$C_{15}H_{24}O_2$			
Aubergone	n.s.g.	229(3.89)	138-1209-78
2(3H)-Benzofuranone, 6-ethylhexahydro-6-methyl-3-methylene-7-(1-methylethyl)-, [3aS-(3aα,6α,7β,7aβ)]-	EtOH	209(3.93)	2-0539-78
2(4H)-Benzofuranone, 6-ethyl-5,6,7,7a-tetrahydro-3,6-dimethyl-7-(1-methylethyl)-	EtOH	216(4.18)	2-0539-78
Cyclodeca[b]furan-2(3H)-one, decahydro-6,10-dimethyl-3-methylene-	EtOH	215(3.95)	2-0539-78
Cyclodeca[b]furan-2(4H)-one, 5,6,7,8,9-10,11,11a-octahydro-3,6,10-trimethyl-	EtOH	218(4.10)	2-0539-78
2,5-Cyclohexadien-1-one, 2,6-bis(1,1-dimethylethyl)-4-hydroxy-4-methyl-	MeOH	236(3.997),270s(3.297)	149-0683-78A
Ethanone, 1-(2-butyl-5-pentylfuranyl)-	C_6H_{12} MeOH	273(3.76) 280(3.74)	44-4081-78 44-4081-78
4β,5α-Eudesm-1-en-3-one, 11-hydroxy-	n.s.g.	228(3.76)	138-1209-78
2(3H)-Naphthalenone, 4,4a,5,6,7,8-hexahydro-1-methoxy-4a-methyl-7-(1-methylethyl)-, (4aR-trans)-	EtOH	257(3.93)	35-1263-78
Spiro[4.5]dec-6-en-8-one, 7-methoxy-1-methyl-4-(1-methylethyl)-	EtOH	266(3.90)	23-1628-78
$C_{15}H_{24}O_2S$			
Benzenesulfinic acid, 4-methyl-, octyl ester	hexane	201(4.09),224(3.96), 254(3.52)	118-0441-78
$C_{15}H_{24}O_3$			
2,5-Cyclohexadien-1-one, 2,6-bis(1,1-dimethylethyl)-4-hydroperoxy-4-methyl-	MeOH	236(3.997),270(3.299), 369(1.364)	149-0683-78A
1,2-Cyclohexanediol, 2-(5-hydroxy-3-methyl-3-penten-1-ynyl)-1,3,3-trimethyl-, cis-erythro	EtOH	229(4.10)	39-1511-78C
cis-threo	EtOH	229(4.10)	39-1511-78C
trans-threo	EtOH	229.5(4.11)	39-1511-78C
2-Cyclohexen-1-one, 3-(3-acetoxybutyl)-2,4,4-trimethyl-	EtOH	247(4.15)	33-2328-78
1-Cyclopentene-1-octanoic acid, 2-methyl-5-oxo-, methyl ester	MeOH	237(4.01?)	2-0280-78
2-Cyclopenten-1-one, 5-acetoxy-2-(2-ethylhexyl)-	EtOH	233(3.99)	107-0155-78
2-Cyclopenten-1-one, 5-acetoxy-2-octyl-	EtOH	231(3.94)	107-0155-78
3(2H)-Furanone, 2,4-bis(1,1-dimethylethyl)-2-(2-oxopropyl)-	EtOH	271(4.2)	88-3597-78
2-Furanpropanoic acid, 3,4-dimethyl-5-pentyl-, methyl ester	MeOH	225(3.87)	31-0299-78
3-Octanone, 1-(1,4-dioxaspiro[4.4]non-6-ylidene)-, (E)-	MeOH	210(3.27)(end abs.)	39-0209-78C
1-Octen-3-one, 1-(1,4-dioxaspiro[4.4]non-6-yl)-, (E)-	MeOH	232.5(4.03)	39-0209-78C
$C_{15}H_{24}O_4$			
1-Nonanone, 1-(3-furanyl)-6,7-dihydroxy-4,8-dimethyl-	MeOH	252(3.41)	102-0317-78
2-Penten-4-yn-1-ol, 3-methyl-5-(1,2-dihydroxy-2,6,6-trimethylcyclohexyl)-, acetate	EtOH	229(4.11)	39-1511-78C
Unidentified compd.	EtOH	244(3.66)	39-1606-78C
$C_{15}H_{24}S_4$			
3,3'(4H,4'H)-Spirobi[2H,6H-cyclopenta-	CH Cl	232(3.05)	49-1017-78

Compound	Solvent	$\lambda_{max}(\log \epsilon)$	Ref.
[b][1,4]dithiepin], octahydro-, [3(5'aS*,8'aS*),5α,8β]- (cont.)			49-1017-78
$C_{15}H_{25}FN_4O_4$ 4-Pyrimidinamine, 1-β-D-arabinofurano-syl-5-fluoro-1,2-dihydro-N-propyl-2-(propylimino)-, monohydrochloride	pH 1	220(4.24),253(4.22), 285s(3.93)	44-4200-78
$C_{15}H_{25}N$ Cyclohexanamine, N-(4,4,6-trimethyl-2-cyclohexen-1-ylidene)-	C_6H_{12}	224(4.36),282(2.48)	33-1025-78
$C_{15}H_{25}NO_6$ Propanedioic acid, [[[2-(1,1-dimethyl-ethoxy)-2-oxoethyl]methylamino]meth-ylene]-, diethyl ester	EtOH	282(4.26)	94-2224-78
$C_{15}H_{25}N_2O$ Methanaminium, N-[9-(dimethylamino)-5-ethoxy-2,4,6,8-nonatetraenyli-dene]-N-methyl-, tetrafluoroborate	CH_2Cl_2	600(5.18)	70-0339-78
$C_{15}H_{26}N_2O_4$ 1,5-Pentanediaminium, N,N'-bis(1-form-yl-2-hydroxyethenyl)-N,N,N',N'-tetra-methyl-, dihydroxide, bis(inner salt)	EtOH	257(4.69)	73-1261-78
$C_{15}H_{26}O_3$ 1,2-Cyclohexanediol, 2-(5-hydroxy-3-methyl-1,3-pentadienyl)-1,3,3-tri-methyl-, [1α,2β,2(1E,3E)]-(±)-	EtOH	236(4.17)	39-1511-78C
[1α,2β,2(1E,3Z)]-(±)-	EtOH	237.5(4.15)	39-1511-78C
2-Cyclohexen-1-one, 3-(3,5-dihydroxy-3-methylpentyl)-2,4,5-trimethyl-	EtOH	249(4.21)	44-4220-78
$C_{15}H_{27}BrSn$ Stannacyclohexa-2,5-diene, 4-(2-bromo-ethyl)-1,1-dibutyl-	C_6H_{12}	213(3.48)	101-0247-78E
$C_{15}H_{27}N$ 1-Propanamine, N-(2,4-diethyl-2,4-octa-dienylidene)-	EtOH	226(3.85)	23-0302-78
Pyridine, 3,5-diethyl-1,2-dihydro-1,2-dipropyl-	EtOH	350(3.40)	23-0302-78
$C_{15}H_{27}N_3O_2$ 1H-1,4-Diazepine-5,7(2H,6H)-dione, 3-(dimethylamino)-2,2-dimethyl-6,6-bis(1-methylethyl)-	CH_2Cl_2	259(3.87)	33-3050-78
4H-Imidazole-2-acetic acid, 5-(dimeth-ylamino)-α,α-diethyl-4,4-dimethyl-, ethyl ester	CH_2Cl_2	275.5(3.99)	33-3050-78
$C_{15}H_{28}Si$ Silane, (5,8-dimethyl-2,4,9-decatrien-yl)trimethyl-	EtOH	236(4.12)	65-1073-78
$C_{15}H_{29}N$ 2,4-Octadien-1-amine, 2,4-diethyl-N-propyl-	EtOH	220(3.79)	23-0302-78

Compound	Solvent	$\lambda_{max}(\log \epsilon)$	Ref.
$C_{15}H_{30}O_2Sn_2$			
Stannane, triethyl[3-oxo-3-[(triethylstannyl)oxy]-1-propenyl]-	MeCN	215(3.86)	65-1244-78
$C_{15}H_{30}Si_3$			
Silane, 1,3-hexadien-5-yne-1,3,6-triyltris[trimethyl-	EtOH	247(3.81),255(3.93), 286(4.29)	70-2160-78

Compound	Solvent	$\lambda_{max}(\log \epsilon)$	Ref.
$C_{16}H_5BrCl_4O_3$			
Spiro[1,3-benzodioxole-2,1'(2'H)-naphthalen]-2'-one, 6'-bromo-4,5,6,7-tetrachloro-	EtOH	218(4.72),248(4.48), 307(3.77),330(3.73)	2-0668-78
Spiro[1,3-benzodioxole-2,2'(1'H)-naphthalen]-1'-one, 6'-bromo-4,5,6,7-tetrachloro-	EtOH	220(4.64),248(4.54), 306(3.97)	2-0668-78
$C_{16}H_5Cl_4NO_5$			
Spiro[1,3-benzodioxole-2,1'(2'H)-naphthalen]-2'-one, 4,5,6,7-tetrachloro-6'-nitro-	EtOH	220(4.68),238(4.65), 243(4.62),295(4.15), 301(4.14)	2-0668-78
$C_{16}H_6Cl_2O_3$			
Anthra[9,1-bc]pyran-2,7-dione, 3,6-dichloro-	CHCl$_3$	529(4.01),566(4.07), 600(3.89)	104-1433-78
$C_{16}H_7BrO_3$			
Anthra[9,1-bc]pyran-2,7-dione, 3-bromo-	CHCl$_3$	543(4.02),580(4.08), 631(3.83)	104-1433-78
$C_{16}H_7ClO_3$			
Anthra[9,1-bc]pyran-2,7-dione, 3-chloro-	CHCl$_3$	513(3.82),545(3.97), 585(3.82)	104-1433-78
Anthra[9,1-bc]pyran-2,7-dione, 6-chloro-	CHCl$_3$	543(4.03),580(4.09), 630(3.84)	104-1433-78
$C_{16}H_8F_4N_2O_4$			
Benzene, 1,1'-(1,2,3,4-tetrafluoro-1,3-butadiene-1,4-diyl)bis[4-nitro-	ether	335(4.45)	104-0191-78
$C_{16}H_8N_2$			
9,10-Anthracenedicarbonitrile	C$_6$H$_{12}$	394(4.15)	61-1068-78
$C_{16}H_8N_2O_2S_2$			
Benzo[1",2":5,6;5",4":5',6']bis[1,4]-oxathiino[3,2-b:3',2'-b']dipyridine	EtOH	215(4.02),243(4.07), 304(3.69),424(3.04)	4-0101-78
$C_{16}H_8N_4S_2$			
[1,4]Dithiino[2,3-b:5,6-b']diquinoxaline	EtOH	215(4.338),253(4.47), 283(3.98)	2-0683-78
$C_{16}H_8N_8$			
Cycloocta[1,2-d:4,3-d':5,6-d":8,7-d"']-tetrapyrimidine	CHCl$_3$	263(4.17)	24-1330-78
$C_{16}H_8O_2S_2$			
Benzo[b]thiophen-3(2H)-one, 2-(3-oxo-benzo[b]thien-2(3H)-ylidene)-	CHCl$_3$	541(4.19)	24-3233-78
$C_{16}H_8O_3$			
Anthra[9,1-bc]pyran-2,7-dione	CHCl$_3$	532(3.95),565(4.00), 602s(3.75)	104-1433-78
$C_{16}H_8O_4$			
Anthra[9,1-bc]pyran-2,7-dione, 6-hydroxy-	CHCl$_3$	481(3.89),506(4.01), 544(3.87)	104-1433-78
Cycloocta[1,2-b:4,3-b':5,6-b":8,7-b"']-tetrafuran	CHCl$_3$	241s(4.06),267(4.43)	24-1330-78

Compound	Solvent	λ_{max}(log ϵ)	Ref.
$C_{16}H_8S_4$			
Cyclooctal[1,2-b:4,3-b':5,6-b":8,7-b"']-tetrathiophene	$CHCl_3$	242(4.04),278(4.26)	24-1330-78
Cyclooctal[1,2-c:3,4-c':5,6-c":7,8-c"']-tetrathiophene	$CHCl_3$	242(4.41),258(4.34)	24-1330-78
$C_{16}H_9BrN_2O_6$			
Furan, 2-[2-[5-(4-bromophenyl)-2-furanyl]-1-nitroethenyl]-5-nitro-, (E)-	dioxan	295(4.23),440(4.38)	73-3252-78
$C_{16}H_9ClFN_3$			
3H-1,4-Benzodiazepine-2-carbonitrile, 7-chloro-5-(2-fluorophenyl)-	isoPrOH	215(4.46),240s(4.31), 322s(3.58),338(3.58)	44-0936-78
$C_{16}H_9ClN_2O_3S_2$			
Benzamide, 4-chloro-2-nitro-N-(5-phenyl-3H-1,2-dithiol-3-ylidene)-	$CHCl_3$	265s(4.2),310(4.1), 495(4.4)	5-0387-78
$C_{16}H_9ClN_2O_3S_3$			
Sulfilimine, N-(4-chloro-2-nitrobenzoyl)-S-(5-phenyl-3H-1,2-dithiol-3-ylidene)-	$CHCl_3$	330(4.2),510(4.3)	5-0387-78
$C_{16}H_9ClN_2O_6$			
Furan, 2-[2-[5-(2-chlorophenyl)-2-furanyl]-1-nitro-2-ethenyl]-5-nitro-, (E)-	dioxan	295(4.19),411(4.35)	73-3252-78
Furan, 2-[2-[5-(4-chlorophenyl)-2-furanyl]-1-nitro-2-ethenyl]-5-nitro-, (E)-	dioxan	300(4.20),435(4.34)	73-3252-78
$C_{16}H_9ClO_3$			
Benzoic acid, 2-(4-chloro-1-oxo-1H-inden-2-yl)-	n.s.g.	217(4.72),269(4.49)	80-1465-78
$C_{16}H_9Cl_2NOS_2$			
Benzamide, 2,4-dichloro-N-(5-phenyl-3H-1,2-dithiol-3-ylidene)-	$CHCl_3$	277(4.2),310s(4.1), 395(4.3)	5-0387-78
Benzamide, 2,6-dichloro-N-(5-phenyl-3H-1,2-dithiol-3-ylidene)-	$CHCl_3$	310(4.1),382(4.2)	5-0387-78
$C_{16}H_9Cl_2NOS_3$			
Sulfilimine, N-(2,4-dichlorobenzoyl)-S-(5-phenyl-3H-1,2-dithiol-3-ylidene)-	$CHCl_3$	320(4.2),525(4.3)	5-0387-78
Sulfilimine, N-(2,6-dichlorobenzoyl)-S-(5-phenyl-3H-1,2-dithiol-3-ylidene)-	$CHCl_3$	330(4.0),500(4.1)	5-0387-78
$C_{16}H_9Cl_2NO_2$			
1,2-Naphthalenedione, 4-[(2,4-dichlorophenyl)amino]-	MeOH	239(4.30),284(4.17), 335(3.77),447(3.56)	78-1377-78
$C_{16}H_9F_4NO_2$			
Benzene, 1-nitro-4-(1,2,3,4-tetrafluoro-4-phenyl-1,3-butadienyl)-, (E,E)-	ether	330(4.48)	104-0191-78 +104-0939-78
$C_{16}H_9NO_2$			
3H-Dibenzo[f,ij]isoquinoline-2,7-dione	50% EtOH	325(3.9),362(4.20), 390s(3.6)	103-1116-78
$C_{16}H_9NO_5$			
4H,8H-Pyrano[2',3':5,6][1]benzopyrano-	MeOH	225(4.4),237(4.41),	88-2911-78

Compound	Solvent	$\lambda_{max}(\log \epsilon)$	Ref.
[3,4-c]pyridine-4,8-dione, 5-hydroxy-2-methyl- (cont.)		251(4.46),256(4.45), 292(4.07),318(4.11)	88-2911-78
$C_{16}H_9N_3O_8$			
Furan, 2-[2-nitro-2-(5-nitro-2-furanyl)ethenyl]-5-(4-nitrophenyl)-	dioxan	345(4.60),431(4.11)	73-3252-78
$C_{16}H_{10}$			
Anthracene, 9-ethynyl-	EtOH	258(5.18),250s(4.91), 325(3.08),342(3.49), 355(3.85),378(4.05), 400(4.05)	35-7041-78
$C_{16}H_{10}BrNO_2$			
1,2-Naphthalenedione, 4-[(4-bromophenyl)amino]-	MeOH	243(4.37),282(4.12), 330(3.82),470(3.66)	78-1377-78
$C_{16}H_{10}Br_2N_2O_2$			
2(1H)-Quinoxalinone, 3-[1-bromo-2-(4-bromophenyl)-2-oxoethyl]-	EtOH	358(3.95)	103-0336-78
$C_{16}H_{10}ClNO_2$			
1,4-Naphthalenedione, 2-chloro-3-(phenylamino)-	C_6H_{12}	468(3.62)	39-1083-78C
	EtOH	482(--)	39-1083-78C
$C_{16}H_{12}Cl_2O_2$			
2-Butene-1,4-dione, 1,4-bis(4-chlorophenyl)-	n.s.g.	272(4.32)	18-1839-78
$C_{16}H_{10}Cl_4$			
Benzene, 1,1'-(2,3,4,4-tetrachloro-1,3-butadienylidene)bis-	MeOH	282(4.96)	150-0582-78
$C_{16}H_{10}F_4$			
Benzene, 1,1'-(1,2,3,4-tetrafluoro-1,3-butadiene-1,4-diyl)bis-, (E,E)-	EtOH	300(4.51)	104-0939-78
$C_{16}H_{10}N_2O_2$			
3H-Indol-3-one, 2-(1,3-dihydro-3-oxo-2H-indol-2-ylidene)-1,2-dihydro-	$C_2H_2Cl_4$	605(4.30)	24-3233-78
$C_{16}H_{10}N_2O_3S_2$			
Benzamide, 2-nitro-N-(5-phenyl-3H-1,2-dithiol-3-ylidene)-	CHCl$_3$	260(4.1),310(4.1), 390(4.3)	5-0387-78
$C_{16}H_{10}N_2O_3S_3$			
Sulfilimine, N-(2-nitrobenzoyl)-S-(5-phenyl-3H-1,2-dithiol-3-ylidene)-	CHCl$_3$	330(4.2),510(4.3)	5-0387-78
$C_{16}H_{10}N_2O_4$			
1H-Indene-1,3(2H)-dione, 2-[[(2-nitrophenyl)amino]methylene]-	EtOH	399(4.39)	97-0256-78
1H-Indene-1,3(2H)-dione, 2-[[(4-nitrophenyl)amino]methylene]-	EtOH	390(4.42)	97-0256-78
5(4H)-Oxazolone, 4-[(4-nitrophenyl)methylene]-2-phenyl-	toluene	380(4.52)	103-0120-78
$C_{16}H_{10}N_2O_6$			
Furan, 2-[2-nitro-2-(5-nitro-2-furanyl)ethenyl]-5-phenyl-, (E)-	dioxan	295(4.16),427(4.29)	73-3252-78

Compound	Solvent	$\lambda_{max}(\log \epsilon)$	Ref.
$C_{16}H_{10}N_4OS$ Thiazolo[4,5-b]quinoxalin-2(3H)-imine, 3-benzoyl-	EtOH	215(4.462),258(4.444), 278(4.416)	2-0683-78
$C_{16}H_{10}N_6S$ 5H-1,2,4-Triazolo[3',4':2,3][1,3,4]-thiadiazino[5,6-b]quinoxaline	EtOH	217(4.43),238(4.36), 268(4.99)	2-0307-78
$C_{16}H_{10}N_8$ 5,5':4',4":5",5"'-Quaterpyrimidine	$CHCl_3$	241(4.30),255s(4.19)	24-1330-78
$C_{16}H_{10}N_8Ni$ Nickel, (6,13-dimethyl-1,4,8,11-tetra-azacyclotetradeca-2,4,7,9,11,14-hex-aene-2,3,9,10-tetracarbonitrilato-(2-)-N^1,N^4,N^8,N^{11}]-, (SP-4-1)-	EtOH	270(3.91),333(4.04), 390(3.89),530(3.74)	24-2919-78
$C_{16}H_{10}N_8NiO_4$ Nickel, (7,16-dihydro-7,16-dinitrodi-pyrido[2,3-b:2',3'-i][1,4,8,11]tetra-azacyclotetradecinato(2-)-N^5,N^9,N^{14}-N^{18}]-, (SP-4-1)-	DMF	404(4.69)	64-1012-78B
$C_{16}H_{10}O$ 17-Oxabicyclo[12.2.1]heptadeca-1(16),2,4,10,12,14-hexaene-6,8-diyne, (E,E,Z,Z)-	EtOH	228(4.11),281(4.92), 296(4.69),536(2.58)	88-4929-78
$C_{16}H_{11}Br$ Anthracene, 9-(1-bromoethenyl)-	EtOH	255(5.10),320(3.08), 335(3.48),352(3.80), 370(3.93)	35-7041-78
Anthracene, 9-(2-bromoethenyl)-	EtOH	255(4.14),333(3.42), 350(3.75),367(3.93), 386(3.88)	35-7041-78
$C_{16}H_{11}BrN_2OS$ 4(5H)-Thiazolone, 5-[(4-bromophenyl)-methylene]-2-(phenylamino)-	MeOH dioxan 70% dioxan	238(4.23),349(4.44) 238(4.25),337(4.41) 237(4.25),346(4.44)	104-0997-78 104-0997-78 104-0997-78
$C_{16}H_{11}BrN_2O_2$ 2(1H)-Quinoxalinone, 3-(1-bromo-2-oxo-2-phenylethyl)-	EtOH	353(3.98)	103-0336-78
2(1H)-Quinoxalinone, 3-[2-(4-bromophen-yl)-2-oxoethylidene]-3,4-dihydro-	EtOH	443(4.53)	103-0336-78
$C_{16}H_{11}BrN_2O_3S$ Acetic acid, [[3-(4-bromophenyl)-3,4-dihydro-4-oxo-2-quinazolinyl]thio]-	MeOH	276(4.16),315(3.60)	2-0678-78
$C_{16}H_{11}BrN_2S$ 2(1H)-Pyrimidinethione, 6-(3-bromophen-yl)-4-phenyl-	C_6H_{12} EtOH	257(4.62),320(4.18) 261(4.24),298(4.32), 414(3.55)	94-1298-78 94-1298-78
$C_{16}H_{11}Br_3N_2O_4$ Cyclobuta[1,2-c:3,4-c']dipyrrole-1,3,4,6(2H,5H)-tetrone, 3a,3b,6a-tribromotetrahydro-2,5-dimethyl-6b-phenyl-	$CHCl_3$	242(4.60),256(4.44), 282(4.19),378(2.98)	24-2677-78

Compound	Solvent	λ_{max}(log ϵ)	Ref.
$C_{16}H_{11}Cl$ Anthracene, 9-(1-chloroethenyl)-	EtOH	254(5.19),320(3.18), 333(3.51),346(3.81), 367(3.97),386(3.91)	35-7041-78
$C_{16}H_{11}ClFN_3O$ 3H-1,4-Benzodiazepine-2-carboxaldehyde, 7-chloro-5-(2-fluorophenyl)-, oxime	isoPrOH	232(4.50),270s(4.20), 319(3.76)	44-0936-78
$C_{16}H_{11}ClFN_3O_2$ 1H-1,4-Benzodiazepine, 7-chloro-5-(2- fluorophenyl)-2,3-dihydro-2-(nitro- methylene)-	isoPrOH	223(4.45),367(4.40)	44-0936-78
$C_{16}H_{11}ClN_2O$ 1-Naphthalenol, 4-[(4-chlorophenyl)- azo]-	pyridine	423(<u>4.3</u>)	44-3882-78
$C_{16}H_{11}ClN_2O_2$ 2(1H)-Quinoxalinone, 3-[2-(4-chloro- phenyl)-2-oxoethylidene]-3,4-dihydro-	EtOH	440(4.50)	103-0336-78
$C_{16}H_{11}ClN_2O_3S$ Acetic acid, [[3-(3-chlorophenyl)-3,4- dihydro-4-oxo-2-quinazolinyl]thio]-	MeOH	276(4.16),316(3.62)	2-0678-78
$C_{16}H_{11}ClN_2S$ 2(1H)-Pyrimidinethione, 4-(3-chloro- phenyl)-6-phenyl-	C_6H_{12}	236s(4.39),257(4.53), 322(4.05)	4-0105-78
	MeOH	242s(4.31),263(4.33), 296(4.32),412(3.34)	4-0105-78
2(1H)-Pyrimidinethione, 4-(4-chloro- phenyl)-6-phenyl-	C_6H_{12}	262(4.47),294s(3.87), 323(4.06)	4-0105-78
	EtOH	247(4.27),270(4.32), 291s(4.31),406(3.34)	4-0105-78
$C_{16}H_{11}ClO$ 1H-Inden-1-one, 2-[(3-chlorophenyl)- methylene]-, cation	H_2SO_4	415(4.37)	65-1644-78
1H-Inden-1-one, 2-[(4-chlorophenyl)- methylene]-, cation	H_2SO_4	432(4.40)	65-1644-78
$C_{16}H_{11}ClOS$ Ethanone, 1-(8-chlorodibenzo[b,f]thie- pin-3-yl)-	MeOH	225(4.46),243(4.55), 255s(4.44),280(4.31)	73-0471-78
$C_{16}H_{11}ClO_4$ Benzeneacetic acid, 2-carboxy-α-[(2- chlorophenyl)methylene]-, cis	n.s.g.	216(4.36),265(4.12), 286(3.96)	80-1465-78
$C_{16}H_{11}Cl_3O_5$ 9H-Xanthen-9-one, 2,4,5-trichloro-1-hy- droxy-3,6-dimethoxy-8-methyl-	MeOH	210(4.06),247s(4.38), 252(4.39),278s(3.77), 320(4.06),360(3.60)	78-2491-78
	MeOH-AlCl$_3$	210(4.41),251(4.44), 280s(4.33),350(4.33), 415(4.06)	78-2491-78
$C_{16}H_{11}FN_2O_2$ 2(1H)-Quinoxalinone, 3-[2-(4-fluoro- phenyl)-2-oxoethylidene]-3,4-dihydro-	EtOH	440(4.52)	103-0336-78

Compound	Solvent	$\lambda_{max}(\log \epsilon)$	Ref.
$C_{16}H_{11}FOS$			
Ethanone, 1-(7-fluoro-9-methylene-9H-thioxanthen-2-yl)-	MeOH	259(4.23),322(4.05)	73-0471-78
$C_{16}H_{11}N$			
1H-Dibenz[e,g]indole	C_6H_{12}	250(4.71),255(4.89), 287(4.15),298(3.87)	44-0381-78
	EtOH and heptane	207(4.38),252(4.71), 258(4.93),290(4.24)	103-0851-78
Propanenitrile, 2-(1H-phenalen-1-ylidene)-	MeOH	253(4.457),280s(--), 295s(--),399(4.313)	24-1824-78
$C_{16}H_{11}NO$			
9-Anthracenecarbonitrile, 10-methoxy-	C_6H_{12}	390(4.00)	61-1068-78
2,5-Cyclohexadien-1-one, 4-(1-naphthalenylimino)-	C_6H_{12}	270(4.10),295(3.96), 340(3.45),500(3.43)	48-0557-78
	EtOH	275(4.09),310(3.86), 335(3.49),510(3.40)	48-0557-78 +80-0617-78
	ether	270(4.30),310(4.04), 340(3.65),490(3.57)	48-0557-78
	$CHCl_3$	310(4.04),335(3.65), 505(3.60)	48-0557-78
	CCl_4	315(4.00),335(3.62), 505(3.61)	48-0557-78
2,5-Cyclohexadien-1-one, 4-(2-naphthalenylimino)-	C_6H_{12}	270(4.36),290(4.18), 312(3.95),470(3.57)	48-0557-78
	EtOH	270(4.39),290(4.20), 315(3.98),490(3.64)	48-0557-78 +80-0617-78
	ether	265(4.38),285(4.23), 315(3.98),470(3.64)	48-0557-78
	$CHCl_3$	270(4.43),315(4.00), 485(3.71)	48-0557-78
	CCl_4	295(4.19),315(4.00), 480(3.60)	48-0557-78
$C_{16}H_{11}NOS$			
Phenanthro[9,10-d]thiazole, 6-methoxy-	EtOH	247s(4.71),254(4.84), 278(4.25),291(4.09), 315(3.83)	39-0685-78C
$C_{16}H_{11}NO_2$			
1H-Indene-1,3(2H)-dione, 2-[(phenylamino)methylene]-	EtOH	360(4.53)	97-0256-78
1,2-Naphthalenedione, 4-(phenylamino)-	MeOH	243(4.48),282(4.27), 338s(3.86),459(3.89)	78-1377-78
1,4-Naphthalenedione, 2-(phenylamino)-	C_6H_{12}	451(3.68)	39-1083-78C
	EtOH	472(--)	39-1083-78C
5(4H)-Oxazolone, 2-phenyl-4-(phenylmethylene)-	toluene	360(4.59)	103-0120-78
2-Quinolinecarboxaldehyde, 4-phenoxy-	DMF	333(3.38)	97-0138-78
$C_{16}H_{11}NO_3$			
1H-Inden-1-one, 2,3-dihydro-2-[(4-nitrophenyl)methylene]-, cation	H_2SO_4	400(4.49)	65-1644-78
Spiro[2H-1-benzopyran-2,2'-[2H-1,3]-benzoxazin]-4'(3'H)-one	isoPrOH	255(3.93),257(3.83), 305(3.96)	103-0122-78
$C_{16}H_{11}NO_6$			
Pyrido[1,2-a]indole-6,10-dione, 7,9-diacetoxy-	MeOH	204(4.39),238(4.30), 280(3.64),313(3.65), 398(3.95)	83-0960-78

Compound	Solvent	$\lambda_{max}(\log \epsilon)$	Ref.
$C_{16}H_{11}N_3$			
9H-Dibenzo[c,e][1,2,3]triazolo[1,5-a]-azepine, 9-methylene-	MeOH	233(4.43)	44-0066-78
3,5-Pyridinedicarbonitrile, 2-methyl-6-(2-phenylethenyl)-, (E)-	EtOH	205(4.62),228s(4.23), 337(4.10)	73-0434-78
$C_{16}H_{11}N_3O$			
Ethanone, 1-[1,2,4]triazolo[1,5-f]phen-anthridin-2-yl-	EtOH	240(4.76),247(4.85), 278s(4.18),297s(3.52), 312(3.51),326(3.54)	142-1577-78A
8H-Phthalazino[1,2-b]quinazolin-8-one, 5-methyl-	n.s.g.	231.8(4.557)	2-0689-78
$C_{16}H_{11}N_3O_3$			
1-Naphthalenol, 4-[(4-nitrophenyl)azo]-	pyridine	475(4.5)	44-3882-78
Pyridinium, 1-(5-formyl-1,2,3,6-tetra-hydro-2,6-dioxo-1-phenyl-4-pyrimi-dinyl)-, hydroxide, inner salt	MeOH	246(4.34),295(4.04), 328(3.97)	103-1372-78
perchlorate	MeOH	251(4.09),292(4.98), 330(3.60)	103-1372-78
$C_{16}H_{11}N_3O_3S$			
4-Thiazolidinone, 5-[(3-nitrophenyl)-methylene]-2-(phenylimino)-	MeOH	234(4.29),336(4.38)	104-0997-78
	dioxan	230(4.33),325(4.32)	104-0997-78
	70% dioxan	232(4.34),325(4.33)	104-0997-78
4-Thiazolidinone, 5-[(4-nitrophenyl)-methylene]-2-(phenylimino)-	MeOH	258(4.14),370(4.37)	104-0997-78
	dioxan	365(4.33)	104-0997-78
	70% dioxan	261(4.14),371(4.35)	104-0997-78
$C_{16}H_{11}N_3O_4$			
2(1H)-Quinoxalinone, 3,4-dihydro-3-[2-(4-nitrophenyl)-2-oxoethyli-dene]-	EtOH	439(4.31)	103-0336-78
ion	EtOH	463(4.87)	103-0336-78
Quinoline, 2-[(2,4-dinitrophenyl)meth-yl]-	EtOH	228(4.3),304(3.8), 315(3.7)	104-1847-78
	EtOH-NaOH	240(4.30),315(3.88), 340s(3.64),515(3.95)	104-1847-78
$C_{16}H_{11}N_5S$			
Thiazolo[4,5-b]quinoxalin-3(2H)-amine, 2-imino-N-(phenylmethylene)-	EtOH	214(4.635),260(4.372), 278(4.406)	2-0683-78
$C_{16}H_{11}N_7O_2S$			
Benzaldehyde, 4-nitro-, 1H-[1,3,4]thia-diazino[5,6-b]quinoxalin-3-ylhydra-zone	EtOH	217(4.64),268(4.26)	2-0307-78
$C_{16}H_{12}$			
Phenanthrene, 1-ethenyl-	MeOH	239(4.37),258(4.59), 300(4.15)	35-3819-78
$C_{16}H_{12}BrClN_4$			
Benzaldehyde, 4-bromo-, (4-chloro-1-phthalazinyl)methylhydrazone	dioxan	299(4.15),359(4.31)	103-0436-78
$C_{16}H_{12}BrNO_2$			
Benzoic acid, 4-bromo-, (3-phenyl-2H-azirin-2-yl)methyl ester	C_6H_{12}	242(4.16)	35-4481-78

Compound	Solvent	λ_{max} (log ϵ)	Ref.
$C_{16}H_{12}BrNO_3$			
1H-Indole, 1-[(6-bromo-1,3-benzodioxol-5-yl)carbonyl]-2,3-dihydro-	EtOH	255(4.39),293(4.28)	87-0199-78
$C_{16}H_{12}BrN_3$			
2-Pyrimidinamine, 4-(3-bromophenyl)-6-phenyl-	EtOH	337(4.13),353(4.49)	94-1298-78
$C_{16}H_{12}BrN_3O_4$			
2H-1,5-Benzodiazepin-2-one, 8-bromo-1,3-dihydro-4-(4-methoxyphenyl)-	EtOH	240s(4.31),260(4.28),274(4.27),330(4.20)	103-0455-78
7-nitro-	70% H_2SO_4	216(4.16),369(4.32)	103-0455-78
$C_{16}H_{12}Br_2$			
Anthracene, 9-(1,2-dibromoethyl)-	EtOH	258(5.08),348(3.66),367(3.79),380(3.88),400(3.78)	35-7041-78
Phenanthrene, 3,6-dibromo-9,10-dimethyl-	hexane	235(4.49),252(4.67),259(4.72),282(4.28),286s(--),297(4.09),309(4.13),329(2.80),347(2.90),364(2.90)	104-0924-78
$C_{16}H_{12}ClFN_4O$			
3H-1,4-Benzodiazepin-2-amine, 7-chloro-5-(2-fluorophenyl)-N-methyl-N-nitroso-	isoPrOH	231(4.49),300(3.96),340s(3.75)	44-0936-78
$C_{16}H_{12}ClNO_2$			
Isoxazole, 3-(4-chlorophenyl)-4,5-dihydro-4-(phenoxymethylene)-	EtOH	252(3.88),285(3.70)	22-0415-78
$C_{16}H_{12}ClN_3O_2$			
1H-1,4-Benzodiazepine, 7-chloro-2,3-dihydro-2-(nitromethylene)-5-phenyl-	isoPrOH	224(4.46),260s(4.06),364(4.42)	44-0936-78
3H-1,4-Benzodiazepine-2-carboxaldehyde, 7-chloro-5-phenyl-, oxime, 4-oxide	isoPrOH	250(4.49),291(4.33),350s(3.62)	44-0936-78
$C_{16}H_{12}ClN_3O_3$			
1H-1,4-Benzodiazepine, 7-chloro-2,3-dihydro-2-(nitromethylene)-5-phenyl-, 4-oxide	isoPrOH	235(4.43),315(4.26),366(4.29)	44-0936-78
$C_{16}H_{12}ClN_3O_4$			
2H-1,5-Benzodiazepin-2-one, 8-chloro-1,3-dihydro-4-(4-methoxyphenyl)-	EtOH	240s(4.34),260(4.28),275(4.20),330(4.20)	103-0455-78
7-nitro-	70% H_2SO_4	217(4.15),365(4.32)	103-0455-78
$C_{16}H_{12}ClN_5O_2$			
Benzaldehyde, 2-nitro-, (4-chloro-2-methyl-1(2H)-phthalazinylidene)-hydrazone	dioxan	280(4.11),410(4.20)	103-0567-78
Benzaldehyde, 4-nitro-, (4-chloro-2-methyl-1(2H)-phthalazinylidene)-hydrazone	dioxan	291(4.15),430(4.26)	103-0567-78
$C_{16}H_{12}Cl_2$			
Stilbene, 2,6-dichloro-2'-ethenyl-	MeOH	225(4.38),241s(4.32),285(4.26)	35-3819-78
$C_{16}H_{12}Cl_2OS$			
Ethanone, 1-[7-chloro-9-(chloromethyl)-	MeOH	241(4.53),263(3.91),	73-0471-78

418                    $C_{16}H_{12}Cl_2OS-C_{16}H_{12}N_2O$

Compound	Solvent	$\lambda_{max}(\log \epsilon)$	Ref.
9H-thioxanthen-2-yl)- (cont.)		312(4.12)	73-0471-78
Ethanone, 1-(8,10-dichloro-10,11-dihydrodibenzo[b,f]thiepin-3-yl)-	MeOH	252(4.33),295s(3.50)	73-0471-78
$C_{16}H_{12}Cl_2O_5$			
9H-Xanthen-9-one, 2,4-dichloro-1-hydroxy-3,6-dimethoxy-8-methyl-	MeOH	210(4.30),243(4.53), 273(4.23),308(4.30), 360(3.84)	78-2491-78
	MeOH-AlCl$_3$	210(4.39),243(4.50), 273s(4.31),286(4.32), 335(4.31),412(3.90)	78-2491-78
9H-Xanthen-9-one, 2,5-dichloro-1-hydroxy-3,6-dimethoxy-8-methyl-	MeOH	214(4.37),251(4.69), 279(3.98),317(4.42), 355s(3.77)	78-2491-78
	MeOH-AlCl$_3$	215(4.62),254(4.69), 270(4.60),348(4.61), 405s(4.14)	78-2491-78
$C_{16}H_{12}F_2$			
Naphthalene, 6-fluoro-1-(4-fluorophenyl)-1,2-dihydro-	MeOH	259(3.96),269s(3.87), 289(3.41),298(3.29)	73-1760-78
$C_{16}H_{12}F_2O$			
1(2H)-Naphthalenone, 7-fluoro-4-(4-fluorophenyl)-3,4-dihydro-	MeOH	243(4.04),297(3.46)	73-1760-78
$C_{16}H_{12}N_2$			
2H-Azirine, 3,3'-[1,1'-biphenyl]-2,2'-diylbis-	MeOH	242(4.31)	44-0066-78
3,5'-Bi-1H-indole	EtOH	217(4.75),231(4.76), 254(4.68)	103-0173-78
5,5'-Bi-1H-indole	EtOH	215(4.41),249(4.72), 301s(3.92)	103-0173-78
5,10-Diazabenzo[b]biphenylene, 2,3-dimethyl-	EtOH	226(3.89),239(4.05), 247(4.00),271s(3.91), 279(4.02),296(3.64), 368s(3.42),384(3.73), 405(3.76)	78-0495-78
Propanedinitrile, (1,4-dihydro-1,4-ethano-7H-benzocyclohepten-7-ylidene)-	hexane	260(4.14),392s(4.45), 398(4.48),405(4.48), 417(4.39),425(4.35)	88-0569-78
	MeOH	228(4.10),262(4.15), 406(4.53),416(4.53), 423s(4.52)	88-0569-78
Propanedinitrile, [1-methyl-2-(1-naphthalenyl)ethylidene]-	C_6H_{12}	273(3.83),283(3.88), 294(3.76),324(3.04)	39-0995-78B
Pyridazine, 3,6-diphenyl-	EtOH	277(4.51)	33-0589-78
5H-Quindoline, 5-methyl- (cryptolepine)	MeOH	224(4.11),246(3.87), 275(4.41),283(4.43), 355s(4.02),370(4.33), 410(3.28),433(3.29)	36-0433-78
	EtOH-KOH	214(4.40),230s(4.01), 297s(4.38),307(4.48), 368(3.50),386(4.08)	36-0433-77
$C_{16}H_{12}N_2O$			
1-Naphthalenol, 4-(phenylazo)-	pyridine	414(4.3)	44-3882-78
2-Naphthalenol, 1-(phenylazo)-	MeOH	228(4.5),307(3.8), 415(4.0),475(4.2)	19-0655-78
2-Propen-1-one, 3-(1H-indazol-3-yl)-1-phenyl-	EtOH	207(4.39),266(4.02), 351(4.17)	103-0771-78

Compound	Solvent	$\lambda_{max}(\log \epsilon)$	Ref.
1H-Pyrazole-4-carboxaldehyde, 1,3-di-phenyl-	EtOH	248(4.42)	22-0415-78
$C_{16}H_{12}N_2OS$			
4(5H)-Thiazolone, 2-(phenylamino)-5-(phenylmethylene)-	MeOH	234(4.25),333(4.47)	104-0997-78
	dioxan	235(4.26),326(4.36)	104-0997-78
	70% dioxan	235(4.19),331(4.38)	104-0997-78
$C_{16}H_{12}N_2O_2$			
Benzo[3,4]cyclobuta[1,2-b]quinoxaline, 7,8-dimethoxy-	EtOH	230(4.04),256(4.27), 259(4.28),283(3.93), 290(3.93),299s(3.75), 403(3.98),422(4.06)	78-0495-78
1,2-Naphthalenedione, 4-[(4-aminophen-yl)amino]-	MeOH	252(4.39),282(4.14), 500(3.66)	78-1377-78
Quinoline, 2-[(4-nitrophenyl)methyl]-	EtOH	272(4.2),300(4.0), 316(3.9)	104-1847-78
2(1H)-Quinoxalinone, 3,4-dihydro-3-(2-oxo-2-phenylethylidene)-	EtOH	441(4.46)	103-0336-78
$C_{16}H_{12}N_2O_3$			
1(2H)-Isoquinolinone, 5-nitro-2-(phen-ylmethyl)-	20% MeCN	261(4.06),313(3.83), 372(3.69)	44-1132-78
1-Naphthalenemethanamine, α-[(5-nitro-2-furanyl)methylene]-	EtOH	265(4.34),480(4.37)	4-0555-78
$C_{16}H_{12}N_2O_4S$			
Benzenesulfonic acid, 4-[(2-hydroxy-1-naphthalenyl)azo]-, sodium salt	H_2O	484(4.38)	40-0271-78
$C_{16}H_{12}N_2O_5$			
Ethanone, 2-[3-(4-nitrophenyl)-1,4,2-dioxazol-5-yl]-1-phenyl-	MeOH	255(4.42),272(4.39)	97-0057-78
$C_{16}H_{12}N_2S$			
2(1H)-Pyrimidinethione, 4,6-diphenyl-	C_6H_{12}	257(4.51),319(4.08)	4-0105-78
	EtOH	241s(4.26),266(4.32), 297(4.39),404(3.28)	4-0105-78
$C_{16}H_{12}N_4$			
Dibenzo[c,e]azocine, 8-azido-5-methyl-	MeOH	280(4.05)	44-0066-78
Pyrido[3,2-c]pyridazine, 4-(2-methyl-1H-indol-3-yl)-	EtOH	214(4.79),275(4.17), 286(4.10)	103-1267-78
$C_{16}H_{12}N_4O_2$			
2-Pyrimidinamine, 4-(3-nitrophenyl)-6-phenyl-	EtOH	245(4.60),341(4.10)	94-1298-78
$C_{16}H_{12}N_4O_4S$			
2H-Naphtho[1,2-d]triazole-7-sulfonic acid, 2-(3-aminophenyl)-9-hydroxy-	n.s.g.	234(4.72),274(4.44), 358(4.36)	56-1355-78
2H-Naphtho[1,2-d]triazole-7-sulfonic acid, 2-(4-aminophenyl)-9-hydroxy-	n.s.g.	236(4.58),288(4.15), 366(4.34)	56-1355-78
2H-Naphtho[1,2-d]triazole-8-sulfonic acid, 2-(3-aminophenyl)-6-hydroxy-	n.s.g.	236(4.53),262(4.34), 356(4.19)	56-1355-78
2H-Naphtho[1,2-d]triazole-8-sulfonic acid, 2-(4-aminophenyl)-6-hydroxy-	n.s.g.	236(4.55),284(4.23), 366(4.38)	56-1355-78
$C_{16}H_{12}N_4O_7S_2$			
2H-Naphtho[1,2-d]triazole-4,7-disulfon-ic acid, 2-(3-aminophenyl)-9-hydroxy-	n.s.g.	238(4.68),270(4.42), 362(4.31)	56-1355-78

Compound	Solvent	$\lambda_{max}(\log \epsilon)$	Ref.
2H-Naphtho[1,2-d]triazole-4,7-disulfon- ic acid, 2-(4-aminophenyl)-9-hydroxy-	n.s.g.	238(4.65),270(4.24), 374(4.38)	56-1355-78
2H-Naphtho[1,2-d]triazole-5,7-disulfon- ic acid, 2-(3-aminophenyl)-9-hydroxy-	n.s.g.	234(4.59),274(4.40), 360(4.27)	56-1355-78
2H-Naphtho[1,2-d]triazole-5,7-disulfon- ic acid, 2-(4-aminophenyl)-9-hydroxy-	n.s.g.	236(4.61),274(4.17), 372(4.38)	56-1355-78
$C_{16}H_{12}N_6S$ Benzaldehyde, 1H-[1,3,4]thiadiazino- [5,6-b]quinoxalin-3-ylhydrazone	EtOH	215(4.412),222(4.3), 241(4.18)	2-0307-78
$C_{16}H_{12}N_8$ 1,4,8,11-Tetraazacyclotetradeca-2,4,6- 9,11,13-hexaene-2,3,9,10-tetracarbo- nitrile, 6,13-dimethyl-	DMF	270(4.15),330(4.30), 350(4.28),360(4.25), 368(4.25),400(3.84), 430(3.68)	24-2919-78
$C_{16}H_{12}N_8O_4$ Dipyrido[2,3-b:2',3'-i][1,4,8,11]tetra- azacyclotetradecine, 5,14-dihydro- 7,16-dinitro-	DMF	339(4.22),382s(4.17), 397(4.21),418(4.18)	64-1012-78B
disodium salt	DMF	395(4.21),419(4.25)	64-1012-78B
$C_{16}H_{12}O$ 9(10H)-Anthracenone, 10-ethylidene-	C_6H_{12}	271(4.25),333(3.74), 346(3.73)	22-0442-78
Benzobicyclo[3.1.0]hex-2-en-4-one, 6-phenyl-, endo	MeOH	252(4.01)	35-3819-78
4H-Cyclohepta[b]furan, 4-(2,4,6-cyclo- heptatrien-1-ylidene)-	EtOH	241(4.29),294(3.77), 355(4.22),450s(3.15)	142-0287-78C
6H-Cyclohepta[c]furan, 6-(2,4,6-cyclo- heptatrien-1-ylidene)-	EtOH	248s(4.43),250(4.45), 368(4.48)	142-0287-78C
1H-Inden-1-one, 2,3-dihydro-2-(phenyl- methylene)-, cation	H_2SO_4	421(4.50)	65-1644-78
Spiro[anthracene-9(10H),1'-cycloprop- an]-10-one	C_6H_{12}	262(4.21),317(3.70), 328(3.68)	22-0442-78
$C_{16}H_{12}OS_2$ 2(1H)-Naphthalenone, 1-(5,6-dihydro- 4H-cyclopenta-1,3-dithiol-2-ylidene)-	MeCN	278(4.77),325(4.18), 365(3.84),490(4.71)	97-0385-78
$C_{16}H_{12}OS_3$ Anthra[1,9-cd]-1,2-dithiole, 6-methoxy- 10-(methylthio)-	EtOH	220(4.36),240(4.33), 292(4.18),408(3.6), 480(3.76)	2-0673-78
$C_{16}H_{12}OSe_3$ Anthra[1,9-cd]-1,2-diselenole, 6-meth- oxy-10-(methylseleno)-	EtOH	220(4.48),252(4.60), 320(3.84)	2-0673-78
$C_{16}H_{12}O_2$ 1,4-Anthracenedione, 2,3-dimethyl-	CH_2Cl_2	277(4.67),295(4.52), 403(3.68)	150-5538-78
9-Anthracenol, acetate	EtOH	253(4.99)	87-0026-78
2-Butene-1,4-dione, 1,4-diphenyl-	n.s.g.	262(4.30)	18-1839-78
2-Butynal, 4-hydroxy-4,4-diphenyl-	hexane	218s(4.11),256s(3.23)	118-0307-78
Dispiro[oxirane-2,9'(10'H)-anthracene- 10',2"-oxirane], trans	EtOH	227(4.03),263(2.76), 270(2.81),277(2.72)	107-0379-78
Propanal, 2-(3H-naphtho[2,1-b]pyran- 3-ylidene)-, (E)-	EtOH	406(4.30)	48-0497-78

Compound	Solvent	$\lambda_{max}(\log \epsilon)$	Ref.
$C_{16}H_{12}O_2S$			
Methanone, phenyl(5-phenyl-1,3-oxathi-ol-2-yl)-	EtOH	241(4.21),283s(3.83), 313(3.94),325s(3.86), 400(3.14)	39-1547-78C
$C_{16}H_{12}O_3$			
3(2H)-Benzofuranone, 2-[(4-methoxyphen-yl)methylene]-, (E)-	EtOH	245(3.99),287s(3.28), 374(4.16)	33-2646-78
(Z)-	EtOH	245(5.13),286s(4.37), 374(5.35)	33-2646-78
4H-1-Benzopyran-4-one, 2-(2-methoxy-phenyl)-	EtOH	232(4.18),306(4.22)	22-0043-78
4H-1-Benzopyran-4-one, 2-(4-methoxy-phenyl)-	EtOH	252(4.10),317(4.44)	22-0043-78
$C_{16}H_{12}O_3S$			
1H-2-Benzopyran-1-one, 4-hydroxy-3-[(4-methylphenyl)thio]-	MeOH	293(3.96)	24-2859-78
	MeOH-base	329(3.78)	24-2859-78
Dibenz[b,e]thiepin-3-carboxylic acid, 6,11-dihydro-11-oxo-, methyl ester	dioxan	246(4.43),280(4.00), 371(3.53)	44-4892-78
$C_{16}H_{12}O_4$			
9,10-Anthracenedione, 3-ethoxy-1-hy-droxy-	EtOH	232s(4.49),245(4.60), 268s(4.47),288(4.53), 416(4.01)	12-1335-78
9(10H)-Anthracenone, 10-acetyl-1,8-di-hydroxy-	EtOH	223(4.20),256(4.08), 385(3.70)	87-0026-78
4H-1-Benzopyran-4-one, 5-hydroxy-7-methoxy-2-phenyl-	EtOH	276(4.32),320(4.10)	22-0043-78
4H-1-Benzopyran-4-one, 7-hydroxy-2-(4-methoxyphenyl)-	EtOH	252(4.01),324(4.41)	22-0043-78
9,10-Phenanthrenedione, 1,3-dimethoxy-	MeOH	224(4.36),264(4.39), 288s(3.97),406(3.85)	12-2259-78
$C_{16}H_{12}O_5$			
9,10-Anthracenedione, 1,4-dihydroxy-2-(1-hydroxyethyl)-	MeOH	203(4.29),228s(--), 246(4.55),281(3.95), 325s(--),453s(--), 477(3.93),510s(--), 560(2.51)	24-3823-78
4H-1-Benzopyran-4-one, 5,7-dihydroxy-2-(4-methoxyphenyl)-	EtOH	278(4.26),328(4.21)	22-0043-78
4H-1-Benzopyran-4-one, 7-hydroxy-3-(3-hydroxy-4-methoxyphenyl)-	EtOH	249(4.36),261s(4.34), 292(4.17)	102-1405-78
4H-1-Benzopyran-4-one, 5-hydroxy-2-(4-hydroxyphenyl)-7-methoxy- (genkwan-in)	MeOH	268(3.99),338(4.12)	39-1572-78C
	MeOH-NaOAc	270(4.09),340(4.09)	39-1572-78C
	MeOH-NaOAc-H_3BO_3	270(4.11),340(4.11)	39-1572-78C
	MeOH-AlCl₃	278(4.06),300(3.99), 342(4.11),380(3.96)	39-1572-78C
	MeOH-NaOMe	265(--),298s(--), 388(--)	39-1572-78C
Naphtho[2,3-b]furan-3-carboxylic acid, 4,9-dihydro-2-methyl-4,9-dioxo-, ethyl ester	dioxan	249(4.52),291(3.88), 339(3.54),372(3.39)	5-0140-78
$C_{16}H_{12}O_5S$			
1H-2-Benzopyran-1,4(3H)-dione, 3-[(4-methylphenyl)sulfonyl]-	MeOH	297(3.42),305(--)	24-2859-78
	MeOH-base	335(3.64+)(changing)	24-2859-78

Compound	Solvent	$\lambda_{max}(\log \epsilon)$	Ref.
$C_{16}H_{12}O_6$			
2,5-Cyclohexadiene-1,4-dione, 2-acetyl-3-(3-acetyl-4-hydroxyphenoxy)-	MeCN	223(4.34),250(4.34), 346(3.62),440s(2.93)	138-1033-78
Pterocarpan, 3,4-dihydroxy-8,9-(methylenedioxy)-, (-)-	EtOH	235(3.92),312(3.80)	102-1419-78
$C_{16}H_{12}O_6S$			
1H-2-Benzopyran-1,4(3H)-dione, 3-[(4-methoxyphenyl)sulfonyl]-	MeOH	300(3.53)	24-2859-78
	MeOH-base	333(3.65+)(changing)	24-2859-78
$C_{16}H_{12}O_7$			
4H-1-Benzopyran-4-one, 3,5,7-trihydroxy-2-(3-hydroxy-4-methoxyphenyl)-(tamarixetin)	MeOH	258(4.24),264s(4.20), 360(4.22)	39-1572-78C
	MeOH-NaOMe	268(4.06),370(3.91)	39-1572-78C
	MeOH-NaOAc	258(3.90),268(3.88), 362(3.82)	39-1572-78C
	MeOH-AlCl	276(4.32),310s(3.79), 440(4.27)	39-1572-78C
	MeOH-AlCl$_3$-HCl	274(4.10),300s(3.79), 355s(3.93),400(3.99)	39-1572-78C
	MeOH-NaOAc-H$_3$BO$_3$	270(4.56),358(3.58)	39-1572-78C
$C_{16}H_{13}BF_4O_4$			
Naphtho[2,1-b]pyrylium, 2-methoxy-3-(methoxycarbonyl)-, tetrafluoroborate	MeCN	241(4.70),289s(3.72), 300(3.87),313(3.94), 332(3.78),345(3.84), 460(3.10)	12-2259-78
$C_{16}H_{13}BrCl_2O$			
Benzene, 1-(bromomethyl)-2-chloro-4-[2-(2-chloro-4-methoxyphenyl)-ethenyl]-, (E)-	EtOH	229(4.19),319s(4.49), 327(4.50)	87-0889-78
$C_{16}H_{13}BrN_2$			
1H-Pyrazole, 5-(3-bromophenyl)-1-methyl-3-phenyl-	EtOH	253(4.46)	94-1298-78
$C_{16}H_{13}BrN_2O_2$			
Benzoic acid, 4-bromo-, (3-hydroxy-3-phenyl-2-propenylidene)hydrazide	benzene	320(3.21),360(2.82)	64-1527-78B
	MeOH	245(4.30),327(4.00), 400(3.90)	64-1527-78B
	MeOH-NaOMe	243(4.20),295s(--), 410(4.40)	64-1527-78B
	dioxan	320(3.63),340(3.52)	64-1527-78B
	CHCl$_3$	250(4.10),295(3.18), 350s(--),390s(--)	64-1527-78B
	C$_6$H$_{12}$	260(--),265(--)	64-1527-78B
$C_{16}H_{13}BrN_2O_3$			
1H-Pyrazole, 1-acetyl-4-(4-bromophenyl)-3-(2-furanylcarbonyl)-4,5-dihydro-	EtOH	217(4.16),291(4.12), 330(4.17)	104-1830-78
$C_{16}H_{13}BrN_4O$			
Benzaldehyde, 4-bromo-, (3,4-dihydro-4-oxo-1-phthalazinyl)methylhydrazone	dioxan	324(4.28)	103-0436-78
2H-1,5-Benzodiazepin-2-one, 8-bromo-1,3-dihydro-4-methyl-3-(phenylazo)-	HOAc	214(4.51),235(4.43), 298(3.76),385(4.55)	103-1270-78
$C_{16}H_{13}Cl$			
Benzene, 1-chloro-2-[2-(2-ethenylphen-	MeOH	218(4.26),234s(4.24),	35-3819-78

Compound	Solvent	$\lambda_{max}(\log \epsilon)$	Ref.
yl)ethenyl]-, (E)- (cont.)		243s(4.26),250(4.27), 298(4.36)	35-3819-78
1,3-Methano-1H-indene, 2-(2-chlorophen-yl)-2,3-dihydro-, (1α,2α,3α)-	MeOH	252(2.85),260(2.96), 266(3.08),274(3.08)	35-3819-78
$C_{16}H_{13}ClFN_3O$			
1H-1,4-Benzodiazepine-2-carboxaldehyde, 7-chloro-5-(2-fluorophenyl)-2,3-di-hydro-	isoPrOH	237(4.40),270s(3.89), 368(3.51)	44-0936-78
$C_{16}H_{13}ClFN_3O_2$			
2,5-Methano-1H-1,4-benzodiazepine, 7-chloro-5-(2-fluorophenyl)-2,3,4,5-tetrahydro-10-nitro-	isoPrOH	258(4.09),318(3.46)	44-0936-78
$C_{16}H_{13}ClN_2$			
1H-Pyrazole, 5-(3-chlorophenyl)-1-meth-yl-3-phenyl-	EtOH	253(4.45)	94-1298-78
1H-Pyrazole, 5-(4-chlorophenyl)-1-meth-yl-3-phenyl-	EtOH	256(4.56)	4-0385-78
$C_{16}H_{13}ClN_2O_2$			
Benzoic acid, 4-chloro-, (3-hydroxy-3-phenyl-2-propenylidene)hydrazide	benzene	325(3.00),385(2.60)	64-1527-78B
	MeOH	243(4.28),325(4.00), 400(3.98)	64-1527-78B
	MeOH-NaOMe	242(4.20),293s(--), 408(4.44)	64-1527-78B
	dioxan	245(4.32),320(4.01), 360s(--)	64-1527-78B
	CHCl$_3$	250(4.03),315(3.29), 360(3.15)	64-1527-78B
	C$_6$H$_{12}$	247(--),255(--)	64-1527-78B
$C_{16}H_{13}ClN_4$			
Benzaldehyde, (4-chloro-2-methyl-1(2H)-phthalazinylidene)hydrazone	dioxan	280(4.04),303(4.06), 384(4.38)	103-0567-78
	MeCN	203s(--),217(--), 279(4.04),300(4.10), 378(4.26)	103-0567-78
Benzaldehyde, (4-chloro-1-phthalazin-yl)methylhydrazone	dioxan	296(4.11),356(4.26)	103-0436-78
	MeCN	219(4.63),291(4.09), 357(4.19)	103-0436-78
	crystal	220(--),298(--), 364(--)	103-0436-78
1(2H)-Phthalazinone, 4-chloro-, (1-phenylethylidene)hydrazone	dioxan	293(4.27),374(4.27)	103-0567-78
	MeCN	211(4.55),291(4.30), 369(4.26)	103-0567-78
$C_{16}H_{13}ClN_4O$			
Benzaldehyde, 4-hydroxy-, (4-chloro-1-phthalazinyl)methylhydrazone	dioxan	284(4.18),364(4.17)	103-0436-78
2H-1,5-Benzodiazepin-2-one, 8-chloro-1,3-dihydro-4-methyl-3-(phenylazo)-	HOAc	210(4.51),235(4.42), 295(3.66),380(4.56)	103-1270-78
$C_{16}H_{13}ClO_2S$			
Ethanone, 1-(8-chloro-10,11-dihydro-10-hydroxydibenzo[b,f]thiepin-2-yl)-	MeOH	237(4.17),256s(4.02), 304(4.03)	73-0471-78
$C_{16}H_{13}ClO_4$			
Benzenepropanoic acid, α-(2-carboxy-phenyl)-2-chloro-	n.s.g.	214(4.41),275(3.95)	80-1465-78

Compound	Solvent	$\lambda_{max}(\log \epsilon)$	Ref.
$C_{16}H_{13}ClO_5$			
Benzoic acid, 3-(5-chloro-2-methoxy-benzoyl)-4-hydroxy-, methyl ester	MeOH	205(3.20),231(3.49), 259s(3.07),329(2.60)	39-0876-78C
Benzoic acid, 4-(5-chloro-2-methoxy-benzoyloxy)-, methyl ester	MeOH	211(3.67),234(3.50), 313(2.81)	39-0876-78C
9H-Xanthen-9-one, 5-chloro-1,3,8-tri-methoxy-	MeOH	236(4.40),242(4.38), 307(4.10),336(3.76)	119-0143-78
9H-Xanthen-9-one, 7-chloro-1,3,8-tri-methoxy-	MeOH	241(4.37),246(4.39), 302(4.03),334(3.58)	119-0143-78
$C_{16}H_{13}FOS$			
9H-Thioxanthen-9-one, 6-fluoro-2-(1-methylethyl)-	MeOH	258(4.68),285s(3.79), 297s(3.53),362(3.79)	73-2656-78
$C_{16}H_{13}F_2NO$			
1H-2-Benzazepin-1-one, 8-fluoro-5-(4-fluorophenyl)-2,3,4,5-tetrahydro-	MeOH	264(3.35),271(3.41)	73-1760-78
2H-1-Benzazepin-2-one, 8-fluoro-5-(4-fluorophenyl)-1,3,4,5-tetrahydro-	MeOH	233(4.04),264(3.47), 271(3.48)	73-1760-78
$C_{16}H_{13}IN_2O_2$			
Benzoic acid, 4-iodo-, (3-hydroxy-3-phenyl-2-propenylidene)hydrazide	benzene	320(3.21),380(3.00)	64-1527-78B
	MeOH	248(4.00),324(3.60), 400(3.90)	64-1527-78B
	MeOH-NaOMe	250(4.00),310s(--), 412(4.16)	64-1527-78B
	dioxan	250(4.14),318(3.74), 355(3.57)	64-1527-78B
	CHCl₃	260(4.15),320s(--), 360(3.39),400s(--)	64-1527-78B
$C_{16}H_{13}N$			
2H-Azirine, 3-(2'-ethenyl[1,1'-biphen-yl]-2-yl)-	C_6H_{12}	300(3.31)	44-0381-78
2H-Azirine, 2-(2-ethenylphenyl)-3-phen-yl-	MeOH	245(4.18)	35-2181-78
1H-Cycloprop[c]isoquinoline, 1a,7b-di-hydro-1a-phenyl-	MeOH	220(3.26),250(2.83), 260(2.68)	35-2181-78
$C_{16}H_{13}NO$			
Phenol, 4-(6-methyl-3-quinolinyl)-	MeOH	267(4.45),295(4.12), 336(3.90)	95-0914-78
	MeOH-HCl	247(4.37),273(3.50)	95-0914-78
	MeOH-NaOH	228(4.67),283(4.39), 308(4.22),321(4.22)	95-0914-78
$C_{16}H_{13}NOS$			
Benzenamine, 4-methyl-N-(5-phenyl-1,3-oxathiol-2-ylidene)-	EtOH	223(4.31),295(4.29)	94-3017-78
Benzenepropanenitrile, β-oxo-2-[(phen-ylmethyl)thio]-	EtOH	204(4.42),238(4.34), 263(3.93),270(3.90), 352(3.44)	103-0141-78
Benzo[b]thiophene, 5-nitroso-4-(2-phen-ylethyl)-	CHCl₃	755(1.20)	39-0692-78C
Isothiazole, 3-(4-methoxyphenyl)-4-phen-yl-	EtOH	283(4.07)	39-0685-78C
Isothiazole, 3-(4-methoxyphenyl)-5-phen-yl-	EtOH	264(4.50),278(4.46)	39-0685-78C
Thiazole, 4-(4-methoxyphenyl)-5-phenyl-	EtOH	239(4.27),289(4.02)	39-0685-78C
Thiazole, 5-(4-methoxyphenyl)-2-phenyl-	EtOH	334(4.39)	39-0685-78C

Compound	Solvent	$\lambda_{max}(\log \epsilon)$	Ref.
$C_{16}H_{13}NO_2$			
2H-Azirine-2-methanol, 3-phenyl-, benzoate	C_6H_{12}	243(4.15)	35-4481-78
Benzamide, N-phenyl-2-(2-propynyloxy)-	EtOH	234s(3.99),241s(3.97), 270(4.09),280s(4.07), 297s(3.90)	33-0716-78
Benzene, 1,1'-[3-(nitromethyl)-1-propyne-1,3-diyl]bis-	n.s.g.	240(4.39),251(4.31)	104-0676-78
Benzene, 1-nitro-4-(4-phenyl-1,3-butadienyl)-	EtOH	360(4.58)	104-0939-78
4H-1-Benzopyran-4-one, 2-methyl-3-(phenylamino)-	EtOH	239(4.19),263(4.02), 275s(3.93),287s(3.77), 293s(3.76),303s(3.62), 348(3.12)	33-0716-78
	12M HCl	232(4.41),253s(4.01), 274(3.73),302(3.80)	33-0716-78
4H-1,3-Benzoxazin-4-one, 2-ethenyl-2,3-dihydro-3-phenyl-	EtOH	241(3.91),246s(3.87), 300s(3.58)	33-0716-78
1H-Indole-2-carboxylic acid, 5-phenyl-, methyl ester	EtOH	206(4.42),261(4.63), 301s(4.22)	103-0173-78
1H-Isoindol-1-one, 2-ethenyl-2,3-dihydro-3-hydroxy-3-phenyl-	EtOH	206(4.47),269(3.86), 280(3.86)	40-0404-78
$C_{16}H_{13}NO_2S$			
Thiopyrano[2,3-b]pyrrol-2(7H)-one, 3-acetyl-7-methyl-6-phenyl-	EtOH	290(4.18),420(4.09)	104-0589-78
$C_{16}H_{13}NO_3$			
1,4-Acridinedione, 9-ethyl-2-methoxy-	dioxan	228(4.45),281(4.46), 377s(3.2)	150-4901-78
1,4-Acridinedione, 2-methoxy-7,9-dimethyl-	dioxan	230(4.42),286(4.45)	150-4901-78
[1,3]Dioxolo[4,5-j]phenanthridine, 5-acetyl-5,6-dihydro-	MeOH	213(4.7),240(4.5), 284(4.1),315(4.1)	5-0608-78
Ethanone, 1-phenyl-2-(3-phenyl-1,4,2-dioxazol-5-yl)-	MeOH	248(4.26),270s(4.15)	97-0057-78
Galanthan-7-one, 2,3,4,12-tetradehydro-9,10-[methylenebis(oxy)]-, (±)-	EtOH	225(4.54),237s(4.28), 274(3.54),306(3.73)	1-0098-78
$C_{16}H_{13}NO_3S_2$			
4-Thiazolidinone, 5-phenyl-2-[(phenylsulfonyl)methylene]-	MeOH	252(3.88),325(4.13)	5-0473-78
	pH 13	302(4.16),372(3.83)	5-0473-78
$C_{16}H_{13}NO_4S$			
5H-Thiazolo[3,2-a]pyridine-7-carboxylic acid, 2,3-dihydro-5-oxo-, methyl ester	MeOH	210(4.6),258(4.5), 335(4.2)	28-0385-78B
$C_{16}H_{13}NO_5$			
1,3-Dioxolo[4,5-b]acridin-10(5H)-one, 4,11-dimethoxy-	EtOH	217(4.33),278(4.74), 395(3.59)	102-0166-78
4H-Pyrano[3,2-c]quinoline-3-carboxylic acid, 5,6-dihydro-6-methyl-2,5-dioxo-, ethyl ester	MeOH	350(4.06),384(4.08)	49-1075-78
$C_{16}H_{13}NS_2$			
Thiazolium, 5-mercapto-4-methyl-2,3-diphenyl-, hydroxide, inner salt	CH_2Cl_2	264(4.24),474(4.00)	5-0029-78
$C_{16}H_{13}N_3$			
[2,2'-Bipyridin]-4-amine, N-phenyl-	neutral	220(3.92),282(4.11)	39-1215-78B

Compound	Solvent	$\lambda_{max}(\log \epsilon)$	Ref.
[2,2'-Bipyridin]-4-amine, N-phenyl-, monocation	H_2O	245(3.91),283(4.05)	39-1215-78B
5H-Dibenzo[c,e][1,2,3]triazolo[1,5-a]-azepine, 4b,9-dihydro-9-methylene-	C_6H_{12}	247(4.10),300(3.56)	44-0066-78
3,5-Pyridinedicarbonitrile, 1,4-dihydro-2-methyl-6-(2-phenylethenyl)-, (E)-	EtOH	205(4.31),224s(4.14), 297(4.47),371(3.50)	73-0434-78
$C_{16}H_{13}N_3O$ 2-Quinoxalinecarboxaldehyde, 3-[(phenylmethyl)amino]-	CHCl$_3$	451(3.50)	97-0177-78
1,2,4-Triazin-6(1H)-one, 1-methyl-3,5-diphenyl-	MeOH	233(4.10),238(4.10), 275(4.44),360(3.89)	4-1271-78
$C_{16}H_{13}N_3O_2$ 1H-Pyrazole, 1-methyl-5-(3-nitrophenyl)-3-phenyl-	EtOH	252(4.48)	94-1298-78
2-Quinoxalinecarboxaldehyde, 3-[(4-methoxyphenyl)amino]-	CHCl$_3$	480(3.37)	97-0177-78
$C_{16}H_{13}N_3O_3$ 1H-1,2-Diazepine, 1-benzoyl-4-(2,5-dioxo-1-pyrrolidinyl)-	MeOH	228(3.88),253s(3.78), 350s(2.53)	33-2887-78
1H-1,2-Diazepine, 1-benzoyl-6-(2,5-dioxo-1-pyrrolidinyl)-	MeOH	225(4.04),260s(3.81), 350s(2.48)	33-2887-78
Glycine, N-(4-oxo-2-phenyl-3(4H)-quinazolinyl)-	MeOH	228(4.60),282(4.20), 319(3.85)	80-1085-78
1H-Indole, 1-acetyl-3-[(4-nitrophenyl)-amino]-	EtOH	206(4.41),229(4.37), 293(3.79),308(3.85), 390(4.27)	103-0757-78
Pyridinium, 1-(benzoylamino)-3-(2,5-dioxo-1-pyrrolidinyl)-, hydroxide, inner salt	EtOH	326(3.59)	33-2887-78
$C_{16}H_{13}N_3O_4$ Benzoic acid, 4-nitro-, (3-hydroxy-3-phenyl-2-propenylidene)hydrazide	MeOH	252(4.26),290(4.08), 320(4.10)	64-1527-78B
	MeOH-NaOMe	260(4.21),370s(--), 455(4.33)	64-1527-78B
	dioxan	250(3.97),310(3.74), 360s(--)	64-1527-78B
	CHCl$_3$	260(4.20),370(3.30), 390(3.12),400s(--)	64-1527-78B
	C_6H_{12}	235(--),260(--), 280(--)	64-1527-78B
$C_{16}H_{13}N_3O_7S_2$ 2,7-Naphthalenedisulfonic acid, 5-amino-4-hydroxy-3-(phenylazo)-, disodium salt	H_2O	531(4.53)	40-0271-78
$C_{16}H_{13}N_5O_2$ Benzaldehyde, 4-nitro-, (2-methyl-1(2H)-phthalazinylidene)hydrazone	dioxan	289(4.15),435(4.33)	103-0567-78
$C_{16}H_{13}N_5O_3$ Benzaldehyde, 4-nitro-, (3,4-dihydro-4-oxo-1-phthalazinyl)methylhydrazone	dioxan	255(4.26),374(4.30)	103-0436-78
2H-1,5-Benzodiazepin-2-one, 1,3-dihydro-4-methyl-7-nitro-3-(phenylazo)-	HOAc	208(4.35),250(4.19), 300(4.11),384(4.43)	103-1270-78

Compound	Solvent	$\lambda_{max}(\log \epsilon)$	Ref.
$C_{16}H_{13}N_5O_6$ 3H-Pyrazole, 4,5-dihydro-3-methyl-4,5-dinitro-3-(4-nitrophenyl)-5-phenyl-	EtOH	262(4.19)	103-0261-78
$C_{16}H_{13}N_5S$ 1H-1,3,4-Thiadiazino[5,6-b]quinoxalin-3-amine, N-(phenylmethyl)-	EtOH	212(4.306),286(4.228), 312(5.04)	2-0307-78
$C_{16}H_{13}O_4$ Naphtho[2,1-b]pyrylium, 2-methoxy-3-(methoxycarbonyl)-, tetrafluoroborate	MeCN	241(4.70),289s(3.72), 300(3.87),313(3.94), 332(3.78),345(3.84), 460(3.10)	12-2259-78
$C_{16}H_{14}$ Anthracene, 9,10-dimethyl-	C_6H_{12}	377(4.05)	61-1068-78
Benzene, 1-ethenyl-2-(2-phenylethenyl)-, cis	MeOH	220s(4.34),258(4.26), 287s(3.93)	35-3819-78
trans	MeOH	220s(4.29),250(4.31), 298(4.38)	35-3819-78
Cyclobutene, 1,2-diphenyl-	MeCN	288s(4.18),297(4.23), 304s(4.22),322s(3.93)	24-3608-78
	ether-EtOH at -190°	286s(--),299(--), 312(--)	24-3608-78
$C_{16}H_{14}Br_2O_5$ Naphtho[1,2-d]-1,3-dioxole-5-acetic acid, 4,7-dibromo-2-methoxy-2-methyl-, methyl ester	MeOH	252(4.84),290s(3.52), 302(3.60),315(3.58), 346(3.49),353(3.52)	12-2259-78
$C_{16}H_{14}ClNO_3$ Benzenepropanamide, α-(4-chlorophenyl)-4-methoxy-β-oxo-	EtOH	211(4.32),237(4.37)	18-3389-78
$C_{16}H_{14}ClNO_6$ 4H-1,3-Benzoxazinium, 2-methyl-4-oxo-3-(phenylmethyl)-, perchlorate	MeCN	235(4.05),294(3.48)	103-0122-78
$C_{16}H_{14}Cl_2O$ Benzene, 2-chloro-4-[2-(2-chloro-4-methoxyphenyl)ethenyl]-1-methyl-, (E)-	EtOH	213(4.40),258(4.21), 310(4.48),321s(4.45)	87-0889-78
$C_{16}H_{14}Cl_2O_2$ 9-Phenanthrenecarboxylic acid, 1,3-dichloro-5,6,7,8-tetrahydro-, methyl ester	MeOH	296(4.06),332s(3.43), 344(3.42)	44-0980-78
$C_{16}H_{14}F_2N_2O_2$ Benzenamine, 4-[1,2-difluoro-2-(4-nitrophenyl)ethenyl]-N,N-dimethyl-, (E)-	ether	420(4.31)	104-0939-78
$C_{16}H_{14}F_3NO_2$ Spiro[cyclopentane-1,3'-[3H]indol]-2-one, 1',2'-dihydro-1'-methyl-2'-(3,3,3-trifluoro-2-oxopropylidene)-	n.s.g.	372(4.31)	77-0779-78
$C_{16}H_{14}FeO_3$ Iron, cyclopentadienyl(5,6-dihydro-4-oxo-7-oxycarbonylazulenyl)-	EtOH	223(4.25),267(4.05), 294s(3.91)	128-0273-78

Compound	Solvent	$\lambda_{max}(\log \epsilon)$	Ref.
$C_{16}H_{14}NO_2$			
2,1-Benzisoxazolium, 3-[2-(4-hydroxy-phenyl)ethenyl]-1-methyl-, perchlorate	MeCN	465(4.49)	44-1233-78
4H-1,3-Benzoxazinium, 2-methyl-4-oxo-3-(phenylmethyl)-, perchlorate	MeCN	235(4.05),294(3.48)	103-0122-78
$C_{16}H_{14}N_2$			
1H-Imidazole, 5-methyl-2,4-diphenyl-	MeOH	217(4.09),272s(4.08), 301(4.30)	33-0286-78
Indeno[1,2-c]pyrazole, 2,3,3a,4-tetra-hydro-2-phenyl-	MeOH	250(4.14),345(4.18)	44-1664-78
Propanedinitrile, (1,2,3,4-tetrahydro-1,4-ethano-7H-benzocyclohepten-7-yli-dene)-	hexane	259(4.05),389s(4.41), 396(4.44),403(4.45), 412s(4.35),422(4.30)	88-0569-78
	MeOH	235(4.05),261(4.14), 404(4.51),414(4.51), 420s(4.50)	88-0569-78
1H-Pyrazole, 1-methyl-3,5-diphenyl-	EtOH	251(3.49)	4-0385-78
1H-Pyrazole, 4-methyl-1,3-diphenyl-	EtOH	285(4.37)	22-0415-78
3H-Pyrazole, 3-methyl-3,5-diphenyl-	hexane	232(4.27),283(3.60), 366(2.30)	44-0034-78
Quinazoline, 4,6-dimethyl-2-phenyl-	EtOH	254(4.40),262(4.44), 288(3.98),326(3.56), 341(3.39)	94-2866-78
Quinazoline, 4,7-dimethyl-2-phenyl-	EtOH	247(4.41),258(4.47), 288(3.90),305(3.93), 318(3.92)	94-2866-78
Quinazoline, 4-ethyl-2-phenyl-	EtOH	262(4.45),269(4.51), 294(4.02),328(3.54), 343(3.38)	94-2866-78
Quinoxaline, 2,6-dimethyl-3-phenyl-	EtOH	243(4.27),328(3.77)	94-2866-78
Quinoxaline, 3,6-dimethyl-2-phenyl-	EtOH	244(4.40),329(3.92)	94-2866-78
Quinoxaline, 2-ethyl-3-phenyl-	EtOH	238(4.41),322(3.89)	94-2866-78
$C_{16}H_{14}N_2O$			
Benzeneacetonitrile, α-[[(4-methoxy-phenyl)amino]methylene]-	EtOH	344(4.34)	78-2229-78
[1]Benzopyrano[4,3-c]pyrazole, 2,3,3a,4-tetrahydro-2-phenyl-	EtOH	255(4.11),304(3.69), 363(4.24)	44-1664-78
1H-Indole, 5-nitroso-4-(2-phenylethyl)-	CHCl₃	725(1.54)	39-0692-78C
Phenol, 2-(1-methyl-4-phenyl-1H-pyra-zol-3-yl)-	EtOH	254(3.95),293(3.60)	114-0457-78C
Phenol, 2-(1-methyl-4-phenyl-1H-pyra-zol-5-yl)-	EtOH	249(3.98),266(3.70)	114-0457-78C
1-Pyrazoline, 4,5-epoxy-3-methyl-3,5-diphenyl-	hexane	365(2.38)	44-0034-78
photoproduct	hexane	268(4.22)	44-0034-78
$C_{16}H_{14}N_2OS$			
2H-1-Benzothiopyran-2-one, 3-amino-4-[(phenylmethyl)amino]-	EtOH	237(4.51),279(4.07), 372(4.06)	103-0033-78
Methanone, [5-phenyl-2-(methylthio)-imidazo[1,2-a]pyridin-3-yl]phenyl-	EtOH	250(4.30),265(4.34), 277(4.30),345(3.79)	95-0631-78
$C_{16}H_{14}N_2O_2$			
1,4-Acridinedione, 2-(dimethylamino)-9-methyl-	dioxan	228(4.51),283(4.32), 334s(4.0),417(3.42)	150-4901-78
1,4-Acridinedione, 2-(ethylamino)-9-methyl-	dioxan	227(4.45),283(4.51), 300s(4.3),330s(4.1), 426(3.55)	150-4901-78

Compound	Solvent	$\lambda_{max}(\log \epsilon)$	Ref.
Benzoic acid, (3-hydroxy-3-phenyl-2-propenylidene)hydrazide	benzene	320(3.21),345s(--), 400s(--)	64-1527-78B
	MeOH	235(4.30),250(4.25), 332(4.13),400s(--)	64-1527-78B
	MeOH-NaOMe	245(4.20),395(4.38)	64-1527-78B
	dioxan	245(3.89),320(3.74), 378(3.18)	64-1527-78B
	CHCl$_3$	250(3.87),315(3.21), 360(--),400s(--)	64-1527-78B
	C_6H_{12}	250(--)	64-1527-78B
1,3,4-Oxadiazol-2(3H)-one, 3,5-bis(4-methylphenyl)-	EtOH	240(4.16),274(4.30), 290(4.27)	33-1477-78
4(1H)-Quinazolinone, 2-(2,6-dimethyl-phenoxy)-	dioxan	243(3.79),250(3.85), 259(3.93),265(3.87), 298(3.48),308(3.63), 320(3.53)	4-1033-78
Sydnone, 3,4-bis(4-methylphenyl)-	EtOH	242(4.28),342(4.10)	33-1477-78
$C_{16}H_{14}N_2O_3$			
Acetamide, N-[2-[2-(2-nitrophenyl)eth-enyl]phenyl]-	benzene	350s(3.53)	33-2813-78
3H-Indolizino[8,7-b]indole-5-carboxylic acid, 2,5,6,11-tetrahydro-3-oxo-, methyl ester, (S)-	MeOH	231(4.40),308(4.33), 321(4.29)	78-1457-78
1H-Pyrazole, 1-acetyl-3-(2-furanylcarb-onyl)-4,5-dihydro-4-phenyl-	EtOH	280s(4.27),293(4.13), 332(4.18)	104-1830-78
$C_{16}H_{14}N_2O_3S$			
1,4,3-Oxathiazin-2-amine, 6-(4-methyl-phenyl)-N-phenyl-, 4,4-dioxide	EtOH	262(4.30)	18-1805-78
2H-1,2,4-Thiadiazin-3(4H)-one, 5-(4-methylphenyl)-4-phenyl-, 1,1-dioxide	EtOH	252(4.06)	18-1805-78
$C_{16}H_{14}N_2O_4$			
2,4,6(1H,3H,5H)-Pyrimidinetrione, 5-[(2-ethyl-5-methylbenzofuranyl)-methylene]-	MeOH	257(4.05),383(3.67)	83-0714-78
$C_{16}H_{14}N_2O_5$			
Benzoic acid, 4,4'-azoxybis-, dimethyl ester	MeOH	265(4.01),330(4.24)	39-0171-78C
$C_{16}H_{14}N_2O_6$			
3-Nortricyclanol, 6-methyl-5-methyl-ene-, 3,5-dinitrobenzoate	EtOH	208(4.51)	33-0732-78
$C_{16}H_{14}N_2S$			
3H-1,5-Benzodiazepine, 2-(methylthio)-4-phenyl-	MeOH	253.5(4.46)	87-0952-78
$C_{16}H_{14}N_4$			
3H-Indole, 3-azido-3,5-dimethyl-2-phen-yl-	EtOH	228(4.20),237(4.25), 246(4.31),252(4.32), 288(4.03),297(4.11), 327(4.23)	94-2866-78
3H-Indole, 3-azido-3,6-dimethyl-2-phen-yl-	EtOH	228(3.99),234(4.03), 248(4.16),315(3.99)	94-2866-78
3H-Indole, 3-azido-3-ethyl-2-phenyl-	EtOH	227(4.19),233(4.22), 242(4.24),247(4.23), 314(4.16)	94-2866-78

Compound	Solvent	$\lambda_{max}(\log \epsilon)$	Ref.
2-Pyridinecarbonitrile, 4,4'-(1,4-but-anediyl)bis-	EtOH	265(3.81)	78-0331-78
$C_{16}H_{14}N_4O$			
Benzaldehyde, (3,4-dihydro-4-oxo-1-phthalazinyl)methylhydrazone	dioxan	320(4.24)	103-0436-78
2H-1,5-Benzodiazepin-2-one, 1,3-di-hydro-4-methyl-3-(phenylazo)-	HOAc	205(4.50),230(4.44), 300(3.70),384(4.61)	103-1270-78
Ethanone, 1-[5-(2-aminophenyl)-1-phen-yl-1H-1,2,4-triazol-3-yl]-	EtOH	217(4.45),252s(4.10), 322(3.60)	142-1577-78A
$C_{16}H_{14}N_4O_2$			
Benzaldehyde, 4-hydroxy-, (3,4-dihydro-4-oxo-1-phthalazinyl)methylhydrazone	dioxan	288(4.18),320(4.11)	103-0436-78
$C_{16}H_{14}N_4O_3$			
1,2,4-Triazolidine-3,5-dione, 1-[2-methoxy-2-(2-pyridinyl)ethenyl]-4-phenyl-	MeOH	201(4.0),240(3.9), 284(3.7)	103-0534-78
$C_{16}H_{14}N_4O_8$			
[4,4]-Bipyridazine]-3,3',6,6'-tetra-carboxylic acid, tetramethyl ester	CH_2Cl_2	230(4.32),328(2.85)	83-0728-78
$C_{16}H_{14}N_6O_2$			
Pyrido[2,3-b]pyrazin-2(1H)-one, 3-(1,4-dihydro-1-methyl-2-oxopyrido[2,3-b]-pyrazin-3(2H)-ylidene)-3,4-dihydro-1-methyl-	pH -1.0	281s(3.79),292s(3.85), 323(3.94),440s(4.35), 466(4.57),497(4.57)	24-1763-78
$C_{16}H_{14}N_8$			
1H-1,2,4-Triazol-5-amine, 1-(1-methyl-1H-tetrazol-5-yl)-N,3-diphenyl-	EtOH	290(4.41)	103-0920-78
$C_{16}H_{14}O$			
Anthracene, 9-ethoxy-	C_6H_{12}	369(3.94)	61-1068-78
Benzeneacetaldehyde, α-(1-phenylethyli-dene)-, (E)-	MeOH	292(4.12)	44-0034-78
(Z)-	MeOH	275(3.98)	44-0034-78
4H-1-Benzopyran, 2-(4-methylphenyl)-	MeOH	244(4.34),281(3.73)	18-1175-78
Phenanthrene, 9-methoxy-10-methyl-	n.s.g.	223(4.21),248(4.62), 255(4.71),271(4.21), 278(4.06),288(3.95), 300(4.01),322(2.43), 337(2.73),353(2.81)	138-1223-78
$C_{16}H_{14}OS_2$			
9,11-Dithiatricyclo[11.3.1.1³,⁷]octa-deca-1(17),3,5,7(18),13,15-hexaen-2-one	MeCN	247s(4.26),254(4.29), 261s(4.10),287s(3.60)	24-2547-78
$C_{16}H_{14}O_2$			
Anthracene, 9,10-dimethoxy-	C_6H_{12}	381(3.95)	61-1068-78
Benzeneacetic acid, α-(1-phenylethyli-dene)-, isomer I	MeOH	256(4.09)	44-0034-78
4H-1-Benzopyran, 6-methoxy-2-phenyl-	MeOH	245(4.38),296(3.65)	18-1175-78
4H-1-Benzopyran, 7-methoxy-2-phenyl-	MeOH	237(4.28),279(3.74)	18-1175-78
2H-1-Benzopyran-3(4H)-one, 4-methyl-4-phenyl-	MeOH	278(3.32),305s(2.69)	44-0303-78
Phenanthrene, 1,3-dimethoxy-	MeOH	230(4.48),247(4.60), 261s(4.52),276s(4.17),	12-2259-78

Compound	Solvent	$\lambda_{max}(\log \epsilon)$	Ref.
Phenanthrene, 1,3-dimethoxy- (cont.)		305s(3.97),312(3.00), 341(3.30),356(3.30)	12-2259-78
2-Propen-1-one, 1-(4-methoxyphenyl)- 3-phenyl-	hexane	196s(4.69),225s(4.21), 308s(4.52)	65-0340-78
2-Propen-1-one, 3-(4-methoxyphenyl)- 1-phenyl-	hexane	189s(4.69),234(4.13), 260(3.90),330(4.45)	65-0340-78
$C_{16}H_{14}O_2S$			
2-Butenoic acid, 2-mercapto-4,4-diphen- yl-	EtOH	210(4.58),238(4.02), 312(4.16)	56-2455-78
	NaHCO₃	215(3.37),240(4.13), 330(4.14)	56-2455-78
Diphenacyl sulfide	EtOH	206(4.31),246(4.30)	39-0955-78C
$C_{16}H_{14}O_3$			
1,8-Anthracenediol, 9,10-dihydro-, monoacetate	EtOH	258(4.49),328(4.18), 278(3.60)	87-0026-78
2H-1-Benzopyran-6-ol, 7-methoxy- 2-phenyl-	EtOH	220(4.28),236(4.37), 328(3.80)	78-0057-78
4H-1-Benzopyran-6-ol, 7-methoxy- 2-phenyl-	EtOH	244(2.70),298(4.16)	78-0057-78
2,5-Cyclohexadiene-1,4-dione, 2-meth- oxy-5-(3-phenyl-2-propenyl)-	EtOH	257(4.45),368(3.24)	78-0057-78
Cyclopenta[b]furo[3,2-g][1]benzopyran- 5(6H)-one, 7,8-dihydro-2,10-dimethyl-	MeOH	246(4.82),290(4.19), 298(4.11),320(4.03)	42-0468-78
Obtusaquinone	EtOH	206(4.24),253(3.85), 268(3.86),399(4.64)	102-1395-78
$C_{16}H_{14}O_4$			
Benzoic acid, 2-(3-ethyl-2-hydroxy- benzoyl)-	EtOH	215(4.48),262(4.14), 338(3.68)	56-0629-78
Benzoic acid, 2-(3-ethyl-4-hydroxy- benzoyl)-	EtOH	206(4.45),295(4.16)	56-0629-78
4H-1-Benzopyran-4-one, 2,3-dihydro- 7-hydroxy-3-(4-methoxyphenyl)-	EtOH	282(4.06),313(3.83)	102-0593-78
	EtOH-NaOH	260(3.46),287s(3.73), 341(4.35)	102-0593-78
	EtOH-NaOAc	262(3.81),287(3.72), 343(4.30)	102-0593-78
2,5-Cyclohexadiene-1,4-dione, 2-(3,4- dihydro-7-methoxy-2H-1-benzopyran- 3-yl)-	EtOH	230(4.12),249(4.20), 283(3.58),289(3.54), 315(2.78)	102-1423-78
Marmelide (5λ.4ε)	EtOH	225(4.30),252(4.13), 261(4.10),269(4.09), 308(?)	25-0848-78
3H-Naphtho[2,1-b]pyran-3-carboxylic acid, 2-methoxy-, methyl ester	MeOH	244(4.68),306(3.77), 318(3.85),350(3.83)	12-2259-78
Propanedioic acid, 7H-benzocyclohepten- 7-ylidene-, dimethyl ester	MeCN	227(4.31),296(4.25), 306(4.24),372(4.16)	78-0533-78
$C_{16}H_{14}O_5$			
Apiumetin acetate	MeOH	245(3.86),330(4.37)	102-2135-78
1,2-Azulenedicarboxylic acid, 3-acet- yl-, dimethyl ester	dioxan	237(4.46),270(4.33), 281(4.58),297(4.53), 306(4.59),331(3.76), 366(3.88),373(3.89), 514(2.88)	5-0376-78
1,3-Benzenediol, 4-(5,6-dimethoxy- 2-benzofuranyl)-	EtOH	293s(3.96),300s(3.99), 330(4.47),343(4.40)	102-0593-78
	EtOH-NaOH	355(4.42),370s(4.53)	102-0593-78
1,3-Benzenediol, 5-(4,6-dimethoxy- 2-benzofuranyl)- (moracin A)	EtOH	217(4.43),304(4.39), 313(4.47),326(4.32)	88-0797-78

Compound	Solvent	λ_{max}(log ϵ)	Ref.
5-Benzofuranol, 2-(3-hydroxy-5-methoxy-phenyl)-6-methoxy- (moracin B)	EtOH	218(4.47),285(4.09), 294(4.14),325(4.44), 337(4.40)	88-0797-78
Benzoic acid, 2-methoxy-, 4-(methoxy-carbonyl)phenyl ester	MeOH	213(4.10),243(4.23), 300(3.56)	39-0876-78C
Benzoic acid, 3-methoxy-, 4-(methoxy-carbonyl)phenyl ester	MeOH	214(4.41),243(4.29), 303(3.48)	39-0876-78C
4H-1-Benzopyran-4-one, 2,3-dihydro-5,7-dihydroxy-2-(4-methoxyphenyl)-	MeOH	289(4.25)	98-0278-78
4H-1-Benzopyran-4-one, 2,3-dihydro-7-hydroxy-3-(3-hydroxy-4-methoxy-phenyl)-	EtOH	282(4.31),315(4.07)	102-0593-78
	EtOH-NaOH	300s(4.34),340(4.63)	102-0593-78
	EtOH-NaOAc	260(4.12),280(4.07), 342(4.45)	102-0593-78
5H-Cyclopenta[b]furo[3,2-g][1]benzo-pyran-5(6H)-one, 7,8-dihydro-4,10-dimethoxy-	EtOH	249(4.65),282(3.61), 330(3.60)	39-0726-78C
6-Isoxanthenone, 2,3,7-trihydroxy-9-propyl-	pH 7.5-9	499(3.45)	140-0699-78
Trichoclin	MeOH	219(4.39),248(4.35), 262(4.12),299(4.05)	102-0143-78
Vesticarpan, (+)-	EtOH	226(4.07),288(3.57)	102-1413-78
$C_{16}H_{14}O_6$			
9,10-Anthracenedione, 1-acetoxy-1,4,4a,9a-tetrahydro-5,8-dihydroxy-	MeOH	207(4.21),226(4.25), 249(4.29),400(3.51)	150-3831-78
2H,8H-Benzo[1,2-b:3,4-b']dipyran-2,9(10H)-dione, 10-acetoxy-8,8-dimethyl-	EtOH	213(4.13),248(3.59), 261(3.55),324(4.11)	2-0184-78
Benzoic acid, 2-hydroxy-, 1,2-ethane-diyl ester	MeOH	238(4.26),306(3.94)	121-0661-78
4H-1-Benzopyran-4-one, 2,3-dihydro-5,7-dihydroxy-2-(3-hydroxy-4-meth-oxyphenyl)- (hesperetin)	MeOH	288(4.31)	98-0278-78
2H-Pyran-2,6(3H)-dione, 3,5-diacetyl-4-(4-methoxyphenyl)-	EtOH and EtOH-NaOH	280(4.34),370(4.35)	2-0196-78
	EtOH-HCl	330(4.17)	2-0196-78
	EtOAc	332(4.12)	2-0196-78
$C_{16}H_{14}O_7S_3$			
3(2H)-Thiophenone, dihydro-2,4-bis[[5-(methylsulfonyl)-2-furanyl]methyl-ene]-	EtOH	286(3.93),353(4.02), 443(3.92)	133-0189-78
$C_{16}H_{14}S_2$			
Benzo[b]cycloocta[e][1,4]dithiin, 2,3-dimethyl-	dioxan	292(4.11),344(3.52)	77-0057-78
$C_{16}H_{15}BrN_2O$			
3H-Pyrazol-4-ol, 3-bromo-4,5-dihydro-5-methyl-3,5-diphenyl-	MeOH	346(2.00)	44-0034-78
$C_{16}H_{15}BrN_2O_5$			
1H-Pyrrolo[1,2-a]indole-9-carbonitrile, 7-acetoxy-6-bromo-2,3,5,6,7,8-hexa-hydro-7-methoxy-6-methyl-5,8-dioxo-	EtOH	244(4.58),322(4.37)	94-3815-78
$C_{16}H_{15}BrN_4O$			
Benzoic acid, 4-methyl-, [[(4-bromo-phenyl)hydrazono]ethylidene]hydra-zide	EtOH	204(4.3),249(4.15), 274(4.07),316s(3.95), 370(4.63)	136-0089-78

Compound	Solvent	$\lambda_{max}(\log \epsilon)$	Ref.
$C_{16}H_{15}BrN_4O_2$ Benzoic acid, 4-methoxy-, [[(4-bromo-phenyl)hydrazono]ethylidene]hydrazide	EtOH	204(4.29),217s(3.98), 271(4.18),316s(4.01), 360(4.67)	136-0089-78E
$C_{16}H_{15}ClFN_3$ 2,5-Methano-1H-1,4-benzodiazepin-10-am-ine, 7-chloro-5-(2-fluorophenyl)-2,3,4,5-tetrahydro-, (2α,5α,10S*)-	isoPrOH	262(3.97),267(3.97), 317(3.41)	44-0936-78
$C_{16}H_{15}ClN_2O$ Benzoic acid, [1-(4-chlorophenyl)ethyl-idene]methylhydrazide	EtOH	295(4.32)	34-0345-78
$C_{16}H_{15}ClN_2O_4S$ Benzoic acid, 2-[[5-[(2-chloroethenyl)-3-nitro-2-thienyl]amino]-, 1-methyl-ethyl ester, (Z)-	MeOH	213(4.31),268(4.03), 335(4.12),407(3.90)	150-5101-78
Benzoic acid, 5-chloro-2-[(5-ethenyl-3-nitro-2-thienyl)amino]-, 1-methyl-ethyl ester	MeOH	214(4.31),271(4.02), 319(4.03),425(3.96)	150-5101-78
$C_{16}H_{15}ClN_2O_7$ 2,4(1H,3H)-Pyrimidinedione, 1-[5-O-(4-chlorobenzoyl)-β-D-arabino-furano-syl]-	MeOH	241(4.35),262s(4.08)	44-0350-78
$C_{16}H_{15}ClO$ Benzene, 2-chloro-4-methoxy-1-[2-(4-methylphenyl)ethenyl]-, (E)-	EtOH	293(4.47),300s(4.43), 320s(4.21),331s(4.07)	87-0889-78
$C_{16}H_{15}ClO_2$ 9-Phenanthrenecarboxylic acid, 3-chlo-ro-5,6,7,8-tetrahydro-, methyl ester	MeOH	241(4.68),285(3.83), 301(3.65),340s(3.52)	44-0980-78
$C_{16}H_{15}F_2N$ 1H-1-Benzazepine, 8-fluoro-5-(4-fluoro-phenyl)-2,3,4,5-tetrahydro-	MeOH	246(3.87),282s(3.33)	73-1760-78
Benzenamine, 4-(1,2-difluoro-2-phenyl-ethenyl)-N,N-dimethyl-	EtOH	338(4.46)	104-0939-78
$C_{16}H_{15}F_3O_2$ 9-Phenanthrenecarboxylic acid, 5,6,7,8-9,10-hexahydro-3-(trifluoromethyl)-	MeOH	225(4.10),275(3.76)	44-0980-78
$C_{16}H_{15}IN_4O_2$ Benzoic acid, 4-methoxy-, [[(4-iodo-phenyl)hydrazono]ethylidene]hydrazide (5λ,4ε)	EtOH	204(4.33),216s(4.19), 270(3.98),319s(?), 371(4.67)	136-0089-78E
$C_{16}H_{15}NO$ 9(10H)-Acridinone, 2-ethyl-N-methyl-	EtOH	257(4.71),297(3.45), 307s(3.04),389(3.89), 408(3.95)	35-5604-78
Benzo[b]cyclohept[e][1,4]oxazine, 7-(1-methylethyl)-	MeOH	207(4.23),215(4.24), 263(4.31),270(4.29), 415(3.96)	18-3316-78
	MeOH-HCl	206(4.22),226(4.33), 265s(4.25),275(4.28), 317(3.84),443(3.84)	18-3316-78

Compound	Solvent	λ_{max}(log ϵ)	Ref.
Benzo[b]cyclohept[e][1,4]oxazine, 8-(1-methylethyl)-	MeOH	214(4.24),263(4.31), 270(4.27),418(3.99)	18-3316-78
	MeOH-HCl	221(4.27),265(4.32), 273(4.46),320(3.80), 436(3.82)	18-3316-78
Benzo[b]cyclohept[e][1,4]oxazine, 9-(1-methylethyl)-	MeOH	207(4.17),215(4.17), 263(4.31),270(4.29), 414(3.95)	18-3316-78
	MeOH-HCl	205(4.13),227(4.20), 265s(4.24),275(4.28), 316(3.83),443(3.82)	18-3316-78
2-Buten-1-one, 1-phenyl-3-(phenyl-amino)-	C_6H_{12}	356(4.33)	48-0945-78
	EtOH	360(4.39)	48-0945-78
1H-Indole, 2-(2-methoxyphenyl)-3-methyl-	EtOH	221(4.27),307(3.96)	120-0096-78
Phenol, 4-(3,3-dimethyl-3H-indol-2-yl)-	EtOH	233(4.18),240s(4.13), 249s(3.16),326(4.36)	120-0096-78
2-Propen-1-one, 1-phenyl-3-[(phenyl-methyl)amino]-	C_6H_{12}	331(4.12)	48-0945-78
	EtOH	339(4.34)	48-0945-78
$C_{16}H_{15}NOSe$ 2-Propene-1-selone, 3-[(4-methoxyphen-yl)amino]-1-phenyl-, (Z)-	benzene	495(4.25)	104-0315-78
$C_{16}H_{15}NO_2$ 9(10H)-Acridinone, 10-ethyl-2-methoxy-	MeOH	305(3.318)	4-0149-78
Benzenepropanamide, 4-methyl-β-oxo-α-phenyl-	EtOH	205(4.38),236(4.34)	18-3389-78
Benzo[b]cyclohept[e][1,4]oxazin-10(11H)-one, 6-(1-methylethyl)-	MeOH	207(4.36),230(4.45), 257(4.34),269(4.33), 286(4.28),410s(3.76), 470(3.98)	18-3316-78
	MeOH-HCl	207(4.34),228(4.49), 260(4.28),270(4.31), 288(4.37),325s(3.82), 460(3.93)	18-3316-78
Benzo[b]cyclohept[e][1,4]oxazin-10(11H)-one, 7-(1-methylethyl)-	MeOH	208(4.39),228(4.45), 257(4.35),268(4.34), 285(4.25),420s(3.79), 480(3.97)	18-3316-78
	MeOH-HCl	208(4.37),228(4.48), 260(4.28),270(4.29), 286(4.36),325(3.85), 480(3.93)	18-3316-78
Benzo[b]cyclohept[e][1,4]oxazin-10(11H)-one, 8-(1-methylethyl)-	MeOH	205(4.34),233(4.45), 258(4.35),270(4.35), 283(4.22),415s(3.76), 480(3.97)	18-3316-78
	MeOH-HCl	205(4.34),230(4.46), 260s(4.32),270(4.35), 285(4.36),322s(3.80), 473(3.94)	18-3316-78
Phenanthridine, 5-acetyl-5,6-dihydro-8-methoxy-	MeOH	218(4.5),236(4.4), 271(4.1),302(3.7)	5-0608-78
$C_{16}H_{15}NO_2S$ Benzenesulfonamide, N-(6,6a-dihydro-cycloprop[a]inden-1a(1H)-yl)-	MeOH	265(3.70),275(3.52)	35-2181-78
$C_{16}H_{15}NO_3$ 9(10H)-Acridinone, 1,3-dimethoxy-N-methyl-	EtOH	210(4.22),225(4.21), 260(4.64),269(4.66),	83-0135-78

Compound	Solvent	$\lambda_{max}(\log \epsilon)$	Ref.
(cont.)		290(4.13),380(3.93)	83-0135-78
Benzenepropanamide, 4-methoxy-β-oxo-α-phenyl-	EtOH	206(4.18),235(4.27)	18-3389-78
Galanthan-7-one, 2,3-didehydro-9,10-[methylenebis(oxy)]-	EtOH	223(4.45),305(3.80)	1-0098-78
4-Isoxazolol, 4,5-dihydro-3-(4-methoxyphenyl)-5-phenyl-	MeOH	277(4.29)	142-0187-78C
4-Isoxazolol, 4,5-dihydro-5-(4-methoxyphenyl)-3-phenyl-	MeOH	266(4.21)	142-0187-78C
Pyridinium, 4-[2-(4-hydroxyphenyl)ethenyl]-1-(2-methoxy-2-oxoethyl)-, hydroxide, inner salt, (E)-	EtOH	520(2.58)	56-1265-78
	acetone	600(2.00)	56-1265-78
	CHCl₃	625(3.00)	56-1265-78
hydrochloride	MeOH	402(4.67)	56-1265-78
$C_{16}H_{15}NO_4$			
Acetamide, N-[(4-acetyl-5-methyl-3-oxo-2(3H)-furanylidene)methyl]-N-phenyl-	EtOH	228(4.08),285(4.01),339(4.20)	4-0155-78
9(10H)-Acridinone, 1-hydroxy-2,3-dimethoxy-10-methyl- (same in acid or base)	EtOH	230(4.18),275(4.66),400(3.77)	83-0135-78
Carbamic acid, (2-acetoxyphenyl)-, phenylmethyl ester	hexane	233(4.30)	44-0775-78
Carbonic acid, 2-(acetylamino)phenyl phenylmethyl ester	hexane	239(4.08)	44-0775-78
$C_{16}H_{15}NO_4S_3$			
1,4,8-Trithia-6-azaspiro[4.5]deca-6,9-diene-9,10-dicarboxylic acid, 7-phenyl-, dimethyl ester	CH₂Cl₂	254.5(4.34)	18-0301-78
$C_{16}H_{15}NO_5$			
9(10H)-Acridinone, 1,5-dihydroxy-2,3-dimethoxy-10-methyl-	EtOH	272(4.60),321(4.32),335s(4.26),405(3.48)	102-2125-78
	EtOH-NaOH	273(4.65),298s(4.47),338(4.31)	102-2125-78
$C_{16}H_{15}N_2S$			
Methanaminium, N-[3-isothiocyanato-3-(2-naphthalenyl)-2-propenylidene]-N-methyl-, perchlorate	MeCN	221(4.62),249s(4.15),285(4.10),337s(4.30),364(4.39)	97-0334-78
1,3-Thiazin-1-ium, 2-(dimethylamino)-4-(2-naphthalenyl)-, perchlorate	MeCN	241(4.37),259s(4.33),277s(4.20),289(4.18),350(4.35),390s(4.06)	97-0334-78
$C_{16}H_{15}N_3$			
Benzenepropanenitrile, α-[(4-methylphenyl)hydrazono]-, anti	EtOH	324(4.33)	118-0445-78
syn	EtOH	341(4.17)	118-0445-78
2-Quinoxalinamine, N-(3,5-dimethylphenyl)-	EtOH	275(4.40),285(4.40),384(3.96)	78-0981-78
$C_{16}H_{15}N_3O$			
Benzenepropanenitrile, α-[(4-methoxyphenyl)hydrazono]-, anti	EtOH	334(4.24)	118-0445-78
syn	EtOH	349(4.24)	118-0445-78
1,2,4-Triazine, 2,5-dihydro-3-methoxy-5,6-diphenyl-	MeOH	220s(4.13),291(3.86)	142-0093-78B
1,2,4-Triazin-6(1H)-one, dihydro-1-methyl-3,5-diphenyl-	MeOH	227(4.18),307(3.79)	4-1271-78
1,2,4-Triazin-6(1H)-one, 2,5-dihydro-2-methyl-3,5-diphenyl-	MeOH	225(4.18),311(3.76)	4-1271-78
	MeOH-H₂SO₄	230(4.10),278(3.71)	4-1271-78
	pH 12	226s(4.31),303(3.62)	4-1271-78

Compound	Solvent	$\lambda_{max}(\log \epsilon)$	Ref.
1,2,4-Triazin-6(1H)-one, 2,5-dihydro- 2-methyl-3,5-diphenyl- (cont.)	H_2O	227(4.26),303(3.81)	4-1271-78
	MeCN	224(4.23),327(3.63)	4-1271-78
	MeCN-H_2SO_4	231(4.14),276(3.77)	4-1271-78
	$CHCl_3$	241(4.02),331(3.49)	4-1271-78
$C_{16}H_{15}N_3O_2$ Butanamide, 3-oxo-N-phenyl-2-(phenyl- hydrazono)-	EtOH	204(3.03),240(1.46), 250(1.40),265(1.00), 374(3.04)	104-0521-78
$C_{16}H_{15}N_3O_2S$ 1,3,5-Triazin-2(1H)-one, tetrahydro- 6-(2-hydroxyphenyl)-1-methyl-3-phenyl- 4-thioxo-	EtOH	269.5(4.23)	4-1193-78
$C_{16}H_{15}N_3O_3$ 2,4,6(1H,3H,5H)-Pyrimidinetrione, 1,3- dimethyl-5-(1-methyl-4(1H)-quinolin- ylidene)-	dioxan	486(4.2)	83-0561-78
	$HClO_4$-HOAc	320(4.0)	83-0561-78
2,4,6(1H,3H,5H)-Pyrimidinetrione, 1,3- dimethyl-5-(2-methyl-4(1H)-quinolin- ylidene)-	MeOH	420(4.2)	83-0561-78
1,3,5-Triazine-2,4(1H,3H)-dione, dihy- dro-6-(2-hydroxyphenyl)-1-methyl- 3-phenyl-	EtOH	216s(4.15),276(3.48), 282s(3.47)	4-1193-78
$C_{16}H_{15}N_3O_5$ 1H-Pyrazole-4-carboxylic acid, 5-[2- (1,3-dihydro-1,3-dioxo-2H-isoindol- 2-yl)ethyl]-2,3-dihydro-3-oxo-, ethyl ester	MeOH	220(4.67),240(4.12), 263(3.79)	1-0056-78
isomer	MeOH	220(4.58),240(4.05), 260(3.58)	1-0056-78
$C_{16}H_{15}N_3S$ 1,2,4-Triazine, 2,5-dihydro-3-(methyl- thio)-5,6-diphenyl-	MeOH	230s(4.14),300(3.93)	142-0093-78B
$C_{16}H_{15}N_5O$ 9H-Imidazo[1,2-a]purin-9-one, 1,4-di- hydro-4,6-dimethyl-1-(phenylmethyl)-	90% MeOH- pH 1	229(4.59),232(4.59), 282(4.00)	44-1644-78
	90% MeOH- pH 7	229(3.54),233(4.58), 266(3.85),304(3.82)	44-1644-78
	90% MeOH- pH 12	227(4.52),230(4.56), 265(3.83),306(3.88)	44-1644-78
$C_{16}H_{15}N_7O_5S_2$ 5-Thia-1-azabicyclo[4.2.0]oct-2-ene- 2-carboxylic acid, 7-[[(3-isoxazo- lylthio)acetyl]amino]-3-[2-(1-methyl- 1H-tetrazol-5-yl)ethenyl]-8-oxo-, (6R-trans)-	MeOH-pH 6.8	324(4.37)	94-1803-78
$C_{16}H_{15}N_8$ 1,2,4,5-Tetrazin-1(2H)-yl, 3,4-dihydro- 4-(1-methyl-1H-tetrazol-5-yl)-2,6-di- phenyl-	benzene	320(3.90),381(3.89), 670(3.52)	103-0800-78
$C_{16}H_{16}$ Ethene, 1,1-bis(2-methylphenyl)-	15M H_2SO_4	332(4.21),444(4.53)	39-0892-78B

Compound	Solvent	$\lambda_{max}(\log \epsilon)$	Ref.
$C_{16}H_{16}BrNO_4$ 1H-Pyrrole-3,4-dicarboxylic acid, 2-bromo-5-phenyl-, diethyl ester	EtOH	276(4.13)	39-1588-78C
$C_{16}H_{16}BrNO_6$ 1H-Pyrrolo[1,2-a]indole-9-carboxalde-hyde, 7-acetoxy-6-bromo-2,3,5,6,7,8-hexahydro-7-methoxy-6-methyl-5,8-di-oxo-	EtOH	254(4.60),322(4.36)	94-3815-78
$C_{16}H_{16}BrNO_8$ 3aH-Indole-2,3,3a,4-tetracarboxylic acid, 1-bromo-1,7a-dihydro-, ester	MeOH	287(3.75)	44-3727-78
$C_{16}H_{16}Br_3NO_8$ 3aH-Indole-2,3,3a,4-tetracarboxylic acid, 1,6,7-tribromo-1,6,7,7a-tetra-hydro-, tetramethyl ester	MeOH	220(4.04)(end abs.)	44-3727-78
$C_{16}H_{16}ClNO_2$ 2H-1-Benzopyran-2-one, 3-chloro-5,6,7,8-tetrahydro-4-(methyl-phenylamino)-	EtOH	243(4.13),302s(3.86), 322(3.93),366(3.87)	4-0511-78
$C_{16}H_{16}ClNO_4$ 1H-Pyrrole-3,4-dicarboxylic acid, 2-chloro-5-phenyl-, diethyl ester	EtOH	275(4.12)	39-1588-78C
$C_{16}H_{16}ClN_3O_3$ Propanehydrazonic acid, 2-methyl-N-(4-nitrophenyl)-, 4-chlorophenyl ester	xylene EtOH	367(4.37) 250(4.12),386(4.28)	18-0512-78 18-0512-78
Propanoic acid, 2-methyl-, 2-(4-chloro-phenyl)-2-(4-nitrophenyl)hydrazide	xylene EtOH	360(4.20) 255(3.86),364(4.19)	18-0512-78 18-0512-78
$C_{16}H_{16}Cl_2N_4O_7$ 1H-Imidazo[4,5-d]pyridazine, 4,7-di-chloro-1-(2,3,5-tri-O-acetyl-β-D-ribofuranosyl)-	pH 1,7,11	249(3.75)	4-0001-78
$C_{16}H_{16}Cl_2O_2$ 9-Phenanthrenecarboxylic acid, 1,3-di-chloro-5,6,7,8,9,10-hexahydro-, methyl ester	MeOH	224(4.31),230(4.33), 237(4.27),249(4.04), 274(3.89)	44-0980-78
$C_{16}H_{16}FeO_4$ Pentanedioic acid, 3-ferrocenylmeth-ylene-, (E)-	EtOH	244(4.14),292(4.17)	128-0273-78
$C_{16}H_{16}N_2$ Benzenamine, N-[1-methyl-3-(phenylami-no)-2-propenylidene]-, monoperchlor-ate	MeOH	238(3.97),367(4.58)	39-1453-78C
Bicyclo[8.2.2]tetradeca-5,10,12,13-tetraene-4,7-dicarbonitrile, cis	EtOH	223(3.83),269(2.62), 277(2.57)	35-1806-78
trans	EtOH	223(3.82),269(2.65), 277(2.60)	35-1806-78
2,4-Heptadien-6-ynal, 5-methyl-, (5-methyl-2,4-heptadien-6-yn-ylidene)hydrazone, (?,?,E,E,Z,Z)-	THF	231(4.06),239s(3.99), 247s(3.80),280s(3.81), 292(3.88),322s(4.35), 341s(4.63),357(4.76), 373(4.76),392(4.57)	18-2112-78

Compound	Solvent	λ_{max}(log ϵ)	Ref.
Pyridazine, 3,4,5,6-tetrahydro-3,6-diphenyl-	PhEt	385(2.81)	24-0596-78
$C_{16}H_{16}N_2O$			
Acetamide, N-[2-[2-(2-aminophenyl)ethenyl]phenyl]-	benzene	346(3.92)	33-2813-78
Benzeneacetic acid, (1-phenylethylidene)hydrazide	EtOH	287(4.23)	34-0345-78
Benzo[g]quinazolin-4(3H)-one, 3-methyl-2-(1-methylethyl)-	EtOH	262(4.66),272(4.62),332(3.73),348(3.70)	2-1067-78
Benzo[h]quinazolin-4(3H)-one, 3-methyl-2-(1-methylethyl)-	EtOH	250(4.69),297(4.03),310(4.07),332(3.76),348(3.78)	2-1067-78
2,5-Cyclohexadien-1-one, 4-diazo-2-(1,1-dimethylethyl)-6-phenyl-	MeOH	288(4.00),370(4.51)	70-2455-78
4H-1,2,5-Oxadiazine, 5,6-dihydro-5-(4-methylphenyl)-3-phenyl-	MeOH	246(4.19)	56-1827-78
2(1H)-Quinazolinone, 1-ethyl-3,4-dihydro-4-phenyl-	MeOH	254(3.89)	32-0591-78
$C_{16}H_{16}N_2OS$			
4H-Thiopyrano[2,3-b]furan-5-carbonitrile, 4-(dimethylamino)-5,6-dihydro-2-phenyl-	C_6H_{12}	306(4.47)	104-0589-78
$C_{16}H_{16}N_2O_2$			
Benzamide, N-methyl-N-[6-(methylamino)-7-oxo-1,3,5-cycloheptatrien-1-yl]-	MeOH	253(4.25),345(4.04),420(4.20)	18-2338-78
Benzeneacetic acid, [(4-methoxyphenyl)methylene]hydrazide	EtOH	291(4.40),304s(4.34)	34-0345-78
Benzeneacetic acid, 2-(phenylacetyl)hydrazide	EtOH	255(2.44)	4-0385-78
Benzoic acid, 2-acetyl-2-(4-methylphenyl)hydrazide	EtOH	229(4.29)	33-1477-78
Benzoic acid, [1-(4-methoxyphenyl)ethylidene]hydrazide	EtOH	282(4.32),291s(4.25)	34-0345-78
Benzoic acid, 2-(1-oxopropyl)-2-phenylhydrazide	EtOH	229(4.22)	33-1477-78
Benzo[a]pyrrolo[3,4-f]quinolizine-1,12-dione, 2,3,5,6,10b,11-hexahydro-3-methyl-	MeOH	243(4.10),308(4.10)	30-0232-78
1H-Isoindole-1,3(2H)-dione, 2-(9-azabicyclo[6.1.0]non-4-en-9-yl)-, (1α,4Z,8α)-	EtOH	215(4.16),237(4.39),295(3.06),304(2.99),325s(2.68)	33-0795-78
4H-1,2,5-Oxadiazine, 5,6-dihydro-5-(4-methoxyphenyl)-3-phenyl-	MeOH	247(4.28)	56-1827-78
1H-Pyrazole-5-carboxylic acid, 1-(4-pentynyl)-3-phenyl-	EtOH	234(4.41)	44-1664-78
Pyrano[2,3-c]pyrazol-6(1H)-one, 1,3,4-trimethyl-5-(phenylmethyl)-	EtOH	318(4.16)	95-0335-78
Pyrano[2,3-c]pyrazol-6(2H)-one, 2,3,4-trimethyl-5-(phenylmethyl)-	EtOH	309.9(4.21)	95-0335-78
$C_{16}H_{16}N_2O_2S$			
1,3-Cyclohexanedione, 2-[(2-benzothiazolylamino)methylene]-5,5-dimethyl-	EtOH	367(4.36)	97-0256-78
$C_{16}H_{16}N_2O_3$			
Benzo[a]pyrrolo[3,4-f]quinolizine-1,12-dione, 2,3,5,6,10b,11-hexahydro-8-methoxy-, (R)-	MeOH	243(4.15),305(4.14)	30-0232-78

Compound	Solvent	λ_{max}(log ϵ)	Ref.
3-Pyridinecarboxamide, N-[2-(2,4-di-methoxyphenyl)ethenyl]-, (E)-	n.s.g.	266(3.31),367(3.50)	88-2723-78
$C_{16}H_{16}N_2O_3S$ 1H-Isoindole-1,3(2H)-dione, 2-(7-oxa-3-thia-1-azaspiro[5.5]undec-1-en-2-yl)-, (±)-	EtOH	220(4.69),293(3.37), 238s(4.15)	87-0895-78
$C_{16}H_{16}N_2O_4$ 1H-Pyrazole-4,5-dicarboxylic acid, 3-phenyl-1-(2-propenyl)-	EtOH	236(4.26)	44-1664-78
2-Pyridinecarboxylic acid, 4,4'-(1,4-butanediyl)bis-	EtOH	264(3.86)	78-0331-78
$C_{16}H_{16}N_2O_5$ 2,4,6(1H,3H,5H)-Pyrimidinetrione, 5-[3-hydroxy-3-(2-methoxyphenyl)-2-propen-ylidene]-1,3-dimethyl-	MeOH	392(4.4)	83-0503-78
$C_{16}H_{16}N_2O_5S$ 2,2'-Anhydro-2-mercapto-1-(3,5-di-O-acetyl-β-D-ribofuranosyl)benzimid-azole	MeOH	249(4.02),282(4.01), 290(4.04)	4-0657-78
5-Thia-1-azabicyclo[4.2.0]oct-2-ene-2-carboxylic acid, 3-methyl-8-oxo-7-[(phenoxyacetyl)amino]-, (6R-trans)	EtOH	260(3.68),266(3.69), 273(3.66),300(3.77)	35-8214-78
$C_{16}H_{16}N_2O_7S$ 4-Oxa-1-azabicyclo[4.2.0]oct-2-ene-2-carboxylic acid, 3-(acetoxymethyl)-8-oxo-7-[(2-thienyl)acetyl]amino]-	EtOH	272(3.94)	23-1335-78
$C_{16}H_{16}N_4$ Benzenamine, N-(4,5-dihydro-5,5-dimeth-yl-4-phenyl-3H-1,2,4-triazol-3-yli-dene)-, (E)-	EtOH	230(4.69),338(3.93), 395(3.88)	23-2194-78
(Z)-	EtOH	235(4.11),345(3.67), 410(3.38)	23-2194-78
$C_{16}H_{16}N_4O$ 3-Cyclohexene-1-carboxamide, 4-amino-1,3-dicyano-N-(2-methylphenyl)-	MeOH	260(4.12)	42-0281-78
3-Cyclohexene-1-carboxamide, 4-amino-1,3-dicyano-N-(phenylmethyl)-	MeOH	260(4.04)	42-0281-78
$C_{16}H_{16}N_4OS$ Acridinium, 3,9-diamino-7-ethoxy-, thiocyanate	MeOH	213(4.1),274(4.5), 295(4.3),308s(--), 377(4.2),418(4.0), 439(4.0)	97-0108-78
$C_{16}H_{16}N_4O_2$ 1,2-Ethanediamine, N-methyl-N'-(1-ni-tro-9-acridinyl)-	pH 1	216(4.342),272(4.608), 332(3.677),436(3.917)	56-2125-78
	pH 7	276(4.209),395(3.601)	56-2125-78
Pyrazolo[3,4-d][1,2]diazepine, 2-acet-yl-6-benzoyl-2,3,3a,6-tetrahydro-8-methyl-	EtOH	331(4.26)	39-1297-78C

Compound	Solvent	$\lambda_{max}(\log \epsilon)$	Ref.
$C_{16}H_{16}N_4O_5$			
Propanehydrazonic acid, 2-methyl-N-(4-nitrophenyl)-, 4-nitrophenyl ester	xylene	362(4.21)	18-0512-78
	EtOH	255s(3.66),325(3.71), 383(4.12)	18-0512-78
Propanoic acid, 2-methyl-, 2,2-bis(4-nitrophenyl)hydrazide	xylene	380(4.18)	18-0512-78
	EtOH	266(3.90),372(4.40)	18-0512-78
$C_{16}H_{16}N_4O_6$			
Benzo[1,2-c:4,5-c']dipyrazole-3,7-diol, 2,6-diacetyl-, diacetate	EtOH	242(4.27)	48-0991-78
$C_{16}H_{16}N_8$			
Formazan, 5-methyl-5-(2-methyl-2H-tet-razol-5-yl)-1,3-diphenyl-	EtOH	244(4.26),275(4.28), 365(4.69)	103-0800-78
1,2,4,5-Tetrazine, 1,2,5,6-tetrahydro-1-(2-methyl-2H-tetrazol-5-yl)-3,5-diphenyl-	EtOH	245(4.30),305(4.15)	103-0800-78
$C_{16}H_{16}O$			
1-Butanone, 1,4-diphenyl-	n.s.g.	228(4.18)	44-1616-78
Methanone, phenyl(2,4,6-trimethylphenyl)-	hexane	243(4.10),272(3.42), 338(1.82),350(1.82)	104-0582-78
Oxacycloheptadeca-2,4,6,8,10,12,14,16-octaene (isomer mixture)	ether	304(4.57),405(3.83)	24-0099-78
isomer	ether	315(4.76),407(3.85)	24-0099-78
4-Oxatetracyclo[9.6.0.0^{2,10}.0^{3,5}]heptadeca-6,8,12,14,16-pentaene, exo	ether	224(3.85),250(3.32)	24-0099-78
6-Oxatetracyclo[9.6.0.0^{2,10}.0^{5,7}]heptadeca-3,8,12,14,16-pentaene, endo	ether	249s(3.26)	24-0099-78
Phenanthrene, 9,10-dihydro-7-methoxy-1-methyl-	EtOH	278(4.47)	77-0526-78
1-Propanone, 2-methyl-1,2-diphenyl-	C_6H_{12}	335(2.03),349(1.81), 364(1.46)	112-0751-78
$C_{16}H_{16}OS$			
Benzo[c]thiophene-1-ol, 1,3-dihydro-3,3-dimethyl-1-phenyl-	hexane	263s(3.08),269s(2.97)	104-0582-78
$C_{16}H_{16}OS_2$			
2,5-Dithia[6.1]metabenzenophan-13-one	MeCN	250(4.04),291(3.41)	24-2547-78
$C_{16}H_{16}O_2$			
Benzene, 1,3-dimethoxy-5-(2-phenyleth-enyl)-, (E)-	n.s.g.	265(4.66)	39-0739-78C
Benzene, 2,4-dimethoxy-1-(2-phenyleth-enyl)-, (E)-	MeOH	238(4.14),295s(4.28), 303(4.28),335(4.38)	12-2259-78
Cyclooctatetraene tricyclic dimer bis-epoxide	ether	224(3.96)	24-0099-78
2,4-Ethanobiscyclopropa[4,5]cyclopenta-[1,2,3-cd:1',2',3'-gh]pentalene-5,6-dione, tetradecahydro-	C_6H_{12}	230.0(3.61)	35-5589-78
	EtOH	407.4(1.22)	35-5589-78
	CH_2Cl_2	412.0(1.23)	35-5589-78
1-Isobenzofuranol, 1,3-dihydro-3,3-di-methyl-1-phenyl-	hexane	257s(2.78),262(2.84), 268(2.73)	104-0582-78
2(4aH)-Phenanthrenone, 9,10-dihydro-7-methoxy-4a-methyl-	EtOH	233(4.71),272(3.85)	44-1580-78
Phenol, 4-[3-(4-methoxyphenyl)-1-prop-enyl]-	EtOH	210(3.70),262(3.69)	39-0088-78C
Spiro[cyclobutane-1,3'(4'H)-[2H]naph-tho[2,3-b][1,4]dioxepin	$CHCl_3$	246(4.17),284(3.66), 315(3.12),324(3.12), 329(3.17)	56-1913-78

Compound	Solvent	$\lambda_{max}(\log \epsilon)$	Ref.
Spiro[2,5-cyclohexadiene-1,1'(2'H)-naphthalen]-4-one, 3',4'-dihydro-5'-methoxy-	EtOH	225(4.30)	22-0202-78
Spiro[2,5-cyclohexadiene-1,1'(2'H)-naphthalen]-4-one, 3',4'-dihydro-7'-methoxy-	EtOH	232(4.31)	22-0202-78
$C_{16}H_{16}O_3$			
3-Buten-2-one, 4-(6,7-dimethoxy-2-naphthalenyl)-	MeOH	223(4.85),268(4.85),342(4.84)	83-0328-78
2,5-Cyclohexadiene-1,4-dione, 2-methoxy-5-(1-methyl-2-phenylethyl)-, (S)-	EtOH	207(4.11),264(4.10),325(2.96)	102-1395-78
Cyclopenta[b]furo[3,2-g][1]benzopyran-5(6H)-one, 2,3,7,8-tetrahydro-2,10-dimethyl-	MeOH	249(4.18),300(4.18)	42-0468-78
2,5-Furandione, 3-(1-ethylpropylidene)-dihydro-4-(phenylmethylene)-, (E)-	toluene	330(4.04)	39-0571-78C
Obtusafuran, (2R,3R)-	EtOH	235s(3.47),305(3.66)	102-1395-78
$C_{16}H_{16}O_4$			
2H-1-Benzopyran-7-ol, 3,4-dihydro-3-(2-hydroxy-4-methoxyphenyl)-, (+)-(vestitol)	EtOH	206(4.66),228(4.05),285(3.76)	102-1413-78
1H-Naphtho[2,1-b]pyran-3-carboxylic acid, 2,3-dihydro-2-methoxy-, methyl ester	MeOH	232(4.82),258s(3.42),267(3.56),277(3.65),288(3.54),318(3.18),332(3.27)	12-2259-78
1-Propanone, 3-phenyl-1-(2,4,6-trihydroxy-3-methylphenyl)-	EtOH	293(4.27)	102-2015-78
	EtOH-NaOH	329(4.37)	102-2015-78
Spiro[1,3-dioxane-5,3'(4'H)-[2H]naphtho[2,3-b][1,4]dioxepin]	CHCl₃	246(4.19),268s(--),276(3.73),285(3.74),296s(--),314(3.55),328(3.34)	56-1913-78
$C_{16}H_{16}O_5$			
1,2-Benzenediol, 3-(3,4-dihydro-7-hydroxy-2H-1-benzopyran-3-yl)-6-methoxy-	EtOH	220(4.31),285(3.53)	102-1413-78
2,5-Furandione, 3-[(3,5-dimethoxyphenyl)methylene]dihydro-4-(1-methylethylidene)-, (E)-	toluene	336(3.92)	39-0571-78C
	C₆H₁₁Me	329(4.18)	39-0571-78C
	CCl₄	332(4.07)	39-0571-78C
	o-C₆H₄Cl₂	336(4.11)	39-0571-78C
1-Naphthalenecarboxaldehyde, 5,8-dihydro-2-hydroxy-3-methoxy-6-methyl-4-(1-methylethyl)-5,8-dioxo-	EtOH-HCl	266(4.36),302s(--),362(3.40)	102-1297-78
	EtOH-NaOH	239(4.38),323(4.07),528(3.48)	102-1297-78
	CHCl₃	261s(--),268(4.41),305(4.05),378(3.36)	102-1297-78
Naphtho[2,3-c]furan-1,3-dione, 4,9-dihydro-5,7-dimethoxy-4,4-dimethyl-	hexane	230(3.75),248s(3.53),279(3.36)	39-0571-78C
2H-Pyran-5-carboxylic acid, 4-(4-methoxyphenyl)-6-methyl-2-oxo-, ethyl ester	n.s.g.	246(4.05),307(4.09)	2-0200-78
$C_{16}H_{16}O_6$			
1-Benzoxepin-5(2H)-one, 3,7-bis(acetoxymethyl)-	EtOH	218(4.36),263(4.05),325(3.30)	39-1490-78C
3-Furancarboxylic acid, 5-acetoxy-4,5-dihydro-5-methyl-4-oxo-2-phenyl-, ethyl ester	EtOH	218(3.97),250(3.69),300(4.16)	118-0291-78

Compound	Solvent	$\lambda_{max}(\log \epsilon)$	Ref.
$C_{16}H_{16}O_7$			
9,10-Anthracenedione, 1,2,3,4-tetrahydro-2,3,5,8-tetrahydroxy-6-methoxy-2-methyl-, cis	MeOH	204(4.48),226(4.53), 299(3.81),475s(--), 498(3.80),534(3.64)	150-3831-78
1,4-Naphthalenedione, 6-(1-acetoxyethyl)-5-hydroxy-2,7-dimethoxy-	MeOH	220(4.51),257s(4.14), 263(4.15),306(3.97), 425(3.55)	31-1257-78
Spiro[1-benzoxepin-5(2H),2'-[1,3]dioxolane]-3,7-dicarboxylic acid, dimethyl ester	EtOH	229(4.22),265(3.79)	39-1490-78C
$C_{16}H_{16}O_7S$			
Propanedioic acid, [(6-methoxy-3-oxo-2(3H)-benzofuranylidene)(methylthio)methyl]-, dimethyl ester	EtOH	256(4.01),350(4.31)	142-0399-78A
$C_{16}H_{16}O_9$			
α-D-Glucopyranosiduronic acid, 4-methyl-2-oxo-2H-1-benzopyran-7-yl-	MeOH	247(3.37),288(3.94), 317(4.14)	136-0023-78F
β-D-Glucopyranosiduronic acid, 4-methyl-2-oxo-2H-1-benzopyran-7-yl-	MeOH	246(3.34),290(3.92), 317(4.11)	136-0023-78F
$C_{16}H_{16}S_2$			
2,4-Dithia[5.1]metabenzenophane	MeCN	245(4.08),255(4.19), 265(4.08)	24-2547-78
$C_{16}H_{17}ClN_2O$			
1H-1,4-Benzodiazepin-2-one, 7-chloro-5-(1-cyclohexen-1-yl)-2,3-dihydro-1-methyl-	10% DMF-pH 1.65	284(4.19),345(3.79)	86-0209-78
$C_{16}H_{17}ClN_2O_4$			
Benzenamine, N-[1-methyl-3-(phenylamino)-2-propenylidene]-, monoperchlorate	MeOH	238(3.97),367(4.58)	39-1453-78C
$C_{16}H_{17}ClN_2O_5S$			
Benzoic acid, 5-chloro-2-[[5-(1-hydroxyethyl)-3-nitro-2-thienyl]amino]-, 1-methylethyl ester	MeOH	209(4.33),275(4.22), 415(4.13)	150-5101-78
$C_{16}H_{17}ClO$			
Naphthalene, 1-(2-chloro-2-propenyl)-6-methoxy-2,3-dimethyl-	EtOH	233(4.65)	39-0110-78C
$C_{16}H_{17}Cl_2N_7$			
2,4-Pteridinediamine, 6-[[(3,4-dichlorophenyl)(1-methylethyl)amino]methyl]-	MeOH-HCl MeOH-KOH	249(4.36),337(4.02) 262(4.59),377(3.91)	87-0331-78 87-0331-78
$C_{16}H_{17}Cl_2N_7O$			
2,4-Pteridinediamine, 6-[[(3,4-dichlorophenyl)(1-methylethyl)amino]methyl]-, 8-oxide	MeOH-HCl MeOH-KOH	260(4.63),361(3.90) 269(4.66),400(3.94)	87-0331-78 87-0331-78
$C_{16}H_{17}IN_2$			
1H-Imidazolium, 4,5-dihydro-3-methyl-1,2-diphenyl-, iodide	EtOH	206(4.40),220(4.40), 270(4.08)	39-0545-78B
$C_{16}H_{17}N$			
Azacycloheptadeca-2,4,6,8,10,12,14,16-octaene	ether	261(4.28),353(4.70), 445s(3.67),532s(3.20),	24-0084-78

Compound	Solvent	$\lambda_{max}(\log \epsilon)$	Ref.
(cont.) Azacycloheptadeca-2,4,6,8,10,12,14,16- octaene, "external"	ether	570(2.95),646(2.30) 250(4.24),350(4.76), 439(3.72),515s(3.08), 560s(2.85),616s(2.34)	24-0084-78 24-0084-78
4-Azatetracyclo[9.6.0.0^{2,10}.0^{3,5}]hepta- deca-6,8,12,14,16-pentaene	MeOH	230s(3.83),267s(3.20)	24-0084-78
6-Azatetracyclo[9.6.0.0^{2,10}.0^{5,7}]hepta- deca-3,8,12,14,16-pentaene	MeOH	255(3.30)	24-0084-78
Benzenamine, N,N-dimethyl-4-(2-phenyl- ethenyl)-	EtOH	333(4.46)	104-0939-78
2H-Pyrrole, 3,4-dihydro-3,3-dimethyl- 5-(4-phenyl-1-buten-3-ynyl)-	MeOH	204(4.17),227(4.09), 300(4.49),338(4.44)	83-0977-78
$C_{16}H_{17}NO$ 4H-Indol-4-one, 1,5,6,7-tetrahydro- 6,6-dimethyl-2-phenyl-	MeOH	215(2.86),240(2.72), 280(2.72),310(3.55)	48-0863-78
$C_{16}H_{17}NO_2$ Benzenebutanoic acid, α-amino-γ-phenyl-, hydrochloride	EtOH	215(4.30),266(2.80)	56-2455-78
Benzoic acid, 2-[[4-(1-methylethyl)- phenyl]amino]-	EtOH	286(4.20),354(3.90)	35-5604-78
2,4,6-Cycloheptatrien-1-one, 2-[(2-hy- droxyphenyl)amino]-4-(1-methylethyl)-	MeOH	208(4.45),235(4.28), 255(4.18),343(3.88), 400(4.07)	18-3316-78
	MeOH-HCl	208(4.25),245(4.32), 257(4.30),378(3.92)	18-3316-78
	MeOH-NaOH	235(4.85),340s(3.62), 410(4.00)	18-3316-78
2,4,6-Cycloheptatrien-1-one, 2-[(2-hy- droxyphenyl)amino]-5-(1-methylethyl)-	MeOH	208(4.38),236(4.42), 346(3.96),409(4.07)	18-3316-78
	MeOH-HCl	241(4.41),382(3.96)	18-3316-78
	MeOH-NaOH	236(4.64),346(3.65), 418(4.00)	18-3316-78
2,4,6-Cycloheptatrien-1-one, 2-[(2-hy- droxyphenyl)amino]-6-(1-methylethyl)-	MeOH	236(4.28),250(4.28), 345(3.87),400(4.08)	18-3316-78
	MeOH-HCl	245(4.20),255(4.19), 383(3.88)	18-3316-78
	MeOH-NaOH	235(4.27),410(3.88)	18-3316-78
1,4-Naphthalenedione, 2-(hexahydro-1H- azepin-1-yl)-	EtOH	241(4.19),278(4.35), 469(3.72)	119-0037-78
$C_{16}H_{17}NO_4$ 1-Azabicyclo[3.2.0]hept-2-ene-2-carbox- ylic acid, 6-(1-hydroxyethyl)-7-oxo-, phenylmethyl ester	EtOH	276(3.89)	35-8004-78
3H-Cyclobut[b]indole-1,2-dicarboxylic 2a,7b-dihydro-3,7b-dimethyl-, dimethyl ester	EtOH	251(3.61),302(3.14), 435(2.79)	88-2979-78
3-Pyridinecarboxylic acid, 5-ethoxy- 1,6-dihydro-6-oxo-2-phenyl-, ethyl ester	EtOH	290(4.20)	142-0161-78A
1H-Pyrrole-3,4-dicarboxylic acid, 2-phenyl-, diethyl ester	EtOH	277(4.11)	39-1588-78C
$C_{16}H_{17}NO_4S$ 2-Butenedioic acid, 2-[2-(phenylmeth- ylene)-3-thiazolidinyl]-, dimethyl ester	MeOH	209(4.3),296(4.3), 376(3.9)	28-0385-78B

Compound	Solvent	$\lambda_{max}(\log \epsilon)$	Ref.
$C_{16}H_{17}NO_5$			
Acetamide, N-[(4-acetyl-2,5-dihydro-3-hydroxy-5-oxo-2-furanyl)methyl]-N-(4-methylphenyl)-	EtOH	230(4.24),265(4.22)	4-0327-78
2-Piperidinone, 4-(5,7-dihydroxy-2-methyl-4-oxo-4H-1-benzopyran-6-yl)-1-methyl-	MeOH	205(4.35),225(4.16), 251(4.22),257(4.23), 295(3.77),318(3.66)	88-2911-78
1H-Pyrrole-2,4-dicarboxylic acid, 3-hydroxy-1-phenyl-, diethyl ester	EtOH	232(4.44),262(4.06)	94-2224-78
$C_{16}H_{17}NO_6$			
Acetamide, N-[(4-acetyl-2,5-dihydro-3-hydroxy-5-oxo-2-furanyl)methyl]-N-(4-methoxyphenyl)-	EtOH	230(4.32),265(4.22)	4-0327-78
3-Quinolineacetic acid, 1-acetoxy-1,4-dihydro-α-methoxy-2-methyl-4-oxo-, methyl ester	EtOH	240(4.30),280s(3.47), 292(3.59),325(3.96)	95-0802-78
$C_{16}H_{17}NO_8$			
3H-Indole-2,3,3,4-tetracarboxylic acid, 1,2-dihydro-, tetramethyl ester	MeOH	220(4.23),321(4.20)	44-3727-78
3aH-Indole-2,3,3a,4-tetracarboxylic acid, 1,7a-dihydro-, tetramethyl ester	MeOH	272(4.10),300s(3.76)	44-3727-78
3aH-Indole-2,3,3a,4-tetracarboxylic acid, 3,7a-dihydro-, tetramethyl ester	MeOH	293(3.77),333s(3.13)	44-3727-78
$C_{16}H_{17}N_2$			
1H-Imidazolium, 4,5-dihydro-3-methyl-1,2-diphenyl-, iodide	EtOH	206(4.40),220(4.40), 270(4.08)	39-0545-78B
$C_{16}H_{17}N_3$			
Benzenamine, N-[1-methyl-2-(phenylazo)-1-propenyl]-, monoperchlorate	CHCl$_3$	409(4.26)	39-1023-78C
1,10-Phenanthrolin-4-amine, N,N-diethyl-	neutral	213(4.38),240(4.00), 263(4.42),284s(4.35), 320(4.10),340s(4.14), 352(4.16)	39-1215-78B
	monocation	208(4.36),225(4.34), 256(4.37),278(4.41), 291(4.24),320(4.14), 355(4.11)	39-1215-78B
$C_{16}H_{17}N_3O_2S$			
Benzoic acid, 2-[[(4-ethoxyphenyl)amino]thioxomethyl]hydrazide	MeOH	238(4.30)	80-0397-78
nickel complex	MeOH	270(4.40)	80-0397-78
Propanehydrazonothioic acid, N-(4-nitrophenyl)-, 4-methylphenyl ester	EtOH	275s(3.939),386s(4.456)	146-0864-78
$C_{16}H_{17}N_3O_3$			
Piperazine, 1-[2-(5-nitro-2-furanyl)-ethenyl]-4-phenyl-, (E)-	EtOH	249(4.35),277(4.38), 500(4.47)	103-0250-78
Propanehydrazonic acid, 2-methyl-N-(4-nitrophenyl)-, phenyl ester	xylene	376(4.30)	18-0512-78
	EtOH	250(4.20),390(4.45)	18-0512-78
Propanoic acid, 2-(4-methylphenyl)-2-(4-nitrophenyl)hydrazide	xylene	360(4.23)	18-0512-78
	EtOH	250s(3.91),366(4.23)	18-0512-78
Propanoic acid, 2-methyl-, 2-(4-nitrophenyl)-2-phenylhydrazide	xylene	360(4.18)	18-0512-78
	EtOH	250(3.88),366(4.22)	18-0512-78

Compound	Solvent	$\lambda_{max}(\log \epsilon)$	Ref.
Pyrazole, 3-[(N-acetyl-N-phenylamino)-acetyl]-4-acetyl-5-methyl-	EtOH	219(4.08),245(3.91)	4-0155-78
1,3,5-Triazine-2,4(1H,3H)-dione, dihydro-6-(2-hydroxy-1-naphthalenyl)-1,3,5-trimethyl-	EtOH	237(4.88),258s(3.51), 266(3.66),279(3.77), 290(3.70),322(3.46), 334(3.54)	4-1193-78
Urea, N-(2,3-dihydro-2-methyl-3-oxo-1H-naphth[1,2-e][1,3]oxazin-1-yl)-N,N'-dimethyl-	EtOH	227(4.79),266s(3.67), 283(3.66),287(3.63), 307(3.04),321(3.04), 321(3.15)[sic]	4-1193-78
$C_{16}H_{17}N_3O_5$			
Glycine, N-(N,1-diformyl-L-tryptophyl)-, methyl ester	MeOH-HCl	300(3.70)	63-1643-78
$C_{16}H_{17}N_5O$			
Acetamide, N-[(3,4-dimethyl-1H-pyrazolo[3,4-d]pyridazin-7-yl)methyl]-N-phenyl-	EtOH	216(4.21),280(3.60)	4-0155-78
3-Formazancarboxamide, 1,5-bis(2-methylphenyl)-	EtOH	268(3.91),295(4.02), 435(4.41)	104-1956-78
3-Formazancarboxamide, 1,5-bis(4-methylphenyl)-	EtOH	265(3.85),298(3.74), 450(4.17)	104-1956-78
$C_{16}H_{17}N_5O_2$			
3,4-Pyridinedicarboxylic acid, 2-methyl-6-(2-phenylethenyl)-, dihydrazide	EtOH	235(5.00),287(5.23), 336(5.30)	106-0782-78
$C_{16}H_{17}N_5O_3$			
3-Formazancarboxamide, 1,5-bis(4-methoxyphenyl)-	EtOH	237(4.14),310(3.87), 377(4.25),472(4.02)	104-1956-78
$C_{16}H_{17}N_5O_3S$			
2,4(1H,3H)-Pyrimidinedione, 1,3-dimethyl-5-[[6-methyl-2-(methylthio)-4-oxopyrido[2,3-d]pyrimidin-8(4H)-yl]-methyl]-	pH 1	266(3.93),271(4.35), 290(4.12),350(4.19)	44-0828-78
	pH 7	266(3.93),274(4.46), 366(4.11)	44-0828-78
	pH 11	266(3.90),274(4.47), 366(4.13)	44-0828-78
$C_{16}H_{17}N_5O_7S_2$			
5-Thia-1-azabicyclo[4.2.0]oct-2-ene-2-carboxylic acid, 3-(acetoxymethyl)-7-[[(2-amino-4-thiazolyl)(methoxyimino)acetyl]amino]-8-oxo-, (E)-	EtOH	237(4.32),305(3.48)	78-2233-78
(Z)-	EtOH	238(4.20),252(4.16), 295(3.78)	78-2233-78
$C_{16}H_{17}N_5O_{10}$			
2,4(1H,3H)-Pyrimidinedione, 1-[2-deoxy-2-[(2,4-dinitrophenyl)amino]-β-D-glucopyranosyl]-	EtOH	257(4.13),345(4.11)	136-0185-78C
$C_{16}H_{17}N_7$			
2,4-Pteridinediamine, 6-[(3,4-dihydro-1(2H)-quinolinyl)methyl]-	MeOH-HCl MeOH-KOH	248(4.33),337(4.02) 260(4.53),375(3.88)	87-0331-78 87-0331-78
$C_{16}H_{17}N_7O$			
2,4-Pteridinediamine, 6-[(3,4-dihydro-1(2H)-quinolinyl)methyl]-, 8-oxide	MeOH-HCl MeOH-KOH	258(4.64),360(3.91) 268(4.64),339(3.95)	87-0331-78 87-0331-78

Compound	Solvent	λ_{max}(log ϵ)	Ref.
C$_{16}$H$_{18}$BrN$_2$O$_2$			
Pyridinium, 1-(4-bromobutyl)-4-[(4-nitrophenyl)methyl]-, bromide	H$_2$O	215s(4.16),260s(4.13) 265(4.15),274s(4.09)	48-0133-78
deprotonated	H$_2$O	552(4.52)	48-0133-78
C$_{16}$H$_{18}$F$_6$NO$_5$P			
1,4,2-Oxazaphosphole, 2,2,2,3-tetrahydro-2,2,2-trimethoxy-5-[2-(2-propenyloxy)phenyl]-3,3-bis(trifluoromethyl)-	C$_6$H$_{12}$	239(2.91),293(2.58)	35-3494-78
C$_{16}$H$_{18}$Fe			
Ferrocene, 1,1':2,2'-bis(1,3-propanediyl)-	CH$_2$Cl$_2$	439(2.45)	101-0077-78L
cation	EtOH	252(--),630(2.62)	101-0077-78L
Ferrocene, 1,1':3,3'-bis(1,3-propanediyl)-	CH$_2$Cl$_2$	434(2.26)	101-0077-78L
cation	EtOH	246(--),690(2.56)	101-0077-78L
C$_{16}$H$_{18}$NO$_5$PS$_3$			
Phosphorothioic acid, O-ethyl O-phenyl S-[(2,3,5,7-tetrahydro-2-methyl-5,7-dioxo-6H-1,4-dithiino[2,3-c]pyrrol-6-yl)methyl] ester	EtOH	210(4.14),264(3.97), 421(3.54)	73-1093-78
C$_{16}$H$_{18}$N$_2$			
9H-Pyrido[3,4-b]indole, 1-pentyl-	MeOH	236(4.64),241(4.62), 251s(4.45),283s(4.08), 289(4.31),338(3.77), 351(3.78)	95-0898-78
C$_{16}$H$_{18}$N$_2$O			
Benzamide, N-methyl-N-[2-(phenylamino)ethyl]-	EtOH	206(4.43),246(4.16)	39-0545-78B
2,5-Cyclohexadien-1-one, 4-[[4-(diethylamino)phenyl]imino]-	C$_6$H$_{12}$	270(4.43),295(4.26), 335(4.03),560(4.44)	48-0557-78
	EtOH	275(4.15),315(3.97), 350(3.85),620(4.51)	48-0557-78 +80-0617-78
	ether	273(4.20),300(4.02), 340(3.83),570(4.31)	48-0557-78
	CCl$_4$	295(4.00),340(3.75), 565(4.12)	48-0557-78
13H-Isoquino[2,1-a][1,8]naphthyridin-13-one, 1,2,3,4,6,7,11b,12-octahydro-	EtOH	238(4.00),309(4.28)	142-1717-78A
10H-Pyrido[3,2-b][1,4]benzoxazine, 10-(3-methylbutyl)-	C$_6$H$_{12}$	233(4.62),237(4.62), 342(4.08)	95-0585-78
4(1H)-Pyrimidinone, 2-methyl-5-(4-pentenyl)-6-phenyl-	EtOH	231(4.04),281(3.68)	39-1293-78C
C$_{16}$H$_{18}$N$_2$OS			
Acetamide, N-(4-piperidino-2H-1-benzothiopyran-2-ylidene)-	EtOH	251(4.13),282(4.82), 387(4.37)	103-0507-78
C$_{16}$H$_{18}$N$_2$OS$_2$			
Piperidinium, 1-[4-(acetylamino)-5-phenyl-1,3-dithiol-2-ylidene]-, hydroxide, inner salt	dioxan	253s(4.12),353(4.18), 367(4.17),435(3.71)	24-2021-78
C$_{16}$H$_{18}$N$_2$O$_2$			
1H-Isoindole-1,3(2H)-dione, 2-(9-azabicyclo[6.1.0]non-9-yl)-, cis	EtOH	215(4.17),237(4.40), 294(3.06),304(2.99), 325s(2.68)	33-0795-78

Compound	Solvent	$\lambda_{max}(\log \epsilon)$	Ref.
6H-Pyrido[3,4-d]azepin-6-one, 2-benz-oyl-1,2,3,4,7,8,9,9a-octahydro-, (±)-	isoPrOH	224(4.31)	35-0571-78
7H-Pyrido[4,3-c]azepin-7-one, 2-benz-oyl-1,2,3,4,6,8,9,9a-octahydro-, (±)-	isoPrOH	245(4.23)	35-0571-78
8H-Pyrido[1,2-a]pyrazin-8-one, 1,2,3,4-tetrahydro-9-hydroxy-2-methyl-1-(phenylmethyl)-	EtOH	287(3.92)	12-0187-78
$C_{16}H_{18}N_2O_3$			
5,8-Isoquinolinedione, 7-methoxy-6-methyl-3-piperidino-	EtOH	255(4.46),325(4.09), 488(3.81)	94-2175-78
1H-Pyrazole-4-carboxylic acid, 3-acet-yl-5-methyl-1-(phenylmethyl)-, ethyl ester	n.s.g.	216(4.11),243(3.85)	44-2665-78
1H-Pyrazole-4-carboxylic acid, 5-acet-yl-3-methyl-1-(phenylmethyl)-, ethyl ester	n.s.g.	214(3.95),250(3.83)	44-2665-78
1H-Pyrazole-4-carboxylic acid, 1,3-di-methyl-5-(phenylacetyl)-, ethyl ester	n.s.g.	212(3.97),235(3.81)	44-2665-78
1H-Pyrazole-4-carboxylic acid, 1,5-di-methyl-3-(phenylacetyl)-, ethyl ester	n.s.g.	212(4.14),243(3.83)	44-2665-78
3-Pyridinecarboxamide, N-[2-(2,4-di-methoxyphenyl)ethyl]-	n.s.g.	216(4.14),271(3.73)	88-2723-78
$C_{16}H_{18}N_2O_3S$			
Benzoic acid, 2-[(5-acetyl-3-amino-2-thienyl)amino]-5-methyl-, ethyl ester	MeOH	219(4.37),252(4.09), 380(3.96)	150-5101-78
$C_{16}H_{18}N_2O_4S$			
Pyridinium, 1-[3-cyano-4-ethoxy-1-(eth-oxycarbonyl)-2-(methylthio)-4-oxo-2-butenyl]-, hydroxide, inner salt	EtOH	295(3.73),379(4.04)	95-1412-78
$C_{16}H_{18}N_2O_5$			
5,8-Isoquinolinediol, 3-(dimethylami-no)-7-methoxy-, diacetate	EtOH	214(4.28),248(4.58), 291(4.26),299(4.29)	4-0569-78
2,4,6(1H,3H,5H)-Pyrimidinetrione, 5-[3-(2-methoxyphenyl)-3-oxopropyl]-1,3-dimethyl-	MeOH	305(3.6)	83-0503-78
$C_{16}H_{18}N_2O_5S$			
2-Propenoic acid, 2-[[4-oxo-3-[(phen-oxyacetyl)amino]-2-azetidinyl]thio]-, ethyl ester, (2R-cis)-	EtOH	269(4.26),273(4.26)	35-8214-78
$C_{16}H_{18}N_2O_7$			
1H-Indole-1-ethanol, β-[2-(acetyloxy)-ethoxy]-6-nitro-, acetate. (R)-	EtOH	247(3.95),316(3.83), 358(3.72)	136-0017-78E
$C_{16}H_{18}N_2S_2$			
Carbamodithioic acid, ethylphenyl-, 2-(2-pyridinyl)ethyl ester	n.s.g.	335(1.95)	97-0381-78
$C_{16}H_{18}N_4O_4$			
1H-Pyrazole-3-carboxylic acid, 4,5-di-hydro-5-oxo-1-phenyl-4-[(tetrahydro-2-furanyl)hydrazono]-, ethyl ester	MeOH	243(4.05),320(4.16)	142-0199-78B
$C_{16}H_{18}N_4O_7$			
6H-Purine-6-thione, 1,7-dihydro-1-	pH 1	229(3.99),321(4.24)	4-0929-78

Compound	Solvent	$\lambda_{max}(\log \epsilon)$	Ref.
(2,3,5-tri-0-acetyl-α-D-ribofurano- syl)- (cont.) 6H-Purine-6-thione, 1,7-dihydro-1- (2,3,5-tri-0-acetyl-β-D-ribofurano- syl)-	pH 7 pH 13 pH 1 pH 7 pH 13	235(4.02),321(4.33) 238(4.08),321(4.34) 230(4.06),320(4.28) 237(4.04),321(4.36) 238(3.95),321(4.32)	4-0929-78 4-0929-78 4-0929-78 4-0929-78 4-0929-78
$C_{16}H_{18}N_4O_8$ Pyrazolo[4,3-d]pyrimidine, 2-(2,3,5- tri-0-acetyl-β-D-ribofuranosyl)-7- hydroxy-	MeOH	217(4.19),261s(3.66), 277(3.79)	104-0601-78
$C_{16}H_{18}N_8O_4$ 1H-Purine-2,6-dione, 8,8'-(1,2-ethane- diyl)bis[3,7-dihydro-1,3-dimethyl-	pH 7.0 pH 13.0 MeOH	211(4.75),275(4.40) 217(4.58),280(4.49) 212(4.67),275(4.42)	24-0982-78 24-0982-78 24-0982-78
$C_{16}H_{18}O$ Benzocyclooctene, 7-(1,1-dimethyleth- oxy)- 6-Benzocyclooctenol, 6-(3-buten-1-yn- yl)-5,6,7,8,9,10-hexahydro- 1H-Inden-1-one, 2,3,3a,4,7,7a-hexahy- dro-7a-methyl-5-phenyl-, cis-(+)- trans-(±)- 1H-Inden-1-one, 2,3,3a,6,7,7a-hexahy- dro-7a-methyl-5-phenyl-, cis-(+)- Naphthalene, 1,2,3,4-tetrahydro-6-meth- oxy-2-methyl-1-methylene-2-(2-prop- ynyl)-	C_6H_{12} EtOH EtOH EtOH EtOH EtOH	240(4.08) 214(4.15),223(4.11), 234(4.00) 212(3.97),245(4.10), 290s(2.30),302s(1.65) 217(3.95),243(4.09), 289s(2.26) 218s(3.96),247(4.12), 290s(2.26),303s(1.30) 260(3.96)	88-0667-78 44-1050-78 22-0343-78 22-0343-78 22-0343-78 39-0110-78C
$C_{16}H_{18}OS$ 4H-Cyclopenta[5,6]naphtho[2,1-b]thio- phene-6-ol, 5,5a,6,7,9,10-hexahydro- 5a-methyl-, cis-(±)-	EtOH	227(4.54),262(3.89), 301f(3.30)	142-0207-78B
$C_{16}H_{18}O_2$ 4,7-Ethano-1H-indene-8,9-dione, 3a,4,7,7a-tetrahydro-5,6-dimethyl- 1-(1-methylethylidene)-, (3aα,4α,7α- 7aα)- 3β,3aβ-Methanobenz[e]inden-2-one, 1,3,3a,4,5,9b-hexahydro-7-methoxy- 9bα-methyl- 1-Naphthalenepropanol, α-methyl-, acetate Spiro[2-cyclohexene-1,1'(2'H)-naphtha- len]-4-one, 3',4'-dihydro-5'-methoxy- Spiro[2-cyclohexene-1,1'(2'H)-naphtha- len]-4-one, 3',4'-dihydro-7'-methoxy-	C_6H_{12} benzene EtOH MeOH EtOH EtOH	252(4.12),462(2.21) 461.3(1.92) 279(3.20) 224(4.91),272(3.74), 282(3.81),291s(3.64) 225(4.28) 225(4.15)	5-0440-78 5-0440-78 39-0110-78C 83-0328-78 22-0202-78 22-0202-78
$C_{16}H_{18}O_2S$ A-Nor-1-thiaestra-2,5,7,9-tetraene- 6,17-diol, (17β)- A-Nor-1-thiaestra-2,5(10),9(11)-trien- 6-one, 17-hydroxy-, (17β)- A-Nor-3-thiaestra-1,5(10),9(11)-trien- 6-one, 17-hydroxy-, (17β)-	EtOH EtOH EtOH	230(4.51),265(4.02), 315f(3.78) 249(4.09),271(3.98), 279(3.89),321(3.96) 235(3.93),241s(3.90), 297(4.05),318(3.99)	142-0207-78B 142-0207-78B 142-0207-78B
$C_{16}H_{18}O_3$ 1-Butanone, 1-(6,7-dimethoxy-2-naphtha-	MeOH	253(4.67),258(4.68),	83-0328-78

Compound	Solvent	$\lambda_{max}(\log \epsilon)$	Ref.
lenyl)- (cont.) Cycloprop[1,2]indeno[5,6-c]furan-1,3- dione, 3a,4b,5,5a,7,7a-hexahydro- 4b,5,5a,6-tetramethyl-	n.s.g.	307(4.20) 264(4.10)	83-0328-78 35-0860-78
Dehydropterosin B, acetyl-	EtOH	244(4.53),250(4.63), 335(3.42)	94-2365-78
3,5-Hexadienoic acid, 2-(methoxymeth- ylene)-3-methyl-6-phenyl-, methyl ester, (E,E,E)- (strobilurin A)	MeOH	224s(4.20),229(4.25), 237(4.19),285s(4.30), 293(4.34),301s(4.33)	24-2779-78
Naphtho[2,3-c]furan-1,3-dione, 3a,4,9,9a-tetrahydro-5,6,7,8- tetramethyl-	CHCl₃	274(2.48)	35-0860-78
$C_{16}H_{18}O_3Se$ 2-Cyclohexene-1-carboxylic acid, 1-methyl-4-oxo-5-(phenylseleno)-, ethyl ester	MeOH	216(4.26),233(3.97)	24-1944-78
$C_{16}H_{18}O_4$ 3-Furancarboxylic acid, 4,5-dihydro- 2-methyl-5-[(4-methylphenyl)methyl]- 4-oxo-, ethyl ester	EtOH	214(4.00),264(3.92)	4-0327-78
1,4-Naphthalenedione, 5,7-diethoxy- 2,3-dimethyl-	EtOH	269(4.24),279s(4.16), 410(3.52)	12-1353-78
$C_{16}H_{18}O_5$ 2H,8H-Benzo[1,2-b:3,4-b']dipyran-2-one, 9-acetoxy-3,4,9,10-tetrahydro-8,8-di- methyl-, (R)-	EtOH	223(4.02),278(3.26), 286(3.26)	2-0179-78
3-Furancarboxylic acid, 4,5-dihydro-5- [(4-methoxyphenyl)methyl]-2-methyl- 4-oxo-, ethyl ester	EtOH	222(4.01),263(4.05)	4-0327-78
5,8-Methano-2H-indeno[1,2-b]furan-3- acetic acid, 3,3a,4,4a,5,8,8a,9b- octahydro-α,2-dioxo-, ethyl ester	MeOH	278(3.80)	87-0815-78
$C_{16}H_{18}O_6$ Pentalenolactone G, methyl ester	MeOH	238(3.84)	88-0923-78
2-Pentenedioic acid, 4-acetyl-3-(4- methoxyphenyl)-, 5-ethyl ester	n.s.g.	280(4.20)	2-0200-78
Spiro[1,3-dioxolane-2,1'(2'H)-naphtha- lene]-7',8'-dicarboxylic acid, 3',4'- dihydro-, dimethyl ester	MeCN	202(4.33),240(3.88), 275s(2.79),286s(2.59)	88-2145-78
$C_{16}H_{18}O_7$ Spiro[1-benzoxepin-6(7H),2'-[1,3]diox- olane-4,5-dicarboxylic acid, 8,9-di- hydro-, dimethyl ester	MeCN	205(4.48),295(3.63)	88-2145-78
Spiro[1,3-dioxolane-2,8'(5'H)-[2H- 2,4a]epoxynaphthalene]-3',4'-di- carboxylic acid, 6',7'-dihydro-, dimethyl ester	MeCN	205(3.86),230s(3.47), 285(2.84)	88-2145-78
$C_{16}H_{18}O_8$ 2H-1-Benzopyran-2-one, 7-(β-D-galacto- furanosyloxy)-4-methyl-	MeOH	247(3.38),290(3.91), 318(4.13)	136-0023-78F
2H-1-Benzopyran-2-one, 7-(α-D-galacto- pyranosyloxy)-4-methyl-	MeOH	248(3.34),289(3.94), 317(4.12)	136-0023-78F
2H-1-Benzopyran-2-one, 7-(β-D-galacto- pyranosyloxy)-4-methyl-	MeOH	248(3.28),289(3.95), 317(4.16)	136-0023-78F

Compound	Solvent	$\lambda_{max}(\log \epsilon)$	Ref.
2H-1-Benzopyran-2-one, 7-(α-D-gluco-pyranosyloxy)-4-methyl-	MeOH	248(3.33),289(3.92), 318(4.13)	136-0023-78F
2H-1-Benzopyran-2-one, 7-(β-D-gluco-pyranosyloxy)-4-methyl-	MeOH	247(3.36),289(3.94), 318(4.14)	136-0023-78F
$C_{16}H_{18}O_9$			
Chlorogenic acid, TiCl₄ complex	acetone	450(3.32)	98-0973-78
$C_{16}H_{18}Sn$			
Stannane, 9H-fluoren-9-yltrimethyl-	heptane	230(4.64),251(4.60), 293(4.09),300(4.06), 334(2.62),355(2.86), 375(2.83)	104-0207-78
$C_{16}H_{19}BrN_2O_2$			
Pyridinium, 1-butyl-4-[(4-nitrophenyl)-methyl]-, bromide	H_2O	215s(4.12),260s(4.10), 265(4.12),273s(4.07)	48-0133-78
deprotonated	H_2O	560(4.47)	48-0133-78
$C_{16}H_{19}BrN_2O_6$			
1H-Pyrrolo[1,2-a]indole-5,8-dione, 9-[[(aminocarbonyl)oxy]methyl]-2,3,6,7-tetrahydro-7,7-dimethoxy-6-methyl-	EtOH	250(4.06),290(3.71), 342(4.00)	94-3815-78
$C_{16}H_{19}ClO$			
Naphthalene, 2-(2-chloro-2-propenyl)-6-methoxy-2-methyl-1-methylene-	EtOH	262(4.05)	39-0110-78C
$C_{16}H_{19}Cl_2N_2O_3PS$			
Phosphorohydrazidothioic acid, [[5-(3,4-dichlorophenyl)-2-furanyl]-methylene]-, O-ethyl O-(1-meth-ylethyl) ester	dioxan	340(4.53)	73-2643-78
$C_{16}H_{19}NO$			
2H-Pyrrole-5-ethanol, 3,4-dihydro-3,3-dimethyl-α-(phenylethynyl)-	MeOH	205(4.47),242(4.40), 253(4.36)	83-0977-78
$C_{16}H_{19}NOS$			
2H-1,2-Thiazine, 3,6-dihydro-3,6-bis(1-methylethylidene)-2-phenyl-, 1-oxide	EtOH	283(4.24),316(4.11)	39-1568-78C
$C_{16}H_{19}NO_2$			
Benzamide, N-[3-(4-oxo-2-cyclohexen-1-yl)propyl]-	EtOH	227(4.30)	94-0620-78
2H-Benz[g]indole-3-carboxylic acid, 3,3a,4,5-tetrahydro-2,2-dimethyl-, methyl ester	MeOH	245(4.27),270(3.48)	35-3494-78
1H-Indole-3-acetic acid, 2-(3-methyl-2-butenyl)-, methyl ester	EtOH	277(4.53),284(3.88), 291(3.83)	78-0929-78
1H-Indole-3-propanoic acid, 2-(3-meth-yl-2-butenyl)-	EtOH	230(4.15),284(3.23), 291(3.16)	78-0929-78
1,2-Naphthalenedione, 4-(hexylamino)-	MeOH	238(4.27),275(4.23), 300s(4.00),325s(3.64), 445(3.68)	78-1377-78
2-Pentenoic acid, 5-[2-(2,2-dimethyl-2H-azirin-3-yl)phenyl]-, methyl ester, (E)-	MeOH	245(4.58),275(3.61)	35-3494-78
$C_{16}H_{19}NO_3$			
1,3-Cyclohexanedione, 2-[[(2-methoxy-	EtOH	364(4.35)	97-0256-78

Compound	Solvent	$\lambda_{max}(\log \epsilon)$	Ref.
phenyl)amino]methylene]-5,5-dimethyl- 1,3-Cyclohexanedione, 2-[[(4-methoxy- phenyl)amino]methylene]-5,5-dimethyl-	EtOH	359(4.43)	97-0256-78 97-0256-78
1H-Pyrrole-2-carboxylic acid, 3-hy- droxy-1-(phenylmethyl)-, 1,1-dimethyl ester	EtOH	265(4.20)	94-2224-78
2(1H)-Quinolinone, 8-hydroxy-4-methoxy- 1-methyl-3-(3-methyl-2-butenyl)- (glycosolone)	MeOH	213(4.57),237(4.27), 251(4.34),259(4.43), 286(3.89),295(3.85), 333(3.47),346(3.43)	25-0272-78
$C_{16}H_{19}NO_4$			
3-Pyridinecarboxylic acid, 5-ethoxy- 1,4,5,6-tetrahydro-6-oxo-2-phenyl-, ethyl ester	EtOH EtOH-KOH	284(4.02) 325(4.03)	142-0161-78A 142-0161-78A
1H-Pyrrole-3-carboxylic acid, 4,5-dihy- dro-5-hydroxy-1,2-dimethyl-4-oxo- 5-(phenylmethyl)-, ethyl ester	EtOH	211(4.04),249(4.19), 326(3.83)	4-1215-78
1H-Pyrrole-3-carboxylic acid, 4,5-dihy- dro-5-hydroxy-2,5-dimethyl-4-oxo- 1-(phenylmethyl)-, ethyl ester	EtOH	209(3.85),246(4.15), 320(3.96)	118-0291-78
1H-Pyrrole-3-carboxylic acid, 4,5-dihy- dro-5-hydroxy-2-methyl-5-[(4-methyl- phenyl)methyl]-4-oxo-, ethyl ester	EtOH	218(4.00),242(4.18), 306(3.81)	4-1215-78
2(1H)-Quinolinone, 4-hydroxy-6,8-di- methoxy-3-(3-methyl-2-butenyl)-	EtOH	223(4.87),253(4.43), 288(4.0),292(3.80), 320(3.65)	142-1433-78A
$C_{16}H_{19}NO_5$			
2,4-Pentadienamide, 5-(6-methoxy-1,3- benzodioxol-5-yl)-N-(2-methoxyphen- yl)-, (E,E)-	n.s.g.	370(4.41)	78-1979-78
1H-Pyrrole-3-carboxylic acid, 4,5-di- hydro-5-hydroxy-5-[(4-methoxyphenyl)- methyl]-2-methyl-4-oxo-, ethyl ester	EtOH	230(4.15),243(4.18), 284(3.72),306(3.79)	4-1215-78
$C_{16}H_{19}NO_8$			
α-D-Xylofuranoside, methyl 3,5-O-(1- methylethylidene)-, 4-nitrobenzoate	EtOH	258(4.09)	136-0257-78H
$C_{16}H_{19}N_2$			
Pyridinium, 2-[2-[4-(dimethylamino)- phenyl]ethenyl]-1-methyl-, iodide	EtOH	460(4.57)	4-0017-78
Pyridinium, 4-[2-[4-(dimethylamino)- phenyl]ethenyl]-1-methyl-, iodide	EtOH	480(4.62)	4-0017-78
$C_{16}H_{19}N_2O_2$			
Pyridinium, 1-butyl-4-[(4-nitrophen- yl)methyl]-, bromide	H_2O	215s(4.12),260s(4.10), 265(4.12),273s(4.07)	48-0133-78
deprotonated	H_2O	560(4.47)	48-0133-78
$C_{16}H_{19}N_3O$			
10H-Pyrido[3,2-b][1,4]benzoxazine- 10-propanamine, N,N-dimethyl-	C_6H_{12}	233(4.91),237(4.90), 343(4.36)	95-0585-78
$C_{16}H_{19}N_3O_2$			
1H-1,4-Benzodiazepin-2-one, 5-(1-cyclo- hexenyl)-2,3-dihydro-1-methyl-7-ni- tro-	10% DMF- pH 1.65	281(4.39)	86-0209-78

Compound	Solvent	$\lambda_{max}(\log \epsilon)$	Ref.
$C_{16}H_{19}N_3O_4$ 1,2,4-Triazine-6-acetic acid, 5-methoxy-3-(4-methoxyphenyl)-α-methyl-, ethyl ester	EtOH	217(4.22),289(4.32)	114-0091-78B
$C_{16}H_{19}N_3O_5$ α-D-Glucopentofuranose, 1,2-O-(1-methylethylidene)-5-C-(1-phenyl-1H-1,2,3-triazol-4-yl)-	MeOH	245(4.26)	136-0357-78H
β-L-Idopentofuranoside, 1,2-O-(1-methylethylidene)-5-C-(1-phenyl-1H-1,2,3-triazol-4-yl)-	MeOH	245(4.25)	136-0357-78H
$C_{16}H_{19}N_3O_6S$ 4-Oxazolecarboxylic acid, 2-[1-[[[2-(dimethoxymethyl)-4-thiazolyl]carbonyl]amino]-1-propenyl]-5-methyl-, methyl ester	MeOH	247(4.38)	88-2791-78
$C_{16}H_{19}N_3O_7$ 1H-Imidazole-4-carboxylic acid, 5-(cyanomethyl)-1-(2,3-di-O-acetyl-5-deoxy-β-D-ribofuranosyl)-, methyl ester	pH 1 pH 7 pH 11	225(4.01) 235(4.03) 238(4.01)	87-1212-78 87-1212-78 87-1212-78
1H-Imidazole-5-carboxylic acid, 4-(cyanomethyl)-1-(2,3-di-O-acetyl-5-deoxy-β-D-ribofuranosyl)-, methyl ester	pH 1 pH 7 pH 11	234(3.93) 242(4.04) 242(4.04)	87-1212-78 87-1212-78 87-1212-78
Uridine, 6-(cyanomethyl)-2',3'-O-(1-methylethylidene)-, 5'-acetate	MeOH pH 14	257(4.03) 332.5(--)	94-2657-78 94-2657-78
$C_{16}H_{19}N_5O_2$ 9H-Purin-6-amine, 9-[2-methoxy-3-(phenylmethoxy)propyl]-, ($\pm$)-	pH 2 and 12	262(4.08)	73-3444-78
$C_{16}H_{19}N_5O_6$ Riboflavin, 8-amino-8-demethyl-	n.s.g.	312(4.30),501(4.63)	69-1942-78
$C_{16}H_{19}N_7O$ 2,4-Pteridinediamine, 6-[[ethyl(4-methoxyphenyl)amino]methyl]-	MeOH-HCl MeOH-KOH	248(4.29),337(4.02) 257(4.55),375(3.92)	87-0331-78 87-0331-78
$C_{16}H_{19}N_7O_2$ 2,4-Pteridinediamine, 6-[[ethyl(4-methoxyphenyl)amino]methyl]-, 8-oxide	MeOH-HCl MeOH-KOH	263(4.47),362(3.85) 268(4.59),400(3.92)	87-0331-78 87-0331-78
$C_{16}H_{20}BrN_2O_3PS$ Phosphorohydrazidothioic acid, [[5-(4-bromophenyl)-2-furanyl]methylene]-, O-ethyl O-(1-methylethyl) ester	dioxan	341(4.58)	73-2643-78
$C_{16}H_{20}ClN_2O_3PS$ Phosphorohydrazidothioic acid, [[5-(2-chlorophenyl)-2-furanyl]methylene]-, O-ethyl O-(1-methylethyl) ester	dioxan	334(4.60)	73-2643-78
Phosphorohydrazidothioic acid, [[5-(3-chlorophenyl)-2-furanyl]methylene]-, O-ethyl O-(1-methylethyl) ester	dioxan	338(4.60)	73-2643-78
$C_{16}H_{20}NO$ Pyridinium, 4-[(2,6-dimethyl-4H-pyran-4-ylidene)methyl]-1,2,6-trimethyl-, perchlorate	MeNO$_2$ CH$_2$Cl$_2$	437(4.64) 450(4.75)	103-0601-78 103-0601-78

Compound	Solvent	$\lambda_{max}(\log \epsilon)$	Ref.
$C_{16}H_{20}NOP$			
1H-Phosphorino[3,4-b]indole, 2-(1,1-di-methylethyl)-2,9-dihydro-4-methyl-, 2-oxide	CHCl₃	280(3.88),331(3.67)	139-0257-78B
$C_{16}H_{20}N_2O$			
4(3H)-Pyridinone, 2-(dimethylamino)-3,3,5-trimethyl-6-phenyl-	EtOH	248(4.09),398(3.94)	27-0468-78
4(3H)-Pyridinone, 2-(dimethylamino)-3,3,6-trimethyl-5-phenyl-	EtOH	222(4.12),385(4.08)	27-0468-78
2H-Pyrrol-2-one, 5-[[4-(dimethylamino)-phenyl]methylene]-4-ethyl-1,5-dihydro-3-methyl-, (Z)-	EtOH	262(4.02),280s(3.78), 396(4.42)	49-0183-78
2H-Pyrrol-2-one, 3-ethenyl-5-[(4-ethyl-3,5-dimethyl-1H-pyrrol-2-yl)methyl-ene]-1,5-dihydro-4-methyl-	MeOH	438(4.60)	4-1117-78
$C_{16}H_{20}N_2OS$			
Azepino[4,3-b][1,4]benzothiazine, 1-ethoxy-11-ethyl-3,4,5,11-tetrahydro-	EtOH	253(4.10),288(3.66)	103-0306-78
1H-[1,2]Oxathiolo[3,2-e][1,2,3]thiadia-zole-7-S^IV, 3,4-bis(1-methylethyl)-1-phenyl-	C₆H₁₂	234(4.03),274(3.90), 441(4.43)	39-0195-78C
$C_{16}H_{20}N_2O_2$			
Acetamide, N-methyl-N-[4-(1-methyl-1H-indol-3-yl)-4-oxobutyl]-	EtOH	245(3.09),304(3.08)	150-0470-78
$C_{16}H_{20}N_2O_3$			
5,8-Etheno-10H-pyrrolo[1,2-e][1,5,8]-oxadiazacyclotetradecine-1,12(2H,11H)-dione, 3,4,14,15,16,16a-hexahydro-, (S)-	MeOH	223s(3.79),271(2.75), 276(2.71)	35-8202-78
$C_{16}H_{20}N_2O_4$			
Hydrazinecarboxylic acid, [4-(benzoyl-oxy)cyclohexylidene]-, ethyl ester	EtOH	228.5(4.39)	39-0045-78C
$C_{16}H_{20}N_2O_4S$			
Spiro[2H-1-benzopyran-2,2'-thiazoli-dine], 8-methoxy-3,3',4',4'-tetra-methyl-6-nitro-	EtOH	215(4.38),222s(4.36), 256(4.43),274(4.26), 354(4.12)	4-1439-78
$C_{16}H_{20}N_2O_5$			
Spiro[2H-1-benzopyran-2,2'-oxazoli-dine], 8-methoxy-3,3',4',4'-tetra-methyl-6-nitro-	EtOH	218(4.66),244s(4.58), 250(4.61),258s(4.53), 277(4.17),343(4.13), 380(3.85)	4-1439-78
$C_{16}H_{20}N_2O_6$			
2,4-Hexadienoic acid, 2-acetoxy-6-(2,5-dihydro-2,3-dimethyl-5-oxo-1H-pyra-zol-1-yl)-6-oxo-, 1-methylethyl ester, (Z,E)-	EtOH	274(4.5)	5-1734-78
$C_{16}H_{20}N_2O_9S$			
1H-Imidazole-4-carboxylic acid, 2,3-di-hydro-2-thioxo-1-(2,3,5-tri-O-acetyl-β-D-ribofuranosyl)-, methyl ester	MeOH	269(4.21),300s(3.92)	94-3322-78

Compound	Solvent	$\lambda_{max}(\log \epsilon)$	Ref.
$C_{16}H_{20}N_2O_{10}$			
1H-Imidazole-4-carboxylic acid, 2,3-di-hydro-2-oxo-1-(2,3,5-tri-O-acetyl-β-D-ribofuranosyl)-, methyl ester	MeOH	267(4.10)	94-3322-78
$C_{16}H_{20}N_2S$			
1H-Imidazole, 1-[2-(cyclopentylthio)-ethyl]-4-phenyl-	EtOH	258(4.20)	4-0307-78
$C_{16}H_{20}N_2S_2$			
1H-[1,2]Dithiolo[5,1-e][1,2,3]thiadia-zole-7-S^{IV}, 3,4-bis(1-methylethyl)-1-phenyl-	C_6H_{12}	235(4.40),255s(4.20), 292(4.05),474(4.14)	39-0195-78C
$C_{16}H_{20}N_3$			
2H-Pyrrolo[3,4-b]quinoxalinium, 2-(1,1-dimethylethyl)-4-ethyl-, tetrafluoro-borate	MeOH	242(4.44),264(4.41), 364(4.23),372(4.28), 597(3.47)	88-4671-78
$C_{16}H_{20}N_3O_4P$			
2,3-Phosphorindione, 1-(1,1-dimethyl-ethyl)-1,6-dihydro-5-methyl-, 2-[(4-nitrophenyl)hydrazono]-, 1-oxide	$CHCl_3$	237(3.96),310s(3.71), 392(4.42),416s(4.37)	139-0027-78B
$C_{16}H_{20}N_3O_5PS$			
Phosphorohydrazidothioic acid, [[5-(2-nitrophenyl)-2-furanyl]methylene]-, O-ethyl O-(1-methylethyl) ester	dioxan	318(4.41)	73-2643-78
Phosphorohydrazidothioic acid, [[5-(3-nitrophenyl)-2-furanyl]methylene]-, O-ethyl O-(1-methylethyl) ester	dioxan	334(4.53)	73-2643-78
Phosphorohydrazidothioic acid, [[5-(4-nitrophenyl)-2-furanyl]methylene]-, O-ethyl O-(1-methylethyl) ester	dioxan	387(4.66)	73-2643-78
$C_{16}H_{20}N_4$			
2-Tetrazene, 1,4-dimethyl-1,4-bis(2-methylphenyl)-	MeCN	303(4.30)	40-0150-78
2-Tetrazene, 1,4-dimethyl-1,4-bis(4-methylphenyl)-	MeCN	347(4.44)	40-0150-78
$C_{16}H_{20}N_4O_2$			
2-Tetrazene, 1,4-bis(4-methoxyphenyl)-1,4-dimethyl-	MeCN	347(4.43)	40-0150-78
$C_{16}H_{20}N_4O_2S_4$			
5,17-Dioxa-2,8,14,20-tetrathia-10,24-25,26-tetraazatricyclo[19.3.1.1^{9,13}]-hexacosa-1(25),9,11,13(26),21,23-hexa-ene	EtOH	208(3.99),254(4.61), 303(4.15)	44-3362-78
$C_{16}H_{20}N_4O_4$			
2-Pyrimidinamine, 1-β-D-arabinofurano-syl-1,4-dihydro-4-imino-N-(phenyl-methyl-, monohydrochloride	pH 1	224(4.35),270s(3.72)	94-3244-78
2,4(1H,3H)-Quinazolinedione, 1-[3-(3,4-dihydro-5-methyl-2,4-dioxo-1(2H)-pyrimidinyl)propyl]-5,6,7,8-tetra-hydro-	H_2O	273(4.26)	56-1035-78

Compound	Solvent	λ_{max} (log ϵ)	Ref.
$C_{16}H_{20}N_4O_4S$			
Thieno[2,3-d:4,5-d']dipyrimidine-2,4,5,7(3H,6H)-tetrone, 1,3,6,8-tetraethyl-1,8-dihydro-	EtOH	245(3.94),304(3.91)	44-1677-78
$C_{16}H_{20}N_4O_6$			
2,5,8,14,17,20-Hexaoxa-10,24,25,26-tetraazatricyclo[19.3.1.1^{9,13}]hexacosa-1(25),9,11,13(26),21,23-hexaene	EtOH	218(4.04),260(3.95)	44-3362-78
$C_{16}H_{20}O$			
Cyclooctanone, 2-[(4-methylphenyl)methylene]-, (E)-	hexane	286(4.24)	54-0305-78
Naphthalene, 1,2,3,4-tetrahydro-6-methoxy-1,2-dimethyl-2-(2-propynyl)-	EtOH	277(3.30)	39-0110-78C
$C_{16}H_{20}OS$			
A-Nor-1-thiaestra-2,5(10),9(11)-trien-17β-ol	EtOH	285(4.06)	142-0207-78B
A-Nor-3-thiaestra-1,5(10),9(11)-trien-17β-ol	EtOH	227(4.19),233(4.18),252(4.05),260s(3.99)	142-0207-78B
2-Propanone, 1-[1,2,3,5,6,7-hexahydro-8-(methylthio)-s-indacen-4-yl]-	MeOH	223s(4.25),270s(3.82),283(3.68),293(3.62)	73-1732-78
$C_{16}H_{20}OSe$			
2-Cyclohexen-1-one, 4-methyl-6-(phenylseleno)-4-propyl-	MeOH	224(4.19),231(4.2)	24-1944-78
$C_{16}H_{20}O_2$			
2H-Benz[e]inden-2-one, 1,3,3a,4,5,9b-hexahydro-7-methoxy-3a,9b-dimethyl-, cis	EtOH	277(3.15)	39-0110-78C
trans	EtOH	277(3.30)	39-0110-78C
Bicyclo[2.2.2]octan-6-one, 2-(4-methoxyphenyl)-5-methyl-	n.s.g.	225(4.18),276(3.31)	88-2929-78
Bicyclo[8.2.2]tetradeca-5,10,12,13-tetraene-4-carboxylic acid, methyl ester, cis	EtOH	226(3.79),277(2.57),284(2.52)	35-1806-78
trans	EtOH	222(3.86),266s(2.51),271(2.60),278(2.56)	35-1806-78
2-Cyclohexen-1-one, 2-[2-(4-methoxyphenyl)ethyl]-3-methyl-	EtOH	227(4.25)	44-4598-78
2-Cyclopropene-1-carboxylic acid, 2-[4-(1,1-dimethylethyl)phenyl]-3-methyl-, methyl ester	EtOH	257(3.96)	104-2144-78
1H-Indene-5-ethanol, 2,4,6-trimethyl-, acetate	EtOH	220(4.33),225(4.35),264(4.01)	94-2365-78
3β,3aβ-Methanobenz[e]inden-2β-ol, 1,3,3a,4,5,9b-hexahydro-7-methoxy-9bα-methyl-	EtOH	279(3.31)	39-0110-78C
Naphthalene, 6-butyl-2,3-dimethoxy-	MeOH	233(5.01),266(3.71),268(3.70),275(3.62),297(3.14),311(3.45),317(3.36),325(3.66)	83-0328-78
$C_{16}H_{20}O_2S$			
2-Cyclohexen-1-one, 3-ethoxy-5,5-dimethyl-2-(phenylthio)-	MeOH	252(4.33),279s(3.43)	44-2676-78
A-Nor-1-thiaestra-2,5(10)-dien-6-one, 17β-hydroxy-	EtOH	220(4.24),224(4.24),256(4.08),278s(3.62)	142-0207-78B

Compound	Solvent	λ_{max}(log ϵ)	Ref.
A-Nor-3-thiaestra-1,5(10)-dien-6-one, 17β-hydroxy-	EtOH	277(4.08)	142-0207-78B
$C_{16}H_{20}O_3$			
2-Naphthalenepropanol, 6,7-dimethoxy-α-methyl-	MeOH	233(5.00),258(3.73), 266(3.75),275(3.68), 297(3.19),311(3.45), 317(3.35),325(3.66)	83-0328-78
Pterosin B acetate	EtOH	216(4.54),258(4.15), 301(3.38)	94-2365-78
$C_{16}H_{20}O_4$			
1,2-Benzenedicarboxylic acid, 4-methyl-3-(3-methyl-2-butenyl)-, dimethyl ester	EtOH	242(3.94),283(3.18)	2-1062-78
1H-Inden-1-one, 6-(2-acetoxyethyl)-2,3-dihydro-3-hydroxy-2,5,7-tri-methyl-, (2S-trans)	EtOH	216(4.55),257(4.17), 298(3.30)	94-2365-78
1,4-Naphthalenediol, 5,7-diethoxy-2,3-dimethyl-	EtOH	226s(4.41),250(4.31), 293(3.63),312(3.62), 343s(3.61),350(3.63)	12-1353-78
$C_{16}H_{20}O_5$			
7-Oxabicyclo[2.2.1]heptane-2-carboxylic acid, 3-(7-ethoxy-7-oxo-1,3,5-hepta-trienyl)-, (1α,2α,3α(1E,3E,5E),4α]-(±)-	EtOH	305(3.65)	39-0980-78C
$C_{16}H_{20}O_6$			
4H-1-Benzopyran-4-one, 5-acetoxy-3-(acetoxymethyl)-6-ethyl-5,6,7,8-tetrahydro-	MeOH	215(3.95),250(3.92)	119-0111-78
$C_{16}H_{21}BrNO_3$			
Pyrrolidinium, 1-[2-(acetoxyethyl)-1-[2-(4-bromophenyl)-2-oxoethyl]-, bromide	EtOH	267(4.22)	145-0595-78
$C_{16}H_{21}BrN_2O_4$			
β-Alanine, N-nitroso-N-pentyl-, 2-(4-bromophenyl)-2-oxoethyl ester	EtOH	256(4.30)	94-3909-78
Butanoic acid, 4-(butylnitrosoamino)-, 2-(4-bromophenyl)-2-oxoethyl ester	EtOH	256.5(4.41)	94-3909-78
Butanoic acid, 4-[(1,1-dimethylethyl)-nitrosoamino]-, 2-(4-bromophenyl)-2-oxoethyl ester	EtOH	256.5(4.25)	94-3909-78
$C_{16}H_{21}BrN_2O_5$			
Butanoic acid, 4-(butylnitrosoamino)-3-hydroxy-, 2-(4-bromophenyl)-2-oxo-ethyl ester	EtOH	256.5(4.20)	94-3909-78
$C_{16}H_{21}ClO$			
Bicyclo[2.2.1]heptan-2-ol, 2-(4-chloro-phenyl)-1,7,7-trimethyl-, exo	C_6H_{12}	240(3.23),256(2.91), 262(2.95),269(2.95), 277(2.87)	12-1223-78
$C_{16}H_{21}ClO_2$			
2-Naphthalenol, 1-(2-chloro-2-propen-yl)-1,2,3,4-tetrahydro-6-methoxy-1,2-dimethyl-	EtOH	277(3.28)	39-0110-78C

Compound	Solvent	$\lambda_{max}(\log \epsilon)$	Ref.
$C_{16}H_{21}Cl_5O$			
1,3-Cyclopentadiene, 1,2,3,4,5-penta-chloro-5-(7-methoxy-1,1,5-trimethyl-5-heptenyl)-	C_6H_{12}	314(3.26)	44-3209-78
$C_{16}H_{21}FN_2O_{10}S$			
2,4(1H,3H)-Pyrimidinedione, 1-[3,4-di-O-acetyl-6-deoxy-6-fluoro-2-O-(meth-ylsulfonyl)-β-D-glucopyranosyl)-5-methyl-	MeOH	265(3.97)	48-0157-78
$C_{16}H_{21}N$			
Methanamine, N-(1-phenyl-3,8-nonadien-ylidene)-	heptane	278(4.02)	70-2284-78
$C_{16}H_{21}NO$			
Azocine, 3-ethenyl-2-(2-furanyl)-1,2,3,4,5,8-hexahydro-1-(2-propenyl)-	heptane	229(4.43)	70-2284-78
$C_{16}H_{21}NO_3$			
Benzenepropanamide, α-acetoxy-N,N-dieth-yl-β-methylene-	EtOH	235(4.00)	44-0954-78
$C_{16}H_{21}NO_4$			
Cyclohexanone, 2-[1-(2-methoxyphenyl)-2-nitroethyl]-4-methyl-	EtOH	209(4.18),274(3.43), 278(3.43)	2-0405-78
Cyclohexanone, 2-[1-(4-methoxyphenyl)-2-nitroethyl]-4-methyl-	EtOH	208(4.13),226(4.18), 277(3.36),283(3.31), 353(3.36)	2-0405-78
Cyclohexanone, 2-[1-(2-methoxyphenyl)-2-nitroethyl]-5-methyl-	EtOH	212(4.03),274(3.26), 279(3.25)	2-0405-78
Cyclohexanone, 2-[1-(4-methoxyphenyl)-2-nitroethyl]-5-methyl-	EtOH	212(4.11),227(4.18), 277(3.37),283(3.27)	2-0405-78
$C_{16}H_{21}NO_5$			
Cyclohexanone, 2-[1-(3,4-dimethoxyphen-yl)-2-nitroethyl]-	EtOH	209(4.25),232(3.94), 280(3.46)	2-0405-78
Cyclohexanone, 2-[1-(4-hydroxy-3-meth-oxyphenyl)-2-nitroethyl]-3-methyl-	EtOH	211(4.17),283(3.47)	2-0405-78
Cyclohexanone, 2-[1-(4-hydroxy-3-meth-oxyphenyl)-2-nitroethyl]-4-methyl-	EtOH	210(4.29),283(3.49)	2-0405-78
Cyclohexanone, 2-[1-(4-hydroxy-3-meth-oxyphenyl)-2-nitroethyl]-5-methyl-	EtOH	211(4.17),283(3.48)	2-0405-78
$C_{16}H_{21}NO_6$			
3-Cyclohexene-1,1,3-tricarboxylic acid, 4-cyano-, triethyl ester	n.s.g.	229(3.93)	128-0097-78
4-Cyclohexene-1,1,3-tricarboxylic acid, 4-cyano-, triethyl ester	n.s.g.	245s(2.81)	128-0097-78
$C_{16}H_{21}N_3O$			
2-Propyn-1-one, 1-[4-(dimethylamino)-phenyl]-3-(4-methyl-1-piperazinyl)-	MeCN	298(4.08),356(4.59)	33-1609-78
$C_{16}H_{21}N_3OS$			
Azepino[4,3-b][1,4]benzothiazin-1(2H)-one, 11-[2-(dimethylamino)ethyl]-3,4,5,11-tetrahydro-	EtOH	253(4.10),286(3.71)	103-0306-78
$C_{16}H_{21}N_3O_2$			
3-Pyridinecarboxylic acid, 5-cyano-	EtOH	212(3.84),263(3.68),	73-2024-78

Compound	Solvent	$\lambda_{max}(\log \epsilon)$	Ref.
1-cyclohexyl-1,4-dihydro-4-imino-6-methyl-, ethyl ester (cont.)		375(4.54)	73-2024-78
$C_{16}H_{21}N_3O_3$			
Hydrazinecarboxamide, 2-[(2,3,3a,4,5,8-9,11a-octahydro-10-methyl-3-methyl-ene-2-oxocyclodeca[b]furan-6-yl)-methylene]-, [3aS-(3aR*,6E,10E,11aS*)]-	isoPrOH	212(4.26),268(4.11)	42-1142-78
1H-Indole-2-carboxylic acid, 2-(2-cya-noethyl)-2,3,4,5,6,7-hexahydro-3-im-ino-4-oxo-, 1,1-dimethylethyl ester	EtOH	258(4.02),317(3.87)	94-3080-78
$C_{16}H_{21}N_3O_4S$			
Benzenesulfonamide, N-[3-(3,4-dihydro-3,6-dimethyl-2,4-dioxo-1(2H)-pyrimi-dinyl)propyl]-4-methyl-	CHCl₃	262(4.16)	70-0149-78
Benzenesulfonamide, N-[3-(3,6-dihydro-3,4-dimethyl-2,6-dioxo-1(2H)-pyrimi-dinyl)propyl]-4-methyl-	CHCl₃	300(4.27)	70-0149-78
$C_{16}H_{21}N_3O_6$			
Glutamine, N-[4-(methylamino)benzoyl-alanyl]-	MeOH	302(3.50)	87-1165-78
Glutamine, N-[4-(methylamino)benzoyl]-sarcosyl]-	MeOH	382(4.09)	87-1165-78
$C_{16}H_{21}N_5O_9$			
Pyrido[3,2-c]pyridazine-1,2-dicarbox-ylic acid, 3-[1,2-bis(methoxycarbo-nyl)hydrazino]-3,4-dihydro-4-meth-oxy-, dimethyl ester	MeOH	210(4.1),250(3.98), 300(3.70)	103-0534-78
$C_{16}H_{21}OP$			
Phosphorin, 4-(1,1-dimethylethyl)-1,1-dihydro-1-methoxy-1-phenyl-	EtOH	325(3.61),350(3.69)	89-0528-78
$C_{16}H_{22}$			
Bicyclo[8.2.2]tetradeca-5,10,12,13-tetraene, 5,6-dimethyl-, cis	EtOH	231(3.80),278(2.58), 285(2.52)	35-1806-78
trans	EtOH	222(3.89),274(2.64), 281(2.59)	35-1806-78
Cyclohexene, 1,1'-(1,3-butadiene-1,4-diyl)bis-	hexane	246(4.42),274(4.19), 291(4.36),310(4.26)	78-1323-78
Tricyclo[10.4.0.0⁶,¹¹]hexadeca-1,3,5-triene	n.s.g.	208(4.13),244(3.69), 252(3.66),269(3.66)	78-1323-78
$C_{16}H_{22}Cl_3N$			
Carbonimidic dichloride, [2-(6-chloro-3,4,4a,5,6,7,8,8a-octahydro-5,5,8a-trimethyl-2-naphthalenyl)ethenyl]-, [4aR-[2(Z),4aα,6β,8aβ]]-	MeOH	293(4.00)	88-1395-78
$C_{16}H_{22}Cl_3NO$			
Carbonimidic dichloride, [2-(6-chloro-1,4,4a,5,6,7,8,8a-octahydro-7-hy-droxy-5,5,8a-trimethyl-2-naphtha-lenyl)ethenyl]-, [4aR-[2(Z),4aα-6β,7β,8aβ]]-	EtOH	294(4.04)	88-1395-78
Carbonimidic dichloride, [2-(6-chloro-3,4,4a,5,6,7,8,8a-octahydro-7-hy-droxy-5,5,8a-trimethyl-2-naphtha-	MeOH	293(4.00)	88-1391-78

Compound	Solvent	$\lambda_{max}(\log \epsilon)$	Ref.
lenyl)ethenyl]-, (E)- (cont.)			88-1391-78
(Z)-	MeOH	288(4.34)	88-1395-78
$C_{16}H_{22}Cl_4N_2Pt_2$			
Platinum, di-μ-chlorodichlorobis(2,4,6-trimethylpyridine)di-	$CHCl_3$	270(4.08)	101-0357-78A
$C_{16}H_{22}Cl_4O$			
1,3-Cyclopentadiene, tetrachloro(7-methoxy-1,1,5-trimethyl-5-hepten-yl)-, (E)-	C_6H_{12}	285(3.26)	44-3209-78
2H-2,4a-Methanonaphthalene, 2,3,4,9-tetrachloro-1,5,6,7,8,8a-hexahydro-1-(methoxymethyl)-5,5,8a-trimethyl-	C_6H_{12}	225(3.70)(end abs.)	44-3209-78
$C_{16}H_{22}N_2$			
Acetaldehyde, 2-(2,7-octadienyl)-2-phenylhydrazone	n.s.g.	279(4.32)	39-0543-78C
Diazene, bicyclo[2.2.1]hept-1-yl)(1-methyl-1-phenylethyl)-, trans	isooctane	366(1.46)	35-0920-78
Diazene, (2-ethenyl-1-methyl-6-hepten-yl)phenyl-	n.s.g.	266(3.96),400(2.24)	39-0543-78C
Diazene, (1-methyl-3,8-nonadienyl)phen-yl-	n.s.g.	266(3.96),400(2.24)	39-0543-78C
$C_{16}H_{22}N_2O$			
2H-Pyrrol-2-one, 5-[(3,4-diethyl-5-methyl-1H-pyrrol-2-yl)methylene]-1,5-dihydro-3,4-dimethyl-	$CHCl_3$	235(3.71),270(3.81), 415(4.27)	64-0924-78B
$C_{16}H_{22}N_2O_4S$			
Phenol, 2-methoxy-4-nitro-6-[2-(3,4,4-trimethyl-2-thiazolidinyl)-1-propen-yl]-, (Z)-	EtOH	210(4.48),238(4.47), 279(4.14),348(4.18), 468(3.60)	4-1439-78
sodium salt	EtOH	212(4.48),237(4.49), 279(4.14),348(4.12), 463(4.05)	4-1439-78
$C_{16}H_{22}N_2O_8$			
Uridine, 2',3'-O-(4-ethoxy-1-methyl-4-oxobutylidene)-, (R)-	MeOH	261(3.98)	64-0056-78C
$C_{16}H_{22}N_4$			
Piperidine, 1-[(cyanoimino)[3-[1-(di-methylamino)ethylidene]-1,4-cyclo-pentadien-1-yl]methyl]-	MeOH	235(3.80),273s(4.05), 343(4.47)	83-0369-78
$C_{16}H_{22}N_4O_5$			
1H-Imidazole-4-carbonitrile, 5-amino-1-[2,3:5,6-bis-O-(1-methylethyli-dene)-α-D-mannofuranosyl]-	MeOH	242(4.05)	39-1381-78C
β-anomer	MeOH	245(4.06)	39-1381-78C
$C_{16}H_{22}O$			
Bicyclo[2.2.1]heptan-2-ol, 1,7,7-tri-methyl-2-phenyl-, endo	C_6H_{12}	243(2.23),248(2.33), 253(2.42),259(2.48), 265(2.35)	12-1223-78
5-Hepten-3-one, 2,4,4-trimethyl-6-phen-yl-, (E)-	EtOH	248(4.01)	35-1791-78
6-Hepten-3-one, 2,4,4-trimethyl-6-phen-yl-	EtOH	247(4.09)	35-1791-78

Compound	Solvent	$\lambda_{max}(\log \epsilon)$	Ref.
5-Hexen-3-one, 2,2,4,4-tetramethyl-6-phenyl-, (E)-	C_6H_{12}	254(4.26),286(3.56), 294(3.51)	39-0155-78B
$C_{16}H_{22}OS$			
A-Nor-1-thiaestra-2,5(10)-dien-17-ol, (9β,17β)-	EtOH	236(3.36),245s(3.33)	142-0207-78B
(17β)-	EtOH	235(3.77),245s(3.69)	142-0207-78B
A-Nor-3-thiaestra-1,5(10)-dien-17-ol, (17β)-	EtOH	237(3.79)	142-0207-78B
$C_{16}H_{22}O_2$			
2-Cyclohexen-1-one, 4-(octahydro-1-hydroxy-7a-methyl-5H-inden-5-ylidene)-	EtOH	305.5(4.20)	39-0024-78C
isomer	EtOH	305.5(4.15)	39-0024-78C
$C_{16}H_{22}O_2S$			
A-Nor-1-thiaestr-3(5)-en-2-one, 17-hydroxy-, (17β)-	EtOH	233(4.01),264s(3.27)	142-0207-78B
A-Nor-3-thiaestr-1(10)-en-2-one, 17-hydroxy-, (5α,17β)-	EtOH	232(4.12),264s(3.34)	142-0207-78B
$C_{16}H_{22}O_2S_2$			
2H-Thiopyran-3(4H)-one, 2,2'-(1,2-ethanediylidene)bis[dihydro-4,4-dimethyl-, (Z,Z)-	C_6H_{12}	216(3.80),242(3.95), 420(4.20)	24-3246-78
	EtOH	242(3.95),426(4.19)	24-3246-78
	$CHCl_3$	435(4.20)	24-3246-78
$C_{16}H_{22}O_3$			
6H-Cycloprop[e]azulen-6-one, 4-acetoxy-1,1a,2,3,4,4a,5,7b-octahydro-1,1,4-trimethyl-, [1aR-(1aα,4α,4aα,7bα)]-	EtOH	241(4.07)	44-0343-78
5H-Inden-5-one, 3-(1-acetoxy-2-methylpropyl)-1,2,3,6,7,7a-hexahydro-7a-methyl-	pentane	290(4.29)	33-0626-78
Strobilurin A, tetrahydro-	MeOH	216(3.78),241(3.81)	24-2779-78
$C_{16}H_{22}O_4$			
Benzoic acid, 2-cyclohexyl-3,6-dihydroxy-, 1-methylethyl ester	MeOH	219(4.37),266(4.09), 303(3.71)	95-1607-78
	MeOH-base	217(4.19),244(4.01), 306(4.29)	95-1607-78
2(3H)-Naphthalenone, 3-acetoxy-4,4a,5-6,7,8-hexahydro-4a-hydroxy-4-methyl-6-(1-methylethenyl)-, [3R-(3α,4β,4aβ,6α)]-	EtOH	220(4.08)	138-1205-78
$C_{16}H_{22}O_{11}$			
Cyclopenta[c]pyran-4-carboxylic acid, 1-(β-D-glucopyranosyloxy)-1,4a,5,7a-tetrahydro-5-hydroxy-7-(hydroxymethyl)- (10-deacetylasperulosidic acid)	MeOH	234(3.9)	32-0013-78
$C_{16}H_{23}Cl_2NO_2$			
Benzenamine, N,N-bis(2-chloroethyl)-4-(5,5-dimethyl-1,3-dioxan-2-yl)-	EtOH	261(4.37),293(3.23)	104-1004-78
$C_{16}H_{23}IN_2O$			
8-Isoquinolineethanaminium, 5,6-dihydro-3-methoxy-N,N,N-trimethyl-β-methylene-, iodide	MeOH	263(4.11)	44-0966-78

Compound	Solvent	$\lambda_{max}(\log \epsilon)$	Ref.
$C_{16}H_{23}NO$			
Azocine, 2-(2-furanyl)-1,2,3,4,5,8-hex-ahydro-1,6-dimethyl-3-(1-methylethen-yl)-	heptane	226(4.33)	70-2284-78
$C_{16}H_{23}NO_3$			
2-Oxetanone, 3-(1-methylethylidene)-4-(tetrahydro-2,2,6,6-tetramethyl-pyrrolo[1,2-b]isoxazol-3(2H)-yli-dene)-, (Z)-	pentane	231(3.64),289(3.56)	12-1757-78
2-Pyrrolidinone, 1-(8-hydroxy-1-oxo-2,4,6-dodecatrienyl)-, (E,E,E)-(±)-	EtOH	324(4.43)	18-2077-78
$C_{16}H_{23}N_2O$			
8-Isoquinolineethanaminium, 5,6-dihy-dro-3-methoxy-N,N,N-trimethyl-β-methylene-, iodide	MeOH	263(4.11)	44-0966-78
$C_{16}H_{23}N_3$			
Cyclohexanone, (2-amino-2-methyl-1-phenylpropylidene)hydrazone	hexane	211(4.34),278(3.26)	103-0278-78
$C_{16}H_{23}N_3O_4$			
1H-Pyrrolo[3,4-d]pyrimidine-7-carbox-ylic acid, 3-butyl-2,3,4,6-tetrahy-dro-5-methyl-2,4-dioxo-	EtOH	240(4.51),271(4.25)	94-3080-78
$C_{16}H_{23}N_5O_5$			
9H-Purin-6-amine, 9-[5-O-(3,3-dimethyl-1-oxobutyl)-β-D-arabinofuranosyl]-	MeOH	259(4.17)	87-1218-78
9H-Purin-6-amine, 9-[5-O-(1-oxohexyl)-β-D-arabinofuranosyl]-	MeOH	259(4.16)	87-1218-78
$C_{16}H_{23}N_5O_6$			
1H-Purine-2,6-dione, 7-[4-amino-4-de-oxy-2,3-O-(1-methylethylidene)-α-D-talopyranosyl]-3,7-dihydro-1,3-di-methyl-	MeOH	272.5(3.89)	136-0073-78C
Zeatin, 9-β-D-glucofuranosyl-	EtOH	268(4.10)	12-1095-78
	EtOH-HOAc	267(4.13)	12-1095-78
	EtOH-NH3	268(4.10)	12-1095-78
Zeatin, O-β-D-glucopyranosyl-	pH 1	274.7(4.21)	12-1291-78
	pH 13	275(4.22),282s(4.11)	12-1291-78
	EtOH	269.5(4.23)	12-1291-78
Zeatin, 7-β-D-glucopyranosyl-	EtOH	277(4.15)	12-1095-78
	EtOH-HOAc	282(4.27)	12-1095-78
	EtOH-NH3	277(4.17)	12-1095-78
Zeatin, 9-β-D-glucopyranosyl-	EtOH	268(4.26)	12-1095-78
	EtOH-HOAc	267(4.27)	12-1095-78
	EtOH-NH3	268(4.27)	12-1095-78
$C_{16}H_{24}F_3N_6O_5P$			
Acetamide, 2,2,2-trifluoro-N-6-[[9-[3-(phosphonooxy)propyl]-9H-purin-6-yl]amino]hexyl]-	pH 5.6	259(4.21)	5-0118-78
$C_{16}H_{24}N_2O$			
Acetaldehyde, [3,5-bis(1,1-dimethyleth-yl)-4-oxo-2,5-cyclohexadien-1-yli-dene]hydrazone	n.s.g.	308(4.48)	70-1179-78

Compound	Solvent	λ_{max}(log ϵ)	Ref.
2,5-Cyclohexadien-1-one, 4-(1-aziridin-ylimino)-2,6-bis(1,1-dimethylethyl)-	n.s.g.	295(3.24)	70-1179-78
$C_{16}H_{24}N_2O_2$ 4H-Pyran-4-one, 3,5-bis[3-(dimethyl-amino)-2-propenylidene]tetrahydro-2-methyl-	EtOH	500(4.88)	70-0102-78
$C_{16}H_{24}N_4$ 9-Azabicyclo[6.1.0]non-4-ene, 9,9'-azo-bis-, [1α,4Z,8α,9[E(1'R*,4'Z,8'S*)]]-	EtOH EtOH-CF₃COOH	209s(3.33),252(3.94) 220(4.25),248s(4.01)	33-0795-78 33-0795-78
$C_{16}H_{24}N_4O_6$ 1H-Imidazole-4-carboxamide, 5-amino-1-[2,3:5,6-bis-O-(1-methylethyli-dene)-α-mannofuranosyl]- β-anomer	MeOH MeOH	266(4.06) 265(4.08)	39-1381-78C 39-1381-78C
$C_{16}H_{24}N_4O_8$ 1,4-Cyclohexadiene-1,4-dicarboxylic acid, 2,5-bis[2-(ethoxycarbonyl)-hydrazino]-, dimethyl ester	EtOH	240(4.23),262(4.06)	48-0991-78
$C_{16}H_{24}N_6O_5$ Adenosine, N-[(pentylamino)carbonyl]-	pH 1 pH 13 70% EtOH	277(4.40) 270(4.32),277(4.27), 298(4.19) 269(4.37),276(4.29)	36-0569-78 36-0569-78 36-0569-78
$C_{16}H_{24}O$ 2,4,6-Cycloheptatrien-1-one, 2,4,7-tris(1-methylethyl)-	MeOH	238(4.31),328(3.80)	35-1778-78
$C_{16}H_{24}OS_2$ 1,3-Dithiane-2-methanol, α-(2-methyl-propyl)-2-(phenylmethyl)-	n.s.g.	223s(--),242s(--), 260(2.95),267(2.85)	33-3087-78
$C_{16}H_{24}O_2$ 5,9-Methanobenzocycloocten-4(1H)-one, 2,3,5,6,7,8,9,10-octahydro-5-hydroxy-2,7,9-trimethyl- lower melting isomer	EtOH EtOH	249(3.96) 248(3.94)	44-3653-78 44-3653-78
$C_{16}H_{24}O_3$ 8H-Indeno[1,2-c]furan-8-one, 1,3,3a,3b-4,5,6,8a-octahydro-1-methoxy-3,3,3b,4-tetramethyl-, (1α,3aα,3bα-4α,8aα)- Pentalenic acid, methyl ester Propanoic acid, 2,2-dimethyl-, 3-(1,1-dimethylethyl)-4-methoxyphenyl ester	pet ether MeOH EtOH	248(3.77),345(1.53) 227(3.73) 202(1.27),230(0.84), 278(0.39)	83-0511-78 88-4411-78 117-0079-78
$C_{16}H_{24}O_5$ 3-Furancarboxylic acid, 2,5-dihydro-2,4-dimethyl-2-(5-methyl-3-oxo-hexyl)-5-oxo-	EtOH	229(4.04)	78-0955-78
$C_{16}H_{24}O_7$ Epi-rhododendrin	EtOH	225(3.86),280(3.24), 287s(--)	95-0041-78

Compound	Solvent	$\lambda_{max}(\log \epsilon)$	Ref.
$C_{16}H_{24}O_9$			
Cyclopenta[c]pyran-4-carboxylic acid, 1-(α-D-glucopyranosyloxy)-1,4a,5,6-7,7a-hexahydro-, methyl ester, [1R-(1α,4aα,7aα)]-	MeOH	238(4.01)	24-2441-78
Cyclopenta[c]pyran-4-carboxylic acid, 1-(β-D-glucopyranosyloxy)-1,4a,5,6-7,7a-hexahydro-, methyl ester, [1S-(1α,4aα,7aα)]-	MeOH	237(4.10)	24-2423-78
α-D-Glucopyranose, 2-O-[1,4a,5,6,7,7a-hexahydro-4-(methoxycarbonyl)-cyclopenta[c]pyran-1-yl]-, [1S-(1α,4aα,7aα)]-	MeOH	237(4.02)	24-2423-78
$C_{16}H_{25}AsO_9S$			
β-D-Galactopyranose, 1-thio-2,3,4,6-tetraacetate 1-(dimethyl-arsinite)	MeOH	223(3.61)	136-0069-78E
$C_{16}H_{25}AsO_9Se$			
β-D-Galactopyranose, 1-seleno-, 2,3,4,6-tetraacetate 1-(dimethyl-arsinite)	MeOH	237(3.69)	136-0069-78E
$C_{16}H_{25}ClO$			
5,9-Methanobenzocycloocten-4(1H)-one, 5-chlorodecahydro-2,7,9-trimethyl-	EtOH	294(1.60)	44-3653-78
$C_{16}H_{25}ClO_4$			
6-Tetradecynoic acid, 9-chloro-8-hy-droxy-3-oxo-, ethyl ester. (R*,S*)-(±)-	EtOH-NaOH	275(4.32)	22-0131-78
$C_{16}H_{25}NO$			
2-Cyclohexen-1-one, 3-[[2-(1-cyclohex-en-1-yl)ethyl]amino]-5,5-dimethyl-	MeOH	292(4.78)	44-4420-78
$C_{16}H_{25}NO_2$			
Benzenamine, 4-(5,5-dimethyl-1,3-diox-an-2-yl)-N,N-diethyl-	EtOH	268(4.35),295(3.36)	104-1004-78
$C_{16}H_{25}NO_4$			
1,4-Cyclohexadiene-1-carboxylic acid, 3-[[(1,1-dimethylethoxy)carbonyl]-amino]-, 1,1-dimethylethyl ester, (±)-	MeOH	284(3.79)	44-1448-78
3,5-Pyridinedicarboxylic acid, 1,4-di-hydro-2,6-dimethyl-4-propyl-, diethyl ester	EtOH	233(4.29),268(4.15), 358(3.92)	103-1226-78
3,5-Pyridinedicarboxylic acid, 1-ethyl-1,4-dihydro-2,4,6-trimethyl-, diethyl ester	EtOH	235(4.11),263(3.94), 351(3.83)	103-1226-78
$C_{16}H_{25}NO_6$			
Glycine, N-[2,2-bis(ethoxycarbonyl)eth-enyl]-, 1,1-dimethylethyl ester	EtOH	221(4.21),279(4.47)	94-2224-78
$C_{16}H_{25}N_3O_4$			
2,3-Pyrrolidinedicarboxamide, 1-cyclo-hexyl-N,N'-diethyl-4,5-dioxo-	EtOH	239(4.04)	4-1463-78

Compound	Solvent	$\lambda_{max}(\log \epsilon)$	Ref.
$C_{16}H_{25}N_3O_6$ Heptanamide, N-(1-β-D-arabinofuranosyl-1,2-dihydro-2-oxo-4-pyrimidinyl)-	isoPrOH	216(4.23),248(4.18), 303(3.91)	94-0981-78
$C_{16}H_{25}N_3O_6S$ 2-Thiazolebutanoic acid, α-(acetylamino)-4-[[(2-ethoxy-2-oxoethyl)amino]-carbonyl]-4,5-dihydro-, ethyl ester	50% EtOH-HCl	267(3.73)	44-1624-78
$C_{16}H_{26}NO_5PS_3$ Phosphorothioic acid, O,O-bis(1-methylpropyl) S-[(2,3,5,7-tetrahydro-2-methyl-5,7-dioxo-6H-1,4-dithiino[2,3-c]pyrrol-6-yl)methyl] ester	EtOH	214(3.63),264(3.98), 419(3.47)	73-1093-78
Phosphorothioic acid, O,O-dibutyl S-[(2,3,5,7-tetrahydro-2-methyl-5,7-dioxo-6H-1,4-dithiino[2,3-c]-pyrrol-6-yl)methyl] ester	EtOH	213(3.48),263(3.92), 418(3.51)	73-1093-78
$C_{16}H_{26}N_2O_7$ 2,5,8,11,14,17,20-Heptaoxa-22,25-diaza-bicyclo[19.3.1]pentacosa-1(25),21,23-triene	EtOH	236(5.28),280(4.60)	44-3362-78
$C_{16}H_{26}O_2$ 5,9-Methanobenzocycloocten-4(1H)-one, decahydro-5-hydroxy-2,7,9-trimethyl-	EtOH	294(1.40)	44-3653-78
2(1H)-Naphthalenone, 5-(1,1-dimethylethoxy)octahydro-4a-methyl-1-methylene-, [4aS-(4aα,5α,8aβ)]-	hexane	220(3.70)	33-2397-78
$C_{16}H_{26}O_3$ 2-Cyclohexen-1-one, 3-(3-acetoxybutyl)-2,4,4,5-tetramethyl-	EtOH	247(4.07)	33-2328-78
3(2H)-Furanone, 2,4-bis(1,1-dimethylethyl)-2-(2-oxobutyl)-	EtOH	270(4.2)	88-3597-78
$C_{16}H_{26}O_6S_2$ Heptos-2-ulofuranose, 3,4:6,7-bis-O-(1-methylethylidene)-, cyclic 1,3-propanediyl mercaptal	MeOH	230(3.96),245(3.87)	94-2782-78
$C_{16}H_{27}IN_2O$ Methanaminium, N-[[3-[3-(dimethylamino)-2-propenylidene]-2-ethoxy-1-cyclohexen-1-yl]methylene]-N-methyl-, iodide	EtOH	550(5.11)	70-0339-78
$C_{16}H_{27}NO$ 5,9-Methanobenzocycloocten-4(1H)-one, 5-aminodecahydro-2,7,9-trimethyl-	EtOH	295(1.36)	44-3653-78
$C_{16}H_{27}NSi$ Benzenamine, N-[2,2-dimethyl-1-(trimethylsilyl)propylidene]-2,6-dimethyl-	C_6H_{12}	214(4.14),244(3.96), 282(3.05),348(2.34)	23-2286-78
	isoPrOH	212(4.07),242(3.95), 281(3.07),343(2.29)	23-2286-78
$C_{16}H_{27}N_2O$ Methanaminium, N-[9-(dimethylamino)-5-ethoxy-4-methyl-2,4,6,8-nonatetra-	CH_2Cl_2	650(4.95)	70-0339-78

Compound	Solvent	$\lambda_{max}(\log \epsilon)$	Ref.
$C_{16}H_{27}N_2O$			
Methanaminium, N-[9-(dimethylamino)-5-ethoxy-4-methyl-2,4,6,8-nonatetraenylidene]-N-methyl-, tetrafluoroborate	CH_2Cl_2	650(4.95)	70-0339-78
Methanaminium, N-[[3-[3-(dimethylamino)-2-propenylidene]-2-ethoxy-1-cyclohexen-1-yl]methylene]-N-methyl-, iodide	EtOH	550(5.11)	70-0339-78
tetrafluoroborate	EtOH	550(5.09)	70-0339-78
$C_{16}H_{28}$			
7-Hexadecen-9-yne, (E)-	EtOH	226(4.41)	39-1631-78C
7-Hexadecen-9-yne, (Z)-	EtOH	226.5(4.25)	39-1531-78C
7-Pentadecyne, 9-methylene-	EtOH	224(4.06),233(4.00)	39-1631-78C
$C_{16}H_{28}N_2O_4$			
1,6-Hexanediaminium, N,N'-bis(1-formyl-2-hydroxyethenyl)-N,N,N',N'-tetramethyl-, dihydroxide bis(inner salt)	EtOH	257(4.74)	73-1261-78
$C_{16}H_{28}N_4$			
9-Azabicyclo[6.1.0]nonane, 9,9'-azobis-, [1α,8α,9[E(1'R*,8'S*)]]-	EtOH	253(3.95)	33-0795-78
	EtOH-CF_3CO-OH	218(4.25),250s(4.02)	33-0795-78
$C_{16}H_{28}N_4O_2P$			
1,3,2-Dioxaphosphorinanium, 2-(diethylamino)-2-[[4-(dimethylamino)phenyl]azo]-4-methyl-, tetrafluoroborate	n.s.g.	280(3.87),464(4.67)	123-0190-78
$C_{16}H_{28}O_4$			
2-Cyclobuten-1-one, 3,4,4-tributoxy-	BuOH	248(4.12),289(1.95)	33-1784-78
$C_{16}H_{30}$			
4,6-Decadiene, 2,2,3,8,9,9-hexamethyl-, (E,E)-	heptane	210s(4.07),220s(4.34), 227(4.47),233(4.50), 240s(4.29)	78-2015-78
(E,Z)-	heptane	210s(3.94),220s(4.24), 228s(4.40),234(4.42), 243s(4.27)	78-2015-78
$C_{16}H_{32}Br_2N_2Ni$			
Nickel, dibromo[N,N'-1,2-ethanediylidenebis(2,4-dimethyl-3-pentanamine)-N,N']-, (T-4)-	hexane	<u>555(3.7),690(3.7)</u>	64-1381-78B
$C_{16}H_{32}GeSn$			
Germane, (1,1-dibutylstannacyclohexa-2,5-dien-4-yl)trimethyl-	C_6H_{12}	222(3.52)	101-0247-78E
$C_{16}H_{32}P_2Pt$			
Platinum, diethynylbis(triethylphosphine)-, cis	ether	246(4.04),285(3.84)	101-0101-78B
trans	ether	258(3.83),267(3.81), 304(3.85)	101-0101-78B
$C_{16}H_{32}SiSn$			
Silane, (1,1-dibutylstannacyclohexa-2,5-dien-4-yl)trimethyl-	C_6H_{12}	224(3.46)	101-0247-78E

Compound	Solvent	$\lambda_{max}(\log \epsilon)$	Ref.
$C_{16}H_{32}Sn_2$			
Stannacyclohexa-2,4-diene, 1,1-dibutyl-6-(trimethylstannyl)-	C_6H_{12}	235(3.62)	101-0247-78E
$C_{16}H_{36}MoO_4$			
Molybdenum, tetra-tert-butoxy-	C_6H_{12}	600(1.60)	35-2744-78

Compound	Solvent	$\lambda_{max}(\log \epsilon)$	Ref.
$C_{17}H_8Cl_4O_4$ Spiro[1,3-benzodioxole-2,1'(2'H)-naphthalen]-2'-one, 4,5,6,7-tetrachloro-6'-methoxy-	EtOH	218(4.71),257(3.38), 310(3.79),328(3.82)	2-0668-78
$C_{17}H_9BrCl_3NO_6S$ Furan, 2-[2-[5-(4-bromophenyl)-2-furanyl]-1-(trichloromethyl)sulfonyl]ethenyl]-5-nitro-	EtOH	203(4.36),260(4.22), 330(4.33),414(4.29)	73-1618-78
$C_{17}H_9ClFe_2O_6$ Iron, hexacarbonyl[μ-[(1,8,9,10-η-:2,3,4,5-η)-9-chloro-10-methyl-bicyclo[6.2.0]deca-1,3,5,7,9-pentaene]]di-	C_6H_{12}	270s(4.28),340(4.08), 420s(3.58)	5-1379-78
$C_{17}H_9Cl_3N_2O_8S$ Furan, 2-[2-(5-nitro-2-furanyl)-2-[(trichloromethyl)sulfonyl]ethenyl]-5-(4-nitrophenyl)-, (E)-	EtOH	203(4.20),233(4.15), 328(4.35),415(4.26)	73-1618-78
$C_{17}H_9Cl_4NO_6S$ Furan, 2-[2-[5-(4-chlorophenyl)-2-furanyl]-1-[(trichloromethyl)sulfonyl]ethenyl]-5-nitro-	EtOH	203(4.34),255(4.27), 330(4.34),414(4.30)	73-1618-78
$C_{17}H_9NO_4$ 8H-Benzo[g]-1,3-benzodioxolo[6,5,4-de]-quinolin-8-one, 12-hydroxy-	EtOH	248(4.14),260s(3.92), 274(4.04),311(3.39)	102-0837-78
	EtOH-NaOH	226(4.36),248s(4.30), 254s(4.26),260s(4.19), 272s(3.92),301(4.08)	102-0837-78
	EtOH-NaOH-HCl	249s(4.06),254(4.11), 257(4.11),273s(3.67), 293(3.98),328s(3.29)	102-0837-78
$C_{17}H_{10}ClFN_4O_2$ 4H-[1,2,4]Triazolo[1,5-a][1,4]benzodiazepine-2-carboxylic acid, 8-chloro-6-(2-fluorophenyl)-, sodium salt	MeOH	225(4.52)	33-0848-78
$C_{17}H_{10}Cl_3NO_5S$ Furan, 2-[2-(1-naphthalenyl)-1-[(trichloromethyl)sulfonyl]ethenyl]-5-nitro-	EtOH	219(4.85),243(4.09), 319(4.15),378(3.88)	73-1618-78
$C_{17}H_{10}Cl_3NO_6S$ Furan, 2-[2-(5-nitro-2-furanyl)-2-[(trichloromethyl)sulfonyl]ethenyl]-5-phenyl-	EtOH	203(4.33),247(4.28), 328(4.33),412(4.29)	73-1618-78
$C_{17}H_{10}N_2O_2$ 2-Naphthalenecarbonitrile, 1,4-dihydro-1,4-dioxo-3-(phenylamino)-	C_6H_{12}	440(3.60)	39-1083-78C
	EtOH	440(--)	39-1083-78C
1H-Pyrrole-2,5-dione, 1-(2-acridinyl)-	EtOH	254(4.92),258(4.95), 360(4.08)	94-0596-78
1H-Pyrrole-2,5-dione, 1-(9-acridinyl)-	EtOH	251(5.20),343s(3.89), 360(4.09),382s(3.67)	94-0596-78
$C_{17}H_{10}N_2O_2S$ Benzenamine, N-2H-naphtho[1,8-bc]thien-	dioxan	392(4.20)	93-1354-78

Compound	Solvent	$\lambda_{max}(\log \epsilon)$	Ref.
2-ylidene-4-nitro- (cont.)			93-1354-78
$C_{17}H_{10}N_4O_4$ Benzeneacetonitrile, 2,4-dinitro- α-2(1H)-quinolinylidene-	EtOH	292(4.1),390(3.7), 525(4.0)	104-1847-78
	EtOH-NaOH	265(4.21),315s(3.95), 395(3.61),565(4.30)	104-1847-78
$C_{17}H_{10}O$ 7H-Benz[de]anthracen-7-one	CH_2Cl_2	232(4.60),237(4.56), 242(4.51),254(4.46), 276s(4.06),285(4.07), 310(4.03),338s(3.66), 355s(3.86),389(4.13), 407s(4.06)	18-2674-78
$C_{17}H_{10}OS_4$ 2-Propanone, 1,3-bis(3H-1,2-benzodi- thiol-3-ylidene)-, (Z,Z)-	dioxan	243(4.43),281(4.19), 294(4.08),324(3.81), 413(3.46),463(3.84), 493(4.34),528(4.62)	78-2175-78
$C_{17}H_{11}ClO_2$ 5H-Benzocyclohepten-5-one, 6-(4-chloro- phenyl)-7-hydroxy-	MeOH	249(4.53),339(3.89)	83-0600-78
$C_{17}H_{11}ClO_4$ 2(5H)-Furanone, 4-benzoyl-5-(2-chloro- phenyl)-3-hydroxy-	H_2O EtOH EtOH-NaHCO$_3$ CHCl$_3$	253(3.67),343(3.73) 256(3.88),348(3.94) 250(3.92),348(4.06) 276(4.04)	4-0737-78 4-0737-78 4-0737-78 4-0737-78
$C_{17}H_{11}ClO_7$ 2-Anthracenecarboxylic acid, 6-chloro- 9,10-dihydro-3,5,8-trihydroxy-1-meth- yl-9,10-dioxo-, methyl ester	CHCl$_3$	240(4.39),281(4.44), 470(4.01)	39-1041-78C
$C_{17}H_{11}Cl_2NO_3$ 6H-1,3-Oxazinium, 2,5-bis(4-chloro- phenyl)-4-hydroxy-3-methyl-6-oxo-, hydroxide, inner salt	MeCN	213s(4.31),259(4.43), 370(3.79)	5-1655-78
$C_{17}H_{11}N$ 8-Azabenzo[4,5]cyclohepta[1,2,3-de]- naphthalene	heptane	215(4.8),225(4.70), 235(4.60),270(4.56), 279(4.51),308(4.04), 352(3.8)	118-0205-78
Benzo[c]phenanthridine	EtOH	254(4.9),262(5.04), 310s(4.09),343(3.86), 360(3.9)	118-0205-78
Benzo[k]phenanthridine	EtOH	215(4.23),226(3.78), 259(4.20),272(4.35), 281(4.23),310(3.70), 340(2.74),355(3.00), 376(3.00)	118-0205-78
$C_{17}H_{11}NOS$ Thiazolo[3,2-a]quinolinium, 1-hydroxy- 2-phenyl-, hydroxide, inner salt	MeOH	270(3.84),306(3.92), 482(4.14)	44-2700-78

Compound	Solvent	$\lambda_{max}(\log \epsilon)$	Ref.
$C_{17}H_{11}NOS_2$			
4H-1-Benzothiopyran-3-carbonitrile, 4-oxo-2-[(phenylmethyl)thio]-	EtOH	246(3.22),286(3.40), 318s(3.10),332(3.17)	78-0725-78
4H-Thiazolo[2,3-c][1,4]benzothiazinium, 1-hydroxy-4-(phenylmethylene)-, hydroxide, inner salt	MeOH	270(4.42),440(3.93)	2-0678-78
$C_{17}H_{11}NO_2$			
1H-Dibenz[e,g]indole-2-carboxylic acid	heptane and EtOH	206(4.31),244(4.40), 254(4.64),262(4.82), 268(4.71),229(4.25), 318(4.22),399(3.62), 354(3.67)	103-0851-78
$C_{17}H_{11}NO_3$			
Anthra[9,1-bc]pyran-2,7-dione, 3-(methylamino)-	CHCl$_3$	538(4.03)	104-1433-78
3H-Dibenz[f,ij]isoquinoline-2,7-dione, 1-hydroxy-3-methyl-	50% EtOH	290f(3.8),490(4.28)	103-1116-78
Naphth[1,2,3-cd]indole-1-carboxylic acid, 2,6-dihydro-2-methyl-6-oxo-	HOAc	320(4.1),365(3.7), 440(4.1)	103-0641-78
	98% H$_2$SO$_4$	375(4.3),450(3.7), 550(4.1)	103-0641-78
$C_{17}H_{11}NO_3S_2$			
4H-Thiazolo[2,3-c][1,4]benzothiazinium, 1-hydroxy-4-(phenylmethylene)-, hydroxide, inner salt, 5,5-dioxide	MeCN	236(4.23),270(4.02), 358(4.33)	2-0678-78
$C_{17}H_{11}NO_6$			
2(5H)-Furanone, 4-benzoyl-3-hydroxy-5-(3-nitrophenyl)-	H$_2$O	262(4.10),342(4.05)	4-0737-78
	EtOH	257(4.24),347(3.97)	4-0737-78
	EtOH-NaHCO$_3$	256(4.23),347(4.16)	4-0737-78
	CHCl$_3$	268(4.29)	4-0737-78
$C_{17}H_{11}N_3O_2$			
Benzeneacetonitrile, 4-nitro-α-2(1H)-quinolinylidene-	EtOH	290(4.0),400(3.7), 480(4.1)	104-1847-78
	EtOH-NaOH	325(4.12),410(3.89), 587(4.51)	104-1847-78
Benz[cd]indol-2-imine, N-(4-nitrophenyl)-	dioxan	404(4.20)	93-1354-78
$C_{17}H_{11}N_3O_3$			
Benzonitrile, 4-[(5-nitro-1-oxo-2(1H)-isoquinolinyl)methyl]-	20% MeCN	261(4.07),313(3.84), 372(3.67)	44-1132-78
$C_{17}H_{12}ClNO_2$			
5(4H)-Oxazolone, 4-[1-(4-chlorophenyl)-ethylidene]-2-phenyl-, (Z)-	EtOH	203(4.45),250(4.17), 350(4.42)	118-0832-78
$C_{17}H_{12}ClNO_3$			
6H-1,3-Oxazinium, 2-(4-chlorophenyl)-4-hydroxy-3-methyl-6-oxo-5-phenyl-	MeCN	212s(4.35),257(4.38), 370(3.68)	5-1655-78
$C_{17}H_{12}ClN_4O_6Rh$			
Rhodium, dicarbonylchloro[2-[(2,4-dinitrophenyl)methyl]-1-methyl-1H-benzimidazole-N^3]-, (SP-4-3)-	EtOH	250(4.35),275s(4.23), 283s(4.12),330(3.43)	104-0189-78

Compound	Solvent	$\lambda_{max}(\log \epsilon)$	Ref.
$C_{17}H_{12}F_3NO_2$ Benzoic acid, 4-(trifluoromethyl)-, (3-phenyl-2H-azirin-2-yl)methyl ester	C_6H_{12}	243(4.08)	35-4481-78
$C_{17}H_{12}F_3NO_4$ 1H-Benzo[de]quinoline-2,3-dicarboxylic acid, 4,5,6-trifluoro-1-methyl-, dimethyl ester	MeOH	234(4.42),342(4.26), 422(3.23)	39-0837-78C
$C_{17}H_{12}N_2$ Benz[cd]indol-2-imine, N-phenyl-	dioxan	389(4.07)	93-1354-78
$C_{17}H_{12}N_2O$ 2-Oxazoleacetonitrile, 4,5-diphenyl-	MeOH	224(4.28),282(4.10)	33-0286-78
Pyrido[1,2-a]indol-1-ol, 10-(2-pyridinyl)-	MeOH	266(4.57),320(3.86), 348(3.96),424(3.92), 438(3.87)	145-1056-78
$C_{17}H_{12}N_2O_2$ 2H-Anthra[1,2-d]imidazole-6,11-dione, 2,2-dimethyl-	dioxan	315s(3.76),405(3.36)	104-0381-78
Benzoic acid, 4-cyano-, (3-phenyl-2H-azirin-2-yl)methyl ester	C_6H_{12}	242(4.00)	35-4481-78
2,5-Pyrrolidinedione, 1-(2-acridinyl)-	EtOH	254(4.94),342s(3.89), 360(4.00)	94-0596-78
$C_{17}H_{12}N_2O_2S$ 2(1H)-Pyrimidinethione, 4-(1,3-benzodioxol-5-yl)-6-phenyl-	C_6H_{12}	238(4.21),253(4.17), 330(4.12)	4-0105-78
	EtOH	234s(4.33),299(4.37), 345(3.98),415s(3.63)	4-0105-78
$C_{17}H_{12}N_2O_4$ 5(4H)-Oxazolone, 4-[1-(3-nitrophenyl)-ethylidene]-2-phenyl-, (E)-	EtOH	202(4.55),328(4.12)	118-0832-78
(Z)-	EtOH	202(4.35),250(4.32), 341(4.37)	118-0832-78
5(4H)-Oxazolone, 4-[1-(4-nitrophenyl)-ethylidene]-2-phenyl-, (E)-	EtOH	204(4.33),247(4.20), 335(4.33)	118-0832-78
(Z)-	EtOH	203(4.36),246(4.16), 348(4.32)	118-0832-78
$C_{17}H_{12}N_2O_5$ 6H-1,3-Oxazinium, 4-hydroxy-3-methyl-2-(4-nitrophenyl)-6-oxo-5-phenyl-, hydroxide, inner salt	MeCN	255(4.63),295s(4.05), 390(3.74)	5-1655-78
$C_{17}H_{12}N_2O_6$ Furan, 2-[2-[5-(4-methylphenyl)-2-furanyl]-1-nitroethenyl]-5-nitro-, (E)-	dioxan	300(4.20),440(4.30)	73-3252-78
$C_{17}H_{12}N_4$ Pyrido[3,2-c]pyridazin-4-amine, N-1-naphthalenyl-	EtOH	254s(4.17),370(4.03)	103-0663-78
Pyrido[3,2-c]pyridazin-4-amine, N-2-naphthalenyl-	EtOH	254(4.22),369(4.08)	103-0663-78
$C_{17}H_{12}N_4O_2$ 1H-Purine-2,6-dione, 3,7-dihydro-1,3-diphenyl-	pH 6.0	220(4.11),279(4.11)	24-0982-78
	pH 13.0	220(4.27),278(4.18)	24-0982-78

Compound	Solvent	$\lambda_{max}(\log \epsilon)$	Ref.
$C_{17}H_{12}N_4O_5$ 1,4-Naphthalenedione, 2-methyl-, 4-[(2,4-dinitrophenyl)hydrazone	acid base	405(4.45) 630(4.94)	36-0258-78 36-0258-78
$C_{17}H_{12}N_6OS$ 4(1H)-Pteridinone, 2,3-dihydro-3-meth- yl-6,7-di-2-pyridinyl-2-thioxo-	DMF-pH 4.0 DMF-pH 9.0	285s(4.32),304(4.42), 373(4.14) 282(4.18),331(4.42), 395(3.97)	24-0971-78 24-0971-78
$C_{17}H_{12}O$ Phenanthro[9,10b]furan, 2-methyl-	EtOH	241(4.56),251(4.71), 258(4.81),282(4.19), 297(4.04),310(4.08)	39-0659-78B
$C_{17}H_{12}O_4$ 2(5H)-Furanone, 4-benzoyl-3-hydroxy- 5-phenyl-	H_2O EtOH EtOH-NaHCO$_3$ CHCl$_3$	254(3.96),377(3.95) 259(3.97),350(3.78) 251(3.96),350(4.07) 273(4.04)	4-0737-78 4-0737-78 4-0737-78 4-0737-78
$C_{17}H_{12}O_5$ 11H-Benzofuro[2,3-b][1]benzopyran- 11-one, 3,8-dimethoxy-	EtOH	244(4.36),342(4.44)	102-1417-78
$C_{17}H_{12}O_6$ 2-Anthraceneacetic acid, 9,10-dihydro- 1,4-dihydroxy-9,10-dioxo-, methyl ester	MeOH	207(4.26),230(4.34), 249(4.55),284(4.01), 318s(--),483(3.96), 515s(--)	24-3823-78
$C_{17}H_{12}O_7$ Laccaic acid D, methyl ester	MeOH	236(3.92),286(3.88), 436(3.75)	102-0895-78
$C_{17}H_{13}AsO$ 4(1H)-Arseninone, 1,2-diphenyl-	EtOH	223(4.29),310(4.13)	88-1471-78
$C_{17}H_{13}BrN_2O$ 1H-Pyrazole, 1-acetyl-3-(3-bromophen- yl)-5-phenyl-	EtOH	243(4.42),272(4.38)	94-1298-78
$C_{17}H_{13}BrN_2OS$ 4(5H)-Thiazolone, 5-[(4-bromophenyl)- methylene]dihydro-3-methyl-2-(phen- ylimino)- 4(5H)-Thiazolone, 5-[(4-bromophenyl)- methylene]-2-(methylphenylamino)-	MeOH MeOH dioxan 70% dioxan	238(4.21),336(4.44) 343(4.44) 338(4.41) 348(4.43)	104-0997-78 104-0997-78 104-0997-78 104-0997-78
$C_{17}H_{13}BrN_2O_2$ 4-Isoquinolinecarboxylic acid, 3-[(4- bromophenyl)amino]-, methyl ester	EtOH	228(4.37),270(4.52), 317(4.50),405(3.84)	138-0677-78
$C_{17}H_{13}BrO_3$ 1H-Indene-1-carboxylic acid, 4-(4-bro- mobenzoyl)-2,3-dihydro-	EtOH	262(4.32)	94-1776-78
$C_{17}H_{13}ClFN_3O_2$ 1H-1,4-Benzodiazepine, 7-chloro-5-(2- fluorophenyl)-2,3-dihydro-2-(1-nitro-	isoPrOH	226(4.45),390(4.43)	44-0936-78

Compound	Solvent	$\lambda_{max}(\log \epsilon)$	Ref.
ethylidene)- (cont.)			44-0936-78
$C_{17}H_{13}ClN_2$			
1H-Indole-2-carbonitrile, 3-[(2-chloro-phenyl)methyl]-1-methyl-	EtOH	290(4.28)	103-1204-78
$C_{17}H_{13}ClN_2O$			
1H-Pyrazole, 1-acetyl-3-(3-chlorophen-yl)-5-phenyl-	EtOH	243(4.51),273(4.46)	94-1298-78
Pyridazine, 3-(4-chlorophenyl)-6-(2-methylphenoxy)-	MeOH	251(4.49)	2-1000-78
$C_{17}H_{13}ClN_2O_3$			
Sydnone, 4-[2-(4-chlorophenyl)ethenyl]-3-(4-methoxyphenyl)-, (E)-	benzene	371(4.42)	48-0071-78
	EtOH	361(4.41)	48-0071-78
(Z)-	benzene	360(4.12)	48-0071-78
	EtOH	350(4.12)	48-0071-78
$C_{17}H_{13}ClN_2S$			
Pyrimidine, 4-(3-chlorophenyl)-2-(meth-ylthio)-6-phenyl-	EtOH	264(4.59),333(3.95)	4-0105-78
Pyrimidine, 4-(4-chlorophenyl)-2-(meth-ylthio)-6-phenyl-	EtOH	268(4.62),332(4.02)	4-0105-78
2(1H)-Pyrimidinethione, 4-(4-chloro-phenyl)-1-(4-methylphenyl)-	MeCN	226s(4.14),300(4.68), 418(3.43)	97-0334-78
$C_{17}H_{13}ClO$			
1(2H)-Naphthalenone, 2-[(3-chlorophen-yl)methylene]-3,4-dihydro-, cation	H_2SO_4	442(4.30)	65-1644-78
1(2H)-Naphthalenone, 2-[(4-chlorophen-yl)methylene]-3,4-dihydro-, cation	H_2SO_4	459(4.37)	65-1644-78
$C_{17}H_{13}ClO_3$			
1H-Indene-1-carboxylic acid, 4-(4-chlo-robenzoyl)-2,3-dihydro-	EtOH	260(4.29)	94-1153-78
1H-Indene-1-carboxylic acid, 6-(4-chlo-robenzoyl)-2,3-dihydro-	EtOH	261(4.26)	94-1153-78
$C_{17}H_{13}ClO_4$			
Benzeneacetic acid, 2-carboxy-α-[(2-chlorophenyl)methylene]-, α-methyl ester, (Z)-	n.s.g.	216(4.36),273(4.07), 293(4.02)	80-1465-78
Benzeneacetic acid, α-[(2-chlorophen-yl)methylene]-2-(methoxycarbonyl)-, (Z)-	n.s.g.	217(4.33),272(4.10), 290(4.07)	80-1465-78
$C_{17}H_{13}Cl_2NO_2$			
1,2-Benzisoxazole, 3-[5-(3,4-dichloro-phenyl)-2-furanyl]-4,5,6,7-tetrahy-dro-	MeOH	217(4.14),317(4.57)	73-0288-78
$C_{17}H_{13}Cl_2NS$			
Butanenitrile, 2-(6,8-dichloro-3H-naph-tho[1,8-bc]thien-3-ylidene)-3,3-di-methyl-	MeOH	226(3.9),249(4.2), 269(4.2),276(4.3), 411(4.2)	5-1974-78
$C_{17}H_{13}Cl_3O_5$			
9H-Xanthen-9-one, 2,4,5-trichloro-1,3,6-trimethoxy-8-methyl-	MeOH	214(4.60),249(4.99), 305(4.64),342s(4.06)	78-2491-78

Compound	Solvent	$\lambda_{max}(\log \epsilon)$	Ref.
$C_{17}H_{13}FO_3$			
1H-Indene-1-carboxylic acid, 4-(4-fluorobenzoyl)-2,3-dihydro-	EtOH	254(4.19)	94-1776-78
$C_{17}H_{13}N$			
9-Anthracenecarbonitrile, 10-ethyl-	C_6H_{12}	386(4.07)	61-1068-78
Benz[cd]indole, 1,2-dihydro-2-phenyl-	n.s.g.	270(3.05),290(3.22), 305(3.27),330(3.0), 350(2.78)	104-1245-78
1-Naphthalenamine, N-(phenylmethylene)-	n.s.g.	270(3.1),290(3.08), 305(2.85),330(2.9), 346(3.08)	104-1245-78
$C_{17}H_{13}NOS$			
Phenanthro[9,10-d]thiazole-2-methanol, α-methyl-	EtOH	254(4.83),261s(4.72), 274s(4.25),284(4.15), 302(4.06),315(4.03)	39-0685-78C
2-Quinolinecarboxaldehyde, 4-[(4-methylphenyl)thio]-	DMF	323(3.91)	97-0138-78
$C_{17}H_{13}NOS_2$			
4-Thiazolidinone, 5-(2,2-diphenylethylidene)-2-thioxo-	EtOH	206(4.58),242(4.04), 274(4.02),400(4.58)	56-2455-78
$C_{17}H_{13}NO_2$			
1,2-Naphthalenedione, 4-(methylphenylamino)-	MeOH	244(4.31),272(4.06), 322(3.81),456(3.83)	78-1377-78
1,2-Naphthalenedione, 4-[(4-methylphenyl)amino]-	MeOH	242(4.40),277(4.16), 330s(3.82),460(3.72)	78-1377-78
5(4H)-Oxazolone, 2-phenyl-4-(1-phenylethylidene)-, (E)-	EtOH	203(4.40),340(4.32)	118-0832-78
(Z)-	EtOH	203(4.35),254(4.13), 345(4.40)	118-0832-78
Phenanthrene, 9-(2-nitro-1-propenyl)-, (E)-	EtOH	214(4.36),253(4.70), 345(3.90)	39-0659-78B
(Z)-	EtOH	296(4.95),340s(4.42)	39-0659-78B
4(1H)-Pyridinone, 1-hydroxy-3,5-diphenyl-	EtOH	240(4.52),297(4.12), 325(3.93)	115-0067-78
2-Quinolinecarboxaldehyde, 4-(4-methylphenoxy)-	DMF	337(3.42)	97-0138-78
$C_{17}H_{13}NO_3$			
1H-Indene-1,3(2H)-dione, 2-[[(2-methoxyphenyl)amino]methylene]-	EtOH	386(4.57)	97-0256-78
1H-Indene-1,3(2H)-dione, 2-[[(4-methoxyphenyl)amino]methylene]-	EtOH	379(4.50)	97-0256-78
1,2-Naphthalenedione, 4-[(4-methoxyphenyl)amino]-	MeOH	245(4.40),279(4.20), 345(3.73),475(3.74)	78-1377-78
1,4-Naphthalenedione, 2-methoxy-3-(phenylamino)-	C_6H_{12}	495(4.51)	39-1083-78C
	EtOH	500(--)	39-1083-78C
1(2H)-Naphthalenone, 3,4-dihydro-2-[(4-nitrophenyl)methylene]-, cation	H_2SO_4	395(4.33)	65-1644-78
6H-1,3-Oxazinium, 4-hydroxy-2-methyl-6-oxo-3,5-diphenyl-, hydroxide, inner salt	MeCN	255(3.98),265s(3.92), 323(3.51)	5-1655-78
6H-1,3-Oxazinium, 4-hydroxy-3-methyl-6-oxo-2,5-diphenyl-, hydroxide, inner salt	MeCN	252(4.25),365(3.64)	5-1655-78
Phenol, 2-(4-phenyl-5-isoxazolyl)-, acetate	EtOH	270(4.03)	114-0069-78B

Compound	Solvent	$\lambda_{max}(\log \epsilon)$	Ref.
Spiro[2H-1-benzopyran-2,2'-[2H-1,3]-benzoxazin]-4'(3'H)-one, 3'-methyl-photoinduced form	isoPrOH	245(4.38),265(4.22), 295(3.91)	103-0122-78
	isoPrOH	370(--),575(--)	103-0122-78
$C_{17}H_{13}NO_4$			
4,7-Ethenofuro[3,4-c]pyridine-1,3,6-(3aH)-trione, 4,5,7,7a-tetrahydro-5-(2-phenylethenyl)-, [3aα,4β,5(E)-7β,7aα]-	MeCN	288(4.34)	78-2617-78
$C_{17}H_{13}N_3$			
Cyclohepta[b]pyrrole-3-carbonitrile, 1,2-dihydro-2-imino-1-(phenylmethyl)-	EtOH	230(4.13),284(4.38), 437(4.12)	138-0677-78
$C_{17}H_{13}N_3O$			
Benz[cd]indazolium, 1-[(4-methoxyphenyl)amino]-, hydroxide, inner salt	EtOH	293(4.32),302(4.25), 352(3.98),531(4.39), 554(4.35)	44-2508-78
Benz[cd]indazolium, 8-methoxy-1-(phenylamino)-, hydroxide, inner salt	EtOH	295(4.27),305(4.20), 348(3.83),498(4.18), 526(4.20),560(4.02)	44-2508-78
$C_{17}H_{13}N_3O_2$			
1H-Indole-2-carbonitrile, 1-methyl-3-[(3-nitrophenyl)methyl]-	EtOH	291(4.35)	103-1204-78
1H-Indole-2-carbonitrile, 1-methyl-3-[(4-nitrophenyl)methyl]-	EtOH	290(4.48)	103-1204-78
Isoquinolinium, 2-[3,3-dicyano-1-(ethoxycarbonyl)-2-propenyl]-, hydroxide, inner salt	EtOH	235(4.43),276(3.37), 333(4.33),460(3.00)	95-1412-78
[1,2,4]Triazolo[1,5-f]phenanthridine-2-carboxylic acid, ethyl ester	EtOH	241s(4.81),247(4.89), 278s(4.12),299(3.22), 311(3.48),326(3.60)	142-1577-78A
$C_{17}H_{13}N_3O_3$			
1H-Pyrazole, 1-acetyl-3-(3-nitrophenyl)-5-phenyl-	EtOH	243(4.41),269s(4.37)	94-1298-78
$C_{17}H_{13}N_3O_3S$			
4-Thiazolidinone, 3-methyl-5-[(3-nitrophenyl)methylene]-2-(phenylimino)-	MeOH	229(4.38),325(4.35)	104-0997-78
4-Thiazolidinone, 3-methyl-5-[(4-nitrophenyl)methylene]-2-(phenylimino)-	MeOH	258(4.14),360(4.24)	104-0997-78
4(5H)-Thiazolone, 2-(methylphenylamino)-5-[(3-nitrophenyl)methylene]-	MeOH	329(4.36)	104-0997-78
	dioxan	238(4.26),321(4.36)	104-0997-78
	70% dioxan	235(4.25),330(4.36)	104-0997-78
4(5H)-Thiazolone, 2-(methylphenylamino)-5-[(4-nitrophenyl)methylene]-	MeOH	357(4.42)	104-0997-78
	dioxan	361(4.38)	104-0997-78
	70% dioxan	364(4.36)	104-0997-78
$C_{17}H_{13}N_3O_4$			
Pyridine, 1,4-dihydro-4-[(2,4-dinitro-1-naphthalenyl)methylene]-1-methyl-	acetone	335(4.1),535(4.2), 607(4.5)	104-1847-78
Quinoline, 1,4-dihydro-4-[(2,4-dinitrophenyl)methylene]-1-methyl-	EtOH	240(4.3),275(4.0), 598(4.4)	104-1847-78
$C_{17}H_{13}N_3O_5$			
Benzamide, 2-nitro-N-(2,3,4,5-tetrahydro-2,5-dioxo-1H-1-benzazepin-3-yl)-	n.s.g.	227(4.60),252(4.17), 305(3.57)	12-0439-78
Sydnone, 3-(4-methoxyphenyl)-4-[2-(4-nitrophenyl)ethenyl]-, (E)-	benzene	398(4.49)	48-0071-78
	EtOH	389(4.48)	48-0071-78

Compound	Solvent	$\lambda_{max}(\log \epsilon)$	Ref.
Sydnone, 3-(4-methoxyphenyl)-4-[2-(4-nitrophenyl)ethenyl]-, (Z)-	benzene EtOH	383(4.23) 372(4.24)	48-0071-78 48-0071-78
$C_{17}H_{13}N_3O_6S$ 2-Pyridinecarboxylic acid, 6-(2-acetyl-4-oxazolyl)-5-[4-(methoxycarbonyl)-2-thiazolyl]-, methyl ester	MeOH	255(4.32),285s(--)	88-2791-78
$C_{17}H_{13}N_5$ Pyrazolo[5,1-c][1,2,4]triazin-7-amine, 3,4-diphenyl-	EtOH	281(4.45),353(3.48)	39-0885-78C
$C_{17}H_{13}N_5O$ 1H-Pyrazolo[3,4-d]pyrimidine-4-methanol, 1-phenyl-α-3-pyridinyl-	EtOH	244(4.58),260s(4.19), 268s(4.08),278s(3.83), 300(3.55)	95-1274-78
$C_{17}H_{13}N_5O_4$ 1,2,3-Triazolo[4',5':3,4]cyclopenta-[1,2-d]pyridazine-4,8-diol, 2,5-di-hydro-5-phenyl-, diacetate	MeOH	216(4.48),299(4.54)	56-0511-78
$C_{17}H_{13}N_7O_2S$ Benzaldehyde, 4-nitro-, (1-methyl-1H-[1,3,4]thiadiazino[5,6-b]quinoxalin-3-yl)hydrazone	EtOH	213(4.32),215(4.41), 250(4.32)	2-0307-78
$C_{17}H_{13}O$ Pyrylium, 2,6-diphenyl-	MeOH MeOH-NaOMe	277(4.21),400(4.40) 243(--),353(--)(changing)	44-4112-78 44-4112-78
$C_{17}H_{14}$ Cyclopentadiene, 1,2-diphenyl- Cyclopropene, 3-ethenyl-1,2-diphenyl- Cycloprop[a]indene, 1,1a,6,6a-tetrahy-dro-6-methylene-1-phenyl-, endo exo Spiro[cyclopropane-1,1'-[1H]indene], 2-phenyl-, anti syn	EtOH EtOH C_6H_{12} tert-BuOH tert-BuOH EtOH	233(4.30),308(3.96) 227(4.44),236(4.33), 315(4.51),332(4.42) 238(4.38),252(4.30), 286(3.71),301(3.59) 210(4.39),239(4.43), 290(3.46),300(3.35) 210(4.39),239(4.43), 290(3.46),300(3.35) 235(4.43),284(3.39), 296(3.31)	44-1481-78 44-1481-78 44-3283-78 44-3283-78 44-3283-78 44-3283-78
$C_{17}H_{14}Br_2$ 5,10-Methano-5H-dibenzo[a,d]cyclohept-ene, 11-bromo-10-(bromomethyl)-10,11-dihydro-, endo exo	hexane hexane	221(4.42),254(3.50), 300(2.26) 210(4.37),254(3.15), 300(2.30)	44-1756-78 44-1756-78
$C_{17}H_{14}ClN_3O$ Benz[cd]indazol-1(2H)-amine, 3(or 5)-chloro-N-(4-methoxyphenyl)- Imidazo[1,5,4-ef][1,5]benzodiazepin-6(7H)-one, 9-chloro-4,5-dihydro-2-(phenylmethyl)- Imidazo[4,5,1-jk][1,4]benzodiazepin-7(4H)-one, 9-chloro-5,6-dihydro-2-(phenylmethyl)-	EtOH MeOH MeOH	295(4.07),354(4.41), 476(3.73) 271s(4.13),276(4.17), 294(3.92),304(3.87) 275(3.87),313(3.86)	44-2508-78 111-0053-78 111-0053-78

Compound	Solvent	λ_{max} (log ϵ)	Ref.
3H-Pyrazol-3-one, 4-[[(4-chlorophenyl)-amino]methylene]-2,4-dihydro-5-methyl-2-phenyl-	C_6H_{11}Me	251s(4.30),255(4.30), 343(4.57)	48-0521-78
	PrOH	253(4.27),342(4.44)	48-0521-78
	PrCN	256(4.30),339(4.49)	48-0521-78
$C_{17}H_{14}ClN_3S$			
1H-Imidazol-1-amine, N-[(2-chlorophenyl)methylene]-2-(methylthio)-4-phenyl-	MeOH	281(4.53),356(4.02)	18-1846-78
1H-Imidazol-1-amine, N-[(2-chlorophenyl)methylene]-2-(methylthio)-5-phenyl-	MeOH	264(4.36)	18-1846-78
1H-Imidazol-1-amine, N-[(4-chlorophenyl)methylene]-2-(methylthio)-4-phenyl-	MeOH	283(4.57),349(4.12)	18-1846-78
1H-Imidazol-1-amine, N-[(4-chlorophenyl)methylene]-2-(methylthio)-5-phenyl-	MeOH	269(4.42)	18-1846-78
1H-Imidazol-1-amine, 4-(4-chlorophenyl)-2-(methylthio)-N-(phenylmethylene)-	MeOH	283(4.55),345(4.16)	18-1846-78
$C_{17}H_{14}Cl_2O_5$			
9H-Xanthen-9-one, 2,4-dichloro-1,3,6-trimethoxy-8-methyl-	MeOH	209(4.11),248(4.44), 285(3.94),304(4.03), 335(3.69)	78-2491-78
9H-Xanthen-9-one, 2,5-dichloro-1,3,6-trimethoxy-8-methyl-	MeOH	216(4.66),250(4.92), 277s(4.30),304(4.56), 332s(4.16)	78-2491-78
$C_{17}H_{14}F_2O_2$			
Benzoic acid, 4-(1,2-difluoro-2-phenylethenyl)-, ethyl ester, (E)-	EtOH	314(4.49)	104-0939-78
$C_{17}H_{14}N_2$			
1H-Indole-2-carbonitrile, 1-methyl-3-(phenylmethyl)-	EtOH	290(4.28)	103-1204-78
1H-Perimidine, 2,3-dihydro-1-phenyl-	MeOH	208(4.52),240(4.55), 365(4.12)	103-1156-78
1H-Pyrazole, 1-ethenyl-3,5-diphenyl-	MeOH	250(4.26),272(4.18), 278(4.18)	95-0095-78
$C_{17}H_{14}N_2O$			
Diazene, (4-methoxy-1-naphthalenyl)-phenyl-	benzene	400(3.99)	44-3882-78
	MeOH	393(4.07)	44-3882-78
	EtOH	395(4.04)	44-3882-78
	acetone	395(4.01)	44-3882-78
	HOAc	395(4.28)	44-3882-78
	CHCl$_3$	398(4.04)	44-3882-78
	pyridine	405(4.06)	44-3882-78
1,4-Naphthalenedione, mono(methylphenylhydrazone)	benzene	460(3.85)	44-3882-78
	MeOH	460(4.03)	44-3882-78
	EtOH	462(3.99)	44-3882-78
	acetone	452(3.94)	44-3882-78
	HOAc	469(4.03)	44-3882-78
	CHCl$_3$	461(3.93)	44-3882-78
	pyridine	460(3.96)	44-3882-78
1-Naphthalenol, 4-[(4-methylphenyl)-azo]-	pyridine	413(4.3)	44-3882-78
6H-Pyrido[4,3-b]carbazol-7-ol, 5,11-dimethyl-	EtOH	245(4.49),288(4.71), 333(3.63)	88-1261-78
	EtOH-HCl	255(4.43),285(4.23), 306(4.39)	88-1261-78

Compound	Solvent	$\lambda_{max}(\log \epsilon)$	Ref.
$C_{17}H_{14}N_2OS$			
Acetamide, N-[4-(phenylimino)-4H-1-benzothiopyran-2-yl]-	EtOH	275s(4.27),349(4.20), 364s(4.18)	103-0507-78
Phthalazine, 4-methyl-1-phenacylthio-	n.s.g.	242.5(4.3522)	2-0689-78
4(1H)-Pyrimidinethione, 4-(4-methoxyphenyl)-6-phenyl-	EtOH	248(4.27),297(4.50), 326s(4.23),450(3.68)	4-0105-78
	C_6H_{12}	233(4.41),258f(4.50), 268f(4.32),326(4.36)	4-0105-78
4-Thiazolidinone, 3-methyl-2-(phenylimino)-5-(phenylmethylene)-	MeOH	233(4.21),330(4.36)	104-0997-78
	dioxan	235(4.23),324(4.33)	104-0997-78
	70% dioxan	235(4.17),327(4.33)	104-0997-78
4(5H)-Thiazolone, 2-(methylphenylamino)-5-(phenylmethylene)-	MeOH	329(4.38)	104-0997-78
	dioxan	323(4.33)	104-0997-78
	70% dioxan	233(4.00),330(4.36)	104-0997-78
$C_{17}H_{14}N_2O_2$			
Cycloprop[a]inden-1a(1H)-amine, 6,6a-dihydro-N-[(4-nitrophenyl)methylene]-	C_6H_{12}	306(3.99)	35-2181-78
4-Isoquinolinecarboxylic acid, 3-(phenylamino)-, methyl ester	EtOH	225(4.34),266(4.51), 312(4.37),406(3.77)	138-0677-78
1,4-Naphthalenedione, 2-(methylamino)-3-(phenylamino)-	C_6H_{12}	549(3.43)	39-1083-78C
	EtOH	552(--)	39-1083-78C
1-Naphthalenol, 4-[(4-methoxyphenyl)azo]-	pyridine	418(<u>4.4</u>)	44-3882-78
Propanoic acid, 2-(9-phenanthrylhydrazono)-	heptane and EtOH	211(2.45),225(2.36), 252(2.62),317(1.90)	103-0851-78
1H-Pyrazole, 5-(1,3-benzodioxol-5-yl)-1-methyl-3-phenyl-	EtOH	259(4.58),291s(4.13)	4-0385-78
3H-Pyrazol-3-one, 4-benzoyl-2,4-dihydro-5-methyl-2-phenyl-	heptane	242(4.28),289(4.25)	18-1525-78
	MeOH	236(4.23),273(4.24)	18-1525-78
	EtOH	238(4.21),278(4.18)	18-1525-78
	$CHCl_3$	246(4.19),287(4.25)	18-1525-78
	CCl_4	291(4.24)	
2(1H)-Quinoxalinone, 3,4-dihydro-3-[2-(4-methylphenyl)-2-oxoethylidene]-ion	EtOH	438(4.40)	103-0336-78
	EtOH	484(4.78)	103-0336-78
2(1H)-Quinoxalinone, 3-(1-methyl-2-oxo-2-phenylethyl)-	EtOH	336(3.79)	103-0336-78
$C_{17}H_{14}N_2O_2S$			
[1]Benzothieno[3'.2':5,6]pyrimido[2,1-a]isoindole-6,13-dione, 4b,5,7,8,9-10-hexahydro-	EtOH	231(4.36),342(3.88)	4-0949-78
4-Thiazolidinone, 5-[(4-methoxyphenyl)imino]-2-(phenylimino)-	MeOH	236(4.21),364(4.44)	104-0997-78
	dioxan	237(4.23),350(4.42)	104-0997-78
	70% dioxan	238(4.24),362(4.42)	104-0997-78
$C_{17}H_{14}N_2O_3$			
Benzamide, N-(2,3,4,5-tetrahydro-2,5-dioxo-1H-1-benzazepin-3-yl)-	n.s.g.	229(4.64),312(3.45)	12-0439-78
Benzo[b]cyclopropa[d]pyran-7b(1H)-amine, 1a,2-dihydro-N-[(4-nitrophenyl)methylene]-	MeOH	280(4.37)	35-3494-78
1(2H)-Isoquinolinone, 2-[(4-methylphenyl)methyl]-5-nitro-	20% MeCN	262(4.06),314(3.83), 372(3.70)	44-1132-78
2(1H)-Quinoxalinone, 3,4-dihydro-3-[2-(4-methoxyphenyl)-2-oxoethylidene]-ionic	EtOH	440(4.40)	103-0336-78
	EtOH	500(4.81)	103-0336-78
Sydnone, 3-(4-methoxyphenyl)-4-(2-phenylethenyl)-, (E)-	benzene	367(4.36)	48-0071-78
	EtOH	357(4.35)	48-0071-78

Compound	Solvent	$\lambda_{max}(\log \epsilon)$	Ref.
Sydnone, 3-(4-methoxyphenyl)-4-(2-phenylethenyl)-, (Z)-	benzene EtOH	357(4.08) 348(4.05)	48-0071-78 48-0071-78
$C_{17}H_{14}N_2O_3S$			
Acetic acid, [[3,4-dihydro-3-(4-methylphenyl)-4-oxo-2-quinazolinyl]thio]-	MeOH	276(4.17),315(3.62)	2-0678-78
$C_{17}H_{14}N_2O_4$			
1,2-Benzisoxazole, 4,5,6,7-tetrahydro-3-[5-(2-nitrophenyl)-2-furanyl]-	MeOH	219(4.09),293(4.29)	73-0288-78
1,2-Benzisoxazole, 4,5,6,7-tetrahydro-3-[5-(3-nitrophenyl)-2-furanyl]-	MeOH	219(4.08),310(4.46)	73-0288-78
1,2-Benzisoxazole, 4,5,6,7-tetrahydro-3-[5-(4-nitrophenyl)-2-furanyl]-	MeOH	224(3.98),280(3.95), 364(4.38)	73-0288-78
1(2H)-Isoquinolinone, 2-[(4-methoxyphenyl)methyl]-5-nitro-	20% MeCN	262(4.07),313(3.82), 367(3.65)	44-1132-78
$C_{17}H_{14}N_2O_4S$			
Acetic acid, [[3,4-dihydro-3-(4-methoxyphenyl)-4-oxo-2-quinazolinyl]thio]-	MeOH	276(4.22),315(3.62)	2-0678-78
$C_{17}H_{14}N_2S$			
Pyrimidine, 2-(methylthio)-4,6-diphenyl-	EtOH	264(4.58),327(3.95)	4-0105-78
2(1H)-Pyrimidinethione, 4-(4-methylphenyl)-6-phenyl-	C_6H_{12} EtOH	235s(4.28),260(4.47), 321(4.11) 248(4.25),300(4.48), 408(3.11)	4-0105-78 4-0105-78
$C_{17}H_{14}N_4$			
Pyrido[3,2-c]pyridazine, 4-(1,2-dimethyl-1H-indol-3-yl)-	EtOH	214(4.76),283(4.13), 292(4.10)	103-1261-78
$C_{17}H_{14}N_4O_2$			
2,7-Naphthyridine-4-carbonitrile, 8-amino-2,3,7,8-tetrahydro-2,7-dimethyl-1,3,8-dioxo-6-phenyl-	n.s.g.	247(4.19),284(4.64), 324(3.90),362(3.99)	39-0554-78C
$C_{17}H_{14}N_4O_3$			
Imidazo[4,5,1-jk][1,4]benzodiazepin-7(4H)-one, 5,6-dihydro-9-nitro-2-(phenylmethyl)-	MeOH	260(4.44),310(4.05)	111-0053-78
3H-Pyrazol-3-one, 2,4-dihydro-5-methyl-4-[[(2-nitrophenyl)amino]methylene]-2-phenyl-	$C_6H_{11}Me$ PrOH PrCN	253(4.28),324s(4.07), 340(4.09) 255(4.41),334(4.34), 370(4.18) 257(4.30),331(4.13), 369(4.11)	48-0521-78 48-0521-78 48-0521-78
3H-Pyrazol-3-one, 2,4-dihydro-5-methyl-4-[[(4-nitrophenyl)amino]methylene]-2-phenyl-	$C_6H_{11}Me$ PrOH PrCN	247(4.00),364(4.31) 251(4.06),370(4.30) 249(4.02),369(4.29)	48-0521-78 48-0521-78 48-0521-78
$C_{17}H_{14}N_4O_3S$			
Benzamide, N-[5-ethyl-3-(4-nitrophenyl)-1,3,4-thiadiazol-2(3H)-ylidene]-	EtOH	290(4.298),340(4.385)	146-0864-78
$C_{17}H_{14}O$			
2,4,10,12-Cyclotridecatetraene-6,8-diyn-1-ol, 1-ethynyl-5,10-dimethyl-	$CHCl_3$	260(4.19),329(3.63)	150-0454S-78
1H-Inden-1-one, 2,3-dihydro-2-[(4-methylphenyl)methylene]-, cation	H_2SO_4	440(4.54)	65-1644-78

Compound	Solvent	$\lambda_{max}(\log \epsilon)$	Ref.
1(2H)-Naphthalenone, 3,4-dihydro-2-(phenylmethylene)-, cation	H_2SO_4	454(4.45)	65-1644-78
$C_{17}H_{14}OS_2$			
2(1H)-Naphthalenone, 1-(4,5,6,7-tetrahydro-1,3-benzodithiol-2-ylidene)-	MeCN	280(4.53),322(3.87), 362(3.49),480(4.36)	97-0385-78
$C_{17}H_{14}O_2$			
2(5H)-Furanone, 5-methyl-3,4-diphenyl-	MeOH	282(4.08)	35-8247-78
2(5H)-Furanone, 5-methyl-3,5-diphenyl-	MeOH	260(4.30)	35-8247-78
1H-Inden-1-one, 2,3-dihydro-2-[(4-methoxyphenyl)methylene]-, cation	H_2SO_4	469(4.51)	65-1644-78
9-Phenanthrenecarboxylic acid, 3-methyl-, methyl ester	MeOH	211(4.41),235(4.42), 256(4.65),305(4.12)	78-0769-78
$C_{17}H_{14}O_3$			
4H-1-Benzopyran-4-one, 7-methoxy-2-methyl-3-phenyl-	MeOH	245(4.34),295(4.11)	130-0493-78
1H-Indene-1-carboxylic acid, 4-benzoyl-2,3-dihydro-	EtOH	253(4.20)	94-1153-78
1H-Indene-1-carboxylic acid, 6-benzoyl-2,3-dihydro-	EtOH	260(4.16)	94-1153-78
Phenanthro[2,3-d][1,3]dioxole, 2-ethoxy-	n.s.g.	253(4.80)	39-0739-78C
$C_{17}H_{14}O_4$			
Anhydrovariabilin	EtOH	244(4.14),335(4.49), 351(4.44)	102-1417-78
4H-1-Benzopyran-4-one, 2-(3,4-dimethoxyphenyl)-	EtOH	238(4.34),316(4.30)	22-0043-78
4H-1-Benzopyran-4-one, 7-methoxy-2-(4-methoxyphenyl)-	EtOH	252(3.99),320(4.36)	22-0043-78
	EtOH	237(4.28),262(4.43), 303s(3.93)	102-1375-78
4H-1-Benzopyran-4-one, 8-methoxy-2-(4-methoxyphenyl)-	MeOH	231(4.29),270(4.21), 324(4.45)	18-1175-78
Phenanthrene, 7,8-dimethoxy-2,3-(methylenedioxy)-	n.s.g.	258(4.89),283(4.32)	39-0739-78C
Phenanthrene, 7,8-dimethoxy-3,4-(methylenedioxy)-	n.s.g.	258(4.88),271(4.83), 291(4.51)	39-0739-78C
9-Phenanthreneacetic acid, 9,10-dihydro-9-hydroxy-10-oxo-, methyl ester	MeOH	244(4.48),280(3.86), 328(3.44)	12-2259-78
$C_{17}H_{14}O_5$			
9,10-Anthracenedione, 1,3,6-trimethoxy-	$CHCl_3$	280(4.62),291s(4.34), 338(3.72),408(3.64)	12-1335-78
9,10-Anthracenedione, 1,3,7-trimethoxy-	EtOH	228(4.36),271s(4.34), 279(4.38),300(4.26), 367(3.84)	12-1335-78
4H-1-Benzopyran-4-one, 5-hydroxy-6,7-dimethoxy-2-phenyl-	EtOH	249(4.21),272(4.43), 313(4.17)	102-1363-78
Furo[2,3-f]-1,3-benzodioxole, 6-(2,4-dimethoxyphenyl)-	EtOH	230s(4.29),240s(4.16), 282(4.17),330(4.57), 347(4.61)	94-1274-78
Kuhlmannin	EtOH	214(4.40),300(3.81)	102-1383-78
9,10-Phenanthrenedione, 1,3,5-trimethoxy-	MeOH	234(4.42),261(4.33), 278s(4.07),412(3.85)	12-2259-78
$C_{17}H_{14}O_6$			
9,10-Anthracenedione, 1,3,8-trihydroxy-6-(3-hydroxypropyl)-	MeOH	224(4.41),252(4.17), 262(4.14),291(4.17), 437(3.94),455s(3.89)	105-0561-78

Compound	Solvent	$\lambda_{max}(\log \epsilon)$	Ref.
2H-1-Benzopyran-2-one, 3-(2,4-dihy-droxyphenyl)-5,7-dimethoxy-	EtOH	252s(4.10),259(4.13), 353(4.28)	94-0135-78
Pterocarpan, 3-hydroxy-4-methoxy-8,9-(methylenedioxy)-, (-)-	EtOH	229(4.00),283(3.45)	102-1419-78
Pterocarpan, 4-hydroxy-3-methoxy-8,9-(methylenedioxy)-	EtOH	236(3.88),312(3.69)	102-1419-78
$C_{17}H_{15}Br$			
Phenanthrene, 6-bromo-3,9,10-trimethyl-	hexane	225(4.47),232(4.52), 254s(--),260(4.80), 280(4.34),284s(--), 296(4.09),308(4.14), 329(2.82),345(2.96), 362(2.98)	104-0924-78
$C_{17}H_{15}BrN_2O$			
Ethanol, 2-[[4-(1-bromo-4-isoquinolin-yl)phenyl]amino]-	EtOH	213(4.55),220(4.61), 259(4.25)	44-0672-78
$C_{17}H_{15}BrO_5$			
2H-Naphtho[1,2-b]pyran-5,6-dione, 4-acetoxy-3-bromo-3,4-dihydro-2,2-dimethyl-	EtOH	252(4.39),285(3.84), 330(3.28),430(3.17)	2-0035-78
$C_{17}H_{15}ClN_2O_4$			
Benzenebutanoic acid, 2-amino-α-[(3-chlorobenzoyl)amino]-γ-oxo-, (±)-	n.s.g.	228(4.46),256(3.90), 367(3.69)	12-0439-78
Cyclohepta[b]pyrrole-3-carboxylic acid, 2-chloro-6-(1-cyano-2-ethoxy-2-oxo-ethylidene)-1,6-dihydro-, ethyl ester	MeOH	258(4.42),281(4.17), 444(4.56)	18-0667-78
$C_{17}H_{15}ClN_2O_8S$			
Benzoic acid, 4-chloro-, [2,3,3a,9a-tetrahydro-3-[(methylsulfonyl)oxy]-6-oxo-6H-furo[2',3':4,5]oxazolo[3,2-a]pyrimidin-2-yl]methyl ester, [2R-(2α,3β,3aβ,9aβ)]-	MeOH	242(4.36)	44-0350-78
$C_{17}H_{15}ClN_4$			
Cyanamide, (6-chloro-3,4-dihydro-1,3-dimethyl-4-phenyl-2(1H)-quinazolin-ylidene)-	MeOH	218(4.36),245(4.19), 269(4.21)	4-1409-78
Ethanone, 1-phenyl-, (4-chloro-1-phtha-lazinyl)methylhydrazone	dioxan	245s(4.20),352(3.95)	103-0436-78
	MeCN	215(4.68),240s(4.16), 351(3.96)	103-0436-78
	crystal	214(--),250(--), 357(--)	103-0436-78
1(2H)-Phthalazinone, 4-chloro-2-meth-yl-, (1-phenylethylidene)hydrazone, (?,Z)-	dioxan	281(4.11),300(4.04), 377(4.27)	103-0567-78
	MeCN	203s(4.49),215(4.56), 280(4.04),298(4.00), 374(4.22)	103-0567-78
$C_{17}H_{15}ClO_4$			
Benzenepropanoic acid, 2-chloro-α-[2-(methoxycarbonyl)phenyl]-	n.s.g.	217(4.32),275(3.79)	80-1465-78
$C_{17}H_{15}ClO_5$			
9H-Xanthen-9-one, 2-chloro-3,6,8-tri-methoxy-1-methyl-	MeOH	213(4.74),247(4.90), 302(4.62),333(4.40)	78-2491-78

Compound	Solvent	λ_{max}(log ϵ)	Ref.

$C_{17}H_{15}FN_2O_2S$
  Benzenesulfonic acid, 4-methyl-, (2-
  fluoro-3-phenyl-2-cyclobuten-1-yl-
  idene)hydrazide
	EtOH	219(4.51),311(4.54)	44-4873-78

$C_{17}H_{15}FN_2O_4$
  Benzenebutanoic acid, 2-amino-
  α-[(4-fluorobenzoyl)amino]-γ-oxo-, (±)-
	n.s.g.	225(4.49),252(3.94),
		363(3.74)
			12-0439-78

$C_{17}H_{15}FO_2S$
  Benzene, 1-[(2-fluoro-3-phenyl-2-cyclo-
  buten-1-yl)sulfonyl]-4-methyl-
	EtOH	217(4.76),230(4.70),
		257(4.84)
			44-4873-78

$C_{17}H_{15}N$
  2-Naphthalenamine, 1-(phenylmethyl)-,
  hydrochloride
	EtOH	212(4.82),244(5.10),
		274s(3.39),283(4.52),
		292(4.47),345(3.90)
			103-1025-78

Tricyclo[9.3.1.1^{4,8}]hexadeca-1(15),4,6- 8(16),11,13-hexaene-5-carbonitrile	EtOH	202(4.63),222(4.39), 249s(3.92),287(3.07), 296s(3.05)	78-0871-78

$C_{17}H_{15}NO$
  Benzenamine, 3-(ethenyloxy)-N-(3-phen-
  yl-2-propenylidene)-
	EtOH	250(4.30),295(4.46),
		340(4.40)
			70-1829-78

$C_{17}H_{15}NO_2$
  1H-Benzo[a]carbazole-1,4(4aH)-dione,
  5,11,11a,11b-tetrahydro-, (4aα,11aα-
  11bα)-
	EtOH	203(4.3),234(4.5),
		258s(4.1),267s(3.9),
		350(3.6)
			12-1841-78

tetradehydro derivative	EtOH	232(4.56),261(4.44), 290(4.02),305s(3.98), 309s(3.92),318s(3.86), 418(3.70),490s(3.51)	12-1841-78
Benzo[h]furo[2,3-b]quinoline, 5,6-di- hydro-7-methoxy-3-methyl-	EtOH	219(4.40),232s(4.32), 250s(4.00),281s(4.24), 295(4.32),315(4.36)	105-0513-78
Benzo[h]furo[2,3-b]quinolin-7(5H)-one, 6,11-dihydro-3,11-dimethyl-	EtOH	227s(4.11),254(4.23), 291s(4.08),309(4.18), 324s(4.03),332s(3.72)	105-0513-78
	acid	214s(4.31),227s(4.15), 248s(3.93),256s(3.90), 298s(4.00),330(4.20)	105-0513-78
Benzoic acid, 4-methyl-, (3-phenyl-2H- azirin-2-yl)methyl ester	C_6H_{12}	239(4.11)	35-4481-78
1H-Indole-2-carboxylic acid, 5-phenyl-, ethyl ester	EtOH	205(4.38),261(4.59), 301(4.17)	103-0173-78

$C_{17}H_{15}NO_2S$
  Quinoline, 2-methyl-4-[(4-methylphen-
  yl)sulfonyl]-
	EtOH	242(4.35),316(3.62),
		327(3.59)
			95-1503-78

Quinoline, 2-[[(4-methylphenyl)sulfon- yl]methyl]-	EtOH	235(4.61),267(3.51), 273(3.53),292(3.47), 297s(3.43),305(3.50), 318(3.56)	95-1503-78

$C_{17}H_{15}NO_3$
  1,4-Acridinedione, 2-methoxy-5,8,9-tri-
  methyl-
	dioxan	240(4.34),288(4.27),
		328(3.88)
			150-4901-78

Benzoic acid, 4-methoxy-, (3-phenyl- 2H-azirin-2-yl)methyl ester	C_6H_{12}	244(4.00)	35-4481-78

Compound	Solvent	$\lambda_{max}(\log \epsilon)$	Ref.
$C_{17}H_{15}NO_5$			
2-Azabicyclo[2.2.2]oct-7-ene-5,6-dicarboxylic acid, 3-oxo-2-(2-phenylethenyl)-, trans	MeCN	288(4.34)	78-2617-78
$C_{17}H_{15}N_3$			
6H-Pyridazino[4,5-b]carbazole, 5-ethyl-1-methyl-	EtOH	232(4.16),248s(4.04), 289(4.52),312s(3.74), .327s(3.53),405(3.9)	77-0309-78
$C_{17}H_{15}N_3O$			
2-Propanone, 1-(1,3-diphenyl-1H-1,2,4-triazol-5-yl)-	EtOH	250(4.14)	94-1825-78
3H-Pyrazol-3-one, 2,4-dihydro-5-methyl-2-phenyl-4-[(phenylamino)methylene]-	$C_6H_{11}Me$	252s(4.25),257(4.25), 339(4.47),386s(3.88)	48-0521-78
	PrOH	251(4.15),338(4.01)	48-0521-78
	PrCN	259s(4.19),336(4.44)	48-0521-78
4(3H)-Quinazolinone, 3-[(1-methylethylidene)amino]-2-phenyl-	MeOH	227(4.45),272(4.23), 303(4.11)	80-1085-78
$C_{17}H_{15}N_3OS$			
Phenol, 2-[[[2-(methylthio)-4-phenyl-1H-imidazol-1-yl]imino]methyl]-	MeOH	281(4.50),351(4.26)	18-1846-78
Phenol, 2-[[[2-(methylthio)-5-phenyl-1H-imidazol-1-yl]imino]methyl]-	MeOH	267(4.30),335(3.85)	18-1846-78
$C_{17}H_{15}N_3O_2$			
Benzenamine, 4-[(6-methyl-1,3-dioxolo-[4,5-g]quinazolin-8-yl)methyl]-	EtOH	226s(4.55),234(4.60), 285(3.75),323s(3.95), 333(3.98)	78-2557-78
1,3-Dioxolo[4,5-g]quinazolin-7(8H)-amine, 6-methyl-8-methylene-N-phenyl-	C_6H_{12}	226(4.54),248s(4.19), 329(3.81),334s(3.34)	78-2557-78
7H-Pyrimido[4,5-d]azepine-7-carboxylic acid, 2-phenyl-, ethyl ester	EtOH	224(4.46),276(4.64), 336(4.00),425(3.23)	142-0275-78C
$C_{17}H_{15}N_3O_2S$			
Benzenesulfonamide, N-[2,2'-bipyridin]-4-yl-4-methyl-	neutral	229(4.28),281(4.18), 308(4.00)	39-1215-78B
	monocation	229(4.33),248(4.23), 280(4.38),306(4.05)	39-1215-78B
$C_{17}H_{15}N_3O_3$			
Benzamide, 2-amino-N-(2,3,4,5-tetrahydro-2,5-dioxo-1H-1-benzazepin-3-yl)-	n.s.g.	228(4.63),294(3.48), 364(3.76)	12-0439-78
1H-1,4-Benzodiazepine-2,5-dione, 3-[2-(2-aminophenyl)-2-oxoethyl]-3,4-dihydro-	n.s.g.	228(4.63),294(3.48), 364(3.76)	12-0439-78
$C_{17}H_{15}N_3O_4$			
Benzeneacetic acid, 4-nitro-, (3-hydroxy-3-phenyl-2-propenylidene)hydrazide	MeOH	240(4.09),325(3.96)	64-1527-78B
	MeOH-NaOMe	245(4.08),398(4.31)	64-1527-78B
	dioxan	240(3.84),275(3.77), 350(2.84)	64-1527-78B
	C_6H_{12}	245(--),265(--)	64-1527-78B
$C_{17}H_{15}N_3O_6$			
Benzenebutanoic acid, 2-amino-α-[(2-nitrobenzoyl)amino]-γ-oxo-, ($\pm$)-	n.s.g.	226(4.48),255(4.08), 365(3.75)	12-0439-78
Pyrido[3,2-b]-1,5-naphthyridine-3,7-dicarboxylic acid, 1,4,6,9-tetrahydro-4,6-dioxo-, diethyl ester	EtOH	303(4.3),360(4.0)	19-0509-78

Compound	Solvent	λ_{max}(log ϵ)	Ref.
$C_{17}H_{15}N_3S$			
1H-Imidazol-1-amine, 2-(methylthio)-4-phenyl-N-(phenylmethylene)-	MeOH	279(4.49),344(4.07)	18-1846-78
1H-Imidazol-1-amine, 2-(methylthio)-5-phenyl-N-(phenylmethylene)-	MeOH	263(4.45)	18-1846-78
$C_{17}H_{15}N_3S_2$			
Propanedinitrile, [(hexahydro-1,3-benzodithiol-2-ylidene)amino]phenyl]-methylene]-	CH_2Cl_2	230(4.13),277(4.24), 314(4.29),324(4.26)	18-0301-78
$C_{17}H_{15}N_5O_2S$			
[1,2,3]Thiadiazolo[5,1-e][1,2,3]thiadiazole-7-S^{IV}, 1,6-dihydro-3,4-dimethyl-1-(4-nitrophenyl)-6-phenyl-	C_6H_{12}	248(4.14),297(4.01), 342(4.14),513(3.41)	39-0195-78C
$C_{17}H_{15}N_5S$			
1H-1,2,3-Triazol-4-amine, N-(3-ethyl-2(3H)-benzothiazolylidene)-1-phenyl-	EtOH	310(4.48),322s(4.40)	33-0108-78
$C_{17}H_{16}$			
Benzobicyclo[2.1.1]hex-2-ene, 5-methyl-6-phenyl-, endo-exo	MeOH	254(3.25),261(3.28), 267(3.29),274(3.26)	35-3819-78
Benzobicyclo[3.1.0]hex-2-ene, 7-methyl-6-phenyl-, endo	MeOH	253(2.69),259(2.76), 266(2.81),271(2.81), 274s(2.76),279(2.69)	35-3819-78
Stilbene, 2-ethenyl-6-methyl-	MeOH	252(4.39),286(4.25)	35-3819-78
Stilbene, 2-propenyl-, cis-trans	MeOH	222(4.24),232s(4.20), 245s(4.13),299(4.42)	35-3819-78
trans-trans	MeOH	223(4.19),240s(4.17), 254(4.31),300(4.39)	35-3819-78
$C_{17}H_{16}BrNO_3$			
1H-2-Pyridine-4-carboxylic acid, 3-(4-bromophenyl)-2,4a,5,7a-tetrahydro-1-oxo-, ethyl ester	EtOH	285(4.04)	142-0153-78A
$C_{17}H_{16}BrN_3O_2$			
7H-Pyrimido[4,5-d]azepine-7-carboxylic acid, 5-bromo-5,6-dihydro-2-phenyl-, ethyl ester	EtOH	277(4.48),317(4.00)	142-0275-78C
7H-Pyrimido[4,5-d]azepine-7-carboxylic acid, 5-bromo-8,9-dihydro-2-phenyl-, ethyl ester	EtOH	265(4.35),338(4.50)	142-0275-78C
$C_{17}H_{16}BrN_5$			
1H-Pyrazol-5-amine, 4-[(2-bromo-4-methylphenyl)azo]-3-methyl-1-phenyl-	$CHCl_3$	390(4.36),405s(4.28)	103-0885-78
$C_{17}H_{16}Br_2$			
Tetracyclo[9.6.0.0^{2,10}.0^{3,5}]heptadeca-6,8,12,14,16-pentaene, 4,4-dibromo-	ether	226(3.86),254(3.32)	24-0107-78
Tetracyclo[9.6.0.0^{2,10}.0^{5,7}]heptadeca-3,8,12,14,16-pentaene, 6,6-dibromo-	ether	248(3.38)	24-0107-78
$C_{17}H_{16}Br_3NO_5$			
1H-Pyrrole-3,4-dicarboxylic acid, 2-bromo-5-(3,5-dibromo-2-methoxyphenyl)-, diethyl ester	EtOH	285(3.96)	39-1588-78C

Compound	Solvent	$\lambda_{max}(\log \epsilon)$	Ref.
$C_{17}H_{16}ClN_3$			
Imidazo[1,5,4-ef][1,5]benzodiazepine, 9-chloro-4,5,6,7-tetrahydro-2-(phenylmethyl)-	MeOH	234(4.54),280(3.84), 301(3.71)	111-0053-78
Imidazo[4,5,1-jk][1,4]benzodiazepine, 9-chloro-4,5,6,7-tetrahydro-2-(phenylmethyl)-	MeOH	257s(3.80),262(3.84), 286(3.72),295(3.70)	111-0053-78
$C_{17}H_{16}Cl_2N_2O_2$			
Benzoic acid, 4-[[(3,4-dichlorophenyl)-methylamino]methylene]amino]-, ethyl ester	isoPrOH	312(4.47)	117-0097-78
$C_{17}H_{16}Cl_3NO_5$			
1H-Pyrrole-3,4-dicarboxylic acid, 2-chloro-5-(3,5-dichloro-2-methoxyphenyl)-, diethyl ester	EtOH	284(3.83)	39-1588-78C
$C_{17}H_{16}NO$			
Cycloheptatrienylium, (5-acetyl-2,3-dihydro-1H-indol-1-yl)-, perchlorate	MeCN	274(4.22),408(4.41)	150-4801-78
$C_{17}H_{16}N_2$			
1H-Pyrazole, 1-methyl-5-(4-methylphenyl)-3-phenyl-	EtOH	252(4.53)	4-0385-78
Quinazoline, 2-phenyl-4-propyl-	EtOH	254(4.04),262(4.09), 287(3.57),320(3.08), 333(2.93)	94-2866-78
Quinoxaline, 2-phenyl-3-propyl-	EtOH	239(4.88),324(4.36)	94-2866-78
$C_{17}H_{16}N_2O$			
1H-Indole, 1-acetyl-3-[(4-methylphenyl)amino]-	EtOH	208(4.45),233(4.37), 260(4.25),339(3.9)	103-0757-78
1H-Pyrazole, 5-(4-methoxyphenyl)-1-methyl-3-phenyl-	EtOH	257(4.57)	4-0385-78
3H-Pyrazole, 3-(4-methoxyphenyl)-5-methylene-3-phenyl-	EtOH	290(4.10)	22-0401-78
$C_{17}H_{16}N_2OS$			
4-Thiazolidinone, 2-[(3-methylphenyl)-imino]-3-(3-methylphenyl)-	dioxan	274(3.82)	103-0148-78
$C_{17}H_{16}N_2O_2$			
Benzeneacetic acid, (3-hydroxy-3-phenyl-2-propenylidene)hydrazide	MeOH	250(4.09),276s(--), 325(3.75)	64-1527-78B
	MeOH-NaOMe	240(4.08),395(4.31)	64-1527-78B
	dioxan	235(4.75),315(3.07), 340(2.96)	64-1527-78B
	C_6H_{12}	230(--),240(--)	64-1527-78B
Benzoic acid, 4-methyl-, (3-hydroxy-3-phenyl-2-propenylidene)hydrazide	benzene	320(3.20),400s(--)	64-1527-78B
	MeOH	240(4.30),325(4.12)	64-1527-78B
	MeOH-NaOMe	245(4.18),290s(--), 405(4.48)	64-1527-78B
	dioxan	240(4.01),315(3.65), 410s(--)	64-1527-78B
	$CHCl_3$	245(4.22),318(3.55)	64-1527-78B
1H-Indole-2-carboxylic acid, 3-(4-aminophenyl)-, ethyl ester	EtOH	204(4.07),243(4.10), 308(3.82)	103-0173-78
1H-Pyrido[3,2-b]indole-2,9-dione, 3,4,5,6,7,8-hexahydro-7-phenyl-	EtOH	249(3.15),315(3.54)	94-3080-78

Compound	Solvent	$\lambda_{max}(\log \epsilon)$	Ref.
4(1H)-Quinazolinone, 1-methyl-2-(2,6-dimethylphenoxy)-	dioxan	241(3.89),253(3.76), 260(3.72),294(3.64), 301(3.80),313(3.75)	4-1033-78
4(3H)-Quinazolinone, 3-methyl-2-(2,6-dimethylphenoxy)-	dioxan	243(3.74),261(3.93), 266(3.89),300(3.51), 310(3.63),322(3.55)	4-1033-78
$C_{17}H_{16}N_2O_3$			
Benzamide, N-[(4-nitrophenyl)methyl]-2-(2-propenyl)-	MeOH	270(4.05)	35-2181-78
Benzoic acid, 4-methoxy-, (3-hydroxy-3-phenyl-2-propenylidene)hydrazide	benzene	320(2.78)	64-1527-78B
	MeOH	256(4.28),325(4.14), 400s(--)	64-1527-78B
	MeOH-NaOMe	245(4.16),290s(--), 420(4.40)	64-1527-78B
	dioxan	250(4.03),320(3.78), 350s(--),410s(--)	64-1527-78B
	CHCl₃	250(4.04),300s(--), 345(3.20),400s(--)	64-1527-78B
	C_6H_{12}	265(--)	64-1527-78B
$C_{17}H_{16}N_2O_4$			
Benzamide, N-[(4-nitrophenyl)methyl]-2-(2-propenyloxy)-	MeOH	275(4.34)	35-3494-78
Benzenebutanoic acid, 2-amino-α-(benzoylamino)-γ-oxo-, (±)-	n.s.g.	227(4.5),252(3.96), 364(3.72)	12-0439-78
Cyclohepta[b]pyrrole-3-carboxylic acid, 2-(1-cyano-2-ethoxy-2-oxoethylidene)-1,2-dihydro-, ethyl ester	MeOH	259(4.39),292(4.42), 341(3.98),500(4.18)	18-0667-78
1H-Pyrazole, 1-acetyl-3-(2-furanyl-carbonyl)-4,5-dihydro-4-(4-methoxyphenyl)-	EtOH	229(4.28),293(4.17), 332(4.20)	104-1830-78
$C_{17}H_{16}N_2O_5$			
Benzoic acid, 4-[[4-(acetoxymethyl)-phenyl]-ONN-azoxy]-, methyl ester	MeOH	230(3.88),265(3.89), 330(4.22)	39-0171-78C
2-Furancarboxamide, N-[2-(acetylamino)-1,2-di-2-furanylethyl]-	EtOH	250(4.20)	103-1184-78
$C_{17}H_{16}N_2O_5S$			
1(2H)-Naphthalenone, 3,4-dihydro-7-nitro-, O-[(4-methylphenyl)sulfonyl]-oxime, (E)-	MeOH	228(4.59),251(4.54)	24-1780-78
$C_{17}H_{16}N_2S$			
Benzenamine, 4-[2-(2,1-benzisothiazol-3-yl)ethenyl]-N,N-dimethyl-	MeCN	308(4.15),435(4.41)	44-1233-78
$C_{17}H_{16}N_3O_2$			
1,2,4-Triazolidin-1-yl, 2-(1-methyl-1-phenylethyl)-3,5-dioxo-4-phenyl-	benzene	316(3.38),480(2.42) (anom.)	44-0808-78
$C_{17}H_{16}N_4$			
3H-Indole, 3-azido-2-phenyl-3-propyl-	EtOH	227(4.13),233(4.16), 241(4.18),248(4.16), 314(4.07)	94-2866-78
2-Pyridinecarbonitrile, 4,4'-(1,5-pentanediyl)bis-	EtOH	265(3.78)	78-0331-78
$C_{17}H_{16}N_4O$			
2H-1,5-Benzodiazepin-2-one, 1,3-dihy-	HOAc	210(3.45),300(3.62),	103-1270-78

Compound	Solvent	$\lambda_{max}(\log \epsilon)$	Ref.
dro-4,8-dimethyl-3-(phenylazo)- (cont.)		370(4.18)	103-1270-78
$C_{17}H_{16}N_4O_2$			
1H-1,2,4-Triazole-3-carboxylic acid, 5-(2-aminophenyl)-1-phenyl-, ethyl ester	EtOH	217(4.45),255s(4.03), 324(3.65)	142-1577-78A
$C_{17}H_{16}N_4O_3$			
1H-Pyrazolo[3,4-d]pyrimidine-4-acetic acid, α-acetyl-1-phenyl-, ethyl ester	EtOH	229(4.05),260(4.21), 345s(4.45),354(4.51)	95-0089-78
$C_{17}H_{16}N_4O_4$			
Butanamide, N-methyl-2-[(4-nitrophenyl)hydrazono]-3-oxo-N-phenyl-	EtOH	204(4.26),235s(--), 376(4.59)	104-0521-78
$C_{17}H_{16}N_4S$			
[1,2,3]Thiadiazolo[5,1-e][1,2,3]thiadiazolo-7-S^{IV}, 1,6-dihydro-3,4-dimethyl-1,6-diphenyl-	C_6H_{12}	246(4.25),300(4.20), 491(4.32)	39-0195-78C
$C_{17}H_{16}N_8O_7$			
Imidazole-4-carboxamidine, 5-amino-1-(phenylmethyl)-, picrate	pH 1	282(4.17)	94-1929-78
	pH 7	285(4.18)	94-1929-78
	EtOH	291(4.26)	94-1929-78
$C_{17}H_{16}N_8O_{14}$			
Riboflavin, 9-nitro-, 3',4',5'-trinitrate	EtOH	220(4.50),270(4.45), 344(3.81),439(3.99)	104-0406-78
$C_{17}H_{16}O$			
2,5-Cyclohexadien-1-one, 2,6-dimethyl-4-(3-phenyl-2-propenylidene)-	MeOH	260(4.03),406(3.66)	18-3612-78
[4.1]Metabenzenophan-11-one	MeCN	256(3.97),288(3.55), 306s(3.28)	24-2547-78
Tricyclo[9.3.1.1^{4,8}]hexadeca-1(15),4,6-8(16),11,13-hexaene-5-carboxaldehyde	isooctane	206(4.54),267(4.04), 274s(3.98),300s(3.23), 307s(3.20),355s(1.67)	78-0871-78
$C_{17}H_{16}O_2$			
Phenanthrene, 3-ethoxy-6-methoxy-	n.s.g.	254(4.69)	39-0739-78C
Phenanthrene, 6-ethoxy-1-methoxy-	n.s.g.	258(4.71)	39-0739-78C
Tricyclo[9.3.1.1^{4,8}]hexadeca-1(15),4,6-8(16),11,13-hexaene-5-carboxylic acid, (+)-	EtOH	207(4.54),218s(4.45), 251s(3.82),289(3.02), 296s(2.98)	78-0871-78
anion, (-)-	EtOH	246s(3.89),276s(2.83)	78-0871-78
$C_{17}H_{16}O_2S$			
Spiro[6H-dibenzo[f,h][1,5]dioxonin-7(8H),3'-thietane]	EtOH	223(4.22),244(3.92), 292(3.63)	56-1913-78
$C_{17}H_{16}O_3$			
Benzenepentanoic acid, δ-oxo-α-phenyl-	EtOH	240(4.07),278(2.88)	142-0161-78A
1,3-Benzodioxole, 5-[2-(4-ethoxyphenyl)ethenyl]-, (E)-	n.s.g.	328(4.28)	39-0739-78C
4H-1-Benzopyran, 2-(3,4-dimethoxyphenyl)-	MeOH	249(4.24),287(3.91)	18-1175-78
4H-1-Benzopyran, 8-methoxy-2-(4-methoxyphenyl)-	MeOH	251(4.35),279(3.99)	18-1175-78
2,5-Cyclohexadien-1-one, 2,6-dimethoxy-4-(3-phenyl-2-propenylidene)-	MeOH	275s(3.65),424(3.53)	18-3612-78

Compound	Solvent	$\lambda_{max}(\log \epsilon)$	Ref.
2-Propen-1-one, 1-(2-hydroxy-4-methoxy-phenyl)-2-methyl-3-phenyl-	MeOH	235(3.22),345(4.39)	130-0493-78
Spiro[6H-dibenzo[f,h][1,5]dioxonin-7(8H),3'-oxetane]	EtOH	222(4.24),243(3.94), 280(3.63)	56-1913-78
$C_{17}H_{16}O_4$			
1,3-Benzodioxole, 5-[2-(2,3-dimethoxy-phenyl)ethenyl]-, (E)-	n.s.g.	299(4.29),326(4.34)	39-0739-78C
Kuhlmannene	EtOH	221(4.34),330(3.38)	102-1383-78
$C_{17}H_{16}O_5$			
6H-Benzofuro[3,2-c][1]benzopyran, 6a,11a-dihydro-3,9-dimethoxy-(+)-variabilin	EtOH	228(4.03),286(3.63)	102-1417-78
2,5-Cyclohexadiene-1,4-dione, 2-(3,4-dihydro-7-methoxy-2H-1-benzopyran-3-yl)-5-methoxy-, (S)-	EtOH	225(4.05),266(4.20), 357(2.98)	102-1423-78
Dalbergione, 4'-hydroxy-3,4-dimethoxy-, (R)-	MeOH	230(4.20),264(4.02)	102-1383-78
5H-Furo[3,2-b]xanthen-5-one, 6,7,8,9-tetrahydro-4,11-dimethoxy-	EtOH	248(4.65),282(3.59), 332(3.62)	39-0726-78C
α-Lapachone, 4-acetoxy-	EtOH	223s(4.32),245(4.54), 251(4.58),284(4.40)	2-0035-78
β-Lapachone, 4'-acetoxy-	MeOH	252(4.60),280(4.13), 335(3.49),384(3.44), 415(3.50)	2-0035-78
3H-Naphtho[2,1-b]pyran-3-carboxylic acid, 2,7-dimethoxy-, methyl ester	MeOH	244(4.64),312s(3.76), 323(3.88),346s(3.89), 351(3.93)	12-2259-78
$C_{17}H_{16}O_6$			
6H-Benzofuro[3,2-c][1]benzopyran-2,10-diol, 6a,11a-dihydro-3,9-dimethoxy-, (6aS,11aS)-	EtOH	230(4.22),294(3.83)	102-1405-78
Benzoic acid, 3,5-dimethoxy-2-(4-methoxybenzoyl)-	EtOH	283(4.12)	12-1335-78
2,5-Cyclohexadiene-1,4-dione, 2-(3,4-dihydro-7-hydroxy-8-methoxy-2H-1-benzopyran-3-yl)-5-methoxy-	EtOH	230s(4.01),264(4.13), 360(3.00)	102-1405-78
$C_{17}H_{16}O_7$			
5H-Benzocyclohepten-5-one, 2,3,4,6-tetrahydroxy-1,7-bis(3-hydroxy-1-propenyl)-, (E,E)- (fomentariol)	MeOH	224(4.21),285(4.29), 330(4.58),460(3.55)	24-3939-78
9H-Xanthen-9-one, 7-hydroxy-1,3,5,6-tetramethoxy-	MeOH	245(4.42),275(3.95), 308s(3.64),320(3.56), 335(3.55),373s(3.80)	102-2119-78
$C_{17}H_{16}O_7S_3$			
2H-Thiopyran-3(4H)-one, dihydro-2,4-bis[[5-(methylsulfonyl)-2-furanyl]-methylene]-	EtOH	287(4.38),337(4.31), 394(4.21)	133-0189-78
4H-Thiopyran-4-one, tetrahydro-3,5-bis-[[5-(methylsulfonyl)-2-furanyl]methylene]-	EtOH	208(4.02),262(4.15), 350(4.51)	133-0189-78
$C_{17}H_{16}O_8S_2$			
4H-Pyran-4-one, tetrahydro-3,5-bis-[[5-(methylsulfonyl)-2-furanyl]-methylene]-	EtOH	261(4.18),368(4.52)	133-0189-78

Compound	Solvent	$\lambda_{max}(\log \epsilon)$	Ref.
$C_{17}H_{16}S_4$ 3,3'(4H,4'H)-Spirobi[2H-1,5-benzodithiepin]	dioxan	226(4.33),254(4.39)	49-1017-78
$C_{17}H_{17}BF_4O_2$ 2-Propenylium, 1,3-bis(4-methoxyphenyl)-, tetrafluoroborate	$C_2H_4Cl_2$	578(5.01)	104-1197-78
$C_{17}H_{17}Br$ Tetracyclo[9.6.0.0^{2,10}.0^{3,5}]heptadeca-6,8,12,14,16-pentaene, 4-bromo-	ether	227s(3.85),254s(3.31)	24-0107-78
Tricyclo[9.3.1.1^{4,8}]hexadeca-1(15),4,6-8(16),11,13-hexaene, 5-(bromomethyl)-	isooctane	202(4.52),208s(4.50), 222s(4.41),248s(3.97), 293s(2.70)	78-0871-78
$C_{17}H_{17}Br_2NO_5$ 1H-Pyrrole-3,4-dicarboxylic acid, 2-bromo-5-(5-bromo-2-methoxyphenyl)-, diethyl ester	EtOH	275(3.94),305(3.84)	39-1588-78C
1H-Pyrrole-3,4-dicarboxylic acid, 2-(3,5-dibromo-2-methoxyphenyl)-, diethyl ester	EtOH	284(4.0)	39-1588-78C
$C_{17}H_{17}ClN_2O_4$ Tricyclo[3.3.1.1^{3,7}]decane-1-carboxamide, N-(4-chloro-2-nitrophenyl)-2-oxo-	MeOH	242(4.35)	4-0705-78
$C_{17}H_{17}ClN_2O_9S$ 2,4(1H,3H)-Pyrimidinedione, 1-[5-0-(4-chlorobenzoyl)-3-0-(methylsulfonyl)-β-D-arabinofuranosyl]-	MeOH	243(4.45),263s(4.12)	44-0350-78
$C_{17}H_{17}ClN_4O_5$ Imidazo[4,5-d]pyridazine, 4-benzyloxy-7-chloro-3-β-D-ribofuranosyl-	pH 1,7,11	248(3.82)	4-0001-78
$C_{17}H_{17}Cl_2NO_5$ 1H-Pyrrole-3,4-dicarboxylic acid, 2-chloro-5-(5-chloro-2-methoxyphenyl)-, diethyl ester	EtOH	274(3.91),304(3.84)	39-1588-78C
1H-Pyrrole-3,4-dicarboxylic acid, 2-(3,5-dichloro-2-methoxyphenyl)-, diethyl ester	EtOH	283(4.0)	39-1588-78C
$C_{17}H_{17}NO$ 9(10H)-Acridinone, N-methyl-2-(1-methylethyl)-	EtOH	257(4.70),297(3.45), 308s(3.08),389(3.88), 407(3.95)	35-5604-78
Benzenamine, 3-(ethenyloxy)-N-(3-phenyl-2-propenyl)-	EtOH	250(4.50)	70-1829-78
2-Propen-1-one, 1-[4-(dimethylamino)-phenyl]-3-phenyl-	hexane	196s(4.53),220(4.15), 292(4.29),354(4.41)	65-0340-78
2-Propen-1-one, 3-[4-(dimethylamino)-phenyl]-1-phenyl-	hexane	254s(4.25),384(4.53)	65-0340-78
Tricyclo[9.3.1.1^{4,8}]hexadeca-1(15),4,6-8(16),11,13-hexaene-5-carboxamide	EtOH	208(4.56),250s(3.68), 279s(2.96)	78-0871-78
$C_{17}H_{17}NO_2$ Acetamide, N-(3',4'-dihydro-4-oxospiro[2,5-cyclohexadiene-1,1'(2'H)-naph-	EtOH	250(4.44)	22-0202-78

Compound	Solvent	$\lambda_{max}(\log \epsilon)$	Ref.
thalen-6'-yl)- (cont.)			22-0202-78
Acetamide, N-(3',4'-dihydro-4-oxospiro-[2,5-cyclohexadiene-1,1'(2'H)-naphthalen-7'-yl)-	EtOH	242(4.45)	22-0202-78
9(10H)-Acridinone, 2-methoxy-10-propyl-	MeOH	302(3.322)	4-0149-78
1,3-Cyclohexanedione, 2-(1H-indol-3-yl-methylene)-5,5-dimethyl-	EtOH	429(4.526)	103-1088-78
	EtOH-NaOEt	442(3.000)	103-1088-78
$C_{17}H_{17}NO_2S$			
Benzenesulfonamide, N-(6,6a-dihydro-1-methylcycloprop[a]inden-1a(1H)-yl)-, endo	MeOH	268(3.26),277(3.15)	35-2181-78
exo	MeOH	265(3.11),275(3.00)	35-2181-78
$C_{17}H_{17}NO_3$			
Benzenepropanamide, 4-methoxy-α-(4-methylphenyl)-β-oxo-	EtOH	205(4.38),236(4.34)	18-3389-78
Phenanthridine, 5-acetyl-5,6-dihydro-8,9-dimethoxy-	MeOH	218(4.5),242(4.4),283(4.1),306(4.1)	5-0608-78
1H-2-Pyrindine-4-carboxylic acid, 2,5,6,7-tetrahydro-1-oxo-3-phenyl-, ethyl ester	EtOH	273(4.00),304(3.89)	142-0161-78A
2H-Pyrrol-2-one, 1,5-dihydro-4-hydroxy-3-(1-oxo-2,4-hexadienyl)-5-(phenyl-methyl)-	EtOH-H$_2$SO$_4$	354(4.48),368s(4.42)	88-3173-78
	EtOH-KOH	258(4.24),283(4.27),333(4.31)	88-3173-78
2(1H)-Quinolinone, 3,4-dihydro-7-meth-oxy-6-(phenylmethoxy)-	MeOH	220(4.33),265(4.02),292s(3.80)	39-0440-78C
$C_{17}H_{17}NO_4$			
Acetamide, N-[(4-acetyl-5-methyl-3-oxo-2(3H)-furanylidene)methyl]-N-(4-meth-ylphenyl)-	EtOH	226(4.19),284(4.01),340(4.20)	4-0155-78
9(10H)-Acridinone, 1,2,3-trimethoxy-10-methyl-	EtOH	213(4.52),270(4.77),403(3.94)	102-0166-78
2-Butenedioic acid, 2-(methyl-1-naph-thalenylamino)-, dimethyl ester, (E)-	MeOH	276(4.26),290s(4.21)	39-0837-78C
Ethanone, 1-[3-(3,5-dimethyl-4-isoxa-zolyl)-6-hydroxy-2,5-dimethyl-7-benzofuranyl]-	EtOH	239(4.15),287(3.73),365(3.41)	94-0526-78
1-Oxadethiaceph-3-em-4-carboxylic acid, 2-ethylidene-3-methyl-, phenylmethyl ester, (Z)-	EtOH	310(4.29)	39-1450-78C
$C_{17}H_{17}NO_5$			
9(10H)-Acridinone, 1-hydroxy-2,3,5-tri-methoxy-10-methyl-	EtOH	266(4.52),276s(4.34),283s(4.52),316(4.08),337s(3.96),350(3.82),410(3.76)	102-2125-78
	EtOH-NaOMe	266(4.52),284s(4.26),318(3.92),336s(3.73),418(3.46)	102-2125-78
	EtOH-AlCl$_3$	267s(4.52),279s(4.16),289(4.02),350(3.92),468(3.76)	102-2125-78
3-Furancarboxylic acid, 5-[(acetylphen-ylamino)methylene]-4,5-dihydro-2-methyl-4-oxo-, ethyl ester	EtOH	211(4.17),283(4.06),337(4.18)	4-0155-78
$C_{17}H_{17}NO_6S$			
1H-Azepine-2,5-dicarboxylic acid, 1-[(4-methylphenyl)sulfonyl]-,	EtOH	355(3.15)	2-0547-78

Compound	Solvent	$\lambda_{max}(\log \epsilon)$	Ref.
dimethyl ester (cont.)			2-0547-78
1,4-Benzenedicarboxylic acid, 2-[[(4-methylphenyl)sulfonyl]amino]-, dimethyl ester	EtOH	327(3.58)	2-0547-78
$C_{17}H_{17}NS$			
2H-Isoindole, 1-(phenylthio)-2-propyl-	EtOH	330(3.85),343s(--)	44-2886-78
$C_{17}H_{17}NSn$			
9H-Fluorene-9-carbonitrile, 9-(trimethylstannyl)-	heptane	205(4.87),209(4.87), 220(4.51),227(4.21), 263(4.47),274(4.31), 289(3.64),301(3.58)	104-0207-78
	HMPTA	340(4.26),407(3.20), 432(3.38),458(3.38)	104-0207-78
$C_{17}H_{17}N_2OP$			
3H-Pyrazole, 5-(diphenylphosphinyl)-3,3-dimethyl-	MeOH	276(3.16),355(2.21)	150-4248-78
$C_{17}H_{17}N_3$			
3H-1,5-Benzodiazepin-2-amine, N,N-dimethyl-4-phenyl-	n.s.g.	258(4.52),346(3.66)	87-0952-78
1,10-Phenanthroline, 4-piperidino-	neutral	212(4.41),230(4.42), 268(4.41),315s(4.10)	39-1215-78B
	monocation	222(4.38),268(4.36), 305s(4.17),352(4.10)	39-1215-78B
$C_{17}H_{17}N_3O$			
Pyrrolo[1,2-c]pyrimidine-5-carboxaldehyde, 1-(dimethylamino)-3-methyl-6-phenyl-	n.s.g.	240(4.54),367(4.24)	44-3544-78
Pyrrolo[1,2-c]pyrimidine-7-carboxaldehyde, 1-(dimethylamino)-3-methyl-6-phenyl-	n.s.g.	246(4.48),272(4.16), 330(3.70),407(4.05)	44-3544-78
1,2,4-Triazine, 2,5-dihydro-6-methoxy-2-methyl-3,5-diphenyl-	MeOH	220s(4.20),283(3.41)	4-1271-78
1,2,4-Triazine, 4,5-dihydro-3-methoxy-4-methyl-5,6-diphenyl-	MeOH	228s(4.13),305(3.89)	142-0093-78B
1,2,4-Triazin-6(1H)-one, 2,5-dihydro-1,2-dimethyl-3,5-diphenyl-	MeOH MeOH-H_2SO_4	234(4.32),265s(3.59) 241(4.17),275s(3.77)	4-1271-78 4-1271-78
1H-1,2,4-Triazole-5-ethanol, α-methyl-1,3-diphenyl-	EtOH	250(4.35)	94-1825-78
$C_{17}H_{17}N_3O_2$			
Butanamide, N-methyl-3-oxo-N-phenyl-2-(phenylhydrazono)-	EtOH	208(4.38),235(4.29), 350(4.40)	104-0521-78
Butanamide, 2-(methylphenylhydrazono)-3-oxo-N-phenyl-	EtOH	208(4.26),240(4.13), 268s(--),335(4.09)	104-0521-78
7H-Pyrimido[4,5-d]azepine-7-carboxylic acid, 5,6-dihydro-2-phenyl-, ethyl ester	EtOH	268(4.50),316(4.29)	142-0275-78C
7H-Pyrimido[4,5-d]azepine-7-carboxylic acid, 8,9-dihydro-2-phenyl-, ethyl ester	EtOH	325(4.40)	142-0275-78C
$C_{17}H_{17}N_3O_3$			
2-Propen-1-one, 3-(2,2-dimethylhydrazino)-1-(3-nitrophenyl)-3-phenyl-	EtOH	226(4.34),347(4.26)	94-1298-78
2,4,6(1H,3H,5H)-Pyrimidinetrione, 1,3,5-trimethyl-5-(2-methyl-4-quin-	dioxan HClO_4-HOAc	318(3.9) 325(4.0)	83-0561-78 83-0561-78

Compound	Solvent	$\lambda_{max}(\log \epsilon)$	Ref.
olinyl)- (cont.)			83-0561-78
$C_{17}H_{17}N_3O_4$			
Benzenebutanoic acid, 2-amino-α-[(2-aminobenzoyl)amino]-γ-oxo-, ($\pm$)-	n.s.g.	216(4.52),225(4.54), 250(4.14),350(3.80)	12-0439-78
Benzoic acid, 2-(2-methyl-1-oxopropyl)-1-(4-nitrophenyl)hydrazide	EtOH	307(4.213)	146-0864-78
$C_{17}H_{17}N_3O_5$			
1H-Pyrrolo[3,4-b]pyridine-3-carboxylic acid, 4,5,6,7-tetrahydro-2,6-dimethyl-4-(3-nitrophenyl)-7-oxo-, methyl ester	EtOH	219(4.36),258(4.24), 335(3.59)	95-0448-78
$C_{17}H_{17}N_3S$			
1,2,4-Triazine, 4,5-dihydro-4-methyl-3-(methylthio)-5,6-diphenyl-	MeOH	235s(4.17),310(3.90)	142-0093-78B
$C_{17}H_{17}N_4O_9P$			
2,4(1H,3H)-Pteridinedione, 6-phenyl-1-(5-O-phosphono-β-D-ribofuranosyl)-	pH 1.0	218s(4.193),273(4.33), 349(3.945)	5-1780-78
	pH 7.0	218s(4.196),272(4.369), 348(3.959)	5-1780-78
	pH 13.0	263(4.435),353(4.041)	5-1780-78
2,4(1H,3H)-Pteridinedione, 7-phenyl-1-(5-O-phosphono-β-D-ribofuranosyl)-	pH 1.0	222(4.358),250s(3.989), 278s(3.875),347(4.316)	5-1780-78
	pH 7.0	223(4.308),250s(3.925), 278s(3.801),348(4.296)	5-1780-78
	pH 13	230s(4.243),250(4.121), 351(4.265)	5-1780-78
$C_{17}H_{17}N_5O_5$			
9H-Purin-6-amine, 9-β-D-arabinofuranosyl-, 5'-benzoate	MeOH	230(4.31),259(4.20)	87-1218-78
$C_{17}H_{17}N_5O_8$			
3-Pyrroline, 2,5-dimethyl-N-(2-pyridyl-N-oxide)-, picrate, trans	MeOH	245(4.25),275(3.86), 335(3.54)	78-2533-78
$C_{17}H_{17}N_6O_9P$			
Guanosine, cyclic 3',5'-[(2-nitrophenyl)methyl phosphate]	MeOH	256(4.26)	31-0014-78
$C_{17}H_{17}O_2$			
2-Propenylium, 1,3-bis(4-methoxyphenyl)-, tetrafluoroborate	$C_2H_4Cl_2$	578(5.01)	104-1197-78
$C_{17}H_{18}$			
1-Butene, 3-methyl-1,3-diphenyl-, cis	EtOH	240(3.91),243(3.91), 253(3.81),265(3.50)	35-4146-78
trans	EtOH	250(4.33),260(4.17), 283(3.23),293(3.08)	35-4146-78
1,3,5,7,9,11,13,15-Cycloheptadecaoctaene	ether	225(4.15),260s(4.42), 276(4.43),362(3.54)	24-0107-78
isomer 18	ether	225(4.20),260s(4.46), 276(4.48),362(3.54)	24-0107-78
Cyclopropane, 1,1-dimethyl-2,3-diphenyl-, cis	EtOH	260(2.95),267(2.89), 275(2.69)	35-4146-78
trans	EtOH	254(2.77),257(2.79), 260(2.83),263(2.83), 265(2.83),270(2.70), 274(2.56)	35-4146-78

Compound	Solvent	$\lambda_{max}(\log \epsilon)$	Ref.
[4.1]Metabenzenophane	MeCN	267(2.95),282(2.78)	24-2547-78
Tricyclo[9.3.1.1^{4,8}]hexadeca-1(15),4,6-8(16),11,13-hexaene, 5-methyl-	EtOH	211(4.49),228s(4.06), 272(2.73),275(2.73), 280s(2.69)	78-0871-78
$C_{17}H_{18}AsN$			
1H-1,3-Benzazarsole, 1-(1-methylprop-yl)-2-phenyl-	MeOH	248(4.16),318(4.08)	101-0001-78K
$C_{17}H_{18}BrNO_5$			
1H-Pyrrole-3,4-dicarboxylic acid, 2-(5-bromo-2-methoxyphenyl)-, diethyl ester	EtOH	275(4.04),305(3.96)	39-1588-78C
$C_{17}H_{18}BrN_3O_3$			
7H-Pyrimido[4,5-d]azepine-7-carboxylic acid, 5-bromo-5,6,8,9-tetrahydro-6-hydroxy-2-phenyl-, ethyl ester	EtOH	267(4.35)	142-0275-78C
$C_{17}H_{18}ClNO_2$			
Cyclohepta[b]pyran-2(5H)-one, 3-chloro-6,7,8,9-tetrahydro-4-(methylphenyl-amino)-	EtOH	241(4.16),303s(3.83), 326(3.96)	4-0511-78
$C_{17}H_{18}ClNO_5$			
1H-Pyrrole-3,4-dicarboxylic acid, 5-chloro-2-(2-methoxyphenyl)-, diethyl ester	EtOH	274(3.9),296(3.85)	39-1588-78C
1H-Pyrrole-3,4-dicarboxylic acid, 2-(5-chloro-2-methoxyphenyl)-, diethyl ester	EtOH	273(4.03),304(3.96)	39-1588-78C
$C_{17}H_{18}ClN_3$			
1,2,4-Triazine, 3-(4-chlorophenyl)-2,3,4,5-tetrahydro-5,5-dimethyl-6-phenyl-	hexane	217(4.31),271(3.91)	103-0278-78
$C_{17}H_{18}Cl_2N_2$			
1H-Benzimidazole, 5-chloro-2-(2-chloro-tricyclo[3.3.1.1^{3,7}]dec-1-yl)-	MeOH	250(3.77),284(3.92), 292(3.91)	4-0705-78
$C_{17}H_{18}FeO_4$			
Ferrocene, [2-(carboxymethyl)-3-ethoxy-3-oxo-1-propenyl]-, (E)-	EtOH	247(4.06),294(4.09)	128-0325-78
(Z)-	EtOH	248(3.96),293(4.03)	128-0325-78
$C_{17}H_{18}INO_3$			
Morphine, 1-iodo-	MeOH	216(4.43),245s(3.81), 288s(3.32),293(3.33)	44-0737-78
$C_{17}H_{18}NO$			
3H-Indolium, 2-(2-hydroxyphenyl)-1,3,3-trimethyl-, iodide	EtOH	233s(4.07),242s(3.91), 286(3.89),325s(3.53)	120-0096-78
3H-Indolium, 2-(4-hydroxyphenyl)-1,3,3-trimethyl-, iodide	EtOH	220(4.28),238s(3.94), 282s(3.47),335(4.10)	120-0096-78
$C_{17}H_{18}NO_3$			
Pyridinium, 1-(2-ethoxy-2-oxoethyl)-4-[2-(4-hydroxyphenyl)ethenyl]-	EtOH	520(2.74)	56-1265-78
	acetone	600(2.09)	56-1265-78
(inner salt)	CHCl$_3$	625(3.02)	56-1265-78
bromide, (E)-	MeOH	406(4.39)	56-1265-78

Compound	Solvent	$\lambda_{max}(\log \epsilon)$	Ref.
chloride, (E)- (cont.)	MeOH	403(4.48)	56-1265-78
Pyridinium, 4-[2-(4-hydroxyphenyl)eth-	EtOH	520(2.83)	56-1265-78
enyl]-1-(2-methoxy-1-methyl-2-oxo-	acetone	600(3.24)	56-1265-78
ethyl)- (inner salt)	CHCl$_3$	625(3.43)	56-1265-78
bromide	MeOH	404(4.46)	56-1265-78

$C_{17}H_{18}NP$

3H-1,3-Benzazaphosphole, 2-(1,1-dimeth- ylethyl)-3-phenyl-	CCl$_4$	258(3.95),305(3.38)	44-0731-78

$C_{17}H_{18}N_2$

Diazene, [2-methyl-1-(phenylmethylene)- propyl]phenyl-, (Z,E)-	hexane	238(4.18),350(4.17), 475(2.54)	49-1081-78

$C_{17}H_{18}N_2O$

Acetic acid, 2-methyl-2-phenyl-1-(2- phenylethenyl)hydrazide	decalin	226(4.32),238(4.25), 284(4.25)	33-1364-78
2H-Benzimidazol-2-one, 1-(2,3,3a,4,5,7a- hexahydro-2,5-methano-1H-inden-7-yl)- 1,3-dihydro-	MeOH	211(4.46),231s(--), 283(3.78)	4-0705-78
6,10:8,11a-Dimethano-11aH-benzo[b]cy- clooocta[e][1,4]diazepin-12(7H)-one, 6,8,9,10,11,13-hexahydro-	MeOH	228(4.53),265(3.70), 310(3.67)	4-0705-78
Ethanone, 1-(1,2,6,7,12,12b-hexahydro- indolo[2,3-a]quinolizin-3-yl)-	EtOH	221(4.60),310(4.61)	1-0077-78
Ethanone, 1-(1,4,6,7,12,12b-hexahydro- indolo[2,3-a]quinolizin-3-yl)-	EtOH	226(4.66),273(3.92), 281(3.93),289(3.87)	1-0216-78
Ethanone, 1-(3,4,6,7,12,12b-hexahydro- indolo[2,3-a]quinolizin-1-yl)-	EtOH	225(4.68),272(3.93), 281(3.94),290(3.89)	1-0216-78
2,5-Methanobenz[b]indeno[4,5-e][1,4]di- azepin-6(1H)-one, 2,3,3a,4,5,5a,7- 12b-octahydro-	MeOH MeOH-HCl	221(4.54) 233(4.44)	4-0705-78 4-0705-78
Tricyclo[3.3.1.1^{3,7}]decanone, 1-(1H- benzimidazol-2-yl)-	MeOH	244(3.79),275(3.91), 282(3.98)	4-0705-78

$C_{17}H_{18}N_2O_2$

Benzeneacetic acid, [1-(4-methoxyphen- yl)ethylidene]hydrazide	EtOH	298(4.31)	34-0345-78
Benzoic acid, 2-(4-methylphenyl)-2-(1- oxopropyl)hydrazide	EtOH	228(4.29)	33-1477-78
Benzoic acid, 4-methyl-, 2-acetyl-2-(4- methylphenyl)hydrazide	EtOH	237(4.33)	33-1477-78
Benzoic acid, 4-[[(methylphenylamino)- methylene]amino]-, ethyl ester	isoPrOH	313(4.42)	117-0097-78
7-Ergolene-8-carboxylic acid, methyl ester	CH$_2$Cl$_2$	223(4.51),292(4.39)	87-0754-78
2H-3,7-Methanoazacycloundecino[5,4-b]- indole-2,9(4H)-dione, 1,5,6,7,8,10- hexahydro-, (±)-	EtOH	286(--),293(3.72)	44-4859-78
isomer	EtOH	242(4.04),316(4.28)	44-4859-78

$C_{17}H_{18}N_2O_3$

Phenylalanine, N-[4-(methylamino)- benzoyl]-	EtOH	299(4.10)	87-1162-78
4H-Pyrido[1,2-a]pyrimidine-9-carboxylic acid, 1,6,7,8-tetrahydro-4-oxo-2- phenyl-, ethyl ester	MeOH	227(4.18),262(4.09), 295s(4.19),310(4.25)	24-2813-78

$C_{17}H_{18}N_2O_4$

Ethanone, 1-[3-(3,5-dimethyl-4-isoxa- zolyl)-6-hydroxy-2,5-dimethyl-7-	EtOH	236(4.35),269s(3.97), 277(4.03),286s(3.94),	94-0526-78

Compound	Solvent	λ_{max} (log ϵ)	Ref.
benzofuranyl]-, oxime (cont.)		332(3.55)	94-0526-78
2,5-Methano-1H-indene-6-carboxamide, octahydro-N-(2-nitrophenyl)-7-oxo-	MeOH	234(4.26)	4-0705-78
1H-Pyrazole-4,5-dicarboxylic acid, 1-(3-butenyl)-3-phenyl-, dimethyl ester	EtOH	234(4.34)	44-1664-78
2-Pyridinecarboxylic acid, 4,4'-(1,5-pentanediyl)bis-	EtOH	264(3.82)	78-0331-78
2H-Pyrido[3,4-e]-1,3-oxazine-4,5(3H,6H)-dione, 7-methoxy-2,2,6-trimethyl-3-phenyl-	MeOH	309(4.4)	120-0101-78
Tricyclo[3.3.1.1^{3,7}]decane-1-carboxamide, N-(2-nitrophenyl)-2-oxo-	MeOH	238(4.27),348(3.49)	4-0705-78
$C_{17}H_{18}N_2O_4S$			
5-Thia-1-azabicyclo[4.2.0]oct-2-ene-2-carboxylic acid, 3-methyl-8-oxo-7-[(phenylacetyl)amino]-, methyl ester	n.s.g.	262(3.86)	88-5219-78
$C_{17}H_{18}N_2O_4S_2$			
4-Thiazolecarboxylic acid, 4,5-dihydro-2-(6-hydroxy-2-benzothiazolyl)-5,5-dimethyl-, 1-ethoxyethenyl ester	EtOH	267(3.90),335(4.21)	44-2366-78
$C_{17}H_{18}N_2O_6$			
Benzamide, N-(1,6-dihydro-6-oxo-1-β-D-ribofuranosyl-3-pyridinyl)-	pH 1	270(4.06)	87-0427-78
	pH 11	270(4.05)	87-0427-78
Uridine, 5-(2-phenylethenyl)-, (E)-	H_2O	309(4.28)	35-8106-78
$C_{17}H_{18}N_2O_6S_2$			
5-Thia-1-azabicyclo[4.2.0]oct-2-ene-2-carboxylic acid, 3-(acetoxymethyl)-8-oxo-7-[(2-thienylacetyl)amino]-, methyl ester, (6R-trans)-	n.s.g.	237(4.12)	88-5219-78
$C_{17}H_{18}N_2O_7$			
1H-Inden-1-one, 5-[(3,5-dinitrobenzoyloxy)octahydro-7a-methyl-, (3aα,5α,7aβ)-(±)-	EtOH	226(4.33)	39-0015-78C
(3aα,5β,7aα)-(±)-	EtOH	230(4.31)	39-0015-78C
2,4(1H,3H)-Pyrimidinedione, 1-[3-O-(4-methylbenzoyl)-β-D-arabinofuranosyl]-	MeOH	242(4.41),265s(4.19)	44-0350-78
$C_{17}H_{18}N_2O_8$			
1H-Indole, 1-(2,3-di-O-acetyl-α-L-arabinofuranosyl)-6-nitro-	EtOH	211(4.19),250(3.81), 315(3.75),357(3.64)	136-0017-78E
$C_{17}H_{18}N_2S_2$			
9,12-Dithiatricyclo[12.3.1.1^{3,7}]nonadeca-1(18),3,5,7(19),14,16-hexaen-2-one, hydrazone	MeCN	281.6(3.98)	24-2547-78
$C_{17}H_{18}N_2S_3$			
Spiro[1,3-benzodithiole-2,4'-[4H-1,3]-thiazine]-5'-carbonitrile, 3a,4,5,5'-6,6',7,7a-octahydro-2'-phenyl-	EtOH	248(4.26)	18-0301-78
$C_{17}H_{18}N_2Se$			
2-Propene-1-selone, 3-[[4-(dimethylamino)phenyl]amino]-1-phenyl-, (Z)-	benzene	520(4.29)	104-0315-78

Compound	Solvent	λ_{max}(log ϵ)	Ref.
$C_{17}H_{18}N_4O$			
Cyclohexanol, 2-(1-phenyl-1H-pyrazolo-3,4-d]pyrimidin-4-yl)-	EtOH	243(4.61),270s(3.88), 300(3.62)	95-1274-78
$C_{17}H_{18}N_4O_2$			
2-Pentanone, 4-hydroxy-4-methyl-1-(1-phenyl-1H-pyrazolo[3,4-d]pyrimidin-4-yl)-	EtOH	223(4.01),263(4.21), 346(4.61),360(4.66)	95-0891-78
1,3-Propanediamine, N-methyl-N'-(1-nitro-9-acridinyl)-	pH 1	217(4.273),273(4.506), 329(3.581),436(3.838)	56-2125-78
	pH 7	272(4.280),401(3.622)	56-2125-78
2H-Pyrazolo[3,4-d]pyrimidine-4,6(5H,7H)-dione, 3-ethyl-5,7-dimethyl-2-(1-phenylethenyl)-	n.s.g.	245(4.44),282(3.92)	4-0359-78
$C_{17}H_{18}N_4O_{12}P_2$			
2,4(1H,3H)-Pteridinedione, 1-[5-O-[hydroxy(phosphonooxy)phosphinyl]-β-D-ribofuranosyl]-6-phenyl-	pH 1.0	215s(4.207),272(4.332),	5-1788-78
	pH 7.0	215s(4.215),272(4.332), 349(3.940)	5-1788-78
	pH 13.0	263(4.412),353(4.021)	5-1788-78
2,4(1H,3H)-Pteridinedione, 1-[5-O-[hydroxy(phosphonooxy)phosphinyl]-β-D-ribofuranosyl]-7-phenyl-	pH 1.0	223(4.337),250s(3.969), 275s(3.862),347(4.324)	5-1788-78
	pH 7.0	223(4.356),250s(3.993), 278s(4.017),348(4.344)	5-1788-78
	pH 13.0	230s(4.255),250(4.121), 288s(3.863),351(4.288)	5-1788-78
$C_{17}H_{18}N_6O_5$			
Adenosine, N-[(phenylamino)carbonyl]-	EtOH	280(4.43)	39-0131-78C
$C_{17}H_{18}N_8O_3$			
Glycine, N-[4-[[(2,4-diamino-6-pteridinyl)methyl]methylamino]benzoyl]-	pH 1	243(4.22),305(4.27)	87-1162-78
$C_{17}H_{18}N_8O_{12}$			
Adenine, 9-β-D-glucopyranosyl-, picrate	MeOH	212(4.39),255(4.24)	136-0301-78C
$C_{17}H_{18}O$			
Benzene, 1,3-dimethyl-2-[(3-phenyl-2-propenyl)oxy]-, (E)-	MeOH	257(4.29)	18-3612-78
Phenol, 2,6-dimethyl-4-(3-phenyl-2-propenyl)-, (E)-	MeOH	258(4.31)	18-3612-78
Tricyclo[9.3.1.1^{4,8}]hexadeca-1(15),4,6-8(16),11,13-hexaene-5-methanol	EtOH	212(4.66),230s(4.18), 270(2.75)	78-0871-78
9H-Xanthene, 1,4,5,8-tetramethyl-	EtOH	248(3.95),250(3.93), 280(3.78),286(3.70)	32-0079-78
$C_{17}H_{18}OS$			
Benzo[c]thiophene, 1,3-dihydro-1-methoxy-3,3-dimethyl-1-phenyl-	hexane	256s(3.06),262s(3.00), 269s(2.86)	104-0582-78
Benzo[c]thiophene-1-ol, 1,3-dihydro-3,3-dimethyl-1-(2-methylphenyl)-	hexane	256s(3.12),262(3.10), 270(2.95)	104-0582-78
Benzo[c]thiophene-1-ol, 1,3-dihydro-3,3-dimethyl-1-(3-methylphenyl)-	hexane	262(3.06),271s(2.96), 276(2.82)	104-0582-78
Benzo[c]thiophene-1-ol, 1,3-dihydro-3,3-dimethyl-1-(4-methylphenyl)-	hexane	256s(3.12),262s(3.07), 270s(2.93)	104-0582-78
$C_{17}H_{18}O_2$			
Benzene, 1-ethoxy-4-[2-(2-methoxyphenyl)ethenyl]-, (E)-	n.s.g.	305(4.26),325(4.26)	39-0739-78C

Compound	Solvent	$\lambda_{max}(\log \epsilon)$	Ref.
Benzene, 1-ethoxy-4-[2-(4-methoxyphen-yl)ethenyl]-, (E)-	n.s.g.	324(4.20)	39-0739-78C
Benzofuran, 2-(4-methoxyphenyl)-3,5-di-methyl-	EtOH	228(4.27),276s(3.55), 282(3.64),293s(3.48)	33-2646-78
4,7-Ethano-1H-indene-8,9-dione, 1-cy-clohexylidene-3a,4,7,7a-tetrahydro-, (3aα,4α,7α,7aα)-	MeCN	248(4.11),452(1.83)	5-0440-78
Isobenzofuran, 1,3-dihydro-1-methoxy-3,3-dimethyl-1-phenyl-	hexane	258s(2.80),262(2.88), 269(2.78)	104-0582-78
1-Isobenzofuranol, 1,3-dihydro-3,3-di-methyl-1-(2-methylphenyl)-	hexane	256(2.96),262(3.06), 269(2.99),283(2.22)	104-0582-78
	dioxan	257s(2.90),262(2.98), 269(2.93)	104-0582-78
	CH_2Cl_2	257s(2.99),262(3.07), 269(3.00)	104-0582-78
1-Isobenzofuranol, 1,3-dihydro-3,3-di-methyl-1-(3-methylphenyl)-	hexane	257(2.99),262(3.08), 269(3.06)	104-0582-78
1-Isobenzofuranol, 1,3-dihydro-3,3-di-methyl-1-(4-methylphenyl)-	hexane	256(3.08),262(3.13), 269(3.05),277s(2.58)	104-0582-78
$C_{17}H_{18}O_2S$			
Benzo[c]thiophene-1-ol, 1,3-dihydro-1-(2-methoxyphenyl)-3,3-dimethyl-	hexane	266s(3.28),273(3.34), 281(3.29)	104-0582-78
$C_{17}H_{18}O_3$			
1-Isobenzofuranol, 1,3-dihydro-1-(2-methoxyphenyl)-3,3-dimethyl-	hexane	262(3.13),269(3.25), 278(3.19)	104-0582-78
Mucronustyrene	MeOH	255(4.22),283s(3.63), 294(3.26)	102-1389-78
2,4-Pentanedione, 3-[(2-ethyl-5-methyl-3-benzofuranyl)methylene]-	MeOH	249(4.18),288(3.85), 327(3.89)	83-0714-78
Phenol, 2-[3-(3,4-dimethoxyphenyl)-2-propenyl]-	EtOH	218(3.88),265(3.69)	39-0088-78C
Phenol, 2,4-dimethoxy-5-(3-phenyl-2-propenyl)-	EtOH	251(4.30),291(3.88)	102-1375-78
Phenol, 2,5-dimethoxy-4-(3-phenyl-2-propenyl)-	EtOH	251(4.30),286(3.87), 294(3.93)	102-1375-78
Phenol, 2,6-dimethoxy-3-(3-phenyl-2-propenyl)-, (E)-	EtOH	256(4.02),282(3.56)	102-1389-78
$C_{17}H_{18}O_4$			
Mucronulastyrene	MeOH	240(4.27),284(3.86)	102-1389-78
1-Propanone, 3-phenyl-1-(2,4,6-tri-hydroxy-3,5-dimethylphenyl)-	EtOH	296(4.24)	102-2015-78
	EtOH-NaOH	338(4.32)	102-2015-78
$C_{17}H_{18}O_5$			
2H-1-Benzopyran-6-carboxylic acid, 5-hydroxy-4-methyl-8-(3-methyl-2-but-enyl)-2-oxo-, methyl ester	n.s.g.	252(3.70),340(2.54)	2-1039-78
2H-1-Benzopyran-7-ol, 3,4-dihydro-3-(3-hydroxy-2,4-dimethoxyphenyl)-, (±)- (mucronulatol)	EtOH	225(4.17),282(3.73), 289(3.6)	102-1405-78
(S)-	EtOH	225(4.19),282(3.73), 290(3.62)	102-1405-78
2,3-Dehydrocacalol, 4-hydroxy-3-meth-oxy-O-methyl-1-oxo-	ether	225(4.08),303(3.96)	102-1161-78
2,5-Furandione, 3-[(3,5-dimethoxyphen-yl)methylene]dihydro-4-(1-methyl-propylidene)-, (E,Z)-	toluene	347(4.08)	39-0571-78C
Naphtho[2,3-c]furan-1,3-dione, 4-ethyl-4,9-dihydro-5,7-dimethoxy-4-methyl-	hexane	228(3.84),249(3.56), 278(3.57)	39-0571-78C

Compound	Solvent	$\lambda_{max}(\log \epsilon)$	Ref.
2,7-Phenanthrenediol, 9,10-dihydro-1,3,5-trimethoxy-	MeOH	275(4.28)	102-1067-78
Propanedioic acid, [(2-ethyl-5-methyl-3-benzofuranyl)methylene]-, dimethyl ester	MeOH	248(4.16),288(3.95), 314(4.05)	83-0714-78
$C_{17}H_{18}O_6$			
2H,6H-Benzo[1,2-b:5,4-b']dipyran-2-one, 7-acetoxy-7,8-dihydro-6-methoxy-8,8-dimethyl-	EtOH	220(4.16),246(3.55), 248(3.37),324(4.12)	102-1805-78
4H-1-Benzopyran-4-one, 5,6,7,8-tetrahydro-5,6,7,8-tetrahydroxy-2-(2-phenylethyl)- (agarotetrol)	EtOH	242(4.03)	88-3921-78
3-Furancarboxylic acid, 5-acetoxy-4,5-dihydro-5-methyl-2-(4-methylphenyl)-4-oxo-, ethyl ester	EtOH	221(4.08),258(3.68), 309(4.26)	118-0291-78
Mucronucarpan, dihydro-	EtOH	225(4.26),298(3.80)	102-1405-78
$C_{17}H_{18}O_7$			
3-Furancarboxylic acid, 5-acetoxy-4,5-dihydro-2-(4-methoxyphenyl)-5-methyl-4-oxo-, ethyl ester	EtOH	230(4.04),332(4.37)	118-0291-78
2H-Furo[2,3-h]-1-benzopyran-2-one, 6-(2,3-dihydroxy-3-methylbutoxy)-5-methoxy-, (R)-	EtOH	223(4.58),254(4.64), 306(4.28)	105-0149-78
$C_{17}H_{18}O_8$			
1,2,3-Benzenetriol, 4-(3-acetoxy-1-propenyl)-, triacetate, (E)-	MeOH	211(4.43),252(4.30)	24-3939-78
$C_{17}H_{18}S_2$			
2,5-Dithia[6.1]metabenzenophane	MeCN	245(3.71),254(3.74), 264(3.63),277s(2.88)	24-2547-78
$C_{17}H_{19}ClN_2O$			
6,10:8,11a-Dimethano-11aH-benzo[b]cyclooctaalpha[e][1,4]diazepin-12(5aH)-one, 2-chloro-5,6,7,8,9,10,11,13-octahydro-	MeOH	231(4.61),268(3.75), 315(3.67)	4-0705-78
$C_{17}H_{19}ClN_2OS$			
Chlorpromazine sulfoxide	MeOH	239(4.55),277(4.12), 300(3.98),343(3.80)	133-0248-78
Chlorpromazine, 3-hydroxy- (reduced)	3M H_2SO_4	252(4.34),280(3.68), 307(3.58)	44-5006-78
cation radical	3M HCl	228(4.39),256(4.21), 280(4.62),348(3.43), 389(3.43),581(3.97)	44-5006-78
10H-Phenothiazin-7-ol, 2-chloro-10-[3-(dimethylamino)propyl]-(reduced)	3M H_2SO_4	253(4.39),283(3.60), 308(3.66)	44-5006-78
cation radical	5M HCl	226(4.42),254(4.20), 283(4.70),349(3.46), 568(4.00)	44-5006-78
$C_{17}H_{19}ClN_2O_2$			
Tricyclo[3.3.1.1^{3,7}]decane-1-carboxamide, N-(2-amino-4-chlorophenyl)-2-oxo-	MeOH	245(3.94),300(3.58)	4-0705-78
$C_{17}H_{19}ClN_2O_2S$			
Chlorpromazine, 3,7-dihydroxy- (reduced)	5M HCl	225(4.44),274(4.17),	44-5006-78

Compound	Solvent	$\lambda_{max}(\log \epsilon)$	Ref.
(cont.)		281(4.23)	44-5006-78
Chlorpromazine, 3,7-dihydroxy-, cation radical	5M HCl	242(4.46),282(4.67), 357(3.68),398(3.50), 623(4.10)	44-5006-78
$C_{17}H_{19}ClN_2O_4$			
Tricyclo[3.3.1.1^{3,7}]decane-1-carboxylic acid, 2-[(4-chloro-2-nitrophenyl)amino]-	MeOH	244(4.33),252(3.67)	4-0705-78
$C_{17}H_{19}ClN_2S$			
Chlorpromazine	n.s.g.	255(4.51),305(3.58)	44-5006-78
cation radical	3M H_2SO_4	217(4.44),268(4.66), 277(4.70),325(3.44), 374(2.86),525(4.10)	44-5006-78
$C_{17}H_{19}ClO_4$			
Strobilurin B	MeOH	225(4.73),279(4.45), 287(4.49),303(4.48), 315s(4.43)	24-2779-78
$C_{17}H_{19}ClO_6$			
Pyrylium, 4-[3-(2,6-dimethyl-4H-pyran-4-ylidene)-1-propenyl]-2,6-dimethyl-, perchlorate	HOAc-HClO$_4$	585(4.651)	83-0236-78
$C_{17}H_{19}Cs$			
Cesium, [bis(2,4-dimethylphenyl)methyl]-	n.s.g.	448(4.58)	35-8271-78
$C_{17}H_{19}FN_2O$			
6,10:8,11a-Dimethano-11aH-benzo[b]cycloocta[e][1,4]diazepin-12(5aH)-one, 2-fluoro-5,6,7,8,9,10,11,13-octahydro-	MeOH	225(4.51),260(3.63), 312(3.70)	4-0705-78
$C_{17}H_{19}FN_2O_4$			
Tricyclo[3.3.1.1^{3,7}]decane-1-carboxylic acid, 2-[(4-fluoro-2-nitrophenyl)amino]-	MeOH	234(4.34),446(3.82)	4-0705-78
$C_{17}H_{19}FN_2S$			
6,10:8,11a-Dimethano-11aH-benzo[b]cycloocta[e][1,4]diazepin-12(5aH)-thione, 2-fluoro-5,6,7,8,9,10,11,13-octahydro-	MeOH	276(3.91),308(3.75), 358(4.07)	4-0705-78
$C_{17}H_{19}IN_2$			
Pyrimidinium, 3,4,5,6-tetrahydro-1-methyl-2,3-diphenyl-, iodide	EtOH	220(4.39)	39-0545-78B
$C_{17}H_{19}N$			
Azacycloheptadeca-2,4,6,8,10,12,14,16-octaene, 1-methyl-	ether	255(4.34),354(4.85), 420s(3.80),585(3.04), 630(2.59)	24-0084-78
$C_{17}H_{19}NO$			
6-Azaestra-1,3,5(10),9(11)-tetraen-17-one, (±)-	EtOH	230(4.26),238(3.68)	65-0841-78
2-Butanone, 4-[1,9(or 3,9)-dimethyl-9H-pyrrolo[1,2-a]indol-9-yl]-	EtOH	215(3.96),232s(3.91), 252s(3.54),315(3.57)	44-1226-78

Compound	Solvent	$\lambda_{max}(\log \epsilon)$	Ref.
$C_{17}H_{19}NO_2$			
Acetamide, N-(3',4'-dihydro-4-oxospiro-[2-cyclohexene-1,1'(2'H)-naphthalen]-5'-yl-	EtOH	238(4.24)	22-0202-78
Acetamide, N-(3',4'-dihydro-4-oxospiro-[2-cyclohexene-1,1'(2'H)-naphthalen]-6'-yl-	EtOH	239(4.37)	22-0202-78
Acetamide, N-(3',4'-dihydro-4-oxospiro-[2-cyclohexene-1,1'(2'H)-naphthalen]-7'-yl-	EtOH	242(4.37)	22-0202-78
Benzoic acid, 2-[[4-(1,1-dimethylethyl)phenyl]amino]-	EtOH	287(4.19),352(3.87)	35-5604-78
Benzo[h]quinolin-2(1H)-one, 3-ethyl-5,6-dihydro-4-methoxy-8-methyl-	EtOH	240s(3.97),260(3.78), 270(3.63),282(3.50), 302(3.46),338s(4.09), 352(4.13)	105-0513-78
Bicyclo[8.2.2]tetradeca-5,10,12,13-tetraene-4-carboxylic acid, 7-cyano-, methyl ester, trans	EtOH	220(3.86),270(2.57), 277(2.54)	35-1806-78
Quinoline, 1,2,3,4-tetrahydro-7-methoxy-6-(phenylmethoxy)-	MeOH	220(4.33),265(4.02), 292s(3.80)	39-0440-78C
$C_{17}H_{19}NO_3$			
1-Azulenecarboxylic acid, 2-hydroxy-6-pyrrolidino-, ethyl ester	MeOH	219(4.39),241(4.03), 345(4.83),406(4.12), 438(3.98)	18-3087-78
7-Isoquinolinol, 1,2,3,4-tetrahydro-4-(4-hydroxyphenyl)-6-methoxy-2-methyl- (±-cherylline)	EtOH	227s(4.17),285(3.62), 295s(3.45)	35-1548-78
$C_{17}H_{19}NO_4$			
1-Azulenecarboxylic acid, 2-hydroxy-6-morpholino-, ethyl ester	MeOH	221(4.45),338(4.76), 400(4.08)	18-3087-78
1H-2-Pyrindine-4-carboxylic acid, 2,4a,5,6,7,7a-hexahydro-7a-hydroxy-1-oxo-3-phenyl-, ethyl ester	EtOH EtOH-KOH	282(4.00) 321(3.94)	142-0161-78A 142-0161-78A
Pyrroledicarboxylic acid, 5-methyl-2-phenyl-, diethyl ester	EtOH	283(4.2)	39-1588-78C
Pyrrolidine, 1-[5-(6-methoxy-1,3-benzodioxol-5-yl)-1-oxo-2,4-pentadienyl]-, (E,E)- (okolasin)	EtOH	247(4.14),260(4.12), 280(4.15),300(4.18), 310(4.15),368(4.27)	5-0573-78
	n.s.g.	374(4.34)	78-1979-78
$C_{17}H_{19}NO_5$			
3-Furancarboxylic acid, 5-[(acetylphenylamino)methyl]-4,5-dihydro-2-methyl-4-oxo-, ethyl ester	EtOH	214(4.12),261(3.98)	4-0155-78 4-0327-78
5H-Inden-5-one, octahydro-7a-methyl-1-[(4-nitrobenzoyl)oxy]-, (1α,3aβ,7aα)-(±)-	EtOH	258(4.14)	39-0045-78C
Morpholine, 4-[5-(6-methoxy-1,3-benzodioxol-5-yl)-1-oxo-2,4-pentadienyl]-, (E,E)-	n.s.g.	373(4.41)	78-1979-78
1H-Pyrrole-2,4-dicarboxylic acid, 3-hydroxy-1-(phenylmethyl)-, diethyl ester	EtOH	230(4.48),262(4.16)	94-2224-78
1H-Pyrrole-3,4-dicarboxylic acid, 2-(2-methoxyphenyl)-, diethyl ester	EtOH	274(4.07),293(4.02)	39-1588-78C
$C_{17}H_{19}N_2$			
Pyrimidinium, 3,4,5,6-tetrahydro-1-	EtOH	220(4.39)	39-0545-78B

Compound	Solvent	$\lambda_{max}(\log \epsilon)$	Ref.
methyl-2,3-diphenyl-, iodide (cont.)			39-0545-78B
$C_{17}H_{19}N_3$			
Pyridazino[4,5-b]carbazole, 5-ethyl-7,8,9,10-tetrahydro-1-methyl-	EtOH	220(3.94),241s(3.92), 280(4.33),324(3.24), 390(3.29)	77-0309-78
$C_{17}H_{19}N_3O_2$			
7H-Pyrimido[4,5-d]azepine-7-carboxylic acid, 5,6,8,9-tetrahydro-2-phenyl-, ethyl ester	EtOH	261(4.38)	142-0275-78C
$C_{17}H_{19}N_3O_2S$			
Propanehydrazonothioic acid, 2-methyl-N-(4-nitrophenyl)-, 4-methylphenyl ester	EtOH	275s(4.762),384(5.268)	146-0864-78
$C_{17}H_{19}N_3O_3$			
Acetamide, N-[2-(4-acetyl-5-methyl-1H-pyrazol-3-yl)-2-oxoethyl]-N-(4-methyl-phenyl)-	EtOH	219(4.13),243(3.94)	4-0155-78
Propanehydrazonic acid, 2-methyl-N-(4-nitrophenyl)-, 4-methylphenyl ester	xylene	370(4.40)	18-0512-78
	EtOH	250(4.20),392(4.46)	18-0512-78
Propanoic acid, 2-methyl-, 2-(4-methyl-phenyl)-2-(4-nitrophenyl)hydrazide	xylene	360(4.20)	18-0512-78
	EtOH	250(3.88),368(4.21)	18-0512-78
$C_{17}H_{19}N_3O_4$			
1H-Pyrazole-4-carboxylic acid, 3-[(acetylphenylamino)acetyl]-5-methyl-, ethyl ester	EtOH	213(4.13),230(4.09)	4-0155-78
$C_{17}H_{19}N_5O_5$			
3H-1,2,3-Triazolo[4,5-b]pyridin-5-amine, 7-(phenylmethoxy)-3-β-D-ribo-furanosyl-	pH 1	252(4.05),295(4.21)	44-4910-78
	pH 11	263(4.13),296(4.22)	44-4910-78
$C_{17}H_{19}N_5O_7S_2$			
5-Thia-1-azabicyclo[4.2.0]oct-2-ene-2-carboxylic acid, 3-(acetoxymethyl)-7-[[(2-amino-4-thiazolyl)(ethoxy-imino)acetyl]amino]-8-oxo-, (E)-(Z)-	EtOH-HCl	244(4.26),300(3.77)	78-2233-78
	EtOH	237(4.23),255(4.17), 295(3.79)	78-2233-78
$C_{17}H_{19}N_5O_8$			
Riboflavin, 9-nitro-	EtOH	222(4.42),271(4.49), 345(3.81),442(4.00)	104-0406-78
$C_{17}H_{19}O_2$			
Pyrylium, 4-[3-(2,6-dimethyl-4H-pyran-4-ylidene)-1-propenyl]-2,6-dimethyl-, perchlorate	HOAc-HClO₄	585(4.651)	83-0236-78
$C_{17}H_{20}BrN_3O_{11}$			
1H-Imidazole, 5-bromo-4-nitro-1-(2,3,5,6-tetra-O-acetyl-β-D-glucofuranosyl)-	pH 1	304(3.94)	12-1095-78
	EtOH	295(3.91)	12-1095-78
	EtOH-NH₃	309(4.00)	12-1095-78
1H-Imidazole, 5-bromo-4-nitro-1-(2,3,5,6-tetra-O-acetyl-β-D-glucopyranosyl)-	pH 1	303(3.85)	12-1095-78
	aq NH₃	303(3.85)	12-1095-78
	EtOH	292.5(3.84)	12-1095-78

Compound	Solvent	$\lambda_{max}(\log \epsilon)$	Ref.
$C_{17}H_{20}N_2$			
9H-Pyrido[3,4-b]indole, 1-hexyl-	MeOH	236(4.60),241(4.58), 251s(4.41),283s(4.04), 289(4.28),338(3.73), 351(3.75)	95-0898-78
$C_{17}H_{20}N_2O$			
Benzamide, N-methyl-N-[3-(phenylamino)-propyl]-	EtOH	206(4.50),244(4.22)	39-0545-78B
2,5-Cyclohexadien-1-one, 4-[[4-(diethylamino)phenyl]imino]-3-methyl-	C_6H_{12}	274(--),295(4.20), 335(3.97),558(4.31)	48-0557-78
	EtOH	275(4.22),303(4.04), 365(3.83),620(4.40)	48-0557-78 +80-0617-78
	ether	272(4.28),295(4.08), 350(3.83),370(4.25)	48-0557-78
	CCl_4	305(3.98),335(3.83), 565(4.20)	48-0557-78
6,10:8,11a-Dimethano-11aH-benzo[b]cycloocta[e][1,4]diazepin-12(5aH)-one, 5,6,7,8,9,10,11,13-octahydro-	MeOH	228(4.54),266(3.70), 305(3.67)	4-0705-78
2,4-Hexadienamide. N-[1-(dimethylamino)-2-phenyl-2-propenylidene]-	n.s.g.	263(4.42),278s(4.34)	33-0815-78
1H-Isoindol-1-one, 3-(dimethylamino)-3a,4,5,7a-tetrahydro-5-methyl-3a-phenyl-	n.s.g.	245.5(4.29)	33-0815-78
Propanedinitrile, [3,5-bis(1,1-dimethylethyl)-4-oxo-2,5-cyclohexadien-1-ylidene]-	n.s.g.	335(4.51)	70-1313-78
Et_3N complex	heptane	511(2.50)	70-1035-78
	benzene	510(2.66)	70-1035-78
	dioxan	510(2.72)	70-1035-78
	MeCN	513(2.73)	70-1035-78
	$CHCl_3$	512(2.63)	70-1035-78
piperidine complex	$CHCl_3$	514(2.61)	70-1035-78
$C_{17}H_{20}N_2OS$			
Promazine sulfoxide	MeOH	232(4.47),274(4.12), 300(4.97),343(3.80)	133-0248-78
Promethazine sulfoxide	MeOH	233(4.47),273(4.07), 297(3.92),338(3.79)	133-0248-78
$C_{17}H_{20}N_2OS_3$			
Spiro[1,3-benzodithiole-2,4'][4H-1,3]-thiazine]-5'-carboxamide, 3a,4,5,5'-6,6',7,7a-octahydro-2'-phenyl-	EtOH	246(4.13)	18-0301-78
$C_{17}H_{20}N_2O_2$			
Indolo[2,3-a]quinolizine-3α-carboxylic acid, 1,2,3,4,6,7,12,12b-octahydro-, methyl ester	EtOH	205s(4.26),226(4.61), 284(3.96),291(3.90)	78-0437-78
3β-	EtOH	205(4.15),227(4.45), 285(3.82),292(3.78)	78-0437-78
2,5-Methanobenz[b]indeno[4,5-e][1,4]-diazepin-6(1H)-one, 2,3,3a,4,5,5a,7-12,12a,12b-decahydro-12a-hydroxy-	MeOH	223(4.54),255s(--), 309(3.71)	4-0705-78 4-0705-78
Tricyclo[3.3.1.1³,⁷]decane-1-carboxamide, N-(2-aminophenyl)-2-oxo-	MeOH	292(3.44)	4-0705-78
$C_{17}H_{20}N_2O_3$			
Acetamide, N-[4-(1-acetyl-1H-indol-3-yl)-4-oxobutyl]-N-methyl-	EtOH	257s(3.74),300(4.11)	150-0470-78

Compound	Solvent	λ_{max}(log ϵ)	Ref.
C$_{17}$H$_{20}$N$_2$O$_3$S			
4H-Indol-4-one, 1,5,6,7-tetrahydro-1,3-dimethyl-, O-[(4-methylphenyl)sulfonyl]oxime, (E)-	MeOH	222(4.42),261(4.25)	24-1780-78
C$_{17}$H$_{20}$N$_2$O$_4$			
8H-Pyrido[1,2-a]pyrazin-8-one, 1-[(3,4-dimethoxyphenyl)methyl]-9-hydroxy-	EtOH	244(4.20),287(4.07)	12-0187-78
Tricyclo[3.3.1.1^{3,7}]decane-1-carboxylic acid, 2-[(2-nitrophenyl)amino]-	MeOH	236(4.31),435(3.76)	4-0705-78
C$_{17}$H$_{20}$N$_2$O$_4$S			
Methyl benzyl-5-epi-penicillinate	EtOH	208(3.80)	39-0668-78C
C$_{17}$H$_{20}$N$_2$O$_5$			
1H-Pyrrole-3-carboxylic acid, 5-[(acetylphenylamino)methyl]-4,5-dihydro-5-hydroxy-2-methyl-4-oxo-, ethyl ester	EtOH	211(3.97),238(4.20), 307(3.78)	4-1215-78
Pyrrolo[2,3-b]indol-3a(1H)-ol, 1,8-diacetyl-2,3,8,8a-tetrahydro-5-methoxy-, acetate	EtOH	251(4.16),299(3.47)	142-0385-78A
C$_{17}$H$_{20}$N$_2$O$_6$S			
1-Azetidineacetic acid, α-(1-methylethylidene)-2-oxo-3-[(phenylacetyl)amino]-4-sulfino-, α-methyl ester, (3R-trans)-	EtOH	208(4.15),225s(3.85)	39-1366-78C
5-Oxa-4-thia-1-azabicyclo[4.2.0]octane-2-carboxylic acid, 3,3-dimethyl-8-oxo-7-[(phneylacetyl)amino]-, methyl ester, 4-oxide, [2S-(2α,6β,7β)]-	EtOH	210(3.97)	39-0668-78C
4-Thia-1-azabicyclo[3.2.0]heptane-2-carboxylic acid, 3,3-dimethyl-7-oxo-6-[(phenylacetyl)amino]-, methyl ester, 4,4-dioxide	EtOH	211(4.18)	39-1366-78C
epimer	EtOH	206(3.97)	39-1366-78C
C$_{17}$H$_{20}$N$_2$O$_7$			
1H-Indole-1-ethanol, β-(2-acetoxy-1-methylethoxy)-6-nitro-, acetate, (R)-	EtOH	247(3.98),315(3.86), 360(3.75)	136-0017-78E
(S)-	EtOH	247(3.98),315(3.87), 360(3.75)	136-0017-78E
C$_{17}$H$_{20}$N$_4$O$_4$			
1H-Cyclopentapyrimidine-2,4(3H,5H)-dione, 1,1'-(1,3-propanediyl)bis[6,7-dihydro-	H$_2$O	278(4.32)	56-1035-78
1H-Purine-2,6-dione, 3,7-dihydro-7-[2-hydroxy-3-(phenylmethoxy)propyl]-1,3-dimethyl-, (±)-	pH 2,7,12	274(4.00)	73-2054-78
C$_{17}$H$_{20}$N$_4$O$_5$			
Riboflavin, 5'-deoxy-	n.s.g.	373(4.03),445(4.10)	69-1942-78
C$_{17}$H$_{20}$N$_4$O$_6$			
1-Cyclohexene-1-acetic acid, 2-[[(2,4-dinitrophenyl)hydrazono]methyl]-, ethyl ester	EtOH	261(4.2),292(3.95), 380(4.45)	23-0424-78
Riboflavin	EtOH	223(4.40),270(4.44), 358(3.77),446(3.92)	104-0406-78

Compound	Solvent	$\lambda_{max}(\log \epsilon)$	Ref.
$C_{17}H_{20}N_4O_6S$			
Riboflavin, 8-demethyl-8-(methylthio)-	n.s.g.	475(4.55)	69-1942-78
$C_{17}H_{20}N_4O_7$			
Riboflavin 5-oxide	n.s.g.	327(3.87),369(3.95), 462(3.97)	69-1942-78
$C_{17}H_{20}N_4O_7S$			
1H-Pyrazolo[4,3-d]pyrimidine, 7-(methylthio)-1-(2,3,5-tri-O-acetyl-β-D-ribofuranosyl)-	MeOH	218(4.14),256(3.57), 304s(3.91),310(3.97), 320s(3.85)	104-0601-78
2H-Pyrazolo[4,3-d]pyrimidine, 7-(methylthio)-2-(2,3,5-tri-O-acetyl-β-D-ribofuranosyl)-	MeOH	218(4.23),261(3.63), 267s(3.62),311(4.00), 320(4.06),334(3.91)	104-0601-78
$C_{17}H_{20}N_4O_9S$			
Riboflavin, 5'-(hydrogen sulfate)	n.s.g.	373(4.02),445(4.10)	69-1942-78
$C_{17}H_{20}N_{10}O_2$			
2H-Pyrimido[4",5":2',3'][1,4]diazepino-[6',5':3,4]pyrrolo[1,2-f]pteridine-1,12-dione, 3,10-diamino-5,6,6a,6b-7,8-hexahydro-5,8,14-trimethyl-, trans	pH -2.0	270(4.30),305s(3.83), 465(4.39)	24-3385-78
	pH 0.7	226(4.56),271(4.17), 330s(3.75),482(4.45)	24-3385-78
	pH 7.0	220(4.60),278(4.14), 330s(3.74),488(4.49)	24-3385-78
	pH 9.5	226(4.55),268(4.20), 325s(3.70),498(4.50)	24-3385-78
	pH 12.0	269(4.25),340(3.74), 510(4.54)	24-3385-78
2H-Pyrimido[4",5":2',3'][1,4]diazepino-[6',5':3,4]pyrrolo[1,2-f]pteridine-1,12-dione, 3,10-diamino-5,6,6a,6b-7,8-hexahydro-6,7,14-trimethyl-(6,7-dimethyldrosopterin)	pH -1.5	262(4.22),458(4.30)	24-3385-78
	pH 0.8	260(4.10),322s(3.64), 472(4.33)	24-3385-78
	pH 5.0	273(4.07),320(3.75), 481(4.40)	24-3385-78
	pH 8.8	267(4.06),322s(3.66), 486(4.38)	24-3385-78
	pH 10.0	265(4.11),330(3.68), 497(4.41)	24-3385-78
	pH 13.0	265(4.14),333(3.70), 504(4.42)	24-3385-78
$C_{17}H_{20}O$			
1(2H)-Naphthalenone, 3,4,4a,5,8,8a-hexahydro-8aβ-methyl-6-phenyl-	EtOH	217s(3.95),244(4.09), 289s(2.33)	22-0350-78
1(2H)-Naphthalenone, 3,4,4a,7,8,8a-hexahydro-8aβ-methyl-6-phenyl-	EtOH	218s(3.96),247(4.12), 288s(2.33)	22-0350-78
$C_{17}H_{20}O_2$			
5H-Benzocycloheptene-5-acetic acid, 9-(1-methylethyl)-, methyl ester	EtOH	271(3.82)	18-1450-78
7H-Benzocycloheptene-7-acetic acid, 5-(1-methylethyl)-, methyl ester	EtOH	227(4.55),255s(3.85)	18-1450-78
7H-Cyclobut[a]indene-7-acetic acid, 2a,7a-dihydro-2a-(1-methylethyl)-, methyl ester, (2aα,7β,7aα)-	EtOH	262(2.89),269(3.07), 276(3.10)	18-1450-78
1H-Cyclopropa[a]naphthalene-1-acetic acid, 1a,7b-dihydro-3-(1-methylethyl)-, methyl ester	EtOH	225(4.22),231(4.21), 238s(4.08),278(3.70), 307s(3.21)	18-1450-78
isomer	EtOH	225(4.29),231(4.28), 238s(4.13),278(3.77), 307s(3.35)	18-1450-78

Compound	Solvent	$\lambda_{max}(\log \epsilon)$	Ref.
Phenol, 2,2'-methylenebis[4,6-dimethyl-	EtOH	279(3.97),287(3.91)	32-0079-78
$C_{17}H_{20}O_3$			
Benzene, 1,4-dimethoxy-2,3-dimethyl-5-(phenylmethoxy)-	EtOH	283(3.40)	23-0517-78
1-Phenanthrenecarboxylic acid, 1,2,3,4-9,10-hexahydro-6-methoxy-1-methyl-	EtOH	273(4.1)	44-4598-78
2-Phenanthrenecarboxylic acid, 1,2,3,4-9,10-hexahydro-6-methoxy-2-methyl-, (±)-	EtOH	273(4.1)	44-4598-78
$C_{17}H_{20}O_3Se$			
2-Cyclohexene-1-carboxylic acid, 1-ethyl-4-oxo-5-(phenylseleno)-, ethyl ester	MeOH	219(4.16),234(4.1)	24-1944-78
$C_{17}H_{20}O_4$			
Phenol, 2-[2-(3,4,5-trimethoxyphenyl)-ethyl]-	EtOH	215(4.32),275(3.51), 281s(3.44)	102-1179-78
$C_{17}H_{20}O_5$			
Spiro[1,3-dioxolane-2,6'(4'H)-naphtho-[2,3-b]furan-4',9'(5'H)-dione, 4'a,7',8',8'a-tetrahydro-3',4'a,5'-trimethyl-	EtOH	243(3.74),302(3.91)	94-3704-78
$C_{17}H_{20}O_5Si$			
9,10-Anthracenedione, 1,4,4a,9a-tetra-hydro-5,8-dihydroxy-1-[(trimethyl-silyl)oxy]-, (1α,4aα,9aα)-	MeOH	209(4.21),229(4.20), 254(3.99),389(3.79)	150-3831-78
$C_{17}H_{20}O_6$			
3-Furancarboxylic acid, 5-(3,4-dimeth-oxyphenyl)methyl]-4,5-dihydro-2-meth-yl-4-oxo-, ethyl ester	EtOH	216(4.17),264(4.00)	4-0327-78
$C_{17}H_{20}Sn$			
Stannane, trimethyl(9-methyl-9H-fluor-en-9-yl)-	heptane	213(4.79),234(4.68), 264(4.43),295(4.14), 302(4.04)	104-0207-78
$C_{17}H_{21}BF_4N_2$			
Methylium, bis[4-(dimethylamino)phen-yl]-, tetrafluoroborate	$C_2H_4Cl_2$	614(5.18)	104-1197-78
$C_{17}H_{21}ClN_2O_2$			
Tricyclo[3.3.1.1^{3,7}]decane-1-carboxam-ide, N-(2-amino-4-chlorophenyl)-2-hydroxy-	MeOH	244(3.90),300(3.56)	4-0705-78
$C_{17}H_{21}ClN_2O_4$			
1H-Indolo[2,3-a]quinolizin-5-ium, 1-ethyl-2,3,4,6,7,12-hexahydro-, per-chlorate	MeOH	363(4.2095)	114-0429-78B
$C_{17}H_{21}IN_2O_5$			
3-Isoquinolinaminium, 5,8-diacetoxy-7-methoxy-N,N,N-trimethyl-, iodide	EtOH	218(4.50),240(4.59)	4-0569-78
$C_{17}H_{21}N$			
5H-Benzo[b]carbazole, 6,6a,7,8,9,10-	n.s.g.	228(4.45),284(3.78),	44-3388-78

Compound	Solvent	$\lambda_{max}(\log \epsilon)$	Ref.
10a,11-octahydro-10a-methyl-, trans (cont.)		291(3.74)	44-3388-78
$C_{17}H_{21}NO$			
4(1H)-Quinolinone, 2,3,5,6,7,8-hexahydro-5-methyl-1-(phenylmethyl)-	heptane	320(4.0)	103-0416-78
4(1H)-Quinolinone, 2,3,5,6,7,8-hexahydro-8-methyl-1-(phenylmethyl)-	heptane	320(4.0)	103-0416-78
$C_{17}H_{21}NO_2$			
Cocculine	EtOH	230(4.10),285(3.58)	95-0886-78
1H-Indole-3-propanoic acid, 2-(3-methyl-2-butenyl)-, methyl ester	EtOH	221(4.25),285(3.40), 291(3.26)	78-0929-78
2H-Pyrrole-3-carboxylic acid, 3,4-dihydro-2,2-dimethyl-5-[2-(2-propenyl)phenyl]phenyl]-, methyl ester	MeOH	240(3.85)	35-2181-78
$C_{17}H_{21}NO_3$			
2-Azabicyclo[2.2.2]oct-5-en-3-one, 7-acetyl-2-(5,5-dimethyl-3-oxo-1-cyclohexen-1-yl)-, endo	EtOH	289(4.21)	78-2617-78
exo	EtOH	288(4.18)	78-2617-78
2(1H)-Pyridinone, 1-[5,5-dimethyl-3-oxo-2-(3-oxobutyl)-1-cyclohexen-1-yl]-	EtOH	231(4.15),306(3.76)	78-2617-78
2H-Pyrrole-3-carboxylic acid, 3,4-dihydro-2,2-dimethyl-5-[2-(2-propenyloxy)phenyl]-, methyl ester	MeOH	255(4.18),285(3.78)	35-3494-78
$C_{17}H_{21}NO_4$			
2,4-Pentadienamide, N,N-diethyl-5-(6-methoxy-1,3-benzodioxol-5-yl)-, (E,E)-	n.s.g.	370(4.36)	78-1979-78
Photocrinamine	MeOH	235s(3.54),295(3.66)	88-1199-78
1H-Pyrrole-3-carboxylic acid, 4,5-dihydro-5-hydroxy-1,2-dimethyl-5-[(4-methylphenyl)methyl]-4-oxo-, ethyl ester	EtOH	215(4.01),248(4.11), 326(3.78)	4-1215-78
2(1H)-Quinolinone, 4,6,8-trimethoxy-3-(3-methyl-2-butenyl)-	EtOH	235(4.6),282(4.0), 292(3.95),343(3.74)	142-1433-78A
$C_{17}H_{21}NO_8$			
5,6,7,8-Indolizinetetracarboxylic acid, 1,2,3,8a-tetrahydro-8a-methyl-, tetramethyl ester	MeOH	228(4.15),288(4.08), 386(3.80)	83-0977-78
$C_{17}H_{21}N_2$			
1H-Indolo[2,3-a]quinolizin-5-ium, 1-ethyl-2,3,4,6,7,12-hexahydro-, perchlorate	MeOH	363(4.2095)	114-0429-78B
Methylium, bis[4-(dimethylamino)phenyl]-, tetrafluoroborate	$C_2H_4Cl_2$	614(5.18)	104-1197-78
$C_{17}H_{21}N_2O$			
3-Isoquinolinaminium, 5,8-diacetoxy-7-methoxy-N,N,N-trimethyl-, iodide	EtOH	218(4.50),240(4.59)	4-0569-78
$C_{17}H_{21}N_3O$			
3H-Pyrazol-3-one, 4-(cyclohexylamino)-methylene]-2,4-dihydro-5-methyl-2-phenyl-	$C_6H_{11}Me$	258(4.27),298(4.39), 337(3.60)	48-0521-78
	PrOH	258(4.22),298(4.38), 333(3.87)	48-0521-78

Compound	Solvent	$\lambda_{max}(\log \epsilon)$	Ref.
(cont.)	PrCN	260(4.21),296(4.38), 326s(3.66)	48-0521-78
$C_{17}H_{21}N_3O_2$ 5,11-Etheno-1H,5H-cyclohepta[d][1,2,4]- triazolo[1,2-a]pyridazine-1,3(2H)-di- one, 2-ethyl-5a,6,10a,11-tetrahydro- 5a,7-dimethyl-	EtOH	248(4.01)	89-0943-78
$C_{17}H_{21}N_3O_{11}$ 1H-Imidazole, 2-nitro-1-(2,3,4,6-tetra- O-acetyl-β-D-glucopyranosyl)-	MeOH	219(3.64),309(3.81)	44-4784-78
$C_{17}H_{21}N_5O_6$ Riboflavin, 8-demethyl-8-(methylamino)-	n.s.g.	305(4.02),488(4.61)	69-1942-78
$C_{17}H_{21}N_5O_{10}S$ 9H-Purin-2-amine, 6-[(methylsulfonyl)- oxy]-9-(2,3,5-tri-O-acetyl-β-D-ribo- furanosyl)-	pH 1 pH 13 50% EtOH	255(4.22),272s(4.08) 264(4.14) 255(4.24),280(4.07)	94-0240-78 94-0240-78 94-0240-78
$C_{17}H_{22}Br_2O_2$ 7,13,15-Hexadecatrien-5-ynoic acid, 14,16-dibromo-, methyl ester	EtOH	212(4.21),225(4.24), 250(3.80),264(3.64), 281(3.39)	88-3637-78
$C_{17}H_{22}NOP$ 1H-Phosphorino[3,4-b]indole, 2-(1,1-di- methylethyl)-2,4a-dihydro-4,4a-di- methyl-, 2-oxide	CHCl$_3$	228(4.42),262(3.96)	139-0257-78B
$C_{17}H_{22}N_2$ 1-Azabicyclo[2.2.2]octane, 5-ethyl- 2-(1H-indol-2-yl)-, [1S-(1α,2β,4α- 5β)]- [2S-(2α,4α,5β)]-	MeOH MeOH	218(4.58),268(3.93), 281(3.89),289(3.74) 218(4.60),271(3.98), 281(3.94),290(3.79)	35-0589-78 35-0589-78
$C_{17}H_{22}N_2O$ Pyrrolo[2,3-b]indole, 1-acetyl-1,2,3- 3a,8,8a-hexahydro-3a-(3-methyl-2- butenyl)-	EtOH	242(3.92),296(3.43)	78-0929-78
$C_{17}H_{22}N_2OS$ 7-Oxa-3-thia-1-azaspiro[5.5]undec-1-en- 2-amine, N-(3-phenyl-2-propenyl)-, (±)-	EtOH	253(4.34)	87-0895-78
$C_{17}H_{22}N_2O_2$ 1-Oxa-5-azaspiro[5.5]undec-2-en-4-one, 2-methyl-5-(methylphenylamino)- Tricyclo[3.3.1.1^{3,7}]decane-1-carboxam- ide, N-(2-aminophenyl)-2-hydroxy-	EtOH MeOH	236(4.26),280s(3.48) 291(3.48)	33-1364-78 4-0705-78
$C_{17}H_{22}N_2O_3$ Cyclo[3-[4-(2-methylaminoethyl)phen- oxy]propanoyl-L-prolyl]	MeOH	270(2.71),276(2.69)	35-8202-78
$C_{17}H_{22}N_2O_4$ 5H-Pyrido[3,2-b]indole-5-propanoic acid, 1,2,3,4,6,7,8,9-octahydro-	EtOH	251(4.16),305(3.52)	94-3080-78

Compound	Solvent	$\lambda_{max}(\log \epsilon)$	Ref.
1-methyl-2,9-dioxo-, ethyl ester (cont.)			94-3080-78
$C_{17}H_{22}N_2O_8S$			
2H-Benzimidazole-2-thione, 1,3-dihydro-1,3-di-β-D-ribofuranosyl-	MeOH	252(4.20),302s(4.26), 311(4.30)	4-0657-78
$C_{17}H_{22}N_2O_9S_3$			
β-D-Glucopyranoside, 3-(methylthio)-1,2,4-thiadiazol-5-yl 1-thio-, 2,3,4,6-tetraacetate	n.s.g.	249(4.05),286(3.54)	106-0764-78
$C_{17}H_{22}N_4O_4$			
2-Pyrimidinamine, 1-β-D-arabinofuranosyl-1,4-dihydro-4-imino-N-(2-phenylethyl)-, monohydrochloride	pH 1	220(4.37),275s(3.79)	94-3244-78
$C_{17}H_{22}N_4O_7$			
Inosine, 2',3'-O-(5-carboxy-1-methylpentylidene)-, (R)-	MeOH	245(4.02)	136-0155-78C
$C_{17}H_{22}O$			
2-Cyclopenten-1-one, 2,3,5-tris(1-methylethyl)-	EtOH	222(3.82),271(4.11)	35-1799-78
$C_{17}H_{22}OS_4$			
2-Propanone, 1,3-bis[5-(1,1-dimethylethyl)-3H-1,2-dithiol-3-ylidene)-, (Z,Z)-	dioxan	245(4.54),344(3.72), 359(3.71),379(3.62), 481(4.57),517(4.82)	78-2175-78
$C_{17}H_{22}O_2$			
2H-Benz[e]indene, 2-ethoxy-3,3a,4,5-tetrahydro-7-methoxy-3a-methyl-, trans	EtOH	267(4.00)	39-0110-78C
$C_{17}H_{22}O_2S_2$			
Spiro[1,3-dithiolane-2,6'(4'H)-naphtho[2,3-b]furan]-9'(5'H)-one, 4'a,7',8'-8'a-tetrahydro-3',4'a,5'-trimethyl-, (4'aα,5'α,8'aα)-(±)-	EtOH	283(4.17)	94-3704-78
(4'aα,5'α,8'β)-(±)-	EtOH	279.5(4.14)	94-3704-78
$C_{17}H_{22}O_4$			
Dispiro[5.0.5.1]tridecane-1,5,8,12-tetrone, 3,3,10,10-tetramethyl-	EtOH	282(2.46)	44-3070-78
Haegeanolide acetate	EtOH	210(4.27)	102-1059-78
Naphtho[2,3-b]furan-2,6(4H,8H)-dione, 9a-ethoxy-4a,8a,9,9a-tetrahydro-3,4a,5-trimethyl-, [4aS-(4aα,5α-8aβ,9aβ)]-	n.s.g.	225(4.24),325(1.68)	73-1113-78
Spiro[cyclohexan-1,1'-[1H]indene]-4,7'-diol, 2',3'-dihydro-5'-methoxy-, 4-acetate, cis	MeOH	273(3.06),277(3.06), 281(3.08)	94-3641-78
$C_{17}H_{22}O_5$			
Hiyodorilactone C	EtOH	210(4.16)	138-1345-78
Lanuginolide, dehydro-	EtOH	215(3.82)	42-1152-78
Santamarine, acetylepoxy-	MeOH	210(3.92)(end abs.)	36-0347-78
Spiro[1,3-dioxolane-2,6'(4'H)-naphtho[2,3-b]furan]-4'-one, 4'a,5',7',8'-8'a,9'-hexahydro-9'-hydroxy-3',4'a-5'a-trimethyl-, (4'aα,5'α,8'aα,9'β)-	EtOH	265.5(3.57)	94-3704-78

Compound	Solvent	λ_{max}(log ϵ)	Ref.
Spiro[1,3-dioxolane-2,6'(4'H)-naphtho- [2,3-b]furan]-9'(5'H)-one, 4'a,7',8'- 8'a-tetrahydro-4'-hydroxy-3',4'a,5'- trimethyl-, (4'α,4'aβ,5'β,8'aβ)-(±)-	EtOH	285.5(4.16)	94-3704-78
$C_{17}H_{23}BF_2N_2O$ 1H-Pyrrole, 1-(difluoroboryl)-3-ethyl- 5-[(3-ethyl-5-methoxy-4-methyl-2H- pyrrol-2-ylidene)methyl]-2,4-dimethyl-	EtOH	528(4.75)	49-0183-78
$C_{17}H_{23}BrN_2O_4$ Butanoic acid, 4-(nitrosopentylamino)-, 2-(4-bromophenyl)-2-oxoethyl ester	EtOH	256(4.31)	94-3909-78
$C_{17}H_{23}ClN_2O_4$ Pyridinium, 1,2,6-trimethyl-4-[(1,2,6- trimethyl-4(1H)-pyridinylidene)meth- yl]-, perchlorate	MeCN	477(5.170)	83-0226-78
$C_{17}H_{23}ClO_4$ Strobilurin B, tetrahydro-	MeOH	210(3.75),230(3.77), 276(3.11),288(3.06)	24-2779-78
$C_{17}H_{23}Cl_6N_3O_2S$ Isodysidenin	MeOH	238(3.52)	88-1519-78
$C_{17}H_{23}NO$ Benzo[f]pyrrolo[1,2-a]quinoline, 1,2,3,4a,5,6,10b,11,12,12a-deca- hydro-8-methoxy-, hydrochloride	EtOH	222(3.93),278(3.29), 287(3.29)	78-1023-78
$C_{17}H_{23}NO_4$ Pyrrolidine, 1-acetyl-2-[2-(4-acetoxy- 3-methoxyphenyl)ethyl]-, (±)-	EtOH	215(3.93),275(3.26), 282(3.23)	33-1200-78
$C_{17}H_{23}NO_5$ Cyclohexanone, 2-[1-(3,4-dimethoxyphen- yl)-2-nitroethyl]-3-methyl-	EtOH	212(4.35),232(4.06), 283(3.52)	2-0405-78
Cyclohexanone, 2-[1-(3,4-dimethoxyphen- yl)-2-nitroethyl]-4-methyl-	EtOH	209(4.35),231(4.07), 280(3.64)	2-0405-78
Cyclohexanone, 2-[1-(3,4-dimethoxyphen- yl)-2-nitroethyl]-5-methyl-	EtOH	212(4.34),233(4.07), 282(3.53)	2-0405-78
$C_{17}H_{23}NO_7$ Propanedioic acid, [1-(1,2-dihydro-4- hydroxy-6-methyl-2-oxo-3-pyridinyl)- 4-methyl-2-oxopentyl]-, dimethyl ester	MeOH	285(3.84)	142-0417-78A
$C_{17}H_{23}N_2$ Pyridinium, 1,2,6-trimethyl-4-[(1,2,6- trimethyl-4(1H)-pyridinylidene)meth- yl]-, perchlorate	CH_2Cl_2 $MeNO_2$	485(5.11) 481(5.04)	103-0601-78 103-0601-78
$C_{17}H_{23}N_3O_2$ 1H-1,4-Diazepine-5,7(2H,6H)-dione, 3-(dimethylamino)-6-ethyl-2,2- dimethyl-6-phenyl-	CH_2Cl_2	267(4.03)	33-3050-78
$C_{17}H_{23}N_3O_3S$ Benzenesulfonamide, N-[3-[3-(aminocarb-	$CHCl_3$	242(3.3),353(3.6)	24-2594-78

Compound	Solvent	$\lambda_{max}(\log \epsilon)$	Ref.
onyl)-1(4H)-pyridinyl]propyl]-N,4-dimethyl- (cont.)			24-2594-78
$C_{17}H_{23}N_3O_6$ D-Ribitol, 1-deoxy-1-[(3,4-dimethoxy-phenyl)(1,2,3,6-tetrahydro-2,6-dioxo-4-pyrimidinyl)amino]-	MeOH	277(4.31)	4-0489-78
$C_{17}H_{23}N_5O_5$ 9H-Purin-6-amine, 9-[2,3:5,6-bis-O-(1-methylethylidene)-α-D-mannofuranos-yl]-	MeOH	259(4.15)	39-1381-78C
β-	MeOH	260(4.14)	39-1381-78C
$C_{17}H_{23}N_5O_9$ Pyrido[3,2-c]pyridazine-1,2-dicarbox-ylic acid, 3-[1,2-bis(methoxycarbo-nyl)hydrazino]-4-ethoxy-3,4-dihydro-, dimethyl ester	MeOH	205(4.0),250(3.93), 300(3.70)	103-0534-78
$C_{17}H_{23}OP$ Phosphorin, 4-(1,1-dimethylethyl)-1-ethoxy-1,1-dihydro-1-phenyl-	$CHCl_3$	335(3.53),358(3.60)	89-0528-78
$C_{17}H_{24}N_2$ Diazene, (1,1-dimethyl-3,8-nonadienyl)-phenyl-	n.s.g.	263(3.94),414(2.16)	39-0543-78C
Diazene, (2-ethenyl-1,1-dimethyl-6-hep-tenyl)phenyl-	n.s.g.	263(3.93),414(2.16)	39-0543-78C
2-Propanone, 2-(2,7-octadienyl)-2-phen-ylhydrazone	n.s.g.	253(3.99),280(3.53)	39-0543-78C
$C_{17}H_{24}N_2O$ 1-Azabicyclo[2.2.2]octane-2-carboxam-ide, 5-ethyl-N-(2-methylphenyl)-, (R)-	MeOH	242(3.94)	35-0589-78
hydrochloride	MeOH	239(3.92)	35-0589-78
(S)-	MeOH	241(3.94)	35-0589-78
1H-Pyrrole, 3-ethyl-5-[(3-ethyl-5-meth-oxy-4-methyl-2H-pyrrol-2-ylidene)-methyl]-2,4-dimethyl-, (Z)-	EtOH	412(4.50)	49-0183-78
$C_{17}H_{24}N_2O_2$ Piperidine, 1-[1-(4-nitrophenyl)cyclo-hexyl]-	n.s.g.	255(4.00)	111-0017-78
hydrochloride	n.s.g.	260(3.93)	111-0017-78
$C_{17}H_{24}N_2O_2S$ Benzenesulfonic acid, 4-methyl-, [2-(1-methylethylidene)cycloheptylidene]hy-drazide	MeOH	225(4.11),274s(3.21)	44-0604-78
$C_{17}H_{24}N_2O_5$ 1H-Indole-1-acetic acid, 3-amino-2-[(1,1-dimethylethoxy)carbonyl]-4,5,6,7-tetrahydro-4-oxo-, ethyl ester	EtOH	254(4.51),286(4.06)	94-3080-78
1H-Indole-2-acetic acid, 2-[(1,1-di-methylethoxy)carbonyl]-2,3,4,5,6,7-hexahydro-3-imino-4-oxo-, ethyl ester	EtOH	260(3.96),317(3.88)	94-3080-78

Compound	Solvent	$\lambda_{max}(\log \epsilon)$	Ref.
$C_{17}H_{24}N_3O_4P$ Methanimidic acid, N-[4-(diethylphos-phinyl)-1-(phenylmethyl)-1H-imida-zol-5-yl]-, ethyl ester	EtOH	216(3.98),260(3.68)	130-0421-78
$C_{17}H_{24}N_5OP$ 1H-Imidazo[4,5-c][1,5,2]diazaphosphor-in-1-amine, 2,5-dihydro-5-(phenyl-methyl)-N,2-dipropyl-, 1-oxide	EtOH	218(4.04),272(3.89)	130-0421-78
$C_{17}H_{24}N_5O_2P$ 4H-Imidazo[4,5-d]-1,2,3-diazaphosphor-in-4-one, 1,2,3,7-tetrahydro-7-(phen-ylmethyl)-3-propyl-2-(propylamino)-, 2-oxide	EtOH	216(4.00),265(3.79)	130-0421-78
$C_{17}H_{24}N_6O_5$ Adenosine, N-[(cyclohexylamino)carbo-nyl]-	EtOH	270(4.35),276s(4.27)	39-0131-78C
$C_{17}H_{24}O$ Bicyclo[2.2.1]heptan-2-ol, 1,7,7-tri-methyl-2-(4-methylphenyl)-, exo	C_6H_{12}	251(2.75),256(2.76), 263(2.74),271(2.58)	12-1223-78
$C_{17}H_{24}O_2$ Bicyclo[2.2.1]heptan-2-ol, 2-(4-meth-oxyphenyl)-1,7,7-trimethyl-, exo	C_6H_{12}	236(3.43),273(3.46), 278(3.51),284(3.47)	12-1223-78
6,7-Dinor-5,8-seco-D-homoestra-4,9-di-en-3-one, 17aβ-hydroxy-, (±)-	EtOH	221(3.78),307(4.28)	39-0024-78C
Hex-5-en-3-one, 6-(4-methoxyphenyl)-2,4,4,5-tetramethyl-	EtOH	261(4.14)	35-1791-78
2-Penten-4-yn-1-ol, 3-methyl-5-(2,6,6-trimethylcyclohex-1-enyl)-, acetate, cis	EtOH	274(4.04)	39-1511-78C
$C_{17}H_{24}O_3$ Bicyclo[7.2.0]undec-4-en-3-one, 2-acet-oxy-4,11,11-trimethyl-8-methylene-cis	MeOH	238(3.93)	94-2535-78
cis	MeOH	237(3.53)	94-2543-78
2-Penten-4-yn-1-ol, 3-methyl-5-(2,2,6-trimethyl-7-oxabicyclo[4.1.0]hept-1-yl)-, acetate, (Z)-(±)-	EtOH	233(4.14)	39-1511-78C
$C_{17}H_{24}O_4$ Azuleno[4,5-b]furan-2(4H)-one, 4-acet-oxy-5,6,6a,7,8,9,9a,9b-octahydro-3,6,9-trimethyl-	EtOH	218(4.15)	2-0539-78
epimer	EtOH	218(4.12)	2-0539-78
2-Penten-4-yn-1-ol, 3-methyl-5-(1,2-di-hydroxy-2,6,6-trimethylcyclohexyl)-, acetate, cis-threo	EtOH	229(4.11)	39-1511-78C
$C_{17}H_{24}O_5$ 2-Cyclohexene-1-carboxylic acid, 1,3-dimethyl-2-[2-(2-methyl-1,3-dioxolan-2-yl)ethenyl]-4-oxo-, ethyl ester, (E)-	EtOH	274(4.08)	104-0660-78
$C_{17}H_{24}O_{11}$ Cyclopenta[c]pyran-4-carboxylic acid, 1-(β-D-glucopyranosyloxy)-1,4a,7,7a-	MeOH	238(3.9)	32-0013-78

Compound	Solvent	$\lambda_{max}(\log \epsilon)$	Ref.
tetrahydro-7-hydroxy-7-(hydroxymeth-yl)-, methyl ester (cont.)			32-0013-78
$C_{17}H_{25}BrO_4$			
Bicyclo[7.2.0]undec-4-en-3-one, 2-acet-oxy-8-(bromomethyl)-8-hydroxy-4,11,11-trimethyl-	MeOH	237(3.99)	94-2535-78
$C_{17}H_{25}N$			
Piperidine, 1-(1-phenylcyclohexyl)-	n.s.g.	240(3.45)	111-0017-78
hydrochloride	n.s.g.	260(3.48)	111-0017-78
$C_{17}H_{25}NO$			
Phenol, 4-(1-piperidinocyclohexyl)-, hydrochloride	n.s.g.	275(3.48)	111-0017-78
$C_{17}H_{25}NO_2$			
11-Azaestr-4-en-3-one, 17β-hydroxy-, (+)-	EtOH	237.5(4.09)	65-1529-78
5H-1,2-Benzoxazin-5-one, 7,8a-bis(1,1-dimethylethyl)-8,8a-dihydro-3-methyl-	EtOH	244(--),312(--)	39-0755-78C
$C_{17}H_{25}NO_3$			
2-Pyrrolidinone, 1-(8-hydroxy-6-methyl-1-oxo-2,4,6-dodecatrienyl)-, (E,E,E)-	EtOH	320(4.42)	18-2077-78
2H-Quinolizin-2-ol, 4-(3,4-dimethoxy-phenyl)octahydro- (lasubine I)	MeOH	279.5(3.5)	94-2515-78
stereoisomer (lasubine II)	MeOH	279.5(3.6)	94-2515-78
$C_{17}H_{26}N_2$			
Benzenamine, 4-(1-piperidinocyclohex-yl)-	n.s.g.	240(5.40)	111-0017-78
hydrochloride	n.s.g.	288(3.28)	111-0017-78
$C_{17}H_{26}N_2O$			
Bicyclo[4.1.0]hept-3-en-2-one, 5-(1-az-iridinylimino)-1,3-bis(1,1-dimethyl-ethyl)-	n.s.g.	272(3.88)	70-1179-78
Cycloheptanone, 2,7-bis[3-(dimethylami-no)-2-propenylidene]-	EtOH	460(4.65)	70-0102-78
2,5-Cyclohexadien-1-one, 2,6-bis(1,1-dimethylethyl)-4-[(2-methyl-1-azir-idinyl)imino]-	n.s.g.	300(4.30)	70-1179-78
Cyclohexanone, 2,6-bis[3-(dimethylami-no)-2-propenylidene]-3-methyl-	EtOH	510(4.93)	70-0102-78
Cyclohexanone, 2,6-bis[3-(dimethylami-no)-2-propenylidene]-4-methyl-	EtOH	510(4.86)	70-0102-78
$C_{17}H_{26}N_2O_2$			
4H-Pyran-4-one, 3,5-bis[3-(dimethylami-no)-2-propenylidene]tetrahydro-2,2-dimethyl-	EtOH	500(4.83)	70-0102-78
$C_{17}H_{26}N_2O_4S$			
Phenol, 2-methoxy-4-nitro-6-[2-(tetra-hydro-3,4,4,6-tetramethyl-2H-1,3-thiazin-2-yl)ethyl]-	EtOH	218(4.44),239(4.34), 324s(4.16),339(4.21), 473(3.43)	4-1439-78
$C_{17}H_{26}N_4O$			
7H-Indazol-7-one, 4-(1-aziridinylimi-no)-6,7a-bis(1,1-dimethylethyl)-	n.s.g.	272(3.68)	70-1179-78

Compound	Solvent	$\lambda_{max}(\log \epsilon)$	Ref.
3,3a,4,7a-tetrahydro- (cont.)			70-1179-78
$C_{17}H_{26}N_6O_5$ Adenosine, N-[(hexylamino)carbonyl]-	pH 1 pH 13 70% EtOH	277(4.37) 269(4.31),277(4.25), 299(4.15) 270(4.35),276(4.27)	36-0569-78 36-0569-78 36-0569-78
$C_{17}H_{26}O_2$ 2,4,6,10-Dodecatetraen-1-ol, 3,7,11- trimethyl-, acetate	EtOH	281(4.67)	44-4769-78
$C_{17}H_{26}O_2S$ Benzenesulfinic acid, 4-methyl-, 5- methyl-2-(1-methylethyl)cyclohexyl ester	hexane	201(4.23),224(4.06), 256(3.63)	118-0441-78
$C_{17}H_{26}O_3$ Cyclohexanol, 1-(5-acetoxy-3-methyl- 3-penten-1-ynyl)-2,2,6-trimethyl-, cis	EtOH	228(4.08)	39-1511-78C
$C_{17}H_{26}O_4$ 1,2-Cyclohexanediol, 2-(5-acetoxy-3- methyl-3-penten-1-ynyl)-1,3,3-tri- methyl-, cis-erythro	n.s.g.	229.5(4.11)	39-1511-78C
$C_{17}H_{26}O_5$ Naphtho[2,3-b]furan-2(4H)-one, 4a,5,6- 7,8,8a,9,9a-octahydro-8a-hydroxy- 4,9a-dimethoxy-3,4a,5-trimethyl- (farformolide B)	MeOH	224(3.91)	95-1592-78
$C_{17}H_{26}S$ Benzene, 2-(ethenylthio)-1,3,5-tris(1- methylethyl)-	heptane	270(3.32)	64-0197-78B
$C_{17}H_{27}BrO$ 2,4-Cyclopentadien-1-one, 2-bromo- 3,4,5-tris(1,1-dimethylethyl)-	C_6H_{12}	434(2.44)	89-0519-78
$C_{17}H_{27}N$ Cyclohexanamine, N-[4,4-dimethyl-6-(2- propenyl)-2-cyclohexen-1-ylidene]-	C_6H_{12}	225(4.32),297(2.51)	33-1025-78
$C_{17}H_{27}NO_3$ 5H-1,2-Benzoxazin-5-one, 7,8a-bis(1,1- dimethylethyl)-4,4a,8,8a-tetrahydro- 4a-hydroxy-3-methyl-	EtOH	235(4.26)	39-0755-78C
$C_{17}H_{27}N_2O$ Methanaminium, N-[3-[3-[3-(dimethylami- no)-2-propenylidene]-2-ethoxy-1-cy- clopenten-1-yl]-2-propenylidene]-N- methyl-, iodide	CHCl_3	640(5.11)	70-0339-78
Methanaminium, N-[3-[3-[3-(dimethylami- no)-2-propenylidene]-2-methoxy-1-cy- clohexen-1-yl]-2-propenylidene]-N- methyl-, methyl sulfate	EtOH	650(5.17)	70-0339-78

Compound	Solvent	$\lambda_{max}(\log \epsilon)$	Ref.
$C_{17}H_{27}N_3O$ 1H-Benzimidazole-1-butanamine, N,N-diethyl-5-methoxy-δ-methyl-	EtOH pH 1	250(4.18),292(4.04) 285(4.10)	4-0297-78 4-0297-78
$C_{17}H_{27}N_3O_2$ 5-Oxazolamine, 4-ethyl-N-(4-ethyl-2-propyl-5-oxazolyl)-N-methyl-2-propyl-	n.s.g.	217(3.81)	70-0963-78
$C_{17}H_{27}N_3O_6$ Octanamide, N-(1-β-D-arabinofuranosyl-1,2-dihydro-2-oxo-4-pyrimidinyl)-	isoPrOH	216(4.23),248(4.18), 303(3.92)	94-0981-78
$C_{17}H_{28}O$ 2H-Cyclopentacyclododecen-2-one, 1,4,5,6,7,8,9,10,11,12,13,13a-dodecahydro-1,3-dimethyl-, cis trans	EtOH EtOH	242(3.96) 242(4.14)	35-1799-78 35-1799-78
$C_{17}H_{28}O_3$ 3(2H)-Furanone, 2,4-bis(1,1-dimethylethyl)-2-(3-methyl-2-oxobutyl)-	EtOH	271(4.1)	88-3597-78
$C_{17}H_{28}S_4$ 3,3'(4H,4'H)-Spirobi[2H-1,5-benzodithiepin], dodecahydro-, [3(5'aR*,9'aR*)-5aα,9aα]-	CH_2Cl_2	232(2.98)	49-1017-78
$C_{17}H_{29}BrN_2O_2$ 1,4-Pentanediamine, N^4-(2-bromo-4,5-dimethoxyphenyl)-N^1,N^1-dimethyl-	EtOH	247(4.15),312(3.75)	4-0297-78
$C_{17}H_{29}NO$ 4(1H)-Quinolinone, 1-butyl-2,3,5,6,7,8-hexahydro-5-methyl-8-(1-methylethyl)-	heptane	<u>225(4.0),285s(3.2)</u>	103-0416-78
$C_{17}H_{29}N_2O$ Methanaminium, N-[9-(dimethylamino)-5-ethoxy-2,8-dimethyl-2,4,6,8-nonatetraenylidene]-N-methyl-, tetrafluoroborate	CH_2Cl_2	630(5.20)	70-0339-78
$C_{17}H_{29}N_3O_5SSi$ 1,2,4-Triazin-5(2H)-one, 6-[5-O-[(1,1-dimethylethyl)dimethylsilyl]-2,3-O-(1-methylethylidene)-β-D-ribofuranosyl]-3,4-dihydro-3-thioxo-	pH 1 pH 13 MeOH	213(3.87),269(4.15) 226(4.21),258(4.06), 313(3.51) 213(3.57),272(4.08)	88-1829-78 88-1829-78 88-1829-78
$C_{17}H_{30}N_2O$ 2-Cyclohexen-1-one, 3-(ethylamino)-2-[(hexahydro-1H-azepin-1-yl)methyl]-5,5-dimethyl-, 2:1 perchlorate	MeOH	293(4.70)	49-1295-78
$C_{17}H_{30}O_2$ 10,12-Hexadecadienoic acid, methyl ester, (E,Z)-	C_6H_{12}	233(4.42)	101-0159-78M
$C_{17}H_{33}NO_2$ 1-Dodecanaminium, N-(1-formyl-2-hydroxyethenyl)-N,N-dimethyl-, hydroxide, inner salt	C_6H_{12} H_2O	255(4.32) 257(4.46)	73-1261-78 73-1261-78

Compound	Solvent	λ_{max} (log ϵ)	Ref.
$C_{18}H_8O_2S_4$			
4H-Indeno[1,7-cd]-1,2-dithiol-4-one, 3-(3H-1,2-benzodithiol-3-tlidene-acetyl)-	dioxan	242s(--),253s(--), 256(4.41),278s(--), 306(4.21),331s(--), 340(4.03),370s(--), 493(4.38)	78-2175-78
$C_{18}H_8O_5$			
2H,9H-Furo[3,2-g:4,5-g']bis[1]benzopyran-2,9-dione	CHCl$_3$	253(4.33),285(4.17), 325(4.10)	25-0954-78
$C_{18}H_9N_3O_2S$			
Benzonitrile, 2-(2H-naphtho[1,8-bc]thien-2-ylideneamino)-5-nitro-	dioxan	390(4.17)	93-1354-78
$C_{18}H_{10}$			
Benzo[3,4]cyclobuta[1,2-a]biphenylene	EtOH	219(4.64),231(4.48), 272(4.66),276(4.62), 284(4.85),301(4.34), 308(4.42),316(4.80), 354(3.59),374(3.62), 394(3.50)	88-1005-78
$C_{18}H_{10}Br_2O$			
Cyclopenta[b][1]benzopyran, 1,3-dibromo-2-phenyl-	THF	233(--),355(4.46), 496(2.96)	103-1075-78
$C_{18}H_{10}ClNO_2$			
Benz[b]acridine-6,11-dione, 2-chloro-12-methyl-	dioxan	234(4.50),290(4.61)	150-4901-78
1H-Pyrrole-2,5-dione, 3-chloro-4-(9-phenanthrenyl)-	MeOH	254(4.76),283(4.01), 294(4.04),363(3.58)	24-2677-78
$C_{18}H_{10}Cl_3NO_8S$			
Benzoic acid, 4-[5-[2-(5-nitro-2-furanyl)-2-[(trichloromethyl)sulfonyl]ethenyl]-2-furanyl]-	EtOH	203(4.23),227(4.09), 256(4.21),327(4.31), 412(4.30)	73-1618-78
$C_{18}H_{10}N_2$			
Benzo[f]benzo[3,4]cyclobuta[1,2-b]quinoxaline	EtOH	226(4.26),254(4.34), 270s(4.24),280(4.27), 321s(3.95),328(4.08), 381s(3.77),399(4.07), 421(4.16)	78-0495-78
Benzo[g]benzo[3,4]cyclobuta[1,2-b]quinoxaline	EtOH	257(4.51),267(4.48), 283(4.51),294(4.62), 317s(4.58),325(4.65), 378(4.03),396(4.06), 417(3.80)	78-0495-78
Biphenyleno[2,1-c]cinnoline	EtOH	212(4.58),231(4.46), 264(4.26),285(4.44), 294(4.49),321(4.37), 331(4.40)	88-1005-78
Phenanthro[1,10,9-cde]cinnoline	EtOH	208(3.70),214(3.76), 232(3.62),252(3.84), 259(3.83),273(3.99), 292(3.21),303(3.27), 311(3.23),361(3.10)	88-1005-78
$C_{18}H_{10}N_4$			
Pyridazino[3,4-c:4,5-c']diquinoline	dioxan	234(4.34),280(4.24),	39-1126-78C

Compound	Solvent	$\lambda_{max}(\log \epsilon)$	Ref.
(cont.)		304(4.28)	39-1126-78C
$C_{18}H_{10}N_4O$ Pyridazino[3.4-c:6,5-c']diquinoline, 3-oxide	dioxan	236(4.41),301(4.38), 410(3.44),434(3.36)	39-1126-78C
$C_{18}H_{10}N_4O_2$ Benzonitrile, 2-(benz[cd]indol-2-ylami- no)-5-nitro-	dioxan	439(4.39)	93-1354-78
2,3-Naphthalenedicarbonitrile, 5-amino- 1,4-dihydro-1,4-dioxo-8-(phenylamino)-	CH_2Cl_2	347(3.42),710s(3.54), 770(3.66)	150-2319-78
Pyridazino[3,4-c:6,5-c']diquinoline, 3,4-dioxide	dioxan	251(4.49),313(4.38), 338(4.38),356(4.18), 384(3.89)	39-1126-78C
$C_{18}H_{10}N_4O_4$ 4,4'-Biquinoline, 3,3'-dinitro-	dioxan	261(4.65),304(4.20)	39-1126-78C
$C_{18}H_{10}N_8Ni$ Nickel, [7,16-dihydrodipyrido[2,3-b- 3',2'-i][1,4,8,11]tetraazacyclotetra- decine-7,16-dicarbonitrilato(2-)- $N^5,N^6,N^{14},N^{18}]$-, (SP-4-2)-	DMF	398s(4.92),415(4.16)	64-1012-78B
$C_{18}H_{10}O$ Cyclopenta[cd]pyran-3(4H)-one	CH_2Cl_2	251(4.26),271s(4.02), 284(4.16),340s(3.88), 351(4.06),372(3.84), 395(4.02)	88-4491-78
$C_{18}H_{10}O_2$ 19,20-Dioxatricyclo[14.2.1.1^{4,7}]eicosa- 1(18),2,4,6,8,14,16-heptaene-10,12- diyne, (Z,Z,E)-	CCl_4	335(4.67),345(4.86), 375s(3.76),398(3.95), 408(3.99),432(3.77), 459s(3.48),470(3.49), 506(2.99),522(2.69)	88-4929-78
$C_{18}H_{10}O_5$ Kuhlmannistyrene	EtOH	253(4.38),290(4.03)	102-1383-78
$C_{18}H_{10}S_2$ Naphthaceno[5,6-cd]-1,2-dithiole	MeCN	214(6.47),236(6.60), 297(6.99),402s(5.29), 425(5.63),561(5.63)	34-0182-78
$C_{18}H_{11}Br_2ClFNO_2$ 3-Quinolinecarboxylic acid, 6-chloro- 2-(dibromomethyl)-4-(2-fluorophen- yl)-, methyl ester	isoPrOH	248(4.70),290s(3.76), 319(3.56),344(3.45)	4-0687-78
$C_{18}H_{11}ClFN_3O$ 4H-Imidazo[1,5-a][1,4]benzodiazepine- 1-carboxaldehyde, 8-chloro-6-(2- fluorophenyl)-	isoPrOH	215(4.56),250s(4.18), 294(4.05)	44-0936-78
$C_{18}H_{11}ClN_2O$ 2,5-Cyclohexadien-1-one, 3-chloro- 4-diazo-2,6-diphenyl-	MeOH	295(4.13),369(4.47)	70-2455-78
$C_{18}H_{11}ClN_2S$ Quinoline, 6-chloro-4-phenyl-2-(4-thia- zolyl)-	dioxan	347(3.33)	97-0400-78

Compound	Solvent	$\lambda_{max}(\log \epsilon)$	Ref.
$C_{18}H_{11}Cl_2NO_2$			
9H-Phenanthro[9',10':3,4]cyclobuta-[1,2-c]pyrrole-9,11(10H)-dione, 8c,11a-dichloro-8b,8c,11a,11b-tetrahydro-, anti-anti	MeCN	224(4.21),277(4.01)	24-2677-78
syn-syn	MeCN	224(4.27),277(4.07)	24-2677-78
$C_{18}H_{11}NOS$			
Methanone, naphtho[2,1-d]thiazol-2-yl-phenyl-	n.s.g.	286(3.95),326(3.94),372(3.48)	42-0702-78
$C_{18}H_{11}NO_2$			
Benz[b]acridine-6,11-dione, 12-methyl-	dioxan	232(4.52),289(4.63),312s(4.1)	150-4901-78
1H-Indene-1,3(2H)-dione, 2-(3-indolyl-methylene)-	EtOH	448(4.727)	103-1088-78
	EtOH-NaOEt	472(4.700)	103-1088-78
$C_{18}H_{11}NO_3$			
Cyclopenta[b][1]benzopyran, 1-nitro-2-phenyl-	THF	246(4.28),380(4.49),500(3.68)	103-1075-78
$C_{18}H_{11}NO_4$			
8H-Benzo[g]-1,3-benzodioxolo[6,5,4-de]-quinolin-8-one, 12-methoxy-	CHCl$_3$	249(4.64),276(4.21),290s(4.00),312(3.60)	102-0837-78
	CHCl$_3$-HCl	229(4.28),259(4.28),294(4.22),328(3.61)	102-0837-78
$C_{18}H_{11}N_3$			
12H-Dibenzo[b,f]quinolizine-7-carbo-nitrile, 12-imino-	EtOH	238(4.50),279(4.07),301(4.01),314(4.09),398(3.95),417(4.03),438(4.04),464(3.77)	44-3536-78
$C_{18}H_{11}N_5O_3$			
Methanone, (2-nitrophenyl)(1-phenyl-1H-pyrazolo[3,4-d]pyrimidin-4-yl)-	EtOH	254(4.56),300s(3.90),345(3.45)	95-1274-78
Methanone, (3-nitrophenyl)(1-phenyl-1H-pyrazolo[3,4-d]pyrimidin-4-yl)-	EtOH	250(4.57),258s(4.54),300s(3.85),360s(3.23)	95-1274-78
Methanone, (4-nitrophenyl)(1-phenyl-1H-pyrazolo[3,4-d]pyrimidin-4-yl)-	EtOH	252s(4.51),262(4.53),365s(3.20)	95-1274-78
$C_{18}H_{11}N_5O_6$			
3H-Pyrazol-3-one, 4-(2-benzoxazolyl-methylene)-2-(2,4-dinitrophenyl)-2,4-dihydro-5-methyl-	n.s.g.	240(4.32),360(4.42)	48-0857-78
$C_{18}H_{12}$			
Acenaphthylene, 1-phenyl-	C_6H_{12}	234(4.77),256(4.16),275(4.06),286s(3.97),316s(4.05),328(4.16),348(3.89),413(3.25),437s(3.18),462s(2.96)	64-0663-78B
$C_{18}H_{12}BrClFN_3$			
4H-Imidazo[1,5-a][1,4]benzodiazepine, 3-bromo-8-chloro-6-(2-fluorophenyl)-1-methyl-	isoPrOH	215s(4.92),242s(4.65),265s(4.32),307(3.36)	44-0936-78
$C_{18}H_{12}ClN_3O$			
Pyrrolo[2,1-a]phthalazine-2-carboxam-ide, 3-(4-chlorophenyl)-	EtOH	257(4.39),293(4.91),372(4.13),450(3.91)	118-0206-78

Compound	Solvent	$\lambda_{max}(\log \epsilon)$	Ref.
$C_{18}H_{12}Cl_2N_4O$ 4H-Imidazo[1,5-a][1,4]benzodiazepine- 3-carboxamide, 8-chloro-6-(2-chloro- phenyl)-	isoPrOH- DMF	270(4.13),290s(4.02)	4-0577-78
$C_{18}H_{12}Cl_3NO_6S$ Furan, 2-[2-[5-(4-methylphenyl)-2-fur- anyl]-1-[(trichloromethyl)sulfonyl]- ethenyl]-5-nitro-	EtOH	203(4.34),245(4.17), 273(4.19),340(4.18), 440(4.30)	73-1618-78
$C_{18}H_{12}Cl_3NO_7S$ Furan, 2-[2-[5-(4-methoxyphenyl)-2-fur- anyl]-1-[(trichloromethyl)sulfonyl]- ethenyl]-5-nitro-	EtOH	203(4.37),250(4.27), 332(4.30),420(4.32)	73-1618-78
$C_{18}H_{12}CoN_6O_4$ Cobalt, [7,16-dihydro-7,16-dinitrodi- benzo[b,i][1,4,8,11]tetraazacyclo- tetradecinato(2-)-N^5,N^9,N^{14},N^{18}]-, (SP-4-1)-	DMF	419(4.59)	64-1012-78B
$C_{18}H_{12}CuN_6O_4$ Copper, [7,16-dihydro-7,16-dinitrodi- benzo[b,i][1,4,8,11]tetraazacyclo- tetradecinato(2-)-N^5,N^9,N^{14},N^{18}]-, (SP-4-1)-	DMF	380(4.75),413s(4.43)	64-1012-78B
$C_{18}H_{12}F_6$ Phenanthrene, 9,10-dimethyl-3,6-bis- (trifluoromethyl)-	hexane	227(4.43),248s(--), 254(4.88),273(4.20), 280s(--),293(4.09), 305(4.11),336(2.71), 353(2.52)	104-0924-78
$C_{18}H_{12}Fe_2O_6$ Iron, hexacarbonyl[μ-(1,8,9,10-η:2,3,4- 5-η)-9,10-dimethylbicyclo[6.2.0]deca- 1,3,5,7,9-pentaene]di-	EtOH	280(4.28),347(4.11), 430s(3.59)	5-1379-78
$C_{18}H_{12}I_2$ 1,1':3',1"-Terphenyl, 4,4'-diiodo-	EtOH	208(4.60),269(4.51)	47-2093-78
$C_{18}H_{12}NOP$ 5,10[1',2']-Benzenophenophosphazine, 10-oxide	EtOH	220(3.98),238s(3.67)	24-1798-78
$C_{18}H_{12}NP$ 5,10[1',2']-Benzenophenophosphazine	EtOH $CHCl_3$	217(4.31),245s(3.35) 248(3.57),279s(2.94)	24-1798-78 24-1798-78
$C_{18}H_{12}N_2$ Benzaldehyde, 2-ethynyl-, [(2-ethynyl- phenyl)methylene]hydrazone	THF	232(4.50),248s(4.41), 255s(4.30),302s(4.42), 315(4.47),332s(4.39), 349s(4.05)	18-2112-78
Benzo[f]quinazoline, 3-phenyl-	EtOH	217(4.61),244(4.38), 261(4.50),283(4.75), 341(3.60),356(3.56)	103-1025-78
$C_{18}H_{12}N_2O$ 2,5-Cyclohexadien-1-one, 4-diazo-2,6-	MeOH	304(4.08),376(4.47)	70-2455-78

Compound	Solvent	$\lambda_{max}(\log \epsilon)$	Ref.
diphenyl- (cont.)			70-2455-78
$C_{18}H_{12}N_2OS$ Naphtho[2,1-d]thiazole-2-carboxamide, N-phenyl-	n.s.g.	288(4.16),318(4.31), 349(4.11)	42-0702-78
$C_{18}H_{12}N_2O_2$ Pyrazino[1,2-a:4,3-a']diindole-13,14- dione, 6,7-dihydro-	$CHCl_3$	570(4.28)	24-3233-78
$C_{18}H_{12}N_2O_4$ Benzene, 1-[2,4-cyclopentadien-1-yli- dene(4-nitrophenyl)methyl]-4-nitro-	CH_2Cl_2	231(4.15),279s(--), 328(4.41)	5-1139-78
[3,5'-Bi-1H-indole]-2,2'-dicarboxylic acid	EtOH	208(4.49),233(4.47), 296(4.25)	103-0173-78
5(4H)-Oxazolone, 4-[3-(4-nitrophenyl)- 2-propenylidene]-2-phenyl-	toluene	410(4.73)	103-0120-78
$C_{18}H_{12}N_2O_6S_2$ 1,2-Benzisothiazol-3(2H)-one, 2,2'- (1,3-butadiene-1,4-diyl)bis-, 1,1,1',1'-tetraoxide	THF	210(4.80),233(4.72), 270(4.78)	40-0404-78
$C_{18}H_{12}N_2O_8$ Benzoic acid, 4-[5-[2-nitro-2-(5-nitro- 2-furanyl)ethenyl]-2-fur-nyl]-, methyl ester, (E)-	dioxan	307(4.04),428(4.15)	73-3252-78
$C_{18}H_{12}N_2S$ Thieno[3,4-d]pyridazine, 5,7-diphenyl-	CH_2Cl_2	388(4.0)	150-5538-78
$C_{18}H_{12}N_4$ Tricyclo[6.2.2.0^{1,4}]dodeca-2,4,6,11- tetraene-9,9,10,10-tetracarbonitrile, 2,3-dimethyl-	$CHCl_3$	290s(3.88),302(4.05), 312(4.06),324(3.86)	5-1379-78
$C_{18}H_{12}N_4O$ Pyrido[2,3-b]pyrido[2,3-d]pyrimidin- 5-one, 6,7,8,9-tetrahydro-9-methyl- 10-(phenylmethyl)-	EtOH	242(4.36),256(3.95), 364(4.01)	103-1261-78
$C_{18}H_{12}N_4O_2$ 1,2-Ethenediol, 1,2-di-1,8-naphthyri- din-2-yl-	$CHCl_3$	290(4.08),318(3.84), 470(4.06),575s(0.08), 625s(0.04)	97-0020-78
$C_{18}H_{12}N_6NiO_4$ Nickel, 7,16-dihydro-7,16-dinitrodi- benzo[b,i](1,4,8,11]tetraazacyclo- tetradecinato(2-)-N^5,N^9,N^{14},N^{18}]-, (SP-4-1)-	DMF	378s(3.83),394(3.94)	64-1012-78B
$C_{18}H_{12}N_8$ 1,4,8,11-Tetraaza[14]annulene-6.13-di- carbonitrile	DMF	338s(4.31),351(4.38), 385s(4.12),405(4.30), 425(4.33)	64-1012-78B
disodium salt	DMF	350(4.22),406(4.19), 423(4.19)	64-1012-78B
$C_{18}H_{12}O$ Cyclopenta[b][1]benzopyran, 2-phenyl-	THF	470(2.70)	103-1070-78

Compound	Solvent	$\lambda_{max}(\log \epsilon)$	Ref.
Cyclopenta[b][1]benzopyran, 2-phenyl- (cont.)	THF	260(4.39),365(4.44), 470(2.67)	103-1075-78
Oxirane, 1-pyrenyl-	MeOH	232(4.64),241(4.91), 254(4.06),264(4.42), 275(4.72),312(4.10), 325(--),342(4.69)	44-3425-78
$C_{18}H_{12}O_2$			
Benzo[b]naphtho[2,3-e]oxepin, 6,13-di- hydro-6,13-epoxy-	THF	233(4.91),266(3.92), 275(3.98),280(3.99), 291(3.94),308(2.98), 332(2.82)	78-0113-78
2,5-Cyclohexadiene-1,4-dione, 2,6-di- phenyl-	decane	400(3.52)	70-2215-78
Naphtho[2',3':3,4]cyclobuta[1,2-b][1,4]- benzodioxin, 5a,11b-dihydro-	ether	230(4.36),276(4.06), 284(4.04),294(3.86), 307(3.48),314(3.19), 321(3.51)	78-0113-78
$C_{18}H_{12}O_2S$			
1H,3H-Thieno[3,4-c]furan-1-one, 4,6-di- phenyl-	CH_2Cl_2	350(4.25)	150-5538-78
3,4-Thiophenedicarboxaldehyde, 2,5-di- phenyl-	CH_2Cl_2	311(4.02)	150-5538-78
$C_{18}H_{12}O_2Se_2$			
Spiro[benzo[b]selenophene-2(3H),2'- [2H][1]benzoselenopheno[3,2-b]py- ran-3-one, 3',4'-dihydro-	C_6H_{12}	247(--),258s(--), 291(--),297s(--), 311(--),358s(--), 376s(--),401(--), 421(--)	22-0241-78
$C_{18}H_{12}O_3$			
7H-Benz[de]anthracen-7-one, 6,8-di- hydroxy-4-methyl-	$CHCl_3$	281(3.95),331(3.84), 357(3.57),375(3.70), 439(4.05),459(4.07)	94-3792-78
4H-1-Benzopyran-4-one, 2-phenyl-7-(2- propynyloxy)-	MeOH	255(4.18),315(4.24)	107-0251-78
Furano[2",3":7,8]flavone, 5"-methyl-	MeOH	247(4.10),275(4.21), 300(4.15)	107-0251-78
1H-Indene-1,3(2H)-dione, 2-(1-oxo-3- phenyl-1-propenyl)-	pH 1	248(4.38),315(3.95), 390(4.63)	65-0142-78
	pH 13	239(4.48),304(4.32), 360(4.40)	65-0142-78
	$CHCl_3$	390(4.60)	65-0142-78
$C_{18}H_{12}O_5$			
6,11[3',4']-Furanodibenzo[b,f][1,4]di- oxocin-14,16-dione, 6,11,13,17-tetra- hydro-, (6α,11α,13R*,17S*)-	ether	266s(3.01),270(3.05), 276(3.01)	78-0073-78
$C_{18}H_{13}BrN_2O$			
10H-Pyrido[3,2-b][1,4]benzoxazine, 3-bromo-10-(phenylmethyl)-	MeOH	234(4.58),350(4.01)	95-0585-78
$C_{18}H_{13}BrN_2OS$			
Ethanethioic acid, S-[4-(3-bromophenyl)- 6-phenyl-2-pyrimidinyl] ester	EtOH	259(4.56),298s(4.13)	94-1298-78
$C_{18}H_{13}BrN_8$			
Naphtho[2,1-e]pyrazolo[5,1-c]triazine,	MeOH	224(4.04),256(4.19),	103-0313-78

Compound	Solvent	λ_{max}(log ϵ)	Ref.
3-[(4-bromo-5-methyl-1H-pyrazol-3-yl)azo]-2-methyl- (cont.)	HOAc	346(4.04),453(3.86) 365(4.19),385(4.15), 460(4.11)	103-0313-78 103-0313-78
	50% HOAc-HCl	380(4.07),475(4.16)	103-0313-78
	70% MeOH- Me₄NOH	370(4.16),480(3.90)	103-0313-78
$C_{18}H_{13}ClFN_3$			
4H-Imidazo[1,5-a][1,4]benzodiazepine, 8-chloro-6-(2-fluorophenyl)-1-methyl-	isoPrOH	220(4.48),240s(4.30)	44-0936-78
6H-Imidazo[1,5-a][1,4]benzodiazepine, 8-chloro-6-(2-fluorophenyl)-1-methyl-	isoPrOH	218s(4.31),255(4.06), 265s(4.05),267(4.04), 288s(3.75)	44-0936-78
$C_{18}H_{13}ClFN_3O$			
4H-Imidazo[1,5-a][1,4]benzodiazepin-4-ol, 8-chloro-6-(2-fluorophenyl)-1-methyl-	isoPrOH	220s(4.54),241s(4.33), 260s(4.03),305s(3.18)	44-0936-78
4H-Imidazo[1,5-a][1,4]benzodiazepine-1-methanol, 8-chloro-6-(2-fluoro-phenyl)-	isoPrOH	215s(4.52),240s(4.40), 305s(3.20)	44-0936-78
4H-Imidazo[1,5-a][1,4]benzodiazepine, 8-chloro-6-(2-fluorophenyl)-1-methyl-, 2-oxide	isoPrOH	227(4.54),245s(4.48), 270s(4.11),315s(3.44)	44-0936-78
$C_{18}H_{13}ClFN_3O_2$			
4H-Imidazo[1,5-a][1,4]benzodiazepine, 8-chloro-6-(2-fluorophenyl)-1-meth-yl-, 2,5-dioxide	isoPrOH	219(4.43),241s(4.25), 267(4.40),308s(4.03)	44-0936-78
4H-Imidazo[1,5-a][1,4]benzodiazepine-1-methanol, 8-chloro-6-(2-fluoro-phenyl)-4-hydroxy-	isoPrOH	216(4.56),240s(4.37), 305s(3.11)	44-0936-78
$C_{18}H_{13}ClN_2O$			
10H-Pyrido[3,2-b][1,4]benzoxazine, 3-chloro-10-(phenylmethyl)-	MeOH	234(4.41),350(3.85)	95-0585-78
$C_{18}H_{13}ClN_2OS$			
Ethanethioic acid, S-[4-(3-chlorophen-yl)-6-phenyl-2-pyrimidinyl] ester	C_6H_{12}	260(4.57),287(4.20), 305s(4.13)	4-0105-78
$C_{18}H_{13}ClN_2O_2$			
1H-Indole-2-carboxylic acid, 3-[(2-chlorophenyl)cyanomethyl]-1-methyl-	EtOH	295(4.30)	103-1204-78
$C_{18}H_{13}ClO_2$			
5H-Benzocyclohepten-5-one, 6-(4-chloro-phenyl)-7-methoxy-	MeOH	233(4.27),272(4.41)	83-0600-78
7H-Benzocyclohepten-7-one, 6-(4-chloro-phenyl)-5-methoxy-		248(4.36),338(3.76)	83-0600-78
$C_{18}H_{13}ClO_4$			
2(5H)-Furanone, 4-benzoyl-5-(2-chloro-phenyl)-3-methoxy-	EtOH	263(3.98)	4-0737-78
$C_{18}H_{13}ClO_6$			
9,10-Anthracenedione, 2-acetyl-6-chlo-ro-5,8-dihydroxy-1-methoxy-3-methyl-	EtOH	233(4.40),257(4.39), 475(3.96)	44-1435-78
$C_{18}H_{13}ClO_7$			
2-Anthracenecarboxylic acid, 6-chloro-	$CHCl_3$	269s(4.46),276(4.51),	39-1041-78C

Compound	Solvent	$\lambda_{max}(\log \epsilon)$	Ref.
9,10-dihydro-5,8-dihydroxy-3-methoxy-1-methyl-9,10-dioxo-, methyl ester (cont.)		305(4.00),470(4.01)	39-1041-78C
$C_{18}H_{13}Cl_2HgN_3$			
Mercury, [1,3-bis(2-chlorophenyl)-1-triazenato-N^3]phenyl-	toluene	399(2.55)	64-1091-78B
Mercury, [1,3-bis(3-chlorophenyl)-1-triazenato-N^3]phenyl-	toluene	394(2.55)	64-1091-78B
Mercury, [1,3-bis(4-chlorophenyl)-1-triazenato-N^3]phenyl-	toluene	402(2.56)	64-1091-78B
$C_{18}H_{13}Cl_2N_3O_2$			
1,2-Propanedione, 1-[7-chloro-5-(2-chlorophenyl)-3H-1,4-benzodiazepin-2-yl]-, 1-oxime	isoPrOH	241(4.49),315s(3.78)	4-0577-78
$C_{18}H_{13}Cl_2N_3O_3$			
3H-1,4-Benzodiazepine-2-acetic acid, 7-chloro-5-(2-chlorophenyl)-α-(hydroxyimino)-, methyl ester	isoPrOH	242(4.51),275s(4.10), 323(3.72),337s(3.71)	4-0577-78
$C_{18}H_{13}Cl_3O_6$			
9H-Xanthen-9-one, 1-acetoxy-2,4,5-trichloro-3,6-dimethoxy-8-methyl-	MeOH	215s(4.47),248(4.88), 291s(4.24),307(4.39), 343s(3.93)	78-2491-78
$C_{18}H_{13}Cl_4N_2O$			
Quinolinium, 3-(aminocarbonyl)-2-(dichloromethyl)-1-[(2,6-dichlorophenyl)methyl]-, chloride	CH_2Cl_2	238(3.7),309(3.9)	5-1536-78
$C_{18}H_{13}Cl_5N_2O$			
4-Isoquinolinecarboxamide, 2-[(2,6-dichlorophenyl)methyl]-1,2-dihydro-1-(trichloromethyl)-	$CHCl_3$	242(3.9),263s(3.8), 307(3.8),327(3.8)	5-1536-78
3-Quinolinecarboxamide, 1-[(2,6-dichlorophenyl)methyl]-1,2-dihydro-2-(trichloromethyl)-	$CHCl_3$	241(3.9),252(3.9), 290s(3.3),381(3.2)	5-1536-78
$C_{18}H_{13}F_2HgN_3$			
Mercury, [1,3-bis(2-fluorophenyl)-1-triazenato-N^3]phenyl-	toluene	396(2.55)	64-1091-78B
$C_{18}H_{13}HgN_5O_4$			
Mercury, [1,3-bis(2-nitrophenyl)-1-triazenato]phenyl-	toluene	423(2.60)	64-1091-78B
Mercury, [1,3-bis(3-nitrophenyl)-1-triazenato]phenyl-	toluene	369s(--)	64-1091-78B
Mercury, [1,3-bis(4-nitrophenyl)-1-triazenato]phenyl-	toluene	441(2.60)	64-1091-78B
$C_{18}H_{13}NO$			
Benz[cd]isoindolo[2,1-a]indol-8(1H)-one, 2,3-dihydro-	EtOH	235(4.46),242s(4.43), 277(4.49),294(4.17), 307(4.22),370(3.95)	94-0630-78
$C_{18}H_{13}NO_2$			
Benzo[6,7]cyclohept[1,2-b]indole-6,7-dione, 5,12-dihydro-12-methyl-	MeOH	244(4.40),250(4.45), 269(4.51),291(4.37), 363(4.12),379(4.18)	18-3579-78

Compound	Solvent	$\lambda_{max}(\log \epsilon)$	Ref.
Benzo[6,7]cyclohept[1,2-b]indol-7(12H)-one, 6-hydroxy-12-methyl-	MeOH	233(4.34),312(4.67), 422(3.86)	18-3579-78
1H-Dibenz[e,g]indole-3-carboxylic acid, methyl ester	C_6H_{12}	254(4.44),261(4.58), 286(3.75),293(3.75), 306(3.54)	44-0381-78
Naphthalene, 1-[2-(4-nitrophenyl)ethenyl]-	EtOH	368.5(4.35)	48-0071-78
9H-Phenanthro[9',10':3,4]cyclobuta[1,2-c]pyrrole-9,11(10H)-dione, 8b,8c,11a-11b-tetrahydro-, anti-anti	MeOH	267(4.05),309(3.29)	24-2677-78
syn-syn	MeOH	265(3.85),309(3.24)	24-2677-78
$C_{18}H_{13}NO_4$			
1H-Isoindole-1,3(2H)-dione, 2-(tetrahydro-5-oxo-3-phenyl-2-furanyl)-	THF	220(4.68),294(3.25)	40-0404-78
1H-Isoindole-1,3(2H)-dione, 2-(tetrahydro-5-oxo-2-phenyl-3-furanyl)-	THF	220(4.67),293(3.27)	40-0404-78
$C_{18}H_{13}NO_4S$			
3-Thiophenecarboxylic acid, 2-(4-nitrophenyl)-5-phenyl-, methyl ester	dioxan	252(4.19),350(4.18)	24-2028-78
$C_{18}H_{13}NO_6$			
2(5H)-Furanone, 4-benzoyl-3-methoxy-5-(3-nitrophenyl)-	EtOH	260(4.29)	4-0737-78
2(5H)-Furanone, 3-hydroxy-4-(4-methoxybenzoyl)-5-(2-nitrophenyl)-	EtOH	290(4.00),347(3.96)	4-0737-78
$C_{18}H_{13}N_3$			
1,10-Phenanthrolin-4-amine, N-phenyl-	neutral	213(4.38),268(4.60), 290s(4.29),295s(4.33), 335(4.28)	39-1215-78B
	monocation	218(4.33),241(4.42), 271(4.02),337(4.22)	39-1215-78B
$C_{18}H_{13}N_3O$			
Pyrrolo[2,1-a]phthalazine-2-carboxamide, 3-phenyl-	EtOH	246(4.18),284(4.66), 366(3.9),446(3.72)	118-0206-78
$C_{18}H_{13}N_3O_2$			
3,5-Pyrazolidinedione, 4-(1H-indol-3-ylmethylene)-1-phenyl-	0.1% NaOH	453(4.590)	103-1088-78
	EtOH	416(4.520)	103-1088-78
	EtOH-NaOEt	452(4.828)	103-1088-78
sodium salt	n.s.g.	418(4.402)	103-1088-78
3H-Pyrazol-3-one,, 4-(2-benzoxazolylmethylene)-2,4-dihydro-5-methyl-2-phenyl-	n.s.g.	210(4.71),250(4.76)	48-0857-78
Pyrimido[4,5-b]quinoline-2,4(3H,10H)-dione, 10-methyl-3-phenyl-	EtOH	220(4.47),265(4.51), 321(3.86),397(3.97)	88-3469-78
$C_{18}H_{13}N_3O_3$			
4(3H)-Quinazolinone, 3-[(dihydro-5-oxo-2(3H)-furanylidene)amino]-2-phenyl-	MeOH	227(4.50),272(4.19), 316(3.71)	80-1085-78
$C_{18}H_{13}N_3O_4$			
1H-Indole-2-carboxylic acid, 3-[cyano-(3-nitrophenyl)methyl]-1-methyl-	EtOH	297(4.20)	103-1204-78
1H-Indole-2-carboxylic acid, 3-[cyano-(4-nitrophenyl)methyl]-1-methyl-	EtOH	297(4.35)	103-1204-78

Compound	Solvent	$\lambda_{max}(\log \epsilon)$	Ref.
$C_{18}H_{13}N_5$			
Pyridinium, 1-[3-(dicyanomethyl)-5-phenyl-1H-pyrazol-4-yl]-4-methyl-, hydroxide, inner salt	EtOH	227(4.50),259(4.52), 460(2.95)	95-1412-78
$C_{18}H_{13}N_5O$			
1H-Pyrazolo[3,4-d]pyrimidine-4-carbox-amide, N,1-diphenyl-	EtOH	250(4.47),275s(3.94), 320(3.90)	95-1274-78
$C_{18}H_{14}BrN_3O$			
Acetamide, N-[4-(3-bromophenyl)-6-phenyl-2-pyrimidinyl]-	EtOH	253(4.60),317(4.23)	94-1298-78
$C_{18}H_{14}ClNO$			
1-Naphthalenecarbonitrile, 5-(4-chloro-benzoyl)-1,2,3,4-tetrahydro-	EtOH	259(4.28)	94-1511-78
$C_{18}H_{14}ClNO_4$			
6H-1,3-Oxazinium, 2-(4-chlorophenyl)-4-hydroxy-5-(4-methoxyphenyl)-3-methyl-6-oxo-, hydroxide, inner salt	MeCN	256(4.40),382(3.68)	5-1655-78
6H-1,3-Oxazinium, 5-(4-chlorophenyl)-4-hydroxy-2-(4-methoxyphenyl)-3-methyl-6-oxo-, hydroxide, inner salt	MeCN	215s(4.26),266(4.30), 291(4.22),365(3.89)	5-1655-78
$C_{18}H_{14}ClN_5$			
7H-Imidazo[4,5-e]-1,2,4-triazine, 3-(4-chlorophenyl)-7-ethyl-6-phenyl-	EtOH	265(4.26),289s(4.00), 338(3.99)	94-3154-78
$C_{18}H_{14}Cl_2N_2O$			
2-Propanone, 1-[7-chloro-5-(2-chloro-phenyl)-1,3-dihydro-2H-1,4-benzodi-azepin-2-ylidene]-, (Z)-	isoPrOH	216(4.47),240s(4.15), 285(3.65),330(4.40)	4-0577-78
$C_{18}H_{14}Cl_2O_6$			
9H-Xanthen-9-one, 1-acetoxy-2,4-di-chloro-3,6-dimethoxy-8-methyl-	MeOH	210(4.19),247(4.57), 282(4.05),303(4.18), 335s(3.85)	78-2491-78
9H-Xanthen-9-one, 1-acetoxy-2,5-di-chloro-3,6-dimethoxy-8-methyl-	MeOH	214(4.17),248(4.56), 272(3.84),303(4.19), 330(3.82)	78-2491-78
$C_{18}H_{14}F_4N_2O_2$			
Benzenamine, N,N-dimethyl-4-[1,2,3,4-tetrafluoro-4-(4-nitrophenyl)-1,3-butadienyl]-, (E,E)-	ether	408(4.42)	104-0939-78
$C_{18}H_{14}NP$			
9H-Phosphorino[3,4-b]indole, 9-methyl-4-phenyl-	EtOH	231(4.30),272(4.38), 327(4.01),385(3.32)	139-0257-78B
$C_{18}H_{14}N_2$			
1H-Perimidine, 1-methyl-2-phenyl-	EtOH	<u>340(4.2)</u>	103-1156-78
1H-Perimidine, 2-methyl-1-phenyl-	MeOH	207(4.62),238(4.45), 340(4.17),400(2.98)	103-1156-78
Tricyclo[8.2.2.$2^{4,7}$]hexadeca-4,6,10,12-13,15-hexaene-5,11-dicarbonitrile, pseudo-ortho	EtOH	232s(4.07),276(3.57), 306(3.11),327s(2.72)	24-0523-78
pseudo-para	EtOH	327[sic](4.19),306(3.28), 326s(2.62)	24-0523-78

Compound	Solvent	λ_{max}(log ϵ)	Ref.
Tricyclo[9.3.1.1⁴,⁸]hexadeca-1(15),4,6-8(16),11,13-hexaene-5,14-dicarbonitrile	EtOH	202(4.53),218(4.54), 238s(4.32),285(3.35), 292s(3.29)	78-0871-78
$C_{18}H_{14}N_2O$			
1H-1,2-Diazepine, 1-benzoyl-4-phenyl-	MeOH	228(4.00),274(3.95), 300s(3.88)	33-2887-78
1H-1,2-Diazepine, 1-benzoyl-6-phenyl-	MeOH	228(4.11),278s(4.06), 370s(2.70)	33-2887-78
Pyridinium, 1-(benzoylamino)-3-phenyl-, hydroxide, inner salt	EtOH	303(3.74)	33-2887-78
10H-Pyrido[3,2-b][1,4]benzoxazine, 10-(phenylmethyl)-	C_6H_{12}	233(4.49),338(3.91)	95-0585-78
Pyrido[1,2-a]indole, 1-methoxy-10-(2-pyridinyl)-	MeOH	261(4.70),319(3.97), 332(3.97),408(3.80), 429(3.81),456(3.63)	145-1056-78
Pyrido[1,2-a]indole, 3-methoxy-10-(2-pyridinyl)-	MeOH	260(4.62),268(4.63), 310(3.98),350(4.02), 416(3.75),440(3.70)	145-1056-78
$C_{18}H_{14}N_2OS$			
Ethanethioic acid, S-(4,6-diphenyl-2-pyrimidinyl) ester	C_6H_{12}	260(4.55),290s(4.14), 304(4.07)	4-0105-78
$C_{18}H_{14}N_2O_2$			
1H-Indole-2-carboxylic acid, 3-(cyanophenylmethyl)-1-methyl-	EtOH	295(4.20)	103-1204-78
3H-Indol-3-one, 2-(1,3-dihydro-1-methyl-3-oxo-2H-indol-2-ylidene)-1,2-dihydro-1-methyl-	CH_2Cl_2	647(4.34)	24-3233-78
1H-Isoindole-1,3(2H)-dione, 2-[(2-methyl-1H-indol-5-yl)methyl]-	CHCl₃	276(3.84),281(3.83)	103-0856-78
1H-Pyrano[3,4-c]pyridin-1-one, 5-ethyl-3-(1H-indol-2-yl)-	EtOH	230(4.22),269(3.86), 382(4.39)	78-1363-78
$C_{18}H_{14}N_2O_2S$			
Butanamide, N-(6-oxo-6H-anthra[9,1-cd]isothiazol-7-yl)-	DMF	435(4.00)	2-1007-78
Pyrimidine, 4-(1,3-benzodioxol-5-yl)-2-(methylthio)-6-phenyl-	EtOH	265(4.43),288s(4.19), 332(4.13)	4-0105-78
$C_{18}H_{14}N_2O_3$			
Acetamide, N-[4-(3,4-dihydro-3,4-dioxo-1-naphthalenyl)amino]phenyl]-	MeOH	249(4.50),275(4.22), 345(3.79),476(3.76)	78-1377-78
Pyridinium, 1-(1-benzoyl-3-cyano-4-methoxy-4-oxo-2-butenyl)-, hydroxide, inner salt	EtOH	252(4.08),364(4.63)	95-1412-78
$C_{18}H_{14}N_2S_2$			
Benzenesulfenamide, N,N'-3,5-cyclohexadiene-1,2-diylidenebis-	n.s.g.	438(4.10),510(4.13)	139-0041-78B
$C_{18}H_{14}N_4$			
[4,4'-Biquinoline]-3,3'-diamine	EtOH	248(4.27),277(3.59), 362(3.58)	39-1126-78C
$C_{18}H_{14}N_4O_2$			
Pyridinium, 1-[3-(1-cyano-2-methoxy-2-oxoethyl)-5-phenyl-1H-pyrazol-4-yl]-, hydroxide, inner salt	EtOH	241(4.50),265(4.55), 465(3.40)	95-1412-78

Compound	Solvent	$\lambda_{max}(\log \epsilon)$	Ref.
$C_{18}H_{14}N_4O_2S$ 1H-Pyrazolo[3,4-d]pyrimidine, 4-[(4-methylphenyl)sulfonyl]-1-phenyl-	EtOH	251(4.59),320(3.29)	95-0089-78
$C_{18}H_{14}N_4O_3$ Acetamide, N-[4-(3-nitrophenyl)-6-phenyl-2-pyrimidinyl]-	EtOH	252(4.50),317(4.00)	94-1298-78
$C_{18}H_{14}N_6O_2$ 5-Isoxazolamine, 4,4'-azobis[3-phenyl-	50% dioxan	272s(3.93),300s(3.83), 363(3.88),413(4.23), 430(4.15)	33-0108-78
$C_{18}H_{14}N_6O_4$ Dibenzo[b,i][1,4,8,11]tetraazacyclotetradecine, 5,14-dihydro-7,16-dinitro-	DMF	354(4.72),394(4.59)	64-1012-78B
$C_{18}H_{14}N_6O_6$ Ethanone, 1,1'-[1,4-dihydro-1,4-bis(4-nitrophenyl)-1,2,4,5-tetrazine-3,6-diyl]bis-	EtOH	240(4.08),262(4.08), 394(4.36)	70-2227-78
$C_{18}H_{14}N_8$ Naphtho[2,1-e]pyrazolo[5,1-c][1,2,4]-triazine, 2-methyl-1-[(5-methyl-1H-pyrazol-3-yl)azo]-	MeOH	222(4.25),250(4.26), 257(4.48),357(4.26), 455(4.08)	103-0313-78
	HOAc	345(4.33),460(3.11)	103-0313-78
	50% HOAc-HCl	360(4.30),470(4.17)	103-0313-78
	70% MeOH-Me₄NOH	365(4.28),400(4.23), 490(4.04)	103-0313-78
$C_{18}H_{14}N_8Ni$ Nickel, [6,13-diethyl-1,4,8,11-tetra-azacyclotetradeca-2,4,7,9,11,14-hexaene-2,3,9,10-tetracarbonitrilato(2-)-N^1,N^4,N^8,N^{11}]-, (SP-4-1)-	EtOH	270(3.87),330(4.36), 390(4.08),530(3.86)	24-2919-78
$C_{18}H_{14}O$ 6,11:7,10-Dimethanobenzo[3,4]cyclobuta-[1,2]cyclodecen-13-one, 6,7,10,11-tetrahydro-, anti	C_6H_{12}	243(4.60),324(3.79), 339(3.76)	12-1585-78
$C_{18}H_{14}O_2$ 9-Anthracenemethanol, α-methylene-, acetate	EtOH	255(5.15),322(3.09), 330(3.26),347(3.75), 365(3.91),384(3.85)	35-7041-78
$C_{18}H_{14}O_2S$ 3-Thiophenecarboxaldehyde, 4-(hydroxymethyl)-2,5-diphenyl-	CH_2Cl_2	261(4.34),310s(3.95)	150-5538-78
$C_{18}H_{14}O_3$ 2H-Furo[2,3-h]-1-benzopyran-2-one, 8,9-dihydro-9-methyl-8-phenyl-	MeOH	260(3.69),325(4.28)	2-0579-78
$C_{18}H_{14}O_4$ 1,4-Anthracenediol, diacetate	CH_2Cl_2	319(3.0),333(3.45), 349(3.7),368(3.8), 387(3.75)	150-5538-78
9,10-Anthracenediol, diacetate	C_6H_{12}	371(4.03)	61-1068-78

Compound	Solvent	$\lambda_{max}(\log \epsilon)$	Ref.
2(5H)-Furanone, 4-benzoyl-3-hydroxy-5-(4-methylphenyl)-	H_2O	257(4.03),330(4.10)	4-0737-78
	EtOH	258(4.06),350(3.88)	4-0737-78
	EtOH-NaHCO$_3$	251(4.06),350(4.14)	4-0737-78
	CHCl$_3$	267(4.11)	4-0737-78
2(5H)-Furanone, 4-benzoyl-3-methoxy-5-phenyl-	EtOH	262(4.04)	4-0737-78
1H-Indene-1,3(2H)-dione, 2-[(2,3-dimethoxyphenyl)methylene]-	C_6H_{12}	347(4.43)	146-0144-78
	acetone	346(--)	146-0144-78
1H-Indene-1,3(2H)-dione, 2-[(2,4-dimethoxyphenyl)methylene]-	C_6H_{12}	399(4.52),419(4.49)	146-0144-78
	acetone	425(--)	146-0144-78
1H-Indene-1,3(2H)-dione, 2-[(2,5-dimethoxyphenyl)methylene]-	C_6H_{12}	430(4.13)	146-0144-78
	acetone	431(--)	146-0144-78
1H-Indene-1,3(2H)-dione, 2-[(2,6-dimethoxyphenyl)methylene]-	C_6H_{12}	357(4.22)	146-0144-78
	acetone	380(--)	146-0144-78
1H-Indene-1,3(2H)-dione, 2-[(3,4-dimethoxyphenyl)methylene]-	C_6H_{12}	398(4.48),419s(--)	146-0144-78
	acetone	412(--)	146-0144-78
1H-Indene-1,3(2H)-dione, 2-[(3,5-dimethoxyphenyl)methylene]-	C_6H_{12}	357(4.53)	146-0144-78
	acetone	350(--)	146-0144-78
1,3-Phenanthrenediol, diacetate	MeOH	247s(4.72),254(4.81), 276(4.11),286(4.03), 297(4.11),335(2.89), 349(2.92)	12-2259-78

$C_{18}H_{14}O_5$

Compound	Solvent	$\lambda_{max}(\log \epsilon)$	Ref.
9(10H)-Anthracenone, 1,8-diacetoxy-	MeOH	210(4.04),258(4.04), 270(3.85),310(2.85), 380(2.65)	87-0026-78
1H-Cyclopenta[b]anthracene-5,10-dione, 2,3-dihydro-1,4,11-trihydroxy-3-methyl-	MeOH	203(4.83),225s(--), 249(4.63),284(3.94), 320s(--),457s(--), 475(3.97),507s(--)	24-3823-78
2(5H)-Furanone, 4-benzoyl-3-hydroxy-5-(4-methoxyphenyl)-	H_2O	248(4.04),341(4.17)	4-0737-78
	EtOH	254(3.99),399(3.99)	4-0737-78
	EtOH-NaHCO$_3$	248(4.02),342(4.09)	4-0737-78
	CHCl$_3$	268(4.11)	4-0737-78

$C_{18}H_{14}O_6$

Compound	Solvent	$\lambda_{max}(\log \epsilon)$	Ref.
4H-1-Benzopyran-4-one, 2-(1,3-benzodioxol-5-yl)-6,7-dimethoxy-	dioxan	322(3.71)	2-1125-78
Naphtho[2,3-b]furan-3-carboxylic acid, 2-(2-oxopropyl)-4,9-dihydro-4,9-dioxo-, ethyl ester	dioxan	251(4.56),293(3.39), 335(3.62),380s(3.34)	5-0140-78

$C_{18}H_{15}AsO$

Compound	Solvent	$\lambda_{max}(\log \epsilon)$	Ref.
4(1H)-Arseninone, 1-methyl-2,6-diphenyl-	EtOH	231(4.26),300(4.19), 333(4.25)	88-1471-78
4(1H)-Arseninone, 2-(4-methylphenyl)-1-phenyl-	EtOH	223(4.10),313(4.00)	88-1471-78
4(1H)-Arseninone, 2-phenyl-1-(phenylmethyl)-	EtOH	228(4.14),320(3.90)	88-1471-78

$C_{18}H_{15}BrN_2O_2$

Compound	Solvent	$\lambda_{max}(\log \epsilon)$	Ref.
1H-Pyrazole-1-carboxylic acid, 5-(3-bromophenyl)-3-phenyl-, ethyl ester	EtOH	262(4.46)	94-1298-78

$C_{18}H_{15}BrO$

Compound	Solvent	$\lambda_{max}(\log \epsilon)$	Ref.
Ethanone, 2-bromo-1-(9,10-dimethyl-2-anthracenyl)-	MeOH	245(4.47),270(4.56), 338(3.39),355(3.53), 375(3.57),426(3.58)	44-3425-78

Compound	Solvent	$\lambda_{max}(\log \epsilon)$	Ref.
$C_{18}H_{15}ClFN_3$			
3H-Imidazo[1,5-a][1,4]benzodiazepine, 8-chloro-6-(2-fluorophenyl)-3a,4-dihydro-1-methyl-	isoPrOH	213(4.57),250s(4.06), 280s(3.57)	44-0936-78
$C_{18}H_{15}ClFN_3O$			
Methanone, [2-[5-(aminomethyl)-2-methyl-1H-imidazol-1-yl]-5-chlorophenyl]-(2-fluorophenyl)-, dihydrochloride	pH 1	215s(4.43),258(4.08), 290s(3.67)	44-0936-78
$C_{18}H_{15}ClN_2O_3$			
1H-Pyrido[3,2-b]indole-2,9-dione, 5-(4-chlorobenzoyl)-3,4,5,6,7,8-hexahydro-	EtOH	266(4.40)	94-3080-78
$C_{18}H_{15}ClN_4O_2$			
Benzamide, N-benzoyl-4-chloro-N-(4,5-dimethyl-1H-1,2,3-triazol-1-yl)-	MeOH	247(4.24)	4-1255-78
Benzoic acid, anhydride with 4-chloro-N-(4,5-dimethyl-1H-1,2,3-triazol-1-yl)benzenecarboximidic acid, (Z)-	MeOH	243(4.19),283(4.25)	4-1255-78
$C_{18}H_{15}ClO_3$			
1H-Indene-1-carboxylic acid, 4-(4-chloro-3-methylbenzoyl)-2,3-dihydro-	EtOH	261(4.17)	94-1776-78
1-Naphthalenecarboxylic acid, 5-(4-chlorobenzoyl)-1,2,3,4-tetrahydro-	EtOH	258(4.29)	94-1511-78
$C_{18}H_{15}ClO_4$			
Benzeneacetic acid, α-[(2-chlorophenyl)methylene]-2-(methoxycarbonyl)-, methyl ester, (Z)-	n.s.g.	219(4.50),275(4.28), 290(4.23)	80-1465-78
$C_{18}H_{15}Cl_2N_3O_2$			
Acetic acid, amino[7-chloro-5-(2-chlorophenyl)-1,3-dihydro-2H-1,4-benzodiazepin-2-ylidene]-, methyl ester, (Z)-	isoPrOH	217s(4.49),280s(3.88), 327(4.42),400s(3.28)	4-0577-78
$C_{18}H_{15}F_4N$			
Benzenamine, N,N-dimethyl-4-(1,2,3,4-tetrafluoro-4-phenyl-1,3-butadien-yl)-, (E,E)-	EtOH	348(4.51)	104-0939-78
$C_{18}H_{15}HgN_3$			
Mercury, (1,3-diphenyl-1-triazenato)-phenyl-	toluene	395(2.54?)	64-1091-78B
$C_{18}H_{15}IN_2O$			
Quinolinium, 2-(2-benzoxazolyl)-1,4-dimethyl-, iodide	EtOH	249(4.39),289(3.94), 299(3.95),369(4.22)	4-0017-78
Quinolinium, 4-(2-benzoxazolyl)-1,2-dimethyl-, iodide	EtOH	342(4.29),370(4.29)	4-0017-78
Quinolinium, 6-(2-benzoxazolyl)-1,2-dimethyl-, iodide	EtOH	251(4.35),297(4.49), 347(4.03),363(3.99)	4-0017-78
Quinolinium, 6-(2-benzoxazolyl)-1,4-dimethyl-, iodide	EtOH	242(4.39),295(4.50), 349(4.05)	4-0017-78
$C_{18}H_{15}IN_2S$			
Benzothiazolium, 3-methyl-2-(4-methyl-2-quinolinyl)-	EtOH	286(4.10),321(4.13), 333(4.18),346(4.15)	4-0017-78

Compound	Solvent	$\lambda_{max}(\log \epsilon)$	Ref.
Quinolinium, 4-(2-benzothiazolyl)-1,2-dimethyl-, iodide	EtOH	336(4.18),360(4.12)	4-0017-78
Quinolinium, 6-(2-benzothiazolyl)-1,2-dimethyl-, iodide	EtOH	252(4.43),299(4.50), 347(4.10)	4-0017-78
Quinolinium, 6-(2-benzothiazolyl)-1,4-dimethyl-, iodide	EtOH	245(4.41),300(4.47), 351(4.06)	4-0017-78
$C_{18}H_{15}KSi$			
Potassium, (triphenylsilyl)-	THF	262s(--),363(4.11)	120-0052-78
$C_{18}H_{15}NO$			
Formamide, N-[1-(phenylmethyl)-2-naphthalenyl]-	EtOH	220s(4.58),240(4.54), 267(4.39),278(4.39)	103-1025-78
Naphthalene, 2-nitroso-1-(2-phenylethyl)-	CHCl$_3$	765(1.54)	39-0692-78C
1-Naphthalenecarbonitrile, 5-benzoyl-1,2,3,4-tetrahydro-	EtOH	250(4.19)	94-1511-78
4(1H)-Pyridinone, 1-methyl-3,5-diphenyl-	EtOH	235(4.46),295(4.14), 305(4.14)	115-0067-78
	CCl$_4$	300(3.89),320(3.90)	115-0067-78
$C_{18}H_{15}NO_2$			
Benz[cd]indolo[2,1-a]indol-8(1H)-one, 2,3,12b,12c-tetrahydro-12b-hydroxy-	EtOH	230(4.05),237s(4.03), 284(3.28),297(3.73)	94-0630-78
1H-Isoindole-1,3(2H)-dione, 2-(5,6,7,8-tetrahydro-1-naphthalenyl)-	EtOH	274(3.33),293(3.29), 300(3.27)	94-0630-78
1,2-Naphthalenedione, 4-[(2,4-dimethylphenyl)amino]-	MeOH	240(4.33),277(4.14), 335s(3.70),455(3.64)	78-1377-78
1,2-Naphthalenedione, 4-(ethylphenylamino)-	MeOH	246(4.38),278(4.08), 322(3.80),476(3.88)	78-1377-78
5(4H)-Oxazolone, 4-[1-(3-methylphenyl)-ethylidene]-2-phenyl-, (Z)-	EtOH	203(4.40),250(4.13), 348(4.39)	118-0832-78
5(4H)-Oxazolone, 4-[1-(4-methylphenyl)-ethylidene]-2-phenyl-, (E)-	EtOH	202(4.40),353(4.32)	118-0832-78
(Z)-	EtOH	202(4.42),251(4.13), 354(4.41)	118-0832-78
2-Propenoic acid, 3-[2'-(2H-azirin-3-yl)[1,1'-biphenyl]-2-yl]-, methyl ester, (E)-	C$_6$H$_{12}$	272(4.22)	44-0381-78
Pyrene, 10b,10c-dihydro-10b,10c-dimethyl-2-nitro-	C$_6$H$_{12}$	288(3.73),342(4.55), 378(4.24),484(4.10), 562(3.07),617(3.14)	44-3475-78
$C_{18}H_{15}NO_2S$			
Phenanthro[9,10-d]thiazole-2-methanol, 6-methoxy- -methyl-	EtOH	248s(4.70),256(4.81), 262s(4.69),280(4.28), 291(4.22),316(3.92)	39-0685-78C
$C_{18}H_{15}NO_3$			
8H-Anthra[9,1-ef][1,4]oxazepin-8-one, 2,3-dihydro-6-methoxy-2-methyl-	MeOH	260(4.30)	20-0911-78
6H-1,3-Oxazinium, 4-hydroxy-3-methyl-2-(4-methylphenyl)-6-oxo-5-phenyl-, hydroxide, inner salt	MeCN	211s(4.35),258(4.37), 365(3.73)	5-1655-78
6H-1,3-Oxazinium, 4-hydroxy-3-methyl-6-oxo-5-phenyl-2-(phenylmethyl)-, hydroxide, inner salt	MeCN	255(4.15),302s(3.78), 330s(3.74)	5-1655-78
5(4H)-Oxazolone, 4-[1-(4-methoxyphenyl)ethylidene]-2-phenyl-, (Z)-	EtOH	201(4.23),251(3.95), 278(3.91),374(4.26)	118-0832-78

Compound	Solvent	$\lambda_{max}(\log \epsilon)$	Ref.
$C_{18}H_{15}NO_4$			
8H-Anthra[9,1-ef][1,4]oxazepin-8-one,	MeOH-HCl	265(4.48)	20-0911-78
3,12b-dihydro-7-hydroxy-6-methoxy-	MeOH-NaOH	315(3.60)	20-0911-78
2-methyl-			
6H-1,3-Oxazinium, 4-hydroxy-2-(4-meth-	MeCN	214s(4.31),226(4.15),	5-1655-78
oxyphenyl)-3-methyl-6-oxo-5-phenyl-,		266(4.27),288(4.21),	
hydroxide, inner salt		365(3.84)	
Spiro[2H-1-benzopyran-2,2'-[2H-1,3]ben-	isoPrOH	260(4.11),270s(4.05),	103-0122-78
zoxazin]-4'(3'H)-one, 5-methoxy-		295(3.68)	
3'-methyl-			
photoinduced form	isoPrOH	415(--),545(--)	103-0122-78
$C_{18}H_{15}N_2O$			
Quinolinium, 2-(2-benzoxazolyl)-1,4-di-	EtOH	249(4.39),289(3.94),	4-0017-78
methyl-, iodide		299(3.95),369(4.22)	
Quinolinium, 4-(2-benzoxazolyl)-1,2-di-	EtOH	342(4.29),370(4.29)	4-0017-78
methyl-, iodide			
Quinolinium, 6-(2-benzoxazolyl)-1,2-di-	EtOH	251(4.35),297(4.49),	4-0017-78
methyl-, iodide		347(4.03),363(3.99)	
Quinolinium, 6-(2-benzoxazolyl)-1,4-di-	EtOH	242(4.39),295(4.50),	4-0017-78
methyl-, iodide		349(4.05)	
Quinolizinium, 2,3-bis(methoxycarbonyl)-	EtOH	227(4.27),245s(4.21),	150-2850-78
4-(2-pyridinyl)-, perchlorate, per-		338s(3.96),349(4.02)	
chlorate ($C_{18}H_{15}N_2O_4$)			
$C_{18}H_{15}N_2S$			
Benzothiazolium, 3-methyl-2-(4-methyl-	EtOH	286(4.10),321(4.13),	4-0017-78
2-quinolinyl)-, iodide		333(4.18),346(4.15)	
Quinolinium, 4-(2-benzothiazolyl)-1,2-	EtOH	336(4.18),360(4.12)	4-0017-78
dimethyl-, iodide (see $C_{18}H_{15}IN_2S$)			
$C_{18}H_{15}N_3$			
Propanedinitrile, [[3-[(dimethylamino)-	MeOH	295(4.30),439(4.59)	83-0369-78
methylene]-1,4-cyclopentadien-1-yl]-			
phenylmethylene]-			
$C_{18}H_{15}N_3O$			
1H-Indole, 1-acetyl-3-(1H-indol-3-yl-	EtOH	205(4.47),241(4.41),	103-0757-78
amino)-		263s(4.43),312(4.22)	
$C_{18}H_{15}N_3OS$			
Benzoic acid, 2-[(1-naphthalenylamino)-	MeOH	300(3.56)	80-0397-78
thioxomethyl]hydrazide			
nickel complex	MeOH	318(4.41)	80-0397-78
$C_{18}H_{15}N_3O_2S$			
1H-Pyrimido[5,4-b][1,4]benzothiazine-	pH 13	259(4.45),283s(4.04),	5-0193-78
2,4(3H,10H)-dione, 3-methyl-10-(phen-		325s(3.45)	
ylmethyl)-	MeOH	248(4.31),282(4.05),	5-0193-78
		340s(3.34)	
	12M HCl	251(4.31),289(3.89),	5-0193-78
		372(3.45)	
$C_{18}H_{15}N_3O_3$			
3H-Pyrazol-3-one, 2,4-dihydro-5-methyl-	benzene	410(3.03)	104-1644-78
4-[1-(3-nitrophenyl)ethylidene]-	EtOH	206(4.32),255(4.38),	104-1644-78
2-phenyl-		284(4.30)	
3H-Pyrazol-3-one, 2,4-dihydro-5-methyl-	benzene	415(2.97)	104-1644-78
4-[1-(4-nitrophenyl)ethylidene]-	EtOH	208(4.30),250(4.25),	104-1644-78
2-phenyl-		286s(4.30)	
geometric isomer	benzene	405s(2.65)	104-1644-78

Compound	Solvent	$\lambda_{max}(\log \epsilon)$	Ref.
(cont.)	EtOH	208(4.35),250(4.30), 286s(4.31)	104-1644-78
enol form	benzene	415(2.95)	104-1644-78
	EtOH	208(4.38),250(4.30), 286s(4.30)	104-1644-78
$C_{18}H_{15}N_3O_4$			
5(4H)-Oxazolone, 4-[[4-(dimethylamino)- phenyl]methylene]-2-(4-nitrophenyl)-	toluene	340(4.08)	103-0120-78
5(4H)-Oxazolone, 2-[4-(dimethylamino)- phenyl]-4-[(4-nitrophenyl)methylene]-	toluene	345(4.20)	103-0120-78
1H-Pyrazole-1-carboxylic acid, 5-(3-ni- trophenyl)-3-phenyl-, ethyl ester	EtOH	260(4.41)	94-1298-78
$C_{18}H_{15}N_3O_5$			
2,4,6(1H,3H,5H)-Pyrimidinetrione, 5- (2,3-dihydro-1-methyl-2-oxo[1]benzo- pyrano[2,3-b]pyrrol-4(1H)-ylidene)- 1,3-dimethyl-	HOAc	336(3.98),448(4.39)	83-0018-78
$C_{18}H_{15}N_3O_5S$			
Diazene, [3,4-dimethoxy-5-(phenylthio)- 2-furanyl](4-nitrophenyl)-	dioxan	236(4.20),416(4.44)	33-1033-78
$C_{18}H_{15}N_5$			
7H-Imidazo[4,5-e]-1,2,4-triazine, 7-ethyl-3,6-diphenyl-	EtOH	259(3.87),270s(3.78), 337(3.66)	94-3154-78
$C_{18}H_{15}N_5O_2S$			
2H-[1,2,3]Thiadiazolo[4,5,1-hi][1,2,3]- benzothiadiazole-3-S^{IV}, 4,6,7,8-tet- rahydro-2-(4-nitrophenyl)-	C_6H_{12}	249(4.13),263(4.12), 303(4.00),350(4.17), 538(4.43)	39-0195-78C
$C_{18}H_{15}N_5O_4$			
Benzamide, N-benzoyl-N-(4,5-dimethyl- 1H-1,2,3-triazol-1-yl)-4-nitro-	MeOH	252(4.24)	4-1255-78
Benzoic acid, anhydride with N-(4,5-di- methyl-1H-1,2,3-triazol-1-yl)-4-nitro- benzenecarboximidic acid, (Z)-	MeOH	237(4.24),295(4.21)	4-1255-78
$C_{18}H_{15}O_2$			
Pyrylium, 4-methoxy-2,6-diphenyl-	MeOH	274(4.39),355(4.41)	44-4112-78
	MeOH-NaOMe	237(--),320(--)	44-4112-78
$C_{18}H_{16}$			
Cyclopentadiene, 2-methyl-1,3-diphenyl-	EtOH	237(4.16),288(3.92)	44-1481-78
1,3-Cyclopentadiene, 3-methyl-1,2-di- phenyl-	EtOH	220(4.16),298(3.92)	44-1481-78
Cyclopropene, 1,3-diphenyl-3-(2-propen- yl)-	EtOH	260(4.09)	44-1481-78
Cyclopropene, 3-ethenyl-3-methyl- 1,2-diphenyl-	EtOH	227(4.34),315(4.49), 332(4.42)	44-1481-78
Cyclopropene, 3-ethenyl-1-methyl- 2,3-diphenyl-	EtOH	260(4.20)	44-1481-78
Cyclopropene, 3-(2-propenyl)-1,2-di- phenyl-	EtOH	225(4.26),228(4.28), 310(4.38),318(4.45), 336(4.32)	44-1481-78
Dibenzotricyclo[4.3.1.0^{3,7}]deca-4,8-di- ene	MeOH	239s(2.75),246s(3.0), 251(3.19),258(3.30), 264(3.19)	35-3819-78
1H-Indene, 3-ethenyl-1-methyl-2-phenyl-	EtOH	232(4.26),303(4.20)	44-1481-78

Compound	Solvent	$\lambda_{max}(\log \epsilon)$	Ref.
1H-Indene, 3-ethenyl-2-methyl-1-phenyl-	EtOH	270(3.86)	44-1481-78
1H-Indene, 2-phenyl-3-(2-propenyl)-	EtOH	227(4.08),293(4.27)	44-1481-78
Stilbene, 2,2'-diethenyl-	MeOH	245(4.43),300(4.34)	35-3819-78
$C_{18}H_{16}BrN$			
Benzocycloheptenylium, 7-(methylphenyl-amino)-, bromide	MeCN	230(3.21),296(4.38), 380(3.08)	78-0533-78
$C_{18}H_{16}BrNO$			
4H,8H-Pyrido[3,2,1-de]phenanthridin-8-one, 11-bromo-5,6-dihydro-4,6-di-methyl-	EtOH	236s(4.65),244(4.69), 268(4.31),305s(3.69), 317s(3.73),329(3.83), 343(3.78)	33-1246-78
geometric isomer	EtOH	236s(4.65),243(4.68), 267(4.32),305s(3.69), 317s(3.74),328(3.82), 343(3.77)	33-1246-78
$C_{18}H_{16}BrNO_4$			
Cyclopenta[3,4]cyclobuta[1,2-b]pyrrole-3a(1H)-carboxylic acid, 6b-(4-bromo-phenyl)-2,3,3b,4,6a,6b-hexahydro-2,3-dioxo-, ethyl ester, (3aα,3bα,6aα,6bα)-	EtOH	230(4.34),260s(3.89)	142-0153-78A
4,7-Methano-3aH-indole-3a-carboxylic acid, 7a-(4-bromophenyl)-1,2,3,4,7,7a-hexahydro-2,3-dioxo-, ethyl ester, (3aα,4β,7β,7aα)-	EtOH	228(4.34)	142-0153-78A
$C_{18}H_{16}ClF$			
Cyclobuta[3",4"]benzo[1",2":3',4']cy-clobuta[1',2':3,4]cyclobuta[1,2]cy-cloocteneene, 2-chloro-1-fluoro-2a,4a-4b,4c,10a,10b,10c,10d-octahydro-	MeOH	252(3.21)	5-1368-78
1,3,5,7,9,11,13,15,17-Cyclooctadeca-nonaene, 1-chloro-2-fluoro-	MeOH	279(3.93),358(3.89), 373(5.32),410(3.89), 453(4.16)	5-1368-78
$C_{18}H_{16}ClF_2NS$			
1-Propanamine, 3-(2-chloro-3,6-difluo-ro-9H-thioxanthen-9-ylidene)-N,N-di-methyl-, hydrochloride	MeOH	270(4.16),295s(3.82), 320s(3.47)	73-2656-78
$C_{18}H_{16}ClF_3$			
Cyclobuta[3",4"]benzo[1",2":3',4']cy-clobuta[1'.2':3,4]cyclobuta[1,2]cy-cloocteneene, 2-chloro-1,1,2-trifluoro-1,2,2a,4a,4b,4c,10a,10b,10c,10d-dec-ahydro-	MeOH	252(3.21)	5-1368-78
1,4-Ethenobenzo[3',4']cyclobuta[1',2'-3,4]cyclobuta[1,2]cycloocteneene, 2-chloro-2,3,3-trifluoro-1,2,3,4,4a-4b,4c,10a,10b,10c-decahydro-	MeOH	252(3.21)	5-1368-78
$C_{18}H_{16}ClN$			
9H-Carbazole, 9-(3-chlorophenyl)-1,2,3,4-tetrahydro-	EtOH	220(4.71),270(4.27), 292s(4.17),300s(4.13)	103-1093-78
9H-Carbazole, 1,2,3,4-tetrahydro-9-phenyl-7-chloro-	EtOH	242(4.78),260s(4.29), 290(4.18),299(4.29), 327(3.68)	103-1093-78

Compound	Solvent	$\lambda_{max}(\log \epsilon)$	Ref.
$C_{18}H_{16}ClNO_6$ 2H-Cyclohepta[gh]pyrrolizine-1,2,5-tri- carboxylic acid, 5-ethyl 1,2-dimethyl ester	CHCl$_3$	255(4.51),273(4.15), 285s(4.03),372(4.17), 392(4.13),416(3.70), 443(3.74),472(3.71), 507(3.53),545(3.06)	39-0429-78C
$C_{18}H_{16}ClN_3OS$ 1H-Imidazol-1-amine, 4-(2-chlorophen- yl)-N-[(4-methoxyphenyl)methylene]- 2-(methylthio)-	MeOH	293(4.39),338(4.34)	18-1846-78
1H-Imidazol-1-amine, 4-(4-chlorophen- yl)-N-[(4-methoxyphenyl)methylene]- 2-(methylthio)-	MeOH	292(4.49),344(4.36)	18-1846-78
$C_{18}H_{16}Cl_2F_2$ Cyclobuta[3",4"]benzo[1",2":3',4']cy- clobuta[1',2':3,4]cyclobuta[1,2]cy- cloocteene, 2,2-dichloro-1,1-difluoro- 1,2,2a,4a,4b,4c,10a,10b,10c,10d-deca- hydro-	MeOH	252(3.21)	5-1368-78
1,4-Ethenobenzo[3',4']cyclobuta[1',2'- 3,4]cyclobuta[1,2]cycloocteene, 2,2- dichloro-3,3-difluoro-1,2,3,4,4a,4b- 4c,10a,10b,10c-decahydro-	EtOH	254(3.25)	5-1368-78
$C_{18}H_{16}Cl_2N_2O_{12}$ Quinolizinium, 2,3-bis(methoxycarbon- yl)-4-(2-pyridinyl)-, perchlorate, perchlorate	EtOH	227(4.27),245s(4.21), 338s(3.96),349(4.02)	150-2850-78
$C_{18}H_{16}F_2$ Cyclobuta[3",4"]benzo[1",2":3',4']cy- clobuta[1',2':3,4]cyclobuta[1,2]cy- cloocteene, 1,2-difluoro-2a,4a,4b,4c- 10a.10b,10c.10d-octahydro-	MeOH	252(3.20)	5-1368-78
1,3,5,7,9,11,13,15,17-Cyclooctadeca- nonaene, 1,2-difluoro-	EtOH	276(3.96),354(4.67), 368(5.32),409(3.32), 448(4.03)	5-1368-78
$C_{18}H_{16}F_6N_2O_5S$ 3H-6-Oxa-1-thia-3,9-diazaspiro[4.4]non- 2-en-7-one, 8-(1-methylethyl)-2-phen- yl-9-(trifluoroacetyl)-, trifluoro- acetate	ether	292(3.90)	78-0611-78
$C_{18}H_{16}N$ Benzocycloheptenylium, 7-(methylphenyl- amino)-, bromide	MeCN	230(3.21),296(4.38), 380(3.08)	78-0533-78
$C_{18}H_{16}N_2O$ Ethanone, 1-(4,5-dihydro-4-methylene- 5,5-diphenyl-1H-pyrazol-3-yl)-	EtOH	240(3.70),345(3.89)	22-0401-78
Ethanone, 1-(5-methyl-1,3-diphenyl-1H- pyrazol-4-yl)-	EtOH	243(4.30),288(3.82)	22-0415-78
1(4H)-Naphthalenone, 4-[[4-(dimethyl- amino)phenyl]imino]-	C_6H_{12}	250(4.22),270(4.31), 325(3.92),540(4.07)	48-0557-78
	EtOH	255(4.30),265(4.27), 308(4.06),595(4.14)	48-0557-78 +80-0617-78
	ether	255(4.29),270(4.29), 302(4.06),560(4.07)	48-0557-78

Compound	Solvent	$\lambda_{max}(\log \epsilon)$	Ref.
(cont.)	CCl$_4$	310(4.00),550(4.02)	48-0557-78
3H-Pyrazol-3-one, 2,4-dihydro-5-methyl-	benzene	400(3.01)	104-1644-78
2-phenyl-4-(1-phenylethylidene)-, (E)-	EtOH	208(4.30),255(4.27),	104-1644-78
		395(4.18)	
6H-Pyrido[4,3-b]carbazole, 7-methoxy-	EtOH	244(4.54),287(4.81),	88-1261-78
5,11-dimethyl-		333(3.66)	
	EtOH-HCl	254(4.53),284(4.38),	88-1261-78
		304(4.53)	
$C_{18}H_{16}N_2OS$			
Pyrimidine, 4-(4-methoxyphenyl)-2-(meth-	EtOH	263(4.39),287s(4.16),	4-0105-78
ylthio)-6-phenyl-		333(4.09)	
2(1H)-Pyrimidinethione, 1-(2-methoxy-	MeCN	300(4.64),410(3.41)	97-0334-78
phenyl)-4-(4-methylphenyl)-			
2(1H)-Pyrimidinethione, 1-(3-methoxy-	MeCN	302(4.66),413(4.43)	97-0334-78
phenyl)-4-(4-methylphenyl)-			
$C_{18}H_{16}N_2O_2$			
Cyclohepta[b]pyrrole-3-carboxylic acid,	EtOH	245(4.25),286(4.41),	138-0677-78
1,2-dihydro-2-imino-1-(phenylmethyl)-,		420(4.08)	
methyl ester			
3H-Indole, 3,3-dimethyl-2-[2-(3-nitro-	EtOH	290(4.1),342(4.3)	103-0985-78
phenyl)ethenyl]-			
4-Isoquinolinecarboxylic acid, 3-[(4-	EtOH	225(4.34),266(4.51),	138-0677-78
methylphenyl)amino]-, methyl ester		312(4.36),411(3.77)	
4-Isoquinolinecarboxylic acid, 3-	EtOH	249(4.71),401(3.77)	138-0677-78
[(phenylmethyl)amino]-, methyl ester			
1,2-Naphthalenedione, 4-[[4-(dimethyl-	MeOH	262(4.50),285(4.05),	78-1377-78
amino)amino]-		520(3.77)	
5(4H)-Oxazolone, 4-[[4-(dimethylamino)-	toluene	470(4.74)	103-0120-78
phenyl]methylene]-2-phenyl-			
1H-Pyrazole-3-carboxylic acid, 4,5-di-	EtOH	293(3.87),334(4.04)	22-0401-78
hydro-4-methylene-5,5-diphenyl-,			
dimethyl ester			
2(1H)-Quinoxalinone, 3-(1-benzoyl-	EtOH	340(3.99)	103-0336-78
propyl)-			
2(1H)-Quinoxalinone, 3-[2-(2,5-dimeth-	EtOH	430(4.30)	103-0336-78
ylphenyl)-2-oxoethylidene]-3,4-dihy-			
dro-			
2(1H)-Quinoxalinone, 3-[2-(4-ethylphen-	EtOH	440(4.40)	103-0336-78
yl)-2-oxoethylidene]-3,4-dihydro-			
$C_{18}H_{16}N_2O_2S$			
[1]Benzothieno[3',2':5,6]pyrimido[2,1-	EtOH	227(4.32),340(3.82)	4-0949-78
a]isoindole-6,13-dione, 4b,5,7,8,9-			
10-hexahydro-4b-methyl-			
2(1H)-Pyrimidinethione, 1,4-bis(4-meth-	MeCN	220(4.35),322(4.53),	97-0334-78
oxyphenyl)-		409(3.44)	
1H-Pyrrole, 2,5-dimethyl-1-[4-[(4-ni-	EtOH	210(4.60),250(4.28),	111-0093-78
trophenyl)thio]phenyl]-		263(4.20),335(4.34)	
4-Thiazolidinone, 5-[(4-methoxyphenyl)-	MeOH	238(4.15),350(4.41)	104-0997-78
methylene]-3-methyl-2-(phenylimino)-	dioxan	238(4.14),350(4.41)	104-0997-78
	70% dioxan	237(4.15),356(4.39)	104-0997-78
4(5H)-Thiazolone, 5-[(4-methoxyphenyl)-	MeOH	242(4.02),363(4.44)	104-0997-78
methylene]-2-(methylphenylamino)-	dioxan	241(4.03),310(4.14),	104-0997-78
		352(4.41)	
	70% dioxan	363(4.43)	104-0997-78
$C_{18}H_{16}N_2O_3$			
1,7-Dioxa-2,8-diazaspiro[4.4]nona-2,8-	EtOH	252(4.34)	22-0415-78
diene, 6-methoxy-3,9-diphenyl-			

Compound	Solvent	$\lambda_{max}(\log \epsilon)$	Ref.
4-Isoquinolinecarboxylic acid, 3-[(4-methoxyphenyl)amino]-, methyl ester	EtOH	228(4.33),266(4.50), 310(4.32),413(3.76)	138-0677-78
Phenol, 2-[2-(3,3-dimethyl-3H-indol-2-yl)ethenyl]-4-nitro-	EtOH	320s(4.3),355(4.3)	103-0985-78
Phenol, 4-[2-(3,3-dimethyl-3H-indol-2-yl)ethenyl]-2-nitro-	EtOH	355(4.3)	103-0985-78
$C_{18}H_{16}N_2O_4$			
Butanedioic acid, bis(2-pyridinylmethylene)-, dimethyl ester	EtOH	255(4.21),295(4.12), 390(3.56),520(3.75)	150-2850-78
$C_{18}H_{16}N_2S$			
Pyrimidine, 4-(4-methylphenyl)-2-(methylthio)-6-phenyl-	EtOH	265(4.46),327(3.92)	4-0105-78
$C_{18}H_{16}N_4$			
Cycloprop[a]indene-3,3,4,4-tetracarbonitrile, 1,1a,2,5,6,6a-hexahydro-1,1a,6a-trimethyl-6-methylene-	$CHCl_3$	260(3.99)	35-0860-78
Cycloprop[a]indene-3,3,4,4-tetracarbonitrile, 1,1a,5,6a-tetrahydro-1,1a,6,6a-tetramethyl-	$CHCl_3$	285(3.83)	35-0860-78
2,2,3,3-Naphthalenetetracarbonitrile, 1,4-dihydro-5,6,7,8-tetramethyl-	$CHCl_3$	274(2.61)	35-0860-78
$C_{18}H_{16}N_4O_4$			
Ethanone, 1-(3,4-dihydro-1-naphthalenyl)-, 2,4-dinitrophenylhydrazone	C_6H_{12}	258(4.21),356(4.32)	56-0529-78
	$CHCl_3$	375(4.28)	56-0529-78
$C_{18}H_{16}N_4O_8$			
2(5H)-Furanone, 3,4-dimethoxy-5-(4-nitrophenoxy)-, (4-nitrophenyl)hydrazone	dioxan	282(4.26),387(4.48)	33-1033-78
$C_{18}H_{16}N_4S$			
2H-[1,2,3]Thiadiazolo[4,5,1-hi][1,2,3]benzothiadiazole-3-S^{IV}, 4,6,7,8-tetrahydro-2,4-diphenyl-	C_6H_{12}	247(4.26),306(4.15), 519(4.33)	39-0195-78C
$C_{18}H_{16}N_8$			
1,4,8,11-Tetraazacyclotetradeca-2,4,6-9,11,13-hexaene-2,3,9,10-tetracarbonitrile, 6,13-diethyl-	$CHCl_3$	330(4.76),350(4.73), 360(4.73),368(4.69), 400(4.13),430(3.90)	24-2919-78
$C_{18}H_{16}O$			
Anthracene, 9-(1-ethoxyethenyl)-	EtOH	253(5.18),315(3.11), 328(3.43),344(3.76), 362(3.94),381(3.90)	35-7041-78
5H-Benzocyclohepten-5-one, 6,7,8,9-tetrahydro-6-(phenylmethylene)-, cation	H_2SO_4	417(4.46)	65-1644-78
Cycloprop[a]indene, 1,1a,6,6a-tetrahydro-1-(4-methoxyphenyl)-6-methylene-, endo	C_6H_{12}	242(4.39),254(4.32), 280(3.85),305(3.58)	44-13283-78
exo	C_6H_{12}	243(4.43),248(4.41), 267(4.19),295(3.74)	44-3283-78
Furan, 2,5-dimethyl-3,4-diphenyl-	EtOH	226(4.45)	104-1894-78
1(2H)-Naphthalenone, 3,4-dihydro-2-[(4-methylphenyl)methylene]-, cation	H_2SO_4	467(4.56)	65-1644-78

Compound	Solvent	$\lambda_{max}(\log \epsilon)$	Ref.
2-Oxabicyclo[3.2.0]hept-6-ene, 6,7-di-phenyl-	MeCN	297(4.21),304s(4.19), 322s(4.10)	24-3608-78
Oxirane, (9,10-dimethyl-2-anthracenyl)-	MeOH	262(5.28),357(3.72), 376(3.91),397(3.88)	44-3425-78
Spiro[cyclopropane-1,1'-[1H]indene], 2-(4-methoxyphenyl)-, trans	EtOH	218(4.36),246(4.39), 290(3.64),301(3.50)	44-3283-78
$C_{18}H_{16}OS$			
Benzo[b]naphtho[1,2-d]thiophen-8(9H)-one, 10,11-dihydro-10,10-dimethyl-	EtOH	233(4.52),260s(4.16), 278s(4.05),330(4.20)	42-1232-78
Benzo[b]naphtho[2,1-d]thiophen-10(7H)-one, 8,9-dihydro-8,8-dimethyl-	EtOH	223(4.28),270s(4.40), 280(4.54),327(4.24), 370s(3.54)	42-1232-78
$C_{18}H_{16}OS_2$			
2(1H)-Naphthalenone, 1-(5,6,7,8-tetra-hydro-4H-cyclohepta-1,3-dithiol-2-ylidene)-	MeCN	286(3.65),325(3.83), 502(4.35)	97-0385-78
$C_{18}H_{16}O_2$			
2-Butene-1,4-dione, 1,4-bis(4-methyl-phenyl)-	n.s.g.	275(4.26)	18-1839-78
2,5-Dioxabicyclo[4.2.0]oct-3-ene, 7,8-diphenyl-, (1α,6α,7α,8β)-	MeCN	247s(2.79),253(2.68), 254s(2.67),259(2.71), 262(2.67),264s(2.61), 269(2.51)	24-3624-78
2,5-Dioxabicyclo[4.2.0]oct-7-ene, 7,8-diphenyl-, cis	MeCN	294(4.22),299s(4.21), 318s(3.97)	24-3608-78
1(2H)-Naphthalenone, 3,4-dihydro-2-[(4-methoxyphenyl)methylene]-, cation	H_2SO_4	491(4.63)	65-1644-78
2,4-Pentadienoic acid, 4,5-diphenyl-, methyl ester, (E,E)-	EtOH	225(4.00),313(4.52)	24-3665-78
Pyreno[1,10a-b:8,8a-b']bisoxirene, 1a,7a,10b,10c-tetrahydro-10b,10c-dimethyl-	EtOH	207(4.01),223(3.94), 230(4.03),239(4.17), 249(3.86),261(3.92), 272(4.00),303(3.55), 317(3.72),333(3.89)	44-3475-78
Tricyclo[9.3.1.1^{4,8}]hexadeca-1(15),4,6-8(16),11,13-hexaene-5,14-dicarboxalde-hyde, (-)-	isooctane	212(4.48),222s(4.38), 260(4.33),269s(4.29), 297(3.56),307s(3.48), 345s(2.12)	78-0871-78
$C_{18}H_{16}O_2S$			
3,4-Thiophenedimethanol, 2,5-diphenyl-	CH_2Cl_2	295(4.21)	150-5538-78
$C_{18}H_{16}O_3$			
1-Cyclopentene-1-acetic acid, 2-(1-naphthalenyl)-5-oxo-, methyl ester	EtOH	293(4.1)	42-0580-78
2(5H)-Furanone, 4-(4-methoxyphenyl)-5-methyl-3-phenyl-	MeOH	307(4.28)	35-8247-78
2(5H)-Furanone, 5-(4-methoxyphenyl)-5-methyl-3-phenyl-	MeOH	263(4.11)	35-8247-78
1H-Indene-1-carboxylic acid, 2,3-di-hydro-4-(4-methylbenzoyl)-	EtOH	260(4.26)	94-1153-78
1H-Indene-1-carboxylic acid, 2,3-di-hydro-6-(4-methylbenzoyl)-	EtOH	263(4.27)	94-1153-78
1-Naphthalenecarboxylic acid, 5-benz-oyl-1,2,3,4-tetrahydro-	EtOH	250(4.18)	94-1511-78
2-Propenoic acid, 3-(2,3-diphenyl-oxiranyl)-, methyl ester	EtOH	237s(4.18)	24-3665-78

Compound	Solvent	$\lambda_{max}(\log \epsilon)$	Ref.
$C_{18}H_{16}O_3S$			
2-Anthracenecarboxylic acid, 9,10-di-hydro-9-methyl-9-(methylthio)-10-oxo-, methyl ester	MeOH	272(4.32),313s(3.62)	44-4892-78
2-Anthracenecarboxylic acid, 10-meth-oxy-9-(methylthio)-, methyl ester	MeOH	265(4.89),285(4.83), 331(3.23),352(3.48), 372(3.69),396(3.81), 417(3.82)	44-4892-78
$C_{18}H_{16}O_4$			
9,10-Anthracenedione, 1,3-diethoxy-	EtOH	232s(4.56),240(4.57), 284(4.54),404(3.88)	12-1335-78
2H-1-Benzopyran-2-one, 5,7-dimethoxy-8-(phenylmethyl)-	MeOH	260(3.94),320(4.02)	2-0584-78
2-Butene-1,4-dione, 1,4-bis(4-methoxy-phenyl)-	n.s.g.	241(4.26)	18-1839-78
1H-Indene-1-carboxylic acid, 2,3-dihy-dro-4-(4-methoxybenzoyl)-	EtOH	288(4.24)	94-1153-78
9-Phenanthreneacetic acid, 9,10-dihy-dro-9-hydroxy-10-oxo-, ethyl ester	MeOH	244(4.47),280(3.85), 329(3.42)	12-2259-78
Tricyclo[9.3.1.1^{4,8}]hexadeca-1(15),4,6-8(16),11,13-hexaene-5,14-dicarboxylic acid, (-)-	EtOH	210(4.50),248(4.13), 287(3.38)	78-0871-78
dianion	EtOH	265(3.37),279s(3.03), 290(2.84)	78-0871-78
$C_{18}H_{16}O_5$			
9,10-Anthracenedione, 1,3-diethoxy-5-hydroxy-	EtOH	231(4.42),252s(4.17), 283(4.31),409(3.96)	12-1335-78
9,10-Anthracenedione, 1,3-diethoxy-7-hydroxy-	EtOH	228(4.42),281(4.47), 300(4.32),372(3.89)	12-1335-78
9,10-Anthracenedione, 1,3-diethoxy-8-hydroxy-	EtOH	224(4.49),245s(4.18), 277s(4.29),283(4.32), 431(3.98)	12-1335-78
9,10-Anthracenedione, 1,4-dihydroxy-2-(3-hydroxybutyl)-	MeOH	207(4.21),230s(--), 249(4.56),284(3.95), 318s(--),458s(--), 483(3.94),515s(--), 562(2.87)	24-3823-78
9,10-Anthracenedione, 1,3,6-trimethoxy-8-methyl-	EtOH	218(4.53),278(4.54), 400(3.67)	39-1041-78C
4H-1-Benzopyran-4-one, 2-(3,4,5-tri-methoxyphenyl)-	EtOH	236(4.28),311(4.44)	22-0043-78
2H-Naphtho[1,2-b]pyran-5,6-diol, 2-methyl-, diacetate	MeOH	219(4.9),262(4.8), 271(4.9)	24-1284-78
2H-Naphtho[2,3-b]pyran-5,10-diol, 2-methyl-, diacetate	MeOH	253(4.8),267s(--), 283(4.4),294(4.4), 305(4.4)	24-1284-78
Phenanthrene, 2,7,8-trimethoxy-3,4-(methylenedioxy)-	n.s.g.	267.5(4.78)	39-0739-78C
Phenanthrene, 4,7,8-trimethoxy-2,3-(methylenedioxy)-	n.s.g.	264(4.81)	39-0739-78C
2-Propen-1-one, 1-(4,6-dimethoxy-1,3-benzodioxol-5-yl)-3-phenyl-, (E)-(helilandin A)	ether	288(4.12)	102-1935-78
$C_{18}H_{16}O_6$			
9,10-Anthracenedione, 2-(1,3-dihydroxy-butyl)-1,4-dihydroxy-, (R*,R*)-	MeOH	202(4.35),229s(--), 246(4.59),251s(--), 284(3.97),321s(--), 456s(--),478(3.98),	24-3823-78

Compound	Solvent	$\lambda_{max}(\log \epsilon)$	Ref.
(cont.)		510s(--),560(2.22)	24-3823-78
9,10-Anthracenedione, 2-(1,3-dihydroxy-butyl)-1,4-dihydroxy-, (R*,S*)-	MeOH	203(4.33),231s(--), 247(4.57),253s(--), 283(3.90),323s(--), 458s(--),480(3.90), 512s(--)	24-3823-78
9,10-Anthracenedione, 1,8-dihydroxy-3-(2-hydroxypropyl)-6-methoxy-	MeOH	224(4.65),252(4.37), 267(4.34),291(4.35), 431(4.18),445s(4.13)	105-0561-78
4H-1-Benzopyran-4-one, 7-hydroxy-2-(3,4,5-trimethoxyphenyl)-	EtOH	234(4.25),310(4.36)	22-0043-78
Naphtho[1,2-b]furan-3-carboxylic acid, 5-acetoxy-4-hydroxy-2-methyl-, ethyl ester	dioxan	250(5.08),280s(4.15), 327(3.35),343(3.19)	5-0140-78
$C_{18}H_{16}O_7$			
9,10-Anthracenedione, 5,8-diacetoxy-1,4,4a,9a-tetrahydro-1-hydroxy-, (1α,4aα,9aα)-	MeOH	223(4.39),250(3.85), 315(3.52)	150-3831-78
4H-1-Benzopyran-4-one, 5,7-dihydroxy-2-(3,4,5-trimethoxyphenyl)-	EtOH	270(4.20),328(4.23)	22-0043-78
4H-1-Benzopyran-4-one, 7-hydroxy-3-(3-hydroxy-2,4-dimethoxyphenyl)-8-methoxy-	EtOH	225(4.42),252(4.49), 303s(3.98)	102-1401-78
6-Deoxy-2-epicryptosporin diacetate	MeOH	218s(--),243(4.6), 250(4.6),270(4.4), 280s(--),327(3.5)	24-1284-78
4H-Naphtho[2,3-b]pyran-4-one, 5,10-diacetoxy-2,3-dihydro-3-hydroxy-2-methyl-, cis	MeOH	221(4.7),252(4.8), 286(4.1),296(4.2), 306(4.1)	24-1284-78
$C_{18}H_{16}O_8$			
Cyclopenta[b][1]benzopyran-3a(1H)-acetic acid, 6,8-diacetoxy-2,3-dihydro-1-oxo-	MeOH	217(3.99),284(3.88)	94-2600-78
1,3,5,8-Naphthalenetetrol, tetraacetate	EtOH	228(4.82),291(3.85), 327(3.23)	12-2099-78
$C_{18}H_{17}BrN_2O_3$			
Hydrazinecarboxylic acid, [3-(3-bromophenyl)-3-oxo-1-phenylpropylidene]-, ethyl ester	EtOH	280(4.35)	94-1298-78
$C_{18}H_{17}BrN_2S$			
1H-Imidazole, 1-[2-[[(4-bromophenyl)-methyl]thio]ethyl]-4-phenyl-	EtOH	257(4.28)	4-0307-78
$C_{18}H_{17}BrN_{10}O$			
1H-Pyrazole-4,5-dione, 3-methyl-1-phenyl-4-[[4-(4-bromo-5-methyl-1H-pyrazol-3-yl)azo]-5-methyl-1H-pyrazol-3-yl]hydrazone]	MeOH	225(4.12),245(4.46), 329(4.30),360(3.81), 410(4.21)	103-0313-78
	HOAc	340(3.97),420(3.93)	103-0313-78
	50% HOAc-HCl	335(4.05),410(3.96)	103-0313-78
	70% MeOH-Me₄NOH	349(4.15),480(3.83)	103-0313-78
$C_{18}H_{17}ClFNS$			
1-Propanamine, 3-(2-chloro-3-fluoro-9H-thioxanthen-9-ylidene)-N,N-dimethyl-	MeOH	230(4.45),270(4.11), 285s(3.92),322(3.45)	73-2656-78

Compound	Solvent	$\lambda_{max}(\log \epsilon)$	Ref.
$C_{18}H_{17}ClF_2$ Cyclobuta[3",4"]benzo[1",2":3',4']cy- clobuta[1',2':3,4]cyclobuta[1,2]cy- clooctene, 2-chloro-1,1-difluoro- 1,2,2a,4a,4b,4c,10a,10b,10c,10d- decahydro-	MeOH	252(3.22)	5-1368-78
$C_{18}H_{17}ClN_2O_3S$ Benzenesulfonamide, 4-chloro-N-(4,5-di- hydro-1'-methylspiro[furan-3(2H),2'- [2H]indol]-3'(1'H)-ylidene)-	EtOH	208(4.51),228(4.37), 281(3.70),468(3.54)	39-1471-78C
$C_{18}H_{17}ClN_2O_4$ Propanedioic acid, [3-(4-chlorophenyl)- 2-cyano-2-propenyl](2-cyanoethyl)-	EtOH	222(4.28),274(4.38)	2-0924B-78
$C_{18}H_{17}ClO_6$ Benzo[1,2-b:4,5-b']difuran-3,7-dicarb- oxylic acid, 4-chloro-2,6-dimethyl-, diethyl ester	MeOH	229(4.58),292(4.35), 303(4.37)	4-0043-78
$C_{18}H_{17}Cl_2F_2NS$ 1-Propanamine, 3-(2-chloro-3,6-diflu- oro-9H-thioxanthen-9-ylidene)-N,N- dimethyl-, hydrochloride	MeOH	270(4.16),295s(3.82), 320s(3.47)	73-2656-78
$C_{18}H_{17}F$ Cyclobuta[3",4"]benzo[1",2":3',4']cy- clobuta[1',2':3,4]cyclobuta[1,2]cy- clooctene, 1-fluoro-2a,4a,4b,4c,10a- 10b,10c,10d-octahydro-	MeOH	252(3.20)	5-1368-78
1,3,5,7,9,11,13,15,17-Cyclooctadeca- nonaene	MeOH	338(4.75),348(4.94), 367(5.42),408(3.41), 449(3.69)	5-1368-78
$C_{18}H_{17}FN_2OS_2$ 4-Thiazolidinone, 3-ethyl-5-[(1-ethyl- 6-fluoro-2(1H)-quinolinylidene)eth- ylidene]-2-thioxo-	EtOH	542(4.85),572(4.79)	103-0060-78
$C_{18}H_{17}N$ 2H-Pyrrole, 2,2-dimethyl-3,5-diphenyl-	EtOH	248(4.25)	88-4511-78
$C_{18}H_{17}NO$ Ethanone, 1-[1-methyl-3-(phenylmethyl)- 1H-indol-2-yl]-	EtOH	209(4.40),240(4.19), 309(4.21)	39-1471-78C
1H-Inden-1-one, 2-[[4-(dimethylamino)- phenyl]methylene]-2,3-dihydro-, cation	H_2SO_4	398(4.49)	65-1644-78
Isoxazole, 3-phenyl-5-(2,4,6-trimethyl- phenyl)-	HOAc HOAc–H_2SO_4	253(3.85) 295(3.87)	103-0264-78 103-0264-78
Isoxazole, 5-phenyl-3-(2,4,6-trimethyl- phenyl)-	HOAc HOAc–H_2SO_4	265(4.41) 300(4.41)	103-0264-78 103-0264-78
4H,8H-Pyrido[3,2,1-de]phenanthridin- 8-one, 5,6-dihydro-4,6-dimethyl-, cis?	EtOH	230s(4.59),235(4.61), 240s(4.55),255s(4.21), 264(4.31),305s(3.71), 315s(3.72),326(3.86), 340(3.81)	33-1246-78
trans	EtOH	230s(4.59),235(4.63), 240s(4.58),253s(4.20), 263(4.31),303s(3.71),	33-1246-78

Compound	Solvent	$\lambda_{max}(\log \epsilon)$	Ref.
(cont.)		315s(3.74),326(3.88), 340(3.83)	33-1246-78
$C_{18}H_{17}NOS$ 4H-Thiopyran-4-one, 3-[(diphenylamino)-methylene]tetrahydro-, (E)-	EtOH	232(3.99),282(3.99), 354(4.30)	4-0181-78
$C_{18}H_{17}NOSe$ 1(2H)-Naphthaleneselone, 3,4-dihydro-2-[[(4-methoxyphenyl)amino]methyl-ene]-	benzene	525(4.28)	104-0315-78
$C_{18}H_{17}NO_2$ 1H-Isoindol-1-one, 3-ethyl-2,3-dihydro-3-hydroxy-2-(2-phenylethenyl)-	THF	228(4.27),266(4.17), 320(4.34)	40-0404-78
$C_{18}H_{17}NO_3$ Acetamide, N-(10-acetoxy-9,10-dihydro-9-phenanthrenyl)-, trans	EtOH	220(4.72),273(4.30), 285(4.08)	44-0397-78
4H-1-Benzopyran-4-one, 6-[(dimethylami-no)methyl]-5-hydroxy-2-phenyl-	n.s.g.	270(4.42),285(4.33), 340(3.74)	103-0497-78
4H-1-Benzopyran-4-one, 8-[(dimethylami-no)methyl]-5-hydroxy-2-phenyl-	n.s.g.	270(4.42),285(4.33), 340(3.74)	103-0497-78
4H-1-Benzopyran-4-one, 8-[(dimethylami-no)methyl]-7-hydroxy-2-phenyl-	n.s.g.	260(4.43),310(4.24)	103-0497-78
2-Butenoic acid, 2-cyano-3-(6-methoxy-2-naphthalenyl)-, ethyl ester, (E)-	EtOH	277(4.26)	95-0146-78
(Z)-	EtOH	273(4.13)	95-0146-78
$C_{18}H_{17}NO_4$ Benzo[5,6]cycloocta[1,2-f]-1,3-benzo-dioxol-5,11-imin-8-ol, 5,6,11,12-tetrahydro-9-methoxy-, (±)-	EtOH EtOH-NaOH	295(3.86) 300(3.88)	78-0241-78 78-0241-78
Codeinone, 10-oxo-	EtOH	237(3.25),287(3.05), 324s(2.67)	23-2467-78
Cyclopenta[3,4]cyclobuta[1,2-b]pyrrole-3a(1H)-carboxylic acid, 2,3,3b,4,6a-6b-hexahydro-2,3-dioxo-6b-phenyl-, ethyl ester, (3aα,3bα,6aα,6bα)-	EtOH	252(3.60)	142-0153-78A
$C_{18}H_{17}NO_7$ 8-Azabicyclo[3.2.1]octane-3-acetalde-hyde, 3-acetoxy-8-(benzoyloxy)-6,7-dioxo-, endo	$CHCl_3$	497(1.40)	44-4765-78
$C_{18}H_{17}NO_9$ 3,4-Furandicarboxylic acid, 2,3-dihy-dro-5-(3-methoxy-3-oxo-1-propenyl)-2-(4-nitrophenyl)-, dimethyl ester, cis	MeCN	262(4.28)	24-3665-78
3,4-Furandicarboxylic acid, 2,5-dihy-dro-2-(3-methoxy-3-oxo-1-propenyl)-5-(4-nitrophenyl)-, dimethyl ester, trans	MeCN	262(4.16)	24-3665-78
$C_{18}H_{17}NS$ Butanenitrile, 3,3-dimethyl-2-(2-meth-yl-3H-naphtho[1,8-bc]thien-3-ylidene)-	MeOH	252(4.282),271(4.265), 402(4.155)	24-1824-78
$C_{18}H_{17}N_3$ Pyrazinamine, N,N-dimethyl-3,5-diphenyl-	EtOH	231(4.21),309(4.22),	94-1322-78

Compound	Solvent	$\lambda_{max}(\log \epsilon)$	Ref.
(cont.)		369(3.96)	94-1322-78
Pyrazinamine, N,N-dimethyl-3,6-diphenyl-	EtOH	236(4.26),270(4.18), 373(4.07)	94-1322-78
Pyrazinamine, N,N-dimethyl-5,6-diphenyl-	EtOH	228(4.23),295(4.24), 364(3.94)	94-1322-78
$C_{18}H_{17}N_3O$			
3H-Pyrazol-3-one, 2,4-dihydro-5-methyl-4-[(methylphenylamino)methylene]-2-phenyl-	$C_6H_{11}Me$	257s(4.13),260(4.15), 322(4.41),372s(3.72)	48-0521-78
	PrOH	256(4.06),310(4.22)	48-0521-78
	PrCN	262(4.09),318(4.31), 365s(3.97)	48-0521-78
3H-Pyrazol-3-one, 2,4-dihydro-5-methyl-2-phenyl-4-[[(phenylmethyl)amino]-methylene]-	$C_6H_{11}Me$	258(4.28),300(4.45), 338(3.46)	48-0521-78
	PrOH	258(4.21),298(4.39), 333(3.84)	48-0521-78
	PrCN	262(4.22),297(4.41), 328s(3.61)	48-0521-78
$C_{18}H_{17}N_3OS$			
1H-Imidazol-1-amine, N-[(2-methoxyphenyl)methylene]-2-(methylthio)-4-phenyl-	MeOH	278(4.44),349(4.24)	18-1846-78
1H-Imidazol-1-amine, N-[(2-methoxyphenyl)methylene]-2-(methylthio)-5-phenyl-	MeOH	267(4.42),332(4.13)	18-1846-78
1H-Imidazol-1-amine, N-[(4-methoxyphenyl)methylene]-2-(methylthio)-4-phenyl-	MeOH	291(4.46),342(4.34)	18-1846-78
1H-Imidazol-1-amine, N-[(4-methoxyphenyl)methylene]-2-(methylthio)-5-phenyl-	MeOH	281(4.44),306(4.38)	18-1846-78
1H-Imidazol-1-amine, 4-(4-methoxyphenyl)-2-(methylthio)-N-(phenylmethylene)-	MeOH	294(4.56),352(4.10)	18-1846-78
4-Thiazolidinone, 5-[[4-(dimethylamino)phenyl]methylene]-2-(phenylimino)-	MeOH	240(4.14),423(4.54)	104-0997-78
	dioxan	243(4.22),403(4.56)	104-0997-78
	70% dioxan	241(4.16),420(4.53)	104-0997-78
$C_{18}H_{17}N_3O_2$			
1H-Imidazole-4-carboxylic acid, 1-phenyl-5-(phenylamino)-, ethyl ester	EtOH	210(4.32),248(4.38), 291(3.86)	118-0741B-78
3H-Pyrazol-3-one, 2,4-dihydro-4-[[(4-methoxyphenyl)amino]methylene]-5-methyl-2-phenyl-	$C_6H_{11}Me$	254s(4.34),257(4.34), 348(4.44)	48-0521-78
	PrOH	254(4.32),360(4.43)	48-0521-78
	PrCN	257(4.33),351(4.42)	48-0521-78
Pyrimido[4,5-b]quinoline-2,4(3H,6H)-dione, 7,8,9,10-tetrahydro-10-methyl-3-phenyl-	EtOH	214(4.27),274(4.06), 377(4.02)	88-3469-78
$C_{18}H_{17}N_3O_2S$			
Phenol, 2-[1-[[(4-methoxyphenyl)methylene]amino]-2-(methylthio)-1H-imidazol-4-yl]-	MeOH	258(4.19),300(4.43), 352(4.34)	18-1846-78
Pyrazolo[4,3-d][1]benzazepine, 2,4,5,6-tetrahydro-6-[(4-methylphenyl)sulfonyl]-	n.s.g.	237(4.25)	39-0862-78C
$C_{18}H_{17}N_3O_5$			
Hydrazinecarboxylic acid, [3-(3-nitrophenyl)-3-oxo-1-phenylpropylidene]-, ethyl ester	EtOH	276(4.36),284s(4.34)	94-1298-78
$C_{18}H_{17}N_3O_5S$			
2(5H)-Furanone, 3,4-dimethoxy-5-(phen-	dioxan	211(4.28),287s(3.88),	33-1033-78

Compound	Solvent	$\lambda_{max}(\log \epsilon)$	Ref.
ylthio)-, (4-nitrophenyl)hydrazone (cont.)		400(4.50)	33-1033-78
$C_{18}H_{17}N_3O_6$			
2(5H)-Furanone, 3,4-dimethoxy-5-phen-oxy-, (4-nitrophenyl)hydrazone	dioxan	273(3.93),394(4.37)	33-1033-78
$C_{18}H_{17}N_5O_7$			
3-Isoxazolecarboxylic acid, 4-[[3-(eth-oxycarbonyl)-2,5-dihydro-5-oxo-1-phenyl-1H-pyrazol-4-yl]hydrazono]-4,5-dihydro-5-oxo-, ethyl ester	MeOH	255(4.20),515(4.34)	142-0199-78B
$C_{18}H_{17}OP$			
Phosphorin, 1,1-dihydro-1-methoxy-1,4-diphenyl-	EtOH	303(4.31),352s(3.78)	89-0528-78
$C_{18}H_{18}$			
Anthracene, 9-(1,1-dimethylethyl)-	C_6H_{12}	372(3.82)	61-1068-78
Benzobicyclo[2.1.1]hex-2-ene, 2,2-di-methyl-5-phenyl-, endo	MeOH	256s(2.77),262(2.90), 269(2.98),275(2.94)	35-3819-78
exo	MeOH	254s(3.00),260(3.09), 267(3.15),274(3.13)	35-3819-78
[2,2,2](1,2,4)(1,2,5)Cyclophane	isooctane	235s(4.00),286(2.81), 294(2.72)	44-1041-78
[2,2,2](1,2,4)(1,3,5)Cyclophane	isooctane	225s(4.00),300(2.58)	44-1041-78
Phenanthrene, 3,6,9,10-tetramethyl-	hexane	228(4.41),251(4.67), 258(4.77),277(4.28), 282s(--),290s(--), 303(4.07),325(2.78), 341(2.93),358(2.95)	104-0924-78
$C_{18}H_{18}BrNO_3$			
1H-2-Pyrindine-4-carboxylic acid, 3-(4-bromophenyl)-2,4a,5,7a-tetrahydro-2-methyl-1-oxo-, ethyl ester, cis	EtOH	288(3.99)	142-0153-78A
$C_{18}H_{18}BrN_5$			
1H-Pyrazol-5-amine, 4-[(2-bromo-4-meth-ylphenyl)azo]-3-methyl-1-(4-methyl-phenyl)-	$CHCl_3$	390(4.78)	103-0885-78
$C_{18}H_{18}Br_2$			
Tricyclo[9.3.1.1^{4,8}]hexadeca-1(15),4,6-8(16),11,13-hexaene, 5,14-bis(bromo-methyl)-	isooctane	224(4.55),250s(4.29), 285s(3.28)	78-0871-78
$C_{18}H_{18}ClN_5OS$			
Thiazolidine, 3-benzoyl-4-[(6-chloro-9H-purin-9-yl)methyl]-2,2-dimethyl-, (R)-	EtOH EtOH-HCl EtOH-NaOH	265(3.79) 264.8(3.79) 265(3.80)	23-0326-78 23-0326-78 23-0326-78
$C_{18}H_{18}F_2O_2$			
2-Pentanone, 3,5-difluoro-4-hydroxy-1-phenyl-4-(phenylmethyl)-	C_6H_{12}	258(3.87)	22-0129-78
$C_{18}H_{18}FeO_3$			
Iron, (η^5-2,4-cyclopentadien-1-yl)-[(1,2,3,3a,8a-η)-5-(ethoxycarbonyl)-1,6,7,8-tetrahydro-8-oxo-1-azulenyl]-	EtOH	269(4.21),294(4.06)	128-0273-78

Compound	Solvent	$\lambda_{max}(\log \epsilon)$	Ref.
$C_{18}H_{18}N_2$			
Dibenzo[a,e]cyclooctene-5,12:6,11-di-imine, 5,6,11,12-tetrahydro-13,14-dimethyl-, $(5\alpha,6\beta,11\beta,12\alpha)$-	hexane	281(3.65),285(3.89)	33-0444-78
4,11-Etheno-5,10-imino-2H-benzo[5,6]-cycloocta[1,2-c]pyrrole, 4,5,10,11-tetrahydro-2,14-dimethyl-, $(4\alpha,5\alpha,10\alpha,11\alpha)$-	hexane	283(3.17),290(3.13)	33-0444-78
1H-Imidazole, 1-ethyl-4-methyl-2,5-di-phenyl-	MeOH	207(4.21),278(4.12)	33-0286-78
1H-Imidazole, 1-ethyl-5-methyl-2,4-di-phenyl-	MeOH	207(4.36),217s(4.16), 266(4.23)	33-0286-78
2H-Pyrrol-5-amine, 2,2-dimethyl-3,4-di-phenyl-	EtOH	252(4.28)	88-4511-78
$C_{18}H_{18}N_2O_2$			
Benzenamine, N,N-dimethyl-4-[4-(4-nitro-phenyl)-1,3-butadienyl]-, (E,E)-	EtOH	442(4.67)	104-0939-78
Ethanone, 2-[3-ethyl-5-(hydroxymethyl)-4-pyridinyl]-1-(1H-indol-2-yl)-	EtOH	224(4.10),236(4.01), 273(3.68),310(4.27)	78-1363-78
1H-Pyrido[3,2-b]indole-2,9-dione, 3,4,6,7,8,9-hexahydro-5-(phenyl-methyl)-	EtOH	255(4.27),310(3.52)	94-3080-78
Tricyclo[9.3.1.1^{4,8}]hexadeca-1(15),4,6-8(16),11,13-hexaene-5,14-dicarboxam-ide	EtOH	214(4.58),245s(4.04), 279s(3.08)	78-0871-78
$C_{18}H_{18}N_2O_3$			
2-Propen-1-one, 1-(1,3-benzodioxol-5-yl)-3-(2,2-dimethylhydrazino)-3-phenyl-	60% DMF	258(4.33),363(4.48)	4-0385-78
$C_{18}H_{18}N_2O_4$			
Acetamide, N-[[6-(acetylamino)-1,3-benzodioxol-5-yl]methyl]-N-phenyl-	MeOH	246s(4.3),292(4.0)	5-0608-78
5,6-Diazatetracyclo[8.6.0.0^{2,9}.0^{3,8}]-hexadeca-3,6,11,13,15-pentaene-4,7-dicarboxylic acid, dimethyl ester methanol adduct	dioxan	255(3.91)	5-1368-78
5,6-Diazatricyclo[8.6.0.0^{2,9}]hexadeca-3,6,8,11,13,15-hexaene-4,7-dicarbox-ylic acid, dimethyl ester	dioxan MeOH	275(3.48) 260(3.67),380(3.48)	5-1368-78 5-1368-78
Propanedioic acid, (2-cyanoethyl)(2-cy-ano-3-phenyl-2-propenyl)-, monoethyl ester	EtOH	274(4.14)	2-0924B-78
1H-Pyrazole-4,5-dicarboxylic acid, 1-(4-pentynyl)-3-phenyl-, dimethyl ester	EtOH	232(4.36)	44-1664-78
2,4,6(1H,3H,5H)-Pyrimidinetrione, 5-[(2-ethyl-5-methyl-3-benzofuranyl)-methylene]-1,3-dimethyl-	MeOH	254(4.08),382(3.69)	83-0714-78
$C_{18}H_{18}N_2O_5$			
Glycine, N-[4-[methyl[(phenylmethoxy)-carbonyl]amino]benzoyl]-	EtOH	260(4.25)	87-1162-78
$C_{18}H_{18}N_2O_6$			
Phenol, 4,4'-(1,2-diethyl-1,2-ethene-diyl)bis[2-nitro-, cis	MeOH	222(4.3),269(4.2), 368(3.6)	83-0184-78
trans	MeOH	265s(4.0),363(3.5)	83-0184-78

Compound	Solvent	$\lambda_{max}(\log \epsilon)$	Ref.
$C_{18}H_{18}N_2S$			
Benzenamine, N-[2-(1,1-dimethylethyl)-4H-3,1-benzothiazin-4-ylidene]-	EtOH	215(4.6),242(4.1), 260(4.1),312(3.8)	139-0209-78B
Benzenamine, 4-[[4-(2,5-dimethyl-1H-pyrrol-1-yl)phenyl]thio]-	EtOH	208(4.48),265(4.30)	111-0093-78
Benzenamine, N-(4-phenyl-3-propyl-2(3H)-thiazolylidene)-, monohydrochloride	EtOH	230s(4.24),265(4.01), 290(4.05)	94-3017-78
1H-Imidazole, 4-phenyl-1-[2-[(phenyl-methyl)thio]ethyl]-	EtOH	257(4.26)	4-0307-78
$C_{18}H_{18}N_3O_3$			
2,1-Benzisoxazolium, 3-[2-[4-(dimethyl-amino)phenyl]ethenyl]-1-methyl-5-nitro-, perchlorate	MeCN	635(4.82)	44-1233-78
$C_{18}H_{18}N_4$			
2,4'-Bi-1H-benzimidazole, 5,5',6,6'-tetramethyl-	EtOH	220(4.45),250(3.83), 290(4.20)	12-2675-78
$C_{18}H_{18}N_4O$			
Benzaldehyde, (4-ethoxy-1-phthalazin-yl)methylhydrazone	dioxan	295s(4.15),345(4.19)	103-0436-78
2H-1,5-Benzodiazepin-2-one, 1,3-dihydro-4,7,8-trimethyl-3-(phenylazo)-	HOAc	210(4.42),235(4.36), 300(3.68),385(4.51)	103-1270-78
Pyrimido[4,5-b]quinolin-2(3H)-one, 4,6,7,8,9,10-hexahydro-4-imino-10-methyl-3-phenyl-	EtOH	214(4.28),280(4.01), 387(4.03)	88-3469-78
Urea, (3-cyano-5,6,7,8-tetrahydro-1-methyl-2(1H)-quinolinylidene)phenyl-	EtOH	235(4.15),294(4.30), 384(3.90)	88-3469-78
$C_{18}H_{18}N_4OS$			
[1,2,3]Thiadiazolo[5,1-e][1,2,3]thia-diazole-7-S^{IV}, 1,6-dihydro-1-(4-meth-oxyphenyl)-3,4-dimethyl-6-phenyl-	C_6H_{12}	246(4.27),306(4.23), 500(4.32)	39-0195-78C
$C_{18}H_{18}N_4O_2$			
8-Azabicyclo[3.2.1]oct-3-ene-6-carbo-nitrile, 8,8'-(1,2-ethanediyl)bis[2-oxo-	CHCl$_3$	246(2.92),253(2.74), 357(2.13)	150-1182-78
$C_{18}H_{18}N_6O_6$			
1H-Pyrrole-2-carboxylic acid, 1-methyl-4-[[[1-methyl-4-[[(1-methyl-4-nitro-1H-pyrrol-2-yl)carbonyl]amino]-1H-pyrrol-2-yl]carbonyl]amino]-	DMF	297(4.55)	78-2389-78
$C_{18}H_{18}N_8O_6$			
1H-Purine-2,6-dione, 7,7'-(1,4-dioxo-1,4-butanediyl)bis[3,7-dihydro-1,3-dimethyl-	CH$_2$Cl$_2$	300(4.12)	36-1045-78
$C_{18}H_{18}O$			
Ethanone, 1-tricyclo[9.3.1.1^{4,8}]hexa-deca-1(15),4,6,8(16),11,13-hexaen-5-yl-	EtOH	207(4.50),223s(4.26), 260(3.90),299s(3.13), 345s(1.76)	78-0871-78
[5.1]Metabenzenophan-12-one	MeCN	246s(4.40),255(4.53), 264(4.41),284(3.61)	24-2547-78
$C_{18}H_{18}O_2$			
Acenaphtho[1,2-b][1,4]dioxin, 8,9-di-hydro-8,8,9,9-tetramethyl-	C_6H_{12}	230(4.70),306(3.94), 318(3.94),450(2.84)	35-6276-78

Compound	Solvent	$\lambda_{max}(\log \epsilon)$	Ref.
Anthracene, 9,10-diethoxy-	C_6H_{12}	383(3.95)	61-1068-78
Benzene, 1,1'-[1,2-ethanediylbis(oxy-2,1-ethenediyl)bis-	EtOH	261(4.57),267s(4.53), 287s(3.90),295s(3.67)	24-3624-78
geometric isomer	EtOH	258(4.56),269s(4.46), 290s(3.59)	24-3624-78
Benzoic acid, 2-[2-(2,4,6-trimethyl-phenyl)ethenyl]-, (E)-	MeOH	283(4.21)	35-3819-78
[1,1'-Biphenyl]-2,2'-diol, 5,5'-di-2-propenyl-	n.s.g.	293(3.90)	100-0442-78
	NaOH	320(4.03)	100-0442-78
[1,1'-Biphenyl]-2,4'-diol, 3',5-di-2-propenyl- (honokiol)	n.s.g.	295(3.85)	100-0442-78
	NaOH	305(4.10)	100-0442-78
2,5-Cyclohexadien-1-one, 4-[3-(4-meth-oxyphenyl)-2-propenylidene]-	MeOH	273(4.20),442(4.32)	18-3612-78
2,5-Dioxabicyclo[4.2.0]octane, 7,8-di-phenyl-, 7-syn-8-anti	MeCN	244s(2.34),249s(2.48), 254s(2.60),256s(2.62), 259(2.67),262(2.67), 296(2.51)	24-3624-78
[3.3]Metacyclophane-6,9-quinone	MeCN	255(4.1),317(3.5)	138-1319-78
Spiro[cyclobutane-1,7'(8'H)-[6H]diben-zo[f,h][1,5]dioxonin]	EtOH	224(4.21),244(3.92), 282(3.64)	56-1913-78
Tetracyclo[9.6.0.0^{2,10}.0^{3,5}]heptadeca-6,8,12,14,16-pentaene-4-carboxylic acid, endo	ether	228s(3.90),259s(3.30)	24-0107-78
exo	ether	228s(3.88),259s(3.30)	24-0107-78
Tetracyclo[9.6.0.0^{2,10}.0^{5,7}]heptadeca-3,8,12,14,16-pentaene-6-carboxylic acid, exo	ether	250(3.30)	24-0107-78
Tricyclo[9.3.1.1^{4,8}]hexadeca-1(15),4,6-8(16),11,13-hexaene-5-carboxylic acid, methyl ester	EtOH	205(4.57),220s(4.42), 249s(3.94),291(3.07)	78-0871-78
$C_{18}H_{18}O_2S$			
2-Butenoic acid, 2-(ethylthio)-4,4-di-phenyl-	EtOH	210(4.61),234(4.06), 300(4.18)	56-2455-78
	NaHCO$_3$	234(4.19),302(4.01)	56-2455-78
2H-Phenanthro[1,2-c]thiopyran-1(4H)-one, 5,6,11,12-tetrahydro-8-methoxy-	EtOH	400(4.27)	39-0576-78C
$C_{18}H_{18}O_3$			
1(2H)-Naphthalenone, 3,4-dihydro-7-methoxy-3-(4-methoxyphenyl)-, (R)-	CHCl$_3$	255(4.02),278s(3.38), 285s(3.33),322(3.52)	12-1011-78
Phenanthrene, 6-ethoxy-1,2-dimethoxy-	n.s.g.	261(4.75),268(4.72)	39-0739-78C
Phenanthrene, 6-ethoxy-2,3-dimethoxy-	n.s.g.	255(4.76)	39-0739-78C
$C_{18}H_{18}O_3S_3$			
3(2H)-Thiophenone, 2,4-bis[[5-(ethyl-thio)-2-furanyl]methylene]dihydro-	EtOH	212(4.10),276(4.12), 407(4.26),472(4.44)	133-0189-78
$C_{18}H_{18}O_4$			
Benzenepropanoic acid, β-hydroxy-α-(2-oxo-2-phenylethyl)-, methyl ester	EtOH	243(4.11),278(3.09)	24-3665-78
4H-1-Benzopyran, 2-(3,4-dimethoxyphen-yl)-7-methoxy-	MeOH	249(4.19),280(4.00)	18-1175-78
1,4-Butanedione, 1,4-bis(2-hydroxy-5-methylphenyl)-	EtOH	257(4.33),340(4.90)	102-1439-78
2,5-Cyclohexadien-1-one, 2,6-dimethoxy-4-[3-(4-methoxyphenyl)-2-propenyli-dene]-	MeOH	245(4.01),280s(3.80), 300(3.80),456(4.62)	18-3612-78
Spiro[2H-1,5-benzodioxepin-3(4H),5'-[1,3]dioxane], 2-phenyl-	CHCl$_3$	276(3.32),282s(--)	56-1913-78

Compound	Solvent	$\lambda_{max}(\log \epsilon)$	Ref.
$C_{18}H_{18}O_5$			
1,3-Benzodioxole, 6-[2-(2,3-dimethoxy-phenyl)ethenyl]-4-methoxy-	n.s.g.	344(4.20)	39-0739-78C
1,3-Benzodioxole, 4-[2-(3,4,5-trimeth-oxyphenyl)ethenyl]-	n.s.g.	310(4.30)	39-0739-78C
Cyclohepta[b]furo[3,2-g][1]benzopyran-5(6H)-one, 7,8,9,10-tetrahydro-4,12-dimethoxy-	EtOH	248(4.70),282(3.49), 331(3.62)	39-0726-78C
2-Propen-1-one, 1-(2-hydroxy-3,4,6-tri-methoxyphenyl)-3-phenyl-	EtOH	340(4.17)	102-1363-78
	EtOH-AlCl$_3$	370(--)	102-1363-78
$C_{18}H_{18}O_5Se$			
2-Oxaspiro[4.5]dec-6-ene-3-carboxylic acid, 1,8-dioxo-9-(phenylseleno)-, ethyl ester	MeOH	218(4.1)	24-1944-78
$C_{18}H_{18}O_6$			
Benzo[1,2-b:4,5-b']difuran-3,7-dicarb-oxylic acid, 2,6-dimethyl-, diethyl ester	MeOH	228(4.47),285(4.23), 292(4.28),297(4.30), 303(4.38)	4-0043-78
Benzoic acid, 3,5-dimethoxy-2-(4-meth-oxybenzoyl)-, methyl ester	EtOH	285(4.24),310(3.98)	12-1335-78
4H-1-Benzopyran-4-one, 1,3-dihydro-2-(4-hydroxyphenyl)-5,6,7-trimethoxy-	EtOH	275(4.12),320s(3.54)	102-1807-78
4H-1-Benzopyran-4-one, 2,3-dihydro-7-hydroxy-5,6,8-trimethoxy-2-phenyl-	EtOH	282(4.7)	102-0587-78
	EtOH-NaOH	296(4.6)	102-0587-78
4H-1-Benzopyran-4-one, 7-ethoxy-2,3-di-hydro-5,6-dihydroxy-8-methoxy-2-phen-yl-	EtOH	292(4.1)	42-1198-78
2-Propen-1-one, 1-(2,4-dihydroxy-3,5,6-trimethoxyphenyl)-3-phenyl-	EtOH	328(4.6)	42-1198-78
	EtOH-NaOH	288(--)	42-1198-78
	EtOH-NaOAc	328(--)	42-1198-78
2-Propen-1-one, 3-(4-hydroxyphenyl)-1-(2-hydroxy-3,4,6-trimethoxyphenyl)-	EtOH	239s(4.05),371(4.50)	102-1363-78
	EtOH-NaOMe	431(--)	102-1363-78
	EtOH-AlCl$_3$	411(--)	102-1363-78
2-Propen-1-one, 3-(4-hydroxyphenyl)-1-(6-hydroxy-2,3,4-trimethoxyphenyl)-	EtOH	235(3.95),370(4.23)	102-1807-78
$C_{18}H_{18}O_7$			
Benzoic acid, 2-hydroxy-, oxy-di-2,1-ethanediyl ester	MeOH	238(4.26),306(3.93)	121-0661-78
2,5-Cyclohexadiene-1,4-dione, 2-(3,4-dihydro-7-hydroxy-8-methoxy-2H-1-benzopyran-3-yl)-3,5-dimethoxy-, (S)-	EtOH	225(4.30),287(4.23), 360(2.85)	102-1423-78
2,5-Cyclohexadiene-1,4-dione, 5-(3,4-dihydro-7-hydroxy-8-methoxy-2H-1-benzopyran-3-yl)-2,3-dimethoxy-	MeOH	266(4.10),385(3.06)	142-0085-78B
3,4-Furandicarboxylic acid, 2,5-dihy-dro-5-(3-methoxy-3-oxo-1-propenyl)-2-phenyl-, dimethyl ester, trans	n.s.g.	228(4.18)(end abs.?)	24-3665-78
Rutaretin diacetate	MeOH	225(3.65),245(3.35), 290(3.40),320(3.76)	102-2135-78
$C_{18}H_{18}S_3$			
2,11,20-Trithia[3.3.3](1,2,4)cyclophane	isooctane	271(3.50)	44-1041-78
2,11,20-Trithia[3.3.3](1,2,4)(1,2,5)-cyclophane	isooctane	256(3.77)	44-1041-78
2,11,20-Trithia[3.3.3](1,2,4)(1,3,5)-cyclophane	isooctane	255(3.70)	44-1041-78

Compound	Solvent	λ_{max}(log ϵ)	Ref.
$C_{18}H_{19}Br$			
Benzene, 2-[2-[2-(bromomethyl)phenyl]-ethenyl]-1,3,5-trimethyl-, (E)-	MeOH	235s(4.23),290(4.21)	35-3819-78
$C_{18}H_{19}BrN_2O_3$			
Acetamide, N-[[2-(acetylamino)-5-bromo-4-methoxyphenyl]methyl]-N-phenyl-	MeOH	217(4.5),253(4.1), 285(3.6)	5-0608-78
$C_{18}H_{19}ClN_2OS$			
Dibenzo[b,f]thiepin, 8-chloro-10,11-di-hydro-10-(4-hydroxypiperazino)-, maleate	MeOH	266(3.98)	73-3092-78
$C_{18}H_{19}ClN_4O$			
5H-Dibenzo[b,e]-1,4-diazepin-2-ol, 8-chloro-11-(4-methylpiperazino)-	MeOH	258(4.26),310(4.21), 343s(3.33)	73-0309-78
$C_{18}H_{19}N$			
3-Benzazocine, 1,2,3,4-tetrahydro-3-methyl-6-phenyl-	MeOH	253(4.10)	83-1029-78
1H-Phenanthro[9,10-b]azirine, 1-butyl-1a,9b-dihydro-	C_6H_{12}	239(3.82),271(3.99), 277(4.01),281(4.00), 288(3.68),295(3.67), 306(3.58)	44-0397-78
$(C_{18}H_{19}N)_n$			
9H-Carbazole, 9-ethenyl-3-(1-methyl-propyl)-, (S)-, homopolymer	CH_2Cl_2	230(4.60),252s(4.11), 265(4.18),290s(3.95), 298(4.08),325s(3.38), 337(3.54),350(3.60)	126-1929-78
$C_{18}H_{19}NO$			
9(10H)-Acridinone, 2-(1,1-dimethyleth-yl)-10-methyl-	EtOH	256(4.73),296(3.45), 307s(3.11),388(3.89), 406(3.96)	35-5604-78
2-Propenenitrile, 3-[3-(bicyclo[2.2.1]-hept-5-en-2-ylcarbonyl)bicyclo[2.2.1]-hept-5-en-2-yl]-, [1α,2α(Z),3α-(1R*,2S*,4R),4α]-	n.s.g.	222(4.08)	33-1427-78
$C_{18}H_{19}NO_2$			
9(10H)-Acridinone, 10-butyl-2-methoxy-	MeOH	300(3.243)	4-0149-78
7H-Indeno[4,5-h]isoquinolin-7-one, 5,6,6a,8,10,11-hexahydro-2-methoxy-6a-methyl-	MeOH	298(4.45)	44-0966-78
2-Propen-1-one, 3-[4-(dimethylamino)-phenyl]-1-(4-methoxyphenyl)-	hexane	230s(4.17),250s(4.05), 320s(4.33),355s(4.55)	65-0340-78
$C_{18}H_{19}NO_3$			
4-Isoquinolinecarboxylic acid, 1,2,4a-5,6,8a-hexahydro-1-oxo-3-phenyl-, ethyl ester	EtOH EtOH-KOH	287(4.00) 330(3.92)	142-0161-78A 142-0161-78A
Pyridinium, 1-(2-ethoxy-1-methyl-2-oxo-ethyl)-4-[2-(4-hydroxyphenyl)ethen-yl]-, hydroxide, inner salt, (E)-hydrobromide	EtOH acetone CHCl₃ MeOH	520(3.22) 600(3.35) 625(3.72) 403(4.06)	56-1265-78 56-1265-78 56-1265-78 56-1265-78
Pyridinium, 1-(3-ethoxy-3-oxopropyl)-4-[2-(4-hydroxyphenyl)ethenyl]-, hydroxide, inner salt, (E)-hydrobromide	EtOH acetone CHCl₃ MeOH	520(3.43) 600(3.45) 625(3.74) 394(4.54)	56-1265-78 56-1265-78 56-1265-78 56-1265-78

Compound	Solvent	$\lambda_{max}(\log \epsilon)$	Ref.
Pyridinium, 4-[2-(4-hydroxyphenyl)eth- enyl]-1-(2-oxo-2-propxyethyl)-, hydroxide, inner salt, (E)- hydrochloride	EtOH acetone CHCl$_3$ MeOH	520(3.03) 600(3.33) 625(3.71) 401(4.68)	56-1265-78 56-1265-78 56-1265-78 56-1265-78
1H-2-Pyrindine-4-carboxylic acid, 2,4a,5,7a-tetrahydro-2-methyl-1- oxo-3-phenyl-, ethyl ester	EtOH	286(4.00)	142-0153-78A
1H-2-Pyrindine-4-carboxylic acid, 2,5,6,7-tetrahydro-2-methyl-1- oxo-3-phenyl-, ethyl ester	EtOH	273(4.02),305(3.86)	142-0161-78A
$C_{18}H_{19}NO_4$			
4H-Dibenzo[de,g]quinoline-2,10,11-triol, 5,6,6a,7-tetrahydro-1-methoxy-6-meth- yl-, (S)- (glaufine)	n.s.g.	217(4.60),274(4.21), 308(3.84)	105-0699-78
1H,4H-3a,8b-Ethenocyclopent[b]indole- 9,10-dicarboxylic acid, 2,3-dihydro- 4-methyl-, dimethyl ester	EtOH	249(3.68),304(3.18), 438(2.88)	88-2979-78
6,8-Indolizinedicarboxylic acid, 1,2,3,8a-tetrahydro-8a-phenyl-, dimethyl ester	MeOH	207(4.17),221s(4.25), 270(3.90)	83-0977-78
$C_{18}H_{19}NO_4S$			
3,4-Thiophenedicarboxylic acid, 2-phen- yl-5-pyrrolidino-, dimethyl ester	dioxan	226(4.31),325(4.18)	24-2021-78
$C_{18}H_{19}NO_5$			
9(10H)-Acridinone, 1,2,3,5-tetrameth- oxy-10-methyl-	EtOH	267(4.88),308(4.42), 324s(4.01),390(3.67)	102-2125-78
3-Furancarboxylic acid, 5-[[acetyl[(4- methylphenyl)amino]methylene]-4,5- dihydro-2-methyl-4-oxo-, ethyl ester	EtOH	213(4.20),281(4.04), 338(4.19)	4-0155-78
$C_{18}H_{19}NO_5S$			
3,4-Thiophenedicarboxylic acid, 2-mor- pholino-5-phenyl-, dimethyl ester	dioxan	230(4.31),311(4.05)	24-2021-78
$C_{18}H_{19}NO_6$			
3-Furancarboxylic acid, 5-[[acetyl(4- methoxyphenyl)amino]methylene]-4,5- dihydro-2-methyl-4-oxo-, ethyl ester	EtOH	214(4.17),278(4.06), 340(4.20)	4-0155-78
$C_{18}H_{19}NO_7S_2$			
2(3H)-Benzothiazolethione, 3-(2,3,5- tri-O-acetyl-β-D-ribofuranosyl)-	EtOH	228(4.15),238(4.03), 326(4.35)	4-0657-78
$C_{18}H_{19}NO_8S$			
2(3H)-Benzoxazolethione, 3-(2,3,5-tri- O-acetyl-β-D-ribofuranosyl)-	EtOH	255(3.91),260(3.88), 296(4.40)	4-0657-78
$C_{18}H_{19}N_2O$			
2,1-Benzisoxazolium, 3-[2-[4-(dimethyl- amino)phenyl]ethenyl]-1-methyl-, perchlorate	MeCN	584(4.77)	44-1233-78
$C_{18}H_{19}N_2OP$			
3H-Pyrazole, 4-(diphenylphosphinyl)- 3,3,5-trimethyl-	MeOH	259(3.57),358(2.28)	150-4248-78
3H-Pyrazole, 5-(diphenylphosphinyl)- 3,3,4-trimethyl-	MeOH	273(3.42),341(2.32)	150-4248-78

Compound	Solvent	$\lambda_{max}(\log \epsilon)$	Ref.
$C_{18}H_{19}N_2S$			
2,1-Benzisothiazolium, 3-[2-[4-(dimeth-ylamino)phenyl]ethenyl]-1-methyl-, perchlorate	MeCN	592(4.58)	44-1233-78
$C_{18}H_{19}N_3$			
1,10-Phenanthrolin-4-amine, N-cyclohex-yl-	neutral	225(4.38),248(4.35), 271(4.33),290s(4.23), 329(4.22)	39-1215-78B
	monocation	210(4.42),250(4.32), 270(4.33),290(4.24), 332(4.17)	39-1215-78B
$C_{18}H_{19}N_3O$			
Ethanol, 2-[methyl(4-phenyl-3H-1,5-ben-zodiazepin-2-yl)amino]-	n.s.g.	258(4.55),346(3.68)	87-0952-78
$C_{18}H_{19}N_3O_2$			
Butanamide, N-methyl-2-(methylphenyl-hydrazono)-3-oxo-N-phenyl-	EtOH	206(4.38),235(4.23), 346(4.34)	104-0521-78
1H-[1,2,4]Triazolo[1,2-a]pyridazine-1,3(2H)-dione, 5,8-dihydro-5,8-bis-(1-methylethylidene)-2-phenyl-	EtOH	222(4.32),276(4.28)	39-1568-78C
$C_{18}H_{19}N_3O_3$			
1H-Indol-3-amine, 1-acetyl-N-[[3-hy-droxy-5-(hydroxymethyl)-2-methyl-4-pyridinyl]methyl]-	EtOH	208(4.58),246(4.09), 290(3.97),312(3.89)	103-0757-78
2,4,6(1H,3H,5H)-Pyrimidinetrione, 1,3-diethyl-5-(2-methyl-4(1H)-quinolin-ylidene)-	dioxan HOAc-HClO₄	465(3.9) 366(3.8)	83-0561-78 83-0561-78
2,4,6(1H,3H,5H)-Pyrimidinetrione, 1,3-diethyl-5-(4-methyl-2(1H)-quinolin-ylidene)-	dioxan HOAc-HClO₄	408(4.4) 390(4.4)	83-0561-78 83-0561-78
$C_{18}H_{19}N_3O_4$			
Benzoic acid, 4-nitro-, 2-benzoyl-1-(1,1-dimethylethyl)hydrazide	EtOH	267(4.11)	33-1477-78
1H-Pyrrolo[3,4-d]pyrimidine-7-carbox-ylic acid, 2,3,4,6-tetrahydro-5-meth-yl-2,4-dioxo-3-phenyl-, 1,1-dimethyl-ethyl ester	EtOH	240(4.58),272(4.28)	94-3080-78
$C_{18}H_{19}N_3O_4S$			
1(2H)-Quinolinecarboxylic acid, 2-[(ethoxycarbonyl)-2-thiazolyl-amino]-, ethyl ester	MeOH	230(4.53),263(3.90)	4-0655-78
$C_{18}H_{19}N_3S$			
6H-1,3,4-Thiadiazin-2-amine, N-(1-meth-ylethyl)-5,6-diphenyl-	isoPrOH	260(4.03),333(3.98)	73-1227-78
3(2H)-Thiazolamine, 2-[(1-methylethyl)-imino]-4,5-diphenyl-	isoPrOH	228(4.08),282(3.82), 333(3.85)	73-1227-78
$C_{18}H_{19}N_4$			
Isoquinolinium, 1-[[4-(dimethylamino)-phenyl]azo]-2-methyl-	n.s.g.	570.5(4.63)	27-0323-78
Quinolinium, 2-[[4-(dimethylamino)phen-yl]azo]-1-methyl-	n.s.g.	591(5.00)	27-0323-78

Compound	Solvent	$\lambda_{max}(\log \epsilon)$	Ref.
$C_{18}H_{19}N_4O_3P$ Urea, N-[5,6-dihydro-6-oxo-5-(phenyl-methyl)-1,4,2-diazaphosphorin-2(1H)-yl]-N'-(phenylmethyl)-	EtOH	222(4.08),249(4.00)	130-0421-78
$C_{18}H_{19}N_4S$ Acridinium, 3,6-bis(dimethylamino)-, thiocyanate	MeOH	231(4.2),271(4.7), 292(4.4),495(5.1)	97-0108-78
$C_{18}H_{19}N_5$ 2-Propanone, 1-(3-methyl-1-phenyl-1H-1,2,4-triazol-5-yl)-, phenylhydrazone	EtOH	274(4.27)	94-1825-78
$C_{18}H_{19}N_5O_4S$ Benzenesulfonamide, 4-[[4-cyano-5-(methoxyimino)-1,3-pentadienyl]-amino]-N-(3,4-dimethyl-5-isoxazolyl)-	MeOH	401(4.72)	83-0433-78
$C_{18}H_{19}N_5O_7S_2$ 5-Thia-1-azabicyclo[4.2.0]oct-2-ene-2-carboxylic acid, 3-(acetoxymethyl)-7-[[(2-amino-4-thiazolyl)[(2-propen-yloxy)imino]acetyl]amino]-8-oxo-, (E)-	EtOH	237(4.34),305(2.48)	78-2233-78
(Z)-	EtOH	236(4.26),254(4.17), 295(3.82)	78-2233-78
$C_{18}H_{20}$ Ethene, 1,1-bis(2,4-dimethylphenyl)-	15M H_2SO_4 $MeSO_3H$	338(4.23),485(4.55) 338(4.11),480(4.48)	39-0892-78B 39-0892-78B
[5.1]Metabenzenophane	MeCN	264(3.17),274s(3.04)	24-2547-78
Naphthacene	isooctane	229(4.81),236(4.97), 258(4.16),274(3.48), 283(3.54),294(3.46), 308s(3.18),316(2.95), 323s(2.63),331(3.15)	35-6174-78
Tricyclo[9.3.1.1^{4,8}]hexadeca-1(15),4,6-8(16),11,13-hexaene, 5,14-dimethyl-	EtOH	212(4.57),214s(4.56), 232s(4.13),276(2.80), 282(2.73)	78-0871-78
$C_{18}H_{20}BrNO_3$ Pyridinium, 1-(2-ethoxy-1-methyl-2-oxo-ethyl)-4-[2-(4-hydroxyphenyl)ethen-yl]-, bromide, (E)-	MeOH	403(4.06)	56-1265-78
Pyridinium, 1-(3-ethoxy-3-oxopropyl)-4-[2-(4-hydroxyphenyl)ethenyl]-, bromide, (E)-	MeOH	394(4.54)	56-1265-78
$C_{18}H_{20}ClNO_2$ 2H-Cycloocta[b]oyran-2-one, 3-chloro-5,6,7,8,9,10-hexahydro-4-(methyl-phenylamino)-	EtOH	242(4.15),303s(3.82), 325(3.92),376s(3.68)	4-0511-78
$C_{18}H_{20}ClNO_2S$ 2H,5H-[1]Benzothiopyrano[4,3-b]pyran-2-one, 4-[bis(1-methylethyl)amino]-3-chloro-	EtOH	242(4.36),270(3.60), 340(3.99),380(3.97)	4-0181-78
$C_{18}H_{20}ClNO_3$ Pyridinium, 4-[2-(4-hydroxyphenyl)eth-enyl]-1-(2-oxo-2-propoxyethyl)-, chloride, (E)-	MeOH	401(4.68)	56-1265-78

Compound	Solvent	$\lambda_{max}(\log \epsilon)$	Ref.
$C_{18}H_{20}ClN_3O_4S_2$			
Sporidesmin H	MeOH	216(4.37),252(4.05), 290(3.81)	39-1476-78C
$C_{18}H_{20}ClN_5O_3$			
Pyrrolo[3',4':4,5]pyrrolo[3,4-b]indole-1,3,5(2H)-trione, 2,5a-diamino-9-chloro-4-cyclohexyl-3a,4,5a,6-tetra-hydro-	EtOH	239(3.94),258(3.92), 315(3.45)	4-1463-78
$C_{18}H_{20}F_3N_3O_{10}$			
2,4(1H,3H)-Pyrimidinedione, 1-[3,4,6-triacetoxy-2-deoxy-2-[(trifluoro-acetyl)amino]-β-D-glucopyranosyl]-	EtOH	257(4.08)	136-0185-78C
$C_{18}H_{20}FeO_4$			
Ferrocene, [4-carboxy-2-(ethoxycarbo-nyl)-1-butenyl]-, (E)-	EtOH	248(4.12),295(4.21)	128-0273-78
Ferrocene, [2-(carboxymethyl)-3-ethoxy-1-methyl-3-oxo-1-propenyl]-, (E)-	EtOH	236s(4.02),281(3.91)	128-0325-78
(Z)-	EtOH	277(3.83)	128-0325-78
Ferrocene, [2-(ethoxycarbonyl)-4-meth-oxy-4-oxo-1-butenyl]-, (E)-	EtOH	248(4.10),294(4.16)	128-0325-78
(Z)-	EtOH	246(4.06),294(4.08)	128-0325-78
$C_{18}H_{20}INO_3$			
1-Iodocodeine	EtOH	217(4.50),247(3.83), 286(3.18)	44-0737-78
$C_{18}H_{20}N_2$			
1-Butene, 3,3-dimethyl-1-phenyl-2-(phen-ylazo)-	hexane	236(4.04),286(3.95), 486(2.08)	49-1081-78
$C_{18}H_{20}N_2O_2$			
Benzamide, N,N'-1,4-butanediylbis-	EtOH	226(4.42)	105-0637-78
Benzoic acid, 4-[[(ethylphenylamino)-methylene]amino]-, ethyl ester	isoPrOH	313(4.44)	117-0097-78
Benzoic acid, 4-methyl-, 2-(4-methyl-phenyl)-2-(1-oxopropyl)hydrazide	EtOH	236(4.35)	33-1477-78
1H-Indole-2-methanol, α-[(3-ethyl-5-(hydroxymethyl)-4-pyridinyl]-methyl-, (±)-	EtOH	268(4.06),273(4.06), 282(3.96),290(3.86)	78-1363-78
2H-3,7-Methanoazacycloundecino[5,4-b]-indole-2,9(4H)-dione, 1,5,6,7,8,10-hexahydro-10-methyl-, (±)-	EtOH	283(3.74)	44-4859-78
isomer	EtOH	242(4.00),311(4.04)	44-4859-78
$C_{18}H_{20}N_2O_3$			
Acetamide, N-[[2-(acetylamino)-5-meth-oxyphenyl]methyl]-N-phenyl-	MeOH	215(4.5),250s(--), 280(3.3)	5-0608-78
Benzoic acid, 4-[[[(2-methoxyphenyl)-methylamino]methylene]amino]-, ethyl ester	isoPrOH	312(4.44)	117-0097-78
Benzoic acid, 4-[[[(4-methoxyphenyl)-methylamino]methylene]amino]-, ethyl ester	isoPrOH	312(4.42)	117-0097-78
2H-3,7-Methanoazacycloundecino[5,4-b]-indole-2,9(4H)-dione, 1,5,6,7,8,10-hexahydro-13-methoxy-, (±)-	EtOH	324(4.20)	44-4859-78
isomer	EtOH	277(3.68),307(3.74)	44-4859-78

Compound	Solvent	$\lambda_{max}(\log \epsilon)$	Ref.
1H-Pyrrolo[3,4-b]pyridine-3-carboxylic acid, 4,5,6,7-tetrahydro-2,6-dimethyl-7-oxo-4-phenyl-, ethyl ester	EtOH	223(4.25),258(4.09), 342(3.65)	95-0448-78
$C_{18}H_{20}N_2O_4$			
1,6-Naphthyridine-3-carboxylic acid, 1,2,5,6,7,8-hexahydro-4-hydroxy-2-oxo-6-(phenylmethyl)-, ethyl ester	EtOH	225(4.46),318(3.99)	142-0267-78C
1H-Pyrazole-4,5-dicarboxylic acid, 1-(4-pentenyl)-3-phenyl-, dimethyl ester	EtOH	232(4.33)	44-1664-78
2H-Pyrido[3,4-e]-1,3-oxazine-4,5(3H,6H)-dione, 3-ethyl-7-methoxy-2,2-dimethyl-6-phenyl-	MeOH	308(4.4)	120-0101-78
Pyrido[2',1':3,4]pyrazino[1,2-b]isoquinolin-2(9H)-one, 6,7,14,14a-tetrahydro-1-hydroxy-11,12-dimethoxy-	EtOH	218(4.10),289(4.0)	12-0187-78
$C_{18}H_{20}N_2O_6$			
2-Butenoic acid, 2-(acetylamino)-4-[2-(1,3-dihydro-1,3-dioxo-2H-isoindol-2-yl)ethoxy]-, ethyl ester, (Z)-	EtOH	220(4.69),233s(4.32), 241s(4.20),294(3.29), 300(3.26)	44-3713-78
3-Butenoic acid, 2-(acetylamino)-4-[2-(1,3-dihydro-1,3-dioxo-2H-isoindol-2-yl)ethoxy]-, ethyl ester, cis	EtOH	218(4.63),239s(3.99), 292(3.27),300s(3.24)	44-3713-78
trans	EtOH	219(4.69),239(4.05), 293(3.31),300(3.27)	44-3713-78
$C_{18}H_{20}N_2O_6$			
Isoquinoline, 3-(5,8-diacetoxy-4-morpholino-7-methoxy)-	EtOH	248(4.61),291s(4.24), 295(4.24)	4-0569-78
Phenol, 4,4'-(1,2-diethyl-1,2-ethanediyl)bis[2-nitro-, (R*,S*)-	MeOH	220(4.5),276(4.1), 363(3.7)	83-0184-78
$C_{18}H_{20}N_2O_7$			
Carbamic acid, (1,6-dihydro-6-oxo-1-β-D-ribofuranosyl-3-pyridinyl)-, phenylmethyl ester	pH 1 pH 11	234(4.13),315(3.67) 237(4.12),315(3.67)	87-0427-78 87-0427-78
$C_{18}H_{20}N_2O_7S$			
2H-Benzimidazole-2-thione, 1,3-dihydro-1-(2,3,5-tri-O-acetyl-β-D-ribofuranosyl)-	MeOH	249(3.89),294s(4.28), 306(4.40)	4-0657-78
$C_{18}H_{20}N_2O_7S_2$			
5-Thia-1-azabicyclo[4.2.0]oct-2-ene-2-carboxylic acid, 3-(acetoxymethyl)-7-methoxy-8-oxo-7-[(2-thienylacetyl)-amino]-, methyl ester	n.s.g.	236(4.12)	88-5219-78
$C_{18}H_{20}N_2O_9$			
2,2'-Anhydro-1-β-D-(3',5'-di-O-acetoacetylarabinofuranosyl)-6-methyluracil	H_2O	224(3.87),255(3.92)	103-0901-78
isomer	H_2O pH 13	218(3.85),272(3.85) 276(4.64)	103-0901-78 103-0901-78
2,2'-Anhydro-3-β-D-(3',5'-di-O-acetoacetylarabinofuranosyl)-6-methyluracil	H_2O pH 13	217(3.79),270(3.79) 276(4.54)	103-0901-78 103-0901-78
isomer	H_2O pH 13	215(3.83),272(3.83) 276(4.60)	103-0901-78 103-0901-78

Compound	Solvent	$\lambda_{max}(\log \epsilon)$	Ref.
$C_{18}H_{20}N_4O_2$			
1,2-Ethanediamine, N-(1-methylethyl)-N'-(1-nitro-9-acridinyl)-	pH 1	218(4.338),273(4.610), 330(3.645),437(3.909)	56-2125-78
	pH 7	250(4.210),268(4.270), 401(3.581)	56-2125-78
1,2-Ethanediamine, N-(1-nitro-9-acridinyl)-N'-propyl-	pH 1	218(4.324),273(4.566), 329(3.637),437(3.902)	56-2125-78
	pH 7	273(4.276),405(3.526)	56-2125-78
$C_{18}H_{20}N_4O_4$			
4H,8H-Benzo[1,2-c:4,5-c']diisoxazole-4,8-dione, 3,7-dipiperidino-	MeCN	217(4.2),312(4.4), 351(4.2)	24-3346-78
$C_{18}H_{20}N_4O_5$			
3H-Imidazo[4,5-b]pyridin-5-amine, 7-(phenylmethoxy)-3-β-D-ribofuranosyl-	pH 1	246(3.97),298(4.10)	4-0839-78
	pH 11	250(3.95),288(4.01)	4-0839-78
$C_{18}H_{20}N_4O_9S$			
β-D-Glucopyranuronic acid, 1-deoxy-1-(1,6-dihydro-6-thioxo-9H-purin-9-yl)-, methyl ester, 2,3,4-triacetate	MeOH	276.5(4.27)	106-0250-78
$C_{18}H_{20}N_4O_{11}$			
1H-Imidazole-5-carbonitrile, 4-nitro-1-(2,3,5,6-tetra-O-acetyl-β-D-glucofuranosyl)-	pH 1	288(3.74)	12-1095-78
	aq NH₃	293(3.76)	12-1095-78
	EtOH	283(3.73)	12-1095-78
1H-Imidazole-5-carbonitrile, 4-nitro-1-(2,3,4,6-tetra-O-acetyl-β-D-glucopyranosyl)-	pH 1	289(3.80)	12-1095-78
	aq NH₃	292(3.80)	12-1095-78
	EtOH	285(3.79)	12-1095-78
$C_{18}H_{20}N_6S_2$			
Hydrazinecarbothioamide, 2,2'-(1,2-dimethyl-1,2-ethanediylidene)bis[N-phenyl-	30% DMF	343(4.63)	96-0140-78
$C_{18}H_{20}N_8O_3$			
DL-Alanine, N-[4-[[(2,4-diamino-6-pteridinyl)methyl]methylamino]benzoyl]-	pH 1	244(4.24),307(4.28)	87-1162-78
β-Alanine, N-[4-[[(2,4-diamino-6-pteridinyl)methyl]methylamino]benzoyl]-	pH 1	245(4.19),300(4.20)	87-1162-78
Glycine, N-[4-[[(2,4-diamino-6-pteridinyl)methyl]methylamino]benzoyl]-N-methyl-	pH 1	246(4.35),297(4.20)	87-1162-78
$C_{18}H_{20}O$			
Benzenemethanol, 2-[2-(2,4,6-trimethylphenyl)ethenyl]-, (E)-	MeOH	220s(4.25),280(4.25)	35-3819-78
$C_{18}H_{20}OS$			
Benzo[c]thiophene, 1,3-dihydro-1-methoxy-3,3-dimethyl-1-(2-methylphenyl)-	hexane	262s(3.16),270s(2.99)	104-0582-78
Benzo[c]thiophene, 1,3-dihydro-1-methoxy-3,3-dimethyl-1-(3-methylphenyl)-	hexane	263s(2.96),270s(2.88), 275s(2.74)	104-0582-78
Benzo[c]thiophene, 1,3-dihydro-1-methoxy-3,3-dimethyl-1-(4-methylphenyl)-	hexane	261s(3.21),269s(3.09)	104-0582-78
$C_{18}H_{20}O_2$			
Acetylacetone, α-(4,6,8-trimethyl-1-azulenyl)-	C_6H_{12}	247(4.48),293(4.70), 339(3.78),349s(3.76),	18-3582-78

Compound	Solvent	λ_{max}(log ϵ)	Ref.
(cont.)		354(3.73),556(2.72), 603s(2.63),660(2.14)	18-3582-78
Benzenemethanol, α-(1,1-dimethylethyl)-, benzoate	isoPrOH	230(4.13),274(2.93)	104-1942-78
2H-Cyclobuta[j]phenanthren-4(5H)-one, 3,3a,6,7-tetrahydro-10-methoxy-3a- methyl-, (3aR*,5aS*)-(±)-	EtOH	260(4.24)	44-4598-78
4,7-Ethano-1H-indene-8,9-dione, 1-cy- clohexylidene-3a,4,7,7a-tetrahydro- 4-methyl-, (3aα,4α,7α,7aα)-	MeCN	249(4.23),450(1.99)	5-0440-78
4,7-Ethano-1H-indene-8,9-dione, 1-cy- clohexylidene-3a,4,7,7a-tetrahydro- 5-methyl-, (3aα,4α,7α,7aα)-	MeCN	247(4.13),253(4.14), 454(2.05)	5-0440-78
Isobenzofuran, 1,3-dihydro-1-methoxy- 3,3-dimethyl-1-(2-methylphenyl)-	hexane	257(2.80),262(2.92), 269(2.87)	104-0582-78
Isobenzofuran, 1,3-dihydro-1-methoxy- 3,3-dimethyl-1-(3-methylphenyl)-	hexane	256(2.90),262(3.02), 269(2.98),273s(2.78)	104-0582-78
Isobenzofuran, 1,3-dihydro-1-methoxy- 3,3-dimethyl-1-(4-methylphenyl)-	hexane	256(3.01),262(3.07), 269(2.99)	104-0582-78
Tricyclo[9.3.1.1^{4,8}]hexadeca-1(15),4,6- 8(16),11,13-hexaene, 5,14-dimethoxy-	isooctane	212(4.48),222s(4.38), 260(4.33),269s(4.29), 297(3.56),307s(3.48), 345s(2.12)	78-0871-78
Tricyclo[9.3.1.1^{4,8}]hexadeca-1(15),4,6- 8(16),11,13-hexaene-5,14-dimethanol	EtOH	214(4.70),233s(4.20), 264s(2.63),274(2.74), 280s(2.66)	78-0871-78
C$_{18}$H$_{20}$O$_2$S			
Benzo[c]thiophene, 1,3-dihydro-1-meth- oxy-1-(2-methoxyphenyl)-3,3-dimethyl-	hexane	267s(3.40),274(3.52), 281(3.48)	104-0582-78
C$_{18}$H$_{20}$O$_2$S$_3$			
Benzenecarbothioic acid, S-[2-(hexahy- dro-1,3-benzodithiol-2-ylidene)-3- oxobutyl] ester	EtOH	238(4.10),273(4.05), 330(4.17)	18-0301-78
C$_{18}$H$_{20}$O$_3$			
Benzene, 1-[2-(4-ethoxyphenyl)ethenyl]- 2,3-dimethoxy-, (E)-	n.s.g.	305(4.43)	39-0739-78C
Benzene, 4-[2-(4-ethoxyphenyl)ethenyl]- 1,2-dimethoxy-, (E)-	n.s.g.	303(4.24),330(4.31)	39-0739-78C
Isobenzofuran, 1,3-dihydro-1-methoxy- 1-(2-methoxyphenyl)-3,3-dimethyl-	hexane	264s(3.24),273(3.38), 280(3.34)	104-0582-78
C$_{18}$H$_{20}$O$_3$S			
Benzene, 1-(1,1-dimethylethyl)-4-[[2- (phenylsulfonyl)ethenyl]oxy]-	dioxan	222(4.19),246(4.28)	99-0194-78
C$_{18}$H$_{20}$O$_4$			
1,3,5-Cyclohexanetrione, 2,2,4-trimeth- yl-6-(1-oxo-3-phenylpropyl)-	MeOH EtOH EtOH-NaOH	350(4.30) 353(4.31) 353(4.33)	102-2011-78 102-2015-78 102-2015-78
1,2-Naphthalenedicarboxylic acid, 5,6,7,8-tetramethyl-, dimethyl ester	EtOH	258(4.71),308(3.67), 316(3.65),345s(3.43)	5-1675-78
1-Propanone, 1-(2,6-dihydroxy-4-methoxy- 3,5-dimethylphenyl)-3-phenyl-	MeOH	223s(4.09),281(4.13), 350(3.56)	102-2011-78
	MeOH-NaOMe	236s(4.17),293(4.07), 396(3.51)	102-2011-78
	EtOH EtOH-NaOH	280(4.21) 296(4.10)	102-2015-78 102-2015-78

Compound	Solvent	λ_{max}(log ϵ)	Ref.
1-Propanone, 1-(2-hydroxy-4,6-dimeth-oxy-3-methylphenyl)-3-phenyl-	EtOH	291(4.37)	102-2015-78
Villostyrene	EtOH	245(4.42),286(3.95)	102-1389-78
$C_{18}H_{20}O_5$			
Ethyne, 1-(5-acetoxy-2,3-O-isopropyli-dene-α-D-ribofuranosyl)-2-phenyl-	EtOH	241(4.23),251(4.11)	94-0898-78
β-	EtOH	240(3.95),251(3.62)	94-0898-78
2,5-Furandione, 3-[(3,5-dimethoxyphen-yl)methylene]-4-(1-ethylpropylidene)-	hexane	262(4.02),268(4.01),331(4.07)	39-0571-78C
dihydro-, (E)-	toluene	338(4.12)	39-0571-78C
Naphtho[2,3-c]furan-1,3-dione, 4,4-di-ethyl-4,9-dihydro-5,7-dimethoxy-	hexane	250(3.61),282(3.42)	39-0571-78C
Phenol, 5-[3-(2-hydroxyphenyl)-2-prop-enyl]-2,3,4-trimethoxy-	EtOH	254(3.81),286(3.49),303(3.34)	102-1379-78
$C_{18}H_{20}O_6$			
Duartin, (-)-	EtOH	225(4.06),278(3.33)	102-1401-78
Ethanone, 1-(2-hydroxy-4-methoxyphen-yl)-2-(2,3,4-trimethoxyphenyl)-	EtOH	228(3.97),266(3.93),317(3.65)	102-1405-78
$C_{18}H_{21}ClN_4O_3S$			
4H-1,4-Thiazin-3-one, 2-(4-chlorophen-yl)-2-hydroxy-6-(2-hydroxyethyl)-4-[(2-methyl-4-aminopyrimidin-5-yl)-methyl]-5-methyl-	EtOH	226(4.39),277(3.92)	94-0722-78
$C_{18}H_{21}Cl_2NO_2S$			
2H,5H-[1]Benzothiopyrano[4,3-b]pyran-2-one, 4-[bis(1-methylethyl)amino]-3,3-dichloro-3,4-dihydro-	EtOH	236(4.25),255s(4.13),340(3.38)	4-0181-78
$C_{18}H_{21}FN_2O_8S$			
2,4(1H,3H)-Pyrimidinedione, 1-(6-deoxy-6-fluoro-2-O-[(4-methylphenyl)sulfon-yl]-β-D-glucopyranosyl]-5-methyl-	MeOH	222(4.20),264(3.99)	48-0157-78
$C_{18}H_{21}N$			
2-Azaspiro[4.4]nona-2,6,8-triene, 1,1,4,4-tetramethyl-3-phenyl-	n.s.g.	242(4.11)	88-0093-78
$C_{18}H_{21}NO$			
4,7-Methano-1H-indene-2-carbonitrile, 1-bicyclo[2.2.1]hept-5-en-2-yl-2,3,3a,4,7,7a-hexahydro-1-hydroxy-	n.s.g.	212(2.74)	33-1427-78
Phenol, 4-[2-(4-aminophenyl)-1-ethyl-1-butenyl]-	EtOH	280(4.10)	104-0725-78
$C_{18}H_{21}NOS_3$			
Spiro[1,3-benzodithiole-2,4'-[4H]thio-pyran]-3'-carboxamide, 2',3',3a,4-5,6,7,7a-octahydro-6'-phenyl-	EtOH	247(4.23)	18-0301-78
$C_{18}H_{21}NO_2$			
2-Azaestra-1,3,5(10),8,14-pentaen-17-ol, 3-methoxy-, (±)-	MeOH	300(4.45)	44-0966-78
2-Azaestra-1,3,5(10),8-tetraen-17-one, 3-methoxy-, (±)-	MeOH	267(4.26)	44-0966-78
4H-Indol-4-one, 1,5,6,7-tetrahydro-1-(2-hydroxyethyl)-6,6-dimethyl-2-phenyl-	MeOH	240(3.93),280(4.00)	48-0863-78

Compound	Solvent	$\lambda_{max}(\log \epsilon)$	Ref.
1,4-Naphthalenedione, 5-(cyclohexyl-amino)-2,3-dimethyl-	C_6H_{12} EtOH	524(3.76) 538(--)	150-2319-78 150-2319-78
$C_{18}H_{21}NO_2S_3$			
Spiro[1,3-benzodithiole-2,4'-[4H-1,3]-thiazine]-5'-carboxylic acid, 3a,4,5,5',6,6',7,7a-octahydro-2'-phenyl-, methyl ester	EtOH	247(4.15)	18-0301-78
$C_{18}H_{21}NO_3$			
1-Azulenecarboxylic acid, 2-hydroxy-6-piperidino-, ethyl ester	MeOH	221(4.41),345(4.73), 409(4.14)	18-3087-78
Butanoic acid, 3-oxo-2-spiro[cyclopent-ane-1,2'-[2H]indol]-3'(1'H)-ylidene-, ethyl ester	EtOH	208(4.42),242(4.44), 293(3.72)	44-3702-78
1,3-Cyclopentanedione, 2-[2-(6,7-dihy-dro-3-methoxy-8(5H)-isoquinolinyli-dene)ethyl]-2-methyl-	MeOH	262(4.26)	44-0966-78
1H-2-Pyrindine-4-carboxylic acid, 2,4a,5,6,7,7a-hexahydro-2-methyl-1-oxo-3-phenyl-, ethyl ester	EtOH and EtOH-KOH	286(4.04)	142-0161-78A
$C_{18}H_{21}NO_4$			
Piperidine, 1-[5-(6-methoxy-1,3-benzo-dioxol-5-yl)-1-oxo-2,4-pentadienyl]-(wisanin)	EtOH	246(4.10),260(4.10), 278(4.14),300(4.17), 309(4.14),365(4.24)	5-0573-78
	n.s.g. n.s.g.	360(4.14) 371(4.42)	78-1979-78 78-1979-78
1H-2-Pyrindine-4-carboxylic acid, 2,4a,5,6,7,7a-hexahydro-7a-hydroxy-2-methyl-1-oxo-3-phenyl-, ethyl ester	EtOH	284(3.97)	142-0161-78A
2H-Pyrrole-3-carboxylic acid, 3,4-di-hydro-2-(4-methoxy-4-oxo-2-butenyl)-2-methyl-5-phenyl-, methyl ester, (E)-	C_6H_{12}	243(4.24)	44-3757-78
(Z)-	C_6H_{12}	243(4.21)	44-3757-78
$C_{18}H_{21}NO_5$			
1,3-Azulenedicarboxylic acid, 2-hy-droxy-6-(dimethylamino)-, diethyl ester	MeOH	230(4.31),265(4.48), 347(4.79),402(4.23), 430(4.17)	18-3087-78
3-Furancarboxylic acid, 5-[[acetyl(4-methylphenyl)amino]methyl]-4,5-dihy-dro-2-methyl-4-oxo-, ethyl ester	EtOH	217(4.15),258(4.00)	4-0327-78
1H-Pyrrole-2,4-dicarboxylic acid, 3-hy-droxy-1-phenyl-, 2-(1,1-dimethyleth-yl) 4-ethyl ester	EtOH	230(4.38),262(4.02)	94-2224-78
$C_{18}H_{21}NO_6$			
2-Butenedioic acid, 2-[(7-methoxy-2,2-dimethyl-2H-1-benzopyran-5-yl)amino]-, dimethyl ester	MeOH	227(4.19),286(4.15), 318(4.11)	24-0439-78
3-Furancarboxylic acid, 5-[[acetyl(4-methoxyphenyl)amino]methyl]-4,5-di-hydro-2-methyl-4-oxo-, ethyl ester	EtOH	220(4.13),262(4.03)	4-0327-78
$C_{18}H_{21}N_3O_2$			
Pyrrolo[3,4-b]indole-1-carboxamide, 2-cyclohexyl-1,2,3,4-tetrahydro-N-methyl-3-oxo-	EtOH	200(4.88),294(4.26)	4-1463-78

Compound	Solvent	$\lambda_{max}(\log \epsilon)$	Ref.
$C_{18}H_{21}N_3O_4$			
7,8-Isoquinolinedione, 6-methyl-3,5-di-morpholino-	EtOH	245(4.33),317(4.98), 427(4.19)	94-2175-78
1H-Pyrazole-4-carboxylic acid, 3-[[acetyl(4-methylphenyl)amino]acetyl]-5-methyl-, ethyl ester	EtOH	212(4.09),231(4.03)	4-0155-78
$C_{18}H_{21}N_3O_5$			
1H-Pyrazole-4-carboxylic acid, 3-[[acetyl(4-methoxyphenyl)amino]acetyl]-5-methyl-, ethyl ester	EtOH	213(4.05),230(4.17)	4-0155-78
$C_{18}H_{21}N_3O_6$			
1-Deazariboflavin	n.s.g.	365(3.60),535(3.83)	69-1942-78
3-Deazariboflavin	n.s.g.	421(4.30)	69-1942-78
5-Deazariboflavin	n.s.g.	338(4.09),397(4.10)	69-1942-78
$C_{18}H_{21}N_5O$			
1H-Pyrazolo[3,4-d]pyrimidine-4-carboxamide, N-hexyl-1-phenyl-	EtOH	250(4.50),270s(4.08), 328(3.29)	95-1274-78
$C_{18}H_{21}N_5O_4$			
1,2-Hydrazinedicarboxylic acid, 1-[2-(1,1-dimethylethyl)-2H-pyrrolo[3,4-b]quinoxalin-1-yl]-, dimethyl ester	MeOH	258(4.71),323(3.65), 330(3.79),338(3.96), 343(4.04),352(4.14), 355(4.12),507(3.42)	88-4671-78
$C_{18}H_{21}N_5O_5$			
7H-Purin-6-amine, 7-β-D-glucofuranosyl-N-(phenylmethyl)-	EtOH	276(4.09)	12-1095-78
	EtOH-NH$_3$	276(4.20)	12-1095-78
	0.1M HOAc	283(4.29)	12-1095-78
7H-Purin-6-amine, 7-β-D-glucopyranosyl-N-(phenylmethyl)-	EtOH	277(4.09)	12-1095-78
	EtOH-NH$_3$	277(4.09)	12-1095-78
	0.1M HOAc	282(4.19)	12-1095-78
9H-Purin-6-amine, 9-β-D-glucofuranosyl-N-(phenylmethyl)-	EtOH	270(4.30)	12-1095-78
	EtOH-NH$_3$	270(4.31)	12-1095-78
	0.1M HOAc	268(4.31)	12-1095-78
9H-Purin-6-amine, 9-α-D-glucopyranosyl-N-(phenylmethyl)-	EtOH	269(4.24)	12-1095-78
	EtOH-NH$_3$	269(4.24)	12-1095-78
	0.1M HOAc	267.5(4.24)	12-1095-78
9H-Purin-6-amine, 9-β-D-glucopyranosyl-N-(phenylmethyl)-	EtOH	267.3(4.30)	12-1095-78
	EtOH-NH$_3$	267.3(4.29)	12-1095-78
	0.1M HOAc	266(4.31)	12-1095-78
6H-Purin-6-imine, 7-β-D-glucopyranosyl-1,7-dihydro-1-(phenylmethyl)-	EtOH	266(3.96)	12-1095-78
	EtOH-NH$_3$	266(4.09)	12-1095-78
	0.1M HOAc	276(4.04)	12-1095-78
$C_{18}H_{21}N_5O_7S_2$			
5-Thia-1-azabicyclo[4.2.0]oct-2-ene-2-carboxylic acid, 3-(acetoxymethyl)-7-[[[(2-amino-4-thiazolyl)[(1-methylethoxy)imino]acetyl]amino]-8-oxo-, (E)-	EtOH	240(4.33),305(3.43)	78-2233-78
(Z)-	EtOH	238(4.24),256(4.16), 295(3.83)	78-2233-78
$C_{18}H_{22}Cl_2NO_9PS_2$			
Phosphoramidic acid, bis[2-[(methylsulfonyl)oxy]ethyl]-, bis(2-chlorophenyl) ester	MeOH	269(3.1271)	65-0720-78

Compound	Solvent	$\lambda_{max}(\log \epsilon)$	Ref.
$C_{18}H_{22}F_2NO_9PS_2$			
Phosphoramidic acid, bis[2-[(methylsul-fonyl)oxy]ethyl]-, bis(4-fluorophen-yl) ester	MeOH	260(3.3927)	65-0720-78
$C_{18}H_{22}NO$			
Pyridinium, 4-[3-(2,6-dimethyl-4H-pyr-an-4-ylidene)-1-propenyl]-1,2,6-tri-methyl-, perchlorate	CH_2Cl_2 MeNO$_2$	518(4.70) 495(4.65)	103-0601-78 103-0601-78
$C_{18}H_{22}N_2$			
9H-Pyrido[3,4-b]indole, 1-heptyl-	MeOH	236(4.60),241(4.58), 251s(4.41),283s(4.04), 289(4.28),338(3.73), 351(3.75)	95-0898-78
$C_{18}H_{22}N_2OS$			
Alimenazine sulfoxide	MeOH	232(4.46),274(4.06), 299(3.94),345(3.78)	133-0248-78
Methylpromazine sulfoxide	MeOH	237(4.51),275(4.10), 299(3.95),342(3.80)	133-0248-78
$C_{18}H_{22}N_2O_2$			
Phenol, 4,4'-(1,2-diethyl-1,2-ethene-diyl)bis[2-amino-, (E)-	MeOH	213(4.6),295(4.0)	83-0184-78
$C_{18}H_{22}N_2O_2S$			
Methopromazine sulfoxide	MeOH	244(4.54),253(4.53), 276(4.11),296(3.84), 335(3.76)	133-0248-78
$C_{18}H_{22}N_2O_4$			
Butanedioic acid, [[[(dimethylamino)-phenylmethylene]amino]methylene]-ethylidene-, dimethyl ester, (E,Z,Z)-	MeOH	328(4.4)	78-0833-78
Piperazine, 1-[5-(6-methoxy-1,3-benzo-dioxol-5-yl)-1-oxo-2,4-pentadienyl]-4-methyl-, (E,E)-	n.s.g.	372(4.48)	78-1979-78
8H-Pyrido[1,2-a]pyrazin-8-one, 1-[(3,4-dimethoxyphenyl)methyl]-1,2,3,4-tet-rahydro-9-hydroxy-2-methyl-	EtOH	285(3.86),310s(3.54)	12-0187-78
1H-Pyrrole-3-propanoic acid, 5-[(4-eth-enyl-2,5-dihydro-3-methyl-5-oxo-1H-pyrrol-2-yl)methyl]-2-formyl-4-meth-yl-, methyl ester	MeOH	312(4.29)	24-0486-78
$C_{18}H_{22}N_2O_5$			
1H-Pyrrole-3-carboxylic acid, 5-[[ace-tyl(4-methylphenyl)amino]methyl]-4,5-dihydro-5-hydroxy-2-methyl-4-oxo-, ethyl ester	EtOH	212(4.01),239(4.23), 306(3.79)	4-1215-78
$C_{18}H_{22}N_2O_6$			
Butanoic acid, 2-(acetylamino)-4-[2-(1,3-dihydro-1,3-dioxo-2H-isoindol-2-yl)ethoxy]-, ethyl ester	EtOH	219(4.61),232s(4.13), 240(3.97),294(3.28), 300s(3.26)	44-3713-78
$C_{18}H_{22}N_4$			
Benzaldehyde, 4-(dimethylamino)-, [[4-(dimethylamino)phenyl]meth-ylene]hydrazone	MeOH	244(4.21),330s(4.28), 393(4.61)	18-3312-78

Compound	Solvent	$\lambda_{max}(\log \epsilon)$	Ref.
Propanedinitrile, [[3-[1-(dimethylami-no)ethylidene]-1,4-cyclopentadien-1-yl]-1-piperidinomethylene]-	MeOH	264(4.12),297(4.19), 390(4.57)	83-0369-78
$C_{18}H_{22}N_4O_2$ 1H-Pyrrolo[3,4-d]pyrimidine-2,4(3H,6H)-dione, 7-[(dimethylamino)methyl]-1,5,6-trimethyl-3-(phenylmethyl)-	EtOH	227(4.48),254(3.93), 290(3.45)	94-3080-78
$C_{18}H_{22}N_4O_4$ 2(3H)-Naphthalenone, 4,4a,5,6,7,8-hexa-hydro-4a,8-dimethyl-, 2,4-dinitro-phenylhydrazone	EtOH	223(4.08),258(4.04), 382(4.28)	12-2527-78
2,4(1H,3H)-Quinazolinedione, 1-[3-(2,3-4,5,6,7-hexahydro-2,4-dioxo-1H-cyclo-pentapyrimidin-1-yl)propyl]-5,6,7,8-tetrahydro-	H_2O	277(4.31)	56-1035-78
$C_{18}H_{22}N_4O_6$ Riboflavin, 6-methyl-	n.s.g.	396(4.28),447(4.06)	69-1942-78
$C_{18}H_{22}N_4O_9$ 1H-Imidazole-5-carbonitrile, 4-amino-1-(2,3,5,6-tetra-O-acetyl-β-D-gluco-furanosyl)-	EtOH pH 1 NH_3	228(3.68),270(3.93) 234(3.91),266(3.86) 228(3.69),264(3.93)	12-1095-78 12-1095-78 12-1095-78
1H-Imidazole-5-carbonitrile, 4-amino-1-(2,3,4,6-tetra-O-acetyl-β-D-gluco-pyranosyl)-	EtOH pH 1 NH_3	270(3.88) 268(3.85) 267(3.88)	12-1095-78 12-1095-78 12-1095-78
$C_{18}H_{22}N_4S$ Propanedinitrile, [[3-(dimethylamino)-(methylthio)methylene]-1,4-cyclopen-tadien-1-yl]-1-piperidinomethylene]-	MeOH	275(4.17),310(4.13), 409(4.50)	83-0369-78
$C_{18}H_{22}N_8O_4$ 1H-Purine-2,6-dione, 8,8'-(1,4-butane-diyl)bis[3,7-dihydro-1,3-dimethyl-	pH 7.0 pH 13.0 MeOH	204(4.73),273(4.35) 214(4.57),279(4.46) 207(4.67),273(4.40)	24-0982-78 24-0982-78 24-0982-78
$C_{18}H_{22}O$ 14β-Estra-1,3,5(10),8-tetraen-3-ol	EtOH	271(4.18)	22-0119-78
$C_{18}H_{22}OS$ A,19-Dinor-1-thiapregna-2,5(10)-dien-20-yn-17-ol, (17α)-	EtOH	235(3.78),245s(3.70)	142-0207-78
$C_{18}H_{22}O_2$ 2H-Cyclobuta[j]phenanthren-4(5H)-one, 1,3,3a,6,7,11b-hexahydro-10-methoxy-3a-methyl-, (3aα,5aS*,11bα)-(±)-	EtOH	278(3.4)	44-4598-78
Estra-4,9-diene-3,17-dione, (-)-	EtOH	301(4.31)	44-1550-78
Estra-5(10),9(11)-diene-3,17-dione, (+)-	EtOH	250(4.30)	44-1550-78
Estra-1,3,5(10)-trien-6-one, 17β-hy-droxy-	EtOH	251(4.07),294(3.31)	94-3567-78
Indan, 1,1-(ethylenedioxy)-3a,4,7,7a-tetrahydro-7aβ-methyl-5-phenyl-, cis	EtOH	217s(3.95),245(4.10), 290s(2.28)	22-0343-78
Indan, 1,1-(ethylenedioxy)-3a,4,7,7a-tetrahydro-7aβ-methyl-5-phenyl-, trans-(±)-	EtOH	218s(3.95),245(4.11), 290s(2.20)	22-0343-78
Indan, 1,1-(ethylenedioxy)-3a,6,7,7a-tetrahydro-7aβ-methyl-5-phenyl-, cis	EtOH	218s(3.95),247(4.13), 290s(2.28)	22-0343-78

Compound	Solvent	$\lambda_{max}(\log \epsilon)$	Ref.
$C_{18}H_{22}O_3$			
2-Butenoic acid, 2-(2,3-dihydro-2,4,6-trimethyl-3-oxo-1H-inden-5-yl)ethyl ester	EtOH	217(4.56),260(4.11), 303(3.27)	94-2365-78
Estra-1,3,5(10)-trien-6-one, 2,17β-di-hydroxy-	EtOH	231(4.07),288(4.12)	94-3567-78
Estra-1,3,5(10)-trien-6-one, 3,17β-di-hydroxy-	EtOH	222(4.30),255(3.93), 327(3.46)	94-3567-78
$C_{18}H_{22}O_4$			
Bicyclo[8.2.2]tetradeca-5,10,12,13-tet-raene-4,7-dicarboxylic acid, dimethyl ester, cis	EtOH	221(3.88),271(2.61), 277(2.56)	35-1806-78
trans	EtOH	222(3.91),265s(2.46), 271(2.58),278(2.54)	35-1806-78
Estra-1,3,5(10)-trien-6-one, 2,3,17β-trihydroxy-	EtOH	236(4.17),281(3.97), 322(3.84)	94-3567-78
9-Phenanthrenecarboxylic acid, 4b,5,6-7,8,10-hexahydro-2,4-dimethoxy-, methyl ester	MeOH	275(3.32),283(3.28)	44-0980-78
$C_{18}H_{22}O_5$			
Pterosin C, diacetate	EtOH	219(4.157),258(4.15), 290(3.24),300(3.25)	94-2365-78
Pterosin G, diacetate	EtOH	216(4.46),258(4.10), 295(3.37),302(3.36)	94-2365-78
$C_{18}H_{23}ClN_2S_2$			
Cyclopentanecarbodithioic acid, 2-imi-no-, 1-[(4-chlorophenyl)amino]cyclo-hexyl ester	EtOH	245(4.10),338(3.80), 408(4.32)	39-0558-78C
$C_{18}H_{23}NO_2$			
2-Azaestra-1,3,5(10)-tetraen-17-ol, 3-methoxy-, (±)-	MeOH	267(4.26)	44-0966-78
Ethanone, 1-(1,2,3,5,6,7-hexahydro-s-indacen-4-yl)-2-morpholino-, hydro-gen maleate	MeOH	257(3.95)	73-0970-78
2H-Pyrrole-3-carboxylic acid, 5-[2-(2-butenyl)phenyl]-3,4-dihydro-2,2-di-methyl-, methyl ester, (E)-	MeOH	230(3.88),240(3.85)	35-2181-78
$C_{18}H_{23}NO_3$			
Coccutrine	EtOH	232(3.90),286(3.29)	95-0886-78
Erysovine, dihydro-	EtOH	232(3.83),299(3.55)	95-0886-78
$C_{18}H_{23}NO_4S$			
1,3-Cyclohexadiene-1-carboxylic acid, 5-[[(4-methylphenyl)sulfonyl]amino]-, 1,1-dimethylethyl ester, (±)-	MeOH	239(4.07),283(3.63)	44-1448-78
$C_{18}H_{23}NO_6$			
Propanedioic acid, [[(2-ethoxy-2-oxo-ethyl)phenylamino]methylene]-, di-ethyl ester	EtOH	228(4.03),292(4.29)	99-2224-78
$C_{18}H_{23}N_5O_6$			
Riboflavin, 8-demethyl-8-(dimethyl-amino)-	n.s.g.	318(4.14),505(4.62)	69-1942-78

Compound	Solvent	λ_{max} (log ϵ)	Ref.
$C_{18}H_{23}N_7O_7$ 1H-Purine-2,6-dione, 7-[4-O-acetyl-6- azido-2,3-O-(1-methylethylidene)-α- D-mannopyranosyl]-3,7-dihydro-1,3- dimethyl-	MeOH	275.5(3.85)	136-0073-78C
$C_{18}H_{23}OP$ Phosphorin, 4-cyclohexyl-1,1-dihydro- 1-methoxy-1-phenyl-	EtOH	329(3.59),358(3.66)	89-0528-78
$C_{18}H_{24}ClN_2O_8PS_2$ Phosphorodiamidic acid, N'-(4-chloro- phenyl)-N,N-bis[2-[(methylsulfonyl)- oxy]ethyl]-, phenyl ester	MeOH	228(5.079),263(4.515)	65-0720-78
$C_{18}H_{24}FN_2O_8PS_2$ Phosphorodiamidic acid, N,N-bis[2-(meth- ylsulfonyl)oxy]ethyl]-N'-phenyl-, 4- fluorophenyl ester	MeOH	227(5.034),267(4.498)	65-0720-78
$C_{18}H_{24}NO_9PS_2$ Phosphoramidic acid, bis[2-[(methyl- sulfonyl)oxy]ethyl]-, diphenyl ester	MeOH	261(2.799)	65-0720-78
$C_{18}H_{24}N_2O_2$ Phenol, 4,4'-(1,2-diethyl-1,2-ethane- diyl)bis[2-amino-	MeOH	213(4.7),229(4.2), 293(3.9)	83-0184-78
$C_{18}H_{24}N_2O_4$ Morpholine, 4-[5-[1-(2-methoxyphenyl)- 2-nitroethyl]-1-cyclopenten-1-yl]-, (R*,S*)-	EtOH	215(4.57),262(4.24), 323(3.73)	2-0405-78
$C_{18}H_{24}N_2O_4S$ Spiro[2H-1-benzopyran-2,2'-[2H-1,3]thi- azine], 3',4',5',6'-tetrahydro-8- methoxy-3,3',5',6',6'-pentamethyl- 6-nitro-	EtOH	229(4.48),258(4.52), 267s(4.43),290(4.27), 357(4.19)	4-1439-78
$C_{18}H_{24}N_2S_2$ Cyclopentanecarbodithioic acid, 2-imi- no-, 1-(phenylamino)cyclohexyl ester	EtOH	235(4.01),336(3.72), 410(4.26)	39-0558-78C
$C_{18}H_{24}N_4O_2$ 1,2,4-Triazine-6-acetic acid, 5-(dieth- ylamino)-α-methyl-3-phenyl-, ethyl ester	EtOH	254(4.39)	114-0091-78B
$C_{18}H_{24}N_4O_5S$ Ethanethioic acid, S-[1-(2-acetoxyeth- yl)-2-[[(4-amino-2-methyl-5-pyrimi- dinyl)methyl](1,2-dioxopropyl)amino]- 1-propenyl] ester	EtOH	233(4.16),278(3.89)	94-0722-78
$C_{18}H_{24}N_4O_6$ Cyclohexanepropanoic acid, 2-[(2,4-di- nitrophenyl)hydrazono]-1-methyl-, ethyl ester	EtOH	365(4.3)	23-0419-78
$C_{18}H_{24}O_3$ 2-Cyclohexen-1-one, 4-(1-acetoxyocta-	EtOH	303.5(4.13)	39-0024-78

Compound	Solvent	$\lambda_{max}(\log \epsilon)$	Ref.
hydro-7a-methyl-5H-inden-5-ylidene)-, (±)- (cont.)			39-0024-78C
Cyclopenta[f][1]benzopyran-7(1H)-one, 2,3,5,6,6a,8-hexahydro-6a-methyl-3-(4-oxopentyl)-, (3S)-	n.s.g.	252(4.27)	44-1550-78
1,4-Phenanthrenedione, 4b,5,6,7,8,8a,9-10-octahydro-3-methoxy-4b,8,8-trimethyl-, (4bS-trans)-	EtOH	272(4.32),370(2.81)	23-0517-78
$C_{18}H_{24}O_4$			
Benzoic acid, 3-hydroxy-4-methoxy-, 1,7,7-trimethylbicyclo[2.2.1]hept-2-yl ester, (1S-endo) (rubaferin)	EtOH	263(3.94),298(3.68)	105-0606-78
Dispiro[5.0.5.1]tridecane-1,5,8,12-tetrone, 3,3,10,10,13-pentamethyl-	EtOH	288(2.48)	44-3070-78
$C_{18}H_{24}O_6$			
4H-1-Benzopyran-4-one, 6-ethyl-5,6,7,8-tetrahydro-5-(1-oxopropoxy)-3-[(1-oxopropoxy)methyl]-	MeOH	215(3.90),254(3.89)	119-0111-78
$C_{18}H_{24}O_{12}$			
Asperulosidic acid	MeOH	233(3.95)	32-0013-78
$C_{18}H_{25}NO$			
Azacyclotetradeca-3,7,11-triene, 14-(2-furanyl)-1-methyl-, (E,E,E)-	heptane	228(4.32)	70-2284-78
Butanenitrile, 2-[3,5-bis(1,1-dimethylethyl)-4-oxo-2,5-cyclohexadien-1-ylidene]-	n.s.g.	310(4.58),323(4.60)	70-1313-78
$C_{18}H_{25}NO_2$			
Apoerysopine, 1,2,3,3a,12b,12c-hexahydro-, dimethyl ether	EtOH	228(3.86),283(3.47)	44-0975-78
2-Azaestra-1,3,5(10)-trien-17-ol, 3-methoxy-, (±)-	MeOH	276(3.57)	44-0966-78
$C_{18}H_{25}NO_3$			
Cyclohexanone, 4-butyl-2-(2-nitro-1-phenylethyl)-	EtOH	211(4.09)	2-0405-78
$C_{18}H_{25}N_2O_8PS_2$			
Phosphorodiamidic acid, N,N-bis[2-[(methylsulfonyl)oxy]ethyl]-N'-phenyl-, phenyl ester	MeOH	230(5.090),257(4.550)	65-0720-78
$C_{18}H_{25}N_3OS$			
Azepino[4,3-b][1,4]benzothiazin-1(2H)-one, 11-[2-(diethylamino)ethyl]-3,4,5,11-tetrahydro-	EtOH	220(4.13),254(4.10), 286(3.73)	103-0306-78
$C_{18}H_{25}N_3O_5$			
Pyrano[2,3-c]pyrazole-2(6H)-acetic acid, 3,4-dimethyl-5-[2-(4-morpholinyl)ethyl]-6-oxo-, ethyl ester	EtOH	306.0(4.21)	95-0335-78
$C_{18}H_{25}N_3P$			
Phosphorus(1+), (4-diazenyl-N,N-dimethylbenzenaminato)diethylphenyl-, (T-4)-, tetrafluoroborate	n.s.g.	278(3.92),470(4.66)	123-0190-78

Compound	Solvent	$\lambda_{max}(\log \epsilon)$	Ref.
$C_{18}H_{25}N_5O_9$ Pyrido[3,2-c]pyridazine-1,2-dicarbox- ylic acid, 3-[1,2-bis(methoxycarbo- nyl)hydrazino]-3,4-dihydro-4-(1- methylethoxy)-, dimethyl ester	MeOH	210(4.03),250(3.90), 305(3.72)	103-0534-78
$C_{18}H_{26}Fe$ Ferrocene, 1,1',2,2',3,3',4,4'-octa- methyl-	EtOH	218($\underline{4.4}$),270s($\underline{3.0}$), 428($\underline{2.1}$)	101-0069-78B
Ferrocinium, octamethyl-, hexafluoro- phosphate	EtOH	215($\underline{4.1}$),276($\underline{4.2}$), 468s(--),620s(--), 760($\underline{2.6}$)	101-0069-78B
$C_{18}H_{26}N_2O_2$ 2,5-Cyclohexadiene-1,4-dione, 2,5-bis- (hexahydro-1H-azepin-1-yl)-	EtOH	228(4.38),379(4.35), 523(2.66)	119-0037-78
$C_{18}H_{26}N_2O_3$ Cyclo[3-[4-[2-(methylamino)ethyl]phen- oxy]propanoyl]-L-leucyl]	MeOH	226s(3.78),275(2.84)	35-8202-78
$C_{18}H_{26}O_2$ Cyclopenta[5,6]naphtho[2,1-b]pyran-7- ol, 2,3,4,6,6a,7,8,9,9a,10,11-dodeca- hydro-2,6a-dimethyl-, [2(R,S),6aS,7S]-	EtOH	248(4.20)	44-1550-78
$C_{18}H_{26}O_4$ Cyclohexanol, 1-(4-methoxyphenyl)-4- [(2-methyl-1,3-dioxolan-2-yl)methyl]-	EtOH	225(4.04),273(3.23), 282(3.18)	39-0024-78C
$C_{18}H_{26}O_5$ 4H-Cyclopent[f]oxacyclotridecin-1,4(6H)- dione, 7,8,9,11a,12,13,14,14a-octahy- dro-13-(methoxymethoxy)-6-methyl- (4-dehydrobrefeldin A)	EtOH	220(3.67)	35-4858-78
epimer	EtOH	224(3.72)	35-4858-78
$C_{18}H_{26}Si_2$ Silane, (bicyclo[4.2.0]octa-2,4-diene- 7,8-diyldi-1,2-ethynediyl)bis[tri- methyl-	hexane	296(4.11)	78-1037-78
Silane, 3,5,7,9-dodecatetraene-1,11- diyne-1,12-diylbis[trimethyl-, (E,E,Z,Z)-	hexane	317(4.48),332(4.78), 348(4.96),368(4.87)	78-1037-78
(Z,E,Z,Z)-	hexane	315(4.47),330(4.76), 346(4.94),365(4.86)	78-1037-78
all-E	hexane	320(4.47),336(4.78), 352(5.01),374(5.05)	78-1037-78
$C_{18}H_{27}NO$ Piperidine, 1-[1-(4-methoxyphenyl)- cyclohexyl]-	n.s.g.	250(3.98)	111-0017-78
hydrochloride	n.s.g.	230(4.11)	111-0017-78
$C_{18}H_{27}NO_3$ 2-Piperidinone, 1-(8-hydroxy-6-methyl- 1-oxo-2,4,6-dodecatrienyl)-, (E,E,E)-	EtOH	317(4.35)	18-2077-78
$C_{18}H_{27}N_3O_{11}$ 4-Pyridinecarboxylic acid hydrazide,	EtOH	223(3.80),263(3.67)	136-0089-78E

Compound	Solvent	$\lambda_{max}(\log \epsilon)$	Ref.
hydrazone with lactose (cont.)			136-0089-78E
$C_{18}H_{27}N_5O_5$ 9H-Purin-6-amine, 9-[5-O-(1-oxooctyl)- β-D-arabinofuranosyl]-	MeOH	259(4.16)	87-1218-78
$C_{18}H_{28}N_2$ Benzenamine, N-methyl-4-[1-(1-piperi- dinyl)cyclohexyl]- hydrochloride	n.s.g. n.s.g.	240(4.11) 245(3.97)	111-0017-78 111-0017-78
$C_{18}H_{28}N_2O$ Cyclohexanone, 2,6-bis[3-(dimethylami- no)-2-propenylidene]-3,4-dimethyl- Cyclohexanone, 2,6-bis[3-(dimethylami- no)-2-propenylidene]-3,5-dimethyl-	EtOH EtOH	510(4.95) 510(4.90)	70-0102-78 70-0102-78
$C_{18}H_{28}N_2O_4$ Pyrrolidinium, 1,1'-(1,4-butanediyl)- bis[1-(1-formyl-2-hydroxyethenyl)]-, dihydroxide, bis(inner salt)	EtOH	257(4.70)	73-1261-78
$C_{18}H_{28}N_6O_5$ Adenosine, N-[(heptylamino)carbonyl]-	pH 1 pH 13 70% EtOH	277(4.38) 269(4.29),277(4.24), 298(4.15) 269(4.35),276(4.28)	36-0569-78 36-0569-78 36-0569-78
$C_{18}H_{29}NO$ Benzene, 1,3,5-tris(1,1-dimethylethyl)-	C_6H_{12}	<u>305(3.5),334(3.68), 750(1.8)</u>	23-2665-78
$C_{18}H_{29}NO_2$ Acetamide, N-(decahydro-2,7,9-trimeth- yl-4-oxo-5,9-methanobenzocycloocten- 5(1H)-yl)-	EtOH	293(1.40)	44-3653-78
$C_{18}H_{29}NO_4$ Butanoic acid, 4-[(8-hydroxy-1-oxo- 2,4,6-dodecatrienyl)amino]-, ethyl ester, (E,E,E)-(±)-	EtOH	295(4.65)	18-2077-78
$C_{18}H_{29}NO_6Si$ 1H-Pyrrole-2,5-dione, 3-[5-O-[(1,1-di- methylethyl)dimethylsilyl]-2,3-O-(1- methylethylidene)-β-D-ribofuranosyl]-	MeOH	221(4.14)	88-1829-78
$C_{18}H_{29}N_2O$ Methanaminium, N-[3-[3-[3-(dimethylami- no)-2-propenylidene]-2-ethoxy-1-cy- clohexen-1-yl]-2-propenylidene]-N- methyl-, tetrafluoroborate iodide	EtOH EtOH	650(5.30) 640(5.36)	70-0339-78 70-0339-78
$C_{18}H_{29}N_2O_2$ Methanaminium, N-[3-[5-[3-(dimethylami- no)-2-propenylidene]-4-ethoxy-5,6-di- hydro-2-methyl-2H-pyran-3-yl]-2-prop- enylidene]-N-methyl-, ethyl sulfate	CHCl$_3$	650(5.20)	70-0339-78

Compound	Solvent	$\lambda_{max}(\log \epsilon)$	Ref.
$C_{18}H_{29}N_3O_2$			
1H-Benzimidazole-1-butanamine, N,N-diethyl-5,6-dimethoxy-δ-methyl-	pH 1	292(4.21)	4-0297-78
1H-Imidazole-4-carboxylic acid, 1-cyclohexyl-5-(cyclohexylamino)-, ethyl ester	EtOH	271(3.87)	118-0741B-78
$C_{18}H_{29}N_7O_4S$			
Adenosine, 3'-[[2-amino-1-oxo-3-(propylthio)propyl]amino]-3'-deoxy-N,N-dimethyl-, (R)-	pH 1	269(4.26)	87-0792-78
	pH 7	276(4.27)	87-0792-78
	pH 13	276(4.27)	87-0792-78
$C_{18}H_{30}N_4O_2$			
Acetamide, N,N'-(2,5-pyrazinediyldi-5,1-pentanediyl)bis-	EtOH	275(3.87),300s(3.12)	88-2217-78
	EtOH-HCl	278(3.87)	88-2217-78
$C_{18}H_{30}O$			
Spiro[4.11]hexadec-3-en-2-one, 1,3-dimethyl-	EtOH	232(4.03)	35-1799-78
$C_{18}H_{30}O_2$			
2,5-Cyclohexadien-1-one, 2,4,6-tris-(1,1-dimethylethyl)-4-hydroxy-	MeOH	240(3.96)	149-0685-78A
$C_{18}H_{30}O_4$			
2,4-Hexadienoic acid, 6-[3-(1-ethoxyethoxy)cyclopentyl]-3-methyl-, ethyl ester	EtOH	264(4.40)	56-0347-78
$C_{18}H_{30}O_5$			
Octadecanoic acid, 9,10,12-trioxo-	MeOH and MeOH-KOH	277(4.00)	2-0275-78
$C_{18}H_{30}Si_2$			
Silane, (3,4,5,6-tetramethyl-3,5-octadiene-1,7-diyne-1,8-diyl)bis[trimethyl-, (Z,Z)-	hexane	241(4.34)	5-1675-78
$C_{18}H_{31}NSi$			
2-Pentanamine, 2,4,4-trimethyl-N-[phenyl(trimethylsilyl)methylene]-	C_6H_{12}	214(3.87),239(3.48), 305(2.45)	23-2286-78
	isoPrOH	212(3.90),239(3.49), 306(2.40)	23-2286-78
$C_{18}H_{32}O$			
3,5-Pentadecadien-2-one, 6,10,14-trimethyl-, α,β-cis	EtOH	298(4.14)	104-1319-78
α,β-trans	EtOH	292(4.24)	104-1319-78
$C_{18}H_{32}O_3Si$			
Cyclohexanol, 1-(5-hydroxy-3-methyl-3-penten-1-ynyl)-2,2,6-trimethyl-6-[(trimethylsilyl)oxy]-	EtOH	230(4.15)	39-1511-78C
$C_{18}H_{34}B_2N_4$			
Boron, bis[μ-(3,5-dimethyl-1H-pyrazolato-N^1:N^2)]tetraethyldi-	hexane	224(4.25)	64-0220-78B

Compound	Solvent	$\lambda_{max}(\log \epsilon)$	Ref.
$C_{19}H_8O_4S_4$ 2H-Pyran-2,4,6(3H,5H)-trione, 3,5-bis- (3H-1,2-benzodithiol-3-ylidene)-	dioxan	243(4.52),275(4.40), 310(4.20),336(4.05), 447(4.24),543(4.66)	78-2175-78
$C_{19}H_9Br_2N_5O_2$ 1,2,3-Triazolo[4',5':3,4]cyclopenta- [1,2-d]pyridazine-4,8-dione, 7-bromo- 5-(3-bromophenyl)-2,5-dihydro-2-phen- yl-	MeOH	256(4.41),400(4.42)	56-0511-78
$C_{19}H_9Cl_2NS$ Benzeneacetonitrile, α-(6,8-dichloro- 3H-naphtho[1,8-bc]thien-3-ylidene)-	MeOH	208(4.4),224(4.4), 252(4.4),367(3.9), 386(4.0),408(3.9)	5-1974-78
$C_{19}H_9F_6NOS$ 1H-Pyrido[2,1-b]benzothiazol-1-one, 2-phenyl-3,4-bis(trifluoromethyl)-	MeOH	224(4.48),250(4.19), 372(4.22),390(4.23)	44-2697-78
$C_{19}H_{10}O$ 1H-Benzo[cd]fluoranthen-1-one	C_6H_{12}	243(4.56),294(3.91), 307(3.94),363(3.82), 379(4.03),403(4.17), 426(4.21),514s(2.67)	88-1539-78
11H-Benzo[cd]fluoranthen-11-one	C_6H_{12}	223(4.47),254(4.47), 278s(4.61),284(4.67), 304(3.99),318(3.96), 351(3.74),369(3.89), 389(3.76),409(3.78), 458(3.29),486(3.21), 523(2.83),534(2.71)	88-1539-78
$C_{19}H_{10}O_2$ 1H-Naphtho[2,1,8-mna]xanthen-1-one	MeOH	211s(4.43),225(4.57), 244(4.57),269(4.00), 280(4.15),299(3.87), 330(3.86),345(3.89), 366(3.64),464s(4.17), 493(4.32),522s(4.23)	88-0637-78
	MeOH-HCl	217(4.55),254(4.55), 271s(4.16),287(3.88), 310(3.89),322(3.91), 349(3.89),416(4.09), 513(4.62)	88-0637-78
$C_{19}H_{11}NO_3$ 1,4-Acridinedione, 2-hydroxy-9-phenyl-	dioxan	224(4.47),288(4.51), 358s(3.5)	150-4901-78
$C_{19}H_{11}NO_7$ 3H-Xanthen-3-one, 2,6,7-trihydroxy- 9-(2-nitrophenyl)-	pH 8.4	523(3.62)	140-0699-78
3H-Xanthen-3-one, 2,6,7-trihydroxy- 9-(3-nitrophenyl)-	pH 8.5	522(4.27)	140-0699-78
3H-Xanthen-3-one, 2,6,7-trihydroxy- 9-(4-nitrophenyl)-	pH 8.4	518(3.60)	140-0699-78
$C_{19}H_{11}NS$ Benzeneacetonitrile, α-5H-naphtho[1,8- bc]thien-5-ylidene-	MeOH	238(4.0),269(4.1), 407(4.4)	5-1974-78

Compound	Solvent	$\lambda_{max}(\log \epsilon)$	Ref.
$C_{19}H_{11}N_3$ 3,5-Pyridinedicarbonitrile, 2,6-di-phenyl-	EtOH	206(4.57),220s(4.30), 281(4.48)	73-0434-78
$C_{19}H_{11}N_3OS_2$ 2,4a-Epithio-4aH-pyrido[2,1-b]benzothi-azole-3,4-dicarbonitrile, 1,2,3,4-tetrahydro-1-oxo-2-phenyl-, (2α,3α,4β,4aα)-	MeOH	407(3.55)	44-2697-78
$C_{19}H_{11}N_5O_2$ 1,2,3-Triazolo[4',5':3,4]cyclopenta-[1,2-d]pyridazine-4,8-dione, 2,5-dihydro-2,5-diphenyl-	MeOH	250(4.42),254s(4.39), 397(4.42)	56-0511-78
$C_{19}H_{12}BrNO_2$ 1H-Pyrrole-2,5-dione, 3-bromo-1-methyl-4-(9-phenanthrenyl)-	MeOH	253(4.86),272(4.04), 285(4.09),375(3.48)	24-2677-78
$C_{19}H_{12}ClNO$ Quinoline, 6-chloro-2-(2-furanyl)-4-phenyl-	dioxan	358(3.57)	97-0400-78
$C_{19}H_{12}ClNS$ Quinoline, 6-chloro-4-phenyl-2-(2-thi-enyl)-	dioxan	359(3.50)	97-0400-78
$C_{19}H_{12}ClN_3O$ 4H-Pyran-3,5-dicarbonitrile, 2-amino-4-(4-chlorophenyl)-6-phenyl-	EtOH	295.6(3.81)	4-0057-78
4H-Pyran-3,5-dicarbonitrile, 2-amino-6-(4-chlorophenyl)-4-phenyl-	EtOH	298(4.40)	4-0057-78
$C_{19}H_{12}Cl_2O_2$ Benzoic acid, 4-chloro-, (4-chlorophen-yl)-2,4-cyclopentadien-1-ylidenemeth-yl ester	EtOH	245(4.33),306(4.31)	1-0463-78
$C_{19}H_{12}Cl_2S$ 3H-Naphtho[1,8-bc]thiophene, 6,8-di-chloro-3-(1-phenylethylidene)-, (E)-	MeOH	221(4.4),261(4.4), 304(3.9),336s(3.7), 381(3.9),400(3.9)	5-1974-78
$C_{19}H_{12}Cl_4O_3$ Spiro[1,3-benzodioxole-2,1'(4'H)-naph-thalen]-4'-one, 4,5,6,7-tetrachloro-2'-(1-methylethyl)-	EtOH	221(4.72),296(3.79), 306(3.75)	2-0668-78
Spiro[1,3-benzodioxole-2,2'(1'H)-naph-thalen]-1'-one, 4,5,6,7-tetrachloro-3'-(1-methylethyl)-	EtOH	221(4.67),240(4.72), 245(4.77),300s(3.82), 305(3.83),375(3.32)	2-0668-78
$C_{19}H_{12}N_2O$ 8H-Cyclohepta[b]quinoxalin-8-one, 7-phenyl-	EtOH	228(4.14),277(4.55)	56-2045-78
12H-Dibenzo[b,f]quinolizine-7-carbo-nitrile, 5-methyl-12-oxo-	EtOH	232s(4.54),236(4.55), 279s(4.12),295(4.11), 306(4.08),386(4.24), 404(4.26),420(4.05)	44-3536-78
$C_{19}H_{12}N_2O_2$ 5H-Indeno[1,2-b]pyridine, 5-[(4-nitro-	n.s.g.	205(4.52),226(4.61),	103-1343-78

Compound	Solvent	$\lambda_{max}(\log \epsilon)$	Ref.
phenyl)methylene]- (cont.)		266(4.36),306(4.31), 346(4.24)	103-1343-78
$C_{19}H_{12}N_4O_3$ 4H-Pyran-3,5-dicarbonitrile, 2-amino-4-(4-nitrophenyl)-6-phenyl-	EtOH	296(4.53)	4-0057-78
4H-Pyran-3,5-dicarbonitrile, 2-amino-6-(3-nitrophenyl)-4-phenyl-	EtOH	296(3.85)	4-0057-78
$C_{19}H_{12}O_2$ Cyclopenta[b][1]benzopyran-1-carbox-aldehyde, 2-phenyl-	THF THF	480(3.39) 255(4.22),300(4.18), 360(4.47),480(3.39)	103-1070-78 103-1075-78
$C_{19}H_{12}O_2Se$ 1H-Indene-1,3(2H)-dione, 2-(2-methyl-4H-1-benzoselenin-4-ylidene)-	dioxan	338(3.272)	83-0170-78
$C_{19}H_{12}O_5$ 3H-Xanthen-3-one, 2,6,7-trihydroxy-9-phenyl-	pH 10.0	552(3.51)	140-0699-78
$C_{19}H_{12}O_7$ Versicolorin A, 6-O-methyl-	MeOH	224(3.88),287(3.86), 316(3.32),442(3.33)	39-0961-78C
$C_{19}H_{12}O_9$ 2(3H)-Benzofuranone, 5,6-dihydroxy-3-(3-hydroxy-4-(4-hydroxy-3-meth-oxyphenyl)-5-oxo-2(5H)-furanyli-dene]-, (E)-	EtOH-HCl	277(3.96),495(3.80)	64-0820-78C
$C_{19}H_{13}BrClN_5O_2$ Benzenecarbohydrazonoyl chloride, 3-bromo-N-[4-[(4-nitrophenyl)-azo]phenyl]-	dioxan	247(4.16),268(4.08), 313(4.02),350(4.08), 441(4.38)	104-2203-78
Benzenecarbohydrazonoyl chloride, 4-bromo-N-[4-[(2-nitrophenyl)-azo]phenyl]-	dioxan	245(4.34),267(4.18), 310(3.95),427(4.57)	104-2203-78
Benzenecarbohydrazonoyl chloride, 4-bromo-N-[4-[(4-nitrophenyl)-azo]phenyl]-	dioxan	246(4.23),267(4.16), 313(4.08),347(4.13), 445(4.40)	104-2203-78
$C_{19}H_{13}Br_2NO_2$ 9H-Phenanthro[9',10':3,4]cyclobuta[1,2-c]pyrrole-9,11(10H)-dione, 8c,11a-di-bromo-8b,8c,11a,11b-tetrahydro-10-methyl-, anti-anti syn-syn	MeOH	213(4.91),272(4.24)	24-2677-78
	MeOH	203(4.65),270(4.07)	24-2677-78
$C_{19}H_{13}ClFNO_3S$ 3-Quinolinecarboxylic acid, 6-chloro-4-(2-fluorophenyl)-2-(methoxythioxo-methyl)-, methyl ester	isoPrOH	215s(4.46),249(4.57), 310s(3.84),340s(3.49), 408(2.76)	4-0687-78
$C_{19}H_{13}ClN_2$ Quinoline, 6-chloro-4-phenyl-2-(1H-pyrrol-2-yl)-	dioxan	366(3.41)	97-0400-78
$C_{19}H_{13}ClN_2S$ Imidazo[1,2-a]pyridine, 2-(4-chloro-	DMF	369(4.65),389(4.59)	33-0129-78

Compound	Solvent	λ_{max}(log ϵ)	Ref.
phenyl)-7-[2-(2-thienyl)ethenyl]- (cont.)			33-0129-78
Quinoline, 6-chloro-2-(2-methyl-4-thia-zolyl)-4-phenyl-	dioxan	348(3.26)	97-0400-78
C$_{19}$H$_{13}$ClO$_6$			
2H,12H-Pyrano[2,3-a]xanthene-2,12-dione, 8-chloro-5,11-dimethoxy-4-methyl-	MeOH	209(4.24),246(4.25), 263(4.03),273(3.99), 320(4.11)	119-0143-78
C$_{19}$H$_{13}$Cl$_2$N$_5$O$_2$			
Benzenecarbohydrazonoyl chloride, 4-chloro-N-[4-(4-nitrophenyl)azo]phen-yl]-	dioxan	243(4.28),268(4.20), 311(4.11),347(4.13), 447(4.45)	104-2203-78
C$_{19}$H$_{13}$Cl$_3$N$_2$O$_7$S			
Acetamide, N-[4-[5-[2-(5-nitro-2-furan-yl)-2-[(trichloromethyl)sulfonyl]eth-enyl]-2-furanyl]phenyl]-	EtOH	206(4.39),247(4.16), 274(4.32),341(4.24), 442(4.33)	73-1618-78
C$_{19}$H$_{13}$F			
Chrysene, 1-fluoro-4-methyl-	hexane	270(5.15)	87-0038-78
Chrysene, 1-fluoro-5-methyl-	hexane	269(4.99)	87-0038-78
Chrysene, 3-fluoro-5-methyl-	hexane	269.5(5.04)	87-0038-78
Chrysene, 6-fluoro-5-methyl-	hexane	269(5.04)	87-0038-78
Chrysene, 7-fluoro-5-methyl-	hexane	271(4.91)	87-0038-78
Chrysene, 9-fluoro-5-methyl-	hexane	268(4.93)	87-0038-78
Chrysene, 11-fluoro-5-methyl-	hexane	266(4.97)	87-0038-78
Chrysene, 12-fluoro-5-methyl-	hexane	268.7(5.11)	87-0038-78
C$_{19}$H$_{13}$N			
Acridine, 9-phenyl-	EtOH	219s(--),254(5.10), 359(4.00)	39-1211-78C
C$_{19}$H$_{13}$NO			
2H-Benzo[a]quinolizin-2-one, 4-phenyl-	EtOH	249(4.49),266(4.47), 322s(4.04),334(4.07), 352(4.03),372(3.91)	95-0198-78
C$_{19}$H$_{13}$NOS$_2$			
3H-Phenothiazin-3-one, 2-[(4-methyl-phenyl)thio]-	CHCl$_3$	265(4.52),457(4.26)	103-0795-78
3H-Phenothiazin-3-one, 7-[(4-methyl-phenyl)thio]-	CHCl$_3$	244(4.38),373(3.92), 523(3.68)	103-0795-78
C$_{19}$H$_{13}$NO$_2$			
Benz[b]acridine-6,11-dione, 2,12-di-methyl-	dioxan	235(4.48),294(4.60)	150-4901-78
Benz[b]acridine-6,11-dione, 4,12-di-methyl-	dioxan	238(4.52),269s(4.4), 294(4.55),376(3.41)	150-4901-78
C$_{19}$H$_{13}$NO$_2$S			
Phenanthridine, 6-(phenylsulfonyl)-	EtOH	223s(4.46),230(4.49), 237(4.48),250(4.31), 307(3.74),322(3.76), 336(3.69)	95-0136-78
C$_{19}$H$_{13}$NO$_3$			
Benz[b]acridine-6,11-dione, 4-methoxy-12-methyl-	dioxan	254(4.61),298(4.42), 397s(3.6)	150-4901-78

Compound	Solvent	$\lambda_{max}(\log \epsilon)$	Ref.
$C_{19}H_{13}NO_5$			
Luguine	dioxan	255(4.44),334(4.36), 345(4.39),384(3.65)	88-2923-78
	dioxan-HCl	250(4.33),269(4.29), 352(4.27),362(4.30), 458(3.80)	88-2923-78
	dioxan-base	255(4.44),334(4.36), 345(4.39),384(3.65)	88-2923-78
$C_{19}H_{13}N_3$			
12H-Dibenzo[b,f]quinolizine-7-carbonitrile, 12-imino-5-methyl-	EtOH	236(4.51),279s(4.12), 301(4.17),311(4.16), 392s(4.02),414(4.08), 438(4.08),464s(3.84)	44-3536-78
3,5-Pyridinedicarbonitrile, 1,4-dihydro-2,6-diphenyl-	EtOH	205(4.54),242(4.34), 264(4.27),365(4.15)	73-0434-78
$C_{19}H_{13}N_3O$			
4H-Pyran-3,5-dicarbonitrile, 2-amino-4,6-diphenyl-	EtOH	298(3.85)	4-0057-78
$C_{19}H_{13}N_3O_2$			
4H-Pyran-3,5-dicarbonitrile, 2-amino-4-(4-hydroxyphenyl)-6-phenyl-	EtOH	298(4.00)	4-0057-78
$C_{19}H_{13}N_5O_3S$			
6H-Anthra[9,1-cd]isothiazol-6-one, 7-[(4,6-dimethoxy-1,3,5-triazin-2-yl)amino]-	DMF	445(4.10)	2-1007-78
$C_{19}H_{14}$			
Chrysene, 5-methyl-	EtOH	271(5.00),287(4.00), 301(3.96),313(4.10), 327(4.08)	44-1656-78
14H-9,10[1',2']-endo-Cyclopentanthracene, 9,10-dihydro-	THF	226(4.58),254(3.72), 267(3.59),275(3.48)	23-0080-78
$C_{19}H_{14}Br_2O_2$			
2H-Pyran-2-one, 3,6-bis(4-bromophenyl)-4,5-dimethyl-	MeCN	248(3.88),325(4.20)	44-2138-78
4H-Pyran-4-one, 3,5-bis(4-bromophenyl)-2,6-dimethyl-	EtOH	229(4.62),261s(4.32)	44-2138-78
$C_{19}H_{14}ClFN_4O_3$			
6H-[1,2,4]Triazolo[1,5-a][1,4]benzodiazepine-2-carboxylic acid, 8-chloro-6-(2-fluorophenyl)-4-methoxy-, methyl ester	MeOH	267(4.00)	33-0848-78
$C_{19}H_{14}ClNO$			
Benzo[h]quinolin-2(1H)-one, 4-(2-chlorophenyl)-5,6-dihydro-	CHCl₃	295(4.23)	48-0097-78
Benzo[h]quinolin-2(1H)-one, 4-(4-chlorophenyl)-5,6-dihydro-	CHCl₃	295(4.23)	48-0097-78
$C_{19}H_{14}ClNO_2S$			
2H,5H-[1]Benzothiopyrano[4,3-b]pyran-2-one, 3-chloro-4-(methylphenylamino)-	EtOH	243(4.51),274(3.87), 301s(3.70),344(4.05), 387(4.11)	4-0181-78

Compound	Solvent	$\lambda_{max}(\log \epsilon)$	Ref.
$C_{19}H_{14}ClN_3O$			
4H-Pyrido[1,2-a]-1,3,5-triazin-4-one, 2-(4-chlorophenyl)-2,3-dihydro-3-phenyl-	C_6H_{12}	238(4.18),356(3.38)	78-0101-78
$C_{19}H_{14}ClN_3O_2$			
1H-Pyrazole, 1-benzoyl-4-(5-chloro-2,1-benzisoxazol-3-yl)-3,5-dimethyl-	EtOH	248(4.07),342(4.07)	94-1141-78
$C_{19}H_{14}ClN_5O_2$			
Benzenecarbohydrazonoyl chloride, N-[4-[(2-nitrophenyl)azo]phenyl]-	dioxan	245(4.18),260(3.95), 307(3.78),425(4.43)	104-2203-78
Benzenecarbohydrazonoyl chloride, N-[4-[(4-nitrophenyl)azo]phenyl]-	dioxan	234(4.19),268(4.08), 312(4.08),341(4.08), 446(4.38)	104-2203-78
$C_{19}H_{14}Cl_2N_2OS_2$			
Hydrazinecarbodithioic acid, [[5-(3,4-dichlorophenyl)-2-furanyl]methylene]-, phenylmethyl ester	dioxan	382(4.57)	73-2643-78
$C_{19}H_{14}Cl_2O_2$			
2H-Pyran-2-one, 3,6-bis(4-chlorophenyl)-4,5-dimethyl-	MeCN	242(4.10),325(4.22)	44-2138-78
4H-Pyran-4-one, 3,5-bis(4-chlorophenyl)-2,6-dimethyl-	EtOH	225(4.49),256(4.14)	44-2138-78
$C_{19}H_{14}F_2O_2$			
2H-Pyran-2-one, 3,6-bis(4-fluorophenyl)-4,5-dimethyl-	MeCN	233(3.97),325(4.02)	44-2138-78
4H-Pyran-4-one, 3,5-bis(4-fluorophenyl)-2,6-dimethyl-	EtOH	255(4.00),299(2.51)	44-2138-78
$C_{19}H_{14}N_2$			
1,1'-Biphenyl, 4-(diazophenylmethyl)-	benzene	529(2.15)	39-1283-78B
2,2'-Spirobi[2H-indene]-5,5'-dicarbonitrile, 1,1',3,3'-tetrahydro-	EtOH	205(4.91),209s(4.86), 237(4.48),242s(4.47), 260s(3.22),268s(3.38), 275(3.53),277s(3.48), 284(3.59)	49-0987-78
$C_{19}H_{14}N_2O$			
Benzene, 1-(diazophenylmethyl)-4-phenoxy-	benzene	531(2.01)	39-1283-78B
Benzo[f]quinazoline, 3-(4-methoxyphenyl)-	EtOH	204s(4.20),227(4.41), 263s(4.05),296(4.57)	103-1025-78
$C_{19}H_{14}N_2O_2$			
2,5-Cyclohexadien-1-one, 4-diazo-3-methoxy-2,6-diphenyl-	MeOH	287(3.84),370(4.38)	70-2455-78
$C_{19}H_{14}N_2O_3$			
1H-1,2-Diazepin-4-ol, 1-benzoyl-, benzoate	MeOH	235(4.36),273s(3.90), 350s(2.76)	33-2887-78
1H-1,2-Diazepin-6-ol, 1-benzoyl-, benzoate	MeOH	233(4.29),270s(3.83), 350s(2.58)	33-2887-78
Pyridinium, 1-(benzoylamino)-3-(benzoyloxy)-, hydroxide, inner salt	EtOH	330(3.72)	33-2887-78
Pyrrolo[2,1-b]quinazolin-9(1H)-one, 3-[(benzoyloxy)methylene]-2,3-dihydro-	EtOH	312(3.96),368(4.23)	142-1729-78A

Compound	Solvent	λ_{max}(log ϵ)	Ref.
$C_{19}H_{14}N_2S$			
Imidazo[1,2-a]pyridine, 2-phenyl-7-[2-(2-theinyl)ethenyl]-	DMF	370(4.63),388(4.58)	33-0129-78
$C_{19}H_{14}N_4$			
Propanedinitrile, [(4,5-dihydro-4-phenyl-1H-pyrazol-3-yl)phenylmethylene]-	EtOH	320(3.63),406(4.39)	78-1163-78
$C_{19}H_{14}N_4OS$			
4(1H)-Pteridinone, 2,3-dihydro-1-methyl-6,7-diphenyl-2-thioxo-	pH 6.0	285s(4.29),306(4.44), 378(4.24)	24-2571-78
	pH 11.0	221(4.47),252(4.23), 290(4.40),383(4.30)	24-2571-78
	MeOH	225s(4.30),280s(4.27), 309(4.46),381(4.23)	24-2571-78
4(1H)-Pteridinone, 2,3-dihydro-3-methyl-6,7-diphenyl-2-thioxo- (in 10% methanol)	pH 3.0	220s(4.38),282s(4.34), 303(4.44),383(4.25)	24-0971-78 +24-2571-78
	pH 10.0	226(4.38),270(4.16), 323(4.50),395(4.08)	24-0971-78
	MeOH	225s(4.33),280s(4.27), 306(4.46),383(4.19)	24-0971-78 +24-2571-78
4(3H)-Pteridinone, 2-(methylthio)-6,7-diphenyl-	pH 5.0	263(4.29),302(4.32), 368(4.18)	24-2571-78
	pH 9.0	258s(4.23),282(4.44), 378(4.18)	24-2571-78
	MeOH	220s(4.35),260(4.30), 295(4.30),365(4.15)	24-2571-78
$C_{19}H_{14}N_4O_2$			
11H-Dibenzo[b,e][1,4]diazepin-11-one, 5,10-dihydro-8-[(4-hydroxyphenyl)-azo]-	DMF	412(4.08)	42-0154-78
$C_{19}H_{14}N_4O_3$			
11H-Dibenzo[b,e][1,4]diazepin-11-one, 8-[(2,4-dihydroxyphenyl)azo]-5,10-dihydro-	DMF	455(4.09)	42-0154-78
$C_{19}H_{14}N_4O_4$			
Methanone, diphenyl-, 2,4-dinitrophenylhydrazone	toluene	382(4.45)	104-2229-78
	EtOH	242s(4.17),293s(3.75), 379(4.40)	104-2229-78
$C_{19}H_{14}N_4S_2$			
2,4(1H,3H)-Pteridinedithione, 3-methyl-6,7-diphenyl-	pH 3.0	230(4.35),252s(4.19), 282(4.32),315(4.45), 425(4.25)	24-0971-78
	pH 10.0	229(4.42),274(4.25), 329(4.53),445(4.10)	24-0971-78
	MeOH	228(4.37),252(4.22), 283(4.31),316(4.46), 424(4.23)	24-0971-78
$C_{19}H_{14}N_6O_5$			
Benzoic acid, 4-nitro-, 2-[4-[(4-nitrophenyl)azo]phenyl]hydrazide	dioxan	265(4.19),407(4.31)	104-2203-78
$C_{19}H_{14}O$			
Chrysene, 5-methoxy-	hexane	265.5(3.85)	87-0038-78

Compound	Solvent	$\lambda_{max}(\log \epsilon)$	Ref.
$C_{19}H_{14}O_2S$			
2-Propynoic acid, 3-phenyl-, 2-benzo-[b]thien-2-ylethyl ester	MeOH	258(4.39)	44-2493-78
$C_{19}H_{14}O_3$			
Benzo[1,2-b:4,5-b']dipyran-2(7H)-one, 4-methyl-9-phenyl-	MeOH	226(3.92),280(3.42), 344(3.40)	2-0856-78
Naphtho[2,3-c]furan-1(3H)-one, 3-(2-methoxyphenyl)-	THF	240(4.92),274(3.86), 283(3.89),297(3.69), 322(3.19),337(3.32)	78-0113-78
$C_{19}H_{14}O_4$			
Acetic acid, (5-oxo-3,4-diphenyl-2(5H)-furanylidene)-, methyl ester	EtOH	237(4.20),290(3.83)	33-3018-78
1H-Indene-1,3(2H)-dione, 2-[3-(3-methoxyphenyl)-1-oxo-2-propenyl]-	pH 1	247(4.48),318(4.00), 408(4.60)	65-0142-78
	pH 7.96	236(4.56),306(4.30), 368(4.46)	65-0142-78
	pH 13	238(4.57),306(4.08), 358(4.23),430(4.32)	65-0142-78
	$CHCl_3$	402(4.51)	65-0142-78
1H-Indene-1,3(2H)-dione, 2-[3-(4-methoxyphenyl)-1-oxo-2-propenyl]-	pH 1	250(4.34),305(3.81), 417(4.70)	65-0142-78
	pH 13	240(4.58),306(4.28), 368(4.56)	65-0142-78
	$CHCl_3$	416(4.64)	65-0142-78
2-Naphthalenecarboxylic acid, 3-(2-methoxybenzoyl)-	THF	232(4.70),307(3.59), 335(3.33)	78-0113-78
$C_{19}H_{14}O_6$			
1,3-Propanedione, 1-(1,3-benzodioxol-5-yl)-3-(4-methoxy-5-benzofuranyl)-	MeOH	240(4.03),290(3.44), 365(4.03)	2-0658-78
$C_{19}H_{14}O_9$			
Benzeneacetic acid, 3,4-dihydroxy-α-[3-hydroxy-4-(4-hydroxy-3-methoxyphenyl)-5-oxo-2(5H)-furanylidene]-	H_2O	259(4.24),383(3.94)	64-0820-78C
	EtOH	263(4.18),403(3.93)	64-0820-78C
$C_{19}H_{14}S_2$			
10H-Thioxanthene, 10-(phenylthio)-	EtOH	500(2.52)(changing to 460(1.47))	142-0383-78C
$C_{19}H_{15}AsO_2$			
4-Arseninol, 2,6-diphenyl-, acetate	EtOH	237(4.23),287(4.54)	88-1471-78
$C_{19}H_{15}BF_4OS$			
Naphtho[1,8-bc]thiolium, 5-ethoxy-2-phenyl-, tetrafluoroborate	CH_2Cl_2	456(4.11)	83-0324-78
$C_{19}H_{15}BrN_2OS_2$			
Hydrazinecarbodithioic acid, [[5-(4-bromophenyl)-2-furanyl]methylene]-, phenylmethyl ester	dioxan	383(4.66)	73-2643-78
$C_{19}H_{15}ClFNO_3$			
3-Quinolinecarboxylic acid, 6-chloro-4-(2-fluorophenyl)-2-(methoxymethyl)-, methyl ester	isoPrOH	210(4.51),235(4.74), 268(3.78),313(3.46), 327(3.51)	4-0687-78
$C_{19}H_{15}ClFN_3$			
4H-Imidazo[1,5-a][1,4]benzodiazepine,	isoPrOH	218(4.51),240s(4.28),	44-0936-78

Compound	Solvent	$\lambda_{max}(\log \epsilon)$	Ref.
8-chloro-6-(2-fluorophenyl)-1,3-di-methyl- (cont.)		265s(3.93)	44-0936-78
6H-Imidazo[1,5-a][1,4]benzodiazepine, 8-chloro-6-(2-fluorophenyl)-1,6-di-methyl-	isoPrOH	228s(4.22),263(3.94), 270(3.93),280s(3.85)	44-0936-78
$C_{19}H_{15}ClFN_5O$ 4H-[1,2,4]Triazolo[1,5-a][1,4]benzodi-azepine-2-carboxamide, 8-chloro-6-(2-fluorophenyl)-N,N-dimethyl-	MeOH	224(4.56),250s(4.32)	33-0848-78
$C_{19}H_{15}ClN_2O$ 1-Naphthaleneacetic acid, [(4-chloro-phenyl)methylene]hydrazide	EtOH	286s(4.39),294(4.38), 308s(4.16)	34-0345-78
$C_{19}H_{15}ClN_2OS_2$ Hydrazinecarbodithioic acid, [[5-(2-chlorophenyl)-2-furanyl]methylene]-, phenylmethyl ester	dioxan	377(4.65)	73-2643-78
Hydrazinecarbodithioic acid, [[5-(4-chlorophenyl)-2-furanyl]methylene]-, phenylmethyl ester	dioxan	382(4.68)	73-2643-78
$C_{19}H_{15}ClN_2S$ 2(1H)-Pyrimidinethione, 4-(4-chloro-phenyl)-6-phenyl-1-(2-propenyl)-	C_6H_{12} EtOH	294(4.51),435(3.31) 298(4.53),404(3.40)	4-0105-78 4-0105-78
2(1H)-Pyrimidinethione, 6-(4-chloro-phenyl)-4-phenyl-1-(2-propenyl)-	EtOH	294(4.59),397(3.60)	4-0105-78
$C_{19}H_{15}ClN_4O_2$ Benzo[g]pteridine-2,4(3H,10H)-dione, 10-[(3-chlorophenyl)methyl]-7,8-di-methyl-	n.s.g.	357(4.28),444(4.08)	69-1942-78
3H-Pyrazol-3-one, 2-(7-chloro-5-phenyl-3H-1,4-benzodiazepin-2-yl)-1,2-dihy-dro-5-methyl-, N-oxide	isoPrOH	220s(4.36),237s(4.27), 298(4.16),367(4.19)	4-0161-78
$C_{19}H_{15}Cl_2NO_2S$ 2H,5H-[1]Benzothiopyrano[4,3-b]pyran-2-one, 3,3-dichloro-3,4-dihydro-4-(methylphenylamino)-	EtOH	242(4.48),350(3.22)	4-0181-78
$C_{19}H_{15}Cl_2N_3O$ Pyridinium, 1-[(2,6-dichlorophenyl)-methyl]-4-[[(phenylamino)carbonyl]-amino]-, hydroxide, inner salt	MeOH CH_2Cl_2	295(3.8),330s(3.57) 340(4.81)	64-0084-78B 64-0084-78B
$C_{19}H_{15}N$ Acridine, 9,10-dihydro-9-phenyl-	EtOH	224s(--),254s(--), 289(4.13),312s(--)	39-1211-78C
10H-Azepino[1,2-a]indole, 11-phenyl-	EtOH	234(4.41),271(4.24), 305s(--),325s(--)	39-1211-78C
$C_{19}H_{15}NO$ 8H-Azepino[1,2-a]indol-8-one, 9,10-di-hydro-11-phenyl-	EtOH	227(4.45),251s(--), 276(4.19),279s(--), 347(4.20)	39-1211-78C
Benzo[h]quinolin-2(1H)-one, 5,6-dihy-dro-4-phenyl-	CHCl$_3$	300(4.13)	48-0097-78
2H-Benzo[a]quinolizin-2-one, 6,7-di-hydro-4-phenyl-	EtOH	257(4.48),282s(4.27)	95-0198-78

Compound	Solvent	$\lambda_{max}(\log \epsilon)$	Ref.
1,4-Ethenoisoquinolin-3(2H)-one, 1,4-dihydro-2-(2-phenylethenyl)-, (E)-	MeCN	297(4.33)	78-2617-78
$C_{19}H_{15}NOS_2$			
[1]Benzopyrano[2,3-b]pyrrole-2,4-dithione, 1,3-dihydro-1,3-dimethyl-3-phenyl-	MeOH	244(4.13),328(4.34), 404(4.36)	83-0018-78
$C_{19}H_{15}NO_2$			
Benzo[6,7]cyclohept[1,2-b]indol-7(12H)-one, 6-methoxy-12-methyl-	MeOH	244(4.30),266(4.12), 295(4.72)	18-3579-78
1H-Dibenz[e,g]indole-2-carboxylic acid, ethyl ester	EtOH and heptane	203(4.24),244(4.41), 254(4.61),262(4.82), 269(4.61),320(4.27), 339(3.99),355(4.06)	103-0851-78
Dibenzo[f,h]isoquinolin-1(2H)-one, 3-ethoxy-	EtOH	242s(4.52),260(4.58), 271s(5.42),346s(3.87), 363(4.04),381s(3.99)	12-0225-78
4,11-Etheno-1H-naphth[2,3-f]isoindole-1,3(2H)-dione, 3a,4,11,11a-tetrahydro-2-methyl-, (3aα,4β,11β,11aα)-	ether	256(3.80),265(3.87), 275(3.87),286(3.65), 302(2.39),306(2.50), 310s(2.26),314(2.22), 319(2.16)	78-0073-78
$C_{19}H_{15}NO_3$			
2-Naphthalenecarboxamide, N-ethyl-1,4-dihydro-1,4-dioxo-	EtOH	214(4.34),248(4.29), 257(4.27),352s(3.48)	104-1962-78
$C_{19}H_{15}NO_4$			
4,11[1',2']-Benzeno-1H-[1,6]benzodioxocino[3,4-c]pyrrole-1,3(2H)-dione, 3a,4,11,11a-tetrahydro-2-methyl-, (3aα,4β,11β,11aα)-	ether	263s(2.94),270(3.00), 277(2.97)	78-0073-78
7,11-Etheno-7H-bisoxireno[1,2:3,4]naphtho[2,3-f]isoindole-8,10(7aH,9H)-dione, 1a,5b,10a,11-tetrahydro-9-methyl-, (1aα,5bα,6aS*,7α,7aβ,10aβ,11α-11aR*)-	THF	250(2.58),265(2.66), 272(2.73),278(2.72)	78-0073-78
Phenol, 2-[(4-hydroxyphenyl)(2-nitrophenyl)methyl]-	EtOH	228(4.33),283(3.64)	39-1211-78C
$C_{19}H_{15}NO_7$			
2(5H)-Furanone, 3-methoxy-4-(4-methoxybenzoyl)-5-(2-nitrophenyl)-	EtOH	296(4.18)	4-0737-78
$C_{19}H_{15}N_2P$			
Cyanamide, (triphenylphosphoranylidene)-	CH_2Cl_2	304(3.74)	39-1237-78C
$C_{19}H_{15}N_3$			
Benzene, 1-azido-2-(diphenylmethyl)-	EtOH	222s(--),254(4.03), 263s(--),280s(--), 289s(--)	39-1211-78C
1,10-Phenanthrolin-4-amine, N-methyl-N-phenyl-	neutral	220(4.35),245(4.41), 277(4.33),330(4.25), 347s(4.21)	39-1215-78B
	monocation	246(4.36),268(4.40), 285(4.35),325(4.30)	39-1215-78B
1,10-Phenanthrolin-4-amine, 1-methyl-N-phenyl-	neutral	225(4.36),305(3.85), 375(3.98)	39-1215-78B

Compound	Solvent	$\lambda_{max}(\log \epsilon)$	Ref.
1,10-Phenanthrolin-4-amine, 1-methyl- N-phenyl- (cont.)	monocation	217(4.40),247(4.38), 268(4.40),283s(4.36), 337(4.28)	39-1215-78B
$C_{19}H_{15}N_3O$			
3H-Pyrazol-3-one, 2,4-dihydro-4-(1H-in- dol-3-ylmethylene)-5-methyl-1-phenyl-	heptane NaOEt	398(4.211),408(4.362) 454(4.641)	103-1088-78 103-1088-78
4H-Pyrido[1,2-a]-1,3,5-triazin-4-one, 2,3-dihydro-2,3-diphenyl-	C_6H_{12}	240(4.18),354(3.41)	78-0101-78
Pyrrolo[2,1-a]phthalazine-2-carboxam- ide, 3-(4-methylphenyl)-	EtOH	255(4.39),293(4.75), 372(3.85),450(3.72)	118-0206-78
$C_{19}H_{15}N_3O_2$			
Pyridinium, 1,3-bis(benzoylamino)-, hydroxide, inner salt	EtOH	324(3.70)	33-2887-78
Pyrrolo[2,1-a]phthalazine-2-carboxam- ide, 3-(4-methoxyphenyl)-	EtOH	282(4.58),293(4.59), 370(3.85),450(3.63)	118-0206-78
$C_{19}H_{15}N_3O_2S$			
Benzenesulfonamide, 4-methyl-N-1,10- phenanthrolin-4-yl-	neutral	220(4.40),238(4.46), 269(4.37),305(4.28)	39-1215-78B
	cation	215(4.41),240s(4.10), 275(4.31),302(4.17), 327s(4.07),393s(4.00)	39-1215-78B
$C_{19}H_{15}N_3O_3S_2$			
Hydrazinecarbodithioic acid, [[5-(2-ni- trophenyl)-2-furanyl]methylene]-, phenylmethyl ester	dioxan	370(4.55)	73-2643-78
Hydrazinecarbodithioic acid, [[5-(3-ni- trophenyl)-2-furanyl]methylene]-, phenylmethyl ester	dioxan	378(4.50)	73-2643-78
Hydrazinecarbodithioic acid, [[5-(4-ni- trophenyl)-2-furanyl]methylene]-, phenylmethyl ester	dioxan	404(4.63)	73-2643-78
$C_{19}H_{15}N_3O_4$			
Pyridine, 4-[(2,4-dinitrophenyl)meth- ylene]-1,4-dihydro-1-(phenylmethyl)-	EtOH	240(4.3),340(3.7), 584(4.3)	104-1847-78
$C_{19}H_{15}N_5O$			
Acetamide, N-(3,4-diphenylpyrazolo[5,1- c][1,2,4]triazin-7-yl)-	EtOH	271(4.57),348(3.46)	39-0885-78C
$C_{19}H_{15}N_5O_3$			
Benzoic acid, 2-[4-[(4-nitrophenyl)- azo]phenyl]hydrazide	dioxan	227(4.16),267(3.95), 410(4.23)	104-2203-78
$C_{19}H_{15}N_7OS$			
6H-Anthra[9,1-cd]isothiazol-6-one, 7- [[4,6-bis(methylamino)-1,3,5-triazin- 2-yl]amino]-	DMF	450(4.08)	2-1007-78
$C_{19}H_{15}OP$			
Phosphine, benzoyldiphenyl-	C_6H_{12}	234(3.39),265(3.32), 391(1.86)	112-0751-78
$C_{19}H_{15}OS$			
Naphtho[1,8-bc]thiolium, 5-ethoxy- 2-phenyl-, tetrafluoroborate	CH_2Cl_2	456(4.11)	83-0324-78

Compound	Solvent	$\lambda_{max}(\log \epsilon)$	Ref.
$C_{19}H_{16}$			
Bicyclo[3.2.0]hepta-2,6-diene, 6,7-di-phenyl-	MeCN	280s(3.95),304(4.18), 311(4.18),327s(3.97)	24-3608-78
Bicyclo[3.1.0]hex-2-ene, 4-methylene-2,6-diphenyl-, endo	EtOH	224(4.33),235(4.36), 298(4.29)	44-3283-78
exo	EtOH	231(4.37),237(4.45), 303(4.34)	44-3283-78
Naphthalene, 1-(1-methyl-2-phenylethen-yl)-, (E)-	EtOH	223(4.79),245s(4.15), 283(4.04)	44-1656-78
(Z)-	EtOH	225(4.87),245s(4.28), 285(3.98)	44-1656-78
Naphthalene, 1-[1-(phenylmethyl)ethen-yl]-	EtOH	225(4.77),283(3.85)	44-1656-78
$C_{19}H_{16}ClNO_4$			
2,4-Pentadienamide, N-(4-chlorophenyl)-5-(6-methoxy-1,3-benzodioxol-5-yl)-, (E,E)-	n.s.g.	380(4.44)	78-1979-78
$C_{19}H_{16}ClN_3O_4$			
1H-Benzimidazolium, 1-(phenylmethyl)-3-(2-pyridinyl)-, perchlorate	EtOH	220(4.48),279(4.40)	73-2046-78
$C_{19}H_{16}FN_2S_2$			
Benzothiazolium, 2-[1-fluoro-3-(3-meth-yl-2(3H)-benzothiazolylidene)-1-prop-enyl]-3-methyl-	EtOH MeNO_2	522(4.94) 523(5.02)	124-0942-78 124-0942-78
$C_{19}H_{16}IN_3$			
1H-Benzimidazolium, 1-(phenylmethyl)-3-(2-pyridinyl)-, iodide	EtOH	220(4.50),270(4.18)	73-2046-78
$C_{19}H_{16}NOP$			
Phenophosphazine, 5,10-dihydro-5-meth-yl-10-phenyl-, 10-oxide	EtOH	226(4.50),275(4.15), 306(3.81),336(3.80)	65-1205-78
$C_{19}H_{16}NO_2P$			
Phenol, 4-(5-methyl-10(5H)-phenophos-phazinyl)-, P-oxide	EtOH	222(4.54),242(4.43), 273(4.27),302(3.88), 337(3.81)	65-1205-78
$C_{19}H_{16}NO_5PS_3$			
Phosphorothioic acid, O,O-diphenyl S-[(2,3,5,7-tetrahydro-5,7-dioxo-6H-1,4-dithiino[2,3-c]pyrrol-6-yl)methyl] ester	EtOH	210(4.18),263(3.95), 420(3.42)	73-1093-78
$C_{19}H_{16}NP$			
9H-Phosphorino[3,4-b]indole, 9-ethyl-4-phenyl-	EtOH	228(4.31),272(4.36), 327(3.97),386s(3.30)	139-0257-78B
$C_{19}H_{16}NPS$			
Phosphinothioic amide, P,P-diphenyl-N-(phenylmethylene)-	C_6H_{12}	260(4.46)	22-0379-78
$C_{19}H_{16}N_2$			
1H-Indole, 3-methyl-2-[(3-methyl-2H-in-dol-2-ylidene)methyl]-, monohydro-bromide	EtOH-H_2SO_4	540(4.61)	83-0954-78

Compound	Solvent	$\lambda_{max}(\log \epsilon)$	Ref.
$C_{19}H_{16}N_2O$			
1-Naphthaleneacetic acid, (phenylmeth- ylene)hydrazide	EtOH	284(4.32),293s(4.31), 302s(4.10)	34-0345-78
10H-Pyrido[3,2-b][1,4]benzoxazine, 8-methyl-10-(phenylmethyl)-	MeOH	233(4.26),340(3.73)	95-0585-78
$C_{19}H_{16}N_2O_2$			
Benzamide, N-[8-(formylmethylamino)- 1-naphthalenyl]-	EtOH	300(4.0)	103-0909-78
1H-Isoindole-1,3(2H)-dione, 2-[(2,3-di- methyl-1H-indol-5-yl)methyl]-	CHCl_3	287(4.0)	103-0856-78
$C_{19}H_{16}N_2O_3$			
Pyridinium, 1-benzoyl-3-cyano-4-ethoxy- 4-oxo-2-butenylide	EtOH	223(4.34),247(4.23), 360(4.77)	95-1412-78
$C_{19}H_{16}N_2O_4$			
1H-Pyrazole-4,5-dicarboxylic acid, 1,3- diphenyl-, dimethyl ester	EtOH	234(4.04)	33-1477-78
$C_{19}H_{16}N_2O_5$			
[1]Benzopyrano[4,3-b]pyrrole-3-carbox- ylic acid, 2,3,3a,4-tetrahydro-2-(4- nitrophenyl)-, methyl ester	MeOH	270(4.54)	35-3494-78
$C_{19}H_{16}N_3$			
1H-Benzimidazolium, 1-(phenylmethyl)- 3-(2-pyridinyl)-, iodide	EtOH	220(4.50),270(4.18)	73-2046-78
perchlorate	EtOH	220(4.48),279(4.40)	73-2046-78
$C_{19}H_{16}N_3PS$			
Phosphine sulfide, (azidophenylmethyl)- diphenyl-	C_6H_{12}	267(3.63)	22-0379-78
$C_{19}H_{16}N_4O$			
4H-Pyrazolo[3,4-d]pyridazin-4-one, 1,5- dihydro-7-methyl-3-phenyl-1-(phenyl- methyl)-	EtOH	213(4.21),260(4.17)	4-0813-78
4H-Pyrazolo[3,4-d]pyridazin-4-one, 2,5- dihydro-7-methyl-3-phenyl2-(phenyl- methyl)-	EtOH	217(4.23),250(4.05)	4-0813-78
$C_{19}H_{16}N_4O_3$			
3,4-Furandiol, tetrahydro-2-(1-phenyl- 1H-pyrazolo[3,4-b]quinoxalin-3-yl)-, [2S-(2α,3β,4β)]-	dioxan	268(4.7),334(4.1), 410(3.8)	136-0079-78H
$C_{19}H_{16}O$			
10H-Phenanthro[9',10':3,4]cyclobuta- [1,2-b]pyran, 8c,11,12,12a-tetra- hydro-	MeCN	247(4.68),255(4.78), 270(4.13),278(4.03), 289(3.97),301(4.07), 321(2.64),329(2.59), 336(2.91),344(2.63), 353(3.03)	24-3608-78
$C_{19}H_{16}O_2$			
4H-Pyran-4-one, 2,6-dimethyl-3,5-di- phenyl-	EtOH	221(4.35),258(4.09)	44-2138-78
2,2'-Spirobi[2H-indene]-5,5'-dicarb- oxaldehyde, 1,1',3,3'-tetrahydro-	EtOH	206(4.60),212s(4.58), 262(4.48),287s(3.43), 295s(3.74)	49-0987-78

Compound	Solvent	λ_{max}(log ϵ)	Ref.
Spiro[2H-cyclopenta[1]phenanthrene-2,2'-[1,3]dioxolane], 1,3-dihydro-	EtOH	214(4.48),223(4.40), 240s(4.23),248(4.70), 255(4.79),271(4.22), 278(4.08),288(3.98), 300(4.06),322(2.90), 328(2.85),336(4.00), 352(2.98)	150-5151-78
Tricyclo[8.5.1.1^{3,8}]heptadeca-1,3,5,7-9,11,14-heptaene-12,14-dicarboxalde-hyde	dioxan	326(4.91),334(4.91), 409(3.72)	89-0956-78
$C_{19}H_{16}O_3$			
2H-1-Benzopyran-2-one, 6-hydroxy-4-methyl-7-(1-phenyl-2-propenyl)-	MeOH	225(4.24),245(4.19), 330(3.87)	2-0856-78
2H-1-Benzopyran-2-one, 7-hydroxy-4-methyl-8-(1-phenyl-2-propenyl)-	MeOH MeOH-NaOH	260(3.53),320(3.88) 250(3.65),375(4.0)	2-0579-78 2-0579-78
2H-1-Benzopyran-2-one, 4-methyl-6-[(3-phenyl-2-propenyl)oxy]-	MeOH	218(4.53),254(4.57), 320(4.46)	2-0856-78
Methanone, (4,5-dihydro-6-methoxy-3H-naphtho[1,8-bc]furan-2-yl)phenyl-	EtOH	207(4.41),257(3.90), 322(4.34)	18-2068-78
Naphtho[2,3-c]furan-1-ol, 1,3-dihydro-3-(2-methoxyphenyl)-	THF	230(4.96),272(3.90), 279(3.91),290(3.61), 305(2.96),317(2.62), 319(2.90)	78-0113-78
$C_{19}H_{16}O_4$			
2H-1-Benzopyran-5-ol, 7-(6-hydroxy-2-benzofuranyl)-2,2-dimethyl-	EtOH	219(4.41),329(4.48), 342(4.62),360(4.55)	138-1239-78
2-Furanacetic acid, 2,5-dihydro-2-methyl-5-oxo-3,4-diphenyl-	EtOH	216s(4.21),264(4.00)	33-3018-78
2(5H)-Furanone, 4-benzoyl-3-methoxy-5-(4-methylphenyl)-	EtOH	260(4.10)	4-0737-78
2,2'-Spirobi[2H-indene]-5,5'-dicarbox-ylic acid, 1,1',3,3'-tetrahydro-	EtOH	207(4.79),211s(4.66), 243(4.36),278(3.52), 287(3.44)	49-0987-78
Spiro[7H-dibenzo[a,c]cyclononene-7,2'-[1,3]dioxolane]-5,9(6H,8H)-dione	EtOH	245(3.68),282s(3.90), 320(2.20)	150-5151-78
$C_{19}H_{16}O_5$			
2(5H)-Furanone, 4-benzoyl-3-methoxy-5-(4-methoxyphenyl)-	EtOH	262(4.13)	4-0737-78
$C_{19}H_{16}O_7$			
Phenanthro[2,3-d][1,3]dioxole-6-carbox-ylic acid, 1,2,3-trimethoxy-	EtOH	258(4.93),284(4.51)	44-3950-78
$C_{19}H_{17}BrN_2$			
1H-Indole, 3-methyl-2-[(3-methyl-2H-indol-2-ylidene)methyl]-, mono-hydrobromide	EtOH-H$_2$SO$_4$	540(4.61)	83-0954-78
$C_{19}H_{17}BrN_2O_2$			
2(1H)-Quinoxalinone, 3-[1-bromo-2-oxo-2-(2,4,6-trimethylphenyl)ethyl]-	EtOH	353(3.98)	103-0336-78
$C_{19}H_{17}BrO_5$			
Egonol, α-bromo-	EtOH	318(4.20)	42-1204-78
$C_{19}H_{17}ClFN_5O$			
4H-[1,2,4]triazolo[1,5-a][1,4]benzodi-azepine-2-carboxamide, 8-chloro-6-(2-	MeOH	250s(4.19)	33-0848-78

Compound	Solvent	$\lambda_{max}(\log \epsilon)$	Ref.
fluorophenyl)-5,6-dihydro-N,N-dimethyl- (cont.)			33-0848-78
$C_{19}H_{17}ClN_2O_6$ 2,1-Benzisoxazolium, 1-methyl-3-[3-(1-methyl-2,1-benzisoxazol-3(1H)-ylidene)-1-propenyl]-, perchlorate	MeCN	615(4.56)	44-1233-78
$C_{19}H_{17}ClO_5$ 9,10-Anthracenedione, 1-chloro-5,7-diethoxy-4-hydroxy-2-methyl-	EtOH	228(4.52),255s(4.16), 279(4.31),445(3.99)	12-1335-78
$C_{19}H_{17}ClO_6$ 9,10-Anthracenedione, 1-chloro-4,5,7,8-tetramethoxy-2-methyl-	EtOH	226(4.52),258(4.33), 400(3.97)	44-1435-78
$C_{19}H_{17}Cl_2N_9O_6$ 1H-Pyrazole-3,5-diamine, 4-[(2,6-dichloro-4-nitrophenyl)azo]-1-(2,4-dinitrophenyl)-N,N,N',N'-tetramethyl-	dioxan	388(4.30)	103-0332-78
$C_{19}H_{17}F_3N_2OS_2$ 4-Thiazolidinone, 3-ethyl-5-[[1-ethyl-6-(trifluoromethyl)-2(1H)-quinolinylidene]ethylidene]-2-thioxo-	EtOH benzene MeOH	537(4.82),570(4.70) 505(--),534(--) 536(--),570(--)	103-0060-78 103-0060-78 103-0060-78
$C_{19}H_{17}F_3N_2OS_3$ 4-Thiazolidinone, 3-ethyl-5-[[1-ethyl-6-[(trifluoromethyl)thio]-2(1H)-quinolinylidene]ethylidene]-2-thioxo-	EtOH	536(4.86),568(4.75)	103-0060-78
$C_{19}H_{17}N$ Benzenamine, 2-(diphenylmethyl)-	EtOH	212(4.40),235s(3.86), 286(3.38)	39-1211-78C
Butanenitrile, 3,3-dimethyl-2-(1H-phenalen-1-ylidene)-	C_6H_{12}	258(4.455),413(4.267)	24-1824-78
$C_{19}H_{17}NO$ 1-Naphthalenecarbonitrile, 1,2,3,4-tetrahydro-5-(4-methylbenzoyl)-	EtOH	260(4.22)	94-1511-78
$C_{19}H_{17}NO_2$ Cyclohept[f]isoindole-1,3(2H,3aH)-dione, 4,7,10,10a-tetrahydro-2-phenyl-	EtOH	261(3.53)	138-0961-78
$C_{19}H_{17}NO_3$ 6H-1,3-Oxazinium, 4-hydroxy-3-methyl-2-(4-methylphenyl)-6-oxo-5-(phenylmethyl)-, hydroxide, inner salt	MeCN	254(4.29),359(3.46)	5-1655-78
Spiro[2H-1-benzopyran-2,2'-[2H-1,3]benzoxazin]-4'(3'H)-one, 3'-(1-methylethyl)-	isoPrOH	267(4.13),275s(3.60), 300(3.79)	103-0122-78
$C_{19}H_{17}NO_4$ 6H-1,3-Oxazinium, 4-hydroxy-2-(4-methoxyphenyl)-3-methyl-6-oxo-5-(phenylmethyl)-, hydroxide, inner salt	MeCN	259(4.17),287(4.08), 346(3.57)	5-1655-78
2,4-Pentadienamide, 5-(6-methoxy-1,3-benzodioxol-5-yl)-N-phenyl-	n.s.g.	374(4.38)	78-1979-78

Compound	Solvent	λ_{max}(log ϵ)	Ref.
$C_{19}H_{17}NO_5$			
2-Azabicyclo[2.2.2]octa-5,7-diene-5,6-dicarboxylic acid, 3-oxo-2-(2-phenylethenyl)-, dimethyl ester, (E)-	EtOH	293(4.26)	78-2617-78
6H-1,3-Oxazinium, 4-hydroxy-2,5-bis(4-methoxyphenyl)-3-methyl-6-oxo-, hydroxide, inner salt	MeCN	218s(4.23),270(4.30), 284s(4.25),377(3.82)	5-1655-78
Stylopine, 13β-hydroxy-	EtOH	232s(3.99),288(3.92)	95-1243-78
$C_{19}H_{17}N_2S_2$			
Benzothiazolium, 3-methyl-2-[3-(3-methyl-2(3H)-benzothiazolylidene)-1-propenyl]-, iodide	n.s.g.	558(5.15)	124-0942-78
	MeNO$_2$	556(5.19)	124-0942-78
$C_{19}H_{17}N_3O$			
Pyridinium, 3-[[(phenylamino)carbonyl]-amino]-1-(phenylmethyl)-, hydroxide, inner salt	MeOH	265(4.99),322(4.46)	64-0084-78B
	CH$_2$Cl$_2$	323(4.04)	64-0084-78B
$C_{19}H_{17}N_3O_2S$			
3H-Pyrimido[5,4-b][1,4]benzothiazine-2,4(4aH,10H)-dione, 3,10-dimethyl-4a-(phenylmethyl)-	MeOH	273(3.96),306(3.78), 334(3.63)	5-0193-78
	12M HCl	297(3.79),363(3.10)	5-0193-78
1H-Pyrimido[5,4-b][1,4]benzothiazinium, 2,3,4,10-tetrahydro-3,10-dimethyl-2,4-dioxo-5-(phenylmethyl)-, hydroxide, inner salt	pH 1	293(3.88),313s(3.85), 355s(2.63)	5-0193-78
	MeOH	266(4.40),315s(3.60)	5-0193-78
3H-Pyrimido[5,4-b][1,4]benzothiazin-2(10H)-one, 3,10-dimethyl-4-(phenylmethoxy)-	MeOH	258(4.78),320s(3.39), 380s(2.90)	5-0193-78
	CF$_3$COOH	276(4.21),303s(3.49), 338s(2.95)	5-0193-78
$C_{19}H_{17}N_3O_3$			
Benzamide, N-[4-[2-(benzoyloxy)ethyl]-1H-imidazol-2-yl]-	EtOH	226(4.47),273s(4.05), 281(4.06),295s(3.97)	35-4208-78
	EtOH-HCl	229(4.42),266(4.21)	35-4208-78
$C_{19}H_{17}N_3O_4$			
Benzamide, N-[2-oxo-2-[(2,3,4,5-tetrahydro-2,5-dioxo-1H-1-benzazepin-3-yl)amino]ethyl]-, (±)-	n.s.g.	227(4.61),252(4.03), 313(3.44)	12-0439-78
Propanoic acid, 3-oxo-3-[(4-oxo-2-phenyl-3(4H)-quinazolinyl)amino]-, ethyl ester	MeOH	233(4.44),270(4.49), 319(4.23)	80-1085-78
$C_{19}H_{17}N_3O_5$			
3H-Pyrazol-3-one, 4-[(2,4-dimethoxy-5-nitrophenyl)methylene]-2,4-dihydro-5-methyl-2-phenyl-	MeOH	245(4.38),375(4.31)	2-0324-78
$C_{19}H_{17}N_3S_2$			
2,5-Pyrrolidinedithione, 1-methyl-3,4-bis[(phenylamino)methylene]-	CHCl$_3$	280(4.24),365(4.37), 405(4.43),520(4.17)	104-2041-78
$C_{19}H_{17}N_5$			
6H-Purin-6-imine, 3,9-dihydro-3,9-bis-(phenylmethyl)-, monoperchlorate	pH 1 and 7	273(4.25)	88-5007-78
$C_{19}H_{18}$			
Bicyclo[3.2.0]hept-6-ene, 6,7-diphenyl-	MeCN	292s(4.13),301(4.16), 309s(4.14),326s(3.92)	24-3608-78

Compound	Solvent	$\lambda_{max}(\log \epsilon)$	Ref.
Cyclopropene, 2-methyl-1,3-diphenyl-3-(2-propenyl)-	EtOH	262(4.23)	44-1481-78
Cyclopropene, 3-methyl-1,2-diphenyl-3-(2-propenyl)-	EtOH	229(4.26),320(4.46), 338(4.33)	44-1481-78
1H-Indene, 1-methyl-2-phenyl-3-(2-propenyl)-	EtOH	226(4.03),230(4.06), 294(4.08)	44-1481-78
1H-Indene, 2-methyl-1-phenyl-3-(2-propenyl)-	EtOH	220(4.25),263(3.84)	44-1481-78

$C_{19}H_{18}ClNO_4$

Compound	Solvent	$\lambda_{max}(\log \epsilon)$	Ref.
Dipyrano[3.4-b:2',3'-h]quinolin-11(1H)-one, 7-chloro-2,3-dihydro-6-methoxy-3,3,9-trimethyl-	MeOH	245(4.37),287(4.58), 340(3.79),435(3.49)	24-0439-78
Phenanthridinium, 7,8,9,10-tetrahydro-5-phenyl-, perchlorate	MeOH	239(5.00),316(4.30)	65-1707-78

$C_{19}H_{18}ClNO_6$

Compound	Solvent	$\lambda_{max}(\log \epsilon)$	Ref.
[1,1'-Biphenyl]-2,4-dicarboxylic acid, 4'-chloro-3-hydroxy-5-methyl-6-nitroso-, diethyl ester	DMF	690(1.99)	104-2189-78

$C_{19}H_{18}ClN_3O_3$

Compound	Solvent	$\lambda_{max}(\log \epsilon)$	Ref.
Spiro[2H-indole-2,5'(4'H)-isoxazole], 3'-(4-chlorophenyl)-1,3-dihydro-1,3,3-trimethyl-5-nitro-	CHCl$_3$	265(4.34),372(4.32)	47-2059-78

$C_{19}H_{18}ClN_9O_6$

Compound	Solvent	$\lambda_{max}(\log \epsilon)$	Ref.
1H-Pyrazole-3,5-diamine, 4-[(2-chloro-4-nitrophenyl)azo]-1-(2,4-dinitrophenyl)-N,N,N',N'-tetramethyl-	dioxan	425(4.37)	103-0332-78

$C_{19}H_{18}N_2OS$

Compound	Solvent	$\lambda_{max}(\log \epsilon)$	Ref.
Azepino[4,3-b][1,4]benzothiazin-1(2H)-one, 3,4,5,11-tetrahydro-11-(phenylmethyl)-	EtOH	252(4.14),286(3.73)	103-0306-78
Benzeneethanethioic acid, S-[2-(4-phenyl-1H-imidazol-1-yl)ethyl] ester	EtOH	258(4.32)	4-0307-78

$C_{19}H_{18}N_2O_2$

Compound	Solvent	$\lambda_{max}(\log \epsilon)$	Ref.
Acetamide, N-(1-acetyl-1H-indol-3-yl)-N-(4-methylphenyl)-	EtOH	205(4.49),232(4.32), 274(3.92),293(3.84), 300(3.89)	103-0757-78
Propanoic acid, 2-(9-phenanthrenylhydrazono)-, ethyl ester	EtOH and heptane	211(4.51),250(4.49), 276(4.29),315(4.15), 573(4.19)	103-0851-78
3H-Pyrazol-3-one, 2,4-dihydro-4-[1-(4-methoxyphenyl)ethylidene]-5-methyl-2-phenyl-, (E)-	benzene EtOH	337(4.11) 208(4.31),355(4.35), 340(3.99)	104-1644-78 104-1644-78
Pyridine, 3-[5-[4-(3-methyl-2-butenyl)-oxy]phenyl]-2-oxazolyl]-	n.s.g. MeOH	250(4.06),261(4.03), 328(4.40) 248(4.10),255(4.03), 325(4.50)	88-2723-78 102-1814-78
2(1H)-Quinoxalinone, 3,4-dihydro-3-[2-oxo-2-(2,4,6-trimethylphenyl)ethylidene]-	EtOH	430(4.32)	103-0336-78

$C_{19}H_{18}N_2O_2S$

Compound	Solvent	$\lambda_{max}(\log \epsilon)$	Ref.
[1]Benzothieno[3,2-e]pyrrolo[1,2-a]pyrimidine-1,5-dione, 2,3,3a,4,6,7,8,9-octahydro-3a-phenyl-	EtOH	225(4.33),260(3.89), 317(3.88)	4-0949-78

Compound	Solvent	λ_{max}(log ϵ)	Ref.
C$_{19}$H$_{18}$N$_2$O$_3$			
3-Pyridinecarboxylic acid, 5-ethyl-4-[2-(1-methyl-1H-indol-2-yl)-2-oxo-ethyl]-	EtOH	236(4.26),279(3.88), 306(4.29)	78-1363-78
C$_{19}$H$_{18}$N$_2$O$_3$S			
Benzenesulfonamide, 4-methyl-N-[3-(1-methyl-2-oxopropylidene)-3H-indol-2-yl]-	CHCl$_3$	259(4.42),265(4.45), 313(3.67),380(3.51)	103-0745-78
C$_{19}$H$_{18}$N$_2$O$_4$			
Phenol, 2-[2-(3,3-dimethyl-3H-indol-2-yl)ethenyl]-6-methoxy-4-nitro-	EtOH	<u>345(4.5),460(3.8)</u>	103-0985-78
C$_{19}$H$_{18}$N$_2$O$_6$			
2-Butenoic acid, 4-[2-[[[(4-nitrophenyl)methyl]amino]carbonyl]phenoxy]-, methyl ester, (E)-	MeOH	275(4.58)	35-3494-78
Propanedioic acid, [3-(1,3-benzodioxol-5-yl)-2-cyano-2-propenyl](2-cyanoethyl)-, monoethyl ester	EtOH	292(4.71),324(4.82)	2-0924B-78
C$_{19}$H$_{18}$N$_4$OS			
2H-[1,2,3]Thiadiazolo[4,5,1-hi][1,2,3]-benzothiadiazole-3-S^{IV}, 4,6,7,8-tetrahydro-2-(4-methoxyphenyl)-4-phenyl-	C$_6$H$_{12}$	248(4.28),312(4.19), 529(4.34)	39-0195-78C
C$_{19}$H$_{18}$N$_4$O$_2$			
Benzamide, N-benzoyl-N-(4,5-dimethyl-1H-1,2,3-triazol-1-yl)-4-methyl-	MeOH	253(4.18)	4-1255-78
Benzoic acid, anhydride with N-(4,5-dimethyl-1H-1,2,3-triazol-1-yl)-4-methylbenzenecarboximidic acid, (Z)-	MeOH	241(4.15),286(4.21)	4-1255-78
1H-Indene-5,5,6,6-tetracarbonitrile, 1-[1-(formyloxy)ethyl]-4,7-dihydro-1,2,3-trimethyl-	n.s.g.	270(3.65)	35-0860-78
2,7-Naphthyridine-4-carbonitrile, 1-amino-2,7-diethyl-2,3,7,8-tetrahydro-3,8-dioxo-6-phenyl-	n.s.g.	250(4.12),287(4.45), 325(3.90),362(4.03)	39-0554-78C
C$_{19}$H$_{18}$N$_4$O$_3$			
Benzamide, N-benzoyl-N-(4,5-dimethyl-1H-1,2,3-triazol-1-yl)-4-methoxy-	MeOH	240(4.26),290(4.23)	4-1255-78
Benzoic acid, anhydride with N-(4,5-dimethyl-1H-1,2,3-triazol-1-yl)-4-methoxybenzenecarboximidic acid, (Z)-	MeOH	233(3.87),304(4.06)	4-1255-78
C$_{19}$H$_{18}$N$_4$O$_4$			
1,2,3,4-Butanetetrol, 1-(1-phenyl-1H-pyrazolo[3,4-b]quinoxalin-3-yl)-, D-	dioxan	269(4.7),334(4.1), 410(3.7)	136-0079-78H
C$_{19}$H$_{18}$N$_4$O$_4$S			
Thiourea, N,N'-bis(3,4-dihydro-3-methyl-2-oxo-2H-1,3-benzoxazin-4-yl)-	EtOH	223s(4.40),256(4.12), 275s(3.86),287s(3.43)	4-1193-78
C$_{19}$H$_{18}$O			
2-Oxabicyclo[3.2.0]hept-6-ene, 1-methyl-6,7-diphenyl-	MeCN	295(4.18),301s(4.17), 321s(3.87)	24-3608-78
2-Oxabicyclo[4.2.0]oct-7-ene, 7,8-diphenyl-	MeCN	288s(4.16),298(4.21), 303s(4.19),321s(3.97)	24-3608-78

Compound	Solvent	$\lambda_{max}(\log \epsilon)$	Ref.
2H-Oxocin, 3,4-dihydro-6,7-diphenyl-	MeCN	233s(4.23),245(4.28), 254s(4.23),280s(3.72)	24-3608-78
$C_{19}H_{18}OSe$			
2-Cyclohexen-1-one, 4-methyl-4-phenyl-6-(phenylseleno)-	MeOH	217(4.27),238s(4.17), 265(3.55)	24-1944-78
$C_{19}H_{18}O_2S$			
2H-Phenanthro[1,2-c]thiopyran-1-ol, 1,12a-dihydro-8-methoxy-12a-methyl-, (1S-trans)-	EtOH	245(4.48),278(4.49), 300s(4.32)	142-0181-78C
2H-Phenanthro[1,2-c]thiopyran-1(11H)-one, 12,12a-dihydro-8-methoxy-12a-methyl-, (S)-	EtOH	244(4.42),271s(4.33), 277(4.33),312(4.34)	39-0576-78C
$C_{19}H_{18}O_3$			
2H-1-Benzopyran-2-one, 5-methoxy-4,7-dimethyl-8-(phenylmethyl)-	MeOH	255(4.11),305(4.28)	2-0574-78
1-Cyclohexene-1-acetic acid, 2-(1-naphthalenyl)-6-oxo-, methyl ester	EtOH	282(4.0)	42-0580-78
1H-Indene-1-carboxylic acid, 4-(3,4-dimethylbenzoyl)-2,3-dihydro-	EtOH	263(4.22)	94-1776-78
1H-Indene-1-carboxylic acid, 4-(4-ethylbenzoyl)-2,3-dihydro-	EtOH	261(4.26)	94-1776-78
1-Naphthalenecarboxylic acid, 1,2,3,4-tetrahydro-5-(4-methylbenzoyl)-	EtOH	260(4.25)	94-1511-78
2,3-Naphthalenedimethanol, α-(2-methoxyphenyl)-	THF	230(5.03),278(3.88), 306(2.61),315(2.39), 320(2.31)	78-0113-78
2H-Phenanthro[1,2-c]pyran-1(11H)-one, 12,12a-dihydro-8-methoxy-12a-methyl-, (S)-	EtOH	221(4.24),240s(4.38), 250s(4.49),258(4.56), 266s(4.52),284(4.21)	39-0576-78C
$C_{19}H_{18}O_4$			
1,3-Benzenediol, 5-(6-hydroxy-2-benzofuranyl)-2-(3-methyl-2-butenyl)- (moracin C)	EtOH	219(4.50),287s(4.20), 296s(4.27),319(4.61), 333(4.54)	138-1239-78
Dipetalolactone	EtOH	222(4.25),244s(4.47), 250(4.53),297(4.46), 308(4.38),343(4.12)	78-1411-78
Harringtonolide	EtOH	242(4.30),310(3.85)	44-1002-78
$C_{19}H_{18}O_5$			
9,10-Anthracenedione, 1,3-diethoxy-8-hydroxy-6-methyl-	EtOH	226(4.55),251s(4.17), 276s(4.32),284(4.35), 434(3.97)	12-1335-78
9,10-Anthracenedione, 1,3-diethoxy-5-methoxy-	EtOH	230(4.42),280(4.33), 395(3.92)	12-1335-78
9,10-Anthracenedione, 1,3-diethoxy-8-methoxy-	EtOH	224(4.48),280(4.39), 348s(3.58),402(3.84)	12-1335-78
4H-1-Benzopyran-4-one, 5,7-dimethoxy-3-(4-methoxyphenyl)-2-methyl-	MeOH	258(4.17),315(4.0)	130-0493-78
Egonol	EtOH	217(4.52),317(4.52)	42-1204-78
Isocrotocaudin	EtOH	218(4.26),254(4.00)	102-1777-78
$C_{19}H_{18}O_6$			
9,10-Anthracenedione, 1,2,4,5-tetramethoxy-7-methyl-	CHCl$_3$	280(3.96),406(3.38)	44-1435-78
Benzo[a]heptalene-7,10-dione, 5,6-dihydro-9-hydroxy-1,2,3-trimethoxy-	isoPrOH	248(4.41),358(4.24)	33-1213-78

Compound	Solvent	$\lambda_{max}(\log \epsilon)$	Ref.
2H-1-Benzopyran-2-one, 8-methoxy-3-(3,4,5-trimethoxyphenyl)-	n.s.g.	335(4.38)	39-0739-78C
9-Phenanthreneacetic acid, 9,10-dihydro-9-hydroxy-1,3-dimethoxy-10-oxo-, methyl ester	MeOH	259(4.55),298s(3.83), 313(3.86),331(3.86)	12-2259-78
1,3-Propanediol, 1-[2-(1,3-benzodioxol-5-yl)-7-methoxy-5-benzofuranyl]- (machicendiol)	EtOH	218(4.30),318(4.56)	42-1204-78
$C_{19}H_{18}O_7$ 9,10-Anthracenedione, 1,2,4,5,7-pentamethoxy-	EtOH	225(4.59),285(4.36), 415(3.85)	44-1435-78
$C_{19}H_{18}O_8$ 4H-1-Benzopyran-4-one, 5,7-dihydroxy-6-methoxy-2-(2,4,5-trimethoxyphenyl)-	EtOH	257(4.20),272(4.13), 360(4.28)	78-1593-78
$C_{19}H_{18}O_{11}$ 9H-Xanthen-9-one, 2-β-D-glucopyranosyl-1,3,6,7-tetrahydroxy-	MeOH	242(4.31),258(4.40), 317(4.11),367(3.99)	105-0449-78
	MeOH-NaOMe	241(--),273(--), 393(--)	105-0449-78
$C_{19}H_{19}BF_4O_2$ 2,4-Pentadienylium, 1,5-bis(4-methoxyphenyl)-, tetrafluoroborate	$C_2H_4Cl_2$	648(4.80)	104-1197-78
$C_{19}H_{19}BrN_2O_2$ Ethanol, 2,2'-[[4-(1-bromo-4-isoquinolinyl)phenyl]imino]bis-	EtOH	213(4.58),221(4.61), 266(4.31)	44-0672-78
$C_{19}H_{19}ClN_2O$ 1,5-Benzodiazocin-2(1H)-one, 8-chloro-3,6-dihydro-1,6,6-trimethyl-4-phenyl-	CHCl$_3$	245(4.15)	47-2039-78
Spiro[2H-indole-2,5'(4'H)-isoxazole], 5-chloro-1,3-dihydro-1,3,3-trimethyl-3'-phenyl-	CHCl$_3$	262(4.34),292s(4.00)	47-2039-78
$C_{19}H_{19}ClN_2O_2$ 3H-1,4-Benzodiazepine, 2-butoxy-7-chloro-5-phenyl-, 4-oxide	isoPrOH	243(4.41),310(4.01)	4-0161-78
$C_{19}H_{19}ClN_2O_4S$ Acetic acid, [3-[[(4-chlorophenyl)sulfonyl]amino]-1,3-dihydro-1,3-dimethyl-2H-indol-2-ylidene]-, methyl ester	EtOH	233(4.26),301(4.05), 336(4.05)	39-1471-78C
$C_{19}H_{19}ClN_2O_5S$ 2,1-Benzisothiazolium, 1-acetyl-3-[2-[4-(dimethylamino)phenyl]ethenyl]-, perchlorate	CH_2Cl_2	660(4.47),705(4.64)	44-1233-78
$C_{19}H_{19}ClO_5$ Benzopyrylium, 4-butyl-2-phenyl-, perchlorate	HOAc-HClO$_4$	390(4.177)	83-0256-78
$C_{19}H_{19}CuN_3O_4$ Copper, [2-[[(2-hydroxyphenyl)methylene]amino]-1-[[[(2-hydroxyphenyl)-methylene]amino]methyl]ethyl methylcarbamato(2-)]-	CH_2Cl_2	238(4.67),275(4.45), 374(4.07),610(2.46)	35-2686-78
	pyridine	377(4.15),617(2.32)	35-2686-78

Compound	Solvent	$\lambda_{max}(\log \epsilon)$	Ref.
(cont.)	THF	232(4.63),247(4.54), 273(4.33),375(3.96), 618(2.44)	35-2686-78
$C_{19}H_{19}F_3N_2O_3$			
1H-Pyrrolo[3,4-b]pyridine-3-carboxylic acid, 4,5,6,7-tetrahydro-2,6-dimeth-yl-7-oxo-4-[3-(trifluoromethyl)phen-yl]-, ethyl ester	EtOH	220s(4.27),256(4.05), 343(3.58)	95-0448-78
$C_{19}H_{19}F_3N_4O_8$			
Acetamide, 2,2,2-trifluoro-N-[2-(2,3,4-tri-O-acetyl-β-D-ribopyranosyl)-2H-benzotriazol-4-yl]-	EtOH	276(4.13),287(4.06)	103-1375-78
$C_{19}H_{19}N$			
6H-Azepino[1,2-a]indole, 7,8,9,10-tetrahydro-11-phenyl-	EtOH	230(4.47),277s(--), 283(4.12)	39-1211-78C
Isoquinoline, 1-[1-(4-phenylbutyl)]-	EtOH	262(3.61),271(3.65), 283(3.54),309(3.43), 322(3.52)	94-2334-78
2H-Pyrrole, 2,2,4-trimethyl-3,5-diphen-yl-	EtOH	240(4.17)	88-4511-78
$C_{19}H_{19}NO$			
1H-Carbazole, 2,3,4,9-tetrahydro-5-methoxy-9-phenyl-	EtOH	224(4.30),244(4.18), 277s(4.08),286(4.12)	103-1093-78
1H-Carbazole, 2,3,4,9-tetrahydro-7-methoxy-9-phenyl-	EtOH	230(4.58),248s(4.29), 290s(4.08)	103-1093-78
1H-Carbazole, 2,3,4,9-tetrahydro-9-(3-methoxyphenyl)-	EtOH	222(4.76),266(4.23), 286(4.21),293s(4.20)	103-1093-78
1(2H)-Naphthalenone, 2-[[4-(dimethyl-amino)phenyl]methylene]-3,4-di-hydro-, cation	H_2SO_4	388(4.35)	65-1644-78
3H-Pyrrol-3-one, 1,2-dihydro-1,2,2-tri-methyl-4,5-diphenyl-	MeCN	225(4.34),276(4.04), 350(3.93)	5-1203-78
3H-Pyrrol-3-one, 1,2-dihydro-1,2,5-tri-methyl-2,4-diphenyl-	MeCN	223(3.89),281(4.00), 337(3.88)	5-1203-78
$C_{19}H_{19}NO_2$			
1,3-Cyclohexanedione, 5,5-dimethyl-2-[(2-naphthalenylamino)methylene]-	EtOH	224(4.52)	97-0256-78
1,4-Ethenoisoquinolin-3(2H)-one, 2-(5,5-dimethyl-3-oxo-1-cyclohexen-1-yl)-1,4-dihydro-	EtOH	298(4.14)	78-2617-78
7H-Indeno[4,5-h]isoquinolin-7-one, 6,6a,10,11-tetrahydro-2-methoxy-5,6a-dimethyl-, (±)-	MeOH	244(4.28),280(4.19), 372(3.91)	44-0966-78
2-Naphthalenamine, N-[(3,4-dimethoxy-phenyl)methyl]-	EtOH	208(4.54),215(4.49), 248(4.71),286(4.14), 294(4.05),336(3.51)	103-0082-78
2-Propen-1-one, 3-morpholino-1,3-di-phenyl-	C_6H_{12} EtOH	333(4.09) 346(3.98)	48-0945-78 48-0945-78
$C_{19}H_{19}NO_2S$			
Quinoline, 3-methyl-2-[1-[(4-methyl-phenyl)sulfonyl]ethyl]-	EtOH	211(4.60),237(4.58), 267(3.56),273(3.56), 295(3.45),300s(3.41), 309(3.54),323(3.61)	95-1503-78

Compound	Solvent	λ_{max}(log ϵ)	Ref.
$C_{19}H_{19}NO_3$ Benzamide, N-(1,1-dimethyl-2,3-dioxo-3-phenylpropyl)-N-methyl-	MeCN	257(4.05)	5-1203-78
$C_{19}H_{19}NO_4$ 4H-Pyrano[2,3-h]quinoline-2-carboxalde-hyde, 1,8-dihydro-5-methoxy-8,8-di-methyl-4-oxo-3-(2-propenyl)-	MeOH	243(4.54),263(4.30), 271(4.32),299(3.65), 335(3.88),338(3.85)	24-0439-78
$C_{19}H_{19}NO_5$ 2-Azabicyclo[2.2.2]oct-7-ene-5,6-di-carboxylic acid, 3-oxo-2-(2-phenyl-ethenyl)-, dimethyl ester, trans	MeCN	287(4.32)	78-2617-78
1,3-Benzodioxol-4-ol, 5-[(5,6,7,8-tet-rahydro-6-methyl-1,3-dioxolo[4,5-g]-isoquinolin-5-yl)methyl]- (ledecorine)	EtOH	240s(3.88),295(3.74)	105-0465-78
Dibenzo[a,i]quinolizine-5,13(1H,6H)-di-one, 2,3-dihydro-8-hydroxy-9,12-di-methoxy-	MeOH	239(4.30),282(3.59)	39-0440-78C
2,4-Pentadienoic acid, 5-[1-(2-ethoxy-2-oxoethylidene)-1,3-dihydro-3-oxo-2H-isoindol-2-yl]-, ethyl ester	THF	215(4.18),287(4.43), 313(4.49)	40-0404-78
Phenanthro[2,3-d][1,3]dioxol-6-amine, 1,2,3-trimethoxy-N-methyl-	EtOH	260(4.78),290(4.57)	44-3950-78
$C_{19}H_{19}NO_6$ 8-Azabicyclo[3.2.1]octane-6,7-dione, 3-acetoxy-8-(benzoyloxy)-3-(2-prop-enyl)-, endo	CHCl$_3$	496(1.65)	44-4765-78
[1,1'-Biphenyl]-2,4-dicarboxylic acid, 3-hydroxy-5-methyl-6-nitroso-, diethyl ester	DMF	690(1.83)	104-2189-78
$C_{19}H_{19}NO_{10}S$ Benzo[b]thiophene-3,4,5,6,7-pentacarb-oxylic acid, 2-amino-, 3-ethyl 4,5,6,7-tetramethyl ester	MeOH	223(4.26),249(4.33), 282(4.34),319(3.85), 362(3.81)	24-0770-78
$C_{19}H_{19}N_2OS$ 2,1-Benzisothiazolium, 1-acetyl-3-[2-[4-(dimethylamino)phenyl]ethenyl]-, perchlorate	CH$_2$Cl$_2$	660(4.47),705(4.64)	44-1233-78
$C_{19}H_{19}N_2O_2$ 2,1-Benzisoxazolium, 1-acetyl-3-[2-[4-(dimethylamino)phenyl]ethenyl]-, perchlorate	MeCN	650(4.45)	44-1233-78
$C_{19}H_{19}N_2O_3$ [1,3]Dioxolo[4,5-f]-2,1-benzisoxazol-ium, 3-[2-[4-(dimethylamino)phenyl]-ethenyl]-1-methyl-, perchlorate	MeCN	530(4.52)	44-1233-78
$C_{19}H_{19}N_3$ 3H-1,5-Benzodiazepine, 2-phenyl-4-pyr-rolidino-	n.s.g.	260(4.54),349(3.64)	87-0952-78
$C_{19}H_{19}N_3O$ 3H-1,5-Benzodiazepine, 2-morpholino-4-phenyl-	n.s.g.	258(4.51),342(3.66)	87-0952-78

Compound	Solvent	$\lambda_{max}(\log \epsilon)$	Ref.
$C_{19}H_{19}N_3OS$			
1H-Imidazol-1-amine, N-[1-(4-methoxyphenyl)ethylidene]-2-(methylthio)-4-phenyl-	MeOH	283(4.53)	18-1846-78
4(5H)-Thiazolone, 5-[[4-(dimethylamino)phenyl]methylene]dihydro-3-methyl-2-(phenylimino)-	MeOH	413(4.54)	104-0997-78
	dioxan	245(4.15),403(4.55)	104-0997-78
	70% dioxan	416(4.54)	104-0997-78
4(5H)-Thiazolone, 5-[[4-(dimethylamino)phenyl]methylene]-2-(methylphenylamino)-	MeOH	281(4.16),426(4.53)	104-0997-78
	dioxan	280(4.15),405(4.53)	104-0997-78
	70% dioxan	282(4.19),425(4.52)	104-0997-78
$C_{19}H_{19}N_3O_2$			
1H-Imidazole-4-carboxylic acid, 1-phenyl-5-[(phenylmethyl)amino]-, ethyl ester	EtOH	213(4.26),226(4.27), 281(4.12)	118-0741B-78
$C_{19}H_{19}N_3O_2S$			
Benzenesulfonamide, N-[2-cyano-1-(1H-indol-3-ylmethyl)ethyl]-4-methyl-, (S)-	MeOH	227(3.75),273(3.79), 281(3.80),289(3.74)	35-0938-78
1H-Imidazol-1-amine, 4-(4-methoxyphenyl)-N-[(4-methoxyphenyl)methylene]-2-(methylthio)-	MeOH	290(4.44),348(4.28)	18-1846-78
$C_{19}H_{19}N_3O_3$			
Propanamide, 3-hydroxy-N-[4-(phenylazo)phenyl]-, methacrylate ester	EtOH	353(4.45)	126-2489-78
Spiro[2H-indole-2,5'(4'H)-isoxazole], 1,3-dihydro-1,3,3-trimethyl-5-nitro-3'-phenyl-	CHCl$_3$	260(4.26),373(4.32)	47-2059-78
$C_{19}H_{19}N_3O_3S$			
Pyrazolo[4,3-d][1]benzazepin-3(2H)-one, 3a,4,5,6-tetrahydro-3a-methyl-6-[(4-methylphenyl)sulfonyl]-	n.s.g.	244(4.11),275(3.92)	39-0862-78C
$C_{19}H_{19}N_3O_5$			
Glycine, benzoyl-DL-kynurenyl-	n.s.g.	227(4.47),252(3.94), 365(3.70)	12-0439-78
DL-Kynurenine, benzoylglycyl-	n.s.g.	227(4.56),252(3.96), 364(3.73)	12-0439-78
$C_{19}H_{19}N_3O_6S$			
Acetic acid, [1,3-dihydro-1,3-dimethyl-3-[[(4-nitrophenyl)sulfonyl]amino]-2H-indol-2-ylidene]-, methyl ester	EtOH	234(4.34),298(4.21), 336(4.17)	39-1471-78C
$C_{19}H_{19}N_5O_5$			
Adenosine, N-benzoyl-2'-deoxy-, 3'-acetate	EtOH	281(4.30)	5-0854-78
$C_{19}H_{19}N_9O_6$			
1H-Pyrazole-3,5-diamine, 1-(2,4-dinitrophenyl)-N,N,N',N'-tetramethyl-4-[(4-nitrophenyl)azo]-	dioxan	408(4.39)	103-0332-78
$C_{19}H_{19}O_2$			
2,4-Pentadienylium, 1,5-bis(4-methoxyphenyl)-, tetrafluoroborate	C$_2$H$_4$Cl$_2$	648(4.80)	104-1197-78

Compound	Solvent	$\lambda_{max}(\log \epsilon)$	Ref.
$C_{19}H_{20}$			
Benzene, 1,1'-(3,3-dimethyl-1,4-penta-dienylidene)bis-	EtOH	248(4.15)	78-1775-78
Benzene, 2-[2-(2-ethenylphenyl)ethen-yl]-1,3,5-trimethyl-	MeOH	224(4.33),252(4.27), 284(4.26)	35-3819-78
Cyclopropane, 1-(2,2-diphenylethenyl)-2,2-dimethyl-	EtOH	267(4.24)	78-1775-78
Cyclopropane, 1-ethenyl-3,3-dimethyl-2,2-diphenyl-	EtOH	226(4.23),254(2.72), 261(2.71),267(2.56)	78-1775-78
2,2'-Spirobi[2H-indene], 1,1',3,3'-tetrahydro-5,5'-dimethyl-	EtOH	215s(4.28),264(3.38), 270(3.52),273(3.57), 279(3.66)	49-0987-78
$C_{19}H_{20}BrNO_3$			
1-Propanone, 1-(4-bromophenyl)-2-hy-droxy-3-morpholino-3-phenyl-	MeOH	259(4.02)	104-1174-78
1-Propanone, 3-(4-bromophenyl)-2-hy-droxy-3-morpholino-1-phenyl-	MeOH	250(3.97)	104-1174-78
$C_{19}H_{20}Br_2N_2O$			
Diaziridinone, bis[1-(4-bromophenyl)-1-methylethyl]-	MeCN	257(2.87),263(2.84), 274(2.58)	44-0922-78
$C_{19}H_{20}ClNO$			
Benzamide, 4-chloro-N-[(4-cyclohexyl)-, (1S-trans)-	MeOH	206(4.35),234(4.13)	44-0355-78
Cyclohexanecarboxamide, N-(4-chloro-phenyl)-2-phenyl-, (1S-trans)-	MeOH	206(4.44),249(4.25)	44-0355-78
$C_{19}H_{20}ClNO_3$			
1-Propanone, 1-(4-chlorophenyl)-2-hy-droxy-3-morpholino-3-phenyl-	MeOH	260(4.00)	104-1174-78
1-Propanone, 3-(4-chlorophenyl)-2-hy-droxy-3-morpholino-1-phenyl-	MeOH	250(3.95)	104-1174-78
$C_{19}H_{20}ClNO_4$			
7H-Pyrano[2,3-c]acridin-7-one, 9-chlo-ro-3,8,9,10,11,12-hexahydro-11-hy-droxy-6-methoxy-3,3-dimethyl-	MeOH	249(4.43),272(4.28), 281(4.28),346(3.65)	24-0439-78
$C_{19}H_{20}N_2$			
1H-Indole, 3-methyl-2-[(4,5,6,7-tetra-hydro-3-methyl-2H-indol-2-ylidene)-methyl]-, monohydrobromide	EtOH	500(4.56)	83-0954-78
$C_{19}H_{20}N_2O$			
Acetic acid, 1-(3,4-dihydro-2-naphtha-lenyl)-2-methyl-2-phenylhydrazide	decalin	233(4.24),291(4.18)	33-1364-78
	EtOH	239(4.16),272(4.26), 346(4.04)	33-1364-78
Benzamide, N-[(2,3,5-trimethyl-1H-in-dol-6-yl)methyl]-	EtOH	235(4.44),283(3.97), 289(3.96)	103-0856-78
Spiro[2H-indole-2,5'(4'H)-isoxazole], 1,3-dihydro-1,3,3-trimethyl-3'-phenyl-	CHCl₃	260(4.30),288s(4.00)	47-2039-78
$C_{19}H_{20}N_2O_2$			
Benzamide, N-[(5-methoxy-2,3-dimethyl-1H-indol-6-ylmethyl]-	EtOH	230(4.48),293(4.01)	103-0856-78
1,2-Diazetidin-3-one, 1-benzoyl-2-(1,1-dimethylethyl)-4-phenyl-	EtOH	220(4.05),269s(3.56), 296s(3.05)	33-1477-78
Estra-1,3,5(10),9(11)-tetraen-17-one, 16-diazo-3-methoxy-	EtOH	213(4.31),261(4.46), 296(3.92)	22-0119-78

Compound	Solvent	$\lambda_{max}(\log \epsilon)$	Ref.
$C_{19}H_{20}N_2O_2$			
2-Propenamide, 2-(acetylamino)-3-phenyl-N-(1-phenylethyl)-	EtOH	217s(4.32),276s(4.32)	70-0957-78
3-Pyridinecarboxamide, N-[2-(2,2-dimethyl-2H-1-benzopyran-6-yl)ethyl]-	n.s.g.	225(4.43),264(3.84), 315(3.32)	88-2723-78
1H-Pyrido[3,2-b]indole-2,9-dione, 3,4,5,6,7,8-hexahydro-1-methyl-5-(phenylmethyl)-	EtOH	251(4.20),305(3.55)	94-3080-78
$C_{19}H_{20}N_2O_2S$			
1H-Pyrido[4,3-b]indole, 2,3,4,5-tetrahydro-5-methyl-2-[(4-methylphenyl)-sulfonyl]-	EtOH	228(4.78),273s(3.91), 283(3.94),292(3.88)	39-1471-78C
$C_{19}H_{20}N_2O_3$			
Benzamide, 4-nitro-N-(2-phenylcyclohexyl)-, (1S-trans)-	MeOH	206(4.29),261(4.07)	44-0355-78
Cyclohexanecarboxamide, N-(4-nitrophenyl)-2-phenyl-, (1S-trans)-	MeOH	206(4.30),316(4.14)	44-0355-78
$C_{19}H_{20}N_2O_4$			
Indolo[2,3-a]quinolizine-1α,3-dicarboxylic acid, 1,2,6,7,12,12bα-hexahydro-, dimethyl ester	EtOH	206(4.37),223(4.36), 292(4.30)	78-2995-78
1β-	EtOH	204s(4.01),224(4.31), 293(4.34)	78-2995-78
Propanedioic acid, (2-cyanoethyl)[2-cyano-3-(4-methylphenyl)-2-propenyl]-, monoethyl ester	EtOH	222(4.00),284(4.27)	2-0924B-78
3,5-Pyridinedicarboxylic acid, 1,4-dihydro-1-[2-(1H-indol-3-yl)ethyl]-, dimethyl ester	EtOH	205s(4.25),224(4.51), 261(4.03),282(3.91), 282(3.91),291(3.83), 392(3.83)	78-2995-78
2H-Pyrido[3,4-e]-1,3-oxazine-4,5(3H,6H)-dione, 7-methoxy-2,2-dimethyl-6-phenyl-3-(2-propenyl)-	MeOH	307(4.4)	120-0101-78
$C_{19}H_{20}N_2O_4S$			
1H-Benzimidazole, 2-[(phenylmethyl)-thio]-1-β-D-ribofuranosyl-	MeOH	252(3.94),258s(3.92), 284(4.12),291(4.14)	24-0996-78
$C_{19}H_{20}N_2O_5$			
DL-Alanine, N-[4-[methyl(phenylmethoxy)carbonyl]amino]benzoyl]-	MeOH	262(4.02)	87-1162-78
Glycine, N-methyl-N-[4-[methyl[(phenylmethoxy)carbonyl]amino]benzoyl]-	EtOH	250(4.29)	87-1162-78
Propanedioic acid, (2-cyanoethyl)[2-cyano-3-(4-methoxyphenyl)-2-propenyl]-, monoethyl ester	EtOH	298(4.22)	2-0924B-78
$C_{19}H_{20}N_2O_9$			
1H-Indole, 6-nitro-1-(2,3,4-tri-O-acetyl-α-D-arabinopyranosyl)-	EtOH	212(4.27),247(3.95), 318(3.89),356(3.80)	136-0017-78E
$C_{19}H_{20}N_2Se$			
1(2H)-Naphthaleneselone, 2-[[[4-(dimethylamino)phenyl]amino]methylene]-3,4-dihydro-	benzene	555(4.31)	104-0315-78
$C_{19}H_{20}N_4O$			
3H-Pyrazol-3-one, 4-[[[4-(dimethylami-	$C_6H_{11}Me$	260(4.34),342s(4.14),	48-0521-78

$C_{19}H_{20}N_4O-C_{19}H_{20}O_2$

Compound	Solvent	$\lambda_{max}(\log \epsilon)$	Ref.
no)phenyl]amino]methylene]-2,4-di- hydro-5-methyl-2-phenyl- (cont.)	 PrOH PrCN	385(4.32) 252(4.17),333s(3.87), 403(4.22) 253(4.21),265s(4.20), 335s(3.90),399(4.26)	48-0521-78 48-0521-78 48-0521-78
$C_{19}H_{20}N_4O_2$ 8-Azabicyclo[3.2.1]oct-3-ene-6-carbo- nitrile, 8,8'-(1,3-propanediyl)bis- [2-oxo-	CHCl$_3$	245(2.94),265(2.68), 353(2.20)	150-1182-78
$C_{19}H_{20}N_4O_2S$ [1,2,3]Thiadiazolo[5,1-e][1,2,3]thiadi- azole-7-S^{IV}, 1,6-dihydro-1,6-bis(4- methoxyphenyl)-3,4-dimethyl-	C_6H_{12}	247(4.27),310(4.27), 505(4.32)	39-0195-78C
$C_{19}H_{20}N_4O_3S$ Benzeneacetamide, N-[(4-amino-2-methyl- 5-pyrimidinyl)methyl]-N-[1-(1,2-oxa- thiolan-3-ylidene)ethyl]-α-oxo-	EtOH	251(4.29)	94-0722-78
$C_{19}H_{20}N_4O_4S$ Benzeneacetamide, N-[(4-amino-2-methyl- 5-pyrimidinyl)methyl]-N-[1-(1,2-oxa- thiolan-3-ylidene)ethyl]-α-oxo-, S- oxide	EtOH	238(4.16),265(4.17)	94-0722-78
$C_{19}H_{20}N_4S$ [1,2,3]Thiadiazolo[5,1-e][1,2,3]thiadi- azole-7-S^{IV}, 3,4-diethyl-1,6-dihydro- 1,6-diphenyl-	C_6H_{12}	245(4.27),265s(4.08), 300(4.25),489(4.33)	39-0195-78C
$C_{19}H_{20}N_6O_6$ 1H-Pyrrole-2-carboxylic acid, 1-methyl- 4-[[[1-methyl-4-[[(1-methyl-4-nitro- 1H-pyrrol-2-yl)carbonyl]amino]-1H- pyrrol-2-yl]carbonyl]amino]-, methyl ester	DMF	298(4.58)	78-2389-78
$C_{19}H_{20}O$ Benzene, [5-[(2-phenylethenyl)oxy]- 1-pentenyl]-	EtOH	247s(4.44),255(4.49), 260s(4.47),280s(3.85), 292(3.57),296s(3.34)	24-3624-78
5-Hexen-3-one, 4-methyl-6,6-diphenyl-	EtOH	224(4.12),255(4.10), 285(3.47),296(3.37)	35-1791-78
2-Oxabicyclo[4.2.0]octane, 7,8-diphen- yl-, all-cis	MeCN	246s(2.34),251s(2.50), 255s(2.61),257s(2.63), 261(2.69),264(2.68), 267s(2.59),271(2.54)	24-3624-78
$C_{19}H_{20}O_2$ 2(3H)-Benzofuranone, 7-(1,1-dimethyl- ethyl)-5-methyl-3-phenyl-	EtOH	266s(2.98),270s(3.06), 279(3.20),285(3.17), 320(2.21)	12-0907-78
1-Benzoxepin-6-ol, 2,5-dihydro-3-meth- yl-8-(2-phenylethyl)-	EtOH	221(3.82),279(3.20)	102-2005-78
[1,1'-Biphenyl]-2-ol, 4'-methoxy-3',5- di-2-propenyl-	n.s.g. + NaOH	259(4.3),293(4.2) 262(4.3),296(4.1)	100-0442-78 100-0442-78
Naphthalene, 1,2-dihydro-6-methoxy- 1-[(4-methoxyphenyl)methyl]-	EtOH	230(4.40),254(3.61), 264(3.68),272s(3.64), 286(3.32),302(3.23),	39-0750-78C

Compound	Solvent	$\lambda_{max}(\log \epsilon)$	Ref.
(cont.) 2,2'-Spirobi[2H-indene]-5,5'-dimethanol, 1,1',3,3'-tetrahydro-	EtOH	310s(3.15) 205s(4.29),215s(4.08), 263(3.27),268s(3.43), 271(3.45),277(3.54)	39-0750-78C 49-0987-78

C$_{19}$H$_{20}$O$_2$S

Compound	Solvent	$\lambda_{max}(\log \epsilon)$	Ref.
16-Thia-D-homoestra-1,3,5(10),6,8,14-hexaen-17aα-ol, 3-methoxy-	EtOH	227(4.39),249(4.32), 272(4.31),280(4.35), 312(4.48),321s(4.44)	39-1252-78C
16-Thia-D-homoestra-1,3,5(10),6,8,14-hexaen-17aβ-ol, 3-methoxy-	EtOH	228(4.39),248(4.34), 272(4.35),280(4.39), 309(4.43),317s(4.40)	39-1252-78C
16-Thia-D-homoestra-1,3,5(10),8,14-pentaen-17a-one, 3-methoxy-	EtOH	320s(4.44),333(4.52), 345s(4.41)	39-0576-78C

C$_{19}$H$_{20}$O$_3$

Compound	Solvent	$\lambda_{max}(\log \epsilon)$	Ref.
2(3H)-Benzofuranone, 7-(1,1-dimethyl-ethyl)-5-methoxy-3-phenyl-	EtOH	292(3.49),330s(2.35)	12-0907-78
1-Benzoxepin-6-ol, 2,5-dihydro-8-[2-(4-hydroxyphenyl)ethyl]-3-methyl-	EtOH	206(4.02),227s(3.59), 275(3.12)	102-2005-78
2,5-Cyclohexadien-1-one, 4-[3-(3,4-di-methoxyphenyl)-2-propenylidene]-2,6-dimethyl-	MeOH	275(4.10),450(4.62)	18-3612-78
2H-Phenanthro[1,2-c]pyran-1-ol, 1,11,12-12a-tetrahydro-8-methoxy-12a-methyl-,	EtOH	220(4.28),247s(4.45), 257(4.60),265(4.62), 282(4.14),292(4.34), 304(4.29)	39-1252-78C
2H-Phenanthro[1,2-c]pyran-1(5H)-one, 6,11,12,12a-tetrahydro-8-methoxy-12a-methyl-, (S)-	EtOH	301s(4.37),314(4.45), 327s(4.30)	39-0576-78C
Phenol, 4-(5,6,7,8-tetrahydro-6,7-di-methylnaphtho[2,3-d]-1,3-dioxol-5-yl)- (attenuol)	EtOH	223s(4.16),287(3.69), 294s(3.63)	31-0422-78

C$_{19}$H$_{20}$O$_3$S

Compound	Solvent	$\lambda_{max}(\log \epsilon)$	Ref.
2H-Phenanthro[1,2-c]thiopyran-1(6H)-one, 5,11,12,12a-tetrahydro-8-meth-oxy-12a-methyl-, 3α-oxide	EtOH	348(4.44)	39-1254-78C
3β-oxide	EtOH	351(4.44)	39-1254-78C

C$_{19}$H$_{20}$O$_3$S$_3$

Compound	Solvent	$\lambda_{max}(\log \epsilon)$	Ref.
2H-Thiopyran-3(4H)-one, 2,4-bis[[5-(ethylthio)-2-furanyl]methylene]-dihydro-	EtOH	218(4.30),272(4.22), 364(4.20),434(4.42)	133-0189-78
4H-Thiopyran-4-one, 3,5-bis[[5-(ethyl-thio)-2-furanyl]methylene]tetrahydro-	EtOH	208(4.03),267(4.09), 408(4.48)	133-0189-78

C$_{19}$H$_{20}$O$_4$

Compound	Solvent	$\lambda_{max}(\log \epsilon)$	Ref.
1,3-Benzenediol, 5-[2-(2,4-dihydroxy-phenyl)ethenyl]-2-(3-methyl-2-but-enyl)-, (E)-	EtOH	220(4.45),240s(4.29), 294s(4.33),303(4.36), 330(4.52)	138-1241-78
Spiro[furan-3(2H),6'-[6H]naphtho[1,8-bc]furan]-2-one, 5-(3-furanyl)-3',4-4',5,5',5'a,7',8'-octahydro-7'-meth-yl-, [5'aS-[5'aα,6'β(R*),7'β]]-(montanin A)	n.s.g.	217(4.03)	88-2025-78

C$_{19}$H$_{20}$O$_4$S

Compound	Solvent	$\lambda_{max}(\log \epsilon)$	Ref.
2H-Phenanthro[1,2-c]thiopyran-1(6H)-one, 5,11,12,12a-tetrahydro-8-meth-oxy-12a-methyl-, 3,3-dioxide, (S)-	EtOH	345(4.44)	39-1254-78C

Compound	Solvent	$\lambda_{max}(\log \epsilon)$	Ref.
$C_{19}H_{20}O_4S_2$			
4H-Pyran-4-one, 3,5-bis[[5-(ethylthio)-2-furanyl]methylene]tetrahydro-	EtOH	207(4.09),261(4.05), 303(4.06),405(4.28)	133-0189-78
$C_{19}H_{20}O_5$			
1,4-Benzenediol, 2,3-dimethyl-5-(phen-ylmethoxy)-, diacetate	EtOH	276(3.26)	23-0517-78
2,5-Cyclohexadien-1-one, 4-[3-(3,4-di-methoxyphenyl)-2-propenylidene]-2,6-dimethoxy-	MeOH	235s(4.12),280(3.93), 300s(3.77),466(4.61)	18-3612-78
Dehydroaguerin B	EtOH	215(4.16)	102-0955-78
2-Propen-1-one, 3-phenyl-1-(2,3,4,6-tetramethoxyphenyl)-	EtOH	292(4.39)	102-1363-78
$C_{19}H_{20}O_6$			
Benzo[1,2-b:4,5-b']difuran-3,7-dicarb-oxylic acid, 2,4,6-trimethyl-, diethyl ester	EtOH	212(4.50),230(4.46), 291(4.28),301(4.30)	4-0043-78
4H-1-Benzopyran-4-one, 2,3-dihydro-7-methoxy-2-(2,3,4-trimethoxyphenyl)-	EtOH	228(4.37),274(4.24), 314(3.95)	102-1405-78
4H-1-Benzopyran-4-one, 2,3-dihydro-5,6,7-trimethoxy-2-(4-methoxyphenyl)-	EtOH	280(4.22),330(3.68)	18-3627-78
4H-1-Benzopyran-4-one, 2,3-dihydro-5,7,8-trimethoxy-2-(4-methoxyphenyl)-	EtOH	286(4.23),332(3.66)	18-3627-78
Oroselol, 9-angeloyloxy-8,9-dihydro-, (8S,9R)-	EtOH	210(4.61),248s(3.78), 258(3.71),322(4.21)	95-0636-78
Spiro[furan-3(2H),1'(2'H)-naphthalene]-5'-carboxylic acid, 5-(3-furanyl)-3',4,4',4'a,5,7',8',8'a-octahydro-2'-methyl-2,4'-dioxo-	EtOH	208(4.42)	102-1967-78
Teucrin H 1	EtOH	215(4.23)	102-1967-78
Teucrin H 4	EtOH	217(4.09)	102-1967-78
$C_{19}H_{20}O_6S_2$			
4H-Pyran-4-one, 3,5-bis[[5-(ethylsul-finyl)-2-furanyl]methylene]tetrahy-dro-	EtOH	211(4.02),261(4.05), 380(4.53)	133-0189-78
$C_{19}H_{20}O_7$			
2,5-Cyclohexadiene-1,4-dione, 2-(3,4-dihydro-7,8-dimethoxy-2H-1-benzopyr-an-3-yl)-3,5-dimethoxy-, (S)-	EtOH	285(4.08),350(2.93)	102-1423-78
$C_{19}H_{21}BrN_2$			
1H-Indole, 3-methyl-2-[(4,5,6,7-tetra-hydro-3-methyl-2H-indol-2-ylidene)-methyl]-, monohydrobromide	EtOH	500(4.56)	83-0954-78
$C_{19}H_{21}BrN_2O_8S$			
Uridine, 5-bromo-2',3'-O-(1-methyleth-ylidene)-, 5'-(4-bromobenzenesulfon-ate)	EtOH	275(4.09)	94-2664-78
$C_{19}H_{21}ClN_2$			
Ibogamine, 9-c-loro-3,4,16,17-tetrade-hydro-9,17-dihydro-, (2α,5β,6α,18β)-	MeOH	227(4.42),276(3.88)	33-0690-78
$C_{19}H_{21}ClN_2O_4$			
1H-1,4-Diazepinium, 2,3-dihydro-1,4-bis(phenylmethyl)-, perchlorate	MeOH	347(4.36)	39-1453-78

Compound	Solvent	$\lambda_{max}(\log \epsilon)$	Ref.
$C_{19}H_{21}ClN_2O_5$			
2,1-Benzisoxazolium, 3-[2-[4-(dimethyl-amino)phenyl]-1-propenyl]-1-methyl-, perchlorate	MeCN	305(4.01),574(4.71)	44-1233-78
$C_{19}H_{21}ClN_4O$			
5H-Dibenzo[b,e][1,4]diazepine, 8-chloro-2-methoxy-11-(4-methylpiperazino)-	MeOH	220(4.44),259(4.29), 310(4.11)	73-0309-78
$C_{19}H_{21}ClN_4O_9$			
7H-Purine, 6-chloro-7-(tetra-O-acetyl-β-D-glucopyranosyl)-	EtOH	248(3.54),282(3.72)	12-1095-78
9H-Purine, 6-chloro-9-(2,3,5,6-tetra-O-acetyl-β-D-glucofuranosyl)-	EtOH	265(3.92)	12-1095-78
9H-Purine, 6-chloro-9-(2,3,4,6-tetra-O-acetyl-β-D-glucopyranosyl)-	EtOH	264(3.94)	12-1095-78
$C_{19}H_{21}FN_4$			
10,14:12,15a-Dimethano-15aH-cycloocta-[c]-1,2,4-triazolo[4,3-a][1,5]benzo-diazepine, 6-fluoro-9,9a,10,11,12,13-14,15-octahydro-3-methyl-, monohydro-chloride	MeOH	257(3.78),322(3.46)	4-0705-78
$(C_{19}H_{21}N)_n$			
9H-Carbazole, 2-ethenyl-9-(2-methylbut-yl)-, (S)-, homopolymer	CH_2Cl_2	240(4.54),267(4.34), 295s(4.15),302(4.28), 320s(3.51),332(3.63), 346(3.68)	126-1929-78
9H-Carbazole, 3-ethenyl-9-(2-methylbut-yl)-, (S)-, homopolymer	CH_2Cl_2	235(4.53),250s(4.32), 267(4.28),293s(4.00), 300(4.18),325s(3.26), 338(3.48),353(3.54)	126-1929-78
$C_{19}H_{21}NO$			
Benzamide, N-(2-phenylcyclohexyl)-, (1S-trans)-	MeOH	206(4.22)	44-0355-78
2-Butenenitrile, 3-[3-(bicyclo[2.2.1]-hept-5-en-2-ylcarbonyl)bicyclo-[2.2.1]hept-5-en-2-yl]-	n.s.g.	224(4.00)	33-1427-78
Cyclohexanecarboxamide, N,2-diphenyl-, (1S-trans)-	MeOH	206(4.39),242(4.12)	44-0355-78
3-Pyrrolidinone, 1,2,2-trimethyl-4,5-diphenyl-	n.s.g.	none above 270 nm	5-1203-78
$C_{19}H_{21}NO_2$			
9(10H)-Acridinone, 2-methoxy-10-pentyl-	MeOH	300(3.470)	4-0149-78
Aporphine, 2,10-dimethoxy-, L-(+)-	EtOH	266(4.10),272(4.12), 298(3.68),310(3.75), 318(3.76)	44-0105-78
Azacycloheptadeca-2,4,6,8,10,12,14,16-octaene-1-carboxylic acid, ethyl ester	ether	246(4.40),318(4.85), 420(3.90)	24-0084-78
isomer?	ether	242(4.34),306(4.86), 408(4.00)	24-0084-78
2-Azaestra-1,3,5(10),8(14),9(11),15-hexaen-17-ol, 3-methoxy-11-methyl-, (±)-	MeOH	257(4.49),263(4.47)	44-0966-78
4-Azatetracyclo[9.6.0.0^{2,10}.0^{3,5}]hepta-deca-6,8,12,14,16-pentaene-4-carbox-ylic acid	C_6H_{12}	230(3.86),260(3.32)	24-0084-78

Compound	Solvent	$\lambda_{max}(\log \epsilon)$	Ref.
6-Azatetracyclo[9.6.0.0^{2,10}.0^{3,7}]hepta-deca-3,8,12,14,16-pentaene-6-carbox-ylic acid, ethyl ester	C_6H_{12}	262s(3.28)	24-0084-78
6-Azatetracyclo[9.6.0.0^{2,10}.0^{3,9}]hepta-deca-4,7,12,14,16-pentaene-6-carbox-ylic acid, ethyl ester	C_6H_{12}	230(4.36),265s(3.30)	24-0084-78

$C_{19}H_{21}NO_3$

Compound	Solvent	$\lambda_{max}(\log \epsilon)$	Ref.
1,3-Cyclopentanedione, 2-[2-(5,6-dihy-dro-3-methoxy-8-isoquinolinyl)-2-propenyl]-2-methyl-	MeOH	261(4.09)	44-0966-78
Estra-1,3,5(10),9(11)-tetraene-16,17-dione, 3-methoxy-, 16-oxime	EtOH	214(4.34),261(4.36)	22-0119-78
	EtOH-NaOH	265(4.35),294(4.32)	22-0119-78
Lirinine, dl-	EtOH	235(3.99),270(4.04),283s(4.02)	2-0421-78
	EtOH-NaOH	236(4.03),272s(4.03),285(4.04)	2-0421-78
1-Propanone, 2-hydroxy-3-morpholino-1,3-diphenyl-	MeOH	245(4.13)	104-1174-78
2-Propenamide, N-[2-methoxy-2-(4-meth-oxyphenyl)ethyl]-3-phenyl-, (E)-	MeOH	217(4.30),223(4.36),275(4.39)	102-1814-78
Pyridinium, 1-(2-butoxy-2-oxoethyl)-4-[2-(4-hydroxyphenyl)ethenyl]-, hydroxide, inner salt, (E)- hydrochloride	EtOH	515(3.05)	56-1265-78
	acetone	600(3.32)	56-1265-78
	CHCl$_3$	625(3.45)	56-1265-78
	MeOH	405(4.62)	56-1265-78
Pyridinium, 1-[1-(ethoxycarbonyl)prop-yl]-4-[2-(4-hydroxyphenyl)ethenyl]-, hydroxide, inner salt, (E)- hydrobromide	EtOH	515(3.70)	56-1265-78
	acetone	610(3.72)	56-1265-78
	CHCl$_3$	630(3.74)	56-1265-78
	MeOH	410(4.41)	56-1265-78
Pyridinium, 1-(4-ethoxy-4-oxobutyl)-4-[2-(4-hydroxyphenyl)ethenyl]-, hydroxide, inner salt, (E)- hydrobromide	EtOH	520(3.68)	56-1265-78
	acetone	600(3.78)	56-1265-78
	CHCl$_3$	625(3.81)	56-1265-78
	MeOH	398(4.50)	56-1265-78
Pyridinium, 4-[2-(4-hydroxyphenyl)eth-enyl]-1-(1-methyl-2-oxo-2-propoxyeth-yl)-, hydroxide, inner salt, (E)- hydrobromide	EtOH	515(3.66)	56-1265-78
	acetone	610(3.69)	56-1265-78
	CHCl$_3$	630(3.86)	56-1265-78
	MeOH	402(4.51)	56-1265-78

$C_{19}H_{21}NO_4$

Compound	Solvent	$\lambda_{max}(\log \epsilon)$	Ref.
Boldine, hydrochloride, (+)-	MeOH	215(--),282(4.18),302(4.17),313(4.12)	12-0313-78
Lindcarpine, N-methyl-	n.s.g.	219(4.62),272(4.20),307(3.81)	105-0699-78
Pallidine, (-)-	EtOH	210(4.23),239(3.79),282(3.52)	100-0169-78
2,6-Phenanthrenediol, 3,5-dimethoxy-8-[2-(methylamino)ethyl]-	MeOH	263(4.74),280s(4.37),304(3.93),317(3.95),345(2.98),363(2.74)	12-0313-78
4H-Pyrano[2,3-h]quinoline-2-carboxalde-hyde, 1,8,9,10-tetrahydro-5-methoxy-8,8-dimethyl-4-oxo-3-(2-propenyl)-	MeOH	245(4.51),262(4.32),272(4.51),324(3.90)	24-0439-78
4H-Pyrano[2,3-h]quinolin-4-one, 1,8-di-hydro-2-(hydroxymethyl)-5-methoxy-8,8-dimethyl-3-(2-propenyl)-	MeOH	245(4.55),270(4.29),276(4.28),307(3.64),341(3.86)	24-0439-78
Thalidine, (-)-	EtOH	2+1[sic](4.56),250s(4.08),291(4.30)	100-0169-78
	EtOH-base	254s(4.11),297(4.27)	100-0169-78

$C_{19}H_{21}NO_4S$

Compound	Solvent	$\lambda_{max}(\log \epsilon)$	Ref.
3,4-Thiophenedicarboxylic acid, 5-phen-yl-2-piperidino-, dimethyl ester	dioxan	217(4.31),320(4.06)	24-2021-78

Compound	Solvent	λ_{max}(log ϵ)	Ref.
$C_{19}H_{21}NO_5$			
Colchiceine, N-deacetyl-, (±)-	MeOH	230(4.49),239(4.47), 248s(4.46),351(4.24), 363s(4.20)	33-1213-78
6-Quinolinol, 1,2,3,4-tetrahydro-1-[(3-hydroxy-4-methoxyphenyl)acetyl]-7-methoxy-	MeOH	253(3.97),287(3.87)	39-0440-78C
$C_{19}H_{21}NO_6S$			
1H-Azepine-2,5-dicarboxylic acid, 1-[(4-methylphenyl)sulfonyl]-, diethyl ester	EtOH	355(3.19)	2-0547-78
$C_{19}H_{21}NO_{10}S$			
Benzo[b]thiophene-3,4,5,6,7-pentacarboxylic acid, 2-amino-4,5-dihydro-, 3-ethyl 4,5,6,7-tetramethyl ester	MeOH	220(4.28),266(3.53), 316(3.71),400(4.27)	24-0770-78
$C_{19}H_{21}N_2$			
1H-1,4-Diazepinium, 2,3-dihydro-1,4-bis(phenylmethyl)-, perchlorate	MeOH	347(4.36)	39-1453-78C
$C_{19}H_{21}N_2O$			
2,1-Benzisoxazolium, 3-[2-[4-(dimethylamino)phenyl]-1-propenyl]-1-methyl-, perchlorate	MeCN	305(4.01),574(4.71)	44-1233-78
$C_{19}H_{21}N_2O_6$			
3,5-Pyridinedicarboxylic acid, 1,4-dihydro-2,6-dimethyl-4-(3-nitrophenyl)-, diethyl ester, anion	DMSO	454(3.76)	103-1226-78
$C_{19}H_{21}N_3O$			
Indolo[2,3-a]quinolizine-6-acetonitrile, 3-ethyl-1,2,3,4,6,7,12,12b-octahydro-2-oxo-, (3R,6S,12bR)-	MeOH	273(3.84),279(3.86), 288(3.79)	35-0938-78
(3R,6S,12bS)-	MeOH	273(3.86),278(3.87), 288(3.78)	35-0938-78
(3S,6S,12bR)-	MeOH	273(3.84),278(3.86), 288(3.79)	35-0938-78
(3S,6S,12bS)-	MeOH	228(3.77),273(3.83), 279(3.84)	35-0938-78
Pericyclivine, 16-cyano-16-de(methoxycarbonyl)-19,20-dihydro-3-hydroxy-	MeOH	272(3.81),282(3.82), 288(3.77),312(3.40)	35-0938-78
Pericyclivine, 16-cyano-16-de(methoxycarbonyl)-19,20-dihydro-15-hydroxy-, 16R-	MeOH	272(3.91),277(3.91), 288(3.81)	35-0938-78
16S-	MeOH	272(3.81),277(3.81), 288(3.71)	35-0938-78
Perivine, 16-cyano-16-de(methoxycarbonyl)-19,20-dihydro-	MeOH	238(4.07),315(4.21)	35-0938-78
$C_{19}H_{21}N_3O_4$			
1H-Pyrrolo[3,4-d]pyrimidine-7-carboxylic acid, 2,3,4,6-tetrahydro-5,6-dimethyl-2,4-dioxo-3-phenyl-, 1,1-dimethylethyl ester	EtOH	243(4.63),273(4.26)	94-3080-78
	EtOH-NaOH	245(--),302(--)	94-3080-78
$C_{19}H_{21}N_3O_5$			
Pyrazolo[4,3-b]pyridine-3-carboxaldehyde, 4-acetyl-4,5,6,7-tetrahydro-, diacetate	n.s.g.	265(3.76)	78-1581-78

Compound	Solvent	$\lambda_{max}(\log \epsilon)$	Ref.
1H-Pyrrolo[3,4-b]pyridine-3-carboxylic acid, 6-ethyl-4,5,6,7-tetrahydro-2-methyl-4-(3-nitrophenyl)-7-oxo-, ethyl ester	EtOH	219(4.36),258(4.25), 335(3.60)	95-0448-78
$C_{19}H_{21}N_3O_7$			
Caerulomycin D	MeOH	231(4.14),267(4.00), 305s(3.52)	23-1836-78
$C_{19}H_{21}N_3S$			
6H-1,3,4-Thiadiazin-2-amine, N-butyl-5,6-diphenyl-, hydrobromide	isoPrOH	262(3.99),334(3.99)	73-1227-78
6H-1,3,4-Thiadiazin-2-amine, N-(1,1-di-methylethyl)-5,6-diphenyl-	isoPrOH	257(4.08),334(4.02)	73-1227-78
3(2H)-Thiazolamine, 2-[(1,1-dimethyl-ethyl)imino]-4,5-diphenyl-	isoPrOH	227(4.20),286(3.95), 340(3.99)	73-1227-78
$C_{19}H_{21}N_5O_5$			
9H-Purin-6-amine, 9-[5-O-(1-oxo-3-phen-ylpropyl)-β-D-arabinofuranosyl]-	MeOH	259(4.17)	87-1218-78
$C_{19}H_{22}BrNO_3$			
Pyridinium, 1-[1-(ethoxycarbonyl)prop-yl]-4-[2-(4-hydroxyphenyl)ethenyl]-, bromide, (E)-	MeOH	410(4.41)	56-1265-78
Pyridinium, 1-(4-ethoxy-4-oxobutyl)-4-[2-(4-hydroxyphenyl)ethenyl]-, bromide	MeOH	398(4.50)	56-1265-78
Pyridinium, 4-[2-(4-hydroxyphenyl)eth-enyl]-1-(1-methyl-2-oxo-2-propoxy-ethyl)-, bromide, (E)-	MeOH	402(4.51)	56-1265-78
$C_{19}H_{22}Br_2N_2O$			
Urea, N,N'-bis[1-(4-bromophenyl)-1-methylethyl]-	MeCN	260(2.77),267(2.81), 275(2.62)	44-0922-78
$C_{19}H_{22}ClNO_4$			
1,2-Dehydroreticulinium chloride	EtOH	250(4.15),309(3.85), 370(3.85)	35-0276-78
$C_{19}H_{22}Fe$			
Ferrocene, 1,1':2,2':3,3'-tris(1,3-pro-panediyl)-	CH₂Cl₂	425(2.49)	101-0077-78L
cation	EtOH	248(--),675(2.51)	101-0077-78L
Ferrocene, 1,1':2,2':4,4'-tris(1,3-pro-panediyl)-	CH₂Cl₂	365(2.18)	101-0077-78L
cation	EtOH	241(4.06),770(2.67)	101-0077-78L
$C_{19}H_{22}FeO_4$			
Ferrocene, [2-(ethoxycarbonyl)-4-meth-oxy-1-methyl-4-oxo-1-butenyl]-, (E)-	EtOH	243(4.12),284(4.10)	128-0325-78
(Z)-	EtOH	234s(4.04),282(3.94)	128-0325-78
Ferrocene, [2-(ethoxycarbonyl)-5-meth-oxy-5-oxo-1-pentenyl]-, (E)-	EtOH	247(4.09),294(4.18)	128-0273-78
$C_{19}H_{22}IN_3$			
1,10-Phenanthrolinium, 4-(cyclohexyl-amino)-1-methyl-, iodide	neutral	232(4.37),248(4.40), 314s(4.26),370(4.23), 376(4.26)	39-1215-78
	monocation	218(4.3),251(4.32), 275(4.31),300s(4.26),	39-1215-78

Compound	Solvent	$\lambda_{max}(\log \epsilon)$	Ref.
(cont.)		332(4.24),350(4.26)	39-1215-78B
$C_{19}H_{22}N$ 3-Benzazocinium, 1,2,3,4-tetrahydro- 3,3-dimethyl-6-phenyl-, iodide	MeOH	260(4.27)	83-1029-78
$C_{19}H_{22}NO_3$ Pyridinium, 1-(2-butoxy-2-oxoethyl)-4- [2-(4-hydroxyphenyl)ethenyl]-, chloride, (E)-	MeOH	405(4.62)	56-1265-78
$C_{19}H_{22}NO_4$ 1,2-Dehydroreticulinium (chloride)	EtOH	250(4.15),309(3.85), 370(3.85)	35-0276-78
$C_{19}H_{22}NTi$ Titanium, bis(η^5-2,4-cyclopentadien- 1-yl)[2-[(dimethylamino)methyl]- phenyl-C,N]-	THF	635(1.65),740(1.68)	35-8068-78
$C_{19}H_{22}NV$ Vanadium, bis(η^5-2,4-cyclopentadien- 1-yl)[2-[(dimethylamino)methyl]- phenyl-C,N]-	THF	434(2.26),514(2.15), 683(2.21)	35-8068-78
$C_{19}H_{22}N_2$ Catharanthine, 18-de(methoxycarbonyl)-	EtOH	226(--),275(3.81), 283(3.87),290(3.85)	33-0690-78
Eburnamenine, (-)-	MeOH	224(4.37),258(4.48), 302(3.90),312(3.94)	33-1682-78
$C_{19}H_{22}N_2O$ Azepino[3,2-b]indol-2(1H)-one, 3,4,5,6- 7,8,9,10-octahydro-6-(phenylmethyl)-	MeOH	209(4.37),253s(3.93), 299s(3.53)	24-0853-78
4H-Carbazol-4-one, 1,2,3,5,6,7,8,9- octahydro-9-(phenylmethyl)-, oxime	MeOH	210(4.27),250(4.15)	24-0853-78
Eburnamonine, (±)-	EtOH-HCl	240(4.32),263(4.01), 292(3.73),301(3.74)	22-0355-78
Ibogamin-18-ol, 3,4-didehydro-	MeOH	275(3.79),282(3.81), 288(3.75)	33-0690-78
Δ^{14}-Isoeburnamine	EtOH	229(4.35),275(3.79), 282(3.80),290(3.70)	102-1452-78
$C_{19}H_{22}N_2O_2$ Acetic acid, (3,4,6,7,12,12b-hexahydro- indolo[2,3-a]quinolizin-1(2H)-yli- dene]-, ethyl ester, (S)-	EtOH	223(4.65),274(3.90), 280(3.90),290(3.82)	22-0355-78
Benzenepropanamide, α-(acetylamino)- N-(1-phenylethyl)-	EtOH	208(4.30),217(3.97), 257(2.57)	70-0957-78
Benzoic acid, 4-[[(methylphenylamino)- methylene]amino]-, butyl ester	isoPrOH	312(4.43)	117-0097-78
3-Pyridinecarboxamide, N-[2-(3,4-di- hydro-2,2-dimethyl-2H-1-benzopyran- 6-yl)ethyl]-	n.s.g.	226(4.20),264(3.82)	88-2723-78
$C_{19}H_{22}N_2O_4$ Acetamide, N-[[2-(acetylamino)-4,5-di- methoxyphenyl]methyl]-N-phenyl-	MeOH	246(3.6),284(4.0)	5-0608-78
2H-Pyrido[3,4-e][1,3]oxazine-4,5(3H,6H)- dione, 7-methoxy-2,2-dimethyl-6-phen- yl-3-propyl-	MeOH	309(4.4)	120-0101-78

Compound	Solvent	$\lambda_{max}(\log \epsilon)$	Ref.
$C_{19}H_{22}N_2O_5S$			
6-Oxabicyclo[3.2.0]heptane-4,4-diaceto-nitrile, 5-methoxy-1-[[[(4-methyl-phenyl)sulfonyl]oxy]methyl]-	dioxan	258(3.23),264(3.32), 269(3.30),275(3.25)	97-0380-78
4-Thia-1-azabicyclo[3.2.0]hept-2-ene-2-carboxylic acid, 3-methyl-7-oxo-6-[(phenoxyacetyl)amino]-, 1,1-di-methylethyl ester, (5R-trans)-	EtOH	263(3.73),268(3.75), 275(3.73),304(3.89)	35-8214-78
$C_{19}H_{22}N_2O_6$			
1,5-Dideazariboflavin	n.s.g.	344(3.62),476(3.98)	69-1942-78
5,8-Isoquinolinediol, 7-methoxy-6-meth-yl-3-morpholino-	EtOH	216(4.42),252(4.63), 290(4.16)	94-2175-78
3,5-Pyridinedicarboxylic acid, 1,4-di-hydro-2,6-dimethyl-4-(3-nitrophen-yl)-, diethyl ester, anion	DMSO	454(3.76)	103-1226-78
$C_{19}H_{22}N_2O_6S$			
1-Azabicyclo[3.2.0]hept-2-ene-2-carbox-ylic acid, 6-(1-hydroxyethyl)-7-oxo-3-[[2-[(phenoxyacetyl)amino]ethyl]-thio]-, sodium salt	H_2O	302(3.89)	35-8004-78
$C_{19}H_{22}N_2O_9$			
1H-Indole, 2,3-dihydro-6-nitro-1-(2,3,4-tri-O-acetyl-α-D-arabinopyranosyl)-	EtOH	252(4.11)	136-0017-78E
1,3-Propanediol, 2-[2-acetoxy-1-(6-ni-tro-1H-indol-1-yl)ethoxy]-, diacetate, (R)-	EtOH	240(3.60),316(3.83), 360(3.40)	136-0017-78E
$C_{19}H_{22}N_3$			
2H-Indazolium, 3-[2-[4-(dimethylamino)-phenyl]ethenyl]-1,2-dimethyl-, per-chlorate	MeCN	428(4.55)	44-1233-78
1,10-Phenanthrolinium, 4-(cyclohexyl-amino)-1-methyl-, iodide	neutral	232(4.37),248(4.40), 314s(4.26),370(4.23), 376(4.26)	39-1215-78B
	monocation	218(4.3),251(4.32), 275(4.31),300s(4.26), 332(4.24),350(4.26)	39-1215-78B
$C_{19}H_{22}N_4O_2$			
1,3-Propanediamine, N-(1-methylethyl)-N'-(1-nitro-9-acridinyl)-	pH 1	217(4.292),273(4.538), 328(3.627),435(3.866)	56-2125-78
	pH 7	273(4.293),402(3.619)	56-2125-78
1,3-Propanediamine, N-(1-nitro-9-acri-dinyl)-N'-propyl-	pH 1	217(4.308),273(4.542), 328(3.643),435(3.885)	56-2125-78
	pH 7	273(4.198),402(3.607)	56-2125-78
$C_{19}H_{22}N_4O_3S$			
2H-1,4-Thiazin-3(4H)-one, 4-[(4-amino-2-methyl-5-pyrimidinyl)methyl]-2-hy-droxy-6-(2-hydroxyethyl)-5-methyl-2-phenyl-	EtOH	231(4.17),278(3.83)	94-0722-78
$C_{19}H_{22}N_4O_4$			
Acetamide, N-[2-[5-methoxy-2-(1,2,3,4-tetrahydro-1,3-dimethyl-2,4-dioxo-5-pyrimidinyl)-1H-indol-3-yl]ethyl]-	MeCN	222(4.34),268(3.94), 306(3.72)	88-2585-78

Compound	Solvent	$\lambda_{max}(\log \epsilon)$	Ref.
$C_{19}H_{22}N_4O_9$			
Acetamide, N-[4,5-dihydro-4-oxo-1-(2,3,5-tri-O-acetyl-β-D-ribofurano-syl)-1H-imidazo[4,5-c]pyridin-6-yl]-	pH 1 pH 7 pH 11	277(4.10),297(4.13) 268(4.12),299(4.09) 223(4.14),285(4.03)	87-1212-78 87-1212-78 87-1212-78
$C_{19}H_{22}N_8O_3$			
Butanoic acid, 2-[[4-[[(2,4-diamino-6-pteridinyl)methyl]methylamino]-benzoyl]amino]-, (±)-	pH 1	245(4.20),308(4.27)	87-1162-78
Butanoic acid, 4-[[4-[[(2,4-diamino-6-pteridinyl)methyl]methylamino]-benzoyl]amino]-	pH 1	245(4.12),300(4.09)	87-1162-78
$C_{19}H_{22}O$			
Estra-1,3,5(10),8,16-pentaene, 3-meth-oxy-	EtOH	279(4.21)	22-0119-78
Estra-1,3,5(10),9(11),16-pentaene, 3-methoxy-	EtOH	264(4.29),298(3.48), 307(3.37)	22-0119-78
$C_{19}H_{22}OS$			
Benzo[c]thiophene-1-ol, 1,3-dihydro-3,3-dimethyl-1-(2,4,6-trimethyl-phenyl)-	hexane	259s(3.11),264s(3.03), 272(2.90),281s(2.52)	104-0582-78
	CH_2Cl_2	264s(3.11),272s(2.97), 283s(2.65)	104-0582-78
	dioxan	264s(2.75),272s(2.72), 281s(2.63)	104-0582-78
$C_{19}H_{22}O_2$			
1,3-Benzenediol, 2-(3-methyl-2-buten-yl)-5-(2-phenylethyl)-	n.s.g.	211(4.13),280(3.88)	102-2115-78
2H-1-Benzopyran-5-ol, 3,4-dihydro-2,2-dimethyl-7-(2-phenylethyl)-	n.s.g.	203(4.27),233s(3.45)	102-2115-78
Estra-1,3,5(10),8-tetraen-17-one, 3-methoxy-	EtOH	279(4.21)	22-0119-78
Estra-1,3,5(10),9(11)-tetraen-17-one, 3-methoxy-	EtOH	213(4.29),263(4.29), 296(3.52)	22-0119-78
Methanone, [2-(1-hydroxy-1-methyleth-yl)phenyl](2,4,6-trimethylphenyl)-	hexane	254(4.00),283s(3.58), 330(2.54)	104-0582-78
$C_{19}H_{22}O_2S$			
16-Thia-D-homoestra-1,3,5(10),8,14-pentaen-17aα-ol	EtOH	318s(4.49),330(4.61), 345(4.49)	39-0576-78C
16-Thia-D-homoestra-1,3,5(10),8,14-pentaen-17aβ-ol	EtOH	317s(4.48),329(4.60), 344(4.48)	39-0576-78C
$C_{19}H_{22}O_3$			
1-Azuleneacetic acid, α-(1-hydroxyeth-ylidene)-4,6,8-trimethyl-, ethyl ester	C_6H_{12}	246(4.54),291(4.64), 339(3.70),349s(3.74), 354(3.76),557(2.70), 604s(2.60),661(2.13)	18-3582-78
1-Benzoxepin-6-ol, 2,3,4,5-tetrahydro-8-[2-(4-hydroxyphenyl)ethyl]-3-meth-yl-	EtOH	207(3.53),277(2.55)	102-2005-78
D-Norestra-1,3,5(10),9(11)-tetraene-16α-carboxylic acid, 3-methoxy-16β-	EtOH	210(4.31),263(4.31), 296(3.50)	22-0119-78
	EtOH	264(4.30),297(3.50)	22-0119-78
16-Oxa-D-homoestra-1,3,5(10),8,14-pentaen-17aα-ol, 3-methoxy-16β-	EtOH	300s(4.40),312(4.49), 325s(4.37)	39-0576-78C
	EtOH	299s(4.36),311(4.45), 323s(4.33)	39-0576-78C

Compound	Solvent	λ_{max}(log ϵ)	Ref.
2-Oxatricyclo[13.2.2.1^{3,7}]eicosa-3,5,7-(20),15,17,18-hexaene-4,12-diol, (±)-(acerogenin A)	EtOH EtOH-NaOH	278(3.378),290s(--) 285-300(--)	94-2805-78 94-2805-78
$C_{19}H_{22}O_3S$			
2H-Phenanthro[1,2-c]thiopyran-1-ol, 1,5,6,11,12,12a-hexahydro-8-methoxy-12a-methyl-, 3α-oxide	EtOH	341.5(4.50)	39-1254-78C
3β-oxide	EtOH	337.5(4.49)	39-1254-78C
9,10-Seco-16-thia-D-homoestra-1,3,5(10)-8(14)-tetraene-9,17a-dione, 3-methoxy-	EtOH	219(4.06),250(3.91), 277s(3.73),280s(3.67)	39-0576-78C
Spiro[naphthalene-1(2H),8'-[2]thiabicyclo[3.3.1]nonane]-4',9'-dione, 3,4-dihydro-6-methoxy-5'-methyl-, [1'S-(1'α,5'α,8'α)]-	EtOH	276(3.26),284(3.23)	39-0576-78C
16-Thia-D-homoestra-1,3,5(10),9,11-tetraen-17a-one, 14-hydroxy-3-methoxy-, (8α,14β)-	EtOH	264.5(4.28)	39-0576-78C
(8β,14α)-	EtOH	265(4.29)	39-0576-78C
2H-Thiopyran-3,5(4H,6H)-dione, 4-[2-(3,4-dihydro-6-methoxy-1(2H)-naphthalenylidene)ethyl]-4-methyl-	EtOH	267(4.26),300s(3.75)	39-0576-78C
$C_{19}H_{22}O_4$			
1,3,5-Cyclohexanetrione, 2,2,4,4-tetramethyl-4-(1-oxo-3-phenylpropyl)-	EtOH EtOH-NaOH	278(4.43) 276(4.44)	102-2015-78 102-2015-78
Mucronulastyrene dimethyl ether	MeOH	240(4.13),282(3.83)	102-1389-78
2H-Pyran-3,5(4H,6H)-dione, 4-[2-(3,4-dihydro-6-methoxy-1(2H)-naphthalenylidene)ethyl]-4-methyl-	EtOH	268(4.20),300s(3.71)	39-0576-78C
$C_{19}H_{22}O_4S$			
2H-Phenanthro[1,2-c]thiopyran-1-ol, 1,5,6,11,12,12a-hexahydro-8-methoxy-12a-methyl-, 3,3-dioxide	EtOH	335.5(4.46)	39-1254-78C
isomer	EtOH	336.5(4.46)	39-1254-78C
$C_{19}H_{22}O_5$			
Mucronulatol dimethyl ether, (±)-	EtOH	227(4.25),282(3.73), 289(3.67)	102-1405-78
Mucronulatol 7-ethyl ether, (±)-	EtOH	226(4.21),283(--), 290(3.53)	102-1405-78
1H-Naphtho[1,8a-c]furan-3,8-dione, 7-[2-(2,5-dihydro-5-oxo-3-furanyl)ethyl]-5,6,6a,7,9,10-hexahydro-7-methyl-	MeCN	211(4.36)	24-2130-78
$C_{19}H_{22}O_5Se$			
2-Cyclohexene-1-acetic acid, 1-(ethoxycarbonyl)-4-oxo-5-(phenylseleno)-, ethyl ester	MeOH	219(4.2),235(4.1)	24-1944-78
$C_{19}H_{22}O_6$			
Eremantholide C	EtOH	266(3.99)	39-1572-78C
3-Furancarboxylic acid, 5-acetoxy-4,5-dihydro-5-methyl-2-[4-(1-methylethyl)phenyl]-, ethyl ester	EtOH	223(4.00),266(3.74), 310(4.31)	118-0291-78
4H,9H-Furo[2',3',4':4,5]naphtho[2,1-c]pyran-4,9-dione, 3,3a,5a,6,10b,10c-hexahydro-3,6-dihydroxy-3a,10b-dimethyl-7-(1-methylethyl)-	EtOH	299(3.83)	142-0123-78B

Compound	Solvent	$\lambda_{max}(\log \epsilon)$	Ref.
$C_{19}H_{22}O_7$			
1H-2-Benzoxacyclotetradecin-1,7(8H)-di-one, 3,4,9,10-tetrahydro-8,9,16-tri-hydroxy-14-methoxy-3-methyl-	MeOH	233(4.58),271(4.06), 314(3.78)	44-2339-78
Ethanone, 1-(2-hydroxy-3,4-dimethoxy-phenyl)-2-(2,3,4-trimethoxyphenyl)-	EtOH	225(4.27),283(4.17), 330s(3.54)	102-1401-78
$C_{19}H_{22}O_8$			
1,1,2-Cyclopropanetricarboxylic acid, 3-(4-methoxybenzoyl)-, 1,1-diethyl 2-methyl ester	EtOH	225(4.12),281(4.28)	73-1727-78
$C_{19}H_{22}O_{10}$			
Aloenin	EtOH	232(3.87),245(3.81), 307(3.91)	18-0842-78
	EtOH-KOH	225(3.98),253(4.14), 353(4.16)	18-0842-78
$C_{19}H_{22}Sn$			
Stannane, trimethyl(1-methyl-3-phenyl-1H-inden-1-yl)-	heptane	237(4.31),261(3.82), 293(3.00)	104-0207-78
	HMPTA	395(3.04),426(3.00)	104-0207-78
$C_{19}H_{23}BF_4N_2$			
2-Propenylium, 1,3-bis[4-(dimethylami-no)phenyl]-, tetrafluoroborate	$C_2H_4Cl_2$	705(5.33)	104-1197-78
$C_{19}H_{23}ClO_5$			
Androst-2-ene-1,4,17-trione, 6-chloro-5,14-dihydroxy-, (5β,6α)-	MeOH	227(3.94),375(1.92)	42-1175-78
$C_{19}H_{23}F_3N_2O$			
Methanone, (1,4,4aβ,5,6,7,8aα-octahy-dro-1,3,4-trimethyl-2-quinoxalinyl)-[4-(trifluoromethyl)phenyl]-, trans	EtOH	282(3.91),335(3.58), 390(3.99)	88-0701-78
$C_{19}H_{23}IN_2O_6$			
Isoquinolinium, 5,8-diacetoxy-7-meth-oxy-2-methyl-3-morpholino-, iodide	EtOH	220(4.52),262(4.44), 304(4.00)	4-0569-78
$C_{19}H_{23}N$			
9H-Carbazole, 2-ethyl-9-(2-methyl-butyl)-, (S)-	CH_2Cl_2	233s(4.57),240(4.68), 248(4.51),267(4.42), 292s(4.08),299(4.28), 320s(3.48),331(3.63), 344(3.62)	126-1929-78
9H-Carbazole, 3-ethyl-9-(2-methyl-butyl)-, (S)-	CH_2Cl_2	235s(4.59),240(4.63), 250s(4.43),268(4.36), 290s(4.04),298(4.26), 325s(3.36),337(3.58), 352(3.63)	126-1929-78
9H-Carbazole, 9-(1-methylethyl)-3-(1-methylpropyl)-, (S)-	CH_2Cl_2	232s(4.54),240(4.58), 249s(4.38),269(4.34), 290s(3.95),298(4.20), 325s(3.34),337(3.51), 351(3.57)	126-1929-78
$C_{19}H_{23}NO$			
Benzenamine, 4-[1-ethyl-2-(4-methoxy-phenyl)-1-butenyl]-, hydrochloride	EtOH	246(4.32)	104-0725-78

Compound	Solvent	$\lambda_{max}(\log \epsilon)$	Ref.
4,7-Methano-1H-indene-2-carbonitrile, 1-bicyclo[2.2.1]hept-5-en-2-yl-2,3,3a,4,7,7a-hexahydro-1-hydroxy-2-methyl-, isomer A	n.s.g.	211(2.84)	33-1427-78
isomer B	n.s.g.	210(2.85)	33-1427-78
$C_{19}H_{23}NO_2$			
2-Azaestra-1,3,5(10),8(14),9(11)-penta-en-17-ol, 3-methoxy-11-methyl-, (±)-	MeOH	242(4.34),247(4.32), 291(3.83)	44-0966-78
5H-Pyrano[2.3-b]quinolin-5-one, 2,3,4-10-tetrahydro-2,2-dimethyl-10-(3-methyl-2-butenyl)-	EtOH	238(4.18),250s(3.95), 317(3.78),329(3.76)	142-0193-78A
$C_{19}H_{23}NO_3$			
Butanoic acid, 2-(1'-methylspiro[cyclo-pentane-1,2'-[2H]indol]-3'(1'H)-yli-dene)-3-oxo-, ethyl ester	EtOH	211(4.48),247(4.51), 317(3.51)	44-3702-78
Erysotrine, hydrochloride	EtOH	280(3.5)	100-0342-78
Pronuciferine, dl-	EtOH	230(4.35),280(3.73)	94-0481-78
$C_{19}H_{23}NO_4$			
3a,8a-Butanofuro[2,3-b]indole-3-carbox-ylic acid, 2,3,3a,8a-tetrahydro-2-hy-droxy-2-methyl-	EtOH	212(4.00),240(3.94), 294(3.36)	44-3702-78
3H-Cyclobut[b]indole-1,2-dicarboxylic acid, 7b-(1,1-dimethylethyl)-2a,7b-dihydro-3-methyl-, dimethyl ester	EtOH	252(3.62),295(3.21), 411(2.81)	88-2979-78
Erythrartine	EtOH	230(8.2),282(3.3)[sic]	100-0342-78
5-Isoquinolinol, 1,2,3,4-tetrahydro-1-[(4-hydroxyphenyl)methyl]-6,7-di-methoxy-2-methyl- (same in acid or base)	MeOH	278(3.77)	44-0580-78
Propanedioic acid, spiro[cyclopentane-1,3'-[3h]indol]-2'(1'H)-ylidene-, diethyl ester	EtOH	204(3.95),232(4.08), 299(3.97),329(4.30)	44-3702-78
4H-Pyrano[2,3-h]quinolin-4-one, 1,8,9-10-tetrahydro-2-(hydroxymethyl)-5-methoxy-8,8-dimethyl-3-(2-propenyl)-	MeOH	246(4.53),261(4.46), 267(4.45),320(3.95)	24-0439-78
3,5-Pyridinedicarboxylic acid, 1,4-di-hydro-2,6-dimethyl-4-phenyl-, diethyl ester, anion	DMSO	465(3.78)	103-1226-78
Reticuline	EtOH	230s(4.09),285(3.76)	105-0360-78
$C_{19}H_{23}NO_5$			
1H-Pyrrole-2,4-dicarboxylic acid, 3-hy-droxy-1-(phenylmethyl)-, 2-(1,1-di-methylethyl) 4-ethyl ester	EtOH	229(4.45),263(4.06)	94-2224-78
1H-Pyrrole-2-propanoic acid, 2-(ethoxy-carbonyl)-2,3-dihydro-3-oxo-1-(phen-ylmethyl)-, ethyl ester	EtOH	332(4.05)	94-3521-78
$C_{19}H_{23}N_2$			
2-Propenylium, 1,3-bis[4-(dimethylami-no)phenyl]-, tetrafluoroborate	$C_2H_4Cl_2$	705(5.33)	104-1197-78
$C_{19}H_{23}N_2O_6$			
Isoquinolinium, 5,8-diacetoxy-7-meth-oxy-2-methyl-3-morpholino-, iodide	EtOH	220(4.52),262(4.44), 304(4.00)	4-0569-78
$C_{19}H_{23}N_3O$			
Perivinol, 16-cyano-16-de(methoxycarbo-	MeOH	274(3.85),282(3.87),	35-0938-78

Compound	Solvent	$\lambda_{max}(\log \epsilon)$	Ref.
nyl)-19,20-dihydro- (cont.)		292(3.81)	35-0938-78
$C_{19}H_{23}N_3O_2$			
Benzoic acid, 4-[[[[4-(dimethylamino)-phenyl]methylamino]methylene]amino]-, ethyl ester	isoPrOH	324(4.34)	117-0097-78
$C_{19}H_{23}N_3O_6$			
1H-Pyrrole-2-carboxylic acid, 4-(cyano-methyl)-2-[[5-(ethoxycarbonyl)-4-hy-droxy-1-methyl-1H-pyrrol-3-yl]meth-yl]-2,3-dihydro-1-methyl-3-oxo-, ethyl ester	EtOH	267(4.27),338(3.87)	94-3521-78
D-Ribitol, 1-deoxy-1-(3,4-dihydro-5,7,8-trimethyl-2,4-dioxopyrimido-[4,5-b]quinolin-10(2H)-yl)-	pH 1	224(4.34),236(4.25),264(4.55),345(4.12)	4-0489-78
	pH 7	225(4.39),250(4.26),271(4.27),334(3.86),394(3.86)	4-0489-78
	pH 13	226(4.43),264(4.55),330(3.99),398(3.96)	4-0489-78
	n.s.g.	335(4.11),394(4.08)	69-1942-78
$C_{19}H_{23}N_5O_3$			
Pyrrolo[3',4':4,5]pyrrolo[3,4-b]indole-1,3,5(2H)-trione, 2,5a-diamino-4-cy-clohexyl-3a,4,5a,6-tetrahydro-6-meth-yl-	EtOH	240(3.87),259(3.87),316(3.35)	4-1463-78
$C_{19}H_{23}N_5O_9$			
Guanosine, 3-methyl-, tetraacetate	MeOH	217(3.98),278(4.22)	88-2579-78
9H-Purin-6-amine, 9-(2,3,4,6-tetra-O-acetyl-β-D-glucopyranosyl)-	MeOH	210(4.46),258(4.26)	136-0301-78C
$C_{19}H_{23}N_7$			
2,4-Pteridinediamine, 6-[[2-(phenyl-methyl)-1-piperidino]methyl]-	MeOH-HCl	248(4.27),338(4.01)	87-0331-78
	MeOH-KOH	260(4.43),373(3.88)	87-0331-78
$C_{19}H_{23}N_7O$			
2,4-Pteridinediamine, 6-[[2-(phenyl-methyl)-1-piperidino]methyl]-, 8-oxide	MeOH-HCl	261(4.41),361(3.79)	87-0331-78
	MeOH-KOH	267(4.50),397(3.84)	87-0331-78
$C_{19}H_{24}N_2$			
14,15-Anhydrodihydrocapuronidine	EtOH	226(3.86),244(3.74),292(3.44)	102-1605-78
1H-Indolo[3,2,1-de]pyrido[3,2,1-ij]-[1,5]naphthyridine, 13a-ethyl-2,3,5,6,12,13,13a,13b-octahydro-, (13aS,13bS)- (vincane)	MeOH	227(4.59),283(3.89)	33-1682-78
9H-Pyrido[3,4-b]indole, 1-octyl-	MeOH	236(4.62),241(4.61),251s(4.42),283s(4.03),289(4.29),338(3.70),351(3.71)	95-0898-78
$C_{19}H_{24}N_2O$			
Epivincanol	MeOH	228(4.53),282(3.93)	33-1682-78
Vincanol	MeOH	195(4.39),227(4.54),281(3.94)	33-1682-78
$C_{19}H_{24}N_2O_2$			
Ibogamine-4,18-diol, (2α,4α,5β,6α,18β)-	MeOH	278(3.87),283(3.90),	33-0690-78

Compound	Solvent	$\lambda_{max}(\log \epsilon)$	Ref.
(cont.) Pyrrolo[2,3-b]indole, 1,8-diacetyl- 1,2,3,3a,8,8a-hexahydro-3a-(3-methyl- 2-butenyl)-	EtOH	290(3.86) 245(4.02),275(3.28), 283(3.22)	33-0690-78 78-0929-78
$C_{19}H_{24}N_2O_2S$ Levomepromazine sulfoxide	MeOH	216(4.29),247(4.54), 277(4.06),295(3.81), 333(3.73)	133-0248-78
$C_{19}H_{24}N_2O_3$ 2-Hexenoic acid, 6-hydroxy-3-[[[2-(1H- indol-3-yl)ethyl]imino]methyl]-, ethyl ester	EtOH	221(4.62),280(3.84), 290(3.76)	22-0355-78
2-Hexenoic acid, 6-hydroxy-3-(2,3,4,9- tetrahydro-1H-pyrido[3,4-b]indol-1- yl)-, ethyl ester, (±)-	EtOH	224(4.59),277(3.85), 282(3.85),292(3.79)	22-0355-78
$C_{19}H_{24}N_2O_4$ Azepino[4,5-b]indole-5,5(2H)-dicarbox- ylic acid, 1,3,4,6-tetrahydro-3-meth- yl-, diethyl ester	EtOH	235(4.53),285(3.90), 293(3.84),340(2.74)	44-3705-78
$C_{19}H_{24}N_2O_6S$ 1-Azetidineacetic acid, 2-acetoxy-α-(1- mercapto-1-methylethyl)-4-oxo-3- [(phenylacetyl)amino]-, methyl ester, [2S-[1(R*),2α,3β]]-	EtOH	213(3.94)	39-0668-78C
$C_{19}H_{24}N_4OS$ 7(4H)-Benzothiazolone, 2-[[4-(diethyl- amino)phenyl]azo]-5,6-dihydro-5,5-di- methyl-	>6M H_2SO_4 0.5M H_2SO_4 pH 2.5-13 >2M NaOH	457(4.66) 592(4.79) 581(4.79) 571(4.78)	73-2289-78 73-2289-78 73-2289-78 73-2289-78
$C_{19}H_{24}N_4O_4$ 2,4(1H,3H)-Quinazolinedione, 1,1'-(1,3- propanediyl)bis[5,6,7,8-tetrahydro-	H_2O	276(4.31)	56-1035-78
$C_{19}H_{24}N_4O_5$ L-Alanine, N-[N-(2-quinoxalinylcarbo- nyl)-D-seryl]-, 1,1-dimethylethyl ester	MeOH	243(4.64),320(3.87), 325(3.87)	18-1501-78
$C_{19}H_{24}N_4O_{12}S$ β-D-Glucopyranuronic acid, 1-deoxy-1- [6-[(6-methyl-β-D-glucopyranurono- syl)thio]-9H-purin-9-yl]-, methyl ester	MeOH	277.5(4.32)	106-0250-78
$C_{19}H_{24}O$ Estra-1,3,5(10),8-tetraene, 3-methoxy- 14β-Estra-1,3,5(10),8-tetraene, 3-meth- oxy- Estra-1,3,5(10),9(11)-tetraene, 3-meth- oxy-	EtOH EtOH EtOH	278(4.22) 273(4.24) 278(4.21)	22-0119-78 22-0119-78 22-0119-78
$C_{19}H_{24}O_2$ 3α,5α-Cyclo-18-nor-D-homoandrost- 13(17a)-ene-6,17-dione	EtOH	240(4.23)	56-2361-78

Compound	Solvent	$\lambda_{max}(\log \epsilon)$	Ref.
Estra-1,3,5(10),8-tetraen-17β-ol, 3-methoxy-	EtOH	278(4.23)	22-0119-78
Estra-1,3,5(10),9(11)-tetraen-17β-ol, 3-methoxy-	MeOH	263(4.30)	94-0171-78
	EtOH	263(4.28),297(3.47)	22-0119-78
18-Nor-5β-androsta-8,11,13-trien-3-one, 11-hydroxy-17ξ-methyl-	EtOH	290(3.57)	39-0163-78C
Phenol, 2,2'-methylenebis[6-(1-methylethyl)-	EtOH	272(3.28)	32-0079-78
$C_{19}H_{24}O_3$			
Estra-1,3,5(10)-trien-6-one, 17β-hydroxy-2-methoxy-	EtOH	226(4.14),278(4.18)	94-3567-78
Estra-1,3,5(10)-trien-6-one, 17β-hydroxy-3-methoxy-	EtOH-CHCl₃	224(4.20),257(3.74), 326(3.29)	94-3567-78
Estra-1,3,5(10)-trien-11-one, 9α-hydroxy-3-methoxy-	EtOH	273(3.10),280(3.08)	78-0393-78
Estra-1,3,5(10)-trien-17-one, 3-hydroxy-11α-methoxy-	EtOH	279(3.28)	111-0313-78
	EtOH-NaOH	242(4.00),296(3.45)	111-0313-78
Phenol, 2,2'-(ethoxymethylene)bis[4,6-dimethyl-	EtOH	282(3.85),288(3.83)	32-0079-78
$C_{19}H_{24}O_3S$			
8,14-Seco-16-thia-D-homoestra-1,3,5(10)-9(11)-tetraen-17a-one, 14α-hydroxy-3-methoxy-	EtOH	266(4.27)	39-0576-78C
14β-hydroxy-	EtOH	266(4.29)	39-0576-78C
$C_{19}H_{24}O_4$			
Estra-1,3,5(10)-trien-6-one, 2,17β-dihydroxy-3-methoxy-	EtOH	235(4.24),281(4.00), 317(3.91)	94-3567-78
Estra-1,3,5(10)-trien-6-one, 3,17β-dihydroxy-2-methoxy-	EtOH	235(4.20),277(3.94), 320(3.74)	94-3567-78
Estra-1,3,5(10)-trien-17-one, 3,16α-dihydroxy-11α-methoxy-	EtOH	279(3.26)	111-0313-78
	EtOH-NaOH	242(4.01),295(3.44)	111-0313-78
$C_{19}H_{24}O_5$			
Benzoic acid, 3,4-dimethoxy-, 3,7-dimethyl-5-oxo-2,6-octadienyl ester (tomentellin)	n.s.g.	257(3.65)	31-0155-78
4H,9H-Furo[2',3',4':4,5]naphtho[2,1-c]-pyran-4,9-dione, 1,2,3,3a,5a,7,10b-10c-octahydro-3-hydroxy-3a,10b-dimethyl-7-(1-methylethyl)-, [3S-(3α,3aβ,5aβ,7β,10bα,10cβ)]-	EtOH	261(4.00)	142-0123-78B
Pterosin A, diacetate	EtOH	218(4.51),260(4.16), 296(3.33),306(3.36)	94-2365-78
Spiro[cyclohexane-1,1'-[1H]indene]-4,7'-diol, 2',3'-dihydro-5'-methoxy-, diacetate	MeOH	273(3.29),280(3.29)	94-3641-78
$C_{19}H_{24}O_6$			
4,16-Dioxatricyclo[11.2.1.1³,⁶]heptadec-6-ene-5,11,14-trione, 17-hydroxy-1-methyl-9-(1-methylethenyl)-	EtOH	217.5(3.95)	12-2049-78
4,16-Dioxatricyclo[11.2.1.1³,⁶]heptadec-6-ene-5,11,14-trione, 17-hydroxy-1-methyl-9-(1-methylethylidene)-	EtOH	217(3.97)	12-2049-78
Eremantholide A	n.s.g.	266(4.00)	39-1572-78C
Spiro[1,3-dioxolane-2,6'(4'H)-naphtho-[2,3-b]furan-4'-one, 9'-acetoxy-4'a,5',7',8',8'a,9'-hexahydro-	EtOH	263.5(3.54)	94-3704-78

Compound	Solvent	$\lambda_{max}(\log \epsilon)$	Ref.
3',4'a,5'-trimethyl- (cont.) Spiro[1,3-dioxolane-2,6'(4'H)-naphtho- [2,3-b]furan-9'(5'H)-one, 4'-acetoxy- 4'a,7',8',8'a-tetrahydro-3',4',5'- trimethyl-	EtOH	282(4.17)	94-3704-78 94-3704-78
$C_{19}H_{24}O_7$ 1H-2-Benzoxacyclotetradecin-1,7(8H)-di- one, 3,4,5,6,9,10-hexahydro-8,9,16- trihydroxy-14-methoxy-3-methyl-	MeOH	232(4.49),270(4.08), 311(3.79)	44-2339-78
1H-2-Benzoxacyclotetradecin-1-one, 3,4,7,8,9,10-hexahydro-7,8,9,16- tetrahydroxy-14-methoxy-3-methyl-	MeOH	236(4.34),272(3.96), 316(3.60)	44-2339-78
$C_{19}H_{25}ClN_2$ 1H-Pyrrole, 2-[2-chloro-3-(4-ethyl-3,5- dimethyl-2H-pyrrol-2-ylidene)-1-prop- enyl]-4-ethyl-3,5-dimethyl-, mono- hydrochloride	EtOH EtOH-NH	571(5.32) 488(4.42)	5-2028-78 5-2028-78
$C_{19}H_{25}NO$ 2-Butenenitrile, 3-[3-(bicyclo[2.2.1]- hept-2-ylcarbonyl)bicyclo[2.2.1]hept- 2-yl]-	n.s.g.	221(3.93)	33-1427-78
$C_{19}H_{25}NO_2$ 2-Azaestra-1,3,5(10),8-tetraen-17-ol, 3-methoxy-11β-methyl-	MeOH	267(4.21)	44-0966-78
19α,20α-Imino-5β,10α-podocarpa-8,11,13- trien-12-ol, N-acetyl-, (±)-	EtOH	282(3.48)	44-4598-78
$C_{19}H_{25}NO_3$ Coccutrine, O-methyl-	EtOH	230(3.95),285(3.50)	95-0886-78
2(1H)-Pyridinone, 1-[2-(3,4-dimethoxy- phenyl)ethyl]-5-(1,1-dimethylethyl)-	EtOH	230(4.18),286(3.70), 312(3.76)	142-0023-78B
$C_{19}H_{25}NO_4$ 2,4-Pentadienamide, N-hexyl-5-(6-meth- oxy-1,3-benzodioxol-5-yl)-, (E,E)-	n.s.g.	368(4.38)	78-1979-78
$C_{19}H_{25}NO_6$ Propanedioic acid, [[(2-ethoxy-2-oxo- ethyl)(phenylmethyl)amino]methyl- ene]-, diethyl ester	EtOH	282(4.25)	94-2224-78
$C_{19}H_{25}NO_8$ 5,6,7,8-Indolizinetetracarboxylic acid, 1,2,3,8a-tetrahydro-2,2,8a-trimeth- yl-, tetramethyl ester	MeOH	227(4.03),289(3.92), 383(3.61)	83-0977-78
$C_{19}H_{25}N_2$ Pyridinium, 1,2,6-trimethyl-4-[3-(1,2,6- trimethyl-4(1H)-pyridinylidene)-1- propenyl]-, perchlorate	CH_2Cl_2 $MeNO_2$	610(5.30) 598(5.11)	103-0601-78 103-0601-78
$C_{19}H_{25}N_3$ Epivincanol, 14-amino-14-deoxy- Vincanol, 14-amino-14-deoxy-	MeOH MeOH	228(4.52),284(3.92) 228(4.50),282(3.89)	33-1682-78 33-1682-78
$C_{19}H_{25}N_4O_{12}P$ Thymidine, 2'-deoxyuridylyl-(5'→3')-	pH 7	263(4.18)	69-4865-78

Compound	Solvent	$\lambda_{max}(\log \epsilon)$	Ref.
$C_{19}H_{25}N_5$			
Propanedinitrile, [[3-[bis(dimethylamino)methylene]-1,4-cyclopentadien-1-yl]-1-piperidinomethylene]-	MeOH	294(4.27),318(4.20), 395(4.55)	83-0369-78
$C_{19}H_{25}N_5O_6$			
Riboflavin, 8-demethyl-8-(propylamino)-	n.s.g.	305(4.00),488(4.61)	69-1942-78
$C_{19}H_{25}OP$			
Phosphorin, 4-cyclohexyl-1-ethoxy-1,1-dihydro-1-phenyl-	EtOH	327(3.54),358(3.61)	89-0528-78
$C_{19}H_{26}ClN_2S_2$			
Thiazolium, 2-[2-[2-chloro-3-[(3-ethyl-2-thiazolidinylidene)ethylidene]-1-cyclopenten-1-yl]ethenyl]-3-ethyl-4,5-dihydro-, iodide	MeOH	70?(5.29)[sic]	104-2046-78
$C_{19}H_{26}N_2$			
Melonine	EtOH	248(3.80),298(3.45)	102-1452-78
1H-Pyrrole, 3-ethyl-5-[3-(4-ethyl-3,5-dimethyl-2H-pyrrol-2-ylidene)-1-propenyl]-2,4-dimethyl-, monohydrobromide	benzene	613(5.32)	5-2028-78
	EtOH	578(5.39)	5-2028-78
	EtOH-NH₃	487(4.84)	5-2028-78
$C_{19}H_{26}N_2O$			
Cinchonamine, dihydro-	MeOH	222(4.63),283(3.99), 292(3.90)	35-0589-78
3-epi-	MeOH	225(4.59),285(3.92), 293(3.88)	35-0589-78
Melonine, N-oxide	EtOH	247(3.82),298(3.31)	102-1452-78
$C_{19}H_{26}N_2OS_2$			
Cyclopentanecarbodithioic acid, 2-imino-, 1-[(4-nitrophenyl)amino]cyclohexyl ester	EtOH	236(4.08),336(3.76), 408(4.30)	39-0558-78C
$C_{19}H_{26}N_2O_3$			
1H-Pyrrole-3-propanoic acid, 5-[(3-ethyl-5-methoxy-4-methyl-2H-pyrrol-2-ylidene)methyl]-2,4-dimethyl-, methyl ester	EtOH	260(4.18),411(4.48)	49-0883-78
$C_{19}H_{26}N_2O_4$			
Morpholine, 4-[6-[1-(2-methoxyphenyl)-2-nitroethyl]-1-cyclohexen-1-yl]-	EtOH	214(4.65),262(4.28), 323(3.77)	2-0405-78
Morpholine, 4-[6-[1-(4-methoxyphenyl)-2-nitroethyl]-1-cyclohexen-1-yl]-, (R*,S*)-	EtOH	231(4.02),352(3.98)	2-0405-78
$C_{19}H_{26}N_2S_2$			
Cyclopentanecarbodithioic acid, 2-imino-, 1-[(4-methylphenyl)amino]cyclohexyl ester	EtOH	236(4.26),336(3.83), 408(4.36)	39-0558-78C
Cyclopentanecarbodithioic acid, 2-imino-, 3-methyl-1-(phenylamino)cyclohexyl ester	EtOH	234(4.12),337(3.80), 408(4.33)	39-0558-78C
Cyclopentanecarbodithioic acid, 2-imino-, 4-methyl-1-(phenylamino)cyclohexyl ester	EtOH	234(4.16),337(3.84), 408(4.35)	39-0558-78C

Compound	Solvent	$\lambda_{max}(\log \epsilon)$	Ref.
$C_{19}H_{26}N_4O_{10}S$			
1H-Purine-2,6-dione, 7-[4-O-acetyl-2,3-O-(1-methylethylidene)-6-O-(methylsulfonyl)-α-D-mannopyranosyl]-3,7-dihydro-1,3-dimethyl-	MeOH	274(3.95)	136-0073-78C
$C_{19}H_{26}O$			
18-Nor-5β-androsta-8,11,13-trien-3α-ol, 17ξ-methyl-	EtOH	270(2.76)	39-0163-78C
18-Nor-5β,17α-pregna-8,11,13-triene-3α,20β-diol, 17β-methyl-	EtOH	270(2.66)	39-0163-78C
4,6,8,10-Tridecatetraen-12-yn-3-one, 11-(1,1-dimethylethyl)-2,2-dimethyl-	EtOH	261(3.77),359s(4.72), 369(4.72)	18-3363-78
$C_{19}H_{26}O_2$			
D-Homoestr-4-ene-3,17a-dione	n.s.g.	240(4.23)	33-2397-78
D-Homo-18-norandrosta-5,13(17a)-dien-17-one, 3β-hydroxy-	EtOH	241(4.23)	56-2361-78
Podocarpa-8,11,13-triene, 13-acetyl-12-hydroxy-	EtOH	263(4.16),340(3.62)	23-0517-78
$C_{19}H_{26}O_3$			
3α,5α-Cyclo-13,17-seco-18-nor-17-methylandrosta-6,13,17-trione	EtOH	210(3.74)	56-2361-78
Estra-1,3,5(10)-triene-3,17β-diol, 11α-methoxy-	EtOH	280(3.27)	111-0313-78
	EtOH-NaOH	241(3.99),296(3.42)	111-0313-78
Estra-1,3,5(10)-triene-9β,11ξ-diol, 3-methoxy-	EtOH	274(3.10),282(3.08)	78-0393-78
$C_{19}H_{26}O_4$			
Dispiro[5.0.5.1]tridecane-1,5,8,12-tetrone, 13-ethyl-3,3,10,10-tetramethyl-	EtOH	288(2.64)	44-3070-78
Estra-1,3,5(10)-triene-3,16α,17β-triol, 11α-methoxy-	EtOH	279(3.25)	111-0313-78
	EtOH-NaOH	240(4.00),296(3.42)	111-0313-78
$C_{19}H_{26}O_5$			
2H-1-Benzopyran-2-one, 3,4-dihydro-5,7-dihydroxy-8-(2-methyl-1-oxobutyl)-4-pentyl-	EtOH	233(4.21),286(4.23), 325(3.88)	102-1783-78
	EtOH-NaOH	212(--),232(--)	102-1783-78
2-Cyclohexene-1-carboxylic acid, 2-(5-ethoxy-3-methyl-5-oxo-1,3-pentadienyl)-1,3-dimethyl-4-oxo-, ethyl ester, trans	EtOH	320(4.44)	104-0660-78
$C_{19}H_{26}O_5Se$			
Undecanedioic acid, 6-oxo-5-(phenylseleno)-, dimethyl ester	MeOH	222(3.99),267(3.09), 310(3.00)	49-0557-78
$C_{19}H_{26}O_7$			
1H-2-Benzoxacyclotetradecin-1,7(8H)-dione, 3,4,5,6,9,10,11,12-octahydro-8,9,16-trihydroxy-14-methoxy-3-methyl-	MeOH	218(4.33),265(4.03), 305(3.64)	44-2339-78
1H-2-Benzoxacyclotetradecin-1-one, 3,4,5,6,7,8,9,10-octahydro-7,8,9,16-tetrahydroxy-14-methoxy-3-methyl-	MeOH	235(4.42),272(4.06), 315(3.74)	44-2339-78
$C_{19}H_{26}O_{10}$			
Miyaginin	MeOH	222(4.06),275(3.26)	95-0366-78

Compound	Solvent	$\lambda_{max}(\log \epsilon)$	Ref.
$C_{19}H_{27}BrO_8$ Propanedioic acid, (bromomethyl)methyl-, bis(5,6-dioxoheptyl) ester	EtOH	272(2.90),422(1.40)	33-1565-78
$C_{19}H_{27}NO$ Butanenitrile, 2-[3,5-bis(1,1-dimethylethyl)-4-oxo-2,5-cyclohexadien-1-ylidene]-3-methyl-	n.s.g.	310(4.60),323(4.61)	70-1313-78
$C_{19}H_{27}NO_2$ 2-Azaestra-1,3,5(10)-trien-17-ol, 3-methoxy-11β-methyl-, (±)-	MeOH	278(3.58)	44-0966-78
$C_{19}H_{27}NO_3$ 2-Azaestra-1,3,5(10)-triene-9,17-diol, 3-methoxy-11-methyl-, (±)-	MeOH	275(3.56)	44-0966-78
2H-Benzo[a]quinolizin-2-one, 9,10-diethoxy-3α-ethyl-1,3,4,6,7,11b-hexahydro-	EtOH	212(4.045),224s(--), 281(3.474)	114-0045-78A
$C_{19}H_{27}NO_4$ Cyclodeca[b]furan-6-carboxaldehyde, 2,3,3a,4,5,8,9,11a-octahydro-10-methyl-3-(morpholinomethyl)-2-oxo-	isoPrOH	208(3.78),230(3.82)	42-1142-78
$C_{19}H_{27}NO_6$ Senkirkine	MeOH	213(3.97)	73-2312-78
$C_{19}H_{27}N_2$ Piperidinium, 1-[2-phenyl-3-(1-piperidinyl)-2-propenylidene]-, perchlorate	MeOH	315(4.32)	39-1453-78C
$C_{19}H_{27}N_2O_8PS_2$ Phosphorodiamidic acid, N,N-bis[2-(methylsulfonyl)oxy]ethyl]-N'-phenyl-, 2-methylphenyl ester	MeOH	227(4.9912),270(4.0969)	65-0720-78
Phosphorodiamidic acid, N,N-bis[2-(methylsulfonyl)oxy]ethyl]-N'-phenyl-, 3-methylphenyl ester	MeOH	228(5.00),271(4.0607)	65-0720-78
Phosphorodiamidic acid, N,N-bis[2-(methylsulfonyl)oxy]ethyl]-N'-phenyl-, 4-methylphenyl ester	MeOH	232(5.0969),266(4.5051)	65-0720-78
$C_{19}H_{27}N_3O_2$ 4H-Imidazole-2-acetic acid, 5-(dimethylamino)-α-ethyl-4,4-dimethyl-α-phenyl-, ethyl ester	CH_2Cl_2	275(2.95)	33-3050-78
$C_{19}H_{27}N_3O_4$ Pyrano[2,3-c]pyrazole-2(6H)-acetic acid, 3,4-dimethyl-6-oxo-5-(piperidinoethyl)-, ethyl ester	EtOH	306.0(4.24)	95-0335-78
$C_{19}H_{27}N_5O_2$ 3,5-Pyridinedicarbonitrile, 1,4-dihydro-4,4-dimethyl-2,6-bis(4-morpholinylmethyl)-	EtOH	205(4.88),217(4.74), 270(4.75),350(4.22)	103-1236-78
$C_{19}H_{28}N_2O_2S$ Benzenesulfonic acid, 4-methyl-, [2-(1,2-dimethylpropylidene)cyclohept-	MeOH	225(4.11),274(3.14)	44-0604-78

Compound	Solvent	$\lambda_{max}(\log \epsilon)$	Ref.
ylidene]hydrazide (cont.)			44-0604-78
$C_{19}H_{28}N_2O_{11}$			
Benzoic acid hydrazide, hydrazone with lactose	EtOH	208s(3.94),225(3.76)	136-0089-78E
$C_{19}H_{28}N_2O_{12}$			
Benzoic acid, 2-hydroxy-, hydrazide, hydrazone with lactose	EtOH	220s(3.83),235(3.94), 300(3.71)	136-0089-78E
Benzoic acid, 4-hydroxy-, hydrazide, hydrazone with lactose	EtOH	215(3.62),255(4.07)	136-0089-78E
$C_{19}H_{28}O$			
Androsta-2,4-dien-17β-ol	EtOH	266(3.78),273(3.76)	44-2715-78
17-Norkaur-9(11)-en-12-one	EtOH	244(4.20),319(2.08)	23-0246-78
Pumiloxide	EtOH	219s(3.92)	105-0286-78
$C_{19}H_{28}OSi$			
Silane, [[6-(1,3-butadienyl)-5,6,7,8,9-10-hexahydro-6-benzocyclooctenyl]-oxy]trimethyl-	EtOH	218(4.43),227(4.40)	44-1050-78
long retention isomer	EtOH	218(4.40),227(4.40)	44-1050-78
$C_{19}H_{28}O_2$			
D-Homoestr-4-en-3-one, 17aβ-hydroxy-	n.s.g.	240(4.23)	33-2397-78
$C_{19}H_{28}O_3$			
5,9-Methanobenzocyclooctene-5(1H)-carb-oxylic acid, 2,3,4,6,7,8,9,10-octa-hydro-2,2,7,7,9-pentamethyl-4-oxo-	EtOH	247(4.00)	78-1251-78
$C_{19}H_{28}O_4$			
Flexilin	n.s.g.	251(4.67)	88-3063-78
$C_{19}H_{28}O_{10}$			
Miyaginin, dihydro-	MeOH	222(4.04),276(3.11)	95-0366-78
$C_{19}H_{29}NO_7$			
1H-Pyrrole-1,2-dipropanoic acid, 2-[((1,1-dimethylethoxy)carbonyl]-2,3-dihydro-3-oxo-, diethyl ester	EtOH	330(3.99)	94-3521-78
$C_{19}H_{30}N_2$			
Benzenamine, N,N-dimethyl-4-(1-piperi-dinocyclohexyl)-	n.s.g.	250(4.11)	111-0017-78
hydrochloride	n.s.g.	250(4.08)	111-0017-78
$C_{19}H_{30}N_5O_3P$			
1H-Imidazole-4-carboxylic acid, 5-[[bis(propylamino)phosphinyl]amino]-1-(phenylmethyl)-, ethyl ester	EtOH	210(4.00),251(3.75)	130-0421-78
$C_{19}H_{30}N_6O_5$			
Adenosine, N-[(octylamino)carbonyl]-	pH 1	270(4.37),277(4.34)	36-0569-78
	pH 13	271(4.21),278(4.21), 298(4.16)	36-0569-78
	70% EtOH	269(4.38),277(4.31)	36-0569-78
$C_{19}H_{30}O_2$			
5,9-Methanobenzocycloocten-4(1H)-one, 2,3,5,6,7,8,9,10-octahydro-5-meth-	EtOH	245(3.91)	78-1251-78

Compound	Solvent	$\lambda_{max}(\log \epsilon)$	Ref.
oxy-2,2,7,7,9-pentamethyl- (cont.)			78-1251-78
$C_{19}H_{30}O_3$			
2H-Cyclopent[h]oxacycloheptadecin-8,16- (or 18)-dione, 1,3,4,5,6,9,10,11,12- 13,14,15,16,17-tetradecahydro-	MeOH	236(3.98)	2-0280-78
$C_{19}H_{30}O_4$			
1-Cyclopentene-1-heptanoic acid, 2-hex- yl-4,5-dioxo-, methyl ester	MeOH and MeOH-KOH	238(4.13)	2-0275-78
$C_{19}H_{30}O_6$			
2-Butenoic acid, 4-[2-(6-hydroxy-1-hep- tenyl)-4-(methoxymethoxy)cyclopent- yl]-4-oxo-, methyl ester	EtOH	224(4.09)	35-4858-78
$C_{19}H_{31}NO_4$			
Butanoic acid, 4-[(8-hydroxy-6-methyl- 1-oxo-2,4,6-dodecatrienyl)amino]-, ethyl ester, (E,E,E)-	EtOH	294(4.60)	18-2077-78
3,5-Pyridinedicarboxylic acid, 1,4-di- hydro-2,4,6-trimethyl-1-pentyl-, diethyl ester	EtOH	235(3.91),264(3.85), 351(3.73)	103-1226-78
$C_{19}H_{31}N_2O$			
Methanaminium, N-[3-[3-[3-(dimethylami- no)-2-propenylidene]-2-ethoxy-1-cy- clohepten-1-yl]-2-propenylidene]-N- methyl-, ethyl sulfate	$CHCl_3$	650(5.06)	70-0339-78
Methanaminium, N-[3-[3-[3-(dimethylami- no)-2-propenylidene]-2-ethoxy-5-meth- yl-1-cyclohexen-1-yl]-2-propenyli- dene]-N-methyl-, ethyl sulfate	$CHCl_3$	665(5.24)	70-0339-78
Methanaminium, N-[3-[3-[3-(dimethylami- no)-2-propenylidene]-2-ethoxy-6-meth- yl-1-cyclohexen-1-yl]-2-propenyli- dene]-N-methyl-, ethyl sulfate	$CHCl_3$	665(5.23)	70-0339-78
$C_{19}H_{31}N_3O_6$			
2(1H)-Pyrimidinone, 4-amino-1-[3-O-(1- oxodecyl)-β-D-arabinofuranosyl]-	isoPrOH	216(4.35),248(4.17), 303(3.90)	94-0981-78
$C_{19}H_{31}N_5$			
3,5-Pyridinedicarbonitrile, 2,6-bis- [(diethylamino)methyl]-1,4-dihydro- 4,4-dimethyl-	EtOH	205(4.92),218(4.73), 271(4.02),350(4.11)	103-1236-78
$C_{19}H_{32}O_3$			
1-Cyclopentene-1-heptanoic acid, 2-hex- yl-5-oxo-, methyl ester	MeOH	240(4.07)	2-0275-78
$C_{19}H_{32}O_4$			
2,4-Hexadienoic acid, 6-[3-(1-ethoxy- ethoxy)cyclopentyl]-3-methyl-, 1-methylethyl ester	EtOH	264(4.59)	56-0347-78
$C_{19}H_{40}Ge_2Sn$			
Germane, (1,1-dibutylstannacyclohexa- 3,5-diene-2,4-diyl)bis[trimethyl-	C_6H_{12}	225(3.48)	101-0247-78E
corresponding silane	C_6H_{12}	230(3.45)	101-0247-78E

Compound	Solvent	$\lambda_{max}(\log \epsilon)$	Ref.
$C_{20}H_3F_{13}$ Anthracene, 1,2,3,4,5,6,7,8-octafluoro- 9-(pentafluorophenyl)-9,10-dihydro-	EtOH	268(3.71)	104-1975-78
$C_{20}H_4Cl_6O_5$ Dinaphtho[2,1-b:1',2'-d]furan-5,9-di- one, 1,3,6,8,11,13-hexachloro-4,10- dihydroxy-	$CHCl_3$	400(4.06),559(4.37)	12-1323-78
$C_{20}H_7F_{10}N_4$ 1,2,4,5-Tetrazin-1(2H)-yl, 3,4-dihydro- 4,6-bis(pentafluorophenyl)-2-phenyl-	benzene	315(4.05),372(3.97), 625(3.67)	103-0218-78
$C_{20}H_8F_{10}N_4$ Formazan, 5-methyl-3,5-bis(pentafluoro- phenyl)-1-phenyl-	benzene	352(4.5)	103-0218-78
$C_{20}H_9NO_2$ 7H,11H-Benz[1,8]indolizino[2,3,4,5,6- defg]acridine-7,11-dione	$CHCl_3$	236(4.59),265(4.71), 320(3.99),390(4.07), 400(4.27),421(4.43)	70-0130-78
	H_2SO_4	205(4.72),242(4.81), 293(3.89),308(3.96), 350(4.50),476(4.32)	70-0130-78
$C_{20}H_{10}Cl_2N_2O_4$ 9,10-Anthracenedione, 1,4-dichloro- 5-nitro-8-(phenylamino)-	dioxan	472(3.73)	40-0082-78
$C_{20}H_{10}Cl_{12}$ Diels-Alder adduct of 5,5'-bicyclopen- tadiene-hexachlorocyclopentadiene	hexane	225(4.22),237(4.33), 246(4.36)	1-0149-78
$C_{20}H_{10}N_4O_9$ Benzofuran, 2,3-bis(2,4-dinitrophenyl)-	n.s.g.	358(4.20)	39-0401-78C
$C_{20}H_{10}O_6$ [2,2'-Binaphthalene]-1,1',4,4'-tetrone, 8,8'-dihydroxy-	dioxan	272(4.06),425(3.66)	102-2042-78
$C_{20}H_{11}BrN_4Ni$ Nickel, [5-bromo-21H,23H-porphinato(2-)- $N^{21},N^{22},N^{23},N^{24}$]-, (SP-4-2)-	CH_2Cl_2	393(5.27),510(4.09), 545(3.98)	150-0690-78
$C_{20}H_{11}NO_2$ 5H,9H-Quino[3,2,1-de]acridine-5,9-di- one	$CHCl_3$	247(4.61),265(4.58), 325(3.59),340(3.74), 450(4.35)	70-0130-78
	H_2SO_4	260(4.76),300(4.23), 350(4.50),400(4.69), 540(4.47)	70-0130-78
$C_{20}H_{12}F_5N_4$ 1,2,4,5-Tetrazin-1(2H)-yl, 3,4-dihydro- 4-(pentafluorophenyl)-2,6-diphenyl-	benzene	385(4.02),675(3.63)	103-0218-78
$C_{20}H_{12}N_2$ [1,1':3',1"-Terphenyl]-4,4'-dicarbo- nitrile	EtOH	206(4.59),259(4.59), 276(4.59)	47-2093-78
[1,1':3',1"-Terphenyl]-4,6'-dicarbo- nitrile	EtOH	208(4.48),258(4.62), 283(4.40)	47-2093-78

Compound	Solvent	$\lambda_{max}(\log \epsilon)$	Ref.
$C_{20}H_{12}N_2O_4$ 1H-Isoindole-1,3(2H)-dione, 2,2'-(1,3-butadiene-1,4-diyl)bis-	THF	212(4.68),287(4.63), 323(4.46)	40-0404-78
$C_{20}H_{12}N_2O_5$ 5(4H)-Oxazolone, 4-[[(4-nitrophenyl)-furanyl]methylene]-2-phenyl-	toluene	445(4.42)	103-0120-78
$C_{20}H_{12}N_2O_6$ 1H-Isoindole-1,3(2H)-dione, 2,2'-(tetrahydro-5-oxo-2,3-furandiyl)bis-	THF	221(4.92),295(3.57)	40-0404-78
$C_{20}H_{12}N_4O_6$ Benzamide, N,N'-3,5-cyclohexadiene-1,2-diylidenebis[4-nitro-	MeCN	258(4.45)	5-1139-78
$C_{20}H_{12}O$ Dinaphtho[2,1-b:1',2'-d]furan	CHCl$_3$	261(4.49),292s(4.04), 301(4.16),328(4.06), 342(4.36),355(4.50)	12-1323-78
$C_{20}H_{12}OS_2$ 9H-Xanthene, 9-(1,3-benzodithiol-2-ylidene)-	CH$_2$Cl$_2$	255(4.44),300(4.01), 370(4.42)	18-2674-78
$C_{20}H_{12}OS_3$ 9H-Thioxanthene, 9-(1,3-benzodithiol-2-ylidene)-, 10-oxide	CH$_2$Cl$_2$	370(4.37)	18-2674-78
$C_{20}H_{12}O_2$ Tribenzo[a,c,f]cyclooctene-9,14-dione	EtOH	230(4.47),285s(3.45), 330s(2.81)	18-1249-78
$C_{20}H_{12}O_2S_3$ 9H-Thioxanthene-, 9-(1,3-benzodithiol-2-ylidene)-, 10,10-dioxide	CH$_2$Cl$_2$ MeCN	402(4.24) 265(4.06),296(3.72), 319(3.76),396(4.32)	18-2674-78 18-2674-78
$C_{20}H_{12}O_3$ Cyclopenta[b][1]benzopyran-1,3-dicarboxaldehyde, 2-phenyl-	THF THF	460(3.40) 286(4.55),360(4.42), 460(3.40)	103-1070-78 103-1075-78
Dinaphtho[1,2-b:2',3'-e][1,4]dioxin-6-ol	n.s.g.	226(4.94),280(3.88), 337(3.67)	112-0835-78
1H-Naphtho[2,1,8-mna]xanthen-1-one	MeOH	221(4.46),248(4.57), 260s(4.45),294s(3.67), 305(3.63),337s(3.63), 350(3.76),376(3.73), 394s(3.56),467s(4.19), 495(4.41),525(4.38)	88-0637-78
	MeOH-HCl	219(4.45),267(4.61), 291s(3.83),314(3.44), 350(3.51),440(3.79), 482s(4.30),516(4.89)	88-0637-78
$C_{20}H_{12}O_4$ 9,10[1',2']-Benzenoanthracene-1,4-dione, 9,10-dihydro-5,8-dihydroxy-	DMSO	430(2.63)	77-0720-78
$C_{20}H_{12}S_2$ 1,3-Benzodithiole, 2-(9H-fluoren-9-yli-	CH$_2$Cl$_2$	251(4.98),298(4.00),	18-2674-78

Compound	Solvent	$\lambda_{max}(\log \epsilon)$	Ref.
dene)- (cont.)		309(4.02),329s(4.07), 388(4.68),402s(4.66)	18-2674-78
	MeCN	249(4.83),273s(3.88), 297s(3.66),308s(3.71), 326s(3.79),385(4.52), 400s(4.49)	18-2674-78
$C_{20}H_{12}S_3$ 9H-Thioxanthene, 9-(1,3-benzodithiol-2-ylidene)-	CH_2Cl_2	244(4.69),271s(4.16), 312s(4.05),363(4.36)	18-2674-78
$C_{20}H_{12}S_4$ 1,3-Benzodithiole, 2,2'-(2,5-cyclohexa-diene-1,4-diylidene)bis-	CS_2	478(4.81),503(4.96)	44-2084-78
$C_{20}H_{13}BrN_4$ 21H,23H-Porphine, 5-bromo-	CH_2Cl_2	400(5.40),497(4.20), 527(3.48),627(2.81)	150-0690-78
$C_{20}H_{13}ClN_2$ Quinoline, 6-chloro-4-phenyl-2-(2-pyri-dinyl)-	dioxan	341(3.26)	97-0400-78
$C_{20}H_{13}ClN_2O$ Benzo[h]quinoline-3-carbonitrile, 4-(2-chlorophenyl)-1,2,5,6-tetrahydro-2-oxo-	$CHCl_3$	267(4.24),395(4.35)	48-0097-78
Benzo[h]quinoline-3-carbonitrile, 4-(4-chlorophenyl)-1,2,5,6-tetrahydro-2-oxo-	$CHCl_3$	270(4.25),397(4.37)	48-0097-78
$C_{20}H_{13}ClN_2O_2$ 1,4-Acridinedione, 5-chloro-9-methyl-2-(phenylamino)-	dioxan	233(4.61),290(4.45), 331(4.24),437(3.68)	150-4901-78
1,4-Acridinedione, 6-chloro-9-methyl-2-(phenylamino)-	dioxan	229(4.64),283(4.53), 327(4.44),442(3.72)	150-4901-78
1,4-Acridinedione, 7-chloro-9-methyl-2-(phenylamino)-	dioxan	231(4.60),288(4.51), 341s(4.2),439(3.66)	150-4901-78
1,4-Acridinedione, 8-chloro-9-methyl-2-(phenylamino)-	dioxan	240(4.55),290(4.42), 340(4.24),439(3.67)	150-4901-78
$C_{20}H_{13}ClO$ 1H-Phenalen-1-one, 2-[(4-chlorophenyl)-methylene]-2,3-dihydro-	octane	254(4.38),290s(--), 312(4.11),358(4.14)	104-2381-78
	MeOH	255(4.33),320(4.16), 370(4.13)	104-2381-78
$C_{20}H_{13}Cl_3O_8$ 9H-Xanthen-9-one, 1,3,6-triacetoxy-2,4,5-trichloro-8-methyl-	MeOH	216(4.30),248(4.66), 282(3.96),397s(3.69), 348(3.69)	78-2491-78
$C_{20}H_{13}N$ Pyrido[4,5-a]fluoranthene, 2-methyl-	$CHCl_3$	270(5.14),305(4.42), 345(3.82),365(4.05), 380(3.77),424(4.25)	103-0875-78
$C_{20}H_{13}NO_2$ 5(4H)-Oxazolone, 4-(2-naphthalenylmeth-ylene)-2-phenyl-	toluene	405(4.43)	103-0120-78

Compound	Solvent	$\lambda_{max}(\log \epsilon)$	Ref.
$C_{20}H_{13}NO_3$			
1,4-Acridinedione, 2-methoxy-9-phenyl-	dioxan	226(4.53),282(4.59), 380s(3.2)	150-4901-78
9,10-Anthracenedione, 1-amino-4-phenoxy-	benzene	473(3.86)	104-2199-78
9,10-Anthracenedione, 1-amino-5-phenoxy-	benzene	466(3.81)	104-2199-78
9,10-Anthracenedione, 2-amino-1-phenoxy-	benzene	415(3.70)	104-2199-78
1H-Phenalen-1-one, 2,3-dihydro-2-[(4-nitrophenyl)methylene]-	MeOH	260(4.09),308(4.21), 370s(--)	104-2381-78
Spiro[2H-1,3-benzoxazine-2,3'-[3H]naphtho[2,1-b]pyran]-4(3H)-one	isoPrOH	295(4.11),310(4.10), 330(3.78),345(3.26)	103-0122-78
$C_{20}H_{13}NO_5$			
Oxysanguinarine	EtOH	241(4.27),281s(4.61), 289(4.70),331(4.17), 348(4.18),370(4.06), 385(4.02)	44-2852-78
$C_{20}H_{13}N_3$			
[1,2,4]Triazolo[1,5-f]phenanthridine, 2-phenyl-	EtOH	246s(4.76),252(4.88), 260s(4.77),285s(4.23), 302s(3.44),317(3.54), 331(3.58)	142-1577-78A
$C_{20}H_{13}N_3O_2S_2$			
2-Thiazoleacetonitrile, 4-hydroxy-α-(4-oxo-5-phenyl-2-thiazolidinylidene)-5-phenyl-	MeOH	249(3.99),292(4.03), 370(4.38)	5-0473-78
$C_{20}H_{14}$			
Anthracene, 9-phenyl-	C_6H_{12}	364(4.05)	61-1068-78
5,8:11,14-Diethenobenzocyclododecene	n.s.g.	307.5(2.13)	89-0046-78
Phenanthrene, 4-phenyl-	hexane	194(4.72),223(4.58), 228(4.58),255(4.56), 296(4.09),334(2.75), 349(2.51)	5-0530-78
Phenanthrene, 9-phenyl-	EtOH	250(4.53),255(4.58), 289(3.84),297(3.90)	35-6679-78
$C_{20}H_{14}BrN$			
3H-Indole, 3-bromo-2,3-diphenyl-	CH_2Cl_2	260(4.03),318(3.70)	94-2866-78
$C_{20}H_{14}ClN$			
3H-Indole, 3-chloro-2,3-diphenyl-	CH_2Cl_2	225(4.22),328(3.88)	94-2866-78
$C_{20}H_{14}ClN_3$			
Imidazo[1,2-a]pyridine, 2-(4-chlorophenyl)-7-[2-(3-pyridinyl)ethenyl]-	DMF	363(4.61),380(4.56)	33-0129-78
[1,2,4]Triazolo[1,5-a]pyridine, 2-(3-chloro-4-(2-phenylethenyl)phenyl]-	DMF	335(4.64)	33-0142-78
[1,2,4]Triazolo[1,5-a]pyridine, 7-[2-(4-chlorophenyl)ethenyl]-2-phenyl-	DMF	330(4.69),347(4.60)	33-0142-7?
$C_{20}H_{14}Cl_2O_8$			
9H-Xant-en-9-one, 1,3,6-triacetoxy-2,5-dichloro-8-methyl-	MeOH	211(4.69),247(4.95), 277(4.41),343(4.17)	78-2491-78
$C_{20}H_{14}Cl_3NO_8S$			
Benzoic acid, 4-[5-[2-(5-nitro-2-furan-	EtOH	203(4.41),229(4.19),	73-1618-78

Compound	Solvent	λ_{max}(log ϵ)	Ref.
y1)-2-[(trichloromethyl)sulfonyl]eth- enyl]-2-furanyl]-, ethyl ester (cont.)		258(4.32),326(4.45), 412(4.41)	73-1618-78
$C_{20}H_{14}Cl_4O_3$ Spiro[1,3-benzodioxole-2,2'(1'H)-naph- thalen]-1'-one, 4,5,6,7-tetrachloro- 3'-(1,1-dimethylethyl)-	EtOH	220(4.73),296(3.81), 304(3.79)	2-0668-78
$C_{20}H_{14}N_2$ Quinazoline, 2,4-diphenyl-	EtOH	265(4.49),307(3.71), 332(3.57)	94-2866-78
Quinoxaline, 2,3-diphenyl-	EtOH	244(4.60),264(4.33), 342(4.08)	94-2866-78
$C_{20}H_{14}N_2O$ Benzo[h]quinoline-3-carbonitrile, 1,2,5,6-tetrahydro-2-oxo-4-phenyl-	CHCl$_3$	268(4.16),395(4.33)	48-0097-78
$C_{20}H_{14}N_2O_2$ 1,4-Acridinedione, 9-methyl-2-(phenyl- amino)-	dioxan	228(4.42),285(4.28), 317(4.13),440(3.64)	150-4901-78
9H-Indeno[2,1-c]pyridine, 3-methyl- 9-[(4-nitro-henyl)methylene]-	n.s.g.	214(4.52),224(4.54), 262(4.52),356(4.16)	103-1343-78
Phenol, 4-benzo[f]quinazolin-3-yl-, acetate	EtOH	219(4.48),246(4.23), 286(4.74),343(3.86), 357(3.83)	103-1025-78
2(1H)-Quinoxalinone, 3,4-dihydro-3- [2-(1-naphthalenyl)-2-oxoethylidene]-	EtOH	445(4.48)	103-0336-78
$C_{20}H_{14}N_2O_3$ 9H-Pyrido[3,4-b]indole-3-carboxylic acid, 1-benzoyl-, methyl ester	EtOH	227(4.01),293(4.02)	95-1635-78
$C_{20}H_{14}N_2O_4$ Cyclobuta[1,2-d:3,4-d']bis[2]benzazep- ine-5,7,12,14(6H,13H)-tetrone, 7a,7b,14a,14b-tetrahydro-, trans insoluble isomer	EtOH	238(2.67),284(1.99)	78-2887-78
	EtOH	240s(2.61),283(1.98)	78-2887-78
4,14:7,11-Diethenocyclododeca[1,2-c- 7,8-c']dipyrrole-1,3,8,10(2H,9H)- tetrone, 5,6,12,13-tetrahydro-	EtOH	223(4.60),270s(3.35), 337(3.60)	24-0523-78
$C_{20}H_{14}N_4$ 3H-Indole, 3-azido-2,3-diphenyl-	EtOH	220(4.19),249(4.13), 318(4.05)	94-2866-78
$C_{20}H_{14}N_4O_2$ 4(1H)-Quinazolinone, 2-[4-(phenylazo)- phenoxy]-	dioxan	244(4.12),256(4.04), 263(4.02),325(4.40)	4-1033-78
$C_{20}H_{14}N_4O_3$ 4H-Pyran-3,5-dicarbonitrile, 2-amino- 6-(4-methylphenyl)-4-(4-nitrophenyl)-	EtOH	301(3.87)	4-0057-78
$C_{20}H_{14}N_4O_4$ 4H-Pyran-3,5-dicarbonitrile, 2-amino- 6-(4-methoxyphenyl)-4-(4-nitrophenyl)-	EtOH	303(4.30)	4-0057-78
$C_{20}H_{14}O$ Benzo[d][15]annulenone, 6,8-bisdehydro- 10-methyl-	ether	223(4.27),262s(4.12), 278s(4.26),299(4.34),	18-1204-78

Compound	Solvent	$\lambda_{max}(\log \epsilon)$	Ref.
(cont.)		377(3.91)	18-1204-78
Benzo[f][15]annulenone, 8,10-bisdehydro-10-methyl-	ether	228(4.40),256(4.27), 274s(4.34),302(4.46), 372s(3.92)	18-1204-78
Isobenzofuran, 1,3-diphenyl-	benzene	305(3.5),315(3.5), 405(3.9)	149-0595-78B
1H-Phenalen-1-one, 2,3-dihydro-2-(phenylmethylene)-	octane	253(4.35),290s(--), 312(4.04),356(4.10)	104-2381-78
	MeOH	255(4.37),307(4.04), 367(4.05)	104-2381-78
Tribenzo[a,c,f]cyclooocten-9(14H)-one	EtOH	235(4.23),255s(3.94), 295s(3.14),325s(2.57)	18-1249-78
$C_{20}H_{14}OS_2$			
2(1H)-Naphthalenone, 1-(4-methyl-5-phenyl-1,3-dithiol-2-ylidene)-	MeCN	297(2.52),325(2.64), 500(4.71)	97-0385-78
$C_{20}H_{14}O_2$			
Benzo[a]pyrene-4,5-diol, 4,5-dihydro-, trans	MeOH	254(4.78),261(4.89), 287(4.17)	44-3462-78
Benzo[a]pyrene-9,10-diol, 9,10-dihydro-, trans	EtOH	230(4.60),242(4.68), 275(4.38),283(4.47), 295s(4.13),337(4.12), 347(4.16),361(4.13)	44-3462-78
Ethanone, 1-(2-phenylcyclopenta[b][1]-pyran-1-yl)-	THF	254(4.16),290(4.16), 355(4.38),486(3.59)	103-1075-78
Methanone, 1,2-phenylenebis[phenyl-	benzene	260(4.0)	149-0595-78B
1,4-Naphthacenedione, 2,3-dimethyl-	CH_2Cl_2	263(5.43),312(4.95), 327(5.25),343(5.42), 360s(3.53),384(3.57), 407(3.63),465(3.85)	150-5538-78
$C_{20}H_{14}O_3$			
Methanone, 7-oxabicyclo[2.2.1]hepta-2,5-diene-2,3-diylbis[phenyl-	EtOH	256(4.11)	78-2305-78
$C_{20}H_{14}O_4$			
[1,1'-Biphenyl]-2,4'-dicarboxylic acid, 5-phenyl-	EtOH	206(4.59),257(4.59)	47-2093-78
$C_{20}H_{14}O_6$			
1H-Furo[3",4":2',3']cycloprop[1',2'-1,2]indeno[5,6-d][1,3]dioxol-3(3aH)-one, 3a-(1,3-benzodioxol-5-yl)-3b,9-dihydro-	EtOH	289(3.74)	94-1592-78
7H-Furo[3".4":1',3']cycloprop[1',2'-2,3]indeno[4,5-d]-1,3-dioxol-9(6H)-one, 9b-(1,3-benzodioxol-5-yl)-9a,9b-dihydro-	EtOH	295(3.90)	94-1592-78
Furo[3',4':6,7]naphtho[1,2-d]-1,3-dioxol-9(6H)-one, 10-(1,3-benzodioxol-5-yl)-7,10-dihydro-	EtOH	289(3.93)	94-3195-78
Furo[3',4':6,7]naphtho[2,3-d]-1,3-dioxol-6(8H)-one, 5-(1,3-benzodioxol-5-yl)-5,9-dihydro-	EtOH	288.5(3.88)	94-3195-78
$C_{20}H_{14}O_6S_2$			
1H-Indene-1,3(2H)-dione, 2,2'-[1,2-ethanediylbis(thio)]bis[2-hydroxy-	MeOH	228(4.96),248(4.35), 280(3.37),288s(3.32), 303s(3.03),353(2.66)	24-3058-78

Compound	Solvent	$\lambda_{max}(\log \epsilon)$	Ref.
$C_{20}H_{14}O_8$			
9,10-Anthracenedione, 1,3,6-triacetoxy-	EtOH	263(4.59),275s(4.28), 330(3.71)	12-1335-78
9,10-Anthracenedione, 1,3,7-triacetoxy-	CHCl$_3$	262(4.59),277s(4.30), 336(3.80)	12-1335-78
$C_{20}H_{14}S$			
Benzo[c]thiophene, 1,3-diphenyl-	CH$_2$Cl$_2$	259(4.30),270s(4.27), 393(4.18)	4-1185-78
$C_{20}H_{14}S_2$			
1,3-Benzodithiole, 2-(diphenylmethylene)-	CH$_2$Cl$_2$	244(4.51),252s(4.43), 329(4.34)	18-2674-78
$C_{20}H_{15}As$			
Acridarsine, 3-methyl-10-phenyl-	THF	397s(3.43),414s(3.57), 435(3.89),461(3.96)	5-0214-78
$C_{20}H_{15}BrN_2O_2$			
Spiro[2H-anthra[1,2-d]imidazole-2,1'-cyclohexane]-6,11-dione, 4-bromo-	dioxan	416(3.56)	104-0381-78
$C_{20}H_{15}Cl$			
1,1'-Biphenyl, 2-[2-(3-chlorophenyl)-ethenyl]-, cis	MeOH	221(4.46),252(4.16), 282(4.13)	39-0915-78B
trans	MeOH	220(4.36),249(4.21), 300(4.36),316s(4.27), 331s(3.94)	39-0915-78B
1,1'-Biphenyl, 2-[2-(4-chlorophenyl)-ethenyl]-, cis	MeOH	222(4.47),250(4.20), 286(4.18)	39-0915-78B
trans	MeOH	224(4.32),254(4.27), 301(4.44),317s(4.38), 335s(4.01)	39-0915-78B
$C_{20}H_{15}ClFN_3O_2$			
4H-Imidazo[1,5-a][1,4]benzodiazepine-4-carboxylic acid, 8-chloro-6-(2-fluorophenyl)-1-methyl-, methyl ester	isoPrOH	220(4.51),240s(4.33), 300s(3.20)	44-0936-78
4H-Imidazo[1,5-a][1,4]benzodiazepine-1-methanol, 8-chloro-6-(2-fluorophenyl)-, acetate	isoPrOH	215(4.61),241s(4.36), 305s(3.11)	44-0936-78
$C_{20}H_{15}ClFN_3O_3$			
4H-Imidazo[1,5-a][1,4]benzodiazepine-1-methanol, 8-chloro-6-(2-fluorophenyl)-, acetate, 5-oxide	isoPrOH	216(4.42),230s(4.39), 257s(4.23),297(4.03)	44-0936-78
$C_{20}H_{15}ClN_2O_2$			
Cyclohepta[b]pyrrole-3-carboxylic acid, 2-chloro-6-(cyanophenylmethylene)-1,6-dihydro-, ethyl ester	MeOH	257(4.46),290(4.25), 400(4.45)	18-0667-78
Spiro[2H-anthra[1,2-d]imidazole-2,1'-cyclohexane]-6,11-dione, 4-chloro-	dioxan	413(3.51)	104-0381-78
$C_{20}H_{15}ClN_2S$			
Quinoline, 6-chloro-2-(2,4-dimethyl-5-thiazolyl)-4-phenyl-	dioxan	348(3.63)	97-0400-78
$C_{20}H_{15}ClN_4$			
Phthalazine, 1-chloro-4-(2,2-diphenyl-hydrazino)-	dioxan	240s(4.20),288(4.22)	103-0575-78

Compound	Solvent	λ_{max}(log ϵ)	Ref.
1(2H)-Phthalazinone, 4-chloro-, diphenylhydrazone	dioxan CHCl$_3$	273(4.31),292(4.02) 273(--),293(--)	103-0575-78 103-0575-78
$C_{20}H_{15}ClO_8$ 9H-Xanthen-9-one, 1,3,6-triacetoxy- 7-chloro-8-methyl-	MeOH	212(4.75),245(4.99), 271(4.57),342(4.19)	78-2491-78
$C_{20}H_{15}IN_6$ 2H-Cyclopenta[d]pyridazine, 5-iodo- 2-methyl-6,7-bis(phenylazo)-	ether	248(4.15),291(4.15), 379(4.36),460(4.32), 474(4.32)	44-0664-78
$C_{20}H_{15}N$ 1H-Indole, 2,3-diphenyl-	MeOH	250(4.42),308(4.24)	44-1230-78
$C_{20}H_{15}NO$ Acridine, 9-(4-methoxyphenyl)-	EtOH	220(4.41),227s(--), 253(5.01),343s(--), 356(3.97)	39-1211-78C
5H-Indeno[1,2-b]pyridine, 5-[(2-methoxy- phenyl)methylene]-	n.s.g.	212(4.46),231(4.56), 289(4.26),340(4.16)	103-1343-78
Phenol, 2-[(3-methyl-9H-indeno[2,1-c]- pyridin-9-ylidene)methyl]-	EtOH	210(4.74),223(4.78), 246(4.56),290(4.08), 340(4.36)	103-0066-78
$C_{20}H_{15}NO_2$ 1,1'-Biphenyl, 2-[2-(4-nitrophenyl)eth- enyl]-, cis	MeOH	235(4.29),323(4.07)	39-0915-78B
trans	MeOH	237(4.32),338(4.09)	39-0915-78B
2-Naphthalenecarboxamide, N-phenyl- 3-(2-propynyloxy)-	EtOH	233s(4.68),237(4.69), 245s(4.60),265s(4.23), 282s(4.20),295s(4.16), 348s(3.34)	33-0716-78
4H-Naphtho[2,3-b]pyran-4-one, 2-methyl- 3-(phenylamino)-	EtOH	248(4.84),265s(4.48), 300(3.90),316(3.70), 358(3.57)	33-0716-78
$C_{20}H_{15}NO_2S$ Phenanthridine, 6-[(4-methylphenyl)sul- fonyl]-	EtOH	229(4.50),237(4.49), 252(4.43),307(3.76), 321(3.70),337(3.59), 352s(3.05)	95-0136-78
$C_{20}H_{15}NO_3$ Benzo[h]quinolin-2(1H)-one, 4-(1,3-ben- zodioxol-5-yl)dihydro-	CHCl$_3$	295(4.24)	48-0097-78
$C_{20}H_{15}NO_3Se$ Selenonium, diphenyl-, 1-nitro-2-oxo- 2-phenylethylide	MeCN	339(3.86)	70-0945-78
$C_{20}H_{15}NO_4$ Dibenzo[de,g]pyrrolo[3,2,1-ij]quinoline- 1,2-dione, 4,5-dihydro-7,8-dimethoxy-	EtOH	257(4.83),324s(4.32), 336(4.40),357s(4.15), 510(3.78)	44-1096-78
Dibenzo[de,g]pyrrolo[3,2,1-ij]quinoline- 1,2-dione, 4,5-dihydro-9,10-dimethoxy-	EtOH	215(4.04),255(4.52), 320(3.82),525(3.08)	44-1096-78
$C_{20}H_{15}NO_5$ 1H,3H-Benz[f]isoquino[8,1,2-hij][3,1]- benzoxazine-1,3-dione, 5,6-dihydro- 8,9-dimethoxy-	EtOH	262(4.58),310(3.95), 324(3.97),378(3.61)	44-1096-78

Compound	Solvent	λ_{max}(log ϵ)	Ref.
1H,3H-Benz[f]isoquino[8,1,2-hij][3,1]-benzoxazine-1,3-dione, 5,6-dihydro-10,11-dimethoxy-	EtOH	240(4.23),261(4.40), 284s(3.91),309s(3.85), 323(3.92),358(3.59), 387(3.53)	44-1096-78
$C_{20}H_{15}NO_6$			
2-Naphthacenecarboxamide, 1,3,10,11,12-pentahydroxy-6-methyl- (6-methylpretetramid)	H_2SO_4-borax	236(4.34),263(4.45), 278(4.42),295s(4.35), 327(4.19),342(4.19), 398(4.22),510(4.24)	39-0145-78C
Sanguinarine, hexahydro-5,8-dioxo-, trans	EtOH	213(4.48),237(4.60), 273(4.02),317(4.07)	44-2852-78
$C_{20}H_{15}NO_7$			
6-Methylpretetramid oxygenation product	H_2SO_4-borax	230(4.11),298(4.62), 421(3.92),520s(3.82), 555(4.00),595(4.09), 640(3.97),694(4.01)	39-0145-78C
isomer	H_2SO_4-borax	273(4.33),294(4.37), 305s(4.34),548(3.98), 575s(3.96),630s(3.77)	39-0145-78C
$C_{20}H_{15}NO_8$			
2-Naphthacenecarboxamide, 6,11,12,12a-tetrahydro-1,3,6,10,12a-pentahydroxy-6-methyl-11,12-dioxo-	H_2SO_4-borax	203(4.35),278(4.39), 290s(4.38),345s(4.23), 394s(4.02),518(4.11)	39-0145-78C
$C_{20}H_{15}N_3$			
Imidazo[1,2-a]pyridine, 2-phenyl-7-[2-(3-pyridinyl)ethenyl]-	DMF	362(4.58),380(4.54)	33-0129-78
1H-1,2,4-Triazole, 1,3,5-triphenyl-	EtOH	243(4.49),284s(3.76)	33-1477-78
2H-1,2,3-Triazole, 2,4,5-triphenyl-	EtOH	219(4.26),297(4.23)	33-1477-78
[1,2,4]Triazolo[1,5-a]pyridine, 2-phenyl-7-(2-phenylethenyl)-	DMF	327(4.64),344(4.54)	33-0142-78
$C_{20}H_{15}N_3O$			
4H-Pyran-3,5-dicarbonitrile, 2-amino-4-(4-methylphenyl)-6-phenyl-	EtOH	297(3.96)	4-0057-78
4H-Pyran-3,5-dicarbonitrile, 2-amino-6-(4-methylphenyl)-4-phenyl-	EtOH	299.2(3.95)	4-0057-78
$C_{20}H_{15}N_3O_2$			
4H-Pyran-3,5-dicarbonitrile, 2-amino-4-(4-methoxyphenyl)-6-phenyl-	EtOH	299.2(4.02)	4-0057-78
$C_{20}H_{15}O_2P$			
Benzaldehyde, 2,2'-(phenylphosphinidene)bis-	CH_2Cl_2	222(4.48),246(4.37), 313s(3.46),362(3.06)	24-0013-78
Spiro[5H-dibenzophosphole-5,8'-[2H,6H]-[1,2]oxaphospholo[4,3,2-hi][2,1]benzoxaphosphole]	EtOH	209(4.76),228s(4.36), 237s(4.22),260s(3.88), 274(4.07)	24-0013-78
$C_{20}H_{16}$			
1,1'-Biphenyl, 2-(1-phenylethenyl)-	C_6H_{12}	245(4.45)	35-6683-78
	C_6H_{12}	210(4.6),235s(4.4), 258s(4.3)	35-6679-78
1,1'-Biphenyl, 2-(2-phenylethenyl)-, cis	MeOH	219(4.46),249(4.22), 284(4.11)	39-0915-78B
trans	MeOH	219(4.38),251(4.22), 298(4.37),312s(4.32), 329s(3.95)	39-0915-78B

Compound	Solvent	$\lambda_{max}(\log \epsilon)$	Ref.
5,8:11,14-Diethenobenzocyclododecene, 9,10-dihydro-	n.s.g.	308(2.11)	89-0046-78
Tribenzo[a,c,f]cyclooctene, 9,14-dihydro-	EtOH	233(3.90),263(3.05), 272(2.87)	18-1249-78
$C_{20}H_{16}ClN$			
Pyridine, 2-[2-(3-chlorophenyl)-1-phenyl-1-propenyl]-, (E)-	MeOH	263(3.88)	87-0225-78
(Z)-	MeOH	270(3.75)	87-0225-78
$C_{20}H_{16}ClNO$			
Benzo[h]quinolin-2(1H)-one, 4-(2-chlorophenyl)-5,6-dihydro-1-methyl-	CHCl$_3$	244(4.18),295(4)	48-0097-78
Benzo[h]quinolin-2(1H)-one, 4-(4-chlorophenyl)-5,6-dihydro-1-methyl-	CHCl$_3$	244(4.23),295(4)	48-0097-78
Phenol, 4-[2-(3-chlorophenyl)-1-(2-pyridinyl)-1-propenyl]-, (E)-	MeOH	243(4.20),270(4.12)	87-0225-78
(Z)-	MeOH	240(4.23),277(4.02)	87-0225-78
$C_{20}H_{16}ClNO_2$			
Cyclopenta[b]pyran-2(5H)-one, 3-chloro-4-(diphenylamino)-6,7-dihydro-	EtOH	249(4.10),278(4.10), 330s(4.02),362(4.08)	4-0511-78
$C_{20}H_{16}ClNO_2S$			
2H,5H-Thiopyrano[4,3-b]pyran-2-one, 3-chloro-4-(diphenylamino)-7,8-dihydro-	EtOH	253(4.02),280(4.19), 320(3.84),374(3.83)	4-0181-78
$C_{20}H_{16}ClNO_6$			
2,1-Benzisoxazolium, 3-[2-(3-hydroxy-2-naphthalenyl)ethenyl]-1-methyl-, perchlorate	MeCN	520(4.52)	44-1233-78
$C_{20}H_{16}Cl_2N_2O_2$			
2,4-Pentanedione, 3-[7-chloro-5-(2-chlorophenyl)-1,3-dihydro-2H-1,4-benzodiazepin-2-ylidene]-	isoPrOH	217(4.52),270s(3.92), 345(4.38)	4-0577-78
$C_{20}H_{16}Cl_2N_2O_3$			
Butanoic acid, 2-[7-chloro-5-(2-chlorophenyl)-1,3-dihydro-2H-1,4-benzodiazepin-2-ylidene]-3-oxo-, methyl ester	isoPrOH	217(4.52),245s(4.24), 331(4.40)	4-0577-78
$C_{20}H_{16}Cl_2N_2O_4$			
Propanedioic acid, [7-chloro-5-(2-chlorophenyl)-1,3-dihydro-2H-1,4-benzodiazepin-2-ylidene]-, dimethyl ester	isoPrOH	214s(4.53),240s(4.23), 271(3.92),313(4.50), 355s(3.62)	4-0577-78
$C_{20}H_{16}NP$			
Acetonitrile, (triphenylphosphoranylidene)-	CH$_2$Cl$_2$	303(3.26)	39-1237-78C
$C_{20}H_{16}N_2$			
Benzenamine, N-(3,7-dimethyl-9H-indeno[2,1-c]pyridin-9-ylidene)-	EtOH	223(4.38),256(4.66), 294(4.10),305(4.14), 380(3.36)	104-2365-78
Diazene, (1,2-diphenylethenyl)phenyl-, (Z,E)-	hexane	248(4.15),347(4.48), 455(2.79)	49-1081-78

Compound	Solvent	$\lambda_{max}(\log \epsilon)$	Ref.
$C_{20}H_{16}N_2O$			
Ethanone, 1-(1,6-dimethylpyrazino[2,1,6-cd:5,4,3-c'd']diindolizin-2-yl)-	EtOH	248(4.33),292(4.52), 405(3.89),432(4.09), 458(4.19)	150-2850-78
$C_{20}H_{16}N_2OS$			
Imidazo[1,2-a]pyridine, 2-(4-methoxyphenyl)-7-[2-(2-thienyl)ethenyl]-	DMF	373(4.62),393(4.58)	33-0129-78
$C_{20}H_{16}N_2O_2$			
Cyclohepta[b]pyrrole-3-carboxylic acid, 2-(cyanophenylmethyl)-, ethyl ester	MeOH	287(4.59),324(3.90), 345(3.58),462(2.97)	18-0667-78
Ethanone, 2,2'-(3,6-pyridazinediyl)bis[1-phenyl-	EtOH	244(4.28),346(3.97)	94-3633-78
Spiro[2H-anthra[1,2-d]imidazole-2,1'-cyclohexane]-6,11-dione	dioxan	405(3.36)	104-0381-78
$C_{20}H_{16}N_2O_3$			
Pyridinium, 4-[2-(4-hydroxyphenyl)ethenyl]-1-[(4-nitrophenyl)methyl]-, hydroxide, inner salt, (E)-	EtOH	540(2.95)	56-1265-78
	acetone	610(3.29)	56-1265-78
	CHCl$_3$	650(3.57)	56-1265-78
hydrochloride	MeOH	402(4.61)	56-1265-78
$C_{20}H_{16}N_2S$			
Imidazo[1,2-a]pyridine, 2-(4-methylphenyl)-7-[2-(2-thienyl)ethenyl]-	DMF	372(4.64),390(4.58)	33-0129-78
$C_{20}H_{16}N_4$			
2-Quinoxalinamine, 3-(4-aminophenyl)-N-phenyl-	EtOH	253(4.45),284(4.47), 388(4.19)	78-0981-78
1,2,4,5-Tetrazine, 1,2-dihydro-1,3,6-triphenyl-	EtOH	236(4.37),264(4.30), 315(3.99)	70-2227-78
$C_{20}H_{16}N_4OS$			
4(1H)-Pteridinone, 1-methyl-2-(methylthio)-6,7-diphenyl-	pH -2.4	262(4.33),303(4.09), 372(4.26)	24-0971-78
	pH 4.0	253(4.29),277(4.40), 370(4.30)	24-0971-78
4(3H)-Pteridinone, 3-methyl-2-(methylthio)-6,7-diphenyl-	pH -3.4	240(4.18),301(4.31), 435(4.23)	24-0971-78
	pH 4.0	263(4.30),300(4.31), 367(4.15)	24-0971-78
$C_{20}H_{16}N_4O_8S_2$			
Benzenesulfonamide, N,N'-(4,5-dimethyl-3,5-cyclohexadiene-1,2-diylidene)-bis[4-nitro-	CH$_2$Cl$_2$	231(4.50),261(4.51), 476(3.56)	5-1146-78
$C_{20}H_{16}N_6$			
2H-Cyclopenta[d]pyridazine, 2-methyl-5,7-bis(phenylazo)-	EtOH	251(4.08),275s(3.90), 348(4.33),469(4.36)	44-0664-78
$C_{20}H_{16}O$			
Chrysene, 6-methoxy-5-methyl-	hexane	271.5(5.05)	87-0038-78
Chrysene, 12-methoxy-5-methyl-	hexane	272(4.94)	87-0038-78
1,3,6,8-Undecatetraen-10-yn-5-one, 1-(2-ethynylphenyl)-9-methyl-, (E,E,E,Z)-	ether	228s(4.24),233(4.24), 255(4.21),263s(4.19), 295s(4.05),352(4.52)	18-1204-78
1,4,6,8-Undecatetraen-10-yn-3-one, 1-(2-ethynylphenyl)-9-methyl-, (E,E,Z,E)-	ether	227(4.23),250s(4.14), 255(4.16),277(4.02), 292(4.05),356(--)	18-1204-78

Compound	Solvent	$\lambda_{max}(\log \epsilon)$	Ref.
$C_{20}H_{16}O_2$			
Naphtho[2',1':4,5]furo[3,2-b]benzofuran, 6b,11a-dihydro-6b,11a-dimethyl-	hexane	213(4.59),240(4.32), 285(3.67),315(3.54), 320(3.42),329(3.58)	88-0511-78
$C_{20}H_{16}O_2S$			
2-Propynoic acid, 3-phenyl-, 3-benzo[b]thien-2-ylpropyl ester	MeOH	257(4.39)	44-2493-78
$C_{20}H_{16}O_2S_2$			
Spiro[2H-[1]benzothieno[3,2-b]pyran-2,2'(3H)-benzo[b]thiophen]-3'-one, 5',8-dimethyl-	C_6H_{12}	241(4.77),256s(4.24), 295(3.55),307(3.64), 350s(3.08),372(3.30), 389(3.35)	22-0241-78
Spiro[2H-[1]benzothieno[3,2-b]pyran-3(4H),2'(3'H)-benzo[b]thiophen]-3'-one, 5',7-dimethyl-	C_6H_{12}	241(4.68),248s(4.57), 261(4.13),297(3.63), 307(3.72),346(3.31), 363(3.42),381(3.40)	22-0241-78
$C_{20}H_{16}O_3$			
2H-Furo[2,3-h]-1-benzopyran-2-one, 8-(1-methylethyl)-9-phenyl-	EtOH	211(4.42),223s(4.36), 261(4.44),306(3.99)	2-0184-78
$C_{20}H_{16}O_4$			
2H,8H-Benzo[1,2-b:3,4-b']dipyran-2,9(10H)-dione, 8,8-dimethyl-10-phenyl-	EtOH	215(4.38),246(3.75), 256(3.69),322(4.10)	2-0184-78
2-Naphthalenecarboxylic acid, 3-(2-methoxybenzoyl)-, methyl ester	THF	230(4.66),310(3.67)	78-0113-78
$C_{20}H_{16}O_5$			
1H-Indene-1,3(2H)-dione, 2-[3-(3,4-dimethoxyphenyl)-1-oxo-2-propenyl]-	pH 1	251(4.34),314(3.98), 428(4.64)	65-0142-78
	pH 13	236(4.48),308(4.15), 374(4.53)	65-0142-78
	$CHCl_3$	428(4.59)	65-0142-78
Naphtho[2,3-c]furan-1(3H)-one, 4-hydroxy-9-(3,4-dimethoxyphenyl)-	EtOH	245(4.38),285s(3.65), 316(3.66),360(3.73)	142-0207-78A
Propanedioic acid, (1-oxo-2-phenyl-1H-inden-3-yl)-, dimethyl ester	EtOH	251(4.53),285s(3.60), 412(3.19)	33-3018-78
sodium derivative	EtOH	250(4.46),285s(4.14), 322s(4.00),370s(3.57), 548(3.84)	33-3018-78
2-Propenoic acid, 3-[2,3-dihydro-3-[(4-methoxyphenyl)methylene]-2-oxo-5-benzofuranyl]-, methyl ester	EtOH	247(4.15),283(4.21), 302s(4.16),362(4.04)	33-2646-78
$C_{20}H_{16}O_6$			
1,5-Anhydro-2,6-di-O-benzoyl-4-deoxy-D-glycero-hex-1-en-3-ulose	EtOH	231(4.48),269(3.86)	136-0433-78H
Spiro[furan-3(2H),6'-[6H]indeno[5,6-d]-[1,3]dioxol]-5(4H)-one, 4-(1,3-benzodioxol-5-yl)-5',7'-dihydro-	EtOH	294.5(3.94)	94-1592-78
$C_{20}H_{16}O_8$			
4H-1-Benzopyran-4-one, 6-acetoxy-7-methoxy-3-(6-methoxy-1,3-benzodioxol-5-yl)-	EtOH	221(4.27),262(4.38), 304(4.04)	102-1419-78
$C_{20}H_{16}O_9$			
Cyclopenta[b]xanthene-3-carboxylic	EtOH	261(4.52),294(4.41),	94-0209-78

Compound	Solvent	$\lambda_{max}(\log \epsilon)$	Ref.
acid, 1,2,3,10-tetrahydro-3,11-dihy-droxy-4,7-dimethoxy-9-methyl-1,10-dioxo- (cont.)		356(3.81)	94-0209-78
$C_{20}H_{16}O_{10}$ 1H-Furo[3,4-b]xanthene-3-carboxylic acid, 3,10-dihydro-3,11-dihydroxy-4,7-dimethoxy-9-methyl-1,10-dioxo-, methyl ester	EtOH	243(4.51),249(4.51), 278(4.26),357(3.88)	94-0209-78
$C_{20}H_{17}AsO_2$ 4-Arseninol, 2-(4-methylphenyl)-6-phenyl-, acetate	EtOH	237(4.32),292(4.60)	88-1471-78
$C_{20}H_{17}BF_4N_2O$ 2H-Pyrido[2,1-b][1,3,4]oxadiazin-5-ium, 3-methyl-6,8-diphenyl-, tetrafluoroborate	EtOH	220(3.94),240(3.95), 315(4.09)	120-0001-78
$C_{20}H_{17}BrN_2$ 1H-Imidazo[1,2-a]pyridin-4-ium, 2-phenyl-1-(phenylmethyl)-, bromide	EtOH	207(4.57),231s(4.25), 289(4.11)	4-1149-78
$C_{20}H_{17}Br_2NO$ 2(1H)-Pyridinone, 3,5-bis(4-bromophenyl)-1,4,6-trimethyl-	MeOH	255(4.01),318(4.02)	44-2138-78
4(1H)-Pyridinone, 3,5-bis(4-bromophenyl)-1,2,6-trimethyl-	EtOH	228(4.16),275(4.16)	44-2138-78
$C_{20}H_{17}ClFN_5O_2$ 6H-[1,2,4]Triazolo[1,5-a][1,4]benzodiazepine]-2-carboxamide, 8-chloro-6-(2-fluorophenyl)-4-methoxy-N,N-dimethyl-	MeOH	267(3.99)	33-0848-78
$C_{20}H_{17}ClN_2O$ 1-Naphthaleneacetic acid, [1-(4-chlorophenyl)ethylidene]hydrazide	EtOH	281(4.36)	34-0345-78
2H-Pyrido[2,1-b][1,3,4]oxadiazin-5-ium, 3-methyl-6,8-diphenyl-, chloride	EtOH	220(3.90),245(3.95), 312(4.11)	120-0001-78
$C_{20}H_{17}ClN_2O_2$ Cyclohepta[b]pyrrole-3-carboxylic acid, 2-chloro-6-(cyanophenylmethyl)-1,6-dihydro-, ethyl ester	MeOH	280(3.89)	18-0667-78
$C_{20}H_{17}ClN_2O_3$ Pyridinium, 4-[2-(4-hydroxyphenyl)ethenyl]-1-[(4-nitrophenyl)methyl]-, chloride	MeOH	402(4.61)	56-1265-78
$C_{20}H_{17}ClN_4O$ 3H-1,4-Benzodiazepine, 7-chloro-2-(3,5-dimethyl-1H-pyrazol-1-yl)-5-phenyl-, 4-oxide	isoPrOH	220s(4.31),250(4.33), 290(4.60),325s(4.04)	4-0161-78
$C_{20}H_{17}ClO_6$ 9,10-Anthracenedione, 2-acetyl-6-chloro-1,5,8-trimethoxy-3-methyl-	EtOH	227(4.41),252(4.36), 281(4.21),410(3.83)	44-1435-78

Compound	Solvent	$\lambda_{max}(\log \epsilon)$	Ref.
$C_{20}H_{17}Cl_2NO$			
2(1H)-Pyridinone, 3,5-bis(4-chlorophenyl)-1,4,6-trimethyl-	MeOH	246(3.81),318(3.83)	44-2138-78
4(1H)-Pyridinone, 3,5-bis(4-chlorophenyl)-1,2,6-trimethyl-	EtOH	225(4.43),275(4.12)	44-2138-78
$C_{20}H_{17}Cl_2NO_2S$			
2H,5H-Thiopyrano[4,3-b]pyran-2-one, 3,3-dichloro-4-(diphenylamino)-3,4,7,8-tetrahydro-	EtOH	250(4.15),293s(3.60)	4-0181-78
$C_{20}H_{17}Cl_3N_4O_2$			
Acetic acid, trichloro-, 1-(5,5,6,6-tetracyano-4,5,6,7-tetrahydro-1,2,3-trimethyl-1H-inden-1-yl)ethyl ester	CHCl$_3$	269(3.68)	35-0860-78
$C_{20}H_{17}F_2NO$			
2(1H)-Pyridinone, 3,5-bis(4-fluorophenyl)-1,4,6-trimethyl-	MeOH	240(3.94),317(3.99)	44-2138-78
4(1H)-Pyridinone, 3,5-bis(4-fluorophenyl)-1,2,6-trimethyl-	EtOH	227(4.38),274(4.18)	44-2138-78
$C_{20}H_{17}NO$			
Acridine, 9,10-dihydro-9-(4-methoxyphenyl)-	EtOH	228(4.3),286(4.15),315s(--)	39-1211-78C
10H-Azepino[1,2-a]indole, 11-(4-methoxyphenyl)-	EtOH	235(4.45),268(4.29),296s(--),325s(--),335s(--)	39-1211-78C
Benzo[h]quinolin-2(1H)-one, 5,6-dihydro-1-methyl-4-phenyl-	CHCl$_3$	246(4.26),295(4.09)	48-0097-78
Benzo[h]quinolin-2(1H)-one, 5,6-dihydro-4-(4-methylphenyl)-	CHCl$_3$	293(4.23)	48-0097-78
1H-Cycloprop[3,4]pyrido[1,2-a]indole, 1a,9b-dihydro-1a-methoxy-9-phenyl-	EtOH	228(4.41),256(4.49),285s(--),314(4.18)	39-1211-78C
2,4a-Methano-1,2,3,4-tetrahydronaphth-[1,8-cd]azepin-5-one, 4-phenyl-	EtOH	245(3.84)	33-1262-78
Pyridinium, 4-[2-(4-hydroxyphenyl)ethenyl]-1-(phenylmethyl)-, hydroxide, inner salt, (E)-hydrochloride	EtOH acetone CHCl$_3$ MeOH	540(3.42) 610(3.51) 630(3.80) 403(4.78)	56-1265-78 56-1265-78 56-1265-78 56-1265-78
$C_{20}H_{17}NO_2$			
Benzo[h]quinolin-2(1H)-one, 5,6-dihydro-4-(4-methoxyphenyl)-	CHCl$_3$	295(4.24)	48-0097-78
Dibenzo[de,g]pyrrolo[3,2,1-ij]quinoline, 4,5-dihydro-7,8-dimethoxy-	EtOH	254(4.67),263(4.85),296(4.11),317(3.97),360s(4.45),376(3.54)	44-1096-78
3-Pyridinecarboxylic acid, 2,5-diphenyl-, ethyl ester	EtOH	273(4.35),373(4.04)	142-0731-78A
3-Pyridinecarboxylic acid, 5,6-diphenyl-, ethyl ester	EtOH	243(4.24),300(3.99)	142-0731-78A
Pyrido[1,2-a]indole, 1-methoxy-10-(2-methoxyphenyl)-	MeOH	264(4.63),310(3.82),323(3.83),338(3.86),408(3.46),432(3.38)	145-1056-78
Pyrido[1,2-a]indole, 3-methoxy-10-(4-methoxyphenyl)-	MeOH	266(4.76),308(4.12),322(4.01),337(3.85),424(3.54)	145-1056-78
$C_{20}H_{17}NO_3$			
Acetic acid, [2,3-dihydro-3-oxo-2-(2-phenylethenyl)-1H-isoindol-1-yli-	THF	218(4.46),250(4.42),280(4.47)	40-0404-78

Compound	Solvent	λ_{max}(log ϵ)	Ref.
dene]-, ethyl ester (cont.)			40-0404-78
Anthra[9,1-bc]pyran-2,7-dione, 3-(but-ylamino)-	CHCl$_3$	542(4.05)	104-1433-78
3-Pyridinecarboxylic acid, 1,6-dihydro-6-oxo-2,5-diphenyl-, ethyl ester	EtOH	280(3.95),328(4.12)	142-0161-78A
$C_{20}H_{17}NO_6$			
4H-Dibenzo[de,g]quinoline-4,5(6H)-di-one, 1,2,9,10-tetramethoxy- (norpon-tevedrine)	EtOH	238(4.60),313(3.99), 325(4.24),478(3.95)	88-2179-78
Sanguinarine, hexahydro-5-hydroxy-8-oxo-	EtOH	219s(4.40),236s(4.17), 290(3.80),318(3.59)	44-2852-78
$C_{20}H_{17}N_2$			
1H-Imidazo[1,2-a]pyridin-4-ium, 2-phen-yl-1-(phenylmethyl)-, bromide	EtOH	207(4.57),231s(4.25), 289(4.11)	4-1149-78
$C_{20}H_{17}N_2O$			
2H-Pyrido[2,1-b][1,3,4]oxadiazin-5-ium, 3-methyl-6,8-diphenyl-, chloride	EtOH	220(3.90),245(3.95), 312(4.11)	120-0001-78
tetrafluoroborate	EtOH	220(3.94),240(3.95), 315(4.09)	120-0001-78
$C_{20}H_{17}N_3$			
Benzenamine, 4-benzo[f]quinazolin-3-yl-N,N-dimethyl-	EtOH	206(4.73),220(4.63), 235(4.67),257(4.60), 300(4.41),325(4.50), 358(4.76)	103-1025-78
$C_{20}H_{17}N_3O$			
Benzene, 1-azido-2-[(4-methoxyphenyl)-phenylmethyl]-	EtOH	230s(--),252(4.01), 264s(--),280s(--), 290s(--)	39-1211-78C
$C_{20}H_{17}N_3O_2$			
Spiro[2H-anthra[1,2-d]imidazole-2,1'-cyclohexane]-6,11-dione, 4-amino-	EtOH	614(3.80)	104-0381-78
9H-1,6[1',2']-endo-[1,2,4]Triazoloben-zo[1,3]cyclopropa[1,2,3-cd]cyclopro-pa[gh]pentalene-9,11(10H)-dione, 1,2,5,6,6a,6b,6c,6d-octahydro-10-phenyl-	CHCl$_3$	292(3.83),310s(3.72)	5-1379-78
$C_{20}H_{17}N_3O_3$			
Pyrimido[4,5-b]quinoline-2,4(3H,10H)-dione, 10-(2-methoxyethyl)-3-phenyl-	EtOH	222(4.58),265(4.60), 322(4.04),397(4.08)	88-3469-78
$C_{20}H_{17}N_4$			
1,2,4,5-Tetrazin-1(2H)-yl, 3,4-dihydro-2,4,6-triphenyl-	heptane	710(3.6)	104-1234-78
$C_{20}H_{17}N_5$			
2H-Benzotriazole, 5,7-dimethyl-2-phen-yl-4-(phenylazo)-	EtOH	224(4.36),270(4.16), 347(4.42),455(3.08)	33-2628-78
$C_{20}H_{17}N_5O$			
2H-Benzotriazol-5-ol, 2-(4-methylphen-yl)-4-[(4-methylphenyl)azo]-	EtOH	231(4.21),282(4.27), 344(4.37),390s(--), 480(4.18),505s(--)	33-2628-78

Compound	Solvent	$\lambda_{max}(\log \epsilon)$	Ref.
$C_{20}H_{17}OP$			
Phosphine oxide, 1,3,5,7-cyclooctatetraen-1-yldiphenyl-	EtOH	257(4.18),266(4.16), 273(4.08)	44-4338-78
Phosphine oxide, diphenyltricyclo-[4.2.0.0^{2,5}]octa-3,7-dien-2-yl-	EtOH	222(4.20),258(3.08), 265(3.11),272(2.97)	44-4338-78
$C_{20}H_{17}OPS$			
9-Phosphapentacyclo[4.3.0.0^{2,5}.0^{3,8} - 0^{4,7}]nonane, 9-phenyl-1-(phenyl-thio)-, 9-oxide	EtOH	258(3.60)	44-4338-78
$C_{20}H_{17}O_2P$			
2H,6H-[1,2]Oxaphospholo[4,3,2-hi][2,1]-benzoxaphosphole, 8,8-dihydro-8,8-di-phenyl-	EtOH	203(4.70),220s(4.42), 253s(2.93),258s(3.13), 266(3.27),272(3.19), 275(3.23)	24-0013-78
$C_{20}H_{17}O_3P$			
1,3-Benzenedimethanol, 2-(5H-dibenzo-phosphol-5-yl)-, P-oxide	EtOH	203(4.70),233s(4.55), 238(4.65),245(4.63), 274s(3.78),283(3.83), 294(3.61),328(3.21)	24-0013-78
$C_{20}H_{17}O_3PS$			
9-Phosphapentacyclo[4.3.0.0^{2,5}.0^{3,8}-0^{4,7}]nonane, 9-phenyl-1-(phenylsul-fonyl)-, 9-oxide	EtOH	259(3.31),266(3.85), 273(3.30)	44-4338-78
$C_{20}H_{17}P$			
Phosphine, 1,3,5,7-cyclooctatetraen-1-yldiphenyl-	EtOH	253s(4.10)	44-4338-78
Phosphine, diphenyltricyclo[4.2.0.0^{2,5}]-octa-3,7-dien-3-yl-	EtOH	251(3.89)	44-4338-78
$C_{20}H_{18}$			
Azulene, 4,6-dimethyl-8-(2-phenyleth-enyl)-	benzene	585(2.76)	104-0243-78
Benzo[3,4]cyclobuta[1,2-1]phenanthrene, 8c,9,10,11,12,12a-hexahydro-	MeCN	223(4.37),239s(4.44), 247(4.69),255(4.79), 261s(4.39),270(4.22), 279(4.02),290(4.00), 302(4.11),321s(2.66), 325(2.75),333s(2.71), 341(3.04),358(3.15)	24-3608-78
Bicyclo[4.2.0]octa-3,7-diene, 7,8-di-phenyl-	MeCN	296(4.13),308s(4.08), 323s(3.81)	24-3608-78
1,3-Cyclohexadiene, 5-phenyl-6-styryl-	EtOH	256(4.30)	78-1323-78
Dibenzo[a,e]cyclooctene, 5,6-didehydro-1,4,7,10-tetramethyl-	hexane	223s(4.21),230s(4.14), 249s(4.07),274(4.82), 282(4.98),294(4.45), 307(4.59),355(3.53), 375(3.54),395(3.60)	88-4269-78
5,12:6,11-Dimethanonaphthacene, 5,5a,6,11,11a,12-hexahydro-	n.s.g.	270-310(c.2.48)	89-0271-78
9,12-Etheno-15,6-metheno-6H-cyclohepta-cyclododecene, 7,8,13,14-tetrahydro-	MeOH	288(4.63),342s(3.44), 348s(3.47),356(3.60), 374(3.39),574(2.21), 600(2.28),626(2.33), 670(2.22),750(1.62)	88-1067-78
2,6-Octadien-4-yne, 3,6-diphenyl-	EtOH	218s(4.30),229(4.37), 241(4.44),252s(4.39)	56-2377-78

Compound	Solvent	$\lambda_{max}(\log \epsilon)$	Ref.
1,3,5,7-Octatetraene, 1,8-diphenyl-, trans-cis-cis-trans	C_6H_{12}	240(4.55),350(4.83), 369(4.88),390(4.75)	78-1323-78
all trans	C_6H_{12}	360(4.82),378(4.98), 400(4.88)	78-1323-78
$C_{20}H_{18}BrNO_4$			
13aαH-Pseudocoptisine, 4-bromo-13α-methyltetrahydro-	EtOH	291(3.98)	44-1992-78
13β-methyl-	EtOH	292(4.00)	44-1992-78
$C_{20}H_{18}BrNO_6$			
3,4-Isoquinolinedione, 1-[(2-bromo-4,5-dimethoxyphenyl)methylene]-1,2-dihydro-6,7-dimethoxy-, (Z)-	EtOH	244(4.27),296(4.26), 402(4.08)	88-2179-78
$C_{20}H_{18}Br_2$			
Benzene, 1,1'-[1,4-bis(1-bromoethyl)-1,2,3-butatriene-1,4-diyl]bis-	C_6H_{12}	253(4.38),328s(3.64), 395(4.52)	56-2377-78
isomer	C_6H_{12}	250(4.22),400(4.53)	56-2377-78
$C_{20}H_{18}Br_4N_4O_4S$			
Acetic acid, 2,2'-thiobis[(2,4-dibromophenyl)hydrazono]-, diethyl ester	EtOH	243(4.23),305(4.08), 373(4.42)	39-0539-78C
$C_{20}H_{18}NO_2P$			
Phenophosphazine, 5,10-dihydro-10-(4-methoxyphenyl)-5-methyl-, 10-oxide	EtOH	222(4.49),243(4.43), 274(4.22),303(3.87), 338(3.81)	65-1205-78
$C_{20}H_{18}NO_5PS_3$			
Phosphorothioic acid, O,O-diphenyl S-(2,3,5,7-tetrahydro-2-methyl-5,7-dioxo-6H-1,4-dithiino[2,3-c]pyrrol-6-yl) ester	EtOH	210(4.11),263(3.90), 420(3.37)	73-1093-78
$C_{20}H_{18}N_2O$			
Ethanone, 1-phenyl-2,2-bis(phenylamino)-	EtOH	246(4.38);286(3.56)	44-3394-78
1-Naphthaleneacetic acid, (1-phenylethylidene)hydrazide	EtOH	276(4.28)	34-0345-78
Phenol, 5-amino-2-(9,10-dihydro-10-methyl-9-acridinyl)-	EtOH	289(4.27)	104-0129-78
$C_{20}H_{18}N_2OS_2$			
Hydrazinecarbodithioic acid, [[5-(4-methylphenyl)-2-furanyl]methylene]-, phenylmethyl ester	dioxan	379(4.57)	73-2643-78
$C_{20}H_{18}N_2O_2$			
1-Naphthaleneacetic acid, [(4-methoxyphenyl)methylene]hydrazide	EtOH	286s(4.33),294(4.39), 305s(4.34)	34-0345-78
5(4H)-Oxazolone, 4-[3-[4-(dimethylamino)phenyl]-2-propenylidene]-2-phenyl-	toluene	500(4.72)	103-0120-78
1H-Phenanthro[9,10-d]imidazole-1-butanoic acid, methyl ester	MeOH	255(4.88)	44-0381-78
$C_{20}H_{18}N_2O_3$			
Benzenemethanamine, N-[2-(5-nitro-2-furanyl)ethenyl]-N-(phenylmethyl)-, (E)-	EtOH	278(4.23),499(4.35)	103-0250-78
Pyrido[3',4':4,5]cyclohept[1,2-b]indol-12(5H)-one, 6-acetoxy-4-ethyl-6,7-di-	EtOH	254(4.16),262(4.12), 271(4.04),278(4.04),	78-1363-78

Compound	Solvent	$\lambda_{max}(\log \epsilon)$	Ref.
hydro-, (±)- (cont.)		342(4.01)	78-1363-78
$C_{20}H_{18}N_2O_3S$ Pyridinium, 1-[1-benzoyl-3-cyano-4-eth-oxy-2-(methylthio)-4-oxo-2-butenyl]-, hydroxide, inner salt	EtOH	260(4.12),337(3.80), 430(4.18)	95-1412-78
$C_{20}H_{18}N_2O_4$ Pyrazinecarboxylic acid, 1-benzoyl-1,2,5,6-tetrahydro-2,5-dimethyl-6-oxo-3-phenyl-	n.s.g.	200(4.65),235(4.34), 263(4.03)	70-0963-78
1H-Pyrazole-4,5-dicarboxylic acid, 1-(4-methylphenyl)-3-phenyl-, dimethyl ester	EtOH	238(4.30),353(3.52)	33-1477-78
$C_{20}H_{18}N_2O_6$ 1,3-Dioxolo[4,5-h]isoquinoline-6-carb-oxamide, 7-(1,3-benzodioxol-5-yl)-6,7,8,9-tetrahydro-N,8-dimethyl-9-oxo-, trans	EtOH	214s(4.44),236s(4.17), 288(3.80),321(3.65)	44-2852-78
$C_{20}H_{18}N_2O_8$ 3H-1,2a-Diazacyclopenta[ef]heptalene-3,4,5,6-tetracarboxylic acid, tetramethyl ester	CHCl$_3$	280(4.01),450(3.99)	39-0429-78C
$C_{20}H_{18}N_2S_2$ Benzenesulfenamide, N,N'-(2,6-dimethyl-2,5-cyclohexadiene-1,4-diylidene)bis-	n.s.g.	280(4.12),468(4.68)	139-0041-78B
Benzenesulfenamide, N,N'-(3,5-dimethyl-3,5-cyclohexadiene-1,2-diylidene)bis-	n.s.g.	435(3.99),500(4.12)	139-0041-78B
$C_{20}H_{18}N_4$ Benzenamine, 2-[2-[2-[(2-aminophenyl)-azo]phenyl]ethenyl]-	benzene	297(4.20),350s(4.09), 430s(3.76)	33-2813-78
$C_{20}H_{18}N_4O$ 4H-Pyrazolo[3,4-d]pyridazin-4-one, 1,5-dihydro-3-methyl-1,7-bis(phen-ylmethyl)-	EtOH	218(4.33),282(3.81)	4-0813-78
4H-Pyrazolo[3,4-d]pyridazin-4-one, 2,5-dihydro-3-methyl-2,7-bis(phen-ylmethyl)-	EtOH	218(4.31),280(3.88)	4-0813-78
Vobasan-17,22-dinitrile, 14,15-didehy-dro-19,20-dihydro-3-oxo-, (20α)-	MeOH	245(3.98),336(4.23)	35-0938-78
$C_{20}H_{18}N_4O_3$ Pyrido[2,3-d]pyrimidine-8(2H)-carboxam-ide, 1,3,4,5,6,7-hexahydro-2,4-dioxo-N,3-diphenyl-	MeOH	223(3.99),291(4.28)	24-2297-78
$C_{20}H_{18}N_6$ 2H-Benzotriazol-5-amine, 2-(4-methyl-phenyl)-4-[(4-methylphenyl)azo]-	EtOH	251(4.41),283(4.14), 354(4.49),428(4.22), 464(4.13)	33-2628-78
$C_{20}H_{18}N_8Ni$ Nickel, [6,13-bis(1-methylethyl)-1,4,8-11-tetraazacyclotetradeca-2,4,7,9,11-14-hexaene-2,3,9,10-tetracarbonitril-ato(2-)-N^1,N^4,N^8,N^{11}]-, (SP-4-1)-	EtOH	270(4.27),330(4.77), 390(4.46),530(4.26)	24-2919-78

Compound	Solvent	$\lambda_{max}(\log \epsilon)$	Ref.
Nickel, [6,13-dipropyl-1,4,8,11-tetra-azacyclotetradeca-2,4,7,9,11,14-hexa-ene-2,3,9,10-tetracarbonitrilato-(2-)-N^1,N^4,N^8,N^{11}]-, (SP-4-1)-	EtOH	270(4.27),330(4.76), 390(4.47),530(4.27)	24-2919-78
$C_{20}H_{18}N_{10}O_6$			
Benzonitrile, 2-[[3,5-bis(dimethylami-no)-1-(2,4-dinitrophenyl)-1H-pyrazol-4-yl]azo]-5-nitro-	dioxan	446(4.39)	103-0332-78
$C_{20}H_{18}O$			
2,4,6,12,14,16-Cycloheptadecahexaene-8,10-diyn-1-one, 2,7,12-trimethyl-	THF	<u>310(4.8)</u>,400s(3.8)	138-1099-78
Ethanone, 1-(10b,10c-dihydro-10b,10c-dimethyl-1-pyrenyl)-, cis	C_6H_{12}	358(4.08),423(3.08), 442(3.00),570(2.00), 616(2.00)	44-3475-78
Ethanone, 1-(10b,10c-dihydro-10b,10c-dimethyl-2-pyrenyl)-, cis	C_6H_{12}	262(4.08),333(4.11), 367(3.97),467(3.42), 570(2.00),617(2.30)	44-3475-78
2(1H)-Pentalenone, 4,5,6,6a-tetrahydro-1,3-diphenyl-, trans	EtOH	231(4.23),266(4.04)	35-1799-78
$C_{20}H_{18}OSi$			
Silane, benzoylmethyldiphenyl-	C_6H_{12}	405(2.26),424(2.38), 446(2.16)	112-0751-78
$C_{20}H_{18}O_2$			
1(2H)-Naphthalenone, 3,4-dihydro-4-(1-1a,6,6a-tetrahydro-6-hydroxycyclo-prop[a]inden-6-yl)-, [1aα,6β(S*),6aα]-	EtOH	252(4.03),280(3.25), 297(3.23)	44-4316-78
$C_{20}H_{18}O_3$			
2H-1-Benzopyran-2-one, 6-methoxy-4-methyl-7-(1-phenyl-2-propenyl)-	MeOH	228(4.01),250(3.82)	2-0856-78
4H-1-Benzopyran-4-one, 7-[(3-methyl-2-butenyl)oxy]-3-phenyl-	MeOH	248(4.53),290(4.21)	2-0973-78
2(5H)-Furanone, 5-methyl-5-(2-oxoprop-yl)-3,4-diphenyl-	EtOH	219s(4.20),265(4.02)	33-3018-78
4H-Furo[2,3-h]-1-benzopyran-4-one, 8,9-dihydro-8,9,9-trimethyl-2-phenyl-	MeOH	252(4.55),310(4.58)	2-0973-78
4H-Furo[2,3-h]-1-benzopyran-4-one, 8,9-dihydro-8,9,9-trimethyl-3-phenyl-	MeOH	241(4.53),256(4.31)	2-0973-78
$C_{20}H_{18}O_4$			
1,4-Anthracenediol, 2,3-dimethyl-, diacetate	CH_2Cl_2	255(5.10),317(3.11), 331(3.47),346(3.74), 364(3.87),384(3.79)	150-5538-78
2H,8H-Benzo[1,2-b:3,4-b']dipyran-2-one, 9,10-dihydro-9-hydroxy-8,8-dimethyl-10-phenyl-	EtOH	212(4.54),247(3.73), 258(3.69),330(4.16)	2-0184-78
4H-1-Benzopyran-4-one, 7-hydroxy-3-[4-hydroxy-3-(3-methyl-2-butenyl)phen-yl]-	MeOH	248(4.41),258s(4.37), 305s(4.00)	18-2398-78
	MeOH-NaOH	255(4.52),331(4.23)	18-2398-78
	MeOH-NaOAc	255(4.55),331(4.11)	18-2398-78
4H-1-Benzopyran-4-one, 5-hydroxy-7-[(3-methyl-2-butenyl)oxy]-3-phenyl-	MeOH	248(4.51),288(4.38)	2-0973-78
3-Cyclobutene-1,2-dicarboxylic acid, 3,4-diphenyl-, dimethyl ester, cis	MeCN	274s(4.07),292(4.20), 300s(4.17),318s(3.85)	24-3608-78
trans	MeCN	294(4.24),300s(4.23), 317s(4.00)	24-3608-78
[2.2.2.2](1,2,4,5)-Cyclophanequinhydrone	MeOH	491(3.11)	89-0756-78

Compound	Solvent	$\lambda_{max}(\log \epsilon)$	Ref.
2-Furanacetic acid, 2,5-dihydro-2-methyl-5-oxo-3,4-diphenyl-, methyl ester	EtOH	216s(4.24),264(4.06)	33-3018-78
4H-Furo[2,3-h]-1-benzopyran-4-one, 8,9-dihydro-5-hydroxy-8,9,9-trimethyl-3-phenyl-	MeOH	261(4.44),285(4.63)	2-0973-78
1H-Indeno[2',1':1,3]cyclopropa[1,2-c]-furan-3(3aH)-one, 3a-(3,4-dimethoxyphenyl)-3b,8-dihydro-	EtOH	273s(3.42),279(3.50), 287s(3.40)	94-1592-78
Isolonchocarpin, 3-hydroxy-	MeOH	264(4.26),300s(3.86)	102-1812-78
$C_{20}H_{18}O_5$			
4H,8H-Benzo[1,2-b:3,4-b']dipyran-4-one, 2,3-dihydro-5-hydroxy-2-(4-hydroxyphenyl)-8,8-dimethyl-, (S)-	MeOH	228(3.94),288(4.08), 337(3.72)	78-3563-78
4H,8H-Benzo[1,2-b:5,4-b']dipyran-4-one, 2,3-dihydro-5-hydroxy-2-(4-hydroxyphenyl)-8,8-dimethyl-, (S)-	MeOH	228(3.94),297(4.28), 338(3.10)	78-3563-78
6-Benzofuranol, 2-(1,3-benzodioxol-5-yl)-5-methoxy-3-methyl-7-(2-propenyl)-	EtOH	254(3.92),290s(4.16), 328(4.51)	44-3717-78
4H-1-Benzopyran-4-one, 3-(3,4-dihydro-3-hydroxy-2,2-dimethyl-2H-1-benzopyran-6-yl)-7-hydroxy- (psoralenol)	EtOH	265(4.29),310s(4.04), 328s(3.54)	102-2046-78
	EtOH-NaOH	260(--),280s(--), 345(--)	102-2046-78
	EtOH-NaOAc	260(--),310s(--), 345(--)	102-2046-78
4H-1-Benzopyran-4-one, 5,7-dihydroxy-3-[4-hydroxy-3-(3-methyl-2-butenyl)-phenyl]-	EtOH	263(4.54),295(4.06), 330(3.72)	18-2398-78
	EtOH-NaOAc	273(3.55),331(4.01)	18-2398-78
4H-1-Benzopyran-4-one, 5,7-dihydroxy-2-(4-hydroxyphenyl)-6-(3-methyl-2-butenyl)-	MeOH	237(4.01),276(4.28), 331(4.25)	78-3569-78
2,4-Hexadienedioic acid, 4-methoxy-2,3-diphenyl-, 6-methyl ester, (?,Z)-	EtOH	215s(4.21),267(4.25), 279s(4.23)	33-3018-78
2-Naphthalenecarboxylic acid, 3,4-dihydro-1,3-dihydroxy-4-oxo-3-(phenylmethyl)-, ethyl ester	n.s.g.	214(4.42),237(4.27), 329(3.95)	1-0347-78
Propanedioic acid, (1-oxo-2,3-diphenyl-2-propenyl)-, dimethyl ester	EtOH	228(3.93),304(4.22)	33-3018-78
α-D-Ribofuranose, 1,2-O-(1,3-diphenyl-2-propynylidene)-	EtOH	242(4.34),250(4.25)	94-0898-78
[1,1':4',1''-Terphenyl]-2',4,4''-triol, 3',6'-dimethoxy-	EtOH	228s(4.64),278(4.62)	98-0632-78
$C_{20}H_{18}O_6$			
9,10-Anthracenedione, 2-acetyl-1,6,8-trimethoxy-3-methyl-	CHCl₃	278(4.32),400(3.73)	44-1435-78
6,11-Ethanodibenzo[b,f][1,4]dioxocin-13,14-dicarboxylic acid, 6,11-dihydro-, dimethyl ester, (6α,11α,13R*-14S*)-	ether	272(3.17),278(3.14)	78-0073-78
Sesamin	EtOH	228(3.98),280(3.52)	42-1204-78
[1,1':4',1''-Terphenyl]-2',3,4,4''-tetrol, 3',6'-dimethoxy-	EtOH	228s(4.71),280(4.60)	98-0632-78
$C_{20}H_{18}O_7$			
2-Anthraceneacetic acid, 9,10-dihydro-1,4-dihydroxy-3-(3-hydroxybutyl)-9,10-dioxo-	MeOH	203(4.36),227s(--), 251(4.55),254s(--), 287(3.84),325s(--), 469s(--),484(3.96), 517(3.77)	24-3823-78

Compound	Solvent	$\lambda_{max}(\log \epsilon)$	Ref.
2-Anthracenecarboxylic acid, 9,10-dihy-dro-3,6,8-trimethoxy-1-methyl-9,10-dioxo-, methyl ester	EtOH	220(4.50),279(4.56), 325(3.85),410(3.68)	39-1041-78C
Naphtho[2,3-b]furan-3-carboxylic acid, 4,5-diacetoxy-2-methyl-, ethyl ester	dioxan	245(4.90),322(4.00), 341(3.88)	5-0140-78
Vesticarpan diacetate	EtOH	230(4.04),284(3.66)	102-1413-78
$C_{20}H_{18}O_8$			
4H-Naphtho[2,3-b]pyran-4-one, 3,5,10-triacetoxy-2,3-dihydro-2-methyl-, cis	MeOH	221(4.7),243(4.7), 251(4.7),283s(--), 294(4.1),305(4.0)	24-1284-78
trans	MeOH	221(4.7),244(4.7), 252(4.7),284s(--), 294(4.1),305(4.0)	24-1284-78
$C_{20}H_{18}O_9$			
9,10-Anthracenedione, 1-(β-D-gluco-pyranosyloxy)-8-hydroxy-	EtOH	220(4.5),253(4.3), 408(3.8)	73-1803-78
9H-Xanthene-2,3-dicarboxylic acid, 1-hydroxy-4,6-dimethoxy-8-methyl-9-oxo-, dimethyl ester	EtOH	245(4.56),251(4.46), 285(4.24),360(3.87)	94-0209-78
$C_{20}H_{18}S$			
Cyclopenta[b]thiopyran, 2,5,6,7-tetra-hydro-2,4-diphenyl-	hexane	220(4.2),234(4.3), 296(3.2)	103-0605-78
Cyclopenta[b]thiopyran, 5,6,7,7a-tetra-hydro-2,4-diphenyl-	hexane	258(4.2),353(3.7)	103-0605-78
$C_{20}H_{18}S_2$			
11H-5,8-Etheno-1,12-metheno-2H-cyclo-hepta[d][1,8]dithiacyclotetradecin, 4,9-dihydro-	MeOH	242(4.24),295(4.55), 363(3.64),381(3.81), 578s(2.32),627(2.43), 672s(2.33),755s(1.90)	88-1067-78
$C_{20}H_{19}BrN_2O$			
1H-Imidazo[1,2-a]pyridin-4-ium, 2,3-di-hydro-2-hydroxy-2-phenyl-1-(phenyl-methyl)-, bromide	EtOH	202(4.44),238(4.18), 327(3.75)	4-1149-78
$C_{20}H_{19}ClFN_3O_2$			
Acetamide, N-[[1-acetyl-7-chloro-5-(2-fluorophenyl)-2,3-dihydro-1H-1,4-ben-zodiazepin-2-yl]methyl]-	isoPrOH	225s(4.41),270s(3.64), 285s(3.40)	44-0936-78
$C_{20}H_{19}ClN_4O$			
1H-Pyrazol-5-ol, 1-(7-chloro-5-phenyl-3H-1,4-benzodiazepin-2-yl)-4,5-di-hydro-3,5-dimethyl-	isoPrOH	233(4.39),250s(4.32), 296(4.38),348(3.72)	4-0161-78
$C_{20}H_{19}ClN_4O_2$			
1H-Pyrazol-5-ol, 1-(7-chloro-5-phenyl-3H-1,4-benzodiazepin-2-yl)-4,5-dihy-dro-3,5-dimethyl-, oxide	isoPrOH	245(4.33),287(4.62), 355s(3.53)	4-0161-78
$C_{20}H_{19}ClN_4O_3$			
3H-1,4-Benzodiazepine, 7-chloro-2-[2-(methoxycarbonylisopropylidene)hy-drazino]-5-phenyl-, 4-oxide	isoPrOH	244(4.37),283(4.48), 350s(3.52)	4-0161-78
$C_{20}H_{19}ClO_6$			
9,10-Anthracenedione, 1-butyl-7-chloro-	EtOH	238(4.58),258(4.07),	44-1435-78

Compound	Solvent	$\lambda_{max}(\log \epsilon)$	Ref.
5,8-dihydroxy-2,4-dimethoxy- (cont.)		287(4.10),296s(4.04), 480(4.08),495(4.07)	44-1435-78
$C_{20}H_{19}HgN_3$			
Mercury, [1,3-bis(2-methylphenyl)-1-triazenato-N]phenyl-	toluene	395(2.56)	64-1091-78B
Mercury, [1,3-bis(4-methylphenyl)-1-triazenato-N]phenyl-	toluene	405(2.55)	64-1091-78B
$C_{20}H_{19}N$			
Cyclopropanecarbonitrile, 2,2-dimethyl-3-(2,2-diphenylethenyl)-, cis	EtOH	227(4.16),263(4.19)	78-1775-78
trans	EtOH	227(4.17),263(4.23)	78-1775-78
2,5-Hexadienenitrile, 4,4-dimethyl-6,6-diphenyl-, cis	EtOH	249(4.14)	78-1775-78
trans	EtOH	251(4.18)	78-1775-78
2-Propenenitrile, 3-(2,2-dimethyl-3,3-diphenylcyclopropyl)-, cis	EtOH	224(4.28),246(4.11)	78-1775-78
trans	EtOH	227(4.29),245(4.21)	78-1775-78
$C_{20}H_{19}NO$			
Benzenamine, 2-[(4-methoxyphenyl)phenylmethyl]-	EtOH	230(4.34),252s(--), 275s(--),286s(--), 374(3.40)	39-1211-78C
4(1H)-Pyridinone, 1,2,6-trimethyl-3,5-diphenyl-	EtOH	236(4.29),275(4.13)	44-2138-78
$C_{20}H_{19}NO_3$			
Cyclobuta[1]phenanthrene-1-carboxylic acid, 2-(aminocarbonyl)-1,2,2a,10b-tetrahydro-, ethyl ester, anti	MeOH	218(4.22),263(4.29), 276(4.19),305(3.62)	24-2677-78
3-Pyridinecarboxylic acid, 1,4,5,6-tetrahydro-6-oxo-2,5-diphenyl-, ethyl ester	EtOH EtOH-KOH	285(3.97) 325(3.99)	142-0161-78A 142-0161-78A
$C_{20}H_{19}NO_3S_4$			
2H-Thiopyran-2,6(3H)-dithione, 3-(dimethylamino)-3-hydroxy-4-(4-methylphenyl)-5-(phenylsulfonyl)-	EtOH	239s(4.12),314(4.45), 380(3.67),463(4.01)	39-1017-78C
$C_{20}H_{19}NO_4$			
1H-Pyrrole-3-carboxylic acid, 4,5-dihydro-5-hydroxy-5-methyl-4-oxo-1,2-diphenyl-, ethyl ester	EtOH	209(4.05),246(4.17), 335(3.98)	118-0291-78
$C_{20}H_{19}NO_5S$			
Pyridinium, 1-(ethoxycarbonyl)-2-(6-methoxy-3-oxo-2(3H)-benzofuranylidene)-2-(methylthio)ethylide	EtOH	274(4.49),330(4.10)	142-0399-78A
$C_{20}H_{19}NO_6$			
Severzinine (dihydrosibiricine)	EtOH	290(4.04)	105-0464-78
$C_{20}H_{19}NO_8$			
Terramycin, 4-de(dimethylamino)-12a-deoxy-	MeOH-HCl	217(4.22),263(4.35), 320(4.19)	39-0145-78C
$C_{20}H_{19}NO_9$			
α-D-Xylofuranoside, methyl 5-benzoate 2-(4-nitrobenzoate)	EtOH	261(4.10)	136-0257-78H

Compound	Solvent	$\lambda_{max}(\log \epsilon)$	Ref.
$C_{20}H_{19}N_2O$			
1H-Imidazo[1,2-a]pyridin-4-ium, 2,3-di-hydro-2-hydroxy-2-phenyl-1-(phenyl-methyl)-, bromide	EtOH	202(4.44),238(4.18), 327(3.75)	4-1149-78
$C_{20}H_{19}N_2S_2$			
Benzothiazolium, 3-methyl-2-[2-methyl-3-(3-methyl-2(3H)-benzothiazolyli-dene)-1-propenyl]-	EtOH	540(5.06)	124-0942-78
$C_{20}H_{19}N_2S_3$			
Benzothiazolium, 3-methyl-2-[2-(methyl-thio)-3-(3-methyl-2(3H)-benzothiazol-ylidene)-1-propenyl]-	EtOH	581(4.79)	124-0942-78
$C_{20}H_{19}N_3O_3$			
1H-Imidazole-4-carboxylic acid, 1-benz-oyl-5-[(phenylmethyl)amino]-, ethyl ester	EtOH	213(4.01),231(4.08), 265s(--)	118-0741B-78
$C_{20}H_{19}N_3O_4$			
Carbamic acid, (6-methyl-8-methylene-1,3-dioxolo[4,5-g]quinazolin-7(8H)-yl)phenyl-, ethyl ester	C_6H_{12}	224(4.65),250s(4.41), 336(4.00)(broad)	78-2557-78
	EtOH	223(4.63),248s(4.35), 340(3.95)	78-2557-78
$C_{20}H_{19}N_3O_6$			
Pyridinium, 1-[1-[2-hydroxy-1-[[(4-ni-trophenyl)methoxy]carbonyl]-1-buten-yl]-4-oxo-2-azetidinyl]-, hydroxide, inner salt	pH 6	272(4.42)	77-0469-78
$C_{20}H_{19}O_3P$			
1,3-Benzenedimethanol, 2-(diphenylphos-phinyl)-	EtOH	204(4.73),223s(4.40), 253s(2.93),259s(3.15), 267(3.30),274(3.31), 281s(3.00)	24-0013-78
$C_{20}H_{19}O_4PS$			
Phosphinic acid, phenyl[8-(phenylsul-fonyl)tricyclo[4.2.0.0^{2,5}]oct-7-en-3-yl]-, syn-exo	EtOH	252(2.66),258(3.01), 264(3.04),273(2.99)	44-4338-78
$C_{20}H_{20}$			
Benzene, 1,1'-(1,2-dicyclopropyl-1,2-ethenediyl)bis-, cis	hexane	254(3.92)	44-1612-78
trans	hexane	242(3.81)	44-1612-78
Bicyclo[4.2.0]oct-7-ene, 7,8-diphenyl-	MeCN	298(4.20),306s(4.18), 325s(3.90)	24-3608-78
Naphthalene, 1,2,3,4-tetramethyl-6-phenyl-	EtOH	218(4.46),260(4.75), 300(3.94)	18-0331-78
$C_{20}H_{20}BrNO_5$			
4H-Pyrano[2,3-h]quinoline-2-carboxylic acid, 3-(2-bromo-2-propenyl)-1,8-di-hydro-5-methoxy-8,8-dimethyl-4-oxo-, methyl ester	MeOH	258(4.43),280(4.11), 350(3.63)	24-0439-78
$C_{20}H_{20}ClNO_5$			
4H-Pyrano[2,3-h]quinoline-2-carboxylic acid, 3-(2-chloro-2-propenyl)-1,8-di-	MeOH	256(4.54),283(4.19), 370(3.74)	24-0439-78

Compound	Solvent	λ_{max}(log ϵ)	Ref.
hydro-5-methoxy-8,8-dimethyl-4-oxo-, methyl ester (cont.)			24-0439-78
$C_{20}H_{20}Cl_2N_2O_3$ 4H-Cyclopent[d]isoxazole, 3-[5-(3,4-di-chlorophenyl)-2-furanyl]-3a,5,6,6a-tetrahydro-6a-(4-morpholinyl)-	MeOH	216(4.33),238(4.30), 327(4.49)	73-0288-78
$C_{20}H_{20}Cl_2N_4O_4$ Butanamide, N-(4-butylphenyl)-2-[(2,6-dichloro-4-nitrophenyl)hydrazono]-3-oxo-	EtOH	204(4.24),221(4.18), 252(4.05),383(4.37)	104-0521-78
$C_{20}H_{20}Fe_2$ Biferrocene cation	MeNO$_2$ CD$_2$Cl$_2$	183.0(2.88) 200.0(2.96)	35-4393-78 35-4393-78
$C_{20}H_{20}N$ Phenanthridinium, 7,8,9,10-tetrahydro-5-(phenylmethyl)-, perchlorate	MeOH	240(4.62),320(4.18)	65-1707-78
$C_{20}H_{20}NO_4$ Columbamine (chloride)	EtOH	229(3.91),268(3.87), 281(3.84),350s(3.82), 430(3.17)	100-0169-78
$C_{20}H_{20}N_2$ 1H-Inden-1-one, 2,3-dihydro-6-methyl-, (2,3-dihydro-6-methyl-1H-inden-1-yli-dene)hydrazone	CH$_2$Cl$_2$	230(3.90),265(3.82), 324s(4.18),342s(4.33), 356s(4.44),367(4.44), 383s(4.32)	5-0440-78
1H-Inden-1-one, 2,3-dihydro-7-methyl-, (2,3-dihydro-7-methyl-1H-inden-1-yli-dene)hydrazone	CH$_2$Cl$_2$	230(3.95),254s(3.63), 282(3.73),353(4.45), 368(4.45),387s(3.63)	5-0440-78
2,4,6-Nonatrien-8-ynal, 7-methyl-, (7-methyl-2,4,6-nonatrien-8-ynyli-dene)hydrazone, (?,?,E,E,Z,Z,E,E)-	THF	260(3.81),271s(3.75), 281s(3.61),312s(3.74), 325s(3.81),396(4.60), 413(4.61),434(4.40)	18-2112-78
$C_{20}H_{20}N_2O$ 1-Butanone, 1-(4,5-dihydro-4-methylene-5,5-diphenyl-1H-pyrazol-3-yl)-	EtOH	240(3.81),343(3.96)	22-0401-78
Formamide, N-[1-[[4-(dimethylamino)-phenyl]methyl]-2-naphthalenyl]-	EtOH	207(4.38),239(4.61), 292(4.00)	103-1025-78
1(4H)-Naphthalenone, 4-[[4-(diethyl-amino)phenyl]imino]-	C$_6$H$_{12}$	270(4.33),305(4.10), 558(4.19)	48-0557-78
	EtOH	270(4.23),315(4.03), 607(4.18)	48-0557-78 80-0617-78
	ether	270(4.32),310(4.11), 572(4.16)	48-0557-78
	CCl$_4$	312(4.02),565(4.12)	48-0557-78
$C_{20}H_{20}N_2O_2$ Dibenzo[b,j][1,7]phenanthroline-8,14-(2H,10H)-dione, 1,3,4,5,9,11,12,13-octahydro-	H$_2$SO$_4$	275(4.95),312(4.25), 325(4.14)	88-3677-78
Dibenzo[b,j][1,10]phenanthroline-5,8-dione, 1,2,3,4,9,10,11,12,13,14-decahydro-	H$_2$SO$_4$	272(4.41),281(4.62), 312(3.60)	88-3677-78
Dibenzo[b,j][4,7]phenanthroline-13,14-dione, decahydro-	H$_2$SO$_4$	252(4.47),302(4.07), 340s(3.88)	88-3677-78

Compound	Solvent	λ_{max}(log ϵ)	Ref.
1H-Pyrazole-5-carboxylic acid, 4,5-di-hydro-5-methyl-4-methylene-1,3-di-phenyl-, ethyl ester	EtOH	245(4.08),375(4.06)	22-0415-78
C$_{20}$H$_{20}$N$_2$O$_3$ 3-Pyridinecarboxylic acid, 5-ethyl-4-[2-(1-methyl-1H-indol-2-yl)-2-oxoethyl]-, methyl ester	EtOH	236(4.60),309(4.57), 384(3.95)	78-1363-78
C$_{20}$H$_{20}$N$_2$O$_5$S Catharanthinic acid, 3,4-dihydroxo-19-thiono-3,4-dihydro-, lactone, (3R,4R)-	EtOH	222(4.57),274(4.25), 278(4.25),281(4.24), 289(4.16)	33-0690-78
C$_{20}$H$_{20}$N$_2$O$_7$S Thiocyanic acid, 1-(2,3,5-tri-O-acetyl-α-L-arabinofuranosyl)-1H-indol-5-yl ester	EtOH	232(4.32),280(3.49)	104-2011-78
Thiocyanic acid, 1-(2,3,4-tri-O-acetyl-α-L-arabinopyranosyl)-1H-indol-5-yl ester	EtOH	234(4.66),275(3.38)	104-2011-78
C$_{20}$H$_{20}$N$_4$O$_2$ Vobasan-17,22-dinitrile, 19,20-dihydro-15-hydroxy-3-oxo-, (R)-	MeOH	230(3.92),315(4.15)	35-0938-78
(S)-	MeOH	240(3.95),316(4.20)	35-0938-78
C$_{20}$H$_{20}$N$_4$O$_2$S 2H-[1,2,3]Thiadiazolo[4,5,1-hi][1,2,3]-benzothiadiazole-3-S^{IV}, 4,6,7,8-tet-rahydro-2,4-bis(4-methoxyphenyl)-	C$_6$H$_{12}$	249(4.28),315(4.22), 535(4.34)	39-0195-78C
C$_{20}$H$_{20}$N$_8$ 1H-Pyrazol-5-amine, 4,4'-azobis[3-meth-yl-1-phenyl-	50% dioxan	247(4.45),270(4.33), 363(4.14),410(4.43), 428s(4.36)	33-0108-78
1,4,8,11-Tetraazacyclotetradeca-2,4,6-9,11,13-hexaene, 6,13-bis(1-methyl-ethyl)-	CHCl$_3$	330(4.71),350(4.67), 360(4.68),368(4.66), 400(4.10),430(3.86)	24-2919-78
1,4,8,11-Tetraazacyclotetradeca-2,4,6-9,11,13-hexaene, 6,13-dipropyl-	CHCl$_3$	330(4.73),350(4.70), 360(4.69),368(4.68), 400(4.10),430(3.87)	24-2919-78
C$_{20}$H$_{20}$N$_8$O$_5$ 4,13-Epoxy-1,3-dioxolo[6,7][1,3]diazo-cino[1,2-e]purine-6,8-diamine, 3a,4,5,6,13,13a-hexahydro-2,2-di-methyl-N^6-[(4-nitrophenyl)methyl-ene]-, [3aR-(3aα,4β,13β,13aα)]-	MeOH	211(4.41),273(4.40)	44-2320-78
C$_{20}$H$_{20}$N$_{10}$O$_4$ 1H-Pyrazole-3,5-diamine, 1-(2,4-dini-trophenyl)-4-(1H-indazol-3-ylazo)-N,N,N',N'-tetramethyl-	dioxan	377(4.74),442s(4.65)	103-0332-78
C$_{20}$H$_{20}$O Furan, 2,5-diethyl-3,4-diphenyl-	EtOH	223(4.32)	104-1894-78
C$_{20}$H$_{20}$O$_2$ Ethanone, 1,1'-tricyclo[9.3.1.1^{4,8}]-hexadeca-1(15),4,6,8(16),11,13-hexa-	EtOH	213(4.40),223s(4.34), 257(4.22),295s(3.50),	78-0871-78

Compound	Solvent	$\lambda_{max}(\log \epsilon)$	Ref.
ene-5,14-diylbis- (cont.) 2,2'-Spirobi[2H-indene]-5-carboxylic acid, 5'-ethyl-1,1',3,3'-tetrahydro-	EtOH	350s(2.09) 205s(4.79),240(4.06), 245s(4.04),263s(3.45), 269(3.48),273(3.50), 278(3.59),287(3.19)	78-0871-78 49-0987-78
$C_{20}H_{20}O_3$			
Furan, 2,5-bis(2-methoxy-5-methylphen- yl)-	EtOH	295(4.01),330(4.44)	102-1439-78
1H-Indene-1-carboxylic acid, 2,3-dihy- dro-4-(2,4,6-trimethylbenzoyl)-	EtOH	252(4.16)	94-1776-78
$C_{20}H_{20}O_4$			
1H-Indeno[1,2-c]furan-1-one, 3a-[(3,4- dimethoxyphenyl)methyl]-3,3a,4,8b- tetrahydro-	EtOH	267(3.32),273(3.48), 281(3.46)	94-1592-78
Spiro[furan-3(2H),2'-[2H]inden]-5(4H)- one, 4-(3,4-dimethoxyphenyl)-1',3'- dihydro-	EtOH	268(3.42),274(3.56), 280s(3.48)	94-1592-78
Tricyclo[9.3.1.1^{4,8}]hexadeca-1(15),4,6- 8(16),11,13-hexaene-5,14-dicarboxylic acid, dimethyl ester, (+)-	EtOH	215(4.54),245s(4.31), 255s(4.26),287(3.46)	78-0871-78
$C_{20}H_{20}O_5$			
9,10-Anthracenedione, 1,3-diethoxy- 6-methoxy-8-methyl-	EtOH	280(4.28),335(3.67), 404(3.45)	12-1363-78
9,10-Anthracenedione, 1,3-diethoxy- 7-methoxy-5-methyl-	EtOH	227(4.36),280(4.43), 290s(4.31),364(3.88)	12-1363-78
9,10-Anthracenedione, 1,3,6-triethoxy-	EtOH	221(4.44),280(4.55), 292s(4.32),337(3.68), 405(3.66)	12-1335-78
9,10-Anthracenedione, 1,3,7-triethoxy-	EtOH	228(4.42),280(4.45), 300s(4.31),359(3.89)	12-1335-78
2H,10H-Benzo[1,2-b:3,4-b']dipyran-10- one, 3,4,8,9-tetrahydro-5-hydroxy- 8-(4-hydroxyphenyl)-2,2-dimethyl-, (S)-	MeOH	272(4.18),328(3.48)	78-3563-78
4H,6H-Benzo[1,2-b:5,4-b']dipyran-4-one, 2,3,7,8-tetrahydro-5-hydroxy-2-(4-hy- droxyphenyl)-8,8-dimethyl-, (S)-	MeOH	235(4.18),298(4.19)	78-3563-78
4H,8H-Benzo[1,2-b:3,4-b']dipyran-4-one, 2,3,9,10-tetrahydro-5-hydroxy-2-(4- hydroxyphenyl)-8,8-dimethyl-, (S)-	MeOH	297(4.19)	78-3563-78
4H-1-Benzopyran-4-one, 2,3-dihydro- 5,7-dihydroxy-6-(3-methyl-2-butenyl)- 2-(4-hydroxyphenyl)- (sophoraflavan- one B)	EtOH EtOH-NaOEt EtOH-NaOH EtOH-NaOAc EtOH-AlCl$_3$	297(4.51) 249(4.77),338(4.77) 339(4.76) 299s(4.38),340(4.47) 314(4.50),395(3.74)	94-3863-78 94-3863-78 94-3863-78 94-3863-78 94-3863-78
Naringenin, 6-C-prenyl-	MeOH	285(4.27),340(3.75)	78-3563-78
Naringenin, 8-C-prenyl-	MeOH	292(4.23),324(3.85)	78-3563-78
$C_{20}H_{20}O_6$			
[2,3'-Bibenzofuran]-6,6'-diol, 2',3'- dihydro-4,4'-dimethoxy-3,3'-dimethyl-	EtOH	262(4.25),290(3.36)	12-1533-78
Colchicine, 7-oxodeacetamido-	MeOH	242(4.46),346(4.20)	33-1213-78
Isocolchicine, 7-oxodeacetamido-	MeOH	245(4.47),342(4.20)	33-1213-78
1-Phenanthrenol, 2,5,6,7-tetramethoxy-, acetate	n.s.g.	263(4.94),275(4.22), 307(4.00)	39-0739-78C
$C_{20}H_{20}O_6S$			
7-Oxabicyclo[4.1.0]heptane-2,3,4-triol,	MeOH	223(4.21),253(3.73),	44-0171-78

Compound	Solvent	$\lambda_{max}(\log \epsilon)$	Ref.
1-[(benzoyloxy)methyl]-5-(phenyl-thio)-, [1H-(1α,2β,3α,4β,5α,6α)]-(cont.)		270(3.42),280(3.19)	44-0171-78
$C_{20}H_{20}O_7$ 4H-1-Benzopyran-4-one, 7,8-dimethoxy-3-(2,3,4-trimethoxyphenyl)-	EtOH	251(4.57),294(3.99)	102-1401-78
$C_{20}H_{20}O_8$ 1,3-Benzodioxole, 5,5'-(4,8-dimethyl-1,2,5,6-tetroxocane-3,7-diyl)bis-	CHCl₃	290(4.0),320s(3.5)	104-2210-78
$C_{20}H_{20}O_{12}$ 9H-Xanthen-9-one, 7-(β-D-glucopyrano-syloxy)-1,5,6-trihydroxy-3-methoxy-	MeOH	260(4.53),298s(4.07), 308(4.11),330(4.21), 373s(3.07)	102-2119-78
	MeOH-NaOAc	265(4.36),355(4.26)	102-2119-78
	MeOH-NaOAc-H₃BO₃	262(4.58),280(4.23), 345(4.10),360(4.28)	102-2119-78
$C_{20}H_{20}S$ Cyclopenta[b]thiopyran, 2,3,5,6,7,7a-hexahydro-2,4-diphenyl-	hexane	207(4.3),213s(4.3), 235s(3.9),258(3.6), 263s(3.6),268s(3.6)	103-0605-78
$C_{20}H_{21}AsO$ 4(1H)-Arseninone, 2-phenyl-1,3,5-tri-2-propenyl-	EtOH	220(4.22),314(3.95)	88-1471-78
$C_{20}H_{21}BrO_5$ 11H-Dibenzo[b,e][1,4]dioxepin-11-one, 7-bromo-3-hydroxy-8-methoxy-1,9-di-methyl-6-(1-methylpropyl)-	EtOH	268(4.11)	39-0395-78C
$C_{20}H_{21}ClN_2OS$ 1-Piperidinecarboximidothioic acid, N-(4-chlorophenyl)-, 2-oxo-2-phenyl-ethyl ester	EtOH	238(4.42),245(4.42), 285s(4.10)	94-3017-78
$C_{20}H_{21}ClN_4$ 3H-1,5-Benzodiazepine, 7-chloro-4-(4-methyl-1-piperazinyl)-2-phenyl-	n.s.g.	262(4.57),356(3.68)	87-0952-78
3H-1,5-Benzodiazepine, 2-(4-chlorophen-yl)-4-(4-methyl-1-piperazinyl)-	n.s.g.	261(4.59),348(3.76)	87-0952-78
$C_{20}H_{21}Cl_2NO_4$ 8H-Pyrano[2,3-h]quinoline-2-carboxylic acid, 4-chloro-3-(2-chloro-2-propen-yl)-9,10-dihydro-5-methoxy-8,8-di-methyl-	MeOH	220(4.43),260(4.65)	24-0439-78
$C_{20}H_{21}FN_4$ 3H-1,5-Benzodiazepine, 2-(4-fluorophen-yl)-4-(4-methyl-1-piperazinyl)-	n.s.g.	259(4.53),341(3.70)	87-0952-78
$C_{20}H_{21}N$ 1H-Phenanthro[9,10-b]azirine, 1-cyclo-hexyl-1a,9b-dihydro-	CHCl₃	242(3.58),269(3.99), 274(4.00),280(3.99), 287(3.86),292(3.69), 305(3.50)	44-0397-78

Compound	Solvent	$\lambda_{max}(\log \epsilon)$	Ref.
$C_{20}H_{21}NO_3$			
4H-1-Benzopyran-4-one, 8-[(diethylamino)methyl]-7-hydroxy-2-phenyl-	n.s.g.	260(4.40),310(4.13)	103-0497-78
$C_{20}H_{21}NO_5$			
2H-Furo[3,2-c]pyrano[2,3-h]quinoline-6-carboxylic acid, 3,4-dihydro-10-methoxy-2,2,8-trimethyl-, methyl ester	MeOH	230(4.39),260(4.50), 272(4.45),398(3.50)	24-0439-78
Phenanthro[2,3-d][1,3]dioxol-6-amine, 1,2,3-trimethoxy-N,N-dimethyl-	EtOH	258(4.87),287(4.64)	44-3950-78
4H-Pyrano[2,3-h]quinoline-2-carboxylic acid, 1,8-dihydro-5-methoxy-8,8-dimethyl-4-oxo-3-(2-propenyl)-, methyl ester	MeOH	255(4.38),282(4.15), 320s(--),361(3.54)	24-0439-78
$C_{20}H_{21}NO_6$			
[1,1'-Biphenyl]-2,4-dicarboxylic acid, 3-hydroxy-4',5-dimethyl-6-nitroso-, diethyl ester	DMF	695(1.83)	104-2189-78
Carpoxidine	EtOH	231(4.16),283(3.77)	95-1243-78
$C_{20}H_{21}N_2$			
Quinolinium, 1-methyl-2-[2-[4-(dimethylamino)phenyl]ethenyl]-, iodide	EtOH	525(4.78)	4-0017-78
Quinolinium, 1-methyl-4-[2-[4-(dimethylamino)phenyl]ethenyl]-, iodide	EtOH	544(4.65)	4-0017-78
$C_{20}H_{21}N_3$			
3H-1,5-Benzodiazepine, 2-phenyl-4-piperidino-	n.s.g.	260(4.54),348(3.66)	87-0952-78
Ibogamine-18-carbonitrile, 3,4-didehydro-, (2α,5β,6α,18β)-	EtOH	222(4.47),275(3.86), 280(3.89),288(3.86)	33-0690-78
$C_{20}H_{21}N_3OS$			
4(5H)-Thiazolone, 5-[[4-(diethylamino)phenyl]methylene]-2-(phenylamino)-	MeOH dioxan 70% dioxan	239(4.23),433(4.68) 245(4.20),413(4.68) 248(4.19),429(4.68)	104-0997-78 104-0997-78 104-0997-78
$C_{20}H_{21}N_3O_2$			
1H-Cyclopropa[3,4]cyclopenta[1,2-d]-[1,2,4]triazolo[1,2-a]pyridazine-1,3(2H)-dione, 5,5b,6,6a,7,8-hexahydro-5b,6,6a-trimethyl-7-methylene-2-phenyl-, (5bα,6α,6aα)-	n.s.g.	255(4.05)	35-0860-78
1H-Cyclopropa[3,4]cyclopenta[1,2-d]-[1,2,4]triazolo[1,2-a]pyridazine-1,3(2H)-dione, 5b,6,6a,8-tetrahydro-5b,6,6a,7-tetramethyl-2-phenyl-, (5bα,6α,6aα)-	n.s.g.	307(4.34)	35-0860-78
$C_{20}H_{21}N_3O_3$			
Pyrimido[4,5-b]quinoline-2,4(3H,6H)-dione, 7,8,9,10-tetrahydro-10-(2-methoxyethyl)-3-phenyl-	EtOH	210(4.32),277(4.05), 380(4.03)	88-3469-78
$C_{20}H_{21}N_3O_5$			
4H-Cyclopent[d]isoxazole, 3a,5,6,6a-tetrahydro-6a-morpholino-3-[5-(2-nitrophenyl)-2-furanyl]-	MeOH	222(4.10),305(4.32)	73-0288-78

Compound	Solvent	$\lambda_{max}(\log \epsilon)$	Ref.
4H-Cyclopent[d]isoxazole, 3a,5,6,6a-tetrahydro-6a-morpholino-3-[5-(3-nitrophenyl)-2-furanyl]-	MeOH	219(4.20),238(4.18), 322(4.48)	73-0288-78
4H-Cyclopent[d]isoxazole, 3a,5,6,6a-tetrahydro-6a-morpholino-3-[5-(4-nitrophenyl)-2-furanyl]-	MeOH	227(4.08),289(4.02), 369(4.38)	73-0288-78
$C_{20}H_{21}N_3S$ 4H-1,3,4-Thiadiazine, 4,5-diphenyl-2-piperidino-	EtOH	210s(4.32),241(4.13), 267(4.30),295s(3.98), 320s(3.80),352s(3.49)	94-3017-78
$C_{20}H_{21}N_9O_4$ Pyrido[2,3-b]pyrazine, 1,2,3,5-tetrahydro-5-methyl-2,3-bis[2-(4-nitrophenyl)hydrazino]-, monohydroiodide	EtOH	264(3.90),374(4.49)	104-0398-78
$C_{20}H_{21}N_9O_6$ DL-Aspartic acid, N-[N-[4-[[(2,4-diamino-6-pteridinyl)methyl]amino]benzoyl]-glycyl]-	pH 1	240s(4.22),290(4.24)	87-1165-78
$C_{20}H_{22}$ Benzene, 1,1'-(3,3,4,4-tetramethyl-1-cyclobutene-1,2-diyl)bis-	MeCN	266s(4.04),277(4.09), 309s(3.89)	24-3608-78
2,2'-Spirobi[2H-indene], 5-ethyl-1,1',3,3'-tetrahydro-5'-methyl-	EtOH	205(4.33),215s(4.28), 261s(3.21),265(3.33), 270(3.48),273(3.54), 279(3.64)	49-0987-78
$C_{20}H_{22}BrNO_5$ 4H-Pyrano[2,3-h]quinoline-2-carboxylic acid, 3-(2-bromo-2-propenyl)-1,8,9-10-tetrahydro-5-methoxy-8,8-dimethyl-4-oxo-, methyl ester	MeOH	226(4.24),255(4.26), 280(4.03),330(3.67)	24-0439-78
$C_{20}H_{22}ClNO_5$ 4H-Pyrano[2,3-h]quinoline-2-carboxylic acid, 3-(2-chloro-2-propenyl)-1,8,9-10-tetrahydro-5-methoxy-8,8-dimethyl-4-oxo-, methyl ester	MeOH	226(4.45),254(4.38), 280(4.16),335(3.77)	24-0439-78
8H-Pyrano[2,3-h]quinoline-2-carboxylic acid, 4-chloro-9,10-dihydro-5-methoxy-8,8-dimethyl-3-(2-oxopropyl)-, methyl ester	MeOH	223(4.72),267(4.91)	24-0439-78
$C_{20}H_{22}N_2O$ Benzamide, N-[(1,2,3,7-tetramethyl-1H-indol-6-yl)methyl]-	CHCl$_3$	238(4.72),292(3.93)	103-0856-78
$C_{20}H_{22}N_2OS$ 1-Piperidinecarboximidothioic acid, N-phenyl-, 2-oxo-2-phenylethyl ester	EtOH	236(4.37),248(4.34), 348(3.72)	94-3017-78
$C_{20}H_{22}N_2O_2$ Benzamide, N-4-[(1-oxo-3-phenyl-2-propenyl)amino]butyl]-, (E)-	EtOH	218(4.31),224(4.34) 273(4.13),300s(3.69)	105-0637-78
Catharanthinic acid	EtOH	224(4.50),275(3.82), 282(3.85),291(3.79)	33-0690-78
1,3-Diazabicyclo[3.1.0]hex-3-ene, 2,6-bis(methoxymethyl)-4,5-diphenyl-, endo	EtOH	243(4.29)	35-4481-78

Compound	Solvent	λ_{max}(log ϵ)	Ref.
1,3-Diazabicyclo[3.1.0]hex-3-ene, 2,6-bis(methoxymethyl)-4,5-diphenyl-, exo	EtOH	243(4.29)	35-4481-78
16-Epipleiocarpamine	MeOH	228(4.22),288(3.75)	95-0950-78
1H-Pyrano[3,4-c]pyridin-1-one, 5-ethyl-5,6,7,8-tetrahydro-3-(1-methyl-1H-indol-2-yl)-7-methyl-, (±)-	EtOH	234(4.26),278(3.90), 365(4.29)	78-1363-78
8αH-Quinidinone	EtOH	210(4.66),240s(4.24), 261s(3.86),295s(3.34), 343(3.70)	35-0576-78
C$_{20}$H$_{22}$N$_2$O$_3$			
[1,4]Dioxino[2,3-g:6,5-h']diisoquino-line, 1,2,3,4,9,10,11,12-octahydro-6-methoxy-2-methyl-	EtOH	215(4.48),246(4.61), 302(3.60)	12-0321-78
Ibogamine-18-carboxylic acid, 3,4-di-hydroxy-, γ-lactone	EtOH	224(4.49),275(3.79), 282(3.83),289(3.79)	33-0690-78
C$_{20}$H$_{22}$N$_2$O$_4$			
Benzoic acid, 4-[[[[4-(ethoxycarbonyl)-phenyl]imino]methyl]methylamino]-, ethyl ester	isoPrOH	323(4.52)	117-0097-78
Indolo[2,3-a]quinolizine-1α,3-dicarbox-ylic acid, 1,2,6,7,12,12bα-hexahydro-2α-methyl-, dimethyl ester	EtOH	206s(4.26),223(4.43), 293(4.41)	78-2995-78
1β-	EtOH	206s(4.31),223(4.44), 293(4.42)	78-2995-78
4-Pyridineacetic acid, 1,4-dihydro-1-[2-(1H-indol-3-yl)ethyl]-3-(methoxy-carbonyl)-, methyl ester, (±)-	EtOH	223(4.52),284(3.86), 292(3.84),347(3.80)	78-2529-78
3,5-Pyridinedicarboxylic acid, 1,4-di-hydro-1-[2-(1H-indol-3-yl)ethyl]-4-methyl-, dimethyl ester	EtOH	205s(4.26),223(4.55), 259(4.07),284(3.90), 291(3.85),374(3.94)	78-2995-78
C$_{20}$H$_{22}$N$_2$O$_5$			
Butanoic acid, 2-[[4-[methyl[(phenyl-methoxy)carbonyl]amino]benzoyl]ami-no]-, (±)-	MeOH	259(4.12)	87-1162-78
Butanoic acid, 4-[[4-[methyl[(phenyl-methoxy)carbonyl]amino]benzoyl]ami-no]-	MeOH	258(4.17)	87-1162-78
Glycine, N-[4-[meethyl[(phenylmethoxy)-carbonyl]amino]benzoyl]-, ethyl ester	EtOH	270(4.21)	87-1162-78
C$_{20}$H$_{22}$N$_2$O$_6$			
1-Propanone, 2-hydroxy-1-(4-methoxy-phenyl)-3-morpholino-3-(4-nitrophen-yl)-	MeOH	256(3.38)	104-1174-78
C$_{20}$H$_{22}$N$_2$O$_7$S			
Thiocyanic acid, 2,3-dihydro-1-(2,3,4-tri-O-acetyl-α-L-arabinopyranosyl)-1H-indol-5-yl ester	EtOH	271(4.20)	104-2011-78
C$_{20}$H$_{22}$N$_2$O$_9$			
1H-Indole, 6-nitro-1-(2,3,4-tri-O-acet-yl-β-D-fucopyranosyl)-	EtOH	212(4.22),247(3.97), 318(3.91),352(3.85)	136-0017-78E
1H-Indole, 6-nitro-1-(2,3,4-tri-O-acet-yl-β-L-rhamnopyranosyl)-	EtOH	210(4.38),248(3.97), 320(3.88),357(3.83)	136-0017-78E
C$_{20}$H$_{22}$N$_4$			
3H-1,5-Benzodiazepine, 2-(4-methylpip-	n.s.g.	258(4.52),342(3.68)	87-0952-78

642    $C_{20}H_{22}N_4$–$C_{20}H_{22}N_4O_6$

Compound	Solvent	λ_{max}(log ϵ)	Ref.
erazino)-4-phenyl- (cont.) 2,6-Pyrazinediamine, N,N,N',N'-tetra-methyl-3,5-diphenyl-	EtOH	242(4.25),332(4.07), 381(4.20)	87-0952-78 94-1322-78
$C_{20}H_{22}N_4O_2$			
Pyrimido[4,5-b]quinolin-2(3H)-one, 4,6,7,8,9,10-hexahydro-4-imino-10-(2-methoxyethyl)-3-phenyl-	EtOH	213(4.32),281(4.04), 387(4.06)	88-3469-78
Urea, [3-cyano-5,6,7,8-tetrahydro-1-(2-methoxyethyl)-2(1H)-quinolinylidene]-phenyl-	EtOH	236(4.18),294(4.29), 385(3.94)	88-3469-78
Vobasan-17,22-dinitrile, 19,20-dihydro-3,15-dihydroxy-, (3α,16R,20α)-	MeOH	274(3.72),283(3.75), 292(3.66)	35-0938-78
(3β,16R,20α)-	MeOH	273(3.74),283(3.76), 292(3.67)	35-0938-78
(3β,20α)-	MeOH	275(3.78),283(3.80), 292(3.70)	35-0938-78
$C_{20}H_{22}N_4O_3$			
Pyrimido[4,5-d]pyrrolo[1,2-a]pyrimidin-7(5H)-one, 1,9a-dihydro-9-(2-hydroxy-ethyl)-8-(2-methoxyphenyl)-2,9a-di-methyl-	EtOH	234s(4.07),288(3.87)	94-0722-78
$C_{20}H_{22}N_4O_4$			
Butanamide, N-(4-butylphenyl)-2-[(2-ni-trophenyl)hydrazono]-3-oxo-	EtOH	207(4.42),233(4.15), 255(4.20),340(4.10), 405(4.23)	104-0521-78
Butanamide, N-(4-butylphenyl)-2-[(4-ni-trophenyl)hydrazono]-3-oxo-	EtOH	205(4.41),235(4.14), 258(4.16),395(4.59)	104-0521-78
$C_{20}H_{22}N_4O_4S$			
Benzeneacetamide, N-[(4-amino-2-methyl-5-pyrimidinyl)methyl]-2-methyl-N-[1-(1,2-oxathiolan-3-ylidene)ethyl]-α-oxo-, S-oxide	EtOH	236(4.14),267(4.14)	94-0722-78
$C_{20}H_{22}N_4O_5$			
6H-Purin-6-one, 1-[2,3-0-(1-methyleth-ylidene)-α-D-ribofuranosyl]-9-(phen-ylmethyl)-	pH 1	252(4.06)	4-0929-78
	pH 7	253(4.06)	4-0929-78
	pH 13	272(4.06)	4-0929-78
β-anomer	pH 1	252(4.06)	4-0929-78
	pH 7	252(4.04)	4-0929-78
	pH 13	271(4.06)	4-0929-78
1H-Pyrazole-3-carboxylic acid, 4,5-di-hydro-4-[(2-hydroxy-4,4-dimethyl-6-oxo-1-cyclohexen-1-yl)hydrazono]-5-oxo-1-phenyl-	MeOH	258(4.19),490(4.23)	142-0199-78B
$C_{20}H_{22}N_4O_5S$			
Benzeneacetamide, N-[(4-amino-2-methyl-5-pyrimidinyl)methyl]-2-methoxy-N-[1-(1,2-oxathiolan-3-ylidene)ethyl]-α-oxo-, S-oxide	EtOH	235s(4.16),268(4.11), 331(3.59)	94-0722-78
$C_{20}H_{22}N_4O_6$			
Acetamide, N-[7-(phenylmethoxy)-1-β-D-ribofuranosyl-1H-imidazo[4,5-b]pyri-din-5-yl]-	pH 1	263(3.74),289(3.88), 301s(3.72)	4-0839-78
	pH 11	253(3.76),284(3.79)	4-0839-78
Acetamide, N-[7-(phenylmethoxy)-3-β-D-ribofuranosyl-1H-imidazo[4,5-b]pyri-	pH 1	263(4.14),284(4.24)	4-0839-78
	pH 11	265(4.19),282(4.19)	4-0839-78

Compound	Solvent	$\lambda_{max}(\log \epsilon)$	Ref.
din-5-yl- (cont.)			4-0839-78
$C_{20}H_{22}O$			
Benzene, 1,1'-(5-methoxy-3,3-dimethyl-1,4-pentadienylidene)bis-, cis	EtOH	250(4.18)	78-1775-78
trans	EtOH	248(4.19)	78-1775-78
Benzene, 1,1'-[3-(2-methoxyethenyl)-2,2-dimethylcyclopropylidene]bis-, cis	EtOH	223(4.31)	78-1775-78
trans	EtOH	225(4.25)	78-1775-78
1,2-Cyclopenteno-9,10-dihydrophenanthrene, 7-methoxy-3',3'-dimethyl-	EtOH	282(4.43)	94-0171-78
$C_{20}H_{22}O_2$			
1-Benzoxepin, 2,5-dihydro-6-methoxy-3-methyl-8-(2-phenylethyl)-	EtOH	206(3.60),216s(3.44)	102-2005-78
1,1'-Biphenyl, 2,4'-dimethoxy-3',5-di-2-propenyl-	n.s.g.	257(4.04),290(3.80)	100-0442-78
Equilenin, 11α-methyl-, methyl ether	EtOH	235(5.83),254(3.79),265(3.76),275(3.76),286(3.57),307(3.01),321(3.26),328(3.25),335(3.38)	94-1533-78
Isoequilenin, 11α-methyl-, methyl ether	MeOH	230(4.84),254(3.94),264(3.83),274(3.82),285(3.60),304(3.00),318(3.29),327(3.25),334(3.38)	94-1533-78
4-Octyne-3,6-diol, 3,6-diphenyl-	EtOH	243s(2.30),248(2.42),252(2.53),258(2.60),264(2.45),268(2.11)	56-2377-78
Tetracyclo[9.6.0.0^{2,10}.0^{3,5}]heptadeca-6,8,12,14,16-pentaene-4-carboxylic acid, ethyl ester, endo	ether	228s(3.82),258s(3.54)	24-0107-78
exo	ether	228s(3.83),258s(3.54)	24-0107-78
Tetracyclo[9.6.0.0^{2,10}.0^{5,7}]heptadeca-3,8,12,14,16-pentaene-6-carboxylic acid, ethyl ester	ether	258(3.30)	24-0107-78
$C_{20}H_{22}O_2S$			
2H-Phenanthro[1,2-c]thiopyran-1(5H)-one, 12a-ethyl-6,11,12,12a-tetrahydro-8-methoxy-, (S)-	EtOH	320s(4.40),332(4.48),345s(4.37)	39-0576-78C
$C_{20}H_{22}O_3$			
2H-Phenanthro[1,2-c]pyran-1(5H)-one, 12a-ethyl-6,10,11,12a-tetrahydro-8-methoxy-, (S)-	EtOH	301s(4.37),314(4.45),327s(4.29)	39-0576-78C
$C_{20}H_{22}O_4$			
1,4-Butanedione, 1,4-bis(2-methoxy-5-methylphenyl)-	EtOH	220(4.14),250(4.60),320(3.07)	102-1439-78
Dipetaline	EtOH	229(4.62),264(4.44),272(4.30),294(4.52),315s(4.29)	78-1411-78
$C_{20}H_{22}O_5$			
Bacchotricuneatin A	MeOH	239(3.37)	44-3339-78
Bacchotricuneatin B	MeOH	240(3.19)	44-3339-78
Bacchotricuneatin C	MeOH	241(3.36)	44-3339-78

Compound	Solvent	λ_{max}(log ϵ)	Ref.
Benzenepentanoic acid, 4-methoxy-γ-(4-methoxyphenyl)-β-methyl-δ-oxo-	MeOH	219(4.19),272(4.19)	5-0726-78
2H,8H-Benzo[1,2-b:3,4-b']dipyran-8-one, 6-(3-hydroxy-3-methyl-1-butenyl)-5-methoxy-2,2-dimethyl- (avicennol)	EtOH	250(4.50),257(4.63), 301(4.27)	78-1411-78
6(2H)-Benzofuranone, 2-(1,3-benzodioxol-5-yl)-3,3a,4,5-tetrahydro-5-methoxy-3-methyl-3a-(2-propenyl)-, (2S)-	MeOH	255(4.10),285(3.58)	102-2038-78
Homoegonol	EtOH	217(4.38),313(4.30)	42-1204-78
Marrubiastrol, 1,2-dehydro-12-oxo-	MeCN	221(4.24),287(4.05)	24-2130-78
$C_{20}H_{22}O_6$			
8H-1,3-Dioxolo[4,5-g][1]benzopyran-8-one, 3a,4,9,9a-tetrahydro-4,9-di-hydroxy-2,2-dimethyl-6-(2-phenyleth-yl)-, [3aS-(3aα,4α,9α,9aα)]-	EtOH	252(3.92)	88-3921-78
Phenol, 2-methoxy-6-[2-(3,4,5-trimeth-oxyphenyl)ethenyl]-, acetate, (E)-	n.s.g.	235(4.36),312(4.45)	39-0739-78C
2-Propen-1-one, 3-(4-methoxyphenyl)-1-(2,3,4,6-tetramethoxyphenyl)-	EtOH	326(4.47)	18-3627-78
Spiro[furan-3(2H),1'(2'H)-naphthalene]-5'-carboxylic acid, 5-(3-furanyl)-3',4,4',4'a,5,7',8',8'a-octahydro-2'-methyl-2,4'-dioxo-, methyl ester, [1'R-[1'α(S*),2'α,4'aα,8'aα]]-	EtOH	210(4.03)	102-1967-78
Triethyleneglycol disalicylate	MeOH	238(4.27),305(3.94)	121-0661-78
$C_{20}H_{22}O_7$			
4H-1-Benzopyran-4-one, 2,3-dihydro-7,8-dimethoxy-3-(2,3,4-trimethoxy-phenyl)-, (+)-	EtOH	282(4.22)	102-1401-78
2-Propen-1-one, 3-(3,4-dimethoxyphen-yl)-1-(6-hydroxy-2,3,4-trimethoxy-phenyl)-	EtOH	260(4.11),375(4.38)	102-1807-78
$C_{20}H_{22}O_8$			
Leprolomin	n.s.g.	224(4.41),288(4.47), 338(3.95)	12-2057-78
$C_{20}H_{22}O_9$			
Benzoic acid, 2,4-dimethoxy-6-[(3,4,5-trimethoxybenzoyl)oxy]-, methyl ester	MeOH	256(3.85),342(3.52)	102-0689-78
$C_{20}H_{22}S$			
Cyclopenta[b]thiopyran, octahydro-2,4-diphenyl-	hexane	191(5.1),206(4.4), 246(2.6),253(2.6), 259(2.7),264(2.5)	103-0605-78
$C_{20}H_{23}ClN_2O_4$			
Piperidinium, 1-[3-(diphenylamino)-2-propenylidene]-, perchlorate	MeOH	230(4.07),344(4.65)	39-1453-78C
$C_{20}H_{23}NO$			
Benzamide, 4-methyl-N-(2-phenylcyclo-hexyl)-, (1S-trans)-	MeOH	206(4.37),234(4.12)	44-0355-78
Benzeneacetamide, N-(2-phenylcyclo-hexyl)-, (1S-trans)-	MeOH	206(4.29),252(2.72), 258(2.76),264(2.66)	44-0355-78
Cyclohexanecarboxamide, N-(4-methyl-phenyl)-2-phenyl-, (1S-trans)-	MeOH	206(4.42),245(4.15)	44-0355-78
Cyclohexanecarboxamide, 2-phenyl-N-(phenylmethyl)-, (1S-trans)-	MeOH	208(4.24),252(2.51), 258(2.58),264(2.46)	44-0355-78

Compound	Solvent	$\lambda_{max}(\log \epsilon)$	Ref.
1H-Indole, 2-[1,1-dimethyl-3-(phenyl-methoxy)propyl]-	MeOH	218(4.71),289(3.85)	88-0539-78
1-Propanone, 1,3-diphenyl-3-piperidino-	H_2O	248(4.12)	19-0819-78
$C_{20}H_{23}NO_2$			
9(10H)-Acridinone, 10-hexyl-2-methoxy-	MeOH	300(3.00)	4-0149-78
Benzamide, 4-methoxy-N-(2-phenylcyclo-hexyl)-, (1S-trans)-	MeOH	206(4.47),250(4.22)	44-0355-78
Cyclohexanecarboxamide, N-(4-methoxy-phenyl)-2-phenyl-, (1S-trans)-	MeOH	206(4.45),249(4.27)	44-0355-78
$C_{20}H_{23}NO_3$			
1-Propanone, 2-hydroxy-1-(4-methylphen-yl)-3-morpholino-3-phenyl-	MeOH	260(3.96)	104-1174-78
1-Propanone, 2-hydroxy-3-(4-methylphen-yl)-3-morpholino-1-phenyl-	MeOH	250(3.89)	104-1174-78
2-Propenamide, N-[2-ethoxy-2-(4-meth-oxyphenyl)ethyl]-3-phenyl-, (E)-	EtOH	218(4.08),224(4.08), 275(4.10)	102-1814-78
$C_{20}H_{23}NO_3S_2$			
4-Piperidinone, 3,5-bis[[5-(ethylthio)-2-furanyl]methylene]-1-methyl-	EtOH	263(4.09),417(4.54)	133-0189-78
$C_{20}H_{23}NO_4$			
2H,5H-Dibenzo[b,d]pyrrolo[1,2-a]azepin-2-one, 6,7,9,10-tetrahydro-3,12,13-trimethoxy-	EtOH	230(4.16),260(3.90), 283(3.79),326(3.62)	44-4464-78
Dibenzo[a,i]quinolizin-13(1H)-one, 2,3,5,6-tetrahydro-8,9,12-trimethoxy-	EtOH	238(4.46),280(3.69)	44-4464-78
Isocorypalmine, (-)-	EtOH	218(3.71),231s(3.67), 286(3.24),295s(3.10)	100-0169-78
2,6-Phenanthrenediol, 8-[2-(dimethyl-amino)ethyl]-3,5-dimethoxy-, hydrio-dide	MeOH	218(--),263(4.80), 280s(4.47),304(4.04), 317(4.06),345(3.12), 363(2.81)	12-0313-78
1-Propanone, 2-hydroxy-1-(4-methoxy-phenyl)-3-morpholino-3-phenyl-	MeOH	275(4.15)	104-1174-78
1-Propanone, 2-hydroxy-3-(4-methoxy-phenyl)-3-morpholino-1-phenyl-	MeOH	240(4.10)	104-1174-78
$C_{20}H_{23}NO_5$			
1,3-Azulenedicarboxylic acid, 2-hy-droxy-6-pyrrolidino-, diethyl ester	MeOH	266(4.44),348(4.77), 402(4.25),431(4.22)	18-3087-78
Isocorydine, N-oxide	EtOH	223(4.39),271(3.95), 306(3.96)	105-0360-78
4H-Pyrano[2,3-h]quinoline-2-carboxylic acid, 1,8,9,10-tetrahydro-5-methoxy-8,8-dimethyl-4-oxo-3-(2-propenyl)-, methyl ester	MeOH	225(4.39),252(4.35), 334(3.70)	24-0439-78
3,5-Pyridinedicarboxylic acid, 4-(4-methoxyphenyl)-2,6-dimethyl-, diethyl ester, anion	DMSO	453(3.76)	103-1226-78
Thalicmidine, N-oxide	EtOH	222(4.49),282(4.04), 306(4.06)	105-0360-78
$C_{20}H_{23}NO_5S_2$			
4-Piperidinone, 3,5-bis[[5-(ethylsul-finyl)-2-furanyl]methylene]-1-methyl-	EtOH	247(4.20),279(4.09), 324(3.93),389(3.65)	133-0189-78
$C_{20}H_{23}NO_6$			
2-Azabicyclo[2.2.2]oct-7-ene-5,6-di-	MeCN	259(3.27)	78-2617-78

Compound	Solvent	$\lambda_{max}(\log \epsilon)$	Ref.
carboxylic acid, 2-(1-methoxy-2-phenylethyl)-3-oxo-, dimethyl ester (cont.)			78-2617-78
1,3-Azulenedicarboxylic acid, 2-hydroxy-6-morpholino-, diethyl ester	MeOH	234(4.35),265(4.43), 348(4.74),405(4.23), 430s(4.10)	18-3087-78
$C_{20}H_{23}NO_7S$			
1H-Indole, 5-(methylthio)-1-(2,3,4-tri-O-acetyl-α-L-arabinopyranosyl)-	EtOH	227(4.44),240(4.23), 251(4.21)	104-2011-78
$C_{20}H_{23}NO_7S_2$			
4-Piperidinone, 3,5-bis[[5-(ethylsulfonyl)-2-furanyl]methylene]-1-methyl-	EtOH	211(4.05),282(4.05), 340(4.13),357(4.32)	133-0189-78
$C_{20}H_{23}N_2$			
Piperidinium, 1-[3-(diphenylamino)-2-propenylidene]-, perchlorate	MeOH	230(4.07),344(4.65)	39-1453-78C
$C_{20}H_{23}N_3$			
1H-Pyrrole, 5-[2-(3,5-dimethyl-1H-pyrrol-2-yl)ethenyl]-2-[(3,5-dimethyl-2H-pyrrol-2-ylidene)methyl]-3-methyl-hydrochloride	EtOH	518(4.63)	5-0289-78
	EtOH	625(4.93)	5-0289-78
$C_{20}H_{23}N_3O_2$			
Butanamide, 2-[(4-butylphenyl)hydrazono]-3-oxo-N-phenyl-	EtOH	204(4.39),230(4.19), 255-270s(--),380(4.56)	104-0521-78
Butanamide, N-(4-butylphenyl)-3-oxo-2-(phenylhydrazono)-	EtOH	204(4.42),245(4.29), 254(4.27),270(4.20), 384(4.53)	104-0521-78
$C_{20}H_{23}N_3O_4$			
1H-Pyrrolo[3,4-d]pyrimidine-7-carboxylic acid, 2,3,4,6-tetrahydro-1,5,6-trimethyl-2,4-dioxo-3-phenyl-, 1,1-dimethylethyl ester	EtOH EtOH-NaOH	243(4.52),250(4.10) 243(--),281(--)	94-3080-78 94-3080-78
$C_{20}H_{23}N_3O_4S$			
Benzenesulfonic acid, 4-methyl-, [1-(1,2,3,3a-tetrahydro-7-hydroxy-5-methylpyrrolo[2,1-b]benzoxazol-8-yl)ethylidene]hydrazide	EtOH	238(4.30),337(4.04), 413(3.41)	44-4472-78
$C_{20}H_{23}N_3O_5S$			
Benzenesulfonic acid, 4-methyl-, [1-(1,3,4,10a-tetrahydro-7-hydroxy-9-methyl-1,4-oxazino[3,4-b]benzoxazol-6-yl)ethylidene]hydrazide	EtOH	227(4.21),316(4.00), 400(3.65)	44-4472-78
$C_{20}H_{23}N_5O$			
9H-Imidazo[1,2-a]purin-9-one, 7-butyl-1,4-dihydro-4,6-dimethyl-1-(phenylmethyl)- (spectra in 10% methanol)	pH 1 pH 6.5	230(4.56),284(3.88) 229(4.47),265(3.79), 315(3.72)	44-1644-78 44-1644-78
	pH 11	230(4.45),266(3.79), 313(3.69)	44-1644-78
$C_{20}H_{23}N_5O_8$			
9H-Imidazo[1,2-a]purin-9-one, 3,4-dihydro-4,6-dimethyl-3-(2,3,5-tri-O-acetyl-β-D-ribofuranosyl)-	MeOH	234(4.24),288(3.72)	88-2579-78

Compound	Solvent	$\lambda_{max}(\log \epsilon)$	Ref.
$C_{20}H_{23}N_7$			
Pyrido[2,3-b]pyrazine, 1,2,3,5-tetrahydro-5-methyl-2,3-bis(2-phenylhydrazino)-, monohydriodide	EtOH	292(4.38),319(4.12), 346(4.20)	104-0398-78
$C_{20}H_{24}$			
1,3,7,9-Cyclododecatetrayne, 5,5,6,6-11,11,12,12-octamethyl-	hexane	237(2.84),248(2.88), 264(2.69),288s(1.48)	35-0692-78
	isooctane	239(2.86),249(2.91), 264(2.71)	44-0168-78
	EtOH	250(3.04),265(2.88)	44-0168-78
[2.2]Metaparacyclophane, 4,6,12,15-tetramethyl-	THF	279(2.72),288(2.73)	18-2668-78
$C_{20}H_{24}BrNO_4$			
3,5-Pyridinedicarboxylic acid, 4-(3-bromophenyl)-1,4-dihydro-1,2,6-trimethyl-, diethyl ester	EtOH	203(4.35),242(4.05), 355(3.64)	103-1226-78
3,5-Pyridinedicarboxylic acid, 4-(4-bromophenyl)-1,4-dihydro-1,2,6-trimethyl-, diethyl ester	EtOH	203(4.35),223(4.26), 244(4.30),353(3.86)	103-1226-78
$C_{20}H_{24}ClN_3OS$			
Prochlorperazine sulfoxide	MeOH	238(4.54),277(4.12), 302(4.01),344(3.82)	133-0248-78
$C_{20}H_{24}NO_2$			
4H-Dibenzo[de,g]quinolinium, 5,6,6a,7-tetrahydro-10,11-dimethoxy-6,6-dimethyl-, (R)-, methyl sulfate	MeOH	214(--),235s(4.07), 271(4.12),305s(3.4)	12-0313-78
$C_{20}H_{24}N_2O$			
Aristoteline	MeOH	290(3.70)	77-0079-78
1H-Cyclobuta[jk]phenanthren-4-ol, 2,3,3a,4,4a,5,9b,9c-octahydro-4-(1H-imidazol-2-yl)-3a,9b-dimethyl-	MeOH	259(2.86),266(2.94), 274(2.94)	33-2843-78
1H-Cyclobuta[jk]phenanthren-4-ol, 2,3,3a,4,4a,5,9b,9c-octahydro-4-(1H-imidazol-4-yl)-3a,9b-dimethyl-	MeOH	259(2.86),266(2.94), 274(2.94)	33-2843-78
Deoxyquinine-deoxyquinidine (mixture)	EtOH	231(4.55),268s(3.52), 279(3.59),290s(3.51), 321(3.63),333(3.68)	35-0576-78
Methanone, 1H-imidazol-2-yl(1,2,3,4,4a-9,10,10a-octahydro-1,4a-dimethyl-1-phenanthrenyl)-, [1R-(1α,4aα,10aα)]-	MeOH	279(3.99)	33-2843-78
Methanone, 1H-imidazol-4-yl(1,2,3,4,4a-9,10,10a-octahydro-1,4a-dimethyl-1-phenanthrenyl)-	MeOH	260(3.98)	33-2843-78
$C_{20}H_{24}N_2O_2$			
Benzoic acid, 4-[[(butylphenylamino)-methylene]amino]-, ethyl ester	isoPrOH	312(4.43)	117-0097-78
2-Propanone, 1-(3-ethenyl-4-piperidinyl)-3-(6-methoxy-4-quinolinyl)-, (3R-cis)-	pH 1	248(4.51),314(3.69), 344(3.72)	35-0576-78
	EtOH	234(4.54),270s(3.52), 278(3.57),290s(3.51), 323s(3.66),332(3.69)	35-0576-78
Quinidine	EtOH	229(4.53),269s(3.54), 279(3.61),288s(3.54), 320(3.67),332(3.72)	35-0576-78

Compound	Solvent	$\lambda_{max}(\log \epsilon)$	Ref.
Quinine	EtOH	230(4.48),268s(3.51), 278(3.55),289s(3.49), 319(3.62),331(3.66)	35-0576-78
Quinoline, 4-[3-[(3-ethenyl-4-piperi-dinyl)methyl]oxiranyl]-6-methoxy-	EtOH	231(4.45),279(3.50), 290s(3.44),320s(3.59), 333(3.65)	35-0576-78
Tricyclo[8.2.2.2^{4,7}]hexadeca-4(16),6-10,12,13-pentaene-5,15-dione, 11,13-bis(dimethylamino)-, pseudo-geminal	C_6H_{12} CH_2Cl_2	531(3.22) 577(3.29)	89-0374-78 89-0374-78
pseudo-ortho-	C_6H_{12} CH_2Cl_2	545(2.17) 595(2.20)	89-0374-78 89-0374-78
$C_{20}H_{24}N_2O_3$ Perivine, 19,20(S)-dihydro-	MeOH	235(4.10),311(4.13)	35-0938-78
16-epi-	MeOH	274(3.65),280(3.66), 288(3.64),312(3.54)	35-0938-78
Vobasan-17-oic acid, 4-demethyl-3-hy-droxy-, methyl ester, (3β)-	MeOH	272(3.92),282(3.92), 287(3.82)	35-0938-78
$C_{20}H_{24}N_2O_3S$ 4H-Carbazol-4-one, 1,2,3,5,6,7,8,9-octahydro-9-methyl-, O-[(4-methyl-phenyl)sulfonyl]oxime	MeOH	221(4.28),259(3.89)	24-0853-78
$C_{20}H_{24}N_2O_4$ 2H-Pyrido[3,4-e]-1,3-oxazine-4,5(3H-6H)-dione, 3-butyl-7-methoxy-2,2-dimethyl-6-phenyl-	MeOH	309(4.4)	120-0101-78
$C_{20}H_{24}N_2O_5$ Furo[3,4-b]pyridine-3-carboxylic acid, 1,4,5,7-tetrahydro-2,5,5,7,7-penta-methyl-4-(3-nitrophenyl)-, methyl ester	EtOH	206(4.38),266(4.01), 328(3.79)	95-0448-78
5,8-Isoquinolinediol, 7-methoxy-6-meth-yl-3-piperidino-, diacetate	EtOH	217(4.46),255(4.63), 294(4.23)	94-2175-78
$C_{20}H_{24}N_2O_5S$ 7-Oxabicyclo[4.2.0]octane-5,5-diaceto-nitrile, 6-methoxy-1-[[[(4-methyl-phenyl)sulfonyl]oxy]methyl]-	MeOH	256(2.77),263(2.84), 268(2.77),273(2.68)	97-0380-78
$C_{20}H_{24}N_2O_8$ Phenol, 4,4'-(2,3-dimethyl-1,3-butane-diyl)bis[2-methoxy-6-nitro-, (R*,S*)-	EtOH	225(4.32),254s(3.87), 296(3.90)	94-0682-78
$C_{20}H_{24}N_2O_9$ 1H-Indole, 2,3-dihydro-6-nitro-1-(2,3,4-tri-O-acetyl-β-D-fuco-pyranosyl)-	EtOH	252(4.19)	136-0017-78E
$C_{20}H_{24}N_4O_3S$ Benzeneacetamide, N-[(4-amino-2-meth-yl-5-pyrimidinyl)methyl]-N-[4-hy-droxy-1-methyl-2-(methylthio)-1-butenyl]-α-oxo-	EtOH	249(4.22),256(4.24)	94-0722-78
2H-1,4-Thiazin-3(4H)-one, 4-[(4-amino-2-methyl-5-pyrimidinyl)methyl]-2-hy-droxy-6-(2-hydroxyethyl)-5-methyl-2-(2-methylphenyl)-	EtOH	231(4.2),277(4.03)	94-0722-78

Compound	Solvent	$\lambda_{max}(\log \epsilon)$	Ref.
$C_{20}H_{24}N_4O_4$			
5H-Furo[2,3-b]pyrrol-5-one, 6-[(4-amino-2-methyl-5-pyrimidinyl)methyl]-hexahydro-4-hydroxy-4-(2-methoxyphenyl)-6a-methyl-	EtOH	221s(4.04),235s(3.88), 277(3.81)	94-0722-78
2H-Pyrrol-2-one, 1-[(4-amino-2-methyl-5-pyrimidinyl)methyl]-1,3-dihydro-3-hydroxy-4-(2-hydroxyethyl)-3-(2-methoxyphenyl)-5-methyl-	EtOH	224(4.19),277(3.99)	94-0722-78
$C_{20}H_{24}N_8O_3$			
DL-Valine, N-[4-[[(2,4-diamino-6-pteridinyl)methyl]methylamino]benzoyl]-	pH 1	244(4.24),308(4.17)	87-1162-78
$C_{20}H_{24}N_{10}O_8S_2$			
3H-1,2,3-Triazolo[4,5-b]pyridin-5-amine, 7,7-dithiobis[3-β-D-ribofuranosyl-	pH 1	282(4.36),319(4.35)	44-4910-78
	pH 11	245(4.26),316(4.40)	44-4910-78
	EtOH	240(4.36),277(4.23), 321(4.28)	44-4910-78
$C_{20}H_{24}OS$			
Benz[c]thiophene, 1,3-dihydro-1-methoxy-3,3-dimethyl-1-(2,4,6-trimethylphenyl)-	hexane	265s(3.07),271s(2.89), 282s(2.52)	104-0582-78
$C_{20}H_{24}O_2$			
2H-1-Benzopyran, 3,4-dihydro-5-methoxy-2,2-dimethyl-7-(2-phenylethyl)-	n.s.g.	208(3.84),232s(3.24)	102-2115-78
Isobenzofuran, 1,3-dihydro-1-methoxy-3,3-dimethyl-1-(2,4,6-trimethylphenyl)-	hexane	256(2.98),262(3.03), 269(3.01),278s(2.74)	104-0582-78
3-Penten-2-one, 3-[3,8-dimethyl-5-(1-methylethyl)-1-azulenyl]-4-hydroxy-, (Z)-	hexane	245(4.42),287(4.65), 307s(4.34),354(3.82), 372(3.76),612(2.69), 666s(2.60)	18-3582-78
Phenol, 3-methoxy-2-(3-methyl-2-butenyl)-5-(2-phenylethyl)-	n.s.g.	211(3.78),230s(3.48), 270(2.63)	102-2115-78
$C_{20}H_{24}O_2S$			
2H-Phenanthro[1,2-c]thiopyran-1-ol, 12a-ethyl-1,5,6,11,12,12a-hexahydro-8-methoxy-, (1R-cis)-	EtOH	316s(4.43),330(4.56), 345(4.45)	39-0576-78C
(1S-trans)-	EtOH	315s(4.44),330(4.57), 344(4.45)	39-0576-78C
$C_{20}H_{24}O_3$			
2H-1-Benzopyran-2-ol, 3,4-dihydro-2,5,7,8-tetramethyl-6-(phenylmethoxy)-, (±)-	EtOH	226(4.08),258(2.95), 265(3.00),278(3.27), 285(3.31)	33-0837-78
1H-Cyclopropa[3,4]benz[1,2-f]azulene-3,5,8(2H)-trione, 1a,3a,4,6,7,9b-hexahydro-1,1,3a,6,9-pentamethyl-, [1aS-(1aα,3aα,6α,9bα)]-	n.s.g.	223(3.98),272(3.87), 318(3.82)	78-0233-78
D-Norestra-1,3,5(10),9(11)-tetraene-16-carboxylic acid, 3-methoxy-, methyl ester, (16α)-	EtOH	214(4.30),264(4.30), 297(3.51)	22-0119-78
(16β)-	EtOH	214(4.30),263(4.30), 297(3.51)	22-0119-78
2H-Phenanthro[1,2-c]pyran-1-ol, 12a-ethyl-1,5,6,11,12,12a-hexahydro-8-methoxy-, (1R-cis)-	EtOH	300s(4.37),312(4.45), 324s(4.33)	39-0576-78C

Compound	Solvent	λ_{max}(log ϵ)	Ref.
$C_{20}H_{24}O_3S$			
2H-Thiopyran-3,5(4H,6H)-dione, 4-[2-(3,4-dihydro-6-methoxy-1(2H)-naphthalenylidene)ethyl]-4-ethyl-	EtOH	267(4.23),300s(3.69)	39-0576-78C
$C_{20}H_{24}O_4$			
8α-Estra-1,4-diene-3,17-dione, 10ξ-acetoxy-	MeOH	246(4.10)	24-0939-78
1,2-Naphthalenedicarboxylic acid, 3,4,5,6,7,8-hexamethyl-, dimethyl ester	EtOH	239(4.35),260(4.66), 326(3.72)	5-1675-78
1H-Naphtho[1,8a-c]furan-3(5H)-one, 7-[2-(2,5-dihydro-5-oxo-3-furanyl)ethenyl]-6,6a,7,8,9,10-hexahydro-7,8-dimethyl-, [6aR-[6aα,7α(E),8β,10aR*]]-	n.s.g.	214(4.07),257(4.25)	24-2130-78
2H-Pyran-3,5(4H,6H)-dione, 4-[2-(3,4-dihydro-6-methoxy-1(2H)-naphthalenylidene)ethyl]-4-ethyl-	EtOH	268(4.20),300s(3.66)	39-0576-78C
$C_{20}H_{24}O_4S$			
A-Nor-1-thiaestra-2,5(10),9(11)-triene-6,17-diol, diacetate, (6α,17β)-	EtOH	283(4.10)	142-0207-78B
$C_{20}H_{24}O_5$			
Ballonigrin, 18-hydroxy-	n.s.g.	259(3.97)	39-1271-78C
Benzene, 1,2,3,4-tetramethoxy-5-[3-(2-methoxyphenyl)-2-propenyl]-, (E)-	EtOH	261(3.76),280(3.72), 304(3.40)	102-1379-78
2-Butenoic acid, 2-methyl-, 2a,3,4,5-5a,6,9b,9c-octahydro-9,9c-dimethyl-2-oxo-2H-naphtho[1,8-bc:3,2-b']difuran-3-yl ester	EtOH	216(4.20)	138-1313-78
1H-Naphtho[1,8a-c]furan-3(5H)-one, 7-[2-(2,5-dihydro-5-oxo-3-furanyl)-2-oxoethyl]-6,6a,7,8,9,10-hexahydro-7,8-dimethoxy-	n.s.g.	211s(4.15),221(4.26)	24-2130-78
$C_{20}H_{24}O_6$			
5-Benzofuranpropanol, 2,3-dihydro-2-(4-hydroxy-3-methoxyphenyl)-3-(hydroxymethyl)-7-methoxy-	MeOH	238(4.10),284(3.72)	94-1619-78
Dibenzo[b,k][1,4,7,10,13,16]hexaoxacyclooctadecin, 6,7,9,10,17,18,20,21-octahydro-	MeCN	250(2.20)	88-0623-78
Euponin	EtOH	210(4.28)(end abs.)	142-0117-78B
3-Furanmethanol, tetrahydro-2-(4-hydroxy-3-methoxyphenyl)-4-[(4-hydroxy-3-methoxyphenyl)methyl]-, [2S-(2α,3β,4β)]-	MeOH	231(3.99),283(3.62)	94-1619-78
$C_{20}H_{24}O_7$			
4-Hexenoic acid, 6-(4-acetoxy-1,3-dihydro-6-methoxy-7-methyl-3-oxo-5-isobenzofuranyl)-4-methyl-, methyl ester, (E)-	MeOH	248(3.99),282(3.35), 290(3.36)	12-0353-78
$C_{20}H_{24}O_{10}$			
Leptophylloside	MeOH	265(3.48),335(3.79)	2-0563-78
	MeOH-NaOMe	280(--),345(--)	2-0563-78
	MeOH-NaOAc	280(--),340(--)	2-0563-78

Compound	Solvent	$\lambda_{max}(\log \epsilon)$	Ref.
$C_{20}H_{24}O_{13}$			
2H-1-Benzopyran-2-one, 7-[(6-O-D-apio-β-D-furanosyl-β-D-glucopyranosyl)-oxy]-6-hydroxy- (diospyroside)	EtOH-H$_2$O	228(4.44),292(4.09), 345(4.21)	88-4783-78
	+base	247(4.49),279(4.08), 307(4.11),395(3.98)	88-4783-78
$C_{20}H_{25}ClO_4$			
1H-Naphtho[1,8a-c]furan-3(5H)-one, 7-[2-chloro-2-(2,5-dihydro-5-oxo-3-furanyl)ethyl]-6,6a,7,8,9,10-hexa-hydro-7,8-dimethyl-, (R)-(S)-	MeCN	212(4.25)	24-2130-78
	MeCN	212(4.23)	24-2130-78
$C_{20}H_{25}FO_5S$			
β-D-arabino-Hept-1-en-3-ulopyranose, 2-deoxy-1-C-fluoro-3,4:5,6-bis(O-1-methylethylidene)-1-S-(4-methyl-phenyl)-1-thio-, (E)-	EtOH	210(3.83),238(3.83), 254s(--)	136-0564-78H
(Z)-	EtOH	210(3.92),238(3.88), 256s(--)	136-0564-78H
$C_{20}H_{25}NO_2$			
2H-Furo[2,3-b]indol-2-one, 3,3a,8,8a-tetrahydro-3a,8-bis(3-methyl-2-but-enyl)-	EtOH	248(3.98),298(3.49)	78-0929-78
$C_{20}H_{25}NO_4$			
6-Aza-9,10-secoestra-1,3,5(10)-triene-9,17-dione, N-acetyl-2β-methoxy-, (±)-	EtOH	225(4.02),267(3.15)	65-0841-78
Gossyrubilone	EtOH	217(4.32),236(4.27), 270(4.34),352(3.90), 400s(--),514(3.43)	102-1297-78
Propanedioic acid, (1'-methylspiro[cy-clopentan-1,3'[3H]indol]-2'(1'H)-yl-idene)-, diethyl ester	EtOH	207(3.79),237(3.91), 302(3.57),342(4.02)	44-3702-78
1H-Pyrrole-3-carboxylic acid, 1-cyclo-hexyl-4,5-dihydro-5-hydroxy-5-methyl-4-oxo-2-phenyl-, ethyl ester	EtOH	208(3.99),245(4.19), 330(3.99)	118-0291-78
$C_{20}H_{25}NO_6$			
α-D-gluco-Pentofuranose, 1,2-O-(1-meth-ylethylidene)-5-C-[3-(2,4,6-trimeth-ylphenyl)-5-isoxazolyl]-	MeOH	222(4.18),272(2.81)	136-0357-78H
β-L-ido-Pentofuranose, 1,2-O-(1-methyl-ethylidene)-5-C-[3-(2,4,6-trimethyl-phenyl)-5-isoxazolyl]-	MeOH	222(4.12),272(2.75)	136-0357-78H
$C_{20}H_{25}N_2O_6$			
Isoquinolinium, 5,8-diacetoxy-7-meth-oxy-2,6-dimethyl-3-morpholino-, iodide	EtOH	220(4.56),271(4.60), 358(3.28)	94-2175-78
$C_{20}H_{25}N_3O_2$			
7,8-Isoquinolinedione, 6-methyl-3,5-di-piperidino-	EtOH	249(4.35),324(3.99), 435(4.26)	94-2175-78
$C_{20}H_{25}N_3O_5$			
1H-Pyrrole-2,4-dicarboxylic acid, 5-methyl-3-[[(phenylamino)carbonyl]-amino]-, 2-(1,1-dimethylethyl) 4-ethyl ester	EtOH	242(4.54),277(4.19)	94-3080-78

Compound	Solvent	$\lambda_{max}(\log \epsilon)$	Ref.
$C_{20}H_{25}N_5O$			
1H-Pyrazolo[3,4-d]pyrimidine-4-carbox-amide, N-(2-ethylhexyl)-1-phenyl-	EtOH	250(4.54),275s(4.05), 329(3.34)	95-1274-78
$C_{20}H_{26}NO$			
Benzenemethanaminium, 4-benzoyl-N,N,N-triethyl-, chloride	pH 8	225(4.32),340(2.30)	149-0007-78B
$C_{20}H_{26}N_2$			
9H-Pyrido[3,4-b]indole, 1-nonyl-	MeOH	236(4.65),241(4.64), 251s(4.46),283s(4.08), 289(4.33),338(3.74), 351(3.75)	95-0898-78
$C_{20}H_{26}N_2O$			
Epivincanol, O-methyl-	MeOH	196(4.34),227(4.49), 278(3.91)	33-1682-78
Vincanol, O-methyl-	MeOH	196(4.36),227(4.50), 278(3.91)	33-1682-78
$C_{20}H_{26}N_2O_2$			
Catharanthinol, 3,4-dihydro-4(S)-hydr-oxy-	MeOH	226(4.51),278(3.86), 283(3.88),290(3.84)	33-0690-78
Quinidine, dihydro-	EtOH	233(4.51),279(3.60), 320(3.65),332(3.71)	35-0589-78
racemic	EtOH	231(4.52),279(3.58), 320(3.64),332(3.70)	35-0589-78
Quinine, dihydro-	EtOH	231(4.43),280(3.49), 333(3.62)	35-0589-78
racemic	EtOH	232(4.54),280(3.59), 321(3.66),333(3.71)	35-0589-78
4-Quinolineethanol, α-[(3(R)-ethenyl-4(S)-piperidinyl)methyl]-6-methoxy-	EtOH	230(4.68),270s(3.60), 279(3.66),289s(3.61), 319(3.72),332(3.77)	35-0576-78
$C_{20}H_{26}N_2O_3$			
Catharanthinol, 3,4-dihydro-3,4-dihy-droxy-, (3R,4S)-	MeOH	224(4.56),275(3.89), 282(3.91),290(3.86)	33-0690-78
Furo[3,4-b]pyridine-3-carboxylic acid, 1,4,5,7-tetrahydro-2,5,5,7,7-penta-methyl-4-(2-pyridinyl)-, ethyl ester	EtOH	259s(3.81),264(3.99), 270(3.92),344(3.72)	95-0448-78
Furo[3,4-b]pyridine-3-carboxylic acid, 1,4,5,7-tetrahydro-2,5,5,7,7-penta-methyl-4-(3-pyridinyl)-, ethyl ester	EtOH	259s(3.81),263(3.83), 269s(3.73),339(3.76)	95-0448-78
Furo[3,4-b]pyridine-3-carboxylic acid, 1,4,5,7-tetrahydro-2,5,5,7,7-penta-methyl-4-(4-pyridinyl)-, ethyl ester	EtOH	259(3.63),261(3.66), 274(3.62),343(3.69)	95-0448-78
$C_{20}H_{26}N_2O_3S$			
Acetic acid, [3-ethyl-4-oxo-5-phenyl-5-piperidino-2-thiazolidinylidene]-, ethyl ester, (Z)-	MeOH	246(3.99),286(4.29)	5-0473-78
$C_{20}H_{26}N_2O_4$			
Butanedioic acid, [[[(diethylamino)-phenylmethylene]amino]methylene]-ethylidene-, dimethyl ester, (E,Z,Z)-	MeOH	330(4.5)	78-0833-78
$C_{20}H_{26}N_2S$			
Vincanol, 14-deoxy-14-(methylthio)-	MeOH	232(4.48),284(3.92)	33-1682-78

Compound	Solvent	$\lambda_{max}(\log \epsilon)$	Ref.
$C_{20}H_{26}N_4O_3$			
Benzoic acid, 4-[[4-cyano-5-(methoxyimino)-1,3-pentadienyl]amino]-, 2-(diethylamino)ethyl ester, (?,Z,E)-	MeOH	402(4.57)	83-0433-78
$C_{20}H_{26}N_4S$			
Propanedinitrile, [[3-(diethylamino)-(methylthio)methylene]-1,4-cyclopentadien-1-yl]-1-piperidinylmethylene]-	MeOH	275(4.12),311(4.16), 416(4.47)	83-0369-78
$C_{20}H_{26}O$			
Estra-1,3,5(10),8-tetraene, 3-methoxy-17β-methyl-	EtOH	278(4.22)	22-0119-78
Phenanthro[3,2-b]furan, 1,2,3,4,4a,5,6-11b-octahydro-4,4,8,11b-tetramethyl-, (4aS-trans)-	EtOH	253(4.13),286(3.74), 295(3.71)	23-0517-78
$C_{20}H_{26}O_2$			
1H-Cyclopropa[3,4]benz[1,2-f]azulene-5,8-dione, 1a,3a,4,4a,6,7,7a,9b-octahydro-1,1,3a,6,9-pentamethyl-, [1aS-(1aα,3aα,6α,9bα)]-	n.s.g.	200(3.63),260(3.71)	78-0233-78
Estra-1,3,5(10),9(11)-tetraen-17β-ol, 3-methoxy-17α-methyl-	MeOH	263(4.29)	94-0171-78
Estra-4,9,11-trien-3-one, 17β-hydroxy-2β,17α-dimethyl-	EtOH	340(4.46)	78-2729-78
$C_{20}H_{26}O_3$			
1H-Cyclopropa[3,4]benz[1,2-f]azulene-5,8-dione, 1a,2,3,3a,4,6,7,9b-octahydro-3-hydroxy-1,1,3a,6,9-pentamethyl-	n.s.g.	218(4.16),256(3.98), 263(4.14),332(3.66)	78-0233-78
Cycloroyleanone, dehydro-	EtOH	285(4.32),380(2.60)	23-0517-78
2H-Cyclotrideca[b]furan-2-one, 6-acetyl-3,3a,4,7,8,11,12,14a-octahydro-9,13-dimethyl-3-methylene-	EtOH	212(4.00),239(3.72)	12-1303-78
Fukujusonorone	EtOH	250(3.95)	100-0001-78
1,4-Phenanthrenedione, 4b,5,6,7,8,8a-hexahydro-3-hydroxy-4b,8,8-trimethyl-2-propyl-, (4bS-trans)-	EtOH	214(4.18),247(3.87), 330(3.82)	23-0733-78
3(4bH)-Phenanthrenone, 5,6,7,8-tetrahydro-4,6-dihydroxy-4b,8,8-trimethyl-2-(1-methylethyl)-, (4bS-trans)-	ether	250s(3.74),263s(3.53), 420(4.16)	33-0709-78
3(4bH)-Phenanthrenone, 5,6,7,8-tetrahydro-4-hydroxy-8-(hydroxymethyl)-4b,8-dimethyl-2-(1-methylethyl)-, (4bS-cis)-	ether	250s(3.76),263s(3.63), 425(4.12)	33-0709-78
$C_{20}H_{26}O_4$			
Estra-1,3,5(10)-trien-6-one, 17β-hydroxy-2,3-dimethoxy-	EtOH	234(4.27),276(4.00), 316(3.85)	94-3567-78
$C_{20}H_{26}O_4S$			
A-Nor-1-thiaestra-2,5(10)-diene-2,17-diol, diacetate, (17β)-	EtOH	237(3.71),258(3.71)	142-0207-78B
A-Nor-3-thiaestra-1,5(10)-diene-2,17-diol, diacetate, (17β)-	EtOH	236(3.75),258(3.70)	142-0207-78B
$C_{20}H_{26}O_5$			
Coleon N, 7,12-bis(O-deacetyl)-	MeOH	233(4.02)	33-0871-78
Marrubiastrol	n.s.g.	212(4.28)	24-2130-78

Compound	Solvent	λ_{max}(log ϵ)	Ref.
$C_{20}H_{26}O_6$			
Ballonigrinolide, 13-hydroxy-	n.s.g.	260(3.88)	102-2132-78
Bis(abeo)-royleanone	MeOH	274(3.89),405(2.23)	33-0871-78
Shikodonin	EtOH	233.5(3.95)	35-0628-78
$C_{20}H_{26}O_7$			
Chaparrinone	MeOH	220(3.90),238(3.93)	100-0584-78
Cyclopenta[c]pyran-4-carboxylic acid, 1,1'-oxybis[1,4a,5,6,7,7a-hexahydro-, dimethyl ester, (1S,4aS,7aR)-	MeOH	238(4.34)	24-2423-78
$C_{20}H_{26}O_8$			
Chaparrinone, 6α-hydroxy-	MeOH	242(4.05)	100-0578-78
$C_{20}H_{26}O_{10}$			
2H-1-Benzopyran-2-one, 6-(2,3-dihydroxy-3-methylbutyl)-7-β-D-glucopyranosyloxy)-, (R)-	MeOH	222(4.27),251(3.48), 294(3.91),326(4.10)	102-0139-78
2H-1-Benzopyran-2-one, 6-[2-(β-D-glucopyranosyloxy)-3-hydroxy-3-methylbutyl]-7-hydroxy-, (R)-	MeOH	223(4.09),247(3.57), 257(3.50),331(4.08)	102-0139-78
2H-1-Benzopyran-2-one, 6-[3-(β-D-glucopyranosyloxy)-2-hydroxy-3-methylbutyl]-7-hydroxy-, (R)-	MeOH	222(4.14),247s(3.55), 256s(3.46),332(4.11)	102-0139-78
$C_{20}H_{27}ClO_2$			
Androst-4-ene-17-carboxaldehyde, 17-chloro-3-oxo-, (17α)-	EtOH	240(4.26)	23-0119-78
$C_{20}H_{27}Cl_2NO_6$			
1,3-Dioxane-5,5-dicarboxylic acid, 2-[4-[bis(2-chloroethyl)amino]phenyl]-, diethyl ester	EtOH	262(4.47),292(3.37)	104-1004-78
$C_{20}H_{27}NO$			
4(1H)-Quinolinone, 2,3,5,6,7,8-hexahydro-5-methyl-8-(1-methylethyl)-1-(phenylmethyl)-	heptane	<u>233(4.0)</u>	103-0416-78
$C_{20}H_{27}NO_4Si$			
1H-2-Pyridine-4-carboxylic acid, 2,4a,5,6,7,7a-hexahydro-1-oxo-3-phenyl-7a-[(trimethylsilyl)oxy]-, ethyl ester	EtOH EtOH-KOH	285(4.01) 325(3.98)	142-0161-78A 142-0161-78A
$C_{20}H_{27}NO_5$			
19-Nor-17α-pregn-4-ene-3,20-dione, 17β-(nitrooxy)-	MeOH	239(4.22)	24-3086-78
$C_{20}H_{27}NO_6$			
Propanedioic acid, [[[2-(1,1-dimethylethoxy)-2-oxoethyl]phenylamino]methylene]-, diethyl ester	EtOH	227(3.93),294(4.27)	94-2224-78
$C_{20}H_{27}N_5O_3$			
2,3-Pyrrolidinedicarboxamide, 1-cyclohexyl-N,N-dimethyl-5-oxo-4-(phenylhydrazono)-	EtOH	230(4.11),293s(4.07), 328(4.47)	4-1463-78
Pyrrolo[3,4-b]indole-1,8b(1H)-dicarboxamide, 3a-amino-2-cyclohexyl-2,3,3a-4-tetrahydro-N,N-dimethyl-3-oxo-	EtOH	245(3.58),297(3.35)	4-1463-78

Compound	Solvent	$\lambda_{max}(\log \epsilon)$	Ref.
$C_{20}H_{27}N_5O_8$			
1H-Purine-2,6-dione, 7-[4-0-acetyl-6-(acetylamino)-6-deoxy-2,3-0-(1-methylethylidene)-α-D-mannopyranosyl]-3,7-dihydro-1,3-dimethyl-	MeOH	274.4(4.18)	136-0073-78C
$C_{20}H_{28}ClN_2S_2$			
Thiazolium, 2-[2-[2-chloro-3-[(3-ethyl-2-thiazolidinylidene)ethylidene]-1-cyclohexen-1-yl]ethenyl]-3-ethyl-4,5-dihydro-	MeOH	678(5.38)	104-2046-78
$C_{20}H_{28}NO_9PS_2$			
Phosphoramidic acid, bis[2-[(methyl-sulfonyl)oxy]ethyl]-, bis(2-methyl-phenyl) ester	MeOH	265(2.9542)	65-0720-78
Phosphoramidic acid, bis[2-[(methyl-sulfonyl)oxy]ethyl]-, bis(3-methyl-phenyl) ester	MeOH	264(2.8751)	65-0720-78
Phosphoramidic acid, bis[2-[(methyl-sulfonyl)oxy]ethyl]-, bis(4-methyl-phenyl) ester	MeOH	269(3.1239)	65-0720-78
$C_{20}H_{28}N_2O$			
Quebrachamine, 3-(hydroxymethyl)-, (-)-	MeOH	226(4.49),283(3.88), 291(3.85)	114-0167-78A
$C_{20}H_{28}N_2O_2$			
Cleavamine, N_a,18β-bis(methoxycarbo-nyl)-3β,4β-epoxydihydro-, reduction product	EtOH	228(4.51),275s(3.79), 283(3.84),291(3.81)	23-0062-78
Oxazole, 2,2'-(2-butyl-1,3-phenylene)-bis[4,5-dihydro-4,4-dimethyl-	C_6H_{12}	275(3.11)	44-0727-78
$C_{20}H_{28}N_2O_4$			
Morpholine, 4-[6-[1-(4-methoxyphenyl)-2-nitroethyl]-4-methyl-1-cyclohexen-1-yl]-	EtOH	227(4.22),277(3.57), 283(3.21),350(3.50)	2-0405-78
$C_{20}H_{28}N_2O_5$			
Morpholine, 4-[6-[1-(3,4-dimethoxyphen-yl)-2-nitroethyl]-1-cyclohexen-1-yl]-	EtOH	211(4.29),280(3.45)	2-0405-78
Phenol, 2-methoxy-4-[1-[4-methyl-2-(4-morpholinyl)-2-cyclohexen-1-yl]-2-nitroethyl]-	EtOH	208(4.38),226(4.10), 282(3.47),363(3.33)	2-0405-78
$C_{20}H_{28}N_2O_{10}$			
Uridine, 2'-0-(tetrahydro-4-methoxy-2H-pyran-4-yl)-, 5'-(4-oxopentanoate)	EtOH	261(4.00)	54-0073-78
$C_{20}H_{28}O$			
Phenanthro[3,2-b]furan, 1,2,3,4,4a,5,6-8,9,11b-decahydro-4,4,8,8b-tetrameth-yl-	EtOH	292(3.71)	23-0517-78
$C_{20}H_{28}O_2$			
Estra-4,9-dien-3-one, 17β-hydroxy-2α,17α-dimethyl-	EtOH	306(4.32)	78-2729-78
2β-	EtOH	304(4.32)	78-2729-78
Estra-5(10),9(11)-dien-3-one, 17β-hy-droxy-2β,17α-dimethyl-	EtOH	240(4.30)	78-2729-78

Compound	Solvent	$\lambda_{max}(\log \epsilon)$	Ref.
Ethanone, 1-[3-[2-acetyl-2-methyl-5-(1-methylethyl)cyclopentyl]-4-methyl-phenyl]-	EtOH	257.5(3.66)	77-0198-78
Podocarpa-8,11,13-triene, 13-acetyl-12-methoxy-	EtOH	258(4.03),318(3.62)	23-0517-78
Veadeiroic acid	EtOH	211(4.66),242(3.87)	102-1671-78
$C_{20}H_{28}O_3$			
2-Butenoic acid, 2-methyl-, 2,3,3a,4,7-8-hexahydro-3a,6-dimethyl-1-(1-methylethyl)-3-oxo-4-azulenyl ester	n.s.g.	216(4.12)	2-0004-78
1H-Cyclopropa[3,4]benz[1,2-f]azulene-5,8-dione, 1a,2,3,3a,4,4a,6,7,7a,9b-decahydro-3-hydroxy-1,1,3a,6,9-pentamethyl-	n.s.g.	203(3.56),261(2.78)	78-0233-78
1-Naphthalenecarboxaldehyde, 5-[2-(3-furanyl)ethyl]-3,4,4a,5,6,7,8,8a-octahydro-7-hydroxy-5,6,8a-trimethyl-	MeOH	217(4.03)	44-3339-78
A-Norandrost-3(5)-ene-3-carboxylic acid, 17-oxo-, methyl ester	n.s.g.	232(4.08)	39-0743-78C
Podocarpa-8,11,13-trien-16-ol, 12-methoxy-, acetate	EtOH	280(3.43)	23-0517-78
$C_{20}H_{28}O_4$			
Allocyafrin B_4	n.s.g.	239(4.21)	23-2113-78
Cyafrin B_4	n.s.g.	241(4.04)	23-2113-78
Xeniolide A	MeOH	268(4.19)	88-4833-78
Xeniolide B	MeOH	241(4.20)	88-4833-78
$C_{20}H_{28}O_5$			
Farformolide A	MeOH	225(4.13)	95-1592-78
Furanoeremophilane, 3β-angeloyloxy-10β-hydroxy-	MeOH	218.5(4.28)	18-3335-78
$C_{20}H_{28}O_6$			
Neurolenin A	n.s.g.	208(4.15),235s(3.78), 305(1.88)	44-4352-78
$C_{20}H_{28}O_7$			
1H-Inden-1-one, 6-[2-(β-D-glucopyranosyloxy)ethyl]-2,3-dihydro-2,5,7-trimethyl- (pteroside B)	EtOH	218(4.40),260(4.04), 305(3.18)	94-2365-78
$C_{20}H_{28}O_8$			
1H-Inden-1-one, 6-[2-(β-D-glucopyranosyloxy)ethyl]-2,3-dihydro-3-hydroxy-2,5,7-trimethyl- (pteroside C)	EtOH	219(4.53),260(4.16), 302(3.30)	94-2365-78
$C_{20}H_{29}ClN_4$			
Pyrrolo[3',4':4,5]pyrrolo[3,4-b]indol-5a(6H)-amine, 9-chloro-4-cyclohexyl-1,2,3,3a,4,5-hexahydro-2,6-dimethyl-	EtOH	261(4.15),321(3.49)	4-1463-78
$C_{20}H_{29}ClO_2$			
Androst-4-en-3-one, 6β-chloro-17β-hydroxy-16β-methyl-	EtOH	239(4.22)	94-1718-78
$C_{20}H_{29}N_5O_2$			
3,5-Pyridinedicarbonitrile, 1,4-dihydro-1,4,4-trimethyl-2,6-bis(morpholinomethyl)-	EtOH	207(4.55),221(4.40), 278(4.30),347(4.02)	103-1236-78

Compound	Solvent	$\lambda_{max}(\log \epsilon)$	Ref.
$C_{20}H_{30}$			
Bicyclo[4.2.0]octa-1,3,5,7-tetraene, 7,8-bis(1,1-dimethylethyl)-2,3,4,5-tetramethyl-	hexane	237(4.43),243(4.44), 305s(2.37),317(2.42), 335s(2.37),372(2.43)	5-1675-78
Cyclobutene, 3,4-bis(3,3-dimethyl-1-butynyl)-1,2,3,4-tetramethyl-, trans	hexane	none above 210 nm	5-1675-78
5,7-Dodecadiene-3,9-diyne, 2,2,5,6,7,8-11,11-octamethyl-, (E,E)-	hexane	241(4.28)	5-1675-78
(Z,Z)-	hexane	229(4.28)	5-1675-78
Naphthalene, decahydro-2-(1,5-dimethyl-2,4-hexadien-1-ylidene)-4a-methyl-8-methylene-	EtOH	275(4.53),285(4.63), 297(4.50)	12-0163-78
Pentalene, 1,3,5-tris(1,1-dimethyleth-yl)-	hexane	212(4.34),274(3.41), 337(3.55),594(1.87)	24-0932-78
$C_{20}H_{30}Br_2O$			
Cyclohexanol, 2-bromo-4-[3-(3-bromo-2,2-dimethyl-6-methylenecyclohexyl)-1-methylene-2-propenyl]-1-methyl-	ether	236(3.92)	88-2453-78
$C_{20}H_{30}FN_3O$			
Adamantano[2,1-b][1,5]benzodiazepin-12-one, 13-(3-dimethylaminopropyl)-2-fluoro-5,5a,12,13-tetrahydro-, maleate	MeOH	317(3.66)	4-0705-78
$C_{20}H_{30}N_2O_{11}$			
Benzoic acid, 3-methyl-, hydrazide, hydrazone with lactose	EtOH	210(3.73),230(3.83)	136-0089-78E
$C_{20}H_{30}N_2O_{12}$			
Benzoic acid, 4-methoxy-, hydrazide, hydrazone with lactose	EtOH	215(3.53),253(4.00)	136-0089-78E
$C_{20}H_{30}N_4O_4$			
Pyrano[2,3-c]pyrazole-2(6H)-acetamide, N-butyl-3,4-dimethyl-5-[2-(4-morpho-linyl)ethyl]-6-oxo-	EtOH	308.2(4.14)	95-0335-78
$C_{20}H_{30}N_8O_{14}$			
1,2,4-Triazine-3,5(2H,4H)-dione, 6,6'-(1,2-ethanediyldiimino)bis[2-β-D-glucopyranosyl-	pH 1 H_2O pH 13	309(4.05) 309(4.05) 298(3.96)	103-1273-78 103-1273-78 103-1273-78
$C_{20}H_{30}O$			
3,7,11,15-Cembratetraene, 13-oxo-, (3E,7E,11E)-	hexane	231(3.91)	44-2127-78
(3E,7E,11Z)-	hexane	234(3.51)	44-2127-78
Furan, 3-[2-(decahydro-5,5,8a-trimeth-yl-2-methylene-1-naphthalenyl)eth-yl]-, [1S-(1α,4aβ,8aα)]- (lambertiane)	EtOH	219s(3.4)	105-0286-78
Veadeirol	EtOH	208(4.70),270(2.79)	102-1671-78
$C_{20}H_{30}OSi$			
Silane, [[5,6,7,8,9,10-hexahydro-6-(1,3-pentadienyl)-6-benzocyclooctenyl]-oxy]trimethyl-	EtOH	235(4.28)	44-1050-78
long-retention-time isomer	EtOH	239(4.34)	44-1050-78
$C_{20}H_{30}O_2$			
Bicyclo[8.1.0]undeca-2,5-dien-4-one,	MeOH	212(3.73),240s(3.57),	88-4155-78

$C_{20}H_{30}O_2-C_{20}H_{32}Cl_2PdS_2$

Compound	Solvent	$\lambda_{max}(\log \epsilon)$	Ref.
7-hydroxy-3,7,11-trimethyl-11-(4-methyl-3-pentenyl)- (cont.)		282(3.00)	88-4155-78
Cembra-7,11,15-triene, 3,4-epoxy-13-oxo-, (1S*,3S*,4S*,7E,11E)-(11Z)-	hexane	234(3.95)	44-2127-78
	hexane	236(3.48)	44-2127-78
Cembra-7,11,15-triene, 3,4-epoxy-14-oxo-, (1S*,3S*,4S*,7E,11E)-	hexane	217(3.03)	44-2127-78
1H-Cyclopent[f]azulene-2-carboxalde-hyde, 3a,4,4a,5,6,7,7a,8,9,9a-deca-hydro-1-hydroxy-3,4a,9a-trimethyl-7-(1-methylethenyl)-	MeOH	257.5(4.06)	78-1551-78
Estr-4-en-3-one, 17β-hydroxy-2α,17α-di-methyl-	EtOH	240(4.21)	78-2729-78
Estr-4-en-3-one, 17β-hydroxy-2β,17α-di-methyl-	EtOH	241(4.20)	78-2729-78
5-Oxatricyclo[9.1.0.0⁴,⁶]dodec-9-en-8-one, 4,9,12-trimethyl-12-(4-methyl-3-pentenyl)- (epoxydilophone)	MeOH	254(2.91)	88-4155-78
5,13-Tetradecadien-9-ynoic acid, 5,13-dimethyl-2-(1-methylethylidene)-, methyl ester, (E)-	MeOH	222(3.48)	35-4268-78

$C_{20}H_{30}O_2S$
Cyclohexanone, 3,3'-thiobis[2-methyl-5-(1-methylethenyl)-	n.s.g.	<u>224(2.9)</u>	64-1535-78B

$C_{20}H_{30}O_3$
Atis-16-en-19-oic acid, ent-7α-hydroxy-	EtOH	210.5(3.43)	102-1637-78
Bacchotricuneatin D	MeOH	217(end absorption)	44-3339-78
1-Phenanthrenecarboxylic acid, 7-ethyl-1,2,3,4,4a,5,8,9,10,10a-decahydro-6-methoxy-1,4a-dimethyl-, [1S-(1α,4aα,10aβ)]-	EtOH	273(2.95),280(2.98)	39-0084-78C
1-Phenanthrenecarboxylic acid, 7-ethyl-1,2,3,4,4a,6,7,8,8a,9,10,10a-dodeca-hydro-1,4a-dimethyl-6-oxo-, methyl ester, [1S-(1α,4aα,7β,8aα,10aβ)]-	EtOH	240(3.17)	39-0084-78C

$C_{20}H_{30}O_4$
Cyafrin A₄	n.s.g.	230s(3.18)	23-2113-78
Vaginatin	n.s.g.	217(4.09)	2-0004-78

$C_{20}H_{31}CoN_4O_5$
Cobalt(III), [[2-(1,7-dioxo-8-hydroxy-imino-2-oxanonyl-7-oximato)-2-(3-[H]-1,7-dioxo-7-hydroxyimino-2-oxanonyl-8-oximato)propyl]pyridinio-	EtOH	456(3.17)	33-1565-78

$C_{20}H_{32}$
1,3,7,11-Cyclotetradecatetraene, 4,8,12-trimethyl-1-(1-methylethyl)-, (all-E)-	EtOH	244s(4.27),251(4.33), 258s(4.23)	12-2707-78

$C_{20}H_{32}Br_2NiS_2$
Nickel, dibromo[(5a,5b,10a,10b-η)-1,2,4,5,5a,5b,6,7,9,10-decahydro-1,1,5,5,6,6,10,10-octamethylcyclo-buta[1,2-d:3,4-d']bisthiepin]-	CHCl₃	260(4.02),308(3.56), 500(3.04)	5-0431-78

$C_{20}H_{32}Cl_2PdS_2$
Palladium, dichloro[(5a,5b,10a,10b-η)-1,2,4,5,5a,5b,6,7,9,10-decahydro-	CH₂Cl₂	228(3.87),251(4.03), 273(3.81),340(3.48)	5-0431-78

Compound	Solvent	$\lambda_{max}(\log \epsilon)$	Ref.
1,1,5,5,6,6,10,10-octamethylcyclo-buta[1,2-d:3,4-d']bisthiepin]- (cont.)			5-0431-78
$C_{20}H_{32}N_6O_5$			
Adenosine, N-[(nonylamino)carbonyl]-	pH 1	270(4.35),277(4.32)	36-0569-78
	pH 13	271(4.18),278(4.19)	36-0569-78
	70% EtOH	269(4.38),276(4.30)	36-0569-78
$C_{20}H_{32}O$			
7-Cyclopentacycloundecenol, 1,2,4,7,8-11,12,12a-octahydro-6,10,12a-trimeth-yl-1-(1-methylethyl)-, (3Z,5E,9E,1R*-7S*,12aR*)-	EtOH	207(4.04)	12-2039-78
(3Z,5Z,9E,1R*,7R*,12aR*)-	EtOH	207(4.01)	12-2039-78
1,3,7-Cyclotetradecatriene, 11,12-epoxy-4,8,12-trimethyl-1-(1-methylethyl)-	EtOH	243s(4.32),249(4.34),255s(4.24)	12-2707-78
3,5-Heptadien-2-ol, 6-(decahydro-4a-methyl-8-methylene-2-naphthalenyl)-2-methyl-, [2α(3E,5E),4aβ,8aα]-(+)-	EtOH	235(4.36),241(4.40),250(4.26)	12-0163-78
	EtOH-acid	275(4.53),285(4.63),297(4.50)	12-0163-78
3,5-Heptadien-2-ol, 6-(4-ethenyl-4-meth-yl-3-(1-methylethenyl)cyclohexyl]-2-methyl- (fuscol)	n.s.g.	234s(4.47),239(4.50),246s(4.40)	88-3641-78
Labda-8(20),13-dien-15-al, (13E)-	EtOH	241(4.20)	105-0286-78
Labda-8(20),13-dien-15-al, (13Z)-	EtOH	240(4.13)	105-0286-78
$C_{20}H_{32}O_2$			
Anhydroverticillol diepoxide	EtOH	206(3.80)	78-2349-78
Atis-16-ene, ent-7α,19-dihydroxy-	EtOH	207.5(3.59)	102-1637-78
Eleganolone	EtOH	240(4.32)	102-1003-78
2,4-Heptadiene-1,6-diol, 2-(decahydro-4a-methyl-8-methylene-2-naphthalen-yl)-6-methyl-	EtOH	239(4.31),247s(4.18)	12-0163-78
5,9-Methanobenzocycloocten-4(1H)-one, 5-ethoxy-2,3,5,6,7,8,9,10-octahydro-2,2,7,7,9-pentamethyl-	n.s.g.	245(3.93)	78-1251-78
2-Pentenal, 3-methyl-5-[1,4,4a,5,6,7-8,8a-octahydro-5-(hydroxymethyl)-2,5,8a-trimethyl-1-naphthalenyl]-, [1S-[1α(E),4aβ,5α,8aα]]-	EtOH	238.5(4.15)	102-0281-78
$C_{20}H_{32}O_3$			
2-Pentenoic acid, 3-methyl-5-[1,4,4a-5,6,7,8,8a-octahydro-5-(hydroxymeth-yl)-2,5,8a-trimethyl-1-naphthalenyl]-	EtOH	218(4.09)	102-0281-78
$C_{20}H_{32}O_4$			
1-Hexanone, 1-[2-hydroxy-4,6-dimethoxy-3-(3-methylbutyl)phenyl]-5-methyl-	EtOH	288(4.10)	142-1355-78A
$C_{20}H_{32}S_2$			
Cyclobuta[1,2-d:3,4-d']bisthiepin, 1,2,4,5,6,7,9,10-octahydro-1,1,5,5,6,6,10,10-octamethyl-	C_6H_{12}	<u>205(3.9),305s(1.9),370s(1.3)</u>	5-0431-78
$C_{20}H_{33}NO_4$			
Pentanoic acid, 5-[(8-hydroxy-6-methyl-1-oxo-2,4,6-dodecatrienyl)amino]-, ethyl ester, (E,E,E)-(±)-	EtOH	295(4.61)	18-2077-78
1-Pyrrolidineheptanoic acid, 2-oxo-5-(2-oxooctylidene)-, methyl ester, (E)-	n.s.g.	272(4.4525)	107-0219-78

Compound	Solvent	$\lambda_{max}(\log \epsilon)$	Ref.
$C_{20}H_{33}N_2O$			
Methanaminium, N-[3-[2-butoxy-3-[3-(di-methylamino)-2-propenylidene]-1-cy-clohexen-1-yl]-2-propenylidene]-N-methyl-, iodide	EtOH	650(5.17)	70-0339-78
Methanaminium, N-[3-[3-[3-(dimethylam-ino)-2-propenylidene]-2-ethoxy-4,6-dimethyl-1-cyclohexen-1-yl]-2-prop-enylidene]-N-methyl-, ethyl sulfate	CHCl₃	665(5.20)	70-0339-78
Methanaminium, N-[3-[3-[3-(dimethylam-ino)-2-propenylidene]-2-ethoxy-5,6-dimethyl-1-cyclohexen-1-yl]-2-prop-enylidene]-N-methyl-, ethyl sulfate	CHCl₃	665(5.22)	70-0339-78
$C_{20}H_{33}N_6O_8P$			
5'-Adenylic acid, N-[(nonylamino)carbo-nyl]-, barium salt	pH 1	270(4.34),277(4.32)	36-0569-78
	pH 13	270(4.24),277(4.22),298(3.97)	36-0569-78
	70% EtOH	269(4.37),277(4.29)	36-0569-78
$C_{20}H_{34}N_2O_3$			
1-Heptanamine, N-heptyl-N-[2-(5-nitro-2-furanyl)ethenyl]-, (E)-	EtOH	273(4.20),509(4.44)	103-0250-78
$C_{20}H_{34}N_6O_{11}P_2$			
Adenosine, N-[8-(acetylamino)octyl]-, 5'-(trihydrogen diphosphate)	H₂O	267(4.24)	87-1137-78
$C_{20}H_{34}O$			
5,9-Methanobenzocyclooctene, 5-ethoxy-1,2,3,4,5,6,7,8,9,10-decahydro-2,2,7,7,9-pentamethyl-	n.s.g.	206(3.73)	78-1251-78
Verticillol, (±)-	EtOH	208(3.98)	78-2349-78
$C_{20}H_{34}O_3$			
2(1H)-Naphthalenone, octahydro-8-(hy-droxymethyl)-4-(5-hydroxy-3-methyl-3-pentenyl)-3,4a,8-trimethyl-, [3S-[3α,4β(E),4aβ,8β,8aα]]- (villenolone)	EtOH	282.5(1.70)	102-0281-78
2,6-Nonadienoic acid, 8-(3-methoxy-3-methylcyclopentyl)-3,8-dimethyl-, ethyl ester, (2E,6Z)-	EtOH	219.5(4.13)	56-0581-78
Verticillol diepoxide	EtOH	205(2.18)	78-2349-78
$C_{20}H_{34}O_4$			
Villenatriolone	EtOH	301(1.61)	102-0281-78
$C_{20}H_{34}O_4Si$			
Pent-cis-2-en-4-yn-1-ol, 5-[1-hydroxy-2,6,6-trimethyl-2-[(trimethylsilyl)-oxy]cyclohexyl]-3-methyl-, acetate, erythro	EtOH	229.5(4.11)	39-1511-78C
threo	EtOH	229.5(4.11)	39-1511-78C
$C_{20}H_{36}P_2Pt$			
Platinum, di-3-buten-1-ynylbis(trieth-ylphosphine)-, (SP-4-2)-, cis	ether	242(4.45),247(4.45),264s(4.13),300(4.16)	101-0101-78B
trans	ether	260(4.38),286(4.01),323(4.24)	101-0101-78B

Compound	Solvent	λ_{max}(log ϵ)	Ref.

$C_{20}H_{37}N_3P$
  Phosphorus(1+), tributyl(4-diazenyl-
  N,N-dimethylbenzenaminato)-,
  (T-4)-, tetrafluoroborate
	n.s.g.	278(3.96),460(4.63)	123-0190-78

$C_{20}H_{38}B_2N_4$
  Boron, tetraethylbis[μ-(3,4,5-trimeth-
  yl-1H-pyrazolato-N^1:N^2)]di-
	hexane	219(4.25),250(3.78)	64-0220-78B

$C_{20}H_{44}MoO_4$
  Molybdenum, tetrakis(2,2-dimethylprop-
  oxy)-
	C_6H_{12}	640(2.51)	35-2744-78

Compound	Solvent	$\lambda_{max}(\log \epsilon)$	Ref.
$C_{21}H_7Cl_5O_6$ Dibenzo[a,kl]xanthene-3,9-dione, 2,5,8,11,13-pentachloro-4,10-di-hydroxy-1-methoxy-	CHCl$_3$	317(4.04),342s(4.04), 375(4.24),410s(4.04), 658(4.11)	12-1323-78
$C_{21}H_{11}BrN_4$ Acenaphth[1,2-e]imidazo[1,2-b][1,2,4]-triazine, 9-(4-bromophenyl)-	DMF	320(4.28),340(4.24), 430(4.07)	103-1277-78
$C_{21}H_{11}ClN_4$ Acenaphth[1,2-e]imidazo[1,2-b][1,2,4]-triazine, 9-(4-chlorophenyl)-	DMF	321(4.48),342(4.45), 431(4.26)	103-1277-78
$C_{21}H_{11}NO_2S$ 8H,12H-Benzo[1,9]quinolizino[3,4,5,6,7-klmn]phenothiazine-8,12-dione, 10-methyl-	CHCl$_3$ H$_2$SO$_4$	264(4.61),323(3.70), 350(3.60),507(3.90) 210(4.54),259(4.49), 344(4.15),460(4.15)	70-0130-78 70-0130-78
$C_{21}H_{11}NO_4S$ 8H,12H-Benzo[1,9]quinolizino[3,4,5,6,7-klmn]phenothiazine-8,12-dione, 10-methyl-, 4,4-dioxide	CHCl$_3$ H$_2$SO$_4$	260(4.47),363(3.91), 418(4.15) 206(4.59),260(4.65), 300(4.18),390(4.31), 480(4.28)	70-0130-78 70-0130-78
$C_{21}H_{12}Br_2OS_4$ 2-Propanone, 1,3-bis[5-(4-bromophenyl)-3H-1,2-dithiol-3-ylidene]-, (Z,Z)-	dioxan	255(4.80),313(4.50), 380(3.74),525(4.52), 563(4.56)	78-2175-78
$C_{21}H_{12}Cl_2N_4O$ [1,2,4]Triazolo[1,5-a]pyridine, 6-chloro-2-[4-[2-(5-chloro-2-benzoxazolyl)-ethenyl]phenyl]-	DMF	345(4.76),357(4.77)	33-0142-78
$C_{21}H_{12}Cl_3NO$ 1,2-Benzisoxazole, 6-chloro-3-[3-chloro-4-[2-(4-chlorophenyl)ethenyl]-phenyl]-	DMF	330(4.66)	33-2904-78
$C_{21}H_{12}N_2O_2S$ Benzamide, N-(6-oxo-6H-anthra[9,1-cd]-isothiazol-7-yl)-	DMF	435(4.06)	2-1007-78
$C_{21}H_{12}N_2O_3S$ 2,4,6(1H,3H,5H)-Pyrimidinetrione, 5-(2-phenyl-5H-naphtho[1,8-bc]thien-5-yli-dene)-	MeOH	334(3.97)	83-0324-78
$C_{21}H_{12}N_4$ Acenaphth[1,2-e]imidazo[1,2-b][1,2,4]-triazine, 9-phenyl-	DMF	321(4.53),324(4.49), 432(4.33)	103-1277-78
$C_{21}H_{12}N_4O_4S$ 6H-Anthra[1,9-bc]thiophen-6-one, 2,4-dinitrophenylhydrazone	MeOH	430(4.06)	83-0324-78
$C_{21}H_{12}O_6$ [2,2'-Binaphthalene]-1,1',4,4'-tetrone, 8-hydroxy-8'-methoxy-	MeOH	255(4.44),412(3.93)	102-2042-78

Compound	Solvent	$\lambda_{max}(\log \epsilon)$	Ref.
$C_{21}H_{13}ClN_4O$			
[1,2,4]Triazolo[1,5-a]pyridine, 2-[4-[2-(2-benzoxazolyl)ethenyl]phenyl]-6-chloro-	DMF	343(4.75),356(4.76)	33-0142-78
[1,2,4]Triazolo[1,5-a]pyridine, 2-[4-[2-(5-chloro-2-benzoxazolyl)ethenyl]phenyl]-	DMF	342(4.75),355(4.76)	33-0142-78
$C_{21}H_{13}Cl_2NO$			
1,2-Benzisoxazole, 6-chloro-3-[3-chloro-4-(2-phenylethenyl)phenyl]-	DMF	328(4.56)	33-2904-78
1,2-Benzisoxazole, 6-chloro-3-[4-[2-(4-chlorophenyl)ethenyl]phenyl]-	DMF	330(4.68)	33-2904-78
$C_{21}H_{13}Cl_3N_2$			
1H-Imidazole, 2,4,5-tris(4-chlorophenyl)-	EtOH	233(4.38),311(4.50)	24-1464-78
$C_{21}H_{13}N$			
9-Anthracenecarbonitrile, 10-phenyl-	C_6H_{12}	386(4.08)	61-1068-78
$C_{21}H_{13}NO_2$			
5H-Naphtho[2,3-a]carbazole-5,13(12H)-dione, 12-methyl-	CHCl$_3$	257(4.29),279(4.38),326(4.29),400(3.86),480(3.68)	12-1841-78
5H,9H-Quino[3,2,1-de]acridine-5,9-dione, 7-methyl-	CHCl$_3$	247(4.33),265(4.33),325(3.44),340(3.61),450(4.34)	70-0130-78
	H_2SO_4	210(4.46),260(4.60),300(3.95),350(4.24),540(4.19)	70-0130-78
$C_{21}H_{13}NO_2S$			
9-Anthracenecarbonitrile, 10-(phenylsulfonyl)-	C_6H_{12}	404(4.02)	61-1068-78
$C_{21}H_{13}NO_4$			
2-Benzopyrylium, 4-hydroxy-3-(4-nitrophenyl)-1-phenyl-, hydroxide, inner salt	EtOH	566(4.41)	103-0493-78
6H-Indeno[1,2-b]oxiren-6-one, 1a,6a-dihydro-6a-(4-nitrophenyl)-1a-phenyl-	EtOH	<u>267(4.3),315(3.8),350s(3.0)</u>	103-0493-78
$C_{21}H_{14}$			
4H-Cyclopenta[def]phenanthrene, 8-phenyl-	C_6H_{12}	230(4.47),257(4.72),304(4.07),330(3.15)	35-6679-78
$C_{21}H_{14}ClNO$			
1,2-Benzisoxazole, 3-[4-[2-(4-chlorophenyl)ethenyl]phenyl]-	DMF	330(4.66)	33-2904-78
1,2-Benzisoxazole, 6-chloro-3-[4-(2-phenylethenyl)phenyl]-	DMF	328(4.64)	33-2904-78
1,2-Benzisoxazole, 6-[2-(4-chlorophenyl)ethenyl]-3-phenyl-	DMF	327(4.61)	33-2904-78
2,1-Benzisoxazole, 3-[4-[2-(4-chlorophenyl)ethenyl]phenyl]-	DMF	378(4.65)	33-2904-78
$C_{21}H_{14}Cl_2N_2$			
1H-Imidazole, 2-(2,6-dichlorophenyl)-4,5-diphenyl-	EtOH	288(4.16)	40-1449-78

Compound	Solvent	$\lambda_{max}(\log \epsilon)$	Ref.
$C_{21}H_{14}N_2O_3$ Benzo[h]quinoline-3-carbonitrile, 4-(1,3-benzodioxol-5-yl)-1,2,5,6-tetrahydro-2-oxo-	CHCl$_3$	254(4.25),395(4.39)	48-0097-78
$C_{21}H_{14}N_2O_3S$ Pyrrolo[3',4':3,4]pyrido[2,1-b]benzothiazole-1,3,5(2H)-trione, 2-ethyl-4-phenyl-	MeOH	232(3.98),340(3.95), 417(3.71)	44-2697-78
$C_{21}H_{14}N_4O$ [1,2,4]Triazolo[1,5-a]pyridine, 2-[4-[2-(2-benzoxazolyl)ethenyl]phenyl]-	DMF	340(4.72),353(4.72)	33-0142-78
$C_{21}H_{14}O$ 14H-Dibenzo[a,j]xanthene	EtOH	221(4.67),256(4.65), 271(4.48),381(3.48)	32-0079-78
Methanone, 4-phenanthrenylphenyl-	EtOH	208(4.62),221(4.48), 252(4.79),285(4.22), 296(4.15),349(3.03)	35-6679-78
$C_{21}H_{14}OS_4$ 2-Propanone, 1,3-bis(5-phenyl-3H-1,2-dithiol-3-ylidene)-	dioxan	247(4.69),304(4.45), 383(3.79),520(4.64), 557(4.74)	78-2175-78
$C_{21}H_{14}O_2$ 2-Benzopyrylium, 4-hydroxy-1,3-diphenyl-, hydroxide, inner salt	EtOH	545(4.27)	103-0493-78
6H-Indeno[1,2-b]oxiren-6-one, 1a,6a-dihydro-1a,6a-diphenyl-	EtOH	<u>255(3.8),305s(2.8), 355s(2.0)</u>	103-0493-78
$C_{21}H_{14}O_6$ 1-Naphthacenecarboxylic acid, 6,11-dihydro-5,12-dihydroxy-2-methyl-6,11-dioxo-, methyl ester	MeOH	203(4.36),265(4.80), 424s(--),452(4.00), 480(4.21),513(4.20), 546(3.60)	24-3823-78
$C_{21}H_{15}Br$ Phenanthrene, 7-(bromomethyl)-4-phenyl-	MeOH	204(4.46),227(4.50), 264(4.69),290s(4.22)	78-0769-78
$C_{21}H_{15}BrO$ 2-Propen-1-one, 3-bromo-1,2,3-triphenyl-	EtOH	256(4.37)	44-0232-78
higher melting isomer	EtOH	252(4.38)	44-0232-78
$C_{21}H_{15}ClN_2$ 1H-Imidazole, 2-(2-chlorophenyl)-4,5-diphenyl-	EtOH	312(4.34)	40-1449-78
Imidazo[1,2-a]pyridine, 7-[2-(2-chlorophenyl)ethenyl]-2-phenyl-	DMF	365(4.57)	33-0129-78
Imidazo[1,2-a]pyridine, 7-[2-(3-chlorophenyl)ethenyl]-2-phenyl-	DMF	363(4.59),382(4.54)	33-0129-78
Imidazo[1,2-a]pyridine, 7-[2-(4-chlorophenyl)ethenyl]-2-phenyl-	DMF	363(4.62),383(4.58)	33-0129-78
Imidazo[1,2-a]pyridine, 2-(4-chlorophenyl)-7-(2-phenylethenyl)-	DMF	360(4.62),380(4.57)	33-0129-78
$C_{21}H_{15}ClN_2O$ Benzo[h]quinoline-3-carbonitrile, 4-(2-	CHCl$_3$	267(4.24),350(4.38)	48-0097-78

Compound	Solvent	$\lambda_{max}(\log \epsilon)$	Ref.
chlorophenyl)-5,6-dihydro-2-methoxy-Benzo[h]quinoline-3-carbonitrile, 4-(2-chlorophenyl)-1,2,5,6-tetrahydro-1-methyl-2-oxo-	CHCl$_3$	242(4.13),392(4.28)	48-0097-78 48-0097-78
$C_{21}H_{15}ClN_4O_3$ 1H-Pyrazolo[3,4-b]pyridine-1-carboxamide, 4-(1,3-benzodioxol-5-yl)-6-(4-chlorophenyl)-3-methyl-	EtOH	237(3.96),255(3.91), 315(3.36)	2-0332-78
$C_{21}H_{15}ClO$ 2-Propen-1-one, 3-chloro-1,2,3-triphenyl-	EtOH	256(4.33)	44-0232-78
high melting isomer	EtOH	257(3.97),280s(3.72)	44-0232-78
$C_{21}H_{15}ClO_4Se$ 1-Benzoseleninium, 2,4-diphenyl-, perchlorate	HOAc-HClO$_4$	416(4.393)	83-0170-78
$C_{21}H_{15}Cl_3FN_5O_3$ 9H,13aH-Oxazolo[3,2-d][1,2,4]triazolo-[1,5-a][1,4]benzodiazepine-7-carboxamide, 2,12,12-trichloro-13a-(2-fluorophenyl)-11,12-dihydro-N,N-dimethyl-11-oxo-	MeOH	252(4.30)	33-0848-78
$C_{21}H_{15}Cl_4N_5O_3$ 9H,13aH-Oxazolo[3,2-d][1,2,4]triazolo-[1,5-a][1,4]benzodiazepine-7-carboxamide, 2,12,12-trichloro-13a-(2-chlorophenyl)-11,12-dihydro-N,N-dimethyl-11-oxo-	MeOH	252(4.21)	33-0848-78
$C_{21}H_{15}N$ 5H-Indeno[1,2-b]pyridine, 5-(3-phenyl-2-propenylidene)-	n.s.g.	210(4.84),240(5.13), 250(4.58),281(4.64), 307(4.56),316(4.6), 374(5.23)	103-1343-78
$C_{21}H_{15}NO$ 1,2-Benzisoxazole, 3-[4-(2-phenylethenyl)phenyl]-	DMF	327(4.63)	33-2904-78
1,2-Benzisoxazole, 3-phenyl-6-(2-phenylethenyl)-	DMF	324(4.55)	33-2904-78
2(1H)-Quinolinone, 1,4-diphenyl-	EtOH	225(4.79),280(4.30), 330(4.08)	44-0954-78
$C_{21}H_{15}NOS$ Thiazolium, 4-hydroxy-2,3,5-triphenyl-, hydroxide, inner salt	dioxan	224s(4.27),275(4.18), 300s(4.01),494(4.19)	24-3178-78
$C_{21}H_{15}NO_2$ 6-Phenanthridinemethanol, benzoate	EtOH	211(4.45),220(4.47), 246(4.68),251s(4.66), 270s(4.05),291(3.82), 299s(3.78),332(3.16), 347(3.09)	95-1503-78
$C_{21}H_{15}NO_3$ 9,10-Anthracenedione, 1-(methylamino)-4-phenoxy-	benzene	515(3.88)	104-2199-78

Compound	Solvent	$\lambda_{max}(\log \epsilon)$	Ref.
9,10-Anthracenedione, 1-(methylamino)-5-phenoxy-	benzene	505(3.88)	104-2199-78
9,10-Anthracenedione, 2-(methylamino)-1-phenoxy-	benzene	438(3.81)	104-2199-78
Spiro[2H-1,3-benzoxazine-2,3'-[3H]naphtho[2,1-b]pyran]-4(3H)-one, 2'-methyl-	isoPrOH	270(3.94),295(4.06), 310(4.03),330(3.70), 344(3.73)	103-0122-78
Spiro[2H-1,3-benzoxazine-2,3'-[3H]naphtho[2,1-b]pyran]-4(3H)-one, 3-methyl-	isoPrOH	298(4.13),310(4.11), 330(3.70),343(3.72)	103-0122-78
photoinduced form	isoPrOH	400(--),520(--), 530(--)	103-0122-78
$C_{21}H_{15}NO_5S$ 1H-Pyrido[2,1-b]benzothiazine-3,4-dicarboxylic acid, 1-oxo-2-phenyl-, dimethyl ester	MeOH	310(4.24),360(4.18), 375(4.25)	44-2697-78
$C_{21}H_{15}NS_2$ Acridine, 9-(1,3-benzodithiol-2-ylidene)-9,10-dihydro-10-methyl-	CH_2Cl_2	238(4.77),263s(4.26), 314(4.18),395(4.30)	18-2674-78
$C_{21}H_{15}N_3O$ Benz[cd]indazolium, 1-[(2-methoxy-1-naphthalenyl)amino]-, hydroxide, inner salt	EtOH	265(4.27),326(3.91), 521(3.99)	44-2508-78
Naphtho[1,8-de]triazine, 2-(2-methoxy-1-naphthalenyl)-	EtOH	341(4.10),356(4.15), 584(2.93),633(2.90), 693(2.71)	44-2508-78
$C_{21}H_{15}N_3O_2$ 1H-Imidazole, 2-(2-nitrophenyl)-4,5-diphenyl-	EtOH	370(3.36)	40-1449-78
1H-Imidazole, 2-(4-nitrophenyl)-4,5-diphenyl-	EtOH	386(4.28)	40-1449-78
$C_{21}H_{15}N_3O_4$ Quinoline, 4-[(2,4-dinitro-1-naphthalenyl)methylene]dihydro-1-methyl-	acetone	374(4.0),637(4.1)	104-1847-78
$C_{21}H_{15}N_7O_2$ Formazan, 5-(4-nitrophenyl)-3-phenyl-1-(1-phthalazinyl)-	C_6H_{12}	280(4.27),397s(4.07), 465(4.25)	103-0450-78
	EtOH	270(4.49),400(4.27), 475(4.10)	103-0450-78
	EtOH-KOH	590(3.80)	103-0450-78
$C_{21}H_{15}Se$ 1-Benzoseleninium, 2,4-diphenyl-, perchlorate	HOAc-HClO$_4$	416(4.393)	83-0170-78
$C_{21}H_{16}$ 9H-Fluorene, 4-(1-phenylethenyl)-	C_6H_{12}	213(4.14),240(3.79), 258(3.90),268(3.84), 280(3.65),288(3.45), 300(3.30)	35-6679-78
Phenanthrene, 2-methyl-5-phenyl-	MeOH	204(4.57),223(4.66), 356(4.67),277s(4.39), 297(4.11)	78-0769-78
$C_{21}H_{16}ClNO_2$ Benzo[h]quinolin-2(1H)-one, 3-acetyl-	CHCl$_3$	247(4.08),372(4.15)	48-0097-78

Compound	Solvent	$\lambda_{max}(\log \epsilon)$	Ref.
4-(2-chlorophenyl)-5,6-dihydro- (cont.)			48-0097-78
Benzo[h]quinolin-2(1H)-one, 3-acetyl- 4-(4-chlorophenyl)-5,6-dihydro-	CHCl$_3$	245(4.17),372(4.28)	48-0097-78
$C_{21}H_{16}ClNO_2S$			
Dibenzo[b,f]thiepin-10-ol, 8-chloro- 10,11-dihydro-, phenylcarbamate	MeOH	268.5(4.00)	73-2427-78
$C_{21}H_{16}ClN_3$			
[1,2,4]Triazolo[1,5-a]pyridine, 7-[2- (4-chlorophenyl)ethenyl]-2-(4-meth- ylphenyl)-	DMF	331(4.68),347(4.58)	33-0142-78
[1,2,4]Triazolo[1,5-a]pyridine, 2-[3- chloro-4-(2-phenylethenyl)phenyl]- 6-methyl-	DMF	338(4.65)	33-0142-78
$C_{21}H_{16}Cl_2O_4$			
2-Cyclohexen-1-one, 6-acetyl-5-(1,3- benzodioxol-5-yl)-3-(3,4-dichloro- phenyl)-	EtOH	292(3.99),377(4.05)	2-0984-78
$C_{21}H_{16}CrO_4S$			
Chromium, tricarbonyl[(2,3,4,5,6-η)-1- methyl-3,5-diphenyl-1H-thiopyran 1- oxide], anti	MeOH	212(4.6),245s(4.3), 282s(4.1),330(3.8), 413(3.7)	24-1709-78
syn	MeOH	212(4.6),259(4.3), 331(3.8),402(3.6)	24-1709-78
$C_{21}H_{16}MoO_4S$			
Molybdenum, tricarbonyl[(2,3,4,5,6-η)- 1-methyl-3,5-diphenyl-1H-thiopyran 1-oxide]	MeOH	211(4.6),275(4.3), 327(3.9),404(3.8)	24-1709-78
$C_{21}H_{16}N_2$			
1H-Imidazole, 2,4,5-triphenyl-	EtOH	304(4.36)	40-1449-78
Imidazo[1,2-a]pyridine, 2-[4-(2-phenyl- ethenyl)phenyl]-	DMF	349(4.69)	33-0129-78
Imidazo[1,2-a]pyridine, 2-phenyl-7-(2- phenylethenyl)-	DMF	359(4.59),379(4.56)	33-0129-78
1H-Pyrazole, 1,3,5-triphenyl-	EtOH	253(4.58)	78-1571-78
	H$_2$SO$_4$	280(4.34)	78-1571-78
$C_{21}H_{16}N_2O$			
Benzo[h]quinoline-3-carbonitrile, 5,6- dihydro-2-methoxy-4-phenyl-	CHCl$_3$	267(4.19),350(4.46)	48-0097-78
Benzo[h]quinoline-3-carbonitrile, 1,2,5,6-tetrahydro-1-methyl-2- oxo-4-phenyl-	CHCl$_3$	243(4.23),390(4.11)	48-0097-78
Benzo[h]quinoline-3-carbonitrile, 1,2,5,6-tetrahydro-4-(2-methyl- phenyl)-2-oxo-	CHCl$_3$	267(4.25),395(4.38)	48-0097-78
Isoindolo[2,1-a]quinazolin-5(6H)-one, 6a,11-dihydro-6a-phenyl-	EtOH	223(4.55),262(3.82), 268(3.79),346(3.44)	4-1141-78
Phenol, 4-(4,5-diphenyl-1H-imidazol- 2-yl)-	EtOH	298(4.31)	40-1449-78
Phenol, 2-(1,4-diphenyl-1H-pyrazol- 5-yl)-	EtOH	241(4.25),267(4.20)	114-0457-78C
$C_{21}H_{16}N_2O_2$			
1,4-Acridinedione, 2-(dimethylamino)- 9-phenyl-	dioxan	224(4.52),285(4.40), 338s(4.0),425(3.0)	150-4901-78

Compound	Solvent	$\lambda_{max}(\log \epsilon)$	Ref.
1,4-Acridinedione, 5,9-dimethyl-2-(phenylamino)-	dioxan	234(4.59),291(4.44), 321s(3.1),408(3.70), 430(3.70)	150-4901-78
1,4-Acridinedione, 7,9-dimethyl-2-(phenylamino)-	dioxan	237(4.49),291(4.50), 314s(4.4),435(3.69)	150-4901-78
1,4-Acridinedione, 2-(ethylamino)-9-phenyl-	dioxan	225(4.55),284(4.51), 320s(4.2),428(3.55)	150-4901-78
1,4-Acridinedione, 9-ethyl-2-(phenyl-amino)-	dioxan	228(4.61),285(4.51), 319(4.36),439(3.72)	150-4901-78
Benzo[h]quinoline-3-carbonitrile, 1,2,5,6-tetrahydro-4-(4-methoxy-phenyl)-2-oxo-	CHCl$_3$	267(4.28),395(4.41)	48-0097-78

$C_{21}H_{16}N_2O_3$

Compound	Solvent	$\lambda_{max}(\log \epsilon)$	Ref.
1,4-Acridinedione, 5-methoxy-9-methyl-2-(phenylamino)-	dioxan	244(4.54),268(4.25), 305(4.23)	150-4901-78
1,4-Acridinedione, 7-methoxy-9-methyl-2-(phenylamino)-	dioxan	237(4.58),309(4.59), 406(3.73),433(3.73)	150-4901-78
1,4-Acridinedione, 2-[(4-methoxyphenyl)-amino]-9-methyl-	dioxan	228(4.59),283(4.52), 333(4.25),450(3.64)	150-4901-78
Sydnone, 3-(4-methoxyphenyl)-4-[2-(1-naphthalenyl)ethenyl]-, (E)-	benzene	379(4.37)	48-0071-78
	EtOH	371(4.36)	48-0071-78
Yohimban, 5,6-dihydro-5-(methoxycarbo-nyl)-21-oxo-	EtOH	219(4.16),273(3.51), 285(3.43),295(3.51), 341(4.13),363(4.10), 381(4.03)	95-1635-78

$C_{21}H_{16}N_2O_3S_2$

Compound	Solvent	$\lambda_{max}(\log \epsilon)$	Ref.
1H-4,11a-Epithiopyrrolo[3',4':3,4]pyri-do[2,1-b]benzothiazole-1,3,5(2H,4H)-trione, 2-ethyl-3a,11b-dihydro-4-phenyl- (3aα,4α,11aα,11bα)-	MeOH	220(4.63)	44-2697-78

$C_{21}H_{16}N_2O_4$

Compound	Solvent	$\lambda_{max}(\log \epsilon)$	Ref.
9H-Pyrido[3,4-b]indole-3-carboxylic acid, 1-(4-methoxybenzoyl)-, methyl ester	EtOH	221(3.76),293(3.77)	95-1635-78

$C_{21}H_{16}N_2S$

Compound	Solvent	$\lambda_{max}(\log \epsilon)$	Ref.
Benzenamine, N-(3,4-diphenyl-2(3H)-thi-azolylidene)-	EtOH	228s(4.35),272s(4.01), 298(4.08)	94-3017-78
2(1H)-Pyrimidinethione, 1-(4-methyl-phenyl)-4-(2-naphthalenyl)-	MeCN	274(4.53),312(4.56), 348s(3.91),420(3.33)	97-0334-78
2(1H)-Pyrimidinethione, 4-(4-methyl-phenyl)-1-(2-naphthalenyl)-	MeCN	303(4.67),411(3.45)	97-0334-78

$C_{21}H_{16}N_4OS$

Compound	Solvent	$\lambda_{max}(\log \epsilon)$	Ref.
7H,11H-[1,3]Thiazino[2,3-b]pteridin-11-one, 8,9-dihydro-2,3-diphenyl-	pH -3.0	220s(4.28),263(4.32), 303(4.16),370(4.23), 435s(3.05)	24-0971-78
	pH 3.0	220(4.34),263(4.27), 308(4.35),372(4.17)	24-0971-78

$C_{21}H_{16}N_4O_5$

Compound	Solvent	$\lambda_{max}(\log \epsilon)$	Ref.
1H-Pyrazole-3-carboxylic acid, 4,5-di-hydro-4-[(3-hydroxy-1-oxo-1H-inden-2-yl)hydrazono]-5-oxo-1-phenyl-, ethyl ester	MeOH	245(4.41),460(4.26)	142-0199-78B

$C_{21}H_{16}N_6$

Compound	Solvent	$\lambda_{max}(\log \epsilon)$	Ref.
Formazan, 3,5-diphenyl-1-(1-phthalazin-	C$_6$H$_{12}$	282(4.36),410(4.33)	103-0450-78

Compound	Solvent	$\lambda_{max}(\log \epsilon)$	Ref.
y1)- (cont.)	EtOH	287(4.31),438(4.33)	103-0450-78
	EtOH-KOH	531(4.11)	103-0450-78
	H_2SO_4	551(3.93)	103-0450-78
	dioxan	285(4.49),292s(4.49), 410s(4.53),425s(4.56), 445(4.60)	103-0450-78
$C_{21}H_{16}O_2$			
Methanone, bicyclo[2.2.1]hepta-2,5-diene-2,3-diylbis[phenyl-	EtOH	250(4.21)	78-2305-78
1H-Phenalen-1-one, 2,3-dihydro-2-[(4-methoxyphenyl)methylene]-	octane	245(4.23),258(4.24), 295s(--),333(4.15), 362(4.25)	104-2381-78
	MeOH	256(4.43),338s(--), 380(4.17)	104-2381-78
$C_{21}H_{16}O_4$			
5H-3,1,4:4a,5,7-Diethanylylidenecyclobuta[cd]pentaleno[1',6':2,3,4]cyclobut[1,2-g]indene-2,8,10,12(1H,6H)-tetrone, decahydro-	EtOH	220(4.1),300(3.0), 390s(1.8)	104-2131-78
$C_{21}H_{16}O_4SW$			
Tungsten, tricarbonyl[(2,3,4,5,6-η)-1-methyl-3,5-diphenyl-1H-thiopyran 1-oxide]	MeOH	210(4.6),216s(4.6), 270(4.3),325(4.0), 400(3.8)	24-1709-78
$C_{21}H_{16}O_6$			
2-Dibenzofurancarboxaldehyde, 1,7-dihydroxy-4,8-dimethoxy-3-phenyl- (penioflavin)	EtOH	266(4.56),298(4.36), 314s(4.21),372(3.81)	1-0075-78
Furo[3',4':6,7]naphtho[2,3-d]-1,3-dioxol-6(8H)-one, 5-(3,4-dimethoxyphenyl)- (4-deoxyisodiphyllin)	EtOH	207(4.56),224(4.46), 258(4.73),310(4.06), 353(3.76)	78-1011-78
$C_{21}H_{16}O_7$			
Anthra[1,2-b]furan-3-carboxylic acid, 2,3,6,11-tetrahydro-5-hydroxy-6,11-dioxo-2-(2-oxopropyl)-, methyl ester, trans	MeOH	204(4.47),224(4.30), 248(4.47),270(--), 317s(--),453(3.70)	24-3823-78
Furo[3',4':6,7]naphtho[2,3-d]-1,3-dioxol-6(8H)-one, 5-(3,4-dimethoxyphenyl)-9-hydroxy- (isodiphyllin)	EtOH	207(4.51),230(4.47), 267(4.60),313(3.92), 325(3.92),364(3.70)	78-1011-78
$C_{21}H_{16}O_8$			
9,10-Anthracenedione, 1,4,5-triacetoxy-2-methyl- (islandicin triacetate)	MeOH	250(4.57),270s(4.15), 342(3.81)	44-1627-78
$C_{21}H_{17}BrO$			
Benzeneethanol, β-(bromophenylmethylene)-α-phenyl-	EtOH	230(4.27),261s(3.82)	44-0232-78
$C_{21}H_{17}ClN_2O_3S$			
Benzenesulfonic acid, 4-[3-(4-chlorophenyl)-4,5-dihydro-5-phenyl-1H-pyrazol-1-yl]-, sodium salt	H_2O	365(4.42)	40-0271-78
$C_{21}H_{17}ClO$			
Benzeneethanol, β-(chlorophenylmethylene)-α-phenyl-	EtOH	245s(3.98)	44-0232-78

Compound	Solvent	$\lambda_{max}(\log \epsilon)$	Ref.
$C_{21}H_{17}FeN_5O_3$			
Iron, tricarbonyl[9-[(2,3,4,5-η)-2,4-cyclohexadien-1-yl]-N-(phenylmethyl)-9H-purin-6-amine]-	aq HOAc EtOH EtOH-NH$_3$	274(4.26) 273(4.34) 273(4.36)	12-1095-78 12-1095-78 12-1095-78
$C_{21}H_{17}N$			
1H-Phenanthr[9,10-b]azirine, 1a,9b-dihydro-1-(phenylmethyl)-	C_6H_{12}	225(4.10),239(3.86), 272(3.96),275(3.97), 281(3.95),288(3.80), 295(3.64),305(3.49)	44-0397-78
$C_{21}H_{17}NO$			
9H-Indeno[2,1-c]pyridine, 9-[(2-methoxyphenyl)methylene]-3-methyl-, trans	n.s.g.	213(4.8),232(4.88), 258(4.69),286(4.2), 347(4.53)	103-1343-78
Phenanthro[9,10-c]isoxazole, 1,3,3a,11b-tetrahydro-1-phenyl-	MeOH	266(4.23),301(3.23)	44-0381-78
2-Propen-1-one, 1,3-diphenyl-3-(phenylamino)-	C_6H_{12} EtOH	376(4.36) 377(4.40)	48-0945-78 48-0945-78
$C_{21}H_{17}NO_2$			
Benzo[h]quinolin-2(1H)-one, 3-acetyl-5,6-dihydro-4-phenyl-	CHCl$_3$	245(4.21),370(4.25)	48-0097-78
5H-Indeno[1,2-b]pyridine, 5-[(3,4-dimethoxyphenyl)methylene]-	n.s.g.	242(4.0),301(3.62), 311(3.64),364(3.72)	103-1343-78
5H-Naphtho[2,3-a]carbazole-5,13(5aH)-dione, 6,12,12aα,12bα-tetrahydro-12-methyl-, (5aα,12aα,12bα)-	CHCl$_3$	256(4.19),263s(4.12), 294s(3.61),310(3.39), 358(3.43)	12-1841-78
Oxiranecarboxamide, N,N,3-triphenyl-, (E)-	EtOH	233(4.16)	44-0954-78
2-Propen-1-one, 3-(hydroxyphenylamino)-1,2-diphenyl-	MeOH CCl$_4$ DMSO	<u>350(4.1)</u> <u>300s(4.0)</u>,412(4.33) 338(4.19)	39-1113-78C 39-1113-78C 39-1113-78C
Pyrrolo[1,2-a]quinoline-3-carboxylic acid, 1-phenyl-, ethyl ester	MeOH	227s(4.42),240s(4.40), 285(4.16),370(4.0), 420(3.75)	44-2700-78
2(1H)-Quinolinone, 3,4-dihydro-3-hydroxy-1,4-diphenyl-	EtOH	252(4.00)	44-0954-78
$C_{21}H_{17}NO_2S$			
Phenanthridine, 6-[[(4-methylphenyl)-sulfonyl]methyl]-	EtOH	210(4.46),224(4.47), 249s(4.62),254(4.62), 296(3.85),305s(3.80), 336(3.30),352(3.27)	95-1503-78
$C_{21}H_{17}NO_2S_3$			
Methanone, (7-methyl-1,4,8-trithia-6-azaspiro[4.5]deca-6,9-diene-9,10-diyl)bis[phenyl-	CH$_2$Cl$_2$	259.5(4.47)	18-0301-78
$C_{21}H_{17}NO_3$			
5(4H)-Oxazolone, 4-[(2-ethyl-5-methyl-3-benzofuranyl)methylene]-2-phenyl-	MeOH	256(4.15),268(4.09), 382(4.37)	83-0714-78
Phenol, 4-(3-methoxypyrido[1,2-a]indol-10-yl)-, acetate	MeOH	269(4.55),310(4.00), 324(3.98),340(3.89), 420(3.48)	145-1056-78
$C_{21}H_{17}NO_4$			
2-Azabicyclo[3.2.0]hept-6-ene-5-carboxylic acid, 3,4-dioxo-1,7-diphenyl-, ethyl ester, cis	EtOH dioxan	257(4.26),328s(3.75), 414(3.57) 254(4.26),383(2.85),	142-0731-78A 142-0731-78A

Compound	Solvent	λ_{max} (log ϵ)	Ref.
(cont.)		403(2.81)	142-0731-78A
2H-Azepine-4-carboxylic acid, 3-hydr-oxy-2-oxo-6,7-diphenyl-, ethyl ester	EtOH	232s(4.26),325(4.04), 395(3.95)	142-0731-78A
	dioxan	227(4.28),286(3.97), 377(3.73)	142-0731-78A
2H-Azepine-6-carboxylic acid, 3-hydr-oxy-2-oxo-4,7-diphenyl-, ethyl ester	EtOH	375(3.79)	142-0731-78A
	dioxan	230(4.27),328(4.09), 365(4.04)	142-0731-78A
[1,3]Benzodioxolo[5,6-c]phenanthridine, 2-ethoxy-1-methoxy-	EtOH	243(4.57),257(4.56), 277(4.70),324(4.19), 364s(3.47),386(3.40)	94-0514-78
$C_{21}H_{17}NO_5$			
Sanguinarine, dihydro-6-(hydroxymethyl)-	MeOH	212(4.24),235(4.39), 284(4.42),322(4.02), 350s(3.54)	102-0839-78
$C_{21}H_{17}N_3$			
2H-1,2,3-Triazole, 2-(4-methylphenyl)-4,5-diphenyl-	EtOH	218(4.42),298(4.47)	33-1477-78
$C_{21}H_{17}N_3O$			
Imidazo[1,2-a]pyridine, 2-(4-methoxy-phenyl)-7-[2-(3-pyridinyl)ethenyl]-	DMF	367(4.56),383(4.53)	33-0129-78
[1,2,4]Triazolo[1,5-a]pyridine, 7-[2-(4-methoxyphenyl)ethenyl]-2-phenyl-	DMF	339(4.63)	33-0142-78
$C_{21}H_{17}N_3O_2$			
4H-Pyran-3,5-dicarbonitrile, 2-amino-4-(4-methoxyphenyl)-6-(4-methylphen-yl)-	EtOH	300(3.97)	4-0057-78
1H-Pyrazole, 4,5-dihydro-3-(4-nitro-phenyl)-1,5-diphenyl-	n.s.g.	445(4.33)	135-0946-78
$C_{21}H_{17}N_3O_7S$			
4-Thia-1-azabicyclo[3.2.0]hept-2-ene-2-carboxylic acid, 7-oxo-6-[(phen-oxyacetyl)amino]-, (4-nitrophenyl)-methyl ester, (5R-trans)-	EtOH	264(4.08),267(4.09), 274(4.04),310(3.90)	35-8214-78
$C_{21}H_{17}N_5O_3S$			
6H-Anthra[9,1-cd]isothiazol-6-one, 7-[(4,6-diethoxy-1,3,5-triazin-2-yl)-amino]-	DMF	445(4.09)	2-1007-78
$C_{21}H_{17}PS_2$			
Phosphorane, 1,3-dithiol-2-ylidene-triphenyl-	n.s.g.	274(3.84),299(4.02), 310(3.99),322(3.98), 364(3.18),460(2.41)	44-0369-78
$C_{21}H_{18}$			
1,1'-Biphenyl, 2-[2-(3-methylphenyl)-ethenyl]-, cis	MeOH	221s(4.55),248(4.29), 284(4.18)	39-0915-78B
trans	MeOH	220(4.40),250(4.23), 301(4.34),316s(4.28), 331s(3.90)	39-0915-78B
1,1'-Biphenyl, 2-[2-(4-methylphenyl)-ethenyl]-, cis	MeOH	220(4.47),251(4.10), 285(4.15)	39-0915-78B
trans	MeOH	224(4.39),253(4.27), 303(4.45),317s(4.41), 333s(4.11)	39-0915-78B

$C_{21}H_{18}$–$C_{21}H_{18}N_2O_3$

Compound	Solvent	$\lambda_{max}(\log \epsilon)$	Ref.
1-Propene, 1,1,3-triphenyl-	EtOH	250(4.25),263(4.12), 269(3.94),285(2.97)	35-4146-78
1,6:8,17:10,15-Trimethano[18]annulene, syn-syn	C_6H_{12}	258s(3.80),349(5.29), 373s(4.42),422(3.92), 589s(2.30),596s(2.34), 605(2.36),614s(2.34), 625s(2.31),637s(2.18)	89-0956-78
$C_{21}H_{18}BrN_5S_2$			
Thiazolium, 3,4-diphenyl-2-[2-[(2-pyridinylamino)thioxomethyl]hydrazino]-, bromide	EtOH	255(4.14),307(4.40)	104-0181-78
$C_{21}H_{18}ClNO_8$			
3H-2a-Azacyclopenta[ef]heptalene-3,4,5,6-tetracarboxylic acid, 2-chloro-, tetramethyl ester	CHCl$_3$	320(3.56),467(3.76)	39-0429-78C
$C_{21}H_{18}N_2$			
Benzenamine, 4-(9-acridinyl)-N,2-dimethyl- (triiodide)	EtOH	259(4.99),290(4.67), 359(4.59),538(3.85)	30-0130-78
Benzenamine, N-(1,3-dimethyl-9H-indeno[2,1-c]pyridin-9-ylidene)-4-methyl-	EtOH	230(4.35),256(4.55), 292(3.99),304(3.99), 329(3.61),390(3.20), 420(3.22)	104-2365-78
Benzenamine, N-[1-phenyl-3-(phenylamino)-2-propenylidene]-	EtOH	245(4.15),292(3.92), 352(4.47)	48-0659-78
monoperchlorate	EtOH	379(4.44)	48-0659-78
Diazene, phenyl[2-phenyl-1-(phenylmethyl)ethenyl]-	hexane	240(4.28),348(4.27), 468(2.60)	49-1081-78
5H-Dibenzo[b,g][1,5]diazonine, 6,13-dihydro-11-phenyl-	EtOH	244(3.98),268(3.89), 270(3.89)	4-1141-78
Isoindolo[2,1-a]quinazoline, 5,6,6a,11-tetrahydro-6a-phenyl-	EtOH	242(3.84),260(3.74), 268(3.53)	4-1141-78
1H-Pyrazole, 4,5-dihydro-1,3,5-triphenyl-	n.s.g.	315s(--),360(4.30)	135-0946-78
$C_{21}H_{18}N_2O$			
1H-Pyrazol-4-ol, 4,5-dihydro-1,3,5-triphenyl-, cis	EtOH	346(4.23)	78-1571-78
Pyrrolo[3,4-b]pyridin-7-one, 5,6-dihydro-2-methyl-4-phenyl-6-(phenylmethyl)-	EtOH	258.5(4.17)	95-0448-78
$C_{21}H_{18}N_2OS$			
2(10H)-Phenazinone, 10-ethyl-3-[(4-methylphenyl)thio]-	CHCl$_3$	236(3.50),429(3.32), 510(3.05)	103-0795-78
$C_{21}H_{18}N_2O_2$			
1H-Isoindole-1,3(2H)-dione, 2-[(2,3,4-9-tetrahydro-1H-carbazol-6-yl)methyl]-	CHCl$_3$	287(3.70)	103-0856-78
8-Quinolinol, 2,2'-(1-methylethylidene)bis-	C_6H_{12}	254(4.94),312(3.83), 322s(3.77)	18-3489-78
	M HCl	267(4.82),328(3.92), 372(3.49)	18-3489-78
	pH 13	264(4.79),338(3.84), 354s(3.77)	18-3489-78
$C_{21}H_{18}N_2O_3$			
4H-Pyran-3-carboxylic acid, 2-amino-5-cyano-4,6-diphenyl-, ethyl ester	EtOH	296(3.78)	4-0057-78

Compound	Solvent	$\lambda_{max}(\log \epsilon)$	Ref.
$C_{21}H_{18}N_2O_4$			
Benzenebutanoic acid, 2-amino-α-[(2-naphthalenylcarbonyl)amino]-γ-oxo-, ($\pm$)-	n.s.g.	231(4.74),365(3.68)	12-0439-78
$C_{21}H_{18}N_2O_4S$			
Berberine thiocyanate	EtOH	228(4.51),266(4.45), 350(3.51),430(3.76)	2-1100-78
Spiro[2H-anthra[1,2-d]imidazole-2,1'-cyclohexane]-6,11-dione, 4-(methylsulfonyl)-	dioxan	391(3.49)	104-0381-78
$C_{21}H_{18}N_4O_3$			
1H-Purine-2,6-dione, 3,7-dihydro-1,3-diphenyl-8-(tetrahydro-2-furanyl)-	pH 6.0 pH 13.0	275(4.19) 278(4.22)	24-0982-78 24-0982-78
$C_{21}H_{18}N_4O_5S$			
6H-1,3,4-Thiadiazine-5-acetic acid, 2-(benzoylamino)-α-[(benzoyloxy)-imino]-, ethyl ester	EtOH	258(4.34?),351(4.32?)	4-0401-78
$C_{21}H_{18}N_5S_2$			
Thiazolium, 3,4-diphenyl-2-[2-[(2-pyridinylamino)thioxomethyl]hydrazino]-, bromide	EtOH	255(4.14),307(4.40)	104-0381-78
$C_{21}H_{18}N_6O$			
Formamide, N-[2-(4-methylphenyl)-4-[(4-methylphenyl)azo]-2H-benzotriazol-5-yl]-	EtOH	248(4.47),279(4.27), 374(4.48),465(3.33)	33-2628-78
$C_{21}H_{18}N_6O_6$			
1H-Isoindole-1,3(2H)-dione, 2-(3a,4,8-9,13,13a-hexahydro-2,2-dimethyl-8-oxo-4,13-epoxy-1,3-dioxolo[6,7][1,3]-diazocino[1,2-e]purin-6(5H)-yl)-, [3aR-(3aα,4β,13β,13aα)]-	MeOH	256(4.33),279s(4.13)	44-2320-78
$C_{21}H_{18}O$			
1,1'-Biphenyl, 2-[2-(3-methoxyphenyl)-ethenyl]-, cis	MeOH	226s(4.27),248(4.07), 287(3.94)	39-0915-78B
trans	MeOH	225(4.41),245(4.23), 299(4.22),320s(4.03), 335s(3.64)	39-0915-78B
1,1'-Biphenyl, 2-[2-(4-methoxyphenyl)-ethenyl]-, cis	MeOH	223s(4.47),257(4.14), 289(4.19)	39-0915-78B
trans	MeOH	264(4.42),295(4.51), 321s(4.44),335s(4.18)	39-0915-78B
1,1'-Biphenyl, 3'-methoxy-2-(2-phenyl-ethenyl)-, cis	MeOH	246(4.16),275(4.14)	39-0922-78B
trans	MeOH	251(4.18),301(4.39)	39-0922-78B
1,1'-Biphenyl, 4-methoxy-2-(2-phenyl-ethenyl)-, cis	MeOH	218(4.51),256(4.27), 283(4.12)	39-0922-78B
trans	MeOH	220(4.42),258(4.33), 298(4.33)	39-0922-78B
1,1'-Biphenyl, 4'-methoxy-2-(2-phenyl-ethenyl)-, cis	MeOH	221(4.45),259(4.32), 282(4.23)	39-0922-78B
trans	MeOH	223(4.33),263(4.35), 300(4.39)	39-0922-78B
1,1'-Biphenyl, 5-methoxy-2-(2-phenyl-ethenyl)-, cis	MeOH	225(4.47),296(4.11)	39-0922-78B

Compound	Solvent	$\lambda_{max}(\log \epsilon)$	Ref.
1,1'-Biphenyl, 5-methoxy-2-(2-phenyl-ethenyl)-, trans	MeOH	221(4.37),249(4.17), 312(4.41)	39-0922-78B
Phenanthrene, 9,10-dihydro-2-methoxy-9-phenyl-	MeOH	280(4.29)	39-0922-78B
Phenanthrene, 9,10-dihydro-3-methoxy-9-phenyl-	MeOH	262(4.15),270(4.08), 307(3.76)	39-0922-78B
Phenanthrene, 9,10-dihydro-6-methoxy-9-phenyl-	MeOH	261(4.20),305(3.79)	39-0922-78B
Phenanthrene, 9,10-dihydro-7-methoxy-9-phenyl-	MeOH	279(4.26)	39-0922-78B
Phenanthrene, 9,10-dihydro-8-methoxy-9-phenyl-	MeOH	270(4.16),285(3.95)	39-0922-78B
1-Propanone, 1,2,2-triphenyl-	C_6H_{12}	355(2.36),350(2.18), 366(1.64)	112-0751-78
$C_{21}H_{18}O_2$			
Benzoic acid, 4-methyl-, 2,4-cyclopen-tadien-1-ylidene(4-methylphenyl)-methyl ester	EtOH	242(4.32),313(4.27)	1-0463-78
$C_{21}H_{18}O_3$			
2H,8H-Benzo[1,2-b:3,4-b']dipyran-2-one, 8,8-dimethyl-10-(4-methylphenyl)-	EtOH	218(4.50),240s(4.20), 287s(4.02),298(4.07), 333(3.97)	2-0184-78
2,5-Furandione, 3-(diphenylmethylene)-dihydro-4-(1-methylpropylidene)-, (Z)-	benzene	354(3.96)	39-0571-78C
2H-Furo[2,3-h]-1-benzopyran-2-one, 8-(1-methylethyl)-9-(4-methylphenyl)-	EtOH	211(4.4),225s(4.32), 260(4.38),306(3.93)	2-0184-78
Naphtho[2,3-c]furan-1,3-dione, 4-ethyl-3a,4-dihydro-4-methyl-9-phenyl-, cis	EtOH	233(4.18),310(4.05)	39-0571-78C
trans	EtOH	235(4.10),311(4.03)	39-0571-78C
$C_{21}H_{18}O_3S_2$			
Benzo[b]thiophen-3(2H)-one, 2-[(3-acet-oxy-5-methylbenzo[b]thien-2-yl)meth-yl]-5-methyl-	EtOH	233(4.44),261(3.95), 297(3.42),306(3.41), 372(2.95)	22-0241-78
$C_{21}H_{18}O_4$			
2H,8H-Benzo[1,2-b:3,4-b']dipyran-2,9(10H)-dione, 8,8-dimethyl-10-(4-methylphenyl)-	EtOH	216(4.38),246(3.82), 256(3.73),322(4.09)	2-0184-78
Benzoic acid, 4-methoxy-, 2,4-cyclo-pentadien-1-ylidene(4-methoxyphen-yl)methyl ester	EtOH	265(4.31),330(4.29)	1-0463-78
2H-1-Benzopyran-2-one, 7-acetoxy-4-methyl-8-(1-phenyl-1-propenyl)-	MeOH	245(4.18),320(4.14)	2-0579-78
	MeOH-NaOH	235(4.26),375(4.27)	2-0579-78
2H-Furo[2,3-h]-1-benzopyran-2-one, 9-(4-methoxyphenyl)-8-(1-methylethyl)-	EtOH	212(4.4),228s(4.34), 262(4.33),306(3.92)	2-0184-78
$C_{21}H_{18}O_5$			
2H,8H-Benzo[1,2-b:3,4-b']dipyran-2,9(10H)-dione, 10-(4-methoxy-phenyl)-8,8-dimethyl-	EtOH	214(4.40),247(3.94), 255(3.89),322(4.09)	2-0184-78
4H,8H-Benzo[1,2-b:5,4-b']dipyran-4-one, 5-hydroxy-2-(4-methoxyphenyl)-8,8-di-methyl-	MeOH	225(4.08),254(4.41), 332(4.44)	78-3569-78
2H-1-Benzopyran-6-carboxylic acid, 5-hydroxy-4-methyl-2-oxo-8-(3-phenyl-2-propenyl)-, methyl ester	MeOH	251(3.84),305(4.12)	2-0856-78
2H-1-Benzopyran-6-carboxylic acid, 4-methyl-2-oxo-5-[(3-phenyl-2-propen-	MeOH	250(4.14)	2-0856-78

Compound	Solvent	λ_{max}(log ϵ)	Ref.
yl)oxy]-, methyl ester (cont.)			2-0856-78
Leiocarpin	MeOH	205(4.69),227(4.78), 290(4.00),305(4.06)	39-0137-78C
Naphtho[2,3-c]furan-1(3H)-one, 9-(3,4-dimethoxyphenyl)-4-methoxy-	EtOH	243(4.66),289(3.76), 306(3.77),353(3.78)	142-0207-78A
$C_{21}H_{18}O_6$			
2H-Anthra[1,2-b]pyran-5-acetic acid, 3,4,7,12-tetrahydro-6-hydroxy-2-methyl-7,12-dioxo-, methyl ester	MeOH	203(4.62),225(4.48), 249(4.65),277(4.14), 314s(--),467(4.00)	24-3823-78
Desmodol	EtOH	240(4.49),277(4.37), 284(4.35),347(4.26)	94-2411-78
Furo[3',4':6,7]naphtho[2,3-d]-1,3-dioxol-6(8H)-one, 5-(3,4-dimethoxyphenyl)-8a,9-dihydro-	EtOH	247(4.30),352(4.09)	78-1011-78
1-Naphthacenecarboxylic acid, 1,2,3,4-6,11-hexahydro-5,12-dihydroxy-2-methyl-6,11-dioxo-, methyl ester, cis	MeOH	205(4.39),251(4.65), 287(4.02),331s(--), 460s(--),480(4.06), 517(3.80),570(3.30)	150-4762-78
trans	MeOH	206(4.38),252(4.65), 287(3.98),331s(--), 460s(--),480(4.02), 518(3.82),569(3.15)	150-4762-78
Naphtho[2,3-c]furan-1(3H)-one, 9-(1,3-benzodioxol-5-yl)-4,9-dihydro-6,7-dimethoxy-	EtOH	286(3.91)	94-3195-78
Naphtho[2,3-c]furan-1(3H)-one, 9-(1,3-benzodioxol-5-yl)-4,9-dihydro-7,8-dimethoxy-	EtOH	285(3.66)	94-3195-78
$C_{21}H_{18}O_7$			
2-Anthraceneacetic acid, 9,10-dihydro-1,4-dihydroxy-9,10-dioxo-3-(3-oxo-butyl)-, methyl ester	MeOH	204(4.41),227s(--), 248(4.63),254s(--), 281(3.98),315s(--), 458s(--),482(4.01), 515(3.84),565(3.16)	24-3823-78
Furo[3',4':6,7]naphtho[2,3-d]-1,3-dioxole-5,8-dione, 9-(3,4-dimethoxyphenyl)-5a,6,8a,9-tetrahydro-	EtOH	206(4.54),235(4.39), 268(3.95),320(3.80)	78-1011-78
1-Naphthacenecarboxylic acid, 1,2,3,4-6,11-hexahydro-4,5,12-trihydroxy-2-methyl-6,11-dioxo-, methyl ester, cis	MeOH	207(4.39),251(4.64), 286(4.00),331s(--), 459s(--),485(4.00), 519(3.82),565(3.00)	150-4762-78
trans	MeOH	206(4.34),250(4.59), 287(3.98),330s(--), 460s(--),484(4.00), 520(3.80),568(2.90)	150-4762-78
$C_{21}H_{18}O_9$			
Cyclopenta[b]xanthene-7-carboxylic acid, 1,2,3,10-tetrahydro-3,11-dihydroxy-4,7-dimethoxy-9-methyl-1,10-dioxo-, methyl ester	EtOH	260(4.55),295(4.36), 357(3.88)	94-0209-78
$C_{21}H_{19}BrO_6$			
2(5H)-Furanone, 3-(1,3-benzodioxol-5-ylbromomethyl)-4-[(3,4-dimethoxyphenyl)methyl]-	MeOH	230(4.24),255(4.25), 315(3.74),395(3.50)	42-1201-78
$C_{21}H_{19}ClN_2O_4$			
Benzenamine, N-[1-phenyl-3-(phenylami-	EtOH	379(4.44)	48-0659-78

Compound	Solvent	$\lambda_{max}(\log \epsilon)$	Ref.
no)-2-propenylidene]-, monoperchlorate (cont.)			48-0659-78
$C_{21}H_{19}ClN_4O_2$			
2H-1,4-Benzodiazepin-2-one, 7-chloro-1,3-dihydro-5-phenyl-, (3-oxo-1-cyclohexen-1-yl)hydrazone, 4-oxide	isoPrOH	219(4.34),239(4.35), 330(4.66)	4-0161-78
2-Pentanone, 5-(8-chloro-6-phenyl-4H-[1,2,4]triazolo[4,3-a][1,4]benzodiazepin-1-yl)-, N-oxide	isoPrOH	227(4.43),258(4.19), 308(4.04)	4-0161-78
$C_{21}H_{19}ClO_7$			
2H-1-Benzopyran-2-one, 8-(3-chloro-2,6-dimethoxybenzoyl)-5,7-dimethoxy-4-methyl-	MeOH	218(4.22),315(3.97)	119-0143-78
$C_{21}H_{19}I_3N_2$			
Acridinium, 9-(4-amino-3-methylphenyl)-10-methyl-, triiodide	EtOH	259(4.99),290(4.67), 359(4.59),538(3.85)	30-0130-78
$C_{21}H_{19}NO$			
Benzo[h]quinoline-2(1H)-one, 5,6-dihydro-1-methyl-4-(4-methylphenyl)-	$CHCl_3$	244(4.22),295(4.05)	48-0097-78
4H-Indol-4-one, 1,5,6,7-tetrahydro-2-phenyl-1-(phenylmethyl)-	MeOH	240(4.39),285(4.12)	48-0863-78
$C_{21}H_{19}NO_2$			
10H-Azepino[1,2-a]indole, 8-methoxy-11-(4-methoxyphenyl)-	EtOH	227(4.38),257(4.21), 352(4.02)	39-1211-78C
Benzo[h]quinolin-2(1H)-one, 5,6-dihydro-4-(4-methoxyphenyl)-1-methyl-	$CHCl_3$	245(4.21),297(4.05)	48-0097-78
1H-Cyclopropa[3,4]pyrido[1,2-a]indole, 1a,9b-dihydro-1a-methoxy-9-(4-methoxyphenyl)-	EtOH	229(4.35),258(4.47), 313(4.16)	39-1211-78C
5H-Indeno[1,2-b]pyridine, 5-[(3,4-dimethoxyphenyl)methyl]-	n.s.g.	212(4.52),252(3.9), 284(4.0),308(4.12)	103-1343-78
$C_{21}H_{19}NO_3$			
Benzamide, N-(4-methoxyphenyl)-N-(1-methyl-4-oxo-2,5-cyclohexadien-1-yl)-	EtOH	231(4.62),276(4.04)	88-1983-78
3-Pyridinecarboxylic acid, 6-methoxy-2,5-diphenyl-, ethyl ester	EtOH	297(4.25)	142-0161-78A
$C_{21}H_{19}NO_4$			
Benzene, 1-[bis(4-methoxyphenyl)methyl]-2-nitro-	EtOH	230(4.39),268(4.04)	39-1211-78C
$C_{21}H_{19}NO_7$			
1,3-Dioxolo[4,5-h]isoquinoline-6-acetic acid, 7-(1,3-benzodioxol-5-yl)-6,7,8-9-tetrahydro-8-methyl-9-oxo-, methyl ester, trans	EtOH	215s(4.44),235s(4.20), 286(3.77),320(3.64)	44-2852-78
$C_{21}H_{19}N_2S$			
Quinolinium, 1-methyl-2-[3-(3-methyl-2(3H)-benzothiazolylidene)-1-propenyl]-	EtOH	582(5.12)	4-0017-78
Quinolinium, 1-methyl-4-[3-(3-methyl-2(3H)-benzothiazolylidene)-1-propenyl]-	EtOH	630(5.14)	4-0017-78

Compound	Solvent	$\lambda_{max}(\log \epsilon)$	Ref.
$C_{21}H_{19}N_3O$			
2H-Pyrrolo[3,4-d]pyrimidin-2-one, 1-ethyl-1,6-dihydro-5-methyl-4,7-diphenyl-	pH 7.38	276(4.18),305s(--), 385(3.15)	32-0591-78
$C_{21}H_{19}N_3O_2$			
Benzene, 1-azido-2-[bis(4-methoxyphenyl)methyl]-	EtOH	230(4.44)	39-1211-78C
2-Butenoic acid, 2-cyano-3-(4,5-dihydro-4-phenyl-3H-pyrazol-3-yl)-4-phenyl-, methyl ester	EtOH	280(3.70),330(3.01)	78-1163-78
2-Butenoic acid, 2-cyano-4-(4,5-dihydro-4-phenyl-3H-pyrazol-3-yl)-3-phenyl-, methyl ester	EtOH	288(3.95)	78-1163-78
Spiro[2H-anthra[1,2-d]imidazole-2,1'-cyclohexane]-6,11-dione, 4-(methylamino)-	EtOH	620(3.85)	104-0381-78
$C_{21}H_{19}N_3O_4S$			
Acetic acid, [4-benzoyl-2-(benzoylamino)-4H-1,3,4-thiadiazin-5(6H)-ylidene]-, ethyl ester, (E)-	EtOH	258(4.30?),295(4.30?)	4-0401-78
2H-1,3,4-Thiadiazine-5-acetic acid, 3-benzoyl-2-(benzoylimino)-, ethyl ester	EtOH	255(4.28?),265(4.30?)	4-0401-78
$C_{21}H_{19}N_5$			
2H-Benzotriazole, 5,7-dimethyl-4-[(4-methylphenyl)azo]-2-phenyl-	EtOH	270(4.10),348(4.38), 450(3.11)	33-2628-78
2H-Benzotriazole, 5,7-dimethyl-2-(4-methylphenyl)-4-(phenylazo)-	EtOH	275(4.19),349(4.43), 450(3.08)	33-2628-78
$C_{21}H_{19}N_5O$			
2H-Benzotriazole, 5-methoxy-2-(4-methylphenyl)-4-[(4-methylphenyl)azo]-	EtOH	230(4.36),280(4.12), 355(4.41),465(3.30)	33-2628-78
$C_{21}H_{19}N_7OS$			
6H-Anthra[9,1-cd]isothiazol-6-one, 7-[[4,6-bis(dimethylamino)-1,3,5-triazin-2-yl]amino]-	DMF	460(4.05)	2-1007-78
6H-Anthra[9,1-cd]isothiazol-6-one, 7-[[4,6-bis(ethylamino)-1,3,5-triazin-2-yl]amino]-	DMF	445(4.15)	2-1007-78
$C_{21}H_{19}N_7O_5$			
1H-Isoindole-1,3(2H)-dione, 2-(8-amino-3a,4,13,13a-tetrahydro-2,2-dimethyl-4,13-epoxy-1,3-dioxolo[6,7][1,3]diazocino[1,2-e]purin-6(5H)-yl)-, [3aR-(3aα,4β,13β,13aα)]-	MeOH	215(4.81),269(4.31)	44-2320-78
$C_{21}H_{19}O_3P$			
Ethanone, 1-[5-(diphenylphosphinyl)-2-hydroxy-4-methylphenyl]-	EtOH	230(4.77),260(4.08), 320(3.57)	95-0503-78
$C_{21}H_{20}$			
Tricyclo[8.5.1.1^{3,8}]heptadeca-1,3,5,7-9,11,14-heptaene, 12,14-diethenyl-	dioxan	228(4.15),323(5.05), 337s(4.88),406(3.76)	89-0956-78
$C_{21}H_{20}BrNO$			
1H-Inden-1-one, 5-bromo-2-(cyclohexyl-	MeOH	280(4.72),560(2.90)	4-1281-78

Compound	Solvent	$\lambda_{max}(\log \epsilon)$	Ref.
amino)-3-phenyl- (cont.)			4-1281-78
$C_{21}H_{20}F_3NO_5$			
Benzo[6,7]cyclohept[1,2,3-ij]isoquino-line-1,10-diol, 4,5,6,6a,7,8-hexahy-dro-2,11-dimethoxy-6-(trifluoroacet-yl)-, (±)-	EtOH	236(4.32),284(3.82)	44-4076-78
4H-Dibenzo[de,g]quinolin-1-ol, 5,6,6a-7-tetrahydro-2,9,10-trimethoxy-6-(trifluoroacetyl)-, (±)-	EtOH	222(4.56),282(4.03), 306(4.11)	44-4076-78
Spiro[7H-benzo[de]quinoline-7,1'-[2,5]-cyclohexadien]-4'-one, 1,2,3,8,9,9a-hexahydro-6-hydroxy-3',5-dimethoxy-1-(trifluoroacetyl)-, cis-(±)-	EtOH	243s(--),290(3.78)	44-4076-78
$C_{21}H_{20}NO_5PS_3$			
Phosphorothioic acid, O-ethyl O-phenyl S-[(2,3,5,7-tetrahydro-5,7-dioxo-2-phenyl-6H-1,4-dithiino[2,3-c]pyrrol-6-yl)methyl] ester	EtOH	212(4.41),260(4.08), 417(3.54)	73-1093-78
$C_{21}H_{20}N_2O_2$			
Benz[a]indolo[3,2-h]quinolizine-4-carb-oxylic acid, 5,6,8,9,14,14b-hexahy-dro-, methyl ester	EtOH	226(4.68),283(3.99), 291(3.92)	95-0850-78
1,3,5-Cycloheptatrien-1-amine, N-(2-methoxyphenyl)-7-[(2-methoxyphenyl)-imino]-	MeOH	213(4.89),236(4.89), 285(4.54),370(3.60), 415(3.68)	18-2185-78
	MeOH-HCl	215(4.89),252(4.88), 415(3.75)	18-2185-78
Ethanone, 1-(4-acetyl-3-methyl-1-(phen-ylmethyl)-1H-pyrazol-5-yl]-2-phenyl-	n.s.g.	214(4.12),260(3.92)	44-2665-78
Ethanone, 1-(4-acetyl-5-methyl-1-(phen-ylmethyl)-1H-pyrazol-3-yl]-2-phenyl-	n.s.g.	222(4.18),258(3.93)	44-2665-78
1H-Isoindole-1,3(2H)-dione, 2-[(1,2,3-7-tetramethyl-1H-indol-6-yl)methyl]-	CHCl₃	234(4.74),292(3.95)	103-0856-78
1-Naphthaleneacetic acid, [1-(4-meth-oxyphenyl)ethylidene]hydrazide	EtOH	286(4.37)	34-0345-78
$C_{21}H_{20}N_2O_3$			
1H-Pyrazole-4-carboxylic acid, 3-acet-yl-5-phenyl-1-(phenylmethyl)-, ethyl ester	n.s.g.	230(4.17)	44-2665-78
1H-Pyrazole-4-carboxylic acid, 5-acet-yl-3-phenyl-1-(phenylmethyl)-, ethyl ester	n.s.g.	213(4.11),240(4.24)	44-2665-78
$C_{21}H_{20}N_2O_4$			
1,7-Dioxa-2,8-diazaspiro[4.4]nona-2,8-diene-6-carboxylic acid, 6-methyl-3,9-diphenyl-, ethyl ester	EtOH	257(4.34)	22-0415-78
2H-Pyrrole-3-carboxylic acid, 3,4-di-hydro-2-(4-nitrophenyl)-5-[2-(2-propenyl)phenyl]-, methyl ester, cis	C₆H₁₂	255(4.13)	35-2181-78
trans	C₆H₁₂	225(4.12)	35-2181-78
$C_{21}H_{20}N_2O_4S$			
Spiro[2H-anthra[1,2-d]imidazole-2,1'-cyclohexane]-6,11-dione, 1,3-dihydro-4-(methylsulfonyl)-	EtOH	526(4.13)	104-0381-78

Compound	Solvent	$\lambda_{max}(\log \epsilon)$	Ref.
$C_{21}H_{20}N_2O_5$ 2H-Pyrrole-3-carboxylic acid, 3,4-di- hydro-2-(4-nitrophenyl)-5-[2-(2-pro- penyloxy)phenyl]-, methyl ester	n.s.g.	270(4.07)	35-3494-78
$C_{21}H_{20}N_4O_3S$ Pyrazolo[3,4-d][1,2]diazepine, 6-benz- oyl-2,3,3a,6-tetrahydro-8-methyl-2- [(4-methylphenyl)sulfonyl]-	EtOH	328(4.13)	39-1297-78C
$C_{21}H_{20}N_4O_4S$ Benz[cd]indole-6-sulfonamide, N,N-di- ethyl-2-[(4-nitrophenyl)amino]-	dioxan	402(4.29)	93-1354-78
$C_{21}H_{20}N_8O_{12}$ 2,3-Pentanedione, 4,5-diacetoxy-, 2,3- bis(2,4-dinitrophenylhydrazone)	$CHCl_3$	390(4.43),437(4.43)	136-0117-78H
$C_{21}H_{20}O_2$ Ethanone, 1,1'-(1,1',3,3'-tetrahydro- 2,2'-spirobi[2H-indene]-5,5'-diyl)- bis-	EtOH	208(4.66),211s(4.65), 217s(4.59),258(4.49), 283s(3.72),291s(3.64)	49-0987-78
2H-Pyran-2-one, 4,5-dimethyl-3,6-bis(4- methylphenyl)-	MeCN	242(4.13),325(4.22)	44-2138-78
4H-Pyran-4-one, 2,6-dimethyl-3,5-bis(4- methylphenyl)-	EtOH	221(4.34),261(4.03), 297(2.93)	44-2138-78
$C_{21}H_{20}O_3$ Benzeneacetic acid, α-(2-methyl-4-oxo- 1-phenyl-2-pentenylidene)-, methyl ester	EtOH	246(4.19),288(3.98)	33-3018-78
2,2'-Spirobi[2H-indene]-5-carboxylic acid, 5'-acetyl-1,1',3,3'-tetra- hydro-, methyl ester	EtOH	206(4.68),211s(4.63), 249(4.38),287(3.63), 310s(2.35)	49-0987-78
$C_{21}H_{20}O_3Se$ 2-Cyclohexene-1-carboxylic acid, 4-oxo- 1-phenyl-5-(phenylseleno)-, ethyl ester	MeOH	218(4.23),265(3.5)	24-1944-78
$C_{21}H_{20}O_4$ 2H,8H-Benzo[1,2-b:3,4-b']dipyran-2-one, 9,10-dihydro-9-hydroxy-8,8-dimethyl- 10-(4-methylphenyl)-	EtOH	212(4.52),247(3.67), 258(3.64),331(4.15)	2-0184-78
4H-1-Benzopyran-4-one, 3-methoxy-7-[(3- methyl-2-butenyl)oxy]-2-phenyl- $(2\lambda,3\epsilon)$	MeOH	266(4.57),310(4.55), ?(4.58)	2-0973-78
Hemileiocarpin	MeOH	245(4.09),287(2.97)	39-0137-78C
2,2'-Spirobi[2H-indene]-5,5'-dicarbox- ylic acid, 1,1',3,3'-tetrahydro-, dimethyl ester	EtOH	207(4.76),211s(4.67), 243(4.43),278(3.55), 287(3.51)	49-0987-78
$C_{21}H_{20}O_5$ Acacetin, 6-C-prenyl-	MeOH	220(4.10),250(4.42), 330(4.34)	78-3569-78
1,3-Benzodioxol-5-ol, 6-(3,4-dihydro- 8,8-dimethyl-2H,8H-benzo[1,2-b:3,4- b']dipyran-3-yl)-, (S)- (leiocin)	MeOH	207(4.65),227(4.55), 280(4.12),292(4.11)	39-0137-78C
2H,8H-Benzo[1,2-b:3,4-b']dipyran-2-one, 9,10-dihydro-9-hydroxy-10-(4-methoxy- phenyl)-8,8-dimethyl-	EtOH	212(4.52),250(3.79), 260(3.77),330(4.15)	2-0184-78

Compound	Solvent	$\lambda_{max}(\log \epsilon)$	Ref.
[3,6'-Bi-4H-1-benzopyran]-4-one, 2',3'-dihydro-3'-hydroxy-7-methoxy-2',2'-dimethyl-	EtOH	265(4.5),312s(4.05)	102-2046-78
[3,6'-Bi-4H-1-benzopyran]-4-one, 2',3'-dihydro-7-hydroxy-3'-methoxy-2',2'-dimethyl-	EtOH	262(4.39),310s(3.94)	102-2046-78
2-Furanacetic acid, 2,5-dihydro-2-methoxy-α-methyl-5-oxo-3,4-diphenyl-, methyl ester	EtOH	222(4.19),238s(3.93), 294(4.03)	33-3018-78

$C_{21}H_{20}O_6$

Compound	Solvent	$\lambda_{max}(\log \epsilon)$	Ref.
9,10-Anthracenedione, 2-acetyl-6,8-diethoxy-1-hydroxy-3-methyl-	EtOH	231(4.45),278s(4.33), 286(4.35),435(3.99)	12-1363-78
9,10-Anthracenedione, 1,4-dihydroxy-2-(2,2,6-trimethyl-1,3-dioxan-4-yl)-, cis	MeOH	202(4.36),231s(--), 247(4.56),251s(--), 271s(--),321(3.64), 461s(--),478(3.80), 508s(--),563s(--)	24-3823-78
trans	MeOH	202(4.18),230s(--), 246(4.40),251s(--), 282(3.79),318s(--), 456s(--),477(3.40), 508s(--),564(2.35)	24-3823-78
Benzeneacetic acid, 2-hydroxy-5-(3-methoxy-3-oxo-1-propenyl)-α-[(4-methoxyphenyl)methylene]-, methyl ester	EtOH	230(4.23),307(4.50)	33-2646-78
4H-1-Benzopyran-4-one, 7-hydroxy-8-(3-hydroxy-3-methyl-2-oxobutyl)-5-methoxy-2-phenyl-	MeOH	212(4.55),268(4.36), 312(4.07)	119-0047-78
2(3H)-Furanone, 3-(1,3-benzodioxol-5-ylmethylene)-4-[(3,4-dimethoxyphenyl)methyl]dihydro- (suchilactone)	MeOH	233(4.35),282(4.09), 335(4.34)	42-1201-78
Furo[3',4':6,7]naphtho[2,3-d]-1,3-dioxol-6(5aH)-one, 5-(3,4-dimethoxyphenyl)-5,8,8a,9-tetrahydro-	EtOH	211(4.34),235(3.92), 288(3.68)	78-1011-78
Isosuchilactone	MeOH	233(4.22),282(4.09), 333(4.22)	42-1201-78
Leiocinol	MeOH	208(4.41),287(3.86), 297(3.85)	39-0137-78C

$C_{21}H_{20}O_7$

Compound	Solvent	$\lambda_{max}(\log \epsilon)$	Ref.
2-Anthraceneacetic acid, 9,10-dihydro-1,4-dihydroxy-3-(3-hydroxybutyl)-9,10-dioxo-, methyl ester	MeOH	204(4.34),228s(--), 250(4.62),253s(--), 288(3.99),324s(--), 460s(--),483(3.97), 516(3.81),566(3.11)	34-3823-78
9,10-Anthracenedione, 2-acetyl-1,5,6,8-tetramethoxy-3-methyl-	EtOH	227(4.41),252(4.36), 281(4.21),410(3.83)	44-1435-78
4H-1-Benzopyran-4-one, 3-[2,4-dihydroxy-5-methoxy-3-(3-methyl-2-butenyl)phenyl]-5,7-dihydroxy- (piscerythrone)	EtOH	267(4.38),295s(4.22), 335(3.88)	138-0879-78
1,2-Phenanthrenediol, 5,6,7-trimethoxy-, diacetate	n.s.g.	260(4.91)	39-0739-78C
2,3-Phenanthrenediol, 4,7,8-trimethoxy-, diacetate	n.s.g.	264(4.64)	39-0739-78C
3,4-Phenanthrenediol, 2,7,8-trimethoxy-, diacetate	n.s.g.	263(4.62)	39-0739-78C

$C_{21}H_{20}O_8$

Compound	Solvent	$\lambda_{max}(\log \epsilon)$	Ref.
2-Anthracenecarboxylic acid, 9,10-dihydro-3,5,6,8-tetramethoxy-1-methyl-	EtOH	224(4.46),266(4.46), 280(4.41),415(3.88)	39-1041-78C

Compound	Solvent	λ_{max}(log ϵ)	Ref.
9,10-dioxo-, methyl ester (cont.)			39-1041-78C
Mucronucarpan, diacetate, (6aS,11aS)-	EtOH	230(4.08),285(3.92)	102-1405-78
Podophyllotoxin, 4'-demethyl-	EtOH	290(3.60)	100-0497-78
C$_{21}$H$_{20}$O$_{10}$			
Conyzorigun	n.s.g.	272(4.28),338(4.34)	77-0152-78
Vitexin (as O-glucoside)	n.s.g.	273(4.21),305s(4.10),	102-0824-78
		335(4.19)	
C$_{21}$H$_{20}$O$_{12}$			
Hyperoside	EtOH	255(4.37),305s(3.97),	95-1395-78
		365(4.30)	
Isoquercitrin	EtOH	259(4.34),305s(3.94),	95-1395-78
		362(4.29)	
C$_{21}$H$_{20}$S			
2H-1-Benzothiopyran, 5,6,7,8-tetrahy-dro-2,4-diphenyl-	hexane	252(4.2),360(3.9)	103-0605-78
5H-1-Benzothiopyran, 6,7,8,8a-tetrahy-dro-2,4-diphenyl-	hexane	225(4.3),237(4.2),298(3.1)	103-0605-78
Cyclopenta[b]thiopyran, 4,5,6,7-tetra-hydro-2-phenyl-4-(phenylmethyl)-	hexane	189(5.0),203s(4.7),240(4.3),297(3.3)	103-0605-78
C$_{21}$H$_{21}$BF$_4$O$_2$			
Cyclopentylium, 2,5-bis[(4-methoxyphen-yl)methylene]-, tetrafluoroborate	C$_2$H$_4$Cl$_2$	664(4.75)	104-1197-78
2,4,6-Heptatrienylium, 1,7-bis(4-meth-oxyphenyl)-, tetrafluoroborate	C$_2$H$_4$Cl$_2$	717(4.70)	104-1197-78
C$_{21}$H$_{21}$ClN$_4$O$_5$S			
Benzenesulfonamide, N-(3-chloro-7,7a-10a,11-tetrahydro-9,9-dimethyl-7,11-epoxy-5H,6H,9H-8,10-dioxa-2,5,11a-triazacyclopenta[6,7]cyclooct[1,2,3-cd]inden-5-ylidene)-4-methyl-,[7R-(7α,7aβ,10aβ,11α)]-	EtOH-HCl EtOH-pH 7	224(4.35),223s(4.34),262(4.01),348(4.32) 224(4.35),233s(4.34),262s(4.00),348(4.33)	87-0112-78 87-0112-78
C$_{21}$H$_{21}$ClO			
Dispiro[cyclopropane-1,3'(9'H)-[4,10]-etheno[5,8]methanobenz[f]azulene-13',1"-cyclopropan]-9'-one, 10'-chloro-3'a,4',4'a,5',8',8'a,10',10'a-octahydro-	MeOH	225s(3.70),254(3.26)	18-1257-78
C$_{21}$H$_{21}$Cl$_3$N$_2$O			
1(2H)-Isoquinolinone, 4-[4-[bis(2-chlo-roethyl)amino]phenyl]-2-(2-chloro-ethyl)-	EtOH	210(4.65),224(4.43),266(4.43)	44-0672-78
C$_{21}$H$_{21}$F$_3$N$_4$O$_4$			
Butanamide, N-(4-butylphenyl)-2-[[4-ni-tro-2-(trifluoromethyl)phenyl]hydra-zono]-3-oxo-	EtOH	206(4.58),250(4.16),390(4.50)	104-0521-78
C$_{21}$H$_{21}$N			
1-Azaspiro[4.5]deca-1,3-diene, 2,4-di-phenyl-	EtOH	248(4.29)	88-4511-78
C$_{21}$H$_{21}$NO			
1H-Inden-1-one, 2-(cyclohexylamino)-3-phenyl-	MeOH	265(4.76),525(4.13)	4-1281-78

Compound	Solvent	$\lambda_{max}(\log \epsilon)$	Ref.
$C_{21}H_{21}NO_2$			
Benzenamine, 2-[bis(4-methoxyphenyl)-methyl]-	EtOH	230(4.39),278(3.79)	39-1211-78C
1,3-Cyclohexanedione, 2-[([1,1'-biphen-yl]-4-ylamino)methylene]-5,5-dimeth-yl-	EtOH	364(4.54)	97-0256-78
Cyclopropanecarboxylic acid, 1-cyano-2-methyl-2-phenyl-3-(1-phenylethyl)-, methyl ester	EtOH	213(4.05),252(2.16), 260(2.38),264(2.38)	44-0626-78
$C_{21}H_{21}NO_3$			
3-Pyridinecarboxylic acid, 1,4,5,6-tet-rahydro-1-methyl-6-oxo-2,5-diphenyl-, ethyl ester	EtOH and EtOH-KOH	286(4.02)	142-0161-78A
$C_{21}H_{21}NO_4$			
1H-Pyrrole-3-carboxylic acid, 4,5-di-hydro-5-hydroxy-5-methyl-1-(4-meth-ylphenyl)-4-oxo-2-phenyl-, ethyl ester	EtOH	210(4.11),246(4.23), 335(4.06)	118-0291-78
1H-Pyrrole-3-carboxylic acid, 4,5-di-hydro-5-hydroxy-5-methyl-2-(4-meth-ylphenyl)-4-oxo-1-phenyl-, ethyl ester	EtOH	210(4.17),247(4.24), 338(4.09)	118-0291-78
1H-Pyrrole-3-carboxylic acid, 4,5-di-hydro-5-hydroxy-5-methyl-4-oxo-2-phenyl-1-(phenylmethyl)-, ethyl ester	EtOH	210(4.10),243(4.18), 328(3.98)	118-0291-78
$C_{21}H_{21}NO_5$			
1H-Pyrrole-3-carboxylic acid, 4,5-di-hydro-5-hydroxy-1-(4-methoxyphenyl)-5-methyl-4-oxo-2-phenyl-, ethyl ester	EtOH	210(4.16),243(4.30), 334(4.06)	118-0291-78
1H-Pyrrole-3-carboxylic acid, 4,5-di-hydro-5-hydroxy-2-(4-methoxyphenyl)-5-methyl-4-oxo-1-phenyl-, ethyl ester	EtOH	210(4.14),242(4.21), 333(4.18)	118-0291-78
$C_{21}H_{21}NO_6$			
Furo[3,4-e]-1,3-benzodioxol-8(6H)-one, 6-[[6-[2-(dimethylamino)ethyl]-1,3-benzodioxol-5-yl]methyl]-, (±)-	EtOH	238s(2.56),292(2.43), 313(2.40)	78-0635-78
Peshawarine, (-)-	EtOH	228(4.96),245(4.32), 293(3.78),333(3.69)	78-0635-78
$C_{21}H_{21}NO_{11}S$			
α-D-Xylofuranoside, methyl 5-benzoate 3-methanesulfonate 2-(4-nitrobenzo-ate)	EtOH	231(4.20),257(4.12)	136-0257-78H
$C_{21}H_{21}N_2S_2$			
Benzothiazolium, 3-ethyl-2-[3-(3-ethyl-2(3H)-benzothiazolylidene)-1-propen-yl]-	EtOH	559(5.20)	124-0942-78
$C_{21}H_{21}N_3O$			
2H-Pyrrolo[3,4-d]pyrimidin-2-one, 1-ethyl-1,3,4,6-tetrahydro-5-methyl-4,7-diphenyl-	MeOH	297(4.04)	32-0591-78
$C_{21}H_{21}N_3O_2S$			
Benz[cd]indole-6-sulfonamide, N,N-di-ethyl-2-(phenylamino)-	dioxan	395(4.25)	93-1354-78

Compound	Solvent	$\lambda_{max}(\log \epsilon)$	Ref.
$C_{21}H_{21}N_3O_3$ Pyrazinecarboxylic acid, 1-benzoyl- 1,2,5,6-tetrahydro-2,5-dimethyl- 6-(methylimino)-3-phenyl-	n.s.g.	200(4.70),217(4.25), 224(4.23),236(3.92)	70-0963-78
$C_{21}H_{21}N_3O_7$ 2H-Pyrano[3',2':5,6][1,4]dioxino[2,3- c]pyridine-7-carbonitrile, 4-acetoxy- 3,4,4a,10a-tetrahydro-3,4a-dimethyl- 2-methyl-9-(2-pyridinyl)-, [2S- (2α,3β,4α,4aα,10aα)]-	MeOH	229(4.17),257(3.84), 292(3.50)	23-1836-78
$C_{21}H_{21}N_3O_8$ DL-Aspartic acid, N-[N-[4-[[(phenyl- methoxy)carbonyl]amino]benzoyl]- glycyl]-	EtOH	268(4.39)	87-1165-78
$C_{21}H_{21}N_7O_{14}$ Riboflavin, 9-nitro-, 2',3'-diacetate dinitrate	EtOH	220(4.50),270(4.46), 342(3.79),439(4.00)	104-0406-78
$C_{21}H_{22}$ Bicyclo[5.2.0]non-8-ene, 8,9-diphenyl- Pyrene, 10b-butyl-10b,10c-dihydro-10c- methyl-, trans	MeCN C_6H_{12}	296(4.17),324s(3.80) 238(3.80),273(2.89), 324(4.53),340(5.00), 343(5.04),358(4.40), 379(4.63),383(4.72), 420(3.49),441(3.63), 463(3.79),478(3.85), 485(3.80),533(2.81), 542(2.77),575(2.63), 581(2.76),592(2.98), 605(3.15),618(3.26), 634(3.28),647(3.32), 652(3.32)	24-3608-78 44-0727-78
$C_{21}H_{22}BrNO_4$ 6H-Benzo[a][1,3]benzodioxolo[5,6-g]- quinolizine, 4-bromo-5,8,14,14a- tetrahydro-2,3-dimethoxy-14-methyl- isomer	EtOH EtOH	288(3.88) 288(3.88)	44-1992-78 44-1992-78
$C_{21}H_{22}BrNO_7$ 3,4-Isoquinolinedione, 1-[(2-bromo- 4,5-dimethoxyphenyl)methyl]-1,2- dihydro-1,6,7-trimethoxy-	EtOH	213(4.29),245(4.02), 292(3.69),340(3.45)	88-2179-78
$C_{21}H_{22}NOP$ 1H-Phosphorino[3,4-b]indole, 2-(1,1-di- methylethyl)-2,9-dihydro-4-phenyl-, 2-oxide	CHCl$_3$	234(4.37),254(4.24), 345(3.63)	139-0257-78B
$C_{21}H_{22}N_2$ 1-Azaspiro[4.5]deca-1,3-dien-2-amine, 3,4-diphenyl-	EtOH	253(4.19)	88-4511-78
$C_{21}H_{22}N_2O$ 1-Butanone, 1-(4,5-dihydro-5-methyl- 4-methylene-1,3-diphenyl-1H-pyrazol- 5-yl)-	EtOH	250(4.07),380(4.03)	22-0415-78

Compound	Solvent	$\lambda_{max}(\log \epsilon)$	Ref.
4(1H)-Pyridinone, 2-(dimethylamino)-3,3-dimethyl-5,6-diphenyl-	EtOH	250(4.33),290s(3.93), 410(4.10)	27-0468-78
$C_{21}H_{22}N_2OS$			
Azepino[4,3-b][1,4]benzothiazine, 1-ethoxy-3,4,5,11-tetrahydro-11-(phenylmethyl)-	EtOH	252(4.20),283(3.76)	103-0306-78
$C_{21}H_{22}N_2O_2$			
1H-Pyrazole-5-carboxylic acid, 4-ethylidene-4,5-dihydro-5-methyl-1,3-diphenyl-, diethyl ester, (E)-	EtOH	247(4.14),365(4.20)	22-0415-78
$C_{21}H_{22}N_2O_3$			
Catharanthine, 19-oxo-	MeOH	222(--),276(3.92), 283(3.95),292(3.89)	33-0690-78
Tabersonine, 3-oxo-	MeOH	297(4.12),331(4.23)	94-1182-78
$C_{21}H_{22}N_2O_4$			
Buxomeline	EtOH	297(4.95),330(4.21)	102-1452-78
Vincadifformine, 14,15-epoxy-3-oxo-	MeOH	222(4.18),297(3.99), 328(4.11)	94-1182-78
$C_{21}H_{22}N_3O_4P$			
3-Phosphorinol, 1-(1,1-dimethylethyl)-1,6-dihydro-2-[(4-nitrophenyl)azo]-5-phenyl-, 1-oxide	CHCl$_3$	241(4.11),320(3.97), 407s(4.51),425(4.52)	139-0027-78B
$C_{21}H_{22}N_3S_2$			
Benzothiazolium, 2-[2-amino-3-(3-ethyl-2(3H)-benzothiazolylidene)-1-propenyl]-3-ethyl-	EtOH	471(4.76)	124-0942-78
$C_{21}H_{22}N_4O_4S$			
Carbamic acid, dimethyl-, 2,4,5,6-tetrahydro-6-[[(4-methylphenyl)sulfonyl]-pyrazolo[4,3-d][1]benzazepin-3-yl ester	n.s.g.	240(4.30),265s(4.13)	39-0862-78C
$C_{21}H_{22}N_8O_5$			
1H-Pyrrole-2-carboxamide, N-[5-[[(2-cyanoethyl)amino]carbonyl]-1-methyl-1H-pyrrol-3-yl]-1-methyl-4-[[(1-methyl-4-nitro-1H-pyrrol-2-yl)carbonyl]amino]-	DMF	295(4.44)	78-2389-78
$C_{21}H_{22}O$			
Ethanone, 1-(5'-ethyl-1,1',3,3'-tetrahydro-2,2'-spirobi[2H-inden]-5-yl)-, (R)-	EtOH	212s(4.52),258(4.17), 278(3.75),293s(3.34)	49-0987-78
$C_{21}H_{22}O_2$			
2,2'-Spirobi[2H-indene]-5-carboxylic acid, 5'-ethyl-1,1',3,3'-tetrahydro-, methyl ester	EtOH	212s(4.61),243(4.17), 269(3.80),273(3.54), 279(3.63),287(3.24)	49-0987-78
$C_{21}H_{22}O_3$			
1H-Indene-1-carboxylic acid, 4-[4-(1,1-dimethylethyl)benzoyl]-2,3-dihydro-	EtOH	261(4.29)	94-1776-78
1H-Inden-1-one, 6-[2-(benzoyloxy)ethyl]-2,3-dihydro-2,5,7-trimethyl-, (R)-	EtOH	230s(4.12),259(4.11), 302(3.27)	94-2365-78

Compound	Solvent	$\lambda_{max}(\log \epsilon)$	Ref.
$C_{21}H_{22}O_3S$			
2H-Phenanthro[1,2-c]thiopyran-1-ol, 1,5,6,12a-tetrahydro-8-methoxy-12a-methyl-, acetate, (1R-cis)-	EtOH	228(4.35),243(4.32), 276(4.35),282(4.35), 306(4.27)	142-0181-78C
2H-Phenanthro[1,2-c]thiopyran-1-ol, 1,11,12,12a-tetrahydro-8-methoxy-12a-methyl-, acetate	EtOH	228(4.38),248(4.35), 270(4.38),279(4.43), 308(4.43),318(4.40)	39-1252-78C
$C_{21}H_{22}O_4$			
Glabridin, 4'-O-methyl-	MeOH	281(4.07),290s(3.96)	142-1533-78A
2-Propen-1-one, 1-(2-hydroxy-4-methoxy-3-(3-methyl-2-butenyl)phenyl]-3-(4-hydroxyphenyl)- (4-hydroxyderricin)	EtOH	243s(4.14),365(4.23)	95-0210-78
$C_{21}H_{22}O_6$			
Benzeneacetic acid, 2-hydroxy-5-(3-methoxy-3-oxo-1-propenyl)-α-[(4-methoxyphenyl)methylene]-, methyl ester, dihydro deriv.	EtOH	222(4.29),300(4.28), 306(4.28)	33-2646-78
4H-1-Benzopyran-4-one, 2,3-dihydro-5,7-dihydroxy-3-[4-hydroxy-2-methoxy-3-(3-methyl-2-butenyl)phenyl]-, (-)-	EtOH	291(4.43),330s(4.02)	94-3863-78
	EtOH-NaOAc	302s(4.23),331(4.49)	94-3863-78
	EtOH-AlCl₃	312(4.43),380(3.72)	94-3863-78
$C_{21}H_{22}O_7$			
1,2-Benzenediol, 5-[2-(2,3-dimethoxyphenyl)ethenyl]-3-methoxy-, diacetate	n.s.g.	301(4.20)	39-0739-78C
1,2-Benzenediol, 3-[2-(3,4,5-trimethoxyphenyl)ethenyl]-; diacetate	n.s.g.	305(4.24)	39-0739-78C
2H-1-Benzopyran-7-ol, 3-(3-acetoxy-2,4-dimethoxyphenyl)-3,4-dihydro-, acetate	EtOH	224(4.25),278(3.65), 284(3.62)	102-1405-78
$C_{21}H_{22}O_8$			
Agarotetrol diacetate	MeOH	213(4.17),250(4.12)	88-3921-78
Aloin, 11-deoxy-, (+)-	EtOH	260s(3.79),268(3.87), 296(3.89),358(4.03)	94-3792-78
4H-1-Benzopyran-4-one, 5,6,7-trimethoxy-2-(2,4,5-trimethoxyphenyl)-	MeOH	252(4.32),305(4.11), 350(4.32)	78-1593-78
$C_{21}H_{22}O_9$			
Cassialoin	EtOH	260s(3.78),268(3.91), 300(3.91),365(4.08)	94-3792-78
$C_{21}H_{22}S$			
2H-1-Benzothiopyran, 3,5,6,7,8,8a-hexahydro-2,4-diphenyl-	hexane	209(4.3),212s(4.3), 238(3.9),258s(3.7), 263s(3.6)	103-0605-78
$C_{21}H_{23}BrN_5O_6PS_2$			
Benzenesulfonic acid, 4-methyl-, 2,2'-[[(4-bromobenzoyl)amino]phosphinylidene]dihydrazide	MeOH	229(4.4092)	65-0720-78
$C_{21}H_{23}ClN_2OS$			
Ethanone, 1-[8-chloro-10,11-dihydro-10-(4-methyl-1-piperazinyl)]dibenzo[b,f]thiepin-2-yl-	MeOH	243(4.28),256s(4.20), 313(3.93)	73-0471-78
$C_{21}H_{23}ClN_5O_6PS_2$			
Benzenesulfonic acid, 4-methyl-, 2,2'-[[(4-chlorobenzoyl)amino]phosphin-	MeOH	228(4.5366)	65-0720-78

Compound	Solvent	$\lambda_{max}(\log \epsilon)$	Ref.
ylidene]dihydrazide (cont.)			65-0720-78
$C_{21}H_{23}FN_5O_6PS_2$			
Benzenesulfonic acid, 4-methyl-, 2,2'-[[(4-fluorobenzoyl)amino]phosphin-ylidene]dihydrazide	MeOH	228(4.3978)	65-0720-78
$C_{21}H_{23}NO_4$			
Benzo[e][1,3]dioxolo[4,5-k][3]benzaze-cine, 5,6,7,8-tetrahydro-3,4-dimeth-oxy-6-methyl-, (E)-	EtOH	295(3.64)	39-0642-78B
Isoquinoline, 3-(6-ethenyl-1,3-benzo-dioxol-5-yl)-1,2,3,4-tetrahydro-7,8-dimethoxy-2-methyl-	EtOH	262(3.25),300(3.69)	39-0642-78B
$C_{21}H_{23}NO_5$			
Canadaline, (±)-	EtOH	226(4.39),262(3.82), 288(3.90),331(3.42)	78-0635-78
$C_{21}H_{23}NO_6$			
Isoquinoline, 6,7,8-trimethoxy-1-(3,4,5-trimethoxyphenyl)-	MeOH	243(4.51),303(4.78), 328(3.73)	142-0001-78A
Methanone, (3,4-dihydro-6,7-dimethoxy-1-isoquinolinyl)[2-(hydroxymethyl)-4,5-dimethoxyphenyl]-	EtOH	234(4.10),283(3.97), 316(3.78)	142-1233-78A
$C_{21}H_{23}N_3OS$			
4(5H)-Thiazolone, 5-[[4-(diethylamino)-phenyl]methylene]dihydro-3-methyl-2-(phenylimino)-	MeOH	422(4.71)	104-0997-78
	dioxan	251(4.18),414(4.70)	104-0997-78
	70% dioxan	233(4.24),423(4.71)	104-0997-78
4(5H)-Thiazolone, 5-[[4-(diethylamino)-phenyl]methylene]-2-(methylphenyl-amino)-	MeOH	283(4.33),435(4.72)	104-0997-78
	dioxan	282(4.27),415(4.70)	104-0997-78
	70% dioxan	284(4.28),434(4.69)	104-0997-78
$C_{21}H_{23}N_3O_2$			
Pericyclivine, 15(R)-acetoxy-16(R)-cya-no-16-de(carbomethoxy)-19,20(S)-di-hydro-	MeOH	270(3.92),279(3.92), 286(3.81)	35-0938-78
$C_{21}H_{23}N_3O_5$			
1,2-Benzisoxazole, 3a,4,5,6,7,7a-hexa-hydro-7a-morpholino-3-[5-(3-nitro-phenyl)-2-furanyl]-	MeOH	217(4.09),244(4.07), 320(4.46)	73-0288-78
1,2-Benzisoxazole, 3a,4,5,6,7,7a-hexa-hydro-7a-morpholino-3-[5-(4-nitro-phenyl)-2-furanyl]-	MeOH	227(3.98),293(4.00), 370(4.39)	73-0288-78
$C_{21}H_{23}N_3O_8$			
Caerulomycin D monoacetate	MeOH	231(4.15),266(3.97), 305s(3.37)	23-1836-78
$C_{21}H_{23}N_3O_{11}$			
1H-Benzotriazole-4,7-dione, 5(or 6)-methyl-1-(2,3,4,6-tetra-O-acetyl-β-D-glucopyranosyl)-	EtOH	246(4.42)	111-0155-78
$C_{21}H_{24}$			
2,2'-Spirobi[2H-indene], 5,5'-diethyl-1,1',3,3'-tetrahydro-, (R)-	EtOH	215s(4.31),264(3.39), 269(3.52),272(3.56), 278(3.66)	49-0987-78

Compound	Solvent	$\lambda_{max}(\log \epsilon)$	Ref.
$C_{21}H_{24}BrNO_6$			
2-Butenedioic acid, 2-[(2-bromo-2-pro-penyl)(7-methoxy-2,2-dimethyl-2H-1-benzopyran-5-yl)amino]-, dimethyl ester	MeOH	225(4.26),230(4.03), 275(4.37),320(3.99)	24-0439-78
$C_{21}H_{24}ClNO_6$			
2-Butenedioic acid, 2-[(2-chloro-2-pro-penyl)(7-methoxy-2,2-dimethyl-2H1-benzopyran-5-yl)amino]-, dimethyl ester	MeOH	282(4.35)	24-0439-78
$C_{21}H_{24}F_3NO_3$			
Furo[3,4-b]pyridine-3-carboxylic acid, 1,4,5,7-tetrahydro-2,5,5,7,7-penta-methyl-4-[3-(trifluoromethyl)phen-yl]-, methyl ester	EtOH	261(3.60),266s(3.58), 274s(3.45),337(3.79)	95-0448-78
$C_{21}H_{24}N_2$			
Benzenamine, N-[3-(cyclohexylimino)-1-propenyl]-N-phenyl-, monoperchlor-ate	MeOH	230(4.21),342(4.77)	39-1453-78C
$C_{21}H_{24}N_2O$			
Urea, N,N'-bis(2-methyl-1-phenyl-1-pro-penyl)-	EtOH	247(4.21)	33-0589-78
$C_{21}H_{24}N_2O_2$			
Ibogamin-18-ol, 3,4-didehydro-, acetate	MeOH	273(3.77),280(3.79), 288(3.72)	33-0690-78
$C_{21}H_{24}N_2O_3$			
4H-1-Benzopyran-4-one, 6,8-bis[(dimeth-ylamino)methyl]-5-hydroxy-2-phenyl-	n.s.g.	275(4.37),285(4.29), 340(3.79)	103-0497-78
Cleavamine, 18β-(methoxycarbonyl)-5-oxo-	EtOH	223(4.42),276(3.81), 283(3.83),291(3.77)	33-0690-78
$C_{21}H_{24}N_2O_4$			
Catharanthine, 3,4-dihydro-3-hydroxy-19-oxo-, (3S,4R)-	EtOH	222(4.53),277(3.90), 283(3.94),292(3.90)	33-0690-78
1(2H)-Isoquinolinone, 4-[4-[bis(2-hy-droxyethyl)amino]phenyl]-2-(2-hy-droxyethyl)-	EtOH	210(4.54),224(4.31), 266(4.31)	44-0672-78
$C_{21}H_{24}N_2O_5$			
DL-Alanine, N-[4-[methyl(phenylmeth-oxy)carbonyl]amino]benzoyl]-, ethyl ester	EtOH	260(4.19)	87-1162-78
β-Alanine, N-[4-[methyl(phenylmethoxy)-carbonyl]amino]benzoyl]-, ethyl ester	EtOH	260(4.19)	87-1162-78
Furo[3,4-b]pyridine-3-carboxylic acid, 5,7-dihydro-2,5,5,7,7-pentamethyl-4-(2-nitrophenyl)-, ethyl ester	EtOH	271.8(3.97)	95-0448-78
Furo[3,4-b]pyridine-3-carboxylic acid, 5,7-dihydro-2,5,5,7,7-pentamethyl-4-(3-nitrophenyl)-, ethyl ester	EtOH	264.2(4.06)	95-0448-78
Glycine, N-methyl-N-[4-[methyl[(phenyl-methoxy)carbonyl]amino]benzoyl]-, ethyl ester	EtOH	250(4.14)	87-1162-78
DL-Valine, N-[4-[methyl[(phenylmeth-oxy)carbonyl]amino]benzoyl]-	MeOH	258(4.13)	87-1162-78

Compound	Solvent	λ_{max}(log ϵ)	Ref.
$C_{21}H_{24}N_2O_6S$ α-D-Mannopyranoside, methyl 2,3-dide-oxy-2,3-[[(4-methylphenyl)sulfonyl]-hydrazono]-4,6-O-(phenylmethylene)-, (R)-	EtOH	231(4.20)	136-0101-78E
$C_{21}H_{24}N_4$ 3H-1,5-Benzodiazepine, 2-(4-methylphen-yl)-4-(4-methyl-1-piperazinyl)-	n.s.g.	261(4.59),345(3.81)	87-0952-78
$C_{21}H_{24}N_4O$ 3H-1,5-Benzodiazepine, 2-(4-methoxy-phenyl)-4-(4-methyl-1-piperazinyl)-	n.s.g.	266(4.58),338(3.90)	87-0952-78
$C_{21}H_{24}N_4S$ [1,2,3]Thiadiazolo[5,1-e][1,2,3]thia-diazole-7-S^{IV}, 1,6-dihydro-3,4-bis-(1-methylethyl)-1,6-diphenyl-	C_6H_{12}	245(4.27),266s(4.06), 300(4.25),485(4.32)	39-0195-78C
$C_{21}H_{24}N_5O_6PS_2$ Benzenesulfonic acid, 4-methyl-, 2,2'-[(benzoylamino)phosphinylidene]di-hydrazide	MeOH	229(4.5705)	65-0720-78
$C_{21}H_{24}N_{10}O_9$ Adenosine, N,N''-carbonylbis-	pH 1 pH 13 EtOH	265(4.19),292(4.54) 268s(4.08),274(4.10), 322(4.57) 266(4.35),284(4.56), 298(4.56)	39-0131-78C 39-0131-78C 39-0131-78C
$C_{21}H_{24}O_3S$ 16-Thia-D-homoestra-1,3,5(10),8,14-pen-taen-17aα-ol, 3-methoxy-, acetate 17aβ-	EtOH EtOH	315s(4.50),329(4.62), 344(4.50) 315s(4.48),328(4.59), 343(4.49)	39-0576-78C 39-0576-78C
$C_{21}H_{24}O_4$ 16-Oxa-D-homoestra-1,3,5(10),8,14-pen-taen-17aα-ol, 3-methoxy-, acetate 17aβ-	EtOH EtOH	300s(4.40),311(4.48), 324s(4.35) 299s(4.39),311(4.47), 321s(4.35)	39-0576-78C 39-0576-78C
$C_{21}H_{24}O_4S$ 1,2-Cyclopenteno-9,10-dihydrophenan-threne-6-sulfonic acid, 7-methoxy-3',3'-dimethyl-, methyl ester 2H-Phenanthro[1,2-c]thiopyran-1-ol, 1,5,6,11,12,12a-hexahydro-8-methoxy-12a-methyl-, acetate, 3-oxide, α- β-	EtOH EtOH EtOH	282.5(4.41) 343(4.50) 340(4.49)	94-0171-78 39-1254-78C 39-1254-78C
$C_{21}H_{24}O_5$ Benzenepentanoic acid, 4-methoxy-γ-(4-methoxyphenyl)-β-methyl-δ-oxo-, methyl ester Cacalohastin-3β-ol, 14-(angeloyloxy)-	MeOH ether	219(4.20),274(4.20) 287(3.59)	5-0726-78 24-3140-78
$C_{21}H_{24}O_6$ 6(2H)-Benzofuranone, 2-(1,3-benzodiox-ol-5-yl)-3,3a,4,5-tetrahydro-5,7-di-	MeOH	239s(3.64),268(4.15), 285s(4.01)	102-2038-78

Compound	Solvent	λ_{max}(log ϵ)	Ref.
methoxy-3-methyl-3a-(2-propenyl)-, (2S,3S,3aR,5R)- (cont.)			102-2038-78
2-Furanol, 3-(1,3-benzodioxol-5-ylmethyl)-4-[(3,4-dimethoxyphenyl)methyl]-tetrahydro-	MeOH	229(3.93),282(3.68)	88-0457-78
C$_{21}$H$_{24}$O$_7$			
Diasin	CH$_2$Cl$_2$	240(2.40)	102-1773-78
C$_{21}$H$_{24}$O$_7$Si			
9,10-Anthracenedione, 5,8-diacetoxy-1,4,4a,9a-tetrahydro-1-[(trimethylsilyl)oxy]-	MeOH	223(4.51),310(3.46)	150-3831-78
C$_{21}$H$_{24}$S			
2H-1-Benzothiopyran, octahydro-2,4-diphenyl-	hexane	191(5.2),206(4.4), 247s(2.6),253(2.7), 259(2.7),265(2.5)	103-0605-78
C$_{21}$H$_{25}$BF$_4$N$_2$			
2,4-Pentadienylium, 1,5-bis[4-(dimethylamino)phenyl]-, tetrafluoroborate	C$_2$H$_4$Cl$_2$	800(5.51)	104-1197-78
C$_{21}$H$_{25}$NO			
Cyclohexanecarboxamide, 2-phenyl-N-(2-phenylethyl)-, (1S-trans)-	MeOH	208(4.29),252(2.65), 258(2.70),264(2.60)	44-0355-78
C$_{21}$H$_{25}$NO$_4$			
1-Phenanthrenethanamine, 3,4,6,7-tetramethoxy-N-methyl-	MeOH	263(4.98),280s(4.53), 307(4.21),320(4.20), 344(3.45),362(3.25)	12-0313-78
C$_{21}$H$_{25}$NO$_5$			
1,3-Azulenedicarboxylic acid, 2-hydroxy-6-piperidino-, diethyl ester	MeOH	232(4.35),266(4.48), 351(4.77),402(4.32), 431(4.25)	18-3087-78
Benzenemethanol, 2-[(3,4-dihydro-6,7-dimethoxy-1-isoquinolinyl)methyl]-4,5-dimethoxy-	EtOH	235(4.40),283(4.13), 318(3.92)	142-1233-78A
C$_{21}$H$_{25}$NO$_6$			
2-Butenedioic acid, 2-[(7-methoxy-2,2-dimethyl-2H-1-benzopyran-5-yl)-2-propenylamino]-, dimethyl ester	MeOH	283(4.36)	24-0439-78
5H-Isoindole-5,5-dicarboxylic acid, 2-benzoyloctahydro-3-oxo-, diethyl ester, trans	n.s.g.	231(3.88)	128-0097-78
Peshawarinediol	EtOH	225(4.21),288(3.63)	78-0635-78
C$_{21}$H$_{25}$N$_2$O			
2,1-Benzisoxazolium, 3-[2-[4-(dimethylamino)phenyl]ethenyl]-1-(1,1-dimethylethyl)-, perchlorate	MeCN	590(4.99)	44-1233-78
C$_{21}$H$_{25}$N$_3$O$_2$			
Butanamide, 2-[(4-butylphenyl)hydrazono]-N-methyl-3-oxo-N-phenyl-	EtOH	204(4.42),233(4.26), 353(4.40)	104-0521-78
Butanamide, N-(4-butylphenyl)-2-[(2-methylphenyl)hydrazono]-3-oxo-	EtOH	212(3.90),246(4.23), 270(4.09),384(4.50)	104-0521-78
Butanamide, N-(4-butylphenyl)-2-[(4-methylphenyl)hydrazono]-3-oxo-	EtOH	208(4.01),249(4.18), 270(4.14),384(4.55)	104-0521-78

Compound	Solvent	$\lambda_{max}(\log \epsilon)$	Ref.
Butanamide, N-(4-butylphenyl)-2-(meth- ylphenylhydrazono)-3-oxo-	EtOH	205(4.52),243(4.39), 250-285s(--),335(4.37)	104-0521-78
$C_{21}H_{25}N_3O_3$ Pyrano[2,3-c]pyrazol-6(2H)-one, 3,4-di- methyl-5-[2-(4-morpholinyl)ethyl]- 2-(phenylmethyl)-	EtOH	308.6(4.24)	95-0335-78
$C_{21}H_{25}N_5O_{10}$ Acetamide, N-[9-(2,3,4,6-tetra-O-acet- yl-β-D-glucopyranosyl)-9H-purin-6- yl]-	MeOH	214(4.35),272(4.28)	136-0301-78C
$C_{21}H_{26}BrNO_4$ 3,5-Pyridinedicarboxylic acid, 4-(4- bromophenyl)-1-ethyl-1,4-dihydro- 2,6-dimethyl-, diethyl ester	EtOH	203(4.34),222(4.25), 244(4.28),351(3.81)	103-1226-78
$C_{21}H_{26}ClN_3O_2S$ Perphenazine sulfoxide	MeOH	238(4.54),277(4.12), 302(4.00),344(3.81)	133-0248-78
$C_{21}H_{26}N_2O$ 2,5-Cyclohexadien-1-one, 2,6-bis(1,1- dimethylethyl)-4-[(phenylazo)methyl- ene]-	n.s.g.	356(4.62)	70-1179-78
$C_{21}H_{26}N_2O_2$ Coronaridine	MeOH	231(4.28),285(3.98), 295(3.88)	102-0835-78
Epivincadine, 6,7-dehydro-	MeOH	226(4.57),285(3.94), 292(3.91)	114-0167-78A
Vincadine, 6,7-dehydro-	MeOH	226(4.58),284(3.90), 292(3.86)	114-0167-78A
$C_{21}H_{26}N_2O_3$ Catharanthine, 3,4-dihydro-3,4-dihy- droxy-, (3R,4R)-	EtOH	226(4.40),277(3.71), 283(3.80),292(3.71)	33-0690-78
Corynantheine, demethyldihydro-	EtOH	227(4.62),284s(3.96), 290(3.86)	95-0950-78
	EtOH-NaOH	227(4.64),278(4.30)	95-0950-78
Hirsutine, demethyl-	EtOH	226(4.57),274(4.00), 283s(3.97),291(3.87)	95-0950-78
	EtOH-NaOH	227(4.58),276(4.38)	95-0950-78
Pseudoyohimbine	EtOH	245(4.03),355(4.35)	5-1096-78
$C_{21}H_{26}N_2O_5$ Furo[3,4-b]pyridine-3-carboxylic acid, 1,4,5,7-tetrahydro-2,5,5,7,7-penta- methyl-4-(3-nitrophenyl)-	EtOH	208(4.38),266(4.02), 329(3.80)	95-0448-78
$C_{21}H_{26}N_2O_6$ 3,5-Pyridinedicarboxylic acid, 1-ethyl- 1,4-dihydro-2,6-dimethyl-4-(4-nitro- phenyl)-, diethyl ester	EtOH	202(4.49),241(4.20), 265s(4.03),330s(3.64)	103-1226-78
$C_{21}H_{26}N_2O_8$ Propanedioic acid, (acetylamino)[2-[2- (1,3-dihydro-1,3-dioxo-2H-isoindol- 2-yl)ethoxy]ethyl]-, diethyl ester	EtOH	220(4.62),240s(3.63), 293(2.96)	44-3713-78

Compound	Solvent	$\lambda_{max}(\log \epsilon)$	Ref.
$C_{21}H_{26}N_4O_2$			
3-Aza-A-homopregna-4a,16-dieno[3,4-d]-tetrazole-6,20-dione	MeOH	250(3.84)	2-0095-78
1,3-Propanediamine, N-(3-methylbutyl)-N'-(1-nitro-9-acridinyl)-	pH 1	218(4.281),273(4.517), 329(3.606),435(3.858)	56-2125-78
	pH 7	273(4.173),402(3.539)	56-2125-78
$C_{21}H_{26}N_8O_3$			
L-Leucine, N-[4-[[(2,4-diamino-6-pteridinyl)methyl]methylamino]benzoyl]-	pH 1	245(4.25),308(4.32)	87-1162-78
$C_{21}H_{26}O$			
9H-Xanthene, 1,8-dimethyl-4,5-bis(1-methylethyl)-	EtOH	250(3.39),257(3.95), 277(3.75),285(3.71)	32-0079-78
$C_{21}H_{26}O_2$			
Benzene, 1,3-dimethoxy-2-(3-methyl-2-butenyl)-5-(2-phenylethyl)-	n.s.g.	207(3.63),239(2.84)	102-2115-78
$C_{21}H_{26}O_3$			
1-Azuleneacetic acid, α-(1-hydroxyethylidene)-3,8-dimethyl-5-(1-methylethyl)-, ethyl ester	hexane	247(4.51),287(4.61), 306s(4.30),354(3.79), 372(3.78),612(2.62), 666s(2.51)	18-3582-78
2H-1-Benzopyran, 3,4-dihydro-2-methoxy-2,5,7,8-tetramethyl-6-(phenylmethoxy)-, (±)-	EtOH	227(4.08),257(2.97), 263(3.03),278(3.24), 285(3.28)	33-0837-78
Estr-4-en-3-one, 17β-ethynyl-17α-(formyloxy)-	MeOH	240(4.23)	24-3086-78
Etiojerva-4,12-diene-3,11-dione, 17,20-epoxy-17α-ethyl-, (20S)-	MeOH	253(4.23)	18-0234-78
	dioxan	254(4.18)	18-0234-78
17-epimer	MeOH	253(4.23)	18-0234-78
	dioxan	257(4.20)	18-0234-78
Etiojerva-4,12-diene-3,11-dione, 17β-formyl-17α-methyl-	dioxan	250(4.26)	18-0234-78
17-epimer	dioxan	250(4.26)	18-0234-78
Etiojerva-4,12-diene-3,11,20-trione, 17α-ethyl-	dioxan	249(4.18)	18-0234-78
17β-	dioxan	251(4.20)	18-0234-78
2,4,6,8-Nonatetraenoic acid, 9-(4-methoxy-2,3,6-trimethylphenyl)-3,7-dimethyl-, cis	EtOH	243(3.91),354(4.46)	33-2697-78
di-cis	EtOH	248(4.03),351(4.48)	33-2697-78
18-Nor-5β,17α-pregna-8,11,13-triene-3,20-dione, 11-hydroxy-17β-methyl-	EtOH	293(3.55)	39-0163-78C
$C_{21}H_{26}O_4$			
Etiojerv-4-ene-3,11,20-trione, 12α,13α-epoxy-17α-ethyl-	EtOH	234(4.11)	18-0234-78
12β,13β-	EtOH	233(3.85)	18-0234-78
Mucronulastyrene diethyl ether	MeOH	239(4.18),283(3.65)	102-1389-78
Phenol, 3-[2-(3-hydroxy-5-methoxyphenyl)ethyl]-6-methoxy-2-(3-methyl-2-butenyl)- (canniprene)	EtOH	206(4.63),224s(4.06), 277s(3.38),281(3.34)	88-4711-78
Pregna-1,4-diene-3,20-dione, 11β,18-epoxy-21-hydroxy-	MeOH	243(4.16)	150-3551-78
$C_{21}H_{26}O_4S$			
8,14-Seco-16-thia-D-homoestra-1,3,5(10)-9(11)-tetraen-17a-one, 14α-acetoxy-	EtOH	266(4.28)	39-0576-78C
14β-	EtOH	266(4.22)	39-0576-78C

Compound	Solvent	$\lambda_{max}(\log \epsilon)$	Ref.
$C_{21}H_{26}O_5$			
Mucronulatol, diethyl ether, (±)-	EtOH	227(4.24),283(3.59), 289(3.56)	102-1405-78
8,14-Seco-16-oxa-D-homoestra-1,3,5(10)-9(11)-tetraen-17a-one, 14α-acetoxy-3-methoxy-	EtOH	266(4.28)	39-0576-78C
14β-	EtOH	267(4.27)	39-0576-78C
$C_{21}H_{26}O_6$			
1,6,10-Dodecatrien-8-yne-1,4-diol, 3-(acetoxymethylene)-7,11-dimethyl-, [S-(Z,E,E)]- (caulerpenyne)	EtOH	252(4.52),265s(4.45), 280s(4.23)	88-3593-78
$C_{21}H_{26}O_7$			
Epipterosin L triacetate	EtOH	218(4.50),260(4.10), 303(3.35)	94-2365-78
Ethanone, 2-(3-ethoxy-2,4-dimethoxy-phenyl)-1-(4-ethoxy-2-hydroxy-3-methoxyphenyl)-	EtOH	225(4.35),285(4.22), 330(3.48)	102-1401-78
Pterosin L triacetate	EtOH	218(4.44),260(4.06), 304(3.20)	94-2365-78
$C_{21}H_{26}O_8$			
Longipilin	MeOH	213(4.32)	102-2131-78
$C_{21}H_{26}S_2$			
3,11-Dithiatricyclo[11.3.1.1^{5,9}]octa-deca-1(17),5,7,9(18),13,15-hexaene, 17-butyl-18-methyl-	EtOH	285(2.56)	44-0727-78
$C_{21}H_{27}NO_4$			
Androst-4-en-3-one, 17α-ethynyl-17β-(nitrooxy)-	MeOH	240(4.24)	24-3086-78
Phenol, 4-[(5-ethoxy-1,2,3,4-tetrahy-dro-6,7-dimethoxy-2-methyl-1-iso-quinolinyl)methyl]-	MeOH	283(3.61)	44-0580-78
1H-Pyrrole-3-carboxylic acid, 1-cyclo-hexyl-4,5-dihydro-5-hydroxy-2-methyl-2-(4-methylphenyl)-4-oxo-, ethyl ester	EtOH	210(3.94),247(4.14), 330(3.97)	118-0291-78
1H-Pyrrole-3-carboxylic acid, 1-cyclo-hexyl-4,5-dihydro-5-hydroxy-2-methyl-4-oxo-5-(phenylmethyl)-, ethyl ester	EtOH	210(3.95),250(4.13), 326(3.85)	4-1215-78
$C_{21}H_{27}NO_5$			
1H-Pyrrole-3-carboxylic acid, 1-cyclo-hexyl-4,5-dihydro-5-hydroxy-2-(4-methoxyphenyl)-5-methyl-4-oxo-, ethyl ester	EtOH	226(4.09),247(4.06), 330(3.95)	118-0291-78
2H-Pyrrole-2,2-dipropanoic acid, 1,3-dihydro-3-oxo-1-(phenylmethyl)-, diethyl ester	EtOH	327(4.08)	94-3521-78
1H-Pyrrole-2-propanoic acid, 2-[(1,1-dimethylethoxy)carbonyl]-2,3-dihydro-3-oxo-1-(phenylmethyl)-, ethyl ester	EtOH	330(4.07)	94-3521-78
$C_{21}H_{27}NO_6$			
1,2-Benzenedimethanol, 4,5-dimethoxy-α-(1,2,3,4-tetrahydro-6,7-dimethoxy-1-isoquinolinyl)-	EtOH	235(4.21),285(3.80)	142-1233-78A

Compound	Solvent	$\lambda_{max}(\log \epsilon)$	Ref.
$C_{21}H_{27}N_2O_4$ 3,5-Pyridinedicarboxylic acid, 4-[4-(dimethylamino)phenyl]-1,4-dihydro-2,6-dimethyl-, diethyl ester, anion	DMSO	454(3.72)	103-1226-78
$C_{21}H_{27}N_2O_5$ Isoquinolinium, 5,8-diacetoxy-7-methoxy-2,6-dimethyl-3-piperidino-, iodide	EtOH	219(4.51),267(4.52), 298(3.81),363(3.38)	94-2175-78
$C_{21}H_{27}N_3O_2$ Epivincamine, 14-amino-14-deoxy- Vincamine, 14-amino-14-deoxy-	MeOH MeOH	226(4.52),282(3.92) 228(4.52),282(3.90)	33-1682-78 33-1682-78
$C_{21}H_{27}N_3S_2$ 2H-1,3,5-Thiadiazine-2-thione, 6-(dicyclohexylamino)-4-phenyl-	CH_2Cl_2	248(4.04),303(4.34), 418(3.90)	48-0647-78
$C_{21}H_{27}N_5O_{10}$ 1H-Purine-2,6-dione, 3,7-dihydro-1,3-dimethyl-7-[3,4,6-tri-O-acetyl-2-(acetylamino)-2-deoxy-β-D-glucopyranosyl]-	MeOH	212(4.29),275(3.94)	136-0301-78C
$C_{21}H_{27}O_7P$ α-D-arabino-Hexofuranoside, methyl 6-[bis(phenylmethoxy)phosphinyl]-5,6-dideoxy-	MeOH	252(2.61),257(2.70), 263(2.64),268(2.48)	136-0349-78H
$C_{21}H_{28}NO_5PS_3$ Phosphorothioic acid, O,O-bis(1-methylpropyl) S-[(2,3,5,7-tetrahydro-5,7-dioxo-2-phenyl-6H-1,4-dithiino[2,3-c]pyrrol-6-yl)methyl] ester	EtOH	213(4.18),263(4.00), 419(3.47)	73-1093-78
Phosphorothioic acid, O,O-dibutyl S-[(2,3,5,7-tetrahydro-5,7-dioxo-2-phenyl-6H-1,4-dithiino[2,3-c]pyrrol-6-yl)methyl] ester	EtOH	213(4.16),262(3.97), 419(3.45)	73-1093-78
$C_{21}H_{28}N_2$ 9H-Pyrido[3,4-b]indole, 1-decyl-	MeOH	236(4.58),241(4.57), 251s(4.39),283s(4.02), 289(4.26),338(3.70), 351(3.72)	95-0898-78
$C_{21}H_{28}N_2OS$ Vincanol, 14-deoxy-14-[(2-hydroxyethyl)thio]-	MeOH	198(4.41),231(4.45), 284(3.91)	33-1682-78
$C_{21}H_{28}N_2O_2$ Epivincadine, (±)-	MeOH	226(4.5),285(3.96), 292(3.93)	114-0167-78A
Piperidine, 1-[1,5-dioxo-5-phenyl-3-(1-piperidinyl)-2(or 3)-pentenyl]-	EtOH	238(4.04),260(2.46), 285(4.09)	104-1446-78
$C_{21}H_{28}N_{10}O_2$ 2H-Pyrimido[4'',5'':2',3'][1,4]diazepino[6',5':3,4]pyrrolo[1,2-f]pteridine-1,12-dione, 3,10-bis(dimethylamino)-5,6,6a,6b,7,8-hexahydro-5,8,14-trimethyl-, trans $(N^3,N^3,N^{10},N^{10},5,8$-hexamethyldrosopterin)	pH −1.5 pH 0.0 pH 7.0	236(4.51),280s(4.02), 468(4.28),585(4.23) 234(4.55),290s(3.95), 485(4.36),583(3.96) 235(4.59),282s(4.01), 330s(3.81),498(4.51)	24-3385-78 24-3385-78 24-3385-78

Compound	Solvent	$\lambda_{max}(\log \epsilon)$	Ref.
(cont.)	pH 10.5	226(4.49),275(4.03), 330s(3.73),521(4.55)	24-3385-78
	pH 13.0	273(4.08),346(3.75), 529(4.57)	24-3385-78
$C_{21}H_{28}O$			
19-Norpregna-1,3,5(10),17(20)-tetraene, 3-methoxy-, (Z)-	EtOH	232s(4.00),237s(3.76), 276(3.34),286(3.41)	35-3435-78
$C_{21}H_{28}O_2$			
Etiojerva-4,12-diene-3,20-dione, 17α-ethyl-	MeOH	238(4.08)	18-0234-78
$C_{21}H_{28}O_2S$			
Benzene, [[3-methyl-5-(2,6,6-trimethyl-2-cyclohexen-1-yl)-2,4-pentadienyl]-sulfonyl]-	EtOH	217(4.30),243(4.45)	44-4769-78
$C_{21}H_{28}O_3$			
Etiojerva-4,12-diene-3,20-dione, 17α-ethyl-11α-hydroxy-11β-	MeOH	238(4.04)	18-0234-78
D(17a)-Homo-C,18-dinorpregn-4-ene-3,20-dione, 13,17a-epoxy-17a-methyl-, (13α,17α,17aα)-	MeOH / EtOH	238(4.04) / 237(4.08)	18-0234-78 / 18-0234-78
D-Homo-18-norandrosta-5,13(17a)-dien-17-one, 3β-acetoxy-	EtOH	240(4.18)	56-2361-78
D-Homo-B-nor-8-isoestra-1,3,5(10)-trien-17aβ-ol, 3-methoxy-, acetate	EtOH	280(3.32),290(3.27)	104-2321-78
9-iso-	EtOH	284(3.56),290(3.51)	104-2321-78
$C_{21}H_{28}O_4$			
Benzo[d]xanthene-9,12-dione, 1,2,3,4-4a,5,6,7,7a,8-decahydro-10-hydroxy-4,4,7,7a-tetramethyl-	EtOH	290(4.02),510(2.38)	12-2685-78
Etiojerv-4-ene-3,20-dione, 12α,13α-epoxy-17α-ethyl-11α-hydroxy-	MeOH	239(4.04)	18-0234-78
Etiojerv-4-ene-3,20-dione, 12β,13β-epoxy-17α-ethyl-11β-hydroxy-	MeOH	238(3.90)	18-0234-78
D(17a)-Homo-C,18-dinorpregn-5-ene-3,20-dione, 11,13-epoxy-17a-hydroxy-17a-methyl-	EtOH	238(4.18)	18-0234-78
19-Norpregn-4-ene-3,20-dione, 17-(formyloxy)-	MeOH	239(4.25)	24-3086-78
Podocarpa-8,11,13-triene, 13-acetyl-12-(carboxymethoxy)-	EtOH	255(4.07),315(3.63)	23-0517-78
$C_{21}H_{28}O_5$			
Pregn-4-en-21-al, 11β,17α-dihydroxy-3,20-dioxo-	MeOH	242(4.23),279(4.12)	44-3405-78
$C_{21}H_{28}O_7$			
4H,9H-Furo[2',3',4':4,5]naphtho[2,1-c]-pyran-4,9-dione, 3-acetoxy-1,2,3,3a-5a,6,6a,7,10b,10c-decahydro-6-hydroxy-3a,10b-dimethyl-7-(1-methylethyl)-, [3S-(3α,3aβ,5aβ,6β,6aβ,7β,10bα-10cβ)]-	EtOH	216(3.95)	142-0123-78B
$C_{21}H_{28}O_9$			
Plagiochiline B	n.s.g.	207.5(3.71)	102-1794-78

Compound	Solvent	$\lambda_{max}(\log \epsilon)$	Ref.
C$_{21}$H$_{29}$ClO$_2$			
Estra-4,6-dien-3-one, 6-chloro-17β-hydroxy-16β-(1-methylethyl)-	EtOH	284(4.33)	94-1718-78
C$_{21}$H$_{29}$NO$_4$			
11-Azaestr-4-en-3-one, 17-acetoxy-11-acetyl-, (17β)-	EtOH	238(4.12)	65-1529-78
6-Aza-9,10-seco-9,17-bis(ethylenedioxy)-estra-1,3,5(10)-triene, (±)-	EtOH	249(4.16),288(3.32)	65-0841-78
C$_{21}$H$_{29}$NO$_5$			
17α-Pregn-4-ene-3,20-dione, 17β-(nitrooxy)-	MeOH	240(4.25)	24-3086-78
C$_{21}$H$_{29}$NO$_6$			
Propanedioic acid, [[[2-(1,1-dimethylethoxy)-2-oxoethyl](phenylmethyl)-amino]methylene]-, diethyl ester	EtOH	283(4.35)	94-2224-78
C$_{21}$H$_{29}$N$_3$			
Eburnamenin-14-amine, 14,15-dihydro-N,N-dimethyl-, (3α,14α,16α)-	MeOH	228(4.54),282(3.93)	33-1682-78
C$_{21}$H$_{29}$N$_3$O$_6$			
2-Azetidinium, N,N,N-triethyl-1-[2-hydroxy-1-[[(4-nitrophenyl)methoxy]-carbonyl]-1-butenyl]-4-oxo-, hydroxide, inner salt	pH 6	273.5(4.47)	77-0469-78
C$_{21}$H$_{29}$N$_5$O$_3$			
1H-Pyrrole-2,3-dicarboxamide, 1-cyclohexyl-2,5-dihydro-N,N'-dimethyl-4-(2-methyl-2-phenylhydrazino)-5-oxo-	EtOH	235(4.20),295(3.85)	4-1463-78
Pyrrolo[3,4-b]indole-1,8b(1H)-dicarboxamide, 3a-amino-2-cyclohexyl-2,3,3a,4-tetrahydro-N,N',4-trimethyl-3-oxo-	EtOH	257(3.66),310(3.29)	4-1463-78
C$_{21}$H$_{29}$N$_7$O$_{14}$P$_2$			
Adenosine, 5'-(trihydrogen diphosphate), 5'→5'-ester with 1,2-dihydro-1-β-D-ribofuranosyl-3-pyridinecarboxamide	80% MeCN	345(3.79)	90-0703-78
C$_{21}$H$_{30}$O$_2$			
Androsta-4,6-dien-3β-ol, acetate	EtOH	239(4.38)	78-0209-78
Estra-4,6-dien-3-one, 17β-hydroxy-16β-(1-methylethyl)-	EtOH	281(4.43)	94-1718-78
Pimara-6,8(14),15-trienoic acid, methyl ester	n.s.g.	234(3.99),241(4.03), 249(3.88)	23-2156-78
Pimara-7,9(11),15-trienoic acid, methyl ester	n.s.g.	236(5.33?),242(5.33?), 251(4.15)	23-2156-78
Pregn-4-ene-3,20-dione	EtOH	240(4.23)	2-0253-78
C$_{21}$H$_{30}$O$_3$			
Estr-1-en-3-one, 17β-acetoxy-9-methyl-, (5α,9β,10α)-	MeOH	235(3.96)	39-0808-78C
(5β,9β,10α)-	MeOH	232(3.98)	39-0808-78C
Estr-4-en-3-one, 6α,7α-epoxy-17β-(hydroxy-16β-(1-methylethyl)-	EtOH	235(4.23)	94-1718-78
19-Norpregn-4-ene-3,20-dione, 17-methoxy-	MeOH	240(4.24)	24-3086-78

$C_{21}H_{30}O_3-C_{21}H_{32}$

Compound	Solvent	$\lambda_{max}(\log \epsilon)$	Ref.
1-Phenanthrenecarboxylic acid, 7-ethenyl-1,2,3,4,4a,4b,5,6,7,8,10,10a-dodecahydro-1,4a,7-trimethyl-8-oxo-, methyl ester	n.s.g.	243(3.78)	23-2156-78
4H-Phenanthro[3,4-b]oxet-4-one, 1,5,6-6a,7,8,9,10,10a,10c-decahydro-10c-hydroxy-7,7,10a-trimethyl-3-(1-methylethyl)-	EtOH	248(4.07),316(3.26)	23-0733-78
Pregn-4-ene-3,20-dione, 14α-hydroxy-	n.s.g.	246(4.16)	102-0578-78
$C_{21}H_{30}O_4$			
Allocyafrin B₄, methyl acetal	MeOH	235(3.98)	23-2113-78
Cyafrin B₄, methyl acetal	MeOH	242(4.14)	23-2113-78
Cyclotetradeca[b]furan-10-carboxylic acid, 2,4,5,8,9,12,13,15a-octahydro-5-hydroxy-3,6,14-trimethyl-, methyl ester, (5R*,6E,10Z,14E,15aR*)-(+)-(sarcoglaucol)	MeOH	215(4.00)	20-0459-78
D(17a)-Homo-C,18-dinorpregn-5-ene-3,20-dione, 13,17a-dihydroxy-17a-methyl-, (13ξ,17α)-	MeOH	239(4.00)	18-0234-78
1-Naphthalenecarboxylic acid, 5-[2-(2,5-dihydro-2-oxo-3-furanyl)ethyl]-decahydro-1,4a-dimethyl-6-methylene-, methyl ester (nivenolide methyl ester)	EtOH	214(3.80)	102-0574-78
$C_{21}H_{30}O_8$			
1H-Inden-1-one, 6-[2-(β-D-glucopyranosyloxy)ethyl]-2,3-dihydro-2-(hydroxymethyl)-2,5,7-trimethyl- (pteroside A)	EtOH	216(4.37),260(4.25), 305(3.46)	94-2365-78
$C_{21}H_{31}BrN_2O_5$			
L-altro-D-allo-2-Trideculose, 4,7¹:9,10-dianhydro-1,3,7,8,11,13-hexadeoxy-7-(hydroxymethyl)-11-methyl-, (2-bromophenyl)hydrazone	EtOH	278(4.16)	39-0561-78C
$C_{21}H_{31}ClO_2$			
Androst-4-en-3-one, 6β-chloro-16β-ethyl-17β-hydroxy-	EtOH	240(4.22)	94-1718-78
$C_{21}H_{31}N_5$			
3,5-Pyridinedicarbonitrile, 1,4-dihydro-4,4-dimethyl-2,6-bis(1-piperidinylmethyl)-	EtOH	218(4.60),221(4.51), 227(4.11),345(4.02)	103-1236-78
$C_{21}H_{31}N_5O_2$			
3,5-Pyridinedicarbonitrile, 4,4-diethyl-1,4-dihydro-2,6-bis(4-morpholinylmethyl)-	EtOH	205(4.72),218(4.83), 260(4.55),340(4.21)	103-1236-78
$C_{21}H_{31}O_5P$			
6H-Dibenzo[b,d]pyran-1-ol, 6a,7,10,10a-tetrahydro-6,6,9-trimethyl-3-pentyl-, dihydrogen phosphate, disodium salt	EtOH	275(3.18),283(3.21)	87-1079-78
$C_{21}H_{32}$			
1H-Indene, octahydro-4,4,7a-trimethyl-1-(3-methyl-2-cyclopenten-1-ylidene)-5-(1-methylethylidene)-, trans	MeOH	245(4.00)	35-4268-78

Compound	Solvent	$\lambda_{max}(\log \epsilon)$	Ref.
$C_{21}H_{32}N_4O_3$ Pyrano[2,3-c]pyrazole-2(6H)-acetamide, N-butyl-3,4-dimethyl-6-oxo-5-[2-(1-piperidinyl)ethyl]-	EtOH	308.5(4.23)	95-0335-78
$C_{21}H_{32}N_4O_5$ 1(2H)-Pyrimidinebutanamide, N-cyclohexyl-N-[(cyclohexylamino)carbonyl]-3,4-dihydro-β-hydroxy-2,4-dioxo-	EtOH	266(4.05)	126-0905-78
$C_{21}H_{32}O$ 9β-Estr-5(10)-en-6-one, 4,4,9-trimethyl-	n.s.g.	255(3.97)	39-1537-78C
$C_{21}H_{32}O_2$ 5,14-Pentadecadien-9-ynoic acid, 5,14-dimethyl-2-(1-methylethylidene)-, methyl ester, (E)-	MeOH	224s(3.90)	35-4268-78
Retinoic acid, 7,8-dihydro-, methyl ester	n.s.g.	315(4.57)	104-0666-78
$C_{21}H_{32}O_3$ 5α-Pregn-16-en-20-one, 3β,11β-dihydroxy-	EtOH	240(3.93)	39-1594-78C
Retinoic acid, 5,6-epoxy-5,6,7,8-tetrahydro-, methyl ester	n.s.g.	320(4.39)	104-0666-78
Retinoic acid, 9,10-epoxy-7,8,9,10-tetrahydro-, methyl ester	n.s.g.	272(4.32)	104-0666-78
$C_{21}H_{32}O_4$ Prostaglandin A_2, methyl ester	MeOH	217(4.01)	44-4377-78
Retinoic acid, 5,6:9,10-diepoxy-5,6,7,8,9,10-hexahydro-, methyl ester	n.s.g.	270(4.20)	104-0666-78
Retinoic acid, 5,6-dihydro-5,6-dihydroxy-, methyl ester, threo	EtOH	347(4.70)	39-1511-78C
$C_{21}H_{32}O_{14}$ Ulmoside	EtOH	204(3.6)	32-0017-78
$C_{21}H_{33}NO_2$ 23-Norcon-20(22)-enin-18-ol, 22-oxide	$CHCl_3$	231(4.01)	78-2639-78
$C_{21}H_{34}N_2O_3$ Carbamic acid, [2-(heptyloxy)phenyl]-, 2-(1-piperidinyl)ethyl ester, monohydrochloride	50% MeOH	206(4.46),236(4.01), 280(3.49)	106-0297-78
$C_{21}H_{34}N_6O_5$ Adenosine, N-[(decylamino)carbonyl]-	pH 1 pH 13 70% EtOH	270(4.37),277(4.34) 269(4.12),277(4.12), 298(4.06) 270(4.39),276(4.31)	36-0569-78 36-0569-78 36-0569-78
$C_{21}H_{34}O_2$ 5,9-Methanobenzocycloocten-4(1H)-one, 2,3,5,6,7,8,9,10-octahydro-2,2,7,7,9-pentamethyl-5-(1-methylethoxy)-	n.s.g.	245(3.91)	78-1251-78
5,9-Methanobenzocycloocten-4(1H)-one, 2,3,5,6,7,8,9,10-octahydro-2,2,7,7,9-pentamethyl-5-propoxy-	n.s.g.	245(3.92)	78-1251-78

Compound	Solvent	λ_{max}(log ϵ)	Ref.
2-Pentenoic acid, 3-methyl-5-(1,2,3,4-4a,7,8,8a-octahydro-1,2,4a,5-tetra-methyl-1-naphthalenyl)-, methyl ester	EtOH	222(4.11)	94-0079-78
$C_{21}H_{35}N_3O_6$			
2(1H)-Pyrimidinone, 4-amino-1-[3-O-(1-oxododecyl)-β-D-arabinofuranosyl]-	isoPrOH	216(4.23),248(4.16), 303(3.91)	94-0981-78
$C_{21}H_{35}N_6O_8P$			
5'-Adenylic acid, N-[(decylamino)carbo-nyl]-, barium salt	pH 1	270(4.34),277(4.33)	36-0569-78
	pH 13	270(4.25),277(4.22), 298(3.95)	36-0569-78
	70% EtOH	269(4.36),277(4.28)	36-0569-78
$C_{21}H_{36}O$			
2,4-Cyclopentadien-1-one, 2,3,4,5-tet-rakis(1,1-dimethylethyl)-	C_6H_{12}	219(4.32),425(2.27)	89-0519-78
$C_{21}H_{38}N_2O$			
Cyclobutanol, 1-(1H-imidazol-2-yl)-2-tetradecyl-	THF	231(3.32)	33-2823-78
Cyclobutanol, 1-(1H-imidazol-4-yl)-2-tetradecyl-	THF	231(3.32)	33-2823-78
$C_{21}H_{40}N_2O$			
1H-Imidazole-2-methanol, α-heptadecyl-	THF	231(3.10)	33-2823-78
1H-Imidazole-4-methanol, α-heptadecyl-	THF	232(3.03)	33-2823-78
$C_{21}H_{41}NO_2$			
1-Hexadecanaminium, N-(1-formyl-2-hy-droxyethenyl)-N,N-dimethyl-, hydroxy, inner salt	EtOH	257(4.40)	73-1261-78
$C_{21}H_{41}N_3O_5Si_2$			
Furo[2',3':4,5]oxazolo[3,2-a]-1,3,5-triazin-4-one, 8-[[(1,1-dimethyleth-yl)dimethylsilyl]oxy]-7-[[[(1,1-di-methylethyl)dimethylsilyl]oxy]meth-yl]-2,3,5a,7,8,8a-hexahydro-3-methyl-	ether	227(3.60)	44-0529-78

Compound	Solvent	$\lambda_{max}(\log \epsilon)$	Ref.
$C_{22}H_{12}N_2$			
Dibenzo[f,h]benzo[3,4]cyclobuta[1,2-b]-quinoxaline	EtOH	223(4.10),247s(4.55), 253(4.58),276s(4.17), 287(4.14),298(4.00), 357(3.56),403(3.99), 425(4.08)	78-0495-78
$C_{22}H_{12}N_2O$			
Benz[cd]indole, 2,2'-oxybis-	MeOH	251(4.24),258(4.26), 280s(3.73),380s(4.06), 401(4.09),430s(3.83), 470s(3.50)	104-2070-78
[1(2H),2'-Bibenz[cd]indol]-2-one	MeOH	250(4.43),320(3.94), 343(4.00),360(4.04), 402(4.06)	104-2070-78
$C_{22}H_{13}N_3O_3$			
4(3H)-Quonazolinone, 3-[(3-oxo-1(3H)-isobenzofuranylidene)amino]-2-phenyl-	MeOH	226(4.65),275(4.17), 315(3.82)	80-1085-78
$C_{22}H_{14}$			
Pyrene, 1-phenyl-	EtOH	204(4.55),235(4.62), 243(4.69),267(4.39), 277(4.61),341(4.49), 375(2.84)	5-0528-78
$C_{22}H_{14}Br_2N_2O$			
1(4H)-Naphthalenone, 2-[(4-bromophenyl)amino]-4-[(4-bromophenyl)imino]-	C_6H_{12}	250(4.12),280(4.20), 340(3.58),460(3.66)	48-0557-78
	EtOH	250(4.40),285(4.46), 306(4.20),335(3.88), 460(3.90)	48-0557-78 +80-0617-78
	ether	250(4.38),280(4.45), 306(4.20),332(3.88), 460(3.90)	48-0557-78
	CCl_4	307(4.11),340(3.76), 465(3.81)	48-0557-78
$C_{22}H_{14}ClN_3$			
Quinoline, 2-(1H-benzimidazol-2-yl)-6-chloro-4-phenyl-	dioxan	358(3.76)	97-0400-78
$C_{22}H_{14}N_2O_5$			
Furan, 3-(2,4-dinitrophenyl)-2,5-di-phenyl-	n.s.g.	322(4.46),376(4.24)	39-0401-78C
$C_{22}H_{14}N_4$			
5,5'-Bi-5H-cyclopenta[2,1-b:3,4-b']di-pyridine	MeOH	250(3.76),300(4.20), 306(4.21),312(4.31)	142-0849-78A
5,5'-Bi-5H-cyclopenta[2,1-b:4,3-c']di-pyridine	MeOH	249(3.78),289(4.08), 302(4.09)	142-0849-78A
$C_{22}H_{14}N_4O_2$			
4(3H)-Quinazolinone, 3-[(1-oxo-1H-iso-indol-3-yl)amino]-2-phenyl-	MeOH	228(4.65),236(4.56), 287(4.06),318(3.77)	80-1085-78
$C_{22}H_{14}N_4O_4$			
4(1H)-Quinazolinone, 2,2'-[1,4-phenyl-enebis(oxy)]bis-	dioxan	246(3.82),260(3.92), 266(3.89),300(3.48), 307(3.60),320(3.51)	4-1033-78

Compound	Solvent	$\lambda_{max}(\log \epsilon)$	Ref.
$C_{22}H_{14}O$			
4H-Cyclohepta[b]furan, 4-(9H-fluoren-9-ylidene)-	MeOH	238(4.74),290s(3.94), 382(4.10),460s(3.69)	142-0287-78C
6H-Cyclohepta[c]furan, 6-(9H-fluoren-9-ylidene)-	MeOH	230s(4.54),249(4.68), 274s(4.27),390(4.36), 425s(4.29)	142-0287-78C
$C_{22}H_{14}O_2$			
Phenanthro[9,10-c]furan-1(3H)-one, 3-phenyl-	EtOH	231(4.61),253(4.60), 261(4.63),278(4.09), 306(3.99),338(3.23), 355(3.16)	35-8247-78
$C_{22}H_{14}O_4$			
Cyclopenta[ef]pentaleno[2,1,6-kla]heptalene-6,7-dicarboxylic acid, dimethyl ester	dioxan	231(4.45),270(4.47), 287(4.69),323(4.25), 350(4.13),386(3.87), 401(3.81),423(3.76), 449(3.64),493(3.33), 558(3.27),594(3.52), 661(2.85),792(2.55)	89-0763-78
$C_{22}H_{14}O_5$			
4H-1-Benzopyran-4-one, 7-(benzoyloxy)-3-(4-hydroxyphenyl)-	EtOH	258(3.86)	18-2398-78
	EtOH-NaOAc	258(4.04)	18-2398-78
$C_{22}H_{14}O_6$			
4H-1-Benzopyran-4-one, 7-(benzoyloxy)-5-hydroxy-3-(4-hydroxyphenyl)-	EtOH	259(4.67),305(3.81), 332(3.73)	18-2398-78
	EtOH-NaOAc	259(4.69),305(4.05), 336(3.94)	18-2398-78
[1,2'-Binaphthalene]-5,5',8,8'-tetrone, 1',4-dihydroxy-2,3'-dimethyl-	EtOH	217(4.67),253(4.40), 430(3.89)	98-0869-78
[2,2'-Binaphthalene]-1,1',4,4'-tetrone, 8,8'-dimethoxy-	MeOH	250(4.70),400(4.68)	102-2042-78
Isobisjuglone, methyl-	dioxan	265(2.47),295(1.83), 370(1.29),490(1.53)	102-2042-78
$C_{22}H_{14}O_{14}$			
1,4-Naphthalenedione, 6,6'-ethylidene-bis[2,3,5,7,8-pentahydroxy-	MeOH	266(4.44),342(4.14), 482(4.06),528(3.91)	105-0371-78
$C_{22}H_{14}S_2$			
1,3-Benzodithiole, 2-(5H-dibenzo[a,d]cyclohepten-5-ylidene)-	CH_2Cl_2	232(4.73),273(4.39), 299(4.37),340s(4.15)	18-2674-78
$C_{22}H_{15}ClN_2O_2$			
Imidazo[1,2-a]pyridine, 7-[2-(1,3-benzodioxol-5-yl)ethenyl]-2-(4-chlorophenyl)-	DMF	367(4.69),385(4.66)	33-0129-78
$C_{22}H_{15}ClN_4O$			
[1,2,4]Triazolo[1,5-a]pyridine, 6-chloro-2-[4-[2-(5-methyl-2-benzoxazolyl)-ethenyl]phenyl]-	DMF	343(4.74),357(4.75)	33-0142-78
[1,2,4]Triazolo[1,5-a]pyridine, 6-chloro-2-[4-[2-(6-methyl-2-benzoxazolyl)-ethenyl]phenyl]-	DMF	344(4.73),358(4.73)	33-0142-78
$C_{22}H_{15}ClN_4O_2$			
[1,2,4]Triazolo[1,5-a]pyridine, 6-chlo-	DMF	353(4.67),365(4.67)	33-0142-78

Compound	Solvent	λ_{max} (log ϵ)	Ref.
ro-2-[4-[2-(5-methoxy-2-benzoxazol-yl)ethenyl]phenyl]- (cont.)			33-0142-78
$C_{22}H_{15}ClN_6$ [1,2,4]Triazolo[1,5-a]pyridine, 2-[3-chloro-4-[2-(2-phenyl-2H-1,2,3-triazol-4-yl)ethenyl]phenyl]-	DMF	337(4.71)	33-0142-78
$C_{22}H_{15}Cl_2NO_2$ 1,2-Benzisoxazole, 6-chloro-3-[3-chloro-4-[2-(4-methoxyphenyl)ethenyl]-phenyl]-	DMF	345(4.57)	33-2904-78
$C_{22}H_{15}Cl_3N_2$ 2H-Imidazole, 2,4,5-tris(4-chlorophenyl)-2-methyl-	EtOH	224(4.49),269(4.19)	24-1464-78
$C_{22}H_{15}NO_2$ 5H-Naphtho[2,3-a]carbazole-5,13(12H)-dione, 7,12-dimethyl-	CHCl$_3$	253(4.50),276(4.64), 327(4.42),400(3.93), 480(3.76)	12-1841-78
5(4H)-Oxazolone, 4-([1,1'-biphenyl]-4-ylmethylene)-2-phenyl-	toluene	390(4.65)	103-0120-78
$C_{22}H_{15}NO_3$ 6H-1,3-Oxazinium, 4-hydroxy-6-oxo-2,3,5-triphenyl-, hydroxide, inner salt	MeCN	255(4.27),374(3.47)	5-1655-78
$C_{22}H_{15}N_3O_2$ 1H-Indene-1,3(2H)-dione, 2-[[[4-(phenylazo)phenyl]amino]methylene]-	EtOH	390(4.66)	97-0256-78
$C_{22}H_{16}$ Anthracene, 9-(4-ethenylphenyl)-	CH$_2$Cl$_2$	258(5.01),332(3.26), 349(3.60),367(3.84), 387(3.79)	117-0177-78
Benzo[a]pyrene, 1,3-dimethyl-	n.s.g.	399.3(4.48)	149-0083-78B
Benzo[a]pyrene, 1,6-dimethyl-	n.s.g.	401.8(4.51)	149-0083-78B
Benzo[a]pyrene, 2,3-dimethyl-	n.s.g.	395.0(4.43)	149-0083-78B
Benzo[a]pyrene, 3,6-dimethyl-	n.s.g.	403.3(4.36)	149-0083-78B
Benzo[a]pyrene, 4,5-dimethyl-	n.s.g.	390.5(4.43)	149-0083-78B
Phenanthrene, 4-(1-phenylethenyl)-	C$_6$H$_{12}$	223(4.04),255(4.31), 287(3.71),299(3.11)	35-6679-78
$C_{22}H_{16}Br_2$ Phenanthrene, 2,7-dibromo-9,10-dihydro-9-methyl-10-methylene-9-phenyl-	hexane	223(4.47),228s(--), 246s(--),252(4.41), 296(4.31)	104-0341-78
	H$_2$SO$_4$	282s(--),289s(--), 298(4.39),336(3.85), 381(3.49),502s(--), 606(3.63)	104-0341-78
Phenanthrene, 3,6-dibromo-9,10-dihydro-9-methyl-10-methylene-9-phenyl-	hexane	233s(--),245(4.60), 287(4.04)	104-0341-78
	H$_2$SO$_4$	264(4.30),271s(--), 289s(--),297s(--), 366(4.24),470s(--), 557(3.49)	104-0341-78

Compound	Solvent	$\lambda_{max}(\log \epsilon)$	Ref.
$C_{22}H_{16}ClNO$			
1,2-Benzisoxazole, 3-[4-[2-(4-chloro-phenyl)ethenyl]phenyl]-5-methyl-	DMF	330(4.69)	33-2904-78
$C_{22}H_{16}ClNO_2$			
1,2-Benzisoxazole, 6-chloro-3-[4-[2-(4-methoxyphenyl)ethenyl]phenyl]-	DMF	342(4.62)	33-2904-78
$C_{22}H_{16}Cl_2N_3$			
Pyrimidinium, 1-(4-chlorophenyl)-4-[(4-chlorophenyl)amino]-5-phenyl-, per-chlorate	CH_2Cl_2	324(4.31)	97-0063-78
$C_{22}H_{16}N_2$			
2-Propenal, 3-(2-ethynylphenyl)-, [3-(2-ethynylphenyl)-2-propenylidene]-hydrazone, (?,?,E,E)-	THF	290s(3.92),339s(4.65), 353(4.72),370s(4.68), 390s(4.38)	18-2112-78
$C_{22}H_{16}N_2O$			
1(4H)-Naphthalenone, 2-(phenylamino)-4-(phenylimino)-	C_6H_{12}	250(4.31),280(4.40), 305(4.05),330(3.83), 465(3.85)	48-0557-78
	EtOH	250(4.32),280(4.38), 295(4.21),340(3.78), 470(3.83)	48-0557-78 +80-0617-78
	ether	250(4.29),280(4.36), 295(4.21),330(3.79), 465(3.83)	48-0557-78
	CCl_4	295(4.32),335(3.83), 460(3.88)	48-0557-78
3H-Pyrazole-4-carboxaldehyde, 3,3,5-triphenyl-	heptane	220(4.22),301(3.88)	103-0879-78
	EtOH	218(4.17),297(3.68)	103-0879-78
3H-Pyrazole-5-carboxaldehyde, 3,3,4-triphenyl-	heptane	225(4.16),304(3.80)	103-0879-78
	EtOH	211(4.18),296(3.65)	103-0879-78
$C_{22}H_{16}N_2O_2$			
1H-Dibenz[e,g]indole, 2,3-dihydro-2-(4-nitrophenyl)-	C_6H_{12}	255(4.62),324(3.80)	44-0381-78
Imidazo[1,2-a]pyridine, 7-[2-(1,3-ben-zodioxol-5-yl)ethenyl]-2-phenyl-	DMF	367(4.68),385(4.66)	33-0129-78
4,11-Imino-1H-naphth[2,3-f]isoindole-1,3(2H)-dione, 3a,4,11,11a-tetra-hydro-2-phenyl-, (3aα,4β,11β,11aα)-	EtOH	265(3.87),275(3.80), 286(3.57),310(2.97), 325(3.08)	44-4469-78
1-Naphthalenecarboxylic acid, 2-(1-naphthalenylcarbonyl)hydrazide	EtOH	222(5.00),282(4.17), 289s(4.15)	4-0385-78
2(1H)-Quinoxalinone, 3-(2-[1,1'-biphen-yl]-4-yl-2-oxoethylidene)-3,4-dihydro-	EtOH	448(4.55)	103-0336-78
ion	EtOH	496(4.83)	103-0336-78
$C_{22}H_{16}N_2O_3$			
Benzo[h]quinoline-3-carbonitrile, 4-(1,3-benzodioxol-5-yl)-1,2,5,6-tetra-hydro-1-methyl-2-oxo-	$CHCl_3$	247(4.16),392(4.19)	48-0097-78
1(4H)-Naphthalenone, 2-[(4-hydroxyphen-yl)amino]-4-[(4-hydroxyphenyl)imino]-	EtOH	250(4.35),275(4.35), 295(4.27),320(3.98), 510(3.86)	48-0557-78 +80-0617-78
	ether	250(4.33),275(4.37), 293(4.29),325(3.98), 500(3.88)	48-0557-78

Compound	Solvent	$\lambda_{max}(\log \epsilon)$	Ref.
$C_{22}H_{16}N_3P$ 2-Butenedinitrile, 2-[(triphenylphos- phoranylidene)amino]-	CH_2Cl_2	305(4.16)	39-1237-78C
$C_{22}H_{16}N_4$ Tricyclo[10.2.2.2^{5,8}]octadeca-5,7,12- 14,15,17-hexaene-2,2,11,11-tetra- carbonitrile	CH_2Cl_2	262(2.83),267(2.76), 297(2.18)	35-1806-78
$C_{22}H_{16}N_4O$ [1,2,4]Triazolo[1,5-a]pyridine, 2-[4- [2-(5-methyl-2-benzoxazolyl)ethen- yl]phenyl]-	DMF	343(4.73),357(4.73)	33-0142-78
[1,2,4]Triazolo[1,5-a]pyridine, 2-[4- [2-(6-methyl-2-benzoxazolyl)ethen- yl]phenyl]-	DMF	342(4.72),357(4.72)	33-0142-78
$C_{22}H_{16}N_4O_2$ [1,2,4]Triazolo[1,5-a]pyridine, 2-[4- [2-(5-methoxy-2-benzoxazolyl)ethen- yl]phenyl]-	DMF	353(4.66)	33-0142-78
$C_{22}H_{16}N_4O_3$ 7H-Benz[de]anthracen-7-one, 3-[(4,6-di- methoxy-1,3,5-triazin-2-yl)amino]-	DMF	425(4.68)	2-0106-78
$C_{22}H_{16}N_4O_6$ Benzamide, N,N-(4,5-dimethyl-3,5-cyclo- hexadiene-1,2-diylidene)bis[4-nitro-	MeCN	257.5(4.50)	5-1139-78
$C_{22}H_{16}N_6$ 1H-Benzimidazole, 1,1',1''-methylidyne- tris-	$CHCl_3$	257(4.00),280(4.07)	25-0126-78
[1,2,4]Triazolo[1,5-a]pyridine, 2-phen- yl-7-[2-(2-phenyl-2H-1,2,3-triazol- 4-yl)ethenyl]-	DMF	330(4.72)	33-0142-78
$C_{22}H_{16}N_6O_2$ Benzoic acid, 4-[3-phenyl-5-(1-phthala- zinyl)-1-formazano]-	EtOH	276(4.08),290(4.13), 330(4.09),460(4.24)	103-0450-78
	EtOH-KOH	541(3.42)	103-0450-78
	dioxan	290(4.35),460(4.42)	103-0450-78
$C_{22}H_{16}O$ 1-Cyclopropene-1-carboxaldehyde, 2,3,3- triphenyl-	EtOH	228(4.08),266(3.82), 308(3.86),347(3.18)	103-0879-78
$C_{22}H_{16}O_2$ 8H,15H-Dinaphtho[1,8-bc:1',8'-gh][1,5]- dioxecin	$CHCl_3$	198(4.2)	39-1385-78C
2(5H)-Furanone, 3,4,5-triphenyl-	EtOH	285(4.09)	35-8247-78
2(5H)-Furanone, 3,5,5-triphenyl-	EtOH	267(4.11)	35-8247-78
Spiro[naphthalene-2(1H),2'(3'H)-[1H]- phenalen]-1-one, 8-hydroxy-	$CHCl_3$	293(4.13),373(4.01)	39-1385-78C
$C_{22}H_{16}O_3$ 2-Benzopyrylium, 4-hydroxy-1-(4-meth- oxyphenyl)-3-phenyl-, hydroxide, inner salt	EtOH	553(4.28)	103-0493-78
2-Benzopyrylium, 4-hydroxy-3-(4-meth- oxyphenyl)-1-phenyl-, hydroxide,	EtOH	552(4.34)	103-0493-78

Compound	Solvent	$\lambda_{max}(\log \epsilon)$	Ref.
inner salt (cont.)			
6H-Indeno[1,2-b]oxiren-6-one, 1a,6a-di- hydro-1a-(4-methoxyphenyl)-6a-phenyl-	EtOH	<u>270s(4.0),345s(2.9)</u>	103-0493-78 103-0493-78
6H-Indeno[1,2-b]oxiren-6-one, 1a,6a-di- hydro-6a-(4-methoxyphenyl)-1a-phenyl-	EtOH	<u>275(4.7),360s(2.3)</u>	103-0493-78
$C_{22}H_{16}O_4$			
9,10[1',2']-Benzenoanthracene-1,4-dione, 9,10-dihydro-5,8-dimethoxy-	MeCN	411(2.65)	77-0720-78
4H-1-Benzopyran-4-one, 3-(4-hydroxy- phenyl)-7-(phenylmethoxy)-	EtOH EtOH-NaOAc	262(4.43),306s(4.01) 262(4.44),306s(4.02)	18-2398-78 18-2398-78
$C_{22}H_{16}O_6$			
Glabratephrinone	MeOH	220(4.34),249(4.28), 255s(4.28),307(4.38)	78-1405-78
$C_{22}H_{17}BrN_2O_2$			
Benzoic acid, [3-(3-bromophenyl)-3-oxo- 1-phenylpropylidene]hydrazide	EtOH	292(4.18)	94-1298-78
$C_{22}H_{17}BrN_2O_3$			
Hydrazinecarboxylic acid, [3-(3-bromo- phenyl)-3-oxo-1-phenylpropylidene]-, phenyl ester	EtOH	280(4.25)	94-1298-78
$C_{22}H_{17}BrOS$			
2-Propen-1-one, 1-(3-bromophenyl)- 3-phenyl-3-[(phenylmethyl)thio]-	EtOH	269(4.14),344(4.34)	94-1298-78
$C_{22}H_{17}ClN_2O$			
Imidazo[1,2-a]pyridine, 2-(4-chloro- phenyl)-7-[2-(4-methoxyphenyl)eth- enyl]-	DMF	364(4.69),383(4.65)	33-0129-78
$C_{22}H_{17}ClN_2O_2$			
Benzoic acid, [3-(4-chlorophenyl)-3- oxo-1-phenylpropylidene]hydrazide	EtOH EtOH EtOH	220(4.46),292(4.36) 292(4.36) 293(4.31)	4-0385-78 34-0345-78 94-1298-78
$C_{22}H_{17}ClOS$			
2-Propen-1-one, 1-(3-chlorophenyl)- 3-phenyl-3-[(phenylmethyl)thio]-	EtOH	258(4.11),342(4.25)	4-0105-78
2-Propen-1-one, 1-(4-chlorophenyl)- 3-phenyl-3-[(phenylmethyl)thio]-	EtOH	266(4.35),339(4.27)	4-0105-78
$C_{22}H_{17}Fe_2NO_{10}$			
Iron, hexacarbonyl[μ-[η^2,η^3-dimethyl 2-[(2-methyl-1-phenyl-1-propenyl)- amino]-2-butenedioato(2-)]]di-	EtOH	327(3.95),440s(3.38)	33-0589-78
$C_{22}H_{17}N$			
9H-Indeno[2,1-c]pyridine, 3-methyl- 9-(3-phenyl-2-propenylidene)-, cis	EtOH	208(4.29),215(4.26), 244(4.53),264(4.28), 275(4.09),388(4.58)	103-0066-78
trans	EtOH	209(4.70),240(4.94), 264(4.64),275(4.50), 389(4.99)	103-0066-78
$C_{22}H_{17}NO$			
1,2-Benzisoxazole, 5-methyl-3-[4-(2- phenylethenyl)phenyl]-	DMF	328(4.65)	33-2904-78

Compound	Solvent	$\lambda_{max}(\log \epsilon)$	Ref.
5H-Indeno[1,2-b]pyridine, 5-[2-(2-prop-enyloxy)phenyl]methylene]-	n.s.g.	212(4.56),230(4.66), 310(4.25),340(4.14)	103-1343-78
3(2H)-Isoquinolinone, 1,4-dihydro-1-phenyl-4-(phenylmethylene)-	EtOH	309(3.98)	5-1103-78
Phenol, 2-(5H-indeno[1,2-b]pyridin-5-ylidenemethyl)-6-(2-propenyl)-	n.s.g.	216(5.21),219(5.28), 309(4.87),332(4.67)	103-1343-78
3H-Pyrrol-3-one, 1,2-dihydro-2,4,5-tri-phenyl-	MeOH	254(4.31),385(3.84)	18-3312-78
$C_{22}H_{17}NOS$			
1,2-Benzisoxazole, 3-[4-[2-[4-(methyl-thio)phenyl]ethenyl]phenyl]-	DMF	347(4.67)	33-2904-78
$C_{22}H_{17}NO_2$			
1,2-Benzisoxazole, 3-[4-[2-(4-methoxy-phenyl)ethenyl]phenyl]-	DMF	340(4.63)	33-2904-78
1,2-Benzisoxazole, 6-[2-(4-methoxyphen-yl)ethenyl]-3-phenyl-	DMF	340(4.58)	33-2904-78
1,2-Benzisoxazole, 3-(4-methoxyphenyl)-6-(2-phenylethenyl)-	DMF	323(4.58)	33-2904-78
2,1-Benzisoxazole, 3-[4-[2-(4-methoxy-phenyl)ethenyl]phenyl]-	DMF	387(4.66)	33-2904-78
1H-Isoindol-1-one, 2,3-dihydro-3-hy-droxy-3-phenyl-2-(2-phenylethenyl)-	THF	212(4.59),219(4.52), 287(4.24),320(4.34)	40-0404-78
4H-1,3-Oxazin-4-one, 2,3-dihydro-2,3,6-triphenyl-	EtOH	208(4.07),303(3.89)	103-0224-78
$C_{22}H_{17}NO_3$			
9,10-Anthracenedione, 1-(dimethylami-no)-4-phenoxy-	benzene	511(3.71)	104-2199-78
9,10-Anthracenedione, 1-(dimethylami-no)-5-phenoxy-	benzene	502(3.72)	104-2199-78
9,10-Anthracenedione, 2-(dimethylami-no)-1-phenoxy-	benzene	448(3.70)	104-2199-78
$C_{22}H_{17}NO_3S$			
2-Propen-1-one, 1-(3-nitrophenyl)-3-phenyl-3-[(phenylmethyl)thio]-	EtOH	239(4.48),346(4.30)	94-1298-78
$C_{22}H_{17}NO_4$			
Pyrido[1,2-a]indol-3-ol, 10-(4-acetoxy-phenyl)-, acetate	MeOH	264(4.59),302(4.00), 321(3.94),337(3.76), 416(3.43),444(3.36)	145-1056-78
Pyrrolo[2,1-a]isoquinoline-2,3-dicarb-oxylic acid, 1-phenyl-, dimethyl ester	MeOH	227s(4.39),277(4.03), 350(4.06)	44-2700-78
$C_{22}H_{17}NS_2$			
Acridine, 9-(1,3-benzodithiol-2-yli-dene)-10-ethyl-9,10-dihydro-	CH_2Cl_2	238(4.73),263s(4.26), 314(4.18),399(4.28)	18-2674-78
$C_{22}H_{17}N_3O$			
4(3H)-Quinazolinone, 2-phenyl-3-[(1-phenylethylidene)amino]-	MeOH	236(4.68),287(4.39), 330(3.79)	80-1085-78
$C_{22}H_{17}N_3O_2$			
4(3H)-Quinazolinone, 3-[(benzoylmeth-yl)amino]-2-phenyl-	MeOH	235(4.76),287(4.44), 314(4.20)	80-1085-78
$C_{22}H_{17}N_3O_3$			
14H-Quinoxalino[2,3-b]phenoxazine-14-	MeOH	262(4.81),311(3.92),	64-0912-78C

Compound	Solvent	$\lambda_{max}(\log \epsilon)$	Ref.
propanoic acid, methyl ester (cont.)		479(4.23)	64-0912-78C
$C_{22}H_{17}N_3O_4$			
Benzoic acid, 3-[(3-nitrophenyl)-3-oxo-1-phenylpropylidene]hydrazide	EtOH	283(4.27),294s(4.26),292(4.26)[sic]	94-1298-78
$C_{22}H_{17}N_3O_5$			
Hydrazinecarboxylic acid, [3-(3-nitro-phenyl)-3-oxo-1-phenylpropylidene]-,phenyl ester	EtOH	278(4.40)	94-1298-78
$C_{22}H_{18}$			
Benz[a]aceanthrylene, 8,12b-dihydro-8,8-dimethyl-	n.s.g.	470s(--),500(3.49),528(3.40)	44-0598-78
Cyclopropene, 1-methyl-2,3,3-triphenyl-	EtOH	261(4.20)	44-1493-78
1,3,9,11-Cyclotridecatetraene-5,7-di-yne, 13-(2,4,6-cycloheptatrien-1-yl-idene)-4,9-dimethyl-	n.s.g.	280(4.4),325(4.0),459(4.1)	88-2795-78
1H-Indene, 1-methyl-2,3-diphenyl-	EtOH	234(4.24),303(4.12)	44-1493-78
1H-Indene, 3-methyl-1,2-diphenyl-	EtOH	230(4.20),295(4.22)	44-1481-78
	EtOH	294(4.24)	44-1493-78
Phenanthrene, 9,10-dihydro-2-(2-phenyl-ethenyl)-	hexane	193(4.47),208(4.45),335(4.68),348(4.51)	5-2105-78
Tetracyclo[4.4.0.0^{2,5}.0^{7,10}]deca-3,8-diene, 1,6-diphenyl-	MeOH	254(3.09),261(4.08),267(2.96),272(2.83)	88-1183-78
$C_{22}H_{18}BrClN_2O_4$			
Pyridinium, 1-[[1-[(4-bromophenyl)meth-yl]-4(1H)-quinolinylidene]methyl]-,perchlorate	EtOH	258(4.27),365(3.91),495(4.05)	124-0071-78
$C_{22}H_{18}ClNO_2$			
Benzo[h]quinolin-2(1H)-one, 3-acetyl-4-(2-chlorophenyl)-5,6-dihydro-1-methyl-	CHCl$_3$	250(4.39),374(4.25)	48-0097-78
Benzp[h]quinolin-2(1H)-one, 3-acetyl-4-(4-chlorophenyl)-5,6-dihydro-1-methyl-	CHCl$_3$	250(4.40),372(4.28)	48-0097-78
$C_{22}H_{18}ClN_2$			
Pyridinium, 1-[[1-[(4-chlorophenyl)-methyl]-4(1H)-quinolinylidene]-methyl]-, perchlorate	EtOH	260(4.32),360(3.98),494(4.10)	124-0071-78
$C_{22}H_{18}ClN_2O$			
1H-Isoindolium, 2-(4-chlorophenyl)-3-[(1-ethyl-2(1H)-pyridinylidene)-methyl]-1-oxo-, iodide	EtOH	220(4.58),240(4.14)	18-2415-78
$C_{22}H_{18}ClN_3O_2$			
1H-1,4-Benzodiazepine-2-methanol, 7-chloro-2,3-dihydro-1-nitroso- ,5-di-phenyl-	isoPrOH	225s(4.43),251(4.18)	4-0855-78
diastereomer	isoPrOH	218(4.57),252(4.23)	4-0855-78
$C_{22}H_{18}ClN_3O_4$			
Quinolinium, 4-formyl-1-phenyl-, hydra-zone, perchlorate	MeOH	490(4.25)	65-1635-78
$C_{22}H_{18}ClN_3O_6$			
Pyridinium, 1-[[1-[1-[(4-nitrophenyl)-	EtOH	265(4.31),355(3.83),	124-0071-78

Compound	Solvent	$\lambda_{max}(\log \epsilon)$	Ref.
methyl]-4(1H)-quinolinylidene]meth-yl]-, perchlorate (cont.)		488(4.07)	124-0071-78
$C_{22}H_{18}ClN_5O_4S$ Benz[cd]indole-6-sulfonamide, 2-[(2-chloro-6-cyano-4-nitrophenyl)amino]-N,N-diethyl-	dioxan	447(4.45)	93-1354-78
$C_{22}H_{18}Cl_2FN_5O_4$ 9H,13aH-Oxazolo[3,2-d][1,2,4]triazolo-[1,5-a][1,4]benzodiazepine-7-carbox-amide, 2,12-dichloro-13a-(2-fluoro-phenyl)-11,12-dihydro-9-methoxy-N,N-dimethyl-11-oxo-	MeOH	252(4.20)	33-0848-78
$C_{22}H_{18}Fe_2$ Ferrocenium, 1,1"-(1,2-ethynediyl)bis-	CH_2Cl_2 $C_6H_5NO_2$	162.0(2.69) 158.2(2.71)	35-4393-78 35-4393-78
$C_{22}H_{18}IN_3O_3$ 1H-Isoindolium, 3-[(1-ethyl-2(1H)-pyri-dinylidene)methyl]-2-(4-nitrophenyl)-1-oxo-, iodide	EtOH	218(4.57),240s(4.42), 310s(4.24),320(4.30)	18-2415-78
$C_{22}H_{18}N_2$ 1H-Imidazole, 2-(4-methylphenyl)-4,5-diphenyl-	EtOH	304(4.37)	40-1449-78
1H-Imidazole, 4-methyl-1,2,5-triphenyl-	EtOH	287(4.21)	94-3798-78
Imidazo[1,2-a]pyridine, 7-[2-(3-methyl-phenyl)ethenyl]-2-phenyl-	DMF	361(4.60),380(4.56)	33-0129-78
Imidazo[1,2-a]pyridine, 2-(4-methyl-phenyl)-7-(2-phenylethenyl)-	DMF	362(4.61),380(4.58)	33-0129-78
Indeno[1,2-c]pyrazole, 1,3a,4,8b-tetra-hydro-1,3-diphenyl-	EtOH	238(3.87),353(3.96)	33-1477-78
3H-Pyrazole, 4,5-dihydro-3,3-diphenyl-5-(phenylmethylene)-	EtOH	315(4.40)	22-0401-78
$C_{22}H_{18}N_2O$ Benzo[h]quinoline-3-carbonitrile, 1,2,5,6-tetrahydro-1-methyl-4-(4-methylphenyl)-2-oxo-	$CHCl_3$	268(4.27),392(4.36)	48-0097-78
1H-Imidazole, 2-(4-methoxyphenyl)-4,5-diphenyl-	EtOH	300(4.33)	40-1449-78
4H-Imidazol-4-one, 3,5-dihydro-5-meth-yl-2,3,5-triphenyl-	EtOH	227.5(4.41)	94-3798-78
Imidazo[1,2-a]pyridine, 2-[4-[2-(4-methoxyphenyl)ethenyl]phenyl]-	DMF	354(4.72),372(4.65)	33-0129-78
Imidazo[1,2-a]pyridine, 7-[2-(2-meth-oxyphenyl)ethenyl]-2-phenyl-	DMF	365(4.61),383(4.55)	33-0129-78
Imidazo[1,2-a]pyridine, 7-[2-(3-meth-oxyphenyl)ethenyl]-2-phenyl-	DMF	361(4.61),381(4.57)	33-0129-78
Imidazo[1,2-a]pyridine, 7-[2-(4-meth-oxyphenyl)ethenyl]-2-phenyl-	DMF	363(4.67),383(4.63)	33-0129-78
Imidazo[1,2-a]pyridine, 2-(4-methoxy-phenyl)-7-(2-phenylethenyl)-	DMF	364(4.58),380(4.55)	33-0129-78
1H-Pyrazole, 4,5-dihydro-4-(phenoxy-methylene)-1,3-diphenyl-	EtOH	262(4.21),380(4.19)	22-0415-78
3H-Pyrazole, 4,5-dihydro-5-methylene-3-(phenoxyphenyl)-	EtOH	293(4.24)	22-0401-78

Compound	Solvent	λ_{max} (log ϵ)	Ref.
$C_{22}H_{18}N_2O_2$			
1,4-Acridinedione, 5,9-dimethyl-2-[(2-methylphenyl)amino]-	dioxan	233(4.63),290(4.46), 320s(4.4),405(3.66)	150-4901-78
1,4-Acridinedione, 7,9-dimethyl-2-[(4-methylphenyl)amino]-	dioxan	232(4.62),290(4.40), 317(4.36),445(3.67)	150-4901-78
1,4-Acridinedione, 5,6,9-trimethyl-2-(phenylamino)-	dioxan	237(4.58),265(4.21), 295(4.39),328(4.38), 407(3.69)	150-4901-78
1,4-Acridinedione, 5,7,9-trimethyl-2-(phenylamino)-	dioxan	237(4.62),268s(4.2), 298(4.46),413(3.72)	150-4901-78
1,4-Acridinedione, 5,8,9-trimethyl-2-(phenylamino)-	dioxan	241(4.56),265(4.25), 300(4.39),417(3.82)	150-4901-78
Benzoic acid, (3-oxo-1,3-diphenylprop-ylidene)hydrazide	EtOH	287(4.39)	4-0385-78 +34-0345-78
Benzo[h]quinoline-3-carbonitrile, 5,6-dihydro-2-methoxy-4-(4-methoxyphenyl)-	CHCl$_3$	267(4.25),350(4.43)	48-0097-78
Benzo[h]quinoline-3-carbonitrile, 1,2,5,6-tetrahydro-4-(4-methoxy-phenyl)-1-methyl-2-oxo-	CHCl$_3$	270(4.23),392(4.34)	48-0097-78
Diol from caulerpin reduction	MeOH	217(4.56),253(4.40), 299(4.31)	150-1683-78
$C_{22}H_{18}N_2O_4$			
Cyclobuta[1,2-d:3,4-d']bis[2]benzaze-pine-5,7,12,14(6H,13H)-tetrone, 7a,7b,14a,14b-tetrahydro-6,12-dimethyl-	EtOH	239(2.42),280(1.67)	78-2887-78
6H,8H-Phenanthro[9,10-d]pyrrolo[1,2-c]-imidazole-9,10-dicarboxylic acid, dimethyl ester	MeOH	242(4.59)	44-0381-78
$C_{22}H_{18}N_2O_6$			
1H-Indene-1,3(2H)-dione, 2,2'-(1,4-pip-erazinediyl)bis[2-hydroxy-	MeOH	229(4.95),248(4.33), 280(3.25),289(3.20), 303(2.83),355(2.27)	24-3058-78
$C_{22}H_{18}N_2S$			
Imidazo[1,2-a]pyridine, 7-[2-[4-(meth-ylthio)phenyl]ethenyl]-2-phenyl-	DMF	370(4.72),389(4.68)	33-0129-78
$C_{22}H_{18}N_3$			
Pyrimidinium, 1,5-diphenyl-4-(phenyl-amino)-, perchlorate	CH$_2$Cl$_2$	321(4.27)	97-0063-78
$C_{22}H_{18}N_3O_2$			
Pyridinium, 1-[[1-[1-[(4-nitrophenyl)-methyl]-4(1H)-quinolinylidene]-methyl]-, perchlorate	EtOH	265(4.31),355(3.83), 488(4.07)	124-0071-78
$C_{22}H_{18}N_3O_3$			
1H-Isoindolium, 3-[(1-ethyl-2(1H)-pyri-dinylidene)methyl]-2-(4-nitrophenyl)-1-oxo-, iodide	EtOH	218(4.57),240s(4.42), 310s(4.24),320(4.30)	18-2415-78
$C_{22}H_{18}N_4$			
Propanedinitrile, [(1-ethyl-3-methyl-4-phenyl-1H-pyrazol-5-yl)phenyl-methylene]-	EtOH	322(4.10)	78-1163-78
$C_{22}H_{18}N_4O_4$			
2,4(3H,8H)-Pteridinedione, 8-(2-acet-	MeOH	273(4.20),368s(3.64),	24-2586-78

Compound	Solvent	$\lambda_{max}(\log \epsilon)$	Ref.
oxyethyl)-6,7-diphenyl- (cont.)		428(3.98)	24-2586-78
Tryptoquivaline F	MeOH	226(4.52),232s(4.49), 255s(4.09),265(4.05), 276s(3.91),290s(3.49), 303(3.45),315(3.34)	94-0111-78
Tryptoquivaline J	MeOH	226(4.61),231s(4.58), 253s(4.21),264s(4.11), 275s(3.99),290s(3.62), 302(3.59),310(3.49)	94-0111-78
$C_{22}H_{18}N_4O_5$			
Tryptoquivaline E	MeOH	226(4.51),232s(4.47), 254s(4.20),266s(4.07), 276s(3.92),291s(3.54), 303(3.48),315(3.42)	94-0111-78
Tryptoquivaline H	MeOH	226(4.52),232s(4.49), 255s(4.09),266s(4.05), 276s(3.93),291s(3.56), 303(3.49),315(3.40)	94-0111-78
$C_{22}H_{18}N_6$			
Formazan, 5-(4-methylphenyl)-1-(1-phtha-lazinyl)-3-phenyl-	C_6H_{12}	280(4.30),407(4.39)	103-0450-78
	EtOH	288(4.31),414(4.32), 444(4.35)	103-0450-78
	EtOH-KOH	526(3.78)	103-0450-78
	dioxan	293(4.13),440(4.29)	103-0450-78
$C_{22}H_{18}N_6O$			
7H-Benz[de]anthracen-7-one, 3-[[4,6-bis(methylamino)-1,3,5-triazin-2-yl]amino]-	DMF	458(4.71)	2-0106-78
Formazan, 5-(4-methoxyphenyl)-1-(1-phthalazinyl)-3-phenyl-	C_6H_{12}	282(4.26),330(4.29), 410(4.45)	103-0450-78
	EtOH	285(4.34),335(4.38), 425s(4.46),440(4.49)	103-0450-78
	EtOH-KOH	529(4.05)	103-0450-78
$C_{22}H_{18}N_6O_7S_2$			
2,7-Naphthalenedisulfonic acid, 4-amino-3-[(4-aminophenyl)azo]-5-hydroxy-6-(phenylazo)-, disodium salt	H_2O	601(4.63)	40-0271-78
$C_{22}H_{18}O$			
1H,3H-3a,8:9,13b-Diethenodibenzo[3,4-7,8]cycloocta[1,2-c]furan, 8,9-dihydro-, endo	EtOH	<u>265(4.0)</u>	88-2815-78
exo	EtOH	<u>275(3.1),282(3.1)</u>	88-2815-78
2H-Indene, 2-(4-methoxyphenyl)-1-phenyl-	MeOH	210(4.78),233(4.57), 310(4.32)	78-0891-78
Naphthalene, 1,1'-[oxybis(methylene)]-bis-	EtOH	<u>278f(4.2),315(2.8)</u>	88-2815-78
$C_{22}H_{18}OS$			
2-Propen-1-one, 1,3-diphenyl-3-[(phenylmethyl)thio]-	EtOH	260(4.09),339(4.29)	4-0105-78
$C_{22}H_{18}O_2$			
Methanone, bicyclo[2.2.2]octa-2,5-dien-2,3-diylbis[phenyl-	EtOH	256(4.18)	78-2305-78

$C_{22}H_{18}O_2S_2-C_{22}H_{19}IN_2O$

Compound	Solvent	$\lambda_{max}(\log \epsilon)$	Ref.
$C_{22}H_{18}O_2S_2$ 1,3-Benzodithiole, 2-[bis(4-methoxy- phenyl)methylene]-	CH_2Cl_2	247(4.55),256s(4.49), 326(4.35)	18-2674-78
$C_{22}H_{18}O_4$ Octacyclo[8.7.2.2^{20,21}.0^{1,10}.0^{3,8}- 0^{12,16}.0^{13,19}.0^{15,18}]heneicosa- 5,20-diene-4,7,11,17-tetrone	n.s.g.	<u>233(3.9),300s(3.0),</u> <u>375s(1.7)</u>	104-2131-78
$C_{22}H_{18}O_5$ 4H-1-Benzopyran-4-one, 2,3-dihydro-5,7- dihydroxy-6-[(4-hydroxyphenyl)methyl]- 2-phenyl-	MeOH	290(4.50),325(3.71)	100-0156-78
4H-1-Benzopyran-4-one, 2,3-dihydro-5,7- dihydroxy-8-[(4-hydroxyphenyl)methyl]- 2-phenyl-	MeOH	282(4.25),328(3.82)	100-0156-78
$C_{22}H_{18}O_6$ 4H-Furo[2,3-h]-1-benzopyran-4-one, 9-(2-hydroxy-2-methyl-1-oxopropyl)- 5-methoxy-2-phenyl-	MeOH	217(4.58),230(4.62), 274(4.60),321(3.97)	119-0047-78
Glabratephrinol	MeOH	214(4.55),247(4.40), 257(4.43),311(4.43)	78-1405-78
$C_{22}H_{18}O_8$ Cyclopenta[b]xanthene-3-carboxylic acid, 1,10-dihydro-4,7,11-trimethoxy- 9-methyl-1,10-dioxo-, methyl ester	EtOH	228(4.38),261(4.25), 290(4.23),415(3.88)	94-0209-78
$C_{22}H_{18}O_9$ 4H-1-Benzopyran-4-one, 7-acetoxy- 3-(2,3-diacetoxy-4-methoxyphenyl)-	EtOH	218(4.50),248(4.52), 305(3.91)	102-1413-78
$C_{22}H_{18}S$ 3H-Naphtho[1,8-bc]thiophene, 3-(3,4-di- hydro-1(2H)-naphthalenylidene)-2- methyl-	MeOH	258(4.293),334(3.764), 410(3.889)	24-1824-78
$C_{22}H_{19}As$ Acridarsine, 10-(2,4,6-trimethylphen- yl)-	THF	386s(2.93),406(3.61), 429(3.95),455(4.08)	142-0299-78C
$C_{22}H_{19}BrN_2$ Quinolinium, 1-[(4-bromophenyl)methyl]- 4-(pyridiniomethyl)-, diperchlorate	EtOH	233(4.84),320(4.11)	124-0071-78
$C_{22}H_{19}ClN_2$ Quinolinium, 1-[(4-chlorophenyl)meth- yl]-4-(pyridiniomethyl)-, diperchlor- ate	EtOH	233(4.66),320(3.95)	124-0071-78
$C_{22}H_{19}ClN_2O_4$ Pyridinium, 1-[[1-(phenylmethyl)-4(1H)- quinolinylidene]methyl]-, perchlorate	EtOH	264(4.15),360(3.79), 492(4.02)	124-0071-78
$C_{22}H_{19}Cl_2N_3O_{10}$ Quinolinium, 1-[(4-nitrophenyl)methyl]- 4-(pyridiniomethyl)-, diperchlorate	EtOH	233(4.71),322(4.07)	124-0071-78
$C_{22}H_{19}IN_2O$ 1H-Isoindolium, 3-[(1-ethyl-2(1H)-pyri-	EtOH	240(4.76),275(4.57)	18-2415-78

Compound	Solvent	$\lambda_{max}(\log \epsilon)$	Ref.
dinylidene)methyl]-1-oxo-2-phenyl-, iodide (cont.)			18-2415-78
Quinolinium, 2-[3-(1,2-dihydro-2-oxo-3H-indol-3-ylidene)-1-propenyl]-1-ethyl-, iodide	MeOH	518s(3.81),560(3.99), 602s(3.81)	64-0209-78B
$C_{22}H_{19}IN_2O_3$			
Quinolinium, 2-[[2-(carboxymethyl)-2,3-dihydro-3-oxo-1H-isoindol-1-ylidene]methyl]-1-ethyl-, iodide	EtOH	242(4.85),554(4.11), 584(4.18)	18-2415-78
$C_{22}H_{19}N$			
Pyridine, 5-methyl-4-phenyl-2-(4-phenyl-1,3-butadienyl)-	EtOH	205(4.64),240(4.41), 316(4.52),342(4.74)	103-0066-78
$C_{22}H_{19}NO$			
1H-Phenalen-1-one, 2-[[4-(dimethylamino)phenyl]methylene]-2,3-dihydro-	MeOH	258(4.19),316(3.91), 368s(--),444(4.34)	104-2381-78
$C_{22}H_{19}NO_2$			
Benzenamine, 4-(7-methoxy-3-phenyl-2H-1-benzopyran-2-yl)-	EtOH	247(3.71),337(3.58)	39-0088-78C
Benzenepropanamide, α-hydroxy-β-methylene-N,N-diphenyl-	EtOH	238(4.00)	44-0954-78
Benzo[h]quinolin-2(1H)-one, 3-acetyl-5,6-dihydro-1-methyl-4-phenyl-	CHCl$_3$	250(4.41),372(4.17)	48-0097-78
Benzo[h]quinolin-2(1H)-one, 3-acetyl-5,6-dihydro-4-(4-methylphenyl)-	CHCl$_3$	245(4.15),372(4.20)	48-0097-78
5H-Naphtho[2,3-a]carbazole-5,13(5aH)-dione, 6,12,12a,12b-tetrahydro-7,12-dimethyl-, (5aα,12aα,12bα)-	CHCl$_3$	253(4.47),274s(4.16), 305s(3.83),350(3.83)	12-1841-78
Oxiranecarboxamide, 3-methyl-N,N,3-triphenyl-, cis	EtOH	233(4.09)	44-0954-78
trans	EtOH	238(4.21)	44-0954-78
2(1H)-Quinolinone, 3,4-dihydro-3-hydroxy-4-methyl-1,4-diphenyl-	EtOH	246(4.00)	44-0954-78
geometric isomer	EtOH	258(4.20)	44-0954-78
$C_{22}H_{19}NO_3$			
Acetamide, N-(6-acetoxy-5,6-dihydro-5-chrysenyl)-, trans	MeCN	256(4.68),266(4.84), 294(4.06),305(4.13), 317(4.03),340(2.83)	44-0397-78
Benzo[h]quinolin-2(1H)-one, 3-acetyl-5,6-dihydro-4-(4-methoxyphenyl)-	CHCl$_3$	245(4.22),370(4.26)	48-0097-78
$C_{22}H_{19}NO_4$			
2H-Azepine-4-carboxylic acid, 3-methoxy-2-oxo-6,7-diphenyl-, ethyl ester	dioxan	253(4.23),300(4.09)	142-0731-78A
2H-Azepine-6-carboxylic acid, 3-methoxy-2-oxo-4,7-diphenyl-, ethyl ester	dioxan	255(4.30),300s(4.02), 350s(3.87)	142-0731-78A
3H-Azepine-6-carboxylic acid, 2-methoxy-3-oxo-4,7-diphenyl-, ethyl ester	dioxan	232(4.22),280(4.00), 355(4.12)	142-0731-78A
3H-Phenanthro[9,10,1-ija]quinolizin-3-one, 5,6-dihydro-2,8,9-trimethoxy-	EtOH	247s(4.48),267(4.70), 320(4.11),344(4.15), 360(4.15),379(4.20)	44-1096-78
3H-Phenanthro[9,10,1-ija]quinolizin-3-one, 5,6-dihydro-2,10,11-trimethoxy-	EtOH	257s(4.61),268(4.70), 302(4.20),320(4.08), 346(3.92),365(3.28), 384(4.00)	44-1096-78
2,3-Pyridinedicarboxylic acid, 5,6-diphenyl-, 3-ethyl 2-methyl ester	EtOH	247(4.24),300s(4.00)	142-0731-78A

Compound	Solvent	$\lambda_{max}(\log \epsilon)$	Ref.
2,5-Pyridinedicarboxylic acid, 3,6-di-phenyl-, 5-ethyl 2-methyl ester	EtOH	270(4.29)	142-0731-78A
$C_{22}H_{19}N_2$ Pyridinium, 1-[[1-(phenylmethyl)-4(1H)-quinolinylidene]methyl]-, perchlorate	EtOH	264(4.15),360(3.79), 492(4.02)	124-0071-78
$C_{22}H_{19}N_2O$ 1H-Isoindolium, 3-[(1-ethyl-2(1H)-pyri-dinylidene)methyl]-1-oxo-2-phenyl-, iodide	EtOH	240(4.76),275(4.57)	18-2415-78
Quinolinium, 2-[3-(1,2-dihydro-2-oxo-3H-indol-3-ylidene)-1-propenyl]-1-ethyl-, iodide	MeOH	518s(3.81),560(3.99), 602s(3.81)	64-0209-78B
$C_{22}H_{19}N_2O_3$ Quinolinium, 2-[[2-(carboxymethyl)-2,3-dihydro-3-oxo-1H-isoindol-1-ylidene]methyl]-1-ethyl-, iodide	EtOH	242(4.85),554(4.11), 584(4.18)	18-2415-78
$C_{22}H_{19}N_3O$ [1,2,4]Triazolo[1,5-a]pyridine, 7-[2-(4-methoxyphenyl)ethenyl]-2-(4-meth-ylphenyl)-	DMF	340(4.65)	33-0142-78
$C_{22}H_{19}N_3O_2$ Quinolinium, 1-[(4-nitrophenyl)methyl]-4-(pyridiniomethyl)-, diperchlorate	EtOH	233(4.71),322(4.07)	124-0071-78
$C_{22}H_{19}N_3O_5$ 4H-Pyran-3-carboxylic acid, 2-amino-5-cyano-6-(4-methylphenyl)-4-(4-ni-trophenyl)-, ethyl ester	EtOH	302(4.02)	4-0057-78
$C_{22}H_{19}N_3O_6$ 4H-Pyran-3-carboxylic acid, 2-amino-5-cyano-6-(4-methoxyphenyl)-4-(4-ni-trophenyl)-, ethyl ester	EtOH	305(4.23)	4-0057-78
$C_{22}H_{19}N_3O_7S$ 4-Thia-1-azabicyclo[3.2.0]hept-2-ene-2-carboxylic acid, 3-methyl-7-oxo-6-[(phenoxyacetyl)amino]-, (4-nitro-phenyl)methyl ester, (5R-trans)-	EtOH	264(4.08),267(4.09), 274(4.05),304(3.94)	35-8214-78
$C_{22}H_{19}N_5O_4S$ Benz[cd]indole-6-sulfonamide, 2-[(2-cy-ano-4-nitrophenyl)amino]-N,N-diethyl-	dioxan	436(4.47)	93-1354-78
$C_{22}H_{19}N_5O_8S_2$ Benzenesulfonamide, N-[1,3-dihydro-1,3-dimethyl-3-[[(4-nitrophenyl)sulfon-yl]amino]-2H-indol-2-ylidene]-4-nitro-	EtOH	210(4.36),270(4.29), 295s(3.95)	39-1471-78C
$C_{22}H_{20}$ Azulene, 4,6-dimethyl-8-(4-phenyl-1,3-butadienyl)-	benzene	600(2.88)	104-0243-78
$C_{22}H_{20}ClNO_2$ Cyclohepta[b]pyran-2(5H)-one, 3-chloro-4-(diphenylamino)-6,7,8,9-tetrahydro-	EtOH	253(4.01),280(4.19), 330(3.94),379(3.87)	4-0511-78

Compound	Solvent	$\lambda_{max}(\log \epsilon)$	Ref.
$C_{22}H_{20}Cl_2FN_3O_6$			
Propanedioic acid, [(chloroacetyl)amino][[4-chloro-2-(2-fluorobenzoyl)-phenyl]azo]-, diethyl ester	MeOH	243(4.25),289(5.14)	33-0848-78
$C_{22}H_{20}IN_5O_4$			
Quinolinium, 4-[[2-(2,4-dinitrophenyl)-2,4-dihydro-5-methyl-3H-pyrazol-3-ylidene]methyl]-1-ethyl-, iodide	n.s.g.	230(4.71),360(4.76)	48-0857-78
$C_{22}H_{20}N_2$			
Benzenamine, 4-methyl-N-(3,5,8-trimethyl-9H-indeno[2,1-c]pyridin-9-ylidene)-	EtOH	250(4.41),257(4.45),291(3.71),302(3.79),314(3.83),380(3.08),415(3.03)	104-2365-78
Benzenamine, 2,2'-(1,2-phenylenedi-2,1-ethenediyl)bis-	EtOH	279(4.35),349(4.24)	33-2813-78
Quinolinium, 1-(phenylmethyl)-4-(pyridiniomethyl)-, diperchlorate	EtOH	237(4.58),319(3.98)	124-0071-78
$C_{22}H_{20}N_2O$			
1H-Imidazol-5-ol, 4,5-dihydro-4-methyl-1,2,4-triphenyl-	EtOH	230(4.22),265(3.88)	94-3798-78
$C_{22}H_{20}N_2O_2$			
Benzenebutanoic acid, α-oxo-γ-phenyl-, phenylhydrazone	EtOH	207(4.67),245(3.89),302(3.98),340(4.19)	56-2455-78
Indolo[2,3-a]quinolizin-1-ol, 2,3,4,6-7,12-hexahydro-, benzoate monoperchlorate	EtOH-HCl	235(4.32),367(4.33)	22-0355-78
	EtOH	234(4.26),366(4.29)	22-0355-78
Pyrano[2,3-c]pyrazol-6(1H)-one, 1,5-bis(phenylmethyl)-3,4-dimethyl-	EtOH	317.3(4.11)	95-0335-78
Pyrano[2,3-c]pyrazol-6(2H)-one, 2,5-bis(phenylmethyl)-3,4-dimethyl-	EtOH	309.7(4.26)	95-0335-78
$C_{22}H_{20}N_2O_3$			
4H-Indol-4-one, 1,5,6,7-tetrahydro-6,6-dimethyl-1-(3-nitrophenyl)-2-phenyl-	MeOH	210(3.55),245(3.71),280(3.39)	48-0863-78
4H-Indol-4-one, 1,5,6,7-tetrahydro-6,6-dimethyl-1-(4-nitrophenyl)-2-phenyl-	MeOH	235(4.26),270(4.28)	48-0863-78
4H-Pyran-3-carboxylic acid, 2-amino-5-cyano-4-(4-methylphenyl)-6-phenyl-, ethyl ester	EtOH	296(3.81)	4-0057-78
4H-Pyran-3-carboxylic acid, 2-amino-5-cyano-6-(4-methylphenyl)-4-phenyl-, ethyl ester	EtOH	297(3.83)	4-0057-78
$C_{22}H_{20}N_2O_4$			
[3,5'-Bi-1H-indole]-2,2'-dicarboxylic acid, diethyl ester	EtOH	210(4.38),232(4.57),300(4.45)	103-0173-78
[5,5'-Bi-1H-indole]-2,2'-dicarboxylic acid, diethyl ester	EtOH	225s(4.45),265(4.7),305(4.6)	103-0173-78
Butanedioic acid, (1H-phenanthro[9,10-d]imidazol-1-ylmethyl)-, dimethyl ester	MeOH	255(4.99)	44-0381-78
4H-Pyran-3-carboxylic acid, 2-amino-5-cyano-4-(4-methoxyphenyl)-6-phenyl-, ethyl ester	EtOH	300(3.90)	4-0057-78

Compound	Solvent	$\lambda_{max}(\log \epsilon)$	Ref.
$C_{22}H_{20}N_3S_2$			
Benzothiazolium, 2-[2-cyano-3-(3-ethyl-2(3H)-benzothiazolylidene)-1-propenyl]-3-ethyl-	EtOH	617(5.12)	124-0942-78
$C_{22}H_{20}N_4$			
2-Quinoxalinamine, 3-(4-amino-2-methylphenyl)-N-(3-methylphenyl)-	EtOH	254(4.46),281(4.42), 384(4.11)	78-0981-78
2,3-Quinoxalinediamine, N,N'-bis(3-methylphenyl)-	EtOH	246(4.39),274(4.47), 354s(4.16),368(4.21), 384s(4.11)	78-0981-78
2,3-Quinoxalinediamine, N,N'-bis(4-methylphenyl)-	EtOH	247(4.42),274(4.50), 356s(4.20),369(4.25), 386s(4.13)	78-0981-78
$C_{22}H_{20}N_4O_2$			
Benzo[1,2-c:4,5-c']dipyrazole, 2,4,6,8-tetrahydro-3,7-dimethoxy-2,6-diphenyl-	EtOH	242(4.21),260(4.19)	48-0991-78
2,3-Quinoxalinediamine, N,N'-bis(3-methoxyphenyl)-	EtOH	246(4.43),273(4.47), 355s(4.16),368(4.23), 383s(4.12)	78-0981-78
$C_{22}H_{20}N_4O_3$			
1H-Pyrazolo[3,4-b]quinoxaline, 3-(2,3-O-isopropylidene-β-D-erythro-furanosyl)-1-phenyl-	MeOH	268(4.5),336(4.0), 408(3.5)	136-0079-78H
$C_{22}H_{20}N_4O_4S$			
Hydrazine, 1-(1,3-dihydro-3,3-dimethyl-1-phenylbenzo[c]thien-1-yl)-2-(2,4-dinitrophenyl)-	toluene EtOH	345(4.24) 258(3.88),349(4.14)	104-2229-78 104-2229-78
$C_{22}H_{20}N_4O_5$			
Hydrazine, 1-(1,3-dihydro-3,3-dimethyl-1-phenyl-1-isobenzofuranyl)-2-(2,4-dinitrophenyl)-	toluene EtOH	345(4.20) 259(3.96),348(4.19)	104-2229-78 104-2229-78
$C_{22}H_{20}N_4O_8$			
3-Pyridinecarboxylic acid, 1,4,5,6-tetrahydro-1-(4-nitrobenzoyl)-2-[(4-nitrobenzoyl)amino]-, ethyl ester	MeOH	231(4.30),268(4.46), 336(3.93)	24-2297-78
$C_{22}H_{20}N_5O_4$			
Quinolinium, 4-[[2-(2,4-dinitrophenyl)-2,4-dihydro-5-methyl-3H-pyrazol-3-ylidene]methyl]-1-ethyl-, iodide	n.s.g.	230(4.71),360(4.76)	48-0857-78
$C_{22}H_{20}N_6O_5S$			
Benzoic acid, 4-[[[(2-amino-1,4-dihydro-4-oxo-6-pteridinyl)methyl][(4-methylphenyl)sulfonyl]amino]methyl]-	pH 1 pH 13	237(4.50),325(3.79) 240(4.48),255(4.37), 363(3.81)	87-0673-78 87-0673-78
$C_{22}H_{20}O_2$			
2,2'a(2'H)-Bi-3H-naphtho[1,8-bc]furan, 4,4',5,5'-tetrahydro-	EtOH	208(4.61),215s(4.53), 256(4.22),286(3.54)	18-2068-78
4-Cyclopentene-1,3-dione, 4-(1-methylethenyl)-2,2-bis(phenylmethyl)-	EtOH	288(3.97),390s(1.96)	78-1567-78
1-Propanone, 2-(4-methoxyphenyl)-1,2-diphenyl-	MeOH	213(3.98),230(3.87), 250(3.60),275(3.11)	78-0891-78

Compound	Solvent	$\lambda_{max}(\log \epsilon)$	Ref.
$C_{22}H_{20}O_3$			
2,5-Furandione, tetrahydro-3-(diphenyl-methylene)-4-(1-ethylpropylidene)-	toluene	355(3.95)	39-0571-78C
Naphtho[2,3-c]furan-1,3-dione, 4,4-di-ethyl-3,4-dihydro-9-phenyl-	toluene	306(3.91)	39-0571-78C
$C_{22}H_{20}O_4$			
Spiro[1,3-dioxane-5,3'(4'H)-[2H]naph-tho[2,3-b][1,4]dioxepin], 2-phenyl-	CHCl$_3$	246(4.20),268s(--), 278(3.72),285(3.72), 296s(--),314(3.55), 328(3.34)	56-1913-78
$C_{22}H_{20}O_5$			
2H,8H-Benzo[1,2-b:3,4-b']dipyran-2-one, 9-acetoxy-9,10-dihydro-8,8-dimethyl-10-phenyl-	EtOH	211(4.51),248(3.69), 257(3.66),330(4.12)	2-0184-78
5H-Furo[3,2-g][1]benzopyran-5-one, 4-acetoxy-2,3-dihydro-2,3,3-tri-methyl-6-phenyl-	MeOH	254(4.77),298(4.68)	2-0973-78
$C_{22}H_{20}O_6$			
2H,6H-Benzo[1,2-b:5,4-b']dipyran-6-one, 7-(2,4-dimethoxyphenyl)-5-hydroxy-2,2-dimethyl-	MeOH	265(4.64),308(3.80)	2-0613-78
2H,6H-Benzo[1,2-b:5,4-b']dipyran-6-one, 7-(3,4-dimethoxyphenyl)-5-hydroxy-2,2-dimethyl-	MeOH	208(4.03),276(4.07)	44-3446-78
4H,8H-Benzo[1,2-b:3,4-b']dipyran-4-one, 3-(2,4-dimethoxyphenyl)-5-hydroxy-8,8-dimethyl-	MeOH	255(4.52),266(4.58), 305(3.91)	2-0613-78
2-Butenoic acid, 4-(1,3-dimethoxy-10-oxo-9(10H)-phenanthrenylidene)-3-methoxy-, methyl ester, (Z,Z)-	MeOH	247(4.51),263s(4.43), 350(4.16)	12-2259-78
4H-Furo[2.3-h]-1-benzopyran-4-one, 8,9-dihydro-9-(2-hydroxy-2-methyl-1-oxopropyl)-5-methoxy-2-phenyl-, (S)-	MeOH	213(4.50),266(4.53), 325(4.02)	119-0047-78
1-Naphthacenecarboxylic acid, 2-ethyl-1,2,3,4,6,11-hexahydro-5,12-dihydr-oxy-6,11-dioxo-, methyl ester, cis	MeOH	205(4.35),251(4.59), 285(3.92),330s(--), 460s(--),482(3.96), 568(3.18)	150-4762-78
trans	MeOH	206(4.33),251(4.57), 285(3.97),330s(--), 460s(--),482(3.94), 518(3.83),569(3.16)	150-4762-78
$C_{22}H_{20}O_7$			
1-Naphthacenecarboxylic acid, 2-ethyl-1,2,3,4,6,11-hexahydro-4,5,12-tri-hydroxy-6,11-dioxo-, methyl ester, (1r,2c,4c)-	MeOH	206(4.36),251(4.62), 287(4.03),330s(--), 460s(--),484(4.07), 520(3.90),568(2.90)	150-4762-78
(1r,2t,4c)-	MeOH	206(4.38),251(4.65), 287(4.06),331s(--), 460s(--),483(4.10), 520(3.92),568(2.83)	150-4762-78
Podophyllotoxin	EtOH	288(3.54)	100-0497-78
$C_{22}H_{20}O_8$			
1-Naphthacenecarboxylic acid, 1,2,3,4-6,11-hexahydro-2,5,12-trihydroxy-4-methoxy-2-methyl-6,11-dioxo-, methyl ester	MeOH	203(4.39),231s(--), 248(4.59),282(3.97), 317s(--),462s(--), 485(4.01),510s(--)	24-3823-78

Compound	Solvent	λ_{max}(log ϵ)	Ref.
isomer	MeOH	203(4.39),227s(--), 229(4.65),253s(--), 286(3.96),317s(--), 454s(--),478(4.03), 513(3.84),558(2.71)	24-3823-78
isomer	MeOH	200(4.35),222s(--), 246(4.64),250s(--), 283(3.90),317s(--), 454s(--),474(3.99), 509(3.90),557(2.68)	24-3823-78
Naphtho[2,3-b]furan-3-carboxylic acid, 4,9-diacetoxy-2-(2-oxopropyl)-, ethyl ester	dioxan	245(4.84),322(3.97), 342(3.85)	5-0140-78
$C_{22}H_{20}O_9$ Cyclopenta[b]xanthene-3-carboxylic acid, 1,2,3,10-tetrahydro-3-hydroxy-4,7,11- trimethoxy-9-methyl-1,10-dioxo-, methyl ester	EtOH	260(4.45),289(4.40), 335(3.80)	94-0209-78
$C_{22}H_{20}O_{10}$ 4H-1-Benzopyran-4-one, 3-(1,3-benzodi- oxol-5-yl)-7-(β-D-glucopyranosyloxy)-	EtOH	222(4.04),263(3.93), 292(3.77)	106-0235B-78
$C_{22}H_{21}N$ Benzenamine, N-(diphenylmethylene)- 2,4,6-trimethyl-	THF	242(3.62),275(3.29), 360(2.92)	35-4886-78
$C_{22}H_{21}NO$ 4H-Indol-4-one, 1,5,6,7-tetrahydro- 6,6-dimethyl-1,2-diphenyl-	MeOH	243(4.01),275(4.08)	48-0863-78
$C_{22}H_{21}NO_2$ 4H-Indol-4-one, 1,5,6,7-tetrahydro- 1-(2-hydroxyphenyl)-6,6-dimethyl- 2-phenyl-	MeOH	240(3.72),275(3.81)	48-0863-78
4H-Indol-4-one, 1,5,6,7-tetrahydro- 1-(3-hydroxyphenyl)-6,6-dimethyl- 2-phenyl-	MeOH	245(3.50),275(3.37)	48-0863-78
$C_{22}H_{21}NO_3$ Acetamide, N-[2-[hydroxy(4-methoxyphen- yl)phenylmethyl]phenyl]-	EtOH	212(4.48),231(4.26), 245s(--),271s(--), 282s(--)	39-1211-78C
$C_{22}H_{21}NO_4$ 3,5-Pyridinedicarboxylic acid, 1,4-di- hydro-1,2,6-trimethyl-, diphenyl ester	EtOH	204(4.35),240(4.26), 264s(4.09),367(3.89)	103-1226-78
$C_{22}H_{21}NO_5$ [1,3]Benzodioxolo[5,6-c]phenanthridine- 13-methanol, 12,13-dihydro-1,2-di- methoxy-12-methyl- (bocconoline)	MeOH	212(4.22),228(4.33), 283(4.45),320(3.96), 350s(3.46)	102-0839-78
	EtOH	229(4.55),284(4.67), 321s(4.19)	94-0166-78
$C_{22}H_{21}N_2O_2S_2$ Benzothiazolium, 2-[2-carboxy-3-(3-eth- yl-2(3H)-benzothiazolylidene)-1-prop- enyl]-3-ethyl-	EtOH	565(5.15)	124-0942-78

Compound	Solvent	λ_{max}(log ϵ)	Ref.
$C_{22}H_{21}N_3$			
Benzenamine, 4-benzo[f]quinazolin-3-yl-N,N-diethyl-	EtOH	208(4.52),235(4.45), 255(4.42),299(4.18), 368(4.60)	103-1025-78
$C_{22}H_{21}N_3O_2$			
Spiro[2H-anthra[1,2-d]imidazole-2,1'-cyclohexane]-6,11-dione, 4-(dimethylamino)-	EtOH	631(3.95)	104-0381-78
$C_{22}H_{21}N_3O_6S_2$			
5-Thia-1-azabicyclo[4.2.0]oct-2-ene-2-carboxylic acid, 3-[[(aminocarbonyl)oxy]methyl]-8-oxo-7-[(2-thienylacetyl)amino]-, phenylmethyl ester, (6R-trans)-	n.s.g.	237(4.08)	88-5219-78
$C_{22}H_{21}N_5O_2$			
Ethanone, 1,1'-(1,3a,7,7a,8,8a-hexahydro-1,7-diphenylpyrazolo[3',4':4,5]-pyrrolo[2,1-c]-1,2,4-triazole-3,5-diyl)bis-	EtOH	242(4.12),298(3.20), 364(4.34)	4-1485-78
Ethanone, 1,1'-(1,4a,7,7a,8,8a-hexahydro-1,5-diphenylpyrazolo[4',3':4,5]-pyrrolo[2,1-c]-1,2,4-triazole-3,7-diyl)bis-	EtOH	241(4.32),285(3.46), 294(3.51),364(4.42)	4-1485-78
$C_{22}H_{21}N_7O_4S$			
Benzoic acid, 4-[[[(2,4-diamino-6-pteridinyl)methyl][(4-methylphenyl)sulfonyl]amino]methyl]-	pH 1	240(4.53),279(3.70), 339(3.87)	87-0673-78
	pH 13	236(4.44),257(4.28), 372(3.72)	87-0673-78
$C_{22}H_{21}P$			
Phosphine, (diphenylmethylene)(2,4,6-trimethylphenyl)-	THF	254(3.20),268(3.01), 324(2.84)	35-4886-78
$C_{22}H_{22}Cl_3N_3O_4$			
Acetic acid, trichloro-, 1-(2,3,6,9-tetrahydro-6,7,8-trimethyl-1,3-dioxo-2-phenyl-1H,5H-cyclopenta[d][1,2,4]-triazolo[1,2-a]pyridazin-6-yl)ethyl ester	CHCl₃	282(3.80)	35-0860-78
$C_{22}H_{22}F_3NO_5$			
Benzo[6,7]cyclohept[1,2,3-ij]isoquinolin-1-ol, 4,5,6,6a,7,8-hexahydro-2,10,11-trimethoxy-6-(trifluoroacetyl)-	EtOH	265(4.05),296(3.93)	44-2521-78
Benzo[6,7]cyclohept[1,2,3-ij]isoquinolin-2-ol, 4,5,6,6a,7,8-hexahydro-1,10,11-trimethoxy-5-(trifluoroacetyl)-	EtOH	267(4.09),289(3.99)	44-2521-78
2H-Dibenzo[2,3:4,5]cyclohepta[1,2-b]-pyrrol-2-one, 5,6,7,7a,8,9-hexahydro-3,11,12-trimethoxy-7-(trifluoroacetyl)-	EtOH	235(4.23),259(4.07), 284(3.88),340(3.73)	44-2521-78
12H-Dibenzo[d,f]quinolin-12-one, 1,2,3-3a,4,5-hexahydro-7,8,11-trimethoxy-3-(trifluoroacetyl)-	EtOH	243(4.33),286(3.67)	44-2521-78

$C_{22}H_{22}F_3NO_6-C_{22}H_{22}N_2O_5S$

Compound	Solvent	λ_{max}(log ϵ)	Ref.
$C_{22}H_{22}F_3NO_6$			
Benzo[6,7]cyclohept[1,2,3-ij]isoquino-line-1,11-diol, 4,5,6,6a,7,8-hexahy-dro-2,10,12-trimethoxy-6-(trifluoro-acetyl)-	EtOH	260(3.95),296(3.77)	44-4076-78
4H-Dibenzo[de,g]quinolin-1-ol, 5,6,6a,7-tetrahydro-2,9,10,11-tetramethoxy-6-(trifluoroacetyl)-	EtOH	235s(--),276(4.09)	44-4076-78
$C_{22}H_{22}F_3N_2$			
Quinolinium, 2-[2-[4-(dimethylamino)-phenyl]ethenyl]-1-ethyl-6-(tri-fluoromethyl)-, iodide	EtOH	556(4.70)	103-0060-78
$C_{22}H_{22}F_3N_2O_2S$			
Quinolinium, 2-[2-[4-(dimethylamino)-phenyl]ethenyl]-1-ethyl-6-[(tri-fluoromethyl)sulfonyl]-, iodide	EtOH	586(4.43)	103-0060-78
$C_{22}H_{22}F_3N_2S$			
Quinolinium, 2-[2-[4-(dimethylamino)-phenyl]ethenyl]-1-ethyl-6-[(tri-fluoromethyl)thio]-, iodide	EtOH	558(4.58)	103-0060-78
$C_{22}H_{22}IN_3$			
Quinolinium, 4-[(2,4-dihydro-5-methyl-2-phenyl-3H-pyrazol-3-ylidene)meth-yl]-1-ethyl-, iodide	n.s.g.	210(4.64),240(4.63)	48-0857-78
$C_{22}H_{22}N_2$			
Benzenamine, 4-(9,10-dihydro-10-methyl-9-acridinyl)-N,N-dimethyl-	EtOH	270(5.59)	104-0129-78
$C_{22}H_{22}N_2O$			
4H-Indol-4-one, 1-(2-aminophenyl)-1,5,6,7-tetrahydro-6,6-dimethyl-2-phenyl-	MeOH	245(3.65),275(3.42)	48-0863-78
Tricyclo[9.3.1.1^{4,8}]hexadeca-1(15),4,6-8(16),11,13-hexaene-5,7-dicarboni-trile, 13-methoxy-6,15,16-tri-methyl-, anti	THF	218(4.69),245(4.30), 338(3.08)	44-3470-78
syn	THF	232(4.57),292(3.43), 346(2.72)	44-3470-78
$C_{22}H_{22}N_2O_3$			
1H-Pyrazole-4-carboxylic acid, 3-meth-yl-5-(phenylacetyl)-1-(phenylmethyl)-, ethyl ester	n.s.g.	211(4.15),241(3.82)	44-2665-78
1H-Pyrazole-4-carboxylic acid, 5-meth-yl-3-(phenylacetyl)-1-(phenylmethyl)-, ethyl ester	n.s.g.	217(4.15),236(3.87)	44-2665-78
$C_{22}H_{22}N_2O_4$			
3-Pyridinecarboxylic acid, 1-benzoyl-2-(benzoylamino)-1,4,5,6-tetrahydro-, ethyl ester	MeOH	228(4.18),268(4.03), 304(4.06),315(4.08)	24-2297-78
$C_{22}H_{22}N_2O_5S$			
Ibogamine-18-carboxylic acid, 3,4-[carbonothioylbis(oxy)]-19-oxo-, methyl ester, (2α,3α,5β,6α,18β)-	MeOH	223(4.23),277(3.88), 283(3.93),292(3.88)	33-0690-78

Compound	Solvent	$\lambda_{max}(\log \epsilon)$	Ref.
$C_{22}H_{22}N_2O_6$			
3-Butenoic acid, 3-benzoyl-4-(3-nitro-phenyl)-2-oxo-, piperidine salt	H_2O	262(4.12),342(4.11)	4-0737-78
	EtOH	256(4.20),347(4.10)	4-0737-78
	$CHCl_3$	260(4.18),341(4.04)	4-0737-78
$C_{22}H_{22}N_2O_6S_2$			
Benzenesulfonamide, N,N'-(4,5-dimethyl-3,5-cyclohexadiene-1,2-diylidene)bis-[4-methoxy-	CH_2Cl_2	232(4.54),475(3.65)	5-1146-78
$C_{22}H_{22}N_3$			
Quinolinium, 4-[(2,4-dihydro-5-methyl-2-phenyl-3H-pyrazol-3-ylidene)-methyl]-1-ethyl-, iodide	n.s.g.	210(4.64),240(4.63)	48-0857-78
$C_{22}H_{22}N_4O_3$			
Pyrido[2,3-d]pyrimidine-8(2H)-carbox-amide, 1,3,4,5,6,7-hexahydro-2,4-di-oxo-N,3-bis(phenylmethyl)-	MeOH	215(3.93),284(4.21)	24-2297-78
$C_{22}H_{22}N_8Ni$			
Nickel, 6,13-dibutyl-1,4,8,11-tetraaza-cyclotetradeca-2,4,7,9,11,14-hexaene-2,3,9,10-tetracarbonitrilato(2-)-N^1,N^4,N^8,N^{11}-, (SP-4-1)-	EtOH	270(4.34),330(4.81), 390(4.53),530(4.33)	24-2919-78
$C_{22}H_{22}O_2$			
Spiro[9-oxabicyclo[6.2.0]decane-10,9'-(10'H)-phenanthren]-10'-one	$CHCl_3$	245(4.40),281(3.43), 327(3.46)	18-2052-78
$C_{22}H_{22}O_3$			
Phenol, 2,2',2''-methylidynetris[4-methyl-	EtOH	282(3.90)	32-0079-78
$C_{22}H_{22}O_5$			
2H,10H-Benzo[1,2-b:3,4-b']dipyran-10-one, 3,4-dihydro-5-methoxy-8-(4-meth-oxyphenyl)-2,2-dimethyl-	MeOH	223(4.01),256(3.98), 324(4.27)	78-3569-78
4H-1-Benzopyran-4-one, 5-hydroxy-7-methoxy-2-(4-methoxyphenyl)-6-(3-methyl-2-butenyl)-	MeOH	221(4.12),251(4.19), 325(4.32)	78-3569-78
$C_{22}H_{22}O_6$			
4H,8H-Benzo[1,2-b:3,4-b']dipyran-4-one, 3-(2,4-dimethoxyphenyl)-9,10-dihydro-5-hydroxy-8,8-dimethyl-	MeOH	258(4.48),288(4.19)	2-0613-78
2H-1-Benzopyran-2-one, 3-(2,4-dihydr-oxyphenyl)-5,7-dimethoxy-6-(3-methyl-2-butenyl)-	EtOH	250(4.06),257(4.03), 356(4.28)	94-0135-78
4H-1-Benzopyran-4-one, 3-(2,4-dimeth-oxyphenyl)-5,7-dihydroxy-6-(3-methyl-2-butenyl)-	MeOH	259(4.45),288(4.09)	2-0613-78
4H-1-Benzopyran-4-one, 3-(2,4-dimeth-oxyphenyl)-5,7-dihydroxy-8-(3-methyl-2-butenyl)-	MeOH	266(4.62),290(4.29)	2-0613-78
4H-1-Benzopyran-4-one, 3-(3,4-dimeth-oxyphenyl)-5,7-dihydroxy-6-(3-methyl-2-butenyl)-	MeOH	218(4.18),254(4.09)	2-0611-78 +44-3446-78
4H-1-Benzopyran-4-one, 3-(3,4-dimeth-oxyphenyl)-5-hydroxy-7-(3-methyl-2-butenyloxy)-	MeOH	255(4.14)	2-0973-78 +44-3446-78

Compound	Solvent	$\lambda_{max}(\log \epsilon)$	Ref.
1,4-Butanedione, 1,4-bis(2-acetoxy-5-methylphenyl)-	EtOH	243(4.77),290(3.51)	102-1439-78
4H-Furo[2,3-h]-1-benzopyran-4-one, 9-(1,2-dihydroxy-2-methylpropyl)-8,9-dihydro-5-methoxy-2-phenyl-	MeOH	215(4.57),270(4.59), 333(4.01)	119-0047-78
4H-Furo[2,3-h]-1-benzopyran-4-one, 3-(3,4-dimethoxyphenyl)-8,9-dihydro-5-hydroxy-8,8,9-trimethyl-	MeOH	248(4.54),288(4.38)	2-0973-78
Phenol, 2,2',2"-methylidynetris[4-methoxy-	EtOH	294(2.91)	32-0079-78
$C_{22}H_{22}O_8$			
Benzoic acid, 2-[(3,4-dihydro-7-hydroxy-2,2,5-trimethyl-4-oxo-2H-1-benzopyran-6-yl)oxy]-3-formyl-4-hydroxy-6-methyl-, methyl ester	EtOH	226(4.45),236(4.43), 277(4.40),332(3.94)	142-0437-78C
2H-Phenaleno[1,9-bc]pyran-2,8(4H)-dione, 10-(2-acetoxy-1-methylethyl)-5,5a-dihydro-6,7,9-trihydroxy-3,5a-dimethyl- (edulon A)	EtOH	240s(4.04),281s(3.67), 313(3.77),486(3.85)	33-1969-78
$C_{22}H_{22}O_{11}$			
4H-1-Benzopyran-4-one, 3-(β-D-glucopyranosyloxy)-5,7-dihydroxy-2-(4-hydroxyphenyl)-6-methyl-	MeOH	270(4.09),335(4.03)	102-0787-78
$C_{22}H_{22}S$			
4H-1-Benzothiopyran, 5,6,7,8-tetrahydro-2-phenyl-4-(phenylmethyl)-	hexane	190(4.9),203s(4.5), 240(4.2),292(3.3)	103-0605-78
$C_{22}H_{22}Sn$			
Stannane, trimethyl(9-phenyl-9H-fluoren-9-yl)-	heptane	260(3.97),295(3.58), 303(3.52)	104-0207-78
$C_{22}H_{23}BF_4O_2$			
Cyclohexylium, 2,6-bis[(4-methoxyphenyl)methylene]-, tetrafluoroborate	$C_2H_4Cl_2$	638(4.65)	104-1197-78
$C_{22}H_{23}ClN_2O_5S$			
Dibenzo[b,f]thiepin, 8-chloro-10,11-dihydro-10-(4-hydroxypiperazino)-, maleate	MeOH	266(3.98)	73-3092-78
$C_{22}H_{23}NO$			
Benzenemethanol, 4-(dimethylamino)-α-(4-methylphenyl)-α-phenyl-	MeCN	472(4.61)	104-1750-78
1H-Inden-1-one, 2-(cyclohexylamino)-5-methyl-3-phenyl-	MeOH	272(4.68),550(2.90)	4-1281-78
2(1H)-Pyridinone, 1,4,6-trimethyl-3,5-bis(4-methylphenyl)-	MeOH	240(4.11),315(4.07)	44-2138-78
4(1H)-Pyridinone, 1,2,6-trimethyl-3,5-bis(4-methylphenyl)-	EtOH	236s(4.34),276(4.14)	44-2138-78
$C_{22}H_{23}NO_4$			
1H-Pyrrole-3-carboxylic acid, 4,5-dihydro-5-hydroxy-5-methyl-1,2-bis(4-methylphenyl)-4-oxo-, ethyl ester	EtOH	212(4.08),248(4.13), 337(3.98)	118-0291-78
1H-Pyrrole-3-carboxylic acid, 4,5-dihydro-5-hydroxy-5-methyl-2-(4-methylphenyl)-4-oxo-1-(phenylmethyl)-	EtOH	210(4.11),246(4.19), 328(3.98)	118-0291-78

Compound	Solvent	$\lambda_{max}(\log \epsilon)$	Ref.
1H-Pyrrole-3-carboxylic acid, 4,5-di-hydro-5-hydroxy-2-methyl-4-oxo-1,5-bis(phenylmethyl)-, ethyl ester	EtOH	210(4.09),248(4.13), 326(3.87)	4-1215-78
Spiro[6H-indeno[5,6-d]-1,3-dioxole-6,1'(2'H)-isoquinoline], 3',4',5,7-tetrahydro-6',7'-dimethoxy-2'-methyl-5-methylene-	EtOH	244(4.01),284(4.15), 296(4.14)	88-3419-78
$C_{22}H_{23}NO_5$			
1H-Pyrrole-3-carboxylic acid, 4,5-di-hydro-5-hydroxy-1-(4-methoxyphenyl)-5-methyl-2-(4-methylphenyl)-4-oxo-, ethyl ester	EtOH	209(4.18),244(4.27), 290(4.24),330(4.18)	118-0291-78
1H-Pyrrole-3-carboxylic acid, 4,5-di-hydro-5-hydroxy-2-(4-methoxyphenyl)-5-methyl-1-(4-methylphenyl)-4-oxo-, ethyl ester	EtOH	210(4.16),244(4.21), 335(4.19)	118-0291-78
1H-Pyrrole-3-carboxylic acid, 4,5-di-hydro-5-hydroxy-2-(4-methoxyphenyl)-5-methyl-4-oxo-1-(phenylmethyl)-, ethyl ester	EtOH	208(4.08),246(4.14), 326(4.04)	118-0291-78
$C_{22}H_{23}NO_6$			
1H-Pyrrole-3-carboxylic acid, 4,5-di-hydro-5-hydroxy-1,2-bis(4-methoxy-phenyl)-5-methyl-4-oxo-, ethyl ester	EtOH	210(4.20),234(4.13), 330(4.18)	118-0291-78
$C_{22}H_{23}NO_8$			
5,6,7,8-Indolizinetetracarboxylic acid, 1,2,3,8a-tetrahydro-8a-phenyl-, tetramethyl ester	MeOH	206(4.25),228(4.13), 287(4.01),384(3.74)	83-0977-78
2,3,4,5-Pyridinetetracarboxylic acid, 1,6-dihydro-6-phenyl-1-(2-propenyl)-, tetramethyl ester	MeOH	228(4.4),278(4.4), 382(4.1)	78-0833-78
$C_{22}H_{23}N_2S_2$			
Benzothiazolium, 3-ethyl-2-[3-(3-ethyl-2(3H)-benzothiazolylidene)-2-methyl-1-propenyl]-	EtOH	545(5.13)	124-0942-78
$C_{22}H_{23}N_3$			
1,3-Benzenediamine, 4-(9,10-dihydro-10-methyl-9-acridinyl)-N',N'-di-methyl-	EtOH	291(4.38)	104-0129-78
$C_{22}H_{23}N_3O$			
2H-Pyrrolo[3,4-d]pyrimidin-2-one, 1-ethyl-1,3,4,6-tetrahydro-5,6-dimeth-yl-4,7-diphenyl-	MeOH	288(3.66)	32-0591-78
$C_{22}H_{23}N_3O_4$			
1H-Indole-2-carboxylic acid, 3-[4-[(2-ethoxy-1-methyl-2-oxoethylidene)hy-drazino]phenyl]-, ethyl ester, anti	EtOH	210(4.38),241(4.45), 301(4.35),354(4.41)	103-0173-78
syn	EtOH	210(4.31),241(4.42), 303(4.22),366(4.37)	103-0173-78
$C_{22}H_{23}N_3O_8$			
DL-Aspartic acid, N-[N-[4-[methyl-[(phenylmethoxy)carbonyl]amino]-benzoyl]glycyl]-	MeOH	260(4.08)	87-1165-78

Compound	Solvent	λ_{max}(log ϵ)	Ref.
L-Glutamic acid, N-[N-[4-[[(phenylmeth-oxy)carbonyl]amino]benzoyl]glycyl]-	MeOH	266(4.37)	87-1165-78
$C_{22}H_{23}N_5O_{13}$ 2,4(1H,3H)-Pyrimidinedione, 1-[3,4,6-tri-O-acetyl-2-deoxy-2-[(2,4-dinitro-phenyl)amino]-β-D-glucopyranosyl]-	EtOH	257(4.08),340(4.01)	136-0185-78C
$C_{22}H_{23}O_2$ Cyclohexylium, 2,6-bis[(4-methoxyphen-yl)methylene]-, tetrafluoroborate	$C_2H_4Cl_2$	638(4.65)	104-1197-78
$C_{22}H_{24}BrNO_6$ 8H-Pyrano[2,3-h]quinoline-2-carboxylic acid, 4-acetoxy-3-(2-bromo-2-propen-yl)-9,10-dihydro-5-methoxy-8,8-di-methyl-, methyl ester	MeOH	262(4.48),300(3.42), 360(3.23)	24-0439-78
$C_{22}H_{24}Br_4N_7$ 1H-Benzimidazolium, 4,7-dibromo-2-[3-(4,7-dibromo-1,3-diethyl-1,3-dihydro-2H-benzimidazol-2-ylidene)-1-triazen-yl]-1,3-diethyl-, tetrafluoroborate	EtOH	359(4.35),414(4.31) (photochromic)	33-2958-78
$C_{22}H_{24}ClNO_3S$ Benzo[f]quinoline, 3-(chloromethylene)-1,2,3,4,4a,5,6,10b-octahydro-8-meth-oxy-4-[(4-methylphenyl)sulfonyl]-, endo	CH_2Cl_2	278(3.48),287(3.28)	78-1027-78
exo	CH_2Cl_2	277(3.34),287(3.26)	78-1027-78
Benzo[f]quinoline, 3-(chloromethyl)-1,4,4a,5,6,10b-hexahydro-8-methoxy-4-[(4-methylphenyl)sulfonyl]-	EtOH	222(4.34),276(3.52), 287(3.32)	78-1027-78
$C_{22}H_{24}ClNO_6$ 8H-Pyrano[2,3-h]quinoline-2-carboxylic acid, 4-acetoxy-3-(2-chloro-2-prop-enyl)-9,10-dihydro-5-methoxy-8,8-di-methyl-, methyl ester	MeOH	260(4.76)	24-0439-78
$C_{22}H_{24}Cl_3NO_3S$ Benzo[f]quinoline, 1,2,3,4,4a,5,6,10b-octahydro-8-methoxy-4-[(4-methylphen-yl)sulfonyl]-3α-(trichloromethyl)-	EtOH	229(4.22),265(3.27), 271(3.27),275(3.32), 286(3.26)	78-1027-78
$C_{22}H_{24}NOP$ 1H-Phosphorino[3,4-b]indole, 2-(1,1-di-methylethyl)-2,9-dihydro-9-methyl-4-phenyl-, 2-oxide	$CHCl_3$	236(4.39),257s(4.23), 346(3.75)	139-0257-78B
9H-Phosphorino[3,4-b]indole, 2-(1,1-di-methylethyl)-2,2-dihydro-2-methoxy-4-phenyl-	$CHCl_3$	251(4.43),285(4.31), 332(3.80),377(4.01)	139-0257-78B
$C_{22}H_{24}N_2O$ Propanedinitrile, (3-methoxyestra-1,3,5(10)-trien-17-ylidene)-	MeOH	232(4.29),242(4.20), 278(3.36),281(3.32), 287(3.29)	24-3094-78
$C_{22}H_{24}N_2O_2$ 1H-Pyrazole-5-carboxylic acid, 4,5-di-hydro-5-methyl-4-(1-methylethylidene)-	EtOH	250(4.16),365(4.00)	22-0415-78

Compound	Solvent	$\lambda_{max}(\log \epsilon)$	Ref.
1,3-diphenyl- (cont.) 2H-Pyrrol-2-one, 5,5'-[(2,5-dimethyl- 1,4-phenylene)dimethylidyne]bis[1,5- dihydro-, (Z,Z)-	EtOH	248(4.00),355(4.29)	22-0415-78 49-1191-78
$C_{22}H_{24}N_2O_3$ Spiro[2H-indole-2,5'(4'H)-isoxazole]- 5-carboxylic acid, 1,3-dihydro-1,3,3- trimethyl-3'-phenyl-, ethyl ester	CHCl$_3$	265s(4.18),304(4.49)	47-2039-78
$C_{22}H_{24}N_2O_3S$ 4H-Carbazol-4-one, 1,2,3,9-tetrahydro- 2,2,9-trimethyl-, O-[(4-methylphen- yl)sulfonyl]oxime	MeOH	223(4.51),252(4.06), 267(4.11),303(4.19), 312s(4.11)	24-0853-78
$C_{22}H_{24}N_2O_8$ Tetracycline, hydrochloride	neutral	248(4.08),276(4.20), 363(4.23)	1-0131-78
	anion	248(4.11),272(4.20), 287s(4.18),369(4.23)	1-0131-78
	dianion	247(4.11),269(4.15), 288s(4.08),380(4.23)	1-0131-78
	trianion	238(4.13),245(4.15), 267(4.15),288s(4.02), 380(4.23)	1-0131-78
	cation	220(4.10),270(4.28), 358(4.18)	1-0131-78
$C_{22}H_{24}N_4O_4$ 3-Pyridinecarboxylic acid, 1,4,5,6- tetrahydro-1-[(phenylamino)carbonyl]- 2-[[(phenylamino)carbonyl]amino]-	MeOH	229(4.02),293(4.19)	24-2297-78
2H-Pyrrol-2-one, 4-(2-acetoxyethyl)- 1-[(4-amino-2-methyl-5-pyrimidinyl)- methyl]-1,5-dihydro-3-(2-methoxy- phenyl)-5-methylene-	EtOH	235(4.23),270(4.21), 277(4.22)	94-0722-78
$C_{22}H_{24}N_8$ 1,4,8,11-Tetraazacyclotetradeca- 2,4,6,9,11,13-hexaene-2,3,9,10- tetracarbonitrile, 6,13-dibutyl-	CHCl$_3$	330(4.77),350(4.73), 360(4.73),368(4.71), 400(4.15),430(3.92)	24-2919-78
$C_{22}H_{24}OS$ 2H-1-Benzothiopyran, 3,5,6,7,8,8a-hexa- hydro-4-(4-methoxyphenyl)-2-phenyl-	hexane	210(4.3),223s(4.3), 277s(3.5),285(3.4)	103-0605-78
$C_{22}H_{24}O_3$ Tricyclo[9.3.1.1^{4,8}]hexadeca-1(15),4,6- 8(16),11,13-hexaene-5,7-dicarboxalde- hyde, 13-methoxy-6,15,16-trimethyl-, anti	THF	217(4.58),253(3.42), 352(3.23)	44-3470-78
syn	THF	212(4.40),240(4.20), 360(2.64)	44-3470-78
$C_{22}H_{24}O_4$ Benzeneacetic acid, 2-(2,3-dihydro- 1-hydroxy-2,4,6-trimethyl-3-oxo-1H- inden-5-yl)ethyl ester, (1S-trans)-	EtOH	216(4.58),257(4.14), 299(3.34)	94-2365-78
1-Benzoxepin-4-carboxylic acid, 2,5-di- hydro-6-methoxy-3-methyl-8-(2-phenyl- ethyl)-, methyl ester	EtOH	215(3.65)	102-2005-78

Compound	Solvent	$\lambda_{max}(\log \epsilon)$	Ref.
Tricyclo[8.2.2.2^{4,7}]hexadeca-4,6,10,12-13,15-hexaene-5,11-dicarboxylic acid, diethyl ester, pseudo-ortho	EtOH	208(5.03),280(3.80), 327(2.81)	24-0523-78
pseudo-para	EtOH	211(4.73),242(4.15), 306(3.26),330s(2.57)	24-0523-78
Tricyclo[8.2.2.2^{4,7}]hexadeca-4,6,10,12-13,15-hexaene-5,12-dicarboxylic acid, diethyl ester, pseudo-gem.	EtOH	207(5.15),238(4.56), 305(3.68),333s(2.88)	24-0523-78
pseudo-meta	EtOH	207(4.85),238(4.33), 304(3.34)	24-0523-78
$C_{22}H_{24}O_5$			
1-Benzoxepin-7-carboxylic acid, 2,3-di-hydro-3-hydroxy-6-methoxy-3-methyl-8-(2-phenylethyl)-, methyl ester	EtOH	203(3.90),206s(3.83), 220(3.65),265(3.39)	102-2005-78
1-Benzoxepin-9-carboxylic acid, 2,3-di-hydro-3-hydroxy-6-methoxy-3-methyl-8-(2-phenylethyl)-, methyl ester	EtOH	207(3.82),215(3.80), 263(3.42),275s(3.26)	102-2005-78
1,5-Methano-9H-furo[3,4-f][1]benzopyr-an-9,10-dione, 1,3,3a,4,5,7,8,9b-octahydro-9b-hydroxy-1,4,4,5-tetra-methyl-7-phenyl-	EtOH	207(4.67),285(4.49), 329(3.96)	88-0429-78
	EtOH-NaOMe	207(4.78),280(4.73), 303s(4.66),383(4.76)	88-0429-78
Naringenin, 6-C-prenyl-4',7-di-O-meth-yl-	MeOH	295(4.30),335(3.67)	78-3563-78
Naringenin, 8-C-prenyl-4',7-di-O-meth-yl-	MeOH	297(4.31),337(3.78)	78-3563-78
Oxireno[c][1]benzoxepin-4-carboxylic acid, 1a,2,8,8a-tetrahydro-7-methoxy-1a-methyl-5-(2-phenylethyl)-, methyl ester	EtOH	205(3.35),215s(3.14)	102-2005-78
Oxireno[c][1]benzoxepin-6-carboxylic acid, 1a,2,8,8a-tetrahydro-7-methoxy-1a-methyl-5-(2-phenylethyl)-, methyl ester	EtOH	216(3.58)	102-2115-78
$C_{22}H_{24}O_6$			
1-Propanone, 1-(2,6-diacetoxy-4-meth-oxy-3,5-dimethylphenyl)-3-phenyl-	MeOH	248(3.68)	102-2011-78
$C_{22}H_{24}O_8$			
Duartin diacetate, (-)-	EtOH	225(4.38),275(3.43)	102-1401-78
Lirionol	MeOH	242(4.33),291(4.04)	102-0779-78
	MeOH-NaOMe	262(4.08),371(4.34)	102-0779-78
$C_{22}H_{24}O_{10}$			
4H-1-Benzopyran-4-one, 5-(β-D-gluco-pyranosyloxy)-2,3-dihydro-7-hydroxy-2-(4-methoxyphenyl)-	MeOH	272(4.14)	102-2119-78
$C_{22}H_{24}O_{11}$			
9H-Xanthen-9-one, 2-β-D-glucopyranosyl-1-hydroxy-3,5,6-trimethoxy-	MeOH	245(4.43),280s(3.98), 312(4.16),335(3.55)	102-2119-78
$C_{22}H_{24}Si$			
Silane, [4-(diphenylmethyl)phenyl]tri-methyl-	n.s.g.	214(4.33),226(4.30)	39-0751-78B
$C_{22}H_{25}Cl_5O$			
Benzene, [[[3,7-dimethyl-7-(1,2,3,4,5-pentachloro-2,4-cyclopentadien-1-yl)-2-octenyl]oxy]methyl]-	C_6H_{12}	313(3.26)	44-3209-78

Compound	Solvent	$\lambda_{max}(\log \epsilon)$	Ref.
$C_{22}H_{25}FN_4$ 3H-1,5-Benzodiazepine, 2-(4-fluorophen- yl)-7,8-dimethyl-4-(4-methyl-1-piper- azinyl)-	n.s.g.	263(4.53),345(3.76)	87-0952-78
$C_{22}H_{25}NO$ Benzeneacetonitrile, α-[3,5-bis(1,1-di- methylethyl)-4-oxo-2,5-cyclohexadien- 1-ylidene]-	n.s.g.	355(4.50)	70-1313-78
Ethanone, 1-(1,2,3,5,6,7-hexahydro-s- indacen-4-yl)-2-[methyl(phenylmeth- yl)amino]-, hydrogen maleate	MeOH	257(3.97)	73-0970-78
$C_{22}H_{25}NO_6$ 1H-2-Benzopyran-1-one, 3-[6-[2-(dimeth- ylamino)ethyl]-1,3-benzodioxol-5-yl]- 3,4-dihydro-7,8-dimethoxy-	EtOH	230s(4.38),247s(4.04), 292(3.60),317(3.31)	78-0635-78
Spiro[3H-2-benzopyran-3,1'(2'H)-iso- quinolin]-4(1H)-one, 3',4'-dihydro- 6,6',7,7'-tetramethoxy-2'-methyl-	EtOH	238(4.35),285(4.14), 312(3.85)	142-1233-78A
$C_{22}H_{25}N_3O_8$ D-Ribitol, 1-deoxy-1-(3,4-dihydro-7,8- dimethyl-2,4-dioxopyrimido[4,5-b]- quinolin-10(2H)-yl)-2,3:4,5-bis- O-(methoxymethylene)-	MeOH	226(4.58),263(4.48), 330(4.04),400(4.13)	4-0489-78
$C_{22}H_{25}N_3P$ Phosphorus(1+), (4-diazenyl-N,N-dimeth- ylbenzenaminato)ethyldiphenyl-, (T-4)-, tetrafluoroborate	n.s.g.	278(3.98),480(4.61)	123-0190-78
$C_{22}H_{25}N_4O_2P$ 2,3,4(1H)-Phosphorintrione, 1-(1,1-di- methylethyl)-5-methyl-2,4-bis(phen- ylhydrazone), P-oxide	CHCl$_3$	266(3.93),347(4.28), 404(4.04),498(4.35)	139-0027-78B
$C_{22}H_{26}ClNO_3S$ Benzo[f]quinoline, 1,2,3,4,4a,5,6,10b- octahydro-3α-(chloromethyl)-4-(4- methylphenylsulfonyl)-8-methoxy- 3β-	EtOH EtOH	229(4.20),265s(3.32), 272s(3.30),276(3.36), 286(3.28) 226(4.22),277(3.40), 287(3.34)	78-1027-78 78-1027-78
$C_{22}H_{26}Cl_4O$ Benzene, [[[3,7-dimethyl-7-(tetrachlo- rocyclopentadienyl)-2-octenyl]oxy]- methyl]-	C$_6$H$_{12}$	302(3.26)	44-3209-78
2H-2,4a-Methanonaphthalene, 2,3,4,9- tetrachloro-1,5,6,7,8,8a-hexahydro- 5,5,8a-trimethyl-1-[(phenylmethoxy)- methyl]-, (1α,2α,4aβ,8aα,9R*)-	C$_6$H$_{12}$	225(2.89)(end abs.)	44-3209-78
$C_{22}H_{26}F_3NO_3$ Furo[3,4-b]pyridine-3-carboxylic acid, 1,4,5,7-tetrahydro-2,5,5,7,7-penta- methyl-4-[3-(trifluoromethyl)phenyl]-, ethyl ester	EtOH	260s(3.61),266s(3.58), 274s(3.45),337(3.80)	95-0448-78
$C_{22}H_{26}F_3N_3O_2S$ Fluphenazine sulfoxide	MeOH	215(4.44),235(4.39)	133-0248-78

Compound	Solvent	$\lambda_{max}(\log \epsilon)$	Ref.
Fluphenazine sulfoxide (cont.)		277(4.12),305(3.99), 350(3.79)	133-0248-78
$C_{22}H_{26}Fe$ Ferrocene, 1,1':2,2':3,4:4',5'-tetra- kis(1,3-propanediyl)- cation	CH$_2$Cl$_2$ EtOH	421(2.48) 260(--),675(2.58)	101-0077-78L 101-0077-78L
$C_{22}H_{26}NO_6$ Isoquinolinium, 3,4-dihydro-1-[2-(hy- droxymethyl)-4,5-dimethoxybenzoyl]- 6,7-dimethoxy-2-methyl-	EtOH	241(4.34),292(4.12), 333(3.95)	142-1233-78A
$C_{22}H_{26}N_2O_2S_4Sn$ Morpholine, 4,4'-[(diphenylstannylene)- bis(thiocarbonothioyl)]bis-	CHCl$_3$	255(4.27),287(3.57)	90-0399-78
$C_{22}H_{26}N_2O_4$ Aspidophytine	EtOH	222(4.46),256(3.77), 304(3.38)	23-1052-78
Benzeneacetic acid, α-(1-oxopropoxy)- 2-benzoyl-1-(1,1-dimethylethyl)hy- drazide	EtOH	220(4.13),252s(3.56), 259s(3.47),266s(3.40), 278s(3.27)	33-1477-78
Benzoic acid, 4-[[[[4-(butoxycarbonyl)- phenyl]methylamino]methylene]amino]-, ethyl ester	isoPrOH	323(4.57)	117-0097-78
5,1-[Epoxy(phenylmetheno)nitrilo]tetra- hydrosantonin, (1S,5R)-	EtOH	235(4.05)	44-0536-78
β-isomer	EtOH	235(4.03)	44-0536-78
2H-Pyrido[3,4-e]-1,3-oxazine-4,5(3H,6H)- dione, 3-cyclohexyl-7-methoxy-2,2-di- methyl-6-phenyl-	MeOH	309(4.4)	120-0101-78
4-Pyridineacetic acid, α-ethyl-1,4-di- hydro-1-[2-(1H-indol-3-yl)ethyl]- 3-(methoxycarbonyl)-, methyl ester	EtOH	223(4.50),284(3.90), 292(3.86),347(3.78)	78-2529-78
Vallesiachotamine, 19,20-dihydro-20- deformyl-20-(methoxycarbonyl)-, (±)-	EtOH	207s(4.29),225(4.49), 293(4.48)	78-2529-78
$C_{22}H_{26}N_2O_5$ Aspidophytine oxide	EtOH	228(4.39),296(3.66)	23-1052-78
Butanoic acid, 2-[[4-[methyl[(phenyl- methoxy)carbonyl]amino]benzoyl]- amino]-, ethyl ester, (±)-	MeOH	260(4.08)	87-1162-78
Butanoic acid, 4-[[4-[methyl[(phenyl- methoxy)carbonyl]amino]benzoyl]- amino]-, ethyl ester	MeOH	260(4.10)	87-1162-78
L-Leucine, N-[4-[methyl[(phenylmeth- oxy)carbonyl]amino]benzoyl]-	EtOH	260(4.28)	87-1162-78
$C_{22}H_{26}N_4$ 3H-1,5-Benzodiazepine, 7,8-dimethyl- 2-(4-methylpiperazino)-4-phenyl-	n.s.g.	260(4.52),350(3.77)	87-0952-78
$C_{22}H_{26}N_4O_2$ 9-Acridinamine, N-(cyclohexylaminopro- pyl)-1-nitro-	pH 1 pH 7	218(4.242),273(4.493), 329(3.581),435(3.814) 271(4.375),406(3.649)	56-2125-78 56-2125-78
$C_{22}H_{26}N_4O_4$ 2H-Pyrrol-2-one, 4-(2-acetoxyethyl)- 1-[(4-amino-2-methyl-5-pyrimidinyl)-	EtOH	218(4.38),281(3.91)	94-0722-78

Compound	Solvent	$\lambda_{max}(\log \epsilon)$	Ref.
methyl]-1,5-dihydro-3-(2-methoxy-phenyl)-5-methyl- (cont.)			94-0722-78
$C_{22}H_{26}N_4O_5$ 2H-Pyrrol-2-one, 4-(2-acetoxyethyl)-1-[(4-amino-2-methyl-5-pyrimidinyl)-methyl]-1,3-dihydro-3-hydroxy-3-(2-methoxyphenyl)-5-methyl-	EtOH	224(4.18),277(3.96)	94-0722-78
$C_{22}H_{26}N_4O_8$ Uridine, 5-[3-[2-(acetylamino)ethyl]-5-methoxy-1H-indol-2-yl]-	EtOH	226(4.05),271(3.82), 296(3.70),325(3.52)	88-2585-78
$C_{22}H_{26}N_5O_7PS_2$ Benzenesulfonic acid, 4-methyl-, 2,2'-[[(4-methoxybenzoyl)amino]phosphin-ylidene]dihydrazide, P-oxide	MeOH	227(4.4748),258(4.3118)	65-0720-78
$C_{22}H_{26}N_8O_8S_2$ 3H-Imidazo[4,5-b]pyridin-5-amine, 7,7'-dithiobis[3-β-D-ribofuranosyl-	pH 1 pH 11 EtOH	270(4.16),324(4.20) 267(4.19),308(4.29) 267(4.24),319(4.25)	4-0839-78 4-0839-78 4-0839-78
$C_{22}H_{26}O_2$ 2(3H)-Benzofuranone, 5,7-bis(1,1-di-methylethyl)-3-phenyl-	EtOH	277(3.11),283(3.10)	12-0907-78
$C_{22}H_{26}O_3$ 2H-1-Benzopyran-2-acetaldehyde, 3,4-di-hydro-2,5,7,8-tetramethyl-6-(phenyl-methoxy)-	EtOH	227(4.08),258(3.01), 264(3.02),280(3.32), 288(3.37)	33-0837-78
2,5-Cyclohexadiene-1,4-dione, 2-[3,7-dimethyl-8-(4-methyl-2-furanyl)-2,6-octadienyl]-5-methyl-, (E,E)-	EtOH	222(4.23),253(4.31), 258(4.27),264(4.23)	12-0157-78
$C_{22}H_{26}O_3S$ Benzene, 1-methoxy-2,3,5-trimethyl-4-[3-methyl-5-(phenylsulfonyl)-1,3-pentadienyl]-	isoPrOH	217(4.39),232(4.35), 290(4.23)	44-4769-78
$C_{22}H_{26}O_4$ 2H-1-Benzopyran-2-acetic acid, 3,4-di-hydro-2,5,7,8-tetramethyl-6-(phenyl-methoxy)-, (±)-	EtOH	228(4.10),258(2.97), 265(2.99),281(3.30), 287(3.34)	33-0837-78
Estra-1,3,5(10),8,14-pentaen-17β-ol, 1,3-dimethoxy-, acetate	MeOH	227(4.18),240(4.10), 294(4.28),303(4.36), 320(4.37)	24-0939-78
2H-Pyran-2-one, 6-ethyltetrahydro-5,6-bis(4-methoxyphenyl)-4-methyl-	MeOH	226(4.33),275(3.72), 283(3.67)	5-0726-78
$C_{22}H_{26}O_5$ 1-Benzoxepin-9-carboxylic acid, 2,3,4-5-tetrahydro-3-hydroxy-6-methoxy-3-methyl-8-(2-phenylethyl)-, methyl ester	EtOH	213(3.94)	102-2005-78
8α-Estra-1,3,5(10)-trien-17-one, 1,3-diacetoxy-	MeCN	265(1.64),272(2.70)	24-0939-78
$C_{22}H_{26}O_7$ Dispiro[furan-3(2H),1'(5'H)-naphthalen-5',2"-oxirane]-2,4'(4'aH)-dione,	n.s.g.	211(3.74)	105-0215-78

Compound	Solvent	$\lambda_{max}(\log \epsilon)$	Ref.
4'a-(acetoxymethyl)-5-(3-furanyl)- octahydro-2'-methyl- (cont.)			105-0215-78
$C_{22}H_{26}O_9$ Magnolenin L aglycone	EtOH	300(4.00)	100-0056-78
$C_{22}H_{27}AsN_2$ 1H-1,3-Benzazarsole, 2-(1,1-dimethyl- 1-[2,2-dimethyl-1-(phenylimino)pro- pyl]-	MeOH	238(4.51),266s(4.01), 282s(3.87),320(3.68)	101-0001-78K
$C_{22}H_{27}NO_2$ 2H-1-Benzopyran-2-amine, 7-methoxy- N,N-bis(1,1-dimethylethyl)-3-phenyl-	EtOH	246(3.85),325(3.56)	39-0088-78C
$C_{22}H_{27}NO_3$ Furo[3,4-b]pyridine-3-carboxylic acid, 5,7-dihydro-2,5,5,7,7-pentamethyl- 4-(3-methylphenyl)-, ethyl ester	EtOH	275.6(3.79)	95-0448-78
$C_{22}H_{27}NO_4$ Benzo[6,7]cyclohept[1,2,3-ij]isoquino- line, 4,5,6,6a,7,8-hexahydro-1,2,10- 11-tetramethoxy-6-methyl-, hydro- chloride, (±)-	EtOH	266(4.11),289(3.95)	44-2521-78
$C_{22}H_{27}NO_5$ Furo[3,4-b]pyridine-3-carboxylic acid, 4-(1,3-benzodioxol-5-yl)-1,4,5,7- tetrahydro-2,5,5,7,7-pentamethyl-, ethyl ester	EtOH	292(3.79),332(3.84)	95-0448-78
Spiro[3H-2-benzopyran-3,1'(2'H)-iso- quinoline], 1,3',4,4'-tetrahydro- 6,6',7,7'-tetramethoxy-2'-methyl-	EtOH	286(3.92)	142-1233-78A
Thalicsimidine	n.s.g.	282(4.06),302(4.02), 311s(3.97)	24-0554-78
$C_{22}H_{27}NO_7$ 2,4-Pyrrolidinedione, 3-[1-hydroxy- 4-methyl-6-(1,2,7-trimethyl-5-oxo- 3,9,10-trioxatricyclo[4.3.1.0^{2,4}]- dec-8-yl)-2,4-heptadienylidene]-	EtOH-H$_2$SO$_4$ EtOH-KOH	353(4.51),366s(4.48) 287(4.21),331(4.22)	88-3173-78 88-3173-78
$C_{22}H_{27}N_3O_2$ Butanamide, N-(4-butylphenyl)-2-[(2,4- dimethylphenyl)hydrazono]-3-oxo-	EtOH	210(3.95),240(4.23), 270(4.09),389(4.51)	104-0521-78
Pyrano[2,3-c]pyrazol-6(2H)-one, 3,4-di- methyl-2-(phenylmethyl)-5-[2-(1-pip- eridinyl)ethyl]-, monohydrochloride	EtOH	310.0(4.25)	95-0335-78
$C_{22}H_{27}N_7O_8$ 1,2-Hydrazinedicarboxylic acid, 1,1'- [2-(1,1-dimethylethyl)-2H-pyrrolo- [3,4-b]quinoxaline-1,3-diyl]bis-, tetramethyl ester	MeOH	260(4.70),330(3.81), 339(3.99),344(4.06), 355(4.14),358(4.16), 509(3.54)	88-4671-78
$C_{22}H_{28}Cl_2O_6$ 3-Furancarboxylic acid, 5,5'-(2,3-di- chloro-1,3-butadiene-1,4-diyl)bis- [4,5-dihydro-2,4,4-trimethyl-, dimethyl ester	MeOH	249(4.57)	25-0274-78

Compound	Solvent	$\lambda_{max}(\log \epsilon)$	Ref.
$C_{22}H_{28}NO_5$			
Isoquinolinium, 3,4-dihydro-1-[[2-(hy-droxymethyl)-4,5-dimethoxyphenyl]-methyl]-6,7-dimethoxy-2-methyl-, iodide	EtOH	245(4.23),290(3.77), 314(3.83),368(3.87)	142-1233-78A
$C_{22}H_{28}N_2O_2S$			
Epivincamine, 14-deoxy-14-(methylthio)-	MeOH	226(4.42),277(3.92)	33-1682-78
$C_{22}H_{28}N_2O_3$			
2,3-Anhydro-4-deacetoxy-N-demethyl-6,7-dihydrovindoline	MeOH	244(4.13),324(4.26)	35-4220-78
4-Quinolineethanol, α-[(3-ethenyl-4-piperidinyl)methyl]-6-methoxy-, acetate, [3R-(3α,4α(S*)]]-	EtOH	235(4.57),270s(3.57), 279(3.62),288s(3.54), 321(3.69),333(3.73)	35-0576-78
$C_{22}H_{28}N_2O_3S$			
4H-Carbazol-4-one, 1,2,3,5,6,7,8,9-octahydro-2,2,9-trimethyl-, O-[(4-methylphenyl)sulfonyl]oxime	MeOH	221(4.29),259(3.92)	24-0853-78
$C_{22}H_{28}N_2O_5$			
Benzoic acid, 4-(dimethylamino)-, oxydi-2,1-ethanediyl ester	MeOH	228(4.17),310(4.73)	121-0661-78
Yohimbine oxindole, 9-methoxy-, (7β,16β,17β,20α)-	EtOH	221(4.37),248(3.55), 290(3.35)	78-3341-78
$C_{22}H_{28}N_2O_8$			
Benzene, 1,1'-(2,3-dimethyl-1,4-butane-diyl)bis[3,4-dimethoxy-5-nitro-, (R*,S*)-	EtOH	222(4.43),265(3.66), 330(3.23)	94-0682-78
$C_{22}H_{27}N_4$			
Diazene, 1,1'-(1,2-diphenyl-1,2-ethene-diyl)bis[2-(1,1-dimethylethyl)-	EtOH	249(4.12),303(3.99)	33-1477-78
$C_{22}H_{28}N_4O_6$			
1H-Imidazole-4-carboxylic acid, 5-[[[[[2,3-O-(1-methylethylidene)-β-D-ribofuranosyl]amino]methylene]amino]-1-(phenylmethyl)-, ethyl ester	EtOH	280(3.83)	4-0929-78
$C_{22}H_{28}N_7$			
1H-Benzimidazolium, 2-[3-(1,3-diethyl-1,3-dihydro-2H-benzimidazol-2-yli-dene)-1-triazenyl]-1,3-diethyl-, tetrafluoroborate	EtOH mixt.	370s(4.26),435(4.43) (photochromic)	33-2958-78
$C_{22}H_{28}O_2$			
1-Dibenzofuranol, 4,6-bis(1,1-dimethyl-ethyl)-2,8-dimethyl-	EtOH	229(4.62),262(4.16), 272(4.08),282(4.16), 306(3.76),318(3.89)	12-0907-78
$C_{22}H_{28}O_3$			
1,4-Benzenediol, 2-[3,7-dimethyl-8-(4-methyl-2-furanyl)-2,6-octadienyl]-5-methyl-, (E,E)-	EtOH	216(4.23),222(4.15), 293(3.58)	12-0157-78
2(3H)-Benzofuranone, 7-(1,1-dimethyl-ethyl)-3-[5-(1,1-dimethylethyl)-3-methyl-2-furanyl]-5-methyl-	C_6H_{12}	213(4.30),279(3.24), 287(3.21)	12-0907-78

Compound	Solvent	$\lambda_{max}(\log \epsilon)$	Ref.
2H-1-Benzopyran-2-ethanol, 3,4-dihydro-2,5,7,8-tetramethyl-6-(phenylmeth-oxy)-, (±)-	EtOH	203(4.36),256(2.62), 263(2.62),280(2.93), 286(2.98)	33-0837-78
Microferin	EtOH	260(4.22)	105-0487-78
Pregn-4-en-20-yn-3-one, 17-(formyloxy)-	MeOH	240(4.23)	24-3086-78
22-Spiroxa-4,6-diene-3,20-dione	EtOH	282(4.42)	33-3068-78
Tricyclo[9.3.1.1^{4,8}]hexadeca-1(15),4,6-8(16),11,13-hexaene-5,7-dimethanol, 13-methoxy-6,15,16-trimethyl-	THF	275(3.40),297(3.15)	44-3470-78
$C_{22}H_{28}O_4$			
2H-Benzo[a]xanthene-8,11-dione, 1,3,4-4a,5,6,6a,12b-octahydro-10-methoxy-4,4,6a,12b-tetramethyl-	MeOH	266(4.08),280s(4.02), 295s(3.97),310s(3.87), 496(2.85)	12-2685-78
1H-Cyclopropa[3,4]benz[1,2-f]azulene-5,8-dione, 3-acetoxy-1a,2,3,3a,4,6,7-9b-octahydro-1,1,3a,6,9-pentamethyl-	n.s.g.	217(3.15),263(4.16), 330(3.76)	78-0233-78
Spiro[1,3-dioxolane-2,1'(2'H)-naphtha-lene], 3',7',8',8'a-tetrahydro-5',8'a-dimethyl-6'-[(phenylmethoxy)-methoxy]-, (±)-	EtOH	244(4.13)	39-1461-78C
$C_{22}H_{28}O_5$			
Benzene, 1-ethoxy-5-[3-(2-ethoxyphen-yl)-2-propenyl]-2,3,4-trimethoxy-	EtOH	258(3.99),280(3.61), 304(3.32)	102-1379-78
1H-Cyclopenta[a]cyclopropa[f]cycloun-decene-4,7-dione, 10-acetoxy-1a,4a-7a,10,11,11a-hexahydro-4a-hydroxy-1,1,3,6,9-pentamethyl-	n.s.g.	278(3.36)	78-0233-78
Estra-1,3,5(10)-trien-6-one, 17β-acet-oxy-2,3-dimethoxy-	EtOH	233(4.27),277(4.00), 317(3.85)	94-3567-78
$C_{22}H_{28}O_6$			
Coleon N, 12-O-deacetyl-	MeOH	235(4.04)	33-0871-78
2-Cyclohexen-1-one, 4-methoxy-6-[1-methyl-2-oxo-2-(3,4,5-trimethoxy-phenyl)ethyl]-6-(2-propenyl)-, [4R-(4α,6α(S*)]]-	EtOH	280(3.95)	44-0586-78
Dibenzo[a,c]cyclooctene-3,10-diol, 5,6,7,8-tetrahydro-1,2,11,12-tetra-methoxy-6,7-dimethyl-	EtOH	214(4.70),248(4.15), 276(3.53)	94-0682-78
Duartin diethyl ether, (-)-	EtOH	225(4.46),278(3.43)	102-1401-78
Marrubiastrol acetate	MeCN	210(4.28)	24-2130-78
$C_{22}H_{28}O_7$			
Coleon D, 16-O-acetyl-	ether	267(3.69),318(3.89), 385(3.66)	33-0871-78
Hiyodorilactone B	EtOH	210(4.48)	138-1345-78
$C_{22}H_{28}O_8$			
Hiyodorilactone A	EtOH	210(4.17)(end abs.)	138-1345-78
$C_{22}H_{28}O_9$			
Magnolenin C aglycone reduction product	EtOH	270(3.75)	100-0056-78
$C_{22}H_{28}O_{12}$			
Cornoside tetraacetate	EtOH	228(3.96)	94-2111-78
$C_{22}H_{29}BrN_4P$			
Phosphorus(1+), [(4-bromophenyl)azo]-phenyldi-1-piperidinyl-, (T-4)-	n.s.g.	330(4.32),500(2.28)	123-0190-78

Compound	Solvent	$\lambda_{max}(\log \epsilon)$	Ref.
$C_{22}H_{29}Br_2N_5O_2$			
3-Azaspiro[5.5]undeca-1,4-diene-1,5-di-carbonitrile, 2,4-bis(bromo-4-morpho-linylmethyl)-	EtOH	205(4.72),219(4.73), 280(4.51),350(4.07)	103-1236-78
$C_{22}H_{29}ClO_3$			
Benzenemethanol, 3-[[3-chloro-5-(1,1-dimethylethyl)-2-hydroxyphenyl]meth-yl]-5-(1,1-dimethylethyl)-2-hydroxy-	dioxan	283(3.70)	49-0767-78
$C_{22}H_{29}FN_4P$			
Phosphorus(1+), [(3-fluorophenyl)azo]-phenyldi-1-piperidinyl-, (T-4)-, tetrafluoroborate	n.s.g.	305(4.13),500(2.20)	123-0190-78
$C_{22}H_{29}NO_3$			
Furo[3.4-b]pyridine-3-carboxylic acid, 1,4,5,7-tetrahydro-2,5,5,7,7-penta-methyl-4-(4-methylphenyl)-, ethyl ester	EtOH	231(4.03),277(3.08), 334(3.83)	95-0448-78
$C_{22}H_{29}NO_4$			
Laudanosinemethine	MeOH	215(4.42),330(4.26)	36-0473-78
methiodide	MeOH	223(4.32),331(4.26)	36-0473-78
3,5-Pyridinedicarboxylic acid, 1,4-di-hydro-2,6-dimethyl-4-phenyl-1-prop-yl-, diethyl ester	EtOH	205(4.24),243(4.20), 353(3.81)	103-1226-78
1H-Pyrrole-3-carboxylic acid, 1-cyclo-hexyl-4,5-dihydro-5-hydroxy-2-methyl-5-[(4-methylphenyl)methyl]-4-oxo-, ethyl ester	EtOH	214(3.98),248(4.15), 326(3.84)	4-1215-78
$C_{22}H_{29}N_2O_3$			
Diploceline cation	MeOH	220(4.54),270(3.86), 280(3.82),289(3.68)	102-1447-78
$C_{22}H_{29}N_3O_2$			
Epivincamine, 14-deoxy-14-(methylamino)-	CH_2Cl_2	228(4.48),282(3.88)	33-1682-78
Vincamine, 14-deoxy-14-(methylamino)-	CH_2Cl_2	233(4.50),284(3.94)	33-1682-78
$C_{22}H_{29}N_5O_2$			
Phosphorus(1+), [(4-nitrophenyl)azo]-phenyldi-1-piperidinyl-, tetrafluo-roborate	n.s.g.	295(4.22),505(2.26)	123-0190-78
$C_{22}H_{29}N_7O_5$			
Purine, 6-(dimethylamino)-9-[(1R,2S,4R-5R)-2,4,5-trihydroxy-3-(4-methoxy-phenyl)-L-alanylamino]cyclopentyl-(1R,2S,4S,5S)-	pH 1	269(4.32)	18-0855-78
	H_2O	276(4.34)	18-0855-78
	pH 13	276(4.34)	18-0855-78
	pH 1	269(4.26)	18-0855-78
	H_2O	276(4.30)	18-0855-78
	pH 13	276(4.36)	18-0855-78
(1S,2R,4R,5R)-	pH 1	268(4.43)	18-0855-78
	H_2O	276(4.28)	18-0855-78
	pH 13	275(4.32)	18-0855-78
(1S,2R,4S,5S)-	pH 1	269(4.38)	18-0855-78
	H_2O	275(4.40)	18-0855-78
	pH 13	276(4.40)	18-0855-78
$C_{22}H_{30}N_2O$			
Acetamide, N-[2-[1,2-bis(3-methyl-2-	EtOH	230(4.48),287(3.85),	78-0929-78

Compound	Solvent	$\lambda_{max}(\log \epsilon)$	Ref.
butenyl)-1H-indol-3-yl]ethyl]- (cont.)		295(3.84)	78-0929-78
Aspidospermidine, 16-methoxy-1-methyl-3-methylene-, (2β,5α,12β,19α)-	EtOH	254(3.90),307(3.73)	35-4220-78
$C_{22}H_{30}N_2O_3$			
Vindoline, 4-deacetoxy-N-demethyl-3(R)-deoxy-6,7-dihydro-	MeOH	245(3.80),301(3.75)	35-4220-78
3(S)-	EtOH	208(4.54),245(3.73), 301(3.72)	35-4220-78
$C_{22}H_{30}N_2O_5$			
1H-Pyrrole-3-propanoic acid, 2-[(1,1-dimethylethoxy)carbonyl]-5-[(4-ethenyl-2,5-dihydro-3-methyl-5-oxo-1H-pyrrol-2-yl)methyl]-4-methyl-, methyl ester	CHCl₃	284(4.21)	24-0486-78
$C_{22}H_{30}N_2O_6$			
1H-Pyrrole-3-propanoic acid, 5-[[1,5-dihydro-4-(1-hydroxyethyl)-3-methyl-5-oxo-2H-pyrrol-2-ylidene]methyl]-2-[(1,1-dimethylethoxy)carbonyl]-4-methyl-, methyl ester	MeOH	255s(--),260(4.32), 385(4.51),404s(--)	24-0486-78
zinc complex	MeOH	265(4.32),270s(--), 425s(--),440(4.54)	24-0486-78
$C_{22}H_{30}N_2S$			
Vincanol, 14-deoxy-14-[(1-methylethyl)-thio]-	MeOH	197(4.43),231(4.47), 284(3.92)	33-1682-78
$C_{22}H_{30}N_4$			
2,2'-Bipyridine, 5,5'-bis(1-methyl-2-piperidinyl)-	EtOH	237(3.5),290(4.3)	103-0988-78
$C_{22}H_{30}N_4O_4$			
Cyclic(N-methyl-L-alanyl-L-leucyl-α,β-didehydro-N-methylphenylalanylglycyl) (tentoxin)	MeOH	282(4.32)	44-0296-78
(Z)-	MeOH	278(4.29)	44-0296-78
$C_{22}H_{30}N_8O_4$			
1,8-Purine-2,6-dione, 8,8'-(1,8-octanediyl)bis[3,7-dihydro-1,3-dimethyl-	pH 7.0	205(4.64),272(4.36)	24-0982-78
	pH 13.0	215(4.50),277(4.42)	24-0982-78
	MeOH	206(4.63),272(4.36)	24-0982-78
$C_{22}H_{30}O_2$			
3(4H)-Dibenzofuranone, 2,6-bis(1,1-dimethylethyl)-4a,9b-dihydro-8,9b-dimethyl-	EtOH	220(4.30),303(3.76)	12-0907-78
$C_{22}H_{30}O_3$			
22-Spirox-4-ene-3,20-dione	EtOH	240(4.21)	33-3068-78
$C_{22}H_{30}O_4$			
Benzene, 1,1'-(1,1,2,2-tetramethyl-1,2-ethanediyl)bis[3,4-dimethoxy-	EtOH	229(4.26),280(3.80), 285s(3.72)	94-0682-78
Benzo[d]xanthene-9,12-dione, 1,2,3,4-4a,5,6,7,7a,8-decahydro-10-methoxy-4,4,7,7a-tetramethyl-	hexane	278(4.24),285(4.25), 390(2.69)	12-2685-78
2H-Benzo[a]xanthene-8,11-dione, 1,3,4-4a,5,6,6a.12,12a,12b-decahydro-10-	EtOH	300(4.19),440(2.51)	12-2685-78

Compound	Solvent	λ_{max}(log ϵ)	Ref.
methoxy-4,4,6a,12b-tetramethyl- (cont.)			12-2685-78
2,5-Cyclohexadiene-1,4-dione, 2-hydr-oxy-5-methoxy-3-[(octahydro-2,5,5,8a-tetramethyl-1(2H)-naphthalenyli-dene)methyl]- (spongiaquinone)	EtOH EtOH-base	282(4.26),450(2.95) 286(4.29),530(3.24)	12-2685-78 12-2685-78
2,5-Cyclohexadiene-1,4-dione, 2-hydr-oxy-5-methoxy-3-[(1,2,3,4,4a,7,8,8a-octahydro-1,2,4a,5-tetramethyl-1-naphthalenyl)methyl]-	EtOH EtOH-base	225(3.82),285(4.31), 422(2.86) 218(4.37),243s(4.11), 289(4.87),525(3.68)	12-2685-78 12-2685-78
1H-Cyclopropa[3,4]benz[1,2-f]azulene-5,8-dione, 3-acetoxy-1a,2,3,3a,4,4a-6,7,7a,9b-decahydro-1,1,3a,6,9-penta-methyl-	n.s.g.	203(3.48),260(3.74)	78-0233-78
Estra-1,3,5(10)-trien-17-ol, 1,3-di-methoxy-, acetate, (8α,17β)-	MeOH	223(3.94),279(3.29), 283(3.30)	24-0939-78
2(1H)-Naphthalenone, 8-(acetoxymethyl)-4-[2-(3-furanyl)ethyl]-3,4,4a,5,6,8a-hexahydro-3,4,8a-trimethyl-, [3S-(3α,4β,4aβ,8aα)]-	MeOH	278(1.58)	44-3339-78
Spiro[benzofuran-2(3H),1'(2'H)-naphtha-lene]-4,7-dione, 3',4',4'a,5',6',7'-8',8'a-octahydro-5-methoxy-2',5',5'a-8'a-tetramethyl-, [1'S-(1'α,2'α,4'aβ-8'aα)]-	EtOH	285(4.34),406(2.57)	12-2685-78
Tridachione	n.s.g.	257(3.78)	35-1002-78
$C_{22}H_{30}O_4S$			
2(3H)-Naphthalenone, 5-(1,1-dimethyl-ethoxy)-4,4a,5,6,7,8-hexahydro-4a-methyl-1-[(phenylsulfonyl)methyl]-, (4aS,5S)-	n.s.g.	219(4.08),252(4.12)	33-2397-78
$C_{22}H_{30}O_5$			
Akiferidin	n.s.g.	265(3.9),300(3.6)	105-0617-78
2H-Cyclotrideca[b]furan-2-one, 5-acet-oxy-6-acetyl-3,3a,4,5,6,7,8,11,12,14a-decahydro-9,13-dimethyl-3-methylene-, (9E,13E)-	EtOH	219(4.18)	12-1303-78
Rubaferidin	EtOH	261(4.19)	105-0606-78
$C_{22}H_{30}O_6$			
2-Cyclohexen-1-one, 6-[2-hydroxy-1-methyl-2-(3,4,5-trimethoxyphenyl)-ethyl]-4-methoxy-6-(2-propenyl)-, [4R-[4α,6α(1S*,2R*)]]- (megaphone)	EtOH	269(2.70),279s(2.52)	44-0586-78
Phenol, 3,3'-(2,3-dimethyl-1,4-butane-diyl)bis[5,6-dimethoxy-, (R*,S*)-	EtOH	208(4.81),225s(4.30), 270(3.15)	94-0682-78
$C_{22}H_{30}O_7$			
Coleon R, 6,12-bis(O-deacetyl)-	MeOH	235(3.95)	33-0871-78
$C_{22}H_{30}O_8$			
Neurolenin B	n.s.g.	207(4.20),235s(c.3.79), 305(1.88)	44-4352-78
$C_{22}H_{30}O_9$			
Coleon Y, 3-O-deacetyl-3-O-formyl-	MeOH	233(3.98)	33-0871-78
$C_{22}H_{31}BrN_2O_5$			
1H-Pyrrole-3-propanoic acid, 5-[[4-(2-bromoethyl)-2,5-dihydro-3-methyl-5-	CHCl$_3$	282(4.25)	24-0486-78

Compound	Solvent	λ_{max} (log ϵ)	Ref.
oxo-1H-pyrrol-2-yl]methyl]-2-[(1,1-dimethylethoxy)carbonyl]-4-methyl-, methyl ester, (±)- (cont.)			24-0486-78
$C_{22}H_{31}NO_3$			
Benzeneethanamine, 2-ethoxy-4-methoxy-6-[2-(4-methoxyphenyl)ethyl]-N,N-dimethyl-	MeOH	206(4.84),225(4.55), 278(3.72),284(3.69)	100-0257-78
$C_{22}H_{31}NO_4$			
Laudanosinemethine, dihydro-	MeOH	225(4.27),280(3.73)	36-0473-78
Phenol, 3-[2-(dimethylamino)ethyl]-2-ethoxy-6-methoxy-4-[2-(4-methoxyphenyl)ethyl]-	pH 12	230s(4.05),255s(3.40), 277(3.36),285(3.34), 297s(3.00)	100-0257-78
	MeOH	205(4.42),225s(4.06), 277(3.30),285(2.98)	100-0257-78
$C_{22}H_{31}NO_5$			
Dispiro[1,3-dioxolane-2,1'-[1H]indene-5'(4'H),2"-[1,3]dioxolane]-4'-methanamine, hexahydro-N-(4-methoxyphenyl)-7'a-methyl-, (3'aα,4'α,7'aβ)-(±)-	EtOH	245(4.11),312(3.34)	65-0841-78
$C_{22}H_{32}N_2O_4$			
Aspidospermidine-3,4-diol, 3-(hydroxymethyl)-16-methoxy-1-methyl-, (2β,3β,4α,5α,12β,19α)-	EtOH	213(4.44),250(3.77), 302(3.65)	33-1554-78
(2β,3β,4β,5α,12β,19α)-	EtOH	213(4.52),251(3.88), 303(3.74)	33-1554-78
$C_{22}H_{32}N_2O_6$			
1H-Pyrrole-3-propanoic acid, 5-[[2,5-dihydro-4-(2-hydroxyethyl)-3-methyl-5-oxo-1H-pyrrol-2-yl]methyl]-2-[(1,1-dimethylethoxy)carbonyl]-4-methyl-, methyl ester, (±)-	CHCl$_3$	282(4.29)	24-0486-78
$C_{22}H_{32}N_4O_2$			
1,1'-Bipyridinium, 2,2',6,6'-tetramethyl-4,4'-dimorpholino-, bis-(tetrafluoroborate)	EtOH	235(3.62),315(4.12)	120-0001-78
$C_{22}H_{32}O$			
6a,12b-Propano-12a,6b-propenocyclobuta-[1,2:3,4]dicycloocten-15-one, 1,2,3-4,5,6,7,8,9,10,11,12-dodecahydro-(6aα,6bβ,12aβ,12bα)-	CHCl$_3$	304(1.64)	44-3776-78
Retinal, 10,14-dimethyl-	hexane	375(4.69)	88-0869-78
13-cis	hexane	360(4.48)	88-0869-78
$C_{22}H_{32}O_2$			
6a,12b:6b,12a-Dipropanocyclobuta[1,2:3,4]dicyclooctene-13,18-dione, dodecahydro-, (6aα,6bα,12aα,12bα)-	CHCl$_3$	304(1.86)	44-3776-78
(6aα,6bβ,12aβ,12bα)-	CHCl$_3$	320(2.19)	44-3776-78
Vitamin A acetate	EtOH	325(4.69)	44-4769-78
$C_{22}H_{32}O_4$			
8H-Cyclopropa[3,4]benz[1,2-f]azulen-8-one, 3-acetoxy-1,1a,2,3,3a,4,4a-5,6,7,7a,9b-dodecahydro-5-hydroxy-	n.s.g.	263(3.68)	78-0233-78

Compound	Solvent	$\lambda_{max}(\log \epsilon)$	Ref.
1,1,3a,6,9-pentamethyl- (cont.)			78-0233-78
Dilopholone, acetoxy-	MeOH	248(3.54),299(3.38)	88-4155-78
Epidilopholone, acetoxy-	MeOH	218(3.90),242(3.80), 285(3.20)	88-4155-78
Isospongiaquinone, dihydro-	EtOH	286(4.22),424(2.73)	12-2685-78
$C_{22}H_{32}O_4S$			
2(1H)-Naphthalenone, 5-(1,1-dimethyl-ethoxy)-4a-methyl-1-[(phenylsulfon-yl)methyl]-, (1R,4aS,5S,8aR)-	n.s.g.	217(3.97),265(3.06), 271(3.02)	33-2397-78
(1S,4aS,5S,8aS)-	n.s.g.	216(3.99),264(2.92), 271(2.86)	33-2397-78
$C_{22}H_{33}ClO_2$			
Androst-4-en-3-one, 6β-chloro-17β-hy-droxy-16β-(1-methylethyl)-	EtOH	240(4.23)	94-1718-78
$C_{22}H_{34}N_4O_5$			
1(2H)-Pyrimidinebutanamide, N-cyclohex-yl-N-[(cyclohexylamino)carbonyl]-3,4-dihydro-β-hydroxy-5-methyl-2,4-dioxo-	EtOH	268(4.04)	126-0905-78
$C_{22}H_{34}O$			
6a,12b:6b,12a-Dipropanocyclobuta[1,2-3,4]dicycloocten-13-one, dodecahy-dro-, (6aα,6bβ,12aβ,12bα)-	$CHCl_3$	304(1.70)	44-3776-78
$C_{22}H_{34}O_2$			
2,4,6,10,14-Hexadecapentaen-1-ol, 3,7,11,15-tetramethyl-, acetate, (all-E)	EtOH	280(4.54)	44-4769-78
$C_{22}H_{34}O_3$			
Pregn-20-ene-20-carboxaldehyde, 3,21-dihydroxy-, (3β,5α,20Z)-	EtOH	253(4.1)	23-0424-78
$C_{22}H_{34}O_5$			
2H-Cyclotrideca[b]furan-2-one, 5-acet-oxy-6-acetyl-3,3a,4,5,6,7,8,9,10,11-12,14a-dodecahydro-3,9,13-trimethyl-, (E)-	EtOH	205(3.90)	12-1303-78
$C_{22}H_{34}O_6$			
Megaphone, tetrahydro-	EtOH	270(2.98),279s(2.87)	44-0586-78
$C_{22}H_{34}O_{12}$			
α-D-Glucopyranoside, 6-O-acetyl-β-D-fructofuranosyl 1',2:4,6-bis-O-(1-methylethylidene)-, 3-acetate	MeOH	229(3.80),279(3.34)	94-2111-78
$C_{22}H_{36}O_2$			
5,9-Methanobenzocycloocten-4(1H)-one, 5-butoxy-2,3,5,6,7,8,9,10-octahydro-2,2,7,7,9-pentamethyl-	n.s.g.	245(3.92)	78-1251-78
$C_{22}H_{38}O_3$			
1-Cyclopenteneoctanoic acid, 2-octyl-5-oxo-, methyl ester (or isomer)	MeOH	238(3.88)	2-0280-78

Compound	Solvent	λ_{max}(log ϵ)	Ref.
$C_{23}H_{10}Br_2O_4S_4$ 2H-Pyran-2,4,6(3H,5H)-trione, 3,5-bis-[5-(4-bromophenyl)-3H-1,2-dithiol-3-ylidene]-	dioxan	242(4.42),288(4.26), 333(4.46),415s(--), 430(4.04),500s(--), 518(4.63)	78-2175-78
$C_{23}H_{11}ClN_2O_2$ Naphtho[1',2',3':4,5]quino[2,1-b]quin-azoline-5,10-dione, 3-chloro-	n.s.g.	440(3.81)	2-0103-78
$C_{23}H_{12}N_2O_2$ Naphtho[1',2',3':4,5]quino[2,1-b]quin-azoline-5,10-dione	n.s.g.	438(3.92)	2-0103-78
$C_{23}H_{12}O_4S_4$ 2H-Pyran-2,4,6(3H,5H)-trione, 3,5-bis-(5-phenyl-3H-1,2-dithiol-3-ylidene)-	dioxan	240(4.22),288(4.11), 323(4.25),360s(--), 408s(--),425(3.93), 490s(--),513(4.52)	78-2175-78
$C_{23}H_{14}ClNO$ Quinoline, 2-(2-benzofuranyl)-6-chloro-4-phenyl-	dioxan	365(3.60)	97-0400-78
$C_{23}H_{14}FeO_3$ Iron, tricarbonyl[9-[(2,3,4,5-η)-2,4,6-cycloheptatrien-1-ylidene]-9H-fluor-ene]-	C_6H_{12}	253(4.33),275(4.20), 285(4.12),366(3.88), 454(4.15)	88-1225-78
$C_{23}H_{14}Fe_2N_2O_7$ Iron, [μ-[N,N'-bis(1-phenylethenyl)ur-eato(2-)-N,N':N,N']]hexacarbonyldi-	EtOH	330(4.13)	33-0589-78
$C_{23}H_{14}N_4O_4S$ 5H-Naphtho[1,8-bc]thiophen-5-one, 2-phenyl-, 2,4-dinitrophenylhydrazone	MeOH	484(4.24)	83-0324-78
$C_{23}H_{15}ClFN_3$ 4H-Imidazo[1,5-a][1,4]benzodiazepine, 8-chloro-6-(2-fluorophenyl)-1-phenyl-	isoPrOH	217s(4.59),250s(4.25), 275s(4.09)	4-0855-78
$C_{23}H_{15}N$ Cyclobuta[b]quinoline, 1,2-diphenyl-	EtOH	282(5.62),444(3.35)	44-4128-78
$C_{23}H_{15}NO_2$ Benzonitrile, 4-(2,5-dihydro-5-oxo-2,4-diphenyl-2-furanyl)-	EtOH	268(4.19)	35-8247-78
Benzonitrile, 4-(2,5-dihydro-5-oxo-2,4-diphenyl-3-furanyl)-	MeOH	276(4.01)	35-8247-78
Benzonitrile, 4-(2,5-dihydro-5-oxo-3,4-diphenyl-2-furanyl)-	MeOH	276(4.09)	35-8247-78
3H-Naphth[2,3-e]indole-6,11-dione, 3-methyl-4-phenyl-	CHCl$_3$	247s(4.46),250(4.48), 291(4.51),306(4.43), 392(3.70),420s(3.64), 440s(3.61)	138-0323-78
$C_{23}H_{15}N_3$ 3,5-Pyridinedicarbonitrile, 2,6-bis(2-phenylethenyl)-, (E,E)-	EtOH	205(4.59),226s(4.35), 254(4.56)	73-0434-78

Compound	Solvent	$\lambda_{max}(\log \epsilon)$	Ref.

$C_{23}H_{15}N_4P$
  Ethenetricarbonitrile, [(triphenyl-
  phosphoranylidene)amino]-
	CH_2Cl_2	233(4.27),268(3.90),
		276(3.85),303(4.13)
			39-1237-78C

$C_{23}H_{16}FeO_3$
  Iron, tricarbonyl[(1,2,3,4-η)-7-(di-
  phenylmethylene)-1,3,5-cyclohepta-
  triene]-
	C_6H_{12}	232(4.27),263(4.20),
		333(4.09),394(3.87)
			88-1225-78

$C_{23}H_{16}N_2O_3$
  Pyrrolo[2,1-a]isoquinoline-3-carboxylic
  acid, 2-hydroxy-1-(1-isoquinolinyl)-,
  methyl ester
	EtOH	221(4.82),284(4.79),
		350(3.95),370(3.99)
			95-0198-78

$C_{23}H_{16}N_4O_2$
  11H-Dibenzo[b,e][1,4]diazepin-11-one,
  5,10-dihydro-8-[(2-hydroxy-1-naph-
  thalenyl)azo]-
	DMF	475(4.12)	42-0154-78

11H-Dibenzo[b,e][1,4]diazepin-11-one,
  5,10-dihydro-8-[(4-hydroxy-1-naph-
  thalenyl)azo]-
	DMF	430(4.08)	42-0154-78

1H-Purine-2,6-dione, 3,7-dihydro- 1,3,8-triphenyl-	pH 6.0	220(4.30),240(4.38), 310(4.38)	24-0982-78
	pH 13.0	220(4.33),312(4.43)	24-0982-78

$C_{23}H_{16}N_6O_2$
  7H-Pyrano[2",3":4,5:5",4":4',5']dipyr-
  rolo[2,3-b:2',3'-b']diquinoxalin-
  7-one, 5b,7a,14,15-tetrahydro-14,15-
  dimethyl-
	DMSO	260(5.3),302(5.1),
		373s(--),391s(--),
		522(4.9)
			24-2376-78

$C_{23}H_{16}O$
  2H-Cyclopenta[1]phenanthren-2-one,
  1,3-dihydro-1-phenyl-
	MeOH	206(4.69),219s(4.48),
		248s(4.71),255(4.83),
		264s(4.44),276(4.25),
		286(4.14),297s(4.06),
		299(4.12)
			54-0197-78

$C_{23}H_{16}O_2$
  1,2-Naphthalenedione, 4-(diphenylmeth-
  yl)-
	heptane	248(4.81),254(4.44),
		306(3.42),386(3.34)
			104-1986-78

	EtOH	215(4.40),253(4.39), 325(3.49),403(3.31)	104-1986-78
	EtOH-KOH	250(4.45),280s(4.00), 483(3.94)	104-1986-78
	CF_3COOH- H_2SO_4	299(3.96),392(4.16), 592(4.20)	104-1986-78
1(4H)-Naphthalenone, 4-(diphenylmeth- ylene)-2-hydroxy-	heptane	236(4.33),266(4.03), 319(3.78),416(4.20)	104-1986-78
	EtOH	211(4.35),237(4.32), 263(4.02),307s(3.79), 421(4.19)	104-1986-78
	EtOH-KOH	250(4.47),284(4.02), 483(4.02)	104-1986-78
	CF_3COOH- H_2SO_4	393(4.18),587(4.31)	104-1986-78

$C_{23}H_{16}O_8$
  4H-1-Benzopyran-4-one, 7-(benzoyloxy)-
  3-(2,4-dihydroxy-5-methoxyphenyl)-
  5-hydroxy-
	EtOH	299(4.04)	138-0879-78

Compound	Solvent	$\lambda_{max}(\log \epsilon)$	Ref.
$C_{23}H_{17}ClN_2O_2$			
1H,3H-Oxazolo[3,4-a][1,4]benzodiazepin-1-one, 8-chloro-3a,4-dihydro-3,6-diphenyl-	isoPrOH	212s(4.74),260s(3.96)	4-0855-78
$C_{23}H_{17}ClN_6$			
[1,2,4]Triazolo[1,5-a]pyridine, 2-[3-chloro-4-[2-(5-methyl-2-phenyl-2H-1,2,3-triazol-4-yl)ethenyl]phenyl]-	DMF	343(4.68)	33-0142-78
[1,2,4]Triazolo[1,5-a]pyridine, 2-[3-chloro-4-[2-(2-phenyl-2H-1,2,3-triazol-4-yl)ethenyl]phenyl]-6-methyl-	DMF	337(4.71)	33-0142-78
$C_{23}H_{17}ClN_6O$			
[1,2,4]Triazolo[1,5-a]pyridine, 2-[3-chloro-4-[2-[2-(4-methoxyphenyl)-2H-1,2,3-triazol-4-yl]ethenyl]phenyl]-	DMF	344(4.67)	33-0142-78
$C_{23}H_{17}N$			
Cyclobuta[b]quinoline, 1,2-dihydro-1,2-diphenyl-, cis	EtOH	234(4.77),306(4.07), 312(4.04),320(4.11)	44-4128-78
trans	EtOH	235(4.75),306(4.07), 312(4.05),320(4.12)	44-4128-78
3H-Naphth[2,3-e]indole, 3-methyl-4-phenyl-	EtOH	207(4.46),223s(4.51), 237(4.63),271(4.75), 288(4.69),302s(4.48), 360s(3.79),376(3.91), 396(3.80)	138-0323-78
$C_{23}H_{17}NO_2$			
Benzamide, N-methyl-N-(1-oxo-2-phenyl-1H-inden-3-yl)-	MeCN	263(3.54),423(3.22)	5-1203-78
4,11[1',2']-Benzeno-1H-naphth[2,3-f]-isoindole-1,3(2H)-dione, 3a,4,11,11a-tetrahydro-2-methyl-	THF	234(4.10),257(3.85), 266(3.92),276(3.88), 287(3.64),305(2.75), 313(2.45),319(2.71)	78-0113-78
4,13-Etheno-1H-anthra[2,3-f]isoindole-1,3(2H)-dione, 3a,4,13,13a-tetrahydro-2-methyl-	THF	263(4.27),295(2.80), 308(3.14),323(3.46), 338(3.80),355(3.85), 375(3.77)	78-0113-78
$C_{23}H_{17}NO_3$			
Spiro[2H-1-benzopyran-2,2'-[2H-1,3]-benzoxazin]-4'(3'H)-one, 3'-(phenylmethyl)-	isoPrOH	268(4.20),298(3.85)	103-0122-78
phootoinduced form	isoPrOH	385(--),480(--)	103-0122-78
$C_{23}H_{17}NO_4$			
2,6-Etheno-2H-bisoxireno[1,2:3,4]anthra-[2,3-f]isoindole-3,5(2aH,4H)-dione, 5a,6,7a,13b-tetrahydro-4-methyl-, (1aR,2α,2aβ,5aβ,6α,6aS,7aα,13bα)-	THF	236(4.70),252(3.65), 262(3.73),271(3.76), 281(3.80),293(3.62), 312(2.85),319(2.82), 327(2.75)	78-0113-78
6H-1,3-Oxazinium, 4-hydroxy-2-(4-methoxyphenyl)-6-oxo-3,5-diphenyl-, hydroxide, inner salt	MeCN	224s(4.24),262(4.27), 303(4.15),375(3.78)	5-1655-78
$C_{23}H_{17}N_3$			
3,5-Pyridinedicarbonitrile, 1,4-dihydro-2,6-bis(2-phenylethenyl)-, (E,E)-	EtOH	205(4.56),225s(4.32), 310(4.71),364s(3.71)	73-0434-78

Compound	Solvent	$\lambda_{max}(\log \epsilon)$	Ref.
$C_{23}H_{17}N_3$			
1H-Pyrrole-3-carbonitrile, 2,5-diphenyl-4-(phenylamino)-	EtOH	210(4.28),243(4.32), 298(4.32)	88-1163-78
$C_{23}H_{17}N_3O$			
Benz[cd]indazolium, 1-([1,1'-biphenyl]-2-ylamino)-8-methoxy-, hydroxide, inner salt	EtOH	298(4.34),308(4.23), 505(4.13),534(4.15), 574(3.93)	44-2508-78
$C_{23}H_{17}N_3O_4$			
Quinoline, 4-[(2,4-dinitrophenyl)methylene]-1,4-dihydro-1-(phenylmethyl)-	EtOH	244(4.3),280(4.0), 590(4.4)	104-1847-78
$C_{23}H_{18}$			
Cyclopentadiene, 1,2,3-triphenyl-	EtOH	246(4.48),305(4.04)	44-1481-78
	EtOH	239(4.38),310(3.91)	44-1493-78
Cyclopropene, 3-ethenyl-1,2,3-triphenyl-	EtOH	227(4.47),314(4.37), 330(4.32)	44-1481-78
	EtOH	227(4.31),313(4.36), 328(4.31)	44-1493-78
1H-Indene, 1-(diphenylmethylene)-4-methyl-	C_6H_{12}	209(4.48),218s(4.33), 250(4.35),291(4.17), 351(4.16)	5-0440-78
1H-Indene, 1-(diphenylmethylene)-5-methyl-	C_6H_{12}	212(4.16),218s(4.11), 248(4.17),288(4.05), 348(4.08)	5-0440-78
1H-Indene, 1-(diphenylmethylene)-6-methyl-	C_6H_{12}	207(4.48),217s(4.33), 248(4.32),288(4.19), 342(4.17)	5-0440-78
1H-Indene, 1-(diphenylmethylene)-7-methyl-	C_6H_{12}	224(4.21),253(4.28), 296(3.98),364(4.11)	5-0440-78
$C_{23}H_{18}Cl_2O_5$			
9H-Xanthen-9-one, 2,5-dichloro-3,6-dimethoxy-8-methyl-1-(phenylmethoxy)-	MeOH	212(4.47),251(4.71), 304(4.40),335s(3.94)	78-2491-78
$C_{23}H_{18}N_2O_2$			
Imidazo[1,2-a]pyridine, 7-[2-(1,3-benzodioxol-5-yl)ethenyl]-2-(4-methylphenyl)-	DMF	366(4.66),387(4.64)	33-0129-78
3aH-Phenanthro[9,10-d]imidazole, 3a-methoxy-2-(4-methoxyphenyl)-	EtOH	317(4.30)	40-1449-78
Spiro[imidazolidine-2,2'-[2H]indene]-1',3'-dione, 1,3-diphenyl-	MeOH	230(4.65),246(4.59), 284s(3.63),290s(3.58), 300s(3.54),343(2.56), 410s(2.39),475s(2.19)	24-3058-78
$C_{23}H_{18}N_2O_3$			
1,7-Dioxa-2,8-diazaspiro[4.4]nona-2,8-diene, 6-phenoxy-3,9-diphenyl-	EtOH	255(4.37)	22-0415-78
Imidazo[1,2-a]pyridine, 7-[2-(1,3-benzodioxol-5-yl)ethenyl]-2-(4-methoxyphenyl)-	DMF	368(4.66),388(4.64)	33-0129-78
$C_{23}H_{18}N_4O$			
[1,2,4]Triazolo[1,5-a]pyridine, 2-[4-[2-(5,6-dimethyl-2-benzoxazolyl)-ethenyl]phenyl]-	DMF	347(4.72),361(4.73)	33-0142-78
[1,2,4]Triazolo[1,5-a]pyridine, 2-[4-[2-(5,7-dimethyl-2-benzoxazolyl)-ethenyl]phenyl]-	DMF	344(4.72),358(4.72)	33-0142-78

Compound	Solvent	$\lambda_{max}(\log \epsilon)$	Ref.
$C_{23}H_{18}N_6$			
[1,2,4]Triazolo[1,5-a]pyridine, 2-(4-methylphenyl)-7-[2-(2-phenyl-2H-1,2,3-triazol-4-yl)ethenyl]-	DMF	331(4.69)	33-0142-78
[1,2,4]Triazolo[1,5-a]pyridine, 7-[2-(5-methyl-2-phenyl-2H-1,2,3-triazol-4-yl)ethenyl]-2-phenyl-	DMF	337(4.66),348(4.66)	33-0142-78
$C_{23}H_{18}O$			
9H-Benzocycloheptadecen-9-one, 16,17-18,19-tetradehydro-8,15-dimethyl-, (E,E,E,Z,E)-	THF	300(4.6),320(4.5)	138-1099-78
9H-Benzocycloheptadecen-9-one, 16,17-18,19-tetradehydro-10,15-dimethyl-, (E,E,E,Z,E)-	THF	300(4.7),405(3.7)	138-1099-78
$C_{23}H_{18}OS_4$			
2-Propanone, 1,3-bis(4-methyl-5-phenyl-1,2-dithiol-3-ylidene)-	dioxan	248(4.62),370(3.78), 495(4.58),528(4.78)	78-2175-78
$C_{23}H_{18}O_2$			
4,6-Etheno-1H-cycloprop[f]isobenzofuran-1-one, 4,4a,5,5a,6,6a-hexahydro-3,6a-diphenyl-	EtOH	230(3.70),242(3.54), 260(3.40)	78-2305-78
2(5H)-Furanone, 4-(4-methylphenyl)-3,5-diphenyl-	MeOH	295(4.13)	35-8247-78
2(5H)-Furanone, 5-(4-methylphenyl)-3,4-diphenyl-	MeOH	284(4.09)	35-8247-78
2(5H)-Furanone, 5-(4-methylphenyl)-3,5-diphenyl-	MeOH	266(4.16)	35-8247-78
2(5H)-Furanone, 4-methyl-3,5,5-triphenyl-	MeOH	256(4.68)	35-8247-78
1H-Indene-2-carboxylic acid, 1,3-diphenyl-, methyl ester	MeOH	242(4.20),300(4.11)	104-0701-78
Methanone, bicyclo[3.2.2]nona-2,6,8-triene-3,4-diylbis[phenyl-	EtOH	250(4.30),274(3.94), 320(2.60)	78-2305-78
Methanone, tricyclo[3.2.2.0^{2,4}]nona-6,8-diene-6,7-diylbis[phenyl-, (1α,2β,4β,5α)-	EtOH	255(4.13),290(3.80), 350(2.70)	78-2305-78
3-Oxabicyclo[3.1.0]hexan-2-one, 1,4,4-triphenyl-	MeOH	260(2.99)	35-8247-78
2-Propen-1-one, 1-[5-(1-hydroxy-3-phenyl-2-propenylidene)-1,3-cyclopentadien-1-yl]-3-phenyl-	MeOH	262(4.121),310(4.531), 440(4.121)	73-0938-78
Spiro[1H-cyclohepta[c]furan-1,1'-cyclotrideca-2,4,10,12-tetraene-6,8-diyn]-3(3aH)-one, 5',10'-dimethyl-	EtOH	258(4.38),316(3.90)	88-2795-78
$C_{23}H_{18}O_3$			
2(5H)-Furanone, 4-(3-methoxyphenyl)-3,5-diphenyl-	MeOH	284(4.26)	35-8247-78
2(5H)-Furanone, 5-(3-methoxyphenyl)-3,4-diphenyl-	MeOH	280(4.10)	35-8247-78
2(5H)-Furanone, 5-(3-methoxyphenyl)-3,5-diphenyl-	MeOH	267(4.14)	35-8247-78
2(5H)-Furanone, 5-(4-methoxyphenyl)-3,4-diphenyl-	EtOH	283(4.11)	35-8247-78
2(5H)-Furanone, 5-(4-methoxyphenyl)-3,5-diphenyl-	EtOH	268(4.13)	35-8247-78

Compound	Solvent	$\lambda_{max}(\log \epsilon)$	Ref.
$C_{23}H_{18}O_3S$			
2-Propen-1-one, 1-(1,3-benzodioxol-5-yl)-3-phenyl-3-[(phenylmethyl)thio]-	EtOH	276(3.84),349(4.38)	4-0105-78
$C_{23}H_{19}BrN_2O_2$			
Benzeneacetic acid, [3-(3-bromophenyl)-3-oxo-1-phenylpropylidene]hydrazide	EtOH	283(4.31),292s(4.30)	94-1298-78
$C_{23}H_{19}Br_3O_5$			
19-Norpregna-1,3,5,7,9,11,14-heptaene, 4,11,15-tribromo-3-methoxy-17,20-20,21-bis[methylenebis(oxy)]-	EtOH	269(4.62),283(4.57), 292(4.55),363(3.88), 378(3.81)	88-0639-78
$C_{23}H_{19}ClN_2O_2$			
Benzeneacetic acid, [1-(4-chlorophenyl)-3-oxo-3-phenylpropylidene]hydrazide	EtOH	300(4.36)	34-0345-78
Benzeneacetic acid, [3-(3-chlorophenyl)-3-oxo-1-phenylpropylidene]hydrazide	EtOH	284(4.38),293(4.36)	94-1298-78
Benzeneacetic acid, [3-(4-chlorophenyl)-3-oxo-1-phenylpropylidene]hydrazide	EtOH	220(4.52),283(4.45), 293s(4.44)	4-0385-78
	EtOH	283(4.45),293s(4.44)	34-0345-78
4H-Imidazol-4-one, 3-(2-chlorophenyl)-3,5-dihydro-5-methyl-5-phenyl-2-(4-methoxyphenyl)-	EtOH	265(4.18)	94-3798-78
$C_{23}H_{19}ClN_2O_4$			
1,4'-Bipyridinium, 2-methyl-4,6-diphenyl-, perchlorate	EtOH	220(3.89),310(4.05)	120-0001-78
$C_{23}H_{19}ClN_2O_4S_2$			
1H-Cyclopenta[b]quinoxaline, 6(and 7)-chloro-4,9-dihydro-4,9-bis(phenylsulfonyl)-	C_6H_{12}	218(4.61),229(4.60), 260(4.17),267(4.16), 274s(4.08),298(3.61)	5-1129-78
$C_{23}H_{19}ClO_6$			
Pyrylium, 2,6-dimethyl-4-[(2-phenyl-4H-1-benzopyran-4-ylidene)methyl]-, perchlorate	HOAc-HClO4	523(4.534)	83-0236-78
$C_{23}H_{19}Cl_2N_4O_9P$			
2,4(1H,3H)-Pteridinedione, 6,7-bis(4-chlorophenyl)-1-(5-O-phosphono-β-D-ribofuranosyl)-	pH 1.0	278(4.253),361(4.179)	5-1780-78
	pH 7.0	278(4.265),362(4.196)	5-1780-78
	pH 13.0	270(4.344),362(4.225)	5-1780-78
$C_{23}H_{19}NO$			
3H-Pyrrol-3-one, 1,2-dihydro-2-(4-methylphenyl)-4,5-diphenyl-	MeOH	254(4.33),383(3.84)	18-3312-78
$C_{23}H_{19}NO_2$			
1,2-Benzisoxazole, 3-[4-[2-(4-ethoxyphenyl)ethenyl]phenyl]-	DMF	342(4.63)	33-2904-78
1,2-Benzisoxazole, 3-[4-[2-(4-methoxyphenyl)ethenyl]phenyl]-5-methyl-	DMF	340(4.64)	33-2904-78
3-Butene-1,2-dione, 4-(methylamino)-1,3,4-triphenyl-, (Z)-	MeCN	237(4.17),254(4.18), 483(3.92)	5-1203-78
3H-Pyrrol-3-one, 1,2-dihydro-2-hydroxy-1-methyl-2,4,5-triphenyl-	MeCN	302(3.87)	5-1203-78
3H-Pyrrol-3-one, 1,2-dihydro-2-(4-methoxyphenyl)-4,5-diphenyl-	MeOH	243(4.35),250s(4.34), 383(3.88)	18-3312-78

Compound	Solvent	$\lambda_{max}(\log \epsilon)$	Ref.
$C_{23}H_{19}NO_3$			
Benzeneacetic acid, α-(benzoylmethyl-amino)phenylmethylene-, (Z)-	MeCN	225(4.21),256(3.98), 317(3.83)	5-1203-78
1,2-Benzisoxazole, 3-(4-methoxyphenyl)-6-[2-(4-methoxyphenyl)ethenyl]-	DMF	340(4.60)	33-2904-78
$C_{23}H_{19}NO_4$			
1H-Benz[f]indole-3-carboxylic acid, 4,9-dihydro-2-methyl-4,9-dioxo-1-(phenylmethyl)-, ethyl ester	dioxan	390(3.60)	5-0129-78
1H-Benz[g]indole-3-carboxylic acid, 4,5-dihydro-2-methyl-4,5-dioxo-1-(phenylmethyl)-, ethyl ester	dioxan	455(3.46)	5-0129-78
$C_{23}H_{19}NS$			
Butanenitrile, 3,3-dimethyl-2-(2-phenyl-5H-naphtho[1,8-bc]thien-5-ylidene)-	MeOH	245(4.122),302(4.122), 417(4.298)	24-1824-78
$C_{23}H_{19}N_2$			
1,4'-Bipyridinium, 2-methyl-4,6-di-phenyl-, perchlorate	EtOH	220(3.89),310(4.05)	120-0001-78
$C_{23}H_{19}N_3O$			
3-Pyrazolidinone, 5-imino-4-(9-methyl-9H-fluoren-9-yl)-1-phenyl-	EtOH	262(4.23),302(3.82)	115-0305-78
$C_{23}H_{19}N_3O_4$			
Benzeneacetic acid, [3-(3-nitrophenyl)-3-oxo-1-phenylpropylidene]hydrazide	EtOH	282(4.34),301s(4.13)	94-1298-78
$C_{23}H_{19}N_5O_{11}$			
1,2,4-Triazine-3,5(2H,4H)-dione, 2-[2-deoxy-3,5-bis-0-(4-nitrobenzoyl)-β-D-erythro-pentofuranosyl]-6-methyl-	pH 1.5	266(4.40)	136-0175-78C
	pH 12.7	267(4.43)	136-0175-78C
	EtOH	261(4.51)	136-0175-78C
$C_{23}H_{19}O_2$			
Pyrylium, 2,6-dimethyl-4-[(2-phenyl-4H-1-benzopyran-4-ylidene)methyl]-, perchlorate	HOAc-HClO₄	523(4.534)	83-0236-78
$C_{23}H_{20}$			
Cyclopropene, 2-methyl-1,3-diphenyl-3-(phenylmethyl)-	EtOH	264(4.02)	44-1481-78
Cyclopropene, 3-methyl-1,2-diphenyl-3-(phenylmethyl)-	EtOH	230(4.28),322(4.39), 339(4.27)	44-1481-78
1H-Indene, 1-(diphenylmethylene)-2,3-dihydro-4-methyl-	C_6H_{12}	207(4.51),231(4.22), 306(4.00)	5-0440-78
1H-Indene, 1-(diphenylmethylene)-2,3-dihydro-5-methyl-	C_6H_{12}	238s(4.19),300s(4.23), 315(4.25)	5-0440-78
1H-Indene, 1-(diphenylmethylene)-2,3-dihydro-6-methyl-	C_6H_{12}	206(4.49),230(4.21), 304s(4.13),317(4.18)	5-0440-78
1H-Indene, 1-(diphenylmethylene)-2,3-dihydro-7-methyl-	C_6H_{12}	209(4.48),239(4.31), 313(4.26)	5-0440-78
1H-Indene, 1-methyl-2-phenyl-3-(phenyl-methyl)-	EtOH	229(4.18),296(4.23)	44-1481-78
1H-Indene, 2-methyl-1-phenyl-3-(phenyl-methyl)-	EtOH	265(3.97)	44-1481-78
$C_{23}H_{20}INO_4$			
Glaucine ethiodide, (+)-	MeOH	220(--),282(4.32), 303(4.32)	12-0313-78

Compound	Solvent	$\lambda_{max}(\log \epsilon)$	Ref.
$C_{23}H_{20}N_2$			
Indeno[1,2-c]pyrazole, 1,3a,4,8b-tetra- hydro-1-(4-methylphenyl)-3-phenyl-	EtOH	240(4.14),356(4.14)	33-1477-78
$C_{23}H_{20}N_2O$			
1H-Imidazole, 2-(4-methoxyphenyl)- 4-methyl-1,5-diphenyl-	EtOH	288(4.30)	94-3798-78
Imidazo[1,2-a]pyridine, 7-[2-(4-ethoxy- phenyl)ethenyl]-2-phenyl-	DMF	364(4.68),383(4.64)	33-0129-78
Imidazo[1,2-a]pyridine, 7-[2-(4-meth- oxyphenyl)ethenyl]-2-(4-methylphenyl)-	DMF	364(4.67),384(4.63)	33-0129-78
3H-Pyrazole, 4,5-dihydro-5-(1-phenoxy- ethylidene)-3,3-diphenyl-	EtOH	295(4.09)	22-0401-78
$C_{23}H_{20}N_2O_2$			
Benzamide, N-[4-methoxy-2-(3-methyl- 3H-indol-3-yl)phenyl]-	EtOH	235(4.34),270(4.16), 285(4.11)	88-1983-78
Benzeneacetic acid, (3-oxo-1,3-diphen- ylpropylidene)hydrazide	EtOH	283(4.39),293s(4.35)	4-0385-78 +34-0345-78
Benzoic acid, [3-(4-methylphenyl)-3- oxo-1-phenylpropylidene]hydrazide	EtOH	219(4.50),292(4.42)	4-0385-78
4H-Imidazol-4-one, 3,5-dihydro-2-(4- methoxyphenyl)-5-methyl-3,5-diphenyl-	EtOH	260(4.34)	94-3798-78
Imidazo[1,2-a]pyridine, 7-[2-(2,3-di- methoxyphenyl)ethenyl]-2-phenyl-	DMF	361(4.60),381(3.54)	33-0129-78
Imidazo[1,2-a]pyridine, 7-[2-(2,4-di- methoxyphenyl)ethenyl]-2-phenyl-	DMF	370(4.67),387(4.61)	33-0129-78
Imidazo[1,2-a]pyridine, 7-[2-(2,5-di- methoxyphenyl)ethenyl]-2-phenyl-	DMF	373(4.60)	33-0129-78
Imidazo[1,2-a]pyridine, 7-[2-(3,4-di- methoxyphenyl)ethenyl]-2-phenyl-	DMF	366(4.69),386(4.66)	33-0129-78
Imidazo[1,2-a]pyridine, 7-[2-(3,5-di- methoxyphenyl)ethenyl]-2-phenyl-	DMF	361(4.61),380(4.57)	33-0129-78
Imidazo[1,2-a]pyridine, 2-(4-methoxy- phenyl)-7-[2-(4-methoxyphenyl)ethen- yl]-	DMF	366(4.64),385(4.62)	33-0129-78
3H-Pyrazole, 4,5-dihydro-3-(4-methoxy- phenyl)-5-(phenoxymethylene)-3-phenyl-	EtOH	293(4.24)	22-0401-78
$C_{23}H_{20}N_2O_2S_2$			
3H-Phenothiazin-3-one, 7-[(4-methyl- phenyl)thio]-2-morpholino-	CHCl$_3$	264(4.76),491(4.19)	103-0795-78
$C_{23}H_{20}N_2O_3$			
Benzamide, N,N'-(2-oxo-3,5,7-cyclohep- tatriene-1,3-diyl)bis[N-methyl-	MeOH	282(4.00),355(3.81)	18-2338-78
Benzoic acid, [1-(4-methoxyphenyl)- 3-oxo-3-phenylpropylidene]hydrazide	EtOH	286s(4.41),293(4.42)	34-0345-78
Benzoic acid, [3-(4-methoxyphenyl)- 3-oxo-1-phenylpropylidene]hydrazide	EtOH	292(4.41)	4-0385-78
$C_{23}H_{20}N_2S$			
Benzenamine, 4-methyl-N-[3-(4-methyl- phenyl)-4-phenyl-2(3H)-thiazolyli- dene]-	EtOH	230s(4.37),275s(4.06), 298(4.11)	94-3017-78
$C_{23}H_{20}N_3O$			
Pyrimidinium, 5-(4-methoxyphenyl)- 1-phenyl-4-(phenylamino)-, per- chlorate	CH$_2$Cl$_2$	292(4.10),321(4.19)	97-0063-78

Compound	Solvent	$\lambda_{max}(\log \epsilon)$	Ref.
$C_{23}H_{20}N_4O_2$			
Ethanone, 1,1'-(5-methyl-1,1'-diphenyl-[4,4'-bi-1H-pyrazole]-3,3'-diyl)bis-	EtOH	242(4.55),276s(4.25)	4-0293-78
Ethanone, 1,1'-(5-methyl-1,1'-diphenyl-[4,5'-bi-1H-pyrazole]-3,3'-diyl)bis-	EtOH	232(4.46),252s(4.37)	4-0293-78
$C_{23}H_{20}N_4O_5$			
3,4-Furandiol, tetrahydro-2-(1-phenyl-1H-pyrazolo[3,4-b]quinoxalin-3-yl)-, diacetate	MeOH	267(4.7),334(4.1), 408(3.7)	136-0079-78H
Tryptoquivaline G	MeOH	226(4.54),232s(4.50), 253s(4.24),265s(4.07), 275s(3.91),291s(3.57), 302(3.48),315(3.40)	94-0111-78
$C_{23}H_{20}N_4O_5S$			
4(1H)-Pteridinone, 2,3-dihydro-6,7-diphenyl-3-β-D-ribofuranosyl-2-thioxo-	10% MeOH-pH 4	230s(4.28),280s(4.17), 310(4.46),383(4.22)	24-2571-78
	10% MeOH-pH 9.0	227(4.37),270(4.06), 330(4.50),398(4.05)	24-2571-78
	MeOH	225s(4.34),288s(4.18), 314(4.47),386(4.17)	24-2571-78
4(1H)-Pteridinone, 6,7-diphenyl-2-(β-D-ribofuranosylthio)-	pH 3.0	222s(4.35),259(4.28), 300(4.32),366(4.19)	24-2571-78
	pH 8.0	221(4.39),278(4.42), 374(4.18)	24-2571-78
	MeOH	220s(4.36),262(4.28), 295(4.32),367(4.16)	24-2571-78
$C_{23}H_{20}O$			
1H-Indene, 1-ethoxy-1,2-diphenyl-	EtOH	243s(4.38),251(4.35), 314(4.31),327(4.32), 342(4.10)	44-0232-78
1H-Indene, 1-ethoxy-2,3-diphenyl-	EtOH	242(4.45),315(4.11)	44-0232-78
1H-Indene, 3-ethoxy-1,2-diphenyl-	EtOH	235(4.15),242(4.10), 302s(4.34),313(4.35)	44-0232-78
$C_{23}H_{20}OS$			
2-Propen-1-one, 1-(4-methylphenyl)-3-phenyl-3-[(phenylmethyl)thio]-	EtOH	267(4.12),337(4.31)	4-0105-78
$C_{23}H_{20}O_2S$			
2-Propen-1-one, 1-(4-methoxyphenyl)-3-phenyl-3-[(phenylmethyl)thio]-	EtOH	283s(4.00),339(4.41)	4-0105-78
$C_{23}H_{20}O_3$			
4H-1-Benzopyran, 7-methoxy-2-[4-(phenylmethoxy)phenyl]-	MeOH	249(4.39),277(4.08)	18-1175-78
2-Naphthalenol, 1,1'-(ethoxymethylene)-bis-	EtOH	236(4.71),271(4.13), 281(4.09),293(3.95), 336(3.63)	32-0079-78
$C_{23}H_{20}O_5$			
9H-Xanthen-9-one, 3,6-dimethoxy-1-methyl-8-(phenylmethoxy)-	MeOH	212(4.56),244(4.67), 302(4.38),327(4.12)	78-2491-78
$C_{23}H_{20}O_6$			
Penioflavin dimethyl ether	EtOH	263(4.58),308(4.23)	1-0075-78
$C_{23}H_{20}O_{10}$			
4H-1-Benzopyran-4-one, 5,6,7-triacet-	EtOH	313(4.04),327(3.63)	18-3627-78

Compound	Solvent	$\lambda_{max}(\log \epsilon)$	Ref.
oxy-2-(4-acetoxyphenyl)-2,3-dihydro-4H-1-Benzopyran-4-one, 5,7,8-triacet-oxy-2-(4-acetoxyphenyl)-2,3-dihydro-	EtOH	264(4.03),322(3.52)	18-3627-78 18-3627-78
$C_{23}H_{20}S$ Spiro[1H-indene-1,2'-thiirane], 2,3-di-hydro-7-methyl-3',3'-diphenyl-	C_6H_{12}	212(4.55),238s(4.16), 267s(3.32),273(3.24), 282(3.11)	5-0440-78
$C_{23}H_{21}Br$ Benzene, 1-bromo-4-(2,2-dimethyl-3,3-diphenylcyclopropyl)-	EtOH	269(3.15),273(3.03), 284(2.67)	35-4146-78
Benzene, 1-bromo-4-(1,1-dimethyl-3,3-diphenyl-2-propenyl)-	EtOH	250(4.34)	35-4146-78
$C_{23}H_{21}ClFN_5O_5$ 9H,13aH-Oxazolo[3,2-d][1,2,4]triazolo-[1,5-a][1,4]benzodiazepine-7-carbox-amide, 2-chloro-13a-(2-fluorophenyl)-11,12-dihydro-9,12-dimethoxy-N,N-di-methyl-11-oxo-. (9α,12α,13aβ)-	MeOH	252(4.19)	33-0848-78
(9α,12β,13aβ)-	MeOH	252(4.22)	33-0848-78
$C_{23}H_{21}ClN_2O_2$ 1H-Imidazol-5-ol, 1-(2-chlorophenyl)-4,5-dihydro-2-(4-methoxyphenyl)-4-methyl-4-phenyl-	EtOH	253(4.26)	94-3798-78
$C_{23}H_{21}ClN_2O_4$ Pyridinium, 1-[[1-[(4-methylphenyl)-methyl]-4(1H)-quinolinylidene]-methyl]-, perchlorate	EtOH	262(4.05),365(3.56), 510(3.81)	124-0071-78
$C_{23}H_{21}Cl_2N_3O_3$ Acetic acid, [7-chloro-5-(2-chlorophen-yl)-1,3-dihydro-2H-1,4-benzodiazepin-2-ylidene][(1-methyl-3-oxo-1-buten-yl)amino]-, methyl ester	isoPrOH	324(4.59)	4-0577-78
$C_{23}H_{21}IN_2O$ Quinolinium, 2-[[1,5-dihydro-1-(4-meth-ylphenyl)-5-oxo-2H-pyrrol-2-ylidene]-methyl]-1-ethyl-, iodide	EtOH	240(4.02),320(3.61)	18-2415-78
Quinolinium, 2-[3-(1,2-dihydro-2-oxo-3H-indol-3-ylidene)-2-methyl-1-pro-penyl]-1-ethyl-, iodide	MeOH	510(4.22),560(4.21)	64-0209-78B
$C_{23}H_{21}IN_2O_2$ 1H-Isoindolium, 3-[(1-ethyl-2(1H)-pyri-dinylidene)methyl]-2-(4-methoxyphen-yl)-1-oxo-, iodide	EtOH	215(4.83),245s(4.55), 270(4.45),330(4.40)	18-2415-78
Quinolinium, 2-[[1,5-dihydro-1-(4-meth-oxyphenyl)-5-oxo-2H-pyrrol-2-ylidene]-methyl]-1-ethyl-, iodide	EtOH	240(4.98),323(4.57)	18-2415-78
$C_{23}H_{21}NO_3$ Benzenamine, 4-[7-methoxy-3-(4-methoxy-phenyl)-2H-1-benzopyran-2-yl)-	EtOH	242(3.67),328(3.81)	39-0088-78C
Benzoic acid, 3-(4,5,6,7-tetrahydro-6,6-dimethyl-4-oxo-2-phenyl-1H-indol-1-yl)-	MeOH	220(4.16),245(4.16), 280(3.90)	48-0863-78

Compound	Solvent	$\lambda_{max}(\log \epsilon)$	Ref.
Benzoic acid, 4-(4,5,6,7-tetrahydro-6,6-dimethyl-4-oxo-2-phenyl-1H-indol-1-yl)-	MeOH	248(4.50),275(4.24)	48-0863-78
Benzo[h]quinolin-2(1H)-one, 3-acetyl-5,6-dihydro-4-(4-methoxyphenyl)-1-methyl-	CHCl$_3$	250(4.41),372(4.16)	48-0097-78
$C_{23}H_{21}NO_4$			
2,3-Pyridinedicarboxylic acid, 5,6-diphenyl-, diethyl ester	EtOH	247(4.19),300s(3.96)	142-0731-78A
$C_{23}H_{21}N_2$			
Pyridinium, 1-[[1-[(4-methylphenyl)-methyl]-4(1H)-quinolinylidene]-methyl]-, perchlorate	EtOH	262(4.05),365(3.56),510(3.81)	124-0071-78
$C_{23}H_{21}N_2O$			
1H-Isoindolium, 3-[(1-ethyl-2(1H)-pyridinylidene)methyl]-2-(4-methylphenyl)-1-oxo-, iodide	EtOH	218(4.89),240s(4.74),325(4.06)	18-2415-78
Quinolinium, 2-[[1,5-dihydro-1-(4-methylphenyl)-5-oxo-2H-pyrrol-2-ylidene]-methyl]-1-ethyl-, iodide	EtOH	240(4.02),320(3.61)	18-2415-78
Quinolinium, 2-[3-(1,2-dihydro-2-oxo-3H-indol-3-ylidene)-2-methyl-1-propenyl]-1-ethyl-, iodide	MeOH	510(4.22),560(4.21)	64-0209-78B
$C_{23}H_{21}N_2OP$			
3H-Pyrazole, 4-(diphenylphosphinyl)-3,3-dimethyl-5-phenyl-	MeOH	296(3.46),355s(2.55)	150-4248-78
$C_{23}H_{21}N_2O_2$			
1H-Isoindolium, 3-[(1-ethyl-2(1H)-pyridinylidene)methyl]-2-(4-methoxyphenyl)-1-oxo-, iodide	EtOH	215(4.83),245s(4.55),270(4.45),330(4.40)	18-2415-78
Pyridinium, 1-ethyl-2-[4-(4-nitrophenyl)-2-phenyl-1,3-butadienyl]-, iodide	EtOH	482(4.39)	146-0341-78
Quinolinium, 2-[[1,5-dihydro-1-(4-methoxyphenyl)-5-oxo-2H-pyrrol-2-ylidene]methyl]-1-ethyl-, iodide	EtOH	240(4.98),323(4.57)	18-2415-78
$C_{23}H_{21}N_2PS$			
3H-Pyrazole, 4-(diphenylphosphinothioyl)-3,3-dimethyl-5-phenyl-	MeOH	315s(3.60)	150-4248-78
3H-Pyrazole, 5-(diphenylphosphinothioyl)-3,3-dimethyl-4-phenyl-	MeOH	307s(3.52)	150-4248-78
$C_{23}H_{21}N_3$			
1H-1,2,3-Triazole, 2,4,5-tris(4-methylphenyl)-	EtOH	223(4.38),302(4.42)	33-1477-78
$C_{23}H_{21}N_3O_2$			
1,2,4-Triazin-3(2H)-one, 4,5-dihydro-4-(4-methoxyphenyl)-2-(2-methylphenyl)-6-phenyl-	EtOH	297(4.68),336(4.72)	5-2033-78
$C_{23}H_{21}N_3O_4$			
1,3,5-Triazine-2,4(1H,3H)-dione, dihydro-6-(2-hydroxy-4-methoxyphenyl)-1-methyl-3,6-diphenyl-	EtOH	230s(3.66),277(3.13),284(3.09)	4-1193-78

Compound	Solvent	$\lambda_{max}(\log \epsilon)$	Ref.
$C_{23}H_{21}N_4O_8P$			
2,4(1H,3H)-Pteridinedione, 1-(2-deoxy-5-O-phosphono-β-D-erythro-pento-furanosyl)-6,7-diphenyl-	pH 1.0	220(4.459),273(4.207), 359(4.152)	5-1780-78
	pH 7.0	220(4.464),274(4.212), 360(4.155)	5-1780-78
	pH 13.0	265(4.316),359(4.185)	5-1780-78
$C_{23}H_{21}N_4O_9P$			
2,4(1H,3H)-Pteridinedione, 6,7-diphenyl-1-(5-O-phosphono-β-D-arabinofurano-syl)-	pH 1.0	220s(4.413),273(4.201), 358(4.107)	5-1780-78
	pH 7.0	221(4.417),273(4.179), 359(4.117)	5-1780-78
	pH 13	265(4.272),359(4.158)	5-1780-78
2,4(1H,3H)-Pteridinedione, 6,7-diphenyl-1-(5-O-phosphono-β-D-ribofuranosyl)-	pH 1.0	220(4.439),273(4.188), 358(4.143)	5-1780-78
	pH 7.0	220s(4.450),273(4.199), 359(4.158)	5-1780-78
	pH 13.0	265(4.305),359(4.188)	5-1780-78
$C_{23}H_{21}N_5$			
2-Propanone, 1-(1,3-diphenyl-1H-1,2,4-triazol-5-yl)-, phenylhydrazone	EtOH	259(4.48)	94-1825-78
$C_{23}H_{21}N_5O_5S$			
6H-Anthra[9,1-cd]isothiazol-6-one, 7-[[4,6-bis(2-methoxyethoxy)-1,3,5-triazin-2-yl]amino]-	DMF	470(4.03)	2-1007-78
$C_{23}H_{21}N_5O_6$			
2,4,6(1H,3H,5H)-Pyrimidinetrione, 5-[[4-(hexahydro-1,3-dimethyl-2,4,6-trioxo-5-pyrimidinylidene)-1,4-di-hydro-2-methyl-3-quinolinyl]meth-ylene]-1,3-dimethyl-	MeOH	435(3.6)	83-0561-78
$C_{23}H_{21}N_7$			
Formazan, 5-[4-(dimethylamino)phenyl]-1-(1-phthalazinyl)-3-phenyl-	C_6H_{12}	263(4.51),370(3.86), 470(3.30)	103-0450-78
	EtOH	255(4.57),275(4.62), 370(4.19),462(3.66), 485s(3.64),516s(3.60)	103-0450-78
	EtOH-KOH	547(3.11)	103-0450-78
$C_{23}H_{22}$			
Benzene, 2-(2,2-diphenylethenyl)-1,3,5-trimethyl-	THF	267(3.06),273(3.02), 290(2.70)	35-4886-78
1-Butene, 3-methyl-1,1,3-triphenyl-	EtOH	250(4.16)	35-4146-78
Cyclopropane, 3,3-dimethyl-1,1,2-tri-phenyl-	EtOH	262(3.06),274(2.72)	35-4146-78
$C_{23}H_{22}ClNO_2$			
2H-Cycloocta[b]pyran-2-one, 3-chloro-4-(diphenylamino)-5,6,7,8,9,10-hexa-hydro-	EtOH	253(4.02),282(4.21), 328(3.95),380(3.87)	4-0511-78
$C_{23}H_{22}N$			
Pyridinium, 2-(2,4-diphenyl-1,3-buta-dienyl)-1-ethyl-, iodide	EtOH	500(2.86)	146-0341-78
$C_{23}H_{22}N_2$			
4,11-Etheno-1H-cyclopenta[b]phenazine,	C_6H_{12}	240(4.58),292s(3.72),	5-0440-78

Compound	Solvent	$\lambda_{max}(\log \epsilon)$	Ref.
1-cyclohexylidene-3a,4,11,11a-tetra- hydro-, (3aα,4β,11β,11aα)- (cont.)		297s(3.78),304s(3.90), 310s(3.94),316(4.00)	5-0440-78
Quinolinium, 1-[(4-methylphenyl)methyl]- 4-(pyridiniomethyl)-, diperchlorate	EtOH	238(4.64),317(4.00)	124-0071-78

$C_{23}H_{22}N_2O_2$

Benzenepentanamide, δ-oxo-N-phenyl- β-(phenylamino)-	n.s.g.	246(4.57)	104-1446-78
1H-Imidazol-5-ol, 4,5-dihydro-2-(4- methoxyphenyl)-4-methyl-1,4-diphenyl-	EtOH	257.5(4.38)	94-3798-78

$C_{23}H_{22}N_2O_3$

1H-Pyrrolo[3,4-b]pyridine-3-carboxylic acid, 4,5,6,7-tetrahydro-2-methyl-7- oxo-4-phenyl-6-(phenylmethyl)-, meth- yl ester	EtOH	259(4.12),341(3.63)	95-0448-78

$C_{23}H_{22}N_2O_4$

4H-Pyran-3-carboxylic acid, 2-amino- 5-cyano-4-(4-methoxyphenyl)-6-(4- methylphenyl)-, ethyl ester	EtOH	299(3.92)	4-0057-78
2H-Pyrido[3,4-e]-1,3-oxazine-4,5(3H,6H)- dione, 7-methoxy-2,2-dimethyl-6-phen- yl-3-(phenylmethyl)-	MeOH	307(4.4)	120-0101-78

$C_{23}H_{22}N_2O_4S$

5-Thia-1-azabicyclo[4.2.0]oct-2-ene-2- carboxylic acid, 8-oxo-7-[(phenyl- acetyl)amino]-3-(phenylmethyl)-, methyl ester	n.s.g.	257(3.85)	88-5219-78

$C_{23}H_{22}N_2O_5S$

5-Thia-1-azabicyclo[4.2.0]oct-2-ene-2- carboxylic acid, 8-oxo-3-(phenoxy- methyl)-7-[(phenylacetyl)amino]-, methyl ester	n.s.g.	262(3.90)	88-5219-78

$C_{23}H_{22}N_2O_6S_2$

5-Thia-1-azabicyclo[4.2.0]oct-2-ene-2- carboxylic acid, 3-(acetoxymethyl)- 8-oxo-7-[(2-thienylacetyl)amino]-, phenylmethyl ester	n.s.g.	236(4.12)	88-5219-78

$C_{23}H_{22}N_3O$

Pyridinium, 2-(2-benzoxazolyl)-4-[2- [4-(dimethylamino)phenylethenyl]- 1-methyl-, iodide	EtOH	531(4.59)	4-0017-78
Pyridinium, 2-(2-benzoxazolyl)-6-[2- [4-(dimethylamino)phenylethenyl]- 1-methyl-, iodide	EtOH	508(4.60)	4-0017-78
Pyridinium, 4-(2-benzoxazolyl)-2-[2- [4-(dimethylamino)phenylethenyl]- 1-methyl-, iodide	EtOH	526(4.55)	4-0017-78

$C_{23}H_{22}N_3PS$

Phosphonium, [3-[(aminothioxomethyl)- hydrazono]-1-butenyl]triphenyl-, hydroxide, inner salt	DMF	275(3.863),344(3.910), 515(3.674)	65-1998-78

$C_{23}H_{22}N_3S$

Pyridinium, 4-(2-benzothiazolyl)-2-[2-	EtOH	521(4.34)	4-0017-78

Compound	Solvent	$\lambda_{max}(\log \epsilon)$	Ref.
[4-(dimethylamino)phenyl]ethenyl]- 1-methyl-, iodide (cont.)			4-0017-78
$C_{23}H_{22}N_4O_4S$			
Hydrazine, 1-[1,3-dihydro-3,3-dimethyl- 1-(2-methylphenyl)benzo[c]thien-1- yl]-2-(2,4-dinitrophenyl)-	toluene EtOH	345(4.20) 256(3.89),350(4.05)	104-2229-78 104-2229-78
Hydrazine, 1-[1,3-dihydro-3,3-dimethyl- 1-(3-methylphenyl)benzo[c]thien-1- yl]-2-(2,4-dinitrophenyl)-	toluene EtOH	345(4.19) 258(3.90),350(4.08)	104-2229-78 104-2229-78
Hydrazine, 1-[1,3-dihydro-3,3-dimethyl- 1-(4-methylphenyl)benzo[c]thien-1- yl]-2-(2,4-dinitrophenyl)-	toluene EtOH	345(4.21) 257(4.02),350(4.11)	104-2229-78 104-2229-78
$C_{23}H_{22}N_4O_5$			
Hydrazine, 1-[1,3-dihydro-3,3-dimethyl- 1-(3-methylphenyl)-1-isobenzofuran- yl]-2-(2,4-dinitrophenyl)-	toluene EtOH	345(4.16) 259(3.92),352(4.16)	104-2229-78 104-2229-78
Hydrazine, 1-[1,3-dihydro-3,3-dimethyl- 1-(4-methylphenyl)-1-isobenzofuran- yl]-2-(2,4-dinitrophenyl)-	toluene EtOH	345(4.16) 254s(4.04),352(4.15)	104-2229-78 104-2229-78
Methanone, [2-(1-hydroxy-1-methyleth- yl)phenyl](2-methylphenyl)-, 2,4- dinitrophenylhydrazone	toluene EtOH	382(4.33) 248s(4.12),293s(3.78), 382(4.29)	104-2229-78 104-2229-78
Methanone, [2-(1-hydroxy-1-methyleth- yl)phenyl](3-methylphenyl)-, 2,4- dinitrophenylhydrazone	toluene EtOH	383(4.28) 249s(4.04),293s(3.75), 384(4.28)	104-2229-78 104-2229-78
$C_{23}H_{22}N_4O_5S$			
Hydrazine, 1-[1,3-dihydro-1-(2-methoxy- phenyl)-3,3-dimethylbenzo[c]thien-1- yl]-2-(2,4-dinitrophenyl)-	toluene EtOH	348(4.29) 261(3.85),351(4.05)	104-2229-78 104-2229-78
Hydrazine, 1-[1,3-dihydro-1-(3-methoxy- phenyl)-3,3-dimethylbenzo[c]thien-1- yl]-2-(2,4-dinitrophenyl)-	toluene EtOH	345(4.21) 262(3.75),348(4.02)	104-2229-78 104-2229-78
Hydrazine, 1-[1,3-dihydro-1-(4-methoxy- phenyl)-3,3-dimethylbenzo[c]thien-1- yl]-2-(2,4-dinitrophenyl)-	toluene EtOH	346(4.19) 257s(3.63),350(4.14)	104-2229-78 104-2229-78
$C_{23}H_{22}N_4O_6$			
9,10-Anthracenedione, 1-[[4,6-bis(2- methoxyethoxy)-1,3,5-triazin-2-yl]- amino]-	DMF	460(4.50)	2-0106-78
Hydrazine, 1-[1,3-dihydro-1-(2-methoxy- phenyl)-3,3-dimethyl-1-isobenzofuran- yl]-2-(2,4-dinitrophenyl)-	toluene EtOH	347(4.21) 264(4.00),349(4.18)	104-2229-78 104-2229-78
Hydrazine, 1-[1,3-dihydro-1-(3-methoxy- phenyl)-3,3-dimethyl-1-isobenzofuran- yl]-2-(2,4-dinitrophenyl)-	toluene EtOH	343(4.17) 262(4.05),350(4.22)	104-2229-78 104-2229-78
Hydrazine, 1-[1,3-dihydro-1-(4-methoxy- phenyl)-3,3-dimethyl-1-isobenzofuran-	toluene EtOH	346(4.18) 260(4.01),350(4.13)	104-2229-78 104-2229-78
$C_{23}H_{22}N_4O_{12}P_2$			
2,4(1H,3H)-Pteridinedione, 1-[5-O-[hy- droxy(phosphonooxy)phosphinyl]-β-D- arabinofuranosyl]-6,7-diphenyl-	pH 1.0	220(4.413),273(4.173), 358(4.107)	5-1788-78
	pH 7.0	221(4.417),273(4.176), 359(4.118)	5-1788-78
	pH 13.0	265(4.272),359(4.146)	5-1788-78
2,4(1H,3H)-Pteridinedione, 1-[5-O-[hy- droxy(phosphonooxy)phosphinyl]-β-D- ribofuranosyl]-6,7-diphenyl-	pH 1.0	220(4.447),273(4.199), 358(4.152)	5-1788-78
	pH 7.0	220(4.471),273(4.223)	5-1788-78

Compound	Solvent	$\lambda_{max}(\log \epsilon)$	Ref.
(cont.)		358(4.146)	5-1788-78
	pH 13.0	264(4.305),359(4.196)	5-1788-78
$C_{23}H_{22}O$			
Benzene, 1,1',1"-(3-ethoxy-1-propene-1,2,3-triyl)tris-	EtOH	260(4.20)	44-0232-78
$C_{23}H_{22}O_2$			
A-Nor-3-oxaestra-1,5,7,9-tetraen-17-one, 7-methyl-2-phenyl-	hexane	213(4.38),236(4.16), 242(4.12),248(4.01), 260(3.95),278s(4.11), 288s(4.27),298(4.39), 311(4.50),328(4.37)	2-0418-78
$C_{23}H_{22}O_2S_2$			
Benzene, 1-[[2-(ethylthio)-1,2-diphenylethenyl]sulfonyl]-4-methyl-, (E)-	EtOH	202(4.66),235s(4.20), 294(4.13)	2-1086-78
(Z)-	EtOH	203(4.57),235s(4.22), 301(4.00)	2-1086-78
$C_{23}H_{22}O_4S_2$			
Benzene, 1-[[2-(ethylsulfonyl)-1,2-diphenylethenyl]sulfonyl]-4-methyl-, (E)-	EtOH	203(4.56),220s(4.40), 244(4.13)	2-1086-78
(Z)-	EtOH	203(4.56),241(4.09)	2-1086-78
$C_{23}H_{22}O_5$			
2H,8H-Benzo[1,2-b:3,4-b']dipyran-2-one, 9-acetoxy-9,10-dihydro-8,8-dimethyl-10-(4-methylphenyl)-	EtOH	211(4.54),247(3.69), 257(3.64),329(4.10)	2-0184-78
4,6-Heptadien-3-one, 1-(3,4-diacetoxyphenyl)-7-phenyl-	MeOH	323(4.91)	78-3005-78
$C_{23}H_{22}O_6$			
2H,8H-Benzo[1,2-b:3,4-b']dipyran-2-one, 9-acetoxy-9,10-dihydro-10-(4-methoxyphenyl)-8,8-dimethyl-	EtOH	211(4.51),249(3.79), 258(3.74),330(4.13)	2-0184-78
Desmodol dimethyl ether	EtOH	240(4.55),279(4.39), 318(4.19),338(4.27)	94-2411-78
$C_{23}H_{22}O_8$			
Dalbinol	MeOH	235(4.29),242s(4.24), 292(4.36)	102-1442-78
$C_{23}H_{22}O_{12}$			
Quercitrin, 3"-O-acetyl-	MeOH	260(4.33),357(4.18)	94-3580-78
Quercitrin, 4"-O-acetyl-	MeOH	260(4.33),357(4.18)	94-3580-78
$C_{23}H_{22}O_{13}$			
4H-1-Benzopyran-4-one, 3-[(6-O-acetyl-β-D-glucopyranosyl)oxy]-2-(3,4-dihydroxyphenyl)-5,7-dihydroxy-	MeOH	257(4.33),265s(--), 360(4.26)	105-0146-78
$C_{23}H_{23}BF_4O_2$			
2,4,6,8-Nonatetraenylium, 1,9-bis(4-methoxyphenyl)-, tetrafluoroborate	$C_2H_4Cl_2$	785(4.97)	104-1197-78
$C_{23}H_{23}BrO_5$			
19-Norpregna-1,3,5,7,9,11-hexaene, 4-bromo-3-methoxy-17,20:20,21-bis[methylenebis(oxy)]-	EtOH	240(4.77),277(3.71), 290(3.76),306(3.80), 322(3.90),348(3.76)	88-0639-78

Compound	Solvent	$\lambda_{max}(\log \epsilon)$	Ref.
$C_{23}H_{23}BrO_6$			
19-Norpregna-1,3,5,7,9-pentaene, 4-bromo-11,12-epoxy-3-methoxy-17,20:20,21-bis[methylenebis(oxy)]-	EtOH	240(4.88),281(3.71), 293(3.82),348(3.54)	88-0639-78
$C_{23}H_{23}ClN_3PS$			
Phosphonium, [3-[(aminothioxomethyl)hydrazono]-1-butenyl]triphenyl-, chloride	EtOH or DMF	268(4.025),339(4.267)	65-1998-78
$C_{23}H_{23}ClO_2$			
3-Butenoic acid, 4-(4-chlorophenyl)-2-cyclohexylidene-4-(4-methylphenyl)-, (E)-	EtOH	269(2.8808)	2-0502-78
$C_{23}H_{23}ClO_3$			
3-Butenoic acid, 4-(4-chlorophenyl)-2-cyclohexylidene-4-(4-methoxyphenyl)-, (E)-	EtOH	230(4.2878)	2-0502-78
$C_{23}H_{23}NO$			
4H-Indol-4-one, 1,5,6,7-tetrahydro-6,6-dimethyl-1-(4-methylphenyl)-2-phenyl-	MeOH	244(4.09),272(3.89)	48-0863-78
4H-Indol-4-one, 1,5,6,7-tetrahydro-6,6-dimethyl-2-phenyl-1-(phenylmethyl)-	MeOH	242(4.22),290(3.86)	48-0863-78
$C_{23}H_{23}NO_2$			
4H-Indol-4-one, 1,5,6,7-tetrahydro-1-(2-methoxyphenyl)-6,6-dimethyl-2-phenyl-	MeOH	212(4.23),245(4.11), 275(4.00)	48-0863-78
4H-Indol-4-one, 1,5,6,7-tetrahydro-1-(3-methoxyphenyl)-6,6-dimethyl-2-phenyl-	MeOH	218(4.58),245(4.32), 275(4.24)	48-0863-78
4H-Indol-4-one, 1,5,6,7-tetrahydro-1-(4-methoxyphenyl)-6,6-dimethyl-2-phenyl-	MeOH	218(4.36),245(4.39), 274(4.24)	48-0863-78
$C_{23}H_{23}NO_4$			
3,5-Pyridinedicarboxylic acid, 1-ethyl-1,4-dihydro-2,6-dimethyl-, diphenyl ester	EtOH	204(4.31),240(4.20), 265s(4.07),374(3.85)	103-1226-78
$C_{23}H_{23}NO_5$			
1H-Pyrrole-3-carboxylic acid, 5-acetoxy-5-methyl-4-oxo-2-phenyl-1-(phenylmethyl)-, ethyl ester	EtOH	212(4.00),239(4.16), 326(4.02)	118-0291-78
$C_{23}H_{23}NSi$			
3H-1,3-Benzazasilole, 2-(1,1-dimethylethyl)-3,3-diphenyl-	CCl_4	252(3.92),256(3.92), 307(3.64)	44-0731-78
$C_{23}H_{23}N_3$			
2-Acridinamine, 9-[4-(dimethylamino)phenyl]-N,N-dimethyl-	EtOH	244(4.53),281(4.68), 464(3.66)	39-1211-78C
	EtOH-HCl	240(4.44),300(4.59), 392(3.66),576(3.83)	39-1211-78C
3-Acridinamine, 9-[4-(dimethylamino)phenyl]-N,N-dimethyl-	EtOH	241(4.19),283(4.26), 455(3.37)	39-1211-78C
	EtOH-HCl	241(4.31),296(4.28),	39-1211-78C

Compound	Solvent	$\lambda_{max}(\log \epsilon)$	Ref.
(cont.)		360(3.62),378(3.69), 492(3.99)	39-1211-78C
$C_{23}H_{23}N_3O_4S$ 2,3,5-Triazabicyclo[2.2.2]octa-2,7-di-ene-7,8-dicarboxylic acid, 5-methyl-4-(methylthio)-1,6-diphenyl-, di-methyl ester	MeOH	225s(4.08),303(4.08)	142-0093-78B
$C_{23}H_{23}N_3O_5$ 2,3,5-Triazabicyclo[2.2.2]octa-2,7-di-ene-7,8-dicarboxylic acid, 4-methoxy-5-methyl-1,6-diphenyl-, dimethyl ester	MeOH	296(4.02)	142-0093-78B
$C_{23}H_{23}N_3O_6S$ 4-Oxa-2,6-diazabicyclo[3.2.0]hept-2-ene-6-acetic acid, α-(1-mercapto-1-meth-ylethyl)-7-oxo-3-(phenylmethyl)-, (4-nitrophenyl)methyl ester	EtOH	208(4.14),262(3.91)	39-0668-78C
1,4-Thiazepine-3-carboxylic acid, 2,3,4,7-tetrahydro-2,2-dimethyl-7-oxo-6-[(phenylacetyl)amino]-, (4-nitrophenyl)methyl ester, (S)-	EtOH	211(4.20),260(4.15), 316(3.93)	39-0668-78C
$C_{23}H_{23}N_3O_7S$ 4-Oxa-2,6-diazabicyclo[3.2.0]hept-2-ene-6-acetic acid, α-(1-mercapto-1-meth-ylethyl)-7-oxo-3-(phenylmethoxy)-, (4-nitrophenyl)methyl ester, [1S-[1α,5α,6(R*)]]-	EtOH	210(4.23),264(4.10), 268(4.10),275s(4.04)	39-0668-78C
4-Thia-1-azabicyclo[3.2.0]heptane-2-carboxylic acid, 3,3-dimethyl-7-oxo-6-[(phenoxyacetyl)amino]-, (4-nitro-phenyl)methyl ester, [2S-(2α,5β,6β)]-	EtOH	210(4.23),264(4.02), 268(4.03),275s(3.94)	39-0668-78C
1,4-Thiazepine-3-carboxylic acid, 2,3,4,7-tetrahydro-2,2-dimethyl-7-oxo-6-[(phenoxyacetyl)amino]-, (4-nitrophenyl)methyl ester, (S)-	EtOH	206(4.16),262(4.05), 315(3.69)	39-0668-78C
$C_{23}H_{23}N_3O_8S$ 5-Oxa-4-thia-1-azabicyclo[4.2.0]octane-2-carboxylic acid, 3,3-dimethyl-8-oxo-7-[(phenylacetyl)amino]-, (4-nitrophenyl)methyl ester, 4-oxide, (2S,6S,7S)-	EtOH	210(4.34),265(4.05)	39-0668-78C
$C_{23}H_{23}N_3PS$ Phosphonium, [3-[(aminothioxomethyl)-hydrazono]-1-butenyl]triphenyl-, chloride	EtOH or DMF	268(4.025),339(4.267)	65-1998-78
$C_{23}H_{23}N_5O_9$ Carbamic acid, N-[9-(2,3,5-tri-O-acet-yl-β-D-ribofuranosyl)-9H-purin-6-yl]-, phenyl ester	EtOH	268(4.28)	39-0131-78C
$C_{23}H_{23}O_2$ 2,4,6,8-Nonatetraenylium, 1,9-bis(4-methoxyphenyl)-, tetrafluoroborate	$C_2H_4Cl_2$	785(4.97)	104-1197-78

Compound	Solvent	$\lambda_{max}(\log \epsilon)$	Ref.
$C_{23}H_{23}O_8P$ 4H,8aH-Phospholo[1,2-a]phosphorin-1,2,3,8a-tetracarboxylic acid, 7-methyl-4-phenyl-, tetramethyl ester	EtOH	223(4.30),262s(3.96), 269s(4.00),275(4.00), 275(4.00),332(4.10)	4-1319-78
$C_{23}H_{24}$ Tricyclo[4.3.2.0^{1,6}]undeca-3,10-diene, 10,11-diphenyl-	MeCN	226(4.23),286(4.02), 293s(4.01),310s(3.93), 330s(3.59)	89-0758-78
$C_{23}H_{24}F_3NO_5$ Benzo[6,7]cyclohept[1,2,3-ij]isoquinoline, 4,5,6,6a,7,8-hexahydro-1,2,10-11-tetramethoxy-6-(trifluoroacetyl)-, (±)-	EtOH	267(4.07),289(3.98)	44-2521-78
$C_{23}H_{24}F_3NO_6$ Benzo[6,7]cyclohept[1,2,3-ij]isoquinolin-1-ol, 4,5,6,6a,7,8-hexahydro-2,10,11,12-tetramethoxy-6-(trifluoroacetyl)-, (±)-	EtOH	258(4.11),296(3.72)	44-4076-78
$C_{23}H_{24}N_2$ Benzenamine, 4-(9,10-dihydro-10-methyl-9-acridinyl)-N-ethyl-N-methyl-hydriodide	EtOH	272(4.20)	104-0129-78
	EtOH	277(4.26)	104-0129-78
4,11-Etheno-1H-cyclopenta[b]phenazine, 3a,4,11,11a-tetrahydro-4,7,8-trimethyl-1-(1-methylethylidene)-	C_6H_{12}	213(4.58),248(4.62), 297s(3.61),304s(3.73), 310s(3.82),317(3.93), 322(3.93),330(3.87), 368s(2.93),395s(2.52)	5-0440-78
$C_{23}H_{24}N_2O_4$ Arenine	n.s.g.	244(4.49),291(3.78), 319(3.57),333(3.57)	105-0348-78
$C_{23}H_{24}O_2$ 3-Butenoic acid, 2-cyclohexylidene-4-(4-methylphenyl)-4-phenyl-	EtOH	265(3.7634)	2-0502-78
$C_{23}H_{24}O_3$ 3-Butenoic acid, 2-cyclohexylidene-4-(4-methoxyphenyl)-4-phenyl-	EtOH	231(4.7067)	2-0502-78
2,2'-Spirobi[2H-indene]-5-carboxylic acid, 5'-acetyl-6'-ethyl-1,1',3,3'-tetrahydro-, methyl ester	EtOH	206(4.69),212s(4.66), 247(4.35),278s(3.56), 288(3.56),305s(3.27)	49-0987-78
$C_{23}H_{24}O_4$ Tricyclo[8.5.1.1^{3,8}]heptadeca-1,3,5,7-9,11,14-heptaene-12,14-dicarboxylic acid, diethyl ester	dioxan	317(4.91),339s(4.84), 397(3.75)	89-0956-78
$C_{23}H_{24}O_5$ 1-Benzoxepin-6-ol, 8-[2-(4-acetoxyphenyl)ethyl]-2,5-dihydro-3-methyl-, acetate	EtOH	204(3.33),206(3.31), 260(2.54)	102-2005-78
$C_{23}H_{24}O_6$ 2H,10H-Benzo[1,2-b:3,4-b']dipyran-10-one, 9-(2,4-dimethoxyphenyl)-3,4-dihydro-5-methoxy-2,2-dimethyl-	MeOH	264(4.40),292(4.05)	2-0613-78

Compound	Solvent	λ_{max}(log ϵ)	Ref.
4H-1-Benzopyran-4-one, 2-(2,4-dimeth-oxyphenyl)-5-hydroxy-7-methoxy-8-(3-methyl-2-butenyl)-	MeOH	267(4.64),293(4.23)	2-0613-78
4H-1-Benzopyran-4-one, 3-(3,4-dimeth-oxyphenyl)-5-hydroxy-7-methoxy-6-(3-methyl-2-butenyl)-	MeOH	216(3.83),280(4.05)	44-3446-78
Dihydrodesmodol dimethyl ether	EtOH	245(4.19),255(4.13), 282(4.26),339(4.29)	94-2411-78
4H-Furo[2,3-h]-1-benzopyran-4-one, 3-(3,4-dimethoxyphenyl)-8,9-dihydro-5-methoxy-8,9,9-trimethyl-	MeOH	238(4.39),291(4.31)	2-0973-78
$C_{23}H_{24}O_7$			
2(3H)-Furanone, 4-[(3,4-dimethoxyphen-yl)methylene]dihydro-3-[(3,4,5-tri-methoxyphenyl)methylene]-	EtOH	261(4.04),271(4.04), 357(3.83)	94-3186-78
Naphtho[2,3-c]furan-1(3H)-one, 4,9-di-hydro-7,8-dimethoxy-9-(3,4,5-tri-methoxyphenyl)-	EtOH	276s(3.44)	94-3186-78
$C_{23}H_{24}O_9$			
2-Propenoic acid, 3-(4-hydroxy-3-meth-oxyphenyl)-, 2-hydroxy-1,3-propane-diyl ester	MeOH	230(3.7),290(3.8), 310(4.1)	102-1673-78
$C_{23}H_{24}O_{11}$			
Tectorigenin, 7-O-methyl-, 4'-O-β-D-glucopyranoside	MeOH	268(4.20),330(3.60)	2-0641-78
	MeOH-NaOAc	268(--),330(--)	2-0641-78
	MeOH-AlCl$_3$	280(--),330(--)	2-0641-78
$C_{23}H_{25}ClN_4O_5S$			
Ethanethioic acid, S-[1-[2-(acetoxy-ethyl)-2-[[(4-amino-2-methyl-5-pyr-imidinyl)methyl][(4-chlorophenyl)-oxoacetyl]amino]-1-propenyl] ester	EtOH	237s(4.17),274(4.24)	94-0722-78
$C_{23}H_{25}NO$			
Mahanimbine	EtOH	223(4.55),239(4.06), 288(4.61),330(3.90), 343(3.42)	42-0308-78
Murrayazoline	EtOH	245(4.69),307(4.16)	42-0308-78
$C_{23}H_{25}NO_4$			
1H-Pyrrole-3-carboxylic acid, 4,5-di-hydro-5-hydroxy-5-methyl-2-[4-(1-methylethyl)phenyl]-4-oxo-1-phenyl-, ethyl ester	EtOH	210(4.20),246(4.23), 284(3.91),335(4.9)	118-0291-78
1H-Pyrrole-3-carboxylic acid, 4,5-di-hydro-5-hydroxy-2-methyl-5-[(4-meth-ylphenyl)methyl]-4-oxo-1-(phenyl-methyl)-, ethyl ester	EtOH	214(4.15),248(4.16), 326(3.88)	4-1215-78
$C_{23}H_{25}NO_5$			
1H-Pyrrole-3-carboxylic acid, 4,5-di-hydro-5-hydroxy-5-[(4-methoxyphenyl)-methyl]-2-methyl-4-oxo-1-(phenyl-methyl)-, ethyl ester	EtOH	210(4.10),228(4.09), 248(4.11),330(3.86)	4-1215-78
$C_{23}H_{25}NO_8$			
5,6,7,8-Indolizinetetracarboxylic acid, 1,2,3,8a-tetrahydro-8a-(phenylmethyl)-,	MeOH	208(4.16),232(4.13), 288(3.98),396(3.67)	83-0977-78

Compound	Solvent	$\lambda_{max}(\log \epsilon)$	Ref.
tetramethyl ester (cont.)			83-0977-78
$C_{23}H_{25}N_2$			
Methylium, bis[4-(dimethylamino)phenyl]phenyl-, triiodide	EtOH	317(4.34),427(4.30), 622(5.01)	64-1520-78B
	acetone	423(4.38),622(5.03)	64-1520-78B
$C_{23}H_{25}N_3$			
Ethanone, 2-[(2,4-dimethylphenyl)amino]-1-phenyl-, (2-methylphenyl)hydrazone, (E)-	EtOH	247(4.36),295(4.08)	5-2033-78
(Z)-	EtOH	241(4.29),293(3.94), 338(4.13)	5-2033-78
2-Propanone, 1-[(4-methylphenyl)amino]-1-phenyl-, (2-methylphenyl)hydrazone, (E)-	EtOH	248(4.32),274(4.29)	5-2033-78
$C_{23}H_{25}N_3O_5$			
Glycine, N-[N-(phenylmethoxy)carbonyl]-L-tryptophyl]-, ethyl ester	isoPrOH	275(3.75),282(3.79), 290(3.73)	63-1617-78
$C_{23}H_{25}N_3O_8$			
L-Glutamic acid, N-[N-[4-[methyl[(phenylmethoxy)carbonyl]amino]benzoyl]glycyl]-	MeOH	260(4.11)	87-1165-78
$C_{23}H_{25}N_7O_7$			
1H-Purine-2,6-dione, 7-[6-azido-4-O-benzoyl-6-deoxy-2,3-O-(1-methylethylidene)-α-D-mannopyranosyl]-3,7-dihydro-1,3-dimethyl-	MeOH	274(3.94)	136-0073-78C
1H-Purine-2,6-dione, 7-[4-azido-6-O-benzoyl-2,3-O-(1-methylethylidene)-α-D-talopyranosyl]-3,7-dihydro-1,3-dimethyl-	MeOH	274(3.93)	136-0073-78C
$C_{23}H_{25}O_8P$			
α-D-arabino-Hex-5-enofuranoside, methyl 5,6-dideoxy-6-(diphenoxyphosphinyl)-, 2,3-diacetate	MeOH	256(2.22),261(2.26), 267(2.22)	136-0349-78H
$C_{23}H_{26}$			
Benzene, 1,1'-[3-(1,1-dimethylethyl)-1-(2-methyl-1-propenyl)-2-cyclopropene-1,2-diyl]bis-	EtOH	204(4.63),263(4.30)	44-1493-78
Cyclopentadiene, 1-(1,1-dimethylethyl)-5,5-dimethyl-2,3-diphenyl-	EtOH	282(3.72)	44-1493-78
Cyclopentadiene, 2-(1,1-dimethylethyl)-5,5-dimethyl-1,3-diphenyl-	EtOH	235(3.83)	44-1493-78
Cyclopentadiene, 3-(1,1-dimethylethyl)-5,5-dimethyl-1,2-diphenyl-	EtOH	271(3.87)	44-1493-78
1H-Indene, 1-(1,1-dimethylethyl)-3-(2-methyl-1-propenyl)-2-phenyl-	EtOH	236(4.28),299(4.16)	44-1493-78
$C_{23}H_{26}F_3NO_5$			
Isoquinoline, 1-[2-(3,4-dimethoxyphenyl)ethyl]-1,2,3,4-tetrahydro-6,7-dimethoxy-2-(trifluoroacetyl)-, (±)-	EtOH	228s(4.30),282(3.84), 286s(3.83)	44-2521-78
$C_{23}H_{26}F_3NO_6$			
Acetamide, 2,2,2-trifluoro-N-[2-	EtOH	235s(4.25),285(3.84)	44-2521-78

Compound	Solvent	$\lambda_{max}(\log \epsilon)$	Ref.
[4,4',5,5'-tetramethoxy-2'-(3-oxopro-pyl)[1,1'-biphenyl]-2-yl]ethyl]-(cont.)			44-2521-78
$C_{23}H_{26}NOP$ 9H-Phosphorino[3,4-b]indole, 2-(1,1-di-methylethyl)-2,2-dihydro-2-methoxy-9-methyl-4-phenyl-	CHCl$_3$	255(4.44),285(4.34), 333(3.81),375(4.05)	139-0257-78B
$C_{23}H_{26}N_2$ 2H-Indeno[2,1-c]pyridine-2-propanamine, 1,3-dihydro-N,N-dimethyl-9-phenyl-	MeOH	243(4.43),258(4.42), 310(3.71),320(3.61)	87-0340-78
$C_{23}H_{26}N_2O$ 17,21-Cyclo-19-norpregna-1,3,5(10)-tri-ene-20,20-dicarbonitrile, 3-methoxy-	MeOH	220(3.89),229(3.81), 279(3.26),287(3.24)	24-3094-78
$C_{23}H_{26}N_2O_4S$ Benzenesulfonamide, N-[1,3-dihydro-3,3-bis(3-oxobutyl)-2H-indol-2-yli-dene]-4-methyl-	CHCl$_3$	278(4.12),294s(3.99)	103-0745-78
Benzenesulfonic acid, 4-methyl-, [1-(6,7-dimethoxy-2-naphthalenyl)but-ylidene]hydrazide	MeOH	256(4.68),301(4.38)	83-0328-78
$C_{23}H_{26}N_2O_6$ Cleavamine, N$_a$,18β-bis(methoxycarbo-nyl)-5-oxo-3β,4β-epoxydihydro-	EtOH	227(4.38),259(4.06), 265s(4.05),281s(3.78), 293(3.67)	23-0062-78
$C_{23}H_{26}N_3OS$ 1,3-Thiazin-1-ium, 4-[4-(dimethylami-no)phenyl]-6-(4-methylphenyl)-2-mor-pholino-, perchlorate	CH$_2$Cl$_2$	243(4.23),268s(4.13), 328(3.91),405(4.08), 538(4.76)	97-0334-78
$C_{23}H_{26}N_4O_4$ 2-Pyrimidinamine, 1-β-D-arabinofurano-syl-1,4-dihydro-N-(phenylmethyl)-4-[(phenylmethyl)imino]-, hydrochloride	pH 1	227s(4.34),247s(4.28)	94-3244-78
$C_{23}H_{26}N_4O_5S$ Ethanethioic acid, S-[1-[2-(acetoxyeth-yl)-2-[[(4-amino-2-methyl-5-pyrimi-dinyl)methyl](oxophenylacetyl)amino]-1-propenyl] ester	EtOH	243(4.22),262(4.19)	94-0722-78
$C_{23}H_{26}N_4O_8$ 1H-Purine-2,6-dione, 7-[6-O-benzoyl-2,3-O-(1-methylethylidene)-α-D-mannopyranosyl]-3,7-dihydro-1,3-dimethyl-	MeOH	274(3.97)	136-0073-78C
$C_{23}H_{26}N_6O_5$ 1H-Imidazo[1,2-a]purine-7-butanoic acid, 4,9-dihydro-α-[(methoxycarbonyl)ami-no]-4,6-dimethyl-9-oxo-1-(phenyl-methyl)-, methyl ester, (±)-	MeOH	237(4.38),261(3.69), 313(3.51)	44-1644-78
$C_{23}H_{26}O_2S$ 2H-1-Benzothiopyran, 4-(3,4-dimethoxy-phenyl)-3,5,6,7,8,8a-hexahydro-2-	hexane	203(4.6),230s(4.1), 264(3.5),285(3.6),	103-0605-78

Compound	Solvent	$\lambda_{max}(\log \epsilon)$	Ref.
phenyl- (cont.)		289(3.6),295s(3.5), 301s(3.4)	103-0605-78
$C_{23}H_{26}O_3$			
Carda-1,4,6,20(22)-tetraenolide, 3-oxo-	MeOH	298.5(4.11)	44-2334-78
$C_{23}H_{26}O_5$			
Benzofuran, 2-(2,4-dimethoxyphenyl)- 4,6-dimethoxy-5-(3-methyl-2-butenyl)-	EtOH	251s(3.94),288s(4.17), 297s(4.26),310s(4.42), 322(4.60),338(4.54)	44-0135-78
Naringenin, 8-C-prenyltrimethyl-	MeOH	272(4.15),285(4.23), 328(3.81)	78-3563-78
2-Propen-1-one, 1-[2-hydroxy-4,6-dimeth- oxy-3-(3-methyl-2-butenyl)phenyl]- 3-(4-methoxyphenyl)-	MeOH	223(3.82),362(4.18)	78-3563-78
2-Propen-1-one, 1-[6-hydroxy-2,4-dimeth- oxy-3-(3-methyl-2-butenyl)phenyl]- 3-(4-methoxyphenyl)-	MeOH	225(3.96),363(4.06)	78-3563-78
$C_{23}H_{26}O_6$			
2-Butenoic acid, 2-methyl-, (6-acetoxy- 5,6-dihydro-9-methoxy-3,5-dimethyl- naphtho[2,3-c]furan-4-yl)methyl ester	ether	288(3.61)	24-3140-78
$C_{23}H_{26}O_7$			
9,10-Anthracenedione, 1-butyl- 2,4,5,7,8-pentamethoxy-	EtOH	227(4.47),262(4.18), 288(4.16),400(3.86)	44-1435-78
$C_{23}H_{26}O_{11}$			
4H-1-Benzopyran-4-one, 2-(3,4-dimeth- oxyphenyl)-5-(β-D-glucopyranosyl- oxy)-2,3-dihydro-	MeOH	275(4.25),318(3.40)	102-2119-78
$C_{23}H_{26}O_{12}$			
9H-Xanthen-9-one, 2-β-D-glucosyl-1-hy- droxy-3,5,6,7-tetramethoxy-	MeOH	242(4.08),258(4.50), 310(3.92),355(3.77)	102-2119-78
$C_{23}H_{27}BF_4N_2$			
Cyclopentylium, 2,5-bis[4-(dimethylami- no)phenyl]methylene-, tetrafluoro- borate	$C_2H_4Cl_2$	820(5.27)	104-1197-78
2,4,6-Heptatrienylium, 1,7-bis[4-(di- methylamino)phenyl]-, tetrafluoro- borate	$C_2H_4Cl_2$	880(5.53)	104-1197-78
$C_{23}H_{27}Cl_4N_2O_{11}P$			
3'-Uridylic acid, 2'-O-(tetrahydro- 4-methoxy-2H-pyran-4-yl)-, 2-chloro- phenyl 2,2,2-trichloroethyl ester	EtOH	260(4.01)	78-1999-78
$C_{23}H_{27}NO_3$			
17,21-Cyclo-19-norpregna-1,3,5(10)-tri- ene-20-carboxylic acid, 20-cyano- 3-methoxy-, (20S)-	MeOH	219(3.94),229(3.86), 273(3.19),278(3.31), 287(3.28)	24-3094-78
19-Norpregna-1,3,5(10),17(20)-tetraen- 21-oic acid, 20-cyano-3-methoxy-, methyl ester, (17E)-	MeOH	232(4.29),277(3.39), 287(3.29)	24-3094-78
	MeOH	239(4.46)	24-1533-78
2-Propenamide, N-[2-methoxy-2-[4-[(3- methyl-2-butenyl)oxy]phenyl]ethyl]- 3-phenyl-, (E)-	MeOH	217(4.31),224(4.35), 276(4.38)	102-1814-78

$C_{23}H_{27}NO_6$–$C_{23}H_{28}N_2O$

Compound	Solvent	$\lambda_{max}(\log \epsilon)$	Ref.
$C_{23}H_{27}NO_6$			
Solidaline	MeOH	233(4.03),281(3.51), 316(3.42)	23-0383-78
	MeOH-HCl	248(4.02),318(3.92), 367(3.42)	23-0383-78
$C_{23}H_{27}N_2$			
Cyclopentylium, 2,5-bis[[4-(dimethyl-amino)phenyl]methylene]-, tetra-fluoroborate	$C_2H_4Cl_2$	820(5.27)	104-1197-78
2,4,6-Heptatrienylium, 1,7-bis[4-(di-methylamino)phenyl]-, tetrafluoro-borate	$C_2H_4Cl_2$	880(5.53)	104-1197-78
$C_{23}H_{27}N_3O_2$			
2,13-Methano-2H-2,6,10-benzotriaza-cyclopentadecine-1,7-dione, 3,4,5,6-8,9,10,11,12,13-decahydro-9-phenyl-	n.s.g.	254s(3.77),265s(3.69)	88-3893-78
2,13-Methano-2H-2,7,11-benzotriaza-cyclopentadecine-1,10(3H,11H)-dione, 4,5,6,7,8,9,12,13-octahydro-8-phenyl-	n.s.g.	253s(3.75),263s(3.65)	88-3893-78
$C_{23}H_{27}N_3O_5$			
1H-Pyrrolo[3,4-b]pyridine-3-carboxylic acid, 6-cyclohexyl-4,5,6,7-tetrahy-dro-2-methyl-4-(3-nitrophenyl)-7-oxo-, ethyl ester	EtOH	220(4.36),258(4.27), 335(3.62)	95-0448-78
$C_{23}H_{27}N_4O_2P$			
2,3,6(1H)-Phosphorintrione, 1-(1,1-di-methylethyl)-4,5-dimethyl-2,6-bis-(phenylhydrazono)-, 1-oxide	EtOH	232(4.15),250(4.15), 370(4.39),455(4.43)	139-0027-78B
	CHCl$_3$	235(4.13),255(4.13), 370(4.36),460(4.41)	139-0027-78B
$C_{23}H_{27}N_9O_6$			
L-Glutamic acid, N-[N-[4-[[(2,4-di-amino-6-pteridinyl)methyl]methyl-amino]benzoyl]-DL-alanyl-	pH 1	245(4.25),309(4.31)	87-1165-78
L-Glutamic acid, N-[N-[4-[[(2,4-di-amino-6-pteridinyl)methyl]methyl-amino]benzoyl]-N-methylglycyl]-	pH 1	246(4.21),290(4.03)	87-1165-78
$C_{23}H_{27}O_8P$			
α-D-arabino-Hexofuranoside, methyl 2,3-di-O-acetyl-5,6-dideoxy-6-(O,O-diphen-ylphosphono)-	MeOH	257(3.02),263(3.10), 268(3.04)	136-0349-78H
$C_{23}H_{28}Br_2N_2$			
2H-Indeno[2,1-c]pyridine-2-propanamine, 1,3-dihydro-N,N-dimethyl-9-phenyl-, dihydrobromide	MeOH	243(4.43),258(4.42), 310(3.71),320(3.61)	87-0340-78
$C_{23}H_{28}ClNO$			
17,21-Cyclo-19-norpregna-1,3,5(10)-tri-ene-20-carbonitrile, 20-(chloromethyl)-3-methoxy-, (20S)-	MeOH	218(3.93),228(3.85), 273(3.18),278(3.30), 287(3.28)	24-3094-78
$C_{23}H_{28}N_2O$			
Phenanthro[1,10-ef]benzimidazol-7(1H)-one, 2,3,3a,4,7a,8,12b,12c-octahydro-3a,12b-dimethyl-10-(1-methylethyl)-	MeOH	265(4.25)	33-2843-78

Compound	Solvent	λ_{max}(log ϵ)	Ref.
$C_{23}H_{28}N_2O_3$			
Aspidospermidine-3-carboxylic acid, 3,4,6,7-tetradehydro-16-methoxy-1-methyl-, methyl ester	EtOH	252(3.91),307(3.74)	23-2560-78
1H-Pyrrolo[3,4-b]pyridine-3-carboxylic acid, 6-cyclohexyl-4,5,6,7-tetrahydro-2-methyl-7-oxo-4-phenyl-, ethyl ester	EtOH	224(4.24),258(4.11), 341(3.66)	95-0448-78
$C_{23}H_{28}N_2O_4$			
Aspidospermidine-3-carboxylic acid, 6,7-didehydro-3,4-epoxy-16-methoxy-1-methyl-, methyl ester	MeOH	258(3.90),312(3.62)	33-1554-78
Catharanthine, 3-acetoxy-3,4-dihydro-, (3R,4R)-	EtOH	227(4.36),277(3.70), 283(3.80),292(3.70)	33-0690-78
10H-3,7-Methanoazacycloundecino[5,4-b]-indole-9,10-dicarboxylic acid, 5-ethyl-1,2,4,7,8,9-hexahydro-, dimethyl ester	EtOH	228(4.37),262(4.11), 268(4.09),283(3.80), 294(3.66)	23-0062-78
$C_{23}H_{28}N_2O_5$			
10H-3,13-Methanooxireno[9,10]azacyclo-undecino[5,4-b]indole-10,11-dicarboxylic acid, 1a-ethyl-1a,2,4,5,11-12,13,13a-octahydro-, dimethyl ester, [1aS-(1aR*,11S*,13R*,13aS*)]-	EtOH	227(4.32),262(4.09), 268(4.08),283s(3.76), 294(3.66)	23-0062-78
DL-Valine, N-[4-[methyl(phenylmethoxy)-carbonyl]amino]benzoyl]-, ethyl ester	EtOH	260(4.26)	87-1162-78
$C_{23}H_{28}N_4O_6$			
1H-Pyrrolo[3,4-b]pyridine-3-carboxylic acid, 4,5,6,7-tetrahydro-2-methyl-6-[2-(4-morpholinyl)ethyl]-4-(3-nitrophenyl)-7-oxo-, ethyl ester	EtOH	218(4.12),259(4.02), 335(3.38)	95-0448-78
$C_{23}H_{28}N_4O_{10}S$			
Theophylline, 7-(4-O-benzoyl-2,3-O-iso-propylidene-6-O-mesyl-α-D-mannopyran-osyl)-	MeOH	274(4.04)	136-0073-78C
$C_{23}H_{28}O$			
1-Hepten-4-one, 6-methyl-3-(1-methyl-ethyl)-1,1-diphenyl-	EtOH	257(4.11),288(3.54), 297(3.51)	35-1791-78
$C_{23}H_{28}O_2S$			
2H-1-Benzothiopyran, 4-(3,4-dimethoxy-phenyl)octahydro-2-phenyl-	hexane	202(4.4),230(3.6), 260s(--),277(3.1), 281(3.1),287(3.0)	103-0605-78
$C_{23}H_{28}O_3$			
Carda-1,4,6-trienolide, 3-oxo-, (14α)-	MeOH	299(4.12)	44-2334-78
Carda-1,4,20(22)-trienolide, 3-oxo-, (14α)-	MeOH	218(4.40)	44-2334-78
Carda-4,6,20(22)-trienolide, 3-oxo-, (14α)-	MeOH	283(4.42)	44-2334-78
2-Dibenzofurancarboxaldehyde, 4,6-bis-(1,1-dimethylethyl)-9-methoxy-8-methyl-	EtOH	226(4.46),248s(4.45), 262(4.55),310s(3.74)	12-0907-78
$C_{23}H_{28}O_4$			
Benzoic acid, 2,4-dimethoxy-3-(3-methyl-2-butenyl)-6-(2-phenylethyl)-,	n.s.g.	212(3.83),260(3.02)	102-2115-78

Compound	Solvent	$\lambda_{max}(\log \epsilon)$	Ref.
methyl ester (cont.)			102-2115-78
2H-1-Benzopyran-2-acetic acid, 3,4-di-hydro-2,5,7,8-tetramethyl-6-(phenyl-methoxy)-, methyl ester, (±)-	EtOH	203(4.78),227(4.09), 283(3.33),289(3.28)	33-0837-78
Carda-4,20(22)-dienolide, 3,11-dioxo-, (14α)-	MeOH	223(4.37)	44-2334-78
Carda-1,4,20(22)-trienolide, 11-hy-droxy-3-oxo-, (11β,14α)-	MeOH	220(4.32),244s(4.19)	44-2334-78
$C_{23}H_{28}O_5$			
Δ¹-Aldosterone, 18-deoxy-, acetate	MeOH	242(4.19)	150-3551-78
Δ¹-Isoaldosterone, 18-deoxy-, acetate	MeOH	242(4.15)	150-3551-78
$C_{23}H_{28}O_6$			
Benzo[3,4]cycloocta[1,2-f][1,3]benzodi-oxole, 5,6,7,8-tetrahydro-1,2,3,13-tetramethoxy-6,7-dimethyl- (gomisin N)	EtOH	217(4.73),251s(4.14), 278s(3.61)	94-3257-78
Estra-1,3,5(10)-trien-6-one, 2,17β-di-acetoxy-3-methoxy-	EtOH	224(4.35),259(3.99), 315(3.60)	94-3567-78
Estra-1,3,5(10)-trien-6-one, 3,17β-di-acetoxy-2-methoxy-	EtOH	227(4.38),273(4.20), 299s(3.95)	94-3567-78
$C_{23}H_{28}O_7$			
Megaphyllone acetate	EtOH	275(3.10),284s(3.01)	44-0586-78
$C_{23}H_{28}O_8$			
Benzo[3,4]cycloocta[1,2-f][1,3]benzo-dioxole-5,6-diol, 5,6,7,8-tetrahydro-1,2,3,13-tetramethoxy-6,7-dimethyl-	EtOH	220(4.59),254s(4.02), 283s(3.49),294s(3.37)	94-3257-78
Dibenzo[a,c]cyclooctene-1,4-dione, 5,6,7,8-tetrahydro-7-hydroxy-2,3,10,11,12-pentamethoxy-6,7-dimethyl-	EtOH	216(4.59),241s(4.16), 274(4.01),365(3.44)	94-0328-78
$C_{23}H_{28}O_9$			
Melcanthin A	MeOH	226(3.66)	44-4984-78
$C_{23}H_{28}O_{10}$			
Melcanthin B	MeOH	222(4.26)	44-4984-78
$C_{23}H_{29}BrO_4$			
Pregna-14,16-dien-20-one, 3β-acetoxy-5α-bromo-6β,19-epoxy-	MeOH	206(3.81),306(4.11)	44-3946-78
$C_{23}H_{29}NO$			
1H-Indene-5-acetamide, 2,3-dihydro-α-methyl-2-(1-methylethyl)-N-(1-phenylethyl)-	n.s.g.	263(3.09),270(3.21), 277(3.27)	94-1486-78
isomer	n.s.g.	263(3.06),271(3.18), 277(3.24)	94-1486-78
$C_{23}H_{29}NO_3$			
2H-1-Benzopyran-2-amine, 7-methoxy-3-(4-methoxyphenyl)-N,N-bis(1-methylethyl)-	EtOH	245(3.58),328(3.71)	39-0088-78C
$C_{23}H_{29}NO_4$			
1-Phenanthreneethanamine, N-ethyl-3,4,6,7-tetramethoxy-N-methyl-, hydriodide	MeOH	217(--),263(5.02), 280s(4.57),308(4.26), 320(4.26),344(3.51),	12-0313-78

Compound	Solvent	$\lambda_{max}(\log \epsilon)$	Ref.
(cont.)		362(3.36)	12-0313-78
$C_{23}H_{29}NO_6$			
1,3-Benzodioxole-5-ethanamine, 6-(3,4-dihydro-1,7,8-trimethoxy-1H-2-benzopyran-3-yl)-N,N-dimethyl-	EtOH	231(3.81),287(3.57)	78-0635-78
$C_{23}H_{29}NO_8$			
5,6,7,8-Indolizinetetracarboxylic acid, 8a-ethenyl-1,2,3,8a-tetrahydro-3,3-dimethyl-1-(1-methylethylidene)-, tetramethyl ester	MeOH	203(4.20),226(4.14), 282(3.73),359(3.78)	83-0977-78
$C_{23}H_{29}N_3O_2$			
Aspidospermidine-3-carboxamide, 3,4,6,7-tetradehydro-16-methoxy-N,1-dimethyl-, $(2\beta,5\alpha,12\beta,19\alpha)$-	EtOH	250(3.95),305(3.73)	23-2560-78
Butanamide, N-(4-butylphenyl)-3-oxo-2-[(2,4,6-trimethylphenyl)hydrazono]-	EtOH	208(4.18),249(4.16), 263(4.18),384(4.40)	104-0521-78
Celabenzene	n.s.g.	257s(3.15),263s(2.98)	88-3893-78
$C_{23}H_{29}N_3O_3$			
1H-Pyrrolo[3,4-b]pyridine-3-carboxylic acid, 4,5,6,7-tetrahydro-2-methyl-7-oxo-4-phenyl-6-(2-pyrrolidinoethyl)-, ethyl ester	EtOH	224(4.23),259(4.10), 342(3.63)	95-0448-78
1,16-Cyclo-3,4-secocorynan-17-oic acid, 4-cyano-3-ethoxy-, methyl ester	MeOH	229(4.49),275(3.85), 287(3.77)	95-0950-78
$C_{23}H_{29}N_3O_4$			
1H-Pyrrolo[3,4-b]pyridine-3-carboxylic acid, 4,5,6,7-tetrahydro-2-methyl-6-(2-morpholinoethyl)-7-oxo-4-phenyl-, ethyl ester	EtOH	224(4.23),259(4.11), 342(3.65)	95-0448-78
$C_{23}H_{29}N_3S$			
6H-1,3,4-Thiadiazin-2-amine, 5,6-diphenyl-N-(1,1,3,3-tetramethylbutyl)-	isoPrOH	259(4.11),334(3.99)	73-1227-78
3(2H)-Thiazolamine, 4,5-diphenyl-2-[(1,1,3,3-tetramethylbutyl)imino]-	isoPrOH	229(4.17),282(3.91), 342(3.92)	73-1227-78
$C_{23}H_{29}N_5O_4S$			
Benzeneacetamide, N-[(4-amino-2-methyl-5-pyrimidinyl)methyl]-N-(4-hydroxy-1-methyl-2-(morpholinothio)-1-buten-yl-α-oxo-	EtOH	239(4.24),252s(4.19)	94-0722-78
$C_{23}H_{30}ClN_3O_3$			
1H-Azecino[5,4-b]indole-6-acetic acid, α-chloro-3-cyano-8-ethoxy-5-ethyl-2,3,4,5,6,7,8,9-octahydro-, methyl ester	EtOH	225(4.47),284(3.81), 293(3.74)	95-0950-78
$C_{23}H_{30}NO_5$			
Isoquinolinium, 1,2,3,4-tetrahydro-1-[[2-(hydroxymethyl)-4,5-dimethoxyphenyl]methylene]-6,7-dimethoxy-2,2-dimethyl-, iodide	EtOH	218(4.68),325(4.26)	142-1233-78A
$C_{23}H_{30}N_2O$			
Methanone, 1H-imidazol-2-yl[1,2,3,4,4a-	MeOH	277(4.21)	33-2843-78

Compound	Solvent	λ_{max} (log ϵ)	Ref.
9,10,10a-octahydro-1,4a-dimethyl-7-(1-methylethyl)-1-phenanthrenyl]-, [1R-(1α,4aβ,10aα)]- (cont.)			33-2843-78
Methanone, 1H-imidazol-2-yl[4b,5,6,7,8-8a,9,10-octahydro-4b,8-dimethyl-2-(1-methylethyl)-9-phenanthrenyl]-	MeOH	277(4.23)	33-2843-78
Methanone, 1H-imidazol-4-yl[4b,5,6,7,8-8a,9,10-octahydro-4b,8-dimethyl-2-(1-methylethyl)-9-phenanthrenyl]-	MeOH	270(4.23)	33-2843-78
$C_{23}H_{30}N_2O_3$			
2,3-Anhydro-4-deacetoxy-6,7-dihydro-vindoline	EtOH	205(4.78),240(4.30), 334(4.22)	35-4220-78
3.4-Anhydro-4-deacetoxy-6,7-dihydro-vindoline	EtOH	212(4.53),253(3.82), 307(3.62)	35-4220-78
Aspidospermidine-3-carboxylic acid, 6,7-didehydro-16-methoxy-1-methyl-, methyl ester	EtOH	252(3.71),304(3.55)	23-2560-78
$C_{23}H_{30}N_2O_3S$			
Eburnamenine-14-carboxylic acid, 14,15-dihydro-14-[(2-hydroxyethyl)thio]-, methyl ester, (3α,14α,16α)- (hydrate)	MeOH	227(4.42),276(3.94)	33-1682-78
(3α,14β,16α)-	MeOH	225(4.43),275(3.94)	33-1682-78
$C_{23}H_{30}N_2O_4$			
Aspidospermidine-3-carboxylic acid, 3-hydroxy-16-methoxy-1-methyl-, methyl ester	EtOH	252(3.83),306(3.77)	23-2560-78
Aspidospermidine-3-carboxylic acid, 4-hydroxy-16-methoxy-1-methyl-, methyl ester	MeOH	251(3.84),302(3.66)	33-1554-78
Aspidospermidine-3-carboxylic acid, 16-methoxy-1-methyl-4-oxo-, methyl ester	EtOH	212(4.39),252(3.88), 304(3.59)	35-4220-78
Isoquinoline, 1,2,3,4-tetrahydro-6,7-dimethoxy-2-methyl-8-[(1,2,3,4-tetra-hydro-6-methoxy-2-methyl-7-isoquino-linyl)oxy]-	EtOH	208(4.48),229(4.20), 255(4.14),287(3.75)	12-0321-78
$C_{23}H_{30}N_2O_5$			
Aspidospermidine-3-carboxylic acid, 3-hydroxy-16-methoxy-1-methyl-4-oxo-, methyl ester, (2β,3β,5α,12β,19α)-	EtOH	213(4.49),248(3.81), 303(3.67)	33-1554-78 +35-4220-78
1H-Pyrrole-3-carboxylic acid, 5-[(acet-ylphenylamino)methyl]-1-cyclohexyl-4,5-dihydro-5-hydroxy-2-methyl-4-oxo-, ethyl ester (hydrate)	EtOH	209(3.90),248(4.14), 332(3.85)	4-1215-78
$C_{23}H_{30}N_4O_4$			
Cyclic(glycyl-D-prolyl-L-leucyl-α,β-di-dehydro-N-methylphenylalanyl)	MeOH	282(4.28)	44-0296-78
L-prolyl isomer	MeOH	282(4.28)	44-0296-78
$C_{23}H_{30}N_6O_8$			
Adenosine, N-[(cyclohexylamino)carbo-nyl]-, 2',3',5'-triacetate	EtOH	270(4.34),276(4.27)	39-0131-78C
$C_{23}H_{30}O$			
9H-Xanthene, 4,5-bis(1,1-dimethyleth-yl)-1,8-dimethyl-	EtOH	255(3.94),280(3.70)	32-0079-78

Compound	Solvent	$\lambda_{max}(\log \epsilon)$	Ref.
$C_{23}H_{30}O_2$			
Dibenzofuran, 4,6-bis(1,1-dimethyleth-yl)-1-methoxy-2,8-dimethyl-	EtOH	229(4.63),252(4.01), 260(4.20),276s(3.99), 284(4.26),302(3.75), 314(3.81)	12-0907-78
$C_{23}H_{30}O_3$			
Carda-1,4-dienolide, 3-oxo-, (14α)-	MeOH	243(4.20)	44-2334-78
Carda-4,6-dienolide, 3-oxo-, (14α)-	MeOH	283(4.42)	44-2334-78
Carda-4,20(22)-dienolide, 3-oxo-, (14α)-	MeOH	224(4.34)	44-2334-78
Carda-8,14,20(22)-trienolide, 3-hy-droxy-	EtOH	217(4.51),244(4.43)	94-3023-78
2,4,6,8-Nonatetraenoic acid, 9-(4-meth-oxy-2,3,6-trimethylphenyl)-3,7-di-methyl-, (Z,E,E,E)-	EtOH	244(4.04),362(4.59)	33-2697-78
(Z,Z,E,E)-	EtOH	250(4.08),356(4.48)	33-2697-78
(Z,Z,Z,E)-	EtOH	249(4.08),348(4.42)	33-2697-78
1-Nonen-6-ol, 6,7-bis(4-methoxyphenyl)-	MeOH	227(4.34),275(3.68), 282(3.56)	5-0726-78
19-Norpregna-1,3,5(10),17(20)-tetraen-21-oic acid, 3-methoxy-, ethyl ester	MeOH	222(4.45),273(3.21), 278(3.32),287(3.29)	24-3094-78
18-Nor-5α-pregna-8,11,13-trien-20-one, 3β-acetoxy-12-methyl-	EtOH	206(4.54),269(2.96)	39-0076-78C
$C_{23}H_{30}O_4$			
2(3H)-Benzofuranone, 7-(1,1-dimethyl-ethyl)-3-[5-(1,1-dimethylethyl)-5-methoxy-3-methyl-2(5H)-furanylidene]-5-methyl-	EtOH	225s(3.94),258(4.03), 269(4.07),351(4.13)	12-0907-78
Benzoic acid, 4-hydroxy-3-methoxy-, 1,3a,4,5,6,7-hexahydro-3,8-dimethyl-5-(1-methylethyl)-6-azulenyl ester (microferinin)	EtOH	266(4.02),296(3.23)	105-0487-78
Card-4-enolide, 3,11-dioxo-, (14α)-	MeOH	237(4.16)	44-2334-78
Etiojerva-5,12-diene-3,11-dione, 17,20-epoxy-17α-ethyl-, 3-ethylene acetal	MeOH	251(4.15)	18-0243-78
17,20-epimer	MeOH	261(4.04)	18-0243-78
Etiojerva-4,12-diene-3,20-dione, 17α-ethyl-11α-hydroxy-, acetate	MeOH	236(4.00)	18-0234-78
11β-	MeOH	237(4.00)	18-0234-78
19-Norpregna-1,3,5(10)-trien-21-oic acid, 3-methoxy-20-oxo-, ethyl ester	MeOH	219(4.02),229(3.89), 273(3.20),278(3.30), 287(3.27)	24-1533-78
Pregna-5,14,16-trien-20-one, 3β-acet-oxy-19-hydroxy-	MeOH	207(3.84),310(4.05)	44-3946-78
$C_{23}H_{30}O_5$			
Carda-4,20(22)-dienolide, 14β,19-di-hydroxy-3-oxo-	MeOH	222(4.36)	44-3946-78
Etiojerv-4-ene-3,20-dione, 11α-acetoxy-12α,13α-epoxy-17α-ethyl-	EtOH	238(4.00)	18-0234-78
12β,13β-epoxy-	EtOH	237(4.00)	18-0234-78
Etiojerv-4-ene-3,20-dione, 11β-acetoxy-12α,13α-epoxy-17α-ethyl-	EtOH	237(4.16)	18-0234-78
12β,13β-epoxy-	EtOH	237(4.08)	18-0234-78
$C_{23}H_{30}O_6$			
Dibenzo[a,c]cycloocten-6-ol, 5,6,7,8-tetrahydro-1,2,3,10,11-pentamethoxy-6,7-dimethyl-	EtOH	213(4.63),254(4.14), 282(3.76),293s(3.67)	94-0328-78

$C_{23}H_{30}O_7-C_{23}H_{32}N_2O_3$

Compound	Solvent	$\lambda_{max}(\log \epsilon)$	Ref.
$C_{23}H_{30}O_7$			
Gomisin H	EtOH	219(4.81),248s(4.32), 276(3.72),285s(3.65)	94-0328-78
Talassin A	EtOH	224(4.56),250(4.49)	105-0377-78
$C_{23}H_{30}O_8$			
1,1,2-Cyclopropanetricarboxylic acid, 3-[4-(pentyloxy)benzoyl]-, 1,1-di-ethyl 2-methyl ester	EtOH	223(3.98),282(4.28)	73-1727-78
$C_{23}H_{31}ClO_2$			
3'H-Cycloprop[1,2]androsta-1,4,6-trien-3-one, 6-chloro-1,2-dihydro-17-hy-droxy-16-(1-methylethyl)-, (1β,2β,16β,17β)-	EtOH	283(4.26)	94-1718-78
$C_{23}H_{31}NO$			
Acetamide, N-(1,2,3,4,4a,9,10,11,12,12a-decahydro-4a,7,8-trimethyl-2-chrysen-yl)-, [2S-(2α,4aα,12aα)]-	EtOH	275(4.10)	39-0163-78C
Androsta-1,3,5-trien-17-one, 3-pyrroli-dino-	EtOH-pyrro-lidine	283s(4.24),290(4.25), 299s(4.22)	22-0033-78
$C_{23}H_{31}NO_4$			
3,5-Pyridinedicarboxylic acid, 1-butyl-1,4-dihydro-2,6-dimethyl-4-phenyl-, diethyl ester	EtOH	205(4.19),243(4.17), 350(3.78)	103-1226-78
3,5-Pyridinedicarboxylic acid, 1,4-di-hydro-2,6-dimethyl-1-(2-methylprop-yl)-4-phenyl-, diethyl ester	EtOH	205(4.33),240(4.19), 344(3.76)	103-1226-78
$C_{23}H_{31}NO_7$			
1,1,3-Cyclohexanetricarboxylic acid, 4-[(benzoylamino)methyl]-, triethyl ester	n.s.g.	231(3.88),248s(3.75)	128-0097-78
$C_{23}H_{31}N_3O_2S$			
Eburnamenine-14-carboxylic acid, 14-[(2-aminoethyl)thio]-14,15-dihydro-, methyl ester	MeOH	225(4.38),276(3.90)	33-1682-78
$C_{23}H_{31}N_3O_3$			
1H-Pyrrolo[3,4-b]pyridine-3-carboxylic acid, 6-[2-[(1,1-dimethylethyl)ami-no]ethyl]-4,5,6,7-tetrahydro-2-meth-yl-7-oxo-4-phenyl-, ethyl ester	EtOH	224(4.23),259(4.11), 341(3.65)	95-0448-78
$C_{23}H_{32}N_2O$			
Eburnamenine, 14-butoxy-14,15-dihydro-(O-butylepivincanol)	MeOH	228(4.53),282(3.94)	33-1682-78
$C_{23}H_{32}N_2O_3$			
Furo[3,4-b]pyridine-3-carboxylic acid, 4-[4-(dimethylamino)phenyl]-1,4,5,7-tetrahydro-2,5,5,7,7-pentamethyl-, ethyl ester	EtOH	203(3.54),249(4.13), 320(3.90)	95-0448-78
Vindoline, 4-deacetoxy-3(R)-deoxy-6,7-dihydro-	EtOH	212(4.42),256(3.80), 307(3.65)	35-4220-78
3(S)-	EtOH	212(4.35),253(3.69), 305(3.53)	35-4220-78

Compound	Solvent	$\lambda_{max}(\log \epsilon)$	Ref.
$C_{23}H_{32}N_2O_4$			
Aspidospermidine-3-carboxylic acid, 3-hydroxy-16-methoxy-1-methyl-, methyl ester, (2β,5α,12β,19α)-	EtOH	252(3.73),305(3.64)	23-2560-78
$C_{23}H_{32}N_2O_5$			
Aspidospermidine-3-carboxylic acid, 3,4-dihydroxy-16-methoxy-1-methyl-, methyl ester	EtOH	213(4.46),254(3.76), 305(3.63)	33-1554-78
$C_{23}H_{32}N_2S$			
Eburnamenine, 14-(butylthio)-14,15-di-hydro-	MeOH	198(4.42),230(4.47), 283(3.92)	33-1682-78
$C_{23}H_{32}N_4O_7$			
L-Leucine, N-[N-[N-[[(1-carboxy-2-phen-ylethenyl)methylamino]carbonyl]gly-cyl]-N-methyl-L-alanyl]-, (Z)-	MeOH	277.5(4.16)	44-0296-78
$C_{23}H_{32}O_3$			
Card-4-enolide, 3-oxo-, (14α)-	MeOH	243(4.20)	44-2334-78
Fexerinin	EtOH	265(3.99),297(3.71)	105-0495-78
$C_{23}H_{32}O_4$			
Acetic acid, [(2-acetyl-4b,5,6,7,8,8a-9,10-octahydro-4b,8,8-trimethyl-3-phenanthrenyl)oxy]-, ethyl ester, (4bS-trans)-	EtOH	255(4.00),312(3.54)	23-0517-78
Card-20(22)-enolide, 8,14-epoxy-3-hy-droxy-	n.s.g.	215(4.29)	94-3023-78
Etiojerva-5,12-diene-3,11-dione, 17α-ethyl-20-hydroxy-, 3-ethylene acetal, (20R)-	MeOH	255(4.15)	18-0243-78
Etiojerva-5,12-diene-3,11-dione, 17β-ethyl-20-hydroxy-, 3-ethylene acetal, (20S)-	MeOH	258(4.00)	18-0243-78
20-epimer	MeOH	255(4.20)	18-0243-78
Isospongiaquinone methyl ether	MeOH	288(4.11),400(2.64)	12-2685-78
Pregn-4-ene-17-carboxylic acid, 3,20-dioxo-, methyl ester	EtOH	240(4.2)	23-0410-78
$C_{23}H_{32}O_5$			
1,4-Phenanthrenedione, 3-(acetoxymeth-oxy)-4b,5,6,7,8,8a,9,10-octahydro-4b,8,8-trimethyl-2-(1-methylethyl)-, (4bS-trans)-	EtOH	273(4.15),363(2.52)	23-0733-78
Xeroferin	EtOH	267(4.03),297(3.97)	105-0499-78
$C_{23}H_{32}O_6$			
Rubaferinin	EtOH	265(4.01),295(3.79)	105-0606-78
Strophanthidin	MeOH	217(4.19)	44-3946-78
$C_{23}H_{33}NO_2$			
17,21-Cyclo-19-norpregna-1,3,5(10)-tri-ene-20-methanol, 20-(aminomethyl)-3-methoxy-, (20S)-	MeOH	218(3.94),229(3.84), 273(3.20),278(3.30), 287(3.27)	24-3094-78
$C_{23}H_{34}O_3$			
Pregn-4-ene-3,20-dione, 17-ethoxy-	MeOH	241(4.22)	24-3086-78

Compound	Solvent	$\lambda_{max}(\log \epsilon)$	Ref.
$C_{23}H_{34}O_4$			
Cyafrin A acetonide	MeOH	236(3.80)	23-2113-78
5α-Pregn-16-en-20-one, 3β-acetoxy-11β-hydroxy-	EtOH	240(3.94)	39-1594-78C
5β-Pregn-16-en-20-one, 3α-acetoxy-11β-hydroxy-	EtOH	240(3.93)	39-1594-78C
$C_{23}H_{34}O_5$			
Prostaglandin A₂, 15-acetate, methyl ester, (+)-	MeOH	217(4.00)	44-4377-78
$C_{23}H_{34}O_6$			
Benzoic acid, 4-hydroxy-3-methoxy-, 4,9-dihydroxy-2,5,5,9-tetramethyl-2-cycloundecen-1-yl ester (fexeridin)	EtOH	267(4.04),297(3.80)	105-0495-78
$C_{23}H_{35}NO_3$			
2H-Pyrrol-2-one, 4-[(3β,5β,14β,17β)-3,14-dihydroxyandrostan-17-yl]-1,5-dihydro-	EtOH	214(4.23)	33-0977-78
$C_{23}H_{36}N_2OS_4$			
Urea, bis[2,3-bis[(1,1-dimethylethyl)-thio]-2-cyclopropen-1-ylidene]-	EtOH	327(4.53)	18-3653-78
$C_{23}H_{36}N_2O_2$			
Tryptophan, 2,5,7-tris(1,1-dimethyl-ethyl)-	50% EtOH-HCl	274(3.90),283s(--),293s(--)	63-1637-78
	isoPrOH	226(4.58),273(3.90),282(3.87),292s(--)	63-1637-78
$C_{23}H_{36}O_2$			
D-Homoestr-4-en-3-one, 17a-(1,1-dimeth-ylethoxy)-, (9β,10α,17aβ)-	n.s.g.	242(4.23)	33-2397-78
(17aβ)-	n.s.g.	241(4.23)	33-2397-78
$C_{23}H_{36}O_3$			
2(3H)-Phenanthrenone, 8-(1,1-dimethyl-ethoxy)-4,4a,4b,5,6,7,8,8a,9,10-deca-hydro-8a-methyl-1-(3-oxobutyl)-, [4aS-(4aα,4bβ,8α,8aα)]-	n.s.g.	251(4.18)	33-2397-78
Pregn-20-ene-20-carboxaldehyde, 3-hy-droxy-21-methoxy-, (3β,5α,20Z)-	EtOH	253(4.2)	23-0424-78
$C_{23}H_{36}O_4S_2$			
1-Penten-3-one, 1-[6-[4-(1,3-dithiolan-2-yl)butyl]-1,4-dioxaspiro[4.4]non-7-yl]-5-ethoxy-4,4-dimethyl-, [6R-[6α,7β(E)]]-	EtOH	234(4.08)	5-1739-78
$C_{23}H_{39}N_3O_6$			
Tetradecanamide, N-(1-β-D-arabino-furanosyl-1,2-dihydro-2-oxo-4-pyrimidinyl)-	isoPrOH	216(4.17),248(4.16),303(3.90)	94-0981-78

Compound	Solvent	$\lambda_{max}(\log \epsilon)$	Ref.
$C_{24}H_8Cl_6O_7$ Dinaphtho[2,1-b:1',2'-d]furan-5,9-dione, 4,10-diacetoxy-1,3,6,8,11,13-hexachloro-	$CHCl_3$	360(3.84),440s(4.22), 470(4.30),498s(4.23)	12-1323-78
$C_{24}H_9Cl_2NO_4$ 1(2H)-Acenaphthylenone, 5,6-dichloro-2-(6-nitro-2-oxo-1(2H)-acenaphthylenylidene)-	$C_6H_5NO_2$	418(0.78),492(0.62)	104-1971-78
$C_{24}H_9I_2NO_4$ 1(2H)-Acenaphthylenone, 5,6-diiodo-2-(6-nitro-2-oxo-1(2H)-acenaphthylenylidene)-	$C_6H_5NO_2$	420(2.60),510(1.70)	104-1971-78
$C_{24}H_9N_3O_8$ 1(2H)-Acenaphthylenone, 5,6-dinitro-2-(6-nitro-2-oxo-1(2H)-acenaphthylenylidene)-	$C_6H_5NO_2$	426(3.20),460(2.31)	104-1971-78
$C_{24}H_{10}N_2O_6$ 1(2H)-Acenaphthylenone, 5-nitro-2-(6-nitro-2-oxo-1(2H)-acenaphthylenylidene)-	$C_6H_5NO_2$	418(0.46),460(0.30), 522(0.24)	104-1971-78
$C_{24}H_{11}NO_4$ 1(2H)-Acenaphthylenone, 5-nitro-2-(2-oxo-1(2H)-acenaphthylenylidene)-	$C_6H_5NO_2$	422(1.68),490(1.60)	104-1971-78
$C_{24}H_{12}S_6$ Cyclododeca[1,2-b:4,3-b':5,6-b":8,7-b"':9,10-b"":12,11-b""']hexathiophene	$CHCl_3$	252(4.42)	24-1330-78
$C_{24}H_{14}BrN_3O_2$ Benzo[f]quinoline, 3-(4-bromophenyl)-8-nitro-1-(3-pyridinyl)-	EtOH	219(4.42),267(4.57), 286(4.65),351(3.87), 369(3.87)	103-1337-78
$C_{24}H_{14}Br_2N_2$ Benzo[f]quinazoline, 1,3-bis(4-bromophenyl)-	EtOH	207(4.53),221(4.58), 267(4.51),291(4.74)	103-0082-78
$C_{24}H_{14}Cl_2N_2$ Benzo[f]quinazoline, 1,3-bis(4-chlorophenyl)-	EtOH	205(4.48),221(4.54), 265(4.43),288(4.65)	103-0082-78
$C_{24}H_{14}Cl_2S$ 3H-Naphtho[1,8-bc]thiophene, 6,8-dichloro-3-(diphenylmethylene)-	MeOH	220s(4.4),253(4.3), 266s(4.3),305(3.9), 322(3.8),337(3.9), 400(3.9),420s(3.9)	5-1974-78
$C_{24}H_{14}N_2O_2$ Naphtho[1',2',3':4,5]quino[2,1-b]quinazoline-5,10-dione, 8-methyl-	n.s.g.	445(3.78)	2-0103-78
Naphtho[1',2',3':4,5]quino[2,1-b]quinazoline-5,10-dione, 16-methyl-	n.s.g.	430(3.85)	2-0103-78
$C_{24}H_{14}N_2O_3$ Naphtho[1',2',3':4,5]quino[2,1-b]quin-	n.s.g.	447(3.61)	2-0103-78

Compound	Solvent	$\lambda_{max}(\log \epsilon)$	Ref.
azoline-5,10-dione, 6-methoxy- (cont.)			2-0103-78

$C_{24}H_{14}O_3$
  Naphtho[2,3-c]furan-5,8-dione, 1,3-di-
    phenyl-
	CH_2Cl_2	280(4.72),480(4.60),
		547s(3.8)
			150-5538-78

$C_{24}H_{14}O_5$
  Naphtho[2,3-c]furan-4,9-dione, 5,8-di-
    hydroxy-1,3-diphenyl-
	CH_2Cl_2	361(3.72),474(4.00)	150-5538-78

$C_{24}H_{14}S_2$
  1,3-Benzodithiole, 2-(7H-benz[de]anthra-
    cen-7-ylidene)-
	CH_2Cl_2	248s(4.66),272s(4.38),
		318(4.40),424(4.49)
			18-2674-78

$C_{24}H_{15}BrS$
  5H-Naphtho[1,8-bc]thiophene, 2-bromo-
    5-(diphenylmethylene)-
	C_6H_{12}	327(4.021),378s(--),
		379(4.308)
			24-1824-78

$C_{24}H_{15}ClN_2S$
  Quinoline, 6-chloro-4-phenyl-2-(2-phen-
    yl-4-thiazolyl)-
	dioxan	352(3.34)	97-0400-78

$C_{24}H_{15}NO$
  2H-Cyclopenta[b]quinolin-2-one, 1,3-di-
    phenyl-
	EtOH	269(6.5),477(5.9),
		552(5.6)
			44-4128-78

$C_{24}H_{15}NO_2$
  5(4H)-Isoxazolone, 4-(9-anthracenyl-
    methylene)-3-phenyl-
	toluene	460(4.08)	103-0120-78

$C_{24}H_{15}NO_2S$
  Methanone, pyrrolo[2,1-b]benzothiazole-
    2,3-diylbis[phenyl-
	MeOH	205(4.62),252(4.56),
		285(4.34),335(4.19)
			44-2697-78

$C_{24}H_{15}NO_3$
  Cyclopenta[b][1]benzopyran, 3-nitro-
    1,2-diphenyl-
	THF	265(4.79),345(4.59),
		460(2.84)
			103-1075-78

$C_{24}H_{15}N_3O_2$
  Benzo[f]quinoline, 8-nitro-3-phenyl-
    1-(3-pyridinyl)-
	EtOH	213(4.52),261(4.61),
		282(4.66),350(3.81),
		366(3.87)
			103-1337-78

$C_{24}H_{16}$
  Phenanthrene, 9-(4-phenyl-1-buten-
    3-ynyl)-, cis
	MeOH	222(4.47),245s(4.63),
		253(4.75),268(4.49),
		284(4.46),329(4.10)
			54-0197-78

trans	MeOH	221(4.50),223s(4.49), 247(4.54),254(4.60), 272(4.48),284(4.49), 333(4.43)	54-0197-78

$C_{24}H_{16}BrN_3$
  Benzo[f]quinolin-8-amine, 3-(4-bromo-
    phenyl)-1-(3-pyridinyl)-
	EtOH	276(4.65),347(4.24),
		384(4.08)
			103-1337-78

$C_{24}H_{16}Br_2N_2$
  Benzo[f]quinazoline, 1,3-bis(4-bromo-
    phenyl)dihydro-
	EtOH	210(4.60),228(4.60),
		278(4.25),290(4.11),
		358(3.90)
			103-0082-78

Compound	Solvent	$\lambda_{max}(\log \epsilon)$	Ref.
$C_{24}H_{16}ClNO_2S$ 2H,5H-[1]Benzothiopyrano[4,3-b]pyran-2-one, 3-chloro-4-(diphenylamino)-	EtOH	245(4.44),272(4.29), 307s(3.85),345(4.04), 390(4.13)	4-0181-78
$C_{24}H_{16}ClNS$ Quinoline, 6-chloro-2-(3-methylbenzo-[b]thien-2-yl)-4-phenyl-	dioxan	355(3.75)	97-0400-78
$C_{24}H_{16}ClN_3$ [1,2,4]Triazolo[1,5-a]pyridine, 2-[3-chloro-4-[2-(1-naphthalenyl)ethenyl]-phenyl]-	DMF	348(4.53)	33-0142-78
[1,2,4]Triazolo[1,5-a]pyridine, 2-[3-chloro-4-[2-(2-naphthalenyl)ethenyl]-phenyl]-	DMF	345(4.69)	33-0142-78
$C_{24}H_{16}ClN_3O$ 4H-Imidazo[1,5-a][1,4]benzodiazepin-4-one, 8-chloro-1-methyl-3,6-diphenyl-	isoPrOH	233(4.54),277(4.26), 352(3.72)	4-0855-78
$C_{24}H_{16}Cl_2N_2$ Benzo[f]quinazoline, 1,3-bis(4-chloro-phenyl)dihydro-	EtOH	208(4.58),234(4.58), 278(4.18),290(4.08), 353(3.92)	103-0082-78
$C_{24}H_{16}F_6$ Phenanthrene, 9,10-dihydro-9-methyl-10-methylene-3,6-bis(trifluoromethyl)-	hexane	221(4.44),225(4.44), 244(4.42),281(4.02)	104-0341-78
	H_2SO_4	255s(--),262(4.24), 305s(--),332s(--), 388(2.96),513s(--), 552(3.10)	104-0341-78
$C_{24}H_{16}N_2$ Benzo[f]quinazoline, 1,3-diphenyl-	EtOH	216(4.59),261(4.53), 284(4.72)	103-0082-78
5,5'-Bi-5H-indeno[1,2-b]pyridine	MeOH	254(3.54),283(3.55), 308(3.77),325(3.47)	142-0849-78A
$C_{24}H_{16}N_2OS$ Benzoic acid, (2-phenyl-5H-naphtho-[1,8-bc]thien-5-ylidene)hydrazide	MeOH	422(4.38),490(4.26)	83-0324-78
$C_{24}H_{16}N_2O_2$ Phenol, 2,2'-benzo[f]quinazoline-1,3-diylbis-	EtOH	208(4.56),222(4.58), 286(4.60),333(4.24)	103-0082-78
Phenol, 4,4'-benzo[f]quinazoline-1,3-diylbis-	EtOH	206(4.48),228(4.55), 303(4.67)	103-0082-78
$C_{24}H_{16}N_3P$ 1-Propene-1,1,2-tricarbonitrile, 3-(triphenylphosphoranylidene)-	CH_2Cl_2	417(4.42)	39-1237-78C
$C_{24}H_{16}N_4O_2$ 11H,18H-Cyclobuta[1",2":3,4;3",4"-5',6']dipyrido[2,1-b][2',1'-b']-diquinazoline-11,18-dione	MeCN	223(4.591),278(4.190), 289(4.223),302(4.201), 322(4.170),337(4.009)	104-2218-78
	CH_2Cl_2	281(4.228),292(4.267), 305(4.250),324(4.220), 338(4.099)	104-2218-78

Compound	Solvent	$\lambda_{max}(\log \epsilon)$	Ref.
4(3H)-Quinazolinone, 3-[(1,5-dihydro-5-oxo-1-phenyl-2H-pyrrol-2-ylidene)-amino]-2-phenyl-	MeOH	216(4.72),235(4.75), 287(4.45)	80-1085-78
$C_{24}H_{16}N_4O_4$			
2-Naphthalenecarboxylic acid, 4-[(10,11-dihydro-11-oxo-5H-dibenzo[b,e][1,4]-diazepin-8-yl)azo]-1-hydroxy-	DMF	428(4.08)	42-0154-78
2-Naphthalenecarboxylic acid, 4-[(10,11-dihydro-11-oxo-5H-dibenzo[b,e][1,4]-diazepin-8-yl)azo]-3-hydroxy-	DMF	425(4.10)	42-0154-78
$C_{24}H_{16}O$			
Cyclopenta[b][1]benzopyran, 1,2-diphen-yl-	THF THF	512(2.91) 270(4.39),365(4.37), 512(2.91)	103-1070-78 103-1075-78
Cyclopenta[b][1]benzopyran, 2,3-diphen-yl-	EtOH	256(4.59),362(4.43), 510(2.89)	103-1075-78
$C_{24}H_{16}O_2$			
9H-Phenanthro[9',10':2,3]cyclopropa-[1,2-c]pyran-9-one, 8b,8c-dihydro-8c-phenyl-	MeOH	216(4.56),240(4.29), 250(4.18),269(4.03), 278(4.03),282(4.04), 295(3.65),307(3.51)	54-0197-78
$C_{24}H_{16}O_3$			
Naphtho[2,3-c]furan-5,8-dione, 6,7-di-hydro-1,3-diphenyl-	CH_2Cl_2	418(3.56),489(3.55)	150-5538-78
$C_{24}H_{16}O_3S$			
Cyclopenta[b][1]benzopyran, 2-phenyl-1-(phenylsulfonyl)-	THF	250(4.21),280(4.21), 360(4.46),470(3.37)	103-1075-78
$C_{24}H_{16}O_4$			
5H-Furo[3,2-g][1]benzopyran-5-one, 4-hydroxy-2(or 3)-methyl-3,6(or 2,6)-diphenyl-	n.s.g.	263(4.44),310(4.33)	2-1124-78
$C_{24}H_{16}S$			
5H-Naphtho[1,8-bc]thiophene, 5-(diphen-ylmethylene)-	MeOH	266(4.2),384(4.3)	5-1974-78
$C_{24}H_{16}S_4$			
2,2'-Bithiophene, 5,5''-(1,2-cyclopent-anediyldi-2,1-ethynediyl)bis-, cis (cardopatine)	EtOH	242(4.33),340(4.82)	102-2097-78
trans	EtOH	242(4.33),340(4.82)	102-2097-78
$C_{24}H_{17}Cl_2NO_2S$			
2H,5H-[1]Benzothiopyrano[4,3-b]pyran-2-one, 3,3-dichloro-4-(diphenylami-no)-3,4-dihydro-	EtOH	240(4.45),262s(4.27), 290s(4.06),340(3.42)	4-0181-78
$C_{24}H_{17}Cl_2NO_2Sn$			
Stannane, (2,6-dichloro-4-nitrophenyl)-triphenyl-	benzene	312(3.91)	70-1573-78
$C_{24}H_{17}N_3$			
Benzo[f]quinolin-8-amine, 3-phenyl-1-(3-pyridinyl)-	EtOH	269(4.61),346(4.17), 370(4.02)	103-1337-78

Compound	Solvent	λ_{max}(log ϵ)	Ref.
[1,2,4]Triazolo[1,5-a]pyridine, 7-[2-(1-naphthalenyl)ethenyl]-2-phenyl-	DMF	350(4.49)	33-0142-78
[1,2,4]Triazolo[1,5-a]pyridine, 7-[2-(2-naphthalenyl)ethenyl]-2-phenyl-	DMF	338(4.73)	33-0142-78
C$_{24}$H$_{17}$N$_3$O$_2$			
3,5-Pyrazolidinedione, 4-(1H-indol-3-ylmethylene)-1,2-diphenyl-	0.1% NaOH	445(4.740)	103-1088-78
	EtOH	435(4.641)	103-1088-78
	EtOH-NaOEt	445(4.830)	103-1088-78
sodium salt	n.s.g.	442(4.733)	103-1088-78
C$_{24}$H$_{17}$O			
[1,1':3',1"-Terphenyl]-2'-yloxy, 5'-phenyl-	MeCN	535(3.78),750(3.48)	70-1304-78
C$_{24}$H$_{18}$			
Cyclotetradeca[a]naphthalene, 11,12,13-14-tetrahydro-10,15-dimethyl-, (E,E,Z,Z)-	CH$_2$Cl$_2$	250(4.10),333(4.65), 396(3.93),420s(3.60)	88-2719-78
Cyclotetradeca[b]naphthalene, 10,11,12-13-tetrahydro-9,14-dimethyl-, (E,E,Z,Z)-	CH$_2$Cl$_2$	248(4.24),261(4.22), 337(4.66),400(3.65)	88-2719-78
7,10:13,16-Diethenocyclotrideca[de]-naphthalene, 11,12-dihydro-	C$_6$H$_{12}$	301(4.06)	88-1459-78
7,14-Ethanodibenz[a,h]anthracene, 7,14-dihydro-, (7R)-	EtOH	232(4.99),284(4.05), 325(3.40)	18-0265-78
Naphthalene, 2-(2-[1,1'-biphenyl]-2-yl-ethenyl)-, cis	MeOH	245(4.60),268(4.54), 275s(4.51),301(4.43),	39-0915-78B
trans	MeOH	252(4.54),272(4.62), 282(4.61),315(4.57), 340s(4.24)	39-0915-78B
Pyrene, 4,5-dihydro-4-methyl-5-methyl-ene-4-phenyl-	hexane	227(4.69),238s(--), 274(4.44),296s(--)	104-0341-78
	H$_2$SO$_4$	290(4.15),304s(--), 316s(--),456s(--), 506(3.88),636(3.28)	104-0341-78
	CF$_3$COOH	289(4.14),304s(--), 314s(--),456s(--), 507(3.89),643(3.26)	104-0341-78
C$_{24}$H$_{18}$Br$_2$N$_4$O			
2H-Benzimidazol-2-one, 5-[4,5-bis(4-bromophenyl)-1H-imidazol-2-yl]-1,3-dihydro-1,3-dimethyl-	EtOH	213(4.64),320(4.52)	103-0425-78
radical	toluene	556(--),748(--)	103-0425-78
C$_{24}$H$_{18}$ClN$_3$			
4H-Imidazo[1,5-a][1,4]benzodiazepine, 8-chloro-1-methyl-3,6-diphenyl-	isoPrOH	215s(4.66),247(4.49), 270s(4.35),340s(3.68)	4-0855-78
C$_{24}$H$_{18}$Cl$_2$N$_2$O$_6$			
2(5H)-Furanone, 4,4'-(azinodiethyli-dyne)bis[5-(2-chlorophenyl)-3-hy-droxy-, [R*,S*-(E,E)]-	EtOH	250(4.10),318(3.94), 463(4.15)	142-1041-78A
C$_{24}$H$_{18}$N$_2$			
Benzo[f]quinazoline, dihydro-1,3-di-phenyl-	EtOH	204(4.72),239(4.68), 278(4.3),287(4.28), 345(4.22)	103-0082-78
Tribenzo[d,h,l][1,2]diazacyclotetra-decine	DMF	284(4.59),350s(4.30)	33-2813-78

Compound	Solvent	$\lambda_{max}(\log \epsilon)$	Ref.
$C_{24}H_{18}N_2O_2$			
Benzamide, N-(2-oxo-4,6-diphenyl-1(2H)-pyridinyl)-	EtOH	225(4.00),245(4.11), 330(3.53)	120-0001-78
$C_{24}H_{18}N_2O_3$			
Furo[2,3-b]quinoxaline, 4,9-dibenzoyl-3a,4,9,9a-tetrahydro-	CH$_2$Cl$_2$	268(4.11)	5-1129-78
$C_{24}H_{18}N_2O_4$			
Cycloocta[1,2-b:5,6-b']diindole-6,13-dicarboxylic acid, 5,12-dihydro-, dimethyl ester (caulerpin)	MeOH	221(4.63),270(4.25), 315(4.43)	150-1683-78
5H-Cyclopenta[d]pyridazine-1,4-dicarboxylic acid, 5-(diphenylmethylene)-, dimethyl ester	MeCN	237(4.27),316(3.99), 381(4.14)	78-2509-78
	MeCN-HCl	245(4.23),300(3.89), 355(4.21),460(4.28)	78-2509-78
$C_{24}H_{18}N_2O_6$			
Cycloocta[1,2-b:5,6-b']diindole-6,13-dicarboxylic acid, 5,6,7,12-tetrahydro-6-hydroxy-7-oxo-, dimethyl ester	MeOH	229(4.39),240(4.39), 303(4.42),379(3.79)	150-1683-78
1H-Indole-2-acetic acid, 3-[2-(3-formyl-1H-indol-2-yl)-3-methoxy-3-oxo-1-propenyl]-α-oxo-, methyl ester	MeOH	222(4.39),244(4.37), 299(4.13),350(4.08)	150-1683-78
Propanoic acid, 3-[2-[1,3-dihydro-3-(3-methoxy-2,3-dioxopropylidene)-2H-indol-2-ylidene]-1,2-dihydro-3H-indol-3-ylidene]-2-oxo-, methyl ester	MeOH	288(4.15),430(3.70), 650(4.10)	150-1683-78
	MeOH-acid	265(4.07),313(3.99), 506(4.10),600(4.38)	150-1683-78
$C_{24}H_{18}N_2O_8$			
Pyrazino[2,1,6-cd:5,4,3-c'd']diindolizine-1,2,6,7-tetracarboxylic acid, tetramethyl ester	EtOH	244(4.43),285(4.48), 338(4.08),350s(4.00), 406s(3.80),430(4.10), 456(4.19)	150-2850-78
$C_{24}H_{18}N_4O_2$			
1H-Purine-2,6-dione, 3,7-dihydro-1,3-diphenyl-8-(phenylmethyl)-	pH 6.0	275(4.21)	24-0982-78
	pH 13.0	282(4.21)	24-0982-78
$C_{24}H_{18}O$			
2-Cyclopenten-1-one, 3,4-diphenyl-5-(phenylmethylene)-, (E)-	CHCl$_3$	321(4.39)	39-0989-78C
$C_{24}H_{18}O_2$			
Benzaldehyde, 2,2'-(1,2-phenylenedi-2,1-ethenediyl)bis-	DMF	257(4.49),293s(4.26), 340(4.20)	33-2813-78
4,7-Ethano-1H-indene-8,9-dione, 1-(diphenylmethylene)-3a,4,7,7a-tetrahydro-, (3aα,4α,7α,7aα)-	C$_6$H$_{12}$ benzene	239(3.66),292(4.26) 459(1.99)	5-0440-78 5-0440-78
Spiro[2H-pyran-4,9'-phenanthren]-2-one, 3,4,9',10'-tetrahydro-3-phenyl-	MeOH	207(4.66),270(4.16), 296(3.41)	54-0197-78
$C_{24}H_{18}O_3$			
2H-1-Benzopyran-2-one, 7-hydroxy-4-phenyl-8-(1-phenyl-2-propenyl)-	MeOH MeOH-NaOH	240(4.13),335(4.04) 260(4.25),385(4.19)	2-0579-78 2-0579-78
2H-Furo[2,3-h]-1-benzopyran-2-one, 8,9-dihydro-9-methyl-4,8-diphenyl-	MeOH	240(3.72),330(3.71)	2-0579-78
$C_{24}H_{18}O_4$			
4H-1-Benzopyran-4-one, 5-hydroxy-	n.s.g.	264(4.53),300(4.31)	2-1124-78

Compound	Solvent	$\lambda_{max}(\log \epsilon)$	Ref.
3-phenyl-7-[(3-phenyl-2-propenyl)-oxy]- (cont.)			2-1124-78
$C_{24}H_{18}O_5$			
α-Indomycinone	MeOH-dioxan	242(4.54),282(4.38), 418(3.99)	78-0761-78
$C_{24}H_{18}O_{10}$			
Furo[3,2-b]furan-2,5-dione, 3,6-bis(3-acetoxy-4-methoxyphenyl)-	acetone	413(4.47)	64-0820-78C
Furo[3.2-b]furan-2,5-dione, 3,6-bis(4-acetoxy-3-methoxyphenyl)-	acetone	396(4.30)	64-0820-78C
$C_{24}H_{18}S_3$			
[1,2]Dithiolo[1,5-b][1,2]dithiole-7-S^{IV}, 3-methyl-2,4,5-triphenyl-	dioxan	250(4.68),273(4.67), 504(4.19)	104-2267-78
1-Propanethione, 2-(4,5-diphenyl-1,3-dithiol-2-ylidene)-1-phenyl-	dioxan	228s(4.52),276(4.37), 330(4.01),481(4.56), 587(2.82)	104-2267-78
$C_{24}H_{19}BF_4N_2O_2$			
Pyridinium, 1-(benzoylamino)-2-hydroxy-4,6-diphenyl-, tetrafluoroborate	EtOH	220(3.83),257(3.90), 315(4.06)	120-0001-78
$C_{24}H_{19}BrN_4Ni$			
Nickel, [2-bromo-5,10,15,20-tetramethyl-21H,23H-porphinato(2-)-N^{21},N^{22}-N^{23},N^{24}]-, (SP-4-2)-	$CHCl_3$	304(4.05),337(4.00), 423(5.23),545(4.11)	5-0238-78
$C_{24}H_{19}BrN_4O$			
2H-Benzimidazol-2-one, 5-[4-(4-bromophenyl)-5-phenyl-1H-imidazol-2-yl]-1,3-dihydro-1,3-dimethyl-	EtOH	216(4.66),320(4.54)	103-0425-78
radical	toluene	546(--),727(--)	103-0425-78
$C_{24}H_{19}ClN_4O$			
2H-Benzimidazol-2-one, 5-[4-(4-chlorophenyl)-5-phenyl-1H-imidazol-2-yl]-1,3-dihydro-1,3-dimethyl-	EtOH	219(4.64),320(4.54)	103-0425-78
radical	toluene	548(--),726(--)	103-0425-78
$C_{24}H_{19}ClN_6$			
[1,2,4]Triazolo[1,5-a]pyridine, 2-[3-chloro-4-[2-(5-methyl-2-phenyl-2H-1,2,3-triazol-4-yl)ethenyl]phenyl]-6-methyl-	DMF	345(4.70)	33-0142-78
$C_{24}H_{19}NO_3Pb$			
Plumbane, (4-nitrophenoxy)triphenyl-	benzene	345(4.22)	70-2243-78
	DMSO	390(--),434(--)	70-2243-78
$C_{24}H_{19}NO_3Sn$			
Stannane, (4-nitrophenoxy)triphenyl-	benzene	308(4.04)	70-1573-78
	benzene	318(4.06)	70-2243-78
	DMSO	370(--),434(--)	70-2243-78
$C_{24}H_{19}N_2O_2$			
Pyridinium, 1-(benzoylamino)-2-hydroxy-4,6-diphenyl-, tetrafluoroborate	EtOH	220(3.83),257(3.90), 315(4.06)	120-0001-78

Compound	Solvent	λ_{max} (log ϵ)	Ref.
$C_{24}H_{19}N_2O_3P$			
Ethenamine, 2-(5-nitro-2-furanyl)-N-(triphenylphosphoranylidene)-	MeOH	497(4.39)	73-3404-78
$C_{24}H_{19}N_3O_3S$			
4-Thiazoleacetic acid, α-(hydroxyimino)-2-[(triphenylmethyl)amino]-, (E)-	EtOH	229s(4.16),316(3.34)	78-2233-78
(Z)-	EtOH-HCl	274(4.13)	78-2233-78
$C_{24}H_{19}N_5O_3$			
2H-Benzimidazol-2-one, 1,3-dihydro-1,3-dimethyl-5-[4-(4-nitrophenyl)-5-phenyl-1H-imidazol-2-yl]-	EtOH	220(4.64),318(4.38), 394(4.10)	103-0425-78
radical	toluene	480(--),552s(--)	103-0425-78
$C_{24}H_{20}$			
Cyclobuta[a]naphthalene, 2a,3,4,8b-tetrahydro-1,2-diphenyl-	MeCN	276s(4.07),295(4.17) 325s(3.74)	24-3608-78
Cyclopropene, 1,2,3-triphenyl-3-(2-propenyl)-	EtOH	228(4.46),316(4.43), 333(4.34)	44-1481-78
1H-Indene, 1,2-diphenyl-3-(2-propenyl)-	EtOH	228(4.31),295(4.29)	44-1481-78
Naphthalene, 1,8-bis(3-methylphenyl)-	C_6H_{12}	301(4.09)	88-1459-78
Naphthalene, 1,8-bis(4-methylphenyl)-	C_6H_{12}	299(4.16)	88-1459-78
Naphthalene, 2,3-bis(phenylmethyl)-	CH_2Cl_2	265(4.0),270(4.0), 281(4.0),290s(3.8)	4-1185-78
$C_{24}H_{20}Br_4O_6$			
19-Norpregna-1,3,5,7,9,11,14-heptaene, 4,11,15,16-tetrabromo-3,16-dimethoxy-17,20:20,21-bis[methylenebis(oxy)]-	MeOH	268(4.63),283(4.60), 292(4.62),363(3.94)	88-0639-78
$C_{24}H_{20}ClIN_2O$			
Pyridinium, 2-[2-(4-chlorophenyl)-3-(1,2-dihydro-2-oxo-3H-indol-3-ylidene)-1-propenyl]-1-ethyl-, iodide	MeOH	534(3.63)	64-0209-78B
$C_{24}H_{20}ClN_3O_3$			
Quinolinium, 3-acetyl-6-hydroxy-1-(4-hydroxyphenyl)-4-[(phenylhydrazono)-methyl]-, chloride	MeOH	475(3.70)	65-1635-78
$C_{24}H_{20}ClN_3O_5$			
Quinolinium, 3-acetyl-1-phenyl-4-[(phenylhydrazono)methyl]-, perchlorate	MeOH	509(4.16)	65-1635-78
$C_{24}H_{20}NP$			
Benzenamine, N-(triphenylphosphoranylidene)-	CH_2Cl_2	305(3.70)	39-1237-78C
$C_{24}H_{20}N_2O$			
1(4H)-Naphthalenone, 2-[(4-methylphenyl)amino]-4-[(4-methylphenyl)imino]-	C_6H_{12}	245(4.05),280(4.14), 300(3.96),335(3.54), 470(3.63)	48-0557-78
	EtOH	250(4.38),275(4.43), 305(4.00),330(3.93), 480(3.90)	48-0557-78 +80-0617-78
	ether	250(4.35),280(4.42), 305(4.20),330(3.94), 470(3.92)	48-0557-78
	CCl_4	295(4.24),335(3.79), 475(3.82)	48-0557-78

Compound	Solvent	$\lambda_{max}(\log \epsilon)$	Ref.
$C_{24}H_{20}N_2O_2$			
1-Naphthaleneacetic acid, 2-(1-naphthalenylacetyl)hydrazide	EtOH	224(5.07),276(4.03), 282(4.11),290s(3.95), 312(2.86)	4-0385-78
Quinoxaline, 1,4-dibenzoyl-2-ethenyl-1,2,3,4-tetrahydro-	CH_2Cl_2	271.5(4.10)	5-1129-78
$C_{24}H_{20}N_2O_2S$			
4-Thiazoleacetic acid, 2-[(triphenylmethyl)amino]-	EtOH-HCl	272(3.98)	78-2233-78
$C_{24}H_{20}N_2O_3$			
1(4H)-Naphthalenone, 2-[(4-methoxyphenyl)amino]-4-[(4-methoxyphenyl)imino]-	C_6H_{12}	247(--),275(--), 340(--),490(--)	48-0557-78
	EtOH	245(4.35),275(4.37), 334(3.86),500(3.83)	48-0557-78 +80-0617-78
$C_{24}H_{20}N_2O_4$			
2H-Cyclopenta[d]pyridazine-1,4-dicarboxylic acid, 5-(diphenylmethyl)-, dimethyl ester	MeCN	295(4.42),335(3.63), 500(2.87)	78-2509-78
$C_{24}H_{20}N_2O_6$			
2(5H)-Furanone, 4,4'-(azinodiethylidyne)bis[3-hydroxy-5-phenyl-, [R*,S*-(E,E)]-	EtOH	253(4.16),317(3.80), 455(4.01)	142-1041-78A
$C_{24}H_{20}N_2O_7$			
2,4,6(1H,3H,5H)-Pyrimidinetrione, 5-[3-(2-hydroxyphenyl)-3-oxo-1-(4-oxo-4H-1-benzopyran-3-yl)propyl]-1,3-dimethyl-	MeOH	309(3.9)	83-0503-78
$C_{24}H_{20}N_2S_2$			
Benzenamine, N,N'-dithiobis[N-phenyl-	n.s.g.	<u>310(4.5)</u>	23-1080-78
$C_{24}H_{20}N_3O$			
Quinolinium, 3-acetyl-1-phenyl-4-[(phenylhydrazono)methyl]-, perchlorate	MeOH	509(4.16)	65-1635-78
$C_{24}H_{20}N_3OS$			
Benzothiazolium, 2-[3-[2-(2-benzoxazolyl)-1-methyl-4(1H)-pyridinylidene]-1-propenyl]-3-methyl-, iodide	EtOH	601(4.92)	4-0017-78
Benzothiazolium, 2-[3-[4-(2-benzoxazolyl)-1-methyl-2(1H)-pyridinylidene]-1-propenyl]-3-methyl-, iodide	EtOH	589(4.65)	4-0017-78
Benzothiazolium, 2-[3-[6-(2-benzoxazolyl)-1-methyl-2(1H)-pyridinylidene]-1-propenyl]-3-methyl-, iodide	EtOH	568(4.23)	4-0017-78
$C_{24}H_{20}N_3O_3$			
Quinolinium, 3-acetyl-6-hydroxy-1-(4-hydroxyphenyl)-4-[(phenylhydrazono)methyl]-, chloride	MeOH	475(3.70)	65-1635-78
$C_{24}H_{20}N_3S_2$			
Benzothiazolium, 2-[3-[4-(2-benzothiazolyl)-1-methyl-2(1H)-pyridinylidene]-1-propenyl]-3-methyl-, iodide	EtOH	585(4.71)	4-0017-78

Compound	Solvent	$\lambda_{max}(\log \epsilon)$	Ref.
$C_{24}H_{20}N_4Ni$ Nickel, [5,10,15,20-tetramethyl-21H,23H- porphinato(2-)-$N^{21},N^{22},N^{23},N^{24}$]-, (SP-4-1)-	CHCl$_3$	300(4.06),332(3.97), 418(5.27),539(4.09)	5-0238-78
$C_{24}H_{20}N_4O$ 2H-Benzimidazol-2-one, 5-(4,5-diphenyl- 1H-imidazol-2-yl)-1,3-dihydro-1,3-di- methyl- radical	EtOH toluene	224(4.84),322(4.69) 542(--),716(--)	103-0425-78 103-0425-78
$C_{24}H_{20}N_4O_3$ 7H-Benz[de]anthracen-7-one, 3-[(4,6-di- ethoxy-1,3,5-triazin-2-yl)amino]-	DMF	425(4.70)	2-0106-78
$C_{24}H_{20}N_4O_4$ [4,4'-Bi-1H-pyrazole]-5-carboxylic acid, 3,3'-diacetyl-1,1'-diphenyl-, methyl ester	EtOH	232(4.48),270s(4.32)	4-0293-78
$C_{24}H_{20}N_4O_5$ Tryptoquivaline F, acetate	MeOH	217s(4.59),226(4.57), 232(4.55),241s(4.40), 252s(4.35),265(4.33), 276s(4.29),303s(3.77), 315s(3.49)	94-0111-78
Tryptoquivaline J, acetate	MeOH	226(4.57),231s(4.53), 253s(4.14),265(4.02), 276(3.95),304(3.56), 315(3.48)	94-0111-78
$C_{24}H_{20}N_4O_6$ Tryptoquivaline H, acetate	MeOH	217s(4.49),226(4.52), 231s(4.49),254s(4.06), 265s(3.97),275s(3.89), 290s(3.49),302(3.42), 315(3.32)	94-0111-78
$C_{24}H_{20}N_4Pd$ Palladium, [5,10,15,20-tetramethyl- 21H,23H-porphinato(2-)-N^{21},N^{22},N^{23}- N^{24}]-, (SP-4-1)-	CHCl$_3$	418(5.28),494(3.30), 533(4.09),566(3.48)	5-0238-78
$C_{24}H_{20}OSe$ 2-Cyclohexen-1-one, 4,4-diphenyl- 6-(phenylseleno)-	MeOH	213(4.4),233(4.1)	24-1944-78
$C_{24}H_{20}O_2$ 1H-Indene-1-carboxylic acid, 2,3-di- phenyl-, ethyl ester	MeOH	239(4.40),303(4.21)	104-0701-78
$C_{24}H_{20}O_2S_4$ Ethanone, 2,2'-(1,2-cyclobutanediyl)- bis[1-[2,2'-bithiophen]-5-yl-	EtOH	244(3.94),252s(3.84), 349(4.54)	102-2097-78
$C_{24}H_{20}O_4$ 2H-1-Benzopyran-2-one, 5,7-dimethoxy- 3-phenyl-8-(phenylmethyl)-	MeOH	255(3.02),320(3.35)	2-0584-78
2H-1-Benzopyran-2-one, 5,7-dimethoxy- 4-phenyl-8-(phenylmethyl)-	MeOH	260(4.09),320(4.14)	2-0584-78

Compound	Solvent	$\lambda_{max}(\log \epsilon)$	Ref.
1,4-Naphthacenediol, 2,3-dimethyl-, diacetate	CH$_2$Cl$_2$	282(5.35),375(3.0), 396(3.34),419(3.61), 446(3.87),476(3.86)	150-5538-78
1(2H)-Naphthalenone, 4-ethoxy-2-(4-eth-oxy-1-oxo-2(1H)-naphthalenylidene)-	CH$_2$Cl$_2$	281(4.28),320(4.03), 640(4.15)	88-2289-78
$C_{24}H_{20}O_5$			
α-Indomycinone, dihydro-	MeOH-dioxan	240(4.60),272(4.45), 416(3.94)	78-0761-78
$C_{24}H_{20}O_6$			
β-Indomycinone	MeOH-dioxan	239(4.57),267(4.35), 417(3.93)	78-0761-78
$C_{24}H_{20}O_7$			
Glabratephrin	MeOH	218(4.22),247(4.23), 256(4.23),309(4.27)	78-1405-78
$C_{24}H_{21}ClN_2O_4$			
Pyridinium, 2-methyl-1-(4-methyl-2-pyridinyl)-4,6-diphenyl-, perchlorate	EtOH	220(3.95),315(4.11)	120-0001-78
Pyridinium, 2-methyl-1-(5-methyl-2-pyridinyl)-4,6-diphenyl-, perchlorate	EtOH	220(3.93),312(4.08)	120-0001-78
$C_{24}H_{21}ClN_2O_7S$			
2(1H)-Pyrimidinone, 1-[5-0-(4-chloro-benzoyl)-3-0-(4-methylbenzoyl)-β-D-arabinofuranosyl]-3,4-dihydro-4-thi-oxo-	MeOH	238(4.69),328(4.40)	44-0350-78
$C_{24}H_{21}ClN_2O_8$			
2,4(1H,3H)-Pyrimidinedione, 1-[5-0-(4-chlorobenzoyl)-2-0-(4-methylbenzoyl)-β-D-arabinofuranosyl]-	MeOH	240(4.60),261s(4.21)	44-0350-78
2,4(1H,3H)-Pyrimidinedione, 1-[5-0-(4-chlorobenzoyl)-2-0-(4-methylbenzoyl)-β-D-xylofuranosyl]-	MeOH	240(4.66),262(4.28)	44-0350-78
$C_{24}H_{21}FeN_6O_6$			
Iron, tris(4-acetyl-4,5-dihydro-2-meth-yl-5-oxo-1H-pyrrole-3-carbonitrilato-0,0')-	CH$_2$Cl$_2$	240(4.39),255s(4.29), 296(4.19),340s(3.91), 560(3.58)	44-3821-78
$C_{24}H_{21}IN_2O$			
Pyridinium, 2-[3-(1,2-dihydro-2-oxo-3H-indol-3-ylidene)-2-phenyl-1-pro-penyl]-1-ethyl-, iodide	MeOH	538(4.17)	64-0209-78B
$C_{24}H_{21}N$			
Benzonitrile, 4-(2,2-dimethyl-3,3-di-phenylcyclopropyl)-	EtOH	225(4.38),276(4.02), 285(3.55)	35-4146-78
Benzonitrile, 4-(1,1-dimethyl-3,3-di-phenyl-2-propenyl)-	EtOH	235(4.47),250(4.41)	35-4146-78
$C_{24}H_{21}NO$			
1,2-Benzisoxazole, 3-[4-[2-[4-(1-meth-ylethyl)phenyl]ethenyl]phenyl]-	DMF	332(4.65)	33-2904-78
3H-Pyrrol-3-one, 1,2-dihydro-1,2-di-methyl-2,4,5-triphenyl-	MeCN	225(4.20),280(4.08), 354(3.96)	5-1203-78
3H-Pyrrol-3-one, 1,2-dihydro-2,2-di-methyl-1,4,5-triphenyl-	MeCN	262(3.16),355(3.03)	5-1203-78

Compound	Solvent	$\lambda_{max}(\log \epsilon)$	Ref.
$C_{24}H_{21}NO_3$			
Benzenepropanamide, α-acetoxy-β-methylene-N,N-diphenyl-	EtOH	238(4.34)	44-0954-78
2(1H)-Quinolinone, 3-acetoxy-3,4-dihydro-4-methyl-1,4-diphenyl-, (E)-	EtOH	255(4.11)	44-0954-78
(Z)-	EtOH	246(4.00)	44-0954-78
Spiro[2H-indene-2,3'-[3H]indeno[1,2-b]pyridine]-1,3,5'(2'H)-trione, 1'-bromo-1',4'-dihydro-, perchlorate (2:1)	MeOH	224(5.06),258(4.71), 268(4.72)	49-1295-78
$C_{24}H_{21}NS$			
2-Azabicyclo[3.1.0]hexane-3-thione, 2-methyl-1,4,5-triphenyl-, (1α,4α,5α)-	MeCN	286(4.17)	5-1222-78
2(1H)-Pyridinethione, 5,6-dihydro-1-methyl-3,4,6-triphenyl-	MeCN	310(4.14)	5-1222-78
$C_{24}H_{21}N_2$			
Pyridinium, 2-methyl-1-(4-methyl-2-pyridinyl)-4,6-diphenyl-, perchlorate	EtOH	220(3.95),315(4.11)	120-0001-78
Pyridinium, 2-methyl-1-(5-methyl-2-pyridinyl)-4,6-diphenyl-, perchlorate	EtOH	220(3.93),312(4.08)	120-0001-78
$C_{24}H_{21}N_2O$			
Pyridinium, 2-[3-(1,2-dihydro-2-oxo-3H-indol-3-ylidene)-2-phenyl-1-propenyl]-1-ethyl-, iodide	MeOH	538(4.17)	64-0209-78B
$C_{24}H_{21}N_3$			
Benzenamine, N-[5-(3,5-dimethylphenyl)-1-phenyl-4(1H)-pyrimidinylidene]-	CH₂Cl₂	292(4.29),328s(4.11)	97-0063-78
$C_{24}H_{21}N_3O$			
Benzenamine, N,N-dimethyl-4-[2-[4-(5-phenyl-1,3,4-oxadiazol-2-yl)phenyl]-ethenyl]-, (E)-	n.s.g.	297(4.49),385(4.46)	135-0946-78
3-Pyrazolidinone, 4-(9-ethyl-9H-fluoren-9-yl)-5-imino-1-phenyl-	EtOH	267(4.20),305(3.84)	115-0305-78
$C_{24}H_{21}N_3O_6S$			
4-Thia-2,6-diazabicyclo[3.2.0]heptane-6-acetic acid, 2-formyl-α-(1-methylethenyl)-7-oxo-3-(phenylmethylene)-, (4-nitrophenyl)methyl ester, [1R-[1α,5α,6(R*)]]-	EtOH	280(4.33)	88-4703-78
$C_{24}H_{21}N_3O_7S$			
4-Thia-2,6-diazabicyclo[3.2.0]heptane-6-acetic acid, 2-formyl-α-(1-methylethenyl)-7-oxo-3-(phenylmethylene)-, (4-nitrophenyl)methyl ester, 4-oxide, [1R-[1α,5α,6(R*)]]-	EtOH	268(4.25)	88-4703-78
$C_{24}H_{22}$			
Bicyclo[4.2.0]octa-1,3,5,7-tetraene, 2,3,4,5-tetramethyl-7,8-diphenyl-	hexane	201(4.66),218s(4.38), 278(4.60),468(3.25)	5-1675-78
Phenanthrene, 9,10-dihydro-3,6,9-trimethyl-10-methylene-9-phenyl-	hexane	222(4.47),244(4.39), 250s(--),288(3.99)	104-0341-78
	H₂SO₄	272(4.41),289s(--), 341(4.16),381s(--), 543(3.71)	104-0341-78
	CF₃COOH	269(4.44),288s(--),	104-0341-78

Compound	Solvent	$\lambda_{max}(\log \epsilon)$	Ref.
(cont.)		338(4.22),388(3.71), 543(3.78)	104-0341-78
$C_{24}H_{22}Br_4N_4O_8$ 1-Oxa-2-azaspiro[4.5]deca-2,6,8-triene-3-carboxamide, N,N'-1,4-butanediyl-bis[7,9-dibromo-8-methoxy-10-oxo-(tetradehydroaerothionin)	dioxan	257(4.30),348(3.74)	5-0066-78
$C_{24}H_{22}ClNO_{10}$ 3H-2a-Azacyclopenta[ef]heptalene-1,3,4,5,6-pentacarboxylic acid, 2-chloro-, 1-ethyl 3,4,5,6-tetra-methyl ester	CHCl$_3$	257s(4.37),448(3.85)	39-0429-78C
$C_{24}H_{22}ClN_3O_3S$ Benzenesulfonamide, 4-chloro-N-[1,2-di-hydro-2-[1-(hydroxyimino)ethyl]-1-methyl-2-(phenylmethyl)-3H-indol-3-ylidene]-	EtOH	230(4.51),290(4.03), 473(4.04)	39-1471-78C
$C_{24}H_{22}Cl_2$ Pentacyclo[4.4.0.0^{2,5}.0^{3,8}.0^{4,7}]decane, 2,5-bis(4-chlorophenyl)-3,4-dimethyl-	C$_6$H$_{12}$	256(4.43),274s(4.28), 305s(2.89)	88-0357-78
$C_{24}H_{22}Cl_2F_2O_4$ Cyclobuta[3",4"]benzo[1",2":3',4']cy-clobuta[1',2':3,4]cyclobuta[1,2]cy-clobutene, 2,2-dichloro-1,1-diflu-orodecahydro-, acetylenedicarboxylic acid methyl ester adduct	MeOH	240s(3.53)	5-1368-78
isomer (5a adduct)	MeOH	235(3.51)	5-1368-78
$C_{24}H_{22}F_8$ 1,1'-Biphenyl, 3,3'-dicyclohexylocta-fluoro-	EtOH	212(4.09),229(3.92), 266(3.28)	104-1537-78
1,1'-Biphenyl, 3,4'-dicyclohexylocta-fluoro-	EtOH	214(4.08),234(4.10), 268(3.34)	104-1537-78
1,1'-Biphenyl, 4,4'-dicyclohexylocta-fluoro-	EtOH	211(4.20),238(5.37), 274(3.58)	104-1537-78
$C_{24}H_{22}F_{12}$ Bi-2,5-cyclohexadien-1-yl, 4,4'-dicyclo-hexyl-1,1',2,2',3,3',4,4',5,5',6,6'-dodecafluoro-	EtOH	216(3.48),265(1.53)	104-1537-78
isomeric mixture	EtOH	217(4.81),266(2.83)	104-1537-78
$C_{24}H_{22}NO$ Pyridinium, 1,2,6-trimethyl-4-[(2-phen-yl-4H-1-benzopyran-4-ylidene)methyl]-, perchlorate	HOAc	453(4.485)	83-0226-78
$C_{24}H_{22}N_2$ Imidazo[1,2-a]pyridine, 7-[2-[4-(1-methylethyl)phenyl]ethenyl]-2-phenyl-	DMF	362(4.66),380(4.61)	33-0129-78
1H-Indole, 3,3'-(1,3-cyclohexadiene-1,4-diyl)bis[2-methyl-	EtOH	231(4.57),291(4.12), 349(4.23)	2-0731-78
	HClO$_4$	227(4.23),273(4.12), 363(4.23)	2-0731-78

Compound	Solvent	$\lambda_{max}(\log \epsilon)$	Ref.
$C_{24}H_{22}N_2O$			
3H-Pyrrol-3-one, 2-[4-(dimethylamino)-phenyl]-1,2-dihydro-4,5-diphenyl-	MeOH	261(4.35),400(4.09)	18-3312-78
$C_{24}H_{22}N_2O_2$			
Benzeneacetic acid, [3-(4-methylphenyl)-3-oxo-1-phenylpropylidene]hydrazide	EtOH	218s(4.45),284(4.40), 291s(4.39)	4-0385-78
4H-Imidazole, 4-ethoxy-2-(4-methoxy-phenyl)-4,5-diphenyl-	EtOH	306(4.38)	40-1449-78
Quinoxaline, 1,4-dibenzoyl-2-ethyl-1,2,3,4-tetrahydro-	C_6H_{12}	220s(4.45),273(4.06)	5-1129-78
$C_{24}H_{22}N_2O_2S_3$			
Spiro[1,3-benzodithiole-2,4'(4'aH)-pyr-rolo[3,4-e][1,3]thiazine]-5',7'-(6'H,7'aH)-dione, 3a,4,5,6,7,7a-hexahydro-2',6'-diphenyl-	EtOH	249(4.30),309(3.57)	18-0301-78
$C_{24}H_{22}N_2O_3$			
Benzeneacetic acid, [1-(4-methoxyphen-yl)-3-oxo-3-phenylpropylidene]hydra-zide	EtOH	303(4.42)	34-0345-78
Benzeneacetic acid, [3-(4-methoxyphen-yl)-3-oxo-1-phenylpropylidene]hydra-zide	EtOH	283(4.43),292s(4.41)	34-0345-78
Imidazo[1,2-a]pyridine, 2-phenyl-7-[2-(3,4,5-trimethoxyphenyl)ethenyl]-	DMF	365(4.66),384(4.62)	33-0129-78
Quinoxaline, 1,4-dibenzoyl-2-ethoxy-1,2,3,4-tetrahydro-	CH_2Cl_2	270(4.10)	5-1129-78
$C_{24}H_{22}N_2O_4$			
Caulerpin, tetrahydro-	MeOH	224(4.72),286(4.49), 294(4.45)	150-1683-78
$C_{24}H_{22}N_3$			
Pyrimidinium, 5-(3,5-dimethylphenyl)-1-phenyl-4-(phenylamino)-, perchlorate	CH_2Cl_2	324(4.21)	97-0063-78
Pyrimidinium, 1-(4-methylphenyl)-4-[(4-methylphenyl)amino]-5-phenyl-, per-chlorate	CH_2Cl_2	328(4.26)	97-0063-78
$C_{24}H_{22}N_3O_2$			
Pyrimidinium, 1-(2-methoxyphenyl)-4-[(2-methoxyphenyl)amino]-5-phenyl-, perchlorate	CH_2Cl_2	355(4.22)	97-0063-78
Pyrimidinium, 1-(3-methoxyphenyl)-4-[(3-methoxyphenyl)amino]-5-phenyl-, perchlorate	CH_2Cl_2	243(4.16),320(4.19)	97-0063-78
Pyrimidinium, 1-(4-methoxyphenyl)-4-[(4-methoxyphenyl)amino]-5-phenyl-, perchlorate	CH_2Cl_2	286(4.07),341(4.21)	97-0063-78
$C_{24}H_{22}N_4O_2$			
Acetamide, N-[2-[2-[2-[[2-(acetylami-no)phenyl]azo]phenyl]ethenyl]phenyl]-	benzene	302(4.29),394(3.89)	33-2813-78
$C_{24}H_{22}N_4O_2Zn$			
Zinc, bis[1,5-dihydro-3,4-dimethyl-5-(2-pyridinylmethylene)-2H-pyrrol-2-onato-N^1,N^5]-, (T-4)-	EtOH DMSO	385(4.43) 396(--)	49-0883-78 49-0883-78

Compound	Solvent	$\lambda_{max}(\log \epsilon)$	Ref.
$C_{24}H_{22}N_6O$			
7H-Benz[de]anthracen-7-one, 3-[[4,6-bis(dimethylamino)-1,3,5-triazin-2-yl]amino]-	DMF	435(4.56)	2-0106-78
7H-Benz[de]anthracen-7-one, 3-[[4,6-bis(ethylamino)-1,3,5-triazin-2-yl]amino]-	DMF	460(4.66)	2-0106-78
$C_{24}H_{22}N_6O_6$			
1H-Pyrazole-3-carboxylic acid, 4-[[3-(ethoxycarbonyl)-1,5-dihydro-5-oxo-1-phenyl-4H-pyrazol-4-ylidene]hydrazino]-2,5-dihydro-5-oxo-1-phenyl-, ethyl ester	MeOH	255(4.13),520(4.02)	142-0199-78B
$C_{24}H_{22}O_2$			
7,14-Ethanodibenz[a,h]anthracene-4,11-(1H,7H)-dione, 2,3,8,9,10,14-hexahydro-, (7R)-	EtOH	222(4.51),261(4.45)	18-0265-78
$C_{24}H_{22}O_4$			
Fulvene, 2,6-diacetoxy-3,4-dimethyl-1,6-diphenyl-	n.s.g.	244(4.18),318(3.84)	33-0266-78
2-Propen-1-one, 1-[2,4-dimethoxy-4-(phenylmethoxy)phenyl]-3-phenyl-	EtOH	234(4.17),306(4.33), 362(4.00)	102-1375-78
2-Propen-1-one, 1-[2,5-dimethoxy-4-(phenylmethoxy)phenyl]-3-phenyl-	EtOH	232(4.12),303(4.36), 365(4.06)	102-1375-78
2-Propen-1-one, 1-[4,5-dimethoxy-2-(phenylmethoxy)phenyl]-3-phenyl-	EtOH	235(4.19),308(4.32), 362(4.03)	102-1375-78
$C_{24}H_{22}O_5$			
α-Indomycinone, tetrahydro-	MeOH-dioxan	241(4.68),267(4.40), 416(3.94)	78-0761-78
$C_{24}H_{22}O_8$			
9,10-Anthracenedione, 1,4-diacetoxy-2-(3-acetoxybutyl)-	MeOH	202(4.48),230s(--), 248(4.48),265s(--), 334(3.72)	24-3823-78
$C_{24}H_{22}O_9$			
9H-Xanthen-9-one, 3-hydroxy-4-[2-hydroxy-2-(4-hydroxy-3-methoxyphenyl)-1-(hydroxymethyl)ethoxy]-2-methoxy-	MeOH	230(4.52),281(3.88), 312(3.83),354(3.63)	102-2040-78
	MeOH-NaOAc	213(4.58),281(3.98), 378(4.02)	102-2040-78
$C_{24}H_{22}O_{13}$			
1,2,3-Benzenetriol, 5-(2,4,6-triacetoxyphenoxy)-, triacetate	MeCN	210(4.47),229s(4.20), 269(3.33)	102-0579-78
$C_{24}H_{23}FN_4O_4S_3$			
1,3-Dithiolylium, 4,5-dicyano-2-(9-julolidinyl)-, fluorosulfonate, p-anisidine complex	MeCN	520(4.15),610(4.66)	44-0678-78
$C_{24}H_{23}N$			
2-Azabicyclo[3.1.0]hexane, 2-methyl-1,4,5-triphenyl-	MeCN	248(3.98)	5-1222-78
Pyridine, 1,2,3,6-tetrahydro-1-methyl-2,4,5-triphenyl-	MeCN	245(3.98)	5-1222-78

$C_{24}H_{23}NO_2-C_{24}H_{24}FeO_4$

Compound	Solvent	$\lambda_{max}(\log \epsilon)$	Ref.
$C_{24}H_{23}NO_2$ 2-Propenal, 3-(phenylmethoxy)-2-[bis-(phenylmethyl)amino]-	EtOH	235(4.10),325(3.63)	73-1261-78
$C_{24}H_{23}N_2$ Pyridinium, 1-[[1-[(4-ethylphenyl)methyl]-4(1H)-quinolinylidene)methyl]-, perchlorate	EtOH	260(4.15),365(3.79), 498(4.02)	124-0071-78
$C_{24}H_{23}N_3O$ 2H-Imidazol-2-one, 1,3-dihydro-4-methyl-1-(4-methylphenyl)-3-[(2-methylphenyl)amino]-5-phenyl-	EtOH	275(4.13)	5-2033-78
2H-Imidazol-2-one, 1-(2,4-dimethylphenyl)-1,3-dihydro-3-[(2-methylphenyl)-amino]-4-phenyl-	EtOH	281(4.03)	5-2033-78
1,2,4-Triazin-3(2H)-one, 4-(2,4-dimethylphenyl)-4,5-dihydro-2-(2-methylphenyl)-6-phenyl-	EtOH	241(4.54),337(4.26)	5-2033-78
$C_{24}H_{23}N_3O_3$ 5H-Pyrrolo[3,4-b]pyridine-3-carboxylic acid, 4-(3-aminophenyl)-6,7-dihydro-2-methyl-7-oxo-6-(phenylmethyl)-, ethyl ester	EtOH	224(4.24),259(4.10), 343(3.64)	95-0448-78
$C_{24}H_{23}N_3O_5$ 1H-Pyrrolo[3,4-b]pyridine-3-carboxylic acid, 4,5,6,7-tetrahydro-2-methyl-4-(3-nitrophenyl)-7-oxo-6-(phenylmethyl)-, ethyl ester	EtOH	259(4.26),336(3.58)	95-0448-78
$C_{24}H_{23}N_3O_7S$ 4-Thia-1-azabicyclo[3.2.0]hept-2-ene-2-carboxylic acid, 3-(4-nitrophenyl)-7-oxo-6-[(phenoxyacetyl)amino]-, 1,1-dimethylethyl ester, (5R-trans)-	hexane	262(4.20),267(4.20), 348(3.72)	35-8214-78
$C_{24}H_{23}N_3O_{11}$ 1H-Naphtho[2,3-d]triazole-4,9-dione, 1-(2,3,4,6-tetra-O-acetyl-β-D-gluco-pyranosyl)-	EtOH	245(4.39),268(4.09), 334(3.49)	111-0155-78
$C_{24}H_{23}N_5$ 2-Propanone, 1-[1-phenyl-3-(phenyl-methyl)-1H-1,2,4-triazol-5-yl]-, phenylhydrazone	EtOH	274(4.30)	94-1825-78
$C_{24}H_{24}$ Pentacyclo[4.4.0.0^{2,5}.0^{3,8}.0^{4,7}]decane, 2,5-dimethyl-3,4-diphenyl-	C_6H_{12}	251(4.34),280s(3.90), 295s(2.62)	88-0357-78
$C_{24}H_{24}Cl_2N_2O_8$ Quinolinium, 1-[(4-ethylphenyl)methyl]-4-(pyridiniomethyl)-, diperchlorate	EtOH	238(4.61),317(4.06)	124-0071-78
$C_{24}H_{24}FeO_4$ Ferrocene, [2-(ethoxycarbonyl)-4-methoxy-4-oxo-1-phenyl-1-butenyl]-, (E)-	EtOH	243(3.81),291(3.60)	128-0325-78
(Z)-	EtOH	244(4.38),293(4.19)	128-0325-78

Compound	Solvent	$\lambda_{max}(\log \epsilon)$	Ref.
$C_{24}H_{24}NO$			
Pyridinium, 1-ethyl-2-[4-(4-methoxy-phenyl)-2-phenyl-1,3-butadienyl]-, iodide	EtOH	530(3.63)	146-0341-78
$C_{24}H_{24}N_2$			
Benzenamine, 2,4,6-trimethyl-N-(1,4,6-trimethyl-9H-indeno[2,1-c]pyridin-9-ylidene)-, (E)-	EtOH	256(4.72),300(4.06), 312(4.14),334s(3.70), 440(2.96)	104-2365-78
Benzenamine, 2,4,6-trimethyl-N-(1,4,7-trimethyl-9H-indeno[2,1-c]pyridin-9-ylidene)-, (E)-	EtOH	207(4.76),251(4.67), 258(4.72),297(4.04), 308(4.07),322s(3.7), 444(2.78)	104-2365-78
4,11-Etheno-1H-cyclopenta[b]phenazine, 1-cyclohexylidene-3a,4,11,11a-tetra-hydro-13-methyl-, (3aα,4β,11β,11aα)-	C_6H_{12}	240(4.58),255s(4.45), 265s(4.27),292s(3.72), 297s(3.78),304s(3.90), 310s(3.94),316(4.00)	5-0440-78
Quinolinium, 1-[(4-ethylphenyl)methyl]-4-(pyridiniomethyl)-, diperchlorate	EtOH	238(4.61),317(4.06)	124-0071-78
$C_{24}H_{24}N_2O$			
Acetic acid, 2-methyl-2-phenyl-1-[2-phenyl-1-(phenylmethyl)ethenyl]hy-drazide	decalin	243(4.36),280(4.07)	33-1364-78
3-Cyclohexen-1-ol, 1,4-bis(2-methyl-1H-indol-3-yl)-	EtOH	228(4.636),292(4.146), 352(4.136)	2-0731-78
	$HClO_4$	238(4.326),272(4.51), 360(4.21)	2-0731-78
$C_{24}H_{24}N_2O_3$			
1H-Pyrrolo[3,4-b]pyridine-3-carboxylic acid, 4,5,6,7-tetrahydro-2-methyl-7-oxo-4-phenyl-6-(phenylmethyl)-, ethyl ester	EtOH	222s(4.26),259(4.11), 343(3.61)	95-0448-78
$C_{24}H_{24}N_2O_4$			
2H-Pyrido[3,4-e][1,3]oxazine-4,5(3H,6H)-dione, 7-methoxy-2,2-dimethyl-6-phen-yl-3-(2-phenylethyl)-	MeOH	309(4.4)	120-0101-78
$C_{24}H_{24}N_2S$			
1H-Pyrrole, 1,1'-(thiodi-4,1-phenyl-ene)-	EtOH	207(4.60),260(4.28), 280(4.18)	111-0093-78
$C_{24}H_{24}N_3PS$			
Phosphonium, [3-[(aminothioxomethyl)-hydrazono]-2-methylenebutyl]tri-phenyl-, hydroxide, inner salt	EtOH or DMF	288(2.810),342(3.152), 459(3.841)	65-1998-78
$C_{24}H_{24}N_4$			
2,3-Quinoxalinediamine, N,N'-bis(3,5-dimethylphenyl)-	EtOH	248(4.34),275(4.41), 354s(4.10),369(4.14), 384s(4.03)	78-0981-78
$C_{24}H_{24}N_4O_6$			
Tryptoquivaline E, acetate	MeOH	217s(4.52),226(4.53), 232s(4.50),255s(4.19), 266s(4.04),276s(3.91), 290s(3.56),302(3.45), 315(3.36)	94-0111-78

Compound	Solvent	$\lambda_{max}(\log \epsilon)$	Ref.
$C_{24}H_{24}N_4O_8$			
Pentanedial, 2-(benzoylhydrazono)-4-[1-(1-formyl-2-hydroxyethoxy)-2-oxoethoxy]-, 1-(benzoylhydrazone)	EtOH	240(3.81),265(3.83), 340(3.33)	136-0089-78E
$C_{24}H_{24}N_5O_4P_3$			
1,3,5,2,4,6-Triazatriphosphorine, 2,2-diamino-2,2,4,4,6,6-hexahydro-4,4,6,6-tetraphenoxy-	MeOH	210(4.33),263(3.50)	121-0945-78
$C_{24}H_{24}N_5P_3S_4$			
1,3,5,2,4,6-Triazatriphosphorine, 2,2-diamino-2,2,4,4,6,6-hexahydro-4,4,6,6-tetrakis(phenylthio)-	MeOH	239(4.99),321(3.98)	121-0945-78
$C_{24}H_{24}N_8O_3$			
L-Phenylalanine, N-[4-[[(2,4-diamino-6-pteridinyl)methyl]methylamino]-benzoyl]-	pH 1	245(4.20),308(4.23)	87-1162-78
$C_{24}H_{24}N_{10}Ni$			
Nickel, [1,14,15,20-tetrahydro-1,3,7-12,14,17-hexamethyldibenzo[c,j]di-pyrazolo[4,3-g:3',4'-m][1,2,5,6,9-12]hexaazacyclotetradecinato(2-)-N^5,N^{10},N^{15},N^{20}]-, (SP-4-3)-	MeOH	295(3.60),340(3.17), 400s(3.36),455(3.57), 615(3.20)	103-0885-78
$C_{24}H_{24}O$			
Benzene, 1-(2,2-dimethyl-3,3-diphenyl-cyclopropyl)-3-methoxy-	EtOH	256(3.11),263(3.20), 270(3.33),277(3.45), 285(3.44)	35-4146-78
Benzene, 1-(2,2-dimethyl-3,3-diphenyl-cyclopropyl)-4-methoxy-	EtOH	273(3.32),282(3.32), 291(3.20)	35-4146-78
Benzene, 1-(1,1-dimethyl-3,3-diphenyl-2-propenyl)-3-methoxy-	EtOH	251(4.19),280(3.66), 294(2.56)	35-4146-78
Benzene, 1-(1,1-dimethyl-3,3-diphenyl-2-propenyl)-4-methoxy-	EtOH	250(4.21),277(3.86), 285(3.65)	35-4146-78
Benzene, 1,1',1"-[3-(1-methylethoxy)-1-propene-1,2,3-triyl]tris-	EtOH	260(4.21)	44-0232-78
Cyclobutanone, 3-(3-methyl-1,2-butadi-enyl)-4-(1-methylethylidene)-2,2-di-phenyl-	C_6H_{12}	224(4.22),245(4.16)	39-1568-78C
3-Cyclohexen-1-one, 2,5-bis(1-methyl-ethylidene)-6,6-diphenyl-	C_6H_{12}	248(4.05),295(4.15)	39-1568-78C
$C_{24}H_{24}O_2S_2$			
Benzene, 1-[[1,2-diphenyl-2-(propyl-thio)ethenyl]sulfonyl]-4-methyl-, (E)-	EtOH	202(4.60),235s(4.13), 294(4.05)	2-1086-78
(Z)-	EtOH	203(4.60),234s(4.23), 303(3.96)	2-1086-78
Benzene, 1-methyl-4-[[2-[(1-methyleth-yl)thio]-1,2-diphenylethenyl]sul-fonyl]-, (E)-	EtOH	203(4.63),235s(4.23), 297(4.11)	2-1086-78
(Z)-	EtOH	203(4.61),235s(4.26), 310(3.98)	2-1086-78
$C_{24}H_{24}O_4$			
2,2'a(2'H)-Bi-3H-naphtho[1,8-bc]furan, 4,4',5,5'-tetrahydro-6,6'-dimethoxy-	EtOH	215s(4.54),260(4.20), 291(3.90)	18-2068-78
2,2'a(2'H)-Bi-3H-naphtho[1,8-bc]furan, 4,4',5,5'-tetrahydro-8,8'-dimethoxy-	EtOH	216(4.67),249(4.21), 357(4.19),280(3.59),	18-2068-78

Compound	Solvent	$\lambda_{max}(\log \epsilon)$	Ref.
(cont.)		290(3.48)	18-2068-78
$C_{24}H_{24}O_4S_2$			
Benzene, 1-[[1,2-diphenyl-2-(propylsulfonyl)ethenyl]sulfonyl]-4-methyl-, (E)-	EtOH	203(4.63),219s(4.45), 245s(4.17)	2-1086-78
(Z)-	EtOH	202(4.60),243s(4.10)	2-1086-78
Benzene, 1-methyl-4-[[2-[(1-methylethyl)sulfonyl]-1,2-diphenylethenyl]sulfonyl]-, (E)-	EtOH	203(4.60),220s(4.42), 240s(3.60)	2-1086-78
(Z)-	EtOH	203(4.60),246(4.15)	2-1086-78
$C_{24}H_{24}O_5$			
2H,6H-Benzo[1,2-b:5,4-b']dipyran-2-one, 7,8-dihydro-4-hydroxy-8,8-dimethyl-3-(3-oxo-1-phenylbutyl)-	dioxan	332(4.54)	2-0375-78
$C_{24}H_{24}O_6$			
Desmodol trimethyl ether	EtOH	214(4.43),240(4.60), 272(4.37),278(4.36), 335(4.36)	94-2411-78
$C_{24}H_{24}O_6S_2$			
Spiro[1,3-dithiane-2,5'(8'H)-furo[3',4'-6,7]naphtho[2,3-d][1,3]dioxol]-8'-one, 9'-(3,4-dimethoxyphenyl)-5'a,6',8'a,9'-tetrahydro-	EtOH	210(4.55),285(3.71)	78-1011-78
$C_{24}H_{24}O_7$			
4H-Furo[2,3-h]-1-benzopyran-4-one, 9-(1-acetoxy-2-hydroxy-2-methylpropyl)-8,9-dihydro-5-methoxy-2-phenyl-, [R-(R*,S*)]-	MeOH	213(4.57),268(4.57), 325(4.02)	119-0047-78
$C_{24}H_{24}O_8$			
Tricyclo[8.2.2.2^{4,7}]hexadeca-4,6,10,12-13,15-hexaene-5,6,11,12-tetracarboxylic acid, tetramethyl ester	EtOH	228(3.93),248s(3.80)	88-0969-78
$C_{24}H_{24}O_8S$			
7-Oxabicyclo[4.1.0]heptane-2,3,4-triol, 1-[(benzoyloxy)methyl]-5-(phenylthio)-, 2,3-diacetate	MeOH	222(4.27),251(3.70), 279(3.16)	44-0171-78
7-Oxabicyclo[4.1.0]heptane-2,3,5-triol, 1-[(benzoyloxy)methyl]-4-(phenylthio)-, 2,3-diacetate, [1R-(1α,2β,3α,4α,5β,6α)]-	MeOH	254(3.83)	44-0171-78
$C_{24}H_{24}O_{13}$			
4H-1-Benzopyran-4-one, 3-[(6-O-acetyl-β-D-glucopyranosyl)oxy]-5,7-dihydroxy-2-(4-hydroxy-3-methoxyphenyl)-	MeOH	255(4.28),265s(--), 357(4.19)	105-0146-78
	MeOH-NaOMe	273(--),333(--), 418(--)	105-0146-78
	MeOH-NaOAc	276(--),323(--), 386(--)	105-0146-78
$C_{24}H_{24}S_4$			
2,2'-Bithiophene, 5,5"-(1,2-cyclobutanediyldi-2,1-ethanediyl)bis-, trans	EtOH	244(3.77),308(4.11)	102-2097-78

Compound	Solvent	$\lambda_{max}(\log \epsilon)$	Ref.
$C_{24}H_{25}ClN_3PS$ Phosphonium, [3-[(aminothioxomethyl)hy- drazono]-2-methylenebutyl]triphenyl-, chloride	EtOH	268(4.191),334(4.096)	65-1998-78
$C_{24}H_{25}NO$ Ethanone, 1-(1,2,3,5,6,7-hexahydro-s- indacen-4-yl)-2-[(phenylmethyl)-2- propenylamino]-, (Z)-2-butenedioate	MeOH	253(3.88),310(3.43)	73-0970-78
$C_{24}H_{25}NO_4$ 3,5-Pyridinedicarboxylic acid, 1,4-di- hydro-1,2,6-trimethyl-, bis(4-methyl- phenyl) ester	EtOH	205(4.26),218s(4.21), 239(4.14),267(4.04), 368(3.78)	103-1226-78
$C_{24}H_{25}NO_6$ Corynoline, 6-ax'-acetonyl-	n.s.g.	237(3.30),288(3.20)	94-1880-78
3,5-Pyridinedicarboxylic acid, 1,4-di- hydro-1,2,6-trimethyl-, bis(3-meth- oxyphenyl) ester	EtOH	202(4.20),221(4.19), 238s(3.97),272(3.88), 367(3.58)	103-1226-78
3,5-Pyridinedicarboxylic acid, 1,4-di- hydro-1,2,6-trimethyl-, bis(4-meth- oxyphenyl) ester	EtOH	203(4.33),225(4.37), 238s(4.28),268(4.18), 367(3.90)	103-1226-78
$C_{24}H_{25}NO_{11}$ 1H-Indene-1,3(2H)-dione, 2-[[(1,2,4,6- tetra-O-acetyl-β-D-glucopyranosyl)- amino]methylene]-	EtOH	238(4.39),283(4.15), 294(4.43),326(4.42)	111-0155-78
1,4-Naphthalenedione, 2-[(2,3,4,6-tet- ra-O-acetyl-β-D-glucopyranosyl)ami- no]-	EtOH	254(4.18),284(3.89), 292(3.88),330(3.40)	111-0155-78
$C_{24}H_{25}N_2O_2S_2$ Benzothiazolium, 1-ethyl-2-[2-(ethoxy- carbonyl)-3-[(1-ethyl-2-benzothiazo- lylidene)-1-propenyl]-	EtOH	563(4.14)	124-0942-78
$C_{24}H_{25}N_3$ 1,2,4-Triazine, 4-(2,4-dimethylphenyl)- 2,3,4,5-tetrahydro-2-(2-methylphen- yl)-6-phenyl-	EtOH	311(4.18)	5-2033-78
$C_{24}H_{25}N_3O_6$ Acetamide, N-(1-acetyl-1H-indol-3-yl)- N-[[3-acetoxy-5-(acetoxymethyl)-2- methyl-4-pyridinyl]methyl]-	EtOH	204(4.46),238(4.21), 267(4.11),290(3.81), 301(3.83)	103-0757-78
$C_{24}H_{25}N_5O_2$ 2,3-Butanedione, 1-[1-methyl-5-[2-oxo- 1-(phenylhydrazono)propyl]-1H-pyrrol- 2-yl]-, 2-(phenylhydrazone)	EtOH	235(4.39),284(3.95), 292(4.02),343(4.49)	4-0293-78
Ethanone, 1-[5-[(3-acetyl-1-phenyl-1H- pyrazol-5-yl)methyl]-4,5-dihydro- 4-(methylamino)-1-phenyl-1H-pyrazol- 3-yl]-, cis	EtOH	241(4.36),360(4.31)	4-0293-78
1,2-Propanedione, 1,1'-(1,2-dimethyl- 1H-pyrrole-3,4-diyl)bis-, 1,1'- bis(phenylhydrazone)	EtOH	234(4.38),287s(3.84), 294(3.98),347(4.48)	4-0293-78
1,2-Propanedione, 1,1'-(1,2-dimethyl- 1H-pyrrole-3,5-diyl)bis-, 1,1'- bis(phenylhydrazone)	EtOH	236(4.45),286s(3.92), 295(3.99),353(4.59)	4-0293-78

Compound	Solvent	$\lambda_{max}(\log \epsilon)$	Ref.
1H-Pyrrolo[2,3-c:4,5-c']dipyrazole, 3,7-diacetyl-3a,4,4a,5,7a,7b-hexahydro-4,4a-dimethyl-1,5-diphenyl-	EtOH	238(4.25),300s(3.98), 344(4.44)	4-0293-78
Spiro[2-pyrazoline-5,5'(1H')-pyrrolo-[3,2-c]pyrazole], 3,3'-diacetyl-3'a,4',6',6'a-tetrahydro-4'-methyl-1,1'-diphenyl-	EtOH	239(4.31),289s(3.63), 300s(3.75),360(4.31)	4-0293-78
$C_{24}H_{25}O_8P$ 4H,8aH-Phospholo[1,2-a]phosphorin-1,2,3-8a-tetracarboxylic acid, 6,7-dimethyl-4-phenyl-, tetramethyl ester	EtOH	223(4.29),269s(3.95), 277s(4.00),289s(4.00), 331(4.08)	4-1319-78
$C_{24}H_{26}$ 7,14-Ethanodibenz[a,h]anthracene, 1,2,3,4,7,8,9,10,11,14-decahydro-	EtOH	218(4.58),263(2.95), 271(2.95),280(2.95)	18-0265-78
$C_{24}H_{26}Br_4N_4O_8$ 1-Oxa-2-azaspiro[4.5]deca-2,6,8-triene-3-carboxamide, N,N'-1,4-butanediyl-bis[7,9-dibromo-10-hydroxy-8-methoxy-, cis-cis (aerothionin)	EtOH	237(4.49),284(4.24)	5-0066-78
$C_{24}H_{26}N_2$ Benzenamine, 4-(9,10-dihydro-10-methyl-9-acridinyl)-N,N-diethyl-	EtOH	276(4.60)	104-0129-78
hydriodide	EtOH	278(4.33)	104-0129-78
perchlorate	EtOH	280(4.25)	104-0129-78
4,11-Etheno-1H-cyclopenta[b]phenazine, 3a,4,11,11a-tetrahydro-7,8,12,13-tetramethyl-1-(1-methylethylidene)-, (3aα,4β,11β,11aα)-	C_6H_{12}	213(4.61),244(4.57), 246(4.57),333s(3.90)	5-0440-78
$C_{24}H_{26}N_2O$ Phenol, 5-(diethylamino)-2-(9,10-dihydro-10-methyl-9-acridinyl)-, monohydriodide	EtOH	282(4.26)	104-0129-78
$C_{24}H_{26}N_2O_4$ Carbamic acid, [1,3-dihydro-3,3-bis(3-oxobutyl)-2H-indol-2-ylidene]-, phenylmethyl ester	CHCl$_3$	280(4.20),285(4.19), 294s(4.11)	103-0745-78
1,4-Cyclohexadiene-1,4-dicarboxylic acid, 2,5-bis[(phenylmethyl)amino]-, dimethyl ester	EtOH	290(4.36)	48-0991-78
3-Pyrrolidinepropanoic acid, α-(1-aminoethylidene)-4,5-dioxo-β-phenyl-1-(phenylmethyl)-, ethyl ester	EtOH	284.4(4.24)	95-0448-78
$C_{24}H_{26}N_2O_5$ 1H-Pyrrole-3-carboxylic acid, 5-[(acetylphenylamino)methyl]-4,5-dihydro-5-hydroxy-2-methyl-4-oxo-1-(phenylmethyl)-, ethyl ester	EtOH	210(4.11),246(4.12), 330(3.81)	4-1215-78
$C_{24}H_{26}N_8Ni$ Nickel, [6,13-dipentyl-1,4,8,11-tetraazacyclotetradeca-2,4,7,9,11,14-hexaene-2,3,9,10-tetracarbonitrilato-(2-)-N^1,N^4,N^8,N^{11}]-, (SP-4-1)-	EtOH	270(4.39),330(4.86), 390(4.58),530(4.53)	24-2919-78

Compound	Solvent	$\lambda_{max}(\log \epsilon)$	Ref.
$C_{24}H_{26}O_2$			
Cyclohexanone, 4-methyl-2-(3-oxo-1,5-diphenyl-4-pentenyl)-	EtOH	227(4.00),295(4.28)	2-0428-78
Cyclohexanone, 5-methyl-2-(3-oxo-1,5-diphenyl-4-pentenyl)-	EtOH	212(4.15),295(4.34)	2-0428-78
Spiro[5.5]undecane-1,9-dione, 3-methyl-7,11-diphenyl-	EtOH	214(4.18),225(3.83), 260(2.57),266(2.53)	2-0428-78
$C_{24}H_{26}O_3$			
3-Butenoic acid, 2-cyclohexylidene-4-(4-methoxyphenyl)-4-(4-methyl-phenyl)-	EtOH	232(4.5922)	2-0502-78
1,3-Cyclopentanedione, 2-methyl-2-[2-(4,5,6,7-tetrahydro-6,6-dimethyl-2-phenylbenzofuran-4-ylidene)ethyl]-	MeOH	205(4.35),234(3.87), 244(3.70),265s(2.62), 284s(4.02),298(4.18), 310(4.27),325(4.15)	2-0418-78
$C_{24}H_{26}O_4$			
3-Butenoic acid, 2-cyclohexylidene-4,4-bis(4-methoxyphenyl)-	EtOH	250(4.120),297(3.9138)	2-0502-78
Cyclopentanone, 2-[1,5-bis(2-methoxy-phenyl)-3-oxo-4-pentenyl]-	EtOH	233(4.02),283(4.18), 335(4.00)	2-0428-78
Cyclopentanone, 2-[1,5-bis(4-methoxy-phenyl)-3-oxo-4-pentenyl]-	EtOH	228(4.21),328(4.30)	2-0428-78
Spiro[4.5]decane-1,8-dione, 6,10-bis(2-methoxyphenyl)-	EtOH	230(3.74),277(3.64), 283(3.64)	2-0428-78
Spiro[4.5]decane-1,8-dione, 6,10-bis(4-methoxyphenyl)-	EtOH	235(4.22),278(3.41), 284(3.37)	2-0428-78
$C_{24}H_{26}O_6$			
2H-1-Benzopyran-2-one, 3-(2,4-dimeth-oxyphenyl)-5,7-dimethoxy-6-(3-methyl-2-butenyl)-	EtOH	247s(4.08),256s(4.02), 290s(3.75),348(4.29)	94-0135-78
4H-1-Benzopyran-4-one, 3-(2,4-dimeth-oxyphenyl)-5,7-dimethoxy-8-(3-methyl-2-butenyl)-	EtOH	264(4.65)	2-0613-78
$C_{24}H_{26}O_7$			
2,4-Cyclohexadien-1-one, 3,5-diacetoxy-2-(1-acetoxy-3-phenyl-1-propenyl)-4,6,6-trimethyl-	MeOH	212(4.32),218s(4.28), 242s(4.11)	102-2011-78
$C_{24}H_{26}O_{12}$			
2H-1-Benzopyran-2-one, 4-methyl-7-[(2,3,5,6-tetra-O-acetyl-β-D-galactofuranosyl)oxy]-	MeOH	247(3.35),287(3.91), 316(4.10)	136-0023-78F
2H-1-Benzopyran-2-one, 4-methyl-7-[(2,3,4,6-tetra-O-acetyl-α-D-galactopyranosyl)oxy]-	MeOH	247(3.35),287(3.94), 317(4.11)	136-0023-78F
β-	MeOH	247(3.41),287(3.95), 317(4.12)	136-0023-78F
2H-1-Benzopyran-2-one, 4-methyl-7-[(2,3,4,6-tetra-O-acetyl-α-D-glucopyranosyl)oxy]-	MeOH	246(3.38),285(3.95), 315(4.11)	136-0023-78F
β-	MeOH	246(3.42),287(3.95), 317(4.11)	136-0023-78F
$C_{24}H_{26}O_{13}$			
4H-1-Benzopyran-4-one, 2-[4-(β-D-gluco-pyranosyloxy)-3-methoxyphenyl]-5,7-dihydroxy-6,8-dimethoxy-	EtOH EtOH-NaOMe	282(4.35),335(4.24) 283(4.45),309s(4.25), 383(4.13)	142-0053-78B 142-0053-78B

Compound	Solvent	$\lambda_{max}(\log \epsilon)$	Ref.
$C_{24}H_{27}B$ Boron, phenylbis(2,4,6-trimethylphen-yl)-	C_6H_{12}	319(4.08)	147-0177-78A
$C_{24}H_{27}NO_2$ Benzo[f]pyrrolo[1,2-a]quinolin-3(2H)-one, 1,4a,5,6,10b,11,12,12a-octahydro-8-methoxy-1-(4-methylphenyl)-	EtOH	277(3.31),287(3.28)	78-1023-78
$C_{24}H_{27}NO_3$ Isotriphyophylline	EtOH	234(4.66),307(4.02), 321(3.94),336(3.90)	28-0129-78B
$C_{24}H_{27}NO_4$ 1H-Pyrrole-3-carboxylic acid, 4,5-di-hydro-5-hydroxy-5-methyl-2-[4-(1-methylethyl)phenyl]-4-oxo-1-(phen-ylmethyl)-, ethyl ester	EtOH	210(4.15),246(4.10), 330(3.94)	118-0291-78
$C_{24}H_{27}N_3$ 2-Butanone, 1-[(4-methylphenyl)amino]-1-phenyl-, (2-methylphenyl)hydrazone	EtOH	247(4.29),275(4.31)	5-2033-78
$C_{24}H_{27}N_3O_8$ L-Glutamic acid, N-[N-methyl-N-[4-[methyl[(phenylmethoxy)carbonyl]-amino]benzoyl]glycyl]-	MeOH	249(4.02)	87-1165-78
L-Glutamic acid, N-[N-[4-[methyl-[(phenylmethoxy)carbonyl]amino]-benzoyl]-DL-alanyl]-	MeOH	260(4.11)	87-1165-78
$C_{24}H_{28}B_2F_8S_4$ 1,3-Benzodithiol-1-ium, 2,2'-(1,10-dec-anediyl)bis-, bis(tetrafluoroborate)	MeCN-H_2SO_4	253(4.02),257s(4.15), 258(4.16),262(4.19), 314(4.20),335s(3.98)	39-1133-78C
$C_{24}H_{28}Br_2CoP_4$ Cobalt, dibromotetrakis(phenylphos-phine)-	CH_2Cl_2-$PhPH_2$	617(2.12),665s(--), 700(2.12)	24-1420-78
$C_{24}H_{28}N_2O$ Ethanone, 1-(1,2,3,5,6,7-hexahydro-s-indacen-4-yl)-2-(4-phenyl-1-pipera-zinyl)-, hydrochloride	MeOH	249(4.24)	73-0970-78
3H-Pyrrol-3-one, 2-(12-azabicyclo-[9.2.1]tetradeca-11(14),13-dien-13-ylmethylene)-5-phenyl-	MeOH	415(4.36),488(4.39), 510(4.51)	49-0137-78
$C_{24}H_{28}N_2O_5S$ Aspidospermidine-3-carboxylic acid, 3,4-[carbonothioylbis(oxy)]-6,7-di-dehydro-16-methoxy-1-methyl-, methyl ester, (2β,3β,4β,5α,12β,19α)-	EtOH	235(4.28),300(3.64)	23-2560-78
$C_{24}H_{28}N_2O_6$ Aspidospermidine-3-carboxylic acid, 3,4-[carbonylbis(oxy)]-6,7-didehydro-16-methoxy-1-methyl-, methyl ester, (2β,3β,4β,5α,12β,19α)-	MeOH	241(3.77),297(3.69)	33-1554-78
3,4-Furandiol, 2-(1,3-diphenyl-2-imida-zolidinyl)tetrahydro-5-methoxy-, 3,4-	MeOH	292(3.65),353(4.53)	136-0349-78H

Compound	Solvent	$\lambda_{max}(\log \epsilon)$	Ref.
diacetate, [2R-(2α,3β,4α,5β)]- (cont.)			136-0349-78H
$C_{24}H_{28}N_2S_2$ 2H-Isoindole, 1,1'-[1,2-ethanediylbis-(thio)]bis[2-propyl-	EtOH	333(4.16),346s(--)	44-2886-78
$C_{24}H_{28}N_4O_4$ 3-Pyridinecarboxylic acid, 1,4,5,6-tet-rahydro-1-[[(phenylmethyl)amino]carb-onyl]amino]-, ethyl ester	MeOH	232(4.29),291(4.37)	24-2297-78
$C_{24}H_{28}N_4O_{10}S$ 1H-Purine-2,6-dione, 7-[6-O-benzoyl-2,3-O-(1-methylethylidene)-4-O-(methylsulfonyl)-α-D-mannopyranosyl]-3,7-dihydro-1,3-dimethyl-	MeOH	274(3.96)	136-0073-78C
$C_{24}H_{28}N_8$ 1,4,8,11-Tetraazacyclotetradeca-2,4,6,9,11,13-hexaene, 6,13-dipentyl-	CHCl$_3$	330(4.78),350(4.76), 360(4.75),368(4.73), 400(4.18),430(3.95)	24-2919-78
$C_{24}H_{28}N_9P_3$ 1,3,5,2,4,6-Triazatriphosphorine, 2,2-diamino-2,2,4,4,6,6-hexahydro-4,4,6-6-tetrakis(phenylamino)-	MeOH	234(4.67),277(3.96)	121-0945-78
$C_{24}H_{28}N_{10}O_9$ Adenosine, 2',3'-O-(1-methylethyli-dene)-N-[[(9-β-D-ribofuranosyl-9H-purin-6-yl)amino]carbonyl]-	EtOH	267(4.33),285(4.59), 292(4.58)	39-0131-78C
$C_{24}H_{28}O_3$ 1,4-Phenanthrenedione, 4b,5,6,7,8,8a-9,10-octahydro-4b,8,8-trimethyl-3-(phenylmethoxy)-, (4bS-trans)-	EtOH	272(4.23),360(2.70)	23-0517-78
$C_{24}H_{28}O_4$ Cauferidin	EtOH	222(4.33),236(4.30), 291(3.94),325(4.15)	105-0271-78
$C_{24}H_{28}O_5$ Benzenebutanoic acid, 5-[1-ethyl-2-(4-methoxyphenyl)-1-butenyl]-2-methoxy-γ-oxo-	MeOH	236(4.38),320(3.39)	5-0726-78
2H-1-Benzopyran-2-one, 7-[(1,4,4a,5,6-7,8,8a-octahydro-4-hydroxy-2,5,5,8a-tetramethyl-6-oxo-1-naphthalenyl)-methoxy]-, [1S-(1α,4α,4aβ,8aα)]-(ferocaulin)	EtOH	217(4.04),243(3.45), 253(3.22),297(3.84), 326(4.10)	105-0267-78
[1S-(1α,4β,4aβ,8aα)]-	EtOH	216(4.26),242(3.76), 253(3.60),295(3.97), 325(4.20)	105-0267-78
Ferocaulidin	EtOH	218(4.31),234(4.18), 297(4.00),325(4.22)	105-0267-78
$C_{24}H_{28}S_4$ 1,3-Benzodithiol-1-ium, 2,2'-(1,10-dec-anediyl)bis-, bis(tetrafluoroborate)	MeCN-H$_2$SO$_4$	253(4.02),257s(4.15), 258(4.16),262(4.19), 314(4.20),335s(3.98)	39-1133-78C

Compound	Solvent	λ_{max}(log ϵ)	Ref.
$C_{24}H_{29}NO_3$			
17,21-Cyclo-19-norpregna-1,3,5(10)-triene-20-carbonitrile, 20-[(formyloxy)-methyl]-3-methoxy-, (20S)-	MeOH	219(3.93),229(3.83), 273(3.18),279(3.30), 287(3.28)	24-3094-78
17,21-Cyclo-19-norpregna-1,3,5(10)-triene-20-carboxylic acid, 20-cyano-3-methoxy-, methyl ester, (20S)-	MeOH	218(3.96),229(3.86), 273(3.20),278(3.26), 287(3.29)	24-3094-78
19-Norpregna-1,3,5(10),17(20)-tetraen-21-oic acid, 20-cyano-3-methoxy-, ethyl ester, (17E)-	MeOH	231(4.30),276(3.40), 287(3.29)	24-1533-78
$C_{24}H_{29}NO_{12}$			
Menisdaurin pentaacetate	EtOH	255.5(4.34)	94-1677-78
$C_{24}H_{29}N_2$			
Cyclohexylium, 2,6-bis[[4-(dimethylamino)phenyl]methylene]-, tetrafluoroborate	$C_2H_4Cl_2$	780(5.27)	104-1197-78
$C_{24}H_{30}$			
Benzo[1,2:3,4:5,6]triscyclooctene, 1,2,5,6,7,8,11,12,13,14,17,18-dodecahydro-	hexane	214(4.54),232s(4.03), 274(2.40),280s(2.26)	33-1663-78
Pyrene, 2,7-bis(1,1-dimethylethyl)-1,9,10,10a-tetrahydro-	EtOH	250(4.38),258(4.67), 268(4.73),303(3.86), 316(3.98),329(3.86), 350(2.80)	1-0720-78
$C_{24}H_{30}ClN_3O_4$			
1H-Azecino[5,4-b]indole-6-acetic acid, 13b-chloro-3-cyano-8-ethoxy-5-ethyl-α-formyl-2,3,4,5,6,7,8,9-octahydro-, methyl ester	EtOH	221(4.66),273(4.37), 290s(4.07)	95-0950-78
	EtOH-HCl	221(4.60),245s(4.09), 283s(3.85),290(3.77)	95-0950-78
$C_{24}H_{30}CuN_2O_2$			
Copper, [[6,6'-[1,2-ethanediylbis(nitrilomethylidyne)]bis[2-methyl-5-(1-methylethenyl)-2-cyclohexen-1-onato]](2-)-N,N',O,O']-	benzene	280(4.11),550(2.63), 650s(2.30)	65-1474-78
$C_{24}H_{30}N_2O_4$			
Indolo[2,3-a]quinolizine-2β-acetic acid, 3-(tert-butoxycarbonyl)-1,2,6,7,12,12b-hexahydro-4-methyl-, methyl ester	MeOH	225(4.43),298(4.53)	24-1540-78
Indolo[2,3-a]quinolizine-3-carboxylic acid, 2β-(acetoxymethyl)-1,2,6,7,12-12b-hexahydro-4-methyl-, tert-butyl ester	MeOH	225(4.50),295(4.48)	24-1540-78
$C_{24}H_{30}N_2O_5$			
L-Leucine, N-[4-[methyl(phenylmethoxy)-carbonyl]amino]benzoyl]-, ethyl ester	MeOH	260(4.09)	87-1162-78
$C_{24}H_{30}N_2O_5S$			
Aspidospermidine-3-carboxylic acid, 3,4-[carbonothioylbis(oxy)]-16-methoxy-1-methyl-, methyl ester, (2β,3β,4β,5α,12β,19α)-	EtOH	208(4.53),233(4.33), 298(3.69)	35-4220-78

Compound	Solvent	λ_{max}(log ϵ)	Ref.
$C_{24}H_{30}N_4O_2$ Pyrano[2,3-c]pyrazol-6(2H)-one, 3,4-di-methyl-5-[2-[4-(4-methylphenyl)-1-piperazinyl]ethyl]-2-(2-propenyl)-	EtOH	249(4.14),306(4.26)	95-0335-78
$C_{24}H_{30}N_4O_3$ Pyrano[2,3-c]pyrazol-6(2H)-one, 5-[2-[4-(2-methoxyphenyl)-1-piperazinyl]-ethyl]-3,4-dimethyl-2-(2-propenyl)-	EtOH	246(3.95),307(4.24)	95-0335-78
$C_{24}H_{30}N_4O_6S_2$ Benzenesulfonamide, N,N'-[(2,4-dioxo-1,3(2H,4H)-pyrimidinediyl)di-3,1-propanediyl]bis[4-methyl-	$CHCl_3$	302(4.26)	70-0149-78
$C_{24}H_{30}O_2$ 1,3-Benzenediol, 2-(3,7-dimethyl-2,6-octadienyl)-5-(2-phenylethyl)-	EtOH	225(3.98),280(3.58)	102-2005-78
$C_{24}H_{30}O_3$ Cauferin	EtOH	217(3.96),241(3.50), 296(3.80),325(3.95)	105-0271-78
$C_{24}H_{30}O_4$ 2H-1-Benzopyran-2-ethanol, 3,4-dihydro-2,5,7,8-tetramethyl-6-(phenylmeth-oxy)-, acetate, (±)-	EtOH	280(3.32),286(3.37)	33-0837-78
$C_{24}H_{30}O_5$ Azuleno[5,6-g]benzofuran-4,5-diol, 2,3,7,8,9,9a,10,11-octahydro-2,6,9,9-tetramethyl-, diacetate	EtOH	251(3.91),298(3.59)	23-0733-78
Benzenepentanoic acid, 4-methoxy-γ-(4-methoxyphenyl)-β-methyl-δ-oxo-, 1,1-dimethylethyl ester	MeOH	220(4.20),274(4.21)	5-0726-78
8H-Dipyrano[3,2-b:3',2'-g][1]benzopyr-an-8-one, 9-(1,1-dimethyl-2-propen-yl)-2,3,4,4a,12,12a-hexahydro-2-(1-hydroxy-1-methylethyl)-4a-methyl-(clausmarin A)	EtOH	225(4.12),248(3.59), 250(3.49),298(3.89), 332(4.83)	77-0281-78
$C_{24}H_{30}O_6$ Benzeneethanol, β-[[6-hydroxy-2,4-di-methoxy-3-(3-methyl-2-butenyl)phen-yl]methylene]-2,4-dimethoxy-	EtOH	267(3.81),286(3.76)	94-0135-78
8α-Estra-1,3,5(10)-triene-1,3,17β-tri-ol, triacetate	MeOH	266(2.68),273(2.63)	24-0939-78
$C_{24}H_{30}O_{10}$ Fasciculide B	EtOH	212(4.18),285(3.52)	112-0267-78
$C_{24}H_{30}O_{13}$ Cornoside, pentaacetate	EtOH	238(3.92)	94-2111-78
$C_{24}H_{31}ClO_2$ 2H-1-Benzopyran-5-ol, 2-(4-chloro-4-methylpentyl)-3,4-dihydro-2-methyl-7-(2-phenylethyl)-	EtOH	211(4.29),233s(3.78), 285(3.24)	102-2005-78
$C_{24}H_{31}NO_3$ 19-Norpregna-1,3,5(10)-trien-21-oic	MeOH	278(3.31),287(3.29)	24-1533-78

Compound	Solvent	$\lambda_{max}(\log \epsilon)$	Ref.
acid, 20-cyano-3-methoxy-, ethyl ester (cont.)			24-1533-78
$C_{24}H_{31}NO_7$			
α-D-Galactopyranose, 1,2:3,4-bis-O-(1-methylethylidene)-6-C-[3-(2,4,6-tri-methylphenyl)-5-isoxazolyl]-, (R)-	MeOH	222(4.14),272(3.07)	136-0357-78H
(S)-	MeOH	222(4.13),272(2.95)	136-0357-78H
1H-Pyrrole-2,4-dicarboxylic acid, 2-(3-ethoxy-3-oxopropyl)-2,3-dihydro-3-oxo-1-(phenylmethyl)-, 2-(1,1-dimeth-ylethyl) 4-ethyl ester	EtOH	248(4.14),317(4.06)	94-3521-78
2H-Pyrrole-2,2-dipropanoic acid, 4-(ethoxycarbonyl)-1,3-dihydro-3-oxo-1-(phenylmethyl)-, diethyl ester	EtOH	245(4.19),313(4.06)	94-3521-78
$C_{24}H_{31}N_3O_2S$			
1-Butanone, 1-[10-[3-(4-methyl-1-piper-azinyl)propyl]-10H-phenothiazin-2-yl]-, S-oxide	MeOH	204(4.32),252(4.53), 311(3.80),369(3.62)	133-0248-78
$C_{24}H_{31}N_3O_3$			
1H-Pyrrolo[3,4-b]pyridine-3-carboxylic acid, 4,5,6,7-tetrahydro-2-methyl-7-oxo-4-phenyl-6-[2-(1-piperidinyl)eth-yl]-, ethyl ester	EtOH	223(4.24),259(4.11), 342(3.65)	95-0448-78
1H-Pyrrolo[3,4-b]pyridine-3-carboxylic acid, 4,5,6,7-tetrahydro-2-methyl-7-oxo-4-phenyl-6-[3-(1-pyrrolidinyl)-propyl]-, ethyl ester	EtOH	224(4.24),259(4.10), 342(3.64)	95-0448-78
$C_{24}H_{31}N_5O_7$			
D-Valine, N-[N-[4-methyl-2-nitro-3-(phenylmethoxy)benzoyl]-L-threonyl]-, hydrazide	MeOH	209(4.876)	87-0607-78
$C_{24}H_{31}N_8O_{16}P$			
Adenosine, uridylyl-(3'→5')-N-[[(1-car-boxy-2-hydroxypropyl)amino]carbonyl]-, (S)-	pH 2	261s(4.35),269(4.37), 276s(4.36)	19-0021-78
	pH 12	261s(4.34),270(4.37), 278(4.30),301(4.01)	19-0021-78
	MeOH	260s(4.39),269(4.41), 276s(4.34)	19-0021-78
$C_{24}H_{32}$			
Pyrene, 2,7-bis(1,1-dimethylethyl)-1,2,3,4,5,12-hexahydro-	EtOH	230(4.68),236(4.89), 282(3.72),292(3.68), 314(3.28),328(3.28)	1-0720-78
Pyrene, 2,7-bis(1,1-dimethylethyl)-1,2,3,6,7,8-hexahydro-	EtOH	229(4.63),235(4.80), 288(3.96),298(4.06), 310(3.90),316(3.81), 330(3.54)	1-0720-78
Pyrene, 2,7-bis(1,1-dimethylethyl)-1,4,5,9,10,11-hexahydro-	EtOH	324(4.04),338(4.12), 353(3.96)	1-0720-78
$C_{24}H_{32}N_2O$			
Methanone, (1-methyl-1H-imidazol-2-yl)-[4b,5,6,7,8,8a,9,10-octahydro-4b,8-dimethyl-2-(1-methylethyl)-9-phenan-threnyl]-	MeOH	274(4.24)	33-2843-78

$C_{24}H_{32}N_2O_2-C_{24}H_{32}O_4$

Compound	Solvent	$\lambda_{max}(\log \epsilon)$	Ref.
$C_{24}H_{32}N_2O_2$			
Benzoic acid, 4-[[(octylphenylamino)-methylene]amino]-, ethyl ester	isoPrOH	314(4.45)	117-0097-78
$C_{24}H_{32}N_2O_5$			
1H-Pyrrole-3-carboxylic acid, 5-[[acetyl(4-methylphenyl)amino]methyl]-1-cyclohexyl-4,5-dihydro-5-hydroxy-2-methyl-4-oxo-, ethyl ester	EtOH	210(4.02),246(4.13), 333(3.81)	4-1215-78
$C_{24}H_{32}N_2O_6$			
Benzoic acid, 4-(dimethylamino)-, 1,2-ethanediylbis(oxy-2,1-ethanediyl) ester	MeOH	228(--),310(4.75)	121-0661-78
1H-Pyrrole-3-carboxylic acid, 5-[[acetyl(4-methoxyphenyl)amino]methyl]-1-cyclohexyl-4,5-dihydro-5-hydroxy-2-methyl-4-oxo-, ethyl ester	EtOH	211(4.02),234(4.18), 334(3.82)	4-1215-78
$C_{24}H_{32}N_2O_6S$			
Aspidospermidine-3-carboxylic acid, 6,7-didehydro-N-methoxy-1-methyl-5-[(methylsulfonyl)oxy]-, methyl ester, (2β,3α,4β,5α,12β,19α)-	MeOH	250(3.85),301(3.69)	33-1554-78
$C_{24}H_{32}N_2O_7S$			
Aspidospermidine-3-carboxylic acid, 6,7-didehydro-3-hydroxy-16-methoxy-1-methyl-4-[(methylsulfonyl)oxy]-, methyl ester	MeOH	252(3.84),304(3.72)	33-1554-78
Aspidospermidine-3-carboxylic acid, 6,7-didehydro-4-hydroxy-16-methoxy-1-methyl-3-[(methylsulfonyl)oxy]-, methyl ester	MeOH	248(3.86),302(3.76)	33-1554-78
$C_{24}H_{32}N_4O_3$			
1H-Pyrrolo[3,4-b]pyridine-3-carboxylic acid, 4,5,6,7-tetrahydro-2-methyl-6-[2-(4-methyl-1-piperazinyl)ethyl]-7-oxo-4-phenyl-, ethyl ester	EtOH	223(4.23),259(4.10), 341(3.65)	95-0448-78
$C_{24}H_{32}N_6O_4$			
1,2-Hydrazinedicarboxylic acid, 1-[[4-(dimethylamino)phenyl][[4-(dimethylamino)phenyl]methylene]hydrazono]-methyl]-, diethyl ester	MeOH	243(4.26),325s(4.28), 388(4.61)	18-3312-78
$C_{24}H_{32}O_3$			
Androsta-1,4,6-trien-3-one, 17β-acetoxy-16β-(1-methylethyl)-	EtOH	222(3.99),252(3.98), 296(4.02)	94-1718-78
Cyclopropa[4,5]card-20(22)-enolide, 3',4-dihydro-3-oxo-, (4α,5S,14α)-	MeOH	220(4.24)(end abs.)	44-2334-78
$C_{24}H_{32}O_4$			
2H-1-Benzopyran-2-one, 7-[[(decahydro-6-hydroxy-1,2,4a,5-tetramethyl-1-naphthalenyl)methoxy]- (fecarpin)	EtOH	220(4.14),243(3.58), 290(3.78),327(3.98)	105-0487-78
Oxepino[2,3-b]benzofuran, 2,9-bis(1,1-dimethylethyl)-4,7-diethoxy-	C_6H_{12}	214(4.37),266(4.07), 303(3.80),399(3.54)	12-1061-78
Tetradecanoic acid, 5,8-dihydro-5,8-dioxo-1-naphthalenyl ester	EtOH	227s(4.12),242(4.19), 248s(4.19),257s(4.03),	87-0026-78

Compound	Solvent	$\lambda_{max}(\log \epsilon)$	Ref.
(cont.)		338(3.45)	87-0026-78
$C_{24}H_{32}O_5$			
Clausmarin A, dihydro-	EtOH	223(4.21),246(3.58), 258(3.48),298(3.88), 333(4.39)	77-0281-78
21,24-Dinorchola-5,22(23)-dien-20-one, 3-acetoxy-16β,23-epoxy-17-hydroxy-	EtOH	271(3.92)	70-0160-78
Isospongiaquinone acetate	MeOH	278(3.97),366(2.75)	12-2685-78
$C_{24}H_{32}O_6$			
Allocyafrin B₄, O,O-diacetyl-	MeOH	235(4.08)	23-2113-78
D-Homo-17-oxaandrosta-5,14-dien-16-one, 17a-acetoxy-3,19-epoxy-3-methoxy-4,4-dimethyl-, (3β,13α)-	n.s.g.	232(4.06)	33-3087-78
$C_{24}H_{32}O_7$			
Megaphone acetate	EtOH	269(3.02),279s(2.83)	44-0586-78
$C_{24}H_{32}O_8$			
Aceroside I	EtOH and EtOH-NaOH	265(3.374),270s(--)	94-2805-78
Spiro[cyclopropane-1,2'(1'H)-phenanthrene]-1',4'(3'H)-dione, 7',9'-diacetoxy-4'b,5',6',7',8',8'a,9',10'-octahydro-3',10'-dihydroxy-2,4'b,8',8'-tetramethyl-	MeOH	231(4.10)	33-0871-78
$C_{24}H_{32}O_9$			
Coleon Y	MeOH	235(3.99)	33-0871-78
$C_{24}H_{33}NO_4$			
3,5-Pyridinedicarboxylic acid, 1,4-dihydro-2,6-dimethyl-1-pentyl-4-phenyl-, diethyl ester	EtOH	205(4.19),243(4.17), 349(3.75)	103-1226-78
$C_{24}H_{33}NO_5$			
3,5-Pyridinedicarboxylic acid, 1-butyl-1,4-dihydro-4-(4-methoxyphenyl)-2,6-dimethyl-, diethyl ester	EtOH	202(4.31),226(4.20), 257s(4.06),284s(3.42), 349(3.73)	103-1226-78
$C_{24}H_{33}NO_7$			
Dihydroanhydro-1,14-didehydrooxodelcosine	MeOH	300(2.11)	18-0248-78
$C_{24}H_{33}N_3O_3S$			
Ethanol, 2-[2-[4-[2-methyl-3(10H)-phenothiazin-10-yl]propyl]-1-piperazinyl]ethoxy]-, S-oxide	MeOH	231(4.46),276(4.09), 300(3.98),345(3.80)	133-0248-78
$C_{24}H_{34}N_2O_3$			
Aspidospermidine-3-methanol, 16-methoxy-1-methyl-, acetate	EtOH	212(4.54),252(3.88), 304(3.75)	35-4220-78
Piperidine, 1-[5-(4-methoxyphenyl)-3-methyl-4-(3-methyl-2-butenyl)-6-nitro-2-cyclohexen-1-yl]-	n.s.g.	236(3.61),267(3.18)	2-1062-78
$C_{24}H_{34}O_2$			
1,3-Benzenediol, 2-(3,7-dimethyloctyl)-5-(2-phenylethyl)-	EtOH	207(3.28)	102-2005-78

Compound	Solvent	$\lambda_{max}(\log \epsilon)$	Ref.
$C_{24}H_{34}O_3$			
Androsta-4,6-dien-3-one, 17β-acetoxy-16β-(1-methylethyl)-	EtOH	282(4.37)	94-1718-78
Cyclopropa[4,5]card-20(22)-enolide, 3',4-dihydro-3-hydroxy-, (3β,4α,5S-14α)-	MeOH	220(4.20)(end abs.)	44-2334-78
$C_{24}H_{34}O_3S_2$			
2-Phenanthrenecarboxaldehyde, 7-acetoxy-1-(1,3-dithian-2-ylidenemethyl)-1,2,3,4,4a,4b,5,6,7,8,10,10a-dodecahydro-2,4b-dimethyl-, [1R-(1α,2β,4aβ,4bα,7α,10aα)]-	n.s.g.	260(3.98)	33-3087-78
$C_{24}H_{34}O_4$			
5H-Inden-5-one, 4-[2-(3,5-dimethoxyphenyl)ethyl]-1-(1,1-dimethylethoxy)-1,2,3,6,7,7a-hexahydro-7a-methyl-, (1S-cis)-	MeOH	226(4.09),250(4.01), 280(3.43)	24-0939-78
$C_{24}H_{34}O_5$			
Pregn-4-ene-3,20-dione, 17,21-dihydroxy-16α-methyl-, 21-acetate	n.s.g.	241(4.20)	80-0921-78
Pregn-4-ene-3,20-dione, 11β-hydroxy-17,21-[(1-methylethylidene)bis(oxy)]-	MeOH	242(4.21)	44-3405-78
$C_{24}H_{35}ClO_3$			
Pregn-20-ene-20-carboxaldehyde, 3-acetoxy-21-chloro-, (3β,5α)-	EtOH	247(4.1)	23-0424-78
$C_{24}H_{35}N_5P$			
Phosphorus(1+), (4-diazenyl-N,N-dimethylbenzenaminato)phenyldi-1-piperidinyl-, (T-4)-, tetrafluoroborate	n.s.g.	277(3.97),463(4.65)	123-0190-78
$C_{24}H_{36}ClNO_4$			
Carbamic acid, [(17-chloro-3-oxoandrost-4-en-17-yl)methoxymethyl]-, ethyl ester	EtOH	240(4.20)	23-0119-78
$C_{24}H_{36}O_3$			
Androst-4-en-3-one, 17β-acetoxy-16β-(1-methylethyl)-	EtOH	239(4.23)	94-1718-78
2H-1-Benzopyran-2-one, 7-hydroxy-4-pentadecyl-	EtOH	326(4.18)	51-0652-78
anion	EtOH	376(4.32)	51-0652-78
$C_{24}H_{36}O_3S_2$			
Androst-5-en-16-one, 3-acetoxy-17-hydroxy-, cyclic 16-(1,3-propanediyl mercaptole), (3β,17β)-	n.s.g.	225(2.90),250(2.95), 330(1.95)	33-3087-78
1H-Benz[e]indene-7-acetaldehyde, 3-acetoxy-6-(1,3-dithian-2-ylidene)methyl)dodecahydro-3a,6-dimethyl-, [3S-(3α,3aα,5aβ,6β,7α,9aα,9bβ)]-	n.s.g.	213(3.81),259(3.76)	33-3087-78
$C_{24}H_{36}O_7$			
Tetrahydromegaphone acetate	EtOH	269(2.94),278(2.82)	44-0586-78
$C_{24}H_{36}O_8$			
Oleandomycin diacetylanhydroglycone	MeOH	234.0(4.12)	35-6733-78

Compound	Solvent	λ_{max}(log ϵ)	Ref.
$C_{24}H_{37}NO_7$			
Dihydroanhydrooxodelcosine	MeOH	307(1.94)	18-0248-78
$C_{24}H_{38}N_2O_6$			
Propanoic acid, 3-[4-[2-[[2-[[(1,1-di-methylethoxy)carbonyl]methylamino]-4-methyl-1-oxopentyl]methylamino]eth-yl]phenoxy]-, (S)-	MeOH	277(3.20),283(3.13)	35-8202-78
$C_{24}H_{38}O_3$			
Benzoic acid, 2-(8-heptadecenyl)-6-hy-droxy-, (Z)- (merulinic acid C)	MeOH	210(4.28),243(3.71), 309(3.50)	64-0912-78C
$C_{24}H_{38}O_3S_2$			
5α-Androstan-2-one, 17β-acetoxy-3β-hy-droxy-, cyclic 2-(1,3-propanediyl mercaptole)	n.s.g.	228s(--),240(3.08)	33-3087-78
$C_{24}H_{38}O_4$			
Benzoic acid, 2-(8-heptadecenyl)-4,6-dihydroxy-, (Z)-	MeOH	225(4.39),259(3.99), 299(3.62)	64-0912-78C
Benzoic acid, 2-hydroxy-6-(16-hydroxy-8-heptadecenyl)-, (Z)-	MeOH	210(4.28),243(3.71), 309(3.50)	64-0912-78C
$C_{24}H_{38}O_5$			
Ardisianone	EtOH	268(4.12),363(3.01)	18-0943-78
$C_{24}H_{38}O_{10}Se_2$			
α-D-Galactopyranose, 6,6'-diselenobis-[6-deoxy-1,2:3,4-bis-O-(1-methyleth-ylidene)-	MeOH	312(2.40)	136-0069-78E
$C_{24}H_{38}Si_2$			
Silane, (bicyclo[4.2.0]octa-2,4-diene-7,8-diyldi-1,2-ethynediyl)bis[tri-ethyl-, (1α,6α,7α,8β)-	hexane	296(4.10)	78-1037-78
Silane, 3,5,7,9-dodecatetraene-1,11-di-yne-1,12-diylbis[triethyl-, (all-E)-	hexane	322(4.46),338(4.78), 355(5.01),375(5.05)	78-1037-78
(E,E,Z,Z)-	hexane	317(4.49),333(4.77), 349(4.95),369(4.89)	78-1037-78
$C_{24}H_{40}O_3$			
1,3-Cyclohexanedione, 2-(1-oxo-9-octa-decenyl)-, (Z)-	MeOH	204s(3.86),233(4.16), 273(4.17)	77-1075-78
$C_{24}H_{40}O_4$			
1,3-Cyclohexanedione, 4-hydroxy-2-(1-oxo-9-octadecenyl)-	MeOH	233(4.07),273(4.04)	77-1075-78

Compound	Solvent	$\lambda_{max}(\log \epsilon)$	Ref.
$C_{25}H_{14}N_2O_3$			
Chryseno[6,5-d]oxazole, 2-(2-nitro-phenyl)-	dioxan	297(4.16),340(4.08), 365(4.09),384(4.12)	48-0986-78
Chryseno[6,5-d]oxazole, 2-(4-nitro-phenyl)-	dioxan	297(4.27),394(4.44)	48-0986-78
$C_{25}H_{15}BrN_3O_2P$			
Phosphonium, triphenyl-, 2-bromo-3,4-dicyano-5-nitro-2,4-cyclopentadien-1-ylide	MeCN	222(4.54),393(4.08)	18-2605-78
$C_{25}H_{15}Cl_2NO$			
1,2-Benzisoxazole, 6-chloro-3-[3-chloro-4-[2-(1-naphthalenyl)ethenyl]-phenyl]-	DMF	347(4.44)	33-2904-78
1,2-Benzisoxazole, 6-chloro-3-[3-chloro-4-[2-(2-naphthalenyl)ethenyl]-phenyl]-	DMF	291(4.34),338(4.62)	33-2904-78
$C_{25}H_{15}NOS$			
Thiocyanic acid, 2,3-diphenylcyclopenta[b][1]benzopyran-3-yl ester	THF	277(4.53),360(4.38), 486(3.30)	103-1075-78
$C_{25}H_{15}NO_4$			
Chryseno[5,6-d]-1,3-dioxole, 2-(4-nitrophenyl)-	dioxan	333(4.06),347(4.16), 376(3.48),396(3.47)	48-0986-78
Spiro[chrysene-5(6H),2'-oxiran]-6-one, 3'-(2-nitrophenyl)-, cis	dioxan	350(3.90)	48-0986-78
Spiro[chrysene-6(5H),2'-oxiran]-5-one, 3'-(2-nitrophenyl)-	dioxan	346(3.86)	48-0986-78
Spiro[chrysene-6(5H),2'-oxiran]-5-one, 3'-(4-nitrophenyl)-	dioxan	348.5(3.97)	48-0986-78
$C_{25}H_{15}N_3O_3$			
Acetamide, N-(5,10-dihydro-5,10-dioxo-naphtho[1',2',3':4,5]quino[2,1-b]-quinazolin-4-yl)-	n.s.g.	435(3.76)	2-0103-78
Acetamide, N-(5,10-dihydro-5,10-dioxo-naphtho[1',2',3':4,5]quino[2,1-b]-quinazolin-6-yl)-	n.s.g.	435(3.77)	2-0103-78
$C_{25}H_{15}N_4P$			
1-Propene-1,1,2,3-tetracarbonitrile, 3-(triphenylphosphoranylidene)-	CH_2Cl_2	230(4.65),268(3.80), 276(3.64),401(3.80), 405(4.28)	39-1237-78C
$C_{25}H_{16}ClNO$			
1,2-Benzisoxazole, 6-chloro-3-[4-[2-(1-naphthalenyl)ethenyl]phenyl]-	DMF	342(4.52)	33-2904-78
1,2-Benzisoxazole, 6-chloro-3-[4-[2-(2-naphthalenyl)ethenyl]phenyl]-	DMF	292(4.35),337(4.70)	33-2904-78
$C_{25}H_{16}N_2O$			
8H-Cyclohepta[b]quinoxalin-8-one, 7,9-diphenyl-	EtOH	228(4.25),277(4.63), 302(4.55),390(4.20)	56-2045-78
$C_{25}H_{16}N_2O_2$			
1,4-Acridinedione, 9-phenyl-2-(phenyl-amino)-	dioxan	227(4.58),284(4.49), 316s(4.2),438(3.69)	150-4901-78
Naphtho[1',2',3':4,5]quino[2,1-b]quin-azoline-5,10-dione, 7,8-dimethyl-	n.s.g.	450(3.84)	2-0103-78

Compound	Solvent	$\lambda_{max}(\log \epsilon)$	Ref.
$C_{25}H_{16}N_2O_3$			
8H-Cyclohepta[b]quinoxalin-8-one, 7,9-diphenyl-, 5,10-dioxide	EtOH	257(4.33),339(4.55), 413(3.81)	56-2045-78
$C_{25}H_{16}N_2O_3S_2$			
1H-4,11a-Epithiopyrrolo[3',4':3,4]pyrido[2,1-b]benzothiazole-1,3,5(2H,4H)-dione, 3a,11b-dihydro-2,4-diphenyl-, (3aα,4α,11aα,11bα)-	MeOH	267(4.50)	44-2697-78
$C_{25}H_{16}N_4OS$			
[1,2,4]Triazolo[1,5-a]pyridine, 2-[4-[2-[5-(2-benzoxazolyl)-2-thienyl]-ethenyl]phenyl]-	DMF	395(4.81)	33-0142-78
$C_{25}H_{16}N_4O_2$			
[1,2,4]Triazolo[1,5-a]pyridine, 2-[4-[5-(2-benzoxazolyl)-2-furanyl]ethenyl]phenyl]-	DMF	387(4.81),408(4.62)	33-0142-78
$C_{25}H_{16}O_2$			
2-Benzopyrylium, 4-hydroxy-1-(1-naphthalenyl)-3-phenyl-, hydroxide, inner salt	EtOH	542(4.26)	103-0493-78
Cyclopenta[b][1]benzopyran-3-carboxaldehyde, 1,2-diphenyl-	THF	490(3.10)	103-1070-78
	THF	260(4.46),305(4.72), 370(4.38),490(3.10)	103-1075-78
6H-Indeno[1,2-b]oxiren-6-one, 1a,6a-dihydro-1a-(1-naphthalenyl)-6a-phenyl-	EtOH	267(4.0),360s(2.3)	103-0493-78
6H-Indeno[1,2-b]oxiren-6-one, 1a,6a-dihydro-6a-(1-naphthalenyl)-1a-phenyl-	EtOH	275(4.5),355s(2.2)	103-0493-78
$C_{25}H_{16}O_4S_4$			
2H-Pyran-2,4,6(3H,5H)-trione, 3,5-bis-(4-methyl-5-phenyl-3H-1,2-dithiol-3-ylidene)-	dioxan	244(4.51),305(4.29), 353s(--),433(4.13), 501s(--),526(4.63)	78-2175-78
$C_{25}H_{16}O_8$			
Kolaflavanone	MeOH	292(4.51),329(3.84)	39-0532-78C
$C_{25}H_{16}O_{12}$			
Spiro[naphtho[2,3-b]furan-2(3H),2'(5'H)-oxireno[d]benzo[1,2-b:5,4-c']dipyran]-5,5',8-trione, 1'a,9'b-dihydro-3,4,4',9-tetrahydroxy-7-methoxy-7'-methyl-	CHCl$_3$	315(3.93),362(3.80), 378s(3.74),484s(3.57), 513(3.80),550s(3.57)	78-0399-78
	H$_2$SO$_4$	410(3.73),494s(3.83), 528(3.96),570(3.88)	78-0399-78
$C_{25}H_{17}Br$			
Benzo[c]phenanthrene, 11-(bromomethyl)-1-phenyl-	MeOH	232(4.70),285s(4.64), 292(4.67),325s(4.01)	78-0769-78
$C_{25}H_{17}ClN_2$			
Imidazo[1,2-a]pyridine, 2-(4-chlorophenyl)-7-[2-(1-naphthalenyl)ethenyl]-	DMF	374(4.61)	33-0129-78
Imidazo[1,2-a]pyridine, 2-(4-chlorophenyl)-7-[2-(2-naphthalenyl)ethenyl]-	DMF	369(4.73),387(4.69)	33-0129-78
$C_{25}H_{17}ClN_2S$			
Quinoline, 6-chloro-2-(4-methyl-2-phenyl-5-thiazolyl)-4-phenyl-	dioxan	365(3.77)	97-0400-78

Compound	Solvent	$\lambda_{max}(\log \epsilon)$	Ref.
$C_{25}H_{17}F_6N_4S_4$ Benzo[1,2-d:4,3-d']bisthiazolium, 3-ethyl-2-[3-[3-ethyl-7-(trifluoromethyl)benzo[1,2-d:4,3-d']bisthiazol-2(3H)-ylidene]-1-propenyl]-7-(trifluoromethyl)-, perchlorate	EtOH	586(5.15)	124-0637-78
$C_{25}H_{17}N$ 9H-Indeno[2,1-c]pyridine, 9-methylene-1,3-diphenyl-	EtOH	205(4.82),260(4.80), 328(4.20),396(2.50)	103-0997-78
$C_{25}H_{17}NO$ 1,2-Benzisoxazole, 3-[4-[2-(1-naphthalenyl)ethenyl]phenyl]-	DMF	343(4.51)	33-2904-78
1,2-Benzisoxazole, 3-[4-[2-(2-naphthalenyl)ethenyl]phenyl]-	DMF	337(4.71)	33-2904-78
1,2-Benzisoxazole, 6-[2-(1-naphthalenyl)ethenyl]-3-phenyl-	DMF	342(4.45)	33-2904-78
1,2-Benzisoxazole, 6-[2-(2-naphthalenyl)ethenyl]-3-phenyl-	DMF	337(4.66)	33-2904-78
$C_{25}H_{17}N_3O_3$ Benzo[f]quinoline, 3-(4-methoxyphenyl)-8-nitro-1-(3-pyridinyl)-	EtOH	218(4.61),277(4.70), 289(4.73),354(4.16), 369(4.18)	103-1337-78
$C_{25}H_{18}$ Benzo[c]phenanthrene, 11-methyl-1-phenyl-	MeOH	205(4.65),229(4.66), 250s(4.29),282(4.58), 289(4.59),310s(4.14)	78-0769-78
Diindeno[1,2,3-cd:1',2',3'-ij]azulene, 4,6,11-trimethyl-	dioxan	231(3.79),245s(3.69), 275s(3.87),285s(3.99), 309(4.42),348(3.33), 367(3.27),388(3.22), 411s(2.46),426(2.45), 464(1.59),502(1.78), 621(2.25),833(1.83), 928(1.81),1111(1.41)	89-0763-78
$C_{25}H_{18}ClN_3$ [1,2,4]Triazolo[1,5-a]pyridine, 2-[3-chloro-4-[2-(1-naphthalenyl)ethenyl]-phenyl]-6-methyl-	DMF	348(4.56)	33-0142-78
[1,2,4]Triazolo[1,5-a]pyridine, 2-[3-chloro-4-[2-(2-naphthalenyl)ethenyl]-phenyl]-6-methyl-	DMF	347(4.71)	33-0142-78
$C_{25}H_{18}N_2$ Imidazo[1,2-a]pyridine, 7-[2-(1-naphthalenyl)ethenyl]-2-phenyl-	DMF	375(4.60)	33-0129-78
Imidazo[1,2-a]pyridine, 7-[2-(2-naphthalenyl)ethenyl]-2-phenyl-	DMF	368(4.70),386(4.66)	33-0129-78
$C_{25}H_{18}N_2OS$ 2(10H)-Phenazinone, 3-[(4-methylphenyl)thio]-10-phenyl-	CHCl$_3$	452(3.27),519(2.96)	103-0795-78
$C_{25}H_{18}N_2O_2$ Phenol, 2,2'-benzo[f]quinazoline-1,3-diylbis-, monomethyl deriv.	EtOH	205(4.63),220(4.63), 283(4.64),323(4.30)	103-0082-78

Compound	Solvent	$\lambda_{max}(\log \epsilon)$	Ref.
$C_{25}H_{18}N_2S$ Imidazo[1,2-a]pyridine, 2-[1,1'-biphenyl]-4-yl-7-[2-(2-thienyl)ethenyl]-	DMF	374(4.71),395(4.66)	33-0129-78
$C_{25}H_{18}N_3$ Methylium, tri-1-H-indol-3-yl-, perchlorate	MeCN	500(7.02)[sic]	103-0345-78
$C_{25}H_{18}N_4O_3$ 4(3H)-Quinazolinone, 3-[[1,5-dihydro-1-(4-methoxyphenyl)-5-oxo-2H-pyrrol-2-ylidene]amino]-2-phenyl-	MeOH	214(4.76),235(4.79), 287(4.50),316(4.27)	80-1085-78
$C_{25}H_{18}N_8O$ 2-Propanone, 1,1-bis(1-phenyl-1H-pyrazolo[3,4-d]pyrimidin-4-yl)-	EtOH	246(4.52),341s(4.29), 357(4.32)	95-0089-78
$C_{25}H_{18}O_{11}$ Furo[3,2-b]furan-2,5-dione, 3-(3-acetoxy-4-methoxyphenyl)-6-(3,4-diacetoxyphenyl)-	acetone	400(4.41)	64-0820-78C
Furo[3,2-b]furan-2,5-dione, 3-(4-acetoxy-3-methoxyphenyl)-6-(3,4-diacetoxyphenyl)-	acetone	386(4.51)	64-0820-78C
$C_{25}H_{18}S$ 3H-Naphtho[1,8-bc]thiophene, 3-(diphenylmethylene)-2-methyl-	C_6H_{12}	296(4.006),328s(--), 340(4.141),433(4.188)	24-1824-78
$C_{25}H_{19}ClN_2O_2$ 1H-Cyclopenta[b]quinoxaline, 4,9-dibenzoyl-6(or 7)-chloro-3a,4,9,9a-tetrahydro-	MeCN	220(4.57),270(4.32)	5-1129-78
$C_{25}H_{19}Cs$ 1,1'-Biphenyl, 4-(diphenylmethyl)-, cesium deriv.	n.s.g.	573(4.65)	35-8271-78
$C_{25}H_{19}NO$ 9H-Indeno[2,1-c]pyridin-9-ol, 9-methyl-1,3-diphenyl-	EtOH	208(4.64),260(4.6), 314(3.63)	103-0997-78
$C_{25}H_{19}NO_3$ 2,5-Pyrrolidinedione, 1-(3-oxo-1,2,3-triphenyl-1-propenyl)-	EtOH	258(4.28)	44-0232-78
$C_{25}H_{19}N_2O$ 2H-Pyrido[2,1-b][1,3,4]oxadiazin-5-ium, 3,6,8-triphenyl-, bromide	EtOH	220(4.07),257(4.06), 288(4.07),335(4.06)	120-0001-78
$C_{25}H_{19}N_3$ [1,2,4]Triazolo[1,5-a]pyridine, 2-(4-methylphenyl)-7-[2-(1-naphthalenyl)ethenyl]-	DMF	351(4.54)	33-0142-78
[1,2,4]Triazolo[1,5-a]pyridine, 2-(4-methylphenyl)-7-[2-(2-naphthalenyl)ethenyl]-	DMF	338(4.72),353(4.63)	33-0142-78
$C_{25}H_{19}N_3O$ Benzo[f]quinolin-8-amine, 3-(4-methoxyphenyl)-1-(3-pyridinyl)-	EtOH	280(4.55),340(4.24), 384(3.92)	103-1337-78

Compound	Solvent	$\lambda_{max}(\log \epsilon)$	Ref.
$C_{25}H_{19}N_5$			
Benzenamine, N,N'-(3-phenyl-2H-pyrido-[1,2-a][1,3,5]triazine-2,4(3H)-diyl-idene)bis-	MeOH	279(4.49),389(4.02)	78-0101-78
$C_{25}H_{20}NO_5PS_3$			
Phosphorothioic acid, O,O-diphenyl S-[(2,3,5,7-tetrahydro-5,7-dioxo-2-phenyl-6H-1,4-dithiino[2,3-c]pyrrol-6-yl)methyl] ester	EtOH	213(4.41),261(4.03),417(3.50)	73-1093-78
$C_{25}H_{20}N_2$			
Benzo[f]quinazoline, 1,2-dihydro-2-methyl-1,3-diphenyl-	EtOH	210(4.56),239(4.49),277(4.26),288(4.28),343(4.59)	103-1019-78
$C_{25}H_{20}N_2O_2$			
1H-Cyclopenta[b]quinoxaline, 4,9-di-benzoyl-3a,4,9,9a-tetrahydro-	CH_2Cl_2	271.5(4.11)	5-1129-78
$C_{25}H_{20}O$			
Benz[2,3]azuleno[7,8,1-kla]-s-indacen-1(3aH)-one, 4,12b-dihydro-6,11,12b-trimethyl-	dioxan	221(4.36),283(4.84),356(3.87),371(3.83),380s(3.68),461(3.24),482(3.33),504s(3.23),650s(2.37),688(2.39),752s(2.29),860s(1.76)	89-0763-78
$C_{25}H_{20}OSi$			
Silane, benzoyltriphenyl-	C_6H_{12}	406(2.35),425(2.46),446(2.25)	112-0751-78
$C_{25}H_{20}O_2$			
4,7-Ethano-1H-indene-8,9-dione, 1-(di-phenylmethylene)-3a,4,7,7a-tetrahy-dro-4-methyl-	C_6H_{12} benzene	239(4.26),293(4.38) 453.5(1.91)	5-0440-78
4,7-Ethano-1H-indene-8,9-dione, 1-(di-phenylmethylene)-3a,4,7,7a-tetrahy-dro-5-methyl-	benzene	459(1.99)	5-0440-78
$C_{25}H_{20}O_2S$			
Ethanone, 1-[2-(4-methoxyphenyl)-4,5-diphenyl-3-thienyl]-	dioxan	255(4.35),313(4.19)	24-2028-78
Spiro[4H-dinaphtho[2,1-f:1',2'-h][1,5]-dioxonin-5(6H),3'-thietane]	$CHCl_3$	244(4.63),296(4.02),330(3.81)	56-1913-78
$C_{25}H_{20}O_3$			
Spiro[4H-dinaphtho[2,1-f:1',2'-h][1,5]-dioxonin-5(6H),3'-oxetane]	$CHCl_3$	244(4.63),296(4.04),330(3.83)	56-1913-78
$C_{25}H_{20}O_8$			
Penioflavin diacetate	EtOH	264(4.45),309(3.88),320s(3.82)	1-0075-78
$C_{25}H_{20}S_4$			
3,3'(4H,4'H)-Spirobi[2H-naphtho[1,8-bc]-1,5-dithiocin]	EtOH	246(4.38),250(4.29),270(4.00)	49-1017-78
$C_{25}H_{21}ClO_5S$			
Thiopyrylium, 2-[(2,6-dimethyl-4H-pyr-an-4-ylidene)methyl]-4,6-diphenyl-,	HOAc-HClO$_4$	561(3.898)	83-0256-78

Compound	Solvent	$\lambda_{max}(\log \epsilon)$	Ref.
perchlorate (cont.)			83-0256-78
$C_{25}H_{21}N$ 1H-Indeno[2,1-c]pyridine, 2,3-dihydro-9-phenyl-2-(phenylmethyl)-, hydrobromide	MeOH	243(4.47),254(4.45), 310(3.75),322(3.62)	87-0340-78
$C_{25}H_{21}NO$ 2,5-Cyclohexadien-1-one, 4-(methylamino)-2,4,6-triphenyl-	EtOH	225(4.49),280(3.75)	5-1634-78
$C_{25}H_{21}N_3O_2S$ 10H-Pyrimido[5,4-b][1,4]benzothiazine, 10-methyl-2,4-bis(phenylmethoxy)-	12M HCl	251(4.27),293(3.90), 372(3.45)	5-0193-78
	MeOH	252(4.58),299(3.90)	5-0193-78
3H-Pyrimido[5,4-b][1,4]benzothiazine-2,4(4aH,10H)-dione, 10-methyl-3,4a-bis(phenylmethyl)-	12M HCl	300(3.90),360(3.21)	5-0193-78
	MeOH	248(4.20),272(3.98), 305(3.83),335(3.68)	5-0193-78
10H-Pyrimido[5,4-b][1,4]benzothiazin-4(3H)-one, 10-methyl-2-(phenylmethoxy)-3-(phenylmethyl)-	12M HCl	252(4.22),292(3.86), 368(3.48)	5-0193-78
	MeOH	258(4.57),320s(3.45), 380s(3.08)	5-0193-78
$C_{25}H_{21}N_3O_3S$ 4-Thiazoleacetic acid, α-(methoxyimino)-2-[(triphenylmethyl)amino]-, (E)-	EtOH	245s(4.25),321(3.40)	78-2233-78
(Z)-	EtOH	234(4.32),295(3.81)	78-2233-78
$C_{25}H_{22}$ Benzene, 1,1',1"-(3,5-dimethyl-1,3-cyclopentadiene-1,2,4-triyl)tris-	isooctane	230(4.253),253s(--), 322(4.22)	44-4090-78
Benzene, 1,1',1"-(5,5-dimethyl-1,3-cyclopentadiene-1,2,3-triyl)tris-	EtOH	242(4.52),300(3.51)	44-1493-78
Benzene, 1,1',1"-[3-(2-methyl-1-propenyl)-1-cyclopropene-1,2,3-triyl]tris-	EtOH	228(4.43),313(4.41), 331(4.31)	44-1493-78
1H-Indene, 3-(2-methyl-1-propenyl)-1,2-diphenyl-	EtOH	305(4.16)	44-1493-78
$C_{25}H_{22}N_2O$ 4H-Indol-4-one, 1,5,6,7-tetrahydro-6,6-dimethyl-2-phenyl-1-(6-quinolinyl)-	MeOH	246(4.12),275(3.86), 295(3.65)	48-0863-78
$C_{25}H_{22}N_2O_2$ Spiro[imidazolidine-2,2'-[2H]indene]-1',3'-dione, 1,3-bis(3-methylphenyl)-	MeOH	230(4.67),250(4.58), 287s(3.66),300s(3.54), 346s(2.55),411(2.40), 476s(2.20)	24-3058-78
Spiro[imidazolidine-2,2'-[2H]indene]-1',3'-dione, 1,3-bis(4-methylphenyl)-	MeOH	230(4.68),250(4.60), 290s(4.58),307s(3.37), 351(2.54),414s(2.42), 483s(2.23)	24-3058-78
$C_{25}H_{22}N_2O_4$ 2H-Cyclopenta[d]pyridazine-1,4-dicarboxylic acid, 5-(diphenylmethyl)-2-methyl-, dimethyl ester	MeCN	273(4.32),290(4.39), 330(3.82),455(2.80)	78-2509-78
2H-Cyclopenta[d]pyridazine-1,4-dicarboxylic acid, 7-(diphenylmethyl)-2-methyl-, dimethyl ester	MeCN	284(4.46),350(3.61), 460(2.66)	78-2509-78

Compound	Solvent	$\lambda_{max}(\log \epsilon)$	Ref.
Spiro[imidazolidine-2,2'-[2H]indene]- 1',3'-dione, 1,3-bis(4-methoxyphenyl)-	MeOH	230(4.69),245s(4.60), 296(3.57),306s(3.56), 417s(2.43),487s(2.24)	24-3058-78
$C_{25}H_{22}N_4NiO$ Nickel, [5-(methoxymethyl)-10,15,20- trimethyl-21H,23H-porphinato(2-)- $N^{21},N^{22},N^{23},N^{24}$]-, (SP-4-2)-	$CHCl_3$	299(4.10),334(4.04), 417(5.27),538(4.10)	5-0238-78
$C_{25}H_{22}N_4O$ 2H-Benzimidazol-2-one, 1,3-dihydro- 1,3-dimethyl-5-[4-(4-methylphenyl)- 5-phenyl-1H-imidazol-2-yl]- radical	EtOH	221(4.63),320(4.48)	103-0425-78
	toluene	540(--),710(--)	103-0425-78
$C_{25}H_{22}N_4O_2$ 2H-Benzimidazol-2-one, 1,3-dihydro- 5-[4-(4-methoxyphenyl)-5-phenyl- 1H-imidazol-2-yl]-1,3-dimethyl- radical	EtOH	216(4.64),321(4.49)	103-0425-78
	toluene	540(--),714(--)	103-0425-78
$C_{25}H_{22}N_4O_2S$ Benzaldehyde, 4-nitro-, [3-(1-methyl- ethyl)-4,5-diphenyl-2(3H)-thiazol- ylidene]hydrazone	isoPrOH	292(4.15),338(4.03), 472(4.28)	73-1227-78
3(2H)-Thiazolamine, 2-[(1-methylethyl)- imino]-N-[(4-nitrophenyl)methylene]- 4,5-diphenyl-	isoPrOH	287(4.24),427(4.18)	73-1227-78
$C_{25}H_{22}N_4O_6$ Tryptoquivaline G acetate	MeOH	218s(4.54),226(4.55), 232s(4.53),255s(4.24), 265s(4.09),276s(3.96), 290s(3.62),303(3.52), 315(3.42)	94-0111-78
$C_{25}H_{22}O_2$ 1H-Indene, 1-[bis(4-methoxyphenyl)meth- ylene]-4-methyl-	C_6H_{12}	208(4.54),230(4.32), 244s(4.31),293(4.18), 306s(4.12),367(4.21)	5-0440-78
1H-Indene, 1-[bis(4-methoxyphenyl)meth- ylene]-5-methyl-	C_6H_{12}	209(4.52),232(4.32), 242s(4.29),293(4.19), 304s(4.11),367(4.31)	5-0440-78
1H-Indene, 1-[bis(4-methoxyphenyl)meth- ylene]-6-methyl-	C_6H_{12}	230(4.36),292(4.22), 301(4.16),358(4.26)	5-0440-78
1H-Indene, 1-[bis(4-methoxyphenyl)meth- ylene]-7-methyl-	C_6H_{12}	207(4.53),235(4.29), 251s(4.28),290(4.19), 306s(4.11),378(4.25)	5-0440-78
4,7-Methanobenzo[3,4]cyclobuta[1,2-b]- furan-8-one, 3a,3b,4,7,7a,7b-hexahy- dro-4,7-dimethyl-5,6-diphenyl-, $(3a\alpha,3b\beta,4\beta,7\beta,7a\beta,7b\alpha)-$	hexane	222(4.30),258(4.00)	44-0315-78
6,9-Methano-3-benzoxepin-10-one, 5a,6,9,9a-tetrahydro-6,9-dimethyl- 7,8-diphenyl-, $(5a\alpha,6\alpha,9\alpha,9a\alpha)-$	hexane	222(4.27),257(3.96)	44-0315-78
$C_{25}H_{22}O_4$ 2H-1-Benzopyran-2-one, 5,7-dimethoxy- 4-methyl-3-phenyl-8-(phenylmethyl)-	MeOH	260(3.84),320(3.99)	2-0584-78

Compound	Solvent	$\lambda_{max}(\log \epsilon)$	Ref.
$C_{25}H_{22}O_6$			
3H,7H,8H-Bis[1]benzopyrano[4,3-b:6',5'-e]pyran-7-one, 6,11-dihydroxy-3,3-dimethyl-8-(2-methyl-1-propenyl)-(cyclomorusin)	MeOH	223(4.45),255(4.38), 283(4.43),383(4.19)	94-1394-78
	MeOH-NaOMe	270(4.47),409(4.41)	94-1394-78
	MeOH-AlCl₃	229(4.51),265(4.35), 285(4.41),379(4.24), 429(3.84)	94-1394-78
3H,7H-Pyrano[2',3':7,8][1]benzopyrano-[3,2-d][1]benzoxepin-7-one, 8,9-di-hydro-6,12-dihydroxy-3,3-dimethyl-9-(1-methylethenyl)-	EtOH	219(4.45),235(4.45), 278(4.47),334(4.16)	94-1394-78
	EtOH-NaOMe	271(4.50),309(4.35), 351(4.31),389(4.27)	94-1394-78
$C_{25}H_{22}O_8$			
Amorphigenin, 8'-acetoxy-6a,12a-dehydro-	MeOH	235(4.36),280(4.26), 310(4.17)	102-1442-78
4H-1-Benzopyran-4-one, 5-hydroxy-2-(4-hydroxy-3-methoxyphenyl)-6,8-dimethoxy-7-(phenylmethoxy)-	EtOH	257(4.21),278(4.27), 349(4.36)	142-0053-78B
	EtOH-NaOAc	267(4.32),421(4.53)	142-0053-78B
$C_{25}H_{22}O_{11}$			
2-Propen-1-one, 3-[4-(acetyloxy)phenyl]-1-[2,3,4,6-tetrakis(acetyloxy)phenyl]-	EtOH	313(4.47)	18-3627-78
$C_{25}H_{23}ClN_2O_8$			
2,4(1H,3H)-Pyrimidinedione, 1-[5-O-(4-chlorobenzoyl)-3-O-(4-methylbenzoyl)-β-D-arabinofuranosyl]-3-methyl-	MeOH	240(4.66),262(4.23)	44-0350-78
$C_{25}H_{23}NO$			
2H-Pyrrole-5-ethanol, 3,4-dihydro-3,3-dimethyl-α-(phenylethynyl)-β-(3-phenyl-2-propynylidene)-	MeOH	206(4.20),234(4.01), 240(4.02),248(4.00), 306(4.34),321(4.06)	83-0977-78
3H-Pyrrol-3-one, 1,2-dihydro-2-[4-(1-methylethyl)phenyl]-4,5-diphenyl-	MeOH	255(4.32),378(3.81)	18-3312-78
3H-Pyrrol-3-one, 2-ethyl-1,2-dihydro-1-methyl-2,4,5-triphenyl-	MeCN	225(4.14),280(4.01), 357(3.88)	5-1203-78
$C_{25}H_{23}NS$			
2-Azabicyclo[3.1.0]hexane-3-thione, 2,6-dimethyl-1,4,5-triphenyl-	MeCN	284(4.13)	5-1222-78
2-Azabicyclo[3.1.0]hexane-3-thione, 2-methyl-1-(4-methylphenyl)-4,5-diphenyl-	MeCN	287(4.17)	5-1222-78
$C_{25}H_{23}N_3O_2S$			
Benz[cd]indole-6-sulfonamide, N,N-diethyl-2-(naphthalenylamino)-	dioxan	424(4.30)	93-1354-78
$C_{25}H_{23}N_3O_3$			
Methanone, [4,5-dihydro-3-(1-methylethyl)-1-(4-nitrophenyl)-5-phenyl-1H-pyrazol-4-yl]phenyl-	EtOH	245(4.760),390(4.394)	146-0864-78
$C_{25}H_{23}O_2P$			
Phosphonium, triphenyl-, 1-acetyl-4-oxo-2-pentenylide	EtOH	220(4.47),356(4.53)	95-0503-78
$C_{25}H_{23}O_3P$			
Phosphonium, triphenyl-, 1-(methoxy-carbonyl)-4-oxo-2-pentenylide	EtOH	227(4.49),352(4.55)	95-0503-78

Compound	Solvent	$\lambda_{max}(\log \epsilon)$	Ref.
$C_{25}H_{24}N_2O_2$ 1H-Inden-1-one, 2,3-dihydro-7-methyl-, [bis(4-methoxyphenyl)methylene]hydrazone	C_6H_{12}	227s(4.42),248s(4.22), 280(4.31),327(4.42)	5-0440-78
$C_{25}H_{24}N_2O_3S$ Benzenesulfonamide, 4-methyl-N-[3-(3-oxo-1-phenylbutyl)-1H-indol-2-yl]-	$CHCl_3$	280(4.12),294(3.98)	103-0745-78
$C_{25}H_{24}N_2O_4S_2$ 1H-Cyclopenta[b]quinoxaline, 3a,4,9,9a-tetrahydro-6,7-dimethyl-4,9-bis(phenylsulfonyl)-	C_6H_{12}	216(4.40),259s(3.79), 266s(3.77),273s(3.71), 295s(3.38)	5-1129-78
$C_{25}H_{24}N_2O_5$ 1H-Cyclopenta[d]pyridazine-1,4-dicarboxylic acid, 5-(diphenylmethylene)-2,5-dihydro-1-methoxy-, dimethyl ester	MeOH	207(4.07),238(3.85), 275(3.93)	78-2509-78
L-Phenylalanine, N-[4-[methyl[(phenylmethoxy)carbonyl]amino]benzoyl]-	EtOH	260(4.21)	87-1162-78
$C_{25}H_{24}N_2O_5S_2$ Spiro[2H-benzimidazole-2,1'-cyclopentan]-2'-one, 1,3-dihydro-5,6-dimethyl-1,3-bis(phenylsulfonyl)-	MeCN	209(4.55),267s(3.76), 274s(3.73),301(3.68)	5-1161-78
$C_{25}H_{24}N_2O_{11}$ 1H-Benz[f]indazole-4,9-dione, 1-(2,3,4-6-tetra-O-acetyl-β-D-glucopyranosyl)-	EtOH	244(4.32),270(4.05)	111-0155-78
2H-Benz[f]indazole-4,9-dione, 2-(2,3,4-6-tetra-O-acetyl-β-D-glucopyranosyl)-	EtOH	245(4.68)	111-0155-78
$C_{25}H_{24}N_4$ 3H-1,5-Benzodiazepine, 2-phenyl-4-(4-phenylpiperaziny1)-	n.s.g.	256(4.64),346(3.67)	87-0952-78
$C_{25}H_{24}N_4O_4$ Pyrazolidine, 2-(2,4-dinitrophenyl)-3-ethenyl-3-methyl-5-phenyl-1-(phenylmethyl)-	MeCN	238(4.17),386(4.24)	78-0903-78
$C_{25}H_{24}N_4O_6S_2$ Benzenesulfonamide, N-[1-methyl-1'-[(4-methylphenyl)sulfonyl]spiro[3H-indole-3,3'-pyrrolidin]-2(1H)-ylidene]-4-nitro-	EtOH	224(4.74),270(4.27)	39-1471-78C
$C_{25}H_{24}O_2$ 1H-Indene, 1-[bis(4-methoxyphenyl)methylene]-2,3-dihydro-4-methyl-	C_6H_{12}	210(4.55),247(4.30), 255s(4.24),316(4.32)	5-0440-78
1H-Indene, 1-[bis(4-methoxyphenyl)methylene]-2,3-dihydro-5-methyl-	C_6H_{12}	211(4.55),247(4.26), 255s(4.24),310(4.29), 320(4.31)	5-0440-78
1H-Indene, 1-[bis(4-methoxyphenyl)methylene]-2,3-dihydro-6-methyl-	C_6H_{12}	210(4.54),249(4.29), 323(4.28)	5-0440-78
1H-Indene, 1-[bis(4-methoxyphenyl)methylene]-2,3-dihydro-7-methyl-	C_6H_{12}	211(4.59),254(4.40), 290s(4.15),318(4.35)	5-0440-78
$C_{25}H_{24}O_2S$ Spiro[1H-indene-1,2'-thiirane], 2,3-di-	C_6H_{12}	210(4.63),229s(4.40),	5-0440-78

Compound	Solvent	$\lambda_{max}(\log \epsilon)$	Ref.
3,3'-bis(4-methoxyphenyl)-4-methyl- (cont.)		241(4.38),246s(4.34), 273(3.74),280s(3.70), 287s(3.55),323(2.30), 338(2.19)	5-0440-78
$C_{25}H_{24}O_3$ 1,2,4-Ethanylylidene-1H-cyclobuta[cd]- pentalen-5(1aH)-one, hexahydro-7,7- bis(4-methoxyphenyl)-	EtOH	230(4.23),250(3.96), 275(3.65),286(3.56)	104-2336-78
$C_{25}H_{24}O_6$ 4H,8H-Benzo[1,2-b:3,4-b']dipyran-4-one, 8,8-dimethyl-2-(2,4-dihydroxyphenyl)- 5-hydroxy-3-(3-methyl-2-butenyl)- (morusin)	EtOH	206(4.49),220s(4.43), 270(4.60),300s(4.00), 320s(3.90),350(3.81)	94-1394-78
	EtOH-NaOMe	267(4.59),371(4.04)	94-1394-78
	EtOH-AlCl$_3$	279(4.63),338(3.89), 415(3.79)	94-1394-78
	EtOH-NaOAc- H$_3$BO$_3$	270(4.62),300s(4.04), 320s(3.97),350s(3.87)	94-1394-78
6H,7H-[1]Benzopyrano[4,3-b][1]benzopyr- an-7-one, 3,8,10-trihydroxy-11-(3- methyl-2-butenyl)-6-(2-methyl-1-pro- penyl)- (pomiferin)	MeOH	280(4.40),310(4.51)	64-1547-78B
4H-1-Benzopyran-4-one, 5,7-dihydroxy- 2-(5-hydroxy-2,2-dimethyl-2H-1-benzo- pyran-6-yl)-3-(3-methyl-2-butenyl)- (kuwanon B)	MeOH	234(4.49),260(4.38), 280s(4.10),330s(4.00)	94-1453-78
	MeOH-NaOMe	268(4.44),310(4.02), 359(4.09)	94-1453-78
	MeOH-AlCl$_3$	214(4.52),269(4.44), 376(3.94)	94-1453-78
4H-1-Benzopyran-4-one, 5,7-dihydroxy- 2-(5-hydroxy-2,2-dimethyl-2H-1-benzo- pyran-8-yl)-3-(3-methyl-2-butenyl)- (kuwanon A)	EtOH	208(4.49),261(4.26), 283s(4.11),325s(3.93)	94-1453-78
	EtOH-NaOMe	268(4.35),327(4.11), 370s(4.00)	94-1453-78
	EtOH-AlCl$_3$	208(4.54),269(4.33), 373(3.81)	94-1453-78
$C_{25}H_{24}O_7$ Acacetin, 6-C-prenyl-, diacetate	MeOH	242(3.92),251(4.21), 311(4.24)	78-3569-78
3H,7H-Pyrano[2',3':7,8][1]benzopyrano- [3,2-d][1]benzoxepin-7-one, 8,9-di- hydro-6,12-dihydroxy-9-(1-hydroxy- 1-methylethyl)-3,3-dimethyl-, (±)-	MeOH	218(4.49),234(4.49), 278(4.51),334(4.24)	94-1394-78
	MeOH-NaOMe	265(4.56),393(4.42)	94-1394-78
	MeOH-AlCl$_3$	224(4.53),261(4.41), 284(4.54),360(4.33), 417(4.02)	94-1394-78
$C_{25}H_{24}O_8$ 4H-Furo[2,3-h]-1-benzopyran-4-one, 9-(1,2-diacetoxy-2-methylpropyl)- 8,9-dihydro-5-hydroxy-2-phenyl-, [R-(R*,S*)]-	MeOH	214(4.49),273(4.48), 315(3.88)	119-0047-78
Kuwanon B photoproduct 2c	MeOH	210(4.38),268(4.32), 290s(4.04),350(4.14)	94-1453-78
	MeOH-NaOMe	233(4.35),276(4.40), 377(4.07)	94-1453-78
	MeOH-AlCl$_3$	210(4.48),278(4.26), 360(4.08),396(4.08)	94-1453-78
3H,7H-Pyrano[2',3':7,8][1]benzopyrano- [3,2-d][1]benzoxepin-7-one, 8,9-di- hydro-9-(1-hydroperoxy-1-methyleth- yl)-6,12-dihydroxy-3,3-dimethyl-	EtOH	280(4.46),335(4.42)	94-1431-78
	EtOH-NaOMe	267(4.46),397(4.34)	94-1431-78
	EtOH-AlCl$_3$	283(4.39),320s(4.15), 356(4.18),370(4.42)	94-1431-78

$C_{25}H_{25}Br_3O_7-C_{25}H_{26}O_2S_2$

Compound	Solvent	$\lambda_{max}(\log \epsilon)$	Ref.
$C_{25}H_{25}Br_3O_7$ 19-Norpregna-1,3,5,7,9,14-hexaene, 4,12,15-tribromo-3,11,16-trimethoxy-17,20:20,21-bis[methylenebis(oxy)]-	EtOH	244s(4.39),264(4.71), 305(4.04),312(4.02), 335(3.65),354(3.45)	88-0639-78
$C_{25}H_{25}NO_3$ Benzoic acid, 4-(4,5,6,7-tetrahydro-6,6-dimethyl-4-oxo-2-phenyl-1H-indol-1-yl)-, ethyl ester	MeOH	215(3.46),255(3.54), 280(4.31)	48-0863-78
$C_{25}H_{25}N_3O$ 2H-Imidazol-2-one, 4-ethyl-1,3-dihydro-1-(4-methylphenyl)-3-[(2-methylphenyl)amino]-5-phenyl-	EtOH	275(4.17)	5-2033-78
$C_{25}H_{25}N_3O_2$ Spiro[2H-anthra[1,2-d]imidazole-2,1'-cyclohexane]-6,11-dione, 4-piperidino-	EtOH	640(3.94)	104-0381-78
$C_{25}H_{25}N_5O_4$ 6aH-Pyrrolo[2,3-c:5,4-c']dipyrazole-6a-carboxylic acid, 3,4-diacetyl-1,3a,3b-6,7,7a-hexahydro-7-methyl-1,6-diphenyl-, methyl ester	EtOH	238(4.34),290s(3.92), 299s(3.97),350(4.47)	4-0293-78
$C_{25}H_{26}$ Tricyclo[4.3.0^{2,5}]undeca-3,10-diene, 3,10-dimethyl-4,11-diphenyl-, (1α,2α,5α,6α)-	C_6H_{12}	252(4.26),268s(4.18), 297s(2.56)	88-0357-78
$C_{25}H_{26}N_2O_3$ L-Phenylalanine, 2-(benzoylamino)-3-phenylpropyl ester, monohydrobromide, (S)-	MeOH	215(4.01),226(4.07)	78-2791-78
$C_{25}H_{26}N_2O_5S_2$ Benzenesulfonamide, N-[1-(1H-indol-3-ylmethyl)-2-[[(4-methylphenyl)sulfonyl]oxy]ethyl]-4-methyl-, (S)-	MeOH	272(3.90),280(3.89), 289(3.82)	35-0938-78
$C_{25}H_{26}N_4O_5$ 3H-Imidazo[4,5-b]pyridin-5-amine, 7-(phenylmethoxy)-N-(phenylmethyl)-3-β-D-ribofuranosyl-	pH 1 pH 11	250s(4.05),309(4.15) 255(4.06),260(4.06), 300(4.11)	4-0839-78 4-0839-78
Methanone, [2-(1-hydroxy-1-methylethyl)phenyl](2,4,6-trimethylphenyl)-, 2,4-dinitrophenylhydrazone	toluene EtOH	375(4.47) 245s(4.20),293s(3.81), 374(4.37)	104-2229-78 104-2229-78
$C_{25}H_{26}O_2$ Phenol, 2,5-dimethyl-4-(1,4,5,8-tetramethyl-9H-xanthen-9-yl)-	EtOH	240(4.40),282(3.80), 290(3.77)	32-0079-78
$C_{25}H_{26}O_2S_2$ Benzene, 1-[[2-(butylthio)-1,2-diphenylethenyl]sulfonyl]-4-methyl-, (E)- (Z)-	EtOH EtOH	203(4.58),235s(4.15), 295(4.11) 203(4.59),235s(4.25), 303(4.03)	2-1086-78 2-1086-78
Benzene, 1-methyl-4-[[2-[(2-methylpropyl)thio]-1,2-diphenylethenyl]sulfonyl]-, (E)-	EtOH	203(4.62),234s(4.22), 295(4.14)	2-1086-78

Compound	Solvent	$\lambda_{max}(\log \epsilon)$	Ref.
Benzene, 1-methyl-4-[[2-[(2-methylprop-yl)thio]-1,2-diphenylethenyl]sulfon-yl]-, (Z)-	EtOH	203(4.54),235s(4.21), 303(4.01)	2-1086-78
$C_{25}H_{26}O_4$			
Hispaglabridin B	MeOH	280(4.17),290s(4.11), 309(3.67)	142-1533-78A
6,10-Methano-4H,6H-pyrano[3,2-i][1]ben-zoxocin-4-one, 2,3,7,8,9,10-hexahydro-5-hydroxy-10-methyl-7-(1-methylethyl-idene)-2-phenyl-	MeOH	229(4.11),300(4.18), 349s(3.37)	102-0517-78
	MeOH-NaOH	238(3.98),251s(3.92), 299(4.08),364(3.74)	102-0517-78
$C_{25}H_{26}O_4S_2$			
Benzene, 1-[[2-(butylsulfonyl)-1,2-di-phenylethenyl]sulfonyl]-4-methyl-, (E)-	EtOH	203(4.64),220s(4.47), 244(4.18)	2-1086-78
(Z)-	EtOH	203(4.61),243(4.14)	2-1086-78
Benzene, 1-methyl-4-[[2-[(2-methylprop-yl)sulfonyl]-1,2-diphenylethenyl]sul-fonyl]-, (E)-	EtOH	202(4.59),226(4.36), 243s(4.16)	2-1086-78
(Z)-	EtOH	202(4.64),242(4.17)	2-1086-78
$C_{25}H_{26}O_5$			
4H,8H-Benzo[1,2-b:3,4-b']dipyran-4-one, 2,3-dihydro-5-hydroxy-2-(4-hydroxy-phenyl)-8,8-dimethyl-6-(3-methyl-2-butenyl)-, (S)-	MeOH	282(4.53),296(4.23), 322(3.87)	78-3563-78
4H-1-Benzopyran-4-one, 5,7-dihydroxy-2-(4-hydroxyphenyl)-6,8-bis(3-methyl-2-butenyl)-	MeOH	270(4.22),330(4.24)	78-3569-78
Cajaflavanone	MeOH	225(4.11),275(4.03), 300(3.84),345(3.21), 360(3.39),380(3.39)	102-2045-78
$C_{25}H_{26}O_6$			
Kuwanon C	EtOH	210(4.63),265(4.49), 315(4.06)	94-1453-78
	EtOH-NaOMe	276(4.56),320s(4.10), 373(4.20)	94-1453-78
	EtOH-AlCl$_3$	275(4.55),335(4.02), 387(3.93)	94-1453-78
Kuwanon D	EtOH	213(4.59),227s(4.07), 290(4.30),320s(3.86)	142-0745-78A
Morusin, 12,13-dihydro-	MeOH	264(4.25),303s(3.83), 320(3.83)	94-1394-78
	MeOH-AlCl$_3$	276(4.31),333(3.79), 370(3.72)	94-1394-78
$C_{25}H_{26}O_7$			
Oxydihydromorusin	MeOH	206(4.38),225(4.26), 242(4.26),270(4.44), 300s(3.82),350s(3.64)	94-1453-78
	MeOH-NaOMe	270(4.41),370(3.81)	94-1453-78
	MeOH-AlCl$_3$	206(4.42),227(4.34), 279(4.47),336(3.76), 409(3.61)	94-1453-78
1H,7H-Pyrano[2',3':7,8][1]benzopyrano-[3,2-d][1]benzoxepin-7-one, 2,3,5,9-tetrahydro-6,12-dihydroxy-9-(1-hy-droxy-1-methylethyl)-3,3-dimethyl-	EtOH	255s(4.22),274(4.34), 300s(4.03),350(4.25)	94-1431-78
	EtOH-AlCl$_3$	265s(4.14),285(4.34), 300s(4.14),362(4.25), 400(4.11)	94-1431-78

Compound	Solvent	$\lambda_{max}(\log \epsilon)$	Ref.
$C_{25}H_{26}O_8$			
1H,7H-Pyrano[2',3':7,8][1]benzopyrano-[3,2-d][1]benzoxepin-7-one, 2,3,8,9-tetrahydro-9-(1-hydroperoxy-1-methyl-ethyl)-6,12-dihydroxy-3,3-dimethyl-	EtOH	256s(4.15),274(4.24), 290s(4.00),350(4.18)	94-1431-78
	EtOH-AlCl$_3$	265s(4.10),285(4.29), 363(4.20),402(4.05)	94-1431-78
$C_{25}H_{26}O_{10}$			
Agarotetrol tetraacetate	EtOH	250(3.79)	88-3921-78
$C_{25}H_{27}NO_4$			
3,5-Pyridinedicarboxylic acid, 1-ethyl-1,4-dihydro-2,6-dimethyl-, bis(4-methylphenyl) ester	EtOH	206(4.36),218s(4.25), 239(4.26),266(4.16), 372(3.89)	103-1226-78
$C_{25}H_{27}NO_6$			
3,5-Pyridinedicarboxylic acid, 1-ethyl-1,4-dihydro-2,6-dimethyl-, bis(4-methoxyphenyl) ester	EtOH	203(4.31),225(4.33), 238s(4.21),268(4.13), 376(3.82)	103-1226-78
$C_{25}H_{27}N_2$			
Pyridinium, 2-[4-[4-(dimethylamino)-phenyl]-2-phenyl-1,3-butadienyl]-1-ethyl, iodide	EtOH	550(3.11)	146-0341-78
$C_{25}H_{27}N_3O_2$			
Pyrrolo[3,4-b]indole-1-carboxamide, 2-cyclohexyl-1,2,3,4-tetrahydro-N-methyl-3-oxo-4-(phenylmethyl)-	EtOH	229(4.52),301(4.28)	4-1463-78
$C_{25}H_{27}N_3PS$			
Phosphonium, [3-[(aminothioxomethyl)-hydrazono]-2-methylenepentyl]tri-phenyl-, chloride	EtOH	268(4.181),333(4.089)	65-1998-78
$C_{25}H_{27}N_5O_{12}$			
Riboflavin, 9-nitro-, 2',3',4',5'-tet-raacetate	EtOH	223(4.51),271(4.56), 343(3.82),442(4.06)	104-0406-78
$C_{25}H_{27}N_7OS$			
6H-Anthra[9,1-cd]isothiazol-6-one, 7-[[4,6-bis(butylamino)-1,3,5-tria-zin-2-yl]amino]-	DMF	460(4.06)	2-1007-78
$C_{25}H_{28}N_2O_5$			
1H-Pyrrole-3-carboxylic acid, 5-[[acet-yl(4-methylphenyl)amino]methyl-4,5-dihydro-5-hydroxy-2-methyl-4-oxo-1-(phenylmethyl)-, ethyl ester	EtOH	211(4.19),244(4.18), 331(3.86)	4-1215-78
$C_{25}H_{28}N_2O_6$			
1H-Pyrrole-3-carboxylic acid, 5-[[acet-yl(4-methoxyphenyl)amino]methyl]-4,5-dihydro-5-hydroxy-2-methyl-4-oxo-1-(phenylmethyl)-, ethyl ester	EtOH	211(4.16),235(4.24), 331(3.84)	4-1215-78
$C_{25}H_{28}N_2O_8$			
Aspidospermidine-3-carboxylic acid, 4-acetoxy-3,6-epoxy-1-formyl-16-methoxy-8-oxo-, methyl ester, (2β,3β,4β,5α,6β,12β,19α)-	MeOH	219(4.41),248(3.94), 297(3.79)	33-1554-78

Compound	Solvent	$\lambda_{max}(\log \epsilon)$	Ref.
$C_{25}H_{28}N_4O_7$			
2-Cyclohexene-1-carboxylic acid, 4-[(2,4-dinitrophenyl)hydrazono]-3-[2-(4-methoxyphenyl)ethyl]-2-methyl-, ethyl ester, (±)-	EtOH	227(4.38)	44-4598-78
$C_{25}H_{28}N_4O_8$			
Benzoic acid, 3,4-dimethoxy-, 5-[(2,4-dinitrophenyl)hydrazono]-3,7-dimethyl-2,6-octadienyl ester	n.s.g.	365(3.86)	31-0155-78
$C_{25}H_{28}N_4O_{10}$			
Riboflavin, 2',3',4',5'-tetraacetate	EtOH	223(4.57),270(4.53), 354(3.91),446(4.08)	104-0406-78
$C_{25}H_{28}O_3$			
Phenol, 2,2',2"-methylidynetris[4,5-dimethyl-	EtOH	283(3.84)	32-0079-78
Phenol, 2,2',2"-methylidynetris[4,6-dimethyl-	EtOH	281(3.84),285(3.83)	32-0079-78
$C_{25}H_{28}O_4$			
4H-1-Benzopyran-4-one, 2,3-dihydro-7-hydroxy-2-(4-hydroxyphenyl)-6,8-bis(3-methyl-2-butenyl)-	MeOH	230(4.21),275(4.02)	2-1126-78
Cyclohexanone, 2-[1,5-bis(2-methoxyphenyl)-3-oxo-4-pentenyl]-	EtOH	216(4.24),287(4.08), 337(3.90)	2-0428-78
Cyclohexanone, 2-[1,5-bis(4-methoxyphenyl)-3-oxo-4-pentenyl]-	EtOH	217(4.25),325(4.33)	2-0428-78
Hispaglabridin A	MeOH	281(4.05),290s(3.95), 312(3.41)	142-1533-78A
6,10-Methano-4H,6H-pyrano[3,2-i][1]-benzoxocin-4-one, 2,3,7,8,9,10-hexahydro-5-hydroxy-10-methyl-7-(1-methylethyl)-2-phenyl-	MeOH	217(4.33),234s(4.09), 301(4.23),348s(3.63)	102-0517-78
2-Propen-1-one, 1-[2,4-dihydroxy-3,5-bis(3-methyl-2-butenyl)phenyl]-3-(4-hydroxyphenyl)-	MeOH	230(3.95),370(4.03)	2-1126-78
Spiro[5.5]undecane-1,9-dione, 7,11-bis-(2-methoxyphenyl)-	EtOH	231(3.79),277(3.68), 283(3.66)	2-0428-78
Spiro[5.5]undecane-1,9-dione, 7,11-bis-(4-methoxyphenyl)-	EtOH	213(3.57),230(4.27), 278(3.44),284(3.41)	2-0428-78
Xanthoangelol	EtOH	225s(4.26),368(4.57)	95-0210-78
$C_{25}H_{28}O_5$			
4H-1-Benzopyran-4-one, 2,3-dihydro-5,7-dihydroxy-2-(4-hydroxyphenyl)-6,8-bis(3-methyl-2-butenyl)-	MeOH	289(4.23),340(3.48)	78-3563-78
4H,6H,10H-Benzo[1,2-b:3,4-b':5,6-b"]-tripyran-4-one, 2,3,7,8,11,12-hexahydro-2-(4-hydroxyphenyl)-6,6,10,10-tetramethyl-, (S)-	MeOH	224(3.84),272(4.23), 322(3.63)	78-3563-78
$C_{25}H_{28}O_6$			
Morusin, 10,11,12,13-tetrahydro-	MeOH	263(4.22),315(3.79)	94-1394-78
	MeOH-AlCl₃	276(4.17),330(3.75)	94-1394-78
$C_{25}H_{28}O_7$			
Kuwanon E	EtOH	212(4.56),289(4.26), 320s(3.73)	142-1295-78A
	EtOH-NaOMe	326(4.16),497(4.17)	142-1295-78A

Compound	Solvent	$\lambda_{max}(\log \epsilon)$	Ref.
Kuwanon E (cont.)	EtOH-NaOAc	214(4.66),295s(3.97), 327(4.49)	142-1295-78A
	EtOH-AlCl$_3$	212(4.55),220s(4.19), 308(4.33),374(3.57)	142-1295-78A
$C_{25}H_{29}NO_3$ 24-Norchola-1,4,6,20(22)-tetraene-23-nitrile, 21-acetoxy-3-oxo-, (20Z)-	MeOH	299(4.11)	44-2334-78
Triphyophylline, N-methyl-	EtOH	231(4.72),306(4.09), 321(3.90),336(3.83)	28-0129-78B
Triphyophylline, 8-methyltetradehydro-	EtOH	230(4.97),312(4.20), 324(4.21),338(4.06)	28-0129-78B
$C_{25}H_{29}NO_5$ 2-Propenoic acid, 3-(4-hydroxyphenyl)-, octahydro-4-(3-hydroxy-4-methoxy-phenyl)-2H-quinolizin-2-yl ester	MeOH	228(4.3),290s(4.3), 298s(4.3),313(4.4)	102-0305-78
$C_{25}H_{29}N_2$ 2,4,6,8-Nonatetraenylium, 1,9-bis[4-(dimethylamino)phenyl]-, tetra-fluoroborate	C$_2$H$_4$Cl$_2$	980(5.58)	104-1197-78
$C_{25}H_{29}N_3O_8$ DL-Aspartic acid, N-[N-[4-[[(phenyl-methoxy)carbonyl]amino]benzoyl]-glycyl]-, diethyl ester	EtOH	268(4.48)	87-1165-78
$C_{25}H_{30}NOP$ 9H-Phosphorino[3,4-b]indole, 2-(1,1-di-methylethyl)-2-ethoxy-9-ethyl-2,2-di-hydro-4-phenyl-	EtOH	222(4.62),252(4.82), 286(4.70),330(3.97), 373(4.33)	139-0257-78B
$C_{25}H_{30}N_2$ 2H-Indeno[2,1-c]pyridine-2-propanamine, N,N-diethyl-1,3-dihydro-9-phenyl-, dihydrobromide	MeOH	244(4.45),259(4.44), 312(3.74),323(3.63)	87-0340-78
$C_{25}H_{30}N_3O$ 12-Azabicyclo[9.2.1]tetradeca-1(14),11-diene, 13-[(3-methoxy-5-phenyl-1H-pyrrol-2-yl)methylene]-, monohydro-chloride	MeOH	280(4.06),497(4.69)	49-0137-78
$C_{25}H_{30}N_2O_4$ Lidocaine 3-hydroxy-2-naphthoate	MeOH	271(3.72),282(3.73), 293(3.45),353(3.37)	94-0936-78
$C_{25}H_{30}N_2O_7$ Aspidospermidine-3-carboxylic acid, 4-acetoxy-3,6-epoxy-16-methoxy-1-methyl-8-oxo-, methyl ester	EtOH	214(4.57),256(3.82), 308(3.70)	35-4220-78
$C_{25}H_{30}N_3$ Methylium, tris[4-(dimethylamino)phen-yl]-, iodide	EtOH	250(4.34),304(4.59), 360(4.15),590(4.08)	64-1520-78J
	95% EtOH	590(5.03)	64-1520-78J
	acetone	423(4.38),590(5.06)	64-1520-78J
$C_{25}H_{30}N_4$ 3H-1,5-Benzodiazepine, 2-[1,4'-bipip-	n.s.g.	259(4.54),344(3.69)	87-0952-78

Compound	Solvent	$\lambda_{max}(\log \epsilon)$	Ref.
eridin]-1'-yl-4-phenyl- (cont.)			87-0952-78
$C_{25}H_{30}N_4O_4$			
2-Pyrimidinamine, 1-β-D-arabinofurano-syl-1,4-dihydro-N-(2-phenylethyl)-4-[(2-phenylethyl)imino]-, monohydro-chloride	pH 1	225s(4.32),250s(4.22)	94-3244-78
$C_{25}H_{30}N_4O_8$			
Uridine, 5-[3-[2-(acetylamino)ethyl]-5-methoxy-1H-indol-2-yl]-2',3'-O-(1-methylethylidene)-	MeCN	219(4.33),274(3.91), 307(3.86),330(3.82)	88-2585-78
$C_{25}H_{30}N_4P$			
Phosphorus(1+), (4-diazenyl-N,N-dimeth-ylbenzenaminato)diphenyl-1-piperidin-yl-, (T-4)-, tetrafluoroborate	n.s.g.	277(3.99),470(4.58)	123-0190-78
$C_{25}H_{30}O_4$			
1-Propanone, 3-phenyl-1-[3,4,5,6-tetra-hydro-7,9-dihydroxy-2-methyl-5-(1-methylethyl)-2,6-methano-2H-1-benz-oxocin-8-yl]-	MeOH	212(4.28),234s(4.11), 298(4.17),351s(3.61)	102-0517-78
$C_{25}H_{30}O_5$			
1-Anthracenecarboxaldehyde, 5,8,8a,9-10,10a-hexahydro-2,3-dihydroxy-7,10a-dimethyl-8-(3-methyl-2-butenyl)-4-(1-methylethyl)-9,10-dihydro-	CHCl₃	245s(--),272(4.49), 351(3.59)	98-0115-78
1-Anthracenecarboxaldehyde, 5,8,8a,9-10,10a-hexahydro-2,3-dihydroxy-10a-methyl-4-(1-methylethyl)-6-(4-meth-yl-3-pentenyl)-9,10-dioxo-, cis	CHCl₃	272(4.58),308s(--), 356(3.67)	102-0151-78
Heliocide H₄	CHCl₃	245s(--),272(4.45), 351(3.54)	98-0115-78
$C_{25}H_{30}O_5Se_2$			
Undecanedioic acid, 6-oxo-5,7-bis(phen-ylseleno)-, dimethyl ester	MeOH	227(4.27),301(3.33)	49-0557-78
$C_{25}H_{30}O_6$			
Tetrahydrokuwanon C	EtOH	261(4.43),303(3.94), 332(3.90)	142-1355-78A
$C_{25}H_{31}ClO_2$			
Androst-4-en-3-one, 6β-chloro-17β-hy-droxy-16β-phenyl-	EtOH	240(4.22)	94-1718-78
$C_{25}H_{31}NO$			
Cyclohexanecarboxamide, 2-phenyl-N-(2-phenylcyclohexyl)-, [1α(1S*,2R*),2β]-[1S-[1α(1R*,2S*)2β]]-	MeOH	208(4.28),252(2.52), 258(2.61),264(2.49)	44-0355-78
	MeOH	208(4.29),252(2.56), 258(2.64),264(2.53)	44-0355-78
$C_{25}H_{31}NO_3$			
24-Norchola-1,4,20(22)-triene-23-ni-trile, 21-acetoxy-3-oxo-, (20Z)-	MeOH	220(4.36)	44-2334-78
24-Norchola-4,6,20(22)-triene-23-ni-trile, 21-acetoxy-3-oxo-	MeOH	282.5(4.42)	44-2334-78

Compound	Solvent	λ_{max}(log ϵ)	Ref.
$C_{25}H_{31}NO_4$			
24-Norchola-4,20(22)-diene-23-nitrile, 21-acetoxy-3,11-dioxo-, (20Z)-	MeOH	231(4.44)	44-2334-78
24-Norchola-1,4,20(22)-triene-23-nitrile, 21-acetoxy-11-hydroxy-3-oxo-, (11β,20Z)-	MeOH	218(4.40)	44-2334-78
$C_{25}H_{31}NO_6S$			
Dispiro[cyclohexane-1,2'-thiirane-3',5'-[5H]indene]-1",4-diol, octahydro-7"a-methyl-, 4-acetate 1"-(4-nitrobenzoate)	EtOH	258(4.17)	39-0045-78C
$C_{25}H_{31}NO_8$			
Carbamic acid, [2-[6-(3,4-dihydro-1,7,8-trimethoxy-1H-2-benzopyran-3-yl)-1,3-benzodioxol-5-yl]ethyl]methyl-, ethyl ester	EtOH	230(3.96),285(3.70)	78-0635-78
$C_{25}H_{31}N_3O_8$			
D-Valine, N-[N-[4-methyl-2-nitro-3-(phenylmethoxy)benzoyl]-L-threonyl]-, methyl ester	MeOH	210(4.479)	87-0607-78
$C_{25}H_{32}CuN_2O_2$			
Copper, [[6,6'-[1,3-propanediylbis(nitrilomethylidyne)]bis[2-methyl-5-(1-methylethenyl]-2-cyclohexen-1-onato]](2-)-N,N',O,O']-	benzene	310(4.08),390(4.14), 760(1.95)	65-1474-78
$C_{25}H_{32}CuN_2O_3$			
Copper, [[2,2'-[(2-hydroxy-1,3-propanediyl)bis(nitrilomethylidyne)]bis[4-(1,1-dimethylethyl)phenolato]](2-)-$N^2,N^{2'},O^1,O^{1'}$]-	CH_2Cl_2-DMSO pyridine DMSO-THF	380(4.06),610(2.43) 384(4.06),616(2.42) 277(4.45),380(4.09), 619(2.49)	35-2686-78 35-2686-78 35-2686-78
$C_{25}H_{32}N_2O_5$			
Aspidospermidine-3-carboxylic acid, 4-acetoxy-6,7-didehydro-16-methoxy-1-methyl-, methyl ester	MeOH	247(3.94),300(3.76)	33-1554-78
$C_{25}H_{32}N_2O_6$			
Aspidospermidine-3-carboxylic acid, 4-acetoxy-3,6-epoxy-16-methoxy-1-methyl-, methyl ester	EtOH EtOH	214(4.44),257(3.81), 308(3.66) 256(3.79),308(3.69)	33-1554-78 39-0757-78
$C_{25}H_{32}N_2O_7$			
Aspidospermidine-3-carboxylic acid, 4-acetoxy-3-hydroxy-16-methoxy-1-methyl-10-oxo-, methyl ester	EtOH	211(4.65),249(3.93), 302(3.74)	35-4220-78
$C_{25}H_{32}N_4O_4$			
Pyrano[2,3-c]pyrazole-2(6H)-acetic acid, 3,4-dimethyl-5-[2-[4-(4-methylphenyl)-1-piperazinyl]ethyl]-6-oxo-, ethyl ester	EtOH	249(4.14),304(4.26)	95-0335-78
$C_{25}H_{32}N_4O_5$			
Pyrano[2,3-c]pyrazole-2(6H)-acetic acid, 5-[2-[4-(2-methoxyphenyl)-1-piperazinyl]ethyl]-3,4-dimethyl-6-	EtOH	246(3.95),305(4.24)	95-0335-78

Compound	Solvent	$\lambda_{max}(\log \epsilon)$	Ref.
oxo-, ethyl ester (cont.)			95-0335-78
$C_{25}H_{32}N_4O_6S_2$			
Benzenesulfonamide, N,N'-[(6-methyl-2,4-dioxo-1,3(2H,4H)-pyrimidinediyl)di-3,1-propanediyl]bis[4-methyl-	CHCl$_3$	300(4.32)	70-0149-78
$C_{25}H_{32}N_4O_7$			
D-Valinamide, N-[4-methyl-2-nitro-3-(phenylmethoxy)benzoyl]-L-threonyl-N-methyl-	MeOH	218(4.751)	87-0607-78
$C_{25}H_{32}O_3$			
16,23-Cyclo-24-norchola-4,20(22)-dien-21-al, 17-acetyl-3-oxo-, (16β,17α)-	EtOH	241(4.34)	70-1214-78
16,23-Cyclo-21-norchola-4,22-dien-24-al, 17-acetyl-3-oxo-, (16β,17α)-	EtOH	241(4.49)	70-1214-78
$C_{25}H_{32}O_5$			
Carda-4,20(22)-dienolide, 11-acetoxy-3-oxo-, (11β,14α)-	MeOH	227(4.33)	44-2334-78
$C_{25}H_{32}O_6$			
Andilesin	n.s.g.	end absorption	77-0533-78
Carda-4,20(22)-dienolide, 19-acetoxy-14β-hydroxy-3-oxo-	MeOH	227(4.38)	44-3946-78
$C_{25}H_{32}O_7$			
1-Pentanone, 1-(6,7,9,10,17,18,20,21-octahydrodibenzo[b,k][1,4,7,10,13,16]-hexaoxacyclooctadecin-2-yl)-	MeOH	303(3.92)	35-0648-78
	MeOH-LiOAc	303(3.91)	35-0648-78
	MeOH-NaOAc	300(3.89)	35-0648-78
	MeOH-KOAc	299(3.88)	35-0648-78
$C_{25}H_{32}O_9$			
Chaparrinone, 6α-senecioyloxy-	MeOH	228(4.34)	87-1186-78 +100-0578-78
$C_{25}H_{33}ClO_2$			
16,24-Cyclo-21-norchola-4,6-dien-3-one, 17-acetyl-6-chloro-, (16β,17α)-	CHCl$_3$	287(4.42)	70-0384-78
$C_{25}H_{33}ClO_3$			
3'H-Cycloprop[1,2]androsta-1,4,6-trien-3-one, 17-acetoxy-6-chloro-1,2-dihydro-16-(1-methylethyl)-, (1β,2β,16β-17β)-	EtOH	283(4.26)	94-1718-78
$C_{25}H_{33}NO_3$			
24-Norchola-4,20(22)-diene-23-nitrile, 21-acetoxy-3-oxo-	MeOH	232.5(4.42)	44-2334-78
$C_{25}H_{34}N_2O_3$			
Acetamide, N-[(3β,5β,17α)-20-(acetoxyimino)-17-methyl-18-norpregna-8,11,13-trien-3-yl]-	EtOH	208(4.16)	39-0163-78C
$C_{25}H_{34}N_2O_4$			
2H-Pyrido[3,4-e][1,3]oxazine-4,5(3H,6H)-dione, 7-methoxy-2,2-dimethyl-3-nonyl-6-phenyl-	MeOH	309(4.4)	120-0101-78

Compound	Solvent	$\lambda_{max}(\log \epsilon)$	Ref.
$C_{25}H_{34}N_2O_6$			
Vindoline	EtOH	250(3.87),304(3.74)	39-0757-78C
$C_{25}H_{34}N_2O_9S_2$			
Aspidospermidine-3-carboxylic acid, 6,7-didehydro-16-methoxy-1-methyl-3,4-bis[(methylsulfonyl)oxy]-, methyl ester	MeOH	242(3.85),297(3.74)	33-1554-78
$C_{25}H_{34}N_4$			
13H-8,13-Methano-7H-benzo[d]pyrrolo-[3,2,1-kl][1,3]benzodiazocine-4,13-diethanamine, 4,5-dihydro-N,N,N',N'-tetramethyl-	EtOH	250(4.16),300(3.81)	39-1671-78C
$C_{25}H_{34}O_2$			
16,24-Cyclo-21-norchola-4,6-dien-3-one, 17-acetyl-, (16β,17α)-	$CHCl_3$	285(4.46)	70-0384-78
$C_{25}H_{34}O_3$			
Carda-3,5,20(22)-trienolide, 3-ethoxy-, (14α)-	MeOH	224.5(4.39)	44-2334-78
16,24-Cyclo-21-norchol-4-en-3-one, 17-acetyl-6,7-epoxy-, (6α,7α,16β,17β)-	$CHCl_3$	241(4.27)	70-0384-78
3'H-Cycloprop[1,2]androsta-1,4,6-trien-3-one, 17-acetoxy-1,2-dihydro-16-(1-methylethyl)-, (1β,2β,16β,17β)-	EtOH	280(4.26)	94-1718-78
$C_{25}H_{34}O_3S_2$			
3H-3,10b-Ethano-1H-naphtho[1,2-c]pyran-8-carboxaldehyde, 7-(1,3-dithian-2-ylidenemethylene)-4,6,6a,7,8,9,10,10a-octahydro-3-methoxy-4,4,8-trimethyl-, [3S-(3α,6aα,7α,10aβ,10bβ)]-	n.s.g.	254(3.93)	33-3087-78
$C_{25}H_{34}O_4$			
3'H-Cycloprop[1,2]androsta-1,4-dien-3-one, 17-acetoxy-6,7-epoxy-1,2-dihydro-16-(1-methylethyl)-, (1β,2β,6α,7α,16β,17β)-	EtOH	233(4.22)	94-1718-78
6H-3a,7-Methanoazulene-6,9-dione, 1,2,3,7,8,8a-hexahydro-4-hydroxy-1,1-dimethyl-7-(3-methyl-2-butenyl)-2-(1-methylethenyl)-5-(2-methyl-1-oxopropyl)-	EtOH EtOH-base	236(3.88),280(4.09) 245s(3.86),270(4.17), 283s(4.12)	39-1633-78C 39-1633-78C
2,4,6,8-Nonatetraenoic acid, 9-(4-methoxy-2,3,6-trimethylphenyl)-3,7-dimethyl-, 2-hydroxybutyl ester, (Z,E,E,E)-	EtOH	244(4.06),363(4.54)	33-2697-78
$C_{25}H_{34}O_5$			
Etiojerva-5,12-diene-3,11,20-trione, 17α-ethyl-, 3,20-bis(ethylene acetal)	MeOH	257(4.00)	18-0234-78
epimer	MeOH	259(4.11)	18-0234-78
$C_{25}H_{34}O_6$			
2-Butenoic acid, 2-methyl-, 4,4a,5,6-7,8,8a,9-octahydro-8a-hydroxy-3,4a,5-trimethyl-9-[(3-methyl-1-oxo-2-butenyl)oxy]naphtho[2,3-b]furan-6-yl ester	MeOH	220(4.57)	18-3335-78

Compound	Solvent	λ_{max}(log ϵ)	Ref.
C$_{25}$H$_{35}$NO$_4$			
Azacyclotridecane, 1-[5-(6-methoxy-1,3-benzodioxol-5-yl)1-oxo-2,4-pentadienyl]-, (E,E)-	n.s.g.	370(4.35)	78-1979-78
3,5-Pyridinedicarboxylic acid, 1-hexyl-1,4-dihydro-2,6-dimethyl-4-phenyl-, diethyl ester	EtOH	205(4.16),243(4.15), 348(3.73)	103-1226-78
C$_{25}$H$_{35}$NO$_5$			
3,5-Pyridinedicarboxylic acid, 1,4-dihydro-4-(4-methoxyphenyl)-2,6-dimethyl-1-pentyl-, diethyl ester	EtOH	202(4.28),227(4.17), 259s(4.00),284s(3.34), 348(3.70)	103-1226-78
C$_{25}$H$_{36}$O$_2$			
1-Naphthalenecarboxylic acid, tetradecyl ester	hexane	298.5(3.82)	95-0774-78
C$_{25}$H$_{36}$O$_3$			
12'-Apo-β,ψ-carotenal, 5,6-dihydro-5,6-dihydroxy-, (5R,6R)-(±)-	benzene	298(3.91),395(--), 415(4.76),437(--)	39-1511-78C
[1,1'-Biphenyl]-2,2'-diol, 3,3',5-tris-(1,1-dimethylethyl)-5'-methoxy-	C$_6$H$_{12}$	234(4.26),267(3.85), 317s(3.46),360s(3.20)	12-1061-78
C$_{25}$H$_{36}$O$_3$S$_2$			
Androsta-5,15-dien-16-one, 3,19-epoxy-17-hydroxy-3-methoxy-4,4-dimethyl-, cyclic 1,3-propanediyl mercaptole, (3β,13α)-	n.s.g.	260(3.45),330(2.00)	33-3087-78
3H-3,10b-Ethano-1H-naphtho[1,2-c]pyran-8-carboxaldehyde, 7-(1,3-dithian-2-ylidenemethyl)-4,6,6a,7,8,9,10,10a-octahydro-3-methoxy-4,4,8-trimethyl-	n.s.g.	260(3.83)	33-3087-78
isomer	n.s.g.	258(3.98),330(2.60)	33-3087-78
C$_{25}$H$_{36}$O$_4$			
2H-1-Benzopyran-2-one, 7-hydroxy-3-(1-oxohexadecyl)-	EtOH	365(4.34)	51-0652-78
anion	EtOH	430(4.65)	51-0652-78
Cyclohexanone, 2,2'-methylenebis[6-(1-cyclohexen-1-yl)-6-hydroxy-, cis-cis	C$_6$H$_{12}$	297(2.255)	104-0067-78
cis-trans	C$_6$H$_{12}$	293(2.22)	104-0067-78
trans-trans	C$_6$H$_{12}$	289(2.25)	104-0067-78
16,24-Cyclo-21-norchol-4-en-3-one, 17-acetyl-22,23-dihydroxy-, (16β,17α)-	EtOH	243(4.18)	70-1214-78
Gummiferolic acid	EtOH	218(3.91)	102-1637-78
6H-3a,7-Methanoazulene-6,9-dione, 1,2,3,7,8,8a-hexahydro-4-hydroxy-1,1-dimethyl-7-(3-methyl-2-butenyl)-2-(1-methylethyl)-5-(2-methyl-1-oxopropyl)-	EtOH	235(3.88),280(4.12)	39-1633-78C
	EtOH-base	245s(3.88),270(4.19), 282s(4.13)	39-1633-78C
Pregna-3,5-dien-20-one, 21-acetoxy-3-ethoxy-	MeOH	239(4.26)	44-2334-78
C$_{25}$H$_{36}$O$_5$			
6H-3a,7-Methanoazulene-6,9-dione, 1,2,3,7,8,8a-hexahydro-4-hydroxy-2-(1-hydroxy-1-methylethyl)-1,1-dimethyl-7-(3-methyl-2-butenyl)-5-(2-methyl-1-oxopropyl)-	MeOH-HCl	232(4.09),280(3.96)	20-0459-78
	MeOH-NaOH	267(4.13)	20-0459-78
Pregna-4,17(20)-dien-3-one, 11β-hydroxy-3-methoxy-17,21-[(1-methyleth-	MeOH	242(4.20)	44-3405-78

Compound	Solvent	$\lambda_{max}(\log \epsilon)$	Ref.
ylidene)bis(oxy)]- (cont.)			44-3405-78
Pregna-4,20-dien-3-one, 11β-hydroxy-21-methoxy-17,20-[(1-methylethyli-dene)bis(oxy)]-	MeOH	242(4.22)	44-3405-78
5β-Pregn-16-en-20-one, 3α,11β-diacetoxy-	EtOH	238(3.98)	39-1594-78C
$C_{25}H_{36}O_6$			
6H-3a,7-Methanoazulene-6,9-dione, 1,2,3-7,8,8a-hexahydro-2-(1-hydroperoxy-1-methylethyl)-4-hydroxy-1,1-dimethyl-7-(3-methyl-2-butenyl)-5-(2-methyl-1-oxopropyl)-, exo	MeOH-HCl MeOH-NaOH	234(3.97),282(4.03) 273(4.09)	20-0459-78 20-0459-78
$C_{25}H_{37}NO_4$			
2H-Pyrrol-2-one, 4-[(3β,5β,14β,17β)-3-acetoxy-14-hydroxyandrostan-17-yl]-1,5-dihydro-	EtOH	213(4.23)	33-0977-78
$C_{25}H_{38}O_2$			
26,27-Dinorcholesta-5,22-dien-24-one, 3-hydroxy-, (3β,22E)-	MeOH	225(4.18)	33-1470-78
$C_{25}H_{38}O_3$			
Ceralbic acid	n.s.g.	213(4.13)	25-0584-78
$C_{25}H_{38}O_3S_2$			
Androst-5-en-16-one, 3,19-epoxy-17-hy-droxy-3-methoxy-4,4-dimethyl-, cyclic 1,3-propanediyl mercaptole, (3β,13α)-	n.s.g.	228s(--),253(2.85), 326(2.48)	33-3087-78
(3β,17β)-	n.s.g.	217s(--),250(3.08), 310(1.92)	33-3087-78
$C_{25}H_{38}O_4$			
6H-3a,7-Methanoazulene-6,9-dione, 1,2-3,7,8,8a-hexahydro-4-hydroxy-1,1-di-methyl-7-(3-methylbutyl)-2-(1-methyl-ethyl)-5-(2-methyl-1-oxopropyl)-	EtOH EtOH-base	235(3.91),280(4.12) 245s(3.84),272(4.19), 285s(4.12)	39-1633-78C 39-1633-78C
Pregn-20-ene-20-carboxaldehyde, 3-acet-oxy-21-methoxy-, (3β,5α,20E)-	EtOH	275(3.95)	23-0424-78
(3β,5α,20Z)-	EtOH	253(4.2)	23-0424-78
$C_{25}H_{38}O_6S$			
Prosta-5,13-dien-1-oic acid, 15-acet-oxy-11-(acetylthio)-9-oxo-, methyl ester, (5Z,11α,13E,15S)-	MeOH	232(3.70)	44-4377-78
$C_{25}H_{40}N_2O$			
2,5-Cyclohexadien-1-one, 2,6-bis(1,1-dimethylethyl)-4-(di-1-piperidinyl-methylene)-	n.s.g.	419(4.36)	70-1035-78
$C_{25}H_{40}O_2$			
Ceralbol	n.s.g.	207(3.76)	31-0421-78
$C_{25}H_{43}N_3O_6$			
Hexadecanamide, N-(1-β-D-arabinofurano-syl-1,2-dihydro-2-oxo-4-pyrimidinyl)-	isoPrOH	216(4.14),248(4.14), 303(3.89)	94-0981-78
$C_{25}H_{44}N_{14}O_7$			
Capreomycin 1B	pH 1 H_2O	268(4.36) 268(4.35)	78-0921-78 78-0921-78

Compound	Solvent	$\lambda_{max}(\log \epsilon)$	Ref.
Capreomycin 1B (cont.)	pH 13	290(4.16)	78-0921-78
$C_{25}H_{44}N_{14}O_8$			
Capreomycin 1A	pH 1	269(4.38)	78-0921-78
	H_2O	268(4.38)	78-0921-78
	pH 13	287(4.20)	78-0921-78
$C_{25}H_{46}N_2O_2$			
1H-Imidazole-2-methanol, α-heptadecyl- α-(tetrahydro-2-furanyl)-	THF	228(3.05)	33-2823-78
isomer	THF	230(3.00)	33-2823-78

Compound	Solvent	$\lambda_{max}(\log \epsilon)$	Ref.
$C_{26}H_{14}BrNO_2$ Methanone, (4-bromophenyl)chryseno-[6,5-d]oxazol-2-yl-	dioxan	297(4.27),350(4.16), 363(4.21),381(4.34)	48-0986-78
$C_{26}H_{14}Br_2$ Phenanthro[3,4-c]phenanthrene, 1,8-di-bromo-	MeOH	232(4.69),248(4.66), 256(4.66),268(4.69), 293s(4.09),305(4.22), 317(4.44),329(4.48), 352(4.20)	54-0265-78
$C_{26}H_{15}Br$ Phenanthro[3,4-c]phenanthrene, 1-bromo-	MeOH	231(4.63),255(4.71), 265(4.69),290(4.15), 304s(4.24),316(4.43), 327(4.44),349s(4.13)	54-0265-78
$C_{26}H_{15}BrO_3$ Spiro[chrysene-5(6H),2'-oxiran]-6-one, 3'-(4-bromophenyl)hydroxymethylene]-	dioxan	300(4.13),362(3.99), 382(3.99),403(4.06)	48-0986-78
$C_{26}H_{15}NO_2$ Methanone, chryseno[6,5-d]oxazol-2-yl-phenyl-	dioxan	295(4.36),333(4.21), 346(4.31),359(4.28), 378(4.37)	48-0986-78
Phenanthro[3,4-c]phenanthrene, 1-nitro-	MeOH	229s(4.53),248(4.70), 267s(4.42),305s(4.15), 330(4.20),353s(4.07), 385(3.36)	54-0265-78
Phenanthro[3,4-c]phenanthrene, 4-nitro-	MeOH	220(4.48),226(4.51), 248(4.57),265(4.40), 301s(4.09),320(4.15), 353s(3.95),380(3.75)	54-0265-78
$C_{26}H_{16}$ 9H-Fluorene, 9-(9H-fluoren-9-ylidene)-	C_6H_{12}	455(4.11)	18-3373-78
$C_{26}H_{16}BrN_2OP$ Phosphonium, triphenyl-, 2-bromo-3,4-dicyano-5-formyl-2,4-cyclopentadien-1-ylide	MeCN	232(4.61),254(4.57), 316(4.18)	18-2605-78
$C_{26}H_{16}Br_2N_6O_4$ 1,2,4,5-Tetrazine, 1,4-bis(2-bromo-4-nitrophenyl)-1,4-dihydro-3,6-di-phenyl-, cation radical perchlorate	n.s.g.	263(4.63),290(4.52), 385(4.41),420(4.26), 555(3.48)	70-2227-78
$C_{26}H_{16}ClN_3O$ [1,2,4]Triazolo[1,5-a]pyridine, 2-[3-chloro-4-[2-(3-dibenzofuranyl)ethen-yl]phenyl]-	DMF	359(4.79)	33-0142-78
$C_{26}H_{16}ClN_3O_4$ 9,10-Anthracenedione, 1-chloro-5-nitro-4,8-bis(phenylamino)-	dioxan	529(4.16)	40-0082-78
9,10-Anthracenedione, 1-chloro-8-nitro-4,5-bis(phenylamino)-	dioxan	535.5(4.18)	40-0082-78
$C_{26}H_{16}N_2O_2S$ 1,2-Benzisoxazole, 3-[4-[2-[5-(2-benz-oxazolyl)-2-thienyl]ethenyl]phenyl]-	DMF	390(4.77)	33-2904-78

Compound	Solvent	$\lambda_{max}(\log \epsilon)$	Ref.
$C_{26}H_{16}N_2O_3$ 1,2-Benzisoxazole, 3-[4-[2-[5-(2-benz- oxazolyl)-2-furanyl]ethenyl]phenyl]-	DMF	383(4.76)	33-2904-78
$C_{26}H_{16}N_2O_4$ 5(4H)-Oxazolone, 4,4'-(1,4-phenylenedi- methylidyne)bis[2-phenyl-	toluene	435(4.85)	103-0120-78
$C_{26}H_{16}N_4O_8$ Benzene, 1,1',1'',1'''-(1,2-ethenediyli- dene)tetrakis[4-nitro-	MeCN	270(4.40),418(1.90)	70-2227-78
$C_{26}H_{16}O_3$ Spiro[chrysene-5(6H),2'-oxiran]-6-one, 3'-(hydroxyphenylmethylene)-	dioxan	301(4.13),362(4.00), 383(4.01),404(4.07)	48-0986-78
$C_{26}H_{17}ClN_6$ 2H-Benzotriazole, 2-[3-chloro-4-[2-(4- [1,2,4]triazolo[1,5-a]pyridin-2-yl- phenyl)ethenyl]phenyl]-	DMF	366(4.80)	33-0142-78
$C_{26}H_{17}N$ Phenanthro[3,4-c]phenanthren-1-amine	MeOH	228(3.61),248(3.65), 270(3.55),290s(3.27), 321(3.27),333(3.31), 347s(3.18),375(2.87), 430s(2.20)	54-0265-78
$C_{26}H_{17}N_3O$ [1,2,4]Triazolo[1,5-a]pyridine, 7-[2- (3-dibenzofuranyl)ethenyl]-2-phenyl-	DMF	350(4.81),365(4.72)	33-0142-78
$C_{26}H_{17}N_3O_4$ 9,10-Anthracenedione, 1-nitro-4,5-bis- (phenylamino)-	dioxan	542(4.09)	40-0082-78
9,10-Anthracenedione, 4-nitro-1,5-bis- (phenylamino)-	dioxan	530.5(4.24)	40-0082-78
$C_{26}H_{18}$ Anthracene, 9,10-diphenyl-	C_6H_{12}	372(4.15)	61-1068-78
$C_{26}H_{18}Br_2N_4$ 1,2,4,5-Tetrazine, 1,4-bis(4-bromo- phenyl)-1,4-dihydro-3,6-diphenyl- cation radical perchlorate	EtOH	227(4.54),275(4.26), 329(4.25)	70-2227-78
	MeCN	227s(4.63),278(4.36), 340(4.20),454(3.54), 623(3.42)	70-2227-78
1,2,4,5-Tetrazine, 3,6-bis(4-bromo- phenyl)-1,4-dihydro-1,4-diphenyl- cation radical perchlorate	EtOH	242(4.45),276(4.41), 340(3.04)	70-2227-78
	MeCN	276(4.47),313(4.36), 453(3.70),590(3.40)	70-2227-78
$C_{26}H_{18}ClNO_2$ Benzo[h]quinolin-2(1H)-one, 3-benzoyl- 4-(2-chlorophenyl)-5,6-dihydro-	$CHCl_3$	247(4.47),374(4.42)	48-0097-78
Benzo[h]quinolin-2(1H)-one, 3-benzoyl- 4-(4-chlorophenyl)-5,6-dihydro-	$CHCl_3$	250(4.44),374(4.25)	48-0097-78
$C_{26}H_{18}ClN_3$ [1,2,4]Triazolo[1,5-a]pyridine, 2-[4- (2-[1,1'-biphenyl]-4-ylethenyl)-3-	DMF	350(4.75)	33-0142-78

Compound	Solvent	$\lambda_{max}(\log \epsilon)$	Ref.
chlorophenyl]- (cont.)			33-0142-78
$C_{26}H_{18}Cl_2N_4$			
1,2,4,5-Tetrazine, 1,4-bis(3-chloro-phenyl)-1,4-dihydro-3,6-diphenyl-cation radical perchlorate	EtOH	272(4.20),335(4.14)	70-2227-78
	MeCN	286(4.44),340s(4.27), 412(3.80),580(3.44)	70-2227-78
1,2,4,5-Tetrazine, 3,6-bis(4-chloro-phenyl)-1,4-dihydro-1,4-diphenyl-cation radical perchlorate	EtOH	236(4.44),275(4.34), 340(3.04)	70-2227-78
	MeCN	276(4.42),312s(4.36), 430(3.67),580(3.40)	70-2227-78
$C_{26}H_{18}N_2$			
9H-Fluoren-9-one, 9H-fluoren-9-ylhydra-zone	benzene	350(4.34)	18-2431-78
5H-Indeno[1,2-b]pyridine, 5-methylene-, dimer	EtOH	206(4.5),256(4.0), 312(4.3)	103-0997-78
$C_{26}H_{18}N_2O_2$			
9,10-Anthracenedione, 1,4-bis(phenyl-amino)-	dioxan	396(3.90),550s(3.91), 597(4.18),635(4.19)	40-0082-78
$C_{26}H_{18}N_4OS$			
[1,2,4]Triazolo[1,5-a]pyridine, 2-[4-[2-[5-(5-methyl-2-benzoxazolyl)-2-thienyl]ethenyl]phenyl]-	DMF	398(4.82)	33-0142-78
$C_{26}H_{18}N_6$			
2H-Benzotriazole, 2-[4-[2-(4-[1,2,4]-triazolo[1,5-a]pyridin-2-ylphenyl)-ethenyl]phenyl]-	DMF	365(4.86)	33-0142-78
$C_{26}H_{18}N_6O_4$			
1,2,4,5-Tetrazine, 1,4-dihydro-1,4-bis(4-nitrophenyl)-3,6-diphenyl-cation radical perchlorate	EtOH	225(--),240(--), 261(--),414(--)	70-2227-78
	n.s.g.	263(4.48),290(4.32), 350(4.30),560(3.28)	70-2227-78
1,2,4,5-Tetrazine, 1,4-dihydro-3,6-bis(3-nitrophenyl)-1,4-diphenyl-cation radical perchlorate	EtOH	265(--),333(--)	70-2227-78
	n.s.g.	263(4.33),333s(3.90), 400s(3.48),500(3.20)	70-2227-78
1,2,4,5-Tetrazine, 1,4-dihydro-3,6-bis(4-nitrophenyl)-1,4-diphenyl-cation radical perchlorate	EtOH	265(4.32),302(4.45), 406(3.74)	70-2227-78
	n.s.g.	267(4.61),305(4.55), 405(4.00),550(3.49)	70-2227-78
$C_{26}H_{18}O$			
9H-Dibenzo[a,g]cycloheptadecen-9-one, 18,19,20,21-tetradehydro-8-methyl-, (all-E)-	THF	<u>300(4.6)</u>,310(<u>4.7</u>)	138-1099-78
$C_{26}H_{18}O_2$			
Benzo[b]benzo[3,4]cyclobuta[1,2-e]-[1,4]dioxin, 4b,10a-dihydro-4b,10a-diphenyl-	ether	261(3.40),266(3.54), 272(3.60)	88-0687-78
Benzofuro[3,2-b]benzofuran, 4b,9b-di-hydro-4b,9b-diphenyl-	ether	282(3.85)	88-0687-78
Benzo[c]phenanthrene-5-carboxylic acid, 12-phenyl-, methyl ester	MeOH	231(4.60),292(4.63)	78-0769-78
6,11-Epoxydibenz[b,e]oxepin, 6,11-di-hydro-6,11-diphenyl-	ether	279(3.49),286(3.48)	88-0687-78

Compound	Solvent	$\lambda_{max}(\log \epsilon)$	Ref.
Ethanone, 1-(1,2-diphenylcyclopenta[b]-[1]benzopyran-3-yl)-	THF	287(4.55),370(4.25), 480(3.05)	103-1075-78
$C_{26}H_{18}O_2S$ Naphtho[2,3-c]thiophene-5,8-dione, 6,7-dimethyl-1,3-diphenyl-	CH_2Cl_2	276(4.70),445(3.97), 490s(3.80)	150-5538-78
$C_{26}H_{18}O_3$ Naphtho[2,3-c]furan-5,8-dione, 6,7-dimethyl-1,3-diphenyl-	CH_2Cl_2	279(4.83),474(4.16), 530s(3.96)	150-5538-78
$C_{26}H_{18}O_3S_3$ 3(2H)-Thiophenone, dihydro-2,4-bis[[5-(phenylthio)-2-furanyl]methylene]-	EtOH	233(4.13),280(4.06), 400(4.04),465(4.20)	133-0189-78
$C_{26}H_{18}O_7S_3$ 3(2H)-Thiophenone, dihydro-2,4-bis[[5-(phenylsulfonyl)-2-furanyl]methylene]-	EtOH	297(4.19),358(4.31), 448(4.21)	133-0189-78
$C_{26}H_{19}ClFN_3O_2$ 4H-Imidazo[1,5-a][1,4]benzodiazepine-3-carboxylic acid, 8-chloro-6-(2-fluorophenyl)-1-phenyl-	isoPrOH	217s(4.59),250s(4.25), 275s(4.09)	4-0855-78
$C_{26}H_{19}ClN_2O_2$ 1-Naphthalenecarboxylic acid, [3-(4-chlorophenyl)-3-oxo-1-phenylpropylidene]hydrazide	EtOH	220(4.45),282(4.42), 293s(4.40)	4-0385-78
$C_{26}H_{19}FeNO_3$ 6H-1,3-Oxazinium, 2-ferrocenyl-4-hydroxy-6-oxo-3,5-diphenyl-, hydroxide, inner salt	MeCN	255(4.39),361(4.05), 480(3.52)	5-1655-78
$C_{26}H_{19}NO$ 1,2-Benzisoxazole, 5-methyl-3-[4-[2-(1-naphthalenyl)ethenyl]phenyl]-	DMF	343(4.52)	33-2904-78
1,2-Benzisoxazole, 5-methyl-3-[4-[2-(2-naphthalenyl)ethenyl]phenyl]-	DMF	337(4.71)	33-2904-78
$C_{26}H_{19}NOS_3$ 3H-Phenothiazin-3-one, 2,7-bis[(4-methylphenyl)thio]-	CHCl$_3$	261(3.92?),465(3.92?)	103-0795-78
$C_{26}H_{19}NO_2$ 1,2-Benzisoxazole, 3-(4-methoxyphenyl)-6-[2-(1-naphthalenyl)ethenyl]-	DMF	342(4.48)	33-2904-78
1,2-Benzisoxazole, 3-(4-methoxyphenyl)-6-[2-(2-naphthalenyl)ethenyl]-	DMF	336(4.67)	33-2904-78
Benzo[h]quinolin-2(1H)-one, 3-benzoyl-5,6-dihydro-4-phenyl-	CHCl$_3$	250(4.38),374(4.20)	48-0097-78
$C_{26}H_{19}N_3$ Imidazo[1,2-a]pyridine, 2-[1,1'-biphenyl]-4-yl-7-[2-(3-pyridinyl)ethenyl]-	DMF	368(4.66),385(4.61)	33-0129-78
[1,2,4]Triazolo[1,5-a]pyridine, 2-[1,1'-biphenyl]-4-yl-7-(2-phenylethenyl)-	DMF	329(4.73),346(4.68)	33-0142-78
$C_{26}H_{20}ClN_2O$ Quinolinium, 2-[[2-(2-chlorophenyl)-2,3-dihydro-3-oxo-1H-isoindol-1-yli-	EtOH	219(5.65),237(5.71), 400(3.74),490(3.62),	18-2415-78

Compound	Solvent	$\lambda_{max}(\log \epsilon)$	Ref.
dene]methyl]-1-ethyl-, iodide (cont.)		540(3.69)	18-2415-78
Quinolinium, 2-[[2-(4-chlorophenyl)-2,3-dihydro-3-oxo-1H-isoindol-1-yli-dene]methyl]-1-ethyl-, iodide	EtOH	213(5.47),243(4.97), 520(3.76),557(3.95)	18-2415-78
$C_{26}H_{20}N_2$			
Imidazo[1,2-a]pyridine, 2-(4-methyl-phenyl)-7-[2-(1-naphthalenyl)ethenyl]-	DMF	376(4.61)	33-0129-78
Imidazo[1,2-a]pyridine, 2-(4-methyl-phenyl)-7-[2-(2-naphthalenyl)ethenyl]-	DMF	370(4.70),389(4.67)	33-0129-78
2,4-Pentadienal, 5-(2-ethynylphenyl)-, [5-(2-ethynylphenyl)-2,4-pentadien-ylidene]hydrazone, (?,?,E,E,E,E)-	THF	276s(4.16),368s(4.73), 389(4.83),405s(4.79), 430s(4.18)	18-2112-78
$C_{26}H_{20}N_2O$			
Benzo[f]quinazoline, 4-acetyl-3,4-di-hydro-1,3-diphenyl-	EtOH	210s(4.48),225(4.67), 253(4.36),278(4.49), 288(4.43),345(4.20)	103-0082-78
Imidazo[1,2-a]pyridine, 2-(4-methoxy-phenyl)-7-[2-(1-naphthalenyl)ethenyl]-	DMF	378(4.60)	33-0129-78
Imidazo[1,2-a]pyridine, 2-(4-methoxy-phenyl)-7-[2-(2-naphthalenyl)ethenyl]-	DMF	371(4.68),390(4.65)	33-0129-78
4H-Pyrido[1,2-a]pyrimidin-4-one, 2,3-dihydro-2,3,3-triphenyl-	C_6H_{12}	252(4.19),369(3.62)	78-0101-78
$C_{26}H_{20}N_2O_2$			
Benzo[f]quinazoline, 1,3-bis(2-methoxy-phenyl)-	EtOH	206(4.61),218(4.63), 271(4.52),301(4.30)	103-0082-78
Benzo[f]quinazoline, 1,3-bis(4-methoxy-phenyl)-	EtOH	205(4.47),226(4.44), 270(4.11),301(4.56)	103-0082-78
1-Naphthalenecarboxylic acid, (3-oxo-1,3-diphenylpropylidene)hydrazide	EtOH	283(4.46),293s(4.41)	4-0385-78
$C_{26}H_{20}N_2O_2S$			
Spiro[2H-anthra[1,2-d]imidazole-2,1'-cyclohexane]-6,11-dione, 4-(phenyl-thio)-	dioxan	502(3.88)	104-0381-78
$C_{26}H_{20}N_2O_4S$			
Spiro[2H-anthra[1,2-d]imidazole-2,1'-cyclohexane]-6,11-dione, 4-(phenyl-sulfonyl)-	dioxan	396(3.58)	104-0381-78
$C_{26}H_{20}N_3O_3$			
Quinolinium, 2-[[2,3-dihydro-2-(4-ni-trophenyl)-3-oxo-1H-isoindol-1-yli-dene]methyl]-1-ethyl-, iodide	EtOH	215(5.40),236(4.92), 530s(4.07),556(4.19)	18-2415-78
$C_{26}H_{20}N_4$			
1,2,4,5-Tetrazine, 1,4-dihydro-1,3,4,6-tetraphenyl-	EtOH	272(4.31),333(4.15)	70-2227-78
cation radical chloride	MeCN	284(4.20),312s(4.15), 333s(3.95),430(3.58), 585(3.40)	70-2227-78
cation radical perchlorate	MeCN	276(4.30),333(4.11), 423(3.52),582(3.32)	70-2227-78
$C_{26}H_{20}N_6$			
2H-Cyclopenta[d]pyridazine, 2-methyl-6-phenyl-5,7-bis(phenylazo)-	ether	248s(4.26),295(4.26), 384(4.61),480(4.34)	44-0664-78

Compound	Solvent	$\lambda_{max}(\log \epsilon)$	Ref.
$C_{26}H_{20}N_8O$			
2-Butanone, 1,1-bis(1-phenyl-1H-pyra-zolo[3,4-d]pyrimidin-4-yl)-	EtOH	247(4.54),340(4.30), 356(4.31)	95-0089-78
$C_{26}H_{20}O$			
Ethanone, 1-(5'-phenyl[1,1':3',1"-ter-phenyl]-4-yl)-	EtOH	205(4.60),249(4.41), 273(4.26)	47-2093-78
Ethanone, tetraphenyl-	C_6H_{12}	330(2.35),338(2.33), 370(1.90)	112-0751-78
$C_{26}H_{20}O_4$			
Naphtho[2,3-c]furan-1(3H)-one, 3,3-bis-(2-methoxyphenyl)-	THF	241(4.84),275(4.07), 282(4.07),297(3.73), 324(3.17),339(3.28)	78-0113-78
$C_{26}H_{21}NO$			
Benzaldehyde, O-(triphenylmethyl)oxime, (E)-	EtOH	260(4.26)	44-1890-78
(Z)-	EtOH	251(4.24)	44-1890-78
$C_{26}H_{21}NO_3$			
Benzenebutanamide, γ-(2-oxo-5-phenyl-3(2H)-furanylidene)-N-phenyl-	EtOH	208(4.48),247(4.54), 365(4.36)	12-2031-78
$C_{26}H_{21}NO_{11}$			
2-Naphthacenecarboxamide, 6,11,12,12a-tetrahydro-1,3,6,10,12a-pentahydroxy-6-methyl-11,12-dioxo-, triacetate	MeOH-HCl	243(4.41),261(4.40), 280s(4.32),415(3.78)	39-0145-78C
$C_{26}H_{21}N_2O$			
1H-Isoindolium, 3-[(1-ethyl-2(1H)-pyri-dinylidene)methyl]-2-(1-naphthalen-yl)-1-oxo-, iodide	EtOH	215(5.01),285(4.37)	18-2415-78
Quinolinium, 2-[[1,5-dihydro-1-(1-naph-thalenyl)-5-oxo-2H-pyrrol-2-ylidene]-methyl]-1-ethyl-, iodide	EtOH	221(4.96),238(4.49), 280(4.19),320(4.14)	18-2415-78
Quinolinium, 2-[(2,3-dihydro-3-oxo-2-phenyl-1H-isoindol-1-ylidene)meth-yl]-1-ethyl-, iodide	EtOH	215(5.19),235s(4.94), 545(3.60),582(3.70)	18-2415-78
$C_{26}H_{21}N_6O_5$			
Pyridinium, 2-[[4-(2-benzoxazolylmeth-ylene)-2-(2,4-dinitrophenyl)-2,4-di-hydro-5-methyl-3H-pyrazol-3-ylidene]-methyl]-1-ethyl-, iodide	n.s.g.	220(4.68),350(4.53)	48-0857-78
$C_{26}H_{21}OPS_2$			
9-Phosphapentacyclo[4.3.0.0^{2,5}.0^{3,8}.-0^{4,7}]nonane, 9-phenyl-1,8-bis(phen-ylthio)-, 9-oxide	EtOH	261(3.84)	44-4338-78
$C_{26}H_{21}O_3P$			
Ethanone, 1-[2-(diphenylphosphinyl)-5-hydroxy[1,1'-biphenyl]-4-yl]-	EtOH	230(4.72),274(4.00)	95-0503-78
Phosphonium, (5-acetyl-2,4-dihydroxy-phenyl)triphenyl-, hydroxide, inner salt	EtOH	226(4.62),268(4.12), 330(4.36)	95-0503-78
$C_{26}H_{21}O_5PS_2$			
9-Phosphapentacyclo[4.3.0.0^{2,5}.0^{3,8}.-0^{4,7}]nonane, 9-phenyl-1,8-bis(phenyl-	EtOH	262(3.22),267(3.37), 273(3.27)	44-4338-78

Compound	Solvent	$\lambda_{max}(\log \epsilon)$	Ref.
sulfonyl)-, 9-oxide (cont.)			44-4338-78
$C_{26}H_{22}$			
7,14-Ethanodibenz[a,h]anthracene, 7,14-dihydro-2,9-dimethyl-, (7R)-	EtOH-0.2% dioxan	226(4.98),237(4.99), 287(4.07)	18-0265-78
$C_{26}H_{22}ClN_3$			
3H-1,5-Benzodiazepine, 2-[4-(4-chloro-phenyl)-3,6-dihydro-1(2H)-pyridinyl]-4-phenyl-	n.s.g.	259(4.69)	87-0952-78
$C_{26}H_{22}Cl_2N_2O_6$			
2(5H)-Furanone, 4,4'-(azinodiethyli-dyne)bis[5-(2-chlorophenyl)-3-meth-oxy-	EtOH	308(4.40),330(4.35)	142-1041-78A
geometric isomer	EtOH	310(4.31),330(4.32)	142-1041-78A
$C_{26}H_{22}N_2O_2$			
Benzo[f]quinazoline, dihydro-1,3-bis(2-methoxyphenyl)-	EtOH	220(4.71),225(4.35), 282(4.37),334(3.88)	103-0082-78
Benzo[f]quinazoline, dihydro-1,3-bis(4-methoxyphenyl)-	EtOH	208(4.73),220(4.68), 255(4.38),278(4.51), 340(4.00)	103-0082-78
$C_{26}H_{22}N_2O_4$			
Caulerpin, di-N-methyl-	MeOH	224(4.72),286(4.49), 294(4.45)	150-1683-78
$C_{26}H_{22}N_2O_6$			
5H-Cyclopenta[d]pyridazine-1,4-dicarb-oxylic acid, 5-[bis(4-methoxyphenyl)-methylene]-, dimethyl ester	MeCN	249(4.28),290(3.74), 360(4.25),440(4.30)	78-2509-78
$C_{26}H_{22}N_4$			
Tetrabenzo[c,f,i,m][1,2,5,8]tetraaza-cyclotetradecine, 6,7,8,10-tetrahy-dro-	benzene	325s(4.14),460(3.52)	33-2813-78
$C_{26}H_{22}O$			
9,10-Epoxyanthracene, 1,2,4a,9,9a,10-hexahydro-9,10-diphenyl-	MeCN	242s(2.42),247s(2.57), 251(2.74),254s(2.74), 257(2.89),264(2.91), 268s(2.60),271(2.68)	150-1301-78
$C_{26}H_{22}O_2$			
4,7-Ethano-1H-indene-8,9-dione, 1-(di-phenylmethylene)-3a,4,7,7a-tetrahy-dro-5,6-dimethyl-	C_6H_{12}	233(4.15),291(4.26), 462(2.19)	5-0440-78
	benzene	462.0(1.78)	5-0440-78
Spiro[cyclobutane-1,5'(6'H)-[4H]dinaph-tho[2,1-f:1',2'-h][1,5]dioxonin]	CHCl₃	244(4.66),293(4.05), 330(3.88)	56-1913-78
$C_{26}H_{22}O_3S$			
3-Thiophenecarboxylic acid, 2-(4-meth-oxyphenyl)-4,5-diphenyl-, ethyl ester	dioxan	255(4.36),314(4.27)	24-2028-78
3-Thiophenecarboxylic acid, 5-(4-meth-oxyphenyl)-2,4-diphenyl-, ethyl ester	dioxan	231(4.33),252(4.32), 315(4.19)	24-2028-78
$C_{26}H_{22}O_4$			
4,7-Ethano-1H-indene-8,9-dione, 1-[bis-(4-methoxyphenyl)methylene]-3a,4,7,7a-tetrahydro-	benzene MeCN	455.6(1.80) 255(4.34),299(4.26)	5-0440-78 5-0440-78

Compound	Solvent	$\lambda_{max}(\log \epsilon)$	Ref.
Spiro[4H-dinaphtho[2,1-f:1',2'-h][1,5]-dioxonin-5(6H),5'-[1,3]dioxane]	CHCl$_3$	246(4.46),273(3.92), 302(4.00),332(3.82)	56-1913-78
$C_{26}H_{22}O_6$ 4H-1-Benzopyran-4-one, 3-(3,4-dimeth-oxyphenyl)-5-hydroxy-7-[(3-phenyl-2-propenyl)oxy]-	n.s.g.	256(4.67),310(4.63)	2-1124-78
$C_{26}H_{22}O_9$ Licoisoflavone B triacetate	EtOH	228(4.55),241s(4.55), 245s(4.57),250(4.59), 255(4.54),261s(4.37), 305(3.87)	94-0144-78
$C_{26}H_{23}NO$ 2,4-Cyclohexadien-1-one, 6-(dimethyl-amino)-2,3,6-triphenyl-	EtOH	235(4.28),285s(3.89), 349(3.73)	5-1634-78
2,4-Cyclohexadien-1-one, 6-(dimethyl-amino)-2,4,6-triphenyl-	EtOH	244(4.36),295s(3.70), 347(3.58)	5-1634-78
2,4-Cyclohexadien-1-one, 6-(dimethyl-amino)-2,5,6-triphenyl-	EtOH	247(4.23),370(3.91)	5-1634-78
2,5-Cyclohexadien-1-one, 4-(dimethyl-amino)-2,4,6-triphenyl-	MeOH	224(4.46),275(3.84)	5-1634-78
4H-Indol-4-one, 1,5,6,7-tetrahydro-6,6-dimethyl-1-(2-naphthalenyl)-2-phenyl-	MeOH	215(4.56),242(4.40), 278(4.30)	87-1165-78
$C_{26}H_{23}NO_3$ 7-Benzoyldehydronuciferine	EtOH	257(4.66),323(4.04), 373s(3.51)	44-1096-78
$C_{26}H_{23}N_2$ Benzo[f]quinazolinium, 1,4-dihydro-2,4-dimethyl-1,3-diphenyl-, iodide	EtOH	209(4.6),240(4.5), 275(4.3),288(4.3), 350(3.9)	103-1019-78
$C_{26}H_{23}N_3$ 3H-1,5-Benzodiazepine, 2-(3,6-dihydro-4-phenyl-1(2H)-pyridinyl)-4-phenyl-	n.s.g.	259(4.66),345(3.66)	87-0952-78
$C_{26}H_{23}N_3O$ 2H-Pyrrolo[3,4-d]pyrimidin-2-one, 1,3,4,6-tetrahydro-5-methyl-4,7-diphenyl-1-(phenylmethyl)-	MeOH	296(4.05)	32-0591-78
$C_{26}H_{23}N_3O_2$ 2-Naphthalenamine, 1-[[4-(dimethylami-no)phenyl]methyl]-N-[(4-nitrophenyl)-methylene]-	EtOH	207(4.60),231(4.58), 257(4.47),288(4.41), 388(3.98)	103-1025-78
$C_{26}H_{23}N_3O_3S$ 4-Thiazoleacetic acid, α-(ethoxyimino)-2-[(triphenylmethyl)amino]-, (E)-	EtOH	250(4.27),323(3.41)	78-2233-78
(Z)-	EtOH	234(4.28),293(3.81)	78-2233-78
4-Thiazoleacetic acid, α-(hydroxyimi-no)-2-[(triphenylmethyl)amino]-, ethyl ester, (E)-	EtOH	235s(4.21),308(3.48)	78-2233-78
(Z)-	EtOH	226s(4.40),264(3.93), 300(3.63)	78-2233-78

Compound	Solvent	$\lambda_{max}(\log \epsilon)$	Ref.
$C_{26}H_{23}N_4O$			
Pyridinium, 2-[[4-(2-benzoxazolylmethylene)-2,4-dihydro-5-methyl-2-phenyl-3H-pyrazol-3-ylidene]methyl]-1-ethyl-, iodide	n.s.g.	214(4.76),250(4.81)	48-0857-78
$C_{26}H_{23}O_3PS$			
Phosphine oxide, diphenyl[8-(phenylsulfonyl)tricyclo[4.2.0.0^{2,5}]oct-7-en-3-yl]-, syn	EtOH	258(3.06),265(3.20), 273(3.12),553(2.85)	44-4338-78
$C_{26}H_{24}$			
[3.3](2,6)Naphthalenophane	C_6H_{12}	217(5.3),270f(4.0), 300(3.2),315(3.2)	88-1425-78
$C_{26}H_{24}ClNO_4S$			
Thiopyrylium, 2,4-diphenyl-6-[(1,2,6-trimethyl-4(1H)-pyridinylidene)methyl]-, perchlorate	EtOH	507(4.295)	83-0256-78
$C_{26}H_{24}ClN_3$			
3H-1,5-Benzodiazepine, 2-(4-chlorophenyl)-4-(4-phenylpiperidino)-	n.s.g.	262(4.65),354(3.72)	87-0952-78
3H-1,5-Benzodiazepine, 7-chloro-2-phenyl-4-(4-phenylpiperidino)-	n.s.g.	263(4.61),358(3.68)	87-0952-78
3H-1,5-Benzodiazepine, 2-[4-(4-chlorophenyl)piperidino]-4-phenyl-	n.s.g.	260(4.61),346(3.73)	87-0952-78
$C_{26}H_{24}ClN_3O_5$			
Quinolinium, 3-acetyl-6-methyl-1-(4-methylphenyl)-4-[(phenylhydrazono)methyl]-, perchlorate	MeOH	508(4.32)	65-1635-78
$C_{26}H_{24}ClN_3O_7$			
Quinolinium, 3-acetyl-6-methoxy-1-(4-methoxyphenyl)-4-[(phenylhydrazono)methyl]-, perchlorate	MeOH	509(4.13)	65-1635-78
$C_{26}H_{24}FN_3$			
3H-1,5-Benzodiazepine, 2-(4-fluorophenyl)-4-(4-phenylpiperidino)-	n.s.g.	262(4.57),346(3.68)	87-0952-78
$C_{26}H_{24}N_2O_3$			
Carbamic acid, [3-(3-oxo-1-phenylbutyl)-1H-indol-2-yl]-, phenylmethyl ester	CHCl$_3$	283(4.11),295(4.09)	103-0745-78
Quinazoline, 4-[(3,4-dihydro-7-methoxy-2,2-dimethyl-2H-1-benzopyran-5-yl)oxy]-2-phenyl-	MeOH	256(4.53),284(4.26)	24-0439-78
$C_{26}H_{24}N_2O_4$			
L-Tryptophan, N-[(phenylmethoxy)carbonyl]-, phenylmethyl ester	isoPrOH	274(3.77),282(3.81), 290(3.75)	63-1617-78
$C_{26}H_{24}N_2O_6$			
2H-Cyclopenta[d]pyridazine-1,4-dicarboxylic acid, 5-[bis(4-methoxyphenyl)methyl]-, dimethyl ester	MeCN	226(4.53),293(4.40), 330(3.67),500(2.77)	78-2509-78
2(5H)-Furanone, 4,4'-(azinodiethylidyne)bis[3-methoxy-5-phenyl-	EtOH	305(4.40),325(4.37)	142-1041-78A
geometric isomer	EtOH	304(4.43),327(4.29)	142-1041-78A

Compound	Solvent	$\lambda_{max}(\log \epsilon)$	Ref.
$C_{26}H_{24}N_2O_9S$			
Sulfamide, (2,3,5-tri-O-benzoyl-D-ribo-furanosyl)-	EtOH	231(4.51)	4-0477-78
$C_{26}H_{24}N_3O$			
Quinolinium, 3-acetyl-6-methyl-1-(4-methylphenyl)-4-[(phenylhydrazono)-methyl]-, perchlorate	MeOH	508(4.32)	65-1635-78
$C_{26}H_{24}N_3O_3$			
Quinolinium, 3-acetyl-6-methoxy-1-(4-methoxyphenyl)-4-[(phenylhydrazono)-methyl]-, perchlorate	MeOH	509(4.13)	65-1635-78
$C_{26}H_{24}N_4O$			
2H-Benzimidazol-2-one, 5-[4,5-bis(4-methylphenyl)-1H-imidazol-2-yl]-1,3-dihydro-1,3-dimethyl-radical	EtOH	218(4.77),323(4.62)	103-0425-78
	toluene	540(--),704(--)	103-0425-78
2-Cyclohexen-1-one, 2-[[2-[2-[2-[(2-am-inophenyl)azo]phenyl]ethenyl]phenyl]-amino]-	benzene	305(4.21),325s(4.16),428s(3.74)	33-2813-78
$C_{26}H_{24}N_4O_2$			
2H-Benzimidazol-2-one, 5-[4-(4-ethoxy-phenyl)-5-phenyl-1H-imidazol-2-yl]-1,3-dihydro-1,3-dimethyl-radical	EtOH	219(4.62),321(4.48)	103-0425-78
	toluene	539(--),713(--)	103-0425-78
$C_{26}H_{24}N_4O_2S$			
Benzaldehyde, 4-nitro-, (3-butyl-4,5-diphenyl-2(3H)-thiazolylidene)hydra-zone	isoPrOH	295(4.10),336(4.00),459(4.35)	73-1227-78
3(2H)-Thiazolamine, 2-[(1,1-dimethyl-ethyl)imino]-N-[(4-nitrophenyl)meth-ylene]-4,5-diphenyl-	isoPrOH	290(4.25),431(4.14)	73-1227-78
$C_{26}H_{24}N_4O_3$			
2H-Benzimidazol-2-one, 5-[4,5-bis(4-methoxyphenyl)-1H-imidazol-2-yl]-1,3-dihydro-1,3-dimethyl-radical	EtOH	218(4.76),322(4.57)	103-0425-78
	toluene	542(--),708(--)	103-0425-78
$C_{26}H_{24}N_4O_5$			
7H-Benz[de]anthracen-7-one, 3-[[4,6-bis(2-methoxyethoxy)-1,3,5-triazin-2-yl]amino]-	DMF	460(4.80)	2-0106-78
$C_{26}H_{24}N_6O_4S$			
4-Thiazolidinecarboxylic acid, 3-benz-oyl-5-[6-(benzoylamino)-9H-purin-9-yl]-2,2-dimethyl-, methyl ester, (4R-trans)-	EtOH	232(4.21),262s(4.11),281(4.33)	23-0326-78
	EtOH-HCl	285(4.33)	23-0326-78
$C_{26}H_{24}O_4$			
Tricyclo[4.2.2.0^{2,5}]deca-3,7-diene-3,8-dicarboxylic acid, 4,7-diphen-yl-, dimethyl ester	C_6H_{12}	274(4.25),293s(4.17),309s(3.93)	88-0357-78
$C_{26}H_{24}O_6$			
Cyclobuta[1,2-a:4,3-a']dinaphthalene-	EtOH	235(4.47),243s(4.42),	12-0085-78

Compound	Solvent	λ_{max}(log ϵ)	Ref.
2,7-dicarboxylic acid, 6a,6b,12b,12c-tetrahydro-6,11-dimethoxy-, dimethyl ester (cont.)		253s(4.30),312(4.49)	12-0085-78
$C_{26}H_{24}O_8$			
4H-1-Benzopyran-4-one, 5,7-diacetoxy-2-(4-acetoxyphenyl)-6-(3-methyl-2-butenyl)-	MeOH	240(3.98),281(4.01), 321(4.20)	78-3569-78
$C_{26}H_{24}O_9$			
Demethylisoglycyrin triacetate	EtOH	255(4.04),328(4.31)	94-0135-78
Licoisoflavanone triacetate	EtOH	224(4.71),244s(4.11), 250s(4.20),256(4.28), 261(4.27),313(3.82)	94-0144-78
$C_{26}H_{25}Br$			
[2.2]Paracyclo(4,8)[2.2]metaparacyclo-phane, 6-bromo-	THF	285(2.79),294(2.45), 305s(2.32)	18-2668-78
$C_{26}H_{25}NO$			
3H-Pyrrol-3-one, 2-ethyl-1,2-dihydro-1-methyl-2-(4-methylphenyl)-4,5-di-phenyl-	MeCN	224(4.17),282(4.03), 356(3.96)	5-1203-78
$C_{26}H_{25}NO_5$			
Deacetamidocolchiceine, 7-benzylimino-	EtOH	247(4.56),356(4.42), 370s(4.34),408s(3.94)	33-1213-78
Dibenzo[a,i]quinolizine-6,13(1H,5H)-di-one, 2,3-dihydro-9,12-dimethoxy-8-(phenylmethoxy)-	MeOH	242(4.55),282(4.03), 317(3.91)	39-0434-78C
C-Homoooxoerysodienone, O-benzyl-	MeOH	240(4.28),282(3.61)	39-0434-78C
$C_{26}H_{25}NO_8$			
5H-[1,3]Dioxolo[6,7]isoquino[2,1-b][2]-benzazocine-14,15-dicarboxylic acid, 7,8-dihydro-3,4-dimethoxy-, dimethyl ester	EtOH	280(4.25),375(3.76)	2-1100-78
$C_{26}H_{25}NS$			
2-Azabicyclo[3.1.0]hexane-3-thione, 2,6-dimethyl-1-(4-methylphenyl)-4,5-diphenyl-	MeCN	286(4.18)	5-1222-78
$C_{26}H_{25}N_2O$			
Pyridinium, 2-[3-(1,2-dihydro-2-oxo-3H-indol-3-ylidene)-2-(4-ethylphen-yl)-1-propenyl]-1-ethyl-, iodide	MeOH	534(3.98)	64-0209-78B
$C_{26}H_{25}N_3$			
3H-1,5-Benzodiazepine, 2-phenyl-4-(4-phenylpiperidino)-	n.s.g.	260(4.53),348(3.65)	87-0952-78
$C_{26}H_{25}N_3O_5$			
8(5H)-Quinolinone, 2-[5,6-dimethyl-4-(2,3,4-trimethoxyphenyl)-2-pyri-dinyl]-5-imino-6-methoxy-	n.s.g.	261(4.53),309(4.43)	88-4775-78
$C_{26}H_{25}N_3O_6$			
5,8-Quinolinedione, 7-amino-2-[5,6-di-methyl-4-(2,3,4-trimethoxyphenyl)-2-pyridinyl]-6-methoxy-	n.s.g.	260(4.69),317(4.65), 484(3.34)	88-4775-78

Compound	Solvent	$\lambda_{max}(\log \epsilon)$	Ref.
$C_{26}H_{25}N_3P$			
Phosphorus(1+), (4-diazenyl-N,N-dimethylbenzenaminato)triphenyl-, (T-4)-, tetrafluoroborate	n.s.g.	277(4.06),490(4.63)	123-0190-78
$C_{26}H_{25}N_5$			
Pyrido[2,3-b]pyrazine, 1,2,3,5-tetrahydro-5-methyl-2,3-bis(2-methyl-1H-indol-3-yl)-, monohydriodide (same spectrum for perchlorate)	EtOH	279(4.30),289(4.20), 351(4.11)	104-0398-78
$C_{26}H_{26}$			
Bicyclo[8.2.2]tetradeca-5,10,12,13-tetraene, 4,7-diphenyl-, trans	EtOH	262(3.00),269(3.00), 279(2.68)	35-1806-78
Triple-layered [2.2]metaparacyclophane	THF	285(2.77),293(2.72)	18-2668-78
$C_{26}H_{26}N_2O_2$			
7,8-Diazatricyclo[4.3.2.0^{2,5}]undeca-3,8,10-triene-7-carboxylic acid, 3,10-dimethyl-4,11-diphenyl-, ethyl ester	C_6H_{12}	253(4.30),278s(3.93), 294s(2.92)	88-0357-78
$C_{26}H_{26}N_3$			
Pyrimidinium, 5-(3,5-dimethylphenyl)-1-(4-methylphenyl)-4-[(4-methylphenyl)amino]-, perchlorate	CH_2Cl_2	332(4.27)	97-0063-78
$C_{26}H_{26}N_3O_2$			
Pyrimidinium, 5-(3,5-dimethylphenyl)-1-(4-methoxyphenyl)-4-[(4-methoxyphenyl)amino]-, perchlorate	CH_2Cl_2	287(4.08),341(4.21)	97-0063-78
$C_{26}H_{26}N_4O$			
1(4H)-Naphthalenone, 2-[[4-(dimethylamino)phenyl]amino]-4-[[4-(dimethylamino)phenyl]imino]-	C_6H_{12}	255(4.42),285(4.40), 332(4.00),560(3.89)	48-0557-78
	EtOH	260(4.31),285(4.20), 332(3.85),565(3.72)	48-0557-78 +80-0617-78
	ether	260(4.46),282(4.40), 335(4.05),565(3.95)	48-0557-78
	CCl$_4$	332(4.05),570(3.92)	48-0557-78
$C_{26}H_{26}N_4O_2$			
[4,4'-Bipyridine]-3,3'-dicarboxamide, 1,1',4,4'-tetrahydro-1,1'-bis(phenylmethyl)-	MeOH	268(3.80),348(3.86)	44-3420-78
isomer	MeOH	276(3.78),354(3.91)	44-3420-78
$C_{26}H_{26}N_4O_4$			
Pyrazolidine, 2-(2,4-dinitrophenyl)-3-methyl-3-(1-methylethenyl)-5-phenyl-1-(phenylmethyl)-	MeCN	236(4.26),388(4.29)	78-0903-78
$C_{26}H_{26}O_2$			
1H-Indene-8,9-diol, 1-(diphenylmethylene)-3a,4,7,7a-tetrahydro-5,6-dimethyl-4,7-ethano-, cis-cis	MeOH	241(4.02),295(4.23)	5-0440-78
cis-trans	MeOH	242(3.97),293(4.08)	5-0440-78
trans-trans	MeOH	241(3.99),294(4.23)	5-0440-78
$C_{26}H_{26}O_5$			
4H,8H-Benzo[1,2-b:3,4-b']dipyran-4-one,	MeOH	263(4.40),284(3.94),	78-3569-78

Compound	Solvent	λ_{max} (log ϵ)	Ref.
5-hydroxy-2-(4-methoxyphenyl)-8,8-di-methyl-6-(3-methyl-2-butenyl)- (cont.)		330(3.99)	78-3569-78
4H,8H-Benzo[1,2-b:5,4-b']dipyran-4-one, 5-hydroxy-2-(4-methoxyphenyl)-8,8-di-methyl-10-(3-methyl-2-butenyl)-	MeOH	265(4.36),289(3.72), 331(3.79)	78-3569-78
Nitiducarpin	MeOH	205(4.20),230(4.38), 292(3.80),307(3.81)	39-0137-78C
$C_{26}H_{26}O_6$			
4H,8H-Benzo[1,2-b:3,4-b']dipyran-4-one, 5-hydroxy-2-(2-hydroxy-4-methoxyphen-yl)-8,8-dimethyl-5-(3-methyl-2-buten-yl)- (4'-O-methylmorusin)	MeOH	270(4.47),300(3.92), 320(3.77)	94-1431-78
	MeOH-NaOMe	270(4.47),300(3.88), 365(3.88)	94-1431-78
4H,8H-Benzo[1,2-b:3,4-b']dipyran-4-one, 5-hydroxy-2-(4-hydroxy-2-methoxyphen-yl)-8,8-dimethyl-5-(3-methyl-2-buten-yl)-	MeOH	270(4.47),300(3.93), 320(3.80)	94-1431-78
	MeOH-NaOMe	270(4.47),300(3.89), 365(4.02)	94-1431-78
$C_{26}H_{26}O_7$			
4-Benzofurancarboxaldehyde, 2-(3,4-di-methoxyphenyl)-3-(3,5-dimethoxyphen-yl)-2,3-dihydro-6-methoxy-	EtOH	277(4.02),346(3.54)	94-3050-78
3H,7H-Pyrano[2',3':7,8][1]benzopyrano-[3,2-d][1]benzoxepin-7-one, 8,9-di-hydro-6-hydroxy-9-(1-hydroxy-1-methyl-ethyl)-12-methoxy-3,3-dimethyl-	EtOH	207(4.46),221(4.46), 238(4.46),282(4.61), 336(4.20)	94-1394-78
	EtOH-AlCl₃	208(4.49),232(4.42), 249s(4.36),291(4.53), 354(4.26)	94-1394-78
$C_{26}H_{26}O_8$			
4H-1-Benzopyran-4-one, 5,7-diacetoxy-3-(2,4-dimethoxyphenyl)-6-(3-methyl-2-butenyl)-	MeOH	263(4.42),287(4.20)	2-0613-78
4H-Furo[2,3-h]-1-benzopyran-4-one, 9-(1,2-diacetoxy-2-methylpropyl)-8,9-dihydro-5-methoxy-2-phenyl-(polystachin)	MeOH	215(4.43),269(4.41), 326(3.93)	119-0047-78
Glycyrin diacetate	EtOH	246s(4.16),255s(4.04), 345(4.43)	94-0135-78
Morusin hydroperoxide methyl ether	MeOH	215(4.25),237(4.25), 279(4.27),337(3.99)	94-1431-78
	MeOH-NaOMe	286(4.34),387(3.55)	94-1431-78
	MeOH-AlCl₃	225(4.26),284(4.29), 359(4.07),424(3.72)	94-1431-78
$C_{26}H_{26}O_{10}$			
Edulon A, di-O-acetyl-	EtOH	232(3.93),279(3.78), 290s(3.75),347s(4.09), 374(4.11)	33-1969-78
$C_{26}H_{27}BrN_2O_6S$			
1-Azabicyclo[3.2.0]hept-2-ene-2-carb-oxylic acid, 6-(1-hydroxyethyl)-7-oxo-3-[[2-[(phenoxyacetyl)amino]-ethyl]thio]-, (4-bromophenyl)methyl ester, [5R-[5α,6α(R*)]]-	EtOH	320(3.90)	35-8004-78
$C_{26}H_{27}NO_3$			
Nuciferine, 7-(α-hydroxybenzyl)-, hydro-chloride	EtOH	243s(4.36),275(4.51)	44-1096-78

Compound	Solvent	$\lambda_{max}(\log \epsilon)$	Ref.
$C_{26}H_{27}NO_3S_2$ 4-Piperidinone, 3,5-bis[[5-(ethylthio)-2-furanyl]methylene]-1-(phenylmethyl)-	EtOH	209(4.30),262(4.11), 420(4.50)	133-0189-78
$C_{26}H_{27}NO_5S_2$ 4-Piperidinone, 3,5-bis[[5-(ethylsulfinyl)-2-furanyl]methylene]-1-(phenylmethyl)-	EtOH	210(4.30),258(4.04), 376(4.55)	133-0189-78
$C_{26}H_{27}NO_7S_2$ 4-Piperidinone, 3,5-bis[[5-(ethylsulfonyl)-2-furanyl]methylene]-1-(phenylmethyl)-	EtOH	244(4.14),286(4.06), 360(3.46)	133-0189-78
$C_{26}H_{27}NO_9$ Benzo[3,4]cycloocta[1,2-f][1,3]benzodioxole-6,7-dicarboxylic acid, 8-(dimethylamino)-1,2,3-trimethoxy-, dimethyl ester	EtOH	248s(5.32),302(5.16), 344(4.88)	44-3950-78
$C_{26}H_{27}N_3O$ 2H-Imidazol-2-one, 4-ethyl-1,3-dihydro-3-[methyl(2-methylphenyl)amino]-1-(4-methylphenyl)-5-phenyl-	EtOH	273(4.12)	5-2033-78
$C_{26}H_{27}N_5O_{10}$ Benzamide, N-[9-(2,3,4,6-tetra-O-acetyl-β-D-glucopyranosyl)-9H-purin-6-yl]-	MeOH	210(4.41),278(4.33)	136-0301-78C
$C_{26}H_{28}N_2$ 4,11-Etheno-1H-cyclopenta[b]phenazine, 1-cyclohexylidene-3a,4,11,11a-tetrahydro-	C_6H_{12}	213(4.67),247(4.70), 292s(3.69),298s(3.76), 305s(3.88),310(3.96), 318(4.07),322(4.06), 331(4.00)	5-0440-78
$C_{26}H_{28}N_2O_3S$ 4H-Carbazol-4-one, 1,2,3,5,6,7,8,9-octahydro-9-(phenylmethyl)-, O-[(4-methylphenyl)sulfonyl]oxime	MeOH	211(4.38),217(4.35), 257(4.11)	24-0853-78
$C_{26}H_{28}N_2O_6$ 1H-Pyrrole-3-propanoic acid, 5-[[3-ethylidene-4-methyl-5-oxo-2-pyrrolidinylidene]-2-oxo-2-(phenylmethoxy)ethyl]-2-formyl-4-methyl-, methyl ester, (?,E)-(±)-	MeOH	234(4.57),309(4.74), 350(4.22)	44-0283-78
Thienamycin, N-benzyloxycarbonyl-, benzyl ester	MeOH	317(4.07)	35-6491-78
$C_{26}H_{28}N_2O_8S_2$ 1H-Indole, 5,5'-dithiobis[1-α-L-arabinopyranosyl-	EtOH	230(4.59),242(4.55)	104-2011-78
$C_{26}H_{28}N_4$ 1,1'-Bipyridinium, 2,2',6,6'-tetramethyl-4,4'-bis(phenylamino)-, bis(tetrafluoroborate)	EtOH	220(3.86),318(4.03)	120-0001-78
dichloride	EtOH	220(3.78),317(3.99)	120-0001-78

Compound	Solvent	$\lambda_{max}(\log \epsilon)$	Ref.
$C_{26}H_{28}N_4O_3$			
Pyrrolo[3',4':4,5]pyrrolo[3,4-b]indole-1,3,5(2H)-trione, 5a-amino-4-cyclo-hexyl-3a,4,5a,6-tetrahydro-2-methyl-6-(phenylmethyl)-	EtOH	243s(3.83),260(3.88), 312(3.40)	4-1463-78
$C_{26}H_{28}N_6O_9$			
Benzamide, N-[9-[3,4,6-tri-O-acetyl-2-(acetylamino)-2-deoxy-β-D-gluco-pyranosyl]-9H-purin-6-yl]-	MeOH	211(4.41),281(4.37)	136-0301-78C
$C_{26}H_{28}O_2$			
Tricyclo[4.2.2.0^{2,5}]deca-3,7-diene, 3,8-bis(4-methoxyphenyl)-4,7-dimethyl-	C_6H_{12}	255(4.44),270s(4.32), 310s(3.01)	88-0357-78
$C_{26}H_{28}O_2S_2$			
Benzene, 1-methyl-4-[[2-[(3-methylbut-yl)thio]-1,2-diphenylethenyl]sulfon-yl]-, (E)-	EtOH	203(4.57),235s(4.17), 300(4.06)	2-1086-78
(Z)-	EtOH	203(4.56),235s(4.24), 310(3.97)	2-1086-78
Benzene, 1-methyl-4-[[2-(pentylthio)-1,2-diphenylethenyl]sulfonyl]-, (E)-	EtOH	203(4.62),235s(4.19), 295(4.12)	2-1086-78
(Z)-	EtOH	202(4.61),234s(4.25), 304(4.00)	2-1086-78
$C_{26}H_{28}O_4S_2$			
Benzene, 1-methyl-4-[[2-[(3-methylbut-yl)sulfonyl]-1,2-diphenylethenyl]-sulfonyl]-, (E)-	EtOH	203(4.66),220s(4.46), 245s(4.17)	2-1086-78
(Z)-	EtOH	202(4.65),244s(4.14)	2-1086-78
Benzene, 1-methyl-4-[[2-(pentylsulfon-yl)-1,2-diphenylethenyl]sulfonyl]-, (E)-	EtOH	203(4.61),219s(4.44), 244(4.15)	2-1086-78
(Z)-	EtOH	203(4.54),245(4.09)	2-1086-78
$C_{26}H_{28}O_5$			
Acacetin, 6,8-di-C-prenyl-	MeOH	226(4.38),283(4.36), 329(4.29)	78-3569-78
4H,6H,10H-Benzo[1,2-b:3,4-b':5,6-b"]-tripyran-4-one, 7,8,11,12-tetrahydro-2-(4-methoxyphenyl)-6,6,10,10-tetra-methyl-	MeOH	261(3.94),280(4.20), 328(4.15)	78-3569-78
Nitiducol	MeOH	207(4.76),283(3.64), 307(4.91)	39-0137-78C
Nitidulan	MeOH	205(4.54),227(4.53), 280(4.13),290(4.13)	39-0137-78C
$C_{26}H_{28}O_6$			
Benzofuran, 2-(3,4-dimethoxyphenyl)-3-(3,5-dimethoxyphenyl)-2,3-dihydro-6-methoxy-4-methyl-	EtOH	209(4.96),230s(4.46), 279s(3.91),282(3.98), 289s(3.83)	94-3050-78
$C_{26}H_{28}O_7$			
4-Benzofuranmethanol, 2-(3,4-dimethoxy-phenyl)-3-(3,5-dimethoxyphenyl)-2,3-dihydro-6-methoxy-	EtOH	234(4.44),278s(3.80), 283(3.84),299s(3.55)	94-3050-78
$C_{26}H_{28}O_8$			
Butanedioic acid, (1,2-diethyl-1,2-eth-enediyl)di-4,1-phenylene ester	MeOH	234(4.24)	5-0726-78

Compound	Solvent	$\lambda_{max}(\log \epsilon)$	Ref.
$C_{26}H_{28}O_9$ Malaphyllin	EtOH	223(4.39),262(4.41), 293(3.93)	105-0377-78

$C_{26}H_{29}BrO_3$
  Phenol, 2-[(3-bromo-2-hydroxy-5-methyl-
  phenyl)methyl]-6-[[5-(1,1-dimethyl-
  ethyl)-2-hydroxyphenyl]methyl]-4-
  methyl-
	dioxan	279s(3.88),286(3.91)	49-0767-78

$C_{26}H_{29}CuN_4O_5$
  Copper, [2,5-dihydro-3-[[[2-[[(2-hy-
  droxyphenyl)methylene]amino]-1-[[[(2-
  hydroxyphenyl)methylene]amino]meth-
  yl]ethoxy]carbonyl]amino]-2,2,5,5-
  tetramethyl-1H-pyrrol-1-yloxyato(2-)]-
	CH_2Cl_2
	
	pyridine
		232(4.76),274(4.45),
		373(4.09),610(2.41)
		377(4.06),610(2.29)
			35-2686-78
			
			35-2686-78

$C_{26}H_{29}NO_5$
  Cryogenine
  Lythrine
	MeOH
	MeOH
		260(4.1),285(4.2)
		261(4.1),284(4.2)
			94-2515-78
			94-2515-78

$C_{26}H_{29}N_2O$
  Pyrylium, 2,6-bis[2-[4-(dimethylamino)-
  phenyl]ethenyl]-4-methyl-, perchlor-
  ate
	EtOH	670(4.673)	83-0236-78

$C_{26}H_{30}N_2$
  1H-Indeno[2,1-c]pyridine, 2,3-dihydro-
  9-phenyl-2-[3-(1-piperidinyl)propyl]-
  1H-Indole, 2,3-dihydro-2-(1H-indol-3-
  yl)-3,3-bis(3-methyl-2-butenyl)-
	MeOH
	
	EtOH
		244(4.44),258(4.42),
		311(3.75),323(3.64)
		220s(4.48),278(4.03),
		288(3.97)
			87-0340-78
			
			78-0929-78

$C_{26}H_{30}N_2O_2$
  1H-Pyrrol-3-ol, 2-(12-azabicyclo[9.2.1]-
  tetradeca-1(14),11-dien-13-ylidene-
  methyl)-5-phenyl-, acetate
	MeOH	415(4.33),488(4.31),
		509(4.48)
			49-0137-78

$C_{26}H_{30}N_2O_3S$
  Benzenesulfonic acid, 4-methyl-, (3-
  methoxyestra-1,3,5(10),8-tetraen-
  17-ylidene)hydrazide, (±)-
  DL-Leucine, N-benzoyl-, 4-(2-methyl-
  propyl)-2-phenyl-5-thiazolyl ester
	EtOH
	
	ether
		275(4.26)
		
		307(4.04)
			22-0119-78
			
			78-0611-78

$C_{26}H_{30}N_2O_4$ Mepivacaine 3-hydroxy-2-naphthoate	MeOH	271(3.73),282(3.74), 293(3.45),353(3.38)	94-0936-78

$C_{26}H_{30}N_4O_2Zn$
  Zinc, bis[2-[(3-ethyl-5-methoxy-4-meth-
  yl-2H-pyrrol-2-ylidene)methyl]-1H-
  pyrrolato-N^1,N^2]-, (T-4)-
	MeOH
	DMSO
		458(4.70)
		459(--)
			49-0883-78
			49-0883-78

$C_{26}H_{30}N_4O_5$
  8H-Oxazolo[4,5-b]phenoxazine-4,6-di-
  carboxamide, N,N,N',N'-tetraethyl-
  2,9,11-trimethyl-8-oxo-
	CHCl₃	263(4.36),276(4.37),
		295s(3.83),382(4.12),
		495(4.04)
			4-0129-78

$C_{26}H_{30}O_4$
  Cyclohexanone, 2-[1,5-bis(2-methoxy-
  phenyl)-3-oxo-4-pentenyl]-4-methyl-
	EtOH	227(4.10),284(4.06),
		335(3.86)
			2-0428-78

Compound	Solvent	$\lambda_{max}(\log \epsilon)$	Ref.
Cyclohexanone, 2-[1,5-bis(2-methoxy-phenyl)-3-oxo-4-pentenyl]-5-methyl-	EtOH	229(4.07),285(4.22), 336(4.04)	2-0428-78
Cyclohexanone, 2-[1,5-bis(4-methoxy-phenyl)-3-oxo-4-pentenyl]-4-methyl-	EtOH	229(4.31),326(4.39)	2-0428-78
Cyclohexanone, 2-[1,5-bis(4-methoxy-phenyl)-3-oxo-4-pentenyl]-5-methyl-	EtOH	228(4.20),324(4.28)	2-0428-78
Heminitidulan	MeOH	205(4.58),228(4.59), 280(4.14)	39-0137-78C
Spiro[5.5]undecane-1,9-dione, 7,11-bis-(2-methoxyphenyl)-3-methyl-	EtOH	227(4.05),276(3.72), 283(3.71)	2-0428-78
Spiro[5.5]undecane-1,9-dione, 7,11-bis-(2-methoxyphenyl)-4-methyl-	EtOH	228(3.99),277(3.72), 283(3.71)	2-0428-78
Spiro[5.5]undecane-1,9-dione, 7,11-bis-(4-methoxyphenyl)-3-methyl-	EtOH	232(4.24),277(3.47), 284(3.41)	2-0428-78
Spiro[5.5]undecane-1,9-dione, 7,11-bis-(4-methoxyphenyl)-4-methyl-	EtOH	230(4.27),277(3.42), 283(3.37)	2-0428-78
$C_{26}H_{30}O_5$			
Nitidulin	MeOH	283(4.01),295(4.02)	39-0137-78C
$C_{26}H_{30}O_6$			
Ferocaulicin	EtOH	219(4.37),235(4.27), 295(4.01),324(4.20)	105-0267-78
Phenol, 2-[2-(3,4-dimethoxyphenyl)-1-(3,5-dimethoxyphenyl)ethyl]-5-methoxy-3-methyl-	EtOH	209(4.88),230s(4.44), 276s(3.85),279(3.86), 286s(3.75)	94-3050-78
$C_{26}H_{31}NO_6$			
Lythridine	MeOH	292.0(3.9)	94-2515-78
$C_{26}H_{31}N_3O_8$			
DL-Aspartic acid, N-[N-[4-[methyl-[(phenylmethoxy)carbonyl]amino]-benzoyl]glycyl]-, diethyl ester	EtOH	262(4.16)	87-1165-78
L-Glutamic acid, N-[N-[4-[[(phenyl-methoxy)carbonyl]amino]benzoyl]-glycyl]-, diethyl ester	EtOH	268(4.53)	87-1165-78
$C_{26}H_{31}OP$			
Phosphine oxide, phenylbis(2,3,5,6-tetramethylphenyl)-	C_6H_{12}	287(3.58),295(3.60)	65-1394-78
$C_{26}H_{31}P$			
Phosphine, phenylbis(2,3,5,6-tetra-methylphenyl)-	C_6H_{12}	199(4.92),295(4.09)	65-1394-78
$C_{26}H_{32}BN$			
Benzenamine, 4-[bis(2,4,6-trimethyl-phenyl)boryl]-N,N-dimethyl-	C_6H_{12}	352(4.69)	147-0177-78A
$C_{26}H_{32}N_2$			
1H-Indole, 1,1'-(1,4-butanediyl)bis-[2,3-dihydro-3,3-dimethyl-2-methyl-ene-	n.s.g.	255s(4.79),296(4.34)	47-2039-78
$C_{26}H_{32}N_2O_2S_2$			
2,3-Butanediol, 1,4-bis[(2-propyl-2H-isoindol-1-yl)thio]-, (R*,R*)-	EtOH	332(4.16),345s(--)	44-2886-78
$C_{26}H_{32}N_3OP$			
Phosphorin, 4-cyclohexyl-2-[[4-(dimeth-	EtOH	360(3.93),514(4.47)	118-0846-78

Compound	Solvent	$\lambda_{max}(\log \epsilon)$	Ref.
ylamino)phenyl]azo]-1,1-dihydro-1-methoxy-1-phenyl- (cont.)			118-0846-78
$C_{26}H_{32}N_4O_4$ 5H-Oxazolo[4,5-b]phenoxazine-4,6-dicarboxamide, N,N,N',N'-tetraethyl-2,9,11-trimethyl-	CHCl$_3$	253(4.53),351(3.95)	4-0129-78
$C_{26}H_{32}N_6O_8$ L-Threonine, N-[[[9-[2,3-O-(1-methylethylidene)-β-D-ribofuranosyl]-9H-purin-6-yl]amino]carbonyl]-O-(phenylmethyl)-, methyl ester	pH 2.0 pH 12.0 MeOH	270(4.30),277(4.32) 271(4.22),278(4.26), 300(4.05) 270(4.35),277s(4.28)	19-0021-78 19-0021-78 19-0021-78
$C_{26}H_{32}O_5$ 1-Anthracenecarboxaldehyde, 5,8,8a,9,10-10a-hexahydro-2-hydroxy-3-methoxy-6,10a-dimethyl-5-(3-methyl-2-butenyl)-4-(1-methylethyl)-9,10-dioxo-(heliocide B₄)	EtOH-HCl EtOH-NaOH CHCl$_3$	266(4.55),298s(--), 364(3.45) 255s(--),266(4.36), 291(4.36),339(4.24) 268(4.53),352(3.56)	102-1297-78 102-1297-78 102-1297-78
1-Anthracenecarboxaldehyde, 5,8,8a,9,10-10a-hexahydro-2-hydroxy-3-methoxy-7,10a-dimethyl-5-(3-methyl-2-butenyl)-4-(1-methylethyl)-9,10-dioxo-(heliocide B₁)	EtOH-HCl EtOH-NaOH CHCl$_3$	265(4.47),348(3.54) 276(4.31),339(3.99), 385(3.80) 267(4.49),350(3.58)	102-1297-78 102-1297-78 102-1297-78
$C_{26}H_{32}O_6$ Benzene, 1,2,4-trimethoxy-5-[2-[2-(2,4,5-trimethoxyphenyl)-3-cyclohexen-1-yl]ethenyl]- (alflabene)	EtOH	261(4.25),300(3.97), 305(3.95),310(3.94)	88-2297-78
Feterin	EtOH	216(3.97),253(3.29), 324(3.98)	105-0263-78
$C_{26}H_{32}O_8$ Olivetoric acid	n.s.g.	218(4.47),272(4.27), 308(4.10)	12-1041-78
$C_{26}H_{33}NO_5S$ Benzo[f]quinoline-3-propanol, 1,2,3,4-4a,5,6,10b-octahydro-8-methoxy-4-[(4-methylphenyl)sulfonyl]-, acetate	EtOH	222(4.19),277(3.34), 287(3.30)	78-1023-78
$C_{26}H_{34}Cl_4N_2O_2$ Phenol, 4,4'-(1,2-diethyl-1,2-ethenediyl)bis[2-[bis(2-chloroethyl)amino]-, (E)-	MeOH	208(4.4),228s(4.3), 285s(3.9)	83-0184-78
$C_{26}H_{34}CuN_2O_2$ Copper, [[6,6'-[1,4-butanediylbis(nitrilomethylidyne)]bis[2-methyl-5-(1-methylethenyl)-2-cyclohexen-1-onato]](2-)-N,N',O,O']-	benzene	312(4.06),385(4.08), 660s(2.15)	65-1474-78
$C_{26}H_{34}N_4$ Pyrrolo[3',4':4,5]pyrrolo[3,4-b]indol-5a(6H)-amine, 4-cyclohexyl-1,2,3,3a-4,5-hexahydro-2-methyl-6-(phenylmethyl)-	EtOH	255(4.15),307(3.70)	4-1463-78
$C_{26}H_{34}N_4O_6S_2$ 1,3(2H,4H)-Pyrimidinedibutanesulfonam-	CHCl$_3$	268(3.97)	70-0149-78

Compound	Solvent	$\lambda_{max}(\log \epsilon)$	Ref.
ide, N,N'-dimethyl-2,4-dioxo-N,N'-di- phenyl- (cont.)			70-0149-78

$C_{26}H_{34}N_4O_8$
  D-Valinamide, N-[4-methyl-2-nitro-
  3-(phenylmethoxy)benzoyl]-L-threonyl-
  N-(2-hydroxyethyl)-
	MeOH	216(4.673)	87-0607-78

$C_{26}H_{34}O_3$
  Estra-1,3,5(10),8-tetraen-17-ol, 3-
  methoxy-, 17β-cyclohexanecarboxylate
	EtOH	278(4.22)	22-0119-78

$C_{26}H_{34}O_4$
  Strongylophorine-2
	MeOH	210(3.94),220(3.89),
		  228(3.83),297(3.68)
			20-0917-78

$C_{26}H_{34}O_5$
  19-Norpregna-1,3,5(10),16-tetraene-
  21,21-dicarboxylic acid, 3-methoxy-
  20-methyl-, dimethyl ester
	EtOH	220s(3.72),278(3.13),
		  287(3.11)
			35-3435-78

$C_{26}H_{34}O_6$
  Benz[f]azulene-5,6,8-triol, 1,2,3,9,10-
  10a-hexahydro-1,1,4-trimethyl-7-(1-
  methylethyl)-, triacetate, (S)-
	EtOH	252(4.04)	23-0733-78

Benz[f]azulene-5,6,8-triol, 1,2,3,9,10-
  10a-hexahydro-1,1,4-trimethyl-7-prop-
  yl-, triacetate, (S)-
	EtOH	254(4.08)	23-0733-78

21,24-Dinorchola-5,22(23)-dien-20-one, 3,17α-diacetoxy-16β,23-epoxy-	EtOH	271(3.91)	70-0160-78

$C_{26}H_{34}O_8$
  Phenol, 3,3'-(2,3-dimethyl-1,4-butane-
  diyl)bis[5,6-dimethoxy-, diacetate,
  (R*,S*)-
	EtOH	205(4.79),223s(4.22),
		  274(3.45)
			94-0682-78

$C_{26}H_{34}O_9$ Cressolide	MeOH	210(3.84)	20-0075-78

$C_{26}H_{34}O_{11}$
  β-D-Glucopyranoside, [2,3-dihydro-2-(4-
  hydroxy-3-methoxyphenyl)-5-(3-hydr-
  oxypropyl)-7-methoxy-3-benzofuran-
  yl]methyl, cis
	MeOH	237(4.10),283(3.63)	94-1619-78

β-D-Glucopyranoside, [tetrahydro-2-(4-
  hydroxy-3-methoxyphenyl)-4-[(4-hy-
  droxy-3-methoxyphenyl)methyl]-3-
  furanyl]methyl, (2α,3β,4β)-(-)-
	MeOH	231(3.97),283(3.57)	94-1619-78

$C_{26}H_{35}N_3O_{17}$ β-D-Lactosyl azide, heptaacetate	MeOH	273(1.81)	78-1427-78
β-D-Maltosyl azide, heptaacetate	MeOH	272(1.70)	78-1427-78

$C_{26}H_{36}Cl_4N_2O_2$
  Phenol, 4,4'-(1,2-diethyl-1,2-ethanedi-
  yl)bis[2-[bis(2-chloroethyl)amino]-,
  (R*,S*)-
	MeOH	216(4.4),287(3.8)	83-0184-78

$C_{26}H_{36}N_2O_7$
  Benzoic acid, 4-(dimethylamino)-, oxy-
  bis(2,1-ethanediyloxy-2,1-ethanediyl)
  ester
	MeOH
	ester
		228(4.15),310(4.75)	121-0661-78

Compound	Solvent	$\lambda_{max}(\log \epsilon)$	Ref.
$C_{26}H_{36}N_7$ 1H-Benzimidazolium, 2-[3-(1,3-diethyl-1,3-dihydro-4,7-dimethyl-2H-benzimidazol-2-ylidene)-1-triazenyl]-1,3-diethyl-4,7-dimethyl-, tetrafluoroborate	EtOH mixt.	356(4.30),413(4.32) (photochromic)	33-2958-78
$C_{26}H_{36}N_7O_4$ 1H-Benzimidazolium, 2-[3-(1,3-diethyl-1,3-dihydro-4,7-dimethoxy-2H-benzimidazol-2-ylidene)-1-triazenyl]-1,3-diethyl-4,7-dimethoxy-, tetrafluoroborate	EtOH mixt.	356(4.33),410(4.36)	33-2958-78
$C_{26}H_{36}O_2$ 16,24-Cyclo-21-norchola-4,6-dien-3-one, 17-acetyl-6-methyl-, (16β,17α)-	CHCl₃	288(4.45)	70-0384-78
$C_{26}H_{36}O_4$ Cyclopropa[4,5]card-20(22)-enolide, 3-acetoxy-3',4-dihydro-, (3β,4α,5S-14α)-	MeOH	220(4.23)(end abs.)	44-2334-78
1,2-Naphthalenedicarboxylic acid, 3,4-bis(1,1-dimethylethyl)-5,6,7,8-tetramethyl-, dimethyl ester	EtOH	236(4.20),276(4.44), 368(3.69)	5-1675-78
Strongylophorine-3	MeOH	211(3.94),221(3.90), 228(3.83),297(3.68)	20-0917-78
$C_{26}H_{36}O_5$ Benzenehexanoic acid, ε-ethyl-δ-hydroxy-4-methoxy-δ-(4-methoxyphenyl)-, 1,1-dimethylethyl ester	MeOH	228(4.44),275(4.06), 281(4.04)	5-0726-78
A-Homo-4-oxacarda-1,14-dienolide, 7-hydroxy-4a,4a,8-trimethyl-3-oxo-, (5α,7α,13α,17α,20S)-	MeOH	206(4.25)	88-3901-78
$C_{26}H_{37}NO_8S_3$ L-threo-α-D-galacto-Octopyranoside, methyl 6-(acetylamino)-6,8-dideoxy-7-S-[2-[(phenylmethyl)thio]ethyl]-1,7-dithio-, 2,3,4-triacetate	n.s.g.	254s(2.73),259(2.60), 266(2.40)	39-0967-78C
$C_{26}H_{38}N_2O_6$ Phenol, 4,4'-(1,2-ethyl-1,2-ethanediyl)bis[2-[bis(2-hydroxyethyl)amino]-, (E)-	MeOH	208(4.4),229s(4.3), 283s(3.9)	83-0184-78
$C_{26}H_{38}N_8O_2S_4$ Formamide, N,N'-[dithiobis[1-methyl-2-[2-(methylthio)ethyl]-2,1-ethenediyl]]bis[N-[(4-amino-2-methyl-5-pyrimidinyl)methyl]-	pH 2 1% NaHCO₃	243(2.61) 233(2.61),278(2.25)	94-3675-78 94-3675-78
$C_{26}H_{38}N_8O_4S_4$ Formamide, N,N'-[dithiobis[1-methyl-2-[2-(methylsulfinyl)ethyl]-2,1-ethenediyl]]bis[N-[(4-amino-2-methyl-5-pyrimidinyl)methyl]-	pH 2	243(2.57)	94-3675-78
$C_{26}H_{38}O_2$ Hexadecanoic acid, 1-naphthalenyl ester	hexane	298.5(3.83)	95-0774-78

Compound	Solvent	$\lambda_{max}(\log \epsilon)$	Ref.
24-Norcholesta-1,4,6-trien-3-one, 25-hydroxy-	EtOH	226(4.08),260(3.97), 304(4.11)	87-1025-78
$C_{26}H_{38}O_3$			
Cyclopent[c]azulene-1-methanol, 10-(ben-zoyloxy)dodecahydro-1,3,3,5a,9-penta-methyl-, [1R-(1α,3aα,5aβ,9β,10α-10aR*)]-	n.s.g.	222(4.08)	25-0584-78
26,27-Dinorergosta-dien-3-one, 1α,2α-epoxy-24-hydroxy-	EtOH	296(4.31)	87-1025-78
$C_{26}H_{38}O_4$			
Gummiferolic acid, ethyl ester	EtOH	214(3.96)	102-1637-78
$C_{26}H_{38}O_5$			
Pregn-20-ene-20-carboxaldehyde, 3,21-diacetoxy-, (3β,5α,20Z)-	EtOH	245(4)	23-0424-78
$C_{26}H_{40}N_2O_6$			
Phenol, 4,4'-(1,2-diethyl-1,2-ethanedi-yl)bis[2-[bis(2-hydroxyethyl)amino]-, (R*,S*)-	MeOH	216(4.5),280(4.2)	83-0184-78
$C_{26}H_{40}O_6S$			
5β,14β-Card-20(22)-enolide, 14-hydroxy-3β-[[(1-methylethyl)sulfonyl]oxy]-	MeOH	215(4.19)	106-0706-78
5β,14β-Card-20(22)-enolide, 14-hydroxy-3β-[(propylsulfonyl)oxy]-	MeOH	217(4.17)	106-0706-78
$C_{26}H_{42}O_2$			
A-Norcholestane-3,6-dione	hexane	277(4.00)	12-0113-78
$C_{26}H_{42}O_3$			
Ophiobol-2-en-21-oic acid, 7,8-epoxy-, methyl ester	n.s.g.	210(3.46)	25-0584-78
$C_{26}H_{42}O_6$			
Ardisianol	EtOH	230s(3.74),283(3.52)	18-0943-78
$C_{26}H_{42}O_8S_2$			
Prosta-5,13-dien-1-oic acid, 15-acet-oxy-11-(acetylthio)-9-[(methylsul-fonyl)oxy]-, methyl ester, (5Z,9β,11α,13E,15S)-	MeOH	233(3.70)	44-4377-78
$C_{26}H_{43}NO$			
A-Norcholest-5-en-3-one, 6-amino-	EtOH	305s(3.97),322(4.08)	12-0113-78
$C_{26}H_{48}O_5$			
2H-Pyran-2-one, 5,6-dihydro-6-(2,4,6-trihydroxyheneicosyl)-	MeOH	207(3.89)	102-1327-78
$C_{26}H_{48}O_6$			
2H-Pyran-2-one, 5,6-dihydro-6-(2,4,6,10-tetrahydroxyheneicosyl)-	MeOH	208(3.97)	102-1327-78
$C_{26}H_{49}BrN_4OP$			
Phosphorus(1+), [(4-bromophenyl)azo]-butoxybis(N-butyl-1-butanaminato)-, (T-4)-, tetrafluoroborate	n.s.g.	333(4.22),500(2.31)	123-0190-78

Compound	Solvent	$\lambda_{max}(\log \epsilon)$	Ref.
$C_{27}H_{13}ClN_2O_2$ Benzo[g]naphtho[1',2',3':4,5]quino-[2,1-b]quinazoline-5,10-dione, 3-chloro-	n.s.g.	470(3.66)	2-0103-78
$C_{27}H_{14}N_2O_2$ Benzo[g]naphtho[1',2',3':4,5]quino-[2,1-b]quinazoline-5,10-dione	n.s.g.	470(3.65)	2-0103-78
$C_{27}H_{15}Cl_2NO_2$ 1,2-Benzisoxazole, 6-chloro-3-[3-chlo-ro-4-[2-(3-dibenzofuranyl)ethenyl]-phenyl]-	DMF	355(4.72)	33-2904-78
$C_{27}H_{16}ClNO_2$ 1,2-Benzisoxazole, 6-chloro-3-[4-[2-(3-dibenzofuranyl)ethenyl]phenyl]-	DMF	352(4.80)	33-2904-78
$C_{27}H_{16}Cl_2N_4O$ 2H-Benzotriazole, 2-[4-[2-[4-(6-chloro-1,2-benzisoxazol-3-yl)phenyl]ethen-yl]-3-chlorophenyl-	DMF	359(4.80)	33-2904-78
$C_{27}H_{17}BrO_2$ 2H-Pyran-2-one, 6-(3-bromophenyl)-3-(1-naphthalenyl)-4-phenyl-	EtOH	267(4.34),348(4.23)	4-0759-78
$C_{27}H_{17}ClN_2O$ Imidazo[1,2-a]pyridine, 2-(4-chloro-phenyl)-7-[2-(3-dibenzofuranyl)-ethenyl]-	DMF	375(4.79),395(4.72)	33-0129-78
$C_{27}H_{17}ClN_4O$ 2H-Benzotriazole, 2-[4-[2-[4-(1,2-benz-isoxazol-3-yl)phenyl]ethenyl]-3-chlo-rophenyl]-	DMF	359(4.78)	33-2904-78
[1,2,4]Triazolo[1,5-a]pyridine, 2-[4-[2-[4-(1,2-benzisoxazol-3-yl)phenyl]-ethenyl]-3-chlorophenyl]-	DMF	348(4.75)	33-2904-78
[1,2,4]Triazolo[1,5-a]pyridine, 2-[4-[2-[4-(2-benzoxazolyl)-2-chlorophen-yl]ethenyl]phenyl]-	DMF	367(4.79)	33-0142-78
[1,2,4]Triazolo[1,5-a]pyridine, 7-[2-[4-(2-benzoxazolyl)-2-chlorophenyl]-ethenyl]-2-phenyl-	DMF	363(4.81)	33-0142-78
[1,2,4]Triazolo[1,5-a]pyridine, 2-[4-[2-[4-(5-chloro-2-benzoxazolyl)phen-yl]ethenyl]phenyl]-	DMF	367(4.92),385(4.75)	33-0142-78
[1,2,4]Triazolo[1,5-a]pyridine, 6-chlo-ro-2-[4-[2-(5-phenyl-2-benzoxazolyl)-ethenyl]phenyl]-	DMF	347(4.76),360(4.78)	33-0142-78
[1,2,4]Triazolo[1,5-a]pyridine, 6-chlo-ro-2-[4-[2-(6-phenyl-2-benzoxazolyl)-ethenyl]phenyl]-	DMF	357(4.78)	33-0142-78
$C_{27}H_{17}ClO_2$ 2H-Pyran-2-one, 6-(3-chlorophenyl)-3-(1-naphthalenyl)-4-phenyl-	EtOH	267(4.44),349(4.32)	4-0759-78
$C_{27}H_{17}Cl_2NO$ 1,2-Benzisoxazole, 3-[4-(2-[1,1'-bi-	DMF	345(4.69)	33-2904-78

$C_{27}H_{17}Cl_2NO-C_{27}H_{18}N_2O_2$

Compound	Solvent	λ_{max} (log ϵ)	Ref.
phenyl]-4-ylethenyl)-3-chlorophenyl]-6-chloro- (cont.)			33-2904-78
$C_{27}H_{17}NO_2$			
1,2-Benzisoxazole, 3-[4-[2-(3-dibenzo-furanyl)ethenyl]phenyl]-	DMF	352(4.80)	33-2904-78
1,2-Benzisoxazole, 6-[2-(3-dibenzofur-anyl)ethenyl]-3-phenyl-	DMF	349(4.77)	33-2904-78
5H-Naphtho[2,3-c]carbazole-8,13-dione, 5-methyl-6-phenyl-	CHCl	246(4.60),268(4.51), 301(4.23),332s(4.38), 344(4.43),417(3.81), 440s(3.76)	138-0323-78
$C_{27}H_{17}NO_4$			
Benzamide, N-(9,10-dihydro-9,10-dioxo-1-phenoxy-2-anthracenyl)-	benzene	376(3.85)	104-2199-78
Benzamide, N-(9,10-dihydro-9,10-dioxo-4-phenoxy-1-anthracenyl)-	benzene	441(3.76)	104-2199-78
Benzamide, N-(9,10-dihydro-9,10-dioxo-5-phenoxy-1-anthracenyl)-	benzene	419(3.87)	104-2199-78
$C_{27}H_{17}N_2O_2P$			
Phosphonium, triphenyl-, 3,4-dicyano-2,5-diformyl-2,4-cyclopentadien-1-ylide	MeCN	269(4.61),278(4.69), 327(3.92)	18-2605-78
$C_{27}H_{18}$			
8H-Cyclopent[a]acenaphthylene, 7,9-di-phenyl-	EtOH	258(4.2),287(4.02), 329(3.65),397(4.01)	104-1532-78
$C_{27}H_{18}BrNO$			
Phenol, 2-[2-[3-(4-bromophenyl)benzo-[f]quinolin-1-yl]ethenyl]-	EtOH	262(4.64),290(4.76), 350(4.32)	103-0418-78
$C_{27}H_{18}ClNO$			
1,2-Benzisoxazole, 3-[4-(2-[1,1'-bi-phenyl]-4-ylethenyl)phenyl]-6-chloro-	DMF	344(4.76)	33-2904-78
$C_{27}H_{18}ClN_3O$			
[1,2,4]Triazolo[1,5-a]pyridine, 2-[3-chloro-4-[2-(3-dibenzofuranyl)ethen-yl]phenyl]-6-methyl-	DMF	360(4.79)	33-0142-78
$C_{27}H_{18}N_2O$			
Benzaldehyde, 3-(11-phenyldibenzo[b,f]-[1,4]diazocin-6-yl)-	CHCl$_3$	245(4.75),345(3.62)	150-4944-78
Benzaldehyde, 4-(11-phenyldibenzo[b,f]-[1,4]diazocin-6-yl)-	CHCl$_3$	262(4.57),272s(4.52), 350(3.60)	150-4944-78
Dibenzo[b,f][1,4]diazocine-1-carbox-aldehyde, 6,11-diphenyl-	CHCl$_3$	245s(4.46),257(4.63), 355(3.72),365s(3.70)	150-4944-78
Dibenzo[b,f][1,4]diazocine-2-carbox-aldehyde, 6,11-diphenyl-	CHCl$_3$	265(4.57),284s(4.41), 340(3.86)	150-4944-78
Dibenzo[b,f][1,4]diazocine-8-carbox-aldehyde, 6,11-diphenyl-	CHCl$_3$	245s(4.54),256(4.63), 335(3.56)	150-4944-78
Imidazo[1,2-a]pyridine, 7-[2-(3-diben-zofuranyl)ethenyl]-2-phenyl-	DMF	377(4.78),396(4.75)	33-0129-78
$C_{27}H_{18}N_2O_2$			
Benzo[f]quinoline, 1-[2-(4-nitrophen-yl)ethenyl]-3-phenyl-	EtOH	277(4.02),344(3.73)	103-0418-78

Compound	Solvent	$\lambda_{max}(\log \epsilon)$	Ref.
$C_{27}H_{18}N_2O_2S$ 1,2-Benzisoxazole, 3-[4-[2-[5-(5-meth-yl-2-benzoxazolyl)-2-thienyl]ethen-yl]phenyl]-	DMF	393(4.78)	33-2904-78
$C_{27}H_{18}N_4O$ [1,2,4]Triazolo[1,5-a]pyridine, 2-[4-[2-[4-(2-benzoxazolyl)phenyl]ethen-yl]phenyl]-	DMF	363(4.92),385(4.74)	33-0142-78
[1,2,4]Triazolo[1,5-a]pyridine, 2-[4-[2-(5-phenyl-2-benzoxazolyl)ethenyl]-phenyl]-	DMF	346(4.75),359(4.76)	33-0142-78
[1,2,4]Triazolo[1,5-a]pyridine, 2-[4-[2-(6-phenyl-2-benzoxazolyl)ethenyl]-phenyl]-	DMF	356(4.77)	33-0142-78
[1,2,4]Triazolo[1,5-a]pyridine, 2-[4-[2-(3-phenyl-1,2-benzisoxazol-6-yl)-ethenyl]phenyl]-	DMF	344(4.79)	33-2904-78
$C_{27}H_{18}O_2$ Benzeneacetaldehyde, α-[(2-phenylcyclo-penta[b][1]benzopyran-1-yl)methyl-ene]-	THF THF	550(2.92) 235(4.10),330(4.40), 375(4.52),550(2.92)	103-1070-78 103-1075-78
Cyclopenta[b][1]benzopyran-1-carbox-aldehyde, 2-phenyl-3-(2-phenyleth-enyl)-, cis	THF THF	536(3.44) 237(4.56),325(4.61), 375(4.21),536(3.44)	103-1070-78 103-1075-78
trans	THF THF	570(3.43) 240(4.41),325(4.69), 375s(4.20),570(3.43)	103-1070-78 103-1075-78
2H-Pyran-2-one, 3-(1-naphthalenyl)-4,6-diphenyl-	EtOH	264(4.40),351(4.28)	4-0759-78
$C_{27}H_{19}$ Methylium,9-anthracenyldiphenyl-	SO_2	780(4.07)	39-0488-78C
$C_{27}H_{19}BrN_2O$ 2(1H)-Pyridinone, 1-amino-6-(3-bromo-phenyl)-3-(1-naphthalenyl)-4-phenyl-	EtOH	257s(4.40),340(4.26)	4-0759-78
$C_{27}H_{19}Cl$ Anthracene, 9-chloro-10-(diphenylmeth-ylene)-9,10-dihydro-	n.s.g.	245(4.41),309(4.05)	39-0488-78C
$C_{27}H_{19}ClN_2$ Imidazo[1,2-a]pyridine, 7-(2-[1,1'-bi-phenyl]-4-ylethenyl)-2-(4-chloro-phenyl)-	DMF	370(4.74),390(4.67)	33-0129-78
Imidazo[1,2-a]pyridine, 7-[2-(4-chloro-phenyl)ethenyl]-2,3-diphenyl-	DMF	369(4.57),385(4.53)	33-0129-78
$C_{27}H_{19}ClN_2O$ 2(1H)-Pyridinone, 1-amino-6-(3-chloro-phenyl)-3-(1-naphthalenyl)-4-phenyl-	EtOH	252s(4.35),339(4.12)	4-0759-78
$C_{27}H_{19}ClN_6O$ 2H-Benzotriazole, 2-[3-chloro-4-[2-(4-[1,2,4]triazolo[1,5-a]pyridin-2-yl-phenyl)ethenyl]phenyl]-5-methoxy-	DMF	372(4.83)	33-0142-78
$C_{27}H_{19}ClO_4S$ Thiopyrylium, 2-[2-(3-benzothiophenyl)-	HOAc-HClO₄	505(3.696)	83-0256-78

Compound	Solvent	λ_{max}(log ϵ)	Ref.
ethenyl]-4,6-diphenyl-, perchlorate (cont.)			83-0256-78
$C_{27}H_{19}F_6N_4S_4$			
Benzo[1,2-d:4,3-d']bisthiazolium, 3-ethyl-2-[5-[3-ethyl-7-(trifluoromethyl)benzo[1,2-d:4,3-d']bisthiazol-2(3H)-ylidene]-1,3-pentadienyl]-7-(trifluoromethyl)-, perchlorate	EtOH	686(5.28)	124-0637-78
$C_{27}H_{19}N$			
Benzo[f]quinoline, 3-phenyl-1-(2-phenylethenyl)-	EtOH	259(4.57),290(4.80), 340(4.20)	103-0418-78
5H-Naphtho[2,3-c]carbazole, 5-methyl-6-phenyl-	EtOH	205(4.71),219(4.67), 232(4.54),251s(4.62), 260s(4.67),265(4.68), 290(4.72),310(4.52), 324(4.51),341(3.69), 358(3.92),377(4.08), 403(3.72),427(3.64)	138-0323-78
$C_{27}H_{19}NO$			
1,2-Benzisoxazole, 3-[4-(2-[1,1'-biphenyl]-4-ylethenyl)phenyl]-	DMF	343(4.76)	33-2904-78
1,2-Benzisoxazole, 6-(2-[1,1'-biphenyl]-4-ylethenyl)-3-phenyl-	DMF	342(4.72)	33-2904-78
1,2-Benzisoxazole, 3-[1,1'-biphenyl]-4-yl-6-(2-phenylethenyl)-	DMF	325(4.62)	33-2904-78
2,1-Benzisoxazole, 7-[4-(2-[1,1'-biphenyl]-4-ylethenyl)phenyl]-	DMF	327(4.30),385(4.74)	33-2904-78
Phenol, 2-[2-(3-phenylbenzo[f]quinolin-1-yl)ethenyl]-	EtOH	255(4.54),284(4.63), 345(4.11)	103-0418-78
$C_{27}H_{19}NO_2$			
1,2-Benzisoxazole, 3-[4-[2-(4-phenoxyphenyl)ethenyl]phenyl]-	DMF	336(4.67)	33-2904-78
Phenol, 2-[2-[3-(2-hydroxyphenyl)benzo[f]quinolin-1-yl]ethenyl]-	EtOH	258(4.59),284(4.63), 325(4.23),355(4.09), 373(4.17)	103-0418-78
$C_{27}H_{19}NO_3S$			
1,2-Benzisoxazole, 3-[4-[2-[4-(phenylsulfonyl)phenyl]ethenyl]phenyl]-	DMF	337(4.73)	33-2904-78
$C_{27}H_{19}NO_4$			
Benzo[h]quinolin-2(1H)-one, 4-(1,3-benzodioxol-5-yl)-3-benzoyl-5,6-dihydro-	CHCl$_3$	247(4.40),374(4.22)	48-0097-78
$C_{27}H_{19}N_3O$			
[1,2,4]Triazolo[1,5-a]pyridine, 7-[2-(3-dibenzofuranyl)ethenyl]-2-(4-methylphenyl)-	DMF	352(4.81),367(4.72)	33-0142-78
$C_{27}H_{19}N_3O_2$			
[1,2,4]Triazolo[1,5-a]pyridine, 7-[2-(3-dibenzofuranyl)ethenyl]-2-(4-methoxyphenyl)-	DMF	352(4.82),368(4.72)	33-0142-78
$C_{27}H_{20}$			
Anthracene, 9-(diphenylmethylene)-9,10-	n.s.g.	257(5.04),320(3.11)	39-0488-78C

Compound	Solvent	$\lambda_{max}(\log \epsilon)$	Ref.
dihydro- (cont.)		335(3.52),352(3.85), 370(4.05),390(4.04)	39-0488-78C
Phenanthrene, 4-(9H-fluoren-9-ylidene)-1,2,3,4-tetrahydro-	hexane	218s(4.18),262(4.62), 355(4.19)	54-0249-78
$C_{27}H_{20}ClNO_2$			
Benzo[h]quinolin-2(1H)-one, 3-benzoyl-4-(2-chlorophenyl)-5,6-dihydro-1-methyl-	CHCl₃	250(4.42),370(4.22)	48-0097-78
Benzo[h]quinolin-2(1H)-one, 3-benzoyl-4-(4-chlorophenyl)-5,6-dihydro-1-methyl-	CHCl₃	250(4.43),370(4.25)	48-0097-78
$C_{27}H_{20}ClN_3$			
[1,2,4]Triazolo[1,5-a]pyridine, 2-[4-(2-[1,1'-biphenyl]-4-ylethenyl)-3-chlorophenyl]-6-methyl-	DMF	350(4.76)	33-0142-78
$C_{27}H_{20}Cl_2S$			
3H-Naphtho[1,8-bc]thiophene, 6,8-dichloro-3-[phenyl(2,4,6-trimethylphenyl)methylene]-	MeOH	274(4.4),336s(3.7), 406(4.2),421(4.3)	5-1974-78
$C_{27}H_{20}N_2$			
Dibenzo[b,f][1,4]diazocine, 1-methyl-6,11-diphenyl-	CHCl₃	252(4.58),347(3.54)	150-4944-78
Dibenzo[b,f][1,4]diazocine, 2-methyl-6,11-diphenyl-	CHCl₃	254(4.60),345(3.57)	150-4944-78
Dibenzo[b,f][1,4]diazocine, 8-methyl-6,11-diphenyl-	CHCl₃	252(4.59),342(3.56)	150-4944-78
Dibenzo[b,f][1,4]diazocine, 6-(3-methylphenyl)-11-phenyl-, (±)-	CHCl₃	240s(4.45),252(4.55), 345(3.26)	150-4944-78
Dibenzo[b,f][1,4]diazocine, 6-(4-methylphenyl)-11-phenyl-, (±)-	CHCl₃	240s(4.38),257(3.56), 345(3.62)	150-4944-78
Imidazo[1,2-a]pyridine, 2-[4-(2-[1,1'-biphenyl]-4-ylethenyl)phenyl]-	DMF	361(4.83)	33-0129-78
Imidazo[1,2-a]pyridine, 7-(2-[1,1'-biphenyl]-4-ylethenyl)-2-phenyl-	DMF	370(4.73),388(4.66)	33-0129-78
Imidazo[1,2-a]pyridine, 2-[1,1'-biphenyl]-4-yl-7-(2-phenylethenyl)-	DMF	365(4.68),383(4.63)	33-0129-78
Imidazo[1,2-a]pyridine, 2,3-diphenyl-7-(2-phenylethenyl)-	DMF	365(4.53),383(4.49)	33-0129-78
$C_{27}H_{20}N_2O$			
Benzenamine, N-(3,4,5-triphenyl-2(3H)-oxazolylidene)-	MeOH	282(4.27),310s(4.22)	39-1113-78C
2-Imidazolin-5-one, 1,2,4,4-tetraphenyl-	EtOH	227(4.42),260s(3.83), 267s(3.75)	94-3798-78
Imidazo[1,2-a]pyridine, 7-[2-(4-phenoxyphenyl)ethenyl]-2-phenyl-	DMF	363(4.66),383(4.61)	33-0129-78
3-Pyridinecarbonitrile, 1,2-dihydro-1-methyl-2-oxo-5,6-diphenyl-4-(1-phenylethenyl)-	MeCN	245(4.30),359(4.11)	5-1222-78
2(1H)-Pyridinone, 1-amino-3-(1-naphthalenyl)-4,6-diphenyl-	EtOH	252(4.35),340(4.15)	4-0759-78
$C_{27}H_{20}N_4O$			
1H-Pyrazolo[3,4-b]pyridin-4-ol, 5-(3,4-dihydro-1-isoquinolinyl)-1,3-diphenyl-	EtOH	275(4.50)	2-0161-78

Compound	Solvent	$\lambda_{max}(\log \epsilon)$	Ref.
$C_{27}H_{20}N_4OS$ [1,2,4]Triazolo[1,5-a]pyridine, 2-[4-[2-[5-(5,6-dimethyl-2-benzoxazolyl)-2-thienyl]ethenyl]phenyl]-	DMF	400(4.82)	33-0142-78
$C_{27}H_{20}O$ 9-Anthracenemethanol, α,10-diphenyl-	n.s.g.	338(3.51),355(3.83), 375(3.96),395(3.93)	39-0488-78C
$C_{27}H_{20}OS$ 1,3-Oxathiole, 2,2,4,5-tetraphenyl-	EtOH	343(3.88)	44-3730-78
$C_{27}H_{20}O_3S_3$ 2H-Thiopyran-3(4H)-one, dihydro-2,4-bis[[5-(phenylthio)-2-furanyl]-methylene]-	EtOH	207(4.50),234(4.37), 298(4.25),429(4.30)	133-0189-78
4H-Thiopyran-4-one, tetrahydro-3,5-bis-[[5-(phenylthio)-2-furanyl]methylene]-	EtOH	208(4.47),233(4.36), 312(4.37),424(4.37)	133-0189-78
$C_{27}H_{20}O_4S_2$ 4H-Pyran-4-one, tetrahydro-3,5-bis[[5-(phenylthio)-2-furanyl]methylene]-	EtOH	208(4.43),233(4.34), 301(4.13),418(4.29)	133-0189-78
$C_{27}H_{20}O_5S_3$ 4H-Thiopyran-4-one, tetrahydro-3,5-bis-[[5-(phenylsulfinyl)-2-furanyl]meth-ylene]-	EtOH	220(4.36),274(4.21), 368(4.40)	133-0189-78
$C_{27}H_{20}O_6S_2$ 4H-Pyran-4-one, tetrahydro-3,5-bis[[5-(phenylsulfinyl)-2-furanyl]methylene]-	EtOH	212(4.24),267(4.12), 380(4.49)	133-0189-78
$C_{27}H_{20}O_7S_3$ 2H-Thiopyran-3(4H)-one, dihydro-2,4-bis[[5-(phenylsulfonyl)-2-furanyl]-methylene]-	EtOH	215(4.37),297(4.27), 346(4.43),400(4.17)	133-0189-78
4H-Thiopyran-4-one, tetrahydro-3,5-bis-[[5-(phenylsulfonyl)-2-furanyl]meth-ylene]-	EtOH	215(4.31),279(4.22), 361(4.43)	133-0189-78
$C_{27}H_{20}O_8S_2$ 4H-Pyran-4-one, tetrahydro-3,5-bis[[5-(phenylsulfonyl)-2-furanyl]methylene]-	EtOH	214(4.30),240(4.21), 266(4.23),376(4.51)	133-0189-78
$C_{27}H_{21}BrN_2O_2$ 1-Naphthaleneacetic acid, [3-(3-bromo-phenyl)-3-oxo-1-phenylpropylidene]-hydrazide	EtOH	284(4.53),293(4.51)	94-1298-78
$C_{27}H_{21}BrN_4O_7$ Methylium, (3-bromophenyl)[4-(dimeth-ylamino)phenyl]phenyl-, picrate	MeCN	447(4.34)	104-1750-78
Methylium, (4-bromophenyl)[4-(dimeth-ylamino)phenyl]phenyl-, picrate	MeCN	455(4.45)	104-1750-78
$C_{27}H_{21}ClN_2O_2$ 1-Naphthaleneacetic acid, [1-(4-chloro-phenyl)-3-oxo-3-phenylpropylidene]-hydrazide	EtOH	290(4.39),298s(4.38)	34-0345-78
1-Naphthaleneacetic acid, [3-(3-chloro-phenyl)-3-oxo-1-phenylpropylidene]-	EtOH	283(4.45),294(4.43)	94-1298-78

Compound	Solvent	$\lambda_{max}(\log \epsilon)$	Ref.
hydrazide (cont.)			94-1298-78
1-Naphthaleneacetic acid, [3-(4-chloro-phenyl)-3-oxo-1-phenylpropylidene]-	EtOH	220(4.51),283(4.47), 293s(4.45)	4-0385-78
hydrazide	EtOH	283(4.47),293s(4.45)	34-0345-78
$C_{27}H_{21}ClN_4O_7$			
Methylium, (4-chlorophenyl)[4-(dimeth-ylamino)phenyl]phenyl-, picrate	9.5M HCl	456(4.45)	104-1750-78
	MeCN	456(4.45)	104-1750-78
Methylium, (3-chlorophenyl)[4-(dimeth-ylamino)phenyl]phenyl-, picrate	MeCN	449(4.36)	104-1750-78
$C_{27}H_{21}NO_2$			
Benzo[h]quinolin-2(1H)-one, 3-benzoyl-5,6-dihydro-1-methyl-4-phenyl-	CHCl$_3$	250(4.40),373(4.3)	48-0097-78
Benzo[h]quinolin-2(1H)-one, 3-benzoyl-5,6-dihydro-4-(4-methylphenyl)-	CHCl$_3$	247(4.40),374(4.21)	48-0097-78
$C_{27}H_{21}NO_3$			
Benzo[h]quinolin-2(1H)-one, 3-benzoyl-5,6-dihydro-4-(4-methoxyphenyl)-	CHCl$_3$	247(4.48),372(4.18)	48-0097-78
$C_{27}H_{21}NO_4S$			
5H-Naphtho[2,3-a]carbazole-5,13(5aH)-dione, 6,12,12a,12b-tetrahydro-12-[(4-methylphenyl)sulfonyl]-, (5aα,12aα,12bα)-	CHCl$_3$	258(4.37),312s(3.97), 342s(3.54)	12-1841-78
$C_{27}H_{21}N_3$			
Propanedinitrile, [3-(methylamino)-2,3-diphenyl-1-(1-phenylethenyl)-2-propenylidene]-, (Z)-	MeCN	254(4.45),274(4.03), 413(3.86)	5-1222-78
[1,2,4]Triazolo[1,5-a]pyridine, 7-(2-[1,1'-biphenyl]-4-ylethenyl)-2-(4-methylphenyl)-	DMF	344(4.76)	33-0142-78
$C_{27}H_{21}N_3O_4$			
1-Naphthaleneacetic acid, [3-(3-nitro-phenyl)-3-oxo-1-phenylpropylidene]hy-drazide	EtOH	282(4.46),293s(4.42)	94-1298-78
$C_{27}H_{21}N_3O_7S$			
4-Thia-1-azabicyclo[3.2.0]hept-2-ene-2-carboxylic acid, 7-oxo-6-[(phenoxy-acetyl)amino]-3-phenyl-, (4-nitro-phenyl)methyl ester, (5R-trans)-	EtOH	262(4.29),267(4.29), 273(4.24),324(3.81)	35-8214-78
$C_{27}H_{21}N_5O_9$			
Methylium, [4-(dimethylamino)phenyl]-(3-nitrophenyl)phenyl-, picrate	MeCN	439(3.95)	104-1750-78
Methylium, [4-(dimethylamino)phenyl]-(4-nitrophenyl)phenyl-, picrate	MeCN	435(3.85)	104-1750-78
$C_{27}H_{22}F_3N_3$			
3H-1,5-Benzodiazepine, 2-[3,6-dihydro-4-[4-(trifluoromethyl)phenyl]-1(2H)-pyridinyl]-4-phenyl-	n.s.g.	260(4.67),342(3.70)	87-0952-78
$C_{27}H_{22}Fe_2N_2O_7$			
Iron, [μ-[N,N'-bis(2-methyl-1-phenyl-1-propenyl)ureato(2-)-N,N':N,N']]-hexacarbonyldi-, (Fe-Fe)-	EtOH	323(4.27)	33-0589-78

Compound	Solvent	$\lambda_{max}(\log \epsilon)$	Ref.
$C_{27}H_{22}N_2O$			
1H-Imidazol-5-ol, 4,5-dihydro-1,2,4,4-tetraphenyl-	EtOH	240s(4.23),266(3.91), 273s(3.90)	94-3798-78
$C_{27}H_{22}N_2O_2$			
1-Naphthaleneacetic acid, (3-oxo-1,3-diphenylpropylidene)hydrazide	EtOH	283(4.49),293s(4.46)	4-0385-78 +34-0345-78
1-Naphthalenecarboxylic acid, [3-(4-methylphenyl)-3-oxo-1-phenylpropylidene]hydrazide	EtOH	281(4.43),293s(4.40)	4-0385-78
$C_{27}H_{22}N_4O_2$			
1H-Pyrazolo[3,4-b]pyridine-5-carboxamide, 4-hydroxy-1,3-diphenyl-N-(2-phenylethyl)-	EtOH	310(4.26)	2-0161-78
$C_{27}H_{22}N_4O_7$			
Methylium, [4-(dimethylamino)phenyl]diphenyl-, picrate	MeCN	460(4.54)	104-1750-78
$C_{27}H_{22}O$			
2H-1-Benzopyran, 3,4-dihydro-2,2,4-triphenyl-	MeOH	277(3.45),284(3.45)	44-0303-78
Phenol, 2-(2,2,3-triphenylcyclopropyl)-	MeOH	277(3.60),283s(3.56)	44-0303-78
$C_{27}H_{22}O_5$			
4H-1-Benzopyran-4-one, 7-(benzoyloxy)-3-[4-hydroxy-3-(3-methyl-2-butenyl)phenyl]-	EtOH EtOH-NaOAc	258(4.51) 258(4.55)	18-2398-78 18-2398-78
$C_{27}H_{22}O_6$			
4H-1-Benzopyran-4-one, 7-(benzoyloxy)-5-hydroxy-3-[4-hydroxy-3-(3-methyl-2-butenyl)phenyl]-	EtOH EtOH-NaOAc	262(4.64),334s(3.91) 262(4.64),334s(3.92)	18-2398-78 18-2398-78
$C_{27}H_{22}O_7S$			
Pyrylium, 2-[2-(3-hydroxyphenyl)ethenyl]-4,6-diphenyl-, carboxymethylsulfonate	EtOH	472(4.411)	83-0256-78
Pyrylium, 2-[2-(4-hydroxyphenyl)ethenyl]-4,6-diphenyl-, carboxymethylsulfonate	EtOH	525(4.284)	83-0256-78
$C_{27}H_{22}O_8S$			
Pyrylium, 2-[2-(3,4-dihydroxyphenyl)-ethenyl]-4,6-diphenyl-, carboxymethylsulfonate	EtOH	556(4.404)	83-0256-78
$C_{27}H_{22}O_9$			
D-Allonic acid, 2,5-anhydro-, tribenzoate	EtOH	230(4.36),275(3.37)	94-0898-78
$C_{27}H_{22}S$			
5H-Naphtho[1,8-bc]thiophene, 5-[phenyl-(2,4,6-trimethylphenyl)methylene]-, (Z)-	MeOH	267(4.2),384(4.3)	5-1974-78
$C_{27}H_{23}ClN$			
Quinolinium, 2-[4-(4-chlorophenyl)-2-phenyl-1,3-butadienyl]-1-ethyl-, iodide	EtOH	245(5.68),320(4.85), 400(4.09),555(5.21)	146-0341-78

Compound	Solvent	$\lambda_{max}(\log \epsilon)$	Ref.
$C_{27}H_{23}F_6N_2$ Quinolinium, 1-ethyl-2-[3-[1-ethyl-6-(trifluoromethyl)-2(1H)-quinolinyl-idene]-1-propenyl]-6-(trifluoro-methyl)-, iodide	EtOH	608(5.20)	103-0060-78
$C_{27}H_{23}NO_2$ Acetamide, N-methyl-N-(4-oxo-3,5,6-tri-phenylbicyclo[3.1.0]hex-2-en-6-yl)-, exo	C_6H_{12}	222(4.47),267(3.76)	5-1634-78
Acetamide, N-methyl-N-(4-oxo-1,3,5-tri-phenyl-2,5-cyclohexadien-1-yl)-	EtOH	224(4.42),286(3.77)	5-1634-78
Acetamide, N-methyl-N-(6-oxo-1,2,5-tri-phenyl-2,4-cyclohexadien-1-yl)-	EtOH	240(4.16),262s(4.09), 342(3.97),383s(3.79)	5-1634-78
$C_{27}H_{23}N_2O$ Quinolinium, 2-[[2,3-dihydro-2-(2-meth-ylphenyl)-3-oxo-1H-isoindol-1-yli-dene]methyl]-1-ethyl-, iodide	EtOH	215(5.13),245(4.66), 520s(4.07),554(4.30)	18-2415-78
Quinolinium, 2-[[2,3-dihydro-2-(4-meth-ylphenyl)-3-oxo-1H-isoindol-1-yli-dene]methyl]-1-ethyl-, iodide	EtOH	215(5.67),240(5.44), 540(4.76),584(4.93)	18-2415-78
$C_{27}H_{23}N_2O_2$ Quinolinium, 2-[[2,3-dihydro-2-(2-meth-oxyphenyl)-3-oxo-1H-isoindol-1-yli-dene]methyl]-1-ethyl-, iodide	EtOH	215(5.90),238(5.28), 540(4.00),582(4.30)	18-2415-78
Quinolinium, 2-[[2,3-dihydro-2-(4-meth-oxyphenyl)-3-oxo-1H-isoindol-1-yli-dene]methyl]-1-ethyl-, iodide	EtOH	215(5.48),238(5.06), 548(3.28),590(3.34)	18-2415-78
Quinolinium, 1-ethyl-2-[4-(4-nitrophen-yl)-2-phenyl-1,3-butadienyl]-, iodide	EtOH	240(5.23),390(4.89), 500(4.41)	146-0341-78
$C_{27}H_{23}N_3O_3S$ 4-Thiazoleacetic acid, α-[(2-propenyl-oxy)imino]-2-[(triphenylmethyl)ami-no]-, (E)-	EtOH	249(4.23),323(3.45)	78-2233-78
(Z)-	EtOH	234(4.30),295(3.80)	78-2233-78
$C_{27}H_{23}O_2P$ Phosphonium, (5-acetyl-2-hydroxy-4-methylphenyl)triphenyl-, hydroxide, inner salt	EtOH	240s(4.31),316(4.37)	95-0503-78
Phosphonium, (5-acetyl-4-hydroxy-2-methylphenyl)triphenyl-, hydroxide, inner salt	EtOH	233(4.67),285(4.31), 350(3.80)	95-0503-78
$C_{27}H_{23}O_3P$ Phosphonium, (5-acetyl-4-hydroxy-2-methoxyphenyl)triphenyl-, hydroxide, inner salt	EtOH	232(4.61),245s(4.54), 269(4.24),277(4.24), 340(3.85)	95-0503-78
$C_{27}H_{24}BrNO_3$ 2,5-Pyrrolidinedione, 1-(3-bromo-1-eth-oxy-1,2,3-triphenyl-2-propenyl)-	MeCN	266s(3.69)	44-0232-78
$C_{27}H_{24}F_3N_3$ 3H-1,5-Benzodiazepine, 2-phenyl-4-[4-[3-(trifluoromethyl)phenyl]-1-pip-eridinyl]-	n.s.g.	260(4.56),347(3.66)	87-0952-78
3H-1,5-Benzodiazepine, 2-phenyl-4-[4-	n.s.g.	259(4.56),344(3.66)	87-0952-78

$C_{27}H_{24}F_3N_3-C_{27}H_{24}O_3$

Compound	Solvent	$\lambda_{max}(\log \epsilon)$	Ref.
[4-(trifluoromethyl)phenyl]-1-piperidinyl]- (cont.)			87-0952-78
$C_{27}H_{24}IO_3P$			
Phosphonium, (5-acetyl-4-hydroxy-2-methoxyphenyl)triphenyl-, iodide	EtOH	233(4.79),269(4.20), 275s(4.16),304s(3.79)	95-0503-78
$C_{27}H_{24}N$			
Quinolinium, 2-(2,4-diphenyl-1,3-butadienyl)-1-ethyl-, iodide	EtOH	245(5.46),325(4.70), 560(4.92)	146-0341-78
$C_{27}H_{24}NO$			
Quinolinium, 1-ethyl-2-[4-(4-hydroxyphenyl)-2-phenyl-1,3-butadienyl]-, iodide	EtOH	450(4.58),580(4.13)	146-0341-78
$C_{27}H_{24}N_2O_6S$			
Benzo[g]-1,3-benzodioxolo[5,6-a]quinolizinium, 5,6-dihydro-9,10-dimethoxy-13-[[(4-methylphenyl)sulfonyl]amino]-, hydroxide, inner salt	EtOH	233(4.46),265(4.31), 297(4.02),348(4.13), 433(3.87)	2-1100-78
$C_{27}H_{24}N_2O_7$			
2,4(1H,3H)-Pyrimidinedione, 1-[2-deoxy-3,5-bis-O-(4-methylbenzoyl)-α-D-erythro-pentofuranosyl]-5-ethynyl-	EtOH	240(4.57),285(4.09)	39-1263-78C
Uridine, 2'-deoxy-5-ethynyl-, 3',5'-bis(4-methylbenzoate)	EtOH	240(4.49),285(4.04)	39-1263-78C
$C_{27}H_{24}N_3O$			
Quinolinium, 4-(2-benzoxazolyl)-2-[2-[4-(dimethylamino)phenyl]ethenyl]-1-methyl-, iodide	EtOH	587(4.70)	4-0017-78
Quinolinium, 6-(2-benzoxazolyl)-2-[2-[4-(dimethylamino)phenyl]ethenyl]-1-methyl-, iodide	EtOH	563(4.86)	4-0017-78
Quinolinium, 6-(2-benzoxazolyl)-4-[2-[4-(dimethylamino)phenyl]ethenyl]-1-methyl-, iodide	EtOH	591(4.63)	4-0017-78
$C_{27}H_{24}N_3S$			
Quinolinium, 4-(2-benzothiazolyl)-2-[2-[4-(dimethylamino)phenyl]ethenyl]-1-methyl-, iodide	EtOH	573(4.71)	4-0017-78
Quinolinium, 6-(2-benzothiazolyl)-2-[2-[4-(dimethylamino)phenyl]ethenyl]-1-methyl-, iodide	EtOH	560(4.86)	4-0017-78
Quinolinium, 6-(2-benzothiazolyl)-4-[2-[4-(dimethylamino)phenyl]ethenyl]-1-methyl-, iodide	EtOH	589(4.62)	4-0017-78
$C_{27}H_{24}O_3$			
5,10-Epoxybenzocyclooctene-6-carboxylic acid, 5,6,9,10-tetrahydro-9-methyl-5,10-diphenyl-, methyl ester	MeCN	247s(2.67),249s(2.72), 252(2.80),255s(2.87), 258(2.94),264(3.00), 271(2.84)	150-1301-78
2-Propenoic acid, 3-(1,2,3,4-tetrahydro-3-methyl-1,4-diphenyl-1,4-epoxynaphthalen-2-yl)-, methyl ester	MeCN	257(3.24),264(3.19), 271(3.01)	150-1301-78

Compound	Solvent	$\lambda_{max}(\log \epsilon)$	Ref.
$C_{27}H_{24}O_3P$ Phosphonium, (5-acetyl-4-hydroxy-2-methoxyphenyl)triphenyl-, iodide	EtOH	233(4.79),269(4.20), 275s(4.16),304s(3.79)	95-0503-78
$C_{27}H_{24}O_4$ 4,7-Ethano-1H-indene-8,9-dione, 1-[bis-(4-methoxyphenyl)methylene]-3a,4,7,7a-tetrahydro-4-methyl-	benzene MeCN	451.3(1.65) 255(4.28),300(4.29)	5-0440-78 5-0440-78
4,7-Ethano-1H-indene-8,9-dione, 1-[bis-(4-methoxyphenyl)methylene]-3a,4,7,7a-tetrahydro-5-methyl-	CH_2Cl_2 MeCN	258(4.30),302(4.33), 452(2.22) 255(4.23),298(4.26), 452(2.13)	5-0440-78 5-0440-78
$C_{27}H_{24}O_7$ 3H,7H-Pyrano[2',3':7,8][1]benzopyrano-[3,2-d][1]benzoxepin-7-one, 12-acet-oxy-8,9-dihydro-6-hydroxy-3,3-dimeth-yl-9-(1-methylethenyl)-, (±)-	EtOH EtOH-AlCl$_3$	232(4.54),280(4.60), 325s(4.01) 233(4.48),285(4.59), 346(4.13),427(3.82)	94-1394-78 94-1394-78
$C_{27}H_{25}NO$ 1-Azaspiro[4.5]dec-2-en-4-one, 1,2,3-triphenyl-	MeCN	262(3.11),354(2.98)	5-1203-78
$C_{27}H_{25}NO_5$ 1H-Isoindole-1,3(2H)-dione, 2-[[1-[2-(3,4-dihydro-6-methoxy-1-naphthalen-yl)ethyl]-2,5-dioxocyclopentyl]methyl]-	EtOH	212(4.30),228(4.26), 238(4.16),278(3.78), 295(3.93)	150-5001-78
$C_{27}H_{25}NO_6$ 1H-Benz[f]indole-3-carboxylic acid, 4,9-diacetoxy-2-methyl-1-(phenyl-methyl)-, ethyl ester	dioxan	258(4.87),355(4.10), 371(4.10)	5-0129-78
$C_{27}H_{25}NO_{11}S$ α-D-Xylofuranoside, methyl 5-benzoate 2-(4-nitrobenzoate) 3-(4-methylbenz-enesulfonate)	EtOH	226(4.45),258(4.16)	136-0257-78H
$C_{27}H_{25}NS_2$ Benzenemethanamine, α-(diphenylmethyl-ene)-N-(hexahydro-1,3-benzodithiol-2-ylidene)-	EtOH	236(4.15),305(3.91)	18-0301-78
$C_{27}H_{25}NS_3$ Spiro[1,3-benzodithiole-2,2'(5'H)-thia-zole], 3a,4,5,6,7,7a-hexahydro-4',5',5'-trimethyl-, α-	EtOH	253(4.11),281(3.96)	18-0301-78
β-	EtOH	253(4.11),281(3.96)	18-0301-78
Spiro[1,3-benzodithiole-2,4'(5'H)-thia-zole], 3a,4,5,6,7,7a-hexahydro-2',5',5'-triphenyl-, α-	EtOH	253(4.21),310(3.53)	18-0301-78
β-	EtOH	253(4.21),305(3.53)	18-0301-78
$C_{27}H_{25}N_3$ 3H-1,5-Benzodiazepine, 2-(3,6-dihydro-4-(4-methylphenyl)-1(2H)-pyridinyl]-4-phenyl-	n.s.g.	258(4.68),342(3.70)	87-0952-78
$C_{27}H_{25}N_3O_3S$ 4-Thiazoleacetic acid, α-(methoxyimino)-2-[(triphenylmethyl)amino]-, ethyl	EtOH	230s(4.34),304(3.48)	78-2233-78

Compound	Solvent	$\lambda_{max}(\log \epsilon)$	Ref.
ester, (E)- (cont.)			78-2233-78
(Z)-	EtOH	236s(4.30),300(3.72)	78-2233-78
$C_{27}H_{25}O_3P$			
Phosphonium, triphenyl-, 1,3-diacetyl-4-oxo-2-pentenylide	EtOH	221(4.56),268(4.00), 275(4.03),305(4.02), 398(4.29)	95-0503-78
$C_{27}H_{25}O_4P$			
Phosphonium, triphenyl-, 3-acetyl-1-(methoxycarbonyl)-4-oxo-2-pentenylide	EtOH	268(3.92),275(3.96), 291(3.91),361(4.35)	95-0503-78
$C_{27}H_{26}$			
Azulene, 1-[2-(6,8-dimethyl-4-azulenyl)ethenyl]-4,6,8-trimethyl-	benzene	468(4.57)	104-0243-78
$C_{27}H_{26}BrN_3O_8$			
1H-Indole-3-propanamide, α-(acetylamino)-6-bromo-N-[2-(3,4,5-triacetoxyphenyl)ethenyl]-	MeOH	227(4.77),290(4.57)	88-2541-78
$C_{27}H_{26}ClN_2S_2$			
Benzothiazolium, 2-[2-[2-chloro-3-[(3-ethyl-2(3H)-benzothiazolylidene)ethylidene]-1-cyclopenten-1-yl]ethenyl]-3-ethyl-, perchlorate	MeOH	821(5.40)	104-2046-78
$C_{27}H_{26}N_2O_2$			
2H-Pyrrol-2-one, 5,5'-[methylenebis(4,1-phenylenemethylidyne)]bis[1,5-dihydro-3,4-dimethyl-, (E,E)-	EtOH	234(4.14),246(4.09), 275(4.21),337(4.50), 358s(4.23)	49-1191-78
(E,Z)-	EtOH	235(4.24),246(4.16), 335(4.67),355s(4.50)	49-1191-78
(Z,Z)-	EtOH	<u>225(4.2),335(4.6)</u>	49-1191-78
$C_{27}H_{26}N_2O_6$			
2H-Cyclopenta[d]pyridazine-1,4-dicarboxylic acid, 5-[bis(4-methoxyphenyl)-methyl]-2-methyl-, dimethyl ester	MeCN	228(4.47),288(4.40), 334(3.72),460(2.82)	78-2509-78
2H-Cyclopenta[d]pyridazine-1,4-dicarboxylic acid, 7-[bis(4-methoxyphenyl)-methyl]-2-methyl-, dimethyl ester	MeCN	226(4.46),283(4.51), 352(3.60),462(2.67)	78-2509-78
$C_{27}H_{26}N_4O_6$			
Tryptoquivaline I	MeOH	235(4.50),250s(4.33), 292(3.98),321s(3.79)	94-0111-78
$C_{27}H_{26}O_8$			
4H-1-Benzopyran-4-one, 5,6-dimethoxy-7-(phenylmethoxy)-2-(2,4,5-trimethoxyphenyl)-	EtOH	252(4.31),304(4.10), 350(4.29)	78-1593-78
3H,7H-Pyrano[2',3':7,8][1]benzopyrano-[3,2-d][1]benzoxepin-7-one, 12-acetoxy-8,9-dihydro-6-hydroxy-9-(1-hydroxy-1-methylethyl)-3,3-dimethyl-,	EtOH	211(4.49),233(4.48), 279(4.50),325s(4.02)	94-1394-78
	EtOH-NaOMe	269(4.44),309(4.29), 353(4.26),387(4.22)	94-1394-78
(±)-	EtOH-AlCl₃	210(4.48),233(4.45), 285(4.50),347(4.08), 420(3.71)	94-1394-78

Compound	Solvent	$\lambda_{max}(\log \epsilon)$	Ref.
$C_{27}H_{26}O_{10}$ 12H-Benzo[b]xanthene-7,10,11-triol, 3,6,8-trimethoxy-1-methyl-, triacetate	EtOH	239(4.60),269(4.57), 320(3.83)	94-0209-78
$C_{27}H_{27}Br_3O_6$ 5H-Tribenzo[a,d,g]cyclononene, 1,6,11- tribromo-10,15-dihydro-2,3,7,8,12,13- hexamethoxy-	$CHCl_3$	246(3.95),291(3.62)	88-1699-78 +102-1791-78
$C_{27}H_{27}N$ Benzo[4,5]pentaleno[2,1,6-kla]heptalen- 2-amine, N,N-diethyl-3,6,11-trimethyl-	dioxan	272s(4.52),300s(4.72), 312s(4.77),322(4.80), 352s(4.42),390s(4.00), 436s(3.91),454(4.00), 516(3.37),557(3.19), 647s(2.00),709(2.16), 784(2.21),880s(2.00), 1042s(1.18)	89-0763-78
$C_{27}H_{27}NO_3$ 1H,7aH-Chryseno[12a,1-e][1,3]oxazin- 7a-ol, 6,7,7b,8,9,15-hexahydro-11- methoxy-3-phenyl-, [7aR-(7aα,7bβ- 15aS*)]-	EtOH	262(3.60)	150-5001-78
$C_{27}H_{27}N_3$ 3H-1,5-Benzodiazepine, 2-(4-methyl- phenyl)-4-(4-phenyl-1-piperidinyl)-	n.s.g.	263(4.62),346(3.74)	87-0952-78
3H-1,5-Benzodiazepine, 2-phenyl-4-[4- (phenylmethyl)-1-piperidinyl]-	n.s.g.	260(4.53),346(3.66)	87-0952-78
$C_{27}H_{27}N_3O$ 3H-1,5-Benzodiazepine, 2-(4-methoxy- phenyl)-4-(4-phenyl-1-piperidinyl)-	n.s.g.	267(4.61),340(3.88)	87-0952-78
3H-1,5-Benzodiazepine, 2-[4-(4-methoxy- phenyl)-1-piperidinyl]-4-phenyl-	n.s.g.	260(4.56),346(3.65)	87-0952-78
$C_{27}H_{27}N_3O_2$ Acetamide, N-[5-ethyl-2,3-dihydro-3-(4- methylphenyl)-2-oxo-4-phenyl-1H-imid- azol-1-yl]-N-(2-methylphenyl)-	EtOH	265(4.04)	5-2033-78
$C_{27}H_{27}N_3O_3$ Diquino[2,3-a:2',3'-c]acridine- 6,12,18(2H,8H,14H)-trione, 1,3,4,5- 7,9,10,11,13,15,16,17-dodecahydro-	H_2SO_4	277(4.83),313(4.20), 372(3.78)	88-3677-78
$C_{27}H_{27}N_4O_2P$ 2,3,4(1H)-Phosphorintrione, 1-(1,1-di- methylethyl)-5-phenyl-, 2,4-dinitro- phenylhydrazone, 1-oxide	$CHCl_3$	260(4.07),353(4.33), 404(4.11),500(4.35)	139-0027-78B
$C_{27}H_{27}N_5$ Pyrido[2,3-b]pyrazine, 5-ethyl-1,2,3,5- tetrahydro-2,3-bis(2-methyl-1H-indol- 3-yl)-, hydriodide	EtOH	280(4.21),289(4.11), 355(4.20)	104-0398-78
$C_{27}H_{27}N_7OS$ 6H-Anthra[9,1-cd]isothiazol-6-one, 7- [(4,6-di-1-piperidinyl-1,3,5-triazin-	DMF	460(4.04)	2-1007-78

Compound	Solvent	$\lambda_{max}(\log \epsilon)$	Ref.
2-yl)amino]- (cont.)			2-1007-78
$C_{27}H_{27}N_7O_8S$			
L-Glutamic acid, N-[4-[[[(2-amino-1,4-dihydro-4-oxo-6-pteridinyl)methyl]-[(4-methylphenyl)sulfonyl]amino]-methyl]benzoyl]-	pH 1 pH 13	240(4.60),325(3.88) 245(4.44),367(3.88)	87-0673-78 87-0673-78
$C_{27}H_{28}NO_4PS$			
2-Butenedioic acid, 2-[[(diphenylphos-phinothioyl)phenylmethyl]amino]-, diethyl ester	C_6H_{12}	315(3.99)	22-0379-78
$C_{27}H_{28}N_2O_3$			
Piperidine, 1-benzoyl-3-ethenyl-4-[[3-(6-methoxy-4-quinolinyl)oxiranyl]-methyl]-	EtOH	222(4.57),232(4.58), 269(3.58),278(3.58), 290s(3.51),323s(3.64), 334(3.69)	35-0576-78
Piperidine, 1-benzoyl-3-ethenyl-4-[3-(6-methoxy-4-quinolinyl)-2-oxoprop-yl]-, (3R-cis)-	EtOH	220(4.69),233(4.73), 266s(3.73),278s(3.63), 323s(3.77),332(3.81)	35-0576-78
$C_{27}H_{28}N_2O_5$			
L-Phenylalanine, N-[4-[methyl[(phenyl-methoxy)carbonyl]amino]benzoyl]-, ethyl ester	EtOH	260(4.25)	87-1162-78
$C_{27}H_{28}N_2O_6$			
α-D-galacto-Hexodialdo-1,5-pyranose, 6-C-(1,3-diphenyl-1H-pyrazol-5-yl)-1,2:3,4-bis-O-(1-methylethylidene)-	MeOH	250(4.74),296(3.91)	136-0357-78H
$C_{27}H_{28}N_2O_7$			
1H-Cyclopenta[d]pyridazine-1,4-dicarb-oxylic acid, 5-[bis(4-methoxyphenyl)-methylene]-2,5-dihydro-1-methoxy-, dimethyl ester	MeOH	207(4.04),244(3.92), 298(3.83)	78-2509-78
$C_{27}H_{28}N_8O_7S$			
L-Glutamic acid, N-[4-[[[(2,4-diamino-6-pteridinyl)methyl][(4-methylphen-yl)sulfonyl]amino]methyl]benzoyl]-	pH 1 pH 13	242(4.55),283(3.70), 338(3.89) 239(4.50),374(3.78)	87-0673-78 87-0673-78
$C_{27}H_{28}O_4$			
4,7-Ethano-1H-indene-8,9-diol, 1-[bis-(4-methoxyphenyl)methylene]-3a,4,7,7a-tetrahydro-5-methyl-, cis-cis	MeOH	254(4.19),300(4.30)	5-0440-78
cis-trans	MeOH	254(4.23),300(4.30)	5-0440-78
trans-trans	MeOH	254(4.26),300(4.33)	5-0440-78
$C_{27}H_{28}O_6$			
Auriculasin 3',4'-dimethyl ether	MeOH	218(4.25),274(4.34)	44-3446-78
Kuwanon A dimethyl ether	MeOH	231(4.47),260(4.39), 325(3.99)	94-1453-78
	MeOH-NaOMe MeOH-AlCl₃	270(4.40),362(3.68) 269(4.46),320(3.95), 376(3.90)	94-1453-78 94-1453-78
Kuwanon B dimethyl ether	EtOH	213(4.57),235(4.60), 261(4.51),320s(4.08)	94-1453-78
	EtOH-NaOMe	267(4.76),358(3.86)	94-1453-78

Compound	Solvent	$\lambda_{max}(\log \epsilon)$	Ref.
Kuwanon B dimethyl ether (cont.)	EtOH–AlCl$_3$	215(4.62),271(4.54), 315s(4.06),375(3.82)	94-1453-78
Morusin dimethyl ether	EtOH	209(4.62),234(4.46), 273(4.60),300s(3.75), 345s(3.15)	94-1394-78
	EtOH–AlCl$_3$	209(4.60),230(4.49), 285(4.63),330s(3.63), 415(3.67)	94-1394-78
Pomiferin 3',4'-dimethyl ether	MeOH	224(4.18),278(4.28)	44-3446-78
$C_{27}H_{28}O_8$			
Morusin hydroperoxide dimethyl ether	EtOH	229(4.50),239(4.46), 278(4.47),360(4.14)	94-1431-78
	EtOH–AlCl$_3$	219(4.59),278(4.47), 360(4.14)	94-1431-78
$C_{27}H_{29}NO_4$			
Benzamide, N-[[1-[2-(3,4-dihydro-6-methoxy-1-naphthalenyl)ethyl]-2,6-dioxocyclohexyl]methyl]-	EtOH	220(4.15),276(3.75)	150-5001-78
Benzamide, N-[[1-[2-(3,4-dihydro-6-methoxy-1(2H)-naphthalenylidene)-ethyl]-2,6-dioxocyclohexyl]methyl]-	EtOH	234(3.90),264(3.20)	150-5001-78
Benzamide, N-(14-hydroxy-3-methoxy-17a-oxo-D-homoestra-1,3,5(10),8-tetraen-18-yl)-	EtOH	233(4.30),263(3.70)	150-5001-78
Benzamide, N-(14-hydroxy-3-methoxy-17a-oxo-D-homoestra-1,3,5(10),9(11)-tetraen-18-yl)-	EtOH	234(4.60),263(3.82)	150-5001-78
$C_{27}H_{30}N_2O$			
2,5-Cyclohexadiene-1,4-dione, 2,6-bis-(1,1-dimethylethyl)-, 4-[(diphenyl-methylene)hydrazone]	n.s.g.	360(4.43)	70-1179-78
$C_{27}H_{30}N_8O_8$			
Sacculatal bis(2,4-dinitrophenylhydra-zone)	EtOH	358(4.58)	102-0153-78
$C_{27}H_{30}O_5$			
2H,10H-Benzo[1,2-b:3,4-b']dipyran-10-one, 3,4-dihydro-5-methoxy-8-(4-meth-oxyphenyl)-2,2-dimethyl-6-(3-methyl-2-butenyl)-	MeOH	225(4.26),272(4.38), 319(4.32)	78-3569-78
4H-1-Benzopyran-4-one, 5-hydroxy-7-methoxy-2-(4-methoxyphenyl)-6,8-bis(3-methyl-2-butenyl)-	MeOH	229(3.32),288(4.35), 325(4.36)	78-3569-78
Parviflorin C	ether	254(4.32),272s(3.72), 282s(3.16),423(4.05)	33-0709-78
Parviflorin D	ether	253(4.39),272s(3.79), 282s(3.17),423(4.14)	33-0709-78
$C_{27}H_{30}O_6$			
Benzene, 2-[2-(3,4-dimethoxyphenyl)-1-(3,5-dimethoxyphenyl)ethenyl]-1,5-dimethoxy-3-methyl-, (E)-	EtOH	291s(4.23),321(4.35)	94-3050-78
(Z)-	EtOH	242s(4.40),285(4.20), 299s(4.17)	94-3050-78
4H-1-Benzopyran-4-one, 3-(3,4-dimeth-oxyphenyl)-5,7-dihydroxy-6,8-bis(3-methyl-2-butenyl)-	MeOH	210(3.80),240(4.10)	44-3446-78

Compound	Solvent	$\lambda_{max}(\log \epsilon)$	Ref.
Cyclotriveratrylene	hexane	232(4.52),290(3.98)	102-1791-78
Parviflorin E	ether	254(4.17),292(3.77), 420(3.91)	33-0709-78
Parviflorin F	ether	254(4.18),290s(3.79), 297s(3.76),420(3.94)	33-0709-78
5H-Tribenzo[a,d,g]cyclononene, 10,15- dihydro-2,3,7,8,12,13-hexamethoxy-	hexane	232(4.52),290(3.98)	88-1699-78
$C_{27}H_{30}O_{14}$ Rhoifolin	MeOH	270(4.23),334(4.30)	95-0366-78
$C_{27}H_{31}NO_3$ D-Homoestra-1,3,5(10),8-tetraen-17a- one, 14-hydroxy-3-methoxy-18-[(phen- ylmethyl)amino]-	EtOH	270(3.40)	150-5001-78
$C_{27}H_{31}NO_9$ 13-Deoxydaunorubicin, hydrochloride	MeOH	243(4.50),251(4.42), 284(4.01),478(4.00), 495(4.00),529(3.82)	87-0280-78
$C_{27}H_{31}NO_{10}$ 13-Deoxyadriamycin, hydrochloride	MeOH	237(4.47),249(4.38), 286(3.91)	87-0280-78
$C_{27}H_{31}NO_{12}$ 1H-Isoindole-1,3(2H)-dione, 2-[3-methyl- 4-[(2,3,4,6-tetra-O-acetyl-β-D-gluco- pyranosyl)oxy]-2-butenyl]-, (E)-	EtOH	221(4.63),241(3.97), 294(3.26)	12-1291-78
$C_{27}H_{31}N_3O_2$ Eburnamenine-14-carboxylic acid, 14,15- dihydro-14-(phenylamino)-, methyl ester	MeOH	227(4.56),281(3.97)	33-1682-78
$C_{27}H_{31}N_3O_{10}$ D-Ribitol, 1-deoxy-1-(3,4-dihydro- 5,7,8-trimethyl-2,4-dioxopyrimido- [4,5-b]quinolin-10(2H)-yl)-, 2,3,4,5-tetraacetate	MeOH	271(4.45),330(4.00), 397(4.09)	4-0489-78
$C_{27}H_{31}O$ Phenoxy, 2,6-bis(1,1-dimethylethyl)- 4-(diphenylmethyl)-	MeCN	410(3.6),630(2.7)	70-1304-78
$C_{27}H_{32}N_2O_8$ Aspidospermidine-3-carboxylic acid, 3,4-diacetoxy-6,7-didehydro-16-meth- oxy-1-methyl-8-oxo-, methyl ester, (2β,3β,4β,5α,12β,19α)-	EtOH	214(4.66),251(3.95), 302(3.76)	33-1554-78
$C_{27}H_{32}N_2O_9$ Aspidospermidine-3-carboxylic acid, 3,4-diacetoxy-6,7-didehydro-19-hy- droxy-16-methoxy-1-methyl-8-oxo-, methyl ester, (2β,3β,4β,5α,12β,19α)-	EtOH	214(4.64),249(3.94), 300(3.70)	33-1554-78
$C_{27}H_{32}N_{10}O_9$ Adenosine, N,N''-carbonylbis[2',3'-O-(1- methylethylidene)-	EtOH	267(4.31),285(4.55), 292(4.53)	39-0131-78C

Compound	Solvent	$\lambda_{max}(\log \epsilon)$	Ref.
$C_{27}H_{32}O$ Phenol, 2,6-bis(1,1-dimethylethyl)-4-(diphenylmethyl)-	C_6H_{12}	270(3.3),286(3.3)	70-1304-78
$C_{27}H_{32}O_6$ Tricoccin S_7	MeOH	252(4.25)	88-3699-78
$C_{27}H_{33}B$ Borane, tris(2,4,6-trimethylphenyl)-	MeOH-EtOH	330(4.2)	147-0177-78A
$C_{27}H_{33}N_3O_5$ Glycine, N-[1-(1,1-dimethylethyl)-N-[(phenylmethoxy)carbonyl]-L-tryptophyl]-, ethyl ester	isoPrOH	278s(--),288(3.71), 293s(--)	63-1617-78
$C_{27}H_{33}N_3O_6$ 1H-Pyrrole-3-carboxylic acid, 5-[3-(ethoxycarbonyl)-2-[[4-(ethoxycarbonyl)-3,5-dimethyl-1H-pyrrol-2-yl]methylene]-4-methyl-2H-pyrrol-5-yl]-2,4-dimethyl-, ethyl ester	EtOH EtOH-HCl	514(4.46) 592(4.64)	5-2024-78 5-2024-78
$C_{27}H_{33}N_3O_8$ L-Glutamic acid, N-[N-[4-[methyl(phenylmethoxy)carbonyl]amino]benzoyl]-glycyl]-, diethyl ester	MeOH	261(4.09)	87-1165-78
$C_{27}H_{33}N_3O_9$ D-Valine, N-[O-acetyl-N-[4-methyl-2-nitro-3-(phenylmethoxy)benzoyl]-L-threonyl]-, methyl ester	MeOH	213(4.638)	87-0607-78
$C_{27}H_{33}OP$ Phosphine, (4-methoxy-2,6-dimethylphenyl)bis(2,4,6-trimethylphenyl)-	heptane	311(4.38)	65-1732-78
$C_{27}H_{33}O_2P$ Phosphine oxide, (4-methoxy-2,6-dimethylphenyl)bis(2,4,6-trimethylphenyl)-	heptane	244(4.46),278(3.44), 288(3.37)	65-1732-78
$C_{27}H_{34}Cl_4N_2O_4$ Benzenamine, 4,4'-(2,4,8,10-tetraoxaspiro[5.5]undecane-3,9-diyl)bis[N,N-bis(2-chloroethyl)-	EtOH	262(4.76),291(3.91)	104-1004-78
$C_{27}H_{34}N_2O_7$ Aspidospermidine-3-carboxylic acid, 3,4-diacetoxy-6,7-didehydro-16-methoxy-1-methyl-, methyl ester	EtOH	212(4.52),249(3.80), 303(3.68)	33-1554-78
$C_{27}H_{34}N_2O_8$ Aspidospermidine-3-carboxylic acid, 3,4-diacetoxy-16-methoxy-1-methyl-8-oxo-, methyl ester	EtOH	213(4.57),250(3.81), 304(3.79)	33-1554-78
Aspidospermidine-3-carboxylic acid, 3,4-diacetoxy-16-methoxy-1-methyl-10-oxo-, methyl ester	EtOH	212(4.64),247(3.87), 300(3.72)	33-1554-78
$C_{27}H_{34}O_6$ Carda-5,14,20(22)-trienolide, 3β,19-diacetoxy-	MeOH	217(4.10)	44-3946-78

Compound	Solvent	$\lambda_{max}(\log \epsilon)$	Ref.
$C_{27}H_{34}O_7$			
Tricoccin S_8	MeOH	254(4.41)	88-3699-78
Tricoccin S_{19}	MeOH	252(4.48)	88-3699-78
$C_{27}H_{34}O_8$			
Olivetoric acid, 4-O-methyl-	n.s.g.	223(4.12),269(4.16), 309(4.00)	12-1041-78
$C_{27}H_{34}O_{10}$			
Chaparrinone, 6α-senecioyloxy-, mono-acetate	MeOH	228(4.33),240s(4.25)	87-1186-78 +100-0578-78
$C_{27}H_{34}O_{11}$			
Arctiin	MeOH	226(4.67),276(4.45)	100-0638-78
$C_{27}H_{35}NO_6$			
Nicotinoylisolineolon	EtOH	264(3.60)	100-0001-78
$C_{27}H_{36}CuN_2O_2$			
Copper, [[6,6'-[1,5-pentanediylbis(nitrilomethylidyne)]bis[2-methyl-5-(1-methylethenyl)-2-cyclohexen-1-onato]]-(2-)-N,N',O,O']-	benzene	310(4.10),375(4.14), 645s(2.18)	65-1474-78
$C_{27}H_{36}N_2O_7$			
Aspidospermidine-3-carboxylic acid, 3,4-diacetoxy-16-methoxy-1-methyl-, methyl ester	EtOH	211(4.59),247(3.87), 301(3.75)	33-1554-78
$C_{27}H_{36}N_4O_6S_2$			
1,3(2H,4H)-Pyrimidinedibutanesulfonamide, N,N',6-trimethyl-2,4-dioxo-N,N'-diphenyl-	CHCl$_3$	266(3.96)	70-0149-78
$C_{27}H_{36}N_4O_{12}$			
Pyrano[2,3-c]pyrazole-2-acetic acid, 2,6-dihydro-3,4-dimethyl-5-[2-(N-methylpiperazino)ethyl]-6-oxo-, ethyl ester, dimaleate	H$_2$O	206(4.54),310(4.22)	95-0335-78
$C_{27}H_{36}O_7$			
Akiferidinin	n.s.g.	265(3.9),300(3.6)	105-0617-78
$C_{27}H_{36}O_8$			
Tetrahydroverrucarin J	EtOH	218(4.20)	33-1975-78
$C_{27}H_{37}F_3N_8O_{12}P_2$			
Pyridinium, 3-(aminocarbonyl)-1-[5-O-[hydroxy[[hydroxy[3-[6-[[6-[(trifluoroacetyl)amino]hexyl]amino]-9H-purin-9-yl]propoxy]phosphinyl]oxy]-phosphinyl]-β-D-ribofuranosyl]-, hydroxide, inner salt	pH 9.5	267(4.32)	5-0118-78
$C_{27}H_{37}NO_2$			
10aH-Benzo[b]cyclohepta[d]furan-10a-carboxamide, 2,4-bis(1,1-dimethylethyl)-N,N-diethyl-8-methyl-	hexane	213s(4.44),233s(4.15), 266(3.60),297(3.64)	88-1229-78
$C_{27}H_{37}NO_3$			
24-Norchola-3,5,20(22)-triene-23-ni-	MeOH	232.5(4.43)	44-2334-78

Compound	Solvent	$\lambda_{max}(\log \epsilon)$	Ref.
trile, 21-acetoxy-3-ethoxy-, (20Z)- (cont.)			44-2334-78
$C_{27}H_{38}$ 19-Norergosta-3,5,7,9,22-pentaene	n.s.g.	267(4.07)	39-1533-78C
$C_{27}H_{38}N_4O_3$ 3-Aza-A-homospirost-4a-eno[3,4-d]tetra- zol-6-one, (25R)-	MeOH	262(4.37)	2-0095-78
$C_{27}H_{38}O_2$ Azafrinal	EtOH	250(4.24),288(4.08), 315(4.08),445(4.91)	33-0822-78
8,10,12,14,16,18,20-Docosaheptaene-2,7- dione, 6,6,10,14,19-pentamethyl-, (all-E)-	EtOH	250(3.93),318(4.06), 433(4.85)	33-0822-78
$C_{27}H_{38}O_4$ Spirost-4-ene-3,6-dione, (25R)-	MeOH	252(4.87)	2-0095-78
Strongylophorine-1	MeOH	207(4.04),221(3.78), 229(3.86),295(3.59)	20-0917-78
$C_{27}H_{38}O_7$ Octyl mycophenolate acetate	MeOH	248(4.04),285(3.37)	12-0353-78
$C_{27}H_{38}O_{12}$ Roseoside tetraacetate	EtOH	236(3.94)	94-2111-78
$C_{27}H_{40}$ 19-Norergosta-3,5,7,9-tetraene	n.s.g.	267(4.05)	39-1533-78C
$C_{27}H_{40}O$ Azafrinol	EtOH	231(4.14),277(3.87), 289(3.97),356s(4.57), 374(4.85),396(5.04), 420(5.01)	33-0822-78
$C_{27}H_{40}O_2$ Hexadecanoic acid, 1-naphthalenylmethyl ester	hexane	298.5(3.83)	95-0774-78
$C_{27}H_{40}O_3$ Cholest-en-19-oic acid, 3β-hydroxy- 7-oxo-, δ-lactone	EtOH	249(3.84)	73-1142-78
$C_{27}H_{40}O_4$ Chola-5,22-dien-24-oic acid, 3-acetoxy-, methyl ester, (3β,20R,22E)-	EtOH	210(4.21)	44-1442-78
(3β,20S,22E)-	EtOH	218(4.11)	44-1442-78
Pregn-4-ene-3,20-dione, 17-[(2,2-di- methyl-1-oxopropoxy)methyl]-	EtOH	237(4.1)	23-0410-78
$C_{27}H_{40}O_5$ Linarienone	EtOH	219(4.11),245(4.04)	94-0079-78
$C_{27}H_{41}NO_2$ Cholest-4-eno[6,5,4-cd]isoxazol-3-one	EtOH	263(3.98)	12-0113-78
Solasodenone	MeOH	242(4.18),310(2.14)	102-1070-78
$C_{27}H_{41}NO_3$ 5'H-Norcholestano[6,5-c]isoxazole-	EtOH	225(3.67)	12-0113-78

Compound	Solvent	$\lambda_{max}(\log \epsilon)$	Ref.
3,5'-dione, (5S)- (cont.)			12-0113-78
$C_{27}H_{42}N_4$			
Cyclopenta[5,6]naphtho[2,1-d]tetrazolo-[1,5-a]azepine, 1-(1,5-dimethylhex-yl)-1,2,3,3a,3b,11,12,12a,12b,13,14-14a-dodecahydro-12a,14a-dimethyl-	EtOH	287(4.47)	2-0559-78
$C_{27}H_{42}N_4O$			
Cyclopenta[5,6]naphtho[1,2-e]tetrazolo-[1,5-a]azepin-5(1H)-one, 1-(1,5-di-methylhexyl)-2,3,3a,3b,4,11,12,12a-12b,13,14,14a-dodecahydro-12a,14a-dimethyl-	EtOH	243(4.1)	2-0559-78
Cyclopenta[5,6]naphtho[2,1-d]tetrazolo-[1,5-a]azepin-5(1H)-one, 1-(1,5-di-methylhexyl)-2,3,3a,3b,4,11,12,12a-12b,13,14,14a-dodecahydro-12a,14a-dimethyl-	EtOH	260(4.38)	2-0559-78
$C_{27}H_{42}N_8$			
1H-Cyclopenta[g]tetrazolo[1,5-c]tetra-zolo[1',5':1,7]azepino[4,5-a][3]-benzazepine, 1-(1,5-dimethylhexyl)-2,3,3a,3b,4,14,15,15a,15b,16,17,17a-dodecahydro-15a,17a-dimethyl-	EtOH	245(4.4)	2-0559-78
3,6-Diaza-AB-bishomocholest-4a-eno-[3,4-d:6,7-d]bistetrazole	MeOH	246(4.31)	2-0095-78
3,7-Diaza-AB-bishomocholest-4a-eno-[3,4-d:7,6-d]bistetrazole	MeOH	250(4.08)	2-0095-78
3,7a-Diaza-AB-bishomo-5-cholesteno-[3,4-d:7a,7-d]bistetrazole	EtOH	243(4.18)	2-0617-78
4,6-Diaza-AB-bishomocholest-4a-eno-[4,3-d:6,7-d]bistetrazole	MeOH	240(4.21)	2-0095-78
$C_{27}H_{42}O$			
Cholesta-4,6-dien-3-one	MeOH	285(4.320)	2-0545-78
19-Nor-9,10-secoergosta-5,7,22-trien-1-one, (5E,7E,22E)-	EtOH	313(4.37)	39-0590-78C
(5Z,7E,22E)-	EtOH	310(4.21)	39-0590-78C
$C_{27}H_{42}O_2$			
Cholesta-2,4-dien-6-one, 4-hydroxy-	EtOH	320(3.89)	12-0171-78
5α-Cholesta-8,14-dien-23-one, 3β-hy-droxy-	MeOH	248(4.26)	88-2609-78
Cholest-4-en-3-one, 1α,2α-epoxy-	EtOH	246.5(4.14)	94-2933-78
$C_{27}H_{42}O_3$			
5α-Cholest-7-ene-3,6-dione, 5-hydroxy-	EtOH	250(4.20)	12-2069-78
Cholest-4-en-3-one, 1α,2α-epoxy-25-hy-droxy-	EtOH	246.5(4.11)	94-2933-78
$C_{27}H_{42}O_4$			
5α-Chol-22-en-24-oic acid, 3-acetoxy-, methyl ester, (3β,20S,22E)-	EtOH	219(3.81)	44-1442-78
6,7-Secocholest-4-en-6-oic acid, 3,7-dioxo-	EtOH	238(4.07)	12-0171-78
$C_{27}H_{42}O_5$			
6,7-Secocholest-4-ene-6,7-dicarboxylic acid, 3-oxo-	EtOH	235(3.95)	12-0171-78

Compound	Solvent	$\lambda_{max}(\log \epsilon)$	Ref.
$C_{27}H_{42}O_6S$			
Card-20(22)-enolide, 3-[(butylsulfonyl)oxy]-14-hydroxy-, (3β,5β)-	MeOH	216(4.21)	106-0706-78
Card-20(22)-enolide, 14-hydroxy-3-[[(2-methylpropyl)sulfonyl]oxy]-, (3β,5β)-	MeOH	215(4.19)	106-0706-78
$C_{27}H_{43}FO_2$			
Cholesta-5,7-diene-1α,3β-diol, 25-fluoro-	EtOH	262s(--),272(--), 282(4.08),293(--)	13-0453-78B
$C_{27}H_{43}NO$			
Cholest-4-eno[6,5,4-cd]isoxazole	EtOH	231(3.90)	12-0097-78
$C_{27}H_{43}NO_2$			
Cholesta-3,5-diene, 6-nitro-	EtOH	267(4.16),318(3.56)	12-0097-78
Cholest-4-ene-3,6-dione, 3-oxime	EtOH	285(4.20)	12-0097-78
Cholest-4-eno[6,5,4-cd]isoxazol-2-ol, (3β)-	EtOH	234(3.74)	12-0097-78
$C_{27}H_{43}NO_3$			
A-Norcholestano[6,5-c]isoxazol-5'-one, 3-hydroxy-, (3β,5S)-	EtOH	217(3.79)	12-0113-78
$C_{27}H_{43}N_5O$			
1H-Cyclopenta[g]tetrazolo[1',5':1,7]azepino[4,5-a][3]benzazepin-5(6H)-one, 1-(1,5-dimethylhexyl)-2,3,3a,3b,4,12-13,13a,13b,14,15,15a-dodecahydro-13a,15a-dimethyl-	EtOH	243(4.1)	2-0559-78
3,6-Diaza-AB-bishomocholest-4a-eno-[6,7-d]tetrazol-4-one	MeOH	247(4.25)	2-0095-78
$C_{27}H_{44}$			
Cholesta-3,5-diene	EtOH	228(4.26),235(4.31), 244(4.12)	44-2715-78
19-Nor-9,10-secoergosta-5,7,22-triene, (7E,22E)-	EtOH	243(4.50),252(4.57), 261(4.40)	39-0590-78C
$C_{27}H_{44}N_2$			
2'H-Cholest-4-eno[4,5,6-cd]pyrazole	n.s.g.	226.5(3.71)	4-0023-78
$C_{27}H_{44}N_2O_2$			
3,7a-Diaza-AB-bishomochloest-5(or 4a)-ene-4,7-dione	MeOH	217(4.24)	2-0617-78
$C_{27}H_{44}O$			
Cholest-8(14)-en-15-one	C_6H_{12}	255(4.12)	2-0257-78
Cholest-13(17)-en-12-one	C_6H_{12}	255(4.30)	2-0257-78
Cholest-13(17)-en-16-one	C_6H_{12}	244(4.08)	2-0257-78
A-Homo-B-norcholest-4a-en-3-one	MeOH	290(2.33)	44-4622-78
$C_{27}H_{44}O_2$			
Cholesta-5,7-diene-1α,3β-diol	EtOH	272(--),283(4.10), 294(--)	94-2933-78
Cholestane-4,6-dione	EtOH	292(4.0)	2-0559-78
	EtOH	295(4.00)	12-0113-78
Vitamin D_3, 1α-hydroxy-	EtOH	265.5(4.23)	94-2933-78
$C_{27}H_{44}O_3$			
Cholesta-5,7-diene-1α,3β,25-triol	EtOH	264(--),272(--), 282(4.05),294(--)	94-2933-78

Compound	Solvent	$\lambda_{max}(\log \epsilon)$	Ref.
Cholestane-4,6-dione, 3β-hydroxy-	EtOH	296(3.98)	12-0113-78
5α-Cholest-7-en-6-one, 3β,5-dihydroxy-	EtOH	249(4.17)	12-2069-78
5,6-Secocholest-3-en-6-oic acid, 5-oxo-	EtOH	230(4.00)	12-0171-78
Vitamin D₃, 1α,25-dihydroxy-	EtOH	264.5(4.22)	94-2933-78
$C_{27}H_{44}O_4$			
5α-Cholest-7-en-6-one, 3β,5,14α-tri- hydroxy-	EtOH	254(4.14)	12-2069-78
2-Propenoic acid, 3-(3,4-dihydroxyphen- yl)-, octadecyl ester, (E)-	EtOH	248(4.11),301s(4.03), 332(4.13)	94-3863-78
6,7-Secocholest-4-en-6-oic acid, 3β-hy- droxy-7-oxo-	EtOH	215(3.88)	12-0171-78
$C_{27}H_{44}O_5$			
6,7-Secocholest-4-ene-6,7-dicarboxylic acid, 3β-hydroxy-	EtOH	212(3.87)	12-0171-78
$C_{27}H_{44}O_9$			
Cholest-7-en-6-one, 1,2,3,5,14,20,22,25- octahydroxy-, (1β,2β,3β,5β,22R)-	EtOH	240(4.10)	105-0388-78
$C_{27}H_{45}NO$			
Cholest-5-en-4-one, 6-amino-	EtOH	321(4.09)	12-0113-78
$C_{27}H_{45}NO_2$			
Cholest-5-en-4-one, 6-amino-3β-hydroxy-	EtOH	326(4.08)	12-0113-78
$C_{27}H_{45}NO_3$			
Cholestane, 4α,5α-epoxy-6β-nitro-	EtOH	280(1.56)	12-0097-78
$C_{27}H_{46}N_6O_5$			
Adenosine, N-[(hexadecylamino)carbonyl]-	pH 1	270(4.34),276(4.31)	36-0569-78
	pH 13	261(4.18),282s(3.88)	36-0569-78
	70% EtOH	269(4.35),275(4.28)	36-0569-78
$C_{27}H_{46}O_9$			
Methyl isopseudomonate A	EtOH	221(4.06)	39-0561-78C

Compound	Solvent	$\lambda_{max}(\log \epsilon)$	Ref.
$C_{28}H_{14}Cl_6O_9$			
Dinaphtho[2,1-b:1',2'-d]furan-4,5,9,10-tetrol, 1,3,6,8,11,13-hexachloro-, tetraacetate	$CHCl_3$	277(4.50),286(4.53), 314(3.97),325(4.08), 362(4.07),379(4.36), 398(4.51)	12-1323-78
$C_{28}H_{15}Cl_3N_2O_2$			
1,2-Benzisoxazole, 6-chloro-3-[3-chloro-4-[2-[4-(6-chloro-1,2-benzisoxazol-3-yl)phenyl]ethenyl]phenyl]-	DMF	340(4.70)	33-2904-78
1,2-Benzisoxazole, 6-chloro-3-[3-chloro-4-[2-[2-(4-chlorophenyl-6-benzoxazolyl]ethenyl]phenyl]-	DMF	357(4.76)	33-2904-78
$C_{28}H_{16}$			
Dibenzo[def,pqr]tetraphenylene	C_6H_{12}	254(4.94),273s(4.66), 294s(4.40),304s(4.32), 317s(4.14),346s(3.02), 361s(2.96)	44-2484-78
$C_{28}H_{16}Cl_2N_2O_2$			
1,2-Benzisoxazole, 3-[4-[2-[4-(1,2-benzisoxazol-3-yl)phenyl]ethenyl]-3-chlorophenyl]-6-phenyl-	DMF	340(4.69)	33-2904-78
1,2-Benzisoxazole, 3-[4-[2-[4-(2-benzoxazolyl)-2-chlorophenyl]ethenyl]phenyl]-6-chloro-	DMF	356(4.75)	33-2904-78
1,2-Benzisoxazole, 6-chloro-3-[4-[2-[4-(5-chloro-2-benzoxazolyl)phenyl]ethenyl]phenyl]-	DMF	361(4.88)	33-2904-78
1,2-Benzisoxazole, 6-chloro-3-[3-chloro-4-[2-[2-(2-phenyl-6-benzoxazolyl)ethenyl]phenyl]-	DMF	360(4.79)	33-2904-78
$C_{28}H_{16}Cl_2S$			
3H-Naphtho[1,8-bc]thiophene, 6,8-dichloro-3-(1-naphthalenylphenylmethylene)-	MeOH	216(4.7),272(4.5), 312(4.1),324(4.2), 406(3.9)	5-1974-78
$C_{28}H_{16}D_4O_4$			
6,17[1',2']-Benzeno-7,8,16-[1,3]butadien[1]yl[4]ylidenedibenzo[b,h]-[1,4,7,10]tetraoxacyclotetradecin-9-d, 6,7,16,17-tetrahydro-6,16,17-d$_3$, (6α,7β,16β,17α)-	ether	267(3.74),274(3.74), 308(3.79)	78-0083-78
$C_{28}H_{16}N_2O_2$			
Benzo[g]naphtho[1',2',3':4,5]quino-[2,1-b]quinazoline-5,10-dione, 8-methyl-	n.s.g.	478(3.61)	2-0103-78
Benzo[g]naphtho[1',2',3':4,5]quino-[2,1-b]quinazoline-5,10-dione, 18-methyl-	n.s.g.	460(3.71)	2-0103-78
Naphtho[2,3-g]phthalazine-6,11-dione, 1,4-diphenyl-	$CHCl_3$	309(4.63),374(3.63)	150-5538-78
$C_{28}H_{16}N_2O_3$			
Benzo[g]naphtho[1',2',3':4,5]quino-[2,1-b]quinazoline-5,10-dione, 6-methoxy-	n.s.g.	480(3.60)	2-0103-78

Compound	Solvent	λ_{max} (log ϵ)	Ref.
$C_{28}H_{16}N_4O_9$ Furan, 3,4-bis(2,4-dinitrophenyl)-2,5-diphenyl-	n.s.g.	305(4.31),380(3.94)	39-0401-78C
$C_{28}H_{16}O_2S$ Anthra[2,3-c]thiophene-5,10-dione, 1,3-diphenyl-	CH_2Cl_2	277(4.63),319(4.65), 460(4.13)	150-5538-78
$C_{28}H_{16}O_3$ Anthra[2,3-c]furan-5,10-dione, 1,3-diphenyl-	CH_2Cl_2	285(4.85),500(4.13)	150-5538-78
$C_{28}H_{16}O_3Se$ 1H-Indene-1,3(2H)-dione, 2-[2,3-dihydro-2-(2-methyl-4H-1-benzoselenin-4-ylidene)-3-oxo-1H-inden-1-ylidene]-	HOAc-HClO$_4$	692(3.436)	83-0170-78
$C_{28}H_{16}O_4$ 9,10-Anthracenedione, 2,3-dibenzoyl-	CHCl$_3$	260(4.80),330(3.84)	150-5538-78
$C_{28}H_{16}O_4S$ Anthra[2,3-c]thiophene-5,10-dione, 6,9-dihydroxy-1,3-diphenyl-	CH_2Cl_2	282(4.77),318(4.63), 496(4.32)	150-5538-78
$C_{28}H_{16}O_6$ 4,11-Epoxyanthra[2,3-c]furan-1,3,6,9-tetrone, 4,11-dihydro-4,11-diphenyl-	CH_2Cl_2	332(3.4)	150-5538-78
$C_{28}H_{16}S_4$ 1,3-Benzodithiole, 2,2'-(9,10-anthracenediylidene)bis-	CH_2Cl_2	237(4.34),273s(3.77), 352s(3.78),395s(3.89), 416(3.97)	18-2674-78
$C_{28}H_{17}ClN_2O_2$ 1,2-Benzisoxazole, 3-[4-[2-[4-(2-benzoxazolyl)-2-chlorophenyl]ethenyl]phenyl]-	DMF	360(4.79)	33-2904-78
1,2-Benzisoxazole, 3-[4-[2-[4-(2-benzoxazolyl)phenyl]ethenyl]phenyl]-6-chloro-	DMF	360(4.86)	33-2904-78
1,2-Benzisoxazole, 3-[4-[2-[4-(5-chloro-2-benzoxazolyl)phenyl]ethenyl]phenyl]-	DMF	362(4.87)	33-2904-78
$C_{28}H_{17}Cl_2N_5O$ [1,2,4]Triazolo[1,5-a]pyridine, 2-[4-[2-[2-chloro-4-[5-(4-chlorophenyl)-1,3,4-oxadiazol-2-yl]phenyl]ethenyl]phenyl]-	DMF	360(4.78)	33-0142-78
[1,2,4]Triazolo[1,5-a]pyridine, 7-[2-[2-chloro-4-[5-(4-chlorophenyl)-1,3,4-oxadiazol-2-yl]phenyl]ethenyl]-2-phenyl-	DMF	356(4.78)	33-0142-78
$C_{28}H_{18}$ 15H-Dibenzo[a,g]cyclotridecene, 15-(2,4,6-cycloheptatrien-1-ylidene)-5,6,7,8-tetrahydro-	n.s.g.	<u>225(4.7),290(4.5), 350(4.2),422(4.2)</u>	88-2795-78
Dibenzo[b,i]triptycene	C_6H_{12}	229(5.05),243(4.80), 250(5.01),269(4.39), 278(4.27),289(3.93),	39-1019-78B

Compound	Solvent	$\lambda_{max}(\log \epsilon)$	Ref.
Dibenzo[b,i]triptycene (cont.)		302(3.10),310(3.37), 316(3.15),321(3.23), 325(3.70)	39-1019-78B
$C_{28}H_{18}ClN_3O$ [1,2,4]Triazolo[1,5-a]pyridine, 2-[4-[2-[4-(5-chloro-2-benzofuranyl)phenyl]ethenyl]phenyl]-	DMF	371(4.94),390(4.77)	33-0142-78
$C_{28}H_{18}ClN_5O$ [1,2,4]Triazolo[1,5-a]pyridine, 2-[4-[2-[2-chloro-4-(5-phenyl-1,3,4-oxadiazol-2-yl)phenyl]ethenyl]phenyl]-	DMF	361(4.77)	33-0142-78
[1,2,4]Triazolo[1,5-a]pyridine, 2-[4-[2-[4-(3-(4-chlorophenyl)-1,2,4-oxadiazol-5-yl]phenyl]ethenyl]phenyl]-	DMF	356(4.84)	33-0142-78
[1,2,4]Triazolo[1,5-a]pyridine, 2-[4-[2-[4-(5-(4-chlorophenyl)-1,2,4-oxadiazol-2-yl]phenyl]ethenyl]phenyl]-	DMF	359(4.90)	33-0142-78
[1,2,4]Triazolo[1,5-a]pyridine, 7-[2-[2-chloro-4-(5-phenyl-1,3,4-oxadiazol-2-yl)phenyl]ethenyl]-2-phenyl-	DMF	357(4.77)	33-0142-78
[1,2,4]Triazolo[1,5-a]pyridine, 7-[2-[4-[3-(4-chlorophenyl)-1,2,4-oxadiazol-5-yl]phenyl]ethenyl]-2-phenyl-	DMF	349(4.79)	33-0142-78
$C_{28}H_{18}Cl_2N_6$ [1,2,4]Triazolo[1,5-a]pyridine, 2-[3-chloro-4-[2-[4-(4-chloro-5-phenyl-2H-1,2,3-triazol-2-yl)phenyl]ethenyl]phenyl]-	DMF	359(4.82)	33-0142-78
[1,2,4]Triazolo[1,5-a]pyridine, 2-(3-chloro-4-[2-[4-(5-chloro-2-phenyl-2H-1,2,3-triazol-4-yl)phenyl]ethenyl]phenyl]-	DMF	350(4.77)	33-0142-78
$C_{28}H_{18}N_2O_2$ 1,2-Benzisoxazole, 6-[2-[4-(1,2-benzisoxazol-3-yl)phenyl]ethenyl]-3-phenyl-	DMF	339(4.73)	33-2904-78
1,2-Benzisoxazole, 3-[4-[2-[4-(2-benzoxazolyl)phenyl]ethenyl]phenyl]-	DMF	360(4.87)	33-2904-78
1,2-Benzisoxazole, 3,3'-(1,2-ethenediyldi-4,1-phenylene)bis-	DMF	342(4.76)	33-2904-78
2,1-Benzisoxazole, 3-[4-[2-[4-(1,2-benzisoxazol-3-yl)phenyl]ethenyl]-phenyl]-	DMF	382(4.73)	33-2904-78
$C_{28}H_{18}N_6O_4$ 4(3H)-Quinazolinone, 3-[[3-(4-nitrophenyl)hydrazono]-1(3H)-isobenzofuranylidene]amino]-2-phenyl-	MeOH	226(4.78),232(4.76), 275(4.31),311(3.96)	80-1085-78
$C_{28}H_{18}O$ Hexahelicene, 5-acetyl-	MeOH	228(4.06),247(4.16), 268(4.02),304s(3.58), 320(3.78),329(3.80), 348s(3.60),370s(3.30)	54-0265-78
Hexahelicene, 8-acetyl-	MeOH	226(4.38),255(4.52), 264s(4.46),300s(4.01), 313s(4.18),323(4.24),	54-0265-78

Compound	Solvent	$\lambda_{max}(\log \epsilon)$	Ref.
(cont.)		347s(3.95),368s(3.47)	54-0265-78
$C_{28}H_{18}O_4$ 2H-Pyran-2-one, 6-(1,3-benzodioxol-5-yl)-3-(1-naphthalenyl)-4-phenyl-	EtOH	269(4.33),377(4.33)	4-0759-78
$C_{28}H_{18}O_7$ Naphtho[2,3-c]furan-4,9-dione, 5,8-di-acetoxy-1,3-diphenyl-	CH_2Cl_2	350(4.1),420(3.96)	150-5538-78
$C_{28}H_{18}S$ 5H-Naphtho[1,8-bc]thiophene, 5-(1-naph-thalenylphenylmethylene)-	MeOH	215(4.8),270(4.3), 375(4.3)	5-1974-78
$C_{28}H_{19}ClN_4O$ [1,2,4]Triazolo[1,5-a]pyridine, 2-[4-[2-[4-(1,2-benzisoxazol-3-yl)phenyl]-ethenyl]-3-chlorophenyl]-6-methyl-	DMF	350(4.76)	33-2904-78
[1,2,4]Triazolo[1,5-a]pyridine, 2-[4-[2-[4-(5-chloro-7-methyl-2-benzoxa-zolyl)phenyl]ethenyl]phenyl]-	DMF	367(4.92),386(4.69)	33-0142-78
$C_{28}H_{19}ClN_4O_2$ 2H-Benzotriazole, 2-[4-[2-[4-(1,2-benz-isoxazol-3-yl)phenyl]ethenyl]-3-chlo-rophenyl]-5-methoxy-	DMF	368(4.81)	33-2904-78
$C_{28}H_{19}ClN_6$ [1,2,4]Triazolo[1,5-a]pyridine, 2-[3-chloro-4-[2-(2,5-diphenyl-2H-1,2,3-triazol-4-yl)ethenyl]phenyl]-	DMF	347(4.63)	33-0142-78
[1,2,4]Triazolo[1,5-a]pyridine, 2-[3-chloro-4-[2-[4-(2-phenyl-2H-1,2,3-triazol-4-yl)phenyl]ethenyl]phenyl]-	DMF	356(4.83)	33-0142-78
[1,2,4]Triazolo[1,5-a]pyridine, 2-[3-chloro-4-[2-(4-phenyl-2H-1,2,3-triazol-2-yl)phenyl]ethenyl]phenyl]-	DMF	358(4.84)	33-0142-78
$C_{28}H_{19}NO_2$ 1,2-Benzisoxazole, 3-[4-[2-(3-dibenzo-furanyl)ethenyl]phenyl]-5-methyl-	DMF	352(4.80)	33-2904-78
$C_{28}H_{19}NO_3$ 1,2-Benzisoxazole, 6-[2-(3-dibenzofur-anyl)ethenyl]-3-(4-methoxyphenyl)-	DMF	350(4.78)	33-2904-78
$C_{28}H_{19}N_3O$ [1,2,4]Triazolo[1,5-a]pyridine, 2-[4-[2-[4-(2-benzofuranyl)phenyl]ethen-yl]phenyl]-	DMF	371(4.91),390(4.76)	33-0142-78
[1,2,4]Triazolo[1,5-a]pyridine, 7-[2-[4-(2-benzofuranyl)phenyl]ethenyl]-2-phenyl-	DMF	368(4.87)	33-0142-78
[1,2,4]Triazolo[1,5-a]pyridine, 2-phen-yl-7-[2-(2-phenyl-6-benzofuranyl)eth-enyl]-	DMF	365(4.80)	33-0142-78
$C_{28}H_{19}N_5O$ [1,2,4]Triazolo[1,5-a]pyridine, 2-[4-[2-[4-(3-phenyl-1,2,4-oxadiazol-5-yl)phenyl]-	DMF	355(4.84)	33-0142-78

Compound	Solvent	$\lambda_{max}(\log \epsilon)$	Ref.
[1,2,4]Triazolo[1,5-a]pyridine, 2-[4- [2-[4-(5-phenyl-1,3,4-oxadiazol-2- yl)phenyl]ethenyl]phenyl]-	DMF	357(4.89)	33-0142-78
[1,2,4]Triazolo[1,5-a]pyridine, 2-phen- yl-7-[2-[4-(3-phenyl-1,2,4-oxadiazol- 5-yl)phenyl]ethenyl]-	DMF	348(4.79)	33-0142-78
$C_{28}H_{20}$ [2.2](3,3')-Biphenylophane-1,15-diene	C_6H_{12}	213(4.74),243s(4.51), 282s(4.28)	44-2484-78
Hexahelicene, 2,15-dimethyl-	CH_2Cl_2	257(4.74),307s(4.39), 316(4.48),328(4.37), 348s(4.09),391(2.92), 414(2.76)	78-2565-78
$C_{28}H_{20}BrNO$ Benzo[f]quinoline, 1-[2-(4-bromophen- yl)ethenyl]-3-(4-methoxyphenyl)-	EtOH	260(4.52),297(4.64), 370(3.97)	103-0418-78
Benzo[f]quinoline, 3-(4-bromophenyl)- 1-[2-(4-methoxyphenyl)ethenyl]-	EtOH	260(4.56),295(4.69), 350(4.40)	103-0418-78
$C_{28}H_{20}ClN$ Benzo[f]quinoline, 3-(4-chlorophenyl)- 2-methyl-1-(2-phenylethenyl)-	EtOH	267(4.64),294(4.48), 324(4.01),342(3.97), 359(3.88)	103-0418-78
$C_{28}H_{20}ClN_3$ [1,2,4]Triazolo[1,5-a]pyridine, 2-[3- chloro-4-(2-phenylethenyl)phenyl]- 7-(2-phenylethenyl)-	DMF	335(4.88)	33-0142-78
$C_{28}H_{20}Cl_2O_2$ Phenanthrene, 3-chloro-10-(4-chloro- phenyl)-6-methoxy-9-(4-methoxyphenyl)-	$CHCl_3$	255(4.65),263s(4.64), 284(4.38),307s(4.06), 316(4.09),350(3.28), 368(3.28)	23-1246-78
$C_{28}H_{20}CoN_{10}$ Cobalt, [7,16-dihydro-7,16-bis(phenyl- azo)dipyrido[2,3-b:2',3'-i][1,4,8,11]- tetraazacyclotetradecinato(2-)-N^5,N^9- N^{14},N^{18}]-, (SP-4-1)-	DMF	449(4.75),505s(4.50)	64-1012-78B
$C_{28}H_{20}CuN_{10}$ Copper, [7,16-dihydro-7,16-bis(phenyl- azo)dipyrido[2,3-b:2',3'-i][1,4,8,11]- tetraazacyclotetradecinato(2-)-N^5,N^9- N^{14},N^{18}]-, (SP-4-1)-	DMF	415(4.70),473s(4.44)	64-1012-78B
$C_{28}H_{20}N_2$ 9H-Indeno[2,1-c]pyridine, 9,9'-(1,2- ethanediylidene)bis[3-methyl-, (E,E)-	EtOH	216(4.58),240(4.64), 267(4.45),401(4.57), 420(4.63),448(4.62)	103-0066-78
$C_{28}H_{20}N_2O$ Acetic acid, 9H-fluoren-9-yl-9H-fluo- ren-9-ylidenehydrazide	isooctane	378(3.55)	18-2431-78
Imidazo[1,2-a]pyridine, 7-[2-(3-diben- zofuranyl)ethenyl]-2-(4-methylphenyl)-	DMF	377(4.77),397(4.71)	33-0129-78

Compound	Solvent	$\lambda_{max}(\log \epsilon)$	Ref.
$C_{28}H_{20}N_2O_2$			
Imidazo[1,2-a]pyridine, 7-[2-(1,3-ben- zodioxol-5-yl)ethenyl]-2-[1,1'-bi- phenyl]-4-yl-	DMF	372(4.75),391(4.71)	33-0129-78
Imidazo[1,2-a]pyridine, 7-[2-(3-diben- zofuranyl)ethenyl]-2-(4-methoxyphen- yl)-	DMF	380(4.76),398(4.71)	33-0129-78
$C_{28}H_{20}N_2O_2S$			
1,2-Benzisoxazole, 3-[4-[2-[5-(5,6-di- methyl-2-benzoxazolyl)-2-thienyl]- ethenyl]phenyl]-	DMF	396(4.77)	33-2904-78
$C_{28}H_{20}N_2O_3$			
2(1H)-Pyridinone, 1-amino-6-(1,3-benzo- dioxol-5-yl)-3-(1-naphthalenyl)-4- phenyl-	EtOH	250s(4.36),344(4.29)	4-0759-78
$C_{28}H_{20}N_4$			
5,7b,10,10c-Tetraazacyclopent[hi]acean- thrylene, 4,9-dimethyl-1,6-diphenyl-	CH_2Cl_2	268(4.80),288s(4.55), 410s(3.47),438s(3.87), 460(4.09),486(4.17)	44-3544-78
$C_{28}H_{20}N_4O$			
[1,2,4]Triazolo[1,5-a]pyridine, 2-[4- [2-[4-(5-methyl-2-benzoxazolyl)phen- yl]ethenyl]phenyl]-	DMF	367(4.92),386(4.74)	33-0142-78
[1,2,4]Triazolo[1,5-a]pyridine, 2-[4- [2-[4-(6-methyl-2-benzoxazolyl)phen- yl]ethenyl]phenyl]-	DMF	368(4.92),386(4.73)	33-0142-78
[1,2,4]Triazolo[1,5-a]pyridine, 2-[4- [2-[4-(7-methyl-2-benzoxazolyl)phen- yl]ethenyl]phenyl]-	DMF	365(4.91),385(4.72)	33-0142-78
$C_{28}H_{20}N_4O_2$			
[1,2,4]Triazolo[1,5-a]pyridine, 2-[4- [2-[4-(5-methoxy-2-benzoxazolyl)- phenyl]ethenyl]phenyl]-	DMF	370(4.91)	33-0142-78
[1,2,4]Triazolo[1,5-a]pyridine, 2-[4- [2-[3-(4-methoxyphenyl)-1,2-benzis- oxazol-6-yl]ethenyl]phenyl]-	DMF	344(4.81)	33-2904-78
$C_{28}H_{20}N_6$			
[1,2,4]Triazolo[1,5-a]pyridine, 2- [1,1'-biphenyl]-4-yl-7-[2-(2-phenyl- 2H-1,2,3-triazol-4-yl)ethenyl]-	DMF	332(4.78)	33-0142-78
[1,2,4]Triazolo[1,5-a]pyridine, 2-phen- yl-7-[2-[4-(2-phenyl-2H-1,2,3-tria- zol-4-yl)phenyl]ethenyl]-	DMF	352(4.83)	33-0142-78
[1,2,4]Triazolo[1,5-a]pyridine, 2-phen- yl-7-[2-[4-(4-phenyl-2H-1,2,3-tria- zol-2-yl)phenyl]ethenyl]-	DMF	354(4.84)	33-0142-78
$C_{28}H_{20}N_{10}Ni$			
Nickel, [7,16-dihydro-7,16-bis(phenyl- azo)dipyrido[2,3-b:2',3'-i][1,4,8,11]- tetraazacyclotetradecinato(2-)-N^5,N^9- N^{14},N^{18}]-, (SP-4-1)-	DMF	410(4.80),456s(4.62)	64-1012-78B
$C_{28}H_{20}O$			
Spiro[anthracene-9(10H),1'-cyclopropan]-	C_6H_{12}	269(4.26),323(3.77)	22-0442-78

Compound	Solvent	$\lambda_{max}(\log \epsilon)$	Ref.
10-one, 2',2'-diphenyl- (cont.)			22-0442-78
$C_{28}H_{20}O_2$			
9(10H)-Anthracenone, 10-(2,2-diphenyl-ethenyl)-10-hydroxy-	MeCN	259(4.45)	22-0442-78
2H-Pyran-2-one, 6-(4-methylphenyl)-3-(1-naphthalenyl)-4-phenyl-	EtOH	264(4.44),357(4.33)	4-0759-78
$C_{28}H_{20}O_3$			
9(10H)-Anthracenone, 10-(2,2-diphenyl-ethenyl)-10-hydroperoxy-	MeCN	260(4.43)	22-0442-78
2H-Pyran-2-one, 6-(4-methoxyphenyl)-3-(1-naphthalenyl)-4-phenyl-	EtOH	271(4.39),372(4.35)	4-0759-78
$C_{28}H_{20}O_4$			
6,18[1',2']-Benzeno-6H-dibenzo[2,3:6,7]-[1,4]dioxocino[5,6-c][1,6]benzodioxo-cin, 17a,18-dihydro-, (6α,6aS*,17aβ-18α)-	ether	274(3.79),281(3.81), 305(3.78),314(3.77), 325s(3.58)	78-0083-78
6,15[1',2']:7,14[1",2"]-Dibenzenodiben-zo[b,h][1,4,7,10]tetraoxacyclododecin, 6,7,14,15-tetrahydro-, (6α,7β,14β,15α)-	THF	265s(3.51),270(3.61), 277(3.61)	78-0083-78
$C_{28}H_{21}ClN_3O$			
Benzo[f]quinolinium, 2-acetyl-4-(4-chlorophenyl)-1-[(phenylhydrazono)-methyl]-, perchlorate	MeOH	500(3.94)	65-1635-78
$C_{28}H_{21}N$			
Benzo[f]quinoline, 1-(1-methyl-2-phen-ylethenyl)-3-phenyl-	EtOH	250(4.70),282(4.68), 345(3.83),362(3.72)	103-0418-78
Benzo[f]quinoline, 2-methyl-3-phenyl-1-(2-phenylethenyl)-	EtOH	263(4.59),294(4.37), 325(3.92),341(3.89), 357(3.85)	103-0418-78
$C_{28}H_{21}NO$			
Benzo[f]quinoline, 3-(4-methoxyphenyl)-1-(2-phenylethenyl)-	EtOH	255(4.46),299(4.66), 360(4.00)	103-0418-78
1,2-Benzisoxazole, 3-[4-(2-[1,1'-bi-phenyl]-4-ylethenyl)phenyl]-5-methyl-	DMF	343(4.76)	33-2904-78
$C_{28}H_{21}NO_2$			
1,2-Benzisoxazole, 6-(2-[1,1'-biphen-yl]-4-ylethenyl)-3-(4-methoxyphenyl)-	DMF	342(4.74)	33-2904-78
1,2-Benzisoxazole, 3-[1,1]-biphenyl]-4-yl-6-[2-(4-methoxyphenyl)ethenyl]-	DMF	342(4.62)	33-2904-78
Phenol, 2-[2-[3-(4-methoxyphenyl)benzo-[f]quinolin-1-yl]ethenyl]-	EtOH	262(4.53),289(4.64), 350(4.23),370(4.13)	103-0418-78
1,3-Propanedione, 1,3-diphenyl-2-[phen-yl(phenylamino)methylene]-	hexane MeOH	270(4.24),329(3.9) 329(--)	48-0945-78 48-0945-78
$C_{28}H_{21}N_3$			
[1,2,4]Triazolo[1,5-a]pyridine, 2-phen-yl-7-[2-[4-(2-phenylethenyl)phenyl]-ethenyl]-	DMF	369(4.87)	33-0142-78
$C_{28}H_{21}N_3O$			
3-Pyrazolidinone, 5-imino-1-phenyl-4-(9-phenyl-9H-fluoren-9-yl)-	EtOH	270(4.27),305(3.69)	115-0305-78

Compound	Solvent	$\lambda_{max}(\log \epsilon)$	Ref.
$C_{28}H_{21}N_3O_2S$ 4H-1,3,4,5-Thiatriazine, 4-(1,2-diphen-ylethenyl)-2,6-diphenyl-, 1,1-diox-ide, (E)-	EtOH	230(4.63),275s(--), 330(4.64),375s(--)	88-3781-78
$C_{28}H_{21}N_4O_3$ Benzo[f]quinolinium, 2-acetyl-4-(4-ni-trophenyl)-1-[(phenylhydrazono)meth-yl]-, perchlorate	MeOH	503(4.06)	65-1635-78
$C_{28}H_{21}N_6O_6$ Benzoxazolium, 2-[[4-(2-benzoxazolyl-methylene)-2-(2,4-dinitrophenyl)-2,4-dihydro-5-methyl-3H-pyrazol-3-ylidene]methyl]-3-ethyl-, iodide	n.s.g.	220(4.69),360(4.53)	48-0857-78
$C_{28}H_{21}OS$ Pyrylium, 2-[(2-methyl-4H-1-benzothio-pyran-4-ylidene)methyl]-4,6-diphen-yl-, perchlorate	HOAc-HClO₄	625(4.453)	83-0226-78
$C_{28}H_{21}SSe$ 1-Benzoseleninium, 4-[(4,6-diphenyl-2H-thiopyran-2-ylidene)methyl]-, perchlorate	HOAc-HClO₄	613(4.090)	83-0170-78
$C_{28}H_{21}S_2$ Thiopyrylium, 2-[2-methyl-4H-1-benzo-thiopyran-4-ylidene)methyl]-4,6-di-phenyl-, perchlorate	HOAc-HClO₄	625(4.453)	83-0226-78
$C_{28}H_{22}$ Anthracene, 9,10-bis(phenylmethyl)-	C_6H_{12}	377(4.15)	61-1068-78
Azulene, 4-[2-(9-anthracenyl)ethenyl]-6,8-dimethyl-	benzene	600(2.83)	104-0243-78
Indeno[2,1-a]indene, 5,5a,10,10a-tetra-hydro-5,5a-diphenyl-	EtOH	262(3.26),267(3.39), 273(3.41)	44-4984-78
$C_{28}H_{22}BrN_2O$ Quinolinium, 2-[2-(4-bromophenyl)-3-(1,2-dihydro-2-oxo-3H-indol-3-ylidene)-1-propenyl]-1-ethyl-, iodide	MeOH	448(3.88),470(3.89), 546(3.89),582(3.95)	64-0209-78B
$C_{28}H_{22}ClN_2O$ Quinolinium, 2-[2-(4-chlorophenyl)-3-(1,2-dihydro-2-oxo-3H-indol-3-ylidene)-1-propenyl]-1-ethyl-, iodide	MeOH	450(4.09),468(4.09), 550(4.20),580(4.33)	64-0209-78B
$C_{28}H_{22}ClN_3O_5$ Benzo[f]quinolinium, 2-acetyl-4-phenyl-1-[(phenylhydrazono)methyl]-, per-chlorate	MeOH	507(3.88)	65-1635-78
$C_{28}H_{22}ClN_3O_6$ Benzo[f]quinolinium, 2-acetyl-4-(4-hy-droxyphenyl)-1-[(phenylhydrazono)-methyl]-, perchlorate	MeOH	492(4.19)	65-1635-78
$C_{28}H_{22}Cl_2O_2$ Benzene, 1,1'-[bis(4-chlorophenyl)eth-enylidene]bis[4-methoxy-	CHCl₃	256(4.44),298(4.18), 330(4.22)	23-1246-78

Compound	Solvent	$\lambda_{max}(\log \epsilon)$	Ref.
$C_{28}H_{22}N_2$			
Imidazo[1,2-a]pyridine, 7-(2-[1,1'-biphenyl]-4-ylethenyl)-2-(4-methylphenyl)-	DMF	373(4.74),391(4.69)	33-0129-78
$C_{28}H_{22}N_2O$			
Imidazo[1,2-a]pyridine, 7-(2-[1,1'-biphenyl]-4-ylethenyl)-2-(4-methoxyphenyl)-	DMF	375(4.71),392(4.67)	33-0129-78
Imidazo[1,2-a]pyridine, 2-[1,1'-biphenyl]-4-yl-7-[2-(4-methoxyphenyl)ethenyl]-	DMF	368(4.73),387(4.69)	33-0129-78
Imidazo[1,2-a]pyridine, 7-[2-(4-methoxyphenyl)ethenyl]-2,3-diphenyl-	DMF	368(4.60),386(4.56)	33-0129-78
3-Pyridinecarbonitrile, 1,2-dihydro-1-methyl-6-(4-methylphenyl)-2-oxo-5-phenyl-4-(1-phenylethenyl)-	MeCN	244(4.25),361(4.08)	5-1222-78
3-Pyridinecarbonitrile, 1,2-dihydro-1-methyl-2-oxo-5,6-diphenyl-4-(1-phenyl-1-propenyl)-, (E)-	MeCN	240(4.31),359(4.12)	5-1222-78
2(1H)-Pyridinone, 1-amino-6-(4-methylphenyl)-3-(1-naphthalenyl)-4-phenyl-	EtOH	251s(4.38),340(4.19)	4-0759-78
$C_{28}H_{22}N_2O_2$			
2(1H)-Pyridinone, 1-amino-6-(4-methoxyphenyl)-3-(1-naphthalenyl)-4-phenyl-	EtOH	249s(4.38),343(4.24)	4-0759-78
$C_{28}H_{22}N_2O_4$			
1-Naphthaleneacetic acid, [3-(1,3-benzodioxol-5-yl)-3-oxo-1-phenylpropylidene]hydrazide	EtOH	282(4.35),293s(4.29), 315(4.05)	4-0385-78
$C_{28}H_{22}N_3O$			
Benzo[f]quinolinium, 2-acetyl-4-phenyl-1-[(phenylhydrazono)methyl]-, perchlorate	MeOH	507(3.88)	65-1635-78
$C_{28}H_{22}N_3OS$			
Quinolinium, 4-(2-benzoxazolyl)-1-methyl-2-[3-(3-methyl-2(3H)-benzothiazolylidene)-1-propenyl]-, iodide	EtOH	613(4.78)	4-0017-78
Quinolinium, 6-(2-benzoxazolyl)-1-methyl-2-[3-(3-methyl-2(3H)-benzothiazolylidene)-1-propenyl]-, iodide	EtOH	600(5.16)	4-0017-78
Quinolinium, 6-(2-benzoxazolyl)-1-methyl-4-[3-(3-methyl-2(3H)-benzothiazolylidene)-1-propenyl]-, iodide	EtOH	657(5.14)	4-0017-78
$C_{28}H_{22}N_3O_2$			
Benzo[f]quinolinium, 2-acetyl-4-(4-hydroxyphenyl)-1-[(phenylhydrazono)methyl]-, perchlorate	MeOH	492(4.19)	65-1635-78
$C_{28}H_{22}N_3S_2$			
Quinolinium, 4-(2-benzothiazolyl)-1-methyl-2-[3-(3-methyl-2(3H)-benzothiazolylidene)-1-propenyl]-, iodide	EtOH	609(4.86)	4-0017-78
Quinolinium, 6-(2-benzothiazolyl)-1-methyl-2-[3-(3-methyl-2(3H)-benzothiazolylidene)-1-propenyl]-, iodide	EtOH	601(5.17)	4-0017-78

Compound	Solvent	$\lambda_{max}(\log \epsilon)$	Ref.
Quinolinium, 6-(2-benzothiazolyl)-1-methyl-4-[3-(3-methyl-2(3H)-benzothiazolylidene)-1-propenyl]-, iodide	EtOH	658(5.15)	4-0017-78
$C_{28}H_{22}N_4O$			
2H-Benzimidazol-2-one, 1,3-dihydro-1,3-dimethyl-5-[4-(1-naphthalenyl)-5-phenyl-1H-imidazol-2-yl]-radical	EtOH	222(4.56),320(4.51)	103-0425-78
	toluene	540(--),736(--)	103-0425-78
2H-Benzimidazol-2-one, 1,3-dihydro-1,3-dimethyl-5-[4-(2-naphthalenyl)-5-phenyl-1H-imidazol-2-yl]-radical	EtOH	227(4.99),322(4.77)	103-0425-78
	toluene	532(--),738(--)	103-0425-78
1H-Pyrazolo[3,4-b]pyridin-4-ol, 5-(3,4-dihydro-3-methyl-1-isoquinolinyl)-1,3-diphenyl-	EtOH	275(4.42)	2-0161-78
$C_{28}H_{22}N_6$			
Dibenzo[b,i][1,4,8,11]tetraazacyclotetradecine, 5,14-dihydro-7,16-di-4-pyridinyl-	DMF	326(4.61),385(4.71), 419(4.22),437(4.19)	64-1012-78B
$C_{28}H_{22}N_{10}$			
Dipyrido[2,3-b:2',3'-i][1,4,8,11]tetraazacyclotetradecine, 5,14-dihydro-7,16-bis(phenylazo)-	DMF	393(4.77),418(4.76), 440s(4.70)	64-1012-78B
$C_{28}H_{22}O_2$			
Phenanthrene, 3-methoxy-10-(4-methoxyphenyl)-9-phenyl-	ether	260(4.72),279s(4.40), 302s(3.96),340(2.94)	23-1246-78
$C_{28}H_{22}O_4$			
1-Naphthaleneacetic acid, α-[3-(4-methoxyphenyl)-3-oxo-1-phenylpropylidene]-	EtOH	277(4.37)	4-0759-78
$C_{28}H_{22}O_5$			
Naphtho[2,3-c]furan-5,8-diol, 4,9-dihydro-1,3-diphenyl-, diacetate	CH_2Cl_2	333(4.4),351s(4.1)	150-5538-78
$C_{28}H_{22}O_7$			
1,2-Benzenediol, 4-[3-(3,5-dihydroxyphenyl)-2,3-dihydro-6-hydroxy-4-[2-(4-hydroxyphenyl)ethenyl]-2-benzofuranyl]- (scirpusin A)	EtOH	224s(4.56),290s(4.27), 311s(4.36),323(4.39)	94-3050-78
$C_{28}H_{22}O_8$			
1,2-Benzenediol, 4-[2-[2-(3,4-dihydroxyphenyl)-3-(3,5-dihydroxyphenyl)-2,3-dihydro-6-hydroxy-4-benzofuranyl]ethenyl]- (scirpusin B)	EtOH	225s(4.55),290(4.17), 308s(4.21),331(4.32)	94-3050-78
[10,10'-Bi-2H,8H-benzo[1,2-b:3,4-b']dipyran]-2,2',9,9'(10H,10'H)-tetrone, 8,8,8',8'-tetramethyl-	EtOH	208(4.69),323(4.33)	2-0179-78
$C_{28}H_{23}BrN_3PS$			
Phosphonium, [3-[(aminothioxomethyl)-hydrazono]-3-(4-bromophenyl)-1-propenyl]triphenyl-, hydroxide, inner salt	DMF	268(4.199),320(3.903), 519(3.572)	65-1998-78

Compound	Solvent	$\lambda_{max}(\log \epsilon)$	Ref.
$C_{28}H_{23}Cl_2N_3O_4S_2$ Benzenesulfonamide, 4-chloro-N-[3-[[(4-chlorophenyl)sulfonyl]amino]-1,3-dihydro-1-methyl-3-(phenylmethyl)-2H-indol-2-ylidene]-	EtOH	226(4.54),289(4.09), 305(3.96)	39-1471-78C
$C_{28}H_{23}NO_3$ Benzo[h]quinolin-2(1H)-one, 3-benzoyl-5,6-dihydro-4-(4-methoxyphenyl)-1-methyl-	$CHCl_3$	250(3.94),374(3.74)	48-0097-78
$C_{28}H_{23}NO_3S_2$ 4-Piperidinone, 1-methyl-3,5-bis[[5-(phenylthio)-2-furanyl]methylene]-	EtOH	208(4.43),235(4.33), 265(4.27),410(4.52)	133-0189-78
$C_{28}H_{23}NO_5S_2$ 4-Piperidinone, 1-methyl-3,5-bis[[5-(phenylsulfinyl)-2-furanyl]methylene]-	EtOH	260(4.16),348(4.51)	133-0189-78
$C_{28}H_{23}NO_7S_2$ 4-Piperidinone, 1-methyl-3,5-bis[[5-(phenylsulfonyl)-2-furanyl]methylene]-	EtOH	217(4.37),263(4.34), 299(4.20),370(4.12)	133-0189-78
$C_{28}H_{23}N_2O$ Quinolinium, 2-[3-(1,2-dihydro-2-oxo-3H-indol-3-ylidene)-2-phenyl-1-propenyl]-1-ethyl-, iodide	MeOH	400(3.92),545(4.49), 581(4.66)	64-0209-78B
$C_{28}H_{23}N_2OP$ 3H-Pyrazole, 5-(diphenylphosphinyl)-3-methyl-3,4-diphenyl-	MeOH	302(3.46),365s(2.78)	150-4248-78
$C_{28}H_{23}N_3$ Propanedinitrile, (3,4-dihydro-1,5-dimethyl-3,4,6-triphenyl-2(1H)-pyridinylidene)-, trans	MeCN	254(4.07),328(4.29)	5-1222-78
Propanedinitrile, [1-[2-(methylamino)-1,2-diphenylethenyl]-2-phenyl-2-butenylidene]-, (Z,E)-	MeCN	240(4.36),271(4.12), 413(3.85)	5-1222-78
Propanedinitrile, [3-(methylamino)-3-(4-methylphenyl)-2-phenyl-1-(1-phenylethenyl)-2-propenylidene]-, (Z)-	MeCN	246(4.26),352(3.44), 414(3.67)	5-1222-78
$C_{28}H_{23}N_3OS$ 1,4-Naphthalenedione, mono[3-(1-methylethyl)-4,5-diphenyl-2(3H)-thiazolylidene]hydrazone	isoPrOH	249(4.33),308(4.18), 543(4.56)	73-1227-78
$C_{28}H_{23}N_3O_2$ 1,2,4-Triazolidine-3,5-dione, 4-phenyl-1-[5-phenyl-1-(1-phenylethenyl)hexa-1,3,4-trienyl]-	heptane	222(4.25),248(4.16), 293(3.97)	39-1568-78C
1H-[1,2,4]Triazolo[1,2-a]pyridazine-1,3(2H)-dione, 5,8-dihydro-2-phenyl-5,8-bis(1-phenylethylidene)-, (E,Z)-	C_6H_{12}	217(4.26),286(4.30)	39-1568-78C
$C_{28}H_{23}N_4O_2$ Benzoxazolium, 2-[[4-(2-benzoxazolylmethylene)-2,4-dihydro-5-methyl-2-phenyl-3H-pyrazol-3-ylidene]methyl]-3-ethyl-	n.s.g.	214(4.76),250(4.81)	48-0857-78

Compound	Solvent	$\lambda_{max}(\log \epsilon)$	Ref.
$C_{28}H_{24}BrN_3PS$ Phosphonium, [3-[(aminothioxomethyl)-hydrazono]-3-(4-bromophenyl)-1-propenyl]triphenyl-, chloride	EtOH or DMF	266(4.201),324(3.937)	65-1998-78
$C_{28}H_{24}N_2OS_2$ 2(10H)-Phenazinone, 10-ethyl-3,8-bis-[(4-methylphenyl)thio]-	CHCl₃	318(3.48),461(3.41), 516(3.31)	103-0795-78
$C_{28}H_{24}N_2O_2$ 1H-Cyclopenta[b]quinoxaline, 4,9-dibenzoyl-3a,4,9,9a-tetrahydro-1-(1-methylethylidene)-	CH₂Cl₂	231(4.44),250s(4.40)	5-1139-78
1-Naphthaleneacetic acid, [3-(4-methylphenyl)-3-oxo-1-phenylpropylidene]-hydrazide	EtOH	283(4.42),293s(4.38)	4-0385-78
$C_{28}H_{24}N_2O_3$ 1-Naphthaleneacetic acid, [3-(4-methoxyphenyl)-3-oxo-1-phenylpropylidene]-hydrazide	EtOH	284(4.43),293s(4.40)	4-0385-78 +34-0345-78
$C_{28}H_{24}N_2O_4$ Benzo[f]quinazoline, 1,3-bis(3,4-dimethoxyphenyl)-	EtOH	208(4.66),226(4.58), 307(4.56)	103-0082-78
$C_{28}H_{24}N_3$ Methylium, tris(1-methyl-1H-indol-3-yl)-, perchlorate	MeCN	535(8.02)[sic]	103-0345-78
$C_{28}H_{24}N_3PS$ Phosphonium, [3-[(aminothioxomethyl)hydrazono]-3-phenyl-1-propenyl]triphenyl-, hydroxide, inner salt	DMF	267(4.079),317(3.897), 531(3.417)	65-1998-78
$C_{28}H_{24}N_4O_2$ 1H-Pyrazolo[3,4-b]pyridine-5-carboxamide, 4-hydroxy-N-(1-methyl-2-phenylethyl)-1,3-diphenyl-	EtOH	320(4.12)	2-0161-78
1,2,4,5-Tetrazine, 1,4-dihydro-3,6-bis-(4-methoxyphenyl)-1,4-diphenyl-	EtOH	272(4.43),332(4.11)	70-2227-78
cation radical perchlorate	n.s.g.	227(4.34),278(4.30), 330(4.45),470(3.61), 670(2.28)	70-2227-78
$C_{28}H_{24}N_4O_4S_2$ 1,2,4,5-Tetrazine, 1,4-dihydro-1,4-bis-[(4-methylphenyl)sulfonyl]-3,6-diphenyl-	EtOH	230(4.48),255(4.44)	70-2227-78
$C_{28}H_{24}N_4O_7$ Methylium, [4-(dimethylamino)phenyl]-(3-methylphenyl)phenyl-, picrate	MeCN	463(4.59)	104-1750-78
Methylium, [4-(dimethylamino)phenyl]-(4-methylphenyl)phenyl-, picrate	MeCN	470(4.63)	104-1750-78
$C_{28}H_{24}N_6O_7$ Benzamide, N-[9-[3-(acetylamino)-2,3²-anhydro-5-O-benzoyl-3-C-(carboxymethyl)-3-deoxy-α-D-xylofuranosyl]-9H-purin-6-yl]-	MeOH	202(4.35),227(4.30), 277(4.22)	136-0039-78A

Compound	Solvent	$\lambda_{max}(\log \epsilon)$	Ref.
Benzamide, N-[9-[3-(acetylamino)-2,3 - anhydro-5-O-benzoyl-3-C-(carboxymeth-yl)-3-deoxy-β-D-xylofuranosyl]-9H-purin-6-yl]]-	MeOH	202(4.34),227(4.29), 278(4.20)	136-0039-78A
$C_{28}H_{24}O$			
Benzene, 1-methoxy-2-(2,2,3-triphenyl-cyclopropyl)-	MeOH	274(3.56),281(3.49)	44-0303-78
1-Propene, 3-(2-methoxyphenyl)-1,1,3-triphenyl-	MeOH	256(4.26)	44-0303-78
$C_{28}H_{24}O_4$			
2H,8H-Benzo[1,2-b:3,4-b']dipyran-2-one, 9,10-dihydro-5-hydroxy-4-methyl-8-phenyl-6-(1-phenyl-1-propenyl)-	MeOH	261(4.04),324(3.82)	2-0856-78
2H,8H-Benzo[1,2-b:3,4-b']dipyran-2-one, 9,10-dihydro-5-hydroxy-4-methyl-8-phenyl-6-(3-phenyl-2-propenyl)-	MeOH	254(4.14),321(3.84)	2-0856-78
2H-1-Benzopyran-2-one, 5,7-dihydroxy-4-methyl-6,8-bis(1-phenyl-1-propenyl)-	MeOH	259(3.94),319(3.2)	2-0856-78
2H-1-Benzopyran-2-one, 4-methyl-5,7-bis[(3-phenyl-2-propenyl)oxy]-	MeOH	252(3.62),320(3.14)	2-0856-78
$C_{28}H_{24}O_6$			
5,17:8,14-Diethenocyclododeca[1,2-f-7,8-f']diisobenzofuran-1,3,10,12-tetrone, 3a,4,6,7,9,9a,12a,13,15-16,18,18a-dodecahydro-	MeCN	220s(4.23)	88-0969-78
$C_{28}H_{24}O_8S$			
Pyrylium, 2-[2-(4-hydroxy-3-methoxy-phenyl)ethenyl]-4,6-diphenyl-, carboxymethylsulfonate	EtOH	550(4.291)	83-0256-78
$C_{28}H_{25}NOS_3$			
Spiro[1,3-benzodithiole-2,4'-[4H-1,3]-thiazin]-6'(5'H)-one, 3a,4,5,6,7,7a-hexahydro-2',5',5'-triphenyl-	CH Cl	229(4.24),251(4.26)	18-0301-78
$C_{28}H_{25}NO_4$			
1H-Isoindole-1,3(2H)-dione, 2-[(5,7,8-9,11,12-hexahydro-2-methoxy-7-oxo-6a(6H)-chrysenyl)methyl]-, (S)-	EtOH	223(3.15),312(3.30)	150-5001-78
$C_{28}H_{25}N_3PS$			
Phosphonium, [3-[(aminothioxomethyl)hy-drazono]-3-phenyl-1-propenyl]tri-phenyl-, chloride	EtOH or DMF	268(4.199),332(4.097)	65-1998-78
$C_{28}H_{26}$			
7,14-Ethanodibenz[a,h]anthracene, 7,14-dihydro-2,4,9,11-tetramethyl-, (7R)-	EtOH	238(5.07),293(4.20)	18-0265-78
$C_{28}H_{26}F_3NO_5$			
2H-Dibenzo[2,3:4,5]cyclohepta[1,2-b]-pyrrol-2-one, 5,6,7,7a,8,9-hexahydro-3,11-dimethoxy-12-(phenylmethoxy)-7-(trifluoroacetyl)-	EtOH	235(4.15),258(3.93), 285(3.75),342(3.59)	44-4464-78
12H-Dibenzo[d,f]quinolin-12-one, 1,2,3-3a,4,5-hexahydro-7,8-dimethoxy-11-(phenylmethoxy)-3-(trifluoroacetyl)-	EtOH	235(4.23),258(4.07), 285(3.85),342(3.74)	44-2521-78

Compound	Solvent	$\lambda_{max}(\log \epsilon)$	Ref.
12H-Dibenzo[d,f]quinolin-12-one, 1,2,3-3a,4,5-hexahydro-8,11-dimethoxy-7-(phenylmethoxy)-3-(trifluoroacetyl)-	EtOH	242(4.36),282(3.73)	44-4464-78
$C_{28}H_{26}NO$			
Quinolinium, 1-ethyl-2-[4-(4-methoxy-phenyl)-2-phenyl-1,3-butadienyl]-, iodide	EtOH	500(3.95),550s(4.09), 582(4.30)	146-0341-78
$C_{28}H_{26}N_4O_4$			
[10,10'-Bibenzo[b][1,6]naphthyridine]-1,1',3,3'(2H,2'H,4H,4'H)-tetrone, 5,5',10,10'-tetrahydro-2,2',5,5'-tetramethyl-	pH 4 pH 10	357(4.20) 247(4.19),297(4.29)	150-1325-78 150-1325-78
$C_{28}H_{26}N_5$			
[1,2,3]Triazolo[1,5-a]pyrimidinium, 5-[2-[4-(dimethylamino)phenyl]ethenyl]-7-methyl-2,3-diphenyl-, perchlorate	EtOH	506(4.83)	103-1160-78
$C_{28}H_{26}N_6O_3$			
7H-Benz[de]anthracen-7-one, 3-[(4,6-di-morpholino-1,3,5-triazin-2-yl)amino]-	DMF	460(4.62)	2-0106-78
$C_{28}H_{26}O_2P$			
Phosphonium, (5-acetyl-2-methoxy-4-meth-ylphenyl)triphenyl-, iodide	EtOH	226(4.92),266(4.29), 300s(3.98)	95-0503-78
Phosphonium, (5-acetyl-4-methoxy-2-meth-ylphenyl)triphenyl-, iodide	EtOH	230(4.53),257(3.92), 277s(3.39),303(3.24)	95-0503-78
$C_{28}H_{26}O_4$			
4,7-Ethano-1H-indene-8,9-dione, 1-[bis-(4-methoxyphenyl)methylene]-3a,4,7,7a-tetrahydro-5,6-dimethyl-, (3aα,4α,7α-7aα)-	C_6H_{12} benzene	257(4.25),305(4.28), 463(2.20) 462.0(1.90)	5-0440-78 5-0440-78
(3aα,4β,7β,7aα)-	C_6H_{12}	254(4.35),298(4.33), 459(2.40)	5-0440-78
$C_{28}H_{26}O_{10}$			
[1,1':4',1"-Terphenyl]-2',3,4,4"-tetrol, 3',6'-dimethoxy-, tetraacetate	EtOH	230s(4.66),263(4.62), 303(4.38)	98-0632-78
$C_{28}H_{26}O_{11}$			
12H-Benzo[b]xanthene-6,7,10,11-tetrol, 3,8-dimethoxy-1-methyl-, tetraacetate	EtOH	240(4.67),257(4.64), 312(3.84)	94-0209-78
$C_{28}H_{26}O_{13}$			
9,10-Anthracenedione, 1-hydroxy-8-[(2,3,4,6-tetra-O-acetyl-β-D-gluco-pyranosyl)oxy]-	EtOH	219(4.7),252(4.5), 406(4.0)	73-1803-78
$C_{28}H_{26}O_{14}$			
1,4-Naphthalenedione, 6,6'-ethylidene-bis[5,8-dihydroxy-2,3,7-trimethoxy-	CHCl₃	247(4.59),326(4.21), 483(4.15),509(4.25), 545(4.12)	105-0371-78
$C_{28}H_{26}O_{16}$			
Benzoic acid, 3,4,5-trihydroxy-, 6'-ester with 2-(3,4-dihydroxyphenyl)-3-(β-D-glucopyranosyloxy)-2,3-dihydro-5,7-dihydroxy-4H-1-benzopyran-4-one	EtOH	292(4.37)	138-0259-78

Compound	Solvent	$\lambda_{max}(\log \epsilon)$	Ref.
$C_{28}H_{27}N$			
2-Propanamine, N-(6-acetyl-2,5,6-tri-phenyl-2,4-cyclohexadien-1-ylidene)-	C_6H_{12}	250s(4.00),337(4.20)	89-0465-78
$C_{28}H_{27}NO_4$			
1H-Isoindole-1,3(2H)-dione, 2-(3-meth-oxy-17a-oxo-D-homoestra-1,3,5(10),8-tetraen-18-yl)-	EtOH	220(3.79),280(3.30)	150-5001-78
$C_{28}H_{27}N_2$			
Benzo[f]quinolinium, 2,4-diethyl-1,2-dihydro-1,3-diphenyl-	EtOH	218(4.75),261(4.19), 270(4.33),279(4.37), 325(3.62)	103-1019-78
$C_{28}H_{27}N_3O_3S$			
4-Thiazoleacetic acid, α-(ethoxyimino)-2-[(triphenylmethyl)amino]-, ethyl ester, (E)-	EtOH	231s(4.34),305(3.45)	78-2233-78
$C_{28}H_{28}ClN_2S_2$			
Benzothiazolium, 2-[2-[2-chloro-[3-[(3-ethyl-2(3H)-benzothiazolylidene)-ethylidene]-1-cyclohexen-1-yl]ethen-yl]-3-ethyl-, iodide	MeOH	794(5.42)	104-2046-78
$C_{28}H_{28}N_2O_4S_2$			
Benzo[b]cyclopenta[e][1,4]diazepine, 3a,4,9,10-tetrahydro-6,7,10,11-tetramethyl-4,9-bis(phenylsulfonyl)-	CH_2Cl_2	231(4.37),266s(--), 273s(--)	5-1146-78
1H-Cyclopenta[b]quinoxaline, 3a,4,9,9a-tetrahydro-6,7-dimethyl-1-(1-methyl-ethylidene)-4,9-bis(phenylsulfonyl)-	MeCN	217(4.60)	5-1146-78
$C_{28}H_{28}N_2O_{13}$			
1H-Benz[f]indazole-3-carboxylic acid, 4,9-dihydro-4,9-dioxo-1-(2,3,4,6-tetra-O-acetyl-β-D-glucopyranosyl)-, ethyl ester	EtOH	245(4.68),268(4.02)	111-0155-78
2H-Benz[f]indazole-3-carboxylic acid, 4,9-dihydro-4,9-dioxo-2-(2,3,4,6-tetra-O-acetyl-β-D-glucopyranosyl)-, ethyl ester	EtOH	248(4.70)	111-0155-78
$C_{28}H_{28}N_4O_8$			
5H-Cyclopenta[d]pyridazine-1,4-dicarb-oxylic acid, 5-[1,4-bis(methoxycarb-onyl)-7-(1-methylethyl)-5H-cyclopen-ta[d]pyridazin-5-ylidene]-7-(1-meth-ylethyl)-, dimethyl ester	MeCN	218(4.46),313(4.13), 367(4.20),468(4.07)	78-2509-78
$C_{28}H_{28}N_5O_9P$			
Uridine, 2'-O-[(2-nitrophenyl)methyl]-, 3'-(N,N'-diphenylphosphorodiamidate)	EtOH	231(4.40),262(4.21)	35-4580-78
$C_{28}H_{28}O_8$			
Gibb-3-ene-1,10-dicarboxylic acid, 4a,7-dihydroxy-1-methyl-8-methylene-2-oxo-, 1,4a-lactone, 10-[2-(4-meth-oxyphenyl)-2-oxoethyl] ester	EtOH	225s(4.17),280(4.15)	78-0345-78
Pentacyclo[10.4.2.2^{4,9}.0^{5,8}.0^{13,16}]ei-cosa-4,8,12,16,17,19-hexaene-	EtOH	220(4.56),274s(3.80), 305(3.46),318(3.40)	88-0449-78

Compound	Solvent	$\lambda_{max}(\log \epsilon)$	Ref.
17,18,19,20-tetracarboxylic acid, tetramethyl ester (cont.)			88-0449-78
$C_{28}H_{28}O_{10}$			
Demethyldihydroglycyrin tetraacetate	EtOH	236s(4.12),313(4.25), 325s(4.24)	94-0135-78
Isophysalin, dehydro-	EtOH	248(3.60)	102-1647-78
Physalin B-7-one	EtOH	232(4.12)	102-1641-78
$C_{28}H_{28}Si$			
Silane, trimethyl[4-(triphenylmethyl)-phenyl]-	n.s.g.	216(4.45)	39-0751-78B
$C_{28}H_{29}FN_2O_3S$			
Quinolinium, 2-[2-[4-(dimethylamino)-phenyl]ethenyl]-1-ethyl-6-fluoro-, p-toluenesulfonate	EtOH	536(4.81)	103-0060-78
$C_{28}H_{29}NO_4$			
1H-Isoindole-1,3(2H)-dione, 2-[(17a)-17a-hydroxy-3-methoxy-D-homoestra-1,3,5(10),8-tetraen-18-yl]-	EtOH	228(3.70),274(3.60)	150-5001-78
$C_{28}H_{29}NO_5$			
Benzoic acid, 2-[[[(5,7,8,9,11,12-hexa-hydro-7-hydroxy-2-methoxy-6a(6H)-chrysenyl)methyl]amino]carbonyl]-isomer 20	EtOH	238(3.60),308(3.80), 320(3.40)	150-5001-78
	EtOH	232(3.47),308(3.61), 320(3.56)	150-5001-78
$C_{28}H_{29}N_3$			
3H-1,5-Benzodiazepine, 7,8-dimethyl-2-phenyl-4-(4-phenylpiperidino)-	n.s.g.	265(4.53),350(3.72)	87-0952-78
$C_{28}H_{29}N_3P$			
Phosphorus(1+), (4-diazenyl-N,N-dieth-ylbenzenaminato)triphenyl-, (T-4)-, tetrafluoroborate	n.s.g.	280(4.02),495(4.67)	123-0190-78
$C_{28}H_{29}N_4OP$			
Phosphorin, 4-(1,1-dimethylethyl)-1,1-dihydro-1-methoxy-1-phenyl-2,6-bis-(phenylazo)-	EtOH	271(4.46),306(4.38), 420(3.72),568(4.67)	118-0846-78
$C_{28}H_{29}N_9O_6$			
DL-Aspartic acid, N-[N-[4-[[(2,4-diami-no-6-pteridinyl)methyl]methylamino]-benzoyl]-L-phenylalanyl]-	pH 1	245(4.22),310(4.29)	87-1165-78
$C_{28}H_{30}Br_4N_4O_{10}$			
1-Oxa-2-azaspiro[4,5]deca-2,6,8-triene-3-carboxamide, N,N'-1,4-butanediyl-bis[10-acetoxy-7,9-dibromo-8-methoxy-, [3(5'R,10'S),5α,10β]-	EtOH	242(4.03),285(3.79)	5-0066-78
$C_{28}H_{30}CuN_2O_2$			
Copper, [[6,6'-[1,2-phenylenebis(ni-trilomethylidyne)]bis[2-methyl-5-(1-methylethenyl)-2-cyclohexen-1-onato]]-(2-)-N,N',O,O']-	benzene	375(4.24),420(4.34), 450s(4.21),650s(2.79), 780(2.06)	65-1474-78

Compound	Solvent	$\lambda_{max}(\log \epsilon)$	Ref.
$C_{28}H_{30}N_2O_7S$			
Ibogamine-18-carboxylic acid, 4-hydroxy-3-[[(4-methylphenyl)sulfonyl]-oxy]-19-oxo-, methyl ester, $(2\alpha,3\alpha,5\beta,6\alpha,18\beta)-$	MeOH	223(4.69),278(3.95), 284(3.99),293(3.95)	33-0690-78
$C_{28}H_{30}N_2O_8$			
Propanedioic acid, [9,11-dihydro-12-methoxy-8-(methoxycarbonyl)-9-oxo-indolizino[1,2-b]quinolin-7-yl]-, 1,1-dimethylethyl 1-methylethyl ester	MeOH	222(4.52),252(4.24), 258(4.27),284(3.51), 300(3.61),313(3.72), 360(4.18)	24-3403-78
$C_{28}H_{30}N_6O$			
7H-Benz[de]anthracen-7-one, 3-[[4,6-bis(diethylamino)-1,3,5-triazin-2-yl]amino]-	DMF	430(4.62)	2-0106-78
$C_{28}H_{30}O_4$			
4,7-Ethano-1H-indene-8,9-diol, 1-[bis-(4-methoxyphenyl)methylene]-3a,4,7,7a-tetrahydro-5,6-dimethyl-, cis-cis	MeOH	253(4.16),297(4.17)	5-0440-78
cis-trans	MeOH	255(4.09),300(4.11)	5-0440-78
trans-trans	MeOH	255(4.24),302(4.31)	5-0440-78
$C_{28}H_{30}O_6$			
Morusin trimethyl ether	EtOH	237(4.27),268(4.34), 285s(4.06),330s(3.68)	94-1394-78
	EtOH-AlCl$_3$	235(4.29),268(4.37), 285s(4.11),330s(3.74)	94-1394-78
$C_{28}H_{30}O_8$			
Gibb-3-ene-1,10-dicarboxylic acid, 2,4a,7-trihydroxy-1-methyl-8-methyl-ene-, 1,4a-lactone, 10-[2-(4-methoxy-phenyl)-2-oxoethyl] ester	EtOH	226s(3.98),280(4.09)	78-0345-78
$C_{28}H_{30}O_9$			
Physalin B	EtOH	228(4.00)	102-1641-78
$C_{28}H_{30}O_{10}$			
Anhydrophysalin E	EtOH	228(3.81)	102-1641-78
Isoanhydrophysalin E	EtOH	232(3.95)	102-1641-78
Isophysalin F	EtOH	250(3.91)	102-1647-78
Physalin F	EtOH	218(4.00)	102-1647-78
Physalin G	EtOH	218(4.03)	102-1647-78
Physalin H	EtOH	230(3.81)	102-1641-78
Physalin J	EtOH	220(3.96)	102-1647-78
$C_{28}H_{30}O_{11}$			
Physalin E-7-one	EtOH	228(3.86)	102-1641-78
$C_{28}H_{31}NO_3S_2$			
4-Piperidinone, 1-(phenylmethyl)-3,5-bis[[5-(propylthio)-2-furanyl]meth-ylene]-	EtOH	209(4.32),263(4.12), 418(4.51)	133-0189-78
$C_{28}H_{31}NO_4$			
1-Phenanthreneethanamine, 3,4,6,7-tetramethoxy-N-methyl-N-(phenyl-methyl)-	MeOH	263(4.73),280s(4.27), 307(3.98),320(3.98), 344(3.25),362(3.13)	12-0313-78

Compound	Solvent	$\lambda_{max}(\log \epsilon)$	Ref.
$C_{28}H_{31}NO_5$ Benzoic acid, 2-[[[(17aβ)-17a-hydroxy-3-methoxy-D-homoestra-1,3,5(10),8-tetraen-18-yl]amino]carbonyl]-	EtOH	220(3.49),274(3.60)	150-5001-78
$C_{28}H_{31}NO_{11}$ 1-Naphthacenecarboxylic acid, 4-[(3-amino-2,3,6-trideoxy-α-L-lyxo-hexopyranosyl)oxy]-2-ethyl-1,2,3,4,6,11-hexahydro-2,5,7,12-tetrahydroxy-6,11-dioxo-, methyl ester, hydrochloride, [1R-(1α,2β,4β)]-	MeOH	235(4.53),254(4.32), 296(3.74),492(4.18), 510(4.06),527(4.00)	87-0280-78
$C_{28}H_{32}NO_4$ 4H-Dibenzo[de,g]quinolinium, 5,6,6a,7-tetrahydro-1,2,9,10-tetramethoxy-6-methyl-6-(phenylmethyl)-, bromide	MeOH	218(4.66),283(4.17), 302(4.19)	12-0313-78
$C_{28}H_{32}N_2O_6S$ Ibogamine-18-carboxylic acid, 3-hydroxy-4-[[(4-methylphenyl)sulfonyl]oxy]-, methyl ester, (2α,3α,5β,6α-18β)-	MeOH	223(4.74),277(3.96), 284(4.01),292(3.98)	33-0690-78
Ibogamine-18-carboxylic acid, 4-hydroxy-3-[[(4-methylphenyl)sulfonyl]oxy]-, methyl ester, (2α,3α,5β,6α-18β)-	MeOH	223(4.68),277(3.94), 285(3.99),292(3.97)	33-0690-78
$C_{28}H_{32}N_4O$ 3H-Indazole-4,7-dione, 6,7a-bis(1,1-dimethylethyl)-3a,7a-dihydro-, 4-[(diphenylmethylene)hydrazone]	n.s.g.	325(4.08)	70-1179-78
$C_{28}H_{32}N_4O_8S_2$ Benzenesulfonic acid, 4-methyl-, [[5-[1-methoxy-2-[[(4-methylphenyl)sulfonyl]hydrazono]ethoxy]-2-phenyl-1,3-dioxan-6-yl]methylene]hydrazide, [2R-[2α,4α,5β(S*)]]-	EtOH	231(4.44)	136-0101-78E
$C_{28}H_{32}N_{10}Ni$ Nickel, [1,10,11,20-tetrahydro-3,8,13-18-tetramethyl-1,11-bis(1-methylethyl)dibenzo[c,j]dipyrazolo[3,4-f:3',4'-m][1,2,5,8,9,12]hexaazacyclotetradecinato(2-)-N^4,N^{10},N^{14},N^{20}]-, (SP-4-1)-	MeOH	385(3.90),400s(4.00), 460(4.25),615(3.45)	103-0885-78
Nickel, [1,10,11,20-tetrahydro-3,8,13-18-tetramethyl-1,11-dipropyldibenzo[c,j]dipyrazolo[3,4-f:3',4'-m]-[1,2,5,8,9,12]hexaazacyclotetradecinato(2-)-N^4,N^{10},N^{14},N^{20}]-, (SP-4-1)-	hexane	256(3.53),280(3.56), 292s(3.52),390(4.11), 410(4.16),455(4.40), 628(3.62)	103-0885-78
	CHCl$_3$	266(3.59),294(3.54), 370(4.11),425s(4.08), 460(4.36),610(3.64)	103-0885-78
$C_{28}H_{32}O_2$ 1-Pyrenedodecanoic acid	EtOH	234(4.55),243(4.81), 255(3.98),265(4.33), 276(4.68),313(4.01), 326(4.40),343(4.58)	128-0163-78

Compound	Solvent	$\lambda_{max}(\log \epsilon)$	Ref.
$C_{28}H_{32}O_6$			
Parviflorin B	ether	258(4.25),288(3.89), 297s(3.87),423(4.11)	33-0709-78
$C_{28}H_{32}O_{11}$			
Physalin D	EtOH	227(3.81)	102-1647-78
Physalin E	EtOH	229(3.83)	102-1641-78
$C_{28}H_{33}Cl_4N_2O_{13}P$			
3'-Uridylic acid, 2'-O-(tetrahydro-4-methoxy-2H-pyran-4-yl)-, 2-chloro-phenyl 2,2,2-trichloroethyl ester, 5'-(4-oxopentanoate)	EtOH	259(4.02)	54-0073-78
$C_{28}H_{34}N_2$			
Benzenamine, N,N-dibutyl-4-(9,10-dihy-dro-10-methyl-9-acridinyl)-	EtOH	276(4.54)	104-0129-78
$C_{28}H_{34}N_4O_4$			
Aphelandrin	EtOH	229(4.28),279s(3.52), 284(3.56),291s(3.45)	133-0355-78
$C_{28}H_{34}N_4O_9$			
1H-Pyrrole-2-carboxylic acid, 4-[[4-(cyanomethyl)-2-(ethoxycarbonyl)-2,3-dihydro-1-methyl-3-oxo-1H-pyrrol-2-yl]methyl]-2-[[5-(ethoxycarbonyl)-4-hydroxy-1-methyl-1H-pyrrol-3-yl]-methyl]-2,3-dihydro-1-methyl-3-oxo-, ethyl ester	EtOH	266(4.27),332(4.17)	94-3521-78
$C_{28}H_{34}O_3$			
Phenol, 2,2',2"-methylidynetris[4-(1-methylethyl)-	EtOH	225(3.87),282(3.39)	32-0079-78
$C_{28}H_{34}O_4$			
1,3-Benzenediol, 2-(3,7-dimethyl-2,8-octadienyl)-5-(2-phenylethyl)-, di-acetate	EtOH	206(3.38),217s(3.22)	102-2005-78
$C_{28}H_{34}O_9$			
2-Butenoic acid, 2-methyl-, 5,6,7,8-tetrahydro-6-hydroxy-1,2,3,13-tetra-methoxy-6,7-dimethylbenzo[3,4]cyclo-octa[1,2-f][1,3]benzodioxol-5-yl ester (tigloylgomisin P)	EtOH	217(4.68),255s(3.99), 282(3.51)	94-3257-78
$C_{28}H_{34}O_{10}$			
Coleon C, 11,12,14,16-tetra-O-acetyl-	ether	240s(3.80),264(3.83), 323(3.91)	33-0871-78
$C_{28}H_{34}O_{12}$			
Chaparrinone, 6α-hydroxy-, tetraacetate	MeOH	240(4.05)	100-0578-78
$C_{28}H_{34}O_{15}$			
Hesperidin	MeOH	285(4.32)	98-0278-78
$C_{28}H_{35}NO_6$			
2-Propenoic acid, 3-(3,4-dimethoxyphen-yl)-, 4-(3,4-dimethoxyphenyl)octahy-dro-2H-quinolizin-2-yl ester	MeOH	233(4.7),288(4.1), 324(4.2)	94-2515-78

Compound	Solvent	$\lambda_{max}(\log \epsilon)$	Ref.
Subcosine II	MeOH	233(4.3),287(4.1), 323(4.2)	94-2515-78
$C_{28}H_{35}N_3O_8$			
L-Glutamic acid, N-[N-methyl-N-[4-[methyl[(phenylmethoxy)carbonyl]amino]benzoyl]glycyl]-, diethyl ester	MeOH	250(4.01)	87-1165-78
L-Glutamic acid, N-[N-[4-[methyl(phenylmethoxy)carbonyl]amino]benzoyl]-DL-alanyl]-, diethyl ester	MeOH	260(4.12)	87-1165-78
$C_{28}H_{36}$			
Cyclobuta[1",2":3,4;3",4":3',4']dicyclobuta[1,2:1',2']dibenzene, 4b,4c-8b,8c-tetrahydro-1,2,3,4,4b,4c,5,6,7-8,8b,8c-dodecamethyl-	$CHCl_3$	268s(3.10),277(3.35), 284s(3.37),287(3.56)	5-1675-78
Dibenzo[a,e]cyclooctene, 1,2,3,4,5,6,7-8,9,10,11,12-dodecamethyl-	$CHCl_3$	280(2.9)	5-1675-78
$C_{28}H_{36}N_4O_4$			
Aphelandrine, (+)-	EtOH and EtOH-HCl	228(4.23),277s(3.46), 283(3.51),291s(3.38)	33-2646-78
	EtOH-NaOH	250(4.30),290(3.76)	33-2646-78
16,17-Secoaphelandrine, 16,17-dihydro-	EtOH	225(4.18),277(3.49), 283s(3.46)	33-2646-78
$C_{28}H_{36}N_4O_8S_2$			
Benzenesulfonic acid, 4-methyl-, 2-[[5-[1-methoxy-2-[2-[(4-methylphenyl)sulfonyl]hydrazino]ethoxy]-2-phenyl-1,3-dioxan-6-yl]methyl]hydrazide	EtOH	227(4.31)	136-0101-78E
$C_{28}H_{36}O_2$			
Tetradecanoic acid, 9-anthracenyl ester	EtOH	253(4.99)	87-0026-78
$C_{28}H_{36}O_3$			
10'-Apo-β,ψ-carotenoic acid, 15,15'-didehydro-5,6-epoxy-5,6-dihydro-, methyl ester	pet ether	402(4.87),425(4.79)	39-1511-78C
10'-Apo-β,ψ-carotenoic acid, 15,15'-didehydro-5,8-epoxy-5,8-dihydro-, methyl ester	pet ether	383(4.82),404(4.78)	39-1511-78C
$C_{28}H_{36}O_4$			
9(10H)-Anthracenone, 1,8-dihydroxy-10-(1-oxotetradecyl)-	EtOH	223(4.24),263(4.00), 283(3.94),360(3.94)	87-0026-78
15,15'-Didehydroazafrinone methyl ester	pet ether	405(4.93),430(4.90)	39-1511-78C
$C_{28}H_{36}O_8$			
2-Butenoic acid, 2-methyl-, 5,6,7,8-tetrahydro-7-hydroxy-2,3,10,11,12-pentamethoxy-6,7-dimethyldibenzo[a,c]cycloocten-1-yl ester	EtOH	215(4.72),248s(4.18), 286s(3.14)	94-0328-78
geometric isomer (tigloylgomisin H)	EtOH	215(4.75),248s(4.21), 285s(3.47)	94-0328-78
Confluentic acid	n.s.g.	224(4.34),269(4.28), 305(3.94)	12-1041-78
2,5-Cyclohexadien-1-one, 2-[[3,4-dihydro-5,7-dihydroxy-6-(2-methyl-1-oxopropyl)-2H-1-benzopyran-8-yl]methyl]-3,5-dihydroxy-4,4-dimethyl-6-(2-	C_6H_{12}	231(4.44),283(4.36)	39-1303-78C

Compound	Solvent	$\lambda_{max}(\log \epsilon)$	Ref.
methyl-1-oxopropyl)- (cont.)			39-1303-78C
2,5-Cyclohexadien-1-one, 2-[[3,4-dihydro-5,7-dihydroxy-2,2-dimethyl-8-(2-methyl-1-oxopropyl)-2H-1-benzopyran-6-yl]methyl]-3,5-dihydroxy-4,4-dimethyl-6-(2-methyl-1-oxopropyl)-(dihydrouliginosin B)	C_6H_{12}	232(4.46),298(4.40)	39-1303-78C
2,5-Cyclohexadien-1-one, 3,5-dihydroxy-4,4-dimethyl-2-(2-methyl-1-oxopropyl)-6-[[2,4,6-trihydroxy-3-(3-methyl-2-butenyl)-5-(2-methyl-1-oxopropyl)-phenyl]methyl]-	C_6H_{12}	229(4.53),294(4.42)	39-1303-78C
$C_{28}H_{36}O_{10}$			
Coleon R, 7a-O-acetyl-	MeOH	233(3.91)	33-0871-78
$C_{28}H_{36}O_{13}$			
Acanthoside B	EtOH	270(3.87)	100-0056-78
	EtOH-base	260(--)	100-0056-78
$C_{28}H_{36}O_{14}$			
Magnolenin C	EtOH	310(4.15)	100-0056-78
	EtOH-base	360(--)	100-0056-78
$C_{28}H_{37}ClN_2O_8$			
Maytansinol	EtOH	232(4.51),244s(4.49),252(4.50),281(3.76),288(3.76)	87-0031-78
$C_{28}H_{38}CuN_2O_2$			
Copper, [[6,6'-[1,6-hexanediyl]bis(nitrilomethylidyne)]bis[2-methyl-5-(1-methylethenyl)-2-cyclohexen-1-onato]](2-)-N,N',O,O']-	benzene	310(4.07),380(4.04),650s(2.20)	65-1474-78
$C_{28}H_{38}O_4$			
Asafrinone methyl ester	n.s.g.	402(4.76),425(4.94),452(4.94)	39-1511-78C
$C_{28}H_{38}O_4S$			
2,6,8-Nonatrien-1-ol, 3,7-dimethyl-5-(phenylsulfonyl)-9-(2,6,6-trimethylcyclohexen-1-yl)-, acetate	isoPrOH	215(4.31),248s(4.11),267s(4.17),273(4.18)	44-4769-78
2,6,8-Nonatrien-1-ol, 3,7-dimethyl-5-(phenylsulfonyl)-9-(2,6,6-trimethyl-2-cyclohexen-1-yl)-, acetate	isoPrOH	218(4.36),247(4.48)	44-4769-78
$C_{28}H_{38}O_5$			
3H-Xanthen-3-one, 9-pentadecyl-2,6,7-trihydroxy-	pH 8-9	506(3.08)	140-0699-78
$C_{28}H_{38}O_7$			
Isospongiaquinone leucotriacetate	MeOH	281(3.39)	12-2685-78
Withastramonolide	EtOH	225(4.17)	105-0073-78
$C_{28}H_{38}O_{14}$			
Dihydromagnolenin C	EtOH	270(3.65)	100-0056-78
$C_{28}H_{38}Si_2$			
Silane, 5,7,9,11-hexadecatetraene-1,3,13,15-tetrayne-1,16-diylbis[tri-	hexane	350(4.51),365(4.82),385(5.05),409(5.09)	78-1037-78

Compound	Solvent	$\lambda_{max}(\log \epsilon)$	Ref.
ethyl-, (all-E)- (cont.)			78-1037-78
(E,E,Z,Z)-	hexane	342(4.47),358(4.79), 378(5.00),402(4.97)	78-1037-78
$C_{28}H_{39}ClO_8$			
Physalolactone	MeOH	217(4.07)	42-1175-78
$C_{28}H_{40}O_4$			
Azafrin methyl ester	EtOH	245(4.20),296s(4.08), 309(4.13),424(4.91), 435s(4.90)	33-0822-78
all-trans, (±)-	benzene	418s(--),429(3.91), 453(4.84)	39-1511-78
$C_{28}H_{40}O_4S$			
2,6,10,14-Hexadecatetraen-1-ol, 3,7,11-15-tetramethyl-5-(phenylsulfonyl)-, acetate	EtOH	220(4.20),258(2.94), 265(3.03),272(2.96)	44-4769-78
$C_{28}H_{40}O_7S_2$			
D-Homo-17-oxaandrost-5-en-16-one, 3,15,17a-triacetoxy-, cyclic 16-(1,3-propanediyl mercaptole), (3β,15β)-	n.s.g.	224(--),250(2.88)	33-3087-78
$C_{28}H_{40}O_{10}$			
Strophanthidin β-D-riboside	EtOH	217(4.18)	105-0567-78
$C_{28}H_{40}P_2Pt$			
Platinum, bis(phenylethynyl)bis(triethylphosphine)-, cis	ether	253(4.57),268(4.53), 309(4.44)	101-0101-78B
trans	ether	264(4.23),288(4.13), 328(4.22)	101-0101-78B
$C_{28}H_{41}NO$			
2,5-Cyclohexadien-1-one, 4-[[2,5-bis-(1,1-dimethylethyl)phenyl]imino]-2,5-bis(1,1-dimethylethyl)-	C_6H_{12}	278(4.04),490(3.36)	23-2665-78
$C_{28}H_{42}N_4O_7$			
Glycine, N-[α,β-didehydro-N-[N-[N-[(1,1-dimethylethoxy)carbonyl]-N-methyl-L-alanyl]-L-leucyl]-N-methylphenylalanyl]-, methyl ester, (Z)-	CHCl₃	277(4.26)	44-0296-78
$C_{28}H_{42}O$			
Ergosta-1,4,7-trien-3-one	EtOH	242(4.18)	39-0074-78C
Ergosta-4,7,22-trien-3-one	n.s.g.	242(4.30)	105-0070-78
Isocalciferol₂, 9,11-didehydro-	EtOH	301(4.59),314(4.71), 329(4.61)	39-1014-78C
$C_{28}H_{42}O_2$			
Heptadecanoic acid, 1-naphthalenylmethyl ester	hexane	298.5(3.83)	95-0774-78
1,4-Naphthalenedione, 2,3-dimethyl-6-(4,8,12-trimethyltridecyl)-	hexane	248(4.51),255(4.54), 262(4.43),272(4.43), 330(3.60)	88-2645-78
$C_{28}H_{42}O_4$			
Pregna-5,17(20)-dien-21-oic acid, 3-[(tetrahydro-2H-pyran-2-yl)oxy]-, ethyl ester, (3β)-	EtOH	223(4.26)	39-1282-78C

Compound	Solvent	$\lambda_{max}(\log \epsilon)$	Ref.
Spirost-4-en-3-one, 2-methoxy-, (2α,25R)-	EtOH	243(4.07)	105-0630-78
$C_{28}H_{42}O_5$ Chol-20(22)-en-24-oic acid, 3-acetoxy-21-oxo-, ethyl ester, (3β,5α,20Z)- isomeric mixture	EtOH EtOH	233(4.05),300(3.4) 233(3.95),300(3.25)	23-0424-78 23-0424-78
$C_{28}H_{42}O_6S$ Card-20(22)-enolide, 3-[(cyclopentyl-sulfonyl)oxy]-14-hydroxy-, (3β,5β)-	MeOH	217(4.21)	106-0706-78
$C_{28}H_{42}O_8$ Digitoxigenin β-D-riboside	EtOH	218(4.19)	105-0567-78
$C_{28}H_{44}$ 3-Deoxyvitamin D_2	EtOH-hexane	263.5(4.25)	39-0590-78C
$C_{28}H_{44}O$ 6α,7aα-Cyclo-B-homocholest-4-en-3-one 6β,7aβ-Cyclo-B-homocholest-4-en-3-one Ergosta-1,5,7-trien-3β-ol Ergosta-6,8(14),22-trien-3-ol Ergosta-8,14,22-trien-3-ol	EtOH EtOH EtOH MeOH MeOH	268(4.20) 258(4.02) 262(3.86),270(3.98), 280(4.01),290(3.73) 252(4.37) 252(4.26)	73-1134-78 73-1134-78 39-0074-78C 94-3582-78 94-3582-78
$C_{28}H_{44}O_2$ Cholesta-1,4-dien-3-one, 4-methoxy- 1,5-Cyclocholest-3-en-2-one, 1-methoxy-, (1α,5β,10α)-	C_6H_{12} C_6H_{12}	234(4.01),270(3.82), 352s(2.01) 242(3.65),285(3.13)	78-1509-78 78-1509-78
$C_{28}H_{44}O_3$ 9,10-Secoergosta-5(10),8,22-trien-3β-ol, 6,19-epidioxy-	EtOH	210(4.21)	88-4895-78
$C_{28}H_{44}O_4$ 4H-1-Benzopyran-4-one, 2-heptadecyl-5,7-dimethoxy- 6,7-Secocholest-4-en-6-oic acid, 3,7-dioxo-, methyl ester	EtOH EtOH	230(4.57),245(4.53), 253(4.52),284(4.16) 238(4.05)	94-2407-78 12-0171-78
$C_{28}H_{44}O_5$ Propanedioic acid, [(3β,5α)-3-hydroxy-pregnan-21-ylidene]-, diethyl ester	EtOH	215(3.8)	23-0424-78
$C_{28}H_{44}O_6S$ Card-20(22)-enolide, 14-hydroxy-3-[(pentylsulfonyl)oxy]-	MeOH	217(4.21)	106-0706-78
$C_{28}H_{44}O_7$ 24(28)-Dehydromakisterone A	EtOH	245(4.15)	105-0393-78
$C_{28}H_{46}N_2$ 1'H-Cholest-5-eno[4,5,6-cd]pyrazole, 1'-methyl-	EtOH	231s(3.90)	4-0023-78
$C_{28}H_{46}O_2$ Cholest-2-en-4-one, 3-methoxy- Ergosta-5,7-diene-1α,3β-diol	EtOH EtOH	262(3.77),320s(2.79) 263(3.90),271(4.04), 282(4.07),293(3.84)	78-1509-78 39-0074-78C

Compound	Solvent	$\lambda_{max}(\log \epsilon)$	Ref.
$C_{28}H_{46}O_3$ 5,6-Secocholest-3-en-6-oic acid, 5-oxo-, methyl ester	EtOH	230(4.00)	12-0171-78
$C_{28}H_{52}N_2O_4$ 1,18-Octadecanediaminium, N,N'-bis(1-formyl-2-hydroxyethenyl)-N,N,N',N'-tetramethyl-, dihydroxide, bis(inner salt)	EtOH	257(4.75)	73-1261-78
$C_{28}H_{54}P_2Pd$ Palladium, (1,3-butadiyne-1,4-diyl-C)-bis(tributylphosphine)-, homopolymer	CH_2Cl_2	287(4.22),314(4.19), 342(4.33)	101-0319-78Q
$C_{28}H_{55}N_5OP$ Phosphorus(1+), butoxybis(N-butyl-1-butanaminato)(4-diazenyl-N,N-dimethyl-benzenaminato)-, (T-4)-, tetrafluoroborate	n.s.g.	280(3.89),462(4.61)	123-0190-78
$C_{28}H_{56}P_2Pt$ Platinum, diethynylbis(tributylphosphine)-, cis	CH_2Cl_2	250(4.13),283(3.79)	101-0101-78B
trans	CH_2Cl_2	267(4.04),302(3.90)	101-0101-78B

Compound	Solvent	$\lambda_{max}(\log \epsilon)$	Ref.
$C_{29}H_{16}Cl_3N_3O_2$ 1,2-Benzisoxazole, 6-chloro-3-[4-[2-[2-chloro-4-[5-(4-chlorophenyl)-1,3,4-oxadiazol-2-yl]phenyl]ethenyl]phenyl]-	DMF	355(4.79)	33-2904-78
$C_{29}H_{16}N_2O_2$ Naphtho[1',2',3':4,5]quino[2,1-b]quinazoline-5,10-dione, 12-phenyl-	n.s.g.	430(3.93)	2-0103-78
$C_{29}H_{17}Cl_2NO_2$ 1,2-Benzisoxazole, 6-chloro-3-[3-chloro-4-[2-(2-phenyl-6-benzofuranyl)-ethenyl]phenyl]-	DMF	370(4.72)	33-2904-78
$C_{29}H_{17}Cl_2N_3O_2$ 1,2-Benzisoxazole, 3-[4-[2-[2-chloro-4-[5-(4-chlorophenyl)-1,3,4-oxadiazol-2-yl]phenyl]ethenyl]phenyl]-	DMF	354(4.77)	33-2904-78
1,2-Benzisoxazole, 6-chloro-3-[4-[2-[2-chloro-4-(5-phenyl-1,3,4-oxadiazol-2-yl)phenyl]ethenyl]phenyl]-	DMF	354(4.77)	33-2904-78
1,2-Benzisoxazole, 6-chloro-3-[4-[2-[4-[5-(4-chlorophenyl)-1,3,4-oxadiazol-2-yl]phenyl]ethenyl]phenyl]-	DMF	354(4.85)	33-2904-78
1,2-Benzisoxazole, 6-chloro-3-[4-[2-[4-[3-(4-chlorophenyl)-1,2,4-oxadiazol-5-yl]phenyl]ethenyl]phenyl]-	DMF	350(4.82)	33-2904-78
$C_{29}H_{17}Cl_3N_4O$ 1,2-Benzisoxazole, 6-chloro-3-[3-chloro-4-[2-[4-(4-chloro-5-phenyl-2H-1,2,3-triazol-2-yl)phenyl]ethenyl]-phenyl]-	DMF	353(4.81)	33-2904-78
1,2-Benzisoxazole, 6-chloro-3-[3-chloro-4-[2-[4-(5-chloro-2-phenyl-2H-1,2,3-triazol-4-yl)phenyl]ethenyl]-phenyl]-	DMF	348(4.77)	33-2904-78
$C_{29}H_{17}Cl_5O_{10}$ Dibenzo[a,kl]xanthene-3,4,9,10-tetrol, 2,5,8,11,13-pentachloro-1-methoxy-, tetraacetate	CHCl₃	260s(4.32),290s(4.14), 320(3.93),334(3.98), 345s(3.84),366(3.72), 434(3.68)	12-1323-78
$C_{29}H_{17}N_3O_3$ Acetamide, N-(5,10-dihydro-5,10-dioxo-benzo[g]naphtho[1',2',3':4,5]quino-[2,1-b]quinazolin-4-yl)-	n.s.g.	435(3.63)	2-0103-78
Acetamide, N-(5,10-dihydro-5,10-dioxo-benzo[g]naphtho[1',2',3':4,5]quino-[2,1-b]quinazolin-6-yl)-	n.s.g.	435(3.62)	2-0103-78
$C_{29}H_{18}ClNO_2$ 1,2-Benzisoxazole, 6-chloro-3-[4-[2-(2-phenyl-6-benzofuranyl)ethenyl]phenyl]-	DMF	365(4.80)	33-2904-78
$C_{29}H_{18}ClN_3O_2$ 1,2-Benzisoxazole, 3-[4-[2-[2-chloro-4-[5-phenyl-1,3,4-oxadiazol-2-yl)phenyl]-ethenyl]phenyl]-	DMF	352(4.76)	33-2904-78

Compound	Solvent	$\lambda_{max}(\log \epsilon)$	Ref.
1,2-Benzisoxazole, 3-[4-[2-[4-[3-(4-chlorophenyl)-1,2,4-oxadiazol-5-yl]-phenyl]ethenyl]phenyl]-	DMF	349(4.81)	33-2904-78
1,2-Benzisoxazole, 6-chloro-3-[4-[2-[4-(3-phenyl-1,2,4-oxadiazol-5-yl)phen-yl]ethenyl]phenyl]-	DMF	350(4.81)	33-2904-78
1,2-Benzisoxazole, 3-[4-[2-[4-[5-(4-chlorophenyl)-1,3,4-oxadiazol-2-yl]-phenyl]ethenyl]phenyl]-	DMF	354(4.85)	33-2904-78
1,2-Benzisoxazole, 6-chloro-3-[4-[2-[4-(5-phenyl-1,3,4-oxadiazol-2-yl)-phenyl]ethenyl]phenyl]-	DMF	353(4.84)	33-2904-78
$C_{29}H_{18}Cl_2N_4O$			
1,2-Benzisoxazole, 3-[4-[2-[2-chloro-4-(4-chloro-5-phenyl-2H-1,2,3-tria-zol-2-yl)phenyl]ethenyl]phenyl]-	DMF	354(4.77)	33-2904-78
1,2-Benzisoxazole, 6-chloro-3-[3-chlo-ro-4-[2-[4-(2-phenyl-2H-1,2,3-tria-zol-4-yl)phenyl]ethenyl]phenyl]-	DMF	353(4.78)	33-2904-78
1,2-Benzisoxazole, 6-chloro-3-[3-chlo-ro-4-[2-[4-(4-phenyl-2H-1,2,3-tria-zol-2-yl)phenyl]ethenyl]phenyl]-	DMF	354(4.78)	33-2904-78
$C_{29}H_{18}N_2O_2$			
Benzo[g]naphtho[1',2',3':4,5]quino-[2,1-b]quinazoline-5,10-dione, 7,8-dimethyl-	n.s.g.	480(3.73)	2-0103-78
$C_{29}H_{18}O_2$			
Spiro[1H-cyclohepta[c]furan-1,15'-[15H]dibenzo[a,g]cyclotridecen]-3(8aH)-one, 5',6',7',8'-tetradehydro-	EtOH	223(4.81),259(4.52), 273(4.50),321(4.18), 341(4.16)	88-2795-78
$C_{29}H_{19}ClN_2O_2$			
1,2-Benzisoxazole, 6-chloro-3-[4-[2-[4-(5-methyl-2-benzoxazolyl)phenyl]-ethenyl]phenyl]-	DMF	363(4.88)	33-2904-78
1,2-Benzisoxazole, 6-chloro-3-[4-[2-[4-(6-methyl-2-benzoxazolyl)phenyl]-ethenyl]phenyl]-	DMF	363(4.87)	33-2904-78
1,2-Benzisoxazole, 6-chloro-3-[4-[2-[4-(7-methyl-2-benzoxazolyl)phenyl]-ethenyl]phenyl]-	DMF	362(4.86)	33-2904-78
$C_{29}H_{19}ClN_2O_3$			
1,2-Benzisoxazole, 6-chloro-3-[4-[2-[4-(5-methoxy-2-benzoxazolyl)phenyl]-ethenyl]phenyl]-	DMF	365(4.86)	33-2904-78
$C_{29}H_{19}ClN_4O$			
1,2-Benzisoxazole, 3-[4-[2-[2-chloro-4-(4-phenyl-2H-1,2,3-triazol-2-yl)-phenyl]ethenyl]phenyl]-	DMF	354(4.79)	33-2904-78
[1,2,4]Triazolo[1,5-a]pyridine, 2-[4-[2-[4-[3-(4-chlorophenyl)-5-isoxa-zolyl]phenyl]ethenyl]phenyl]-	DMF	356(4.89),375(4.71)	33-0142-78
$C_{29}H_{19}Cl_2NO$			
1,2-Benzisoxazole, 6-chloro-3-[3-chlo-ro-4-[2-[4-(2-phenylethenyl)phenyl]-	DMF	370(4.80)	33-2904-78

Compound	Solvent	$\lambda_{max}(\log \epsilon)$	Ref.
ethenyl]phenyl]- (cont.)			33-2904-78
1,2-Benzisoxazole, 6-[2-(2-chlorophen-yl)ethenyl]-3-[4-[2-(2-chlorophenyl)-ethenyl]phenyl]-	DMF	333(4.82)	33-2904-78
1,2-Benzisoxazole, 6-[2-(3-chlorophen-yl)ethenyl]-3-[4-[2-(3-chlorophenyl)-ethenyl]phenyl]-	DMF	334(4.86)	33-2904-78
1,2-Benzisoxazole, 6-[2-(4-chlorophen-yl)ethenyl]-3-[4-[2-(4-chlorophenyl)-ethenyl]phenyl]-	DMF	337(4.90)	33-2904-78
$C_{29}H_{19}NO_2$			
1,2-Benzisoxazole, 3-[4-[2-(2-phenyl-6-benzofuranyl)ethenyl]phenyl]-	DMF	365(4.79)	33-2904-78
1,2-Benzisoxazole, 3-phenyl-6-[2-(2-phenyl-6-benzofuranyl)ethenyl]-	DMF	364(4.78)	33-2904-78
$C_{29}H_{19}N_3O_2$			
1,2-Benzisoxazole, 3-[4-[2-[4-(3-phen-yl-1,2,4-oxadiazol-5-yl)phenyl]eth-enyl]phenyl]-	DMF	348(4.80)	33-2904-78
1,2-Benzisoxazole, 3-[4-[2-[4-(5-phen-yl-1,3,4-oxadiazol-2-yl)phenyl]eth-enyl]phenyl]-	DMF	352(4.84)	33-2904-78
$C_{29}H_{20}ClNO$			
1,2-Benzisoxazole, 6-chloro-3-[4-[2-[4-(2-phenylethenyl)phenyl]ethenyl]-phenyl]-	DMF	368(4.87)	33-2904-78
$C_{29}H_{20}ClN_5O_2$			
[1,2,4]Triazolo[1,5-a]pyridine, 2-[4-[2-[2-chloro-4-[5-(4-methoxyphenyl)-1,3,4-oxadiazol-2-yl]phenyl]ethenyl]-phenyl]-	DMF	362(4.83)	33-0142-78
[1,2,4]Triazolo[1,5-a]pyridine, 7-[2-[2-chloro-4-[5-(4-methoxyphenyl)-1,3,4-oxadiazol-2-yl]phenyl]ethenyl]-2-phenyl-	DMF	358(4.79)	33-0142-78
$C_{29}H_{20}NOP$			
2H-Phosphorino[4,3-b]indole, 1,2,3-tri-phenyl-, 2-oxide	EtOH	220s(4.53),255(4.26), 382(3.20)	139-0257-78B
$C_{29}H_{20}N_2O_2$			
1,2-Benzisoxazole, 3-[4-[2-[4-(5-meth-yl-2-benzoxazolyl)phenyl]ethenyl]-phenyl]-	DMF	362(4.87)	33-2904-78
1,2-Benzisoxazole, 3-[4-[2-[4-(6-meth-yl-2-benzoxazolyl)phenyl]ethenyl]-phenyl]-	DMF	362(4.86)	33-2904-78
1,2-Benzisoxazole, 3-[4-[2-[4-(7-meth-yl-2-benzoxazolyl)phenyl]ethenyl]-phenyl]-	DMF	360(4.86)	33-2904-78
$C_{29}H_{20}N_2O_3$			
1,2-Benzisoxazole, 6-[2-[4-(1,2-benz-isoxazol-3-yl)phenyl]ethenyl]-3-(4-methoxyphenyl)-	DMF	339(4.75)	33-2904-78
1,2-Benzisoxazole, 3-[4-[2-[4-(5-meth-oxy-2-benzoxazolyl)phenyl]ethenyl]phenyl]-	DMF	365(4.84)	33-2904-78

Compound	Solvent	$\lambda_{max}(\log \epsilon)$	Ref.
$C_{29}H_{20}N_4O$			
1,2-Benzisoxazole, 3-phenyl-6-[2-[4-(2-phenyl-2H-1,2,3-triazol-4-yl)phenyl]-ethenyl]-	DMF	349(4.81)	33-2904-78
1,2-Benzisoxazole, 6-[2-[4-(4-phenyl-2H-1,2,3-triazol-2-yl)phenyl]ethenyl]-	DMF	352(4.81)	33-2904-78
[1,2,4]Triazolo[1,5-a]pyridine, 2-[4-[2-[4-(3-phenyl-5-isoxazolyl)phenyl]-ethenyl]phenyl]-	DMF	356(4.90),374(4.72)	33-0142-78
[1,2,4]Triazolo[1,5-a]pyridine, 2-[4-[2-[4-(5-phenyl-2-oxazolyl)phenyl]-ethenyl]phenyl]-	DMF	372(4.88)	33-0142-78
$C_{29}H_{20}O$			
2,4-Cyclopentadien-1-one, 2,3,4,5-tetraphenyl-	CH_2Cl_2	262(4.46),340(3.88), 510(3.03)	18-2674-78
	MeCN	260(4.40),338(3.79)	18-2674-78
	MeCN-HBF$_4$	259(4.40),337(3.81), 502(3.09)	18-2674-78
2H-Cyclopenta[1]phenanthren-2-one, 1,3-dihydro-1,3-diphenyl-	MeOH	206(4.75),219s(4.54), 248s(4.72),255(4.83), 264s(4.48),276(4.25), 286(4.15),297(4.07), 299(4.11)	54-0197-78
7H-Dibenzo[a,c]cyclononen-7-one, 5,9-diphenyl-	EtOH	229s(4.45),252(4.12), 261s(4.08),272s(4.06), 309(4.26)	150-5151-78
$C_{29}H_{21}BrN$			
Pyridinium, 1-(2-bromophenyl)-2,4,6-triphenyl-, iodide	EtOH	215(4.72),315(4.66)	73-2046-78
Pyridinium, 1-(4-bromophenyl)-2,4,6-triphenyl-, iodide	EtOH	213(4.09),313(4.32)	73-2046-78
perchlorate	EtOH	210(4.31),313(4.33)	73-2046-78
$C_{29}H_{21}BrN_2O_2$			
2(1H)-Quinoxalinone, 3-[1-(4-bromo-benzoyl)-2,2-diphenylethyl]-	EtOH	347(4.28)	103-0336-78
$C_{29}H_{21}ClN_4O$			
[1,2,4]Triazolo[1,5-a]pyridine, 2-[4-[2-[2-chloro-4-(5,6-dimethyl-2-benzoxazolyl)phenyl]ethenyl]phenyl]-	DMF	371(4.84)	33-0142-78
[1,2,4]Triazolo[1,5-a]pyridine, 7-[2-[2-chloro-4-(5,6-dimethyl-2-benzoxazolyl)phenyl]ethenyl]-2-phenyl-	DMF	368(4.80)	33-0142-78
$C_{29}H_{21}ClN_6$			
[1,2,4]Triazolo[1,5-a]pyridine, 2-[3-chloro-4-[2-[4-(2-phenyl-2H-1,2,3-triazol-4-yl)phenyl]ethenyl]phenyl]-6-methyl-	DMF	357(4.83)	33-0142-78
[1,2,4]Triazolo[1,5-a]pyridine, 2-[3-chloro-4-[2-[4-(4-phenyl-2H-1,2,3-triazol-2-yl)phenyl]ethenyl]phenyl]-6-methyl-	DMF	358(4.84)	33-0142-78
$C_{29}H_{21}Cl_2N_3O_2$			
2H-Imidazo[4,5-d]oxazol-2-one, 3,6-bis-(2-chlorophenyl)-3,3a,6,6a-tetrahydro-6a-methyl-3a,5-diphenyl-	EtOH	236(4.04),256s(3.72)	94-3798-78

Compound	Solvent	$\lambda_{max}(\log \epsilon)$	Ref.
2H-Imidazo[4,5-d]oxazol-2-one, 3,6-bis-(4-chlorophenyl)-3,3a,6,6a-tetrahydro-6a-methyl-3a,5-diphenyl-	EtOH	243(4.55),270s(3.97)	94-3798-78
$C_{29}H_{21}NO$			
1,2-Benzisoxazole, 3-[4-[2-[4-(2-phenylethenyl)phenyl]ethenyl]phenyl]-	DMF	367(4.86)	33-2904-78
1,2-Benzisoxazole, 6-(2-phenylethenyl)-3-[4-(2-phenylethenyl)phenyl]-	DMF	336(4.85)	33-2904-78
1,2-Benzisoxazole, 3-phenyl-6-[2-[4-(2-phenylethenyl)phenyl]ethenyl]-	DMF	366(4.84)	33-2904-78
Pyridinium, 1-(4-hydroxyphenyl)-2,4,6-triphenyl-, hydroxide, inner salt	EtOH-Et$_3$N	305(4.45),468(3.29)	44-3678-78
	EtOH-10%TFE-Et$_3$N	305(4.43),455(3.24)	44-3678-78
	EtOH-50%THF	308(4.34)	44-3678-78
	+ Et$_3$N	306(4.44),419(3.27)	44-3678-78
(and other solvent mixtures)	TFE-Et$_3$N	306(4.45),389(3.36)	44-3678-78
$C_{29}H_{21}N_2O_2$			
Pyridinium, 1-(3-nitrophenyl)-2,4,6-triphenyl-, perchlorate	EtOH	210(4.37),315(4.35)	73-2046-78
$C_{29}H_{21}N_2O_4P$			
Phosphonium, triphenyl-, 2,5-bis(methoxycarbonyl)-3,4-dicyano-2,4-cyclopentadien-1-ylide	MeCN	233(4.63),253(4.65), 276(4.15),292(4.11)	18-2605-78
$C_{29}H_{21}N_3O$			
[1,2,4]Triazolo[1,5-a]pyridine, 7-[2-[4-(2-benzofuranyl)phenyl]ethenyl]-2-(4-methylphenyl)-	DMF	368(4.86)	33-0142-78
[1,2,4]Triazolo[1,5-a]pyridine, 2-[4-[2-[4-(5-methyl-2-benzofuranyl)phenyl]ethenyl]phenyl]-	DMF	372(4.90)	33-0142-78
[1,2,4]Triazolo[1,5-a]pyridine, 2-[4-[2-[4-(6-methyl-2-benzofuranyl)phenyl]ethenyl]phenyl]-	DMF	374(4.89)	33-0142-78
[1,2,4]Triazolo[1,5-a]pyridine, 2-(4-methylphenyl)-7-[2-(2-phenyl-6-benzofuranyl)ethenyl]-	DMF	366(4.81)	33-0142-78
$C_{29}H_{21}N_3O_2$			
[1,2,4]Triazolo[1,5-a]pyridine, 2-[4-[2-[4-(5-methoxy-2-benzofuranyl)-phenyl]ethenyl]phenyl]-	DMF	373(4.91)	33-0142-78
$C_{29}H_{21}N_5O$			
[1,2,4]Triazolo[1,5-a]pyridine, 2-[4-[2-[4-[3-(4-methylphenyl)-1,2,4-oxadiazol-5-yl]phenyl]ethenyl]phenyl]-	DMF	357(4.85)	33-0142-78
[1,2,4]Triazolo[1,5-a]pyridine, 2-[4-[2-[4-[5-(3-methylphenyl)-1,3,4-oxadiazol-2-yl]phenyl]ethenyl]phenyl]-	DMF	359(4.89)	33-0142-78
[1,2,4]Triazolo[1,5-a]pyridine, 7-[2-[4-[3-(4-methylphenyl)-1,2,4-oxadiazol-5-yl]phenyl]ethenyl]-2-phenyl-	DMF	348(4.80)	33-0142-78
$C_{29}H_{21}N_5O_2$			
[1,2,4]Triazolo[1,5-a]pyridine, 2-[4-[2-[4-[3-(4-methoxyphenyl)-1,2,4-oxadiazol-5-yl]phenyl]ethenyl]phenyl]-	DMF	355(4.86)	33-0142-78

Compound	Solvent	$\lambda_{max}(\log \epsilon)$	Ref.
[1,2,4]Triazolo[1,5-a]pyridine, 2-[4-[2-[4-[5-(2-methoxyphenyl)-1,3,4-oxadiazol-2-yl]phenyl]ethenyl]phenyl]-	DMF	358(4.87)	33-0142-78
[1,2,4]Triazolo[1,5-a]pyridine, 2-[4-[2-[4-[5-(3-methoxyphenyl)-1,3,4-oxadiazol-2-yl]phenyl]ethenyl]phenyl]-	DMF	358(4.89)	33-0142-78
[1,2,4]Triazolo[1,5-a]pyridine, 2-[4-[2-[4-[5-(4-methoxyphenyl)-1,3,4-oxadiazol-2-yl]phenyl]ethenyl]phenyl]-	DMF	361(4.90)	33-0142-78
[1,2,4]Triazolo[1,5-a]pyridine, 7-[2-[4-[3-(4-methoxyphenyl)-1,2,4-oxadiazol-5-yl]phenyl]ethenyl]-2-phenyl-	DMF	349(4.81)	33-0142-78
$C_{29}H_{22}$			
Benzene, 1,1',1"-[3-(1-phenylethenyl)-1-cyclopropene-1,2,3-triyl]tris-	EtOH	228(4.45),318(4.32), 332(4.25)	44-1493-78
1,3-Cyclopentadiene, 1,2,3,4-tetraphenyl-	benzene	350(4.09)	104-1532-78
$C_{29}H_{22}N_2O_2$			
2(1H)-Quinoxalinone, 3-(1-benzoyl-2,2-diphenylethyl)-	EtOH	354(3.96)	103-0336-78
$C_{29}H_{22}N_2O_3S$			
Benzenesulfonamide, 4-methyl-N-[3-(oxodiphenylethylidene)-3H-indol-2-yl]-	CHCl$_3$	267(4.43),327(3.79), 395(3.55)	103-0745-78
$C_{29}H_{22}N_3OS$			
Quinolinium, 2-[[2,3-dihydro-3-oxo-2-(4-phenyl-2-thiazolyl)-1H-isoindol-1-ylidene)methyl]-1-ethyl-, iodide	EtOH	242(5.20),540(4.02), 586(4.00)	18-2415-78
$C_{29}H_{22}N_4O$			
[1,2,4]Triazolo[1,5-a]pyridine, 2-[4-[2-[4-(5,6-dimethyl-2-benzoxazolyl)-phenyl]ethenyl]phenyl]-	DMF	369(4.91),387(4.72)	33-0142-78
[1,2,4]Triazolo[1,5-a]pyridine, 2-[4-[2-[4-(5,7-dimethyl-2-benzoxazolyl)-phenyl]ethenyl]phenyl]-	DMF	368(4.91),387(4.73)	33-0142-78
$C_{29}H_{22}N_6$			
[1,2,4]Triazolo[1,5-a]pyridine, 2-[1,1'-biphenyl]-4-yl-7-[2-(5-methyl-2-phenyl-2H-1,2,3-triazol-4-yl)ethenyl]-	DMF	338(4.72),348(4.73)	33-0142-78
[1,2,4]Triazolo[1,5-a]pyridine, 2-(4-methylphenyl)-7-[2-[4-(2-phenyl-2H-1,2,3-triazol-4-yl)phenyl]ethenyl]-	DMF	352(4.86)	33-0142-78
[1,2,4]Triazolo[1,5-a]pyridine, 2-(4-methylphenyl)-7-[2-[4-(4-phenyl-2H-1,2,3-triazol-2-yl)phenyl]ethenyl]-	DMF	354(4.85)	33-0142-78
$C_{29}H_{22}N_6O$			
[1,2,4]Triazolo[1,5-a]pyridine, 2-(4-methoxyphenyl)-7-[2-[4-(4-phenyl-2H-1,2,3-triazol-2-yl)phenyl]ethenyl]-	DMF	356(4.85)	33-0142-78
$C_{29}H_{23}BrN_2$			
Benzenamine, 4-[1-[2-(4-bromophenyl)-ethenyl]benzo[f]quinolin-3-yl-N,N-dimethyl-	EtOH	288(4.17),368(4.75)	103-0418-78

Compound	Solvent	$\lambda_{max}(\log \epsilon)$	Ref.
C$_{29}$H$_{23}$NO			
Benzo[f]quinoline, 3-(4-methoxyphenyl)-2-methyl-1-(2-phenylethenyl)-	EtOH	268(4.64),294(4.53), 342(3.95),360(3.94)	103-0418-78
3H-Pyrrol-3-one, 1,2-dihydro-2-methyl-1,2,4,5-tetraphenyl-	MeCN	262(3.22),362(3.02)	5-1203-78
C$_{29}$H$_{23}$N$_2$			
Pyridinium, 2,4,6-triphenyl-1-(2-pyridinylmethyl)-, perchlorate	EtOH	211(4.67),308(4.68)	73-2046-78
Pyridinium, 2,4,6-triphenyl-1-(4-pyridinylmethyl)-, perchlorate	EtOH	209(4.68),313(4.57)	73-2046-78
C$_{29}$H$_{23}$N$_3$			
[1,2,4]Triazolo[1,5-a]pyridine, 2-(4-methylphenyl)-7-[2-[4-(2-phenylethenyl)phenyl]ethenyl]-	DMF	368(4.87)	33-0142-78
C$_{29}$H$_{23}$N$_3$O$_2$			
2H-Imidazo[4,5-d]oxazol-2-one, 3,3a,6-6a-tetrahydro-6a-methyl-3,3a,5,6-tetraphenyl-	EtOH	236(4.48)	94-3798-78
C$_{29}$H$_{23}$N$_3$O$_9$			
1H-Imidazole, 2-nitro-1-(2,3,5-tri-O-benzoyl-α-D-ribofuranosyl)-	MeOH	230(4.61),275(3.69), 283(3.72),317(3.86)	44-4784-78
1H-Imidazole, 2-nitro-1-(2,3,5-tri-O-benzoyl-β-D-ribofuranosyl)-	MeOH	231(4.64),275(3.72), 283(3.76),305(3.83)	44-4784-78
C$_{29}$H$_{24}$NS			
Pyridinium, 1-methyl-2-[(2-methyl-4H-1-benzothiopyran-4-ylidene)methyl]-4,6-diphenyl-, perchlorate	MeCN	455(3.997)	83-0226-78
C$_{29}$H$_{24}$N$_2$			
Benzenamine, N,N-dimethyl-4-[2-(3-phenylbenzo[f]quinolin-1-yl)ethenyl]-	EtOH	257(4.57),288(4.60), 377(4.24)	103-0418-78
1H-Pyrazole, 4,5-dihydro-3,5-diphenyl-1-[4-(2-phenylethenyl)phenyl]-	n.s.g.	390(4.62)	135-0946-78
C$_{29}$H$_{24}$N$_2$O			
3-Pyridinecarbonitrile, 1,2-dihydro-1-methyl-6-(4-methylphenyl)-2-oxo-5-phenyl-4-(1-phenyl-1-propenyl)-, (E)-	MeCN	242(4.22),361(4.05)	5-1222-78
C$_{29}$H$_{24}$N$_3$O			
Benzo[f]quinolinium, 2-acetyl-4-(4-methylphenyl)-1-[(phenylhydrazono)methyl]-, perchlorate	MeOH	497(4.26)	65-1635-78
C$_{29}$H$_{24}$N$_3$O$_2$			
Benzo[f]quinolinium, 2-acetyl-4-(4-methoxyphenyl)-1-[(phenylhydrazono)methyl]-, perchlorate	MeOH	487(3.90)	65-1635-78
C$_{29}$H$_{24}$N$_5$O			
[1,2,3]Triazolo[1,5-a]pyrimidinium, 7-methyl-5-[3-(3-methyl-2(3H)-benzoxazolylidene)-1-propenyl]-2,3-diphenyl-, perchlorate	EtOH	547(4.91)	103-1160-78

Compound	Solvent	$\lambda_{max}(\log \epsilon)$	Ref.
$C_{29}H_{24}N_5O_5$			
Quinolinium, 4-[[2-(2,4-dinitrophenyl)-2,4-dihydro-4-[(4-hydroxyphenyl)methylene]-5-methyl-3H-pyrazol-3-ylidene]methyl]-1-ethyl-, iodide	n.s.g.	220(4.61)	48-0857-78
$C_{29}H_{24}O$			
Anthracene, 9-(diphenylmethylene)-10-ethoxy-9,10-dihydro-	n.s.g.	242(4.04),297(4.02)	39-0488-78C
Anthracene, 9-(ethoxyphenylmethyl)-10-phenyl-	n.s.g.	338(3.51),356(3.83), 375(3.96),395(3.95)	39-0488-78C
$C_{29}H_{24}O_3S$			
9-Anthracenesulfonic acid, 10-(diphenylmethylene)-9,10-dihydro-	n.s.g.	244(3.89),295(3.78)	39-0488-78C
$C_{29}H_{24}O_4$			
2-Butenoic acid, 2-(1-naphthalenyl)-3-phenyl-4-(4-methoxybenzoyl)-, methyl ester	EtOH	277(4.36)	4-0759-78
$C_{29}H_{25}NO_2S$			
Methanone, [2-phenyl-5-(1-piperidinyl)-3,4-thiophenediyl]bis[phenyl-	CH_2Cl_2	253(4.52),312(4.03), 390(3.45)	24-2021-78
$C_{29}H_{25}NO_3$			
Methanone, [2-phenyl-5-(1-piperidinyl)-3,4-furandiyl]bis[phenyl-	dioxan	247(4.44),315(4.18), 374(3.82)	24-3336-78
$C_{29}H_{25}N_2O$			
Quinolinium, 2-[3-(1,2-dihydro-2-oxo-3H-indol-3-ylidene)-2-(4-methylphenyl)-1-propenyl]-1-ethyl-, iodide	MeOH	425s(3.80),432(3.83), 472(3.83),548(4.07), 581(4.23)	64-0209-78B
$C_{29}H_{25}N_2O_2$			
Quinolinium, 2-[3-(1,2-dihydro-2-oxo-3H-indol-3-ylidene)-2-(4-methoxyphenyl)-1-propenyl]-1-ethyl-, iodide	MeOH	402(3.54),475(3.60), 550(4.00),584(4.19)	64-0209-78B
$C_{29}H_{25}N_3$			
Propanedinitrile, [3,4-dihydro-1,5-dimethyl-6-(4-methylphenyl)-3,4-diphenyl-2(1H)-pyridinylidene]-, trans	MeCN	254(4.02),330(4.16)	5-1222-78
Propanedinitrile, [1-[2-(methylamino)-2-(4-methylphenyl)-1-phenylethenyl]-2-phenyl-2-butenylidene]-, (Z,E)-	MeCN	240(4.30),273(4.10), 413(3.80)	5-1222-78
$C_{29}H_{25}N_3OS$			
1,4-Naphthalenedione, mono[(3-butyl-4,5-diphenyl-2(3H)-thiazolylidene)-hydrazone]	isoPrOH	248(4.36),308(4.21), 538(4.63)	73-1227-78
$C_{29}H_{25}N_3O_7$			
1H-Imidazol-2-amine, 1-(2,3,5-tri-O-benzoyl-α-D-ribofuranosyl)-	MeOH	229(4.62),269(3.46), 275(3.51),282(3.43)	44-4784-78
1H-Imidazol-2-amine, 1-(2,3,5-tri-O-benzoyl-β-D-ribofuranosyl)-	MeOH	229(4.62),274(3.48), 281(3.38)	44-4784-78
β-D-Ribofuranosylamine, N-1H-imidazol-2-yl-, 2,3,5-tribenzoate	MeOH	228(4.66),274(3.51), 281(3.42)	44-4774-78

Compound	Solvent	$\lambda_{max}(\log \epsilon)$	Ref.
$C_{29}H_{26}N_2O_2$ Acetic acid, cyano(3,4-dihydro-1,5-di-methyl-3,4,6-triphenyl-2(1H)-pyridin-ylidene)-, methyl ester, trans	MeCN	259(3.96),339(4.21)	5-1222-78
$C_{29}H_{26}N_2O_2P_2$ 3H-Pyrazole, 4,5-bis(diphenylphosphin-yl)-3,3-dimethyl-	MeOH	261(3.58),370(1.77)	150-4248-78
$C_{29}H_{26}N_3O$ Quinolinium, 4-[(2,4-dihydro-4-[(4-hy-droxyphenyl)methylene]-5-methyl-2-phenyl-3H-pyrazol-3-ylidene]methyl]-1-ethyl-, iodide	n.s.g.	210(4.67),246(4.65), 360(4.23)	48-0857-78
$C_{29}H_{26}N_3OPS$ Phosphonium, [3-[(aminothioxomethyl)hy-drazono]-3-(4-methoxyphenyl)-1-prop-enyl]triphenyl-, hydroxide, inner salt	DMF	267(4.072),316(3.826), 516(2.935)	65-1998-78
$C_{29}H_{26}N_3PS$ Phosphonium, [3-[(aminothioxomethyl)hy-drazono]-3-(4-methylphenyl)-1-propen-yl]triphenyl-, hydroxide, inner salt	DMF	268(4.199),317(3.877), 516(3.076)	65-1998-78
$C_{29}H_{26}N_4O_8S$ 2(1H)-Pteridinethione, 6,7-diphenyl-4-[(2,3,5-tri-O-acetyl-β-D-ribofurano-syl)oxy]-	MeOH	220s(4.36),262s(4.19), 280s(4.28),303(4.39), 373(4.25)	24-2571-78
4(1H)-Pteridinone, 2,3-dihydro-6,7-di-phenyl-2-thioxo-3-(2,3,5-tri-O-acetyl-β-D-ribofuranosyl)-	MeOH	224s(4.33),280s(4.13), 313(4.47),384(4.17)	24-2571-78
4(1H)-Pteridinone, 6,7-diphenyl-2-[(2,3,5-tri-O-acetyl-β-D-ribofurano-syl)thio]-	MeOH	220s(4.39),259(4.32), 284(4.34),365(4.17)	24-2571-78
$C_{29}H_{26}O$ 2,5-Cyclohexadien-1-one, 2-(1,1-dimeth-ylethyl)-4-(diphenylmethylene)-6-phenyl-	benzene	390(4.60)	70-1304-78
$C_{29}H_{26}O_8$ 3H,7H-Pyrano[2',3':7,8][1]benzopyrano-[3,2-d][1]benzoxepin-7-one, 6,12-di-acetoxy-8,9-dihydro-3,3-dimethyl-9-(1-methylethenyl)-, (±)-	EtOH EtOH-AlCl$_3$	219(4.47),232(4.41), 273(4.30),355(3.95) 220(4.52),272(4.33), 355(4.04)	94-1394-78 94-1394-78
$C_{29}H_{26}O_{10}$ Neocercosporin	MeOH	222(4.69),270(4.50), 475(4.37),562(3.93)	95-1553-78
Neosporin	MeOH	222(4.69),270(4.50), 475(4.37),562(3.93)	25-0233-78
$C_{29}H_{26}O_{12}$ 12H-Benzo[b]xanthene-6,7,8,10,11-pentol, 3-methoxy-1-methyl-, pentaacetate	EtOH	242(4.74),300(3.99)	94-0209-78
$C_{29}H_{27}Cl_2N_4OP$ 2(1H)-Phosphorinone, 6-[(4-chlorophen-yl)azo]-4-cyclohexyl-1-phenyl-,	EtOH	265(4.20),293(4.18), 353(4.16),515(4.40)	118-0846-78

Compound	Solvent	λ_{max} (log ϵ)	Ref.
(4-chlorophenyl)hydrazone, 1-oxide			118-0846-78
$C_{29}H_{27}NO_6S$ Pyrylium, 2-[2-[4-(dimethylamino)ethen- yl]-4,6-diphenyl-, carboxymethylsul- fonate	EtOH	645(4.491)	83-0256-78
$C_{29}H_{27}N_3OPS$ Phosphonium, [3-[(aminothioxomethyl)hy- drazono]-3-(4-methoxyphenyl)-1-prop- enyl]triphenyl-, chloride	EtOH or DMF	268(4.097),327(3.917)	65-1998-78
$C_{29}H_{27}N_3O_3S$ 4-Thiazoleacetic acid, α-[(2-propenyl- oxy)imino]-2-[(triphenylmethyl)ami- no]-, ethyl ester, (E)-	EtOH	233s(4.36),305(3.48)	78-2233-78
$C_{29}H_{27}N_3PS$ Phosphonium, [3-[(aminothioxomethyl)hy- drazono]-3-(4-methylphenyl)-1-propen- yl]triphenyl-, chloride	EtOH or DMF	266(4.083),325(3.878)	65-1998-78
$C_{29}H_{27}O$ [1,1'-Biphenyl]-2-yloxy, 3-(1,1-dimeth- ylethyl)-5-(diphenylmethyl)-	MeCN	425(3.6),700(3.3)	70-1304-78
$C_{29}H_{28}N_2O_5$ 2(1H)-Pyrimidinone, 1-[2-deoxy-5-O- [(4-methoxyphenyl)diphenylmethyl]- -D-ribofuranosyl]-	50% EtOH	230(4.15),306(3.76)	33-2579-78
$C_{29}H_{28}O$ [1,1'-Biphenyl]-2-ol, 3-(1,1-dimethyl- ethyl)-5-(diphenylmethyl)-	benzene	295(3.48)	70-1304-78
$C_{29}H_{28}O_3P$ Phosphonium, (5-acetyl-4-ethoxy-2-meth- oxyphenyl)triphenyl-, iodide	EtOH	235(4.82),269(4.23), 304s(3.68)	95-0503-78
$C_{29}H_{28}O_8$ 2H,6H-Benzo[1,2-b:5,4-b']dipyran-2-one, 3,3'-methylenebis[7,8-dihydro-4-hy- droxy-8,8-dimethyl-	dioxan	255(3.61),320(4.63), 340(4.68)	2-0375-78
Kuwanon A diacetate	MeOH	224(4.56),259(4.39), 327(3.99)	94-1453-78
	MeOH-NaOMe	268(4.46),345(4.03)	94-1453-78
	MeOH-AlCl₃	268(4.46),310(3.87), 383(3.86)	94-1453-78
Kuwanon B diacetate	EtOH	233(4.62),262(4.43), 322(4.07)	94-1453-78
	EtOH-NaOMe	269(4.53),359(4.17)	94-1453-78
	EtOH-AlCl₃	225(4.56),270(4.46), 326(3.92),383(3.93)	94-1453-78
Morusin diacetate	EtOH	234(4.27),270(4.54), 354(3.69)	94-1394-78
	EtOH-AlCl₃	279(4.62),33(4.87), 415(3.77)	94-1394-78
$C_{29}H_{28}O_9$ 3H,7H-Pyrano[2',3':7,8][1]benzopyrano- [3,2-d][1]benzoxepin-7-one, 6,12-di-	EtOH	233(4.61),271(4.53), 326(4.18)	94-1394-78

Compound	Solvent	$\lambda_{max}(\log \epsilon)$	Ref.
acetoxy-8,9-dihydro-9-(1-hydroxy-1-methylethyl)-3,3-dimethyl-, (±)-	EtOH-AlCl$_3$	233(4.60),270(4.54), 326(4.21)	94-1394-78
$C_{29}H_{28}O_{10}$			
Ethaneperoxoic acid, 1-(12-acetoxy-8,9-dihydro-6-hydroxy-3,3-dimethyl-7-oxo-1H,7H-pyrano[2',3':7,8][1]benzopyrano-[3,2-d][1]benzoxepin-9-yl)-1-methyl-, ethyl ester	EtOH EtOH-AlCl$_3$	234(4.62),276(4.68) 235(4.65),284(4.70)	94-1431-78 94-1431-78
$C_{29}H_{29}N_3O_3S$			
4-Thiazoleacetic acid, α-(1-methyleth-oxy)imino]-2-[(triphenylmethyl)ami-no]-, ethyl ester, (E)-	EtOH	231s(4.37),302(3.47)	78-2233-78
$C_{29}H_{30}$			
Benzene, 1,1',1"-(3,5-dipropyl-1,3-cy-clopentadiene-1,2,4-triyl)tris-	isooctane	257(4.230),315(4.152)	44-4090-78
$C_{29}H_{30}ClN_2S_2$			
Benzothiazolium, 2-[2-[2-chloro-3-[(3-ethyl-2(3H)-benzothiazolylidene)eth-ylidene]-1-cyclohepten-1-yl]ethenyl]-3-ethyl-	MeOH	790(5.45)	104-2046-78
Benzothiazolium, 2-[2-[2-chloro-3-[(3-ethyl-2(3H)-benzothiazolylidene)eth-ylidene]-5-methyl-1-cyclohexen-1-yl]-ethenyl]-3-ethyl-, iodide	MeOH	795(5.33)	104-2046-78
$C_{29}H_{30}N_2O_2$			
2H-Pyrrol-2-one, 5,5'-[1,3-propanediyl-bis(4,1-phenylenemethylidyne)]bis-[1,5-dihydro-3,4-dimethyl-, (E,E)-(E,Z)-	EtOH	238(4.03),247s(4.06), 275s(4.24),322(4.43)	49-1191-78
	EtOH	237(4.10),246(4.01), 331(4.54),358s(4.29)	49-1191-78
(Z,Z)-	EtOH	235f(4.1),335(4.5)	49-1191-78
$C_{29}H_{30}N_2O_6$			
5,1-(Epoxyimino)tetrahydrosantonin, N,O-dibenzoyl-, (1S,4S,5R)-	MeOH	234(4.40),273(3.68)	44-0536-78
$C_{29}H_{30}N_4$			
Tchibangensine	EtOH	227(4.71),286(4.06), 293(4.10),319(4.19)	102-0539-78
	acid	220(4.70),250(4.12), 274s(3.96),283s(3.91), 291(3.78),358(4.38)	102-0539-78
$C_{29}H_{30}O_2$			
4,8-Etheno-4H-indeno[5,6-d]-1,3-dioxole, 5-(diphenylmethylene)-3a,4a,5,7a,8,8a-hexahydro-2,2,9,10-tetramethyl-, (3aα,4α,4aβ,7aβ,8α,8aα)-	CHCl$_3$	297(4.23)	5-0440-78
(3aα,4β,4aα,7aα,8β,8aα)-	CHCl$_3$	299(4.14)	5-0440-78
$C_{29}H_{30}O_7$			
Morusin, 5-O-acetyl-, dimethyl ether	MeOH	206(4.62),239(4.57), 263(4.49),319(4.01)	94-1394-78
	MeOH-AlCl$_3$	207(4.63),239(4.58), 263(4.51),319(4.02)	94-1394-78

Compound	Solvent	$\lambda_{max}(\log \epsilon)$	Ref.
$C_{29}H_{30}O_8$			
Kuwanon D diacetate	EtOH	211(4.93),278(4.28), 335(3.75)	142-0745-78A
	EtOH-AlCl$_3$	211(4.98),289(4.29), 385(3.91)	142-0745-78A
$C_{29}H_{30}O_{14}$			
Ethanone, 2-(1,3-benzodioxol-5-yl)-1- [2-hydroxy-4-[(2,3,4,6-tetra-O-acet- yl-β-D-glucopyranosyl)oxy]phenyl]-	EtOH	227(4.04),272(4.16), 320(3.84)	106-0235B-78
$C_{29}H_{31}N_9O_6$			
L-Glutamic acid, N-[N-[4-[[(2,4-diami- no-6-pteridinyl)methyl]methylamino]- benzoyl]-L-phenylalanyl]-	pH 1	258(4.45),302(4.45)	87-1165-78
$C_{29}H_{32}N_2O_{11}$			
Dibenzo[a,c]cycloocten-6-ol, 12-(2,4- dinitrophenoxy)-5,6,7,8-tetrahydro- 1,2,3,10,11-pentamethoxy-6,7-dimethyl-	EtOH	212(4.70),250(4.31), 284(4.06),300s(3.98)	94-0328-78
$C_{29}H_{32}N_6O_8$			
L-Threonine, O-(phenylmethyl)-N-[[(9- β-D-ribofuranosyl-9H-purin-6-yl)am- ino]carbonyl]-, phenylmethyl ester	pH 2 pH 12	270(4.31),277(4.30) 270(4.24),279(4.23), 299(4.02)	19-0021-78 19-0021-78
	MeOH	269(4.33),275s(4.25)	19-0021-78
$C_{29}H_{32}O_2$			
11-Dodecenoic acid, 12-(1-pyrenyl)-, methyl ester	hexane	223(4.26),238s(4.50), 244(4.63),268s(4.28), 277(4.46),329s(4.28), 343(4.46),356s(4.37)	128-0163-78
$C_{29}H_{32}O_4$			
Bicyclo[7.2.0]undec-4-ene-2,3-diol, 4,11,11-trimethyl-8-methylene-, dibenzoate	MeOH	228(4.42),273(3.28), 281(3.23)	94-2535-78
isomer 15	MeOH	228(4.51),275(3.28), 280(3.20)	94-2535-78
$C_{29}H_{32}O_9$			
Malaphyll	EtOH	221(4.61),260(4.46), 292(3.95)	105-0377-78
$C_{29}H_{34}N_2O_{14}S$			
2H-Benzimidazole-2-thione, 1,3-dihydro- 1,3-bis(2,3,5-tri-O-acetyl-β-D-ribo- furanosyl)-	MeOH	253(4.03),302s(4.38), 313(4.47)	4-0657-78
$C_{29}H_{34}O_2$			
3(4H)-Dibenzofuranone, 2,6-bis(1,1-di- methylethyl)-4a,9b-dihydro-8,9b-di- methyl-4-(phenylmethylene)-, (4E,4aα,9bα)-	EtOH	231(4.12),290(4.24)	12-0907-78
1-Pyrenedodecanoic acid, methyl ester	hexane	234(4.59),243(4.86), 254(4.02),264(4.39), 276(4.73),312(4.03), 326(4.43),343(4.63)	128-0163-78
$C_{29}H_{34}O_5$			
Cyclopentanone, 2,2'-[1,5-bis(2-meth-	EtOH	214(4.24),275(3.66),	2-0428-78

Compound	Solvent	$\lambda_{max}(\log \epsilon)$	Ref.
oxyphenyl)-3-oxo-1,5-pentanediyl]bis- Cyclopentanone, 2,2'-[1,5-bis(4-meth- oxyphenyl)-3-oxo-1,5-pentanediyl]bis-	EtOH	280(3.65) 227(4.32),279(3.54)	2-0428-78 2-0428-78
$C_{29}H_{34}O_{11}$ Physalin I	EtOH	227(3.86)	102-1641-78

$C_{29}H_{34}O_{16}$
  Tectorigenin, 7-O-methyl-, 4'-O-β-D-
  glucopyranosyl-(1→6)-β-D-glucopyr-
  anoside
	MeOH
	MeOH-NaOAc
	MeOH-AlCl$_3$
		270(4.17),335(3.26)
		270(--),335(--)
		280(--),335(--)
			2-0641-78
			2-0641-78
			2-0641-78

$C_{29}H_{35}BrO_5$
  Benzoic acid, 4-bromo-, 8-[(acetoxy-
  methyl)-2,3,3a,4,5,5a,6,9,10,10a-
  decahydro-3a,5a-dimethyl-1-(1-meth-
  ylethyl)-6-oxocyclohept[e]inden-9-yl
  ester, [3aR-(3aα,5aβ,9β,10aα)]-
	MeOH	208(4.20),246(4.26),
		  282s(2.89)
			23-2113-78

$C_{29}H_{36}N_2O_4$
  Bupivacanine 3-hydroxy-2-naphthoate
	MeOH	271(3.73),282(3.73),
		  293(3.44),353(3.38)
			94-0936-78

$C_{29}H_{36}N_2O_8$
  Aspidospermidine-3-carboxylic acid, 7-
  acetyl-3,4-diacetoxy-7,8-didehydro-
  16-methoxy-1-methyl-, methyl ester,
  (2β,3β,4β,5α,12β,19α)-
	EtOH	212(4.45),246(3.72),
		  316(4.36)
			33-1554-78

$C_{29}H_{36}N_5O_{18}P$
  Coenzyme F$_{420}$
	pH 0.1
	
	pH 4.5
		230(4.60) 250(4.36),
		  267(4.33),375(4.52)
		235(4.62),250(4.38),
		  267(4.41),385s(4.41),
		  395(4.44)
			69-4583-78
			
			69-4583-78

$C_{29}H_{36}O_4$
  Pregn-4-ene-3,20-dione, 17-[(benzoyl-
  oxy)methyl]-
	EtOH	238(4.15)	23-0410-78

$C_{29}H_{36}O_5$
  Cyclohept[e]inden-6(2H)-one, 8-(acet-
  oxymethyl)-9-(benzoyloxy)-3,3a,4,5-
  5a,9,10,10a-octahydro-3a,5a-dimethyl-
  1-(1-methylethyl)-, [3aR-(3aα,5aβ-
  9β,10aα)]-
  Fuselol angelate
	MeOH
	
	n.s.g.
		230(4.23),274s(3.08),
		  282s(2.99)
		
		217(4.35),254(3.33),
		  324(4.15)
			23-2113-78
			
			105-0441-78

$C_{29}H_{36}O_5S$
  Nona-2,4,6-trien-1-ol, 9-(4-methoxy-
  2,3,6-trimethylphenyl)-3,7-dimethyl-
  5-(phenylsulfonyl)-, acetate
	EtOH	217(4.42),238(4.27),
		  292(4.17)
			44-4769-78

$C_{29}H_{36}O_6$ Benzoyllineolon	EtOH	233(4.10),278(3.13)	100-0001-78

$C_{29}H_{36}O_6S_2$
  2(3H)-Naphthalenone, 5-(1,1-dimethyl-
  ethoxy)-4,4a,5,6,7,8-hexahydro-4a-
  methyl-1,3-bis[(phenylsulfonyl)-
	n.s.g.	217(4.31),253(4.10)	33-2397-78

Compound	Solvent	$\lambda_{max}(\log \epsilon)$	Ref.
methyl]-, [3R-(3α,4aβ,5β)]- (cont.)			33-2397-78
$C_{29}H_{38}O_6$ 4H-1-Benzopyran-4-one, 2-(2,4-dimeth-oxyphenyl)-5,7-dimethoxy-3,8-bis(3-methylbutyl)-	EtOH	229(4.68),257(4.65), 291(4.24),318(4.38)	142-1355-78A
$C_{29}H_{40}O_6$ 5α,14α-Bufa-20,22-dienolide, 3β-acet-oxy-23-(ethoxycarbonyl)-	EtOH	322(3.7)	23-0424-78
$C_{29}H_{41}NO_4$ Pregna-5,17(20)-dien-21-oic acid, 20-cyano-3,3-[(2,2-dimethyl-1,3-prop-anediyl)bis(oxy)]-, ethyl ester, (17E)-	MeOH	238(4.09)	24-1533-78
$C_{29}H_{42}O_6$ 25,26,27-Trinor-5α-lanost-8-en-24-oic acid, 3β-acetoxy-7,11-dioxo-	MeOH	272(3.90)	5-0419-78
$C_{29}H_{43}O_2$ Pyrylium, 4-[3-[2,6-bis(1,1-dimethyl-ethyl)-4H-pyran-4-ylidene]-1-propen-yl]-2,6-bis(1,1-dimethylethyl)-, per-chlorate	MeCN	562(4.63),603(5.43)	4-0365-78
$C_{29}H_{44}O$ Cholesta-1,5,7-trien-3-one, 4,4-dimeth-yl-	ether	275(3.78)	35-5626-78
$C_{29}H_{44}O_2$ Cholesta-5,7-dien-3-one, 1α,2α-epoxy-4,4-dimethyl-	ether	261s(3.85),272(3.97), 282(3.97),293s(3.73)	35-5626-78
18-Nor-5α-cholesta-9,11,13-trien-3β-ol, 12-methyl-, acetate	EtOH	205(4.67),225(4.07)	39-0076-78C
Octadecanoic acid, 1-naphthalenylmethyl ester	hexane	298.5(3.83)	95-0774-78
$C_{29}H_{44}O_3$ Cholesta-2,4-dien-6-one, 4-acetoxy-	EtOH	315(3.86)	12-0171-78
Cholesta-3,5-dien-7-one, 1α-acetoxy-	EtOH	280(4.39)	44-0574-78
Cholesta-4,7-dien-6-one, 3β-acetoxy-	EtOH	267(4.25)	12-2069-78
29,30-Dinorlup-12-en-20-oic acid, 11-oxo-, methyl ester	C_6H_{12}	235(4.01)	73-2190-78
$C_{29}H_{44}O_4$ Cholesta-4,7-dien-6-one, 3β-acetoxy-14-hydroxy-	EtOH	264(4.19)	12-2069-78
$C_{29}H_{44}O_6S$ Card-20(22)-enolide, 3-[(cyclohexylsul-fonyl)oxy]-14-hydroxy-, (3β,5β)-	MeOH	217(4.19)	106-0706-78
$C_{29}H_{45}NO_3$ Cholest-4-eno[6,5,4-cd]isoxazol-2-ol, acetate	EtOH	232(3.76)	12-0097-78
Cholest-4-eno[6,5,4-cd]isoxazol-3-one, 1,2-ethanediyl acetal	EtOH	232(3.74)	12-0113-78
$C_{29}H_{45}NO_4$ 5'H-A-Norcholestano[6,5-c]isoxazol-	EtOH	217(3.78)	12-0097-78

Compound	Solvent	$\lambda_{max}(\log \epsilon)$	Ref.
5'-one, 3-acetoxy-, (3β,5S)- (cont.)			12-0097-78
$C_{29}H_{46}O$			
Pregn-4-en-3-one, 20-methyl-21-[3-methyl-2-(1-methylethyl)cyclopropyl]-, (20R)-	EtOH	242(4.19)	88-0837-78
9,10-Secocholesta-5,7,10(19)-trien-3-one, 4,4-dimethyl-, (5E,7E)-	ether	253(4.03),264(4.08), 290s(3.60),310s(3.43), 320s(3.08)	35-5626-78
Stigmasta-4,20(22)-dien-3-one	EtOH	239(4.32)	2-0253-78
$C_{29}H_{46}O_2$			
Cholesta-5,7-dien-3-one, 1α-hydroxy-4,4-dimethyl-	ether	261s(3.87),272(3.98), 282(3.98),293s(3.74)	35-5626-78
Stigmast-4-ene-3,6-dione	EtOH	253(3.97),328(2.01)	106-0082-78
$C_{29}H_{46}O_4$			
Cholestane-4,6-dione, 3β-acetoxy-	EtOH	298(3.95)	12-0113-78
Cholestane-3,4,6-trione, cyclic 3-(1,2-ethanediyl acetal), (5β)-	EtOH	298(3.95)	12-0113-78
Cholest-4-en-3-one, 4-acetoxy-6β-hydroxy-	n.s.g.	245(3.49)	2-0426-78
Cholest-4-en-6-one, 3-acetoxy-4-hydroxy-	EtOH	300(4.2)	2-0559-78
Cholest-4-en-6-one, 3-acetoxy-7α-hydroxy-	EtOH	238(3.79)	12-0171-78
9,11-Secocholest-7-en-11-al, 3β-acetoxy-9-oxo-	EtOH	241(3.86)	56-1713-78
$C_{29}H_{46}O_5$			
6,7-Secocholest-4-ene-6,7-dioic acid, 3-oxo-, dimethyl ester	EtOH	236(3.98)	12-0171-78
6,7-Secocholest-4-en-6-oic acid, 3β-acetoxy-7-oxo-	EtOH	209(3.89)	12-0171-78
9,11-Secocholest-7-en-11-oic acid, 3-acetoxy-9-oxo-, (3β,5α)-	EtOH	241(3.87)	56-1713-78
$C_{29}H_{46}O_6$			
6,7-Secocholest-4-ene-6,7-dioic acid, 3 -acetoxy-	EtOH	212(3.88)	12-0171-78
$C_{29}H_{47}NO_3$			
Cholest-5-en-4-one, 3β-acetoxy-6-amino-	EtOH	327(4.13)	12-0113-78
$C_{29}H_{47}NO_4$			
4-Aza-A-homocholest-5-en-4a-one, 3β-acetoxy-6-hydroxy-	EtOH	300(4.15)	2-0559-78
$C_{29}H_{48}O$			
Previtamin D_3, 4,4-dimethyl-	ether	257(3.90)	35-5626-78
iodine added	ether	256(4.09)	35-5626-78
$C_{29}H_{48}O_2$			
Cholesta-5,7-diene-1α,3α-diol, 4,4-dimethyl-	ether	261s(3.91),272(4.04), 282(4.04),292s(3.78)	35-5626-78
Cholesta-5,7-diene-1α,3β-diol, 4,4-dimethyl-	ether	261s(3.94),272(4.08), 282(4.08),293s(3.85)	35-5626-78
Cholest-2-en-4-one, 3-ethoxy-	EtOH	236(3.69)	78-1509-78
Cholest-4-en-3-one, 4-ethoxy-	MeOH	257(4.06)	78-1509-78
Cholest-4-eno[4,3-b]oxet-3(4'H)-ol, 4'-methyl-	EtOH	220(3.60)	78-1509-78

Compound	Solvent	$\lambda_{max}(\log \epsilon)$	Ref.
3-Epivitamin D_3, 1α-hydroxy-4,4-dimethyl-	ether	257(4.05)	35-5626-78
Vitamin D_3, 1α-hydroxy-4,4-dimethyl-	ether	257(4.06)	35-5626-78
$C_{29}H_{48}O_7$			
2-Propenoic acid, 3-(4-hydroxyphenyl)-, 1-(6,6-dimethoxyhexyl-10,10-dimethyl-, cis	MeOH	225(3.97),299s(4.17), 310.5(4.20)	102-0979-78
trans	MeOH	226(4.03),300s(4.26), 311.5(4.30)	102-0979-78
$C_{29}H_{50}N_6O_5$			
Adenosine, N-[(octadecylamino)carbonyl]-	pH 1	268(4.06)	36-0569-78
	pH 13	261(4.20),283s(3.85)	36-0569-78
	70% EtOH	270(4.30),276(4.22)	36-0569-78
$C_{29}H_{51}N_3O_6$			
2(1H)-Pyrimidinone, 4-amino-1-[3-O-(1-oxoeicosyl)-β-D-arabinofuranosyl]-	isoPrOH	216(4.19),248(4.16), 303(3.90)	94-0981-78

Compound	Solvent	λ_{max} (log ϵ)	Ref.
$C_{30}H_{15}N_8P$ 4-Cyclopentene-1,1,2,2,3,3,4-heptacarb-onitrile, 5-[(triphenylphosphoranyli-dene)amino]-	CH_2Cl_2	238(4.09),268(3.91), 277(3.87),303(3.76)	39-1237-78C
$C_{30}H_{17}Br$ Dinaphth[1,2-a:2',1'-j]anthracene, 9-bromo-	CH_2Cl_2	289(4.48),338(4.89), 393(3.88)	88-1893-78
$C_{30}H_{18}O_2$ [2,2'-Bi-1H-indene]-1,1'-dione, 3,3'-diphenyl-	benzene MeOH	453(3.60) 253(4.72),300s(4.23), 459(3.58)	40-0144-78 40-0144-78
$C_{30}H_{18}O_8$ Asphodelin, (+)-	MeOH MeOH-KOH	258(4.48),290s(--), 435(4.04) 285(--),420(--)	94-1111-78 94-1111-78
$C_{30}H_{19}ClN_2O_2$ 1,2-Benzisoxazole, 3-[4-[2-[4-[3-(4-chlorophenyl)-5-isoxazolyl]phenyl]-ethenyl]phenyl]-	DMF	352(4.84)	33-2904-78
$C_{30}H_{19}Cl_2N_3O_3$ 1,2-Benzisoxazole, 6-chloro-3-[4-[2-[2-chloro-4-[5-(4-methoxyphenyl)-1,3,4-oxadiazol-2-yl]phenyl]eth-enyl]phenyl]-	DMF	355(4.79)	33-2904-78
$C_{30}H_{20}$ Hexacyclo[14.6.2.2^{4,7}.2^{10,13}.1^{3,21}-1^{14,18}]triaconta-1,3(25),4,6,8,10-12,14,16,18(26),19,21,23,27,29-pentadecaene	dioxan	267(4.54),306s(3.44), 325(3.26)	24-0205-78
$C_{30}H_{20}Br_2N_3P$ Phosphonium, triphenyl-, 2,4-dibromo-3-cyano-5-(phenylazo)-2,4-cyclopen-tadien-1-ylide	MeCN	223(4.64),432(4.32)	18-2605-78
$C_{30}H_{20}ClN_3O_2$ 1,2-Benzisoxazole, 6-chloro-3-[4-[2-[4-[3-(4-methylphenyl)-1,2,4-oxadia-zol-5-yl]phenyl]ethenyl]phenyl]-	DMF	349(4.81)	33-2904-78
$C_{30}H_{20}ClN_3O_3$ 1,2-Benzisoxazole, 3-[4-[2-[2-chloro-4-[5-(4-methoxyphenyl)-1,3,4-oxadia-zol-2-yl]phenyl]ethenyl]phenyl]-	DMF	355(4.78)	33-2904-78
1,2-Benzisoxazole, 6-chloro-3-[4-[2-[4-[3-(4-methoxyphenyl)-1,2,4-oxadia-zol-5-yl]phenyl]ethenyl]phenyl]-	DMF	349(4.82)	33-2904-78
1,2-Benzisoxazole, 6-chloro-3-[4-[2-[4-[5-(4-methoxyphenyl)-1,3,4-oxadia-zol-2-yl]phenyl]ethenyl]phenyl]-	DMF	355(4.85)	33-2904-78
$C_{30}H_{20}N_2$ Benzo[b]naphtho[1,2-f][1,4]diazocine, 7,14-diphenyl-, (±)-	$CHCl_3$	240(4.67),252s(4.62), 345(3.53)	150-4944-78

Compound	Solvent	$\lambda_{max}(\log \epsilon)$	Ref.
Benzo[f]quinoline, 1-(2-phenylethenyl)-3-(2-quinolinyl)-	EtOH	257(4.68),287(4.69), 316(4.53),329(4.53), 344(4.45),374(3.88)	103-0418-78
Benzo[f]quinoline, 3-phenyl-1-[2-(2-quinolinyl)ethenyl]-	EtOH	265(4.73),338(4.40)	103-0418-78
Benzo[f]quinoline, 3-phenyl-1-[2-(4-quinolinyl)ethenyl]-	EtOH	272(3.96),340(3.46)	103-0418-78
$C_{30}H_{20}N_2O_2$			
1,2-Benzisoxazole, 3-[4-[2-[4-(3-phenyl-5-isoxazolyl)phenyl]ethenyl]phenyl]-	DMF	351(4.83)	33-2904-78
1,2-Benzisoxazole, 3-[4-[2-[4-(5-phenyl-2-oxazolyl)phenyl]ethenyl]phenyl]-	DMF	367(4.82)	33-2904-78
$C_{30}H_{20}N_3P$			
1-Propene-1,1,2-tricarbonitrile, 3-phenyl-3-(triphenylphosphoranylidene)-	CH_2Cl_2	236(4.18),265(3.79), 269(3.94),297(3.71), 415(4.19)	39-1237-78C
$C_{30}H_{20}N_4O_2$			
5,7b,10,10c-Tetraazacyclopent[hi]acean-thrylene-2,7-dicarboxaldehyde, 4,9-dimethyl-1,6-diphenyl-	CH_2Cl_2	274(4.72),370(4.12), 452s(4.19),467(4.28)	44-3544-78
$C_{30}H_{20}N_8O$			
Ethanone, 1-phenyl-2,2-bis(1-phenyl-1H-pyrazolo[3,4-d]pyrimidin-4-yl)-	EtOH	252s(4.70),258(4.71), 380s(4.58),397(4.74), 417(4.82)	95-0089-78
$C_{30}H_{20}O$			
Cyclopenta[b][1]benzopyran, 1,2,3-tri-phenyl-	THF	285(4.64),365(4.38), 530(3.19)	103-1075-78
$C_{30}H_{20}O_2$			
Benzo[b]naphtho[2',3':3,4]cyclobuta-[1,2-e][1,4]dioxin, 5a,11b-dihydro-5a,11b-diphenyl-	THF	233(4.84),278(3.86), 308(3.41),322(3.31)	78-2263-78
6,13-Epoxybenzo[b]naphtho[2,3-e]oxepin, 6,13-dihydro-7,12-diphenyl-	THF	243(4.72),300(4.06)	78-2263-78
Methanone, (9,10-dihydro-9,10-etheno-anthracene-11,12-diyl)bis[phenyl-	EtOH	262(4.18)	78-2305-78
$C_{30}H_{20}O_3$			
Naphtho[2,3-c]furan-1(3H)-one, 3-(2-hy-droxyphenyl)-4,9-diphenyl-	THF	244(5.00),291(4.17), 305(4.16),330(3.85), 344(3.96)	78-2263-78
7H-Oxireno[b][1]benzopyran, 1a,7a-di-hydro-1a-phenyl-7-(2-phenyl-4H-1-benzopyran-4-ylidene)-	$CHCl_3$	245(4.49),275(4.43), 313(3.99),387(4.07)	24-3957-78
$C_{30}H_{20}O_3S$			
Cyclopenta[b][1]benzopyran, 1,2-diphen-yl-3-(phenylsulfonyl)-	THF	250(--),280(4.59), 360(4.44),470(3.18)	103-1075-78
$C_{30}H_{20}O_4$			
2-Butene-1,4-dione, 2,3-dibenzoyl-1,4-diphenyl-	CCl_4	260-310(4.62)	12-1265-78

Compound	Solvent	$\lambda_{max}(\log \epsilon)$	Ref.
$C_{30}H_{20}O_7$ [1,2'-Bianthracene]-9,9',10(10'H)-tri-one, 1',4,5,8'-tetrahydroxy-2,6'-di-methyl-	MeOH	255(3.76),288(3.74), 290(3.60),304s(--), 372(3.54),435(3.18)	94-1111-78
	MeOH-KOH	270(--),290(--), 380(--),430(--), 510(--)(changing)	94-1111-78
$C_{30}H_{20}O_8$ [2,9'-Bianthracene]-9,10,10'(9'H)-tri-one, 1,4',5',8,9'-pentahydroxy-2',6-dimethyl-	MeOH	228(4.48),263(4.18), 293(3.88),390(3.90), 435(3.84)	94-1111-78
	MeOH-KOH	255(--),300(--), 380(--),520(--)	94-1111-78
$C_{30}H_{20}S_2$ Thieno[3,4-c]thiophene, tetraphenyl-	CH_2Cl_2	256(4.50),295(4.44), 340(3.77),380(3.24), 550(4.07)	44-3893-78
$C_{30}H_{21}BrN_3P$ Phosphonium, triphenyl-, 2-bromo-5-cy-ano-3-(phenylazo)-2,4-cyclopentadien-1-ylide	MeCN	220(4.53),428(4.08)	18-2605-78
$C_{30}H_{21}ClN_2O_2$ 1,2-Benzisoxazole, 6-chloro-3-[4-[2-[4-(5,6-dimethyl-2-benzoxazolyl)-phenyl]ethenyl]phenyl]-	DMF	365(4.87)	33-2904-78
1,2-Benzisoxazole, 6-chloro-3-[4-[2-[4-(5,7-dimethyl-2-benzoxazolyl)-phenyl]ethenyl]phenyl]-	DMF	363(4.86)	33-2904-78
$C_{30}H_{21}ClO_3$ 3-Butenoic acid, 4-(4-chlorophenyl)-2-(9H-fluoren-9-ylidene)-4-(4-meth-oxyphenyl)-, (E)-	EtOH	227(4.72),258(4.62), 266(4.72),286(4.17), 335(4.18)	2-0502-78
$C_{30}H_{21}Fe_2NO_6$ Iron, hexacarbonyl[μ-[$\eta^2:\eta^3$-N-(1,2-di-phenylethenyl)-α-(1-methylethyli-dene)benzenemethanaminato(2-)]]di-, (Fe-Fe)	EtOH	326(4.02),440s(3.37)	33-0589-78
$C_{30}H_{21}NO_2$ 1,2-Benzisoxazole, 5-methyl-3-[4-[2-(2-phenyl-6-benzofuranyl)ethenyl]phenyl]-	DMF	364(4.78)	33-2904-78
Pyridinium, 1-(2-carboxyphenyl)-2,4,6-triphenyl-, hydroxide, inner salt	EtOH	240(4.62),306(4.61)	73-2046-78
Pyridinium, 1-(4-carboxyphenyl)-2,4,6-triphenyl-, hydroxide, inner salt	EtOH	211(4.34),308(4.37)	73-2046-78
$C_{30}H_{21}NO_3$ 1,2-Benzisoxazole, 3-(4-methoxyphenyl)-6-[2-(2-phenyl-6-benzofuranyl)ethen-yl]-	DMF	364(4.78)	33-2904-78
$C_{30}H_{21}N_3$ [1,2,4]Triazolo[1,5-a]pyridine, 2-[1,1'-biphenyl]-4-yl-7-[2-(1-naphthalenyl)-ethenyl]-	DMF	349(4.61)	33-0142-78

Compound	Solvent	$\lambda_{max}(\log \epsilon)$	Ref.
[1,2,4]Triazolo[1,5-a]pyridine, 2-[1,1'-biphenyl]-4-yl-7-[2-(2-naphthalenyl)-ethenyl]-	DMF	340(4.79)	33-0142-78
$C_{30}H_{21}N_3O_2$			
1,2-Benzisoxazole, 3-[4-[2-[4-[3-(4-methylphenyl)-1,2,4-oxadiazol-5-yl]-phenyl]ethenyl]phenyl]-	DMF	348(4.81)	33-2904-78
1,2-Benzisoxazole, 3-[4-[2-[4-[5-(3-methylphenyl)-1,3,4-oxadiazol-2-yl]-phenyl]ethenyl]phenyl]-	DMF	352(4.84)	33-2904-78
$C_{30}H_{21}N_3O_3$			
1,2-Benzisoxazole, 3-[4-[2-[4-[3-(4-methoxyphenyl)-1,2,4-oxadiazol-5-yl]phenyl]ethenyl]phenyl]-	DMF	350(4.81)	33-2904-78
1,2-Benzisoxazole, 3-[4-[2-[4-[5-(2-methoxyphenyl)-1,3,4-oxadiazol-2-yl]phenyl]ethenyl]phenyl]-	DMF	352(4.83)	33-2904-78
1,2-Benzisoxazole, 3-[4-[2-[4-[5-(3-methoxyphenyl)-1,3,4-oxadiazol-2-yl]phenyl]ethenyl]phenyl]-	DMF	352(4.84)	33-2904-78
1,2-Benzisoxazole, 3-[4-[2-[4-[5-(4-methoxyphenyl)-1,3,4-oxadiazol-2-yl]phenyl]ethenyl]phenyl]-	DMF	355(4.84)	33-2904-78
$C_{30}H_{21}N_5O$			
4(3H)-Quinazolinone, 3-[[1,5-dihydro-1-phenyl-5-(phenylimino)-2H-pyrrol-2-ylidene]amino]-2-phenyl-	MeOH	234(4.78),284(4.16), 303(3.92)	80-1085-78
$C_{30}H_{22}$			
7H-Dibenzo[a,c]cyclononene, 7-methyl-ene-5,9-diphenyl-	EtOH	276(5.02),288(5.02), 350s(3.10)	150-5151-78
$C_{30}H_{22}CoN_8$			
Cobalt, [7,16-dihydro-7,16-bis(phenyl-azo)dibenzo[b,i][1,4,8,11]tetraaza-cyclotetradecinato(2-)-N^5,N^9,N^{14},N^{18}]-, (SP-4-1)-	DMF	444(4.60),471s(4.57)	64-1012-78B
$C_{30}H_{22}CuN_8$			
Copper, [7,16-dihydro-7,16-bis(phenyl-azo)dibenzo[b,i][1,4,8,11]tetraaza-cyclotetradecinato(2-)-N^5,N^9,N^{14},N^{18}]-, (SP-4-1)-	DMF	415(4.10)	64-1012-78B
$C_{30}H_{22}NO_2$			
Pyridinium, 1-(2-carboxyphenyl)-2,4,6-triphenyl-, perchlorate	EtOH	241(4.63),307(4.52)	73-2046-78
Pyridinium, 1-(4-carboxyphenyl)-2,4,6-triphenyl-, perchlorate	EtOH	210(4.55),309(4.53)	73-2046-78
$C_{30}H_{22}N_2$			
Pentanedinitrile, 2-(diphenylmethyl-ene)-3,4-diphenyl-	C_6H_{12}	238(4.79),241(4.77), 291(4.90),296(4.58), 302(4.60),308s(4.58), 314s(4.55)	5-0440-78

Compound	Solvent	$\lambda_{max}(\log \epsilon)$	Ref.
$C_{30}H_{22}N_2O_2$			
1,2-Benzisoxazole, 3-[4-[2-[4-(5,6-di-methyl-2-benzoxazolyl)phenyl]ethen-yl]phenyl]-	DMF	365(4.87)	33-2904-78
1,2-Benzisoxazole, 3-[4-[2-[4-(5,7-di-methyl-2-benzoxazolyl)phenyl]ethen-yl]phenyl]-	DMF	362(4.86)	33-2904-78
$C_{30}H_{22}N_4O_2$			
1,2-Benzisoxazole, 3-(4-methoxyphenyl)-6-[2-[4-(2-phenyl-2H-1,2,3-triazol-4-yl)phenyl]ethenyl]-	DMF	349(4.83)	33-2904-78
1,2-Benzisoxazole, 3-(4-methoxyphenyl)-6-[2-[4-(4-phenyl-2H-1,2,3-triazol-2-yl)phenyl]ethenyl]-	DMF	352(4.82)	33-2904-78
[1,2,4]Triazolo[1,5-a]pyridine, 2-[4-[2-[4-[3-(4-methoxyphenyl)-5-isoxa-zolyl]phenyl]ethenyl]phenyl]-	DMF	356(4.89),374(4.70)	33-0142-78
$C_{30}H_{22}N_8Ni$			
Nickel, [7,16-dihydro-7,16-bis(phenyl-azo)dibenzo[b,i][1,4,8,11]tetraaza-cyclotetradecinato(2-)-N^5,N^9,N^{14},N^{18}]-, (SP-4-1)-	DMF	410(4.49),458s(4.31)	64-1012-78B
$C_{30}H_{22}O_2$			
3-Butenoic acid, 2-(9H-fluoren-9-yli-dene)-4-(4-methylphenyl)-4-phenyl-, (Z)-	EtOH	247(4.39),297(4.18),321(3.74),336(3.90)	2-0502-78
$C_{30}H_{22}O_3$			
3-Butenoic acid, 2-(9H-fluoren-9-yli-dene)-4-(4-methoxyphenyl)-4-phenyl-, (Z)-	EtOH	227(4.72),266(4.82),286(4.28),332(4.34)	2-0502-78
$C_{30}H_{22}O_6$			
[9,9'-Bianthracene]-10,10'(9H,9'H)-di-one, 4,4',5,5'-tetrahydroxy-2,2'-di-methyl-	EtOH	257(4.19),268(4.26),363(4.33)	94-3792-78
$C_{30}H_{22}O_7$			
[2,9'-Bianthracene]-9,10'(9'H,10H)-di-one, 1,4',5',8,9'-pentahydroxy-2',6-dimethyl-	MeOH	255s(--),263(3.62),271(3.70),305(3.70),370(3.77)	94-1111-78
changing in base	MeOH-KOH	272(--),306(--),306(--),380(--)	94-1111-78
$C_{30}H_{22}O_{10}$			
3',8"-Binaringenin	MeOH	228(4.64),290(4.52),321(4.43)	102-2140-78
	MeOH-base	227(4.57),322(4.72)	102-2140-78
	MeOH-NaOAc	230(5.12),322(4.74)	102-2140-78
	MeOH-AlCl$_3$	227(4.74),311(4.69),375(3.99)	102-2140-78
Mesuaferrone A (8,8'-binaringenin)	EtOH	224(4.58),294(4.47)	2-0167-78
Neorhustflavanone	MeOH	275s(4.37),296(4.42),322s(4.24)	142-0663-78A
$C_{30}H_{23}$			
Cycloheptatrienylium, [4-(3,5-diphenyl-1,3-cyclopentadien-1-yl)phenyl]-	MeCN	543(4.20)	138-0237-78

Compound	Solvent	$\lambda_{max}(\log \epsilon)$	Ref.
$C_{30}H_{23}Cl_2N_3O_3$			
2H-Imidazo[4,5-d]oxazol-2-one, 3,6-bis-(2-chlorophenyl)-3,3a,6,6a-tetrahydro-5-(4-methoxyphenyl)-6a-methyl-3a-phenyl-	EtOH	257.5(4.28)	94-3798-78
2H-Imidazo[4,5-d]oxazol-2-one, 3,6-bis-(4-chlorophenyl)-3,3a,6,6a-tetrahydro-5-(4-methoxyphenyl)-6a-methyl-3a-phenyl-	EtOH	248(4.53),275s(4.23)	94-3798-78
$C_{30}H_{23}FN_2O_8S$			
2(1H)-Pyrimidinone, 5-fluoro-3,4-dihydro-4-thioxo-1-(2,3,5-tri-O-benzoyl-β-D-xylofuranosyl)-	EtOH	238(3.88),277(3.85), 337(4.07)	65-1725-78
$C_{30}H_{23}FN_2O_9$			
2,4(1H,3H)-Pyrimidinedione, 5-fluoro-1-(2,3,5-tri-O-benzoyl-β-D-xylofuranosyl)-	EtOH	243(3.90),273(4.00)	65-1725-78
$C_{30}H_{23}NO$			
1,2-Benzisoxazole, 5-methyl-3-[4-[2-[4-(2-phenylethenyl)phenyl]ethenyl]phenyl]-	DMF	368(4.87)	33-2904-78
$C_{30}H_{23}NO_2$			
1,2-Benzisoxazole, 3-(4-methoxyphenyl)-6-[2-[4-(2-phenylethenyl)phenyl]ethenyl]-	DMF	366(4.86)	33-2904-78
$C_{30}H_{23}N_2O$			
Quinolinium, 2-[[2,3-dihydro-2-(1-naphthalenyl)-3-oxo-1H-isoindol-1-ylidene]methyl]-1-ethyl-, iodide	EtOH	223(4.81),238(4.70), 450s(4.28),556(4.59)	18-2415-78
Quinolinium, 2-[[2,3-dihydro-2-(2-naphthalenyl)-3-oxo-1H-isoindol-1-ylidene]methyl]-1-ethyl-, iodide	EtOH	213(5.25),241(5.00), 525s(4.10),560(--), 604s(3.93)	18-2415-78
$C_{30}H_{23}N_3O$			
[1,2,4]Triazolo[1,5-a]pyridine, 2-[4-[2-[4-(5,6-dimethyl-2-benzofuranyl)-phenyl]ethenyl]phenyl]-	DMF	375(4.89)	33-0142-78
$C_{30}H_{23}N_3O_2$			
1H-Pyrazol-3-ol, 4,5-dihydro-5-imino-4-(9-methyl-9H-fluoren-9-yl)-1-phenyl-, benzoate	EtOH	270(4.26),308(3.81)	115-0305-78
$C_{30}H_{23}N_6O_5$			
Quinolinium, 2-[[4-(2-benzoxazolylmethylene)-2-(2,4-dinitrophenyl)-2,4-dihydro-5-methyl-3H-pyrazol-3-ylidene]methyl]-1-ethyl-, iodide	n.s.g.	230(4.58),360(4.40)	48-0857-78
Quinolinium, 4-[[4-(2-benzoxazolylmethylene)-2-(2,4-dinitrophenyl)-2,4-dihydro-5-methyl-3H-pyrazol-3-ylidene]methyl]-1-ethyl-, iodide	n.s.g.	220(4.67),350(4.48)	48-0857-78
$C_{30}H_{24}$			
Phenanthro[3,4-c]phenanthrene, 10,12,13,15-tetramethyl-	CH_2Cl_2	243(4.59),263(4.78), 271s(4.68),330(4.32),	78-2565-78

Compound	Solvent	$\lambda_{max}(\log \epsilon)$	Ref.
(cont.)		366s(4.01),406s(2.65), 425(2.37)	78-2565-78
$C_{30}H_{24}ClN_3O_2$ [1,2,4]Triazolo[1,5-a]pyridine, 2-[3-chloro-4-[2-(4-methoxyphenyl)ethenyl]-phenyl]-7-[2-(4-methoxyphenyl)ethen-yl]-	DMF	359(4.92)	33-0142-78
$C_{30}H_{24}Cl_2N_2O_2$ Acridine, 6-chloro-9-(6-chloro-2-meth-oxy-10-methyl-9(10H)-acridinylidene)-9,10-dihydro-2-methoxy-10-methyl-	CH_2Cl_2	241(4.71),302(4.24), 446(4.20)	35-5604-78
$C_{30}H_{24}NO_3Sb$ Antimony, (4-nitrophenolato-O^1)tetra-phenyl-	benzene DMSO	366(4.27) 434(--)	70-2243-78 70-2243-78
$C_{30}H_{24}N_2O$ Benzenecarboximidic acid, N-(2-phenyl-ethenyl)-, anhydride, (E,E)-	EtOH	235(4.25),310(4.43)	35-4481-78
$C_{30}H_{24}N_2O_2$ 1H-Pyrazole-3-carboxylic acid, 4,5-di-hydro-4-methylene-5,5-diphenyl-, diphenylmethyl ester	EtOH	339(4.06)	22-0401-78
2(1H)-Quinoxalinone, 3-[1-(diphenyl-methyl)-2-(4-methylphenyl)-2-oxo-ethyl]-	EtOH	357(4.09)	103-0336-78
$C_{30}H_{24}N_2O_8S$ 2(1H)-Pyrimidinone, 3,4-dihydro-4-thi-oxo-1-(2,3,5-tri-O-benzoyl-β-D-xylo-furanosyl)-	EtOH	238(3.98),279(3.76), 330(3.60)	65-1725-78
$C_{30}H_{24}N_2O_9$ 2,4(1H,3H)-Pyrimidinedione, 1-(2,3,5-tri-O-benzoyl-β-D-xylofuranosyl)-	EtOH	243(3.88),268(3.98)	65-1725-78
$C_{30}H_{24}N_4O$ [1,2,4]Triazolo[1,5-a]pyridine, 2-[4-[2-[4-[5-(1-methylethyl)-2-benzoxa-zolyl]phenyl]ethenyl]phenyl]-	DMF	366(4.92),384(4.75)	33-0142-78
[1,2,4]Triazolo[1,5-a]pyridine, 2-[4-[2-[4-(5-propyl-2-benzoxazolyl)phen-yl]ethenyl]phenyl]-	DMF	367(4.92),387(4.74)	33-0142-78
$C_{30}H_{24}N_8$ Dibenzo[b,i][1,4,8,11]tetraazacyclo-tetradecine, 5,14-dihydro-7,16-bis-(phenylazo)-	DMF	394(4.98),425s(4.90)	64-1012-78B
$C_{30}H_{24}O_8$ Spiro[2H,10H-benzo[1,2-b:3,4-b']dipyr-an-10,10'(8'H)-[2H]dipyrano[2,3-c-2',3'-f][1]benzopyran]-2,2',9(8H)-trione, 11',12'-dihydro-8,8,8',8'-tetramethyl-	EtOH	227(4.37),248(4.12), 259(4.05),298(4.47), 319(4.40)	2-0179-78
$C_{30}H_{24}S$ Thiophene, 2,5-diphenyl-3,4-bis(phenyl-	CH_2Cl_2	297(4.2)	4-1185-78

Compound	Solvent	$\lambda_{max}(\log \epsilon)$	Ref.
methyl)- (cont.)			4-1185-78
$C_{30}H_{24}Se$ Selenophene, 2,5-diphenyl-3,4-bis(phenylmethyl)-	$CHCl_3$	300(4.2)	4-1185-78
$C_{30}H_{25}NO_2S_3$ Methanone, (3a,4,5,6,7,7a-hexahydro-2'-phenylspiro[1,3-benzodithiole-2,4'-[4H-1,3]thiazine]-5',6'-diyl)bis-[phenyl-	CH_2Cl_2	257(4.55)	18-0301-78
$C_{30}H_{25}N_3O_3$ 2H-Imidazo[4,5-d]oxazol-2-one, 3,3a,6-6a-tetrahydro-5-(4-methoxyphenyl)-6a-methyl-3,3a,6-triphenyl-	EtOH	242(4.44),258(4.40)	94-3798-78
$C_{30}H_{25}N_4O$ Quinolinium, 2-[[4-(2-benzoxazolylmethylene)-2,4-dihydro-5-methyl-2-phenyl-3H-pyrazol-3-ylidene]methyl]-1-ethyl-, iodide	n.s.g.	210(4.36),246(4.42)	48-0857-78
Quinolinium, 4-[[4-(2-benzoxazolylmethylene)-2,4-dihydro-5-methyl-2-phenyl-3H-pyrazol-3-ylidene]methyl]-1-ethyl-, iodide	n.s.g.	214(4.65),246(4.69)	48-0857-78
$C_{30}H_{25}N_4OP$ Phosphorin, 1,1-dihydro-1-methoxy-1,4-diphenyl-2,6-bis(phenylazo)-	EtOH	423(3.64),567(4.35)	118-0846-78
$C_{30}H_{25}O_2P$ Phosphonium, triphenyl-, 1-benzoyl-4-oxo-2-pentenylide	EtOH	222(4.55),310s(4.02), 366(4.29)	95-0503-78
$C_{30}H_{26}$ Benzene, 1,1'-[2,3-bis(phenylmethyl)-1,3-butadiene-1,4-diyl]bis-	C_6H_{12}	297(4.4)	4-1185-78
Hexacyclo[14.6.2.2^{4,7}.2^{10,13}.1^3,^{21}-1^{14,18}]triaconta-1,3(25),4,6,10,12-14,16,18(26),21,27,29-dodecaene	dioxan	268(4.59),300s(3.44), 320(3.39)	24-0205-78
Spiro[5H-benzo[cd]pyrene-5,1'-[1H]indene], 2',3,3',4-tetrahydro-5',7',8-trimethyl-	MeOH	238(4.55),247(4.78), 259(4.40),270(4.34), 282(4.63),316(3.90), 329(4.26),345(4.40)	78-2569-78
$C_{30}H_{26}N_2$ Acridine, 9-(2,10-dimethyl-9(10H)-acridinylidene)-9,10-dihydro-2,10-dimethyl-	CH_2Cl_2	273(4.72),293(4.21), 429(4.17)	35-5604-78
$C_{30}H_{26}N_2O_7S$ 1H-Imidazole, 2-(methylthio)-1-(2,3,5-tri-O-benzoyl-β-D-ribofuranosyl)-	MeOH	230(4.63),273(3.53), 281(3.40)	44-4774-78
$C_{30}H_{26}N_2O_9S$ 1H-Imidazole, 2-(methylsulfonyl)-1-(2,3,5-tri-O-benzoyl-β-D-ribofuranosyl)-	MeOH	231(4.60),274(3.45), 282(3.36)	44-4774-78

Compound	Solvent	$\lambda_{max}(\log \epsilon)$	Ref.
$C_{30}H_{26}O_4$			
Phenanthrene, 3,6-dimethoxy-9,10-bis-(4-methoxyphenyl)-	CH_2Cl_2	256(4.73),262s(4.72), 283(4.38),289(4.39), 305s(4.08),314(4.13), 355(3.43),372(3.41)	23-1246-78
$C_{30}H_{26}O_{13}$			
Trifolin, 6"-O-p-coumaroyl-	MeOH	268(4.31),300s(4.37), 315(4.41),360(4.08)	105-0275-78
$C_{30}H_{26}O_{14}$			
Isoquercitrin, 3"-O-p-coumaroyl-	MeOH	257s(4.39),270(4.42), 300s(4.51),314(4.55), 360(4.25)	105-0159-78
$C_{30}H_{26}S_3$			
2,17,32-Trithia[3.3.3](3,4',5)biphenylo<2>phane	$CHCl_3$	263(4.56),300s(3.56)	24-0205-78
isomer	dioxan	264(4.33)	24-0205-78
$C_{30}H_{27}NO_6S$			
Sulfonium, dimethyl-, 2-(benzoyloxy)-3-[4-(benzoyloxy)-3,4-dihydro-3-isoquinolinyl]-1-(methoxycarbonyl)-2-propenylide	EtOH	232(4.59),269(4.30), 307(4.06)	95-0198-78
$C_{30}H_{27}N_2O$			
Quinolinium, 2-[3-(1,2-dihydro-2-oxo-3H-indol-3-ylidene)-2-(4-ethylphenyl)-1-propenyl]-1-ethyl-, iodide	MeOH	401(3.79),548(4.29), 582(4.44)	64-0209-78B
$C_{30}H_{27}N_5O_{11}$			
Carbamic acid, (phenoxycarbonyl)[9-(2,3,5-tri-O-acetyl-β-D-ribofuranosyl)-9H-purin-6-yl]-, phenyl ester	EtOH	256s(3.88),269(4.00)	39-0131-78C
$C_{30}H_{28}N_2O_2$			
Acetic acid, cyano[3,4-dihydro-1,5-dimethyl-6-(4-methylphenyl)-3,4-diphenyl-2(1H)-pyridinylidene]-, methyl ester, trans	MeCN	259(4.06),339(4.23)	5-1222-78
$C_{30}H_{28}N_3O_2PS$			
Benzenesulfonamide, N-[(3,3-dimethyl-4-phenyl-3H-Pyrazol-5-yl)diphenylphosphoranylidene]-4-methyl-	MeOH	350(2.70)	150-4248-78
$C_{30}H_{28}O_{11}$			
1,2-Anthracenedicarboxylic acid, 9,10-dihydro-8-hydroxy-3,6-dimethoxy-7-[2-methoxy-5-(2-methoxyethyl)phenyl]-9,10-dioxo-, dimethyl ester	MeOH-1% HOAc	236(4.25),244s(4.22), 287(4.61),344s(3.79), 426(3.88)	12-2651-78
1,2-Anthracenedicarboxylic acid, 9,10-dihydro-7-[5-(2-hydroxyethyl)-2-methoxyphenyl]-3,6,8-trimethoxy-9,10-dioxo-, dimethyl ester	MeOH	223(4.60),286(4.65), 335s(3.98),356s(3.82)	12-2651-78
$C_{30}H_{28}O_{14}$			
4H-1-Benzopyran-4-one, 3-(1,3-benzodioxol-5-yl)-7-[(2,3,4,6-tetra-O-acetyl-β-D-glucopyranosyl)oxy]-	EtOH	221(4.07),264(3.97), 292(3.83)	106-0235B-78

Compound	Solvent	$\lambda_{max}(\log \epsilon)$	Ref.
$C_{30}H_{29}Cl_2N_4OP$ Phosphorin, 2,6-bis[(4-chlorophenyl)-azo]-4-cyclohexyl-1,1-dihydro-1-methoxy-1-phenyl-	EtOH	275(4.34),307(4.32), 579(4.55)	118-0846-78
$C_{30}H_{29}N_4O_2P$ 2,3,4(1H)-Phosphorintrione, 1-(1,1-dimethylethyl)-5-methyl-, 2,4-bis(1-naphthalenylhydrazone), 1-oxide	$CHCl_3$	275s(4.16),334(4.06), 373(4.08),428(4.18), 516(4.18)	139-0027-78B
$C_{30}H_{30}$ 1,1':4',1":4",1"'-Quaterphenyl, 2,2"',4,4"',6,6"'-hexamethyl-	C_6H_{12}	267.5(4.46)	44-2435-78
$C_{30}H_{30}N_2O_4$ 27,28-Diazaoctacyclo[12.12.2.0^{2,13}-0^{3,12}.0^{4,11}.0^{15,26}.0^{16,25}.0^{17,24}]-octacosa-5,7,9,18,20,22,27-heptaene-1,14-dicarboxylic acid, dimethyl ester	EtOH	250s(3.30)	5-1368-78
5H-Pyrano[3,2-c]quinolin-5-one, 4-[2-(4,9-dihydro-9-methyl-4-oxofuro[2,3-b]quinolin-2(3H)-ylidene)propyl]-2,3,4,6-tetrahydro-2,2,6-trimethyl-(pteledimerin)	MeOH	217(4.58),233(4.64), 252s(4.20),278(3.83), 287(3.89),316(4.06), 329(3.98)	88-3681-78
	MeOH-HCl	218(4.66),235(4.74), 278(3.87),287(4.01), 311(4.06),324s(3.85)	88-3681-78
$C_{30}H_{30}N_6O$ 7H-Benz[de]anthracen-7-one, 3-[(4,6-di-1-piperidinyl-1,3,5-triazin-2-yl)amino]-	DMF	460(4.71)	2-0106-78
$C_{30}H_{30}O_4$ 4,7-Ethano-1H-indene-8,9-diol, 1-(diphenylmethylene)-3a,4,7,7a-tetrahydro-5,6-dimethyl-, diacetate, c,c-	C_6H_{12}	241(3.98),296(4.23)	5-0440-78
c,c-	C_6H_{12}	240(3.99),295(4.22)	5-0440-78
t,c-	C_6H_{12}	241(3.99),294(4.24)	5-0440-78
t,t-	C_6H_{12}	240(3.99),295(4.23)	5-0440-78
$C_{30}H_{31}NO_8$ Benzaldehyde, 2-[(5,6-dihydro-1,2,3,10-tetramethoxy-6-methyl-4H-dibenzo[de-g]quinolin-9-yl)oxy]-4,5-dimethoxy-(dehydrothaliudine)	pH 2	202(4.63),222s(4.40), 235s(4.37),270(4.86), 305s(4.04),319(4.01), 347(3.47),364(3.44)	100-0271-78
	MeOH and MeOH-base	257(4.68),272(4.67), 330(4.24)	100-0271-78
$C_{30}H_{31}N_3O_8$ L-Glutamic acid, N-[N-[4-[methyl[(phenylmethoxy)carbonyl]amino]benzoyl]-L-phenylalanyl]-	MeOH	261(4.12)	87-1165-78
$C_{30}H_{31}N_4OP$ Phosphorin, 4-cyclohexyl-1,1-dihydro-1-methoxy-1-phenyl-2,6-bis(phenylazo)-	EtOH	264(4.33),304(4.25), 418(3.58),569(4.52)	118-0846-78
$C_{30}H_{32}N_4O_2S$ 3(2H)-Thiazolamine, N-[(4-nitrophenyl)-methylene]-4,5-diphenyl-2-[(1,1,3,3-tetramethylbutyl)imino]-	isoPrOH	292(4.17),433(4.07)	73-1227-78

Compound	Solvent	$\lambda_{max}(\log \epsilon)$	Ref.
$C_{30}H_{32}O_4$			
4,8-Etheno-5H-indeno[5,6-d]-1,3-dioxole, 5-[bis(4-methoxyphenyl)methylene]-3a,4,4a,7a,8,8a-hexahydro-2,2,9-trimethyl-, cis-cis	MeOH	254(4.24),300(4.34)	5-0440-78
trans-trans	MeOH	254(3.99),300(4.17)	5-0440-78
$C_{30}H_{32}O_7$			
4H-1-Benzopyran-4-one, 5,7-diacetoxy-2-(4-methoxyphenyl)-6,8-bis(3-methyl-2-butenyl)-	MeOH	230(4.41),266(4.31), 323(4.47)	78-3569-78
$C_{30}H_{32}O_9$			
[2,5'-Bibenzofuran]-3,3'-dimethanol, 2,2',3,3'-tetrahydro-2'-(4-hydroxy-3-methoxyphenyl)-5-(3-hydroxy-1-propenyl)-7,7'-dimethoxy- (herpetotriol)	n.s.g.	276(4.33)	88-4111-78
$C_{30}H_{32}O_{11}$			
Isophysalin F acetate	EtOH	252(3.64)	102-1647-78
$C_{30}H_{33}ClN_2O_9S$			
Thioxanthene, 2-chloro-9-(4-methylpiperazinomethyl)-7-propanoyl-, bis(hydrogen maleate hemihydrate)	MeOH	236(4.25),263(3.95), 314(4.11)	73-0471-78
$C_{30}H_{33}NO_8$			
Benzaldehyde, 4,5-dimethoxy-2-[(5,6,6a-7-tetrahydro-1,2,3,10-tetramethoxy-6-methyl-4H-dibenzo[de,g]quinolin-9-yl)oxy]-, (S)-	pH 2	221(4.61),280(4.30), 300(4.19),312(4.12)	100-0271-78
	MeOH	220(4.62),237s(4.48), 277(4.50),300(4.30), 312(4.30),337(4.01)	100-0271-78
$C_{30}H_{33}N_4O_3P$			
Phosphorin, 4-(1,1-dimethylethyl)-1,1-dihydro-1-methoxy-2,6-bis[(4-methoxyphenyl)azo]-1-phenyl-	EtOH	229(4.43),325(4.34), 412(3.62),570(4.56)	118-0846-78
$C_{30}H_{34}$			
Benzene, 1,1'-[(1,2,3,4-tetramethyl-3-cyclobutene-1,2-diyl)di-2,1-ethynediyl]bis[2,4,6-trimethyl-, trans	hexane	257(4.62),266(4.66), 290(3.14)	5-1675-78
Benzene, 1,1'-(3,4,5,6-tetramethyl-3,5-octadiene-1,7-diyne-1,8-diyl)-bis[2,4,6-trimethyl-, (E,E)-	hexane	234s(4.43),257s(4.18), 296s(4.70),307(4.80)	5-1675-78
(Z,Z)-	hexane	234s(4.40),283(4.57), 307(4.53)	5-1675-78
Bicyclo[4.2.0]octa-1,2,3,7-tetraene, 2,3,4,5-tetramethyl-7,8-bis(2,4,6-trimethylphenyl)-	hexane	234s(4.22),243(4.27), 275(4.46),284s(4.36), 415(3.02)	5-1675-78
$C_{30}H_{34}N_4$			
18-Dehydronigritanin	MeOH	226(4.77),282(4.16), 290(4.09)	32-0615-78
Usambarine	MeOH	226(4.81),275(4.17), 282(4.19),290(4.1)	102-1687-78
$C_{30}H_{34}N_4O$			
1(4H)-Naphthalenone, 2-[[4-(diethylamino)phenyl]amino]-4-[[4-(diethylamino)phenyl]imino]-	C_6H_{12}	270(4.46),287(4.43), 340(4.06),575(3.99)	48-0557-78

Compound	Solvent	$\lambda_{max}(\log \epsilon)$	Ref.
(cont.)	EtOH	270(4.43),290(4.37), 335(4.05),590(3.89)	48-0557-78 +80-0617-78
	ether	270(4.46),285(4.43), 340(4.06),580(3.98)	48-0557-78
	CCl_4	290(4.46),335(4.17), 590(4.01)	48-0557-78
Nigritanin, 18-dehydro-10-hydroxy-	MeOH	226(4.74),281(4.20), 290(4.12),310(3.76)	32-0615-78
$C_{30}H_{34}O_5$			
Apigenin, 7-O-prenyl-6,8-di-C-prenyl-	MeOH	221(3.91),272(4.21), 331(4.34)	78-3569-78
$C_{30}H_{34}O_8$			
Gibb-3-ene-1,10-dicarboxylic acid, 4a- hydroxy-2,7-dimethoxy-1-methyl-8- methylene-, 1,4a-lactone, 10-[2-(4- methoxyphenyl)-2-oxoethyl] ester	EtOH	225s(3.98),280(4.08)	78-0345-78
Gomisin H, benzoyl-	EtOH	217(4.66),250s(4.20), 282s(3.56)	94-0328-78
$C_{30}H_{34}O_{12}$			
Physalin E acetate	EtOH	228(3.95)	102-1641-78
$C_{30}H_{35}NO_8$			
1,1,3-Cyclohexanetricarboxylic acid, 4-[(dibenzoylamino)methyl]-, tri- ethyl ester	n.s.g.	225(4.15),250(4.07)	128-0097-78
$C_{30}H_{36}ClF_3N_2O_9$			
Maytansine, O^3-de[2-(acetylmethylamino)- 1-oxopropyl]-O^3-(trifluoroacetyl)-	EtOH	232(4.54),240s(4.51), 252(4.48),281(3.76), 288(3.76)	87-0031-78
$C_{30}H_{36}N_4$			
Nigritanin	MeOH	226(4.76),282(4.16), 290(4.09)	32-0615-78
$C_{30}H_{36}N_4O$			
Nigritanin, 10-hydroxy-	MeOH	226(4.75),282(4.17), 290(4.10),310(3.76)	32-0615-78
$C_{30}H_{36}N_4O_6$			
18-Nor-5α-pregna-8,11,13-triene-20α- carboxaldehyde, 3β-acetoxy-12-meth- yl-, 2,4-dinitrophenylhydrazone	EtOH	206(4.68),358(4.36)	39-0076-78C
$C_{30}H_{36}O_4$			
1-Cyclododecene-1-carboxylic acid, 2-(2-carboxy-1,2-diphenylethenyl)-, 1-ethyl ester, (?,E)-	EtOH	224(4.26),272s(4.00)	33-3018-78
1-Oxaspiro[4.11]hexadec-3-ene-6-carbox- ylic acid, 2-oxo-3,4-diphenyl-, ethyl ester	EtOH	215s(4.20),270(4.02)	33-3018-78
$C_{30}H_{38}BrClN_2O_9$			
Maytansine, O^3-(bromoacetyl)-O^3-de[2- (acetylmethylamino)-1-oxopropyl]-	EtOH	233(4.40),252(4.35), 280(3.66),289(3.65)	87-0031-78
$C_{30}H_{38}N_2$			
Benzenamine, 4-(9,10-dihydro-10-methyl-	EtOH	276(4.57)	104-0129-78

Compound	Solvent	$\lambda_{max}(\log \epsilon)$	Ref.
9-acridinyl)-N,N-bis(3-methylbutyl)-			104-0129-78
$C_{30}H_{38}N_2O_4$ Vincamine, O-phenyl-	MeOH	222(4.54),272(3.97)	33-1682-78
$C_{30}H_{38}N_4O_9$ L-Proline, 1-[N-[N-[4-methyl-2-nitro- 3-(phenylmethoxy)benzoyl]-L-threon- yl]-D-valyl]-, methyl ester	MeOH	216(4.773)	87-0607-78
$C_{30}H_{38}O_5$ Cyafrin A$_4$ acetonide benzoate	MeOH	231(4.27),280s(2.95)	23-2113-78
$C_{30}H_{38}O_6$ 24-Norchola-1,14,20,22-tetraen-3-one, 6,7-diacetoxy-21,23-epoxy-4,4,8- trimethyl-	n.s.g.	220(4.00)	2-1042B-78
$C_{30}H_{38}O_{12}$ Coleon Y, tri-O-acetyl-	MeOH	234(3.97)	33-0871-78
$C_{30}H_{39}ClN_2O_9$ Maytanacine	EtOH	233(4.48),242s(4.45), 252(4.45),281(3.73), 289(3.73)	87-0031-78
$C_{30}H_{39}OP$ Phosphine oxide, tris(2,3,5,6-tetra- methylphenyl)-	C_6H_{12}	289(3.77),295(3.76)	65-1394-78
$C_{30}H_{39}P$ Phosphine, tris(2,3,5,6-tetramethyl- phenyl)-	C_6H_{12}	205(4.96),298(4.03), 317(4.20)	65-1394-78
$C_{30}H_{42}N_8O_8$ 1,2-Hydrazinedicarboxylic acid, 1,1'- [azinobis[[4-(dimethylamino)phenyl]- methylidyne]]bis-, tetraethyl ester	MeOH	241(4.22),320s(4.25), 382(4.65)	18-3312-78
$C_{30}H_{42}O_2$ 2,5-Cyclohexadien-1-one, 4,4'-(1,2-eth- anediylidene)bis[2,6-bis(1,1-dimeth- ylethyl)-	hexane	243(4.29)	70-1452-78
$C_{30}H_{42}O_4$ ^{n}(17a)-Homo-C,18-dinorcholesta-13,15- 17-triene-3,7,11-trione, 9-hydroxy- 4,4,15,17a-tetramethyl-, (5α,9β)- (same spectrum in base)	EtOH	221(4.42),266(4.13), 320(3.46)	39-0471-78C
$C_{30}H_{42}O_6$ D(17a)-Homo-C,18-dinorchola-13,15,17- trien-24-oic acid, 3-acetoxy-9-hy- droxy-4,4,15,17a-tetramethyl-11-oxo-, methyl ester, (3β,5α,9β)-	EtOH	222(4.18),262(4.02), 312(3.37)	39-0471-78C
$C_{30}H_{44}Br_2O_2$ 5α-Ergosta-7,9(11),14-trien-3β-ol, 22α,23α-dibromo-, acetate	EtOH	228(3.99),236(4.01), 268(3.96)	39-0076-78C

$C_{30}H_{44}O_2-C_{30}H_{48}O_3$

Compound	Solvent	$\lambda_{max}(\log \epsilon)$	Ref.
$C_{30}H_{44}O_2$			
Ergosta-1,3,5,7-tetraen-3-ol, acetate	EtOH	251(4.97),360(4.86)	39-0074-78C
$C_{30}H_{44}O_5$			
Urs-12-en-30-oic acid, 18-hydroxy-3,6-dioxo-	EtOH	208(4.45)	102-1979-78
	EtOH-acid	215(3.98)	102-1979-78
	EtOH-base	210(5.01)	102-1979-78
$C_{30}H_{46}O$			
Ergosta-1,5,7-trien-3-one, 4,4-dimethyl-	EtOH	215(4.04),274(4.03)	39-0074-78C
$C_{30}H_{46}O_2$			
Ergosta-5,7-dien-3-one, 1α,2α-epoxy-4,4-dimethyl-	EtOH	271(3.98),282(3.98)	39-0074-78C
D:B-Friedoolean-5(10)-ene-1,6-dione	EtOH	264(3.99)	18-2702-78
2(5H)-Furanone, 4-[4,8,12-trimethyl-14-(2,6,6-trimethyl-1-cyclohexen-1-yl)-3,7,11-tetradecatrienyl]-, (E,E,E)- (same in acid and base)	MeOH	213(4.26)	35-0307-78
$C_{30}H_{46}O_3$			
2(5H)-Furanone, 5-hydroxy-4-[4,8,12-trimethyl-14-(2,6,6-trimethyl-1-cyclohexen-1-yl)-3,7,11-tetradeca-trienyl]-, (E,E,E)- (hydroxymokupal-ide)	MeOH	211(4.24)	35-0307-78
	MeOH-base	251(3.68)	35-0307-78
$C_{30}H_{46}O_4$			
[2,2'-Bioxepin]-5,5'(2H,2'H)-dione, 4,4',6,6'-tetrakis(1,1-dimethyl-ethyl)-2,2'-dimethyl-	CH_2Cl_2	248(3.90),292(3.75)	88-3557-78
$C_{30}H_{46}O_6$			
Propanedioic acid, [(3β,5α)-3-acetoxy-pregnan-21-ylidene]-, diethyl ester	$CHCl_3$	217(3.5)	23-0424-78
$C_{30}H_{48}$			
Olean-11,13(18)-diene	EtOH	242(4.50),251(4.55), 260(4.36)	39-0384-78C
$C_{30}H_{48}O$			
D:B-Friedoolean-5(10)-en-1-one	EtOH	249(4.28)	18-2702-78
D:B-Friedoolean-5(10)-en-6-one	EtOH	250(4.03)	18-2702-78
$C_{30}H_{48}O_2$			
Ergosta-5,7-dien-3-one, 1α-hydroxy-4,4-dimethyl-	EtOH	273(4.09),282(4.07)	39-0074-78C
D:A-Friedoolean-3-en-2-one, 3-hydroxy-	EtOH	278(3.94)	2-0351-78
	base	320(3.72)	2-0361-78
Oleana-9,12-diene-3β,7α-diol (castanopsin)	MeOH	286(3.9)	102-0575-78
$C_{30}H_{48}O_3$			
Cholest-5-en-7-one, 1α-acetoxy-3α-meth-yl-	EtOH	241(4.09)	44-0574-78
3β-	EtOH	241(4.08)	44-0574-78
Cucurbita-1(10),5-dien-7-one, 3β,11α-dihydroxy-	EtOH	242(3.70),302(4.03)	19-0583-78
Hexadecanoic acid, 2-(2,3-dihydro-2,4,6-trimethyl-3-oxo-1H-inden-5-yl)ethyl ester, (R)-	EtOH	217(4.53),260(4.18), 303(3.34)	94-2365-78

Compound	Solvent	$\lambda_{max}(\log \epsilon)$	Ref.
Lanost-5-ene-7,11-dione, 3β-hydroxy-	EtOH	240.5(4.05)	19-0583-78
5α-Lanost-8-ene-7,11-dione, 3β-hydroxy-	EtOH	273(3.86)	39-0329-78C
$C_{30}H_{48}O_4$			
4H-1-Benzopyran-4-one, 2-heptadecyl-5,7-dimethoxy-6,8-dimethyl-	EtOH	226(4.22),229(4.21), 256(4.01),280(3.65), 310(3.64)	94-2407-78
Hexadecanoic acid, 2-(2,3-dihydro-1-hydroxy-2,4,6-trimethyl-3-oxo-1H-inden-5-yl)ethyl ester	EtOH	217(4.56),258(4.18), 300(3.42)	94-2365-78
D(17a)-Homo-C,18-dinorcholesta-13,15,17-triene-3,7,9,11-tetrol, 4,4,15,17a-tetramethyl-, (3β,5α,7β,9β,11α)-	EtOH	208(4.43),272(2.74), 282(2.74)	39-0471-78C
$C_{30}H_{48}O_5$			
6,7-Secocholest-4-en-6-oic acid, 3β-acetoxy-7-oxo-, methyl ester	EtOH	210(3.89)	12-0171-78
$C_{30}H_{48}O_7$			
Prosta-5,13-dien-1-oic acid, 10-(ethoxycarbonyl)-9-oxo-15-[(tetrahydro-2H-pyran-2-yl)oxy]-, ethyl ester, (8RS,12RS,15SR,5Z,13E)-	EtOH-NaOH	288(4.20)	22-0131-78
$C_{30}H_{50}O$			
Ergosta-5,7-dien-3β-ol, 4,4-dimethyl-	EtOH	273(4.07)	39-0074-78C
Lup-20-en-28β-ol	EtOH	203(3.40)	2-0416-78
$C_{30}H_{50}O_2$			
Cholest-4-en-3-one, 4-(1-methylethoxy)-	C_6H_{12}	253(3.97)	78-1509-78
Ergosta-5,7-diene-1α,3α-diol, 4,4-dimethyl-	EtOH	273(4.01),282(4.01)	39-0074-78C
1α,3β-	EtOH	273(4.01),282(4.01)	39-0074-78C
$C_{30}H_{50}O_3$			
Dammar-24-en-26-al, 3β,20-dihydroxy-, (20S)-	n.s.g.	228(3.75)	39-0360-78C
5α-Lanost-8-en-7-one, 3β,11α-dihydroxy-	EtOH	253(3.98)	39-0329-78C
isomer 13	EtOH	256(4.14)	39-0329-78C
$C_{30}H_{50}O_5$			
6,7-Secocholest-4-en-6-oic acid, 3β-acetoxy-7-hydroxy-, methyl ester	EtOH	210(3.86)	12-0171-78
$C_{30}H_{56}Cl_4N_4Pd_2$			
Palladium, bis[2,3-bis[bis(1-methylethyl)amino]-2-cyclopropen-1-ylidene]di-μ-chlorodichlorodi-	CH_2Cl_2	325(3.61)	89-0457-78

Compound	Solvent	$\lambda_{max}(\log \epsilon)$	Ref.
$C_{31}H_{19}NO_3S$ 1H-Pyrido[2,1-b]benzothiazol-1-one, 3,4-dibenzoyl-2-phenyl-	MeOH	257(4.25),370(4.04)	44-2697-78
$C_{31}H_{20}N_2O$ Benzaldehyde, 3-(7-phenylbenzo[b]naph-tho[1,2-f][1,4]diazocin-14-yl)-	$CHCl_3$	240(4.73),268s(4.41), 350(3.45)	150-4944-78
Benzaldehyde, 4-(7-phenylbenzo[b]naph-tho[1,2-f][1,4]diazocin-14-yl)-	$CHCl_3$	240(4.61),265(4.63), 330s(3.70),360s(3.54)	150-4944-78
$C_{31}H_{20}N_4OS$ [1,2,4]Triazolo[1,5-a]pyridine, 2-[4-[2-[5-(6-phenyl-2-benzoxazolyl)-2-thienyl]ethenyl]phenyl]-	DMF	405(4.85)	33-0142-78
$C_{34}H_{21}NO$ 1,2-Benzisoxazole, 3-[1,1'-biphenyl]-4-yl-6-[2-(1-naphthalenyl)ethenyl]-	DMF	342(4.51)	33-2904-78
1,2-Benzisoxazole, 3-[1,1'-biphenyl]-4-yl-6-[2-(2-naphthalenyl)ethenyl]-	DMF	337(4.70)	33-2904-78
$C_{31}H_{21}NO_4$ Dinaphtho[3,2-c:2',3'-g]indole-5,8,13,16-tetrone, 5b,6,14,15-tetra-hydro-6-(4-methylphenyl)-	EtOH	213(4.06),241(4.29), 255(4.29),263s(4.11), 309(3.57),370(3.81), 417(3.81)	104-1962-78
$C_{31}H_{21}S_2$ Thiopyrylium, 2,4-diphenyl-6-(9H-thio-xanthen-9-ylidenemethyl)-, perchlor-ate	EtOH $CHCl_3$	554(3.528) 568(3.654)	83-0226-78 83-0226-78
$C_{31}H_{22}N_2$ Benzo[b]naphtho[1,2-f][1,4]diazocine, 14-(3-methylphenyl)-7-phenyl-, (±)-	$CHCl_3$	240(4.64),255(4.59), 350(3.52)	150-4944-78
Benzo[b]naphtho[1,2-f][1,4]diazocine, 14-(4-methylphenyl)-7-phenyl-, (±)-	$CHCl_3$	240(4.68),255(4.62), 345(3.57)	150-4944-78
Imidazo[1,2-a]pyridine, 2-[1,1'-biphen-yl]-4-yl-7-[2-(1-naphthalenyl)ethenyl]-	DMF	378(4.66)	33-0129-78
Imidazo[1,2-a]pyridine, 2-[1,1'-biphen-yl]-4-yl-7-[2-(2-naphthalenyl)ethenyl]-	DMF	373(4.76),391(4.72)	33-0129-78
$C_{31}H_{22}N_2O_3$ 1,2-Benzisoxazole, 3-[4-[2-[4-[3-(4-methoxyphenyl)-5-isoxazolyl]phenyl]-ethenyl]phenyl]-	DMF	352(4.84)	33-2904-78
$C_{31}H_{22}O$ 2,5-Cyclohexadien-1-one, 4-(diphenyl-methylene)-2,6-diphenyl-	benzene	405(4.7)	70-1304-78
3-Cyclopenten-1-one, 3,4-diphenyl-2,5-bis(phenylmethylene)-, (E,E)-	$CHCl_3$	322(4.56),352(4.42)	39-0989-78C
(E,Z)-	$CHCl_3$	324(4.54),354(4.36)	39-0989-78C
(Z,Z)-	$CHCl_3$	325(4.51),363(4.34)	39-0989-78C
$C_{31}H_{22}O_3$ Naphtho[2,3-c]furan-1(3H)-one, 3-(2-methoxyphenyl)-4,9-diphenyl-	THF	245(4.82),293(3.94), 305(3.97),331(3.68), 345(3.79)	78-2263-78

Compound	Solvent	$\lambda_{max}(\log \epsilon)$	Ref.
$C_{31}H_{22}O_4$ 2-Naphthalenecarboxylic acid, 3-(2-methoxybenzoyl)-1,4-diphenyl-	THF	246(4.75),282(3.94), 292(3.95),304(3.97), 327(3.49),342(3.57)	78-2263-78
$C_{31}H_{22}O_8$ [2,9'-Bianthracene]-9,10,10'(9'H)-trione, 1,4',5',8-tetrahydroxy-9'-methoxy-2',6-dimethyl-	MeOH	228(4.30),263(3.88), 292(3.58),395(3.22), 435(3.13)	94-1111-78
$C_{31}H_{23}NO_2$ 4,11-Etheno-1H-naphth[2,3-f]isoindole-1,3(2H)-dione, 3a,4,11,11a-tetrahydro-2-methyl-5,10-diphenyl-, (3aα,4β,11β,11aα)-	ether	286(4.03),295(3.99)	88-0687-78
2H-Isoindole, 2-methyl-1-phenyl-3-[2-(3-phenyl-1-isobenzofuranyl)ethenyl]-, 2-oxide	benzene MeOH	358(3.56),491(4.65) 278(--),348(--), 350(--),499(--)	40-0144-78 40-0144-78
$C_{31}H_{23}NO_4$ 7,11-Etheno-7H-bisoxirano[1,2:3,4]naphtho[2,3-f]isoindole-8,10(7aH,9H)-dione, 1a,5b,10a,11-tetrahydro-9-methyl-1a,5b-diphenyl-, (1aα,5bα,6aS*,7α-7aβ,10aβ,11α,11aR*)-	ether	253(2.93),259(3.00), 265(3.03),272(3.04), 279(3.04)	88-0687-78
$C_{31}H_{23}N_2O_2P$ Cyclohepta[b]pyrrole-2-acetic acid, 3-cyano-α-(triphenylphosphoranylidene)-, methyl ester	$CHCl_3$	269(4.34),307(4.31), 318s(4.23),354(3.76), 450(4.22)	18-1573-78
$C_{31}H_{23}O$ [1,1':3',1''-Terphenyl]-2'-yloxy, 5'-(diphenylmethyl)-	MeCN	460(3.95),750(3.70)	70-1304-78
$C_{31}H_{24}$ Fluoranthene, 7,10-dimethyl-8-(4-methylphenyl)-9-phenyl-	C_6H_{12}	297(4.41)	2-0152-78
Spiro[2.4]hepta-4,6-diene, 4,5,6,7-tetraphenyl-	MeCN	248(4.67),303(3.95)	5-1648-78
$C_{31}H_{24}Cl_2N_2O_8S$ 2(1H)-Pyrimidinone, 1-[2,5-bis-O-(4-chlorobenzoyl)-3-O-(4-methylbenzoyl)-β-D-arabinofuranosyl]-3,4-dihydro-4-thioxo-	MeOH	238(4.77),327(4.28)	44-0350-78
$C_{31}H_{24}Cl_2N_2O_9$ 2,4(1H,3H)-Pyrimidinedione, 1-[2,5-bis-O-(4-chlorobenzoyl)-3-O-(4-methylbenzoyl)-β-D-arabinofuranosyl]-	MeOH	241(4.85),267s(3.96)	44-0350-78
$C_{31}H_{24}N_2O_2$ 1,2-Benzisoxazole, 3-[4-[2-[4-[5-(1-methylethyl)-2-benzoxazolyl]phenyl]ethenyl]phenyl]-	DMF	362(4.87)	33-2904-78
1,2-Benzisoxazole, 3-[4-[2-[4-(5-propyl-2-benzoxazolyl)phenyl]ethenyl]phenyl]-	DMF	362(4.87)	33-2904-78

Compound	Solvent	$\lambda_{max}(\log \epsilon)$	Ref.
$C_{31}H_{24}N_4O_7S_2$			
Imidazo[1,2-a]-1,3,5-triazine-2,4(3H-8H)-dithione, 8-(2,3,5-tri-O-benzoyl-α-D-ribofuranosyl)-	MeOH	230(4.58),297(4.48), 312s(4.44)	44-4784-78
β-	MeOH	230(4.61),296(4.49), 315s(4.42)	44-4784-78
$C_{31}H_{24}N_4O_8S$			
Imidazo[1,2-a]-1,3,5-triazin-2(1H)-one, 3,4-dihydro-4-thioxo-1-(2,3,5-tri-O-benzoyl-β-D-ribofuranosyl)-	MeOH	218(4.55),229(4.60), 278s(4.20),283(4.23), 291(4.21)	44-4774-78
Imidazo[1,2-a]-1,3,5-triazin-2(3H)-one, 4,8-dihydro-4-thioxo-8-(2,3,5-tri-O-benzoyl-α-D-ribofuranosyl)-	MeOH	230(4.61),270(4.16), 300(4.05)	44-4784-78
β-	MeOH	230(4.62),271(4.19), 299(4.11)	44-4784-78
$C_{31}H_{24}N_4O_9$			
Imidazo[1,2-a]-1,3,5-triazine-2,4(1H-3H)-dione, 1-(2,3,5-tri-O-benzoyl-β-D-ribofuranosyl)-	MeOH	230(4.64),274(3.56), 282(3.42)	44-4774-78
Imidazo[1,2-a]-1,3,5-triazine-2,4(3H-8H)-dione, 8-(2,3,5-tri-O-benzoyl-α-D-ribofuranosyl)-	MeOH	230(4.67),256s(4.09), 281s(3.58)	44-4784-78
β-	MeOH	231(4.69),255s(4.08), 281s(3.48)	44-4784-78
$C_{31}H_{24}O$			
[1,1':3',1''-Terphenyl]-2'-ol, 5'-(diphenylmethyl)-	benzene	305(3.9)	70-1304-78
$C_{31}H_{24}O_2$			
3-Butenoic acid, 2-(9H-fluoren-9-ylidene)-4,4-bis(4-methylphenyl)-	EtOH	247(4.44),296(4.09), 322(3.56),337(3.79)	2-0502-78
2-Propen-1-one, 3-(1-benzoylcyclopropyl)-1,2,3-triphenyl-, (Z)-	MeCN	249(4.42),290s(3.84), 315s(3.58)	5-1648-78
$C_{31}H_{24}O_3$			
3-Butenoic acid, 2-(9H-fluoren-9-ylidene)-4-(4-methoxyphenyl)-4-(4-methylphenyl)-	EtOH	225(4.55),267(4.60), 287(4.11),333(4.14)	2-0502-78
$C_{31}H_{24}O_4$			
3-Butenoic acid, 2-(9H-fluoren-9-ylidene)-4,4-bis(4-methoxyphenyl)-	EtOH	248(4.43),285(4.05), 340(3.60)	2-0502-78
$C_{31}H_{24}O_7$			
[2,9'-Bianthracene]-9,10'(9'H,10H)-dione, 1,4',5',8-tetrahydroxy-9'-methoxy-2',6-dimethyl-	MeOH	263(3.48),271(3.54), 306(3.54),375(3.65)	94-1111-78
$C_{31}H_{24}O_7S$			
Pyrylium, 2-[2-(2-hydroxy-1-naphthalenyl)ethenyl]-4,6-diphenyl-, carboxymethylsulfonate	EtOH	483(4.348)	83-0256-78
$C_{31}H_{25}ClN_4O$			
[1,2,4]Triazolo[1,5-a]pyridine, 2-[4-[2-[2-chloro-4-[5-(1,1-dimethylethyl)-2-benzoxazolyl]phenyl]ethenyl]phenyl]-	DMF	368(4.83)	33-0142-78

Compound	Solvent	λ_{max}(log ϵ)	Ref.
[1,2,4]Triazolo[1,5-a]pyridine, 7-[2-[2-chloro-4-[5-(1,1-dimethylethyl)-2-benzoxazolyl]phenyl]ethenyl]-2-phenyl-	DMF	366(4.81)	33-0142-78
$C_{31}H_{25}NO_3$			
1,2-Benzisoxazole, 3-[3,4-bis[2-(4-methoxyphenyl)ethenyl]phenyl]-	DMF	318(4.70)	33-2904-78
1,2-Benzisoxazole, 6-[2-(2-methoxyphenyl)ethenyl]-3-[4-[2-(2-methoxyphenyl)-ethenyl]phenyl]-	DMF	347(4.79)	33-2904-78
1,2-Benzisoxazole, 6-[2-(3-methoxyphenyl)ethenyl]-3-[4-[2-(3-methoxyphenyl)-ethenyl]phenyl]-	DMF	339(4.84)	33-2904-78
1,2-Benzisoxazole, 6-[2-(4-methoxyphenyl)ethenyl]-3-[4-[2-(4-methoxyphenyl)-ethenyl]phenyl]-	DMF	350(4.89)	33-2904-78
$C_{31}H_{25}N_5O_8$			
Imidazo[1,2-a]-1,3,5-triazin-4(8H)-one, 2-amino-8-(2,3,5-tri-O-benzoyl-β-D-ribofuranosyl)-	pH 1 pH 7 pH 11	233(4.66),268(4.27) 235(4.60),263s(4.39) 234(4.63),260s(4.31)	87-0883-78 87-0883-78 87-0883-78
$C_{31}H_{26}ClNO_3S$			
1,2-Benzisoxazole, 6-chloro-3-[4-[2-[4-[[4-(1,1-dimethylethyl)phenyl]-sulfonyl]phenyl]ethenyl]phenyl]-	DMF	338(4.74)	33-2904-78
$C_{31}H_{26}Cl_2$			
Cyclopropane, 1,1-bis(4-chlorophenyl)-2,2-dimethyl-3-(2,2-diphenylethenyl)-	EtOH	234(4.46),268(4.28)	35-4131-78
Cyclopropane, 3-[2,2-bis(4-chlorophenyl)ethenyl]-2,2-dimethyl-1,1-diphenyl-	EtOH	273(4.26)	35-4131-78
$C_{31}H_{26}N_2O_3$			
Carbamic acid, [3-(3-oxo-1,3-diphenyl-propyl)-1H-indol-2-yl]-, phenylmethyl ester	CHCl$_3$	283(4.18),295(4.18)	103-0745-78
$C_{31}H_{26}N_2O_8S$			
2(1H)-Pyrimidinone, 3,4-dihydro-5-methyl-4-thioxo-1-(2,3,5-tri-O-benzoyl-β-D-xylofuranosyl)-	EtOH	242(3.88),332(4.17)	65-1725-78
$C_{31}H_{26}N_2O_9$			
2,4(1H,3H)-Pyrimidinedione, 5-methyl-1-(2,3,5-tri-O-benzoyl-β-D-xylo-furanosyl)-	EtOH	243(3.97),268(3.96)	65-1725-78
$C_{31}H_{26}N_4O$			
Ethanone, 1-(1,3,7,9-tetraphenyl-1,2,7,8-tetraazaspiro[4.4]nona-2,8-dien-6-yl)-	EtOH	242(4.54),339(4.47)	22-0415-78
[1,2,4]Triazolo[1,5-a]pyridine, 2-[4-[2-[4-[5-(1,1-dimethylethyl)-2-benz-oxazolyl]phenyl]ethenyl]phenyl]-	DMF	367(4.92),387(4.74)	33-0142-78
$C_{31}H_{26}N_4O_2$			
1,2,7,8-Tetraazaspiro[4.4]nona-2,8-di-ene-6-carboxylic acid, 1,3,7,9-tetra-phenyl-, methyl ester	EtOH	239(4.48),339(4.44)	22-0415-78

Compound	Solvent	$\lambda_{max}(\log \epsilon)$	Ref.
$C_{31}H_{26}O_7$			
4H-1-Benzopyran-4-one, 2-[3,4-bis(phen-ylmethoxy)phenyl]-7-hydroxy-5,8-di-methoxy-	MeOH MeOH-AlCl$_3$ MeOH-NaOAc	270(3.32),335(3.29) 270(--),325(--) 280(--)	2-1079-78 2-1079-78 2-1079-78
$C_{31}H_{27}NO_3S$			
1,2-Benzisoxazole, 3-[4-[2-[4-[[4-(1,1-dimethylethyl)phenyl]sulfonyl]phen-yl]ethenyl]phenyl]-	DMF	337(4.73)	33-2904-78
$C_{31}H_{28}O_4$			
Spiro[7H-dibenzo[a,c]cyclononene-7,2'-[1,3]dioxolane]-5,9-diol, 5,6,8,9-tetrahydro-5,9-diphenyl-	EtOH	253(3.64),257s(3.62), 325s(2.45)	150-5151-78
$C_{31}H_{28}O_{12}$			
1,2-Anthracenedicarboxylic acid, 7-[5-[2-(formyloxy)ethyl]-2-methoxyphen-yl]-9,10-dihydro-3,6,8-trimethoxy-9,10-dioxo-, dimethyl ester	MeOH	221(4.58),285(4.64), 335(4.55),355(4.41)	12-2651-78
$C_{31}H_{28}O_{14}$			
Astragalin, 6"-O-feruloyl-	MeOH	267(4.28),300s(4.36), 330(4.40)	105-0490-78
	MeOH-NaOMe MeOH-NaOAc	275(--),388(--) 276(--),318(--), 378(--)	105-0490-78 105-0490-78
$C_{31}H_{29}NO_6S$			
Pyrylium, 2-[4-[4-(dimethylamino)phen-yl]-1,3-butadienyl]-4,6-diphenyl-, carboxymethylsulfonate	EtOH	700(4.505)	83-0256-78
$C_{31}H_{29}N_6O_4$			
Quinolinium, 4-[[4-[[4-(dimethylamino)-phenyl]methylene]-2-(2,4-dinitrophen-yl)-2,4-dihydro-5-methyl-3H-pyrazol-3-ylidene]methyl]-1-ethyl-, iodide	n.s.g.	210(4.54),240(4.48)	48-0857-78
$C_{31}H_{29}N_7O_3$			
1,2-Propanedione, 1-(3,7-diacetyl-1,3a-4a,5,7a,7b-hexahydro-1,5-diphenyl-4H-pyrrolo[2,3-c:4,5-c']dipyrazol-4-yl)-, 1-(phenylhydrazone), (3aα,4aβ-7aβ,7bα)-	EtOH	239(4.42),298s(3.00), 355(4.69)	4-1485-78
$C_{31}H_{30}BrNO_3$			
Isoquinoline, 1-[[2-bromo-4-(phenyl-methoxy)phenyl]methyl]-1,2,3,4-tetra-hydro-6-methoxy-7-(phenylmethoxy)-	EtOH	206(4.82),231s(4.59), 283(3.94)	2-0421-78
$C_{31}H_{30}O_9$			
4H,8H-Benzo[1,2-b:3,4-b']dipyran-4-one, 5-acetoxy-2-(2,4-diacetoxyphenyl)-8,8-dimethyl-3-(3-methyl-2-butenyl)-	EtOH EtOH-AlCl$_3$	235(4.47),262(4.53), 318(3.86) 235(4.49),262(4.53), 318(3.91)	94-1394-78 94-1394-78
$C_{31}H_{30}O_{10}$			
3H,7H-Pyrano[2',3':7,8][1]benzopyrano-[3,2-d][1]benzoxepin-7-one, 6,12-di-	EtOH	232(4.53),269(4.47), 324(4.11)	94-1394-78

Compound	Solvent	$\lambda_{max}(\log \epsilon)$	Ref.
acetoxy-9-(1-acetoxy-1-methylethyl)-8,9-dihydro-3,3-dimethyl-, (±)- (cont.)	EtOH-AlCl$_3$	232(4.54),269(4.49), 324(4.13)	94-1394-78
$C_{31}H_{30}O_{11}$			
1,2-Anthracenedicarboxylic acid, 9,10-dihydro-3,6,8-trimethoxy-7-[2-methoxy-5-(2-methoxyethyl)phenyl]-9,10-dioxo-, dimethyl ester	MeOH	223(4.57),285(4.63), 336s(3.93),352s(3.82)	12-2651-78
Ethaneperoxoic acid, 1-(6,12-diacetoxy-8,9-dihydro-3,3-dimethyl-7-oxo-3H,7H-pyrano[2',3':7,8][1]benzopyrano[3,2-d][1]benzoxepin-9-yl)-1-methylethyl ester	EtOH	233(4.64),271(4.58), 325(4.24)	94-1431-78
	EtOH-AlCl$_3$	236(4.86),270(4.78), 321(4.42)	94-1431-78
$C_{31}H_{31}N_4$			
Quinolinium, 4-[[4-[[4-(dimethylamino)-phenyl]methylene]-2,4-dihydro-5-methyl-2-phenyl-3H-pyrazol-3-ylidene]-methyl]-1-ethyl-, iodide	n.s.g.	230(4.49),370(4.49), 360(4.08)	48-0857-78
$C_{31}H_{32}N_2O_7$			
Lanciferine	EtOH	236(3.87),280(4.20)	42-1096-78
$C_{31}H_{32}N_2O_8$			
Lanciferine, 10-hydroxy-	EtOH	240(3.89),280(4.19)	42-1096-78
$C_{31}H_{32}N_4O_5$			
8H-Oxazolo[4,5-b]phenoxazine-4,6-di-carboxamide, N,N,N',N'-tetraethyl-9,11-dimethyl-8-oxo-2-phenyl-	CHCl$_3$	253(4.26),282(4.33), 302(4.31),322(4.35), 379(4.19),504(4.18)	4-0129-78
$C_{31}H_{32}O_6$			
4a,9a-(Epoxyethanoxy)-5,9-etheno-5H-indeno[5,6-b]dioxin, 6-[bis(4-methoxyphenyl)methylene]-2,3,5a,6,8a,9-hexahydro-14-methyl-	MeOH	254(4.25),299(4.32)	5-0440-78
4,7-Ethano-1H-indene-8,9-diol, 1-[bis-(4-methoxyphenyl)methylene]-3a,4,7,7a-tetrahydro-5-methyl-, diacetate, c,c-	MeOH	255(4.25),300(4.34)	5-0440-78
t,c-	C$_6$H$_{12}$	255(4.24),300(4.32)	5-0440-78
t,t-	MeOH	255(4.18),300(4.29)	5-0440-78
$C_{31}H_{32}O_8$			
4H-1-Benzopyran-4-one, 5,7-diacetoxy-2-(4-acetoxyphenyl)-6,8-bis(3-methyl-2-butenyl)-	MeOH	254(3.92),320(4.21)	78-3569-78
$C_{31}H_{33}NO_{10}S_3$			
6-Oxabicyclo[3.2.0]heptane-5-carbonitrile, 1,4,4-tris[[[(4-methylphenyl)-sulfonyl]oxy]methyl]-	dioxan	257(3.12),263(3.22), 267(3.21),274(3.16)	97-0380-78
$C_{31}H_{34}ClN_2$			
3H-Indolium, 2-[2-[2-chloro-3-[(1,3-di-hydro-1,3,3-trimethyl-2H-indol-2-yli-dene)ethylidene]-1-cyclopenten-1-yl]-ethenyl]-1,3,3-trimethyl-, perchlorate	MeOH	798(5.48)	104-2046-78

Compound	Solvent	$\lambda_{max}(\log \epsilon)$	Ref.
$C_{31}H_{34}Cl_4N_3O_{12}P$ 3'-Cytidylic acid, N-(4-methoxybenz-oyl)-2'-O-(tetrahydro-4-methoxy-2H-pyran-4-yl)-, 2-chlorophenyl 2,2,2-trichloroethyl ester	EtOH	289(4.45)	78-1999-78
$C_{31}H_{34}N_4O_4$ 5H-Oxazolo[4,5-b]phenoxazine-4,6-di-carboxamide, N,N,N',N'-tetraethyl-9,11-dimethyl-2-phenyl-	CHCl$_3$	253(4.53),288(4.12), 387(4.20)	4-0129-78
$C_{31}H_{34}N_4O_{10}$ D-Valine, N-[N-[4-methyl-2-nitro-3-(phenylmethoxy)benzoyl]-L-threo-nyl]-, (4-nitrophenyl)methyl ester	MeOH	208(4.718)	87-0607-78
$C_{31}H_{34}O_8$ Benzoic acid, 3,4-diacetoxy-, [1,2,3,4-4a,6-hexahydro-1,4a-dimethyl-7-(1-methylethyl)-6-oxo-5-hydroxy-1-phen-anthrenyl]methyl ester, (1S-cis)-	ether	232(4.28),281s(3.23), 418(4.10)	33-0709-78
Benzoic acid, 3,4-diacetoxy-, 1,2,3,4-4a,6-hexahydro-5-hydroxy-1,1,4a-tri-methyl-7-(1-methylethyl)-6-oxo-3-phenanthrenyl ester, (3S-trans)-	ether	232(4.32),281s(3.14), 415(4.10)	33-0709-78
4H-1-Benzopyran-4-one, 5,7-diacetoxy-3-(3,4-dimethoxyphenyl)-6,8-bis(3-methyl-2-butenyl)-	MeOH	206(4.02),254(4.32)	44-3446-78
$C_{31}H_{34}O_9$ Kuwanon E triacetate	EtOH	212(4.70),278(4.16), 334(3.69)	142-1295-78A
	EtOH-NaOMe	325(4.22),495(4.27)	142-1295-78A
	EtOH-AlCl$_3$	212(4.72),301(4.19), 390(3.75)	142-1295-78A
$C_{31}H_{36}N_4O_{18}S$ β-D-Glucopyranuronic acid, 1-deoxy-1-[6-[(2,3,4-tri-O-acetyl-6-methyl-β-D-glucopyranuronosyl)thio]-9H-purin-9-yl]-, methyl ester, 2,3,4-triacetate	MeOH	275(4.31)	106-0250-78
$C_{31}H_{38}N_4O_2$ 21H-Biline-1,19-dione, 7,8,12,13-tetra-ethyl-22,24-dihydro-2,3,7,18-tetra-methyl-, (Z,Z,Z)-	CHCl$_3$	275(4.27),310s(4.38), 370(4.71),650(4.16)	64-0924-78B
$C_{31}H_{38}O_{12}$ Picras-3-ene-2,11,16-trione, 1,12,20-triacetoxy-6-[(3-methyl-1-oxo-2-but-enyl)oxy]-, (1β,6α,12α)-	MeOH	226(4.32),238s(4.27), 245s(4.20)	87-1186-78 +100-0578-78
$C_{31}H_{40}O_2$ 28-Norursa-9(11),12,15,17,19,21-hexaen-3β-ol, acetate	C$_6$H$_{12}$	209(4.22),236(4.35), 265s(--),274(4.19), 285(4.20)	24-1160-78
$C_{31}H_{40}O_3$ Phenol, 2,2',2"-methylidynetris[4-(1,1-dimethylethyl)-	EtOH	278(3.28)	32-0079-78

Compound	Solvent	$\lambda_{max}(\log \epsilon)$	Ref.
Phenol, 2,2',2"-methylidynetris[6-(1,1-dimethylethyl)-	EtOH	263(3.83),278(3.82)	32-0079-78
$C_{31}H_{41}ClN_2O_9$ Maytansinol 3-propanoate	EtOH	233(4.50),252(4.45), 280(3.75),288(3.74)	87-0031-78
$C_{31}H_{41}NO_4$ Veratraman-3-ol, 28-acetyl-4,5,14,15-16,17-hexadehydro-16,23-epoxy-5,6-dihydro-, acetate	MeOH	280(3.45),288(3.47)	150-4523-78
$C_{31}H_{41}N_3O_8$ L-Glutamic acid, N-[N-[4-[methyl[(phenylmethoxy)carbonyl]amino]benzoyl]-L-leucyl]-, diethyl ester	MeOH	259(4.13)	87-1165-78
$C_{31}H_{42}O_{11}$ 1H-2,8a-Methanocyclopenta[a]cyclopropa-[e]cyclodecen-11-one, 5-acetoxy-1a,2-5,5a,6,7,8,9,10,10a-decahydro-5a,6,10-trihydroxy-4-(hydroxymethyl)-1,1,7,9-tetramethyl-, triacetate propanoate	MeOH	211(3.42)	88-2119-78
$C_{31}H_{43}NO_4$ Antibiotic X-14547A	n.s.g.	229(4.10),245(4.38)	35-6784-78
$C_{31}H_{44}O_8$ 14β-Bufa-4,20,22-trienolide, 14-hydroxy-3β-[(4-O-methyl-α-L-rhamnopyranosyl)-oxy]-	MeOH	297(3.79),355(4.65)	145-0493-78
$C_{31}H_{46}O_2$ 28-Noroleana-9(11),12,17-trien-3β-ol, acetate	C_6H_{12}	306s(--),319(4.32), 333s(--)	24-1160-78
28-Norursa-9(11),12,17-trien-3β-ol, acetate	C_6H_{12}	293s(--),308(3.93), 321(4.12),338s(--)	24-1160-78
$C_{31}H_{46}O_4$ 30-Norlup-12-ene-11,20-dione, 29-acetoxy-	C_6H_{12}	237(4.04)	73-2190-78
$C_{31}H_{46}O_9$ Cyasterone, 22-acetyl-	EtOH	240(4.07)	105-0175-78
$C_{31}H_{47}NO_2$ Molliorin C	EtOH	257(4.08)	31-0300-78
$C_{31}H_{48}O_4$ Cholesta-5,7-diene-1α,3β-diol, diacetate	EtOH	263(--),271(--), 282(4.10),294(--)	94-2933-78
3-Epiisomasticadienolalic acid, methyl ester	EtOH	212(4.07)	102-2107-78
$C_{31}H_{48}O_5$ Cholest-4-en-3-one, 4,6β-diacetoxy-	n.s.g.	247(3.48)	2-0426-78
Cholest-5-en-7-one, 1α,3β-diacetoxy-	EtOH	237(4.10)	44-0574-78
$C_{31}H_{50}O_3$ Cholesta-5,7-diene-1α,3α-diol, 4,4-dimethyl-, 3-acetate	ether	261s(3.85),272(3.95), 282(3.95),293s(3.70)	35-5626-78

Compound	Solvent	$\lambda_{max}(\log \epsilon)$	Ref.
Cholesta-5,7-diene-1α,3β-diol, 4,4-di-methyl-, 3-acetate	ether	261s(3.85),272(3.97), 282(3.97),293s(3.66)	35-5626-78
$C_{31}H_{50}O_4$			
Hexadecanoic acid, 2-[2,3-dihydro-2-(hydroxymethyl)-2,4,6-trimethyl-3-oxo-1H-inden-5-yl]ethyl ester, (S)-	EtOH	217(4.57),258(4.20), 304(3.44)	94-2365-78
Isomasticadienediolic acid, methyl ester	EtOH	208(4.04)	102-2107-78
$C_{31}H_{50}O_6S$			
Card-20(22)-enolide, 14-hydroxy-3-[(octylsulfonyl)oxy]-, (3β,5β)-	MeOH	216(4.23)	106-0706-78
$C_{31}H_{55}NO_6$			
4-Hexadecenamide, N-[2-hydroxy-2-(4-hydroxy-3,5,5-trimethyl-6-oxo-1-cyclohexen-1-yl)-1-(methoxymethyl)ethyl]-7-methoxy-9-methyl-	MeOH	235(3.81)	44-4359-78
$C_{31}H_{55}N_3O_6$			
Docosanamide, N-(1-β-D-arabinofuranosyl-1,2-dihydro-2-oxo-4-pyrimidinyl)-	isoPrOH	216(4.21),248(4.18), 303(3.91)	94-0981-78

Compound	Solvent	$\lambda_{max}(\log \epsilon)$	Ref.
$C_{32}H_{20}$			
Benzo[a]aceanthrylene, 7,8-diphenyl-	EtOH	276(4.5),308(4.0), 324(4.0),353(2.9), 370(3.2),432(3.7)	4-1185-78
$C_{32}H_{20}Br_2N_4S_2$			
Disulfide, bis[4-(3-bromophenyl)-6-phenyl-2-pyrimidinyl]	EtOH	263(4.94),320(4.49)	94-1298-78
$C_{32}H_{20}Cl_2N_4S_2$			
Pyrimidine, 2,2'-dithiobis[4-(3-chlorophenyl)-6-phenyl-	EtOH	262(4.66),321(4.23)	4-0105-78
Pyrimidine, 2,2'-dithiobis[4-(4-chlorophenyl)-6-phenyl-	EtOH	268(4.86),323(4.46)	4-0105-78
$C_{32}H_{20}N_2$			
3,5-Cyclopentadiene-1,3-dicarbonitrile, 2-[4-(2,4,6-cycloheptatrien-1-ylidene)-2,5-cyclohexadien-1-ylidene]-4,5-diphenyl-	acetone MeCN DMSO	733(4.23) 687(4.26) 672(4.23)	138-0237-78 138-0237-78 138-0237-78
$C_{32}H_{20}N_2O_2$			
1,2-Benzisoxazole, 3-[4-[2-[4-(2-benzoxazolyl)-1-naphthalenyl]ethenyl]-phenyl]-	DMF	301(4.18),379(4.64)	33-2904-78
$C_{32}H_{20}O_4$			
Methanone, (9-hydroxy-3-phenylbenzo[b]-cyclobut[e]oxepin-1,2-diyl)bis-[phenyl-	MeOH	254(4.64),287(4.30), 326(4.15)	78-2305-78
$C_{32}H_{21}N_3O$			
[1,2,4]Triazolo[1,5-a]pyridine, 2-[1,1'-biphenyl]-4-yl-7-[2-(3-dibenzofuranyl)ethenyl]-	DMF	352(4.87),368(4.77)	33-0142-78
$C_{32}H_{21}N_5O$			
[1,2,4]Triazolo[1,5-a]pyridine, 2-[4-[2-[4-[5-(1-naphthalenyl)-1,3,4-oxadiazol-2-yl]phenyl]ethenyl]phenyl]-	DMF	362(4.88)	33-0142-78
[1,2,4]Triazolo[1,5-a]pyridine, 2-[4-[2-[4-[5-(2-naphthalenyl)-1,3,4-oxadiazol-2-yl]phenyl]ethenyl]phenyl]-	DMF	362(4.92)	33-0142-78
$C_{32}H_{22}N_4O_4$			
9,10-Anthracenedione, 1-nitro-4,5,8-tris(phenylamino)-	dioxan	382s(4.00),426s(3.79), 577s(--),634(4.31), 682(4.34)	40-0082-78
$C_{32}H_{22}N_4S_2$			
Pyrimidine, 2,2'-dithiobis[4,6-diphenyl-	EtOH	263(4.73),320(4.34)	4-0105-78
$C_{32}H_{22}O$			
Cyclopenta[b][1]benzopyran, 1,2-diphenyl-3-(2-phenylethenyl)-, cis	THF THF	542(2.85) 267(4.57),300(4.48), 541(2.85)	103-1070-78 103-1075-78
trans	THF THF	570(3.06) 305(4.59),330(4.58), 570(3.06)	103-1070-78 103-1075-78

Compound	Solvent	$\lambda_{max}(\log \epsilon)$	Ref.
$C_{32}H_{22}O_9$			
9,10-Anthracenedione, 2,2'-(1-hydroxy-3-methyl-1,3-propanediyl)bis[1,4-dihydroxy-	MeOH	203(4.26),227s(--), 248(4.48),252s(--), 281(3.91),317(3.30), 456s(--),488(3.93), 514s(--)	24-3823-78
$C_{32}H_{23}N_3$			
[1,2,4]Triazolo[1,5-a]pyridine, 2-[1,1'-biphenyl]-4-yl-7-(2-[1,1'-biphenyl]-4-ylethenyl)-	DMF	345(4.83)	33-0142-78
$C_{32}H_{23}N_3O_2$			
9,10-Anthracenedione, 1,4,5-tris(phenylamino)-	dioxan	402(3.92),569s(--), 619(4.35),662(4.39)	40-0082-78
$C_{32}H_{24}$			
1,3,9,11-Cyclotridecatetraene-5,7-diyne, 13,13'-(1,2-ethenediylidene)-bis[4,9-dimethyl-	$CHCl_3$	284(4.48),566(4.92)	150-0454S-78
[24]Metacyclophanetetraene (40% cis)	C_6H_{12}	283(4.63)	1-0109-78
[24]Metaparametaparacyclophane	C_6H_{12}	237(4.64),292(4.55)	1-0109-78
[24]Orthoparaorthoparacyclophanediene, c,c,c,t-	C_6H_{12}	232(4.58),250s(--), 313(4.85)	1-0109-78
Unnamed cyclophane	n.s.g.	234(4.67)	89-0268-78
$C_{32}H_{24}N_2O_4$			
[4,4'-Biisoquinoline]-3,3'(2H,2'H)-dione, 6,6'-dimethoxy-2,2'-diphenyl-	DMF	341(4.30),505(4.45)	64-0332-78B
$C_{32}H_{24}N_2O_5$			
Acetamide, N-acetyl-N-[6-(1,3-benzodioxol-5-yl)-3-(1-naphthalenyl)-2-oxo-4-phenyl-1(2H)-pyridinyl]-	EtOH	244(4.40),296s(4.05), 335(4.21)	4-0759-78
$C_{32}H_{24}N_3O$			
Benzo[f]quinolinium, 2-acetyl-4-(2-naphthalenyl)-1-[(phenylhydrazono)methyl]-, perchlorate	MeOH	502(4.20)	65-1635-78
$C_{32}H_{24}N_4O_6$			
Trichotomine dimethyl ester	$CHCl_3$	245(4.51),340(4.48), 351(4.53),620(4.79), 658(4.85)	78-1457-78
$C_{32}H_{24}O_2$			
3-Cyclopenten-1-one, 2-[(4-methoxyphenyl)methylene]-3,4-diphenyl-5-(phenylmethylene)-, (E,Z)-	$CHCl_3$	342(4.28)	39-0989-78C
(Z,Z)-	$CHCl_3$	265(4.29),348(4.48)	39-0989-78C
$C_{32}H_{24}O_2$			
2-Naphthalenecarboxylic acid, 3-(2-methoxybenzoyl)-1,4-diphenyl-, methyl ester	THF	232(4.66),304(4.08)	78-2263-78
$C_{32}H_{24}O_5$			
Anthra[2,3-c]furan-5,10-diol, 4,11-dihydro-1,3-diphenyl-, diacetate	CH_2Cl_2	290(4.24),334(4.39)	150-5538-78

Compound	Solvent	$\lambda_{max}(\log \epsilon)$	Ref.
$C_{32}H_{25}ClN_2O_2$			
1,2-Benzisoxazole, 3-[4-[2-[2-chloro-4-[5-(1,1-dimethylethyl)-2-benzoxazolyl]phenyl]ethenyl]phenyl]-	DMF	362(4.80)	33-2904-78
1,2-Benzisoxazole, 6-chloro-3-[4-[2-[4-[5-(1,1-dimethylethyl)-2-benzoxazolyl]phenyl]ethenyl]phenyl]-	DMF	362(4.86)	33-2904-78
$C_{32}H_{25}NO_2$			
2H-Isoindole, 2-ethyl-1-phenyl-3-[2-(3-phenyl-1-isobenzofuranyl)ethenyl]-, 2-oxide	benzene	490(4.50)	40-0144-78
	MeOH	498(--)	40-0144-78
$C_{32}H_{25}O_2P$			
Phosphonium, (4-acetyl-5-hydroxy[1,1'-biphenyl]-2-yl)triphenyl-, hydroxide, inner salt	EtOH	230(4.64),288(4.29), 354(3.83)	95-0503-78
$C_{32}H_{26}$			
7,10:20,23-Dietheno-17,13-metheno-13H-cycloeicosa[de]naphthalene, 11,12,18,19-tetrahydro-[2₄]Paracyclophanetriene	C_6H_{12}	307.5(4.19)	88-1459-78
	EtOH	225(4.56),283(4.40)	1-0155-78
1,1':2',1'':3'',1''':2''',1''''-Quinquephenyl, 3,3''''-dimethyl-	n.s.g.	237.5(4.73)	89-0268-78
1,1':4',1''-Terphenyl, 2',3'-bis(phenylmethyl)-	CH_2Cl_2	250(4.2)	4-1185-78
7,10:13,16:19,22-Triethenocyclononadeca[de]naphthalene, 11,12,17,18-tetrahydro-	C_6H_{12}	311.5(4.10)	88-1459-78
7,11:14,18:21,25-Trimetheno-11H-cyclodocosa[de]naphthalene, 12,13,19,20-tetrahydro-	C_6H_{12}	306(3.95)	88-1459-78
$C_{32}H_{26}N_2O_2$			
1,2-Benzisoxazole, 3-[4-[2-[4-[5-(1,1-dimethylethyl)-2-benzoxazolyl]phenyl]ethenyl]phenyl]-	DMF	362(4.87)	33-2904-78
4,11-Etheno-1H-cyclopenta[b]phenazine, 1-[bis(4-methoxyphenyl)methylene]-3a,4,11,11a-tetrahydro-, (3aα,4β-11β,11aα)-	C_6H_{12}	238(4.55),241s(4.54), 255s(4.33),308(4.42), 316(4.42)	5-0440-78
$C_{32}H_{26}N_2O_4S_2$			
Acenaphtho[1,2-b]quinoxaline, 6b,7,12-12a-tetrahydro-9,10-dimethyl-7,12-bis(phenylsulfonyl)-	C_6H_{12}	255(3.91),267(3.97), 276(4.01),288(4.01), 298(3.83)	5-1129-78
$C_{32}H_{26}N_4O_6$			
4,8-Iminocyclohepta[c]pyrrole-1,3,5(2H)-trione, 9,9'-(1,2-ethanediyl)bis-[3a,4,8,8a-tetrahydro-2-phenyl-	CH_2Cl_2	247(3.86)	150-1182-78
isomer	CH_2Cl_2	247(3.70)	150-1182-78
$C_{32}H_{26}N_4O_7S_2$			
Imidazo[1,2-a]-1,3,5-triazine-4(1H)-thione, 2-(methylthio)-1-(2,3,5-tri-O-benzoyl-β-D-ribofuranosyl)-	MeOH	232(4.71),257s(4.15), 274(3.95),282(3.91)	44-4774-78
$C_{32}H_{26}N_4O_8S$			
Imidazo[1,2-a]-1,3,5-triazin-2(1H)-one,	MeOH	229(4.67),262(4.19),	44-4774-78

Compound	Solvent	$\lambda_{max}(\log \epsilon)$	Ref.
4-(methylthio)-1-(2,3,5-tri-O-benz-oyl-β-D-ribofuranosyl)- (cont.)		281s(3.72),305(3.45)	44-4774-78
Imidazo[1,2-a]-1,3,5-triazin-2(8H)-one, 4-(methylthio)-8-(2,3,5-tri-O-benz-oyl-α-D-ribofuranosyl)-	MeOH	233(4.73),257s(4.19), 272s(3.88),282s(3.79), 298s(3.46)	44-4784-78
β-	MeOH	233(4.74),255s(4.23), 274s(3.89),283s(3.83), 300(3.52)	44-4784-78
$C_{32}H_{26}O_2$			
2,4,10,12-Cyclotridecatetraene-6,8-diyn-1-ol, 1,1'-(1,2-ethynediylbis[5,10-dimethyl-	CHCl$_3$	263(4.44),329(3.97)	150-0454S-78
$C_{32}H_{26}O_3$			
2-Propenoic acid, 3-(1,2,3,4-tetrahy-dro-1,3,4-triphenyl-1,4-epoxynaph-thalen-2-yl)-, methyl ester	MeCN	256s(3.61),264(3.35), 268s(3.10),271(3.10)	150-1301-78
$C_{32}H_{26}O_4$			
Spiro[4H-dinaphtho[2,1-f:1',2'-h][1,5]-dioxonin-5(6H),5'-[1,3]dioxane], 2'-phenyl-	CHCl$_3$	246(4.50),273(3.96), 302(4.02),332(3.85)	56-1913-78
$C_{32}H_{26}O_{12}$			
[2,2'-Binaphthalene]-1,1',4,4',8,8'-hexol, hexaacetate	MeOH	235(4.80),253s(4.60), 280(4.07),292(4.07)	102-2042-78
$C_{32}H_{27}N$			
Methanamine, 1-(5,9-diphenyl-7H-diben-zo[a,c]cyclononen-7-ylidene)-N,N-di-methyl-	EtOH	250s(4.27),256s(4.24), 265s(4.20),336s(4.22), 356(4.32)	150-5151-78
also qualitative spectra in other solvents	C$_6$H$_{12}$	245(--),255(--), 267(--),335(--), 349(--)	150-5151-78
$C_{32}H_{27}N_5O$			
[1,2,4]Triazolo[1,5-a]pyridine, 2-[4-[2-[4-[3-[4-(1,1-dimethylethyl)phen-yl]-1,2,4-oxadiazol-5-yl]phenyl]eth-enyl]phenyl]-	DMF	355(4.86)	33-0142-78
[1,2,4]Triazolo[1,5-a]pyridine, 2-[4-[2-[4-[5-[4-(1,1-dimethylethyl)phen-yl]-1,3,4-oxadiazol-2-yl]phenyl]eth-enyl]phenyl]-	DMF	360(4.90)	33-0142-78
$C_{32}H_{27}O_3P$			
Phosphonium, triphenyl-, 3-acetyl-1-benzoyl-4-oxo-2-pentenylide	EtOH	221(4.72),269(4.32), 276(4.33),420(4.22)	95-0503-78
$C_{32}H_{28}$			
1,3,5-Cycloheptatriene, 3-[4-(3,4-di-phenyl-1,3-cyclopentadien-1-yl)-2,5-dimethylphenyl]-, cation	MeCN	519(4.04)	138-0237-78
$C_{32}H_{28}N_2O_2$			
1H-Cyclopenta[b]quinoxaline, 4,9-diben-zoyl-1-(dicyclopropylmethylene)-3a,4,9,9a-tetrahydro-	MeCN	265(4.42)	5-1139-78

Compound	Solvent	$\lambda_{max}(\log \epsilon)$	Ref.
$C_{32}H_{28}N_2O_4S_2$			
Benzo[b]cyclopenta[e][1,5]diazepine, tetrahydro-6,7-dimethyl-10-phenyl-4,4-bis(phenylsulfonyl)-	CH_2Cl_2	232(4.34),265s(3.97), 273s(3.90)	5-1161-78
1H-Cyclopenta[b]quinoxaline, 3a,4,9,9a-tetrahydro-6,7-dimethyl-1-(phenyl-methylene)-4,9-bis(phenylsulfonyl)-, (E)-	MeCN	208(4.55),283(4.33)	5-1161-78
$C_{32}H_{28}N_4$			
1,2,17,18-Tetraazacyclodotriaconta-2,4,6,12,14,16,18,20,22,28,30,32-dodecaene-8,10,24,26-tetrayne, 7,12,23,28-tetramethyl-, (?,?,?,?,E,E,E,E,Z,Z,Z,Z)-	THF	232(4.15),269s(4.20), 312s(4.52),339(4.65)	18-2112-78
$C_{32}H_{28}N_4NiO$			
Nickel, [2,6-dimethyl-4-[(10,15,20-tri-methyl-21H,23H-porphin-5-yl)methyl]-phenolato(2-)-$N^{21},N^{22},N^{23},N^{24}$]-, (SP-4-2)-	$CHCl_3$	300(4.07),344(4.02), 420(5.20),541(4.14)	5-0238-78
$C_{32}H_{28}N_4O$			
[1,2,4]Triazolo[1,5-a]pyridine, 2-[4-[2-[4-[5-(1,1-dimethylethyl)-7-meth-yl-2-benzoxazolyl]phenyl]ethenyl]-phenyl]-	DMF	368(4.91),388(4.72)	33-0142-78
[1,2,4]Triazolo[1,5-a]pyridine, 2-[4-[2-[4-[7-(1,1-dimethylethyl)-5-meth-yl-2-benzoxazolyl]phenyl]ethenyl]-phenyl]-	DMF	367(4.91),386(4.72)	33-0142-78
$C_{32}H_{28}N_4O_2$			
1,2,7,8-Tetraazaspiro[4.4]nona-2,8-di-ene-6-carboxylic acid, 1,3,7,9-tetra-phenyl-, ethyl ester	EtOH	242(4.51),324(4.39)	22-0415-78
$C_{32}H_{28}O_4$			
2,3'-Bi-2H-1-benzopyran, 3',4'-dihydro-7-methoxy-2'-(4-methoxyphenyl)-3-phenyl-	EtOH	243(3.75),282(3.53), 330(3.54)	39-0088-78C
$C_{32}H_{28}O_7$			
2-Propen-1-one, 3-(6-methoxy-1,3-benzo-dioxol-5-yl)-1-[4-methoxy-2,3-bis-(phenylmethoxy)phenyl]-	EtOH	245(4.01),314(3.96), 398(4.19)	102-1419-78
$C_{32}H_{28}O_8$			
5,16:8,13-Diethenodibenzo[a,g]cyclodo-decene-2,3,10,11-tetracarboxylic acid, 6,7,14,15-tetrahydro-, tetramethyl ester	EtOH	253(4.74)	88-0969-78
Methanone, [3-(6-methoxy-1,3-benzodiox-ol-5-yl)oxiranyl][4-methoxy-2,3-bis-phenylmethoxy)phenyl]-	EtOH	233(4.26),295(4.03), 293[sic](3.80)	102-1419-78
$C_{32}H_{30}F_2N_2O_3S$			
Quinolinium, 1-ethyl-6-fluoro-2-[3-(1-ethyl-6-fluoro-2(1H)-quinolinyli-dene)-1-propenyl]-, p-toluenesul-fonate	EtOH	612(5.19)	103-0060-78

Compound	Solvent	$\lambda_{max}(\log \epsilon)$	Ref.
$C_{32}H_{30}N_2$			
Acridine, 2-ethyl-9-(2-ethyl-10-methyl-9(10H)-acridinylidene)-9,10-dihydro-10-methyl-	CH Cl	238(4.73),296(4.26), 428(4.14)	35-5604-78
$C_{32}H_{30}N_2O_4$			
L-Phenylalanine, N-benzoyl-, 2-(benzoyl-amino)-3-phenylpropyl ester, (S)-	MeOH	215(4.36),227(4.30)	78-2791-78
$C_{32}H_{30}N_2O_5S_2$			
Spiro[2H-benzimidazole-2,1'-[4]cyclo-penten]-3'-one, 2'-(dicyclopropyl-methylene)-1,3-dihydro-5,6-dimethyl-1,3-bis(phenylsulfonyl)-	MeCN	213(4.73),293(4.17)	5-1161-78
$C_{32}H_{30}N_4O_{10}S$			
2(1H)-Pteridinone,6,7-diphenyl-4-[(2,3,4,6-tetra-O-acetyl-β-D-gluco-pyranosyl)oxy]-	MeOH	225s(4.31),260s(4.19), 280s(4.31),303(4.40), 374(4.26)	24-2571-78
4(1H)-Pteridinone, 2,3-dihydro-6,7-di-phenyl-3-(2,3,4,6-tetra-O-acetyl-β-D-glucopyranosyl)-2-thioxo-	pH 9.0	223(4.38),265(4.06), 331(4.48),391(4.01)	24-2571-78
	MeOH	225s(4.37),280s(4.16), 314(4.47),381(4.19)	24-2571-78
4(1H)-Pteridinone, 6,7-diphenyl-2-[(2,3,4,6-tetra-O-acetyl-β-D-gluco-pyranosyl)thio]-	MeOH	220s(4.37),260s(4.28), 280(4.31),366(4.12)	24-2571-78
$C_{32}H_{30}O_{14}$			
Secalonic acid G	dioxan	240(4.34),265(4.28), 338(4.57)	88-4633-78
$C_{32}H_{31}N_5O_3$			
Pyrrolo[3',4':4,5]pyrrolo[3,4-b]indole-1,3,5(2H)-trione, 5a-amino-4-cyclo-hexyl-3a,4,5a,6-tetrahydro-6-(phen-ylmethyl)-2-(phenylmethylene)amino-	EtOH	262(4.32),297s(4.03)	4-1463-78
$C_{32}H_{32}$			
7,10:19,22-Diethenodibenzo[a,k]cyclo-eicosene, 5,6,11,12,17,18,23,24-octa-hydro-	EtOH	253(2.57),257(2.69), 261(2.73),266(2.84), 274(2.75),288s(--), 296s(--)	88-3777-78
$C_{32}H_{32}Cl_2N_4O_3$			
21H,23H-Porphine-2-carboxylic acid, 7,12-bis(2-chloroethyl)-18-formyl-3,8,13,17-tetramethyl-, ethyl ester	CHCl₃	424(5.28),517(4.10), 556(4.01),589(3.82), 644(3.73)	12-2491-78
$C_{32}H_{32}N_2O_4S_2$			
Benzo[b]cyclopenta[e][1,4]diazepine, 10,10-dicyclopropyl-3a,4,9,10-tetra-hydro-6,7-dimethyl-4,9-bis(phenyl-sulfonyl)-	MeCN	272(3.67)	5-1146-78
1H-Cyclopenta[b]quinoxaline, 1-(dicy-clopropylmethylene)-3a,4,9,9a-tetra-hydro-6,7-dimethyl-4,9-bis(phenylsul-fonyl)-, cis	MeCN	215(4.56),261(4.35)	5-1146-78
Spiro[2H-benzimidazole-2,1'-cyclopent-2'-ene], 5'-(dicyclopropylmethylene)-1,3-dihydro-5,6-dimethyl-1,3-bis(phen-ylsulfonyl)-	MeCN	212(4.64),307(3.87)	5-1161-78

Compound	Solvent	$\lambda_{max}(\log \epsilon)$	Ref.
$C_{32}H_{32}N_2O_6S_2$			
Benzo[b]cyclopenta[e][1,4]diazepin-2(3H)-one, 10,10-dicyclopropyl-3a,4,9,10-tetrahydro-3-hydroxy-6,7-dimethyl-4,9-bis(phenylsulfonyl)-, cis	MeCN	266s(3.87),273s(3.71)	5-1146-78
$C_{32}H_{32}N_4O_5$			
Inosine, 2,3-dihydro-3-methyl-2',3',5'-tris-0-(phenylmethyl)-	EtOH	259(4.07)	88-4047-78
$C_{32}H_{32}O_3$			
1,2,4-[1]Propanyl[3]ylidenepentalene, 1,2,3,3a,4,6a-hexahydro-5,8,8-tris-(4-methoxyphenyl)-	EtOH	227(4.46),278(3.75), 284(3.71)	104-2336-78
$C_{32}H_{32}O_4$			
Cacalohastin, 12-(dehydrocacalohastin-14-yl)-	ether	245(4.80),250(4.81), 290(4.45),323(4.18), 350(4.03)	24-3140-78
$C_{32}H_{32}O_5$			
Furan, tetrahydro-2,3,4,5-tetrakis(4-methoxyphenyl)-	MeCN	276(3.9),284s(--)	35-3526-78
$C_{32}H_{32}O_8$			
5,16:8,13-Diethenodibenzo[a,g]cyclododecene-2,3,10,11-tetracarboxylic acid, 1,4,6,7,9,12,14,15-octahydro-, tetramethyl ester	EtOH	225(4.28)	88-0969-78
$C_{32}H_{33}BrO$			
Phenol, 4-(4-bromophenyl)-2,6-bis[4-(1,1-dimethylethyl)phenyl]-	CH_2Cl_2	251(4.64),255(4.64), 312(3.71)	24-0264-78
$C_{32}H_{33}ClN_4O_3$			
21H,23H-Porphine-2-carboxylic acid, 12-(2-chloroethyl)-7-ethyl-18-formyl-3,8,13,17-tetramethyl-, ethyl ester	$CHCl_3$	425(5.28),519(4.16), 559(4.15),588(3.84), 644(3.68)	12-2491-78
isomer	$CHCl_3$	425(5.26),518(4.08), 557(4.00),590(3.79), 646(3.77)	12-2491-78
$C_{32}H_{34}Cl_4N_5O_{11}P$			
3'-Adenylic acid, N-(4-methoxybenzoyl)-2'-0-(tetrahydro-4-methoxy-2H-pyran-4-yl)-, 2-chlorophenyl 2,2,2-trichloroethyl ester	EtOH	287(4.55)	54-0073-78
$C_{32}H_{34}N_2O_4S_2$			
Spiro[2H-benzimidazole-2,1'-cyclopentane], 2'-(dicyclopropylmethylene)-1,3-dihydro-5,6-dimethyl-1,3-bis-(phenylsulfonyl)-	MeCN	213(4.67),306(3.88)	5-1161-78
$C_{32}H_{34}N_2O_8$			
Lanciferine, 10-methoxy-	EtOH	238(3.92),281(4.18)	42-1096-78
$C_{32}H_{34}O_2$			
Estra-1,3,5(10),9(11)-tetraene, 3,17β-bis(phenylmethoxy)-	EtOH	264(4.26)	111-0313-78

Compound	Solvent	$\lambda_{max}(\log \epsilon)$	Ref.
$C_{32}H_{35}NO_{10}S_3$ 7-Oxabicyclo[4.2.0]octane-6-carboni- trile, 1,5,5-tris[[[(4-methylphenyl)- sulfonyl]oxy]methyl]-	dioxan	257(2.71),262(2.79), 266(2.77),273(2.72)	97-0380-78
$C_{32}H_{35}N_4OP$ Phosphorin, 4-cyclohexyl-1,1-dihydro- 1-methoxy-2,6-bis[(4-methylphenyl)- azo]-1-phenyl-	EtOH	267(4.31),311(4.29), 419(3.60),573(4.54)	118-0846-78
$C_{32}H_{36}Br_2N_2O_4$ Piperidine, 1,1'-[2-[3-(4-bromophenyl)- 3-hydroxy-2-propenylidene]-3-[2-(4- bromophenyl)-2-oxoethyl]-1,5-dioxo- 1,5-pentanediyl]bis-	CHCl$_3$	265(3.92)	104-2026-78
$C_{32}H_{36}N_6O_8$ L-Threonine, N-[[[9-[2,3-O-(1-methyl- ethylidene)-β-D-ribofuranosyl]-9H- purin-6-yl]amino]carbonyl]-O-(phen- ylmethyl)-, phenylmethyl ester	pH 2 pH 12 MeOH	270(4.31),277(4.33) 271(4.23),278(4.26), 299(4.04) 270(4.34),277s(4.27)	19-0021-78 19-0021-78 19-0021-78
$C_{32}H_{37}CuN_3O_4$ Copper, [2-[[[5-(1,1-dimethylethyl)-2- hydroxyphenyl]methylene]amino]-1- [[[[5-(1,1-dimethylethyl)-2-hydroxy- phenyl]methylene]amino]methyl]ethyl phenylcarbamato(2-)]-	CH$_2$Cl$_2$ THF pyridine	234(4.81),279(4.50), 384(4.06),609(2.50) 235(4.79),275(4.40), 385(3.98),618(2.48) 385(4.05),611(2.35)	35-2686-78 35-2686-78 35-2686-78
$C_{32}H_{38}N_2O$ 2'H-Androsta-5,16-dieno[16,17-c]pyra- zol-3-ol, 16,17-dihydro-2',5'-di- phenyl-, (3β,16β,17β)-	n.s.g.	244s(4.12),255(4.16), 316s(3.93),360(4.34)	78-1633-78
$C_{32}H_{38}O_2$ Pentacyclo[11.9.1.1^{2,12}.1^{5,9}.1^{16,20}]- hexacosa-1,5,7,9(26),12(25),13(23),16- 18,20(24)-nonaene, 7,18-dimethoxy- 23,24,25,26-tetramethyl-	THF	211(4.72),238(4.16), 287(3.54)	44-3470-78
Pentacyclo[;9.3.1.1^9,^{13}.0^4,^{16}.0^6,^{18}]- hexacosa-1(25),4,6(18),9,11,13(26)- 16,21,23-nonaene, 11,23-dimethoxy- 5,17,25,26-tetramethyl-	THF	207(4.11),259(3.82), 310(3.00),340(2.70)	44-3470-78
$C_{32}H_{38}O_2S_2$ 15,23-Dithiapentacyclo[11.11.1.1^{2,12}- 1^{5,9}.1^{17,21}]octacosa-1,5,7,9(28),12- (27),13(25),17,19,21(26)-nonaene, 7,19-dimethoxy-25,26,27,28-tetra- methyl-	THF	207(--),247(4.40), 303(--),327(2.81)	44-3470-78
$C_{32}H_{40}N_4$ Isoborrevine	EtOH	226(4.72),288(4.15), 294(4.15)	77-0826-78
$C_{32}H_{40}N_4O_2$ 21H-Bilin-1(22H)-one, 3,8,12,17-tetra- ethyl-19-methoxy-2,7,13,18-tetra- methyl-, (Z,Z,Z)-	CHCl$_3$	270s(4.18),300(4.28), 370(4.62),670(4.10)	64-0924-78B
21H-Bilin-1(22H)-one, 7,8,12,13-tetra- ethyl-19-methoxy-2,3,17,18-tetra-	CHCl$_3$	270(4.17),295(4.23), 370(4.60),670(4.08)	64-0924-78B

Compound	Solvent	$\lambda_{max}(\log \epsilon)$	Ref.
methyl-, (Z,Z,Z)- (cont.)			64-0924-78B
$C_{32}H_{40}N_4O_6$			
Aphelandrine, N(6),N(10)-diacetyl-	EtOH	229(4.31),282(3.57), 290s(3.41)	33-2646-78
	EtOH-NaOH	223(5.18),247(4.28), 287(3.69)	33-2646-78
Cyclo[3-(4-β-aminoethyl)phenoxypropan-oyl-L-prolyl]$_2$	MeOH	224(4.40),277(3.53), 284(3.46)	35-8202-78
$C_{32}H_{40}N_4O_{11}$			
1,4-Dioxa-7-azaspiro[4.4]nonane-8-carb-oxylic acid, 7-[N-[N-[4-methyl-2-ni-tro-3-(phenylmethoxy)benzoyl]-L-threonyl]-D-valyl]-, methyl ester, (S)-	MeOH	209(4.759)	87-0607-78
$C_{32}H_{40}N_8O_8$			
1,5,9,13,17,21,25,29-Octaazapentacyclo-[27.3.1.1^{5,9}.1^{13,17}.1^{21,25}]hexatria-conta-7,14,23,30-tetraene-6,16,22, 32,33,34,35,36-octone, 7,15,23,31-tetramethyl-	50% EtOH	272(4.51)	56-1365-78
$C_{32}H_{40}O_8$			
Tricyclo[8.2.2.2^{4,7}]hexadeca-4,6,10,12-13,15-hexaene-5,6,11,12-tetracarbox-ylic acid, tetrakis(1-methylethyl) ester	EtOH	215(4.61),302(3.33)	24-0523-78
$C_{32}H_{41}ClN_2O_9$			
Maytansine, O^3-de[2-(acetylmethylami-no)-1-oxopropyl]-O^3-(1-oxo-2-buten-yl)-	EtOH	231(4.42),252(4.32), 280(3.65),289(3.65)	87-0031-78
$C_{32}H_{41}ClO_3$			
Phenol, 2-[[3-chloro-5-(1,1-dimethyl-ethyl)-2-hydroxyphenyl]methyl]-4-(1,1-dimethylethyl)-6-[[5-(1,1-di-methylethyl)-2-hydroxyphenyl]methyl]-	dioxan	285(3.95)	49-0767-78
$C_{32}H_{44}N_2O_9$			
Streptolydigin	EtOH	250(4.05),347(4.43), 368s(4.34)	35-4225-78
	EtOH-H$_2$SO$_4$	242(4.00),355(4.57), 368s(4.56)	35-4225-78
	EtOH-KOH	261(4.22),289(4.28), 334(4.35)	35-4225-78
$C_{32}H_{44}O_6S$			
24-Norchol-16-en-23-oic acid, 3,3-[1,2-ethanediylbis(oxy)]-22-(phenylsul-fonyl)-, methyl ester, (5α)-	EtOH	218(3.96),252(2.93), 258(2.93),266(3.05), 273(3.05)	35-3435-78
$C_{32}H_{46}N_2O_8$			
Lycotonine, anthranoyl-	EtOH	219(4.44),249(3.90), 342(3.73)	95-1376-78
$C_{32}H_{47}ClO_3$			
Oleana-9(11),12-dien-28-oyl chloride, 3β-acetoxy-	C_6H_{12}	283(3.95)	24-1160-78

Compound	Solvent	$\lambda_{max}(\log \epsilon)$	Ref.
Ursa-9(11),12-dien-28-oyl chloride, 3β-acetoxy-	C_6H_{12}	282(3.99)	24-1160-78
$C_{32}H_{48}O_4$			
D(17a)-Homo-C,18-dinorcholesta-13,15,17-trien-11-one, 3-acetoxy-9-hydroxy-4,4,15,17a-tetramethyl-, (3β,5α,9β)-	EtOH	218(4.45),262(4.11), 314(3.40)	39-0471-78C
Mokupalide, acetoxy-	MeOH	210(4.24)	35-0307-78
	MeOH-base	210(4.27),251(3.63)	35-0307-78
Spiro[naphthalene-1(2H),2'(4'H)-[1H]-phenalene]-2,4'-dione, 5,7'-bis(1,1-dimethylethoxy)-3,3',4,4a,5,5',6,6'-6'a,7,7',8',9',9'a-tetradecahydro-4a,6'a-dimethyl-	n.s.g.	256(4.08)	33-2397-78
$C_{32}H_{50}O_3$			
Neoilexonol acetate	MeOH	251(4.07)	95-0249-78
$C_{32}H_{50}O_4$			
2,5-Cyclohexadien-1-one, 4-[2,4-bis-(1,1-dimethylethyl)-5-ethoxyphen-oxy]-2,4-bis(1,1-dimethylethyl)-5-ethoxy-	C_6H_{12}	239(4.23),279(3.83), 284(3.81)	12-1061-78
9β-Lanost-5-ene-7,11-dione, 3β-acetoxy-	EtOH	239(3.81),255(3.78)	19-0583-78
Lanost-5-en-7-one, 3β-acetoxy-9α,11α-epoxy-	EtOH	238(3.95)	19-0583-78
$C_{32}H_{52}O_2$			
Lanosta-7,9(11)-dien-3β-ol, acetate	EtOH	236(4.13),244(4.22), 253(4.03)	19-0591-78
$C_{32}H_{52}O_6$			
6,7-Secocholest-4-en-6-oic acid, 3β,7-diacetoxy-, methyl ester	EtOH	212(3.90)	12-0171-78
$C_{32}H_{55}N_{10}O_{19}P$			
Amikacin, 4'-O-adenylyl-	H_2O	260(4.09)	88-3917-78
$C_{32}H_{56}P_2Pt$			
Platinum, di-1,3-butadiynylbis(tributyl-phosphine), cis	CH_2Cl_2	238(4.37),276(4.05), 301(4.19)	101-0101-78B
trans	CH_2Cl_2	241(4.45),251s(4.32), 268(4.01),287(3.95), 318(4.39)	101-0101-78B
$C_{32}H_{64}N_4Ni$			
Nickel, bis[N,N'-1,2-ethanediylidene-bis[2,4-dimethyl-3-pentanamine]-N,N']-, (T-4)-	hexane	500(3.8),690(3.3)	64-1381-78B

Compound	Solvent	$\lambda_{max}(\log \epsilon)$	Ref.
$C_{33}H_{18}N_2O_2$ Benzo[g]naphtho[1',2',3':4,5]quino[2,1-b]quinazoline-5,10-dione, 18-phenyl-	n.s.g.	480(3.73)	2-0103-78
$C_{33}H_{20}O_9$ [2,2':7',2"-Ternaphthalene]-1,1',1",4-4',4"-hexone, 8,8',8"-trihydroxy-6,6',6"-trimethyl-	EtOH	218(4.96),250(4.59), 435(4.16)	88-3889-78
$C_{33}H_{21}ClN_4O$ [1,2,4]Triazolo[1,5-a]pyridine, 2-[4-[2-[2-chloro-4-(6-phenyl-2-benzoxazolyl)phenyl]ethenyl]phenyl]-	DMF	373(4.87)	33-0142-78
[1,2,4]Triazolo[1,5-a]pyridine, 7-[2-[2-chloro-4-(6-phenyl-2-benzoxazolyl)phenyl]ethenyl]-2-phenyl-	DMF	371(4.84)	33-0142-78
$C_{33}H_{21}NO_2$ 1,2-Benzisoxazole, 3-[1,1'-biphenyl]-4-yl-6-[2-(3-dibenzofuranyl)ethenyl]-	DMF	351(4.80)	33-2904-78
$C_{33}H_{21}N_3$ Benzo[f]quinoline, 3-(2-quinolinyl)-1-[2-(2-quinolinyl)ethenyl]-	EtOH	275(4.78),315(4.59), 330(4.62),345(4.60), 378(4.06)	103-0418-78
Benzo[f]quinoline, 3-(2-quinolinyl)-1-[2-(4-quinolinyl)ethenyl]-	EtOH	275(4.69),320(4.54), 331(4.57),344(4.50), 378(3.96)	103-0418-78
Benzo[f]quinoline, 3-(2-quinolinyl)-1-[2-(6-quinolinyl)ethenyl]-	EtOH	259(4.77),287(4.71), 319(4.58),330(4.62), 344(4.61),374(3.90)	103-0418-78
$C_{33}H_{21}N_3O_2$ 1,2-Benzisoxazole, 3-[4-[2-[4-[5-(1-naphthalenyl)-1,3,4-oxadiazol-2-yl]-phenyl]ethenyl]phenyl]-	DMF	357(4.84)	33-2904-78
$C_{33}H_{22}N_2O$ Imidazo[1,2-a]pyridine, 2-[1,1'-biphen-yl]-4-yl-7-[2-(3-dibenzofuranyl)eth-enyl]-	DMF	380(4.82),400(4.76)	33-0129-78
$C_{33}H_{22}N_2O_2$ 1,2-Benzisoxazole, 3-[4-[2-[4-(5-meth-yl-2-benzoxazolyl)-1-naphthalenyl]-ethenyl]phenyl]-	DMF	300(4.20),382(4.66)	33-2904-78
$C_{33}H_{22}N_4O$ [1,2,4]Triazolo[1,5-a]pyridine, 2-[4-[2-(3-[1,1'-biphenyl]-4-yl-1,2-benz-isoxazol-6-yl)ethenyl]phenyl]-	DMF	345(4.83)	33-2904-78
[1,2,4]Triazolo[1,5-a]pyridine, 2-[4-[2-[4-(5-phenyl-2-benzoxazolyl)phen-yl]ethenyl]phenyl]-	DMF	369(4.94),389(4.77)	33-0142-78
[1,2,4]Triazolo[1,5-a]pyridine, 2-[4-[2-[4-(6-phenyl-2-benzoxazolyl)phen-yl]ethenyl]phenyl]-	DMF	373(4.94)	33-0142-78
$C_{33}H_{22}O_2$ Benzeneacetaldehyde, α-[(1,2-diphenyl-cyclopenta[b][1]benzopyran-3-yl)-	THF THF	540(2.97) 255(4.32),285(4.34),	103-1070-78 103-1075-78

Compound	Solvent	$\lambda_{max}(\log \epsilon)$	Ref.
methylene]-, (Z)- (cont.)		393(4.45),530(3.08)	103-1075-78
$C_{33}H_{23}$ Methyl, diphenyl(10-phenyl-9-anthracenyl)-	SO_2	820(4.08)	39-0488-78C
$C_{33}H_{23}Cl$ Anthracene, 9-chloro-10-(diphenylmethylene)-9,10-dihydro-9-phenyl-	n.s.g.	240(4.08),260(3.94), 290(3.84)	39-0488-78C
$C_{33}H_{23}NO$ 1,2-Benzisoxazole, 3-[1,1'-biphenyl]-4-yl-6-(2-[1,1'-biphenyl]-4-ylethenyl)-	DMF	343(4.76)	33-2904-78
$C_{33}H_{23}NO_3S$ 1,2-Benzisoxazole, 3-[4-[2-[4-([1,1'-biphenyl]-4-ylsulfonyl)phenyl]ethenyl]phenyl]-	DMF	340(4.75)	33-2904-78
$C_{33}H_{23}OS$ Pyrylium, 2,4-diphenyl-6-[(2-phenyl-4H-1-benzothiopyran-4-ylidene)methyl]-, perchlorate	HOAc-HClO₄	635(4.015)	83-0226-78
Thiopyrylium, 2,4-diphenyl-6-[(2-phenyl-4H-1-benzopyran-4-ylidene)methyl]-, perchlorate	HOAc-HClO₄	600(4.392)	83-0226-78
$C_{33}H_{23}S_2$ Thiopyrylium, 2,4-diphenyl-6-[(2-phenyl-4H-1-benzothiopyran-4-ylidene)methyl]-, perchlorate	HOAc-HClO₄	644(3.959)	83-0226-78
$C_{33}H_{24}$ Anthracene, 9-(diphenylmethyl)-10-phenyl-	n.s.g.	343(3.09),358(3.46), 378(3.68),398(3.67)	39-0488-78C
$C_{33}H_{24}N_2$ Imidazo[1,2-a]pyridine, 2-[1,1'-biphenyl]-4-yl-7-(2-[1,1'-biphenyl]-4-ylethenyl)-	DMF	375(4.78),393(4.73)	33-0129-78
Imidazo[1,2-a]pyridine, 7-(2-[1,1'-biphenyl]-4-ylethenyl)-2,3-diphenyl-	DMF	376(4.68)	33-0129-78
$C_{33}H_{24}O$ 9-Anthracenemethanol, α,α,10-triphenyl-	n.s.g.	268(4.04),345(3.45), 361(3.76),381(3.94), 401(3.90)	39-0488-78C
$C_{33}H_{24}O_3$ 2H-Cyclopenta[1]phenanthren-2-one, 1,3-dihydro-1,3-bis[(4-methoxyphenyl)methylene]-, (E,Z)-	CHCl₃	259(4.57),279(4.40), 373(4.53)	39-0989-78C
$C_{33}H_{25}N_2OP$ 3H-Pyrazole, 5-(diphenylphosphinyl)-3,3,4-triphenyl-, P-oxide	MeOH	245(3.00),380(1.68)	150-4248-78
$C_{33}H_{26}ClN_3O_2$ 1,2-Benzisoxazole, 6-chloro-3-[4-[2-[4-[3-[4-(1,1-dimethylethyl)phenyl]-	DMF	349(4.82)	33-2904-78

Compound	Solvent	$\lambda_{max}(\log \epsilon)$	Ref.
1,2,4-oxadiazol-5-yl]phenyl]ethenyl]-phenyl]- (cont.)			33-2904-78
1,2-Benzisoxazole, 6-chloro-3-[4-[2-[4-[5-[4-(1,1-dimethylethyl)phenyl]-1,3,4-oxadiazol-2-yl]phenyl]ethenyl]-phenyl]-	DMF	354(4.86)	33-2904-78
$C_{33}H_{26}N_2O_7$ 1H-Benzimidazole, 1-(2,3,5-tri-O-benzoyl-β-D-ribofuranosyl)-	MeOH	230(4.62),265s(3.71), 273(3.80),280(3.77)	24-0996-78
$C_{33}H_{26}N_2O_7S$ 2H-Benzimidazole-2-thione, 1,3-dihydro-1-(2,3,5-tri-O-benzoyl-β-D-ribofuranosyl)-	MeOH	227(4.73),258s(3.96), 284s(3.96),300s(4.36), 307(4.46)	24-0996-78
$C_{33}H_{27}NO$ Methanone, (4-methylphenyl)phenyl-, O-(triphenylmethyl)oxime, (E)-	EtOH	263.5(4.16)	44-1890-78
(Z)-	EtOH	266.5(4.20)	44-1890-78
$C_{33}H_{27}N_3O_2$ 1,2-Benzisoxazole, 3-[4-[2-[4-[3-[4-(1,1-dimethylethyl)phenyl]-1,2,4-oxadiazol-5-yl]phenyl]ethenyl]phenyl]-	DMF	349(4.80)	33-2904-78
1,2-Benzisoxazole, 3-[4-[2-[4-[5-[4-(1,1-dimethylethyl)phenyl]-1,3,4-oxadiazol-2-yl]phenyl]ethenyl]phenyl]-	DMF	353(4.85)	33-2904-78
$C_{33}H_{28}$ Spiro[2.4]hepta-4,6-diene, 5,6-bis(4-methylphenyl)-4,7-diphenyl-	MeCN	247(4.57),300s(3.87)	5-1648-78
$C_{33}H_{28}FeN_5O_4$ Quinolinium, 4-[[2-(2,4-dinitrophenyl)-4-(ferrocenylmethylene)-2,4-dihydro-5-methyl-3H-pyrazol-3-ylidene)methyl]-1-ethyl-, iodide	n.s.g.	210(4.77),248(4.73)	48-0857-78
$C_{33}H_{28}NO_4P$ Cyclohepta[b]pyrrole-2-acetic acid, 3-(ethoxycarbonyl)-α-(triphenylphosphoranylidene)-, methyl ester	CHCl$_3$	272(4.47),309(4.25), 354(3.81),448(4.19)	18-1573-78
$C_{33}H_{28}N_2$ 4,11-Etheno-1H-cyclopenta[b]phenazine, 1-(diphenylmethylene)-3a,4,11,11a-tetrahydro-7,8,13-trimethyl-	C_6H_{12}	243(4.71),246s(4.69), 298s(4.87),303(4.87), 310s(4.86),318s(4.86), 323s(4.84),332s(4.78)	5-0440-78
$C_{33}H_{28}N_2O_2$ 1,2-Benzisoxazole, 3-[4-[2-[4-[5-(1,1-dimethylethyl)-7-methyl-2-benzoxazolyl]phenyl]ethenyl]phenyl]-	DMF	362(4.86)	33-2904-78
1,2-Benzisoxazole, 3-[4-[2-[4-[7-(1,1-dimethylethyl)-5-methyl-2-benzoxazolyl]phenyl]ethenyl]phenyl]-	DMF	363(4.86)	33-2904-78
$C_{33}H_{28}N_4O$ [1,2,4]Triazolo[1,5-a]pyridine, 2-[4-[2-[4-(5-cyclohexyl-2-benzoxazolyl)-	DMF	366(4.92),384(4.74)	33-0142-78

Compound	Solvent	$\lambda_{max}(\log \epsilon)$	Ref.
phenyl]ethenyl]phenyl]- (cont.)			33-0142-78
$C_{33}H_{28}N_4O_6$ 4,8-Iminocyclohepta[c]pyrrole-1,3,5(2H)- trione, 9,9'-(1,3-propanediyl)bis- [3a,4,8,8a-tetrahydro-2-phenyl-	CHCl$_3$	247(3.84)	150-1182-78
$C_{33}H_{28}O_2$ Spiro[2.4]hepta-4,6-diene, 5,6-bis(4- methoxyphenyl)-4,7-diphenyl-	MeCN	248(4.53)	5-1648-78
$C_{33}H_{28}O_2P$ Phosphonium, (4-acetyl-5-methoxy[1,1'- biphenyl]-2-yl)triphenyl-, iodide	EtOH	224(4.53)	95-0503-78
$C_{33}H_{28}O_4$ 12H-Benzo[a]xanthene, 9-methoxy-5-(4- methoxyphenyl)-6-[(4-methoxyphenyl)- methyl]-	EtOH	258(4.07),288s(3.62), 330(2.93)	39-0088-78C
$C_{33}H_{29}ClN_2O_9$ 2,4(1H,3H)-Pyrimidinedione, 1-[5-O-(4- chlorobenzoyl)-2,3-bis-O-(4-methyl- benzoyl)-β-D-arabinofuranosyl]-3- methyl-	MeOH	242(4.82),262(4.25)	44-0350-78
$C_{33}H_{30}Cl_2O_2$ Benzene, 1,1'-[5,5-bis(4-chlorophenyl)- 3,3-dimethyl-1,4-pentadienylidene]- bis[4-methoxy-	EtOH	247(4.52),263(4.48)	35-4131-78
$C_{33}H_{30}FeN_3$ Quinolinium, 1-ethyl-4-[[4-(ferrocenyl- methylene)-2,4-dihydro-5-methyl-2- phenyl-3H-pyrazol-3-ylidene]methyl]-, iodide	n.s.g.	210(4.60),340(4.28)	48-0857-78
$C_{33}H_{30}N_2O_4S_2$ 1H-Cyclopenta[b]quinoxaline, 3a,4,9,9a- tetrahydro-6,7-dimethyl-1-[(4-methyl- phenyl)methylene]-4,9-bis(phenylsul- fonyl)-, (E)-	MeCN	210(4.61),288(4.39)	5-1161-78
$C_{33}H_{30}N_2O_5S_2$ Benzo[b]cyclopenta[e][1,4]diazepine, tetrahydro-10-(4-methoxyphenyl)-6,7- dimethyl-4,9-bis(phenylsulfonyl)-	MeCN	265s(4.02),273s(3.97)	5-1161-78
$C_{33}H_{30}N_4O$ 1-Butanone, 1-(1,3,7,9-tetraphenyl- 1,2,7,8-tetraazaspiro[4.4]nona-2,8- dien-6-yl)-	EtOH	242(4.32),342(4.27)	22-0415-78
$C_{33}H_{30}N_4O_2$ 1,2,7,8-Tetraazaspiro[4.4]nona-2,8-di- ene-6-carboxylic acid, 6-methyl- 1,3,7,9-tetraphenyl-, ethyl ester	EtOH	242(4.52),324(4.39), 358(4.41)	22-0415-78
$C_{33}H_{30}O_5$ 2,3'-Bi-2H-1-benzopyran, 3',4'-dihydro- 7-methoxy-2',3-bis(4-methoxyphenyl)-	EtOH	243(3.64),280(3.50), 327(3.27)	39-0088-78C

Compound	Solvent	$\lambda_{max}(\log \epsilon)$	Ref.
2,3'-Bi-2H-1-benzopyran, 2'-(3,4-di-methoxyphenyl)-3',4'-dihydro-7-meth-oxy-3-phenyl-	EtOH	245(3.76),285(3.46), 330(3.68)	39-0088-78C
$C_{33}H_{31}BrO_8$			
Benzoic acid, 4-bromo-, [2-(3,4-dimeth-oxyphenyl)-3-(3,5-dimethoxyphenyl)-2,3-dihydro-6-methoxy-4-benzofuran-yl]methyl ester	EtOH	235(4.54),277s(3.91), 283(3.96),297s(3.68)	94-3050-78
$C_{33}H_{31}N_3$			
3H-1,5-Benzodiazepine, 2-[4-(diphenyl-methyl)-1-piperidinyl]-4-phenyl-	n.s.g.	260(4.55),346(3.65)	87-0952-78
$C_{33}H_{32}N_2O_5$			
L-Phenylalanine, N-[N-[(phenylmethoxy)-carbonyl]-L-phenylalanyl]-, phenyl-methyl ester	MeOH	211(4.21),236s(3.93)	102-0552-78
$C_{33}H_{32}O_2$			
Cyclopropane, 1,1-bis(3-methoxyphenyl)-2,2-dimethyl-3-(2,2-diphenylethenyl)-	EtOH	276(4.28),283(4.24)	35-4131-78
Cyclopropane, 3-[1,1-bis(3-methoxyphen-yl)ethenyl]-2,2-dimethyl-1,1-diphenyl-	EtOH	274(4.20)	35-4131-78
1,4-Pentadiene, 1,1-bis(3-methoxyphen-yl)-3,3-dimethyl-5,5-diphenyl-	EtOH	250(4.29)	35-4131-78
1,4-Pentadiene, 1,1-bis(4-methoxyphen-yl)-3,3-dimethyl-5,5-diphenyl-	EtOH	245(4.49)	44-1997-78
$C_{33}H_{33}NO_5$			
Dibenz[d,f]azecin-8(5H)-one, 6,7,9,10-tetrahydro-2,13-dimethoxy-3,12-bis-(phenylmethoxy)-	MeOH	236(4.45),284(4.42)	39-0434-78C
Quinoline, 1,2,3,4-tetrahydro-7-meth-oxy-1-[[4-methoxy-3-(phenylmethoxy)-phenyl]acetyl]-6-(phenylmethoxy)-	MeOH	250s(4.11),286(3.90)	39-0440-78C
$C_{33}H_{33}N_5O_{10}$			
Benzamide, N-[3,7-dihydro-3-(phenyl-methyl)-7-(2,3,4,6-tetra-O-acetyl-β-D-glucopyranosyl)-6H-purin-6-yl-idene]-	EtOH	236(4.11),334(4.24)	12-1095-78
$C_{33}H_{34}Cl_2N_4O_2$			
21H,23H-Porphine-2-carboxylic acid, 7,12-bis(2-chloroethyl)-18-ethenyl-3,8,13,17-tetramethyl-, ethyl ester	CHCl₃	413(5.31),511(4.11), 550(4.09),579(3.90), 635(3.48)	12-2491-78
$C_{33}H_{34}Cl_2N_4O_3$			
21H,23H-Porphine-2-carboxylic acid, 18-acetyl-7,12-bis(2-chloroethyl)-3,8,13,17-tetramethyl-, ethyl ester	CHCl₃	419(5.29),512(4.14), 549(3.95),585(3.81), 638(3.66)	12-2491-78
$C_{33}H_{34}NO_4$			
Isoquinolinium, 3,4-dihydro-6-methoxy-1-[[4-methoxy-3-(phenylmethoxy)phen-yl]methyl]-2-methyl-7-(phenylmeth-oxy)-, chloride	EtOH	247(4.11),309(3.83), 362(3.86)	35-0276-78
iodide	EtOH	247(4.11),309(3.83), 362(3.86)	35-0276-78

Compound	Solvent	$\lambda_{max}(\log \epsilon)$	Ref.
$C_{33}H_{35}ClN_4O_2$			
21H,23H-Porphine-2-carboxylic acid, 7-(2-chloroethyl)-18-ethenyl-12-ethyl-3,8,13,17-tetramethyl-, ethyl ester	$CHCl_3$	412(5.31),512(4.06), 552(4.15),578(3.92), 636(3.33)	12-2491-78
21H,23H-Porphine-2-carboxylic acid, 12-(2-chloroethyl)-18-ethenyl-7-ethyl-3,8,13,17-tetramethyl-, ethyl ester	$CHCl_3$	412(5.26),511(4.12), 549(4.01),579(3.86), 634(3.60)	12-2491-78
$C_{33}H_{35}ClN_4O_3$			
21H,23H-Porphine-2-carboxylic acid, 18-acetyl-7-(2-chloroethyl)-12-ethyl-3,8,13,17-tetramethyl-, ethyl ester	$CHCl_3$	419(5.29),514(4.12), 551(3.94),585(3.81), 637(3.61)	12-2491-78
$C_{33}H_{35}N_5Ni$			
Nickel, [2,7,12,17-tetraethyl-3,8,13,18-tetramethyl-21H,23H-porphine-5-carbonitrilato(2-)-$N^{21},N^{22},N^{23},N^{24}$]-, (SP-4-2)-	$CHCl_3$	305(4.04),406(5.21), 550(3.93),588(4.31)	39-0366-78C
$C_{33}H_{35}N_5O_4$			
Adenosine, N,N-dimethyl-2',3',5'-tris-O-(phenylmethyl)-	EtOH	291(4.27)	88-4047-78
$C_{33}H_{35}N_8O_9P$			
Guanosine, N-(2-methyl-1-oxopropyl)-2'-O-[(2-nitrophenyl)methyl]-, 3'-(N,N-diphenylphosphorodiamidate)	EtOH	233(4.49),261(4.34), 283s(4.20)	35-4580-78
$C_{33}H_{36}N_4NiO_2$			
Nickel, [2,7,12,17-tetraethyl-3,8,13,18-tetramethyl-21H,23H-porphin-5-yl formato(2-)-$N^{21},N^{22},N^{23},N^{24}$]-, (SP-4-2)-	$CHCl_3$	400(5.43),521(4.25), 556(4.50)	39-0366-78C
$C_{33}H_{36}N_4O_4$			
21H,23H-Porphine-2-carboxylic acid, 18-acetyl-7-ethyl-12-(2-hydroxyethyl)-3,8,13,17-tetramethyl-, ethyl ester	$CHCl_3$	419(5.21),513(4.05), 550(3.89),585(3.74), 639(3.62)	12-2491-78
$C_{33}H_{37}N_3O_8$			
DL-Aspartic acid, N-[N-[4-[methyl[(phenylmethoxy)carbonyl]amino]benzoyl]-L-phenylalanyl]-, diethyl ester	MeOH	251(4.10)	87-1165-78
$C_{33}H_{38}N_4NiO$			
Nickel, [2,7,12,17-tetraethyl-5-methoxy-3,8,13,18-tetramethyl-21H,23H-porphinato(2-)-$N^{21},N^{22},N^{23},N^{24}$]-, (SP-4-2)-	$CHCl_3$	404(5.30),524(4.09), 559(4.19)	39-0366-78C
$C_{33}H_{39}S_3$			
Cyclopropenylium, tris[[4-(1,1-dimethylethyl)phenyl]thio]-, perchlorate	MeCN	232(4.51),290(4.30)	18-3653-78
$C_{33}H_{42}N_4O_6$			
Aphelandrine, N(6),N(10)-diacetyl-O(34)-methyl-	EtOH	228(4.31),276s(3.48), 281(3.53),288s(3.39)	33-2646-78

Compound	Solvent	$\lambda_{max}(\log \epsilon)$	Ref.
$C_{33}H_{42}O_4$ Spiro[3,5-cyclohexadiene-1,6'-dibenzo- [d,f][1,3]dioxepin]-2-one, 2',4,10'- tris(1,1-dimethylethyl)-5-methoxy- 4',8'-dimethyl-	C_6H_{12}	218(4.53),255(4.21), 286s(3.73)	12-1069-78
$C_{33}H_{42}O_5$ Spiro[3,5-cyclohexadiene-1,6'-dibenzo- [d,f][1,3]dioxepin]-2-one, 3',4,10'- tris(1,1-dimethylethyl)-2',5-dimeth- oxy-8'-methyl-	C_6H_{12}	217(4.53),256(4.23), 297(3.96)	12-1069-78
$C_{33}H_{44}N_2O_4$ 4H-Benzo[a]quinolizine, 2-[(6,7-dieth- oxy-3,4-dihydro-1-isoquinolinyl)meth- yl]-9,10-diethoxy-3-ethyl-1,6,7,11bα- tetrahydro-, (±)-	EtOH	232(4.28),282(4.03), 310(3.82)	114-0055-78A
dihydrochloride	H_2O	213(4.17),242(4.14), 290(3.83),307(3.87), 353(3.87)	114-0055-78A
$C_{33}H_{46}N_2O$ 8H-Benzo[b]cyclopenta[7,8]phenanthro- [1,10-ef][1,4]diazepin-8-one, 3-(1,5- dimethylhexyl)-1,2,3,3a,4,5,5a,5b,6- 7,9,15,15a,15b-tetradecahydro-3a,5b- dimethyl-, [3R-[3α(R*),3aα,5aβ,5bα- 15aα,15bβ]]-	EtOH	280(3.96),291(3.97), 305(3.95)(plus shoulders)	4-0023-78
$C_{33}H_{46}N_2O_4$ Emetan, 1',2'-didehydro-6',7',10,11- tetraethoxy-, (±)-	pH 1	213(4.20),251(4.40), 290(3.85),309(3.93), 348(3.84)	114-0045-78A
	EtOH	231(4.38),281(3.93), 310(3.71)	114-0045-78A
compd. with N-acetyl-L-leucine	EtOH	232(4.31),260(3.91), 280(4.01),3-8(3.80)	114-0045-78A
Emetan, 2,3-didehydro-6',7',10,11- tetraethoxy-, dihydrochloride, (1'α)-	H_2O	213(4.21),230(4.11), 282(3.77)	114-0055-78A
(1'β)-	H_2O	213(4.21),230(4.11), 282(3.77)	114-0055-78A
$C_{33}H_{48}N_2$ 1'H-Cholest-5-eno[4,5,6-cd]pyrazole, 1'-phenyl-	EtOH	261(4.26)	4-0023-78
$C_{33}H_{48}N_2O$ 8H-Benzo[b]cyclopenta[7,8]phenanthro- [1,10-ef][1,4]diazepin-8-one, 3-(1,5- dimethylhexyl)-1,2,3,3a,4,5,5a,5b,6- 7,8a,9,14b,15,15a,15b-hexadecahydro- 3a,5b-dimethyl-, [3R-[3α(R*),3aα,5aβ- 5bα,8aα,14bβ,15aα,15bβ]]-	EtOH	252(3.91),275(3.83), 282(3.82)	4-0023-78
$C_{33}H_{48}N_2O_4$ Emetan, 6',7',10,11-tetraethoxy-, (1'β)- dihydrochloride	EtOH H_2O	226(4.15),285(3.83) 211(4.21),230(4.08), 282(3.75)	114-0045-78A 114-0045-78A
$C_{33}H_{48}O_6$ 30-Norlup-12-ene-11,20-dione, 3β,29-di-	C_6H_{12}	238(4.05)	73-2190-78

Compound	Solvent	$\lambda_{max}(\log \epsilon)$	Ref.
acetoxy- (cont.)			73-2190-78
$C_{33}H_{50}N_2$			
1H-Benzo[b]cyclopenta[7,8]phenanthro-[1,10-ef][1,4]diazepine, 3-(1,5-di-methylhexyl)-2,3,3a,4,5,5a,5b,6,7,8-8a,9,14b,15,15a,15b-hexadecahydro-3a,5b-dimethyl-, [3R-[3α(R*),3aα-5aβ,5bα,8aα,14bβ,15aα,15bβ]]-	EtOH	294(3.16),321(3.33)	4-0023-78
Cholest-4-en-6-one, phenylhydrazone	EtOH	236s(3.92),314(4.40), 321(4.38),328s(4.11)	4-0023-78
$C_{33}H_{50}O_8Si_2$			
α-D-Glucopyranoside, methyl 3,6-bis-O-[(1,1-dimethylethyl)dimethylsilyl]-, dibenzoate	n.s.g.	210(3.88),220(4.18), 230(4.37),240(4.11) (end absorptions)	33-1832-78
$C_{33}H_{54}N_2O$			
1-Tridecanone, 1-(3-dodecyl-2-quinoxa-linyl)-	EtOH	244(4.54),306(3.79), 322(3.78)	24-1019-78
$C_{33}H_{58}N_2O_2$			
Spiro[5α-cholestane-3,2'-imidazolidine]-1'β,3'α-diylbis(oxy), 4',4',5',5'-tetramethyl-	tert-BuOH	236(3.63)	35-0934-78

Compound	Solvent	λ_{max} (log ϵ)	Ref.
$C_{34}H_{20}O_2S_4$ 5H-Naphtho[2,1-c][1,2]dithiol-5-one, 4-[(4,5-diphenyl-3H-1,2-dithiol-3-ylidene)acetyl]-1-phenyl-	dioxan	239(4.71),443(4.28), 488(4.37),517(4.55) (plus shoulders)	78-2175-78
$C_{34}H_{21}ClN_2O_2$ 1,2-Benzisoxazole, 3-[4-[2-[2-chloro-4-(6-phenyl-2-benzoxazolyl)phenyl]-ethenyl]phenyl]-	DMF	367(4.84)	33-2904-78
1,2-Benzisoxazole, 6-chloro-3-[4-[2-[4-(5-phenyl-2-benzoxazolyl)phenyl]eth-enyl]phenyl]-	DMF	365(4.91)	33-2904-78
1,2-Benzisoxazole, 6-chloro-3-[4-[2-[4-(6-phenyl-2-benzoxazolyl)phenyl]eth-enyl]phenyl]-	DMF	368(4.91)	33-2904-78
$C_{34}H_{21}F_3O_2$ 3-Buten-2-one, 4-(2,3-diphenylcyclo-penta[b][1]benzopyran-3-yl)-1,1,1-trifluoro-3-phenyl-, (Z)-	THF THF	530(3.08) 240(4.48),275(4.55), 370(4.57),540(2.97)	103-1070-78 103-1075-78
$C_{34}H_{22}ClN_5O$ [1,2,4]Triazolo[1,5-a]pyridine, 7-[2-[4-(5-[1,1'-biphenyl]-4-yl-1,3,4-oxadiazol-2-yl)-2-chlorophenyl]-ethenyl]-2-phenyl-	DMF	358(4.83)	33-0142-78
$C_{34}H_{22}N_2$ Naphtho[2,3-g]phthalazine, 1,4,6-tri-phenyl-	CH_2Cl_2	294(4.94),376(--), 398(--),442(3.64), 465(3.81),494(3.74)	4-1185-78
$C_{34}H_{22}N_2O$ Furo[3,4-g]phthalazine, 1,4,6,8-tetra-phenyl-	CH_2Cl_2	240(4.52),293(4.55), 357(4.21),552(3.95)	4-0793-78
$C_{34}H_{22}N_2O_2$ 1,2-Benzisoxazole, 6-[2-[4-(1,2-benz-isoxazol-3-yl)phenyl]ethenyl]-3-[1,1'-biphenyl]-4-yl-	DMF	340(4.78)	33-2904-78
1,2-Benzisoxazole, 3-[1,1'-biphenyl]-4-yl-6-[2-(2-phenyl-6-benzoxazolyl)-ethenyl]-	DMF	355(4.80)	33-2904-78
1,2-Benzisoxazole, 3-[4-[2-[4-(5-phenyl-2-benzoxazolyl)phenyl]ethenyl]phenyl]-	DMF	364(4.89)	33-2904-78
1,2-Benzisoxazole, 3-[4-[2-[4-(6-phenyl-2-benzoxazolyl)phenyl]ethenyl]phenyl]-	DMF	368(4.87)	33-2904-78
$C_{34}H_{22}N_2S$ Thieno[3,4-g]phthalazine, 1,4,6,8-tet-raphenyl-	$CHCl_3$ $CHCl_3$-HCl	297(4.70),527(4.00) 300(4.50),333(4.50), 606(3.80)	4-0793-78 4-0793-78
$C_{34}H_{22}N_4O_4S_2$ Pyrimidine, 2,2'-dithiobis[4-(1,3-ben-zodioxol-5-yl)-6-phenyl-	EtOH	241(4.58),265(4.52), 338(4.43)	4-0105-78
$C_{34}H_{22}OS$ [1]Benzothiepino[4',5':3,4]cyclopenta-[1,2-b][1]benzopyran, 6-phenyl-14-(2-phenylethenyl)-, (E)-	THF THF	434(4.11),565(2.79) 285(4.40),378(4.57), 385(4.59),434(4.11),	103-1070-78 103-1075-78

Compound	Solvent	$\lambda_{max}(\log \epsilon)$	Ref.
(cont.)		565(2.79)	103-1075-78
$C_{34}H_{22}O_2$			
Anthracene, 2,3-dibenzoyl-9-phenyl-	CH_2Cl_2	303(4.63),394(3.82)	4-1185-78
$C_{34}H_{22}O_2S$			
Benzo[c]thiophene, 5,6-dibenzoyl-1,3-diphenyl-	C_6H_{12}	223(4.4),271(4.3), 390(4.2)	4-1185-78
	$CHCl_3$	299(4.60),426(4.00)	4-1185-78
$C_{34}H_{22}O_3$			
Isobenzofuran, 5,6-dibenzoyl-1,3-diphenyl-	CH_2Cl_2	265s(4.53),280(4.57), 330(4.12),385(4.05), 440(4.02)	4-0793-78
$C_{34}H_{22}O_4$			
Ethanedione, 1,1'-[1,1':3',1"-terphenyl]-4,4'-diylbis[2-phenyl-	dioxan	307.5(4.58)	47-2093-78
Ethanedione, 1,1'-[1,1':3',1"-terphenyl]-4,4"-diylbis[2-phenyl-	dioxan	303(4.64)	47-2093-78
Ethanedione, 1,1'-[1,1':3',1"-terphenyl]-4,6'-diylbis[2-phenyl-	dioxan	280(4.57)	47-2093-78
$C_{34}H_{23}ClN_6$			
[1,2,4]Triazolo[1,5-a]pyridine, 2-[3-chloro-4-[2-[4-(2,5-diphenyl-2H-1,2,3-triazol-4-yl)phenyl]ethenyl]-phenyl]-	DMF	348(4.78)	33-0142-78
[1,2,4]Triazolo[1,5-a]pyridine, 2-[3-chloro-4-[2-[4-(4,5-diphenyl-2H-1,2,3-triazol-2-yl)phenyl]ethenyl]-phenyl]-	DMF	361(4.85)	33-0142-78
$C_{34}H_{23}N_3O$			
[1,2,4]Triazolo[1,5-a]pyridine, 7-[2-[4-(2-benzofuranyl)phenyl]ethenyl]-2-[1,1'-biphenyl]-4-yl-	DMF	368(4.91)	33-0142-78
[1,2,4]Triazolo[1,5-a]pyridine, 2-[1,1'-biphenyl]-4-yl-7-[2-(2-phenyl-6-benzofuranyl)ethenyl]-	DMF	367(4.86)	33-0142-78
[1,2,4]Triazolo[1,5-a]pyridine, 2-[4-[2-[4-(5-phenyl-2-benzofuranyl)phenyl]ethenyl]phenyl]-	DMF	373(4.93),392(4.76)	33-0142-78
$C_{34}H_{23}N_5O$			
[1,2,4]Triazolo[1,5-a]pyridine, 2-[4-[2-[4-(3-[1,1'-biphenyl]-4-yl-1,2,4-oxadiazol-5-yl)phenyl]ethenyl]phenyl]-	DMF	356(4.87)	33-0142-78
[1,2,4]Triazolo[1,5-a]pyridine, 2-[4-[2-[4-(5-[1,1'-biphenyl]-4-yl-1,3,4-oxadiazol-2-yl)phenyl]ethenyl]phenyl]-	DMF	363(4.94)	33-0142-78
[1,2,4]Triazolo[1,5-a]pyridine, 7-[2-[4-(3-[1,1'-biphenyl]-4-yl-1,2,4-oxadiazol-5-yl)phenyl]ethenyl]-2-phenyl-	DMF	348(4.83)	33-0142-78
$C_{34}H_{24}$			
1,3,9,11-Cyclotridecatetraene-5,7-diyne, 13,13'-(1,2,3-butatriene-1,4-diylidene)bis[4,9-dimethyl-	$CHCl_3$	277(4.65),604(5.03)	150-0454S-78

Compound	Solvent	$\lambda_{max}(\log \epsilon)$	Ref.
$C_{34}H_{24}N_2$			
3,5-Cyclopentadiene-1,3-dicarbonitrile, 2-[4-(2,4,6-cycloheptatrien-1-ylidene)-2,5-dimethyl-2,5-cyclohexadien-1-ylidene]-4,5-diphenyl-	MeCN DMSO acetone	609(3.73) 605(3.62) 630(--)	138-0237-78 138-0237-78 138-0237-78
$C_{34}H_{24}N_4O$			
[1,2,4]Triazolo[1,5-a]pyridine, 2-[4-[2-[4-[5-(phenylmethyl)-2-benzoxazolyl]phenyl]ethenyl]phenyl]-	DMF	367(4.93),389(4.75)	33-0142-78
$C_{34}H_{24}N_6$			
[1,2,4]Triazolo[1,5-a]pyridine, 2-[1,1'-biphenyl]-4-yl-7-[2-[4-(2-phenyl-2H-1,2,3-triazol-4-yl)-phenyl]ethenyl]-	DMF	353(4.91)	33-0142-78
[1,2,4]Triazolo[1,5-a]pyridine, 2-[1,1'-biphenyl]-4-yl-7-[2-[4-(4-phenyl-2H-1,2,3-triazol-2-yl)-phenyl]ethenyl]-	DMF	355(4.90)	33-0142-78
$C_{34}H_{24}O$			
Cyclopenta[b][1]benzopyran, 1,2-diphenyl-3-(4-phenyl-1,3-butadienyl)-, (E,E)-	THF THF	580(2.65) 270(4.55),340(4.77), 580(2.65)	103-1070-78 103-1075-78
Cyclopenta[b][1]benzopyran, 2-phenyl-1,3-bis(2-phenylethenyl)-	THF THF	610(2.87) 264(4.31),335(4.78), 610(2.87)	103-1070-78 103-1075-78
$C_{34}H_{25}N$			
Pyridinium, 5,9-diphenyl-7H-dibenzo[a,c]cyclononen-7-ylide	MeCN	248(4.02),280s(3.46), 350s(2.38),555(2.48)	150-5151-78
$C_{34}H_{25}N_3$			
[1,2,4]Triazolo[1,5-a]pyridine, 2-[1,1'-biphenyl]-4-yl-7-[2-[4-(2-phenylethenyl)phenyl]ethenyl]-	DMF	369(4.91)	33-0142-78
$C_{34}H_{25}O_2$			
1-Benzopyrylium, 4-[2-methyl-3-(2-phenyl-4H-benzopyran-4-ylidene)-1-propenyl]-2-phenyl-, perchlorate	CH_2Cl_2	665(4.56),728(5.00)	4-0365-78
$C_{34}H_{26}NO$			
Pyridinium, 1-methyl-2,4-diphenyl-6-[(2-phenyl-4H-1-benzopyran-4-ylidene)methyl]-, perchlorate	EtOH	450(4.217)	83-0226-78
$C_{34}H_{26}NS$			
Pyridinium, 1-methyl-2,4-diphenyl-6-[(2-phenyl-4H-1-benzothiopyran-4-ylidene)methyl]-, perchlorate	EtOH	467(4.289)	83-0226-78
$C_{34}H_{26}N_2$			
Phthalazine, 6,7-bis(phenylmethyl)-1,4-diphenyl-	CH_2Cl_2	302(4.00)	4-0793-78
$C_{34}H_{26}N_2O_4$			
5,13-Epoxyfuro[3',4':6,7]naphtho[2,3-g]phthalazine-6,12-dione, 5,5a,12a,13-tetrahydro-1,4,5,13-tetramethyl-8,10-	CH_2Cl_2	295(4.50),430(3.98), 507(3.85)	150-5538-78

Compound	Solvent	$\lambda_{max}(\log \epsilon)$	Ref.
diphenyl- (cont.)			150-5538-78
$C_{34}H_{26}N_4O_2S_2$ Pyrimidine, 2,2'-dithiobis[4-(4-methoxyphenyl)-6-phenyl-	EtOH	232(4.65),264(4.78), 292s(4.70),326(4.68)	4-0105-78
$C_{34}H_{26}N_4S_2$ Pyrimidine, 2,2'-dithiobis[4-(4-methylphenyl)-6-phenyl-	EtOH	267(4.82),322(4.48)	4-0105-78
$C_{34}H_{26}O_2$ 2,4,10,12-Cyclotridecatetraene-6,8-diyn-1-ol, 1,1'-(1,3-butadiyne-1,4-diyl)bis[5,10-dimethyl-	CHCl$_3$	265(4.52),329(3.97)	150-0454S-78
Ethanone, 1,1'-[1,1':3',1"-terphenyl]-4,4'-diylbis[2-phenyl-	dioxan	265(4.45),289(4.62)	47-2093-78
Ethanone, 1,1'-[1,1':3',1"-terphenyl]-4,4"-diylbis[2-phenyl-	dioxan	285(4.68)	47-2093-78
Ethanone, 1,1'-[1,1':3',1"-terphenyl]-4,6'-diylbis[2-phenyl-	EtOH	211(4.61),267(4.59)	47-2093-78
$C_{34}H_{26}O_7$ Ethyne, 1-phenyl-2-(2,3,5-tri-O-benzoyl-α-D-ribofuranosyl)-	EtOH	230(4.94),252(4.54), 272(3.87),283(3.79)	94-0898-78
β-	EtOH	231(4.67),251(4.27), 272(3.60),282(3.46)	94-0898-78
$C_{34}H_{27}NO_3S_2$ 4-Piperidinone, 1-(phenylmethyl)-3,5-bis[[5-(phenylthio)-2-furanyl]methylene]-	EtOH	207(4.54),214(4.28), 263(4.17),405(4.54)	133-0189-78
$C_{34}H_{27}NO_5S_2$ 4-Piperidinone, 1-(phenylmethyl)-3,5-bis[[5-(phenylsulfinyl)-2-furanyl]-methylene]-	EtOH	211(4.50),297(4.27), 395(3.88)	133-0189-78
$C_{34}H_{27}NO_7S_2$ 4-Piperidinone, 1-(phenylmethyl)-3,5-bis[[5-(phenylsulfonyl)-2-furanyl]-methylene]-	EtOH	211(4.45),263(4.35), 392(3.29)	133-0189-78
$C_{34}H_{27}N_3O$ Methanone, phenyl[1,2,3,4-tetrahydro-1,2,4,5-tetraphenyl-1,2,4-triazin-3-yl)-	EtOH	243(4.54),285(4.11)	44-3394-78
$C_{34}H_{28}N_2O_2P_2$ 3H-Pyrazole, 4,5-bis(diphenylphosphinyl)-3-methyl-3-phenyl-	MeOH	258(3.55),380(1.94)	150-4248-78
$C_{34}H_{28}N_2O_3$ 1H-Cyclopenta[b]quinoxaline, 4,9-dibenzoyl-3a,4,9,9a-tetrahydro-1-[1-(4-methoxyphenyl)ethylidene]-, (E)-	MeCN	270(4.43)	5-1139-78
(Z)-	MeCN	272(4.38)	5-1139-78
$C_{34}H_{28}N_2O_6$ 2,4-Hexadienedinitrile, 3,4-dihydroxy-2,5-bis[3-methoxy-4-(phenylmethoxy)-phenyl]-	EtOH	268(4.36),359(4.18)	64-0820-78C

Compound	Solvent	λ_{max} (log ϵ)	Ref.
2,5-Hexadienedinitrile, 3,4-dihydroxy-2,5-bis[4-methoxy-3-(phenylmethoxy)-phenyl]-	EtOH	267(4.39),359(4.22)	64-0820-78C
$C_{34}H_{28}N_2O_7S$ 1H-Benzimidazole, 2-(methylthio)-1-(2,3,5-tri-O-benzoyl-β-D-ribofuranosyl)-	MeOH	227(4.72),258s(4.07), 275s(4.05),282(4.16), 291(4.10)	24-0996-78
$C_{34}H_{28}N_{10}Ni$ Nickel, [1,10,11,20-tetrahydro-3,8,13-18-tetramethyl-1,11-diphenyldibenzo-[c,j]dipyrazolo[3,4-f:3',4'-m]-[1,2,5,8,9,12]hexaazacyclotetradecinato(2-)-N^4,N^{10},N^{14},N^{20}]-, (SP-4-1)-	$CHCl_3$	312(4.49),395(4.29), 428s(4.39),460(4.62), 628(3.80)	103-0885-78
$C_{34}H_{30}$ Fluoranthene, 7,10-diethyl-8-(4-ethylphenyl)-9-phenyl-	C_6H_{12}	275(4.53),284(4.48), 295(4.34)	2-0152-78
$C_{34}H_{30}N_2$ 4,11-Etheno-1H-cyclopenta[b]phenazine, 1-(diphenylmethylene)-3a,4,11,11a-tetrahydro-7,8,12,13-tetramethyl-	C_6H_{12}	240(4.54),243(4.58), 245(4.56),298(4.37), 304(4.37),317(4.35), 330(4.16)	5-0440-78
$C_{34}H_{30}N_2O_2$ 4,11-Etheno-1H-cyclopenta[b]phenazine, 1-[bis(4-methoxyphenyl)methylene]-3a,4,11,11a-tetrahydro-7,8-dimethyl-	C_6H_{12}	244(4.71),247(4.71), 305s(4.52),310(4.53), 318(4.54),322s(4.51), 329s(4.41)	5-0440-78
$C_{34}H_{30}N_2O_4$ Benzenebutanamide, γ-[1,2-dihydro-1-(4-methoxyphenyl)-2-oxo-5-phenyl-3H-pyrrol-3-ylidene]-N-(4-methoxyphenyl)-	EtOH	210(4.31),253(4.47), 400(3.75)	12-2031-78
Benzenebutanamide, γ-[1,2-dihydro-5-(4-methoxyphenyl)-2-oxo-1-phenyl-3H-pyrrol-3-ylidene]-4-methoxy-N-phenyl-	EtOH	210(4.48),230(4.46), 274(4.32),402(3.92)	12-2031-78
$C_{34}H_{30}N_2O_5$ 1,2-Dehydroapateline	MeOH MeOH-NaOH	288s(--),335(3.46) 292s(--),337(3.56)	12-2077-78 12-2077-78
$C_{34}H_{30}O_5$ 12H-Benzo[a]xanthene, 3,9-dimethoxy-5-(4-methoxyphenyl)-6-[(4-methoxyphenyl)methyl]-	EtOH	260(4.08),290s(3.68), 350(3.07)	39-0088-78C
12H-Benzo[a]xanthene, 5-(3,4-dimethoxyphenyl)-9-methoxy-6-[(2-methoxyphenyl)methyl]-	EtOH	240(3.80),249s(3.86), 275s(3.55),328(2.71)	39-0088-78C
$C_{34}H_{30}O_{17}$ [1,1'-Biphenyl]-2,2',4,4',6,6'-hexol, 3-(3,5-diacetoxyphenoxy)-, hexaacetate	MeCN	226s(4.46),270s(3.52)	83-0393-78
$C_{34}H_{30}S_2$ 1,4:18,21-Dietheno-13,9-metheno-5H,9H,17H-naphtho[1,8-qr][1,11]dithiacyclotetracosin, 7,8,14,15-tetra-	$CHCl_3$	307(4.17)	88-1459-78

Compound	Solvent	$\lambda_{max}(\log \epsilon)$	Ref.
hydro- (cont.)			88-1459-78
6H,17H-10,13-Etheno-1,5:18,22-dimeth-	$CHCl_3$	304(4.13)	88-1459-78
enonaphtho[1,8-qr][1,10]dithiacyclo-			
pentacosin, 8,9,14,15-tetrahydro-			
1,4:9,12:17,20-Trietheno-5H,16H-naph-	$CHCl_3$	303(4.10)	88-1459-78
tho[1,8-pq][1,10]dithiacyclotricosin,			
7,8,13,14-tetrahydro-			
$C_{34}H_{32}Br_2$			
1,2,3,5,7,11,13-Cyclotetradecaheptaen-	THF	231(4.17),239(4.22),	18-3351-78
9-yne, 1,8-bis(4-bromophenyl)-4,11-		269(4.15),315(4.48),	
bis(1,1-dimethylethyl)-		332(4.66),348s(4.90),	
		361(5.28),514(4.76),	
		574(2.94),628(3.36)	
$C_{34}H_{32}Cl_2$			
1,2,3,5,7,11,13-Cyclotetradecaheptaen-	THF	230(4.14),238(4.21),	18-3351-78
9-yne, 1,8-bis(4-chlorophenyl)-4,11-		268(4.16),332(4.68),	
bis(1,1-dimethylethyl)-		360(5.29),512(4.77),	
		570(2.73),623(3.33)	
$C_{34}H_{32}N_2O_5$			
Apateline	MeOH	283(3.5),305s(--)	12-2077-78
	MeOH-NaOH	297(3.6)	12-2077-78
Norapateline, N-methyl-	MeOH	282(3.72)	12-2539-78
	MeOH-NaOH	302(3.95)	12-2539-78
$C_{34}H_{32}N_2O_5S_2$			
Benzenesulfonamide, N-[4,5-dimethyl-2-	MeCN	340(4.21)	5-1146-78
[(phenylsulfonyl)amino]phenyl]-N-[5-			
[1-(4-methoxyphenyl)ethylidene]-1,3-			
cyclopentadien-1-yl]-			
Benzo[b]cyclopenta[e][1,4]diazepine,	MeCN	228(4.58),266s(3.96),	5-1146-78
3a,4,9,10-tetrahydro-10-(4-methoxy-		273s(3.90)	
phenyl)-6,7,10-trimethyl-4,9-bis-			
(phenylsulfonyl)-, cis			
trans	MeCN	265(4.00),273(3.92)	5-1146-78
1H-Cyclopenta[b]quinoxaline, 3a,4,9,9a-	MeCN	268(4.37)	5-1146-78
tetrahydro-1-[1-(4-methoxyphenyl)eth-			
ylidene]-6,7-dimethyl-4,9-bis(phenyl-			
sulfonyl)-, (1E,3aα,9aα)-			
Spiro[2H-benzimidazole-2,1'-[3]cyclo-	MeCN	212(4.63),236s(4.38)	5-1146-78
pentene], 1,3-dihydro-2'-[1-(4-meth-			
oxyphenyl)ethylidene]-5,6-dimethyl-			
1,3-bis(phenylsulfonyl)-			
$C_{34}H_{32}N_4O$			
1-Butanone, 1-(6-methyl-3,7,9-triphen-	EtOH	242(4.56),300(4.45),	22-0415-78
yl-1,2,7,8-tetraazaspiro[4.4]nona-		347(4.49)	
2,8-dien-6-yl)-			
$C_{34}H_{32}O_6$			
2,3'-Bi-2H-1-benzopyran, 2'-(3,4-di-	EtOH	240(3.94),272(3.80),	39-0088-78C
methoxyphenyl)-3',4'-dihydro-7-meth-		322(3.27)	
oxy-3-(4-methoxyphenyl)-			
$C_{34}H_{34}$			
[14]Annulene, 1,14-di-tert-butyl-7,10-	THF	228s(4.10),239(4.16),	18-3345-78
diphenyl-1,8-bisdehydro-		278(4.11),315s(4.35),	
		331s(4.58),347s(4.90),	
		360(5.33),480s(4.27),	

Compound	Solvent	$\lambda_{max}(\log \epsilon)$	Ref.
(cont.)		507(4.57),573(2.73), 624(3.26)	18-3345-78
[14]Annulene, 3,7-di-tert-butyl-10,14-diphenyl-1,8-bisdehydro-	THF	275s(4.11),315s(4.30), 335s(4.57),352s(3.94), 367(5.27),474s(4.29), 499(4.56),574(2.55), 624(3.10)	18-3345-78
[14]Annulene, 7,14-di-tert-butyl-3,10-diphenyl-1,8-bisdehydro-	THF	226s(4.09),234(4.16), 260(4.13),270s(4.03), 315s(4.45),332s(4.61), 345s(4.88),357(5.29), 478s(3.10),508(4.72), 571(2.77),623(3.21)	18-3345-78
$C_{34}H_{34}N_2$ Acridine, 9,10-dihydro-10-methyl-2-(1-methylethyl)-9-[10-methyl-2-(1-methylethyl)-9(10H)-acridinylidene]-	EtOH	235(4.74),264s(4.41), 285s(4.26),422(4.15)	35-5604-78
$C_{34}H_{34}N_2O_7S_2$ Benzo[b]cyclopenta[e][1,4]diazepin-2(3H)-one, 3-acetoxy-10,10-dicyclopropyl-3a,4,9,10-tetrahydro-6,7-dimethyl-4,9-bis(phenylsulfonyl)-, cis	MeCN	265s(3.88),273s(3.74), 346(2.52)	5-1146-78
$C_{34}H_{34}O_7$ Scirpusin A hexamethyl ether	EtOH	228s(4.57),290(4.25), 310s(4.39),322(4.42)	94-3050-78
$C_{34}H_{35}BrO_2$ [1,1':3',1"-Terphenyl]-2'-ol, 5'-(4-bromophenyl)-4,4"-bis(1,1-dimethylethyl)-, acetate	CH_2Cl_2	252(4.73),255(4.74)	24-0264-78
$C_{34}H_{35}N_5O_4$ 21H,23H-Porphine-2,18-dipropanoic acid, 8-cyano-3,7,12,13,17-pentamethyl-, dimethyl ester	$CHCl_3$	406(5.07),512(4.01), 550(4.24),571(3.99), 621(3.08)	104-0794-78
$C_{34}H_{36}N_2O_2$ 4H-Indol-4-one, 1,1'-(1,2-ethanediyl)-bis[1,5,6,7-tetrahydro-6,6-dimethyl-2-phenyl-	MeOH	240(3.66),280(3.48)	48-0863-78
$C_{34}H_{36}O_4$ Oxepino[2,3-b]benzofuran, 2,9-bis(1,1-dimethylethyl)-4,7-bis(phenylmethoxy)-	C_6H_{12}	212(4.56),234s(3.99), 291(3.92),298s(3.91), 330s(3.29)	12-1061-78
$C_{34}H_{37}N_3O_{11}S$ Daunorubicin 13-tosylhydrazone, hydrochloride	MeOH	234(4.62),252(4.46), 291(3.90)	87-0280-78
$C_{34}H_{37}N_3O_{12}S$ Adriamycin 13-tosylhydrazone, hydrochloride	MeOH	234(4.58),252(4.43), 291(3.92)	87-0280-78
$C_{34}H_{38}N_2O_2$ 2'H-Androsta-5,16-dieno[16,17-c]pyrazol-3-ol, 2',5'-diphenyl-, acetate, (3β)-	dioxan	222(4.25),235(4.08), 285(4.36)	78-1633-78

$C_{34}H_{38}N_2O_2-C_{34}H_{45}CuN_4O_5$

Compound	Solvent	$\lambda_{max}(\log \epsilon)$	Ref.
2'H-Androsta-5,16-dieno[17,16-c]pyrazol-3-ol, 2',5'-diphenyl-, acetate, (3β)-	dioxan	220(4.32),235(4.20), 276(4.39),282s(4.37), 290s(4.28)	78-1633-78
$C_{34}H_{38}N_4Ni$ Nickel, [5-ethenyl-2,7,12,17-tetraethyl-3,8,13,18-tetramethyl-21H,23H-porphinato(2-)-$N^{21},N^{22},N^{23},N^{24}$]-, (SP-4-2)-	CHCl$_3$	404(5.21),530(4.02), 566(4.21)	39-1660-78C
$C_{34}H_{38}N_4O_4$ 21H,23H-Porphine-13,17-dipropanoic acid, 2,8,12,18-tetramethyl-, diethyl ester	CHCl$_3$	399(5.27),496(4.16), 529(3.92),566(3.80), 619(3.61)	12-0639-78
$C_{34}H_{38}O_{15}$ 2-Propen-1-one, 1-[2-acetoxy-4-methoxy-3-(2,3,4,6-tetra-O-acetyl-β-D-glucopyranosyl)phenyl]-3-methoxy-2-(4-methoxyphenyl)-	n.s.g.	230(4.31),257(4.30)	12-2699-78
$C_{34}H_{39}N_3O_8$ L-Glutamic acid, N-[N-[4-[methyl[(phenylmethoxy)carbonyl]amino]benzoyl]-L-phenylalanyl]-, diethyl ester	MeOH	260(4.11)	87-1165-78
$C_{34}H_{39}N_5O_4$ 21H-Biline-8,12-dipropanoic acid, 2-cyano-10,24-dihydro-1,3,7,13,17,18-hexamethyl-, dimethyl ester, dihydrobromide	CHCl$_3$-1%HBr	452(4.57),518(5.05)	104-0794-78
$C_{34}H_{40}N_2O_5$ Piperidine, 1-[[6-(4-methoxybenzoyl)-5-(4-methoxyphenyl)-2-(1-piperidinylcarbonyl)-1,5-cyclohexadien-1-yl]-acetyl]-	EtOH	300(4.2),360(4.38)	104-2026-78
$C_{34}H_{40}O_2$ Benzoic acid, 4-[4,8,11-tris(1,1-dimethylethyl)-1,2,3,5,7,11,13-cyclotetradecaheptaen-9-yn-1-yl]-, methyl ester	THF	235(4.05),255s(4.01), 268(4.06),305s(4.35), 326s(4.63),340s(4.89), 355(5.21),491(4.57), 561(2.84),612(3.26)	18-3351-78
$C_{34}H_{41}N_6OP$ Phosphorin, 4-cyclohexyl-2,6-bis[[4-(dimethylamino)phenyl]azo]-1,1-dihydro-1-methoxy-1-phenyl-	EtOH	244(4.38),367(4.28), 599(4.57)	118-0846-78
$C_{34}H_{44}N_6O_{12}$ Actinocylbis(threonyl-D-valine)	MeOH	240(4.669),448(4.367)	87-0607-78
$C_{34}H_{44}O_2$ Phenol, 4-[4,5-bis(1,1-dimethylethyl)-1,8-dimethyl-9H-xanthen-9-yl]-2-(1,1-dimethylethyl)-5-methyl-	EtOH	245(4.25),277(3.75), 286(3.74)	32-0079-78
$C_{34}H_{45}CuN_4O_5$ Copper, [3-[[[2-[[[5-(1,1-dimethyleth-	CH$_2$Cl$_2$	229(4.81),250(4.72),	35-2686-78

Compound	Solvent	λ_{max}(log ϵ)	Ref.
yl)-2-hydroxyphenyl]methylene]amino]-1-[[[[5-(1,1-dimethylethyl)-2-hydroxymethylene]amino]methyl]ethoxy]carbonyl]amino]-2,5-dihydro-2,2,5,5-tetramethyl-1H-pyrrol-1-yloxyato(2-)]-(cont.)	THF pyridine	279(4.53),384(4.10), 608(2.51) 233(4.79),250(4.62), 275(4.41),384(4.00), 618(2.49) 386(4.06),610(2.35)	35-2686-78 35-2686-78 35-2686-78
C$_{34}$H$_{46}$ 1,2,3,5,7,9,13,15,17-Cyclooctadecanonaen-11-yne, 1,4,10,13-tetrakis(1,1-dimethylethyl)-	THF	217(4.05),255s(3.77), 268(3.92),342s(4.65), 356(4.98),372(5.65), 464s(3.68),499(4.03), 539(4.28),607(2.00), 646s(1.95),669(2.07), 681(2.07),720s(2.14), 751(2.50)	18-3359-78
C$_{34}$H$_{46}$O$_3$ Cholesta-3,5-dien-7-one, 19-(benzoyloxy)-	EtOH	230(4.33),278(4.45)	73-1142-78
C$_{34}$H$_{48}$N$_2$ Cholest-4-eno[4,5,6-de]pyrimidine, 2'-phenyl-	EtOH	257(4.17)	4-0023-78
C$_{34}$H$_{48}$N$_{10}$O$_{10}$ Actinocylbis(threonyl-D-valine hydrazide)	MeOH	240(4.519),443(4.284)	87-0607-78
C$_{34}$H$_{48}$O$_2$ 2,4,6,11,13,15-Cyclooctadecahexaene-8,17-diyne-1,10-diol, 1,7,10,11-tetrakis(1,1-dimethylethyl)- lower melting isomer	EtOH EtOH	230(4.32),281(4.94), 291(5.05),328(4.14) 230(4.26),281(4.88), 291(4.99),328(4.09)	18-3359-78 18-3359-78
C$_{34}$H$_{48}$O$_6$ 14(13→12)-Abeo-5α-cholesta-12,14,16-trien-21-oic acid, 3β-acetoxy-9β-hydroxy-4,4,15,24-tetramethyl-11-oxo-, methyl ester	EtOH	220(4.23),261(3.93), 312(3.28)	39-0471-78C
C$_{34}$H$_{48}$O$_{10}$ Tangulic acid, 11-oxodiacetyl-	EtOH	256(4.12)	42-1169-78
C$_{34}$H$_{50}$N$_2$ Cholest-5-eno[4,5,6-de]pyrimidine, 3',4-dihydro-2'-phenyl-	EtOH	230(4.67),260s(4.45)	4-0023-78
C$_{34}$H$_{50}$N$_2$O$_2$ Phenol, 2-[[[(3β,5α,22β,25S)-spirosolan-3-yl]imino]methyl]- Phenol, 4-[[[(3β,5α,22β,25S)-spirosolan-3-yl]imino]methyl]-	EtOH EtOH	224(4.20),255(4.11), 316(3.61),400(3.03) 220(4.07),271(4.27), 384(3.62)	88-0159-78 88-0159-78
C$_{34}$H$_{50}$O$_2$ Cholest-2-en-4-one, 3-(phenylmethoxy)- Cholest-4-eno[4,3-b]oxet-3(4'H)-ol, 4'-phenyl-	EtOH EtOH	262(3.76) 224(3.96)	78-1509-78 78-1509-78

Compound	Solvent	$\lambda_{max}(\log \epsilon)$	Ref.
$C_{34}H_{50}O_5$			
D(17a)-Homo-C,18-dinorcholesta-13,15,17-trien-11-one, 3,9-diacetoxy-4,4,15-17a-tetramethyl-, (3β,5α,9β)-	EtOH	215(4.41),260(3.99), 310(3.30)	39-0471-78C
$C_{34}H_{50}O_6$			
D(17a)-Homo-C,18-dinorcholesta-13,15,17-trien-11-one, 3,7-diacetoxy-9-hydroxy-4,4,15,17a-tetramethyl-, (3β,5α,7α-9β)-	EtOH	218(4.43),260(4.06), 310(3.40)	39-0471-78C
(3β,5α,7β,9β)-	EtOH	217(4.38),264(4.05), 315(3.35)	39-0471-78C
$C_{34}H_{50}O_9$			
22-Acetylcyasterone 2,3-acetonide	EtOH	238(4.14)	105-0175-78
$C_{34}H_{52}O_5$			
Lanosta-5,8-dien-7-one, 3β,11α-diacetoxy-	EtOH	254(4.12)	19-0583-78
19-Norlanosta-1(10),5-dien-7-one, 3β,11α-diacetoxy-9-methyl-, (3β,9β,11α)-	EtOH	238(3.79),296(4.05)	19-0583-78
$C_{34}H_{54}O_5$			
Dammar-13(17)-en-16-one, 3,11-diacetoxy-, (3β,8α,9β,11α,14β)-	EtOH	242(4.15)	19-0591-78
19-Norlanost-5-en-7-one, 3,11-diacetoxy-, (3β,9β,10α,11α)-	EtOH	244(4.06)	19-0583-78

Compound	Solvent	$\lambda_{max}(\log \epsilon)$	Ref.
$C_{35}H_{20}F_{18}O$ 4,7-Methano-1,2,3-metheno-1H-cyclopenta[3',4'[cyclobuta[1',2':3,4]cyclobuta[1,2]benzen-8-one, 2,3,3a,3b,3c-4,7,7a,7b,7c-decahydro-4,7-dimethyl-5,6-diphenyl-1,2,3,3a,7c,9-hexakis-(trifluoromethyl)-	EtOH	260(3.89)	12-0221-78
$C_{35}H_{20}O_4S_4$ 2H-Pyran-2,4,6(3H,5H)-trione, 3,5-bis-(4,5-diphenyl-3H-1,2-dithiol-3-ylidene)-	dioxan	243(4.53),278s(--), 333s(--),433(4.03), 514s(--),536(4.58)	78-2175-78
$C_{35}H_{21}Cl_2N_3O_2$ 1,2-Benzisoxazole, 3-[4-[2-[4-(5-[1,1'-biphenyl]-4-yl-1,3,4-oxadiazol-2-yl)-2-chlorophenyl]ethenyl]phenyl]-6-chloro-	DMF	355(4.83)	33-2904-78
$C_{35}H_{22}ClN_3O_2$ 1,2-Benzisoxazole, 3-[4-[2-[4-(5-[1,1'-biphenyl]-4-yl-1,3,4-oxadiazol-2-yl)-2-chlorophenyl]ethenyl]phenyl]-	DMF	355(4.82)	33-2904-78
1,2-Benzisoxazole, 3-[4-[2-[4-(3-[1,1'-biphenyl]-4-yl-1,2,4-oxadiazol-5-yl)-phenyl]ethenyl]phenyl]-6-chloro-	DMF	350(4.84)	33-2904-78
1,2-Benzisoxazole, 3-[4-[2-[4-(5-[1,1'-biphenyl]-4-yl-1,3,4-oxadiazol-2-yl)-phenyl]ethenyl]phenyl]-6-chloro-	DMF	357(4.90)	33-2904-78
$C_{35}H_{22}Cl_2N_4O$ 1,2-Benzisoxazole, 6-chloro-3-[3-chloro-4-[2-[4-(2,5-diphenyl-2H-1,2,3-triazol-4-yl)phenyl]ethenyl]phenyl]-	DMF	347(4.72)	33-2904-78
1,2-Benzisoxazole, 6-chloro-3-[3-chloro-4-[2-[4-(4,5-diphenyl-2H-1,2,3-triazol-2-yl)phenyl]ethenyl]phenyl]-	DMF	357(4.80)	33-2904-78
$C_{35}H_{23}ClN_4O$ 1,2-Benzisoxazole, 3-[4-[2-[2-chloro-4-(4,5-diphenyl-2H-1,2,3-triazol-2-yl)phenyl]ethenyl]phenyl]-	DMF	357(4.80)	33-2904-78
[1,2,4]Triazolo[1,5-a]pyridine, 2-[4-[2-[2-chloro-4-(4,5-diphenyl-2-oxazolyl)phenyl]ethenyl]phenyl]-	DMF	372(4.80)	33-0142-78
$C_{35}H_{23}NO_2$ 1,2-Benzisoxazole, 3-[1,1'-biphenyl]-4-yl-6-[2-(2-phenyl-6-benzofuranyl)-ethenyl]-	DMF	365(4.80)	33-2904-78
$C_{35}H_{23}N_3O_2$ 1,2-Benzisoxazole, 3-[4-[2-[4-(3-[1,1'-biphenyl]-4-yl-1,2,4-oxadiazol-5-yl)-phenyl]ethenyl]phenyl]-	DMF	350(4.84)	33-2904-78
1,2-Benzisoxazole, 3-[4-[2-[4-(5-[1,1'-biphenyl]-4-yl-1,3,4-oxadiazol-2-yl)-phenyl]ethenyl]phenyl]-	DMF	357(4.89)	33-2904-78

Compound	Solvent	λ_{max}(log ϵ)	Ref.
$C_{35}H_{24}N_2O_2$			
1,2-Benzisoxazole, 3-[4-[2-[4-[5-(phen-ylmethyl)-2-benzoxazolyl]phenyl]eth-enyl]phenyl]-	DMF	362(4.87)	33-2904-78
$C_{35}H_{24}N_2O_3$			
1,2-Benzisoxazole, 3-[4-[2-[4-[5-(phen-ylmethoxy)-2-benzoxazolyl]phenyl]eth-enyl]phenyl]-	DMF	365(4.85)	33-2904-78
$C_{35}H_{24}N_4O$			
1,2-Benzisoxazole, 3-[1,1'-biphenyl]-4-yl-6-[2-[4-(2-phenyl-2H-1,2,3-tri-azol-4-yl)phenyl]ethenyl]-	DMF	350(4.85)	33-2904-78
1,2-Benzisoxazole, 3-[1,1'-biphenyl]-4-yl-6-[2-[4-(4-phenyl-2H-1,2,3-tri-azol-2-yl)phenyl]ethenyl]-	DMF	352(4.85)	33-2904-78
1,2-Benzisoxazole, 6-[2-[4-(4,5-diphen-yl-2H-1,2,3-triazol-2-yl)phenyl]eth-enyl]-3-phenyl-	DMF	355(4.83)	33-2904-78
[1,2,4]Triazolo[1,5-a]pyridine, 2-[4-[2-[4-(3-[1,1'-biphenyl]-4-yl-5-is-oxazolyl]phenyl]-	DMF	356(4.91),373(4.72)	33-0142-78
[1,2,4]Triazolo[1,5-a]pyridine, 2-[4-[2-[4-(4,5-diphenyl-2-oxazolyl)phen-yl]ethenyl]phenyl]-	DMF	371(4.85)	33-0142-78
$C_{35}H_{25}NO$			
1,2-Benzisoxazole, 3-[1,1'-biphenyl]-4-yl-6-[2-[4-(2-phenylethenyl)phen-yl]ethenyl]-	DMF	366(4.86)	33-2904-78
$C_{35}H_{25}NO_2$			
4,13-Etheno-1H-anthra[2,3-f]isoindole-1,3(2H)-dione, 3a,4,13,13a-tetrahy-dro-2-methyl-5,12-diphenyl-, (3aα,4β,13β,13aα)-	THF	262(4.69),269(4.87), 315(3.32),330(3.65), 346(3.92),364(4.07), 381(4.01)	78-2263-78
$C_{35}H_{25}NO_2S_2$			
Benzeneacetic acid, α-phenyl-, 2,2-di-phenyl-1-(2-thioxo-3(2H)-benzothia-zolyl)ethenyl ester	CHCl₃	246(4.38),276s(4.15), 336(4.27)	44-4961-78
$C_{35}H_{25}NO_4$			
4,13[1',2']-Benzeno-1H-naphtho[2',3'-2,3][1,4]dioxocino[6,7-c]pyrrole-1,3(2H)-dione, 3a,4,13,13a-tetra-hydro-2-methyl-6,11-diphenyl-	THF	241(4.73),271(3.81), 296(4.06)	78-2263-78
2,6-Etheno-2H-bisoxireno[1,2:3,4]anthra-[2,3-f]isoindole-3,5(2aH,4H)-dione, 5a,6,7a,13a-tetrahydro-4-methyl-8,13-diphenyl-, (1aR*,2α,2aβ,5aβ,6α,6aS*-7aα,13bα)-	THF	245(4.85),299(4.04)	78-2263-78
2,6-Etheno-2H-bisoxireno[1,2:3,4]anthra-[2,3-f]isoindole-3,5(2aH,4H)-dione, 5a,6,7a,13b-tetrahydro-4-methyl-7a-13b-diphenyl-, (1aR*,2α,2aβ,5aβ,6α-6aS*,7aα,13bα)-	ether	235(4.85),240(4.88), 257(3.66),265(3.70), 275(3.75),284(3.74), 297(3.55),320(2.87), 328(2.95)	78-2263-78
C H O			
Pyrylium, 2-[(2,6-diphenyl-4H-pyran-	MeCN	228(4.42),268(4.42),	4-1225-78

Compound	Solvent	$\lambda_{max}(\log \epsilon)$	Ref.
4-ylidene)methyl]-4,6-diphenyl-, perchlorate		347(4.58),560(4.68), 596(4.80)	4-1225-78
Pyrylium, 2-[(4,6-diphenyl-2H-pyran-2-ylidene)methyl]-4,6-diphenyl-, perchlorate	MeCN	285(4.52),335(4.66), 585(4.54),628(4.58)	4-1225-78
Pyrylium, 4-[(2,6-diphenyl-4H-pyran-2-ylidene)methyl]-2,6-diphenyl-, perchlorate	MeCN	230(4.33),268(4.42), 379(4.43),543(5.08)	4-1225-78
$C_{35}H_{27}O_2$ 1-Benzopyrylium, 2-phenyl-4-[2-[(2-phenyl-4H-benzopyran-4-ylidene)methyl]-1-butenyl]-, perchlorate	CH_2Cl_2	727(4.78)	4-0365-78
$C_{35}H_{28}NO_2PS$ 2-Butene-1,4-dione, 2-[[(diphenylphosphinothioyl)phenylmethyl]amino]-1,4-diphenyl-	C_6H_{12}	255(4.14),315(4.11)	22-0379-78
$C_{35}H_{28}N_2O_5$ Alkaloid, m. 250-260°d.	MeOH MeOH-NaOH	290s(4.0),340s(3.7) 354(4.2)	12-2077-78 12-2077-78
$C_{35}H_{28}N_2O_7$ 1H-Pyrazole, 4-(2,3,5-tri-O-benzoyl-α-D-ribofuranosyl)-3-phenyl- β-	EtOH EtOH	231(4.54),275(3.75), 283(3.69) 231(4.68),275(3.77), 282(3.69)	94-0898-78 94-0898-78
$C_{35}H_{28}O$ Anthracene, 9-ethoxy-9,10-dihydro-9-phenyl-10-(diphenylmethylene)-	n.s.g.	238(4.1),290(4.0)	39-0488-78C
$C_{35}H_{30}N_2O_2$ Benzonitrile, 4,4'-[3-[2,2-bis(4-methoxyphenyl)ethenyl]-2,2-dimethylcyclopropylidene]bis- 1,4-Pentadiene, 1,1-bis(4-cyanophenyl)-5,5-bis(4-methoxyphenyl)-3,3-dimethyl-	EtOH EtOH	250(4.63),272(4.35) 237(4.53),248(4.50), 265(4.48)	35-4131-78 35-4131-78
$C_{35}H_{30}N_2O_5$ Alkaloid, m. 247-258°d.	MeOH MeOH-NaOH	290s(3.8),337(3.6) 227(3.7),295s(3.8)	12-2077-78 12-2077-78
$C_{35}H_{31}ClO_8$ 2H,6H-Benzo[1,2-b:5,4-b']dipyran-2-one, 3,3'-[(4-chlorophenyl)methylene]bis-[7,8-dihydro-4-hydroxy-8,8-dimethyl-	dioxan	335(4.7)	2-0375-78
$C_{35}H_{31}N_4O_2P$ 2,3,4(1H)-Phosphorintrione, 1-(1,1-dimethylethyl)-5-phenyl-, 2,4-bis(1-naphthalenylhydrazone)-, 1-oxide	$CHCl_3$	288s(4.10),342(4.20), 380(4.21),512(4.32)	139-0027-78B
$C_{35}H_{32}N_2O_2$ 4,11-Etheno-1H-c-clopenta[b]phenazine, 1-[bis(4-methoxyphenyl)methylene]-3a,4,11,11a-tetrahydro-7,8,13-trimethyl-, (3aα,4β,11β,11aα)-	C_6H_{12}	211(4.72),244(4.57), 247(4.57),312(4.44), 318(4.44),325s(4.41), 332s(4.31)	5-0440-78

Compound	Solvent	$\lambda_{max}(\log \epsilon)$	Ref.
$C_{35}H_{32}N_2O_4S_2$ 1H-Benzimidazole, 2,3-dihydro-5,6-di-methyl-2,2-bis(4-methylphenyl)-1,3-bis(phenylsulfonyl)-	MeCN	213(4.77),266s(3.80), 274s(3.77),307(3.86)	5-1161-78
$C_{35}H_{32}N_2O_5$ 1,2-Dehydrotelobine	MeOH and MeOH-NaOH	287s(--),336(3.65)	12-2077-78
$C_{35}H_{32}O_6$ 12H-Benzo[a]xanthene, 5-(3,4-dimethoxy-phenyl)-3,9-dimethoxy-6-(2-methoxy-phenyl)methyl]-	EtOH	256(4.70),283s(3.80), 291s(3.72),355(3.14)	39-0088-78C
$C_{35}H_{32}O_8$ 2H,6H-Benzo[1,2-b:5,4-b']dipyran-2-one, 3,3'-(phenylmethylene)bis[7,8-dihydro-4-hydroxy-8,8-dimethyl-	dioxan	330(4.72)	2-0375-78
$C_{35}H_{33}NO$ 1,2-Benzisoxazole, 6-[2-[4-(1-methyl-ethyl)phenyl]ethenyl]-3-[4-[2-[4-(1-methylethyl)phenyl]ethenyl]phenyl]-	DMF	344(4.88)	33-2904-78
$C_{35}H_{33}N_6O_9P$ Cytidine, N-benzoyl-2'-O-[(2-nitrophen-yl)methyl]-, 3'-(N,N'-diphenylphos-phorodiamidate)	EtOH	232(4.51),262(4.50), 303(4.11)	35-4580-78
$C_{35}H_{34}N_2OP$ Methylium, bis[4-(dimethylamino)phen-yl][4-(diphenylphosphinyl)phenyl]-, perchlorate	EtOH	262(4.16),318(4.39), 430(4.21),636(4.75)	65-2383-78
$C_{35}H_{34}N_2O_5$ Apateline, N-methyl- Telobine	MeOH MeOH-NaOH MeOH and MeOH-NaOH	280(3.61) 290(3.72) 278(3.66),305s(--)	12-2539-78 12-2539-78 12-2077-78
$C_{35}H_{34}N_2O_6$ Gilletine	MeOH	237(4.34),274s(3.33), 290(3.41),301s(3.36)	142-0995-78A
$C_{35}H_{34}N_2P$ Methylium, bis[4-(dimethylamino)phenyl]-[4-(diphenylphosphino)phenyl]-, per-chlorate	EtOH	261(4.15),314(4.15), 454(4.16),628(4.78)	65-2383-78
$C_{35}H_{36}O_4$ Cyclopropane, 1,1-bis(3-methoxyphenyl)-3-[2,2-bis(3-methoxyphenyl)ethenyl]-2,2-dimethyl-	EtOH	277(4.29),283s(4.25)	35-4131-78
1,4-Pentadiene, 1,1,5,5-tetrakis(3-methoxyphenyl)-3,3-dimethyl-	EtOH	262(4.25)	35-4131-78
1,4-Pentadiene, 1,1,5,5-tetrakis(4-methoxyphenyl)-3,3-dimethyl-	EtOH	244(4.50),264(4.44)	44-1997-78
$C_{35}H_{36}O_8$ Scirpusin B heptamethyl ether	EtOH	228s(4.67),286(4.22), 308s(4.28),330(4.40),	94-3050-78

Compound	Solvent	$\lambda_{max}(\log \epsilon)$	Ref.
(cont.)		350s(4.19)	94-3050-78
$C_{35}H_{38}NO_4$ 1-Phenanthrenethanaminium, 3,4,6,7-tetramethoxy-N-methyl-N,N-bis(phenylmethyl)-, bromide	MeOH	266(4.91),283s(4.48), 309(4.16),322(4.17), 345(3.52),363(3.34)	12-0313-78
$C_{35}H_{38}N_2$ 1,4-Pentadiene, 1,1-bis[4-(dimethylamino)phenyl]-3,3-dimethyl-5,5-diphenyl-	EtOH	262(4.42)	44-1997-78
$C_{35}H_{38}N_4NiO$ Nickel, [3-(2,7,12,17-tetraethyl-3,8,13,18-tetramethyl-21H,23H-porphin-5-yl)-2-propenalato(2-)-N^{21},N^{22}-N^{23},N^{24}]-, (SP-4-2)-	$CHCl_3$	408(4.89),436(4.94), 536(3.79),560(3.94), 592(3.78)	39-1660-78C
$C_{35}H_{38}N_4O_4$ 9,13-Phorbinedipropanoic acid, 3,4-didehydro-3,4,8,14,18-pentamethyl-, dimethyl ester	$CHCl_3$	401(5.35),501(4.18), 536(3.65),564(3.83), 618(3.63)	65-1080-78
$C_{35}H_{38}N_4O_6$ Biliverdin dimethyl ester Biliverdin XIIIα dimethyl ester	EtOH EtOH	377(4.74),665(4.19) 375(4.60),655(4.14)	5-1990-78 5-1990-78
$C_{35}H_{38}N_6O_{11}$ L-Threonine, O-(phenylmethyl)-N-[[[9-(tri-O-acetyl-β-D-ribofuranosyl)-9H-purin-6-yl]amino]carbonyl]-, phenylmethyl ester	pH 2 pH 12 MeOH	271(4.31),278(4.32) 272(4.23),280(4.24), 300(4.09) 270(4.33),277s(4.27)	19-0021-78 19-0021-78 19-0021-78
$C_{35}H_{39}N_3O_2$ 2'H-Androsta-5,16-dieno[17,16-c]pyrazole-17(16H)-carbonitrile, 3-acetoxy-2',5'-diphenyl-, (3β,16β,17β)-	dioxan	241(4.25),293(3.74), 340(4.20)	78-1633-78
$C_{35}H_{40}N_4NiO$ Nickel, [3-(2,7,12,17-tetraethyl-3,8,13,18-tetramethyl-21H,23H-porphin-5-yl)-2-propen-1-olato(2-)-$N^{21},N^{22},N^{23},N^{24}$]-, (SP-4-2)-	$CHCl_3$	404(5.15),530(3.98), 566(4.17)	39-1660-78C
$C_{35}H_{40}O_8$ Ohchinal	EtOH	225(4.17),275(3.27), 282(3.26)	138-0331-78
$C_{35}H_{42}N_2O_3$ 2'H-Androsta-5,16-dieno[17,16-c]pyrazol-3-ol, 16,17-dihydro-17-methoxy-2',5'-diphenyl-, acetate, (3β,16β,17β)-	dioxan	240s(4.25),250(4.27), 304(3.87),344(4.38)	78-1633-78
$C_{35}H_{42}N_4NiO$ Nickel, [2,7,12,17-tetraethyl-3,8,13,18-tetramethyl-21H,23H-porphine-5-propanolato(2-)-$N^{21},N^{22},N^{23},N^{24}$]-, (SP-4-2)-	$CHCl_3$	338(4.06),410(5.14), 538(3.92),574(4.00)	39-1660-78C
$C_{35}H_{42}O_{16}$ 1-Propanone, 1-[2-acetoxy-4-methoxy-3-(tetraacetoxy-β-D-glucopyranosyl)-	n.s.g.	266(4.15)	12-2699-78

Compound	Solvent	$\lambda_{max}(\log \epsilon)$	Ref.
phenyl]-3,3-dimethoxy-2-(4-methoxy-phenyl)- (cont.)			12-2699-78
$C_{35}H_{44}N_2O_6$ 9,10-Secoergosta-1(10),5,7,9(11),22-pentaen-3-ol, 3,5-dinitrobenzoate	EtOH	301(4.50),314(4.63), 328(4.53)	39-1014-78C
$C_{35}H_{44}N_4O_6$ 22H-Biline-8,12-dipropanoic acid, 3-ethyl-1,4,5,15,16,19,21,24-octahydro-2,7,13,17-tetramethyl-1,19-dioxo-18-vinyl-, dimethyl ester, (4R,16R)-	CHCl$_3$	452(4.37)	24-0486-78
hydrochloride	CHCl$_3$	498(4.96)	24-0486-78
zinc complex	MeOH	510(4.99)	24-0486-78
(4R,16S)-	CHCl$_3$	452(4.37)	24-0486-78
hydrochloride	CHCl$_3$	496(4.98)	24-0486-78
zinc complex	MeOH	510(5.00)	24-0486-78
$C_{35}H_{44}O_6$ Pregnane-3,11,17,20-tetrol, 3,20-di-benzoate, (3α,5β,11β,20R)-	EtOH	230(4.48)	39-0163-78C
$C_{35}H_{48}ClN_3O_{10}$ Maytansine methyl ether	EtOH	232(4.41),240(4.38), 253(4.37),280(3.71), 288(3.71)	87-0031-78
$C_{35}H_{48}O_4$ 9,10-Secoergosta-5(10),7,22-trien-3β-ol, 6,19-epidioxy-, benzoate	EtOH	207(4.36),227(4.19), 273(3.02),281(2.98)	88-4895-78
C-6 epimer	EtOH	209(4.30),229(4.18), 274(3.09),281(3.04)	88-4895-78
$C_{35}H_{50}O_{11}$ 22-Acetylcyasterone 2,3-diacetate	EtOH	242(4.03)	105-0175-78
$C_{35}H_{52}O_6$ Diacetyl methyl dehydromaslinate	EtOH	241(4.12),252(4.18), 260(4.21)	42-1169-78
$C_{35}H_{52}O_8$ Diacetyl methyl acutangulate, 11-oxo-	EtOH	256(4.23)	42-1169-78

Compound	Solvent	$\lambda_{max}(\log \epsilon)$	Ref.
$C_{36}H_{16}O_2S_2$			
Pyreno[1,10-bc]thiopyran-5(4H)-one, 4-(5-oxopyreno[1,10-bc]thiopyran-4(5H)-ylidene)-, cis	$C_6H_5NO_2$	584(4.22)	110-1540-78
trans	$C_6H_5NO_2$	728(4.52)	110-1540-78
Pyreno[10,1-bc]thiopyran-3(4H)-one, 4-(3-oxopyreno[10,1-bc]thiopyran-4(3H)-ylidene)-, cis	$C_6H_5NO_2$	543(4.45)	110-1540-78
trans	$C_6H_5NO_2$	668(4.65)	110-1540-78
$C_{36}H_{23}ClN_2O_2$			
1,2-Benzisoxazole, 3-[4-[2-[2-chloro-4-(4,5-diphenyl-2-oxazolyl)phenyl]-ethenyl]phenyl]-	DMF	368(4.74)	33-2904-78
$C_{36}H_{23}NO$			
[2,2'-Bi-1H-inden]-1-one, 3,3'-diphen-yl-1'-(phenylimino)-	benzene	425(3.70)	40-0144-78
$C_{36}H_{24}Cl_2N_4O_8S_2$			
Benzo[b]cyclopenta[e][1,4]diazepine, 6,7-dichloro-3a,4,9,10-tetrahydro-10,10-bis(4-nitrophenyl)-4,9-bis-(phenylsulfonyl)-	MeCN	266(4.37)	5-1146-78
1H-Cyclopenta[b]quinoxaline, 1-[bis(4-nitrophenyl)methylene]-6,7-dichloro-3a,4,9,9a-tetrahydro-4,9-bis(phenyl-sulfonyl)-, cis	MeCN	228(4.57),255(4.43), 315(4.26)	5-1146-78
Spiro[2H-benzimidazole-2,1'-[2]cyclo-pentene], 5'-[bis(4-nitrophenyl)meth-ylene]-5,6-dichloro-1,3-dihydro-1,3-bis(phenylsulfonyl)-	MeCN	216(4.68),282(4.41)	5-1161-78
$C_{36}H_{24}N_2O_2$			
1,2-Benzisoxazole, 3-[4-[2-[4-(3-[1,1'-biphenyl]-4-yl-5-isoxazolyl)phenyl]-ethenyl]phenyl]-	DMF	351(4.86)	33-2904-78
1,2-Benzisoxazole, 3-[4-[2-[4-(4,5-di-phenyl-2-oxazolyl)phenyl]ethenyl]-phenyl]-	DMF	367(4.79)	33-2904-78
$C_{36}H_{24}O_3$			
Methanone, (1,4-dihydro-1,4-diphenyl-1,4-epoxynaphthalene-2,3-diyl)bis-[phenyl-	EtOH	258(4.15)	78-2305-78
$C_{36}H_{24}S_2$			
1,3-Benzodithiole, 2-(2,3,4,5-tetra-phenyl-2,4-cyclopentadien-1-yli-dene)-	CH_2Cl_2	247s(4.45),256(4.50), 274s(4.39),435(4.60)	18-2674-78
	MeCN	245s(4.33),253(4.38), 272s(4.24),430(4.45)	18-2674-78
	MeCN-HBF$_4$	232s(4.43),314s(3.75), 532(4.44)	18-2674-78
$C_{36}H_{25}Cl$			
7H-Dibenzo[a,c]cyclononene, 7-[(4-chlo-rophenyl)methylene]-5,9-diphenyl-	EtOH	215(4.59),224s(4.53), 240(4.48),265s(4.38), 335(3.92),346(3.95)	150-5151-78
$C_{36}H_{25}NO$			
Methanone, phenyl(3,4,5,6-tetraphenyl-	CHCl$_3$	253.5(4.65)	39-1315-78C

Compound	Solvent	$\lambda_{max}(\log \epsilon)$	Ref.
2-pyridinyl)- (cont.)			39-1315-78C
$C_{36}H_{25}N_3O_9$ 1H-Naphtho[2,3-d]triazole-4,9-dione, 1-(2,3,5-tri-O-benzoyl-β-D-ribo-furanosyl)-	EtOH	231(4.62),270(4.11), 333(3.43)	111-0155-78
$C_{36}H_{26}$ 7H-Dibenzo[a,c]cyclononene, 7-(2,4,6-cycloheptatrien-1-ylidene)-5,9-di-phenyl-	C_6H_{12}	245(4.32),263s(4.23), 275s(4.13),290s(3.98), 300s(3.61),325s(2.70), 399(2.51)	150-5151-78
	EtOH	245(4.32),265s(4.23), 275s(4.15),290s(3.91), 300s(3.63),325s(2.58), 401(2.51)	150-5151-78
	CH_2Cl_2	245(4.36),265(4.26), 270s(4.21),293s(3.90)	150-5151-78
7H-Dibenzo[a,c]cyclononene, 5,9-diphen-yl-7-(phenylmethylene)-	EtOH	224(4.42),240(4.58), 271s(4.24),308s(3.85), 325s(3.81),340(3.84), 400s(3.06)	150-5151-78
$C_{36}H_{26}Cl_2N_2O_4S_2$ Benzo[b]cyclopenta[e][1,4]diazepine, 6,7-dichloro-3a,4,9,10-tetrahydro-10,10-diphenyl-4,9-bis(phenylsul-fonyl)-	CH_2Cl_2	231(4.61)	5-1146-78
1H-Cyclopenta[b]quinoxaline, 6,7-di-chloro-1-(diphenylmethylene)-3a,4,9-9a-tetrahydro-4,9-bis(phenylsulfon-yl)-, cis	MeCN	230(4.52),267(4.32), 273s(4.30)	5-1146-78
Spiro[2H-benzimidazole-2,1'-[2]cyclo-pentene], 5,6-dichloro-5'-(diphenyl-methylene)-1,3-dihydro-1,3-bis(phen-ylsulfonyl)-	MeCN	209(4.58),221(4.58), 320(3.86)	5-1161-78
Spiro[2H-benzimidazole-2,1'-[3]cyclo-pentene], 5,6-dichloro-2'-(diphenyl-methylene)-1,3-dihydro-1,3-bis(phen-ylsulfonyl)-	MeCN	275(4.24)	5-1161-78
$C_{36}H_{26}N_2O_5$ 5,6-Diazaspiro[3.4]octa-5,7-diene-7,8-dicarboxylic acid, 1,2-bis(diphenyl-methylene)-3-oxo-, dimethyl ester	$CHCl_3$	335(4.35),415(3.70)	18-0649-78
$C_{36}H_{26}O_5$ 1H-Benzo[a]cyclobuta[d]cycloheptene-8,9-dicarboxylic acid, 2-(diphenyl-methylene)-2,3-dihydro-3-phenyl-, dimethyl ester	CHCl	320(4.16),343(4.22)	18-0649-78
$C_{36}H_{27}NO_9$ 1H-Indene-1,3(2H)-dione, 2-[[(2,3,5-tri-O-benzoyl-β-D-ribofuranosyl)-amino]methylene]-	EtOH	234(4.68),283(4.15), 294(4.38),328(4.39)	111-0155-78
1,4-Naphthalenedione, 2-[(2,3,5-tri-O-benzoyl-β-D-ribofuranosyl)amino]-	EtOH	257(4.30),283(4.08), 333(3.56)	111-0155-78
$C_{36}H_{28}O$ 7H-Dibenzo[a,c]cyclononene-7-methanol,	EtOH	248(4.32),265s(4.20),	150-5151-78

Compound	Solvent	λ_{max} (log ϵ)	Ref.
α,5,9-triphenyl- (cont.) 1,4-Epoxynaphthalene, 1,2,3,4-tetrahydro-1,2,4-triphenyl-3-(2-phenylethenyl)- (also at low temperature)	MeCN	290s(3.79),297s(3.48) 258(4.41),284s(3.61), 294(3.31)	150-5151-78 150-1301-78
$C_{36}H_{28}O_4$ 1,4-Naphthacenediol, 8,9-dimethyl-6,11-diphenyl-, diacetate	CH_2Cl_2	386(3.0),413(3.3), 434(3.70),464(4.0), 496(4.0)	150-5538-78
2-Propen-1-one, 3,3'-(3,4-dibenzoyl-1,2-cyclobutanediyl)bis[1-phenyl-	EtOH	250(4.29),330s(3.23)	44-2056-78
$C_{36}H_{32}N_2O_8$ Cancentrine, 10-oxo-	EtOH	268(3.35),284s(3.34), 324s(2.90),436(2.65)	23-2467-78
1H-Indole, 1-[2,3-di-O-acetyl-5-O-(triphenylmethyl)-α-L-arabinofuranosyl)-6-nitro-	EtOH	212(4.56),234s(4.12), 250(3.95),315(3.86), 357(3.78)	136-0017-78E
β-	EtOH	211(4.56),234s(4.08), 250(3.86),315(3.72), 357(3.60)	136-0017-78E
$C_{36}H_{32}N_4O_5$ 21-Phorbinecarboxylic acid, 3,4-didehydro-9,14-diethenyl-3-(3-methoxy-3-oxo-1-propenyl)-4,8,13,18-tetramethyl-20-oxo-, methyl ester	$CHCl_3$	440(5.22),536(4.05), 580(4.09),598(4.10), 656(3.15)	12-2491-78
$C_{36}H_{32}O_4$ Heptacyclo[28.2.2.2^{14,17}.0^{4,25}.0^{6,27}-0^{9,20}.0^{11,22}]hexatriaconta-4,6(27),9-11(22),14(36),16,20,25,30(34),32-decaene-15,31,33,35-tetrone	$CHCl_3$	415(3.70)	89-0757-78
stereoisomer	$CHCl_3$	417(3.70)	89-0757-78
$C_{36}H_{34}$ Bi-1,3,5,7,9,11,13,15,17-cyclooctadecanonaen-1-yl	benzene	380(5.22),400s(5.01), 425s(4.78),466(4.67), 510s(4.10)	88-4567-78
Fluoranthene, 8-(4-ethylphenyl)-9-phenyl-7,10-dipropyl-	C_6H_{12}	270(4.38),295(4.54)	2-0156-78
$C_{36}H_{34}N_2O_2$ 4,11-Etheno-1H-cyclopenta[b]phenazine, 1-[bis(4-methoxyphenyl)methylene]-3a,4,11,11a-tetrahydro-7,8,12,13-tetramethyl-	C_6H_{12}	212(4.86),241s(4.57), 243(4.62),246(4.62), 255s(4.48),312(4.47), 319(4.48)	5-0440-78
$C_{36}H_{34}N_4O_5$ 21-Phorbinecarboxylic acid, 3,4-didehydro-9-ethenyl-14-ethyl-3-(3-methoxy-3-oxo-1-propenyl)-4,8,13,18-tetramethyl-20-oxo-, methyl ester, (E)-	$CHCl_3$	437(5.19),535(3.93), 583(4.16),592(4.18), 649(3.13)	12-2491-78
21-Phorbinecarboxylic acid, 3,4-didehydro-14-ethenyl-9-ethyl-3-(3-methoxy-3-oxo-1-propenyl)-4,8,13,18-tetramethyl-20-oxo-, methyl ester, (E)-	$CHCl_3$	440(5.15),532(4.03), 578(3.97),597(4.00), 653(3.24)	12-2491-78
21H,23H-Porphine-2-propanoic acid, 7,12-diethenyl-18-(3-methoxy-3-oxo-1-propenyl)-3,8,13,17-tetramethyl--oxo-, methyl ester, (E)-	$CHCl_3$	429(5.22),521(4.15), 558(4.07),591(3.88), 646(3.72)	12-2491-78

Compound	Solvent	λ_{max} (log ϵ)	Ref.
$C_{36}H_{36}$ Quadruple-layered metaparacyclophane	THF	286(3.25)	18-2668-78
$C_{36}H_{36}FeN_5O_5$ Iron, [dimethyl 7,12-diethenyl-3,8,13-17-tetramethyl-21H,23H-porphine-2,18-dipropanoato(2-)-N^{21},N^{22},N^{23}-N^{24}]nitrosyl-, (SP-5-13)-	n.s.g.	398(4.91),487(4.05), 546(4.02),570(4.04)	39-0974-78C
$C_{36}H_{36}N_2O_5$ Telobine, N-methyl-	MeOH and MeOH-NaOH	275(3.83),305s(--)	12-2077-78
$C_{36}H_{36}N_2O_6$ Gilletine, N-methyl-	MeOH	237s(4.39),277s(3.39), 289(3.44),304s(3.37)	142-0995-78A
$C_{36}H_{36}N_4O_5$ 21H,23H-Porphine-13-propanoic acid, 3-ethenyl-8-ethyl-17-(3-methoxy-3-oxo-1-propenyl)-2,7,12,18-tetramethyl-β-oxo-, methyl ester, (E)-	CHCl$_3$	426(5.20),519(4.06), 558(4.08),587(3.94), 641(3.61)	12-2491-78
$C_{36}H_{36}N_9P_3$ 1,3,5,2,4,6-Triazatriphosphorine, 2,2,4,4,6,6-hexahydro-2,2,4,4,6,6-hexakis(phenylamino)-	MeOH	237(4.88),276(3.87)	121-0945-78
$C_{36}H_{36}O_8$ 5,16:8,13-Diethenodibenzo[a,g]cyclododecene-2,3,10,11-tetracarboxylic acid, 6,7,14,15-tetrahydro-, tetraethyl ester	EtOH	230s(4.69),253(4.83), 345(3.84)	88-0969-78
$C_{36}H_{36}O_{15}$ α-Naphthocyclinone methyl ester dimethyl ether	EtOH	228(4.65),247s(--), 332(3.77),427(3.63)	44-1438-78
	EtOH-NaOH	232(4.65),276s(--), 378(4.08),555(3.73)	44-1438-78
	CHCl$_3$	249s(--),286s(--), 328(3.72),438(3.53)	44-1438-78
$C_{36}H_{38}$ 1,2,3,5,7,11,13-Cyclotetradecaheptaen-9-yne, 1,8-bis(1,1-dimethylethyl)-4,11-bis(4-methylphenyl)-	THF	238(4.16),264(4.14), 331(4.62),361(5.28), 513(4.74),571(2.83), 624(3.20)	18-3351-78
$C_{36}H_{38}N_2$ Acridine, 2-(1,1-dimethylethyl)-9-[2-(1,1-dimethylethyl)-10-methyl-9(10H)-acridinylidene]-9,10-dihydro-10-methyl-	CH$_2$Cl$_2$	238(4.68),292(4.24), 426(4.15)	35-5604-78
$C_{36}H_{38}N_2O_6$ Tubocurine, (+)-	MeOH	225(4.52),281(3.86)	36-1204-78
$C_{36}H_{38}N_2O_8S_2$ Benzo[b]cyclopenta[e][1,4]diazepine-2,3-diol, 10,10-dicyclopropyl-2,3-3a,4,9,10-hexahydro-6,7-dimethyl-	MeCN	266s(3.69),273s(3.50)	5-1146-78

Compound	Solvent	$\lambda_{max}(\log \epsilon)$	Ref.
4,9-bis(phenylsulfonyl)-, diacetate, (2α,3α,3aα)- (cont.)			5-1146-78
$C_{36}H_{38}N_4O_4$			
21H,23H-Porphine-13,17-dipropanoic acid, 3,7-diethenyl-2,8,12,18-tetramethyl-, dimethyl ester	CHCl$_3$	407(5.23),505(4.15), 541(4.05),575(3.83), 630(3.72)	12-0639-78
Protoporphyrin XIII, dimethyl ester	CHCl$_3$	407(5.23),505(4.16), 541(4.06),575(3.84), 630(3.74)	12-0639-78
$C_{36}H_{38}N_4O_5$			
21H,23H-Porphine-13,17-dipropanoic acid, 3-acetyl-2,7,12,18-tetramethyl-8-ethenyl-, dimethyl ester	CHCl$_3$	414(5.24),512(4.08), 551(4.07),580(3.90), 635(3.37)	12-0365-78
$C_{36}H_{38}N_4O_6$			
Diacetyldeuteroporphyrin 2, methyl ester	CHCl$_3$	521(3.92),564(4.28), 588(4.06),646(3.23)	23-2430-78
Diacetyldeuteroporphyrin 9, methyl ester	CHCl$_3$	519(4.17),553(3.91), 588(3.85),642(3.58)	23-2430-78
$C_{36}H_{38}O_2$			
1,2,3,5,7,11,13-Cyclotetradecaheptaen-9-yne, 1,8-bis(1,1-dimethylethyl)-4,11-bis(4-methoxyphenyl)-	THF	240(4.16),267(4.20), 316(4.48),334(4.32), 367(5.25),524(4.78), 625(3.35)	18-3351-78
1,2,3,5,7,11,13-Cyclotetradecaheptaen-9-yne, 1,8-bis(1,1-dimethylethyl)-5,12-dimethoxy-4,11-diphenyl-	THF	236(4.04),253s(4.04), 264(4.05),321s(4.47), 344s(3.95),356(5.25), 492(4.54),590(3.84), 637(3.66)	18-3345-78
$C_{36}H_{40}Cl_2N_4O_4$			
21H,23H-Porphine-13,17-dipropanoic acid, 2,8-bis(2-chloroethyl)-3,7,12,18-tetramethyl-, dimethyl ester	CHCl$_3$	400(5.28),499(4.14), 532(3.99),567(3.87), 621(3.67)	12-0639-78
21H,23H-Porphine-13,17-dipropanoic acid, 3,7-bis(2-chloroethyl)-2,8,12,18-tetramethyl-, dimethyl ester	CHCl$_3$	401(5.25),498(4.17), 533(3.99),567(3.83), 622(3.70)	12-0639-78
$C_{36}H_{40}N_4NiO_2$			
Nickel, [methyl 3-(2,7,12,17-tetraethyl-3,8,13,18-tetramethyl-21H,23H-porphin-5-yl)-2-propenoato(2-)-$N^{21},N^{22},N^{23},N^{24}$]-, (SP-4-2)-	CHCl$_3$	404(5.08),530(3.97), 566(4.13)	39-1660-78C
$C_{36}H_{40}N_4O_4$			
21H,23H-Porphine-13,17-dipropanoic acid, 3-ethenyl-8-ethyl-2,7,12,18-tetramethyl-, dimethyl ester	CHCl$_3$	403(5.23),502(4.12), 539(4.07),570(3.85), 624(3.59)	12-0365-78
21H,23H-Porphine-13,17-dipropanoic acid, 8-ethenyl-3-ethyl-2,7,12,18-tetramethyl-, dimethyl ester	CHCl$_3$	403(5.24),502(4.12), 539(4.08),570(3.85), 624(3.60)	12-0365-78
$C_{36}H_{40}N_4O_5$			
21H,23H-Porphine-13,17-dipropanoic acid, 3-acetyl-8-ethyl-2,7,12,18-tetramethyl-, dimethyl ester	CHCl$_3$	410(5.28),511(4.01), 550(4.16),575(3.97), 634(3.17)	12-0365-78

Compound	Solvent	$\lambda_{max}(\log \epsilon)$	Ref.
21H,23H-Porphine-13,17-dipropanoic acid, 8-acetyl-3-ethyl-2,7,12,18-tetramethyl-, dimethyl ester	CHCl$_3$	410(5.27),511(4.00), 550(4.16),575(3.97), 634(3.12)	12-0365-78
21H,23H-Porphine-13,17-dipropanoic acid, 3-ethenyl-8-(1-hydroxyethyl)-2,7,12,18-tetramethyl-, dimethyl ester	CHCl$_3$	404(5.24),502(4.16), 537(4.02),572(3.82), 626(3.66)	12-0365-78
21H,23H-Porphine-13,17-dipropanoic acid, 8-ethenyl-3-(1-hydroxyethyl)-2,7,12,18-tetramethyl-, dimethyl ester	CHCl$_3$	404(5.24),502(4.15), 537(4.02),572(3.82), 626(3.67)	12-0365-78
$C_{36}H_{40}N_4O_6$ Deuteroporphyrin IX, 2-(ethoxycarbonyl)-4-methyl-, dimethyl ester	CHCl$_3$	408(5.06),511(3.99), 550(4.15),576(3.94), 637(3.11)	104-0794-78
21H,23H-Porphine-13,17-dipropanoic acid, 3-acetyl-8-(2-hydroxyethyl)-2,7,12,18-tetramethyl-, dimethyl ester	CHCl$_3$	410(5.27),510(4.03), 549(4.13),576(3.92), 634(3.21)	12-0365-78
$C_{36}H_{41}ClN_2O_3$ Pregn-5-eno[17,16-c]pyrazol-20-one, 3-acetoxy-2'-(2-chlorophenyl)-2',16-dihydro-5'-phenyl-, (3β,16β)-	dioxan	235(4.25),275(3.99), 308(4.25),316(4.30), 355(4.25),362(4.26)	78-1633-78
Pregn-5-eno[17,16-c]pyrazol-20-one, 3-acetoxy-2'-(4-chlorophenyl)-2',16-dihydro-5'-phenyl-, (3β,16β)-	dioxan	245(4.19),252(4.19), 312(4.10),323(4.11), 372(4.14)	78-1633-78
Pregn-5-eno[17,16-c]pyrazol-20-one, 3-acetoxy-5'-(2-chlorophenyl)-2',16-dihydro-2'-phenyl-, (3β,16β)-	dioxan	246(4.23),300s(4.00), 310(4.03),364(4.08)	78-1633-78
Pregn-5-eno[17,16-c]pyrazol-20-one, 3-acetoxy-5'-(3-chlorophenyl)-2',16-dihydro-2'-phenyl-, (3β,16β)-	dioxan	251(4.21),229s(3.78), 313(3.92),324(3.97), 376(4.19)	78-1633-78
Pregn-5-eno[17,16-c]pyrazol-20-one, 3-acetoxy-5'-(4-chlorophenyl)-2',16-dihydro-2'-phenyl-, (3β,16β)-	dioxan	244(4.19),252(4.19), 310(4.10),322(4.11), 371(4.14)	78-1633-78
$C_{36}H_{42}$ 1,1':3',1"-Terphenyl, 4,4"-bis(1,1-dimethylethyl)-5'-[4-(1,1-dimethylethyl)phenyl]-	CH$_2$Cl$_2$	259(4.94)	24-0264-78
$C_{36}H_{42}N_2O_3$ Pregn-5-eno[17,16-c]pyrazol-20-one, 3-acetoxy-2',16-dihydro-2',5'-diphenyl-, (3β,16β)-	dioxan	246(4.30),298(3.94), 310(4.03),322(4.05), 372(4.18)	78-1633-78
$C_{36}H_{42}N_4O_2$ 2-Propenoic acid, 3-(2,7,12,17-tetraethyl-3,8,13,18-tetramethyl-21H,23H-porphin-5-yl)-, methyl ester	CHCl$_3$	406(5.09),506(4.00), 540(3.71),576(3.68), 630(3.20)	39-1660-78C
$C_{36}H_{42}N_6O_4Zn$ Zinc, [2,3,7,8,12,13,17,18-octaethyl-5,10-dinitro-21H,23H-porphinato(2-)-$N^{21},N^{22},N^{23},N^{24}$]-, (SP-4-2)-	n.s.g.	405(5.25),537(4.17), 574(4.23)	39-0768-78C
$C_{36}H_{42}O$ [1,1':3',1"-Terphenyl]-2'-ol, 4,4'-bis-(1,1-dimethylethyl)-5'-[4-(1,1-di-	CH$_2$Cl$_2$	251(4.68),255(4.68), 308(3.72)	24-0264-78

Compound	Solvent	$\lambda_{max}(\log \epsilon)$	Ref.
methylethyl)phenyl]- (cont.)			24-0264-78
$C_{36}H_{42}O_2$ [1,1':3',1"-Terphenyl]-2,2'-diol, 4,4"-bis(1,1-dimethylethyl)-5'-[4-(1,1-dimethylethyl)phenyl]-	CH_2Cl_2	254(4.49),308(3.53)	24-0264-78
$C_{36}H_{43}ClFeN_5O_2$ Iron, chloro[2,3,7,8,12,13,17,18-octaethyl-5-nitro-21H,23H-porphinato(2-)-$N^{21},N^{22},N^{23},N^{24}$]-, (SP-5-13)-	n.s.g.	377(4.91),509(3.89), 537(3.88),637(3.54)	39-0974-78C
$C_{36}H_{43}N_5O_2Zn$ Zinc, [2,3,7,8,12,13,17,18-octaethyl-5-nitro-21H,23H-porphinato(2-)-$N^{21},N^{22},N^{23},N^{24}$]-, (SP-4-2)-	n.s.g.	406(5.29),540(4.16), 578(4.18)	39-0974-78C
$C_{36}H_{44}FeN_5O$ Iron, nitrosyl[2,3,7,8,12,13,17,18-octaethyl-21H,23H-porphinato(2-)-$N^{21},N^{22},N^{23},N^{24}$]-, (SP-5-12)-	n.s.g.	389(4.92),478(4.01), 531(3.97),557(3.97)	39-0974-78C
$C_{36}H_{44}IIrN_4$ Iridium, iodo[2,3,7,8,12,13,17,18-octaethyl-21H,23H-porphinato(2-)-$N^{21},N^{22},N^{23},N^{24}$]-, (SP-5-12)-	$CHCl_3$	396(5.22),512(4.13), 541(4.39)	101-0317-78P
$C_{36}H_{44}N_4OV$ Vanadium, [2,3,7,8,12,13,17,18-octaethyl-21H,23H-porphinato(2-)-$N^{21},N^{22}N^{23}-N^{24}$]oxo-, (SP-5-12)-	benzene	408(5.06),534(4.20), 572(4.56)	78-0379-78
	HOAc	404(5.46),532(4.18), 570(4.51)	78-0379-78
	CF_3COOH	387(5.04),470(3.64), 513(3.82),541(3.75), 562s(3.72),613s(3.64)	78-0379-78
	pyridine	408(5.41),423(5.05), 536(4.15),572(4.41)	78-0379-78
$C_{36}H_{44}N_4O_6$ 21H-Biline-8,12-dipropanoic acid, 2-(ethoxycarbonyl)-10,24-dihydro-1,3,7,13,17,18-hexamethyl-, dimethyl ester, dihydrobromide	$CHCl_3$-1%HBr	456(4.57),520(5.16)	104-0794-78
$C_{36}H_{44}O_8$ 5,16:8,13-Diethenodibenzo[a,g]cyclododecene-2,3,10,11-tetracarboxylic acid, 1,2,3,4,6,7,9,10,11,12,14,15-dodecahydro-, tetraethyl ester	EtOH	228(3.95)	88-0969-78
$C_{36}H_{45}IrN_4$ Iridium, hydro[2,3,7,8,12,13,17,18-octaethyl-21H,23H-porphinato(2-)-$N^{21},N^{22},N^{23},N^{24}$]-, (SP-5-31)-	benzene	393(5.23),502(4.15), 531(4.51)	101-0317-78P
$C_{36}H_{46}N_4OV$ Vanadium, [2,3,7,8,12,13,17,18-octaethyl-2,3-dihydro-21H,23H-porphinato(2-)-$N^{21},N^{22},N^{23},N^{24}$]oxo-, [(SP-5-12)-(trans)]-	$CHCl_3$	409(5.06),506(3.62), 541(3.69),574(3.85), 632(4.52)	78-0379-78

Compound	Solvent	$\lambda_{max}(\log \epsilon)$	Ref.
$C_{36}H_{46}O_7$ Prosta-5,13-dien-1-oic acid, 15,15-[1,2-ethanediylbis(oxy)]-9,11-dihydroxy-, 2-[1,1'-biphenyl]-4-yl-2-oxoethyl ester, (5Z,9α,11α,13E)-	n.s.g.	284(4.37)	87-0443-78
$C_{36}H_{48}N_2O_{10}$ Lycaconitine	EtOH EtOH-NaOH	231(4.00),277(2.00) 223(--),228s(--), 253(--),261s(--), 312(--)	95-1376-78 95-1376-78
$C_{36}H_{48}N_6$ [3,7':3',7"-Ter-1H-indole]-3,3',3"-(2H,2'H)-triethanamine, N,N,N',N',N"-N"-hexamethyl-	EtOH EtOH-HCl	245s(4.00),289s(3.93), 293(3.96) 263s(3.78),270(3.82), 282(3.80),291(3.65)	39-1671-78C 39-1671-78C
$C_{36}H_{48}N_6O_{12}$ D-Valine, N-[N-[[2-amino-9-[[[2-hydroxy-1-[[[1-(methoxycarbonyl)-2-methylpropyl]amino]carbonyl]propyl]amino]carbonyl]-4,6-dimethyl-3-oxo-3H-phenoxazin-1-yl]carbonyl]-L-threonyl]-, methyl ester, [1S-[1R*(S*),2S*]]-	MeOH	241(4.567),443(4.381)	87-0607-78
$C_{36}H_{48}N_8O_8$ 1,5,9,13,17,21,25,29-Octaazapentacyclo-[27.3.1.1^{5,9}.1^{13,17}.1^{21,25}]hexatriaconta-7,14,23,30-tetraene-6,16,22-32,33,34,35,36-octone, 7,15,23,31-tetraethyl-	50% EtOH	272(4.50)	56-1365-78
$C_{36}H_{48}O_8$ Tricyclo[8.2.2.2^{4,7}]hexadeca-4,6,10,12-13,15-hexaene-5,6,11,12-tetracarboxylic acid, tetrakis(1,1-dimethylethyl) ester	EtOH	216(4.63),302(3.32)	24-0523-78
Tricyclo[8.2.2.2^{4,7}]hexadeca-4,6,10,12-13,15-hexaene-5,6,11,12-tetracarboxylic acid, tetrakis(2-methylpropyl) ester	EtOH	215(4.61),302(3.35)	24-0523-78
$C_{36}H_{50}ClN_3O_{10}$ Maytansine, O^9-ethyl-	EtOH	233(4.45),242s(4.42), 253(4.45),280(3.73), 288(3.75)	87-0031-78
$C_{36}H_{50}N_6$ [3,7':3',7"-Ter-3H-indole]-3,3',3"-triethanamine, 1,1',1",2,2',2"-hexahydro-N,N,N',N',N",N"-hexamethyl-	EtOH	246(4.16),300(3.88)	39-1671-78C
$C_{36}H_{50}N_8O_{10}$ D-Valinamide, N-[[2-amino-9-[[[2-hydroxy-1-[[[2-methyl-1-[(methylamino)-carbonyl]propyl]amino]carbonyl]propyl]amino]carbonyl]-4,6-dimethyl-3-oxo-3H-phenoxazin-1-yl]carbonyl]-L-threonyl-N-methyl-, [1S-[1R*(S*),2S*]]-	MeOH	240(4.569),443(4.383)	87-0607-78

Compound	Solvent	$\lambda_{max}(\log \epsilon)$	Ref.
$C_{36}H_{51}NO_4$ 9,10-Secocholesta-5,7,10(19)-trien-3-ol, 4,4-dimethyl-, 4-nitrobenzoate, (3β,5E,7E)-	ether	256(4.34)	35-5626-78
$C_{36}H_{52}N_4O_6$ 2,16-Dioxa-6,9,20,23-tetraazatricyclo-[24.2.2.2^{12,15}]dotriaconta-12,14,26-28,29,31-hexaene-5,8,19,22-tetrone, 9,23-dimethyl-7,21-bis(2-methylpropyl)-, [7S-(7R*,21R*)]-	MeOH	224(4.40),276(3.51), 283(3.43)	35-8202-78
$C_{36}H_{52}N_{12}O_4$ Tetrapyrido[4,3-b:4',3'-f:4",3"-j-4''',3'''-n][1,5,9,13]tetraazacyclo-hexadecine-6,12,18,24(2H,8H,14H,20H)-tetrone, 2,8,14,20-tetrakis(3-aminopropyl)-4a,5,10a,11,11a,17,22a,23-octahydro-	CH_2Cl_2	340(4.1)	24-2594-78
$C_{36}H_{52}O_8$ Urs-12-ene-27,28-dioic acid, 3-acetoxy-, dianhydride with acetic acid, (3β)-	EtOH	246(4.02)	42-1169-78
$C_{36}H_{57}FeN_6O_6$ Iron, tris[1-hydroxy-3,6-bis(2-methylpropyl)-2(1H)-pyrazinonato-O,O']-	EtOH	227(4.69),315(4.26), 410(3.64)	94-1320-78

Compound	Solvent	$\lambda_{max}(\log \epsilon)$	Ref.
$C_{37}H_{25}NO$			
1,2-Benzisoxazole, 6-[2-(1-naphthalen-yl)ethenyl]-3-[4-[2-(1-naphthalenyl)-ethenyl]phenyl]-	DMF	353(4.80)	33-2904-78
1,2-Benzisoxazole, 6-[2-(2-naphthalen-yl)ethenyl]-3-[4-[2-(2-naphthalenyl)-ethenyl]phenyl]-	DMF	350(4.95)	33-2904-78
$C_{37}H_{26}$			
12H-Indeno[1,2-k]fluoranthene, 7,13-bis(4-methylphenyl)-	$CHCl_3$	288(4.56),418(3.94)	39-0989-78C
$C_{37}H_{26}N_2O_9$			
1H-Benz[f]indazole-4,9-dione, 1-(2,3,5-tri-O-benzoyl-β-D-ribofuranosyl)-	EtOH	234(4.74),270(3.88)	111-0155-78
2H-Benz[f]indazole-4,9-dione, 2-(2,3,5-tri-O-benzoyl-β-D-ribofuranosyl)-	EtOH	239(4.76)	111-0155-78
$C_{37}H_{27}ClN_4O$			
1,3,4-Oxadiazole, 2-[4-[2-[4-[3-(4-chlorophenyl)-4,5-dihydro-5-phenyl-1H-pyrazol-1-yl]phenyl]ethenyl]-phenyl]-5-phenyl-	n.s.g.	308(4.25),420(4.70)	135-0946-78
$C_{37}H_{27}F_2N_4S_4$			
Benzo[1,2-d:4,3-d']bisthiazolium, 3-ethyl-2-[5-[3-ethtl-7-(4-fluorophen-yl)benzo[1,2-d:4,3-d']bisthiazol-2(3H)-ylidene]-1,3-pentadienyl]-7-(4-fluorophenyl)-, dichloride	EtOH	692(5.18)	124-0637-78
$C_{37}H_{27}NO_2$			
2H-Isoindole, 1-phenyl-3-[2-(3-phenyl-1-isobenzofuranyl)ethenyl]-2-(phen-ylmethyl)-, 2-oxide	benzene MeOH	482(4.45) 485(--)	40-0144-78 40-0144-78
$C_{37}H_{27}N_5O_3$			
1,3,4-Oxadiazole, 2-[4-[2-[4-[4,5-di-hydro-3-(4-nitrophenyl)-5-phenyl-1H-pyrazol-1-yl]phenyl]ethenyl]phenyl]-5-phenyl-	n.s.g.	290(4.52),385(4.56), 468(4.58)	135-0946-78
$C_{37}H_{27}O_2$			
Pyrylium, 4-[3-(2,6-diphenyl-4H-pyran-4-ylidene)-1-propenyl]-2,6-diphenyl-, perchlorate	CH_2Cl_2 $MeNO_2$	628(4.78),685(5.43) 625(4.81),680(5.38)	4-0365-78 4-0365-78
$C_{37}H_{28}N_4O$			
1,3,4-Oxadiazole, 2-[4-[2-[4-(4,5-di-hydro-3,5-diphenyl-1H-pyrazol-1-yl)-phenyl]ethenyl]phenyl]-5-phenyl-	n.s.g.	305(4.29),415(4.69)	135-0946-78
$C_{37}H_{28}O$			
3-Cyclopenten-1-ol, 1,3,4-triphenyl-2,5-bis(phenylmethylene)-, (E,E)-	$CHCl_3$	342(4.42)	39-0989-78C
(E,Z)-	$CHCl_3$	356(4.50)	39-0989-78C
(Z,Z)-	$CHCl_3$	348(4.45)	39-0989-78C
7H-Dibenzo[a,c]cyclononene, 7-[(4-meth-oxyphenyl)methylene]-5,9-diphenyl-	EtOH	240s(4.19),260s(4.15), 335s(4.30),348(4.38)	150-5151-78

Compound	Solvent	$\lambda_{max}(\log \epsilon)$	Ref.
$C_{37}H_{29}Cl_2N_3O$ Pyrrolo[2,3-d]pyridazine, 7-(4-chloro-benzoyl)-5-cyclohexyl-1-(4-chloro-phenyl)-4,6-diphenyl-	CHCl$_3$	244(4.57),300(4.15)	4-0241-78
$C_{37}H_{30}ClN_3O$ Pyrrolo[2,3-d]pyridazine, 7-benzoyl-4-(4-chlorophenyl)-5-cyclohexyl-1,6-diphenyl-	CHCl$_3$	244(4.65),300(4.23)	4-0241-78
$C_{37}H_{31}N_3O$ Pyrrolo[2,3-d]pyridazine, 7-benzoyl-5-cyclohexyl-1,4,6-triphenyl-	CHCl$_3$	244(4.57),300(4.11)	4-0241-78
$C_{37}H_{31}N_4O_2S_4$ Benzo[1,2-d:4,3-d']bisthiazolium, 3-ethyl-2-[3-[3-ethyl-7-(4-methoxyphen-yl)benzo[1,2-d:4,3-d']bisthiazol-2(3H)-ylidene]-1-propenyl]-7-(4-methoxyphenyl)-, diiodide	EtOH	602(5.16)	124-0637-78
$C_{37}H_{34}O_{10}$ 4H-1-Benzopyran-4-one, 5,8-diacetoxy-5,6,7,8-tetrahydro-6,7-bis(4-methoxy-benzoyl)-2-(2-phenylethyl)-	MeOH	211(4.67),258(4.64)	88-3921-78
$C_{37}H_{34}O_{14}$ Neocercosporin tetraacetate	MeOH	220(4.68),275(4.48), 475(4.40)	95-1553-78
Neosporin tetraacetate	MeOH	220(4.68),275(4.48), 475(4.40)	25-0233-78
$C_{37}H_{38}N_2O_6$ Cocsulinine, O,O-dimethyl-	MeOH	237s(4.57),276s(3.46), 291(3.51),301s(3.45)	142-0995-78A
$C_{37}H_{40}Cl_4N_5O_{13}P$ 3'-Adenylic acid, N-(4-methoxybenzoyl)-2'-O-(tetrahydro-4-methoxy-2H-pyran-4-yl)-, 2-chlorophenyl 2,2,2-tri-chloroethyl ester, 5'-(4-oxopent-anoate)	EtOH	287(4.54)	54-0073-78
$C_{37}H_{40}N_2O_6$ Johnsonine	MeOH MeOH-NaOH	281(3.82) 284(3.85)	12-2539-78 12-2539-78
$C_{37}H_{40}N_2O_7$ Thaligosidine	MeOH pH 12	275(3.72),283(3.72) 275(3.75),284(3.76), 310s(3.22)	44-0580-78 44-0580-78
$C_{37}H_{40}N_4NiO$ Nickel, [5-(2,7,12,17-tetraethyl-3,8,13,18-tetramethyl-21H,23H-porph-in-5-yl]-2,4-pentadienolato(2-)-N^{21},N^{22},N^{23},N^{24}]-, (SP-4-2)-	CHCl$_3$	400(4.67),430(4.55), 566(3.87),594(3.83)	39-1660-78C
$C_{37}H_{40}N_4O_7$ Haplophytine	EtOH	220(4.69),265(4.16), 305(3.65)	23-1052-78

Compound	Solvent	$\lambda_{max}(\log \epsilon)$	Ref.
$C_{37}H_{40}N_4O_8$ Haplophytine oxide	EtOH	231(4.57),292(3.95)	23-1052-78
$C_{37}H_{44}ClIrN_4O$ Iridium, carbonylchloro[2,3,7,8,12,13-17,18-octaethyl-21H,23H-porphinato-(2-)-$N^{21},N^{22},N^{23},N^{24}$]-, (OC-6-42)-	$CHCl_3$	402(5.23),516(4.16), 547(4.52)	101-0317-78P
$C_{37}H_{44}N_2O_4$ Pregn-5-eno[17,16-c]pyrazol-20-one, 3-acetoxy-2',16-dihydro-5'-(4-methoxyphenyl)-2'-phenyl-, (3β,16β)-	dioxan	250(4.28),300(4.09), 310(4.15),319(4.13), 368(4.15)	78-1633-78
$C_{37}H_{44}N_4NiO_2$ Nickel, [2,3,7,8,12,13,17,18-octaethyl-21H,23H-porphin-5-yl formato(2-)-$N^{21},N^{22},N^{23},N^{24}$]-, (SP-4-2)-	$CHCl_3$	402(5.48),524(4.06), 559(4.29)	39-0366-78C
$C_{37}H_{44}O$ 1,1':3',1"-Terphenyl, 4,4"-bis(1,1-dimethylethyl)-5'-[4-(1,1-dimethylethyl)phenyl]-2'-methoxy-	CH_2Cl_2	254(4.65)	24-0264-78
$C_{37}H_{44}O_6$ 18-Nor-5β,17α-pregn-13-ene-3α,11β,20β-triol, 17β-methyl-, 11-acetate 3,20-dibenzoate	EtOH	230(4.36)	39-0163-78C
$C_{37}H_{46}N_4NiO$ Nickel, [2,3,7,8,12,13,17,18-octaethyl-5-methoxy-21H,23H-porphinato(2-)-$N^{21},N^{22},N^{23},N^{24}$]-, (SP-4-2)-	$CHCl_3$	405(5.32),525(4.15), 560(4.25)	39-0366-78C
$C_{37}H_{47}FeN_4O$ Iron, methoxy[2,3,7,8,12,13,17,18-octaethyl-21H,23H-porphinato(2-)-$N^{21},N^{22},N^{23},N^{24}$]-, (SP-5-12)-	n.s.g.	394(5.04),460s(4.22), 582(4.05)	39-0974-78C
$C_{37}H_{52}ClN_3O_9S$ Maytansine, 9-dehydroxy-9-(propylthio)-	EtOH	234(4.39),245(4.37), 255(4.41),280(3.71), 289(3.69)	87-0031-78
$C_{37}H_{52}ClN_3O_{10}$ Maytanbutine methyl ether	EtOH	233(4.49),240(4.45), 253(4.45),280(3.78), 288(3.77)	87-0031-78
$C_{37}H_{54}O_2$ Lanosta-7,9(11)-dien-3β-ol, benzoate	EtOH	236(4.33),244(4.25), 255(3.98)	19-0591-78
$C_{37}H_{67}NO_{12}$ 10-Epierythromycin B	n.s.g.	280(1.75)	44-2351-78

Compound	Solvent	$\lambda_{max}(\log \epsilon)$	Ref.
$C_{38}H_{23}BrN_4Ni$ Nickel, [5-bromo-10,15,20-triphenyl-21H,23H-porphinato(2-)-N^{21},N^{22},N^{23}-N^{24}]-, (SP-4-2)-	CH_2Cl_2	415(4.56),530(4.42)	150-0690-78
$C_{38}H_{23}N_5NiO_2$ Nickel, [5-nitro-10,15,20-triphenyl-21H,23H-porphinato(2-)-N^{21},$N^{22}N^{23}$-N^{24}]-, (SP-4-2)-	CH_2Cl_2	413(5.11),532(4.01), 572(3.85)	150-0690-78
$C_{38}H_{24}N_2S$ [2]Benzothieno[5,6-g]phthalazine, 1,4,7,9-tetraphenyl-	CH_2Cl_2	274(4.66),322(4.94), 398(3.79),656(3.73)	4-0793-78
$C_{38}H_{24}N_4$ Phthalazino[6,7-g]phthalazine, 1,4,7,10-tetraphenyl-	CH_2Cl_2	278(4.97),386(3.64), 426s(3.85),441(3.91), 462s(3.78)	4-0793-78
	CH_2Cl_2-HCl	440(3.83),465(4.02), 492(4.00)	4-0793-78
$C_{38}H_{24}N_4Ni$ Nickel, [5,10,15-triphenyl-21H,23H-porphinato(2-)-N^{21},N^{22},N^{23},N^{24}]-, (SP-4-2)-	CH_2Cl_2	406(5.43),521(4.27), 551(3.75)	150-0690-78
$C_{38}H_{24}O_2S$ Naphtho[2,3-c]thiophene, 6,7-dibenzoyl-1,3-diphenyl-	CH_2Cl_2	258(4.67),290(4.56), 325(4.72),424(3.87), 536(3.91)	4-0793-78
$C_{38}H_{24}O_3$ Naphtho[2,3-c]furan, 6,7-dibenzoyl-1,3-diphenyl-	CH_2Cl_2	240(4.62),285s(4.47), 297(4.56),312(4.47), 370s(4.06),430(3.77), 572(3.76)	4-0793-78
$C_{38}H_{24}O_4$ Naphthalene, 2,3,6,7-tetrabenzoyl-	CH_2Cl_2	267(4.91),345(4.68), 356s(3.66)	4-0793-78
$C_{38}H_{26}Cl_2N_2O_2$ 1H-Cyclopenta[b]quinoxaline, 4,9-di-benzoyl-1-[bis(4-chlorophenyl)-methylene]-3a,4,9,9a-tetrahydro-	MeCN	283(4.36)	5-1139-78
$C_{38}H_{26}Cl_2N_4O_9S_2$ Spiro[2H-benzimidazole-2,1'-[4]cyclo-penten]-3'-one, 2'-[bis(4-chloro-phenyl)methylene]-1,3-dihydro-5,6-dimethyl-1,3-bis[(4-nitrophenyl)-sulfonyl]-	MeCN	234.5(4.66)	5-1161-78
$C_{38}H_{26}N_4$ 21H,23H-Porphine, 5,10,15-triphenyl-	CH_2Cl_2	411(5.69),507(4.25), 542(3.71),583(3.71), 641(3.34)	150-0690-78
$C_{38}H_{26}N_4O_6$ 1H-Cyclopenta[b]quinoxaline, 4,9-di-benzoyl-1-[bis(4-nitrophenyl)meth-	MeCN	259(4.46),331(4.24)	5-1139-78

Compound	Solvent	$\lambda_{max}(\log \epsilon)$	Ref.
ylene]-3a,4,9,9a-tetrahydro- (cont.)			5-1139-78
$C_{38}H_{26}O_3$ Naphtho[2,3-c]furan, 6,7-dibenzoyl- 4,9-dihydro-1,3-diphenyl-	CH_2Cl_2	333(4.42)	4-0793-78
$C_{38}H_{26}O_{12}$ Asphodelin tetraacetate	MeOH	262(4.63),344(3.99)	94-1111-78
$C_{38}H_{27}N_3O_2$ 1H-[1,2,4]Triazolo[1,2-a]pyridazine- 1,3(2H)-dione, 5,8-bis(diphenyl- methylene)-5,8-dihydro-2-phenyl-	EtOH	241(4.42),350(4.40)	39-1568-78C
$C_{38}H_{27}OS_2$ Pyrylium, 2-methyl-4,6-bis[(2-phenyl- 4H-1-benzothiopyran-4-ylidene)meth- yl]-, perchlorate	HOAc-HClO₄	590(4.508)	83-0236-78
$C_{38}H_{27}O_3$ Pyrylium, 4-methyl-2,6-bis[(2-phenyl- 4H-1-benzopyran-4-ylidene)methyl]-, perchlorate	HOAc-HClO₄	556(4.435),655(4.223)	83-0236-78
$C_{38}H_{28}Cl_2N_2O_5S_2$ Spiro[2H-benzimidazole-2,1'-[4]cyclo- penten]-3'-one, 2'-[bis(4-chlorophen- yl)methylene]-1,3-dihydro-5,6-dimeth- yl-1,3-bis(phenylsulfonyl)-	MeCN	267(4.23),320s(4.09)	5-1161-78
$C_{38}H_{28}Cl_2N_4O_8S_2$ Benzo[b]cyclopenta[e][1,4]diazepine, 10,10-bis(4-chlorophenyl)-3a,4,9,10- tetrahydro-6,7-dimethyl-4,9-bis[(4- nitrophenyl)sulfonyl]-	MeCN	247(4.51)	5-1146-78
1H-Cyclopenta[b]quinoxaline, 1-[bis(4- chlorophenyl)methylene]-3a,4,9,9a- tetrahydro-6,7-dimethyl-4,9-bis[(4- nitrophenyl)sulfonyl]-, cis	MeCN	252(4.24),281(4.47)	5-1146-78
Spiro[2H-benzimidazole-2,1'-[2]cyclo- pentene], 5'-[bis(4-chlorophenyl)- methylene]-1,3-dihydro-5,6-dimethyl- 1,3-bis[(4-nitrophenyl)sulfonyl]-	MeCN	252(4.60)	5-1161-78
Spiro[2H-benzimidazole-2,1'-[3]cyclo- pentene], 2'-[bis(4-chlorophenyl)- methylene]-1,3-dihydro-5,6-dimethyl- 1,3-bis[(4-nitrophenyl)sulfonyl]-	MeCN	252(4.58),290(4.36)	5-1161-78
$C_{38}H_{28}N_2O_2$ 1H-Cyclopenta[b]quinoxaline, 4,9-dibenz- oyl-1-(diphenylmethylene)-3a,4,9,9a- tetrahydro-	CH_2Cl_2	231(4.49),284(4.39)	5-1139-78
$C_{38}H_{28}N_4O_2$ 9,10-Anthracenedione, 1,4,5,8-tetrakis- (phenylamino)-	dioxan	620(--),680(4.38), 741(4.52)	40-0082-78
$C_{38}H_{28}N_4O_9S_2$ Spiro[2H-benzimidazole-2,1'-[4]cyclo- penten]-3'-one, 2'-[bis(4-nitrophen- yl)methylene]-1,3-dihydro-5,6-dimeth-	MeCN	286(4.49)	5-1161-78

Compound	Solvent	$\lambda_{max}(\log \epsilon)$	Ref.
yl-1,3-bis(phenylsulfonyl)- (cont.)			5-1161-78
$C_{38}H_{28}O$			
Spiro[2.4]hepta-4,6-diene, 1-benzoyl-4,5,6,7-tetraphenyl-	MeCN	245(4.70)	5-1648-78
$C_{38}H_{28}O_3$			
2-Propen-1-one, 3-(1,2-dibenzoylcyclopropyl)-1,2,3-triphenyl-, [1α,1(Z)-2α]-	MeCN	251(4.60),280s(4.10), 313s(3.76)	5-1648-78
[1α,1(Z),2β]-	MeCN	250(4.59),280s(3.95), 316s(3.44)	5-1648-78
$C_{38}H_{28}O_4$			
Naphtho[2,3-c]-1(3H)-furanone, 3,3-bis-(2-methoxyphenyl)-4,9-diphenyl-	THF	246(4.83),283(4.06), 305(3.96),329(3.58), 343(3.68)	78-2263-78
$C_{38}H_{29}N_2O_3P$			
Cyclohepta[b]pyrrole-2-acetic acid, 3-cyano-8-(4-methoxyphenyl)-α-(triphenylphosphoranylidene)-, methyl ester	CHCl$_3$	266(4.16),312(4.18), 360s(3.75),383s(3.70), 464(3.98)	18-1573-78
$C_{38}H_{29}N_3O$			
Oxazole, 2-[4-[2-[4-(4,5-dihydro-3,5-diphenyl-1H-pyrazol-1-yl)phenyl]-ethenyl]phenyl]-5-phenyl-, (E)-	n.s.g.	325(4.24),415(4.76)	135-0946-78
$C_{38}H_{29}O_2$			
Pyrylium, 4-[3-(2,6-diphenyl-4H-pyran-4-ylidene)-1-methyl-1-propenyl]-2,6-diphenyl-, perchlorate	MeCN	640(4.76),698(5.32)	4-0365-78
Pyrylium, 4-[3-(2,6-diphenyl-4H-pyran-4-ylidene)-1-(2-methyl-1-propenyl)]-2,6-diphenyl-	CH$_2$Cl$_2$	650(4.71),696(5.28)	4-0365-78
$C_{38}H_{30}$			
7,10:25,28-Dietheno-13,17:18,22-dimethenocyclopentacosa[de]naphthalene, 11,12,23,24-tetrahydro-	C$_6$H$_{12}$	310(4.14)	88-1459-78
7,11:14,18:19,23:26,30-Tetramethenocycloheptacosa[de]naphthalene, 12,13,24,25-tetrahydro-	C$_6$H$_{12}$	307(4.24)	88-1459-78
$C_{38}H_{30}Cl_2N_2O_4S_2$			
Benzo[b]cyclopenta[e][1,4]diazepine, 10,10-bis(4-chlorophenyl)-3a,4,9,10-tetrahydro-6,7-dimethyl-4,9-bis(phenylsulfonyl)-	CH$_2$Cl$_2$	231(4.54)	5-1146-78
1H-Cyclopenta[b]quinoxaline, 1-[bis(4-chlorophenyl)methylene]-3a,4,9,9a-tetrahydro-6,7-dimethyl-4,9-bis(phenylsulfonyl)-, cis	MeCN	274(4.32),287(4.33)	5-1146-78
$C_{38}H_{30}Cl_2N_2O_6S_2$			
1H-Cyclopenta[b]quinoxaline, 1-[bis(4-methoxyphenyl)methylene]-6,7-dichloro-3a,4,9,9a-tetrahydro-4,9-bis(phenylsulfonyl)-	CH$_2$Cl$_2$	231(4.60),260(4.30)	5-1161-78
2,5-Methano-1,6-benzodiazocine, 11-[bis(4-methoxyphenyl)methylene]-	MeCN	255(4.57)	5-1161-78

Compound	Solvent	$\lambda_{max}(\log \epsilon)$	Ref.
8,9-dichloro-1,2,5,6-tetrahydro-1,6-bis(phenylsulfonyl)- (cont.)			5-1161-78
$C_{38}H_{30}N_2$ Diazene, bis[2-(diphenylmethyl)phenyl]-	EtOH	257(4.66),268s(4.64), 287(4.54),310(4.40), 342(4.28),390(3.76)	39-1211-78C
$C_{38}H_{30}N_2O_4$ Spiro[anthracene-2(1H),2'(1'H)-naphthalene]-1',4',9,10(3'H)-tetrone, 3,4-dihydro-1-[(4-methylphenyl)amino]-3'-[[(4-methylphenyl)amino]methylene]-	EtOH	207(3.98),236(4.13), 301(3.78),331(3.85), 419(3.70),484(3.91), 517s(3.83)	104-1962-78
$C_{38}H_{30}N_4O$ 1,3,4-Oxadiazole, 2-[4-[2-[4-[4,5-dihydro-3-(4-methylphenyl)-5-phenyl-1H-pyrazol-1-yl]phenyl]ethenyl]phenyl]-5-phenyl-, (E)-	n.s.g.	305(4.28),418(4.78)	135-0946-78
$C_{38}H_{30}N_4O_2$ 1,3,4-Oxadiazole, 2-[4-[2-[4-[4,5-dihydro-3-(4-methoxyphenyl)-5-phenyl-1H-pyrazol-1-yl]phenyl]ethenyl]phenyl]-5-phenyl-, (E)-	n.s.g.	310(4.40),418(4.76)	135-0946-78
$C_{38}H_{30}N_4O_8S_2$ Benzo[b]cyclopenta[e][1,4]diazepine, 3a,4,9,10-tetrahydro-6,7-dimethyl-10,10-bis(4-nitrophenyl)-4,9-bis(phenylsulfonyl)-	CH_2Cl_2	231(4.56),267s(4.47), 274s(4.45)	5-1146-78
1H-Cyclopenta[b]quinoxaline, 1-[bis(4-nitrophenyl)methylene]-3a,4,9,9a-tetrahydro-6,7-dimethyl-4,9-bis(phenylsulfonyl)-	CH_2Cl_2	231(4.53),250(4.45), 327(4.30)	5-1146-78
Spiro[2H-benzimidazole-2,1'-[2]cyclopentene], 5'-[bis(4-nitrophenyl)methylene]-1,3-dihydro-5,6-dimethyl-1,3-bis(phenylsulfonyl)-	MeCN	288(4.39)	5-1161-78
Spiro[2H-benzimidazole-2,1'-[3]cyclopentene], 2'-[bis(4-nitrophenyl)methylene]-1,3-dihydro-5,6-dimethyl-1,3-bis(phenylsulfonyl)-	CH_2Cl_2	231(4.44),258(4.37), 303(4.30)	5-1161-78
$C_{38}H_{30}O_2$ 3-Cyclopenten-1-ol, 2-[(4-methoxyphenyl)methylene]-1,3,4-triphenyl-5-(phenylmethylene)-, (Z,Z)-	CHCl$_3$	290(4.32),356(4.43)	39-0989-78C
$C_{38}H_{31}ClN_2O_6S_2$ 1H-Cyclopenta[b]quinoxaline, 1-[bis(4-methoxyphenyl)methylene]-7-chloro-3a,4,9,9a-tetrahydro-4,9-bis(phenylsulfonyl)-	MeCN	258(4.42)	5-1161-78
2,5-Methano-1,6-benzodiazocine, 11-[bis(4-methoxyphenyl)methylene]-8-chloro-1,2,5,6-tetrahydro-1,6-bis(phenylsulfonyl)-	MeCN	257(4.40),263(4.40)	5-1161-78
$C_{38}H_{31}N$ Benzenamine, 4-[(5,9-diphenyl-7H-di-	EtOH	220s(4.59),230s(4.48),	150-5151-78

Compound	Solvent	$\lambda_{max}(\log \epsilon)$	Ref.
benzo[a,c]cyclononen-7-ylidene)methyl]-N,N-dimethyl- (cont.)		250s(4.28),280(4.18), 350s(4.36),386(4.57)	150-5151-78
$C_{38}H_{32}Cl_2N_2O_4S_2$ Spiro[2H-benzimidazole-2,1'-cyclopentane], 2'-[bis(4-chlorophenyl)methylene]-1,3-dihydro-5,6-dimethyl-1,3-bis(phenylsulfonyl)-	MeCN	237s(4.44),304(3.82)	5-1161-78
$C_{38}H_{32}N_2O_4S_2$ Benzo[b]cyclopenta[e][1,4]diazepine, 3a,4,9,9a-tetrahydro-6,7-dimethyl-10,10-diphenyl-4,9-bis(phenylsulfonyl)-	CH_2Cl_2	231(4.46)	5-1146-78
1H-Cyclopenta[b]quinoxaline, 1-(diphenylmethylene)-3a,4,9,9a-tetrahydro-6,7-dimethyl-4,9-bis(phenylsulfonyl)-, cis	CH_2Cl_2	231(4.44),285(4.26)	5-1146-78
Spiro[2H-benzimidazole-2,1'-[3]cyclopentene], 2'-(diphenylmethylene)-1,3-dihydro-5,6-dimethyl-1,3-bis(phenylsulfonyl)-	MeCN	280(4.26)	5-1161-78
$C_{38}H_{34}N_2O_4S_2$ Spiro[2H-benzimidazole-2,1'-cyclopentane], 2'-(diphenylmethylene)-1,5-dihydro-5,6-dimethyl-1,3-bis(phenylsulfonyl)-	MeCN	305(3.82)	5-1161-78
$C_{38}H_{38}$ 1,2,3,5,7,9,13,15,17-Cyclodecanonaen-11-yne, 1,10-bis(1,1-dimethylethyl)-4,13-diphenyl-	THF	240s(4.11),250(4.15), 267s(4.08),282(4.21), 367s(4.71),389s(5.23), 401(5.52),534s(4.25), 574(4.68),673s(2.18), 701(2.36),785(2.99)	18-3359-78
$C_{38}H_{38}O_4$ Benzoic acid, 4,4'-[4,11-bis(1,1-dimethylethyl)-1,2,3,5,7,11,13-cyclotetradecaheptaen-9-yne-1,8-diyl]bis-, dimethyl ester	THF	233(4.24),276(4.18), 317(4.51),334(4.66), 351(4.84),365(5.27), 525(4.79),578(3.19), 616(3.47)	18-3351-78
Heptacyclo[28.2.2.2^{14,17}.0^{4,25}.0^{6,27}-0^{9,20}.0^{11,22}]hexatriaconta-4,6(27)-9,11(22),14(36),16,20,25,30,32,33-undecaene-15,35-dione, 31,33-dimethoxy-, pseudo-genminal-	n.s.g.	450(3.40)	89-0757-78
pseudo-ortho-	n.s.g.	447(3.40)	89-0757-78
$C_{38}H_{38}S_2$ Anthracene, 9,9'-[1,8-octanediylbis-(thiomethylene)]bis-	$CHCl_3$	350(4.06),368(4.18), 387(4.08)	24-3519-78
$C_{38}H_{40}N_2O_{14}S_2$ 1H-Indole, 5,5'-dithiobis[1-(2,3,4-tri-O-acetyl-α-L-arabinopyranosyl)-	EtOH	230(4.62),243(4.60)	104-2011-78
$C_{38}H_{42}N_2O_6$ Tubocurine, O,O-dimethyl-, (+)-	MeOH	225(4.55),280(3.80)	36-1204-78

Compound	Solvent	$\lambda_{max}(\log \epsilon)$	Ref.
$C_{38}H_{42}N_2O_7$			
Thaligosine	pH 12	282(3.88),305s(3.64)	44-0580-78
	MeOH	282(3.86)	44-0580-78
Thaligosinine	pH 12	284(4.11),307s(3.53)	44-0580-78
	MeOH	282(3.90)	44-0580-78
$C_{38}H_{42}N_4NiO_2$			
Nickel, [methyl 5-(2,7,12,17-tetraeth-yl-3,8,13,18-tetramethyl-21H,23H-porphin-5-yl)-2,4-pentadienoato-(2-)-$N^{21},N^{22},N^{23},N^{24}$]-, (SP-4-2)-	CHCl$_3$	404(4.84),533(3.81),563(3.97)	39-1660-78C
$C_{38}H_{42}N_4O_6$			
21H,23H-Porphine-13,17-dipropanoic acid, 3-(1-acetoxyethyl)-8-ethenyl-2,7,12,18-tetramethyl-, dimethyl ester	CHCl$_3$	405(5.24),502(4.16),537(4.02),573(3.84),626(3.70)	12-0365-78
21H,23H-Porphine-13,17-dipropanoic acid, 8-(1-acetoxyethyl)-3-ethenyl-2,7,12,18-tetramethyl-, dimethyl ester	CHCl$_3$	405(5.24),502(4.16),537(4.01),573(3.82),626(3.68)	12-0365-78
$C_{38}H_{42}N_4O_7$			
Deuteroporphyrin IX, 2-(ethoxycarbonyl-acetyl)-4-methyl-, dimethyl ester	CHCl$_3$	408(5.04),511(3.88),555(4.02),558(3.86),636(2.95)	104-0794-78
$C_{38}H_{43}N_2O_6$			
Berbamanium, 12-hydroxy-6,6',7-trimeth-oxy-2,2',2'-trimethyl-, iodide	EtOH	282(3.87)	105-0353-78
$C_{38}H_{44}N_2O_7$			
Thalirugine	pH 12	280(3.86),285(3.87),310s(3.45)	44-0580-78
	MeOH	280(3.81)	44-0580-78
$C_{38}H_{44}N_4O_6$			
21H,23H-Porphine-8,13,17-tripropanoic acid, 3-ethyl-2,7,12,18-tetramethyl-, trimethyl ester	CHCl$_3$	400(5.22),498(4.15),533(3.99),567(3.82),621(3.69)	12-0365-78
$C_{38}H_{44}O_2$			
[1,1':3',1"-Terphenyl]-2'-ol, 4,4"-bis-(1,1-dimethylethyl)-5'-[4-(1,1-di-methylethyl)phenyl]-, acetate	CH$_2$Cl$_2$	251(4.71),255(4.72)	24-0264-78
$C_{38}H_{44}O_4$			
Pentacyclo[19.3.1.1^3,^{7}.1^9,^{13}.1^{15},^{19}]-octacosa-1(25),3,5,7(28),9,11,13(27)-15,17,19(26),21,23-dodecaene-25,26-27,28-tetrol, 5,11-bis(1,1-dimethyl-ethyl)-17,23-dimethyl-	dioxan	280(4.03),287s(3.92)	49-0767-78
$C_{38}H_{44}O_9$			
Ohchinin acetate	EtOH	279(4.34)	138-0331-78
$C_{38}H_{45}BrN_4Ni$			
Nickel, [5-(2-bromoethenyl)-2,3,7,8,12-13,17,18-octaethyl-21H,23H-porphin-ato(2-)-$N^{21},N^{22},N^{23},N^{24}$]-, (SP-4-2-cis)-	CHCl$_3$	346(4.22),407(5.30),531(4.06),568(4.30)	39-0366-78C

Compound	Solvent	$\lambda_{max}(\log \epsilon)$	Ref.
trans (cont.)	$CHCl_3$	347(4.16),407(5.23), 531(4.02),568(4.21)	39-0366-78C
$C_{38}H_{46}ClN_4Rh$ Rhodium, (2-chloroethenyl)[2,3,7,8,12-13,17,18-octaethyl-21H,23H-porphin-ato(2-)-$N^{21},N^{22},N^{23},N^{24}$]-, [SP-5-31-(E)]-	$CHCl_3$	395(5.23),512(4.23), 544(4.74)	101-0329-78P
$C_{38}H_{46}FeN_2O_9$ Iron, tricarbonyl[(6,7,8,9-η)-9,10-secoergosta-5(10),6,8,22-tetraen-3-yl 3,5-dinitrobenzoate, (α)-stereoisomer	EtOH	210(4.56)	39-1014-78C
	EtOH	210(4.73)	39-1014-78C
$C_{38}H_{46}O_2$ 2,2'-Dinorrhodoxanthin	$CHCl_3$	482(5.04),504(5.12), 525(5.03)	33-0242-78
$C_{38}H_{47}N_4ORh$ Rhodium, [2,3,7,8,12,13,17,18-octaethyl-21H,23H-porphinato(2-)-$N^{21},N^{22}N^{23}$-N^{24}](2-oxoethyl)-, (SP-5-31)-	$CHCl_3$	386(5.29),512(4.25), 544(4.65)	101-0329-78P
$C_{38}H_{48}N_4Ni$ Nickel, [2,3,5,7,8,12,13,17,18-nonaethyl-21H,23H-porphinato(2-)-N^{21},N^{22},N^{23}-N^{24}]-, (SP-4-2)-	$CHCl_3$	303(4.02),343(4.10), 411(5.18),537(3.94), 574(4.04)	39-0366-78C
$C_{38}H_{48}O_2$ 2,2'-Dinorrhodoxanthin, 4,4'-dihydro-	$CHCl_3$	454(4.97),481(5.13), 514(5.07)	33-0242-78
$C_{38}H_{49}IrN_4$ Iridium, ethyl[2,3,7,8,12,13,17,18-octaethyl-21H,23H-porphinato(2-)-$N^{21},N^{22},N^{23},N^{24}$]-, (SP-5-31)-	$CHCl_3$	388(5.24),500(4.24), 530(4.48)	101-0317-78P
$C_{38}H_{50}$ 1,2,3,5,7,9,11,15,17,19,21-Cyclodocosa-undecaen-13-yne, 1,4,12,15-tetrakis-(1,1-dimethylethyl)-	THF	222(4.02),248(4.08), 261s(3.95),277s(3.96), 290(4.08),303(4.23), 317s(3.98),377s(3.89), 395(5.17),413(5.48), 512s(3.82),556(4.02), 591(4.05),712s(1.92), 757s(1.86),782(1.89), 853s(1.82),895(2.03)	18-3363-78
$C_{38}H_{52}O_2$ 2,4,6,8,13,15,17,19-Cyclodocosaoctaene-10,21-diyne-1,12-diol, 1,9,12,20-tetrakis(1,1-dimethylethyl)- lower melting isomer	EtOH	249(4.35),292s(4.70), 307(5.08),320(5.14), 341s(4.54),359(4.44)	18-3363-78
	EtOH	249(4.35),292s(4.66), 307(5.04),320(5.20), 341s(4.53),359(4.42)	18-3363-78
$C_{38}H_{54}Br_2$ Benzene, 1,1'-(2,3-dibromo-1,4-diundec-ylidene-2-butene-1,4-diyl)bis-	C_6H_{12}	243(4.36)	56-2377-78

Compound	Solvent	$\lambda_{max}(\log \epsilon)$	Ref.
$C_{38}H_{54}ClN_3O_{10}$ Maytansine, $N^{2'}$-deacetyl-9-dehydroxy-9-ethoxy-$N^{2'}$-(2-methyl-1-oxopropyl)-	EtOH	233(4.44),242(4.40), 253(4.41),280(3.73), 288(3.71)	87-0031-78
$C_{38}H_{54}N_8O_{12}$ D-Valinamide, N-[[2-amino-9-[[[2-hydroxy-1-[[[1-[[(2-hydroxyethyl)amino]-carbonyl]-2-methylpropyl]amino]carbonyl]propyl]amino]carbonyl]-4,6-dimethyl-3-oxo-3H-phenoxazin-1-yl]-carbonyl]-L-threonyl-N-(2-hydroxyethyl)-, [1S-[1R*(S*),2S*]]-	MeOH	242(4.568),443(4.364)	87-0607-78
$C_{38}H_{54}O_{15}$ β-D-Glucopyranoside, 2-(3,4-dimethoxyphenyl) 3-O-(6-deoxy-2,3,4-tri-O-methyl-α-L-mannopyranosyl)-2,6-di-O-methyl-, 4-[3-(3,4-dimethoxyphenyl)-2-propenoate]	EtOH	232(4.18),287(4.07), 324(4.18)	94-2111-78
$C_{38}H_{56}N_2O_4S$ Olean-12-en-28-oic acid, 3-[[(4-methylphenyl)sulfonyl]hydrazone], methyl ester	EtOH	218s(--),272(4.46)	115-0001-78
$C_{38}H_{56}O_6S$ Olean-12-en-28-oic acid, 16-hydroxy-3-[[(4-methylphenyl)sulfonyl]oxy]-, methyl ester, (3β,16α)-	EtOH	210s(--),223(4.28)	102-2107-78
$C_{38}H_{58}O_2$ 13-Hexacosyne-12,15-diol, 12,15-diphenyl-	EtOH	243s(2.15),248(2.32), 252(2.45),258(2.52), 264(2.40),268s(2.06)	56-2377-78

Compound	Solvent	λ_{max}(log ϵ)	Ref.
$C_{39}H_{24}$ 4,9'-Bi-9H-fluorene, 9-(9H-fluoren-9-ylidene)-	C_6H_{12}	477(4.24)	18-3373-78
$C_{39}H_{26}$ 9,2':7',9"-Ter-9H-fluorene	C_6H_{12}	268f(4.6),290(4.5), 296(4.4),302(4.4), 310(4.2)	18-3373-78
9,2':9',9"-Ter-9H-fluorene	C_6H_{12}	268(4.6),305(4.2)	18-3373-78
9,4':9',9"-Ter-9H-fluorene	C_6H_{12}	230s(4.8),272(4.5), 292(4.1),302(4.1)	18-3373-78
9,9':9',9"-Ter-9H-fluorene	C_6H_{12}	225s(4.7),265(4.5), 292(4.0),303(4.0)	18-3373-78
$C_{39}H_{26}N_2O_2$ 5,6-Diazaspiro[3.4]octa-5,7-dien-1-one, 7-benzoyl-2,3-bis(diphenylmethylene)-	$CHCl_3$	325(4.41),427(3.74)	18-0649-78
$C_{39}H_{26}O_2$ 1H-Benzo[a]cyclobuta[d]cyclohepten-1-one, 8-benzoyl-2-(diphenylmethylene)-2,3-dihydro-3-phenyl-	CHCl	245(4.40),329(4.24)	18-0649-78
$C_{39}H_{27}O_2$ 1-Benzopyrylium, 2-phenyl-4-[2-phenyl-3-(2-phenyl-4H-1-benzopyran-4-ylidene)-1-propenyl]-, perchlorate	CH_2Cl_2	660(4.65),728(5.11)	4-0365-78
$C_{39}H_{30}N_4O$ 1,3,4-Oxadiazole, 2-[4-[2-[4-[4,5-di-hydro-5-phenyl-3-(2-phenylethenyl)-1H-pyrazol-1-yl]phenyl]ethenyl]-phenyl]-5-phenyl-	n.s.g.	432(5.00)	135-0946-78
$C_{39}H_{30}O_7S$ Pyrylium, 2-[3-(4,6-diphenyl-2H-pyran-2-ylidene)-1-propenyl]-4,6-diphenyl-, carboxymethylsulfonate	acetone	725(4.674)	83-0256-78
$C_{39}H_{31}O_2$ Pyrylium, 4-[2-[(2,6-diphenyl-4H-pyran-4-ylidene)methyl]-1-butenyl]-2,6-di-phenyl-, perchlorate	MeCN	640(4.54),688(5.10)	4-0365-78
Pyrylium, 4-[3-(2,6-diphenyl-4H-pyran-4-ylidene)-1-methyl-1-butenyl]-2,6-diphenyl-, perchlorate	MeCN	675(4.76),723(5.11)	4-0365-78
$C_{39}H_{32}NO_4P$ Cyclohepta[b]pyrrole-2-acetic acid, 3-(ethoxycarbonyl)-8-phenyl-α-(tri-phenylphosphoranylidene)-, methyl ester	$CHCl_3$	264(4.35),305(4.30), 365(3.84),458(4.18)	18-1573-78
$C_{39}H_{32}N_4O_4$ Carbamic acid, [(phenylmethylene)di-1H-indole-3,2-diyl]bis-, bis(phenylmeth-yl) ester	$CHCl_3$	283(4.40),295(4.39)	103-0745-78
$C_{39}H_{32}O_{15}$ 2-Propenoic acid, 3-(4-hydroxyphenyl)-,	MeOH	268(4.49),300s(4.71),	105-0275-78

Compound	Solvent	$\lambda_{max}(\log \epsilon)$	Ref.
3',6'-diester with 3-(β-D-galacto-pyranosyloxy)-5,7-dihydroxy-2-(4-hydroxyphenyl)-4H-1-benzopyran-4-one, (E,E)- (cont.)		315(4.76),360s(4.21)	105-0275-78
$C_{39}H_{32}O_{16}$ 2-Propenoic acid, 3-(4-hydroxyphenyl)-, 3',6'-diester with 2-(3,4-dihydroxy-phenyl)-3-(β-D-glucopyranosyloxy)-5,7-dihydroxy-4H-1-benzopyran-4-one, (E,E)-	MeOH MeOH-NaOMe	270(4.45),302s(4.70), 315(4.75),360s(4.30) 274(--),368(--)	105-0335-78 105-0335-78
$C_{39}H_{34}ClN_2O_2$ Benzoxazolium, 2-[2-[2-chloro-3-[(3-ethyl-5-phenyl-2(3H)-benzoxazolyli-dene)ethylidene]-1-cyclopenten-1-yl]ethenyl]-3-ethyl-5-phenyl-	MeOH	751(5.45)	104-2046-78
$C_{39}H_{34}N_2O_6S_2$ 1H-Cyclopenta[b]quinoxaline, 1-[bis(4-methoxyphenyl)methylene]-3a,4,9,9a-tetrahydro-6-methyl-4,9-bis(phenyl-sulfonyl)-	MeCN	255(4.36),295(4.18)	5-1161-78
1H-Cyclopenta[b]quinoxaline, 1-[bis(4-methoxyphenyl)methylene]-3a,4,9,9a-tetrahydro-7-methyl-4,9-bis(phenyl-sulfonyl)-	MeCN	255(4.41),295(4.29)	5-1161-78
2,5-Methano-1,6-benzodiazocine, 11-[bis(4-methoxyphenyl)methylene]-1,2,5,6-tetrahydro-8-methyl-1,6-bis(phenylsulfonyl)-	MeCN	265(4.36)	5-1161-78
$C_{39}H_{34}O_8$ 4H-1-Benzopyran-4-one, 5,8-dimethoxy-2-[3-methoxy-4,5-bis(phenylmethoxy)-phenyl]-7-(phenylmethoxy)-	MeOH	275(4.06),330(3.98)	2-1079-78
$C_{39}H_{39}N_5O_5$ 9H-Purin-6-amine, 9-[2,3,4,6-tetrakis-O-(phenylmethyl)-β-D-glucopyranosyl]-picrate	MeOH MeOH	214(4.47),260(4.18) 210(4.73),254(4.31)	136-0301-78C 136-0301-78C
$C_{39}H_{42}N_4NiO$ Nickel, [2,7,12,17-tetraethyl-3,8,13-18-tetramethyl-α-phenyl-21H,23H-por-phine-5-methanolato(2-)-N^{21},N^{22},N^{23}-N^{24}]-, (SP-4-2)-	CHCl$_3$	346(4.18),408(5.21), 536(3.96),578(4.18)	39-1660-78C
$C_{39}H_{44}N_2O_6$ Tubocurinemethine, O,O-dimethyl-	MeOH	225(4.66),284(4.34), 303s(4.24)	36-0473-78
$C_{39}H_{44}N_2O_7$ Thaliracebine	MeOH	278(3.90)	100-0257-78
$C_{39}H_{46}N_2O_6$ Tubocurinemethine, dihydro-O,O-dimeth-yl-	MeOH	225(4.70),280(3.95)	36-0473-78

Compound	Solvent	$\lambda_{max}(\log \epsilon)$	Ref.
$C_{39}H_{46}N_2O_7$ Thaliruginine	pH 12 MeOH	281(3.97),309s(3.09) 281(3.90)	44-0580-78 44-0580-78
$C_{39}H_{46}N_2O_8$ Thalirugidine	pH 12 MeOH	281(3.94) 278(3.82)	44-0580-78 44-0580-78
$C_{39}H_{46}N_4Ni$ Nickel, [5,6,10,11,15,16,22,22-octaeth-yl-4,7:14,17-diimino-2,21-methano-9,12-nitrilo-12H-1-benzazacyclonona-decinato(2-)-N^1,N^{23},N^{24},N^{25}]-, (SP-4-2)-	CHCl$_3$	364(3.95),416(4.84), 500(3.70),566(3.62), 625(3.99),677(4.56)	39-1660-78C
$C_{39}H_{46}N_4NiO$ Nickel, [3-(2,3,7,8,12,13,17,18-octa-ethyl-21H,23H-porphin-5-yl)-2-prop-enalato(2-)-$N^{21},N^{22},N^{23},N^{24}$]-, (SP-4-2)-	CHCl$_3$	337(4.22),409(4.79), 434(4.80),540s(3.81), 571(3.99),602s(3.88)	39-1660-78C
$C_{39}H_{46}N_4O_2$ 13'-Peribogamine	n.s.g.	235(4.55),287(4.08), 295(4.03)	36-0272-78
$C_{39}H_{48}IrN_5$ Iridium, (2-cyanoethyl)[2,3,7,8,12,13-17,18-octaethyl-21H,23H-porphinato-(2-)-$N^{21},N^{22},N^{23},N^{24}$]-, (SP-5-31)-	CHCl$_3$	390(5.26),500(4.12), 531(4.50)	101-0317-78P
$C_{39}H_{48}N_2O_9$ Kidamycin	MeOH	216s(4.58),243(4.67), 270s(4.52),434(3.93)	78-0761-78
$C_{39}H_{48}N_4Ni$ Nickel, [2,3,7,8,12,13,17,18-octaethyl-5-(1-propenyl)-21H,23H-porphinato-(2-)-$N^{21},N^{22},N^{23},N^{24}$]-, (SP-4-2)-	CHCl$_3$	408(5.23),530(4.05), 566(4.22)	39-0366-78C
Nickel, [2,7,12,17-tetraethyl-5-(1-hep-tenyl)-3,8,13,18-tetramethyl-21H,23H-porphinato(2-)-$N^{21},N^{22},N^{23},N^{24}$]-, (SP-4-2)-	CHCl$_3$	406(5.21),529(4.00), 564(4.17)	39-1660-78C
$C_{39}H_{48}N_4NiO$ Nickel, [3-(2,3,7,8,12,13,17,18-octa-ethyl-21H,23H-porphin-5-yl)-2-propen-1-olato(2-)-$N^{21},N^{22},N^{23},N^{24}$]-, (SP-4-2)-	CHCl$_3$	404(5.14),530(3.97), 567(4.16)	39-1660-78C
$C_{39}H_{48}N_4O$ 2-Propenal, 3-(2,3,7,8,12,13,17,18-octaethyl-21H,23H-porphin-5-yl)-	CHCl$_3$	383s(4.79),410(5.02), 510(4.02),543(3.77), 581(3.83),632(3.60)	39-1660-78C
$C_{39}H_{48}N_4O_3$ Demethylcapuvosine	EtOH	233(4.57),279(4.05), 287(4.09),294(4.05)	102-1605-78
$C_{39}H_{50}N_6O_3$ [3,7':3',7"-Ter-1H-indole]-1,1',1"-tri-	EtOH	252(4.46),284(3.82)	39-1671-78C

Compound	Solvent	$\lambda_{max}(\log \epsilon)$	Ref.
carboxaldehyde, 3,3',3"-tris[2-(di-methylamino)ethyl]-2,2',2",3,3',3"-hexahydro- (cont.)			39-1671-78C
nitrated derivative	EtOH	245(4.23),330(4.23)	39-1671-78C
$C_{39}H_{52}FN_3O_6$			
4a,13b-Etheno-1H,9H-benzo[c]cyclopenta-[h][1,2,4]triazolo[1,2-a]cinnoline-1,3(2H)-dione, 6,8-diacetoxy-11-(5-fluoro-1,5-dimethylhexyl)-5,6,7,8-8a,8b,10,10a,11,12,13,13a-dodecahy-dro-8a,10a-dimethyl-2-phenyl-	EtOH	254(3.62)	13-0453-78B
$C_{39}H_{52}N_4$			
Melonine, 1,1'-methylenebis-, dihydro-chloride	EtOH	248(3.82),298(3.37)	102-1452-78
$C_{39}H_{54}O_6$			
Lup-20(29)-en-28-oic acid, 2-hydroxy-3-[[3-(4-hydroxyphenyl)-1-oxo-2-propenyl]oxy]-, [2α,3β(E)]-	MeOH	312(4.46)	94-1798-78
[2α,3β(Z)]-	MeOH	310(4.22)	94-1798-78
Lup-20(29)-en-28-oic acid, 3-hydroxy-2-[[3-(4-hydroxyphenyl)-1-oxo-2-propenyl]oxy]-, [2α(E),3β]-	MeOH	308(4.95)	94-1798-78
$C_{39}H_{56}ClN_3O_9P$			
Maytansine, $N^{2'}$-deacetyl-9-deoxy-$N^{2'}$-(2-methyl-1-oxopropyl)-9-(propyl-thio)-	EtOH	233(4.41),245(4.38), 255(4.41),281(3.71), 289(3.69)	87-0031-78
$C_{39}H_{56}O_2$			
Lup-20(29)-en-3-ol, 3-phenyl-2-propen-oate, [3β(E)]-	MeOH	279(4.34)	95-0249-78
$C_{39}H_{68}O_6$			
Tetradecanoic acid, 2-[(1-oxo-2,4,6-octatrienyl)oxy]-1,3-propanediyl ester, (E,E,E)-	EtOH	307(4.54)	12-2085-78
$C_{39}H_{69}N_6NiO_6$			
Nickel(2+), tris(1,2,3,4,4a,5,6,7-octa-hydro-1-hydroxy-2,2,4a,7,7-pentameth-yl-1,8-naphthyridine 8-oxide-0,0')-, diperchlorate	CH₂Cl₂	269(4.22),338(3.96), 488(3.34),630(3.10), 813(3.06)	33-2851-78

Compound	Solvent	λ_{max} (log ϵ)	Ref.
$C_{40}H_{24}O_2S$ Anthra[2,3-c]thiophene-5,10-dione, 1,3,6,9-tetraphenyl-	CH_2Cl_2	277(4.45),322(4.41), 462(3.89)	150-5538-78
$C_{40}H_{27}N_6S_3$ Benzothiazolium, 2-[[bis[(3-phenyl- 2(3H)-benzothiazolylidene)amino]- methylene]amino]-3-phenyl-	EtOH	401(3.81)	18-0535-78
$C_{40}H_{28}ClN_3$ [1,2,4]Triazolo[1,5-a]pyridine, 7-(2- [1,1'-biphenyl]-4-ylethenyl)-2-[4- (2-[1,1'-biphenyl]-4-ylethenyl)-3- chlorophenyl]-	DMF	360(4.94)	33-0142-78
$C_{40}H_{28}N_2$ 1H-Indole, 5-(2,3-diphenyl-3H-indol- 3-yl)-2,3-diphenyl-	MeOH	257(4.67),314(4.49)	44-1230-78
$C_{40}H_{28}O$ Cyclopenta[b][1]benzopyran, 3-(2,4-di- phenyl-1,3-butadienyl)-1,2-diphenyl-, (E,E)-	THF THF	550(2.84) 235(4.41),275(4.42), 350(4.35),550(2.84)	103-1070-78 103-1075-78
$C_{40}H_{30}N_2O_3$ 1H-Pyrazole-3-propanoic acid, 5-benz- oyl-α,β-bis(diphenylmethylene)-, methyl ester	$CHCl_3$	252(4.47),290(4.35), 335s(4.04)	18-0649-78
$C_{40}H_{30}N_2O_{11}$ 2H-Benz[f]indazole-3-carboxylic acid, 4,9-dihydro-4,9-dioxo-2-(2,3,5-tri- O-benzoyl-β-D-ribofuranosyl)-, ethyl ester	EtOH	232(4.73)	111-0155-78
$C_{40}H_{30}N_4O_4$ [4,4'-Bispiro[2H-anthra[1,2-d]imidazole- 2,1'-cyclohexane]]-6,6'',11,11''-tetr- one	dioxan	500(4.22)	104-0381-78
$C_{40}H_{30}N_4O_6$ 1H-Cyclopenta[b]quinoxaline, 3a,4,9,9a- tetrahydro-1-(diphenylmethylene)-6,7- dimethyl-4,9-bis(4-nitrobenzoyl)-	CH_2Cl_2	231(4.56),250(4.55), 285(4.47)	5-1139-78
$C_{40}H_{30}N_4O_8$ 1H-Cyclopenta[b]quinoxaline, 1-[bis(4- methoxyphenyl)methylene]-3a,4,9,9a- tetrahydro-4,9-bis(4-nitrobenzoyl)-	CH_2Cl_2	231(4.51),255(4.54), 285(4.41)	5-1139-78
$C_{40}H_{30}O_2$ 1,2,3,5,7,11,13-Cyclotetradecaheptaen- 9-yne, 1,8-bis(4-methoxyphenyl)- 4,11-diphenyl-	THF	245(4.29),261s(4.31), 283(4.43),330(4.23), 347(4.39),395(5.44), 563(4.72),665(3.68)	18-3351-78
$C_{40}H_{30}O_{13}$ [2,9'-Bianthracene]-9,10,10'(9'H)-tri- one, 1,4',5',8,9'-pentaacetoxy-2',6- dimethyl-	MeOH	263(4.75),345(3.75)	94-1111-78

Compound	Solvent	$\lambda_{max}(\log \epsilon)$	Ref.
$C_{40}H_{32}N_2O_4$ 1H-Cyclopenta[b]quinoxaline, 4,9-dibenzoyl-1-[bis(4-methoxyphenyl)methylene]-3a,4,9,9a-tetrahydro-	CH_2Cl_2	231(4.51),260(4.45), 290(4.37)	5-1139-78
$C_{40}H_{34}Cl_2N_2O_6S_2$ Benzo[b]cyclopenta[e][1,4]diazepine, 10,10-bis(4-chlorophenyl)-3a,4,9,10-tetrahydro-4,9-bis[(4-methoxyphenyl)sulfonyl]-6,7-dimethyl-	MeCN	243.5(4.68)	5-1146-78
1H-Cyclopenta[b]quinoxaline, 1-[bis(4-chlorophenyl)methylene]-3a,4,9,9a-tetrahydro-4,9-bis[(4-methoxyphenyl)sulfonyl]-6,7-dimethyl-, cis	MeCN	243(4.68)	5-1146-78
$C_{40}H_{34}N_4O_4$ [4,4"-Bispiro[2H-anthra[1,2-d]imidazole-2,1'-cyclohexane]]-6,6",11,11"-tetrone, 1,1",3,3"-tetrahydro-	EtOH EtOH-NaOH $MeNO_2$ $MeNO_2$-$HClO_4$	558(4.11) 650(3.83) 541(4.27) 580(4.30),610(4.30)	104-0381-78 104-0381-78 104-0381-78 104-0381-78
$C_{40}H_{34}N_4O_{10}S_2$ 1H-Cyclopenta[b]quinoxaline, 1-[bis(4-methoxyphenyl)methylene]-3a,4,9,9a-tetrahydro-6,7-dimethyl-4,9-bis[(4-nitrophenyl)sulfonyl]-	MeCN	255(4.57),300(4.37)	5-1161-78
$C_{40}H_{34}O_{13}$ Scirpusin A hexaacetate	EtOH	214(4.61),292s(4.48), 310(4.49)	94-3050-78
$C_{40}H_{34}O_{16}$ 2-Propenoic acid, 3-(4-hydroxy-3-methoxyphenyl)-, 6'-ester with 5,7-dihydroxy-2-(4-hydroxyphenyl)-3-[[3-O-[3-(4-hydroxyphenyl)-1-oxo-2-propenyl]-β-D-glucopyranosyl]oxy]-4H-1-benzopyran-4-one	MeOH	270(4.49),300s(4.66), 320(4.69)	105-0490-78
$C_{40}H_{36}$ 1,3-Cyclopentadiene, 1,2,3,4,5-pentakis(phenylmethyl)-	isooctane	276(3.92)	44-4090-78
$C_{40}H_{36}N_2O_4S_2$ 1H-Cyclopenta[b]quinoxaline, 1-[bis(4-methylphenyl)methylene]-3a,4,9,9a-tetrahydro-6,7-dimethyl-4,9-bis(phenylsulfonyl)-, cis	MeCN	274(4.24),290(4.27)	5-1146-78
Spiro[2H-benzimidazole-2,1'-[3]cyclopentene], 2'-[bis(4-methylphenyl)methylene]-1,3-dihydro-5,6-diphenyl-1,3-bis(phenylsulfonyl)-	MeCN	288(4.32)	5-1161-78
$C_{40}H_{36}N_2O_6S_2$ 2,5-Methano-1,6-benzodiazocine, 11-[bis(4-methoxyphenyl)methylene]-1,2,5,6-tetrahydro-8,9-dimethyl-1,6-bis(phenylsulfonyl)-	CH_2Cl_2	232(4.52),267(4.38)	5-1161-78

Compound	Solvent	λ_{max} (log ϵ)	Ref.
$C_{40}H_{36}N_4$			
1,2,21,22-Tetraazacyclotetraconta-2,4-6,8,14,16,18,20,22,24,26,28,34,36,38-40-hexadecaene-10,12,30,32-tetrayne, 9,14,29,34-tetramethyl-, (?,?,?,?-E,E,E,E,Z,Z,Z,Z,Z,E,E,E,E)-	THF	238s(4.00),258s(4.17), 381(4.99),468s(3.98)	18-2112-78
$C_{40}H_{38}N_2O_6S_2$			
2,5-Methano-1,6-benzodiazocine, 11-[bis(4-methoxyphenyl)methylene]-1,2,3,4,5,6-hexahydro-8,9-dimethyl-1,6-bis(phenylsulfonyl)-	CH_2Cl_2	232(4.51),267(4.30)	5-1161-78
$C_{40}H_{42}N_4Ni$			
Nickel, [2,7,12,17-tetraethyl-3,8,13-18-tetramethyl-5-(2-phenylethenyl)-21H,23H-porphinato(2-)-N^{21},N^{22},N^{23}-N^{24}]-, (SP-4-2)-	$CHCl_3$	405(5.09),528(3.93), 566(4.11)	39-1660-78C
$C_{40}H_{42}O_{12}$			
Lappaol F	MeOH	282(4.03)	88-3035-78
$C_{40}H_{44}N_2O_7$			
4H-Dibenzo[de,g]quinoline, 5,6,6a,7-tetrahydro-1,2,11-trimethoxy-6-methyl-4H-dibenzo[de,g]quinolin-9-yl)oxy]-	EtOH	278(4.51),304(4.47)	44-0105-78
$C_{40}H_{44}N_2O_8$			
Spiro[2,5-cyclohexadiene-1,7'(1'H)-cyclopent[ij]isoquinolin]-4-one, 2',3'-8',8'a-tetrahydro-5,5',6'-trimethoxy-1'-methyl-2-[(5,6,6a,7-tetrahydro-1,2,10-trimethoxy-6-methyl-4H-dibenzo[de,g]quinolin-9-yl)oxy]-	EtOH	232s(4.63),258(4.39), 263(4.40),275(4.32), 304(4.18)	44-0105-78
diastereomer	EtOH	232s(4.67),258(4.42), 263(4.41),275(4.30), 304(4.2)	44-0105-78
$C_{40}H_{44}N_4O_4$			
Ervafoline	EtOH	252(3.86),306(3.95), 326(4.03)	88-3707-78
$C_{40}H_{46}N_2O_8$			
Pennsylvanine, (+)-	EtOH	282(4.23),304(4.16), 320s(4.01)	100-0169-78
Thalmelatine, (+)-	EtOH	282(4.12),315s(3.86)	100-0169-78
$C_{40}H_{46}N_4NiO_2$			
Nickel, [2,3,7,8,12,13,17,18-octaethyl-10-(3-oxo-1-propenyl)-21H,23H-porphine-5-carboxaldehydato(2-)-N^{21},N^{22},N^{23},N^{24}]-, (SP-4-2)-	$CHCl_3$	338(4.19),441(4.82), 660(3.99)	39-1660-78C
$C_{40}H_{46}N_4O_8$			
21H,23H-Porphine-2,6-dipropanoic acid, 3,7-bis(ethoxycarbonyl)-1,4,5,8-tetramethyl-, diethyl ester	$CHCl_3$	518(3.96),561(4.33), 586(4.03),641(3.42)	23-2430-78
$C_{40}H_{46}O_4$			
[1,1':3',1"-Terphenyl]-2,2'-diol, 4,4"-	CH_2Cl_2	252(4.66)	24-0264-78

Compound	Solvent	$\lambda_{max}(\log \epsilon)$	Ref.
bis(1,1-dimethylethyl)-5'-[4-(1,1-di-methylethyl)phenyl]-, diacetate (cont.)			24-0264-78
[1,1':3',1"-Terphenyl]-2',4'-diol, 4,4'-bis(1,1-dimethylethyl)-5'-[4-(1,1-dimethylethyl)phenyl]-, diacetate	CH_2Cl_2	243(4.82)	24-0264-78
$C_{40}H_{46}O_{14}$			
Lappaol H	MeOH	282(4.01)	88-3035-78
$C_{40}H_{47}N_2O_8$			
Thalirabine, iodide	pH 2	209(4.77),273(3.81), 280(3.78)	100-0257-78
	pH 12	276(3.90),283(3.91)	100-0257-78
	MeOH	207(4.99),276(3.82), 283(3.80)	100-0257-78
$C_{40}H_{48}N_2O_2$			
Capuvosidine	EtOH	228(4.61),280(3.99), 286(3.99),295(3.86)	102-1605-78
	EtOH-acid	230(--),283(--), 295(--),340(--)	102-1605-78
$C_{40}H_{48}N_4Ni$			
Nickel, [5-(1,3-butadienyl)-2,3,7,8,12-13,17,18-octaethyl-21H,23H-porphin-ato(2-)-$N^{21},N^{22},N^{23},N^{24}$]-, (SP-4-2)-	$CHCl_3$	338(4.18),408(5.12), 534(4.02),568(4.17)	39-1660-78C
$C_{40}H_{48}N_4O_2$			
Capuvoisidine	EtOH	228(4.61),280(3.99), 286(3.99),295(3.86)	42-1099-78
	EtOH-acid	230(--),283(--), 295(--),340(--)	42-1099-78
$C_{40}H_{48}O_2$			
Oxepino[2,3-b]benzofuran, 2,8-bis(1,1-dimethylethyl)-4,7-bis[4-(1,1-dimeth-ylethyl)phenyl]-	C_6H_{12}	248(4.64),339(3.80)	12-1061-78
$C_{40}H_{50}N_4O_2$			
20-Phorbinecarboxylic acid, 3,4,20,21-tetradehydro-3,4,8,9,13,14,18,19-octahydro-18,19-dihydro-, methyl ester ($6\lambda,7\epsilon$)	$CHCl_3$	415s(4.94),430(5.16), 503(3.76),573(3.84), 652(4.16),704s(3.82), ?(4.61)	39-1660-78C
2-Propenoic acid, 3-(2,3,7,8,12,13,17-18-octaethyl-21H,23H-porphin-5-yl)-, methyl ester	$CHCl_3$	408(5.16),509(4.05), 542(3.79),579(3.76), 631(3.38)	39-1660-78C
Vobasan-17-oic acid, 3-(5-ethyl-1,4,5-6,7,8,9,10-octahydro-2H-3,7-methano-azacycloundecino[5,4-b]indol-12-yl)-, methyl ester (dihydrocapuvoisidine)	EtOH	232(4.63),288(4.15), 295(4.11)	42-1099-78 +102-1605-78
$C_{40}H_{51}N_4O_2Rh$			
Rhodium, (4-hydroxy-1-oxobutyl)[2,3,7-8,12,13,17,18-octaethyl-21H,23H-por-phinato(2-)-$N^{21},N^{22},N^{23},N^{24}$]-, (SP-5-31)-	$CHCl_3$	394(5.25),510(4.13), 544(4.70)	101-0329-78P
$C_{40}H_{52}N_6O_{14}$			
D-Valine, N-[O-acetyl-N-[[9-[[[2-(acet-yloxy)-1-[[[1-(methoxycarbonyl)-2-	MeOH	240(4.698),446(4.296)	87-0607-78

Compound	Solvent	$\lambda_{max}(\log \epsilon)$	Ref.
methylpropyl]amino]carbonyl]propyl]-amino]carbonyl]-2-amino-4,6-dimethyl-3-oxo-3H-phenoxazin-1-yl]carbonyl]-L-threonyl]-. methyl ester (cont.)			87-0607-78
$C_{40}H_{56}O$			
β,β-Carotene, 5,6-epoxy-5,6-dihydro-	hexane	266(4.32),442s(5.01), 445(5.15),472(5.07)	33-0822-78
$C_{40}H_{56}O_2$			
β,β-Carotene, 15,15'-didehydro-5,6-di-hydro-5,6-dihydroxy-, (5R,6R)-(±)-	benzene	416s(--),436(5.22), 465(5.13)	39-1511-78C
	hexane	404(--),423(--), 450(--)	39-1511-78C
$C_{40}H_{56}O_8$			
Tricyclo[8.2.2^{4,7}]hexadeca-4,6,10,12-13,15-hexaene-5,6,11,12-tetracarbox-ylic acid, tetrakis(2,2-dimethyl-propyl) ester	EtOH	213(4.63),301(3.36)	24-0523-78
$C_{40}H_{58}O_2$			
β,β-Carotene, 5,6-dihydro-5,6-dihydr-oxy-, (5R,6R)- (5R,6R)-(±)-	EtOH	266(4.42),425(5.02), 448(5.19),476(5.15)	33-0822-78
	benzene	432s(--),456(5.29), 484(5.23)	39-1511-78C
(5R,6R,9'-cis)-	EtOH	269(4.15),330(4.17), 417(4.77),440(4.89), 468(4.80)	33-0822-78
$C_{40}H_{60}O_4$			
β,β-Carotene, 5,5',6,6'-tetrahydro-5,5',6,6'-tetrahydroxy-, all-trans	benzene	423(4.85),450(5.03), 479(5.03)	39-1511-78C
$C_{40}H_{64}P_2Pt$			
Platinum, bis(phenylethynyl)bis(tri-butylphosphine), cis	CH_2Cl_2	252(4.58),267(4.54), 307(4.47)	101-0101-78B
trans	CH_2Cl_2	263(4.55),280(4.48), 322(4.46)	101-0101-78B

Compound	Solvent	$\lambda_{max}(\log \epsilon)$	Ref.
$C_{41}H_{27}O_2$ 3H-Cyclopenta[b][1]benzopyrylium, 3-(2,6-diphenyl-4H-1-benzopyran-4-ylidene)-1,2-diphenyl-, perchlorate	CHCl₃	532(3.89),566(4.07), 653(3.87),714(3.98)	103-0373-78
$C_{41}H_{28}CoN_4O_2$ Cobalt, [ethyl 10,15,20-triphenyl-21H,23H-porphine-5-carboxylato(2-)-N²¹,N²²,N²³,N²⁴]-, (SP-4-2)-	benzene	407(5.33),525(4.09)	150-0690-78
$C_{41}H_{28}O_2$ 1H-Cyclopenta[b][1]benzopyrylium, 3-(2,6-diphenylpyrylium-4-yl)-1,2-diphenyl-, diperchlorate	CHCl₃	565(3.54),655(3.57)	103-0373-78
$C_{41}H_{29}NO$ 1,2-Benzisoxazole, 6-(2-[1,1'-biphenyl]-4-ylethenyl)-3-[4-(2-[1,1'-biphenyl]-4-ylethenyl)phenyl]-	DMF	352(4.95)	33-2904-78
1,2-Benzisoxazole, 3-[3,4-bis(2-[1,1'-biphenyl]-4-ylethenyl)phenyl]-	DMF	320(4.87)	33-2904-78
$C_{41}H_{29}N_3O_4$ 3,5-Pyridinedicarbonitrile, 2,6-bis-[(2,3-dihydro-1,3-dioxo-2-phenyl-1H-inden-2-yl)methyl]-1,4-dihydro-4,4-dimethyl-	EtOH	208(4.01),230(4.87), 257(3.67),370(3.09)	103-1226-78
$C_{41}H_{29}O_2$ 1-Benzopyrylium, 4-[3-(2,6-diphenyl-4H-pyran-4-ylidene)-2-phenyl-1-propenyl]-2-phenyl-, perchlorate	CH₂Cl₂	645(4.75),704(5.29)	4-0365-78
$C_{41}H_{42}NO_6P$ 1H-Pyrrole-3-propanoic acid, 2-[(1,1-dimethylethoxy)carbonyl]-4-methyl-5-[2-oxo-2-(phenylmethoxy)-1-(triphenylphosphoranylidene)ethyl]-, methyl ester	MeOH	267(4.29),275(4.31), 294(4.44)	44-0283-78
$C_{41}H_{42}N_4O_6$ 21H,23H-Porphine-2,18-dipropanoic acid, 3,7,8,13,17-pentamethyl-12-[(phenylmethoxy)carbonyl]-, dimethyl ester	CHCl₃	407(5.06),510(3.99), 549(4.15),574(3.94), 634(3.15)	104-0794-78
$C_{41}H_{44}N_4Ni$ Nickel, [2,7,12,17-tetraethyl-3,8,13-18-tetramethyl-5-[2-(4-methylphenyl)-ethenyl]-21H,23H-porphinato(2-)-N²¹-N²²,N²³,N²⁴]-, (SP-4-2)-	CHCl₃	404(5.10),530(3.98), 564(4.14)	39-1660-78C
$C_{41}H_{45}NO_{18}$ Alatamine	EtOH	233(4.39),272(3.72)	78-1915-78
$C_{41}H_{46}N_4O_3$ Macrocarpamine	EtOH	230(4.58),254(4.39), 284s(3.93),291(3.88)	33-0337-78
$C_{41}H_{48}N_2O_8$ Thalicarpine, (+)-	EtOH	207(4.50),279(4.06), 303(3.70)	100-0169-78

Compound	Solvent	$\lambda_{max}(\log \epsilon)$	Ref.

$C_{41}H_{48}N_2O_9$
  Thaliadanine
    same spectrum in acid or base
	MeOH	281(4.33),302(4.18),
		312(4.11)
			100-0271-78

$C_{41}H_{50}I_2N_2O_8$ Thalirabine dimethiodide	MeOH	212(5.01),275(3.84), 283(3.80)	100-0257-78
$C_{41}H_{50}N_2O_8$ Thalirugidine, O,O-dimethyl-	MeOH	278(3.96)	44-0580-78
$C_{41}H_{50}N_2O_{11}$ Hedamycin	MeOH	213(4.53),244(4.67), 264s(4.47),434(3.95)	78-0761-78

$C_{41}H_{53}NO_2$
  2,5-Cyclohexadien-1-one, 4,4'-[3-(3,5-
    diethyl-4-imino-2,5-cyclohexadien-1-
    ylidene)-1,2-cyclopropanediylidene]-
    bis[2,6-bis(1,1-dimethylethyl)-
	C_6H_{12}	280(4.17),500s(4.25),
		539(5.84)
			44-1577-78

$C_{41}H_{53}N_5O_5$
  Vincaleukoblastine, 3-(aminocarbonyl)-
    O^4-deacetyl-3,18'-dide(methoxycarbo-
    nyl)-
	MeOH	213(4.70),225s(4.57),
		260(4.15),287(4.06),
		295(4.07)
			87-0088-78

$C_{41}H_{54}N_6O_5$
  Vincaleukoblastin-23-oic acid, 4-de-
    acetyl-18'-de(methoxycarbonyl)-,
    hydrazide
	MeOH	213(4.73),224s(4.57),
		258(4.11),287(4.05),
		294(4.05)
			87-0088-78

$C_{41}H_{55}NO_2$
  2,5-Cyclohexadien-1-one, 4-[2-(4-amino-
    3,5-diethylphenyl)-3-[3,5-bis(1,1-di-
    methylethyl)-4-hydroxyphenyl]-2-cy-
    clopropen-1-ylidene]-2,6-bis(1,1-di-
    methylethyl)-
	MeOH	320(4.36),430(4.49)	44-1577-78

Phenol, 4,4'-[3-(3,5-diethyl-4-imino-
    2,5-cyclohexadien-1-ylidene)-1-cyclo-
    propene-1,2-diyl]bis[2,6-bis(1,1-di-
    methylethyl)-
	MeOH	280(4.36),330(4.12),
		420(4.59)
			44-1577-78

Compound	Solvent	λ_{max}(log ϵ)	Ref.
$C_{42}H_{24}Cl_6N_4$			
1,1'-Bi-1H-imidazole, 2,2',4,4',5,5'-hexakis(4-chlorophenyl)-	EtOH	222(4.74),298(4.62)	24-1464-78
1H-Imidazole, 2,4,5-tris(4-chlorophenyl)-1-[2,4,5-tris(4-chlorophenyl)-2H-imidazol-2-yl]-	EtOH	231(4.70),279(4.53)	24-1464-78
$C_{42}H_{24}O_5$			
Anthra[2,3-c]furan-5,10-dione, 7,8-di-benzoyl-1,3-diphenyl-	CH_2Cl_2	275(4.37),520(4.11)	150-5538-78
$C_{42}H_{28}$			
7H-Dibenzo[a,c]cyclononene, 7-(9H-fluo-ren-9-ylidene)-5,9-diphenyl-	EtOH	248(4.64),269s(4.45),280s(4.34),307(4.11),320s(4.09),336s(4.02),412(4.52)	150-5151-78
	C_6H_{12}	250(4.65),270s(4.48),280s(4.42),308(4.13),322(4.12),336s(4.06),414(4.56)	150-5151-78
7H-Dibenzo[a,c]cyclononene, 7-(1H-phen-alen-1-ylidene)-5,9-diphenyl-	C_6H_{12}	245(4.48),280s(4.23),300s(3.93),330(3.74),345(3.75),400s(3.04),430s(3.20),459(3.48),510(3.46)	150-5151-78
	EtOH	245(4.45),270s(4.22),300s(3.81),326s(3.62),350(3.53),400s(3.04),430s(3.18),459(3.32),508(3.36)	150-5151-78
	MeCN	245(4.53),280s(4.23),300s(3.85),326(3.79),350(3.71),400(3.30),430s(3.42),460s(3.63),506(3.75)	150-5151-78
$C_{42}H_{28}N_2$			
Benzenamine, N,N'-(3,3'-diphenyl[2,2'-bi-1H-indene]-1,1'-diylidene)bis-	benzene	421(3.84)	40-0144-78
$C_{42}H_{30}$			
1,2,3,5,7,9,13,15,17-Cyclooctadecanona-en-11-yne, 1,4,10,13-tetraphenyl-	THF	261(4.40),293(4.42),411(5.03),431(4.71),572(4.31),615(4.69),735(2.43),826(3.19)	18-3359-78
Fluoranthene, 8-(3,4-dimethylphenyl)-7,9,10-triphenyl-	C_6H_{12}	297(4.63),330(4.01)	2-0152-78
$C_{42}H_{30}N_4$			
1H-Imidazole, 2,4,5-triphenyl-1-(2,4,5-triphenyl-4H-imidazol-4-yl)-	EtOH	227(4.59),270(4.42)	24-1464-78
$C_{42}H_{34}N_4O_8$			
1H-Cyclopenta[b]quinoxaline, 1-[bis(4-methoxyphenyl)methylene]-3a,4,9,9a-tetrahydro-6,7-dimethyl-4,9-bis(4-nitrobenzoyl)-, cis	CH_2Cl_2	231(4.62),256(4.64),288(4.50)	5-1139-78
$C_{42}H_{34}O_4$			
1,2,3,5,7,11,13-Cyclotetradecaheptaen-9-yne, 1,4,8,11-tetrakis(4-methoxy-	THF	248(4.34),285(4.53),333(4.28),350(4.41),	18-3351-78

Compound	Solvent	$\lambda_{max}(\log \epsilon)$	Ref.
phenyl- (cont.)		408(5.47),576(4.76), 665(3.36)	18-3351-78
$C_{42}H_{36}O_{15}$ 1,2-Benzenediol, 4-[2-[6-(acetyloxy)-2-[3,4-bis(acetyloxy)phenyl]-3-[3,5-bis(acetyloxy)phenyl]-2,3-dihydro-4-benzofuranyl]ethenyl]-, diacetate (scirpusin B heptaacetate)	EtOH	303s(4.33),310(4.34)	94-3050-78
$C_{42}H_{37}N_5O_9$ Benzoic acid, 4-methyl-, 2',3',5'-triester with N-(1,4-dihydro-4-oxo-1- -D-ribofuranosylimidazo[1,2-a]-1,3,5-triazin-2-yl)-4-methylbenzamide	dioxan	239(4.72),265s(4.30), 305(4.34)	44-4774-78
$C_{42}H_{38}N_4O_4$ [4,4"-Bispiro[2H-anthra-[1,2-d]imidazole-2,1'-cyclohexane]-6,6",11,11"-tetrone, 1,1",3,3"-tetrahydro-3,3"-dimethyl-	EtOH EtOH-NaOH $MeNO_2$ $MeNO_2-HClO_4$	567(4.40) 567(4.40) 558(4.37) 625(4.67)	104-0381-78 104-0381-78 104-0381-78 104-0381-78
$C_{42}H_{42}ClFeN_6O_{10}$ Heminbis(L-aspartic acid)	7:3 $CHCl_3$-EtOH	403(5.12),479(4.22), 538(4.13),597(3.88)	104-0786-78
$C_{42}H_{44}Ir_2N_4O_6$ Iridium, hexacarbonyl[μ-(2,3,7,8,12,13-17,18-octaethyl-21H,23H-porphinato-(2-)-$N^{21},N^{22},N^{23},N^{24}$]di-	$CHCl_3$	350(4.63),388(4.92), 445(4.55),520(4.05), 548(4.08)	101-0317-78P
$C_{42}H_{44}O_{22}$ 9,10-Anthracenedione, 1,8-bis(2,3,4,6-tetra-O-acetyl-β-D-glucopyranosyl-oxy)-	ether	216(4.4),252(4.4), 351(3.7)	73-1803-78
$C_{42}H_{46}S_2$ Anthracene, 9,9'-[1,12-dodecanediylbis-(thiomethylene)]bis-	$CHCl_3$	350(4.09),368(4.23), 387(4.12)	24-3519-78
$C_{42}H_{50}N_4O_5$ 13'-Perivoacangine	n.s.g.	227(4.62),286(4.18), 294(4.18)	36-0272-78
$C_{42}H_{54}Cl_2O_2Pd_2$ Palladium, di-μ-chlorobis[(16,17,20-η^3)-3-methoxy-19-norpregna-1,3,5(10)-17(20)-tetraene]di-	EtOH	222s(4.45),248s(4.13), 278(3.55),287(3.54)	35-3435-78
$C_{42}H_{55}N_4ORh$ Rhodium, [2,3,7,8,12,13,17,18-octaethyl-21H,23H-porphinato(2-)-N^{21},N^{22},N^{23}-N^{24}](1-oxohexyl)-, (SP-5-31)-	benzene	397(5.31),513(4.20), 545(4.81)	101-0329-78P
$C_{42}H_{57}IrN_4$ Iridium, hexyl[2,3,7,8,12,13,17,18-octaethyl-21H,23H-porphinato(2-)-$N^{21},N^{22},N^{23},N^{24}$]-, (SP-5-31)-	benzene	391(5.32),500(4.26), 530(4.54)	101-0317-78P
$C_{42}H_{57}N_4O_2Rh$ Rhodium, (2,2-diethoxyethyl)[2,3,7,8-	benzene	387(5.12),397(5.15),	101-0329-78P

Compound	Solvent	$\lambda_{max}(\log \epsilon)$	Ref.
12,13,17,18-octaethyl-21H,23H-porph-inato(2-)-N^{21},N^{22},N^{23},N^{24}]-, (SP-5-31)- (cont.)		513(4.19),544(4.76)	101-0329-78P
$C_{42}H_{62}O_5$ Tetradecanoic acid, 9,10-dihydro-9-oxo-1,8-anthracenediyl ester	CH_2Cl_2	228(3.79),270(4.03), 310(3.40)	87-0026-78
$C_{43}H_{30}$ Fluoranthene, 8-(2,3-dihydro-1H-inden-5-yl)-7,9,10-triphenyl-	C_6H_{12}	297(4.58),330(3.97)	2-0152-78
$C_{43}H_{30}NO_4$ Pyrylium, 4-[3-(2,6-diphenyl-4H-pyran-4-ylidene)-1-[2-(4-nitrophenyl)-1-propenyl]-2,6-diphenyl-, perchlorate	CH_2Cl_2	638(4.70),695(5.39)	4-0365-78
$C_{43}H_{31}O_2$ Pyrylium, 2-[3-(2,6-diphenyl-4H-pyran-4-ylidene)-2-phenyl-1-propenyl]-4,6-diphenyl-, perchlorate	CH_2Cl_2	736(5.05)	4-0365-78
Pyrylium, 4-[3-(2,6-diphenyl-4H-pyran-4-ylidene)-2-phenyl-1-propenyl]-2,6-diphenyl-, perchlorate	CH_2Cl_2	635(4.79),690(5.49)	4-0365-78
$C_{43}H_{32}N_4O$ 1,3,4-Oxadiazole, 2-[4-[2-[4-(3-[1,1'-biphenyl]-4-yl-4,5-dihydro-5-phenyl-1H-pyrazol-1-yl)phenyl]ethenyl]phen-yl]-5-phenyl-, (E)-	n.s.g.	425(4.93)	135-0946-78
$C_{43}H_{34}N_4O_2$ 1,2,7,8-Tetraazaspiro[4.4]nona-2,8-di-ene-6-carboxylic acid, 1,3,7,9-tetra-phenyl-, diphenylmethyl ester	EtOH	240(4.41),342(4.28)	22-0415-78
$C_{43}H_{38}Cl_2$ Tetracyclo[6.2.1.0^{1,8}.0^{3,6}]undeca-2,4,6,9-tetraene, 11,11-dichloro-2,7-bis(1,1-dimethylethyl)-4,5,9,10-tetraphenyl-	$CHCl_3$	254(4.53),320s(3.86)	138-0657-78
$C_{43}H_{48}ClN_5OZn$ Zinc, [4-chloro-N-(2,3,7,8,12,13,17,18-octaethyl-21H,23H-porphin-5-yl)benz-amidato(2-)-N^{21},N^{22},N^{23},N^{24}]-, SP-4-2)-	CH_2Cl_2	412(5.64),504(3.43), 540(4.28),578(4.20)	18-1444-78
$C_{43}H_{48}N_4Ni$ Nickel, [3-ethenyl-2,7,8,12,13,17,18-heptaethyl-5-(phenylmethyl)-21H,23H-porphinato(2-)-N^{21},N^{22},N^{23},N^{24}]-, (SP-4-2)-	$CHCl_3$	345(4.19),411(5.28), 534(4.07),569(4.22)	39-0366-78C
$C_{43}H_{48}N_4O_3V$ Vanadium, [1-(3,7,8,12,13,17,18-hepta-ethyl-21H,23H-porphin-2-yl)ethyl benzoato(2-)-N^{21},N^{22},N^{23},N^{24}]-, (SP-5-13)-	$CHCl_3$	411(5.45),535(4.10), 574(4.36)	78-0379-78

Compound	Solvent	$\lambda_{max}(\log \epsilon)$	Ref.
$C_{43}H_{48}N_6O_3Zn$ Zinc, [3-nitro-N-(2,3,7,8,12,13,17,18-octaethyl-21H,23H-porphin-5-yl)benz-amidato(2-)-$N^{21},N^{22},N^{23},N^{24}$]-, (SP-4-2)-	CH_2Cl_2	412(5.59),505(3.40), 541(4.27),579(4.21)	18-1444-78
Zinc, [4-nitro-N-(2,3,7,8,12,13,17,18-octaethyl-21H,23H-porphin-5-yl)benz-amidato(2-)-$N^{21},N^{22},N^{23},N^{24}$]-, (SP-4-2)-	CH_2Cl_2	411(5.60),503(3.53), 541(4.27),579(4.19)	18-1444-78
$C_{43}H_{49}NO_{19}$ Wilfordine	EtOH	232(4.35),271(3.64)	78-1915-78
$C_{43}H_{49}N_5OZn$ Zinc, [N-(2,3,7,8,12,13,17,18-octaeth-yl-21H,23H-porphin-5-yl)benzamidato-(2-)-$N^{21},N^{22},N^{23},N^{24}$]-, (SP-4-2)-	CH_2Cl_2	412(5.62),504(3.42), 540(4.29),578(4.20)	18-1444-78
$C_{43}H_{50}N_4NiO$ Nickel, [2,3,7,8,12,13,17,18-octaethyl-α-phenyl-21H,23H-porphine-5-methanol-ato(2-)-$N^{21},N^{22},N^{23},N^{24}$]-, (SP-4-2)-	$CHCl_3$	307(4.04),354(4.16), 412(5.21),539(3.93), 579(4.15)	39-0366-78C
$C_{43}H_{51}N_5O$ Benzamide, N-(2,3,7,8,12,13,17,18-octa-ethyl-21H,23H-porphin-21-yl)-	CH_2Cl_2	401(5.03),530(3.87), 566(4.11),612(3.42)	18-1444-78
$C_{43}H_{52}N_4O_5$ Voacamine	n.s.g.	226(4.60),289(4.13), 295(4.15)	36-0271-78
$C_{43}H_{52}N_4O_6$ Epivoacorine	n.s.g.	226(4.63),288(4.17), 295(4.18)	36-0271-78
$C_{43}H_{52}N_5ORh$ Rhodium, [2,3,7,8,12,13,17,18-octaeth-yl-21H,23H-porphinato(2-)-N^{21},N^{22}-N^{23},N^{24}](2-oxoethyl)(pyridine)-, (OC-6-14)-	$CHCl_3$	352(4.51),407(5.24), 522(4.27),552(4.49)	101-0329-78P
$C_{43}H_{55}ClO_4$ Phenol, 2-[[3-chloro-5-(1,1-dimethyl-ethyl)-2-hydroxyphenyl]methyl]-4-(1,1-dimethylethyl)-6-[[5-(1,1-di-methylethyl)-3-[[5-(1,1-dimethyleth-yl)-2-hydroxyphenyl]methyl]-2-hy-droxyphenyl]methyl]-	dioxan	284(4.01)	49-0767-78
$C_{43}H_{55}N_5O_7$ Vindesine	MeOH	214(4.73),266(4.24), 288(4.14),296(4.10)	87-0088-78
$C_{43}H_{56}O_{17}$ Chukrasin C, 3-deacetyl-	EtOH	208(3.88),268(3.96)	33-1814-78
$C_{43}H_{56}O_{18}$ Chukrasin A, deacetyl-	EtOH	208(3.87),268(3.92)	33-1814-78
$C_{43}H_{57}NO_2$ 2,5-Cyclohexadien-1-one, 4,4'-[3-[4-	C_6H_{12}	280(4.17),500s(4.57),	44-1577-78

Compound	Solvent	$\lambda_{max}(\log \epsilon)$	Ref.
imino-3,5-bis(1-methylethyl)-2,5-cyclohexadien-1-ylidene]-1,2-cyclopropanediylidene]bis[2,6-bis(1,1-dimethylethyl)- (cont.)		539(4.85)	44-1577-78
$C_{43}H_{57}N_5O_7$ Vindesine, 6,7-dihydro-	MeOH	214(4.67),266(4.11), 288s(4.08),296(4.06)	87-0088-78
$C_{43}H_{59}NO_2$ Phenol, 4,4'-[3-[4-imino-3,5-bis(1-methylethyl)-2,5-cyclohexadien-1-ylidene]-1-cyclopropene-1,2-diyl]bis[2,6-bis(1,1-dimethylethyl)-	MeOH	280(4.26),330(4.07), 420(4.54)	44-1577-78
$C_{44}H_{26}O_{12}$ [2,2':7',2":7",2"'-Quaternaphthalenyl]-1,1",1"',4,4",4"',5',8'-octone, 1',8,8",8"'-tetrahydroxy-3',6,6",6"'-tetramethyl-	EtOH	219(4.83),245(4.54), 435(4.13)	88-3889-78
$C_{44}H_{27}N_3O_2$ 7H-Isoindolo[5,6-g]phthalazine-7,9(8H)-dione, 1,4,6,8,10-pentaphenyl-	CH_2Cl_2	260(4.66),295(4.59), 383(4.03),399(4.03)	4-0793-78
$C_{44}H_{28}N_2SSe$ 21-Selena-23-thiaporphyrin, tetraphenyl-	n.s.g.	441(5.37),520(4.47), 551s(3.70),630(2.41), 692(3.63)	88-1885-78
$C_{44}H_{28}N_2STe$ 21-Tellura-23-thiaporphyrin, tetraphenyl-	n.s.g.	445(4.96),508s(4.10), 543s(3.87),625(3.52), 668(3.45)	88-1885-78
$C_{44}H_{28}N_2Se_2$ 21,23-Diselenaporphyrin, tetraphenyl-	n.s.g.	447(5.27),528(4.41), 632(3.36),694(3.50)	88-1885-78
$C_{44}H_{28}N_4$ Tetrabenzo[f,1,v,b$_1$][1,2,17,18]tetraazacyclodotriacontine, 5,6,7,8,25,26-27,28-octadehydro-, (?,?,?,?,E,E,E,E)-	THF	287s(4.78),303s(4.82), 333(4.92)	18-2112-78
$C_{44}H_{29}N_3O_2S$ 6,10-Epithio-6H-isoindolo[5,6-g]phthalazine-7,9(6aH,8H)-dione, 9a,10-dihydro-1,4,6,8,10-pentaphenyl-	CH_2Cl_2 CH_2Cl_2-HCl	278(4.30),300s(4.20) 300s(4.30)	4-0793-78 4-0793-78
5,9[3',4']-Thiopheno-5H-pyrrolo[3,4-g]phthalazine-6,8(5aH,7H)-dione, 8a,9-dihydro-1,4,7,11,13-pentaphenyl-, (5α,5aα,8aα,9α)-	CH_2Cl_2 CH_2Cl_2-HCl	262(4.50),300s(4.30) 280(4.50)	4-0793-78 4-0793-78
(5α,5aβ,8aβ,9α)-	CH_2Cl_2 CH_2Cl_2-HCl	260(4.50),300s(4.20) 280(4.50)	4-0793-78 4-0793-78
$C_{44}H_{29}N_3O_3$ 6,10-Epoxy-6H-isoindolo[5,6-g]phthalazine-7,9(6aH,8H)-dione, 9a,10-dihydro-1,4,6,8,10-pentaphenyl-	CH_2Cl_2	298(4.09)	4-0793-78

Compound	Solvent	$\lambda_{max}(\log \epsilon)$	Ref.
$C_{44}H_{29}N_5O_2$ 21H,23H-Porphine, 2-nitro-5,10,15,20-tetraphenyl-	$CHCl_3$	425(5.24),534(4.12), 559(3.53),600(3.49), 661(3.84)	39-0768-78C
$C_{44}H_{30}$ 7H-Dibenzo[a,c]cyclononene, 7-(5H-dibenzo[a,d]cyclohepten-5-ylidene)-5,9-diphenyl-	C_6H_{12}	225(4.62),252(4.34), 289(4.36),338(4.27), 343s(4.24),350(4.13)	150-5151-78
	CH_2Cl_2	245(4.36),290(4.43), 336(4.34),345s(4.27), 352s(4.18)	150-5151-78
$C_{44}H_{31}N_3O_2$ 6H-Isoindolo[5,6-g]phthalazine-7,9-(6aH,8H)-dione, 9a,10-dihydro-1,4,6,8,10-pentaphenyl-	CH_2Cl_2	234(4.77),301(4.12)	4-0793-78
$C_{44}H_{32}$ Fluoranthene, 7,8,10-triphenyl-9-(5,6,7,8-tetrahydro-2-naphthalenyl)-	C_6H_{12}	298(4.65),330(4.00)	2-0152-78
$C_{44}H_{32}N_4O$ 21H,23H-Porphin-2-ol, 2,3-dihydro-5,10,15,20-tetraphenyl-	benzene	419(5.33),517(4.26), 542(4.22),592(3.89), 645(4.59)	35-6228-78
$C_{44}H_{32}N_4O_2$ 21H-Bilin-1(19H)-one, 23,24-dihydro-19-(hydroxyphenylmethylene)-5,10,15-triphenyl-	$CHCl_3$	345(4.57),565(4.37), 587s(4.33)	39-0768-78C
	$CHCl_3$-TFA	375(4.44),431(4.23), 585(4.44)	39-0768-78C
21H,23H-Porphine-5,15-diol, 5,15-dihydro-5,10,15,20-tetraphenyl-	$CHCl_3$	326(4.28),427(4.69), 491(3.15)	39-0768-78C
	$CHCl_3$-TFA	387(4.57),461(5.23), 516(3.82),557(3.84)	39-0768-78C
$C_{44}H_{32}N_4O_8S$ 2(1H)-Pteridinethione, 6,7-diphenyl-4-[(2,3,5-tri-O-benzoyl-β-D-ribofuranosyl)oxy]-	MeOH	228(4.75),262s(4.26), 282s(4.32),302(4.38), 373(4.24)	24-2571-78
4(1H)-Pteridinone, 2,3-dihydro-6,7-diphenyl-2-thioxo-3-(2,3,5-tri-O-benzoyl-β-D-ribofuranosyl)-	MeOH	229(4.78),275(4.21), 314(4.47),383(4.18)	24-2571-78
4(1H)-Pteridinone, 6,7-diphenyl-2-[(2,3,5-tri-O-benzoyl-β-D-ribofuranosyl)thio]-	MeOH	228(4.78),281(4.40), 366(4.16)	24-2571-78
$C_{44}H_{32}N_4O_9$ 2,4(1H,3H)-Pteridinedione, 6,7-diphenyl-3-(2,3,5-tri-O-benzoyl-β-D-ribofuranosyl)-	MeOH	229(4.83),273(4.30), 364(4.21)	24-2586-78
$C_{44}H_{33}O_3$ Pyrylium, 4-[3-(2,6-diphenyl-4H-pyran-4-ylidene)-2-(4-methoxyphenyl)-1-propenyl]-2,6-diphenyl-, perchlorate	CH_2Cl_2	635(4.75),693(5.48)	4-0365-78
$C_{44}H_{40}N_2O_{13}$ Isochelocardin, hydrochloride	MeOH	226(4.56),273(4.70), 307(4.31),438(3.98)	44-2855-78

Compound	Solvent	$\lambda_{max}(\log \epsilon)$	Ref.
$C_{44}H_{41}Cl$ Tetracyclo[6.2.1.0^{1,8}.0^{3,6}]undeca-2,4,6,9-tetraene, 11-chloro-2,7-bis(1,1-dimethylethyl)-11-methyl-4,5,9,10-tetraphenyl-	CHCl$_3$	251(4.38),323s(3.81)	138-0657-78
$C_{44}H_{41}ClO$ Tetracyclo[6.2.1.0^{1,8}.0^{3,6}]undeca-2,4,6,9-tetraene, 11-chloro-2,7-bis(1,1-dimethylethyl)-11-methoxy-4,5,9,10-tetraphenyl-	CHCl$_3$	260(4.50),319s(4.03)	138-0657-78
$C_{44}H_{48}F_{12}N_{12}O_8$ Acetamide, N,N',N'',N'''-[(4a,5,10a,11-16a,17,22a,23-octahydro-6,12,18,24-tetraoxotetrapyrido[4,3-b'4',3'-f-4'',3''-j:4''',3'''-n][1,5,9,13]tetra-azacyclohexadecine-2,8,14,20(6H,12H-18H,24H)-tetrayl)tetra-3,1-propane-diyl]tetrakis[2,2,2-trifluoro-	CHCl$_3$	335(4.1)	24-2594-78
$C_{44}H_{48}N_6OZn$ Zinc, [4-cyano-N-(2,3,7,8,12,13,17,18-octaethyl-21H,23H-porphin-5-yl)benz-amidato(2-)-N^{21},N^{22},N^{23},N^{24}]-, (SP-4-2)-	CH$_2$Cl$_2$	412(5.54),504(3.34), 540(4.28),578(4.21)	18-1444-78
$C_{44}H_{50}ClFeN_6O_6S_2$ Heminbis-L-methionine	7:3 CHCl$_3$-EtOH	405(5.06),504(4.01), 540(3.97),635(3.74)	104-0786-78
$C_{44}H_{50}ClN_4Rh$ Rhodium, (2-chloro-2-phenylethenyl)-[2,3,7,8,12,13,17,18-octaethyl-21H,23H-porphinato(2-)-N^{21},N^{22},N^{23}-N^{24}]-, [SP-5-31-(Z)]-	benzene	394(5.05),513(4.06), 546(4.67)	101-0329-78P
$C_{44}H_{50}N_4Ni$ Nickel, [2,3,7,8,12,13,17,18-octaethyl-5-(2-phenylethenyl)-21H,23H-porphin-ato(2-)-N^{21},N^{22},N^{23},N^{24}]-, (SP-4-2)-	CHCl$_3$	335(4.22),407(5.19), 532(4.07),568(4.22)	39-1660-78C
$C_{44}H_{50}N_8$ Quadrigemine A	EtOH	244(4.30),304(4.04)	39-1671-78C
	EtOH-HCl	234(4.29),294(3.96)	39-1671-78C
Quadrigemine B	EtOH	245(4.30),304(4.00)	39-1671-78C
	EtOH-HCl	236(4.29),295(3.92)	39-1671-78C
$C_{44}H_{51}N_4ORh$ Rhodium, [2,3,7,8,12,13,17,18-octaeth-yl-21H,23H-porphinato(2-)-N^{21},N^{22}-N^{23},N^{24}](phenylacetyl)-, (SP-5-31)-	benzene	397(5.21),512(4.15), 544(4.72)	101-0329-78P
$C_{44}H_{51}N_5O_2Zn$ Zinc, [4-methoxy-N-(2,3,7,8,12,13,17-18-octaethyl-21H,23H-porphin-5-yl)-benzamidato(2-)-N^{21},N^{22},N^{23},N^{24}]-, (SP-4-2)-	CH$_2$Cl$_2$	412(5.66),504(3.38), 540(4.23),578(4.12)	18-1444-78

Compound	Solvent	$\lambda_{max}(\log \epsilon)$	Ref.
$C_{44}H_{52}N_4O_5$ Vincaleukoblastine, 4-de(acetyloxy)- 3,3',4,4'-tetrahydro-3,4'-dideoxy-	EtOH	263(4.40),287(4.27), 295(4.24),311(3.98)	23-2560-78
$C_{44}H_{53}N_5O_4$ Vincaleukoblastine, 4-de(acetyloxy)- 3,3',4,4'-tetrahydro-3-de(methoxy- carbonyl)-3,4'-dideoxy-3-[(methyl- amino)carbonyl]-	EtOH	267(4.38),287(4.26), 296(4.17),312(3.96)	23-2560-78
$C_{44}H_{54}N_4O_5$ Vincaleukoblastine, 4-de(acetyloxy)- 3,3',4,4'-tetrahydro-3,4'-dideoxy- 6,7-dihydro-	EtOH	268(4.31),284(4.21), 292(4.14),313(3.87)	23-2560-78
$C_{44}H_{54}N_4O_6$ Vincaleukoblastine, 4-de(acetyloxy)- 3',4'-didehydro-4'-deoxy-, (3ξ)-	EtOH	272(4.27),286(4.18), 295(4.07),317(3.80)	23-2560-78
$C_{44}H_{55}N_5O_4$ Vincaleukoblastine, 4-de(acetyloxy)- 3,3',4,4'-tetrahydro-3-de(methoxy- carbonyl)-3,4'-dideoxy-6,7-dihydro- 3-[(methylamino)carbonyl]-	EtOH	261(4.25),287(4.11), 296(4.07),312(3.93)	23-2560-78
$C_{44}H_{56}O_4$ Pentacyclo[19.3.1.1^{3,7}.1^{9,13}.1^{15,19}]- octacosa-1(25),3,5,7(28),9,11,13(27), 15,17,19(26),21,23-dodecaene-25,26- 27,28-tetrol, 5,11,17,23-tetrakis- (1,1-dimethylethyl)-	dioxan	279(4.06),287s(3.95)	49-0767-78
$C_{44}H_{57}IrN_4$ Iridium, [(1,5,6-η)-1,5-octadienyl]- [2,3,7,8,12,13,17,18-octaethyl- 21H,23H-porphinato(2-)-N^{21},N^{22},N^{23}- N^{24}]-	CHCl$_3$	383(5.10),388(5.17), 498(4.10),529(4.55)	101-0317-78P
$C_{44}H_{60}N_{12}O_{12}$ Carbamic acid, [(4a,5,10a,11,16a,17- 22a,23-octahydro-6,12,18,24-tetraoxo- tetrapyrido[4,3-b:4',3'-f:4",3"-j- 4"',3"'-n][1,5,9,13]tetraazacyclo- hexadecine-2,8,14,20(6H,12H,18H,24H)- tetrayl)tetra-2,1-ethanediyl]tetra- kis-, tetraethyl ester	CHCl$_3$	334(4.5)	24-2594-78
$C_{44}H_{64}P_2Pt$ Platinum, bis[[(4-ethynylphenyl)ethyn- yl]bis(tributylphosphine)-, cis	CH$_2$Cl$_2$	287(4.70),308(4.75)	101-0101-78B
trans	CH$_2$Cl$_2$	279s(3.38),297(4.59), 338(4.73)	101-0101-78B
$C_{45}H_{34}N_4O_2$ 21H,23H-Porphin-5-ol, 5,15-dihydro- 15-methoxy-5,10,15,20-tetraphenyl-	CHCl$_3$	325(4.28),428(4.69), 490(3.20)	39-0768-78C
$C_{45}H_{37}N_3O_3S$ 4-Thiazoleacetic acid, α-[(triphenyl- methoxy)imino]-2-[(triphenylmethyl)- amino]-, ethyl ester, (E)-	EtOH	238s(4.42),307(3.48)	78-2233-78

Compound	Solvent	$\lambda_{max}(\log \epsilon)$	Ref.
(Z)- (cont.)	EtOH	241s(4.36),298(3.76)	78-2233-78
$C_{45}H_{40}ClN_4O_{12}P$ 3'-Uridylic acid, 5'-O-[(4-methoxyphenyl)diphenylmethyl]-2'-O-[(2-nitrophenyl)methyl]-, 4-chlorophenyl 2-cyanoethyl ester	MeOH	233(4.27),260(4.15)	35-8210-78
$C_{45}H_{41}ClO_2$ Tetracyclo[6.2.1.0^{1,8}.0^{3,6}]undeca-2,4,6,9-tetraen-11-ol, 11-chloro-2,7-bis(1,1-dimethylethyl)-4,5,9,10-tetraphenyl-, acetate	CHCl$_3$	252(4.54),325s(3.96)	138-0657-78
$C_{45}H_{42}N_8O_{12}$ 9H-Purin-6-amine, 9-[2,3,4,6-tetrakis-O-(phenylmethyl)-β-D-glucopyranosyl]-, picrate	MeOH	210(4.73),254(4.31)	136-0301-78C
$C_{45}H_{44}O_2$ Tetracyclo[6.2.1.0^{1,8}.0^{3,6}]undeca-2,4,6,9-tetraene, 2,7-bis(1,1-dimethylethyl)-11,11-dimethoxy-4,5,9,10-tetraphenyl-	CHCl$_3$	248(4.53),321s(3.96)	138-0657-78
$C_{45}H_{51}N_5O_{14}$ 21H,23H-5-Azaporphine-2,8,12,17-tetrapropanoic acid, 3,7,18-tris(2-methoxy-2-oxoethyl)-13-methyl-, tetramethyl ester	CH$_2$Cl$_2$	386(5.06),504(3.81), 540(4.24),563(4.11), 613(4.21)	39-0871-78C
$C_{45}H_{52}I_2N_4O_{14}$ 21H-Biline-3,8,13,17-tetrapropanoic acid, 10,23-dihydro-1,19-diiodo-2,7,18-tris(2-methoxy-2-oxoethyl)-12-methyl-, tetramethyl ester, dihydrobromide	CH$_2$Cl$_2$	468(4.41),548(4.96)	39-0871-78C
$C_{45}H_{52}N_4O_{10}$ Vincaleukoblastine, 1-demethyl-4'-deoxy-3',4'-epoxy-19'-oxo-, (3'α,4'α)-	EtOH	212(4.70),224s(4.58), 252(4.10),279(4.09), 286(4.11),294(4.07), 314s(3.75)	23-2560-78 +142-0201-78A
$C_{45}H_{52}N_4O_{14}$ 3,7,12,17-Corrintetrapropanoic acid, 1,2,3,7,8,12,13,17,18,19-decadehydro-21,23-dihydro-2,13,18-tris(2-methoxy-2-oxoethyl)-8-methyl-, tetramethyl ester	CH$_2$Cl$_2$	396(5.08),410(4.98), 544(4.32),555(4.32), 595(4.32)	39-0871-78C
$C_{45}H_{56}N_4O_{14}$ 3,7,12,17-Corrintetrapropanoic acid, 1,2,3,7,8,12,13,17,18,19-decadehydro-5,10,15,21,22,23-hexahydro-2,13,18-tris(2-methoxy-2-oxoethyl)-8-methyl-, tetramethyl ester	MeOH	275(3.92)	39-0871-78C
$C_{45}H_{56}O_{19}$ 6-Dehydrochukrasin A	EtOH	202(3.99),268(3.85)	33-1814-78

Compound	Solvent	$\lambda_{max}(\log \epsilon)$	Ref.

$C_{45}H_{58}O_6$
  Chola-20(22),23-diene-3,7,11-triol,
  4,4,14-trimethyl-24,24-diphenyl-,
  triacetate, (3β,5α,7β,11α,20E)-
	CH_2Cl_2	310(4.43)	5-0419-78

$C_{45}H_{58}O_{18}$ Chukrasin C	EtOH	207(3.83),268(4.01)	33-1814-78
$C_{45}H_{58}O_{19}$ Chukrasin A	EtOH	208(3.71),268(3.91)	33-1814-78

$C_{45}H_{60}O_6$
  Chol-23-ene-3β,7β,11α-triol, 4,4,14-
  trimethyl-24,24-diphenyl-, triacetate
	CH_2Cl_2	254(4.22)	5-0419-78

$C_{45}H_{60}O_{17}$ 3-Deacetylchukrasin B	EtOH	208(3.83),268(3.94)	33-1814-78

$C_{45}H_{66}N_4$
  Cyclopropenylium, 1,1'-(2,3-diphenyl-
  2-cyclopropen-1-ylidene)bis[2,3-bis-
  [bis(1-methylethyl)amino]-, diper-
  chlorate
	CH_2Cl_2	330(4.22)	89-0446-78

$C_{46}H_{29}F_3N_4O_3$
  21H-Bilin-1(4H)-one, 19-benzoyl-
  5,10,15-triphenyl-4-(trifluoroacetyl)-
	$CHCl_3$
	
	$CHCl_3$-TFA
		337(4.37),396(4.30),
		547(4.43),586(4.48)
		353(4.38),573(4.37)
			39-0768-78C
			
			39-0768-78C

$C_{46}H_{30}N_2O_3$
  5,6-Diazaspiro[3.4]octa-5,7-dien-1-one,
  7,8-dibenzoyl-2,3-bis(diphenylmethyl-
  ene)-
	$CHCl_3$	320(4.42),432(3.70)	18-0649-78

$C_{46}H_{30}O_3$
  1H-Benzo[a]cyclobuta[d]cyclohepten-
  1-one, 8,9-dibenzoyl-2-(diphenyl-
  methylene)-2,3-dihydro-3-phenyl-
	$CHCl_3$	258(4.48),330(4.33)	18-0649-78

$C_{46}H_{33}N_3O_3$
  Spiro[cyclopropane-1,10'-[5,8]methano-
  [1H][1,2,4]triazolo[1,2-a]pyridaz-
  ine]-1',3'(2'H)-dione, 2-benzoyl-
  5',8'-dihydro-2',5',6',7',8'-
  pentaphenyl-
	MeCN	247(4.43),280s(4.04)	5-1648-78

$C_{46}H_{34}N_2$
  Phthalazine, 1,4,5,8-tetraphenyl-6,7-
  bis(phenylmethyl)-
	C_6H_{12}	248(4.7),313(4.2)	4-1185-78

$C_{46}H_{34}O$
  Isobenzofuran, 1,3,4,7-tetraphenyl-
  5,6-bis(phenylmethyl)-
	CH_2Cl_2	330(3.90),388(4.25)	4-1185-78

$C_{46}H_{34}O_2$
  p-Terphenyl, 2',3'-dibenzoyl-5',6'-bis-
  (phenylmethyl)-
	EtOH	250s(4.60)	4-1185-78

$C_{46}H_{34}O_4$ 1,2,3,4,5,7,9,15,17-Cyclooctadecanona-	THF	282(4.50),305(4.48),	18-3351-78

Compound	Solvent	$\lambda_{max}(\log \epsilon)$	Ref.
ene-11,13-diyne, 1,6,10,15-tetrakis-(4-methoxyphenyl)- (cont.)		329(4.35),358(4.37), 381(4.56),453(5.46), 670(5.06),784(3.59)	18-3351-78
$C_{46}H_{36}N_4O_{10}$			
2,4(3H,8H)-Pteridinedione, 6,7-diphenyl-8-[2-[(2,3,5-tri-O-benzoyl-β-D-ribofuranosyl)oxy]-	MeOH	229(4.73),273(4.29), 282(4.28),290s(4.21), 380s(3.68),430(4.02)	24-2586-78
2,4(3H,8H)-Pteridinedione, 8-(2-hydroxyethyl)-6,7-diphenyl-3-(2,3,5-tri-O-benzoyl-β-D-ribofuranosyl)-	MeOH	229(4.69),273(4.20), 282(4.18),290s(4.16), 380s(3.83),430(3.80)	24-2586-78
$C_{46}H_{36}O$			
Isobenzofuran, 4,7-dihydro-1,3,4,7-tetraphenyl-5,6-bis(phenylmethyl)-	CH_2Cl_2	324(4.46)	4-1185-78
$C_{46}H_{42}BN$			
Methylium, [4-(dimethylamino)phenyl]-(4-methylphenyl)phenyl-, tetraphenylborate	MeCN	470(4.66)	104-1750-78
$C_{46}H_{43}N_5O_6$			
Benzamide, N-[9-[2,3,4,6-tetrakis-O-(phenylmethyl)-α-D-glucopyranosyl]-9H-purin-6-yl]-	MeOH	210(4.74),280(4.31)	136-0301-78C
β-	MeOH	209(4.74),280(4.34)	136-0301-78C
$C_{46}H_{52}N_4NiO_{14}$			
Nickel, [tetramethyl 1,2,3,7,8,12,13-17,18,19-decadehydro-1,22-dihydro-2,7,18-tris(2-methoxy-2-oxoethyl)-1,12-dimethyl-3,8,13,17-corrintetrapropanoato(2-)-$N^{21},N^{22},N^{23},N^{24}$]-, (SP-4-2)-	CH_2Cl_2	325(4.23),468(4.50), 804(3.99)	39-0871-78C
$C_{46}H_{52}N_4O_{11}$			
Vincaleukoblastine, 4'-deoxy-3',4'-epoxy-19',22-dioxo-, (3'α,4'α)-	EtOH	217(4.78),258(4.24), 280(4.17),288(4.23), 296(4.27)	23-2560-78
$C_{46}H_{54}N_4O_9$			
Vincaleukoblastine, 3',4'-didehydro-4'-deoxy-19'-oxo-	EtOH	213(5.17),262(4.63), 283(4.53),293(4.47)	23-0062-78
Vincaleukoblastine, 3',4'-didehydro-4'-deoxy-22-oxo-	EtOH	218(4.69),255(4.20), 296(4.20)	23-2560-78
$C_{46}H_{54}N_4O_{10}$			
Leurosine, 5'-oxo-	EtOH	215(4.70),263(4.15), 283(4.06),290(4.04), 307(3.91)	23-0062-78
Leurosine, 19'-oxo-	EtOH	214(4.66),262(4.13), 284(4.05),294(4.00), 309s(3.74)	23-0062-78
Vincaleukoblastine, 4'-deoxy-3',4'-epoxy-22-oxo-	EtOH	219(4.68),253(4.20), 262(4.15),289(4.18), 296(4.22)	23-2560-78
$C_{46}H_{55}N_5O_8$			
Vincristine, 3',4'-dehydro-	n.s.g.	218(4.69),255(4.20), 296(4.20)	142-0201-78A

Compound	Solvent	λ_{max}(log ϵ)	Ref.
$C_{46}H_{56}N_4O_9$			
Vinblastine, 3',4'-dehydro-, N-oxide	EtOH	213(4.98),265(4.46), 283(4.33),292(4.24), 306(3.99)	23-0062-78
Vincaleukoblastine, 4'-deoxy-22-oxo-	EtOH	218(4.59),248(4.05), 294(4.07)	23-2560-78
isomer	EtOH	218(4.70),250(4.21), 292(4.12)	23-2560-78
$C_{46}H_{56}N_4O_{10}$			
Catharinine	EtOH	220(4.69),268(4.15), 288(4.10),296(4.04)	78-0677-78
Leurosidine, 19'-oxo-	EtOH	214(4.66),262(4.11), 285(4.04),294(4.00), 309s(3.74)	23-0062-78
Vincaleukoblastine, 19'-oxo-	EtOH	213(4.69),259(4.14), 285(4.02),294(3.99), 311s(3.74)	23-0062-78
$C_{46}H_{56}N_4O_{11}$			
Vincaleukoblastine, 3'α-hydroxy-19'-oxo-	EtOH	213(5.01),261(4.45), 283(4.39),293(4.33)	23-0062-78
$C_{46}H_{58}N_4O_8$			
Vincaleukoblastine, 3',4'-didehydro- 4'-deoxy-6,7-dihydro-	EtOH	213(4.70),255(4.15), 287(4.10),295(4.08), 307(3.80)	23-2560-78
$C_{46}H_{58}N_4O_9$			
Leurosine, 6,7-dihydro-	EtOH	213(4.68),260(4.12), 284(4.03),294(4.00), 310(3.75)	23-2560-78
$C_{46}H_{58}N_4O_{10}$			
Vincaleukoblastine, 3'α-hydroxy-	EtOH	212(5.00),258(4.49), 285(4.40),294(4.37)	23-0062-78
$C_{46}H_{58}N_8O_{16}$			
L-Proline, 1-[N-[N-[[2-amino-9-[[[2-hy- hydroxy-1-[[[1-[[2-(methoxycarbonyl)- 4-oxo-1-pyrrolidinyl]carbonyl]-2-meth- ylpropyl]amino]carbonyl]propyl]amino]- carbonyl]-4,6-dimethyl-3-oxo-3H-phen- oxazin-1-yl]carbonyl]-L-threonyl]-D- valyl]-4-oxo-, methyl ester	MeOH	240(4.578),443(4.410)	87-0607-78
$C_{46}H_{62}N_8O_{14}$			
Actinomycindioic D acid, 4^A-de(N-meth- ylglycine)-4^B-de(N-methylglycine)-5^A- de(N-methyl-L-valine)-5^B-de(N-methyl- L-valine)-, dimethyl ester	MeOH	240(4.540),443(4.357)	87-0607-78
$C_{47}H_{29}CoN_6$			
Cobalt, (dicyanomethyl)[5,10,15,20- tetraphenyl-21H,23H-porphinato(2-)- $N^{21},N^{22},N^{23},N^{24}$]-, (SP-5-31)-	benzene	412(5.03),552(4.20), 550(3.92),620(3.12)	101-0109-78N
$C_{47}H_{33}CoN_4O$			
Cobalt, (2-oxopropyl)[5,10,15,20-tetra- phenyl-21H,23H-porphinato(2-)-N^{21},N^{22}- N^{23},N^{24}]-, (SP-5-31)-	benzene	412(5.15),527(4.12)	101-0109-78N

Compound	Solvent	$\lambda_{max}(\log \epsilon)$	Ref.
$C_{47}H_{33}N_5NiO_2$ Nickel, [ethyl 2,7,12,18-tetraphenyl-17,22,23,24,25-pentaazapentacyclo-$[17.2.1.1^3,^6.1^8,^{11}.1^{13},^{16}]$pentacosa-1,3(25),4,6,8,10,12,14,16(23),18,20-undecaene-17-carboxylato(2-)-$N^{22},N^{23},N^{24},N^{25}]-$, (SP-4-2)-	CH_2Cl_2	451(4.82),587(3.72), 705(4.13)	35-4733-78
$C_{47}H_{35}N_5O_2$ Carbamic acid, (5,10,15,20-tetraphenyl-21H,23H-porphin-2-yl)-, ethyl ester	CH_2Cl_2	419(4.40),515(4.36), 548(3.79),590(3.84), 647(3.53)	35-4733-78
17,22,23,24,25-Pentaazapentacyclo[17.2-$1.1^3,^6.1^8,^{11}.1^{13},^{16}]$pentacosa-1,3(25)-4,6,8,10,12,14,16(23),18,20-undeca-ene-17-carboxylic acid, 2,7,12,18-tetraphenyl-, ethyl ester	CH_2Cl_2	453(4.91),515(3.78), 530(3.69),600(3.85), 651(3.97),729(4.08)	35-4733-78
$C_{47}H_{36}BrP$ Phosphorane, bromo(5,9-diphenyl-7H-di-benzo[a,c]cyclononen-7-yl)triphenyl-	MeCN	224s(5.13),259(4.47), 266(4.48),272(4.26), 300(4.50),435(3.37)	150-5151-78
$C_{47}H_{42}N_{12}O_{21}$ Adenosine, N,N''-carbonylbis[N-(4-nitro-benzoyl)-, 2',2''',3',3''',5',5'''-hexa-acetate	EtOH	270(4.57)	39-0131-78C
$C_{47}H_{52}N_2OS_4$ Urea, bis[2,3-bis[[4-(1,1-dimethyleth-yl)phenyl]thio]-2-cyclopropen-1-yli-dene]-	MeCN	238(4.01),247(4.27)	18-3653-78
$C_{47}H_{53}N_5O_{16}$ 21H,23H-5-Azaporphine-2,8,12,17-tetra-propanoic acid, 3,7,13,18-tetrakis-(2-methoxy-2-oxoethyl)-, tetramethyl ester	CH_2Cl_2	391(4.92),500(3.87), 538(4.27),564(3.87), 614(4.27)	39-0871-78C
$C_{47}H_{54}I_2N_4O_{16}$ 21H-Biline-3,8,13,17-tetrapropanoic acid, 10,23-dihydro-1,19-diiodo-2,7,12,18-tetrakis(2-methoxy-2-oxo-ethyl)-, tetramethyl ester, dihydro-bromide	CH_2Cl_2	469(4.70),547(4.97)	39-0871-78C
$C_{47}H_{54}N_4O_{16}$ 3,7,12,17-Corrintetrapropanoic acid, 1,2,3,7,8,12,13,17,18,19-decadehydro-21,23-dihydro-2,8,13,18-tetrakis(2-methoxy-2-oxoethyl)-, tetramethyl ester	CH_2Cl_2	399(4.98),412(4.88), 546(4.20),556(4.20), 595(4.20)	39-0871-78C
$C_{47}H_{54}O_{24}$ Acteoside nonaacetate	EtOH	283(4.26)	94-2111-78
$C_{47}H_{55}CoN_4O_{14}$ Cobalt(1+), [tetramethyl octadehydro-2,7,18-tris(2-methoxy-2-oxoethyl)-1,12,19-trimethyl-3,8,13,17-corrin-tetrapropanoato-$N^{21},N^{22},N^{23},N^{24}]-$,	CH_2Cl_2	281(4.33),356(4.28), 500(4.13),580(3.71)	39-0871-78C

Compound	Solvent	$\lambda_{max}(\log \epsilon)$	Ref.
bromide, (SP-4-3)- (cont.)			39-0871-78C
$C_{47}H_{55}N_4NiO_{14}$ Nickel(1+), [tetramethyl octadehydro- 2,7,18-tris(2-methoxy-2-oxoethyl)- 1,12,19-trimethyl-3,8,13,17-corrin- tetrapropanoato-$N^{21},N^{22},N^{23},N^{24}$]-, bromide	CH_2Cl_2	278(4.47),356(4.41), 568(4.11)	39-0871-78C
$C_{47}H_{58}N_4O_{16}$ 3,7,12,17-Corrintetrapropanoic acid, 1,2,3,7,8,12,13,17,18,19-decadehydro- 5,10,15,21,22,23-hexahydro-2,8,13,18- tetrakis(2-methoxy-2-oxoethyl)-, tetramethyl ester	MeOH	275(3.93)	39-0871-78C
$C_{47}H_{60}N_8$ Quadrigemine B Hofmann base reduction product	EtOH EtOH-HCl	240(4.03),296(3.68) 238(3.97),294(3.62)	39-1671-78C 39-1671-78C
$C_{47}H_{60}O_{19}$ Chukrasin D	EtOH	207(3.81),269(3.99)	33-1814-78
$C_{47}H_{62}O_{18}$ Chukrasin B	EtOH	207(3.81),267(4.01)	33-1814-78
$C_{48}H_{28}N_2O_2$ Naphthaceno[2,3-g]phthalazine-8,13-di- one, 1,4,7,14-tetraphenyl-	CH_2Cl_2	253(4.58),293(4.89), 340(4.59),475(4.07), 506(3.74)	4-0793-78
$C_{48}H_{28}O_4$ Pentacene-5,14-dione, 9,10-dibenzoyl- 6,13-diphenyl- 1H-Indene-1,3(2H)-dione, 2,2'-(2,3,5,6- tetraphenyl-2,5-cyclohexadiene-1,4- diylidene)bis-	CH_2Cl_2 MeCN	255(4.71),316(4.85), 425(4.08),445(4.08) 254(4.53),363(3.93), 495(3.33)	4-0793-78 78-1985-78
$C_{48}H_{30}N_6P_2$ 3,5-Octadienedinitrile, 3,4,5,6-tetra- cyano-1,8-bis(triphenylphosphoranyl- idene)-	CH_2Cl_2	485(4.34)	39-1237-78C
$C_{48}H_{30}O_4$ 1H-Indene-1,3(2H)-dione, 2,2'-(4',5'- diphenyl[1,1':2',1"-terphenyl]- 3',6'-diyl)bis-	MeCN	282(4.12)	78-1985-78
$C_{48}H_{30}O_5$ 6,13-Epoxypentacene-5,14-dione, 9,10- dibenzoyl-5a,6,13,13a-tetrahydro- 6,13-diphenyl- 5,12[3',4']Furanonaphthacene-6,11-di- one, 2,3-dibenzoyl-5,5a,11,12-tetra- hydro-14,16-diphenyl-	CH_2Cl_2 CH_2Cl_2	263(4.86),347(3.63) 245(4.76),332(4.56)	4-0793-78 4-0793-78
$C_{48}H_{34}N_4NiO_2$ Nickel, [methyl 19a,19b-dihydro-1-meth- yl-3,8,13,18-tetraphenyl-1H,20H,22H- cyclopropa[b]porphinecarboxylato(2-)- $N^{20},N^{21},N^{22},N^{23}$]-	CH_2Cl_2	412(4.98),505(3.77), 570s(3.85),608(4.28)	78-2295-78

Compound	Solvent	$\lambda_{max}(\log \epsilon)$	Ref.
Nickel, [methyl 19-methyl-2,7,12,17-tetraphenyl-22,23,24,25-tetraaza-pentacyclo[16.3.1.1^{3,6}.1^{8,11}.1^{13,16}]-pentacosa-1,3(25),4,6,8,10,12,14,16-(23),17,20-undecaene-19-carboxylato-(2-)-N^{22},N^{23},N^{24},N^{25}]-	CH$_2$Cl$_2$	437(5.09),525(3.65), 643(4.18)	78-2295-78
$C_{48}H_{35}ClN_4NiO_2$			
Nickel, chloro(methyl α-methyl-5,10,15-20-tetraphenyl-21H,23H-porphine-21-acetato-N^{21},N^{22},N^{23},N^{24})-	CH$_2$Cl$_2$	445(4.93),550(3.36), 625(3.91),680(3.81)	78-2295-78
$C_{48}H_{36}$			
[2$_6$](Orthopara)$_3$cyclophanehexaene, all cis	C$_6$H$_{12}$	229(4.67),312(4.60)	1-0109-78
$C_{48}H_{36}N_4$			
Hexabenzo[b,f,j,p,t,x][1,4,15,18]tetra-azacyclooctacosine	dioxan	287(4.91),377(4.55)	33-2813-78
$C_{48}H_{36}N_4O_8$			
5H-Cyclopenta[d]pyridazine-1,4-dicarb-oxylic acid, 7-(diphenylmethyl)-5-[7-(diphenylmethyl)-1,4-bis(methoxy-carbonyl)-5H-cyclopenta[d]pyridazin-5-ylidene]-, dimethyl ester	MeCN	219(4.79),315(4.00), 460(4.02)	78-2509-78
[5,5'-Bi-2H-cyclopenta[d]pyridazine]-1,1',4,4'-tetracarboxylic acid, 7,7'-bis(diphenylmethyl)-, tetramethyl ester	MeCN	267(4.51),272(4.52), 290(4.50),350(4.02), 498(3.14)	78-2509-78
$C_{48}H_{38}N_4O_{11}$			
2,4(3H,8H)-Pteridinedione, 8-(2-acet-oxyethyl)-6,7-diphenyl-3-(2,3,5-tri-O-benzoyl-β-D-ribofuranosyl)-	MeOH	229(4.71),273(4.29), 282(4.28),290s(4.23), 431(4.04)	24-2586-78
$C_{48}H_{38}N_8Ni_2$			
Nickel, [μ-[5,10,15,20-tetramethyl-2-[(10,15,20-trimethyl-21H,23H-porphin-5-yl)methyl]-21H,23H-porphinato(4-)-N^{21},N^{22},N^{23},N^{24},N$^{21'}$,N$^{22'}$,N$^{23'}$,N$^{24'}$]-di-	CHCl$_3$	302(4.28),333(4.22), 425(5.29),431(5.30), 542(4.37)	5-0238-78
corresponding palladium chelate	CH$_2$Cl$_2$	423s(5.29),431(5.32), 497(3.08),537(4.38), 568s(3.71)	5-0238-78
$C_{48}H_{39}N_4O_3P$			
Phosphonic acid, (17-methyl-2,7,12,18-tetraphenyl-22,23,24,25-tetraazapen-tacyclo[17.2.1.1^{3,6}.1^{8,11}.1^{13,16}]pen-tacosa-1(22),2,4,6,8(24),9,11,13,15-18,20-undecaen-17-yl)-, dimethyl ester	CH$_2$Cl$_2$	442(4.80),605s(3.95), 635(3.97),700(4.09)	78-2295-78
nickel chelate	CH$_2$Cl$_2$	445(4.85),585(3.76), 685(4.06)	78-2295-78
$C_{48}H_{44}$			
Dibenzosemibullvalene, 2,3,4,5,9,10,11-12-octamethyl-1,6,8,13-tetraphenyl-	CHCl$_3$	300s(3.38),320s(2.97)	5-1675-78

Compound	Solvent	$\lambda_{max}(\log \epsilon)$	Ref.
$C_{48}H_{46}N_4O_8S_4$ Quinoxalino[2',3':4,5]cyclopenta[1,2-b][1,5]benzodiazepine, 5,5a,5b,6,11-11a,13,14-octahydro-2,3,8,9,13,13-hexamethyl-5,6,11,14-tetrakis(phenylsulfonyl)-, (5aα,5bβ,11aβ)-	CH_2Cl_2	231(4.57),267s(4.03), 273s(3.92)	5-1146-78
$C_{48}H_{54}N_4NiO_{16}$ Nickel, [tetramethyl 1,2,3,7,8,12,13-17,18,19-decadehydro-1,22-dihydro-2,7,12,18-tetrakis(2-methoxy-2-oxoethyl)-1-methyl-3,8,13,17-corrintetrapropanoato(2-)-$N^{21},N^{22},N^{\angle 3},N^{24}$]-, (SP-4-2)-	MeOH	320(4.52),796(3.96)	39-0871-78C
$C_{48}H_{56}$ Bicyclo[14.14.2]dotriaconta-1,3,5,7,8-9,11,13,15,17,19,21,25,27,29-pentadecaene-23,31-diyne, 7,10,22,25-tetrakis(1,1-dimethylethyl)-	THF	267(4.37),292s(4.52), 303(4.60),326(4.02), 368s(4.39),427s(5.01), 446(5.31),466(5.59), 567s(4.05),604(4.33), 653(4.33),740(2.95), 825(2.90),945(2.69)	88-1483-78
$C_{48}H_{58}N_8$ Quadrigemine B Hofmann base	EtOH EtOH-HCl EtOH-NaOH	283(3.89) 230(4.34),292(3.90) 260(4.21)	39-1671-78C 39-1671-78C 39-1671-78C
$C_{48}H_{68}N_{12}O_{12}$ Carbamic acid, [(4a,5,10a,11,16a,17-22a,23-octahydro-6,12,18,24-tetraoxotetrapyrido[4,3-b:4',3'f:4",3"-j-4'",3'"-n][1,5,9,13]tetraazacyclohexadecine-2,8,14,20(6H,12H,18H,24H)-tetrayl)tetra-3,1-propanediyl]tetrakis-, tetraethyl ester	CH_2Cl_2	336(4.2)	24-2594-78
$C_{49}H_{32}O_5$ 1H-Indene-1,3(2H)-dione, 2-[6'-(2,3-dihydro-1,3-dioxo-1H-inden-2-yl)-4',5'-diphenyl[1,1':2',1"-terphenyl]-3'-yl]-2-methoxy-	MeCN	228(4.98),282(4.14)	78-1985-78
1H-Indene-1,3(2H)-dione, 2-[4-(1,3-dihydro-1,3-dioxo-2H-inden-2-ylidene)-1-methoxy-2,3,5,6-tetraphenyl-2,5-cyclohexadien-1-ylidene]-	MeCN	225(4.88),260(4.56), 282(4.58),450(3.29)	78-1985-78
1H-Indene-1,3(2H)-dione, 2,2'-(5-methoxy-2,3,5,6-tetraphenyl-2-cyclohexene-1,4-diylidene)bis-	MeCN	254(4.53),360(3.92), 490(3.33)	78-1985-78
$C_{49}H_{36}N_4NiO$ Nickel, [ethyl 19-methyl-2,7,12,17-tetraphenyl-22,23,24,25-tetraazapentacyclo[16.3.1.1^{3,6}.1^{8,11}.1^{13,16}]-pentacosa-1,3(25),4,6,8,10,12,14,16-(23),17,20-undecaene-19-carboxylato-(2-)-$N^{22},N^{23},N^{24},N^{25}$]-	CH_2Cl_2	437(5.00),525(3.79), 612(4.05),645(4.13)	78-2295-78

Compound	Solvent	λ_{max}(log ϵ)	Ref.
$C_{49}H_{37}ClN_4NiO_2$			
Nickel, chloro(ethyl α-methyl-5,10,15-20-tetraphenyl-21H,23H-porphine-21-acetato-$N^{21},N^{22},N^{23},N^{24}$)-	CH_2Cl_2	455(5.01),527(3.86),560(3.95),630(3.90),685(3.76)	78-2295-78
$C_{49}H_{38}N_4O_2$			
22,23,24,25-Tetraazapentacyclo[16.3.1-$1^3,^6.1^8,^{11}.1^{13},^{16}$]pentacosa-1,3(25)-4,6,8,10,12,14,16(23),17,20-undeca-ene-19-carboxylic acid, 19-methyl-2,7,12,17-tetraphenyl-, ethyl ester	CH_2Cl_2	430(5.27),535(4.07),577(3.95),627(3.78),687(3.88)	78-2295-78
22,23,24,25-Tetraazapentacyclo[17.2.1-$1^3,^6.1^8,^{11}.1^{13},^{16}$]pentacosa-1(22),2-4,6,8(24),9,11,13,15,18,20-undecaene-17-carboxylic acid, 17-methyl-2,7,12-18-tetraphenyl-, ethyl ester	benzene	442(4.75),515(3.48),670(4.16)	78-2295-78
nickel chelate	CH_2Cl_2	447(4.93),580(3.82),667(4.16)	78-2295-78
$C_{49}H_{40}O_5$			
1H-Indene-1,3-diol, 2-[4-(1,3-dihydro-1,3-dihydroxy-2H-inden-2-ylidene)-1-methoxy-2,3,5,6-tetraphenyl-2,5-cy-clohexadien-1-yl]-2,3-dihydro-	MeCN	264(4.23),288(4.11)	78-1985-78
1H-Indene-1,3-diol, 2,2'-(5-methoxy-2,3,5,6-tetraphenyl-2-cyclohexene-1,4-diylidene)bis[2,3-dihydro-	MeCN	258(4.48)	78-1985-78
$C_{49}H_{41}N_5$			
21H,23H-Porphine, 2,3-dihydro-2-(1-methyl-2-pyrrolidinyl)-5,10,15,20-tetraphenyl-	benzene	421(5.31),519(4.22),546(4.09),600(3.80),654(4.61)	35-6228-78
$C_{49}H_{57}N_4NiO_{16}$			
Nickel(1+), [tetramethyl octadehydro-2,7,12,18-tetrakis(2-methoxy-2-oxo-ethyl)-1,19-dimethyl-3,8,13,17-corr-intetrapropanoato-$N^{21},N^{22},N^{23},N^{24}$]-, bromide, (SP-4-3)-	CH_2Cl_2	277(4.38),357(4.34),423(3.99),569(4.01)	39-0871-78C
$C_{49}H_{64}O_{19}$			
Chukrasin E	EtOH	207(3.82),267(3.99)	33-1814-78
$C_{49}H_{67}Br_2NO_{14}$			
Oleandomycin, 4',11-bis(4-bromobenz-oate)	MeOH	243(4.27)	35-6733-78
$C_{50}H_{29}NO_3$			
Spiro[6a,16c-ethenodibenz[4,5:6,7]-as-indaceno[2,1-c]isoquinoline-7(16H)-2'-[2H]indene]-1',3',16-trione, 5-methyl-17,18-diphenyl-	MeCN	230(4.99),262(4.66),378(2.13)	78-1985-78
$C_{50}H_{31}NO_4$			
Spiro[6a,16c-ethenodibenz[4,5:6,7]-as-indaceno[2,1-c]isoquinoline-7(16H)-2'-[2H]indene]-1',3',16-trione, 11b,16a-dihydro-11b-hydroxy-5-methyl-17,18-diphenyl-	MeCN	228(4.92),260(4.67),285(4.66)	78-1985-78

Compound	Solvent	$\lambda_{max}(\log \epsilon)$	Ref.
$C_{50}H_{32}N_2O_2$ Benzo[g]phthalazine, 7,8-dibenzoyl- 1,4,6,9-tetraphenyl-	CH_2Cl_2	271(4.78),375s(3.97), 383(3.98)	4-0793-78
$C_{50}H_{32}N_2S$ [2]Benzothieno[5,6-g]phthalazine, 1,4,6,7,9,10-hexaphenyl-	CH_2Cl_2	270(4.46),322(4.91), 394(3.79),630s(3.82), 664(3.92)	4-0793-78
$C_{50}H_{32}N_4$ Phthalazino[6,7-g]phthalazine, 1,4,5,7- 10,12-hexaphenyl-	CH_2Cl_2	293(4.87),404(3.73), 450s(3.99),470(4.03), 490s(3.89)	4-0793-78
	CH_2Cl_2-HCl	290(4.82),408(3.93), 465s(3.83),495(4.02), 524(4.01)	4-0793-78
$C_{50}H_{32}O_2S$ Naphtho[2,3-c]thiophene, 6,7-dibenzoyl- 1,3,4,9-tetraphenyl-	CH_2Cl_2	258(4.72),328(4.61), 426(3.87),534(3.94)	4-0793-78
Naphtho[2,3-c]thiophene, 6,7-dibenzoyl- 1,3,5,8-tetraphenyl-	CH_2Cl_2	245(4.66),263(4.66), 299(4.73),390s(3.77), 544(4.03)	4-0793-78
$C_{50}H_{34}O_3$ Naphtho[2,3-c]furan, 6,7-dibenzoyl-4,9- dihydro-1,3,4,9-tetraphenyl-	CH_2Cl_2	327(4.40)	4-0793-78
Naphtho[2,3-c]furan, 6,7-dibenzoyl-4,9- dihydro-1,3,5,8-tetraphenyl-	CH_2Cl_2	336(4.40)	4-0793-78
$C_{50}H_{34}O_5$ 1H-Indene-1,3(2H)-dione, 2-[6'-(2,3-di- hydro-1,3-dioxo-1H-inden-2-yl)-4',5'- diphenyl[1,1':2',1"-terphenyl]-3'- yl]-2-ethoxy-	MeCN	228(4.98),282(4.14)	78-1985-78
1H-Indene-1,3(2H)-dione, 2-[4-(2,3-di- hydro-1,3-dioxo-1H-inden-2-yl)-4-eth- oxy-2,3,5,6-tetraphenyl-2,5-cyclohex- adien-1-ylidene]-	MeCN	226(4.86),264(4.58), 284(4.59),450(3.21)	78-1985-78
1H-Indene-1,3(2H)-dione, 2,2'-(5-eth- oxy-2,3,5,6-tetraphenyl-2-cyclohex- ene-1,4-diylidene)bis-	MeCN	254(4.64),362(3.98), 490(3.34)	78-1985-78
$C_{50}H_{38}N_4O_4$ 22,23,24,25-Tetraazapentacyclo[17.2.1- $1^{3,6}.1^{8,11}.1^{13,16}$]pentacosa-1(22),2- 4,6,8(24),9,11,13,15,18,20-undecaene- 17-acetic acid, 17-(methoxycarbonyl)- 2,7,12,18-tetraphenyl-, methyl ester nickel chelate	CH_2Cl_2	447(4.84),515(3.62), 555s(3.77),672(4.27)	78-2295-78
	CH_2Cl_2	447(4.93),580(3.87), 672(4.17)	78-2295-78
$C_{50}H_{45}N_5$ 21H,23H-Porphine-2-methanamine, N,N-di- ethyl-2,3-dihydro-α-methyl-5,10,15,20- tetraphenyl-	benzene	421(5.29),519(4.23), 547(4.08),600(3.81), 655(4.60)	35-6228-78
$C_{50}H_{52}N_4O_5V$ Vanadium, [2,3,7,8,12,13,17,18-octaeth- yl-21H,23H-porphine-5,15-diyl dibenz- oato(2-)-$N^{21},N^{22},N^{23},N^{24}$]oxo-, (SP-5-12)-	$CHCl_3$	419(5.45),541(4.18), 577(4.03)	78-0379-78

Compound	Solvent	$\lambda_{max}(\log \epsilon)$	Ref.
$C_{50}H_{60}N_4O_{12}$ [7,8'-Biaspidospermidine]-3,3'-dicarboxylic acid, 4,4'-diacetoxy-7,8-didehydro-3,6:3',6'-diepoxy-16,16'-dimethoxy-1,1'-dimethyl-, dimethyl ester	EtOH	249(4.25),309(4.05)	39-0757-78C
$C_{51}H_{32}CuN_6O_3$ Copper, [4-nitro-N-(5,10,15,20-tetraphenyl-21H,23H-porphin-21-yl)benzamidato(2-)]-	CH_2Cl_2	429(5.33),558(4.03), 592(4.06)	35-4733-78
$C_{51}H_{32}N_6NiO_3$ Nickel, [4-nitro-N-(5,10,15,20-tetraphenyl-21H,23H-porphin-21-yl)benzamidato(2-)]-	CH_2Cl_2	420(5.10),560(4.06), 585s(3.95)	35-4733-78
$C_{51}H_{34}N_4NiO_2S$ Nickel, [4-methyl-N-(5,10,15,20-tetraphenyl-21H,23H-porphin-21-yl)benzenesulfonamidato(2-)-$N^{21},N^{22},N^{23},N^{24}$]-, (SP-4-2)-	CH_2Cl_2	423(4.35),538(4.15), 575(3.91)	35-4733-78
$C_{51}H_{34}N_6O_3$ Benzamide, 4-nitro-N-(5,10,15,20-tetraphenyl-21H,23H-porphin-21-yl)-	CH_2Cl_2	427(5.27),505s(3.50), 544(4.04),583(4.02), 640(3.89)	35-4733-78
$C_{51}H_{35}Cl_2Hg_2N_5O_2S$ Mercury, dichloro[μ-[4-methyl-N-(5,10-15,20-tetraphenyl-21H,23H-porphin-21-yl)benzenesulfonamidato(2-)-$N^1:N^{23}$]]di-	CH_2Cl_2	450(5.38),520s(3.76), 550(3.99),604(4.18), 660(3.85)	35-4733-78
$C_{51}H_{35}CuN_5O_2S$ Copper, [4-methyl-N-(5,10,15,20-tetraphenyl-21H,23H-porphin-21-yl)benzenesulfonamidato(2-)]-	CH_2Cl_2	432(5.25),560(3.96), 593(4.06)	35-4733-78
$C_{51}H_{35}N_5NiO_2S$ Nickel, [4-methyl-N-(5,10,15,20-tetraphenyl-21H,23H-porphin-21-yl)benzenesulfonamidato(2-)-N^1,N^{22},N^{23},N^{24}]-, (SP-4-3)-	CH_2Cl_2	425(5.10),566(4.07), 585s(4.03)	35-4733-78
$C_{51}H_{35}N_5O_2SZn$ Zinc, [4-methyl-N-(5,10,15,20-tetraphenyl-21H,23H-porphin-21-yl)benzenesulfonamidato(2-)-N^1,N^{22},N^{23},N^{24}]-, (SP-4-3)-	CH_2Cl_2	433(5.23),520s(3.45), 560s(3.80),590s(4.05), 612(4.13)	35-4733-78
$C_{51}H_{36}N_4O_2S$ 21H,23H-Porphine, 2-[(4-methylphenyl)-sulfonyl]-5,10,15,20-tetraphenyl-	CH_2Cl_2	425(5.31),525(4.03), 562(3.49),605(3.48), 663(3.72)	35-4733-78
$C_{51}H_{37}N_5O_2S$ Benzenesulfonamide, 4-methyl-N-(5,10-15,20-tetraphenyl-21H,23H-porphinato-(2-)-$N^{21},N^{22},N^{23},N^{24}$]-, (SP-5-21)-	CH_2Cl_2	432(5.34),510(3.36), 548(3.97),585(4.00), 640(3.90)	35-4733-78

Compound	Solvent	$\lambda_{max}(\log \epsilon)$	Ref.
$C_{51}H_{66}O_3$ 2,5-Cyclohexadien-1-one, 4-[2-[3',5'-bis(1,1-dimethylethyl)-4'-hydroxy-[1,1'-biphenyl]-4-yl]-3-[3,5-bis-(1,1-dimethylethyl)-4-hydroxyphenyl]-2-cyclopropen-1-ylidene]-2,6-bisO(1,1-dimethylethyl)-	C_6H_{12}	275(4.57),315(4.56), 371(4.55),392(4.79), 395s(4.34),405(4.65)	44-1573-78
perchlorate	C_6H_{12}	232(4.09),360(4.79)	44-1573-78
dianion	C_6H_{12}	254(4.08),318(4.08), 432(4.77)	44-1573-78
$C_{52}H_{32}$ 4,9'-Bi-9H-fluorene, 9-[4,9'-bis-9H-fluoren]-9-ylidene-	C_6H_{12}	490(4.44)	18-3373-78
$C_{52}H_{34}N_2$ Diazene, bis([9,9'-bi-9H-fluoren]-9-yl)-	benzene	355(4.22)	18-2431-78
$C_{52}H_{35}CoN_4O$ Cobalt, (2-oxo-2-phenylethyl)[5,10,15-20-tetraphenyl-21H,23H-porphinato-(2-)-$N^{21},N^{22},N^{23},N^{24}$]-, (SP-5-31)-	benzene	412(5.18),530(4.12), 620(3.08)	101-0109-78N
$C_{52}H_{38}$ Benzene, 1,4-bis(2,4,5-triphenyl-1,4-cyclopentadien-1-yl)-	benzene	355(4.31)	104-1532-78
$C_{52}H_{38}N_2$ Pyridinium, 1,1'-(1,3-phenylene)bis-[2,4,6-triphenyl-, diperchlorate	EtOH	209(4.86),321(4.86)	73-2046-78
Pyridinium, 1,1'-(1,4-phenylene)bis-[2,4,6-triphenyl-, diperchlorate	EtOH	210(4.67),308(4.68)	73-2046-78
$C_{52}H_{46}N_4O_{12}$ [5,5'-Bi-2H-cyclopenta[d]pyridazine]-1,1',4,4'-tetracarboxylic acid, 7,7'-bis[bis(4-methoxyphenyl)methyl]-, tetramethyl ester	MeCN	229(4.81),279(4.57), 295(4.49),350(4.01), 494(3.26)	78-2509-78
$C_{52}H_{56}N_8O_4$ [1,5,9,13]Tetrazacyclohexadecino[2,3-b:6,7-b':10,11-b'':14,15-b''']tetra-quinoline, 7,15,23,31(5H,13H,21H,29H)-tetrone, 5a,6,13a,14,21a,22,29a,30-octahydro-5,13,21,29-tetrapropyl-	CH_2Cl_2	235(4.5),336(4.7), 346(4.7)	5-0188-78
[1,5,9,13]Tetraazacyclohexadecino[3,2-c:7,6-c':11,10-c'':15,14-c''']tetra-quinoline-6,14,22,30(5H,8H,16H,24H)-tetrone, 4b,12b,13,20b,21,28b,29,32-octahydro-8,16,24,32-tetrapropyl-	CH_2Cl_2	243(4.9),299(4.7), 308(4.7),411(4.3)	5-0532-78
$C_{52}H_{63}BrO_{20}$ Chukrasin A, 1'-O-(4-bromobenzoyl)-	EtOH	200(4.58),220s(4.24), 248(4.42)	33-1814-78
$C_{52}H_{68}N_4O_{15}$ 6,15:21,30-Bis(ethanoxyethanoxyethano)-6H,15H,21H,30H-dinaphtho[2,1-f :1',2'-h][1,7,10,16,22,25,31,4,13,19,28]-heptaoxatetraazacyclopentatriacontine-5,16,20,311(4H,17H,19H,32H)-tetrone,	n.s.g.	274(4.23),331(3.70)	33-2407-78

$C_{52}H_{68}N_4O_{15}-C_{55}H_{86}O_{23}$

Compound	Solvent	$\lambda_{max}(\log \epsilon)$	Ref.
7,8,10,11,13,14,22,23,25,26,28,29-dodecahydro- (cont.)			33-2407-78
$C_{52}H_{70}N_4O_4$ 1-Butanone, 1,1',1",1"'-(3,8,13,18-tetrabutyl-21H,23H-porphine-2,7,12,17-tetrayl)tetrakis-	CHCl$_3$	427(5.38),522(4.20), 556(3.95),594(3.85), 648(3.49)	107-0579-78
$C_{52}H_{76}N_4O_{11}$ 6,15:21,30-Bis(ethanoxyethanoxyethano)-6H,15H,21H,30H-dinaphtho[2,1-f$_1$:1',2'-h$_1$][1,7,10,16,22,25,31,4,13,19,28]-heptaoxatetraazacyclopentatriacontine, 4,5,7,8,10,11,13,14,16,17,19,20,22-23,25,26,28,29,31,32-eicosahydro-, (S)-	n.s.g.	277(3.97),333(3.77)	33-2407-78
$C_{52}H_{108}Cl_2P_4Pd_2$ Palladium, μ-1,3-butadiyne-1,4-diyldichlorotetrakis(tributylphosphine)di-	CH$_2$Cl$_2$	263(4.92),276(4.91), 299(4.53)	101-0319-78Q
$C_{53}H_{62}N_4O_{13}S$ Vincaleukoblastine, 3'-[[(4-methylphenyl)sulfonyl]oxy]-19'-oxo-, (3'α,4'α)-	EtOH	265(4.27),272(4.27), 286(4.23),295(4.18), 312s(3.88)	23-0062-78
$C_{53}H_{78}O_{17}$ Condurango glycoside A	EtOH	217(4.20),224(4.18), 280(4.31)	133-0056-78
$C_{54}H_{38}N_4NiO_2$ Nickel, [methyl 2,7,12,18-tetraphenyl-17-(phenylmethyl)-22,23,24,25-tetra-azapentacyclo[17.2.1.1^{3,6}.1^{8,11}-1^{13,16}]pentacosa-1,3(25),4,6,8,10,12-14,16(23),18,20-undecaene-17-carboxylato(2-)-N^{22},N^{23},N^{24},N^{25}]-, (SP-4-2)-	CH$_2$Cl$_2$	447(4.93),582(3.85), 667(4.15)	78-2295-78
$C_{54}H_{50}N_4O_{15}$ Carbobenzoxyisochelocardin acethydrazone	MeOH	224(4.69),272(4.90), 300(4.64),311(4.64), 430(4.08)	44-2855-78
$C_{54}H_{65}BrO_{19}$ Chukrasin B, 1'-O-(4-bromobenzoyl)-	EtOH	198(4.55),248(4.40)	33-1814-78
$C_{54}H_{86}$ 3,3'-Bicholesta-2,5-diene	CHCl$_3$	272(4.55),280(4.61), 292(4.48)	13-0163-78A
3,3'-Bicholesta-3,5-diene	C$_6$H$_{12}$	267(4.45),276(4.53), 290(4.39)	2-0257-78
	CHCl$_3$	298(4.70),312(4.80), 328(4.67)	13-0163-78A
$C_{55}H_{74}MgN_4O_6$ Bacteriochlorophyll a	ether	359(4.81),394(4.58), 540s(3.52),572(4.26), 705s(3.96),768(4.89)	39-1150-78C
$C_{55}H_{86}O_{23}$ Glucoacetyldigoxoside	MeOH	219(4.15)	31-1254-78

Compound	Solvent	$\lambda_{max}(\log \epsilon)$	Ref.
$C_{56}H_{38}O_4$ 9(10H)-Anthracenone, 10,10'-dioxybis- [10-(2,2-diphenylethenyl)-	MeCN	260(4.41)	22-0442-78
$C_{56}H_{62}N_4OPRh$ Rhodium, [2,3,7,8,12,13,17,18-octaeth- yl-21H,23H-porphinato(2-)-N^{21},N^{22},N^{23}- N^{24}](2-oxoethyl)(triphenylphosphine)-, (OC-6-42)-	$CHCl_3$	365(4.51),419(4.94), 530(4.16),560(4.18)	101-0329-78P
$C_{56}H_{70}MgN_4O_6$ Chlorophyll b	ether	429(4.73),453(5.18), 548(3.76),595(4.01), 645(4.67)	39-1150-78C
$C_{56}H_{72}MgN_4O_5$ Chlorophyll a	ether	410(4.95),430(5.12), 535(3.68),579(3.94), 619(4.21),664(4.94)	39-1150-78C
$C_{58}H_{40}$ 7H-Dibenzo[a,c]cyclononene, 7-(5,9-di- phenyl-7H-dibenzo[a,c]cyclononen-7- ylidene)-5,9-diphenyl-	EtOH	248s(4.48),275s(4.26), 295s(3.88),402(4.28)	150-5151-78
7H-Dibenzo[a,c]cyclononene, 5,9-diphen- yl-7-(2,3,4,5-tetraphenyl-2,4-cyclo- pentadien-1-ylidene)-	C_6H_{12}	245(4.64),291s(4.19), 300s(3.90),335s(2.94), 405(2.93)	150-5151-78
$C_{60}H_{64}N_8O_4$ Tetrapyrido[4,3-b:4',3'-f:4",3"-j:4"'- 3"'-n][1,5,9,13]tetraazacyclohexadec- ine-6,12,18,24(2H,8H,14H,20H)-tetrone, 4a,5,10a,11,16a,17,22a,23-octahydro- 2,8,14,20-tetrakis(3-phenylpropyl)-	$CHCl_3$	339(4.5)	24-2594-78
$C_{64}H_{72}N_8O_8$ Tetrapyrido[4,3-b:4',3'-f:4",3"-j:4"'- 3"'-n][1,5,9,13]tetraazacyclohexadec- ine-6,12,18,24(2H,8H,14H,20H)-tetrone, 4a,5,10a,11,16a,17,22a,23-octahydro- 2,8,14,20-tetrakis[3-(2-methoxyphen- yl)propyl]-	$CHCl_3$	339(4.5)	24-2594-78
$C_{64}H_{72}N_8O_{16}$ Tetrapyrido[4,3-b:4',3'-f:4",3"-j:4"'- 3"'-n][1,5,9,13]tetraazacyclohexadec- ine-6,12,18,24(2H,8H,14H,20H)-tetrone, 2,8,14,20-tetrakis[[(3,4-dimethoxy- phenyl)methoxy]methyl]-4a,5,10a,11- 16a,17,22a,23-octahydro-	$CHCl_3$	328(4.1)	24-2594-78
$C_{65}H_{100}O_{27}$ Condurango glycoside C^1	EtOH	217(4.21),223(4.14), 279(4.32)	133-0056-78
$C_{65}H_{106}O_{22}$ Lanosta-9(11),17(20),25-trien-18-oic acid, 3-[(O-6-deoxy-2,3,4-tri-O-meth- yl-β-D-glucopyranosyl-(1→2)-O-[O- 2,3,4,6-tetra-O-methyl-β-D-glucopyran- osyl-(1→2)-2,4,6-tri-O-methyl-β-D-	hexane	253(4.00)	94-3722-78

Compound	Solvent	$\lambda_{max}(\log \epsilon)$	Ref.
glucopyranosyl-(1→4)-3-O-methyl-β-D-xylopyranosyl)oxy]-16-oxo-, methyl ester, (3β)- (cont.)			94-3722-78
Lanosta-9(11),17(20),25-trien-18-oic acid, 16-oxo-3-[(O-2,3,4,6-tetra-O-methyl-β-D-glucopyranosyl-(1→3)-O-2,4,6-tri-O-methyl-β-D-glucopyranosyl-(1→4)-O-6-deoxy-2,3-di-O-methyl-β-D-glucopyranosyl)-(1→2)-3,4-di-O-methyl-β-D-xylopyranosyl)oxy]-, methyl ester, (3β)-	hexane	253(3.98)	94-3722-78
$C_{66}H_{44}N_2O_2S$			
7,10-Epithionaphtho[2,3-g]phthalazine, 8,9-dibenzoyl-7,8,9,10-tetrahydro-1,4,6,7,10,11-hexaphenyl-	CH_2Cl_2	248(4.71),382(3.87)	4-0793-78
$C_{66}H_{69}N_9O_{18}P$			
Adenosine, 5'-O-methoxytrityl-2'-O-acetyluridylyl-3'-(2-cyanoethyl)-5'-2',3'-isopropylidene-N^6-[N-(O-benzyl)threonylcarbonyl]-, benzyl ester	pH 2	235(4.24),262s(4.27), 270(4.36),277s(4.35)	19-0021-78
	pH 12	237(4.43),260s(4.34), 270(4.40),278(4.38), 300(4.08)	19-0021-78
	MeOH	234(4.29),261s(4.38), 269(4.42),277s(4.34)	19-0021-78
$C_{66}H_{104}O_{32}$			
Holotoxin B	MeOH	none above 210 nm	94-3722-78
$C_{67}H_{106}O_{32}$			
Holotoxin A	MeOH	none above 210 nm	94-3722-78
$C_{68}H_{48}Cl_8N_8O_4$			
[1,5,9,13]Tetraazacyclohexadecino[2,3-b:6,7-b':10,11-b":14,15-b"']tetraquinoline-7,15,23,31(5H,13H,21H,29H)-tetrone, 5,13,21,29-tetrakis[(2,6-dichlorophenyl)methyl]-5a,6,13a,14,21a-22,29a,30-octahydro-	CH_2Cl_2	231(4.7),330(4.7), 339(4.7)	5-0188-78
[1,5,9,13]Tetraazacyclohexadecino[3,2-c:7,6-c':11,10-c":15,14-c"']tetraquinoline-6,14,22,30(5H,8H,16H,24H)-tetrone, 8,16,24,32-tetrakis[(2,6-dichlorophenyl)methyl]-4b,12b,13,20b-21,28b,29,32-octahydro-	CH_2Cl_2	243(5.0),298(4.7), 305(4.7),404(4.4)	5-0532-78
$C_{68}H_{56}N_8O_4$			
[1,5,9,13]Tetraazacyclohexadecino[2,3-b:6,7-b':10,11-b":14,15-b"']tetraquinoline-7,15,23,31(5H,13H,21H,29H)-tetrone, 5a,6,13a,14,21a,22,29a,30-octahydro-5,13,21,29-tetrakis(phenylmethyl)-	CH_2Cl_2	235(4.6),336s(4.7), 346(4.7)	5-0188-78
[1,5,9,13]Tetraazacyclohexadecino[3,2-c:7,6-c':11,10-c":15,14-c"']tetraquinoline-6,14,22,30(5H,8H,16H,24H)-tetrone, 4b,12b,13,20b,21,28b,29,32-octahydro-8,16,24,32-tetrakis(phenylmethyl)-	CH_2Cl_2	245(4.9),300(4.6), 307(4.6),407(4.2)	5-0532-78

Compound	Solvent	λ_{max}(log ϵ)	Ref.
$C_{68}H_{60}N_{12}O_{12}$ Tetrapyrido[4,3-b:4',3'-f:4",3"-j:4"'-3"'-n][1,5,9,13]tetraazacyclohexadec-ine-6,12,18,24(2H,8H,14H.20H)-tetrone, 2,8,14,20-tetrakis[3-(1,1-dihydro-1,3-dioxo-2H-isoindol-2-yl)propyl]-4a,5-10a,11,16a,17,22a,23-octahydro-	CH_2Cl_2	335(4.2)	24-2594-78
$C_{68}H_{80}N_8O_{12}$ Tetrapyrido[4,3-b:4',3'-f:4",3"-j:4"'-3"'-n][1,5,9,13]tetraazacyclohexadec-ine-6,12,18,24(2H,8H,14H,20H)-tetrone, 2,8,14,20-tetrakis[3-(3,4-dimethoxy-phenyl)propyl]-4a,5,10a,11,16a,17-22a,23-octahydro-	$CHCl_3$	338(4.5)	24-2594-78
$C_{68}H_{84}N_{12}O_{12}S_4$ Benzenesulfonamide, N,N',N",N"'-[(4a,5-10a,11,16a,17,22a,23-octahydro-6,12-18,24-tetraoxotetrapyrido[4,3-b:4',3'-f:4",3"-j:4"',3"'-n][1,5,9,13]tetra-azacyclohexadecine-2,8,14,20(6H,12H-18H,24H)-tetrayl)tetra-3,1-propane-diyl]tetrakis[N,4-dimethyl-	CH_2Cl_2	335(4.0)	24-2594-78
$C_{72}H_{56}N_4O_{17}$ 2,4(3H,8H)-Pteridinedione, 6,7-diphen-yl-3-(2,3,5-tri-O-benzoyl-β-D-ribo-furanosyl)-8-[2-[(2,3,5-tri-O-benz-oyl-β-D-ribofuranosyl)oxy]ethyl]-	MeOH	229(4.96),273(4.35), 282(4.34),295s(4.26), 380s(3.58),430(3.82)	24-2586-78
$C_{72}H_{82}O_2$ [1,1':3',1"-Terphenyl]-2'-ol, 2-[[4,4"-bis(1,1-dimethylethyl)-5'-[4-(1,1-di-methylethyl)phenyl][1,1':3',1"-ter-phenyl]-2'-yl]oxy]-4,4'-bis(1,1-di-methylethyl)-5'-[4-(1,1-dimethyl-ethyl)phenyl]-	CH_2Cl_2	252(4.95),255(4.95), 308(3.86)	24-0264-78
$C_{74}H_{84}O_3$ [1,1':3',1"-Terphenyl]-2'-ol, 2-[[4,4"-bis(1,1-dimethylethyl)-5'-[4-(1,1-di-methylethyl)phenyl][1,1':3',1"-ter-phenyl]-2'-yl]oxy]-4,4"-bis(1,1-di-methylethyl)-5'-[4-(1,1-dimethyleth-yl)phenyl]-, acetate	CH_2Cl_2	252(4.96),255(4.96)	24-0264-78
$C_{74}H_{122}O_{27}$ Lanosta-9(11),17(20),25-trien-18-oic acid, 16-oxo-3-[(O-2,3,4,6-tetra-O-methyl-β-D-glucopyranosyl-(1→4)-O-[O-(2,3,4,6-tetra-O-methyl-β-D-gluco-pyranosyl-(1→3)-O-2,4,6-tri-O-methyl-β-D-glucopyranosyl-(1→4)-6-deoxy-2,3-di-O-methyl-β-D-glucopyranosyl-(1→2)-3-O-methyl-β-D-xylopyranosyl)oxy]-, methyl ester, (3β)-	hexane	253(3.91)	94-3722-78
$C_{76}H_{86}N_8Ni_2$ Nickel, [μ-[[5,5'-(1,3-butadiyne-1,4-diyl)bis[2,3,7,8,12,13,17,18-octa-	$CHCl_3$	377s(4.55),427(5.05), 457(5.06),484(5.12),	39-0366-78C

Compound	Solvent	$\lambda_{max}(\log \epsilon)$	Ref.
ethyl-21H,23H-porphinato]](4-)-N^{21}-N^{22},N^{23},N^{24},$N^{21'}$,$N^{22'}$,$N^{23'}$,$N^{24'}$]]di-(cont.)		531s(4.29),574s(4.41), 618(4.76)	39-0366-78C
$C_{80}H_{162}Cl_2P_6Pd_2Pt$ Platinum, di-μ-1,3-butadiyne-1,4-diyl-bis[chlorobis(tributylphosphine)pall-adium]bis[tributylphosphine)-	CH_2Cl_2	281(4.52),343(4.76)	101-0319-78Q
$C_{80}H_{162}Cl_2P_6Pt_3$ Platinum, di-μ-1,3-butadiyne-1,4-diyl-dichlorohexakis(tributylphosphine)-tri-	CH_2Cl_2	269(4.49),289(4.46), 336(4.55),360(4.47)	101-0319-78Q
$C_{83}H_{138}O_{32}$ Lanosta-9(11),17(20),25-trien-18-oic acid, 16-oxo-3-[(O-2,3,4,6-tetra-O-methyl-β-D-glucopyranosyl)-(1→3)-O-2,4,6-tri-O-methyl-β-D-glucopyrano-syl)-(1→4)-O-[O-2,3,4,6-tetra-O-meth-yl-β-D-glucopyranosyl)-(1→3)-O-2,4,6-tri-O-methyl-β-D-glucopyranosyl)-(1→4)-6-deoxy-2,3-di-O-methyl-β-D-glucopyranosyl-(1→2)]-3-O-methyl-β-D-xylopyranosyl]oxy]-, methyl ester, (3β)-	hexane	252(3.98)	94-3722-78
$C_{147}H_{147}Au_{11}I_3P_7$ Gold, triiodoheptakis[tris(4-methyl-phenyl)phosphine]undeca-, (31Au-Au)-	EtOH	305(4.87),417(4.55)	35-5085-78
$C_{150}H_{147}Au_{11}N_3P_7$ Gold, tris(cyano-C)heptakis[tris(4-methylphenyl)phosphine]undeca-, (31Au-Au)-	EtOH	303(5.00),415(4.50)	35-5085-78

1-   -78, Acta Chem. Scand. B, 32 (1978)
  0027   S.B. Christensen and P. Krogs-
          gaard-Larsen
  0056   J. Lykkeberg
  0066   G.A. Ulsaker and K. Undheim
  0068   T. Laerum and K. Undheim
  0075   J. Gripenberg
  0077   M. Lounasmaa and M. Puhakka
  0098   O. Møller et al.
  0109   B. Thulin et al.
  0118   K. Torssell and O. Zeuthen
  0131   W.F. Smyth et al.
  0149   C. Moberg
  0152   S.C. Olsen and J.P. Snyder
  0155   K. Ankner et al.
  0216   M. Lounasmaa and M. Puhakka
  0293   T. Olsson and O. Wennerstrom
  0303   K.E. Nielsen and E.B. Pedersen
  0327   P. Krogsgaard-Larsen et al.
  0347   S.C. Sharma and K. Torssell
  0463   O. Wennerstrom and I. Raston
  0469   P. Krogsgaard-Larsen et al.
  0625   C.L. Pedersen et al.
  0720   P.E. Hansen et al.

1-   -78, Acta Chem. Scand. A, 32 (1978)
  0259A   G. Borch et al.

2-   -78, Indian J. Chem. Sect. B, 16 (1978)
  0004   R. Rajendran et al.
  0016   S.A. Nadgouda et al.
  0027   K.R. Ravindranath et al.
  0035   R.B. Gupta and R.N. Khanna
  0095   H. Singh and K.K. Bhutani
  0103   N.R. Ayyangar and D.R. Wagle
  0106   N.R. Ayyangar et al.
  0143   G.P. Dhareshwar and B.D. Hosangadi
  0148   Y. Gopichand et al.
  0151   R.A. Misra
  0152   K. Ghose and A.J. Bhattacharya
  0161   S.S. Chakravorti et al.
  0165   J.P. Saxena et al.
  0167   M.S. Raju et al.
  0179   V.S. Kamat et al.
  0184   V.S. Kamat et al.
  0188   J.S. Patel et al.
  0196   A.K. Ghosal et al.
  0200   A.K. Ghosal et al.
  0226   V.P. Arya et al.
  0253   P. Balakrishnan et al.
  0257   R.V. Almanlu et al.
  0275   K.S. Jadhav et al.
  0280   K.S. Jadhav et al.
  0289   M. Sharma et al.
  0307   I.S. Ismail and R. Jacobi
  0324   P.S. Dhamankar et al.
  0332   M. Abdalla et al.
  0361   S.K. Talapatra et al.
  0375   S.S. Patwardhan and R.N. Usgaonkar
  0405   M.C. Moorjani and G.K. Trivedi
  0416   A. Chatterjee and A. Banerjee
  0418   S.R. Ramadas and S. Padmanbhan
  0421   B.R. Pai et al.
  0426   M.S. Ahmad et al.
  0428   L. Anandan and G.K. Trivedi
  0469   L. Rateb et al.

  0502   A.I. Hashem and E.E. Eid
  0539   M.V. Karwe et al.
  0545   K.A. Joshi and A.B. Kulkarni
  0547   N.R. Ayyangar et al.
  0559   M.S. Ahmad et al.
  0563   P. Sharma et al.
  0574   V.K. Ahluwalia et al.
  0579   V.K. Ahluwalia et al.
  0584   V.K. Ahluwalia et al.
  0611   A.C. Jain et al.
  0613   A.C. Jain et al.
  0617   H. Singh and T.R. Bhardwaj
  0641   U. Khera and S.S. Chhiber
  0643   M. Wij and S. Rangaswami
  0658   G.P. Garg et al.
  0668   T.R. Kasturi and R. Sivaramakrish-
          nam
  0673   N.R. Ayyangar et al.
  0678   P.B. Talukdar et al.
  0683   I.M. Ismail and W. Sauer
  0689   H. Jahine et al.
  0731   S. Manna et al.
  0786   A.K. Sen and S. Ray
  0853   P.H. Ladwa et al.
  0856   A.C. Jain et al.
  0860   S. Krishnappa
  0910   A.M.J. Ali and Z.Y. Al-Saigh
  0924B   M.E. Singh and G. Bagavant
  0970   V. Chandrasekharan et al.
  0973   A.C. Jain et al.
  0984   M.A. El-Hashash et al.
  1000   H. Jahine et al.
  1007   N.R. Ayyangar et al.
  1039   A.C. Jain et al.
  1042B   A.K. Dey et al.
  1062   S.H. Mashraqui and G.K. Trivedi
  1067   B.P. Joshi and B.D. Hosangadi
  1079   S.C. Chhabra et al.
  1086   M.S.R. Naidu and S.G. Peeran
  1100   N. Viswanathan and V. Balakrish-
          nam
  1124   A.C. Jain and D.K. Tuli
  1125   A.C. Jain and R. Khazanchi
  1126   A.C. Jain and R.C. Gupta

4-   -78, J. Heterocyclic Chem., 15 (1978)
  0001   P.D. Cook, P. Dea and R.K. Robins
  0017   E. Barni and P. Savarino
  0023   J. Daunis et al.
  0029   C.D. Weis
  0031   C.D. Weis
  0043   H.L. McPherson and B.W. Ponder
  0057   M. Quinteiro et al.
  0063   R. Friary
  0077   R. Friary
  0081   L. Grehn
  0097   J. Bielawski et al.
  0101   G.E. Martin et al.
  0105   F.G. Baddar et al.
  0113   D.T. Connor et al.
  0115   D.T. Connor et al.
  0127   M.P. Serve et al.
  0129   S.K. Sengupta et al.
  0133   A.M. Kiwan and A.Y. Kassir
  0149   C. Blanchard et al.
  0155   B. Chantegrel and S. Gelin

0161	A. Walser and G. Zenchoff		1281	P. Tarburton et al.
0167	E. Campaigne and R.K. Mehra		1295	P. Demaree et al.
0181	L. Mosti, P. Schenone and G. Menozzi		1319	D.G. Holah, A.N. Hughes and D. Kleemola
0221	C. Ochoa and M. Stud		1349	S. Romani and W. Klotzer
0241	T. Uchida		1387	M. Robba et al.
0253	P. Goya and M. Stud		1409	J. Bristol
0257	V.H. Belgaonkar and R.N. Usgaonkar		1411	M.L. Stein et al.
0293	M. Ruccia, N. Vivona and G. Cusmano		1425	M. Iwao and T. Kuraishi
			1439	M. Maguet and R. Guglielmetti
0297	S.J. Yan et al.		1463	J.P. Yevich et al.
0307	R.F. Fibiger et al.		1471	O. Ceder and B. Hall
0311	Y. Lin et al.		1485	M. Ruccia et al.
0327	B. Chantegrel and S. Gelin		1493	M. Payard et al.
0333	A. Dinner and E. Rickard		1505	S.W. Schneller and R.S. Hosmane
0359	K. Senga et al.		1521	L.H. Morcos Guindi and A.F. Temple
0365	J.A. Van Allan et al.			
0385	F.G. Baddar et al.			
0395	E. Alcalde et al.		5- -78, Ann. Chem. Liebigs (1978)	
0401	E. Campaigne and T.P. Selby		0029	R. Huisgen and T. Schmidt
0463	G.Y. Hajos and A. Messmer		0066	A.R. Forrester et al.
0477	P. Goya and M. Stud		0118	V. Berariu et al.
0489	W.T. Ashton, R.D. Brown and R.L. Tolman		0129	U. Kücklander
			0140	U. Kücklander
0511	F. Evangelisti et al.		0188	W.H. Gündel and H. Berenbold
0529	A.H. Albert et al.		0193	H. Fenner et al.
0551	H. Cairns and A.R. Payne		0214	R.J.M. Weustink et al.
0555	A. Tanaka, T. Usui and S. Yoshina		0227	P.C. Thieme and E. Hädicke
0569	H. Fukumi et al.		0238	B. von Maltzan
0577	A. Walser, T. Flynn and R.I. Fryer		0289	A. Treibs et al.
0609	G.E. Martin and J.C. Turley		0376	K. Hafner et al.
0637	G. Heinisch et al.		0387	G.J. Wentrup and F. Boberg
0655	J.G. Lombardino et al.		0398	H.-J. Kabbe
0657	G. Gosselin et al.		0419	G. Habermehl et al.
0685	G.R. Newkome and S.J. Garbis		0431	A. Krebs et al.
0687	A. Walser and T. Flynn		0440	W. Friedrichsen et al.
0705	J.K. Chakrabarti et al.		0473	G. Satzinger
0711	R.A. Hollins and A.C. Pinto		0528	P. Studt
0715	R. Lazaro and J. Elguero		0530	P. Studt
0737	T. Kurihara et al.		0532	H. Berenbold and W.H. Gündel
0759	Y.A. El-Farkh et al.		0573	H.D. Scharf et al.
0781	K. Senga et al.		0608	K. Krohn et al.
0793	L. Lepage and Y. Lepage		0627	R. Neidlein and K.F. Cepera
0813	S. Gelin and D. Hartmann		0726	K. Krohn and C. Hemme
0839	B.L. Cline, R.P. Panzica and L.B. Townsend		0804	A. Roedig and M. Försch
			0854	H. Köster et al.
0855	A. Walser, R.F. Lauer and R.I. Fryer		1096	G. Toth et al.
			1103	G. Toth et al.
0929	J.L. Montero et al.		1129	W. Friedrichsen and R. Schmidt
0949	M.S. Manhas et al.		1139	W. Friedrichsen and H.-G. Oeser
0987	S.W. Schneller and J.L. May		1146	W. Freidrichsen and H.-G. Oeser
1027	J. de Mendoza and P. Rull		1161	W. Friedrichsen and H.-G. Oeser
1033	M. Hedayatullah and J. Pailler		1203	T. Eicher et al.
1041	G. Maury et al.		1222	T. Eicher et al.
1043	A.W. McGee et al.		1365	S. Penades and G. Schafer
1093	C. Suarez et al.		1368	R. Neuberg et al.
1117	D.A. Lightner and Y.T. Park		1379	M. Magon and G. Schroder
1141	P. Aeberli and W. Houlihan		1406	A. Roedig and M. Forsch
1149	D.E. Kuhla and H.A. Watson, Jr.		1536	W.H. Gündel and H. Berenbold
1153	B. Nedjar et al.		1568	E. Schaumann and S. Grabley
1185	L. Lepage and Y. Lepage		1634	V. Weisskopf and H. Perst
1193	G. Bobowski and J. Shavel, Jr.		1648	W.D. Schröer and W. Friedrichsen
1215	B. Chantegrel and S. Gelin		1655	W. Friedrichsen et al.
1225	G.A. Reynolds and J.A. Van Allan		1675	H. Straub
1255	W. Kehrbach and N.E. Alexandrou		1734	R. Müller and F. Lingens
1271	A. Camparini et al.		1739	W. Bartmann et al.

1749   H.-J. Schäfer et al.
1775   A. Roedig and M. Försch
1780   M. Hattori and W. Pfleiderer
1788   M. Hattori and W. Pfleiderer
1796   M. Hattori et al.
1880   F. Heinrich and W. Luttke
1974   R. Neidlein and G. Humburg
1990   H. Lehner et al.
2024   A. Treibs and D. Grimm
2028   A. Treibs and D. Grimm
2033   H. Gnichtel et al.
2074   H.D. Carnadi et al.
2105   P. Studt

12-   -78, Australian J. Chem., 31 (1978)
0085   T. Teitei et al.
0097   J.T. Pinhey et al.
0113   J.T. Pinhey et al.
0157   J.C. Coll et al.
0163   B.F. Bowden et al.
0171   M.S. Ahmad and I.A. Khan
0179   C.G. Freeman et al.
0187   C.M. Richards and A. Hofmann
0221   R.N. Warrener and E.E. Nunn
0225   J.T. Craig et al.
0313   J.B. Bremner and K.N. Winzenberg
0321   I.R.C. Bick et al.
0353   R.M. Carman
0365   P.S. Clezy et al.
0389   G.B. Barlin and P. Lakshminarayana
0439   D.E. Rivett and F.H.C. Stewart
0627   B.B. Greene and K.G. Lewis
0639   P.S. Clezy et al.
0649   D.J. Brown and P. Waring
0907   F.R. Hewgill and G.B. Howie
1011   J.W. Loder et al.
1041   J.A. Elix and B.A. Ferguson
1061   F.R. Hewgill and G.B. Howie
1069   F.R. Hewgill and G.B. Howie
1081   W.L.F. Armarego and H. Schou
1095   D.E. Cowley et al.
1113   R.N. Warrener et al.
1129   T.S. Lee et al.
1223   J.M. Coxon et al.
1265   J.R. Cannon et al.
1291   C.C. Duke et al.
1303   B.F. Bowden et al.
1323   D.W. Cameron and M.D. Sidell
1335   D.W. Cameron et al.
1353   D.W. Cameron and M.J. Crossley
1363   D.W. Cameron et al.
1493   C.L. Raston et al.
1533   J.K. MacLeod et al.
1553   J.K. MacLeod et al.
1561   D.W. Johnson and L.N. Mander
1569   L. Lombardo and D. Wege
1585   L. Lombardo et al.
1607   M.B. Stringer and D. Wege
1757   G.J. Baxter et al.
1841   R. Bergamasco et al.
1965   R.L. Crumbie et al.
2031   V. Candeloro and J.H. Bowie
2039   B.F. Bowden et al.
2049   B.F. Bowden et al.
2057   J.A. Elix et al.
2069   J.F. Kinnear et al.

2077   I.R.C. Bick and S. Sotheeswaran
2085   I. Addae-Mensah and D.W. Cameron
2099   D.W. Cameron et al.
2259   D.W. Cameron et al.
2271   H.J. Banks et al.
2317   D. St.C. Black and J.E. Doyle
2463   P.J. Newcombe and R.K. Norris
2477   D.J. Freeman et al.
2491   P.S. Clezy and C.J.R. Fookes
2505   D.J. Brown and T. Nagamatsu
2517   W. Cowden and N. Jacobsen
2527   I.A. Southwell
2539   I.R.C. Bick and H.M. Leow
2651   D.W. Cameron et al.
2675   E.R. Cole, G. Crank and E. Lye
2685   R. Kazlauskas et al.
2699   R.A. Eade et al.
2707   B.F. Bowden et al.

13-   -78A, Steroids, 31 (1978)
0163   Y. Kurasawa et al.

13-   -78B, Steroids, 32 (1978)
0453   J.L. Napoli et al.

18-   -78, Bull. Chem. Soc. Japan, 51 (1978)
0179   H. Besso et al.
0234   A. Murai et al.
0243   A. Murai et al.
0248   T. Amiya et al.
0265   N. Harada et al.
0301   M. O-oka et al.
0309   R. Okazaki et al.
0331   A. Oku and A. Matsui
0512   A.S. Shawali et al.
0535   K. Akiba et al.
0649   K. Ueda and F. Toda
0667   N. Abe and T. Nishiwaki
0839   M. Tada et al.
0842   T. Hirata and T. Suga
0855   T. Suami et al.
0909   S. Kamiyama et al.
0943   T. Kusumi et al.
1175   K. Kurosawa and Y. Ashihara
1178   I. Tabushi et al.
1204   J. Ojima and Y. Shiroishi
1249   Y. Shigemitsu et al.
1257   H. Takeshita et al.
1444   K. Ichimura
1450   M. Kato et al.
1501   M. Shin et al.
1525   A. Roy and K. Nag
1573   T. Nishiwaki et al.
1708   M. Yudasaka and H. Hosoya
1723   H. Nakanishi et al.
1793   H. Inoue and M. Hida
1805   K. Hasegawa et al.
1839   K. Tada et al.
1846   C. Yamaazaki
2052   K. Maruyama et al.
2068   T. Horaguchi and T. Abe
2077   A. Ishida and T. Mukaiyama
2112   J. Ojima et al.
2131   T. Asao et al.
2185   T. Nozoe et al.
2264   S.Y. Matsuzaki and A. Kuboyama

2338	K. Kikuchi et al.
2375	Y. Inoue et al.
2398	M.Nakayama et al.
2415	A.E. Abdel-Rahman and Z.H. Khalil
2425	K. Hiraki et al.
2431	N.Eda et al.
2437	M. Ando and S. Emoto
2601	K. Okada et al.
2605	S. Yoneda et al.
2668	H. Horita et al.
2674	K. Akiba et al.
2698	T. Matsuura et al.
2702	E. Akiyama et al.
2718	R. Abu-Eittah and R. Hilal
3027	M. Kondo
3035	K. Tatsuta et al.
3087	K. Kikuchi et al.
3175	K. Fukuyama et al.
3312	M. Takahashi et al.
3316	T. Nozoe and T. Someya
3335	H. Nagano and T. Takahashi
3345	T. Nomoto et al.
3351	S. Akiyama et al.
3359	M. Iyoda et al.
3363	M. Iyoda and M. Nakagawa
3373	M. Minabe et al.
3389	K. Takahashi et al.
3489	Y. Yamamoto et al.
3579	K. Fujimori and K. Yamane
3582	K. Kohara
3612	K. Kurosawa et al.
3627	H. Obara et al.
3653	S. Inoue et al.

19-   -78, Bull. Acad. Polon. Sci., 26 (1978)

0021	J. Boryski and B. Golankiewicz
0509	W. Czuba and J.A. Bajgrowicz
0583	Z. Paryzek
0591	Z. Paryzek and R. Wydra
0655	L. Skulski and D. Maciejewska
0819	S. Kinastowski and J. Grabarkie-wicz-Szczesna
0851	I.Z. Siemion and M. Szkoda
0907	S. Kinastowski and H. Kasprzyk

20-   -78, Bull. soc. chim. Belges, 87 (1978)

0027	P. Meunier et al.
0075	B. Tursch et al.
0309	Y. Van Haberbeke et al.
0459	M. DePotter et al.
0487	M. Albericci et al.
0621	H.J. Meyer et al.
0693	E. Francotte et al.
0911	J. Pecher and A. Waefelaer
0917	J.C. Braekman et al.

22-   -78, Bull. soc. chim. France, Pt. II, (1978)

0033	R. Bucourt and J. Dube
0043	E.M. Gaydou and J.-P. Bianchini
0119	D. Hainaut and R. Bucourt
0129	E. Elkik and H. Assadifar
0131	J. Martel et al.
0202	J.C. Jacquesy and M. Jouannetaud
0241	M. Schaefer et al.
0255	R. Jacquesy and J.F. Patoiseau

0273	R. Faure et al.
0343	G. Amiard and R. Bucourt
0350	G. Amiard and R. Bucourt
0355	G. Costerousse et al.
0379	G. Cauquis and B. Divisia
0401	P. Battioni et al.
0415	P. Battioni et al.
0439	C. Raby and J. Buxeraud
0442	J. Moutet and G. Reverdy
0457	S. Rebuffat et al.
0612	R.M. Dupeyre and A. Rassat
0621	D. Pilo-Veloso and A. Rassat

23-   -78, Can. J. Chem., 56 (1978)

0041	Y. Pepin et al.
0062	J.P. Kutney et al.
0080	D.N. Butler and I. Gupta
0119	H. Driguez et al.
0221	K.-E. Teo et al.
0246	E. Fujita and M. Ochiai
0302	M. Guay and F. Lamy
0326	M. Iwakawa et al.
0383	R.H.F. Manske et al.
0410	D. Mukherjee and C.R. Engel
0419	G. Dionne and C.R. Engel
0424	C.R. Engel and G. Dionne
0517	R.H. Burnell and M. Ringuet
0613	J.A. Findlay et al.
0733	O.E. Edwards and P.-T. Ho
0992	R.J. Crawford et al.
1052	P. Yates et al.
1063	A. Fischer and C.C. Grieg
1080	O. Ho and M. Matsuda
1246	R.H. Mitchell et al.
1319	A.J. Paine and N.H. Werstiuk
1335	T.T. Coway et al.
1429	S. Soth et al.
1593	I. Kurobane et al.
1628	G.L. Lange et al.
1646	J.W. ApSimon et al.
1657	L.J. Madgzinski et al.
1687	P. Deslongchamps et al.
1752	G.W. Schnarr and W.A. Szarek
1758	A. Fischer and K.C. Teo
1796	T.D. Cyr and G.A. Poulton
1836	A.G. McInnes et al.
1970	D.R. Arnold and C.P. Hadjianton-iou
2003	S.N. Bhat et al.
2113	W.A. Ayer et al.
2156	J.W. ApSimon and S.F. Hall
2194	L.M. Cabelkova-Taguchi and J. Warkentin
2286	A.G. Brook et al.
2369	S.J. Loeb et al.
2430	H. Kobayashi et al.
2467	H.L. Holland et al.
2560	J.P. Kutney et al.
2665	L.R.C. Barclay et al.
2681	M.A. Ragan

24-   -78, Chem. Ber., 111 (1978)

0013	D.Hellwinkel and W. Krapp
0084	H. Röttele et al.
0099	G. Schröder et al.
0107	P. Hildenbrand et al.

0205	F. Vögtle and G. Steinhagen	3178	H. Gotthardt and C.M. Weisshuhn
0240	J. Adler et al.	3233	U. Luhmann et al.
0264	K. Dimroth et al.	3246	U. Luhmann and W. Lüttke
0282	F.G. Klärner and M. Wette	3336	H. Gotthardt et al.
0439	S. Biechert et al.	3346	R. Neidlein et al.
0486	A. Gossauer and J.-P. Weller	3385	N. Theobald and W. Pfleiderer
0523	I. Böhm et al.	3403	E. Baxmann and E. Winterfeldt
0554	N. Mollov et al.	3519	K. Wald and H. Wamhoff
0566	W. Tochtermann et al.	3608	G. Kaupp and M. Stark
0596	W. Duismann and C. Rüchardt	3624	G. Kaupp et al.
0770	H. Biere et al.	3665	W. Eberbach and B. Burchardt
0780	W. Sucrow and M. Slopianka	3750	A.Q. Hussein et al.
0791	W. Sucrow et al.	3790	R. Mengel and W. Pfleiderer
0832	H. Bertram and K. Wieghardt	3823	K. Krohn and M. Radeloff
0853	V. Bardakos and W. Sucrow	3912	K. Heyns and R. Hohlweg
0932	P. Bischof et al.	3927	R. Aumann and J. Knecht
0939	U. Eder et al.	3939	W. Steglich and L. Zechlin
0971	I.W. Southon and W. Pfleiderer	3957	A. Schönberg and P. Eckert
0982	H. Fuchs et al.		
0996	I.W. Southon and W. Pfleiderer	25-	-78, Chem. and Ind.(London), (1978)
1006	I.W. Southon and W. Pfleiderer	0092	J.S. Davidson and S.S. Dhami
1019	E. Logemann et al.	0126	H. Singh and P. Singh
1160	C.H. Briekorn and G. Unger	0233	S. Matsueda
1223	C.G. Kreiter and R. Aumann	0272	B.P. Das and D.N. Chowdhury
1284	K. Krohn et al.	0274	K. Ohkata et al.
1330	T. Kauffmann et al.	0584	J.S. Calderon et al.
1420	R. Bartsch et al.	0628	J.R. Merchant and A.S. Gupta
1464	G. Domany et al.	0848	D.K. Chakraborty et al.
1533	G. Haffer et al.	0954	S. Sengupta and S.C. Das
1540	J. Müller and E. Winterfeldt		
1709	L. Weber et al.	27-	-78, Chimia, 32 (1978)
1753	R. Gottlieb and W. Pfleiderer	0323	J. Fabian
1763	R. Gottlieb and W. Pfleiderer	0468	S. Chaloupka and H. Heimgartner
1780	V. Bardakos and W. Sucrow		
1798	D. Hellwinkel et al.	28-	-78A, Compt. rend., 286 (1978)
1824	R. Neidlein and K.F. Cepera	0021	B. Vidal and G. Bastaert
1944	H. Plieninger and W. Gramlich	0047	L. Duhamel and J.Y. Valnot
2021	H. Gotthardt and C.M. Weisshuhn	0521	D. Bouin-Roubaud et al.
2028	H. Gotthardt and C.M. Weisshuhn	0663	H. Galons et al.
2130	R. Tschesche and B. Streuff		
2206	H.D. Scharf and J. Mattay	28-	-78B, Compt. rend., 287 (1978)
2258	H. Dürr and H. Schmitz	0129	M. Lavault and J. Bruneton
2297	H. Wamhoff and L. Lichtenthäler	0201	J. Kister et al.
2376	K. Fickentscher et al.	0353	G. Bonhomme and J. Lemaire
2407	W. Flitsch et al.	0385	C. Rocheville-Divorne and J.-P.
2423	L.F. Tietze and U. Niemeyer		Roggero
2441	L.F. Tietze and P. Marx		
2547	M. Atzmuller and F. Vogtle	30-	-78, Doklady Akad. Nauk S.S.S.R.,
2557	H.-D. Martin et al.		238-243 (1978)
2563	W. Grimme and U. Heinze	0130	O.N. Chupakhin et al.
2571	I.W. Southon and W. Pfleiderer	0232	A.A. Akhrem et al.
2586	K. Ienaga and W. Pfleiderer	0338	E.G. Kovalev
2594	W.H. Gündel and W. Kramer	0350	V.M. Potapov et al.
2677	H. Wamhoff and H.-J. Hupe	0597	S.P. Kolesnikov et al.
2738	A. Roedig and M. Forsch		
2779	G. Schramm et al.	31-	-78, Experientia, 34 (1978)
2813	H. Wamhoff and L. Lichtenthäler	0014	J. Engels and R. Reidys
2859	M. Adler and K. Schank	0155	Y. Asakawa et al.
2919	I. Kohlmeyer et al.	0299	A. Groweiss and Y. Kashman
2925	F. Seela and R. Richter	0300	F. Cafieri et al.
3012	H. Musso	0421	J.S. Calderon et al.
3058	A. Schönberg et al.	0422	B.S. Joshi et al.
3086	H. Hofmeister et al.	1254	Z. Imre and O. Ersoy
3094	K. Annen et al.	1257	G.W. van Eijk and H.J. Roemans
3112	H. Siegel et al.		
3140	F. Bohlmann and C. Zdero		

32-   -78, Gazz. chim. ital., 108 (1978)
0013   A. Bianco et al.
0017   A. Bianco et al.
0057   H. Tornetta et al.
0079   G. Casiraghi et al.
0581   P.L. Beltrame et al.
0591   G. Tarzia and G. Panzone
0615   J.V. Oguakwa et al.
0693   F. Orsini et al.

33-   -78, Helv. Chim. Acta, 61 (1978)
0108   H. Balli and L. Felder
0129   J.P. Pauchard and A.E. Siegrist
0142   J.P. Pauchard and A.E. Siegrist
0242   F. Kienzle and R.E. Minder
0266   S. Pürro et al.
0286   H. Bader and H.-J. Hansen
0337   F. Mayerl and M. Hesse
0401   H.-R. Waespe et al.
0430   P.X. Iten et al.
0444   W. Rettig and J. Wirz
0589   Y. Nakamura et al.
0626   A.K. Dey and H.R. Wolf
0690   J.P. Kutney et al.
0709   P. Rüedi and C.H. Eugster
0716   V. Scherrer et al.
0732   A. Chollet and P. Vogel
0795   L. Hoesch et al.
0815   U. Widmer et al.
0822   W. Eschenmoser and C.H. Eugster
0837   N. Cohen et al.
0844   P. Schiess and M. Heitzmann
0848   R. Heckendorn and A.R. Gagneux
0871   K. Grob et al.
0977   T.W. Güntert et al.
0990   P. Dubs and R. Stüssi
1025   P. Margaretha
1033   P.X. Iten and C.H. Eugster
1200   F. Roessler et al.
1213   M.A. Iorio et al.
1221   C.R. Hutchinson et al.
1246   K. Nagarajan et al.
1257   W. Heller and C. Tamm
1262   D. Berney and K. Schuh
1364   P. Schiess and E. Sendi
1427   H. Schmid et al.
1470   C. Delsetch et al.
1477   M. Marky et al.
1554   J.P. Kutney et al.
1565   H. Flohr et al.
1609   U. Lienhard et al.
1622   D. Seebach et al.
1663   W. Leupin and J. Wirz
1682   P. Pfaffli and H. Hauth
1695   M. von Buren et al.
1755   C. Wentrup
1784   D. Bellus et al.
1814   T. Ragettli and C. Tamm
1832   H.H. Brandstetter and E. Zbiral
1969   G. Buchbauer et al.
1975   W. Breitenstein and C. Tamm
2116   H. Link et al.
2328   R. Kaiser and D. Lamparsky
2397   J. Gutzwiller et al.
2407   J.-M. Lehn et al.
2419   H. Link

2542   J. Bruhn et al.
2579   E. Volz and C. Tamm
2628   H. Balli and S. Gunzenhauser
2646   P. Datwyler et al.
2681   A.P. Alder et al.
2697   G. Engler et al.
2813   C.-P. Ehrensperger et al.
2823   T. Yatsunami and S. Iwasaki
2831   S. Iwasaki
2843   S. Iwasaki
2851   F. Heinzer et al.
2887   H. Fritz et al.
2904   B.S.F.E. de Sousa and A.E. Sie-
        grist
2958   R. Naef and H. Balli
3018   V. Veprek-Bilinski et al.
3050   B. Scholl et al.
3068   P. Wieland
3087   W. Lottenbach and W. Graf
3108   K.L. Brown et al.

34-   -78, J. Chem. Eng. Data, 23 (1978)
0182   P.J. Nigrey and A.F. Garito
0261   G.Y. Sarkis
0345   Y.A. Al-Farkh et al.

35-   -78, J. Am. Chem. Soc., 100 (1978)
0202   D.J. Sandman et al.
0211   M.M. Morrison and D.T. Sawyer
0276   P.R. Borkowski et al.
0285   H. Olsen and J.P. Snyder
0307   M.B. Yunker and P.J. Scheuer
0309   V. Chowdhry and F.H. Westheimer
0571   M.R. Uskokovic et al.
0576   J. Gutzwiller and M.R. Uskokovic
0581   G. Grethe et al.
0589   G. Grethe et al.
0628   I. Kubo et al.
0648   R.R. Hautala and R.H. Hastings
0692   C. Santiago et al.
0860   H. Hogeveen and W.F.J. Huurdeman
0877   E.N. Marvell and C. Lin
0883   E.N. Marvell and T.H.C. Li
0920   N.A. Porter et al.
0926   L. de Vries
0934   J.F.W. Keana et al.
0938   J.P. Kutney et al.
0980   D. Davalian et al.
1002   C. Ireland et al.
1008   T. Miyashi et al.
1172   R.G. Weiss and G.S. Hammond
1235   R.C. Haddon et al.
1263   E. Wenkert et al.
1548   D.J. Hart et al.
1778   H. Takaya et al.
1791   Y. Hayakawa et al.
1799   Y. Hayakawa et al.
1806   T. Tsuji et al.
1876   P.S. Engel et al.
2181   A. Padwa and A. Ku
2248   T. Sasaki et al.
2444   L.J. Magdzinski and Y.L. Chow
2464   R.L. Petty et al.
2586   A.J. Fatiadi et al.
2686   D.L. Dubois et al.
2744   M.H. Chisholm et al.

2959   L.N. Donelsmith et al.
3435   B.M. Trost and T.R. Verhoeven
3494   A. Padwa et al.
3526   E. Steckhan
3730   C. Santiago et al.
3819   M. Sindler-Kulyk and W.H. Laarho-
       ven
4131   H.E. Zimmerman and T.R. Welter
4146   H.E. Zimmerman et al.
4208   M. Braun et al.
4220   J.P. Kutney et al.
4225   V.J. Lee et al.
4268   M.B. Gravestock et al.
4320   L.T. Scott and W.R. Brunsvold
4326   H. Hart and M. Sasaoka
4393   M.J. Powers and T.J. Meyer
4481   A. Padwa et al.
4580   E. Ohtsuka et al.
4733   H.J. Callot et al.
4858   P.A. Bartlett and F.R. Green, III
4886   T.C. Klebach et al.
5085   P.A. Bartlett et al.
5122   M.J. Mirbach et al.
5141   H. Hart and E. Dunkelblum
5170   B.C. Pal
5232   J.M. Ohrt et al.
5472   M.M. Rogic and T.R. Demmin
5564   T. Hino and M. Taniguchi
5589   R. Bartetzko et al.
5604   I. Agranat and Y. Tapuhi
5626   E. Berman et al.
5856   L.A. Paquette and M.R. Detty
6035   H.E. Smith et al.
6171   R.P. Thummel and W. Nutakul
6225   B. Fuchs and M. Pasternak
6228   Y. Harel and J. Manassen
6276   T.-S. Fang and L.A. Singer
6416   W.K. Musker et al.
6483   B. Ganem et al.
6491   G. Albers-Schonberg
6516   E.M. Kosower et al.
6535   L.T. Scott and W.R. Brunsvold
6649   K.A. Horn and G.B. Schuster
6679   R. Koussini et al.
6683   P. Fournier de Violet et al.
6728   J. Froborg and G. Magnusson
6733   H. Ogura et al.
6784   J.W. Westley et al.
7009   P.S. Engel et al.
7041   Z. Rappoport et al.
7055   O. Bensaude et al.
7264   M.S. Wrighton et al.
7364   T. Nylund and H. Morrison
7375   R.L. Chan and T.C. Bruice
7382   R. Kluger and J. Chin
7423   J. Vasilevskis et al.
7629   R.C. Haddon et al.
7661   S. Senda et al.
7670   H.L. Levine and E.T. Kaiser
7758   R.E. Moore et al.
7934   R.V. Hoffman and H. Shechter
8004   D.H. Shih et al.
8006   L.D. Cama and B.G. Christensen
8068   L.E. Manzer
8106   D.E. Bergstrom and M.K. Ogawa

8202   J.C. Lagarias et al.
8210   E. Ohtsuka et al.
8214   I. Ernest et al.
8247   A. Padwa et al.
8271   A. Streitwieser, Jr. et al.

36-   -78, J. Pharm. Sci., 67 (1978)
0258   L.B. Burns et al.
0271   D.G.I. Kingston
0272   D.G.I. Kingston
0334   S.G. Schulman et al.
0347   F.S. El-Feraly and Y.M. Chan
0433   D. Dwuma-Badu et al.
0473   J.A. Naghaway and T.O. Soine
0569   C.I. Hong et al.
1045   N. Bodor et al.
1204   J.A. Naghaway and T.O. Soine

39-   -78B, J. Chem. Soc., Perkin Trans.
       II (1978)
0012   D. Forrest et al.
0155   A.J.A. van der Weerdt et al.
0545   B. Fernandez et al.
0607   P. Beltrame et al.
0642   A.J. Kirby and C.J. Logan
0659   R.A. Humphry-Baker et al.
0751   H.V. Ansell and R. Taylor
0892   Kabir-ud-Din and P.H. Plesch
0915   P.H.G. ophet Veld and W.H. Laar-
       hoven
0922   P.H.G. ophet Veld and W.H. Laar-
       hoven
0995   A.B.B. Ferreira and K. Salisbury
1019   R. Godfrey
1157   J.S. Davies and W.A. Thomas
1215   M.J. Cook et al.
1277   S. Matsui and H. Aida
1283   T. Oshima et al.
1292   K.C. Brown et al.
1338   J.C. Craig et al.

39-   -78C, J. Chem. Soc., Perkin Trans I,
       (1978)
0015   D.J. Humphreys et al.
0024   D.J. Humphreys et al.
0045   D.J. Humphreys et al.
0074   R. Ahmad et al.
0076   R. Edmunds et al.
0084   C.G. DeGrazia and W.B. Whalley
0088   J.O. Oluwadiya and W.B. Whalley
0110   R.A. Packer and J.S. Whitehurst
0131   P.A. Lyon and C.B. Reese
0137   F.R. van Heerden et al.
0145   C.H. Hassall and G.J. Thomas
0159   W. Cocker et al.
0163   A.C. Campbell et al.
0171   N. Claydon and J.F. Grove
0191   R. Purvis et al.
0195   R.M. Christie et al.
0209   R.L. Jones and N.H. Wilson
0232   D. Bryce-Smith et al.
0295   M.J.V. deO. Baptista and D.A.
       Widdowson
0299   A. Albini et al.
0329   Z. Paryzek et al.

0360	T. Biftu and P. Stevenson		1113	F. DeSarlo and G. Renzi
0363	M.G. Barlow et al.		1117	R.M. Acheson et al.
0366	D.P. Anold et al.		1126	S. Kanoktanaporn and J. MacBride
0378	M.G. Barlow et al.		1133	I. Degani and R. Fochi
0384	P. Sengupta et al.		1150	J.K.M. Sanders et al.
0387	B. Lythgoe et al.		1161	C.B. Lee et al.
0395	P. Djura and M.V. Sargent		1194	K.-Y. Chu and J. Griffiths
0401	M.G. Bartle et al.		1198	J.M. Lindley et al.
0419	S. Jordan and R.E. Markwell		1208	R.C. Forster and L.N. Owen
0429	N. Abe et al.		1211	R.N. Carde et al.
0434	E. McDonald and A. Suksamrarn		1237	P.J. Butterfield et al.
0440	E. McDonald and A. Suksamrarn		1252	T. Terasawa and T. Okada
0460	T. Kametani et al.		1254	T. Terasawa and T. Okada
0471	W. Lawrie et al.		1263	P.J. Barr et al.
0483	M.T. Garcia-Lopez et al.		1267	K.D. Croft et al.
0488	M. Nojima et al.		1271	G. Savona et al.
0513	A. Albert		1282	J. Wicha and K.Bal
0532	P.J. Cotterill et al.		1293	L.B. Davies et al.
0539	D.E. Ames et al.		1297	J.R. Frost and J. Streith
0543	R. Baker et al.		1303	T. Meikle and R. Stevens
0554	R.R. Rastogi et al.		1315	T. Uchida
0558	N. Fukada et al.		1366	C.M. Pant and R.J. Stoodley
0561	R.G. Alexander et al.		1381	G. Mackenzie and G. Shaw
0571	P.J. Davey et al.		1385	D.J. Brickwood et al.
0576	T. Terasawa and T. Okada		1450	C.L. Branch et al.
0590	B. Lythgoe et al.		1453	D. Lloyd et al.
0668	W. Baker et al.		1461	A. Ardon-Jiminez and T.G. Halsall
0681	C. Venturello		1471	A.S. Bailey et al.
0685	M. Maeda and M. Kojima		1476	R. Rahman et al.
0692	G. Bartoli et al.		1487	M. Ahmed et al.
0726	T. Watanabe et al.		1490	J.K. Holroyde et al.
0730	P. Crabbe et al.		1511	M. Akhtar et al.
0739	R.M. Letcher and K.M. Wong		1533	T.G.C. Bird et al.
0743	J.R. Hanson et al.		1537	J.R. Bull et al.
0750	M.J. Begley et al.		1547	U.J. Kempe et al.
0755	A.R. Forrester et al.		1568	C. Boan and L. Skattebøl
0757	T. Nabih et al.		1572	P.W. LeQuesne et al.
0762	K. Yamauchi and M. Kinoshita		1580	G.R. Chalkley et al.
0768	B. Evans et al.		1588	J.W. ApSimon et al.
0789	F.J. Lalor et al.		1594	J. Cairns et al.
0804	F. Bondavalli et al.		1606	T.G. Halsall et al.
0808	J.C.A. Boeyens et al.		1621	W.B. Turner
0829	P.J. Kocienski et al.		1631	L. Carlton and G. Read
0837	J.M. Vernon et al.		1633	G.D. John and P.V.R. Shannon
0862	G.R. Proctor and B.M.L. Smith		1660	D.P. Arnold et al.
0871	J. Engel et al.		1671	K.P. Parry and G.F. Smith
0876	R. Graham and J.R. Lewis			
0885	E.J. Gray et al.		40- -78, Nippon Kagaku Kaishi (1978)	
0896	A.R. Mattocks et al.		0075	K. Shibata et al.
0909	M.J. Sasse and R.C. Storr		0082	S. Isakawa and S. Tsuruoka
0913	S. Oae et al.		0144	S. Nan'ya and E. Maekawa
0928	S. Jordan et al.		0150	K. Sugiyama et al.
0937	J.B. Chakrabarti et al.		0271	H. Shosenji et al.
0955	A.J.H. Labuschagne et al.		0404	K. Yanagi and T. Nishiyama
0961	C.P. Gorst-Allman et al.		0489	C. Shibata et al.
0967	B. Bannister		0582	M. Matsumura et al.
0974	R. Bonnett et al.		0827	M. Tsuchiya
0980	T.A. Eggelte et al.		1127	S. Shiraishi et al.
0989	H.G. Heller and R.D. Piggott		1249	M. Hida and S. Kawakami
1006	C.D. Campbell et al.		1449	Y. Sakaino and H. Kakisawa
1014	A.G.M. Barrett et al.		1690	N. Negishi et al.
1017	M. Muraoka et al.			
1023	H. McNab		42- -78, J. Indian Chem. Soc., 55 (1978)	
1041	G. Roberge and P. Brassard		0146	A.K. Koli and W.R. Williams
1066	P.F. Alewood et al.		0154	R. Grover and B.C. Joshi
1083	K.-Y. Chu and J. Griffiths		0242	A.K. Koli and E. McClary

0281	J.S. Shukla and P. Bhatia
0308	P. Bhattacharyya et al.
0468	K.P. Sanghvi and K.N. Trivedi
0580	D. Nasipuri and S.K. Konar
0702	A. Chaudhuri et al.
0847	P.K. Sen et al.
0916	S.C. Shaw et al.
1096	G. Lewin et al.
1099	H.-P. Husson et al.
1114	D.P. Chakraborty, S. Roy and R. Guha
1122	E. Wenkert and A.C. Kryger
1142	S. Krishnan et al.
1152	S.K. Talapatra et al.
1169	A.S.R. Anjaneyulu et al.
1175	A.B. Ray et al.
1198	F.C. Bose and N. Adityachaudhury
1201	S. Ghosal and S. Banerjee
1204	R. Talapatra et al.
1224	R.W.°Gray and A.S. Dreiding
1228	B.S. Thyagarajan et al.
1232	D. Nasipuri et al.

44-   -78, J. Org. Chem., 43 (1978)

0034	L.E. Friedrich et al.
0066	A. Padwa et al.
0076	J.P. Dirlam et al.
0079	W.H. Koster et al.
0105	S.M. Kupchan et al.
0161	L.M. Lerner
0167	F.L. Lam and T.-C. Lee
0168	G. Buchi and K.-C.T. Luk
0171	S.M. Kupchan and W.L. Sunshine
0232	A.S. Monahan
0268	E.V. Crabtree et al.
0283	A. Gossauer and R.-P. Hinze
0289	P.D. Cook and R.K. Robins
0296	D.H. Rich et al.
0303	A. Padwa et al.
0315	A.G. Anastassiou et al.
0343	D. Caine et al.
0350	T. Sasaki et al.
0355	H.C. Price et al.
0367	R.A. Sharma and M. Bobek
0369	N.C. Gonnella and M.P. Cava
0381	A. Padwa et al.
0397	Y. Ittah et al.
0416	M. Mizuno and M.P. Cava
0441	K. Okada and S. Sekiguchi
0469	M. Ichiba et al.
0481	B.A. Otter et al.
0497	A.I. Fetell and H. Feuer
0511	J.D. Bryant and N.J. Leonard
0516	B.N. Holmes and N.J. Leonard
0519	B.C. Clark, Jr. et al.
0529	W. Wierenga and J.A. Woltersom
0536	J.T. Edward and M.J. Davis
0541	J.A. Montgomery and H.J. Thomas
0544	M.A. Templeton and J.C. Parham
0574	W.H. Okamura et al.
0580	W.N. Wu et al.
0586	S.M. Kupchan et al.
0595	S. Yoneda et al.
0598	A. Streitwieser, Jr. and L.L. Nebenzahl
0604	L.E. Friedrich et al.

0610	R.P. Hanzlik and J.M. Hilbert
0626	W. Nagai and Y. Hirata
0664	A.G. Anderson, Jr., et al.
0672	C.N. Filer et al.
0678	N.F. Haley
0689	M. Nakazaki et al.
0700	H.O. House et al.
0727	T.D. Harris et al.
0731	H.M. Walborsky and P. Ronman
0736	E.C. Taylor et al.
0737	A.A. Liebman et al.
0740	J. Wolinsky et al.
0775	E.D. Smith and L. Elrod, Jr.
0808	W.H. Pirkle and P.L. Gravel
0828	A. Srinivasan et al.
0914	M.M. Chan and J.L. Kice
0922	P.E. McGann et al.
0936	A. Walser et al.
0944	Y. Ogata and K. Takagi
0950	D.S. Matteson et al.
0954	G.P. Butke et al.
0962	L.M. Lerner
0966	R.J. Chorvat
0975	H. Iida et al.
0980	R. Srinavasan et al.
0998	Y. Wang and H.P.C. Hogenkamp
1002	J.G. Buta et al.
1041	M. Nakazaki et al.
1050	R.W. Thies and E.P. Seitz
1096	J.M. Saa and M.P. Cava
1114	D. Caine et al.
1118	M.R. Detty and L.A. Paquette
1126	K.E. Wiegers and S.G. Smith
1132	J.W. Bunting et al.
1193	K. Hirota et al.
1197	L.D. Taylor et al.
1226	R.L. Garnick et al.
1230	G.T. Cheek and R.F. Nelson
1233	N.F. Haley
1238	A.I. Fetell and H. Feuer
1248	T. Fujita et al.
1331	S.B. Brown and M.J.S. Dewar
1361	T. Goka et al.
1435	J.L. Grandmaison and P. Brassard
1438	K. Schroder and H.G. Floss
1442	D.J. Vanderah and C. Djerassi
1448	S.P. Singer and K.B. Sharpless
1459	K.G. Taylor and J.B. Simons
1481	A. Padwa et al.
1493	H.E. Zimmerman and S.M. Aasen
1550	M. Rosenberger et al.
1569	R.B. Miller and C.G. Gutierrez
1573	L.A. Wendling and R. West
1577	L.A. Wendling and R. West
1580	L.L. Miller et al.
1586	R.J. Hunadi and G.H. Helmkamp
1602	A.G. Anderson, Jr., et al.
1612	S. Nishida and F. Kataoka
1613	H.H. Sun and K.L. Erickson
1616	L. Mandell et al.
1624	D.K. Minster et al.
1627	M.L. Casey et al.
1644	C.R. Frihart et al.
1656	C.E. Browne et al.
1664	A. Padwa et al.
1677	K. Senga et al.

1756	S.J. Cristol et al.		3420	F.M. Moracci et al.
1890	J.A. Villarreal et al.		3425	N.C. Yang et al.
1900	E.G. Janzen and R. C. Zawalski		3446	A.C. Jain et al.
1912	R.F. Heldeweg et al.		3454	M.D. Higgs and D.J. Faulkner
1992	B.R. Pai et al.		3462	R.E. Lehr et al.
1997	H.E. Zimmerman and W.T. Gruenbaum		3470	D. Kamp and V. Boekelheide
2026	S.F. Krauser and A.C. Watterson, Jr.		3475	D. Kamp and V. Boekelheide
2029	A. Padwa and P.H.J. Carlsen		3478	G.L. Grunewald et al.
2056	H.D. Perlmutter and R.B. Trattner		3536	C.K. Bradsher and I.J. Westerman
2084	M. Sato et al.		3544	R. Buchan et al.
2093	D.R. Morton and R.A. Morge		3653	H.O. House et al.
2125	T.E. Catka and E. Leete		3678	Z. Rappoport et al.
2127	B.N. Ravi and D.J. Faulkner		3702	M.E. Kuehne and R. Haffer
2138	N. Ishibe and S. Yutaka		3705	M.E. Kuehne et al.
2153	H.O. House et al.		3713	D.D. Keith et al.
2167	R.H. Wightman et al.		3717	G. Büchi and P.-S. Chu
2190	J.J. Eisch et al.		3723	N. Cohen et al.
2224	J.P. Snyder et al.		3727	C.K. Lee et al.
2248	I.M. Cuccovia et al.		3730	S. Mataka et al.
2289	A.G. Hortmann et al.		3757	A. Padwa and P.H.J. Carlsen
2291	R.S. Glass and R.J. Swedo		3776	Y. Tobe et al.
2311	S. Daluge and R. Vince		3778	R.J. Chorvat
2320	T. Sasaki et al.		3813	W.J. Lipa et al.
2325	J.C. Parham et al.		3821	J.F. Blount et al.
2334	G.R. Lenz and J.A. Schulz		3848	W.R. Dolbier, Jr., and O.T. Garza
2339	G.A. Ellestad et al.		3882	S. Kishimoto et al.
2351	J. Tadanier et al.		3893	R. Gleiter et al.
2366	E.H. White et al.		3904	A. Nickon et al.
2435	L.-S. Wen et al.		3937	J.A. Secrist and P.S. Liu
2484	D.N. Leach and J.A. Reiss		3946	E. Yoshii et al.
2487	J. Font et al.		3950	G.R. Krow et al.
2493	A.H.A. Tinnemans and D.C. Neckers		3983	G. Büchi et al.
2508	P. Spagnolo et al.		4069	T.C. Kung and C.D. Gutsche
2521	S.M. Kupchan et al.		4076	S.M. Kupchan et al.
2536	R.S. Klein et al.		4081	W.J. Gensler et al.
2557	D. Becker et al.		4090	S.S. Hirsch and W.J. Bailey
2562	D. Becker et al.		4110	I. Arai and G.D. Daves, Jr.
2587	J.H. Keck, Jr., et al.		4112	S. Bersani et al.
2621	S.F. Nelsen et al.		4115	M.T. Edgar et al.
2665	S. Gelin et al.		4125	B.W. Cue, Jr., et al.
2676	G.F. Koser et al.		4128	A.A. Ardakani et al.
2697	K.T. Potts and D.R. Choudhury		4200	A.F. Cook and M.J. Holman
2700	K.T. Potts and D.R. Choudhury		4215	J. Moron and G. Roussi
2715	S.C. Welch and M.E. Walters		4220	F.J. Schmitz et al.
2833	R.H. Smithers		4271	Y. Ittah et al.
2852	M. Shamma and H.H. Tomlinson		4303	G. Doddi et al.
2855	E. Bernstein et al.		4316	H.O. House et al.
2870	J.L. Ruth and D.E. Bergstrom		4338	J.P. Albarella and T.J. Katz
2886	S.S. Simons, Jr., and D.F. Johnson		4352	P.S. Manchand and J.F. Blount
2930	G. Barany et al.		4359	J.S. Mynderse and R.E. Moore
2932	G. Bartoli et al.		4369	H.O. House and T.V. Lee
3044	M. Imazawa and F. Eckstein		4377	A.E. Greene et al.
3070	K. Ohkata et al.		4394	M.V. Lakshmikantham et al.
3116	F.G. Bordwell and J.A. Hautala		4415	L. Tsai et al.
3194	S.M. Waraszkiewicz et al.		4420	F.M. Schell and P.M. Cook
3209	R.S. Glass et al.		4438	R. Busson and H. Vanderhaeghe
3283	H.E. Zimmerman and T.P. Cutler		4447	P.E. Krieger and J.A. Landgrebe
3324	Y. Wang and H.P.C. Hogenkamp		4464	S.M. Kupchan et al.
3339	H. Wagner et al.		4469	D.E. Remy et al.
3348	J.H. Hall and M. Wojciechowska		4472	M. Akiba et al.
3362	G.R. Newcome et al.		4512	M.L. Greenlee et al.
3367	N. Sato		4598	U.R. Ghatak et al.
3374	A.G. Hortmann et al.		4622	V. Dave and E.W. Warnhoff
3388	F.M. Miller and R.A. Lohr, Jr.		4642	M.L. Kaplan et al.
3394	J. Correia		4712	L.A. Paquette et al.
3405	K.B. Sloan et al.		4765	G. Büchi et al.

4769   P.S. Manchand et al.
4774   E.J. Prisbe et al.
4784   E.J. Prisbe et al.
4844   R.W. Morrison, Jr., et al.
4859   R.J. Sundberg et al.
4873   V.T. Yue et al.
4892   J. Ackrell
4910   B.L. Cline et al.
4961   H. Cohn et al.
4966   F.H. Greenberg and Y. Gaoni
4975   H.S. Freeman et al.
4984   N.H. Fischer et al.
4987   L.E. Salisbury
4996   F. Ramirez et al.
5006   M. Neptune and R.L. McCreery

47-   -78, J. Polymer Sci., Polymer Chem.
          Ed., 16 (1978)
0041   H. Uemura et al.
1343   J.L. Nash, Jr., et al.
1367   R.E. Thompson and G.B. Butler
2039   S. Watura et al.
2093   G. Rabilloud and B. Sillion

48-   -78, J. prakt. Chem., 320 (1978)
0071   H.G. Henning et al.
0097   A.H. Moustafa et al.
0133   G.W. Fischer et al.
0157   M. von Janta-Lipinski et al.
0313   R. Mayer et al.
0497   M. Weissenfels and H.J. Hense
0521   W. Freyer
0557   I.M. Issa et al.
0647   H. Hartmann et al.
0659   G.W. Fischer
0677   D. Martin et al.
0857   A.M. Osman et al.
0863   S.R. Ramadas and S. Padmanabhan
0945   H.G. Henning et al.
0986   W.I. Awad and R.R. Al-Sabti
0991   N.R. El-Rayyes and F.H. Al-Hajjar
1029   Y.M. Temerk et al.

49-   -78, Monatsh. Chem., 109 (1978)
0123   O.S. Wolfbeis and G. Uray
0137   H. Berner et al.
0183   H. Falk et al.
0337   F. Wille and W. Schwab
0557   H. Berner et al.
0575   G. Haufe et al.
0767   H. Kämmerer et al.
0883   H. Falk and A. Leodolter
0905   O.S. Wolfbeis
0987   E. Langer et al.
1017   S. Smolinski et al.
1075   H.W. Schmidt and H. Junek
1081   J. Schantl and P. Karpellus
1191   H. Falk et al.
1227   K. Schlögl and R. Schölm
1295   H. Möhrle

51-   -78, Naturwiss., 65 (1978)
0652   H. Alpes and W.G. Pohl

54-   -78, Rec. trav. chim., 97 (1978)
0073   J.H. van Boom and P.M.J. Burgers

0105   J.W. van Straten et al.
0121   D. Paquer et al.
0197   R.J.F.M. van Arendook et al.
0249   B. Feringa and H. Wynberg
0265   P.M. op den Brouw and W.H. Laar-
          hoven
0305   W. Verboom and H.J.T. Bos

56-   -78, Polish J. Chem., 52 (1978)
0037   W.E. Hahn and E. Kozlowska-Gramsz
0347   M. Kozlowska and W. Sobotka
0511   L. Lechowski
0529   S. Mejer and R. Pacut
0581   M. Kozlowska and W. Sobotka
0629   J.K. Ruminski et al.
1035   K. Golankiewicz and L. Celewicz
1085   K. Golankiewicz et al.
1255   E. Wyrzykiewicz et al.
1265   E. Wyrzykiewicz et al.
1355   J. Kraska and J. Sokolowska-
          Gajda
1365   K. Golankiewicz and B. Skalski
1579   L. Strekowski
1713   W.J. Rodewald and Z. Chilmonczyk
1827   T. Kiersznicki and A. Rajca
1913   S. Smolinski et al.
2039   K. Krowicki
2045   W.E. Hahn and M. Jatczak
2125   S. Skonieczny et al.
2233   W. Zielinski
2361   W.J. Rodewald and J.W. Morzycki
2377   W. Jaziobedzki et al.
2455   W. Podkoscielny et al.
2479   Z. Witczak and M. Krolikowska

60-   -78, J. Chem. Soc., Faraday Trans.
          II, (1978)
1909   J. Kroner et al.

61-   -78, Ber. Bunsen Gesell. Phys. Chem.,
          82 (1978)
0396   H. Felbecker et al.
1068   S. Schoof et al.

62-   -78A, Z. phys. Chem. (Leipzig), 259
          (1978)
0762   D. Lohse et al.
1083   M.S.A. Abd El-Mottaleb et al.

63-   -78, Z. physiol. Chem., 359 (1978)
0211   J.P. Lautié and M. Olomucki
0407   J. Müller and G. Pfleiderer
0813   M. Christensen et al.
1617   E. Jaeger et al.
1637   M. Löw et al.
1643   M. Löw et al.
1659   F. Hansske et al.

64-   -78B, Z. Naturforsch., 33b (1978)
0075   N.A. Kassab et al.
0080   I.F. Eckhard et al.
0084   W.H. Gündel
0118   H.G. Grant and L.A. Summers
0197   R.R.M. Brand et al.
0209   Z.H. Khalil
0220   C.E. May et al.

0332 W. Fabian
0663 G. Schaden
0924 H. Falk et al.
0976 S. Kato et al.
1012 C. Reichardt and W. Scheibelein
1020 R. Hansel et al.
1091 P. Peringer
1381 H. tom Dieck et al.
1520 M.J. Volz-de Lecea
1527 L. Rateb et al.
1535 M. Hargreaves et al.

64-   -78C, Z. Naturforsch., 33c (1978)
0056 F. Seela and H. Rosemeyer
0326 E. Schlimme
0807 B.M. Giannetti et al.
0820 H. Besl et al.
0912 M.G. Peter

65-   -78, Zhur. Obshchei Khim., 48 (1978)
      (English translation pagination)
0142 A.P. Apsitis and L.S. Geita
0340 A.A. Sukhorukov et al.
0720 I.P. Titarenko and L.D. Protsenko
0841 E.V. Popova and G.S. Grinenko
0891 A.Y. Deich
1073 V.P. Yur'ev et al.
1080 V.D. Rumyantseva et al.
1201 O.G. Piskunova et al.
1205 O.G. Piskunova et al.
1244 O.A. Zazyadko et al.
1352 E.E. Nifant'ev et al.
1394 G.V. Ratovskii et al.
1474 G.V. Panova et al.
1529 K.A. Pivnitskii and Y.P. Badanova
1635 P.D. Romanko et al.
1644 R.I. Zalewski et al.
1707 B.M. Gutsulyak and Z.L. Nivitskii
1725 E.M. Kaz'mina et al.
1732 N.A. Rozanel'skaya et al.
1817 G.V. Ratovskii et al.
1956 K.O. Kim et al.
1998 L.V. Megera and M.I. Shevchuk
2027 V.F. Traven' et al.
2318 S.A. Chernyuk et al.
2383 N.S. Bychkov et al.

67-   -78, J. Structural Chem., 19 (1978)
0882 D.M. Kizhner et al.

69-   -78, Biochemistry, 17 (1978)
0094 K. Quiggle et al.
1942 C. Walsh et al.
3321 J.V. Staros and J.R. Knowles
4583 L.D. Eirich et al.
4865 F.T. Liu and N.C. Yang

70-   -78, Izvest. Akad. Nauk S.S.S.R.,
      27 (1978)
0091 V.A. Mironov et al.
0102 Z.A. Krasnaya et al.
0107 Z.A. Krasnaya et al.
0130 F.M. Stoyanovich and M.A. Marakatkina
0133 A.N. Volkov et al.
0149 Y.S. Shvetsov et al.

0160 A.V. Kamernitskii et al.
0178 L.G. Abakumova et al.
0215 M.I. Kanishchev et al.
0339 Z.A. Krasnaya et al.
0352 A.K. Margaryan et al.
0384 A.A. Akhrem et al.
0448 A.A. Volod'kin et al.
0550 Y.S. Dol'skaya et al.
0844 L.B. Volodarskii and V.A. Samsonov
0901 Y.L. Frolov et al.
0923 U.M. Dzhemilev et al.
0932 S.A. Khurshudyan et al.
0945 S.A. Shevelev et al.
0957 E.I. Karpeiskaya et al.
0963 G.Y. Kondrat'eva et al.
1035 A.A. Volod'kin et al.
1179 G.F. Bannikov et al.
1214 A.V. Kamernitskii et al.
1220 A.Y. Lazaris and A.N. Egorochkin
1241 M.A. Vardanyan et al.
1287 U.M. Dzhemilev et al.
1304 I.V. Khudyakov et al.
1313 L.I. Kudinova et al.
1339 B.A. Arbusov et al.
1361 S.A. Khurshudyan et al.
1452 L.I. Kudinova et al.
1573 L.M. Epshtein et al.
1674 E.P. Serebryakov et al.
1829 Z.M. Garashchenko and G.G. Skvortsova
2080 V.V. Semenov et al.
2087 V.V. Semenov and S.A. Shevelev
2160 F.G. Yusupova et al.
2215 I.A. Shlyapnikova et al.
2227 V.K. Ivanova et al.
2243 L.M. Epshtein et al.
2284 U.M. Dzhemilev et al.
2455 G.I. Nikiforov et al.

73-   -78, Coll. Czech. Chem. Comm., 43
      (1978)
0057 J. Podlahova
0156 R. Kada et al.
0160 R. Kada et al.
0257 O. Hritzova and P. Kristian
0288 A. Krutosikova et al.
0309 K. Sindelar et al.
0434 V. Skala et al.
0463 I. Srokova et al.
0471 K. Sindelar et al.
0621 R. Kada et al.
0870 F. Povazanec et al.
0938 P. Zahradnik et al.
0970 Z.J. Vejdelek et al.
1093 F. Gregan et al.
1113 J. Jizba et al.
1134 L. Kohout and J. Fajkos
1142 J. Fajkos and J. Joska
1227 E. Bulka et al.
1248 V. Kral and Z. Arnold
1261 M. Svoboda et al.
1276 M. Rajsner et al.
1438 P. Sedmera et al.
1511 J.J.K. Novak
1571 T. Kucerova and J. Mollin

1618	D. Geisbacher et al.
1727	V. Zikan et al.
1732	Z.J. Vejdelek
1747	J.O. Jilek et al.
1760	M. Rajsner et al.
1803	J. Cudlin et al.
1808	J. Cudlin et al.
1917	L. Kniezo et al.
2024	J. Kuthan et al.
2037	R. Kada and J. Kovac
2041	J. Stetinova et al.
2046	A.R. Katritzky et al.
2054	A. Holy et al.
2190	V. Pouzar and A. Vystreil
2289	V. Zelenska et al.
2312	R.J. Jovceva et al.
2395	K. Topek et al.
2415	V. Konecny et al.
2427	V. Bartl et al.
2619	V. Svata et al.
2635	M. Siroky and M. Prochazka
2643	A. Krutosikova et al.
2656	M. Protiva et al.
2711	Y. Koblizkova et al.
2740	O. Exner and N. Motekov
2763	O. Exner and P. Vetesnik
2897	R. Karlicek et al.
3056	J. Kuthan et al.
3092	J. Jilek et al.
3103	A. Holy et al.
3252	I. Srokova et al.
3268	H. Hrebabecky and J. Beranek
3404	D. Vegh et al.
3414	A. Rybar et al.
3444	A. Holy

77-   -78, J. Chem. Soc., Chem. Comm. (1978)

0029	A.G. Schultz and I.-C. Chiu
0057	L. Moegel et al.
0079	M. Bittner et al.
0124	F. Mizuno and Y. Inoue
0152	E.K. Adesogan and A.L. Okunade
0198	A. Matsuo et al.
0281	A. Shoeb et al.
0284	D.E. Bergstrom and K.F. Rash
0309	M.M. Baradarani and J.A. Joule
0377	W.H. Rastetter et al.
0442	T. Miyashi et al.
0469	P.C. Cherry et al.
0526	E. Huq et al.
0528	M. Mellor et al.
0529	E.G. Scovell and J.K. Sutherland
0533	A.W. Dunn et al.
0557	J.M. Mellor et al.
0620	Y. Nakamura et al.
0627	T.J. Simpson
0628	M. Kalyanasundaram
0720	H. Iwamura and K. Makino
0779	A.H. Jackson et al.
0826	F. Tillequin et al.
0850	P.M. Cullis et al.
0869	N. Ikota and B. Ganem
1033	E. Piers and H.E. Morton
1075	A. Mudd

78-   -78, Tetrahedron, 34 (1978)

0041	N.H. Andersen et al.
0057	L. Jurd and J.N. Roitman
0067	T. Sasaki, S. Eguchi and S. Hattori
0073	J. Rigaudy et al.
0083	J. Rigaudy et al.
0101	J. Boedeker and K. Courault
0113	J. Rigaudy and D. Sparfel
0209	M. Fetizon et al.
0233	E.L. Ghisalberti et al.
0241	S.F. Dyke et al.
0331	D.E. Caddy and J.H.P. Utley
0345	E.P. Serebryakov et al.
0379	R. Bonnett et al.
0393	P. Aclinou et al.
0399	K. Eckardt et al.
0437	M. Lounasmaa and M. Hameila
0453	A. Martvon, L. Floch and S. Sekretar
0495	J.W. Barton et al.
0505	Y. Kito, M. Namiki and K. Tsuji
0533	B. Foelisch et al.
0577	E.G. Sundholm
0591	P.M. Adler et al.
0599	S. Krishnappa and Sukh Dev
0611	G.C. Barrett
0635	M. Shamma et al.
0677	R.Z. Andriamialisoa et al.
0725	W.-D. Rudorf
0761	U. Sequin
0769	W.H. Laarhoven et al.
0833	S.D. Worley et al.
0871	H. Keller et al.
0891	H.P. Kraemer and H. Plieninger
0903	G. Cauquis and B. Chabaud
0921	S. Nomoto et al.
0929	V. Bocchi, G. Casnati and R. Marchelli
0951	L. Baiocchi et al.
0955	L. Baiocchi et al.
0981	S.D. Carter and G.W.H. Cheeseman
0989	J. Becher et al.
1011	A.G. Gonzalez et al.
1023	R.P. Loven and W.N. Speckamp
1027	R.P. Loven and W.N. Speckamp
1037	S.J. Harris and D.R.M. Walton
1133	M. Ikehara, T. Maruyama and H. Miki
1163	J. Martelli and R. Carrie
1221	J.I. Okogun et al.
1241	S. Zbaida and E. Breuer
1251	C.R. Duffner and F. Kurzer
1323	E.N. Marvell et al.
1363	H.-P. Husson et al.
1377	I.D. Biggs and J.M. Tedder
1405	R. Vlegaar et al.
1411	R.D.H. Murray and I.T. Forbes
1427	W.A. Szarek et al.
1457	S. Iwadare et al.
1509	A. Enger et al.
1551	J.C. Braekman et al.
1567	D.W. Theobald
1571	A.V. Chapman et al.

1581   G. Jones, J.R. Phipps and P. Raf-
        ferty
1593   S. Ahmad, H. Wagner and S. Razaq
1595   F.R. Hewgill
1633   B. Green, B.L. Jensen and P.L.
        Lalan
1661   F.J. McEnroe and W. Fenical
1707   W. Reineke et al.
1769   Y. Kuwahara et al.
1775   H.E. Zimmerman and R.T. Klun
1845   F.C. Brown, D.G. Morris and A.N.
        Murray
1889   N. Numao et al.
1915   K. Yamada, Y. Shizuri and Y. Hirata
1943   M.B. Gase et al.
1957   Y. Bessiere et al.
1965   H. Sadlo and W. Kraus
1979   S. Linke, J. Kurz and H.-J. Zeiler
1985   T, Eicher et al.
1999   J.H. Van Boom et al.
2015   G. Giacomelli et al.
2027   J. Froborg and G. Magnusson
2077   M.O. Stallard et al.
2175   E.G. Frandsen
2201   K.Y. Geetha et al.
2213   R.S. Sagitullin, S.P. Gromov and
        A.N. Kost
2229   F. Roelants and A. Bruylants
2233   R. Bucourt et al.
2259   D. Bensaude et al.
2263   J. Rigaudy and D. Sparfel
2295   H.J. Callot and E. Schaeffer
2305   S. Lahiri et al.
2321   R. Osman and Y. Shvo
2349   B. Karlsson et al.
2389   M. Bialer et al.
2439   K.B. White and W. Reusch
2449   A. Matsuda, M. Tezuka and T. Ueda
2491   S. Huneck and G. Hofle
2509   W. Friedrichsen and H. von Wallis
2529   M. Lounasmaa et al.
2533   R. Perrone and V. Tortorella
2557   J. Fetter, K. Lempert and J. Møller
2565   J.H. Borkent and W.H. Laarhoven
2569   J.H. Borkent et al.
2609   P.S. Mariano et al.
2617   P.S. Mariano et al.
2627   V.K. Srivastava and L.M. Lerner
2633   A. Matsuda et al.
2639   H. Dadoun et al.
2729   L. Nedelec et al.
2783   K.J. Crowley and S.G. Traynor
2791   N.J. McCorkindale et al.
2797   M. Franck-Neumann and C. Dietrich-
        Buchecker
2861   Y.J. Lee, W.A. Summers and J.G.
        Burr
2887   M.S. Puar and B.R. Vogt
2893   Y. Uchio
2927   T. La Noce and A.M. Giuliani
2967   S. Colombi et al.
2995   M. Lounasmaa et al.
3005   A.C. Bazan, J.M. Edwards and U.
        Weiss
3215   V.M. Karpov et al.
3291   B. Kryczka et al.

3331   P. Baeckström
3341   P.L. Majumder et al.
3563   A.C. Jain, R.C. Gupta and P.D.
        Sarpal
3569   A.C. Jain et al.
3631   H.J. Eggelte et al.

80-   -78, Revue Roumaine Chim., 23 (1978)
0079   S.H. Etaiw et al.
0397   R. Craciuneanu and E. Florean
0617   I.M. Issa et al.
0921   O. Mantsch and M. Buchwald
1085   M. Anwar et al.
1465   A.M. El-Abbady et al.

83-   -78, Arch. Pharm., 311 (1978)
0018   F. Eiden and W. Hirschmuller
0115   H. Fenner and R. Teichmann
0135   R. Hansel and E.M. Cybulski
0153   H. Fenner and A. Motschall
0161   A. Burkartsmaier and E. Mutsch-
        ler
0170   R. Neidlein and I. Körber
0184   H. Hamacher et al.
0236   R. Neidlein and I. Körber
0256   R. Neidlein and I. Körber
0267   K.C. Liu et al.
0287   F. Eiden and C. Herdeis
0294   G. Dannhardt
0303   H. Fenner and R.W. Grauert
0324   R. Neidlein and H. Seel
0328   W. Wiegrebe et al.
0369   K.H. Schlingensief and K. Hartke
0393   K.-W. Glombitza et al.
0433   J. Schnekenburger et al.
0503   F. Eiden et al.
0511   G. Rucker and H.W. Hembeck
0561   F. Eiden et al.
0600   K. Rehse et al.
0714   R. Neidlein and E. Bernhard
0728   G. Seitz and T. Kämpchen
0754   G. Rücker and M. Schikarski
0786   G. Seitz et al.
0848   W. Löwe
0954   C.H. Brieskorn and G. Wittig
0960   K. Görlitzer and E. Engler
0977   G. Dannhardt and R. Obergangs-
        berger
1029   H.W. Bersch et al.

86-   -78, Talanta, 25 (1978)
0209   J. Volk et al.

87-   -78, J. Med. Chem., 21 (1978)
0026   B.L. van Duuren et al.
0031   S.M. Kupchan et al.
0038   S.S. Hecht et al.
0088   C.J. Barnett et al.
0096   C.K. Chu et al.
0106   T.S. Lin and W.H. Prusoff
0109   T.S. Lin and W.H. Prusoff
0112   R.D. Elliott and J.A. Montgomery
0130   T.S. Lin et al.
0199   R. Zee-Cheng et al.
0204   J.A. Beisler
0225   J.M. Kokosa et al.

0280	T.H. Smith et al.		0803	K. Komatsu et al.
0331	D.F. Worth et al.		0837	D. Sica and F. Zollo
0340	C.R. Ellefson et al.		0847	A. Jensen and M.A. Ragan
0344	A. Giner-Sorolla et al.		0869	S.P. Tanis et al.
0427	K. Miyai et al.		0923	H. Seto et al.
0443	W. Skuballa et al.		0961	S.J. Wratten and D.J. Faulkner
0483	J.P. Dirlam and J.E. Presslitz		0969	J. Kleinschroth and H. Hopf
0485	L.A. Mitscher et al.		0977	M. Oda et al.
0520	G.R. Gough et al.		1005	J.W. Barton and R.B. Walker
0578	G. Alonso et al.		1067	Y. Fukazawa et al.
0607	A.K.A. Chowdhury et al.		1071	R. Bloch
0649	M.W. Gemborys et al.		1099	T.C. Klebach et al.
0673	M.G. Nair et al.		1163	S. Robev
0704	W.S. Mungall et al.		1175	G. Märkl and J.B. Rampal
0706	D.K. Parikh and R.R. Watson		1183	H.-D. Martin and M. Hekman
0738	M. Yasumoto et al.		1199	Y. Tsuda et al.
0754	P.L. Stutz et al.		1223	G. Mehta et al.
0781	B. Cavalleri et al.		1225	Z. Goldschmidt and S. Antebi
0792	K.L. Fong and R. Vince		1229	W. Verboom and H.J.T. Bos
0804	E.A. Coats et al.		1261	J. Lallemand et al.
0815	J.M. Cassady et al.		1291	H. Koga et al.
0883	S.-H. Kim et al.		1299	M. Hirama et al.
0889	G.D. Diana et al.		1391	S.J. Wratten et al.
0895	N. Cohen et al.		1395	S.J. Wratten and D.J. Faulkner
0952	C.R. Ellefson et al.		1401	I. Tsujino et al.
0990	S.W. Schneller et al.		1425	M.E. Blank and N.W. Haenel
1025	A. Mourino et al.		1447	K. Ienaga and W. Pfleiderer
1044	C.R. Rasmussen et al.		1459	F. Vögtle and R. Wingen
1079	H. Yoshimura et al.		1471	G. Märkl and J.B. Rampal
1093	J. Hannah et al.		1483	S. Nakatsuji et al.
1137	A. Hampton et al.		1519	C. Charles et al.
1162	D.C. Suster et al.		1539	K. Nakasuji et al.
1165	D.C. Suster et al.		1553	Y. Asakawa et al.
1186	M.E. Wall and M.C. Wani		1699	G. Combaut et al.
1212	P.D. Cook et al.		1751	Y. Tamura et al.
1218	D.C. Baker et al.		1801	L. Tchisambou et al.
			1829	T. Sato et al.
88- -78, Tetrahedron Letters, (1978)			1885	A. Ulman et al.
0093	A. Padwa and F. Nobs		1887	E. Tichineanu et al.
0125	H. Wamhoff and H.-J. Hupe		1893	C.F. Wilcox, Jr., et al.
0143	M. Sato et al.		1983	T. Ohta et al.
0155	Y. Honma and Y. Ban		2025	R.Y. Malakov et al.
0159	G.J. Bird et al.		2119	C.O. Fakunle et al.
0303	U. Eisner et al.		2127	P. Naegeli et al.
0307	C. Rossi and L. Tuttobello		2145	W. Tochtermann and H. Timm
0345	W. Abraham et al.		2179	L. Castedo et al.
0357	T. Mukai and Y. Yamashita		2187	U. Sanyal et al.
0429	X.A. Dominguez		2217	S. Rajappa and R. Sreenivasan
0441	K. Issleib et al.		2235	R.E. Schwartz et al.
0449	P. Blickle and H. Hopf		2251	K.A. Parker and M. Sworin
0457	G. Rucker and B. Langmann		2289	M. Hirama et al.
0475	W. Gramlich and H. Plieninger		2297	I. Mori et al.
0511	M.K. Logani et al.		2299	S. Takano et al.
0539	A.J. Hutchison and Y. Kishi		2331	H. Gotthardt et al.
0543	D.S. Kemp et al.		2387	I.M. Takakis and W.C. Agosta
0569	T. Nakazawa et al.		2445	G. Stork et al.
0599	A.T. Balaban et al.		2453	B.M. Howard and W. Fenical
0623	W. Kogel and G. Schroder		2469	E.A. Wildi and B.K. Carpenter
0637	R.G. Cooke and I.J. Dagley		2507	T. Otsubo et al.
0639	J. Zjawiony and H. Zajac		2541	R.J. Andersen
0645	H. Babsch and H. Prinzbach		2553	L.A. Mitscher et al.
0667	H.N.C. Wong et al.		2557	R.D. Roberts and T.A. Spencer
0687	J. Rigaudy et al.		2579	S. Nakatsuka et al.
0701	J. Szmuszkovicz et al.		2585	I. Saito et al.
0795	J.C. Espie et al.		2609	L. Minale and R. Riccio
0797	M. Takasugi et al.		2645	R.N. Ferguson et al.

2653	A.B. de Oliveira et al.
2675	P.T. Lansbury and R.W. Erwin
2719	T.C. Walsgrove and F. Sondheimer
2723	B.A. Burke and H. Parkins
2731	G. Jones and P. Rafferty
2765	K.A. Reimann and D.M. Piatak
2791	H. Abe et al.
2795	T. Asao et al.
2803	F. Yoneda et al.
2815	R. Todesco et al.
2819	P.H. McCabe and C.R. Nelson
2907	T. Itaya and K. Ogawa
2911	E. Schlittler and U. Spitaler
2923	L. Castedo et al.
2929	U.R. Ghatak et al.
2979	P.D. Davis and D.C. Neckers
3007	H. Karpf and H. Junek
3035	A. Ichihara et al.
3063	A.J. Blackman and R.J. Wells
3097	O. Salcher et al.
3165	R. Kazlauskas et al.
3173	R.C.F. Jones and S. Sumaria
3227	D. Lerdal and C.S. Foote
3287	G. Rousseau et al.
3419	T.-T. Wu et al.
3463	R. Rajee and V. Ramamurthy
3469	A. Lacroix and J.-P. Fleury
3527	L. Ananthasubramanian et al.
3557	A. Nishinaga et al.
3563	H. Kuzuhara et al.
3593	V. Amico et al.
3597	A. Nishinaga et al.
3603	V.L. Mizyuk and A.V. Semenovsky
3637	F.J. Schmitz and Y. Gopichand
3641	Y. Gopichand and F.J. Schmitz
3653	M.J. Robins and S.D. Hawrelak
3677	G. Ege and C. Freund
3681	J. Reisch et al.
3687	A. Konnecke et al.
3699	A. Mondon et al.
3707	A. Henriques et al.
3729	M. Franck-Neumann and J.J. Loh- mann
3777	E. A. Truesdale
3781	B.B. Jarvis et al.
3823	Y. Yokoyama et al.
3889	H. Laatsch
3893	H. Wagner et al.
3901	B. Epe and A. Mondon
3917	S. Toda et al.
3921	E. Yoshii et al.
3931	A.G. Gonzalez et al.
4047	T. Haya and H. Matsumoto
4073	R. Baker et al.
4093	D. Harris et al.
4111	J. Favre-Bonvin et al.
4135	R.S. Sagitullin et al.
4155	R. Kazlauskas et al.
4205	A. Alexakis et al.
4225	R.A. Berger and E. LeGoff
4263	R. Gompper and U. Wolf
4269	M.K. Au et al.
4293	Y. Fujise et al.
4403	T. Sato et al.
4411	H. Seto et al.
4467	D.A. Lightner et al.

4479	G.T. Carter et al.
4491	A. Gold et al.
4511	A. Laurent et al.
4559	T.R. Kasturi and S. Parvathi
4567	I.T. Storie and F. Sondheimer
4629	D.N. Kevill and M.A. Park
4633	I. Kurobane et al.
4671	R. Kreher and G. Use
4703	S.J. Eagle and J. Kitchin
4711	L. Crombie and W.M.L. Crombie
4741	T. Kato and J. Zemlicka
4749	G.L. Lange and F.C. McCarthy
4775	A.S. Kende and P.C. Naegely
4783	P. Forgacs et al.
4805	M. Suzuki and E. Kurosawa
4833	Y. Kashman and A. Groweiss
4857	H. Numan et al.
4895	S. Yamada et al.
4909	S. Takano et al.
4913	T. Hino et al.
4929	H. Ogawa et al.
5007	T. Fujii et al.
5071	J.W. Pavlik and R.M. Dunn
5219	A. Mangia and A. Scandroglio

89- -78, Angew. Chem.(Intl. Ed.), (1978)

0046	N. Jacobsen and V. Boekelheide
0068	R. Kreher and K.J. Herd
0203	E. Stoldt and R. Kreher
0268	F. Vögtle and E. Hammerschmidt
0271	H. Prinzbach et al.
0369	R. Neidlein and E. Bernhard
0374	R. Reimann and H.A. Staab
0446	R. Weiss et al.
0457	R. Weiss and C. Priesner
0465	H.H. Eckhardt and H. Perst
0468	R. Bartetzko and R. Gleiter
0519	G. Maier and S. Pfriem
0528	G. Märkl et al.
0756	H.A. Staab and V. Schwendemann
0757	H.A. Staab and U. Zapf
0758	G. Kaupp and M. Stark
0763	K. Hafner et al.
0853	M. Schäfer-Ridder et al.
0855	H.J. Gölz et al.
0943	J. Frank et al.
0956	W. Wagemann et al.

90- -78, J. Inorg. Nucl. Chem., 40 (1978)

0211	B.J. Sejekan et al.
0399	T.N. Srivastava et al.
0703	M. Hughes and R.H. Prince

93- -78, J. Appl. Chem. U.S.S.R., 51 (1978)

1354	M.A. Mostoslavskii et al.
2264	V.V. Kalmykov et al.

94- -78, Chem. Pharm. Bull. Japan, 26 (1978)

0038	M. Tanaka et al.
0079	I. Kitagawa et al.
0111	M. Yamazaki et al.
0135	T. Kinoshita et al.
0144	T. Saitoh et al.
0166	H. Ishii et al.

0171	T. Miura and M. Kimura		2205	T. Tsuchiye et al.
0209	T. Kato et al.		2224	T. Momose et al.
0240	M. Ikehara and T. Maruyama		2277	I. Yokoe et al.
0245	T. Hirota et al.		2334	T. Miyadera et al.
0328	Y. Ikeya et al.		2340	A. Matsuda et al.
0367	F. Yoneda et al.		2365	M. Fukuoka et al.
0466	H. Takahashi and H. Otomasu		2391	N. Yamaji et al.
0481	Z. Horii et al.		2407	A. Ueno et al.
0514	H. Ishii et al.		2411	A. Ueno et al.
0526	K. Takahashi and M. Takani		2422	H. Takayama and T. Okamoto
0530	T. Watanabe et al.		2435	H. Takayama and T. Okamoto
0596	M. Machida et al.		2449	M. Ikehara and H. Miki
0620	T. Momose et al.		2483	K. Morinaga et al.
0630	M. Terashima et al.		2508	C. Kaneko et al.
0635	S. Kobayashi et al.		2515	K. Fuji et al.
0643	T. Murakami et al.		2535	T. Yoshida et al.
0645	T. Fujii et al.		2543	T. Yoshida et al.
0682	Y. Ikeya et al.		2575	H. Takayama
0722	A. Takamizawa et al.		2600	T. Satake et al.
0898	N. Hamamichi et al.		2635	M. Miyahara et al.
0930	S. Tamura et al.		2657	H. Inoue and T. Ueda
0936	N.I. Nakano et al.		2664	H. Inoue and T. Ueda
0972	T. Momose et al.		2729	H. Takayanagi et al.
0981	M. Akiyama et al.		2765	T. Tsuji and Y. Ohtsuka
0985	M. Ikehara and Y. Takatsuka		2782	H. Ogura et al.
1015	H. Itokawa et al.		2805	M. Nagai et al.
1021	S. Kitagawa and H. Tanaka		2866	Y. Tamura et al.
1026	S. Kitagawa and H. Tanaka		2874	Y. Tamura et al.
1111	A. Yagi et al.		2914	K. Kohashi et al.
1141	T. Kurihara et al.		2933	T. Sato et al.
1153	T. Aono et al.		2990	M. Saneyoshi et al.
1162	K. Uekama et al.		3017	K. Hirai and T. Ishiba
1182	N. Aimi et al.		3023	F. Abe and T. Yamauchi
1274	M. Komatsu et al.		3050	K. Nakajima et al.
1298	Y.A. Al-Farkh et al.		3080	T. Murata et al.
1320	Y. Maebayashi et al.		3154	F. Yoneda et al.
1322	A. Ohta et al.		3186	T. Momose et al.
1375	Y. Ito and Y. Hamada		3195	T. Momose et al.
1394	T. Nomura et al.		3244	K. Kondo et al.
1403	T. Kitagawa et al.		3257	Y. Ikeya et al.
1415	T. Kitagawa et al.		3322	H. Tanaka et al.
1431	T. Nomura et al.		3433	H. Matsumoto and Y. Ohkura
1443	H. Obase et al.		3504	K. Funakoshi et al.
1453	T. Nomura et al.		3521	T. Momose et al.
1486	S. Naruto and A. Terada		3567	A. Nakagawa et al.
1511	T. Aono et al.		3580	N. Tanaka et al.
1533	T. Kurosawa et al.		3582	C. Kaneko et al.
1592	T. Momose et al.		3585	K. Takahashi and M. Takani
1619	T. Satake et al.		3633	A. Ohsawa et al.
1629	S. Kitagawa and H. Tanaka		3641	Y. Shoyama and I. Nishioka
1677	K. Takahashi et al.		3675	K. Hira et al.
1718	G. Goto et al.		3695	T. Hino et al.
1776	T. Aono et al.		3704	K. Yamakawa and T. Satoh
1798	A. Yagi et al.		3722	I. Kitagawa et al.
1803	T. Hashimoto et al.		3792	K. Hata et al.
1825	Y. Yamamoto et al.		3798	Y. Goto and N. Honjo
1832	A. Ueno et al.		3815	T. Kametani et al.
1880	N. Takao et al.		3863	M. Komatsu et al.
1890	T. Tsuchiya and J. Kurita		3891	M. Okada et al.
1896	T. Tsuchiya and J. Kurita		3896	T. Uno et al.
1929	T. Fujii et al.		3905	M. Mochizuki et al.
2027	H. Katayama et al.		3909	M. Okada et al.
2046	A. Ohta et al.		3914	M. Mochizuki et al.
2111	H. Sasaki et al.			
2122	T. Ueda et al.		95- -78, J. Pharm. Soc. Japan, 98 (1978)	
2175	H. Fukumi et al.		0041	T. Inoue et al.

0089   E. Hayashi et al.
0095   E. Hayashi and N. Shimada
0136   E. Hayashi and N. Shimada
0146   T. Kametani et al.
0165   H, Ochi et al.
0198   M. Watanabe et al.
0210   M. Kozawa et al.
0249   K. Ito and J. Lai
0286   S. Yoshina et al.
0335   Y. Sato et al.
0358   T. Takahashi et al.
0366   M. Kanao and H. Matsuda
0448   Y. Sato et al.
0503   S. Furukawa et al.
0585   Y. Hamada and Y. Ito
0623   K. Kurata et al.
0631   K. Kurata et al.
0636   M. Kozawa et al.
0774   Y. Yamano et al.
0802   T. Kurihara et al.
0850   S. Sakai et al.
0886   M. Ju-Ichi et al.
0891   E. Hayashi et al.
0898   K. Wakabayashi et al.
0910   K. Tsuji et al.
0914   T. Yazaki et al.
0929   T. Okano et al.
0950   S. Sakai and N. Shinma
1072   K. Nakashima et al.
1243   C. Tani et al.
1274   S. Suzuki
1285   K. Ito and J. Lai
1376   S. Sakai et al.
1395   A. Sakushima et al.
1402   H. Ichibagase et al.
1412   H. Fujito et al.
1441   T. Kurihara and S. Suzuki
1503   E. Hayashi and N. Shimada
1551   M. Yasumoto et al.
1553   S. Matsueda et al.
1592   K. Ito et al.
1607   M. Yokota et al.
1635   F. Hamaguchi et al.
1658   C. Tani et al.

96-   -78, The Analyst, 103 (1978)
0140   A.G. Asuero and J.M. Cano
0879   A.E. Mahgoub et al.

97-   -78, Z. Chemie, 18 (1978)
0020   M. Weissenfels and B. Ulrici
0057   W. Schroth and O. Peters
0063   J. Liebscher et al.
0108   H. Boland and R. Muller
0138   M. Weissenfels et al.
0177   E. Lippmann et al.
0256   H.-G. Henning and J.-U. Thurner
0334   J. Liebscher and H. Hartmann
0345B   H. Poleschner et al.
0380   R. Cizmarikova and J. Heger
0381   J. Berner and K. Kirmse
0385   E. Fanghanel et al.
0400   G. Sarodnick and G. Kempter

98-   -78, J. Agr. Food Chem., 26 (1978)
0115   R.D. Stipanovic et al.

0195   W.G. Galetto and P.G. Hoffman
0278   C.T. Seitz and R.E. Wingard, Jr.
0632   H.G. Cutler et al.
0869   F.L. Carter et al.
0973   N.A.M. Eskin et al.
1173   R.M. Seifert et al.
1316   G.C. Miller and D.G. Crosby

99-   -78, Theor. Exptl. Chem., 14 (1978)
0194   G.V. Ratovskii et al.

100-   -78, Lloydia, 41 (1978)
0001   Y. Shimizu et al.
0056   K.V. Rao and W.-N. Wu
0156   C.D. Hufford and W.L. Lasswell,
        Jr.
0166   G.A. Cordell et al.
0169   M. Shamma and A.S. Rothenberg
0184   I.H. Bowen and J.R. Lewis
0257   W.-T. Liao et al.
0271   W.-T. Liao et al.
0342   M.M. El-Olemy et al.
0383   H. Rosler et al.
0442   F.S. El-Feraly and W.-S. Li
0497   G.C. Hokanson
0578   M.C. Wani et al.
0584   A.A. Seida et al.
0638   A.A. Omar

101-   -78A, J. Organomet. Chem., 144 (1978)
0357   P. Courtot et al.

101-   -78B, J. Organomet. Chem., 145 (1978)
0069   A.N. Nesmeyanov et al.
0101   K. Sonogashira et al.
0C01   N. Minami et al.

101-   -78C, J. Organomet. Chem., 146 (1978)
0235   F.P. Colonna et al.

101-   -78E, J. Organomet. Chem., 148 (1978)
0247   P. Jutzi and J. Baumgartner
0257   P. Jutzi and J. Baumgartner

101-   -78G, J. Organomet. Chem., 150 (1978)
0101   B. Floris and G. Illuminati
0C25   H. Okinoshima and W.B. Weber

101-   -78J, J. Organomet. Chem., 153 (1978)
0245   H.K. Hofstee et al.
0265   G.E. Herberich et al.

101-   -78K, J. Organomet. Chem., 154 (1978)
0001   J. Heinicke and A. Tzschach

101-   -78L, J. Organomet. Chem., 155 (1978)
0077   M. Hillman et al.

101-   -78M, J. Organomet. Chem., 156 (1978)
0159   G. Zweifel and S.J. Backlund

101-   -78N, J. Organomet. Chem., 157 (1978)
0001   J.Y. Becker and J. Klein
0109   M.E. Kastner and W.R. Scheidt

101-   -78P, J. Organomet. Chem., 159 (1978)
0317   H. Ogoshi et al.
0329   I. Ogoshi et al.

101-   -78Q, J. Organomet. Chem., 160 (1978)
0159   J.A. Connor and G.A. Hudson

0319   K. Sonogashira et al.

102-   -78, Phytochemistry, 17 (1978)
0135   I.H. Suhr et al.
0139   J. Lemmich and S. Havelund
0143   M. Miyakado et al.
0151   R.D. Stipanovic et al.
0153   Y. Asakawa and T. Takemoto
0156   W.H. Hui and M.-M. Li
0166   A. Ahond et al.
0275   V. Bardouille et al.
0279   A.R. de Vivar
0281   B. Rodriguez
0305   Anonymous
0317   L.T. Burka
0328   A. Chatterjee et al.
0338   F. Faini et al.
0447   J.H. Tatum and R.E. Berry
0457   Y. Asakawa et al.
0505   F.M. Dean et al.
0511   M.A. de Alvarenga et al.
0517   I.B. De Alleluia
0539   C. Richard et al.
0552   A.M. Clark and C.D. Hufford
0574   E.T. Rojas and L. Rodriguez-Hahn
0575   P. Pant and R.P. Rastogi
0578   G.E. Gallili
0579   K.W. Glombitza et al.
0583   P. Joseph-Nathan et al.
0587   P.C. Bose and N. Adityachaudhury
0593   A.B. deOliveira et al.
0689   S. Ghosal et al.
0779   C.L. Chen and H.M. Chang
0787   R. Higuchi and D.M.X. Donnelly
0824   A.B. Segelman et al.
0831   B. Adeoti et al.
0834   D.P. Chakraborty et al.
0835   V. Vecchietti et al.
0837   O.R. Gottlieb et al.
0839   H. Itokawa et al.
0895   A. Yagi et al.
0939   R.H. White and L.P. Hager
0955   A.G. Gonzalez et al.
0979   T. Shibuya et al.
1003   C. Francisco et al.
1059   W. Kisiel
1067   R. Sunder et al.
1069   H. Ripperger
1070   G. Adam et al.
1161   F. Bohlmann and C. Zdero
1179   T. Hashimoto and M. Tajima
1297   A.A. Bell et al.
1327   W. Herz and G. Ramakrishnan
1359   J. Trofast et al.
1363   K. Panichpol and P.G. Waterman
1375   M. Gregson et al.
1379   W.D. Ollis et al.
1383   W.D. Ollis et al.
1389   K. Kurosawa et al.
1395   M. Gregson et al.
1401   W.D. Ollis et al.
1405   K. Kurosawa et al.
1413   K. Kurosawa et al.
1417   K. Kurosawa et al.
1419   J.T. Cook et al.
1423   K. Kurosawa et al.

1439   P. Majumder and S. Saha
1442   S.S. Chibber and U. Khera
1447   C.A. Coune and L.J.G. Angenot
1452   A. Rabaron et al.
1605   I. Chandon-Loriaux et al.
1637   M. Pinar et al.
1641   L.R. Row et al.
1647   L.R. Row et al.
1671   R. Pinchin et al.
1673   R. Cooper et al.
1687   L.J.G. Angenot et al.
1731   M.R. Pollard et al.
1773   M.A. De Alvarenga et al.
1777   A. Chatterjee et al.
1783   E.G. Crichton and P.G. Waterman
1791   G. Combaut et al.
1794   Y. Asakawa et al.
1797   M.L. Bittner et al.
1805   G.A. Kuznetsova et al.
1807   R.N. Barua et al.
1812   F. Delle Monache et al.
1814   M.D. Manandhar et al.
1935   F. Bohlmann et al.
1967   E. Gacz-Baitz et al.
2005   Y. Asakawa et al.
2011   T. Uyar et al.
2015   H. Misirlioglu
2021   S. Ohmiya et al.
2023   N.N. Sabri and W.E. Court
2027   S. Hatanaka and S. Kaneko
2038   C.J. Aiba et al.
2040   H. Nielsen and P. Arends
2042   M. Pardhasaradhi and M.H. Babu
2045   S. Bhanumati et al.
2046   J.L. Suri et al.
2047   S.K. Dutta et al.
2097   A. Selva et al.
2107   T. Pozzo-Balbi et al.
2115   Y. Asakawa et al.
2119   S. Ghosal et al.
2125   I.H. Bowen et al.
2131   F.C. Seaman and N.H. Fischer
2132   G. Savona et al.
2135   S.K. Garg et al.
2140   M.A. El-Sohly et al.
2145   M. Sarkar et al.

103-   -78, Khim. Geterosikl. Soedin., 14
            (1978)
0033   V.L. Savel'ev et al.
0037   R.R. Kostikov et al.
0049   L.N. Zhukaaskaite et al.
0060   Z.Y. Krainer et al.
0066   N.S. Prostakov et al.
0082   M.A. Mikhaleva et al.
0109   S.N. Dashkevich
0120   B.M. Krasovitskii et al.
0122   B.S. Luk'yanov et al.
0141   L.A. Zhmurenko et al.
0148   A.P. Engoyan et al.
0173   N.N. Suvorov et al.
0183   V.T. Grachev et al.
0208   R.O. Kochkanyan et al.
0218   O.M. Polumbrik et al.
0224   Y.S. Andreichikov et al.
0226   P.P. Kornuta et al.

0250	D. Vegh et al.		1036	N.N. Suvorov et al.
0261	F.A. Gabitov et al.		1070	Y.N. Porshnev et al.
0264	S.D. Sokolov and G.B. Tikhomirova		1075	B.E. Zaitsev et al.
0278	A.V. Eremeev et al.		1085	V.L. Savel'ev et al.
0306	R.G. Glushkov et al.		1088	V.S. Belezheva et al.
0310	V.G. Granik et al.		1093	N.M. Przheval'skii et al.
0313	V.M. Dziomko and B.K. Berestevich		1116	T.A. Mikhailova et al.
0332	M.V. Gorelik et al.		1123	N.A. Klyuev et al.
0336	Y.S. Andreichikov et al.		1132	V.F. Sedova et al.
0345	T.V. Stupnikova et al.		1137	V.F. Sedova and V.P. Mamaev
0347	A.N. Kost et al.		1139	A.N. Kost et al.
0373	Y.N. Porshnev et al.		1145	V.V. Kuz'nenko and A.F. Pozharskii
0380	A.N. Kost et al.		1152	M. Starshikov and A.F. Pozharskii
0399	B.A. Trofimov et al.		1156	A.F. Pozharskii and N.M. Starshi-
0416	V.M. Potapov et al.			kov
0418	N.S. Kozlov et al.		1160	V.A. Chuiguk and P.D. Medik
0425	V.E. Blokhin et al.		1163	I.N. Zhurkovich and G.I. Gerasi-
0436	B.I. Buzykin et al.			menko
0443	N.N. Smirnova		1184	G.D. Varlamov et al.
0450	Y.A. Sedov and M.A. Chernova		1188	V.K. Daukshas et al.
0455	Z.F. Solomko et al.		1204	N.A. Kogan
0458	A.D. Sinegibskaya et al.		1217	V.S. Velezheva et al.
0465	L.V. Alam and I.Y. Kvitko		1223	R.G. Glushkov et al.
0493	V.M. Zolin et al.		1226	A.E. Sausin' et al.
0497	N.A. Tyukavkina et al.		1236	V.V. Kastron et al.
0503	A.V. Eremeev et al.		1241	I.N. Azerbaev et al.
0507	O.M. Glozman et al.		1261	V.G. Granik et al.
0518	L.B. Shagalov et al.		1267	M.F. Budyka and P.B. Terent'ev
0530	P.B. Terent'ev and N.G. Kotova		1270	L.V. Pribega et al.
0534	P.B. Terent'ev et al.		1273	I.V. Alekseeva et al.
0538	V.M. Potapov et al.		1277	V.P. Kruglenko et al.
0567	B.I. Buzykin et al.		1278	A.N. Kost et al.
0575	B.I. Buzykin et al.		1337	N.S. Kozlov et al.
0601	A.I. Tolmachev et al.		1343	N.S. Prostakov et al.
0605	E.N. Kharlamova et al.		1372	R.O. Kochkanyan et al.
0641	M.V. Kazankov and E.G. Sadvykh		1375	M.M. Kul'chitskii and S.V. Boya-
0663	M.F. Budyka et al.			cheva
0694	B.V. Lopatin et al.			
0702	V.A. Zagorevskii et al.		104-	-78, Zhur. Organ. Khim., 14 (1978)
0738	I.N. Azerbaev et al.		0067	T.I. Akimova and M.N. Tilichenko
0745	R.S. Sagitullin et al.		0106	P.A. Sharbatyan et al.
0757	V.S. Velezheva et al.		0111	V.S. Poplavskii et al.
0771	V.V. Dikopolova and N.N. Suvorov		0129	V.N. Charushin et al.
0784	I.S. Levi et al.		0159	V.A. Loskutov et al.
0795	T.K. Pashkevich et al.		0181	I.M. Bazavova et al.
0800	V.P. Shchipanov et al.		0189	A.V. El'tsov et al.
0843	M.I. Komendatov et al.		0191	A.P. Sevast'yan et al.
0851	V.N. Buyanov et al.		0194	G.I. Migachev et al.
0856	L.G. Yudin et al.		0207	I.P. Beletskaya et al.
0875	N.S. Prostakov et al.		0221	S.A. Andreev et al.
0879	I.N. Domnin et al.		0243	Y.N. Porshnev et al.
0885	V.M. Dziomko et al.		0264	V.A. Mironov et al.
0901	I.A. Drizina et al.		0286	Z.F. Solomko et al.
0909	A.F. Pozharskii et al.		0315	L.E. Nivorozhkin et al.
0920	V.P. Shchipanov and G.D. Kadochni-		0341	G.I. Borodkin et al.
	kova		0377	O.A. Zagulyaeva et al.
0985	I.B. Lazarenko et al.		0381	M.V. Gorelik and H.I. Kuon
0988	A.A. Ziyaev et al.		0390	M.V. Gorelik et al.
0993	G.V. Grishina et al.		0398	A.N. Chupakhin et al.
0997	N.S. Prostakov et al.		0406	Z.I. Litvak and V.M. Berezovskii
1009	R.G. Glushkov et al.		0412	Y.A. Artsybasheva and B.V. Ioffe
1013	R.G. Glushkov and T.V. Stezhko		0521	K.M. Dyumaev et al.
1016	L.P. Prikazchikova et al.		0559	Y.V. Pozdnyakovich and V.D. Shtein-
1019	M.A. Mikhaleva et al.			garts
1025	M.A. Mikhaleva et al.		0566	L.L. Rodina et al.
1032	L.N. Koikov et al.		0576	V.P. Sergutina et al.

0582   D.A. Oparin et al.
0589   A.I. Grigor'eva et al.
0601   O.V. Goryunova et al.
0619   G.I. Migachev et al.
0660   I.M. Yakovleva et al.
0666   L.P. Davydova et al.
0676   T.D. Mechkov et al.
0701   M.I. Komendantov et al.
0717   V.V. Sverev et al.
0725   I.G. Vitenberg
0758   V.B. Piskov and V.P. Kasperovich
0786   A.E. Vasil'ev and L.S. Shishkanova
0794   V.M. Bairamov et al.
0817   V.A. Buevich and V.V. Rudchenko
0829   G.A. Karlivan and R.E. Valter
0832   V.N. Knyazev and V.N. Drozd
0846   S.A. Andreev et al.
0924   G.I. Borodkin et al.
0939   Y.A. Fialkov et al.
0997   S.M. Ramsh et al.
1004   V.V. Belogorodskii and A.L. Remizov
1041   V.N. Drozd and M.V. Grandberg
1089   E.S. Turbanova et al.
1174   S.N. Semenova et al.
1197   V.K. Grif et al.
1218   S.P. Gromov et al.
1225   N.A. Shenberg et al.
1229   S.M. Ramsh et al.
1234   G.M. Polumbrik and I.G. Ryabokon'
1245   V.I. Letunov and N.P. Soldatova
1319   N.I. Zakharova et al.
1353   M. Komendantov et al.
1402   N.K. Genkina et al.
1433   M.V. Gorelik et al.
1446   V.L. Lapteva and N.P. Shusherina
1463   V.S. Velezheva et al.
1532   Y.N. Kreitsberga and O.Y. Neiland
1537   Y.L. Salenko et al.
1544   R.S. Sagitullin et al.
1582   V.A. Ostrovskii et al.
1614   B.A. Trofimov et al.
1617   N.V. Smirnova and M.B. Kolesova
1624   K.L. Muravich-Aleksandr et al.
1644   A.A. Ginesina et al.
1750   O.F. Ginzburg et al.
1830   I.A. Aleksandrova et al.
1847   E.R. Zakhs et al.
1894   M.G. Vinogradov et al.
1908   A.N. Patel
1910   Y. Degutis et al.
1936   V.I. Savin
1942   V.I. Savin
1956   R.G. Dubenko et al.
1962   V.A. Gailite et al.
1971   Y.V. Samusenko et al.
1975   V.M. Vlasov and G.G. Yakobson
1986   V.V. Russkikh et al.
2011   V.I. Mukhanov et al.
2026   V.L. Lapteva et al.
2041   E.A. Panfilova et al.
2046   Y.L. Slominskii et al.
2064   V.N. Drozd and N.V. Grandberg
2070   Y.I. Rozhinskii and V.L. Plakidin
2131   F.Z. Galin et al.

2144   I.N. Domnin et al.
2189   E.Y. Belyaev et al.
2199   Y.E. Gerasimenko et al.
2203   B.I. Buzykin et al.
2208   S.G. Sheiko et al.
2210   N.M. Baranova et al.
2218   N.S. Zefirov et al.
2229   D.A. Oparin et al.
2252   M.L. Shpak et al.
2267   V.N. Drozd et al.
2316   S.A. Andreev et al.
2319   G.A. Lanovaya and V.P. Mikheeva
2321   E.B. Krylova et al.
2333   B.M. Lerman et al.
2336   B.M. Lerman et al.
2361   S.G. Sheiko et al.
2365   N.S. Prostakov et al.
2381   V.D. Orlov et al.
2420   B.A. Trofimov et al.

105-   -78, Khim. Prirodn. Soedin., 14 (1978)
0070   E.V. Popova et al.
0073   R.N. Tursanova et al.
0101   G.A. Tolstikov et al.
0149   N.F. Komissarenko et al.
0156   G.G. Zapesochnaya et al.
0159   S.Z. Ivanova et al.
0175   B.Z. Usmanov et al.
0186   A. Karimov et al.
0190   Y.R. Kushmuradov et al.
0217   Y. Oganesyan et al.
0263   M.E. Perel'son et al.
0267   Z.A. Kuliev and T.K. Khasanov
0271   Z.A. Kuliev and T.K. Khasanov
0275   G.G. Zapesochnaya et al.
0286   V.A. Raldugin et al.
0335   S.Z. Ivanova et al.
0348   V.A. Raldugin et al.
0353   L.I. Topuriya
0358   I.A. Israilov et al.
0360   A. Karimove et al.
0371   E.A. Kol'tsova et al.
0377   V.Y. Bagirov et al.
0388   U. Baltaev et al.
0393   U. Baltaev et al.
0441   A.A. Nabiev et al.
0449   G. Kitanov and K.F. Blinova
0464   T. Irgashev et al.
0465   I.A. Israilov et al.
0487   L.A. Golovina et al.
0490   G.G. Zapesochnaya et al.
0495   K. Bizhanova et al.
0499   K. Bizhanova et al.
0513   I.A. Bessonova et al.
0561   N.K. Utkina and O.B. Maksimov
0567   M.S. Pal'yants and N.K. Abubakirov
0606   L.A. Golovina et al.
0617   A.Y. Kushmuradov et al.
0630   Y.S. Vollerner et al.
0637   E.F. Nesmelova et al.
0699   S.V. Karimova et al.

106-   -78, Die Pharmazie, 33 (1978)
0051   I. Bornschein et al.
0082   H. Ripperger

0185   S. Leistner et al.
0235B   A. Levai and R. Bogner
0250   J. Dehlke
0297   J. Cizmarik et al.
0706   H.J. Siemann et al.
0764   G. Wagner and B. Dietzsch
0782   H.M. Mokhtar and L. Rateb

107-   -78, Synthetic Comm., 8 (1978)
0099   T. Imanishi et al.
0109   A.E. Hauck and C.S. Giam
0155   M.P.L. Caton et al.
0219   A. Barco et al.
0251   A.C. Jain et al.
0353   T.A. Eggelte et al.
0379   T.J. McCarthy et al.
0579   K.S. Chamberlin and E. LeGoff

110-   -78, Russian J. Phys. Chem., 52 (1978)
0044   L.P. Krasnomolova et al.
1540   V.D. Paramonov et al.

111-   -78, European J. Med. Chem., 13 (1978)
0017   A. Kalir et al.
0053   P. Geneste et al.
0093   A. Valla et al.
0155   G. Alonso et al.
0213   R. Royer et al.
0313   G. Azadian-Boulanger et al.
0435   L. Rene et al.
0515   D. Maysinger et al.

112-   -78, Spectroscopy Letters, 11 (1978)
0267   N.K. Narain
0751   E.M. Dexheimer et al.
0835   L. Prajer-Janczewska and K. Rudolf

114-   -78A, Acta Chim. Acad. Sci. Hung., 96
        (1978)
0045   J. Rohaly and C. Szantay
0055   C. Szantay and J. Rohaly
0167   B. Zsadon et al.
0217   T. Veszpremi et al.

114-   -78B, Acta Chem. Acad. Sci. Hung. 97
        (1978)
0069   V. Szabo et al.
0091   J. Nyitrai et al.
0429   G. Kalaus et al.

114-   -78C, Acta Chem. Acad. Sci. Hung., 98
        (1978)
0457   V. Szabo et al.

115-   -78, Egyptian J. Chem., 21 (1978)
0001   A.M. Dawidar et al.
0067   A.A. Hartsuch et al.
0073   A.A. Hartsuch and R.M. Issa
0305   M. Hammad and K. El-Bayouki

116-   -78, Macromolecules, 11 (1978)
0312   D. Bailey et al.
0568   J.A. Moore and J.E. Kelly

117-   -78, Org. Preps. Procedures Intl., 10
        (1978)

0048   S.H. Doss et al.
0079   L.K.T. Lam and K. Farhat
0097   J.A. Virgilio and E. Heilweil
0137   F.G. Schreiber and R. Stevenson
0177   S. Nozaki et al.

118-   -78, Synthesis (1978)
0144   E. Suzuki et al.
0205   J.H. Boyer and J.R. Patel
0206   B.C. Uff and R.S. Budhram
0208   D.T. Connor et al.
0286   A. Faure and G. Descotes
0291   S. Gelin
0307   H. Hauptmann and M. Mader
0374   G. Mehta and D.N. Dhar
0441   N. Furakawa et al.
0445   S. Morooka et al.
0448   S. Gelin and M. Chabannet
0543   P. Schiess and C. Suter
0592   T. Severin and I. Ipach
0633   L. Baiocchi et al.
0741B   J.T. Hunt and P.A. Bartlett
0832   C. Cativiela and E. Melendez
0846   G. Markl and R. Liebl
0900   S. Gelin and C. Deshayes

119-   -78, S. African J. Chem., 31 (1978)
0037   I.R. Green et al.
0047   R. Vlegaar et al.
0075   L.M. du Plessis and J.A.D. Eras-
        mus
0111   A.A. Chalmers et al.
0143   C.P. Gorst-Allman et al.

120-   -78, Pakistan J. Sci. Ind. Research,
        21 (1978)
0001   A.R. Katritzky et al.
0052   M.A. Hamid
0062   B. Robinson and M.U. Zubair
0096   T.S.L. Beswick et al.
0101   A. Butt et al.
0111   M. Younas and S.S. Bokhari

121-   -78, J. Macromol. Sci., 12 (1978)
0661   D. Bailey, D. Tirrell and O. Vogl
0945   M. Kajiwara et al.
1477   T. Otsu and H. Ohnishi

123-   -78, Moscow U. Chem. Bull., 33 (1978)
0190   V.N. Abramov et al.
0701   A.G. Popova et al.

124-   -78, Ukrain. Khim. Zhur., 44 (1978)
0071   I.N. Chernjuk et al.
0398   A.M. Osipov et al.
0637   S.G. Fridman and I.L. Kotova
0838   J.I. Slominskij
0844   R.V. Sendega and T.A. Prolsajlo
0942   A.J. Il'chenko et al.
1064   I.V. Alekseeva et al.
1187   A.I. Il'chenko

126-   -78, Makromol. Chem., 179 (1978)
0047   J. Kumanotani
0131   G. Sanchez et al.
0905   M. Hattori et al.

1803   R. Kerber et al.
1929   E. Chiellini et al.
1999   D. Rohde and G. Wegner
2195   M. Draminski and J. Pitha
2489   C.D. Eisenbach et al.
2845   O. Nuyken and R. Kerber

128-   -78, Croatica Chem. Acta, 51 (1978)
0097   V. Skaric et al.
0163   B. Katusin-Razem
0259   M. Dumic et al.
0273   M. Lacam and V. Rapic
0325   M. Lacam and V. Paric

130-   -78, Bioorg. Chem., 7 (1978)
0189   V. Chowdhry and F. H. Westheimer
0289   J.P. Kutney et al.
0409   G.B. Birrell et al.
0421   P.A. Bartlett et al.
0493   A.C. Jain et al.

133-   -78, Pharm. Acta Helv., 53 (1978)
0056   H. Koch and E. Steinegger
0189   J.K. Sugden et al.
0241   M. Otagiri et al.
0248   B. Kreyenbuhl et al.
0355   H. Bosshardt et al.

135-   -78, J. Appl. Spectroscopy S.S.S.R.,
       28-29 (1978)
0946   B.M. Krasovitskii et al.
1225   N.N. Chipanina et al.

136-   -78A, Carbohydrate Research, 60 (1978)
0039   A. Rosenthal and M. Ratcliffe
0251   H.A. Stuber and B.M. Tolbert
0267   C.R. Nelson and J.S. Gratzl

136-   -78B, Carbohydrate Research, 61 (1978)
0359   K. Wallenfels et al.

136-   -78C, Carbohydrate Research, 62 (1978)
0073   F. Leclercq et al.
0089   W.A. Szarek et al.
0155   H. Rosemeyer and F. Seela
0175   G.T. Shiau and W.H. Prusoff
0185   H.G. Garg and R.W. Jeanloz
0301   N. Pravdic and I. Franjic-Mihalic

136-   -78E, Carbohydrate Research, 64 (1978)
0017   T.D. Miniker et al.
0069   J.R. Daniel and R.A. Zingaro
0089   M.M.A. Rahman et al.
0101   S. Honda et al.

136-   -78F, Carbohydrate Research, 65 (1978)
0023   M. Courtin-Duchateau and A. Vey-
       rieres

136-   -78H, Carbohydrate Research, 67 (1978)
0079   M.A.E. Sallam
0117   T. Okuda et al.
0127   C.H. Lee et al.
0257   F.M. Unger et al.
0349   F.M. Unger et al.
0357   D. Horton and J. Tsai

0433   F. Shafizadeh et al.
0564   J.M.J. Tronchet et al.

137-   -78, Finnish Chem. Letters (1978)
0151   T. Simonen

138-   -78, Chemistry Letters (1978)
0029   H. Ogoshi et al.
0157   T. Miyashi et al.
0217   M. Kasai et al.
0237   K. Takahashi et al.
0259   A. Sakurai and Y. Okumura
0263   Y. Tamaru et al.
0301   K. Naya et al.
0323   K. Maruyama et al.
0331   M. Ochi et al.
0433   K. Yoshihara et al.
0657   F. Toda et al.
0677   K. Kikuchi et al.
0797   K. Shibata et al.
0879   M. Nakayama et al.
0961   M. Oda et al.
1005   A. Murai et al.
1033   K. Maruyama et al.
1099   J. Ojima et al.
1205   N. Katsui et al.
1209   A. Murai et al.
1219   A. Kato et al.
1223   T. Kitamura et al.
1239   M. Takasugi et al.
1241   M. Takasugi et al.
1281   C. Kaneko et al.
1297   T. Sato et al.
1313   C. Kuroda et al.
1319   T. Shinmyozu et al.
1345   T. Takahashi et al.

139-   -78A, P,S and Related Elements, 4
       (1978)
0303   G. Entenmann et al.

139-   -78B, P,S and Related Elements, 5
       (1978)
0027   G. Märkl and G. Habel
0041   D.A. Armitage and M.J. Clark
0191   N. Furukawa et al.
0209   L.LeGrand and N. Lozac'h
0251   D. Barillier
0257   G. Märkl et al.

140-   -78, J. Anal. Chem. S.S.S.R., 33
       (1978)
0032   I.V. Pyatnitskii et al.
0060   V.V. Lukachina and A.T. Pilipenko
0699   V.A. Nazarenko et al.
1305   V.A. Nazarenko et al.

142-   -78A, Heterocycles, 9 (1978)
0001   M. Iorio et al.
0011   S. Nishigaki et al.
0139   M. Uchida et al.
0153   Y. Tsuda et al.
0161   T. Sano et al.
0193   P. Venturella and A. Bellino
0201   J.P. Kutney et al.
0207   T. Momose and K. Kanui

0287	F.L. Lam and J.C. Parham
0385	M. Nakagawa et al.
0399	Y. Tominaga et al.
0417	N.N. Girotra and N.L. Wendler
0635	T. Nomura and T. Fukai
0663	F. Chen et al.
0731	T. Sano et al.
0739	S. Senda et al.
0745	T. Nomura et al.
0793	K. Senga et al.
0849	K. Kloc et al.
0995	D. Dwuma-Badu et al.
1041	T. Kurihara et al.
1233	V. Simanek et al.
1295	T. Nomura and T. Fukai
1355	T. Nomura et al.
1433	P. Venturella et al.
1533	L.A. Mitscher et al.
1577	M. Ruccia et al.
1717	T. Yamazaki et al.
1729	T. Kurihara and Y. Sakamoto
1733	T. Konakahara and Y. Takagi
1741	R. Okazaki et al.

142-   78B, Heterocycles, 10 (1978)
0023	T. Fujii et al.
0053	T. Horie et al.
0085	H. Shibata and S. Shimizu
0093	T. Sasaki et al.
0117	S. Nakajima and K. Kawazu
0123	Y. Hayashi et al.
0199	A. Schmitz et al.
0207	T. Komeno et al.

142-   -78C, Heterocycles, 11 (1978)
0083	V. Scherrer et al.
0105	R. Huisgen and J. Geittner
0181	T. Terasawa and T. Okada
0187	K. Kikuchi et al.
0191	M. Ozaki et al.
0267	H. Yamamoto et al.
0275	H. Yamamoto et al.
0287	T. Asao et al.
0293	T. Miyashi et al.
0299	R.J.M. Weustink et al.
0327	M. Tochtermann and H. Timm
0347	J.P. O'Brien and S. Teitel
0351	Y. Fujise et al.
0383	C.C. Price and J. Follweiler
0387	K. Hafner et al.
0401	S. Iida et al.
0409	R. Kreher and K.J. Herd
0437	G. Buchi and P.G. Williard

145-   -78, Arzneimittel Forsch., 28 (1978)
0493	H. Kubinyi
0595	P. Kourounakis and A.H. Beckett
1056	K. Posselt

146-   -78, J. Appl. Chem. Biotech., 28 (1978)
0144	K.Y. Chu and J. Griffiths
0341	Z.H. Khalil
0864	A.S. Shawali et al.

147-   -78, J. Luminescence, 16 (1978)
| 0177 | M.E. Long |

149-   -78A, Photochem. Photobiol., 27 (1978)
| 0683 | M.J. Thomas and C.S. Foote |

149-   -78B, Photochem. Photobiol., 28 (1978)
0007	G.I. Glover et al.
0083	S.-M. Park
0595	A. Singh et al.

150-   -78, J. Chem. Research(M), (1978)
0189	J.H. Adams and J.R. Lewis
0470	M. Harris and J.A. Joule
0582	E.V. Dehmlow and Naser-Ud-Din
0690	H.J. Callot and E. Schaefer
1182	N. Dennis et al.
1301	G. Kaupp and E. Teufel
1325	W.R. Knappe and F. Rothfelder
1683	B.C. Maiti et al.
2201	M. Pardo et al.
2216	J.-F. Verchere
2319	K.Y. Chu and J. Griffiths
2340	M. Rahimizadeh and E.S. Waight
2850	G.G. Abbot et al.
3272	C. DeGrand et al.
3551	L.J. Chinn et al.
3831	A. Rösner et al.
4248	H. Heydt and M. Regitz
4356	J.M. Tedder and D.J. Woodcock
4523	H. Suginome et al.
4762	K. Krohn
4801	B. Föhlisch and E. Haug
4901	K. Joos et al.
4944	J.M. Ruxer and G. Solladie
5001	G.M. Segal et al.
5101	J.K. Chakrabarti et al.
5151	M. Rabinovitz and A. Gazit
5344	P.M. Collins et al.
5451	R.P. Duffley and R. Stevenson
5538	D. Villesot and Y. Lepage
0454S	L. Lombardo and F. Sondheimer